PRINTED CIRCUITS HANDBOOK

To Ann

ABOUT THE EDITOR

Clyde F. Coombs, Jr. recently retired from Hewlett-Packard, where he had worked as an electronics engineer and manager. One of the most successful editors in professional publishing today, he developed and edited all five previous editions of the *Printed Circuits Handbook,* and also edited the *Electronic Instrument Handbook* and the *Communications Network Test and Measurement Handbook,* three of McGraw-Hill's best-selling technical handbooks.

PRINTED CIRCUITS HANDBOOK

Clyde F. Coombs, Jr. Editor-in-Chief

Sixth Edition

New York Chicago San Francisco Lisbon
London Madrid Mexico City Milan
New Delhi San Juan Seoul
Singapore Sydney Toronto

The McGraw·Hill Companies

Cataloging-in-Publication Data is on file with the Library of Congress

McGraw-Hill books are available at special quantity discounts to use as premiums and sales promotions, or for use in corporate training programs. For more information, please write to the Director of Special Sales, Professional Publishing, McGraw-Hill, Two Penn Plaza, New York, NY 10121-2298. Or contact your local bookstore.

Printed Circuits Handbook, Sixth Edition

1 2 3 4 5 6 7 8 9 0 DOC DOC 0 1 9 8 7

ISBN: 978-0-07-146734-6
MHID: 0-07-146734-3

Sponsoring Editor Steve Chapman
Editorial Supervisor Jody McKenzie
Project Manager Madhu Bhardwaj (International Typesetting and Composition)
Acquisitions Coordinator Alexis Richard
Copy Editors Andy Saff, Julie M. Smith
Proofreader Deepa Pathak
Indexer Clyde F. Coombs, Jr.
Production Supervisor Jean Bodeaux
Composition International Typesetting and Composition
Illustration International Typesetting and Composition

CONTENTS

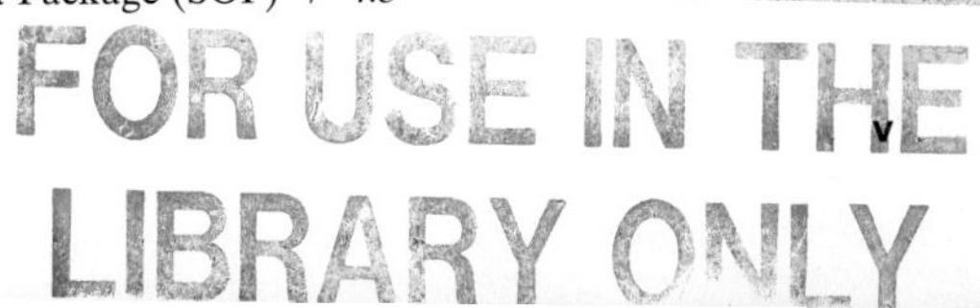

Part 8 Assembly

Part 9 Solderability Technology

Part 10 Solder Materials and Processes

Part 11 Nonsolder Interconnection

Part 12 Quality

Part 15 Flexible Circuits

Chapter 66. Special Constructions of Flexible Circuits 66.1

Chapter 67. Quality Assurance of Flexible Circuits 67.1

LIST OF CONTRIBUTORS

Mudasir Ahmad *Cisco Systems, San Jose, CA,* (CHAPS. 58, 59)

Simon Ang *University of Arkansas, Fayetteville, AR* (CHAP. 4)

Joyce M. Avery *Avery Environmental Services, Saratoga, CA* (CHAP. 60)

David W. Bergman *IPC, Bannockburn, IL* (APPENDIX)

Mark Brillhart *Palo Alto, CA* (CHAPS. 58, 59)

William Brown *University of Arkansas, Fayetteville, AR* (CHAP. 4)

Jody Byrum *Lockheed Martin, Newtown, PA* (CHAP. 41)

Brian F. Conaghan *Parelec, Rocky Hill, NJ* (CHAP. 26)

Don Cullen *MacDermid Inc., Waterbury, CT* (CHAP. 32)

C. D. Dupriest *Lockheed Martin, Dallas, TX (CHAP. 27)*

Darvin Edwards *Texas Instruments, Dallas, TX* (CHAP. 17)

Sylvia Ehrler *Feinmetall, GmbH, Herrenberg, Germany* (CHAP. 12)

Bini Elhanan *Valor Computerized Systems Inc., Yavne, Israel* (CHAP. 18)

Gary M. Freedman *Hewlett-Packard Co, Singapore* (CHAPS. 1, 44–50)

Dennis Fritz *MacDermid Inc., Waterbury, CT* (CHAP. 21)

Judith Glazer *Hewlett-Packard, Palo Alto, CA* (CHAP. 57)

Marshall I. Gurian *Marshall Gurian Consulting, Tempe, AZ* (CHAP. 34)

Terry Haney *Excellon Automation Co., Rancho Dominquez, CA* (CHAP. 25)

Ralph J. Hersey, Jr. *Lawrence Livermore National Laboratory, Livermore, CA* (CHAP. 15)

Happy T. Holden *Mentor Graphics, Longmont, CO* (CHAPS. 2, 19, 20, 22, 23)

Stacy Kalisz Johnson *Agilent Technologies, Gilbert AZ* (CHAP. 53)

Mike Jouppi *Thermal Management Inc., Centennial, CO* (CHAP. 16)

Edward J. Kelley *Isola Group, Chandler, AZ* (CHAPS. 6–11)

Peter G. Moleux *Peter Moleux and Associates, Newton Centre, MA* (CHAP. 60)

Gregory C. Munie *Kester, Itasca, IL* (CHAP. 43)

Hayao Nakahara *N.T. Information, Huntington, NY* (CHAPS. 5, 30, 31)

Robert (Bob) Neves *Microtek Laboratories, Anaheim, CA and Changzhou, China* (CHAP. 51)

Dominique K. Numakura *DKN Research, Haverhill, MA* (CHAPS. 61–67)

Gerard O'Brien *S.T. and S. Company, Bayport, NY* (CHAP. 42)

Stig Oresjo *Agilent Technologies, Loveland, CO* (CHAP. 53)

Kenneth P. Parker *Agilent Technologies, Loveland, CO* (CHAPS. 54, 55)

Mel Parrish *STI Electronics Inc., Madison AL* (CHAP. 52)

Tarak A. Railkar *Texas Instruments, Dallas, TX* (CHAP. 4)

Lee W. Ritchey *3Com, Santa Clara, CA* (CHAPS. 13, 14)

Michael Roesch *Hewlett-Packard Co., Palo Alto, CA* (CHAP. 12)

Gary Roper *One Source Group, Dallas, TX* (CHAP. 35)

John W. Stafford *JWS Consulting PLC, Phoenix, AZ* (CHAP. 3)

Valerie St. Cyr *Teradyne Inc., North Reading, PA* (CHAP. 27)

Laura J. Turbini *University of Toronto, Canada* (CHAP. 43, 56)

Hans Vandervelde *Laminating Company of America (LCOA), Garden Grove, CA* (CHAP. 24)

David A. Vahghan *Taiyo America Inc., Carson City, NV* (CHAP. 33)

Paul T. Vianco *Sandia National Laboratories, Albuquerque, NM* (CHAP. 40)

Jim Watkowski *Mac Dermid Inc., Waterbury, CT* (CHAP. 28, 29)

David J. Wilkie *Everett Charles Technologies, Pomona, CA* (CHAPS. 36–39)

PREFACE

With the implementation of the European Union (EU) directive on Restriction of Hazardous Substances (RoHS), the printed circuit industry has been forced to undergo a revolution in technology that is unprecedented in its history. The result is commonly called lead-free, a term that the reader will find used throughout this book, even though the restrictions are on several materials in addition to lead, as described in Chapter 1. The reason for this emphasis is that the biggest impact on the industry is the elimination of lead in the solder that is used for interconnection of the components and the board.

These changes are not driven by demands from the marketplace or the advancement of technology; rather, they have been legislated for reasons perceived to benefit society in general. Although the net effect, either positive or negative, on the global environment of elimination of lead in printed circuit assemblies has been the source of great debate, the reality is that it has been adopted worldwide and is a fact of life for all those who design, fabricate, and assemble printed circuits.

Tin-lead solder has been the basis for interconnections since the beginning of the use of printed circuits, and all associated materials and processes have been developed with the use of that material in mind. As a result, the abrupt shift to new solder alloys poses many questions about the related or affected processes and technologies that must now adapt and change to meet the lead-free requirements. Past experience or previous literature cannot be relied on to provide answers to these questions. The purpose of this book is to provide this information as specifically and in as much detail as possible, using industry standards, where they exist, or best practice that has a sound technical foundation and has been shown to work.

The most obvious change is in the metallurgy of the solder and in mating surfaces. However, all alternative solder alloys melt at a higher temperature than eutectic solder, and this has required new base materials, new assembly processes, as well as new test, inspection, and reliability criteria and methods. With specific regard to the lead-free alternative alloys, we start with the periodic table of elements and describe what elements are candidates for use, and what the effect each has when incorporated into an alloy. Thus we rely on physics and material science to help define the appropriate alternative for a given application. Similarly, to help the reader choose the best base material for the higher assembly and operating temperatures and higher component speeds, we describe the chemistry of the material.

The issues of defining and predicting reliability of products in this new situation have absorbed a great deal of corporate resources since the EU directive was announced. The predictive models and history were based on tin-lead alloys, and new models are needed for lead-free alloys. For this edition, we have added new material and expanded existing discussions on this subject.

Although the lead-free technology revolution has gotten much more attention, there have also been important evolutionary changes in the technology as the industry continues to meet the need for higher circuit and component density and faster circuits. As a result, printed circuit boards continue to be designed to be ever smaller, or larger, to suit the requirements of specific applications. This book updates the reader on these advancements. In addition, with the continuing global nature of the printed circuit industry, along with the constant pressure to decrease both time to market and time to volume, there is a growing need for international standards covering layout and design information and communications that enable each organizational element in the supply chain to function effectively with minimum manual intervention and delays for design clarification. For the first time, we have added chapters that describe these parts of the overall process. In addition, we have introduced discussions on embedded components and conformal coatings, both important issues in many applications of the technology.

This edition addresses these new elements of the printed circuit processes, both revolutionary and evolutionary, while still maintaining its foundation on the basics of the technology. No matter how sophisticated the leading edge of the technology becomes, at the core of all printed circuits is the plated-through hole in its various forms. This remains one of the most important technical achievements of the twentieth century. Although based on the plated-through hole, printed circuits technology has evolved over the years to be more reliable, efficient, and reproducible, but the process described in the first edition of this book is still recognizable in the sixth. As a result, those new to the technology will still find introductory information, while experienced practitioners will find industry standard methods and best practices to help them with the most recent developments.

As the industry has grown, it has become more specialized. This has created the need to standardize documentation and communication techniques as well as to understand the specific capabilities of all suppliers in the overall value delivery chain. The result is that process capabilities and limitations at each step must be known, the board must be designed with these clearly in mind, and consistent acceptability criteria must be agreed to in advance, before the responsibility for the board passes from designer to fabricator to assembler to end user. This has created a community of people who have not been intimately involved in printed circuit issues before, and who now find a working knowledge of printed circuits critical for performance in their jobs. This book provides information for these people as well. They not only will find the basic information useful in understanding the issues, but also find specific guidelines on the development and management of the value chain for the success of all.

Although the industry's preferred term for the subject of this book is *printed wiring* or *etched wiring,* the term *printed circuits* has passed into the world's languages as representing the process and products described. As a result, we will use the terms interchangeably.

The impact of all these changes in the printed circuit technology is reflected in this book, as over 75 percent of the chapters are either revised or new to this edition. That means that this sixth edition contains the most new information since the first edition.

The continuing cooperation and support of the leadership and staff of the IPC-Association Connecting Electronic Industries (IPC) are acknowledged, not just for this edition, but for all the previous editions. IPC, under the leadership of Ray Pritchard, now retired, and Dieter Bergman, celebrated its 50th anniversary in 2007, and it would be impossible to overemphasize the contribution it has made, not just to the electronics industry but to a world that increasingly relies on electronic products. Special appreciation goes to Jack Crawford for the help in identifying and providing IPC material critical for the preparation of this edition.

Finally, I want to acknowledge and thank the authors, who have given so liberally of their time and skill in preparing the chapters for this book. They have made a great contribution to the literature and to the industry.

Clyde F. Coombs, Jr.
Editor-in-Chief

P · A · R · T · 1

LEAD-FREE LEGISLATION

CHAPTER 1
LEGISLATION AND IMPACT ON PRINTED CIRCUITS

Gary M. Freedman
Hewlett-Packard Corporation, Business Critical Systems, Singapore

1.1 LEGISLATION OVERVIEW

The dramatic rise in output and rapid obsolescence of consumer and commercial electronic goods has prompted the European Union (EU) to draft two pieces of environmental legislation that impact the electronics industry:

- Directive 2002-96-EC 27 Jan 2003 of the European Parliament and of the Council on Waste Electrical & Electronic Equipment, commonly abbreviated as WEEE.
- Directive 2002-95-EC of the European Parliament and of the Council 27 Jan 2003, which restricts the use of certain hazardous substances in electrical and electronic equipment. This directive is also known as RoHS (Restriction of Hazardous Substances).

1.2 WASTE ELECTRICAL AND ELECTRONIC EQUIPMENT (WEEE)

This European Community Directive was fully effective in August 2005. It requires manufacturers selling electrical or electronic equipment to member countries of the European Union to be responsible for end-of-life disposal of their finished goods. The directive attempts to limit the stream of waste materials to landfills, encourage recycling, and pressure equipment designs and materials to be more environmentally friendly. The end results are tariffs for purchasing electronic products to help with disposal costs and encouraging electronic manufacturers to institute take-back programs for recycling. The list of affected goods is very long. Table 1.1 includes examples of affected products.

The directive also mandates that the symbol shown in Fig. 1.1 be placed conspicuously on products signifying separate collection for disposal.

1.3 RESTRICTION OF HAZARDOUS SUBSTANCES (RoHS)

The second piece of legislation has the biggest impact on the electronics industry. The RoHS directive makes it illegal to manufacture or import into the member states of the EU any electrical or electronic equipment that contains restricted materials—materials that have been the

TABLE 1.1 Examples of Products Affected by the WEEE Directive

Category	Examples
Large household appliances	Washers, dryers, refrigerators, freezers, microwave ovens, air conditioners, etc.
Small household appliances	Vacuum cleaners, toasters, irons, clocks, hair dryers, etc.
IT and telecommunications equipment	Computers, printers, monitors, copying equipment, calculators, data storage equipment, projectors, telephones, etc.
Consumer equipment	Radios, televisions, cameras, audio equipment, musical instruments, etc.
Lighting equipment	Light bulbs, lamps, other lighting equipment, etc.
Electrical and electronic tools	Tools for drilling, sawing, sewing, turning, milling, sanding, grinding, cutting, shearing, making holes, punching, folding, bending, riveting, nailing, welding, soldering, etc.
Toys, leisure and sports equipment	Electric trains; car racing sets; video game consoles; computers for biking, diving, running, rowing, etc.; sports equipment with electric or electronic components; slot machines; etc.
Medical devices	Equipment for radiotherapy, cardiology, dialysis, and nuclear medicine; pulmonary ventilators; laboratory equipment for in vitro fertilization, diagnosis, analyzers, freezers, etc.
Monitoring and control instruments	Smoke detectors, thermostats, instruments for measuring and weighing, laboratory equipment, industrial monitoring equipment, etc.
Automatic dispensers	Automatic dispensers for hot drinks, cold drinks, bottles, cans, solid products, money, etc.

FIGURE 1.1 Symbol for separate collection for disposal.

mainstay of electronics manufacturing. The legislation took full effect on July 1, 2006. Its implications are far reaching and apply to all electrical and electronic equipment unless specifically addressed by the main body of the legislation or an exemption. Definitions of terms within the RoHS legislation have been slow to come; some of the exemptions have yet to be fully defined, whereas others have been revised. Imprecision of the legislation delayed early adoption within the electronics industry. As a result, there is no standard industry set of materials and evaluation of alternative materials for manufacturability or for long-term reliability. Chapter 45 on "Soldering Materials" lists and describes alternative materials to tin-lead solder. Chapters 58 and 59 on "Reliability" provide information on methods to predict product life.

1.3.1 RoHS Restricted Materials

The RoHS directive targets six widely used materials (see Table 1.2).

1.3.2 Analysis for RoHS Compliance

The directive focuses on the smallest homogeneous part or material that can be separated from the mass of a component or assembly. As an example, take the case of a plastic packaged integrated circuit (IC) with a copper (Cu) lead-frame, where the lead-frame is plated with some surface finish for solderability. The surface finish would be evaluated for RoHS compliance separate from the whole package or even from the weight of the lead-frame itself. Instead, the weight-percent of (Pb) within the surface finish would be evaluated on the ratio of the weight of any Pb to the weight of the surface finish plated upon the lead-frame. Similarly, any Pb alloyed within the Cu lead-frame would be evaluated based on the total weight of the assayed Pb to the total weight of the Cu lead-frame. Similarly, the IC's encapsulant would be evaluated for the weight of any PBB or PBDE as compared to the weight of the encapsulant in which it is contained or upon which it is coated.

TABLE 1.2 Materials Impacted by "Restriction of Hazardous Substances" Legislation

Substance	Symbol or Abbreviation	Maximum Allowable Value (Wt%)	Uses in Electronics
Hexavalent chromium	Cr^{+6}	0.1	Platings, especially for surface passivation of environmentally unstable metals, pigments, and plastics; colorant as lead chromate; hexavalent chromate finish for plated metal enclosures, fasteners, clips, and screws
Cadmium	Cd	0.01	Platings, ink pigments, paint pigments, batteries, detectors, and thick film circuits
Mercury	Hg	0.1	Pigments, curing catalyst for plastics and foams, switch contacts, batteries PVC stabilizer, and colorant
Lead	Pb	0.1	Solder PWB surface finish; component terminations; surface finish; PVC stabilizer and colorant; plastics colorant as lead chromate; cables and harnesses; sheet metal plating (to reduce risk of Zn whiskers); alloy agent for steel, copper, aluminum, etc.
Polybrominated biphenyls	PBB	0.1	Circuit board flame retardant and other electrical insulators
Polybrominated diphenyl ethers	PBDE	0.1	Circuit board flame retardant and other electrical insulators, and flame retardant for plastics

1.3.3 Analytical Methods for RoHS Compliance

The RoHS legislation does not set forth specific recommendations for analysis, nor is there wide consensus within the electronics community on standard analytical screening methods for determining compliance with RoHS legislation. Simple wet chemistry spot tests can be conducted for presence or absence of Cr^{+6}, Hg, Cd, and Pb. These tests, if conducted properly, have good sensitivity for qualitative analysis. Atomic absorption spectrophotometry (AAS), x-ray fluorescence (XRF), energy dispersive spectroscopy (EDS or EDAX), infrared (IR) or ultraviolet (UV) spectrophotometry, and gas chromatography-mass spectroscopy (GC-MS, that is effective for PBB or PBDE) are but a few of the methods requiring expensive analytical instrumentation. Quantitative data can be gleaned from these and other methods, but capital investment is high, trained personnel are required, and use of traceable calibration standards are a must.

1.3.4 Exceptions and Exclusions

The RoHS legislation exempts numerous applications of the four targeted elemental species. Most are not relevant to the discussion of printed circuit board soldering. Relevant applications are listed in Table 1.3.

TABLE 1.3 RoHS Exemptions as per the Second Amendment of RoHS Legislation*

Lead in high melting temperature type solders (i.e., lead-based alloys containing 85 percent by weight or more lead)
Lead in solders for servers; storage and storage array systems; network infrastructure equipment for switching, signaling, and transmission; and network management for telecommunications
Lead in electronic ceramic parts (e.g., piezoelectronic devices)
Lead as a coating material for the thermal conduction module c-ring
Lead in solders consisting of more than two elements for the connection between the pins and the package of microprocessors with a lead content of more than 80 percent and less than 85 percent by weight
Lead in solders to complete a viable electrical connection between semiconductor die and carrier within integrated circuit Flip Chip packages
Cadmium and its compounds in electrical contacts and cadmium plating except for applications banned under Directive 91-338-EEC

* *Source:* L 280-18 EN *Official Journal of the European Union* 25.10.2005, 2nd Amendment Annex to Directive 2002-95-EC.

It should be noted that even in cases where Pb is removed from a component or solder, that the package must be otherwise RoHS-compliant.

Although counterintuitive, the directive declares an exemption for Pb-based interconnects where the Pb content is greater than 85 percent. This exemption is meant to address specific devices, such as ceramic ball grid arrays (CBGAs), ceramic column grid arrays (CCGAs), flip-chip devices, and other high-Pb interconnection schemes. CBGAs and CCGAs rely on high Pb-content balls or columns, respectively, which melt at higher temperatures than eutectic Sn-Pb and thus will not collapse upon reflow. Since these ceramic devices are heavy, if the solder balls or columns beneath were to collapse, then interconnect-to-interconnect shorting would take place. Several years are typically needed to determine reliability impact of metallurgical changes to IC packages. In the case of these area-array devices, the high-lead content solder connection is an integral part of the package and not just a link between a component lead and bonding pad on a circuit board. Further, these packages are generally used in high-end systems such as telecommunications equipment and powerful computers. The RoHS legislation exempts such equipment from restrictions on Pb until 2010. In the meantime, package manufacturers are working to find and test reliable replacements for Pb-based solders for high-end packages. Some smaller ICs have already made the switch to Pb-free interconnects. Other exclusions or exemptions detailed in the current RoHS directive are Cd and Pb in batteries, Pb in video monitor screens, and Hg in fluorescent light bulbs. It is curious to note that RoHS legislation targets Pb in solders even though solder represents a minor use of Pb (it is estimated to comprise less than 10 percent of world Pb usage). Conversely, the exempted Pb-acid storage battery is the major consumer of Pb accounting for more than 85 percent worldwide Pb usage.

In August 2006, additional exemptions were allowed for the RoHS legislation. As will be discussed in subsequent chapters, lead has been used to reduce or eliminate the occurrence of tin whiskers (metallic dendrites of tin that grow from pure tin surfaces). Metallic whiskers, such as from tin or zinc, are known to pose a reliability risk in terms of electrical shorting between oppositely charged conductors. Therefore, the RoHS legislation has been amended to allow lead in surface finishes of components with pitch $\leq$0.65 mm for tin whisker represion. This applies to NiFe (Alloy 42, also known as Kovar) lead-frames as well as to components with copper lead-frames. Curiously, the exemption does not cover connectors.

1.4 RoHS' IMPACT ON THE PRINTED CIRCUIT INDUSTRY

Never has there been such a sweeping change in electronics manufacturing as imparted by WEEE and RoHS—particularly the latter. Every industry is being impacted either by processes and materials or materials availability.

1.4.1 Components

Generally, ICs, passive devices, and connectors are not rated to survive temperature excursions much above that of eutectic Sn-Pb solder and are rated only to survive a brief thermal cycle. IC makers and component manufacturers have worked to understand the impact of the higher process temperatures required for Pb-free soldering.

Plastic encapsulated components are known to absorb water from the atmosphere. As these components are heated to reflow soldering temperature, the water expands and can cause component cracking. This phenomenon is popularly known as "popcorning" since the plastic package often bulges prior to cracking and remains deformed upon cooling. To counter this, plastic packaged components are "baked out" prior to soldering. The higher temperature regimes of most lead-free solders will exacerbate the issue of popcorning requiring more attention to component drying cycles. In addition, the higher reflow temperature of Pb-free soldering has necessitated revamping of the classification schedule for package moisture sensitivity. This is especially relevant to components that are not hermetically sealed. Also, longer bake-out cycles are likely required.

Many components, such as ICs and passive devices (resistors, capacitors, and so on), may not be suited for the higher process temperatures generally associated with lead-free soldering. The higher temperature may result in changes to electrical characteristics, cracking, melting, or other component damage or degradation. Only components certified by the parts manufacturer or by the user should be considered for Pb-free assembly.

The manufacturer of each component assigns it a moisture sensitivity classification that dictates how long the component can be stored in a normal workplace environment before requiring a controlled bake-out cycle. Detailed information can be found about component moisture sensitivity classification, storage conditions, and bake-out requirements in joint industry standards J-STD-020[1] and J-STD-033.[2]

In addition to changes in moisture sensitivity, component laminates, encapsulants, or overmolding compounds may be prone to degradation as a result of the higher temperature exposures and times. Other negative attributes are associated with the move to Pb-free solders. Most of the Pb-free solders are poorer wetting (slower to spread on a solderable surface), have grainer appearance (which is not necessarily a negative attribute, but different from what is generally accepted as a positive attribute in lead-based solder joints).

With the advent of Pb-free solders as a mainstream technology, electrical components meant for surface mounting are being reformulated or requalified to withstand the higher requisite temperatures of Pb-free processing. Similarly, standards organizations such as the Joint Electron Devices Engineering Council (JEDEC) have respecified the maximum safe temperature regime for components due to the generally higher reflow soldering temperature required for Pb-free soldering. The process engineer must consult component manufacturer's specifications to establish safe reflow parameters for Pb-free soldering.

1.4.2 Workmanship Standards

The most widely used workmanship standard, IPC-A-610, has been rewritten to accommodate the inspection of Pb-free soldered assemblies. The changes were introduced with version D of that document. Inspection personnel require retraining to give proper attention to the changes and renewed inspection criteria.

1.4.3 Laminates

Many of the widely used printed wiring board (PWB) laminates for leaded (Pb'd) solder assembly will not endure the higher temperatures of Pb-free processing. Only laminates with characteristic high-glass transition temperature (T_g) and high-decomposition temperature (T_d) should be investigated for use in the Pb-free process. PWB laminates are more likely to sag during Pb-free reflow. Some may even darken in color or delaminate when exposed to Pb-free soldering temperatures. Careful attention to material selection and testing is a must for Pb-free soldering. Reformulation of circuit board laminates for higher temperature use is required to reduce the possibility of warpage, charring, delamination, sagging, via- or barrel-cracking, and so on. High-temperature PWB laminates are available but are typically more expensive than those used for Sn-Pb processing.

[1] J-STD-020: Joint IPC-JEDEC Standard for Moisture-Reflow Sensitivity Classification for Nonhermetic Solid State Surface Mount Devices, Joint Electron Device Engineering Council (JEDEC), Arlington, VA, USA, and IPC-Association Connecting Electronic Industries, Bannockburn, IL, USA.

[2] J-STD-033: Joint IPC-JEDEC Standard for Handling, Packing, Shipping, and Use of Moisture-Reflow Sensitive Surface Mount Devices, JEDEC, Arlington, VA, USA, and IPC, Bannockburn, IL, USA.

1.4.4 Solder Flux

Some of the fluxes in use for tin-lead soldering may not be compatible with the higher reflow temperatures associated with most Pb-free alloys. Critical materials in the flux may boil off, decompose, or oxidize before thorough fluxing can take place. Also, the flux must remain on metal surfaces to be bonded until reflow occurs since it serves as an oxygen barrier. Were the flux to disappear too soon, reoxidation of the solder and bonding surfaces may occur, which will retard solder wetting. Fluxes for Pb-free solders must be tailored to the composition of the new solder alloys and their inherently higher soldering process temperatures, and to the Pb-free surface finishes on component leads and PWB bonding pads.

1.4.5 Hygiene

Distinct health benefits are associated with the removal of Pb from printed circuit assemblies. Pb's toxicity is well documented, and the most serious problems are associated with direct ingestion of Pb-bearing materials (for example, children ingesting peeling Pb-based paint or lead in drinking water from Pb-soldered plumbing). This accumulating heavy metal dissolves slowly in the human body. Although there is conflicting evidence as to whether Pb from solder can leach into the water table from landfills at a rate that warrants concern, it is in everyone's best interest to explore alternative solder alloys and migrate from Pb-based solders.

Printed circuit board assembly operations will certainly benefit in that the labor force will encounter lower toxicity. Solder dross generated during wave soldering or disturbed while performing wave solder maintenance is a major source of Pb inhalation in manufacturing plants. Airborne dross is also generated at solder fountain operations, hot-air rework, and even during hand soldering. The move to a Pb-free environment will also eliminate concerns for ingested fines from Pb-contaminated hands.

1.4.6 Lead Replacement Cost

In the United States and other countries, some legislative initiatives aimed at restricting the use of Pb have failed. This failure spurred research for solder replacement alloys. Several corporations and universities devised Pb-free solders, some of which were patented. As is the case with any patented item or process, the rights to invention reside with the inventor or sponsoring corporation in the short term, and several of these alloys may not be used freely. In some cases, only slight modifications in alloy composition separate commonly available solder alloys from patented solder compositions. The cost of patent licensing, whether by the end user or solder manufacturer, will be borne largely by the end user.

In almost every case, the cost of Pb-free solder is at minimum 20 percent higher than for Sn-Pb solder. PWB laminates are more costly, as are many of the components. Energy costs go up with the need for higher reflow soldering temperatures. There is significant cost in training of personnel to deal with the change to a Pb-free environment. Separate lines for Pb'd and Pb-free solder may be required.

Both tin and lead are abundant, easy to refine, and thus inexpensive. This is not the case for some of the constituents of lead-free solders being considered, such as indium (In), gallium (Ga), and silver (Ag). Even bismuth (Bi) is a minor constituent of Pb ore and comes largely from Pb refining. In wave soldering, a large volume (hundreds of pounds) of solder is required to fill the wave solder reservoir (solder pot). The cost and availability of some of the lead-free alloys make them impractical for the large volume of the solder pot. Compare Sn-Pb solder at roughly USD $1.50 per pound to Sn-0.7Cu at USD $2.30 and Sn2Ag0.8Cu0.5Sb at USD $4.00 per pound. Granted, with lead being such a heavy element, price per pound is not the best metric; price per unit volume is a more significant measure. Nonetheless, moving away from the Sn-Pb binary eutectic will mean significantly higher manufacturing materials costs. Also, during wave soldering, the molten and flowing wave will tend to dissolve certain surface finishes, which may eventually build up and change the composition of the solder alloy. Different solder alloys have different metals dissolution properties. Therefore, using the same lead-free solder alloy for wave soldering as for reflow may be impractical.

Some of the overt costs and impacts of moving to lead-free solders were mentioned previously in this chapter. There are other potentially hidden costs. Soldering equipment may not have the bandwidth to handle the higher temperatures of some of the pb-free alloys. Reflow ovens may need to be constructed out of different materials to handle long-term effects of the higher temperatures. Fan bushings and polymeric seals to maintain inert atmosphere integrity may not withstand much above the temperature regimes required for Sn-Pb processing. These same considerations apply to wave soldering. Hand soldering irons will require higher temperatures, and more boards will likely have to be scrapped due to localized charring while trying to attain reflow during the hand soldering operation or at repair. Board materials, fluxes, and such may require reformulating, which is also likely to drive costs higher.

1.4.7 Equipment Changes

In some cases, soldering equipment changes are required to accommodate higher process temperatures and contact with materials other than Sn and Pb. Such changes are discussed in more detail in Chap. 47.

1.4.8 Scrapping and Material Mistakes

There will be costs associated with segregating materials (Pb'd versus Pb-free) and associated increased floor space for carrying two inventories (Pb'd and Pb-free). Noncompliant materials will have to be scrapped or sold off to those still using leaded processes. It is anticipated that in the early throes of the conversion to Pb-free, parts inventories occasionally will be inadvertently mixed, which may impact finished goods reliability and also make these products unacceptable for import to Europe after July 1, 2006, due to tenets of the RoHS directive.

1.4.9 Training

Every facet of manufacturing is touched by this change, whether it is the soldering process, solder joint inspection, electrical testing, material inventories, bill-of-materials verification, purchase specifications, inventory separation, and incoming materials inspection. The adaptations of RoHS and Pb-free standards for printed circuit assembly certainly require the most extensive changes that the electronics industry has ever needed to make. Great care has to go into training of personnel at every level of printed circuit assembly (PCA) building and support.

1.4.10 Reliability Testing

The change of materials and solders to accommodate the elimination of Pb necessitates reliability testing of the Pb-free assemblies to ensure that reliability is comparable to that of Sn-Pb assemblies. Reliability test methods, parameters, and models will have to be refined to match material properties of Pb-free assemblies.

1.4.11 In-Circuit Test (ICT)

Since the process temperature for most lead-frees are higher than for eutectic- or near-eutectic Sn-Pb solder, any flux residues associated with No-Clean soldering are more thoroughly baked onto PWB surface metals. This inhibits electrical test probe contact. Even with today's No-Clean solder pastes and with Sn-Pb solders, electrical probing can be a challenge. Often the residues that cover test points necessitate multiple seating cycles of the test probes to penetrate the flux residue.

1.5 LEAD-FREE PERSPECTIVES

The relative benefits of removing or reducing lead remain controversial. Some studies have shown a net negative effect due to the energy materials and processes required for the production and use of replacement lead-free materials. Whatever the long-term environmental impact, the short-term consequences clearly create a technical challenge to ensure reliable, cost-effective printed circuit assemblies. There are neither drop-in replacements for Sn-Pb solder nor industry wide consensus for a single-solder alloy. The alloys that are most promising have significant challenges due to raw materials pricing, limited reliability data, and yet-to-be-determined interactions with other materials. It has taken many years to understand the properties of Sn-Pb solder, its metallurgical behavior, solder joint properties, and soldering methodologies. Any new solders will have to undergo scrutiny to determine key characteristics of process and reliability. Since the paradigm for characterizing solders and reliability has been honed through the millennia-long use of Sn-Pb solder, the development cycle for Pb-free solder should be significantly streamlined. On the other hand, the rapid changeover to Pb-free solder brings with it risks with regard to process effectiveness and control, as well as the long-term reliability of the end product. These risks will require continuous reevaluations of the solder alloy choice, appropriate surface finishes, soldering conditions, materials compatibility (PWB, IC packages, and such), redefinition of solder volume requirements for reliable solder joint strength, and even some equipment changes. As a result, the Chap. 45 on "Solder Materials" contains a detailed discussion of "lead-free" alloys and materials and offer several choices on materials and discussion of their characteristics.

Since nearly all the practical Pb-free alloy choices have a higher melting temperature, a requisite higher energy consumption and energy cost is associated with the new soldering materials. Equipment changes needed for Pb-free processing will be discussed in Chaps. 46 and 47.

Also, as Pb-based solder will remain in use until at least 2010 as per RoHS, another likely mode of failure will be the inadvertent use of Pb-free solder with Pb'd packages or Pb'd solder with Pb-free packages. Although there is some backward and forward compatibility in terms of solders, packages, and surface finishes, there are also some difficulties. Some packages meant for use in the Pb'd process will not tolerate the higher soldering temperatures associated with most Pb-free solders. In the case of a Pb-free ball grid array (BGA) soldered with Pb'd solder, or vice versa, it is already known that there will be a negative impact to the resultant solder joint reliability for the leading Pb-free candidate solders. Increased tensile strength with increased brittleness is commonly reported for the tin-silver-copper system—a leading replacement candidate.

China, Japan, and Korea have legislative initiatives taking shape that are similar to the EU's, WEEE, and RoHS directives. This is the case for several U.S. states also. There is no escaping the fact that Pb-free soldering will be the mainstay of all future electronics interconnections.

P · A · R · T · 2

PRINTED CIRCUIT TECHNOLOGY DRIVERS

CHAPTER 2
ELECTRONIC PACKAGING AND HIGH-DENSITY INTERCONNECTIVITY*

Clyde F. Coombs Jr.
Editor-In-Chief, Los Altos, California

Happy T. Holden
Westwood Associates, Loveland, Colorado

2.1 *INTRODUCTION*

All electronic components must be interconnected and assembled to form a functional and operating system. The design and the manufacture of these interconnections have evolved into a separate discipline called *electronic packaging.* Since the early 1950s, the basic building block of electronic packaging is the printed wiring board (PWB), and it will remain that into the foreseeable future. This book outlines the basic design approaches and manufacturing processes needed to produce these PWBs.

This chapter outlines the basic considerations, the main choices, and the potential trade-offs that must be accounted for in the selection of the interconnection methods for electronic systems. Its main emphasis is on the analysis of potential effects that the selection of various printed wiring board types and design alternatives could have on the cost and performance of the complete electronic product.

2.2 *MEASURING THE INTERCONNECTIVITY REVOLUTION (HDI)*

The continuing increase in component performance and lead density, along with the reduction in package sizes, has required that PWB technology find corresponding ways to increase the interconnection density of the substrate. With the introduction and continued refinement of such packaging techniques as the ball grid array (BGA), chip-scale packaging (CSP), and chip-on-board (COB), traditional PWB technology has approached a point where alternative ways of providing high-density interconnection have had to be developed. (See Chaps. 3 and

* Adapted from Coombs, Clyde F. Jr., *Printed Circuits Handbook* (4th ed.), chap. 1, "Electronic Packaging and Interconnectivity," (McGraw-Hill, New York, 1996.)

4 for detailed discussions of component and packaging technologies.) This has been called at times high-density interconnects (HDI), the interconnection revolution, or the density revolution, because doing the same things in the same way, only smaller, was no longer sufficient.

2.2.1 Interconnect Density Elements

The extent of these interconnect density issues is not always observable, but the chart[1] in Fig. 2.1 can help one define and understand it. The chart portrays the interrelationship between component packaging, surface-mount technology (SMT) assembly, and PWB density. As can be seen, these three elements are interlinked. A change in one has a significant effect on the overall interconnection density. The metrics are as follows:

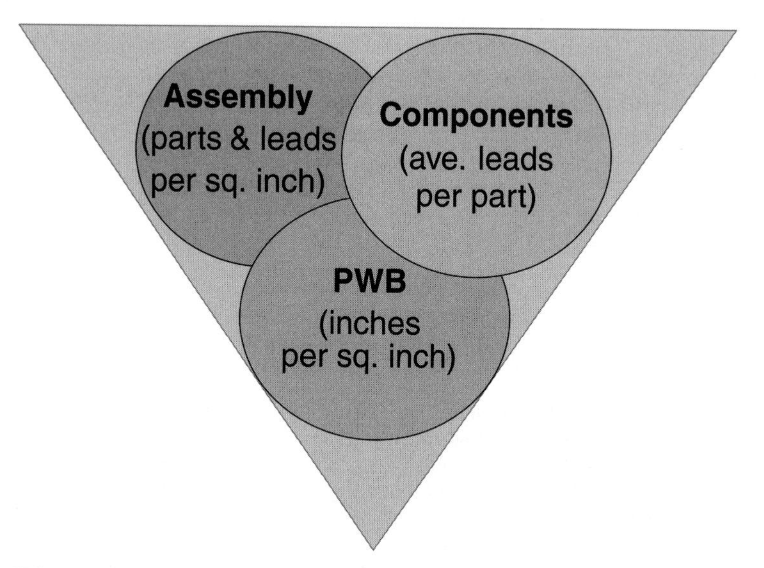

FIGURE 2.1 Representation of the metrics of assembly, component, and PWB technologies and their general relationship to each other.

- *Assembly complexity:* the measure of the difficulty of assembling surface-mounted components in parts per square inch and leads per square inch.
- *Component packaging complexity:* the degree of sophistication of a component, measured by its average leads (I/Os) per part.
- *Printed wiring board density:* the amount of wiring a PWB has as measured by the average length of traces per square inch or the area of that board, including all signal layers. The metric is inches per square inch.

2.2.2 Interconnect Technology Map

To visualize the interrelationships of the three elements, see Fig. 2.2. It shows these elements as axes of a three-dimensional technology map that defines the passage from conventional PWB structures to advanced technologies and shows how changes in just one of the elements can increase or decrease the total density of the entire electronic package.

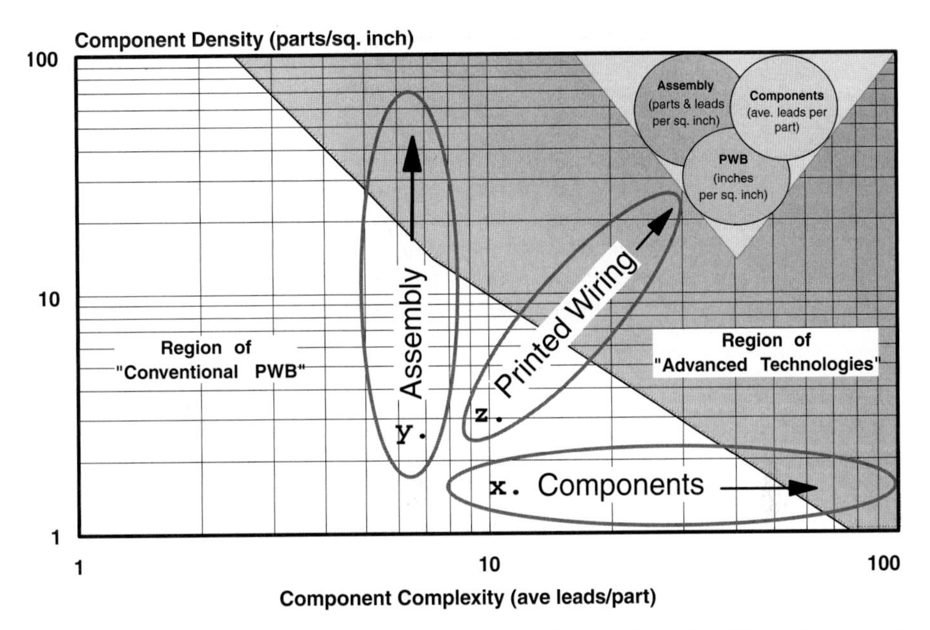

FIGURE 2.2 Component technology map, showing the relationship of assembly, PWB, and component technologies on overall package density and technology.

To describe the component complexity of an assembly, the total component connections (I/Os) include both sides of an assembly, as well as edge fingers, or contacts, which are divided by the total number of components on the assembly. The resulting average leads (I/Os) per part provides the *x*-axis of Fig. 2.2. The horizontal oval shape shows how the component complexity can vary from two leads per part in discrete circuit elements to the very large numbers seen on BGA and application-specific integrated circuits.

When Fig. 2.2 is used to describe surface-mount assemblies, the vertical (*y*-axis) dimension (shown as a vertical oval) indicates how complex it is to assemble the board by number of components per square inch or square centimeter for the surface area of the PWB. This vertical oval can vary from 1 to over 100 parts per square inch. As the parts become smaller and closer together, this number naturally goes up. A second assembly measure is average leads (I/Os) per square inch or square centimeter. This is the *x*-axis value multiplied by the *y*-axis value. (For a further description of these issues, and equations for quantifying them, see Chap. 18.)

The *z*-axis oval in Fig. 2.2 describes the printed wiring board's density. This is the wiring required to connect all the I/Os of the components at the size of the assembly specified, assuming three nodes per net. This axis has the units inches per square inch, or centimeters per square centimeter. A further description of this metric is provided in this chapter and in more detail in Chap. 19.

2.2.3 An Example of the Interconnect Revolution

By charting products of a particular type over time, an analysis will show how the interconnect technology has changed and continues to change, its rate of change, and the direction of these changes. An example is given in Fig. 2.3. This shows how component technology, assembly technology, and PWB technology have led to the evolution of the same computer CPU from:

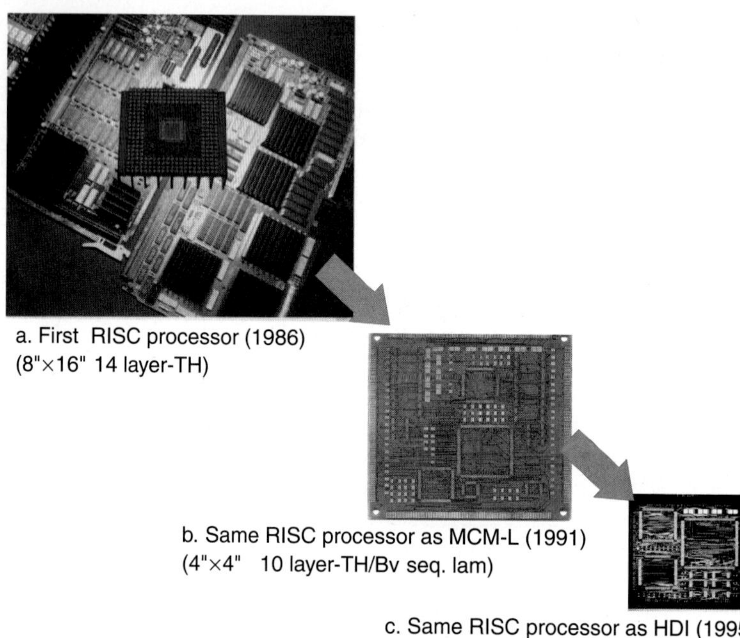

FIGURE 2.3 Example of the same computer CPU board as it used alternative component, assembly, and PWB technologies. (*a*) Size and appearance of each generation. (*b*) Movement of total board density from traditional to HDI. (*c*) HDI.

A 14-layer board through-hole with a surface area of 128 in^2 in 1986 (Fig. 2.3[*a*]) to

A 10-layer surface-mount technology board with a surface area of 16 in^2 in 1991 (Fig. 2.3[*b*]) to

A high-density interconnect board with sequential build-up microvias, buried and blind vias, and a surface area of 4 in^2 in 1995 (Fig. 2.3[*c*])

2.2.4 Region of Advanced Technologies

A second valuable feature of the chart in Fig. 2.2 is the area identified as the region of advanced technologies. This is where calculations and data have shown that it is necessary to have an HDI structure. Therefore, this is the barrier, or wall of HDI: on one side it is most cost effective to use traditional PWB technologies; on the other side it becomes cost effective to use HDI technologies. Continuing beyond this point, HDI becomes necessary.

2.3 HIERARCHY OF INTERCONNECTIONS

To have the proper perspective on where PWBs fit into electronic systems, it will be helpful to describe briefly the packaging hierarchy of electronic systems. Some time ago, the Institute for Interconnecting and Packaging Electronic Circuits (IPC)[2] proposed eight categories of system elements in ascending order of size and complexity, which will be used here to illustrate typical electronic packaging structures. These are as follows:

Category A consists of fully processed active and passive devices. Bare or uncased chips and discrete capacitors, resistors, or their networks are typical examples of this category.

Category B comprises all packaged devices (active and passive) in plastic packages, such as DIPs, TSOPs, QFPs, and BGAs, as well as those in ceramic packages, such as PGAs, and connectors, sockets, and switches. All are ready to be connected to an interconnecting structure.

Category C is substrates that interconnect uncased or bare chips (i.e., the components of category A) into a separable package. Included here are all types of multichip modules (MCMs), chip-on-boards (COBs), and hybrids.

Category D covers all kinds of substrates that interconnect and form assemblies of already packaged components, i.e., those from categories B and C. This category includes all types of rigid PWBs, flexible and rigid-flexible, and discrete-wiring boards.

Category E covers the back planes made by printed wiring and discrete-wiring methods or with flexible circuits, which interconnect PWBs, but not components, from the preceding groups.

Category F covers all intraenclosure connections. Included in this category are harnesses, ground and power distribution buses, RF plumbing, and co-ax or fiberoptic wiring.

Category G includes the system assembly hardware, card racks, mechanical structures, and thermal control components.

Category H encompasses the entire integrated system with all its bays, racks, boxes, and enclosures and all auxiliary and support subsystems.

As seen from the preceding list, PWBs are exactly in the center of the hierarchy and are the most important and universally used element of electronic packaging.

The packaging categories F, G, and H are used mainly in large mainframes, supercomputers, central office switching, and some military systems. Since there is a strong trend toward the use of miniaturized and portable electronic products for the majority of electronic packaging designs, trade-offs are made in the judicious application and selection among the elements of the first five categories. These are discussed in this chapter.

2.4 FACTORS AFFECTING SELECTION OF INTERCONNECTIONS

Selection of the packaging approaches among the various aforementioned elements is dictated not only by the system function, but also by the component types selected and by the operating parameters of the system, such as the clock speeds, power consumption, and heat management methods, and the environment in which the system will operate. This section provides a brief overview of these basic constraints that must be considered for proper packaging design of the electronic system.

2.4.1 Speed of Operation

The speed at which the electronic system operates is a very important technical factor in the design of interconnections. Many digital systems operate at close to 100 MHz and are already reaching beyond that level. The increasing system speed is placing great demands on the ingenuity of packaging engineers and on the properties of materials used for PWB substrates.

The speed of signal propagation is inversely proportional to the square root of the dielectric constant of the substrate materials, requiring designers to be aware of the dielectric properties of the substrate materials they intend to use. The signal propagation on the substrate between chips, the so-called *time of flight,* is directly proportional to the length of the conductors and must be kept short to ensure the optimal electrical performance of a system operating at high speeds.

For systems operating at speeds above 25 MHz, the interconnections must have transmission line characteristics to minimize signal losses and distortion. Proper design of such transmission lines requires careful calculation of the conductor and dielectric separation dimensions and their precise manufacture to ensure the expected accuracy of performance. For PWBs, there are two basic transmission line types:

1. Stripline
2. Microstrip (for details, see Chap. 15)

2.4.2 Power Consumption

As the clock rates of the chips increase and as the number of gates per chip grows, there is a corresponding increase in their power consumption. Some chips require up to 30 W of power for their operation. With that, more and more terminals are required to bring power in and to accommodate the return flow on the ground planes. About 20 to 30 percent of chip terminals are used for power and ground connections. With the need for electrical isolation of signals in high-speed systems operation, the count may go to 50 percent.

Design engineers must provide adequate power and ground distribution planes within the multilayer boards (MLBs) to ensure efficient, low-resistance flow of currents, which may be substantial in boards interconnecting high-speed chips consuming tens of watts and operating at 5 V, 3.3 V, or lower. Proper power and ground distribution in the system is essential for reducing *di/dt* switching interference in high-speed systems, as well as for reducing undesirable heat concentrations. In some cases, separate bus-bar structures have been required to meet such high power demands.

2.4.3 Thermal Management

All the energy that has been delivered to power integrated circuits (ICs) must be efficiently removed from the system to ensure its proper operation and long life. The removal of the heat from a system is one of the most difficult tasks of electronic packaging. In large systems, huge heat-sink structures, dwarfing the individual ICs, are required to air-cool them, and some computer companies have built giant superstructures for liquid cooling of their computer modules. Some computer designs use liquid immersion cooling. Still, the cooling needs of large systems tax the capabilities of existing cooling methods.

The situation is not that severe in smaller, tabletop or portable electronic equipment, but it still requires packaging engineers to ameliorate the hot spots and ensure longevity of operation. Since PWBs are notoriously poor heat conductors, designers must carefully evaluate the method of heat conduction through the board, using such techniques as heat vias, embedded metal slugs, and conductive planes.

2.4.4 Electronic Interference

As the frequency of operation of electronic equipment increases, many ICs, modules, or assemblies can act as generators of radio frequency (RF) signals. Such electromagnetic interference (EMI) emanations can seriously jeopardize the operation of neighboring electronics or even of other elements of the same equipment, causing failures, mistakes, and errors, and must be prevented. There are specific EMI standards defining the permissible levels of such radiation, and these levels are very low.

Packaging engineers, and especially PWB designers, must be familiar with the methods of reducing or canceling this EMI radiation to ensure that their equipment will not exceed the permissible limits of this interference.

2.4.5 System Operating Environment

The selection of a particular packaging approach for an electronic product is also dictated by its end use and by the market segment for which that product is designed. The packaging designer has to understand the major driving force behind the product use. Is it cost driven, performance driven, or somewhere in between? Where will it be used—for instance, under the hood of a car, where environmental conditions are severe, or in the office, where the operating conditions are benign? The IPC[2] has established a set of equipment operating conditions classified by the degree of severity, which are listed in Table 2.1.

2.4.5.1 Cost. The universal digitization of most electronic functions led to the merger of consumer, computer, and communication technologies. This development resulted in the increased appeal of electronics and the need for mass production of many electronic products. Thus, product cost has become the most important criterion in any design of electronic systems. While complying with all the aforementioned design and operation conditions, the design engineer must keep cost as the dominant criterion, and must analyze all potential trade-offs in light of the best cost/performance solution for the product.

TABLE 2.1 Realistic Representative-Use Environments, Service Lives, and Acceptable Cumulative-Failure Probabilities for Surface-Mounted Electronics by Use Categories

	Worst-case use environment						
Use category	T_{min}, °C	T_{max}, °C	ΔT,* °C	t_D, h	Cycles/ year	Years of service	Acceptable failure risk, %
1—Consumer	0	+60	35	12	365	1–3	~1
2—Computers	+15	+60	20	2	1460	~5	~0.1
3—Telecomm	–40	+85	35	12	365	7–20	~0.01
4—Commercial aircraft	–55	+95	20	12	365	~20	~0.001
5—Industrial & automotive	–55	+95	20	12	185		
(passenger compartment)			&40	12	100	~10	~0.1
			&60	12	60		
			&80	12	20		
6—Military	–55	+95	40	12	100		
ground & ship			&60	12	265	~5	~0.1
7—LEO	–40	+85	35	1	8760	5–20	~0.001
Space GEO				12	365		
8—Military *b*	*a*			40	2	365	
avionics *c*	–55	+95	60	2	365	~10	~0.01
			80	2	365		
			&20	1	365		
9—Automotive			60	1	1000		
(under hood)	–55	+125	&100	1	300	~5	~0.1
			&140	2	40		

& = in addition.

* ΔT represents the maximum temperature swing, but does not include power dissipation effects; for power dissipation, calculate ΔT_e.

The importance of the rigorous cost trade-off analysis during the design of electronic products is underscored by the fact that about 60 percent of the manufacturing costs are determined in the first stages of the design process, when only 35 percent of the total design effort has been expended.

Attention to manufacturing and assembly requirements and capabilities (so-called design for manufacturability and assembly [DFM/A]) during product design can reduce assembly costs by up to 35 percent and PWB costs by up to 25 percent.

The elements that must be considered for the most cost-effective electronic packaging designs are:

- Optimization of the PWB design and layout to reduce its manufacturing cost
- Optimization of the PWB design to reduce its assembly cost
- Optimization of the PWB design to reduce testing and repair costs

The following sections provide some guidelines on how to approach such optimization of PWB designs. Basically, the costs of the electronic assemblies are directly related to their complexity and there are a number of measurements relating the effects of various PWB design elements to their costs to guide the design engineer in selection of the most cost-effective approach.

2.5 ICS AND PACKAGES

The most important factors influencing PWB design and layout are the component terminal patterns and their pitches, especially those of ICs and their packages, since these dictate the density of the interconnecting substrates. Thus, this element will be considered first.

Driven by the need for improved cost and performance, the complexity of ICs is constantly increasing. Due to relentless progress in IC technology, the gate density on a chip is increasing by about 75 percent per year, resulting in the growth of IC chip I/O terminals by 40 percent per year, which places ever increasing demands on the methods of their packaging and interconnection.

As a result, the physical size of electronic gears keeps shrinking by 10 to 20 percent per year, while the surface area of substrates is being reduced by about 7 percent per year. This is accomplished by continuously increasing wiring densities and reducing linewidths, which has severely stressed PWB manufacturing methods, reduced processing yields, and increased the costs of the boards.

2.5.1 IC Packages

Since their inception, IC chips have been placed within ceramic or plastic packages. Until about 1980, all IC packages had terminal leads that were soldered into plated through-holes (PTHs) of the PWBs. Since then, an increasing number of IC packages have their terminals made in a form suitable for surface-mounting technology (SMT), which has become the prevailing method of component mounting.

There has been a proliferation of IC package types, both for through-hole assembly as well as for surface mounting, varying in their lead configurations, placement, and pitches. Also, IPC-SM-782[3] provides a good catalog of the available SMT packages and of the PWB footprint formats they require for their assembly.

Basic I/O termination methods of IC packages include the following:

- *Peripheral,* where the terminations are located around the edges of the chip or package
- *Grid-array,* where the terminations are located on the bottom surface of the chip or package

Most IC packages have peripheral terminations at their edges. The practical limit on the peripheral lead pitches on packages is about 0.3 mm, which permits locating, at most, 500 I/Os on a large IC package, as shown in Table 2.2. It has also become evident that, in typical board assembly operations, the yields plummet as the lead pitches go below 0.5 mm.

TABLE 2.2 Various Array Package Body Sizes, Configurations, and Lead Pitches

Body size (mm)	Number of I/Os	Minimum pitch (mm)
8 × 8	24	0.5
9 × 9	68	0.5
10 × 10	144	0.5
13 × 13	154	0.65
23 × 23	168	1.27
23 × 23	208	1.27
23 × 23	217	1.27
23 × 23	240	1.27
23 × 23	249	1.27
27 × 27	225	1.27
27 × 27	256	1.27
27 × 27	272	1.27
27 × 27	292	1.27
27 × 27	300	1.27
27 × 27	316	1.27
31 × 31	304	1.50
31 × 31	329	1.27
31 × 31	360	1.27
31 × 31	385	1.27
35 × 35	313	1.27
35 × 35	352	1.27
35 × 35	388	1.27
35 × 35	420	1.27
35 × 35	456	1.27
37 × 37	676	0.8
42.5 × 52.5	1247	1.0
52.5 × 52.5	2577	1.0

Various area array components come with a large variety of body sizes, numbers of I/Os, and I/O pitches. These components are called chip-scale packages (CSPs), plastic ball grid arrays (PBGAs), ceramic ball grid arrays (CBGAs), plastic pin grid arrays (PPGAs), and ceramic column grid arrays (CCGAs).

It is expected that chips with terminal counts below 150 to 200 will continue to use packages with peripheral leads, if these can be soldered within practical assembly yields. But for IC packages with over 150 to 200 I/Os, it is very attractive to use the grid-array terminations, since in such a case the entire bottom surface area can be utilized for terminations, which makes it possible to place large numbers of I/Os within a limited area.

This consideration has led to the development of a number of area array solder-bumping termination methods for IC and multichip module (MCM) packages, variously called pad grid, land grid, or ball grid arrays (BGAs) with terminal grids set at 1 mm (0.040 in), 1.27 mm (0.050 in), and 1.50 mm (0.060 in), respectively.

Use of grid arrays provides a number of benefits. The most important is the minimal footprint area on the interconnecting substrate, but grid arrays also offer better electrical performance due to low electrical parasitics in high-speed operation, simplified adaptation into SMT component placement lines, and better assembly yields, despite the impossibility of direct visual inspection of the joints.

Due to continuous decrease of the terminal pitches on packages, it is important that PWB designers carefully assess the manufacturing and assembly capabilities of PWB substrates requiring such fine-pitch terminations, to ensure the greatest yields and lowest cost of the product.

2.5.2 Direct Chip Attach

The relentless pressures of size, weight, and volume reduction of electronic products have resulted in a growth in interest in direct chip attach (DCA) methods, where bare IC chips are mounted directly to the substrate. These methods are extensively used on chip-on-board (COB) and multichip module (MCM) assemblies, as shown in Fig. 2.4.

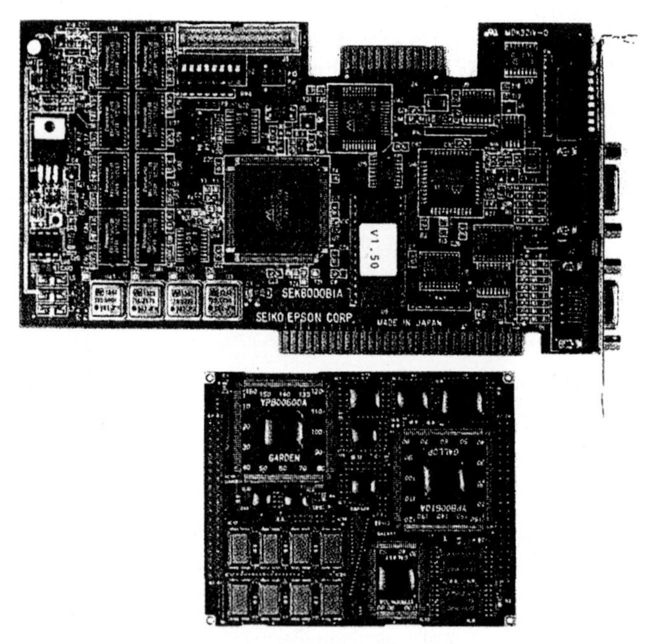

FIGURE 2.4 COB and MCM assemblies.

Three methods of bare chip attachment to the substrate are as follows:

1. *Wire-bonding* is the oldest and the most flexible and widely used method. (More than 96 percent of all chips today are wire-bonded.)
2. *Tape-automated bonding* (TAB) is useful with small I/O pitches and provides the ability to pretest the chips before assembly.
3. *Flip-chipping* is used for its compactness and improved electricals, typical of which is the C4 process of IBM.

The problems of thermal coefficient of expansion (TCE) mismatch between silicon chips that are directly flip-chipped onto a laminate substrate have been effectively eliminated by using a filled epoxy underfill encapsulation technique between the chip and the substrate (see

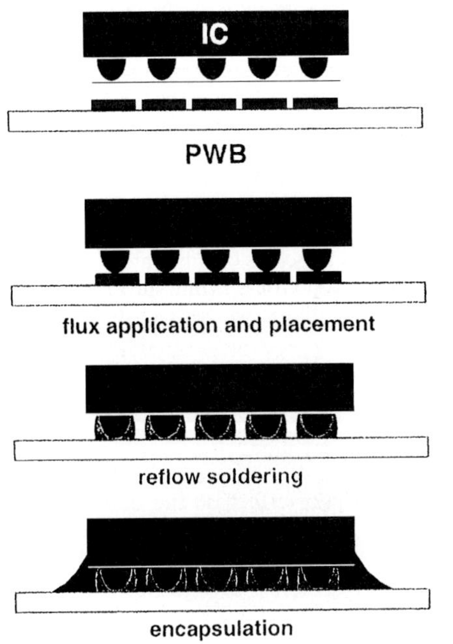

FIGURE 2.5 Underfill between the chip and PWB on flip-chip on board.

Fig. 2.5). This method distributes stresses over the entire area of the chip and thus significantly improves the reliability performance of this assembly method.

While the chips that need area array or flip-chip terminations—i.e., those with high pin counts—are the fastest-growing category of ICs, they still represent only a very small percent of all ICs used. Designers must, therefore, ascertain which of the DCA methods will be the most cost beneficial for a particular application.

2.5.3 Chip-Scale Packages (CSPs)

When mounting unpackaged chips on these interconnecting substrates, it is not always possible to ascertain that only properly operating chips have been assembled. By now, there are a number of methods proposed to solve this known good die (KGD) problem.

As one of the ways to resolve this problem, a number of manufacturers have developed a set of miniature packages, only slightly larger than the chip itself, which protect the chip and redistribute the chip termination to a grid array. These miniature packages permit testing and burning in of chips prior to their final assembly. A typical example of such chip scale packages is shown in Fig. 2.6. There are a number of such packages on the market.

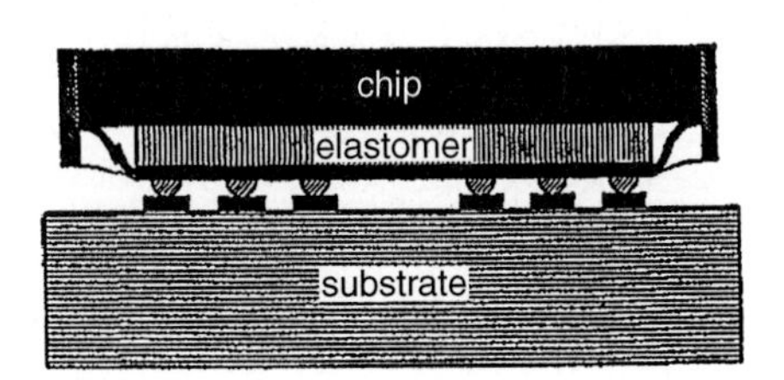

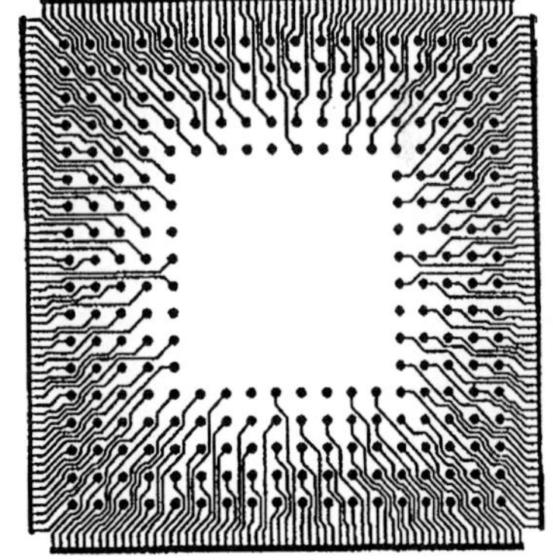

FIGURE 2.6 CSP by Tessera, Inc., San Jose, California.

The designer, however, must analyze the termination pitches of these new CSPs because some use very tight grids, such as 0.5 mm (0.020 in) or less, which need special PWB techniques to permit signal redistribution from these packages to the rest of the board.

In general, the current PWB technology is adequate to provide direct chip terminations if wire-bonding or TAB techniques are used for interconnecting bare chips to the substrate. It requires placing suitable bonding pads spaced by the required pitch in one or two rows around the chip site. While this somewhat reduces the packaging efficiency of the board, it is still an effective method for DCA assemblies.

With grid array, the situation is more difficult because the signals from internal rows of area grid terminations must be routed between the terminals located closer to the edge, which do not permit more than one, or at most two, conductors to pass through. In most cases, these signals from internal rows are brought down into internal layers of MLBs.

The conventional PCB constructions today cannot handle any grid arrays with pitches below 0.020 in, while some flip-chip ball grid arrays go below 0.010-in pitch. In cases when grid distances of the area terminations are below 0.50 mm (0.040 in), special redistribution layers are frequently used, which distribute signals to the conventionally made PTHs in supporting MLB.

Such layers consist of unsupported dielectric layers where small vias or blind holes are formed by laser or plasma etching or are photoformed and then plated using additive or semi-additive metallization processes. While this approach requires some extra area beyond the chip perimeter to complete the signal transfer and increases the costs of substrates, it permits the mounting of flip-chips and CSPs on PWBs. A typical method for forming such redistribution layers, called surface laminar circuit (SLC),[4] has been developed at IBM's Yasu plant.

2.6 DENSITY EVALUATIONS

2.6.1 Component Density Analysis

Because the components and their terminations exert a major effect on the design of the PWB, a number of metrics have been developed to establish the relationships between component density and PWB density. A major analysis of these relationships has been made by H. Holden[5] and some of his charts and derivations are provided here to guide the design engineers during the development of a rational PWB design.

This information is very useful in determining where the designed product will fit in the component density spectrum and what, therefore, is to be expected for PWB density.

Figure 2.7 provides a generalized view of the relationships among the component density, their terminal density, and the necessary wiring density that will be required to accommodate the selected degree of component complexity. The definition of the wiring connectivity W_f is provided.

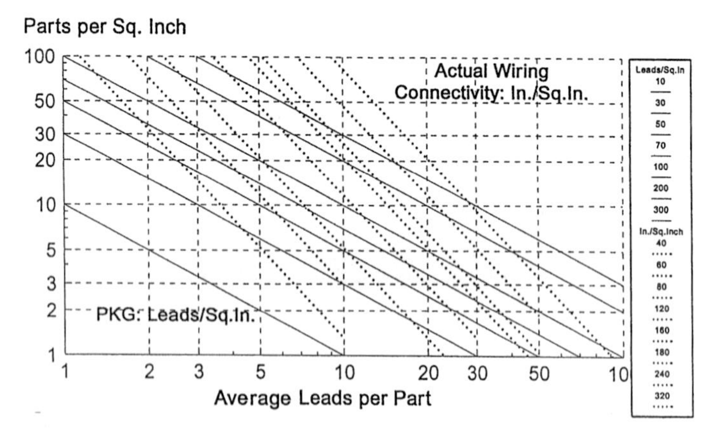

FIGURE 2.7 Plot of general relations between component and wiring density.

2.6.2 PWB Density Metrics

It is essential for the proper design of PWBs to determine the density requirements and then analyze alternative methods of board construction for the most cost-effective design. There are a number of basic terms and equations used for the calculation and analysis of PWB wiring density.

$$W_c = \frac{T*L}{G} \text{ in/in}^2 \tag{2.1}$$

where W_c = wiring capacity
T = tracks per channel
L = number of signal layers
G = channel width

But it is more important to determine the required wiring density that will be sufficient to interconnect all the components on the desired board size. There have been a number of empirically developed equations that permit the calculation of such a wiring demand. The simplest has been developed by Dr. D. Seraphim:[6]

$$W_d = 2.25\, N_t*P \tag{2.2}$$

where W_d = wiring demand
N_t = number of I/Os
P = pitch between packages

2.6.3 Special Metrics for Direct Chip Attach

The assembly of uncased or bare chips on substrates has become popular mostly due to the ability of such assemblies to reduce the area of interconnections. The ideal limit for such assembly would be to place all the chips tightly together, without any space in between. This would result in 100 percent packaging efficiency, a metric measuring the ratio of silicon area to the substrate area. Naturally, such 100 percent efficiency is not achievable, but this metric is still useful in ranking various substrate construction or bare chip attachment methods, as shown in Fig. 2.8.

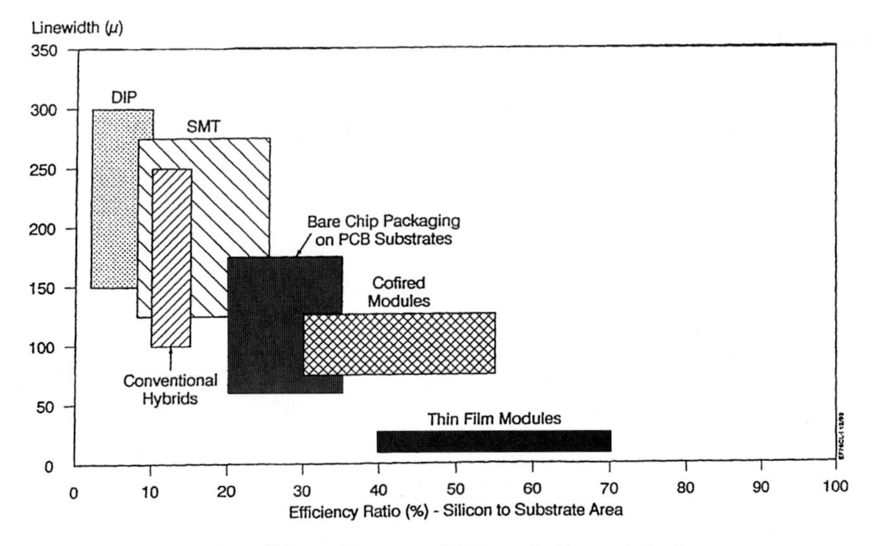

FIGURE 2.8 Packaging efficiency. *(Courtesy of BPA, used with permission.)*

Packaging efficiency of 100 percent is impossible to achieve because all chip-mounting methods require some space around the chips. Even with flip-chips, there must be a distance left between the chips to permit room for the placement tool.

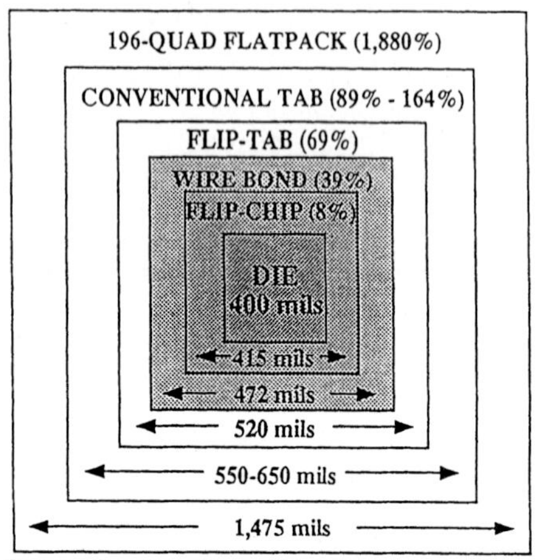

FIGURE 2.9 Chip area required to accommodate bonding methods.

Dr. H. Charles[7] of Johns Hopkins University has listed the dimensions in Table 2.3 for the necessary spacing between the chips (or the total width of the frame around the chips) for various chip attachment methods. These or very similar distances have also been cited by a number of other sources.

Even with the flip-chip mounting, packaging efficiency must be derated to about 90 percent, for wire-bonding to 70 percent and for TABs to about 50 percent, and in some cases much more. A very similar situation is shown graphically in Fig. 2.9. The packaging efficiency deratings shown in Fig. 2.9 are required to accommodate only the wiring bond pads on the substrates. But the mounting of bare chips on PWBs requires additional signal redistribution area to permit placement of larger-diameter PTHs farther out for communication with internal layers. It is evident that the packaging efficiencies on PWBs could be reduced to the range of only 20 to 30 percent, unless special surface signal redistribution layers (as previously mentioned) are used, which are made of unreinforced dielectric material. In such cases, packaging efficiency and the chip-to-chip distances will again be similar to the values cited in Table 2.3.

TABLE 2.3 Spacing Required Between Chips

Attachment method	Chip spacing, mils
Flip-chip	15–20
Wire-bonding	70–80
Flip TAB	100–120
Regular TAB	150–400

It is apparent that direct chip attach on PWBs will result in the significant reduction of the packaging efficiency of such assemblies, except for the fact that components can be mounted on both sides of the PWB substrate. It has been shown that wire-bonding can be done on both sides of a PWB with some special fixturing; also, outer lead bonding (OLB) of TABs can be done on both sides of the PWB substrate. Thus, while single-sided bare chip assembly on PWBs reduces its packaging efficiency to about half that of other types of substrate constructions, the ability to place components on both sides of PWBs brings it back to the same packaging efficiency level as others.

2.7 METHODS TO INCREASE PWB DENSITY

There are three basic ways to increase the connectivity or available conductor capacity of PWBs:[8]

- Reduce hole and pad diameters
- Increase the number of conductive channels between pads by reducing the widths of the conductors
- Increase the number of signal planes

The effect of each approach on manufacturing yields, and thus on board costs, will be discussed in sequence. It should be noted that the last option is the simplest but the most costly solution, and thus should be used only after the methods suitable for resolving the first two conditions have been proven inadequate for achieving the desirable board density.

2.7.1 Effect of Pads on Wiring Density

The major obstacles preventing increase of conductor channel capacity are large pad diameters around the plated through-holes (PTHs), since, at the present state of technology, PWBs still require pads wider than the conductors at their location. These pads reduce the obtainable connectivity of PWB boards and must be accounted for in a proper analysis of interconnection density I_d. For instance, in one design, the reduction of pad diameters from 55 to 25 mils (by 55 percent) doubled the interconnection density, while the reduction in conductor pitch C_p from 18 to 7 mils (by 61 percent) increased it only by 50 percent. It is obvious that the reduction of pad diameters, or their total elimination, could be a more efficient way to increase the wiring capacity of complex PWBs.

The purpose of copper pads surrounding the drilled holes in PWBs is to accommodate any potential layer-to-layer or pattern-to-hole misregistrations and thus prevent any hole breakout outside the copper area of the pads. This misregistration is caused mainly by the instability and movement of the base laminate during its processing through the PWB or multilayer board (MLB) manufacturing steps.

The base material standards specify that such movement be limited to a maximum of 300 ppm, but the actual base material excursions are closer to 500 ppm, producing 10 mils of layer movement within a 20-in distance. For many applications this tolerance is too wide, as it requires at least a 10-mil-wide annular ring around drilled holes, resulting in considerable conductor channel blockage.

Another cause of material instability in MLBs is the excessive material movement that occurs if the laminating temperature exceeds the glass transition temperature T_g of the laminate resin. On the other hand, if the laminating temperature remains below the T_g of the resin, there is minimal dimensional variation of the base material, as the resin is still in its linear expansion phase. This explains the need for use of high-T_g resins in the PWB industry.

The data obtained from the performance of new, more stable unidirectional laminates indicate that the base material movement is reduced, for instance to 200 ppm from 500 ppm, and the requirements for the annular ring width will be reduced to 4 mils from 10 mils.

Table 2.4 illustrates the connectivity gains made possible when a more stable laminate material is used, permitting a reduction in the initial diameters of the pads (as given in the first column) spaced at 2.5 mm (0.100 in) while keeping the conductor pitches constant. The most effective use of the signal plane area is achieved when the pads are eliminated and the z-axis interconnects are confined within the width of the conductors forming the invisible vias.

This derivation is based on actual data obtained from the performance of new, more stable, unidirectional laminates. While MLBs using these new, more dimensionally stable, unidirectional laminates with reduced pad diameters could be manufactured by conventional manufacturing methods, the production of MLBs with invisible vias requires the use of a sequential manufacturing process similar to the SLC process previously described.

PWB manufacturers are reasonably comfortable with the production of boards with 4- or 5-mil-wide conductors, but they still require large pads around plated holes to ensure against hole breakout. This limits the currently available wiring density to about 40 to 60 in/in^2 per plane, as seen from Table 2.4. A technology that will permit PWB manufacturers to fabricate invisible vias could increase the connectivity per PWB signal plane from this current range to the level of 100 to 140 in/in^2. Conductor widths of 0.002 in will offer a PWB of 200 to 250 in/in^2 per signal plane.

TABLE 2.4 Effect of Pad Diameters on Interconnectivity Density

Pad dia, in	Cond pitch, in	I_d @ 500 ppm, in/in^2	I_d @ 200 ppm, in/in^2	I_d @ invisible via, in/in^2
0.055	0.010	20	37	55
0.036	0.018	30	48	55
0.025	0.009	40	96	100
0.025	0.007	60	130	143

TABLE 2.5 Effect of Increased Connectivity on Reduction of Layers

Pad dia, in	Cond pitch, in	I_d @ 500 ppm, layers	I_d @ 200 ppm, layers	I_d @ invisible via, layers
0.055	0.010	10	6	4
0.036	0.018	7	4	4
0.025	0.009	5	2	2
0.025	0.007	4	2	2

Table 2.5 illustrates the most important result of increased connectivity per layer: a reduction in the number of signal layers needed to provide the same wiring density W_d. Table 2.5 was constructed by applying connectivity data from Table 2.4 to a 50-in² MLB with total wiring length of 10,000 in. Note also that the layer count in Table 2.4 has been brought up to the next higher full-layer value, i.e., the calculated 1.4 layers have been recorded as 2 layers.

The major benefit of such a reduction in the layer count is that it can result in a significant reduction of the manufacturing cost while providing the same total interconnection length.

2.7.2 Reduction of Conductor Width

An obvious method of increasing the connectivity of PWBs is to reduce the widths of conductors and spaces and thus increase the number of available wiring channels on each signal plane, as described previously. This is the direction that has been used in the IC and PWB industries for many years. However, it is impossible to decrease conductor widths or spaces indefinitely. The reduction of the conductor width is limited by the current-carrying capacity of thin, small conductors, especially when these conductors are long, as they frequently are on PWBs. There are processing limits to this conductor reduction, since manufacturing yields may plummet if the reduction stretches the process capabilities beyond their normal limits.

There is also a limit to the reduction of the spaces between the conductors, governed mainly by electrical considerations, i.e., by the need to prevent excessive cross talk, to minimize noise, and to provide proper signal propagation conditions and characteristic impedance.

Still, such conductor reductions, if achieved within the described limits, can be an effective path for increasing the PWB density and the reduction of PWB manufacturing costs. As seen from Table 2.6, constructed from cost data derived from the Columbus program of BPA, the reduction of conductor widths from 6 to 3 mils halves the number of signal layers necessary to

TABLE 2.6 Effect of Conductor Widths on Number of Layers and Board Cost for a 6-in × 8-in MLB, with $I_d = 450$ in/in², 65 to 68 Percent Yields

Line-space	Total no. of layers	No. of signal layers	Board cost, %
3–3	8	4	55
4–4	10	6	64
5–6	12	7	77
5–7	14	8	87
6–6	16	8	90
7–8	20	10	100

ensure the same connectivity (while their yields, interconnection density, and board area were kept constant). This reduction in the number of layers can significantly reduce the manufacturing costs of PWB boards.

2.7.3 Effect of Conductor Widths on Board Yields

It is obvious that any successful increase of conductor density I_d in PWBs would be effective only if the processes exist that permit manufacture with reasonable yields. Unfortunately, the yields of thin conductors in PWBs fall rapidly as their widths are reduced below 5 mils, as shown in Fig. 2.10. Therefore, the understanding of manufacturing yields is very important for analysis of the most cost-effective manufacturing process, because the process yields have a major effect on the cost of interconnection substrates.

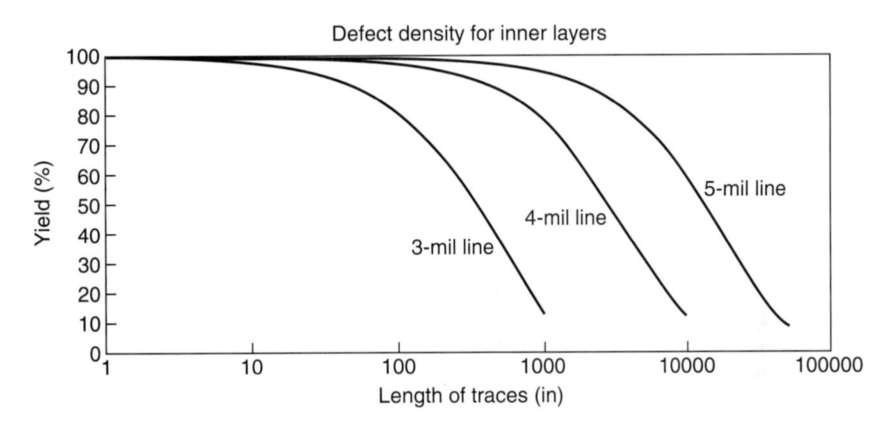

FIGURE 2.10 Board yields vs. conductor width.

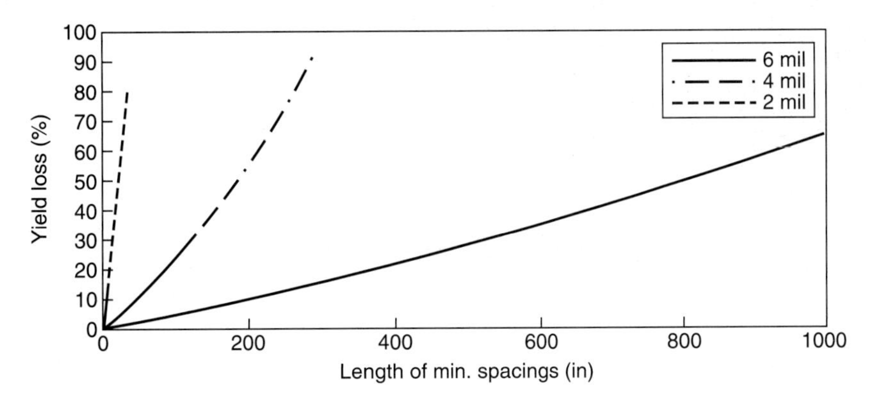

FIGURE 2.11 Yield loss from shorts.

A useful empirical equation for calculating the manufacturing cost is:

$$\text{Cost } C = \frac{(\text{material} + \text{process costs})}{\text{Yield } Y} \tag{2.3}$$

To establish the effect of the interconnection density I_d on the final yield of substrates, the total processing yield can be split into two components: one that depends on the conductor density, i.e., Y_{Id}, and the second, which is a function of the combined yields of the rest of the manufacturing processes:

$$Y_{\text{total}} = Y_{Id} * Y_{\text{proc}} \tag{2.4}$$

In a well controlled manufacturing operation, the process-dependent yields (such as plating) remain fairly constant for a given technology, permitting the yield function to be based solely on the changes in thc conductor widtlıs.

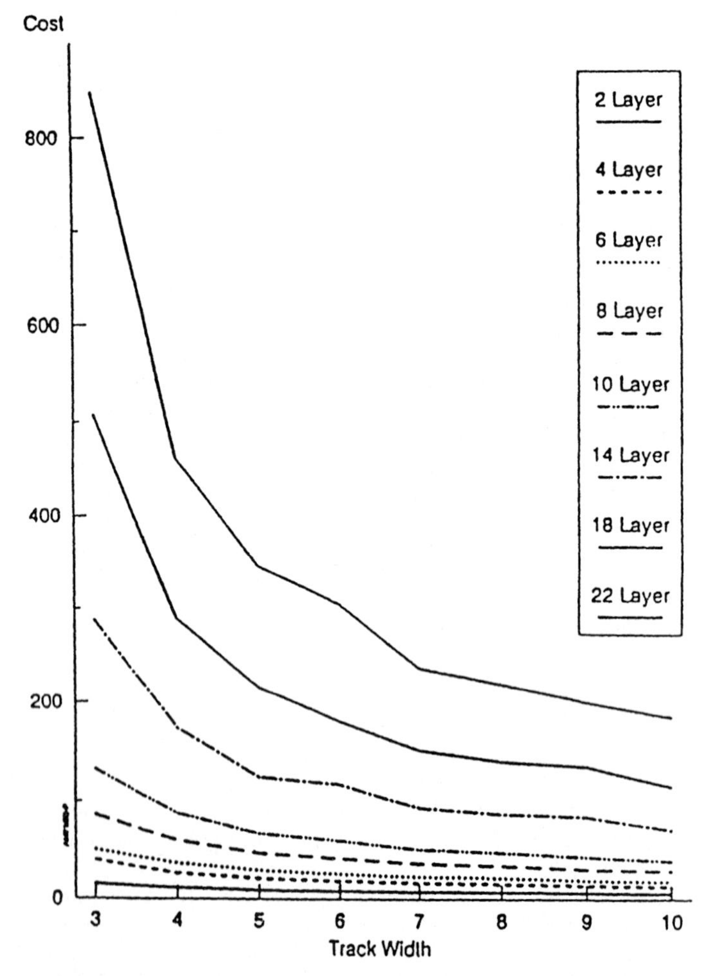

FIGURE 2.12 Cost relationships between number of layers and conductor widths.

As seen from Fig. 2.11, the defects that affect this density-dependent yield function Y_{Id} are conductor opens and shorts between them. It would be reasonable to assume that such defects have a Poisson distribution over the total length TL of conductors of a substrate, with an average defect frequency of v. The yield is the probability of zero defects ($n = 0$) in the total conductor length TL. Thus,

$$Y = (\text{at } n = 0) = e^{(-v^*\text{TL})} \quad \text{(Poisson distribution)} \tag{2.5}$$

As seen from Figs. 2.10 and 2.11, the defect frequency v depends also on the widths of lines and spaces, i.e., on the conductor pitch C_p. With a decrease of C_p, v will increase, but for very large C_p, v should be 0, since Y_{Id} will be 100 percent.

For instance, in the case of a design using invisible pads, where $C_p = 2w$, the interconnection density I_d can be expressed as $I_d = \text{TL/A}$, and I_d is proportional to C_p, i.e., $I_d * C_p = 1$, and $\text{TL} = A/C_p$. Therefore, v in this equation can be empirically expressed as:

$$v = -\ln \frac{Y_0}{\text{TL}_0} * \left(\frac{C_{p0}}{C_p}\right)^b \tag{2.6}$$

where b is an exponent dependent on the technology or process used to form the conductors. This exponent b varies considerably from facility to facility and among various pattern formation methods, and must be empirically determined for each case.

2.7.4 Increase in Number of Conductor Layers

This is the simplest and most straightforward solution: when there is insufficient room on existing layers to place all the necessary interconnecting paths, add a layer. This approach has been widely practiced in the past, but when cost effectiveness of the substrates is of paramount importance, a very careful design analysis must be made to minimize layer counts in MLBs, because there is a significant cost increase with every additional layer in the board. As seen from Table 2.6, calculated for 6- × 8-in MLBs produced in large quantity with yields and conductor density kept constant, there is almost a linear relationship between board costs and layer count.

Table 2.6 also shows that any increase in the number of signal layers in boards operating at frequencies requiring transmission line characteristics will double the total number of layers, due to the need to interleave ground or DC power planes between signal planes.

A typical example of the effect of layer count on the finished MLB yield can be seen from Fig. 2.12, prepared some years ago by BPA. We can see that there is a definite decrease in the manufacturing yields with an increased number of layers in any of the linewidth categories. This is rather a typical situation in board manufacturing because increased complexity and thickness of MLB with a higher number of layers usually leads to a larger number of problems on the production floor.

REFERENCES

1. Toshiba, "New Polymeric Multilayer and Packaging," *Proceedings of the Printed Circuit World Conference V,* Glasgow, Scotland, January 1991.
2. The Institute for Interconnecting and Packaging Electronic Circuits, 7380 N. Lincoln Ave, Lincolnwood, IL 60646.
3. IPC-SM-782, "Surface Mount Design and Land Pattern Standard," The Institute for Interconnecting and Packaging Electronic Circuits.
4. Y. Tsukada et al., "A Novel Solution for MCM-L Utilizing Surface Laminar Circuit and Flip Chip Attach Technology," *Proceedings of the 2d International Conference on Multichip Modules,* Denver, CO, April 1993, pp. 252–259.

5. H. Holden, "Metrics for MCM-L Design," *Proceedings of the IPC National Conference on MCM-L,* Minneapolis, MN, May 1994.
6. D. Seraphim, "Chip-Module-Package Interface," *Proceedings of Insulation Conference,* Chicago, IL, September 1977, pp. 90–93.
7. H. Charles, "Design Rules for Advanced Packaging," *Proceedings of ISHM* 1993, pp. 301–307.
8. G. Messner, "Analysis of the Density and Yield Relationships Leading Toward the Optimal Interconnection Methods," *Proceedings of Printed Circuits World Conference VI,* San Francisco, CA, May 1993, pp. M 19 1–20.

CHAPTER 3
SEMICONDUCTOR PACKAGING TECHNOLOGY

John W. Stafford
JWS Consulting PLC, Phoenix, Arizona

3.1 INTRODUCTION

A revolution has occurred in the electronics industry due to advances in semiconductor design and manufacturing, the packaging of semiconductor die, the packaging of electronics systems, and product physical design.

The driving force in this revolution originates with the advances that have occurred in integrated circuit (IC) technology and the levels of integration obtained. The initial driver to this was the development of a micrometer-level lithographic capability.[1] Semiconductor packaging and printed circuit board (PCB) evolution tracks the advances in IC technology. Figures 3.1 and 3.2 show the trend of chip transistor density and the increase in chip frequency versus time. There is concern that, with the current semiconductor technology advances, by about 2010 we may begin to approach significant process and technology obstacles. As a consequence of being able to put more circuitry on a silicon IC, its package size and the number of package input/output (I/O) pins have increased, as has the wiring density required of the medium that interconnects the packaged ICs.

This continuing thrust for higher levels of integration has forced an ongoing effort for smaller and cheaper means of packaging these ICs so they can be interconnected in a cost-effective manner that does not degrade the electrical performance of the assembled circuit. As a result, high-performance systems require consideration of both the IC design and its packaged format and the design of the interconnect that connects the ICs.

Table 3.1 compares the computing capability available over the years. What has occurred is that there has been a constant increase in functionality (i.e., instructions per second) along with continuous decreases in the cost per instruction. The results of this trend are the ever increasing functionality of portable wireless communications, such as pagers and cellular phones, and their reduction in size to minimum ergonomic standards.

Worldwide in 1998 there were some 308 million cell phone subscribers; the number grew to 475 million in 1999. It is estimated that in 2001 there were 1000 million (i.e., 1 billion) cell phone subscribers. From 2000 to 2001, more than half a billion cellular phones were manufactured, and these are now considered a commodity product. This same technology trend is spawning new wireless products for local area networks. These applications will allow cellular

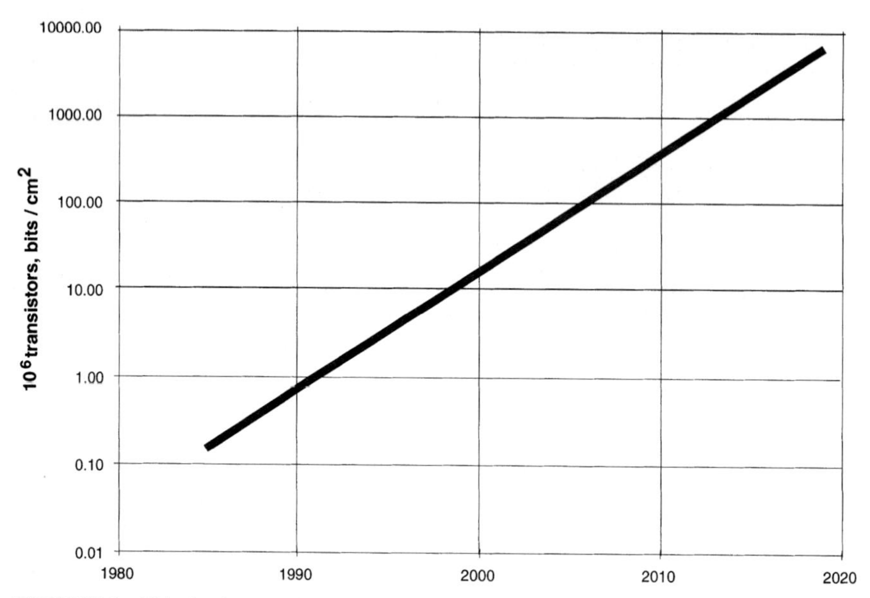

FIGURE 3.1 Chip density trend for logic and DRAM components.

phones, personal computers, etc. to exchange data (including video) using appropriate radio frequency (RF) protocols (Bluetooth operating in the 2.4-GHz frequency range is one such protocol) if they are in proximity to one another.

The requirements of the Internet for increasing functionality and bandwidth and lower cost will require new concepts for electronic physical design as well as the incorporation of optical components and interconnects. These trends will continue indefinitely.

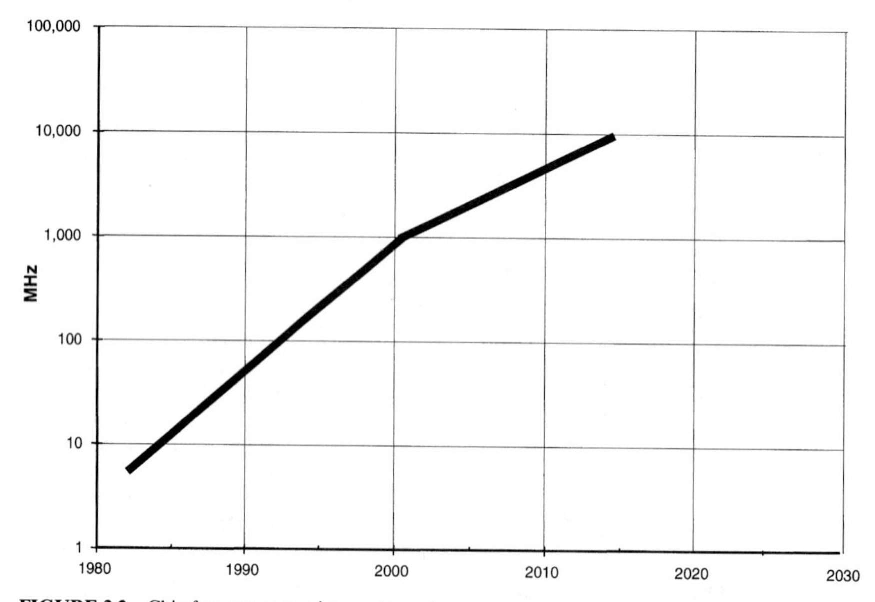

FIGURE 3.2 Chip frequency requirement trend.

TABLE 3.1 Cost Comparison

	Approx. no. of instructions/s	Price	Cents/ instruction
1975 IBM mainframe	10,000,000	$10,000,000	100
1976 Cray 1	160,000,000	$20,000,000	12.5
1979 Digital VAX	1,000,000	$ 200,000	20.0
1981 IBM PC	250,000	$ 3,000	1.2
1984 Sun Microsystems 2	1,000,000	$ 10,000	1.0
1994 Pentium PC	66,000,000	$ 3,000	0.0045

Source: New York Times, April 20, 1994.

3.1.1 Packaging and Printed Circuit Technology Relationships

The packaging density trend lines (package area/die area) are summarized in Fig. 3.3. The area efficiencies of array package concepts are evident and have been maintained over a period of 20 years.

It is apparent, then, that the trends for new semiconductor chip packaging are increasing the number of package I/Os. Area array packages for high- I/O semiconductors have emerged to minimize package size and enhance package electrical performance (i.e., lower lead inductance). The impact of this will be pressure on PCB fabricators to minimize linewidth and space in order to escape from high- I/O area array packages or direct chip attach (DCA) semiconductors. In 1986, for PCB interconnects, 6-mil lines and spaces were nominal; in 1992, 8-mil lines and spaces were nominal; in 1995, 6-mil lines and spaces were nominal, and in 2000 4-mil lines and spaces are nominal to meet the interconnect requirements of commercial hand-held electronics products.

To meet the needs for a finer-pitch circuit board technology for ball grid array (BGA) packages, chip-scale packages (CSPs), and high-density circuit cards, a new circuit board

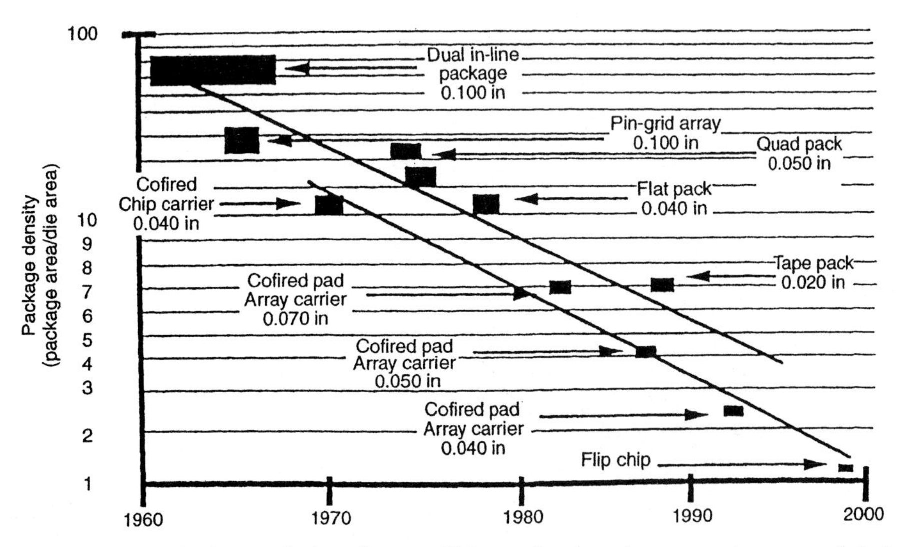

FIGURE 3.3 Packaging trends show the area efficiency of array package concepts over a period of 20 years.

technology called high-density interconnect, capable of 1.58-mil (40-μm) lines and spaces have been developed.

3.1.2 Electronic Packaging Issues and Concerns

Electronic packaging begins where circuit design leaves off.

3.1.2.1 Physical Design and Packaging Issues. The issues involved in electronic physical design and packaging are as follows:

- Selection of appropriate electronic components (i.e., semiconductors, discrete, and passives)
- Mechanical layout and assembly of components, interconnectors, and cases
- Production engineering/technology
- Electrical parameters of interconnects (controlled-impedance design, cross talk, clock skew, signal propagation delay, electromagnetic interference for RF circuits, etc.)
- Thermal conditions (heat dissipation, cooling, etc.)

3.1.2.2 Digital Circuit Design Considerations. Digital circuits should be able to do the following:

- Transfer a complete logic swing in shortest time
- Have the interconnect characteristic impedance designed to equal the load impedance
- Have characteristic impedance that is purely resistive to minimize reflections
- Accommodate clock skew

For the interconnection of digital semiconductor devices, a major issue that must be considered is clock skew, which results from varying the length of clock lines and is a major design consideration for high-speed products.

3.1.2.3 Analog Circuit Design Considerations. Analog circuits should be able to do the following:

- Maximize power transferred from input to output
- Make the driver impedance a complex conjugate on the transmission line

3.1.2.4 Power Issues for Silicon Semiconductors. The following items are power considerations for silicon semiconductors:

- Transistor-transistor logic and complementary metal oxide semiconductor (CMOS) power dissipation depend on frequency and increase dramatically at high frequencies.
- Drain of an emitter collector logic device is independent of frequency for a given load.
- Resistance loading drains the output capacitor of CMOS circuitry.
- Terminating CMOS circuits to control reflection imposes a power penalty.

Gallium arsenide (GaAs) semiconductors are now technologically competitive with silicon,[2] especially for high-speed logic applications. For GaAs logic, power dissipation is independent of frequency, and GaAs circuit operation is unaffected by power supply voltages down to about 1 V.

3.1.2.5 RF Semiconductors. GaAs bipolar CMOS and silicon germanium are semiconductor technologies now being used in RF circuits of many wireless products. Special attention is required in the packaging of RF components to control the electrical parasitics, the thermal resistance for components such as power amplifiers, etc.

3.1.3 Requirements for Electronic Systems

The requirements for electronics systems and products driven by semiconductor technology developments are as follows:

1. The advances in integrated semiconductor technology mean products operate at higher speeds and have higher performance and greater functionality.
2. The reliability and quality of products are givens and are expected to be built in at no cost premium.
3. The volume (i.e., size) of the electronics products is diminishing, and is constrained only by ergonomic requirements and the ability to dissipate heat (i.e., power).
4. The costs of the components and assembly are expected to continuously decrease with time.
5. The time to market impacts all of the preceding items.

3.2 *SINGLE-CHIP PACKAGING*

Prior to 1980, the semiconductor package predominately used was the dual inline package (DIP). The package is rectangular in shape with leads on a 0.100-in pitch along the long sides of the package.

Figure 3.4 shows the various semiconductor formats and package trends. The packages on the left side of Fig. 3.4 are essentially perimeter I/O packages—i.e., DIPs, quad flat packages (QFPs), plastic leaded chip carriers (PLCCs, tape automated bonding (TAB), etc. The package types on the lower right side of Fig. 3.4 represent area array packages such as pin grid array (PGA) packages (an array of pins attached to the package base for electrical connection), land grid array packages (an array of conducting pads on the package base for electrical connection, sometimes called pad array carriers [PACs], or, when the lands have reflowed solder balls attached, ball grid array [BGA] packages), and multichip modules. The pitch of the I/Os of DIPs and PGA packages is 0.1 in, while the I/O pitches of the balance of the parts

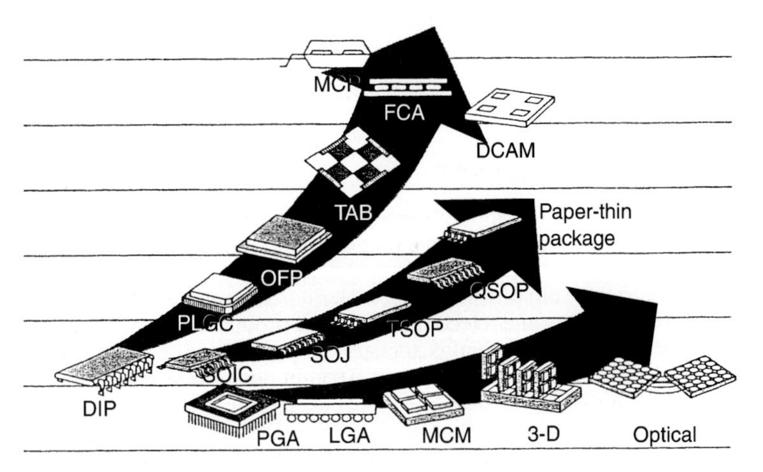

FIGURE 3.4 Integrated circuit packaging trends.

are 0.060 in or less (i.e., 0.050 in, 0.5 mm, 0.4 mm, 0.3 mm). Array-type packaging concepts have emerged to provide higher package electrical performance and/or lower packaging densities (package area/die area).

3.2.1 Dual Inline Packages (DIPs)

Figure 3.5 shows the configuration of a DIP. DIPs are available with a cofired ceramic body with the leads brazed along the long edges or in a postmolded construction where the die is bonded to a lead frame and gold wires interconnect the chip to the lead frame leads prior to molding of a plastic body around the lead frame. DIPs are limited to 64 or fewer I/Os.

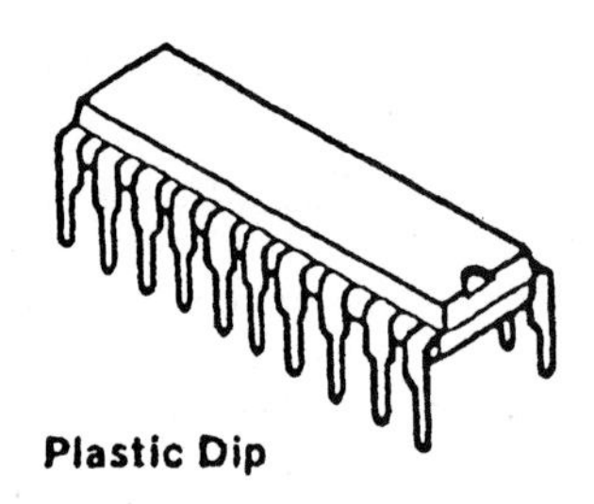

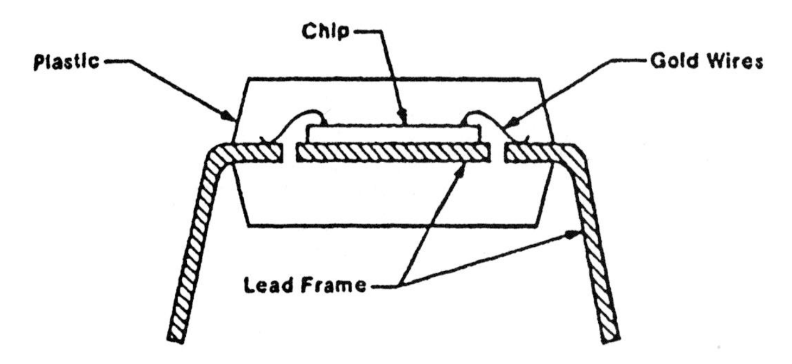

FIGURE 3.5 Dual inline package.

3.2.2 Leadless Ceramic Chip Carriers

To improve the form factor, packages for commercial and military applications were developed called leadless ceramic chip carriers, consisting essentially of the cavity portion of the ceramic hermetic DIP (see Fig. 3.6) with solderable lands printed onto the bottom of the leadless ceramic chip carrier package. These parts were assembled on ceramic substrates and used in both military and telecommunications products. Almost concurrently, leaded versions of the leadless packages begin to appear. The pitches of the leadless parts were 0.040 and 0.050 in, while the leaded parts were on a 0.050-in pitch. Reference 3 discusses these developments in detail. By 1980, thrust-to-quad surface-mount packaging had begun, with the emphasis on leaded plastic quad packages.

Reference 4 shows that by 1993 there was a dramatic swing away from through-hole-mounted parts (i.e., DIPs) to surface-mount packages. In 1993, 50 percent of the semiconductor

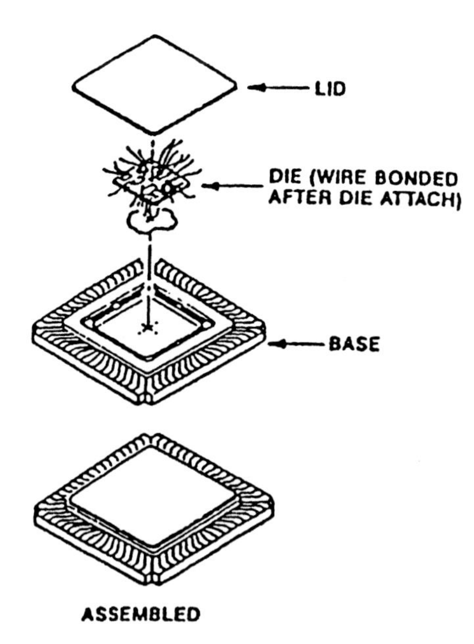

FIGURE 3.6 Leadless ceramic chip carrier.

packages fabricated were through-hole-mounted DIPs. By 2000, this percentage dropped to 30 percent, and by 2005 it is projected to drop to only 15 percent.

3.2.3 Plastic Quad Flat Package (PQFP)

The driver for the surface-mount plastic packages has been the development of the plastic quad flat package (PQFP), which consists of a metal leadframe with leads emanating from all four sides. The leadframe is usually copper, to which the semiconductor die is "die bonded" (usually epoxy die bonded). The I/Os of the die are connected by wire bonds to the leadframe leads. The conventional method of wire-bonding is thermosonic gold ball wedge bonding. A plastic body is then molded around the die and the leads are trimmed and formed. Figure 3.7(*a*) shows a cross-sectional view of a PQFP.

PQFPs have their leads formed in a gull-wing fashion (see Fig. 3.7), while PLCCs have their leads in the shape of a J, which are formed (i.e., folded) underneath the package.

Figure 3.8 shows the lead pitch and pin count limit versus QFP size and lead pitch. QFPs are in production and readily used in the assembly of product with 0.5-mm pitch. Based on molding capability and impact of lead length on electrical performance,

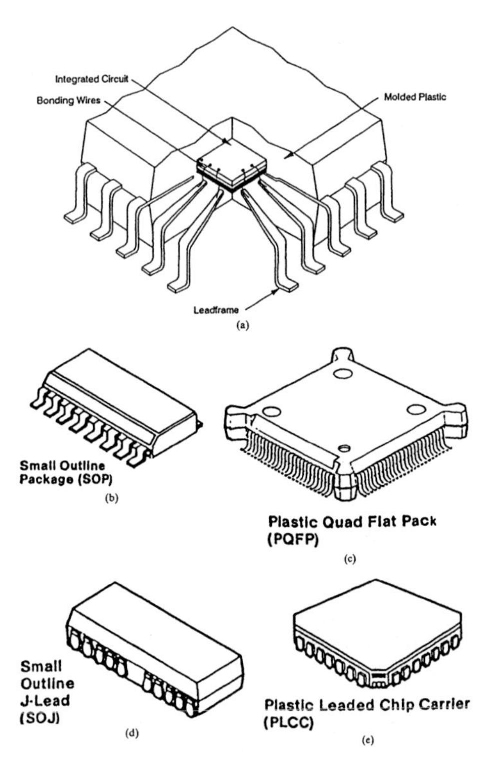

FIGURE 3.7 Surface-mount plastic package types.

a molded body 30 mm on a side is thought to be the practical limit. QFPs with 0.5-mm pitch based on the preceding are limited to around 200 I/Qs. QFPs with 0.4-mm pitch have been implemented.

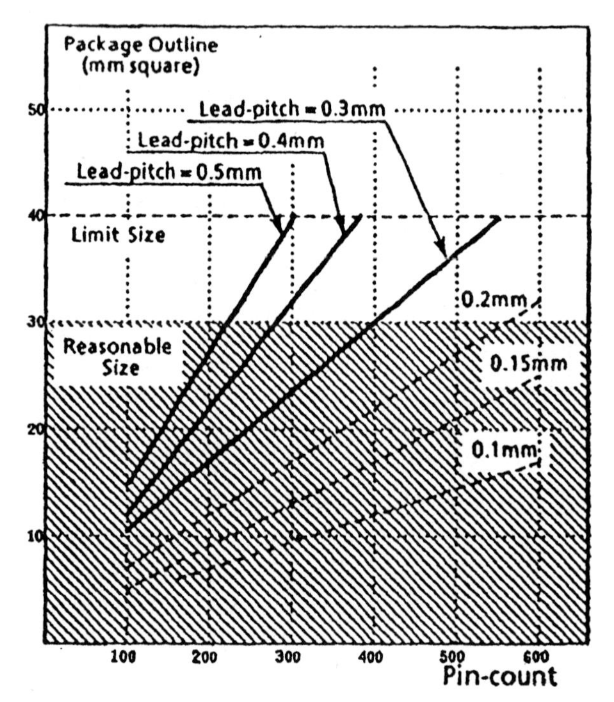

FIGURE 3.8 Lead pitch and pin count limit vs. QFP size and lead pitch.

Figure 3.7 shows the variety of surface-mount plastic packages that have been developed based on PQFP technology.

Ceramic and plastic QFPs, as well as PLCCs, are used to package gate array and standard cell logic and microprocessors. Small-outline IC and small-outline J-lead packages are used to package memory (SRAM and DRAM) as well as linear semiconductors. Pin count for all package types is limited only by molding capability and the demand for ever thinner molded packages.

3.2.4 Pin Grid Array (PGA) and Pad Array Carrier (PAC)

Consider the impact of using a perimeter I/O package versus an area array package. Figure 3.9 illustrates the differences between a perimeter array package (leadless chip carrier) and an area array package (PAC). Figure 3.10 shows the relation between the package area versus I/O for perimeter and area array packages. It is clear from Fig. 3.10 that for semiconductors with more than 100 I/Os, PGA and PAC packages have become increasingly attractive for packaging very-large-scale semiconductor ICs and ultra-large-scale semiconductor ICs. The scalable limit is determined only by fatigue issues of the solder connection or joint.

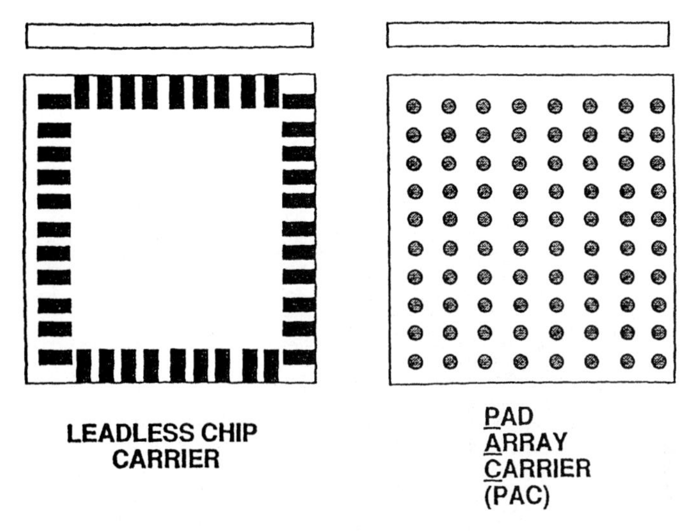

FIGURE 3.9 Perimeter I/O package vs. area array package.

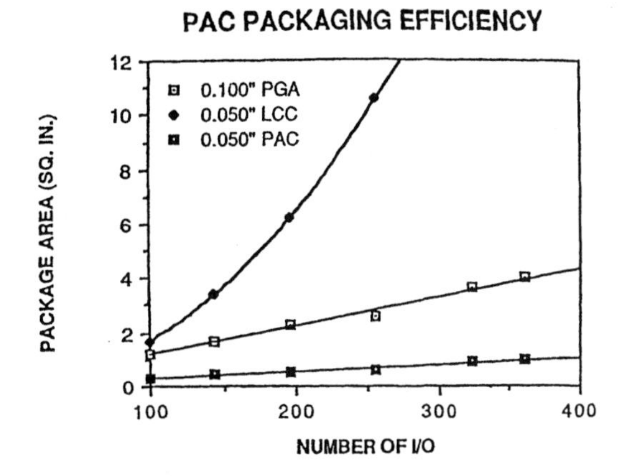

FIGURE 3.10 PAC packaging efficiency

3.2.4.1 Perimeter Array I/O Package Advantages and Disadvantages

- Perimeter array I/O packages at 0.050-in pitch can be readily surface-mount assembled. Both 0.4-mm pitch perimeter I/O QFPs and 0.3-mm pitch perimeter I/O QFPs are now in manufacture. Reviewing Fig. 3.8, which shows QFP size versus lead pitch and pin count, it is apparent that the usefulness of QFPs is limited to around 400 I/Os.
- There still is a controversy in the packaging and assembly community as to whether 0.3-mm pitch QFPs, particularly with a large number of I/Os, are a high-throughput, high-assembly-yield part due to the fragility of the lead (0.015 mm wide) and the possibility of solder shorts between leads.

3.2.4.2 Pad Array and BGA Packages. This disconnect is being addressed first by pad array and BGA package technology. References 5 and 6 provide details of this emerging technology, particularly as it relates to low-cost plastic array packages with reflowed solder balls attached to the package array I/Os. Figure 3.11 shows a ceramic BGA (CBGA) and Fig. 3.12 shows a cross section of the plastic BGA.

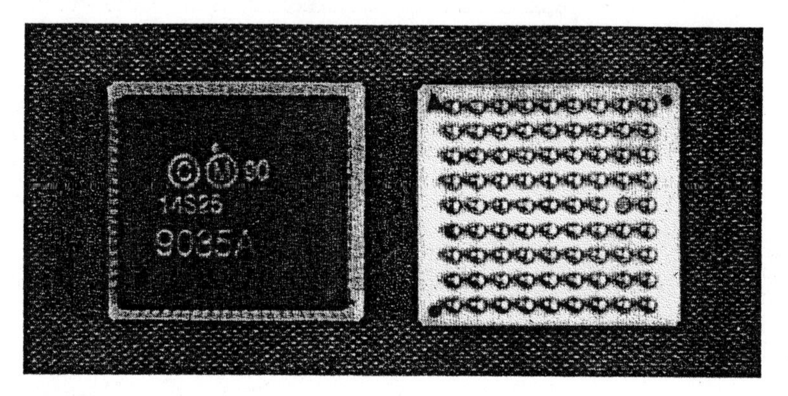

FIGURE 3.11 Ceramic ball grid array.

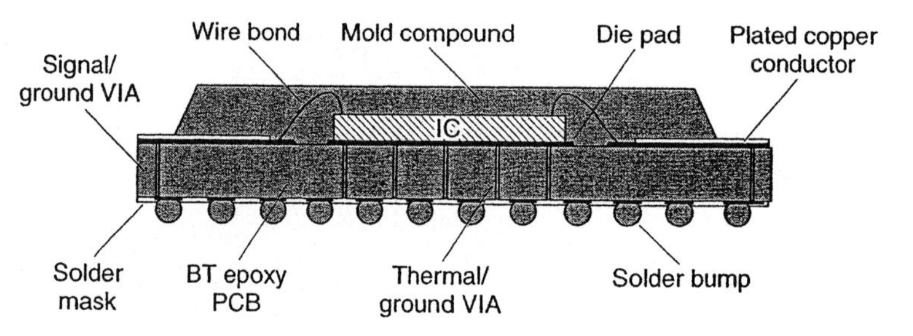

FIGURE 3.12 Ball grid array cross section.

The advantage of ball grid array packages (ceramic or plastic) include the following:

- The package offers a high-density interconnect. Pad array pitches of 1.27, 1.0, 0.8, and 0.5 mm are now commercially available.
- The packages have achieved six-sigma soldering (demonstrated for 1.27- and 1.00-mm I/O pitches) because of the large volume of solder on the I/O pad.
- The package is a low-profile part (package thicknesses as low as 1 mm are now available).
- The package has superior electrical performance in that the total lead length is short, controlled impedance interconnects can be designed in, and low-dielectric constant and low-loss materials can be used for the substrate.
- The package depending on the design has the potential for superior thermal performance.
- The package concept is extendable to multichip packages (MCPs).

Table 3.2 gives the expected ranges of I/O for BGA packages for high-end microprocessors and for dies used in portable products. In 2000, approximately 60 percent of all BGA packages had 1.27- or 1.00-mm pitch and the balance had 0.8- or 0.5-mm pitch. By 2004, 60 percent of all BGA packages will have 0.8- or 0.5-mm pitch, with a small number having less than 0.5-mm pitch.

TABLE 3.2 BGA Pin Counts

Year	2001	2005	2010
Pin count, high-end logic	700–1000	1000–1900	1200–4000
Pin count, portable products	256	312	360

The area efficiency of BGA packages that have a relatively low pad count (i.e., less than 140) has increased in the last several years. This new form of the BGA package, called chip-scale package (CSP), has an area efficiency of around 80 percent (i.e., 80 percent of the package area comprises the silicon die area). The pad pitch on CSPs range from 0.8 to 1 mm. CSPs should be considered a variant of the BGA package technology. Figure 3.13 shows a typical CSP package.

Amkor Tape CSP

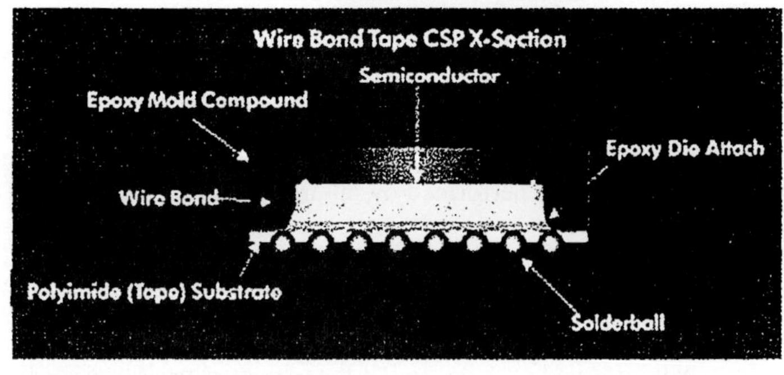

FIGURE 3.13 Typical CSP technology. (*Courtesy of Amkor Inc.*)

A new and smaller package called a wafer-scale package (WSP) has been developed. Figure 3.14 shows such a package. The area efficiency of this package is 100 percent (i.e., 100 percent of the package area comprises the silicon die area). A WSP is a package where all the packaging processes are completed on the silicon wafer, including application of the I/O solder balls. The WSP has relatively low I/O count and has pad pitches ranging from 0.8 to 1 mm. Generally, a redistribution of the chip I/Os uniformly over the die area is required to obtain these pitches. The WSP requires no chip under encapsulation, as is required by direct chip attach (DCA), discussed in Sec. 3.2.5.

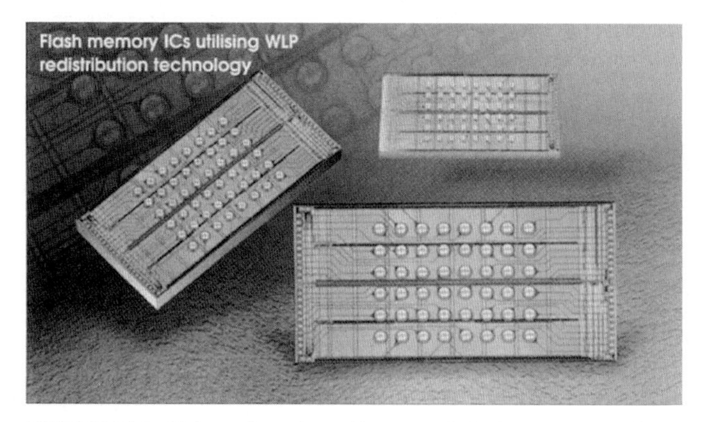

FIGURE 3.14 Wafer-scale package. (*Courtesy of Flipchip Technologies.*)

3.2.5 Direct Chip Attach (DCA)

The next step down from area array packages is DCA. The methods of attaching a semiconductor die directly to an interconnect board (PCB, multilayer ceramic, etc.) are:

- Die bond/wire-bond
- TAB
- Flip-chip bonding

An exhaustive discussion of wire-bond, TAB, and controlled-collapse chip connection (C4) or solder-bumped flip-chip-to-board interconnect technology can be found in Chap. 6 of Ref. 7. Figure 3.15 illustrates these chip interconnect methodologies, which will be discussed in the following text.

3.2.5.1 Die Bonding to Printed Circuit. The preferred method of die bonding and wire-bonding is epoxy die bonding to the interconnect (i.e., PCB or multilayer ceramic) and gold ball-wedge wire-bonding. One of the advantages of gold ball-wedge wire-bonding is that a wedge bond can be performed on an arc around the ball bond. This is not true for wedge-wedge wire-bonding.

Wire bonds can be made with wire diameters as small as 0.8 mil. Thermosonic ball-wedge bonding of a gold wire, shown in Fig. 3.15(*e*), is performed in the following manner:

1. A gold wire protrudes through a capillary.
2. A ball is formed over the end of the wire by capacitance discharge or by passing a hydrogen torch over the end of the gold wire.

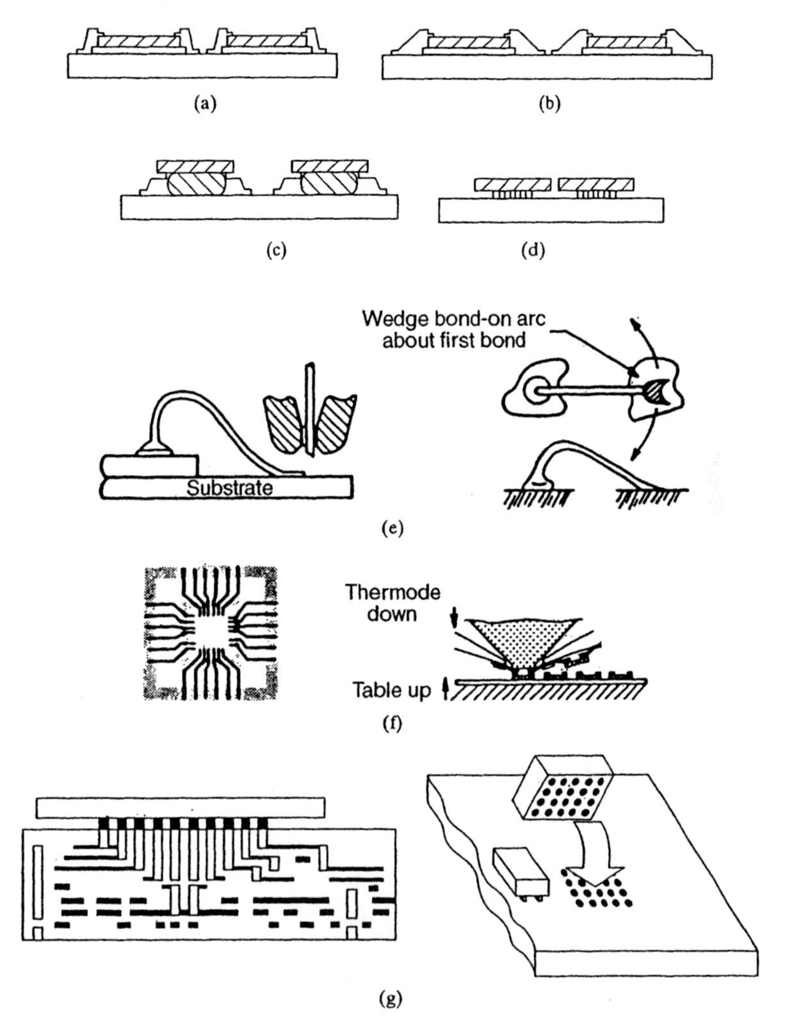

FIGURE 3.15 DCA interconnect methodologies: (*a*) die bond/wire-bond module; (*b*) TAB module; (*c*) flip TAB module; (*d*) flip-chip module; (*e*) thermosonic gold wire-bonding; (*f*) TAB bonding; (*g*) flip-chip bonding.

3. Bonding of the ball is accomplished by simultaneously applying a vertical load to the ball bottomed out on the die bond pad while ultrasonically exciting the capillary (the die and substrate are usually heated to a nominal temperature).
4. The capillary is moved up and over to the substrate or lead bond pad, creating a loop, and, under load and ultrasonic excitation, a bond is made.
5. The wire is clamped relative to the capillary and the capillary moves up, breaking the wire at the bond.

Die bond and wire-bond attach suffer from the problem that this method of chip attach is difficult to repair, particularly if the chip is encapsulated.

3.2.5.2 Tape-Automated Bonding (TAB). TAB, shown in Fig. 3.15(*f*), is more expensive than wire bonding and may require a substantial fan-out from the die to make the outer lead bond. TAB is a process in which chemically etched, prefabricated copper fingers, in the form of a continuously etched tape consisting of repetitive sites, are simultaneously bonded using temperature and pressure to gold or gold-tin eutectic bumps that are fabricated on the I/Os of the die. The outer leads of the TAB-bonded die are excised and simultaneously bonded to tinned pads on the interconnect, using temperature and pressure.

3.2.5.3 Solder-Bumped Dies. The use of solder-bumped dies for packaging electronic systems was pioneered by IBM and was called controlled-collapse chip connection (C4) by IBM. The solder bump composition of the C4 die is approximately 95Pb/5Sn. The C4 dies in the IBM application were attached to multilayer ceramic substrates by reflow soldering, using a flux that required cleaning after reflow. The initial IBM application of C4 technology was for high-end computer packaging. Flip-chip attach shown in Fig. 3.15(*g*), where the die I/Os are solder bumped (usually with 95Pb/5Sn or eutectic Pb/Sn solder) and the chip reflow is attached to its interconnect, has now emerged as a viable packageless technology for consumer commercial products. For the DCA technology, the PCB flip-chip lands are usually solder-finished with a eutectic Pb/Sn solder. A no-clean flux is used for DCA soldering. Once the chip has been solder-attached, it is underencapsulated to provide moisture protection and to enhance the thermal cycling performance of the assembled die. References 8 to 10 discuss some of the emerging developments for direct attach of solder-bumped die to PCBs for commercial product applications. Figure 3.16 shows the DCA application discussed in Ref. 10. In this application, the CBGA packaged microprocessor was solder-bumped and direct chip attached to illustrate the potential savings in PCB real estate.

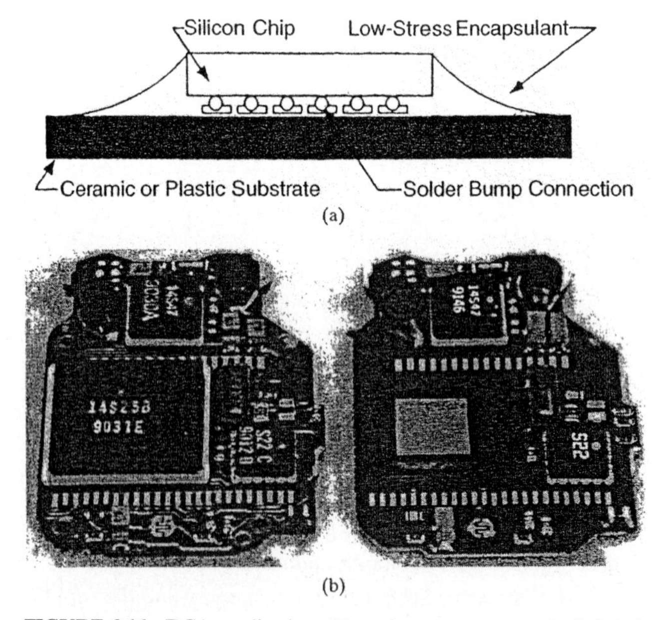

FIGURE 3.16 DCA applications. The microprocessor on the left is in BGA, while the microprocessor on the right is attached by solder bumps to show potential area savings.

3.3 MULTICHIP PACKAGES

MCPs have emerged as a packaging option where performance is an issue or where interconnect area for connecting packaged ICs is at a premium. The performance issue is one of increasing semiconductor performance and the impact of the interconnect medium on the performance of the assembled die. The consequence of this is that the interconnects can no longer be considered to provide an instantaneous electrical connection. It can be shown that for copper conductors on polyimide at a clock frequency of 200 MHz, the maximum allowable length of interconnect is 170 mm, or about 6.8 in. As a reference point, the clock speed for the DEC ALPHA RISC microprocessor was 200 MHz or better in 1995. The clock speed of the Pentium III microprocessor is 500 MHz, and microprocessors with 1 GHz clock speeds are commercially available.

3.3.1 Multichip Versus Single-Chip Packages

MCPs for memory have found broad application in personal computer and laptop computer products. Multiple memory dies (packaged or unpackaged) are assembled on a rectangular interconnect (PCB or multilayer ceramic) with I/Os along one rectangular edge. Such a package is called a single inline package (SIP). The I/Os of the SIP can have DIP leads assembled on them for through-hole attachment onto PCB, or they can have lands on the package to mate with a suitable connector. For through-hole-mounted SIPs, the leads are usually on a 0.100-in pitch. Memory packaged in system-on-a-package or thin system-on-a-package or small-outline J-lead packages are assembled onto SIPs whose substrate is a PCB in appropriate multiples to offer enhanced memory capability. DCA or memory die is also possible. Figure 3.17 shows several SIP configurations. Figure 3.17(*a*) shows an early version of a SIP using a ceramic substrate and leadless ceramic chip carriers with soldered-on leads. Figure 3.17(*b*) shows a typical version of an SIP to mate with an SIP-type connector.

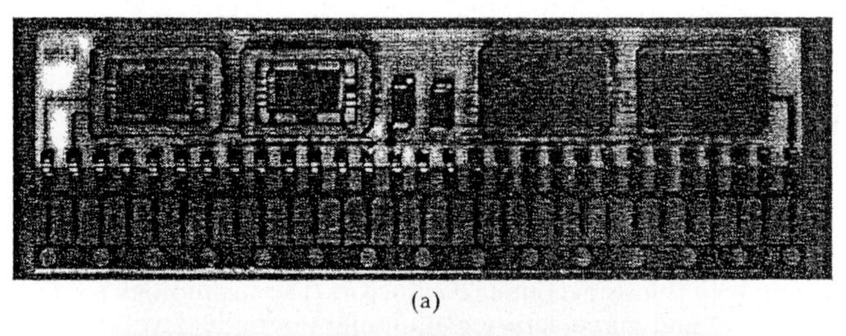

(a)

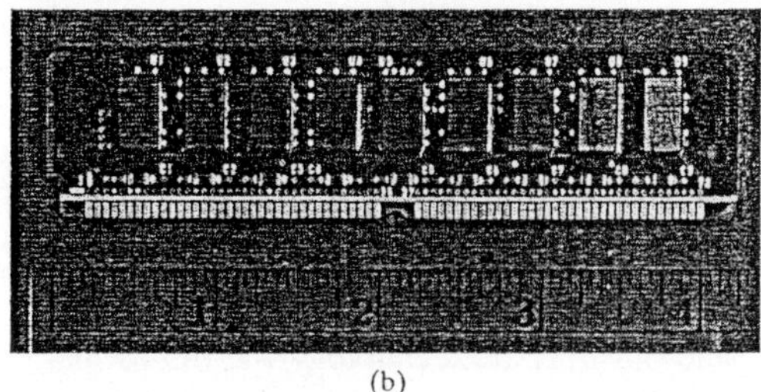

(b)

FIGURE 3.17 SIP configurations.

3.3.2 MCPs Using Printed Circuit Technology

MCPs using PCB technology, but using leads on the substrate, are also being implemented to provide improved product functionality and board space savings. Figure 3.18 shows the concept of the lead-on-substrate MCP. The leads to the substrate can be fabricated integrally with the board during the PCB fabrication or can be soldered on using a suitable high-temperature solder. The lead-on-substrate MCPs can be postmolded or cover-coated with a suitable encapsulant.

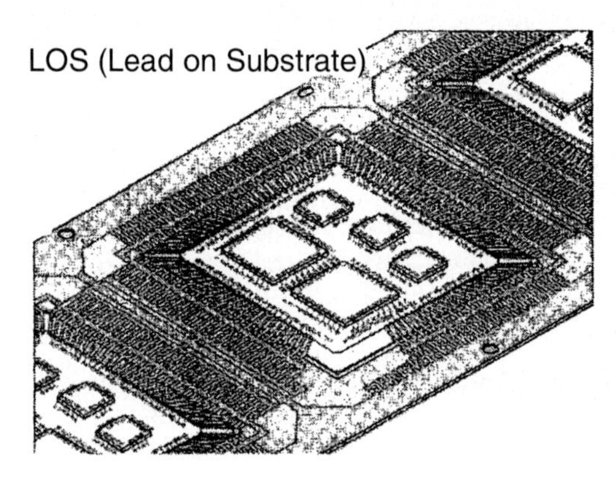

FIGURE 3.18 LOS printed circuit MCP.

The lead-on-substrate MCP could also be implemented using a cofired ceramic package with the leads brazed on and finished by cover coating with a suitable encapsulant or hermetic sealing of the assembled ICs. Chapter 7 of Ref. 7 gives a detailed discussion of the cofired multilayer ceramic package manufacturing processes.

3.3.3 MCPs Using Organic Substrates

The more prevalent form of MCPs uses a multilayer organic interconnect built on a substrate such as silicon, alumina ceramic, or metal composite. The dielectric films are patterned serially one on top of another to produce a multilayer interconnect. The dielectric of choice is polyimide with thin-film copper conductors. The technology for the interconnect in question uses processes and manufacturing equipment initially developed for semiconductor manufacture. This technology has a 1-mil line and space capability. The packages can have the cavity facing up (facing away from the PCB) or down (facing the PCB) if a heat sink is required for heat dissipation.

Figure 3.19 shows wire-bonded die on a silicon substrate packaged in a premolded QFP and a ceramic QFP. A premolded QFP is a lead frame about which a plastic body has been molded and which contains a cavity for a die or a substrate. The method of interconnect from the silicon substrate to the package is wire-bonding. As shown in Fig. 3.15, bare die can be attached to the silicon substrate by die/wire-bonding, TAB bonding, or flip-chip bonding. As with the lead-on-substrate technology, the I/O format for the QFPs that house the multichip substrate follows the standards for QFP packages.

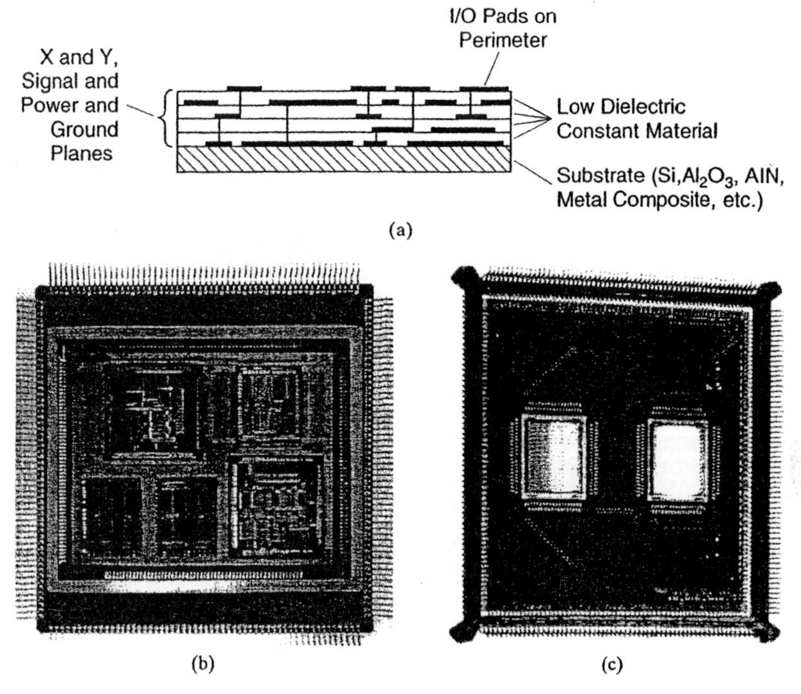

FIGURE 3.19 MCP example.

3.3.4 MCP and PGA

PGA packages can also be used as MCPs containing the type of substrate shown in Fig. 3.19. The PGA packages can have their cavity facing up (facing away from the PCB) or down (facing the PCB) if the package requires a heat sink for power dissipation. The pitch of the pins of the PGA is usually 0.1 in.

3.3.5 Multichip Stacked-Die Packages

A new form of MCP is emerging at the time of this writing. The packaged dies are stacked one on top of each other and are interconnected generally by wire bonds. The dies are thinned in wafer format to thicknesses of 6 mil or less. Figure 3.20 shows the various formats. The substrates for the multichip stacked-die packages can be tape, multilayer organic (FR-4, FR-5, etc.), multilayer organic with high-density interconnect layers, or multilayer ceramic. The preferred package format is a BGA. The initial application of stacked die has been to incorporate memory with logic dies rather than integrate logic and memory on a single silicon die.

3.3.6 MCP and Known Good Die

One of the issues in the application of MCP technology is the availability of known good die. Clearly considering that, for most dies, the package die test yield after wafer test runs

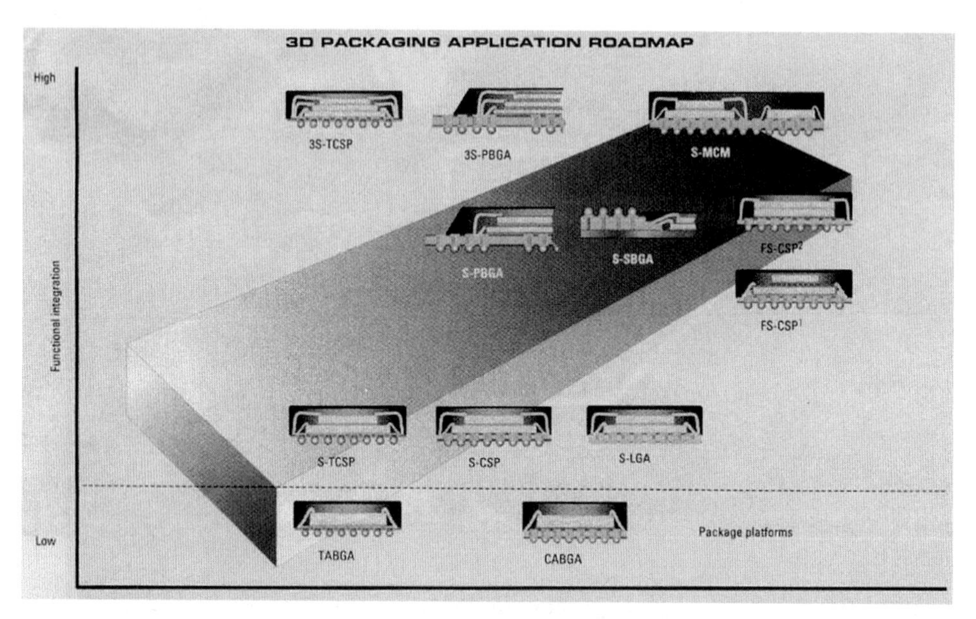

FIGURE 3.20 Formats for multichip stacked die packages.

somewhere around 95 percent, the MCPs are limited to perhaps five IC dies per package at best. The limitation is the resultant yield loss of a finished multichip module package. As an example, for a two-die MCP where each die has an assembled yield of 90 percent the MCP yield would be 81 percent (i.e., 0.9^2) and for a four-chip MCP the yield would be 65.6 percent (i.e., 0.9^4).

3.3.7 System-in-a-Package

An enhanced form of MCP called system-on-a-package or system-in-a-package has emerged. This package incorporates some or all of the multichip packaging technologies previously discussed. Figure 3.21 shows all of the various packaging features that make up a system-on-a-package or system-in-a-package. One feature of this package concept is that it includes not only silicon die but all the passives, etc., required to provide a system function. Passives (i.e., resistors, capacitors, and inductors) can be assembled on the substrate or can be integrated into the substrate itself (i.e., embedded passive substrate technology). Reference 11 provides a detailed overview of the system-on-a-package or system-in-a-package concept. Another example of this concept is the RF front-end HiperLAN module shown in Fig. 3.22.

3.4 OPTICAL INTERCONNECTS

As the performance of semiconductors increases, as measured by clock frequency, the allowable length of interconnect that does not degrade the device performance decreases. The emergence of the Internet and the need for ever increasing bandwidth has become the

(a)

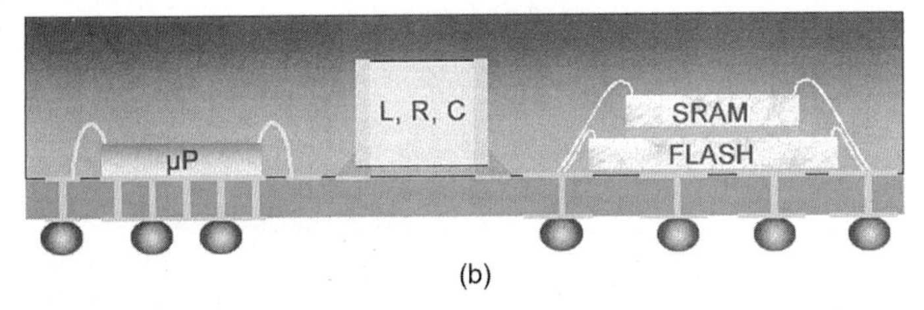

(b)

FIGURE 3.21 (*a*) Packaging technologies features that make up system-in-a-package; (*b*) cross section of system-in-a-package.

INTARSIA RF FRONT-END

- RF front-end for HiperLAN
  - For 5.2 GHz wireless LAN module
  - 12.2 x 12.2 mm wafer-level CSP
- Provides all functions to convert between first IF and 5.2 GHz RF
  - GaAs transceiver
  - GaAs power amplifier
  - T/R
  - LNA transistors
  - Prescaler
  - System RF filters
- Silicon carrier with integrated passives
  - 2 GaAs flip chip
  - 1 Si flip chip
  - 2 wirebonded GaAs discretes

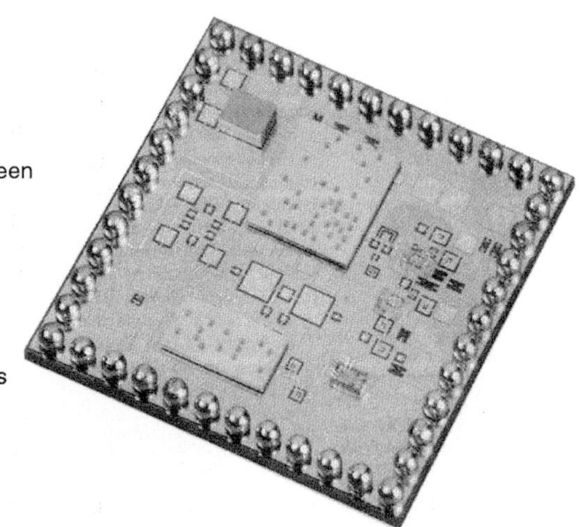

FIGURE 3.22 Example of system-in-a-package.

major driver for optical component and interconnect technology development. For wave-division multiplexing and Ethernet systems, we have reached the copper bandwidth limit of 5 Gbs.

3.4.1 Components and Packages

The demand for increased bandwidth is being driven closer to the end user (i.e., residence or desktop). The issues facing low-cost optoelectronic physical designs are (1) low-cost assembly/processes and low-cost packages for optoelectronic components and (2) the capability for standard automated board assembly of all optoelectronic parts. Figure 3.23 shows a variety of package types used to package light emitters and detectors. The packages are generally of the TO header type or the butterfly DIP or through-hole-mounted DIP type. In general the packages are hermetic. The through-hole-mounted DIP and the butterfly DIP packages are most often used to package edge-emitting lasers. An example of a fiber pigtailed butterfly DIP packaged edge-emitter laser is shown in Fig. 3.24.

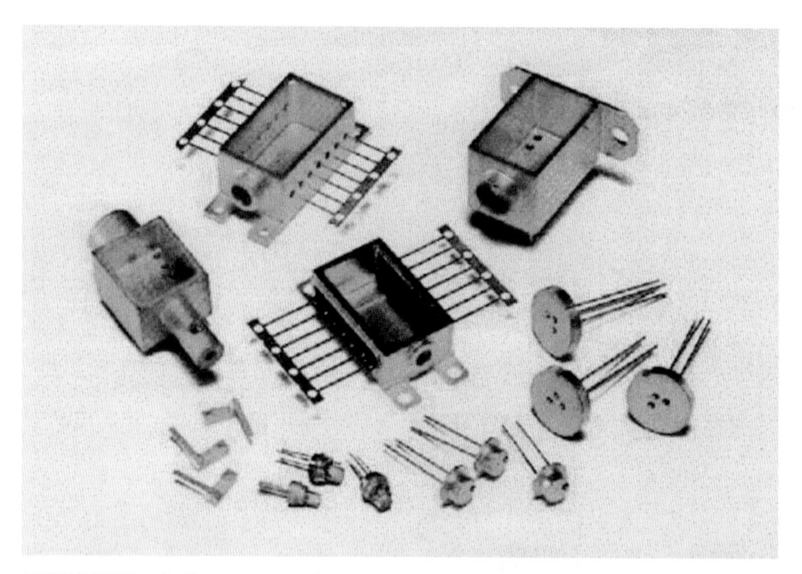

FIGURE 3.23 Package types used to package light emitters and detectors.

FIGURE 3.24 DIP laser 0/7.

3.4.2 Advantages of Optical Interconnects

The advantage of optical interconnects are as follows:

- The signal propagation is independent of the bit rate (up to 20,000 Gbits/s)
- The component is immune to electromagnetic interference and cross talk
- The optical signals can pass through one another (optical noninteraction)
- The components lend themselves readily to multiplexing

Advances in optoelectronic devices are leading to active consideration of local optical signal distribution to:

- Reduce the pin count of packages for complex very-large-scale semiconductor IC chips by replacing I/O complexity with bandwidth
- Carry very-high-bit-rate signals in hybrid circuits and PCB assemblies
- Allow highly complex interconnections to be achieved without the cost and difficulties of a metal-based backplane

3.5 HIGH-DENSITY/HIGH-PERFORMANCE PACKAGING SUMMARY

Where high performance and a high level of interconnects are required, MCPs and multichip modules have emerged as one packaging idea that can provide the performance required. In addition, fiberoptic transmitters and receivers are now commercially available to provide high-performance offboard optical interconnects in place of a hard-wired backplane interconnect using standard PCB technology and cabling. These fiberoptic transmitters and receivers generally use hermetic and nonhermetic custom DIP package formats for packaging. With advances in optoelectronic semiconductors, onboard optical interconnects are also being investigated and demonstrated. Reference 12 discusses one such demonstration of onboard optical interconnects where a polyimeric optical waveguide, including branch and cross-circuits, was fabricated in the polyimide dielectric of a multilayer copper polyimide silicon substrate. Optical waveguide technology that is embedded in PCBs is now being developed.

3.6 ROADMAP INFORMATION

The NEMI *Year 2000 Roadmap*[13] is a good source of detailed roadmap information for IC packaging, PCB technology, optoelectronics, and so on.

REFERENCES

1. D. R. Herriott, R. J. Collier, D. S. Alles, and J. W. Stafford "EBES: A Practical Lithographic System," *IEEE Transactions on Electron Devices,* vol. ED-22, no. 7, July 1975.
2. Ira Deyhimy, Vitesse Semiconductor Corp., "Gallium Arsenide Joins the Giants," *IEEE Spectrum,* February 1995.
3. J. W. Stafford, "Chip Carriers—Their Application and Future Direction," *Proceedings of the International Microelectronics Conference,* Anaheim, CA, February 26–28, 1980, New York, June 17–19, 1980. Also published in *Electronics Packaging and Production,* vol. 20, no. 7, July 1980.
4. Ron Iscoff, "Costs to Package Die Will Continue to Rise," *Semiconductor International,* December 1994, p. 32.

5. Bruce Freyman and Robert Pennisi, Motorola, Inc., "Overmolded Plastic Pad Array Carriers (OMPAC): A Low Cost, High Interconnect Density IC Packaging," *Proceedings of the 41st Electronics Components Technology Conference,* Atlanta, GA, May 1991.

6. Howard Markstein, "Pad Array Improves Density," *Electronics Packaging and Production,* May 1992.

7. Rao R. Tummala and Eugene J. Rymaszewski, *Microelectronics Packaging Handbook,* Van Nostrand Reinhold, New York, 1989.

8. Yutaka Tsukada, Dyuhei Tsuchia, and Yohko Machimoto, IBM Yasu Laboratory, Japan, "Surface Laminar Circuit Packaging," *Proceedings of the 42nd Electronics Components and Technology Conference,* San Diego, CA, May 18–20, 1992.

9. Akiteru Rai, Yoshihisa Dotta, Takashi Nukii, and Tetsuga Ohnishi, Sharp Corporation, "Flip Chip COB Technology on PWB," *Proceedings of the 7th International Microelectronics Conference,* Yokohama, Japan, June 3–5, 1992.

10. C. Becker, R. Brooks, T. Kirby, K. Moore, C. Raleigh, J. Stafford, and K. Wasko, Motorola, Inc., "Direct Chip Attach (DCA), the Introduction of a New Packaging Concept for Portable Electronics," *Proceedings of the 1993 International Electronics Packaging Conference,* San Diego, CA, September 12–15, 1993.

11. William F. Shutler, Alberto Parolo, Stefano Orggioni, and Claudio Dall'Ara, "Examining Technology Options for System On a Package," *Electronics Packaging and Production,* September, 2000.

12. K. W. Jelley, G. T. Valliath, and J. W. Stafford, Motorola, Inc., "1 Gbit/s NRZ Chip to Chip Optical Interconnect," *IEEE Photonics Technology Letters,* vol. 4, no. 10, October 1992.

13. National Electronic Manufacturing Initiative, Inc. (NEMI) *Year 2000 Roadmap,* NEMI, Herndon, VA, 2000.

CHAPTER 4
ADVANCED COMPONENT PACKAGING

Tarak A. Railkar, Ph.D.
Texas Instruments, Inc., Dallas, Texas

Simon Ang, Ph.D.
High Density Electronics Center, Department of Electrical Engineering
University of Arkansas, Fayetteville, Arkansas

William Brown, Ph.D.
High Density Electronics Center, Department of Electrical Engineering
University of Arkansas, Fayetteville, Arkansas

4.1 INTRODUCTION

Packages that contain electronic devices such as integrated circuits (ICs) must, at a minimum, perform the following four basic functions:

- Provide for electrical interconnection (signal, power, and ground) between the various components in the package
- Offer mechanical, electrical, and environmental protection for the devices it contains
- Provide sufficient input/output connections to allow fan-out with other parts of the electronic system of which it is a part
- Provide for dissipation of heat that may be generated by the electronic devices contained within the package when the system is powered

In the simplest terms, an electronic package must provide for circuit support and protection, power distribution, signal distribution, and heat dissipation. These basic requirements must be met if a package contains a single IC, a multiple number of ICs, or a combination of ICs and passive devices.[1]

Driven by the desire, and need, to make electronic systems smaller, faster, cheaper, and less power hungry, semiconductor technology continues its relentless efforts on increasing component and interconnect density, input/output (I/O) capability, and power dissipation, which, in turn, places increasing demands on packaging technology.[2] These advances in ICs drive packaging technology by increasing the number of I/O connections, the operating speeds, and the thermal dissipation requirements. Furthermore, systems packaging interconnections must accommodate passive components required for electromagnetic interference (EMI) reduction, filter circuits, terminations, and impedance matching.[3] Consequently, for high-performance ICs, the package will continue to play an increasingly role in determining performance and cost of the IC. As a result, packaging technologies must continue to improve on the protection provided for the IC, the handling of thermal dissipation, and the routing of more and more signal interconnections, as well as power and ground distribution, through smaller and smaller spaces.[4] Many packages include passive devices in the form of integral passives. The advent of micro-electro-mechanical systems (MEMS), micro-opto-electro-mechanical systems (MOEMS), and even biological packages and applications has further added to the already high demands on packaging. Optoelectronic packages contain not only semiconductor devices but also optical components, such as optical fibers, lens assemblies, and, depending on the application, elements, such as optical multiplexers/demultiplexers. MEMS and MOEMS both include tiny moving parts inside the package. In view of the increased scope, *advanced packaging* is probably a more appropriate terminology than simply *electronic packaging*.

Over the past decade or two, semiconductor technology has evolved to the point where packaging cannot be considered as an afterthought in device design and manufacturing. The package plays an increasingly more integral, sometimes decisive role in the performance of the semiconductor device or devices it contains. The International Technology Roadmap for Semiconductors (ITRS) addresses this fact,[5] stating the following in its assembly and packaging section: "There is an increased awareness in the industry that assembly and packaging is becoming a differentiator in product development. Package design and fabrication are increasingly important to system applications. It is no longer just a means of protecting the integrated circuit (IC), but also a way for the systems designer to ensure form fit and function for today's product—spanning consumer products to high-end workstations." Although the definition of high-density packaging varies across the interconnect pitch to a package that has to be codesigned with the chip, it is understood that high-density packaging is and will continue to be a requirement for high-performance ICs and systems.[5] Figure 4.1 shows the evolution of what have been termed advanced packaging technologies for the past few years in terms of system-level packaging efficiency. It is anticipated that system-on-a-chip (SOC) and system-on-a-package (SOP) technologies will dominate advanced packaging for the next several years.

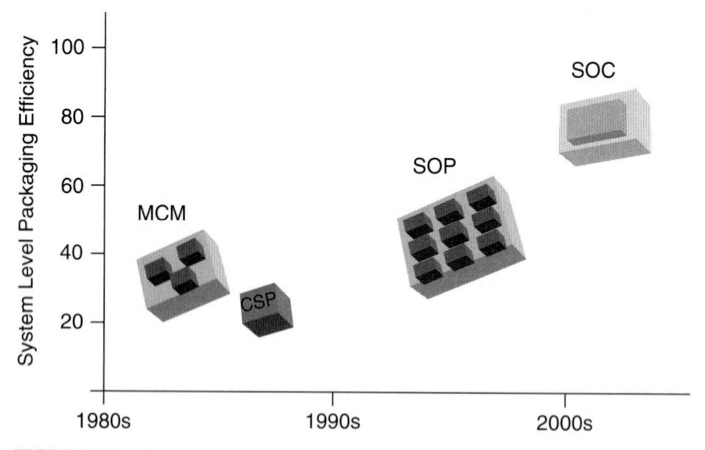

FIGURE 4.1 Evolution of system-level packaging efficiency.

4.2 LEAD-FREE

A new requirement on packages—that of lead-free solder—has initiated a tremendous amount of developmental effort in the past few years. This effort stems from the requirement of packages exported to the European Union (EU) market, to pass Restriction of Hazardous Substances (RoHS) compliance requirements set forth in the Directive 2002/95/EC of the

European Parliament.[6] Origins of the RoHS requirements stem from the need to restrict the use of specific materials found in electrical and electronic products due to their hazardous nature, the risk posed to the environment, their ability to pollute landfills, and the risk posed during its use in component manufacturing and recycling.

The substances banned under RoHS requirements and its maximum allowable trace values are as follows:

- Lead: 0.1 percent by weight (1,000 ppm)
- Cadmium: 0.01 percent by weight (100 ppm)
- Mercury: 0.1 percent by weight (1,000 ppm)
- Hexavalent chromium: 0.1 percent by weight (1,000 ppm)
- Polybrominated biphenyls (PBB) and polybrominated diphenyl ethers (PBDE): 0.1 percent by weight (1,000 ppm) combined

Two other related but different terms sometimes used in this context are the "lead-free" and "green." RoHS is clearly more than "lead-free," but "green" includes absence of antimony (Sb) and halogens (chlorine, bromine, and iodine). However, the eutectic tin-lead solder (Sn63%–Pb37%, or SnPb for short) cannot be used in any of the preceding. This single requirement is bringing about a paradigm shift to the world of advanced packaging. The SnPb solder has been in use in the electronics and related industry for about three decades now. During this time, the entire infrastructure related to package manufacture—beginning from the assembly materials such as substrates, flux, and underfill, and in case of flip-chip, wafer bumping—equipment such as substrate bake and solder reflow ovens and the package reliability requirements such as the JEDEC standards have been centered around the SnPb solder. Changing the requirement to now eliminate Pb from electronic packages translates to a change in all of the preceding. This change stems from an inherent difference in the physical and metallurgical properties—and hence, the processibility and reliability of the two solder materials. SnPb is an eutectic composition of its two parent elements, and has a melting temperature of 183°C. Lead-free solder alloys, almost regardless of their type and composition, have a higher melting point. Further, most of the lead-free compositions are not eutectic in nature, meaning that they do not melt at one temperature. Instead, such compositions have a softening range of temperature. Additionally, the higher melting/softening temperature imposes additional requirements on the assembly materials and on the assembly process. For example, because of the higher solder reflow temperatures for Pb-free solders, the substrates must be able to maintain their integrity and withstand a higher temperature. The flux used in the flip-chip assembly process now needs to be more aggressive in order to be effective on Pb-free solders and must not burn off at higher temperatures that correspond to the higher melting temperatures of Pb-free solders, while simultaneously leaving behind a residue that is not corrosive. The underfill also needs to be suitably selected so that its chemistry is compatible with the flux residue.

Additional challenges are due to the fact that it is difficult to change all components on a system, all at a time, to lead-free. Consequently, the industry must cope with a scenario where a system has mixed metallurgy—partly SnPb based and the others Pb-free—on the system boards.

4.3 SYSTEM-ON-A-CHIP (SOC) VERSUS SYSTEM-ON-A-PACKAGE (SOP)

For decades, the most efficient way to add more functions to an electronic system has been to integrate more functions on a chip either by reducing minimum line width, increasing the size of the chip, or both. Recently, however, it is beginning to be recognized that on-chip integration poses several practical challenges as the die size grows to increasing proportions. With increasing die size, the die-per-wafer yield at the wafer fabrication starts going down, partly due to the increased complexity, but also due the challenges in the wafer manufacturing processes.

To complicate the problem further, off-chip interconnects are beginning to contribute to the challenge of SOC integration. What this means is that interconnects, and not the ICs, are dominating, and often limiting, the system performance and increasing its cost.[7] Consequently, there is mounting support for moving interconnects off the chip and onto the package.[8]

4.3.1 System-on-a-Chip (SOC)

There are many reasons why semiconductor manufacturers continue to push on-chip integration of a system.[9] The reason most often cited is that on-chip integration allows faster interconnection between circuit components simply because of the shorter distances involved. However, it has been pointed out that this may not be the case for all designs. For example, on a large digital IC running at a high clock frequency, a signal traveling on a global interconnection trace may take dozens of clock cycles to reach its destination. In such a case, dividing the device into smaller dice and using high-density interconnects on a package substrate can actually be faster.[10]

Another reason for continuing with on-chip system integration is to avoid the configuration of a multichip package (MCP) because, historically, packages containing a system composed of a multiple number of chips have been larger than those containing a single-chip system.

Although designers will continue to be pressured to integrate as much functionality onto a single chip as possible, there are many factors that must be considered when deciding if SOC is the right approach to the design of an electronic system. Furthermore, as minimum geometries continue to shrink and chip sizes continue to increase due to the incorporation of additional functionality, a point may be reached where it becomes economically impractical, if not impossible, to fabricate SOC. At this point, SOP may provide the only practical alternative to SOC. Now that chip and package are starting to be codesigned, an opportunity exists to develop capabilities for designing the SOP solutions that would cost less and perform better than SOP solutions.[11] Figure 4.2 provides a list of some of the challenges that must be addressed if SOC is to become the standard approach to electronic system integration.

SOC Challenges	SOP Challenges
Fundamental: latency, SiO2 insulation	Design: High speed digital, optical analog, RF
Process complexity	Large-area intelligent manufacturing, cost/yield
SOC design and test	Thermal management
Wafer fab costs and yields	Testing and reliability
Intellectual property for integrated functions	High performance for low cost fab

FIGURE 4.2 Issues that must be addressed prior to full-scale implementation of SOC and SOP.

4.3.2 System-on-a-Package (SOP)

The definition of SOP depends on what is considered to be a system. In fact, much of what is referred to today as an electronic system is in reality a subsystem, meaning that it does not perform an autonomous electronic function. Consequently, SOP has multiple definitions. For example, some consider a multiple number of logic and memory ICs in a single package as an SOP. On the other hand, if the SOP does not contain analog peripheral device drivers, one might argue that the package really doesn't contain a system. Additionally, complete systems usually contain passive devices in addition to the ICs. Passive devices are required to make systems work, and if an entire system is to be placed in a package, then the required passives must go in as well,[3,11] making it what is often referred to an MCM.

Area array packaging is becoming an enabling technology for addressing the needs of customized SOP solutions in order to reduce size, weight, and pincount at the second level of interconnection.[12] This type of packaging technology includes pin grid array (PGA), ball grid array (BGA), and chip-scale packages (CSP), with BGA playing the largest role, primarily

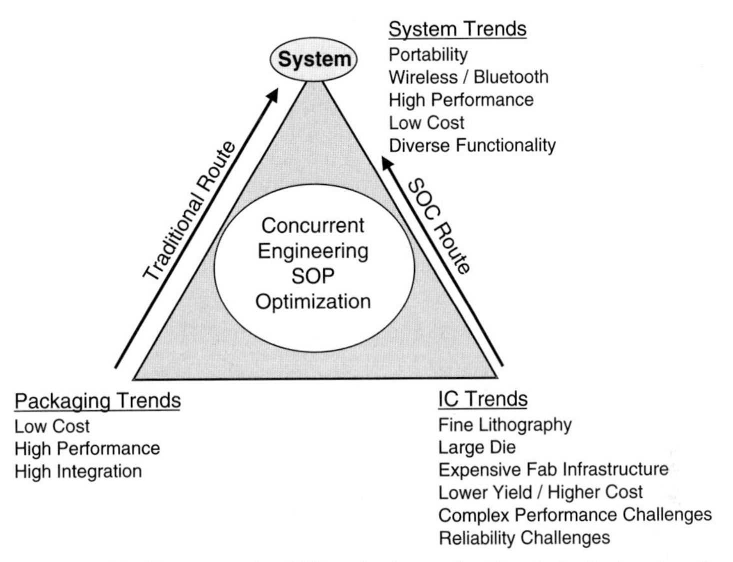

FIGURE 4.3 Future trends of IC, packaging, and system technologies as envisioned by personnel of the Packaging Research Center at The Georgia Institute of Technology. *(Drawing provided courtesy of the Packaging Research Center, the Georgia Institute of Technology.)*

because of its versatility. Compared to conventional leaded packages, BGA packages exhibit improved electrical performance due to a shorter distance between the IC and the solder balls, improved thermal performance by the use of thermal vias or heat dissipation through power and ground planes, reduced handling related lead damage, and increased manufacturing yields due to their self-alignment feature.[13] In fact, BGA has been one of the biggest contributing factors to the proliferation of mobile phones and other wireless communication systems. However, even with the advantages created by area-array packaging, performance, reliability, and cost requirements are challenging currently available area-array electronic package design, primarily by because of materials issues[14] Fig. 4.3 provides an overview of IC, packaging, and system trends that are anticipated for the year 2007 and beyond.

4.4 MULTICHIP MODULES

Just as the name implies, MCM technology mounts multiple, unpackaged ICs (bare die), along with signal conditioning or support circuitry such as capacitors and resistors, to form a system, or a subsystem on a single substrate.[1] In fact, in the late 1980s and early 1990s, MCMs were considered to be the ultimate interconnect and packaging solution, capable of meeting every challenge of the electronics industry. This technology placed ICs in close proximity to each other, thereby enhancing system performance by reducing interconnect delay.[15]

Size is often the primary driver for MCM-based systems. The typical multicomponent discrete assembly provides a silicon-to-board efficiency of <10 percent (actual total die area versus the total printed circuit board area). MCM technology can often increase the silicon-to-board efficiency to 35 or 40 percent with chip and wire assembly processes, and to 50 percent or higher with some of the higher-density processes.[16] Thus, with reduced size and weight, MCMs offer a practical approach to reducing overall system size while providing enhanced performance due to a reduction in the interconnect distance between chips.[1] Multichip modules typically use three to five times less board area than their equivalent discrete solution.[16]

Since a multichip module is, by definition, a single substrate containing two or more ICs, it is unlikely that they are a thing of the past. Instead, they have been renamed from MCMs to MCPs, few chip packages (FCPs), or other similarly descriptive names. These packaging solutions usually contain from about two to six ICs that are packed onto an inexpensive laminate substrate. Although they are multichip modules, they may be called something else. Intel refers to its two-chip package as a CSP, although in fact it is an MCP or MCM .[17] Interestingly, some still believe that when time is an important consideration while reaching the market with a low-cost and reliable product, multichip solutions win. On the other hand, others believe that it's only a matter of time before SOCs take over completely and multichip packaging will not be necessary, and they view multichip solutions only as stopgaps.[18] Nevertheless, regardless of their nomenclature, laminate based multifunctional packages are among the key package types in the industry.

4.5 MULTICHIP PACKAGING

In general, an MCP is any package that contains more than one die. Low-cost solutions to multichip packaging are available for high-volume applications.[11]

Advanced electronic packaging technologies have historically been limited to applications in high-performance products such as supercomputers. Today, they are being moved into lower-cost products through the use of chip-on-board (COB), BGA, or land grid array (LGA) for second-level interconnect assembly, and surface-mount technology (SMT) mass production equipment. Prismark's December 2005 *Report on Semiconductor and Packaging* found that laminate packages constitute 47 percent of the overall packaging volumes followed by lead-frame packages at 39 percent.

Improvement in the silicon efficiency of MCM designs can be accomplished by packaging ICs onto very-thin, compact, and lightweight microcarriers using BGA or LGA connections, such as those shown in Fig. 4.4. Because of the development of lower-cost materials (i.e., low-cost

FIGURE 4.4 Photograph showing three different BGA packages and one LGA package: in the top-left, 1,600-pin FCBGA (1.0 mm pitch, ceramic substrate); in the top-right, 1,140-pin FC-LGA (1.0 mm pitch, build-up substrate); in the bottom-left, 768-pin TAB-BGA (1.0 mm pitch); and in the bottom-right, 672-pin EBGA (1.27 mm pitch). *(Photograph provided courtesy of Fujitsu Microelectronics, Inc.)*

photosensitive materials), large-area processing (as practiced in the fabrication of displays), large-area lithography, and low-cost metallization processes as practiced in printed wiring board (PWB) fabrication, economically viable multichip packaging is realistic.[19]

4.5.1 Few Chip Packaging (FCP)

An increasing number of companies are embracing FCP for technical and business reasons. Although these FCPs are virtually indistinguishable from their single-chip packages, they are a radical departure from the MCMs of the early 1990s. Instead of 10 to 20 die, today's FCPs typically contain two to five die mounted on a laminate substrate in a BGA package. This rebirth can be attributed, at least in part, to improved bare die testing and handling, along with the availability of low-cost, high-performance laminate substrates. Furthermore, there is a growing trend to use FCP as an alternative to SOC, resulting in a system-in-a-package (SIP). This approach is also referred to as SOP.

4.5.2 Chip-Scale Packaging (CSP)

CSP was introduced in Japan in the early 1990s and in the United States soon thereafter as a less expensive alternative to MCMs.[20] According to IPC/JEDEC J-STD-012 definition, CSP is a single-die, direct-surface mountable package with an area of no more than 1.2 times the original die area. CSPs come in many forms—wire-bonded, flip-chip, leaded, and BGA. A typical wire-bonded CSP packaging process (Fig. 4.5) starts with the attachment of the die on

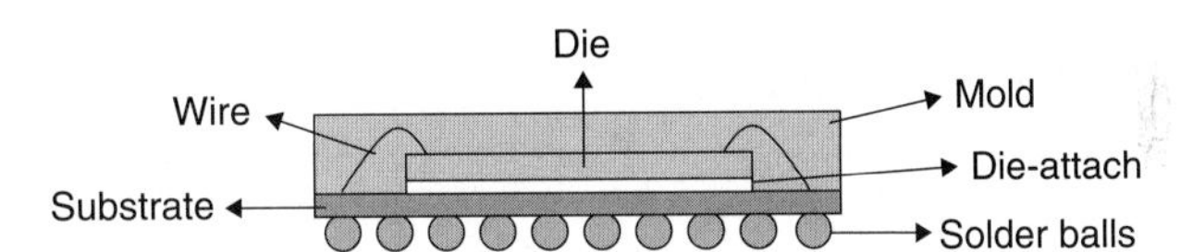

FIGURE 4.5 Schematic of a wire-bonded CSP.

an interposer using an epoxy. Wire-bonding is then performed from the I/O pads of the die to the interposer. These wirebonds must be as low and as close to the die as possible to achieve the minimized package size. A plastic transfer molding process is used to encapsulate the die. Finally, solder balls are formed on the bottom of the interposer. In flip-chip CSPs, the solder-bumped die is bonded onto the interposer as shown in Fig. 4.6. The area-array version of CSPs, also known as fine-pitch ball grid assemblies (FPBGAs), is now widely used in many portable and telecommunication products.

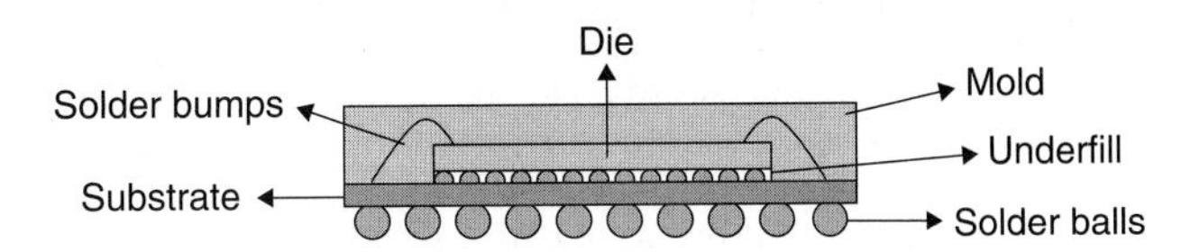

FIGURE 4.6 Schematic of a flip-chip CSP.

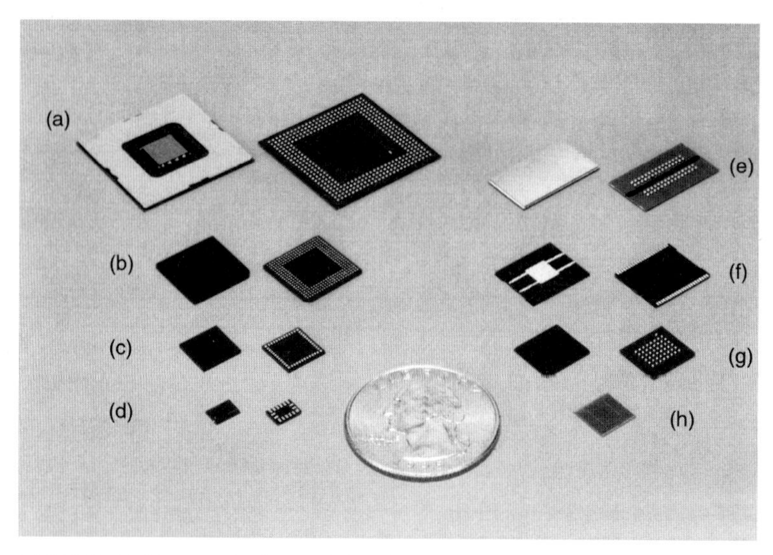

FIGURE 4.7 Photograph of several chip-size packages. On the left are logic CSPs:
Row 1: 352-pin TAB-BGA (0.80 mm pitch)
Row 2: 288-pin FBGA (0.5 mm pitch)
Row 3: 48-pin BCC (0.5 mm pitch)
Row 4: 16-pin BCC (0.65 mm pitch)
On the right are memory CSPs:
Row 1: 60-pin μBGA (0.8 mm pitch)
Row 2: 46-pin SON (0.5 mm pitch)
Row 3: 48-pin FBGA (0.8 mm pitch)
Row 4: 16 M bit flash memory chip
(Photograph provided courtesy of Fujitsu Microelectronics, Inc.)

CSPs have smaller size, lesser weight, and improved electrical performance. CSPs offer the same space and material savings and short signal paths as direct chip attach (DCA) methods such as COB and flip-chip-on-board (FCOB) offer. The advantages of using a chip-scale package over DCA are easier handling, more protection for the chip, and simpler board assembly. Figure 4.7 shows a selection of commercially available CSPs. CSPs are somewhat limited to use in moderate I/O ICs.[20]

4.5.3 Wafer-Level Packaging (WLP)

Wafer-level package is defined as an IC package completely fabricated at the wafer level and assembled with standard SMT. Alternatively, a wafer-level package can be defined as the package in which the die and "package" are fabricated and tested on the wafer prior to singulation. Significant reductions in cost and device form factor can be achieved with WLP, while at the same time increasing the electrical performance. Wafer-level packaging was first investigated by Sandia National Laboratories and Fujitsu in the mid-1990s. Since then, WLP technology has been applied to flash and dynamic random access memory (DRAM) devices. However, widespread market adoption of WLPs has yet to occur. A WLP is not the optimal package for high I/O devices because of the limitations of the SMT assembly process and the wiring constraints of the printed circuit board (PCB) to which it is mounted. WLPs are typically used in devices that are small, with 2 to 36 bumps, such as in power metal-oxide semiconductor field effect transistors (MOSFETs), electronically erasable read-only memory (EEPROM), small logic devices, and transient voltage suppressors. Cellular phone market consumes more than 90 percent of all WLPs produced today. This is because WLPs enable the small form factor and increased functions desired in modern cellular phones. However, there are indications that higher-end DDR2 and DDR3 DRAM devices will be able to take advantage of the benefits offered by WLPs. Several factors must be addressed in adopting WLPs. These factors include I/O density, reliability, burn-in, assembly processes, and cost. It is important to remember that a WLP is not the ultimate package for certain devices.

The benefits of wafer-level packaging can be summarized as follows:

- Smallest IC package size, as it is truly CSP
- Lowest cost per I/O because the interconnections are all done at the wafer level in one set of parallel steps
- Lowest cost of electrical testing, as this is done at the wafer level
- Lowest burn-in cost, as burn-in is done at the wafer level
- Elimination of underfilling with organic materials around the solder joint
- Enhancement of electrical performance because of the short interconnections

According to the latest ITRS roadmap, the pitch of area-array packages is expected to decrease to 100 μm by 2009. Simultaneously, the electrical performance of these interconnections needs to be improved to support data rates in excess of 10 Gbps, while guaranteeing thermomechanical reliability and lowering the cost. These requirements are thus challenging, requiring innovative interconnection designs and technologies. To meet these challenges, nano wafer-level packaging that uses nano materials and structures to bring about unprecedented advances in electrical, mechanical, and thermal properties in the chip-to-package interconnections is being investigated.[21,22]

4.5.4 Three-Dimensional (3-D) Packaging

Another type of multichip packaging gaining wider acceptance is stacked die packaging.[23,24,25] At the present time, most packaging of this type is in the form of stacked memory in CSPs, and the stacking is done only to save space. However, it is a growing trend that can be applied to any electronic system where volume density is of concern. Stacked die packaging is illustrated in the top view of Fig. 4.8, while the bottom view shows how packaged devices can be stacked.

The advantages of stacked die packaging are further enhanced by using thinned die. Thinning die has the potential of enhancing first-level interconnect reliability while minimizing the increase in vertical profile. A silicon die with a thickness <100 μm is quite flexible and can relieve stresses induced by packaging and coefficient of thermal expansion (CTE) mismatch.[26]

The latest in IC packaging is a stacked ball grid array (SDBGA) package for the communication market. It is used to stack various ICs in one package with the resulting savings in real estate on the motherboard and manufacturing cost. Compared to conventional packages, an area savings of up to 70 percent can be realized. Popular SDBGA sizes range from 8 mm × 8 mm to 14 mm × 14 mm and a pin count between 80 and 140. The total package height is typically about 1.4 mm.

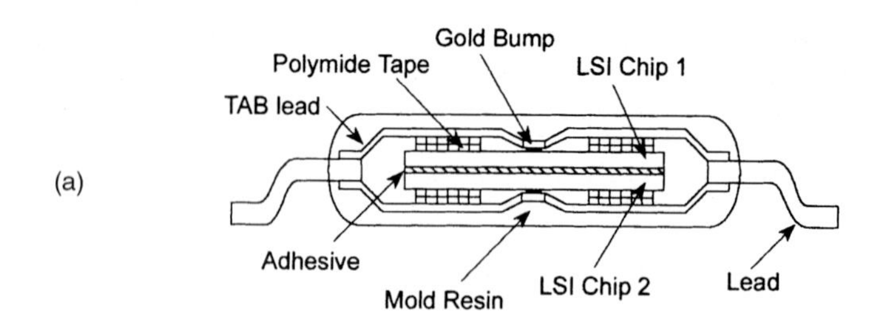

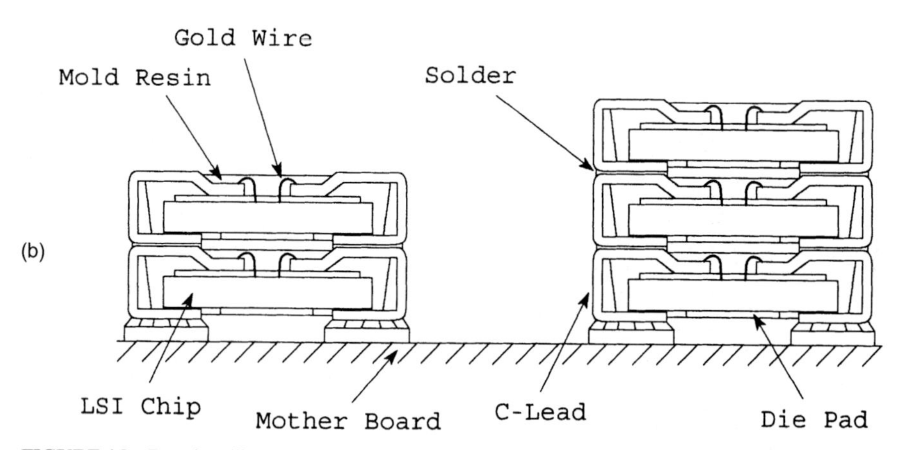

FIGURE 4.8 Drawings illustrating 3-D packaging: at the top, die (or chip) stack packaging; at the bottom, package stacking.

4.6 ENABLING TECHNOLOGIES

The end application and the electrical, thermal, and reliability performance demands it presents on the package often dictate the choice of assembly materials, assembly process, and package test technologies. For packages that involve multiple die or components, the need for known good die cannot be overemphasized. Applications such as cell-phones, where the available space has a premium but which demands an increasing functionality, may require enabling technologies such as wafer thinning along with stacked die or stacked package. However, in view of the relatively small product lifetime, the reliability requirements may be somewhat more forgiving than those for a server application, wherein the long product lifetime, large body size, high electrical signal speeds, and high power dissipation pose a different set of challenges. This section will review some of the key enabling technologies required across different application spaces and will also discuss the resulting impact of lead-free requirements.

4.6.1 Known Good Die (KGD)

When an electronic system is fabricated or an electronic component is assembled, KGD considerations play a very important role in minimizing waste, optimizing the material, and infrastructural resources, and keeping the costs low.[27,28] Packaging and test combined, especially that for high-performance devices, form a higher fraction of the total device cost today than it did a few years ago. Consequently, it makes even more sense to weed out the known reject die and not let it into the assembly process. By doing so, the manufacturer can control the overall cost of packaging while simultaneously using the assembly equipment time only for known good subcomponents. This requirement directly translates into a need for developing adequate test and probe technologies that are suitable for flip-chip as well as wire-bond devices. Shrinking pad pitches and narrowing geometries require the development of advanced wafer probe technologies.

4.6.2 Chip Thinning

As noted in a previous section, one approach to increasing the functional density of an electronic system is to stack ICs vertically. Reducing the thickness of the chips allows for more chips to be stacked within a package, thereby substantially increasing the functional density. In addition to increasing the functionality per unit area of substrate (module), thinning chips also improves their thermal performance, results in a more mechanically reliable device, allows for flexibility so they can conform to a curved surface, and relieves stress induced on a chip due to packaging[26,29] Thinning chips can result also in an improved electrical performance in view of the reduced signal path across two or more chips. However, this is not necessarily always the case, as cross-talk and interference across the chips can also increase due to the reduced distance, thereby potentially degrading the electrical performance. Decoupling capacitors may be needed for adequate isolation; however, a more detailed electrical analysis is beyond the scope of this chapter.

Four techniques that are primarily used for wafer thinning are mechanical grinding, chemical mechanical polishing (CMP), atmospheric downstream plasma (ADP, sometimes used with wet etching), and dry chemical etching (DCE). Factors to be considered when selecting the appropriate thinning technique include the throughput of the time taken to thin a wafer, quality, and defect density in the resulting thinned surface, and cost. Due to its high material removal rate, mechanical grinding continues to be the most popular method for wafer thinning. However, this technique also introduces to the thinned surface a significant amount of mechanical damage that needs to be removed. This is achieved by fine grinding or through CMP, which relies on chemically assisted mechanical polishing of the wafer surface. CMP can deliver very flat and defect-free wafer surface, however, the material removal rate is small—of the order of a few micrometers per minute. Thicknesses routinely achieved vary from 50 to 100 μm, although thicknesses as small as 15 to 25 μm (silicon is flexible in this thickness range) have been demonstrated.[30] Figure 4.9 is a photograph of a silicon wafer that has been thinned to approximately 50 μm using a plasma process that causes bending of the wafer.

FIGURE 4.9 Photograph of a silicon wafer that has been thinned to approximately 50μm by plasma etching. Warpage was caused by the wafer thinning process. *(Photograph provided courtesy of Materials and Manufacturing Research Laboratories, University of Arkansas, Fayetteville AR.)*

4.6.3 Lead-Free Chip Attach Processing

Packaged devices require a larger amount of real estate as compared to the die due to the requirement for a substrate (or "chip carrier").[31] Substrates provide electrical connections from the die bond pads to the external world. Bare die attachment eliminates the need for a substrate, thereby saving valuable module space and reducing interconnect distance, but such system configuration can increase the risk of reliability performance of the system. Direct die attachment can be accomplished by conventional die attach/wire bonding, tape-automated bonding (TAB), and flip-chip processes.

Lead-free requirements impose the following additional requirements:

- Use of suitable lead-free materials in the assembly bill-of-materials
- Compatibility of these materials with lead-free processing temperatures

Performance of the package moisture sensitivity level (MSL) can be impaired by the higher temperature compatibility arising from lead-free requirements.

4.6.3.1 Wire Bond. For most part, transition to lead-free of a wirebond package on a component level is relatively straightforward. Through the use of gold wires, the first level interconnect is already lead-free. Further, most of the metal lead-frame-based packages are lead-free-compliant as long as the mold compound or the die-attach materials are free of lead or similar components and are able to withstand the higher temperatures seen by the package during component reflow. BGA-based wirebond packages, on the other hand, additionally need use of lead-free BGA balls on the package to make it compatible with lead-free requirements.

4.6.3.2 Tape-Automated Bonding (TAB). TAB is rather expensive because it requires a custom tape for each different IC and its construction requires a significant amount of processing. Furthermore, the assembly process is equipment-intensive because of testing of the IC while it is mounted on the tape, excising the IC from the tape and shaping (forming) its leads prior to mounting it onto a substrate, requiring a custom die for gang bonding. However, it is attractive because it provides the opportunity to perform burn-in and die testing prior to final assembly. The TAB process is schematically illustrated in Fig. 4.10.

TAB packages are usually naturally lead-free at the component level. Assembly to the application board does require lead-free solder along with a requirement for the package to be able to withstand the high temperatures associated with lead-free processing.

4.6.3.3 Flip-Chip Bonding. The flip-chip assembly technique is one of the important techniques of die attach to either the carrier (substrate) or directly to the board usually employing either solder bumps or z-axis conductive polymers. Another common mode of attaching silicon dice in the flip-chip configuration is stud bumping, where a gold wire is used in a similar fashion as in wirebonding, but the wire is truncated after establishing contact to the die pads. The resultant stud is then flattened using a coining process. The original flip-chip assembly, pioneered by IBM, eliminates all the packaging area that is not occupied by the die itself, generally resulting in the smallest footprint and the lowest profile. However, wire-bonded devices can be back-ground to reduce the overall package thickness. It is not trivial to perform a similar back-grinding operation on flip-chip devices for two reasons. First, the bumping process prefers to use full thickness wafers due to breakage concerns, and second, it is extremely difficult to back-grind bumped wafers.

In general, flip-chip assemblies can use either grid arrays or peripheral pads for I/O connection at pitches between 0.1 and 0.65 mm, but are limited mostly by routing constraints through substrate manufacturing capabilities. Consequently, most flip-chip devices in production today use a solder bump area-array pitch in the range of about 0.15–0.25 mm. Of all the techniques available, flip-chip assembly offers the highest packaging density and there is increased use of flip-chip attachment. Consequently, many bumping contractors also offer redistribution services. Although redistribution is done on the chip (actually on the wafer), it is considered a packaging operation.

Additional requirements presented by the need for lead-free, RoHS-compliant, "green" packages call for lead-free solder bumps. Devices that use gold stud bumps previously referenced are

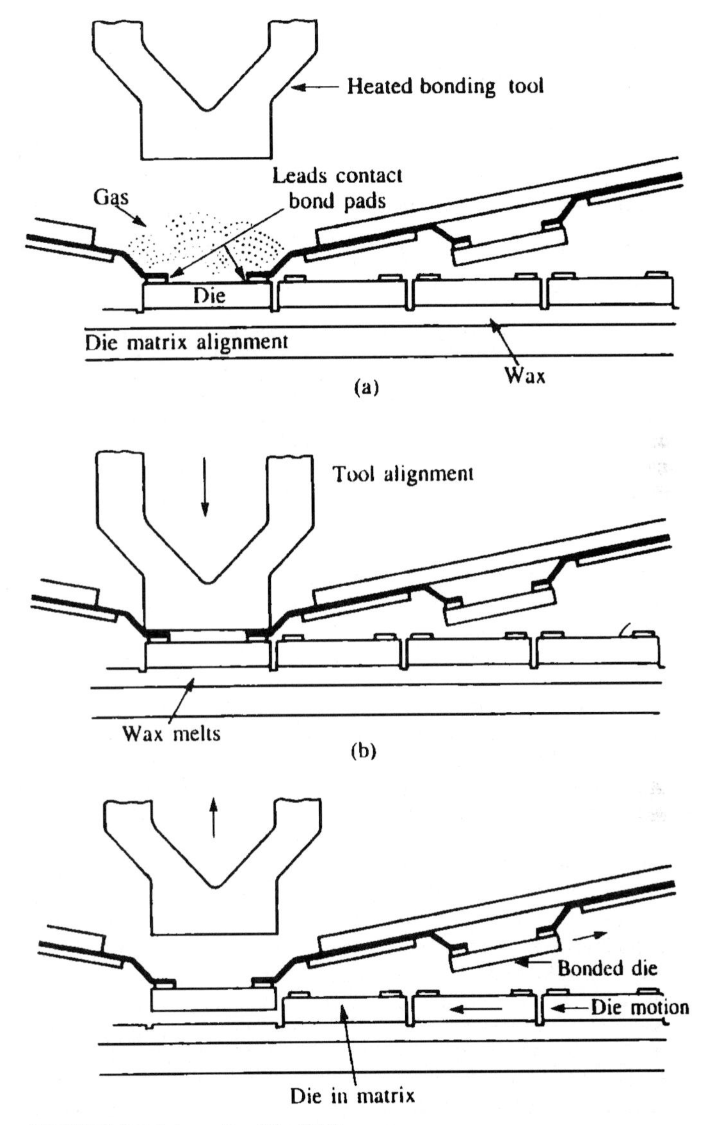

FIGURE 4.10 Schematic of the TAB process.

already lead-free; however, stud bumping is typically limited to a smaller die and those with a small number of I/O count requirements (note that I/O here refers to the total number of connections required across the die and the carrier/substrate). Furthermore, the stud-bump-based lead-free solution has its challenges due to the serial nature of the stud-bumping and die-attach processes. Consequently, a lot of effort continues to go into the development of lead-free solder-based wafer bumping. Lead-free wafer bumping is discussed in more detail in the next section.

4.6.3.4 Wafer Bumping. Wafer bumping is a wafer-level packaging technology that uses solder bumps to form interconnects between the integrated circuit (IC or chip) and the actual package. To create a solderable site with an electromigration barrier, the manufacturer usually first deposits an under-bump metallurgy (UBM) of electroless nickel and gold directly onto the I/O pad of the IC. Other typical UBM consists of a 20–50 nm of evaporated chromium acting as contact and adhesion layer, a diffusion barrier of 100–200 nm of Cr:Cu, a conductive layer of 0.4–0.6 μm of copper, and finally an optional oxide protector of about 100 nm of gold. This is followed by the placement of solder—either by evaporation, electroplating, or screen printing—onto each I/O site and a thermal reflow to form a solder ball as shown in Fig. 4.11.

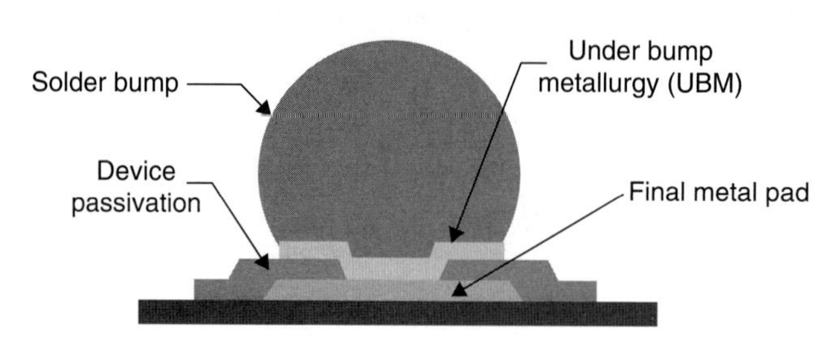

FIGURE 4.11 Schematic of a solder bump and underlying metallurgy.

Flip-chip ball grid array (FCBGA) is a package type that uses solder bumping to yield area-array interconnection. Area-array interconnection ensures signal and power/ground integrity far more superior to conventional peripheral wire-bonding interconnection due to elimination of wirebond inductance and a lower electrical resistance. Solder bumping enables the realization of CSPs. CSPs are well accepted in handheld consumer products such as cellular phones and PDAs, since the package size is of vital importance. However, the ultimate package is a true chip size package, the wafer-level CSPs (WL-CSP). WL-CSP is predicted to enjoy the highest continuous aggregated growth rate (CAGR) of more than 20 percent since it enables high-performance microprocessor chip (MPU), application-specific integrated circuit (ASIC), and memory devices for the high-end computing and networking markets. Bumped die for FCOB assembly is common in electronic manufacturing due to the saving in the printed circuit board real estate.

There are three primary wafer-bumping processes for solder placement: screen printing, electroplating, and evaporation. All these processes have been used in production, but bump pitch, I/O count, start-up cost, and production volume are critical criteria that dictate which process is best for a particular solder placement. Evaporation of solder requires substantial investment in capital equipment since it is performed in a high-vacuum environment. Electroplating of solder yields a finer solder bump pitch, but it is primarily limited to solders with binary alloys such as SnAg and SnCu. Screen printing of solder is the most cost-efficient for solder bump pitch greater than 120 μm.[32]

Solder bumps must undergo a reflow process to create the solder bump structure shown in Fig. 4.11 as a result of its surface tension. Since the composition of the solder may be high-lead, eutectic, or lead-free, the equipment and processes must be capable of handling a wide range of temperature profile variations while maintaining a tight thermal uniformity within each process profile. A typical reflow process undergoes five transitions: pre-heat, flux activation, soak, reflow, and cooldown. The purpose of the pre-heat is to evaporate solvents from the solder paste. A slow thermal ramp-up rate prevents damage due to thermal shock. The time and temperature to evaporate the solvents depends on the solder paste that is applied. As the temperature is ramped up, the flux reacts with the oxide and contaminants on the surfaces to be joined. The time and temperature should be long enough to allow the flux to clean these surfaces fully, but not so long that the flux may be exhausted before soldering takes place. The soak temperature should be approximately 20–40°C below the peak reflow temperature.

Time and temperature depend on the mass and solder materials. During reflow, the assembly is briefly brought to the temperature sufficient to reflow the solder. After reflow, a gradual cooling should be used to prevent thermal shock. Gradual cooling produces a finer grain structure and fatigue-resistant solder joint. During reflow process, care must be exercised to prevent the formation of oxides that degrade subsequent processes. A nitrogen-with-flux or hydrogen-flux-free reflow process has been used to control oxide formation.

There is a trend in the industry toward smaller area-array packages with lead-free bump sizes in the range of 10–15 μm. Metal bumps, including gold, copper, and nickel, are lead-free wafer bumping.[33] In addition, gold and copper have a greater mechanical shear strength compared to lead solders. This helps to relieve residual stresses caused by the mismatch of the CTE between the chip and printed-circuit board materials, and hence strengthens the bump connections.[34]

Gold stud bumps are formed by a modified wire bonder that uses thermosonic energy (150–200°C) to first attach a gold bondwire to the die bond pads and then shear the top of the gold wire without leaving a tail.[35] To achieve a uniform bump height, the gold studs may be flattened or "coined" by pressing any remaining wire tail to the ball or shearing it off. This yields a planar, flat-top gold bump that does not require any coining in a single-step process.

Copper is another candidate material for wafer bumping. Compared to gold, copper costs less but has a good electrical and heat-dissipating performance due to its low electrical resistivity and high thermal conductivity.

Copper studs can be created by bonding a copper bondwire directly onto the aluminum bonding pads on a wafer using a thermosonic bumping machine. However, the ultrasonic power and bonding force required for copper studs are generally higher than those for the gold studs due to the hardness of copper. Also, a reducing gas, usually 5 percent hydrogen in nitrogen, is blown over the end of the copper wire during formation of the ball to prevent oxidation of copper. Copper bumping has many advantages over solder and gold bumping in terms of increased electrical and mechanical performance and reduction of material cost; however, it is relatively new to the flip-chip industry.

4.6.3.5 Chip-on-Board (COB). In COB assembly, a back of a bare (unpackaged) IC is attached directly onto a PWB, wire-bonded, and then encapsulated with a polymer. The die bondpad pitch is generally in the range of about 0.175–0.25 mm and IC placement must be very accurate. For many applications requiring miniaturization, and especially those where space is limited, COB assembly can be the most cost-effective packaging option. It is a mature technology and offers high packaging density, low packaging cost, and fast signal speed because the dies are wire-bonded directly onto a board.[36] The photograph in Fig. 4.12 shows details of a COB-mounted power diode. Usually, as a final step, a glob top is deposited onto the chip to serve as passivation and protection.

FIGURE 4.12 Photograph of chip-on-board. (A COB assembly).

4.6.3.6 Passive Devices: Surface-Mount Components. Since the invention of the IC, almost continuous progress has been made in creating a given functionality in a smaller area. Unfortunately, surface mount (SM), discrete passive (DP) component size, which has seen continuing size reductions from 0805 components (80 mils × 50 mil or 2 mm × 1.25 mm) to 0201 components (20 mil × 10 mil or 0.5 mm × 0.25 mm) with preliminary work being done on 0105 components (10 mil × 5 mil or 0.25 mm × 0.125 mm), has not shrunk accordingly. Furthermore, increasing numbers of DPs are required in newer electronic systems. Consequently, DPs can occupy significant substrate area. In fact, system substrates are now dominated by individually placed DP components. The key challenges presented by DPs are their fairly constant physical size, their increasing numbers, the cost of individual handling and attachment, the parasitic effects of their associated electrical connections, the reliability of solder joints, and the wide range of required values.[3,37,38] Thus, they continue to pose challenges for miniaturization of electronic systems, particularly for the wireless and consumer markets.

Lead-free requirements pose manufacturing challenges, not only in package assembly but also in areas of component manufacturing assembly and of lead-free assembly.

4.6.3.7 Passive Devices: Integrated Passive Components. Recently, much attention has been directed toward the use of integrated passive (IP) devices (also referred to as embedded or integral passive devices). IPs are located within the substrate and are formed during the fabrication of the substrate. Consequently, they require processing steps in addition to those that provide only interconnection. There are no external leads to be individually attached because all IPs are connected to metallizations within the substrate, whether it is laminate, ceramic, or any other material. They do not compete with ICs for surface space, but may complicate routing within inner substrate layers. They have no casing or packaging of their own to contribute weight and volume to the system. The lead length between the IP and an interconnect is usually only mils. Also, since the IP is typically a planar structure, there is much less parasitic inductance, capacitance, and resistance compared to DPs. Furthermore, IPs eliminate pick-and-place operations and two solder joints. A shortcoming of IPs is the fact that they cannot be tested individually so the yield of the IPs directly impacts the yield of the substrates. The arguments for changing to IPs include reduced cost, smaller system size, better electrical performance, and improved system reliability. Figure 4.13 shows schematically the difference between discrete passives, located on the surface of a substrate, and integrated passives, which are located below the surface of a substrate.

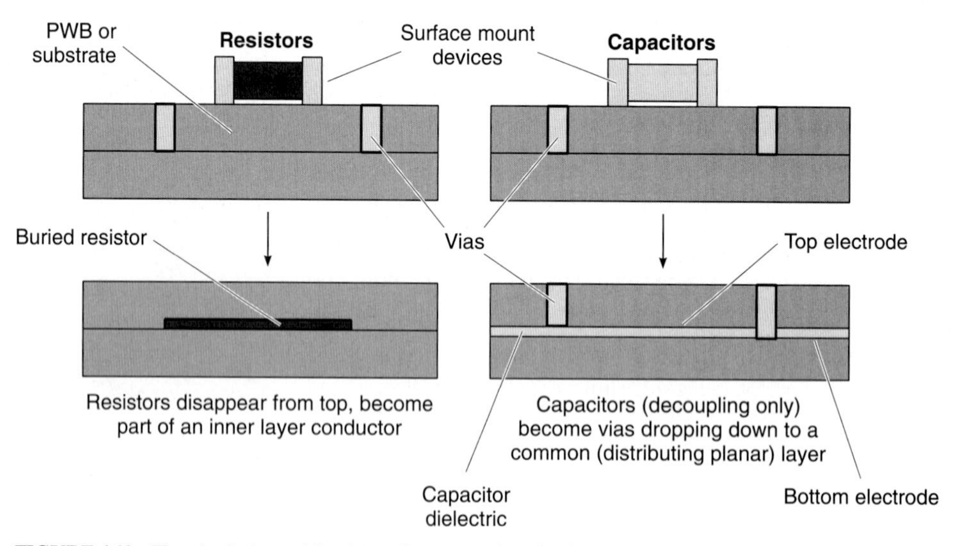

FIGURE 4.13 The physical transition from discrete passives (top) to integrated passives (bottom).

Thin-film passive components rely on deposition and photolithography to define conductive, resistive, and dielectric materials physically that produce a desired electrical response.[3] Integrated resistors are fabricated by depositing and patterning a layer of thin-film or thick-film

TABLE 4.1 Characteristics of Typical Tantalum Nitride and Nichrome Resistors

Resistor material → characteristic ↓	Tantalum nitride	Nichrome
Sheet resistance	20–150 $\Omega/\square$ 100 $\Omega/\square$ (typical)	25–300 $\Omega/\square$ 100–200 $\Omega/\square$ (typical)
Sheet resistance tolerance	±10% of nominal value	±10% of nominal value
Temperature coefficient of resistance	−75 + 50 ppm/°C (typ.) 0 ± 25 ppm/°C (anneal)	0 + 50 ppm/°C 0 + 25 ppm/°(anneal)
Resistor drift (1000 h at 150° in air)	<1000 ppm	<1000 ppm (anneal) <200 ppm (sputter/ann.)
Resistor tolerance after anneal and laser trim	±0.10% standard ±0.03% (bridge trim)	0.10%

resistive material in series with an interconnect line on an insulating substrate. Table 4.1 gives some characteristics of typical tantalum nitride and nichrome resistors.

The integration of capacitors into interconnect substrates is largely driven by three applications: decoupling of ICs from power supplies, analog functions such as those in radio frequency (RF)/wireless, and termination of transmission lines. A wide variety of dielectric materials can be used to build integrated capacitors. There are two broad classifications of dielectrics—paraelectrics and ferroelectrics—and they have very distinct electrical properties, the most significant being the dielectric constant. Some of the dielectric materials that can be used to fabricate integrated capacitors are listed in Table 4.2, along with their composition and dielectric constant.

TABLE 4.2 Dielectric Material Candidates for Application in Integrated Capacitors

Material	Composition	Dielectric constant	Paraelectric (P) or Ferroelectric (F)
Silicon monoxide	SiO	4.5–6.8	F
Silicon dioxide	SiO_2	4–5	F
Silicon nitride	Si_3N_4	6–7	F
Silicon carbide	SiC	20–45	F
Aluminium oxide	Al_2O_3	6–10	F
Aluminium nitride	AIN	8–10	F
Tantalum oxide (a)	Ta_2O_5	25	F
Tantalum oxide (h)	Ta_2O_5	50	F
Titanium oxide	TiO_2	10–100	F
Diamond-like-carbon	sp^2 and sp^3 carbon	4–6	F
BCB	Organic	2.7	F
Polyimide	Organic	3–4.5	F
Barium strontium titanate	$BaSrTiO_3$	up to 1000	p
Lead zirconate titanate	$PbZr_xTi\text{-}_xO_3$	up to 2000	P
BPZT	$Ba_{0.8}Pb_{0.2}(Zr_{0.12}Ti_{0.88})O_3$	up to 3000	P
Barium titanate	$BaTiO_3$	up to 5000	P

Integrated inductors are the easiest of the IPs to fabricate since they are usually spirals of the conductor material. Also, since inductors are magnetic devices, they pose an integration problem not shared by resistors and capacitors: They perform best when there is a sufficient volume of space to allow their magnetic fields to be unimpeded by other structures. As a result, there is a "keep-away" distance required for inductors in order to avoid loss of inductance relative to an isolated structure and to prevent interference with nearby signal lines and ground planes from the inductor's field.

The easiest way to fabricate integrated inductors is simply to form a spiral out of the interconnect material, typically one to eight turns with a total outer diameter of about 0.2 mm, which provides an inductance of around 1–40 mH.[3]

4.6.4 Hybrid SnPb- and Pb-free Systems

In an ideal world, the transition from SnPb to Pb-free could happen overnight. However, in reality, the transition is not likely to happen instantly; instead, this is expected to occur over a period of at least 6 to 12 months. During this time, SnPb and lead-free alloys are likely to coexist on the same printed circuit board assembly (PCBA), which poses potential issues with component soldering and reliability.

When a BGA with lead-free solder is reflowed with SnPb solder paste on the motherboard using a SnPb reflow profile, the lead-free solder balls stay virtually in an unmolten state. However, the lead in the eutectic tin-lead solder paste can diffuse through the grain boundaries in the lead-free BGA solder ball, the extent of which depends on several factors such as the maximum reflow temperature and the time during which this temperature stays over the melting temperature of SnPb solder. This can result in potential solder joint reliability issues, as the Pb-rich grain boundaries have a higher risk of microcracks, which can lead to solder opens.

The industry is also exploring benefits of an approach called backward-compatibility, where the component termination is lead-free, but the on-board solder is Pb containing solder—eutectic SnPb in many cases. Such a joint is neither Pb-free nor RoHS-compliant, but can allow for the component supplier to deliver lead-free components while letting the customer keep the assembly materials and reflow profiles mostly constant.

4.7 ACKNOWLEDGMENT

The first version of this chapter was written for the fifth edition of this handbook by Drs. Simon Ang, Fred Barlow, William Brown, and Tarak A. Railkar.

REFERENCES

1. Brown, William D., and Ulrich, Richard K. (eds.), *Advanced Electronic Packaging, 2nd ed.*, Wiley-IEEE Press, 2006.
2. Thompson, Terrence, "Top 10+ Critical Issues for HDI Manufacturing," *HDI*, August 1999, pp. 34–43.
3. Ulrich, R. K., Brown, W. D., Ang, S. S., Barlow, F. D., Elshabini, A., Lenihan, T. G., Naseem, H. A., Nelms, D. M., Parkerson, J., Schaper, L. W., and Morcan, G., "Getting Aggressive with Passive Devices," *Circuits and Devices,* Vol. 16, No. 5, pp. 17–25, 2000.
4. Baliga, John, "High-Density Packaging: The Next Interconnect Challenge," *Semiconductor International,* February 2000, pp. 91–100.
5. *International Technology Roadmap for Semiconductors,* 1999, available online at http://public.itrs.net
6. Fjelstad, J., and DiStefano, T., "Where Are We Headed?" *Electronic Packaging and Production*, December 1999, pp. 16–20.
7. Maner, K. U., Porter, E. V., Schaper, L. W., Ang, S. S., and Brown, W. D., "The Seamless High Off-Chip Connectivity (SHOCC) Packaging Technology," *Electrochemical Society Proceedings,* Vol. 99–7, 1999, pp. 188–194.

8. Shutler, William F., "Examining Technology Options for System on a Package," *Electronic Packaging and Production*, September 2000, pp. 32–36.
9. Davidson, E., "Large Chip vs. MCM for a High-Performance System," *IEEE Micro*, July/August 1998, pp. 33–41.
10. Rinebold, Kevin, "Few-Chip Packaging: An MCM Renaissance," *HDI*, October 2000, pp. 18–21.
11. Brinton, James B., "SOC Isn't Cutting It Yet. Is Multi-Chip Package a Better Answer Today?" *Semiconductor Business News*, January 28, 2000.
12. "Analog Devices, Technology Background," available online at http://www.analog.com/industry/dsp/bga/background.html.
13. Frear, Darrel R., "Materials Issues in Area-Array Microelectronic Packaging," *Journal of Materials*, Vol. 51, No. 3, 1999, pp. 22–27.
14. Barrett, Kelly, "A Rose and an MCM by Any Other Name . . . . ," Electronic News Online, May 17, 1999, http://www.electronicnews.com/enews/Issue/1999/05171999/20mcmbl.asp.
15. Landers, Thomas L., Brown, William D., Fant, Earnest W., Malstrom, Eric M., and Schmitt, Neil M., *Electronic Manufacturing Processes*, Prentice Hall, 1994.
16. Wong, K. K. H., Kaja, S., and DeHaven, P. W., "Metallization by Plating for High-Performance Multichip Modules," *IBM Journal of Research and Development*, Vol. 42, No. 5, 1998, available online at http://www.research.ibm.com/journal/rd/425/wong.html.
17. Carpenter, Karen, "IBM Packaging Technology Keeping Pace with Semiconductor Roadmap," *IBM Microelectronics MicroNews*, Vol. 4, No. 4, 1998, available online at http://www.chips.ibm.com/micronews/vol4_no4/packaging.html.
18. Woida, Jack, "The Challenge of System on a Chip," *Electronic Packaging and Production*, May 2000, pp. 18–26.
19. Carchon, G., Brebels, S., Vaesen, K., Pieters, P., Nauwelaers, B., and Beyne, E., "Design of Microwave MCM-D CPW Quadrature Couplers and Power Dividers in X-, Ku- and Ka-band," *International Journal of Microcircuits and Electronic Packaging*, Vol. 23, No. 3, 2000, pp. 257–264.
20. Afonso, Sergio, Schaper, Leonard W., Parkerson, James P., Brown, William D., Ang, Simon S., and Naseem, Hameed A., "Modeling and Electrical Analysis of Seamless High Off-Chip Connectivity (SHOCC) Interconnects," *IEEE Transactions on Advanced Packaging*, Vol. 22, No. 3, 1999, pp. 309–320.
21. Davidson, E., "Large Chip vs. MCM for a High Performance System," *IEEE Micro*, July/August 1998, pp. 33–41.
22. Baliga, John, "Making Room for More Performance with Chip Scale Packages," *Semiconductor International*, Vol. 21, No. 11, October 1998, pp. 85–92.
23. Baliga, John, "Ball Grid Arrays: The High Pincount Workhorse," *Semiconductor International*, September 1999, available online at http://www.semiconductor-intl.com/semiconductor/issues/issues/1999/sep99/docs/feature2.asp.
24. "WMU-VLSI, Part 5: System Packaging Styles," available online at http://www.cs.wmich.edu/~vlsi2/Pages/what/part_5.html.
25. Gann, Keith D., "Neo-Stacking Technology," *HDI*, December 1999, pp. 20–22.
26. Chen, Daniel, Hellmold, Steffen, Yee, Mike, and Kilbuck, Kevin, "Stacked Multi-Chip CSP Standards: A Push Forward," *Advanced Packaging*, June/July 2000, pp. 49–53.
27. Francis, David, and Jardine, Linda, "Stackable 3-D Chip-Scale Package Uses Silicon as the Substrate," *Chip Scale Review*, May/June 2000, p. 79.
28. Leseduarte, S., Marco, S., Beyne, E., Van Hoof, R., Marty, A., Pinel, S., Vendier, O., and Coello-Vera, A., "Residual Thermomechanical Stresses in Thinned-Chip Assemblies," *IEEE Transactions on Components and Packaging Technologies*, Vol. 23, No. 4, 2000, pp. 673–679.
29. Rates, Jim, "KGD: A State of the Art Report," *Advanced Packaging*, September 1999, pp. 30–36.
30. Beddingfield, Craig, Ballouli, Walid, Carney, Frank, and Nair, Raj, "Wafer-Level KGD," *Advanced Packaging*, September 1999, pp. 26–30.
31. Fry, Michael A., Kline, Jerry D., Prince, John L., and Tanel, Gary A., "In-Line KGD Test Speeds Flip Chip Assembly," *Electronic Packaging and Production*, February 2001, pp. 36–41.
32. Baliga, J., "Contract Bumping is Increasing," *Semiconductor International*, March 1999, p. 42.

33. Nappi, L., Rapelo, J., Pohjonen, H., Holloway, P., and Ristolainen, E., "Integrating Passive Components with Thin-Film Deposition," *HDI*, July 2000, pp. 38–48.
34. Logan, Elizabeth A., and Young, James L., "Integration Ends Passives' Domination of Wireless Circuits," *Electronic Packaging and Production*, August 2000, pp. 26–30.
35. Hou, Zhenwei, Johnson, R. Wayne, Yaeger, Erin, Konarski, Mark, and Crane, Larry, "Lead-Free Solder for Flip Chip," *HDI*, September 2000, pp. 38–47.
36. Young, Jedediah J., Malshe, Ajay P., Brown, W. D., and Lenihan, Timothy, "Modeling and Analysis of Very Thin Silicon Chips for Conformal Electronics Systems," paper presented at the International Conference and Exhibition on High-Density Interconnect and Systems Packaging, , Santa Clara, CA, April 18–20, 2001.
37. Railkar, Tarak A., and Warren, Robert W., in Brown, William D., and Ulrich, Richard K. (eds.), "Electronic Package assembly" *Advanced Electronic Packaging, 2nd ed.*, Wiley-IEEE Press, 2006.
38. http://europa.eu.int/eur-lex/pri/en/oj/dat/2003/l_037/l_03720030213en00190024.pdf.
39. Malshe, Ajay P., O'Neal, Chad, Singh, Sushila B., Brown, William D., Eaton, William P., and Miller, William M., "Challenges in the Packaging of MEMS," *International Journal of Microcircuits and Electronic Packaging*, Vol. 22, No. 3, 1999, pp. 233–241.
40. O'Neal, Chad B., Malshe, Ajay P., Schmidt, William F., Gordon, Matthew H., Reynolds, Robert R., Brown, William D., Eaton, William P., and Miller, William M., "A Study of the Effects of Packaging Induced Stress on the Reliability of the Sandia MEMS Microengine," paper presented at the Pacific Rim/ASME International Electronic Packaging Technical Conference and Exhibition, Kauai, HI, July 8–13, 2001.

CHAPTER 5
TYPES OF PRINTED WIRING BOARDS

Dr. Hayao Nakahara
N.T. Information Ltd., Huntington, New York

5.1 INTRODUCTION

Since the invention of printed wiring technology by Dr. Paul Eisner in 1936, several methods and processes have been developed for manufacturing printed wiring board (PWBs) of various types. Most of these have not changed significantly over the years; however, some specific trends continue to exert major influences on the types of PWBs required and the processes that create them:

1. Computers and portable telecommunications equipment require higher-frequency circuits, boards, and materials, and also use more functional components that generate considerable amounts of heat that need to be extracted.
2. Consumer products have incorporated digital products into their design, requiring more functionality at ever-lower total cost.
3. Products for all uses continue to get smaller and more functional, driving the total circuit package itself to become more dense, causing the PWBs to evolve to meet these needs.

These trends have led to the larger use of nonorganic base substrates, such as aluminum and soft iron. In addition, alternate ways to create boards have been developed. These will be discussed in this chapter, along with the traditional board structures and processes. The terms *printed wiring board, PWB,* and *board* will be used synonymously. Also, the words *laminate, substrate,* and *panel* will be used interchangeably.

5.2 CLASSIFICATION OF PRINTED WIRING BOARDS

PWBs may be classified in many different ways according to their various attributes. One fundamental structure common to all of them is that they must provide electrical conductor paths which interconnect components to be mounted on them.

5.2.1 Basic PWB Classifications

There are two basic ways to form these conductors:

1. *Subtractive:* In the subtractive process, the unwanted portion of the copper foil on the base substrate is etched away, leaving the desired conductor pattern in place.
2. *Additive:* In the additive process, formation of the conductor pattern is accomplished by adding copper to a bare (no copper foil) substrate in the pattern and places desired. This can be done by plating copper, screening conductive paste, or laying down insulating wire onto the substrate on the predetermined conductor paths.

The PWB classifications given in Fig. 5.1 take into consideration all these factors, i.e., fabrication processes as well as substrate material. The use of this figure is as follows:

- Column 1 shows the classification of PWBs by the nature of their substrate.
- Column 2 shows the classification of PWBs by the way the conductor pattern is imaged.
- Column 3 shows the classification of PWBs by their physical nature.
- Column 4 shows the classification of PWBs by the method of actual conductor formation.

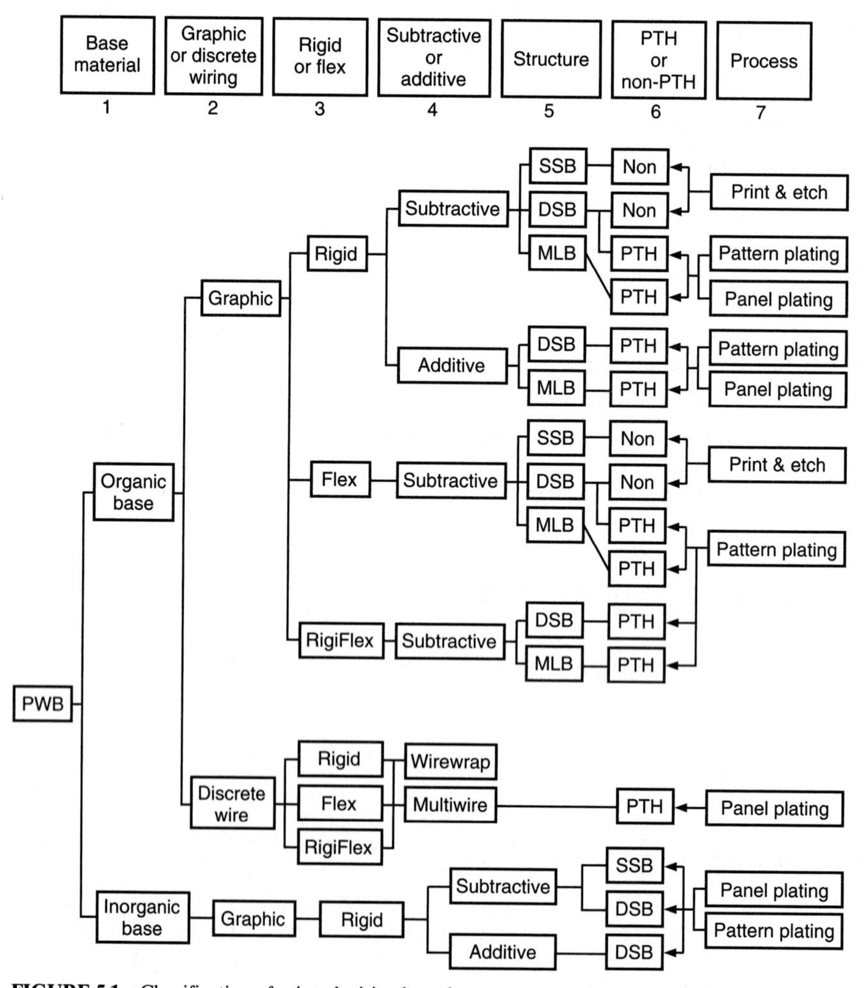

FIGURE 5.1 Classification of printed wiring boards.

- Column 5 shows the classification of PWBs by the number of conductor layers.
- Column 6 shows the classification of PWBs by the existence or absence of plated-through-holes (PTHs).
- Column 7 shows the classification of PWBs by production method.

5.3 ORGANIC AND NONORGANIC SUBSTRATES

One of the major issues that has arisen with the ever-higher speed and functionality of components used in computers and telecommunications is the availability of materials for the PWB substrate that are compatible with these product and process needs. This includes the stresses on substrate material created by more and longer exposure to soldering temperatures during the assembly process, as well as the need to match the coefficient of thermal expansion for components and substrate. The resultant search has found new materials, both organic and nonorganic based. The details of these materials are explained in Chaps. 6 through 11, but this outlines the basic character of the two types of substrate.

5.3.1 Organic Substrates

Organic substrates consist of layers of paper impregnated with phenolic resin or layers of woven or nonwoven glass cloth impregnated with epoxy resin, polyimide, cyanate ester, BT resin, etc. The usage of these substrates depends on the physical characteristics required by the application of the PWB, such as operating temperature, frequency, or mechanical strength.

5.3.2 Nonorganic Substrates

Nonorganic substrates consist mainly of ceramic and metallic materials such as aluminum, soft iron, and copper-invar-copper. The usage of these substrates is usually dictated by the need of heat dissipation, except for the case of soft iron, which provides the flux path for flexible disk motor drives.

5.4 GRAPHICAL AND DISCRETE-WIRE BOARDS

Printed wiring boards may be classified into two basic categories, based on the way they are manufactured:

1. Graphical
2. Discrete-wire

5.4.1 Graphical Interconnection Board

A *graphical PWB* is the standard PWB and the type that is usually thought of when PWBs are discussed. In this case, the image of the master circuit pattern is formed photographically on a photosensitive material, such as treated glass plate or plastic film. The image is then transferred to the circuit board by screening or photoprinting the artwork generated from the master. Due to the speed and economy of making master artwork by laser plotters, this master can also be the working artwork.

Direct laser imaging of the resist on the PWB can also be used. In this case, the conductor image is made by the laser plotter, on the photoresistive material, which is laminated to the board, without going through the intermediate step of creating a phototool. This tends to be somewhat slower than using working artwork as the tool and is not generally applied to mass production. Work continues on faster resists, as well as exposure systems, and this method will undoubtedly continue to emerge.

5.4.2 Discrete-Wire Boards

Discrete-wire boards do not involve an imaging process for the formation of signal conductors. Rather, conductors are formed directly onto the wiring board with insulated copper wire. Wire-wrap® and Multiwire® are the best known discrete-wire interconnection technologies. Because of the allowance of wire crossings, a single layer of wiring can match multiple conductor layers in the graphically produced boards, thus offering very high wiring density. However, the wiring process is sequential in nature and the productivity of discrete-wiring technology is not suitable for mass production. Despite this weakness, discrete-wiring boards are in use for some very high density packaging applications. See Fig. 5.2 for an example of a discrete-wiring board.

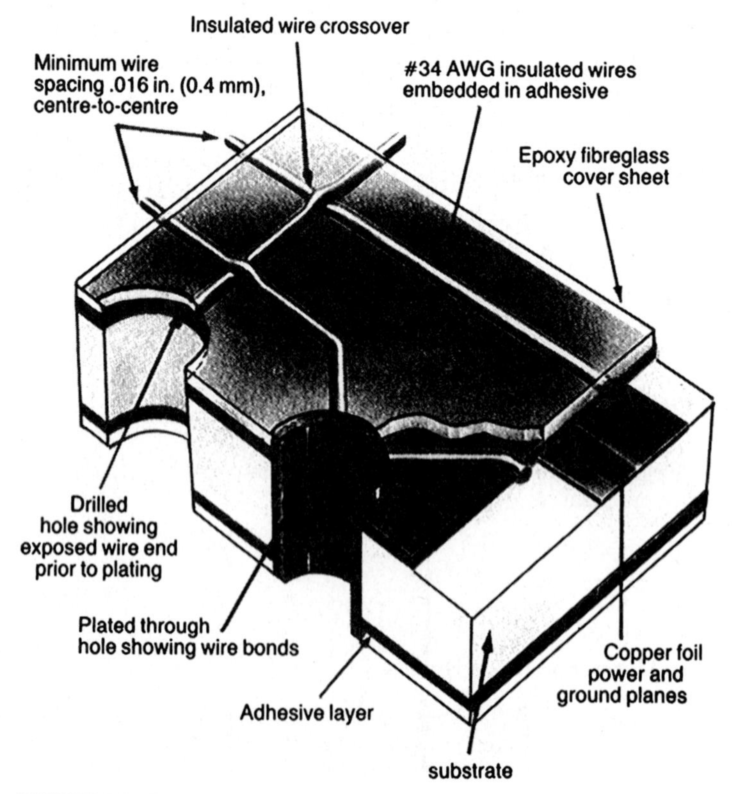

FIGURE 5.2 Example of discrete-wiring board.

5.5 RIGID AND FLEXIBLE BOARDS

Another class of boards is made up of the *rigid* and *flexible* PWBs. Whereas boards are made of a variety of materials, flexible boards generally are made of polyester and polyimide bases. *Rigi-flex boards,* a combination of rigid and flexible boards usually bonded together, have gained wide use in electronic packaging (see Fig. 5.3). Most rigi-flex boards are three-dimensional

FIGURE 5.3 Rigi-flex printed wiring board.

structures that have flexible parts connecting the rigid boards, which usually support components; this packaging is thus volumetrically efficient.

5.6 GRAPHICALLY PRODUCED BOARDS

The majority of boards produced in the world are graphically produced. There are three alternative types:

1. Single-sided boards
2. Double-sided boards
3. Multilayer boards

5.6.1 Single-Sided Boards (SSBs)

Single-sided boards (SSBs) have circuitry on only one side of the board and are often referred to as print-and-etch boards because the etch resist is usually printed on by screen-printing techniques and the conductor pattern is them formed by chemically etching the exposed, and unwanted, copper foil.

5.6.1.1 Typical Single-Sided Board Materials. This method of board fabrication is generally used for low-cost, high-volume, and relatively low functionality boards. In the Far East, for example, the majority of SSBs are made of paper-based substrates for lowest cost, with the most popular grade of paper-based laminate being XPC-FR, which is a flame-retardant phenolic material that is also highly punchable. In Europe, FR-2 grade paper laminate is the most popular substrate for SSBs because it emits less odor than XPC-FR when placed in high-voltage, high-temperature environments, such as inside a television set chassis. In the United States, CEM-1 material, which is a composite of paper and glass impregnated with epoxy resin, is the most popular substrate for SSBs. While not as low cost as XPC-FR or FR-2, CEM-1 has gained popularity because of its mechanical strength and also because of the relative unavailability of paper phenolic laminates.

5.6.1.2 Single-Sided Board Fabrication Process. Given the emphasis on cost and low complexity, SSBs are generally produced in highly automated, conveyorized print-and-etch lines, using the following basic process flow.

Step 1: Cut substrate into appropriate panel size by either sawing or shearing.
Step 2: Place panel in loader which feeds them into the line.
Step 3: Clean panels.
Step 4: Screen panel with ultraviolet curable etch-resist ink.
Step 5: Cure the etch-resist ink.
Step 6: Etch exposed copper.
Step 7: Strip the resist.
Step 8: Apply solder resist.
Step 9: Screen legend.
Step 10: Form holes by drilling or punching.
Step 11: Test for shorts and opens.

The conveyor speed of automated print-and-etch lines ranges from 30 to 45 ft/min. Some lines are equipped with an online optical inspection which enables the elimination of the final electrical open/short test.

As previously noted, after the conductor pattern is generated in the print-and-etch line, holes for component insertion are formed on the panel by punching when the panel is made of paper-based substrate, but must be formed by drilling when the panel is made of glass-based substrate.

5.6.1.3 Process Variations. In some variations, the conductor surface of the PWB gets insulated, exposing only pads, and then conductive paste is screened to form additional conductors on the same side of the board, thus forming double conductive layers on a single side.

Most metal-core PWB consumer applications are made of aluminum substrate, which comes as a copper-clad material. PWBs made of such material do not have through-holes, and components are usually surface-mount types. These circuits are frequently formed into three-dimensional shapes.

5.6.2 Double-Sided Boards

By definition, double-sided boards (DBs) have circuitry on both sides of the boards. They can be classified into two categories:

1. Without through-hole metallization
2. With through-hole metallization

The category of through-hole metallization can be further broken into two types:

1. Plated-through-hole (PTH)
2. Silver-through-hole (STH)

5.6.2.1 Plated-Through-Hole Technology. PTH technology is discussed in some detail in Sec. 5.7; however, some comments are appropriate here.

Metallization of holes by copper plating has been practiced since the mid-1950s. Since PWB substrate is an insulating material, and therefore nonconductive, holes must be metallized first before subsequent copper plating can take place. The usual metallization procedure is to catalyze the holes with palladium catalyst followed by electroless copper plating. Then, thicker plating is done by galvanic plating. Alternately, electroless plating can be used to plate all the way to the desired thickness, which is called *additive plating*.

The biggest change in the manufacturing process of double-sided PTH boards, and also of multilayer boards (MLBs), is the use of *direct metallization* technologies. (See Chap. 30 for full discussion of electroless and direct metallization for through-hole boards.) Here, simply, it eliminates the electroless copper process. The hole wall is made conductive by palladium catalyst, carbon, or polymer conductive film, then copper is deposited by galvanic plating. The elimination of electroless copper, in turn, allows the elimination of environmentally hazardous chemicals, such as formaldehyde, and EDTA, which are two main components of electroless copper-plating solutions.

5.6.2.2 Silver-Through-Hole Technology. STH boards are usually made of paper phenolic materials or composite epoxy paper and glass materials, such as CE-1 or CE-3. After double-sided copper-clad materials are etched to form conductor patterns on both sides of the panel, holes are formed by drilling. Then the panel is screened with silver-filled conductive paste. Instead of silver, copper paste can also be used.

Since STHs have a relatively high electrical resistance compared with PTHs, the application of STH boards is limited. However, because of their economic advantage (the cost of

STH boards is usually one-half to two-thirds that of functionally equivalent PTH boards), their application has spread to high-volume, low-cost products such as audio equipment, floppy disk controllers, car radios, remote controls, etc.

5.6.3 Multilayer Boards (MLBs)

By definition, MLBs have three or more circuit layers (see Fig. 5.4). Main applications of MLBs used to be confined to sophisticated industrial electronic products. Now, however, they are the mainstream of all electronic devices, including consumer products such as portable video cameras, cellular phones, and audio discs.

5.6.2.3 Layer Count. As personal computers and workstations become more powerful, mainframe computers and supercomputers are being replaced in many applications by these smaller machines. As a result, the use of highly sophisticated MLBs, which have layer counts over 70, are being reduced, but the technology to produce them is proven. At the other end of the layer-count spectrum, thin and high-density MLBs with layer counts between 4 and 8 are mainstream. The drive toward thinner MLBs will continue and is made possible by the continuing concurrent advancement of materials and equipment to handle thin core materials.

5.6.2.4 Via and Via Production Technologies. As PWBs have had to address the issues of higher speed, higher density, and the rise of surface-mount components that use both sides, the need to communicate between layers has increased dramatically. At the same time, the space available for vias has decreased, causing a continuing trend toward smaller holes, more holes on the board, and the decline of the use of holes that penetrate the entire board, which use space on all layers. As a result, the use of buried and blind vias has become a standard part of multilayer board technology, driven by the need for this increased package density (Fig. 5.4).

One of the immediate issues that arise from these trends is the problems of drilling and the associated cost of this fabrication step. Printed wiring boards, which once were stacked three high on a drilling machine, must be drilled individually, and the number of holes per board has risen, to accommodate the need for vias. This has caused a major problem for fabricators, who find that a lack of drilling capacity is creating a big demand on funds for additional machines, while the cost of drilling continues to increase dramatically. Therefore, alternate methods for creating vias are being developed. These pressures will be ongoing, and therefore the process listed here, or some equivalent, will undoubtedly become more important as the drive to miniaturization continues and drilling individual holes becomes less and less practical.

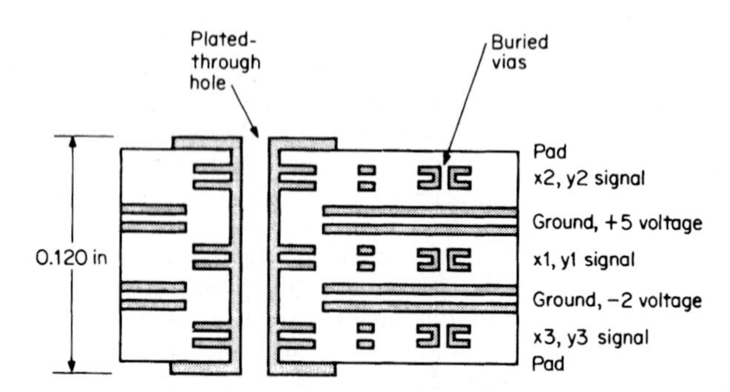

FIGURE 5.4 Cross-section multilayer board with buried via holes. Buried vias are built into each of the double-sided boards that make up the final multilayer structure.

These processes have been developed to mass-produce vias without drills.

Surface Laminar Circuits (SLCs). The most notable MLB technology developed to form vias is the *sequential* fabrication of multilayers without press operations. This is particularly important for surface blind via holes.

The process for fabricating a board using surface laminar circuits is as follows (see Fig. 5.5):

1. Innerlayer ground and power distribution patterns are formed.
2. Panel receives an oxide treatment.
3. Insulating photosensitive resin is coated over the panel by curtain or screen-coating methods.
4. Holes are formed by photoexposure and development.
5. Panel is metallized by usual copper reduction process (consisting of catalyzing and electroless copper plating or by direct metallization process).
6. Thicker deposition of copper is made by continuation of electroless copper plating or galvanic plating.
7. Circuit patterns are formed by dry film tenting process.

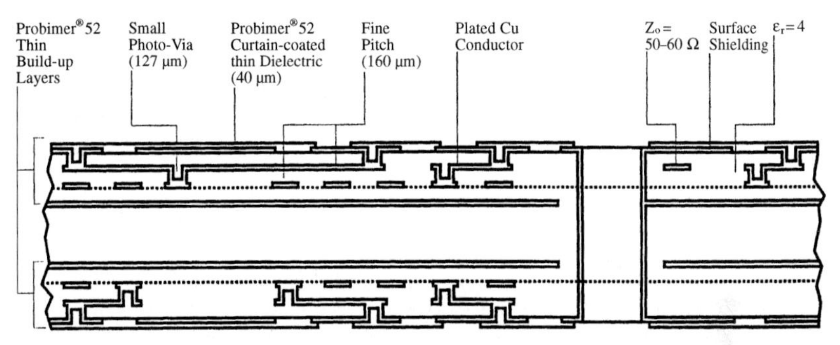

FIGURE 5.5 Example of surface laminar circuit (SLC) board cross section. *(Courtesy IBM Yasu and Ciba-Geigy Limited.)*

DYCOstrate®. A different approach to small via creation has been taken by Dyconex AG of Switzerland. After ground and power patterns are formed on the panel, and the panel is oxide-treated, polyimide-backed copper foil is laminated on the panel. Holes in the copper are formed by a chemical etching process, and the insulating polyimide material underneath the holes is removed by plasma etching. PWBs made in such a way are called DYCOstrate. In other, similar technologies, different dielectric materials are used, and they are removed by alkaline solutions. The rest of the process is similar to that for SLC; that is, holes are metallized and a thick copper deposition is made by electroless or galvanic plating, and the circuit pattern is formed by a tent-and-etch process (see Fig. 5.5).

Drilled Vias. In both SLC and DYCOstrate cases, through-holes can also be made by conventional drilling and plating processes, in addition to the surface blind via holes.

Cost Impact. The manufacturing cost of these sequential technologies is not necessarily directly cheaper than conventional MLB technology, which depends on a laminating press operation. However, since the cost of making standard holes in a board can be as high as 30 percent of the total manufacturing cost, and the creation of holes in these processes is comparatively inexpensive, the overall cost for equivalent functionality can be less. In addition, the fine pattern capability for this process is excellent. For example, an eight-layer conventional structure can often be reduced to a four-layer structure, reducing the total cost for the same packaging density.

5.7 MOLDED INTERCONNECTION DEVICES

Three-dimensional circuit technology was of great interest in the early to mid-1980s. The proponents for this technology, however, realized the mistake of trying to make it directly competitive with conventional flat circuits and have developed a niche where the substrate also offers other functional uses, such as structural support for the product.

Manufacturers of three-dimensional circuits prefer to call them *molded interconnection devices* (MIDs). In many applications of MIDs, the number of components to interconnect the electronic and electrical components can be reduced, thus making the total assembly cost cheaper and the final structure more reliable.

5.8 PLATED-THROUGH-HOLE (PTH) TECHNOLOGIES

In 1953, the Motorola Corporation developed a PTH process called the *Placir* method,[1] in which the entire surface and hole walls of an unclad panel are sensitized with $SnCl_2$ and metallized by spraying on silver with a two-gun spray. Next, the panel is screened with a reverse conductor pattern, using a plating resist ink, leaving metallized conductor traces uncovered. The panel is then plated with copper by an electroplating method. Finally, the resist ink is stripped and the base silver removed to complete the PTH board. One problem associated with the use of silver is the migration caused by silver traces underneath the copper conductors.

The Placir method was the forerunner of the semiadditive process, which is discussed in Chap. 31.

In 1955, Fred Pearlstein[2] published a process involving electroless nickel plating for metallizing nonconductive materials. This catalyzer consists of two steps. First, the panel is sensitized in $SnCl_2$ solution, and then it is activated in $PdCl_2$ solution. This process presented no problem for metallizing nonconductive materials.

At the same time that Pearlstein's paper was published, copper-clad laminates were starting to become popular. Manufacturers of PWBs applied this two-step catalyzing process to making PTHs using copper-clad laminates. This process, however, turned out to be incompatible with the copper surface. A myriad of black palladium particles called *smads* were generated between copper foil and electrolessly deposited copper, resulting in poor adhesion between the electroless copper and the copper foil. These smads and electroless copper had to be brushed off with strong abrasive action before the secondary electroplating process could begin. To overcome this smad problem, around 1960 researchers began attempting to develop better catalysts; the products of their research were the predecessors of modern palladium catalysts.[3]

The mid-1950s was a busy time in the area of electroless copper-plating solutions. Electrolessly deposited nickel is difficult to etch. But since it adheres somewhat better to the base than does electroless copper, research for the development of stable electroless copper-plating solutions was quite natural. Many patent applications for these solutions were filled in the mid-1950s. Among the applicants were P. B. Atkinson, Sam Wein, and a team of General Electric engineers, Luke, Cahill, and Agens. Atkinson won the case, and a patent[4] teaching the use of Cu-EDTA as a complexing agent was issued in January 1964 (the application had been filed in September 1956).

5.8.1 Subtractive and Additive Processes

Photocircuits Corporation was another company engaged throughout the 1950s in the development of chemicals for PTH processes. Copper-clad laminates were expensive, and a major portion of expensive copper foil had to be etched (*subtracted*) to form the desired conductor pattern. The engineers at Photocircuits, therefore, concerned themselves with plating (*adding*) copper conductors wherever necessary on unclad materials for the sake of economy. Their efforts paid off. They were successful in developing not only the essential chemicals for reliable PTH processes but also the fully additive PWB manufacturing technology known as the CC-4* process. This process is discussed in detail in Chap. 31.

With the use of $SnCl_2$-$PdCl_2$ catalysts and EDTA-base electroless copper-plating solutions, the modern PTH processes became firmly established in the 1960s. The process of metallizing hole walls with these chemicals for the subsequent formation of PTHs is commonly called the *copper reduction process.* In the subtractive method, which begins with copper-clad laminates, pattern plating and panel plating are the two most widely practiced methods of making PTH boards. These methods are discussed in the following subsections.

5.8.2 Pattern Plating

In the pattern-plating method, after the copper reduction process, plating resist layers of the reverse conductor image are formed on both sides of the panel by screening resist inks. In most fineline boards, photosensitive dry film is used instead. There are some minor variations in the pattern-plating method (see Fig. 5.6):

1. Catalyzing (preparing the nonconductive surface to cause copper to come out of solution onto that surface)
2. Thin electroless copper (0.00001 in) followed by primary copper electroplating; thick electroless copper (0.0001 in)
3. Imaging (application of a plating resist in the negative of the desired finished circuit)
4. Final electroplating copper
5. Solder plating (as etching resist) 0.0002 or 0.0006 in
6. Stripping plating resist
7. Etching of base copper
8. Solder etching (0.0002-in case); solder reflow (0.0006-in case)
9. Solder mask followed by hot-air solder coater leveler if solder etching is used
10. Final fabrication and inspection

Most manufacturers of DSBs with relatively wide conductors employ thick electroless copper plating. However, thin electroless copper followed by primary electroplating is preferred for boards having fine-line conductors, because a considerable amount of surface is brushed off for better adhesion of dry film. This provides a higher reliability for PTHs. Solder reflow boards had been preferred by many customers, particularly in military and telecommunications applications, until the emergence of hot-air solder coater levelers. Although the solder-over-copper conductors protect the copper from oxidization, solder reflow boards have some limitations. Solder mask is hard to apply over reflowed solder, and it tends to wrinkle and peel off in some areas when the boards go through component soldering. A more serious problem is the solder bridging that occurs when the conductor width and clearance become very small.

In step 9, the entire surface of the board except for the pads is covered by solder mask, and then the board is immersed into the hot-air solder coater leveler, resulting in a thin coating of

* CC-4 is a registered trademark of Kollmorgen Corporation.

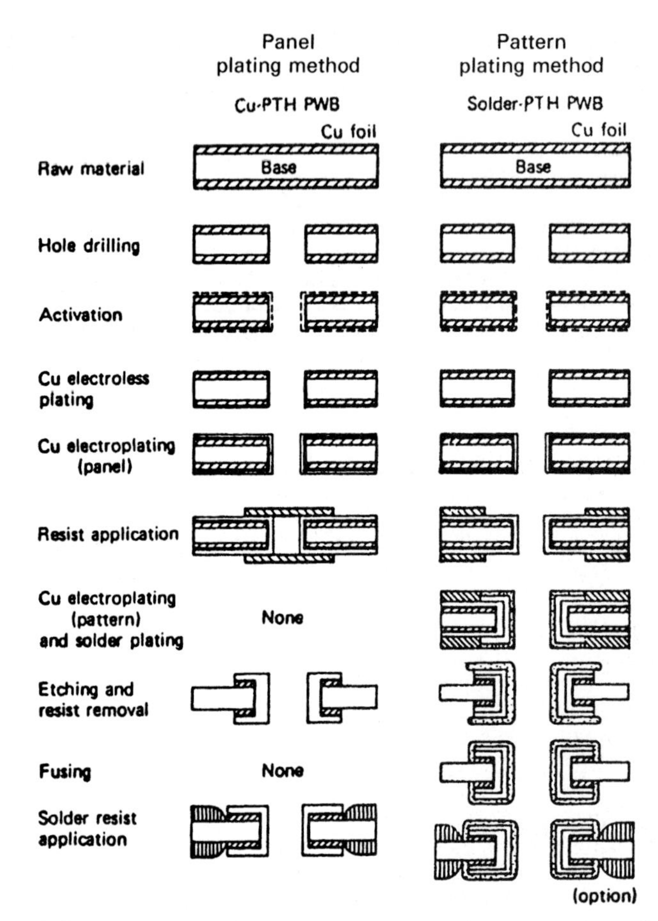

FIGURE 5.6 Key manufacturing steps in panel plating and pattern-plating methods.

solder over the pads and the hole walls. The operation sounds simple, but it requires constant fine-tuning and maintenance of the hot-air solder coater leveler; otherwise, some holes may become heavily clogged with solder and are then useless for component insertion.

One advantage of the pattern-plating method over the panel-plating method is in etching. The pattern-plating method needs to etch only the base copper. The use of ultrathin copper foil (UTC), which is usually ⅛ or ¼ oz thick, offers a real advantage. However, as long as electroplating is used, the pattern-plating method cannot escape from a current distribution problem, regardless of the thickness of the base foils. The panel-plating method by electroplating suffers from the same problem but to a lesser degree. Good current distribution is very difficult to achieve when the boards are not of the same size or type, and particularly if some have large ground planes on the outer faces being plated. When the board has a few holes in an isolated area remote from the bulk of the circuitry, these tend to become overplated, making component lead insertion difficult during assembly. To minimize this current distribution problem, various countermeasures are practiced, such as special anode position, anode masking, agitation, and plating thieves. But none of these offers a decisive solution to the distribution problem, and they are extremely difficult to implement flexibly and effectively in a large plating operation, where a large number of product mixes have to be handled all the time.

Another advantage of the pattern-plating method is its ability to form padless micro-via holes of a diameter ranging from 0.012 to 0.016 in. Micro-via holes enable better usage of conductor channels, thereby increasing the connective capacity of the board.

5.8.3 Panel Plating

In the panel-plating method, there are two variations for finishing the board after the panel is plated with electrolytic copper to the desired thickness. In the *hole-plugging* method, the holes are filled with alkaline-etchable ink to protect the hole walls from being etched; this is used in conjunction with screened etch resist. In the other method, called *tent-and-etch* or simply *tenting,* the copper in the hole is protected from etching by covering the hole or tenting with dry film, which is also used as an etch resist for conductors on the panel surface. The simplified sequence of the panel-plating method is as follows (see Fig. 5.6):

1. Catalyzing
2. Thin electroless copper deposition (0.0001 in)
3. Electroplating copper (0.001 to 0.0012 in)
4. Hole plugging with alkaline-resolvable ink; tenting (dry film lamination)
5. Screen-print etching resist (conductor pattern); photoexpose the panel for conductor pattern
6. Etching copper
7. Stripping etching resist
8. Solder mask
9. Solder coater leveler (optional)
10. Final fabrication and inspection

The panel-plating method is ideal for bare copper board. However, it is a difficult way to make padless via holes, which are becoming more popular. Generally, the conductor width of 0.004 in is considered to be the minimum realizable by this method for mass production.

Although the use of the panel-plating method in the United States and western Europe is limited, nearly 60 percent of the PTH boards in Japan are manufactured by this method.

5.8.4 Additive Plating

Plated-through holes can be formed by additive (electroless) copper deposition, of which there are three basic methods: fully additive, semiadditive, and partially additive. Of these, semiadditive involves pattern electroplating for PTHs with very thin surface copper, but the other two form PTHs solely by electroless copper deposition. The additive process has various advantages over the subtractive process in forming fineline conductors and PTHs of high aspect ratio. A detailed account of the additive process is given in Chap. 31.

5.9 SUMMARY

Modern electronic packaging has become very complex. Interconnections are pushed more into lower levels of packaging. The choice of which packaging technology to use is governed by many factors: cost, electrical requirements, thermal requirements, density requirements, and so on. Material also plays a very important role. All things considered, PWBs still play important roles in electronic packaging.

REFERENCES

1. Robert L. Swiggett, *Introduction to Printed Circuits,* John F. Rider Publisher, Inc., New York, 1956.
2. Private communication with John McCormack, PCK Technology, Division of Kollmorgen Corporation.
3. C. R. Shipley, Jr., U.S. Patent 3,011,920, Dec. 5, 1961.
4. R. J. Zebliski, U.S. Patent 3,672,938, June 27, 1972.

P · A · R · T · 3

MATERIALS

CHAPTER 6
INTRODUCTION TO BASE MATERIALS

Edward J. Kelley
Isola Group, Chandler, Arizona

6.1 INTRODUCTION

The field of base materials can appear deceptively simple. In simple terms, base materials are comprised of just three components: the resin system, the reinforcement, and the conductive foil. However, the variants in each of these components and the many possible combinations of these components make the discussion of base materials much more complex. One of the primary reasons for this complexity is the fact that printed circuits are used in so many differing applications. This results in many different sets of requirements in terms of cost and performance, and therefore many grades of base materials.

Also, because base materials are the most fundamental components of the printed circuit itself, they interact with virtually every other printed circuit manufacturing process. Therefore, not only are the physical and electrical properties of the material critical, but their compatibility with manufacturing processes is also of great importance.

In addition, the advent of the European Union's Restriction of Hazardous Substances (RoHS) directive and the lead-free assembly processes that result are redefining the requirements for base materials. RoHS has a severe impact on all aspects of base materials technology. The impact of lead-free assembly on base materials and a method of selecting materials for lead-free assembly are discussed in Chaps. 10 and 11. Requirements to support circuit densification, reliability, and electrical performance are also critical and will be discussed in Chap. 9. This chapter discusses grades and specifications of base materials, as well as the manufacturing processes used to make them.

6.2 GRADES AND SPECIFICATIONS

The various types of base materials can be classified by the reinforcement type, the resin system used, the glass transition temperature (T_g) of the resin system, as well as many other properties of the material. With the advent of lead-free assembly processes, properties other than T_g are becoming as important, if not more important when selecting a base material. The decomposition temperature (T_d) is one of these properties and will be defined shortly.

The most commonly referenced specification for base materials is IPC-4101, "Specification for Base Materials for Rigid and Multilayer Printed Boards." Historically, the National Electrical Manufacturers Association, NEMA, has also been used in specifying base materials.

6.2.1 NEMA Industrial Laminating Thermosetting Products

One of the first classifications of base materials for printed circuits (and other electrical components) was completed by the National Electrical Manufacturers Association (NEMA). NEMA's industrial laminating thermosetting products standard documents many of the materials used in printed circuits as well as specifications for some of their properties. Historical NEMA grades are outlined in Table 6.1 and cross-referenced with IPC-4101 in Table 6.2. Some of the commonly used materials have been FR-2, CEM-1, CEM-3, and, of course, FR-4.

TABLE 6.1 NEMA Base Material Grades

Grade	Resin	Reinforcement	Flame Retardant
XXXPC	Phenolic	Cotton Paper	No
FR-2	Phenolic	Cotton Paper	Yes
FR-3	Epoxy	Cotton Paper	Yes
FR-4	Epoxy	Woven Glass	Yes
FR-5	Epoxy	Woven Glass	Yes
FR-6	Polyester	Mat Glass	Yes
G-10	Epoxy	Woven Glass	No
CEM-1	Epoxy	Cotton Paper/Woven Glass	Yes
CEM-2	Epoxy	Cotton Paper/Woven Glass	No
CEM-3	Epoxy	Woven Glass/Mat Glass	Yes
CEM-4	Epoxy	Woven Glass/Mat Glass	No
CRM-5	Polyester	Woven Glass/Mat Glass	Yes
CRM-6	Polyester	Woven Glass/Mat Glass	No
CRM-7	Polyester	Mat Glass/Glass Veil	Yes
CRM-8	Polyester	Mat Glass/Glass Veil	No

FR-2 is comprised of multiple plies of paper impregnated with a flame-resistant phenolic resin. It possesses good punching characteristics and is relatively low in cost. It has typically been used in simple applications such as radios, calculators, and toys, where dimensional stability and high performance are not required. FR-3 is also paper-based, but uses an epoxy resin system.

CEM-1 uses a paper-based core with woven glass cloth on the surfaces, both impregnated with an epoxy resin. This enables the material to be punched while realizing improved electrical and physical properties. CEM-1 has been used in both consumer and industrial electronics.

CEM-3, a composite of dissimilar core materials, uses an epoxy resin impregnated, nonwoven fiberglass core with epoxy resin impregnated, woven fiberglass cloth surface sheets. It is higher in cost than CEM-1, but is more suitable for plated through holes. CEM-3 has been used in early home computers, automobiles, and home entertainment products.

FR-4, by far the most commonly used material for printed circuits, is constructed of woven fiberglass cloths impregnated with an epoxy resin or epoxy resin blend. The outstanding electrical, mechanical, and thermal properties of FR-4 have made it an excellent material for a wide range of applications including computers and peripherals, servers and storage networks, telecommunications, aerospace, industrial controls, and automotive applications. FR-4 will be discussed in more detail later.

These grades are also used by ANSI, the American National Standards Institute.

TABLE 6.2 IPC-4101 Base Material Summary

Spec. Sheet #	Reinforcement Type	Resin System	ID Reference*	T_g Range	Other Properties
00	Cellulose Paper	Phenolic	XPC	N/A	UL94 HB
01	Cellulose Paper	Phenolic	XXXPC	N/A	UL94 HB
02	Cellulose Paper	Phenolic, Flame Resistant	FR-1	N/A	UL94 V-1
03	Cellulose Paper	Phenolic, Flame Resistant	FR-2	N/A	UL94 V-1
04	Cellulose Paper	Epoxy, Flame Resistant	FR-3	N/A	UL94 V-1
05	Cellulose Paper	Phenolic, Phosphorus Flame Resistant	FR-2	N/A	UL94 V-1, limits on Br and Cl content
10	Woven E-Glass Surface/ Cellulose Paper Core	Epoxy/Phenolic, Flame Resistant	CEM-1	100°C min.	UL94 V-0
11	Woven E-Glass Surface/ Non-Woven E-Glass Core	Polyester/Vinyl Ester, Flame Resistant	CRM-5	N/A	UL94 V-1, inorganic fillers
12	Woven E-Glass Surface/ Non-Woven E-Glass Core	Epoxy, Flame Resistant	CEM-3	N/A	UL94 V-0, with or without inorganic fillers
14	Woven E-Glass Surface/ Non-Woven E-Glass Core	Epoxy, Flame Resistant	CEM-3	N/A	UL94 V-0, limits on Br and Cl content
20	Woven E-Glass Fabric	Epoxy, Non-Flame Resistant	G-10, MIL-S-13949/ 03–GE/GEN	N/A	UL94 HB
21	Woven E-Glass	Difunctional and Multifunctional Epoxy, Flame Resistant	FR-4, MIL-S-13949/04–GF/ GFN/GFK/GFP/GFM	110°C min.	UL94 V-0
22	Woven E-Glass	Epoxy, Hot Strength Retention, Non-Flame Resistant	G-11 MIL-S-13949/02–GB/ GBN/GBP	135°C–175°C	UL 94 HB
23	Woven E-Glass	Epoxy, Hot Strength Retention, Flame Resistant	FR-5, MIL-S-13949/05–GH/ GHN/GHP	135°C–185°C	UL94 V-1
24	Woven E-Glass	Epoxy/Multifunctional Epoxy, Flame Resistant	FR-4, MIL-S-13949/04–GF/ GFG/GFN	150°C min.	UL94 V-0
25	Woven E-Glass	Epoxy/PPO, Flame Resistant	MIL-S-13949/04–GF/ GFG/GFN	150°C–200°C	UL94 V-1
26	Woven E-Glass	Epoxy/Multifunctional Epoxy, Flame Resistant	FR-4, MIL-S-13949/ 04–GF/GFT	170°C min.	UL94 V-0
27	Unidirectional E-Glass, Cross-Plied	Epoxy/Multifunctional Epoxy, Flame Resistant	N/A	110°C min.	UL94 V-1
28	Woven E-Glass	Epoxy/Non-Epoxy, Flame Resistant	MIL-S-13949/ 04–GFN/GFT	170°C–220°C	UL94 V-1
29	Woven E-Glass	Epoxy/Cyanate Ester, Flame Resistant	MIL-S-13949/ 04–GFN, GFT	170°C–220°C	UL94 V-1

(continued)

TABLE 6.2 IPC-4101 Base Material Summary (Continued)

Spec. Sheet #	Reinforcement Type	Resin System	ID Reference*	T_g Range	Other Properties
30	Woven E-Glass	Bismaleimide/Triazine (BT)/Epoxy	GPY MIL-S-13949/ 26–GIT/GMT	170°C–220°C	UL94 HB
31	N/A	Epoxy/Multifunctional Epoxy	N/A	90°C min.	Non-reinforced film, inorganic fillers, thermal conductivity
32	Woven E-Glass	Epoxy/Multifunctional Epoxy	N/A	90°C min.	Inorganic fillers, thermal conductivity
33	N/A	Epoxy/Multifunctional Epoxy	N/A	150°C min.	Non-reinforced film, inorganic fillers, thermal conductivity
40	Woven E-Glass	Polyimide	GPY MIL-S-13949/10–GI/ GIN/GIJ/GIP/GIL	200°C min.	UL94 HB, with or without inorganic fillers
41	Woven E-Glass	Polyimide	GPY MIL-S-13949/10– GIL/GIP	250°C min.	UL94 HB with or without inorganic fillers
42	Woven E-Glass	Polyimide/Epoxy	GPY MIL-S-13949/10–GIJ	200°C min. 200°C min.	UL94 HB with or without inorganic fillers
50	Woven Aramid	Epoxy/Multifunctional Epoxy	MIL-S-13949/15–AF/ AFN/AFG	150°C–200°C	UL94 V-1
53	Non-Woven Aramid Paper	Polyimide	MIL-S-13949/31–BIN/BIJ	220°C min.	UL94 HB
54	Unidirectional Aramid Fiber, Cross Plied	Cyanate Ester	N/A	230°C min.	UL94 V-1
55	Non-Woven Aramid Paper	Epoxy/Multifunctional Epoxy	MIL-S-13949/22–BF/ BFN/BFG	150°C–200°C	UL94 V-1
58	Non-Woven Aramid Paper	Multifunctional Epoxy/Non-Epoxy, Phosphorus Flame Resistant	N/A	135°C–185°C	UL94 V-0, limits on Br and Cl content
60	Woven Quartz Fiber	Polyimide	MIL-S-13949/19–QIL	250°C min.	UL94 HB
70	Woven S-2 Glass	Cyanate Ester		230°C min.	UL94 V-1
71	Woven E-Glass	Cyanate Ester	MIL-S-13949/29–GCN	230°C min.	UL94 V-1
80	Woven E-Glass Surface/Cellulose Paper Core	Epoxy/Phenolic (catalyzed for additive process), Flame Resistant (Bromine or Antimony)	CEM-1	100°C min.	UL94 V-0, Kaolin or inorganic catalyst

81	Woven E-Glass Surface/ Non-Woven E-Glass Core	Epoxy (catalyzed for additive process), Flame Resistant	CEM-3	N/A	UL94 V-0, Kaolin or inorganic catalyst
82	Woven E-Glass	Epoxy/Multifunctional Epoxy, (catalyzed for additive process), Flame Resistant	FR-4	110°C min.	UL94 V-1, Kaolin or inorganic catalyst
83	Woven E-Glass	Epoxy/Multifunctional Epoxy	FR-4	150°C–200°C	UL94 V-1, Kaolin or inorganic catalyst
90	Woven E-Glass	Polyphenylene Ether, Bromine/ Antimony Flame Retardant	N/A	175°C min.	UL94 V-1
91	Woven E-Glass	Polyphenylene Ether, Flame Retardant	N/A	175°C min.	UL94 V-1
92	Woven E-Glass	Epoxy/Multifunctional Epoxy, Phosphorus Flame Retardant	FR-4	110°C–150°C	UL94 V-1, limits on Br and Cl content
93	Woven E-Glass	Epoxy/Multifunctional Epoxy, Aluminum Hydroxide Flame Retardant	FR-4	110°C–150°C	UL94 V-1, limits on Br and Cl content
94	Woven E-Glass	Epoxy/Multifunctional Epoxy, Phosphorus Flame Retardant	FR-4	150°C–200°C	UL94 V-1, limits on Br and Cl content
95	Woven E-Glass	Epoxy/Multifunctional Epoxy, Aluminum Hydroxide Flame Retardant	FR-4	150°C–200°C	UL94 V-1, limits on Br and Cl content
96	Woven E-Glass	Polyphenylene Ether	N/A	175°C min.	UL94 V-1, limits on Br and Cl content
97	Woven E-Glass	Difunctional, Multifunctional Epoxy	FR-4, MIL-S-13949/ 04–GF/GFN/GFK/ GFP/GFM	110°C min.	UL94 V-0, inorganic fillers
98	Woven E-Glass	Epoxy/Multifunctional Epoxy, Flame Retardant	FR-4, MIL-S-13949/04–GF/GFG/GFN	150°C min.	UL94 V-0, inorganic fillers
99	Woven E-Glass	Epoxy/Multifunctional Epoxy/Modified or Non-Epoxy (max. wt. 5%)	FR-4	150°C min.	UL94 V-0, inorganic fillers, T_d, Z-axis CTE, and time to delamination requirements
101	Woven E-Glass	Epoxy/Multifunctional Epoxy/ Modified or Non-Epoxy (max. wt. 5%)	FR-4	110°C min.	UL94 V-0, inorganic fillers, T_d, Z-axis CTE, and time to delamination requirements

(continued)

TABLE 6.2 IPC-4101 Base Material Summary (Continued)

Spec. Sheet #	Reinforcement Type	Resin System	ID Reference*	T_g Range	Other Properties
121	Woven E-Glass	Difunctional, Multifunctional Epoxy/Modified or Non-Epoxy (max. wt. 5%)	FR-4	110°C min.	UL94 V-0, T_d, Z-axis CTE, and time to delamination requirements
124	Woven E-Glass	Epoxy/Multifunctional Epoxy/Modified or Non-Epoxy (max. wt. 5%)	FR-4	150°C min.	UL94 V-0, T_d, Z-axis CTE, and time to delamination requirements
126	Woven E-Glass	Epoxy/Multifunctional Epoxy/Modified or Non-Epoxy (max. wt. 5%)	FR-4	170°C min.	UL94 V-0, Inorganic Fillers, T_d, Z-axis CTE, and time to delamination requirements, 130°C max. operating temp.
129	Woven E-Glass	Epoxy/Multifunctional Epoxy/Modified or Non-Epoxy (max. wt. 5%)	FR-4	170°C min.	UL94 V-0, T_d, Z-axis CTE, and time to delamination requirements, 130°C max. operating temp.

* ANSI, NEMA, and/or MIL-S-13949 grade.

6.2.2 PC-4101 Specification for Base Materials for Rigid and Multilayer Printed Boards

The most commonly used specification for base materials is IPC-4101. This specification presents a classification scheme and specification sheets for the various materials in use. Table 6.2 summarizes the various materials by specification sheet number. Each specification sheet in IPC-4101[1] includes property requirements for that particular material type. As these specification sheets are updated periodically, it is recommended that the latest revision of this document be reviewed. This is particularly true in light of new requirements for materials that must be compatible with lead-free assembly. Table 6.2 is presented for reference only and is not all-inclusive. UL94 comments in Table 6.2 reference the minimum flammability requirements for that material. Materials may exceed these minimum ratings. Also note that where a non-halogen-based flame retardant is used, it is shown along with the resin system description. Definitions of the UL flammability ratings are given in Chap. 8.

Some of the properties documented in the specification sheets of IPC-4101 include T_g, copper peel strengths at different conditions, volume resistivity, surface resistivity, moisture absorption, dielectric breakdown voltage, permittivity, loss tangent, flexural strengths and arc resistance. These properties will be discussed further in Chap. 8.

IPC-4101 also provides a classification scheme for identification of laminate and prepreg materials; this is described in Secs. 6.5 and 6.6.

6.3 *PROPERTIES USED TO CLASSIFY BASE MATERIALS*

Historically, the T_g has been the most common property used to classify FR-4 base materials, and this is one of the primary properties documented in IPC-4101. Materials with high T_g values were generally viewed as being more reliable. While this question was never as simple as specifying only the T_g, the move toward lead-free assembly has caused a fundamental redefinition of what properties are needed to ensure a given level of reliability. The reason is that the alloys used in lead-free assembly require higher peak temperatures than those used in tin-lead assembly. These higher peak temperatures can approach the point where many base materials begin to decompose. For this reason, the decomposition temperature is becoming another property used to classify base materials.

Other important properties include Z-axis expansion characteristics, moisture absorption, and adhesion characteristics within the material, commonly measured using time-to-delamination tests such as the T260 test. The T_g and its impact on Z-axis expansion are summarized below along with the decomposition temperature. The other properties will be discussed in Chaps. 8 and 10.

6.3.1 Glass Transition Temperature, T_g

The T_g of a resin system is the temperature at which the material transforms from a relatively rigid or "glassy" state, to a more deformable or softened state. This is a thermodynamic change in the material which is reversible as long as the resin system has not been degraded. This is to say that when a material that has been heated above its T_g is cooled back below the T_g, it returns to a more rigid state with essentially the same properties as before. However, if the material is heated to too high a temperature beyond T_g, irreversible changes in properties may result. The temperature at which this will happen varies by material type and is related to resin decomposition, discussed later.

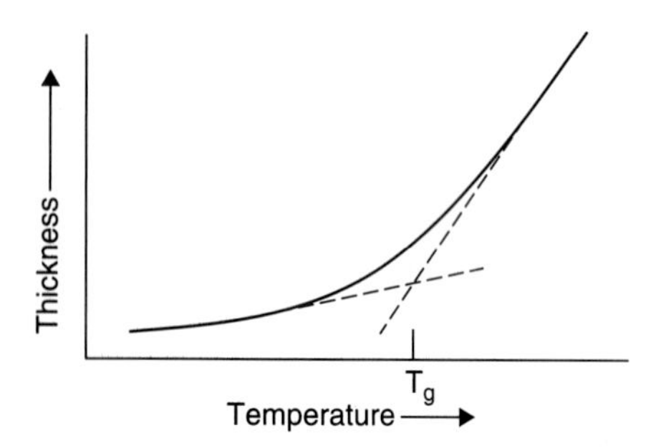

FIGURE 6.1 Glass transition temperature by TMA.

A material is not in a liquid state when it is above the T_g, as some discussions of T_g imply. It is a temperature at which physical changes take place due to the weakening of molecular bonds within the material. It is important to understand T_g since the properties of base materials are different above T_g than below T_g. While the T_g is typically described as being a very precise temperature, this is somewhat misleading, because the physical properties of the material can begin to change as the T_g is approached and some of the molecular bonds are affected. As the temperature increases, more of the bonds become weakened until for all practical purposes, all relevant bonds are affected. This explains the curved line in Fig. 6.1, which is discussed in the following text.

The T_g of a resin system has several important implications. These include:

- The impact on thermal expansion.
- A measure of the degree of cure of the resin system.

6.3.1.1 ***Thermal Expansion.*** All materials will undergo changes in physical dimensions in response to changes in temperature. The rate at which the material expands is much lower below the T_g than above. Thermomechanical analysis (TMA) is a procedure used to measure dimensional changes versus temperature. Extrapolating the linear portions of the curve to the point at which they intersect provides a measure of the T_g (see Fig. 6.1). The slopes of the linear portions of the curve above and below the T_g represent the respective rates of thermal expansion, or as they are typically called, the coefficients of thermal expansion (CTEs). CTE values are important since they influence the reliability of the finished circuit. Other things being equal, less thermal expansion will result in greater circuit reliability as less stress is applied to plated holes.

6.3.1.2 ***Degree of Cure.*** The resin systems used in base materials begin with subcomponents that contain reactive sites on their molecular structures. The application of heat to the resin system and curing agents causes the reactive sites to *cross-link* or bond together. This process of curing the resin system brings about physical changes in the material in proportion to the degree to which the cross-linking occurs, including increases in the T_g. When most of the reactive sites have cross-linked, the material is said to be fully cured and its ultimate physical properties will be established.

Besides TMA, two other thermal analysis techniques are also commonly used for measuring T_g and degree of cure: differential scanning calorimetry (DSC) and dynamic mechanical analysis (DMA).

- DSC measures heat flow versus temperature rather than the dimensional changes measured by TMA. The heat absorbed or given off will also change as the temperature increases through the T_g of the resin system. T_g measured by DSC is often somewhat higher than measurements by TMA.
- DMA measures the modulus of the material versus changes in temperature, and also will give somewhat higher values for T_g.

Materials that are not fully cured can cause problems in circuit manufacturing processes and with finished circuit reliability. For example, an undercured multilayer circuit at the drilling process can result in excessive resin smear across the internal circuit connections in the hole formed. This happens because an undercured resin system, and the resulting lower than normal T_g, will result in excessive softening of the resin when it is subjected to the heat generated during the drilling process. If this smeared resin is not completely removed, it can prevent electrical connection when the hole is plated with copper. In addition, finished circuits in which the resin system is not fully cured can exhibit greater amounts of Z-axis thermal expansion. This adversely affects circuit board reliability due to the increased stress on the plated through holes as the circuit expands.

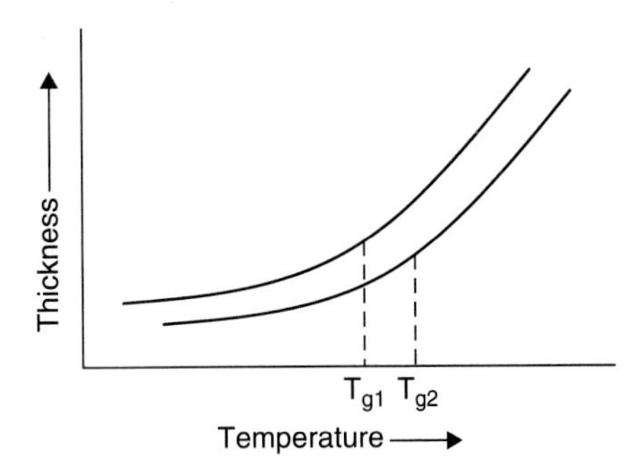

FIGURE 6.2 Measuring the degree of cure by measuring T_g values.

Because increased cross-linking will require greater amounts of energy (heat) to weaken the bonds in the molecular structure of the resin system, measurements of T_g can be used to determine the degree of cure of the resin system. For example, a measure of the degree of cure can be obtained by performing two thermal analyses on the same sample. After the first analysis, the sample is subjected to a thermal cycle designed to promote any additional cross-linking available within the resin system, and then a second thermal analysis is performed. The degree of cure is measured by comparing the difference in measured T_g between the two analyses (see Fig. 6.2). If the material is fully cured, the difference between T_{g2} and T_{g1} will be very small, typically within a few degrees Celsius. Negative differences in these measured values or negative "delta T_g" values where T_{g1} is greater than T_{g2} are also indicative of full cure. However, great care must be exercised when using these methods to assess degree of cure. An understanding of the resin system is important because not all resins will behave the same way when performing sequential tests such as this. The degree of cure for some advanced resin systems is not easily assessed by these techniques. In addition, sample preparation methods, particularly sample drying prior to testing, can also influence results.

6.3.1.3 Advantages and Disadvantages of High T_g Values. Implicit in many discussions of T_g is the assumption that higher values of T_g are always better. This is not always the case. While it is certainly true that higher values of T_g will delay the onset of high rates of thermal expansion for a given resin system, total expansion can differ from material to material. A material with a lower T_g could exhibit less total expansion than a material with a higher T_g, due to differing resin CTE values or by incorporating fillers into the resin system that lower the CTE of the composite material. This is illustrated in Fig. 6.3. Material C exhibits a higher T_g than material A, but material C exhibits more total thermal expansion because its CTE value above T_g is much higher. On the other hand, with the same CTE values above and below the T_g, the higher-T_g material B exhibits less total thermal expansion than material A. Finally, although the T_g values are the same, material B exhibits less total expansion than material C due to a lower CTE value above T_g. This will be discussed further in Chap. 10 in the context of the impact of lead-free assembly temperatures on base materials.

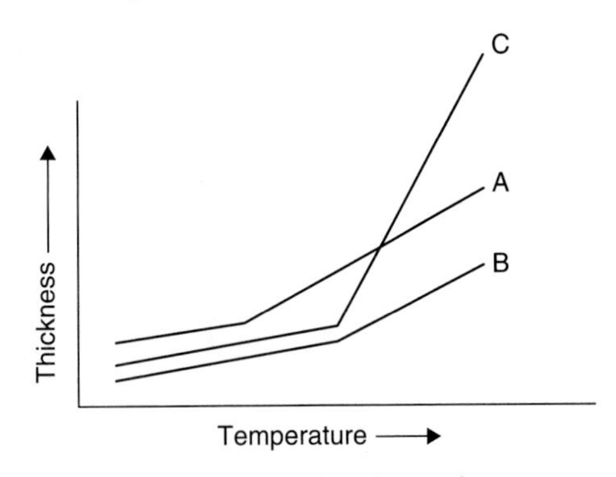

FIGURE 6.3 T_g versus thermal expansion.

There are other considerations as well. Most notably, many standard 140°C T_g FR-4 materials exhibit higher decomposition temperatures, T_d's, than the standard 170°C T_g FR-4s. T_d is an important property for lead-free assembly, with higher values generally preferred. Advanced FR-4 resin systems can combine both high-T_g and high- T_d. This will be discussed in detail in Chap. 10.

In addition, higher T_g resin systems can be harder and more brittle than lower T_g systems. This can adversely affect productivity in the printed circuit manufacturing process. In particular, the drilling process can be affected as slower drilling, reduced drill bit life or reduced stack heights can be required for high-T_g materials.

Lower copper peel strength values and shorter times to delamination can also be correlated with higher T_g values, though other factors influence these properties as well. The time to delamination is a measure of the time it takes for the resin and copper, or resin and reinforcement, to separate or delaminate, and can be correlated to T_d. This will also be discussed in Chaps. 8 and 10, in the context of lead-free assembly. This time-to-delamination test utilizes TMA equipment to bring a sample to a specified temperature and then measures the time it takes for failure to occur. Failure is typically delamination between the resin and copper foil, or resin and glass within the laminate. Temperatures of 260°C (T260) and 288°C (T288) are commonly used for this testing.

6.3.2 Decomposition Temperature, T_d

As a material is heated to higher temperatures, a point is reached where the resin system will begin to decompose. The chemical bonds within the resin system will begin to break down and volatile components will be driven off, reducing the mass of the sample. The decomposition temperature, T_d, is a property which describes the point at which this process occurs. The traditional definition of T_d is the point where 5 percent of the original mass is lost to decomposition. However, 5 percent is a very large number when multilayer PCB reliability is considered, and temperatures where lower levels of decomposition occur are very important to understand, particularly with respect to lead-free assembly. To illustrate this, consider Fig. 6.4.

In Fig. 6.4, you can see curves for two FR-4 materials. The "traditional FR-4," a 140°C T_g material in this case, has a decomposition temperature of 320°C by the 5 percent weight loss definition. The "enhanced FR-4" has a decomposition temperature of 350°C by the 5 percent weight loss definition. Many standard high-T_g FR-4 materials actually have decomposition

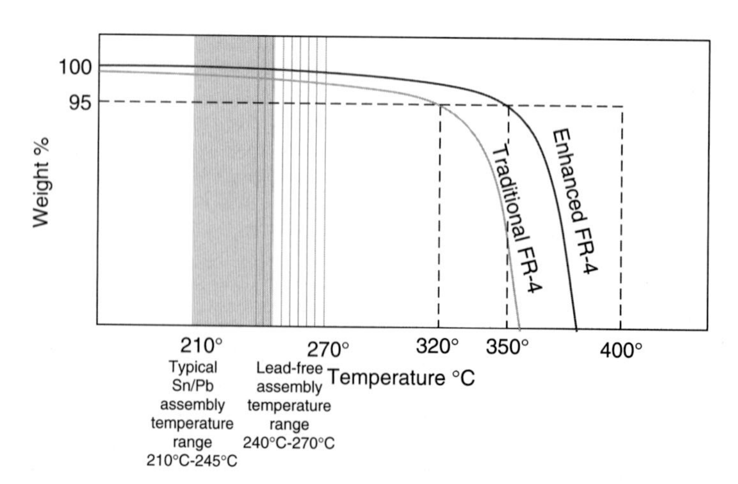

FIGURE 6.4 Illustration of decomposition curves.

temperatures in the range of 290–310°C, while the 140°C T_g FR-4 materials generally have slightly higher T_d values. The shaded regions indicate the peak temperature ranges for standard tin-lead assembly and lead-free assembly. A very common question is, if a PCB will be assembled at 260°C, and the material has a decomposition temperature of 310–320°C, then why wouldn't it be compatible with lead-free assembly?

The answer lies in the level of decomposition in the temperature ranges where assembly will take place. In the tin-lead temperature range, neither material exhibits a significant level of decomposition. However, in the lead-free assembly temperature range, the traditional FR-4 begins to exhibit 1.5–3 percent weight loss. This level of decomposition can compromise long-term reliability or result in defects such as delamination during assembly, particularly if multiple assembly cycles or rework cycles are performed.

6.4 TYPES OF FR-4

As can be seen in Table 6.2, several materials listed are considered FR-4. This begs the question: What is FR-4? In Table 6.2, it can be seen that each material considered FR-4 is flame-resistant, uses a woven fiberglass reinforcement and is primarily an epoxy-based resin system.

6.4.1 The Many Faces of FR-4

So what can be different about these materials? The most obvious difference between the commercially available FR-4 materials is in T_g values. Specification sheet 21 refers to an FR-4 with a minimum T_g of 110°C. More common are FR-4 materials with T_g values between 130°C to 140°C and 170°C to 180°C. Therefore, it is obvious that FR-4 materials can consist of different types of epoxy resins. Indeed, there are many types of epoxy resins that will provide differing T_g values, as well as differences in other properties.

Specification sheets 92-95 list FR-4 materials with non-halogen-based flame retardants, as well as different T_g ranges. Sheets 92 and 94 indicate the use of a phosphorus-based flame retardant, whereas sheets 93 and 95 show the use of aluminum hydroxide. So FR-4 may include different types of flame retardants.

In addition, several FR-4 materials, such as those shown in sheets 99, 101, and 126, show that these materials contain inorganic fillers. These fillers are often used to reduce the amount of Z-axis expansion. Some of these materials also have requirements for T_d, Z-axis expansion and time-to-delamination, which historically were not properties included on the specification sheets. The addition of specification sheets with these properties was partly driven by their relevance to compatibility with lead-free assembly processes.

Finally, specification sheet 82 covers an FR-4 material catalyzed for an additive copper-plating process. A catalyst used for this material is a kaolin clay filler coated with palladium. In short, there are many combinations of resin system, flame retardant and filler that can result in many types of FR-4.

6.4.2 The Longevity of FR-4

FR-4 materials have been the most successful, most commonly used materials in printed circuit manufacturing for many years. Why? Well, as we just discussed above, FR-4 actually encompasses a range of material types, even though they share certain properties and are primarily epoxy-based. The result is that there is often an FR-4 material readily available for the most common end-use applications. For relatively simple applications, the 130°C to 140°C T_g FR-4s have become the material of choice. In higher layer count multilayer circuits, in very thick circuits, as well as those requiring improved thermal properties, the 170°C to 180°C T_g FR-4s have become the material of choice. With the advent of lead-free assembly, T_d and other properties must also be considered along with T_g. Further, as noted above, higher T_g does not always translate to better performance in lead-free assembly applications. In other words, as the range of end-use applications has grown, so has the range of available FR-4 materials.

In addition, the components used in FR-4 materials, particularly woven fiberglass cloth and epoxy resins, provide a very good combination of performance, processability, and cost. The range of woven fiberglass cloth styles available makes it easy to control dielectric and overall circuit thicknesses. As noted earlier, the range of epoxy resin types also makes it relatively simple to tailor material properties to end-use applications. The combination of good electrical, thermal, and mechanical properties of epoxies also explains why they have become the primary type of resin used in printed circuits. Epoxy-based materials are also relatively easy to process through conventional printed circuit manufacturing processes, at least in comparison to some of the other types of materials available. The processability of these materials has helped control the cost of FR-4-based printed circuits.

Last, the development of FR-4 materials took advantage of established manufacturing processes and materials. Weaving processes have been in place for many years. The fundamental process of weaving fiberglass yarns into glass cloth is not so different from weaving textile yarns into cloth fabrics. Using the same basic manufacturing technology avoided additional upfront research and development costs, but more importantly it avoids to some extent highly specialized capital assets and aids in achieving economies of scope for the suppliers of fiberglass cloth. The result is good control of the costs of these materials. Likewise, epoxy resins have been in use in a wide range of applications outside of printed circuits, resulting in a very large installed manufacturing base for these types of materials. This also results in good cost competitiveness for epoxy resins. In summary, the range of FR-4 types and their unique combination of performance, processability and cost characteristics have made them the workhorse of the printed circuit industry.

6.5 LAMINATE IDENTIFICATION SCHEME

As IPC specifications are regularly reviewed and updated, it is recommended that the latest IPC specification be used for actual applications. However, an example of a laminate designation in IPC-4101 is:

L	25	1500	C1/C1	A	A
Material Designator	Specification Sheet Number	Nominal Laminate Thickness	Metal Cladding Type and Nominal Weight/Thickness	Thickness Tolerance Class	Surface Quality Class

- **Material designator** "L" refers to laminate.
- **Specification sheet number** This designation references the specification sheet number as in Table 6.2.
- **Nominal laminate thickness** Four digits identify this variable. It may be specified either over the cladding or over the dielectric. For metric specification, the first digit represents whole millimeters, the second represents tenths of millimeters, and so on. For orders requiring English units, the four digits indicate the thickness in ten-thousandths of an inch (tenths of mils). In the example shown, 1500 is designated for the English usage of 0590.

TABLE 6.3 Metal Cladding Types Summarized in IPC-4101 (IPC-CF-148, IPC-4562, and IPC-CF-152 are Actual Metal Cladding Specifications)[2]

Designation	Metal Cladding Type
A	Copper, wrought, rolled (IPC-4562, grade 5)
B	Copper, rolled (treated)
C	Copper, electrodeposited (IPC-4562, grade 1)
D	Copper, electrodeposited, double treat (IPC-4562, grade 1)
G	Copper, electrodeposited, high ductility (IPC-4562, grade 2)
H	Copper, electrodeposited, high temperature elongation (IPC-4562, grade 3)
J	Copper, electrodeposited, annealed (IPC-4562, grade 4)
K	Copper, wrought, light cold rolled (IPC-4562, grade 6)
L	Copper, wrought, annealed (IPC-4562, grade 7)
M	Copper, wrought, rolled, low temperature annealable (IPC-4562, grade 8)
N	Nickel
O	Unclad
P	Copper, electrodeposited, high temperature elongation, double treat (IPC-4562, grade 3)
R	Copper, reverse treated electrodeposited (IPC-4562, grade 1)
S	Copper, reverse treated electrodeposited, high temperature elongation (IPC-4562, grade 3)
T	Copper, copper foil parameters as dictated by contract or purchase order
U	Aluminum
V	Copper, revers treated electrodeposited, high temperature elongation (IPC-4562, grade 3) for buried capacitance applications
X	As agreed between user and supplier
Y	Copper Invar Copper
Z	Copper, electrodeposited, high temperature elongation, double-treat (IPC-4562, grade 3) for buried capacitance applications

- **Metal cladding type and nominal weight/thickness** Five designators are used to specify cladding type and thickness. The first and fourth indicate the type, the second and fifth indicate thickness, and the third is a slash to differentiate sides of the base material. Table 6.3 lists the types of metal cladding. Table 6.4 lists copper foil weights and thicknesses.
- **Thickness tolerance class** This variable references the thickness tolerance as agreed between user and supplier (see Table 6.5). Classes A, B, and C refer to measurement by micrometer of the base material without cladding. Class D requires measurement by microsection (see Fig. 6.5). Classes K, L, and M refer to measurement by micrometer of the base material with the metal cladding. Class X refers to a requirement agreed upon between user and supplier.
- **Surface quality class** This class is identified by either A, B, C, D, or X (as agreed upon between user and supplier). Specimens are examined with normal or corrected 20/20 vision. The worst 50 mm × 50 mm (1.97 in. × 1.97 in.) is examined at 10 × magnification. Indentations are located visually using normal or corrected 20/20 vision. The longest dimension of each foil indentation in a specimen is measured with a suitable reticule on a minimum 4 × magnifier, with referee inspections at 10 ×. A point value is allocated according to the longest dimension, as shown in Table 6.6. Surface quality class is determined by the total point count of foil indentations within 300 mm × 300 mm (11.81 in. × 11.81 in.), as shown in Table 6.7.

TABLE 6.4 Foil Weights and Thicknesses from IPC-4101 (IPC-CF-148, IPC-4562, and IPC-CF-152 are Actual Metal Cladding Specifications)

Foil Designator	Common Industry Terminology	Area Weight (g/m²)	Nominal Thickness (μm)	Area Weight (oz/ft²)	Area Weight (g/254 in.²)	Nominal Thickness (μm)
E	5 μm	45.1	5.1	0.148	7.4	0.20
Q	9 μm	75.9	8.5	0.249	12.5	0.34
T	12 μm	106.8	12.0	0.350	17.5	0.47
H	1/2 oz	152.5	17.1	0.500	25.0	0.68
M	3/4 oz	228.8	25.7	0.750	37.5	1.01
1	1 oz	305.0	34.3	1	50.0	1.35
2	2 oz	610.0	68.6	2	100.0	2.70
3	3 oz	915.0	102.9	3	150.0	4.05
4	4 oz	1220.0	137.2	4	200.0	5.40
5	5 oz	1525.0	171.5	5	250.0	6.75
6	6 oz	1830.0	205.7	6	300.0	8.10
7	7 oz	2135.0	240.0	7	350.0	9.45
10	10 oz	3050.0	342.9	10	500.0	13.50
14	14 oz	4270.0	480.1	14	700.0	18.90

TABLE 6.5 Base Laminate Thickness Tolerances from IPC-4101

Nominal Thickness of Laminate, mm [in.]	Class A/K	Class B/L	Class C/M	Class D
0.025 to 0.119 mm [0.0009 to 0.0047]	+/– 0.025 mm [+/–0.000984]	+/– 0.018 mm [+/–0.000709]	+/– 0.013 mm [+/–0.000512]	–0.013 +0.025 mm [–0.000512 +0.000984]
0.120 to 0.164 mm [0.0047 to 0.0065]	+/– 0.038 mm [+/–0.00150]	+/– 0.025 mm [+/–0.000984]	+/– 0.018 mm [+/–0.000709]	–0.018 +0.030 mm [–0.000709 +0.00118]
0.165 to 0.299 mm [0.0065 to 0.0118]	+/– 0.050 mm [+/–0.00197]	+/– 0.038 mm [+/–0.000150]	+/– 0.025 mm [+/–0.000984]	–0.025 +0.038 mm [–0.000984 +0.00150]
0.300 to 0.499 mm [0.0118 to 0.0196]	+/–0.064 mm [+/–0.00252]	+/–0.050 mm [+/–0.00197]	+/–0.038 mm [+/–0.00150]	–0.038 +0.050 mm [–0.00150 +0.00197]
0.500 to 0.785 mm [0.0197 to 0.0309]	+/– 0.075 mm [+/–0.00295]	+/– 0.064 mm [+/–0.00252]	+/– 0.050 mm [+/–0.00197]	–0.050 +0.064 mm [–0.00197 +0.00252]
0.786 to 1.039 mm [0.0309 to 0.04091]	+/– 0.165 mm [+/–0.006496]	+/– 0.10 mm [+/–0.00394]	+/– 0.075 mm [+/–0.00295]	N/A
1.040 to 1.674 mm [0.04091 to 0.06594]	+/–0.190 mm [+/–0.007480]	+/–0.13 mm [+/–0.00512]	+/–0.075 mm [+/–0.00295]	N/A
1.675 to 2.564 mm [0.06594 to 0.10094]	+/–0.23 mm [+/–0.00906]	+/– 0.18 mm [+/–0.00709]	+/– 0.10 mm [+/–0.00394]	N/A
2.565 to 3.579 mm [0.10094 to 0.14091]	+/–0.30 mm [+/–0.0118]	+/– 0.23 mm [+/–0.00906]	+/– 0.13 mm [+/–0.00512]	N/A
3.580 to 6.35 mm [0.14094 to 0.250]	+/– 0.56 mm [+/–0.0220]	+/– 0.30 mm [+/–0.0118]	+/– 0.15 mm [+/–0.00591]	N/A

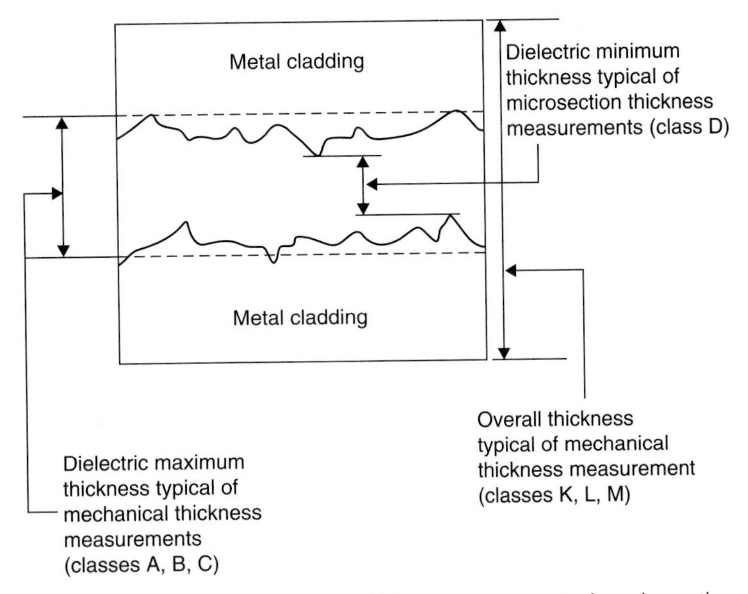

FIGURE 6.5 Dielectric minimum thickness measurements by microsection. *(from IPC-4101)*

TABLE 6.6 Pit and Dent Point Values from IPC-4101

Longest Dimension, mm [in.]	Point Value
0.13–0.25 mm [0.005 to 0.009]	1
0.26–0.50 mm [0.009 to 0.019]	2
0.51–0.75 mm [0.019 to 0.029]	4
0.76–1.00 mm [0.029 to 0.039]	7
>1.00 mm [0.039]	30

TABLE 6.7 Surface Quality Grades from IPC-4101

Surface Quality Class	Maximum Point Count	Other Requirements
Class A	29	
Class B	17	
Class C	5	Longest dimensions </=380 μm [14.96 mil]
Class D	0	Longest dimensions <125 μm [4.92 mil] Resin spots = 0
Class X	As agreed between user and supplier (AABUS)	

6.6 PREPREG IDENTIFICATION SCHEME

An example of a prepreg identification in IPC-4101 is:

P	25	E7628	TW	RE	VC
Material Designator	Specification Sheet	Reinforcement Style Number	Resin Content Method	Flow Parameter Method	Optional Prepreg Method

- **Material designator** P refers to prepreg.
- **Specification sheet number** This designation references the specification sheet number as in Table 6.2.
- **Reinforcement style** The reinforcement type and style of prepreg is indicated by five digits based on the chemical type and style. In the example, E refers to E-glass and 7628 is the glass fabric style.
- **Resin content method** This variable refers to the method used to specify resin content of a prepreg. The two methods are RC, which refers to percent resin content of the prepreg, and TW, which refers to the treated weight of the prepreg. A 00 designator would indicate that no method is specified.
- **Flow parameter method** This variable indicates how much the resin in the prepreg will flow under specified conditions—a critical property. The options are:

  MF: resin flow percent
  SC: scaled flow thickness
  NF: no flow
  RE: rheological flow
  DH: delta H
  PC: percent cure
  00: no method specified

- **Optional prepreg method** Other test methods may also be designated here. These include:

  VC: volatile content
  DY: dicy inspection
  GT: gel time
  00: none specified

The choice of test methods and their required nominal values and tolerances are normally agreed upon between user and supplier.

In addition to properties on the specification sheets, IPC-4101 includes class specifications for other properties as well. These include length and width, bow and twist, and thermal conductivity, among others.

6.7 LAMINATE AND PREPREG MANUFACTURING PROCESSES

While there are a few different processes that can be used to integrate the components that make up printed circuit base materials, the overwhelming majority of materials are manufactured using a common process. In recent years however, new techniques have been developed and continue to be developed. These new processes have been designed to either lower the

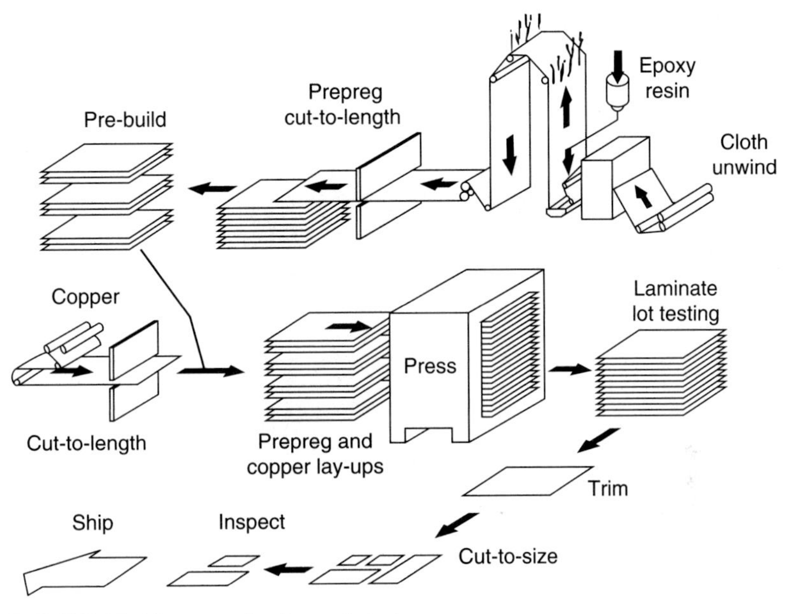

FIGURE 6.6 The conventional manufacturing process.

cost of manufacturing or to improve material performance, or both. Common to all processes is the need to manufacture a copper clad laminate and the bonding sheets or *prepregs* used to manufacture multilayer circuits.

6.7.1 Conventional Manufacturing Processes

The overall conventional manufacturing process is illustrated in Fig 6.6. This process can be subdivided into two processes: prepreg manufacturing and laminate manufacturing. Prepreg is also called *B-stage* or bonding sheets, while laminate is sometimes called *C-stage*. The terms B-stage and C-stage refer to the degree to which the resin system is polymerized or cured. B-stage refers to a state of partial cure. B-stage is designed to melt and continue polymerizing when it is exposed to sufficient temperatures. C-stage refers to a state of "full" cure (typically we never realize *full* cure, in the sense that not all reactive sites on the resin molecules have cross-linked; however, we use the term *full cure* to mean that the overwhelming majority of such sites have reacted and additional exposure to temperature will do little to advance the state of cure).

6.7.2 Prepreg Manufacturing

The first step in most of these processes involves coating a resin system onto the chosen reinforcement, most commonly a woven fiberglass cloth. Rolls of fiberglass cloth or alternative reinforcement type are run through equipment called *treaters*. Figure 6.7 shows some of the resin system components that are mixed and aged prior to application to the fiberglass cloth. Figure 6.8 illustrates the overall treating process. The fiberglass cloth is drawn through a pan containing the resin system and then precise metering rolls help control thickness and also push resin into the yarns of the glass cloth (see Fig. 6.9).

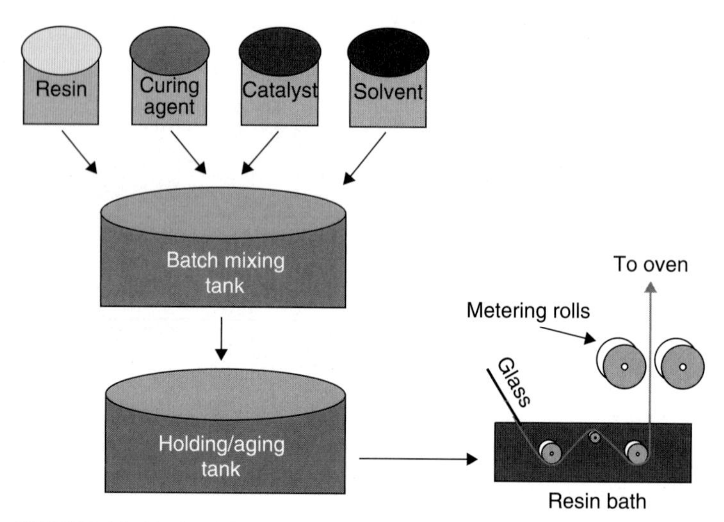

FIGURE 6.7 Resin mixing, aging, and treating.

Next the cloth is pulled through a series of heating zones. These heating zones commonly utilize forced air convection, infrared heating, or a combination of the two. In the first set of zones, solvent used to carry the resin system components is evaporated off. Subsequent zones are dedicated to partially curing the resin system, or B-staging the resin. Finally, the prepreg is then rewound into rolls or cut into sheets.

A number of process controls are required in this operation. The concentrations of the resin system components must be controlled and the viscosity of the resin system maintained within acceptable limits. Tension on the cloth as it is pulled through the treater is also important, for among other reasons, it is vital not to distort the weave pattern of the cloth. Control of the

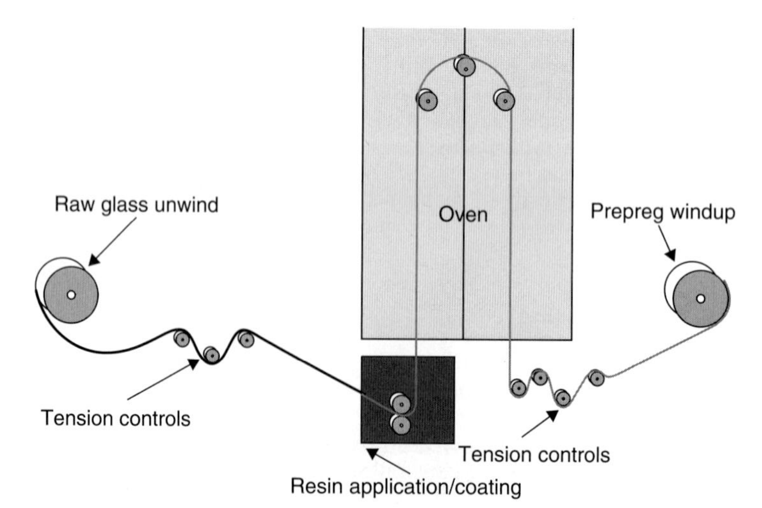

FIGURE 6.8 "Treating" fiberglass cloth with resin.

resin-to-glass ratio or resin content, the degree of cure of the resin and cleanliness are also critical.

Because the resin system at this point is only partially cured, prepregs must typically be stored in temperature- and humidity-controlled environments. Temperature, for obvious reasons, could affect the degree of cure of the resin and therefore its performance in laminate or multilayer circuit pressing. Since moisture can affect the performance of many curing agents and accelerators, not to mention the performance of the resin system during lamination or press cycles, humidity is also important to control during prepreg storage. Absorbed moisture that becomes trapped during lamination cycles can also lead to blisters or delaminations within the laminate or multilayer circuit.

FIGURE 6.9 Prepreg after treating with resin.

6.7.3 Laminate Manufacturing

The process of copper clad laminate manufacturing begins with the prepreg material. Prepregs of certain fiberglass cloth styles and specific resin contents are combined with the desired copper foils to make the finished laminate. First, the prepregs and copper foils are sheeted to the desired size. Figure 6.10 shows an automated copper-sheeting process.

These materials are then laid up in the proper sequence to produce the desired copper clad laminate. Figures 6.11 and 6.12 are examples of an automated build-up process where the prepreg and copper foil materials are combined prior to being pressed together. Several of these individual sandwiches of prepreg and copper foil are stacked on top of each other, typically separated by stainless steel plates, although other separator materials, including aluminum, can also be used. These stacks are then loaded into multi-opening lamination

FIGURE 6.10 Sheeting of rolls of copper foil.

FIGURE 6.11 Automated build-up of copper and prepreg.

FIGURE 6.12 Automated build-up of copper and prepreg.

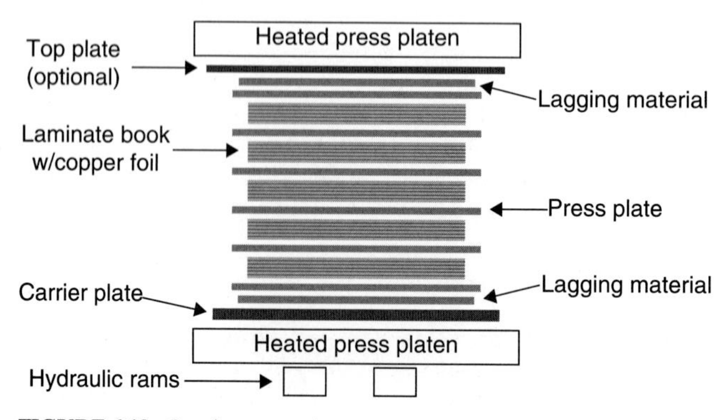

FIGURE 6.13 Laminate pressing.

FIGURE 6.14 Lamination press with multiple openings. *(Photo Courtesy of Polyclad Laminates, Inc.)*

presses (see Figs. 6.13 and 6.14), where pressure, temperature, and vacuum are applied. The specific press cycle used will vary depending upon the particular resin system, the degree of cure of the prepregs and other factors. The presses themselves have many platens that can be heated by steam or hot oil that flows through the platens, or they can also be heated electrically.

Process controls in the lamination/pressing process are also critical. Cleanliness in the manufacturing environment and cleanliness of the steel separator plates are critical in achieving good surface quality and in avoiding embedded foreign material within the laminate. Control of the temperature rise during lamination will provide the desired amount of resin flow, while control of the cooldown rate can impact warp and twist. The length of time that the laminate is above the temperature required to initiate the curing reaction will determine the degree of cure, as will the degree to which the actual temperature exceeds this temperature.

While these descriptions of prepreg and laminate manufacturing give a simple description of the processes used, it is important to understand that there are many variables that influence the quality and performance of the finished products. In addition, many of these variables interact with each other, meaning that a change in one may influence others and may require adjustments to these other process variables. In summary, prepreg and laminate manufacturing is much more complex than it appears at first glance.

6.7.4 Direct Current or Continuous Foil Manufacturing

Continuous foil or direct current manufacturing is an alternative method used to manufacture copper clad laminates (see Fig. 6.15). Prepregs are still used in this process, however, the lay-up and pressing operations are somewhat different. In this process, the copper foils are not sheeted, but kept in rolls. To start, copper foil with the side to be bonded to the prepreg facing up, is laid

FIGURE 6.15 Laminate pressing with continuous copper foil and direct current heating. *(Illustration Courtesy of Cedal)*

down, while still part of a continuous roll. The prepregs are laid up in the proper sequence, and then the roll of copper foil is passed over the prepregs so that it is applied to the top of what will become the finished laminate. Typically, two rolls of foil are used to allow dissimilar copper weights or types to be bonded to either side of the laminate. A separator plate, typically anodized aluminum, is then placed on top of this sandwich, and the process is repeated so that several laminates are stacked up.

At this point, the stack of laminates is loaded into a press and subjected to heat, pressure, and vacuum. However, as opposed to the conventional process, this technique involves applying direct current to the copper foil that runs throughout the stack. The current heats the foil and therefore the prepregs are adjacent to the foils. By controlling the amount of current, you control the temperature of the materials in the stack.

6.7.5 Continuous Manufacturing Processes

Over the years, continuous lamination processes have also been designed. Rather than cut the prepregs and foils into sheets, lay them up and press them as individual laminates, these processes use rolls of prepreg or bare fiberglass cloth and rolls of copper foil that are continuously unwound and fed together into a continuous horizontal press. One process starts with rolls of prepreg. Another starts with the bare, untreated fiberglass cloth, applies a resin to the cloth and is sandwiched with the copper foil as it is fed continuously into a horizontal press. At the back end of the press, sections of the continuous laminate can be cut into sheets, or with thin laminates, rolls of copper clad laminate can be manufactured.

REFERENCES

1. IPC-4101, "Specification for Base Materials for Rigid and Multilayer Printed Boards."
2. IPC-4562, "Metal Foil for Printed Wiring Applications."

CHAPTER 7
BASE MATERIAL COMPONENTS

Edward J. Kelley
Isola Group, Chandler, Arizona

7.1 INTRODUCTION

Although there are many types of base materials, they all contain three components:

- The resin system, including additives
- The reinforcement(s)
- The conductor

Each of these components is important in its own right, and in combination they determine the properties of the base material as well as the relative cost of the material.

Environmental legislation such as the European Union's Restriction of Hazardous Substances (RoHS) directive has a profound impact on all levels of the electronics supply chain, including these components. RoHS restricts the use of lead, which is an element in the solder used for component assembly onto printed circuits. The impact on the base materials and components is primarily the result of higher assembly temperatures that are associated with lead-free assembly. Table 7.1 summarizes the key issues for base material components. RoHS issues will be discussed further in Chap. 10.

The RoHS directive also restricts specific halogen-containing flame retardants. However, most base materials for printed circuits do not contain the restricted flame retardants. Nevertheless, there is growing interest in halogen-free materials, and this will be discussed further in this chapter.

7.2 EPOXY RESIN SYSTEMS

The most successful and widely used resin systems for printed circuit applications are epoxy resin systems. There are many types of epoxy resins, and this class of resin system continues to be the workhorse among printed circuit board materials. This is a result of the combination of good mechanical, electrical, and physical properties and the relatively low cost of epoxies in comparison to the higher-performance resins. In addition, epoxy systems are relatively easy to process, which aids in keeping manufacturing costs down.

TABLE 7.1 Lead-Free Assembly Impact on Base Material Components

Component	Lead-Free Assembly Impact	Potential Solutions	Related Considerations
Resin System	1. Peak assembly temperatures can reach the point where resin decomposition begins. 2. Higher temperatures result in increased thermal expansion and stress on plated holes as a result. 3. Vapor pressure of absorbed moisture much higher at lead-free assembly temperatures; can lead to blistering/delamination. 4. "Phenolic" lead-free compatible materials often not as good for electrical performance, especially Df.	1. Formulate resin system with higher decomposition temperatures. 2. Formulate for lower coefficients of thermal expansion. 3. Evaluate materials for moisture absorption/release characteristics; drying processes in PCB fabrication and/or assembly. 4. Evaluate non-dicy/non-phenolic laminate materials.	1. Reformulation can adversely affect electrical properties and manufacturability. 2. Can also impact mechanical properties and manufacturability. 3. PCB storage conditions during manufacturing and prior to assembly are much more important, especially humidity conditions. 4. Cost/performance trade-off. New materials now available that offer more choices.
Fiberglass Cloth	Thermal and mechanical stress on resin-to-glass bond. Volatilized components can also stress this bond.	Cleanliness, moisture resistance, and choice of proper coupling agent for resin adhesion are more important.	Loss of resin-to-glass adhesion through thermal cycling could impact CAF resistance.
Copper Foil	Thermal and mechanical stress on resin-to-copper bond.	Copper nodularization and roughness, treatments for improved adhesion, i.e. coupling agents.	Roughness will impact signal attenuation, especially at high frequencies[i].

Brist, Gary, Hall, Stephen, Clauser, Sidney, and Liang, Tao, "Non-Classical Conductor Losses Due to Copper Foil Roughness and Treatment," ECWC 10/IPC/APEX Conference, February 2005.

7.2.1 Definition of Epoxy

One of the most common versions of epoxy resin for printed circuit applications is manufactured from reacting epichlorohydrin and bisphenol-A. This reaction is shown in Fig. 7.1. The bromination of the bisphenol-A provides flame retardancy to the finished resin system.

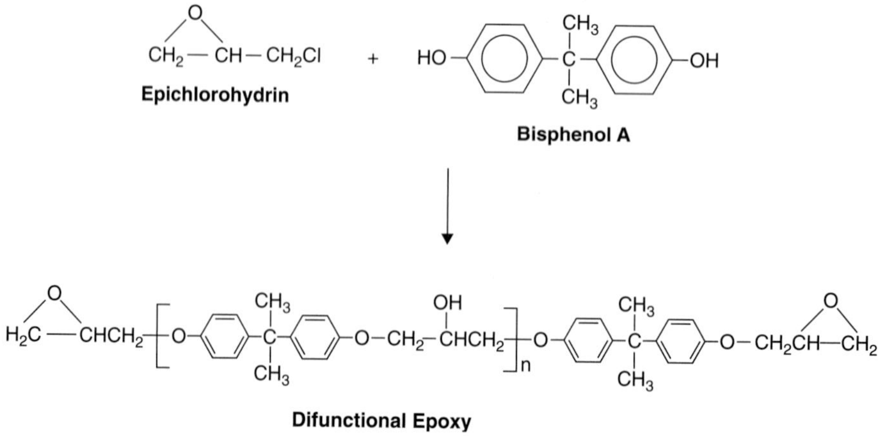

Epichlorohydrin

Bisphenol A

Difunctional Epoxy

FIGURE 7.1 Reaction to form difunctional epoxy resin.

Figure 7.2 shows the structure of tetrabromobisphenol-A, or TBBPA. Figure 7.3 shows a reaction used to manufacture brominated epoxy resins. Brominated epoxy resins such as this are the most common component used to incorporate flame retardancy to the finished product, although non-halogen-based flame retardants can also be used, and are discussed in Sec. 7.4.2.3. The triangular rings on either end of the difunctional epoxy are the epoxide functional groups. In subsequent resin polymerization, these groups react and result in the curing of the resin system. The –OH groups present on the epoxy molecule also react with the epoxide groups, providing cross-linking among the epoxy molecules (see Fig. 7.4).

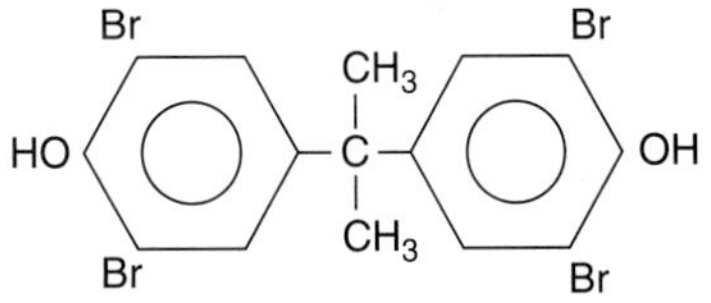

FIGURE 7.2 Tetrabromobisphenol-A (TBBPA).

Difunctional epoxy resin

Tetrabromobisphenol A (TBBPA)

Brominate epoxy resin

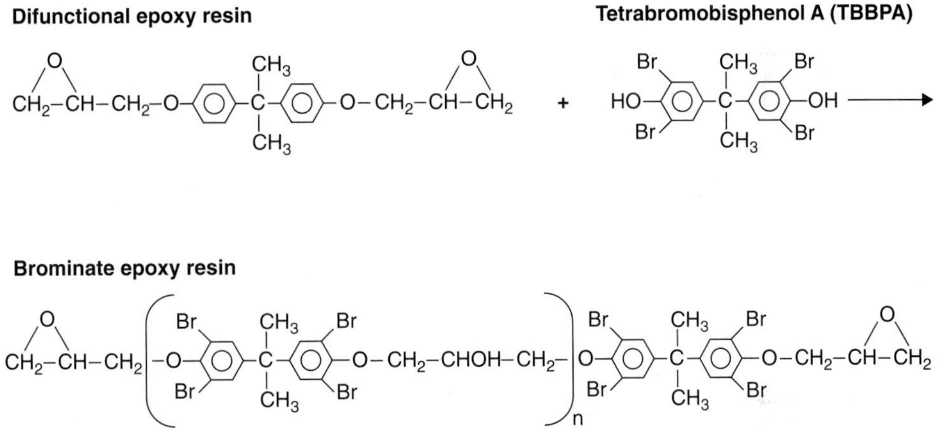

FIGURE 7.3 Brominated difunctional epoxy resin.

7.2.2 Difunctional Epoxies

The epoxy resins shown in Figs. 7.1 and 7.3 are difunctional epoxies. The molecular weight of the epoxy can be varied based on the number of repeating groups shown in the center of the molecule. On either end of the molecule, you see the epoxide functional groups. The name *difunctional epoxy* is derived from the fact that there are two epoxide groups, one on either

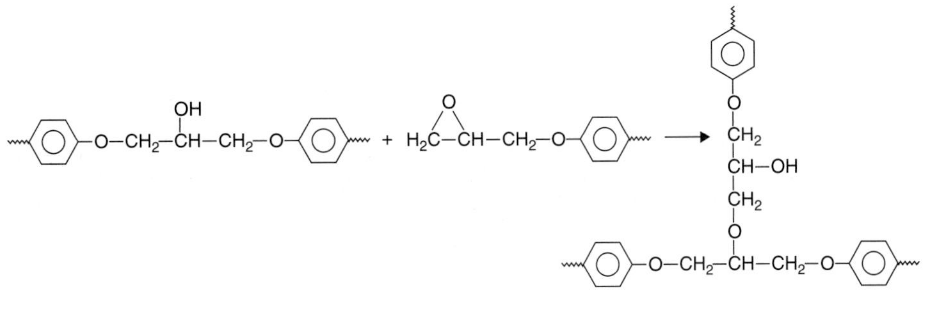

FIGURE 7.4 Cross-linking of –OH and epoxide functional groups.

end of the molecule. The molecular weight, the curing agents used to react the resin and other factors will affect the finished properties of the resin system, including T_g and T_d. The T_g is the temperature at which the resin turns from a rigid or glassy state to a softer, more deformable state. The T_g is important because it affects the thermal and physical properties of the base material and finished circuit board, especially thermal expansion properties. The T_d is the decomposition temperature and will influence thermal reliability of the printed circuit. These properties are discussed in more detail in Chaps. 6 and 8, and again in Chap. 10 as they relate to compatibility with lead-free assembly processes.

Difunctional epoxies can have a range of T_g's, but are typically below 120°C. These epoxies are sometimes used in relatively unsophisticated products such as simple double-sided printed circuits, but are more commonly blended with other epoxies in higher performance systems.

7.2.3 Tetrafunctional and Multifunctional Epoxies

The use of epoxy compounds with more than two epoxide functional groups per molecule results in greater cross-linking when the resin is cured. Among other things, this can result in higher T_g levels. Resin systems with these types of epoxies can exhibit improved thermal and physical properties. However, the curing chemistry will also have an impact on these properties, and often the higher-T_g materials can be harder and more brittle, requiring process adjustments in the printed circuit manufacturing process. Common commercially available laminate materials based on epoxy resin systems can be segmented into a few T_g ranges, 125 to 145°C, 150 to 165°C, and greater than 170°C. There are epoxy systems above190°C, but they are less common. These resin systems are normally blends of difunctional, tetrafunctional, and multifunctional epoxy resins. Figures 7.5 and 7.6 are examples of tetrafunctional and multifunctional epoxies.

A cost-performance trade-off exists among the commonly used epoxy resin systems. In general, resin systems offering higher levels of T_g and T_d will cost more. In addition, the high-T_g systems can incur increased circuit manufacturing costs, due primarily to increased multilayer lamination cycle times and decreases in drilling productivity. However, the improvement in performance is often required to meet design and reliability requirements. Chapter 11 outlines a procedure for selecting cost-effective materials for various printed circuit designs.

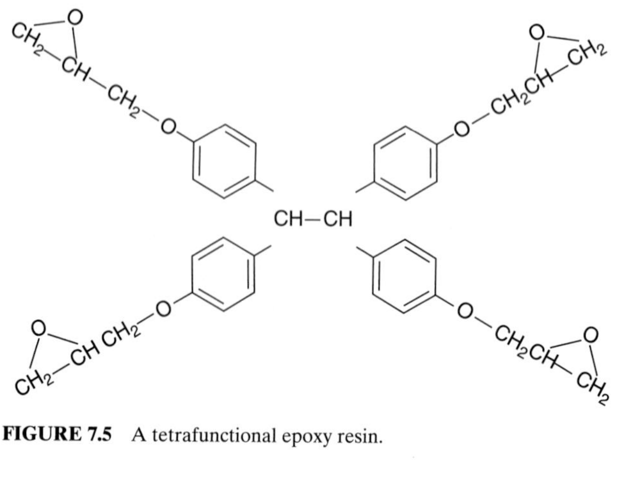

FIGURE 7.5 A tetrafunctional epoxy resin.

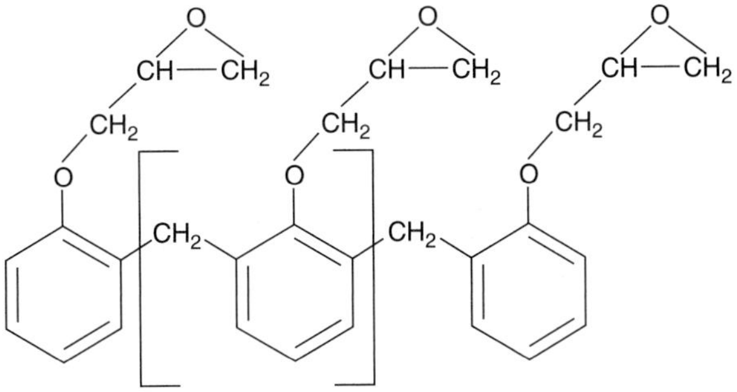

FIGURE 7.6 A multifunctional epoxy phenol novolac resin.

7.3 OTHER RESIN SYSTEMS

Many other resin systems are also available. In choosing a resin system, the circuit board designer and fabricator will have to consider what level of performance is needed, as there is typically a strong cost-performance relationship. This cost-performance relationship is driven not only by the base price of the material itself, but also by the impact the material has on processing costs, both during laminate and prepreg manufacturing, and during printed circuit manufacturing processes.

7.3.1 Epoxy Blends

Blends of epoxy resins with other types of resins have also been developed. These are used when performance demands exceed the capabilities of even the high-T_g/T_d epoxies, but where the costs of the highest performance materials cannot be justified. In many cases, the driving force behind these materials is the need for improved electrical properties versus the standard epoxy offerings. Specifically, improvements in the dielectric constant (permittivity) and dissipation factor (loss tangent) are the properties of interest. Materials with lower values for these properties are needed for circuits that operate at high frequencies.

These blends include epoxy-polyphenylene oxide (PPO) (see Fig. 7.7), epoxy-cyanate ester, and epoxy isocyanurate (see Fig. 7.8). While these materials have been developed to minimize impacts to common printed circuit manufacturing processes, they can affect productivity in multilayer lamination and drilling and can require special desmear and hole wall conditioning processes, depending on the design of the printed circuit and fabrication process used. On the other hand, they typically have less of an impact to these processes when compared to the even higher performance materials.

2,6 Xylenol

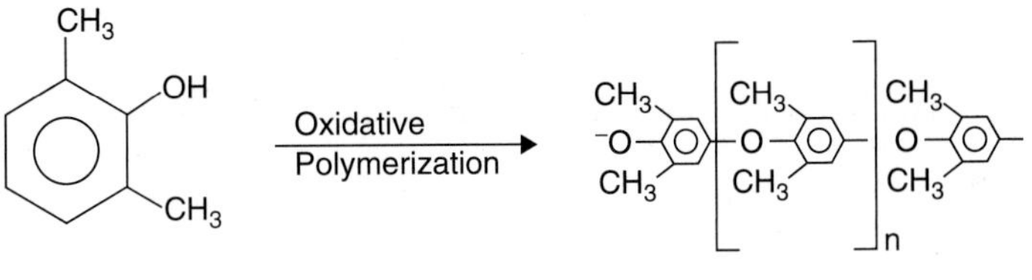

FIGURE 7.7 Polyphenylene oxide (PPO).

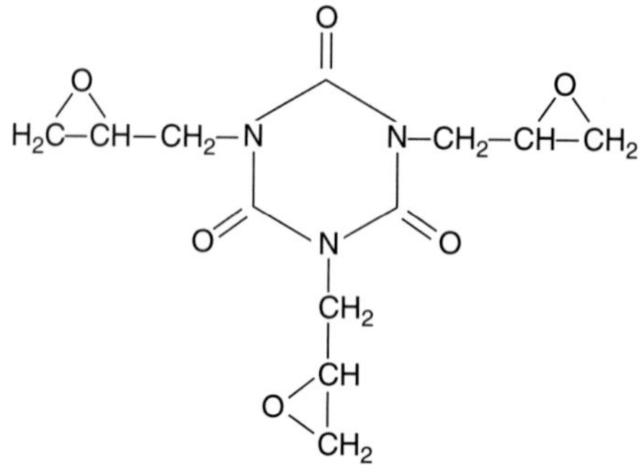

FIGURE 7.8 Epoxy isocyanurate.

These epoxy blends are often used in high-frequency applications including antennas, radio frequency (RF), and wireless communications equipment, and high-speed computing applications.

7.3.2 Bismaleimide Triazine (BT)/Epoxy

Typically, epoxy is added to BT resins to modify the properties of pure BT. Therefore these materials are also considered epoxy blends. BT/epoxies normally have T_g values in the range of 180°C to 220°C and exhibit a good combination of electrical, thermal and chemical resistance properties. BT/epoxy is commonly used in BGA substrates and chip scale packages since it can meet the requirements of specifications for use in semiconductor chip packaging. It is also suitable for high-density multilayers requiring good thermal, electrical, and chemical performance.

The primary disadvantage of BT-based materials is cost. The greater the amount of BT, the higher the cost. BT can also be more brittle than pure epoxy systems, and moisture absorption can also be higher.

7.3.3 Cyanate Ester

Cyanate ester systems possess very high-T_g values, typically in the neighborhood of 250°C, and exhibit very good electrical performance. Thermal stability is also good. However, cyanate esters are relatively expensive and require special processing that can add additional cost to the finished circuit. For these reasons, cyanate ester materials are still used only in niche applications. Blends with epoxies have also been used to modify processability and cost.

7.3.4 Polyimide

When extreme heat resistance is required, polyimide resins offer outstanding performance. With a T_g of 260°C for the purest systems to 220°C for modified or toughened systems, and very high decomposition temperatures, circuits manufactured with polyimide materials exhibit very high levels of thermal reliability. The high T_g of these resin systems also helps to minimize thermal expansion, as most of the thermal expansion occurs pre-T_g, where expansion rates are relatively low. These materials have commonly been used in burn-in boards, aerospace and avionics, down-well oil drilling, and military applications where thermal performance is vital. Again, however, these resin systems are relatively expensive and more difficult to process, generally limiting them to niche applications.

7.3.5 Polytetrafluoroethylene (PTFE,Teflon®)

A variety of PTFE-based products are also available for applications where extremely good electrical properties are required. These materials often require very specialized processing and are relatively expensive. Hybrids of PTFE-based materials and other material types are often used to gain the performance benefits of the PTFE-based materials where needed, while controlling the total circuit cost by using other less costly materials in other layers of the multilayer circuit.

7.3.6 Polyphenylene Ether (PPE)

PPE-based products offer superior electrical properties compared to the epoxy blends discussed above, and also offer excellent thermal performance. This resin system is well suited for RF and wireless communications and high-speed computing applications. While early versions of these resin systems were more difficult to process, improvements in resin formulation and rheology control have been made so that PPE systems can be processed using conventional PCB manufacturing processes with minor adjustments.

7.3.7 Other Resins and Formulations

For many of these resins, several different types of molecular structures can be used to modify the ultimate properties of the base material. In addition, other types of resins are also in use or are being developed to meet the growing requirements for the various applications in which printed circuits are used. Often, these resins are proprietary in nature, and therefore discussions of these products focus on their properties rather than the specific chemistry used to achieve the properties. In addition, fillers are increasingly being used to modify thermal expansion or electrical performance, or both. Fillers will be discussed later in this chapter.

7.4 ADDITIVES

The resin systems discussed above will typically contain a variety of additives that either promote curing of the resin system or modify the properties in some way. Some important types of additives include the following.

7.4.1 Curing Agents and Accelerators

Each resin system contains organic components that must be reacted together to promote polymerization and cross-linking. Curing agents and accelerators are used to promote these reactions. Amine-based curing agents are commonly used to cure epoxy resins. Some of these, such as aliphatic diamines, are used to cure epoxies at room temperature. Others, such as aromatic diamines, require elevated temperatures. Figure 7.9 shows an example of curing epoxy with an aromatic amine. Note the –OH group formed on the new molecule. As shown in Fig. 7.4, the –OH can also cross-link with other epoxide groups.

Historically, the most common curing agent used in epoxy resin systems for printed circuit base materials has been dicyandiamide, or "dicy." Figure 7.10 illustrates the polymerization of

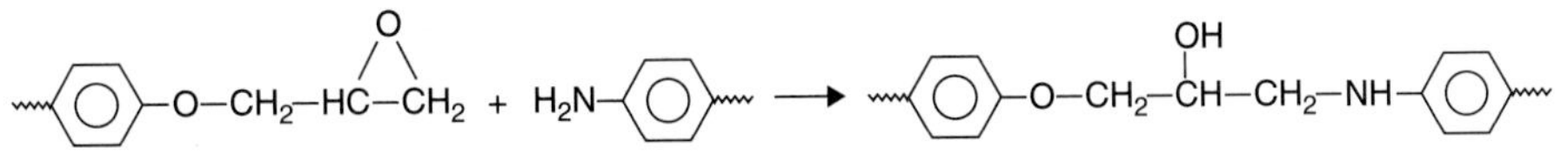

O
OH
O–CH2–HC–CH2 + H2N–
O–CH2–CH–CH2–NH–

FIGURE 7.9 Curing epoxy with an aromatic amine.

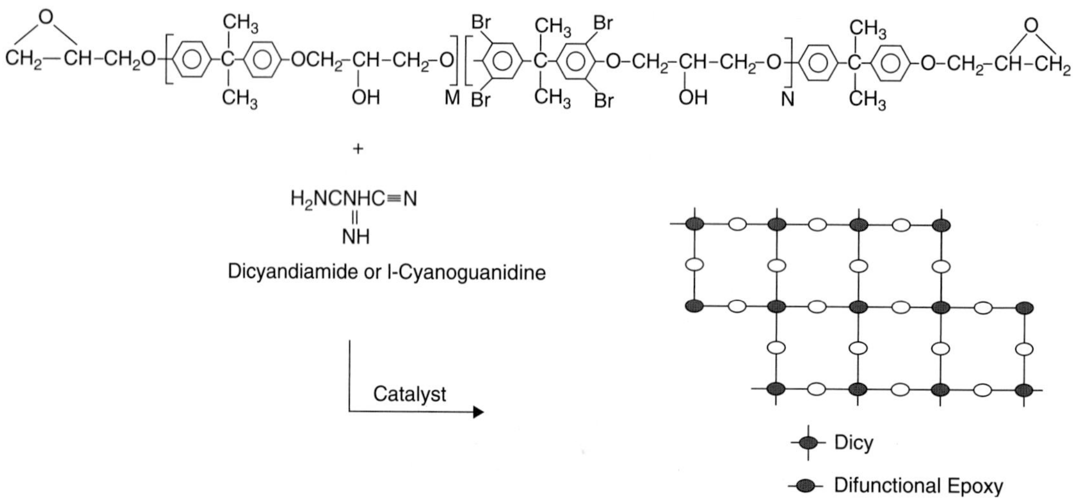

O
CH2—CH-CH2O [CH3 C CH3] OCH2-CH-CH2-O [Br CH3 Br C Br CH3 Br] M O-CH2-CH-CH2-O OH N CH3 C CH3 O-CH2-CH-CH2 O
OH

+

$H_2NCNHC{\equiv}N$
NH
Dicyandiamide or I-Cyanoguanidine

Catalyst

Dicy
Difunctional Epoxy

FIGURE 7.10 Curing epoxy with dicyandiamide (dicy).

epoxy and dicy. However non-dicy systems have been developed to promote faster curing, reduce moisture sensitivity, and improve thermal stability. The choice of curing agent and accelerator is also driven by the resin type. Non-dicy-cured epoxy materials will be discussed further in Chap. 6, as they are often used in lead-free assembly applications. Figure 7.11 illustrates

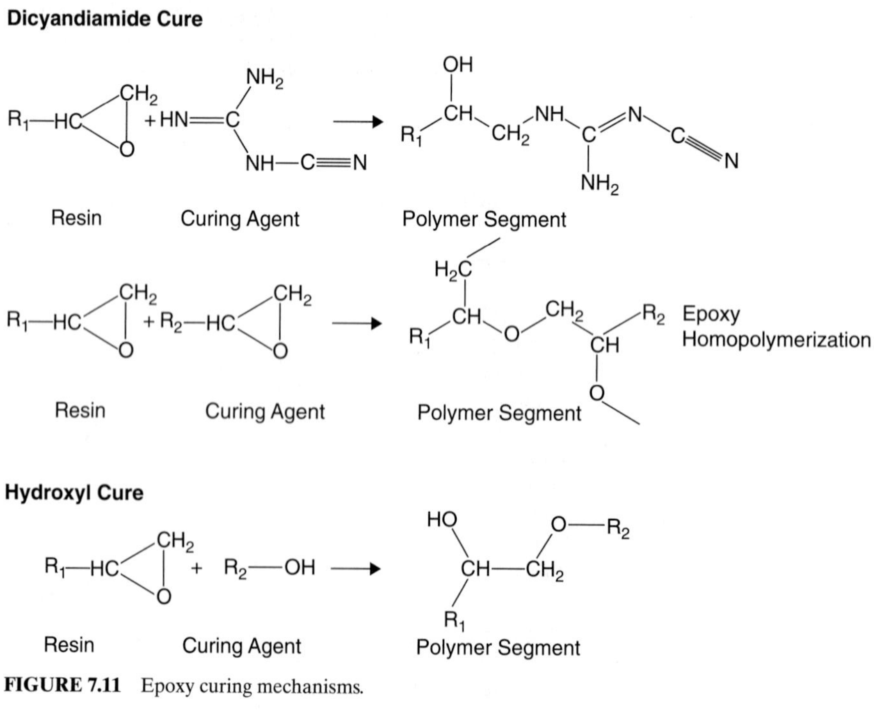

FIGURE 7.11 Epoxy curing mechanisms.

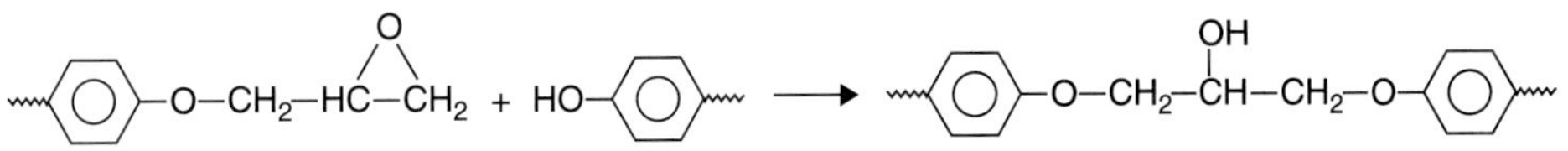

FIGURE 7.12 Phenol curing mechanism.

some of the curing reactions that occur as the resin system polymerizes, and Fig. 7.12 illustrates a phenol curing mechanism.

7.4.2 Flame Retardants

While once the subject of little attention, flame retardants used in the resin system have become a more significant consideration. This is due primarily to legislative initiatives and the increasing focus on the toxicity and environmental impact of some of these compounds. While scientific evidence shows that some of these compounds truly pose a risk, the scientific community also generally agrees that other flame retardants are generally safe. Unfortunately, politics often clouds the decision-making process, and sometimes marketing efforts promoting environmental friendly or "green" products are based more on perception than scientific reality. Some have even argued that specific alternative flame retardant chemistries could even be worse, on balance, than those currently used. Nevertheless, the subject of flame retardants has become, and is expected to remain, an area of continued research.

7.4.2.1 Legislative Issues. The European Union's RoHS and WEEE directives (the WEEE directive addresses waste electronic equipment and recycling requirements) affect not only the lead used in printed circuits, but the flame retardants used in the resin system as well. The RoHS directive restricts the use of specific types of brominated flame retardants. The restricted class of compounds consists of polybrominated biphenyls (PBBs) and polybrominated biphenyl oxides (PBBOs), also called polybrominated diphenyl ethers (PBDEs). The generic structure of these compounds is shown in Fig. 7.13. Specific compounds within these classes of flame retardants can vary in their toxicity, and given the dynamic nature of legislative initiatives, it is important to check the current status of these compounds when making decisions on what materials to use.

In the case of standard epoxy materials used in printed circuits, flame retardancy has commonly been achieved by brominating the epoxy resin. This normally involves manufacturing

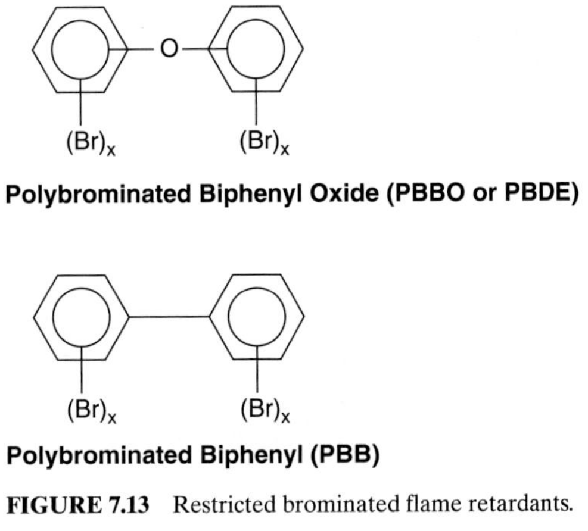

Polybrominated Biphenyl Oxide (PBBO or PBDE)

Polybrominated Biphenyl (PBB)

FIGURE 7.13 Restricted brominated flame retardants.

the epoxy resins with tetrabromobisphenol A (TBBPA), which contains bromine within the chemical backbone (see Fig. 7.2). The RoHS directive does not restrict TBBPA. Because TBBPA is reacted into the epoxy resin itself, it is not available for release into the environment. Under excessive exposure to heat, the bromine is released and retards burning. TBBPA has been used successfully for many years as a flame retardant and is still used in the overwhelming majority of materials. However, while the RoHS directive applies only to specific brominated flame retardants, the WEEE directive may require separation and special handling of materials containing any brominated flame retardant. The separation and special handling relate to concerns about the by-products of incineration, especially if incineration is done at too low a temperature. Furthermore, individual countries continue to look at introducing their own legislative initiatives in regard to these flame retardants, and thus is always important to check the status of these efforts.

7.4.2.2 Flame-Retardant Chemistry. Polymer combustion occurs in a continuous cycle, as highlighted in Fig. 7.14. Heat generated in the flame is transferred back to the polymer surface, producing volatile polymer fragments that constitute fuel for further burning. These fragments diffuse into the flame zone, where they react with oxygen by free-radical chain reactions. This in turn produces more heat and continues the cycle. Flame retardancy is achieved by interrupting this cycle.

There are two basic ways to interrupt this cycle. One method is called solid phase inhibition and involves changes in the polymer substrate. Systems that promote extensive polymer crosslinking at the surface form a carbonaceous char upon heating. The char then insulates the underlying polymer from the heat of the flame, preventing production of new fuel and further burning. Other systems evolve water during heating, cooling the surface and increasing the amount of energy needed to maintain the flame.

The second method is called vapor phase inhibition and involves changes in the flame chemistry. Reactive species are built into the polymer that are transformed into volatile free-radical inhibitors during burning. These materials diffuse into the flame and inhibit the branching radical reactions. As a result, increased energy is required to maintain the flame and the cycle is interrupted. For many materials, both solid and vapor phase inhibition occur.

Polymers, including the various types of epoxy resins, differ in their inherent flammability. The types of resins and curing agents selected can impact the fundamental flammability of the resin system and determine how much additional flame-retardant components are needed to achieve the desired flammability rating. For example, polymers with a high concentration of

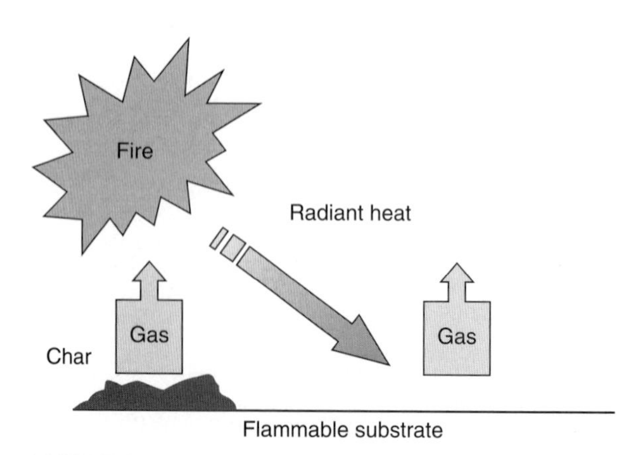

FIGURE 7.14 Polymer combustion.

aromatic groups will generally have higher thermal stability as well as the ability to form char on burning. Additives such as TBBPA or epoxy resins built from TBBPA contain bromine on the molecular backbone. Most organic halogen compounds such as these retard burning by vapor phase inhibition. They typically decompose to yield HBr or HCl (HBr in the case of TBBPA-based materials), which quench chain-branching free radical reactions in the flame. Some halogen acids also catalyze char formation.

7.4.2.3 Halogen-Free Systems. Halogen-free flame retardants and resin systems are also available. Alternatives include phosphorus-based compounds, nitrogen-based flame retardants, inorganic flame retardants, and hydrated fillers. These can be subdivided into reactive components and additives. A reactive component is one that becomes directly incorporated into the resin system itself, such as TBBPA. As with TBBPA, reacted components have the advantage of not being available for release to the environment through leaching or solvent extraction. An example of a reactive, non-halogen flame retardant would be epoxidized phosphorus compounds. The epoxide groups of these compounds make them reactive and result in chemical bonding into the polymer backbone. In contrast, red phosphorus is an inorganic solid that can be dispersed into an epoxy formulation. Hydrates, such as aluminum hydroxide or magnesium hydroxide, decompose and liberate water, which then suppresses the burning process.

In choosing a flame retardant, consideration must be given to the impact on the performance of the resin system and the finished base material. These materials, at the levels required for flame retardancy, can affect the physical properties of the laminate, change rheological properties, and alter cure kinetics of the resin system. Generally, the reactive compounds are preferred since they are bound to the polymer backbone, which prevents release into the environment, and, in comparison to additives or fillers, they seem better suited to obtaining the desired material properties. Table 7.2 summarizes some of the available halogen-free flame retardants.

Organic phosphorus-based flame retardants have become the most common alternative flame retardants used in printed circuit base materials. However, others are also being used and combinations of two or more types of halogen-free flame retardants can also be used synergistically to achieve adequate flammability levels while minimizing adverse impacts on material properties. Some of the adverse impacts that must be evaluated are moisture absorption, copper peel strength degradation, T_g reduction, changes in resin flow, degradation in mechanical or electrical performance, and the impact on Conductive Anodic Filament (CAF) resistance.

TABLE 7.2 Halogen-Free Flame Retardants

Type	Primary Mechanism	Examples	Considerations
Phosphorus	Forms char	Red phosphorus, ammonium polyphosphate, organic phosphorus compounds	Red phosphorus is difficult to process and toxic. Organic phosphorus compounds are a common alternative, but generally more costly and can lower T_g.
Inorganic/Hydrated Fillers	Evolve water, endothermic, may promote char	Alumina trihydrate, magnesium hydroxide, zinc borate, antimony oxides	Inexpensive, but generally require high loadings and can degrade mechanical and other properties. Toxicity of antimony-based retardants is a concern. Can be used with other flame retardants synergistically.
Nitrogen	Intumescent system, generates gas that foams char	Melamine, melamine cyanurate	Impact on material properties and other environmental effects are concerns.

7.4.3 UV Blockers/Fluorescing Aids

Some resins naturally absorb ultraviolet light. Others absorb very little. There are two reasons why this property is important. First, automatic optical inspection (AOI) equipment is often laser-based and relies on the resin system in the base material to fluoresce upon exposure. In this way, the AOI equipment can distinguish between the base laminate and the conductor pattern. The second reason involves photoimaging on very thin circuits, during the solder mask imaging process, for example. This process commonly involves exposure to ultraviolet (UV) light through an image of the desired pattern onto both sides of the circuit, which is coated with a mask or photoresist. The UV light initiates chemical changes to the photoresist, typically polymerization, which makes it less soluble in a developer solution. In very thin circuits, if the base material does not absorb UV light sufficiently, the UV light from one side of the circuit board can pass through to the other side and cause unwanted exposure of the photoresist on the opposite side. Therefore, components that absorb UV light are sometimes added to resin systems that do not sufficiently absorb UV light.

7.5 *REINFORCEMENTS*

While there are also a variety of reinforcements used in base materials, woven fiberglass cloths are by far the most common. Other materials include paper, glass matte, non-woven aramid fibers, non-woven fiberglass, and a variety of fillers. The advantages of woven glass cloths include a good combination of mechanical and electrical properties, a wide range of types for achieving various laminate thicknesses, and good economics.

7.5.1 Woven Fiberglass

The process of manufacturing woven fiberglass cloth begins with melting the various inorganic components required for a particular grade of glass. The molten components travel through a furnace and ultimately flow through specialized bushings to form the individual fiberglass filaments and yarns. These yarns are then used in a weaving process to manufacture the fiberglass cloth. The relative concentrations of the components used affect the chemical, mechanical, and electrical properties of the fiberglass. The compositions of a few fiberglass types are provided in Table 7.3. Figure 7.15 illustrates the process used to manufacture the fiberglass yarns. The yarns are then woven into fiberglass cloths.

TABLE 7.3 Fiberglass Compositions in Percentages

Component	E-Glass	NE-Glass	S-Glass	D-Glass	Quartz
Silicon Dioxide	52–56	52–56	64–66	72–75	99.97
Calcium Oxide	16–25	0–10	0–0.3	0–1	
Aluminum Oxide	12–16	10–15	24–26	0–1	
Boron Oxide	5–10	15–20		21–24	
Sodium Oxide and Potassium Oxide	0–2	0–1	0–0.3	0–4	
Magnesium Oxide	0–5	0–5	9–11		
Iron Oxide	0.05–0.4	0–0.3	0–0.3	0.3	
Titanium Oxide	0–0.8	0.5–5			
Fluorides	0–1.0				

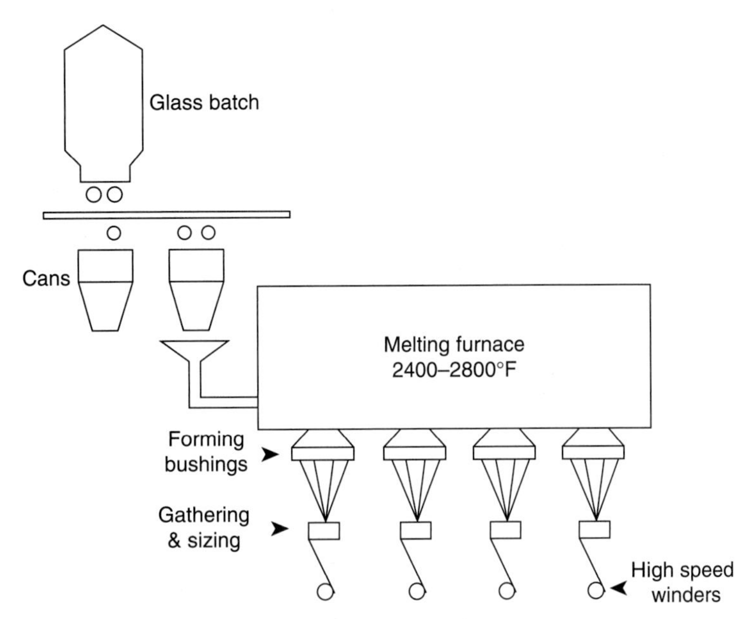

FIGURE 7.15 Illustration of fiberglass yarn manufacture.

E-glass is by far the most commonly used fiberglass for printed circuits. It provides an excellent combination of electrical, mechanical, and chemical properties at a reasonable cost. NE-glass is used in limited quantities and offers improved permittivity (Dk) and loss (Df) properties (see Table 7.4). However, NE-glass is more expensive than E-glass. S-glass provides greater strength but is more difficult to process through the mechanical drilling operation. Other glass types are seldom used.

TABLE 7.4 Comparison of E-Glass and NE-Glass

Property	E-Glass	NE-Glass
CTE, ppm/°C	5.5	3.4
Dielectric constant @ 1 MHz	6.6	4.4
Dissipation factor @ 1 MHz	0.0012	0.0006

Several diameters of filaments are available (see Table 7.5), as well as many different yarn types. D, DE, E, and G filaments are the most commonly used. Combining the various yarn

TABLE 7.5 Glass Filament Diameters

Filament Designation	Nominal Diameter (Microns)	Nominal Diameter (Inches)
B	3.5	0.00015
C	4.5	0.00018
D	5	0.00021
DE	6	0.00025
E	7	0.00028
G	9	0.00037
H	10	0.00043
K	13	0.00051

FIGURE 7.16 Plain weave.

types in the weaving process leads to many different types of cloth styles. While there are also many different weave types, virtually all cloths for printed circuits use a plain weave (see Fig. 7.16). The plain weave consists of yarns interlaced in an alternating fashion, one over and one under every other yarn. This weave pattern provides good fabric stability. Some of the common plain weave cloth styles used for printed circuits are shown in Table 7.6. The fiberglass cloths are also coated with a finish or coupling agent tailored for improving the bond between the glass filaments and the specific resin coated onto the glass. This coupling agent and the resin-to-glass bond that results are important considerations for CAF resistance, which will be discussed in Chap. 9.

7.5.2 Yarn Nomenclature

Because there can be so many types of yarn manufactured from the available grades of glass and filament diameters, special nomenclature systems have been developed. The two systems used are the U.S. System and the TEX/Metric System.

7.5.2.1 The U.S. System. An example of a yarn name in the U.S. system is ECD 450-1/0. This yarn is used in making a 1080 style glass cloth. Each letter and number in the name describes something about the yarn:

- **First letter** This describes the glass composition. Electrical grade, or E-glass, is the most common grade used for manufacturing materials for printed circuits.
- **Second letter** C indicates that the yarn is composed of continuous filaments. S would indicate staple filaments, and T indicates texturized continuous filaments.
- **Third letter** This letter indicates the individual filament diameter (see Table 7.5).

TABLE 7.6 Common Fiberglass Cloth Styles for Printed Circuit Base Materials

Style	Approx. Fiberglass Thickness (In.)	Warp Yarn	Fill Yarn	Count (Ends/In.)	Weight (Oz./Yd.^{2})
104	0.0013	ECD 900-1/0	ECD 1800-1/0	60 × 52	0.55
106	0.0014	ECD 900-1/0	ECD 900-1/0	56 × 56	0.73
1067	0.0014	ECD 900-1/0	ECD 900-1/0	69 × 69	0.71
1080	0.0023	ECD 450-1/0	ECD 450-1/0	60 × 47	1.42
1280	0.0026	ECD 450-1/0	ECD 450-1/0	60 × 60	1.55
1500	0.0052	ECE 110-1/0	ECE 110-1/0	49 × 42	4.95
1652	0.0045	ECG 150-1/0	ECG 150-1/0	52 × 52	4.06
2113	0.0028	ECE 225-1/0	ECD 450-1/0	60 × 56	2.31
2116	0.0038	ECE 225-1/0	ECE 225-1/0	60 × 58	3.22
2157	0.0051	ECE 225-1/0	ECG 75-1/0	60 × 35	4.36
2165	0.0040	ECE 225-1/0	ECG 150-1/0	60 × 52	3.55
2313	0.0029	ECE 225-1/0	ECD 450-1/0	60 × 64	2.38
3070	0.0031	ECDE 300-1/0	ECDE 300-1/0	70 × 70	2.74
3313	0.0033	ECDE 300-1/0	ECDE 300-1/0	60 × 62	2.40
7628	0.0068	ECG 75-1/0	ECG 75-1/0	44 × 32	6.00
7629	0.0070	ECG 75-1/0	ECG 75-1/0	44 × 34	6.25
7635	0.0080	ECG 75-1/0	ECG 50-1/0	44 × 29	6.90

- **First number** This represents 1/100th the normal bare glass yardage in one pound of the basic yarn strand. In the preceding example, multiply 450 by 100, which results in 45,000 yards in one pound.
- **Second number** Here 1/0 represents the number of basic strands in the yarn. The first digit represents the original number of twisted strands. The second digit separated by the diagonal represents the number of strands plied or twisted together. Thus 1/0 means that the yarn is a singles yarn (no or "zero" plying required).

The name may also include a designation indicating the number of turns per inch in the twist of the final yarn, and the direction of the twist. An example would be 3.0S, or three turns per inch with an "S" direction twist. An "S" twist has spirals that run up and to the left. A "Z" twist has spirals that run up and to the right.

7.5.2.2 The TEX/Metric System An example of a yarn name in the TEX/Metric System is EC9 33 1X2.

- **First letter** This designates the glass composition.
- **Second letter** C indicates that the yarn is composed of continuous filaments. T would indicate textured continuous filaments. D indicates staple filament.
- **First number** Here 9 indicates the individual filament diameter expressed in microns.
- **Second number** In this example, 33 represents the nonlinear weight of the bare glass strand expressed in TEX. TEX is the mass in grams per 1,000 meters of yarn.
- **Third number** IX2 indicates yarn construction or the basic number of strands in the yarn. The first digit represents the original number of twisted strands and the second digit, after the X, indicates the number of these strands twisted or plied together.

7.5.3 Fiberglass Cloths

With the possible combinations of glass compositions, filament diameters, yarn types, and the number of different weave patterns available, the number of possible fiberglass cloths can almost be unlimited. The effects that these glass fabrics have on the base material are driven by these variables. In addition, the fabric count, or the number of warp yarns and fill yarns, also helps determine the properties of the fabric and the base material. Warp yarns are those that lie in the length (machine direction) of the fabric, whereas the fill yarns lie across the warp direction. The warp direction is also commonly called the grain direction.

As pointed out earlier, glass cloths used in printed circuits are typically, though not always, made from E-glass, and virtually all use what is called a plain weave. A plain weave results in a fabric constructed so that one warp yarn passes over and under one filling yarn (and vice versa). This weave pattern offers good resistance to yarn slippage and fabric distortion. Common fiberglass cloths used in materials for printed circuits are shown in Table 7.6. Three of the most common are shown in Fig. 7.17. At a given resin content, each style will yield a different nominal thickness. Having the flexibility of many glass styles and thicknesses is important for meeting controlled impedance and overall thickness requirements.

During the manufacture of the filaments and woven fabrics, a variety of surface finishes may be applied to the glass to improve manufacturability, help prevent abrasion and static, and aid in holding the filaments together. The most important of these surface finishes to the laminate and printed circuit manufacturer is the coupling agent applied to the finished fabric. The coupling agent, typically an organosilane compound, aids in the wetting of and adhesion to the resin applied to the glass. It is important for the reliability of the finished

(a) 1080 Cloth

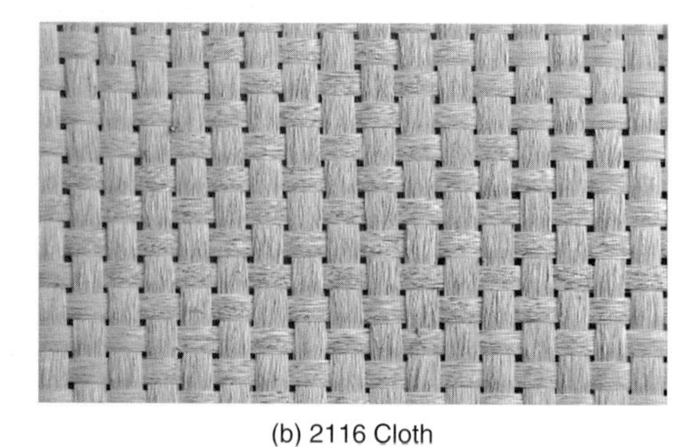

(b) 2116 Cloth

(c) 7628 Cloth

FIGURE 7.17 The three most common fiberglass cloths used in materials for printed circuits.

circuit board that this coupling agent improves the resin to glass adhesion both during the circuit manufacturing process (for example, in mechanical drilling) as well as within the actual end-use environment of the circuit. The coupling agent also plays a role in long-term CAF resistance, which is discussed in Chap. 9. A variety of these compounds are commercially available, with the specific choice primarily driven by the type of resin to be applied to the fiberglass cloth. Figure 7.18 provides examples of silane coupling agents.

$(CH_3O)_3$—Si—R—CH—CH_2 (epoxide ring, O bridging CH and CH_2)

Epoxy Silane

$(CH_3O)_3$—Si—R—NH_2

Amino Silane

GLASS(Si)—O—Si—R—Z
GLASS(Si)—O—Si—R—Z
GLASS(Si)—O—Si—R—Z

Bonding to Glass Cloth **Bonding to Resin**

FIGURE 7.18 Examples of silane coupling agents.

7.5.4 Other Reinforcements

While woven fiberglass cloth makes up the overwhelming majority of the reinforcements used for printed circuit base materials, other types can be used exclusively or in combination with woven glass fabric. These other reinforcements include the following.

7.5.4.1 Glass Matte. Compared to woven cloths, glass matte reinforcement exhibits a more random orientation. Chopped strand matte consists of fiberglass strands that have been chopped into 1- to 2-inch lengths and distributed evenly. Continuous strand matte, as the name implies, consists of continuous strands of fiberglass in a random, spiral orientation. Glass matte is used in the core of CEM-3, which is used in relatively unsophisticated products.

7.5.4.2 Aramid Fiber. As opposed to the inorganic chemistry of fiberglass reinforcements, aramid fibers consist of aromatic polyamide organic compounds, and therefore exhibit different properties. The unique properties of aramid fibers can offer advantages in specific high-performance printed circuits and laminate-based multichip modules (MCM-L). For example, aramid fiber-reinforced materials are sometimes used in microvia applications since they are easily ablated by plasma or laser. Other interesting properties of aramid fibers are their low weight, high strength, and negative coefficient of thermal expansion (CTE) in the axial direction. When combined with the resin system, the resulting composite can offer reduced overall CTEs in the X-Y plane compared to conventional materials.

7.5.4.3 Linear Continuous Filament Fiberglass. A unique process for producing fiberglass-reinforced laminates winds continuous yarns in a linear orientation. The resulting laminate possesses three layers of fiberglass filaments, with the outside layers parallel to each other and the middle layer perpendicular to the outside. With an equal number of linear filaments running in each axis, this type of reinforcement results in a laminate with improved dimensional stability.

7.5.4.4 Paper. Cellulose-based papers can also be used as a reinforcement for base materials. Paper-based reinforcements are also used with other reinforcements such as woven glass, and these materials can also allow punching of through holes rather than drilling. This makes them economical in some high-volume, low-technology applications including consumer electronics such as radios, toys, calculators, and video game systems. Paper is used in FR-2, FR-3, and the core of CEM-1.

7.5.4.5 Fillers. Fillers are small particles that can be added to a resin system to modify the properties of the composite material. These materials range from talc, silica, kaolin clay powders, and tiny hollow glass spheres to a variety of other inorganic materials. These materials are typically used to tailor the properties of the base material for specific uses. For example, kaolin clay powders coated with a layer of palladium and dispersed within the base materials have been used as catalysts for electroless copper plating. Hollow glass microspheres have been used to reduce the dielectric constant of materials. Other fillers are being used to reduce thermal expansion properties and improve reliability, enhance the machinability of materials in the drilling process, alter electrical properties such as Dk and Df, and lower total costs. The use of fillers to reduce z-axis expansion has become quite common, especially as lead-free assembly is adopted. The reduced z-axis expansion helps offset the increased expansion and stress on plated through holes that results from the higher temperatures associated with lead-free assembly. This is discussed further in Chap. 9.

7.5.4.6 Expanded Teflon. Although not typically thought of as a reinforcement, Teflon® that has been "expanded" into a spongelike structure on a microscopic scale is currently being used in applications that require prepregs with very low dielectric constants or loss properties.

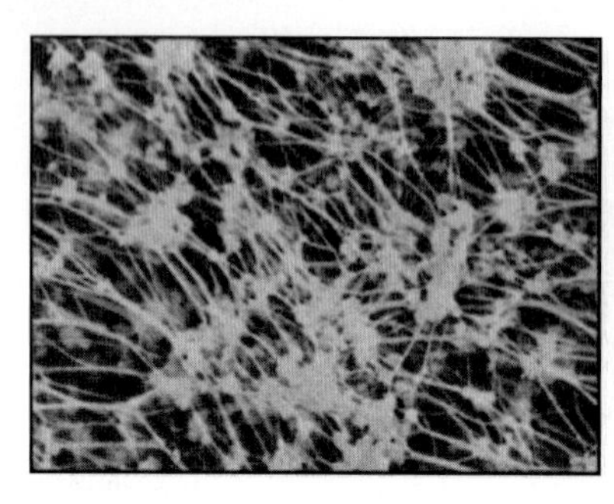

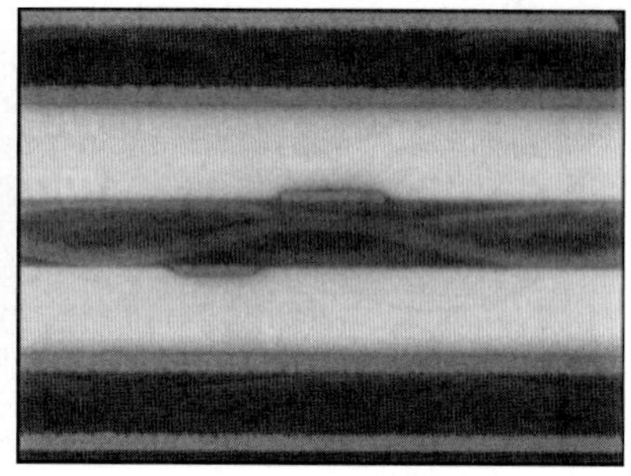

FIGURE 7.19 Expanded Teflon (left) and a PCB cross section showing prepreg using expanded Teflon (right). *(Courtesy of W.L. Gore and Associates.)*

A resin system is applied to the expanded Teflon structure and is made into a b-staged sheet capable of bonding printed circuit layers together. This material can be used to make prepregs for high-frequency applications and is shown in Fig. 7.19.

7.6 CONDUCTIVE MATERIALS

The primary conductive material used in printed circuits is copper foil. However, the trend toward circuit densification has brought about recent developments in copper foil technology as well. In addition, copper foils subsequently plated with other metal alloys are employed to manufacture printed circuits with resistive components buried within a multilayer structure. Copper foil grades are shown in Table 7.7.

TABLE 7.7 Foil Grades from IPC-4562

Grade	Foil Description
1	Standard electrodeposited (STD-Type E)
2	High-ductility electrodeposited (HD-Type E)
3	High-temperature elongation electrodeposited (HTE-Type E)
4	Annealed electrodeposited (ANN-Type E)
5	As rolled-wrought (AR-Type W)
6	Light cold rolled-wrought (LCR-Type W)
7	Annealed-wrought (ANN-Type W)
8	As rolled-wrought low-temperature annealable (LTA-Type W)
9	Nickel, standard electrodeposited
10	Electrodeposited low temperature annealable (LTA-Type E)
11	Electrodeposited annealable (A-Type E)

7.6.1 Electrodeposited Copper Foil

The most common foil used in manufacturing printed circuits is electrodeposited copper foil (ED foil). ED foil is produced through an electrochemical process in which copper feed stock or scrap copper wire is first dissolved in a sulfuric acid solution. The purified copper sulfate/ sulfuric acid solution is then used to electroplate copper onto a cylindrical drum typically made from stainless steel or titanium. Figure 7.20 illustrates the overall process used to make electrodeposited copper foil. This process results in a copper foil with a relatively smooth, shiny side, and a coarser matte side, as illustrated in Fig. 7.21. The shiny side mirrors the surface

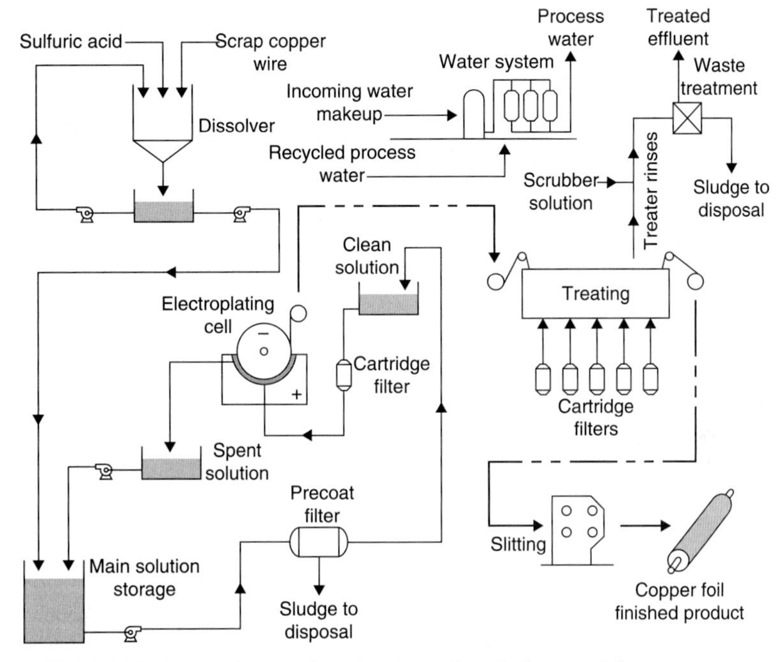

FIGURE 7.20 Process for manufacturing electrodeposited copper foil.

of the plating drum, while the microscopically rough matte side is formed by the copper grain structure. By controlling the plating solution chemistry, the surface condition of the plating drum, and the electroplating parameters, one can modify the properties of the copper foil for various usage environments. For example, mechanical properties such as tensile strength and elongation, as well as the surface profile of the matte side, can all be adjusted through control

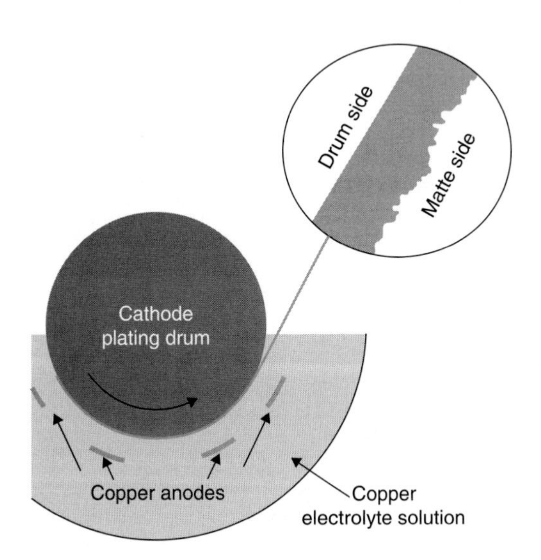

FIGURE 7.21 Electroplating copper foil.

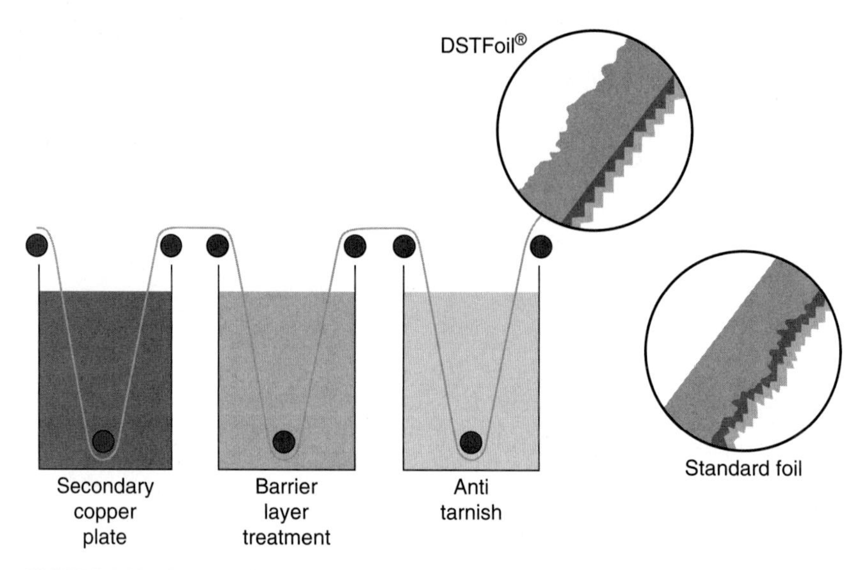

FIGURE 7.22 Treating of copper foil.

of these process variables. The foil produced in this process is then run through a treatment process. This treatment process typically plates copper nodules that further roughen the surface for adhesion, and also applies other metal barrier layers and antitarnish coatings, as shown in Fig. 7.22. Standard copper foil area weights and thicknesses were shown in Table 6.4.

The foils most commonly used in printed circuit manufacturing are Grade 1 and Grade 3 foils. Unlike Grade 1 foil, Grade 3 foil must meet specific elongation requirements at elevated temperatures (180°C). Grade 3 foil—also commonly referred to as high-temperature elongation, or simply "HTE" foil—is widely used in base laminates employed to manufacture multilayer printed circuit boards. The increased ductility at elevated temperatures provides resistance to copper foil cracks when the multilayer circuit is thermally stressed and expands in the z-axis. Changes to the plating bath are made to alter the grain structure of HTE foil. This results in different mechanical properties. Tables 7.8 and 7.9 show the tensile strength and ductility requirements for standard Grade 1 copper foil and high-temperature elongation Grade 3 copper foil. These requirements come from IPC-4562, "Metal Foil for Printed Wiring Applications," which lists requirements for the other grades as well.

The surface profile of the copper foil is also important in printed circuit manufacturing. On one hand, a relatively rough surface profile aids in the bond strength of the foil to the resin system. On the other hand, a rough profile may require longer etching times, which impact productivity and the geometry of the etched features. With longer etching times, propensity to

TABLE 7.8 Tensile and Elongation Properties of Grade 1 Copper Foil

Property @ 23°C	1/2 Ounce	1 Ounce	2 Ounce
Tensile strength:			
kpsi	30	40	40
MPa	207	276	276
Elongation %	2	3	3

TABLE 7.9 Tensile and Elongation Properties of Grade 3 Copper Foil

Property	1/2 Ounce	1 Ounce	2 Ounce
Tensile strength @ 23°C:			
kpsi	30	40	40
MPa	207	276	276
Elongation % @ 23°C	2	3	3
Tensile strength @ 180°C:			
kpsi	15	20	20
MPa	103	138	138
Elongation % @ 180°C	2	2	3

TABLE 7.10 Foil Profile Criteria

Foil Profile Type	Max. Foil Profile (Microns)	Max. Foil Profile (μInches)
S – Standard	N/A	N/A
L – Low profile	10.2	400
V – Very low profile	5.1	200
X – No treatment or roughness	N/A	N/A

form trapezoidal circuit traces increases since there is more time for lateral etching of the conductor. This has obvious implications for manufacturing fine-line circuits in high-yield and controlling impedance properties. Low-profile and very-low-profile characteristics are included in the IPC-4562 specification and are summarized in Table 7.10. Figures 7.23 and 7.24 give an example of profile differences between a standard and low-profile copper foil. In addition, as circuit operating frequencies increase, roughness of the copper foil can also impact signal attenuation. At higher frequencies, more of the electrical signal is conducted near the surface of the conductor. A rougher profile results in a longer path for the signal to travel, which also results in greater attenuation, or loss. As a result, high-performance materials require foils with low profiles that have adequate adhesion to the high-performance resin systems.

Subsequent to the manufacturing of the base copper foil, a variety of surface treatments are typically applied, and these too will vary depending on the usage environments. These treatments fall into four categories.

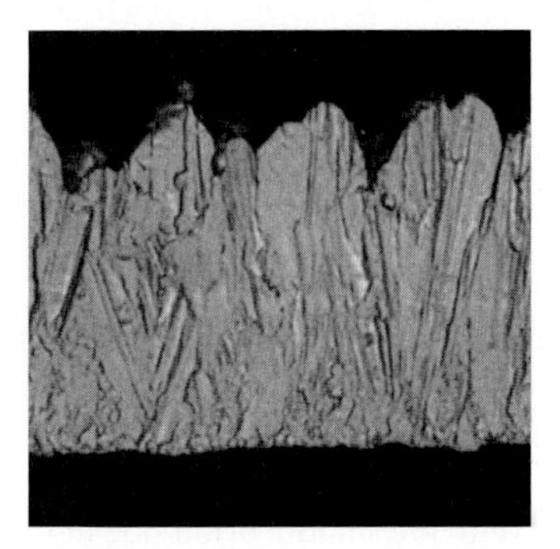
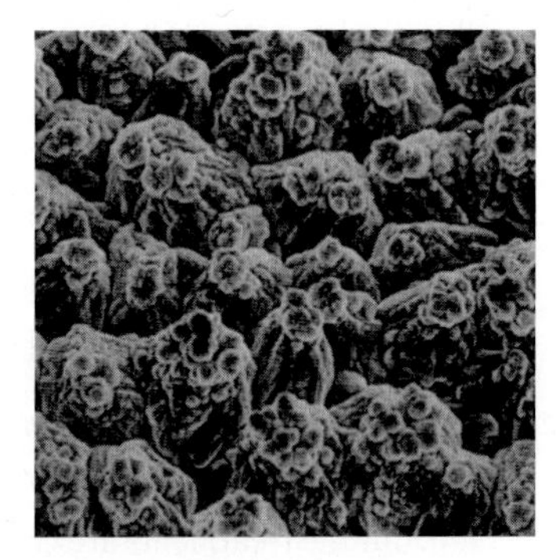

FIGURE 7.23 Cross section and the matte side of standard Grade 1 foil. *(Courtesy of Gould Electronics)*

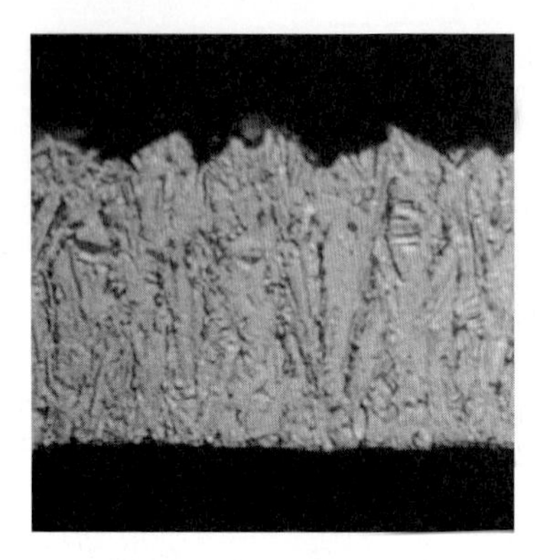
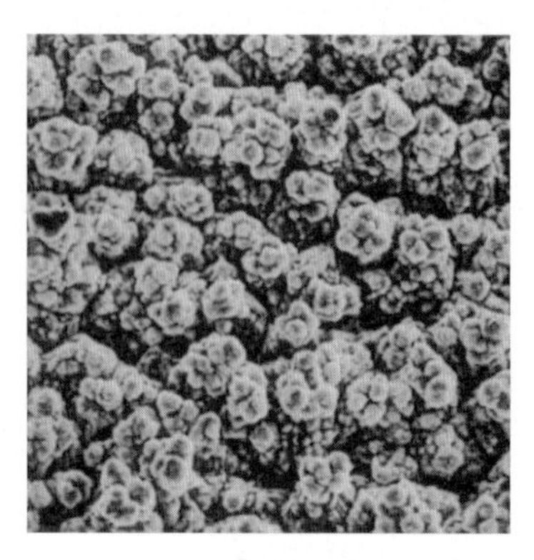

FIGURE 7.24 Cross section and the matte side of low-profile Grade 1 foil. *(Courtesy of Gould Electronics)*

7.6.1.1 Bonding Treatments or Nodularization. This treatment increases the surface area of the foil by plating copper or copper-oxide nodules to the surface of the foil. The increased surface area results in increased foil to resin bond strengths. The thickness of this treatment is relatively small, but can be tailored for adhesion to high-performance resin systems such as polyimides, cyanate esters, and BT. The matte side images in Figs. 7.23 and 7.24 include these nodules.

7.6.1.2 Thermal Barriers. A coating of zinc, nickel, or brass is usually applied over the nodules. This coating can prevent thermal or chemical degradation of the foil to resin bond during manufacture of the laminate, the printed circuit, and the circuit assembly. These coatings typically measure several hundred angstroms in thickness and vary in color due to the specific metal-alloy used, although most treatments are brown, gray, or a yellow mustard color.

7.6.1.3 Passivation and Antioxidant Coatings. In contrast to the other coatings, these treatments are virtually always applied to both sides of the foil. Although many of these treatments are chromium-based, organic coatings can also be utilized. The primary purpose of these treatments is to prevent oxidation of the copper foil during storage and lamination. These coatings are usually less than 100 angstroms thick and are typically removed by the cleaning, etching, or scrubbing processes normally used at the start of printed circuit manufacturing processes.

7.6.1.4 Coupling Agents. The use of coupling agents, primarily silanes such as those used to promote fiberglass to resin adhesion, can also be used on copper foils. These coupling agents can improve the chemical bond between the foil and the resin system and can also be used to help prevent oxidation or contamination.

7.6.2 Drum Side Treated Foils (DSTF) or Reverse Treated Foils (RTF)

Drum side treated (DSTFoil®) or reverse treated foil (RTF) is also an electrodeposited copper foil, but the treatments are coated onto the smooth drum side of the foil rather than the matte side, as with conventional electrodeposited foil (see Figs. 7.25 and 7.26). This results in a very-low-profile surface bonded to the laminate, with the rough matte side facing out. The low

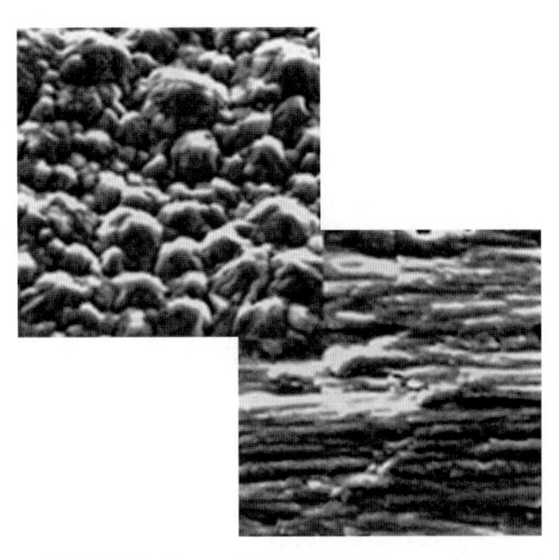

FIGURE 7.25 External sides for RTF (upper photo) and standard (lower photo) foils. *(Photos Courtesy of Gould Electronics)*

surface profile against the laminate aids in the production of fine circuit features on the inner layer, while the matte surface can aid in photoresist adhesion. The low profile against the laminate can also improve electrical performance at high frequencies. In addition, in very thin laminates, the low-surface profile can aid in achieving consistent dielectric thickness with a reduced concern of inadequate dielectric separation between the corresponding points on the tooth profile. These benefits come at the expense of a slight reduction in peel strength.

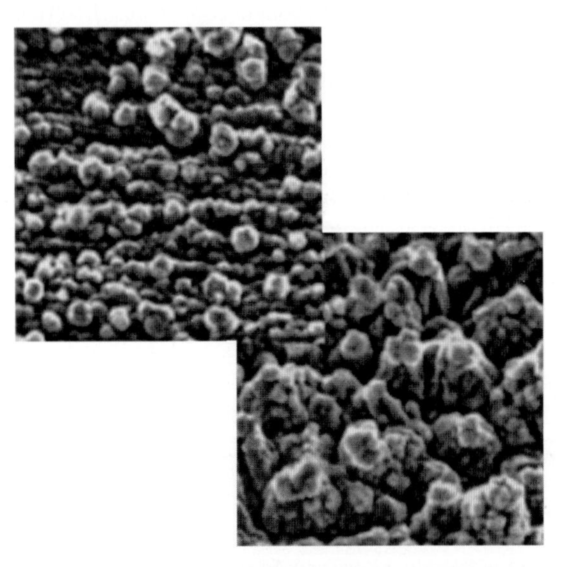

FIGURE 7.26 Sides bonded to laminate for RTF (upper photo) and standard foils (lower photo). *(Photos Courtesy of Gould Electronics)*

7.6.3 Wrought Annealed Copper Foils

Wrought annealed foil is typically used in flexible circuit manufacturing because of its superior ductility. In contrast to an electrodeposition process, the process of creating wrought annealed foil begins with working a slab or ingot of copper through a series of rollers in conjunction with heat cycles to obtain the desired thickness and mechanical properties. The resulting grain structure of wrought annealed copper foil, which is very random compared to the columnar or fine-grain structure of electrodeposited foil, is a significant contributor to the differences in mechanical properties. In addition, wrought annealed foil consists of two very-low-profile sides, so the treatment steps account for any surface roughness. The tensile and elongation requirements for Grade 7 wrought annealed copper are provided in Table 7.11. The requirements for other wrought foils are included in IPC-4562.

TABLE 7.11 Tensile and Elongation Properties of Wrought Annealed Foil

Property	½ Ounce	1 Ounce	2 Ounce
Tensile strength @ 23°C:			
kpsi	15	20	25
MPa	103	138	172
Fatigue ductility % @ 23°C	65	65	65
Elongation % @ 23°C	5	10	20
Tensile strength @ 180°C:			
kpsi	TBD	14	22
MPa		97	152
Elongation % @ 180°C	TBD	6	11

7.6.4 Copper Foil Purity and Resistivity

IPC-4562, "Metal Foil for Printed Wiring Applications," also specifies the purity and resistivity of electrodeposited and wrought copper foils. Electrodeposited copper foils without treatment have a minimum purity of 99.8%, with any silver counted as copper. The value for wrought foils is 99.9%. Table 7.12 provides the resistivity requirements for electrodeposited foil. The maximum resistivity requirements for wrought copper range from 0.155 to 0.160 ohm-gram/m^2, depending on the weight.

TABLE 7.12 Maximum Resistivity of Deposited Foil

Weight Designator	Common Industry Terminology	Maximum Resistivity
E	5 μm	0.181 ohm-gram/m^2
Q	9 μm	0.171 ohm-gram/m^2
T	12 μm	0.170 ohm-gram/m^2
H	½ ounce	0.166 ohm-gram/m^2
M	¾ ounce	0.164 ohm-gram/m^2
1 oz. and over (305 g/m^2)	1 ounce	0.162 ohm-gram/m^2

7.6.5 Other Foil Types

The electrodeposited copper foils described above account for most of the conductive foils used in rigid printed circuits. However, modified versions of these foils are sometimes used in niche applications. These modified versions include double-treated copper foil and resistive foils.

7.6.5.1 Double-Treated Copper Foil. As discussed in the preceding section, the foil surface that is bonded to the base laminate is treated with coatings designed to improve foil-to-resin bond strength and reliability. In double-treated foils, these coatings are also applied to the foil surface that forms the outside laminate surface. It is also possible to have a "reverse-treated" double-treat foil, meaning that the smooth surface is bonded to the laminate with the matte surface facing out, with both sides having been treated.

The advantage of double-treat foil is that it eliminates the oxide or other surface preparation process typically used to prepare the inner-layer circuitry for multilayer lamination. However, no abrasion of this double-treat coating can be tolerated, and removal of any surface contamination becomes difficult. This also makes double-treated foil more sensitive to handling practices in the circuit manufacturing process.

7.6.5.2 Resistive Foils. Other treatments can also be applied to the base foil for use in manufacturing inner-layer circuits with *buried resistors*. This technology can enable the creation of resistors on internal layers of a multilayer circuit, with removal of many of the resistors commonly assembled on the outside of the multilayer circuit. This can improve board reliability and free up space on the outside of the board for active components. These foils typically use a resistive metal alloy coated onto the base foil. The laminate made with this foil can then be sequentially imaged and etched to produce the desired circuit pattern along with resistive components.

REFERENCES

1. IPC-4101, "Specification for Base Materials for Rigid and Multilayer Printed Boards."
2. Polyclad Product Reference Materials.
3. Kelley, Edward, "Meeting the Needs of the Density Revolution with Non-Woven Fiberglass Reinforced Laminates," EIPC/Productronica Conference, November 1999.
4. W.L. Gore Technical Literature.
5. Levchik, Sergei V., Weil, Edward D., "Thermal Decomposition, Combustion and Flame-Retardancy of Epoxy Resins—a Review of the Recent Literature," Polymer International/Society of Chemical Industry, 2004.
6. Nelson, Gordon L., *Fire and Polymers II, Materials and Tests for Hazard Prevention*, American Chemical Society, 1995.
7. IPC-WP/TR-584, "IPC White Paper and Technical Report on Halogen-Free Materials Used for Printed Circuit Boards and Assemblies."
8. Clark-Schwebel Industrial Fabrics Guide.
9. BGF Industries, Inc. Fiberglass Guide.
10. Gould Electronics Product Reference Materials.
11. Kelley, Edward J., and Micha, Richard A., "Improved Printed Circuit Manufacturing with Reverse Treated Copper Foils." IPC Printed Circuits Expo, March 1997.
12. Jawitz, Martin W., *Printed Circuit Materials Handbook,* McGraw-Hill Companies, Inc., 1997.
13. IPC-4562, "Metal Foil for Printed Wiring Applications."
14. Brist, Gary, Hall, Stephen, Clauser, Sidney, and Liang, Tao, "Non-Classical Conductor Losses Due to Copper Foil Roughness and Treatment," ECWC 10/IPC/APEX Conference, February 2005.
15. IPC-4412, "Specification for Finished Fabric Woven from E-Glass for Printed Boards."

CHAPTER 8
PROPERTIES OF BASE MATERIALS

Edward J. Kelley
Isola Group, Chandler, Arizona

8.1 INTRODUCTION

A variety of base material properties are of interest to the printed circuit manufacturer, assembler, and original equipment manufacturer (OEM). These include thermal, physical, mechanical, and electrical properties. This chapter introduces some of the most important properties and also provides some comparisons between material types. Most of the test methods used to evaluate these properties can be found in the IPC Test Methods Manual, IPC-TM-650.

8.2 THERMAL, PHYSICAL, AND MECHANICAL PROPERTIES

Historically, the properties that received the greatest amount of attention were the glass transition temperature, T_g, and the coefficients of thermal expansion, or CTEs, particularly in the z-axis. With the advent of lead-free assembly processes, other properties have increased in importance as well. The most notable is the decomposition temperature, T_d. These properties were described in more detail in Chap. 6, and will be discussed again in Chap. 10, which focuses on the impact of lead-free assembly on base materials. However, some additional information as well as examples of the test data are included here, as well as comparisons of some common material types.

8.2.1 Thermomechanical Analysis T_g and CTEs

Materials undergo changes in physical dimensions in response to changes in temperature. The rates of expansion of fiberglass cloth reinforced materials differ in the respective axes of the material due to the directionality of the reinforcement. The length and width of the laminate, or printed circuit, are termed the X/Y plane, whereas the axis perpendicular to this plane is the z-axis.

Thermal expansion can be measured by thermomechanical analysis (TMA). TMA uses a device that measures a dimension of a sample versus temperature. Depending upon the orientation of the sample in the device, either the x/y CTE or the z-axis CTE can be measured.

Figure 8.1 provides an example of a TMA scan on a high-T_g, filled FR-4 material designed to be compatible with most lead-free assembly applications. The T_g is determined by extrapolating the linear portions of the expansion curve to the point where they intersect. In this case, a T_g of

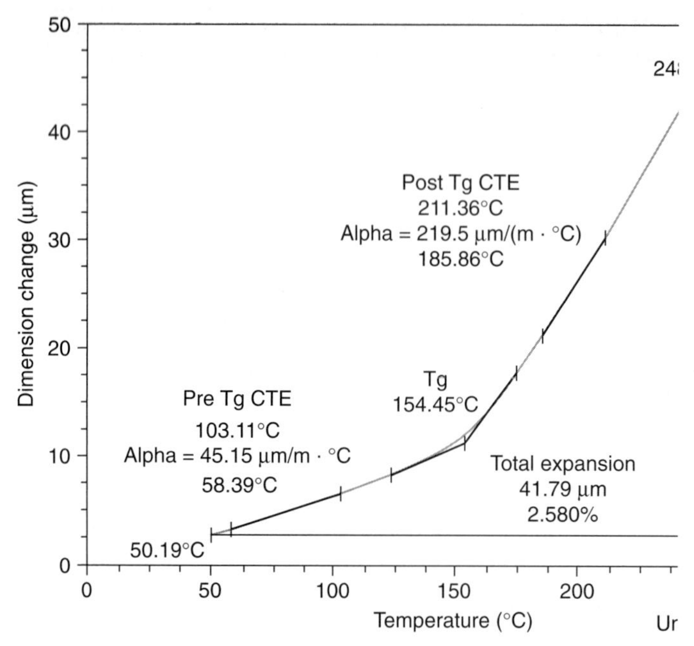

FIGURE 8.1 TMA scan illustrating T_g and CTE determination.

154.45°C is measured. The z-axis CTE values are typically calculated both pre-T_g, also called "alpha 1" (α_1), and post-T_g, also called "alpha 2" (α_2). In this example, the pre-T_g CTE is measured at just over 45 ppm/°C, and the post-T_g CTE is measured at just over 219 ppm/°C. The total expansion from 50°C to 250°C is also shown as a percent expansion value at 2.58 percent.

Thermal expansion in the z-axis can significantly affect the reliability of printed circuits. Since plated through holes run through the z-axis of the printed circuit, thermal expansion and contraction in the base materials causes strain and plastic deformation in the plated through holes and can also deform the copper pads on the surface of the printed circuit. With sufficient stress on the external pads, they can be pulled toward the plated through hole during thermal stress and subsequently appear lifted from the surface upon cooling. These "lifted" or "rotated" pads are an indication of excessive thermal expansion. Thermal cycling over time can fatigue the plated through hole and ultimately cause failure from cracking of the copper plated within the hole or separation of the conductor from the hole wall.

Thermal expansion in the x/y axes is of more importance when discussing the attachment of components to the printed circuit. This is of particular importance when chip scale packages (CSPs) and direct chip attach components are used because the difference in thermal expansion between the printed circuit board and the component can compromise the reliability of the bond between them as they undergo thermal cycles. The x/y CTE can also impact interlaminar adhesion and delamination resistance in copper clad laminates or PCBs. If individual layers of materials with very different x/y CTE properties are adjacent to each other, thermal cycling or thermal excursions can cause enough stress at the interface of these layers to cause a separation or delamination. The thermal excursions experienced in PCB assembly processes can severely stress these interfaces, and the higher temperatures of lead-free assembly result in additional stress. For this reason, more attention should be given to the x/y CTE values of individual layers in a PCB that will experience lead-free assembly. With a given material type, this typically means that the choice of fiberglass cloth styles and resin contents adjacent to each other be examined. Furthermore, hybrid constructions,

TABLE 8.1 T_d and CTE Values at 40% Resin Content of Some Common Base Materials

Material	T_g (°C)	Decomposition Temp. T_d (5% Wt. Loss), °C	Z-Axis Expansion (% from 50°C to 260°C)	X/Y CTE (ppm/°C from–40°C to 125°C)
FR-4 Epoxy	140	315	4.5	13–16
Enhanced FR-4 Epoxy	140	345	4.4	13–16
Enhanced, Filled FR-4	150	345	3.4	13–16
High-T_g FR-4 Epoxy	175	305	3.5	13–16
Enhanced High-T_g FR-4	175	345	3.4	13–16
Enhanced, Filled High-T_g FR-4	175	345	2.8	12–15
BT/Epoxy Blend	190	320	3.3	14–16
PPO/Epoxy	175	345	3.8	15–16
Low D_k/D_f Epoxy Blend-A	200	350	2.8	11–15
Low D_k/D_f Epoxy Blend-B	180	380	3.5	13–15
Advanced Low D_k/D_f	215	363	2.8	13–14
Cyanate Ester	245	375	2.5	11–13
Polyimide	260	415	1.75	12–16
Halogen-Free, Filled High-T_g FR-4	175	380	2.8	13–16

meaning PCBs that utilize different base material types, should be analyzed very carefully, as different base material types can have different CTE values even with a given fiberglass cloth style and resin content.

8.2.1.1 CTE Values. CTE values of some common materials at approximately 40 percent resin content are shown in Table 8.1.

8.2.1.2 Controlling Thermal Expansion. The rate of thermal expansion is a function of the components used in the base material and their relative concentrations. The resin system will have a relatively high coefficient of thermal expansion compared to fiberglass cloth or other types of inorganic reinforcements.

In controlling z-axis expansion, the key factors to consider are the choice of resin system, the resin system T_g, and the resin content of the base materials. Fillers in the resin system, in addition to the fiberglass cloth, can also be used to lower the CTE of the material. Table 8.1 compares the thermal expansion of several commercially available base materials. These values can vary significantly based on the exact resin content of the material or PCB tested. In multilayer PCBs, the amount of copper in the sample will also have a significant impact as the z-axis expansion of copper is very low compared to the resin system.

Note the general difference in CTE values as T_g increases. A higher T_g delays the onset of the more rapid post-T_g rate of expansion. Also note that the filled materials have lower levels of z-axis expansion in comparison to equivalent unfilled materials.

8.2.2 T_g Determination by Other Methods

Besides TMA, two other methods are commonly used to measure T_g. These are differential scanning calorimetry (DSC) and dynamic mechanical analysis (DMA). DSC measures heat flow versus temperature rather than dimensional changes as measured by TMA. The heat absorbed or given off will change as the temperature increases through the T_g of the resin system. T_g measured by DSC is typically higher than measurements by TMA. DMA measures the modulus of the material versus temperature and also typically results in higher measured values of T_g.

As more complex resin systems are developed, and as more blends of resins are used, it becomes more difficult to measure T_g by TMA. For example, if two resins are used that have

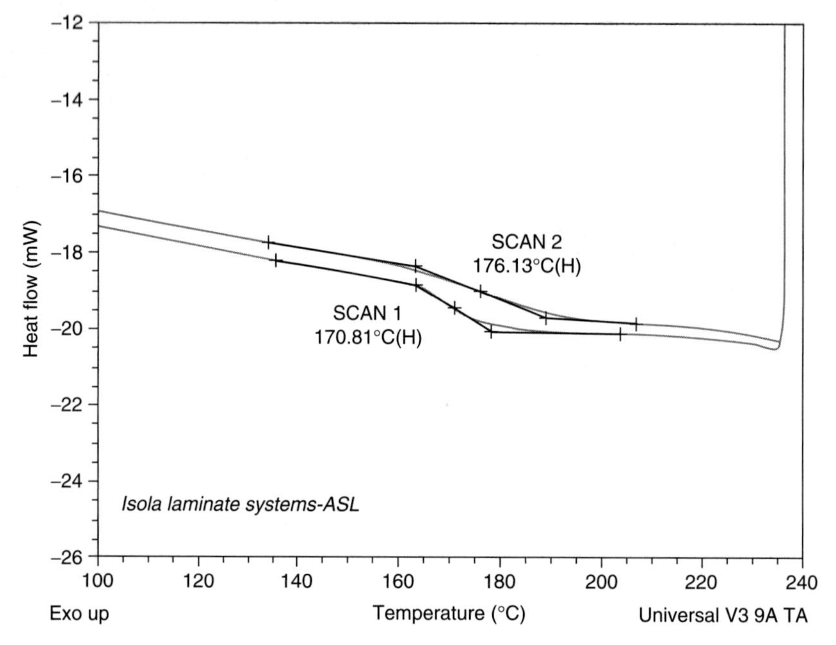

FIGURE 8.2 Measurements of T_g by DSC.

somewhat different TMA T_gs, the combination of the two resins makes it difficult to get a clear transition in expansion rates. For these materials, DSC and especially DMA are the preferred techniques for measuring T_g. Figures 8.2 and 8.3 provide examples of DSC and DMA scans, respectively, and the resulting measurement of T_g.

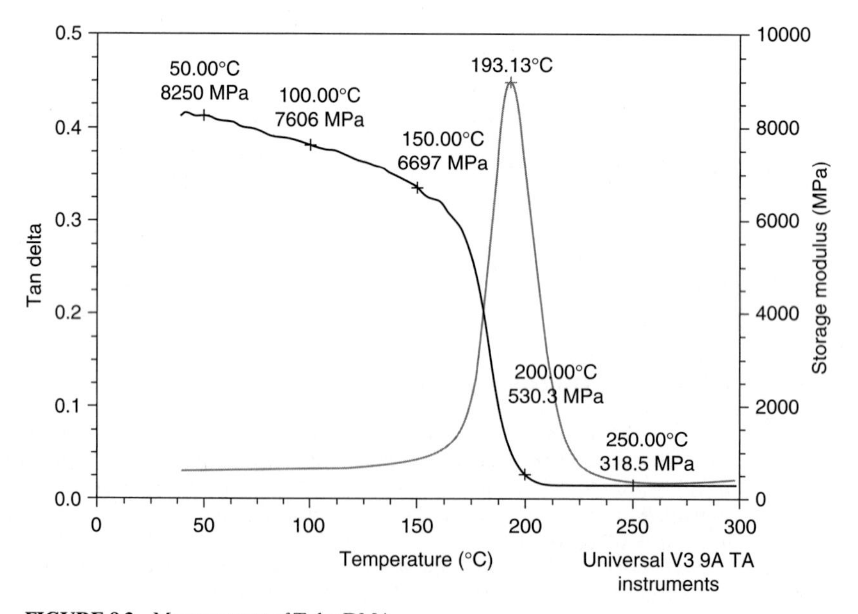

FIGURE 8.3 Measurement of T_g by DMA.

8.2.3 Decomposition Temperature

As a material is heated to higher temperatures, a point is reached where the resin system will begin to decompose. The chemical bonds within the resin system begin to break down and volatile components are driven off, reducing the mass of the sample. The decomposition temperature, T_d, is a property that describes the point at which this process occurs. The traditional definition of T_d is the point where 5 percent of the original mass is lost to decomposition. Table 8.1 shows 5 percent decomposition temperatures for several common materials. However, 5 percent is a very large number when multilayer PCB reliability is considered, and temperatures where lower levels of decomposition occur are very important to understand, particularly with respect to lead-free assembly. To illustrate this, consider Fig. 8.4.

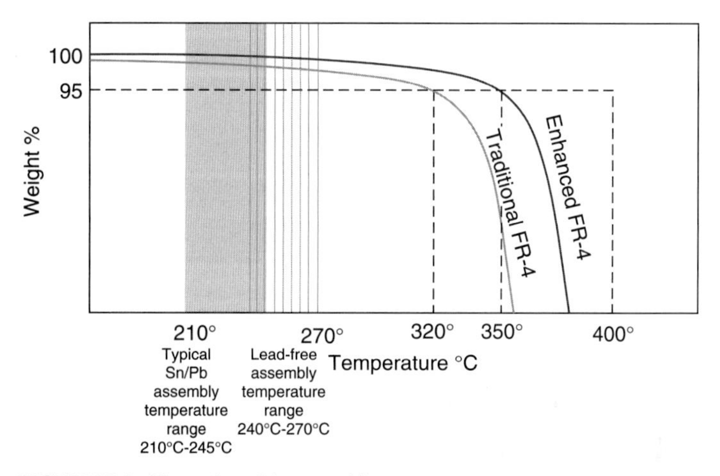

FIGURE 8.4 Illustration of decomposition temperature measurement.

In Fig. 8.4, you can see curves for two FR-4 materials. The "traditional FR-4"—a 140°C T_g material, in this case—has a decomposition temperature of 320°C by the 5 percent weight loss definition. The "enhanced FR-4" has a decomposition temperature of 350°C by the 5 percent weight loss definition. Many standard high-T_g FR-4 materials actually have decomposition temperatures in the range of 290–310°C, whereas the 140°C T_g FR-4 materials generally have slightly higher T_d values. The shaded regions indicate the peak temperature ranges for standard tin-lead assembly and lead-free assembly. A very common question is, if a PCB will be assembled at 260°C, and the material has a decomposition temperature of 310–320°C, then why wouldn't it be compatible with lead-free assembly?

The answer lies in the level of decomposition in the temperature ranges where assembly will take place. In the tin-lead temperature range, neither material exhibits a significant level of decomposition. However, in the lead-free assembly temperature range, the traditional - FR-4 begins to exhibit 1.5-3 percent weight loss. This level of decomposition can compromise long-term reliability or result in defects such as delamination during assembly, particularly if multiple assembly cycles or rework cycles are performed. Figures 8.5 through 8.7 show ThermoGravimetric Analysis(TGA) curves assessing decomposition temperatures for three types of base material.

These curves show different measurements for decomposition temperature. The value traditionally reported is the level at which 5 percent weight loss occurs, or where the mass is 95 percent of the starting sample mass. But as already discussed, this is a very high level of

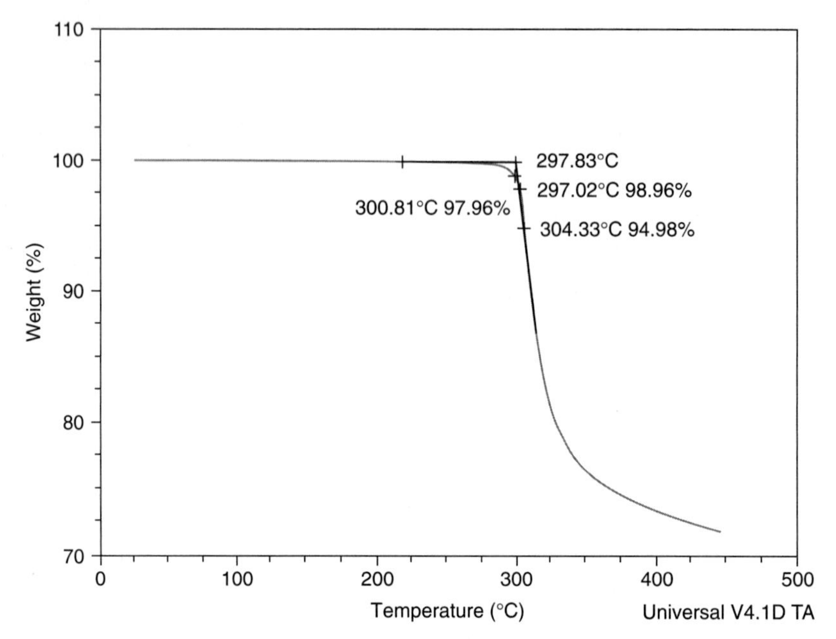

FIGURE 8.5 Decomposition curve for standard high-T_g FR-4.

decomposition when considering PCB reliability. These curves also show 1 percent and 2 percent decomposition temperatures, as well as the "onset" decomposition temperature. The onset temperature extrapolates the linear portions of the curve. The point at which they intersect is reported as the onset temperature.

Note that in Figs. 8.5 and 8.6 decomposition occurs very rapidly, while in Fig. 8.7 decomposition is more gradual, highlighted by the much bigger range between 1 percent, 2 percent, and 5 percent decomposition temperatures. With the material in Fig. 8.7, the 5 percent decomposition temperature would be reported as 405°C, even though some decomposition is beginning around 336°C. This highlights the need to understand not just the 5 percent decomposition temperature, but lower levels as well.

8.2.4 Time to Delamination

Time to delamination refers to a specific test procedure used to measure how long a material will resist blistering or delamination at a specific temperature. The procedure utilizes a thermomechanical analyzer (TMA) in which a sample is heated to the specified temperature. The most common temperature used is 260°C. This test is commonly called a T260 test. Other temperatures are also used, such as 288°C or 300°C. Table 8.2 compares the performance of various material types. When reviewing this table, it is important to note that there can be large variation between similar materials across the manufacturers of these materials. T260 values can be affected by the specific resins and curing agents used in a given material as well as the differences in CTE values between the components in the sample. For example, note the lower T260 values reported for the 175°C epoxy versus the 140°C epoxy. Although the T260 value is shorter, thermal cycling tests typically show improved reliability with the 175°C T_g material because of the lower levels of thermal expansion that result with the higher-T_g material. The enhanced FR-4 materials, which use a different curing chemistry, generally exhibit

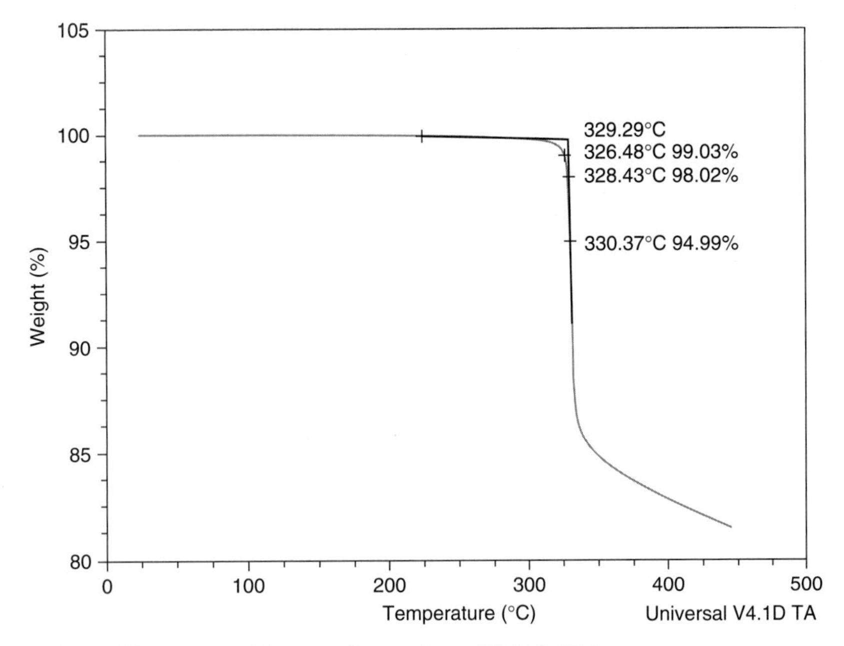

FIGURE 8.6 Decomposition curve for an enhanced high-T_g FR-4.

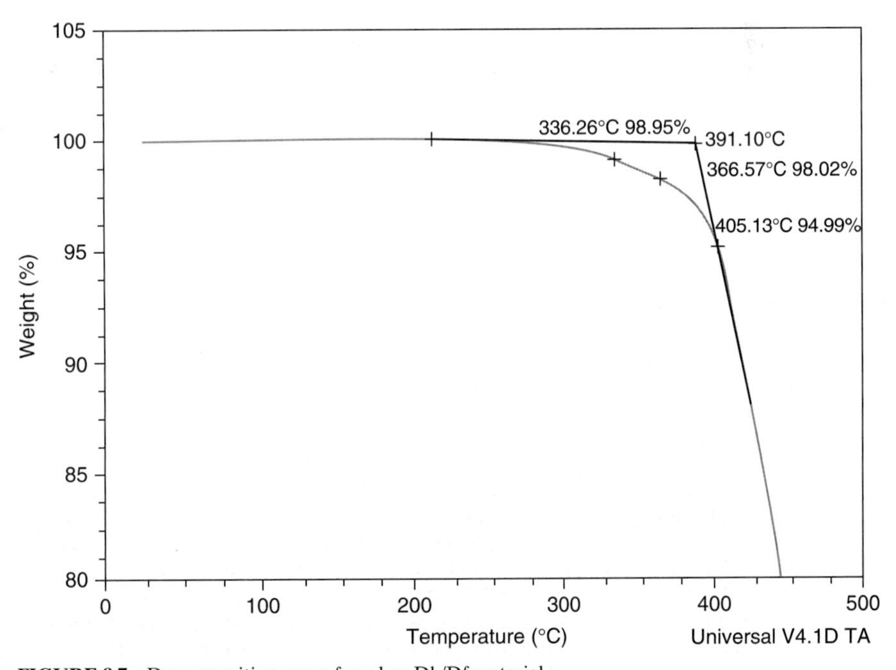

FIGURE 8.7 Decomposition curve for a low Dk/Df material.

Table 8.2 Time to Delamination and Arc Resistance Data for Common Materials

Material	T_g (°C)	T260	T288	Arc Resistance
Standard FR-4 Epoxy	140	8–18	—	65–120
Enhanced FR-4 Epoxy	140	20–30	5–10	75–120
Enhanced, Filled FR-4	150	25–45	6–12	80–120
High-T_g FR-4 Epoxy	175	4–10	—	70–120
Enhanced High-T_g FR-4	175	30+	7–15	70–120
Enhanced, Filled High-T_g FR-4	175	30+	8–16	80–120
BT/Epoxy Blend	190	30+	2–8	100–120
PPO/Epoxy	175	30+	8–20	110–120
Low D_k/D_f Epoxy Blend-A	200	30+	6–12	110–125
Low D_k/D_f Epoxy Blend-B	180	30+	10–20	110–120
Advanced Low D_k/D_f	220	30+	15–35	110
Polyimide	260	30+	30+	120–130
Halogen-Free, Filled High-T_g FR-4	175	20–30	8–12	120–130

better performance in both time-to-delamination tests as well as thermal cycling tests. Figure 8.8 is an example of a TMA scan on a multilayer PCB that evaluates T260 performance. The top curve plots the temperature while the other curve measures sample thickness. Delamination results in a rapid increase in sample thickness and indicates the endpoint for the test. Figure 8.9 is a T288 scan for the same material. Note that times to delamination on PCBs are generally much shorter than times on either copper clad laminate or unclad laminate.

Although time-to-delamination tests have received much attention as lead-free assembly processes have become common, it is important not to focus exclusively on one property or type of measurement when specifying laminate materials for lead-free applications. First, the correlation between time-to-delamination performance and lead-free assembly compatibility is not always apparent. Long T260 or T288 times by themselves do not ensure good reliability in lead-free applications. Conversely, some materials with good but not necessarily great T260 or T288 times have exhibited excellent performance in lead-free applications. So although it is important to consider time-to-delamination performance when specifying materials for lead-free assembly applications, these values should not be looked at exclusively. A balance of several properties, as will be discussed in Chap. 10, is necessary for these applications.

8.2.5 Arc Resistance

In this test technique, a low current arc is placed above the surface of a material. Arc resistance describes the time the material resists tracking, or forming a conductive path, under this condition. Arc resistance values are also shown in Table 8.2. Figure 8.10 illustrates the test technique.

8.2.6 Density

The density of various materials is given in Table 8.3.

8.2.7 Copper Peel Strength

Peel strength testing is the most popular method used to measure the bond between the conductor and the substrate. Peel strengths can be measured in the "as received" condition, after thermal stress, at elevated temperature, and after exposure to processing chemicals. The standard procedure requires that the sample consist of traces or strips of copper foil, or other metal to be tested, and that they be imaged onto the sample using standard printed circuit

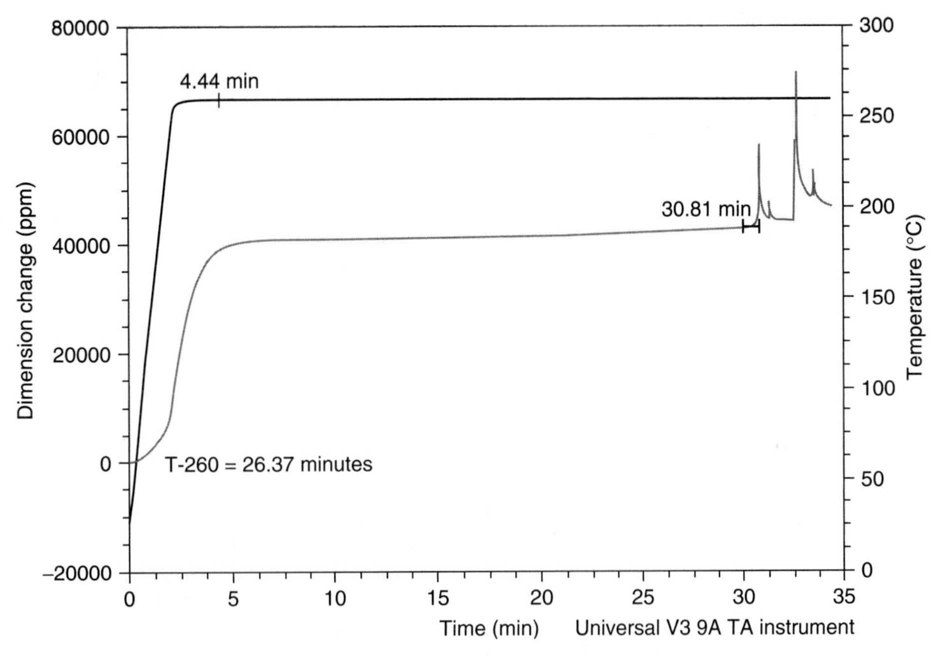

FIGURE 8.8 T260 example for an enhanced high-T_g FR-4.

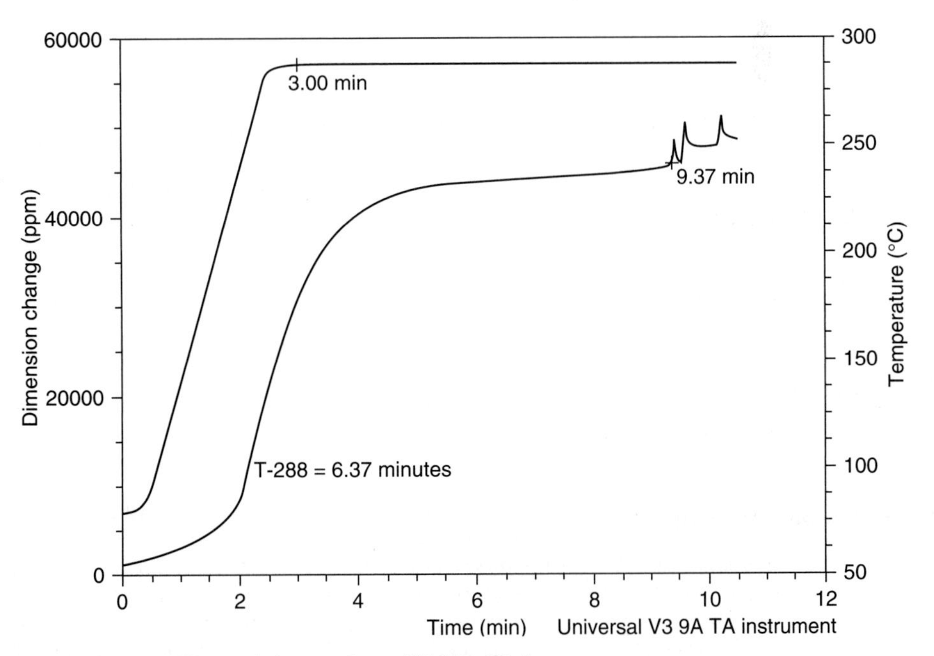

FIGURE 8.9 T288 example for an enhanced high-Tg FR-4.

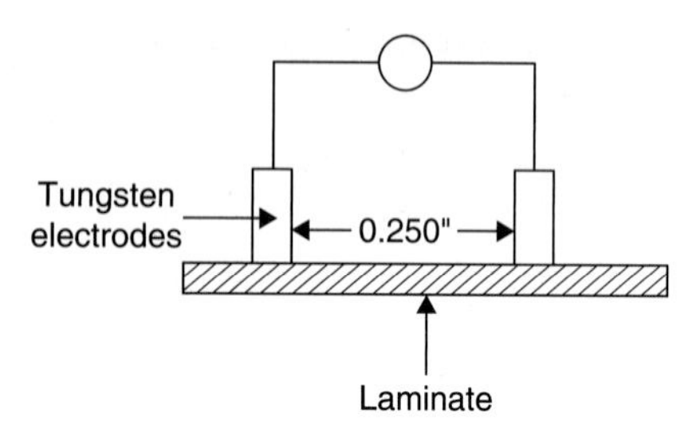

FIGURE 1.10 Arc resistance test technique.

manufacturing processes. The strips should be at least 0.79 mm (0.032 in.) in width for testing after exposure to processing chemistries and 3.18 mm (0.125 in.) in width for the other tests. Metal cladding thickness can influence the measured peel strength and therefore it is accepted practice to use 1 oz. cladding, and if a thinner cladding is used, it may be plated up to 1 oz. thickness.

One end of the strip is peeled back and attached to a load tester such as a tensile strength tester equipped with a load cell. The peel strength is calculated per the formula:

$$\text{lb./in.} = L_m/W_s$$

where L_m = Minimum load, and
W_s = Measured width of the peel strip

TABLE 8.3 Densities of Common Base Material Types

Material	Density (g/cm³)
FR-4 Epoxy	1.79
Filled FR-4 Epoxy	1.97
High-T_g FR-4 Epoxy	1.79
BT/Epoxy Blend	1.77
Low Dk Epoxy Blend	1.77
Cyanate Ester	1.71
Polyimide	1.68
APPE	1.51

For testing after thermal stress, the sample is first floated on solder at 288°C for 10 seconds. For testing after exposure to processing chemistries, the sample is exposed to a series of conditions. First the sample is exposed to an organic stripper at 23°C for 75 seconds. Historically, methylene chloride has been used, but due to environmental concerns, equivalents are now allowed. After drying, the sample is then immersed in a solution of 10 grams/liter sodium hydroxide at 90°C for 5 minutes. The sample is rinsed and then exposed to 10 grams/liter sulfuric acid and 30 grams/liter boric acid at 60°C for 30 minutes. The sample is again rinsed and dried and then immersed in a hot oil bath maintained at 220°C for 40 seconds. Finally, the sample is immersed in a degreaser at 23°C for 75 seconds to remove the oil and then dried. Testing at an elevated temperature may be performed by placing the sample in a hot fluid or in hot air while performing the peel strength measurements. For FR-4 materials, 125°C is commonly used.

Table 8.4 shows 1 oz. copper peel strength values in pounds per inch at various conditions for several common materials.

Note that polyimide, long used for its thermal reliability characteristics, exhibits the lowest peel strength values. This illustrates that the absolute value of peel strength is not always the best indicator of quality or reliability. Measuring and monitoring of peel strengths for a given material type for comparison to its normal values can be used as a process control tool, an inspection criteria, or for failure analysis purposes, but in choosing a material, a higher peel strength does not necessarily imply higher reliability. Last, if drum-side treated foil, also called reverse-treated foil, is used, the peel strength values will be lower because of the lower surface profile.

TABLE 8.4 Peel Strength Values for Common Base Material Types with Standard Copper (lb./in.)

Material	T_g (°C)	Peel Strength after Solder Float	Peel Strength at Elevated Temperature	Peel Strength after Exposure to Chemistry
Standard FR-4 Epoxy	140	9.0	7.0	9.0
Enhanced FR-4 Epoxy	140	8–9	6–7	8–9
Enhanced, Filled FR-4	150	8–9	6–7	8–9
High-T_g FR-4 Epoxy	175	8.5–9	6.5–7.0	8.5–9.0
Enhanced High-T_g FR-4	175	8–9	6–7	8–9
Enhanced, Filled High-T_g FR-4	175	8–9	6–7	8–9
BT/Epoxy Blend	190	8–9	7.5–8.5	8–9
PPO/Epoxy	175	7–8	6–7	7–8
Low Dk/Df Epoxy Blend-A	200	6–7	6–8	7–9
Low Dk/Df Epoxy Blend-B	180	7–8	6–7	7–8
Advanced Low Dk/Df	215	7.0	6.0	7.0
Cyanate Ester	245	8.0	7.5	8.0
Polyimide	260	7.0	6.0	7.0
Halogen Free, Filled High-T_g FR-4	175	7–9	6–8	7–9

8.2.8 Flexural Strength

Flexural strength is a measure of the load that a material will withstand without fracturing when supported at the ends and loaded in the center, as shown in Fig. 8.11. IPC-4101 specifies the minimum flexural strength of various materials, some of which are summarized in Table 8.5.

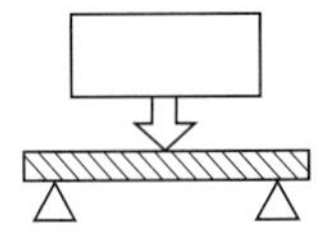

FIGURE 8.11 Flexural strength test.

8.2.9 Water and Moisture Absorption

The ability of a material to resist water absorption, either from the air or when immersed in water, is important for printed circuit reliability. Besides the obvious concerns of moisture causing defects when a material is subjected to thermal excursions, absorbed moisture also affects the ability of a material to resist conductive anodic filament (CAF) formation when a bias is applied to the circuit.

The test method for measuring water absorption for metal-clad base laminates involves immersing a sample in distilled water at 23°C for 24 hours after etching off the metal cladding and drying the sample for 1 hour at 105°C to 110°C and cooling in a dessicator. The sample is

TABLE 8.5 Flexural Strength Requirements of Base Materials

Material Type	Min. Flexural Strength Lengthwise (kg/m²)	Min. Flexural Strength Crosswise (kg/m²)
XXXPC	8.44×10^6	7.39×10^6
CEM-1	2.11×10^7	1.76×10^7
CEM-3	2.32×10^7	1.90×10^7
FR-1	8.44×10^6	7.04×10^6
FR-2	8.44×10^6	7.39×10^6
FR-3	1.41×10^7	1.13×10^7
FR-4	4.23×10^7	3.52×10^7
FR-5	4.23×10^7	3.52×10^7
Polyimide/Woven E-Glass	4.23×10^7	3.17×10^7
Cyanate Ester/Woven E-Glass	3.52×10^7	3.52×10^7

TABLE 8.6 Moisture Absorption and Methylene Chloride Resistance of Common Materials

Material	T_g (°C)	Moisture Absorption (%)	Methylene Chloride Resistance (%)
FR-4 Epoxy	140	0.1	0.7
Filled FR-4 Epoxy	155	0.22	0.42
High-Tg FR-4 Epoxy	180	0.1	0.7
BT/Epoxy Blend	185	<0.5	0.7
Low D_k Epoxy Blend	210	0.1	0.7
Cyanate Ester	250	<0.5	0.32
Polyimide	250	0.35	0.41

weighed after drying, immersed in water under the specified conditions, and weighed again. The water absorption is calculated as follows:

Increase in weight, percent = (wet weight – conditioned weight)/conditioned weight × 100

An additional moisture absorption test measures weight gain after 60 minutes at 15 psi. Table 8.6 shows the moisture absorption of some common material types. These values can vary with resin content.

8.2.10 Chemical Resistance

One common method used to evaluate the chemical resistance of base laminates is to measure absorption of methylene chloride. Similar to water absorption testing, etched samples are exposed to this solvent and the weight gain is measured. The standard procedure starts by etching off the metal cladding of the samples, drying them in an oven for 1 hour at 105°C to 110°C, and measuring the initial weights. The samples are then soaked in methylene chloride at 23°C for 30 minutes, allowing them to dry for 10 minutes, and weighing again. The calculation is as follows:

Change in weight, percent = (final weight – initial weight)/initial weight × 100

Table 8.6 lists methylene chloride resistance for some common materials.

8.2.11 Flammability

Underwriters Laboratories (UL) classifies flammability properties as 94V-0, 94V-1, or 94V-2. Definitions of these classifications are as follows:

- **94V-0** Specimens must extinguish within 10 seconds after each flame application (see Fig. 8.12) and a total combustion of less than 50 seconds after 10 flame applications. No samples are to drip flaming particles or have glowing combustion lasting beyond 30 seconds after the second flame test.
- **94V-1** Specimens must extinguish within 30 seconds after each flame application and a total combustion of less than 250 seconds after 10 flame applications. No samples are to drip flaming particles or have glowing combustion lasting beyond 60 seconds after the second flame test.
- **94V-2** Specimens must extinguish within 30 seconds after each flame application and a total combustion of less than 250 seconds after 10 flame applications. Samples may drip flaming particles, burning briefly, and no specimen will have glowing combustion beyond 60 seconds after the second flame test.

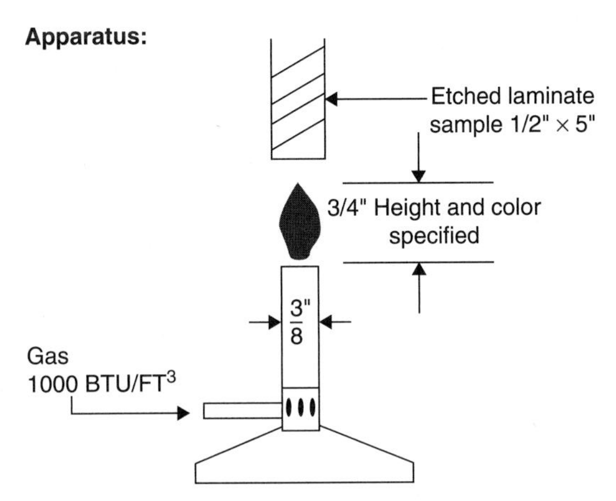

FIGURE 1.12 Flammability testing.

8.3 Electrical Properties

A variety of base material electrical properties are important to understand when designing and manufacturing printed circuits. Some of the most important properties are discussed in this section. As noted earlier in the chapter, the demand for circuits operating at high frequencies requires materials with good permittivity and loss characteristics. These properties will be discussed further in Chap. 4.

8.3.1 Dielectric Constant or Permittivity

The dielectric constant can be defined as the ratio of the capacitance of a capacitor with a given dielectric material to the capacitance of the same capacitor with air as a dielectric, as illustrated in Fig. 8.13. In other words, the dielectric constant is a measure of the ability of a material to store an electric charge. There are actually several test methods used to measure dielectric constant or permittivity and a complete discussion of these methods is beyond the scope of this chapter. However, it should be noted that measuring permittivity at high frequencies can be very difficult and that reported values can vary with the specific test method used. Therefore, when comparing the reported values for different materials, the best comparisons will be those that used the same test method.

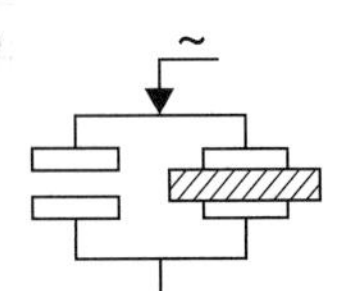

FIGURE 8.13 Dielectric constant.

Furthermore, the dielectric constant is not really a constant. As just implied, the dielectric constant will vary with frequency. It will also vary with temperature and humidity. So besides test method, the frequency, temperature, and humidity conditions must also be considered. Last, even with the same material type, variations in resin content (resin-to-reinforcement ratio) also affect the dielectric constant. These variations will be further discussed in Chap. 10. Table 8.7 shows dielectric constants for some common fiberglass (E-glass) reinforced materials at 50 percent resin content.

8.3.2 Dissipation Factor or Loss Tangent (Tan δ)

The dissipation factor in an insulating material is the ratio of the total power loss in the material to the product of the voltage and current in a capacitor in which the material is a dielectric. Many of the test methods (such as that shown in Fig. 8.14) used for measuring dielectric constant also measure the dissipation factor. Dissipation factor also varies with frequency,

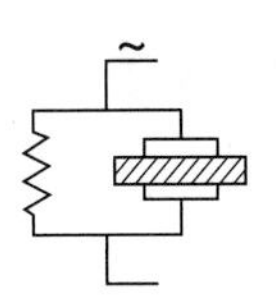

FIGURE 8.14 Dissipation factor.

TABLE 8.7 Dielectric Constants and Dissipation Factors of Common Materials

	Dielectric Constant		Dissipation Factor	
Material	(1 MHz)	(1 GHz)	(1 MHz)	(1 GHz)
Standard FR-4 Epoxy	4.7	4.3	0.025	0.016
Filled FR-4 Epoxy	4.7	4.4	0.023	0.016
High-T_g FR-4 Epoxy	4.7	4.3	0.023	0.018
BT/Epoxy Blend	4.1	3.8	0.013	0.010
Epoxy/PPO	3.9	3.8	0.010	0.011
Low D_k Epoxy Blend	3.9	3.8	0.009	0.010
Cyanate Ester	3.8	3.5	0.008	0.006
Polyimide	4.3	3.7	0.013	0.007
APPE	3.7	3.4	0.005	0.007

resin content, temperature, and humidity. Dissipation factor is discussed in more detail in Chap. 10. Table 8.7 also lists the dissipation factors of some common fiberglass (E-glass) reinforced materials.

8.3.3 Insulation Resistance

The insulation resistance between two conductors or plated holes is the ratio of the voltage to the total current between the conductors. Two measures of electrical resistance are volume and surface resistivities. Since these properties can vary with temperature and humidity, testing is normally performed at two standardized environmental conditions, one involving humidity conditioning, the other involving elevated temperature. Humidity conditioning subjects the sample to 90 percent relative humidity and 35°C for 96 hours (96/35/90). The elevated temperature conditioning normally subjects the sample to 125°C for 24 hours (24/125).

8.3.4 Volume Resistivity

Volume resistivity is the ratio of the DC potential applied to electrodes embedded in a material to the current between them, typically expressed in megohm-centimeters. The measured current flows between electrodes 1 and 3, while stray current flows between electrodes 2 and 3, as shown in Fig. 8.15. Table 8.8 shows the volume resistivity values of some common fiberglass-reinforced material types.

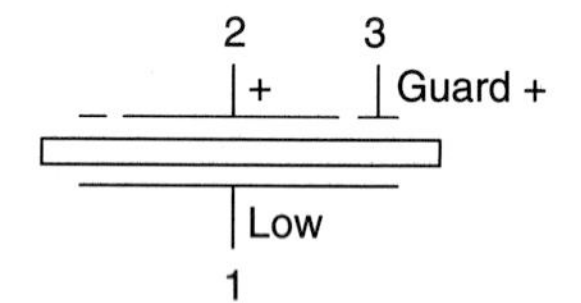

FIGURE 8.15 Circuit for volume resistance.

8.3.5 Surface Resistivity

The surface insulation resistance between two points on the surface on any insulating material is the ratio of the DC potential applied between the two points to the total current between them. For surface resistivity, the measured current flows between electrodes 1 and 2, while stray current flows between electrodes 1 and 3, as shown in Fig. 8.16. Table 8.8 shows the surface resistivity values of some common fiberglass-reinforced materials.

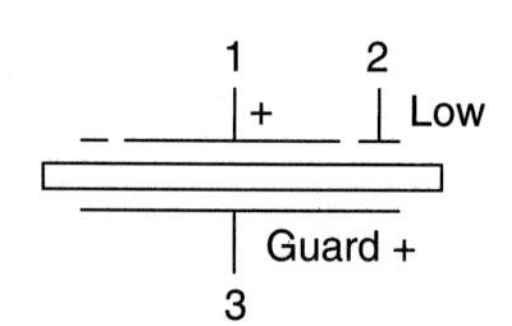

FIGURE 8.16 Circuit for surface resistance.

TABLE 8.8 Additional Electrical Properties of Common Base Materials

Material	Volume Resistivity (96/35/90)	Volume Resistivity (24/125)	Surface Resistivity (96/35/90)	Surface Resistivity (24/125)	Electric Str. (V/mil)
FR-4 Epoxy	10^8	10^7	10^7	10^7	1250
Filled FR-4 Epoxy	10^{11}	10^{10}	10^8	10^9	1250
High-T_g FR-4 Epoxy	10^8	10^7	10^7	10^7	1300
BT/Epoxy Blend	10^7	10^7	10^6	10^7	1200
Low D_k Epoxy Blend	10^8	10^7	10^7	10^7	1200
Cyanate Ester	10^7	10^7	10^7	10^7	1650
Polyimide	10^7	10^7	10^7	10^7	1350

8.3.6 Electric Strength

Electric strength is a measure of the ability of an insulating material to resist electrical breakdown perpendicular to the plane of the material when subjected to short-term, high voltages at standard AC power frequencies of 50 to 60 Hz and is reported in volts per mil (see Fig. 8.17). Results can be affected by moisture content in the sample, so measurements may vary with different preconditioning environments. Unless otherwise noted, measurements are taken at 23°C, after preconditioning for 48 hours in distilled water at 50°C and immersion in ambient temperature distilled water for 30 minutes minimum, 4 hours maximum. Measurements are performed under an oil medium to prevent flashover on a small specimen. The values may decrease with increasing specimen thickness for an otherwise identical material. Table 8.8 compares the electric strength of some common fiberglass-reinforced materials.

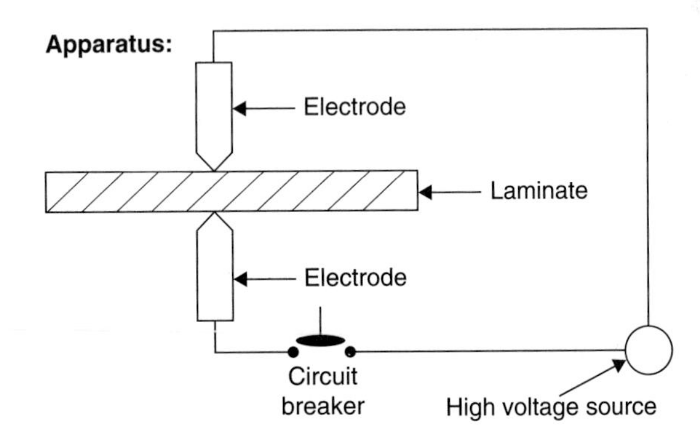

FIGURE 8.17 Dielectric strength.

8.3.7 Dielectric Breakdown

Dielectric breakdown measures the ability of rigid insulating materials to resist breakdown parallel to the laminations (or in the plane of the material) when subjected to extremely high voltages at standard AC power frequencies of 50 to 60 Hz (see Fig. 8.18). As with electric

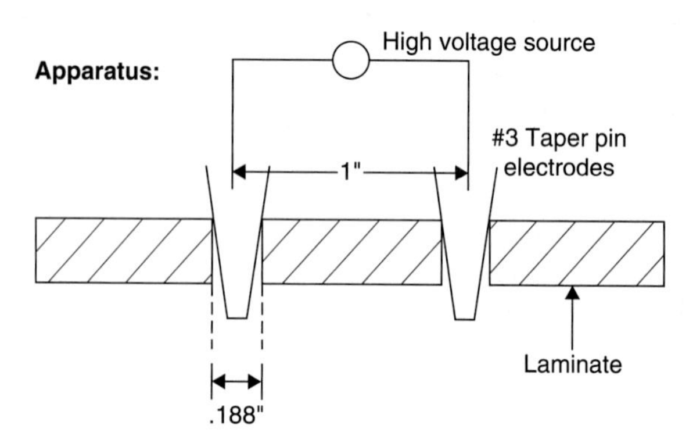

FIGURE 8.18 Dielectric breakdown strength parallel to laminations.

strength, values obtained on most materials are highly dependent on the moisture content and preconditioning method. Unless otherwise noted, measurements are performed at 23°C after preconditioning for 48 hours in distilled water at 50°C followed by immersion in ambient temperature distilled water for 30 minutes minimum, 4 hours maximum. Dielectric breakdown is also performed in an oil medium. Dielectric breakdown for the materials shown in Table 8.8 is normally above 50 kilovolts.

REFERENCES

1. NEMA, "Industrial Laminating Thermosetting Products" Standard, 1998.
2. IPC-4101, "Specification for Base Materials for Rigid and Multilayer Printed Boards."
3. Isola Product Reference Materials.
4. IPC-TM-650 T.

CHAPTER 9

BASE MATERIALS PERFORMANCE ISSUES

Edward J. Kelley
Isola Group, Chandler, Arizana

9.1 INTRODUCTION

As the fundamental building block for printed circuits, base materials must meet the needs of the printed circuit board (PCB) manufacturer, the circuit assembler, and the original equipment manufacturer (OEM). A balance of properties must be achieved that satisfies each member of the supply chain. In some cases, the desires of one member of the supply chain conflict with another. For example, the need for improved electrical performance by the OEM, or improved thermal performance by the assembler, may necessitate the use of resin systems that require longer multilayer press cycles or less productive drilling processes, or both.

Lead-free assembly processes are driving the need for greater thermal reliability. This will be discussed further in Chaps. 10 and 11. Other trends are also driving the need for greater performance. These include:

- Circuit densification
- Higher circuit operating frequencies

The density of IC packaging technologies such as ball grid array (BGA) and chip scale packages continues to increase. In turn, this requires greater interconnection density in the printed circuits onto which they are assembled. The need for increased density in the PCB impacts each of the components of the base material as well as the way in which they are manufactured. To achieve high levels of interconnection, component pitch densities result in smaller and more closely spaced plated through holes and circuit features. As the space between these holes and features decreases, the potential for conductive anodic filament (CAF) failures increases substantially.

Higher operating frequencies can also impact each of the three main components of base materials. Circuits operating at high frequencies are driving the use of materials with low dielectric constants, low dissipation factors, and tighter thickness tolerances. These performance issues and the impact to the printed circuit manufacturing process are discussed in this chapter.

9.2 METHODS OF INCREASING CIRCUIT DENSITY

There are basically three ways to increase printed circuit density:

- Decrease conductor line widths and the spacings between them
- Increase the number of circuit layers in the PCB
- Reduce via and pad sizes.

Decreasing conductor line widths requires very-low-profile copper foils for high yields in circuit etching processes. However, other things being equal, lower profiles result in decreased foil adhesion to the dielectric. Balancing the copper surface profile for both adhesion to the dielectric and the ability to etch fine circuit features, not to mention the impact of surface roughness on electrical performance at high frequencies, is an important consideration. Copper foil manufacturers continue to research methods to improve the chemical adhesion between the foil and the various dielectric materials in use, relying less on a rough surface profile for mechanical adhesion, and allowing for very low profiles for circuit etching and lower conductor losses at high frequencies.

Increasing circuit layer counts have resulted in both greater overall multilayer thicknesses and thinner individual dielectrics, making thickness control and thermal reliability more important than ever. Adding layers to a PCB also demands improved registration capabilities. One of the critical variables in controlling registration is the dimensional stability of the laminate material, which can become more challenging with the thinner laminates generally used as layer counts increase. Reducing via and pad sizes also requires improved laminate dimensional stability for registration of high layer count circuits.

9.3 COPPER FOIL

One obvious method to increase printed circuit functionality is to put more circuitry per unit area of the circuit. Printed circuit densification has driven several improvements in copper foil technology. One of the first developments was high temperature elongation (HTE) foils. Other advances include low- and very-low-profile foils, thin foils, and foils for high-performance resin systems.

9.3.1 HTE Foil

HTE or Class 3 copper foil exhibits improved elongation properties at elevated temperatures as compared to standard electrodeposited or Class 1 copper foil. Typical elongation values for HTE copper foil range from 4–10 percent at 180°C.

The growth in multilayer printed circuits has resulted in HTE becoming the most commonly used foil, since its excellent ductility at elevated temperatures helps prevent inner-layer copper foil cracking. As a printed circuit experiences a thermal cycle, the base materials will expand. The z-axis expansion applies stress to the connection of the inner-layer foil and the plated hole. With HTE foil, the reliability of this connection is improved. This property is particularly important in thicker circuits and high resin content constructions where increased z-axis expansion occurs.

9.3.2 Low-Profile and Reverse-Treated Copper Foils

Three classifications describe the profile of the copper foil surface as shown in Table 9.1.

Copper foil profile is important for etching of fine-line circuits. Figures 7.23 through 7.26 illustrate the difference between standard- and low-profile foils. As can be seen in those photos,

TABLE 9.1 Copper Foil Profiles

Foil Profile Type	Max. Foil Profile (Microns)	Max. Foil Profile (μ Inches)
S – Standard	N/A	N/A
L – Low Profile	10.2	400
V – Very Low Profile	5.1	200
X – No Treatment or Roughness	N/A	N/A

the tooth profile of the standard-profile foil is much more pronounced. The etching of the lower-profile foil results in more control of the geometry of the circuit trace. In addition, in very thin laminates, the large tooth structure of the standard profile foil can result in inconsistent dielectric thickness, making impedance control more difficult, and can even result in electrical failures if the tooth structures from the opposing sides of the laminate protrude sufficiently.

Reverse-treated foils (RTF) take this concept a step further. When copper foil is manufactured, there is a very smooth, shiny side and a rougher matte side. Conventional technology involved treating the matte side and laminating this side to the base material. Reverse-treated foil, as its name implies, involves putting the treatments on the smooth, shiny side of the foil and laminating this side to the base material. This has two important effects. First, the side bonded to the base material has an extremely low profile that aids in etching very fine circuit traces. Second, the rougher matte side, which is now on the surface of the laminate, can improve photoresist adhesion. This enables the removal of surface roughening processes during PCB manufacturing and can also improve inner-layer imaging and etching yields. Figure 9.1 compares laminates made with conventional and RTF foils.

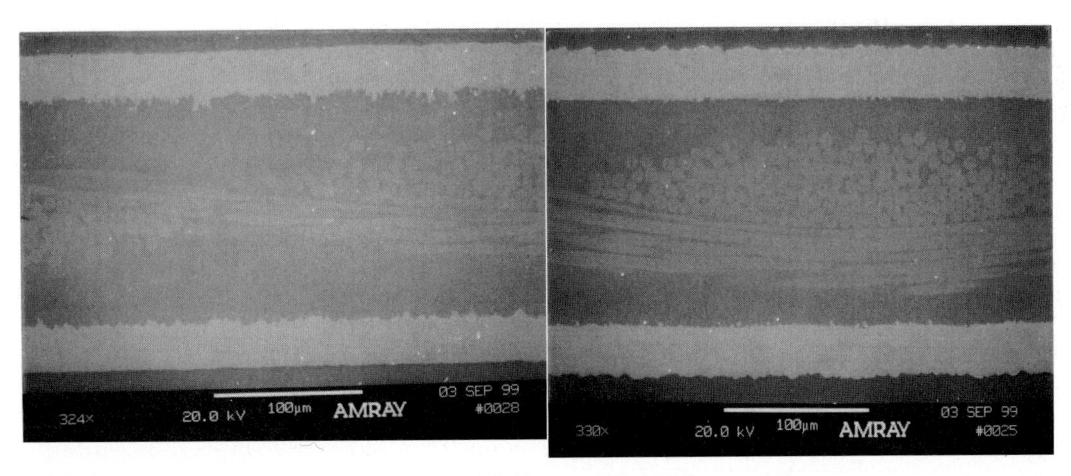

FIGURE 9.1 Comparison of laminates with standard (left) versus RTF (right) copper foils.

9.3.3 Thin Copper Foils

The capability to etch fine-line circuits is also improved through the use of thinner copper foils. Although electrical considerations can limit the use of very thin foils on innerlayer circuits, these thin foils can be used on external layers since the outerlayer process involves plating on top of the foil to the desired overall thickness. For dense, fine-line circuitry, 5.0 micron and 9.0 micron copper foils are sometimes used. Processes to use 3.0 micron copper foil have also been developed.

9.3.4 Foils for High-Performance Resin Systems

Many of the high-performance resin systems such as BT, polyimide, cyanate ester, and even some high-T_g epoxies exhibit lower peel strengths and resistance to undermining of the copper foil when exposed to aggressive chemistries. For these applications, foils with increased nodularization and coupling agents tailored to the resin system are often used. The increased nodularization results in more surface area for mechanical adhesion whereas the specific coupling agent aids in chemically bonding the foil to the resin system.

9.3.5 Copper Roughness and Attenuation

As circuit operating frequencies increase, more of the signal travels in the outermost part of the conductor. The "skin depth"—that is, the region where much of the signal travels—is shown in Fig. 9.2 as a function of frequency. As shown in this graph, the skin depth approaches the average roughness of 0.5 oz. copper foil above 1 GHz. Signal attenuation due to conductor losses related to the roughness of the foil becomes an important factor at these frequencies, and should be considered by the design engineer.

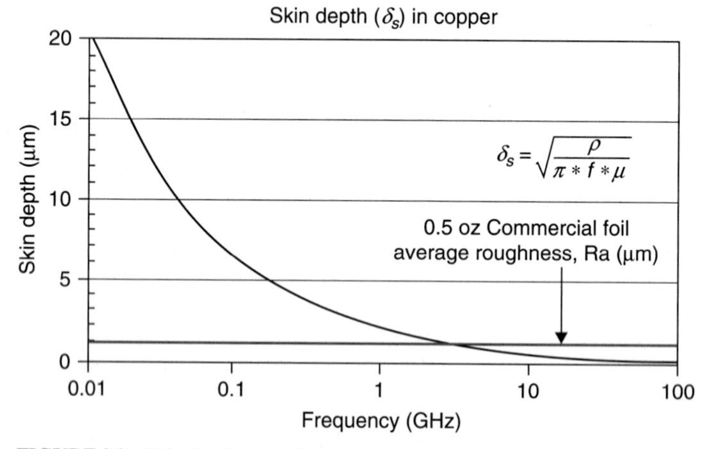

FIGURE 9.2 Skin depth versus frequency.

A study of several different copper foil types correlated roughness and attenuation values.[6] Figures 9.3 and 9.4 show pictures of the foils to highlight their relative roughness differences. Figure 9.5 shows the roughness distributions of these foils. Finally, in Fig. 9.6, the loss values associated with each of these foil types are graphed versus frequency. Up to about 1 GHz, there is very little difference in the observed loss across the several foil types. However, at higher frequencies, the difference becomes much greater, correlating to the roughness of the individual foil types; the greater the roughness, the greater the measured attenuation.

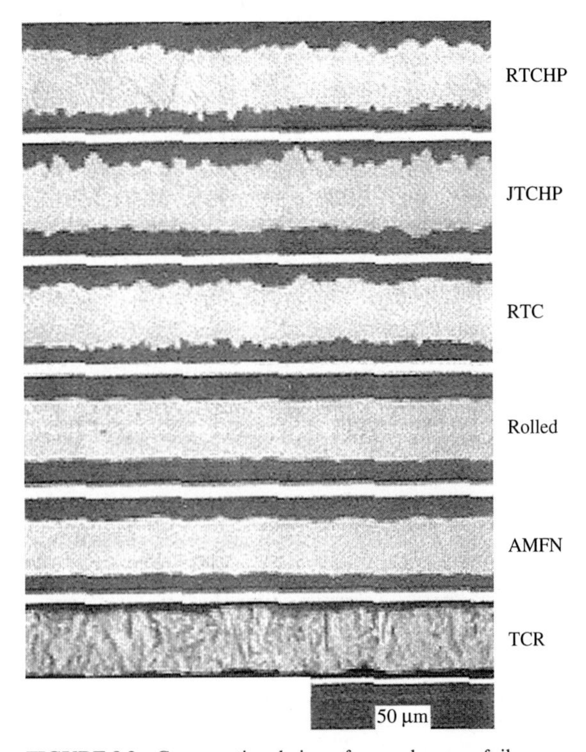

FIGURE 9.3 Cross-sectional view of several copper foils.

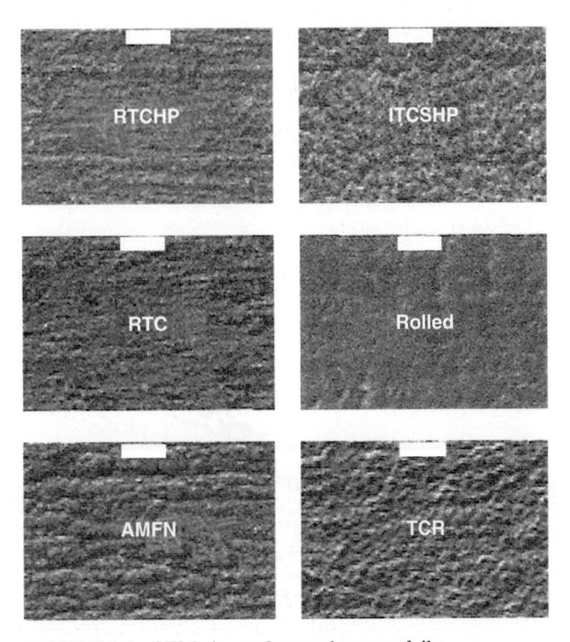

FIGURE 9.4 SEM views of several copper foils.

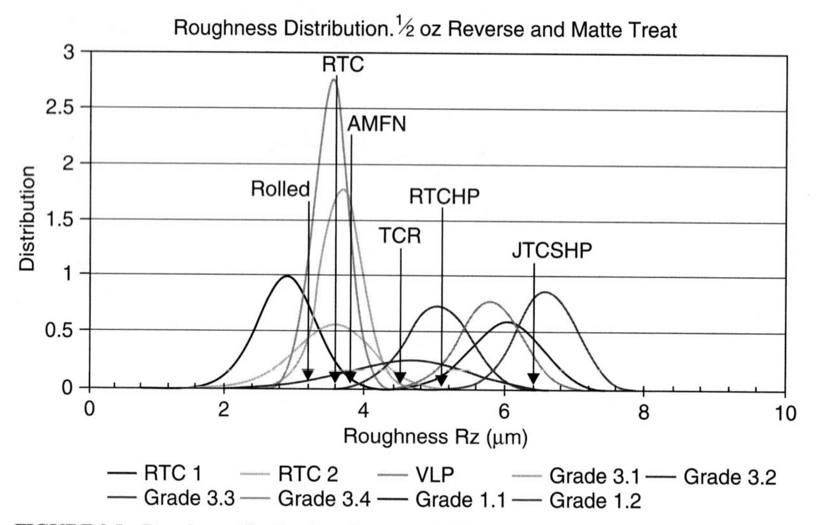

FIGURE 9.5 Roughness distributions for several foil types.

The roughness that results from the oxide/oxide alternative surface preparation process during printed circuit manufacturing is also important. Figures 9.7 and 9.8 compare the roughness obtained from two of these processes. The base copper foil and FR-4 resin system used were held constant. The test vehicles from which these cross sections were taken were also measured for attenuation. This measurement technique was used to calculate an effective dissipation factor, Df, for these material sets. The measured Df for the sample in Fig. 9.7, with the relatively smooth profile, was 0.021 at 1 GHz. The measured Df for the sample in Fig. 9.8, with the rougher profile, was 0.026. Obviously, the rougher profile created by the oxide alternative process in Fig. 9.8 resulted in a significantly higher loss value.

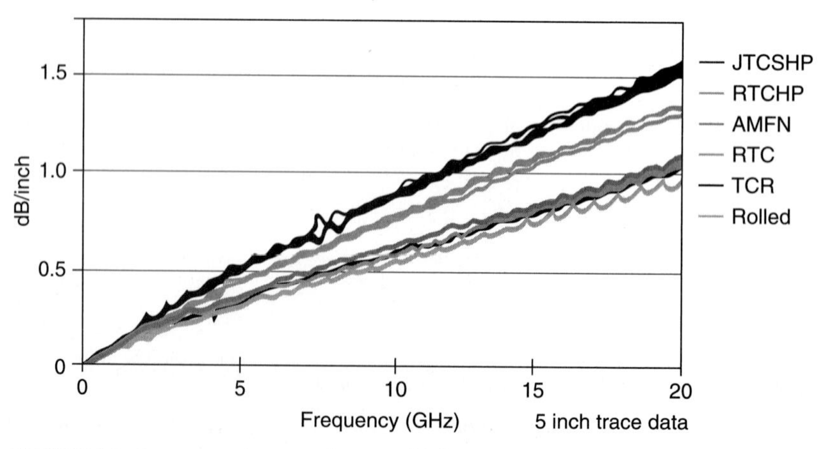

FIGURE 9.6 Loss versus frequency for several foil types.

FIGURE 9.7 Copper profile obtained from process A.

FIGURE 9.8 Copper profile obtained from process B.

9.4 LAMINATE CONSTRUCTIONS

To satisfy requirements for impedance, layer count, and overall PCB thickness, a broad range of laminate dielectric thicknesses are needed. Table 9.2 shows many common laminate dielectric thicknesses, along with typical constructions and resin contents. Individual laminate suppliers may have preferred constructions, so not every construction shown in this section will be available from every laminate supplier. In addition, some of the high-performance resin systems will have slightly different constructions or resin contents in order to target certain performance characteristics, such as dielectric constant.

9.4.1 Single-Ply versus Multiple-Ply Constructions

With dielectrics below 0.0040 in., there is often no choice but to use a single ply of fiberglass cloth to achieve the desired thickness. With dielectrics in the 0.0040 in. to 0.0080 in. range,

TABLE 9.2 Common Laminate Constructions

Laminate Thickness (in.)	Fiberglass Cloth(s)	Approx. Resin (%)
0.0020	1-106	71–72
0.0025	1-1080	53–54
0.0030	1-2113	46–47
0.0035	1-2113	52–53
0.0040	1-2113	53–54
0.0040	1-3070	49–50
0.0040	1-2116	43–44
0.0040	1-106/1-1080	59–60
0.0040	2-106	68–69
0.0040	2-1080	50–51
0.0045	1-2116	50–51
0.0045	2-1080	55–56
0.0050	1-2116	52–53
0.0050	1-2165	47–48
0.0050	1-1652	42–43
0.0050	2-1080	56–57
0.0050	1-2313/1-106	52–53
0.0060	2-2113	46–47
0.0062	1-2157	49–50
0.0062	1-1080/1-2313	53–54
0.0080	1-7628	44–45
0.0080	1-2116/1-2313	49–50
0.0080	2-2116	45–46
0.0090	2-2116	50–51
0.0100	2-1652	42–43
0.0100	2-2165	47–48
0.0120	2-2157	47–48
0.0120	2-1080/1-7628	47–48
0.0140	2-7628	40–41
0.0210	3-7628	40–41

there are both single-ply and multiple-ply options. Above 0.0080 in., multiple plies of cloth are typically required to achieve the desired thickness. Within each range, there are often multiple cloth and resin content combinations that can achieve the desired dielectric thickness. The choice of laminate construction can significantly impact both cost and performance. The choice of single-ply versus multiple-ply construction, when the option is available, is no exception.

As should be expected, a single-ply construction will typically represent a cost savings compared to a multiple-ply construction. The magnitude of this savings will depend on the specific glass styles involved and a host of other parameters. Performance can also be affected and should be considered when specifying the constructions to be used. First, single-ply constructions are often lower in resin content, as can be observed in Table 9.2. Resin content issues are discussed in the following section. The other main benefit of single-ply constructions is dielectric thickness control, beyond resin content considerations. Other things being equal, as with resin contents, tighter thickness tolerances can be achieved using a single-ply construction than a multiple-ply construction since the variation in thickness control with one ply of prepreg will statistically be less than with multiple plies.

9.4.2 Resin Contents

As Table 9.2 illustrates, the same dielectric thickness can be achieved with multiple fiberglass cloth and resin content combinations. Constructions with relatively lower resin contents are often preferred since they result in less z-axis expansion and can therefore improve reliability in many applications. In addition, lower resin contents can also improve dimensional stability, resistance to warpage, and dielectric thickness control. On the other hand, constructions with higher resin contents result in lower dielectric constant values, which are sometimes preferred for electrical performance, which is discussed further in Section 9.8. In addition, a certain minimum resin content is required to ensure adequate resin-to-glass wet-out and to prevent voids from occurring within the laminate. The ability to wet out the glass filaments fully with resin is also important for CAF resistance. In summary, for each glass style, there is an optimal resin content range that balances the various performance requirements.

9.4.3 Laminate Flatness and Flexural Strength

When manufacturing the innerlayer circuit patterns, the flatness and flexural strength of the base material is important for successfully transporting these circuit layers through conveyorized equipment. This is particularly important for very thin laminates. If the laminate is curled, it can become caught or damaged inside this equipment. If the thin laminate with the circuit image sags on the conveyors, similar damage can occur.

For these reasons, it is sometimes preferred to use constructions with relatively high glass contents and use the thickest cloths possible, as this will result in greater flexural strengths, bearing in mind the performance requirements just discussed in Section 9.4.2. In addition, *balanced* constructions, or those that use a symmetrical construction, are normally preferred to avoid curling. Asymmetrical constructions are more prone to curling and causing problems in conveyorized equipment. Consider a 0.0080-in. laminate as an example. Table 9.2 shows three constructions for this thickness, 1-7628, 1-2313/1-2116, and 2-2116. The asymmetrical 1-2313/1-2116 construction will be more prone to curling than the other constructions. One word of caution is warranted, however: Curling is but one property that must be considered. Dimensional stability, thickness control, and other properties must also be considered when choosing a construction.

9.5 PREPREG OPTIONS AND YIELD-PER-PLY VALUES

Just as there are a variety of base laminate constructions, there are a variety of prepreg options. Each fiberglass cloth style can be treated to multiple resin contents and flow values. While the same properties must be considered for prepreg materials as for base laminates, prepregs must also contain sufficient resin to flow and fill in the innerlayer circuitry of a multilayer circuit board. Since the innerlayer circuits can vary in terms of copper thickness and circuit density, a variety of resin content and flow value options are typically required when specifying prepreg styles. Higher resin contents may be needed when filling heavy copper weights and signal layers, while lower resin contents can be used against lighter copper weights and power or ground circuit patterns. Table 9.3 shows some common prepreg styles with various resin contents and thickness yields per ply.

TABLE 9.3 Common Prepreg Styles

Glass Style	Approximate Resin Content (%)	Thickness (in.)*
106	62	0.0015
106	66	0.00175
106	71	0.0020
106	75	0.00225
1080	54	0.00225
1080	57	0.0025
1080	64	0.0030
1080	66	0.00325
2113	50	0.0035
2113	55	0.0040
2116	50	0.0045
2116	52	0.00475
2116	55	0.00525
2165	47	0.0050
2165	50	0.00525
2165	54.0	0.0060
2157	45	0.00575
2157	47	0.0060
2157	48	0.00625
2157	50	0.0065
7628	40	0.0070
7628	42	0.00725
7628	44	0.0075
7628	45	0.00775

*Assumes a given flow value and represents thickness without filling circuitry.

9.6 DIMENSIONAL STABILITY

As circuit layer counts grow and via-to-pad size ratios get tighter, the alignment or registration of the layers of circuitry become extremely important. Although several material and process variables contribute to the capability to achieve layer-to-layer and via-to-innerlayer feature registration, laminate dimensional stability is one of the most important. This is especially true in high-layer-count circuits that use thin laminates, since thinner laminates are generally not as dimensionally stable as thicker laminates. An example can help illustrate this point.

9.6.1 A Model of Printed Circuit Registration Capability

Table 9.4 lists several of the key variables that influence an overall via-to-innerlayer pad registration capability in the printed circuit manufacturing process. The values in the table represent individual process standard deviations using a specific multilayer circuit design. If we make the assumption that these processes are normally distributed and that they are centered on the desired nominal value, we can use additivity of variance to estimate an overall registration capability. In other words, taking the square root of the sum of the squares of each individual process standard deviation gives an overall process standard deviation that can be used to assess capability.

In other words, if we are trying to maintain at least tangency of a drilled hole to an innerlayer pad with a 13.5 mil hole, the pad needs to be almost 27.5 mil in diameter for material type A ($13.5 + 13.8 = 27.3$). The same procedure can be used to calculate clearance or antipad diameters with the desired plated through hole to clearance distance, which is also an important design consideration. It is important to note that items such as dimensional stability and

TABLE 9.4 Multilayer Registration Variables

Process Variable	Process σ Material Type A	Process σ Material Type B
Artwork Plotting	0.33	0.33
Artwork Alignment	0.70	0.70
Post Etch Punch	0.40	0.40
Laminate Stability	1.70	0.90
Drill Set-Up	0.80	0.80
Hole Location	1.00	1.00
Overall σ	*2.30*	*1.79*
Overall +/–3σ	*6.90*	*5.37*
Drill + ? Capability	13.80	10.75

hole location will be significantly impacted by the specific material type, board design, and manufacturing processes used, and that actual registration systems are more complex than described here. The point of this analysis is to show the impact of laminate dimensional stability on printed circuit design. As you can see in this example, improving laminate stability from a standard deviation of 1.70 to 0.90 improves the overall registration capability from drill size + 13.8 mil to drill size + 10.75 mil, or a reduction of over 3 mil in the required diameter of the internal feature. Requirements for increased circuit density require smaller circuit features, such as internal pad sizes. This, in turn, drives the need for better registration capabilities and therefore improved laminate dimensional stability, as this model suggests.

9.6.2 Dimensional Stability Test Methods

A common test method used to evaluate dimensional stability starts with a sample of a copper-clad laminate with scribed targets or holes in the four corners of the sample. Baseline measurements of the distances between these holes are taken prior to conditioning of the sample. One conditioning procedure involves etching the copper cladding off and remeasuring and determining the dimensional movement compared to the baseline. A second method subjects the sample to a thermal cycle, commonly a bake at 150°C for 2 hours. Again, measurements are taken after conditioning and compared to the baseline dimensions. A third method involves first etching the cladding off, measuring, subjecting the sample to the bake cycle, and measuring again. Each of these methods can be used as a process control tool in the laminate manufacturing process.

However, these test methods are of limited value to the printed circuit manufacturer, who etches circuit images on these laminates, combines them with prepreg materials and other laminates, and presses them together under temperature and pressure to form a multilayer circuit. It is the predictability and consistency of laminate movement through the printed circuit manufacturing process, across a variety of circuit patterns and especially through the multilayer lamination cycle, that is of concern for the circuit manufacturer and ultimately for the designer who wants to increase circuit density. Multilayer lamination cycles commonly reach 185°C or higher, and normally exceed the T_g of the base material. Above the T_g, the resin softens and allows tension in the laminate to be released and is also subject to stresses from the surrounding materials and lamination pressure. Most of the movement of the laminate in the circuit manufacturing process occurs during this lamination cycle.

9.6.3 Improving Dimensional Stability

Although many variables in laminate and circuit manufacturing processes can influence dimensional stability, some common techniques involve laminate press cycle optimization, control of laminate resin contents, and the use of higher T_g materials. New materials and process techniques have also been developed to improve dimensional stability.

9.6.3.1 Laminate Manufacturing Process Considerations. Historically, some printed circuit manufacturers had the laminate manufacturer bake the laminate prior to shipment, or would bake it themselves prior to use. The intent was to relieve any stress that may be stored within the laminate prior to use in the circuit manufacturing operation. Although this process may help, the added material handling and cycle time does not usually justify the process. Instead, many laminate manufacturers reduce the lamination pressure at a specified point in their lamination cycle to minimize the stress that becomes stored in the finished product.

New laminate manufacturing techniques may also offer improved dimensional stability. The direct current and continuous manufacturing processes described in Chap. 6 result in consistent thermal profiles from laminate to laminate and can use low lamination pressures. Optimizing these parameters can lead to improved consistency in dimensional stability.

Other process controls in the laminate manufacturing process are also important for dimensional stability. Control of the raw materials, especially the glass cloth, can contribute to dimensional stability. The tension applied to the cloth during the treating process, the heat rise, temperature, and pressure profiles during lamination, as well as laminate stack-up techniques during lamination can all affect dimensional stability. Consistency in these parameters will generally improve the consistency observed during the printed circuit manufacturing process.

9.6.3.2 Impact of Reverse-Treated Foils (RTF). RTF copper foils typically allow the elimination of surface-roughening processes in the innerlayer circuit imaging process. These surface-roughening processes are normally mechanical scrubbing processes that can stretch or distort thin laminates. Much of the distortion induced by these processes is elastic in that the laminates tend to move back toward their original dimensions. This movement takes time, however. So laminates imaged one length of time after scrubbing can potentially move a different amount if imaged after a different length of time after scrubbing. If the laminates are imaged before they have a chance to fully relax after mechanical scrubbing, some movement may occur after imaging, which then distorts the image placed on the laminate. In addition, variation in the mechanical scrubbing process over time can introduce lot-to-lot variation, which adversely impacts registration capabilities. RTF foil, which facilitates the removal of these scrubbing processes because of the improved photoresist adhesion to the matte side of the foil, can therefore result in improvements in registration capability.

9.6.3.3 Fiberglass Cloths and Resin-to-Glass Ratios. Each fiberglass cloth style used in laminates and prepregs has a resin content range that results in sufficient wet-out of the glass cloth, is relatively easy to control, and therefore results in more uniform thickness and more consistent dimensional stability. The specification of laminates and prepregs with resin contents in the desired ranges can result in improvements in dimensional stability and therefore registration capabilities. As a result, having a range of fiberglass cloth styles available is important to be able to achieve a wide range of dielectric thicknesses.

9.6.3.4 Non-Woven Reinforcements. Because of the woven, serpentine geometry of the yarns in fiberglass cloths, they can behave like springs when used as the reinforcement of base materials. During the resin treating process and the laminate pressing operation, the fiberglass cloth is subjected to stresses that can be stored in the laminate as the resin is cured. These stresses can then be released during the circuit manufacturing process, causing dimensional changes.

Non-woven materials can avoid these stresses. In one non-woven material type, short, randomly oriented fibers are treated with the resin system. In a second type, linear strands of fiberglass are layed down in a balanced, cross-plied orientation which resists subsequent stresses.

9.6.3.5 Multilayer Press Cycle Optimization. Most of the dimensional changes in the laminate occur during the multilayer press cycle when the temperature exceeds the T_g of the resin in the laminate. Above the T_g, the resin system in the laminate softens and allows any stored stress in the reinforcement to be released and allows the laminate to be affected by the adjacent materials and the pressure of the lamination cycle.

Understanding the rheology of the resin in the prepreg is important when designing the multilayer lamination cycle. The point at which the resin begins to melt, the point at which it begins to cure, and the relationship between the heat rise and the viscosity profile of the resin are all important. With respect to viscosity, not only is the minimum viscosity achieved important, but the length of time that the resin is below a certain viscosity, allowing the resin to flow and fill the internal circuit features, is also important. With an understanding of these parameters, it is possible to design "kiss" cycles, or "soak" cycles, where pressure and temperature profiles, respectively, are designed to improve performance, including dimensional stability.

In addition, although seldom done in practice, using a resin system in the prepreg materials that can be cured below the T_g of the resin system in the laminate can avoid the softening of the laminate resin system and therefore prevent much of the movement that takes place. The resistance to the use of this technique is usually driven by a desire to keep the resin system same throughout the multilayer PCB.

9.7 HIGH-DENSITY INTERCONNECT/MICROVIA MATERIALS

One method used to increase circuit density is to use blind and buried vias. Rather than placing a via hole completely through the printed circuit board, blind and buried vias go only partly through the multilayer circuit, joining only the layers that require connection. By not extending these vias through the entire multilayer, real estate on the other layers becomes available for additional circuit routing. Buried vias are those that are not visible from the outside of the finished circuit board, and are formed in a subcomposite or copper-clad laminate. Blind vias are those that are visible from the outside of the multilayer circuit but do not go completely through it. By limiting the size of these vias, you can significantly increase interconnection density. Microvia or high-density interconnection (HDI) printed circuit designs utilize these technologies to increase circuit density.

While the materials already discussed are used in blind and buried via applications using conventional processes, additional materials can be used to increase density using more specialized process techniques. The specialized processes used to form microvias include laser ablation, plasma etching, and photoimaging, with laser formation by far the most common.

The resin system will generally ablate much faster in laser drilling processes. Also, plasmas are not effective in etching through fiberglass. As a result, materials that use an alternative reinforcement or do not contain an inorganic reinforcement have been developed for these applications.

For blind via applications, resin-coated copper foil can be used to form the external circuit layer and dielectrics between layers 1 to 2 and n to n-1, using laser or plasma processes to form the vias. Buried vias could be formed in sequential processes. Two basic types of resin-coated copper foil are available. The first type uses one layer of partially cured resin (see Fig. 9.9).

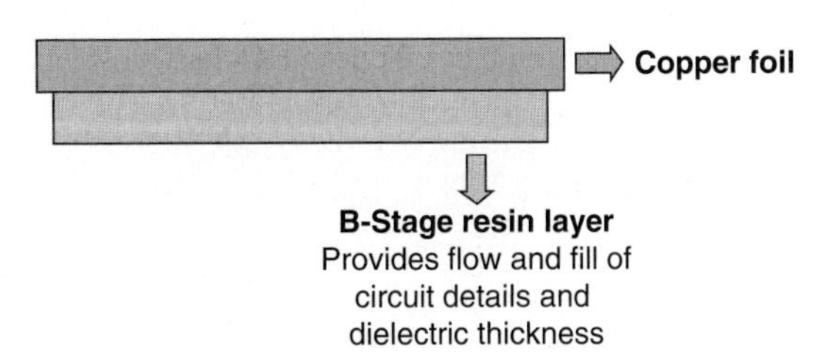

FIGURE 9.9 Resin-coated copper foil with a single layer of B-staged resin.

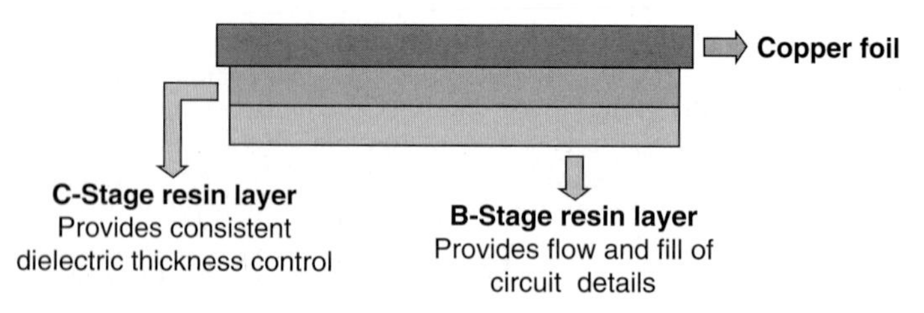

FIGURE 9.10 Resin-coated copper foil with a C-stage and B-stage layer.

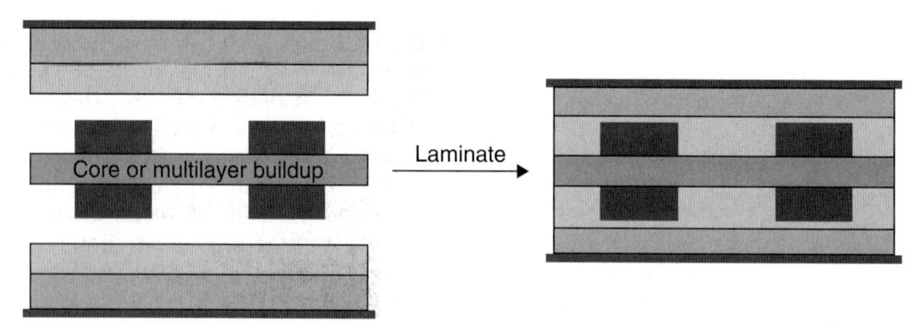

FIGURE 9.11 C-stage plus B-stage resin-coated copper foil laminated to PCB.

This resin-coated foil is then laminated to the rest of the multilayer circuit. A second type of resin-coated copper foil uses two layers of resin (see Figs. 9.10 and 9.11). The first layer is fully cured whereas the second is partially cured. This technique guarantees a minimum dielectric separation between the external foil and the circuitry on the next layer in, since the cured resin layer limits how close the internal circuit layer can get to the external foil.

Another material used in HDI designs utilizes an organic reinforcement that can be laser-ablated or plasma-etched. The most common organic reinforcement used is aramid fiber-based. The aramid fibers are randomly oriented and formed into a sheet that is impregnated with the resin system. In this way, both copper-clad laminates and prepregs can be manufactured and used in multilayer applications. Table 9.5 shows some available thicknesses of Thermount aramid fiber reinforcement with 50 percent resin content. An additional reinforcement that can be used to make a prepreg material is expanded polytetrafluoroethylene (PTFE). This material has a sponge like structure that can also be impregnated with a resin system and used in HDI applications (see Fig. 7.19). Expanded PTFE also has a very low dielectric constant and loss factor.

TABLE 9.5 Commonly Available Thicknesses of Thermount

Thermount Type	Prepreg Thickness	Laminate Thickness
E210	0.0018 in./46 μm	0.0020 in./51 μm
E220	0.0030 in./76 μm	0.0032 in./81 μm
E230	0.0037 in./94 μm	0.0039 in./99 μm

The third process technique used in these applications involves photoimaging a permanent dielectric material in order to form the microvias. These photoimageable dielectrics resemble plating resists but must be able to be catalyzed for subsequent plating operations that will form the external circuit image, and must adhere sufficiently to the rest of the multilayer circuit to provide long-term reliability.

9.8 CAF GROWTH

Conductive anodic filament (CAF) formation is a term used to describe an electrochemical reaction in which conductive paths are formed within a dielectric material due to transport of metal or metal salts through the dielectric. These paths may form between two circuit traces, between two vias, or between a trace and a via, as illustrated in Fig. 9.12. CAF formation between a hole and a plane inside the PCB is also possible, and is similar in concept to hole-to-trace CAF formation. By definition, as circuit density increases, the space between these features decreases. With shorter paths between features, CAF growth becomes a more critical reliability consideration.

For CAF growth to occur, a bias and a path for this filament growth must be present. In fiberglass-reinforced materials, a gap between the resin and the fiberglass filaments is the most common pathway. If the glass is not completely wet-out with resin, or if the bond between the resin system and the fiberglass filaments is insufficient or is compromised, the resulting gap can become such a path. Hollow fiberglass filaments can also provide a path. In addition, there must be a medium in which this electromigration can occur, such as absorbed moisture that allows dissolved ionic species to migrate and promote the electrochemical reaction that leads to CAF. Figure 9.13 is an actual example of CAF formation.

Research on CAF has been done for many years. Some of the key findings of this research include the following:

- CAF is a two-step process consisting of path formation and the electrochemical reaction.
- The filament is usually in the form of a copper salt.

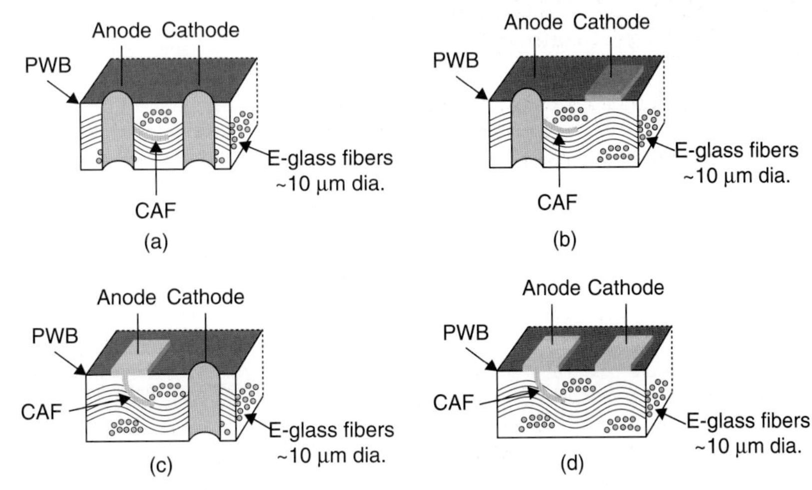

FIGURE 9.12 Pathways for CAF formation: (a) hole to hole; (b) hole to track; (c) track to hole: (d) track to track.

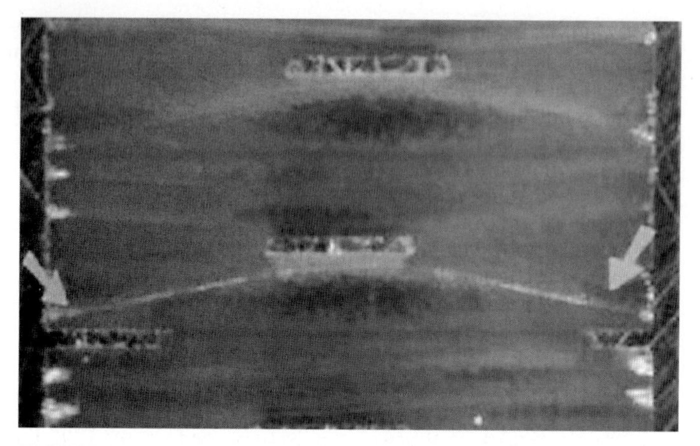

FIGURE 9.13 An example of CAF between two plated through holes.

- Moisture absorption can influence the rate of CAF formation, and there is a minimum threshold for moisture absorption below which CAF is unlikely to occur.
- Temperature can influence the rate of the electrochemical reaction.
- The level of bias can also influence the rate of CAF formation.
- Resin-to-glass wet-out is critical in that it eliminates a potential pathway.
- Beyond glass wet-out, the bond between the resin and the glass is important to ensure that adhesion is not lost due to moisture or thermal stress. Selection of the coupling agent used to improve adhesion between the resin and the glass is critical in this regard, and different resin systems may require different coupling agent types. Figure 9.14 shows examples of silane coupling agents, where "R" represents various chemical species that can be modified for different applications.
- The cleanliness of the glass, both before coating with silane as well as prior to coating with the resin system, is important to ensure proper wetting and bonding of resin to glass.
- The curing agent used in the resin system can influence CAF formation. This may be due to the tendency of some curing agents to absorb moisture, or to the electrochemical nature of the specific curing agent, or both.
- Contaminants on the glass cloth or within the resin system can also accelerate CAF formation. For example, residual hydrolyzable chlorides present in some epoxy resins are known to catalyze electrochemical reactions that can lead to CAF.

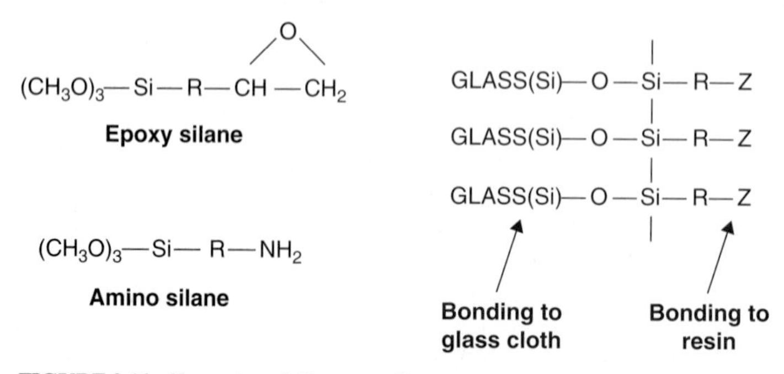

FIGURE 9.14 Examples of silane coupling agents.

- Resin system thermal stability and thermal expansion may influence performance as well, especially as lead-free assembly processes cause PCBs to be subjected to more severe thermal excursions. This is due to the possibility of thermal decomposition of the resin causing reduced resin-to-glass adhesion or even voids within the laminate that can provide pathways for CAF formation. With respect to thermal expansion, the higher temperatures of lead-free assembly result in a greater total stress at the resin-to-glass interface due to the CTE mismatch between them. If this stress is high enough, or the bond strength low enough, separation of the resin from the glass can occur and provide a pathway for CAF as well.

The processing used in printed circuit fabrication can also influence CAF resistance. Some of the factors to consider in PCB fabrication include the following:

- The innerlayer surface must be clean. Adequate rinsing of the innerlayers after chemical processes is important to minimize residual ionic contamination.
- Storage conditions and shelf life of prepreg materials are important to control to ensure adequate flow properties during multilayer lamination.
- Multilayer lamination processing is important to ensure good resin flow and full wetting of resin to glass in the prepreg layers. Vacuum, temperature, and pressure profiles are all important.
- Drilled hole quality is also important. With respect to CAF performance, minimizing resin-to-glass fracturing that can result in excessive wicking of plating chemistries is critical.
- Control of desmear and electroless copper-plating chemistries is also important. Minimizing any further attack of the resin-to-glass interface and plating into the dielectric are the key considerations. Good rinsing to minimize residual ionic contamination from these processes is also important.

9.8.1 CAF Testing

Many different types of test vehicles are used to assess CAF performance, but most include the design features described in Fig. 9.12. Differences between common test vehicles include the distances between features, the thickness and layer counts of the PCBs, and the glass styles and resin contents used to build the PCB. All of these factors influence CAF, if not directly, as a result of the processes used to build the specific test vehicle. In addition, testing parameters such as bias, temperature, and humidity also influence results. Various OEMs will also have different requirements for the time of testing, with typical time requirements ranging from 500 to 1,000 hours.

Figure 9.15 shows an example of CAF found in one test vehicle. In this case, the pathway formed between individual filaments of the glass cloth, either because they had not been sufficiently coated with resin or because resin decomposition, moisture, or other volatile compounds created a void.

9.8.1.1 CAF Test Example #1. One of the common test vehicles used to assess CAF performance consists of a 10 layer PCB, often made of single-ply 2116 style dielectrics in each of the core and prepreg layers. Common test conditions are either 65°C/85 percent relative humidity (RH) or 85°C/85 percent RH, with 10 volts or 100 volts bias. This test vehicle includes design features such as those shown in Fig. 9.12, but the hole-to-hole features are typically the most critically examined. In addition, placement of the holes relative to the glass cloth weave can influence results. When the holes are placed in line with the glass weave, the filaments can bridge the holes, providing a potential CAF pathway. If the holes are placed diagonal to the weave orientation such that no individual filament bridges two holes under test, this potential pathway is eliminated. In practice, holes placed in line with the glass weave are the most relevant for assessing CAF resistance. Hole spacing is also a critical parameter,

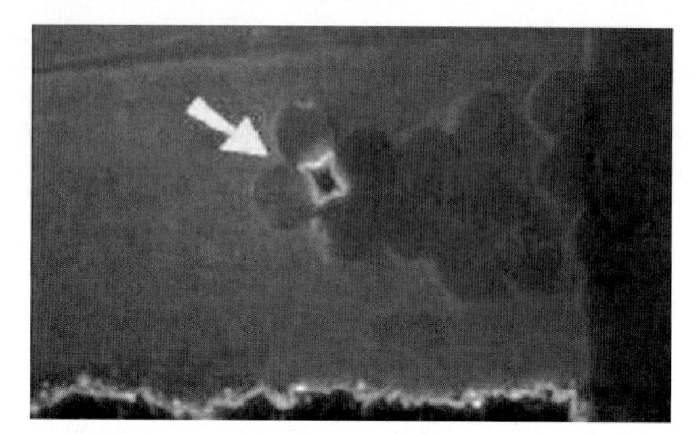

FIGURE 9.15 CAF formation between glass filaments in a PCB test vehicle.

and this particular test vehicle examines holes placed approximately 10, 15, 20, and 25 mil apart. Although the definition of failure can vary slightly from OEM to OEM, the criteria usually consist of some minimum insulation resistance value, and often a maximum decrease in resistance of one order of magnitude, or "a decade drop in resistance." Figure 9.16 provides results for one laminate material processed by one PCB fabricator.

This chart plots average insulation resistance for 25 test coupons (on a log scale in this example) versus time in hours at a 10 volt bias with 65°C/85 percent RH environmental conditions. It also shows the data separately for the 10, 15, 20, and 25 mil hole-to-hole spacings. Other things being equal, average insulation resistance will be higher for holes spaced further apart, as this means there will be more dielectric between features. For this particular test, the average insulation resistance for each spacing did not decrease by an order of magnitude, indicating good results. However, it is also important to examine the individual coupons tested, and a common way to

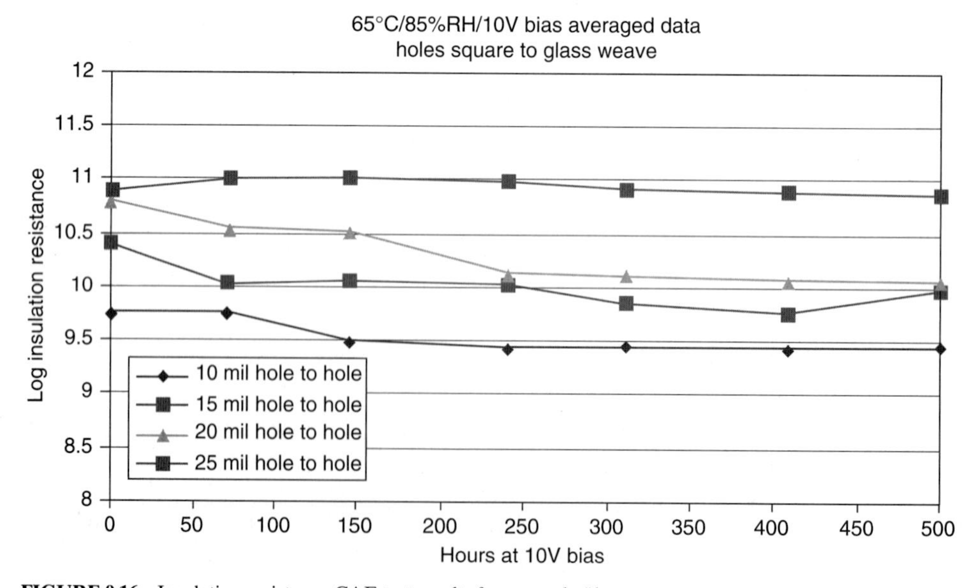

FIGURE 9.16 Insulation resistance CAF test results for example #1.

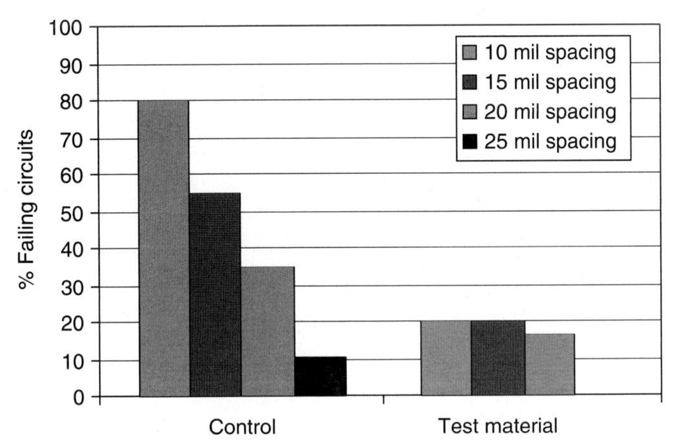

FIGURE 9.17 Percent failing coupons in CAF test example #1.

assess performance in this regard is to calculate the percentage of individual test coupons that pass or fail. Figure 9.17 provides this data for this same test, and shows results for the test material highlighted in Fig. 9.16 versus a control group that utilized a different laminate material.

As is common, the control group shows fewer individual failures as the spacing between holes increase. The test material shows a much lower percentage of individual failures for each spacing.

9.8.1.2 CAF Test Example #2. In a second example, four commonly available FR-4 materials were compared by building the same CAF test vehicle as described in example #1. These materials are all promoted as being "CAF-resistant." The coupons were manufactured by the same PCB fabricator during the same time period. In this evaluation, the test conditions were 100 volts bias, 85°C and 85 percent RH. The same hole-to-hole spacings were evaluated, but for simplicity, Figs. 9.18 and 9.19 show only the data for the 15 and 25 mil spacings, respectively. It is common to use a 1 meg-ohm resistor in series with the test point in these tests, so where the average insulation resistance drops to 10^6 ohm (6 on the y-axis of the chart), this indicates that the resistance has dropped excessively and it is the resistor that is now being

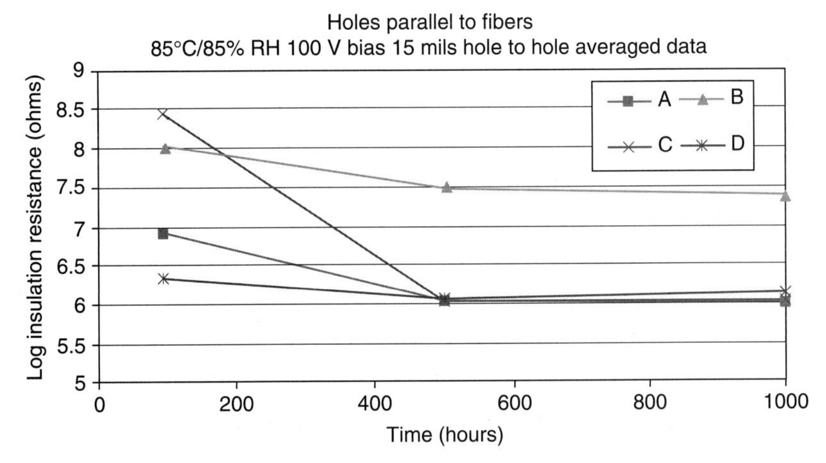

FIGURE 9.18 CAF results for 15 mil hole spacings in example #2.

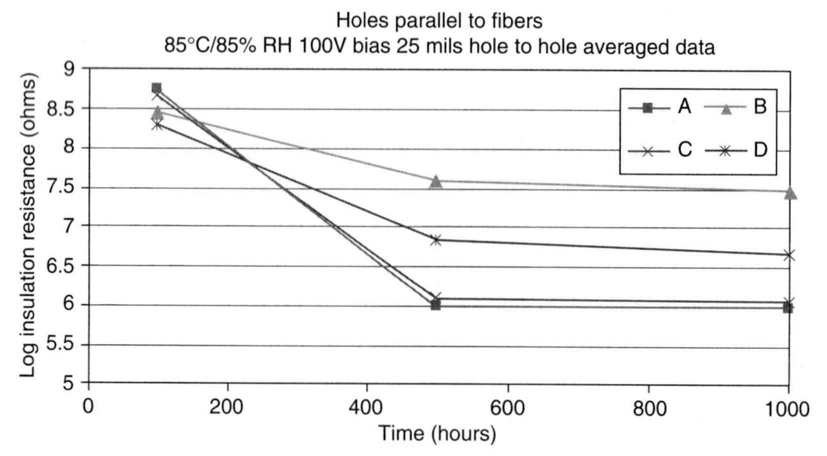

FIGURE 9.19 CAF results for 25 mil hole spacings in example #2.

measured. In this evaluation, materials A, C, and D all drop to this level at 500 hours of testing with 15 mil hole-to-hole spacings. At 25 mil spacings A and C both drop to this level. B and C do not, but C drops in resistance by more than one order of magnitude. Only material B drops by less than one order of magnitude at both spacings. By these criteria, materials A, C, and D would have failed this test. Only material B passed.

9.8.1.3 CAF Test Example #3. In this third example of CAF testing, a different test vehicle design was used, and different temperature, humidity, and bias levels were also used. However, all the test coupons were manufactured at the same time by the same PCB fabricator, and in addition, the same resin system was used in groups 1 through 4. The differences between groups 1 through 4 involve different types of fiberglass cloth and differences in process controls used when treating the glass cloth with the resin system. This is the key point of this third example.

Figure 9.20 shows that while each group used exactly the same laminate material type, differences in glass cloth and treating process controls resulted in different CAF performance levels.

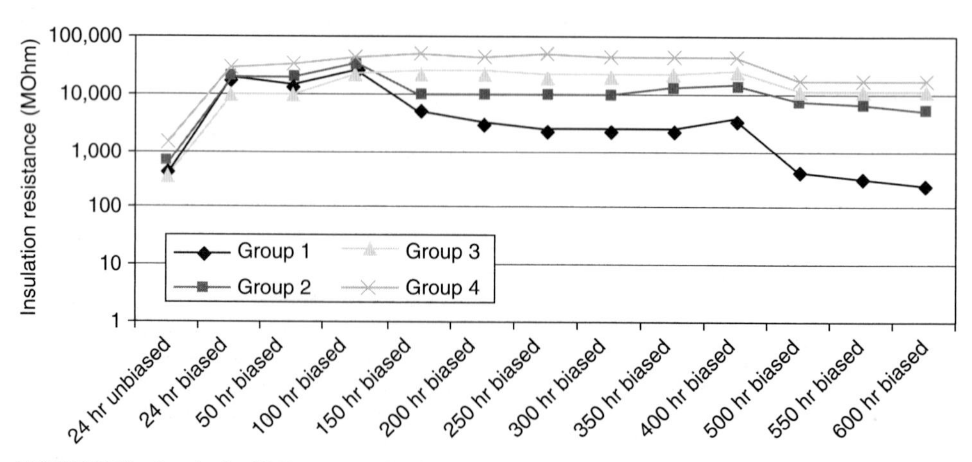

FIGURE 9.20 Results for CAF test example #3.

For each group, the chart shows average insulation resistance for a set of coupons over time. Clearly, the average insulation resistance varied by group, with group 1 showing the most significant decline in resistance. In addition, evaluating the individual coupons in each group provides additional insight into performance differences. In group 1, 40 percent of the coupons failed by 600 hours, with the first failure occurring at 400 hours. In group 2, 20 percent of the coupons failed at 600 hours. Groups 3 through 5 did not have any failures in this test. So even though the resin system was the same for each group, differences in glass cloth and laminate manufacturing processes clearly can have a significant impact on CAF performance.

9.8.2 Choosing a Base Material for CAF Resistance

Some of the key conclusions of these examples and other research on CAF performed in the industry include the following:

- All base materials have some level of CAF resistance, but even materials promoted as "CAF-resistant" can vary widely in actual performance.
- The test vehicle design has a significant impact on results. The closer the circuit features, the more likely that CAF failures will occur. In addition, the other attributes of the test vehicle design can impact the fabrication variables that influence CAF—for example, the thickness of the PCB and drilled hole sizes can lead to high aspect ratios that can be more challenging to manufacture. As a result of these factors, there is not universal agreement on what constitutes a "CAF-resistant" material. This can vary by OEM and application.
- Even one type of base material (resin system type) can vary in performance based on other factors, such as the glass cloth quality and the controls used during laminate and prepreg manufacturing.
- The specific glass styles can also impact performance. Although 7628 glass was the focus of much attention in CAF testing, 106 and 1080 styles can be challenging as well, in that getting full resin-to-glass wetting can be challenging with these styles.

When selecting a material for applications requiring CAF resistance, consider the following:

- It is critical to define appropriately the performance level required and to design a test vehicle and test protocol that reflects the requirements of the application. Failure to do so can result in either underspecifying the material attributes and risking PCB failures or overspecifying the material and paying too much for it.
- For resistance to CAF growth, the resin system and the bonding of the resin system to the fiberglass cloth are key considerations. Coupling agents, typically silanes, are used to promote adhesion between the fiberglass filaments and the resin system. The molecular structure of the coupling agent includes a portion that bonds well to the fiberglass surface and other portions that bond well to the resin system. Specific coupling agents have been developed for specific resin types. Achieving a good bond between the fiberglass filaments and the resin system with an appropriate coupling agent helps prevent this interface from becoming a path for CAF growth.
- The use of expanded glass cloths can facilitate full wet-out of the resin to the glass, but such cloths typically cost more. The important point is that each glass style must be able to achieve good wet-out of all of the glass filaments.
- Low rates of moisture absorption and low levels of ionic contamination within the resin system also aid in resisting CAF growth. Epichlorohydrin is a precursor material used to make epoxy resins. The use of this compound can result in residual hydrolyzable chlorides in the epoxy resin. In the presence of moisture, these hydrolyzable chlorides form ionic species that can promote electromigration. When CAF resistance is a critical consideration, the use of high-purity resins may be required.

- Comparisons of dicy-cured materials with non-dicy cured materials, especially the "phenolic" FR-4 materials developed for lead-free assembly, show that the elimination of dicy can have a positive impact on CAF resistance. This may be due to reduced moisture absorption when dicy is removed, or to the elimination of an electrochemical reaction that may be promoted in the presence of dicy, or to both.
- Lead-free assembly can stress the resin-to-glass bond, and even cause some level of resin decomposition, which can create microvoids within a PCB. These phenomena can make it easier for pathways for CAF formation to be created. The use of more thermally stable resin systems may be required when CAF resistance is needed in products that will experience lead-free assembly.

9.9 ELECTRICAL PERFORMANCE

Base material electrical properties are an important consideration in sophisticated printed circuits operating at high frequencies. High data rates, measured in gigabits per second (Gbps), and high clock speeds make the dielectric constant (D_k) and dissipation factor (D_f) of base materials very important in high-speed digital circuits. Wireless and RF applications operating at very high frequencies also demand very low D_k and D_f values. Moreover, the consistency of these properties over a large frequency range is also important. These properties were defined in Chap. 8 as follows:

- **Dielectric constant/permittivity** This is the ratio of the capacitance of a capacitor with a given dielectric material to the capacitance of the same capacitor with air as a dielectric. It refers to the ability of a material to store an electric charge.
- **Dissipation factor/loss tangent** The property is the ratio of the total power loss in a material to the product of the voltage and current in a capacitor in which the material is a dielectric.

9.9.1 Importance of D_k and D_f

These properties are important because they affect signal transmission in the printed circuit. At low frequencies, a signal path in a printed circuit can typically be represented electrically as a capacitance in parallel with a resistance. However, as frequencies increase, at some point signal paths must be considered transmission lines where the electrical and dielectric properties of the base materials have a greater effect on signal transmission. A full discussion of capacitive versus transmission line environments is beyond the scope of this chapter, but the premise is to determine, for the transmission of a signal pulse of a given rise time, the acceptable length of a conductor before a significant voltage difference is realized along its length. Conductors longer than this critical value are then regarded as transmission lines. Because the velocity of signal propagation is inversely proportional to the square root of the permittivity of the dielectric, a lower permittivity value results in faster signal speeds and a longer rise distance. With a larger rise distance, larger conductor lengths are acceptable before a significant voltage drop is experienced. However, if the ratio of conductor length to rise distance is large enough, signal reflections from a mismatched load impedance may be received back at the source after the pulse has reached its maximum plateau value, and pulse additions that occur under these circumstances may lead to false triggering of a device.

On the other hand, signal attenuation can result in missed signals. One of the causes of signal attenuation is dielectric loss. As the circuit operates, the dielectric medium absorbs energy from the signal. Attenuation of the signal by the dielectric is directly proportional to the square root of permittivity and directly proportional to the loss tangent. In addition, dielectric losses increase as frequencies increase. When a high bandwidth is desired, this effect has a greater impact on the higher-frequency components, and the bandwidth of the propagating pulse decreases and degrades the rise time.

Because the permittivity and loss tangent vary with frequency, and other factors to be discussed later in this chapter, the degree to which these properties vary is also an important circuit design consideration. If these properties vary significantly with frequency, designing a circuit with devices that operate at various frequencies becomes that much more complex. In addition, operating within a given bandwidth becomes that much more difficult as different frequency components experience different dielectric properties, which in turn lead to differences in signal propagation and loss.

Therefore, base materials with low permittivity values and low loss factors are desired for high-speed, high-frequency printed circuits. In addition, consistency of these properties across frequencies is also required. Besides frequency dependence, since operating environments can also vary, the consistency of these properties across environmental conditions is also important and is discussed in the following sections.

9.9.2 High-Speed Digital Basics

Figure 9.21 is a representation of high-speed digital communication that involves sending bits of information coded in waveforms.

The zeros and ones of binary information are coded on the rise time or on both the rise time and fall time. The high voltage represents 1 and the low voltage represents 0. The faster the rise time, the faster the signal. To achieve faster rise times, sinusoidal wave forms are superimposed on one another. The range of frequencies used is called the bandwidth, with the bandwidth given as 0.35/rise time. In short, a faster rise time allows for a greater range of frequencies, or greater bandwidth.

Figure 9.22 provides an example of eye pattern analysis. In this analysis, the height of the central eye opening measures noise margin in the received signal. The width of the signal band at the corner of the eye measures the jitter. The thickness of the signal line at the top and bottom of the eye is proportional to noise and distortion in the receiver output. Transitions between the top and bottom of the eye show the rise and fall times of the signal.

Figure 9.23 illustrates potential signal integrity differences when different base materials are used. The top chart shows an example of a 10 Gbps signal at the source. Note the pattern in this chart. It changes from 0 in the x-axis (0) to its peak value (1). Now look at the chart in the lower-left corner of Fig. 9.23, representing the use of a standard FR-4 material. Note the change in the pattern, particularly the decreased amplitude. When the signal degrades as illustrated

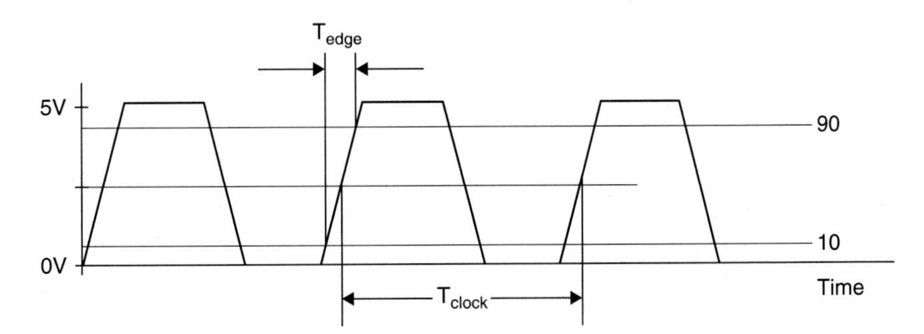

FIGURE 9.21 Digital communication.

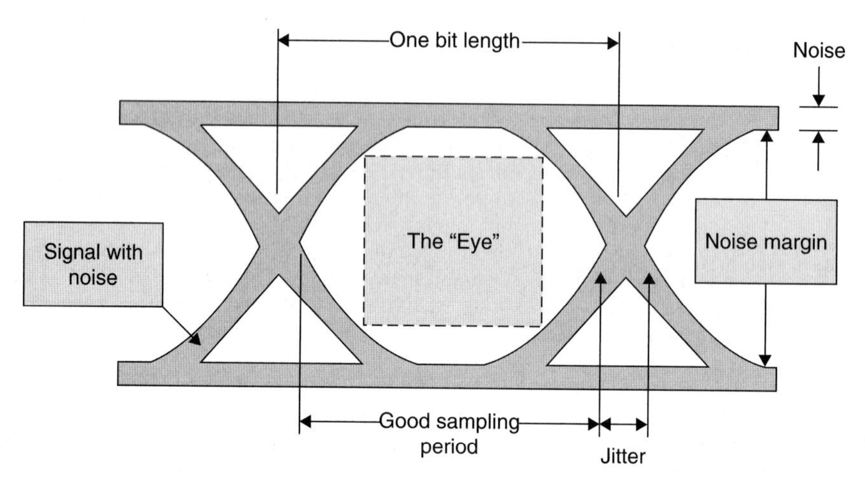

FIGURE 9.22 Eye pattern analysis.

here, the function of the circuit can be significantly impacted. Now examine the chart in the lower-right corner. This is for a low D_k/D_f material. Note the improved amplitude compared to the standard FR-4 material and a pattern much closer to the signal at the source. Figure 9.24 takes this analysis further. The chart in the upper-left corner is the signal at the source. This is as good as it gets. The remaining charts illustrate how the signal is affected with materials of

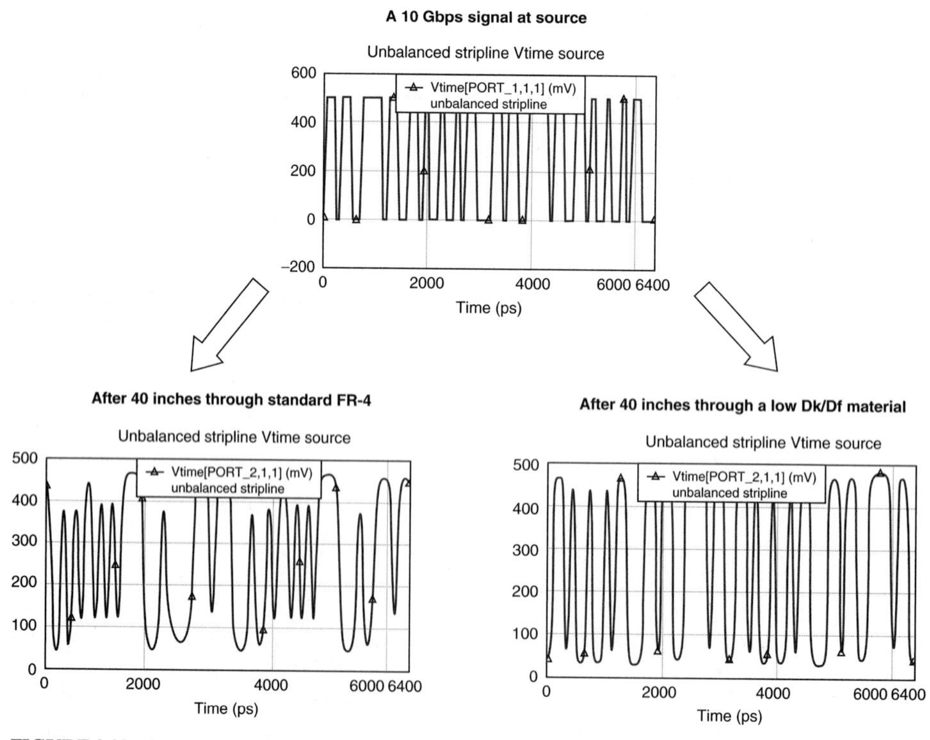

FIGURE 9.23 Substrate influence on signal integrity.

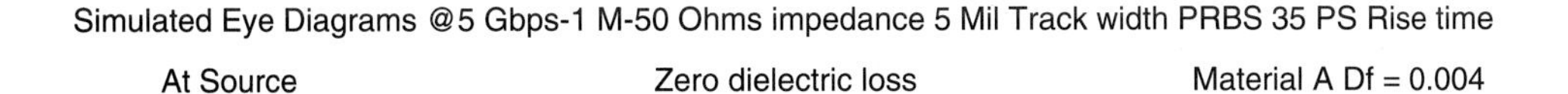

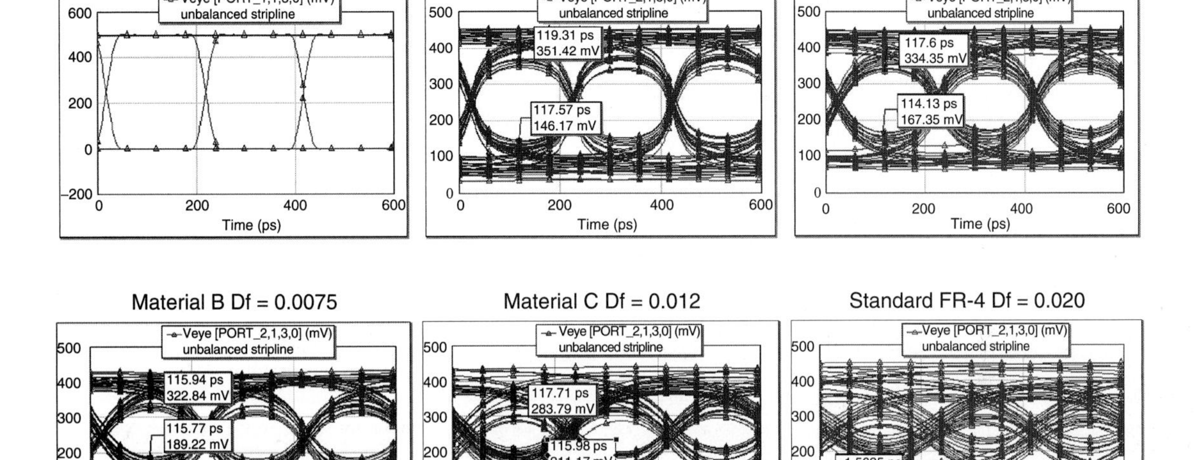

FIGURE 9.24 Simulated eye diagrams at 5 Gbps, 1 M, 50 ohms impedance, 5 mil track width.

increasing D_f. Note the diagram for the standard FR-4 under these conditions. This would represent an unacceptable condition, as the eye is almost completely closed. As the D_f is decreased from the FR-4 level, the pattern improves.

Another way to illustrate the impact of D_f is to graph loss in decibels (dB) per meter versus the D_f. Figure 9.25 provides a simulation of this relationship for a 5 mil line. Because wider line widths are less "lossy" because of reduced skin effect, some of this can be compensated for by using a wider line. However, this negatively impacts circuit density. Figure 9.26 shows

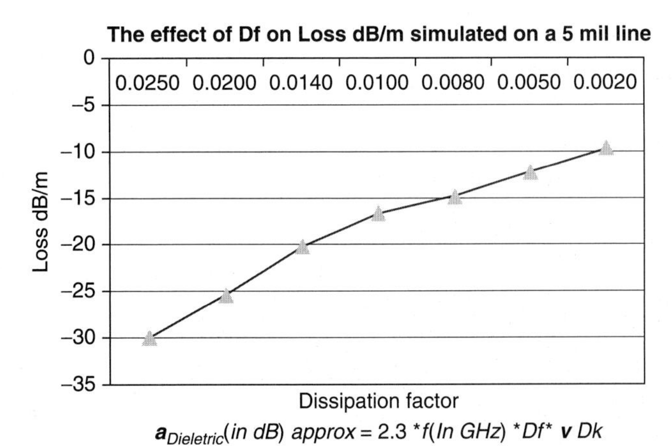

FIGURE 9.25 The effect of D_f on dielectric loss in dB/m simulated on a 5 mil line.

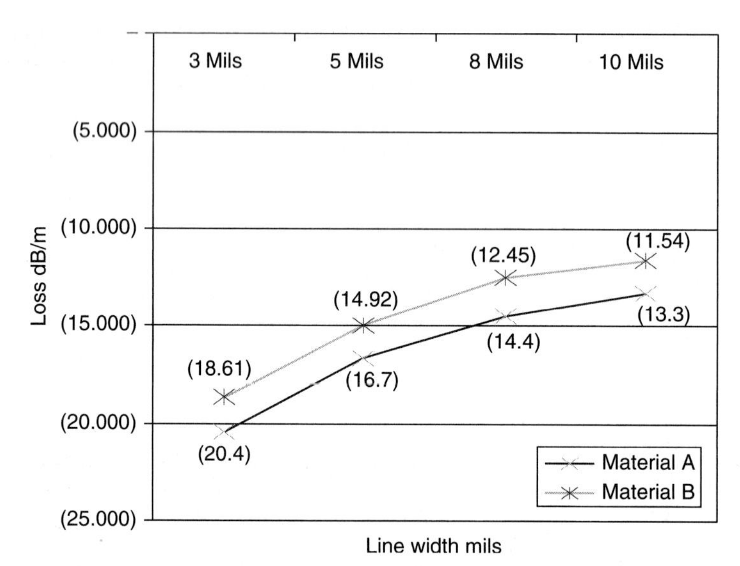

FIGURE 9.26 Effect of line width on loss for two material types.

the impact of line width on loss for two different low D_k/D_f materials. In this case, material B would allow the designer to use a thinner line to increase circuit density while maintaining a given level of loss. For example, to keep the loss in this example below 15 dB/m, material A would require a line width of approximately 8 mils. However, with the same loss requirement, material B enables the line to be reduced to approximately 5 mil. This is a significant benefit in increasing circuit density.

9.9.3 Choosing a Base Material for Electrical Performance

The dielectric constant and dissipation factor of the dielectric material are determined by both the resin system and the reinforcement type. Therefore, each of these should be considered when choosing a material.

A variety of low-dielectric, low-loss resin systems are available for high-speed circuit applications. These include polytetrafluoroethylene (PTFE or Teflon®), cyanate ester, epoxy blends, and allylated polyphenylene ether (APPE). Likewise, a few different reinforcements and fillers are available that can be used to modify the electrical properties of the base material. Although E-glass is still the most commonly used fiberglass reinforcement, it should be noted that others are available. In addition, inorganic fillers are sometimes used to modify electrical properties as well. Table 9.6 provides electrical property data on some of the available fiberglass materials. Table 9.7 provides data on some of the base material composites available.

TABLE 9.6 Dielectric Constants and Dissipation Factors of Common Glass Types

Reinforcement	D_k@ 1 MHz	D_k@ 1 GHz	D_f@ 1 MHz	D_f@ 1 GHz
E-Glass	6.6	6.1	0.0020	0.0035
NE-Glass	4.4	4.1	0.0006	0.0018
S-Glass	5.3	5.2	0.0020	0.0068
D-Glass	3.8	4.0	0.0010	0.0026

TABLE 9.7 Dielectric Constants and Dissipation Factors of Common Resin/Reinforcement Composites

Resin System	Reinforcement	D_k@ 1 MHz	D_k@ 1 GHz	D_f@ 1 MHz	D_f @ 1 GHz
Epoxy	E-Glass	4.4	3.9	0.020	0.018
Epoxy	Aramid Fiber	3.9	3.8	0.024	0.020
Halogen-Free Epoxy	E-Glass	4.3	4.0	0.015	0.013
Epoxy/PPO	E-Glass	3.9	3.9	0.011	0.010
Modified Epoxy	E-Glass	3.9	3.7	0.012	0.012
Epoxy Blend	E-Glass	3.9	3.7	0.009	0.009
Epoxy Blend	SI™-Glass	3.6	3.4	0.008	0.008
Low-Loss Blend	E-Glass	3.8	3.7	0.006	0.007
Very-Low-Loss Blend	E-Glass	3.5	3.4	0.003	0.0036
Cyanate Ester	E-Glass	3.8	3.7	0.008	0.011
Polyimide	E-Glass	4.3	3.9	0.014	0.015
APPE	E-Glass	3.7	3.4	0.005	0.007
PTFE	E-Glass	2.3	2.3	0.0013	0.0009
Hydrocarbon	E-Glass/Ceramic	3.4	3.3	0.0025	0.0024

Table 9.7 lists values for approximately 50 percent resin content. These values change as the resin content varies, with the magnitude of the change depending on the specific resin system. Also, comparable resin system types from various suppliers could differ. For example, since there are many types of epoxies, with a wide range of electrical properties, specific epoxy formulations from different material suppliers can exhibit somewhat different dielectric and loss properties. In addition, the test method used to measure D_k and D_f, especially at higher frequencies, will have a significant impact on the reported value. When comparing materials, it is critical to compare values derived from the same test method, or at least to understand the differences in the test methods. The differences in test methods are beyond the scope of this chapter, but their importance should not be underestimated. The purpose of this table is to highlight the relative differences between various resin systems and reinforcement types.

These values also vary with resin content and frequency. Figure 9.27 illustrates the resin content dependence of the dielectric constant for a few materials. Figure 9.28 shows the impact of resin content on the dissipation factor of the low D_k epoxy blend, and Fig. 9.29 shows the impact of frequency on D_k for a few of these materials. Beyond frequency and resin content effects, the dielectric constant and dissipation factor can vary with temperature and moisture absorption.

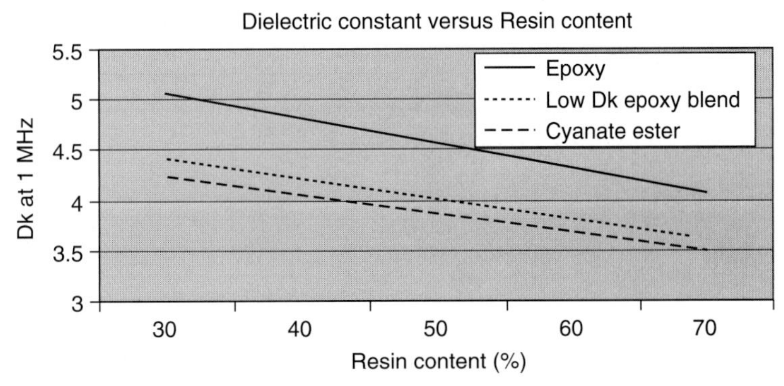

FIGURE 9.27 Dielectric constant versus resin content.

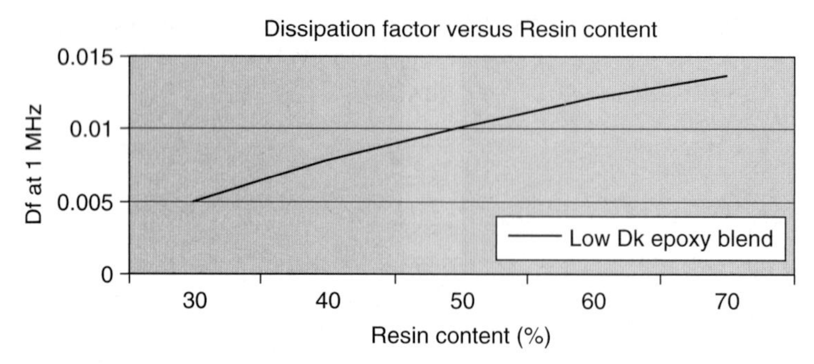

FIGURE 9.28 Dissipation factor versus resin content for a low D_k epoxy blend.

When choosing a material for a specific application, it is important to understand the operating conditions and environment in which the circuit will be used. Some resin systems exhibit less sensitivity to these conditions than others. For example, base material suppliers are continually developing resin systems to meet the demanding electrical properties requirements of high-speed and wireless applications, which include consistent properties over frequency ranges and environmental conditions. In short, different material types can have different responses to changes in frequency, resin content, temperature, and humidity conditions. These responses become more important to understand in high-speed, RF, and wireless applications.

In addition, when selecting a material, it is important not simply to select the material with the lowest dielectric constant or dissipation factor, since there are typically cost-performance trade-offs to be made. In general, the lower the dielectric constant and dissipation factor, the more costly the material, and very often the more difficult it is to process.

In summary, some general relationships include the following:

- The dielectric constant generally decreases with increasing resin content.
- The dissiption factor often increases with increasing resin content.
- The dielectric constant typically drops as frequency increases.
- The dielectric constant and dissipation factor typically increase as water absorption rises.

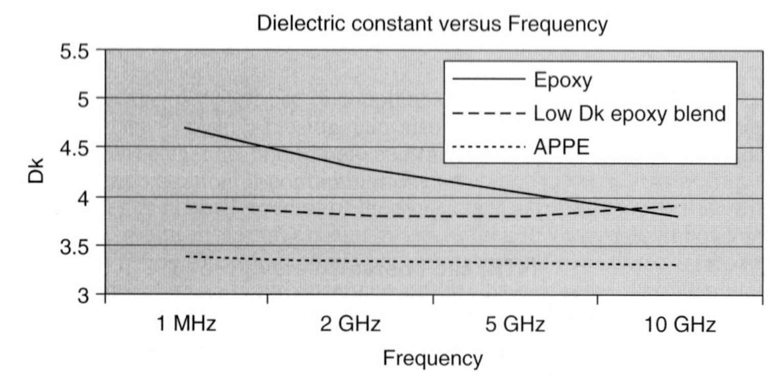

FIGURE 9.29 Dielectric constant versus frequency.

- The dielectric constant of E-glass is only mildly frequency-dependent and therefore lower resin contents result in less variation across frequencies.
- The dissipation factor generally rises with frequency, but may exhibit maxima at certain frequencies.

9.9.4 Electrical Performance of Lead-Free Compatible FR-4 Materials

Most of the FR-4 laminate materials developed for lead-free assembly applications use an alternative resin chemistry in comparison to the conventional dicyandiamide ("dicy") cured FR-4s. The most common alternatives are commonly referred to as "phenolic" or "novolac" cured materials. Although there is some variation between these materials, as a group they tend to exhibit somewhat different electrical performance, particularly with respect to dielectric loss, or D_f. For most applications operating in typical frequency ranges, the differences are not significant. However, as operating frequencies increase toward the higher end of "FR-4 applications" and impedance control becomes more critical, these differences can become very significant. Table 9.8 includes two common dicy-cured FR-4 materials (A and C) and two phenolic FR-4 materials compatible with many lead-free assembly applications (B and D). It also includes a lead-free compatible material (E) designed to improve Dk and D_f performance in comparison to the "phenolic" lead-free compatible materials. Note that different types of measurement systems can result in different measured values for D_k and D_f. Laminate resin contents and other factors also influence these properties. Thus, in Table 9.8 the comparisons between these materials are more important than the absolute values reported for each material. For these comparisons, the same measurement system and resin contents were used.

TABLE 9.8 Properties of Several Base Material Types

Product	Description	Glass Transition Temp., °C	Decomp. Temp., °C	% Expansion, 50–260°C (40% RC)	D_k@ 2 GHz	D_k@ 5 GHz	D_f@ 2 GHz	D_f@ 5 GHz
A	Dicy-cured, 140°C T_g FR-4	140	320	4.2	3.9	3.8	0.021	0.022
B	Phenolic, Mid-T_g FR-4	150	335	3.4	4.0	3.9	0.026	0.027
C	Dicy-cured, High-T_g FR-4	175	310	3.5	3.8	3.7	0.020	0.021
D	Phenolic, High-T_g FR-4	175	335	2.8	3.9	3.8	0.026	0.026
E	Non-Dicy, Non-Phenolic	200	370	2.8	3.7	3.7	0.013	0.014

In Table 9.8, the improved phenolic FR-4 materials are clearly not quite as good in terms of electrical performance at higher frequencies as the conventional dicy-cured materials, particularly with respect to D_f. However, material E exhibits thermal properties at least as good as the phenolic materials. In addition, the electrical performance of material E is even better than the conventional dicy-cured materials, especially in terms of D_f performance. Figures 9.30 and 9.31 provide higher frequency D_f and D_k data for several lead-free compatible FR-4 materials in comparison to a standard dicy-cured, high-T_g FR-4, using a split-post resonant cavity test method. It also includes data for the non-dicy/non-phenolic material. Note that there is a wide range of D_f values for the phenolic materials, and to a lesser extent some variation in D_k values

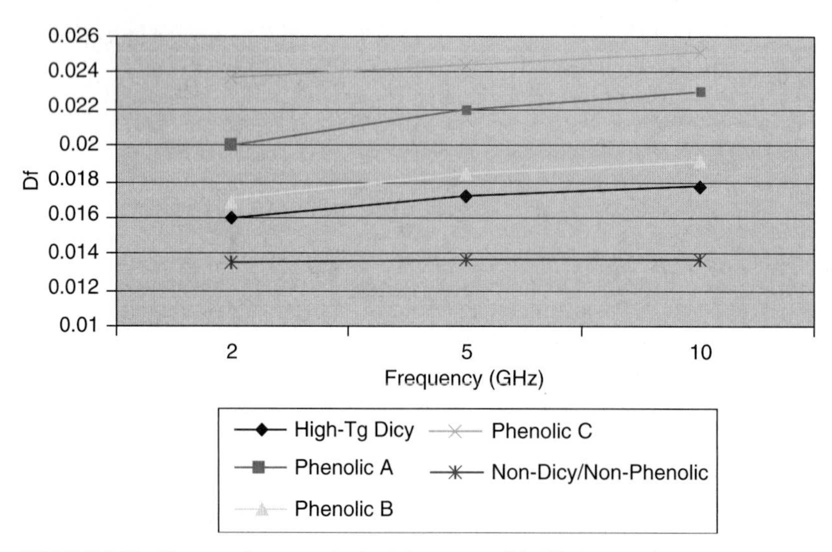

FIGURE 9.30 D_f versus frequency for lead-free compatible FR-4 materials.

as well. In comparison, the non-dicy/non-phenolic material offers lower and more stable D_f properties and a slightly lower D_k.

Furthermore, Table 9.9 provides several laminate constructions for a common phenolic material, along with resin content and D_k data at 2 and 5 GHz. Note how different constructions and resin contents can vary in D_k values. Table 9.10 provides the same construction data with D_f data at 2 and 5 GHz. Tables 9.11 and 9.12 provide this same data for the non-dicy/non-phenolic material with additional data at 10 GHz.

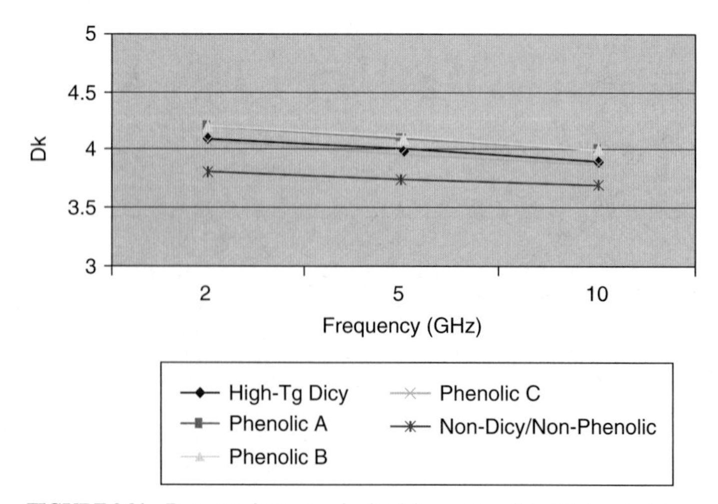

FIGURE 9.31 D_k versus frequency for lead-free compatible FR-4 materials.

TABLE 9.9 Laminate Constructions and D_k Data for a Common Phenolic FR-4

Core Thickness	Standard Construction	Resin Content	Dk@ 2.0 GHz	Dk@ 5.0 GHz
0.0025	1-1080	58	3.72	3.65
0.0030	1-2113	44	4.04	3.98
0.0035	2-106	65	3.58	3.50
0.0040	1-2116	45	4.02	3.96
0.0043	106/1080	60	3.68	3.60
0.0050	1-1652	42	4.10	4.04
0.0053	106/2113	56	3.76	3.69
0.0060	1080/2113	53	3.83	3.76
0.0070	1-7628	41	4.12	4.07
0.0070	2-2113	51	3.87	3.81
0.0080	2-2116	45	4.02	3.96
0.0095	2-2116	52	3.85	3.78
0.0100	2-1652	42	4.10	4.04
0.0120	2-1080/7628	47	3.97	3.91
0.0140	2-7628	41	4.12	4.07
0.0180	2-7628/2116	42	4.10	4.04
0.0210	3-7628	39	4.18	4.12
0.0240	3-7628/2113	41	4.12	4.07
0.0280	4-7628	40	4.15	4.09
0.0310	4-7628/2116	40	4.15	4.09
0.0340	5-7628	40	4.15	4.09
0.0350	5-7628	41	4.12	4.07
0.0390	6-7628	37	4.23	4.18

TABLE 9.10 Laminate Constructions and D_f Data for a Common Phenolic FR-4

Core Thickness	Standard Construction	Resin Content	Df@2.0 GHz	Df@5.0 GHz
0.0025	1-1080	58	0.026	0.027
0.0030	1-2113	44	0.021	0.022
0.0035	2-106	65	0.028	0.029
0.0040	1-2116	45	0.021	0.027
0.0043	106/1080	60	0.026	0.028
0.0050	1-1652	42	0.020	0.021
0.0053	106/2113	56	0.025	0.026
0.0060	1080/2113	53	0.024	0.025
0.0070	1-7628	41	0.020	0.021
0.0070	2-2113	51	0.023	0.024
0.0080	2-2116	45	0.021	0.027
0.0095	2-2116	52	0.024	0.025
0.0100	2-1652	42	0.020	0.021
0.0120	2-1080/7628	47	0.022	0.023
0.0140	2-7628	41	0.020	0.021
0.0180	2-7628/2116	42	0.020	0.021
0.0210	3-7628	39	0.019	0.020
00240	3-7628/2113	41	0.020	0.021
0.0280	4-7628	40	0.020	0.021
0.0310	4-7628/2116	40	0.020	0.021
0.0340	5-7628	40	0.020	0.021
0.0350	5-7628	41	0.020	0.021
0.0390	6-7628	37	0.019	0.020

TABLE 9.11 Laminate Constructions and D_k Data for a Non-Dicy/Non-Phenolic Material

Core Thickness	Standard Construction	Resin Content	Dk@2.0 GHz	Dk@5.0 GHz	Dk@ 10.0 GHz
0.0020	1-106	70	3.40	3.38	3.37
0.0025	1-1080	57	3.67	3.66	3.65
0.0027	1-1080	59	3.63	3.61	3.61
0.0030	1-1080	63	3.54	3.53	3.52
0.0032	1-2113	49	3.87	3.85	3.85
0.0035	1-2113	51	3.82	3.80	3.80
0.0035	2-106	65	3.50	3.48	3.48
0.0040	1-3070	49	3.87	3.85	3.85
0.0040	1-2116	45	3.97	3.96	3.95
0.0043	106/1080	61	3.59	3.57	3.56
0.0045	1-2116	49	3.87	3.85	3.85
0.0050	1-1652	42	4.06	4.04	4.03
0.0053	106/2113	56	3.70	3.68	3.68
0.0060	1080/2113	53	3.77	3.76	3.75
0.0070	1-7628	40	4.12	4.10	4.09
0.0070	2-2113	51	3.82	3.80	3.80
0.0080	2-2116	45	3.97	3.96	3.95
0.0100	2-1652	42	4.06	4.04	4.03
0.0120	2-1080/7628	47	3.92	3.91	3.90
0.0140	2-7628	40	4.12	4.10	4.09
0.0160	2-2116/7628	45	3.97	3.96	3.95
0.0180	2-7628/2116	41	4.09	4.07	4.06
0.0210	3-7628	40	4.12	4.10	4.09
0.0240	2-1652/2-7628	41	4.09	4.07	4.06
0.0280	4-7628	40	4.12	4.10	4.09

TABLE 9.12 Laminate Constructions and D_f Data for a Non-Dicy/Non-Phenolic Material

Core Thickness	Standard Construction	Resin Content	Df@2.0 GHz	Df@5.0 GHz	Df@10.0 GHz
0.0020	1-106	70	0.0151	0.0154	0.0154
0.0025	1-1080	57	0.0136	0.0139	0.0139
0.0027	1-1080	59	0.0138	0.0141	0.0141
0.0030	1-1080	63	0.0143	0.0146	0.0146
0.0032	1-2113	49	0.0121	0.0130	0.0130
0.0035	1-2113	51	0.0130	0.0132	0.0132
0.0035	2-106	65	0.0145	0.0148	0.0148
0.0040	1-3070	49	0.0122	0.0130	0.0130
0.0040	1-2116	45	0.0121	0.0125	0.0125
0.0043	106/1080	61	0.0141	0.0123	0.0123
0.0045	1-2116	49	0.0122	0.0130	0.0130
0.0050	1-1652	42	0.0110	0.0122	0.0122
0.0053	106/2113	56	0.0135	0.0138	0.0138
0.0060	1080/2113	53	0.0132	0.0134	0.0134
0.0070	1-7628	40	0.0104	0.0119	0.0119
0.0070	2-2113	51	0.0130	0.0132	0.0132
0.0080	2-2116	45	0.0123	0.0125	0.0125
0.0100	2-1652	42	0.0111	0.0120	0.0120
0.0120	2-1080/7628	47	0.0125	0.0127	0.0127
0.0140	2-7628	40	0.0114	0.0119	0.0119
0.0160	2-2116/7628	45	0.0123	0.0125	0.0125
0.0180	2-7628/2116	41	0.0115	0.0120	0.0120
0.0210	3-7628	40	0.0114	0.0129	0.0129
0.0240	2-1652/2-7628	41	0.0115	0.0120	0.0120
0.0280	4-7628	40	0.0114	0.0119	0.0119

9.10 ELECTRICAL PERFORMANCE OF LOW D_K/D_F LEAD-FREE COMPATIBLE MATERIALS

However, for performance at very high frequencies, lower D_k/D_f materials are preferred. While low D_k/D_f materials have been available for many years, the advent of lead-free assembly has complicated material selection, and in these applications not only are the laminate D_k and D_f properties critical, but their thermal properties are just as important. Critical material properties for lead-free assembly compatibility will be discussed in a subsequent chapter. Figures 9.32 and 9.33 provide D_f and D_k data for three different low D_k/D_f materials that are also compatible with lead-free assembly.

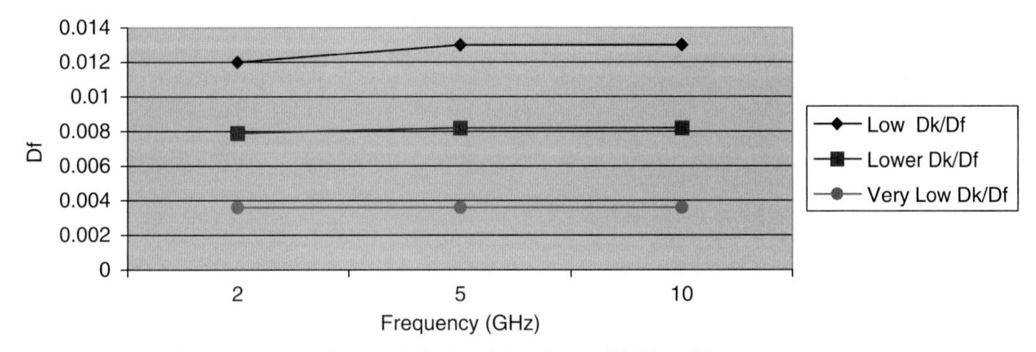

FIGURE 9.32 Df versus Frequency for Low D_k/D_f, Lead-Free Compatible Materials.

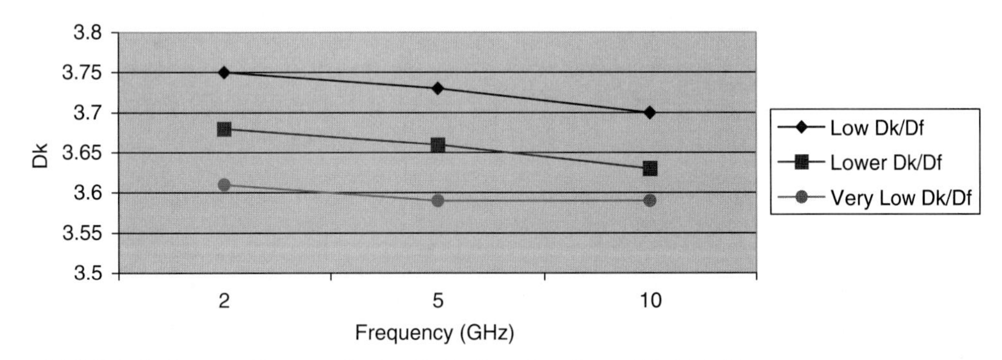

FIGURE 9.33 D_k versus Frequency for Low D_k/D_f, Lead-Free Compatible Materials.

Table 9.13 provides D_k and D_f data for the very low D_k/D_f material by construction and resin content. This sample of constructions illustrates how these properties vary with resin content in addition to frequency. For a full list of constructions available for any product, it is recommended that the manufacturer's data sheets be consulted.

TABLE 9.13 Laminate Constructions and D_k Data For A Low D_k/D_f Material

Thickness (in.)	Construction	Resin Content %	D_k @ 2 GHz	D_k @ 5 GHz	D_k @ 10 GHz	D_f @ 2 GHz	D_f @ 5 GHz	D_f @ 10 GHz
0.0020	1-106	70	3.28	3.24	3.24	0.0058	0.0066	0.0072
0.0027	1-1080	60	3.51	3.47	3.47	0.0059	0.0066	0.0071
0.0030	1-1080	63	3.44	3.40	3.40	0.0059	0.0066	0.0071
0.0035	1-2113	51	3.73	3.70	3.70	0.0060	0.0066	0.0070
0.0040	1-3070	49	3.78	3.75	3.75	0.0060	0.0066	0.0070
0.0043	106/1080	62	3.46	3.42	3.42	0.0059	0.0066	0.0072
0.0050	106/2113	55	3.63	3.59	3.59	0.0060	0.0066	0.0071
0.0060	1080/2113	54	3.65	3.62	3.62	0.0060	0.0066	0.0071
0.0070	2-2113	52	3.70	3.67	3.67	0.0060	0.0066	0.0071
0.0080	2-3070	49	3.78	3.75	3.75	0.0061	0.0066	0.0070

REFERENCES

1. IPC-4101, "Specification for Base Materials for Rigid and Multilayer Printed Boards."
2. Isola Group Product Reference Materials.
3. Clark-Schwebel Industrial Fabrics Guide.
4. BGF Industries, Inc. Fiberglass Guide.
5. Gould Electronics Product Reference Materials.
6. Brist, Gary, Hall, Stephen, Clouser, Sidney, and Liang Tao, "Non-Classical Conductor Losses Due to Copper Foil Roughness and Treatment," ECWC 10/IPC Expo 2005.
7. Kelley, Edward J. and Micha, Richard A., "Improved Printed Circuit Manufacturing with Reverse Treated Copper Foils," IPC Printed Circuits Expo, March 1997.
8. IPC-TM-650 Test Methods Manual.
9. IPC-4562 Metal Foil for Printed Wiring Applications.
10. Kelley, Edward and Christofferson, Owen, "Multilayer Printed Circuits with Exacting Registration Requirements, Achieving Next Generation Design Capabilities," IPC Printed Circuits Expo, May 1998.
11. Kelley, Edward, "Meeting the Needs of the Density Revolution with Non-Woven Fiberglass Reinforced Laminates," EIPC/Productronica Conference, November, 1999.

CHAPTER 10
THE IMPACT OF LEAD-FREE ASSEMBLY ON BASE MATERIALS

Edward J. Kelley
Isola Group, Chandler, Arizona

10.1 INTRODUCTION

The European Union's Restriction of Hazardous Substances (RoHS) directive has had a profound impact on all levels of the electronics industry supply chain, including the base materials used in printed circuit boards (PCBs). While debate over the net environmental benefits of this initiative continues, it is clear that all levels of the supply chain must deal with the implications. The July 2006 deadline for the first group of products that must comply has come and gone, and deadlines for other product groups are established. Indeed, many of the product groups with exemptions that allow for more time to comply are being forced to transition ahead of schedule due to component availability and constraints within the assembly services industry. The product groups granted initial exemptions are generally those with the most stringent reliability requirements, and many involve complex PCB designs that even further complicate the issues with respect to base materials. This chapter will present the key issues related to base materials, important properties with respect to lead-free assembly compatibility, and examples of material evaluation. Further discussion of material selection is presented in Chap.11.

10.2 ROHS BASICS

RoHS restricts the use of:

- Lead (Pb)
- Cadmium (Cd)
- Mercury (Hg)
- Hexavalent chromium (Cr^{VI})
- Polybrominated biphenyl (PBB)
- Polybrominated diphenyl ethers (PBDE)

The first four substances are metals used for a variety of applications, whereas the last two are generally used as flame retardants in plastic materials. It is important to note that these halogenated flame retardants are not generally used in laminate materials for printed circuits. In addition, bromine (Br), the halogen used in these flame retardants, is itself not restricted.

Chapter 7 discussed flame retardants in more detail. The key point to be made here is that the most common flame retardant used in base materials, tetrabromo-bisphenol-A (TBBPA), is not restricted by RoHS. Furthermore, in FR-4 materials, TBBPA is generally reacted into one of the epoxy resins, and thus does not exist as a free-standing molecule, but is incorporated into the molecular backbone of the resin system.

The metals that are restricted are not used in base materials either. So from the standpoint of compliance, most laminate materials are acceptable. The key issue is the restriction on the use of lead, and the implications that this has for assembly of components onto PCBs. Eutectic tin/lead (Sn/Pb) has been the primary solder alloy used in assembly of printed circuits, and has a melting point of 183°C. Peak assembly temperatures using eutectic tin/lead commonly reach 230–235°C. With the elimination of lead from electronics assembly, alternative solder alloys have been developed. These lead-free alloys have higher melting points. Tin/silver/copper (Sn/Ag/Cu) alloys, also called 'SAC' alloys, are the most common lead-free solder alternatives, with the most common of these alloys having melting points around 217°C and peak assembly temperatures of up to 260°C. Assembly rework temperatures can be even higher. It is the increase in temperature to which PCBs are exposed to that is the primary issue with respect to base materials. In other words, the issue for base materials is not compliance with the restriction on the banned substances, but compatibility with the manufacturing processes and temperatures used.

10.3 BASE MATERIAL COMPATIBILITY ISSUES

The question of base material compatibility is very complex. This is the result of several factors. First, PCB design and construction have a significant impact on the base material properties required. Thin, low-layer-count PCBs may have different requirements than very thick, high-layer-count PCBs. Copper weights, aspect ratios, and other design features also have an impact. End-use application and the associated requirement for long-term reliability and electrical performance also impacts the decision-making process. The requirements for a cell phone, video game, or even a computer motherboard are very different than those for high-end servers, telecommunications gear, avionics, and critical medical and automotive electronics. Finally, not all lead-free assembly processes are the same. Some designs experience peak temperatures of around 245°C, whereas others experience peak temperatures of up to 260°C, or even higher in some cases. Some PCBs may experience two to three thermal cycles, others up to five, six, or even more depending on how many reworks are allowed. All of this makes it impossible to recommend one base material type for all applications without either underspecifying the laminate material and risking defects during assembly or later on in the field, or overspecifying the material and paying too much for the material or limiting availability.

Beyond the complexity introduced by the PCB design and construction, there are fundamentally two issues:

- Surviving lead-free assembly processes without defects
- Maintaining the required level of long-term reliability for the given application

The critical base material properties required for each are summarized here, and discussed in more detail later in this chapter. The impact on the base material components is discussed in Section 10.4.

10.3.1 Surviving Lead-Free Assembly without Defects

The primary defects related to base materials that can occur during lead-free assembly include measling, blistering, or delamination. These defects are related to the thermomechanical properties of the base material as well as the interaction of these properties with any absorbed moisture that may be contained within the PCB during assembly, which can be volatilized as the PCB is heated. Fundamentally, these defects represent either the creation of voids within the PCB due to the volatilization of moisture, retained solvents, other organic components including the by-products of resin decomposition, or the loss of adhesion between the fundamental base material components: glass cloth, resin system, and copper. The adhesion loss and subsequent blistering or delamination may be the result of pressure from volatilized components, mismatches between component thermal expansion rates, degradation of the components at the interfaces, or combinations of these factors. Figure 10.1 is an example of delaminations within a PCB after lead-free assembly processing. This PCB used a standard dicyandiamide (dicy) cured, high-T_g FR-4 base material. The left and right-side images in this figure are from the same area of the PCB, but photographed with different lighting to highlight the specific interfaces where delamination occurred.

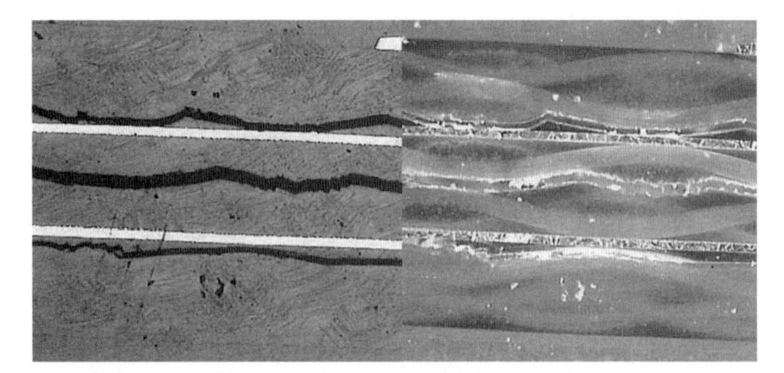

FIGURE 10.1 PCB delamination after lead-free assembly.

The key base material properties that relate to survivability in lead-free assembly processes include:

- Decomposition temperature (T_d)
- Coefficients of thermal expansion (CTE_s)
- Moisture absorption
- Time to delamination

Time to delamination, or T260 time, which is the time it takes for a sample to delaminate when held at an isotherm temperature of 260°C, is not a fundamental base material property, but is a functional test related to the other material properties. In multilayer PCBs, this property can be affected by the quality of the innerlayer copper surface preparation and multilayer lamination quality, among other things.

10.3.2 Lead-Free Assembly and Long-Term Reliability Issues

The second fundamental issue for base materials and PCBs is understanding the impact to long-term reliability assuming a PCB has survived assembly. Other things being equal, the long-term reliability of a PCB assembled in a lead-free process with higher temperatures is not as good as a PCB assembled in a tin-lead assembly process with lower temperatures. The magnitude of the difference is dependent on many factors, but they include the type of base material used. The key properties of the base material with respect to long-term reliability include:

- CTEs
- Glass transition temperature (T_g)
- Decomposition temperature (T_d)
- Moisture absorption

Additionally, the specific construction of the PCB in terms of resin content, the types of fiberglass cloth used, and the adhesion and long-term thermochemical compatibility between the resin system and copper which is plated within vias can have a significant impact on long-term reliability. Some of these factors relate to the impact they have on CTEs, both in the z-axis as well as in the x- and y-axes. For example, using a very high resin content prepreg layer adjacent to a very low resin content layer increases the stress at the interface of these layers due to the differences in thermal expansion that result from the differences in resin-to-glass concentrations. The higher temperatures of lead-free assembly result in a greater absolute level of expansion or stress at these interfaces, which can lead to blistering or delamination. The distribution of the glass filaments within the resin system can have a similar effect. Within a given layer or prepreg, if there are "resin-rich" versus "glass-rich" areas, the differences in thermal expansion between these areas cause stresses that can lead to delamination, or at the interface with a plated-through hole (PTH) will result in points of stress-strain discontinuity resulting from differences in the z-axis expansion values. It is not uncommon to see cracks in the PTH copper begin at these points and ultimately lead to failure.

Section 10.5 discusses the material properties relevant to both surviving lead-free assembly as well as ensuring long-term reliability. Although the test methods used to assess long-term reliability are beyond the scope of this chapter, where relevant, a brief description will be provided to help explain performance differences. However, before discussing these properties in more detail, Section 10.4 presents some of the concerns related to the components used to make base materials.

10.4 THE IMPACT OF LEAD-FREE ASSEMBLY ON BASE MATERIAL COMPONENTS

The higher temperatures of lead-free assembly have an impact on each of the three main components of laminate materials: the resin system, the glass cloth, and the copper foil. Table 10.1 summarizes the key issues for each of these main components.

10.5 CRITICAL BASE MATERIAL PROPERTIES

Substantial work describing the impact of lead-free assembly temperatures on base materials and finished PWBs has been gathered.[2,3,4,5,6,7] These works have identified the critical base material properties that must be considered when selecting a material for lead-free assembly applications, though work on this subject is continuing. Table 10.2 provides a summary of these properties.

A very critical point that must be emphasized with respect to these issues is that although laminate manufacturers can easily make improvements in one property, doing so often can adversely affect other important properties. For example, it is relatively easy to formulate a resin system with a very high time to delaminate in conventional T260 or T288 tests, or to engineer a resin system with a very high decomposition temperature. However, this is often

TABLE 10.1 Primary Lead-Free Issues for Base Material Components

Component	Lead-Free Assembly Impact	Potential Solutions	Related Considerations
Resin System	1. Peak assembly temperatures can reach point where resin decomposition begins.	1. Formulate resin system with higher decomposition temperatures.	1. Reformulation can adversely affect electrical properties and manufacturability.
	2. Higher temperatures result in increased thermal expansion and stress on plated holes as a result.	2. Formulate for lower coefficients of thermal expansion.	2. Can also impact mechanical properties and manufacturability.
	3. Vapor pressure of absorbed moisture is much higher at lead-free assembly temperatures; can lead to blistering/delamination.	3. Evaluate materials for moisture absorption/release characteristics; drying processes in PCB fabrication and/or assembly.	3. PCB storage conditions during manufacturing and prior to assembly are much more important, especially humidity conditions.
	4. "Phenolic" lead-free compatible materials are often not as good for electrical performance, especially D_f.	4. Evaluate non-dicy/non-phenolic laminate materials.	4. Cost/performance trade-off. New materials now available that offer more choices.
Fiberglass Cloth	Thermal and mechanical stress on resin-to-glass bond. Volatilized components can also stress this bond.	Cleanliness, moisture resistance, and choice of proper coupling agent for resin adhesion are more important.	Loss of resin-to-glass adhesion through thermal cycling could impact CAF resistance.
Copper Foil	Thermal and mechanical stress on resin-to-copper bond.	Copper nodularization and roughness, treatments for improved adhesion, i.e. coupling agents.	Roughness impacts signal attenuation, especially at high frequencies.[1]

achieved at the expense of mechanical properties and can make the material more difficult to use successfully in conventional PWB manufacturing processes, or without sacrificing design flexibility. Interactions between these and other material properties such as hardness, modulus, and fracture toughness with the processes used in PCB fabrication can greatly influence the reliability of a PCB during and after assembly. For example, an increase in decomposition temperature with a simultaneous decrease in fracture toughness does little good if the PCB exhibits cracking or fracturing during scoring processes or similar defects during the mechanical drilling of tight pitch holes for PTHs. Therefore, achieving the best balance of properties to meet the needs of the OEM, Electronic Manufacturing Services company (EMS), and PWB manufacturer is absolutely critical.

10.5.1 The Historical Focus on T_g

Chapter 6 explained how and why T_g has historically been one of the properties used to classify base materials, while Chap. 8 provided additional information on the measurement of T_g. In short, the historical focus on T_g as an indicator of reliability is probably the result of its effect on total z-axis expansion. However, because of differences in CTE values, even the relationship

TABLE 10.2 Critical Base Material Properties for Lead-Free Assembly Applications

Property	Definition	Issue
Decomposition Temperature, T_d	Measurement of weight loss from resin system degradation as a function of temperature. T_d is typically defined as the point at which 5 percent of the original mass is lost to decomposition, but other levels can also be reported, e.g. 1 percent, 2 percent, or "onset."	Resin decomposition can result in adhesion loss and delamination. A 5 percent level of decomposition is severe, and intermediate levels are important for assessing reliability since peak temperatures in lead-free assembly can reach onset points of decomposition. A high T_d by the 5 percent definition does not guarantee performance. Conversely, a low T_d by the 5 percent definition is not necessarily bad if the onset temperature of decomposition is high enough.
Glass Transition Temperature, T_g	Thermodynamic change in polymer from a relatively rigid, glassy state, to a softened, more deformable state.	Several properties change as the T_g is exceeded, including the rate at which a material expands versus temperature. Modulus also decreases significantly as T_g is exceeded.
Coefficients of Thermal Expansion, CTEs	Rate of change in physical dimension (in x-, y-, or z-axis) as a function of temperature, expressed as a coefficient of thermal expansion. Can also be expressed as a percentage expansion over a given temperature range.	Resin system CTE values above T_g are much higher than below T_g. Z-axis expansion induces stress on plated vias. The higher temperatures of lead-free assembly result in more total expansion for a given material. Several mature lead-free-compatible materials incorporate inorganic fillers that reduce CTE values. X- and y-axis CTEs are also important for reliability at component and layer interfaces.
Moisture Absorption	Tendency of a material to absorb moisture from the surrounding environment. Can be assessed by more than one method, including water soak or in an increased pressure and humidity environment.	Vapor pressure of water is much higher at lead-free assembly temperatures. Absorbed moisture can volatilize during thermal cycling and cause voiding or delamination. PWBs that initially pass lead-free assembly testing may exhibit defects after storage in an uncontrolled environment as a result of moisture absorption. This should be considered when evaluating materials and PWB designs.
Time to Delamination	Not a fundamental property; measurement of the time for delamination to occur at a specific temperature, e.g. 260°C (T260) or 288°C (T288).	This property is related to decomposition temperature and adhesion between material components. Thermal expansion and moisture absorption can also influence results. In multilayer PWBs, the treatment of the internal copper surfaces is also critical, among other factors.

between T_g and thermal expansion requires further analysis. Figure 10.2 illustrates the relationship between T_g and thermal expansion for the products in Table 10.3.

From Fig. 10.2, products with the same pre- and post-T_g CTE values (A and C, for example; note that above T_g the lines are parallel) differ in total expansion based on their T_g values. For example, the conventional 175°C T_g material (C) exhibits less total expansion than the conventional 140°C T_g material (A) because the onset of the higher post-T_g rate of expansion is delayed by 35°C. However, the 175°C material with a lower CTE value (D) exhibits much less total expansion than the conventional 175°C T_g material even though the T_g values are the same. Furthermore, the 150°C T_g material with reduced CTE values (B) exhibits approximately the same total expansion (3.4 percent) as the conventional 175°C T_g

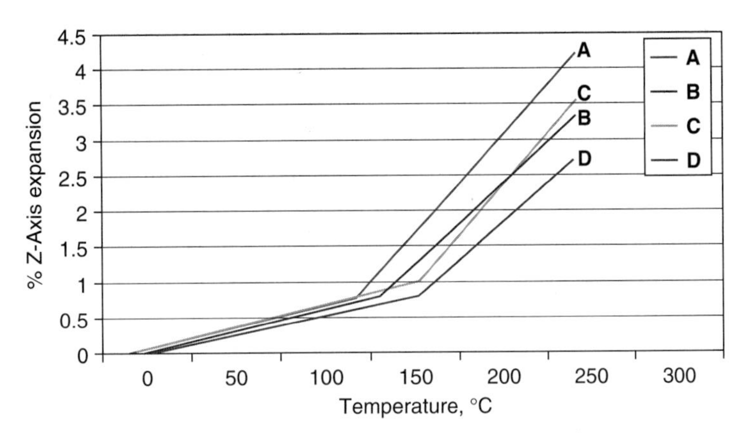

FIGURE 10.2 The impact of T_g and CTE values on total expansion.

material (3.5 percent). However, with a decomposition temperature that is significantly higher, this mid-T_g FR-4 material is much more compatible with lead-free assembly than the conventional 175°C T_g material.

10.5.2 The Importance of Decomposition Temperature

Although glass transition temperature (T_g) and z-axis expansion have been a primary focus of attention over the years, only with the introduction of lead-free assembly has the decomposition temperature (T_d) gained significant attention. The decomposition temperature has always been important in terms of reliability, yet most people have used T_g as a proxy for material reliability. One reason for this is that other things being equal, a higher-T_g results in less total thermal expansion, and therefore less stress on plated vias. What wasn't discussed is that it is common for conventional dicy-cured, high-T_g FR-4 materials to exhibit somewhat lower decomposition temperatures than conventional dicy-cured, 140°C T_g FR-4 materials. This is highlighted by the fact that most conventional 140°C T_g FR-4 materials exhibit longer T260 times than conventional high-T_g FR-4 materials. To highlight the importance of T_d, examine Fig. 10.3.

The traditional dicy-cured, high-T_g FR-4 materials we have become familiar with have T_d values in the range of 290–310°C. Traditional 140°C T_g FR-4 materials are generally somewhat higher, with a typical example of a material with a T_d of 320°C shown in Fig. 10.3. In the typical tin-lead assembly environment, peak temperatures do not reach the point where decomposition is significant for either the traditional or enhanced products. However, in the lead-free assembly environment, peak temperatures reach the point where a small but significant level of decomposition can occur for the conventional materials, but not for the enhanced products.

TABLE 10.3 Properties of Some Common FR-4 Base Materials

Product	Description	Glass Transition Temp. (°C)	Decomposition Temp. (°C)	% Expansion, 50–260°C (40% Resin Content)
A	Conventional Dicy-Cured 140°C T_g FR-4	140	320	4.2
B	Phenolic" Mid-T_g FR-4	150	335	3.4
C	Conventional Dicy-Cured High-T_g FR-4	175	310	3.5
D	Improved High-T_g FR-4	175	335	2.8

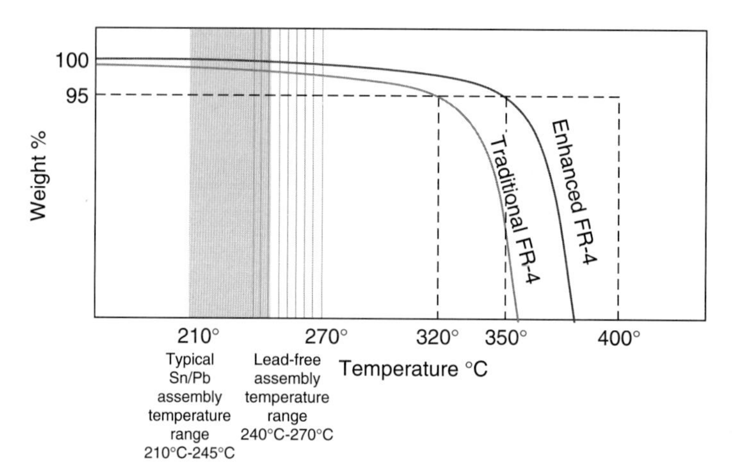

FIGURE 10.3 Decomposition curves for a traditional and improved FR-4 material.

This seemingly small level of decomposition in the conventional products can have extremely significant effects on reliability, especially if multiple thermal cycles are experienced.

On the other hand, as long as the temperatures experienced during assembly and rework do not exceed the temperature at which decomposition begins, there may not be further benefit to higher decomposition temperatures. In addition, the resin system modifications often made to increase T_d can sometimes lead to other problems, such as brittleness or hardness of the resin system that negatively impacts PCB manufacturability (as in drilling, scoring, and routing) or fracture toughness in tight pitch PTH areas. As not all lead-free-compatible materials use the same resin systems, great care should be used in selecting a material that balances the requirements of the OEM, assembler, and PCB fabricator. A simple review of a material data sheet or IPC slash sheet is not sufficient to guarantee success. Involvement with the base material supplier to understand fully the capabilities of a material and the processing required for success is even more important for lead-free applications. To expand on this, examine Figs. 10.4 through 10.6, which were also presented in Chap. 8.

First, note the slope of the decomposition curve in Fig. 10.6. By the 5 percent decomposition definition, this material has a T_d of 405°C. However, looking at 1 percent and 2 percent values, the temperatures are 336°C and 367°C, which are significantly lower than a reported 5 percent value of 405°C. This is not to say that a 1 percent T_d of 336°C is bad. In fact, this material has exhibited exceptional reliability in lead-free applications. However, it does point to the need to understand decomposition levels other than the 5 percent level. In contrast, the material in Fig. 10.5 has a 5 percent T_d of 330°C with 1 percent and 2 percent T_d's of 326°C and 328°C, respectively. In this case, there is a very small difference between the 1 percent, 2 percent, and 5 percent values, and this material has also proven to have excellent performance in a broad range of lead-free assembly applications. Last, the standard high-T_g material in Fig. 10.4 exhibits a 5 percent T_d of 305°C with 1 percent and 2 percent values of 297°C and 301°C. This is also a small range of values, but in this case the absolute values are lower and begin to approach the range of temperatures that can be experienced, at least locally, in a printed circuit board that undergoes lead-free assembly and rework. Materials such as this have not performed with adequate reliability to be considered for anything but the simplest lead-free applications. In addition, if PCBs made with these materials are stored in uncontrolled environments and are allowed to absorb even a modest level of moisture, the combination of lower T_d's with the higher vapor pressure of water at lead-free temperatures can lead to catastrophic defects in assembly or severely degraded long-term reliability.

To take this a step further, examine the materials listed in Table 10.3. These are four FR-4 materials, with material C being the same material as shown if Fig. 10.4, and material D being

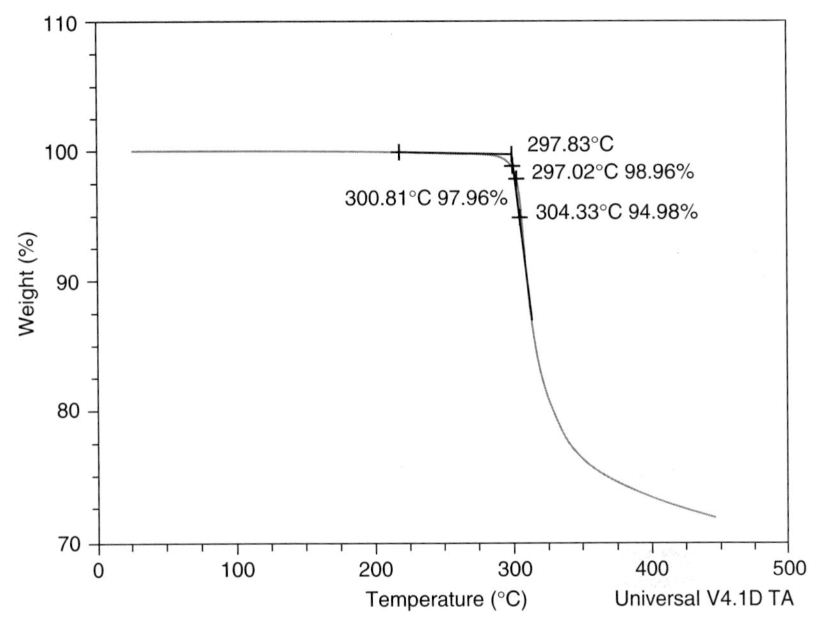

FIGURE 10.4 Decomposition curve for a standard dicy-cured high-T_g FR-4.

the same as the material in Fig. 10.5. The impact of decomposition temperature when these materials are exposed to multiple thermal cycles to different peak temperatures is highlighted in Figs. 10.7 and 10.8. Figure 10.7 graphs cumulative weight loss (decomposition) for these materials when cycled repeatedly to a peak temperature of 235°C. Clearly, there is little impact on

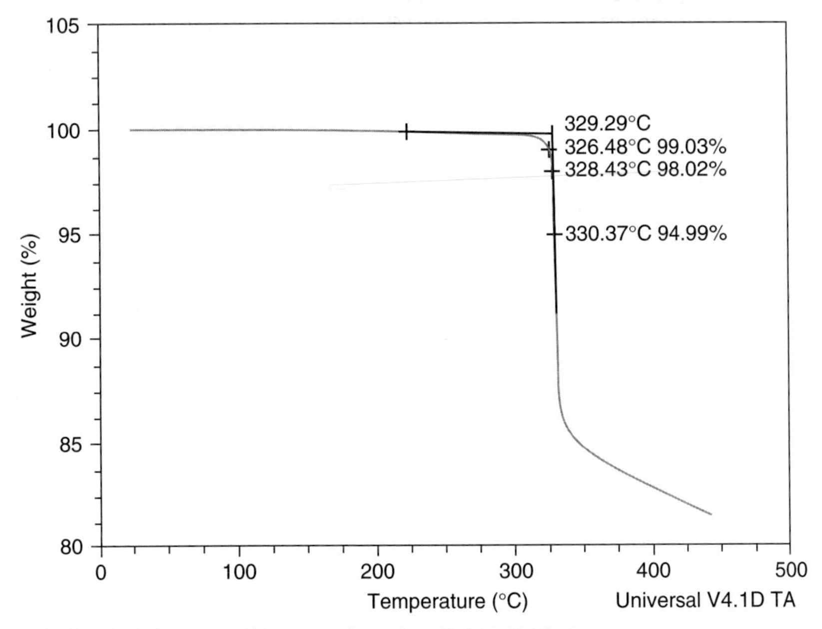

FIGURE 10.5 Decomposition curve for a phenolic high-T_g FR-4.

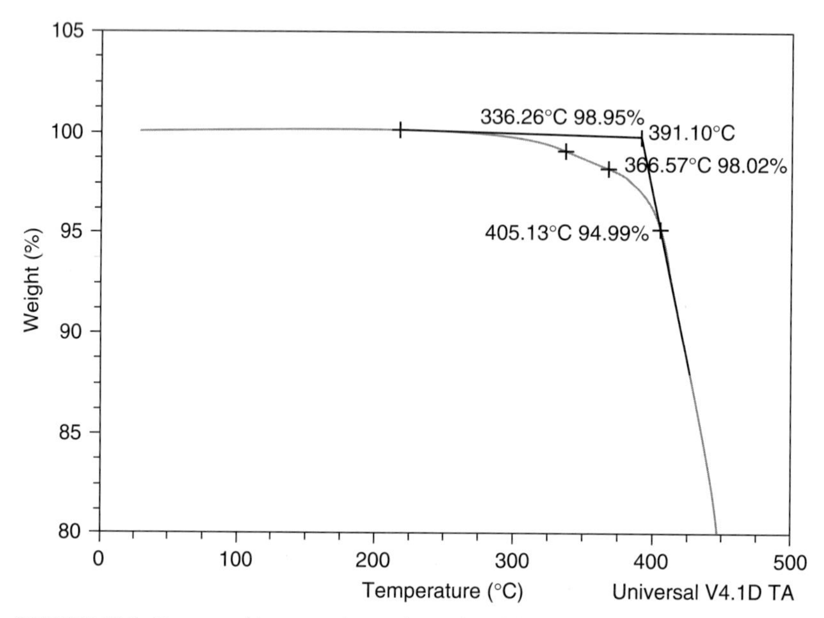

FIGURE 10.6 Decomposition curve for an alternative high-T_g/high-T_d material.

resin decomposition when the peak temperature reaches 235°C. Figure 10.8 presents the same results when the peak temperature is increased to 260°C. The increase in temperature to 260°C has a severe impact on resin decomposition for the traditional FR-4 materials as they experience multiple thermal cycles, especially the conventional high-T_g material (product C). The rapid degradation in material C after only a few thermal cycles highlights why this material is not recommended for lead-free assembly applications. Indeed, as more data are gathered, thermal cycling seems to have a much greater impact than even one cycle to a very high temperature, even if the cycling occurs for a long time. As an example, review Fig. 10.9, which plots decomposition versus time at different isotherm temperatures for a standard dicy-cured high-T_g FR-4 material (C in Figs. 10.7 and 10.8). In Fig. 10.9 it can be seen that relatively low levels of decomposition are experienced for extended time periods even at the 260°C and 275°C isotherm temperatures. This same material, C in Fig. 10.8, exhibits rapid decomposition when cycled up to 260°C and cooled to room temperature multiple times.

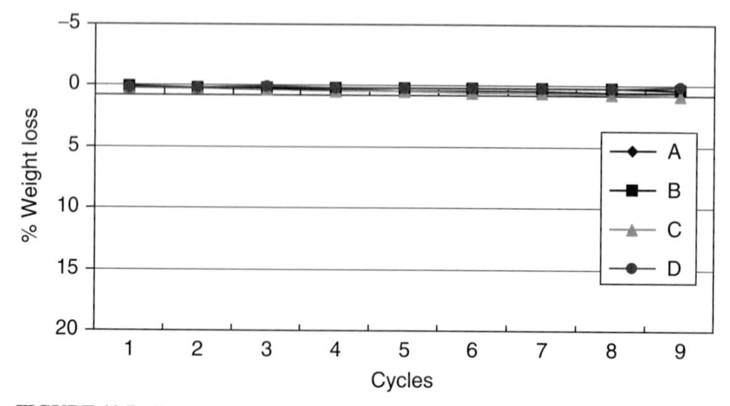

FIGURE 10.7 Decomposition through multiple cycles to 235°C.

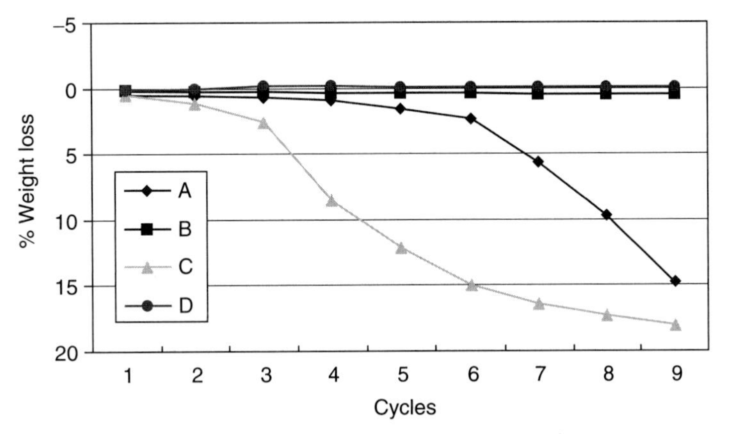

FIGURE 10.8 Decomposition through multiple cycles to 260°C.

10.5.3 Moisture Absorption

As pointed out in Table 10.2, the vapor pressure of water at lead-free assembly temperatures—260°C, for example—is much higher than at eutectic tin-lead assembly temperatures such as 230°C. Figure 10.10 plots the vapor pressure of water, in both mm Hg and psi, versus temperature. At 230°C the vapor pressure of water is near 400 psi. At 260°C, it is close to 700 psi. Therefore, any absorbed moisture within a PCB during assembly can have a much greater impact in lead-free assembly, as the greater pressure stresses the adhesion between the base material components and can also create small voids within the resin system.

This means that much more care must be used in selecting materials for their moisture absorption properties. However, all common base materials absorb some level of moisture.

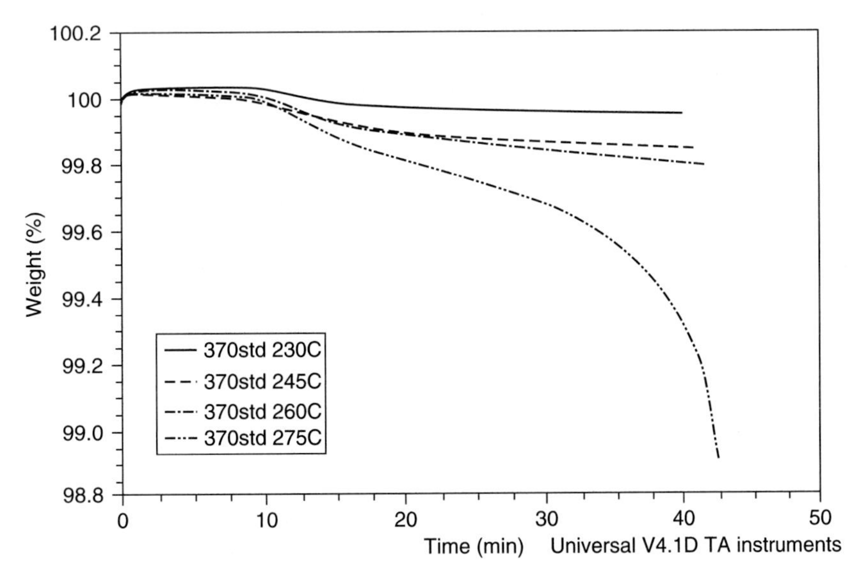

FIGURE 10.9 Decomposition versus time at different isotherm temperatures..

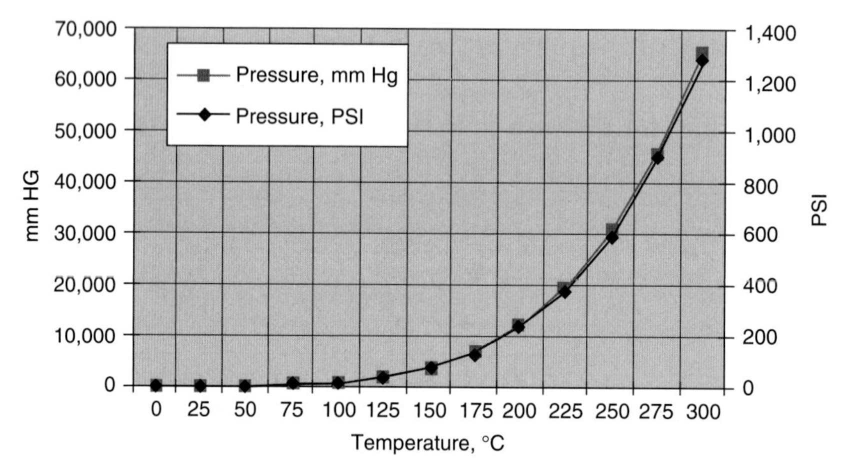

FIGURE 10.10 Vapor pressure of water versus temperature.

So the storage of these materials, both during manufacturing of the PCB as well as prior to assembly, must also be examined. Additional drying or baking steps may be needed in some applications to drive off any absorbed moisture prior to exposure to high temperatures. Additionally, it can be difficult to correlate actual performance with the moisture absorption data commonly found on material data sheets. One reason for this is that it may not be just the moisture absorption that is important, but the rate at which moisture can be driven from the material as well. As an illustration, consider the data in Fig. 10.11.

These materials were conditioned for 60 minutes in 15 psi steam and then solder shocked for 20 seconds at 288°C. Their moisture absorption was measured and compared to their

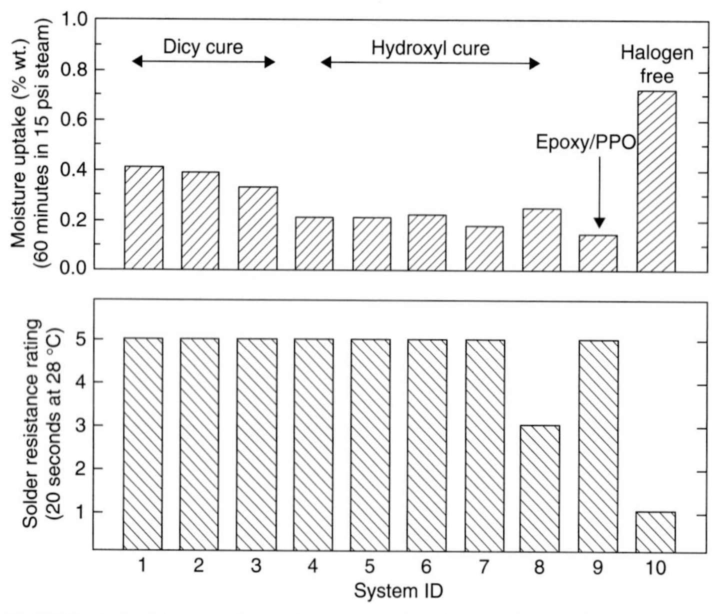

FIGURE 10.11 Moisture absorption versus solder shock resistance for several materials.

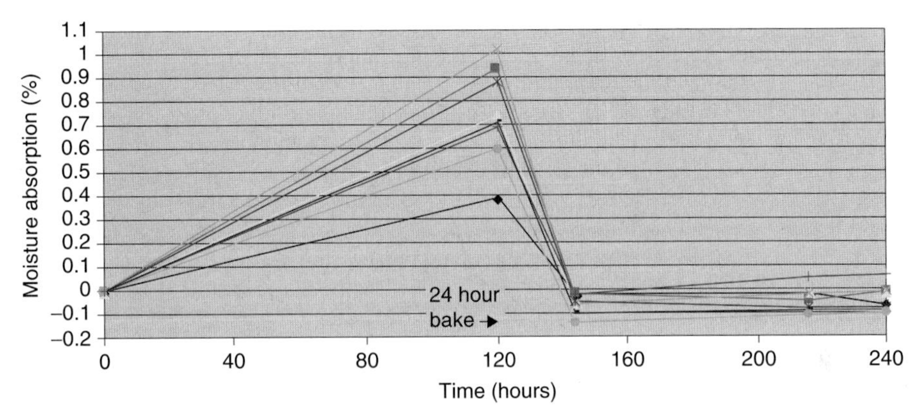

FIGURE 10.12 Percent weight change after a five-day ambient moisture test and 125°C bake.

performance in the solder shock testing. The particular halogen-free material shown in this graph exhibits a very high level of moisture absorption and does indeed exhibit poor solder shock resistance. However, with the exception of only one material, all the other materials exhibited good solder shock resistance in this test, even though their moisture absorption levels ranged from less than 0.2 percent to just over 0.4 percent. The one other material that exhibited some level of defects had an intermediate level of moisture absorption, around 0.3 percent. The point is that it is very difficult to draw conclusions about material performance based on reported levels of moisture absorption. Additional reliability testing relevant to the application being considered is needed to make good decisions on material compatibility.

With respect to moisture absorption and the subsequent drying or baking of PCBs, consider the tests summarized by Figs. 10.12 and 10.13. In each test, several different types of base material were evaluated, including dicy-cured and phenolic-cured FR-4 materials, as well as low D_k/D_f and halogen-free materials. Figure 10.12 shows moisture absorption data for samples that were dried, then soaked in water at room temperature for five days, measured, and then baked at 125°C for 24, 96, and 120 hours. Samples from this test were then submitted to 35°C and 85 percent RH for seven days, measured, and then baked at 125°C for 24 and 48 hours. The results of this second test are shown in Fig. 10.13. Although these graphs show that the different types of materials may absorb somewhat different amounts of moisture, in each case the moisture was

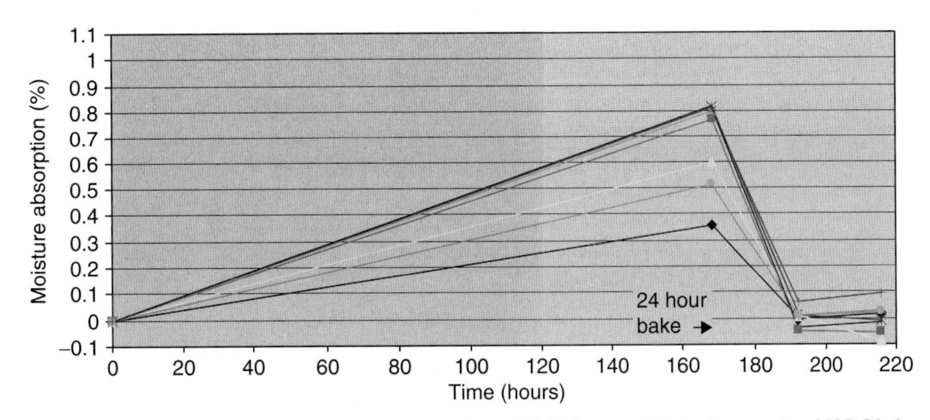

FIGURE 10.13 Percent weight change after seven days at 35°C/85 percent RH moisture test and 125°C bake.

able to be driven off by baking. Beyond this, little correlation could be found between the relative levels of moisture absorption and performance in lead-free assembly applications. Although the materials tested here that showed relatively low levels of moisture absorption have performed well in lead-free assembly applications, some that showed relatively high levels of moisture absorption in these tests have also performed well. The materials in this group that have not done well in lead-free applications fell somewhere in the middle in terms of relative moisture absorption. Thus, although moisture absorption is a critical property, it must also be evaluated within the context of other properties.

10.5.4 Time to Delamination

The time to delamination, and how it is measured, was discussed in further detail in Chap. 8. This test method involves placing a sample of laminate or PCB in a thermomechanical analyzer (TMA) and heating the sample to a specific isothermal temperature, most commonly 260°C (T260) or 288°C (T288). The time it takes for the sample to delaminate is then measured and reported. Figure 10.14 is an example of a complex backplane PCB that was subjected to T260 testing. This PCB used a conventional dicy-cured high-T_g FR-4 material and delaminated before reaching 2 minutes. Note the charring and degradation of the resin system. Alternate resin systems, including alternate types of FR-4 materials, can offer significantly improved performance. Figure 10.15 compares time-to-delamination performance at different temperatures for four different resin systems: two dicy-cured FR-4s and two that use an alternate type of curing mechanism. Note that the materials using an alternative curing mechanism

FIGURE 10.14 Delamination and resin decomposition after T260 testing.

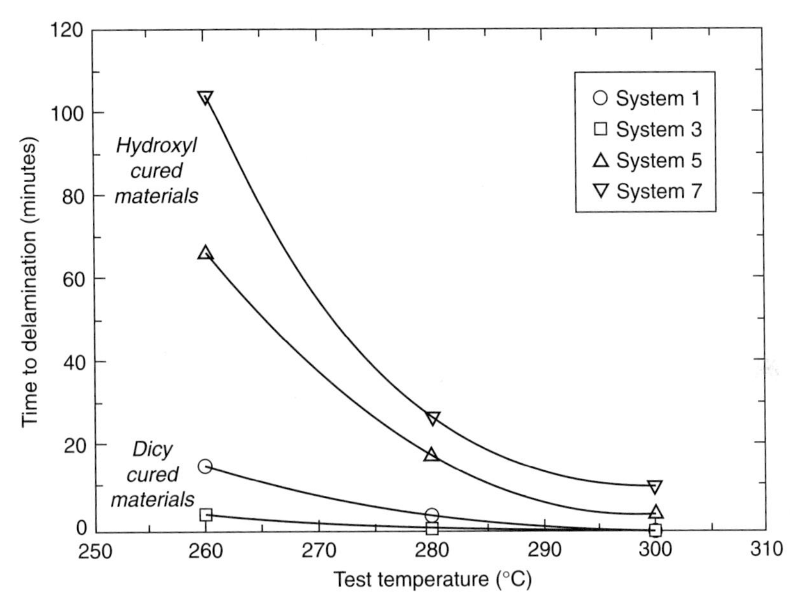

FIGURE 10.15 Time to delamination versus temperature for different resin systems.

offer much longer times to delamination at each temperature, although all systems exhibit much shorter times as the temperature is increased.

Times to delamination also vary based on the type of sample measured. Laminates with the copper foil removed generally exhibit longer times than laminates with copper foil. Clad laminates in turn exhibit longer times than multilayer PCBs. Figure 10.16 provides data on clad versus unclad samples for several material types in T260 and T288 testing. In multilayer PCBs, usually the interface between the resin and the internal copper surfaces will fail first. Time to delamination in multilayer PCBs is affected by the quality of the surface treatment of

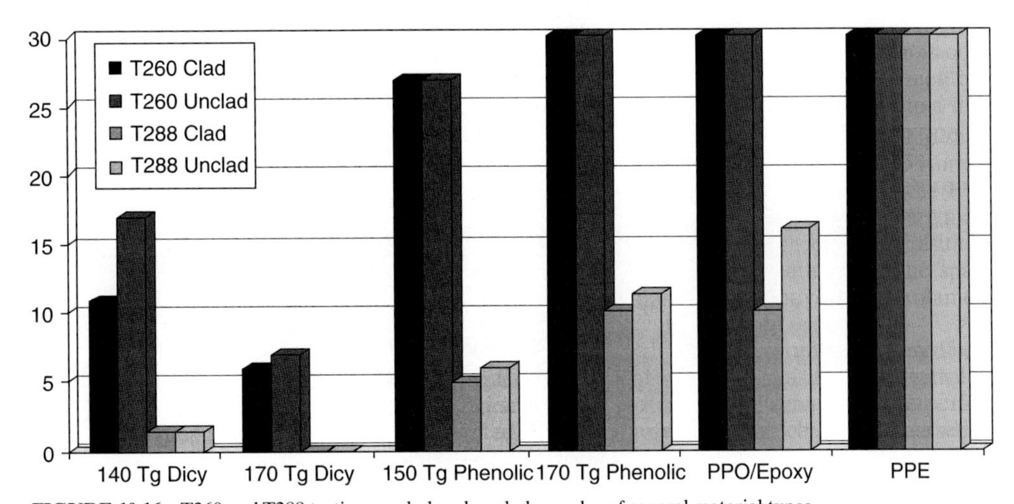

FIGURE 10.16 T260 and T288 testing on clad and unclad samples of several material types.

the internal copper layers, the quality of the multilayer lamination processing, as well as the type and condition of the bonding prepreg.

Time to delamination testing of multilayer PCBs has become a common test for lead-free assembly compatibility. However, several factors beyond the base material type can have an influence on performance. So care needs to be taken in assessing why a particular PCB might exhibit a low time to delamination. On the other hand, if a particular type of base material exhibits low times to delamination, little can be done in PCB manufacturing to improve performance. In short, a given material offers a certain performance entitlement. Whether this performance level is realized in the finished PCB is dependent on several factors, primarily related to PCB manufacturing processing.

10.5.5 The Impact of Lead-Free Assembly on Other Properties

Exposure to assembly temperatures can have a significant effect on the properties of base materials, including the T_g and modulus properties of the material. This is particularly true if the material is exposed to multiple thermal cycles. In addition, the magnitude of the effect is dependent on the specific peak temperature experienced. Figures 10.17 and 10.18 compare the effect of multiple thermal cycles on dynamic mechanical analysis (DMA) T_g for various material types when cycled to 235°C (Fig. 10.17) and 260°C (Fig. 10.18). When cycled repeatedly to 235°C, none of these materials exhibits a significant change in DMA T_g. However, when cycled repeatedly to 260°C, the dicy-cured materials show significant degradation in the measured DMA T_g, especially the high-T_g dicy-cured material.

Figures 10.19 and 10.20 present similar data for measurements of the percent change in DMA modulus through multiple thermal cycles. Note that even when cycled to a peak temperature of 235°C, the high T_g dicy-cured FR-4 material exhibits a consistent decline in DMA modulus. The 140°C T_g dicy-cured material exhibits a decrease only after several cycles.

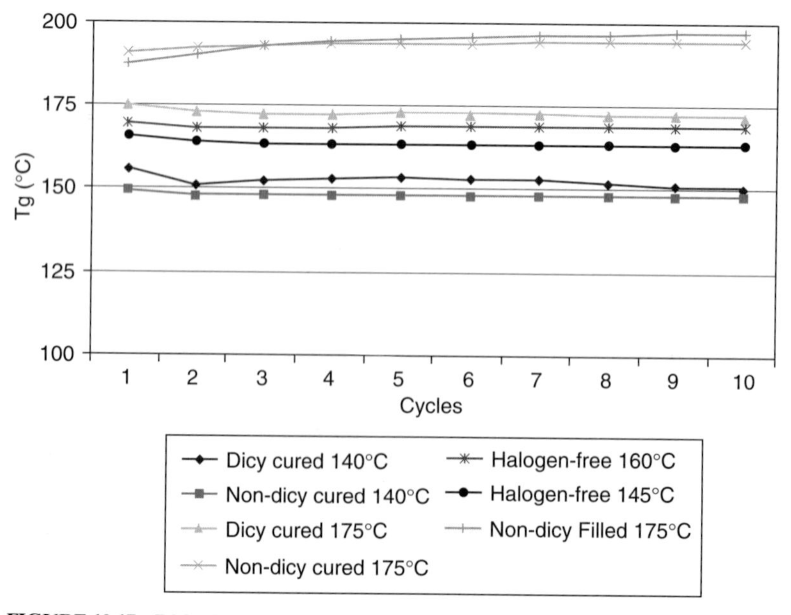

FIGURE 10.17 DMA T_g versus multiple thermal cycles to 235°C.

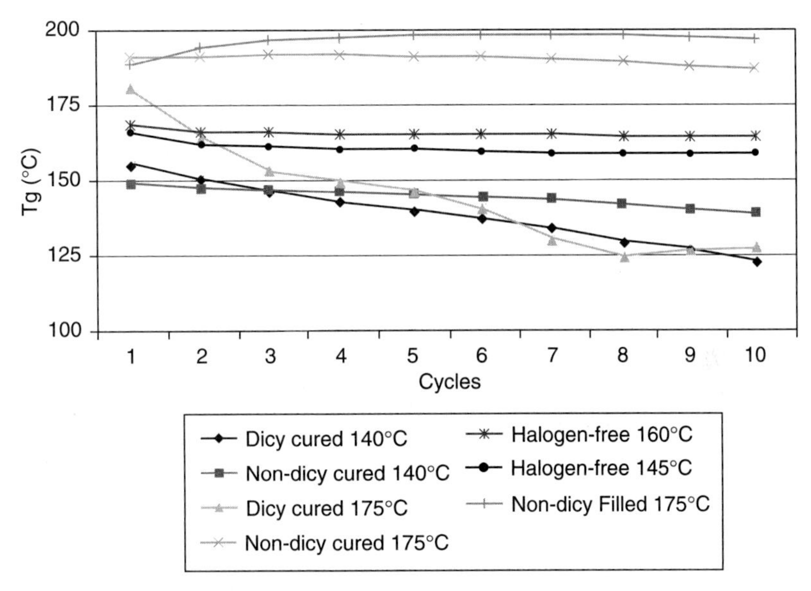

FIGURE 10.18 DMA T_g versus multiple thermal cycles to 260°C.

When cycled to a peak temperature of 260°C, both of the dicy-cured FR-4 materials exhibit rapid declines in DMA modulus. In all of these cases, the non-dicy (phenolic) FR-4 materials show little impact whether cycled to 235°C or 260°C. If anything, they exhibit very slight increases in T_g and modulus.

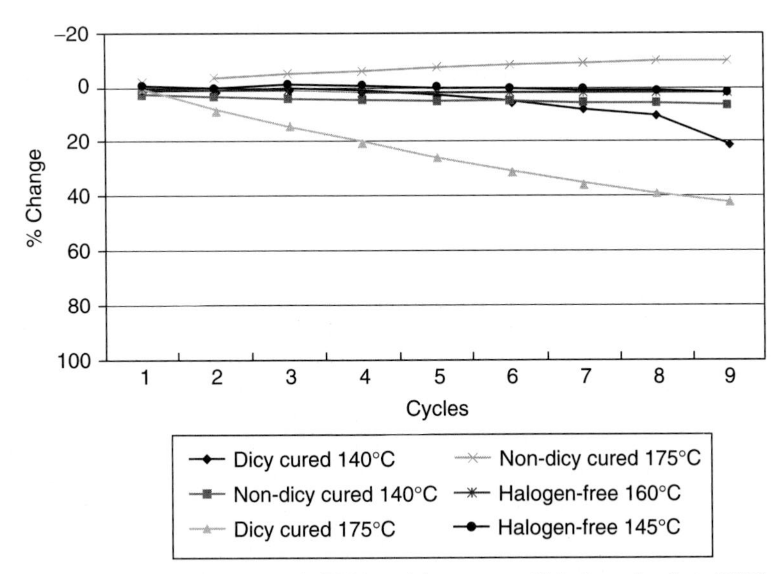

FIGURE 10.19 Percent change in DMA modulus versus multiple thermal cycles to 235°C.

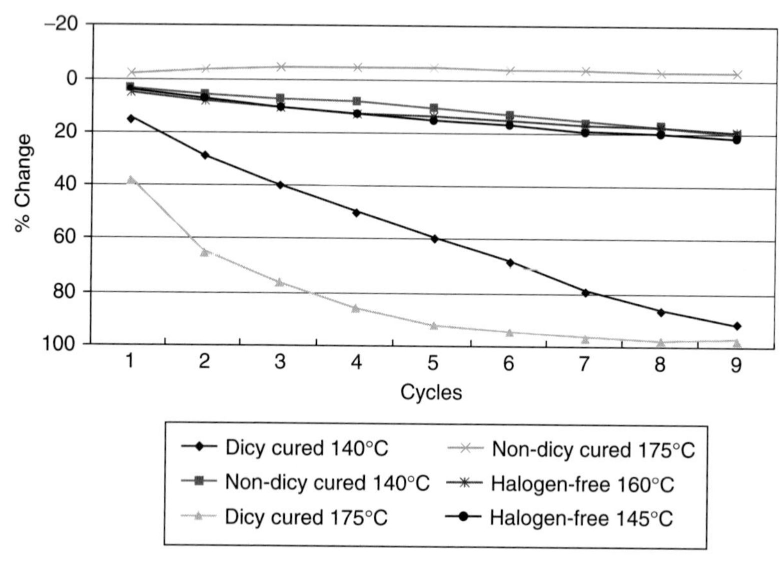

FIGURE 10.20 Percent change in DMA modulus versus multiple thermal cycles to 260°C.

10.6 IMPACT ON PRINTED CIRCUIT RELIABILITY AND MATERIAL SELECTION

In addition to the works already cited, there have been excellent studies on the impact of lead-free assembly, specifically on PWB reliability.[8,9,10] These works present statistical analyses showing the impact of lead-free assembly on PWB reliability and reach important conclusions regarding the base materials. Although there is not perfect agreement among all published works, the differences typically are the result of a different focus—for example, whether the focus is on complex, thick PWBs with stringent reliability requirements versus relatively less complex PWBs with shorter intended field lifetimes or less stringent reliability requirements. Conclusions include:

- A minimum decomposition temperature is critical for lead-free assembly compatibility, although higher T_d's are not *always* better. Trade-offs with other properties, such as manufacturability, fracture toughness, and so on, make achieving the right balance of properties critical.
- T_g and CTE values are important because of the effects on thermal expansion, especially in thicker PWBs.
- Most conventional (dicy-cured) high-T_g FR-4 materials are generally not compatible with lead-free assembly, or can be used successfully only in a very limited range of applications.
- Conventional 140°C T_g materials may still be suitable for PWB designs with limited thickness and reliability requirements, particularly when intermediate peak temperatures are used in assembly. This is largely the result of these materials having slightly higher decomposition temperatures than the higher-T_g equivalents.
- Mid-T_g FR-4 materials with high decomposition temperatures are viable products for many lead-free assembly applications involving intermediate-complexity PWB designs.
- Materials with a sufficiently high-decomposition temperature, high-T_g, and reduced CTE values are suitable for the broadest range of applications, including complex PWBs assembled at 260°C peak temperatures.

- Balancing material properties with PWB manufacturability is critical. Materials that exhibit excellent properties have failed because of difficulties experienced when fabricating the PWB, such as fracturing in drilling, routing or scoring, difficulty in texturing drilled holes for copper plating, resin recession, or hole wall pull-away during thermal stress.

10.6.1 Example of Material Types and Properties versus Assembly Reliability

To highlight a couple of these conclusions, consider the following test. First, multilayer PWBs made from the materials in Table 10.3 were processed through Infra-Red (IR) reflow cycles at different peak temperatures. The PWB was a 10-layer, 0.093-in. (2.6 mm) thick board "designed to fail," meaning the copper weights and patterns, construction, and resin contents were chosen so that the board would be more sensitive to thermal cycles. In addition, the dwell time at the peak temperature was 1.5 minutes. This allowed differences in material performance to be detected more clearly. Figure 10.21 graphs the percentage of boards surviving six reflow cycles without any evidence of blisters, measles, or delamination.

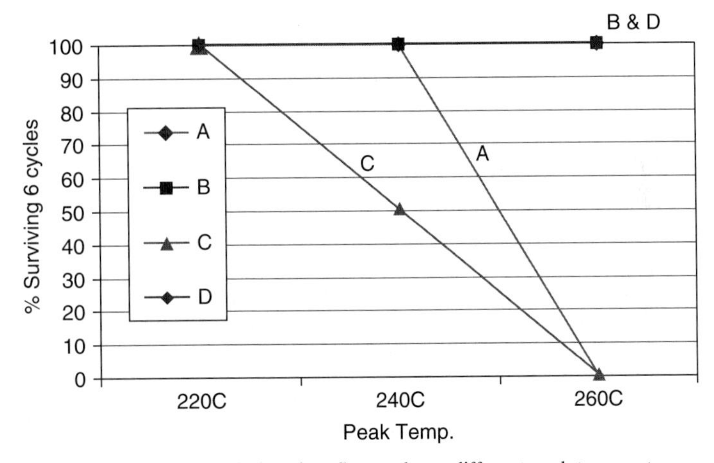

FIGURE 10.21 Survival after six reflow cycles at different peak temperatures.

Notably, the first material to exhibit defects is the conventional dicy-cured, high-T_g FR-4 material. This material began to exhibit defects when the peak temperature reached 240°C. At a peak temperature of 260°C, the conventional dicy-cured FR-4 materials, both the 175°C T_g and the 140°C T_g products, all exhibited evidence of defects. On the other hand, the materials with higher decomposition temperatures, both the 150°C T_g and the 175°C T_g products, all survived six cycles to 260°C.

10.6.2 Example of Material Types and Properties versus Long-Term Reliability

In another test, three high-T_g materials were evaluated through Interconnect Stress Test (IST) testing.[11] This particular test provided insight into the effect of thermal expansion and decomposition temperature on long-term reliability, as assessed by the IST method. The IST test method uses an electric current to heat test coupons that contain a network of plated through holes. The test samples are generally preconditioned several times to simulate exposure to

TABLE 10.4 Materials Evaluated through IST Testing

Product	Description	Glass Transition Temp. (°C)	Decomposition Temp. (°C)	% Expansion, 50–260°C (40% Resin Content)
C	Conventional High-T_g	175	310	3.5
D*	High-T_g/High-T_d	175	335	3.4
D	High-T_g/High-T_d/ Reduced CTE	175	335	2.7

assembly processes, and then cycled back and forth between an elevated temperature, most commonly 150°C, and room temperature. The samples are cycled in this manner until failure occurs, with failure generally being defined as a change in measured resistance of 10 percent. The materials evaluated in this example are described in Table 10.4.

Note that the T_g values of these materials are the same, but differences exist in decomposition temperatures and thermal expansion values. Product D* is similar to product D except that it has a higher level of thermal expansion. Product D* exhibits approximately the same thermal expansion as Product C, but Product D* has a higher decomposition temperature. Product D has both a high decomposition temperature and a very low level of thermal expansion. The PWB tested was a 14-layer, 0.120-in. (3.1 mm) thick multilayer with 0.012-in. (0.30 mm) diameter plated through holes. The average copper plating in the via was 0.8 mil (20.3 micron), although 1.0 mil (25.4 micron) had been requested. Figure 10.22 charts the average number of cycles to failure (10 percent resistance change in the plated via net) for each material type at each preconditioning level: as is (no preconditioning), three cycles to 230°C, six cycles to 230°C, three cycles to 255°C, and six cycles to 255°C.

Clearly, the two materials with improved decomposition temperatures exhibit much better performance than the conventional high-T_g product. Also, in comparing product D to product D*, it appears that the lower thermal expansion of product D does offer improvement in the number of cycles to failure, but this improvement is smaller in comparison to the improvement due to the higher decomposition temperature, at least for this PWB design. The benefit of reduced thermal expansion becomes more important as the thickness of the PWB increases. In addition, the technique used to reduce the CTE values in this case also provides benefits in PWB manufacturability.

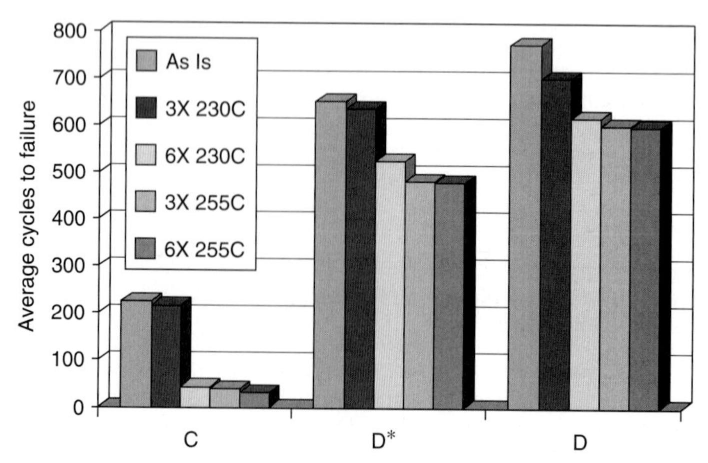

FIGURE 10.22 IST results for three 175°C T_g materials.

10.6.3 Understanding the Potential Impact on Electrical Performance

Most of the FR-4 laminate materials developed for lead-free assembly applications use an alternative resin chemistry in comparison to the conventional dicyandiamide (dicy) cured FR-4s. The most common alternatives are commonly referred to as phenolic or novolac cured materials. Although there is some variation between these materials, as a group they tend to exhibit somewhat different electrical performance, particularly with respect to dielectric loss, or D_f (dissipation factor). For most applications operating in typical frequency ranges, the differences are not significant. However, as operating frequencies increase toward the higher end of FR-4 applications and impedance control becomes more critical, these differences can become very significant. Table 10.5 includes the same materials as Table 10.3, but also shows the differences in dielectric constant (D_k) and dissipation factor (D_f). It also includes a lead-free-compatible material designed to improve D_k and D_f performance in comparison to the phenolic lead-free-compatible materials. Note that different types of measurement systems can result in different measured values for D_k and D_f. Laminate resin contents and other factors also influence these properties. So when examining Table 10.5, the comparisons between these materials are more important than the absolute values reported for each material. For these comparisons, the same measurement system and resin contents were used.

TABLE 10.5 Properties of Several Base Material Types

Product	Description	Glass Transition Temp. (°C)	Decomp. Temp. (°C)	% Expansion, 50–260°C (40% RC)	D_k@ 2 GHz	D_k@ 5 GHz	D_f@ 2 GHz	D_f@ 5 GHz
A	Conventional 140°C T_g FR-4	140	320	4.2	3.9	3.8	0.021	0.022
B	Improved Mid-T_g FR-4	150	335	3.4	4.0	3.9	0.026	0.027
C	Conventional High-T_g FR-4	175	310	3.5	3.8	3.7	0.020	0.021
D	Improved High-T_g FR-4	175	335	2.8	3.9	3.8	0.026	0.026
E	Non-Dicy, Non-Phenolic	200	370	2.8	3.7	3.7	0.013	0.014

From this table it is clear that the improved phenolic FR-4 materials are not quite as good in terms of electrical performance in the range of 2–5 GHz as the conventional dicy-cured materials, particularly with respect to D_f. However, material E exhibits thermal properties at least as good if not better than even the improved materials discussed up to this point. In addition, the electrical performance of material E is even better than the conventional dicy-cured materials, especially in terms of D_f performance. This non-dicy, non-phenolic material is an attractive material for use in applications requiring not only lead-free compatibility, but electrical performance superior to the phenolic materials that have been adopted in existing lead-free designs.

10.7 SUMMARY

While most base materials comply with the RoHS directive, the question of compatibility with lead-free assembly processes is a more complex issue. The material properties that are important for lead-free assembly compatibility include:

- Decomposition temperature (T_d)
- Coefficients of thermal expansion (CTEs)

- Glass transition temperature (T_g) (especially because of the impact on thermal expansion)
- Moisture absorption
- Time-to-delamination performance, which is not a fundamental material property, but is a simple method used to assess thermal stability of a composite material at different temperatures

While laminate manufacturers can easily make improvements in one of these properties, it is not as easy to make improvements without affecting other properties, including properties important for ease of manufacturing in the PCB fabrication process. Achieving the optimal balance of properties that consider the requirements of each level of the supply chain—from OEM to EMS to PCB fabricator—is crucial for success in lead-free assembly applications.

Conventional dicy-cured FR-4 materials, especially the high-T_g dicy-cured materials, are significantly impacted by lead-free assembly temperatures, and are generally not recommended for lead-free applications. Alternative resin systems, especially the phenolic-cured FR-4 materials, have gained widespread acceptance in these applications, and where both thermal performance and electrical performance are critical, non-dicy, non-phenolic materials are also available. Chapter 11 discusses material selection in lead-free applications in more detail.

REFERENCES

1. Brist, Gary, Hall, Stephen, Clauser, Sidney, and Liang, Tao, "Non-Classical Conductor Losses Due to Copper Foil Roughness and Treatment," ECWC 10/IPC/APEX Conference, February 2005.
2. Bergum, Erik, "Application of Thermal Analysis Techniques to Determine Performance of Base Materials through Assembly," IPC Expo Technical Conference Proceedings, Spring 2003.
3. Kelley, Edward, "An Assessment of the Impact of Lead-Free Assembly Processes on Base Material and PCB Reliability," IPC/Soldertec Conference, Amsterdam, June 2004.
4. Hoevel. Dr. Bernd, "Resin Developments Targeting Lead-Free and Low D_k Requirements," EIPC Conference, 2005.
5. Christiansen, Walter, Shirrell, Dave, Aguirre, Beth, and Wilkins, Jeanine, "Thermal Stability of Electrical Grade Laminates Based on Epoxy Resins," IPC Printed Circuits Expo, Anaheim, CA, Spring 2001.
6. Kelley, Ed, Bergum, Erik, Humby, David, Hornsby, Ron, Varnell, William, "Lead-Free Assembly: Identifying Compatible Base Materials for Your Application," IPC/Apex Technical Conference, February 2006.
7. St. Cyr, Valerie A., "New Laminates for High Reliability Printed Circuit Boards." Proceedings of IPC Technical Conference, February, 2006
8. Freda, Michael, and Furlong, Jason, "Application of Reliability/Survival Statistics to Analyze Interconnect Stress Test Data to Make Life Predictions on Complex, Lead-Free Printed Circuit Assemblies," EPC 2004, October 2004.
9. Brist, Gary, and Long, Gary, "Lead-Free Product Transition: Impact on Printed Circuit Board Design and Material Selection," ECWC 10/APEX/IPC Conference, February 2005.
10. Ehrler, Sylvia, "Compatibility of Epoxy-Based PCBs to Lead-Free Assembly," EIPC Winter Conference, 2005, *Circuitree,* June 2005.
11. IST procedure developed and offered through PWB Interconnect Solutions, Inc., www.pwbcorp.com.

CHAPTER 11
SELECTING BASE MATERIALS FOR LEAD-FREE ASSEMBLY APPLICATIONS

Edward J. Kelley
Isola Group, Chandler, Arizona

11.1 INTRODUCTION

Chapter 10 discussed the impact of lead-free assembly on printed circuit boards (PCBs) and base materials, and also discussed many of the critical base material properties relevant to performance in lead-free applications. These properties included:

- Decomposition temperature (T_d)
- Glass transition temperature (T_g)
- Coefficients of thermal expansion (CTEs)
- Moisture absorption
- Time to delamination, such as T260 and T288 tests

Also, the need to achieve the right balance of properties with ease of use in the PCB fabrication process, in addition to the requirements of assemblers and original equipment managers (OEMs), was highlighted.

This is the result of the impact that PCB fabrication can have on the performance of the material and finished PCB during assembly and in end-use application. A given material type offers a certain level of performance in the finished PCB. However, this level of performance may or may not be realized, depending on how the materials were processed during fabrication. This makes it extremely complicated to make general recommendations on what material should be used in a given application. However, this chapter describes an approach that has been used quite successfully in addressing this question. But first, some of the PCB fabrication and assembly issues that can affect performance are outlined.

11.2 PCB FABRICATION AND ASSEMBLY INTERACTIONS[1]

The reliability of the assembled PCB in the end-use application is a function not only of the base materials selected, but how the materials were processed during PCB fabrication and component assembly. Different base material types may require different processing conditions in order to achieve the best results in the finished product. In reviewing these items, understand that these are general considerations that should be evaluated within the context of the specific fabrication and assembly processes and with regard to the specific materials being considered. The specific requirements for long-term reliability must also be considered. So while it is not possible to give exact, optimal processing conditions for all applications, the items that follow represent parameters that should be considered as products are converted to lead-free assembly.

11.2.1 PCB Fabrication Considerations

As just stated, selecting the appropriate base material for a given application is only one of several key issues when converting to lead-free assembly. Several factors related to PCB fabrication are also critical.

11.2.2 Moisture Absorption

As discussed in Chap. 10, a given level of moisture absorption can be much more serious in lead-free assembly applications than it is for conventional tin-lead assembly applications. This is due to the increase in water vapor pressure at the higher temperatures of lead-free assembly. Figure 11.1 is a graph showing the vapor pressure of water versus temperature. At lead-free assembly temperatures, the pressure is much higher, which causes substantially more stress within the laminate and on the interfaces within the PCB, including resin-glass, resin-oxide, resin-copper, and resin-resin interfaces. As a result, additional drying or baking processes may be needed to drive off moisture prior to processes involving thermal cycles.

11.2.2.1 PCB Construction and CTE Values. While all PCB constructions contain some level of CTE mismatch between each of the layers, the total amount of expansion and stress at the interfaces are greater in lead-free applications due to the higher temperatures involved. So even though the CTE values may be the same for a given construction, transitioning a part from standard tin-lead assembly to lead-free assembly results in greater levels of stress within the PCB. In some cases, this increased stress can lead to defects such as delamination, especially if combined with the increased stress resulting from any absorbed moisture. These thermal expansion-related stresses can be most severe in hybrid constructions where two or more different types of base material are used in the same construction, particularly if the

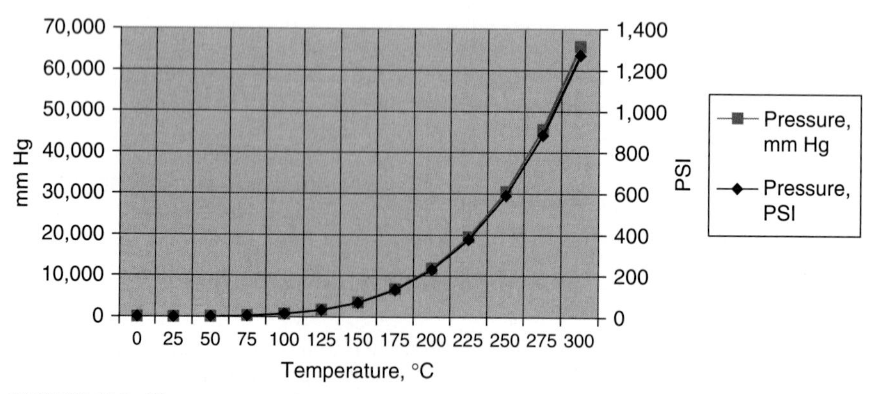

FIGURE 11.1 Vapor pressure of water versus temperature.

TABLE 11.1 X- and Y-axis CTE Values for Some Common Glass Styles

Glass Style	X-axis CTE (ppm/°C)	Y-axis CTE (ppm/°C)
106	22.2	22.2
1080	16.9	19.4
2113	15.3	15.9
2116	14.7	14.9
1652	14.1	14.1
7628	12.1	15.9

CTE values of the two material types are very different. However, even if the same type of material is used throughout the PCB construction, care in selecting the specific glass styles and resin contents should be exercised. This is due to the differences in x-y CTEs between the various glass styles and resin contents that can be used. Adjacent layers with very different CTE values lead to increased levels of stress at higher temperatures, with the possibility of adhesion loss and delamination is the result. Table 11.1 provides x- and y-axis CTE values for some common glass styles to highlight the possible differences. When possible, you should consider choosing glass styles with the smallest differences in CTE values for adjacent layers.

11.2.2.2 Oxide and Oxide Alternative Processes. During PCB fabrication, the internal copper surfaces are generally chemically treated to improve the adhesion between the resin system and the copper layer. Two primary types of processes are used. The first involves an oxidation of the copper surface to micro-roughen the surface, usually followed by a reduction step to improve chemical resistance of this layer in subsequent processes. The second type of process is referred to as an oxide alternative, or oxide replacement process. These processes generally use etching chemistries, which also roughen the copper surface to provide greater surface area for resin bonding, and often use proprietary chemistries designed to further enhance adhesion and chemical resistance. In short, compatibility of the resin system with the oxide or oxide alternative is critical for any application, but becomes even more important in lead-free assembly applications due to the increased stress on the resin to innerlayer copper bond.

Typical validation of this bond includes assessing the peel strength between the oxide or oxide alternative treated copper and the resin system, along with time-to-delamination testing such as T260 and T288 testing. Samples prepared for these tests should use representative prepreg materials and be laminated using the same press cycles used for actual PCB fabrication.

The potential for moisture absorption in these processes is the other critical consideration. Oxide and oxide alternative treatments can be compromised if exposed to high temperatures, or even moderate temperatures for extended times. Oxidation of these surfaces during postdrying or baking can lead to adhesion or chemical resistance problems in subsequent processes. On the other hand, moisture absorbed by the laminate materials during these processes can impact the quality of multilayer lamination and subsequent performance during thermal cycles. In short, it is important to dry the innerlayers as much as possible after these processes while staying within the guidelines provided by the oxide and oxide replacement chemistry supplier.

11.2.2.3 Multilayer Lay-Up Prior to Lamination. In addition to the oxide and oxide alternative processing discussed in the previous section, the hold time between this process and multilayer lay-up and lamination is also important. This is due not only to the possibility of oxidation of the innerlayer surfaces, but to the potential for moisture absorption by the exposed resin surfaces on the innerlayer, for the reasons already discussed. This hold time should be minimized, and the storage of the innerlayers prior to lay-up should be in a temperature- and humidity-controlled environment to reduce the likelihood of absorbing moisture. Typical environmental conditions for this storage are 68°F or 20°C, and a maximum of 50 percent relative humidity (RH), although lower humidity levels are recommended.

Similarly, the prepreg materials used in constructing the multilayer PCB must be stored in a temperature- and humidity-controlled environment. This is due not only to the potential for moisture absorption, but also because the shelf life and performance of the prepreg can be significantly impacted by temperature and humidity conditions prior to use. This can often be a subtle point, but it is a critical one, especially as lead-free applications demand greater performance from the base materials. Again, storage conditions generally should be no worse than 68°F or 20°C and 50 percent RH for typical FR-4 materials, although some high-performance materials can be even more sensitive and perform better if stored in even cooler or drier environments. Some of the high-performance materials may even require vacuum dessication prior to use. This is a fairly common practice for polyimide materials, for example. Finally, storage of the multilayer books prior to lamination should ideally be done in a controlled environment, with the hold time between lay-up and lamination also minimized.

11.2.2.4 Multilayer Lamination Processing. Multilayer lamination cycles must be designed around the specific material being used. However, there are some general recommendations that can be applied to most materials:

- **Pre-lamination vacuum** Expanding on the discussion of moisture absorption, it is important to apply a vacuum to the lamination books prior to ramping up the temperature and curing the resin in the prepreg. This removes moisture that could volatilize during the lamination cycle. Drawing a vacuum can also remove other volatile components such as residual solvents in the prepreg materials. This is an often overlooked factor that can be vitally important. Recommended vacuum times can depend on the actual moisture levels of the materials used, but generally range from 15 to 30 min. It is important during this time that the materials are not heated too much. In addition, as mentioned in the previous section, some materials, such as polyimides, are often vacuum-dessicated prior to lay-up.
- **Heat rise and pressure application** For a given type of resin system, it is important to understand the rheology of the resin in the prepreg as a function of the rate of temperature rise. This, along with the proper application of pressure, ensures adequate filling of the internal circuitry and good wet-out of the resin into the fiberglass filaments. Complete wet-out of the internal circuitry and good resin flow on the oxide or oxide alternative surface are important for reliability in subsequent thermal cycles.
- **Stress relief pressure profiles** It is important to have sufficient pressure on the lamination book, as the resin in the prepregs are heated and begin to flow into the innerlayer circuit patterns. The actual pressure needed can vary with the type of material being used and the level of vacuum achieved during the process. However, after the resin cures to a certain point and stops flowing, a reduction in pressure can help reduce the levels of stress in the multilayer PCB. This can improve reliability in subsequent thermal cycling. The reduction in pressure is typically to the range of 50 to 75 psi.
- **Cool down rate** After the product has been held at the proper temperature for the proper period of time, which will depend on the type of material used, the rate at which the product is cooled is also important. In general, slower cooldown rates will minimize warpage and residual stress in the finished product. However, slower cooldown rates can adversely impact productivity, so a reasonable balance must be identified. In many cases, a slow cooldown through the T_g of the resin system, followed by a somewhat faster cooling rate after the product cools below T_g, represents a good balance of performance and productivity.

11.2.2.5 Drilling. Many of the materials designed for compatibility with lead-free assembly require optimization of drilling parameters. Phenolic and other high-performance resin systems can have higher modulus values and be harder than conventional dicy-cured FR-4 materials.

Undercut drill bits generally provide better hole quality with these materials. Other drill process variables must also be examined, and include:

- Maximum hit counts
- Drill chip loads (feeds and speeds)
- Infeed and retract rates
- Drill resharpenings
- Stack heights
- Vacuum levels
- Entry and backer material types

11.2.2.6 Desmear Processing. Many of the resin systems developed for lead-free assembly compatibility, including the phenolic-type materials, are more chemically resistant than the conventional dicy-cured materials. However, most can still be processed using conventional chemical desmear processes. Adjustments may be needed in the times and/or temperatures used for solvent swelling and permanganate desmear processes, and the supplier of these chemistries should be consulted when transitioning to lead-free compatible materials. The processing guidelines provided by the laminate material supplier will also typically provide recommendations for desmear processing. For some advanced products, plasma desmear may be recommended.

11.2.2.7 Surface Finishes. If a lead-free Hot Air Solder Leveling (HASL) process is used, the higher temperatures required by the lead-free alloys should be considered when selecting the base material. This will be discussed in Section 11.3 as well. The common alternatives to HASL, which are growing in use as lead-free assembly is adopted, typically do not involve a bake process or significant thermal cycle. Although this is good in terms of the reduced thermal cycling exposure to the base material, it also results in one less opportunity for moisture to be removed from the PCB prior to assembly. This should be considered when deciding whether a drying step is needed prior to assembly.

11.2.3 PCB Assembly Considerations

Although a complete discussion of assembly variables is beyond the scope of this chapter, a few key points need to be outlined with respect to material selection and lead-free assembly variables. First, as has been pointed out several times, the storage conditions of the PCB prior to assembly should be examined. Moisture absorption in the PCB prior to assembly can have a significant impact, and in severe cases can lead to delamination within the PCB. Storage in a controlled environment is encouraged, and drying of PCBs prior to assembly may also need to be considered. One of the challenges in drying PCBs prior to assembly is avoiding any negative impact to the surface finish that could adversely affect solderability. Consulting with the supplier of the surface finish supplier is strongly suggested before defining these drying processes. However, Table 11.2 offers some general recommendations for drying steps prior to assembly based on the type of surface finish.

TABLE 11.2 Drying Recommendations Prior to Assembly Based on Surface Finish Type

Final Finish	Temperature	Time (hr)	Comments
Tin	125°C	4	Higher temperature may reduce solderability
Silver	150°C	4	Silver may tarnish, but solderability should not be affected
Nickel/Gold	150°C	4	No issues arise with extended bake on nickel/gold finish
Organic Coating	105°C	2	Extended bake cycles may negatively impact multiple heat cycle assembly processes

Development of the assembly profile is important not just for the obvious requirements of component assembly, but also to ensure that the PCB is not damaged as a result of the thermal exposures. These objectives often compete with each other. Profiles vary with PCB thickness, copper distribution, component densities, and other factors. Defining these profiles to meet both objectives can be a very complex process. But with respect to the impact on the PCB and the base materials, care should be exercised in defining the heat ramp rate, the time the PCB is exposed to the peak temperature, and the cooldown rate. Minimizing temperature gradients across the PCB is important in reducing the stresses that arise from thermal expansion. Unequal copper distribution across the PCB and differences in component mass across the PCB can lead to "hot spots" that can approach decomposition temperatures or create areas of stress resulting from thermal expansion differences. These hot spots or severe temperature gradients can be exacerbated by trying to speed up the process by setting the reflow oven zone temperatures very high. You can reduce these hot spots and thermal gradients by setting up "soak" profiles where the PCB is allowed to stabilize at specific temperatures before heating to the peak temperature.

Cooldown rates must also be controlled for similar reasons. Cooling too fast can lead to significant temperature gradients and induce thermal expansion related stresses that can result in blisters, delamination, or pad cratering, which is fracturing in the PCB base material that propagates underneath a copper pad on the surface of the PCB. Finally, rework procedures must be examined and controlled more strictly in lead-free applications. Control of the rework temperature and time that the PCB is exposed to this temperature is critical. Since rework involves local heating of a specific area of the PCB, the issue of temperature gradients can be even more severe.

For all of these reasons, it is strongly recommended that manufacturers perform studies to assess compatibility among the base material, the PCB fabrication process, and the assembly process. First article qualifications when transitioning specific PCB designs to lead-free assembly are strongly recommended. Success with one PCB design does not necessarily mean that other designs can use the same material or fabrication process successfully. In short, lead-free assembly processes are significantly more demanding on the PCB and base materials used and require extensive engineering work to validate compatibility with the assembly process and the requirements for long-term reliability.

11.3 SELECTING THE RIGHT BASE MATERIAL FOR SPECIFIC APPLICATIONS[2,3]

A very common request from PCB fabricators, electronic manufacturing services (EMS) companies, and OEMs is whether a given material is compatible with lead-free assembly. Although everyone is looking for a simple answer, the solution becomes complex because of the range of PCB designs (board thicknesses, layer counts, aspect ratios, via pitch, and so on) as well as the differences in lead-free assembly processes, such as the specific peak temperature and number of thermal cycles a PCB will experience. Previous sections also outlined how the PCB fabrication process and assembly process can influence compatibility. In addition, people usually want to use the least expensive material that is suitable for a given application. So although it is easy to specify an advanced product for lead-free compatibility, the approach taken here is to balance cost and performance. In an attempt to simplify this discussion, this section describes a material selection tool that has been developed to suggest what materials should be considered for a given application.

This tool is based on data such as that presented in Chap. 10, in the references, as well as empirical results from prototype and production experience with lead-free assembly applications. The experience of several people with "hundreds of years" of combined experience has been leveraged in designing these tools. However, no such tool can address every specific application with 100 percent confidence. In addition, the capabilities of various PCB fabrication

Color code	Application recommendation
Green	Material generally recommended for typical applications of this type
Yellow	Material may be acceptable for applications of this type, but is not generally recommended
Red	Not recommended

FIGURE 11.2 Color code key.

processes can also impact the performance of the finished PCB. So although these tools are based on considerable data and experiences from a number of sources, they are intended to serve as a general guide for typical applications, and as such, it remains the user's responsibility to confirm acceptability of any material recommended. This is particularly true with respect to long-term reliability requirements. For example, the field reliability requirement for a cell-phone PCB is going to be very different for a very complex high-end computer or telecommunications infrastructure PCB.

The intent in developing this tool was to come up with a simple method for dealing with this multitude of variables in PCB design and assembly. Figure 11.2 shows the basic color-coding selected for this. Figure 11.3 shows an example of the actual chart format. In the horizontal axis it divides PCBs into thickness categories, and in the vertical axis it differentiates them by the number of reflow processes.

This format forces definition of a "typical" PCB design for each range of thickness shown on the x-axis. Although this is very difficult and "exceptions to the rules" are evident as soon

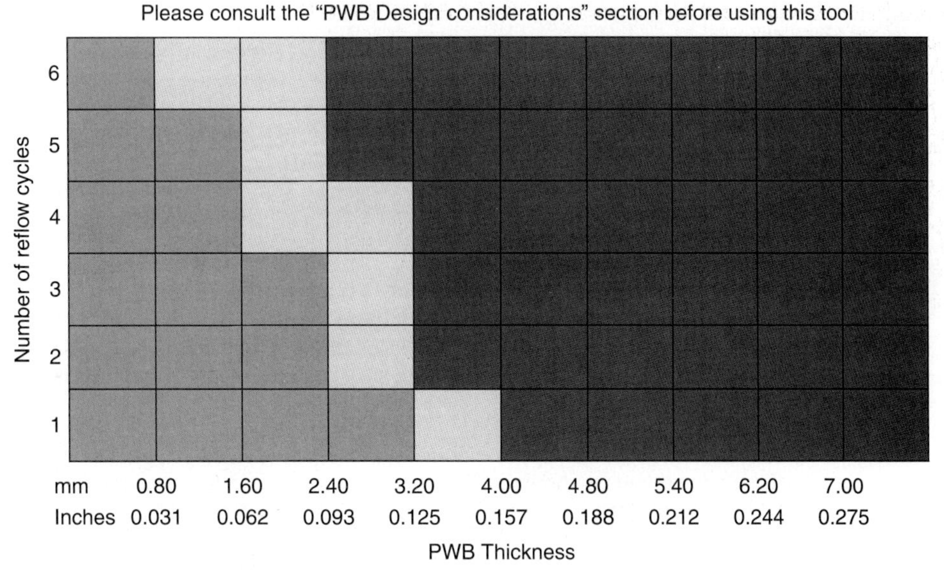

FIGURE 11.3 Example of a chart format.

Layers	2–6	2–8	2–14	2–18	6–22	10–26	10–30	14–34	14–40	14–50
Micro vias	Yes	Yes	No	No	No	No	No	No	No	No
Cu Wt (oz)	<2	<2	<2	<2	<2	<2	<2	<2	<2	<2
RC %	35–55	35–55	35–55	35–55	45–60	45–60	45–60	45–60	45–60	45–60
Aspect ratio	<3:1	<5:1	<8:1	<10:1	<10:1	<10:1	<10:1	<10:1	<10:1	<10:1
Retained Cu %	<50	<25	<25	<25	<25	<25	<25	<25	<25	<25

mm 0.80 1.60 2.40 3.20 4.00 4.80 5.40 6.20 7.00
Inches 0.031 0.062 0.093 0.125 0.157 0.188 0.212 0.244 0.275

PWB Thickness

PTH Cu (μm)	>18	>18	>25	>25	>25	>25	>25	>25	>25	>25
Surface finish	Ni/Au, Silver, Tin, OSP									
Lamination cycles	1	1	1	1	1	1	1	1	1	1
Mixed materials	No	No	No	No	No	No	No	No	No	No
Blind and buried vias	No	No	No	No	No	No	No	No	No	No
External planes	No	No	No	No	No	No	No	No	No	No

mm 0.80 1.60 2.40 3.20 4.00 4.80 5.40 6.20 7.00
Inches 0.031 0.062 0.093 0.125 0.157 0.188 0.212 0.244 0.275

PWB Thickness

FIGURE 11.4 "Typical" PWB features for the selection tool.

as a "typical" definition is provided, these definitions represent fair descriptions of a broad range of products. In addition, you will see that an attempt is made to define a procedure for accommodating the exceptions. Figure 11.4 outlines the "typical" features for PCBs in each thickness range in the charts.

To accommodate the exceptions to these criteria, a method was developed to "adjust" the selection tool based on specific design features or processing conditions. The basic concept for these adjustments is shown in Fig. 11.5, and the specific adjustments are shown in Fig. 11.6.

For the purposes of illustration, the materials outlined in Table 10.5, presented here as Table 11.3.

Figure 11.7 shows the charts for product A with two different assembly temperature ranges, 210 to 235°C for tin-lead assembly and 235 to 260°C for lead-free assembly.

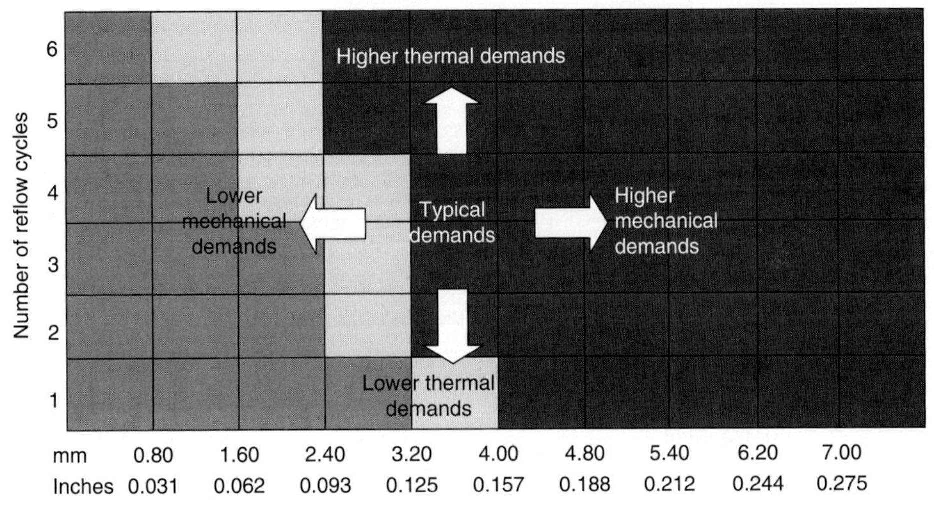

FIGURE 11.5 Concept for adjustments to material selection.

As you can see when these charts are presented side by side, the range of designs for which this product is deemed suitable decreases as the assembly temperature increases. This should be expected based on the earlier discussion of material properties. The other key point is that there may still be a group of products, albeit more limited, where standard 140°C T_g material may be adequate and the most cost-effective option. This may help clarify some of the confusion about whether standard FR-4 materials are lead-free-compatible or not. For specific designs with specific requirements for reliability, the answer is yes. For other designs, applications, or reliability requirements, the answer is no. The value of this tool is that it attempts to define the range of PCB designs where specific materials should be considered.

Figure 11.8 presents similar charts for product C, which is the higher-T_g conventional FR-4 material. This is where the conversation gets very complicated. In Table 11.3, you can see that the decomposition temperature for this product is 310°C, which is the lowest of the materials described. On the one hand, the higher-T_g of this product helps reduce the total amount of z-axis expansion and therefore stress on plated vias. On the other hand, the lower decomposition temperature makes this material the most sensitive to higher assembly temperatures. In fact, the temperature range of 235 to 260°C is a very broad one for this product. Limited success may be seen when assembling at the low end of this range, but as temperatures increase, especially toward the high end of this range, this product is simply not recommended, due to the potential for resin decomposition and resultant defects. In fact, for these reasons and the potential interactions with PCB fabrication and assembly processes, not to mention the potential for moisture absorption to further impact performance, standard high-T_g FR-4 materials should not be considered at all for lead-free applications. Products such as B, D, and E are significantly more robust in these applications and offer attractive cost-performance relationships. In summary, you should exercise extreme caution when considering a conventional high-T_g FR-4 for a lead-free application and should discuss these issues with your laminate material supplier. In contrast, consider the charts for products B, D, and E in Figs. 11.9, 11.10, and 11.11, respectively.

PWB Adjustments for tool reference

Please consult the "PWB Design considerations" section before using this tool

Layers	If greater than typical consider moving up and right on charts
Micro vias	For PWBs thicker than 1.60 mm (0.062 inches) additional evaluation may be required
Cu Wt	If greater than 2 Oz (70 μm) consider moving up and right on charts
RC	If greater than maximum range consider moving right on charts
Aspect ratio	If greater than maximum range consider moving right on charts
Retained Cu	If greater than maximum consider moving up and right on charts

PTH Cu (μm)	If less than typical consider moving up and right on charts
Surface finish	If HASL or reflowed solder consider moving up on chart for each cycle (treat as an additional reflow cycle)
Multiple lamination cycles	If multiple lamination cycles consider moving up on chart for each additional cycle (treat as an additional reflow cycle)
Mixed materials	If mixed material use lowest performing material as reference and consider moving up and right on that chart
Blind and buried vias	If yes consider moving up and right on charts
External planes	If yes consider moving up and right on charts

FIGURE 11.6 Adjustments based on design features or process conditions.

TABLE 11.3 Examples for a Lead-Free Material Selection Tool

Product	Description	Glass Transition Temp. (°C)	Decomp. Temp. (°C)	% Expansion, 50–260°C (40% RC)	D_k@ 2 GHz	D_k@ 5 GHz	D_f@ 2 GHz	D_f@5 GHz
A	Conventional 140°C T_g FR-4	140	320	4.2	3.9	3.8	0.021	0.022
B	Phenolic Mid-T_g FR-4	150	335	3.4	4.0	3.9	0.026	0.027
C	Conventional High-T_g FR-4	175	310	3.5	3.8	3.7	0.020	0.021
D	Phenolic High-T_g FR-4	175	335	2.8	3.9	3.8	0.026	0.026
E	Non-Dicy, Non-Phenolic	200	370	2.8	3.7	3.7	0.013	0.014

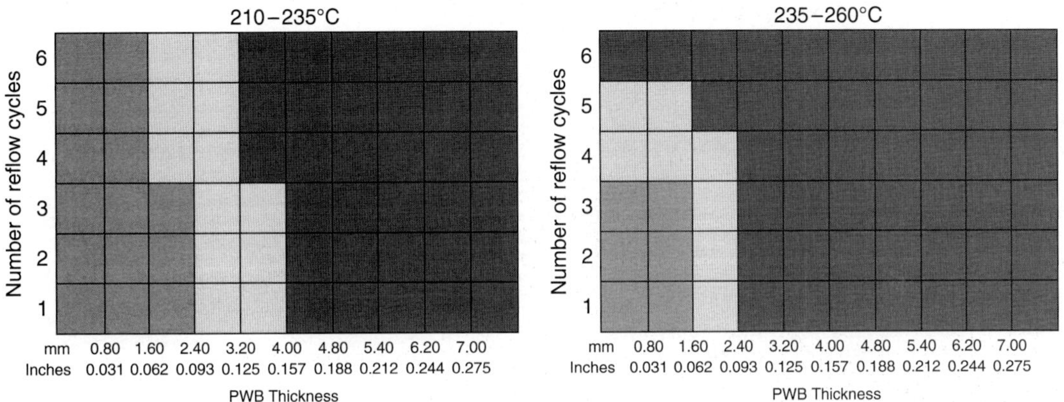

FIGURE 11.7 Charts for product A (140°C T_g conventional FR-4).

Product B, while having a T_g of 150°C, is better suited for lead-free assembly than product C, with its 175°C T_g. This is due the fact that product B has a decomposition temperature approximately 25°C higher than product C. In addition, as shown in Table 11.3, products B and C exhibit approximately the same total amount of thermal expansion in the range from 50 to 260°C. The charts for the different peak temperatures are shown to be the same for product B in Fig. 11.9. This is partly due to the limited experience to date with this product in thick PWB designs. In addition, at lower peak temperatures, the number of reflow cycles that this product

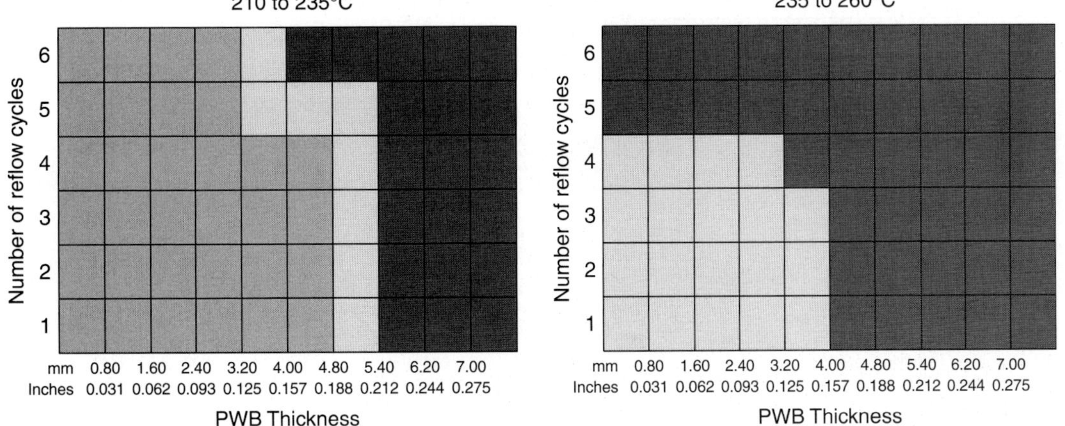

FIGURE 11.8 Charts for product C (175°C T_g conventional FR-4).

Peak reflow temperature

Please consult the "PWB Design considerations" section before using this tool

210–235°C

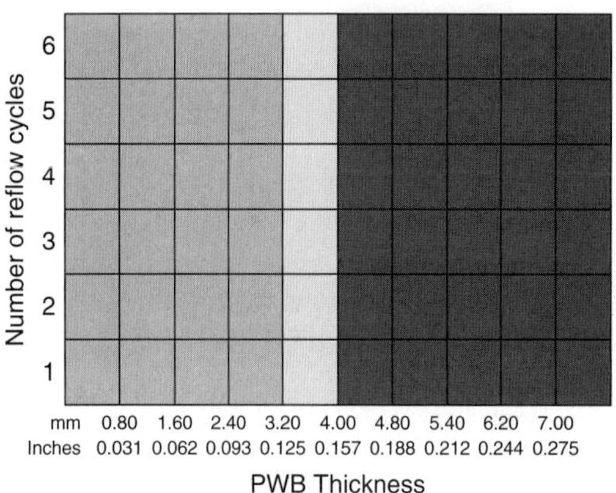

235–260°C

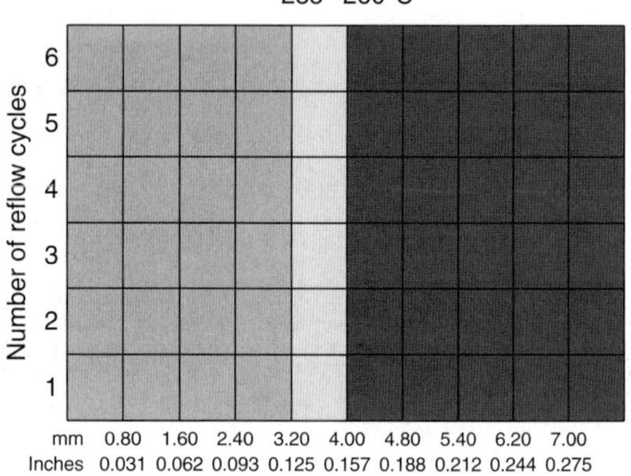

FIGURE 11.9 Charts for product B (phenolic 150°C T_g FR-4).

can withstand can go beyond six for thinner PCBs. If the two charts extended beyond six cycles, more differences would be seen.

Product D, with a high decomposition temperature and very low level of thermal expansion, is suitable for the broadest range of applications. This is illustrated in Fig. 11.10. Product E offers not only a high decomposition temperature and very low thermal expansion, but also significantly improved electrical performance. Combining these recommendations allows us

Peak reflow temperature

Please consult the "PWB Design considerations" section before using this tool

210–235°C

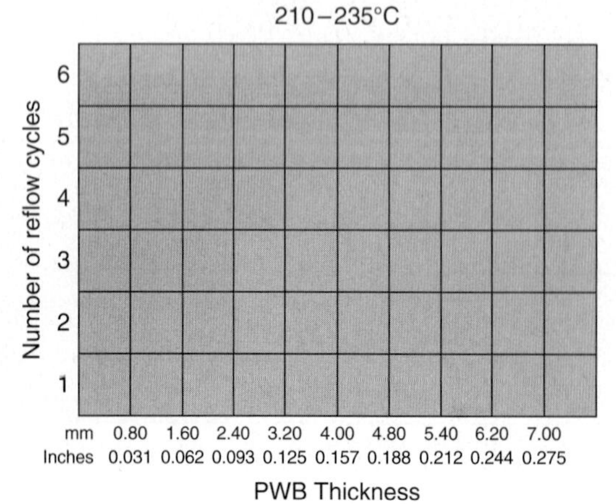

235–260°C

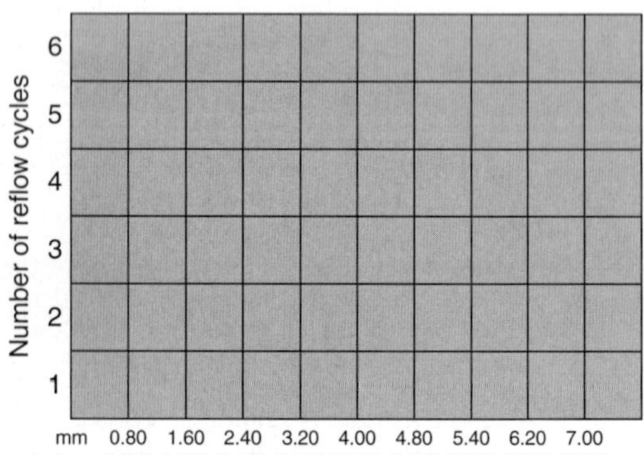

FIGURE 11.10 Charts for product D (phenolic 175°C T_g FR-4).

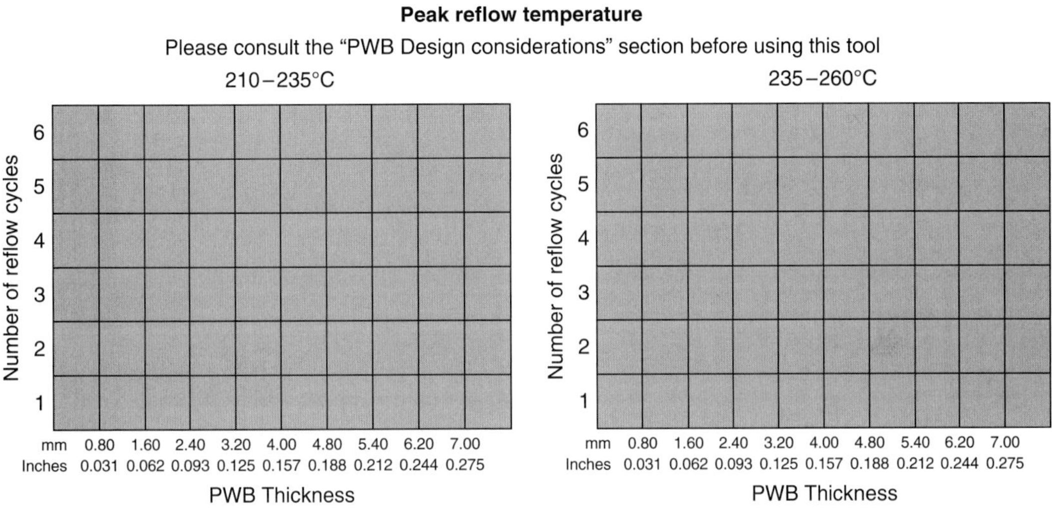

Peak reflow temperature

Please consult the "PWB Design considerations" section before using this tool

210–235°C

Number of reflow cycles

6
5
4
3
2
1

mm 0.80 1.60 2.40 3.20 4.00 4.80 5.40 6.20 7.00
Inches 0.031 0.062 0.093 0.125 0.157 0.188 0.212 0.244 0.275

PWB Thickness

235–260°C

Number of reflow cycles

6
5
4
3
2
1

mm 0.80 1.60 2.40 3.20 4.00 4.80 5.40 6.20 7.00
Inches 0.031 0.062 0.093 0.125 0.157 0.188 0.212 0.244 0.275

PWB Thickness

FIGURE 11.11 Charts for product E (non-dicy, non-phenolic material with improved electrical performance).

to make some general recommendations that also take material cost into consideration. Figure 11.12 summarizes the "cost versus performance" recommendations for materials A through D in lead-free applications. Where improved electrical performance is required, product E should be considered.

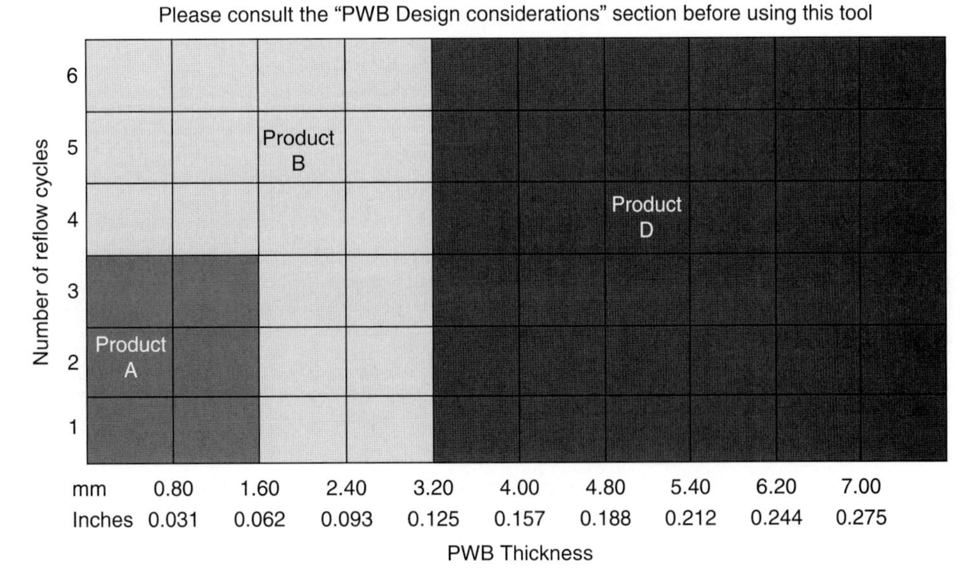

FIGURE 11.12 Summary of recommendations in lead-free assembly for materials A through D.

11.4 EXAMPLE APPLICATION OF THIS TOOL

A simple example helps illustrate the practical application of this tool. This example is based on a real PCB design that was being converted to a lead-free assembly process. Some key features of the PCB as they relate to how it differed from a "typical" PWB as defined in Fig. 11.4, and the recommended adjustments suggested, are shown in Table 11.4.

TABLE 11.4 Example of a PWB Being Converted to Lead-Free Assembly

PWB Design Feature	Attribute	Recommended Adjustment
Thickness	1.60 mm (0.062 inches)	—
Number of Reflows	3	—
Material	Conventional 140°C T_g FR-4	—
Surface Finish	Lead Free HASL (HASL not Typical)	Treat as an additional reflow cycle
Layer Count	10(</=8 Typical)	Consider moving up and right on chart

Figure 11.13 shows the results of making the adjustments suggested, and shows that this material is not recommended for this application. Figure 11.12 recommends that product B be used in this application, and empirical evidence has shown that when the standard 140°C T_g material was used, some level of assembly-related defects was observed. When product B was used, no assembly-related defects were experienced.

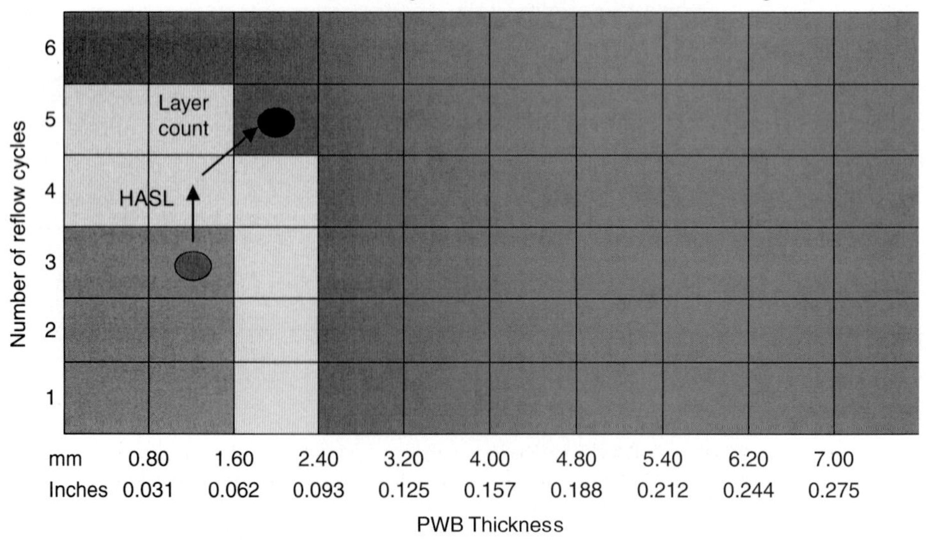

FIGURE 11.13 Example of practical application of the selection tool.

11.5 DISCUSSION OF THE RANGE OF PEAK TEMPERATURES FOR LEAD-FREE ASSEMBLY

As was discussed in Chap. 11 in the context of decomposition temperatures, lead-free assembly temperatures represent a very critical range for many base materials. In the selection charts discussed in Sections 11.3 and 11.4, the lead-free temperature range is given as 235 to 260°C. Although some lead-free applications may fall outside this range, the bulk will likely fall within it. However, with respect to the impact on base materials, 235 to 260°C is a broad range. While this represents a range of only 25°C, for some materials an increase of only 10°C could be the difference between success and failure. Much more data must be gathered before more specific recommendations can be made, but Table 11.5 provides some very general guidance when selecting FR-4 materials such as those discussed in Sections 11.3 and 11.4 for two different ranges of lead-free assembly peak temperatures. All the other variables discussed up to this point must be considered as well, so once again, it is recommended that the user validate the material recommendations suggested here.

TABLE 11.5 General Recommendations by Lead-Free Peak Temperature Range

Category	230 to 245°C Peak Temperatures	246 to 260°C Peak Temperatures
Conventional 140°C T_g FR-4s	May be suitable for PCBs of limited thickness and complexity where a limited number of thermal cycles will be experienced.	Generally not recommended, but may be suitable for low-technology applications where a limited number of thermal cycles will be experienced.
Conventional High-T_g FR-4s	Generally not recommended; even though initial results may be positive, extended storage in an uncontrolled environment prior to assembly could lead to problems.	Not recommended.
Phenolic Mid-T_g FR-4s	Suitable for a broad range of applications because of the higher decomposition temperature and overall thermal expansion being similar to standard high-T_g materials.	Suitable for a broad range of applications because of the higher decomposition temperature and overall thermal expansion being similar to standard high-T_g materials. Consider higher T_g versions for thicker, more complex PCBs.
Phenolic High-T_g FR-4s	Compatible in the broadest range of designs and applications. Recommended when long-term reliability is critical.	Compatible in the broadest range of designs and applications. Recommended when long-term reliability is critical.

11.6 LEAD-FREE APPLICATIONS AND IPC-4101 SPECIFICATION SHEETS

Revision B of the IPC-4101 specification was released in June 2006. The committee responsible for this specification worked hard to add material specification sheets that included some of the properties discussed in Chaps. 10 and 11 in regards to lead-free assembly. Table 11.6 summarizes some of the new specification sheets that include properties such as decomposition temperature, z-axis expansion, and time-to-delamination values.

TABLE 11.6 New Specification Sheets in IPC-4101B

Slash Sheet	Min. T_g	Filled/Unfilled	T_d@5%	T260	T288	T300	Pre-T_g CTE	Post-T_g CTE	CTE% 50–260°C
/101	110	Filled	310	30	5	NA	60	300	4.0
/121	110	NA	310	30	5	NA	60	300	4.0
/99	150	Filled	325	30	5	NA	60	300	3.5
/124	150	NA	325	30	5	NA	60	300	3.5
/126	170	Filled	340	30	15	2	60	300	3.0
/129	170	NA	340	30	15	2	60	300	3.5

It is important to understand that these requirements apply to laminate materials and not the PCBs manufactured from them. Furthermore, the time-to-delamination requirements are for unclad samples. All of the variables outlined in Section 11.2 regarding printed circuit fabrication and assembly must also be considered when specifying a material. It is also critical to understand that no one property will guarantee success in lead-free applications. When faced with uncertainty, the natural tendency is to overspecify each property. Not only could this cause the user to pay too much for a material, but overspecifying specific properties can also lead to functional problems. These problems can be the result of the interaction between advanced materials and the PCB fabrication processes required to manufacture reliable PCBs.

Take decomposition temperature, T_d, as an example. Section 10.2.2 discussed the importance of T_d with respect to lead-free assembly compatibility. Although the 5 percent level of decomposition has been the standard reported value, the point was made that this is a very high level of decomposition from the standpoint of PCB reliability. It is important to understand when decomposition begins and how close this temperature is to actual assembly and rework process temperatures. A material with a reported T_d, based on the 5 percent level, of 325°C may perform just as reliably, or even more reliably, than a material with a 5 percent T_d value of 340°C, depending on what the decomposition characteristics are in the lead-free assembly and rework temperature ranges. Specifying materials with significantly higher T_d values may not improve reliability, and may potentially add cost and difficulty to the manufacturing process if the higher T_d material is more difficult to process—for example, if the lamination times are longer, or if it is harder to drill or score. Although future specification sheets may include T_d values based on lower levels of decomposition, it is very important to understand these properties in some detail and specify the appropriate material for a given application.

In summary, experience to date indicates that specification sheets 99 and 124 are suitable for a very broad range of applications. For lead-free applications, the user should consider these as the baseline. Simple low-technology PCBs or those that do not have stringent long-term reliability requirements may successfully use the materials covered by sheets 101 or 121, whereas very advanced products may need more stringent requirements, such as those covered by sheets 126 and 129. Finally, always check for the most recent revision of IPC-4101, as new specification sheets will be added as more experience is gained and data are gathered.

11.7 ADDITIONAL BASE MATERIAL OPTIONS FOR LEAD-FREE APPLICATIONS

Sections 11.3 through 11.6, including Table 11.6 on the IPC-4101B specification sheets, focused on FR-4 type base materials. These are still the most commonly used material types by far. Chapter 6 touches on the history of FR-4 materials and the development of improved versions of FR-4 as requirements have changed. In addition, several non-FR-4 materials compatible with

TABLE 11.7 Range of Material Types Compatible with Specific Lead-Free Applications

Resin Type	T_g by DSC, °C	T_d	T260 (min.)	T288, (min.)	% Expansion, 50–260°C	D_k@2 GHz	D_k@ 10 GHz	D_f@2 GHz	D_f@ 10 GHz
Filled, Non-Dicy FR-4	140	330	>30	>5	3.0	4.0	3.9	0.020	0.022
Filled, Non-Dicy FR-4	150	330	>30	>5	3.0	4.0	3.9	0.020	0.022
Phenolic, Filled FR-4	150	335	>30	>5	3.4	4.0	3.9	0.020	0.022
Phenolic, Epoxy FR-4	175	340	>30	>10	3.5	4.1	4.0	0.021	0.023
Phenolic, Epoxy FR-4	180	350	>30	>15	3.5	4.0	3.9	0.022	0.024
Phenolic, Filled Epoxy FR-4	180	340	>30	>15	2.8	4.2	4.1	0.020	0.022
Non-Phenolic, Non-Dicy	200	370	>30	>20	2.8	3.7	3.7	0.013	0.014
Halogen Free FR-4	150	380	>30	>5	3.0	4.0	3.9	0.020	0.022
High-T_g, Halogen-Free FR-4	170	390	>30	>15	2.8	4.0	3.9	0.018	0.019
Modified Epoxy, Low D_k/D_f	180	360	>30	>15	3.5	3.6	3.5	0.011	0.012
Low D_k/D_f Blend	225	365	>30	>15	2.8	3.6	3.5	0.007	0.008
Low D_k/D_f Blend/RF	220	350	>30	>10	3.6	3.0–3.6	3.0–3.6	<0.0045	0.0045
Polyimide	260	415	>30	>30	1.5	3.9	3.9	0.017	0.018

lead-free assembly are also available. These include polyimide materials, well known for their thermal reliability and low D_k/D_f materials. Table 11.7 indicates the range of material types compatible with specific lead-free applications and their various properties. As discussed throughout this chapter, not all materials are compatible in all applications. However, the properties included in this table should help identify candidate materials for evaluation.

11.8 SUMMARY

The question of whether a given laminate material is compatible with lead-free assembly can be very complex due to the many variables in PCB design and manufacturing. In addition, there are variations in lead-free assembly processes. Specifically, peak temperatures, dwell times at peak temperatures, heatup rates, cooldown rates, and rework processes can all vary and complicate this decision. As stated earlier, the temperature range from 235 to 260°C is critical for laminate materials; some materials may be moderately successful at the low end of this range but fail at the higher end. Additional work investigating these assembly-related variables is under way. The existing tools for base material selection for lead-free assembly continue to be developed and improved as more data are collected and analyzed.

Critical base material properties with respect to lead-free assembly applications include:

- **The decomposition temperature** Some minimum value is needed, but above that minimum level, higher values are not *always* better.
- **Thermal expansion properties** Improved materials have reduced levels of thermal expansion.

- **Glass transition temperature** A higher glass transition temperature delays the onset of rapid thermal expansion and therefore reduces total expansion within a temperature range.
- **Moisture absorption** In particular, if PWBs are stored in an uncontrolled or humid environment prior to assembly, drying of boards prior to assembly or selection of a material less prone to moisture-related problems should be considered.
- **Time to delamination** T260 and T288 values are a simple way to screen materials for compatibility with lead-free assembly. However, testing at higher temperatures or requiring excessive times may not be necessary.

Balancing these properties with the level of reliability required along with PWB manufacturability and assembly concerns is vital.

REFERENCES

1. Isola "Lead-Free Assembly Compatible PWB Fabrication and Assembly Processing Guidelines."
2. Bergum, Erik J., and Humby, David, "Lead Free Assembly: A Practical Tool for Laminate Materials Selection," IPC-Soldertec, June 2005.
3. Kelley, Edward, Bergum, Erik, Humby, David, Hornsby, Ron, and Varnell, William, "Lead-Free Assembly: Identifying Compatible Base Materials For Your Application," IPC Expo, 2006.

CHAPTER 12
LAMINATE QUALIFICATION AND TESTING

Michael Roesch
Hewlett-Packard Company, Palo Alto, California

Sylvia Ehrler
Feinmetall GmbH, Herrenberg, Germany

12.1 INTRODUCTION

Copper-clad laminate materials (single- or double-sided) are the base material for almost all printed circuit boards (PCBs). The properties of these materials have a large influence on the quality of the final product and therefore need to be understood and tested prior to release/manufacturing.

12.1.1 Impact of RoHS and Lead-Free Requirements

With a change toward RoHS-compliant "lead-free" assembly processes, laminate materials and complete printed circuit boards are facing a set of increased challenges:

- **Laminates** Materials have to survive the increased thermal stress of repeated lead-free assembly cycles. Due to the higher melting temperature of lead-free solders, a peak temperature during reflow of 260°C must realistically be expected (30°C to 40°C higher than eutectic soldering profiles). Most printed circuit boards will be required to withstand at least five such assembly cycles, but that number may increase to six and higher for more complex assemblies. A rule of thumb states that a 10°C increase of temperature doubles the reaction energy starting degradation or decomposition of the resin compound. The increased peak temperature therefore requires new or improved laminate materials with a higher decomposition temperature and enhanced thermal stability. New laminate test methodologies are also needed to assess the impact of the higher assembly temperatures.
- **PCB reliability** There should not be any degradation in the reliability of the printed circuit boards after assembly compared to boards assembled using traditional soldering profiles. When the same laminate material is used, plated through holes will see an increase in thermal expansion stresses after an equal number of assembly cycles with higher peak temperatures. This means that if the same laminate is used, the reliability margin of the PCB has decreased. To achieve comparable reliability of boards after both eutectic or lead-free soldering, a change

in the base material is most likely. Additional PCB reliability concerns include innerlayer bonding, conductive anodic filament (CAF) growth, and dielectric strength, particularly with reduced dielectric thicknesses.

- **PCB performance** In addition to the preceding reliability challenges posed by lead-free assembly processes, it also has to be guaranteed that all other performance characteristics of the printed circuit board stay the same. This includes dielectrical properties such as dielectric constant D_c (which influences impedance), dissipation factor D_f, and thermomechanical properties such as copper peel strength, glass transition temperature, or coefficient of thermal expansion (CTE). These properties should not be affected by the assembly processes applying higher temperatures.

12.1.2 Material Evaluation Process

In almost all cases, the laminate manufacturer will provide a set of data for its specific materials. This datasheet is a good starting point and can certainly be used to evaluate laminates at an early state of a project. Often, though, it will be necessary to perform additional tests in the PCB facility to verify or complement the manufacturer's data to ensure that the laminate materials meet the requirements of manufacturing and assembly processes.

This chapter introduces the most important laminate properties and their characterization methods and has been revised to reflect RoHS-compliant manufacturing process and temperature requirements. It will serve as a quick reference guide for the printed circuit board specialist as well as an introduction to laminate testing for those who are new to the field. Laminate properties according to industry standards such as the National Electrical Manufacturers Association (NEMA) and IPC will be introduced. Then the actual test methods will be discussed in the form of a best-practice guide. This guide will help the PCB specialist make quick decisions during the testing process. Methods to test mechanical, thermomechanical, and electrical properties of printed circuit board materials will be described. The focus will be on how the tests are performed, what the results actually mean in today's manufacturing environment, and how relevant the obtained test data are. In addition to the raw base material testing, the chapter will introduce a best-practice qualification guide to assess interactions between manufacturing processes and laminate properties.

12.2 INDUSTRY STANDARDS

A number of different industry standards can be used to evaluate copper-clad laminate properties. Many of the test methods for similar properties are very comparable, though, and in the end the materials engineer must decide which standard to use as a basis for his or her characterization. In some cases, it may even be necessary to define a new characterization method to qualify a laminate material for a specific customer application or a particular requirement.

12.2.1 IPC-TM-650

This document is the most comprehensive and most widely used compilation of copper-clad laminate test methods. It is administered by IPC (Association Connecting Electronic Industries) and available in print or online.*

*Available at http://www.ipc.org/html/testmethods.htm.

Most of the test methods described in this chapter are based on the test manual IPC-TM-650. It contains industry-approved test techniques and procedures for chemical, mechanical, electrical, and environmental tests on all forms of printed wiring and connectors. An IPC test method is a procedure by which the properties or constituents of a material, an assembly of materials, or a product can be examined. These test procedures do not contain acceptability levels for specific properties.

12.2.2 IPC Specification Sheets

IPC publishes a wide variety of technical specification sheets for PCB and printed circuits assembly (PCA) products. A few are important in this context:

- **IPC 4101, "Specifications for Base Materials for Rigid and Multilayer Printed Boards," Rev. B 06/06** This document contains specification sheets for laminate materials used to manufacture PCBs for commercial and military applications. The minimum requirements for key material properties can be found in these specification sheets.
- **IPC 4103, "Specification for Plastic Substrates, Clad or Unclad, for High Speed/High Frequency Interconnections," 01/02** This document contains specification sheets for Teflon (polytetrafluoroethylene, or PTFE) and other laminate materials used to manufacture PCBs for high-speed or high-frequency applications.
- **IPC 4104, "Specification for High-Density Interconnect (HDI) and Microvia Materials," 05/99** This document contains specification sheets for materials used in HDI and microvia applications.

12.2.3 ASTM

The American Society for Testing and Materials (ASTM) develops standard test methods and specifications that are in some cases related to laminate materials. They will be referenced as appropriate. More information can be found online.*

12.2.4 NEMA

NEMA publishes standards on laminated thermosetting materials. The last published version is the LI 1-1998 Standard for Industrial Laminated Thermosetting Products, which includes the latest information concerning the manufacture, testing, and performance of laminated thermosetting products. Most copper-clad laminate materials are named according to NEMA grades, but it is important to understand that NEMA grades represent only material categories and not one individual material. One of the most commonly known material grades is FR-4, which is often thought of as a specific laminate material. This is not the case, however, as each NEMA grade represents a material class defined by properties that fall into a certain class. Thus, properties of different laminate materials from different suppliers that fall into the same class (FR-4, for example) may not be identical. More information can be found online.** (For a more detailed discussion of FR-4, see Sec. 6.4.)

*Available at http://www.astm.org.
**Available at http://www.nema.org.

TABLE 12.1 NEMA Grade Material Designations and Properties

NEMA grade	Resin system	Reinforcement	Description
XXXPC	Phenolic	Paper	Phenolic paper, punchable
FR-2	Phenolic	Paper	Phenolic paper, punchable, flame-resistant
FR-3	Epoxy	Paper	Epoxy paper, cold punchable, high insulation resistance, flame-resistant
CEM-1	Epoxy	Paper-glass	Epoxy paper core and glass on laminate surface, flame-resistant
CEM-2	Epoxy	Paper-glass	Epoxy paper core and glass on laminate surface
CEM-3	Epoxy	Glass matte	Epoxy non-woven glass core and woven-glass surface, flame-resistant
CEM-4	Epoxy	Glass matte	Epoxy non-woven glass core and woven-glass surface
FR-6	Polyester	Glass matte	Polyester non-woven glass, flame-resistant
G-10	Epoxy	Glass (woven)	Epoxy woven glass, not flame-resistant
FR-4	Epoxy	Glass (woven)	Epoxy woven glass, flame–resistant
G-11	Epoxy	Glass (woven)	High-temp epoxy, woven glass, not flame-resistant
FR-5	Epoxy	Glass (woven)	High-temp epoxy, woven glass, flame-resistant

12.2.5 NEMA Grades

Table 12.1 shows technical and industrial laminate grades as they apply to the most important printed circuit board materials.

12.3 LAMINATE TEST STRATEGY

Laminates contain different reinforcement materials (woven/non-woven glass/organic fibers, expanded PTFE, etc.), resin types (phenolic, epoxy, cyanate ester, polyimide, BT, etc.), resin formulations (blended, functionality, etc.), hardeners (dicyanodiamide [dicy], phenol-novolak, cresol-novolak, p-aminophenol, isocyanurate, etc.), and sometimes filler particles (ceramic or organic). The ratio of all of these individual components can vary widely. To define a test strategy for laminates, it is important to understand the different main components of the materials as well as the conditions during manufacturing, as these will have a large influence on their properties and quality.

The evaluation and qualification of a laminate material can be a quite complex task and the new requirements for lead-free processing have specifically increased the focus on the thermal characteristics of laminate materials. A number of different qualification test methods are available to the materials engineer. However, very often not all of them are relevant or necessary for the final product. At the same time, the materials engineer faces pressure to produce quick results driven by shorter development cycles and increasing product and process requirements. The organization of the test methods described in this chapter follows a proposed best-practice guide aimed to help perform full material evaluations efficiently and at the same time address the new requirements of lead-free processing. The sections allow quick decisions, especially when a number of materials are being evaluated at the same time.

12.3.1 Data Comparison

The first step in every material evaluation is data comparison and assessment. This step may already allow the elimination of some material candidates without performing actual tests.

The first and most important thing to understand is the state of the material that is being considered. Is it an established material formulation that has a solid history at other manufacturing locations, or is it a newly developed material that has not seen mass production yet?

New developments often do not come with complete data sheets that allow an early assessment, whereas the laminator can often provide one for established materials.

Cost is another important factor. The laminator can provide guidance about volume pricing for the new material.

Other considerations are the history of the supplier, in terms of having met previous commitments and whether the material is already in high-volume production. If it is a new material that has not yet been ramped to high volume, if no more than one PWB manufacturer is using it, and if no second source can be identified, then these facts may be a red flag when considering the material for use.

12.3.2 Two-Tier Test Strategy

The qualification procedure proposed in this chapter follows a two-tier test strategy that enables quick decisions during a material qualification program. The first set of tests is easy to perform and focuses on key properties that should be evaluated when testing a new laminate material. Failure to pass any of them will likely result in rejection of a material for manufacturing. If multiple materials are being evaluated, the first tests will allow the easy elimination of the least probable candidates.

The second set of tests is an expanded set of qualification tests that focuses on all key properties that are commonly evaluated for laminates. However, the final decision about a qualification test plan for a new laminate should always be driven by product-level requirements, which means that additional laminate tests as well as board-level tests may need to be added to the set of tests described in the following section.

12.4 INITIAL TESTS

This first set of tests is most commonly performed when starting a material evaluation and qualification project. The tests can be performed easily and give some early indications of whether the material exhibits any major problems.

12.4.1 Surface and Appearance

This test is one of the very first tests to happen as the new material arrives in the shop. The laminate is inspected for its appearance and visual quality. Often the materials engineer will be able to assess the quality of the product without applying specification guidelines. Even packaging and shipping materials should be used to assess the laminator's overall quality control.

IPC-TM-650 method 2.1.2 and method 2.1.5 define procedures for how to inspect the surface and appearance of copper-clad laminates. These standards should be applied only to surface areas that are pertinent to the finished board. More than 90 percent of the copper will usually be removed during manufacturing, and a new material should not be discarded solely for cosmetic reasons. On the other hand, for areas that will contain edge card connector pads or very fine-line traces, these surface standards must be applied.

The specifications categorize the longest permissible dimensions of any copper pits and dents and define a point value-based rating system (as shown in Table 12.2).

TABLE 12.2 Longest Permissible Dimensions

Longest dimension (mm)	Point value
0.13–0.25	1
0.26–0.50	2
0.51–0.75	4
0.76–1.00	7
Over 1.00	30

TABLE 12.3 Material Class

Class	A	B	C	D	X
Total point count	<30	<5 All dents <0.38 mm	<17	0	Requirements agreed upon

The total point value for an area of 300 mm × 300 mm of laminate determines the material class (as shown in Table 12.3).

In addition to copper pits and dents, the specifications contain procedures on how to inspect for scratches, wrinkles, and inclusions.

12.4.2 Copper Peel Strength

The copper peel strength of copper-clad laminates is an indicator for a variety of their performance characteristics. It indicates the bond strength of the copper circuitry of a double-sided printed circuit construction after manufacturing and any successive assembly or repair cycles. To assess the copper bond strength of the outerlayer circuitry of a multilayer construction accurately, it is necessary to take a close look at the exact build-up and copper foil used in the construction and to build specific test vehicles that are similar to the original stack-up. Copper peel strength measurements have grown increasingly important as circuit traces and pads are becoming smaller. In addition, this test allows an assessment of any potential weaknesses in multilayer constructions and identifies the interface that might delaminate first after thermal stress. The higher the initial bond strength of a material, the better. For all characterizations after thermal stress in the temperature range of lead-free assembly, it is also important to look at the decrease of bond strength. This is especially important for applications that require multiple reflow operations and one or more repair cycles.

The test method is specified in IPC-TM-650, method 2.4.8, and the applicable test patterns are specified in IPC TM-650, method 5.8.3. Figure 12.1 shows an actual test pattern using a sample 100 mm × 100 mm coupon with 4-mm-wide copper strips.

The material being tested should be processed using the same fabrication steps that will be applied for the actual product. IPC-TM-650, method 2.4.8, recommends a specimen size of 50.8 mm × 50.8 mm (2.0 in. × 2.0 in.) and a minimum sample size of two for both the warp and fill (X and Y) direction of the laminate. The end of the test strip is peeled off so that a tensile

FIGURE 12.1 Test pattern used for a copper peel strength test.

tester can clamp it. The load cell should be calibrated to account for the weight of the clamp. The copper foil strip is then peeled from the material at a rate of 2 in. per min. The measured value is dependent on the copper thickness. The thicker the copper foil, the more force is required to deform it plastically. The value is also dependent on the angle at which it is being pulled. Therefore, it is always important to specify the thickness of the copper foil tested and to maintain a peeling force radius of 90° during the test. The minimum load is determined as specified and the peel strength is calculated using Eq. 12.1:

$$\text{lb/in} = L_M/W_S \tag{12.1}$$

where L_M = Minimum load
W_S = Measured width of peel strip

Copper bond retention after exposure to soldering or touch-up temperatures is measured as outlined in IPC-TM 650, method 2.4.8, condition B. The test samples are covered with silicone grease before floating them on the solder pot at 288°C for 5 to 20 sec. to avoid solder contamination on the test strip. After allowing the samples to cool down to room temperature and removing the grease, the test is performed in the same way as previously described. It is important to inspect the specimen for any blistering or delaminations prior to testing (see also the discussion of solder shock testing in Section 12.5.2) as this may interfere with the test procedure and may indicate that the laminate material has problems with solder shock resistance.

Another factor that can influence the adhesion strength of the copper foil is exposure to processing chemicals during manufacturing. Details about this test method can be found in IPC-TM 650, method 2.4.8, condition C. This test is usually performed only if specific requirements at the end-product level call for it or if the material during manufacturing exhibits low copper bond retention properties.

Minimum copper peel strength values for unstressed material should be 1 N/mm or higher. The acceptable values after thermal stress are 0.8 N/mm and 0.55 N/mm after process solutions. Measurements taken at elevated temperature (125°C) should yield a minimum value of 0.7 N/mm. The authors recommends using the preceding values as guidelines and validating acceptable minimum values with actual product requirements.

12.4.3 Solder Shock

The solder shock test is one of several methods to assess the thermal resistance of copper-clad laminates. It is easy to perform and represents another key test during the early assessment of a material. There are a number of different methods to choose from, which will be described in detail in Section 12.5.2. During the initial assessment of the material, it is important to choose at least one of the described test methods to make certain that the material meets the minimum requirements, especially if the material is used in higher-temperature lead-free assembly processes. Aside from solder shock testing of bare laminate material, it is also recommended that the PCB engineer consider PCB-level temperature shock as well as repeated reflow testing with a particular focus on resin-reinforcement delaminations. This will ensure that not only the raw material but also the completed PCB will be able to withstand the required temperature regime.

12.4.4 Glass Transition Temperature (T_g)

A further important data point within the initial assessment is the determination of the glass transition temperature (T_g). Different test methods are described in more detail in Section 12.5.2. For the first assessment, one method should be selected. When choosing the method, it is important to make sure that the test procedure is compatible with the material in test (for some materials, the glass transition can be determined only using dynamic mechanical analysis [DMA]). A minimum of two samples should be tested. If a data sheet of the material is available, the measured value should correspond with the manufacturer's data (within the measurement accuracy of the equipment). The values of T_g1 and T_g2 (measured on the same sample) should not be more than 7°C apart. If T_g2 is significantly below the T_g1 measurement, this may be a strong first indication that resin system degradation has started during the measurement procedure. It is recommended to confirm this finding using the T_d measurement procedure.

12.4.5 Thermal Decomposition Temperature (T_d)

The determination of the thermal decomposition temperature (T_d) completes the initial assessment. This is a new test methodology, and both ASTM and IPC test procedures are available. IPC-TM-650, method 2.3.40, is called "Procedure to Determine the Thermal Decomposition Temperature T_d Using Thermo Gravimetric Analysis (TGA)," and the ASTM methodology is called "Standard Test Method for Rapid Thermal Degradation of Solid Electrical Insulating Materials by Thermo Gravimetric Analysis." In both cases, the test methodology is based on thermogravimetric analysis (TGA) of a laminate specimen that is heated with a set ramp rate assessing the onset and process of decomposition. At first, the thermoset laminate resin passes the glass transition point where the polymer changes from a rigid, brittle, and "glasslike" state to a softer and more rubberlike material. As the material is heated further, the three-dimensional cross-linked structure begins to break down and individual bonds in the polymer structure open. As decomposition continues, gaseous decomposition products are set free, which can be detected as a loss of mass of the sample in test. This usually goes hand in hand with visible damage to the sample (delamination and/or strong discoloration) and significant changes in the mechanical properties of the laminate.

A dry test specimen of 2–20 mg is heated at 5°C/min to 150°C and then held at that temperature for 15 min. This ensures that the sample is completely dry and establishes the base weight used to characterize the weight loss. Then the sample is heated at 5°C/min up to a temperature of 800°C and the weight loss is continually recorded. The temperature at which a 5 percent weight loss of the sample is detected is defined as the T_d.

Although this test methodology is a good gauge of the thermal stability of laminate materials, the method and results must be evaluated carefully to ensure that correct conclusions are drawn from the results. For example, the results of this test are dependent on the surface/volume ratio of the specimen in test-specimens of the same weight, but different surface areas will show different results. The sample with the large surface area loses weight at a faster rate and therefore shows a lower T_d. Samples with a large percentage of nondecomposing filler or reinforcement material will also behave differently as the measured weight change is based only on the resin content of the sample. This again will lead to a seemingly higher measured T_d, which may lead to erroneous conclusions. Finally, the process of decomposition is as important as the determined value of T_d and a comparison of the onset of weight loss may also be a valuable parameter to assess when comparing laminate materials using this methodology.

In general, a higher T_d value is desirable for materials that are being selected for lead-free assembly processes as this indicates that the laminate withstands the higher assembly temperatures with an increased margin.

Figure 12.2 shows an example of different T_d measurements for three different laminate materials. It shows the typical decomposition rates for some sample materials and also highlights how the assessment of a material can change if different criteria are chosen. If the IPC value of 5 percent weight loss is chosen, then M3 is superior to M1 and M2. But if the materials engineer decides that the onset of weight loss (decomposition) is critical for the application, then M3 is the worst performing material. The authors recommend using materials with a high onset temperature as well as a high 5 percent decomposition temperature.

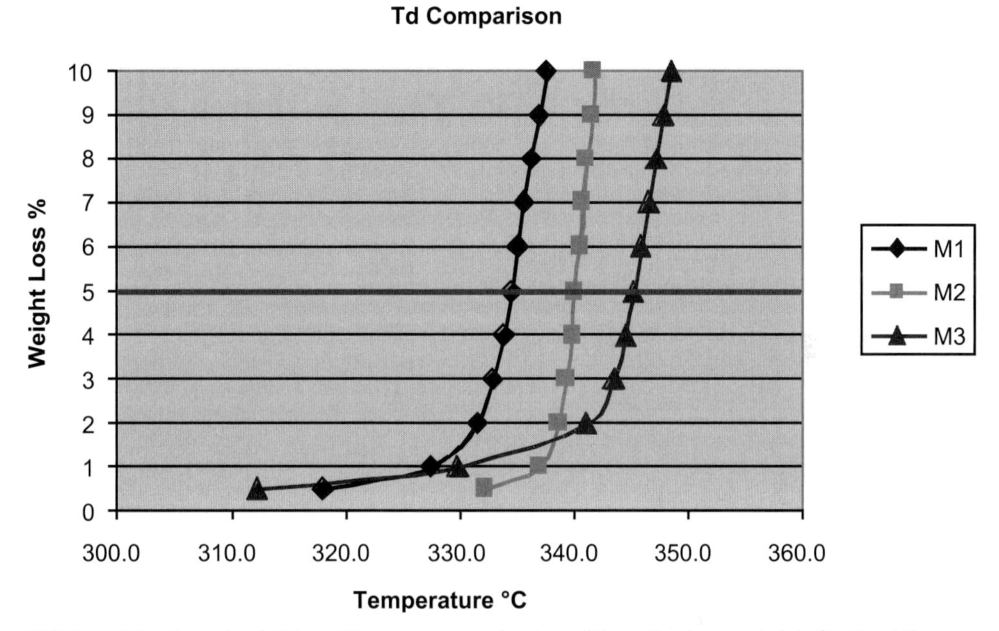

FIGURE 12.2 Example of different T_d measurements for three different laminate materials. (*T_d GraphCourtesy of Helmut Kroener, Multek.*)

12.5 FULL MATERIAL CHARACTERIZATION

The following set of characterization methods summarizes the most commonly used test methods to qualify copper-clad laminates for manufacturing. It is intended to guide the materials engineer through the laminate qualification process. Yet it can only serve as a best-practices guide, as the final set of qualification methods will vary for each material and should always be driven by its specific application.

12.5.1 Mechanical Tests

Commonly used mechanical tests include tensile tests, flexural strength tests, and copper peel strength tests.

12.5.1.1 Modulus and Tensile Strength (Tensile Test). The test method to determine modulus and yield stress for copper-clad laminates described in this section is based on ASTM D 882, "Standard Test Methods for Tensile Properties of Thin Plastic Sheeting." Other relevant industry standards are IPC-TM-650, method 2.4.19 (for flexible laminate materials), and IPC-TM-650, method 2.4.18.3 (for deposited organic free films). Elastic modulus and yield stress are always determined by the same method.

The elastic modulus is the indicator value for the stiffness and strength of a composite material under load, whereas the yield stress indicates the maximum load the material can take in x/y direction before it brakes. The modulus is the more important of the two values. It is used to differentiate materials and their actual strength during qualification testing. The elastic modulus is also one of the key properties of a material that need to be understood when performing computational modeling of packages and printed circuit board assemblies.

In many applications, materials with different properties are combined, which leads to thermomechanical stresses that cause fatigue or interfacial failures. Computational modeling is often used when there is not sufficient time or resources available to perform actual product test, but fast reliability assessments are needed.

The standard test specimen for this test consists of a strip of laminate material 152.4 mm (6 in.) long and 12.7 mm (0.5 in.) wide. A minimum of 10 samples should be prepared and all copper must be removed before testing. It is important that only samples with identical thickness and glass stack-up be tested, as this will allow a meaningful comparison of the materials in the test. The test engineer also must ensure that both the warp and fill direction of the laminate composite are measured and documented separately.

After the samples are cut, they need to be inspected for any fractures, delamination, or roughness along the edges and if necessary they need to be sanded. The cross-sectional area of each specimen is calculated from the width and thickness in the measurement area. Then the individual samples need to be conditioned at 23°C (73.4°F) and 50 percent relative humidity (RH) for 24 hours. Although most specifications recommend conditioning of the material prior to cutting and preparation of the samples, the authors recommend first cutting the specimens to size and then conditioning them. This ensures more uniform properties of the individual samples.

The tests are performed using a standard tension and compression apparatus, which can be operated at a controlled rate of crosshead movement. The load-extension curve should be recorded and the specimen tested until it breaks. If extensometers are used, it is necessary to minimize the stress on the sample at the contact points of specimen and indicator.

The tensile strength is calculated by dividing the maximum load at break by the original minimum cross-sectional area. The unit is megapascals (MPa). Dividing the elongation at break by the initial gauge length and multiplying it by 100 will calculate percent elongation. When extensometers are employed, only the length of the section between them is used; otherwise, the distance between the grips represents the initial gauge length. Young's modulus is calculated by drawing a tangent to the linear portion of the stress-strain curve, selecting any point on this tangent, and dividing the tensile stress at that point by the corresponding strain. The unit is gigapascals (GPa). For all calculations, the values for 10 samples should be recorded and the average calculated.

In addition to the absolute values for the different mechanical properties, the stress-strain curve gives the materials engineer information about the elastic and plastic portion of the deformation in the laminate samples. The elastic (linear) part of the deformation is restorable, whereas plastic changes in the materials are not. Figure 12.3 shows a typical stress-strain curve.

12.5.1.2 Flexural Strength. The test method to determine the flexural strength of copper-clad laminates is based on ASTM D 790 ("Flexural Properties for Un-Reinforced and Reinforced Plastics and Insulating Material") and IPC-TM-650, method 2.4.4B ("Flexural Strength of Laminates at Ambient Temperatures"). The test method described in the preceding section to determine the modulus in tensile mode does have the inherent problem that the glass reinforcement and its properties have a large influence on the results. This may be avoided when testing copper-clad laminates in flexural mode. Here the attributes of the resin system have a greater influence on the outcome of the test. When comparing the test results of different laminate materials, it is important to compare only data that were generated using the same test method.

This test also requires that the copper be completely removed from the laminate material by means of standard etching processes. The specific size (length and width) of the samples and test parameters (test span and speed) are dependent on the thickness of the laminate and can be found in IPC-TM-650, method 2.4.4. See Fig. 12.4 for a typical flexural strength measurement curve.

The samples are conditioned to room temperature. They are then centered on the supports of the standard tension and compression test apparatus with the long axis of the specimen perpendicular to the loading nose and support. The samples are loaded at the defined speed

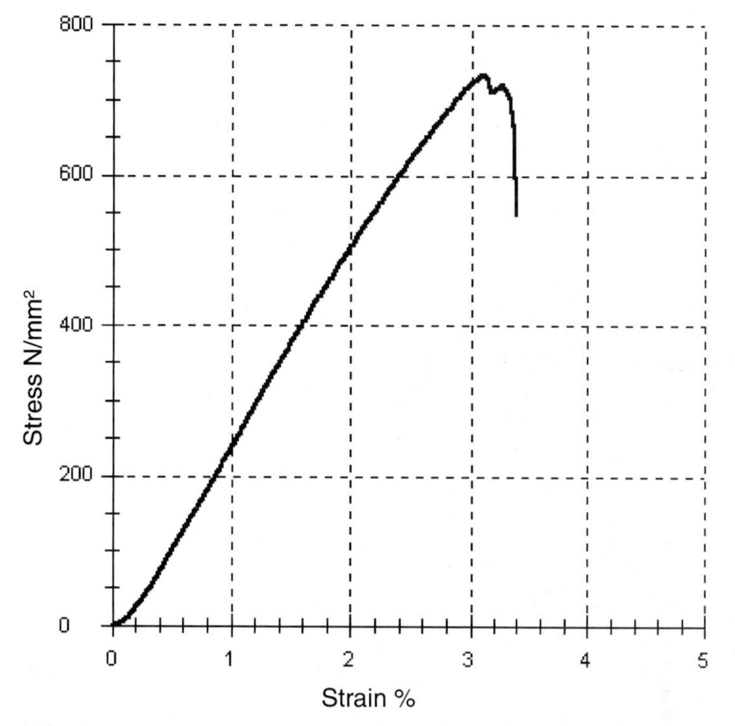

FIGURE 12.3 Typical stress-strain curve.

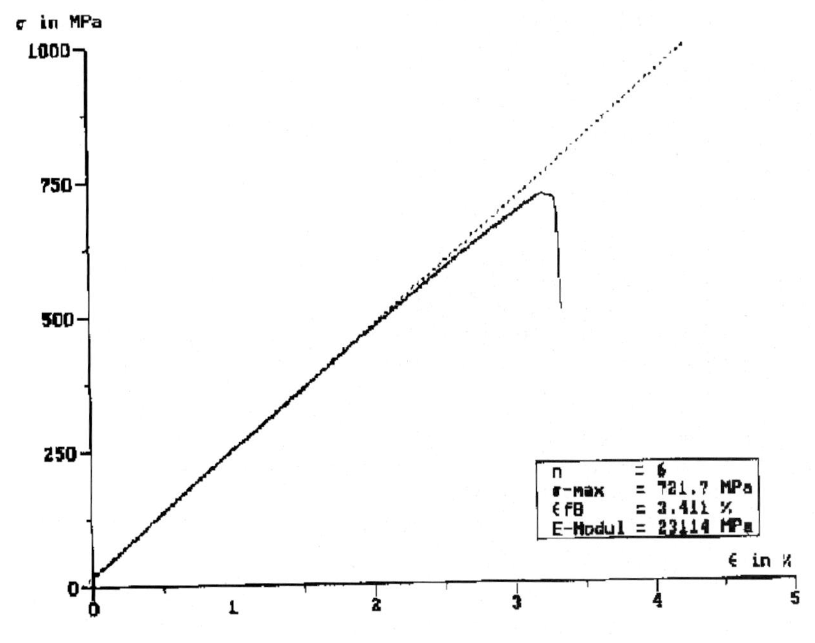

FIGURE 12.4 Typical flexural strength measurement curve.

and tested until they break. The load at breakage should be recorded. Use Eq. 1.2 to calculate the flexural strength.

$$S = 3PL/2bd^2 \tag{12.2}$$

where S = Flexural strength (psi)
P = Load at breakage (lb.)
L = Span (in.)
B = Width of specimen (in.)
D = thickncss (in.)

12.5.1.3 Copper Peel Strength (Before and After Thermal Stress). Test methods are described in Section 12.4.2 and results can be transferred to this point in the evaluation. To complete the data, perform additional tests if necessary after specific thermal stress (solder dip, reflow with different peak temperatures, and/or different dwell times) or after exposure to plating solutions.

12.5.2 Thermomechanical Tests

Properties that are tested thermomechanically include T_g, CTE, thermal resistance, solder dip resistance, and solder float resistance.

12.5.2.1 Glass Transition Temperature T_g (DSC, TMA, and DMA). The resin systems used in today's copper-clad laminates are almost all three-dimensionally cross-linked thermoset polymers. This means that the materials are manufactured (in a curing reaction) into a hard final product and cannot be remelted many times the way thermoplastic polymers can. In many cases, different resin systems are blended to create a material with specific properties. The functionality of the individual resins and their ratio in the blend define the properties of the thermoset. The degree of cure of the hardened thermoset resin system can be described by characteristics such as percent of uncured monomer and heat of final cure or by mechanical properties such as hardness, modulus, and yield stress. But one of the most commonly used characteristics to describe the degree of cure and cross-link density of a thermoset polymer is the glass transition temperature (T_g).

The simplest definition of the T_g of a laminate is to call it the temperature at which the mechanical properties of a laminate change (deteriorate) rapidly. It is at this temperature that the polymer changes from a rigid, brittle, and "glasslike" state to a softer and more rubberlike material. However, this should not be confused with the melting point of a crystalline substance or the extreme softening of a thermoplastic polymer. In a thermoset material at T_g, the relative mobility of the cross-linked polymer molecules changes. Below T_g, they are immobile and the material is rigid; above T_g, the relative mobility of the molecules increases and slight movements are possible, which leads to a loss of mechanical strength in the material. Although all of the methods described in this section enable us to determine a specific glass transition temperature for laminate materials, it also needs to be emphasized that the glass transition in thermoset resins is not a single defined temperature. It usually occurs within a temperature range (for some resin systems, the range can span as much as 100°C), and the T_g is determined within this range.

In addition, the difference between T_g and T_d must be pointed out here. T_g is the temperature at which mechanical properties of the material change, whereas T_d is the temperature where the material starts to irreversibly degrade. If one wants to assess a laminate's capability to withstand high-temperature lead-free processing temperatures, T_d is the more relevant measure. T_g is relevant when assessing loss of mechanical strength which can lead to PCB warpage, bowing, and increased z-axis expansion, which in turn can have a significant impact on PCB reliability.

There are a number of different measurement techniques to determine the T_g. The most commonly used methodologies are thermal analysis techniques. Other techniques are available (such as spectral analysis and electrical characterization), but their use is limited and therefore they will not be discussed here. The three thermal analysis methods that will be discussed in this section are as follows:

- Differential scanning calorimetry (DSC)
- Thermomechanical analysis (TMA)
- Dynamic mechanical analysis (DMA)

All of these techniques measure slightly different property changes in the laminate material and therefore the values that are obtained will be different. They usually will follow this simple rule of thumb:

$$T_g\text{ (DMA)} > T_g\text{ (DSC)} > T_g\text{ (TMA)}$$

12.5.2.1.1 T_g by Thermomechanical Analysis (TMA). Thermomechanical analysis measures dimensional changes in a material as it is heated from room temperature to a preset final temperature. The change in length (width or height) of the specimen with the change in temperature determines the CTE of the material. At T_g, the expansion coefficient of the material changes and it is this property change that is used to determine the T_g by TMA.

The test procedure is specified in IPC-TM-650, method 2.4.24C. The specimen should have a minimum thickness of 0.51 mm and at least two samples taken from random locations in the material need to be tested. Any copper cladding needs to be removed by etching. The treatment of the copper foil leaves a negative imprint on the laminate surface, which can lead to problems during the measurement right below the actual glass transition temperature. Therefore, the surface of the specimen should be sanded lightly prior to testing. The edges of the sample should also be smooth and burr-free. Use care to minimize stress or heat on the specimen.

The samples need to be preconditioned for 2 hours at 105°C (221°F), then cooled to room temperature in a desiccator. The actual measurement should be started at a temperature no higher than 35°C (95°F); an initial temperature of 23°C is recommended. Unless otherwise specified, a scan rate of 10°C (18°F) per min. is commonly used. The temperature ramp needs to be continued at least 30°C (54°F) above the anticipated transition region. The T_g is defined as the temperature at which the two tangent lines for the thermal expansion coefficient intersect. Figure 12.5 shows an example of a typical TMA scan. The material has a T_g of 136.7°C.

It is recommended to retest the same sample after allowing it to cool down to room temperature to determine whether the measured value for T_g2 is comparable to the measured value of T_g1. If T_g1 is significantly lower than T_g2, then this could be an indication that the sample material was not fully cured when it was tested the first time. If T_g1 is significantly higher than T_g2, that could indicate that the sample material has already started to decompose and was approaching T_d. The authors recommend further testing if these measurements differ by more than 5°C.

12.5.2.1.2 T_g by Differential Scanning Calorimetry (DSC). Differential scanning calorimetry measures the difference in heat absorption or emission from a test specimen in comparison to a reference sample (usually nitrogen gas). This makes the technique applicable to determine a variety of property changes in polymer materials. It can detect the exothermic cure reaction, crystallization energy, and residual reactivity in polymers, as well as endothermic melting points. For epoxy-based resin systems, DSC is a well-suited technique as these materials go through a crystalline transition at T_g and this property change can be used to determine the glass transition. For other resin systems that are more amorphous and have a T_g that occurs over a wider range (such as polyimides), the detection may be more difficult by DSC than TMA.

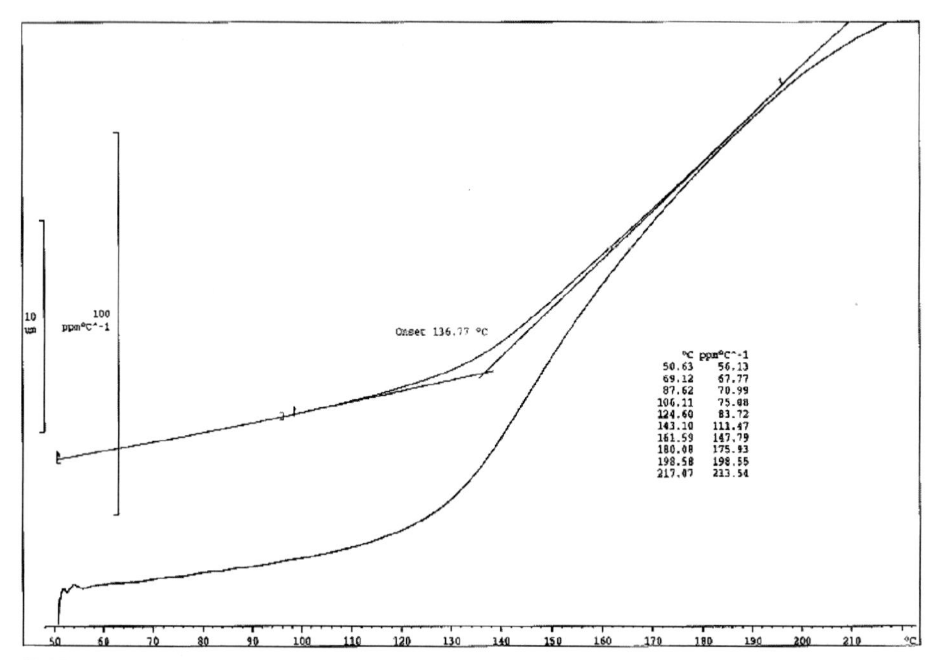

FIGURE 12.5 Typical TMA scan.

The test procedure is specified in IPC-TM-650, method 2.4.25C. The specimen should be a solid piece weighing between 15 mg and 25 mg. In the case of very thin materials, multiple pieces may be used. The specimen should be of the size and configuration that fits within the sample pan of the DSC equipment. All of the specimen preparation needs to take place before preconditioning to avoid any moisture influence prior to testing. Sample edges should be smoothed and burrs removed by light sanding or an equivalent method to achieve proper thermal conduction. Use care to minimize stress or heating of the specimen. Even though the IPC test method allows samples to be tested with copper cladding, the authors recommend testing only without copper cladding. This increases the relative mass of polymer in the DSC sample pan and leads to a better signal and detection of T_g.

The samples need to be preconditioned for 2 hours at 105°C (221°F), then cooled to room temperature in a desiccator for at least a half-hour prior to testing. Place the specimen in a standard aluminum sample pan with an aluminum lid. For reference purposes, a cover lid crimped onto the sample pan should be used. Start the scan at a temperature at least 30° lower than the anticipated onset of T_g. Unless otherwise specified, a scan at a rate of 20°C/min (36°F/min) is recommended and the temperature ramp needs to be continued at least 30°C (54°F) above the anticipated transition region. The glass transition temperature is determined using the heat flow curve. In many cases, the DSC equipment will have applicable software installed to determine T_g. A tangent line is fit onto the curve above the transition region, and a second tangent line is fit below the transition region. The temperature on the curve halfway between the two tangent lines, or 1/2 δCp, is the glass transition point T_g. Figure 12.6 shows an example of a DSC scan where the material has been scanned twice. The T_g of the first scan was 192.9°C. The second scan yielded a slightly higher value of 195.3°C (likely caused by additional curing of the material).

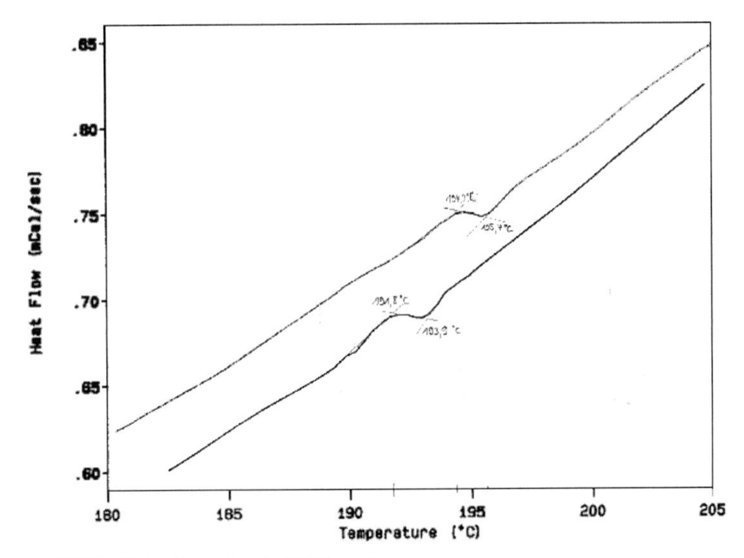

FIGURE 12.6 Example of a DSC result.

12.5.2.1.3 T_g by Dynamic Mechanical Analysis (DMA). Dynamic mechanical analysis applies an oscillatory stress or strain on the sample while the temperature is increased during the experiment. The ability of the material to elastically store the mechanical strain energy changes during the heat cycle as it changes its viscoelastic behavior from glassy to rubbery. This property change determines the glass transition temperature by DMA.

The test procedure is specified in IPC-TM-650, method 2.4.24.2. The test specimen should consist of a strip of laminate material compatible to the measuring equipment. For all samples with woven reinforcement, it is necessary to make sure that the specimens are cut parallel or perpendicular to the woven structure. The analysis is based on an assumption of constant specimen geometry; therefore, the test specimens must be stiff enough not to deform plastically during the experiment. All copper needs to be etched off.

While the IPC-method recommends preconditioning of the samples at 23°C and 50 percent RH for at least 24 hours prior to testing, the authors recommend preconditioning similar to that employed for TMA or DSC measurements. Samples are baked for 2 hours at 105°C (221°F), then cooled to room temperature in a desiccator for at least a half-hour prior to testing. The sample is mounted in the fixture making sure that it is perpendicular to the clamps. Both clamps are tightened using torque screwdrivers to ensure that the sample does not slip during the measurement and that no stresses have built up around the sample that could negatively influence the results. The sample is loaded with a frequency of 1 Hz (6.28 radian/sec) while it is heated in dry nitrogen or dry air at a rate of no faster than 2°C per min. The temperature ramp needs to be continued at least 50°C above the transition region. The glass transition temperature is defined as the temperature corresponding to the maximum in the tan δ versus temperature curves at a frequency of 1 Hz. Tan δ is calculated from Eq. 12.3.

$$\tan \delta = E''/E' \tag{12.3}$$

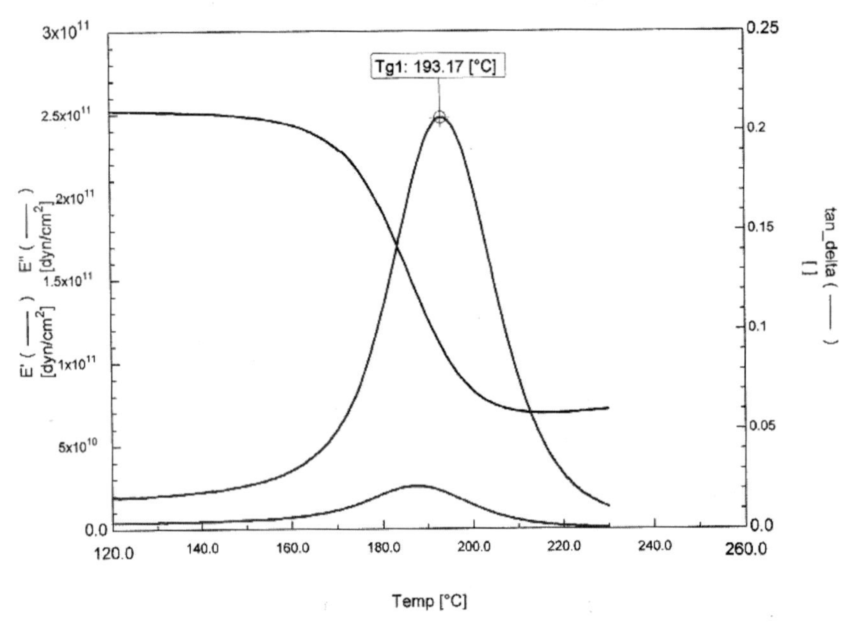

FIGURE 12.7 Typical DMA scan.

where E″ = loss modulus
E′ = storage modulus

Figure 12.7 shows an example of a typical DMA scan. The material has a T_g of 193.17°C.

12.5.2.2 Coefficient of Thermal Expansion (CTE). The CTE describes the property of materials to expand as they are being heated. Because of the reinforced composite construction of most laminates, the CTE in x and y direction is usually different from the CTE in z direction. By convention, the x direction corresponds to the warp direction of the reinforcement, whereas the y direction corresponds to the fill direction of the fabric. The z direction is the direction perpendicular to the plane of the laminate. The difference between the values is caused by the reinforcement, which severely restrains the expansion in x and y, whereas the resin can expand unrestrained in the z direction.

For copper-clad laminate materials used in printed circuit boards, the CTE in x, y, and z direction is of great importance. The x and y direction is critical because of all the components that will be mounted on the final printed circuit board. The greater the mismatch between the laminate material and the soldered components, the higher the risk that temperature changes will lead to solder fatigue and eventually to a reduction in the reliability of the board. The z direction is of equal importance because the expansion of the laminate during thermal cycles will lead to stresses in the copper plating (CTE of 17 ppm/K) of the plated through or buried holes in the board. A low z-axis CTE and high T_g are therefore generally desirable for increased through hole reliability.

A number of different measurement techniques are available to determine the CTE of copper-clad laminates. The most commonly used is TMA following IPC-TM-650, method 2.4.24C, as described in Section 12.2.1. When determining the CTE of a laminate, the temperature scan must start at a temperature sufficiently lower than the specified temperature range for which the CTE is being determined to allow the heat rate to stabilize. The typical scan rate is also 10°C (18°F) per min. and the scan should be continued to at least 250°C (482°F). Because the expansion properties of the material change at T_g, there are usually two CTE values that are being reported: the CTE below T_g is commonly referred to as α_1, and the CTE above T_g as α_2 (see Eq. 12.4).

$$\text{CTE}: \alpha = \Delta L / \Delta T \text{ ppm/K} \tag{12.4}$$

where ΔL = change in length
ΔT = change in temperature
K = degrees Kelvin

In the TMA scan example (Fig. 12.5), the material has a z-axis CTE(α_1) of ~ 65ppm/K and a z-axis CTE(α_2) of ~150 ppm/K.

The CTE in x and y direction can also be measured by TMA, although care must be taken during the preparation of the sample to avoid any influence of reinforcement material on the TMA probe. The results are strongly dependent on the properties of the reinforcement fabric.

Another method to determine the x- and y-axis CTE of laminate materials employs strain gauges. This method is described in IPC-TM-650, method 2.4.41.2. Many details need to be taken into consideration when making use of this method. The strain gauges need to be calibrated for the specific temperature range and the adhesive used to attach them needs to be stable over the whole range. Special care needs to be taken during specimen preparation and gauge attachment. It is recommended to run one heat cycle prior to the actual measurement to remove any residual stresses. More details can be found in IPC-TM-650.

12.5.2.3 Thermal Resistance. The thermal resistance of laminate materials is one of the key properties, especially in light of the lead-free processing conditions. It is the most important indicator of the performance of printed circuit boards during assembly operations. As already mentioned in the Sec. 12.1, most printed circuit boards will be required to withstand at least five reflow cycles with a peak temperature in the range of 260°C, but that number may increase to six and above for more complex assemblies. During exposure to these process temperatures, the laminate cannot delaminate or begin decomposition. Another important factor that can be assessed with thermal resistance tests is the performance of the laminate in printed circuit board applications that expose the board to high-operating temperatures. Several different test methods should be considered when qualifying laminates.

12.5.2.3.1 Solder Dip Resistance. The test method is described in IPC-TM-650, method 2.4.23. The original purpose of this test technique was to assess the solderability of the laminate surface, but today it is often used to assess the ability of laminate materials to withstand the temperatures in a molten solder bath. It is important to note that the original IPC methodology has not been revised in a long time and the current version may not reflect the actual temperature requirements of lead-free processing. The authors still feel that this method is a very good assessment and differentiation technique for laminate materials and recommend that either solder pot temperatures or exposure times are increased if the laminates are intended for use in lead-free assembly processes. Performance factors that are being considered are resistance to softening, loss of surface resin, scorching, delamination, blistering, and measling.

The material will be tested in three different surface configurations: (1) a surface upon which no metal cladding was ever applied (if possible), (2) a surface with the metal cladding removed by standard etching processes, and (3) a surface with metal cladding as received. The sample size for all specimen is 1 $^1/_4$ in. × 1 $^1/_4$ in. thickness and three samples are required for each surface configuration. All samples are tested using the same procedure.

Preclean the samples by immersing them for 15 sec. in 10 percent hydrochloric acid (HCl) (by volume) and then rinse in water. The HCl should be at a temperature of 60°C (140°F). Dry the specimen quickly to avoid excess oxidation of the sample. Dip the sample into a flux agent and allow draining for 60 sec. before proceeding with the solder dip. Stir and then skim the surface of the molten solder with a clean stainless steel paddle to ensure that the solder is of uniform composition and at a temperature of 245°C (473°F). Immerse the specimen edgewise into the molten solder. The insertion and withdrawal should be at a rate of 1 in. per sec. and the dwell time in the solder should be 4 sec. Upon withdrawal, allow the solder to solidify by air-cooling while the specimen is in a vertical position. Thoroughly remove the flux.

The samples are examined for any evidence of discoloration or surface contaminants, loss of surface resin, softness, delamination, interlaminar blistering, or measles. The specimens having metal cladding also need to be examined for blistering or delamination of the metal foil from the laminate material.

12.5.2.3.2 Solder Float Resistance. This test addresses the thermal resistance of the laminate material floating on the solder bath. Because this method subjects the sample to a thermal gradient across the z-axis of the material similar to an actual wave solder operation, the results of this test are particularly important and—as mentioned previously—either solder pot temperatures or exposure times should be increased if the laminates are intended for use in lead-free assembly processes.

The test method follows IPC-TM-650, method 2.4.13. At least two samples need to be tested for each material and they should be taken from random locations. For double-clad laminate, the copper foil needs to be removed from the backside of each specimen by using standard etching processes. The test samples need to be preconditioned in an air circulating oven at 135°C for one hour to remove any excess moisture that could lead to premature failures. After preconditioning, the specimens can be held in a desiccator at room temperature. After attaching the sample to the solder float test fixture, float it for 10 sec., foil side down, on the surface of the molten solder at a temperature of 260°C (for method A) or 288°C (for method B). Then remove the sample and tap the edges to remove any excess solder. Thoroughly clean and visually examine the sample for blistering, delamination, or wrinkling.

In case more than one material is being qualified and no failures are being observed after testing for 10 sec. or if the goal is to determine the weakest point in the laminate and test to failure, then the authors recommend increasing the solder float time incrementally or repeating the test with the same sample until failures are observed.

12.5.2.3.3 T260 (TMA). Aside from testing materials in solder baths, there is one other technique to verify the thermal resistance of laminates. This test method is described in IPC-TM-650, method 2.4.24.1, and is used to determine the time to delamination of laminates and printed boards using a TMA system (see also section 12.5.2.1.1).

The sample requirements are identical to the ones for determination of the T_g by TMA. A minimum of two specimens should be tested, which can be taken from random locations of the material. Materials are tested as received; the metal cladding is not removed. The samples are preconditioned for 2 hours at 105°C (221°F), and then cooled down to room temperature in a desiccator. The TMA temperature ramp should start from an initial temperature no higher than 35°C (95°F) at a scan rate of 10°C/min. After the scan reaches the specified isothermal temperature, hold at that temperature (onset) for 60 min. or until failure. The time to delamination is determined as the time from the onset of the isotherm to failure. Failure is defined as any event or deviation of the data plot where the thickness is shown to have irreversibly changed. On occasion, some materials will delaminate before the isotherm is reached. In this case, the temperature at the time of failure is recorded. For epoxy laminates and similar materials, the recommended isothermal temperature is 260°C (500°F). For materials intended for lead-free processing and for polyimides and other high-temperature materials, increasing the isothermal temperature to 288°C (550°F) is recommended.

12.5.3 Electrical Characterization

Electrical properties include D_c and D_f, surface and volume resistivity, and dielectric strength breakdown.

12.5.3.1 Dielectric Constant (D_c) and Dissipation Factor (D_f). The dielectric constant (also known as permittivity, D_c, and ε_r) is defined as the ratio of the capacitance of a capacitor with a given dielectric (laminate) to the capacitance of the same capacitor with air as the dielectric. It is a measure of the ability of a material to store electrostatic energy and determines the relative speed that an electrical signal will travel within that material. The higher the D_c, the slower the resulting signal propagation speed will be. The signal speed is inversely proportionate to the square root of the dielectric constant. It is not an easy property to measure or to specify, because it depends not only on the material properties and the resin-to-glass ratio, but also on the test method, test signal frequency, and the conditioning of the samples before and during the test. It also tends to shift with temperature. Still, D_c values are important for computer simulations that are used to predict the performance of impedance-controlled high-end multilayer constructions, especially when new laminate materials are considered. See Fig. 12.8 for a schematic representation of the test.

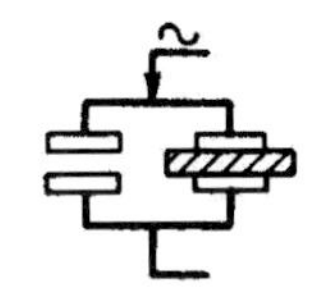

FIGURE 12.8 Dielectric constant test.

Related to the dielectric constant is the dissipation factor (D_f) or loss tangent. This is a measure of the percentage of the total transmitted power that will be lost as electrons dissipate into the laminate material. See Fig. 12.9 for a schematic representation of the test.

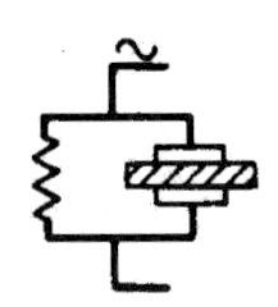

FIGURE 12.9 Dissipation factor test.

Several test methods are specified in IPC-TM-650 to determine the dielectric constant and the dissipation factor (methods 2.5.5, 2.5.5.1, and 2.5.5.2). Both methods 2.5.5 and 2.5.5.2 test laminate material with fully removed copper. At least three samples need to be tested because of the large influence of the dielectric thickness on the determined value. Method 2.5.5.1 employs samples with an etched pattern. All of these measurements are based on the capacitance of the corresponding sample. With this capacitance and the area and thickness of the sample capacitor it is possible to calculate the dielectric constant.

The dissipation factor is determined with methods 2.5.5.1 and 2.5.5.2. In one case, the value can be read from the equipment display (Agilent Technologies Mdl 4271A); in the other, it can be calculated from the measured sample conductance and capacitance and the measuring frequency.

12.5.3.2 Surface and Volume Resistivity. Electrical resistivity of laminate materials is differentiated between specific surface resistance and volume resistance. The surface resistance σ characterizes the electrical resistance between two conductors along the surface of the laminate material. The volume resistance ρ describes the electrical resistance between two layers of conducting copper along the z-axis of the laminate material. The higher the values for both of these electrical properties, the better, as this ensures proper isolation of individual copper conductors in the printed circuit board.

Both resistance values are determined according to IPC-TM-650, method 2.5.17. All resistance measurements are done with equipment capable of measuring up to 10^{12} megohm while applying 500 volts direct current to the test specimen (Agilent Technologies 16008A). The sample's size is 101.6mm × 101.6mm for laminates thicker than 0.51 mm and 50.8 mm × 50.8 mm for laminates below 0.51 mm thickness. The test pattern according to method 2.5.17 should be applied to the samples using standard photo and etch processes.

All measurements are performed by applying 500 volts direct current. The voltage needs to be applied to the samples for 60 sec. before taking the resistance reading to allow the test structures to stabilize. The surface resistance is determined between the outer ring electrode and the inner solid electrode. The volume resistance is determined between the solid front and back electrodes after changing the connecting cables appropriately. The values for volume and surface resistivity can be calculated from the measured resistance values as shown in Eqs. 12.5 and 12.6.

Volume resistivity ρ (meg-ohm-cm)

$$\rho=(R^*A)/T \tag{12.5}$$

where R = Measured resistance in meg-ohms
A = Effective area in square cm
T = Average thickness of specimen in cm

Surface resistivity σ (meg-ohms)

$$\sigma = (R*P)/D \tag{12.6}$$

where R = Measured resistance in meg-ohms
P = Effective perimeter of electrode in cm
D = Width of test gap in cm

12.5.3.3 Dielectric Strength Breakdown. The dielectric strength of a laminate material is its ability to resist electrical breakdown. The dielectric strength defines a specific voltage that the laminate resists for a specified time whereas the dielectric breakdown voltage defines the maximum voltage at which the laminate fails. These properties can be measured perpendicular to the reinforcement (z-axis) or parallel to the reinforcement (x-y axis). The more important value is the z-axis strength because more and more thin prepreg and laminate cores are being used in high-end multilayer applications. The minimum thickness in the laminate is defined as the shortest distance between the copper treatment peaks that needs to resist the desired test voltage. The values for dielectric strength vary with test setup, temperature, humidity, frequency, and wave shape, but are, if tested under controlled conditions, comparable between materials.

The test method following IPC-TM-650, method 2.5.6.2, describes the determination of the perpendicular electric strength of laminates. Four specimens should be tested. The sample size is recommended to be 4 in. × 4 in. with any copper cladding removed. Unless otherwise specified, the samples need to be conditioned for 48 hours in distilled water at 50°C. After that, the samples are immersed in ambient temperature distilled water for a minimum of 30 min. to allow the samples to achieve temperature equilibrium without a substantial change in moisture content. The test is performed at ambient temperature (23°C). Relative humidity is not significant as the tests are performed under oil. Samples are inserted into the high-voltage test equipment and tested to failure at a 500volts/sec. increase. The results are reported in volts/mil.

The test method (see Fig. 12.10) following IPC-TM-650, method 2.5.6, describes the determination of the parallel electric strength of laminates. Four specimens should be tested: two in machine direction, and two in the transverse direction for reinforced materials. The samples size is recommended to be 3 in. long by 2 in. wide with any copper cladding removed. Two holes 0.188 in. in diameter are to be drilled along the centerline of the 3 in. dimension and midway between the edges in the 2 in. dimension, with a spacing of 1 in. center to center. Conditioning and test setup in oil is identical. Electrodes are inserted in the holes and the samples are tested to failure at a 500volts/sec. increase. The results are reported in kilovolts (KV).

FIGURE 12.10 Dielectric strength breakdown

12.5.4 Other Laminate Properties

Other laminate properties to consider include flammability and water absorption.

12.5.4.1 Flammability. The flammability of laminate materials is classified according to Underwriters Laboratories (UL) specifications. All of the tests are performed using a standard test setup under an exhaust hood using a Bunsen burner as a source for the flame. The categories are as follows:

- **UL-94-V-0** Specimens must extinguish within 10 sec. after each flame application and within a total of less than 50 sec. after a total of 10 flame applications. No samples are to drip flaming particles or exhibit glowing combustion that lasts for more than 30 sec. after the second flame test.

- **UL-94-V-1** Specimens must extinguish within 30 sec. after each flame application and within a total of less than 250 sec. after a total of 10 flame applications. No samples are to drip flaming particles or exhibit glowing combustion that lasts for more than 60 sec. after the second flame test.
- **UL-94-V-2** Specimens must extinguish within 30 sec. after each flame application and within a total of less than 250 sec. after a total of 10 flame applications. Samples may drip flaming particles or burn briefly, but no specimen may exhibit glowing combustion that lasts for more than 60 sec. after the second flame test.

In most cases, the laminator will provide results for these tests routinely. In the case of new materials for which the laminator has not supplied any flammability data, it may be a good idea to check this property. This does not require an elaborate test setup, and you can get a first indication by igniting a sample specimen under an exhaust hood using a lighter.

12.5.4.2 Water Absorption. Depending on their specific molecular composition, every laminate material will absorb a certain amount of water. This will happen not only during the many wet processing steps in printed circuit board manufacturing but also as a result of exposure to normal environmental conditions. The absorbed moisture may change the properties of the laminate and increase the risk of blistering and delaminations during high-temperature processes such as reflow soldering.

The test method, according to IPC-TM-650, method 2.6.2.1, determines the amount of water that is absorbed by a laminate material sample when immersed in water for 24 hours. The test is easy to perform and the results for different laminates are readily comparable.

The test samples for this test need to be 2 in. long by 2 in. wide. The thickness is not specified but should not vary widely when more than one material is characterized. The edges of the samples need to be sanded smooth and copper cladding is to be removed using standard etching processes. The samples are preconditioned in a drying oven for 1 hour at 105°C (221°F), cooled down to room temperature in a desiccator, and weighed immediately after removal. Then the samples are placed in distilled water at 23°C. It is important to place the samples on their edge to maximize the laminate area exposed to water. After 24 hours, the samples are removed, dried with a dry cloth, and immediately weighed. The moisture absorption is reported in percent weight increase.

12.5.5 Additional Tests

In addition to the previously described test methods, many more can be found in IPC-4101 and in the test methods manual IPC-TM-650. All of these tests address laminate properties that may have a significant impact on the performance of the final product. The final decision as to which qualification tests to include or exclude always needs to be made on a case-by-case basis depending on the performance requirements for the printed circuit board.

At the same time, there are a number of laminate qualification tests that rarely will be performed in the printed circuit manufacturing facility. In many cases, the laminate supplier will perform a number of qualification tests itself and share the results with its customers. This may in many cases be sufficient, especially after a close relationship between the supplier and the PCB manufacturer has been established.

12.5.6 Prepreg Testing

During the qualification of a new laminate material, it may also become necessary to perform prepreg-specific tests to verify its quality. The most commonly tested properties of prepregs are resin content, flow during mutilayer processing, and gel time. Details about additional properties and all corresponding test methods can be found in IPC-4101.

12.6 CHARACTERIZATION TEST PLAN

Table 12.4 summarizes all of the test procedures.

TABLE 12.4 Laminate Characterization Test Plan

1. Step	Data comparison	Unit	Test result			
	Material available in production volume	yes/no				
	Material cost	$/m^2				
	Material data sheet available	yes/no				
2. Step	**First test runs in small volume**	**Unit**	**Test method**	**Conditioning**	**Data sheet**	**Test result**
	Surface and appearance		IPC-TM-650, method 2.1.2 & 2.1.5	As received		
	Copper peel strength laminate Condition A	N/mm	IPC-TM-650, method 2.4.8	As received		
	Copper peel strength laminate Condition B	N/mm	IPC-TM-650, method 2.4.8	Solder pot 288°C/10s		
	Solder shock laminate 288°C/10s	Pass/fail	IPC-TM-650, method 2.4.23	As received		
	Solder shock laminate 288°C/60s	Pass/fail	IPC-TM-650, method 2.4.23	As received		
	Glass transition temperature (DSC, TMA, or DMA)	°C	IPC-TM-650, as applicable	As received		
	Decomposition temperature (T_d)	°C	IPC-TM-650, method 2.3.40	As received		
	Z-axis expansion laminate TMA	ppm/°C	IPC-TM-650, method 2.4.24	As received		
3. Step	**Material characterization**	**Unit**	**Test method**	**Conditioning**	**Data sheet**	**Test result**
	Dielectric constant (1 MHz)	—	IPC-TM-650, method 2.5.5.2	24h/23°C/50%		
	Dielectric constant (1 MHz)	—	IPC-TM-650, method 2.5.5.2	96h/35°C/90%		
	Dissipation factor (1 MHz)	—	IPC-TM-650, method 2.5.5.2	24h/23°C/50%		
	Dissipation factor (1 MHz)	—	IPC-TM-650, method 2.5.5.2	96h/35°C/90%		
	Surface resistance	MW	IPC-TM-650, method 2.5.17	24h/23°C/50%		
	Surface resistance	MW	IPC-TM-650, method 2.5.17	96h/35°C/90%		
	Volume resistance	MW cm	IPC-TM-650, method 2.5.17	24h/23°C/50%		
	Volume resistance	MW cm	IPC-TM-650, method 2.5.17	96h/35°C/90%		

TABLE 12.4 Laminate Characterization Test Plan *(Continued)*

3. Step	Material characterization	Unit	Test method	Conditioning	Data sheet	Test result
	Dielectric withstanding voltage	V/mil	IPC-TM-650, method 2.5.6	48h in 50°C H_2O		
	Glass transition temperature (DSC)	°C	IPC-TM-650, method 2.4.25C			
	Glass transition temperature (TMA)	°C	IPC-TM-650, method 2.4.24C			
	Glass transition temperature (DMA)	°C	IPC-TM-650, method 2.4.24.2			
	Decomposition temperature T_d	°C	IPC-TM-650, method 2.3.40	As received		
	CTE x,y (α_1)	ppm/K	IPC-TM-650, method 2.4.41.2			
	CTE x,y (α_2)	ppm/K	IPC-TM-650, method 2.4.41.2			
	CTE z (α_1)	ppm/K	IPC-TM-650, method 2.4.24			
	CTE z (α_2)	ppm/K	IPC-TM-650, method 2.4.24			
	Water absorption	%	IPC-TM-650, method 2.6.2.1			
	Flammability		UL-94	—		

12.7 MANUFACTURABILITY IN THE SHOP

During the qualification of a new material, it is also crucial to run the laminate through the production processes to check whether they are compatible. Table 12.5 summarizes all the steps that are necessary to control.

The first process step is innerlayer processing. Here, a change in stiffness of the new laminate material can have a significant impact on the processability in horizontal manufacturing lines. This is especially important for thin cores. A change in the copper quality of the new laminate may impact the adhesion of the innerlayer photoresist and can also affect the copper etch rates. During automatic optical inspection (AOI), it is necessary to verify the contrast between copper circuitry and laminate and adjust the AOI settings if necessary. In case of necessary innerlayer repairs, the settings for the welding process may need adjustment to avoid thermal damage of the laminate and ensure reliable interconnects. The last step during innnerlayer processing is copper blackoxide. Here, it is necessary to verify the compatibility of the employed blackoxide (reduced/nonreduced or alternative) to the base material to obtain a sufficient innerlayer bond strength of the multilayer product. The dimensional stability of the innerlayers must be measured after the multilayer cycle to adjust the innerlayer scaling factors of the new material.

The next step after multilayer lamination and edge routing is drilling. A change in the thermomechanical properties of the laminate may impact the quality of the drilled holes, and adjustments to drill speeds or desmear settings may be necessary. After metallization and plating, the adhesion of the copper in the plated through hole as well as reliable contacts to all innerlayers must be verified.

TABLE 12.5 Summary of Production Steps

Compatibility with innerlayer (IL) process steps
Preclean Resist lamination Exposure Develop, Etch, Strip Punching IL-Automatic optical inspection IL-Repair/outerlayer (OL)-repair Blackoxide/oxide alternatives
Compatibility with multilayer process steps
IL drying Lay up Press cycle Edge routing Dimensional stability check
Compatibility with process steps from drill to electroless copper
Drilling Brush/Pumice/Deburr Desmear/Plasma Desmear/Other E'less copper or similar process
Compatibility with soldermask process step
Compatibility with different finish metallization processes
Electroless Ni/Au Immersion Sn Immersion Ag
Compatibility with routing process

The adhesion of soldermask may also change as well when a new material is introduced. This is especially critical in combination with any metal finish process steps (such as electroless Ni/immersion Au, immersion Sn, and immersion Ag). In addition to having a negative impact on soldermask adhesion, these processes in combination with new laminates can also lead to nonselective metal plating or skip plating. One of the last manufacturing steps that should be evaluated during the introduction of a new laminate material is board routing and scoring. Here the use of a new base material with changed stiffness or different reinforcement may lead to necessary process adjustments.

P · A · R · T · 4

ENGINEERING AND DESIGN

CHAPTER 13
PHYSICAL CHARACTERISTICS OF THE PCB

Lee W. Ritchey
3Com Corp., Santa Clara, California

13.1 CLASSES OF PCB DESIGNS

Printed circuit boards (PCBs) or printed wiring boards (PWBs) can be divided into two general classes which have common characteristics based on their end functions. These two classes have very different materials and design requirements and functions and, as a result, need to be treated differently throughout the design and fabrication processes. The first class contains analog, RF, and microwave PCBs such as are found in stereos, transmitters, receivers, power supplies, automotive controls, microwave ovens, and similar products. The second contains digital-based circuitry such as is found in computers, signal processors, video games, printers, and other products that contain complex digital circuitry. Table 13.1 lists many of the characteristics of each class of PCBs.

TABLE 13.1 Characteristics of RF/Analog versus Digital-Based PCBs

RF, microwave, analog PCB	Digital-based PCB
Low circuit complexity	Very high circuit complexity
Precise matching of impedance often needed	Tolerant of impedance mismatches
Minimizing signal losses essential	Tolerant of lossy materials
Small circuit element sizes often essential	Small circuit element sizes desirable
Only 1 or 2 layers	Many signal and power layers
High feature accuracy needed	Moderate feature accuracy needed
Low/uniform dielectric constants needed	Dielectric constant secondary

13.1.1 Characteristics of Analog, RF, and Microwave PCBs

As can be seen from Table 13.1, the materials, design, and fabrication needs of this class of PCBs are markedly different from those of PCBs commonly referred to as digital.

- Circuit complexity is low because most components used have two, three, or four leads. This is due to the high usage of resistors, transistors, capacitors, transformers, and inductors.

- Traces, pads, and vias often act as inductors, capacitors, and coupling elements in the actual circuit. Their shapes may have a material effect on overall circuit performance. For example, the lead inductance and capacitance in a transistor collector circuit wire may act as the resonant components for an RF amplifier or it may degrade performance if it is unwanted. Figure 13.1 shows the impedance of traces as a function of their capacitance.

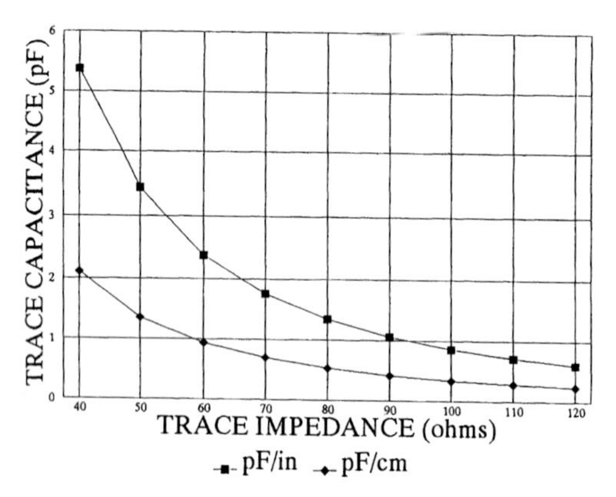

FIGURE 13.1 Trace capacitance vs. trace impedance, based on $L_0 = 8.5$ nH/in. *(Prepared by Ritch Tech, 1992.)*

- Two traces running side by side may be used to couple a signal from one circuit to another as is done in directional couplers of microwave amplifiers. (This same coupling in a digital circuit may result in a signal getting into a neighboring circuit causing a malfunction.)
- A series of conductors running side by side may function as a band-pass filter. Proper performance of filters, as well as most other wideband RF circuits, depends on all frequencies traveling with equal speed through the structures. To the extent that this is not true, frequencies that arrive later distort the signal being processed. This is called *phase distortion.*

Figure 13.2 illustrates the dielectric constant of various PCB materials as a function of frequency. Notice that some materials exhibit a dramatic decline in dielectric constant as frequency increases. The speed with which a signal travels through a dielectric is a function of the dielectric constant. Figure 13.3 illustrates signal velocity as a function of dielectric constant. From these two graphs it can be seen that using a dielectric material with a nonuniform dielectric constant in RF applications may result in severe phase distortion because the higher-frequency components arrive at the output before the lower frequencies.

- A trace in a power supply circuit may be expected to carry several amps without significant heating or voltage drop. Its resistance may even be used as a sense element to detect current flow. Similarly, handling large currents with insufficient copper in a trace may result in a voltage drop that degrades circuit performance. Figure 13.4 illustrates trace resistance of a copper trace as a function of its width and thickness. Figures 13.5 and 13.6 illustrate conductor heating as a function of width, thickness, and current flow.
- PCBs used in consumer electronics tend to share lower circuit complexity with RF and analog PCBs. However, the performance demands are far lower. The need for the lowest possible cost offsets this. To achieve the cost objectives, every effort is made to keep all connections on a single side and to form all holes in a single operation by punching. This

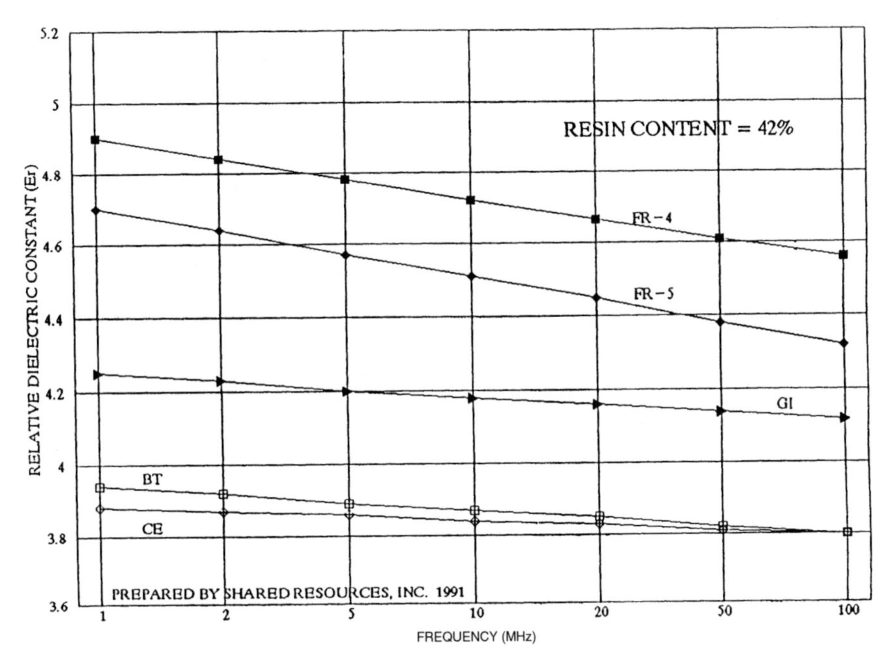

FIGURE 13.2 Dielectric constants versus frequency of various PCB materials.

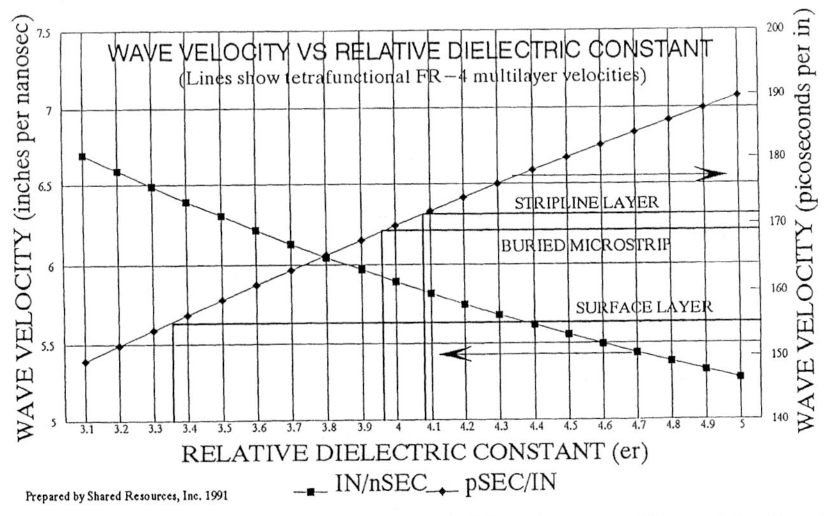

FIGURE 13.3 Signal velocity as a function of dielectric constant. *(Prepared by Shared Resources, Inc., 1991.)*

eliminates both drilling and plating. The substrate material system is often resin impregnated paper, the lowest-cost substrate system for electronic packaging.

Summarizing, successful RF and analog design depends heavily on the properties of the materials used and on the physical shapes of the conductors and their proximity to each other

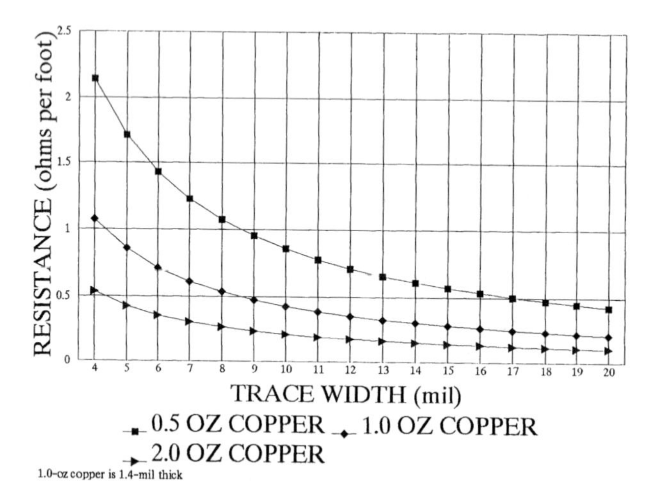

FIGURE 13.4 Trace resistance vs. trace width and thickness. *(Prepared by Ritch Tech.)*

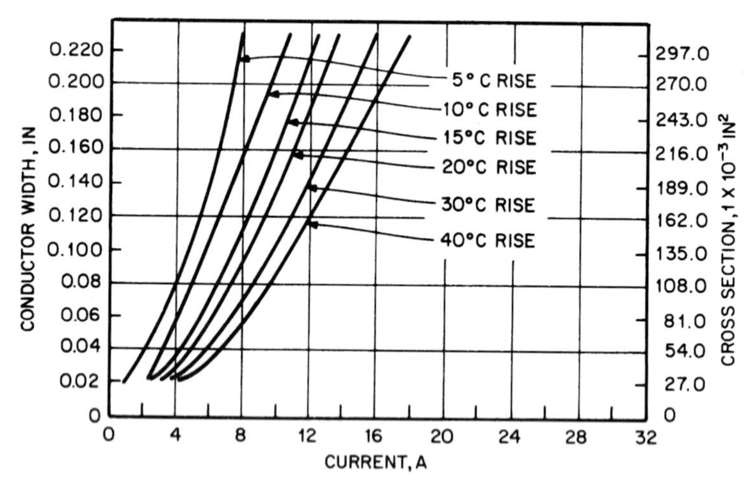

FIGURE 13.5 Temperature rise vs. current for 1-oz copper.

rather than on the ability to handle very large numbers of circuits simultaneously. Hand routing or connecting of the individual parts coupled with manipulating the shapes of individual copper features are essential parts of this design process. For these reasons, the design tools and design team must be chosen to meet these needs. Physical layout tools that provide convenient graphical manipulation of PCB shapes are a must.

13.1.2 Characteristics of Digital-Based PCBs

Compared to RF and analog PCBs, digital-based PCBs have complex interconnections, but are tolerant of rather wide feature size and materials variations.

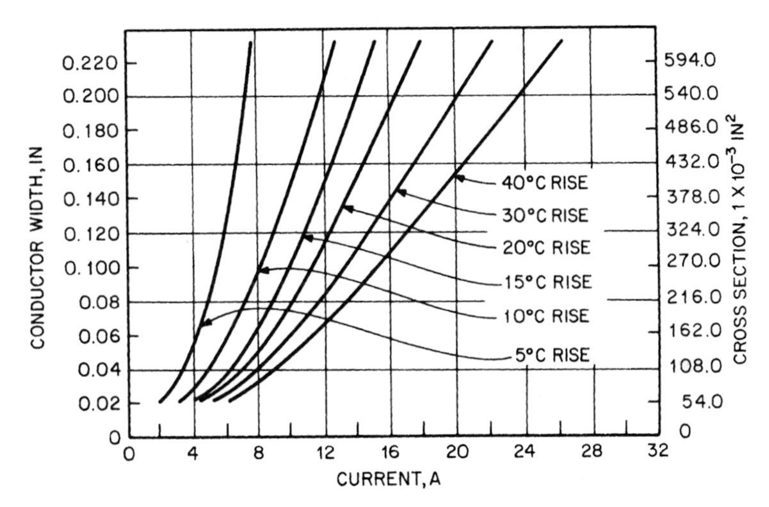

FIGURE 13.6 Temperature rise vs. current for 2-oz copper.

- They are characterized by very large numbers of components, often numbering in the hundreds and sometimes the thousands.
- Digital components often have large numbers of leads, as high as 400 or more. This high lead count stems from logic architectures that have data and address buses as wide as 128 bits or more. To connect PCBs with these wide data buses, digital systems often contain board-to-board connectors with as many as 1,000 pins.
- Digital circuits have increasingly fast edges and low propagation delays to achieve faster performance. Edge rates as fast as 1 ns are now encountered in devices destined for products as common as video games. Table 13.2 lists edge speeds of some commonly used logic families, edge rate being the time required for a logic signal to switch from one logic level to the other (switching speed). Propagation delays, the time required for a signal to travel through a device, are decreasing along with edge rates.

TABLE 13.2 Typical Logic Family Switching Speeds

Logic family	Edge speed, ns	Critical length, in
STD TTL	5.0	14.5
ASTTL	1.9	5.45
FTTL	1.2	3.45
HCTTL	1.5	4.5
10KECL	2.5	7.2
BICMOS	0.7	2.0
10KHECL	0.7	2.0
GaAs	0.3	0.86

- These fast edges and short propagation delays lead to transmission line effects such as coupling, ground bounce, and reflections that can result in improper operation of the resulting PCB. Table 13.2 illustrates the degree to which a fast switching signal will couple into a neighboring line as a function of the edge-to-edge separation and the height of the signal pair above the underlying power plane. The critical length listed in Table 13.2 is the length of parallelism between two traces at which the coupling levels in Fig. 13.7 are reached.

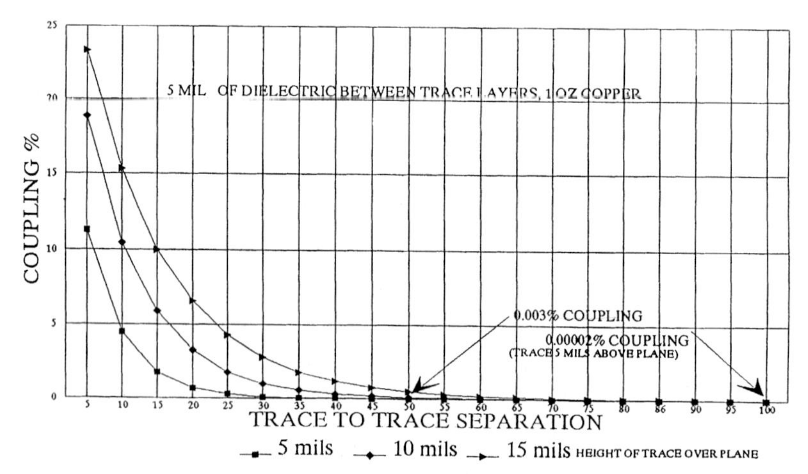

FIGURE 13.7 Trace-to-trace coupling. *(Prepared by Shared Resources, Inc.)*

The digital circuits themselves are designed to function properly with input signals that vary over a relatively wide range of values. Figure 13.8 illustrates the signal levels for a typical logic family, in this case ECL. The smallest output signal from an ECL driver is the difference between VOL_{max} and VOH_{min} or 0.99 V. The smallest input voltage to a device at which the logic part is designed to work properly is the difference between VIL_{max} and VIH_{min} or 0.37 V. The difference between these two levels, the noise margin of 0.62 V is available to counteract losses in the wiring and the dielectric and from other sources such as coupling and reflections. From this it can be seen that digital logic has a high tolerance of losses and higher immunity to noise.

This tolerance of noise and losses makes it possible to have trace features and base materials that introduce substantial losses and distortion while still achieving proper operation. It is this relatively high tolerance of distortion that makes it possible to manufacture economical digital PCBs.

Summary. The large number of connections in digital PCBs generally requires multiple wiring layers to successfully distribute power and interconnect all the devices. As a result, the design task is heavily weighted on the side of successfully making many connections in a limited number of routing layers while obeying transmission line rules. The base materials need to have characteristics that result in a PCB that is economical to fabricate and able to withstand the soldering processes while preserving high-speed performance. Compared to RF PCBs, losses in the dielectric tend to be of little concern for digital PCBs. The actual shapes of conductors, pads, holes, and other features have little effect on performance. (For detailed treatment of these topics, see Howard W. Johnson and Martin Graham, *High Speed Digital Design: A Handbook of Black Magic,* Prentice-Hall, New York, 1993.)

The PCB design system and the design skill set for digital PCBs must be optimized to ensure accuracy in making large numbers of connections while successfully handling the high speed requirements of the system. Achieving this in a reasonable amount of time demands the use of a CAD system that contains an automatic router for use in connecting the wires.

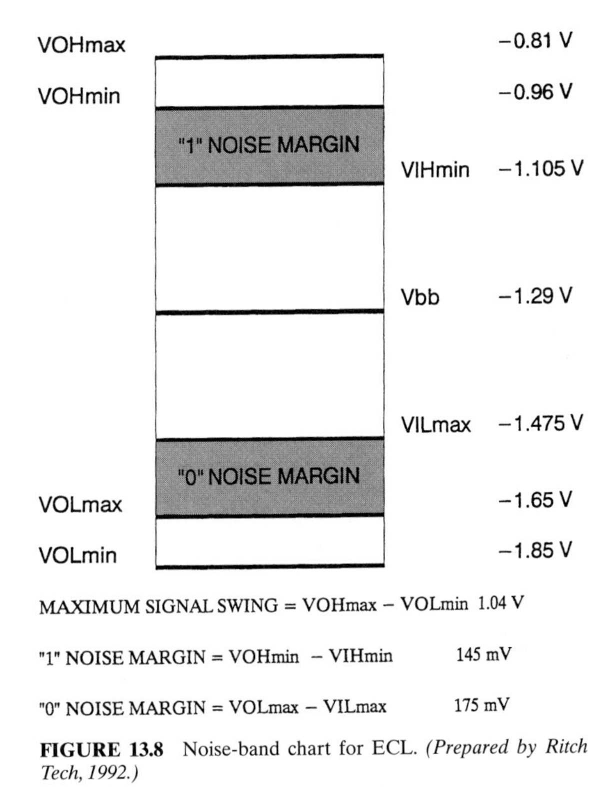

FIGURE 13.8 Noise-band chart for ECL. *(Prepared by Ritch Tech, 1992.)*

13.2 TYPES OF PCBs OR PACKAGES FOR ELECTRONIC CIRCUITS

The range of choices for packaging electronic circuits is quite broad. Some of the parameters that influence the choices are weight, size, cost, speed, ease of manufacture, repairability, and function of the circuit. The more common types are listed as follows with a brief description of their characteristics.

Levels of packaging are often used when referring to how electronic circuits are packaged. The first level of packaging is the housing of an individual component. This is usually an encapsulating coating, a molded case, or a cavity-type package such as a pin grid array (PGA). The second level of packaging is the PCB or substrate on which individual components are mounted. Third-level packaging is any additional packaging beyond these two. Most often third-level packaging takes the form of a multichip module (MCM) that has bare components mounted in it which is itself mounted on a PCB along with other components.

13.2.1 Single- and Double-Sided PCBs

These PCBs have conductor patterns on one or both sides of a base laminate with or without plated through-holes to interconnect the two sides. These are the workhorses of consumer electronics, automotive electronics, and the RF/microwave industry. They are the lowest-cost choice for consumer products. Laminate materials range from resin-impregnated paper for consumer electronics to blends of low-loss Teflon™ for RF applications.

13.2.2 Multilayer PCBs

These PCBs (see Fig. 13.9) have one or more conductor layers (usually power planes) buried inside in addition to having a conductor layer on each outside surface. The inner layers are connected to each other and to the outer layers by plated through-holes or vias. These are the packages of choice for nearly all digital applications ranging from personal computers to supercomputers. Numbers of layers range from 3 to as many as 50 in special applications. Laminate materials are nearly always some type of woven glass cloth impregnated with one of several resin systems. The resin system is chosen to satisfy requirements such as the ability to withstand high temperatures, cost, dielectric constant, or resistance to chemicals.

13.2.3 Discrete-Wire or Multiwire PCBs

This class of PCBs is a variation of the multilayer package. A circuit board is constructed by etching a pair of power layers back to back on a laminate substrate. A layer of partially cured, still sticky laminate is bonded to each side of this power plane structure. Discrete wiring is rolled into this sticky adhesive in patterns that will connect leads or serve as access to surface-mount component pads.

Once all of the wires have been rolled into place, a second layer of laminate is placed over the wires, followed by copper foil sheets. This sandwich is then laminated, drilled, and processed like any multilayer PCB. The resulting PCB has outer layers and power planes much like any multilayer PCB. The principle difference is that the printed wiring signal layers have been replaced by discrete wiring layers. In some cases of very high wiring density, alternating layers of power planes between wiring layers serve as isolation.

Designing a discrete wiring PCB (see Fig. 13.10) involves adding a special discrete-wiring router to a standard PCB design system to generate the files for the machine that rolls the wire into the dielectric. Discrete wiring once provided a faster prototyping alternative to multilayer fabrication. At the present time, both technologies are equally rapid and cost effective during prototyping. However, for modest to high-volume production, multilayer technology is more cost effective than discrete wiring.

13.2.4 Hybrids

These circuits are usually single- or double-sided ceramic substrates with a collection of surface-mount active components and screened-on resistors made from metallic pastes. They are most often found in hearing aids and other miniature devices.

13.2.5 Flexible Circuits

These circuits are made by laminating copper foil onto a flexible substrate such as Kevlar® or Kapton®. They can range from a single conductor layer up to several layers. They are most often used to replace a wiring harness with a flat circuit to save weight or space. Often, the flexible circuit will contain active and passive components. Common applications are cameras, printers, disk drives, avionics, and video tape recorders.

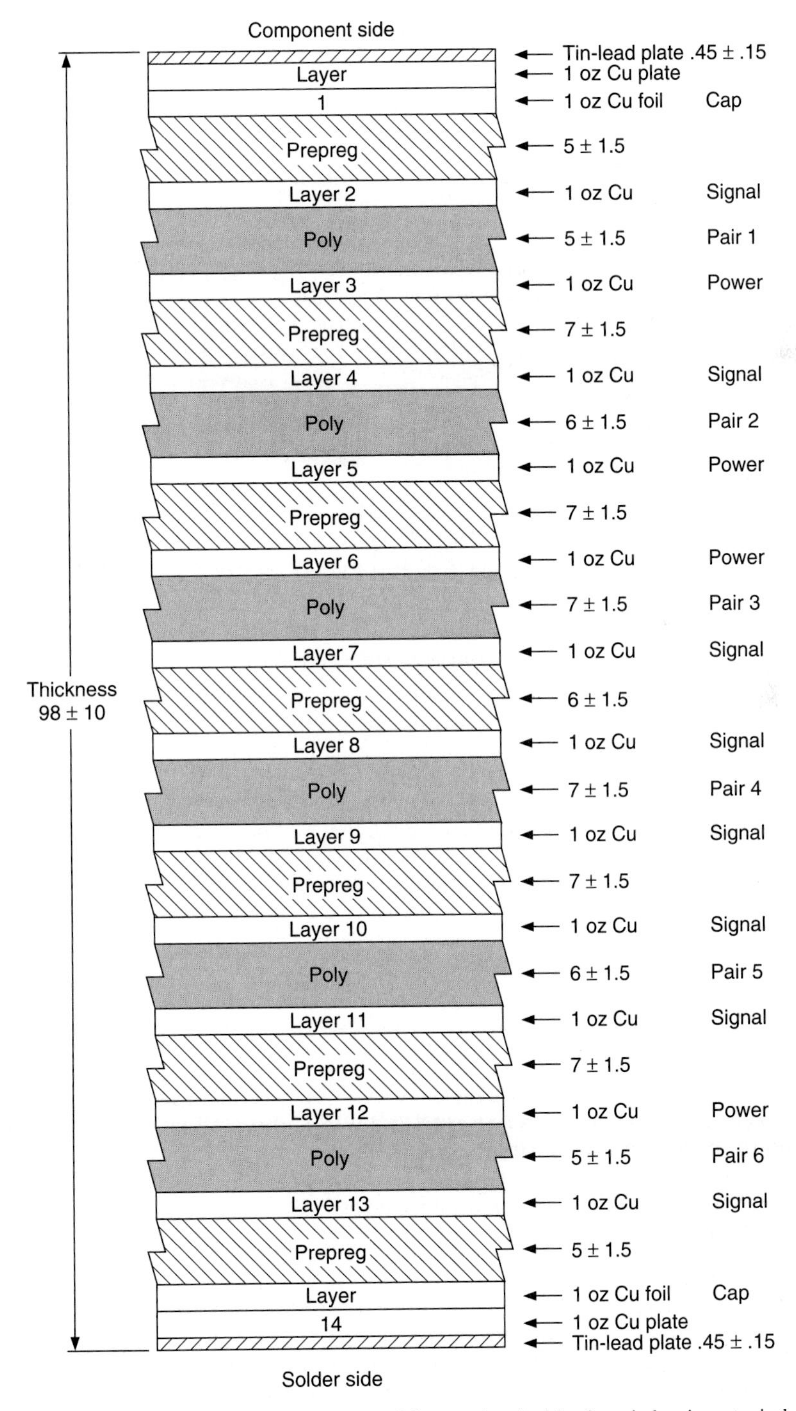

FIGURE 13.9 Cross section of 14-layer multilayer printed wiring board, showing a typical inner layer and prepreg material relationship. In this case, to reduce *z*-axis expansion, the innerlayers are polyimide, while the prepreg material is semicured polyimide. Typical signal, power, and ground layers are also indicated, as well as the thickness of the copper foil for each layer.

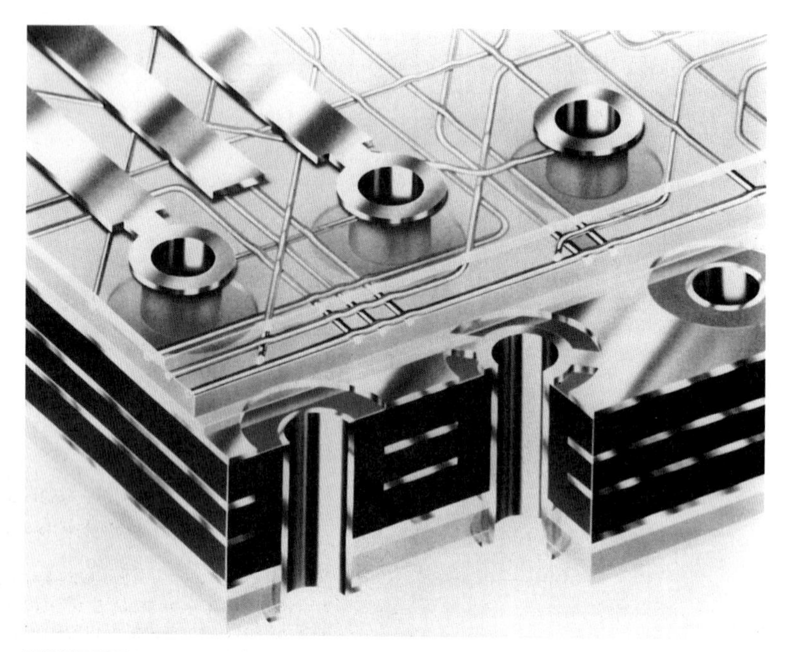

FIGURE 13.10 Cross section of a discrete-wire PCB. *(Courtesy of Icon Industries.)*

13.2.6 Flexible Rigid or Flex-Rigid

As the name suggests, these are combinations of flexible PCBs and rigid PCBs in a single unit. The flexible portion of the circuit is made first and included in the lamination process of the rigid portion of the assembly. This process eliminates wiring harnesses and the associated connectors. Applications include avionics and portable equipment such as laptop computers. As a rule, a flex-rigid assembly is more expensive than an equivalent combination of PCBs and cables.

13.2.7 Backplanes

Backplanes are special cases of multilayer PCBs. They tend to contain large quantities of connectors that have been installed using press fit pins. In addition, backplanes are used to distribute large amounts of dc power to the system. This is accomplished by laminating several power planes inside the backplane and by bolting bus bars onto the outside surfaces. Some applications require that active components, such as surface-mount ICs, be soldered to their surfaces. This greatly increases the difficulty of assembly as a result of the need to solder fine-pitch parts to a large, thick PCB.

13.2.8 MCMs (Multichip Modules)

Multichip modules are essentially miniature PCBs. Miniaturization is achieved by removing components such as ICs from their packages and mounting them directly to the substrate using wire bonds, flip chip, TAB or flip TAB. The motivation for using an MCM is miniaturization, reduction in weight, or a need to get high-speed components as close to each other as possible to achieve high-speed performance goals. MCMs usually represent a third level of packaging in a system between packaged components and the carrier PCB. As a result, this additional level of packaging virtually always results in a more expensive, more complex assembly than the equivalent circuits in standard packages. There are several types of MCM package.

13.2.8.1 MCM-L, Multichip Module Laminate. This version of an MCM is manufactured from very thin laminates and metal layers using the same techniques employed in the manufacture of standard PCBs. Features such as holes, lands, and traces are much finer and require tooling similar to that used to manufacture semiconductors. This is the least expensive MCM type to design, tool, and manufacture. The same design tools and methodologies used for PCBs can be used.

13.2.8.2 MCM-C, Multichip Module Ceramic. This version of an MCM is manufactured by depositing conductor layers on thin layers of uncured ceramic material, punching and backfilling holes for vias, stacking the layers, and firing the total to create a hard ceramic multilayer substrate. This is the second least expensive MCM type to design, tool, and manufacture. It has been the workhorse of IBM's large mainframe computers for at least two decades. The same design tools and methodologies used for PCBs can be used.

13.2.8.3 MCM-D, Multichip Module Deposited. This version of an MCM is manufactured by depositing alternating thin films of organic insulators and thin films of metal conductors on a substrate of silicon, ceramic, or metal. The design and manufacturing techniques used for this technology resemble that used to create integrated circuit metallization. The thermal conductivity of the substrate is quite good. Design and fabrication support for MCM-D is limited.

13.2.8.4 MCM-D/C, Multichip Module Deposited and Cofired. This version of an MCM is a combination of a cofired, multilayer ceramic substrate containing the common wiring for a family of modules and deposited conductor and insulation layers containing the personality wiring. It has all of the problems of each technology it uses plus problems related to mismatches in temperature coefficient of the two materials systems.

13.2.8.5 MCM-Si, Multichip Module Silicon. As the name implies, this MCM technology starts with a silicon substrate like that used to make integrated circuits. Conductor patterns are formed using silicon dioxide (glass) as an insulator and aluminum or a similar metal for the wiring patterns in the same manner as is employed to build an integrated circuit. In fact, the same design tools and fabrication tools used to build ICs are used to build MCM-Si modules.

A significant advantage of MCM-Si is the fact that the substrate is the same material as the ICs that will be attached to it. Therefore, it is thermally matched to the ICs, ensuring reliable contacts over extremes of temperature.

13.2.8.6 Summary of MCM Technologies. MCM packaging may be seen as a way to achieve higher performance from a collection of high-speed ICs than can be accomplished by mounting them onto a PCB or as a way to reduce size and weight. In reality, higher levels of integration nearly always result in a more economical solution. For all but low-volume, specialty applications, such as aerospace electronics and specialty processors for very high performance equipment, this has proven to be true for quite some time. This is likely to continue to be so for some time as semiconductor technology continues to improve the density of functionality that can be placed on a single IC. One need only examine the progression of microprocessor performance to see this phenomenon at work.

When a high-performance product requires integrated circuits made with different processing technologies, such as analog and CMOS or ECL and CMOS, integration does not represent a reasonable alternative to MCMs. Examples of this type of product are high-performance graphics products and video signal processing equipment.

13.3 METHODS OF ATTACHING COMPONENTS

A wide range of methods has evolved for attaching components to PCBs. The methods chosen as well as the combinations of methods chosen for a product have a substantial impact on the final cost, ease of assembly, availability of components, ease of test, and ease of rework. The five basic attachment combinations are: through-hole only, through-hole mixed with surface-mount on one side, surface-mount one side only, surface-mount both sides, and surface-mount both sides with through-hole.

13.3.1 Through-Hole Only

All component leads attach to the PCB by being inserted into holes that pass through the PCB. The components may be secured by wave soldering or by pressing into holes that result in an interference fit (press fit). Assembly involves a component placement operation followed by a wave-soldering operation. This method is still the workhorse of the low-cost consumer electronics industry.

13.3.2 Through-Hole Mixed with Surface-Mount

Components such as connectors and PGAs are attached to the PCB with through-hole technology. All other components are mounted using surface-mount packages. This is the most common method used to assemble electronics products. Assembly is a two-step process that involves placing all surface-mount parts and soldering them in place with a solder reflow system, then inserting all through-hole parts and soldering them in place in a wave-soldering operation. Alternatively, the through-holes may be hand-soldered if the quantity is small.

13.3.3 Surface-Mount, One Side Only

This type of package is made up of only surface-mount parts all mounted on the same side of the PCB. Assembly is a one-step process that involves placing all components and soldering them in place using some form of solder reflow.

13.3.4 Surface-Mount, Both Sides

This type of package contains surface-mount components on both sides. Assembly is a two-step process that involves placing all components on one side and reflow soldering them, followed by placing all components on the other side and reflow soldering them. In addition to more complex assembly operations, designing and testing this type of assembly is much more complex, as parts often share the same area on opposite sides of the PCB, causing conflicts when trying to locate vias and test points.

13.3.5 Surface-Mount, Both Sides with Through-Hole

As the name implies, this package type contains surface-mount parts on both sides, as well as through-hole parts such as connectors. In most cases, the surface-mount components on one side are passives, such as bypass capacitors and resistors that can withstand being passed through wave soldering. Assembly is a three-step process that involves placing the surface-mount components on the primary side and reflow soldering them. Once this is complete, the secondary-side surface-mount components are glued in place, the through-hole components are inserted, and the PCB is sent through wave soldering.

Because of the extra operations and exposure of components on the secondary side to molten solder, this type of assembly is prone to many assembly defects. It is important to note that soldering fine-pitch parts on the secondary side using wave soldering results in excessive solder bridging and should be avoided. In addition, active components, such as ICs, may be damaged by exposure to excessive heating.

13.4 COMPONENT PACKAGE TYPES

Over time, a wide assortment of packages has been developed to house components. Selecting the correct package type for each component is one of the most important parts of the design process. Package types chosen affect ease of design, assembly, test, and rework, as well as product cost and component availability.

13.4.1 Through-Hole

This class of component is characterized by parts that have wire or formed leads. These leads pass through holes drilled or punched in the PCB and are soldered to lands on the back side or to plating in the holes. This is the original package type used for electronic components. A major benefit of through-hole components is the fact that every component lead passes all the way through the PCB. Because of this, there is automatic access to any PCB layer to make connections. Further, every lead is available on the bottom of the PCB, so test tooling is easy to construct. With the advent of surface-mount components, through-hole is used primarily for connectors and pluggable devices such as microprocessors mounted in PGA packages.

Through-hole packages are often preferred for ICs and other components that dissipate large amounts of heat, because of the relative ease with which heat-sinking devices can be fitted to them. In addition, it is much easier to provide a socket for a through-hole device. This eases the task of changing programmable parts and microprocessors when it is necessary to upgrade a system.

Caution: Integrated circuits in through-hole packages are becoming difficult to find as they are displaced by surface-mount equivalents and should be avoided in new designs unless a secure supply of components is available for the production life of the design.

13.4.2 Surface-Mount

This package type is the mainstream choice for packaging electronic components of every type, including connectors. Its principle characteristic is that all connections between a component lead and the PCB or substrate is made with a lap joint to a pad on the surface of the PCB. This has both advantages and disadvantages. On the advantage side, since there are no holes piercing the PCB, wiring space on inner layers and on the reverse side is not consumed with component lead holes. Because of this, it is usually possible to wire a PCB in fewer layers than would be true with through-hole parts. Another and larger benefit is the fact that surface-mount components are always smaller than their through-hole equivalent, making it possible to fit more parts in a given area.

The main disadvantages of surface-mount components stem from the fact that there are no leads to easily grip with instrumentation probes and that there may not be access to the leads from the reverse side for purposes of production testing. This gives rise to the need to add a test pad to most nets on the back side in order to perform production test. It also gives rise to the need for very expensive, complex adapters in order to provide access to leads on processors and other complex devices to probe their inputs and outputs when performing diagnostic work.

Yet another disadvantage of surface-mount parts stems from their small size. It is more difficult to remove heat from SMT packages than it is for their through-hole equivalent. In some cases, such as high-performance processors, the heat generated by the IC is too high to permit proper operation in an SMT package.

13.4.3 Fine Pitch

Fine pitch is a special class of surface-mount components. This class is characterized by lead pitches lower than 0.65 mm (25 mils). These fine lead pitches are usually driven by very high lead count ASICs (160 pins and up) or by the extreme miniaturization requirements of PCMCIA (Personal Computer Memory Card Industry Association) cards, PDAs (personal digital assistants), and other small, high-performance products. The motivation for designating a special fine-pitch-component class of surface-mount parts is the extra difficulty of successfully testing, assembling, and reworking these parts on PCBs, as well as in building PCBs with accurately formed patterns and solder masks to mate with the leads of fine-pitch parts. Fine-pitch parts are the source of most manufacturing defects in a well-run SMT assembly line. The defects stem from lack of coplanarity of the leads, bent leads, insufficient solder on the joints, and poor alignment of the leads to the patterns on the PCB.

Successful manufacture using fine-pitch components involves very tight cooperation among the PCB designer, the PCB fabricator, the component manufacturer, and the PCB assembler/tester. It almost always involves specialized assembly, test, and rework tooling. Design is most often done by convening a series of meetings of the engineering personnel of all these groups to evolve a set of rules, processes, equipment, tooling, and components. These meetings need to start at the product development stage and continue to be held until the proper pad shapes and sizes have been established and the production process is stable.

13.4.4 Press Fit

Press fit is a special form of through-hole technology. Components are fastened to the PCB by deliberately designing an interference fit between the component lead and the plated through-hole in the PCB. The principle application of press-fit technology is the attachment of connectors into backplanes. The reason for this is that early backplanes were built by wire wrapping the signal connections onto the connector pins extending out the back of the backplane. Trying to solder the connector pins to the backplane through this field of pins proved difficult, if not impossible. The solution was press fit.

Successful assembly of a press-fit backplane rests in designing a hole size small enough to create a solid connection with the pins while ensuring that the hole is large enough to permit the insertion of the pin without fracturing the hole barrel.

Caution: Hot-air leveling the solder on a backplane results in a hole with irregular diameter. This irregularity will almost certainly result in damaged hole plating when the insertion operation is done. Be sure to note on the fabrication drawing of a press-fit backplane that hot-air solder leveling is prohibited. Solder must be plated onto the backplane traces and pads and fused using IR reflow or hot-oil reflow to fuse the lead and tin into a solder alloy.

13.4.5 TAB

TAB stands for tape-automated bonding, which is a technique for attaching bare IC die to a printed circuit board. It uses a subminiature lead frame that attaches directly to the bonding pads of an IC at one end, spreads out to a much larger pitch, and attaches to pads on a PCB. The tape in the title describes the method for carrying the parts prior to assembly, which is a tape with the TAB lead frame and IC built into it. The tape is wound onto a reel for handling. The principle application of TAB components is products such as pagers and portable phones that are made in very high volume and can justify the automation involved in attaching TAB parts to substrates.

13.4.6 Flip Chip

Flip-chip technology involves plating raised pillars of metal on the bonding pads of ICs, turning them upside down, and attaching them to a matching pattern on a substrate. The substrates are most often silicon and precision ceramics. From this description, it can be seen that this is a very specialized packaging method. To succeed, a source of tested good bare die with plated-on pillars must be available. This situation occurs almost exclusively in very high performance supercomputers where the ICs have been specially designed for the application or in very high volume applications such as pagers and cellular phones.

13.4.7 BGA

BGA or ball grid array is a relatively new technology that is a cross between pin grid arrays and surface-mount. High-pin-count IC die are mounted on a multilayer substrate made from ceramic or organic material. The die are connected to the substrate using standard wire-bond techniques and encapsulated in epoxy or another form of cover. The bottom side of the substrate contains an array of high-melting-point solder balls which connect to the wire-bond pads through the multilayer substrate. These balls mate with a matching array of pads on the PCB and are reflow soldered during the same operation as all other surface-mount parts. (See Fig. 13.11.)

The appeal of BGA technology is as an alternative to high-pin-count, fine-pitch SMT ICs. As mentioned earlier, assembling high-pin-count, fine-pitch SMT parts is very difficult, owing to the fragile nature of the component leads. BGAs represent a much more robust package during component-level test and during assembly.

As with most technologies, there are some disadvantages to BGAs. Among these are:

- The solder joints are hidden from view, so inspecting them requires x-rays.
- The solder joints are not accessible for rework, so the soldering process must have a very high success rate.
- Part removal requires special tooling.
- The pattern on the PCB surface is an array of pads that require one via each all the way through the PCB, hampering routing.
- All component leads are concentrated in a much smaller area than the equivalent part in a fine-pitch SMT package. This concentrates the wiring in a small area pierced with many vias. As a result, the BGA PCB will likely have more wiring layers than the equivalent SMT PCB.
- The BGA package is more expensive than the equivalent fine-pitch SMT package, by as much as three times.

13.4.8 Wire-Bonded Bare Die

As the name implies, this method of assembly involves attaching bare IC die directly to a substrate using an adhesive or reflowing solder and connecting the bonding pads to pads on the

FIGURE 13.11 Typical ball grid array package. *(Courtesy of Icon Industries.)*

PCB using wire-bond techniques. Virtually all digital watches and many other similar consumer products use this assembly technique. It is very inexpensive when used in very high volume products with only a single IC to connect.

13.5 MATERIALS CHOICES

A wide variety of materials has been developed for use in packaging electronic circuits. These can be divided into three broad classes: reinforced organics, unreinforced organics, and inorganics. These are used primarily for rigid PCBs, flexible and microwave/RF PCBs, and multichip modules, respectively. The following treatment of the available materials concentrates on the principle materials systems used in PCB design along with the properties that warrant their use. See Chaps. 6 and 8 in this handbook for detailed data provided on loss tangents, temperature of coefficient of expansion, glass transition temperature, and other electrical properties.

IPC, the Institute for Interconnecting and Packaging Electronic Circuits, publishes a comprehensive series of standards that list in detail the properties of all types of laminates, resins, foils, reinforcement cloths, and processes that are candidates for the manufacture of PCBs. These standards start with IPC-L-108B and run up to IPC-CF-152. It is recommended that copies of the applicable standards be obtained at the start of a program in order to ensure a thorough understanding of all important characteristics of the materials being considered for a design.

Properties important to the manufacture of PCBs include (see Table 13.3):

- *Glass transition temperature* T_g—The temperature at which the coefficient of thermal expansion in a resin system makes a sharp change in rate from a slow rate of change to a rapid rate of change. A high T_g is important for PCBs that are very thick to guard against barrel cracking or pad fractures during the soldering operation.

- *Coefficient of thermal expansion* T_{CE}—Surface-mount assembly process subjects the printed wiring assembly to more numerous temperature shocks than typical through-hole processes. At the same time, the increase in lead density has caused the designer to use more and more layers, making the board more susceptible to problems concerned with the base material's coefficient of thermal expansion T_{CE}. This can be a particular problem with regard to the z-axis expansion of the material, as this induces stresses in the copper-plated hole, and becomes a reliability concern. Figure 13.12 shows typical z-axis expansion for a variety of printed circuit base laminate materials.

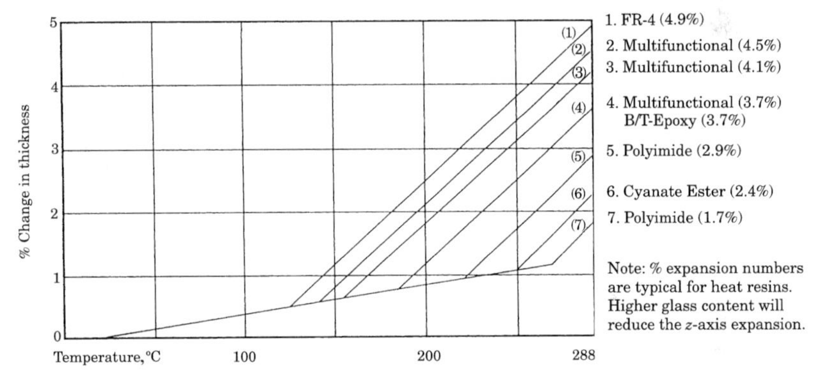

FIGURE 13.12 Typical z-axis expansion via thermal mechanical analysis. *(Courtesy of Nelco International Corp.)*

- *Relative dielectric constant* e_r—This characteristic measures the effect that a dielectric has on the capacitance between a trace and the surrounding structures. This capacitance affects impedance as well as the velocity at which signals travel along a signal line. (See Figs. 13.3 and 13.4) Higher e_r produces lower impedance, higher capacitance, and lower signal velocity.
- *Loss tangent, tan (f), or dissipation factor*—A measure of the tendency of an insulating material to absorb some of the ac energy from electromagnetic fields passing through it. Low values are important for RF applications, but relatively unimportant for logic applications.
- *Electrical strength or dielectric breakdown voltage DBV*—The voltage per unit thickness of an insulator at which an arc may develop *through* the insulator.
- *Water absorption factor WA*—The amount of water an insulating material may absorb when subjected to high relative humidity, expressed as a percent of total weight. Absorbed water increases relative dielectric constant as well as reduces DBV.

13.5.1 Reinforcement Materials

The principal reinforcement for PCB substrate materials is cloth woven from glass fibers. A variation of this glass is cloth made from quartz fibers. The resulting material has a slightly lower dielectric constant than ordinary glass, but at a substantial cost premium and a more difficult drilling cycle. Kevlar is an alternate woven reinforcement that results in a lower-weight material system with a lower dielectric constant, also at a higher cost and higher difficulty in processing.

TABLE 13.3 Properties of Some Common PCB Materials Systems*

	T_g	e_r	tan (f)	DBV, V/mil	WA, %
Std. FR-4 epoxy	125C	4.1	0.02	1100	.14
Multifunctional epoxy	145C	4.1	0.022	1050	.13
Tetrafunctional epoxy	150C	4.1	0.022	1050	.13
BT/epoxy	185C	4.1	0.013	1350	.20
Cyanate ester	245C	3.8	0.005	800	.70
Polyimide	285C	4.1	0.015	1200	.43
Teflon	N.A.	2.2	0.0002	450	0.01

* All with E-glass reinforcement, except Teflon.

The original reinforcement material for PCBs was paper or cardboard in some form. Paper impregnated with a resin system is still in use in consumer applications where lowest possible cost is necessary and where performance is not an issue.

13.5.2 Polyimide Resin Systems

Polyimide resin-based laminates are the workhorse of electronics that must withstand high temperatures in operation or in assembly or repair. Common applications include down-the-hole well-drilling equipment, avionics, missiles, supercomputers, and PCBs with very high layer count. The principle advantage of polyimide is its ability to withstand high temperatures. It has approximately the same dielectric constant as epoxy resin systems. It is more difficult to work with in fabrication, is more costly than FR-4 systems, and absorbs more moisture.

13.5.3 Epoxy-Based Resin Systems

Epoxy resin-based laminates are the workhorse of virtually all consumer and commercial electronic products. There are several variations of this pervasive laminate family, each developed to answer a special need. Among these are standard FR-4, multifunctional epoxy, difunctional epoxy, tetrafunctional epoxy, and BT or bismaleimide triazine blends. Each of these was developed to answer the need for a resin with a successively higher T_g or glass transition temperature. Multifunctional epoxy is the most commonly used form.

13.5.4 Cyanate Ester-Based Resin Systems

This resin system is a recent entrant into the high-performance resin system category. It is said to offer processing characteristics superior to those of the FR-4 blends while offering a higher T_g.

13.5.5 Ceramics

A wide variety of ceramic or alumina substrate materials have been developed for use in hybrids and multichip modules. These materials are the subject of specialized manufacturing processes beyond the scope of this handbook. The reader needing information on this group of materials is advised to contact a major manufacturer of ceramic materials.

13.5.6 Exotic Laminates

Kevlar, Kapton, Teflon, and RO 2800 are materials developed for specialty applications. The first two, in the form of thin films, are commonly used as substrates for flexible circuits. The latter two are the principal dielectrics for microwave and RF circuits. All of these materials can be used with or without reinforcements.

13.5.7 Embedded Components Materials

Specialized materials have been developed to allow construction of passive components such as resistors and capacitors into the PCB structure itself. Most of these materials are patented and available from a very small supplier base.

13.5.7.1 Embedded Resistors. These are formed by plating a very thin film of nickel or other metal onto a copper foil layer and laminating this foil, plated side in, to an FR-4 or other substrate material. To form a resistor, a window is opened in the copper foil, exposing the underlying nickel resistor layer. A resistor of the appropriate value is formed in the resistance material layer. Contact is made with the resulting resistor by etching connecting pads in the copper foil layer and drilling holes through these pads and plating the holes. (See Fig. 13.13.)

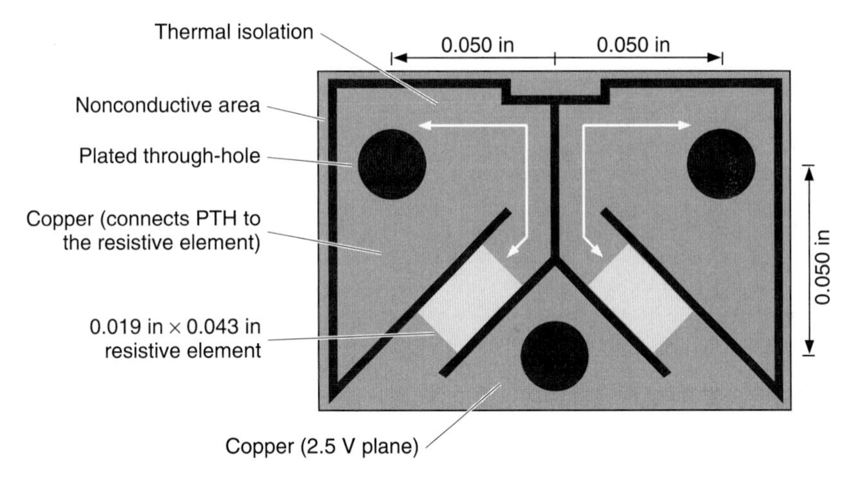

FIGURE 13.13 The two light diagonal shapes in the figure above are terminating resistors for ECL transmission lines. They are formed directly in the VTT powerplane by etching the copper away from the underlying nickel resistive layer. One end of each resistor is connected to the device terminal using a via and the other end is connected directly to the –2.5V plane. *(Courtesy of Ohmega Industries.)*

Resistive material is available in 25- and 100-Ω/square values. The principal application of embedded resistor technology is as terminating resistors for ECL transmission lines and as resistors on flexible circuits in products such as cameras and portable tape and CD players. Practical resistor values range from about 10 to 1000 Ω.

13.5.7.2 Embedded Capacitance. This is formed by placing two copper planes close to each other using very thin dielectrics (1.5 to 2.0 mils). The principal application is in the creation of very high quality, high-frequency capacitance between two power planes. This does indeed result in high-quality capacitance, but usually at a high cost resulting from the need to add a pair of extra planes to a PCB in order to create the capacitance (see Fig. 13.14).

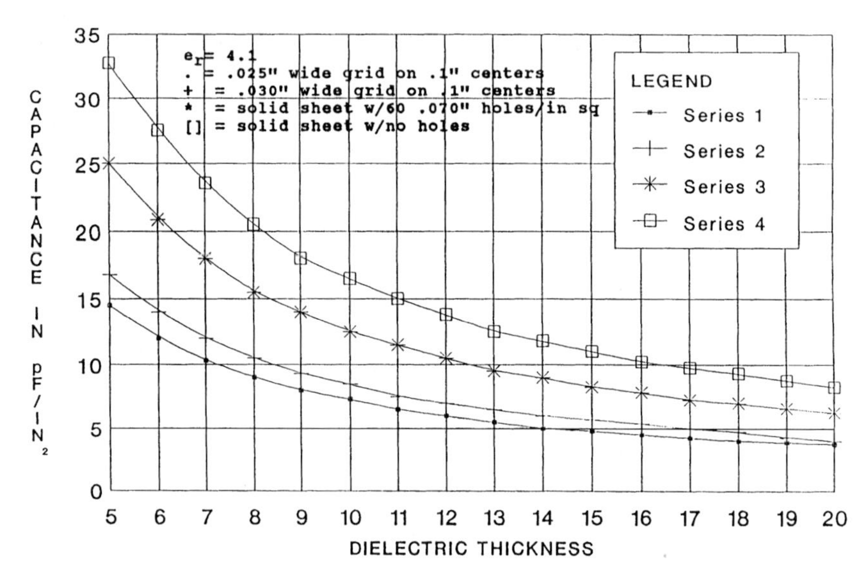

FIGURE 13.14 Capacitance per unit area vs. dielectric thickness.

13.6 FABRICATION METHODS

A wide variety of fabrication methods has been developed to meet the needs of the electronics industry. The following descriptions are quick summaries of each method intended to acquaint the reader with their advantages and disadvantages and likely applications. Detailed treatments are presented elsewhere in this book.

13.6.1 Punch Forming

Punch forming is used in the manufacture of very low cost, single-sided PCBs such as are used in many consumer electronic products. The process involves printing and etching the conductor patterns on one side of a laminate substrate, usually paper-reinforced epoxy. All holes are punched in a single stroke by a die containing a pin and opening for each hole. The PCB outline is formed in a second die that "blanks" it from the panel in which it is processed.

Often, several PCBs will be contained in a single panel sized to travel through the assembly process. After a PCB is punched out of the panel, it is forced back into the vacant hole and is held in place by interlocking fibers along the edges of the PCB. After assembly and testing is complete, each PCB is pressed from the panel. This is known as *crackerboarding* and is aimed at reducing overall manufacturing costs.

13.6.2 Roll Forming

Roll forming is a process used to manufacture flexible circuits in very large quantities. This is the lowest-cost method for the manufacture of flexible circuits. However, it involves substantial tooling, so it is applicable only for very high volume products. Examples of PCBs manufactured using this method are printer head connections, disk drive head connections, and the circuits used in cameras and camcorders. The PCBs can be single- or double-sided.

Roll forming resembles newspaper printing in that the process starts with a large roll of copper-clad laminate that is fed through a long process line containing stations which perform each operation on a continuous basis, starting with printing the conductor pattern, etching it, forming the holes, testing, and blanking from the roll itself. The process can include lamination of a cover insulator layer over the conductors as well.

13.6.3 Lamination

Lamination is the process by which PCBs of more than two layers are formed. The process begins by etching the conductor patterns of the inner layers onto thin pieces of laminate called *details*. These details are then separated by partially cured laminate called *prepreg* and stacked in a *book* with layers of prepreg on the top and bottom and foil sheets on the outside. This stack is placed into a press capable of heating the combination to a temperature that causes the prepreg resin to reach the liquid state. The liquefied resin flows into the voids in the copper patterns to create a *solid* panel upon cooldown. Once cooled, the panel is sent through the drilling and plating operations much like a two-sided PCB.

Note that some materials, such as polyimide, do not have a prepreg form to act as the glue during lamination. In these cases, a special glue sheet must be used during lamination to fasten the individual layers together.

13.6.4 Subtractive Plating

Subtractive plating is a method of forming traces and other conductive patterns on a PCB by first covering a sheet of laminate with a continuous sheet of copper foil. A layer of etch resist is applied such that it covers the copper pattern that is desired. The panel with protective coating is passed through an etcher that removes (subtracts) the unwanted copper, leaving behind the desired patterns. This is the dominant, almost only, method in common usage in the printed circuit industry today.

13.6.5 Additive Plating

As the name implies, this method of forming conductor patterns involves beginning with a bare substrate and plating on the conductor patterns. There are two methods for doing this: electroless plating on areas sensitized to accept electroless copper and electroplating by first applying a very thin coating of electroless copper over the entire surface to act as a conductive path, followed by electroplating to full thickness.

Additive plating is seen as a method for reducing the amount of chemicals required to manufacture PCBs, and this is true. However, the process does not yield copper sufficiently robust to withstand the handling of normal assembly and rework. As a result, it is not commonly available in production.

13.6.6 Discrete Wire

Discrete wire is a method for forming the wiring layers by rolling round wire into a soft insulating material coated onto the outsides of power plane cores. This method is often referred to as *multiwire.* It is available from only a very small number of manufacturers and offers few advantages over conventional multilayer processing. It is described more fully in Sec. 13.2.3.

13.7 CHOOSING A PACKAGE TYPE AND FABRICATION VENDOR

A key part of arriving at a successful design is choosing PCB materials, component-mounting techniques, and fabrication methods that meet the performance needs of the product being designed while achieving the lowest possible costs. Among the decisions that are part of this process are deciding whether to package a product on one large PCB or several smaller PCBs, whether to spread components out to hold the layer count down or increase layers, move components closer together, and design a smaller PCB, whether to package some components in a multichip module that is then mounted on the PCB, as well as other issues.

Part of this decision-making process is arriving at an overall package choice that can be manufactured by the mainstream fabricators and assemblers. Failure to do this will result in excessively high prices and long lead times stemming from the lack of a competitive supplier base from which to choose. At the extreme, where some of the more exotic materials systems are used, there may be as few as one supplier to turn to. In markets where there is substantial price pressure, such as with disk drives and PCs, it is imperative that the design choices be made such that the PCBs can be manufactured at offshore fabricators. Not doing this will place the product at a competitive disadvantage.

13.7.1 Trading Off Number of Layers Against Area

Cost of the bare PCB is often a significant contributor to the overall cost of an assembly. As the number of layers in a PCB increases, the cost increases. A standard practice is to spread components out to make room for the connecting wiring as a way to avoid adding additional wiring layers. As might be expected, there is a point at which PCB size grows to where a smaller PCB with more layers yields a more economical solution. Determining where this breakpoint is requires some knowledge about the PCB fabrication process.

Table 13.4 shows typical costs of four-, six-, and eight-layer 18 in by 24 in standard process panels built at offshore manufacturers. This table can be used to calculate the relative cost of PCBs as layer count is increased to reduce area. While the absolute costs in the table are based on Spring 1995 pricing for Pacific Rim fabricators, the percentage relationships between the costs of panels of various layer counts are a good indicator of relative costs for deciding when to increase layer count and reduce area.

13.7.1.1 Background Information. Multilayer PCBs of six or more layers are normally built on standard 18 in by 24 in panels using pin lamination. Many four-layer PCBs are built offshore using a process called mass lamination with panels sizes of 36 in by 48 in (four times a "standard" panel). The pricing of individual PCBs is based on how many PCBs fit on a stan-

TABLE 13.4 PCB Panel Process Cost vs. Layer Count
Price per panel, $ U.S., Spring 1995

Number of layers	Panels per mo. 100	Panels per mo. 250	Panels per mo. 1000	Panels per mo. 5000
4 mass lam*	$260	$250	$240	$231
4 pin lam†	84	80	77	74
6 pin lam†	113	108	104	100
8 pin lam†	147	140	135	130

To determine the number of PCBs that will fit onto a panel, allow 0.125 in between PCBs and allow 0.75-in margin on all four sides. Net areas are 34.5 in by 46.5 in and 16.5 in by 22.5 in.

For gold plating on connector tips, add up to $2 per PCB.

* 36 in × 48 in panel.

† 18 in × 24 in panel.

dard panel. Therefore, designers need to choose finished PCB sizes with this in mind (assuming the size is negotiable).

The pricing matrix in Table 13.4 is in price per panel based on the following:

- Solder mask over bare copper, silk screen one side only
- Standard multifunctional FR4 laminate
- 1-oz copper inner layers, ½-oz outer layer foil plated up to 1.5 oz nominal
- Thickness accuracy: ±10% overall, ±1.5 mil any dielectric layer
- No controlled impedance requirement or testing
- 1-mil minimum copper plating in holes
- No vias smaller than 13 mils
- Traces and spaces: 7 mil, 7 mil
- Standard delivery: (no acceleration premium)
- Fabrication site: Pacific Rim Fabricators
- No gold plating
- Tested to CAD netlist

Since PCB fabrication is based on standard panels, the cost of each PCB is affected by how much of each panel is used to form the PCBs built on it and the amount of the panel that is scrapped. Clearly, the more of each panel that is usable, the lower the average cost of each PCB will be. As one chooses the size and form factor of each PCB, attention should be paid to how many will fit onto a standard panel in order to minimize the scrap material created.

13.7.2 One PCB vs. Multiple PCBs

One way to keep individual PCB layer counts down is to divide the circuitry among several smaller, simpler PCBs. There is a hidden cost associated with doing this. The hidden cost is spread across several organizations, ranging from the design activity, through manufacturing, and into the sales and service organization. The costs are those associated with handling multiple assemblies, such as managing multiple designs and their documentation, managing the procurement, inventory, testing and stocking of multiple assemblies, and the cost of interconnecting the multiple assemblies. In almost all cases, these costs exceed any savings that might be realized by the creation of multiple assemblies.

CHAPTER 14
THE PCB DESIGN PROCESS

Lee W. Ritchey
3Com Corp., Santa Clara, California

14.1 OBJECTIVE OF THE PCB DESIGN PROCESS

The objective of the PCB design process is to engineer a PCB, including all of its active circuits, that functions properly over all the normal variation in component values, component speeds, materials tolerances, temperature ranges, power supply voltage ranges, and manufacturing tolerances and to produce all of the documentation and data needed to fabricate, assemble, test, and troubleshoot the bare PCB and the PCB assembly. Doing less than this in any area exposes the manufacturer and user of the PCB assembly to excessive yield losses, excessively high manufacturing costs, and unstable performance.

Achieving the objective involves carefully designing a process that matches the end product, selecting design tools with controls and analytical utilities, and selecting a materials system and components that match.

14.2 DESIGN PROCESSES

Figure 14.1 is a flowchart of the major steps in a complete PCB design process, beginning with specification of the desired end product and continuing through to archiving or storing away the design database in a form that permits subsequent design modifications or regeneration of documentation as necessary to support ongoing production. This process takes advantage of all the computer-based tools that have been developed to assure a "right the first time" design. The basic process is the same for either analog or digital PCBs. The differences in the design process for the two classes of PCBs center around the differences in complexity of these two types of circuits, as mentioned in Chap. 13.

14.2.1 The System Specification

The design team begins a new design by creating a *system specification.* This is a list of the functions the design is to perform, the conditions under which it must operate, its cost targets, development schedule, development costs, repair protocols, technologies to be used, weight and size, and other requirements as are appropriate. A rough definition of each of these variables is necessary at the start to permit proper choices of materials, tools, and instrumentation. For example, a project may involve the design of a portable computer that must weigh less

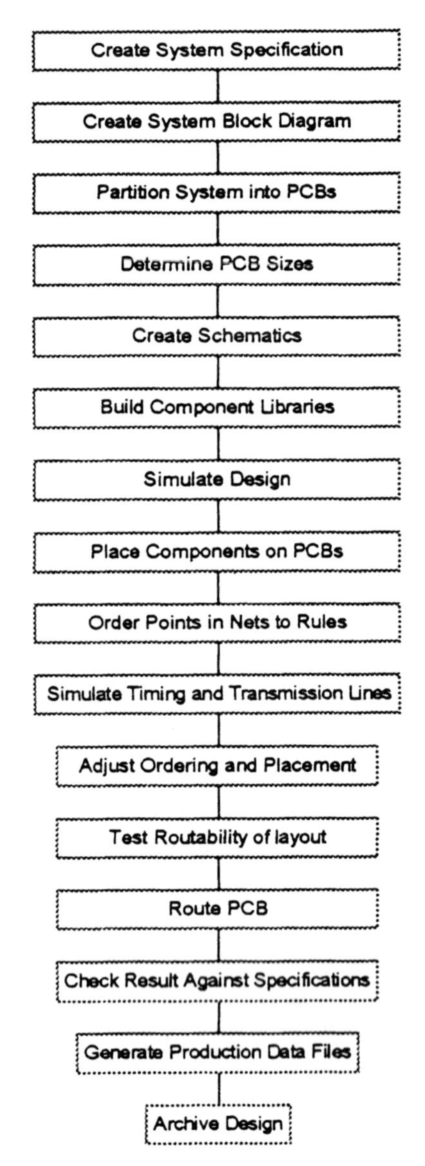

FIGURE 14.1 PCB design process steps.

than 5 lb, fit into a briefcase, operate on batteries for two hours, have a mean time between failures (MTBF) of 200,000 hours or more, cost less than $2000, have 4 Mbytes of memory, 240 Mbytes or more of mass storage, and be MS-DOS compatible. This specification serves as the starting point for a new design.

14.2.2 System Block Diagram

Once the system specification has been completed, a block diagram of the major functions is created, showing how the system is to be partitioned and how the functions link or relate to each other. Figure 14.2 is an example of this partitioning.

14.2.3 Partitioning System into PCBs

Once the major functions are known and the technologies that will be used to implement them are determined, the circuitry is divided into PCB assemblies, grouping functions that must work together onto a single PCB. Usually this partitioning is done where data buses link functions together. Often these buses will be contained on a backplane into which a group of daughter boards are plugged. In the case of a personal computer (PC), partitioning often results in a mother board and several smaller plug-in modules such as memory, display driver, disk controller, and PC card (PCMCIA) interface.

14.2.4 Determining PCB Size

As soon as the amount of circuitry and the technology that each PCB must contain is known, the area and size of each PCB may be estimated. Often, the PCB size is fixed in advance by the end use. For example, a system based on VME or multibus technology will have to use the PCB sizes defined by the standard. In this case, system partitioning and component packaging technology will be dictated by what will fit onto these standard PCB sizes.

The finished cost of a PCB often turns on the number of layers and the quantity that will fit onto the standard manufacturing panel sizes. (For most PCB fabricators, this size is 18 in by 24 in with a usable area of 16.5 in by 22.5 in) Sizing PCBs to utilize all or most of the panel area results in the most cost-effective bare PCB. (See Table 13.4.)

14.2.5 Creating the Schematic

Once the system function, partitioning, and technologies have been determined, the schematic or detailed connections between components can take place. Schematics and block diagrams are normally created on CAE (computer-aided engineering) systems. These systems allow the designer to draw the schematic on a CRT screen or terminal. The data needed by all of the following steps in the design process is generated by the CAE system from this schematic.

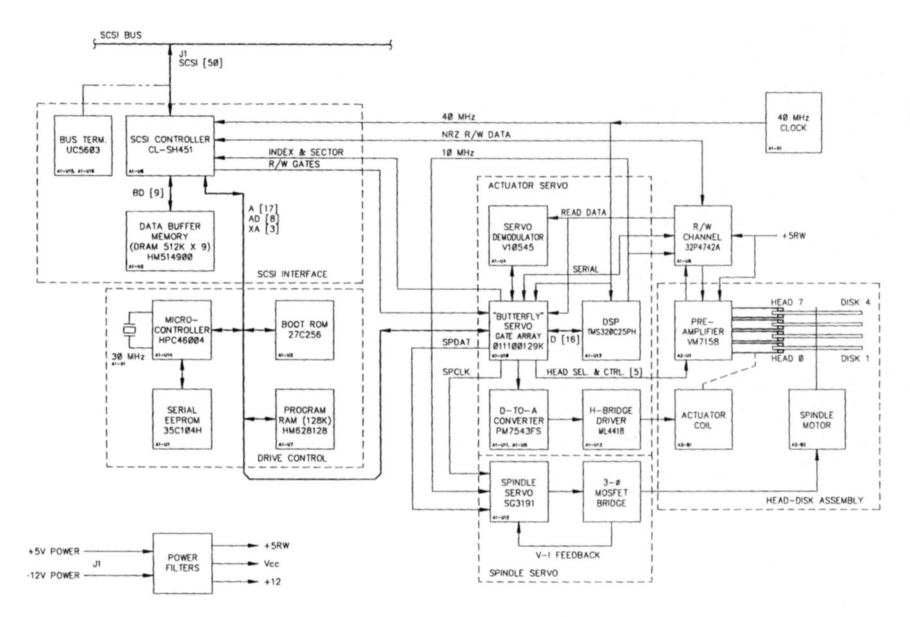

FIGURE 14.2 This is a block diagram of a digital device (in this case a disk drive) that has been segmented to its assembly levels. The dashed lines represent initial partitioning of the entire product to printed circuit assemblies and also shows the expected interface (connector) requirements.

14.2.6 Building Component Libraries

The tools used in the PCB design process must be supplied a variety of information about each part in order to complete each step. This information is entered into a library or set of libraries, one entry per component. Among the pieces of data needed are:

- Type of package that houses the component, e.g., through-hole, QFP, DIP
- Size of component, lead spacing, lead size, pin-numbering pattern
- Function each pin performs, e.g., output, input, power pin
- Electrical characteristics of each pin, e.g., capacitance, output impedance

14.2.7 Simulating Design

To be sure a design will perform its intended function over the intended range of conditions, some form of design verification must be done. These conditions may include component value accuracies, range of component speeds, operating and storage temperature ranges, shock and vibration conditions, humidity ranges, and power supply voltage range. Historically, this has been done by building breadboards and prototypes and subjecting them to rigorous testing. As systems and their operation software have grown more complex, this technique has proved to be inadequate. To solve this problem, simulators have been developed that allow a computer to simulate a function without having to build it. These simulators make it possible to perform tests far quicker, more rigorously, and more completely than any breadboard or prototype could ever be expected to.

Defects discovered by a simulator can be corrected in the simulation model with ease and the tests rerun before any commitment to hardware is made.

14.2.8 Placing Components on PCBs

Upon successful completion of the logical and gross timing simulation process, the actual physical layout can begin. It begins by placing the components of the design on the surface of the PCB in patterns that group logical functions together. Once this is done, the groups of components are located on the PCB surfaces such that functions that interact are adjacent, components that generate heat are cooled properly, components that interface to the outside world are near connectors, and so on. This placement operation can be done manually by the designer using graphics-based tools or automatically by the PCB CAD system.

14.2.9 Sequencing Nets to High-Speed Rules

Most logic families have sufficiently fast rise and fall times and short enough propogation delays to be subject to high-speed problems such as coupling and reflections. To ensure that these high-speed effects do not result in malfunctions, it is necessary to arrange the connections between the loads, terminations, and drivers to control these high-speed phenomena. This arranging of nodes or points in a net is referred to as *sequencing* or *scheduling.* Once the components have been placed on the surface of the PCB, the spatial arrangement of all the nodes on each net is known. At this point, it is possible to determine how to connect the driver to the loads and terminators to form proper transmission lines, ensuring that improper stubs are not created and that the terminator is at the end of the net.

14.2.10 Simulating Timing and Transmission Lines Effects

Upon completion of component placement and sequencing the nodes in each net, it is possible to estimate the length and characteristics of each net. This is possible because the x-y location of each point in a net is known, the order of connections is known, and the fact that the actual wiring must be done in either the x or y direction is also known. This length information can be used to model the high-speed switching characteristics of each net and predict the presence of excessive noise and reflections as well as to estimate the length of time required for signal to travel the length of each line, all before actually routing or building the PCB.

This simulation step makes it possible to detect potential malfunctioning signals prior to routing and take steps to fix the problem while the time invested in the design is still modest.

14.2.11 Adjusting Sequencing and Placement

If the simulations done in Sec. 14.2.10 reveal excessive time delays or reflection problems, the placement may need to be adjusted to move critical parts closer together or add terminations to nets with excessive reflections. By doing this simulation and adjustment, a design can be assured of meeting the "right the first time" goal so important to high-performance designs.

14.2.12 Testing Routability of Placement

At this point, enough analysis has taken place to know that the design will function correctly if routed. However, it may not route in the number of signal layers required by the cost goal. Most CAD systems have tools, such as rats-nest analyzers, that help the designer determine if the design will fit into the allowed signal layers. If it will not, the routability analyzer may give clues on how to revise the component placement to achieve a successful route. Once the placement has been adjusted, the timing and transmission line simulation steps must be repeated to ensure that the set of goals has been met.

14.2.13 Routing PCB

This step involves fitting all the connections into the signal layers in the form of copper traces, following spacing and length rules. It usually involves a combination of hand routing special signals and automatically routing the rest.

14.2.14 Checking Routed Results

After all the connections have been routed into the signal layers, the actual shape and length of each wire is known, as is the layer(s) on which the wires have been routed and which nets are neighbors. This physical data can be loaded into the timing and transmission line analyzers to do a final check that all design goals have been met. Any violations that are detected can be repaired by hand rerouting as necessary. Once this set of checks has been completed and any adjustments made, the final routed result is checked against the schematic netlist to ensure that there are no discrepancies. A final check is performed on the Gerber data to ensure that the line width and spacing rules have been complied with and that there is no solder mask of silk screen on any pad, as well as that traces and other features that must be protected from solder are covered by solder mask.

14.2.15 Generating Manufacturing Files

This step involves generating the photoplotting files, pick-and-place files, bare and loaded board test files, drawings, and bills of material needed to do the actual manufacturing. Typical lists of these files are shown in Tables 14.1 and 14.2.

14.2.16 Archiving Design

Once all the manufacturing data has been created, the design database and all of the manufacturing data files are stored on a magnetic tape or other storage media for future use to incorporate changes and for backup in the event that the files and drawings created for manufacturing are lost or destroyed.

TABLE 14.1 A Typical Collection of Design Files Sent to a PCB Fabricator

File name	File contents
BBBBpCCC.arc*	Arc file of Gerber files containing:
applist.p	List of photoplot apertures for artwork
ly1 thru lyx.ger	Gerber photoplot data for x PCB layers
topmsk.ger	Gerber photoplot data for top solder mask
botmsk.ger	Gerber photoplot data for bottom solder mask
topslk.ger	Gerber photoplot data for top silk screen
botslk.ger	Gerber photoplot data for bottom silk screen
pc_356.out	IPC 356 data for blank PCB netlist testing
name0.rep	Drill allocation report for plated holes
name0.prf	Excellon drill file for all plated holes
name1.rep	Drill allocation report for all nonplated holes
XX.XX.fab	Fabrication drawing in HPGL format, sheet XX of XX

* BBBBpCCC is the part number of the PCB.

TABLE 14.2 A Typical Collection of Design Files Sent to a PCB Assembler

File name	File contents
BBBBaCCC.arc*	Arc file of all assembly data containing:
applist.a	Aperture list for plotting paste mask
readme.asy	Readme file describing assembly
tpstmsk.ger	Gerber photoplot data for top paste mask
bpstmsk.ger	Gerber photoplot data for bottom paste mask
BBBB-CCC.dbg	Mfg. output, data format info.
BBBB-CCC.dip	Mfg. output, *x-y* loc. dip components
BBBB-CCC.log	Mfg. output, component log.
BBBB-CCC.man	Mfg. output, *x-y* loc. manual insert components
BBBB-CCC.smt	Mfg. output, *x-y* loc. top smt components
BBBB-CCC.smb	Mfg. output, *x-y* loc. bottom smt components
BBBB-CCC.unp	Mfg. output, parts not mounted
BBBB-CCC.vcd	Mfg. output
XX_XX.asy	Assembly drawing in HPGL format, sheet XX of XX
XX_XX.fab	Fabrication drawing in HPGL format, sheet XX of XX

* BBBBaCCC is the part number of the assembly.

14.3 DESIGN TOOLS

From the definition of the objective of the PCB design process it can be seen that the process extends from concept all the way through fabrication to assembly and test. Computer-based tools have evolved to automate or improve the speed and accuracy of every step in the process. These tools can be divided into three major groupings based on where they are used:

- CAE (computer-aided engineering) tools
- CAD (computer-aided design) tools
- CAM (computer-aided manufacturing) tools

It is apparent from the names of the tools that they are used in circuit design, physical layout of the PCB, and manufacture of the bare PCB and the PCB assembly.

14.3.1 CAE Tools

CAE tools is the name generally used to refer to the computer-based tools and systems that are employed in the stages of design before the physical layout step or to analyze and evaluate the electrical performance of the final physical layout. These include:

14.3.1.1 Schematic Capture Systems. As the name implies, these tools are used by the design engineer to draw a schematic or circuit diagram. The simplest systems are graphical replacements for the classical drawing board, allowing the engineer to place logic and electronic symbols on the surface of the drawing and connect their terminals with lines. More advanced systems perform substantial error checking such as guarding against multiple uses of the same pin or net name. Failure to connect critical pins such as power pins can be done by using information contained in their component libraries about each part. In addition, these systems can generate netlists to be used by simulators and PCB routers and bills of material for use in the manufacture of the PCB assembly.

14.3.1.2 Synthesizers. Synthesizers are specialized CAE tools that allow a designer to specify the logic functions that a design is expected to perform in the form of logical operations such as dual full adder, 16-bit-wide register, or other macro functions. The synthesizer will extract the equivalent logic circuit functions from a function library and connect them together as specified by the designer to arrive at a complete logic diagram. This synthesized circuit can then be used as part of a bigger design. Some advantages of synthesizers are that all functions of a given type will be implemented the same way and be error free and that time needed to compose a system schematic is reduced by eliminating the labor required to design repeating circuits.

14.3.1.3 Simulators. Simulators are software tools that create computer-based models of a circuit and run them with input test patterns to verify that the circuit will do the intended function when implemented in hardware. Even when run on very large computers, simulators usually run at only a tiny fraction of the speed of the actual circuit. When circuits grow complex, as in a 32-bit microprocessor or digital signal processor, the time required to perform a complete simulation can be very long, sometimes so long that this method of circuit verification becomes impractical. Typical simulation speeds are 1 or 2 s for each machine cycle. A machine cycle may be as little as 2ns or 500 million/s for a system with a 500-MHz clock. This is a 500 million-to-1 slowdown! As circuits have grown more complex and simulation times have become excessive, engineers have been forced to build physical models of a proposed circuit and run actual code against the model as a method of ensuring that the design is accurate. Clearly, this adds both time and cost to the development cycle, both by adding the time needed to build the model and the time needed to locate design errors and fix them. The solution to this problem is circuit emulation.

14.3.1.4 Emulators. Emulators or circuit emulators are collections of programmable logic elements, such as PLAs (programmable logic arrays), that can be configured to represent almost any kind of logic circuit. These emulators are commercially available as standard products from several EDA (electronic design automation) companies. The resulting hardware emulation of a circuit can be operated much faster than a software simulation, sometimes as fast as ¹⁄₁₀₀th of actual final operating speed. Due to this increased speed, the verification of a circuit can take place much faster. In some cases, the emulations are used as substitutes for the actual circuit to verify that the software created to run with the circuit is error free before any commitment is made to final hardware. This technique is used extensively in the design of complex ICs such as microprocessors and custom ASICs. In fact, the Intel Pentium™ microprocessor and its operating system was completely emulated and run successfully on a large hardware emulator prior to making the first silicon.

The use of emulation technology has eliminated the need to iterate or modify designs as errors are detected, saving very large sums of development money as well as development time. In some cases, it has made circuits practical that would not be without it. For example, most supercomputers and other advanced products are built with multilayer ceramic PCBs and MCMs. These packaging technologies do not lend themselves to modification with external wiring to correct errors. As a result, it is necessary to produce entirely new assemblies in order to correct design errors. The same is true for the integrated circuits in the system as well.

14.3.1.5 Circuit Analyzers. Circuit analyzers are tools that examine circuits to ensure that they will perform properly over the range of timing variations in circuits and tolerances of components that are expected to be encountered in normal manufacturing. These analyzers do this by constructing mathematical models of each circuit and then varying the values of each component over their expected tolerance ranges. The behavior of the circuit is calculated by its model and the result compared to preestablished limits. Violations are flagged to alert the design engineer. Among the conditions checked for are freedom from excessive signal coupling from neighboring circuits and that transient behavior such as reflections, overshoot and undershoot, and ringing are within proper limits. This type of analysis is often referred to as worst-case tolerance and timing analysis. It can be run on a design before physical layout begins as well as after layout is completed.

Some examples of circuit analyzers are SPICE and PSPICE. SPICE and PSPICE build mathematical models of each circuit and then perform thousands of complex calculations to predict how the circuit will respond to input signals. Most suppliers of CAD systems also offer their own adaptations of these analysis tools.

14.3.1.6 Impedance Predicting Tools. These tools are used to examine the cross section, trace sizes, and materials properties of a PCB to ensure that the resulting circuit impedance is within allowable limits or to interactively adjust these parameters to achieve a desired final impedance. This is an essential step in the design of the PCB itself. Most suppliers of CAD systems intended for high-speed design supply some form of on-line impedance analysis tools as part of their systems.

14.3.2 CAD Tools

CAD tools are used to turn the electrical circuit described by the schematic into a physical package or PCB. CAD tools are typically operated by PCB designer specialists who are skilled in the areas of PCB manufacture and assembly rather than by electrical engineers. These tools are fed netlists, component lists, wiring rules, and other layout information by schematic capture or CAE tools. In their simplest form, they allow the designer to create the pad patterns for the component leads and PCB shape and then manually connect the component leads with copper traces. The most sophisticated CAD tools can automatically determine the optimum location of each component on the PCB (autoplacement) and then automatically connect (autorouting) all leads while following high-speed layout rules. This is accomplished by providing the CAD tool with a table of rules specifying which components must be located in groups or near connectors as well as by specifying how much space must be maintained between neighboring traces, the maximum length allowed between points on a net, etc.

The outputs of CAD tools are the information files needed to fabricate, assemble, and test the PCB assembly. These are test netlists, photoplotting files, bills of materials, pick-and-place files, and assembly drawings. CAD tools are made up of circuit routers, placement tools, checking tools, and output file generation tools.

14.3.2.1 Placement Tools. Placement tools are used to arrange the components on a PCB surface. Placement tools tend to be part of a complete CAD system rather than a module purchased separately. The inputs to a placement tool are:

- Component list or bill of material
- Netlist or manner in which the components connect to each other
- Shapes, sizes, and spacial arrangement of the component leads
- Shape of the PCB with areas into which components cannot be placed (keepouts)
- Instructions concerning fixed locations for components, such as connectors
- Electrical rules, such as maximum and minimum distance between points on a net
- Thermal rules, such as which parts must be kept apart or near sources of air flow

Placement tools range from completely manual to fully automatic. All have some form of graphical feedback to the designer that assesses the quality of the placement in terms of its ability to be routed or connected in the desired number of signal layers. Most have spacing rules that ensure that the components have enough room between them for successful assembly, rework, and testing.

14.3.2.2 Routers. Routers are the part of a CAD system that makes the physical connections between the components as specified by the netlist. A router operates on the PCB netlist and placement after the placement step has been completed. Routers range from the completely manual, in which case the designer specifies where wires are to be located by using a graphical display and a mouse or light pen, to fully automatic, where a specialized software program takes the netlist, the placement, the spacing rules, and the wiring rules and makes all the decisions necessary to completely connect all components. The principal advantage of manual routing lies in the fact that the designer can tailor every connection to his liking. The principal disadvantage of manual routing is the fact that it is slow and time consuming, often taking several minutes to completely route and check a single net. Automatic or autorouters solve this speed problem. However, the ability to control the detailed shape of each net is limited by the ability of the autorouter to follow wiring rules. Some advanced autorouters are able to comply with very complex wiring rules. A significant problem with autorouters lies in the fact that they may not find ways to successfully route all of the wires. When this occurs, the designer must add more wiring space in the form of more layers or attempt to complete the routing manually. An important feature of a good autorouter is its manual routing option, as it substantially affects the ease with which this often necessary "finishing" operation is completed. Nearly all routers have a suite of checking tools that ensure that the final route matches the netlist and that all of the spacing rules have been followed.

Routers come in several forms and can be purchased as an integral part of a CAD system or as modules to add onto CAD systems. Some router types are:

Gridded Router. This type of router operates by placing wires on a predefined grid pattern. The routing surface is divided into a uniform grid that provides a proper gap between traces when wires are routed on every grid line. It is the first form of both manual and autorouter offered with CAD systems. The primary disadvantages of gridded routers are that it is difficult to manage more than one trace width without losing wiring density and it requires end points of nets to be on the routing grid in order to connect to them successfully. Offgrid component connections typically have to be made by hand and checked by hand.

Gridless Router. This form of router does not depend on a grid to locate wires on a surface. Instead, it places as many wires in a space as will fit and still maintain the spacing rules established by the design engineer to ensure proper electrical performance while optimizing manufacturability. Multiple trace widths on the same layer are handled easily by this type of router. Once the routing job is completed, the router divides up any unused space equally. The advantage of this technique lies in its ability to optimize manufacturability by keeping spaces as large as possible. The disadvantage of this type of router lies in the fact that it usually depends on a given wiring layer being all horizontal or all vertical—a real disadvantage in SMT applications where components do not need through-holes to connect them but a very powerful router for designs with a very regular array of high-pin-count parts, such as big CPUs and massively parallel processors. This router type is the workhorse of very high complexity digital designs where there is a great deal of regularity and a need to achieve predictable spacings and trace lengths for speed and performance reasons.

Shape-based Router. This type of router recognizes shapes already placed in a wiring surface and routes wires to avoid them. Spacing between wires and other objects, such as vias, used to change layers and component pads is maintained as the router places a wire in a space. This router is becoming the workhorse of SMT-based designs.

14.3.2.3 Checking Tools. These tools verify that the routed PCB complies with rules such as spacing between traces and trace and holes by comparing the actual spacings found in the finished artwork to rules provided by the designer. They also ensure that all nets are completely connected and are not connected to objects they should not be, such as other nets and mechanical features on the PCB, by comparing the routed results to data supplied by the CAD system. Some checking tools also check to ensure that transmission line rules are followed and that coupling from neighboring traces is within limits. Checking tools are usually an integral part of a CAD system.

14.3.2.4 Output File Generators. Once a PCB has been routed and all connectivity verified as accurate, the CAD system holds this information in a neutral form specific to the way in which its operating system is built. For this data to be useful in manufacturing, it must be converted into forms usable by other equipment such as photoplotters, testers, assembly equipment, and MRP systems. Output file generating routines do this conversion. Most CAD systems are equipped with a limited set of these when shipped. Additional generators or converters must be ordered as add-ons.

14.3.3 CAM Tools

CAM tools are CAD systems tailored to the needs of the fabrication process. The output of the PCB design process is a set of CAD files that describes each artwork layer of a PCB, the silk screen requirements, drilling requirements, and netlist information. This information must be modified before it can be used to build a PCB. For example, if a fabricator needs to build several copies of a PCB on a single panel, the fabricator will need to add specialized tooling patterns to the artwork and alter trace widths in order to compensate for etching. Initially, these operations were done manually with a high potential for error and significant labor costs. CAM stations or tools allow the fabricator to do all of these operations automatically and rapidly.

CAM stations can check artwork against spacing rules, breakout rules, and connectivity rules and make corrections if necessary. CAM stations can synthesize netlists (the manner in which points in a design are connected) from the Gerber data for use at bare PCB test, in those cases where no netlist was provided by the customer. In cases where a netlist was provided by the customer, the Gerber synthesized netlist can be compared to the CAD generated netlist as a final way of verifying that the artwork does, in fact, match the schematic—one more safeguard against data corruption anywhere in the translation processes.

14.4 SELECTING A SET OF DESIGN TOOLS

The task of selecting the proper set of CAE, CAD, and CAM tools for a project or company is often one of the most critical steps in determining project success. To reduce the problems associated with interfacing tools to each other, it is desirable to choose all of the tools from a single vendor. However, very few vendors are able to offer best-in-class tools for every phase of the design process. Further, it is very difficult for a single CAD/CAE system to handle the wide variety of design types that may be required across a very large process or across an entire company. For example, a CAD system with a powerful autorouter is of very little use to a design team laying out single-sided PCBs with many irregular parts such as transformers and power transistors. Conversely, a system tailored to do power supply layout well will fare very poorly on a CPU design with many high-speed buses. Knowing this, how does one select the "right" tool set?

14.4.1 Specification

The first step in the tool selection process is to characterize the types of design that the tool set will be required to handle. Next, the level of simulation and design rule checking required to ensure "right the first time" designs must be determined in order to select the right level of simulation and checking tools. For example, performing transmission line analysis on a stereo PCB may be an elegant step but it is a waste of resources. Choosing not to do this level of analysis on a high-speed disk drive PCB may result in a design that is unstable throughout its product life.

Key to success with any tool is the availability of designers qualified to operate it. This pool of designers needs to be assessed for each type of candidate system. If a system or tool is chosen that does not have a pool of qualified operators from which to draw, the result will likely be substantial delays while the necessary expertise is developed or designs that are substandard if the learning time is too short.

14.4.2 Supplier Survey

Once these facts are known, a survey of potential tool suppliers must be conducted. Basic elements include:

1. Determining how closely each tool candidate comes to meeting the need and its cost to acquire, set up, and maintain
2. Assessing the long-term viability of the supplier to ensure that the tool does not become an "orphan" should the supplier fail
3. Making a check with other users of the candidate tools to ensure that they perform as advertised

14.4.3 Benchmarking

Representative benchmark designs need to be made on each candidate system to assess how well they are done. Depending on the size of the potential sale and the size of the benchmark design, a vendor may do the benchmark free of charge. If not, one should be prepared to pay for this valuable step in the evaluation process. Only after completing all of these evaluation steps is it possible to make an informed selection. Doing less, such as relying on an outsider to recommend tools or taking a vendor's word for it, carries the risk of owning a tool that delays development or, worse, having to repeat the selection process in the middle of a project.

14.4.4 Multiple Tools

Knowing the wide variety of designs that may be encountered and the fact that virtually all CAD and CAE tools have been designed to be very good at some subset of design types, it is unrealistic to expect that a single set of tools from a single vendor will be able to deal with all problems equally well. A company with a wide range of design types, such as a company engaged in the design of computers and instrumentation, should expect to own more than one set of tools, each set optimized to its set of tasks. Trying to force one set of tools onto all types of problems is certain to result in overtooling simple designs and undertooling complex ones.

14.5 INTERFACING CAE, CAD, AND CAM TOOLS TO EACH OTHER

A major argument for buying all of the CAE, CAD, and CAM tools from a single vendor is to ensure that they all play together. In the past, this was a major concern because each vendor had proprietary data formats and there were no industry standard data formats. IPC, IEEE, and other trade associations have evolved standard forms of data interchange between systems. These have been adopted by suppliers, such that it is relatively easy to interface best-in-class tools from different vendors to each other.

14.6 INPUTS TO THE DESIGN PROCESS

14.6.1 Libraries

Each CAE and CAD tool uses a series of libraries that contain information describing each component that may be used in a design. These range from a simple description of the physical size of the pads and their relative positions to a full logical model that can be exercised in a simulator. Libraries do not usually come as part of a system. They must be purchased separately or developed one part at a time by the user. Libraries in mature systems can be quite large and represent a substantial investment in time to develop them. Unfortunately, libraries are usually unique to a given tool and cannot be transferred easily should a new tool be chosen.

14.6.1.1 Pad Shapes and Physical Features. The most basic library used by a CAD system describes the physical characteristics of a part in a manner that allows the CAD system to create its mounting holes pattern and pads as well as its silk screen outline and solder mask pattern. This library entry will contain a *pad stack* that describes how large the component lead holes are and the size and shape of the pads that will appear in each type of PCB layer. For example, an outer layer pad will need to be large enough to ensure adequate annular ring, an antipad will be needed in a power plane to ensure that the plated-through hole barrel does not touch the power plane, or a thermal pad will be necessary to make a connection to the plane in a manner that still allows reliable soldering. These library entries may contain the unique part numbers used by a company to build a bill of materials, in which case, the CAD system will be able to produce a bill of material in ready-to-use form.

Some physical feature libraries also contain information about the nature of a pin, such as whether it is an input, output, or power pin. This data is used by the checking programs to ensure that the points in a net are ordered properly for high-speed performance or to ensure that a net has the correct kinds of pins in it.

14.6.1.2 Functional Models. CAE tools that simulate the operation of a PCB require a library of models that describe how each part operates logically. These are functional models. Functional models do not contain information about propagation delay or rise times needed to verify that timing rules are complied with. Functional models are often used to configure emulators.

14.6.1.3 Simulation Models. Simulation models are extended versions of functional models. They contain all the functional information as well as detailed information about path delays through a part and rise and fall times. They are used to ensure that worst-case timing conditions result in a properly operating design.

14.6.2 PCB Characteristics

One of the sets of data required by the physical layout system is a description of the PCB or its physical characteristics. This includes its size, number and kinds of layers, thicknesses of insulating layers, copper thicknesses, and areas that are not available for parts or traces.

14.6.3 Spacing and Width Rules

To ensure compliance with manufacturing and transmission line rules, the trace widths and trace spacings for each layer must be entered into the CAD system. This is typically done in tabular form.

14.6.4 Netlists

Netlists describe to the CAD system how the pins of each device connect to each other. Systems that manage routing or layout to high-speed design rules will require netlists that contain instructions on how to handle each net, such as what impedance to use, what spacing to preserve with respect to neighbors, and whether terminations or special ordering is needed.

14.6.5 Parts Lists

Parts lists tell the CAD system what type of library entry to use for each part in the design.

CHAPTER 15
ELECTRICAL AND MECHANICAL DESIGN PARAMETERS

Ralph J. Hersey, Jr.
Lawrence Livermore National Laboratory, Livermore, California

15.1 PRINTED CIRCUIT DESIGN REQUIREMENTS

The electrical characteristics of printed board and multichip electrical connection substrates have become a critical functional product definition, and design requirement, for many electrical and electronic products. Until the late 1980s, most printed board designs were printed wiring designs, in that, with the exception of power and ground distribution, component placement and the arrangement of conductive and nonconductive patterns were not critical for functional electrical requirements. This was particularly true for most digital applications. However, since the late 1980s, electrical signal integrity has become a more serious design consideration in order to meet both functional performance and regulatory compliance requirements.

15.2 INTRODUCTION TO ELECTRICAL SIGNAL INTEGRITY

Electrical signal integrity is a combination of frequency and voltage/current, depending on the application. In low-level analog, very small leakage voltages or currents, thermal instabilities, and electromagnetic couplings can exceed acceptable limits of signal distortion. In a similar manner, most digital components can erroneously switch by application of less than 1 V of combined dc and ac signals.

15.2.1 Drivers for Electrical Signal Integrity

Analog and digital signals are subject to many issues that can cause signal distortion and degrade signal integrity. Some of these concern the transmission of the signal, and some the return paths.

15.2.1.1 Signal Integrity Terms for Both Analog and Digital Signals. Table 15.1 provides a listing of many of the issues that affect the signal transmission.

TABLE 15.1 Representative List of Issues for Both Analog and Digital Signals

Rise time	Thermal offset voltage
Fall time	Thermal offset current
Skew	Low-level amplifier
Jitter	High-impedance amplifier
High skew rate	Charge amplifier
Intermodulation distortion	Integrating amplifier
Harmonic distortion	Wideband amplifier
Phase distortion	Video amplifier
Crossover distortion	Precision amplifier

15.2.1.2 Return Paths. All electrical signals have a signal conductor and a signal return path. Very frequently, the signal conductor is shown on the schematic and the return conductor is neither shown nor mentioned on the schematic/logic drawing. This can also present a problem with PB CAD (computer-aided design) tools. Some CAD tools are "dumb" in that they will automatically route a transmission line as one of the signal conductors but will not route the necessary signal ground on one of the adjacent conductive pattern layers.

15.2.2 Analog Electrical Signal Integrity

The design of some analog printed circuits is a critical balance of all of the known parameters and characteristics of the complete design-through-use product development, manufacturing, assembly, test, and use processes. Analog designs cover all or portions of the complete electromagnetic spectrum, from dc all the way up into the GHz range of frequencies. Active and passive electrical/electronic components and materials have various levels of sensitivity to operating environments and conditions, such as temperature, thermal shock, vibration, voltage, current, electromagnetic fields, and light. In particular, the signal input terminals, voltage connections, and especially analog signal grounds are critical to analog signal integrity.

15.2.2.1 Sensitive Circuitry Isolation. One of the key methods to improve analog signal integrity is to isolate or separate the more critical or sensitive portions of the design. Sensitive circuitry may be susceptible to one or more external forces, such as electromagnetic, voltage, and grounding systems, mechanical shock/vibration, and thermal. Sometimes the more sensitive circuitry is repackaged into a separate function module that provides its own isolation and separation from the offending condition. Isolation and separation can be provided by physical separation, electromagnetic and thermal barriers, improved ground practices and design, power source filtering, signal isolators, shock and vibration dampeners, and elevated or lowered temperature controlled environment.

15.2.2.2 Thermal Electromotive Force. Below a few millivolts, thermal electromotive forces (EMFs) can have a significant impact on low-level analog signal integrity. Thermal EMFs of a nonsymmetrical sequence of various metal junctions (conductors) or symmetrical sequences of various metal junctions operating at different temperatures will generate and inject undesirable voltage (or induce unwanted currents) into the electrical signal path. This thermocouple effect is desirable in the case of temperature measurements. However, in the case of other low-level measurements it is an undesirable characteristic. Therefore, the requirement for low-signal-level PCDs is to ensure that all components and electrical interconnection networks and corresponding electrical terminations (such as soldered, welded, wire-bonded, or conductive adhesive) are symmetrical and isothermal. Electrical components, such as thin/thick-film resistors of different values (resistance), may have resistor elements manufactured from different formulations or compositions of materials and will have a designed-in thermal EMF error due to component selection.

15.2.3 Digital Electrical Signal Integrity

Each digital logic family of integrated circuits has manufacturer-specified electrical operating parameters and signal transfer characteristics, many of which have become industry standards due to multiple sourcing (manufacturing) of the family of components. The electrical signal integrity requirements for digital ICs are primarily the high and low electrical (voltage/current) requirements for the output, input, clock, set, reset, clear, and other signal names; the signal rise/fall times, clock frequency(s) and setup/hold times; and the voltage and ground connections as are necessary for the control and operation of the IC.

The input, output, and electrical signal transfer parameters and characteristics for digital ICs vary from logic or microprocessor family to family. Signal ICs are a large matrix of components, consisting of the semiconductor substrate materials, such as silicon, silicon/germanium, and gallium arsenide, that make up the various types of transistors. As shown in Table 15.2, the large number of digital IC families available create a complex matrix of design issues and requirements.

TABLE 15.2 Typical Digital Logic Rise and Fall Times

Logic family	Typical rise/fall time, ns	Logic family	Typical rise/fall time, ns
STD TTL	5	H	6
L	6.5	HCT	8
S	3.5	AC	3
ALS	1.9	ACT	
FTTL	1.2	10K ECL	0.7
BiCMOS	0.7	100ECL	0.5
		GaAs	0.3

The electrical signal integrity of the rise and fall times of electrical signals is a major driver and concern for high-speed and high-frequency printed circuits. Table 15.2 lists some of the typical rise and fall times of some of the popular digital logic families.

15.3 INTRODUCTION TO ELECTROMAGNETIC COMPATIBILITY

Electromagnetic compatibility (EMC) is a serious design requirement for both functional performance to design and regulatory compliance requirements. EMC encompasses the control and reduction of electromagnetic fields (EMF), electromagnetic interference (EMI), and radio frequency interference (RFI) and covers the whole electromagnetic frequency spectrum from dc to 20 GHz. Worldwide, the electronics industry has had to pay increasing attention to EMC to comply with both national and international standards and regulations. EMC involves major design considerations that ensure proper function within the electronic component, assembly, or system in order to:

1. Limit the emission (radiative or conductive) from one electronic component assembly, or system, to another.
2. Reduce the susceptibility of an electronic component, assembly, or system to external sources of EMF, EMI, or RFI.

There are three keys to EMC:

1. Design the product so that it produces less stray electromagnetic energy.
2. Design the product so that it is less susceptible to stray electromagnetic energy.
3. Design the product to prevent stray electromagnetic energy from entering or leaving the product.

15.4 NOISE BUDGET

Good design requires up-front determination of a *noise budget,* which should be included in the product definition requirements. The noise budget is the summation of all of the dc and ac voltages (or currents) that form a boundary within which the component, assembly, or system is designed to function.

$$e_{\text{noise}} = e_{\text{dc}} + e_{\text{ac}} \tag{15.1}$$

where e_{dc} = dc electrical noise
e_{ac} = ac electrical noise

The *dc noise budget* consists of the voltage settings (preset) of the power supplies, the operating tolerance of the power supply, and the series dc voltage drops of the voltage distribution system.

The *ac noise budget* consists of the effectiveness of the local bypass capacitor, the amount of decoupling between the load, the bulk decoupling capacitor, and the power distributions system, the local voltage drops in the component's voltage/ground conductors, and the component's input voltage tolerance.

As mentioned in Sec. 15.5.3, many of the operating electrical, mechanical, thermal, and environmental parameters and conditions can have a major influence on the noise budget. With a limited focus on digital designs, additional noises that may need to be considered are EMC radiated and conducted emissions from other electromagnetic equipment and thermally generated voltages (thermocouple effect) due to electrical connections with differing layers of metals operating at different temperatures. The following is a list of most of the electrical voltages that should be considered for a noise margin analysis:

Switching noise*	Changes in supply voltage
Cross-talk*	Changes in junction temperature
Impedance mismatch*	Changes in die ground voltage (IR)
Component wirebond (IR)†	Component lead (IR)

15.5 DESIGNING FOR SIGNAL INTEGRITY AND ELECTROMAGNETIC COMPATIBILITY

15.5.1 High Speed and High Frequency

Higher operating speeds and frequencies have had a dramatic effect on electronic packaging technology as a whole, but in particular on PBs and PBAs. Higher operating speeds have

* For digital circuits these three may amount to about 50 to 60 percent of the noise budget.
† (IR) voltage drop (current X resistance).

forced an evolutionary change from traditional PWD methods and practices, which were suitable for lower operating speeds, into the realm of serious PCD. Many of the technical design practices for most digital designs have had to adopt high-frequency analog techniques for design, synthesis, and analysis. In addition, component packaging densities have increased, functional electronic packaging densities have increased, more digital components are of the CMOS family design, and CMOS power dissipation increases with operating frequency, the end result being that thermal management has become more of a concern to PBD and other electronic packaging design personnel.

15.5.2 Leakage Currents and Voltages

Input guarding for stray leakage currents and voltages can be an important design consideration, in particular for some analog but also for some digital designs (especially CMOS). Many analog design requirements are based on thermal, pressure, force, strain, and other sensor technologies that have very low levels and ranges of electrical output voltages or currents. In addition, many of the requirements are for measurement accuracy and precision of a few percent or less. In combination, these requirements can present a challenge to electronic and PBD personnel. Very few sensors are robust in terms of their electrical output parameters; most of them have what is termed in the analog world "low-level" signal output for voltage and/or current. Therefore, many of these sensors require signal conditioning and amplifier circuits to improve the integrity of the electrical signal(s). Many of these signal conditioning and amplifier circuits have high-input-resistance characteristics which cause them to be more susceptible to erroneous signals. As a result, "guarding," as described later, becomes necessary to control stray voltages and currents.

Small, undesired, and unintended leakage currents and voltages can have a significant impact on the electrical signal integrity of some analog, as well as digital, applications. Leakage currents of a few nA or less, and leakage voltages of a few μV or less, can affect the designed functional performance of an electronic assembly. It is nearly impossible, from a practical point-of-view, for a high-input-resistance circuit to distinguish between a desired and undesired electrical signal. Thus, extreme care must be used in the design, manufacturing, and assembly processes of high-input-resistance products. Therefore, suitable PBD concepts must be included in the design to reduce and control the leakage currents and voltages. The following list identifies some of the common causes for leakage currents and voltages in PBAs.

- Insufficient (surface or volume) insulation resistance of base laminate
- Environmental contamination, fingerprints and skin oils, human breath, residual manufacturing and processing chemicals, improperly cured materials, solder fluxes, and surface moisture, such as humidity.
- Surface and subsurface contamination, such as can be found:

  On or in the assembled component

  Between conformal coatings and the surfaces they are protecting

  On, in, or under the solder resist

  Between conductive patterns on or in the electrical interconnection substrate

15.5.2.1 Design Concepts to Control Leakage Currents and Voltages. The primary concept for input guarding is to limit and control undesirable leakage currents and voltages or to prevent their formation in the first place. In theory, the principal is very simple: There are no leakage currents or voltages if there are no differences in electrical potential. In practice this is difficult to achieve and may not be a viable solution. However, by minimizing (the goal being to eliminate) the differences of potential between the critical electrical interconnection networks and components (leads and/or bodies) and all other materials, the formation of undesired leakage currents and voltages can be controlled (minimum effect) or eliminated (maximum effect).

- Form a *Faraday cage* around the critical conductive patterns and components using a combination of conductive patterns (frequently called *input guarding* and *guard rings*) and shielding enclosures.
- Keep all unguarded voltages out of the Faraday cage or protected area.
- Electrically connect the Faraday cage to a low-impedance voltage source that follows the critical (protected) voltage.

15.5.2.2 Guard Rings—A Design Method to Control Leakage Currents and Voltages. The control of electrical leakage to the more critical signal terminals/leads can be optimized through the selection of components, various levels of implementing input guarding into the design, and the selection of the materials for the electrical interconnection substrate.

Some components provide unconnected, unused, or guard terminals/leads adjacent to the input terminals. Care must be exercised for the balance or trim terminals/leads, as in most cases these terminals are connected (internally) in the component directly into the input differential amplifier circuitry of the component; thus, any undesirable leakage currents or voltages to these terminals/leads may result in undesired operation. Some linear operational amplifiers (op-amps) and other linear components are more suitable for input guarding than others; some have two (or more) unused terminals/leads that are used to improve electrical isolation between the protected terminals/leads in the component itself, as well as the component land pattern and electrical interconnection substrate.

The simplest method for providing input guarding to control input leakage currents and voltages is through the use of guard rings of conductive patterns *on all layers of PB conductive patterns* that surround the terminals/leads and associated circuitry. The guard ring is attached to a low-impedance voltage source that best follows the input signal or, as recommended by some analog IC manufacturers in their application notes, to the metal case of the component. As a result, the input terminals of high-input-resistance, low-bias-current, low-offset-voltage Op-amps can be guarded from stray electrical leakage currents and voltages.

15.5.3 Voltage and Ground Distribution Concepts

There are a few main concepts for the distribution of voltage(s) and ground in PBs, their assemblies, and other electronic assemblies. In general, most serious PCDs use one or more ground planes for the common electrical connection(s), for the source(s) of electrical power, and for the reference or return electrical signal path. The keys to good voltage and ground distribution systems are:

- Providing a low-impedance voltage and ground distribution system
- Meeting functional performance product definition design requirements
- Optimizing EMC

Depending on the design, the ground system may also be used for the grounding conductor interconnection(s) for the electrical safety and similar compliance requirements. For electrical signal integrity considerations, it is generally desirable to have separate but parallel electrical interconnection networks for the grounded (signal and power) and grounding (electrical safety) conductors. Like ground, voltage distribution for serious PCDs generally consists of one or more voltage planes (or portions thereof), although for some designs routed conductive patterns or buses may be a functionally acceptable option. A bused voltage and ground system may be acceptable for some designs, but they are generally limited to the PCD with lower operating frequencies and slower rise and fall times. The voltage and ground distribution and the location and type of bypass capacitors can have a significant impact on EMC and electrical signal integrity.

15.5.3.1 Grounding Concepts. Electrical grounding is one of the most important concepts and probably the least understood aspect of electrical signal integrity and EMC. All electrical conductors, including ground, form very subtle, but active, electrical interconnection network that can significantly compromise the product definition's requirements for electrical signal integrity and EMC. In particular, the grounding system is critical to ensure compliance to functional performance and regulatory requirements. Grounding (and voltage distribution) concepts are a matrix of requirements, concepts, concerns, considerations, and practices. In general, there is no universal solution suitable for all applications. Grounding is considered by some to be an art, which can be supported in that some grounding systems are completely unstructured and the reasons that some systems work, while others do not, are not clear. As a result, there has been an ongoing search to find a set of rules that can be used for the design of grounding systems and, unfortunately, many of the rules are conflicting. For example, a modular modem PCMCIA electronic assembly may have a suitable grounding system for normal telecom line operation. Yet, this may be totally inadequate if a 100-1kA electrical current, coupled through the PCMCIA assembly into the personal computer's electrical grounding system, is induced in the telecom line due to a nearby lightning strike or fallen power line. Similarily, a suitable electrical safety grounding system for power line frequencies may not be suitable near a high-power, high-frequency radio, television, or telecommunications transmitter. The following are some of the major concerns and considerations that are involved in good grounding practices:

- Integration of analog and digital signal converters, especially with more than 12 bits of resolution—ground loops and noise
- High-speed and high-frequency operation—ground pull-up and EMC
- High-speed and high-frequency bus line drivers and receivers—major ground pull-up and EMC concerns
- Low-signal-level analog sensors (transducers)
- Length of the conductors in the voltage/grounding system as a considerable portion of the electrical wavelength of one of the frequencies of the signal range of EMC
- When designing a grounding system, consideration of developing a grounding map that identifies all grounding requirements and voltage/current/frequency requirements

15.5.3.2 Grounding Systems. The following is an introduction to several grounding and voltage distribution systems and their electrical characteristics.

Single-Point and Point-Source Grounding Concepts. Single-point grounding is a method whereby the grounding electrical interconnection network is connected to ground at a single point at either the source or load end of the electrical interconnection network. The voltage drops between the various grounding nodes is a function of the interconnection network impedances, operating frequency, and current. Point-source or *star* grounding systems have a single-point grounding location for all electrical loads. The point-source grounding point is connected to another grounding point using a low-impedance bus or grounding conductor.

Multiple-Point Grounding Concepts. A multiple-point grounding system may be in the form of a loop or tree-like structure. In a loop grounding system the voltage drops around the loop may vary, depending on the electrical characteristics of each of the loads attached to the grounding loop. In a tree, the grounding system has good voltage regulation and allows leads to be independently attached or removed from the tree without a significant impact on the remaining loads.

Ground Planes. Ground planes are the grounding system of choice for most serious PCD requirements. Ground planes can improve the electrical signal integrity of the grounding system and EMC, provided all critical conductors are buried (lands only) on outer layer(s) of the printed board assembly.

Separation of Grounds. The identification and separation of natural groupings of grounds into similar requirements increases the assurance of conformance to product definition requirements. Some of the natural groupings are as follows:

- Electrical safety ground
- Power supply ground
- Low-level analog ground
- High-level analog ground
- Digital ground
- I/O ground
- Pulsed power/energy ground

When defining the printed circuit design requirements for a particular assembly, results of an analysis of the grounding elements necessary for proper functional performance must be included.

15.5.3.3 Bypass Capacitors. The selection, location, number, and value of the bypass capacitor(s) for PCDs can affect functional circuit performance. The purpose of the bypass capacitor is to provide the necessary electrical energy to minimize the effect of a component's normal transient switching and load currents during functional operation. The selection, location, and placement of the bypass capacitor can have a significant impact on EMC and a lesser impact on functional performance. One of the keys to EMC management is to prevent or minimize the generation and subsequent radiation of electromagnetic fields in the first place.

15.5.3.4 Voltage and Ground Buses. As operating frequencies and speeds increase, voltage and ground bus distribution systems may function as lump constant shock-excited oscillators. The frequency of oscillation is dependent on the series inductances of the voltage and ground bus system and shunt capacitors. One of the worst cases is when the voltage and ground buses are placed as railroad rails with the bypass capacitors, and digital integrated circuits are alternated in position like railroad ties bridging between the rails.

15.5.3.5 Voltage and Ground Planes. Voltage and ground planes can be an effective means to provide a relatively low resistance and impedance[1] to distribute voltage and ground within a PBD. However, maximum effectiveness—solid metal, with no holes or cutouts for plated through-holes or other necessary features—is an unobtainable condition. Therefore, voltage and ground planes are a compromise in design and requirements due to the necessary holes in the planes for electrical interconnections and component mounting.

Voltage and Ground Plane Resistance. The sheet resistance of solid copper voltage and ground planes is relatively low for most copper foil thicknesses. For 35-μm-thick copper foil, the dc (solid) sheet resistance is less than 1 mΩ/sq. The dc resistance of copper foil is shown in Eq. (15.2). The (solid) sheet resistance for selected copper foils is shown in Table 15.3. Due to the almost infinite number of possible variations in the size, placement, and shape of perforated mesh planes, the following data are presented for informational and comparative purposes only.

$$R_{DC} = \frac{17.2}{t_{\mu m}} \text{ m}\Omega/\text{sq} \qquad (15.2)$$

t_{mm} is in μm

TABLE 15.3 Solid Area Copper Foil Sheet Resistance

Copper thickness, μm	Sheet resistance, mΩ/sq
5	3.44
9	1.911
12	1.433
17	1.012
26	0.662
35	0.491
70	0.246

Resistance models for periodic mesh grid planes[2] is of interest for *uniform* grids, but is of limited use for grid planes that are not periodic and for ac electrical network analysis.

Voltage and Ground Plane Impedance. The impedance of perforated mesh voltage and ground planes is difficult to perform due to the number of variations in the size, placement, and shape of perforated mesh planes.

15.6 MECHANICAL DESIGN REQUIREMENTS

The design intent of PBs and their assemblies is primarily to mount and lend mechanical support to components as well as to provide all of the necessary electrical interconnections. However, PBAs should not be used as a (major) structural member.

There are three general forms of PBAs:

1. *Functional module*—A plug-in and a mechanically mounted PBA. The functional module is in the form of a component, whereby leads or other types of electrical terminals provide both electrical interconnections and mechanical mounting of the module to the next higher level of electronic packaging.
2. *Plug-in module*—This provides all of the necessary electrical interconnections at one or more edge-board connectors. A plug-in module typically plugs into a mother board, or sometimes electrical cables are used. The plug-in module is mechanically supported on one edge by the edge-board connector and on one or more edges by card guides, rails, or a mounting frame.
3. *Mechanically mounted PBA*—This is mounted and/or supported in a mechanical assembly or housing with a number of mechanical fasteners around its periphery (and internally to the PBA for additional support, if required). The most common mechanically mounted PBA is the mother board, such as is frequently used in personal computers. Mechanically mounted PBAs are one of the members of an electrical/electronic assembly and are physically mounted in the assembly using one or more mechanical fasteners, such as screws, clips, and standoffs. A common example of a mechanically mounted PBA is a modular power supply.

All forms of PBAs have many common requirements necessary to meet their product definition requirements, although there will be significant variations in the product definition requirements for PBAs due to their specific form and application requirements. For example, the requirement for PBA flatness due to bow and twist may be different for a plug-in than for a mechanically mounted PBA. The requirements for the number and location of mounting fasteners for a mechanically mounted PBA will be different for a relatively thick MLB with low-mass components than for a simpler PB with high-mass components.

The following list of factors, based primarily on Ginsberg,[3] must be considered and evaluated in the design of a PB and its assembly:

- Configuration of the PB, size, and form factor(s)
- Need for mechanical attachment, mountings, and component types
- Compatibility with EMC and other environmental
- PBA mounting (horizontal or vertical) as a consequence of other factors such as dust and environment
- Environmental factors requiring special attention, such as thermal management, shock and vibration, humidity, salt spray, dust, altitude, and radiation
- Degree of support
- Retention and fastening
- Ease of removal

15.6.1 General Mechanical Design Requirements

The general mechanical design requirements for PBs and PBAs include the methods of dimensioning, mounting, guiding during insertion and removal of plug-in components or assemblies, retention, and extraction. Frequently, the PBA mounting method is predetermined as a design requirement to an established compatibility with existing hardware. In other cases, the printed board designer has a choice in determining which PB mounting method is more suitable after considering such design factors as the following:

- PB size and shape based on form, fit, and function requirements
- Input/output terminations and locations
- Area and volume restrictions
- Accessibility requirements
- Ease of repair/maintenance
- Modularity requirements
- Type of mounting hardware
- Thermal management
- EMC
- Type of circuit in its relationship to other circuits

15.6.1.1 Dimensioning and Tolerancing. The dimensioning and tolerancing system for PBs and PBAs must ensure that the product is appropriately defined for all of the product form, fit, and function requirements for the complete product life cycle, from definition through manufacturing to end use. Dimensioning and tolerancing are critical at least for the design, manufacturing, assembly, inspection, test, and acceptance phases.

Regardless of the dimension and tolerance standards that are used to establish and document the product definition's mechanical design and acceptance requirements,[4] there should be at least two (primary) data reference features in every PBD (see Fig. 15.1). The purpose is to ensure the integrity of a PB's and PBA's datum reference throughout the production and acceptance process. In general, it should be a nonfunctional hole (in the case of PBs) or a surface feature that is used as the primary datum reference for final dimensional measurements and acceptance of the product. The datum reference should not be a machined edge that is formed in the last phases of the manufacturing, fabrication, or assembly process in a secondary machining operation.

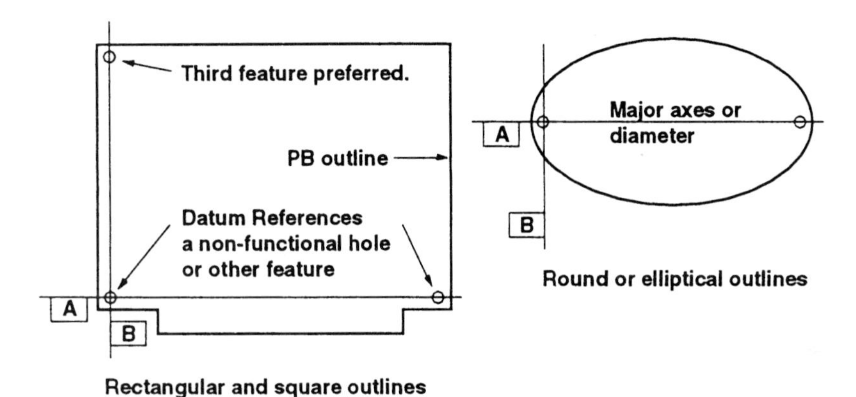

FIGURE 15.1 All printed boards and their assemblies should have their datum references included into the design to ensure the integrity of the mechanical datum references throughout a product's production and acceptance cycles.

15.6.1.2 Mechanically Mounted Printed Board Assemblies (PBAs). PBAs should be mounted to ensure their mechanical (and sometimes electrical grounding) integrity throughout their product life cycle. The following are some of the generally accepted requirements and practices for mechanically mounted PBAs:

- PBAs should be supported within 25 mm of the edge of the PBA on at least three sides.
- As good practice, fabricated printed boards (PBs) having a thickness of about 0.7 to 1.6 mm should be mechanically supported on 100-mm or lower intervals, and PBs thicker than 2.3 mm on less than 1.3-mm intervals.
- Fasteners should not be located on less than the PBs thickness or the fastener's head diameter (whichever is lower) from the edge of the PB.

15.6.1.3 Guides for Printed Board Assemblies. A major advantage of using plug-in printed board assemblies rather than other electronic packaging techniques is the suitability of PBAs for use with mechanical PBA card guides for ease of maintenance, changing configuration, and up-grading function or performance. There are many PBA guide hardware systems that are available either commercially, as industry standard, or proprietarily. The PBA may be predetermined as a design requirement or may be developed based on the size and shape of the PBA, *the degree of dimensional accuracy needed to ensure proper mating alignment with the mating connector system,* and the desired degree of sophistication. Some PBA guide systems contain a built-in locking system that provides mechanical retention and thermal management (conductive).

Caution: Some PBA card guide systems have become somewhat of industry standards and can be obtained or assembled to fit most PBAs. However, not all of the industry-standard-like PBA guide systems are compatible or interchangeable for retaining or extracting PBAs.

15.6.1.4 Retaining Printed Board Assemblies. Quite often, shock, vibration, and normal handling requirements necessitate that the PBA be retained in the equipment by mechanical devices. Some PBA retaining systems are attached as hardware to the PB during assembly; other retaining systems are built into the PBA mounting hardware frequently called a *cage.* The selection of a proper PBA retaining system is important, since the retaining devices may reduce the amount of PB area available for component mounting and interconnections, and can add significantly to the cost of the electronic equipment.

15.6.1.5 Extracting Printed Board Assemblies. A number of unique principles have been developed and applied to solve the various problems of extracting PBAs from their plug-in enclosures. The result has been a proliferation of proprietary and a few industry-standard-like extraction systems. The most common industry-standard extractor is injection-molded plastic hardware that is free to partially rotate when attached to the PBA with a pressed-in pin. Many of these PBA extraction tools use a minimum of PB space, thereby maximizing available PB area for components and conductor routing. They also protect both the PBA and the associated mating connector(s) from damage during the extraction process.

The following should be considered when selecting among the many different types of PBA extraction tools:

- The area of the PBA available for attachment
- The extractor's effect on the PBA-to-PBA mounting pitch
- The need for special provisions in the PBD, such as mounting holes, mounting clearance holes, and notches
- The size of the extractor, especially if the extractor is to be stored in the equipment with which it is used
- The need for an extraction device that is permanently attached to the PBA, usually by riveting
- The need for specially designed considerations, such as load-bearing flanges, in the PBA mounting chassis or cage hardware
- The suitability of the extractor to be used with a variety of PB sizes, shapes, and thicknesses
- The cost of using the extractor, both in piece price and added design costs
- The degree of access required inside the equipment to engage and use the extraction tool

15.6.2 Shock and Vibration

Shock, vibration, flexing, and bowing can be functional performance and reliability concerns for PBAs—more so for larger PBAs. For many PBAs, the worst-case exposure to shock and vibration occurs in nonfunctional or operational usage, during shipping and other forms of transportation from one location to another, or possibly in functional use when the functionally operating product containing PBAs is inadvertently dropped on the floor. Other PBAs are designed to withstand specified levels of shock and vibration in transportation and in use. The design requirements for shock and vibration vary, depending on each family of general requirements. For example, there are nonoperating shock and vibration withstand requirements for vehicular, train, ship, and air transportation for domestic and international shipment for Level 1, 2, or 3 products, which include various procedures and requirements for packaging. The shock and vibration functional design requirements are many and varied, and are very dependent on the application. Some sources of shock and vibration are very obvious, while others are very subtle. The levels and duration of shock and vibration vary significantly in each application: An electronic sensor mounted on a vehicle's axle is different from the radio mounted on the dashboard. There are the differences among ground-based, rack-mounted, industrial control equipment, aircraft, aerospace, and munitions applications. Some vibrations are subtle, low-level continuous, and are frequently caused by electric or gas motor-driven rotating machinery and equipment. Continuous low-level vibrations can induce *mechanical fatigue* in some electrical/electronic equipment.

15.6.2.1 Shock. *Mechanical shock* can be defined as a pulse, step, or transient vibration, wherein the excitation is nonperiodic.[5] Shock is a suddenly applied force or increment of force, by a sudden change in the direction or magnitude of a velocity vector. With few exceptions, shocks are not easily transmitted to electronic equipment by relatively light mounting frames and structures. Most shocks to electronic equipment in the consumer, commercial, and industrial markets are due to dropping during handling or transportation, the exceptions being electronic sensors or equipment mounted to heavier mounting frames, such as vehicle axles and punch-press-like equipment, or military equipment subjected to air dropping or to explosive forces, such as munitions. Most shock impact forces result in a transient type of dampened vibration which is influenced by the natural frequencies of the mounting frame. Generally, shock either results in instantaneous failure or functions as a stress concentrator by reducing the effective strength of the connection or lead for subsequent failure due to additional shock(s) and vibration.

15.6.2.2 Vibration. *Vibration* is a term that describes oscillation in a mechanical system, and is defined by the frequency (or frequencies) of oscillation and amplitude. PBAs that are subjected to extended periods of vibration will often suffer from fatigue failure, which can occur in the form of broken wires or component leads, fractured solder joints, cracked conductive patterns, or broken contacts on electrical connectors. The frequency(s) of vibration, resonances, and amplitude(s) all influence the rate to failure.

Flexing and bowing in PBAs is the result of induced shock and vibration into the PBA. Different PBA mounting methods have differing susceptibilities to shock and vibration. In general, most small PBA functional modules are manufactured as components and are frequently encapsulated with a polymeric potting material into a solid mass, and they therefore have minimal shock and vibration requirements within the module. The plug-in PBA is restrained on one edge by the edge-board connector(s), and to some extent along the two sides of the PBA by mechanical guides. This leaves only one edge of the PBA free to flex from shock and vibration or to bow from residual manufacturing or assembly stress in the PBA. However, a handle along the free edge of the PBA or a restraining bar can be located across the center of the free edge of the PBA and mechanically attached for support at the ends of the restraining bar to the PBA mounting hardware (card cage). Generally, the mating edge-board connector has a molded-plastic body that provides mechanical support to the mated edge-board connector and has sufficient compliance in the electrical contacts to maintain good electrical connections within the connector's performance specifications. Mechanically mounted PBAs can be more of a shock and vibration concern for three main reasons:

1. The PBA can be very large, sometimes as large as 600 mm^2. More frequently, the maximum width is 430 mm (the width of a standard electronic chassis) and less than 600 mm long. Although most are less than 300 mm^2, this still creates a large area and can be a problem if unsupported.
2. The PBA is misused as a mechanical support structure for high-mass components, such as magnetic components (iron-cored transformers and inductors), power supplies, and large (in physical size) function modules.
3. The PBA is not included in the mechanical design definition.

15.6.2.3 Major Shock and Vibration Concerns. These include the following:

- Flexing between PBAs may cause shorts between adjacent PBAs or to the enclosure.
- The fundamental mode is the primary mode of concern because it has the large displacements that cause fatigue damage to solder joints, component leads, and connector contacts.
- Continuous flexing of a PBA will fracture component leads and, more important, surface-mounted component solder joints, due to mechanical fatigue failure. (Mechanically induced flexing or vibration in assembled PBs is used under controlled conditions to induce failures in solder joints for quality and reliability studies.)
- Movement of a PBA within its mechanical guides will be amplified due to shock and vibration resonance or harmonic resonance.

- Fatigue life modeling for PBA mounted components has become significantly more complex due to in-use simultaneous application of both vibration and thermal cycles. Vibration strains and thermal strains should be superposed for more representative modeling.

15.6.2.4 Types of Edge Mounting. The problems, methods of analysis, and minimization of the effects of shock and vibration on PBAs are the same as in other engineering applications, and similar solutions can be used. PBAs are designed and manufactured in a wide range of shapes and sizes, with rectangular being the most common because of the shape of most electrical/electronic equipment, especially for plug-in PBAs. Though PBAs are a multi-degree-of-freedom system, the fundamental mode is of primary importance because it has the large displacements that are the primary cause of fatigue failure in solder joints, component leads, wires, and connector contacts.[6] Most vibrational fatigue damage occurs at the fundamental or natural frequency because displacement is highest and stress is maximized. Edge or boundary conditions are terms that are used to define the method of attachment of the PBA (or, more generically, a panel) to its mounting frame. The term *free edge* is used to define those edges that are not restrained and are free to move and/or rotate along the edge of the PBA out of their normal mounting plane. The terms *supported edge* or *simple support* used to define an edge that is restrained in out-of-plane movements but is allowed rotational movement around the PBA's edge. The terms *fixed edge* or *clamped edge* are used to define an edge that is restrained in both out-of-plane and rotational movements. Illustrations of the definitions for fixed edge, supported edge, and free edge, and their applications to plug-in mounted PBAs, are shown in Fig. 15.2.

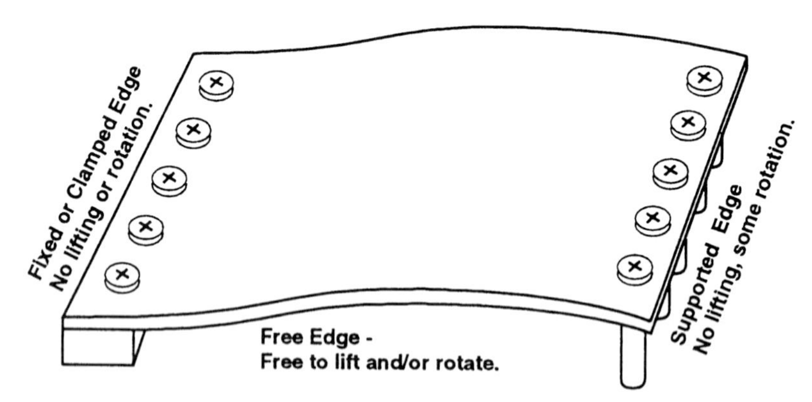

FIGURE 15.2 The method of mounting PBAs contributes to shock and vibration concerns, but can be reduced by good design practices.

15.6.2.5 Board Deflection. The amount of strain on a PBA component is a function of the maximum deflection of the PBA when subjected to shock and vibration. Those components mounted closest to the center of the assembly are subjected to the greatest strain, as illustrated in Fig. 15.3.

A set of empirical maximum deflection (δ) calculations has been developed by Steinberg, and his latest equation[7] has more parameters than previous equations, reflecting the sophistication and requirements for modern PBDs (units adjusted from inches to mm).

$$\delta = \frac{k\, 0.00022\, B}{ct\sqrt{L}} \tag{15.3}$$

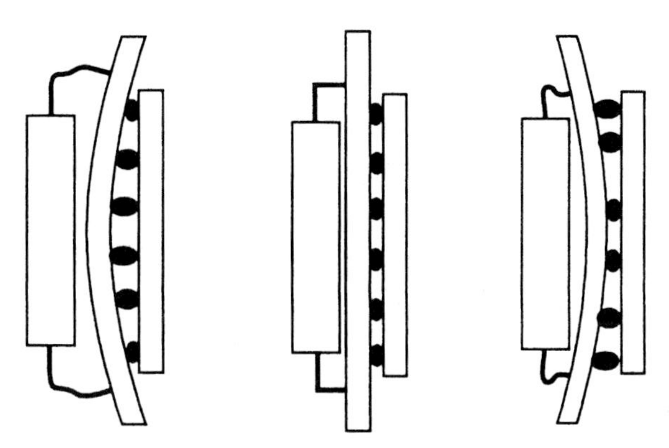

FIGURE 15.3 The PB bends in a PBA during shock and vibration, with the most severe stresses applied to the components mounted centermost on the assembly.

where k = units conversion coefficient; for inches $k = 1$, for mm, $k = \sqrt{25.4}$
B = length of PB edge parallel to component located at center of board (worst case), mm
L = length of component, mm
t = thickness of PB, mm
c = 1.0 for standard DIP
= 1.26 for DIP with side brazed leads
= 1.0 for PGA with four rows of pins (one row extending along the perimeter of each edge)
= 2.25 for CLCC

Analysis of the maximum deflection calculation formula reveals (as expected) that components with some compliance built into their component mounting and electrical terminations (such as the DIP and PGA) can be subjected to about twice the vibrational deflection as an SMT CLCC, provided component size, PB size, and PB thickness are equivalent. The latest equation for maximum deflection calculations is rated for 10 million stress reversals when subjected to harmonic (sinusoidal) vibration, and 20 million stress reversals when subjected to random vibration. It must be understood that this equation is a first approximation for predicting solder joint life. There are many factors that must be included for a more rigorous analysis and prediction. A more thorough discussion is found in Barker.[6]

15.6.2.6 Natural (Fundamental) Resonance of Printed Board Assemblies. The mechanical mounting of PBAs and their components is a key design consideration in the ability of the PBA to withstand shock and vibration. The overall size of the PBA is not a major factor, provided a suitable mechanical support structure is included as a part of the PBA's product definition requirements. There is a large matrix of ways to mount panels (PBAs) using various combinations of free edges, supported edges, fixed edges, and point supports and by calculating fundamental resonance. The following four examples compare the fundamental natural resonances of the same rectangularly shaped PBA using different edge-mounting techniques. Additional formulas for calculating other natural resonances are found in Barker,[6] Steinberg,[7] and others.

In the following examples—demonstrations of the sensitivity of PBAs to their methods of mounting—the same PBA is used for direct comparative purposes (see Fig. 15.4). The following are the design requirements and material parameters that were used for the calculations:

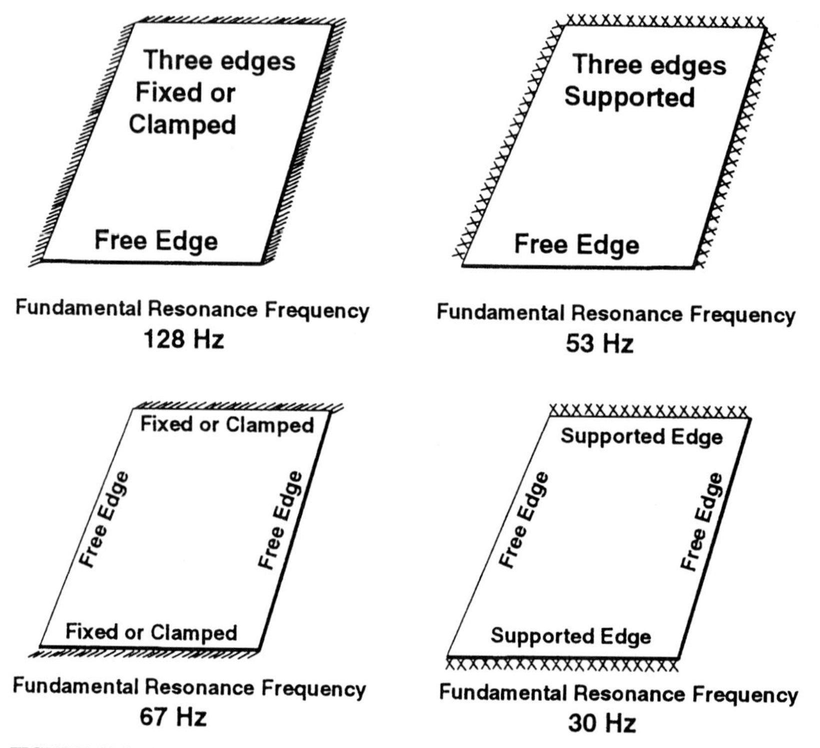

FIGURE 15.4 The same PBA can have a wide range of natural resonances depending on the mounting method, which is why shock and vibration concerns are a critical design consideration.

E	Modulus of elasticity (type GF epoxy-glass)	1.378×10^4 MPa
μ	Poisson's ratio	0.12 (dimensionless)
a	Longer dimension of the PBA	200 mm
b	Narrower dimension of the PBA	150 mm
t	Thickness of the PBA	1.6 mm
W	Weight of the PBA including components	0.25 kg

$$D = \frac{Et^3}{12(1 - \mu^2)} \tag{15.4}$$

where D is the plate bending stiffness.

$$M = \frac{W}{g} \tag{15.5}$$

$$f_n = C_0 P_0 \sqrt{\frac{Dab}{M}} \text{ Hz} \tag{15.6}$$

where f_n is the natural resonant frequency.

15.6.3 Methods of Reinforcement and Snubbers

Printed board assemblies are stiffened using one or more methods to raise the natural resonant frequency sufficiently above the shock and vibration threat. The most obvious is to change the method of retaining or mounting the PBA in the next higher level of assembly. Frequently, though, this may not be an acceptable option due to the resources and schedule changes that may be necessary for extensive redesign. However, some simpler modification in the design may meet requirements. Sometimes changing the plug-in PBA guide from a loose supportive guide to a tighter spring-loaded or clamping-type guide may be sufficient. Other methods are to add ribs or stiffeners, additional single-point mounting locations, or *snubbers* across the surfaces of the PBAs.

REFERENCES

1. Donald R. J. White and Michel Mardiguian, "EMI Control, Methodology and Procedures," *Interference Control Technologies, emf-emi Control,* 4th ed., Gainsville, Va., pp. 5.5–5.6.
2. Ruey-Beei Wu, "Resistance Modeling of Periodically Perforated Mesh Planes in Multilayer Packaging Structures," *IEEE Transactions on Components, Hybrids, and Manufacturing Technology,* vol. 12, no. 3, September 1989, pp. 365–372.
3. G. L. Ginsberg, "Engineering Packaging Interconnection System," Chap. 4 in Clyde F. Coombs, Jr. (ed), *Printed Circuits Handbook* 3d ed., McGraw-Hill, New York, 1988 pp. 4–17.
4. ANSI-YI4.5 "Dimensioning and Tolerancing," American National Standards Institute, New York, (date of current issue).
5. Cyril M. Harris and Charles E. Crede (eds.), *Shock and Vibration Handbook,* vol. 1, McGraw-Hill, New York, 1961, p. 1–2.
6. Donald B. Barker, Chap. 9, in *Handbook of Electronic Packaging Design,* Michael Pecht (ed.), Marcel Dekker, Inc., New York, 1991, p. 550.
7. Dave S. Steinberg, *Vibration Analysis for Electronic Equipment,* 2d ed., John Wiley, New York, 1988.

CHAPTER 16
CURRENT CARRYING CAPACITY IN PRINTED CIRCUITS

Mike Jouppi
Thermal Management Inc.,Centennial, Colorado

16.1 INTRODUCTION

One important consideration when designing electronics is to ensure that the electrical components operate at temperatures that will maintain long life and be reliable. Current carrying capacity of the printed circuit board traces is a part of managing the board temperature, which directly impacts the components. (A *trace* is a copper conductor in a printed circuit board. The terms *conductor* and *trace* are used interchangeably for a printed circuit throughout this chapter. *Track* is another common term for trace or conductor.) Properly sizing the traces for current is necessary to achieve the desired temperature rise at the board level.

It is a difficult task to generalize current carrying capacity for all printed circuits and all applications. The only safe way to simplify this task is to overdesign. As technology requires pushing the limits on current levels, as well as using small trace widths and trace spacing, overdesigning is not a solution.

Several sets of charts for sizing traces are presented in this chapter. These charts are explained and additional charts are presented. Thermal management of printed circuits often requires an accurate estimate of trace heating. This chapter presents information about conductor sizing charts and trace heating information as an aid to printed circuit design.

16.2 CONDUCTOR (TRACE) SIZING CHARTS

Charts are used to define the size of a trace for a given current level and temperature rise. The temperature rise of a trace is defined as the increase in temperature, above the local board temperature, that the trace reaches when current is flowing through it. The board temperature is similar to the ambient temperature when component power dissipations are low, although when the component power is several watts or more, the board temperature is significantly higher than the ambient temperature, especially in a still air or vacuum environment.

16.2.1 Conductor Sizing Chart Development

Charts have been in existence for many years. They have been adequate for most applications, although they have their limitations. The external and internal conductor-sizing charts in IPC 2221, "Generic Standard on Printed Board Design," are the standard for sizing traces and are included as Figs. 16.1 and 16.2, respectively.

The original external conductor chart was developed in 1956. It was assembled from trace heating test data taken from boards of different materials, different thicknesses, and different copper weights, as well as from boards with and without copper planes. The external charts are nonconservative for thin double-sided boards without copper (power, ground, or thermal) planes. "Nonconservative" in this context means that the traces are higher in temperature for a given current level than the charts suggest. The temperature differences are discussed later in the chapter.

The internal conductor chart is conservative for traces with cross-sectional area up to 700 sq. mil. It becomes less conservative at high current levels or with larger cross-sectional areas. The internal conductor-sizing chart represents half the current from the external conductor chart. It does not represent internal conductor test data. The internal trace chart has significant margin and should be used for sizing both internal and external conductors when possible. When that is not possible, a more detailed approach can be taken. The first step is to use new charts that focus on the baseline set of charts.

16.2.2 New Conductor (Trace) Sizing Charts

IPC-2152, "Standard for Determining Current Carrying Capacity in Printed Board Design," is written specifically for the carrying capacity of conductor current. Additional charts are provided that separate the variables that impact the temperature rise of a trace when current is applied.

There is a baseline set of charts for internal and external conductors. The baseline is conservative for most designs, although generally less conservative than the IPC 2221 internal conductor-sizing chart. The difference between the new and old is that attention is given to parameters that affect the temperature rise of a trace; some of these parameters include the board thickness, presence of copper planes, and the environment surrounding the circuit board. This chapter is presented as additional information on the existing charts in IPC 2221, as well as an aid to bridge the gap between the old and new charts that are presented.

16.3 CURRENT CARRYING CAPACITY

Current carrying capacity is commonly defined in terms of the temperature rise in a trace as a result of applying a specific amount of current to a specific size trace. The temperature rise is dependent on:

- The current level
- The board thickness
- The cross-sectional area of the trace
- The thickness of the trace for a given cross-sectional area
- The distance from a trace to copper planes
- The board material
- The environment (still air, forced air, vacuum, etc.)
- Skin effect at high speed (GHz)

16.3.1 Circuit Board Designs

Circuit boards can be extremely different from one design to another. They can vary from a double-sided board the size of a postage stamp to a 40-layer board that is 2 ft. wide and 4 ft. long. The environment in which they operate is also a consideration—for example, whether they are used on earth, in space, or on some other planet.

Because there is such a broad range of printed circuit board (PCB) construction and materials, a single trace-sizing chart cannot be expected to describe the temperature rise of a trace as a function of current for all boards. Fine-line and space, heavy copper, single-layer boards, and multilayer boards all constitute different configurations in which traces with the same cross-sectional area could vary from 10 to 70°C, or more for the same amount of applied current.

16.3.2 Conductor (Trace) Sizing Charts

Five different charts are presented in this chapter. The first has been around since the beginning of the printed circuit industry and is intended for external traces. The second is for sizing internal traces. The third and fourth are for internal and external traces. These are from more recent studies and are referred to as baseline charts. The fifth is used with the baseline charts to account for the heat spreading and cooling effect when copper planes exist in the board. Even with these charts, there are times when charts alone do not offer enough information and analysis tools must be used to solve current carrying capacity problems.

The charts included in this chapter are presented as a guide for sizing copper traces. There are limitations to their use and those limitations are discussed in the trace-sizing guidelines. The trace-sizing charts are used to manage trace heating, which includes sizing parallel conductors and vias. Finally, design problem areas such as "swiss-cheese" effect and environments are discussed at the end of this chapter.

For each set of charts, a discussion of what each chart represents and where it best applies for circuit board design applications is included. The charts included are as follows:

- External and internal charts from IPC 2221, the industry standard for printed circuit board design:
  - The external chart, Fig. 16.1, does not include design margin unless there is at least one internal copper plane in the board, such as a power or ground plane.
  - The internal chart, Fig. 16.2, is conservative, has significant margin, and keeps internal board heating to a minimum; more discussion is included in the design guidelines.
- Baseline charts. These are additional charts for sizing conductors and show an example of the effect of variables that impact trace temperature rise when current is applied. (Additional baseline charts can be obtained from Thermal Man, Inc., that take into account FR-4, BT, copper planes, and board thicknesses of 0.038 to 0.059 in.) Further discussion follows regarding the significance of the copper weight or thickness, board material, board thickness, and copper planes:
  - Internal conductor sizing charts for 1 oz. copper in a 1.78 mm (0.07 in.) thick polyimide board. (It is common to refer to the thickness of copper by weight, e.g., $^1/_4$ oz., $^1/_2$ oz., 1 oz., 2 oz., etc. The weight refers to the weight of a 1 sq. ft. piece of copper at a specific thickness. For example, 1oz. copper is 1 sq. ft. of 0.00135 in. copper. The weight can be calculated using the density of copper, 0.323 lb./in.).[3]
  - External conductor sizing charts for 2 oz. external conductors in a 1.78 mm (0.07 in.) thick polyimide board.
  - Plane chart for a 1.78 mm thick polyimide PWB. This is a chart that describes the reduction in trace temperature rise as a function of the distance from a single copper plane. This chart is used with the internal and external baseline charts.

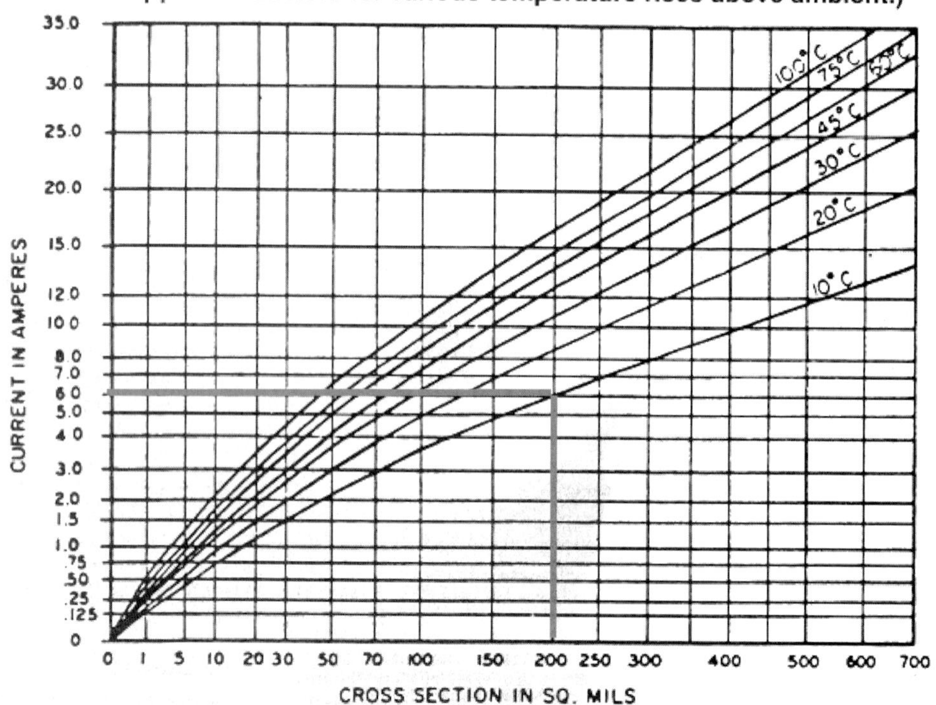

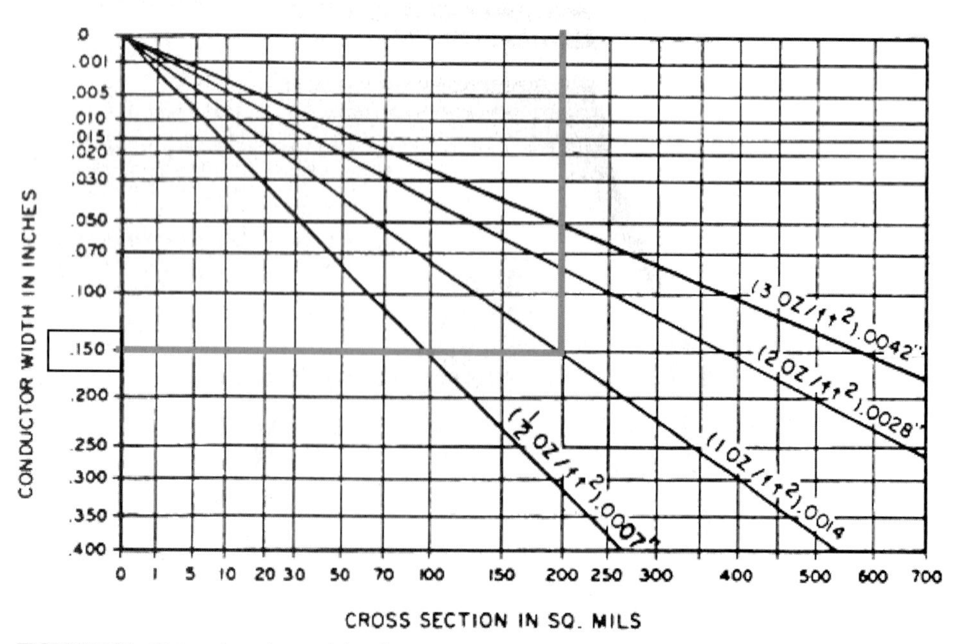

FIGURE 16.1 External conductor sizing chart.

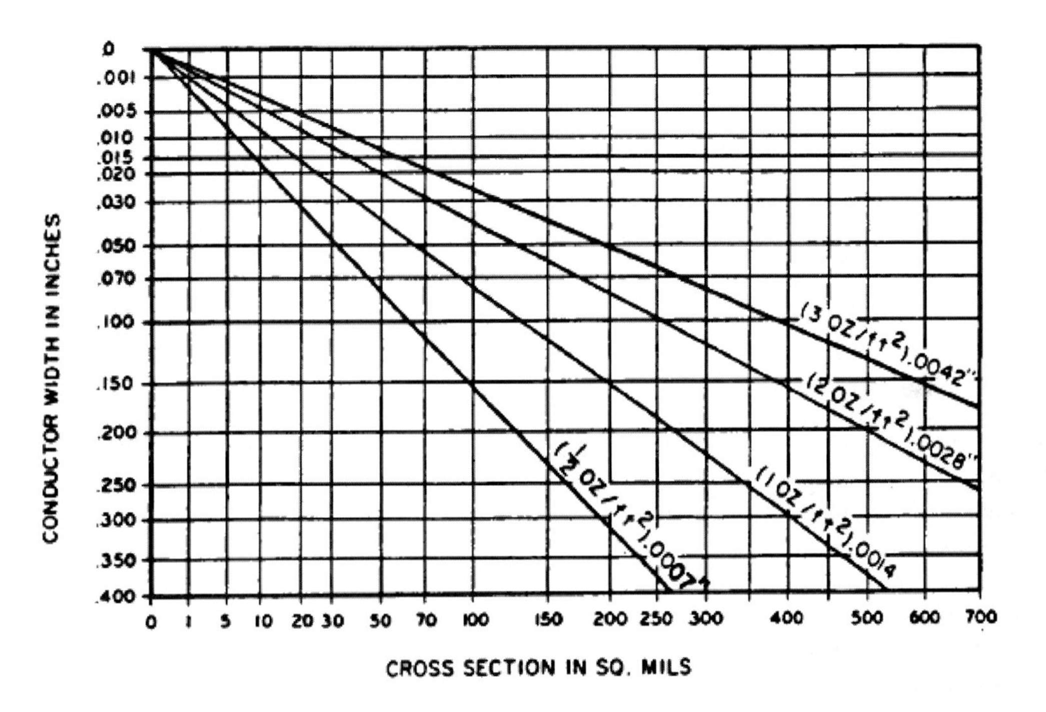

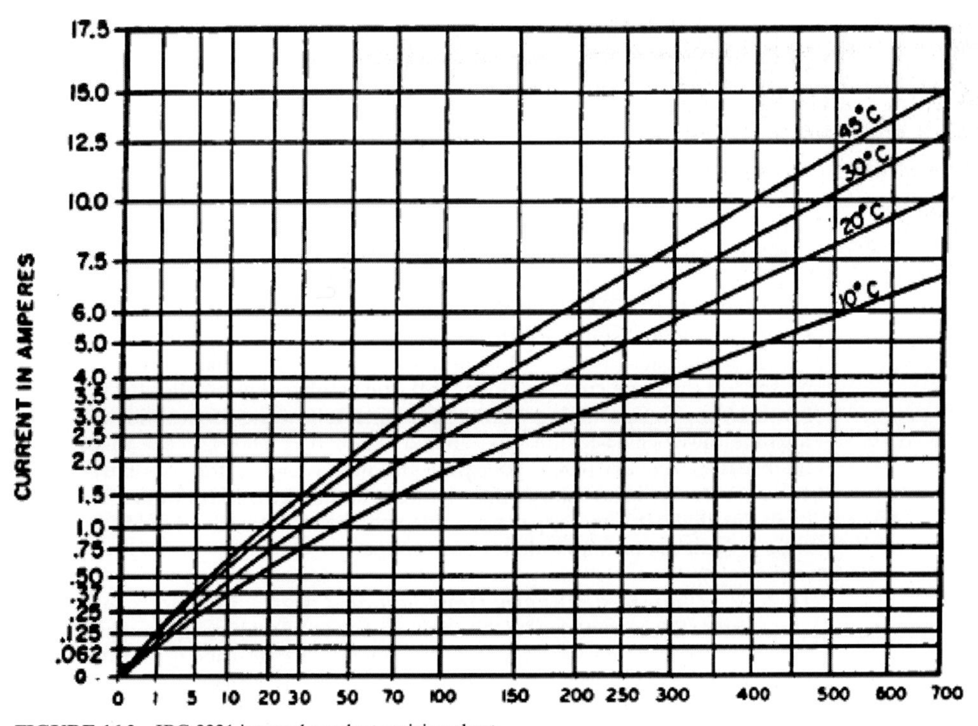

FIGURE 16.2 IPC-2221 internal conductor sizing chart.

16.4 CHARTS

16.4.1 External and Internal Conductor Charts

Figure 16.1 is for sizing external conductors, whereas Fig. 16.2 is for internal conductors; both are from the IPC standard 2221. External refers to the outermost layers of the printed circuit board. Until recent years, internal trace heating data were not broadly published. Figure 16.2 represents a conservative approach to internal trace heating; it does not represent internal trace heating data, although it does represent half the current from the external trace-sizing chart.

16.4.2 Trace Temperature Rise

The trace temperature rise derived from a conductor-sizing chart represents the increase in the trace temperature above the local board temperature surrounding the trace. For example, if the board temperature is 85°C as a result of component heating, and the trace is designed for a 10°C rise, then the trace will be 95°C. The temperature rise is the increase in temperature that the trace experiences when current is applied.

16.4.3 How to Use the Charts

Conductor sizing charts require knowing two of three variables: current, trace temperature rise, or trace size. If you know any two of these variables, you can calculate the other. When a more accurate estimate of the trace temperature rise is needed, board thickness, copper thickness, and board material are also used.

16.4.3.1 Chart Basics:Known Current. When the current and the desired temperature rise are known, then the trace width can be calculated for various trace thicknesses. A 10°C rise is a common temperature rise for a design. The temperature rise should always be minimized. If the designer can manage a 1°C rise or less, then the contribution to board heating will be minimized. Increasing the size of a trace lowers the temperature rise, lowers the voltage drop, decreases component temperatures, and improves the reliability of the product.

The following is an example of using Fig. 16.1:

What is the required size trace that will have a 10°C rise if 6 amps is applied to it? Starting with the top chart, follow the line going across from 6 amps to the curve labeled 10°C. The 10°C curve represents the temperature rise that occurs for a specific size trace at that current level. Next, follow the line going down from the 10°C curve and look at where it intersects the axis labeled "CROSS SECTION IN SQ MILS." The cross section is the cross-sectional area of the trace. The final step is to determine the trace width.

The lower chart in Fig. 16.1 is used to determine the trace width for various copper thicknesses. For the same cross-sectional area, the width will be smaller for thick copper and wider for thin copper. Continuing with the example, follow the vertical line down from 200 sq. mil. into the lower chart to the line labeled "(1 oz/ft^2) 0.0014""; see the discussion of copper thickness in Section 16.4.5. Following the line across to the axis labeled "CONDUCTOR WIDTH IN INCHES" shows that the trace should be 0.15 in. wide for a copper trace that is 0.0014 in. thick. If 3 oz. copper were selected, the trace width would be approximately 0.05 in. wide rather than 0.15 in. wide for 1 oz. copper.

16.4.3.2 Chart Basics:Known Cross-Sectional Area. If a trace size is known, the temperature rise or the current can be determined. The temperature rise can be estimated only in the range between the curves of constant temperature rise. The current can be estimated if the temperature rise is given. Each of these calculations is performed using the method presented in Fig. 16.1.

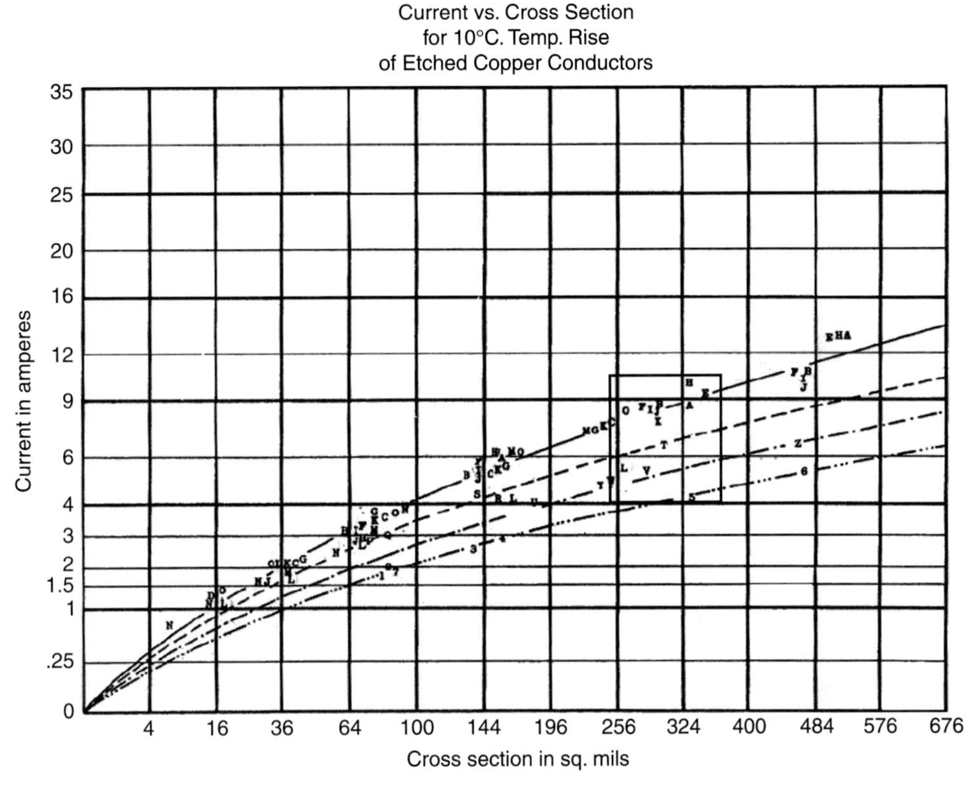

FIGURE 16.3 National Bureau of Standards 10°C chart.[1]

16.4.4 Chart Data

Data collection, to develop conductor-sizing charts, is performed following test procedures defined in IPC-TM-650 2.5.4.1a, "Conductor Temperature Rise Due to Current Changes in Conductors." Understanding these data, test vehicle, and test conditions used to derive conductor charts is important for understanding the limitations of these charts.

The chart shown in Fig. 16.1 was created from external trace heating data from double-sided boards. The internal trace chart, Fig. 16.2, was not developed from internal trace data; it represents half the current defined in the external chart. Further discussion on the internal trace chart is included in Sec. 16.4.5, "Trace-Sizing Design Guidelines."

The external trace chart represents a best-fit line through trace temperature data points from boards with different materials (epoxy and phenolic, different board thicknesses (3.175 mm [0.125 in.], 1.587 mm [0.0625 in.], and 0.794 mm [0.0312 in.]), and different copper thicknesses ($^1/_2$ oz., 1 oz., 2 oz., and 3 oz.), and, most important, from boards with copper planes. Board material, board thickness, copper thickness, and especially the copper planes all have an impact on the temperature rise of the trace. Averaging these variables together skews the results.

The top curve in Fig. 16.3 shows the original 10°C line used to create the 10°C curve in Fig. 16.1. All of the curves in Fig. 16. 3 represent a 10°C rise, but each curve represents a different set of conditions. Going from the top down, the next curve is supposed to represent coatings, although it is a mix of traces with coatings, thin boards, and boards that were dip-soldered. The next curve represents test boards that were dip-soldered. The last curve represents traces that were removed from the board and tested in free air. The boxed area in Fig. 16.3 is increased in size and shown in Fig. 16.4.

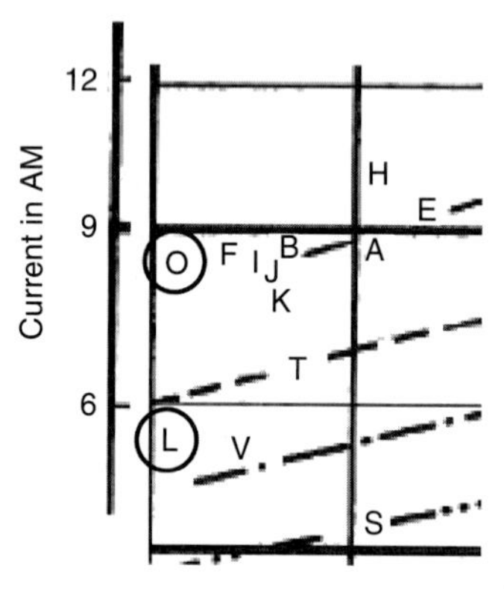

FIGURE 16.4 Influence of copper planes.

The top curve in Fig. 16.3 includes a variety of data points as described previously. In Fig. 16.4, several things can be observed with data points O and L, specifically the influence of board thickness and the influence of a copper plane. O and L are the same size trace in boards that are identical, with one exception: The board with trace O has a copper plane on the back of the board. A difference of 3 amps is observed between the two cases for a 10°C rise.

Data point L represents an external, 1 oz. copper trace on a 0.03 in. thick PWB. A trace on a thin board runs hotter than a thicker board. The trace in the thin board reaches a 10°C rise at a lower current level than the other boards. Data point O also represents an external, 1 oz. copper trace on a 0.794 mm (0.0312 in.) thick PWB, although it has a 1 oz. copper plane on the backside of the board. The trace on the board with the copper plane can have significantly more current before reaching a 10°C rise than the trace on the board without the plane due to the heat-spreading capability of the plane.

As a result of the way the curves in Figs. 16.1 and 16.2 were developed, there are areas where they are conservative and areas where they are not. Guidelines have been prepared to help further define these areas of the curves.

16.4.5 Trace-Sizing Design Guidelines

This section summarizes and expands the design notes from IPC-2221:

- Design charts have been prepared as an aid in *estimating* the temperature rise of a trace above the local board temperature versus current for various cross-sectional areas of etched copper conductors. It is assumed that for normal design, conditions prevail where the conductor surface area is relatively small compared to the adjacent free panel area and that the panel area is approximately 3 in. × 3 in. or greater.
- The curves as presented are not derated. Consideration should be given to allow for variations in etching techniques, conductor width estimates, and copper thickness.
- A standardized process is not used to determine the amount of undercut resulting from the etching process that creates the traces from the layers of copper of a printed circuit board. Circuit board manufacturers have their own techniques for adjusting artwork to provide the desired trace size in a design. Variations exist from one manufacturer to another. The IPC Process Capability, Quality, and Relative Reliability ($PCQR^2$) database is a source for determining the capability of a circuit board manufacturer. Two of the pieces of data collected are trace width and thickness source. This database is a source for finding the cross-sectional area of a trace of a final product compared to the initial artwork. Etch back is primarily a concern for small traces where the aspect ratio (the ratio of width to thickness) is small.
  - The final width of a trace should take into account the undercut that occurs during manufacturing. Consult a board manufacturer for undercut estimates.
  - The IPC $PCQR^2$ database is a source for copper thickness in a final product. The thickness of copper has a minimum allowable value that is less than the assumed thicknesses found in literature (see Sec. 16.7). A random selection of a half-dozen trace sizes from board coupons was collected for four different copper weights and is presented in Table 16.1.
- The temperature rise of a trace should always be kept to a minimum. The trace temperature rise otherwise referred to as the delta T (ΔT) is the temperature rise above the local board temperature surrounding the trace. Components and other conditions drive the board to temperatures above those of the surrounding environment. The temperature rise (ΔT) from the charts describes the temperature rise above the board temperature. For example, if the circuit board is at 75°C and the trace is sized for a 10°C rise, the trace temperature will be 85°C.

TABLE 16.1 Sample Internal Copper Thickness

	1/2 oz. (in.)	1 oz. (in.)	2 oz. (in.)	3 oz. (in.)
Maximum	0.0006	0.0015	0.00260	0.00400
Minimum	0.0006	0.0010	0.00230	0.00380
Median	0.0006	0.0011	0.00245	0.00385
Average	0.0006	0.00117	0.00247	0.00387

- The IPC internal trace-sizing chart is suggested under the following conditions if:
  - The board does not have internal or external copper planes
  - The panel thickness is 0.8 mm (0.315 in.) or less
  - The conductor thickness is 0.108 mm (0.00425 in.) or thicker
- For single conductor applications, a chart may be used directly for determining conductor widths, conductor thickness, cross-sectional area, and current carrying capacity for various temperature rises.
- For groups of similar parallel conductors, if closely spaced, the temperature rise may be found by using an equivalent cross-section and an equivalent current. The equivalent cross-section is equal to the sum of the cross-sections of the parallel conductors, and the equivalent current is the sum of the currents in the conductors. Parallel conductors refer to conductors on all layers of a PWB. Closely spaced is as much as 25.4 mm (1 in.) spacing and less.
- Closely related to parallel conductors are coils. "For applications where etched coils are to be used, the maximum temperature rise may be obtained by using an equivalent cross-section equal to 2n times the cross-section of the conductor, and an equivalent current equal to 2n times the current in the coil, where n is equal to the number of turns."[2]
- The effect of heating due to attachment of power-dissipating parts is not included. Component heating will impact the board temperature rise. The increase in electrical resistance of the trace due to an elevated temperature is a secondary effect on trace temperature rise. The final trace temperature is estimated as the PWB temperature plus the delta T selected for the trace.
- All design guidelines presented are for a still air environment.
- The IPC internal conductor chart is recommended for the following:
  - Flex circuits
  - All trace sizing, which includes external traces when possible
  - Space (vacuum) and high-altitude environments
- The IPC external conductor-sizing chart should be used only when at least one copper plane (power or ground) exists in the PWB.
- For trace sizes outside the range of the chart (over 700 sq. mil), consider the following:
  - Extrapolating the external IPC charts is not recommended.
  - If extrapolation is to be performed, use a baseline chart; however, extrapolating the IPC charts is not recommended.
  - Use a thermal analysis software tool for determining the temperature rise.
- For space (vacuum) environments, consider the following:
  - The internal conductor chart is best for sizing both internal and external traces that must operate in space or vacuum environments.

- When sizing traces and taking into account the copper planes in the board (still air), consider the following:

1. Using charts developed from trace heating data in PWBs with no internal copper and following IPC-TM-650 guidelines is the recommended method for sizing traces. Charts that include the influence of copper planes should be used for determining design margin. A copper plane that is 0.127 mm (0.005 in.) or less from a trace can reduce the trace temperature rise by as much as 70 percent.
2. The presence of copper planes in the printed circuit board provide heat spreading and lower the temperature rise of a trace. When sizing traces using data that takes into account copper planes, the design guidelines change. The heat spreading is significant and heating from all conductors in the area of a plane are mutually influential. Therefore, all traces within the area of the plane must be considered when summing the current following the parallel conductor rule. (The trace must be under or over the plane, not adjacent to the plane, to use charts that consider planes, and the plane must be larger than 3 in. × 3 in. in area.)
3. The distance from a trace to the plane has a significant effect on the temperature rise of the trace. Multiple factors are involved when sizing traces using charts that take into account the heat spreading due to the presence of copper planes:

   a. The trace temperature rise is a function of the size of the plane:

   b. Below 9 sq. in. (3 in. × 3 in.), the impact has not been completely characterized, although the cooling effect is not enough to take the heat spreading into account over the baseline prediction.

   c. From 9 sq. in. to 40 sq. in., the cooling effect from the increased area is improved by approximately 10 percent.

   d. Above 40 sq. in., the effect on trace temperature remains the same.

   e. The trace temperature rise is a function of the total power dissipated by all the traces. There is a diminishing return when the power dissipated by the traces exceeds the capability of the board to spread and dissipate the power.

   f. The thickness of the copper plane has a direct effect on the trace temperature rise.

The next section shows a very specific set of trace heating data. The data set is presented so that the variables that impact the temperature rise of a powered trace are separated. A discussion is included with each data set.

16.5 BASELINE CHARTS

The charts in Figs. 16.5 through 16.8, called *baseline charts*, were developed from current carrying capacity testing using polyimide PWBs with no internal copper planes. Test vehicles were developed and current carrying capacity testing was performed following IPC-TM-650-2.5.4.1a, "Conductor Temperature Rise Due to Current Changes in Conductors."

One internal data set, 1 oz. internal traces, and one external data set, 2 oz. external traces, on a 0.07 in. thick polyimide PWB, are presented. These two data sets are a subset of data collected for $^1/_2$ oz., 1 oz., 2 oz., and 3 oz. copper weights, FR-4, and polyimide boards, and PWBs with board thicknesses of 0.965 mm (0.038 in.), 1.498 mm (0.059 in.), and 1.78 mm (0.07 in.). The other data sets are used to discuss the effect of the other variables with respect to the charts in Figs. 16.5 through 16.8.

- **Board thickness** Traces in the 0.965 mm (0.038 in.) thick boards are approximately 30 to 35 percent higher in temperature than the 1.78 mm (0.07 in.) thick PWB data, and the 1.498 mm (0.059–in.) thick boards are 20 percent higher.

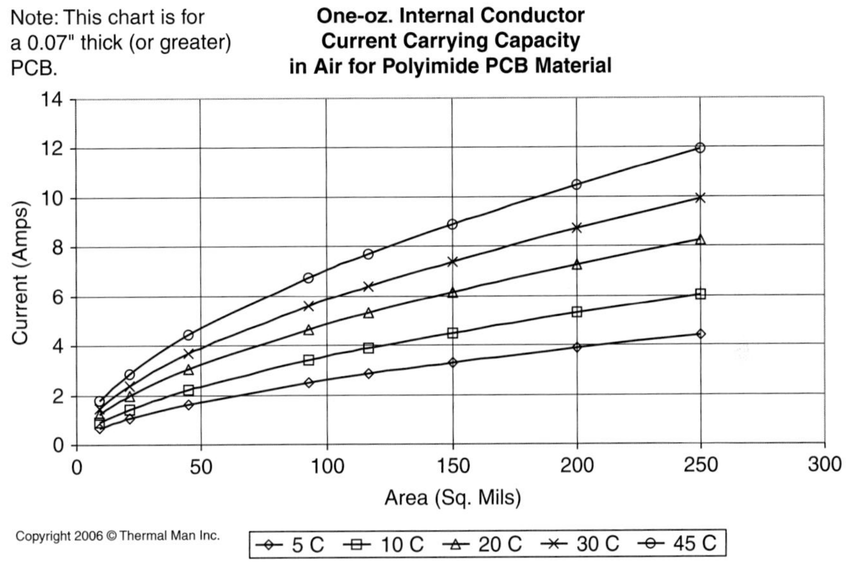

FIGURE 16.5 Baseline (Jouppi) 1 oz. internal conductor chart. [The baseline (Jouppi) charts represent the author's personal data collection.]

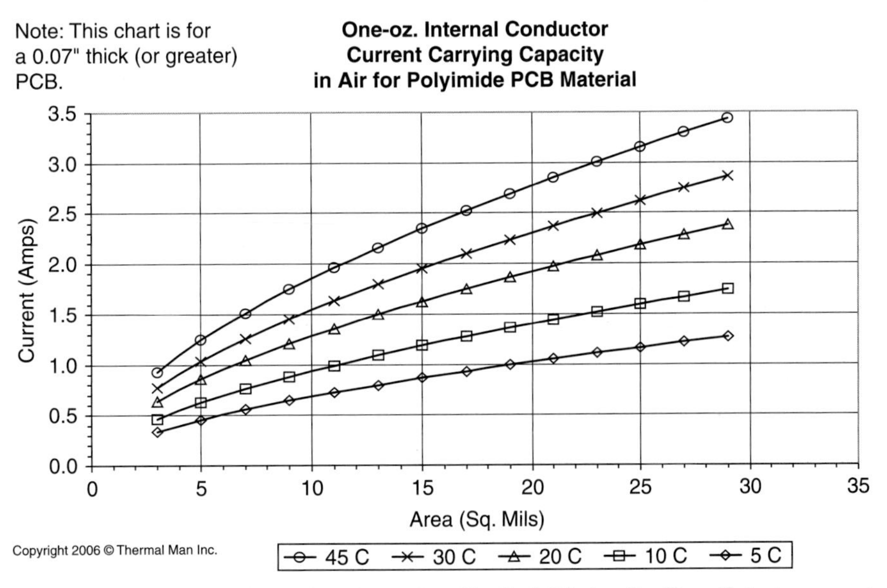

FIGURE 16.6 One-ounce internal conductor chart (fine line). [The baseline (Jouppi) charts represent the author's personal data collection.]

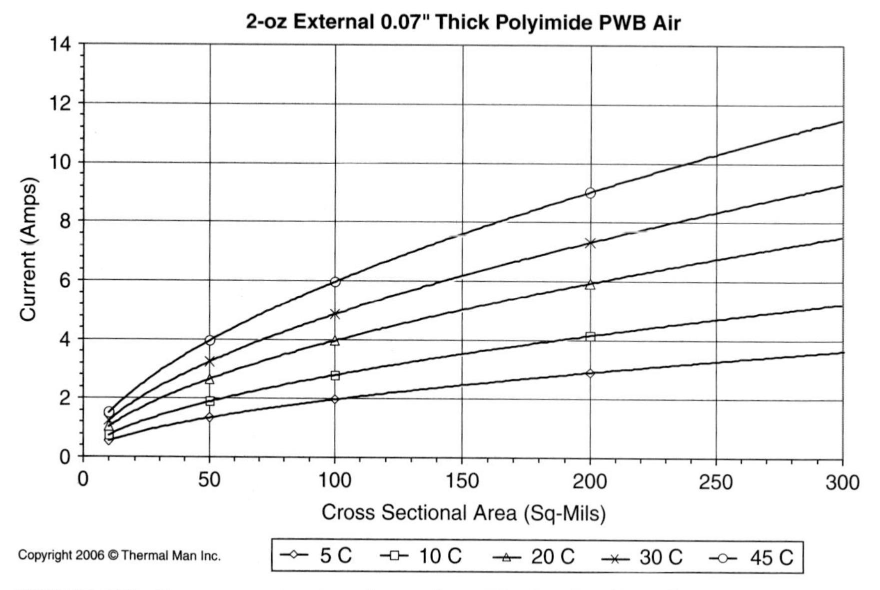

FIGURE 16.7 Two-ounce external conductor chart. (The baseline (Jouppi) charts represent the author's personal data collection.)

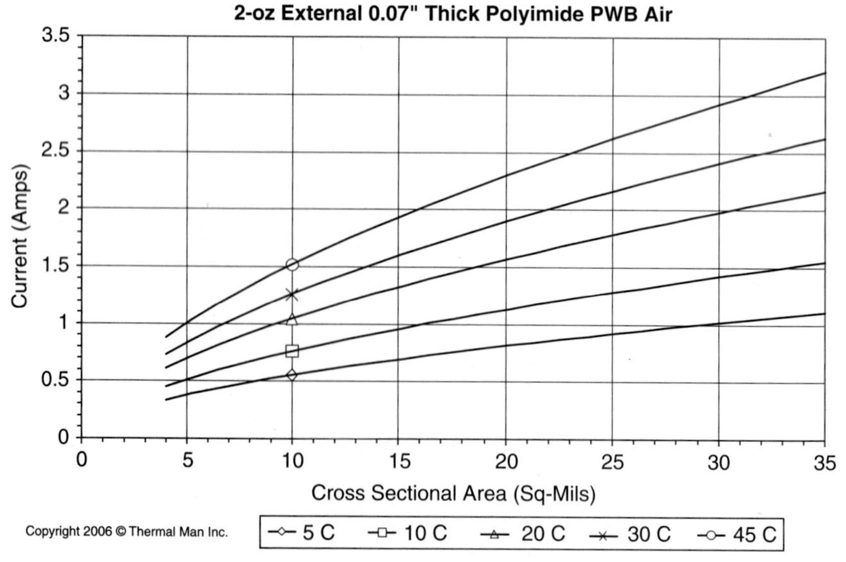

FIGURE 16.8 Two-ounce external conductor chart (fine line). (The baseline (Jouppi) charts represent the author's personal data collection.)

- **Copper weights** Half-ounce copper traces are similar in temperature rise for the same size cross-sectional area as the 1 oz. traces. An increase in temperature rise of 10 to 15 percent over the 1 oz. traces is observed for 2 oz. traces, and 15 to 20 percent increase for 3 oz. traces for the same cross-sectional area. The higher percentage is for a 45°C rise and the lower percentage is for a 10°C rise.
- **Board material** The FR-4 versus polyimide board material properties did not have a significant impact on trace temperature, which is determined primarily by the thermal conductivity of the dielectric laminate material construction. Table 16.2 lists measured thermal conductivity values for each of the test boards. The column labeled kz presents values through the thickness of the board and represents the resin thermal conductivity. The values in columns kx and ky are "in-plane" and the difference is attributed to the influence of the glass fiber.

TABLE 16.2 Dielectric Thermal Conductivity

Material	kx (w/in.-C)	ky (w/in.-C)	kz (w/in.-C)
FR-4	0.0124	0.0124	0.0076
Polyimide	0.0138	0.0138	0.0085

16.5.1 Copper Planes

One factor that has significant impact on the temperature rise of a trace for a given current level and trace size is the influence of copper planes. Whether they are power, ground, or simply thermal planes, the copper planes help spread the heat and lower the temperature rise of what would otherwise be a hot spot.

When designing with baseline charts, the copper planes add margin to the design. Multiple planes have more of an impact on lowering the temperature rise of a trace, although a single plane of 1 oz. copper is the starting point from the baseline configuration.

16.5.2 Single Plane

Trace heating data was collected from testing 1.78-mm (0.07 in.) thick polyimide PWBs and 0.965-mm (0.038 in.) and 1.49-mm (0.059 in.) thick FR-4 PWBs. The testing was performed in air and vacuum. Computer (thermal) models were developed that simulate the steady state and transient (time-dependent) temperature response for trace heating in each test board. The models were correlated to the test data. After correlating the models, copper planes were added to the models and used to develop charts that represent trace temperature rise as a function of current for traces in boards with copper planes. (The finite element models were developed using the thermal analysis software package ANSYS Thermal Analysis System [TAS]. The thermal models were correlated to the trace heating data collected in still air environments.) Figure 16.9 represents some of the results from that work.

Figure 16.9 shows the temperature rise of an internal, 0.010 wide, 1 oz. (0.00135 in.) copper trace (13.5 sq. mil), in a 127 mm (5 in.) wide × 127 mm long × 1.78 mm thick polyimide board, with 1.85 amps. The baseline column represents a trace in a board with no copper planes. The other two columns represent the same trace and current, although now there is a 1 oz. copper plane in the board. The trace is centered in the board and can be above or below the plane. The center column illustrates the temperature rise if the copper plane is 0.508 mm (0.02 in.)

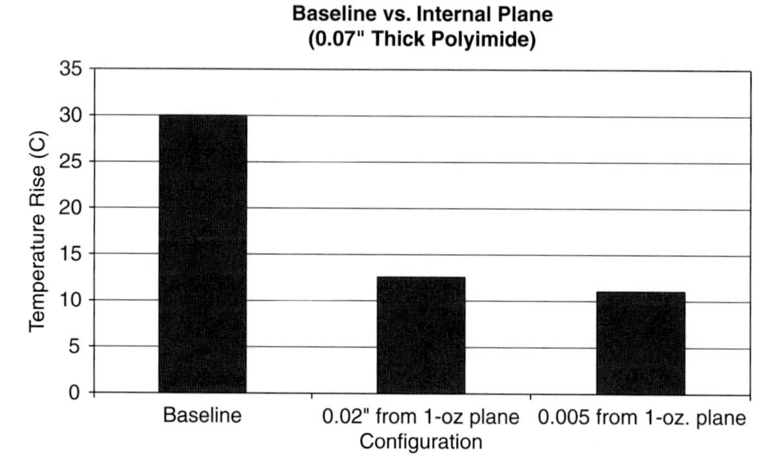

FIGURE 16.9 Copper plane influence on trace temperature.

from the trace. The column on the right shows the temperature rise of the trace when it is 0.127 mm (0.005 in.) from the plane.

Several key points can be observed when looking at this chart and comparing against the IPC-2221 internal conductor-sizing chart.

- The predicted temperature rise of the same trace using the internal conductor-sizing chart in IPC-2221 is 266°C.
  - The internal IPC chart is conservative for sizing conductors in printed circuit board applications.
  - Significant margin exists in trace sizing if copper planes exist.
  - Design guidelines require a full description of their origin.
  - When multiple variables impact the temperature rise of a trace, the impact of each variable requires definition to improve the level of design optimization.
- The presence of a single plane has a significant impact on the temperature rise of a trace.

Trace temperature rise has been characterized with respect to board thickness, copper weight, board material, internal versus external, air versus vacuum, as well as the distance from copper planes. A single copper plane has the most significant impact of all the variables described in this chapter. Guidelines for estimating trace temperature rise when a copper plane is present are as follows:

- Use a baseline chart to calculate the trace temperature for a given current and trace size:
  - The baseline represents the trace temperature rise in a PWB with dielectric only (no internal copper planes).
  - The baseline represents a specific board (dielectric) material and board thickness. (A 1.78 mm thick polyimide PWB, baseline conductor chart is provided in this chapter.)
  - The influence of a copper plane on the baseline trace temperature rise is more significant for thinner boards and boards with lower dielectric thermal conductivity values.
- The distance from a trace to a copper plane must be known. The closer the trace is to the plane, the lower the temperature rise of the trace.
- The thickness of the copper plane must be known. As the amount of copper increases, the temperature rise of the trace decreases.

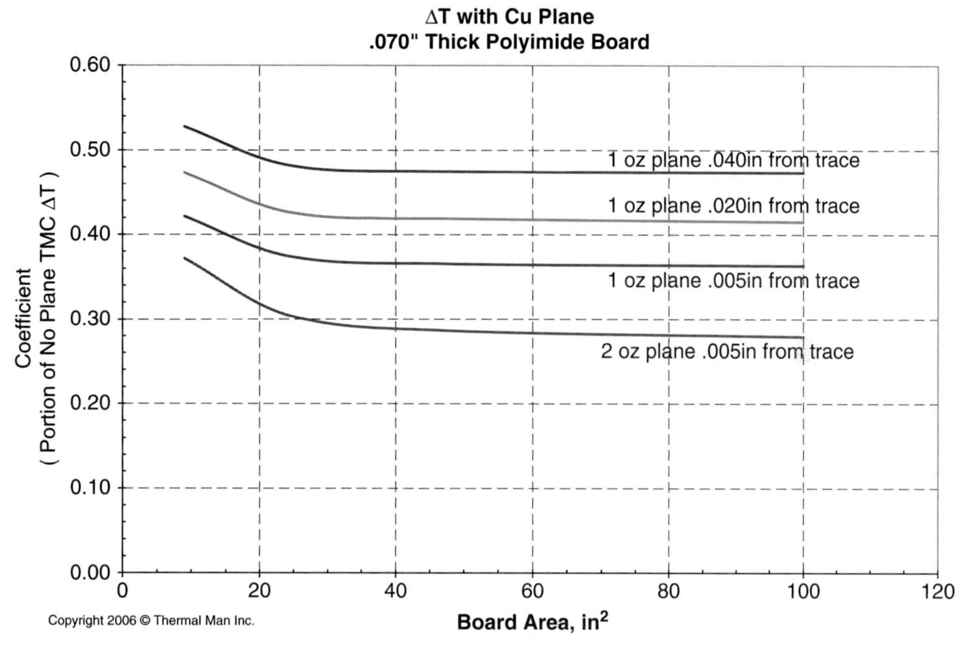

FIGURE 16.10 Distance from trace to copper plane.(The copper planes do not take into account the swiss-cheese effect from vias. The planes are assumed to be continuous solid planes.)

- The number of copper planes is not as significant, although the planes should be continuous with very few etched away areas:
  - The most significant decrease in trace temperature occurs with the introduction of a single plane.
  - As the number of planes increase, the trace temperature rise decreases, but not as much as the influence from the first plane.
- The area that the copper plane covers on a specific layer is important. The plane must completely cover the trace in order for it to impact the temperature of the trace. The size of the internal copper plane impacts the amount of heat spreading that can occur:
  - Below 58 cm^2 (9 sq. in.), the impact has not been completely characterized, although the effect is reduced and is observed on the left hand side of the curves in Fig. 16.10.
  - From 58 cm^2 (9 sq. in.) to 258.1 cm^2 (40 sq. in.), the effect is increased by approximately 10 percent.
  - Above 40 sq. in., the effect remains the same.
- The chart shown in Fig. 16.10 is used to determine the reduction in a trace temperature rise as a function of the distance from the trace to a copper plane.

Figure 16.10 represents estimates of temperature rise as a function of copper plane size and distance from trace to plane in a 0.07–in. (1.78 mm) thick polyimide board. The set of curves in Fig. 16.10 is used with the baseline charts, shown in Figs. 16.5 through 16.8. The first step is to calculate a delta T using the appropriate chart from Figs. 16.5 through 16.8.

The second step is to determine the size of a copper plane in the board and the distance from the trace to the plane. For example, start with a 1 oz. solid copper ground plane in a 127 mm (5 in.) long and 203.2 mm (8 in.) wide board 258.1 mm (40 in.^{2}). Assume that the trace is several dielectric layers from the ground plane. If the distance from the trace to the plane is 0.02–in., then Fig. 16.10 can be used to find the coefficient to be used with the baseline trace temperature rise.

The size of the copper plane, 40 sq. in., is selected on the x-axis of Fig. 16.10. The distance from trace to the plane, 0.02 in., is the second curve from the top in Fig. 16.10. The coefficient is found on the y-axis directly across from the intersection of the curve and copper plane area. The trace temperature rise will bc approximately 0.42 times the temperature rise calculated from Fig. 16.5 or Fig. 16.6.

If the trace were located 0.127 mm (0.005 in.) from a 2 oz. copper plane, the third curve from the top in Fig. 16.10, the temperature rise would be 0.29 times the temperature rise. So, if a 10°C rise were calculated from the baseline chart, the temperature rise would be 4.2°C in the first example and 2.9°C in the second example.

16.5.3 Parallel Conductors

Parallel conductors refer to traces that are parallel to each other as shown in Fig. 16.11. Parallel conductors also refer to traces on adjacent layers that are parallel to each other. The design rule for sizing parallel traces from the design guidelines is stated in this section.

For groups of similar parallel conductors, if they are closely spaced (that is, as much as 25.4 mm [1.0 in.] spacing and less when using a baseline chart), the temperature rise may be found by using an equivalent cross-section and an equivalent current. The equivalent cross-section is equal to the sum of the cross-sections of the parallel conductors, and the equivalent current is the sum of the currents in the conductors.

Example: Determine the maximum current that can be applied to 16 parallel traces that are 0.028 mm (0.0011–in.) wide $^1/_4$ oz. copper (0.00035 in. thick):

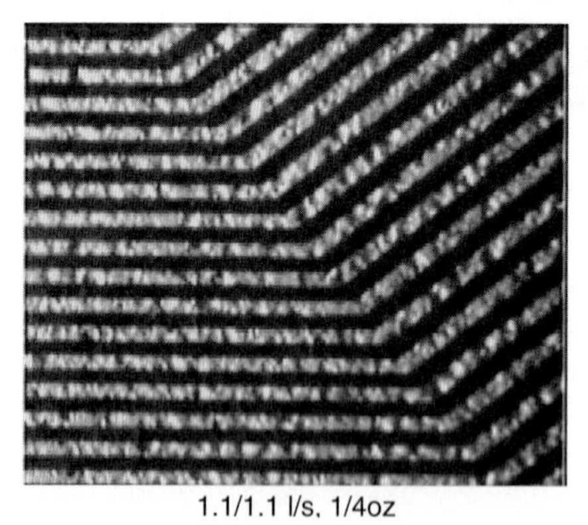

1.1/1.1 l/s, 1/4oz

FIGURE 16.11 Parallel traces. *(Photograph used by permission from Richard Snogren.)*

- Cross-sectional area of one trace = 0.00035 in. × 0.0011 in. = 3.85E-07 in.^{2} = 0.385 sq. mil.
- Cross-sectional area of 16 traces = 3.85 × E-07 in.^{2} × 16 = 6.2 sq. mil.
- Find the current for the total cross-sectional area, 6.2 sq.–mil., and a 10°C rise, using Fig. 16.6 results in 0.7 amps.
- Divide the current by 16 traces, 0.7/16 = 0.0438 amps.
- Round down = 0.04 amps per trace.
- For boards that are less than 0.07 in. thick, the temperature rise will be higher.
- If copper planes are present in the board, then the temperature rise will be lower.
- Trace data for $^1/_4$ oz. copper is not available, so there are assumptions regarding the temperature rise. It is assumed that $^1/_4$ oz. copper will exhibit a similar temperature rise for a defined cross-sectional area as 1 oz. copper.

Closely related to parallel conductors are coils. "For applications where etched coils are to be used, the maximum temperature rises may be obtained by using an equivalent cross-section equal to 2n times the cross-section of the conductor, and an equivalent current equal to 2n times the current in the coil, where n is equal to the number of turns."[2]

Example: Determine the maximum current for a coil with 10 turns and the same size trace as the previous example:

- Trace cross-sectional area = 0.385 sq.–mil. or 3.85E-07 sq. in.
- The number of turns, n, is = 10.
- 2 × n × cross-sectional area = 2 × 10 × 0.385 sq.–mil. = 7.7 sq. mil.
- Equivalent current = 2 × n × current = 0.8 amp.
- Coil current for a 10°C rise = 0.8 amp/2n = 0.04 amp.

16.5.4 Board Thickness

The board thickness is a part of defining a baseline for creating a trace-sizing chart and has a direct impact on conductor heating. The thickness of the board affects the heat transfer path, causing energy to flow away from the trace. As the board thickness increases, the heat transfer path away from the trace increases. When the heat transfer path is increased, the thermal resistance is lower and the temperature rise is lower.

The dielectric material is the first material that begins to transfer heat from the trace. The board material—although poor in terms of thermal conductivity—is better than just air itself. Figure 12 illustrates the impact of board thickness on trace temperature rise. Each category represents the same trace and current level. The x-axis represents the board thickness in in. A 9 percent increase is seen in the trace temperature rise, going from a 0.07 in. thick to 0.059 in. thick board. A 43 percent increase is seen in the trace temperature rise, going from 0.07 in. thick to 0.038 in. thick. And a 260 percent increase is seen in the trace temperature rise going from 0.07 in. thick to a trace in free air. The free air case is equivalent to the IPC internal conductor-sizing chart.

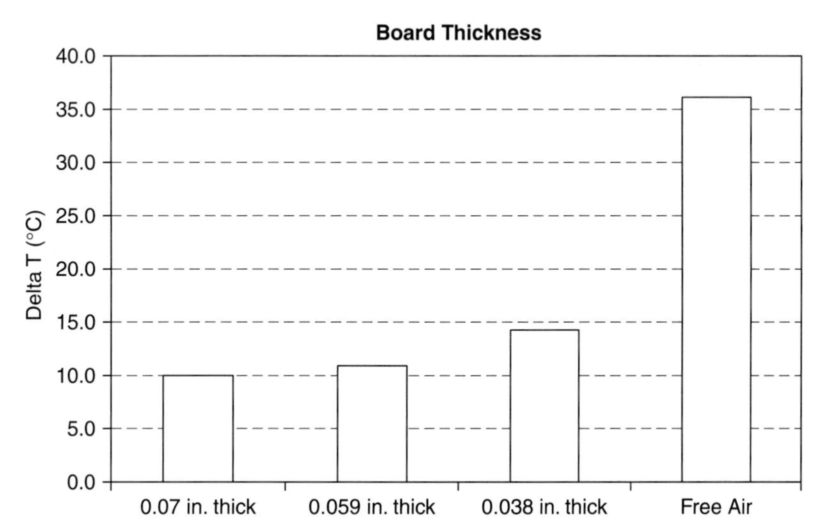

FIGURE 16.12 Temperature rise versus board thickness.

16.5.5 Board Material

The PWB material, or more precisely the PWB material thermal conductivity, has a direct impact on the temperature rise of a trace. The differences between the FR-4 material and polyimide material thermal conductivity, shown in Table 16.3, have approximately a 2 percent impact on trace temperature rise.

The thermal conductivity reported in most data sheets is the z-axis thermal conductivity, which represents the resin in the material. The x- and y-axes are higher due to the woven fibers in the laminate. There is not much information provided on the thermal conductivity of

TABLE 16.3 Material Thermal Conductivity

Material	Kx w/in.-C (w/m-K)	Ky w/in.-C (w/m-K)	Kz w/in.-C (w/m-K)
FR-4	0.0124 (0.488)	0.0124(0.488)	0.0076 (0.299)
Polyimide	0.0138 (0.543)	0.0138 (0.543)	0.0085 (0.335)
Copper OFHC	9.935 (391.2)	9.935 (391.2)	9.935 (391.2)
Air	0.000879 (0.0346)	0.000879 (0.0346)	0.000879 (0.0346)

laminate materials, and the values listed in Table 16.3 for FR-4 and polyimide are measured values from current carrying capacity test board coupons.

The thermal conductivity of copper is almost 1,000 times greater than the dielectric material and is the reason that internal copper planes have such a significant effect on trace temperature rise. The thermal conductivity of air, which is 10 times less than the dielectric, helps explain why external traces generally run hotter than internal traces.

16.5.6 Environments

The term *environment* refers to the surroundings to which the circuit board is exposed and in which it operates. The circuit board may be mounted in an electronics box in a vacuum or surrounded by still air. The board and electronics may be exposed to forced air or it could be immersed in an inert fluid.

It may be necessary to estimate the impact of one environmental condition over another; therefore, it is important to know that the baseline configuration is for a still air environment.

One environment that is not more conservative than still air is an environment in vacuum or in space (see Fig. 16.13). In a vacuum, the internal and external traces run approximately at the same temperature. The internal IPC chart is recommended for sizing traces

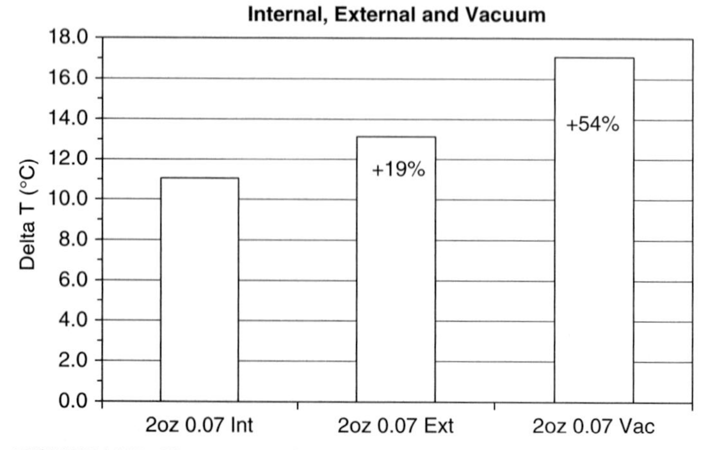

FIGURE 16.13 Vacuum versus baseline environments.

for both internal and external traces in vacuum environments. If the baseline chart is used, the temperature rise should be derated by 55 percent for internal traces and 35 percent for external traces.

16.5.7 Vias

The cross-sectional area of a via should maintain the same cross-sectional area as the trace or be larger than the trace coming into it. If the via has less cross-sectional area than the trace, then multiple vias are required to maintain the cross-sectional area. The cross-sectional area can be calculated based on the barrel diameter and the plating thickness. Consult with the printed circuit board manufacturer to determine the plating thickness. Figure 16.14 illustrates the cross-sectional area of a via.

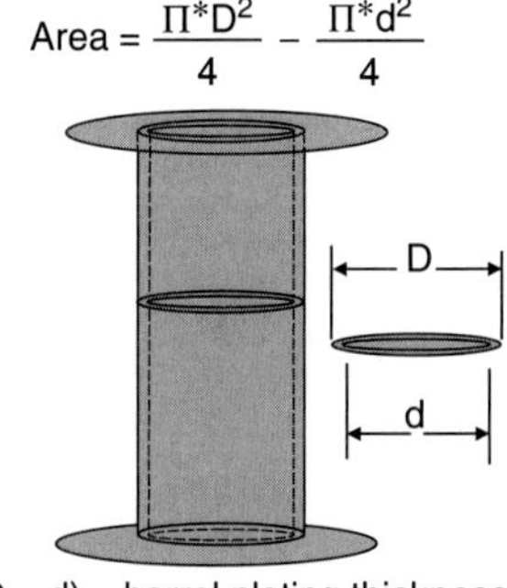

FIGURE 16.14 Via cross-sectional area.

16.5.8 Trace to Via to Plane

If a trace is connected to a via and the via is connected to a plane, the plane will conduct heat away from the via and the via will run cooler than the trace.

16.5.9 Microvias

Microvias respond to current the same as a through-hole via. The cross-sectional area is the parameter to match to a current level and temperature rise.

16.6 ODD-SHAPED GEOMETRIES AND THE "SWISS CHEESE" EFFECT

High current is often applied to copper planes that deliver the current to various locations in the printed circuit. It is not uncommon for these copper planes to be odd-shaped geometries, or "swiss-cheese" (referring to the holes in copper planes resulting from vias, holes, and sections of copper that are etched away from the copper plane). The simple conductor sizing charts become limited in their use for these applications.

A technique for evaluating the temperature distribution in copper planes of odd-shaped geometries, as well as the result of many vias and cutouts, is by using a voltage drop analysis.

16.6.1 Voltage Drop Analysis

Voltage drop in an odd-shaped geometry can be calculated accurately only with numerical techniques. The easiest way is to use software tools (such as ANSYS Thermal Analysis System (TAS) Thermal Modeling Software) that are designed for this type of problem.

There is a direct analogy between thermal resistance and electrical resistance. Because of this, thermal analysis tools can be used to calculate voltage drop rather than temperature drop. The following summarizes the analogy between thermal and voltage drop analysis: (*Mho* is a unit of conductance equal to the reciprocal of an ohm; mho is more properly referred to as siemens = amperes/volts.)

Temperature (degrees)	is analogous to	voltage (volts)
Heat flux (watts)	is analogous to	current (amps)
Thermal conductivity (Watts/(in-C)	is analogous to	electrical conductivity (Mhos/in.)[1]
Resistance (degrees/watt)	is analogous to	resistance (ohms or 1/Mhos)

The voltage drop is calculated based on actual dimension for the problem. If plate elements are used to represent a copper plane in a printed circuit board, the actual PCB dimensions are needed and the actual copper thickness is required.

16.6.2 Voltage Sources

A voltage source is defined as a point where voltage is applied in a circuit. This would typically be a power supply. In a thermal analysis tool, this is analogous to defining a boundary temperature.

16.6.3 Current Source (or Sink)

A point where a defined amount of current is added or removed is a *heat load* in a thermal analysis tool. If a component draws a specified current at a point in the PCB model, the current would be represented as a negative heat load at that point. If a voltage source acts as a constant current source as opposed to a constant voltage source, it would be represented as a positive heat load.

16.6.4 Electrical Conductivity

For geometric elements such as plates, bricks, and tetrahedrons, thermal conductivity should be in units of Mhos/length, where Mhos equals 1 ohm. Units of length must be consistent with those used in the rest of the model.

16.7 COPPER THICKNESS

Copper thickness is assumed 35.6 μm (0.0014 in.) thick for 1 oz. copper or possibly 34.3 μm (0.00135 in.). Two-ounce copper is considered twice those values and half the thickness for $^1/_2$ oz. copper. IPC 2221 specifies the minimum acceptable copper thickness for an internal layer as shown in Table 16.4. The minimums are significantly less than what designers may think and should be considered when pushing current limits on small trace widths. For example, 1 oz. is

TABLE 16.4 Minimum Internal Copper Foil Thickness[3]

Internal conductor thickness after processing		
	Minimum thickness	
Copper foil	(μm)	(in.)
1/8 oz.	3.5	0.000138
1/4 oz.	6	0.000236
3/8 oz.	8	0.000315
1/2 oz.	12	0.000472
1 oz.	25	0.000984
2 oz.	56	0.002205
3 oz.	91	0.003583
4 oz.	122	0.004803

TABLE 16.5 Minimum External Conductor Thickness[4]

External conductor thickness after plating		
	Minimum	
Base copper foil	(μm)	(in.)
1/8 oz.	20	0.000787
1/4 oz.	20	0.000787
3/8 oz.	25	0.000984
1/2 oz.	33	0.001299
1 oz.	46	0.001811
2 oz.	76	0.002992
3 oz.	107	0.004213
4 oz.	137	0.005394

allowed to be a minimum of 0.00098 in. thick rather than 0.0014 in. This significant difference can cause a 50 percent difference in trace temperature rise. Fortunately, copper thickness tends to run more toward a nominal, although attention should be given to the fact that minimum thicknesses do exist and have the potential of being a problem if the limits are being pushed on temperature rise and current level.

External conductors have a thicker minimum (see Table 16.5) due to plating during the final processing steps at the manufacturer. It is always recommended to discuss the details of the board with the circuit board manufacturer. The board manufacturer will know what the typical thickness of the final product will be, which will vary from one board design to another. The final thickness can also be defined on drawings that are sent to the manufacturer.

REFERENCES

1. Hoynes, D. S., "Characterization of Metal-Insulator Laminates, by Progress Report to Navy Bureau of Ships," National Bureau of Standards Report 4283, January 1955–December 1955. (The National Bureau of Standards is now the National Institute of Standards and Technology [NIST].)
2. Ibid., May 1, 1956, p. 25.
3. IPC-2221, "Internal Layer Foil Thickness after Processing, Copper Foil Minimum," Table 10-1.
4. Ibid., Table 10-2.

CHAPTER 17
PCB DESIGN FOR THERMAL PERFORMANCE

Darvin Edwards
Texas Instruments, Dallas, Texas

17.1 INTRODUCTION

The reliability and electrical functionality of electronic components is partially determined by the temperature at which they operate. As such, control of the component temperatures in the system is an important design consideration. Factors that impact device temperatures include the power at which they operate, the air flow surrounding them, heat generation upstream to them, the environment in which the system operates (either indoors or outdoors), the system orientation (either vertical or horizontal), and a variety of printed circuit board (PCB) layout and design properties. These PCB design factors include the design of the copper (Cu) traces contacting the components, the number and area of Cu planes that are connected to them, any thermal vias that might be designed between them and the spreading planes, the proximity of other devices that dissipate power, and any cuts in the thermally conducting layers. Additional PCB features that impact the thermal performance of components include chassis screws, connectors, edge guides, and shields.

To control component temperatures, PCB factors that impact thermal energy flow must be considered in the design phase of the PCB layout. These factors include many complex interactions that make application of simple equations to calculate system temperatures impossible. For example, the historical equation used to calculate component temperatures from a thermal resistance parameter called Theta-ja (θ_{ja}) as shown in Eq. 17.1 is not applicable in modern systems. This is stated clearly in the Joint Electron Device Engineering Council (JEDEC) standard for θ_{ja}.[1] θ_{ja} is not a constant. It is a function of the PCB onto which the component is placed and can vary by a factor of two or more as a function of the PCB design layout. Therefore, if component temperatures are calculated using Eq. 17.1, wildly erroneous estimates may be accepted, which might lead to a system design that fails thermally.

$$T_{junction} = T_{ambient} + (\theta_{ja} * Power) \tag{17.1}$$

where $T_{junction}$ = the temperature of the active portion of the component
$T_{ambient}$ = the temperature of the air at a specified location
θ_{ja} = the thermal resistance of the component as defined by JEDEC
$Power$ = the power of the component

This chapter describes best-design practices that enable the PCB designer to achieve the best possible thermal performance for a given design. As there is no method to calculate analytically the combined impact of the methods described, the designer is encouraged to use sophisticated tools to model the final component temperatures.

17.2 THE PCB AS A HEAT SINK SOLDERED TO THE COMPONENT

The PCB can be considered to be a heat sink that has been soldered to the leads or solder joints of the electronic component. The physical design of the PCB dramatically impacts its efficiency as a heat sink and the temperatures at which the components operate. Figure 17.1 helps to illustrate this point. Here, a packaged component (shown in cross section) is attached to the PCB. Heat is generated by current flowing through electrical resistances on the active surface of the die. This raises the temperature of the surface, resulting in a thermal gradient. Thermal energy (heat) flows from regions of high temperatures to regions of lower temperatures. For the component illustrated in Fig. 17.1, heat flows from the die through the die attach, then through any Cu metallization in the package substrate, then through the solder joints into the PCB. If there are good thermal conduction paths in the PCB, the heat spreads out over a large area of the PCB, allowing the potential for efficient convection and radiation into the environment. If there are few heat transfer paths in the PCB, the component is insulated and the temperature increases.

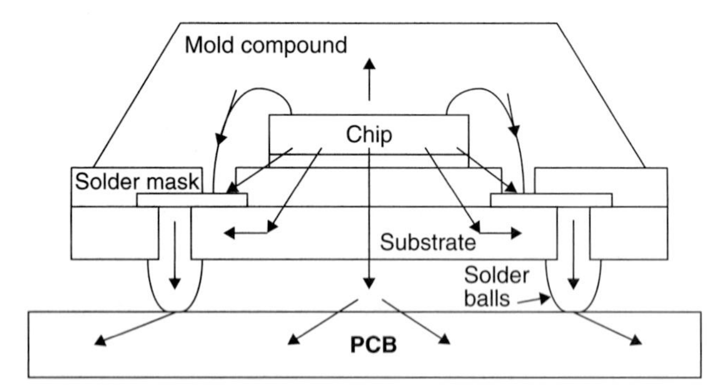

FIGURE 17.1 Cross section of an electronic component on a PCB. The arrows indicate heat conduction paths.

How important is the PCB to the thermal performance of components? Depending on the PCB design, up to 60 to 95 percent of the thermal energy can be dissipated by the PCB. This type of performance can be achieved by PCBs that meet the following criteria:

- Large spreading planes to conduct heat away from the components
- Sparsely populated PCBs with large areas for convection and radiation
- Long traces interconnecting the components, again to conduct heat away from the components
- Sufficient spacing of PCBs in a system rack to enable adequate convection

Failure to meet these conditions will result in reduced thermal dissipation from the PCB, higher component operating temperatures, degraded device reliability, and perhaps even lack of electrical functionality.

17.3 OPTIMIZING THE PCB FOR THERMAL PERFORMANCE

To optimize the PCB's thermal performance, it is important to consider the trace layout, the thermal planes, and the thermal vias. Component spacing on the PCB and the maximum PCB power dissipation (thermal saturation) are also critical.

17.3.1 The Impact of the Trace Layout

The thermal conductivity of a material is a measure of the thermal energy that can flow through the material under an applied temperature gradient. Figure 17.2 shows a plate of material where two sides are held at different temperatures, T_1 and T_2. Experimentally, the one-dimensional thermal energy transferred through this material is found to be governed by Eq. 17.2.

$$q = \frac{kA}{l}(T_1 - T_2) \tag{17.2}$$

where
- q = the heat energy
- A = the area of the plate
- l = the thickness of the plate
- k = the thermal conductivity of the material
- T_1, T_2 = the applied temperatures on the plate's opposing faces

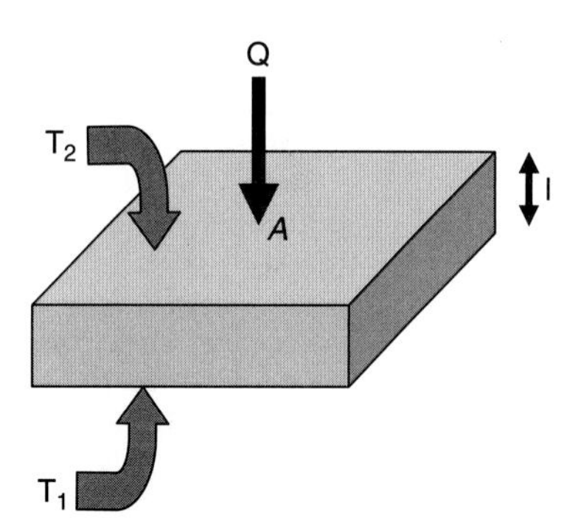

FIGURE 17.2 Typical thermal conductivity test schematic showing a block of material with area A constrained by two different temperatures, T_1 and T_2, on opposing faces. Q indicates the heat flow direction when T_2 is greater than T_1.

Table 17.1 lists typical thermal conductivity values important to PCB and electronic components. Where a range of material properties is given, multiple factors determine the exact thermal conductivity value. These factors can include the filler percentage and composition in a polymer, or in the case of silicon (Si), the doping type and level. Material property measurement or vendor data should be used to determine the thermal conductivity of the specific material of interest. One alloy of Cu is shown to demonstrate that the Cu alloy

TABLE 17.1 Typical Thermal Conductivity Values[2,3]

Material	k (w/m-°C) at 25°C
Air	0.0026
Copper (pure)	386
Copper EFTEC-64T	301
FR-4 (in-plane)	0.6–0.85
FR-4 (out-of-plane)	0.25–0.3
Solder mask material	0.15–0.5
IC encapsulant	0.6–1.0
Silicon	110–149

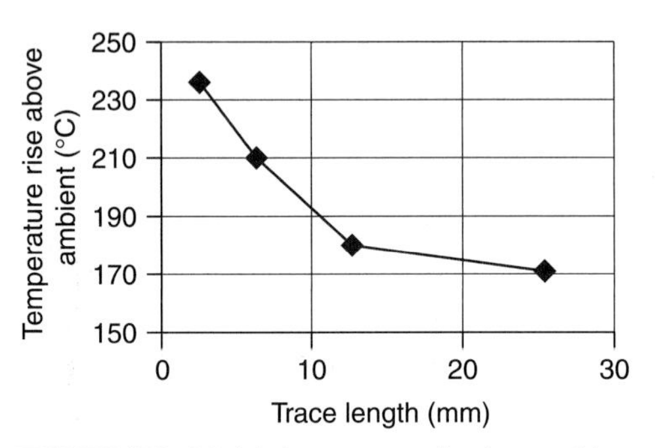

FIGURE 17.3 Modeled temperature rise above ambient thermal performance of an 8-pin SOIC powered at 1 watt as a function of the trace length.

composition can play a role in the thermal conductivity of metals. These material properties are valid at room temperature (~23°C). Due to the temperature-dependent nature of thermal conductivity, further references for these properties should be pursued when operation is expected at temperatures above 85°C or below –25°C.

It is critical to note that the thermal conductivity of Cu is three orders of magnitude greater than the conductivity of most polymers such as PCBs and solder mask materials. This means the majority of the thermal energy will be conducted through the Cu, which implies that the layout of the Cu traces and power planes will be critical to the thermal performance of the PCB as a heat sink. Figure 17.3 shows this graphically. In this model, an eight-pin small-outline integrated circuit (SOIC) package dissipating 1 watt was soldered to a single-layer FR4 PCB of 1.57 mm thickness. The temperature rise above ambient is plotted as a function of the trace length connected to the device pins. As shown, the device temperature changes by ~40 percent for this variation in trace length. The data show that to get the lowest possible temperatures, the longest possible traces should be used to spread heat away from components. Equally important, the widest possible traces should be used. These thermal design optimizations need to be made within the constraints of the electrical performance requirements such as the system time delay budget.

An important feature of thermal management is evident from this chart. After a trace length of about 15 mm, there was little additional improvement as the trace lengthened further—seeking thermal optimization by increasing the trace length had reached a point of diminishing return. Often, once a specific parameter has been optimized, further changes in that parameter lead to little additional efficiency gain. Intuitively, heat flow can be thought of by analogy to fluid flow through pipes of different diameters. The pipe with the smallest diameter limits the fluid flow rate for a given fluid pressure. If the diameter of this constriction is made larger, the next smallest pipe diameter becomes the next constriction. Thermal resistance is analogous to the resistance to fluid flow in this pipe illustration. Once a thermal resistance restriction has been minimized, the thermal conduction "bottle neck" moves to a different portion of the problem.

The PCB trace thickness is another important parameter that interacts with the trace length and width. If the traces are thick, they offer less thermal resistance to heat transfer. If the traces are thin, their thermal resistance is increased and the heat will not spread as far. For best thermal performance, use the thickest possible Cu foil material with the thickest possible electroplating. Unfortunately, the trace thickness is often specified to achieve the best possible etch performance for tight pitch routing, which in turn gives the smallest PCB with the fewest signal layers and lowest cost. Within these limits, ensure that Cu traces are as thick as possible. Use them to provide direct thermal conduction paths to thermal features such as thermal vias, thermal side rails, or thermal conduction screw holes.

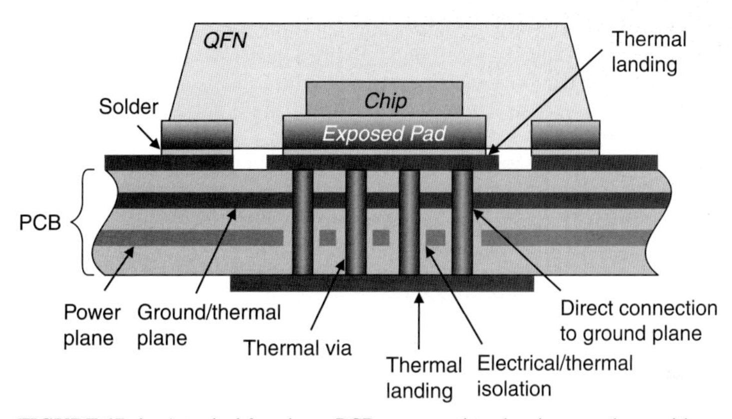

FIGURE 17. 4 A typical four-layer PCB cross section showing a package with an exposed pad soldered to a thermal landing that is in turn connected to the PCB ground plane through thermal vias. The vias are isolated from the power plane for electrical reasons. In this figure, the ground plane would become the thermal spreading plane assuming it was continuous over a large area.

17.3.2 Thermal Planes

Cu planes in the PCB can provide very effective heat spreading for the electronic components. The intent is to conduct the heat over as large an area as possible in order to optimize convection and radiation heat loss from the PCB. Figure 17.4 shows a cross-sectional view of a PCB highlighting features that make good thermal planes. These include:

- A thermal landing (sometimes called a thermal collection plate) or solder land to gather the heat that the device is shedding
- Thermal vias to conduct heat from the thermal landing into a buried plane (most often either the power or ground plane)
- Continuous Cu to spread the heat
- A possible thermal landing on the bottom side of the PCB where an indirect heat sink might be attached
- Isolation areas to ensure that the thermal vias don't short all planes in the PCB

Many integrated circuit (IC) packages are optimized to "dump" power into such thermal planes in the PCB. Figure 17.5 shows images of packages with exposed pads that are intended

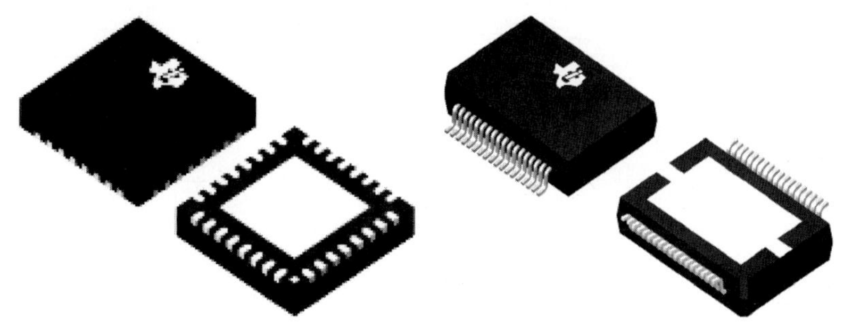

FIGURE 17.5 Two package types with exposed thermal "slugs" on the bottom of the package intended to be soldered to thermal spreading elements on the PCB.

to be soldered to thermal collection plates on the PCB. The die of these packages are glued directly to these exposed pads, giving a very low thermal resistance into the PCB. This thermal resistance can be found from vendor data sheets and might be listed as Theta-jc (θ_{jc}), for junction to case, or as Theta-jp (θ_{jp}), indicating junction to pad resistance. Ball grid array (BGA) packages often include thermal balls that are designed with optimized thermal conduction paths to the die inside the package. These balls should be soldered to thermal collection lands for spreading heat throughout the PCB. Often, thermal ball arrays include both power and ground balls, allowing at least two PCB planes to be used for spreading, thereby increasing the spreading efficiency of the PCB.

Large thermal spreading elements that must be soldered to the PCB can sometimes present manufacturing problems. Improper control of solder paste leading to too much paste under these large thermal pads can result in "floating" the components on a pool of molten solder during reflow. When this happens, continuity yield suffers as the package tends to tilt one way or the other, lifting high side leads off their solder lands. Some users faced with this problem eliminated the solder between the exposed thermal pad and PCB, which led to high operating temperatures and severely degraded reliability in the field. The solution to the floating package problem is to optimize the solder paste volume, not to eliminate the thermal conduction path to the PCB. Always connect thermal management features of electronic components to the appropriate thermal collection pattern, which should then be connected through thermal vias into a thermal plane. Failure to do so will result in devices not running at the expected thermal efficiencies.

Figure 17.6 shows the impact of thermal spreading plane area on the thermal performance of a 12 × 12 mm chip scale package (CSP) BGA-type package with 49 thermal balls. When the thermal plane size is small, the thermal performance is poor, whereas when the thermal plane size is large, the thermal performance improves by a factor of two or more. Often, suppliers of components with exposed thermal attach pads will also provide guidance as to the recommended size and shape of the thermal planes. It is best to ensure that the maximum plane area be close to the component; there shouldn't be a large thermal resistance getting from the component into the plane. Maximize the area of the thermal spreading planes to maximize thermal dissipation.

It is important to point out that the plane must be continuous, that is, with few or no isolation breaks in the Cu area. Since the thermal conductivity of Cu is about 1,000 times higher than that of the FR-4, a 1 mm F-R4 break in a plane offers about the same thermal resistance

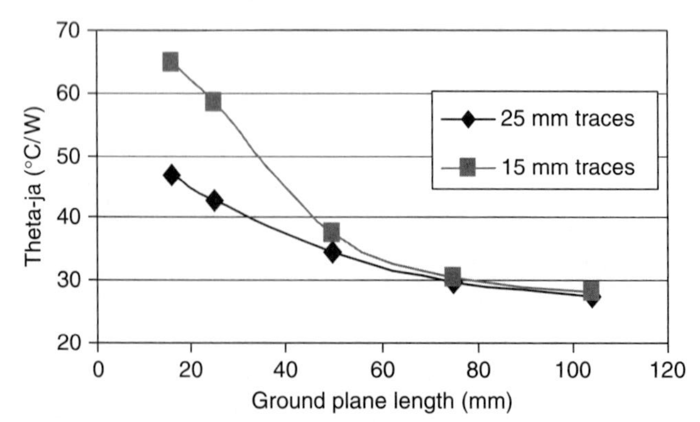

FIGURE 17.6 The effective θ_{ja} of a 12 × 12 mm Chip scale package (CSP) with 49 thermal balls as a function of the (x,y) length of a continuous thermal plane under the package. Two trace lengths were assumed, showing different amounts of coupling of heat from the traces into the plane.

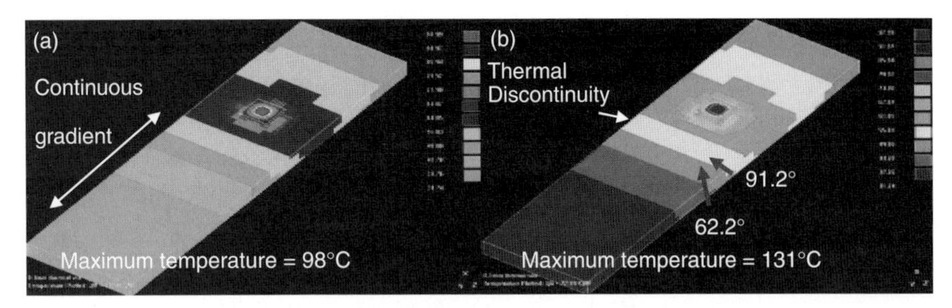

FIGURE 17.7 Thermal contours of a component on an isolated section of PCB in two configurations: (a) with a solid thermal plane over the section of PCB, (b) with a 0.25 mm electrical isolation zone in the thermal plane. The isolation zone increased the component temperature by 33°C.

as a 1,000 mm span of Cu. A model result showing the impact of an electrical path break in a thermal plane is shown in Fig. 17.7. Here, a small 7 × 7 mm device with an exposed solder pad was intended to be cooled by a specific size of PCB area. However, due to electrical constraints, a 0.25 mm electrical noise isolation break was placed halfway down the plane to produce a quiet island. The PCB designer didn't think this would have much impact, but thermally, it cut the plane size by half. As shown in the figure, there was a 29°C temperature drop across this single 0.25 mm break in the thermal plane and the component temperature increased by 33°C. For best thermal performance, don't hamper thermal spreading in a thermal plane with continuous electrical isolation cuts.

Sometimes, the overall thermal performance of a PCB can be enhanced by carefully utilizing isolation cuts in the PCB to thermally segregate components that can afford to run at higher operating temperatures. For example, if a series of power regulator components can operate at junction temperatures of 150°C but are dumping so much heat into the PCB that the digital components' maximum junction temperatures are exceeded, it might be effective to cut the thermal plane between the power regulator components and digital portion of the PCB. This design technique will cause the temperatures of the power components to rise, but will lower the temperatures of the digital components. Of course, care must be taken to ensure the Cu connecting to the power regulation devices can carry the required current, so judicious application of this design technique is advised.

For the best spreading performance, the thickest possible Cu should be used in the construction of the thermal planes. If a thermal plane is to be 0.5 oz. Cu thickness (17.8 μm), it is wise to double this plane with another thermal plane, giving an effective 1 oz. layer of Cu to spread the heat over the PCB surface. Little additional advantage is gained when the thickness of all planes for spreading thermal energy is greater than 2.8 oz. Cu.

17.3.3 Thermal Vias

Thermal vias are simply those that connect the thermal lands or thermal collection plates to the thermal planes. They are formed using the same processing techniques that are used for the PCB's electrical vias. However, unlike electrical vias, the primary intent of a thermal via is that it conducts thermal energy efficiently into the thermal planes. As such, a number of conditions should be met:

- There should be at least one thermal via associated with each thermal ball of a BGA package.
- The density of thermal vias under a thermal landing should be maximized within the limits of what can be allowed without degrading the PCB mechanical integrity.
- If the PCB uses through hole vias, the plating thickness should be maximized to optimize thermal conduction.

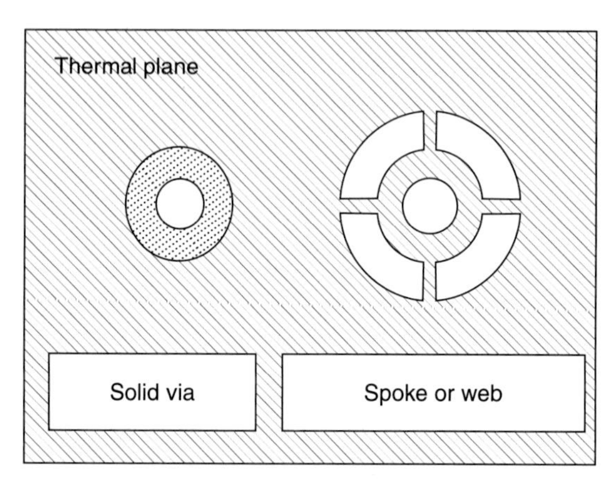

FIGURE 17. 8 An illustration of two possible thermal via connections to the thermal spreading plane. The first via is a solid connection, which is recommended. The second via connection is made with a spoke or web connection to the thermal plane. This is not recommended for a thermal via.

- If the PCB uses build up technology and nested vias are possible, nested vias should be used to connect to the thermal plane.
- If the PCB uses build-up technology but nested vias are not allowed, the shortest possible path between vias should be used to route the thermal energy into the spreading planes.

Figure 17.8 shows the proper way to connect thermal vias to power or ground planes. A web or spoke via connection is not recommended, as either constricts heat flow into the thermal plane. The exit side of the thermal via should have a Cu ring to anchor it to the PCB. Common thermal via dimensions for exposed pad packages are 1.0 to 1.2 mm center-to-center spacing, 0.3 mm drill diameter, with 0.025 mm Cu plating thickness or higher. Through hole thermal vias should be filled with solder, epoxy, or solder mask to avoid wicking solder from the thermal joints or thermal balls that might otherwise create an open or a reliability hazard.

The thermal resistance of an individual thermal via can be easily calculated from the via diameter, the via plating thickness, and the via length between the thermal land and thermal plane using Eq. 17.3.

$$\theta_{via} = \frac{l_{via}}{k * \pi * \left(\left(\frac{diameter_{via}}{2} \right) - \left(\frac{diameter_{via} - (2 * T_{plating})}{2} \right)^2 \right)} \tag{17.3}$$

where

θ_{via} = the via's thermal resistance (°C/W)
k = the thermal conductivity of Cu (W/mm-°C)
$diameter_{via}$ = the via's drill diameter (mm)
$T_{plating}$ = the Cu plating thickness in the via (mm)
l_{via} = the via's length between the thermal land and thermal spreading plane

Substituting some typical values such as a drill diameter of 0.3 mm, a plating thickness of 0.025 mm, a via length of 0.38 mm, and a thermal conductivity of 0.389 W/mm-°C, the typical thermal resistance of a thermal via is found to be 45°C/W. Since a thermal resistance is analogous to an electrical resistance, equations for calculating parallel resistance can be used to calculate the effective thermal resistance of a thermal via array. Using Eq. 17.4 below, a 4 × 4 array of vias is found to give a thermal resistance of 2.8 °C/W. It is evident that the thermal performance of a component can be optimized with a relatively small number of thermal vias (see Eq. 17.4).

$$\frac{1}{R_{total}} = \sum_{i=1}^{n} \frac{1}{R_i} \tag{17.4}$$

where R_{total} = the total resistance of the via group
R_i = the resistance of each individual via

Evaluating the thermal via resistance as a function of trace plating variability, a 0.015 mm thick plated via with the same dimensions as the preceding will have a thermal resistance of 73°C/W. Thus, the thermal performance of a PCB is sensitive to variations in the plating thickness of its thermal vias. To ensure thermal via performance, the plating thickness in the via should be checked. This is usually performed by parallel polishing down from the surface of the PCB rather than through a cross-sectional polishing of the via. When cross-sectioning a via, if sectioning plane does not intersect the exact center of the via, an incorrect plating thickness will be measured; this problem does not occur when parallel polishing into the depth of the PCB.

17.3.4 Component Spacing on the PCB

There is a finite spreading thermal resistance to any PCB as a function of its thickness and the number of planes in the PCB. Eq. 17.5 is a simple equation to calculate the in-plane thermal conductivity of a PCB as a function of the spreading plane thicknesses versus the FR-4 material layers. For example, if a 1.57 mm thick FR-4 board has a single solid thermal spreading plane that is 0.036 mm thick, the effective in-plane thermal conductivity of the plane is 8.9 W/m-°C. This is substantially lower than the thermal conductivity of pure Cu, which is 386 W/m-°C. Therefore, thermal spreading through the PCB with a single plane will be worse than thermal spreading through a plate of Cu with the same thickness as the PCB.

$$k_{effective} = \frac{\sum_i k_i t_i}{\sum_i t_i} \tag{17.5}$$

where $k_{effective}$ = the PCB's effective in-plane thermal conductivity
k_i = the thermal conductivity of the ith layer
t_i = the thickness of the ith layer

The net effect of the limited spreading resistance of a PCB is that component temperatures will rise as they are clustered more tightly on a PCB of a given size and construction. For example, Table 17.2 shows the maximum component temperature of four small devices on a

TABLE 17.2 Case Example of Component Temperature versus PCB Spacing

Component spacing (mm)	Maximum temperature (°C)
50	81.6
25	84.4
12.5	92.9
8	98.4

100 × 100 mm PCB containing two buried planes. As the components were moved closer to each other, the device temperatures increased from 81.6°C to 98.4°C, representing an increase of 30 percent compared to the ambient temperature of 25°C. To make the best use of a PCB as a heat sink spreader, the power of individual components should be distributed as evenly as possible on the PCB to minimize hot spots.

The location of a component with respect to the air flow is also important. As air flows across power sources on a PCB, it picks up heat and increases in temperature. Components in the air flow downstream from high-power devices are heated by the warmer air. Air temperatures can easily rise 10 to 30°C just past a high-powered component such as a microprocessor, power amplifier, or power regulator block. If a component is running too hot, one possible solution is to move it as far upstream in the air flow as possible such that it receives the coolest air.

17.3.5 Thermal Saturation of the PCB

When the power of individual components has been evenly distributed over the PCB surface such that further geometrical rearrangement of the components or further optimization of the trace and thermal plane design is ineffective in cooling the components, thermal saturation of the PCB is said to have occurred. At this point, the power dissipation capability of the PCB becomes the factor limiting the component temperatures. The maximum power dissipation of a PCB with good thermal planes and evenly distributed power sources is in turn limited by a number of geometrical and system-level factors. These include:

- The air stream velocity
- Any air ducting or shrouding around the PCB
- The configuration of heated surfaces around the PCB
- The altitude at which the PCB is operating
- The orientation of the PCB with respect to gravity

Each of these parameters impacts either the convection or radiation from the PCB. Convection is the heat transfer mechanism whereby heat is removed from the PCB through conduction of thermal energy into the gas or fluid that surrounds the PCB. In a natural convection environment, the heating of the air around the PCB causes its density to decrease. In the presence of a gravitational field, the less dense air will rise, carrying away heat with it—fresh, cool air moves in to replace the heated air. In a forced-air environment, the heated air is "blown" away from the surface by cooler air, which in turn becomes heated through the conduction of thermal energy from the PCB. The heat removed from a surface of area A due to convection, either natural or forced, can be represented by the simplified one-dimensional Eq. 17.6.

$$q = hA(T_{surface} - T_{ambient}) \tag{17.6}$$

where
q = heat
h = convection coefficient
A = surface area
$T_{surface}$ = the average temperature of the surface
$T_{ambient}$ = the temperature of the ambient air

Equations to calculate the convection coefficient h are beyond the scope of this chapter. Convection is a function of (1) the PCB size, with smaller PCBs having higher convection efficiencies; (2) the PCB orientation, with vertical PCBs being more efficient; (3) altitude, with lower altitudes being more efficient than higher altitudes; and (4) air ducting, with ducting tending to make the PCB more efficient in forced air environments.[4]

Radiation is the removal of heat from a surface by the emission, or "radiation," of photons. These photons are generated when thermally excited atoms collide with each other, resulting in overlapping electron energy states. Since electrons are fermions, meaning no two electrons can occupy the same energy band, when the energy states of two electrons overlap during the collision, one electron is forced into a higher, empty energy state. This reduces the momentum of the colliding atoms, ensuring that energy is conserved. After collision, the displaced electron decays back to its initial energy state, giving up a photon of energy in the process. At the surface of the object, this photon is emitted, carrying away some of the thermal energy, while in the bulk of the material, this photon is reabsorbed by neighboring atoms. The simplified one-dimensional equation that describes the total energy loss from a surface due to radiation is shown in Eq. 17.7. This equation for radiation neglects heat that the PCB might reabsorb from heated objects such as power supplies around it, but does include an ambient temperature radiation reabsorption.

$$q = \varepsilon \sigma A \left(T^4_{surface} - T^4_{ambient} \right) \qquad (17.7)$$

where
q = heat
ε = the emissivity of the surface
σ = the Stefan-Botzmann constant (5.67×10^{-8} W/m^2-°K^4)
$T_{surface}$ = the PCB surface temperature
$T_{ambient}$ = the ambient temperature

The emissivity of the object is a unitless number that describes how efficiently a surface radiates. It varies from 0, which is a surface with no radiation that is perfectly reflective, to 1, which is a perfectly radiative and absorptive black surface. The emissivity of the surface is a function of both the type of material that composes the surface as well as the roughness of the surface. Typical solder mask material has an emissivity ranging from 0.85 to 0.95. Typical exposed Cu trace has an emissivity ranging from 0.1 to 0.3 depending on the roughness and oxidation condition of the Cu.

It is important to note that the area of the PCB plays a major role in both of these equations. This is intuitive; the larger the PCB, the more heat it can dissipate. In combination, the preceding equations can be used to estimate the power that a given PCB at a given temperature can dissipate. For example, Fig. 17.9 shows the best-case power that can be dissipated by a series of PCBs ranging in size from 10 × 10 cm to 20 × 20 cm. The PCBs were assumed to be oriented horizontally in an unobstructed natural convection environment of 25°C with no significant neighboring radiating surfaces. Convection was allowed to vary as a function of temperature. Most importantly, the power was assumed to be uniformly distributed over the PCB such that there were no thermal gradients on the PCB. This is a major assumption, meaning that the power dissipation figures in the chart are best-case, a condition that no practical PCB will attain. The chart indicates that 50 W can't realistically be dissipated in natural convection from a PCB that is 10 × 10 cm, but it can be dissipated by the 15 × 15 and 20 × 20 cm PCB. The 15 × 15 cm PCB will rise 80°C above the ambient to dissipate 50 W, while the 20 × 20 cm PCB will rise only 50°C to dissipate the same power. This graph has been formatted to look the same as typical heat sink curves that are available from many vendors.

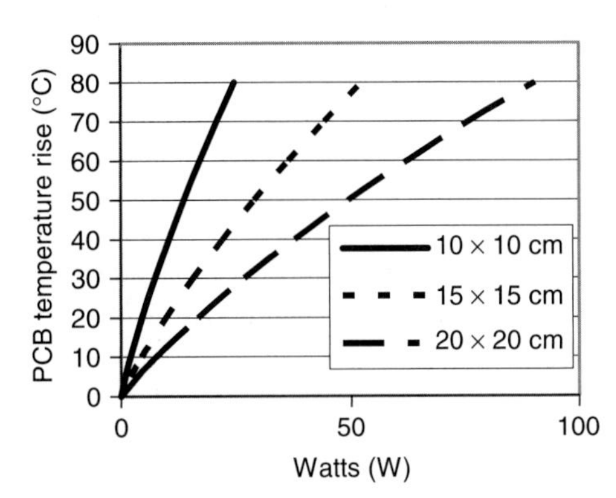

FIGURE 17.9 Temperature rise above 25°C of a horizontal PCB in natural convection as a function of PCB size and power dissipation. Both convection and radiation are dissipating the power.

17.4 CONDUCTING HEAT TO THE CHASSIS

When the PCB is thermally saturated and the component temperatures are still too high to be tolerated at the maximum obtainable system air velocity, other means of conducting heat from the PCB to even larger area system structures are required. The chassis of the system is often the largest surface area structure in the system, is exposed to the ambient air, and often makes a good sink for heat that cannot be dissipated by the PCB alone. Mechanisms to conduct heat into the chassis include chassis screws, gap fillers, connectors, and side rails. Sometimes, radio frequency (RF) shields that are appropriately connected to components can provide additional heat sinking.

17.4.1 Chassis Screws

A properly designed chassis screw thermal sink connection in a PCB includes connection to the thermal spreading plane of the PCB, through hole plating, and a thermal contact area for the screw and stand-off sleeve to clamp onto. This configuration is shown in Fig. 17.10. The thermal screw should be as close to the hot electronic component as possible to minimize thermal resistance into the chassis. As well, the chassis where the thermal screw attaches should have a high thermal conductivity to spread the heat away from the screw. Preferably, the chassis should be made of metal. If a plastic chassis is used, molding a plate of aluminum (Al) into the plastic will improve its thermal heat spreading capability. A plastic chassis alone is usually a poor thermal spreader and may not provide substantial thermal performance enhancement.

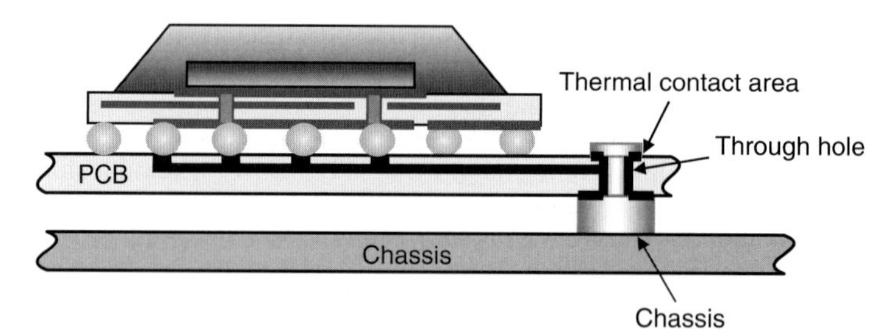

FIGURE 17.10 Schematic of a chassis screw implementation for sinking heat to the chassis. The thermal screw mounts through a plated through via that is shorted to the thermal plane as shown. The thermal plane conducts heat from the thermal balls of the PBGA package.

How much improvement might be expected from using the chassis of an enclosure for cooling? This depends substantially on the enclosure construction and the location of the chassis screw with respect to the hot component. In one example, a device was cooled with a thermal screw that connected between a hard drive PCB and the hard drive enclosure. Connecting the device thermally to the hard disk drive resulted in a 10 percent thermal enhancement.

17.4.2 Gap Fillers

It is sometimes not possible to use screws for conducting heat to the chassis. In other cases, large areas of high-power density might need to be sunk into the chassis. In these instances, another type of thermal conduction path is available that lends itself to conducting heat to the chassis: gap fillers. Gap fillers are thermal compounds that fill the gap between the PCB and chassis whose sole purpose is to conduct heat to the chassis. There are many different types of gap fillers. The most common is a soft, flexible silicone rubber material that has been filled with thermally conducting particles to enhance its bulk thermal conductivity. Sometimes, thermally filled foams are used. When the soft, compliant material is compressed between the PCB and chassis, it conforms to the protruding components on the PCB, making a good thermal connection. Important criteria to consider in selecting elastomeric gap fill materials are the thickness needed, the thermal conductivity required, and the pliancy of the material. A variety of materials with varying compositions are readily available on the market. There is normally a trade-off between the pliancy of the material and the thermal conductivity, with higher thermal conductivity materials exhibiting less compliancy.

Plastic sacks filled with a thermal fluid are sometimes used as gap filler materials. The fluid is usually optimized to produce convection in the gap of interest. The convection in the fluid can give an effective thermal conductivity that is substantially higher than the conductivity of the fluid alone.

17.4.3 Connectors

Connectors can either provide a direct or indirect means of conducting heat from the PCB. Direct conduction occurs in those system configurations where the PCB is plugged into an edge connector or back plane socket, or is held in place with an edge guide. To take advantage of these direct connect thermal features, it is necessary to extend thermal planes into the areas where the connection or clamping occurs. The connector, clamping, or edge guide should have as large a contact area as possible to optimize thermal conduction of heat from the low thermal conductivity PCB material. In some military applications, the PCB is built around a thick Cu core that is clamped by the edge rail of the card cage. The thick Cu core provides very effective thermal conduction from the PCB into the edge rail, which is then cooled by a system of channels containing either moving air or moving water.

Indirect conduction of heat from the PCB can occur when cables are plugged into connectors on the PCB. These indirect conduction paths are much harder to include accurately in system-level thermal analyses, but can provide some margin in a PCB thermal design. Care must be taken, however, to ensure that these plug-in cables don't block critical airflow paths that would lead to overheating.

17.4.4 RF Shields

RF shields are used over sensitive RF and analog circuits to minimize electrical interference with the circuit function or to minimize electrical radiation from the circuit into the surrounding environment. RF shields are usually made of a thin metal which is soldered to ground on the PCB. In most instances, the RF shield is a continuous plate or box that encloses the circuits. Unfortunately, this continuous box creates a dead zone in the air flow directly above the components within the box, degrading natural convection for those components. To enhance the thermal performance of components within RF shields, perforations or meshing of the RF shield cage is recommended. If these perforations are kept less than 1/10 the wavelength of the shielded electromagnetic radiation, the RF shield will function adequately to quiet the circuit and block out extraneous signals while allowing air flow through the shield to cool the interior components.

If additional steps are taken, the shield can be used to spread heat from hot components inside the shield over a larger section of the PCB, thereby maximizing convection and radiation cooling for the heated components. Figure 17.11 shows a schematic of a shield over a stacked package component. It is normally difficult to conduct heat from the top device in such a stack to the PCB, but when the RF shield is brought into contact with the top device, heat can conduct into the shield and down to the PCB. It is suggested that the thermal connection between the shield and the electrical components be made with a thermal epoxy or thermal grease after the shield is soldered in place to avoid issues with mechanical tolerances

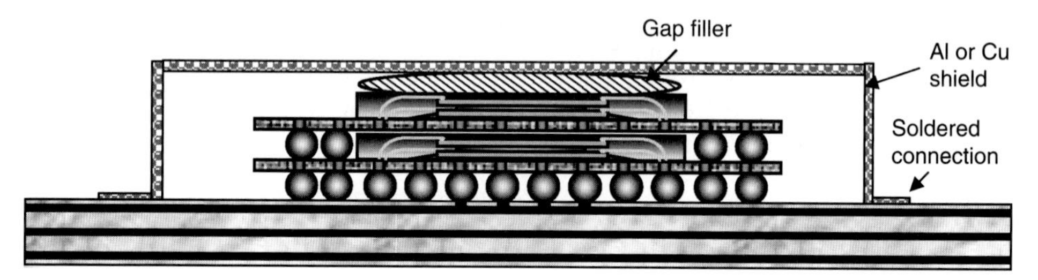

FIGURE 17.11 A stacked package configuration on a PCB with a perforated RF shield that is being used to spread heat from the top package. Thermal gap filler material such as thermal grease is used to conduct the heat from the top package into the RF shield.

or interference between the shield and the components. If the high-powered components to be cooled reside immediately outside the shield, it is possible to conduct heat from them into the shield such that the shield behaves as a heat sink.

17.5 PCB REQUIREMENTS FOR HIGH-POWER HEAT SINK ATTACH

Most IC components with power dissipations of 2.5 W or higher require attached heat sinks to dissipate the power. These heat sinks are often glued or clamped to the components with few special constraints placed on the PCB. If a large number of components requiring heat sinks are placed on the PCB, it is important that the PCB be thick enough to handle the weight of the attached heat sinks, especially if the system environment subjects it to vibration or mechanical shock. High-powered components in the 50–300 W range require special PCB design features to handle the high-loading force of the heat sink against the electronic component. Clamping forces of 20–200 lb. are possible for these high-power heat sinks in order to get the best possible thermal contact between the heat sinks and components. One such high-power heat sink configuration is shown in Fig. 17.12.[5] Here, the heat sink is spring loaded

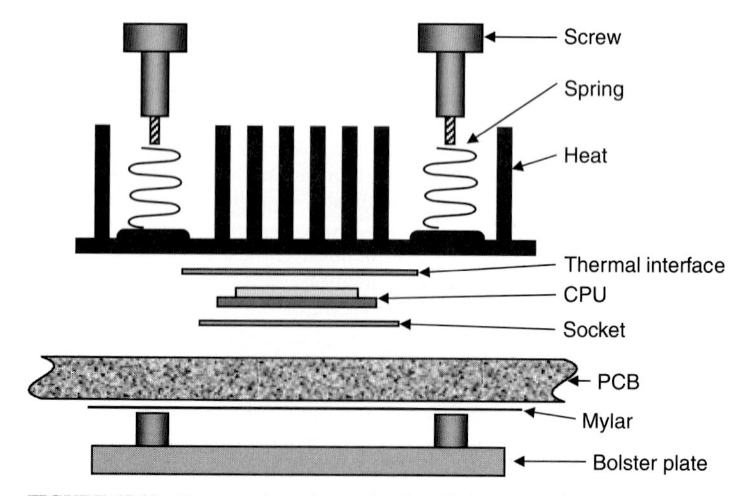

FIGURE 17.12 Cross section of a mechanical design for a high-power heat sink clamped on a microprocessor. The bolster plate on the bottom of the PCB minimizes PCB warpage and breakage.

against the package with tensioned screws which tighten into a bolster plate. The bolster plate provides the rigidity needed to avoid bending and damaging the PCB under the spring loading. A mylar film between the bolster plate and PCB minimizes electrical shorting of the fixture to the PCB. For such heat sink constructions to work, no components can be placed between the PCB and bolster plate.

17.6 MODELING THE THERMAL PERFORMANCE OF THE PCB

Each of the many PCB thermal optimization techniques described in this chapter will have some amount of impact on the thermal performance of the system. Determining the relative impact of a single optimization is difficult to perform analytically. When an entire ensemble of optimization techniques is employed across multiple devices operating at a variety of powers, analytical solutions of the thermal problem become impossible. The simplicity of system-level temperature calculations implied by Eq. 17.1 is misleading unless the effective θ_{ja} of each component in the specific system is known *a priori*. Since the impact of the PCB design on the thermal performance of the system must be determined before design completion and tooling, the question becomes how to quantify the impact of the PCB thermal features. Relayout and redesign of the PCB to fix thermal issues once tooling is generated is prohibitively expensive and leads to noncompetitive design cycle times. The only predictive method currently available that addresses this need is computer simulation, which can incorporate all relevant PCB design features, perform a thermal analysis, and report back the component temperatures to identify problem locations.

Many commercial computer simulation tools exist, ranging in complexity, accuracy, and cost. Computational fluid dynamics (CFD) programs incorporate equations that solve the problems of air flow around the PCB, thermal conduction from the PCB into the air, and thermal conduction within the PCB. Solutions using CFD don't require special knowledge of convection coefficient equations, thereby making the modeling process easier and less prone to convection coefficient calculation errors. Accuracy of CFD analysis can be as high as +/–5 percent. CFD codes tend to be slow, however, taking substantial computer resources to run a reasonably complex model. A step down from CFD are codes that employ user supplied convection coefficients that are applied to the PCB and component surfaces. These solutions can be quite accurate, though no more than +/–10 percent precision can be projected without substantial correlation to experimental data.

Whichever analysis code is selected, it should be capable of reading in data from the user's PCB layout tool and should be capable of incorporating the specific thermal conduction paths that were optimized for component performance. It should be able to handle relatively complex system-level structures and should be capable of inputting a variety of thermal characteristics of the components. Postprocessing identification of PCB regions and components above limit temperatures is a plus. The code should run fast enough to enable multiple design iterations to optimize the PCB layout without slowing down the design cycle.

17.6.1 System-Level Thermal Modeling Phases

System-level thermal analysis normally includes three phases due to the complexity of the task and the difficulty of inputting a fully detailed system design into a tool. The first phase concentrates on airflow optimization through the system to minimize any dead zones in the cabinet. The PCBs, cables, fans, and other air blockages are included in the analysis during this phase. The second phase includes dissipation of power on the PCB by applying power areas to the PCB. During this phase, it is fairly easy to change locations of power sources to optimize the air cooling paths. The final phase is the fully detailed phase, which includes the PCB layout geometries and details of the electrical components. By the time phase three is reached, there should be few remaining major iterations for thermal optimization.

During each phase of thermal modeling, the solids of the components, PCB, and system enclosure must be defined to the code. These solids are meshed by the code, a process through which the objects are broken into tens of thousands of discrete nodal points. The computer calculates thermal parameters such as temperature and heat flow at each of these nodal points as a function of neighboring nodal points rather than trying to solve analytical equations for the thermal fields. If the meshing is fine enough, the calculated solution will be very accurate. If the meshing is too coarse, errors will creep into the model. There is a trade-off between accuracy and run time in computer simulations; models with coarser meshes usually run much faster than models with fine meshes. It is best for users to become familiar with the mesh sensitivity of their chosen modeling tool to optimize run time versus accuracy before running critical analyses.

When performing a CFD system-level analysis to determine temperatures on a PCB, it is important to include all airflow obstructions that might change the convection on the PCB and its components. Common airflow obstructions include cables, RF shields, daughter cards, brackets, air filters, capacitors, transformers, memory single in-line modules (SIMs), power regulators, and hard drives. Failure to include these air blockages in the modeling process can result in systems that run too hot or that suffer from thermal shutdown. External blockages should be accounted for as well. Cabinet vents should not be located where a user might carelessly toss a magazine or CD cover. The impact of accumulated dust over the lifetime of the system should also be considered, as the dust can substantially reduce convection.

From a PCB layout standpoint, the highest velocity air stream on the board may well occur between two tall components if they block the airflow path. The blockage causes the air to be channeled through the gap, resulting in higher air velocities. High-power electronic components can sometimes be more effectively cooled by placing them in this channeled air stream. Locations to avoid for high-power components are the leeward side of air blockages, immediately downstream of high-power components, or on the bottom center of a natural convection-cooled PCB.

17.6.2 Required Component Thermal Parameters

Thermal models of a PCB loaded with detailed models of components in a system can become very complex with many nodes. Such analyses can run for days and weeks, making them impractical. As well, it is often impossible to get detailed models of each component to be used in a system. To address these issues, the industry has developed two levels of component thermal abstraction. The first method reduces the electrical component's thermal behavior to two thermal resistances, θ_{jc}, which represents the thermal resistance from the active portion of the component to the top component surface, and Theta-jb (θ_{jb}), which represents the thermal resistance between the active portion of the component to a point on the PCB at the edge of the component. The component is then represented to the system as shown in Fig. 17.13. The thermal resistance from the top of the component to the ambient, R_a, is calculated either through CFD or from the convection coefficients. It also includes radiation heat losses. Thermal conduction through the PCB to other components and to the air is calculated by the simulation tool. Accuracy of a two-resistor component thermal model temperature delta with respect to the ambient is in the

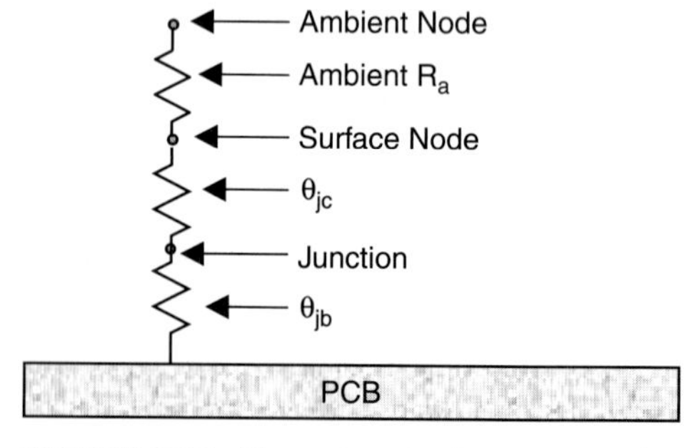

FIGURE 17.13 Thermal conduction representation of an electrical component by two thermal resistances, θ_{jc} and θ_{jb}. The resistance to ambient (R_a) is the result of convective and radiative heat loss.

+/–20 percent range, even when using CFD. The parameters θ_{jc} and θ_{jb} are usually available in the component supplier's data sheets. If not, the supplier should be contacted for the information.

A more complex modeling configuration has been developed to improve thermal estimations from +/–20 percent to +/–5 percent. Called a *compact model*, these resistor network topologies better represent heat flow within the package and from one side of the package to another. Compact models should therefore be used whenever they are available. Often, the system design engineer must ask the component supplier for them since they are not commonly printed in data sheets. Many tools allow the user to mix compact component models with the simpler two-resistor models.[6]

17.6.3 Handling Cu Traces and Power Planes

Due to the complexity of meshing the fine geometries of a PCB trace layout, many if not most modeling tools include the ability to "smear" Cu layers into solid sheets. A smeared layer is a single sheet whose average thermal conductivity is equivalent to the thermal conductivity of the detailed trace layer. Unfortunately, most tools use an incorrect averaging technique to determine the effective thermal conductivity of a layer. The tool asks for the percent Cu coverage in the trace layer. Based on this input, the tool then calculates the thermal conductivity of the trace layer as the weighted average between the Cu conductivity and the insulator conductivity. For example, if a 100 percent coverage of Cu layer has a thermal conductivity of 380 W/m-°C, a 95 percent coverage of Cu would have an effective conductivity of 360 W/m-°C, and a 0 percent Cu coverage layer would have a conductivity of 0.8 W/m-°C. Smearing in this fashion neglects critically important details such as isolation regions that might cut through thermal spreading planes. Therefore, trace smearing using a conductivity averaging scheme should be avoided.

If the modeling tool can't represent the PCB in full complexity, some attempt should be made to perform a better estimation of the thermal conductivity in Cu spreading layers connected to critical components. The overall smeared thermal conductivity of the plate should be replaced with patches of thermal conductivity that account for heat flowing either parallel or perpendicular to the Cu conductors. Eqs. 17.8 and 17.9 can be used to estimate parallel and series thermal conductivities in Cu layers respectively. Figure 17.14 shows graphically what is meant by a parallel thermal conductivity. Here, multiple stripes of thermally conductive materials pass thermal energy from the top to bottom. Each material channel can be considered as a thermal resistor. The thermal resistance of each resistor can be calculated as a function of its

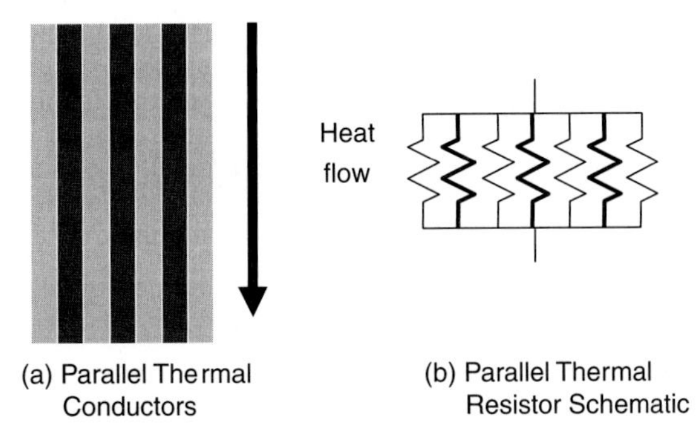

FIGURE 17.14 A diagram to illustrate how heat flowing along parallel conductors can be represented through electrical analogy to electrical resistors in parallel. The parallel resistance representation is used in the derivation of Eq. 17.8.

cross-sectional area (thickness × width) and thermal conductivity, and the effective parallel resistance can be backed out (Eq. 17.8).

$$k_{parallel} = \frac{\sum_i k_i \cdot Area_i}{\sum_i Area_i} \tag{17.8}$$

where $k_{parallel}$ = the effective thermal conductivity of a plate where the heat flows parallel to the conductors
k_i = the thermal conductivity of the ith material
$Area_i$ = the area of the ith material (thickness of stripe × width of strip)

In the case of series thermal resistances, the dimension of interest is the length of each section of material in the direction of the heat flow (see Eq. 17.9 and Fig. 17.14).

$$k_{series} = \frac{\sum_i Length_i}{\sum_i \frac{Length_i}{k_i}} \tag{17.9}$$

where k_{series} = the effective thermal conductivity of a plate where the heat flows perpendicular to the conductors
k_i = the thermal conductivity of the ith material
$Length_i$ = the length of the ith stripe parallel to the heat flow path

REFERENCES

1. JEDS-51.2, "Integrated Circuits Thermal Test Method Environment Conditions–Natural Convection (Still Air)," Section 1.1.
2. The SRC/CINDAS Microelectronics Packaging Materials Database, Purdue University, 1999.
3. Azar, K., and Graebner, J. E., "Experimental Determination of Thermal Conductivity of Printed Wiring Boards," Twelfth IEEE Semiconductor Thermal Measurement and Management Symposium, 1996, pp. 169–182.
4. Holman, J. P., *Heat Transfer*, McGraw-Hill, New York, 1990, pp. 281–368.
5. Lopez, Leoncio D., Nathan, Swami, and Santos, Sarah, "Preparation of Loading Information for Reliability Simulation," *IEEE Transactions on Components and Packaging Technologies,* Vol. 27, No. 4, December 2004, pp. 732–735.
6. Vinke, Heinz, and Lasance, Clemens J. M., "Compact Models for Accurate Thermal Characterization of Electronic Parts," *IEEE Transactions on Components, Packaging, and Manufacturing Technology—Part A*, Vol. 20, No. 4, December 1997, pp. 411–419.

CHAPTER 18
INFORMATION FORMATING AND EXCHANGE

Bini Elhanan
Valor Computerized Systems Ltd., Yavne, Israel

18.1 INTRODUCTION TO DATA EXCHANGE

This chapter presents the elements of data exchange, the data exchange process, its pitfalls and best practices, the most widely used data exchange formats and their main characteristics, and evolution drivers in data exchange.

The electronics manufacturing marketplace is a dynamic domain where supply chains of trading partners compete in a global economy in which time to market and efficiency are key factors for success. Outsourcing is becoming more common; printed circuit board (PCB) fabrication has been outsourced in the 1980s and early 1990s, whereas assembly is following very quickly. Data communication is a vital key to supply chain management. Efficient and accurate data exchange is an extremely important factor in customer supply networks.

The challenge of more efficient and accurate data exchange has only recently begun to be addressed and the future for data exchange improvements holds much potential.

18.1.1 Data Exchange Format Defined

Once the design of a printed wiring board (PWB) is complete, it is necessary to pass the design intent to internal or external suppliers for manufacture and assembly. This is mostly done using electronic data file transfer. A short dictionary entry for a data exchange format would read:

Virtual format that one computer program outputs and another one reads and creates an internal representation thereof that can be manipulated and output. The main criteria for data exchange format quality are its clarity and the accuracy with which it passes the design intent between trading partners.

18.1.2 PCB Design Process Overview

PCB design and manufacturing process (see Fig. 18.1) proceeds as follows:

1. In the *schematic stage,* components and their interconnection pattern (netlist) are defined, simulated, and verified for functionality using schematic capture tools such as Mentor's Design Architect and Cadence's Concept. (The schematic interconnection list is usually called *schematic netlist.* It is an electrical network-based grouping of connection points of

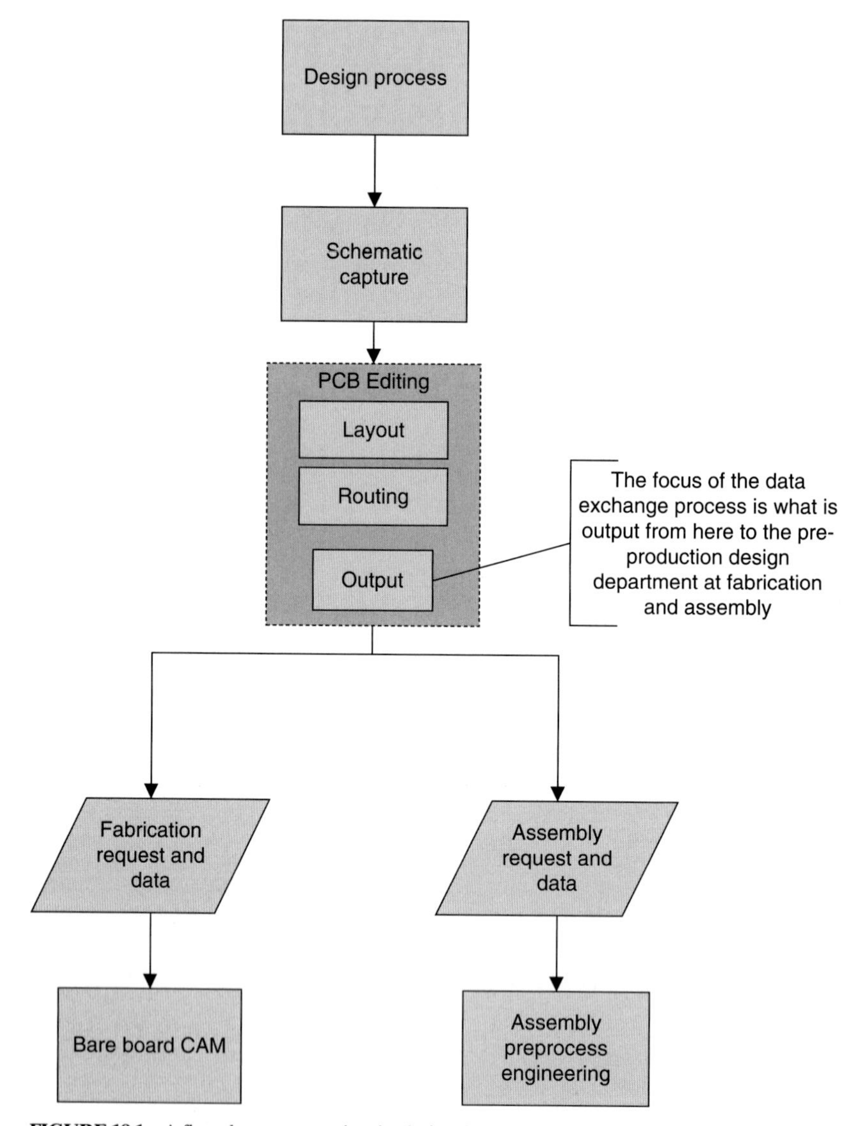

FIGURE 18.1 A flow chart representing the design through manufacturing data exchange process.

the form component ID + pin ID. A physical or PCB-level netlist is the same grouping, whereas the points are described by their physical [geometric] location on the board and the board side[s] from which they can be tested.)

2. The *editing stage* begins with the layout phase. The component shapes (still virtual and within design tools) and the interconnection list are transferred to the target PCB design tool (such as Mentor's Boardstation or Cadence's Allegro) and are manually or (semi-) automatically placed on the PCB as determined by the system's mechanical engineering and manufacturing rules.

3. Still within the PCB editing stage, the layout phase is followed by the routing phase. This routing (manual and automatic) implements the interconnection list in a PCB conductive pattern composed of traces, planes, and vias (vertical conductors).
4. After PCB routing and layout have been completed, the output phase begins, where the design is output from the PCB editing system and sent to bare board manufacturing to produce the actual bare board layers and then laminate, drill, and finish them.
5. When PC boards are ready, the design data together with the boards are sent to assembly houses, where the components from the bill of materials (BOM) are assembled onto the finished bare board at the locations specified by the design.

18.1.3 Board Manufacturing Data

In addition to PCB design data, copious information pertinent to board manufacture is exchanged between partners in the manufacturing and assembly process. Such information can include standard compliance requirements (e.g. ,UL), statements of workmanship (SOW), and delivery and packaging instructions. Modern data exchange formats address some of these comprehensive data exchange issues. The design-to-manufacturing data exchange process should ideally progress to become online collaboration. The drivers of this progression are globalization, the supply chain, and time-to-market pressure.

Historically, the PCB design-to-manufacturing flow began with manually laid-out artworks whose images were printed on copper-clad laminates and developed, etched, and stripped within the same vertically integrated organization. Advances in miniaturization and multilayer board technology led to the need for specialization, and outsourcing of bare board fabrication became the norm. Together with advances in PCB fabrication tools and digital communication, the outsourcing supplier communication advanced from courier-delivered artwork to data tapes and disks and later to electronic transfer over modems and finally over the Internet. Globalization and outsourcing continue to push the envelope of electronic business communication in the PCB industry as well as everywhere around us.

The PCB fabrication tools driving the history of data exchange are mostly imaging tools and placement machines. PCB lithography began with hand-routed artworks and continued with photo plotters, aperture wheels, and computerized numerical control (CNC). Gerber Scientific was the most successful of the early photo-plotter vendors, and hence the Gerber 274D format caught on.[1] Later these were replaced with laser plotters without aperture wheels, but the Gerber data format remained the "lowest common denominator" format. In the PCB assembly arena, robotic assembly machines of varying complexities succeeded hand assembly. This trend led to the need for electronic assembly data, just as the artwork-to-laser-plotter development in board fabrication led to the need for electronic fabrication data.

Bare board fabrication and board assembly share some common data elements but also have distinct data needs. Examples of common elements are the outer circuit and mask layers and through-drill information. Unique data elements are, for example, inner layers for board fabricators and component placement locations and functional component descriptions for assembly houses. All these data elements are described in the next section.

18.2 THE DATA EXCHANGE PROCESS

When a printed circuit assembly (PCA) design (layout, routing, and verification) is complete, the board data must be sent to the PCB manufacturer. An output format is chosen to export the data from the computer-aided design (CAD) system. Historically, the output to bare board manufacturing was in Gerber data format and Excellon drill information and was accompanied by some textual "readme" files. (These files did not contain precise definition of layer stack-up, drill span, or component outlines). These files would be sent—or rather, "thrown over the wall" to the fabricator, sometimes without even being checked. When the board reached the assembly stage, the board assembler would receive Gerber images of the outer layers, x, y locations of the components and a BOM file. These methods were recognized as troublesome because errors and lack of communication necessitated many clarification calls, faxes, and e-mails, and slowed down the process.

18.2.1 Assessing Quality of Data Exchange Formats

Designers and editors, forced to verify the data after output for correctness and manufacturability, began to work with CAD and computer-aided manufacturing (CAM) tool vendors and industry consortia to develop better formats and methods for data exchange. They sought a data exchange format that was explicit, intelligent, optimized, and bidirectional:

- **Explicit** There should be no need for guesswork or reverse engineering of design and no need for external files (see Fig. 18.2).

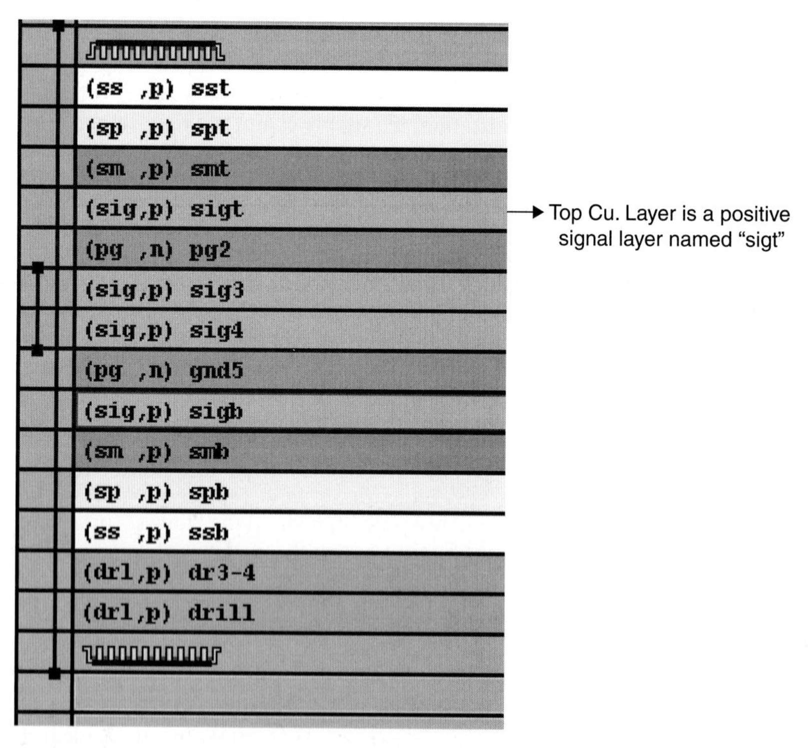

FIGURE 18.2 A CAM system view of a typical six-layer PCA after input from an "intelligent" format that defines the layers' order and characteristics explicitly and unambiguously. If Gerber files were used it would take some time to arrange the layers' order and characteristics manually. The example includes the following:

Silk screen legend files: **sst** (for top) and **ssb** (for bottom)
Solder paste stencil aperture layers: **spt** and **spb**
Solder mask images: **smt** and **smb**
Outer copper layers: **sigt** and **sigb**
Second layer: **pg2** (power layer, which is negative)
Fifth layer: **gnd5** (ground layer, which is also negative)
Inner circuit layers: **sig3** and **sig4**
Drill layers: **drill** (through drill layer) and **dr3-4** (buried via connecting **sig3** and **sig4**)

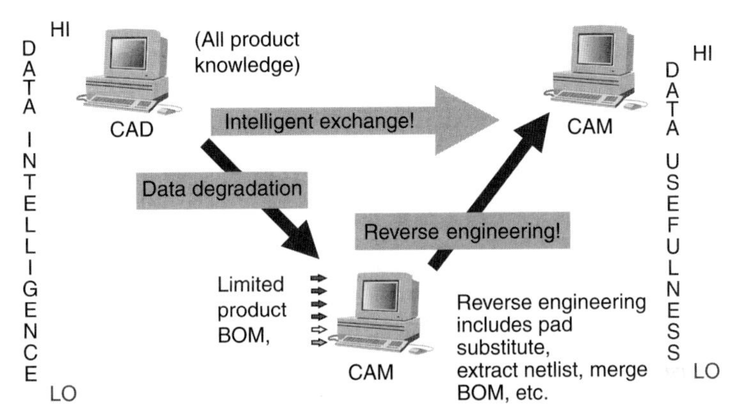

FIGURE 18.3 OEM, CEM, and Fab communication levels. In this figure the y-axis denotes the level of intelligence and the x-axis denotes the data exchange process progression over time. When the formats used carry less design intelligence, it is lost (data degradation). Non-value-added reverse engineering (such as manually rebuilding the board stack-up) is required to rebuild intelligence that is required for manufacturing. If an intelligent exchange format is used, no design data are lost and no reverse engineering is required, resulting in a faster, more accurate, and more efficient process.

- **Intelligent** The format retains CAD information that may help the manufacturer (see Fig. 18.3).
- **Optimized** A surface should be represented by a clear outline and not drawn with overlapping vectors (see Fig. 18.4).
- **Bidirectional** The format should be inherently capable of passing data back and forth, not just passing them one way. This can be achieved through free viewers, annotations tools, and so on.

Other important issues to consider in the context of data exchange are the following:

- **Accountability** Who is responsible if something is misunderstood? If the exchange always succeeds, it is easier to differentiate between design and manufacturing errors.

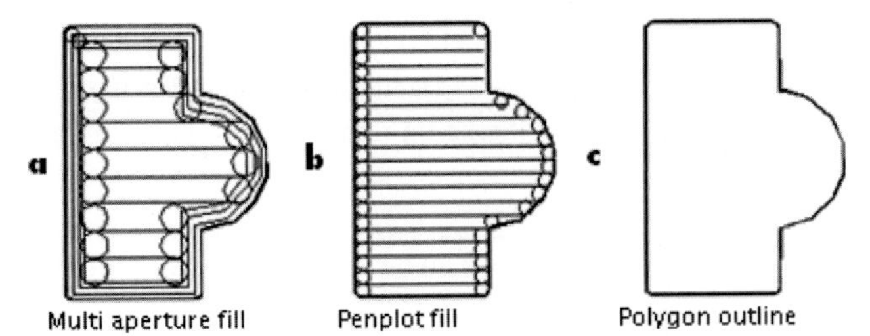

FIGURE 18.4 Optimizing surfaces: (a) a polygon filled by different vector widths; (b) the same polygon as drawn by a pen plotter; (c)the polygon as defined by its external outline. Option c is clearly much more efficient and economical. The outline in (c) is the actual outline of the polygon, whereas (a) and (b) represent approximations.

- **Data ownership** Who can authorize changes? If reverse engineering is performed, subtle design changes may be introduced without OEM authorization.
- **Trust** Is the supplier an ally to whom all available data are provided, or are some design details withheld to protect intellectual property rights?

The next section details all the data elements that designers and manufacturers need to pass among each other.

18.2.2 Elements of Data Exchange: Intelligent Design Data

The goal of a supply chain is short time to market, high quality, and lowest cost, and achieving this goal requires high clarity, efficient communication, and not just a bare-bones data hand-over.

Although a PCB can be manufactured from supplied artworks and drill tapes alone, it is much more efficient to send accurate and verifiable electronic data. The data can include information elements beyond the graphic image and component location information that form the minimum "pure" design data required for improved manufacturing. Among these elements are:

- Mechanical specifications
- Component geometry
- Component tolerances and vendor information
- Standards that the product has to meet

Some of the information belongs in an SOW and need not be repeated in the case of an established partnership. Included may be marking instructions, packaging instructions, and information needed to ensure conformance to standards.

18.2.3 Elements of Data Exchange for Fabrication

This section lists the information required for data exchange for fabrication.

18.2.3.1 Minimum Requirements. These are the minimum required elements of data needed for fabrication:

- Layer graphics (see Fig. 18.5)
- Drill data: locations, spans (for blind and buried vias), tolerances, and plating thickness
- Board outline and rout information
- Stack-up requirements (layer order)

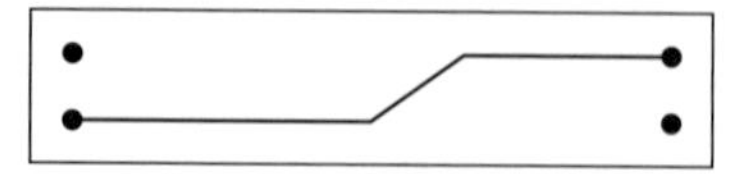

FIGURE 18.5 A layer graphic representing the features of a circuit (or a document) layer represented by computer-readable geometric entities involving x and y coordinates and geometric objects—lines, arcs, circles, and so on.

18.2.3.2 Important Additional Information. These elements are important but not absolutely required for fabrication:

- Netlist information is required to verify that the graphic and drill data truly represent the design. It is highly advisable always to include netlist data so that the graphics can be verified.
- Assembly panelization definition is required to deliver multiple PCBs in a ready-to-assemble array form (four or more cell-phone boards can be assembled together due to their small dimensions).

18.2.3.3 Extra Information. This extra information is not covered by computerized formats, but is useful in fabrication:

- Electrical (impedance) and material requirements (solder mask, legend, etc.). (Impedance requirements refer not only to line width and tight tolerances thereof but also the actual impedance value, in ohms, and the frequency and impedance model to use.)
- Fabrication drawing (extra instructions, dimensions).
- Type of finish and quality requirements (hot air solder leveling [HASL] versus organic solderability preservative [OSP], solder mask).
- Packaging and delivery instructions.
- Testing coupon requirements.

These additional information elements can be provided as well:

- Design for manufacturing (DFM) analysis results and criteria (for example, spacing criteria and violations thereof)
- Component placement information, which can help the manufacturer create solder mask clearances that provide a good fit for soldering

18.2.4 Elements of Data Exchange for Assembly

This section lists the data elements to consider for assembly processes.

18.2.4.1 Common to Fabrication and Assembly. These elements are common to fabrication and assembly:

- Outer layer graphics (for circuit, solder mask, and legend images)
- Drill information (especially through drills and mechanical drills)
- Board dimensions and outline

18.2.4.2 Required for Assembly. These elements are required specifically for assembly:

- Component placement information, including rotations.
- BOM and approved vendor list (AVL) information (AVL is not required if parts are supplied by consignment).
- Mechanical assemblies and their locations (screws, shields and heat sinks, etc.).
- Board electrical schematics for testing (see Fig. 18.6). This information is usually transferred either as an element of a CAD database or as a human- and machine-readable drawing format (usually as a Hewlett Packard Pen Plotter [HPGL] or Data eXchange Format [DXF] file). (Machine-readable in this context means identification of textual labels for nets (signals) and pin and component names. The computer programs cannot understand the connectivity symbolism of the line drawings within schematic drawings.)

18.2.4.3 Extra information. This extra information, not currently covered by electronic formats, must be read and understood:

- Test requirements
- Material instructions
- Delivery and packaging instructions

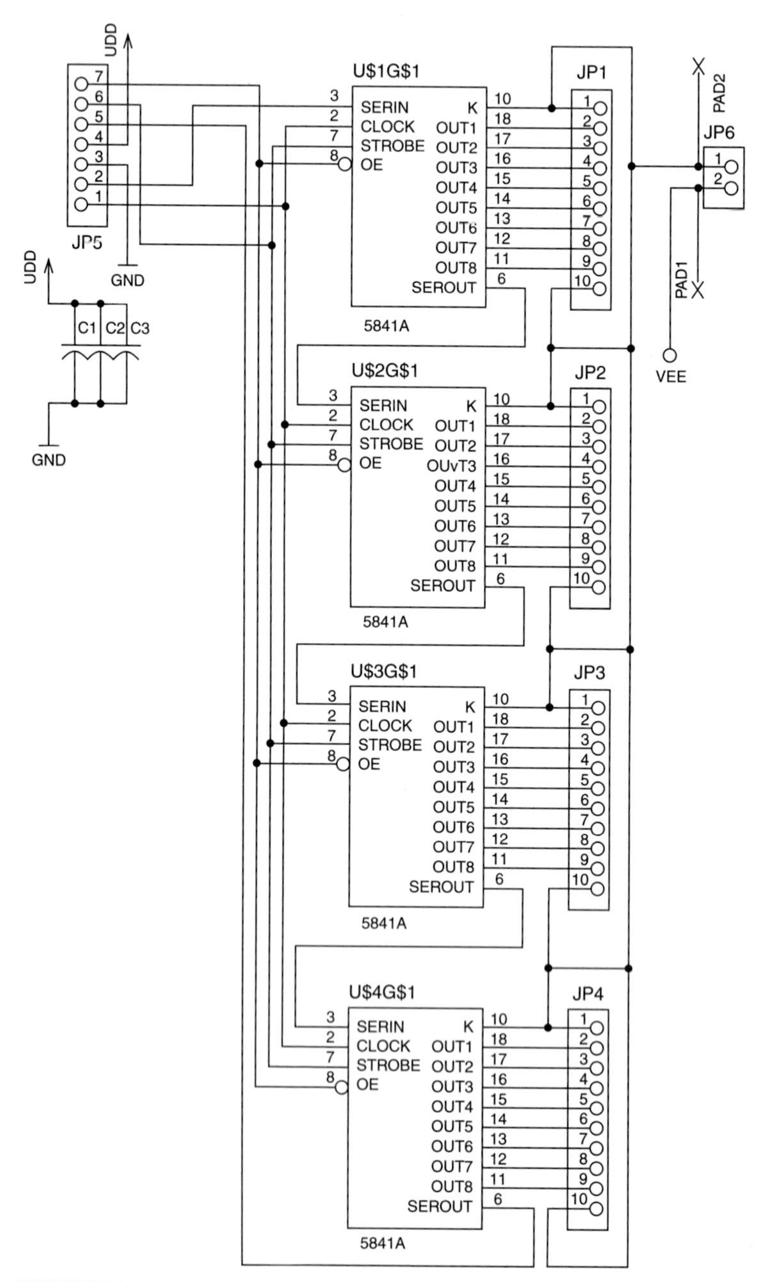

FIGURE 18.6 Product schematic drawing (for components, signals, and interconnections) as sent by the product electrical designer to the PCB editor. There are four 18-pin logical components having SERIN (serial in) and eight output (out1-out8) pins each connected to output connectors (JP1-JP4) and to an input connector JP5 and a ground connection with three capacitors.

18.3 DATA EXCHANGE FORMATS

This section describes the most widely used data exchange formats in the electronics industry according to a common list of characteristics. Advantages and disadvantages of each format are discussed.

18.3.1 Format Types and Their Characteristics

There are several format families differentiated by their origins, properties, and usage. They fall into four categories:

- Historical file collections
- Full-design databases
- Component information (BOM and AVL)
- Formats that support full design to manufacturing

18.3.1.1 Historical File Collections. The historical file collection that is used to transfer design data is composed of three file types:

- **Graphic image formats** Gerber 274D and 274X[2] belong to this group (see Section 18.3.2.1). (There are other formats in this group, most of which are used internally within CAM systems and as archive formats within bare-board fabrication facilities. Pentax format, DaiNippon Screen, and MDA Autoplot belong to this group.) Most of these formats originated as a photo-plotter machine language; they describe single, graphic (drill, drawing), two-dimensional layers. They are usually accompanied by drill files indicating drill locations and diameters. The main drawbacks to graphic image formats are that they represent single layers rather than whole PCBs and that they cannot be verified unless accompanied by a netlist file.
- **Netlist** A netlist presents a list of net numbers, and for each net a list of points by their x and y coordinates, and the surface upon which they can be found (see Fig. 18.7). This is the basic information representing PCB connectivity against which PCB graphics can be verified

```
327LDIN_47    IC52  -U5          A01X+110078Y+030997X0200Y   R180
317LDIN_47    VIA   -      MD0100PA00X+110274Y+031194X0230Y   R180
327LDIN_47    IC10  -J2          A01X+134371Y+039414X0200Y
317LDIN_47    VIA   -      MD0100PA00X+134174Y+039217X0230Y
```

(a)

```
Signal LDIN_47
IC52 U5
IC10 J2
```

(b)

FIGURE 18.7 Example of a netlist: (a) this part represents a physical netlist with x and y coordinates (the first line translates as "Pin U5 of component IC52 is connected to net LDIN_47 and is at location [11.0078, 3.0997]"; (b) this part represents the same netlist but at schematic stage with only component and pin names; the electric signal identified as LDIN_47 connects IC52 pin V5 to IC10 pin J2 through two vias.

and from which a test program for an electrical test machine can be constructed. One type of netlist identifies the locations of the nets that the PCB is designed to implement. Another type of netlist is a CAD software design package schematic netlist that identifies connectivity between components (by component and pin number as an input/output [I/O] pin identifier), but does not identify their physical location.

- **Component information** In addition to being part of the historical file collection, BOM and AVL files are also part of an independent file class that is required even when the more advanced formats are used (see Section 18.3.5).

18.3.1.2 Full-Design Database: CAD Formats. Full-design databases, which are formatted using either the CAD system database format or American Standard Code for Information Interchange (ASCII)* extracts thereof, represent information for the component and their connectivity through the board as hierarchical arrangements of nets and subnets broken into traces, vias, and planes. They are transferred for board assembly and test but not for board fabrication. Because OEMs consider these formats to present a proprietary information security risk, they often prefer to convey inferior quality information using other formats in order to preserve privacy.

18.3.1.3 Explicit Design for Manufacturing Formats. Formats for full design for manufacturing were explicitly built for design-to-manufacturing data exchange and contain nearly all of the required computer-readable data elements in a well-designed, explicit, intelligent layout. Some of these formats originated as DFM/CAM vendor formats whereas others were designed by an industry association committee.

18.3.1.4 Component Information: BOM and AVL. A PCA may be designed for more than one product variant and to support future functionality. Therefore, for almost all designs, some of the reference designators (or actual bare board footprints on which components can be placed) will not be populated for all assembled board orders. BOM files are used to tell the assembler which reference designators should actually be placed on a given board for a specific order.

The separation between electrical functionality of components and their actual procurement leads to the need for an AVL file. CAD databases usually contain part numbers that represent functionality and are internal to the design organization rather than part numbers (including the vendor and catalog number) that customers can order. Component engineers are charged with finding true part numbers to implement the internal ones, and this information is usually organized in spreadsheets or text files called AVLs or approved component lists (ACLs).

18.3.2 Historical File Formats Descriptions

Each format description that follows includes its domain (the elements that are covered), its history and expected future path, distinctive characteristics, and a short example. Format description references can be found at the end of the chapter.

18.3.2.1 Gerber (274D and 274X), Drill and "Readme." Gerber formats are the "lowest common denominator" for passing bare-board fabrication information. They are composed of one file per graphic layer, one file for each drill layer, text files that list the PCB stack-up, and optional drawing files that list critical dimensions and requirements. Inclusion of a netlist file with each Gerber transfer is imperative to ensure correct board fabrication.

* See http://www.lookuptables.com/.

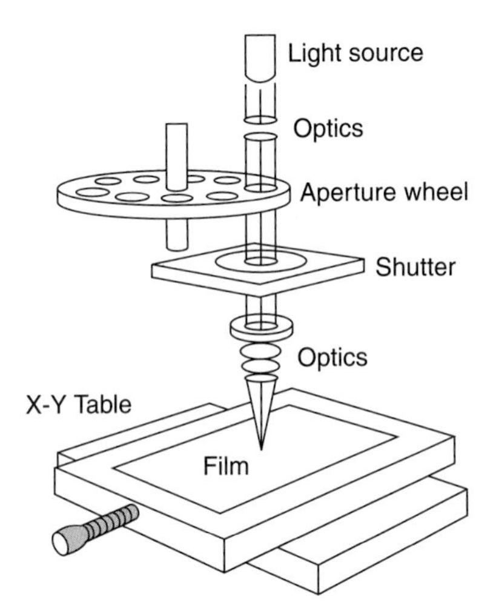

FIGURE 18.8 A photo plotter uses a stationary light source to deliver light through a rotating variable size aperture wheel (which determines the size of the light spot on the table) onto a table that can move in the x and y directions. A shutter shuts or allows light passage as required. Precision optics control the focus of the light spot.

A common problem with both Gerber 274D and drill formats (usually Excellon I and II) is the lack of definition of both units and scale factor of numbers (Excellon is a precision drill machines manufacturer). Each coordinate is given as a collection of digits, and the translator, or user, must define the units (in in. or mm) and the location of the decimal points in the format. Another problem stems from the use of arbitrary numbers ("Dcodes" for aperture numbers on an aperture wheel) to describe the width of lines and size of pads. Accurate translation of these files depends on the correct definition of an aperture wheel, which is described separately. Figure 18.8 shows a photo-plotter schematic that explains the source of the terminology.

The commands that drive this mechanism are x and y coordinate movement commands to position the table, those that turn the shutter on and off, and Dcode commands that rotate the aperture wheel to the required position.

Gerber 274X was designed in the early 1990s (by a team of two engineers—one from Gerber Scientific and another from what used to be the biggest PCB factory in the United States, the AT&T plant in Richmond, Virginia) to overcome some Gerber 274D problems. Gerber 274X permeated slowly at first, but its adoption accelerated and Gerber 274D is now used less frequently.

Both 274D and 274X are used to transfer graphic images of circuit and mask layers and images of documentation layers used for instructions. Other graphic drawing formats are occasionally used to transfer board and drawing data. The most common formats are HPGL (the format of HP pen plotters) and DXF (the AutoCAD exchange format).

Several drill formats originating from drill machine makers other than Excellon (Sieb and Mayer, Posalux, etc.) are sometimes used to transfer drill data or to archive old drill data. They suffer from the same problems of Dcode (explained in Fig. 18.9) and unit definition.

18.3.2.1.1 Gerber 274D. Gerber RS-274-D has been the most common format for describing PCB plot data. This format was originally intended to drive vector photo plotters (see Fig. 18.8), produced by Gerber Systems Corporation (see Fig. 18.9 for an annotated snapshot section of a Gerber file).

Example

`G90*`	G90 indicates absolute coordinates. This means that that each set of coordinates is referenced to the table's origin (0,0). The converse to absolute is incremental; each coordinate is measured relative the previous coordinate value and is set by issuing the G91 command.
`G70*`	G70 indicates inch units.
`G54D10*`	G54, tool select (line 3), is the most commonly encountered Gcode and instructs the plotter to rotate the aperture wheel to the position described by D*xx* immediately following the G54 command. If this code is omitted, the the photo plotter then identifies the next D*xx* command as telling it to select a tool.
`G01X0Y0D02*`	Dcodes are instructions to the photo plotter that, naturally, include the letter *D*. The first three Dcodes control the movement of the x,y table. D02 (D2) tells the plotter to move to the x,y location specified with the shutter closed.
`X450Y330D01*`	D01 (D1): This command tells the plotter to move with the shutter open to the x,y location specified.
`.X455Y300D03*`	D03 (D3): This command tells the plotter to move with the shutter closed to the x,y location specified, then open and close the shutter—known as flashing the exposure.
`G54D11*`	
`Y250D03*`	
`Y200D03*`	
`Y150D03*`	
`X0Y0D02*`	
`M02*`	The *M* in Mcommands stands for miscellaneous; in this case, M02 means stop or end of file.

FIGURE 18.9 An annotated section of a Gerber file.

The basic commands of this format include:

- Selecting the aperture
- Opening and closing the aperture shutter
- Moving the head to a given x, y coordinate

Typically, commands are separated by the asterisk character (*).[3]

18.3.2.1.2 Gerber 274X. Gerber RS-274-X is a common format in use today for describing PCB plot data. The format can be divided into two parts:

- The Gerber part
- The extended part, which includes commands for the following:
  - Standard aperture definition
  - Aperture Macro definition (special symbols)
  - Layer polarity selection

Typically, extended commands begin with % and end with *%.

Unlike Gerber 274D format, Gerber 274X includes definition of coordinates using the FS command (see Fig. 18.10). Reading software uses this information when reading Gerber parts.

The most serious limitation of Gerber 274X is polygon representation. No solution was provided for describing internal cutouts or clearances. Each implementation of Gerber 274X-based CAD output deals with these in its own way. The most problematic way is to use self-intersecting polygons (SIPs) where the "pen" does not leave the board while drawing the internal cutouts, but "swerves" to the inside of the external outline to draw the internal cutout and then returns (see Fig. 18.11).

Example

`%FSLAX23Y23*%`	This command indicates (L)eading zero suppression,(A)bsolute, 2.3, or xx.xxx and yy.yyy decimal point location
`%MOIN*%`	This command specifies inches
`%SFA1.000B1.000*%`	
`%ADD11C,0.00500*%`	These are aperture definition commands defining various diameter circles (5, 8, 10, 20, and 25 mil, respectively)
`%ADD13C,0.00800*%`	
`%ADD14C,0.01000*%`	
`%ADD70C,0.02000*%`	
`%ADD71C,0.02500*%`	
`G54D11*`	Here the Gerber part begins.
`%LPD*%`	This command indicates that digitized data is dark. However, when the entire film is reversed, the digitized data will be clear.

```
G11*G70*G01*D02*G54D10*X-0020000Y-0250000D02*X-0020000Y-0265200D01*
X0010000Y-0255000D02*X0010000Y-0250000D01*X0025200Y-0344800D02*
X0025200Y-0285000D01*X0022800Y-0265200D02*X0022800Y-0270000D01*
```

This is straight Gerber code with multiple commands on a line.

FIGURE 18.10 An annotated Gerber 274X section.

An SIP is a polygon with two nonconsecutive edges (segments or curves) touching each other. SIPs are illegal in CAM systems that define legal polygons as those whose edges intersect only at endpoints of consecutive edges.

Translation of 274X data into a CAM system might fail with a "Self Intersecting Polygon" error because some CAD systems create surfaces using self-intersecting polygons.

SIPs are not mathematically robust, and the following SIP operations are problematic:

- Resizing (enlarge, shrink, scale—especially with differing x and y values)
- Accurate copper calculations (these require unambiguous definition of the copper location)

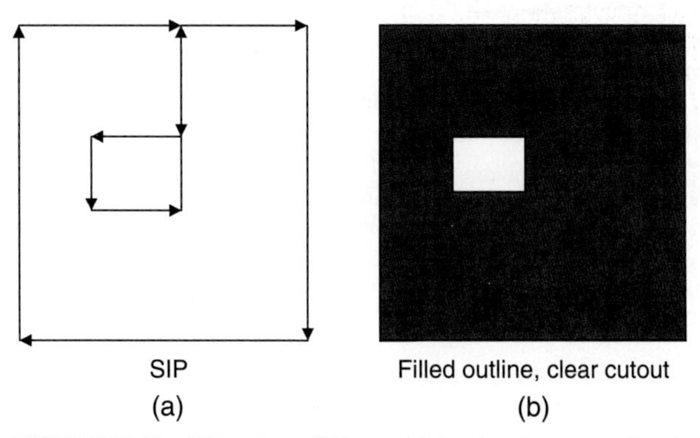

FIGURE 18.11 SIP and non-SIP ways of drawing the same polygon: (a) An SIP: as drawn by self-intersection, with a single path and the pen not removed from the paper; (b) not an SIP with a filled rectangle as an outline and a clear rectangle as the cutout.

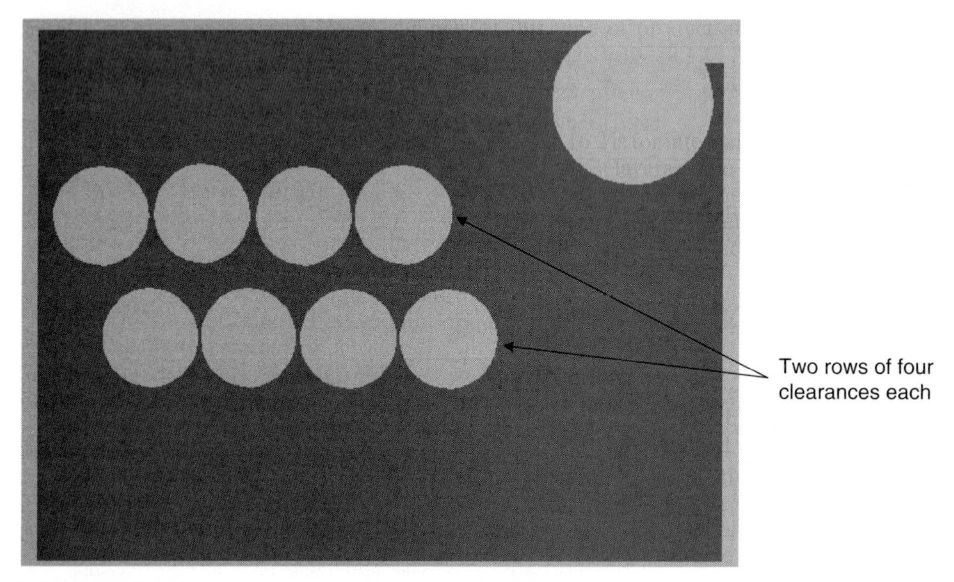

FIGURE 18.12 This surface is composed of one rectangular island, with a circular cutout in the top-right corner representing copper and eight round holes representing clearances.

In most CAM systems, a surface is a mathematical entity that describes planes. It can contain islands (positive polygons) and holes (negative polygons). If the layer is positive, islands represent copper, whereas holes represent non-copper (see Fig. 18.12). Copper planes usually implement power or ground connections, where the clearances provide spacing for data signal vias that pass through the plane but must be electrically disconnected from it.

There are two non-ambiguous ways of drawing planes with cutouts in 274X. The first is to use positive polarity for the islands and negative polarity for the holes. The second is to decompose the plane into positive surfaces that, when combined, yield the outline with the holes.

18.3.2.2 Netlist Formats. A netlist is a set of connection points joined together to form networks. Each point represents a contact point on the surface of a bare board. A connection point can be either a drilled hole or a surface mount pad. All points belonging to one network should be connected to each other through PCB layer circuitry and/or power and ground planes.

*18.3.2.2.1 IPC-D-356.** IPC-D-356 is an ANSI-accepted standard that has become the most widely used standard for transferring netlist information. The IPC-D-356 format is used to transfer netlist information within the PCB design and fabrication community. This information can be used to verify the integrity of the design by netlist extraction from the Gerber graphics and comparison with the IPC-D-356 CAD reference. The information is also used within both the bare board and the assembled board test domains.

Most data in the IPC-D-356 file (see Fig. 18.13) consist of electrical test records of two types:

- Drill records (starting with "317")
- Surface mount pad records (starting with "327")

There are also other general parameter records.

**References* ANSI/IPC-D-356 published by the Institute for Interconnecting and Packaging Electronic Circuits (IPC).

```
C This is a comment field
P UNITS CUST
317A01AUXNEG A10 -1 D0380PA00X+068250Y+002250X0620Y0620
317UN2CAP62PCA0 A10 -2 D0380PA00X+068250Y+001250X0620Y0000
327UN2CAP62PCA0 A9 -1 A01X+066330Y+001500X0800Y0250
327A01AUXPOS A9 -2 A01X+066330Y+002000X0800Y0250
327N/C A9 -3 A01X+066330Y+002500X0800Y0250
327N/C A9 -4 A01X+066330Y+003000X0800Y0250
327N/C A9 -5 A01X+064170Y+003000X0800Y0250
327N/C A9 -6 A01X+064170Y+002500X0800Y0250
```

FIGURE 18.13 A section of an IPC-D-356 file. It contains two drill records ("317") and six surface mount records ("327"). The four bottom solder mask (SMT) points should not be connected, hence their "net name" fields contain "N/C," indicating "Not connected."

Each IPC-D-356 record consists of one line and (for historic reasons) is of a fixed length with a maximum of 80 characters.

The information extracted from translation of IPC-D-356 files to an internal netlist might include solder mask coverage and midpoint flags, along with the dimensions and location of connection points and their groupings into networks. Information in component and pin identification fields is not necessarily extracted for comparison purposes.

There are two amendments to IPC-D-356: IPC-D-356A and IPC-D-356B. Most of the extra information included in these formats is important for bare-board electrical testing and, unless buried passives are used, the information is generally not needed for CAD-to-CAM data consistency verification. If buried passives are used, it is better to transfer IPC-D-356A after verifying readability by the supplier.

18.3.2.2.2 Mentor Graphics Neutral File. Several nonstandard netlist formats can be output by various CAD software suites. All such formats contain connection point records and their grouping into nets. American designers and fabricators occasionally used a stripped-down version of the Mentor CAD format Neutral File to transfer netlist information, and thus it is described here.

The Mentor Graphics Board Station Fablink application produced the Mentor Graphics Neutral File format. It has several sections, each describing one aspect of a board (components, netlist, holes, etc.).

A full neutral file contains these information sections:

- BOARD: board attributes
- NET: net points
- GEOM: packages and pins
- COMP: components and "toeprints"—another name for a pad pattern outline
- HOLE: drilled holes
- PAD: approximate shape of pads

The BOARD and NET sections contain sufficient information to derive a netlist. If the file includes all sections, it is possible to derive components and drills but not full routing (trace) information.

The Fablink user can control which sections are included in the file (see Fig. 18.14).

```
# file: /tmp_mnt/disk/ed8/mj/demo/1/pcb/mfg/neutral_file
# date: Monday June 2, 2005; 18:46:15
#
#############################################
###Panel Added Part Information
#############################################
#############################################
###Board Information
#############################################
BOARD SYSTEST_BOARD OFFSET x:0.0 y:0.0 ORIENTATION 0
B_UNITS Inches
#############################################
###Attribute Information
#############################################
B_ATTR 'MILLING_ORIGIN' 'MILLING 0 0.0 0' -0.4 -4.0
B_ATTR 'DRILL_ORIGIN' '' 0.0 0.0
B_ATTR 'BOARD_DEFINITION_IDENTIFIER' '' 0 0
....
###Nets Information
#############################################
NET /+8VTO10V
N_PIN J1-1 0.5 -3.2 term_1 0
N_PIN CR1-1 -0.2 -2.7 term_1 0
N_PIN W2-3 0.3 -2.5 term_1 0
NET /DATA_BIT_1
N_PIN J2-1 0.8 -1.7 term_1 0
N_PIN U2-3 1.1 -3.1 term_1 0
....
#############################################
###Geometry Information
#############################################
GEOM DIP20
G_PIN 1 0.0 0.0 term_1 Thru 0.033
G_PIN 2 0.0 -0.1 term_1 Thru 0.033
G_PIN 3 0.0 -0.2 term_1 Thru 0.033
G_PIN 4 0.0 -0.3 term_1 Thru 0.033
....
#############################################
###Component Information
```

Comments that denote file section boundaries

FIGURE 18.14 An excerpt from a Mentor Neutral File. Section names and records roles are pretty obvious, because clear key words are used. The number sign, #, denotes a comment.

```
#############################################
COMP C1 PN-150uFCAP 150uFCAP c123_d5 -0.2 -2.0 1 0
C_PIN C1-1 -0.2 -2.0 0 1 0 term_1 /N$577
C_PIN C1-2 0.6 -2.0 0 1 0 term_1 GROUND
....
#############################################
###Hole Information
#############################################
HOLE PTH -0.2 -3.8 0.093
HOLE PTH -0.2 -1.2 0.093
HOLE PTH 3.9 -1.2 0.093
#############################################
###Pad Information
#############################################
PAD VIA VIA_1 Thru 0.03
P_SHAPE VIA_1 PHYSICAL_1 CIRCLE 0.05
P_SHAPE VIA_1 PHYSICAL_2 CIRCLE 0.05
```

FIGURE 18.14 (Continued).

18.3.2.3 Component Placement List. Component placement lists (CPL) are ASCII or spreadsheet files used to transfer component x, y locations if CAD data files are not sent. CPL files are usually minimally formatted into columns. Required columns are the reference designator, x, y locations, and component rotation. In Fig. 18.15, the example CPL file also includes CAD package information.

```
!
! SIDE 1
!
R101    0   R0805          10467C    5065000   880000    90    SMT
R141    0   R0805          10467C    6670000   585000    270   SMT
R127    0   R0805          10467C    6595000   585000    270   SMT
R182    0   R0805          10467C    3850000   1260000   0     SMT
R199    0   R0805          10467C    4015000   1020000   0     SMT
R62     0   R0805          10467C    7885000   810000    180   SMT
R512    0   R0805          10467C    -45000    1065000   90    SMT
R6      0   R0805          10467C    3475000   140000    180   SMT
R214    0   R0805          10467C    3000000   1580000   0     SMT
R76     0   R0805          10467C    7940000   505000    90    SMT
R128    0   R0805          10488C    4275000   895000    0     SMT
R44     0   R0805          10488C    3850000   1415000   0     SMT
R59     0   R0805          10512C    8495000   1035000   90    SMT
M9      0   M_74F153_SO16  105328C   4835000   1305000   270   SMT
```

FIGURE 18.15 An example of a typical simple text tabulated CPL file. The first line translates as "Place resistor R101 at x= 5.065 y=0.88 at 90°. Use part number 10467C. For information, see CAD package R0805. The mounting process is SMT." This example is just a text file listing.

18.3.3 CAD Formats

The leading CAD system vendors have created several formats. When there is a high level of cooperation between customer and supplier, the design customer can send the whole design database from the CAD system on which it was created (and achieve the goal of intelligent exchange as depicted in Fig. 18.3). CAD formats are excellent for testing assembled boards because they include the schematic information. In addition, they include component locations and packages, as well as enough data to derive the graphics of any board layer. They must be accompanied by BOM information and also AVL information (unless parts are consigned) to define which components to place for each board variant and which components to procure.

CAD formats are less suitable for board fabrication, because they lack a flattened WYSIWYG ("what you see is what you get") representation of the circuit layers. Examples of CAD formats include Mentor's Board Station Neutral and Geoms files, as well as Cadence or ORCAD layout files.

When CAD formats are used for data exchange, their format is, by definition, the format of the specific CAD system that was used for layout and routing. The designer must therefore verify that the supplier can read it or use some translation tool to convert the data to a format readable by the supplier, risking conversion errors.

18.3.4 Modern Data Exchange Formats

ODB++ and IPC-2581 are examples of formats that were explicitly designed for the purpose of CAD to CAM data exchange. ODB++ is the internal format for Valor's* tools and was developed early as an exchange format. IPC-2581 is a vendor-independent format developed as a joint effort by Valor and an IPC-sponsored committee to correct the historical problems associated with data exchange.

ODB++ and IPC-2581 are the only formats that can include all the data needed for manufacturing, including BOM and AVL information. They come closest to the idea of intelligent exchange depicted in Fig 18.3.

18.3.4.1 ODB++. The ODB++ data format is a common language used for DFM and CAD/CAM data exchange. It overcomes many data communication obstacles within design/manufacturing supply chains. Data in this powerful open database impart an integrated and accurate physical model of all bare-board, component, and test-related information. It is designed as a simple yet comprehensive description of all entities used in the manufacture and assembly of a PCB.

The Valor Universal Viewer (VUV) is free software that enables ODB++ design data to be viewed graphically on Windows, Sun, or HP-UX workstations.**

ODB++ uses a standard file system structure.[6] ODB++ denotes a job using a simple directory tree that can be transferred between systems with no loss of data (see Fig. 18.16).

The advantages of a directory tree compared to one large file are apparent when a job is being read or saved. The flexible tree structure allows only selected parts of the job to be read or saved, avoiding the overhead of reading and writing a large file.

When a job must be transferred, standard compression utilities can be used to convert a directory tree into a single file.

ODB++ is widely used in the PCB industry and many vendor tools support it, including some CAD systems that offer output to ODB++ format. Description of all ODB++ format elements is beyond the scope of this chapter, but a small excerpt from a layer features file, shown in Fig. 18.17) can exemplify its explicitness and its clarity.

---

* Valor Computerized Systems is a leading provider of CAD, CAM, and DFM tools for the electronics industry. See www.valor.com.

** To obtain a free download of VUV, go to http://www.valor.com/, then select solutions _> Valor Universal Viewer _> Download and follow the simple registration procedure.

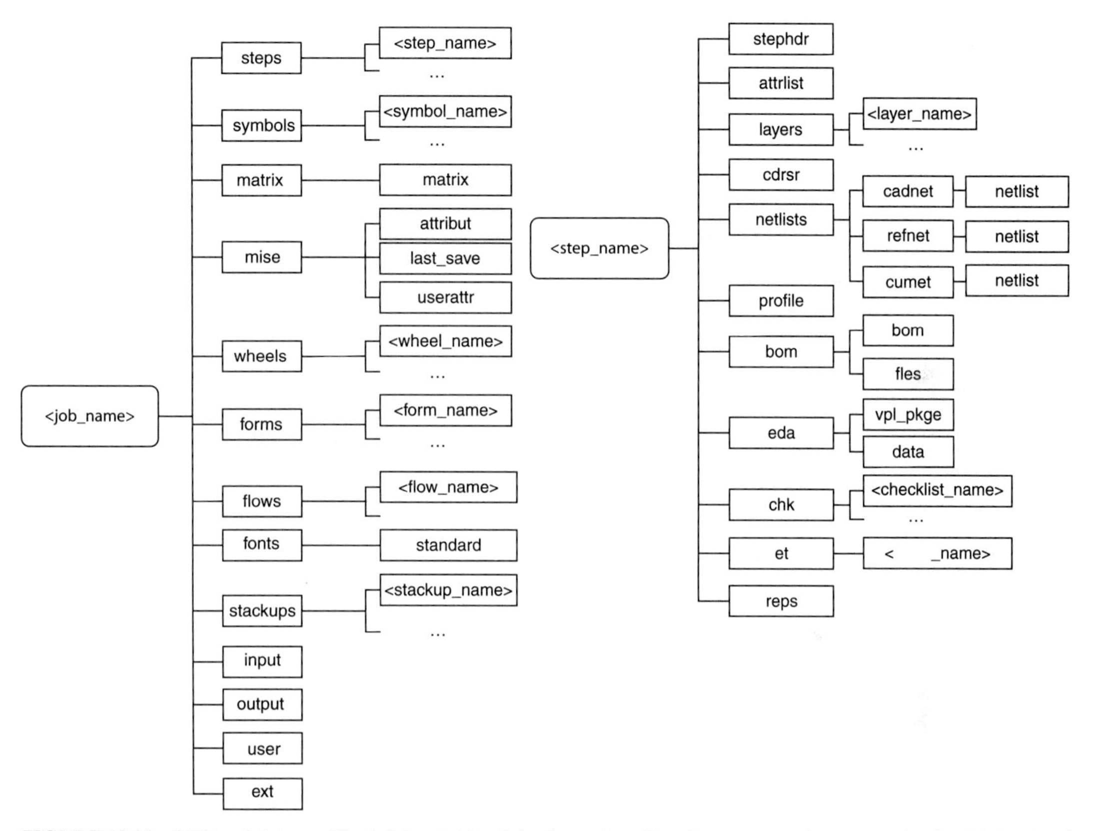

FIGURE 18.16 ODB++ job tree—The left hand side of the figure describes the top-most elements under the Job level and the right hand side describes the top-most elements under the Step level.

18.3.4.2 IPC-2581. IPC-2581, on the part of the IPC and the electronics industry, is a universal vendor-independent data exchange format.* IPC-D-350 was the first such format in the 1970s, followed by IPC-2511 GenCAM in the 1990s; in the new millennium, IPC-2581 evolved.

IPC-2581 (nicknamed "Offspring") is based on ODB++ and is the culmination of the ODB++ and GenCAM convergence project sponsored by the National Electronics Manufacturing Initiative (NEMI). IPC-2581 has inherited some of IPC-2511 GenCAM characteristics.

IPC-2581 is actually a series of IPC-258*x* documents, where x represents 1–9; IPC-2581 consists of the generic requirements (or, in this case, the full format description), whereas IPC-2582 and above are sectional requirements. IPC-2584, for example, contains the sectional requirements for fabrication and explains which generic IPC-2581 elements are required for bare-board fabrication and which elements are optional.

* For an IPC-2581 viewer, Contact IPC at http://www.ipc.org.

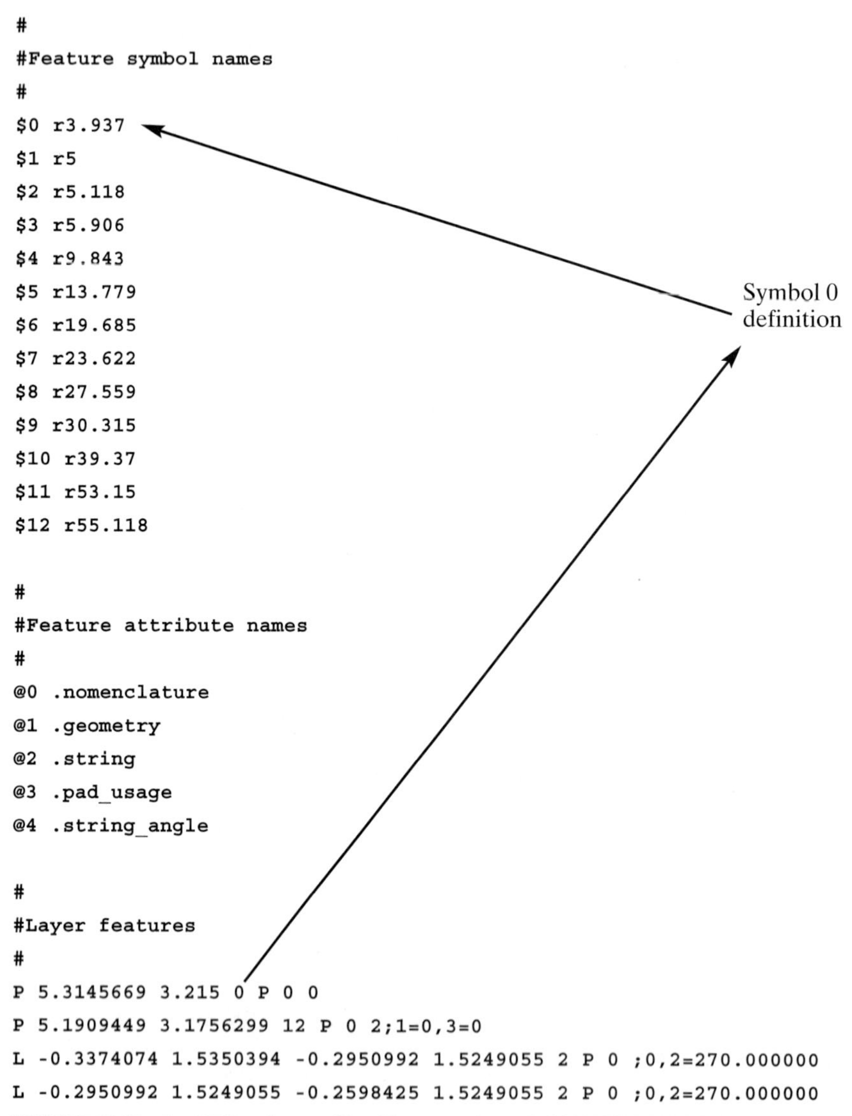


```
#
#Feature symbol names
#
$0 r3.937
$1 r5
$2 r5.118
$3 r5.906
$4 r9.843
$5 r13.779
$6 r19.685
$7 r23.622
$8 r27.559
$9 r30.315
$10 r39.37
$11 r53.15
$12 r55.118

#
#Feature attribute names
#
@0 .nomenclature
@1 .geometry
@2 .string
@3 .pad_usage
@4 .string_angle

#
#Layer features
#
P 5.3145669 3.215 0 P 0 0
P 5.1909449 3.1756299 12 P 0 2;1=0,3=0
L -0.3374074 1.5350394 -0.2950992 1.5249055 2 P 0 ;0,2=270.000000
L -0.2950992 1.5249055 -0.2598425 1.5249055 2 P 0 ;0,2=270.000000
```

FIGURE 18.17 An ODB++ feature file with annotations. "P 5.3145669 3.215 0 P 0 0" indicates a pad at (5.314.., 3.215) with a 3.937 mil diameter. (The first 0 points to symbol 0, clearly represented by "$0 r3.937" in the symbols section. "P 0 0" indicates that the pad has "P" (positive polarity), is not rotated 0°, and has zero attribute values.

IPC-2581 is an all-inclusive, explicit, accurate, and intelligent format containing all data needed for both manufacturing and assembly. The following data elements can be described in a full-mode IPC-2581 file:

- Original schematic capture files
- Original layer/stack instance files
- Original conductor routing files

- BOM and AVL
- Component packages and logical Nets
- Outerlayer copper lands
- Solder mask, solder paste, and legend layers
- Drilling and routing layers
- Documentation layers
- Physical Nets
- Outer copper layers (non-lands)
- Innerlayers
- Nondocument miscellaneous layers
- Instructions and specifications
- Component tuning and programming instructions, program files, and/or references thereof

This format contains information not available in the earlier formats (including details about personnel and the enterprise, material requirements, and component tuning and programming). Most industry tools lack these items in their internal formats, so at this point these sections are usable primarily in human-readable form.

Future evolution and market acceptance of IPC-2581 depend on committee work and vendor implementation and could well take several more years.

18.3.5 Component Information

BOM and AVL files do not usually follow any global standard, but rather conform to the internal standards of each organization. Their format may be simple ASCII or a spreadsheet (see Fig. 18.18). Interpretation of BOM and AVL files requires either manual or semi-automatic parsing followed by conversion into a known supported format. This process might entail creation of templates that recognize various types of BOM and AVL files and translate them semi-automatically.

BOM and AVL files are sometimes combined in a single BOM/AVL file that lists reference designators to be placed and part-ordering information.

Figure 18.19 is a typical AVL file sample as seen in already-tabulated format.

The line numbers are not part of the file. Lines 1 through 5 are header lines and lines 6 through 15 are data lines. There are two types of data lines; line 1 contains column headers for the first line type, and line 3 contains column headers for the second line type. Lines 6 and 12 are lines of the first type, representing functional devices (usually referred to as customer part

ITEM	CPN	GEOM	COUNT	DESCRIPTION	REFERENCE
1	1016975		1	SPROM_ASSY	
2	13r010c0	smd0402	6	cap, 1.0pF, sheet5	C502 C503 C504 C519 C587 C618
3	13r030c0	smd0402	2	cap, 3.0pF, sheet6	C517 C545
4	13r040c0	smd0402	1	cap, 4.0pF, sheet6	C531
5	13r0r5c0	smd0402	1	cap, 0.5pF, sheet5	C591
6	13r101j0	smd0402	3	cap, 100pF, sheet6	C500 C533 C536
7	13r102k1	smd0402	5	cap, 1000pF, sheet5	C518 C571 C573 C581 C617

FIGURE 18.18 A typical spreadsheet-style BOM file containing the customer part number, part information, and reference designators.

```
1 COMPONENT          DESCRIPTION
2
3    VENDOR NAME                          VENDOR ID  VENDOR PART ID
4------------------------------------------------------------------
5
6  XYZ00001          CAP CHIP 1PF 0.25PF 50V COG 0402
7
8     KEMET ELECTRONICS CORPORA       KEMET       C0402C109C5GACTU
9     MURATA ELECTRONICS              MURATA      GRM36COG010C050AQ
10    TEXAS INSTRUMENTS               TI          GRM36COG010C050AQ
11
12 XYZ00002          CAP CHIP 3PF 0.25PF 50V COG 0402
13
14    MURATA ELECTRONICS              MURATA      GRM36COG030C050AQ
15    TEXAS INSTRUMENTS               TI          GRM36C0G030C050AQ
```

FIGURE 18.19 Typical simple text tabulated AVL file excerpt.

numbers [CPNs]). Lines 8 through10 and 14 and 15 are of the second type and represent alternative orderable part numbers (usually referred to as manufacturer part numbers [MPNs]).

Line 6 translates as "Part number XYZ00001 is a 1 Pico Farad chip capacitor of description 'CAP CHIP 1PF 0.25PF 50V COG 0402,'" whereas line 8 translates as "to get the XYZ00001 functionality from Kemet Electronics Corporation, specify catalog number C0402C109C5GACTU."

18.4 DRIVERS FOR EVOLUTION

Data exchange for PCB manufacturing must evolve in two directions:

- Technology-based evolution
- Improved means of communication in a high pace, competitive global economy

With the rapid advance of technology in the PCB industry, data exchange formats must adapt to include new information. For example, none of the CAD formats mentioned here and neither ODB++* nor IPC-2581 support buried resistors and capacitors "out of the box," and therefore buried passive technology development is hindered by data communication errors and costs.

Other technological challenges that need to be addressed are better integrated BOM, AVL, and CAD data packaging coming out of design houses, and tool support for specifications, standards, revision information, and ECO data in an online framework.

With increasing trends toward globalization and supply chains, communication should be more than mere data exchange. The technology exists but is not yet utilized to its fullest extent. Online collaboration tools should replace file transfer. Means of maintaining issue lists and following up on their resolution would ensure that manufacturing requirements are met without compromising design integrity. Means of incrementally issuing design revisions without complete manufacturing reengineering would speed up prototype evolution.

---

* ODB++ version 7 to be released in 2007 supports buried passive resistors and capacitors.

18.5 ACKNOWLEDGMENT

The author thanks Susan Kayesar for her contribution to this chapter.

REFERENCES

1. Dean, Graham, "A Review of Modern Photoplotting Formats," Electronics Manufacturing Technology, http://www.everythingpcb.com/p13447.htm.
2. Document 40101-S00-066A, Mania Barco Corporation. (http://members.optusnet.com.au/~eseychell/rs274xrevd_e.pdf)
3. http://www.artwork.com/gerber/appl2.htm.
4. Mentor Graphics Fablink User's Manual. (www.mentor.com/)
5. ANSI/IPC-D-356, Institute for Interconnecting and Packaging Electronic Circuits, 3000 Lakeside Drive Bannockburn, IL, USA.
6. Valor ODB++ manual, available upon request from Paul Barrow at Paul.Barrow@valor.com; Tel, +972-8-9432430 (ext. 165); Fax, +972-8 – 9432429; Valor Computerized Systems, Ltd., P.O. Box 152, Yavne 70600, Israel.
7. IPC-2581, "Generic Requirements for Printed Board Assembly Products Manufacturing Description Data and Transfer Methodology"; IPC-258x and sectional requirements thereof, Institute for Interconnecting and Packaging Electronic Circuits, 3000 Lakeside Drive, Bannockburn, IL, USA.

CHAPTER 19
PLANNING FOR DESIGN, FABRICATION, AND ASSEMBLY

Happy T. Holden
Mentor Graphics, Longmont, Colorado

19.1 INTRODUCTION

Advances in interconnection technologies have occurred in response to the evolution of component packages, electronic technology, and increasingly complex functions. Therefore, it comes as no surprise that various forms of printed wiring remain the most popular and cost-effective method of interconnections.

Manufacturing, assembly, and test technologies have responded by improvements in their technologies. These increased capabilities have made selection of technologies, design rules, and features so complex that a new function has developed to allow for the prediction and selection of design parameters and performance versus manufacturing costs. This is the planning for design, fabrication, and assembly. This activity has also been called *design for manufacturing and assembly* or sometimes-*predictive engineering*. It is essentially the selection of design features and options that promote cost-competitive manufacturing, assembly, and test practices. Later in this chapter, we will offer a process to define producibility unique to each design or manufacturing process. Section 19.3.4 describes a process to define producibility unique to each design.

The purpose of this chapter is to provide information, concepts, and processes that lead to a thoughtfully and competitively designed printed circuit, ensuring that all pertinent design and layout variables have been considered.

19.1.1 Design Planning and Predicting Cost

Reducing costs to remain competitive is a principle responsibility of product planning. On the average, 75 percent of the recurring manufacturing costs are determined by the design drawing and specifications.[1] This was one of the conclusions found by an extensive study that General Electric conducted on how competitive products were developed. Manufacturing typically determines production setup, material management, and process management costs (see Fig. 19.1), which are a minor part of the overall product cost.

Cost Mainly Determined by Design Drawing and Specifications		Cost Mainly Determined by MANUFACTURING		
		Production Setup Cost	Material Management Cost	Process Management Cost
• Material cost • Purchased part cost • Process cost (Including assembly, adjustment and inspection	75%	13%	6%	6%
	(excluding advertising, sales, administration and design costs)			

FIGURE 19.1 Design determines the majority of the cost of a product.

Time to market along with competitive prices can determine a product's ultimate success. Having the first of a new electronic product on the market has many advantages. By planning the printed wiring board (PWB) layout and taking into consideration aspects and costs of PWB fabrication and assembly, the entire process of design and prototyping can be done with minimum redesign (or respins).

19.1.2 Design Planning and Manufacturing Planning

Electronics is one of the biggest enterprises globally. It is common for design to be done in one hemisphere and manufacturing in another. It is also common for manufacturing to be done in a number of different places simultaneously. An integrated approach must be adopted when the intention is to rationalize fabrication and assembly as part of the entire production system and not as individual entities, as shown in Fig. 19.2. This dispersed manufacturing must be taken into consideration during the design planning and layout process. No finished product is ever better than the original design or the materials from which it is made.

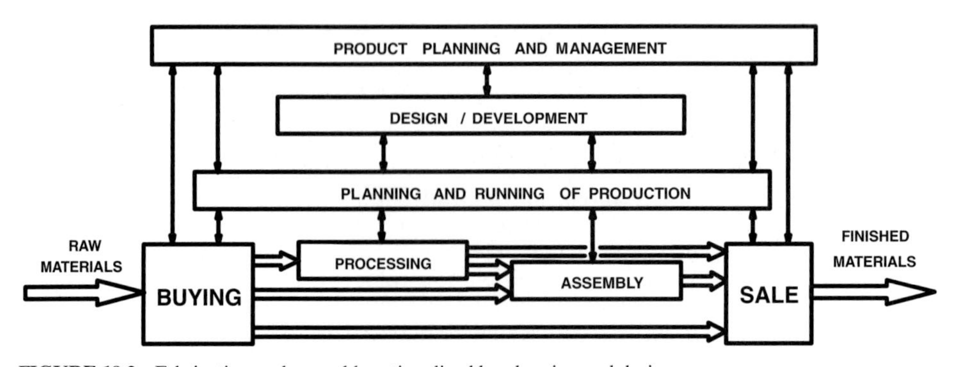

FIGURE 19.2 Fabrication and assembly rationalized by planning and design.

19.2 GENERAL CONSIDERATIONS

The planning process central focus will be the trade-offs between the loss and gain in performance for layout, fabrication, assembly, and test versus the costs in these domains. Therefore, some major considerations will be discussed in the following sections:

- New product design process (Secs. 19.3.1 and 19.3.2)
- The role of metrics (Secs. 19.3.3 and 19.3.4)
- Layout trade-off planning (Sec. 19.4)
- PWB fabrication trade-off planning (Sec. 19.5)
- Assembly trade-off planning (Sec. 19.6)
- Tools for manufacturing audits (Sec. 19.7)

19.2.1 Planning Concepts

Planning for design, fabrication, and assembly (PDFA) is a methodology that addresses all the factors that can impact production and customer satisfaction. Early in the design process, the central idea of PDFA is to make design decisions to optimize particular domains, such as producibility, assemblability, and testability, as well as to fit into a product family, such as custom automated manufacturing. Planning takes place continuously in the electronic design environment (see Fig. 19.3). The data and specifications flow in one direction, from product concept to

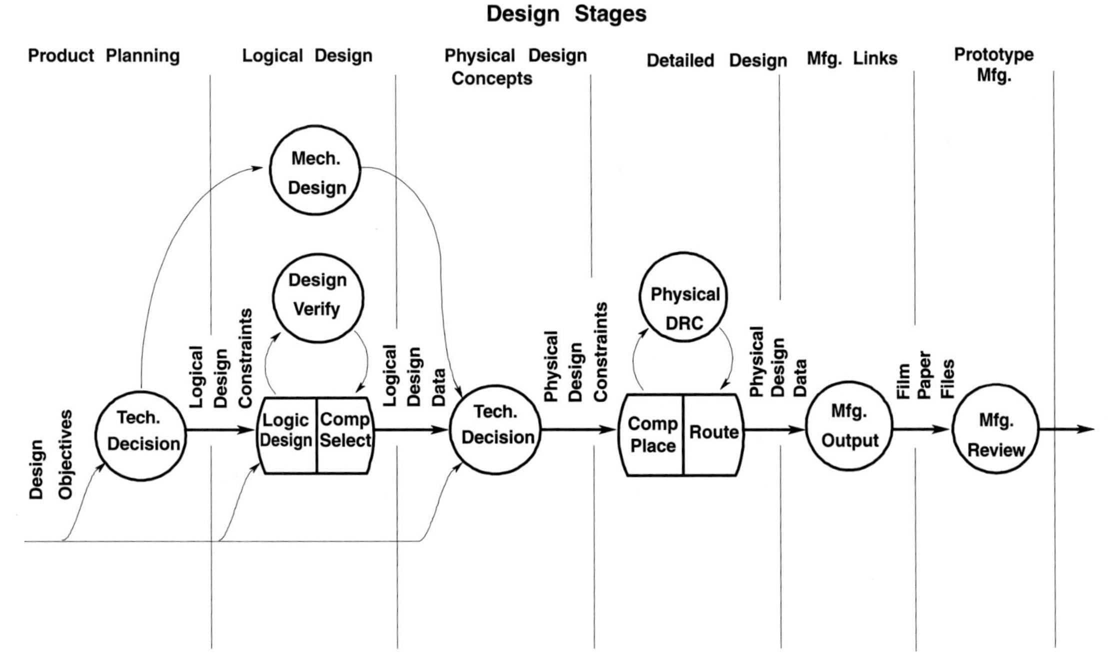

FIGURE 19.3 Electronic design environment.

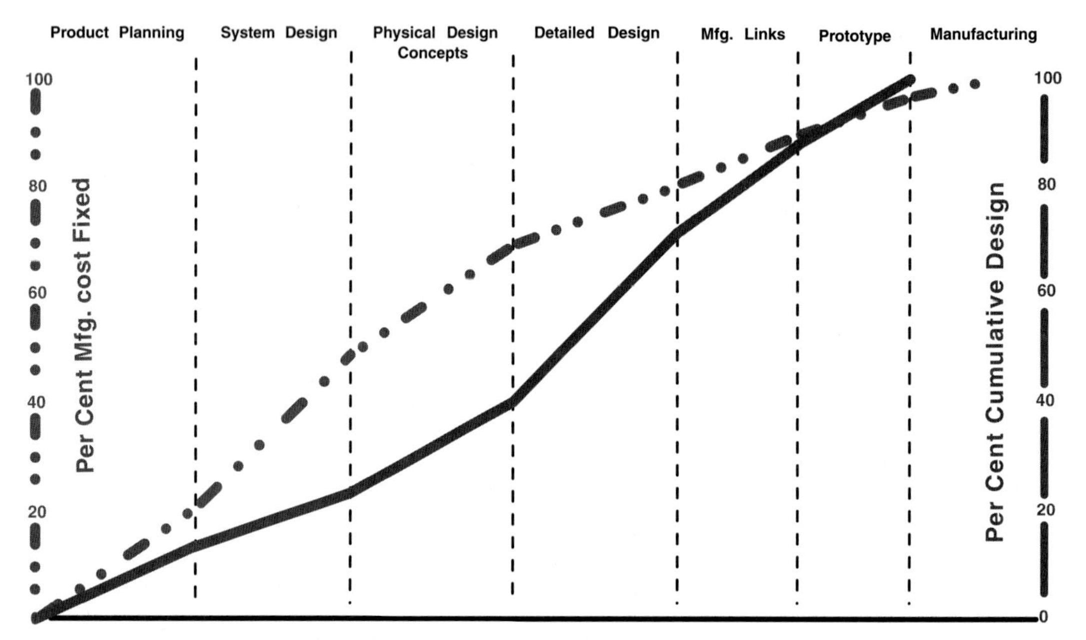

FIGURE 19.4 Design cost accumulation versus intrinsic manufacturing costs.

manufacturing. During the design process, 60 percent of the manufacturing costs are determined in the first stages of design when only 35 percent of the design engineering costs have been expended. The typical response is shown in Fig. 19.4.[1]

19.2.2 Producibility

Producibility is now regarded as an intrinsic characteristic of a modern design. Like the concept of quality in manufacturing, such a characteristic must be built in, not inspected in. Producibility must be designed in; it cannot be a "checkpoint" in the design process or inspected in by tooling.

19.3 NEW PRODUCT DESIGN

The keys to superior producibility in new product design can be found in the expanded design process. One of those keys is the role of metrics or data-based analysis of planning trade-offs.

19.3.1 Expanded Design Process

The new expanded design process that incorporates planning, trade-offs, and manufacturing audits is shown in Table 19.1. The process is made up of 12 separate functions that incorporate the planning and trade-offs sections in this book:

TABLE 19.1 The Expanded Electronics Design Process

Function	Purpose
Specification	User-supplied constraints and ideas formulate executable specifications.
Capture of system description	Technology trade-off analysis: balance of loss and gain in various domains' performance versus cost.
Synthesis	Generation of a netlist from the executable specifications.
Trade-off	Selection of layout, fabrication, and assembly features versus cost.
Physical CAD	Conversion of netlists to system and module layouts.
Simulation	Detailed analysis of design structures to support all other design (CAD) activities.
Design advisor	Continuous display of design rated by performance rules.
Manufacturability audit	Check of design to manufacturing design rules and capabilities.
Tooling	Conversion of module layouts to panel layout.
MFG	Conversion of module layouts to physical products (fabrication or assembly).
PDM database	Enterprise wide database containing all product information (product data management, or PDM) including design files, libraries, manufacturing information and revisions, etc.
Internet	Multi-team designs via access over the internet

This differs from the more conventional design process (as seen in Fig. 19.3) by the inclusion of four important functions:

- The formal technology trade-off analysis during the specification phase
- Detailed trade-off selection of features for layout, fabrication, and assembly
- Design advice during component placement and routing
- Manufacturing audits to review the finished layout for producibility, time to market, and competitiveness

19.3.2 Product Definition

The initial new product design stage is specification and product definition. This key step takes ideas, user requirements, opportunities, and technologies and formulates the executable specifications of a new product. During this operation, the ability of technologists who may lack any manufacturing experience to predict what will happen in manufacturing can affect both time to market and ultimate product costs. Figure 19.5 shows the technology trade-off analysis that requires the balance of loss and gain in various domains performance versus costs. Size and partitioning for integrated circuits (IC) and application-specific integrated circuit (ASIC) must be balanced with overall packaging costs and the resultant electrical performance. All of these factors affect the manufacturing and product cost.

Another definition of this process could be called a "verified design."[2] A verified design is one that was predicted from models or measures that have been correlated to past designs. This is in contrast to the traditional approach, which is a "nonverified design," or trial and error." This is diagrammed in Fig. 19.6. The advantage of the verified design can be a significant reduction in redesigns to achieve the original product objectives.

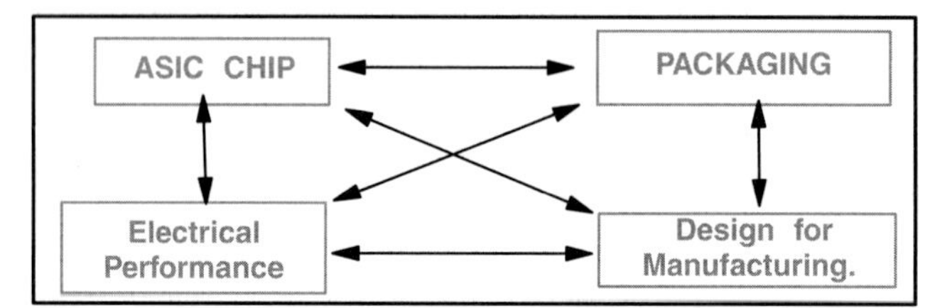

FIGURE 19.5 Specifications determine product partitioning and producibility.

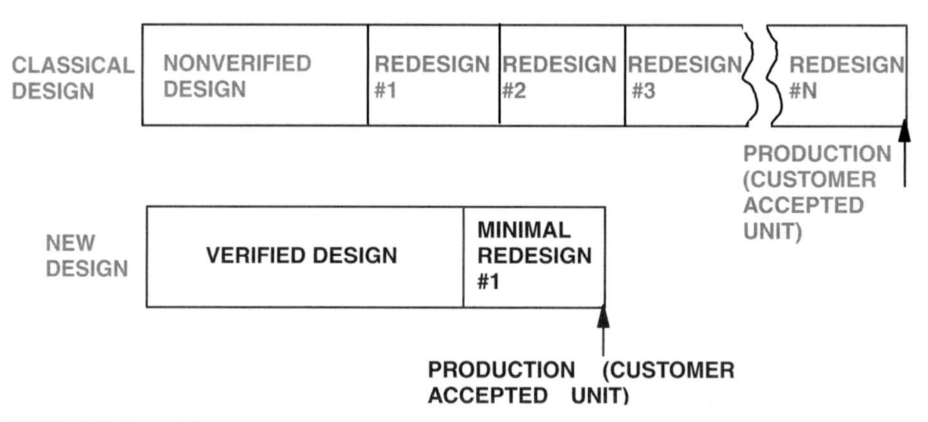

FIGURE 19.6 A design incorporating trade-offs versus traditional design.

19.3.3 Metrics for Predicting and Planning Producibility

Metrics are data and statistically backed measures, such as wiring demand (Wd) (see Section 19.4.1). These measures can be density, connectivity, or, in this context, producibility. These measures are the basis for predicting and planning. When such measures are used in the design process, two categories are applied to a product. Only the metrics can be shared by all in the design team. The non-metrics provide little assistance in design.

Metrics

- **Metrics** Both the product and the process are measured by physical data using statistical process control (SPC) and total quality management (TQM) techniques (predictive engineering process).
- **Figure of merit** Both the product and the process are scored by linear equations developed by consensus expert opinion (expert opinion process).

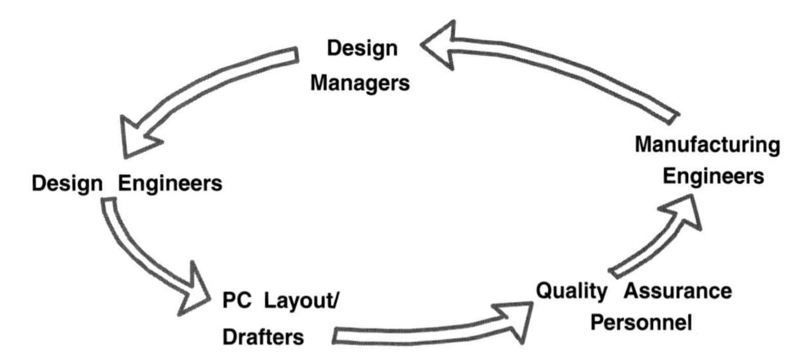

FIGURE 19.7 The benefits of metrics as a common design language.

Non-metrics

- **Opinion** Opinion, albeit from an expert, is applied after or concurrent with design (manufacturing engineering inspection process).
- **No opinions** No attempt to inspect or improve the design is done during the specification, partitioning, or design stage (over the wall process).

Metrics also establish a common language that links manufacturing to design. The producibility scores form a nonopinionated basis allowing a team approach that results in a quality, cost-competitive product (Fig. 19.7).

The strategy in applying these measures is shown in Fig. 19.8. The analysis process is unique to every individual and company, but certain conditions have to be met and considered if the product is going to be successful. If the score meets producibility requirements,

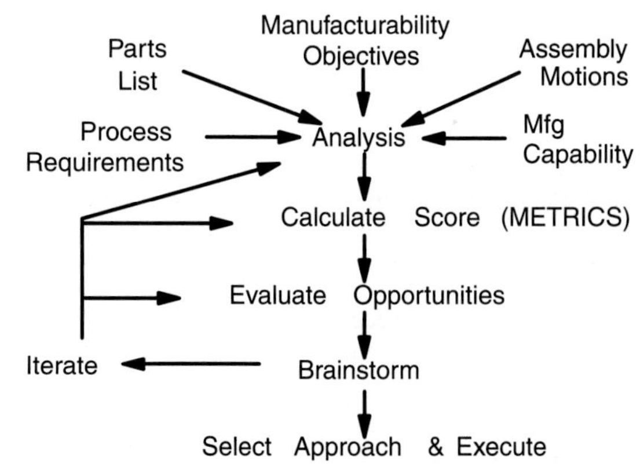

FIGURE 19.8 A process using measures and metrics to obtain a producible product.

then select this approach; if not, then evaluate other opportunities and repeat the process. In the rest of this chapter, measures and metrics are introduced that provide insight for layout, fabrication, and assembly planning.

19.3.4 Non-Metrics

It is always preferable to have metrics when discussing producibility. But if metrics are not available, then the opinions of experts are better than no opinions at all. The problem with opinions is that they are difficult to defend and explain, and when used in relation to producibility, many times they vary with each person. Sometimes, the opinion process is implemented with good intentions, by having experienced production experts review a new design. This is the *expert opinion process* and although it is sometimes successful, it is difficult to replicate and often results in building barriers between manufacturing and design. That is why the figure of merit process is so popular. For a small amount of work by experts, it produces a scoring procedure that can be used and understood by all.

19.3.5 Figure of Merit (FOM) Metric

Metrics are the preferred measures for design planning, but their availability for predicting producibility is often limited. Metrics also can take many months to develop and the amount of experimentation may make them costly. The measure that is much more cost-effective and quicker to develop is the figure of merit. The FOM is the result of one or two days of work by a group of design and manufacturing experts. The process consists of eight steps:

1. Define or identify the new measure to be developed.
2. Determine why the measure was selected. Ensure the relevance of the measure to communication.
3. Survey the customers. Identify what the measure is that customers need to have communicated.
4. Identify needs and expectations and collect data.
5. Brainstorm contributing factors and variables.
6. Determine the major contributors and normalize scores. Use multivote, paired-ranking, ranking voting, or Pareto techniques to verify data if available. These are the coefficients (Cx) of the equation.
7. Construct FOM factor weightings (FWx). Fill in FOM table values for 1, 25, 50, 75, and 100.
8. Construct a linear equation model (coefficient score × FOM weightings).

19.3.6 Figure of Merit Linear Equation

This FOM procedure uses classical TQM techniques to brainstorm, rank, and formulate an equation that will score producibility, assemblability, or any other measure that can be used in design planning. The two factors used in the producibility score are made up of the (1) the coefficient, Cx, and (2) the factor weighting, FWx.

19.3.6.1 Coefficient, Cx. The coefficients in the producibility score are the result of brainstorming all the possible contributors to producibility that can affect the product (Fig. 19.9a). These are grouped into common ideas or factors by such techniques as clouds of affinity or Kay-Jay (Fig. 19.9b). These factors are ranked by voting or other Pareto techniques such as

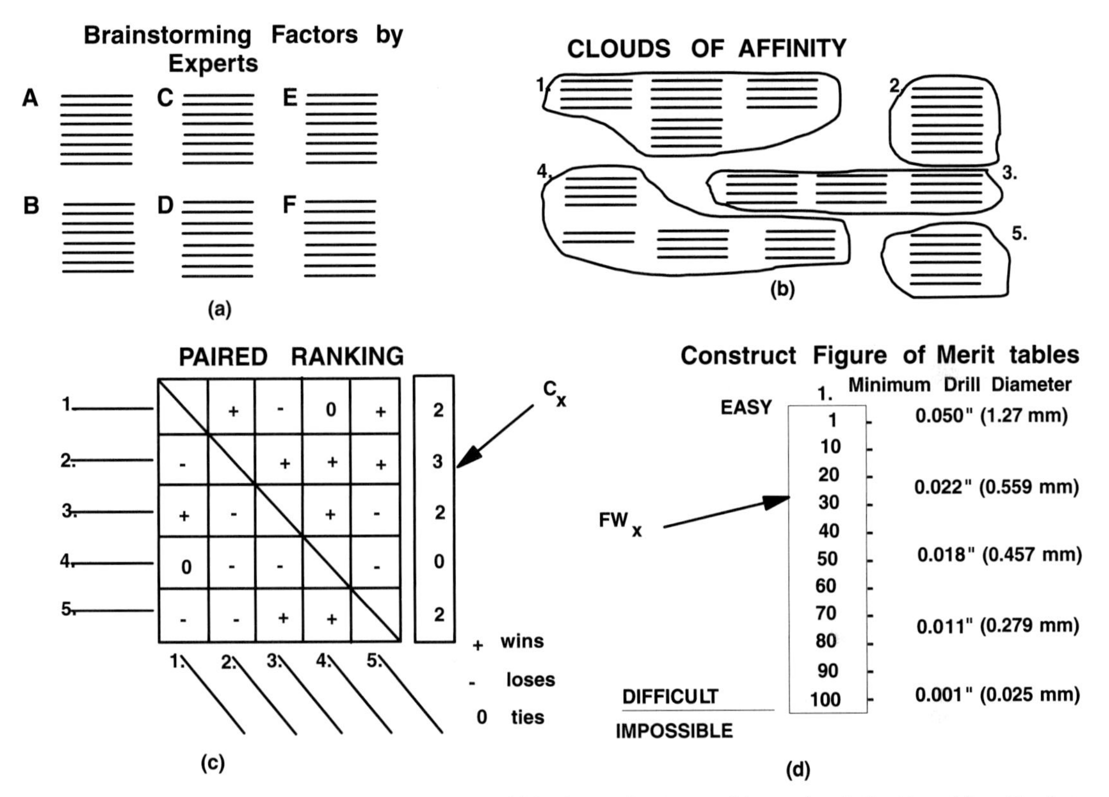

FIGURE 19.9 Elements of the figure of merit process: (a) brainstorming factors; (b) grouping similar ideas; (c) ranking factors; and (d) assigning values to factors.

paired ranking, as seen in Fig 19.9c. Whatever the method of voting, the values are normalized by dividing by the smallest nonzero value. The resulting voting scores form the coefficients, Cx. Those factors with no votes are zeros and drop out of consideration.

19.3.6.2 Factor Weightings, FWx. Each factor that emerges from the ranking process is calibrated by assigning values from one (1) to one hundred (100), as seen in Fig. 19.9d. The 1 factors are easy to manufacture, and the 100 factors are impossible today but may only be very difficult in a few years.

The resulting scoring equation will look like and be used as shown in Eq. 19.1.

$$\text{SCORE} = (C1)\,(FW1) + (C2)\,(FW2) + (C3)\,(FW3) + (Cn)\,(FWn) + \cdots \qquad (19.1)$$

where Cx = coefficients based on ranking
FWx = factor weightings of assigned values (1 – 100)

For example, assume that the producibility of a bare PWB may be scored with the preceding equation if the FOM process established the following factors:

- Size of the substrate: C1 = 1.5
- Number of drilled holes: C2 = 3.0
- Minimum trace width: C3 = 4.0

Now assume that the proposed PWB design specifies the following:

- Size of the substrate: FW1 = 36
- Number of drilled holes: FW2 = 18
- Minimum trace width: FW3 = 31

The producibility SCORE would equal:

$$232 = 54 + 54 + 124 = 1.5 \times 36 + 3.0 \times 18 + 4.0 \times 31$$

The SCORE can be used if it is calibrated based on the prior history of this type of product. Prior products are scored with the FOM linear equation (Eq. 19.1). Those products that went into production smoothly and presented minimum problems determine the minimum SCORE that should be sought. The SCORE from products that were a problem, had to be reengineered, or presented delays in introduction determines the SCORE that the design should exceed if problems occur and producibility is to be obtained.

19.4 LAYOUT TRADE-OFF PLANNING

Predicting density and selecting design rules are two of the primary planning activities for layout. The actual layout of a PWB is covered in Chapters 13, 14, and 15. The selection of design rules not only affects circuit routing but profoundly affects fabrication, assembly, and testing.

19.4.1 Balancing the Density Equation

With the need for more parts on an assembly and the trend to make things smaller for portability or faster speeds, the design process is a challenging one. The process is one of balancing the density equation with considerations for certain boundary conditions such as electrical and thermal performance. Unfortunately, many designers do not realize that there is a mathematical process to determine the routing rules of a printed circuit. Let me briefly explain. The density equation, as seen in Eq. 19.2 and Fig. 19.10, has two parts: the left side, which is the component wiring demand, and the right side, which is the substrate wiring capability.

$$\text{Component PWB Wiring Demand} < \text{PWBs Design Rules \& Construction Wiring Capabilities} \tag{19.2}$$

where PWB Wiring Demand = total connection length required to connect all the parts in a circuit

PWB Wiring Capability = substrate wiring length available to connect all the components

Four conditions can exist between wiring demand and substrate capability:

- **Wiring Demand > Substrate Capability** If the substrate capacity is not equal to the demand, the design can never be finished. There is not enough room for either traces or vias. To correct this, either the substrate has to be bigger or components have to be removed.
- **Wiring Demand = Substrate Capability** Although theoretically optimum, this condition leaves no room for variability and it would take an unacceptable amount of time to complete the design.

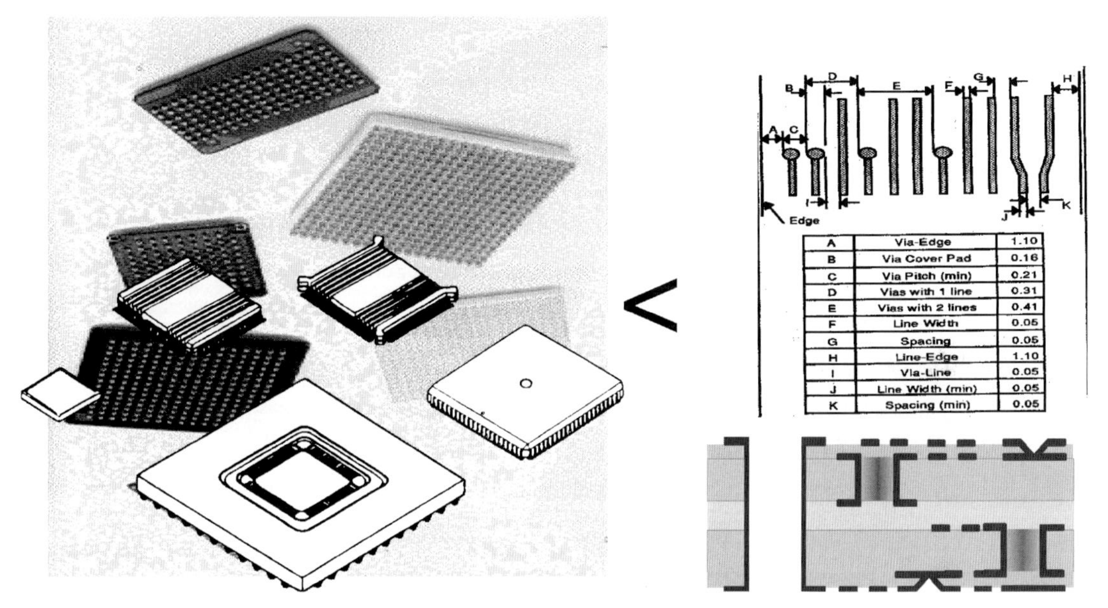

FIGURE 19.10 Balancing the density equations to achieve an optimal layout.

- **Wiring Demand < Substrate Capability** This condition should be your goal. There should be enough extra capacity to complete the design on time with only a minimum of over specification and costs.
- **Wiring Demand << Substrate Capability** This is the condition that usually prevails. By PCB layout, the schedule is tight and timing is all important. Many choose tighter traces or extra layers to help shorten the layout time. This increases the manufacturing costs 15 to 50 percent higher than is necessary, which is sometimes called the sandbag approach. It is unfortunate, as the preceding models would help to create a better-planned environment.

19.4.2 Wiring Demand (W_d)

Wiring demand is the total connection length (in inches) required to connect all the parts in a circuit. If the design specifies an assembly size (in square inches), then a wiring density in inches per square inch or centimeter per square centimeter is created. Models early in the design planning process can estimate the wiring demand. Three cases can control the maximum wiring demand:

- The wiring required to break out from a component such as a flip chip or chip scale package
- The wiring created by two or more components tightly linked, such as a central processing unit (CPU) and its cache or a digital signal processor (DSP) and its input/output (I/O) control
- The wiring demanded by all integrated circuits and discretes collectively

Models are available to calculate the component wiring demand for all three cases; see Sec. 19.4.3. Since it is not always easy to know which case controls a particular design, it is usually to calculate all three cases to see which one is the most demanding and thus controls the layout.

Wiring demand is defined by Eq. 19.3.

$$W_d = W_c \times \varepsilon \text{(in cm/square cm or in./square in.)} \qquad (19.3)^3$$

where W_d = Wiring Demand
W_c = Wiring Capacity
ε = PWB Layout Efficiency (determined in Sec. 19.4.3)

19.4.3 Wiring Capacity (W_c)

Substrate wiring capacity is the wiring length available to connect all the components. It is determined by two factors:

- **Design rules** These rules specify the traces, spaces, via lands, keepouts, and such that make up the surface of the substrate.
- **Structure** This factor determines the number of signal layers and the combination of through and buried vias that permit interconnection between layers and the complex blind, stacked, and variable depth vias available in high-density interconnection (HDI) technologies.

These two factors determine the maximum wiring available on the substrate. To figure the wiring available to meet the demand, multiply the maximum wiring by the layout efficiency. The data are straightforward except for layout efficiency. Layout efficiency expresses what percentage of wiring capacity can be used in the design. Equation 19.4 shows the formula for determining wiring capacity for each signal layer. The total substrate capacity is the sum of all the signal layers.

$$W_c = T \times L/G \quad \text{(in cm/square cm or in./square in.)} \qquad (19.4)$$

where T = number of traces per wiring channel or distance between two via pads
L = number of signal layers
G = wiring channel width or length between centers of the via pads

19.4.4 Layout Efficiency

Layout efficiency is the percentage of capacity from design rules and structure that a designer can deliver on the board. Layout efficiency is the ratio of the actual wiring density that it takes to wire up a schematic versus the maximum wiring density or W_d divided by W_c. Layout efficiencies, for ease of calculations, are typically assumed to be 50 percent. Table 19.2 provides a more detailed selection of efficiencies.

19.4.5 Selecting Design Rules

To calculate a potential set of design rules and signal layers, first the wiring demand (W_d) should be calculated. Wiring models help accomplish this.

TABLE 19.2 Typical Layout Efficiencies

Design scenario	Conditions	Efficiency* (%)
Through hole, rigid	Gridded CAD	6–12
Surface mount/mixed	W/wo back side passives, gridless CAD	8–15
Surface mount/mixed	W/back side actives, gridded CAD	9–18
Surface mount only	W/wo back side passives, gridless CAD	Up to 20
Surface mount/mixed	1-sided blind vias, gridless CAD	Up to 25%
Surface mount/mixed	2-sided blind vias, gridless CAD	Up to 30
Built-up technologies	2-sided micro-blind vias, gridless CAD	Up to 50%

* Determined from analysis of printed circuit designs (actual wiring capacity from the CAD system divided by maximum wiring capacity (Eq. 19.4).

19.4.5.1 Wiring Demand Models. Seven wiring models are reported in the literature, but the first three are used commonly. The three wiring models include:

- Coors, Anderson & Seward Statistical Wiring Length[4]
- Toshiba Technology Map[5]
- Hewlett Packard (HP) Design Density Index[6]

The other four wiring models include:

- Equivalent ICs per square inch[7]
- Rent's Rule[8]
- Section Crossing[9]
- Geometric Analysis[10]

19.4.5.1.1 Coors, Anderson, & Seward Statistical Wiring Length. This wiring demand model is based on a stochastic model of wiring involving all terminals. The probable wire length is calculated based on the distance of a second terminal and the spatial geometry of all other terminals. This is the most recently determined wiring model and represents the most practical approximation of surface mounting technology. Eq. 19.5 presents the mathematical model that results.

$$d = D * N_i/A \text{ (in. per sq. in.)} \tag{19.5}$$

where
D = ave. interconnection distance (in.)
$D = E(x)*G$
$E(x)$ = expectation of occurrence
G = pad placement grid (in.)
N_i = total number of interconnections
A = routing area (sq. in.)

Equation 19.6 for E(9x) is the

$$E(x) = \frac{1}{a}\,\frac{((S-T)(S\,a-2))e\hat{\ }aS + S(2-(S-T)a)e\hat{\ }a(S-T)-2T}{(S-T)e\hat{\ }aS - Se\hat{\ }a(S-T)+T} \tag{19.6}$$

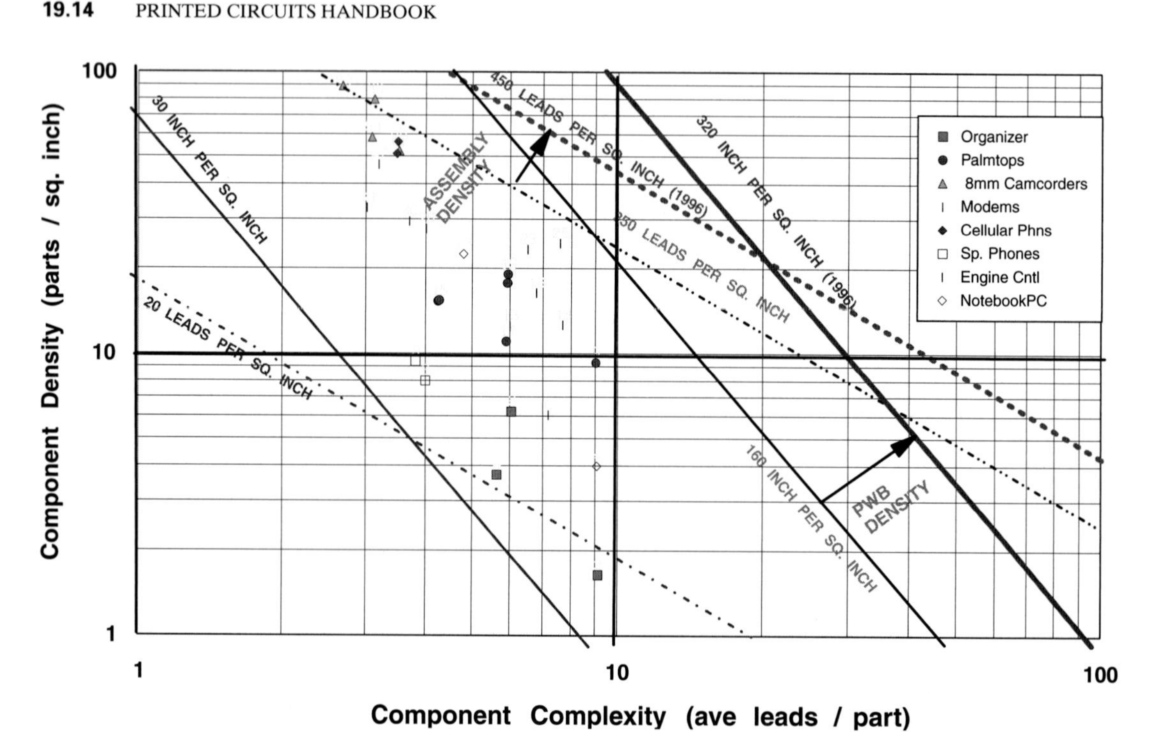

FIGURE 19.11 Packaging technology map.

where $S = M + N$
$T = (M^2 + N^2)^{.5}$
$a = \ln \alpha$
α = empirically derived constant = 0.94
M = board width of grid point = (width/G) +1
N = board length of grid point = (length/G) +1
$N_i = 2*Nt/3$

19.4.5.1.2 Toshiba's Technology Map. The packaging technology map is a simple technique to predict a PWB, Chip-On-Board, or MCM-L wiring demand and its assembly complexity. By plotting components per square inch (or components per square centimeter) against average leads per component on a Log-log graph (see Fig. 19.11), you can calculate the wiring demand W_D in inches per square inch (or centimeters per square centimeters) and assembly complexity in leads per square inch (or leads per square centimeter). Eqs. 19.7 and 19.8 show the equations for these two metrics.

$$\text{Wiring Demand } W_D = \beta \times (\text{comp})^{0.5} \times (\text{leads}) \tag{19.7}$$

where β = wiring coefficient (typically 3.5 but can vary from 2.5–4.0 on average; notes/net is a good approximation)
comp = components per board area in sq. in. or sq. cm.
leads = average leads (connections) per component

$$\text{Assembly Complexity} = (\text{comp}) \times (\text{leads}) \tag{19.8}$$

comp = components per board area in sq. in. or sq. cm.
leads = average leads (connections) per component

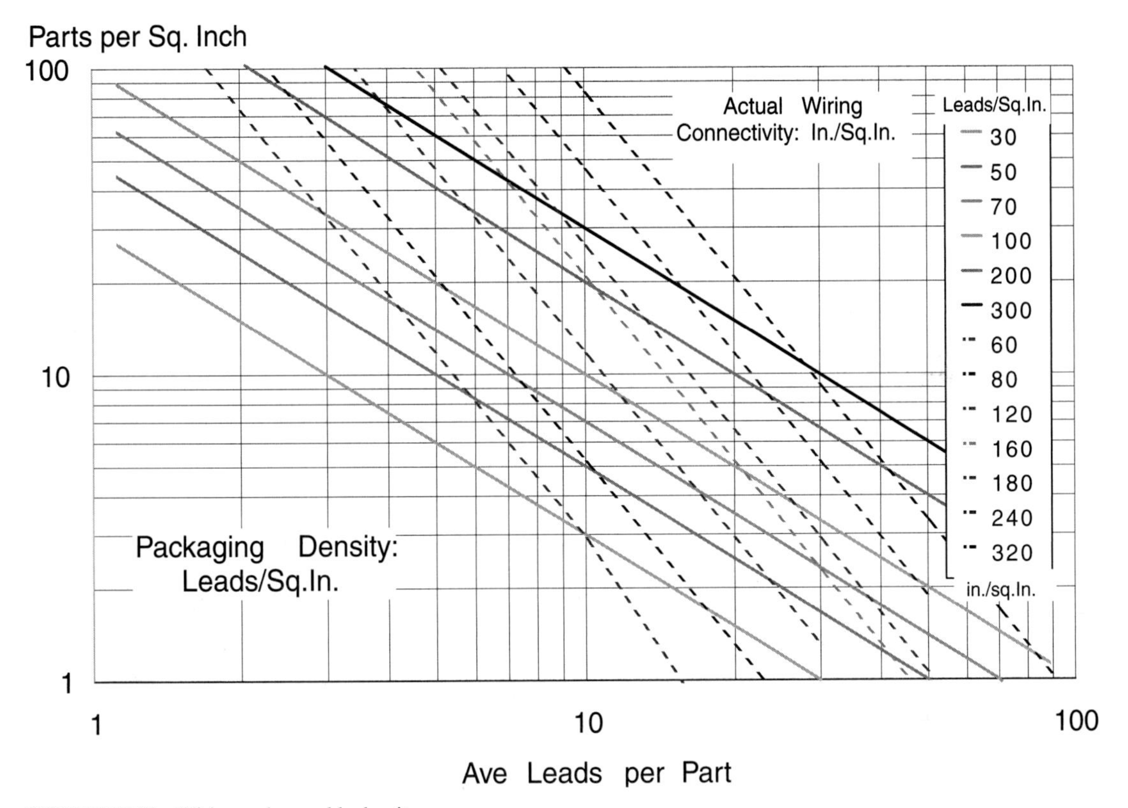

FIGURE 19.12 Wiring and assembly density.

Using these two equations, Fig. 19.12 shows lines of constant wiring demand (in cm. per sq. cm or in. per sq. in.) and assembly complexity (in leads per sq. cm. or sq. in.) that can be plotted on this chart (see Fig. 19.11).

19.4.5.1.3 HP's Design Density Index. Another metric is called the Design Density Index (DDI). It is a correlation of the actual design rules for a PWB compared to the DDI metric. Equation 19.9 gives DDI, and Fig. 19.13 shows a typical calibration chart.

$$\text{DDI} = 13.6 \times (\text{EIC/board area})^{\wedge}1.53 \qquad (19.9)$$

where EIC (equivalent integrated circuits) = total component leads/16
board area = top surface area of a PCB (sq. inch.)

The chart shown in Fig. 9.13 gives a good visual record in PWB layout of a company's efficiency. As various PC boards are charted, their DDIs form a distribution. This distribution is a form of layout efficiency (ε) since at the bottom of the distribution, more EICs are connected than at the top of the distribution.

19.4.5.1.4 Density of Equivalent ICs. EIC per unit area has been a traditional measure of density since the introduction of CAD systems in the early 1970s. A simple measure of the number of electrical connections required per unit area of the board, it remains in use with

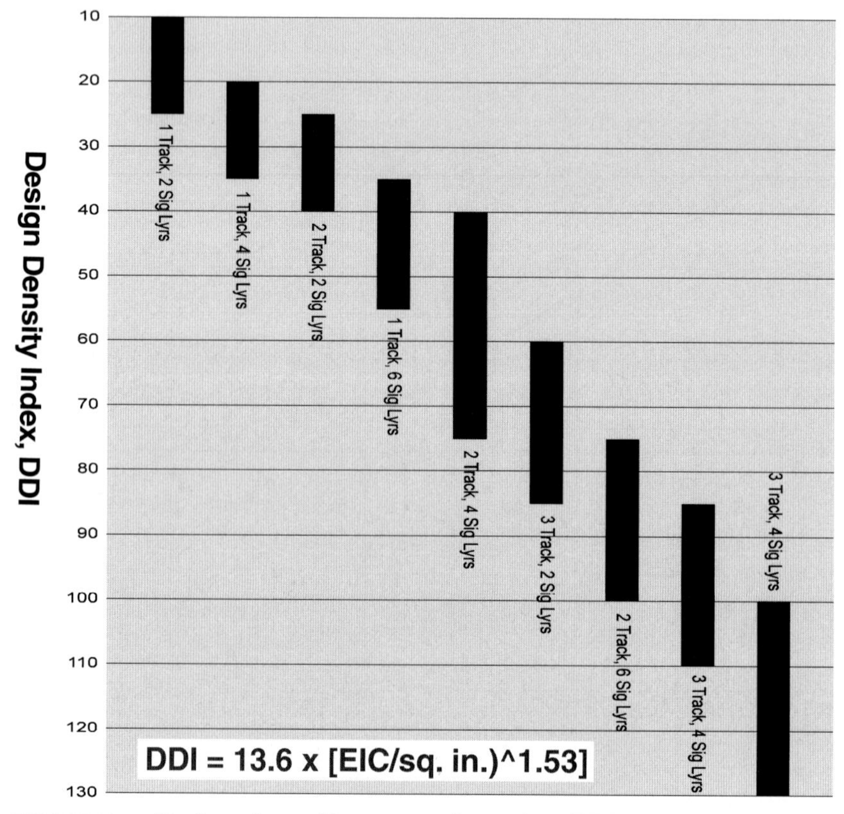

FIGURE 19.13 Design rules and layer count for various DDIs.

surface mounting technology and is usually referred to as EIC density. An EIC is the total number of leads of the components divided by 14 or 16, the old number of pins on a dual-inline package (DIP). Many people also use 20 as a divisor. Equation 19.10 defines EIC density mathematically in EIC/sq. in.

$$\text{EIC Density} = \text{Connections}/16/\text{Board Area} \tag{19.10}$$

where Connections = total component leads
Board Area = top surface board area (in sq. cm. or sq. in.)

19.4.6 Typical Example of Wiring Demand Calculation

As an example, take a typical consumer electronic board with these characteristics:

- Design: consumer PCB, all through hole components
- Components: 86

- Leads: 1,540
- Size: 19.6 in. × 19.5 in. = 57 sq. in.
- EIC/sq. in. = 110/57 sq. in. = 1.93 (using Eq. 19.10)
- DDI = 13.6 × (1.93) $^{1.53}$ = 319.7 ~

Figure19.13 advises two tracks on two signal layers (using Eq. 19.9).

19.5 PWB FABRICATION TRADE-OFF PLANNING

The metrics for PWB and chip-on-board (COB) fabrication involve with trade-offs between the performance objectives and the PWB price. Calculating prices requires the PCB characteristics and manufacturing yield. Manufacturing yield requires producibility estimates. Three items are required to predict PCB prices:

- Fabrication complexity matrix
- Prediction of producibility and first-pass yield
- Relative price as a function of a price index

19.5.1 Fabrication Complexity Matrix

The fabrication complexity matrix is supplied by a PWB fabricator. It relates the various design choices on a PWB to design points. These points are based on the actual prices that a fabricator will charge for these features. They are calculated by dividing the actual costs and the values by the smallest nonzero amount. Typical factors that fabricators can use to price a printed wiring board include the following:

- Size of the board and number that fit on a panel
- Number of layers
- Material of construction
- Trace and space widths
- Total number of holes
- Smallest hole diameter
- Solder mask and component legends
- Final metalization or finish
- Gold-plated edge connectors
- Factors specific to the design

A typical fabrication complexity matrix would look like Table 19.3. This is not a complete matrix, but does show the design factors and the design points assigned to each.

19.5.2 Predicting Producibility

The simple truth about printed circuit boards, multichip modules, and hybrid circuits is that the design factors such as those listed previously can have a cumulative effect on manufacturing yield. These factors all affect producibility. Specifications can be selected that individually may not adversely affect yields but cumulatively can significantly reduce yields. A simple

TABLE 19.3 Example of Fabrication Complexity Matrix

Factors	Pts.	Highest	Pts.	High middle	Pts.	Low middle	Pts.	Lowest
No. of layers	12	8	8	6	4	4	1	2
Trace width (mils)	8	4	5	5–6	3	7–8	1	10
No. of holes	10	5000–8000	5	3000–5000	3	10000–3000	1	>10000

algorithm is available that collects these factors into a single metric, in this case called the complexity index (CI). It is given in Eq. 19.11.

$$\text{Complexity Index} = \frac{(\text{Area})\,(\text{Holes/unit area})^{\wedge 2}\ (\text{no. Layers})^{\wedge 3}}{(\text{Min. trace width})\,(\text{Min.annular ring})\,(\text{Min. hole dia.})} \tag{19.11}$$

where Area = top area of the substrate to be designed
Holes = total number of drilled holes, blind, buried, and through
Holes/unit area = holes divided by board Area
Trace width = the minimum trace width on the substrate
Layers = the total number of layers in the substrate
Annular ring = $^1/_2$ the difference between the via land and the hole dia.
Hole dia. = finished hole size

where Area, number of holes, minimum trace width, number of layers, and minimum tolerance (absolute number) are the factors of the board being designed

19.5.2.1 First-Pass Yield. The first-pass yield equation derives from the Wiebel probability failure equations.[11] Equation 19.12 is of a more general form of the equation typically used to predict ASIC yields by defect density.

$$\text{FPY\%} = \frac{100}{\text{Exp}[(\log \text{CI/A})^{\wedge}\text{B}]} \tag{19.12}$$

where FPY= first-pass yield
CI = complexity index
A,B = constants

To determine the constants A and B in Eq. 19.12, a fabricator will need to characterize their manufacturing process. Selecting a number of printed circuits currently being produced that have various complexity indexes does this, ideally some low, medium, and high. The first-pass yield (at electrical test without repair) of these printed circuits for several production runs is recorded. Any statistical software program[12] that has a model-based regression analysis can now determine A and B from the model (Eq. 19.13).

$$\text{FPY} = f[x] = 100 \div \text{EXP}\{\text{LOG}(\text{complexity} \div \text{PARM}[1])^{\wedge}\text{PARM}[2]\} \tag{19.13}$$

where PARM[1] = A
PARM[2] = B

The first-pass yield will follow the examples in Fig. 19.14. Constant A determines the slope of the inflection of the yield curve, whereas constant B determines the x-axis point of the inflection.

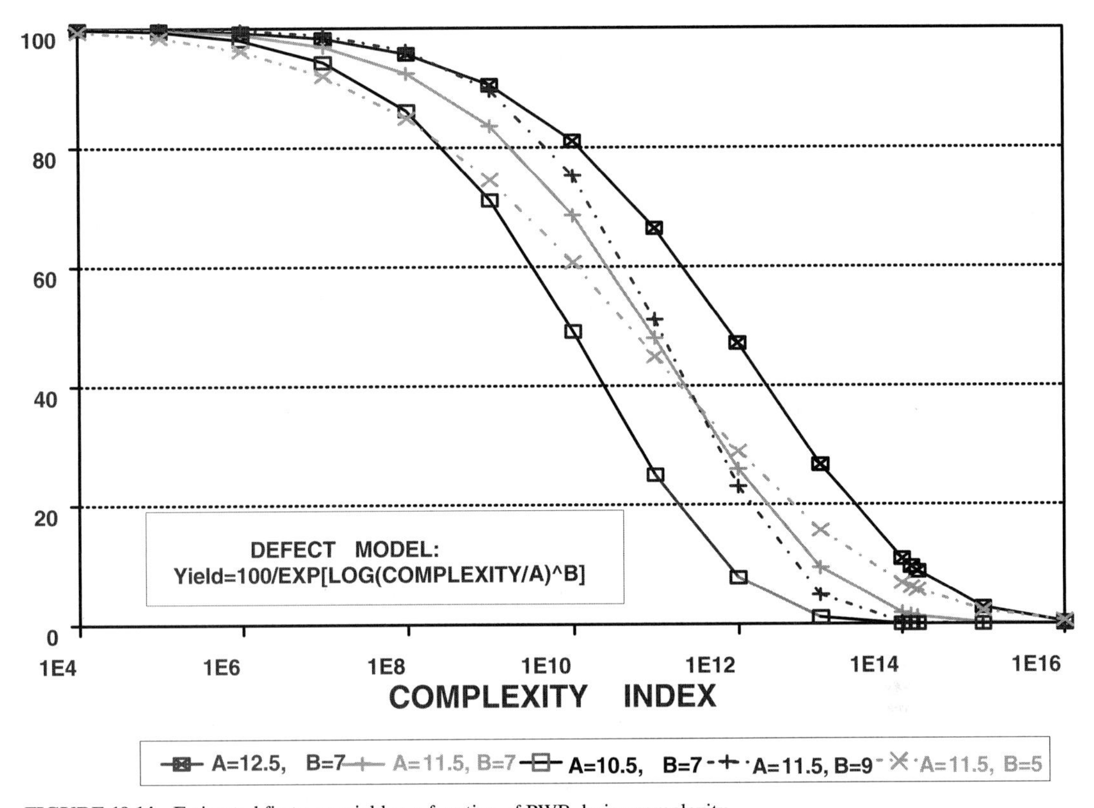

FIGURE 19.14 Estimated first-pass yield as a function of PWB design complexity.

Alternatively, any spreadsheet can be used to determine constants A and B. The [REGR] function in a spreadsheet like Excel™ or Lotus 1-2-3™ is used. The [REGR] function is defined as: (=LINEST(known_y's,known_x's,TRUE,TRUE). To use this function, you must first put the FPY function into the form y = Ax + B. This is done by creating two columns: complexity index (which we will call X1) and yield. A third column is created for {log[log(X1)]}, whereas a fourth column is created for {log[ln(-yield/100)]}. Provide the regression function with column 4 as known_ys and column 3 as known_xs. The regression function will return 10 values; FIT (slope & int.), sig-M (slope & int.), r2, sig-B(slope & int.), F,df (slope & int.), and reg sum sq (slope & int.). The constant B is equal to the FIT (slope) and the constant A is 10^[–FIT(int.)/FIT(slope)]. (Remember, to calculate an array, follow these steps: highlight the array on the spreadsheet; type the array formula, making sure that the cursor is in the edit bar; then press CTRL + SHIFT + ENTER.)

19.5.2.2 Yield Calculation Steps. To calculate the first-pass yield coefficients, there are six steps:

1. Collect design attributes of 10 to 15 currently running boards with various size and layers (see Table 19.4).
2. Collect the first-pass yield information for these selected boards, at least 10 runs (see Table 19.5).

TABLE 19.4 Example of PCB Part Information and Calculated Complexity Index

P/N	PCS SIZE Length (in.)	PCS SIZE Width (in.)	Hole quantity	Hole density	Layers	Thickness (mil.)	Min. Trace width (mil.)	Min. Annular ring (mil.)	Min. Hole size (mil.)	Aspect ratio	Complexity x1
#1	11.500	9.210	4842	45.7	6	82.7	6	4	12	6.89	1.14E+06
#2	11.500	8.900	3217	31.4	8	62.0	5	5	8	7.75	2.01E+06
#3	6.050	1.090	379	57.5	4	46.2	6	5	12	3.85	1.49E+04
#4	5.210	4.330	1538	68.2	6	72.4	6	5	10	7.24	5.46E+05
#5	6.050	1.410	868	101.8	6	50.0	6	5	14	3.57	1.62E+05
#6	7.240	11.460	5274	63.6	8	61.9	5	5	14	4.42	2.17E+06
#7	11.460	6.510	6015	80.6	8	61.9	6	5	8	7.74	8.00E+06
#8	9.600	10.370	4970	49.9	12	47.0	5	5	13	3.62	4.77E+06
#9	11.790	9.490	6034	53.9	10	62.0	5	5	12	5.17	5.60E+06
#10	12.780	4.200	4038	75.2	12	62.0	6	5	10	6.20	1.08E+07
#11	6.050	1.455	554	62.9	6	50.0	6	5	14	3.57	6.40E+04
#12	5.380	5.670	1118	36.7	4	55.1	7	5	18	3.06	1.27E+04
#13	1.250	2.660	220	66.2	4	32.0	7	8	14	2.29	2.72E+03

3. Calculate the board complexity index and average yield.
4. Prepare a spreadsheet of transformed CI (x1) and Yield (Y) (see Table19.6)
5. Calculate regression coefficients (see Table 19.7).
6. Calculate A and B from regression fits.

19.5.2.3 PCB Part Information. Collect design attributes of 10 to 15 currently running boards with various size and layers (see Table 19.3).

19.5.2.4 PCB Production Yield Data. Collect the first-pass yield information for these selected boards, for at least 10 runs (see Table 19.5)

TABLE 19.5 Example of PCB Production Yields from 10 Runs

P/N Run	First pass yield (at Electrical test) 1	2	3	4	5	6	7	8	9	10	Average
#1	92.6	84.9	86.9	82.9	90.0	95.2	90.1	92.4	93.0	92.0	90.0
#2	85.6	85.7	94.2	86.2	86.6	86.7	86.6	89.7	95.6	85.6	88.2
#3	95.6	93.9	93.6	97.8	97.3	95.1	94.2	96.2	96.9	91.6	95.2
#4	92.3	93.3	95.0	92.6	94.5	95.5	93.5	89.7	88.8	89.8	92.5
#5	94.8	97.6	96.4	96.6	94.1	93.2	94.3	93.2	91.8	91.2	94.3
#6	86.8	85.3	88.1	90.8	86.5	87.0	88.0	87.2	86.5	88.9	87.5
#7	90.3	80.9	86.3	87.9	87.1	92.7	87.3	87.6	82.4	88.9	87.1
#8	89.3	87.0	86.6	83.5	91.6	90.1	87.5	86.2	88.1	87.3	87.7
#9	87.2	87.9	86.6	85.7	89.4	87.3	90.6	86.5	85.9	82.9	87.0
#10	86.6	88.1	82.6	84.4	88.0	87.0	79.0	87.0	88.0	80.0	85.1
#11	92.7	92.7	97.2	95.1	96.9	94.3	93.4	95.3	91.6	96.6	94.6
#12	93.8	95.3	98.2	96.3	94.1	95.9	94.2	92.3	94.8	95.6	95.0
#13	99.2	98.6	99.1	98.7	97.9	99.6	99.6	99.1	99.0	99.3	99.0

TABLE 19.6 Example of Excel™ Transformation Setup for Complexity and Yield Data

Sample	Complexity X1	YIELD Y	Log-log (x1)	Log in (−Y/100)	Log-log fit all data	Error all data	Log-log fit avg. only	Error avg. only
#1-1	1144136	92.6	0.78	−1.11	90.6	−2.0	90.1	−2.5
#2-1	2006116	85.6	0.80	−0.81	89.4	3.8	88.9	3.3
#3-1	14909	95.6	0.62	−1.35	97.0	1.4	96.7	1.1
#4-1	546435	92.3	0.76	−1.10	92.0	−0.3	91.5	−0.7
#5-1	162158	94.8	0.72	−1.27	94.1	−0.7	93.6	−1.1
#6-1	2167611	86.8	0.80	−0.85	89.2	2.5	88.7	1.9
#7-1	8002482	90.3	0.84	−0.99	86.1	−4.2	85.6	−4.7
#8-1	4773612	89.3	0.82	−0.95	87.4	−1.9	86.9	−2.4
#9-1	5604279	87.2	0.83	−0.86	87.0	−0.2	86.5	−0.7
#10-1	10848424	86.6	0.85	−0.84	85.3	−1.3	84.8	−1.8
#11-1	64014	92.7	0.68	−1.12	95.4	2.7	95.0	2.3
#12-1	12737	93.8	0.61	−1.19	97.2	3.4	96.9	3.1
#13-1	2716	99.2	0.54	−2.10	98.4	−0.8	98.2	−1.0
#1-2	1144136	84.9	0.78	−0.78	90.6	5.7	90.1	5.2
#2-2	2006116	85.7	0.80	−0.81	89.4	3.7	88.9	3.2
#3-2	14909	93.9	0.62	−1.20	97.0	3.1	96.7	2.8
#4-2	546435	93.3	0.76	−1.16	92.0	−1.3	91.5	−1.7
#5-2	162158	97.6	0.72	−1.62	94.1	−3.5	93.6	−4.0
***	***	***	***	***	***	***	***	***
#11-13	64014	96.6	0.68	−1.46	95.4	−1.2	95.0	−1.6
#12-13	12737	95.6	0.61	−1.35	97.2	1.6	96.9	1.3
#13-13	2716	99.3	0.54	−2.13	98.4	−0.9	98.2	−1.1
				AVG Error	—	**0.4**	—	**0.0**
				STD DEV	—	**4.4**	—	**4.5**

19.5.2.5 Calculating the Board Complexity Index and Average Yield: Regression Analysis Methodology. To determine the constants A and B in Eq. 19.12, you can use any statistical software program that has a model-based regression analysis. The model is shown in Eq. 19.13. Table 19.6 shows an example of an Excel™ spreadsheet setup, and Table 19.7 shows the regression results.

TABLE 19.7 Example of Excel™ Regression Results

	All data		AVG only	
Log-log fit	slope	int.	slope	int.
FIT	3.17	−3.48	3.06	−3.38
Sig-M	0.18	0.13	0.34	0.25
R2,sig-B	0.73	0.19	0.88	0.12
F, df	312.60	115	81.75	11
Reg sum sq	11.01	4.05	1.14	0.15
B	3.17		3.06	
A	12.57		12.66	

TABLE 19.8 Example of One Company's Fabrication Complexity Matrix

Fabrication factors	Pts.	Highest	Pts.	High middle	Pts.	Low middle	Pts.	Lowest
Material of construction	147	Polyimide	88	BT	49	FR-4	40	CEM III
No. of layers	196	12 layers	137	8 layers	89	4 layers	36	2–sided
No. of holes/ panel	270	<20,001	180	10,001–20,000	90	3001 –10000	27	>3000
Min. Trace/spacing	25	2 mil.	10	3–4 mil.	6	5–6 mil.	1	<=6 mil.
Gold tabs	48	3 Sides	32	2 Sides	16	1 Side	0	None
Annular ring	30	>2 mil.	21	2–4 mil.		4–6 mil.	1	<6 mil.
Solder mask	25	2 S Dry Film	17	2 S LPI	7	1 S LPI	5	Screened
Metalization	75	Electroless Ni/Au	69	Selective lead-free solder coat	46	Immersion silver	29	SMOBC/ organic coat
Min. hole dia.	166	<= 8 mil.	84	9–12 mil.	69	13–20 mil.	5	<20 mil.
Controlled impedance tolerance	105	+/–5 %	62	+/–10 %	30	+/–20 %	0	None

Points are per PANEL For board: Divide by no. per PANEL

19.5.3 Example of a Complete PWB Complexity Matrix

This section presents an example of how one company approached this planning process as part of its PWB design for manufacturing program.

19.5.3.1 PWB Fabrication Complexity Matrix (FCM). The PWB fabrication complexity matrix that this company developed is shown in Table 19.8. This FCM is built on a "per panel" basis of 18 in. by 24 in. rather than a "per board" basis. Also, volume is assumed to be in a preset amount.

19.5.3.2 PWB Complexity. Figure 19.14 shows this company's first-pass yield. The curve with A = 11.5 and B = 9.0 was current for six months. Price index is the total points from the complexity matrix divided by the first-pass yield.

19.5.3.3 Relative Costs. The price index data for this company is shown in Fig. 19.15. The price index (PI) can vary from 150, which corresponds to a 70 percent price reduction, to a PI of 1,000 which corresponds to a 275 percent increase in price.

19.5.3.4 PWB Fabrication Example. Let's continue with the consumer electronics board example from Sec. 19.4.5. Table 19.9 lists the initial design characteristics of the printed circuit and the resultant total design points. Table 19.10 shows the calculation of the complexity index, estimated first-pass yield, and resultant price index and price adjustment.

The wiring demand indicates that a 0.007 in trace and a 0.008 in spacing(two track) for a 0.100 in grid (channel) represent more density than is required (see Sec. 19.4.5). A one-track wiring on two signal layers could achieve the required wiring density or a 0.012 in trace

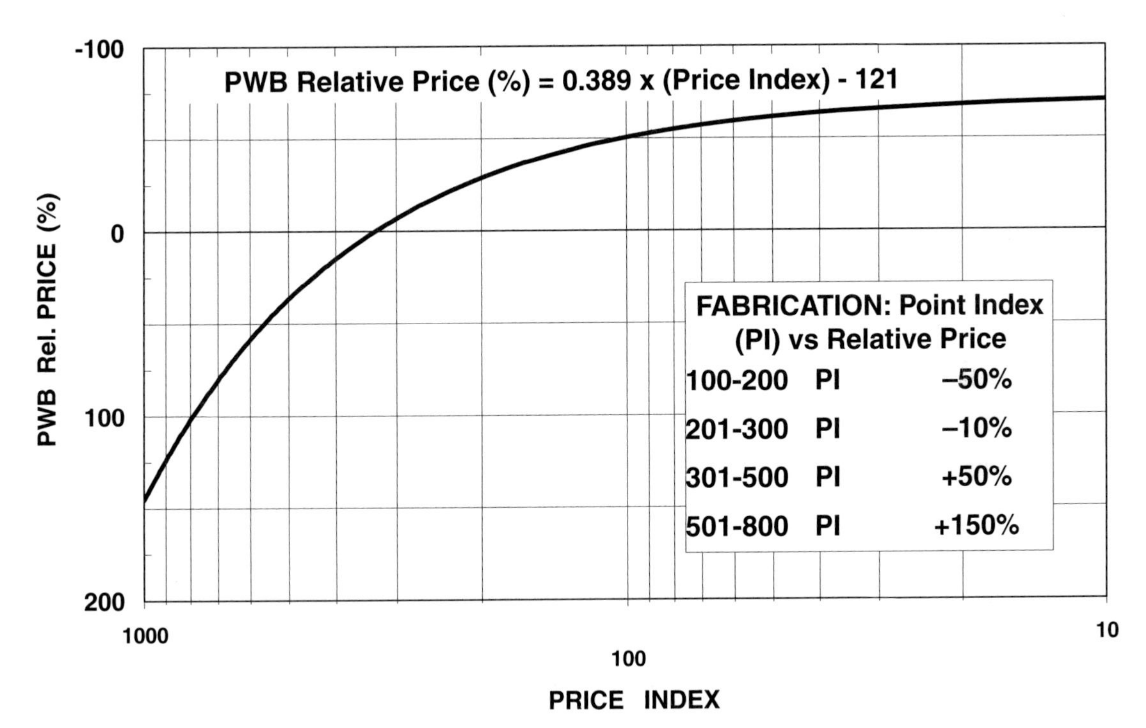

FIGURE 19.15 PWB relative price as a function of the price index.

with a 0.013 in spacing. The size required a panel layout, as shown in Fig. 19.16a, of six per panel. A reduction of 14 sq. in in the board size would allow eight boards on a 18 in × 24 in panel (see Fig. 19.16b).

A simple Gate Array ASIC was employed to reduce the small-scale ICs until 14 sq. in of space was freed up. The final optimized PCB had the fabrication factors seen in Table 19.11. The boards were eight up on a panel (Fig. 19.16b) and had a higher producibility by a 18 percent reduction in its complexity and an overall price reduction of 24.7 percent.

TABLE 19.9 One Company's Fabrication Complexity Makeup

Fabrication factors	PWB design #1	Points	FCM
Mat'l of construction	FR-4	49	FR-4
No. of layers	six	137	Six layers
No. of holes	1655 × 6 = 9930	90	3001–10,000
Min. trace wd.	0.008 in	6	6–8 mil
Gold tabs	None	0	None
Annular ring	0.010 in	1	<6 mil
Solder mask	2-sided LPI	17	2-sided LPI
Metalization	SMOBC/SSC	69	SMOBC/SSC
Min. hole dia.	0.025 in	5	<20 mil
Cont. impedance	None	0	None
TOTAL POINTS =		374	

TABLE 19.10 One Company's Initial Consumer Board Characterisitics

Size	519.0 sq. in.
No. of Layers	Six layers
No. of Holes	1,655
Min. Trace Width	0.008 in.
Tolerance (+/–in)	0.003
Complexity Index	68.06
First-Pass Yield	90.7%
Price Index	412.3
Rel. Price	+ 14.5%

A=35, H=1,000,L=2,T=.01,T_O=.01

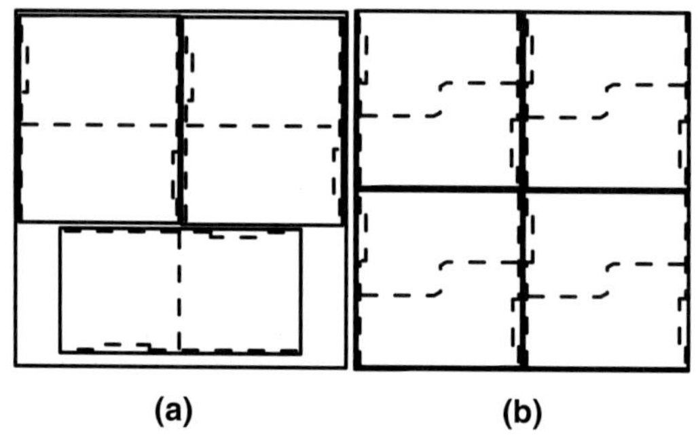

FIGURE 19.16 Fabrication panel layout of the consumer electronics board: (a) the original design; (b) the final layout.

TABLE 19.11 One Company's Final Consumer Board Characteristics

Size	43.6 sq. in.
No. of Layers	Four layers
No. of Holes	1,156
Min. Trace Width	0.012 in.
Tolerance (+/–in.)	0.003
Complexity Index	55.8 (24.0)
First-Pass Yield	96.7%
Price Index	331.9
Rel. Price	–10.2%

A=35, H=1,000,L=2,T=.01, T_O=.01

19.6 ASSEMBLY TRADE-OFF PLANNING

The metrics of assembly trade-offs relates factors of process, component selection, and test to assembly prices. Yields and rework are factored into the points of the Assembly Report Card. The point total provides an estimation of the relative prices of assembly and test.

19.6.1 Assembly Complexity Matrix

The PCB assembler supplies the assembly complexity matrix. This matrix relates various process assembly and test choices that the assembler provides along with component size, orientation, complexity, and various other known qualities to the costs of providing for these design choices. To relate these qualities to the costs, the matrix allocates design points. Typical factors that affect assembly costs are:

- One-pass or two-pass infrared reflow
- Wave solder process
- Manual or automatic parts placement
- Odd-shaped parts
- Part quality level
- Connector placement
- Test coverage
- The ability to diagnose tests
- Assembly stress testing
- Repair equipment compatibility

By collecting all the costs associated with assembly, test, and repair and then normalizing these costs with the smallest nonzero value, a matrix such as shown in Table 19.12 can be produced.

TABLE 19.12 Factors in an Assembly Complexity Matrix

Factors	Pts	Highest	Pts	Middle	Pts	Lowest
Solder Process	35	1 pass IR	20	2 pass IR	0	IR & wave solder
Placement	8	100% auto	5	99%–90% auto	0	>90% auto
Digital Test Coverage	9	<98%	3	98%–90%	0	>90%
Manual Attachment	8	100% Auto	25	Brackets	0	Post solder assembly

19.6.2 Example of an Assembly Complexity Matrix

An example of the assembly complexity matrix is the Assembly Report Card created by IBM-Austin. This complexity matrix has 10 factors that range in points from 0 to 35. The total points can affect the prices from a 30 percent discount to a 30 percent penalty. Table 19.13 illustrates this assembly complexity matrix trade-offs with design points.

19.6.2.1 SMT Assembly Report Card. The IBM Assembly Report Card's[13] 10 assembly trade-off factors are shown in Table 19.13. The report card was the work of many assembly engineers with the help of the accounting department. The report card was introduced in early 1990s; within two years, through the use of the report card, assembly point scores were averaging around 75 instead of the earlier 50, with its Gaussian distribution. Remember, higher scores indicate more producibility. This effect can be seen in Fig. 19.17. The scores are no longer normally distributed.

TABLE 19.13 Assembly Complexity Matrix Defined as an Assembly Report Card

Assembly Factors	Points	A	Points	B	Points	C	Points	D
Assembly process	35	2-pass IR	25	IR/wave B/S passives	20	IR/wave ≤ 5 B/S actives	0	IR/wave >5 B/S actives
• 2-pass IR is lowest-cost process due to low defect/repair level. • Maximize SMT content, some PTH OK. • Back-side SMT attachment with a wave solder process requires the use of adhesive. • Back-side actives have a high defect level when the wave solder process is used.								
Stress tests	15	0 hours	12	≤3 hours in situ	6	≤6 hours in situ and ≤3 hours static	0	<6 hours
• Stress test is a high-cost process step due to cost of chambers, fixtures, and process time. • Stress test can be eliminated by using robust components, design, and process.								
Parts SPQL	10	No high hitters	7	<2 high hitters	4	<4 high hitters	0	>5 high hitters
• Parts SPQL is the quality level of the parts used in the process (shipped product quality level). • High hitters are parts known to have a high defect rate.								
ICT digital test coverage	9	>98% coverage	6	>95% coverage	3	>90% coverage	0	<90% coverage
• The key to high test coverage is the selection of major components that have built-in testability and a PWB design with 1 test point per net.								
Diagnosability	9	≤10 min	6	≤20 min	3	≤30 min	0	>40 min
• Effectiveness of diagnostic tools provided by the card designer influences diagnostic time.								
Placement/ insertion	6	100% auto	4	≥95% auto	2	≥90% auto	0	<90% auto
• Elimination of manual component placement and insertion reduces process time, defects, and process cost.								
Manual attachment	6	100% auto	4	Simple bracket	2	Difficult bracket	0	Post-solder assembly
• Simple bracket = <1 min assembly time • Difficult bracket = >1 min assembly time • Post-solder assembly = manual solder operation								
Connector selection	4	Auto assembly and retention and keyed	2	Manual and retention and keyed	1	Manual and retention	0	Manual
• Manually inserted connectors that are keyed and have retention reduce defects. • Prepackaged SMT connectors that can be automatically placed are considered equivalent to keyed with retention.								
Handling damage exposure	3	No handling list violations	2	<3 handling list violations	1	<5 handling list violations	0	>6 handling list violations
• Single inline packages (SIPs) > 0.5 in high • Memory SIMM connectors with plastic latches • Unshrouded headers over 0.5 in high								
Repair	3	100% auto	2	≥90% auto	1	<90% auto	0	<90% auto and difficult
• Auto repair refers to the use of semiautomated repair tools for the removal of large components and connectors. • A difficult repair is one that takes >10 min.								

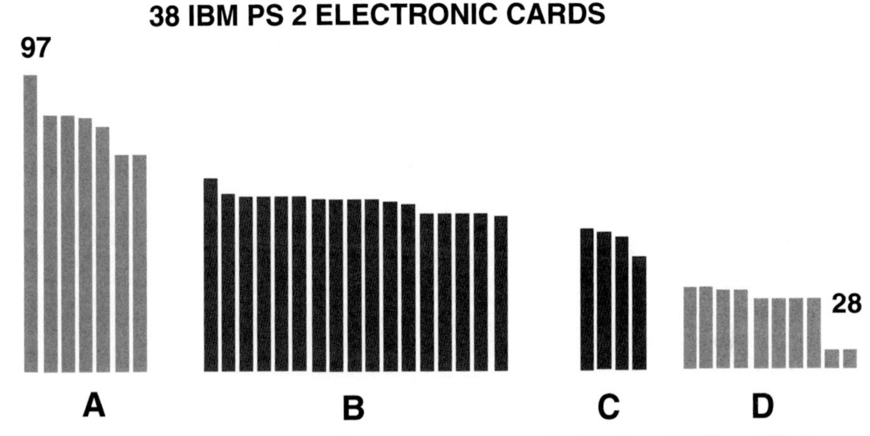

FIGURE 19.17 Improvement in producibility scores as a result of the Assembly Report Card.

REFERENCES

1. General Electric, "Review of DFM Principles," Internal DFM Conference Paper, Charlottesville, VI, 1982.
2. Hawiszczak, Robert, "Integrating Design for Producibility into a CAE Design Environment," NEPCON EAST, June 1989, pp. 3–14.
3. Seraphim, D. P., Lasky, R.C., and Li, C.Y., *Principles of Electronic Packaging,* McGraw-Hill, 1989, pp. 39–52.
4. Coors, G., Anderson, P., and Seward, L.,"A Statistical Approach to Wiring Requirements," *Proceedings of International Electronics Packaging Society (IEPS),* 1990, pp. 774–783.
5. Ohdaira, H., Yoshida, K., and Sassoka, K., "New Polymeric Multilayer and Packaging," Proceedings of Printed Circuit World Conference V, Glasgow, Scotland, reprinted in *Circuit World*, Vol. 17, No. 12, January 1991.
6. Holden, H., "Design Density Index," HP DFM Worksheet, Hewlett Packard, April 1991.
7. IPC-D-275 Task Group, "ANSI/IPC-D-275 Design Standard for Rigid Printed Boards and Rigid Printed Board Assemblies," IPC, September 1991, pp. 50–52.
8. Donath, W., "Placement and Average Interconnection Lengths of Computer Logic," *IEEE Transactions on Circuits and Systems,* No. 4, 1979, pp. 272–277.
9. Sutherland, S. and Oestreicher, D., "How Big Should a Printed Circuit Board Be?" *IEEE Transactions on Computers,* Vol. C-22, No. 5, May 1973, pp. 537–542.
10. Moresco, L., "Electronic System Packaging: The Search for Manufacturing the Optimum in a Sea of Constraints," *IEEE Transactions on Components, Hybrids and Manufacturing Technology,* Vol. 13, 1990, pp. 494–508.
11. Holden, H. T., "PWB Complexity Factor: CI," *IPC Technical Review*, March 1986, p.19.
12. STATGRAPHICS, Ver. 2.6, by Statistical Graphics Corp., 2115 East Jefferson St., Rockville, MD 20852, (301) 984–5123.
13. Hume, H., Komm, R., and Garrison, T., IBM, "Design Report Card: A Method for Measuring Design for Manufacturability," Surface Mount International Conference, September 1992, pp. 986–991.

CHAPTER 20
MANUFACTURING INFORMATION, DOCUMENTATION, AND TRANSFER INCLUDING CAM TOOLING FOR FAB AND ASSEMBLY

Happy T. Holden
Mentor Graphics, Longmont, Colorado

20.1 INTRODUCTION

The manufacturing of printed circuit boards (PCBs) and assemblies (PCAs) begins with the soft-tooling process. This process is the transformation of customer computer-aided design (CAD) data and specifications into the necessary tools required for manufacturing the bare printed circuit board and populating the assembly. The typical tools required for manufacturing printed circuit boards include artwork for photoprinting of inner conductive layers, outer conductive layers, and solder mask patterns. Artwork is also created for screen-printing patterns for nomenclature and via-plugging layers. Additional tools required include drill and routing numerical controlled (NC) programs, electrical testing netlists and fixtures, and CAD reference soft-tools.

The tooling process for assembly requires assembly drawings, bills of materials (BOM), and schematic or logic diagrams, and includes the solder-paste stencil design, assembly array layout, and the resulting computer numeric control (CNC) programs for component placement, in-circuit test program creations, and possibly the creation of a functional test program. Many of these procedures will be covered in subsequent chapters in the "Assembly" section of this Handbook.

During the tooling process, the customer part numbers are analyzed to determine the compatibility of the design features with the manufacturing process capabilities. Additionally, attempts to optimize the manufacturing of the product at the lowest cost are a primary goal. However, the majority of the costs have been defined before the design is transmitted to the manufacturing site by the PCB designer. An early investment in time by the PCB design team and the manufacturing tooling team can result in the most significant savings in overall product cost.

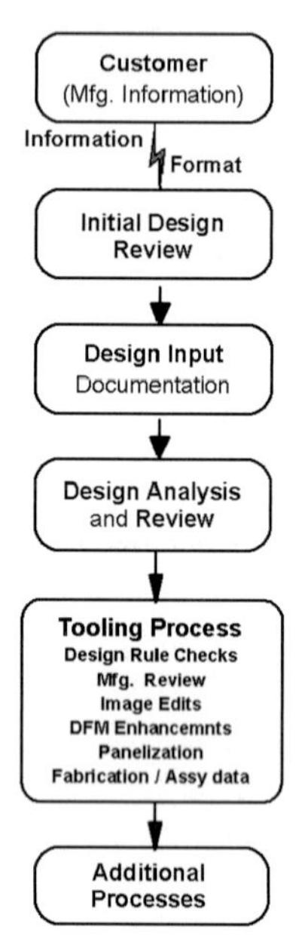

FIGURE 20.1 The soft-tooling process (computer-aided manufacturing [CAM]) for PCB fabrication and assembly.

This chapter describes the PCB tooling process, as defined in Fig. 20.1, including the transfer of information, design reviews, optimization of materials, definition of BOM and routings, tool creation, and additional processes that are required.

20.2 MANUFACTURING INFORMATION

The tooling process begins with the receipt of information from the customer. Unfortunately, although the time required to send information to the manufacturer has decreased to minutes or hours from day(s), a significant issue with the provision of packages to manufacturers is the completeness of the information provided. The Institute for Interconnecting and Packaging Electronic Circuits (IPC) has a new document, IPC-2610, that defines a complete documentation package, including the following:

- IPC-2611, "Generic Requirements for Electronic Product Documentation"
- IPC-2612, "Sectional Requirements for Electronic Diagramming Documentation (Schematic and Logic Descriptions)"
- IPC-2613, "Sectional Requirements for Assembly Documentation (Electronic Printed Board and Module Assembly Descriptions)"
- IPC-2614, "Sectional Requirements for Board Fabrication Documentation (Printed Circuit Board Description Including Embedded Passives)"
- IPC-2615, "Sectional Requirements for Dimensions and Tolerances"
- IPC-2616, "Sectional Requirements for Electrical and Mechanical Part Descriptions (Specification and Source Control Part Descriptions)"
- IPC-2617, "Sectional Requirements for Discrete Wiring Documentation (Wire Harness, Point to Point and Flexible Cable Descriptions)"
- IPC-2618, "Sectional Requirements for Bill of Material Documentation (Complete Listing of Parts, Materials, and Procurement Documents)"

Timely tooling of a part number depends on having the correct information. All features required to exist on the PCB must be defined for the manufacturer. The information is defined via design data, drawings, and textual information. Automated data are encouraged whenever possible.

20.2.1 Required Information

The information and common data formats required to permit the tooling of the PCB include the following:

- Part number information

  Information: This information defines the part number to be built, including revision number, releases, dates, etc.

  Format: This information is typically provided in the part drawing or may be provided as an additional text file.

- Fabrication drawing

  Information: This drawing describes the unpopulated printed board and all features that become part of the board. It may contain specific design requirements such as material requirements, multilayer stack-up diagram, dielectric separation between layers, controlled impedance requirements, solder mask type, nomenclature color, location, size requirements for the fabricator's ID, Underwriter's Laboratories (UL), electrostatic discharge (ESD), country of origin, and dimensional tolerances, as well as testing and electrical performance expectations. The dimensional datum (see Fig. 20.2) should be clearly identified.

  Format: Common formats for drawings are Adobe PDF, HP-GL, HP-GL-II, and PostScript.

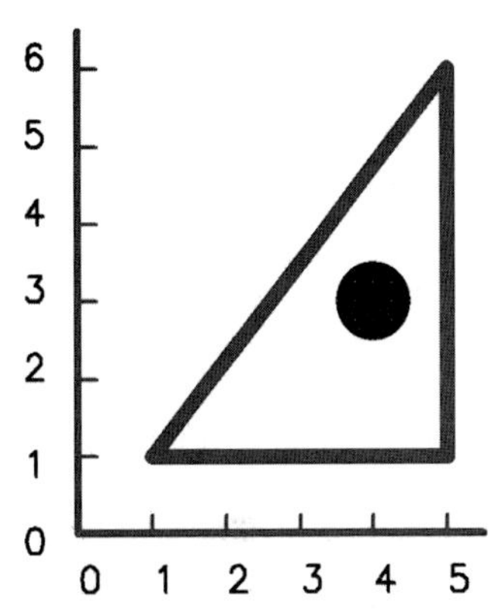

FIGURE 20.2 Datum identification examples.

- Drill drawing

  Information: Although the drill data are provided via data files, this information typically contains only the location of holes and the tool number. The drill drawing identifies the location of the datum reference planes and the coordinate dimensioning system of the board (see Fig. 20.2). The tool number is referenced against the drill drawings to determine the required sizes, plating status, size tolerances, and the total count for verification. This is for plated (TH), non-plated (NPTH), blind/buried vias, and countersink holes.

  Format: Common formats for drawings are Adobe PDF, HP-GL, HP-GL-II, and PostScript.

- Subpanel drawing

  Information: Many assembly operations require boards to be provided in a subpanel form (many parts on one shippable unit). The drawings define the orientation and position of each part, the subpanel dimensions, tooling hole information, special markings, and specific manufacturing processes and tolerances.

  Format: Common formats for drawings are Adobe PDF, HP-GL, HP-GL-II, and PostScript.

- Manufacturing notes

  Information: Manufacturing information is usually included on the fabrication drawing or attached documents. This document details the following:

  - Board details: The type, size and shape of the printed board, bow and twist allowances, overall board thickness requirements including tolerances, tooling hole information, special markings, and specific manufacturing processes and tolerances.
  - Materials: Type class and grade of materials, including color if applicable. Plating and coating material(s), type, thickness, and tolerances. Solder mask and marking inks type, minimum thickness, and permanency.
  - Conductors: Shape and arrangement of both conductors and nonconductor patterns, thickness, dimensions, and tolerances, including conductor width and spacing allowances.
  - Acceptability: Location of quality conformance coupons or circuitry, control documents for acceptability, X-OUTs accepted and assembly-panel concerns.

    Format: Common formats for drawings are Adobe PDF, HP-GL, HP-GL-II, and PostScript.

- Artwork data

  Information: These data consist of files for each circuit layer, coating (e.g., solder mask), marking (i.e., nomenclature), solder paste, hole plugging, and possible test-probing layer.

Format: The data required are usually RS-274X, commonly called Gerber data or ODB++ (Orbitech Data Base). Gerber and ODB++ data are provided as a standard output from most PCB CAD systems. Other possible formats include GenCAM and the new IPC-2581 (see Sec. 18.3.4 "Modern Data Exchange Formats" for more details).

- Aperture list files

  Information: The definitions of the shapes used for drawing are required for each layer provided of artwork data. Special shapes such as thermal pads should specifically define their method of construction.

  Format: This information is usually provided as a text file, although the information may also be defined in the beginning of the individual artwork files, including constructions of complex apertures.

- Drill data

  Information: This information may consist of a single or multiple file(s), and defines the location and tool number used for each hole in the PCB. The files required should define all plated, unplated (which can be combined with plated if fully defined), buried via, and blind via layers.

  Format: Common data files provided are in Excellon format.

- Drill tool files

  Information: This information describes the size, plating status, layer-from, and layer-to (in the case of buried and blind vias) drill-data format and filenames. This information is referenced against the drill drawing.

  Format: This information is usually provided as a text file, although the information may also be defined in the beginning of the individual drill files.

- Special requirements information

  Information: The drawing or a file should describe any special requirements not defined in other information. Typical of such requirements would be the details and image enlargement of "break-away" tabs on a subpanel or details of "hole-plugging" requirements. It is important for the PCB designer not to assume that requirements are understood, but to refer to specifications or clearly define the requirements.

  Format: This information is typically provided in the part drawing or may be provided as an additional text or drawing file.

- Netlist data

  Information: Netlist data define the connectivity of the circuitry.

  Format: This information can be provided from the CAD systems in various formats, or it can be extracted from the drill and artwork data. Contact the PCB manufacturer for compatible formats if the netlist data are to be provided directly. The IPC has defined a neutral format, IPC-356, that provides all the information necessary for netlist and electrical test fixture creation. (See Sec. 18.3.1 and 18.3.2.2 for examples)

In addition, the IPC has defined an alternative neutral standard format for most of the previously defined data. These formats provide simpler processing at the manufacturer. They include IPC-D-350, IPC-2511 (GenCAM), and IPC-2581 (the offspring of ODB++ and GenCAM). These formats can be generated by most PCB CAD systems and processed by most PCB CAM/tooling systems. The PCB customer should review the compatibility of these formats with the PCB manufacturer prior to sending data formatted with them.

20.2.2 Fabrication Information Formats

Formats for fabrication information include Gerber, ODB++, GenCAM, and IPC-2581.

20.2.2.1 Gerber Data. Gerber data format is the oldest artwork format. It was developed with the first NC photo plotters in the early 1970s. It is the "least complete" of data formats in

```
*                   (Always a good idea to start a plot with an asterisk)
G04 *               (Comment; Date: Sun Feb 26 16:48:19 2006)
G04 *               (Comment; Layer 1: Layer_1.gbr )
G54D12*             (G54, tool select Aperture number D12)
D2*                 (D02: move to the x-y location specified with the shutter
                     closed.)
X40000Y30000D03*    (Go to X-4.0000", Y-3.0000, D03 (D3): move to the x-y
                     location specified with the shutter closed; then open and
                     close the shutter-known as flashing the exposure.)
G54D10*             (Select Aperture number D10)
X10000Y10000D02*    (Go to X-1.0000", Y-1.0000, not drawing  a line)
G01X50000D01*       (Go to X-5.0000", Y-1.0000, drawing the base of the triangle)
Y60000*             (Go to X-5.0000", Y-6.0000, still drawing)
X10000Y10000*       (Go to X-1.0000", Y-1.0000, drawing the hypotenuse)
M02*                (End of plot)
```

FIGURE 20.3 Example of a Gerber format program.

that it consists of an aperture wheel definition, an x-y starting coordinate, an "open shutter" command, an x-y ending coordinate, and a "close shutter" command. The aperture wheel must also be defined. Figure 20.3 shows a typical Gerber format program of the Gerber plot in Fig. 20.2.

The Gerber 274-X is the preferred version today, as the apertures are embedded in the format. Figure 20.4 shows the RS-274X 2.5 leading absolute format. For further explanation on Gerber see Sec. 18.3.2.1.

```
GERBER RS-274 X        FIRE 9xxx
                       BARCO BDP
G04 Date:  Sun Feb     G04 Date:  Sun Feb 26     ;
26 16:57:09 2006 *     16:57:39 2006 *           ;        Sun Feb 26 16:57:58 2006
G04 Layer_1:           G04 Layer 1:              ;         Layer 1: Layer_1.gbr
Layer_1.gbr *          Layer_1.gbr *             ;
%FSLAX25Y25*%          G04%FILE="Layer_1.gbr";
%MOIN*%                %*                      U=MIL
%SFA1.000B1.000*%      G04%PAR.%*              X90.0,510.0 Y90.0,610.0
%MIA0B0*%              G04%MODE=A;%*           A10=C,20.00
%IPPOS*%               G04%UNIT=I;%*           A12=C,100.00
%ADD10C,0.02000*%      G04%ZERO=L;%*           A12
                       G04%FORM=2.5;%*         F400.00,300.00
%ADD12C,0.10000*%      G04%IMTP=POSITIVE;%*    A10
%LNLayer_1'.gbr*%      G04%ADRS=Q;%*           M100.00,100.00D500.00,D,600.00D100.00,100.00
%SRX1Y1I0J0*%          G04%MIRR=N;%*
G54D10*                G04%CROS=X;%*
%LPD*%                 G04%EMUL=UP;%*
G54D12*                G04%XSCL=1.000000;%*
X40000Y30000D03*       G04%YSCL=1.000000;%*
```

FIGURE 20.4 Example showing RS-274D 2.5, the leading absolute format.

```
G54D10*
X10000Y10000D02*
G01X50000D01*
Y60000*
X10000Y10000*
M02*
```

```
G04%POEX=11,13;%*
G04%POIN=14,15;%*
G04%MRGE=PAINT;%*
G04%NEXT=="-";%*
G04%NFLG=P;%*
G04%EOP.%*
G04%APR,100000.%*
G04%A10:CIR,2000,X0,Y0.
%*
G04%A12:CIR,10000,X0,Y0
.%*
G04%EOA.%*
G54D10*
G54D12*
X40000Y30000D03*
G54D10*
X10000Y10000D02*
G01X50000D01*
Y60000*
X10000Y10000*
M02*
```

```
IPC D-350
C    Date:  Sun  Feb  26  16:58:20  2006
C    Database:   (Untitled)
C
P    UNITS   CUST
P    IMAGE   NCON POS
P    IMAGE   NCON POS
322D1000L01S          NULL                                      GO    X 00004000Y 00003000
112D0200L01S          NULL                                      GO    X 00001000Y 00001000X
00005000Y     00001000
000                                                             X
00005000Y    00006000X    00001000Y    00001000
999
```

FIGURE 20.4 *(Continued)*

20.2.2.2 ODB++ Format. Orbitech developed ODB++ as a "complete" data description format used for design to manufacturing (DFM) and CAD/CAM data exchange. It contains all the relevant design data for manufacturing and assembly. See Sec. 18.3.4.1 for ODB++ details.

20.2.2.3 GenCAM. Developed by IPC to improve data transfer for manufacturing in the 1980s, GenCAM is an ASCII-based, English-readable, and universal vendor-independent data exchange format. IPC-D-350 was the first such format in the 1970s, followed by IPC-2511/GenCAM in the 1990s, and in the new millennium IPC-2581 evolved.

20.2.2.4 IPC-2581. This is the new IPC coordinated data transfer format. Started in the late 1990s, it is available on some CAD systems. A description is available in Sec. 18.3.4.2.

20.3 INITIAL DESIGN REVIEW

The purpose of the initial design review is to determine the potential fit of the product to the manufacturing facility, determine general cost information, and prepare for tooling. Proper upfront analysis of a product prior to manufacturing or tooling saves time and materials.

It is the responsibility of the manufacturing site to determine the fit of a given product to its capabilities. PCB manufacturing sites should monitor and maintain a list of manufacturing capabilities and a technology roadmap of where the facility is developing additional process capabilities. This list of capabilities defines the acceptability of a product for manufacturing or whether the product is a research and development project.

20.3.1 Design Review

Reviewing the incoming package for design and performance requirements (e.g., line width and spacing, impedance) versus the PCB manufacturing capabilities will define the capability for manufacturing and provide a prediction of the resulting yield. Among the design characteristics that should be reviewed against process capabilities are the following:

Item	Issue
Maximum number of layers	Higher-layer-count products require greater control of processes and tooling tolerances.
Board thickness	Some process or handling equipment may have limitations on board thickness (either too thin or too thick).
Minimum feature width	Finer line widths require better control of artwork and process tolerances. In addition, the relationship of line width to copper weight is significant in providing a well-defined line. Fine lines with lower copper weights are easier to manufacture than fine lines with higher copper weights. For normal impedances, like 50 ohm, thinner lines will require thinner dielectrics.
Minimum feature spacing	Finer feature spacing requires better control of artwork and process tolerances. In ddition, the relationship of feature spacing to copper weight is significant. Fine spaces with lower copper weights are easier to manufacture than fine spaces with higher copper weights. Minimum feature spacings affect electrical cross-talk. Closer spacing to reference ground will minimize this cross-talk.
Minimum finished hole size	Smaller holes require higher manufacturing process capabilities and imply (due to design characteristics) finer tolerances on tooling and internal registration systems. The aspect ratio (AR) of diameter to board thickness is a major process concern.

Maximum aspect ratio	Small holes and thick boards result in difficulty during the plating processes and can result in defective products that may or may not pass electrical testing. Plating high-AR holes requires chemistry and process parameter enhancements.
PCB dimensional tolerances	Fine PCB profile or cutout tolerances may result in punching/blanking requirements versus routing, or changes in the routing parameters or programming.
Feature-to-feature tolerances	The location of features on the PCB relative to other features may require alternative materials or process changes to reduce the tolerances.
Hole size tolerances	Consistency in the plating process versus selected drilling hole size and plating densities has significant impact on the capability to control hole tolerances. Adjustments to the PCB design, drill hole size selection, or process parameters may be required to produce an acceptable product with tighter tolerances.
Impedance requirements	The multilayer stack-up thicknesses and tolerances, along with the copper thickness and line widths, must match the nominal impedance and the upper-and lower-acceptability limit. Often, various impedances cannot be fabricated on the same layer.

20.3.2 Material Requirements

Determination of the bill of materials is required during the initial analysis of the design. The determination of the BOM and other material-processing requirements will define the manufacturing facility's capability to produce and its material cost structure. In addition, the definition of the material requirements will be the basis for the generation of the process traveler requirements.

The primary materials requiring definition are those included in the BOM, including laminates, prepregs, copper foil, solder mask, and gold. The materials may be explicitly defined by the PCB customer (e.g., usage of a specific solder mask),or may be implied in the drawings or specifications provided with the tooling package.

Several factors impact the selection of the raw materials, including the following:

- Customer-defined physical constraints, for example, the definition of the physical dimensions between conductive layers
- Customer specification of electrical properties, for example, the definition of the impedance requirements on certain layers
- Manufacturing process capabilities related to lamination thicknesses and tolerances
- Specification of material dielectric requirements, for example, the usage of FR-4 or polyimide
- Specifications of physical operating parameters, for example, the minimum requirements of the glass transition temperature

The determination of the laminates, prepregs, and copper foils are based on the following:

- Standard constructions for the PCB manufacturer of a defined PCB layer count, final thickness, copper weight, and dielectric spacing.

- Custom constructions based on defined physical constraints (e.g., minimum dielectric spacing). These custom constructions are defined through knowledge of the lamination pressing thickness of materials versus copper circuitry densities and the availability of materials from suppliers.
- Custom constructions based on defined electrical property constraints, such as controlled impedances, cross-talk, or capacitance. These custom constructions are typically defined via equations or software models provided with certain product parameters.

The determination of solder mask is based on the customer specifications and drawings. Once the acceptability of the solder masks is defined, the PCB manufacturer selects the acceptable masks based on either preferred process (due to volume or cost) or the design characteristics' interaction with the solder mask. These design characteristics include the following:

- **Tenting or plugging of vias** A hole plugging epoxy may be preferred over liquid photoimageable solder mask with a secondary via plugging process.
- **Platable area densities/higher external copper weights** Thin solder masks may not be able to ensure coverage of high plating.
- **Secondary processes** Post-solder-mask processes may chemically or mechanically alter the appearance of certain solder masks.

The determination of gold requirements is based on the thickness and area of the gold. These factors can be used to calculate the requirement of gold per PCB product.

20.3.3 Process Requirements

The selection of the proper product routing (or traveler) is critical to the upfront analysis of the product acceptability to manufacturing. Considering a typical multilayer PCB product, the product routings can be broken into two parts: the innerlayer pieces and the outerlayer piece.

The product routings of the innerlayer pieces are fairly standard, and are typically as follows:

Step	Typical process requirements
Layer cleaning	Cleaning of the laminate surface via either mechanical or chemical cleaning processes
Imaging	Coating of the laminate with photoresist material and the exposure of the photoresist with artwork defining the innerlayer pattern
Develop-etch-strip (DES)	Developing of the photoresist, etching of the exposed copper, and stripping of the remaining photoresist
Innerlayer inspection	Inspection of the innerlayer piece to the PCB design intent
Oxide	Coating of the innerlayer piece with an oxide layer prior to lamination

One of the few decision points in innerlayer manufacturing is whether the product will require inspection. This decision can be based on the manufacturing facility's process capabilities and the design of the specific innerlayer piece. For example, if the process capability for innerlayer manufacturing of designs at 0.008 in. lines and spaces is 100 percent yield, and the product has been designed at or above 0.008 in. lines and spaces, then the product may not require inspection. Typically, the design package will identify the design technologies (e.g., line width and spacing); however, these should be confirmed during the design analysis and review stage.

The product routings of the outerlayer pieces define the finished product appearance and, as a result, are more complex. Assuming a pattern plating process, the typical outerlayer process routings for a solder-mask-over-bare-copper (SMOBC)/hot-air-solder-level (HASL) product is as follows:

Step	Typical process requirements
Lamination	Lamination of innerlayer pieces with prepreg and copper foils to create the outerlayer piece
Drilling	Addition of the holes providing pathways for electrical conductivity between outer layer and innerlayer pieces and for other PCB design purposes
Electroless copper plating	Deposition of copper on the surface and in the holes of the product, providing the conductivity necessary for electroplating
Imaging	Coating of the panel with photoresist material, the exposure of the photoresist with artwork defining the outerlayer pattern, and the developing of the outerlayer pattern
Electroplate copper tin plating	The plating of the final circuitry of the product Sacrificial plating over the final circuitry of the product
Strip-etch-strip (SES)	Stripping of the remaining photoresist, etching of the exposed copper, and stripping of the sacrificial plating
Solder mask	Coating of the bare copper product with either dry film or liquid photoimageable solder mask, exposing of the panel with the PCB customer-supplied artwork pattern, developing of the pattern (exposing the sites requiring solder), and curing of the solder mask
Nomenclature	Screening of the nomenclature onto the panel and curing
HASL	Coating of the exposed copper sites with solder, leveling to the customer's requirements of thickness
Depanelization	NC routing of the PCB products from the manufacturing panel
Electrical testing	Electrical testing of the product for conformance to the PCB design
Inspection	Verification of product conformance to specifications prior to shipment to the PCB customer

After the electroplate copper operation, the process may change to meet the surface finish requirements. Some of the alternative steps affecting the outerlayer product routing include the following:

Item	Typical process requirements
Selective gold plating	Before the SES process, the panel is coated with additional resist and the selective sites are exposed. These exposed sites are then plated with nickel and gold. The panel will then proceed through the SES process. The HASL step would be omitted.
NPTH in planes without annular ring	A secondary drilling operation will be required to drill the holes, if the hole is to be located through the copper or the hole/feature size is beyond the process capabilities of tenting. This step may occur at the depanelization step.

Scoring	Additional tooling holes may be required to register the panel to the scoring machine blades. Programming of the scoring machine will be required to set the locations and depth of the cuts. This step may occur prior to NC routing of the product.
Countersink holes	The countersinking process is normally performed prior to depanelization.
Gold finger plating	Nickel and gold are plated over the finger area on connectors. This process typically requires shearing/NC routing separation of the panel to place the fingers at the edge of the remaining pieces, taping the product to expose only the fingers, stripping the solder, plating nickel, plating gold, and removing the tape. These operations occur prior to the depanelization process. Finger-plated products usually require chamfering of the fingers.
Via plugging	Plugging of vias may occur prior to solder mask or after the HASL operations.
Chamfer	After depanelization of the products from the manufacturing panel, the PCB products are processed through a machine to add an angle to the edge of the part or fingers.
Organic coating	Products requiring organic coating would omit the HASL step and be coated prior to inspection.

20.3.4 Multilayer Stack-Up

Multilayer stack-up, defined in Sec. 13.2.2 and Chap. 27, must be carefully specified. This is the function of the stack-up definition (see Figs. 13.9, 27.1, and 27.2), which is usually part of the fabrication drawing. Impedance specifications (from the electrical performance data block) should be compatible with the materials stack-up. Figure 20.5 shows a cross-sectional view of the finished multilayer. Figure 20.6 shows the stack-up code.

20.3.5 Panelization

The selection of the manufacturing panel is one of the most important steps in achieving product profitability. Several factors affect the selection of panel sizes to produce a specific product. These include:

- Material utilization
- Process-specific constraints
- Process limitations

All of these factors impact the selection of the manufacturing panel size and the profitability of the product.

20.3.5.1 Material Utilization. The material manufacturing costs, which correspond to 30 to 40 percent of total costs, are directly related to the square inches of material processed. The material costs can include the following: laminate, prepreg, copper foil, solder mask, photoresists, drill bits, chemicals, and so on. Generally, these materials are consumed relative to the panel area manufactured.

Description	Stock No	Usage	Base	Finish	Er
Liquid Photoimageable Mask	PSR-4000 BL-0	Mask	0.600		4.100
Copper Foil	H oz	Copper	0.600	2.000	
Prepreg 2313	NP-170B	Dielectric	3.700	3.700	3.840
2116 Core	NP-170TL	Copper	1.300	1.300	
2116 Core	NP-170TL	Dielectric	4.000	4.000	4.270
2116 Core	NP-170TL	Copper	1.300	1.300	
Prepreg 2313	NP-170B	Dielectric	3.700	3.700	3.840
2-1080 Core	NP-170TL	Copper	0.600	0.600	
2-1080 Core	NP-170TL	Dielectric	4.000	4.000	3.900
2-1080 Core	NP-170TL	Copper	0.600	0.600	
Prepreg 2313	NP-170B	Dielectric	3.700	3.700	3.840
2-2116	NP-170TL	Copper	1.300	1.300	
2-2116	NP-170TL	Dielectric	10.000	10.000	3.990
2-2116	NP-170TL	Copper	1.300	1.300	
Prepreg 2313	NP-170B	Dielectric	3.700	3.700	3.840
2116 Core	NP-170TL	Copper	0.600	0.600	
2116 Core	NP-170TL	Dielectric	4.000	4.000	4.270
2116 Core	NP-170TL	Copper	0.600	0.600	
Prepreg 2313	NP-170B	Dielectric	3.700	3.700	3.840
2116 Core	NP-170TL	Copper	1.300	1.300	
2116 Core	NP-170TL	Dielectric	4.000	4.000	4.270
2116 Core	NP-170TL	Copper	1.300	1.300	
Prepreg 2313	NP-170B	Dielectric	3.700	3.700	3.840
Copper Foil	H oz	Copper	0.600	2.000	
Liquid Photoimageable Mask	PSR-4000 BL-0	Mask	0.600		4.100

FIGURE 20.5 Stack-up drawing to define thickness, layer order, and tolerances for a multilayer.

The appropriate manufacturing panel size (see Fig. 20.7) should be selected such that the shippable product consumes the highest percentage of the manufactured panel as possible, thus reducing waste material and product costs.

During the design phase of the PCB, the board profile is defined. If the profile's design results in poor manufacturing panel utilization, it will significantly impact the cost of the product. When defining the PCB board profile, the PCB designer should ask the manufacturing site to provide

```
        PRIMARY (Component) Side  Reference
  Plating layer and thickness
 Copper foil thickness, layer_1 reference and name
 Prepreg layer finished thickness and tolerance
 Copper foil thickness, layer_2 reference and name
 Copper-clad-core thickness and tolerance
 Copper foil thickness, layer_3 reference and name
 Prepreg layer finished thickness and tolerance
 :  continue with stackup
 Prepreg layer finished thickness and tolerance
 Copper foil thickness, layer_N reference and name
 Plating layer and thickness
        SECONDARY (Solder)  Side  Reference
```

FIGURE 20.6 Stack-up code listing.

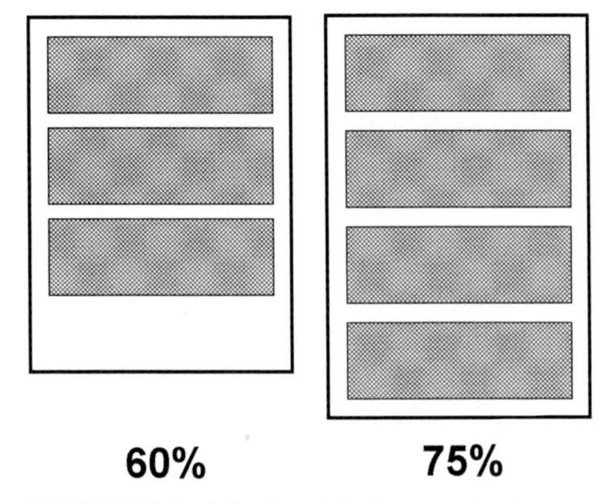

FIGURE 20.7 Selection of the best panel to increase the material utilization of the shippable product.

DFM feedback. Oddly shaped PCB profiles can still result in good material utilization, through nesting of the PCB on the panel.

General limitations in the selection of appropriate panel sizes include the following: minimum spacing between products for depanelization processes (typically 0.100 in.) and minimum panel border-to-product spacing to permit tooling and registration systems (typically 1.0 in.).

In Asia, minimizing materials to minimize costs plays a much more important role than in North America or Europe, where rapid turnaround time may be the singular focus. Table 20.1 shows a typical set of multilayer panel sizes for a large Asian PCB manufacturer.

TABLE 20.1 Typical Multilayer Panel Sizes for a PCB

Type	Panel size (inches)	Useable area (with soldermask)
Standard	12 × 14	10.2" × 12.2"
Standard	12 × 16	10.2" × 14.2"
Standard	12 × 20	10.2" × 18.2"
Standard	13 × 16	11.2" × 14.2"
Standard	13 × 20	11.2" × 18.2"
Standard	14 × 16	12.2" × 14.2"
Standard	14 × 20	12.2" × 18.2"
Standard	14 × 21	12.2" × 19.2"
Standard	14 × 22	12.2" × 20.2"
Standard	14 × 23	12.2" × 21.2"
Standard	14 × 24	12.2" × 22.2"
Standard	16 × 20	14.2" × 18.2"
Standard	16 × 21	14.2" × 19.2"
Standard	16 × 22	14.2" × 20.2"
Standard	16 × 23	14.2" × 21.2"
Standard	16 × 24	14.2" × 22.2"
Standard	17 × 14	15.2" × 12.2"
Standard	18 × 14	16.2" × 12.2"
Standard	18 × 16	16.2" × 14.2"
Standard	18 × 20	16.2" × 18.2"
Standard	18 × 22	16.2" × 20.2"
Standard	18 × 23	16.2" × 21.2"
Standard	18 × 24	16.2" × 22.2"
Chip packaging	12 × 18	10.0" × 16.0"
HDI-microvia	16 × 18	14.0" × 16.0"
Optional		

20.3.5.2 Product Process-Specific Constraints. In the selection of manufacturing panel sizes, there may be processing constraints that restrict the usage of specific panels or limit the method of placing the shippable product on the panel. For example, the manufacturing of products with gold tab plating may require additional spacing between products and restrictions on the rotation and nesting of products. These constraints may force the usage of panel sizes below the optimal material utilization.

20.3.5.3 Process Limitations. Process limitations may require the usage of nonoptimal panel sizes. For example, due to a product's registration requirements, additional tooling may be required, thus reducing the available area for shippable product and reducing the material utilization. Another example is the limitations of some processes to permit larger panel sizes due to the physical processing constraints of the equipment.

The panelization process results in the placement of the PCB single images in the locations defined in the panelization definition step during the initial design review. In addition to the placement of the PCB single images and rotated/nested images (Fig. 20.8), the following features may be added to the panel for manufacturing.

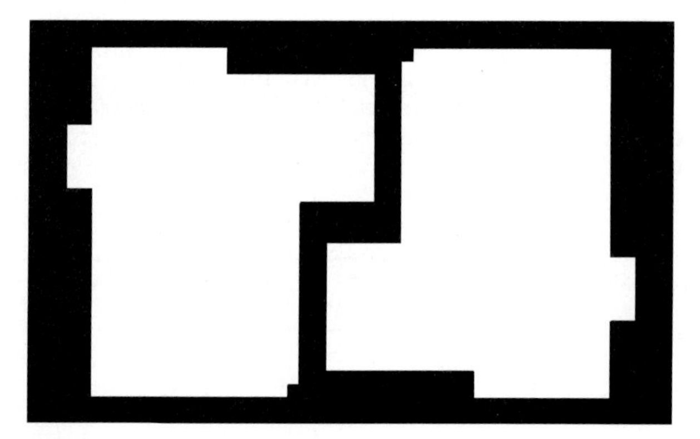

FIGURE 20.8 Nesting and rotation of a multi-image panel.

20.3.6 Initial Design Analysis

After the initial design review, the basics of the design are known and a decision on whether to manufacture the design has to be made. If it is decided to proceed with manufacturing, the cost of the product must be determined. Once the cost has been determined, the product price can be provided to the PCB customer.

The material and process requirements with the manufacturing panel selection, combined with the prediction of yield based on the design review, define the basics of the product costs.

20.3.6.1 Automated Design Analysis. Many automated design analysis software systems exist. You can find them on the Internet by searching on key words such as "Automated Design Analysis," "Computer Aided Manufacturing," or "Design Validation." Figure 20.9 shows a report resulting from a CAM macro executed on an ODB++ design file, "8-Layer."

20.3.6.2 Design Analysis Checksheet. Figure 20.10 shows a design analysis checksheet for determining whether a PCB design is ready for manufacturing.

DATE Sun Mar 12 17:48:51 2006 | AGENT FILE USED

QUOTE REPORT

OPERATOR: H HOLDEN | JOB #: 8-Layer | REV: 1st.

NO LOGO

PCB Fabricator
612 Wheelers Farms Road
Milford CT 06461
USA
203-283-3000

CUSTOMER: XYZ Corporation
HOLDEN
Cust street address
Anywhere CT 00000
USA
PHONE: Cust phone FAX: Cust fax
E-MAIL CustEmail

BOARD INFORMATION

BOARD SIZE: 9.050" X 5.377" | THICKNESS. 0.062mil | BORDER LAYER PRESENT? No
PANEL SIZE. 22.000" X 16.000" | CORE SIZE. 4.000mil | # OF NETS 562
LAMINATE TYPE. NP-170 | COPPER WEIGHT 0 50 oz.

TOTAL (Elec.) layers: 8 | # of INTERNAL layers 2 | # of PLANE layers: 6 | POS 2 | NEG 0

DRILL INFORMATION

# OF DRILL TOOLS / TOTAL COUNT:	10	1768	BLIND/BURIED VIAS?	Yes
SMALLEST DRILL / TOTAL COUNT	13.000mil	1484	# OF DRILL FILES.	1
TIGHTEST DRILL TO TRACE:	0.064mil			

DESIGN ANALYSIS

	Internals	Other (user defined):
SMALLEST TRACE:	6.000mil	6.000mil
SMALLEST AIR GAP:	5.000mil	5.000mil
SMALLEST ANNULAR RING:	N/A	0.000mil
SMALLEST THERMAL ANNULAR RING:	N/A	0.000mil
SMALLEST PAD:	20.000mil	20.000mil

SMT PADS:

	TOP	BOTTOM	TOTAL
# OF PADS.	912	1288	2200
Smallest PITCH:	19.680mil	2.000mil	

GOLD TABS:

# OF FINGERS: TOP	# OF FINGERS: BOTTOM	FINGER SIZE: LENGTH	FINGER SIZE: WIDTH
0	22	600 0 mil	250.0 mil

GOLD THICKNESS 0.000015mil

SOLDERMASK INFO:

TOP: Yes BOTTOM: Yes
VIAS Tented / Plugged: Yes

SILKSCREEN INFO:

TOP Yes BOTTOM Yes
COLOR: White

FIGURE 20.9 Automated design analysis report performed on an ODB++ file.

CHECKLIST

CAD netlist compare
Annular ring error
Padstack checklist
Plane clearance error
Manufacturability analysis, DRC
Thermal leg count violation
Circuit checklist
Unterminated lines
Resist slivers
Copper islands
Solder mask checklist
Sholder short violation
Mask coverage
Mask to via check
Solder paste check
Silkscreen clipping
Teardrop pad addition
Part to part clearance / automation
Quality parts
Hole audit / lead diameter
Part density
Height clearance
Allowable machine span
Part spacing
Drill optimization
Automatic solder mask generator
Bare board test points
In-circuit test point analysis
In-circuit test checklist
Boundary scan audit
Test point management
Design profile
Registration generator
Keepout audit

FIGURE 20.10 A design analysis checksheet.

20.4 DESIGN INPUT

The input of the design into the CAM system is primarily performed by defining the apertures and shapes to be used and then loading Gerber data into the PCB CAM system. Alternatively, most PCB CAM systems accept the IPC-350 format.

The information loaded into the PCB CAM systems includes all artwork layers (e.g., circuitry layers and solder mask layers) and the drill files. Although some PCB CAM systems can accept NC routing files (i.e., board profiling), these files are normally not part of the PCB design system's capabilities to create, and therefore are not provided to the PCB manufacturer.

Before loading the PCB design files, the PCB CAM operator must relate the aperture codes within the design file to physical shapes within the PCB CAM system. These shapes are usually round, square, and rectangular, but may also include complex shapes (e.g., thermal relief pads for innerlayer planes). The complex shapes may need to be created on the CAM system from information provided by the PCB designer. The complete definition of these complex shapes by the PCB designer is critical to the success of the resulting design. Incomplete descriptions could result in nonfunctional designs.

During the loading process of the design files, it is important to review any log files and screen messages for missing aperture definitions or damaged design files. After loading the files, the PCB CAM operator should review the design for any problems with the interpretation of the data or apertures while aligning the data for further processing. The PCB CAM operator also must ensure that the apertures match those defined by the PCB designer and that the design input process has proceeded without failure, since subtle errors may not be discovered within the manufacturing site but only at the PCB customer site, and such errors could result in nonfunctional designs.

20.4.1 Documentation

20.4.1.1 Drill Drawing Example. Figure 20.11 shows an example of a drill drawing.

20.4.1.2 Fabrication Drawings. See Fig. 20.12 for an example of a fabrication drawing.

20.4.1.3 Stack-Up Drawing. Figure 20.13 shows a stack-up drawing.

20.4.1.4 Stack-Up Data Block. Table 20.2 shows an example of a stack-up data block.

20.4.1.5 Manufacturing Notes. Figures 20.11 and 20.12 show examples of manufacturing notes.

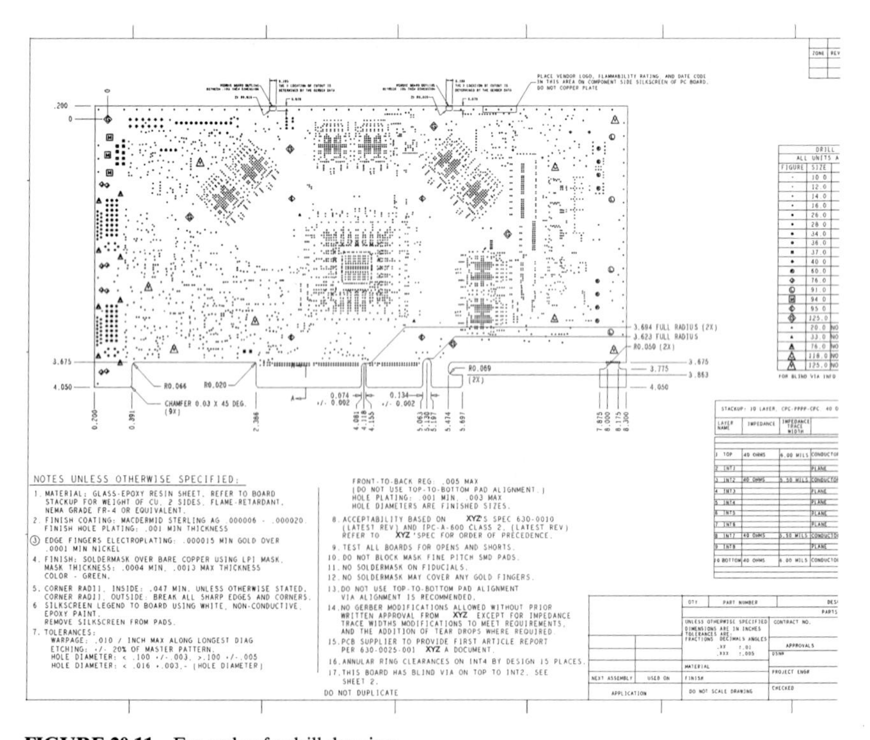

FIGURE 20.11 Example of a drill drawing.

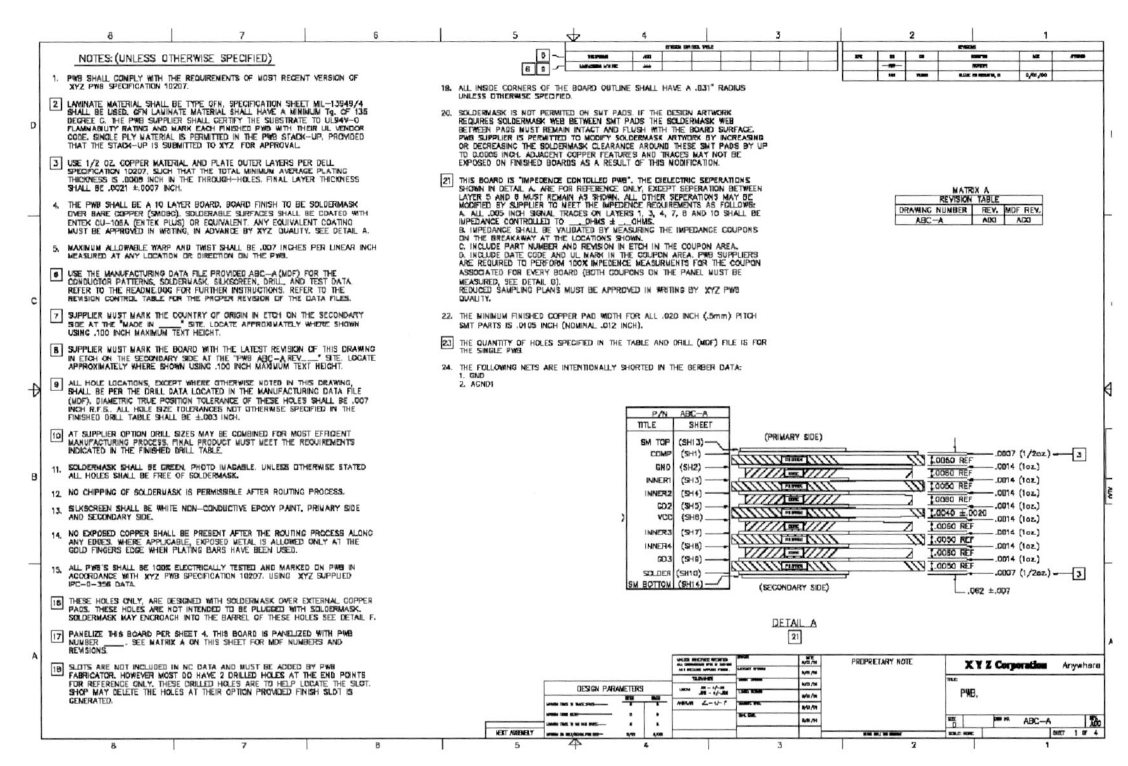

FIGURE 20.12 Example of a fabrication drawing.

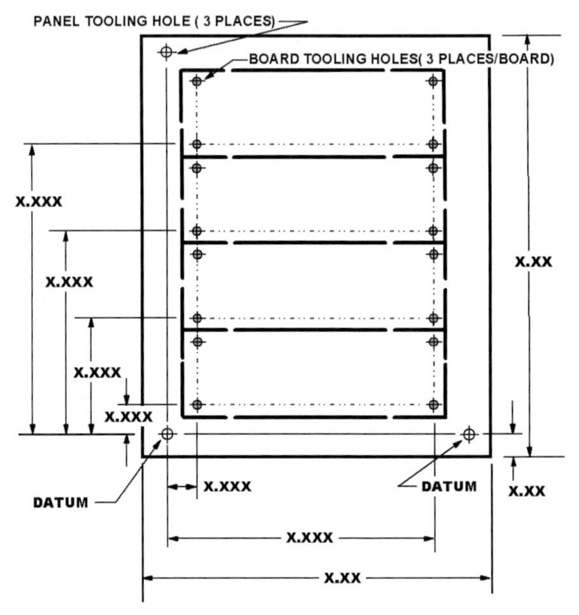

FIGURE 20.13 Example of a stack-up drawing.

TABLE 20.2 Example of a Stack-Up Data Block

P/N	ABC-d	
Title	Sheet	
S/M top	(SH14)	
Component	L1-(SH 1)	→
Ground	L2-(SH 2)	→
Power	L3-(SH 3)	→
Signal_1	L4-(SH 4)	→
Signal_2	L5-(SH 5)	→
Power	L6-(SH 6)	→
Power	L7-(SH 7)	→
Signal_3	L8-(SH 8)	→
Signal_4	L9-(SH 9)	→
Power	L10-(SH 10)	→
Ground	L11-(SH 11)	→
Solder	L12-(SH 12)	→
S/M	(SH 13)	
Bottom		

20.4.1.6 Electrical Performance Data Block. Table 20.3 shows an example of an electrical performance data block.

TABLE 20.3 Example of an Electrical Performance Data Block

Bus	Data	Agp 8x	PCI	DDR2/3 clocks
Type	Single	Single	Differential	Differential
Tolerance	50 ohms ± 10%	55 ohms ± 10%	100 ohms ± 10%	120 ohms ± 10%
Geometry (mils)	Trace/space	Trace/space	Trace/space Space to other pair	Trace/space Space to other pair
Top (Layer 1)	9.5/10 (Ref L2)	7.75/10 (Ref L2)	5.0/6 (Ref L2)	4.0/10 (Ref L2)
PWR (Layer 2)	VCC	VCC	VCC	VCC
GND (Layer 3)	GND	GND	GND	GND
Signal 1 (Layer 4)	4.5/4.5 (Ref L3 & L6)	3.725./4.25 (Ref L3 & L6)	4/10/4/10 (Ref L3 & L6)	3/10/3/10 (Ref L3 & L6)
Signal 2 (Layer 5)	4.5/4.5 (Ref L3 & L6)	3.725./4.25 (Ref L3 & L6)	Not allowed	Not allowed
PWR (Layer 6)	VCC2	VCC2	VCC2	VCC2
PWR (Layer 7)	VDD	VDD	VDD	VDD
Signal 3 (Layer 8)	4.5/4.5 (Ref L7 & L10)	3.725/4.25 (Ref L7 & L10)	Not allowed	Not allowed
Signal 4 (Layer 9)	4.5/4.5 (Ref L7 & L10)	3.725/4.25 (Ref L7 & L10)	4/10/4/10 (Ref L7 & L10)	3/10/3/10 (Ref L7 & L10)
GND (Layer 10)	GND	GND	GND	GND
PWR (Layer 11)	VCC1	VCC1	VCC1	VCC1
Bottom (Layer 12)	9.5/10 (Ref L11)	7.75/10 (Ref L11)	5.0/6 (Ref L11)	4.0/10 (Ref L11)
Measurement requirement	Measure on coupon	Measure on coupon	Measure on coupon	Measure on coupon

20.5 DESIGN ANALYSIS AND REVIEW

Design analysis and review is a refinement of the initial design review performed earlier but is the first step in the CAM (soft-tooling) process. This step focuses on the actual design data submitted for fabrication and assembly. The design analysis and review step consists of the following tasks:

- Design rule checking
- Manufacturability analysis
- Single-image edits
- DFM enhancements
- Panelization
- Fabrication parameters extraction

These tasks provide the final checking of the PCB design requirements against the capabilities of the PCB manufacturer and preparation of the design for manufacturing.

Most fabrication problems occur or are discovered during the initial tooling (CAM) phase. The following are among the causes of such problems:

- Poorly documented designs
- Improperly formatted drill files
- Use of 274-D Gerber with separate aperture tables
- Lack of proper clearance on solder mask openings
- Inadequate drill-to-copper clearance
- Misaligned layers
- Inadequate manufacturing clearances
- Lack of an IPC-D-356 netlist
- Use of positive plane layers instead of negative ones
- Minimal communication between the designer and the fabricator

The best advice for these problems is to fix them during the design phase. Consider the following precautions and considerations:

- Try not to use 274-D with separate aperture tables.
- Because there are no standards for aperture tables, the fabricator must do extra work to accommodate the tables, with increased opportunity for error.
- Use 274-X whenever possible. The apertures are embedded and files are easier for the fabricator to work with. This provides better support for special aperture shapes.
- Better yet, if possible, use "intelligent" formats like GenCAM, DirectCAM, or ODB++. These formats should enable the fabricator to better understand your design. Also, such formats have the netlist information embedded.
- Design rules can be embedded in some formats.

20.6 THE CAM-TOOLING PROCESS

The CAM-tooling process consists of six or more distinct manufacturing engineering activities. These include the following:

- Design rule checking (DRC)
- Manufacturing review
- Single-image edits
- DFM enhancements
- Panelization
- Fabrication and assembly parameters extraction

20.6.1 Design Rule Checking

Design systems lay out circuits to defined rules; however, these systems may fail to adhere to these rules, because of either system failure or manual intervention. The purpose of reviewing the data package, in addition to reviewing the documented design constraints, is to confirm that the product can be produced at the manufacturing site and to the expected manufacturing yields. The analysis of the data describing the circuitry of the PCB to the manufacturing facility's production capabilities is critical to the success of the product.

Table 20.4 defines typical design rule checking that is performed, the reasons for checking, and the method of checking.

20.6.2 Manufacturability Review

The design rule checking results are reviewed against the PCB manufacturing capability matrix, with dispositions being made to the acceptability of violations. The PCB customer may be contacted regarding design violations, to correct and retransmit the design or to permit design changes to be performed by the PCB manufacturer.

Additionally, the factors reviewed during the initial design review are revisited to confirm the results based on design actuals. If differences are noted, product yields and/or cost predictions may need to be altered. During this step of tooling, several concerns are uppermost in the fabricator's mind: Does the board meet the intent of the design? Can it be built? Will it function as planned? Does it pass a master rule set? Can it be improved, that is, be made cheaper or better? Is there component interconnectivity? Does the IPC-D-356 netlist show any shorted nets or broken nets? Were any multiple plane ties not happening? Did a design work-around fool the system—or allow intentional errors to slip through?

20.6.2.1 Bare-Board Analysis. Several physical artwork problems are not discovered by design rule checking and are not strictly speaking the result of design rule violations. Nonetheless, these problems can affect the manufacturability of a board and its yield. Table 20.5 describes some of these issues.

Figure 20.14 shows examples of many of these problems. Most modern CAM-tooling systems can correct these manufacturability issues. Thus the manufacturing notes on the fabrication drawings should acknowledge this and allow for the fabricator to implement corrections. Other, less frequent, artwork errors discovered include:

- Signal integrity and performance issues
- Verification of last-minute changes [life savers and engineering change orders (ECO)]
- "Made in XX" and revision controls

20.6.2.2 Printed Circuit Assembly. In a similar situation, assembly tooling requires the checks and analysis of supplied BOM, machine files, and artwork for the assembly process. These actions are specifically designed to locate potential manufacturability problems during

TABLE 20.4 Design Rule Checking

DRC item	Purpose of checking	Method of checking
Drill layer duplicate coordinates	Duplicate drill holes consume manufacturing capacity and may result in broken drill bits.	Most PCB CAM systems have an explicit check for this problem, provided a given radius of tolerance.
Minimum drill hole-to-hole spacing	Drill holes too close together may result in broken material between the holes.	Most PCB CAM systems have an explicit feature spacing check.
NPTH-to-board edge minimum spacing	Drill holes too close to the board edge may result in broken material at the board edge.	Merge the routing profile onto a copy of the NPTH layer and execute a spacing check of the pad to trace (the route layer would appear as a trace).
Missing drill holes	Missing holes may result in nonfunctional designs or products that cannot be used due to missing mounting or tooling holes.	Check the pad registration against a circuit reference layer, and verify pads without PTH holes.
Extra drill holes	Extra drill holes could result in nonfunctional designs due to cutting of traces or shorts (if the holes become plated).	Check the pad registration against a circuit reference layer, and verify PTH holes without pads.
Board edge-to-board edge minimum spacing	Routing cuts leaving minimal material may result in fractured or broken material between routing.	Check the feature spacing of the routing layer as a design layer, without considering electrical connectivity.
Copper-to-board edge minimum spacing	Tolerances of the routing operation may result in copper exposed at the edges of the product.	Check the feature spacing of the routing layer as a design layer against the circuitry layer.
Minimum annular ring	Most customer specifications define the minimum acceptable annular ring on PCB products. Annular ring is dependent on the design and the registration tolerances of the manufacturing processes.	Most PCB CAM systems have an explicit annular ring check.
Minimum pad-to-pad spacing	In addition to feature spacing and potential shorts, pad spacing affects the capability to test products electrically.	Most PCB CAM systems have an explicit feature-to-feature spacing check.
Minimum pad-to-track spacing	Feature spacing below manufacturing capabilities could result in shorts and poor product yields.	Most PCB CAM systems have an explicit feature-to-feature spacing check.
Minimum track-to-track spacing	Feature spacing below manufacturing capabilities could result in shorts and poor product yields.	Most PCB CAM systems have an explicit feature-to-feature spacing check.
Copper-to-NPTH minimum spacing	NPTH locations too close together may prevent proper tenting of holes and require secondary drilling operations. NPTH too close together may result in damaged features (e.g., cut traces).	Check the feature minimum spacing against a copy of the design layer merged with the NPTH layers.
Minimum line width	Line widths below manufacturing capabilities could result in poor product yields.	Some PCB CAM systems have an explicit check for this problem; others may require review of the apertures used and highlighting of the apertures for visual inspection.
Track termination without pad	Although this may be the design's intent, missing pads may be the result of poor design information or loading failures. These problems can result in nonfunctional designs.	Most PCB CAM systems have an explicit check for this problem.
Padstack alignment	Misaligned padstacks may result in unpredictable annular ring results, incorrect compensations of registration, and product scrap.	Most PCB CAM systems have an explicit pad registration check.

(continued)

TABLE 20.4 Design Rule Checking (Continued)

DRC item	Purpose of checking	Method of checking
Minimum solder mask pad clearance	Solder mask pad clearances below manufacturing capabilities could result in solder mask on the pads and poor product yields.	Check the solder mask layer (as a circuit layer) against the circuitry layer using an annular ring check.
Minimum solder mask edge-to-feature spacing	Solder mask edge to feature spacing below manufacturing capabilities could result in exposed features and poor product yields.	Check the feature minimum spacing against a copy of the design layer merged with the solder mask layer.
Minimum solder mask annular ring for NPTH	NPTH solder mask clearances may need to be larger to prevent ghosting of the solder mask from light diffraction through the product mask.	Check the solder mask layer (as a circuit layer) against the NPTH layer, and verify that there are no matches of the NPTH to the solder mask layer.
Solder-mask-to-board edge minimum clearance	PCB edge clearances may need to be larger to prevent ghosting of the solder mask from light diffraction through the product.	Check the feature minimum spacing against a negative copy of the solder mask layer merged with the routing layer.
Solder mask minimum web	Solder mask webs below manufacturing capabilities could result in solder mask breakdown and poor product yields. This problem could result in a PCB assembler with solder bridging defects.	Use minimum feature-to-feature spacing on the solder mask layer.
Plane-to-board-edge minimum clearance	Spacing below the tolerances of the routing operation may result in exposure of copper at the edges of the product.	Check the feature spacing of the routing layer as a design layer against the plane.
Minimum plane layer annular ring	An annular ring below manufacturing registration and tolerance capabilities could result in open connections and poor product yields.	Most PCB CAM systems have an explicit annular ring check.
Minimum plane layer clearance	Plane clearances below manufacturing registration and tolerance capabilities could result in shorts and poor product yields	Most PCB CAM systems have an explicit annular ring check.
Plane-to-plane isolation	Plane layer isolation typically results from either incorrect designs or interpretation of aperture lists; the result is a nonfunctional product.	Most PCB CAM systems have an explicit layer-to-layer isolation check.
Nomenclature-to NPTH minimum spacing	Nomenclature spacing below manufacturing capabilities may result in nomenclature in the hole, due to misregistration or bleeding of the nomenclature ink.	Check the feature minimum spacing against a copy of the nomenclature layer merged with the NPTH layers.
Nomenclature-to-solder mask minimum spacing	Nomenclature spacing below manufacturing capabilities may result in nomenclature on product features, due to misregistration or bleeding of the nomenclature ink.	Check the feature minimum spacing against a copy of the nomenclature layer merged with each solder mask layer.
Nomenclature-to-feature minimum spacing	Nomenclature spacing below manufacturing capabilities may result in nomenclature on product features, due to misregistration, bleeding of the nomenclature ink, or skipping of the screen over features, resulting in illegible markings.	Check the feature minimum spacing against a copy of the design layer merged with the nomenclature layer.
Nomenclature minimum feature sizes	Nomenclature sizes below manufacturing capabilities could result in illegible nomenclature and poor product yields.	Some PCB CAM systems have an explicit check for minimum line width; others may require review of the apertures used and highlighting of the apertures for visual inspection.

TABLE 20.5 Physical Artwork Issues

Physical artwork issue	Description of issue	Illustration
Acid traps (photo-resist slivers)	Small pieces of photo-resist that have minimum adhesion and can break off, resulting in plating or redeposits	Figure 20.14a
Pin holes	Small openings in planes or pours that can result in loss of adhesion	Figure 20.14b
Copper slivers	Small features that may lose adhesion and break off	Figure. 20.14c
Starved/isolated thermals	Copper legs for thermals that are cut off from the rest of the plane	Figure 20.14d
Soldermask(S/M) slivers (web)	Small S/M features too tiny to image and adhere properly	Figure 20.14e
S/M to trace spacing	Small spacings that create S/M slivers	Figure 20.14f
S/M coverage	Features left open that could result in solder bridging	Figure 20.14g
Solder bridging	Small gaps between soldered features without S/M that could bridge solder	Figure 20.14h
Legend on pads	Silkscreen or legend on surface mount technology (SMT) lands	Figure 20.14i

board assembly and to collect design statistical data for quoting, planning, and allocating resources. Each action covers a specific aspect of board assembly, including component location, fiducial coverage, and testability issues.

These include but are not limited to the following:

- **Fiducial analysis** SMT component placement needs to align the machine to the individual board. By analyzing global and fine-pitch fiducials, you can identify and correct potential problems. Table 20.6a shows the checklist of a typical analysis.

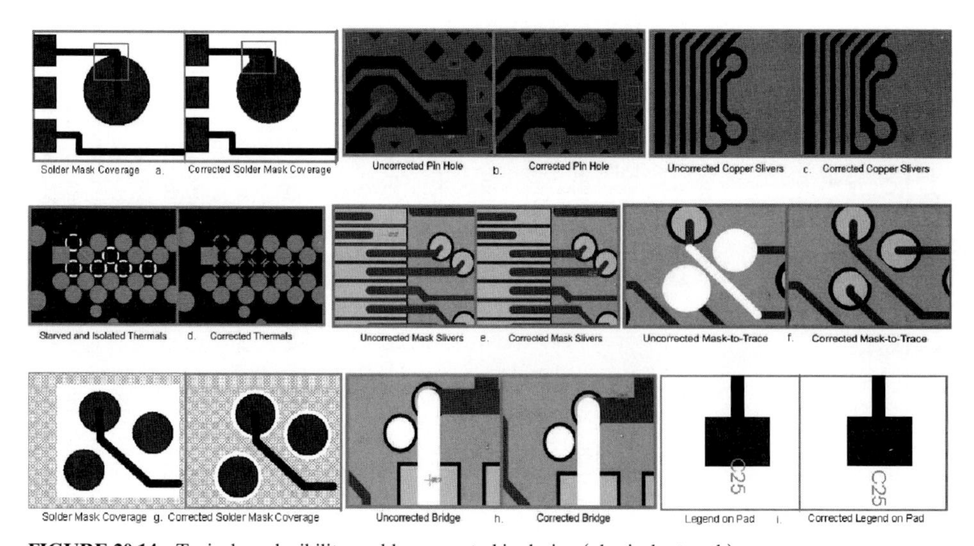

FIGURE 20.14 Typical producibility problems created in design (physical artwork).

TABLE 20.6 Checklist of various assembly and test analyses that need to be performed in Tooling

A. Fiducial analysis	B. Component analysis	C. Padstack analysis	D. Testpoint analysis	E. Solder paste analysis
Global fiducial summary	Comp. max height	Toeprint to toeprint	Toeprint testpoints	Solder paste not on SMD pad
Local fiducial summary	Comp. pitch	Extra solder mask web	Via testpoints	SMD pad without solder paste
Non uniform local fiducial	Comp. rotation	No solder mask web	Other testpoints	Solder paste outside copper
Fiducial to copper	Comp. length Min	Special toeprints	Testpoint to rout	Solder paste not exposed by solder mask
Fiducial to hole	Comp. length max	Via to via center	Testpoint to NPTH	Solder paste to solder mask
Close fiducials	Comp. Spacing	Via to via edge	Testpoint to testpoint	Toe distance min (rect. pins)
Far fiducials	Comp. to toeprint	Via in SMD	Testpoint to toeprint	Heel distance mn (rect. pins)
Fiducial to like pad	Rout spacing	Via to toeprint	Testpoint to capped via	Left Distance min (rect. pins)
Fiducial close to rout	Carrier wall spacing	No thieving pad	Testpoint wrong shape	Right distance min (rect. pins)
Fiducial far from rout	PTH to carrier wall	Toeprint to plane	Testpoint to THMT toeprint	Toe distance max (rect. pins)
Fiducial to background	NPTH to carrier wall	Toeprint annular ring	Testpoint to tooling hole	Heel distance max (rect. pins)
Fiducial to solder mask	Via to carrier wall	Toeprint SM clearance	Testpoint to uncapped via	Left distance max (rect. pins)
Fiducial to solder paste	Label to component	Toeprint to rout	Testpoint to exposed copper	Right distance max (rect. pins)
Fiducials to silk screen	Assembled board thickness	Hole in SMD	Testpoint to unexposed copper	Pin area/pad area min
Fiducial to conveyed edge	Comp. touch silkscreen	Flow solder orientation	Testpoint under component	Large pin area/pad area min
Component center fiducial	Component to conveyed edge	Gold fingers to capped via	Testpoint to component	Pin area/pad area max
Component covers fiducial	SMT component to conveyed edge	Gold fingers to thru TP	Testpoint to component angle	Large pin area/pad area max
Component local fiducial	Other side area keepout	Gold fingers to surface TP	Testpoint to solder mask	Pin width/pad pitch
Component global fiducial	Other Side pins keepout	Gold Fingers to uncapped via	Capped testpoint vias	Pin Pitch/pad pitch

Coarse-pitch component global fiducial	Comp. to gold fingers	Shadowing of inline components	Nets with one testpoint	Direct connection to plane
Fine-pitch component global fiducial	Component to tooling hole	Shadowing of staggered components	Nets without testpoints	No clearance in plane
	Component to mounting hole	Toeprint to v-cut rout	Nets with multiple testpoints	ET pin without PTH
	Hole under component	Toeprint to conveyed edge	Via capped on both sides	MP pin on PTH
	Uncapped via under component	Toeprint to conveyed edge	Uncapped nontestpoint vias	Edge to edge
	No orientation indication	Toeprint to conveyed Edge	Testpoint to conveyed edge	Missing hole
	Component uncommon orientation	Toeprint to break tab rout	Testpoints outside keep-in area	Extra holes
	Wrong orientation indication	Exposed Via to exposed toeprint	Testpoint to silkscreen	Pin too large
	No square pad on THMT pin #1	Irregular exposed toeprint	Testpoints within keep-out area	Pin too small
	Chip edge protection missing	Via to toeprint masked trace	Nets with missing testpoints	Pin touching
	Chip corner protection missing	Testpoints/via Clusters	Nets with necessary testpoints	No pad for pin
	Comp. outside keep-in area	Nonuniform SMD pads size	Nets with extra testpoints	Multiple pads for pin
	Comp. within keep-out area		Net testpoint summary	Pin center to pad heel
	Comp. exceed max height		Net testpoint count report	Pin center to pad toe min
	Comp. under min height			Pin center to pad heel min
	Component to exposed via			Pin center to pad toe max
	Trace under component			Pin diameter vs. hole diameter
	Comp. to reference designator			Large pin right distance max (rect. pins)
	Comp. without reference designator			Large pin toe distance vs. heel distance (rect. pins)

(continued)

TABLE 20.6 Checklist of Various Assembly and Test Analyses That Need to Be Performed in Tooling (Continued)

A. Fiducial analysis	B. Component analysis	C. Padstack analysis	D. Testpoint analysis	E. Solder paste analysis
	Misplaced reference designator			Large pin left distance vs. right distance (rect. pins)
	Wave solder component orientation			Pin diameter/pad diameter min (%)
	Component relative orientation			Pin diameter/pad diameter min (abs)
				Large pin diameter/pad diameter min (%)
				Large pin diameter/pad diameter min (abs)
				Pin diameter/pad diameter max (%)
				Pin diameter/pad diameter max (abs)
				Large pin diameter/pad diameter max (%)
				Large pin diameter/pad diameter max (abs)
				Max pin width/pad width min (%) (rect. pins)

(*Source:* Valor Computerized Systems)

- **Component analysis** SMT components can have a multitude of missing, incorrect, or incomplete characteristics or properties. Component analysis looks for these potential problems, such as one component extending under another component. Table 20.6b provides a typical checklist for such an analysis.
- **Padstack analysis** SMT components have critical toeprints (TPs) and need to be checked with respect to vias, planes, solder mask, fabrication edges, and gold fingers. Table 20.6c shows a checklist for a typical analysis.
- **Testpoint analysis** SMT components need to be tested by in-circuit test (ICT). These analysis check testpoints to vias, surface features, tooling holes, and electrical nets. See Table 20.6d for a typical checklist for testpoint analysis.
- **Solder paste analysis** SMT components have to be soldered correctly. This is the second most important aspect of assembly, just behind placing the right part in the right place. Many checks are necessary to ensure manufacturability, as seen in Table 20.6e.

The benefits of performing a comprehensive and thorough assembly analysis include the following:

- Estimation of cost can be derived from the results of the analysis.
- Manufacturing defects can be identified and corrected prior to construction on the assembly line.
- Such analysis can help detect potential problems so that any necessary reworking can be implemented early in the design process. Each analysis category is governed by a set of engineering reference file (ERF) values, usually unique to each production facility, thereby supporting full job portability between different assembly sites.

The design rule checking results are reviewed against the PCB or assembly manufacturing capability matrix, with dispositions being made to the acceptability of violations. The PCB and assembly customer may be contacted regarding design violations; the designer may need to send corrected designs or to allow design changes to be performed by the PCB manufacturer or assembler.

Additionally, the factors reviewed during the initial design review are revisited to confirm the results based on design actuals. If differences are noted, product yields and/or cost predictions may need to be altered.

When issues are addressed during the fabrication or assembly stage, the design database may not get updated to reflect vital changes, thus compromising the integrity of the design. If the design is respun in the future, this issue might be overlooked, resulting in scrap, or worse, faulty boards.

20.6.3 Single-Image Edits

Almost all designs include information that must be removed before they can be manufactured. Most of these items of information are used as references during the PCB layout phase to assist the designer in understanding the available real estate. However, these references must be removed or modified in order to manufacture the board to design intent. Among these typical items are the following:

Item	**Impact if not removed/modified**
NPTH pads on outerlayers	Plated holes
Routing crop marks	Copper at board edge
Direct contact plane markings of drilled holes	Open circuits
Clip nomenclature withsolder mask data	Nomenclature on pads or in holes
Scaling and compensation	Finished traces or geometries that are the wrong size

20.6.4 DFM Enhancements

The manufacturability of the PCB can be improved through design enhancements. Many of the enhancements can be performed on CAD systems (or through specialized post-CAD software); other enhancements can be performed at the manufacturing site.

Ideally, all design enhancements would be performed at the design site to maintain the consistency of products from multiple manufacturers; however, some of the DFM enhancements can impact the CAD system's capability to provide further design changes. Specifically, the optimization of track-to-feature spacing could result in blocked routing channels, preventing an autorouter from performing. The following are typical DFM enhancements (see Fig. 20.15):

Item	Result from enhancement
Copper balancing	The balancing of copper on the surface of the design will improve the distribution of plating (in the pattern-plating process), preventing areas of overplating due to the isolation of features. This improvement enables the manufacturer to control the plated hole diameters to tolerances and reduce plating heights, which impact solder mask coverage. (See Fig. 20.15a.)
Teardropping	As annular rings decrease due to tighter design specifications, the probability of defects occurring due to registration/annular ring failures increases. The addition of teardropping increases the pad-to-trace junction size, improving the reliability of the connection and improving product yields.(See Fig. 20.15b.)
Removal of nonfunctional pads on innerlayers	Innerlayer shorts result from several factors, among them the proximity of circuitry to other circuitry. The occurrence of shorts can be correlated to the running length of the circuitry versus the running length at some minimum spacing between circuitry (at the process capability of the operation). Reducing the total distance at minimum spacing will reduce the probability of shorts, resulting in improved yields. (See Fig. 20.15c.)
Optimization of track-to-feature spacing	As described in the nonfunctional pads discussion, the occurrence of shorts can be correlated to the running length of the circuitry versus the running length at some minimum spacing between circuitry (at the process capability of the operation). Reducing the total distance at minimum spacing will reduce the probability of shorts, resulting in improved yields. (See Fig. 20.15d.)
Track width optimization	The occurrence of opens can be correlated to the running length of the circuitry versus the running length at minimum track width (at the process capability of the operation). Reducing the total distance at minimum track width will reduce the probability of opens, resulting in improved yields. (See Fig. 20.15e.)

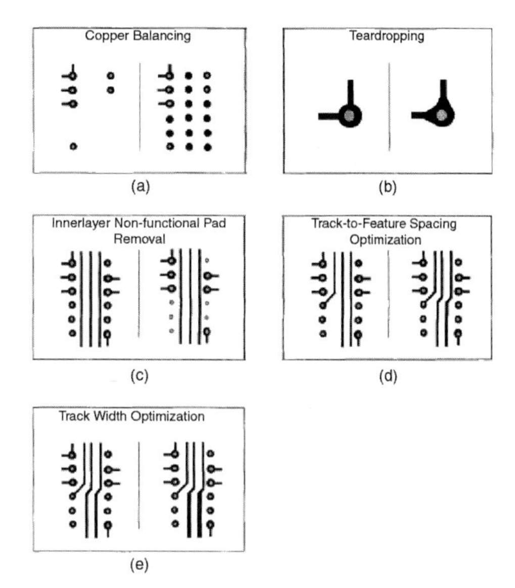

FIGURE 20.15 Typical producibility problems created in design.

20.6.5 Panelization

In the panelization process, the PCB single images are placed in the locations defined in the panelization definition step of the initial design review. In addition, the following features may be added to the panel for manufacturing:

Item	Purpose
Customer coupons	These coupons are provided to the customers for their inspection of the PCB product. These coupons are typically treated as separate PCB single images.
Internal coupons	During the manufacturing process, destructive testing may be required to confirm product quality. These coupons are typically treated as separate PCB single images.
Tooling holes	Holes used for registration of manufacturing panels to artwork, drilling, or routing operations are added to the panel borders.
Outerlayer thieving patterns	Plating thieving patterns are sometimes needed to balance the platable area across the panel.
Innerlayer venting patterns	The addition of copper to innerlayers outside the PCB single image provides more consistency in the lamination process.
Textual markings	Operator aides may be provided outside the PCB single image to improve manufacturing quality.

20.6.6 Fabrication and Assembly Parameter Extraction

Additional manufacturing information must be extracted or calculated from the design file before fabrication or assembly can begin. This information is loaded and stored in several files and databases used to run various fabrication and assembly equipment. In addition to requiring the basic tooling steps already defined, most PCB manufacturing sites also require the designer to provide the following information:

- Raw materials shearing and labeling instructions
- Copper foil thicknesses and etch rates
- Automatic optical inspection files
- Lamination lay-up BOM and order/stack height
- Lamination press cycles, heat input, and pressure settings
- Drill-file location and revision levels
- Plating area calculations and current density settings
- Copper foil thicknesses and etch rates
- Router-file location and revision levels
- Electrical test programs
- Quality control procedures and coupon disposition
- Final packing and shipping instructions

20.6.6.1 AOI Files. The CAD reference files, if required, are generally created at the PCB manufacturing site. These files prepare automated optical inspection (AOI) systems with data that confirm that manufactured pieces match, within tolerances, the PCB design.

20.6.6.2 Lamination Parameters. Each multilayer stack-up and material may require specific lamination processes. Modern multilayer presses have microprocessor-controlled time, temperature, and pressure. These parameters require selection in the tooling process.

20.6.6.3 Drill Files. Drill files and the respective drill bit tools must be archived and available for download to the appropriate drilling machine when production needs them. Drill bit preparation precedes the actual drilling operation in order to create the drill cassette.

20.6.6.4 Plating Coefficients. Plating requires specific direct currents (DC) to be applied to this part. The current density of the appropriate plating bath is multiplied by the plated area on each production rack. After the first actual run, direct measurement of the plated metals is used to adjust the plated area to achieve the correct thicknesses.

20.6.6.5 Routing Files. Router files and the respective router cutting tools have to be archived and available for download to the appropriate routing machine when production needs them. Router tool preparation precedes the actual routing or secondary drilling operation in order to create the final fabrication cassette.

20.6.6.6 Electrical Test Programs and Fixture Creation. An electrical test netlist file may be either created internally from the artwork or prepared from a supplied IPC-D-356 file. This test file has to be compared to the netlist extracted from the artwork supplied. Any difference must be resolved before the actual electrical test file is finalized. In addition, the electrical test fixture must be constructed or sent to an outside contractor to prepare.

20.6.6.7 CAD to Assembly Processes. The assembly tooling operations include CAD interfacing to component data and BOM. These operations include the generation of manufacturing documentation (see Fig. 20.16) and assembly engineering for component machine management, program optimization, and line balancing (see Fig. 20.17).

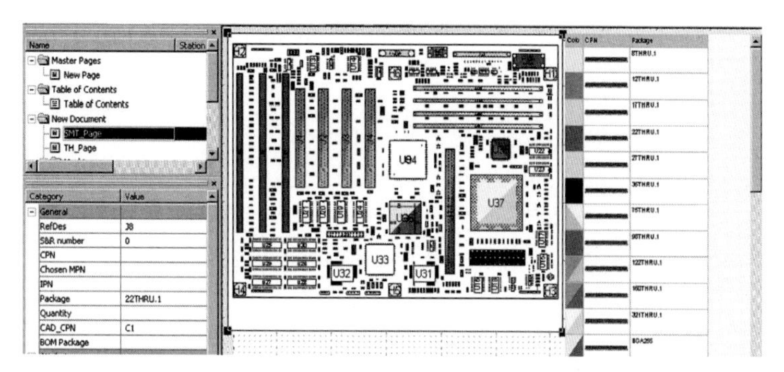

FIGURE 20.16 Assembly manufacturing documentation and instructions creation.

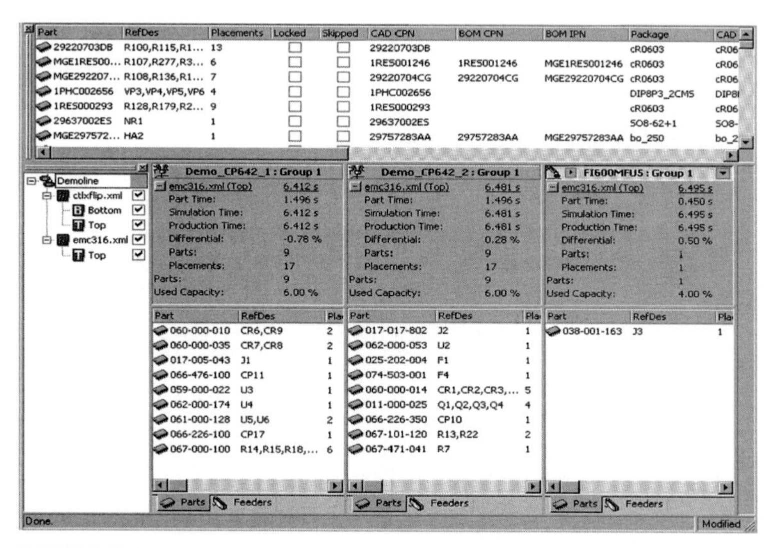

FIGURE 20.17 An assembly BOM control optimizing mixed vendors lines with line balancing.

20.7 ADDITIONAL PROCESSES

Because of variations in the PCB manufacturing process, additional CAM-tooling operations may be required. Variations in manufacturing that may call for additional tooling include the following:

- Rivet lamination versus pin lamination
- Laser drilling
- Multiple panel sizes
- Very large lamination presses (48 in. × 48 in.)
- A very high degree of automation

New processes or PCB product features that may require additional tooling processes include the following:

- Embedded passives
- Embedded waveguides

Embedded passives and embedded waveguides are complicated to implement in production. Alignment and overlap are critical. Proper scaling for the particular embedded materials is especially important, depending on whether the materials are additive or subtractive. Figure 20.18 shows some of the critical tooling dimensions.

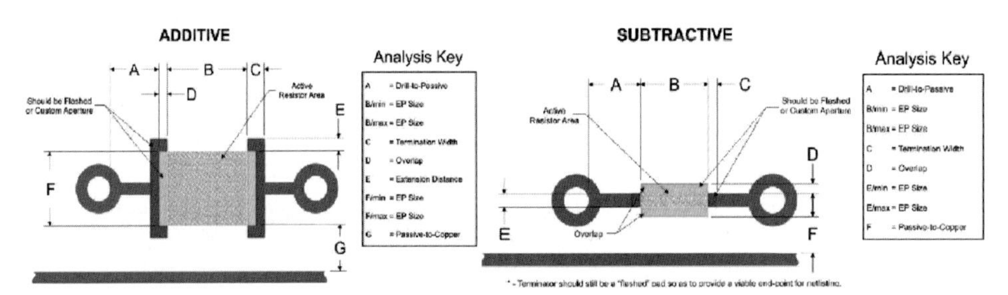

FIGURE 20.18 Critical tooling data for embedded passives.

An important step in all the processes already defined is the necessity for proper information management. Archival of PCB customer-supplied information and the files generated from the tooling process are critical for disaster recovery. Archiving systems available on the market also provide for centralized data management and distribution of the information to the various departments within an organization.

20.7.1 Macros

Macros (or scripts, as they are sometimes called) are used to automate tooling operations. Most modern CAM systems come with the facility to create, edit, debug, and exercise macros. One example of a macro was shown in Fig. 20.9. This macro automatically scans all the CAM information (artwork and NC files) and reports minimum, maximum, and critical geometries.

20.7.2 Design-to-Fab and Assembly Automation

Although a vast number of sophisticated CAD and CAM software packages are available today, experienced operators are still needed to exercise this software. During the era of captive printed circuit manufacturing by original equipment manufacturers (OEMs), internally developed software tools were created that actually automated the entire CAM tooling process. This level of automation was made possible through the development of software that automatically created drill, rout, fabrication, and panel drawings and manufacturing notes from the CAD database. In doing all this, the software captured in a digital header file all the pertinent information on the drawings. The header file could then automatically supply macros that would complete all the tooling. Although this software was never extended to the commercial market, today new automatic blueprint creation software is available that works off of the CAD database.

20.8 ACKNOWLEDGMENT

I would like to thank Jeff Miller, WISE Software Solutions (www.wssi.com), DownStream Technologies (www.downstreamtech.com), and Julian Coates of Valor, Inc. (www.valor.com) for their invaluable contributions to this chapter.

CHAPTER 21
EMBEDDED COMPONENTS

Dennis Fritz
MacDermid Incorporated, Waterbury, Connecticut

21.1 INTRODUCTION

Historically, electrical components have been mounted on the outside of printed circuit boards, either with through hole leads or with terminations all on a single side (surface mount) with the interconnection traces embedded inside the board. Technology, however, has been developed to embed electrical components inside the board also. The earliest components to be embedded were resistors, which are made by etching a pattern on a sheet of resistive material and then connecting them with the rest of the circuit through the standard multilayer process. In addition, capacitors are formed from thin dielectric material between closely spaced copper-foil planes, and inductors are formed by etching coils of the copper conductive foil during innerlayer manufacture. These fabrication techniques have expanded to include the ability to "place" small, discrete passive components inside the board. This allows the normal multilayer board pressing operation to encapsulate these placed components.

21.2 DEFINITIONS AND EXAMPLE

- **Components** These are the electrical units that make up all electrical devices.
- **Passive components** These components—including resistors, capacitors, or inductors—influence electrical flow in a circuit, but do not cause a gain in current or voltage.
- **Active components** In contrast, these components can provide gain in an electronic circuit.
- **Integrated passive arrays** Multiple passive devices may be packaged in either a leaded or surface-mount technology (SMT) package or sometimes in arrays of all the same value. *Integrated passive component* is a general term for multiple passive components that share a substrate and packaging. Most typically, integrated passives are on the surface of a separate substrate that is then placed in an enclosure and surface-mounted on the primary interconnect substrate, in which case they are called *passive arrays* or *passive networks*. This chapter discusses only embedded components.

- **Embedded components** Components formed or inserted inside the primary interconnect substrate are said to be embedded. Embedded components can be either passives or actives (see the preceding definitions).
- **Formed components** A component that the circuit board fabricator manufactures inside the primary interconnect substrate (as opposed to on the surface) is said to be *formed.* Such components are made from raw materials at the same time as the board.
- **Placed embedded components** In contrast to formed components, components (both discrete passives and small actives) can also be "placed" inside the board using conventional placement machines and technologies. This manufacturing technique integrates very well with the "any laycr via" type of contruction.

See Fig. 21.1 for a cross section of a multilayer board with all types of embedded components.

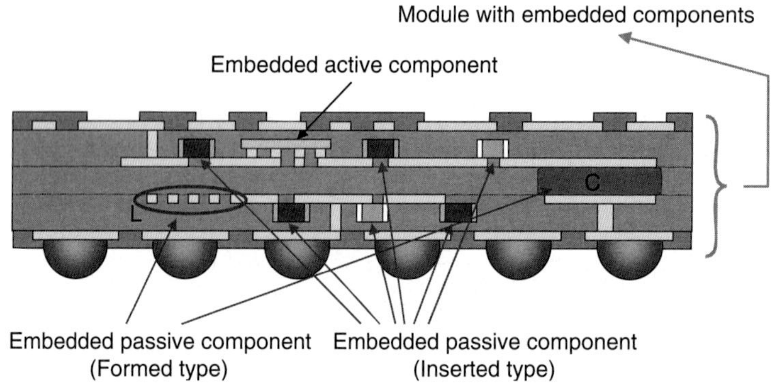

FIGURE 21.1 Cross section of a multilayer board showing the use of all types of embedded components, both passive and active. *(Courtesy of Jisso International Council [JIC].)*

21.3 APPLICATIONS AND TRADE-OFFS

Embedding components is not appropriate for all electronic packaging. A board designer must consider several factors when deciding whether to use embedded components or not. Use of discrete components, whether on the surface or placed inside the board, involves incremental costs in that the more components that are used, the higher the cost. However, the trade-off is that the cost of fabricating one formed embedded component is about the same as that of forming an entire layer of those components. Another factor to consider is that surface-mount components can have the same size and shape for widely differing electrical values. Formed embedded components on a single layer are larger and can increase electrical value. Widely differing electrical values may require more than one layer of embedded formed material.

21.3.1 Advantages

Advantages from embedding components inside the board include:

- **Reduction in board area (enabler)** Reduction in surface area results in more circuits per fabrication panel. Reduction in board size can yield a smaller board for the same functions, as components are now inside the board.

- **Increase in functionality, higher density (enabler)** The surface area is the same, but some other function can be added. For example, the designer can increase cell phone capability by adding a Bluetooth connection to the ear piece receiver.
- **Improvement in performance (enabler)** Higher speeds, shorter paths, and lower noise are possible. For example, cell phones can be enhanced with built-in digital cameras or even video cameras.
- **Improvement in total assembled costs** System cost is key; the higher board price is often offset by a lower bill-of-materials purchased for the system, as fewer standard components are needed.

21.3.2 Disadvantages

Disadvantages of embedding components include:

- Quality can be affected; in particular, meeting tolerances can be more difficult.
  Trimming embedded resistors is slow and expensive.
- Design tools are lacking.
  The design is "parametric" where designer sizes components.
- Technology implementation does not give all values needed.
  Some capacitance values are not obtainable commercially (>100 nF/cm^2).
  Manufacturers of sheet resistor materials make only one decade of values.
- The cost of prototypes can be high.
  Resistors cost the same whether one or thousands are made per layer.
- Test tools are lacking.
  Test voltages may cause dielectric breakdown.
  Charge-up time for capacitors may limit test speed.
- Rework is not possible inside the board.
- Additional capital investment is required.

21.3.3 Principles of Cost Trade-Off

The circuit design tools identify the resistors and capacitors that must be added to make an electronic circuit functional. The tools simulate circuit operation and "size" the resistors and capacitors as appropriate. Whether a resistor is purchased and mounted on the board surface or formed inside the board from a layer of resistive material has little effect on the electrical operation of a circuit. Therefore, if many resistors can be formed inside the board from the same series of chemical steps, the inclusion of perhaps thousands of internal resistors costs no more than adding one internal formed resistor. The critical economic analysis is to find the cost of adding internal formed resistors and then dividing by the number that can be used internally to see whether embedded resistors inside the board make sense.

21.4 DESIGNING FOR EMBEDDED COMPONENT APPLICATIONS

21.4.1 Resistors

Resistors were the first component to be embedded in a multilayer board, making it a truly three-dimensional interconnection and circuit system. While early designs were custom for each applicaton, over time design standards and "best practices" have been developed, which allow for successful inclusion of resistors, and other components, without the need to start from the beginning. The following discussion provides these for resistors, with subsequent sections providing similar information for other components.

21.4.1.1 Resistor Terminology. The designer must size embedded formed passive resistors. This is in contrast to the "library selection" of conventionally manufactured resistors that are selected from a table of available components. The principle of embedded resistors is given by the following formula for a term *ohms per square*; this means that any square area defined in the material will have the same resistance and that by grouping squares appropriately, the designer can define the resulting resistance.

Resistance in any resistor is determined by Eq. 21.1

$$R = \rho \, L/A \qquad \text{(Eq. 21.1)}$$

where R = resistance in ohms
ρ = resistivity in ohm meters (inches)
L = length of the resistor
A = cross-sectional area (resistor thickness times resistor width)

Note: Resistivity (ρ) is a material property. It is a constant at a given temperature and expressed in resistance units (ohms) for area and length. For instance, the resistivity of copper is $\rho = 7.09\ e^{-7}$ ohm in.^{2}/in.

Sheet resistance is a commonly used term to describe raw materials used in embedded formed resistors. These materials are manufactured in sheet form with a constant thickness, giving a uniform *R* value for any square resistor. Therefore, the material is specified by its sheet resistance value, given in ohms/square (Ω/□). The resistance of a specific resistor is simply designed by changing the ratio of length to width. For instance, 100 Ω/□ material forms a 100 ohm resistor in any square configuration. A resistor three squares long by one square wide would be 300 Ω, and a resistor two squares wide by one square long would be 50 Ω.

21.4.1.2 Resistor Design Parameters. The circuit board designer is responsible for sizing the resistor to give the appropriate value. Material suppliers have given their sizing parameters to computer-aided design (CAD) software companies for entry in their component libraries. Also, manufacturing tolerances for each supplier product have been communicated to computer-aided manufacturing (CAM) suppliers. Figure 21.2 gives typical design parameters for sizing both the active embedded formed resistor area and the termination pads for the traditional copper innerlayer upon which the resistor is built.

21.4.2 Capacitor Design

All production formed capacitors use laminate-like materials. Most typically, the capacitor manufacturer starts with an especially flat copper foil on at least one side. This copper foil is precisely coated with the dielectric material: filled polymer. Either the top foil electrode is then laminated or two layers of coated foil are bonded polymer to polymer.

21.4.2.1 Capacitor Terminology. Capacitors are electrical elements that store charge on parallel electrode plates. They are not batteries that operate chemically. Capacitors store more charge with closer parallel plates, and the higher the dielectric constant, the more charge is stored.

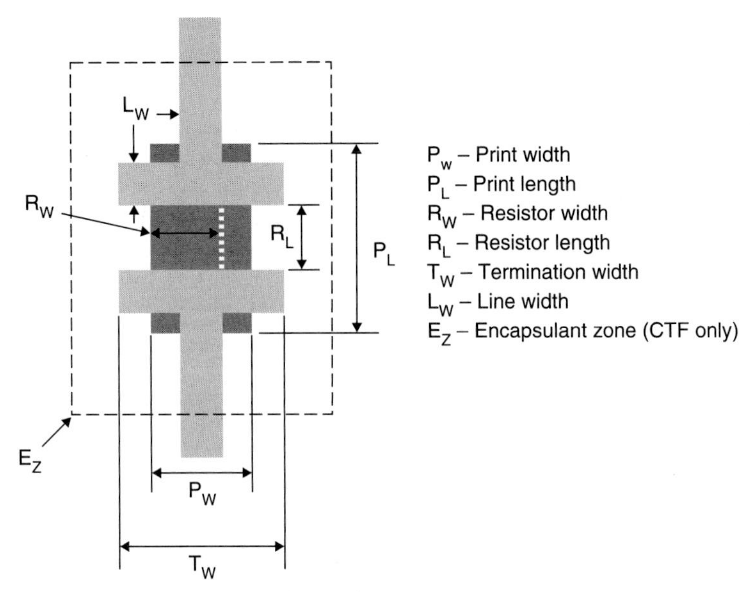

FIGURE 21.2 Resistor terminations.

Again, the value of embedded capacitors must be designed to the correct size. The value of a formed embedded capacitor is given by the Eq. 21.2.

$$C = A \times D_k \times K/t \tag{21.2}$$

where C is the value of the capacitor
A is the area of the formed planar capacitor L × W
D_k is the dielectric constant of the capacitor material
t is the thickness between the capacitor plates
K is a conversion constant = 8.854×10^{-14} Farads/cm

21.4.2.2 Capacitor Design Parameters. Similar design tolerancing for capacitor design parameters (compared to resistors in Sec. 24.4.1) is available for capacitors, as shown in Fig. 21.3.

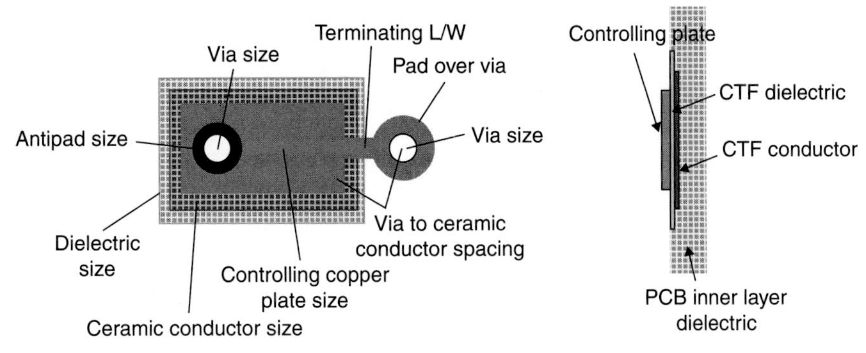

FIGURE 21.3 Capacitor design features.

Even conventional epoxy glass laminate has some capacitance value, and the oldest formed embedded capacitors were carefully controlled thin conventional glass/epoxy laminate layers. Both epoxy resin and E-glass have dielectric constants of about 4, not a particularly large value. However, by eliminating the glass and adding compounds such as barium-titanate to the epoxy or polyimide resin, dielectric constants can be improved to the 10 to 20 range. If really high capacitance values are needed, ceramic or inorganic compositions must be used. Newer technologies demonstrate dielectric constants in the 100 to 2,000 range, capable of forming large value capacitors inside the circuit board. These larger capacitors are useful for decoupling, one of the largest uses of capacitors.

Thickness is the divisor in the capacitance formula. The smaller the dielectric thickness, the higher the device capacitance. Early formed embedded distributed capacitors simply used the readily available 100 micron (4 mil) epoxy/glass innerlayer construction, common in multilayer circuit boards. Then, thinner woven glass allowed 50 micron glass/epoxy layers, doubling the device capacitance. Finally, polymer only or barium-titanate-filled polymer demonstrated thicknesses down to 8 microns, improving capacitance 6 to 20 times the value of the original 100 micron epoxy glass. With the thinner materials comes a higher price tag for precision laminate manufacture and special fabrication/handling techniques. The cost/benefit analysis is a trade-off between these premium materials and the smaller-size/higher-capacitance values of the filled polymers.

21.4.3 Inductors

Inductors use the magnetic effects of electrons traveling in wires to influence the travel speed of other electrical signals. Typically inductors are a coil configuration. No special material technology has been demonstrated for inductors. These spirals of copper circuitry are typically formed similarly to the copper-conductive traces of innerlayers. Again, design software must determine the line width, length of the coil, and the number of "turns" of copper to size the proper inductor.

21.5 MATERIALS

21.5.1 Resistor Materials

Resistors, discrete or formed, are actually very poor electrical conductors when compared to copper. Their function is to restrict the flow of electricity in a circuit. Common compositions of resistors include metal oxides, carbon particles, or small conductive particles separated widely by an organic polymer. Conventionally, the value of a resistor is rated in ohms (Ω), with higher values restricting more current. Also, resistors are sized by tolerance in percentage, allowable maximum current, and parameters that deal with the electrical frequency of operation.

Material suppliers have designated the range of resistance values that are optimum for their material. The designer determines the length and width that is allowable in the design, and then determines which material to specify for manufacture so that the final resistor value is achieved in the allocated circuit board space.

Formed embedded resistors of similar values are designed on one specific layer so that the maximum economic value is achieved for that layer.

Materials are also available that can combine mixed values of resistance on one layer. Polymer thick films (PTF) are applied by successive screen print/cure cycles so that all the desired resistance values can be made on one layer. Also, some sheet capacitance manufacturers laminate a layer of resistor foil as one electrode in their construction. Therefore, a fabricator can mix resistance and capacitance on one sheet type raw material.

21.5.2 Resistor Fabrication Details

Several different raw materials and processing systems are used to fabricate embedded resistors. The most commonly used are listed in this section.

21.5.2.1 Photoprint. Photoprint involves a supplier coating a thin, resistive layer on a copper-foil sheet and selling this to either a laminate supplier or directly to the board fabrication shop. The fabricator uses two imaging and etching steps: the first to image the copper conductors on the layer, and the second to size individual resistors by using a second, different etchant solution, specifically for the resistor composition. Typically, alkaline etchants are used in step one, and acidic etchants are used in step two.

21.5.2.2 Screen or Stencil Print. Screen or stencil print is the operation of adding specific resistive paste compositions directly to the etched innerlayer. Polymer compositions (PTF) are cured directly at normal board exposure temperatures (150–200°C). Ceramic compositions cure at much higher temperatures (900°C), so the sequence of manufacturing operations is changed. Copper foil first has the uncured ceramic paste screened. Then firing of the paste occurs in a nitrogen blanketed oven. The foil, with cured ceramic resistors, is laminated to B-stage epoxy, and the epoxy layer cured to C stage with resistors pressed into the resin. Finally, the innerlayer copper traces are etched, exposing the embedded ceramic resistors.

21.5.2.3 Plating. Plating is an additive process where the copper layer is first etched. Then the layer is catalyzed for electroless plating, and photoresist is applied. Resistor locations are imaged into the resist and the pattern developed. Exposed catalyst in the proper areas then initiates plating of the resistors. The resist is stripped, background catalyst is removed, and the resistors are in the proper locations.

21.5.2.4 Inkjet Printing. Inkjet printing is the additive process of applying appropriate resistor materials (usually PTF-type compositions) directly to etched innerlayers. Resistors are applied one at a time, in contrast to the mass screening of conventional PTF. It is also possible to print several resistor value compositions before a final single cure step.

21.5.2.5 Photoimageable Discrete. In a photoimageable discrete process, resistive compositions have been dispersed in traditional solder mask–like materials, which are applied to the surface of an etched layer. Imaging and development leave the desired geometry resistor after the development step.

21.5.2.6 Resistor Trimming. Resistor trimming is a necessary step. All of the resistors given here have been shown trimmable to very precise values. This is normally done with a modified laser machine, adapted from trimming ceramic surface mount discrete resistors. Either the machine uses a probe card pattern to measure the values of a number of resistors at one time or it uses a flying probe technology from electrical test to measure and trim a single resistors. The flying probe system is used in hybrid circuit manufacture, and the probe card system is standard for arrays of resistors from discrete component suppliers. An interpretation of the laser trim process is shown in Fig. 21.4.

21.5.3 Capacitor Fabrication

Normally, embedded formed polymer capacitors are made from purchased sheet materials. It is difficult for the board fabricator to apply a liquid dielectric formulation precisely to laminate, but one photolithographic capacitor formation technique has been developed. Unneeded dielectric is removed with a developer solution similar to liquid photoimageable solder mask.

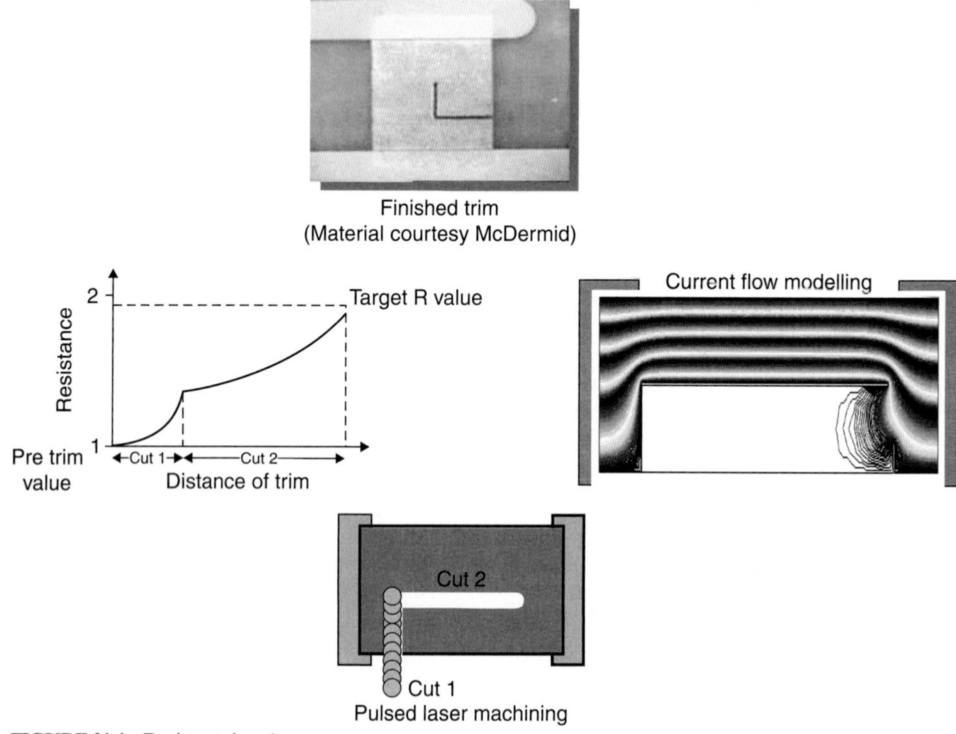

FIGURE 21.4 Resistor trimming.

Embedded formed ceramic capacitors have been demonstrated, but the cure step for the ceramic capacitor composition (900°C) may limit this technology to very large shops that install the "cure on copper foil" technology also needed for ceramic resistors.

21.5.4 Inductor Fabrication Process

Formation of inductors simply involves the design in copper circuitry, and then careful imaging and etching to ensure close tolerance conductors.

21.5.5 Active Integrated Circuit Fabrication

Manufacture of formed embedded active devices has been demonstrated, but not produced commercially. An example is polymer semiconductor radio frequency identification (RFID) tags. Conventional RFID devices, now in mass production, use a tiny silicon integrated circuit, made conventionally and bonded to the antenna. Semiconducting polymers make up the source, drain, and gate of the integrated circuit and are applied by inkjet or screen printing.

21.6 MATERIAL SUPPLY TYPES

21.6.1 Resistor Materials

Formation of the resistor by the designer, working with the board fabricator, involves making a material selection. All resistor values are not possible with each raw material technology. Table 21.1 shows the various forms of the raw resistor material.

TABLE 21.1 Embedded Formed Resistor Supplier Types

Process	Technology
Photoprint (Sheet resistor material)	Thin-film etch (NiCr, Pt, NiP)
Screen or stencil print	Polymer thick-film addition (PTF)
	Ceramic thick-film addition (CTF)
Plating	Selective addition
Inkjetting	Selective addition
Photoimageable discrete	Photopolymer (filled)

Table 21.2 gives the range capability of the various resistor manufacturing technologies.

TABLE 21.2 Embedded Formed Resistor Ranges

Resistance in Ohms	10	100	1,000	10e4	10e5	10e6
Photoprint (NiCr, NiP)	X	X	X			
Photoprint (Pt)		X	X	X		
Screen or stencil print (PTF)	X	X	X	X	X	X
Screen or stencil print (CTF)	X	X	X	X	X	X
Plating	X	X	X			
Inkjetting	X	X	X	X		
Photoimageable discrete	X	X	X	X		

Where materials are sold in foil form or plated, only one value of resistors can be produced per layer. Since the resistive material is consistent across the panel, the final value of the finished resistor is determined by the length-to-width ratio. If the resistive material is made at 100 ohms per square properties, a resistor that is one unit wide and five units long will have a value of 500 ohms. Or, if the resistor is one unit long and five wide, the value will be 20 ohms. Practical resistor ranges evolve from the basic properties of the starting foil material.

Paste compositions, both polymer and ceramic, can have different dispersions of resistive particles, thus giving widely variable resistance values. However, to limit manufacturing steps, selection of certain paste values, usually 10 to 15 times apart, are used. Thus, a design with resistors from 10 ohms to 100,000 ohms could use pastes of 100 Ω/□ 1,000 Ω/□ per square, and perhaps 50,000 Ω/□.

A graphical representation of various formed resistor configurations is shown in Fig. 21.5.

21.6.2 Capacitor Materials

Capacitors are needed in a great range of values. Capacitors may be the conventional "discrete" or a whole capacitance sheet may be used—*distributed capacitance*. Distributed capacitance is quite useful electrically, when high-speed integrated circuits need voltage immediately to turn on or off. The distance to voltage planes may in fact be too long in some designs to allow efficient electrical operation of very fast integrated circuits.

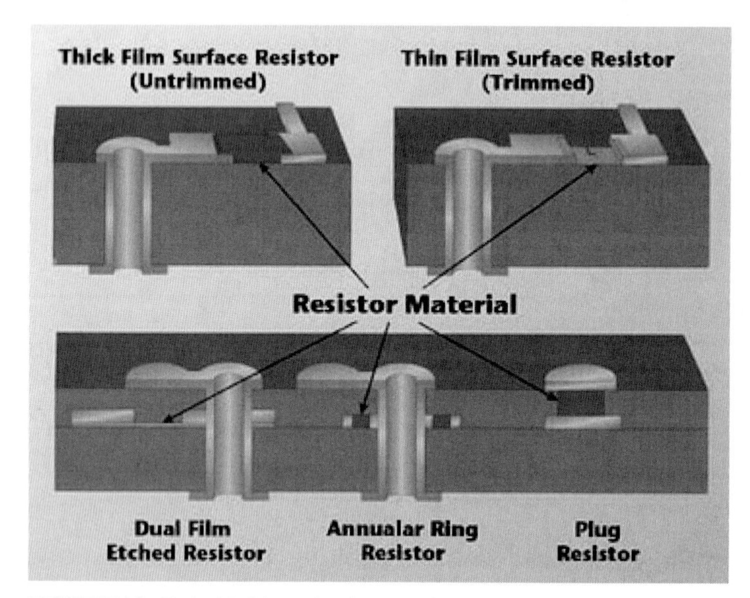

FIGURE 21.5 Embedded formed resistor configurations.

Since the area used to make the formed embedded capacitor can vary, to compare the capabilities of each technology, we must choose a standard area. Therefore, the capacitance at 1 sq. cm is convenient, as shown in Table 21.3.

Capacitor values typically vary from a picofarad to a few microfarads. Thus, low picofarad applications in timing, decoupling, and filtering are the best applications for formed embedded capacitors inside circuit boards.

TABLE 21.3 Embedded Formed Capacitor Technologies

Process	Technology	Dielectric Constant	Micron Thick	Capacitance of a 1 sq. cm Area
Etched copper planes	FR-4/glass, copper clad laminate	4.5	50	78 pF/cm^2
			25	156 pF/cm^2
	Polyimide (filled) copper-clad laminate	11.0	14	700 pF/cm^2
	Polyimide (unfilled) copper-clad laminate	3.5	25	122 pF/ cm^2
	Proprietary dielectric (filled) copper-clad laminate	4.2	24	140 pF/cm^2
			12	310 pF/cm^2
	Epoxy (filled) copper-clad laminate	15	16	850 pF/cm^2
Screen or stencil printed discrete	Ceramic thick-film	1500	25	93 nF/cm^2 (93,000 pF/cm^2)
Photoimageable discrete	Photopolymer (filled)	22	11	1.5 nF/ cm^2 (1,500 pF/cm^2)
Emerging	Sol gel formation	150	0.5	150–300 nF/cm^2

P · A · R · T · 5

HIGH-DENSITY INTERCONNECTION

CHAPTER 22
INTRODUCTION TO HIGH-DENSITY INTERCONNECTION (HDI) TECHNOLOGY

Happy T. Holden
Mentor Graphics, Longmont, Colorado

22.1 INTRODUCTION

The use of more complex components with very high input/output (I/O) counts has pushed the board fabricator to reexamine techniques for creating smaller vias, and many new or redeveloped processes have appeared on the market. These processes include revised methods of creating holes, such as laser drilling, micro-punching, and mass etching; new methods for additively creating dielectric with via holes using photosensitive dielectric materials; and new methods for metallizing the vias such as conductive adhesives and solid post vias. All of these methods share some common traits. They all allow the designer to increase significantly routing density through the use of vias in surface-mount technology (SMT) pads, to reduce size and weight of product, and to improve the electrical performance of the system. These types of boards are generically called high-density interconnects (HDI). An HDI board typically will have, as an average, over 110 to 130 electrical connections per square inch (20 connections per sq. cm) on both sides of the board.

22.2 DEFINITIONS

Printed wiring boards (PWBs) with microvia hole structures are called different names, such as HDI, SBU (sequential build-up), and BUM (build-up multilayer). However, HDI covers a broader range of high-density wiring boards such as extremely high-layer-count multilayer boards (MLBs) without microvia holes. MLBs with microvia holes are not necessarily built sequentially, nor do they necessarily have build-up structures. These definitions are not appropriate for the discussions in this chapter, and therefore we shall address MLBs with microvia holes simply as *microvia hole boards* (all microvia hole boards are essentially multilayer boards).

Some trade and academic organizations define the microvia hole to be a hole of a certain diameter or less. For example, IPC defines a microvia hole as a hole with a diameter equal to or less than 150 μm. However, when a surface blind via (SBV) hole is formed between layer 1 (L1) and layer 3 (L3), the diameter of such a hole is typically 250 μm in order to facilitate reliable plating, but the hole is still considered a microvia hole. Since all microvia holes are essentially SBVs and are normally small in diameter in order to increase circuit density, it seems more appropriate to define the microvia hole as an SBV without limiting its diameter. As long as a hole has SBV structure, it is defined as a microvia hole throughout this chapter.

The first microvia PCB in general production: (a) Hewlett Packard's FINSTRATE was put into production in 1984. It was a copper-core, build-up technology that had direct wire-bonded integrated circuits (ICs). After laminating layers of plasma-metalized polytetrafluoroethylene (PTFE) to the copper core, vias were mechanically drilled through the copper core and then insulated with PTFE. Additional through holes were then drilled along with 5 mil blind vias. (b) The first photodielectric microvia board was produced in volume in Japan by IBM-Yasu. This is the SLC technology with two build-up layers on one side of the four conventional FR-4 layers. These can be seen in the 5th Edition of the PCB Handbook, Fig. 21.1.

22.2.1 HDI Characterization

This generation of printed boards is characterized by very small blind, buried, and through vias made by techniques other than mechanical drilling. To turn blind vias into buried vias, these process techniques are repeated and the layers are "built up," hence the name *build-up* or *sequential build-up circuits (SBU).*

This type of printed circuits actually started in 1980, when researchers started investigating ways to reduce the size of vias. The first innovator is not known, but some of the earliest pioneers include Larry Burgess of MicroPak Laboratories (developer of LaserVia), Dr. Charles Bauer at Tektronixs (who produced photodielectric vias),[1] and Dr. Walter Schmidt at Contraves (who developed plasma-etched vias). Laser-drilled vias were used in mainframe computer multilayers in the late 1970s. These were not as small as the laser-drilled vias today and were produced only in FR-4 with great difficulty and at great cost.

The first production build-up or sequential printed boards appeared in 1984, starting with the Hewlett Packard laser-drilled FINSTRATE computer boards, followed in 1991 in Japan with Surface Laminar Circuits (SLC)[2] by IBM and in Switzerland with DYCOstrate[3] by Dyconex. Figure 21.1 in the 5th Edition of the PCB Handbook shows one of those first Hewlett Packard FINSTRATE boards and one of the first IBM SLC boards.

Since the introduction of SLC technology in 1991 (see Chap. 5; Fig. 5.5), many variations of methods for mass producing HDI wiring boards have been developed and implemented. However, if one technology is to be picked as a winner judged in terms of volume produced, laser-drilling technology is the one. Other methods are still used by a number of PWB manufacturers, but in a much smaller scale.

The purpose of this chapter is to examine a variety of microvia hole formation technologies, structures, and materials.

However, a greater emphasis will be placed on the laser-drilling process (*laser via* hereafter) since it is the most popular process today and it seems its popularity will grow in the future. It must be understood that via hole formation is just one element of fabricating HDI wiring boards. Fabrication of HDI wiring boards with microvia holes involves many new processes not common to conventional board fabrication. Therefore, additional emphasis will be placed upon these new fabrication processes that are common to other microvia technologies.

22.2.2 Advantages and Benefits

Four main factors drive printed boards to require higher wiring densities:

- More components can be placed on both sides of the printed circuit.
- Components are closer together.

- Size and pitch of components are smaller while the number of I/Os is increasing.
- Smaller geometries allow faster transmission of signals and reduce signal crossing delay.

At the same time, enhanced performance is required for faster signal rise times, reduced parasitics, reduction of radio frequency interference (RFI) and electromagnetic interference (EMI), fewer layers, and improved high-temperature performance and reliability. HDI provides all of these advantages and more.

22.2.3 HDI Traditional PCB Comparison

The interactions among printed circuit boards, components, and assemblies are best seen by the packaging technology map (see Chap. 2; Fig. 2.2). Components are characterized by the average I/Os per part, assemblies by the components per sq. cm and I/Os per sq. cm, and the printed circuit by its wiring density in cm per sq. cm. Fig. 22.1 shows the approximate crossover between the traditional printed circuit board and the next generation with microvias.[4]

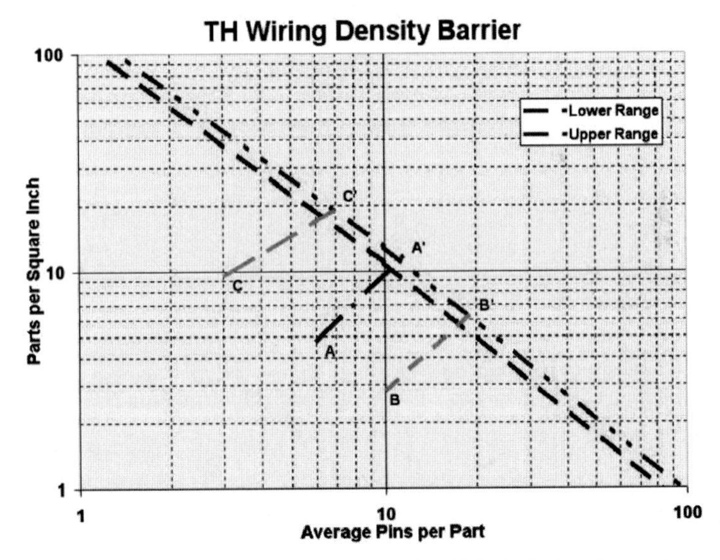

FIGURE 22.1 The through-hole density barrier. If the ends of an electrical net contain an SMT pad connected by a trace and a via hole, then there is only so many such connections (using conventional design rules' spacing) that can be placed in any square area. This *through-hole barrier* is shown crossing this technology roadmap. To cross the barrier (and prevent via starvation), the fabricator must stack vias using blind and buried vias and such.

22.2.4 Design, Cost, and Performance Trade-Offs

The HDI structure is cost-effective for higher-density assemblies shown throughout this chapter. The relative price of HDI versus relative density is depicted in Fig. 22.2. Cost parity (for similar wiring densities) is achieved with a four-layer HDI microvia at approximately an eight-layer, through-hole printed circuit multilayer. Wiring capacities and densities greater than an eight-layer multilayer can be achieved at a lower price with a properly designed HDI substrate. At very high densities, there are no through-hole multilayers that can meet the wiring capacities and density demand necessary, whereas HDI can easily meet the requirements.

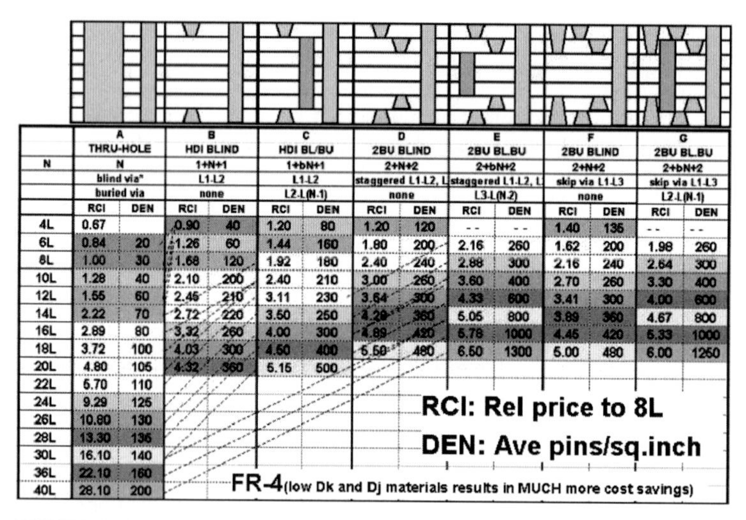

N	A THRU-HOLE		B HDI BLIND		C HDI BL/BU		D 2BU BLIND		E 2BU BL.BU		F 2BU BLIND		G 2BU BL.BU	
	N		1+N+1		1+bN+1		2+N+2		2+bN+2		2+N+2		2+bN+2	
	blind via*		L1-L2		L1-L2		staggered L1-L2, L		staggered L1-L2, L		skip via L1-L3		skip via L1-L3	
	buried via		none		L2-L(N-1)		none		L3-L(N-2)		none		L2-L(N-1)	
	RCI	DEN	RCI	DEN	RCI	DEN	RCI	DEN	RCI	DEN	RCI	DEN	RCI	DEN
4L	0.67		0.90	40	1.20	80	1.20	120	- -	- -	1.40	135	- -	- -
6L	0.84	20	1.26	60	1.44	160	1.80	200	2.16	260	1.62	200	1.98	260
8L	1.00	30	1.68	120	1.92	180	2.40	240	2.88	300	2.16	240	2.64	300
10L	1.28	40	2.10	200	2.40	210	3.00	260	3.60	400	2.70	260	3.30	400
12L	1.55	60	2.45	210	3.11	230	3.64	300	4.33	600	3.41	300	4.00	600
14L	2.22	70	2.72	220	3.50	250	4.29	360	5.05	800	3.89	360	4.67	800
16L	2.89	80	3.32	260	4.00	300	4.89	420	5.78	1000	4.45	420	5.33	1000
18L	3.72	100	4.03	300	4.50	400	5.50	480	6.50	1300	5.00	480	6.00	1250
20L	4.80	105	4.32	360	5.15	500								
22L	5.70	110												
24L	9.29	125												
26L	10.80	130												
28L	13.30	136												
30L	16.10	140												
36L	22.10	160												
40L	28.10	200												

RCI: Rel price to 8L
DEN: Ave pins/sq.inch
FR-4(low Dk and Dj materials results in MUCH more cost savings)

FIGURE 22.2 Relative price and density comparison between conventional through-hole (TH) boards and HDI microvia boards with different structures. The relative cost index (RCI) is actual production pricing of boards normalized to the eight-layer price. The density (DEN) is a measure of the maximum average pins (leads) per sq. in. (for both sides of the PCB). The diagonal lines are equivalent density boards. These are for FR-4 printed circuits.

22.2.5 Specifications and Standards

The Institute for Interconnecting and Packaging Electronic Circuits (IPC) has been working on standards and specifications for high-density interconnecting structures. Three standards and a guideline are the documents that are available from the IPC. The four documents are:

- **IPC-2315** "Design Guide for High Density Interconnecting Structures and Microvias"
- **IPC-2226** "Design Standard for High Density Interconnecting Structures and Microvias"
- **IPC-4104** "Standard for Qualification and Conformance of Materials for High Density Interconnection Structures and Microvias"
- **IPC-6016** "Standard for Qualification and Performance Specification for High Density Interconnection Structures"

The IPC standards that cover HDI design rules, materials, and specifications are:

- **IPC-2226** This specification educates users in microvia formation, selection of wiring density, selection of design rules, interconnecting structures, and material characterization. It is intended to provide standards for use in the design of printed circuit boards utilizing microvia technologies.
- **IPC-4104** This standard identifies materials used for high-density interconnection structures. A series of specifications (slash sheets) are defined for specific available materials. Each sheet outlines engineering and performance data for materials used to fabricate high-density interconnecting structures. These materials include dielectric insulator, conductor, and dielectric/conductor combinations. The slash sheets are provided with letters and numbers for identification purposes. For example, if a user wishes to order from a vendor and reference the specification sheet number 1, the number "1" would be substituted for the "S"

in the preceding designation example (IPC-4104/1). To start the ordering process, one can use the slash sheets in the IPC-4104 document in combination with relevant IPC documents for each material sets (e.g., IPC-CF-148, IPC-MF-150, IPC-4101, IPC-4102, IPC-4103, etc.).

- **IPC-6016** This document contains the general specifications for high-density substrates not already covered by other IPC documents.

22.3 HDI STRUCTURES

The IPC HDI Structure Subcommittee, which is responsible for defining performance requirements, used the following methodology to specify HDI products. Since a microvia can be any shape—including straight wall, positive or negative taper, or cup—determining the methods used for producing microvias can be segmented into three methodologies (A, B, or C), as noted in Table 22.1.

TABLE 22 .1 Ten Process Methods Utilized to Produce Microvias

Hole formation	Microvia process*
Mechanically drilled	A
Dry-etched (plasma)	A
Mechanically punched	A
Abrasive blast	A
Laser-drilled	A
Post-pierced	B
Photo-formed	A
Insulation displacement	B, C
Wet-etched	A
Conductive bonding sheets	C

*A: create hole, then make conductive; B: create conductive via, then add dielectric; C: create conductive via and dielectric simultaneously.

All of these technologies provide approximately the same high-density design rules. These design rules endow build-up technologies with four to eight times the wiring density of conventional, all-drilled through-hole vias. The ten via profiles are depicted in Fig. 22.3.

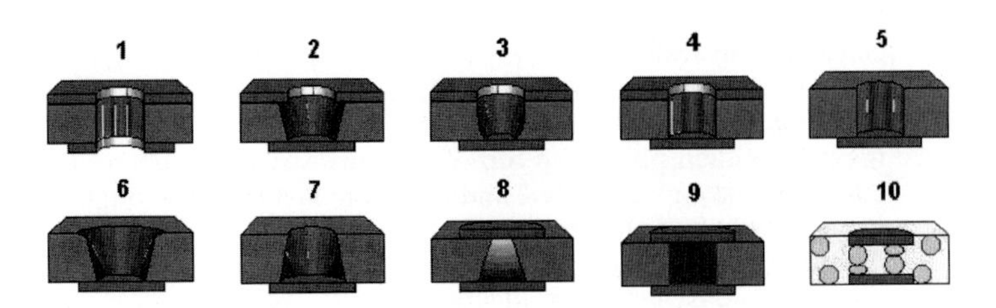

FIGURE 22.3 Representative microvia profiles and the processes that produce them.

The structure of the high-density interconnection is by type: Type I, Type II, Type III, Type IV, Type V, and Type VI (see Table 22.2). However, some of these type constructions are based on which microvia material is to be used. Thus, the following definition applies to all high-density interconnect substrates (HDIS).

TABLE 22.2 Description of the 6 IPC HDI Type Structures of IPC-2226

Type I	1[C]0 or 1[C]1, with through-vias from surface to surface
Type II	1[C]0 or 1[C]1, with through-vias buried in the core and from surface to surface
Type III	≥2[C]≥0, 2 or more HDI layers added to through-via in core or from surface to surface
Type IV	≥2[P]≥0 where P is a passive "substrate with no electrical connecting functions"
Type V	Coreless construction using layer pairs
Type VI	Alternate construction using conductive pastes

[C], printed circuit core; [P], passive substrate core; 0, 1, and 2, number of build-up layers on each side of the core [C] or [P].

The core, defined as [C], can be identified as an A, B, or C type core. Thus, [CA] is a core with internal vias only; redistribution makes contact with the surface. [CB] is core with internal and external (through microvia structures). High-density interconnecting structures make contact with the innerlayers of the core. [CC] is passive core, in which there are no interconnections.

22.3.1 Construction

Type I to Type VI constructions currently describe all known HDI build-up structures, but as the technology evolves, new ones are likely to be created. The notation used is as follows:

x[C]x

where x = 0,1,2,3, etc. (that is, the number of build-up layers on that side of the core)
[C] = a standard laminated core of materials, with or without vias of *n* layers

22.3.1.1 Type I Constructions. This construction (1[C]0 or 1[C]1) describe an HDI in which there are both plated microvias and plated through holes used for interconnection. Type I constructions describe the fabrication of a single microvia layer on either one side (1[C]0) or both sides (1[C]1) of an underlined printed circuit substrate core. The printed circuit core substrate is typically manufactured using conventional printed circuit techniques. The substrate may be rigid or flexible and can have as few as one circuit layer or be as complex as a prefabricated multilayer printed circuit with buried vias. A single layer of dielectric material is then placed on top of the core substrate. Microvias are formed in the dielectric connecting layer 1 to layer 2 and layer n to layer n-1. Through holes are then drilled, connecting layer 1 to layer n. Then the microvias and through holes are metallized or filled with conductive material. Layer 1 and layer n are circuitized and fabrication is completed. Figure 22.4a shows this construction as illustrated in IPC-2226.

22.3.1.2 Type II Constructions. Type II (1[C]0 or 1[C]1) has the same HDI layers as Type I. The difference is the core, [C]. Type II allows through vias to be placed in the core before the HDI layers are applied. The processes are the same except for the cores through vias being

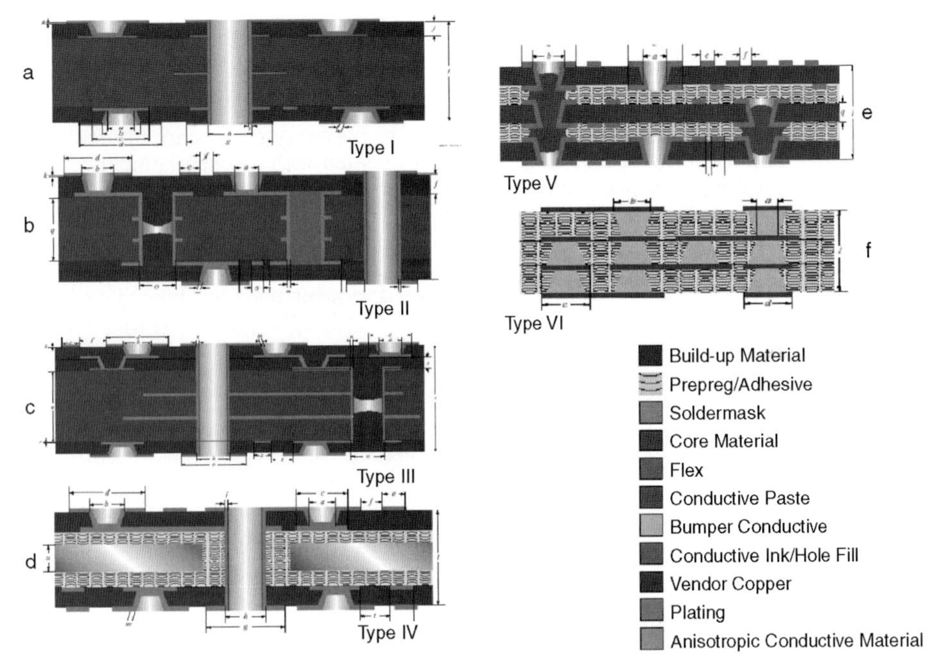

FIGURE 22.4 IPC-2226 microvia structures from Type I to Type VI.

filled before the HDI layers are applied. Figure 22.4b shows this construction as illustrated in IPC-2226.

22.3.1.3 Type III Constructions. This construction (2[C]0) describes an HDI in which there are both plated microvias and plated through holes used for interconnection. Type III constructions describe the fabrication of two microvia layers on either one side (2[C]0) or both sides (2[C]2) of an undrilled or drilled printed circuit substrate core. The printed circuit core substrate is typically manufactured using printed circuit techniques. The substrate may be rigid or flexible and have as few as one circuit layer or be as complex as a prefabricated multilayer printed circuit with buried vias. A single layer of dielectric material is then placed on top of the core substrate. Microvias are formed in the dielectric, connecting layer 2 to layer 3 and layer n-1 to layer n-2. This first microvia layer is either metallized or filled with conductive material and then circuitized. A second layer of dielectric material is then placed on top of this circuitized layer and microvias are formed connecting layer 1 to layer 2 and layer n to layer n-1. Through holes are then drilled connecting layer 1 to layer n. The microvias and through holes are then metallized or filled with conductive material. Layer 1 and layer n are circuitized and fabrication is completed. Figure 22.4c.shows this construction as illustrated in IPC-2226.

22.3.1.4 Type IV Constructions. This construction (1[P]0 or 1[P]1 or >2 [P] > 0) describes an HDI in which the build-up layers are used over an existing drilled and plated passive substrate. The printed circuit or metal core substrate is typically manufactured using conventional printed circuit techniques. This substrate may be rigid or flexible. Figure 22.4d shows this construction as illustrated in IPC-2226.

22.3.1.5 *Type V Constructions.*

The Type V construction describes an HDI where there is no core. Both plated and conductive paste layer pairs are interconnected through a co-lamination process. The multilayer is created with an even number of layers (two-sided flex or rigid layer pairs) laminated together at the same time that the interconnections between the even and odd layer are made. This is not a build-up process; it is essentially a single-lamination-parallel process. Figure 22.4e shows this construction as illustrated in IPC-2226.

Layer pairs are prepared using conventional processes of etching, plating, drilling, and so on, or by conductive paste processes. The layer pairs are laminated together using B-stage resin systems or some other form of dielectric adhesive into which conductive paste vias have been placed.

22.3.1.6 *Type VI Constructions.*

Type VI describes constructions of HDI in which the electrical interconnections and wiring can be formed simultaneously. Another variety forms the electrical interconnections and the mechanical structure simultaneously. The layers may be formed sequentially or co-laminated. The conductive interconnect may be formed by means other than electroplating, such as anisotropic films/pastes, conductive pastes, dielectric piercing posts, and so on. Figure 22.4f shows this construction as illustrated in IPC-2226.

22.3.2 Design Rules

The designer should be aware that not all fabricators have equal capabilities in the areas of fine-pitch imaging, etching, layer-to layer alignment, via formation, and plating. For this reason, the HDI design guide categorizes design rules into two categories: the preferred producibility range and the reduced producibility range. For simplicity of design, this *Handbook* will further divide the design rules into three category ranges: A, B, and C, with A being the easiest to produce and C being the most difficult. If more stringent design standards are selected, the number of fabricators that can produce such a board is limited. Circuits produced with design rules in the A category will be easier to produce, will have higher yields, and therefore can be fabricated at lower cost. To keep costs at a minimum, you should select the design rules most appropriate for the application:

- **Category A** This specification allows conventional HDI processes to be used with relaxed tolerances. It should have the highest yield and lowest cost. It is estimated that 100 percent of HDI fabricators can meet these design rules.
- **Category B** This is the conventional HDI process. It is estimated that 75 percent of HDI suppliers can meet these design requirements under production conditions.
- **Category C** Top-level fabrication shops, representing 20 percent of the total HDI fabricators, can meet these design rules. Panel sizes are often reduced to increase yield, which increases the final cost. Production volumes are presently limited, with special attention required during the production process. These rules require smaller panels and more exotic fabrication techniques. They are generally required only in electronic packages, chip-on-board (COB), flip-chip interposers, or MCM applications. At present, yields are lower and costs are high. It is estimated that fewer than 20 percent of all fabricators can achieve these design rules in limited production or prototype volumes.

Typical HDI design rules are given in Fig. 22.5. This figure bridges the two design categories from the specification IPC-2226, "Design Standards for HDI and Microvias."[5] This is a Type III HDI structure.

Symbol	Feature	Category A micron [mil]	Category B micron [mil]	Category C micron [mil]
a	Microvia hole diameter at target land (as formed)	125 - 200 [5-8]	100 - 200 [4-8]	75 - 250 [3-10]
b	Microvia diameter at capture land (as formed, no plating)	350 / 14	300 / 12	250 / 10
b-1	via in SMT pad width	300 / 12	250 / 10	250 / 10
c	landing pad diameter	350 / 14	300 / 12	250 / 10
d-1	conductor/landing pad spacing on rigid innerlayer	125 / 5	100 / 4	75 / 3
d	pad to pad spacing on innerlayer	125 / 5	75 / 3	75 / 3
e-1	conductor line width on landing layer	125 / 5	100 / 4	75 / 3
e	conductor width on innerlayers	125 / 5	75 / 3	75 / 3
f	min. finished hole size - fhs (plated through-hole)	250 / 10	200 / 8	150 / 6
g,o	surface via pad (fhs + annular ring x 2)min.	fhs + 350 / 14	fhs + 300 / 12	fhs + 250 / 10
	through-hole pad diameter	600 / 24	500 / 20	400 / 16
h	trace spacing on rigid outerlayer	125 / 5	100 / 4	87 / 3.5
I	trace width on rigid outerlayer	125 / 5	100 / 4	87 / 3.5
k	fhs drill hole plane clearance on innerlayer (fhs + annular ring x 2)*	fhs + 700 / 28	fhs + 66 / 24	fhs + 500 / 20
	surface conductor to unplated hole	250 / 10	200 / 8	200 / 8
n	min. unplated hole diameter	350 / 14	300 / 12	300 / 12
m-1	min. HDI dielectric thickness	75 / 3	62.5 / 2.5	50 / 2
ar	Min. aspect ratio, (m-1)/ a	<1.0	1.0	1.0
z	min. board thickness	725 / 29	600 / 24	500 / 20
m	min. core thickness	100 / 4	62.5 / 2.5	50 / 2
p	min.prepreg thickness	100 / 4	62.5 / 2.5	50 / 2
	min. plated thickness	25 / 1	25 / 1	30 / 1.2

FIGURE 22.5 Typical HDI design rules with corresponding dimensional symbols for IPC-2226 figures. Illustrated is an IPC Type III HDI structure for categories A, B, and C. (Source IPC-2226, "Design Standards for HDI and Microvias.")

22.4 DESIGN

HDI and microvias place a new burden on printed circuit design. The various IPC types impose significant changes in the multilayer stack-up from a conventional board. But additionally, microvias can be implemented in many different ways and with different design rules. This section presents just a few of the design issues.

22.4.1 Multiple Layers of Microvia Holes

For interconnecting IC substrates with hundreds of I/O terminals, a single layer of microvia holes may not be sufficient to satisfy interconnection requirements. Double, triple, or even quadruple microvia hole layers may be necessary whether the board is a motherboard or a chip package substrate. When microvia holes are made only between adjacent layers (see Fig. 22.6), all three microvia processes—photovia, plasma via, and laser via—can be used. However, when a design requires microvia holes that must connect beyond adjacent layers—such as L1 and L3 (skipvia)—laser via processing is the only practical option.

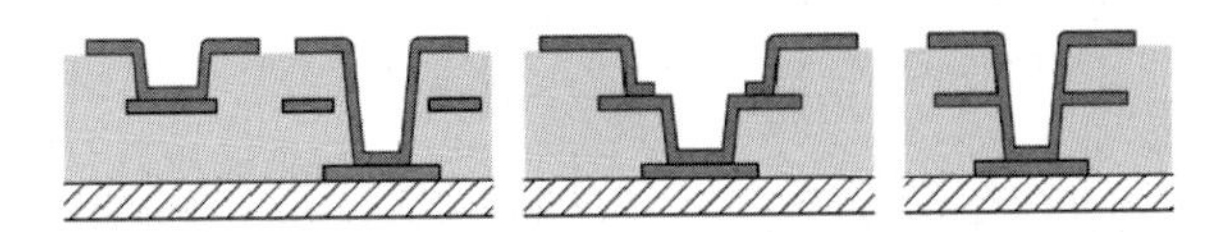

FIGURE 22.6 Microvia connections beyond adjacent layers.

22.4.2 Design Tools

Printed circuit design has become one of the most important functions in the electronic product realization process.[6] The demands on printed circuit design keep mounting. The reasons for this increasing demand include the following:

- To regain time for product schedules
- To lower the cost of fabrication and assembly
- To enable new area-array components, such as chip scale packages (CSPs), micro-BGA, and flip-chip interposers
- To decrease the time to market
- To improve electrical performance for high frequency and to reduce EMI

Bringing the process back under control requires a methodology that involves planning the printed circuit layout process with predictive wiring density models. The other benefits include reducing printed circuit fabrication and assembly costs.

22.4.2.1 Required Features of HDI CAD Toolsets and Autorouters. A partial list of features that computer-aided design (CAD) tools require to design HDI boards includes the following:

- Optimization of a mixed via by a router
- Autorouting of cost budget for mass via generation
- Buried or blind via control on the layer
- Via padstack control (landless)
- Staggered via control
- Pad within pad
- Blind via push/shove during manual routing
- Any angle routing
- Manufacturing process rules at all design phases
- Buried components

Some of these features have already been discussed in previous chapters.

22.4.2.2 Autorouters. Autorouters have been a part of printed circuit design systems for many years. An autorouter will automatically place vias and traces on the printed circuit based on the schematic and part geometries. Elaborate configuration menus drive the appropriate placement of these features. A special autorouter is required for HDI structures, because many HDI processes employ mass via generation as used for SLC technologies. These processes produce all the vias simultaneously and at any desired diameter. Since the cost of vias is rather insignificant, the autorouter should have the capability of achieving a near-zero via cost. When the routing is done, a different but more optimized design results.

22.4.3 Trade-Off Analysis

After a product has been partitioned, the circuit has been designed, and the components have been selected, the physical design must be planned with an eye toward achieving the lowest manufacturing cost while meeting all performance and operating boundary conditions. This is especially true for HDI designs. Conventional through-hole printed circuit design has not changed too much over the years. Finer geometries, more layers, and surface mounting have been added, but the conventional design process remains essentially the same. Microvias and HDI, however, introduce many changes to the design process and require new design rules and layer structures. Experience and history will not help here. The unfortunate truth about printed circuit layout is that there is an almost infinite number of combinations of layer structures and design rules that can satisfy the schematic and bill of materials. With all these choices, especially the new ones that HDI offers, a trade-off tool is required to find the best set of design rules and features that allows the rapid design of a printed circuit board, is producible, and meets all the performance expectations while providing the lowest total manufacturing costs. When used early in the design process, before the actual physical design of the printed circuit, the tools require predictive models that can anticipate cost and performance. Information and processes for planning a HDI design can be found in Chap. 19 of this *Handbook*. Trade-off information of through-hole density and relative costs can be seen in Fig. 22.2.

22.5 DIELECTRIC MATERIALS AND COATING METHODS

This section provides an overview of the dielectric and applied conductive materials used in microvia and via filling. Some of these materials can be used in both IC chip carrier and PWB HDI applications. The discussions are focused on the HDI PWB arena and on materials for which information is readily available. In section 22.5.2, cross-references are made to the relevant material specifications of the IPC/JPCA-4104 specification for HDI and microvia materials. A brief material roadmap discussion is included to illustrate material property trends.

Figure 22.7 shows the compatibility of laser via, photovia, and plasma via methods with four basic surface dielectric structures on which microvia holes are to be formed. Although laser via methods can cope with all four dielectric structures, photovia and plasma via methods are applicable to only one structure, respectively, as shown in the figure. This is one reason why laser via is more widely used today. Another wiring layer is built over the existing microvia holes, which become buried via holes (BVHs).

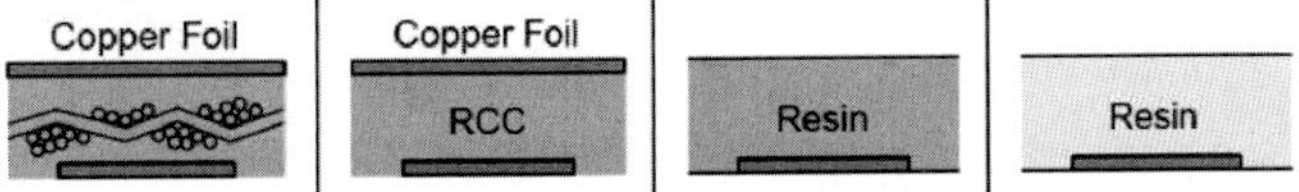

	Standard Configuration Copper Foil	RCC Copper Foil RCC	Thermally Curable Resin Resin	Photoimageable Resin Resin
Photovia	X	X	X	O
Laser Via, CO_2	O	O	O	O
Laser Via, Yag	O	O	O	O
Plasma Via	X	O	X	X

FIGURE 22.7 Compatibility of via hole formation methods with four basic dielectric layer structures.

The specification IPC-4104 will define material requirements for HDI applications. This IPC specification applies only to the *surface HDI layers*, the conventional multilayer core materials are covered by IPC specification IPC-4101B.

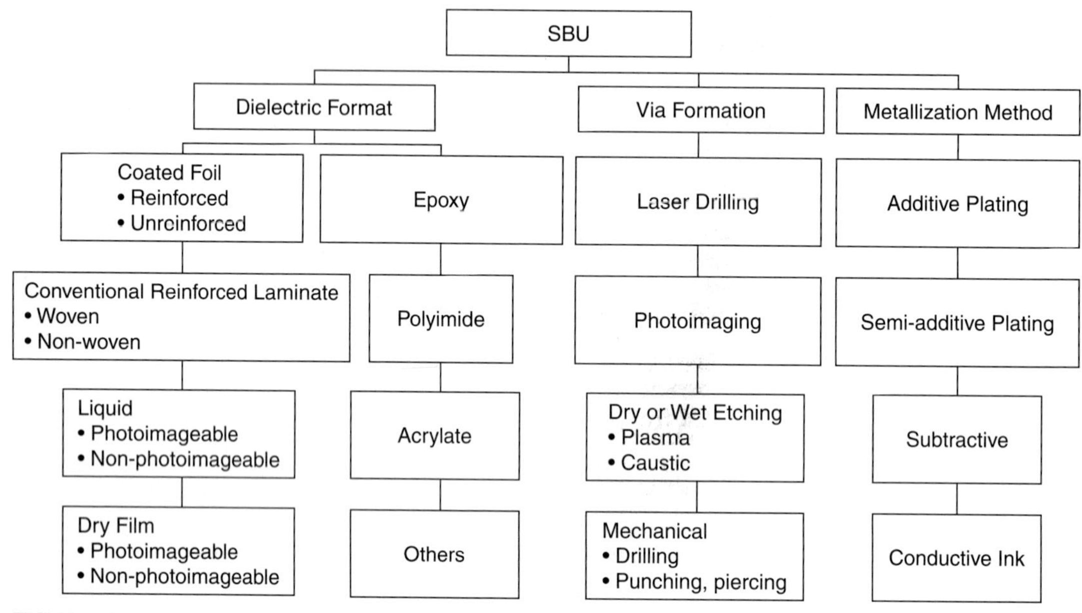

FIGURE 22.8 Material and technology choices for SBU fabrication. (*Courtesy of DuPont.*)

22.5.1 Materials for HDI Microvia Fabrication

Figure 22.8 shows a material and technology selection flowchart for use when choosing dielectric materials. In using the flowchart, you should ask the following questions regarding the dielectric you are considering:

- Will the dielectric use chemistry compatible with current chemistry used by core substrate material?
- Will the dielectric have acceptable plated copper adhesion? (Many original equipment manufacturers [OEMs] want ≥6 lb/in [1.08 kgm/cm] per 1 oz. [35.6 μm] copper.)
- Will the dielectric provide adequate and reliable dielectric spacing between metal layers?
- Will it meet thermal needs?
- Will the dielectric provide a desirable "high" T_g for wire bonding and rework?
- Will it survive thermal shock with multiple SBU layers (i.e., solder floats, accelerated thermal cycles, multiple reflows)?
- Will it have platable, reliable microvias (that is, will it have latitude to ensure good plating to the bottom of the via)?

22.5.1.1 Copper-Clad Dielectric Materials. Due to relative ease of implementation, copper-clad dielectrics are used on a larger scale than unclad dielectrics. Copper-clad dielectrics provide a method that requires the least number of changes in manufacturing flow because they typically use the same dielectric and reinforcements found in standard PWBs. Copper-clad-based materials have a longer history in making blind vias than any other method. This makes many designers, OEMs, and PWB fabricators more comfortable with copper-clad-based materials.

These materials can be nonreinforced or reinforced. The reinforcement can be woven or non-woven and can be aramid, glass, and so on. These dielectrics are suitable for via formation by laser or other mechanical removal methods.

Due to its wide availability and familiarity, FR-4 material is often initially evaluated in laser-drilling microvias. Reinforced with thin 106 or 1080 woven glass (since thicker glass is more difficult to vaporize with lasers, but there is 1086 uniform-woven glass just for laser ablation), one- or two-ply laminates are selected with a resin content close to 70 percent. The "laser drilling" is done by ablation of the via using a conformal mask or a directly focused beam, with either an ultraviolet (UV) Nd:YAG or a CO_2 laser. These materials may also be coated on a copper foil. Typical applications use single-side clad material where the copper clad is used as the outer layer and the C stage is bonded to the subcomposite. These materials are suitable for via formation using methods such as plasma or laser. To meet fine circuitry and smaller via needs, very thin copper is available. Another approach, practiced by many Asian PWB fabricators, is to thin down incoming copper clad by etching and/or planarizing the copper surface precisely.

22.5.1.2 Unclad Dielectric Materials. If the dielectric is reinforced, microvias are formed by laser drilling or other mechanical means. If it is nonreinforced, it can be photoimageable in addition to the previous choices. To add conductivity, subtractive processing is the standard manufacturing practice in the United States and Europe. Semi- or fully additive techniques have been practiced in continuous high volume only in Asia, most notably Japan. In Japan, which leads the world in microvia production, about 22 percent of manufacturing begins with a non-copper-clad dielectric material.

Japan has long accepted the additive manufacturing methodology for creating circuits on the board surface as well as the through-hole connection.

The trade-off between clad and unclad dielectric materials follows a benefit analysis similar to that used for subtractive versus additive processing. Copper-clad-based material is fabricated using standard practices familiar to most shops. However, it innately generates more waste and has a higher cost due to the subtractive nature of the manufacturing sequence. In addition, copper-clad materials demonstrate limitations in fine-line capability similar to those seen with standard manufacturing methods used today.

Unclad dielectrics require considerable resources for optimized plating of copper to achieve reliable and consistent copper peel strengths. Unclad dielectrics and thin copper-clad dielectrics, however, gain in importance as the requirements for line and space geometries reach ≤75 μm and required via sizes are lower than 75 μm in diameter. Figure 22.9 depicts the influence of base copper on the aspect ratio of a via. The base copper increases the aspect ratio, making it more difficult to plate the via, and can cause issues with bottlenecking whereby the via plates shut at the top without sufficient plating in the bottom. Other data show that thinner copper etches more uniformly, which helps the cause for controlled impedance.

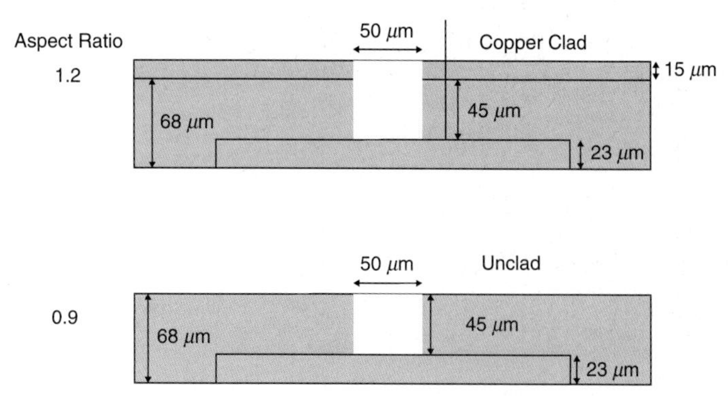

FIGURE 22.9 Impact of 1.2 oz. base copper on the aspect ratio of a 50 μm via. *(Courtesy of Vantico.)*

TABLE 22.3 Utilization of Unclad Dielectrics by Key Market Segments[4]

	Chip package substrates	Cellular, camcorder, PDA	Computer/ telecommunication*
HDI penetration	High	High	Low
Line/space and via diameter (year 2000)	50-μm lines and spaces; 50-μm via diameter	75-μm lines and spaces; 100-μm via diameter	100-μm lines and spaces; 125-μm via diameter
Use of unreinforced dielectrics	High	High	Low
Use of unclad dielectrics	High	Low, but increasing	Very low

*Excluding laptop computers

22.5.1.3 Clad versus Unclad Dielectric Materials. Table 22.3 compares the utilization of clad versus unclad dielectric materials as circuit geometry shrinks. This table shows that as the technology requirements get tougher, the trend toward using unclad dielectrics increases. As line and space requirements get denser, and via diameters get smaller, requirements for chip package substrates to use unclad or thin copper dielectrics increase. In the case of mobile phones and other portable applications, for example, it is foreseen that by the year 2005 lines and spaces will be down to 50 to 75 μm, respectively, and via sizes to 50 μm, with a subsequent high usage of unclad material.

22.5.2 Examples of HDI Microvia Organic Substrates

From a mechanical standpoint, materials may be grouped as reinforced and nonreinforced laminates and prepregs, as in Fig. 22.8. Reinforced materials are generally better in dimensional stability, lower in coefficient of thermal expansion (CTE), and less sensitive to thermal cracking, whereas nonreinforced materials often have a lower dielectric constant (D_k) and may be photoimageable. This can be seen in Fig. 22.10, which shows the D_j versus D_k (at 1 MHz) of various resins and reinforcements.

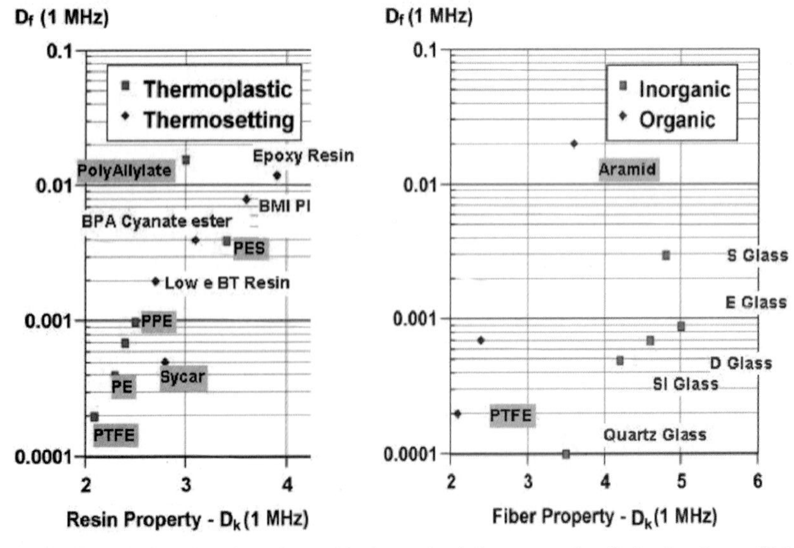

FIGURE 22.10 Dielectric options of resins and reinforcements by dissipation factor (D_j) versus dielectric constant (D_k). *(Source: Holden Consulting.)*

22.5.2.1 Nonreinforced Dielectric Materials. Nonreinforced dielectric materials include resin-coated copper foils, unclad photoimageable dielectric materials, and non-photoimageable dielectric materials.

22.5.2.1.1 Resin-Coated Copper Foils. Resin-coated foils are the most common materials used for build-up multilayer microvia applications. Many product variations are available and fit well within the existing multilayer-manufacturing infrastructure. Epoxy-based coated foils are the most common and have performance properties similar to FR-4 but with no glass reinforcement. Peel strengths, thermal performance, and electrical properties are excellent. A variety of other resin systems are also being developed and starting to be used for coated foil build-up applications. Resin-coated foils come in two general types with either one or two resin layers. One-pass coated foils have a single B-stage layer designed to flow, fill, and provide thickness control at the same time. Two-pass coated foils have a C-stage resin layer adjacent to the foil and a B-stage layer for flow and fill. The fully cured C-stage layer acts as a "stop" during the lamination process, typically enabling better thickness control. Resin-coated foils are available in a variety of thicknesses, yielding between 1.0 mil (25 micron) and 3.0 mil (76 micron) thick, finished dielectric layers. Copper foils most commonly used are $^1/_2$ oz. (18 micron) and $^3/_8$ oz. (13.34 micron), but there is substantial interest in thinner copper foils for improved laser efficiency and better fine-line circuitry definition.

22.5.2.1.2 Other Resins. Since build-up technology is still in its relative infancy but evolving rapidly, many different and diverse approaches are being investigated for the resins used and variations on the via-formation process. Low-D_k/D_j resins such as polyphenylene ether (PPE) are being used to address signal speed and integrity demands in resin-coated foil build-up structures. Another approach is to use additively plateable resin and use the copper as a sacrificial carrier, eliminating the need to laser through copper. An additively plateable resin has the characteristic of high surface adhesion to electroless copper or direct metallization. The sacrificial foil makes it compatible with normal multilayer lamination processes, saving the need for expensive coating machinery, while also protecting the thin dielectric from harm until it is laminated to the printed circuit. This also gives excellent surface topography for good peel strength. The properties of a resin-coated foil are given in Table 22.4. See IPC- 4104 specification sheets 12, 13, and 19 through 22 for more complete material performance information.

TABLE 22.4 General Properties of Resin-Coated Copper (RCC)

	Units	Via foil	Conditioning
Dielectric constant at 1 MHz	—	3.4	C-24/23/50
Dissipation factor at 1 MHz	—	0.0205	C-24/23/50
Electrical strength	V/mil	1776	D-48/50
Insulation resistance	MΩ	2.65×10^5	C-96/35/90
Surface resistivity	MΩ	6.60×10^8	E-24/125
	MΩ	4.71×10^8	C-96/35/90
Volume resistivity	MΩ-cm	7.17×10^7	E-24/125

Source: Allied-Signal

Laminate suppliers should be contacted when special needs arise, as the resulting composite will require understanding of fabrication and design issues.

Resin-coated foils come in two general types:

- One-pass coated foils have a single B-stage layer designed to flow and fill and to provide thickness control, all at the same time.
- Two-pass coated foils have a C-stage resin layer against the foil and a B-stage layer for flow and fill. The fully cured C-stage layer acts as a "stop" during the lamination process, typically enabling better thickness control.

With the exception of the laser process, all the basic technology to produce SBU using resin-coated foil is the same as, and found in, most multilayer PWB production operations.

Another alternative is to use additively platable resins. In this approach, the copper foil acts as a sacrificial carrier. The process is similar to that described earlier, but the first step after lamination is to etch off all the copper chemically. This eliminates the need to laser through copper and makes acquisition of registration targets simpler. The structure left behind by the copper also gives excellent surface topography to the resin for good peel strengths from the subsequent additive plating step. This approach allows for extremely fine feature definition and much greater laser efficiency.

22.5.2.1.3 Unclad Nonreinforced Photoimageable Dielectric Materials. Material chemistry options for this group include epoxies, epoxy blends, polynorborenes, and polyimides. They can be applied as liquid or dry film, negative or positive imaging, and can be solvent- or aqueous-developable. To improve the dielectric's adhesion to copper, most dielectric suppliers require a copper pretreatment with a black oxide or conversion coating (oxide replacement) prior to application of the dielectric. Most of these materials are either epoxy or epoxy- and novolac-based to provide high T_g and good plating, and most provide adhesion values for plated copper of at least 1.1 kg/cm at a 25 μm copper thickness. Typically these dielectrics are metallized using conventional solvent swell and permanganate etching techniques. The liquid materials use only safe solvents (those not known to lead to harmful health effects with prolonged exposure. Typical microvias are seen in Chap. 23; Fig. 23.27. and Chap. 29; Figs. 29.5 and 29.12.

A unique advantage for photoimageable materials is that the speed used to make small or large vias is the same. With the growing need to make embedded reactives (passives), large rectangular areas need to be opened up to place these devices. Currently, this is only economical with photoimaging. Most photoimageable materials are easy and fast to laser-drill because they are nonreinforced. See IPC-4104 specification sheets 1, 2, 7 through 10, and 16 for more complete material performance information. A partial list of available materials is shown in Table 22.5. Properties of photodielectrics can be seen in Tables 22.6 and 22.7.

TABLE 22.5 Partial List of six Photodielectrics available on the Market

Ciba	Probelec™ 81/7081 liquid dielectric
Dupont	ViaLux™ 81 photodielectric dry film
Enthone-OMI	Envision® PDD-9015 photodefinable liquid dielectric
MacDermid	MACu Via-C photodefinable liquid dielectric
Shipley	MultiPosit 9500 CC liquid dielectric
Morton	DynaVia2000™ photoimageable dielectric dry film

TABLE 22.6 Comparison of Four Photoimageable Dielectrics for Electrical and Mechanical Properties *(Courtesy of Holden Consulting)*

Factors	Epoxy	Acrylate	Polyimide
Cost	Excellent (+2)	Excellent (+2)	Poor (−1)
Processibility	Excellent (+2)	Excellent (+2)	Poor (−1)
Performance:			
MIR	Good (+1)	Fair (0)	Fair (0)
D_k	Fair (0)	Fair (0)	Good (+1)
T_g	Fair-good (0)	Poor (−1)	Excellent (+2)
CTE	Good (+1)	Poor (−1)	Excellent (+2)
% Moisture	Good (+1)	Good (+1)	Poor (−1)
Adhesion on lam.Cu	Excellent (+2)	Good (+1)	Poor (−1)
Deposited Cu adhesion	Good (+1)	Good (+1)	Poor (−1)
Overall rating	+10	+5	0

Key: Excellent = +2, Good = +1, Fair = 0, Poor = −1

TABLE 22.7 Partial List of Liquid and Film Dielectrics for HDI Available on the Market

	Product A	Product B	Product C	Product D
Product	L-PID, Neg.	L-PID, Neg.	L-PID, Neg.	DF-PID, Neg.
Material	Epoxy	Epoxy	Polyimide	Epoxy
Insulation resistance		1–10 E+13Ω		8 E+13Ω
MIR		1–4 E+9Ω		1.5 E+11Ω
D_k@ 1 Mhz	3.2	4.0–3.4	2.8	3.4
10 Mhz				4.1
1 Ghz				4.2
Loss factor@1 Mhz	<0.01	0.02	0.004	0.007
1GHz		0.015		0.01
Breakdown voltage (between layers)		>2000		
E-migration	Pass	Pass		Pass
T_g	140–180°C	135°C	300°C	170°C
Peel Strength	>9 N/cm >5 lb/in	14 N/cm 8 lb/in		
TCE (ppm/°C)	60–70	60–70		60–70
Tensile modulus		3000–3500 N/mm^2		4.0 E+5 psi

22.5.2.1.4 Nonphotoimageable, Nonreinforced Dielectric Materials. This group can be laser-drilled, plasma-etched, and/or mechanically treated to form microvias. As stated earlier, many of the photoimageable dielectrics are laser-drillable.

As with the photoimageable group, to improve the dielectric's adhesion to copper, most dielectric suppliers require a copper pretreatment with a black oxide or conversion coating (oxide replacement) process. See IPC-4104 specification sheets 6, 11, 17, and 18 for more complete material performance information. A partial list of materials is seen in Table 22.8.

TABLE 22.8 Partial List of Commercially Available Nonphotoimageable, Nonreinforced Dielectric Materials and Their Suppliers

Osada Ajinonomoto	ABF dry film
Tamura	TBR-25A-3 thermoset ink
Taiyo	HBI-200BC thermal cure ink
MacDermid	MACuVia-L liquid dielectric
Enthone-OMI	Envision® liquid dielectric
3M	Electronic bonding film
B. F. Goodrich	Polynorborene liquid dielectric

22.5.2.2 Reinforced Dielectric Materials. Copper-clad dielectrics for HDI can be reinforced, as in FR-4, or unreinforced, as in coated copper foil. The dielectrics can be epoxy, as in FR-4, or polyimide, cyanate ester, bismalene-triazine (BT), PPE, or PTFE. Reinforcements are typically glass cloth, but there are a variety of glass as well as aramid paper and exotic fibers such as quartz or carbon fiber.

22.5.2.3 Laser-Drillable, Woven-Glass Laminate. A new family of prepregs for HDI consists of uniform-glass prepregs such as 1086 and 1087 glass. This glass is a thinner fabric than the typical 1080 glass cloth; it has more glass plies but they are spread uniformly across

TABLE 22.9 Comparison of Conventional 1080, 1080 LD, and the New 1086 LD Prepregs *(Courtesy of NanYa Plastics)*

Fabric type	Base weight (g/cm²)	Resin content	Warp × Fill (W + F)	Warp	Width	Fill	Width	Thickness	
				mm	mil	mm	mil	mm	mil
1080	48	62	60 × 48	0.19	7.48	0.264	10.39	0.057	2.24
1080 LD	48	62	60 × 48	0.280	11.02	6.44	17.3	0.045	1.77
1086 LD	55	62	60 × 61	0.288	11.3	0.395	155.7	0.045	1.77

the fabric. Table 22.9 describes the differences between the conventional 1080 glass cloth and the 1080 LD (laser-drillable) and the new 1086 LD prepreg. Figure 22.11 shows these two glass prepregs. Table 22.10 compares the performance of RCC, conventional 1080 prepreg (PP), and the new 1086 LD prepreg, whereas Fig. 22.12 shows laser-drilled microvias in the two different prepregs.

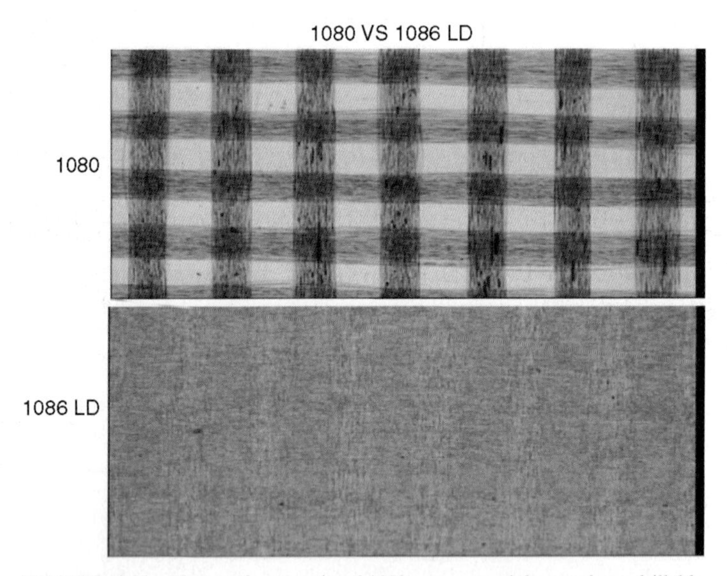

FIGURE 22.11 Photos of conventional 1080 prepreg and the new laser-drillable 1086 LD uniform-glass prepreg. *(Courtesy of NanYa Plastics.)*

22.5.2.4 Aramid-Reinforced, Non-Woven, Non-Glass Laminate. In 1965, scientists at DuPont discovered a method for producing an almost perfect polymer chain extension using the polymer poly-p-benzamide. The key structural feature of this molecule is the paraorientation on the benzene ring, which allows it to form rod-like structures with a simple repeating molecular backbone. The term *aramid* now refers generically to organic fibers in the aromatic polyimide family. Kevlar was the first para-aramid fiber to become familiar, due to its use in bullet-resistant vests and as a lightweight, high-strength structural reinforcement. Aramid-reinforced prepregs and laminates have proven their functionality for a number of years in high-reliability applications and more recently in consumer electronics.

The low CTE of aramid non-woven reinforced prepreg and laminate provides a closer match to the CTE of the silicon chip. Depending on the type of resin and the resin and copper

TABLE 22.10 Comparison of the Performance of RCC, Conventional Prepregs, and the New Laser-Drillable 1086 LD Glass Prepreg *(Courtesy of NanYa Plastics)*

Build-up layer	RCC	Convention P/P	1086 LDP
Dimension stability	Poor	Good	Better
Thickness control	Poor	Good	Good
Surface smooth	Good	Good	Better
Rigidity	Poor	Good	Good
Handling	Good	Good	Good
Drill ability	Excellent	Poor	Good

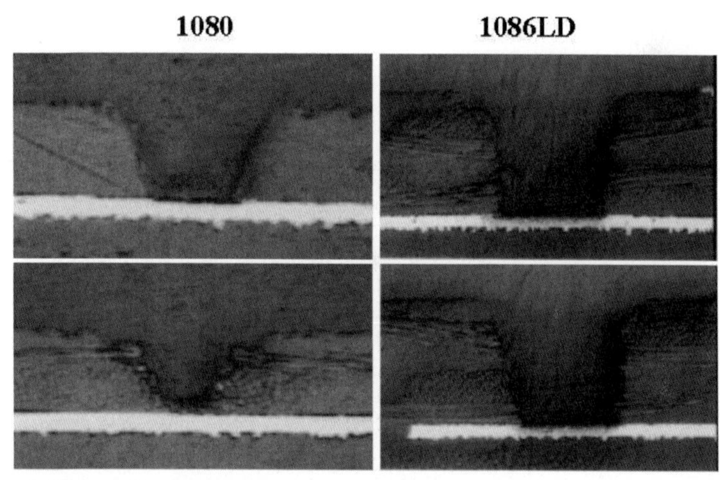

FIGURE 22.12 Photos of cross sections of laser-drilled vias in conventional prepregs and the new, uniform, laser-drillable 1086 LD prepregs. (*Courtesy of NanYa Plastics.*)

content of the laminate and the prepreg, the CTE of the PWB can be tailored to between 10 and 16 ppm/°C (Fig. 22.13). This allows the designer the option of finding a best fit of the CTE of the PWB to the CTE of the components.

Reliability can be designed in by PWB designers, as they know which component packages are used, what the CTE requirements of these packages are, and what the life-expectancy requirements of the electronic equipment are. The ability to tailor in-plane CTE has made non-woven aramid-reinforced PWBs one of the most favorable material options for avionics, satellites, and telecom applications where long life expectancy and high reliability are needed.

In mobile phone applications, where chip scale packages (CSPs) are commonly employed, low-CTE non-woven aramid reinforcement extends solder joint life as much as three times over FR-4 and resin-coated foil (RCF). After more than 1,000 thermal cycles (–40 to +125°C), non-woven aramid-reinforced epoxy resin does not crack, as is common with nonreinforced dielectrics.

Aramid-reinforced laminate and prepreg allow fast microvia hole formation and at the same time maintain the performance characteristic of a smooth surface for fine-line conductor imaging. The ablation speed of non-woven (aramid) laminates and prepregs is close to that achieved when using nonreinforced materials such as resin-coated foil, dry film, or liquid dielectrics. Since aramid laminates are very stable, they allow the fabrication of double-sided, very thin, etched innerlayers, which are then pressed to a multilayer package in a single

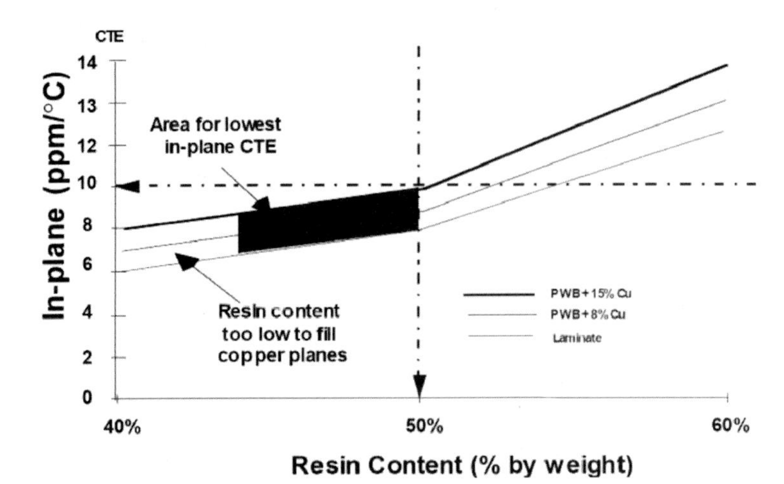

FIGURE 22.13 Impact of copper and resin content on in-plane CTE. *(Source: DuPont.)*

pressing cycle. Thus, these thin innerlayers can be processed in parallel. Aramid-reinforced material is a cost-effective alternative when laser hole formation is used to form single- and multiple-layer interconnects, and is compatible with FR-4 materials used in the core of aramid–FR-4 mixed constructions. Laser drilling of stepped microvia holes is accomplished in one subsequent operation (see Fig. 22.14). This allows the designer to interconnect up to four layers on each side of the inner core without sequential processing—a substantial productivity advantage for the PWB fabricator that results in the lowest possible manufacturing cost. See IPC-4104 specification sheets 5 and 23 for more complete material performance information.

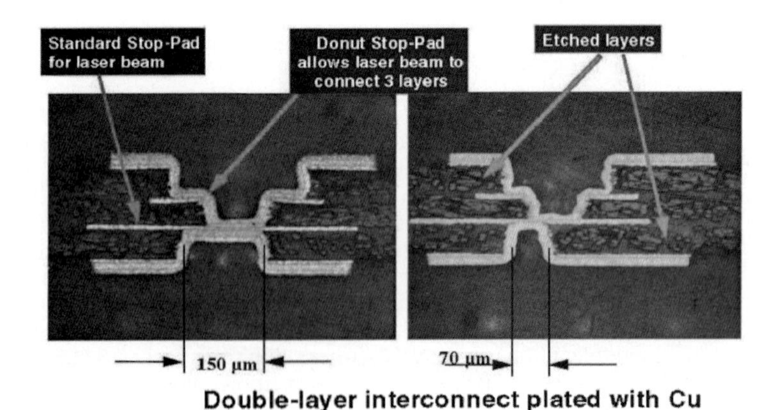

FIGURE 22.14 Photos of conventional and skip-microvias laser-drilled in Thermount materials. *(Courtesy of DuPont.)*

22.5.3 Via Filling

The demand for filling of through holes with epoxy resin or conductive paste has been rising ever since surface-mount technology became widely adopted in the PWB industry.

22.5.3.1 Basics. Following are the traditional examples of the types of applications that require via filling processing capability:

- Preventing acid residue from attacking the Cu plating at the through-hole opening, which can in turn cause an open circuit
- Averting mishandling due to loss of vacuum during board-level assembly or during vacuum-assisted transport in production
- Avoiding blowout of flux and/or solvent residue during assembly and solder reflow
- Stopping flux from dripping from the through holes to the opposite side of the PWB
- Keeping solder resist ink from migrating into through holes during screen printing, which can cause formation of nodules at the through-hole opening during solder plating and/or gold plating
- Enhancing the planarity of solder resist on the surface of filled through holes and the planarity of core layer for SBU PWBs
- Improving the stability of solder paste printing volume for via-in-pad designs
- Preventing the migration of solder into through holes in via-in-pad designs

The filling of through holes with nonconductive resin and conductive materials using dispensing, screen-printing, and roll-coating methods has been extensively evaluated and tested. Of the three basic methods described earlier, screen printing is the most common process. It allows for efficient, selective filling of large numbers of through holes.

22.5.3.2 Via-Filling Materials. Selection of the proper resin-based material is the most important objective when using a screen-printing process to fill through-hole vias in build-up multilayer cores. In particular, the primary issues to be considered when choosing an appropriate fill material are:

- Ease of printing
- Ease of grinding (planarizing)
- Adhesion to hole wall and panel surface

The most commonly used materials for permanently filling through holes include single-cure (thermal) resins, photoimageable dielectrics, conductive pastes, and the dual-cure (UV + thermal) epoxy resin utilized in the Noda screen flat plug process. Following are brief descriptions and comparisons of the characteristics of each of these types of via-fill materials.

22.5.3.2.1 Conventional Prepregs and RCC. In fabricating a base core with a pattern, the core panel is usually panel-plated and the pattern is made with a dry-film tenting process. Some fabricators seem to prefer pattern plating, however. The choice depends on the fabricator's familiarity with these processes. After the pattern is formed, dielectric material is laminated over the core (in the case of prepreg, with copper foil) and the holes can be filled with resins, depending on the plated hole diameter and the thickness of the core. It is generally agreed that when the diameter of plated holes is equal to or less than 0.3 mm *and* the core thickness is equal to 0.6 mm or less, these holes can be filled effectively by the lamination process (although the resin thickness of 80 μm is preferred in the case of RCC). See Fig. 22.15.

When the diameter/thickness conditions are not met, it is necessary to fill the holes using a separate process. A screening process does this from one side of the panel with polyester screen with an oversized hole pattern. After hole filling and curing the resin completely, the fabricator removes excess resin using a belt sander (#600 to #800) or a ceramic brush. Figure 22.16 illustrates the process. This is a tricky but necessary operation for microvia board makers, particularly for photovia processing.

Hole filling is costly but yields a few advantages over simultaneous hole filling by lamination of RCC or prepreg. The edge of the plated holes is well protected; therefore, forming

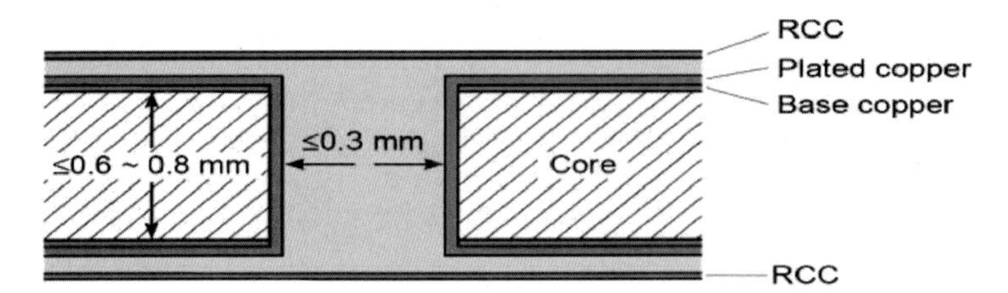

FIGURE 22.15 Conditions under which the through hole can be filled with resin by RCC lamination.

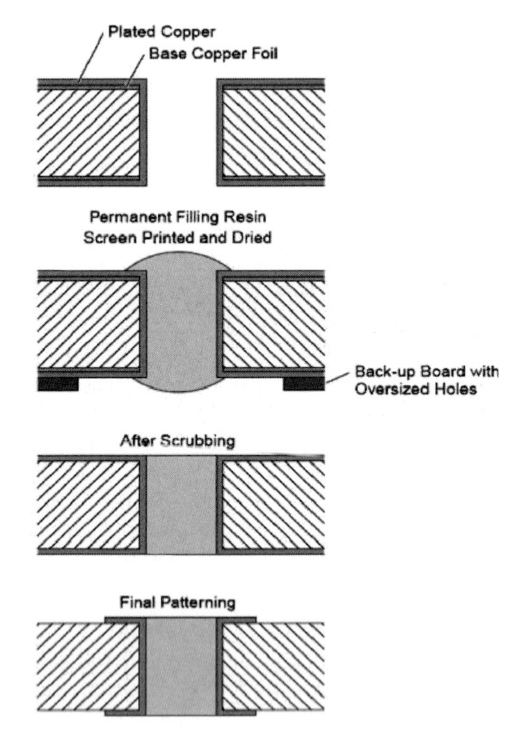

FIGURE 22.16 Hole filling and surface scrubbing.

finer annular rings on the base core by tenting is easier. There is no concavity problem over the hole that can cause difficulties in later processes.

22.5.3.2.2 Dual-Cure (UV + Thermal) Epoxy Resin. The main purpose for using a dual-cure epoxy resin is the stability of the B-cured material and the ease with which it can be removed by grinding during planarization. A homogeneous, semicured state can be reached because the liquid environment within the UV exposure system helps to control the temperature and stabilize the intensity of UV light. Furthermore, because the temperature of the liquid environment is kept low, the expansion of air bubbles that may have been trapped in the fill material at room temperature is prevented.

In formulating epoxy resin via-fill ink, one must consider the adhesion between the fill material and the copper plating in the hole barrel. Normally it may seem suitable to use a material that does not include an acrylic component. However, some suppliers have found it more beneficial to add an acrylic component because of its resistance to moisture absorption. They determined that it would be preferable to improve the ink's adhesion to copper by pretreatment with a black oxide or chemical etching process. This is similar to the approach taken by many dielectric suppliers.

22.5.3.2.3 Photoimageable Dielectrics. An advantage of a dielectric being photoimageable is that unwanted material is washed off during development, thus eliminating the planarization step. Typically, one screen pass is sufficient to fill or plug vias, but two passes can also be used to maximize hole filling. Standard screen-printing equipment is needed with a bed-of-nails or dimple plate to maximize air release out of the plugged holes. If a squeegee is being used, single-sided application is recommended. Two coats are recommended to minimize the effects of material. Cores as small or smaller than 0.030 in. (0.76 mm) require a secondary coating application.

22.5.3.2.4 Conductive Paste. In addition to the previous advantages of via filling stated earlier, conductive via fills also dissipate heat, increase electrical conductivity in the hole, and allow a via-in-pad design option (Fig. 22.17). The latter advantage is another way to recover real estate.

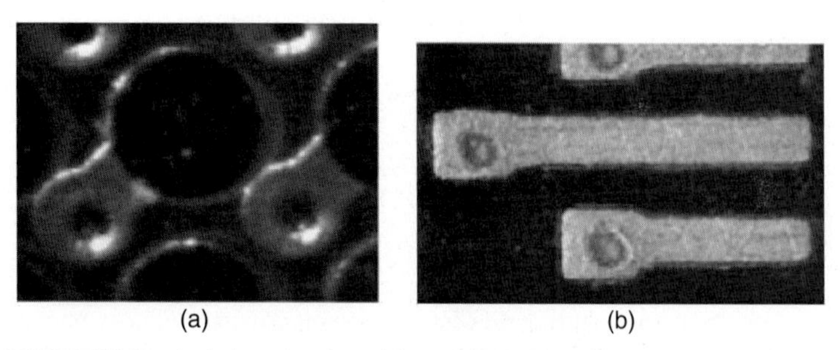

FIGURE 22.17 A via-in-pad option with conductive via fill: (a) traditional "dog-bone" design; (b) via-in-pad design. *(Courtesy of DuPont.)*

One example is a screen-printable paste made of silver, copper, and epoxy. The optimum mix of metal particle shapes and sizes bonded together with epoxy resin produces essentially zero shrinkage. In general, the plated or unplated through hole is filled with paste using a stencil-printing process. After drying and curing, the plugged through holes are planarized and plated to make them solderable. Some typical design rules are an aspect ratio of 1:1 to 6:1 with vacuum assist, a via size of 6 to 25 mil (152 to 635 μm), and a core thickness of 6 to 85 mil (152 to 2159 μm). The product is extremely reliable during assembly.

22.5.3.2.5 Single-Cure (Thermal) Resin. Typically, thermally cured resins exhibit good adhesion strength to copper. Unfortunately, the rate of heat dissipation throughout the filled core material is unstable, making it difficult to achieve a reliable semicured state prior to planarization.

As a result, a rapid and unpredictable rise in temperature is likely to occur; as such, it is imperative to fill the holes completely, leaving no voids, if this type of material is to be used.

22.5.3.2.6 Solder Mask Ink. Using this type of material, it is relatively easy to fill the hole. However, the presence of volatile elements (solvents) and the degree of shrinkage during curing have a significant effect on the stability of the manufacturing process. In the case of small diameter through holes, the solvent residue may remain in craters or dishes at the mouth of the hole opening. Furthermore, the copper adhesion strength of these materials is typically lower than that of other hole fill materials.

22.5.3.2.7 Resin-Coated Foil (RCF). Similar to the potential problems faced with photoimageable dielectrics, there is a divergence between the material characteristics required for hole filling and the characteristics that are desirable for an innerlayer dielectric-copper material. Nonetheless, it is likely that the RCF foil materials will prove to be acceptable for filling vias in the limited range of specific applications that exhibit a low-aspect ratio between hole diameter and core thickness.

Liquid dielectric resin is coated more or less the same way as in photovia. However, there is a fundamental difference between photovia and laser via dielectric. In laser via processing, the resin is fully cured before laser drilling (microvia formation). This is a big advantage over photovia materials since the resin movement after hole formation is much more stable than in the case of photovia processing in which resin is fully cured after holes are formed. This resin (hole) movement makes pattern-imaging registration difficult for photovia processing.

22.6 HDI MANUFACTURING PROCESSES

This section discusses processes that employ non-drilling via-hole formation techniques. Through-via drilling is possible below 0.20 mm (0.008 in.), but cost and practicality discourage this. Below 0.20 mm (0.008 in.), laser and other via-formation processes are more cost-effective. Each of the five major via-hole formation processes used for printed circuits is discussed in the following sections.

- Photovia
- Plasma via
- Laser via
- Solid (paste) via
- Insulation displacement via

The manufacturing process for each microvia technology begins with a base core, which may be a simple double-sided board carrying power and ground planes or a multilayer board carrying some signal pattern in addition to power and ground planes. The core usually has plated through holes (PTHs). These PTHs become BVHs. Such a core is often called an *active core.*

22.6.1 Photovia Process

Prior to dielectric material coating by any of the methods described previously, the copper surface of the base core must be treated by an adhesion promotion process to ensure good adhesion of dielectric material to the copper surface. Today, very few manufacturers use oxide treatment for this purpose. The most popular adhesion promotion treatment is a special etching process offered by many suppliers of chemicals. This step is common to all microvia processes.

Dielectric resin is semicured after coating to eliminate tackiness, and then the hole pattern is exposed by photoexposure processing. The usual photodeveloping process creates microvia holes and the dielectric is fully cured, typically at 160°C for about 1 hour. Then, the panel goes through a permanganate etching process to remove any residual resin at the bottom of the hole and simultaneously create microporous surfaces that act as an anchor and ensure desirable peel strength after copper plating.

The level of peel strength is controversial. Minimum peel strength required for chip package substrates is about 600 g/cm^2, but motherboard users, particularly cell phone makers, demand a minimum of 1.0 kg/m^2 or more in order for cell phone handsets to withstand drop tests. Laser via materials usually yield stronger peel strength because of fillers that can be added to dielectric resin. When etched, these fillers generate a superior microporous surface structure needed for strong peel strength.

After permanganate etching, the panel is catalyzed and metallized in an electroless copper bath and panel-plated galvanically to desired thickness. Some photovia process practitioners roughen the resin surface mechanically by brushing or liquid horning prior to catalyzing. Then, the conductor pattern is formed by dry-film tenting and etching. Some manufacturers prefer to use the pattern-plating method for this purpose. Very few microvia board manufacturers use direct metallization methods for metallizing holes prior to galvanic plating. Several Japanese manufacturers use electroless copper all the way to the desired thickness in panel plating and use a positive electrodeposited (ED) system to achieve fine lines and very small annular rings.

One important step in microvia hole board fabrication when resin is the dielectric choice, whether the process is photovia or laser via, is the removal of residual catalysts (normally palladium) entrapped in a microporous surface that can cause migration. The process used in this step is normally a trade secret.

Photovia processing is now used primarily to fabricate semiconductor chip package substrates because a large number of holes can be formed in one photoexposure and development step. However, as mentioned previously, photovia processing suffers more from material shrinkage than laser via processing after full cure, and hole locations tend to move randomly, which makes subsequent registration for patterning difficult. Because of this problem, photovia users limit the size of the panel to be much smaller than the usual panel size prevalent in motherboard fabrication to about 400 mm × 400 mm. Small hole formation is also difficult with the photovia process. As a result, even makers of package substrates are now resorting more to laser via processes as the laser-drilling speed is being improved. Photovia process users engaged in mass production are found only in Japan today. All photodielectric processes have certain characteristics in general. Table 22.11 lists the typical processing factors. A standard photovia process sequence is described in Fig. 22.18a.

TABLE 22.11 Four Typical Photodielectrics (Three Epoxy and One Polyimide) and the Processing Parameters for Coating, Exposure, Developing, Desmears, and Metallization

	Product A	Product B	Product C	Product D
Product	L-PID, negative	L-PID, negative	L-PID, negative	DF-PID, negative
Material	Epoxy	Epoxy	Polyimide	Epoxy
Preclean	Chemical clean	Pumice jet	Br. oxide	Chemical clean
Apply PID	Curtain coat 150–400 cps	Curtain coat 200–600 cps	Extrusion coat 12,000–25,000 cps	Vac, Iam. 60 s @ 65°C
Thickness	50 μm	50 μm	37 μm	63 μm
Drying	15 h @ 90°C	6 h @ 25/40°C 3 h @ 140°C	5 h @ 25°C 15 h @ 125°C	N/A
Exposure	800–1,200 mJ/cm^2	800–1,600 mJ/cm^2	2,000–3,000 mJ/cm^2	700–1,200 mJ/cm^2
Heat bump	15 h @ 90°C	12 h @ 125°C	N.R.	20 h @ 85°C
Development	Aqueous proprietary 75 min @ 35°C	Organic GBL 60 min @ 30°C	Organic proprietary 150 min @ 30°C	Aqueous proprietary 60 min @ 35°C
Final cure	UV: 1.0 J/cm^2 + 60 h @ 145°C	60 h @ 150°C	120 h @ 175°C	UV: 2 J/cm^2 + 60h @ 150°C
Roughen				
Swell	4 h @ 65°C	4 h @ 75°C	2 h @ 65°C	5 h @ 60°C
Etch	4 h @ 80°C	8 h @ 75°C	1 h @ 75°C	10 h @ 75°C
Neutralize	6 h @ 50°C	6 h @ 50°C	3 h @ 45°C	3 h @ 25°C
Metallize: electroless Cu	0.3–0.5 μm	0.7–1.0 μm	0.3–0.5 μm	0.4 μm
Bake	20 h @ 90°C	60 h @ 150°C	60 h @ 150°C	15 h @ 90°C
Electroplate				

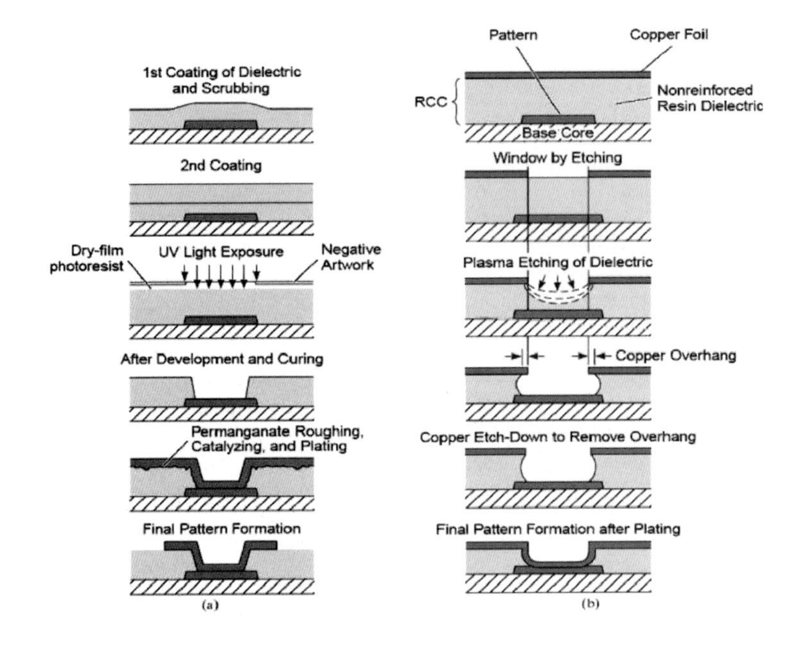

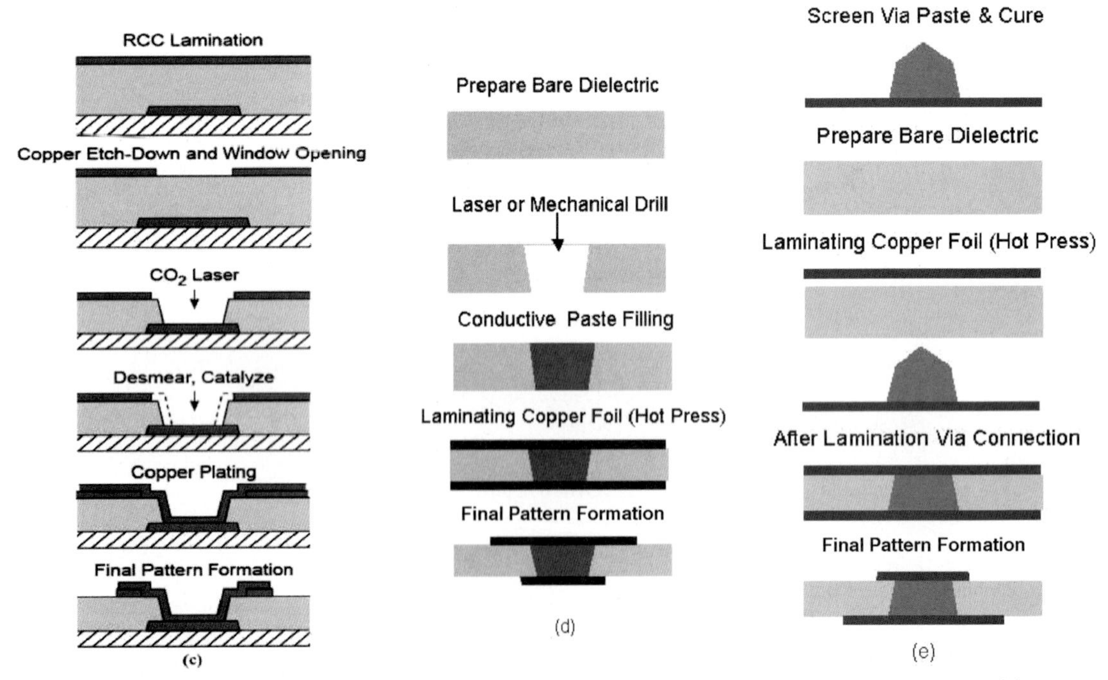

FIGURE 22.18 Standard HDI manufacturing processes: (a) a typical photovia process with liquid resin dielectric; (b) the standard plasma via process with RCC; (c) the semiconformal laser via process; (d) the standard conductive (paste) solid via process; (e) the insulation displacement via formation process.

22.6.2 Plasma Via Process

Plasma via processing was developed by a Swiss PWB maker, Dyconex. Products made with the plasma via process are called DYCOstrate.

There are many variations to the plasma via process, one of which is illustrated in Fig. 22.18b. Today, it is mainly used to fabricate sophisticated flex and flex-rigid wiring boards in small quantities.

First, an opening or window is made through copper foil by a normal etching process. When plasma etching is applied through this window, the shape of holes tends to be like a bowl (as shown in Fig. 22.28b), which is not suitable for reliable plating (although new plasma-etching equipment claims to have solved this problem).[7,8] Another problem is related to how the microvia hole is formed. The copper edge of the window hangs out over the hole, which results in poor reliability after panel plating. Therefore, to ensure reliable plated holes, a secondary etching is necessary to remove this copper overhang. One good thing results from this secondary etching, however: Since surface copper is thinned, formation of finer conductors is made easier. Nevertheless, by the time the panel is ready for plating for subsequent conductor pattern formation, it takes several times longer than other processes in a mass-production environment.

Plasma via processing is effective for forming through holes on flexible materials since holes are formed by plasma etching from both sides of the film and the bowl effect is minimized.

Plasma via etching evolved from the traditional process of plasma desmearing of through holes. Different gas, magnetrons, and equipment fixturing are employed by current plasma via-etching equipment. The plasma is generated in a partial vacuum filled with a mixture of oxygen, nitrogen, and chlorofluoro (CF_4) gasses. Microwave magnetrons create the plasma field and special low-frequency kilowave units help provide rapid etching of organics.

22.6.3 Laser Via Process

Laser via processing is by far the most popular microvia hole formation process. But it is not the fastest via formation process. In Fig. 22.19, the chemical etching of small vias is the fastest, with an estimated rate of 40,000 to 50,000 vias per second. This is also true of plasma via

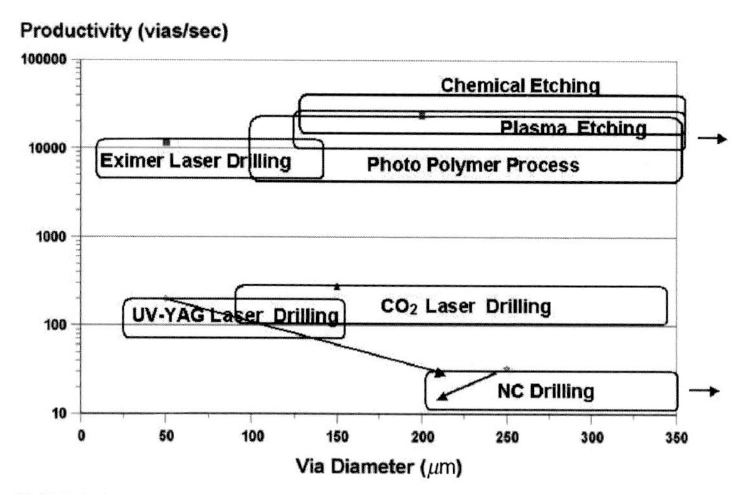

FIGURE 22.19 The various rates of via formation based on diameter and method of via ablation, from chemical, plasma, photo and laser techniques.

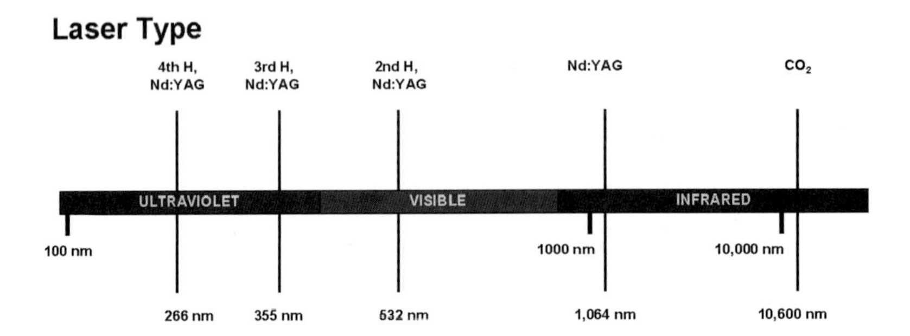

FIGURE 22.20 The wavelengths used by laser via drilling from infrared to ultraviolet region, including the harmonics of Nd:YAG lasers.

formation and photovia formation. These are all *mass-via-formation* processes. Laser drilling is one of the oldest microvia generation techniques.[9] The wavelengths for laser energy are in the infrared and ultraviolet region. Figure 22.20 shows the five major wavelengths used in current laser drills. The absorption curves of many organic dielectrics epoxy, polyimide, matte copper, and glass fibers are shown in Fig. 22.21. Laser drilling requires programming the beam fluence size and energy. High-fluence beams can cut metal and glass, whereas low-fluence beams cleanly remove organics but leave metals undamaged. A beam spot size as small as approximately 20 microns (<1 mil) is used for high-fluence beams and about 100 microns (4 mil) to 350 microns (14 mil) for low-fluence beams.[10,11]

However, there are many variations. For the purpose of drilling microvia holes, there are four laser systems: UV/Yag laser, CO_2 laser, Yag/CO_2, and CO_2/CO_2 combinations. Then there are three dielectric materials: RCC, resin only (dry-film or liquid resin), and reinforced prepreg. Therefore, the number of ways to make microvia holes by laser systems is driven by the permutation of four laser systems and these three dielectric materials.

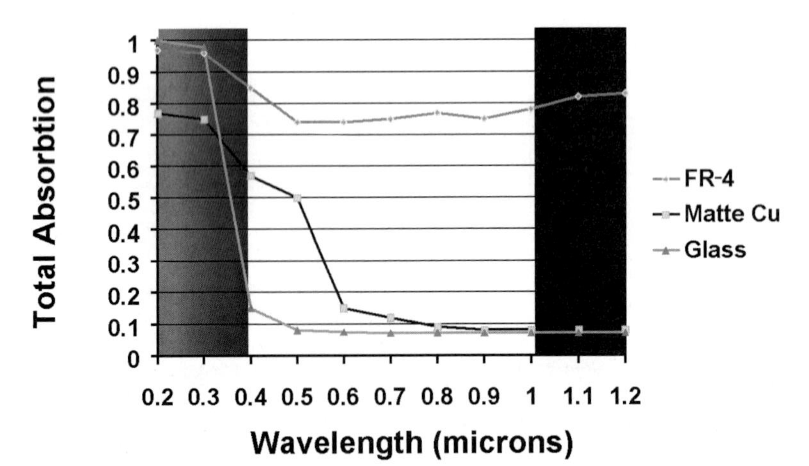

FIGURE 22.21 The absorption wavelength of different materials respond to different wavelength of laser energy. Epoxy, copper and glass of laminate are ablated by the UV wavelength but only the resin is ablated by the CO_2 wavelengths .

There are several factors to observe in laser via processing: position accuracy of lased holes (microvia holes), uneven diameters of holes, dimensional change of the panel after curing dielectric, dimensional change of the panel due to temperature and humidity variations, alignment accuracy of the photoexposure machine, unstable nature of negative artwork, and so on. These should be carefully monitored and are important for all microvia hole processes.

22.6.3.1 ***UV/Yag Laser.*** A Yag laser can penetrate copper; therefore, it is not necessary to pre-etch the window when the choice of dielectric is RCC or a copper-foil/prepreg combination.

Positioning accuracy in relation to the capture pad is good, but there are weaknesses associated with the Yag laser. The speed of drilling (ablation) is slower than the CO_2 laser, particularly when a hole diameter is equal to or larger than 125 μm, because the laser beam is very small. Therefore, it is necessary to do trepanning (spiral drilling pattern), which results in longer drilling time. Another weakness is the Yag laser's sensitivity to the thickness of dielectric material. If the thickness of the RCC's resin is not even, the Yag laser beam, the power of which is adjusted to remove the thicker part of the resin, may damage the capture pad. If adjusted to the thinner part, it may not reach the capture pad.

Although the Yag laser can penetrate copper and glass-reinforced prepreg, the situation is worse in this material than with RCC. The laser power must be adjusted to penetrate through the crossover part of woven glass. Then, when the laser beam hits the part of prepreg where there is now a glass bundle (opening), the beam can really damage the capture pad. To avoid this problem, the operator must reduce the laser power and apply laser ablation for a longer duration, which increases drilling time considerably. Depending on the construction of the dielectric layer through which the Yag laser beam must penetrate, the drilling speed must be adjusted to about 3 to 50 holes per second when glass-reinforced dielectric is used.

When the dielectric is RCC, the speed of the currently available UV/Yag laser is somewhere around 100 to 300 holes per second.

When the dielectric is resin without copper foil, a CO_2 laser is the best since a CO_2 laser can offer much faster drilling speed (a single-head CO_2 laser can drill as many as 25,000 holes per minute, including overhead such as loading a panel onto the drilling table and registering the panel).

The best feature of the UV/Yag laser drilling machine is its ability to drill a very small hole (down to 20 to 30 μm, whereas the smallest hole a CO_2 laser can drill is 50 μm in a mass-production environment) and its superior position accuracy through RCC.

For higher-speed drilling, even Yag laser users are now etching copper foil down to 6 to 9 μm, the process more frequently seen in CO_2 drilling.

22.6.3.2 ***CO_2 and Twin CO_2.*** A CO_2 laser beam cannot penetrate copper foil unless the foil is very thin, less than 5 μm, and the surface is treated to be dark to absorb the CO_2 laser beam (this process is called CO_2 laser direct drilling and will be treated later).

When the surface is resin only, a single-headed CO_2 laser machine can drill holes at the rate of 20,000 to 25,000 per minute depending on hole density and distribution. The drilling speed continues to increase. The denser the hole, the faster the drilling speed because time lost by table motion is minimized. A dual-head CO_2 machine can boost drilling productivity by about 70 percent compared to a single-head machine.

Manufacturers of dielectric resins mix fillers in the resins to enhance peel strength. Some fillers can slow down the drilling speed of a CO_2 laser beam (beam absorption is weaker in this case). For example, when the required dielectric thickness is 60 μm, resin is coated twice, which is a frequent practice to produce a thick coating. The first coating is made with resin without filler (40 μm), and the second coating (20 μm) is resin with filler. Since the laser beam goes through a thin layer of filled resin, the speed degradation is minimized.

The most desirable shape of drilled microvia holes is tapered, which is accomplished by pulse drilling. Normally, two to three pulses are used. When three pulses are applied on the same spot continuously, the capture pad may be overheated and excessive epoxy smear may result. Therefore, when three pulses are to be applied, the entire panel is drilled first by two pulses and then the third pulse is once again applied to all holes. This way, the quality of the holes is ensured, although it takes longer to drill.

When the choice of dielectric is RCC, the opening or window must be formed first on copper foil, as in the case of plasma via. Prior to window formation, an etching process thins the surface copper, usually down to 6 to 7. This process is sometimes called *half-etching*.

Half-etching yields a few benefits. Window formation is made easier, and formation of fine-line conductors is also made easier. H_2O_2/H_2SO_4 etchant is often used for this purpose for even thickness. As mentioned previously, this half-etching procedure is also becoming common for Yag laser users.

There are two variations to CO_2 laser drilling through a window opening. One is called *conformal* and the other *semiconformal* (see Fig. 22.18c).

In the conformal method, a laser beam with a diameter larger than the window opening is used. Straight ablation of dielectric results in a bowl-shaped hole, as in the case of plasma via. Therefore, improved systems use laser beam diameters slightly larger than the window opening, and beams are pulsed to minimize the bowl effect.

In the semiconformal method, a laser beam with a diameter slightly smaller than the window opening is used. This results in a slight step when plated afterward, as shown in Fig. 22.18c. To minimize this step, half-etching is important in semiconformal drilling.

Some manufacturers using CO_2 drilling etch down copper completely after RCC lamination. The dielectric surface below copper is already microporous, which generates good adhesion except prior to the RCC stage. Therefore, some special treatment is applied to the resin surface, such as electrolessly plated nickel layer, prior to copper plating, but such treatment is kept secret.

The RCC approach suffers from misregistration of window opening to capture pad when the surface and capture pad diameter becomes less than 250 μm, which causes serious defects (see Fig. 22.22). Laser via process practitioners have several solutions to this problem. They can use liquid resin or resort to a Yag or Yag/CO_2 laser. Since a Yag laser can break through copper foil, the beam can be directed to the capture pad position accurately. Another choice is to use laser direct imaging (LDI) for window etching in order to benefit from the speed of the CO_2 laser.

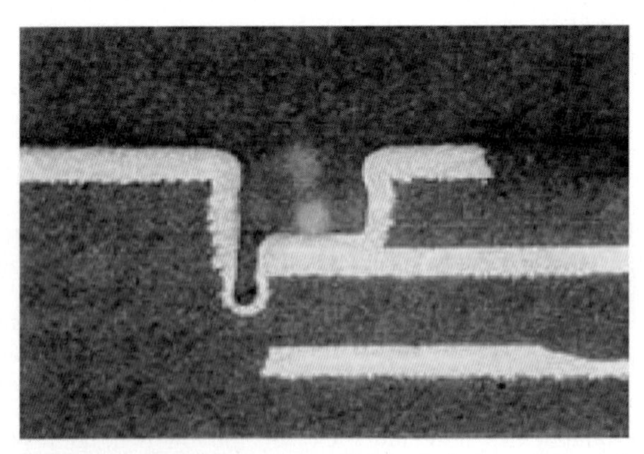

FIGURE 22.22 Misregistration of laser hole to capture pad. (*Courtesy of Ibiden.*)

Another solution is to use CO_2 laser direct drilling. The advantage of using RCC is the pre-established peel strength. Manufacturers go through delicate steps to establish the appropriate peel strength when using liquid resin. The degree of resin cure, permanganate etching, control, and regeneration of the permanganate solution affects the resultant peel strength as well as cost.

In this laser direct drilling method, copper foil is etched down to about 5 μm and its surface goes through oxide treatments such as black oxide or other methods, typically as alternatives to oxide treatment, which is normally an etching process. The copper thickness is further reduced by 1 μm, and the surface color becomes dark, allowing the CO_2 beam to penetrate.

Alternatively, ultrathin copper RCC may be used. Carrier foil, usually copper foil of 70-μm thickness (2 oz.), is coated with a very thin layer (10 to 20 μm) of current conductive release film, and a thin foil of 3 to 5 μm is plated. After such RCC is laminated to the core, the carrier film is peeled off and the surface condition is the same as etched-down copper. Drilling speed in this direct CO_2 laser drilling method is degraded by 30 to 40 percent compared to that of straight resin drilling, but the registration to the capture pad is ensured.

22.6.3.3 Yag/CO_2. This combination is particularly useful for a copper/prepreg structure. The Yag laser offers good position accuracy and cuts through copper, but it tends to leave glass fibers sticking out in the hole after prepreg drilling. And as mentioned before, it is sensitive to dielectric thickness variation. The CO_2 laser beam cuts through glass fiber more cleanly than the Yag. In this combination laser drilling, the power of the Yag laser beam is adjusted to cut through copper and remove a portion of glass fiber and leaves the finishing to the CO_2 laser. The speed of drilling is governed by the speed of the Yag laser and is therefore slower than a pure CO_2 laser. However, this is the preferred laser-drilling machine for fabricating infrastructure types of PWBs, such as ones used for servers, network routers, and base stations that typically have copper/prepreg surface structures. The size of the panels used to fabricate such boards is usually large; therefore, the positioning accuracy of the Yag laser is preferred. Microvia hole boards used by the automotive industry also have this copper/prepreg structure.

The speed of drilling by a Yag/CO_2 machine is much slower than that of a CO_2 laser. However, the number of microvia holes to be drilled for infrastructure boards is normally one magnitude less than that seen in cell phone applications, in which the number of microvia holes can reach as many as 650,000 to 700,000 per $m.^2$ Furthermore, the price of infrastructure boards is much more than that of mass-produced cellular phone motherboards. Therefore, slower speed of drilling, which means higher drilling cost, does not affect the total cost of manufacturing.

Residual glass fibers in the hole that can affect reliability may be removed chemically or mechanically. Mechanical removal is preferred. Blasting with aluminum oxide (about 20 μm diameter particles) is commonly used to remove glass fibers and smears as well. An excimer laser is also used to remove smears in the hole by a sweeping motion through the copper opening. In both cases, it takes about 30 sec. to desmear one side of the panel.

22.6.4 Dry Metallization (Conductive Inks, Conductive Paste, and Insulation Displacement)

Three types of dry metallization are covered: conductive ink, conductive paste, and insulation displacement. Conductive ink describes a single-layer dielectric with microvias formed by photoimaging, laser, or insulation displacement. A conductive paste is used to fill the microvias and act as the conductive path between layers. Surface metallization may be accomplished either by laminating copper foil onto the dielectric surface or by chemical deposition. This process is shown in Fig. 22.18d.

Insulation displacement is a unique process in that the silver conductive paste is screened on copper foil and cured. The conductive paste forms a "pointed" via that penetrates conventional prepregs during lamination and adheres to the copper foil on the reverse side, forming a via. Figure 22.18e shows this via formation process.

Standard activities concentrate on the use of photoimaging and screen printing while overcoming the limitations and expense associated with processes such as plating and etching. The concept is built on the use of a photoimageable dielectric to produce both vias and circuit channels that are then filled with a conductive ink. These approaches eliminate the need for a separate dielectric and clad-metal layer that requires either etching or a combination of plating and etching to produce a circuit. They also eliminate the resist deposition and stripping previously required to define that circuit. The conductive ink technique of metallization virtually eliminates the generation of metal waste streams.

Conventional multilayer printed wiring boards must be subjected to drilling and through-hole plating to create interconnections. These holes represent inefficient use of PWB area. In addition, to connect the printed wiring with part connection lands, some through holes must be provided in areas other than where the lands are located.

REFERENCES

1. Bauer, Charles E., and Bold, William A., Tektronix, Inc., U.S. Patent 4,566,186, "Multilayer Interconnect Circuitry Using Photoimageable Dielectric," January 28, 1985.
2. Tsukada, Y., and Tsuchida, S., "Surface Laminar Circuit, A Low Cost High Density Printed-Circuit Board," *Proceedings of the Surface Mount International Conference and Exposition*, San Jose, CA, September 1992.
3. Schmidt, W., "A Revolutionary Answer to Today's and Future Interconnect Challenges," *Proceedings of the Sixth PC World Conference*, San Francisco, May 1993.
4. Holden, H., "Segmentation of Assemblies: A Way to Predict Printed Circuit Characteristics," *Proceedings of IPC T/MRC*, New Orleans, December 6, 1994.
5. Holden, H., IPC-2315, "Design Guidelines for HDI and Microvias," IPC, 1998, pp. 55.
6. Holden, H., "The Challenge: To Plan Successful Products When Packaging Is So Complicated," *Future Circuits*, Vol. 2, No. 1, 1997, pp.106–109.
7. Seraphim, D. P., Lasky, R. C., and Li, C.Y., *Principles of Electronic Packaging*, McGraw-Hill, 1989, pp. 39–52.
8. Heller, W. R., His, C. G., and Mikhail, W. F., "Wireability: Designing Wiring Space for Chips and Chip Packages," *IEEE Design Test*, August 1984, pp. 43–51.
9. Sweetman, E., "Characteristics and Performance of PHP-92: AT&T's Triazine-Based Dielectric for Polyhic MCMs," *International Journal of Microcircuits and Electronic Packaging,* Vol. 15, No. 4, 1992, pp. 195–204.
10. Gonzalez, Ceferino G., "Materials for Sequential Build-Up (SBU) of HDI-Microvia Organic Substrates,", The Board Authority, June 1999, pp 56–58
11. Circuit Tree HDI Materials, The Board Authority Journals on HDI, June 1999 and April 2000, CircuiTree magazine, BNP Publishing.
12. Bakoglu, H. B., *Circuits, Interconnections and Packaging for VLSI,* Addison Wesley, 1990.
13. Hannemann, R. J., "Introduction: The Physical Architecture of Electronic Systems," *Physical Architecture of VLSI Systems*, R. Hannemann, A. D. Kraus, and M. Pecht (eds.), John Wiley & Sons, 1994, pp. 1–21.
14. Moresco, L., "Electronic System Packaging: The Search for Manufacturing the Optimum in a Sea of Constraints," *IEEE Transactions on Components, Hybrids and Manufacturing Technology*, Vol. 13, 1990, pp. 494–508.
15. Maliniak, D., "Future Packaging Depends Heavily on Materials," *Electronic Design*, January 1992, pp. 83–97.
16. Powell, D., and Weinhold, M., "Laser Ablation of Microvia Holes in Woven Aramid-reinforced PWBs," *Chip Scale Review*, September 1997, pp 38–45.
17. Poulin, D., Reid, J., and Znotins, T. A., "Materials Processing with Excimer Lasers," International Congress on Application of Lasers and Electro-Optics (ICALEO) paper, November 1982.
18. Knudsen, P. D., et al., U.S. Patent 5,262,280, November 16, 1993.
19. Shipley, C. R., U.S. Patent 4,902,610, February 20, 1990.

20. Shipley, C. R., U.S. Patent 5,246,817, September 21, 1993.
21. Sakamoto, Kazunori, Yoshida, Shingo, Fukuoka, Kazuyoshi, and Andô, Daizo, "The Evolution and Continuing Development of ALIVH High-Density Printed Wiring Board," presented at IPC Expo 2000; featured in *Circuit Tree,* May 2000, pp 34–37.
22. Itou, Motoaki, "High-Density PCBs Provide for More Portable Design," http://www.nikkeibp.com/nea/nov99/tech/.
23. "Microvia Substrates: An Enabling Technology for Minimalist Packaging 1998–2008," BPA Group Ltd., 1999, pp. 4–3 to 4–17.
24. Tsukada, Yutaka, et al., "Surface Laminar Circuit and Flip Chip Attach Packaging," *Proceedings of the Seventh IMC,* 1992.
25. Tsukada, Yutaka, *Introduction to Build-Up Printed Wiring Board* (in Japanese), Nikkan Kogyo Shinbun, 1998.
26. Holden, Happy, "Special Construction Printed Wiring Boards," *Printed Circuit Handbook, 4th ed.,* Clyde F. Coombs, Jr. (ed.), McGraw-Hill, 1995, chap. 4.
27. Takahashi, Akio, "Thin Film Laminated Multilayer Wiring Substrate," *JIPC Proceeding,* Vol. 11, No. 7, November 1996, pp. 481–484.
28. Shiraishi, Kazuoki, "Any Layer IVH Multilayer Printed Wiring Board," *JIPC Proceeding,* Vol. 11, No. 7, November 1996, pp. 485–486.
29. Fukuoka, Yoshitaka, "New High Density Printed Wiring Board Technology Named B2it," *JIPC Proceeding,* Vol. 11, No. 7, November 1996, pp. 475–478.
30. Apol, Tim, "Directional Plasma Etching—Straight Sidewalls, No Undercut," *PC Fabrication,* Vol. 20, No.12, December 1997, pp. 38–40.
31. Tsuyama, Koichi, et al., "New Multi-Layer Boards Incorporating IVH: HITAVIA," *Hitachi Chemical Technical Report,* No. 24, 1995-1, pp. 17–20.
32. Tokyo Ohka Company Brochure.

BIBLIOGRAPHY-ADDITIONAL READING

1. Coors, G., Anderson, P., and Seward, L., "A Statistical Approach to Wiring Requirements," *Proceedings of the IEPS*, 1990, pp. 774–783.
2. Sutherland, I. E., and Oestreicher, D., "How Big Should a Printed-Circuit Board Be?" *IEEE Transactions on Computers*, Vol. C-22, No.5, May 1973, pp. 5323–5542.

CHAPTER 23

ADVANCED HIGH-DENSITY INTERCONNECTION (HDI) TECHNOLOGIES

Happy T. Holden
Mentor Graphics, Longmont, Colorado

23.1 INTRODUCTION

Since the introduction of HDI in the mid-1980s (HP-Finstrate, Siemens's MicroWiring, and IBM's Surface Laminar Circuit [SLC] technology; see Figs. 22.1, 22.2, and 22.3), many variations of making high-density interconnect (HDI) wiring boards have been developed and implemented for mass production. However, if one technology is to be judged in terms of volume produced and picked as a winner, laser-drilling technology is the one. Other methods are still used by a number of printed wiring board (PWB) manufacturers, but on a much smaller scale.

The purpose of this chapter is to examine a variety of HDI fabrication processes as seen in Fig. 23.1. However, a greater emphasis will be placed on the laser-drilling process (hereinafter *laservia*) since it is the most popular process today and it seems its popularity will grow in the future. It must be understood that interconnect via hole (IVH) formation is just one element of fabricating HDI wiring boards. The other two important factors are the various dielectric materials and the methods of metallization. Fabrication of HDI wiring boards with microvia holes involves many new processes not common to conventional board fabrication. Therefore, additional emphasis will be placed upon these new fabrication processes that are common to other microvia technologies.

23.2 DEFINITIONS OF HDI PROCESS FACTORS

The 21 different HDI processes shown in Fig. 23.1 have three basic fabrication factors in their process makeup:

- Dielectric materials
- IVH formation
- Methods of metallization

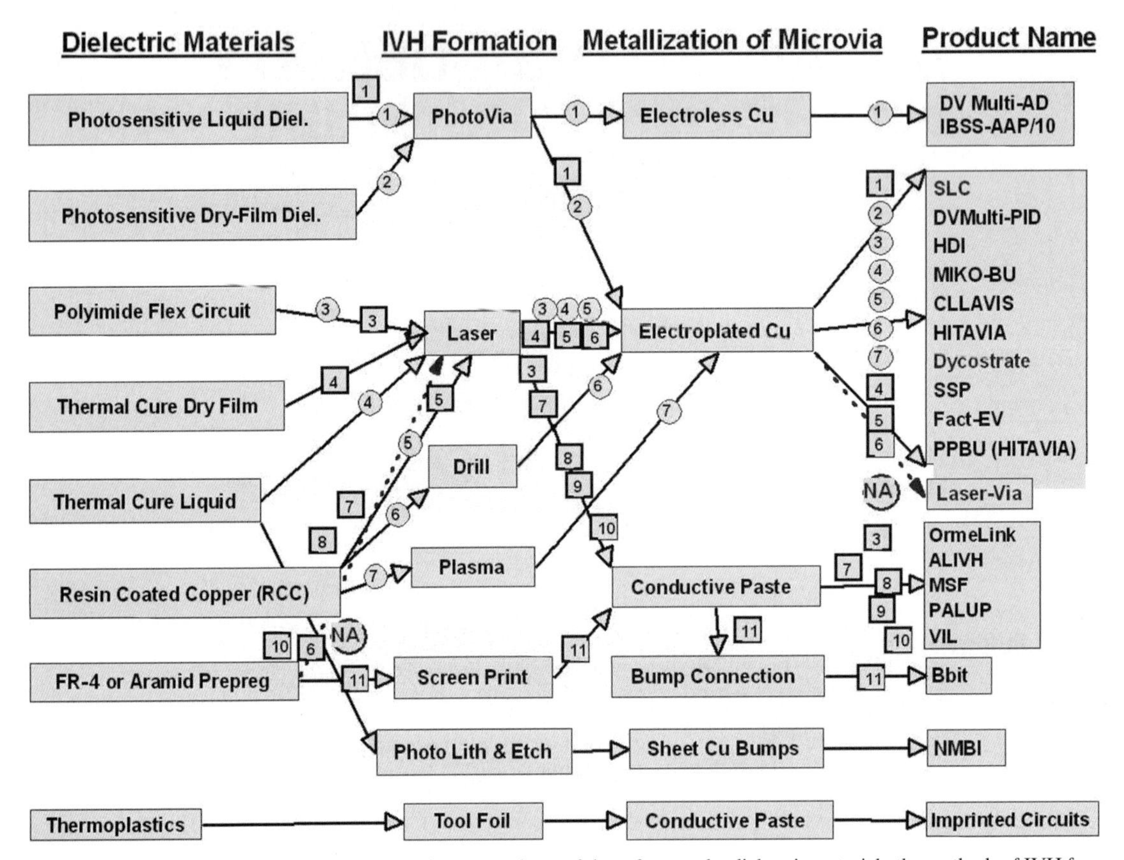

FIGURE 23.1 The HDI technologies in use today are made up of three factors: the dielectric materials, the methods of IVH formation, and the method of metallization of those z-axis via connections. In recent years up until now, 21 different HDI processes have been used.

23.2.1 Dielectric Materials

Eight different general dielectric materials have been used in the current or past HDI processes. IPC slash sheets such as IPC-4101B and IPC-4104 cover many of these, but many are not yet specified by IPC standards. The materials are:

- Photosensitive liquid dielectrics
- Photosensitive dry-film dielectrics
- Polyimide flexible film
- Thermally cured dry films
- Thermally cured liquid dielectrics
- Resin-coated copper (RCC) foil
- Conventional FR-4 cores and prepregs
- Thermoplastics

23.2.2 Interconnect Via Hole Formation

Seven different methods of forming the IVHs have been used in current or past HDI processes. Laser drilling is the most prominent, but the other six come into use also. The methods are:

- Photo processes to define vias in photodielectrics
- Various laser-drilling methods, including UV-Yag, UV-Eximer, and CO_2
- Mechanical drilling
- Plasma drilling
- Screen printing of via pastes
- Photo and etch of solid vias
- Toolfoil

23.2.3 Method of Metallization

Four different methods of metallizing the IVHs have been used in the current or past HDI processes. The methods are:

- Fully additive electroless copper
- Conventional electroless and electroplating copper
- Conductive pastes
- Fabricating solid metal vias

23.3 HDI FABRICATION PROCESSES

The three basic and original HDI processes are laser drilling laminate (Finstrate) and flexible materials (microwiring), photodielectrics with SLC and plasma-etching polyimide (Dycostrate). These and the other 18 HDI processes have been integrated into standard printed circuit manufacturing flows. There is no standard way to organize these processes, but the method of IVH formulation is the most unique characteristic of each process and will be used for this purpose.

23.3.1 Photo-Via Technologies [1,2]

23.1.1 IBSS-AAP10 System. The Interpenetrating Polymer Network Build-Up Structure System (IBSS) is a new photodielectric. This photo-imageable dielectric (PID) is a proprietary photodielectric developed by Ibiden of Japan. It complements Ibiden's older epoxy type of photo-redistribution layer technology, IP-10.

23.3.1.1.1 Structure. IBSS is specifically formulated for integrated circuit and flip-chip packages and has a high T_g and flexibility. The bismalene-triazine(BT) core also provides high-temperature resistance. A typical IBSS structure and its features are illustrated in Fig. 23.2.

23.3.1.1.2 Manufacturing Process. IBSS is a liquid system applied by roller coater to a rigid core, typically BT. The coating is cured and photodeveloping creates the vias. The IBSS requires an aggressive swell-and-etch because of the higher-temperature nature of the photo-dielectric (PD). The metallization is fully additive copper, also proprietary to Ibiden.

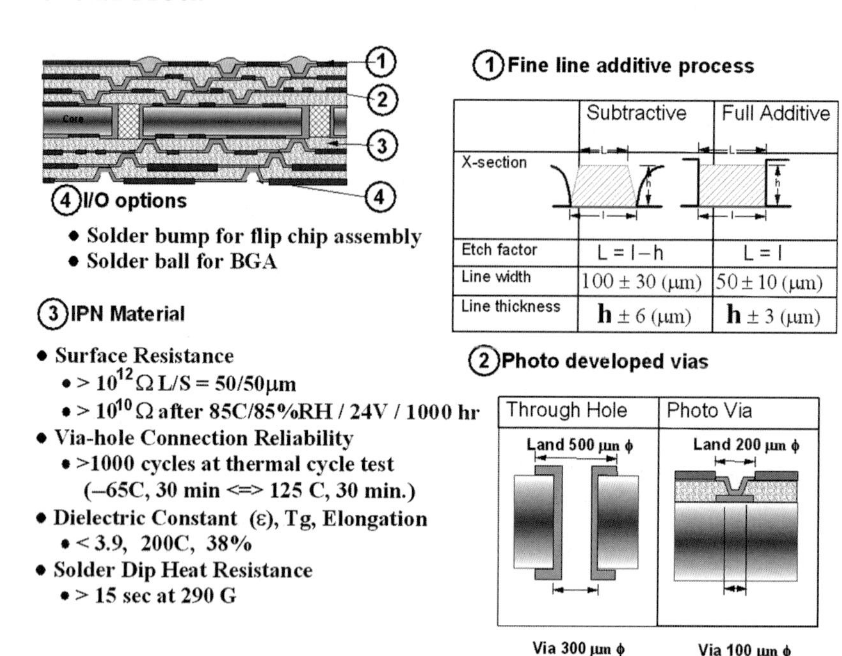

FIGURE 23.2 The IBSS/AAP10 multilayer substrate structure.

The imaging for additive metalization uses a permanent dry-film photoresist. For additional layers of dielectric, an adhesion promoter is applied to the surface copper. Figure 23.3 outlines the IBSS process.

23.3.1.2 Surface Laminar Circuits. SLC technology was the first of the PID microvia, build-up technologies to reach the interconnect market in high volume. Development of SLC technology began during the late 1980s at IBM (Yasu, Japan), and the first product was introduced in 1990.[1] The first flip-chip, direct-chip-attach (FC/DCA) product employing SLC technology was shipped from Yasu in 1992, and in 1995 IBM Endicott, New York, shipped its first printed wiring boards that utilized SLC technology.[2] Used by many original equipment manufacturers (OEMs) in Japan, the SLC process provided the high-density interconnect technology for Sony's first digital Camcorder, the DCR-PC7, seen in Fig. 23.4. This eight-layer build-up structure had two SLC layers on each side of a four-layer core. The assembly density of this was over 612 pins per sq. in. due to the wall-to-wall 0.5 mm. pitch chip scale package (CSP) used.[3]

The most suitable photosensitive dielectric was the photo-imageable solder masks, due to their proven compatibility with printed circuit assembly processes, ability to withstand exposure to service environments, and possession of the necessary via-imaging characteristics. SLC technology was originally developed and implemented with a liquid PID solder mask applied by curtain coating. In 1995, it was also qualified with a dry-film PID with virtually identical electrical properties. The dry-film version did not require the extensive surface grinding processes that the liquid PID process required.

PID technology is based on photoimageable polymeric systems to form blind microvias in dielectric material between layers of circuitry. The use of PIDs allows all microvias on a panel to be formed simultaneously, with no incremental-per-via cost. Its use is particularly advantageous on applications with high densities of vias (e.g., more than 50,000 on an 18 in. × 24 in. panel).

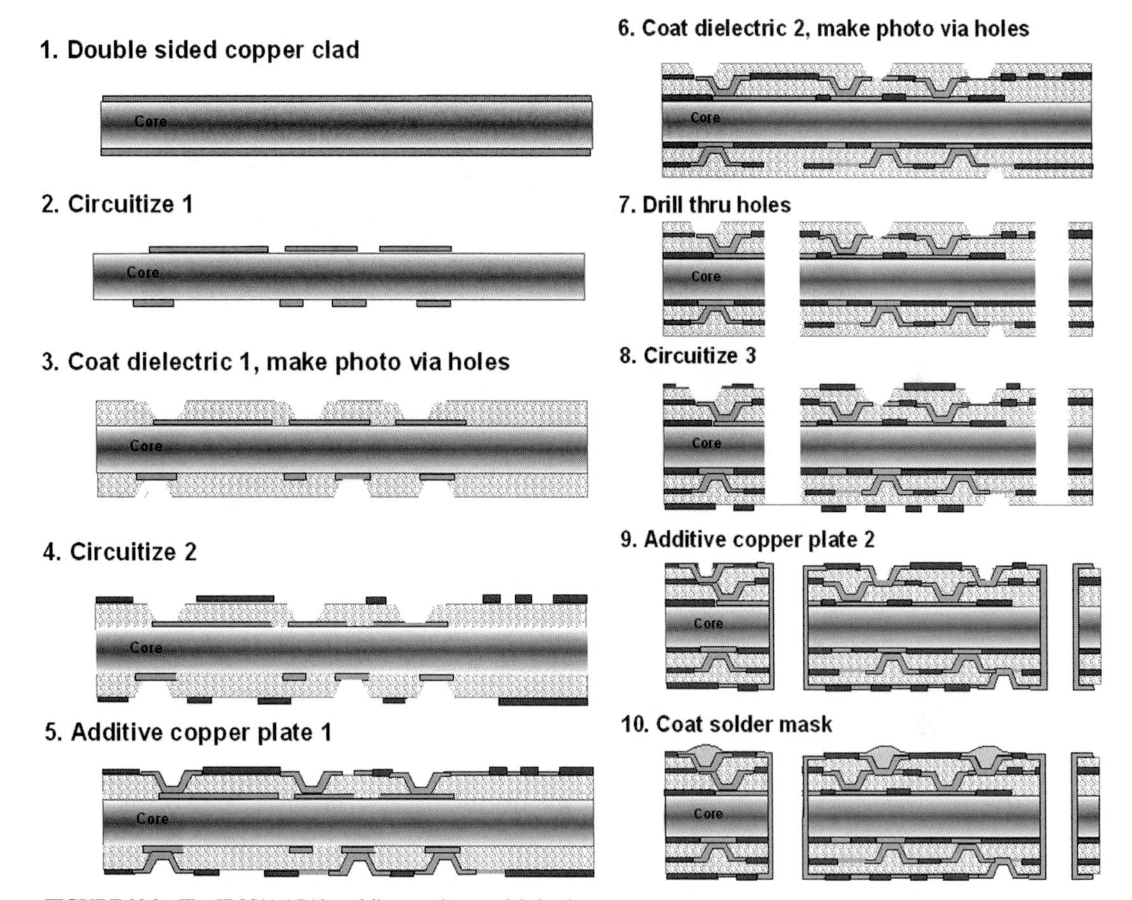

FIGURE 23.3 The IBSS/AAP10 multilayer substrate fabrication process.

23.3.1.2.1 Liquid Versus Dry PID. A typical flowchart of the PID technology fabrication process, including options for multiple build-up layers and a multilayer core, is shown in Fig. 23.5. At this level in the process, PID technology is the same whether a liquid or dry-film PID is used.

Liquid and dry-film PIDs do provide slightly different via wall profiles. The liquid PID has tapered via walls, whereas the dry-film PID has essentially vertical via walls. The tapered via walls provide good plating coverage on the via walls and base. Vertical walls allow for a smaller via top opening, and correspondingly smaller capture land for a given via bottom diameter.[4,5]

23.3.1.2.2 Tools. The use of a liquid PID requires two unique pieces of equipment: a coater of some variety, either curtain, slot, roller coater or screen printer (with associated drying ovens), and a leveling tool (surface sander). A leveling tool is required to planarize the surface of the cured liquid PID to accommodate fine-line photolithography on the surface. Liquid PID provides a conformal coating over underlying circuit features, producing nonuniform planarity. The leveling operation also removes a lip of exposed PID that overhangs the via openings in these processes. Many liquid PIDs have a self-leveling characteristic and do not require any other leveling.

FIGURE 23.4 The Sony DCR-PC7 digital camcorder utilized the SLC process for the 2+4+2 build-up structure and 0.5 mm CSPs to achieve over 600 connections (pins) per sq. in.

The use of a dry-film PID requires only one unique piece of equipment: a vacuum laminator. Vacuum laminators are common at printed circuit fabrication shops; in addition, the capital to obtain one is significantly less than for a curtain or slot coater. The dry-film PID has excellent planarization and does not require a leveling process due to the film's low solvent content, low shrinkage, and vacuum lamination process.[6,7]

PID technologies were originally conceived as an alternative to multilayer lamination as a way to make multilayer PWBs. PID technology builds up each surface signal layer on top of the previous layers in a sequential fashion. The base under all of the outer build-up layers is a conventional double-sided or multilayer circuit board containing voltage, ground planes, and even signal layers. Mechanically drilled and plated holes are used only in the core layers, to make connections to the backside of the laminate base and accommodate mounting pinned components. There is greater wiring density to use plated through holes (PTHs), which extend only through the base printed circuit structure and not through the build-up layers, especially when two or more build-up layers are used. Figure 23.6 provides additional details of the SLC manufacturing process of multilayer printed circuit boards.[8,9]

23.3.1.3 DV Mult-PID. DV Mult-PID is from NEC. This structure uses a PID primarily as an integrated circuit (IC) substrate and for high-density electronic products. It is capable of three build-up layers per side. Figure 23.7 shows a typical NEC high-density board. The use of a dry-film PID requires only one unique piece of equipment: a vacuum laminator. Vacuum laminators are common at printed circuit fabrication shops; in addition, they cost much less than a curtain or slot coater. The dry-film PID has excellent planarization, does not require a leveling process due to the film's low solvent content, low shrinkage, and vacuum lamination. The modified structure of DV Multi-AD uses laminates and RCC for the build-up layers and fully additive electroless plating. Figure 23.8 illustrates this structure and manufacturing process.

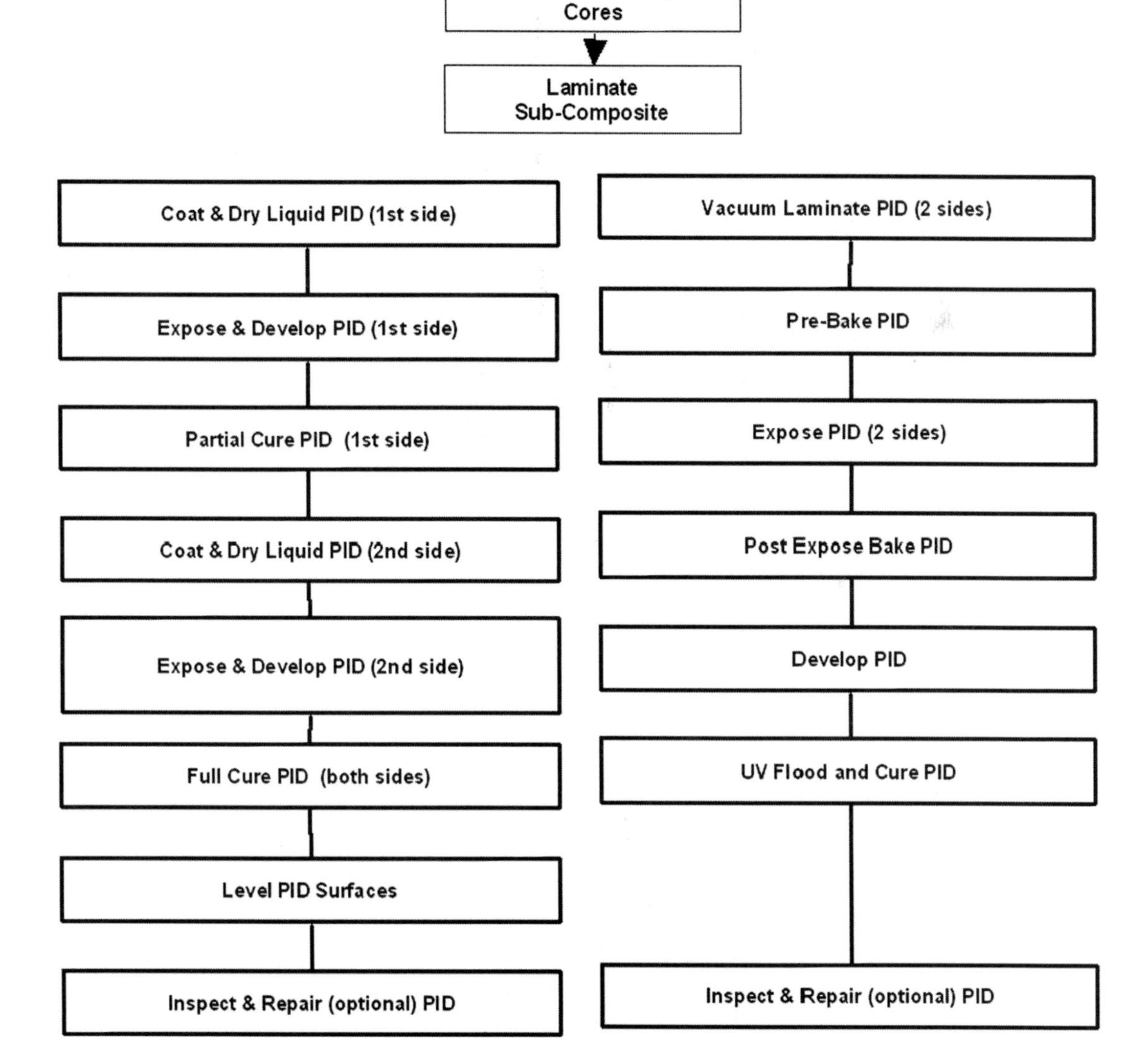

FIGURE 23.5 The manufacturing flows for the two-major SLC manufacturing processes.

23.3.2 Laser Via Technologies[(3–5), [3–6], [7–10]]

Circled processes 3, 4, and 5, as well as squared 3 through 10 in Fig. 23.1, utilize a laser drill via generation technique. This process was first used in the late 1970s to drill small holes in G-10 laminate for buried vias in mainframe computer boards by IBM for its 3081 system and by Burroughs. (Attempts to find pictures of these products proved unfruitful.) The Hewlett

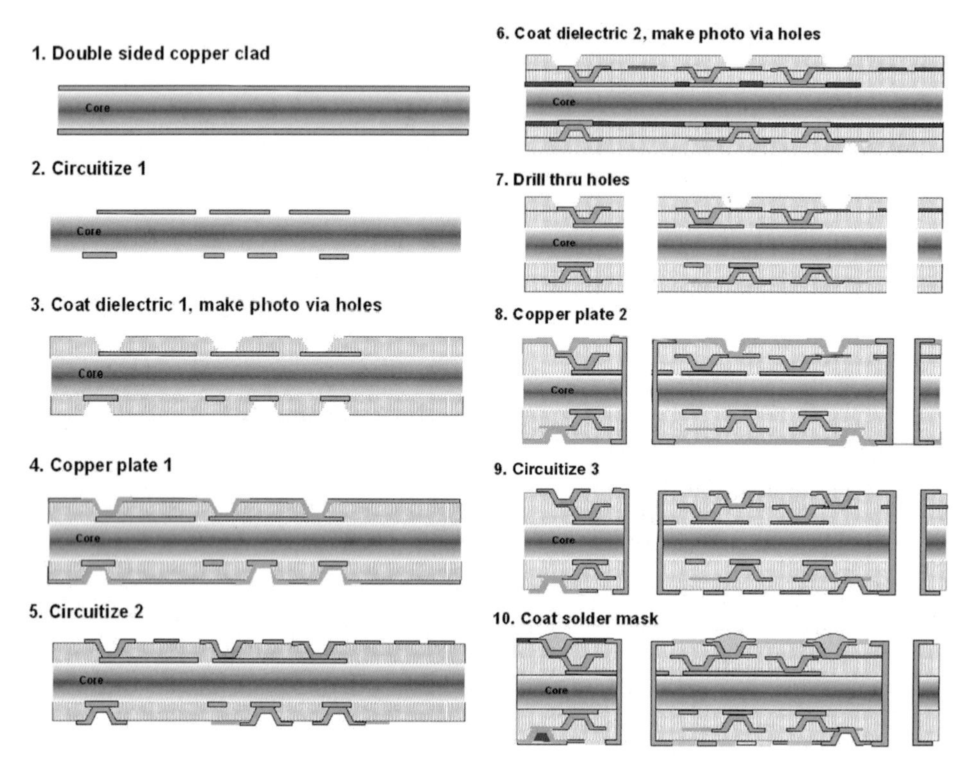

FIGURE 23.6 Additional details of the SLC manufacturing process.

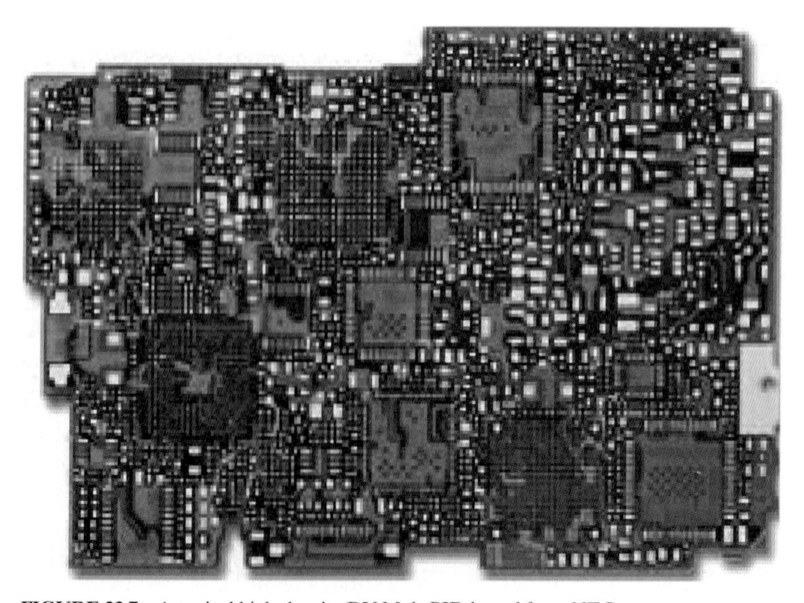

FIGURE 23.7 A typical high-density DV Mult-PID board from NEC.

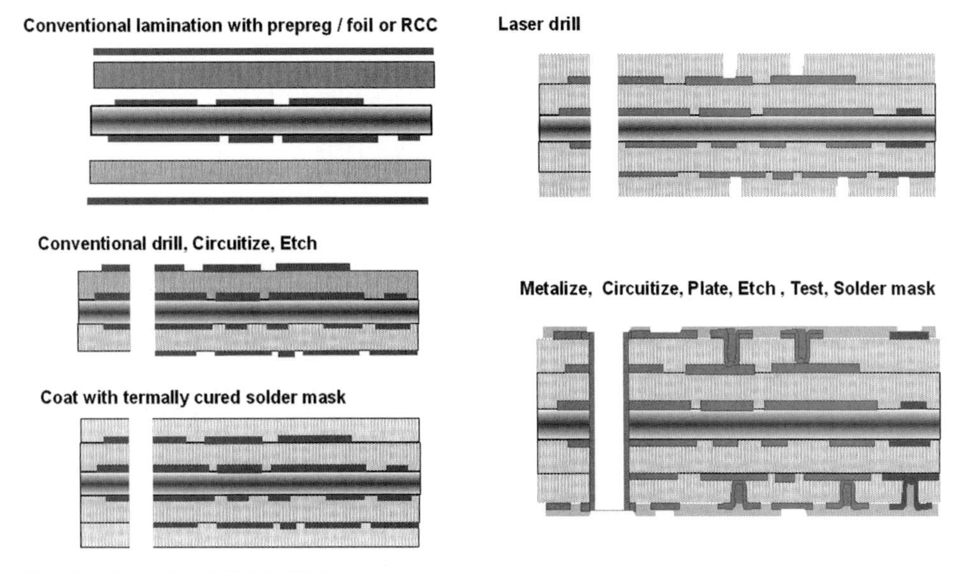

FIGURE 23.8 The DV Mult-PID board manufacturing process.

Packard laser-drilled Finstrate family of boards, first produced in 1983, is shown in Chapter 22. Siemens' laser-drilled microwiring was produced since 1987.

23.3.2.1 High-Density Interconnect (HDI). The HDI process developed by General Electric and now owned by Lockheed Martin is the one most similar to IC processes. This process is what is called a "chips first," as the assembly is done before the substrate is completed and the IC bonding is direct to the substrate. It does not use flip-chip or wire bonding, but can utilize unmodified chips directly. A number of build-up layers and very fine geometry are possible. There are obvious advantages in making multichip modules and in using standard chips. The materials and bonding technologies have proven to be very reliable and suitable for military applications. The disadvantage is the maximum panel that can be sputtered. Future use of PVD/CVD metallization may make this process more cost-effective.

23.3.2.1.1 Structure. The circuit structure of HDI consists essentially of common conventional flex circuits or more advanced microvia flex structures. These are purchased for this process. The polyimide film is usually 25- or 50-micron adhesiveness. The laser ablates away the polyimide to produce blind and through vias. The through vias are stopped by the aluminum bonding pads of the integrated circuits. By doing this, integrated circuits designed for wire bonding can be directly attached. Figure 23.9 shows a typical HDI structure.

23.3.2.1.2 Manufacturing Process. The process starts with a finished single-sided, double-sided, or multilayer polyimide flex circuit. An adhesive is applied to the flex circuit. All the ICs and components are bonded to the polyimide film layer pair flex circuitry and cured. The assembly is turned over and a laser drills down through the flex circuitry to make blind and through vias including opening up the bonding pads on the IC chip. Gold, under-bump metallurgy, or C4 bumps are not required. The panel goes through tungsten sputtering to metalize the vias and seed the top layer. Circuit pattern can be applied, plated, and then etched to complete the circuit. The process is diagrammed in Fig. 23.10.[10,11]

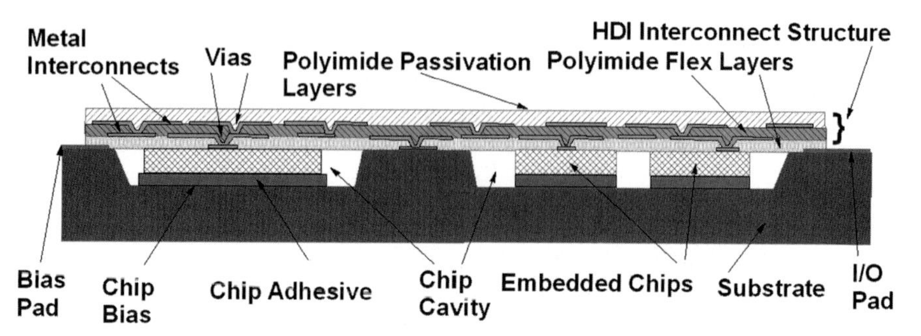

FIGURE 23.9 Structure for the HDI multilayer substrate with lasered blind vias and direct connection to the ICs.

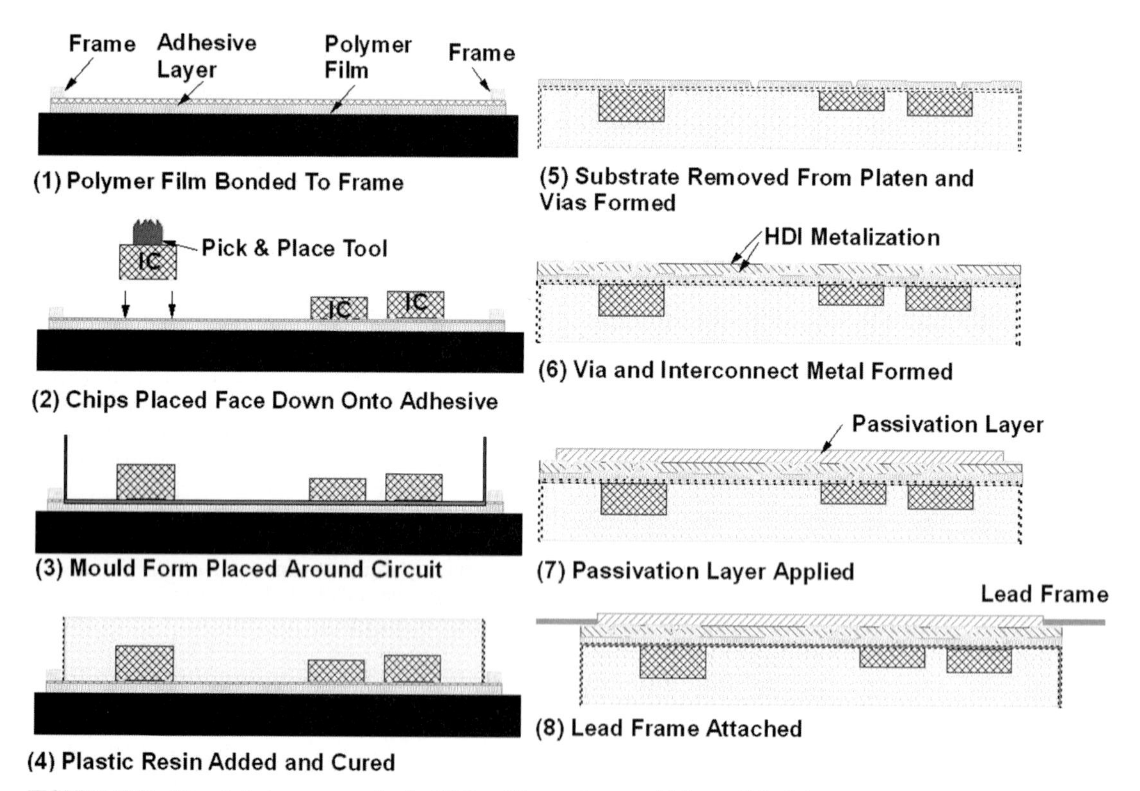

FIGURE 23.10 Manufacturing sequence for the HDI multilayer substrate with lasered blind vias.

23.3.2.2 Meiko-BU. Meiko Circuits of Japan takes a photoresist and coats it onto a stainless steel panel. This starts the process to produce high density interconnect structures (HDIS) in a remarkable way. The advantages of this process is that surface geometries are not determined by etching or full additive metallization, the vias are under the surface lands, and the circuits are all flush with the dielectric, permitting the elimination of solder masks. On the negative side, this is a more expensive process that involves carriers.

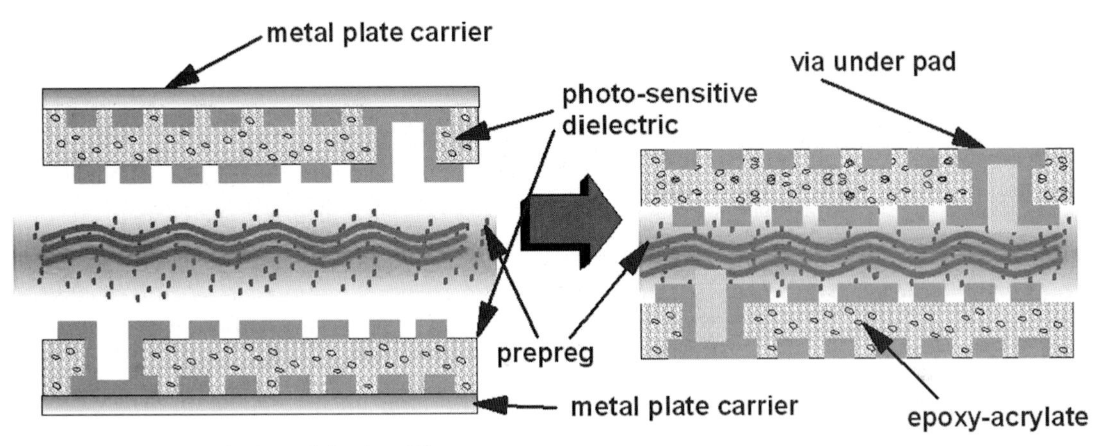

FIGURE 23.11 The carrier-formed circuit multilayer structure.

23.3.2.2.1 Structure. Figure 23.11 shows the structure of Meiko's build-up circuits. These are called carrier-formed because a stainless steel carrier serves as the base for the photodielectric even though they are laser-drilled to form the vias. The resulting structure is similar to other HDIs. The core is still a rigid board and the build-up layers are PID.

23.3.2.2.2 Manufacturing Process. The manufacturing process (see in Fig. 23.12) starts by taking a photoresist and coating it onto a stainless steel panel. The surface pattern is exposed and developed in the photoresist. First gold is plated, then nickel, and finally copper is plated on the panel. When the resist is stripped, the PID is applied over the entire panel and via holes are laser-drilled in the dielectric. Once metalized and plated, photoresists can define the circuitry by etching. The process can be repeated until the circuitry is complete or it can be laminated to FR-4 materials as rigidizers.[12] [13]

23.3.2.3 *CLLAVIS.* The CLLAVIS build-up technology is marketed by CMK of Japan. This laser-drilled microvia technology is the most common of HDI processes. The cross-sectional view in Fig. 23.13 shows the filled buried vias in the multilayer core, as well as the optional filled microvias that can be stacked. This structure is also available with the simpler, unfilled staggered microvias.

23.3.2.3.1 Manufacturing Process. The CLLAVIS manufacturing process is outlined in Fig. 23.14 and is identical to most of the laser-via-build-up technologies used. [14–28]

23.3.2.4 *SSP.* The SSP technology was developed by Ibiden of Japan. It used standard FR-4, copper plating, and laser drilling. The additional step is the application of a thin adhesive to each finished single-sided, pump-plated core. The process sequence is as follows:

1. Start with single-sided copper-clad laminate.
2. Laser drill from the non-copper side.
3. Desmear laser holes and run through the electroless copper process.
4. Plate up bumps on the clad side.
5. Image and circuitize the copper side.
6. Apply a thin adhesive to the unclad side.

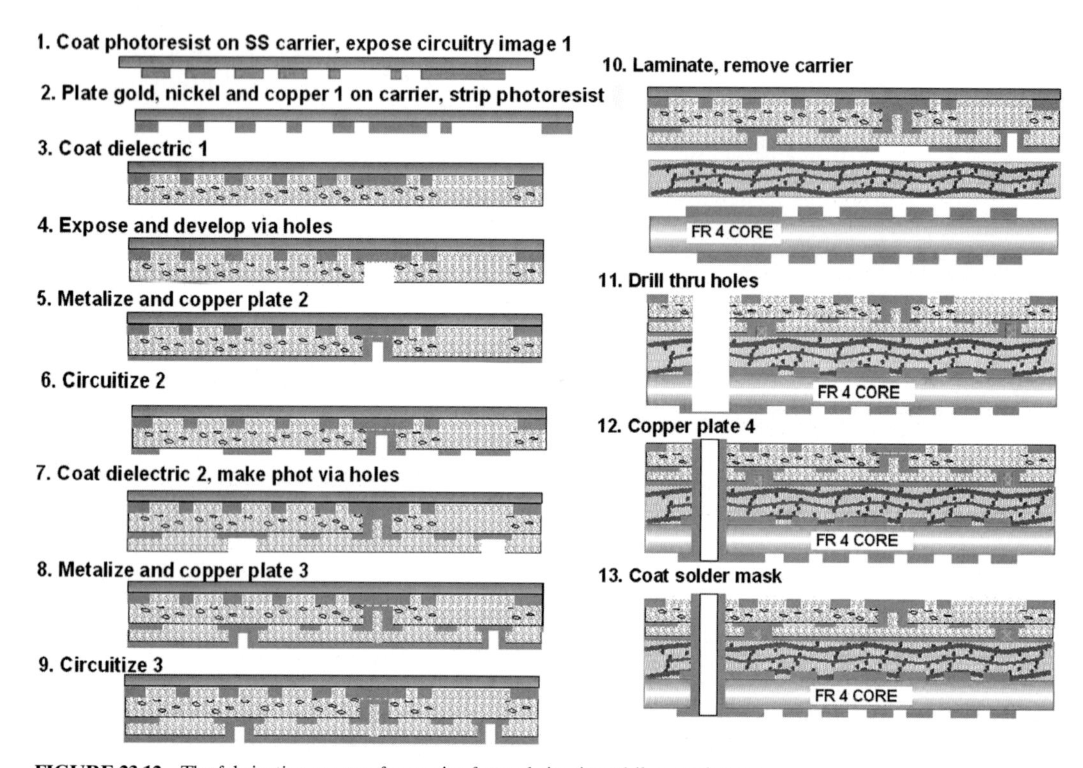

FIGURE 23.12 The fabrication process for carrier-formed circuit multilayer substrate.

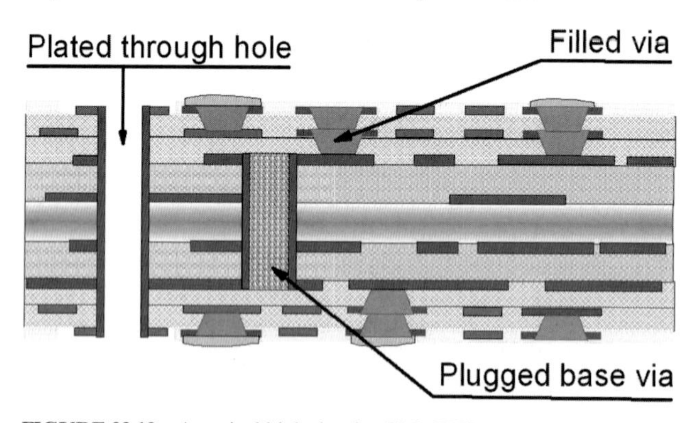

FIGURE 23.13 A typical high-density CLLAVIS.

Repeat steps 1–6 for the remaining cores of the multilayer

7. Lay up the finished cores with copper foil.
8. Vacuum laminate the finished layers.
9. Image and circuitize the outer layers
10. Coat the solder mask and finish.

This process can be seen in Fig. 23.15.

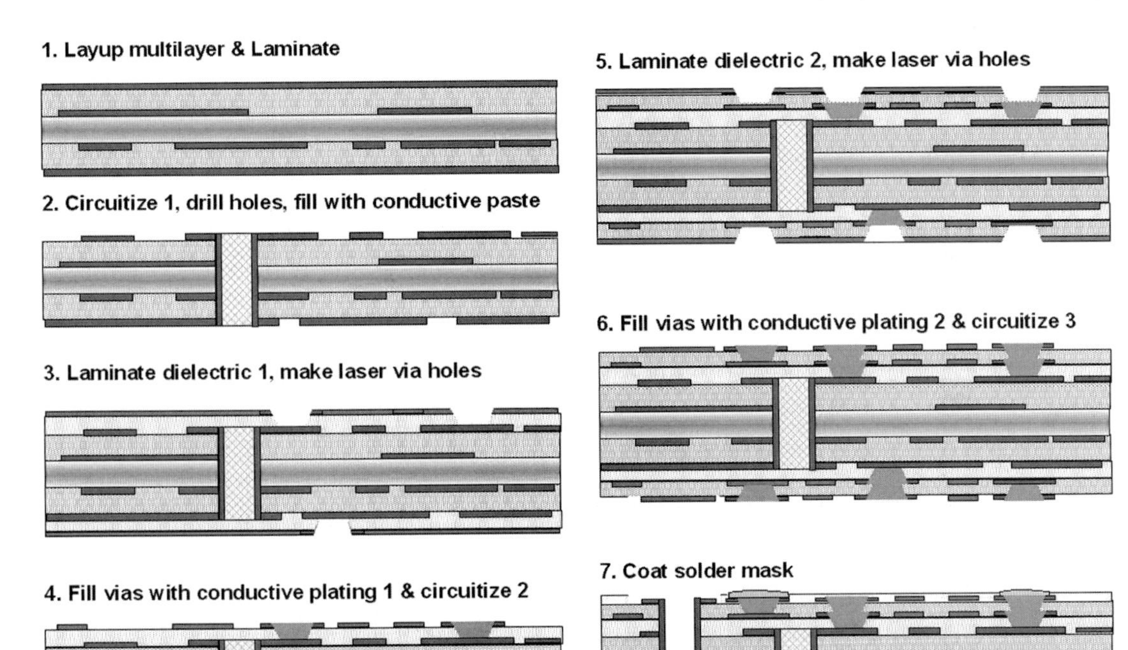

FIGURE 23.14 The typical manufacturing process for CLLAVIS boards.

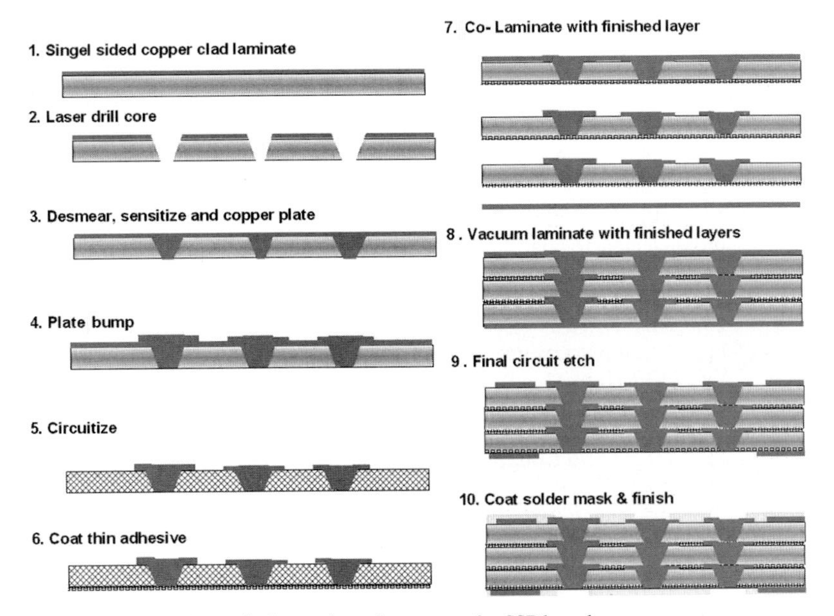

FIGURE 23.15 The typical manufacturing process for SSP boards.

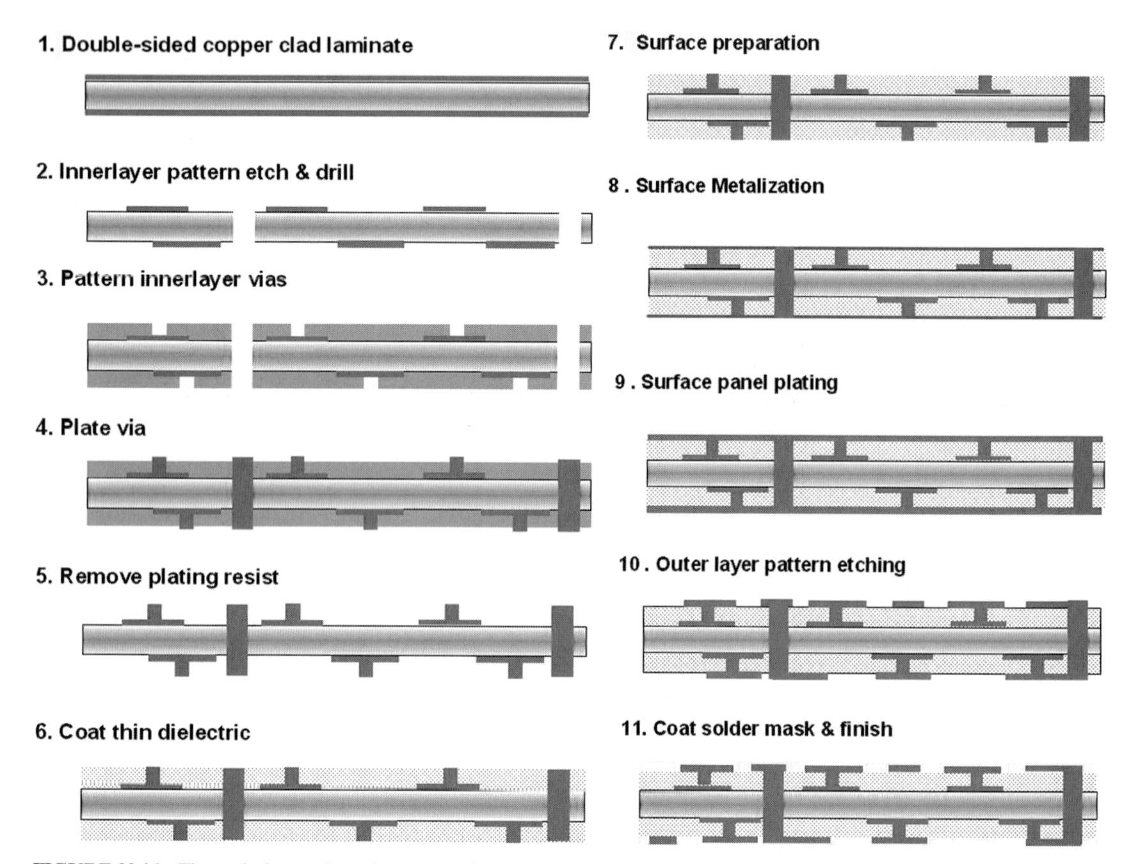

FIGURE 23.16 The typical manufacturing process for FACT-EV boards.

23.3.2.5 ***FACT-EV.*** FACT-EV (Fujikiko Advanced Chemical Technology-Etched Via Post) is from Fujikiko of Japan. As with the SSP process, the via is a solid-plated copper post. In this case, the process uses standard dry-film photoresist to define the posts and a thin-liquid dielectric to coat the plated posts. Unlike SSP, however, the process is sequential and each two sets of layers are processed on the prior layers. The process is outlined in Fig. 23.16.

23.3.2.6 ***PPBU.*** The PPBU (Prepreg Build-Up), from CMK, is a standard sequential lamination process using laser-drilled vias similar to the CLLAVIS process. The diagrams in Fig. 23.17 show two of the structures, a standard 2+4+2 build-up and an advanced 3+2+3 stacked build-up.

23.3.2.7 ***Solid Conductive Via Fill.*** The next group of HDI technologies all utilize metallic copper pastes or a solid sheet of metal to form the via connections. Table 23.1 presents these alternatives to copper plating for forming IVH connections.

23.3.2.8 ***OrmeLink.*** CTS's co-lamination process and Ormet's transient liquid phase sintering (TLPS) process (OrmaLink) are similar to the ALIVH process conductive paste in that it is a via paste copper-tin organometallic matrix that sinters into a solid metallurgical via.

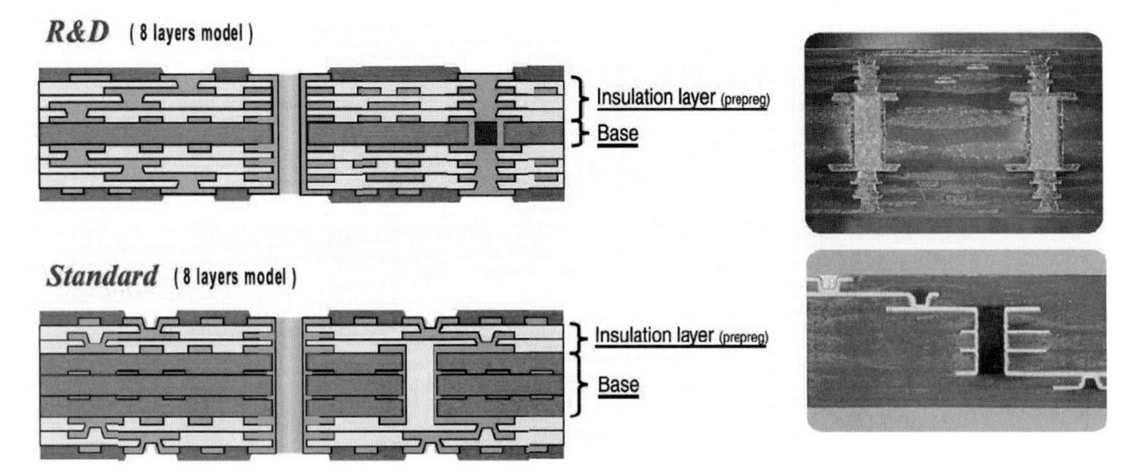

FIGURE 23.17 The typical manufacturing process for FACT-EV boards.

CTS's process is called ViaPly. Past users, in addition to Sheldahl, included Litronics, now called Allied-Signal Substrates, in Costa Mesa, California. Up to four layer pairs have been connected (eight metal layers) using OrmeLink.

23.3.2.8.1 Structure. The Ormet structure is made up of polyimide or FR-4 layer pairs. Different materials can be mixed if a rigid core or heat spreader is required. The conductive paste is a TLPS ink of copper-tin. The structure is shown in Fig. 23.18. Figure 23.19 shows the cross-sections of two finished circuits with lasered-via polyimide layer pairs, with TLPS solid metallurgical vias connecting the layer pairs and FR-4 innerlayer cores with buried TLPS vias.

23.3.2.8.2 Manufacturing Process. The manufacturing process is shown in Fig. 23.20. The microvias are lasered or punched in the polymide adhesive and then filled with the TLPS paste. The structure can now take layer pairs from any other HDIS process (such as Sheldahl's) and turn them into a multilayer structure through sintering. The conductive pastes have to be sintered in a condensing vapor of fluorocarbon at 215°C for 2 min. The structure is then postcured by baking for 40 min. at 175°C. Table 23.2 details the process.

TABLE 23.1 Alternative HDI Technologies

Fabricator	Trade name	IVH process	Metalization
Dyconex	DYCOre	Copper etching	Etched copper bumps
Ormet	OrmeLink	Laser,plasma,or photodielectric	Cu/Sn organo-metallic
Matsushita Comp.	ALIVH	Laser	Copper particles in epoxy
Toshiba	Bbit	Insulation displacement	Silver/epoxy paste
Parelec	PARMOD	Drill, laser, or photodielectric	Metallo-organic decomposition, Cu or Ag
Namics	Unimec	Punch, drill	Silver, palladium, and copper particle pastes
North Corp.	NMBI NMTI (Neo-Manhattan)	Image and etch	Etched copper bump
Denso	PALUP	Laser	Cu organo-metallic
Ibiden	SSP	Laser	Plated copper bumps
Fujikiko	Fact-Ev	Photoprocessing	Plated copper posts

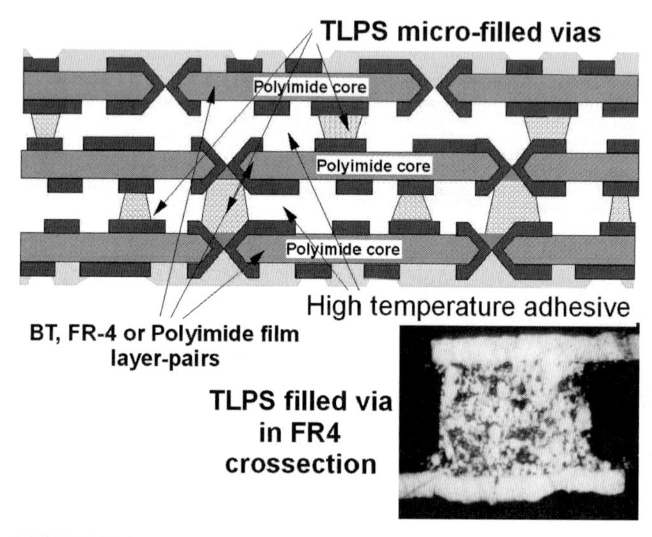

FIGURE 23.18 The co-lamination (OrmeLink) multilayer structure.

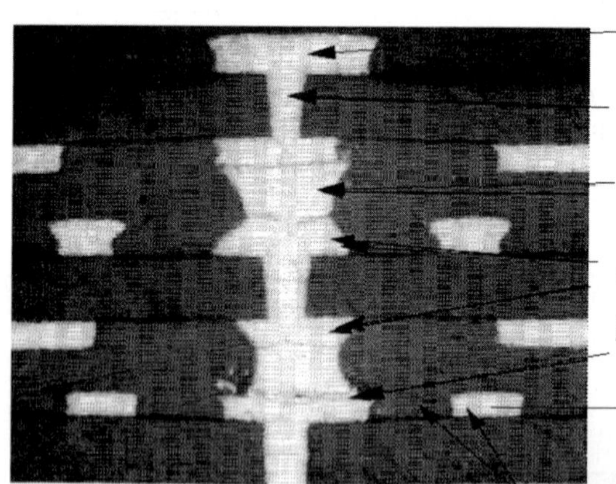

photos courtesy Ormet Technology

FIGURE 23.19 Examples of TLPS cross-sections of three layer pairs with lasered-vias filled with TLPS paste vias and buried vias for FR-4 innerlayers.

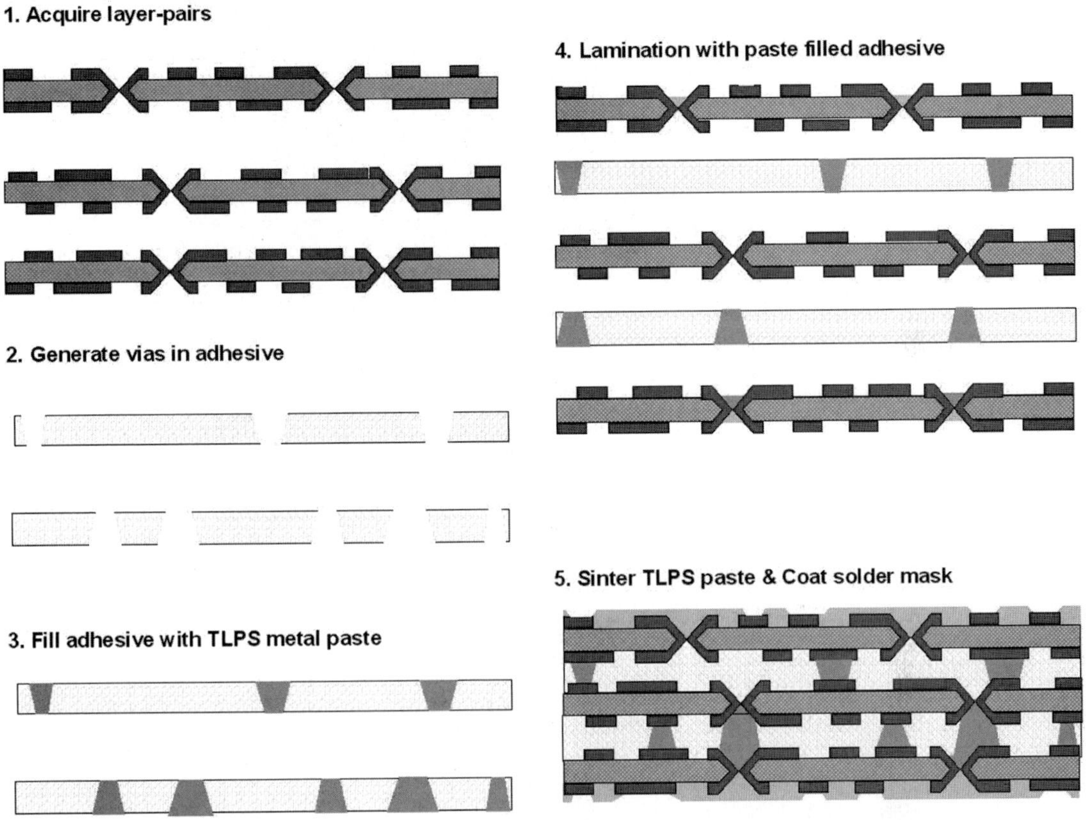

1. Acquire layer-pairs

2. Generate vias in adhesive

3. Fill adhesive with TLPS metal paste

4. Lamination with paste filled adhesive

5. Sinter TLPS paste & Coat solder mask

FIGURE 23.20 The OrmaLink multilayer substrate fabrication process.

TABLE 23.2 Properties and Curing Processes for the Ormet Type of TLPS Conductive Pastes

Property	Curing process
Ormet 2005 series ink	Specification and processing parameter
Electrical conductivity	Bulk 4.0×10^{-3} Ohms-cm. Sheet resistance 10.0×10^{-3} Ohms/sq. in.
Adhesion (Tensile Pull) on Various Materials	
FR-4 ($T_g = 125°C$)	1,300 psi (minimum)
Copper	2,921 psi (average)
Printability	With 230 stainless steel wire mesh and emulsion thickness of 7.5 μm Sintered thickness 28–38 μm 200 μm traces on 400 μm pitch
Cure cycle	30 min. drying at 85°C 2 min. vapor cure at 215°C 40 min. postcure at 175°C

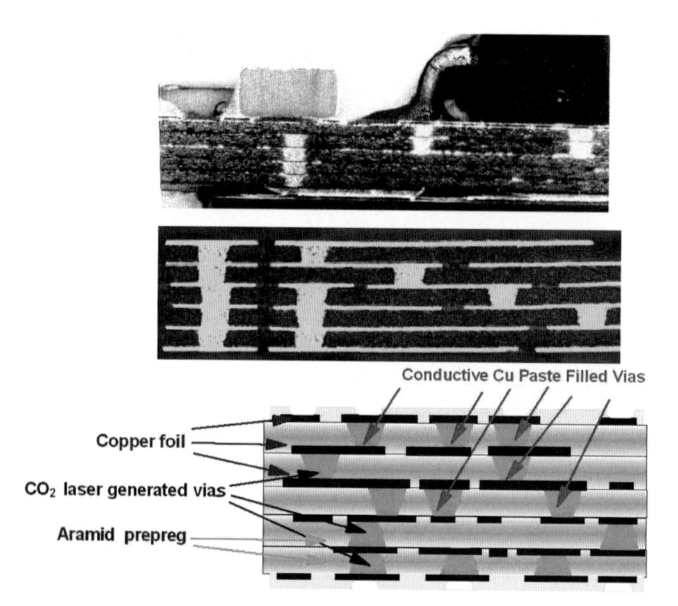

FIGURE 23.21 The parallel-bonded solid via structure of the ALIVH multilayer with cross-sections of examples.

23.3.2.9 ALIVH. The ALIVH (Any Layer Interstitial Via Hole) process has been in development for many years by Matsushita Components of Osaka, Japan. The novel process eliminates additive metallization and plating, but defines all features by subtractive etching of the copper foil. The build-up process is not sequential, though; it uses layer pairs and aramid-epoxy prepreg with copper-paste vias that can be laminated at one time into a three-dimensional structure. CMK (the second largest PCB maker in the world) and other Japanese firms have licensed the process. Six to 10 layers have been laminated in this way.

23.3.2.9.1 Structure. Figure 23.21 shows the structure and cross-sections for an ALIVH product. The PCB consists of laser-produced blind vias. The core material is an epoxy-aramid laminate. The man-made aramid filaments are ideal to be cut with a CO_2 or UV laser. If the DuPont Kevlar filaments are added, then the resulting material will have a very low coefficient of thermal expansion (CTE). This is useful for mounting ceramic packages and for direct attachment of flip-chip integrated circuits. The structure can be as simple as a two-sided PCB or as complex as a many-layered PCB. The vias consist of a copper-epoxy paste that connects the top and bottom copper foil. If used as a prepreg layer without copper, the vias connect the various ALIVH layer pairs into a multilayer structure. This is not a sequential build-up process, but rather a parallel build-up process.

23.3.2.9.2 Manufacturing Process. The ALIVH process is shown in Fig. 23.22. The process starts with an epoxy-aramid B-stage prepreg. The laser cutting of holes can proceed very rapidly. The material is then printed with a conductive paste of copper and epoxy to fill these holes. Copper foil is applied and the structure is laminated to attach the foil and cure the prepreg and conductive pastes that serve as vias. The sheet of material is imaged and etched to provide the various circuits. Registration is less critical because the vias are now *under* the surface lands. Several of these two-sided layer pairs can be produced, inspected, and tested. The two-sided structures can then have additional single layers of B-stage/conductive paste layers with foil laminated to one or both sides, or the B-stage/conductive paste layer can be used to attach a number of layer pairs in one parallel lamination. The outside is imaged, etched, and completed as a normal PCB.

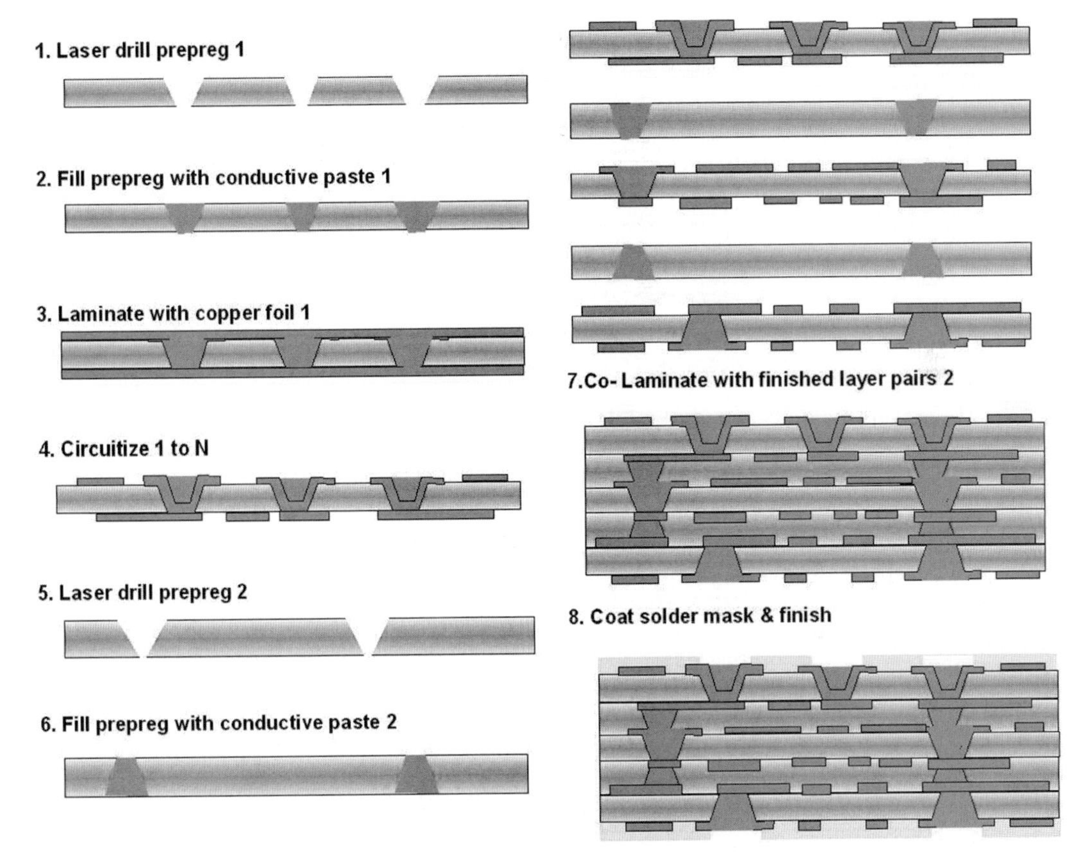

FIGURE 23.22 The ALIVH manufacturing sequence for a multilayer substrate.

An advanced manufacturing process results in a product known as ALIVH-FB. This product has a fine-line, tight via structure suitable for wire-bonding and flip-chip substrates, as shown in Fig. 23.23. The manufacturing process for each microvia technology begins with a base core, which may be a simple double-sided board carrying power and ground planes or a multilayer board carrying some signal pattern in addition to power and ground planes. The core usually has PTHs, which become blind via holes (BVHs). Such a core is often called an *active core*.

In fabricating a base core with a pattern, manufacturers usually panel-plate the core panel and create the pattern through a dry-film tenting process. Some makers seem to prefer pattern plating, however. The choice depends on the fabricator's familiarity with these processes. After the pattern is formed, dielectric material is laminated over the core (in the case of prepreg, with copper foil) and the holes can be filled with resins, depending on the plated hole diameter and the thickness of the core. It is generally agreed that when the diameter of plated holes is equal to or less than 0.3 mm and the core thickness is equal to 0.6 mm or less, these holes can be filled effectively by the lamination process (although the resin thickness of 80 μm is preferred in the case of RCC).

When the diameter and thickness conditions are not met, it is necessary to fill the holes by a separate process. A screening process does this from one side of the panel with a polyester screen that has an oversized hole pattern. After filling the holes the resin is cured completely.

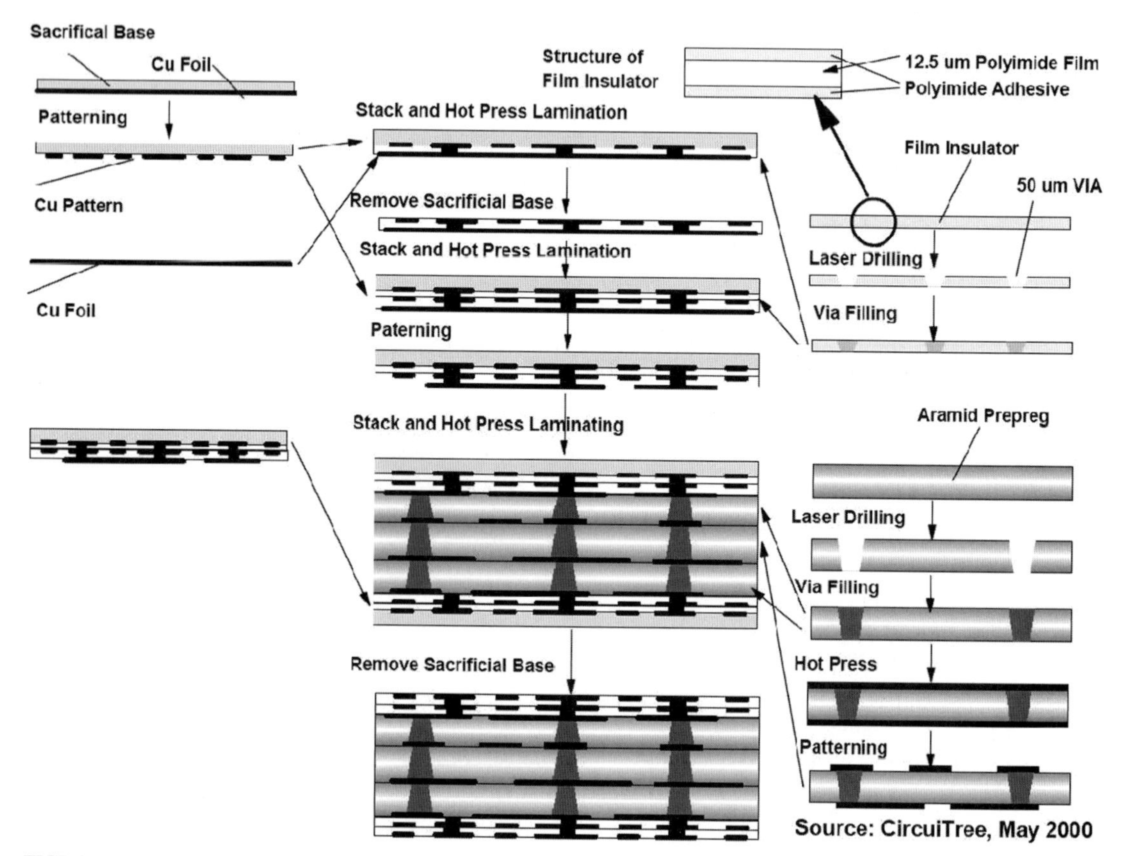

FIGURE 23.23 The ALIVH-FB manufacturing process for a multilayer substrate. *(Illustration Courtesy of CircuiTree.)*

23.3.2.10 MSF. Shinko of Japan developed the MSF HDI technology. It utilizes laser-drilled RCC materials and vias filled with a conductive paste. After testing, these are laminated into the parallel build-up structure. Figure 23.24 shows the typical manufacturing sequence for this HDI technology.

23.3.2.11 PALAP. PALAP (Patterned Prepreg Layup Process) is a process that was developed by a consortium consisting of the Japanese firms Denso, Wako Corporation, Airex, Kyosha, Noda Screen, and O.K. Print. Originally, the process started with copper-clad laminates (CCL), but now utilizes thermoplastics like PEEK resins (polyether-ether ketone) or a new plastic called PAL-CLAD. PAL-CLAD is characterized by the electrical properties and heat resistance of BIAC, a recyclable thermoplastic resin film produced by Japan Gore-Tex, Inc.

The single-lamination process compares favorably to conventional PCB processing, where lamination, curing, and wiring patterning are repeated layer after layer. PALUP boards can be multilayered by pressing together all thermoplastic resin layers, each having wiring patterns, as shown in Fig. 23.25. This significantly improves quality, lowers costs, and shortens delivery times. PALAP boards also offer high-interconnecting reliability by adopting metallic paste for filling vias, and have excellent high-frequency properties due to a low dielectric constant.

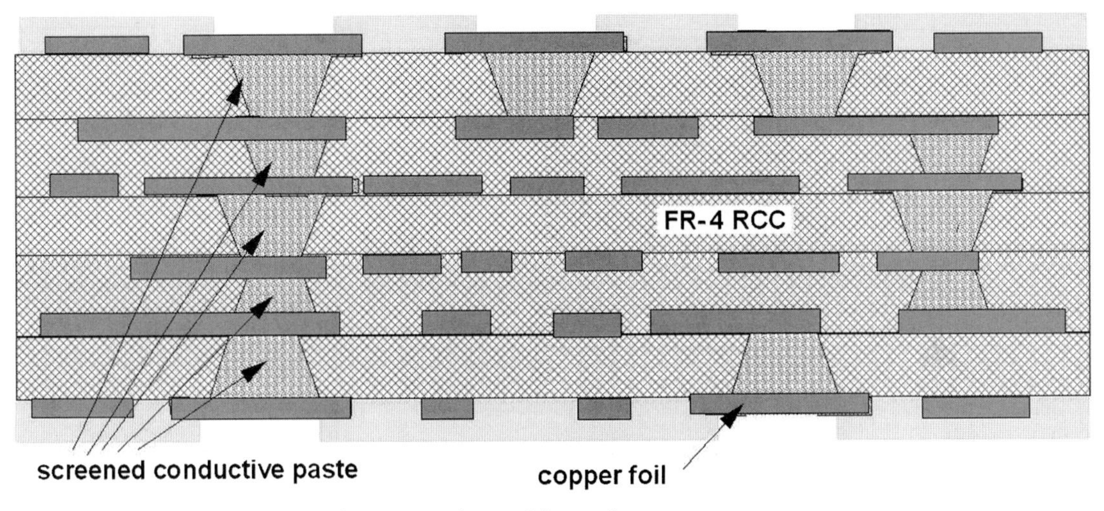

FIGURE 23.24 The MSF manufacturing sequence for a multilayer substrate.

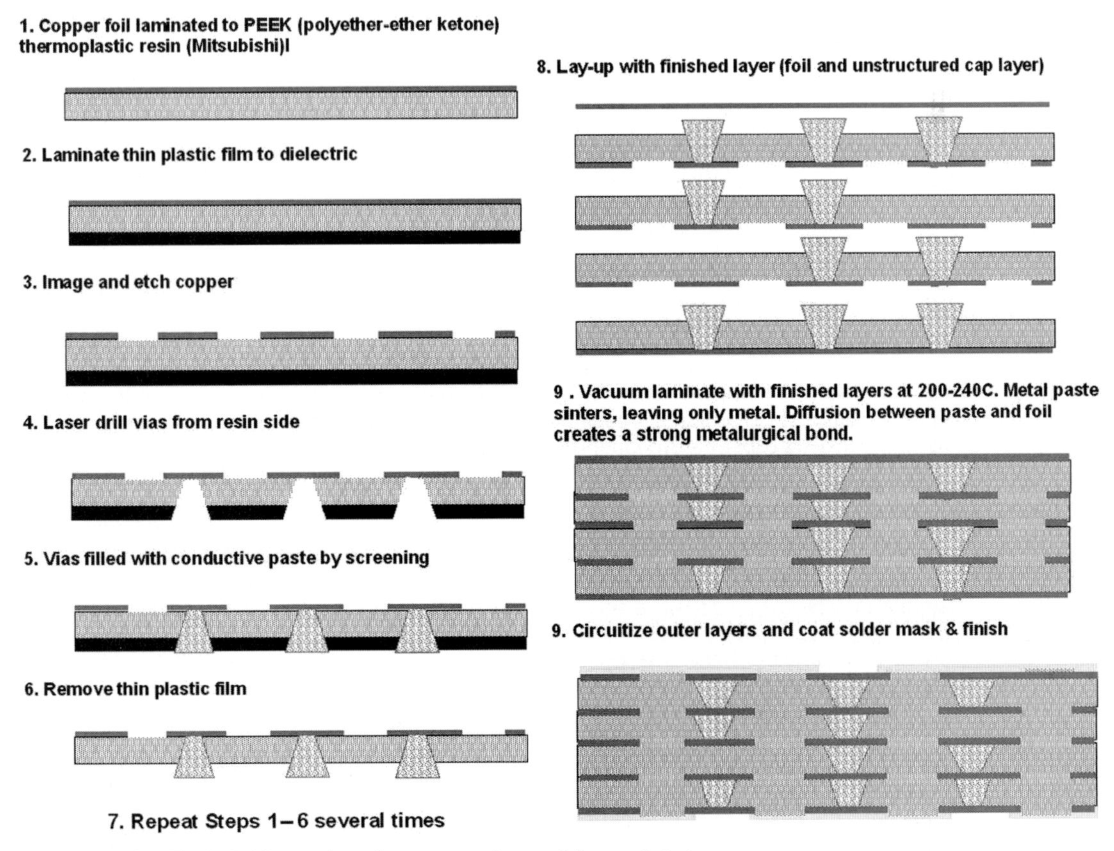

FIGURE 23.25 The PALAP manufacturing sequence for a multilayer substrate.

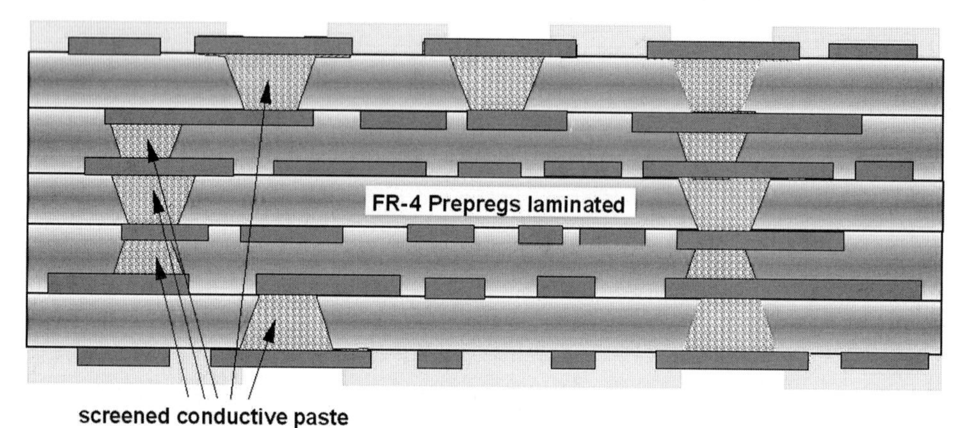

FIGURE 23.26 The VIL manufacturing sequence for a multilayer substrate.

PALAP boards also give due consideration to the environment, one of the major issues requiring action in the area of electronics products, including those used in information technology and automobiles. Because PALAP boards use thermoplastic resin as their base material, they enable material recycling in which only the resin is separated and reused.

23.3.2.12 VIL. The VIL (Victor Interconnected Layers) HDI technology was developed by Victor of Japan. It utilizes FR-4 prepreg materials that are laser-drilled and sequentially laminated after the vias are filled with a conductive paste. Figure 23.26 shows the typical manufacturing sequence for this HDI technology.

23.3.3 Mechanical Drill-Via Technologies (See Fig. 23.6)

Hitachi HITAVIA is an example of the mechanical drill-via technologies available.

23.3.3.1 HITAVIA. The Hitachi HITAVIA technology is the only HDI process developed for use with conventional mechanical drilling. It utilizes RCC or prepreg materials that are mechanically drilled and sequentially laminated after the vias are filled with a conventional electroless copper and copper plating. The typical manufacturing sequence for this HDI technology is similar to that of previously described HDI processes such as CLLAVIS and PPBU.

23.3.4 Plasma-Via Technologies (See Fig. 23.7)

Plasma-via technologies include DYCOstrate, plasma micromilling, and plasma-etched redistribution layers (PERL).

23.3.4.1 DYCOstrate. Starting in 1989, Dr. Walter Schmidt of Contraves in Switzerland developed the plasma-etching process for microvias. Evolving from the traditional plasma desmear process, the plasma etcher for vias was jointly developed by Dyconex (successor to Contraves) and Technic Plasma of Germany. Dyconex trademarked and patented its via-generation process as DYCOstrate. Hewlett Packard licensed the technology in 1993 and developed it for mass, low-cost production. The low-cost production process of Hewlett Packard is called PERL (for plasma-etched redistribution layers).

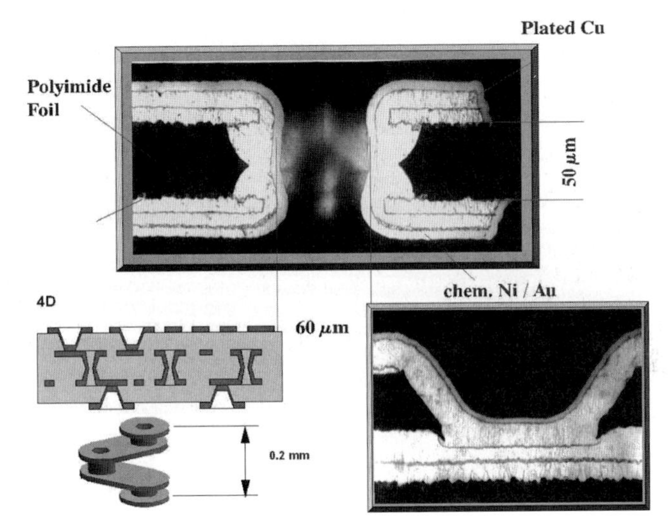

FIGURE 23.27 The DYCOstrate multilayer structure and cross-sections of through and blind vias.

DYCOstrate, along with SLC and Finstrate, is among the most enduring production processes for microvias generation. The DYCOstrate process was first employed for high-reliability military, aerospace, medical, and IC packaging starting in 1991. Since that time, Dyconex has used the process to produce hundreds of different printed boards both in polyimide film and with the new laminates called FR-4 RCC. This Dyconex called DYCOstrate-C.

23.3.4.1.1 Structure. Figure 23.27 shows the structure of DYCOstrate. The core material can be polyimide film with plasma-etched holes, drilled epoxy fiberglass, or other plasma-etchable materials such as liquid crystal polymers. Multiple layers can be built up to increase density with the resulting buried and blind vias, but two-sided DYCOstrate structures are also used in products because of the very high density that 0.075 mm through holes allow.

23.3.4.1.2 Manufacturing Process. The manufacturing process for a DYCOstrate substrate utilizes common printed circuit board techniques. Only the via-generation process is different. To produce a plasma-etched via hole, manufacturers use two or three process steps that replace the conventional mechanical drilling-debur-desmear steps:

1. Define photographically the location and geometry of the via holes.
2. Etch an opening in the copper foil that will serve as the resist mask.
3. For blind vias, reduce the copper-foil thickness to eliminate the copper overhang.

Figure 23.27 shows microsections of a plated through hole and a blind via in polyimide film. The plasma-etch procedure is basically an isotropic process, as indicated by the undercut seen on the through hole; considering the actual dimensions, however, this undercut is too small to cause any plating problems.

When the etching depth is increased, as in the case of a blind via, the resulting undercut is generally too big to allow reliable plating. To overcome this problem, the manufacturer can reduce copper use by etching the copper foil, eliminating the copper overhang, and providing a thinner copper foil for fine-line resolution.

Figure 23.28 shows two common products made with the Dycostrate process (a four-layer flex-multilayer chip-on-board (COB) for a hearing aid) and the PERL process (a six-layer FR-4 COB for a networking module).

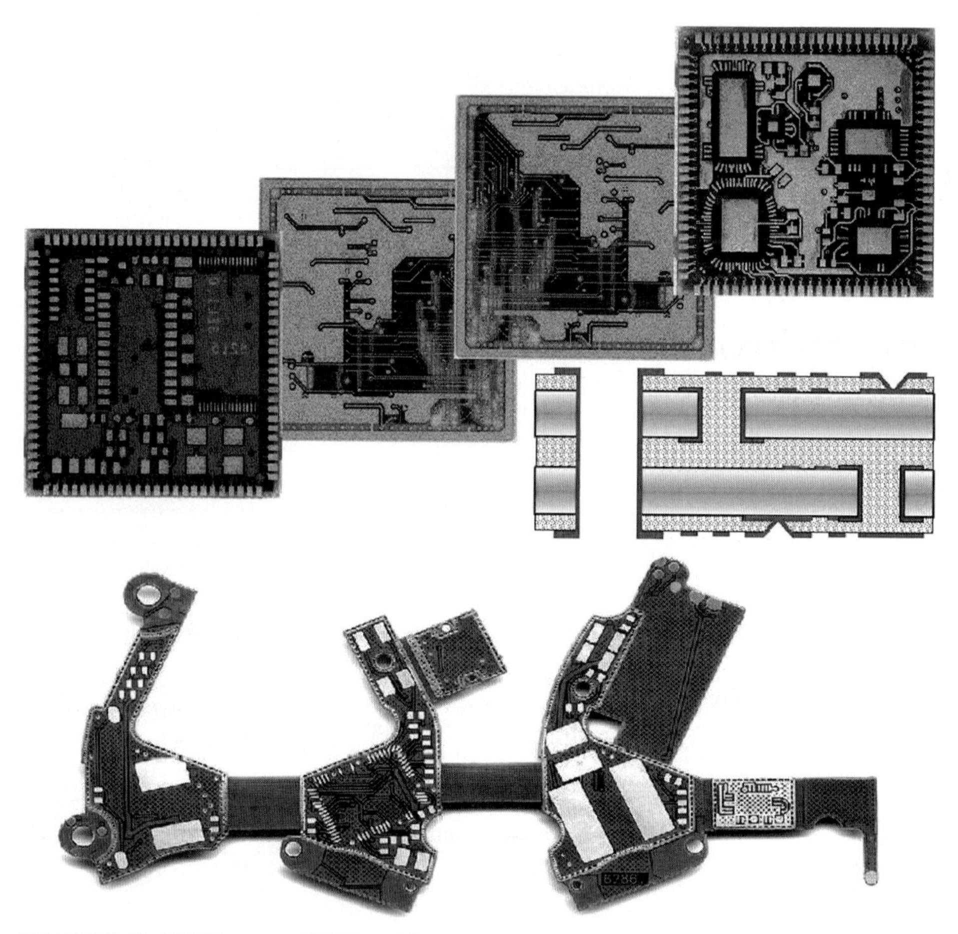

FIGURE 23.28 DYCOstrate and PERL multilayer examples.

The manufacturing process for a DYCOstrate substrate with a standard four-layer build-up is shown in Fig. 23.29. Starting with a prefabricated double-sided DYCOstrate foil, the manufacturer bonds two single-sided copper clad foils to the center-core foil using standard lamination techniques of mass-lam-like process. The resulting four-layer structure is then processed and structured analog to the two-layer foil, producing blind vias instead of through holes.

Further refinements in magnetrons for IC manufacturing provide the opportunity for finer plasma-etched vias. Additionally, Dyconex has demonstrated landless vias to be simple and reliable, offering much higher densities at lower cost and simplifying fabrication registration.

23.3.4.2 Plasma Micromilling. The drilling of simple holes is not the only process in which plasma etching is used. Plasma can be also be used to sculpture the surface of the substrate or to fabricate slots, grooves, stepped windows, and more. Even angled vias or tubular systems can be formed. This is useful since cavities can be formed for wire bonding or recessed

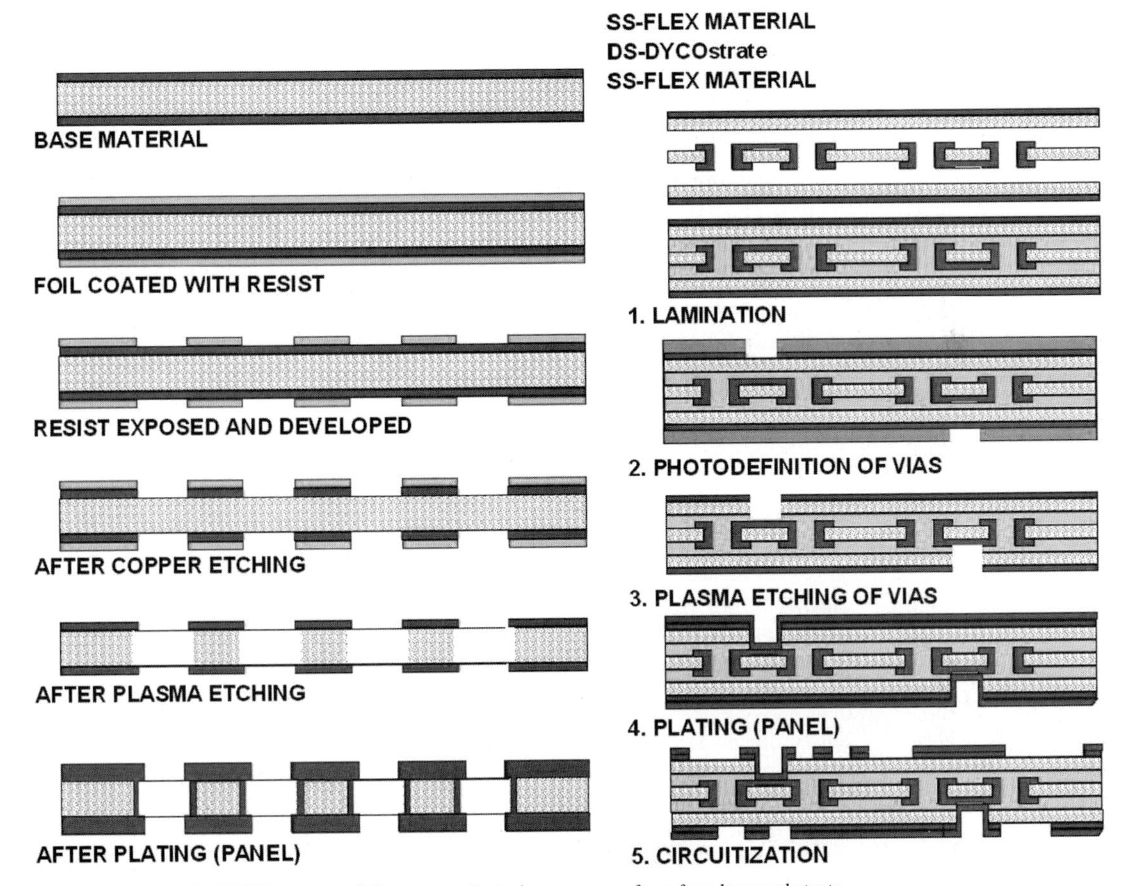

FIGURE 23.29 The DYCOstrate multilayer manufacturing sequence for a four-layer substrate.

mounting of components. A simple definition of the periphery outline by artwork can provide extremely accurate final fabrication, or windows in the final etch step can define areas that will flex.

23.3.4.3 Plasma-Etched Redistribution Layers (PERL). The PERL process was developed by Hewlett Packard to produce microblind and buried vias using the DYCOstrate plasma via process. The materials used are the FR-4 epoxy-coated copper foils (RCC) to replace epoxy-glass prepregs in normal multilayer production. These are copper foils coated with a B-stage epoxy or dual C-stage/B-stage epoxy films.

23.3.4.3.1 Structure. The structure of DYCOstrate PERL is shown in Fig. 23.28. The core material can be standard FR-4 multilayer innerlayers or a through-hole-plated two-sided or multilayer board. Multiple layers can be built up to increase density with the resulting buried and blind vias. Current production ranges from 4 layers to 12 layers with various buried board and buried via constructions. As with DYCOstrate, many materials are plasma-etchable and the resin butter-coated copper foils come in many thicknesses and resin types, such as BT, cyanate ester, and polyphenelene-ether (PPE).

23.3.4.3.2 Manufacturing Process. The manufacturing process for a DYCOstrate PERL substrate utilizes common printed circuit board techniques. Only the via-generation process is different. To produce a plasma-etched via hole requires only two or three process steps that replace the conventional mechanical drilling-debur-desmear steps:

1. Define photographically the location and geometry of the via holes.
2. Etch an opening in the copper foil that will serve as the resist mask.
3. For blind vias, reduce the copper-foil thickness to eliminate the copper overhang.

The manufacturing process for a DYCOstrate PERL substrate with a standard four-layer build-up starts with FR-4 innerlayers or a prefabricated double-sided or multilayer board. Two single-sided copper foils with resin coatings are bonded to the center-core foil using standard lamination techniques. This structure could also be procured through standard mass-lam channels. The resulting four-layer structure is then processed and structured analog to the two-layer DYCOstrate foil, producing blind vias. Through holes can be added or they can be in the preexisting buried board.

23.3.5 Screen-Printed Via Technologies[11]

Toshiba's new Buried Bump Interconnect Technology is a new screen-printed via process.

23.3.5.1 Buried Bump Interconnect Technology (BBIT). Toshiba has developed a new process referred to as BBIT (Buried Bump Interconnection Technology). With this process, a conductive paste is employed to replace via drilling, additive metallization, and via plating. The process has the advantage that hole-producing equipment is not required. Instead, a screened silver-epoxy paste is cured to be pointed like a thumbtack. This paste displaces the glass and epoxy during lamination to connect to the copper on the opposite side of the prepreg. The mass via-generation techniques, the simplification of metallization, and the fact that no drilling equipment is required make this new technique potentially very inexpensive. Currently, the design rules and features are for consumer products, but future uses are for simple plastic ball grid arrays (PBGAs).

23.3.5.1.1 Structure. The BBIT structure appears similar to that of other PCBs employing conductive paste vias. As shown in Fig. 23.30, the structure is similar to ALIVH. The difference is that the BBIT structure uses standard FR-4 materials and a novel way to produce a via.

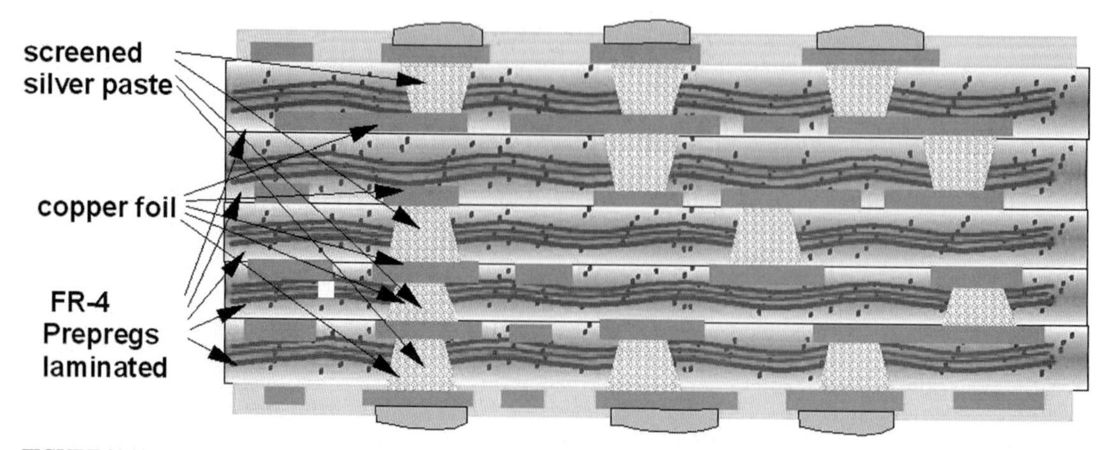

FIGURE 23.30 Structure of the BBIT substrate.

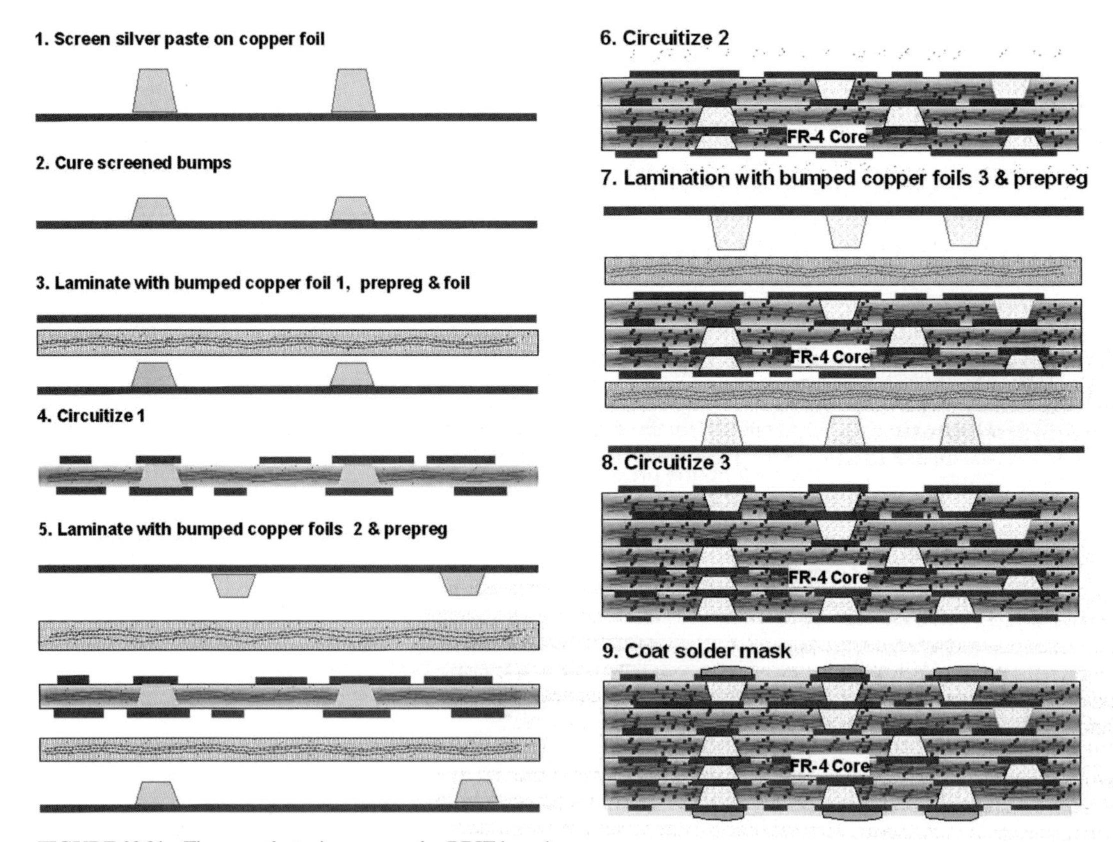

FIGURE 23.31 The manufacturing process for BBIT boards.

23.3.5.1.2 Manufacturing Process. A silver ink is screened on copper foil in the locations where vias are required. This copper is used in the manufacture of standard FR-4 laminate. The silver paste thumbtack-like obstruction forces itself through the glass cloth during lamination to connect itself to the copper foil on the other side of the laminate and to cure the prepreg. The sheet of material is imaged and etched to provide the various circuits. Registration is less critical because the vias are now under the surface lands. Several of these two-sided layer pairs can be produced, inspected, and tested. The two-sided structures can then have additional single layers of B-stage/conductive paste layers with foil laminated to one or both sides, or the B-stage/conductive paste layer can be used to attach a number of layer pairs in one parallel lamination. The outside is imaged, etched, and completed as a normal PCB. This process is seen in Fig. 23.31.

A newer BBIT process has been defined to manufacture wire-bonding and flip-chip substrates. In this BBIT process (shown in Fig. 23.32), the prepreg is laser-drilled so that smaller cream Ag solder can be used to bond to the copper.

23.3.6 Photo-Defined/Etched-Via Technologies

Neo-Manhattan Bump Interconnect is an example of the photo-defined and etched-via technologies available.

23.3.6.1 Neo-Manhattan Bump Interconnect (NMBI). The NMBI technology is one of the newest of the Japanese HDI technologies. It might be considered a "third-generation"

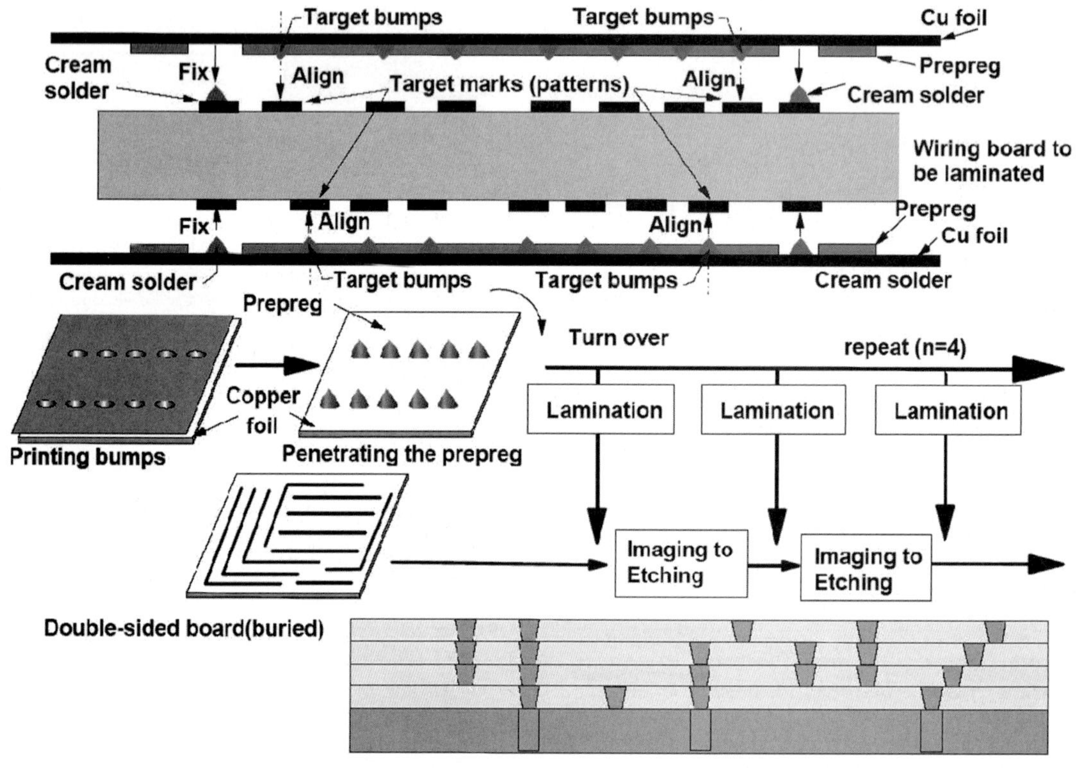

FIGURE 23.32 The manufacturing process for the fine-pitch BBIT substrates.

HDI technology after technologies as ALIVH is considered a second-generation HDI technology. NMBI was pioneered by North Corporation of Japan.

23.3.6.1.1 Structure. Figure 23.33 shows the structure of an NMBI substrate. Also in this figure is an scanning electron microscope (SEM) of the etched-metal bumps that make up NMBI's via interconnects.

23.3.6.1.2 Manufacturing Process. The manufacturing process for NMBI is unusual in that it does not drill the via connections. Instead, a new material is employed that consists of two different coppers (one thicker than the other) that are bonded together. The thicker copper is imaged and etched to form the interconnect vias. This structure is then filled with a film or liquid dielectric. It is bonded to copper foil and both are cured. These copper foils are then imaged and etched to circuitize the layer pair. The layer pairs can be tested and later be stacked with uncured NMBI structures to form the final multilayer. Like the second-generation ALIVH HDI structure, an NMBI structure allows the connection of any layer to any other layer. This manufacturing process is shown in Fig. 23.34.

23.3.7 ToolFoil Technologies

Imprinted circuits are an example of a ToolFoil technology.

23.3.7.1 Imprinted Circuits. Imprint pattering is a similar technique used to manufacture compact discs (CDs). No photoresist, registration, or conventional techniques are employed.

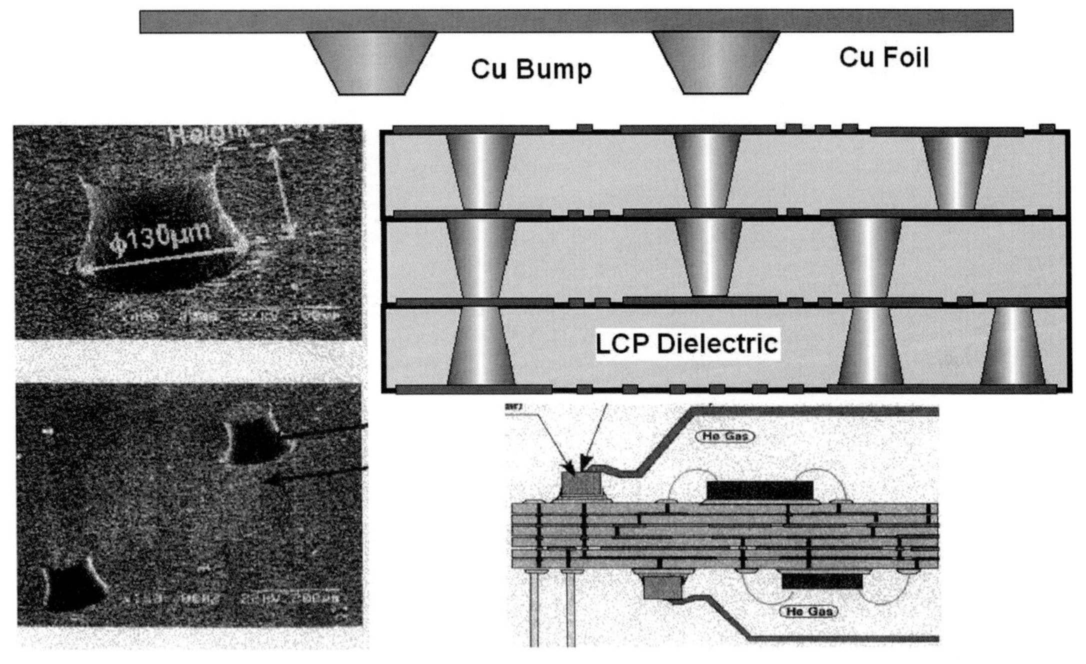

FIGURE 23.33 The Neo-Manhattan Bump Interconnect structure.

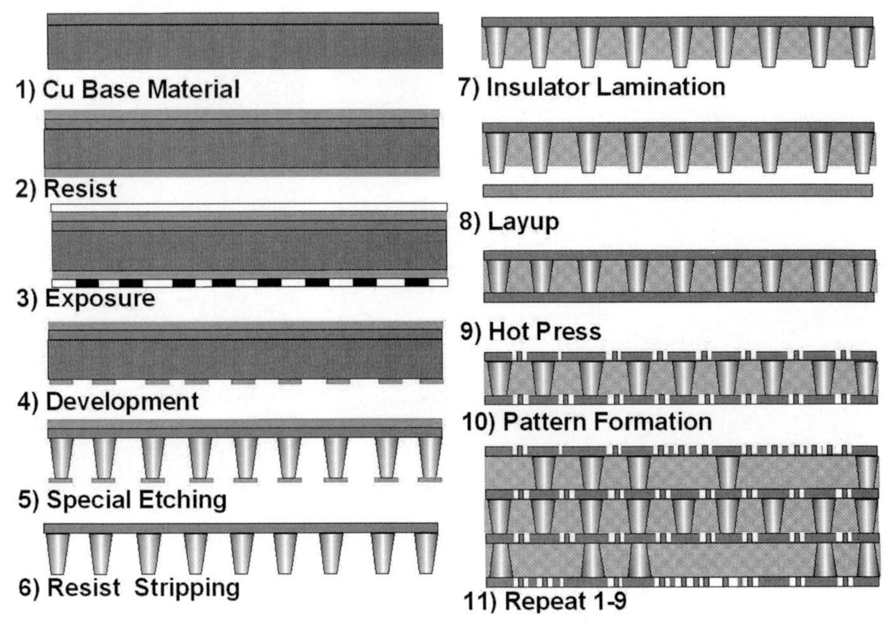

FIGURE 23.34 The manufacturing process for the NMBI substrates.

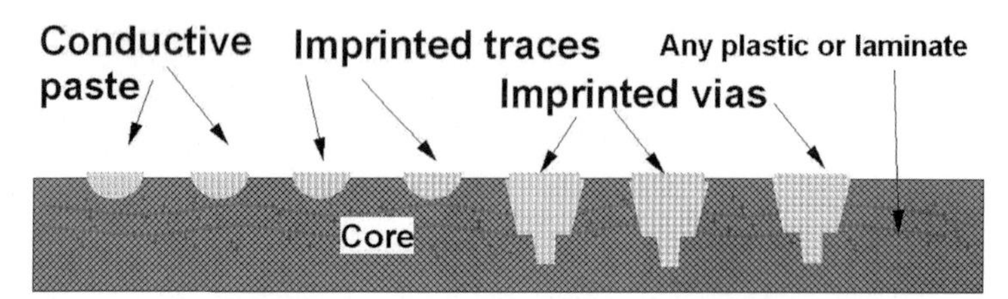

FIGURE 23.35 A simplified diagram of the imprinted printed circuit structure.

Every substrate is a copy of the mold (or ToolFoil, in this case), and each can be done in sequence to make a build-up substrate. The unique characteristic of CDs and DVDs is the millions of 0.5 micron pits or vias. A typical CD includes over 3 km of these. The simple manufacturing process and perfect reproduction of the master tool creates inexpensive and accurate substrates. The patented process is a lab process; actual complexity is still under evaluation.

23.3.7.1.1 Structure. Figure 23.35 shows a diagram of the imprinted circuit structure, whereas Fig. 23.36 shows SEMs of an imprint-patterned circuit. The unique characteristic

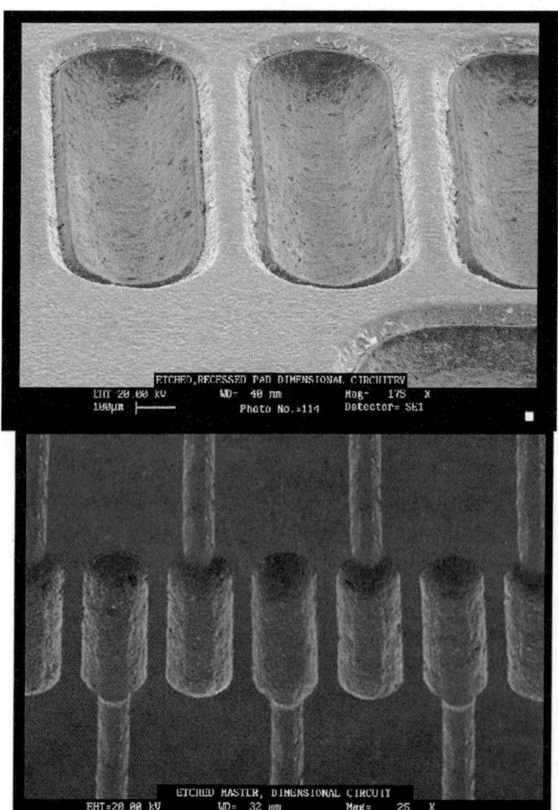

12 mil SMT land with ~ 17 microns of copper

30 mil pitch SMT without the solder paste

25 mil pitch SMT lands with 4 mil traces

FIGURE 23.36 Various views of SEMs of imprinted circuit liquids and traces. *(Courtesy of Dimensional Circuits.)*

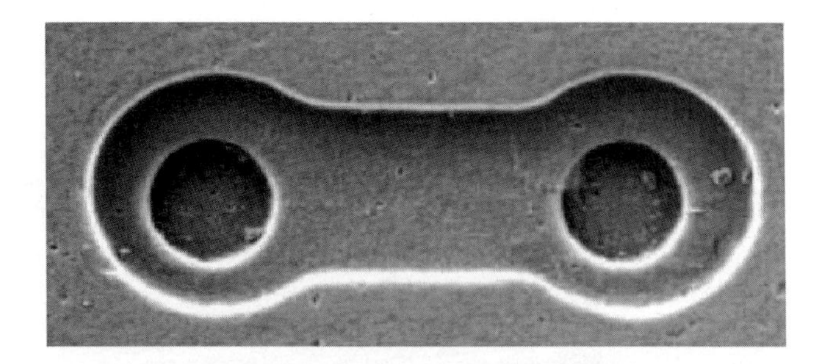

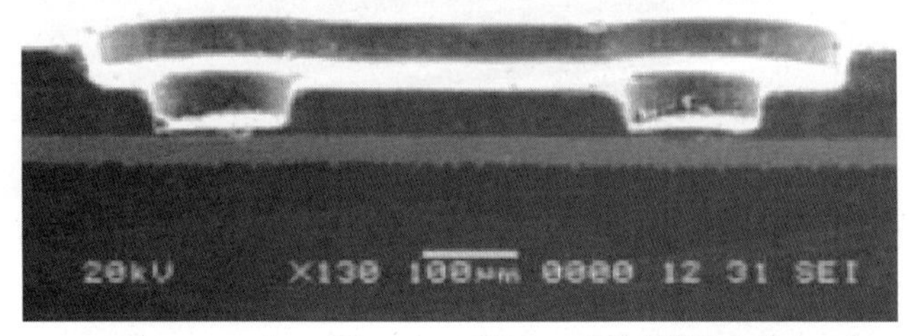

Cross-section of Imprinted Dexter LF 69702.0 dielectric

FIGURE 23.37 SEM examples of an imprinted circuit using Dexter LF69702 epoxy and showing how traces and vias are recessed. *(Courtesy of Dimensional Circuits.)*

of this structure is that all features are embedded in the substrate. The critical dielectric is the part that is imprinted or molded. A long fiber molding compound, such as that used on components' overmolding, is best suited for this task. Such a compound can be used with or without an FR-4 backing. The cross section of an imprint circuit shows the features similar to a CD or DVD, that is, the various impressions. All are metallized but the vias are deeper. This allows the vias to contact the next layer of imprinted circuits, as shown in Fig. 23.37.

23.3.7.1.2 Manufacturing Process. A master of the substrate is machined in copper by a UV laser or by photochemical machining. The vias are deep, and the circuits and pads are shallow. Since only one master tool, called a ToolFoil, is required, time can be taken to make sure the master is perfect. With a laser, it is possible to have perfect registration of lands to via holes or even landless vias. This master is electroformed with nickel and back-filled to make a master tool. A sample ToolFoil is shown in Fig. 23.38.

For production, the mold is filled with a thermoset or thermoplastic resin and then cured. The substrate is additive or semi-additive metallized, and plated thicker with copper. The recesses are filled with an etch resist, and the surface resist is removed to expose the copper surface by polishing or by using an abrasive agent. The exposed copper is etched away and the etch resist is dissolved. The entire manufacturing process does not employ photoresists, exposure, or registration. Because of this, the yield is expected to be very high. Figure 23.39 shows the process for tool generation and production substrate. The process for metallization is shown in Fig. 23.40.

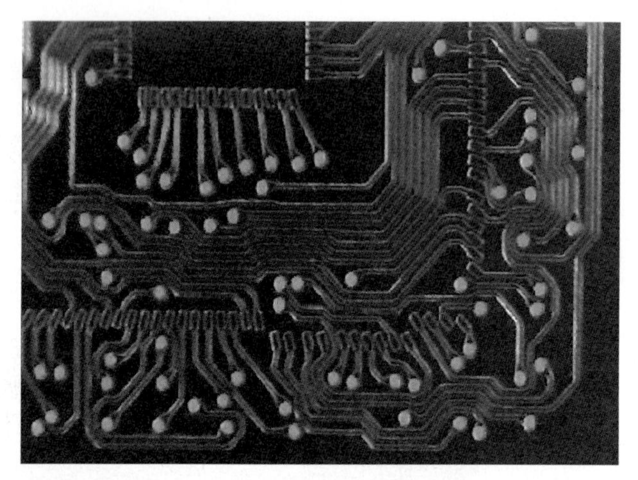

FIGURE 23.38 Example of a ToolFoil for imprinting. *(Courtesy of Dimensional Circuits.)*

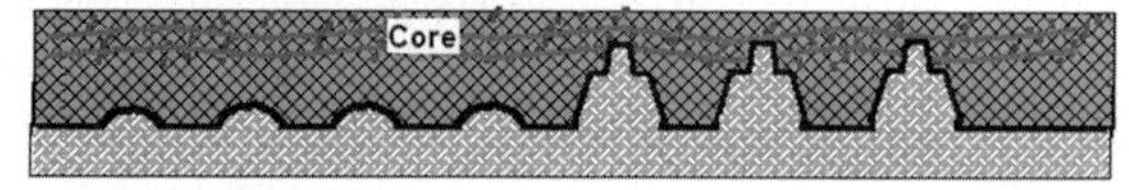

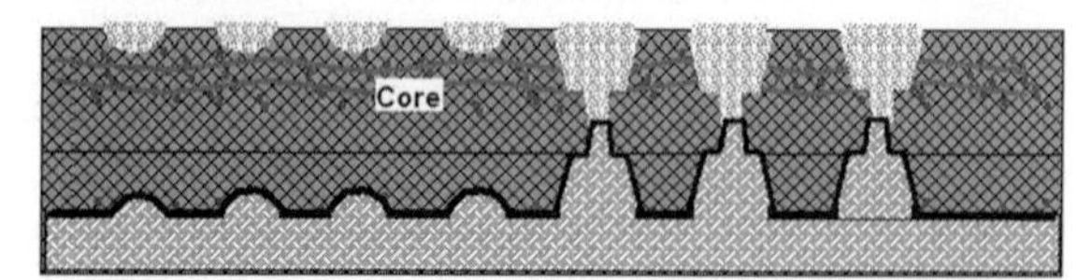

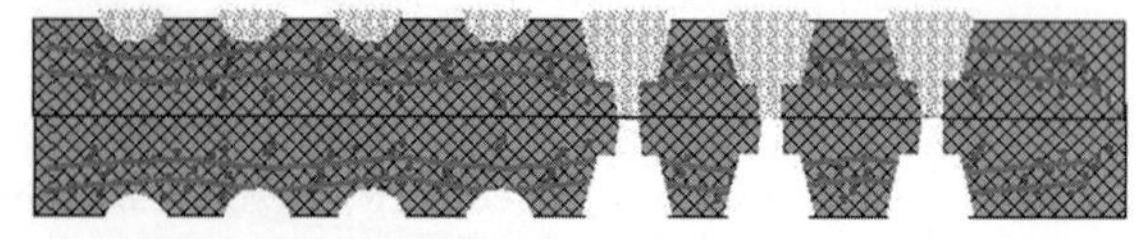

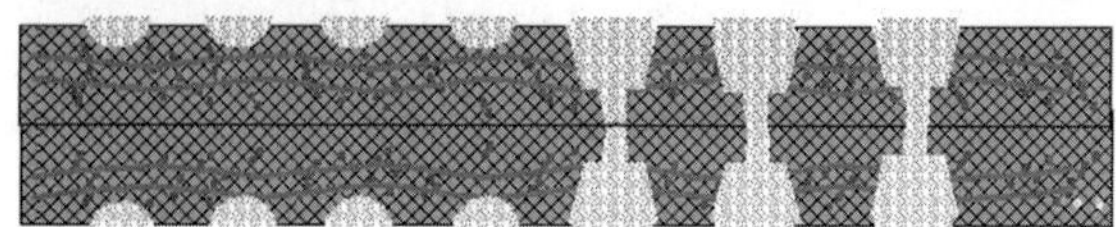

FIGURE 23.39 The manufacturing process for the ToolFoil and using the tool to make imprinted circuits.

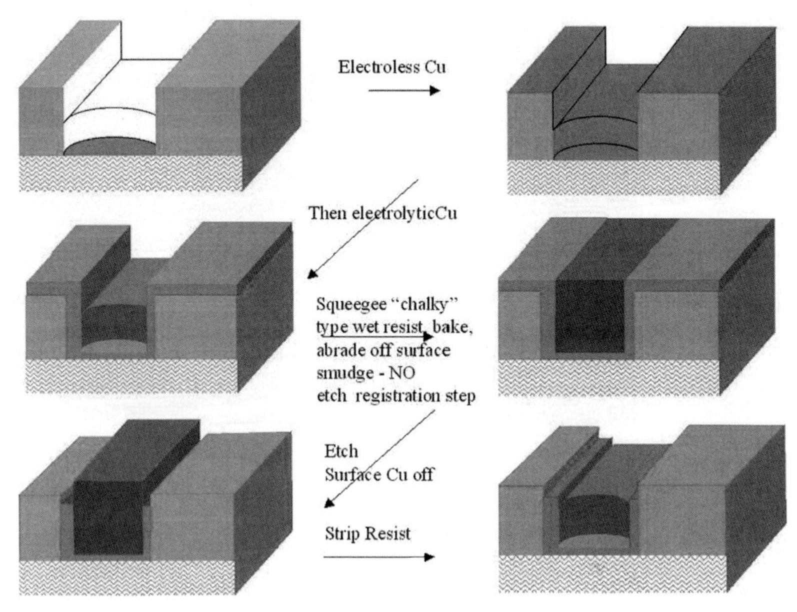

FIGURE 23.40 Methods of metallization using electroless plating, plugging, and etching. *(Courtesy of Dimensional Circuits.)*

23.4 NEXT-GENERATION HDI PROCESSES

HDI and microvias provide a big boost in density due to miniaturization. These technologies will continue to evolve and become smaller, following integrated circuit cell geometries. So the next revolution will be in optical wiring. Currently, optical networking ties together continents and cities, providing the backbone of modern worldwide networking. The fiercely competitive market is what technology will provide "the last mile" on connectivity? Optical wiring is competing for that market.

23.4.1 Printed Optical Waveguides

Although electrical signals can be multiplexed on a single 100 micron wire or trace, many individual laser wavelengths can each carry many multiplexed signals on a single 100 micron waveguide. This results in a 10,000× increase in information handling while not being influenced by magnetic and electrical fields such as electronic signals. These optical fibers can now be mass produced on a printed circuits as polymer optical waveguides, as shown in Fig. 23.41. Much like when the industry transitioned from single-point soldering by individual wires to printed circuits, single-point fiber-optical cabling can now have low-cost printed waveguides.

23.4.1.1 Structures. The optical waveguide structures are shown in Fig. 23.42. The transmitter package provides the multiplexed laser signal sources, through vertical cavity surface

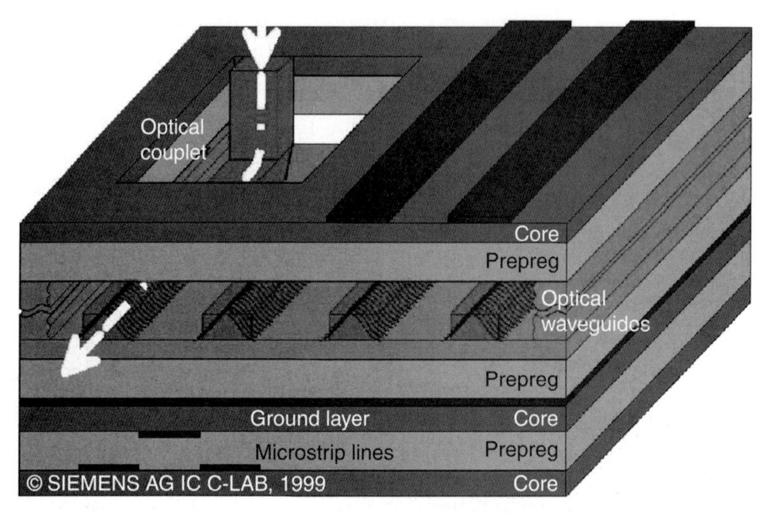

FIGURE 23.41 Structure of a printed optical waveguide. *(Courtesy of Siemens C-Lab.)*

emitting laser (VCSEL) arrays into mirrors at the optical polymer waveguide level. At the receiver package, mirrors direct the laser signal up into photodiode arrays.

23.4.1.2 Components. Figure 23.43 shows what the transmitter and receiver BGA packages will look like, whereas Fig. 23.44 shows what the micro-mirrors, VCSEL, and photo-receptor sensors may look like.

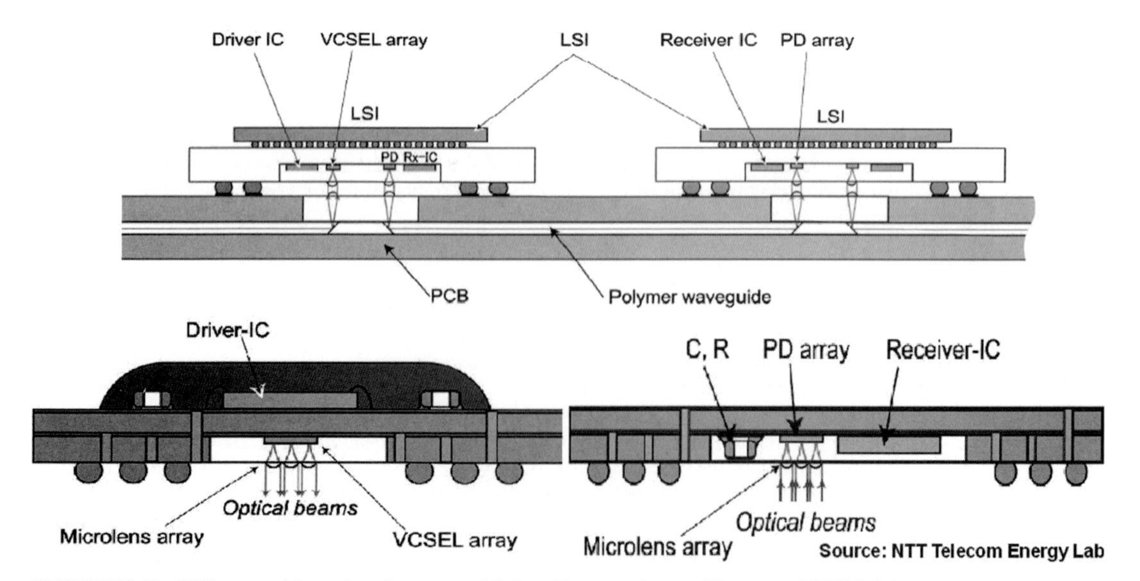

FIGURE 23.42 NTT waveguide-on-board concept with OptoBump packages. *(Courtesy of NTT-Lab.)*

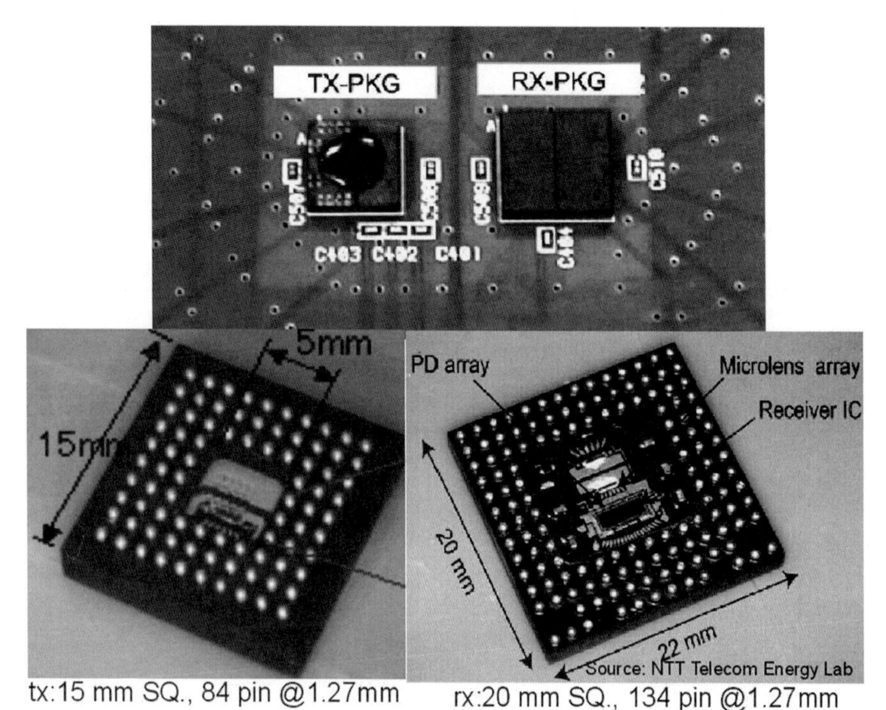

FIGURE 23.43 NTT chip-to-chip optical interconnection, Tx and Rx. *(Courtesy of NTT-Lab.)*

23.4.1.3 Optical Waveguide Materials. The optical materials being considered for integrated waveguides are currently polymers. Polymers have a number of advantages:

- **Stability** Such materials offer high thermal stability and long-term photostability. Bellcore Telecordia compliance (i.e., 1209, 1221) tested for less than 600 hr. at 85°C/85 relative humidity (RH), solder temperature >230°C, and degradation temperature >350°C.
- **Well established** A huge body of data has been gathered over the last 100 years on polymers including all the popular photoresists.
- **Useful** Polymers have unique properties (such as bend radius, modulators, and index tuning) that you cannot get anywhere else. These properties include unique processing options (photolithographic, reactive ion etching [RIE], direct laser writing, molding, and printing).

They also have a number of disadvantages:

- **Unstable** Many have low thermal stability (POF below 80°C) with photodegration (laser dyes < days) and are sensitive to delamination, moisture, and chemicals.
- **Unknown** New materials require new processes, equipment, and experience.
- **Useless** Some polymers have losses of POF of 20 dB/km, whereas optical glass is <0.1 dB/km. The packaging cost of polymers is 80 percent of the cost in devices.

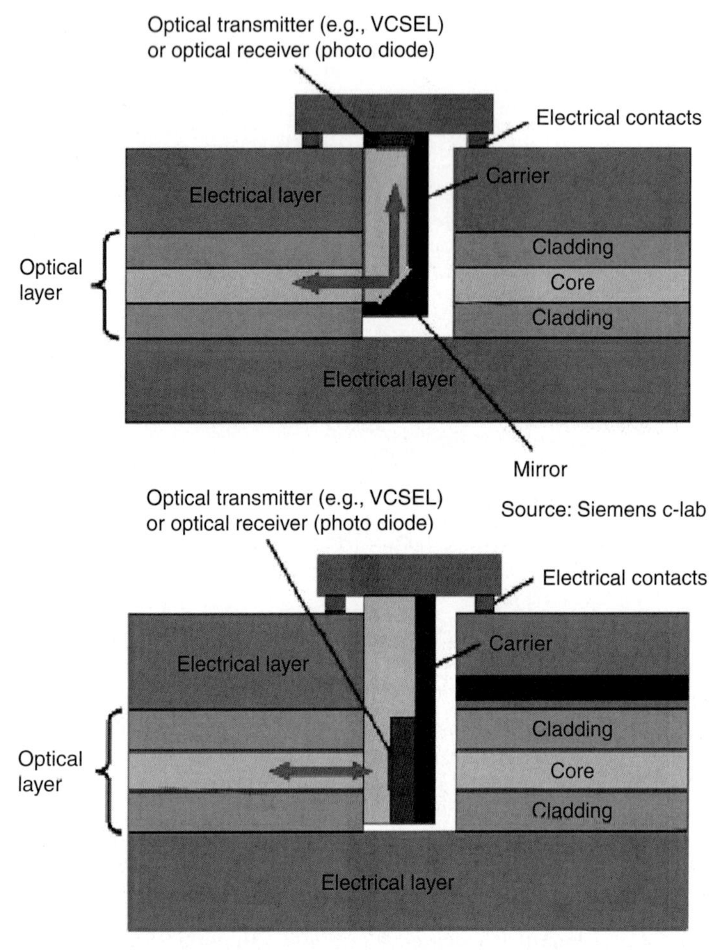

FIGURE 23.44 Various components and mounting schemes for the VCSEL, mirrors, and photo-receptors. *(Courtesy of Siemens C-Lab.)*

Candidate materials are acrylates, halogenated acrylates, cyclobutenes, polyimides, and polysiloxane. Table 23.3 lists many of the most popular materials. The optical losses of the polymers (dB/cm) are in the lower wavelengths near 840 nm.

23.4.1.4 Manufacturing Process. The manufacturing process for polymer optical waveguides is much like using a liquid photoresist. Roller coating or meniscus coating of the liquid polymer is all that is required to apply the optical polymer on a standard multilayer innerlayer. After drying, the optical polymer is exposed through standard photo masks and developed. After a final curing, they are laminated into standard multilayers with prepregs. Figure 23.45 shows the lay-up of the TRUEMODE optical polymer from Terahertz Corporation.

TABLE 23.3 Polymers for Optical Waveguides and Candidate Materials

			Waveguide optical loss (dB/cm)		
Manufacturer	Polymer type	Patterning techniques	840 nm	1300 nm	1550 nm
Allied Signal	Halogenated acrylate	Lithographic, RIE laser	0.01	0.03	0.07
	Acrylate	Lithographic, RIE laser	0.02	0.2	0.5
Dow	Benzocyclobutene	RIE	0.8	1.5	
Chemical	Perfluorocyclobutene	RIE	0.01	0.02	0.03
DuPont	Acrylate (Polyguide) Teflon AF	Lithographic, RIE	0.2	0.6	
Amoco	Fluorinated polyimide	Lithographic,		0.4	1.0
BF goodrich	Polynorbornenes	Lithographic	0.18		
Gen electric	Polyetherimide	RIE, Laser	0.24		
JDSU	Acrylate	RIE			
TeraHertz	Acrylate	Lithographic	0.03	0.4	0.8
NTT	Halogenated acrylate	RIE	0.02	0.07	1.7
	Polysiloxane	RIE	0.17	0.43	
Asahi	Cytop	RIE		0.3	
Nippon paint	Polysilane-photosensitive	Lithographic Photo bleaching	0.1	0.06–0.2	0.04–0.9

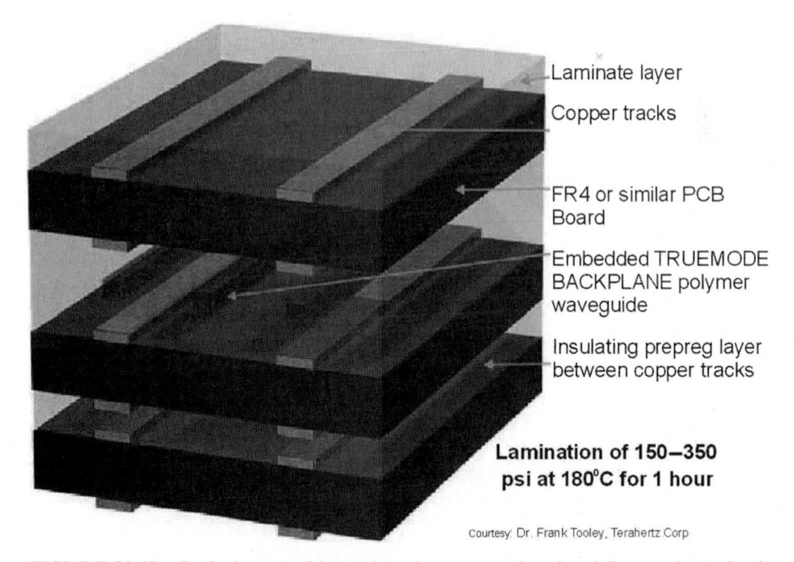

FIGURE 23.45 Optical waveguide stack-up in a conventional multilayer prior to lamination. *(Courtesy of Terahertz Corporation.)*

REFERENCES

1. Sakamoto, Kazunori, Yoshida, Shingo, Fukuoka, Kazuyoshi, and Andô, Daizo,"The Evolution and Continuing Development of ALIVH High-Density Printed Wiring Board," paper presented at IPC Expo 2000; featured in *CircuiTree,* May 2000.
2. Itou, Motoaki, tp://www.nikkeibp.com/nea/nov99/tech/.

3. "Microvia Substrates: An Enabling Technology for Minimalist Packaging 1998–2008," BPA Group, Ltd., 1999, pp. 4-3–4-17.
4. Tsukada, Yutaka, et al., "Surface Laminar Circuit and Flip Chip Attach Packaging," *Proc. 7th IMC,* 1992.
5. Tsukada, Yutaka, *Introduction to Build-Up Printed Wiring Board* (in Japanese), Nikkan Kogyo Shinbun, Tokyo, 1998.
6. Holden, Happy, "Special Construction Printed Wiring Boards," *Printed Circuit Handbook,* 4th ed., Clyde F. Coombs, Jr. (ed.), McGraw-Hill, 1995, chap. 4.
7. Takahashi, Akio, "Thin Film Laminated Multilayer Wiring Substrate," *JIPC Proceeding,* Vol. 11, No. 7, November 1996, pp. 481–484.
8. Shiraishi, Kazuoki, "Any Layer IVH Multilayer Printed Wiring Board," *JIPC Proceeding,* Vol. 11, No. 7, November 1996, pp. 485–486.
9. Fukuoka, Yoshitaka, "New High Density Printed Wiring Board Technology Named B2it," *JIPC Proceeding,* Vol. 11, No. 7, November 1996, pp. 475–478.
10. Apol, Tim, "Directional Plasma Etching—Straight Sidewalls, No Undercut," *PC Fabrication,* Vol. 20, No.12, December 1997, pp. 38–40.
11. Tsuyama, Koichi, et al., "New Multi-Layer Boards Incorporating IVH: HITAVIA," *Hitachi Chemical Technical Report,* No. 24 (1995-1), 1995, pp. 17–20.
12. Tokyo Ohka company brochure.
13. Holden, Happy, "Micro-Via Printed Wiring Boards: The Challenges of the Next Generation of Substrates and Packages," *Future Circuits International, pp 76–79,* Vol. 1, 1997.
14. Eric, Bogatin, "Signal Integrity and HDI Substrates," *Board Authority,* Vol. 1, No. 2, June 1999, pp. 22–26; a PDF copy is available for download at www.Megatest.com.
15. Figure 12—Integrated 3-D Assembly with Thin-Film, Flip Chip, MEMS and PWB.
16. Erben, Christoph, "New Materials in Optoelectronics: Advantages and Challenges for Polymers," Second Optoelectronics Packaging Workshop, Austin, TX, August 21–22, 2001, pp. B1– B22.
17. Schroder, Henning, "Photonic and Optical Wiring—Advantages and Challenges for Polymers," Second Optoelectronics Packaging Workshop, Austin, TX, August 21–22, 2001, pp.B1–B22.
18. Watsun, Jim, "Chip-to-Chip Optical Interconnec—Optical Pipedream?" First Optoelectronics Packaging Workshop, February 21–22, 2001, Austin, TX, pp.B1–B22.
19. Griese, Elmar, "Optical Interconnection Technology for PCB Applications." *PCB Fab,* June 2002, pp. 20–36.
20. Griese, E., Himmler, A., Klimke, K., Koske, A., Kropp, J.-R., Lehmacher, S., Neyer, A., and Süllau, W., "Self-Ligned Coupling of Optical Transmitter and Receiver Modules to Board-Integrated Multimode Waveguides," *Micro- and Nano-Optics for Optical Interconnection and Information Processing, Proceedings of SPIE,* Vol. 4455, 2001, M., pp. 243–250.
21. Griese, E., "A High-Performance Hybrid Electrical-Optical Interconnection Technology for High-Speed Electronic Systems," *IEEE Transactions Advanced Packaging,* Vol. 24, No. 3, August 2001, pp. 375–383.
22. Krabe, D., and Scheel, W., "Optical Interconnects by Hot Embossing for Module and PCB technology: The EOCB Approach," *Proceedings of 49th Electronics Components & Technology Conference,* June 1999, pp. 1164–1166.
23. Krabe, D., Ebling, F., Arndt-Staufenbiel, N., Lang, G., and Scheel, W., "New Technology for Electrical/Optical Systems on Module and Board Level: The EOCB Approach," *Proceedings of 50th Electronics Components & Technology Conference,* May 2000, pp. 970–974.

P · A · R · T · 6

FABRICATION

CHAPTER 24
DRILLING PROCESSES

Hans Vandervelde
Laminating Company of America, Garden Grove, California

24.1 INTRODUCTION

The purpose of through-hole drilling printed circuit boards is twofold: (1) to produce an opening through the board that will permit a subsequent process to form an electrical connection between top, bottom, and internal conductor pathways, and (2) to permit through-the-board component mounting with structural integrity and precision of location.

The quality of a hole drilled through a printed circuit board is measured by its ability to interface with the following processes: plating, soldering, and forming a highly reliable, nondegrading electrical and mechanical connection.

As with any process, the elements of the drilling process are:

- Materials
- Machines
- Methods
- Workers

When workers are properly trained and educated so that they possess a sound understanding of the other three elements, it is possible to drill holes meeting the aforementioned requirements with high productivity, consistency, and yield.

The goals of this chapter are as follows:

- Understanding the drilling process thoroughly
- Recognizing what might go wrong
- Locating where problems could occur
- Detecting whether problems do occur
- Finding root causes of problems
- Correcting undesirable conditions
- Striving for zero discrepancies
- Making improvement a team effort

24.2 MATERIALS

Materials that affect the drilling process are as follows and as shown in the fishbone diagram of Fig. 24.1.

- Laminate material
- Drill bits
- Drill bit rings
- Entry material
- Backup material
- Tooling pins

24.2.1 Laminate Material

A typical circuit board laminate panel consists of three basic components:

- Supporting fibers
- Resin
- Copper layers

The laminate substrate material is constructed of supporting fibers (most commonly a glass fiber weave) and a resin (most commonly an epoxy compound). Finished board thickness may range from 0.010 to 0.300 in or thicker, with the most common panel thickness ranging around $^1/_{16}$ (0.0625 in).

24.2.1.1 Supporting Fibers. Generally, the larger the glass fibers in the weave, the lower the cost of the base material. However, larger fibers are less desirable from a drilling point of view because they are more likely to cause the drill bit to deflect, resulting in decreased hole

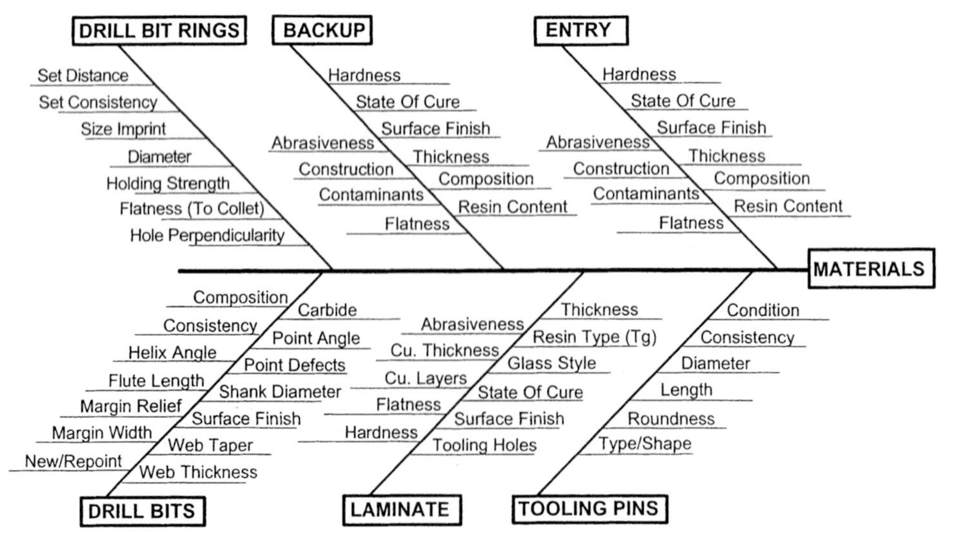

FIGURE 24.1 Materials used in the drilling process, with major variables and special considerations identified.

registration accuracy. In addition, larger fibers adversely affect drilled hole wall quality because they create larger (drilling) voids, defined as tear-out of supporting fibers. In simple terms: the larger the glass fibers, the larger the chunks torn out during drilling and thus the greater the hole wall roughness.

24.2.1.2 Resin. The most common resin system is FR-4 epoxy. However, general drilling considerations (see Fig. 24.1) are the same for all resin systems. The glass transition (T_g) rating of a material is defined as the temperature at which fully cured resin starts to soften. The T_g of FR-4 is typically around 130°C or higher. Many materials have resin systems that include additives or that are different from FR-4 resins (e.g., polyimide) in order to raise the T_g. The reason higher T_g is desired is because it implies a more stable material with lowered z-axis expansion rates. Although higher-T_g material may reduce the extent of hole wall resin smear, the compromise of a higher-T_g product is that the material is more brittle and more abrasive to the drill, resulting in increased tool wear and hole wall roughness. This increased abrasiveness to the drill bit may be offset by reducing the surface speed to reduce the spindle speed, resulting in less frictional heat being generated. Lowering drilling heat reduces drill wear.

24.2.1.3 Copper Layers. Outer and inner copper layers may be of various thicknesses, the most common being what is referred to as $^1/_2$-oz and 1-oz copper. One-ounce copper equates to a thickness of approximately 1.4 mils (0.0014 in). The more copper layers within a laminate, the more balanced the panel from a drilling point of view, meaning reduced occurrences or extents of drilled hole defects such as voids (torn-out fiber bundles). However, in order to compensate for a greater number of copper layers, the chip load (advance per revolution) that determines the infeed rate usually needs to be adjusted to control nail-heading. Increasing the chip load reduces the amount of nail-heading. In addition, more copper layers wear the drill bit at a faster rate and may require lowering the maximum hit count per drill bit.

24.2.2 Drill Bits

24.2.2.1 Materials. Drill bits are made of tungsten carbide because its wear resistance (and relatively low cost) make it the most ideal material for cutting the very abrasive circuit board laminate materials. The compromise of this very hard (carbide) material is that it is also brittle and subject to damage in the form of chips if not handled carefully and correctly.

24.2.2.2 Handling and Inspection. When handling drill bits, do not permit the bits to come in contact with one another and be careful not to touch them to the sides of the tool pods. Modern drilling machines utilize drill cassettes to reduce time per load by eliminating manual tool changes. These cassettes may accommodate 120 or more drill bits and reduce drill bit handling damage. If drill bits are measured to verify diameter using a contact-type measuring device (such as a contact micrometer), take the measurement away from the point to prevent chipping of the cutting corners. Following diameter measurement, inspect the drill bit for damage using a microscope. After use, again use care when removing drill bits from the tool pods or cassettes and remember, if they are intended for repointing, to use the same careful handling practices as when they were new.

24.2.2.3 Geometric Attributes. The geometry of a drill bit very much affects the way it behaves during drilling. (See Fig. 24.2 for attribute nomenclatures.) The land is the area remaining after fluting. In order to reduce the amount of land that creates friction with the hole wall (thus generating heat), drill bits are margin relieved. The amount of land remaining in contact with the hole wall during drilling is referred to as the margin. The wider the margin, the greater the friction area and the higher the drilling temperature, resulting in higher extents of heat-related hole quality defects such as resin smear and plowing (defined as furrows in the resin).

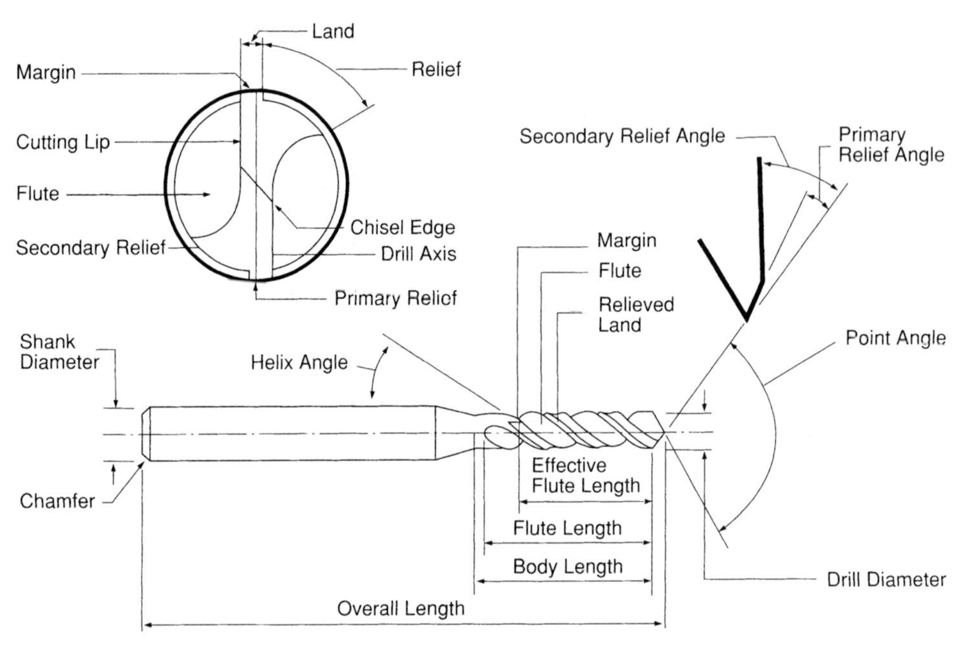

FIGURE 24.2 Drill bit geometry.

The result of increasing the land mass and the web is a smaller flute area. Less flute space implies reduced amounts of available area to remove drilling debris, which again raises the drilling temperature. It is important to understand that some drill designs that are meant to increase strength (particularly in smaller-diameter drill bits, with the intent to reduce breakage) may include what are referred to as partial margin reliefs (see Fig. 24.3). What this means is that when the drill bit is viewed from the point, a relieved margin is seen. However, viewing

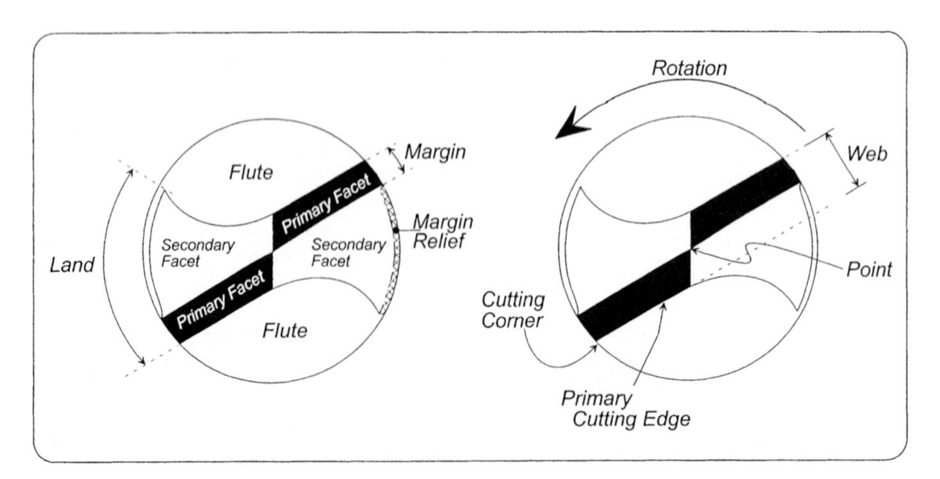

FIGURE 24.3 Drill tip attribute nomenclatures.

this type of design drill bit from the side reveals that the relief extends only part of the way, typically about one-fifth of the total flute length The major drawback of this particular design is that it increases drilling temperatures (documented to be as much as 25 percent or more), resulting in increased heat-related drilled hole defects (such as smear) and drill breakage due to packed margins.

24.2.2.4 Flute Length. Minimum flute length must equal total drilled depth (total laminate thickness + entry thickness + backup penetration depth) plus at least 0.050 in of unused drill flute remaining above the stack at the bottom of the drill stroke to allow debris to be removed by the vacuum system. If debris cannot be removed from the drill flutes during and between drill strokes, the results are extensive hole quality defects and drill breakage.

24.2.2.5 New Drill Bits. It is not necessary to inspect incoming orders of drill bits 100 percent. A procedure proved to be effective is to inspect drill bits according to an acceptable quality level (AQL) method (such as MIL-STD-105). This type of sampling plan allows the user to determine the percentage of drill bits that are inspected to judge an entire lot of drill bits to a predetermined quality level. Using this method, finding a defined quantity of drill bits that do not meet the specification is cause to return all received drill bits to the vendor; if the selected quantity of inspected drill bits meets specification, the entire lot of drill bits is accepted. Inspection criteria might include point geometry defects (refer to Fig. 24.4), damage (chips), diameter (drill and shank), and flute length, as well as correct ring set distance and size imprint (matching both the actual diameter and that labeled on the box).

24.2.2.6 Repointed Drill Bits. Drill bits are typically repointed for reasons of economics. The cost of repointing a tool may be as low as 15 percent of the cost of a new tool. The number of times a drill is repointed varies anywhere from 1 to as many as 10 times or more and typically depends on the drill diameter. The smaller the drill, the fewer times it is normally repointed. The reason is that smaller-diameter holes are more critical and require better hole quality.

There are two methods used to repoint tools.

1. The first is to specify a certain number of times the tool is to be repointed before being discarded. This number typically varies between one to three for smaller-diameter tools depending on hole quality experienced after repointing.
2. The second is to specify a minimum overall length of the tool at which it is discarded. Minimum overall length is determined by calculation based on minimum remaining flute length required to drill the required total drilling depth. This method does not allow determination of how many times a drill is actually repointed because stock removal during each regrind may vary from 0.002 to 0.005 in or more.

Repointed drill bits cannot be expected to perform as well as new drill bits because only the points are refurbished to a quality that may be as good as a new drill bit while the rest of the drill bit, including the critical margin, is not. The condition of the margin is very important because it is the part of the drill bit that finishes the hole wall. A rough margin results in a rough hole wall surface. When inspecting repointed drill bits, examine the sides of the drills for margin damage and/or fused or packed drilling debris remaining from previous use. These drill bits contaminate holes from the very first one drilled and may cause run-out due to an imbalanced condition of the drill bit resulting from the buildup. Drill bits must be repointed to the same point geometry specifications as those that apply to new drill bits. Point inspection criteria, therefore, is the same for repointed drill bits as for new drill bits.

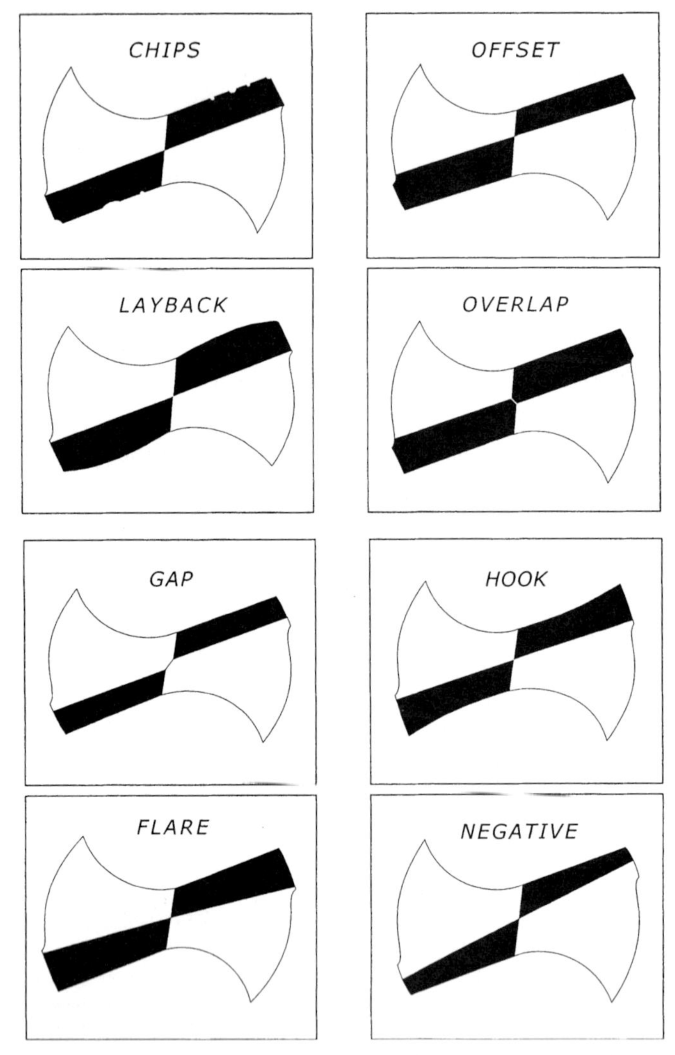

FIGURE 24.4 Drill point defect illustrations. (*Courtesy of LCOA Technical Center.*)

24.2.3 Drill Bit Rings

Drill bit rings are set to a common distance from the point to the back of the ring, thereby allowing a controlled drilling depth. The quality of these rings is as critical as the consistency of drill bit attributes because they can equally affect drill bit performance. Rings that fit loosely on the drill shank have been known to move during tool changes, resulting in insufficient drilling depth. Rings that fit too tightly may crack. Rings that have protruding material ("flash") on the inside diameter may cause improper seating of the drill bit in the collet (of the spindle) or may prevent the drill bit from being properly inserted into the tool pod or cassette, resulting in destructive tool change problems.

Rings have specific color codes assigned to each size and are commonly imprinted with size, diameter, and design or series number. Rings may be either machined or mold-injected. While machined rings are superior because of their consistency and quality, their cost is prohibitive for many drill bit manufacturers. Some of the drawbacks of mold-injected rings that must be monitored are inside diameter, affecting how well the ring fits on the drill shank, and remaining flash.

24.2.4 Entry Material

24.2.4.1 Purpose. The purpose of the entry material is fivefold:

- Centers the drill
- Prevents drill breakage
- Prevents copper burrs
- Avoids contamination of the hole or drill bit
- Prevents pressure footmarks

24.2.4.2 Types. There are many different available types (constructions) of materials used as entry material for printed circuit board drilling, although few are specifically designed and engineered for this purpose. Engineered products are designed to improve hole registration accuracy and reduce drill breakage. The performance qualifications of the most popular materials are discussed in the next section.

Commonly available entry materials, listed in order of performance quality with respect to the five characteristics listed in the preceding section, are:

- Aluminum-clad cellulose core composite
- Solid aluminum (various alloys and thicknesses)
- Solid or melamine-clad phenolic
- Aluminum-clad phenolic

24.2.4.3 Performance. The right entry material will improve drilled hole registration and lower the risk of drill bit breakage by minimizing drill deflection upon contact with the stack. In order for the entry material to function properly, it must be flat and free of pits, dents, and scratches. Warped or twisted material will result in increased extents of entry burrs and drill bit breakage. Surface imperfections and materials that are too hard contribute to drill deflection, resulting in decreased hole registration accuracy and breakage of small-diameter drills.

Phenolic materials or phenolic composites (i.e., aluminum-clad phenolic) often warp and under most drilling conditions contaminate the hole wall, which results in problems with adherence of the plating because desmearing chemicals are not designed to remove phenolic resin. Solid aluminum materials of the correct composition and hardness that are not of an excessive thickness, yet are not too thin, may work satisfactorily with larger-diameter drill bits. However, drilling with solid aluminum materials (0.008 in and thicker) may increase the risk of breakage of smaller-diameter drills. Aluminum-clad cellulose core materials provide a hard surface to prevent burrs yet minimize drill deflection and breakage associated with solid aluminum.

24.2.5 Backup Material

24.2.5.1 Purpose. The purpose of backup material is defined by the following criteria. An ideal backup material will:

- Provide a safe medium for drill stroke termination
- Prevent copper burrs
- Not contaminate the hole or drill bit
- Minimize drilling temperatures
- Improve hole quality

24.2.5.2 Types. Numerous materials are available that are sold as backup material. Few of the materials used as backup materials are actually specifically engineered for circuit board drilling. Many of the popular backup products are composites with a variety of surface coatings or skins bonded to several different core materials. Available backup products include the following:

- Epoxy-paper-clad, wood-core composite utilizing a bonding agent with lubricating properties
- Aluminum-clad, wood-core composite
- Epoxy-paper-clad, wood-core composite
- Melamine-clad, wood-core composite
- Urethane-clad, wood-core composite
- Solid phenolic
- Aluminum-clad phenolic composite
- Plain wood
- Hardboard

24.2.5.3 Performance. Desired qualities in a backup material are minimal thickness variations, flatness (no bow, warp, or twist), no abrasives or contaminants, a smooth surface, low cutting energy (minimizing drilling temperatures), and a surface hardness that supports the laminate copper surface (to prevent burrs) yet does not cause damage or extensive wear to the drill bit.

Backup materials with lubricating properties have been proven to significantly reduce drilling temperatures by as much as 50 percent or more, often resulting in temperatures below the T_g of the laminate product being drilled. This advantage greatly reduces hole wall defects such as roughness, smear, and nail-heading and often allows increased drill stack heights and/or greatly increased drill bit maximum hit counts. The importance of these benefits is significant reduction in drilling cost per hole and improved productivity and yield.

Remember that drilled backup debris exits the stacks by passing through the holes in the laminate material and that therefore contamination (from the backup material) is of great concern. Materials containing phenolic, or composed of solid phenolic, are not suitable for circuit board drilling. Phenolic materials or phenolic composites (i.e., aluminum-clad phenolic) often warp and under most drilling conditions contaminate the hole wall, which results in problems with adherence of the plating because desmearing chemicals are not designed to remove phenolic resin. Hardboard types of materials cannot be maintained to thickness variation tolerances acceptable for circuit board drilling and are a source for a great variety of contaminants (e.g., oils crystallized on the surface for hardening purposes).

24.2.6 Tooling Pins

Seldom is any due attention given to the tooling pins. They come in many shapes and sizes and their cost, with respect to how much they add to the cost of fabricating a circuit board, is insignificant. Yet, quite often, tooling pins are found to be damaged or deformed (e.g., "mushroomed" from being hammered into the stack) or do not fit snugly. Tooling pins that do not hold the stack tightly in place or that allow the stack to move create a large variety of problems from burrs and other hole defects to poor registration or drill bit breakage. These unnecessary problems may be prevented by simply replacing tooling pins when they start to show signs of wear or damage. Use tooling pins that are hardened to minimize wear and deformation, and (ideally) $^3/_{16}$ in in diameter. Pins that are less than $^3/_{16}$ in in diameter (i.e., $^1/_8$ in) do not hold the stack firmly in place during drilling and may result in stack movement.

24.3 MACHINES

Machine variables that affect the drilling process are as follows and as shown in the fishbone diagram of Fig. 24.5.

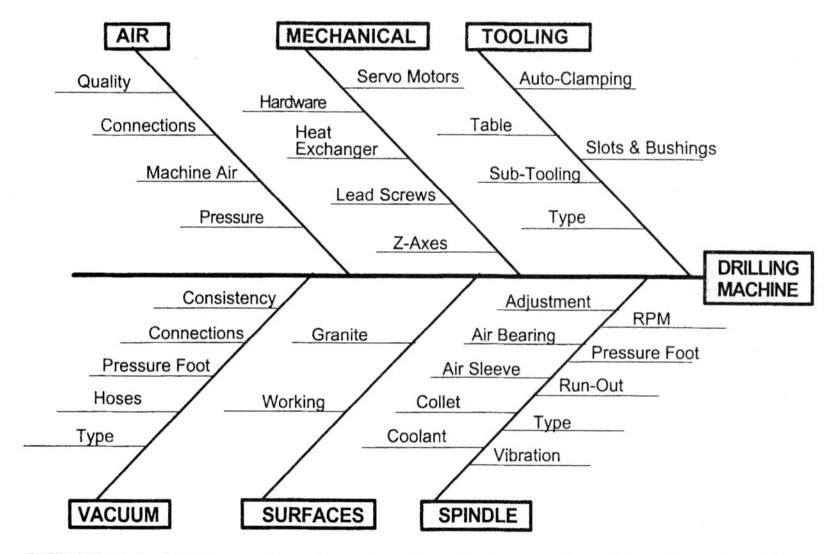

FIGURE 24.5 Drilling machine elements, with major issues and special considerations identified.

24.3.1 Air

24.3.1.1 Quality. Maintaining high levels of air quality is important. Air supplied to the machine and the spindle assemblies needs to be clean and dry. This is accomplished by filters that remove dirt and moisture from the air. In areas where the relative humidity is high, or when using air-bearing-type spindles, an inline air drying and filtering system may be necessary to control air moisture levels and dirt. Moist air causes corrosion of air surfaces (such as the spindle sleeve and other spindle components operating on air) and reduces the life of these components, resulting in increased repair costs. Dirty air affects operation of components relying on air by clogging the channels that supply the air (e.g., to the spindle sleeves and table air shoes). In addition, dirt serves as an abrasive that causes faster wear of machine components, resulting in reduced life cycles between repairs, again increasing downtime and repair costs. Cleaning the machine air filters and purging air compressors and lines on a routine basis may minimize problems resulting from dirt or moisture.

24.3.1.2 Connections and Pressure. Check connections, gauges, hoses, and switches for leaks and wear. Air hoses connected to the pressure foot pistons often crack near the piston connector; by simply bending the hoses in these areas it can be determined whether they are leaking. Other locations that commonly have poor connections and air leaks are the collet air connector on top of the spindle and the air manifold on the side of the spindle sleeve housing that supplies air to the collet and the spindle sleeve. Sufficient air pressure to both the collet and the spindle sleeve is critical and must be maintained adequately for proper operation. The tool table is designed to ride on a bearing of air supplied through the table air shoes. Verify proper operation of the air shoes by slightly rotating them back and forth; the shoes should move freely. If the shoes do not move with ease or do not move at all, the most likely reason is that the air channels are blocked due to trapped drilling debris, meaning the shoes need to be cleaned or replaced. Check the pressure foot air gauges for the correct air pressure; equal pressure must be supplied to each of the spindle stations. Insufficient or unequal air pressure results in burrs and increases the possibility of drill bit breakage.

24.3.2 Vacuum

Effective vacuum is absolutely essential because it greatly affects the resulting drilled hole quality. Heat (drilling temperature) seen by the hole walls depends on how effectively hot drilling chips are removed. Excessive temperature due to inefficient removal of drilling debris causes heat-related hole defects such as smearing and plowing, increases the opportunity for plugged holes, and is a major contributor to drill wear. Check hoses for proper connections, wear, and restrictions and examine the inside of the pressure foot for holes worn inside the connector attaching the vacuum hose.

24.3.3 Tooling

24.3.3.1 Bushings and Slots. Maintain tight tolerances on tool table bushings and slots and replace bushings that are sunken below the tool table surfaces or are worn. Bushings and slots that do not hold tooling pins snugly and allow stack movement during the drill stroke cause burrs, registration problems, and drill breakage.

24.3.3.2 Subtooling. Avoid subtooling plates that overlap across individual tool table stations and fasten the plates securely to the base plates with no separation or gaps. Ensure the subtooling is not warped, does not vary significantly in thickness, and has no surface protrusions (such as broken drill bits) to allow stacks to lay flat and to minimize variation in drill bit penetration depth into the backup material. If your subtooling is other than metal, watch for potential shrinkage or expansion problems indicated by difficulties in fitting the pinned stacks to the existing pinning holes in the plates. If this problem occurs, the results are the same as those caused by bushings and slots that allow stack movement.

24.3.4 Spindles

Spindle assemblies are one of the single most important components of the drilling machine. Proper operation is essential and requires maintenance (e.g., collet and collet seat cleaning) and verification on a regular basis.

24.3.4.1 Collet Maintenance. For mechanical (ball) bearing spindles, cleaning of the collet and the collet seat (inside the spindle) needs to be performed a minimum of once per shift; dirty collets increase drill run-out. When processing certain laminate materials that create greater amounts of drilling dust, and depending on your vacuum system efficiency, a higher cleaning frequency may be necessary to maintain acceptable levels of drill run-out. It is recommended that after cleaning the collets be returned to the same spindle from which they were removed because collets tend to adjust to the respective collet seats. For air-bearing spindles, most manufacturers recommend cleaning the collets only when the run-out measures excessive.

24.3.4.2 Run-Out Measurement. Drill/collet concentricity, or total indicated run-out (TIR), is a measure that indicates how true the assembly rotates. It can be determined while the spindle is running at various speeds (rpm), which is referred to as a dynamic form of measurement. Another method is performed while the spindle is not running and is called a static measurement. Static TIR is determined by rotating a $^1/_8$-in (0.1250-in)-diameter pin installed into the collet by hand while reading movement on a dial indicator placed against the pin at a distance of approximately 0.800 in (simulating the distance of the drill point) from the collet nose (refer to Fig. 24.6). Of course, the pin used to measure TIR must itself be concentric.

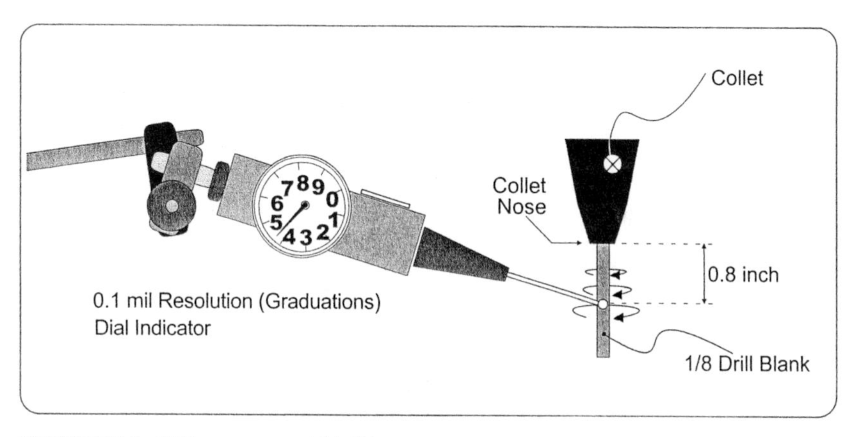

FIGURE 24.6 TIR measurement (static).

Maximum acceptable TIR for drills greater than 0.0200 in in diameter is 0.5 mils (0.0005 in). When using drills 0.0200 in or less in diameter, the maximum TIR needs to be maintained within 0.2 mils (0.0002 in) to prevent drill breakage. If excessive TIR is noted, it is wise to replace the pin and again measure the run-out. Drill bit blanks make ideal pins for measuring TIR and may be acquired from drill bit suppliers. Excessive spindle run-out results in drill breakage, causes burrs and other drilled hole defects, and adversely affects hole registration accuracy. Excessive TIR may often be corrected by simply cleaning the collet and the collet seat or by replacing a worn collet. Measure the run-out of spindles of each machine at a minimum frequency of once per week.

24.3.4.3 Spindle Speed. Maintaining correctly adjusted spindle speed is important because spindles running at speeds other than those displayed on the monitor imply that drill bits are rotating at surface speeds other than those that are desired. When actual rpm is higher than the displayed rpm, it implies higher surface speeds (refer to Sec. 24.4), and vice versa. Higher surface speeds cause greater frictional drill heat, resulting in faster drill wear and greater extents of heat-related hole defects such as smearing and plowing. Noncontact-type tachometers able to measure up to 150,000 rpm or more are available for around $300 (a lot less than the cost of rebuilding just one spindle) and allow measurement of actual spindle rpm in a matter of minutes. Verifying spindle speeds once every 6 months is usually sufficient.

24.3.4.4 Pressure Foot. The pressure foot insert lead distance to the point of the drill bit is set at approximately 0.050 in to give the pressure foot sufficient time to hold the stack flat before the drill bit contacts and enters the stack. If the lead distance is significantly less than the specified distance or if the drill bit point protrudes from the pressure foot, drill bits will break. To ensure the pressure foot assemblies function properly, check pressure foot inserts for wear or damage daily and verify that the pistons and guide rods are not bent and move smoothly.

24.3.4.5 Adjustment. When adjusting the physical *z*-axis height of the spindle, for instance to accommodate a thicker subtooling plate, care must be taken to also adjust the pressure foot so its height remains identical relative to the spindle. The pressure foot adjustment is independent from the spindle and, if not performed correctly, may result in a gap between the bottom of the spindle casing and the top of the pressure foot window. This gap is noticed only while the spindle is engaged and will draw most of the vacuum air, effectively cutting off the vacuum from the pressure foot insert, resulting in a great variety of hole defects and drill bit breakage. If a gap is present, it is usually accompanied by great amounts of drilling debris spewed across the stacks and drilling machine.

24.3.5 Mechanical Factors

24.3.5.1 Heat (Coolant) Exchanger. Drill motor temperatures are reduced by exchanging a fluid through the spindles. The fluid is processed through a heat exchanger that works much like a car radiator. As with a car, it is important to maintain the recommended coolant mixture, proper fill level, and flow rate indicated by the flowmeter. Check the operation of the fan and minimize coolant flow restriction (due to algae buildup) by cleaning filters regularly. Algae growth is promoted through exposure to ultraviolet light and may be controlled by using black hoses instead of clear ones in addition to adding growth inhibitors to the coolant mixture.

24.3.5.2 Hardware. Although simple to perform, checking mechanical connections for loose or worn parts (i.e., causing travel slop of pressure foot assembly movement) is often neglected. It may be accomplished by observing the machine while it is running or checking by hand while the machine is idle.

24.3.5.3 Lead Screws and Servos. The environment and cleanliness of the drill room determine how often lead screws need to be cleaned and lubricated to ensure smooth operation and minimize wear. Most machine manufacturers recommend performing this type of maintenance every 6 months. When lubricating the lead screws, only a light coat of the appropriate grease needs to be applied. Excessive amounts of grease on the lead screws defeat the purpose, as this causes more dirt to be attracted, resulting in faster wear. Constant searching for programmed *x-y* locations indicates problems with the servo motors or lead screws. If this occurs, the displayed *x* or *y* (actual) location displayed on the controller will change continuously while the machine is stopped.

24.3.5.4 z-Axis Stroke. Because machines may have as many as six mechanical connections in the *z* axis, travel slop during the drill stroke or retraction may occur, causing chip loads to vary during the stroke, which results in burrs and other hole defects and drill breakage. Check connections by hand for slop while the machine is idle and observe and listen for improper operation while the machine is running.

24.3.6 Surfaces

A clean drilling room decreases repair costs and the possibility of improper operation of the drilling machines. Wipe surfaces clean using a lint-free cloth or use a vacuum to remove debris.

Never use compressed air to clean the drilling machine because debris will be blown into areas that need to be kept clean (e.g., lead screws).

24.3.6.1 Granite. Granite provides a stable platform that absorbs unwanted vibration. It also offers a smooth and level surface to support the movement of the table. Keep granite surfaces clean to prevent tool table air shoes from becoming clogged and to minimize collection of dirt on the lead screws.

24.3.6.2 Working Surfaces. Maintain clean working surfaces, such as the tool table, to minimize the possibility of dirt particles getting trapped between the stacked laminate, entry, and backup panels, causing stack separation that may result in burrs and possible drill bit breakage.

24.4 METHODS

Drilling parameter variables that affect the drilling process are as follows and as shown in the fishbone diagram of Fig. 24.7.

24.4.1 Surface Speed and Spindle Speed

As holes are getting smaller, higher spindle speeds are required to achieve the desired surface speed that determines throughput. However, higher surface speeds result in higher drilling temperatures that may increase heat-related hole defects such as smearing and plowing.

24.4.1.1 Definition. Surface speed is a measure of how much distance is covered by the drill's diameter while it is rotated by the spindle and is expressed in surface feet per minute (sfm). It is used to calculate spindle speed (rpm) for a given drill diameter. The formula to calculate spindle speed using the desired surface speed is shown in Eq. (24.1).

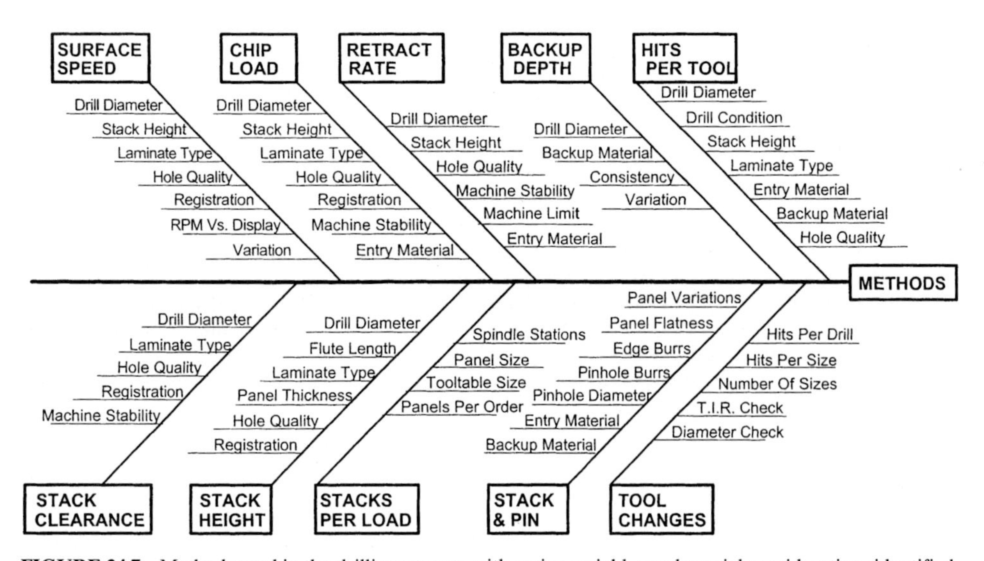

FIGURE 24.7 Methods used in the drilling process, with major variables and special considerations identified.

$$\text{spindle speed (rpm)} = \frac{\text{sfm} \times 12}{\pi \times \text{diam.}} \tag{24.1}$$

where diam. = drill diameter (in)
$\pi = 3.1415$

24.4.1.2 Effects. The higher the surface speed, the higher the spindle speed, and subsequently the higher the frictional heat that is generated, translating into greater extents of heat-related hole defects and drill wear. Conversely, lower surface speeds imply lower spindle speeds and less frictional heat. When more abrasive materials (e.g., materials with higher T_g such as multifunctional FR-4, polyimide, or cyanate ester) are processed or drilled stack height is increased, drilling temperatures increase. To offset the resulting increase in temperature under such conditions or when excessive extents of heat-related hole defects are apparent, decrease the surface speed to lower the spindle speed.

24.4.2 Chip Load and Infeed Rate

24.4.2.1 Definition. Chip load is defined as advance per revolution and is usually expressed in mils (1 mil equals $^1/_{1000}$ of an inch [0.001 in]). It implies the distance the drill travels during the drill stroke per each full revolution of the drill bit. Chip load is used to calculate the infeed rate in inches per minute (ipm).

$$\text{infeed rate (ipm)} = \text{chip load (mils/revolution)} \times \text{rpm} \tag{24.2}$$

24.4.2.2 Effects. Higher chip loads result in greater throughput. However, chip load and infeed rate affect hole registration accuracy, drill breakage, burrs, and mechanical types of drilled hole defects such as voids (tear-out of the supporting fibers) and nail-heading. Faster infeed rates translate into higher (top laminate) entry burrs, lower extents of nail-heading defects and (bottom laminate) exit burrs, increased occurrences of drill bit breakage, and higher extents of drilling voids. Lower infeed rates result in exactly the opposite but lower throughput.

24.4.3 Retract Rate

24.4.3.1 Definition. Retract rate is the speed at which the drill bit exits the stack following the drill stroke and is expressed in inches per minute (ipm). Machine default (maximum) settings vary between manufacturers and may range anywhere from 500 to 1000 ipm.

24.4.3.2 Effects. Higher retract rates imply lower processing times per load. While the maximum retract rate setting may be fine for larger-diameter drill bits, when using drill bits in the diameter range of 0.0250 in (size #72) to 0.0135 in (size #80), retract rates may have to be reduced to 500 ipm or lower to prevent drill breakage. When drilling with sizes smaller than 0.0135 in in diameter, retract rates may have to be reduced even further. The maximum retract rate that may be used with any given drill diameter without causing drill breakage greatly depends on the stability and vacuum system efficiency of the drilling machine as well as the drilled stack height, laminate construction and thickness, type of entry material, use of proper stacking and pinning procedures, and the design characteristics of the drill bit such as flute length, web thickness, and web taper.

24.4.4 Backup Penetration Depth

24.4.4.1 Definition. Backup depth is the distance a drill bit penetrates the backup material at the bottom of the drill stroke. The minimum backup penetration depth setting varies depending on drill diameter and is determined by calculating the drill bit point length (see Fig. 24.8) and adding approximately 0.010 in. As a rule of thumb, backup penetration depth may be set to a distance equal to the drill diameter or 0.040 in, whichever is less.

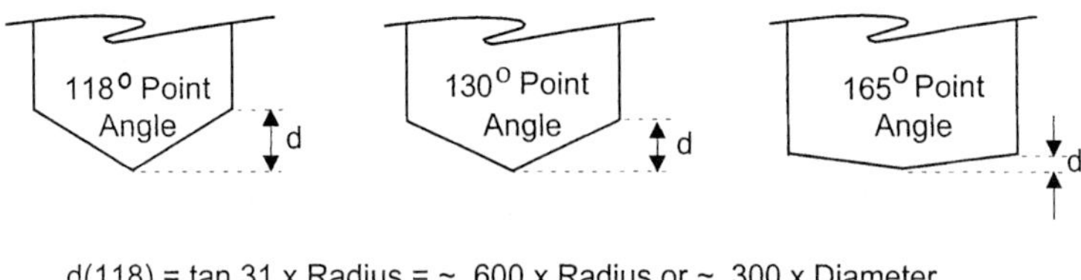

d(118) = tan 31 x Radius = ~ .600 x Radius or ~ .300 x Diameter
d(130) = tan 25 x Radius = ~ .466 x Radius or ~ .233 x Diameter
d(165) = tan 7.5 x Radius = ~ .132 x Radius or ~ .065 x Diameter

FIGURE 24.8 Point length calculation.

24.4.4.2 Effects. Excessive backup penetration depth increases drill wear and the occurrence of breakage of small-diameter drill bits, adversely affects hole quality, and increases process time per load. Insufficient backup penetration depth results in incompletely drilled holes. This implies that thickness variations of the backup material are very important, meaning that minimal variations are a much desired characteristic of the backup and need to be considered when choosing a material suitable for your application.

24.4.5 Hits Per Tool

24.4.5.1 Definition. The maximum hits per tool specified for any given drill size implies the number of drill strokes a drill bit is used for until its expected effective life is expired. Maximum hit count per tool is product and process specific and is affected by laminate material construction, panel thickness, drilled stack height, surface speed, and the type of entry and backup material used. Therefore, no specific number of hits per tool can be arbitrarily specified.

24.4.5.2 Effects. Excessive drill wear caused by excessive maximum hit count increases drilled hole defects and may prevent proper repointing. Conservative maximum hit counts greatly impact drilling cost per hole and increase time per load because of increased numbers of tool changes.

24.4.6 Stack Clearance Height

24.4.6.1 Definition. Stack clearance height is the distance between the point of the drill and the surface of the stack at the top of the drill stroke. Maintain a minimum stack clearance distance of $^1/_8$ in (0.125 in), which implies a space between the bottom of the pressure foot and the top of the stack of 0.075 in, assuming the pressure foot lead distance to the point of the drill is correctly set at 0.050 in. Stack clearance may be adjusted for each load by simply sliding a 0.075-in shim between the pressure foot and the stack and adjusting the upper limit ("UP#") until the pressure foot touches the shim.

24.4.6.2 Effects. The less the stack clearance distance between the drill point and the top of the stack, the shorter the drill stroke and therefore the shorter the processing time per load. Increasing the stack clearance distance allows more time between drill strokes and gives the tool table more time to settle, which may improve hole registration accuracy and prevent small-diameter drill bit breakage. In addition, the greater the time between drill strokes, the more likely drilling debris will be removed from the drill flutes and, consequently, the lower the drilling temperatures, which results in reduced occurrences of drill breakage and lower extents of drilled hole quality defects.

24.4.7 Drilled Stack Height

Material construction (panel thickness, number of copper layers, laminate type, etc.), drill bit diameter, and flute length, as well as hole quality and registration accuracy requirements, all are factors that need to be considered when deciding on appropriate drilled stack heights. A greater number of panels in the drilled stack means higher drilling temperatures, accelerated drill wear, and greater drill deflection, affecting the resulting hole quality and registration accuracy. When using smaller-diameter drill bits, stack heights need to be reduced to prevent drill breakage and to accommodate the shorter flute lengths. As a rule of thumb, the maximum total drilled depth (number of panels × panel thickness + entry thickness + backup penetration depth) that can safely be handled by the drill bit without breakage is approximately 17 times its diameter.

24.4.8 Stacking and Pinning

24.4.8.1 Building the Stack. Inspect all laminate panels and the entry and backup materials for surface damage and remove burrs from the panel edges as well as from the pinning holes. Even though the registration tooling holes on the laminate material may not be used to pin the stacked panels together, it is important to remove any resin buildup remaining around these holes after lamination (a common occurrence). Burrs and raised surface areas do not permit the panels to lay flat, causing gaps resulting in drilled hole registration problems, burrs, hole quality defects, and drill breakage. Reject entry and backup materials with excessive nicks, scratches, and other surface defects as well as those that are warped or twisted.

24.4.8.2 Pinning Procedures. Wipe the surfaces of all laminate panels and the backup material using a lint-free cloth to remove any debris before stacking the panels (allowing intimate contact between the stacked panels). Verify that the pinning holes and pin insertion are perpendicular to the stack and avoid using pins that are damaged or deformed.

24.4.8.3 Installation. Before placing the pinned stacks onto the drilling machine tool table, inspect the surface for burrs or broken drill bits protruding from the table that may prevent the stacks from lying flat. Do not continue if the pin bushings of the tool table are sunken or worn to the point that they do not hold the stack firmly in place. Loose or sunken tooling pin bushings cause movement of the stack during drilling and result in a variety of hole defects, registration problems, and drill breakage, yet are simple to replace at a minimum cost. After stacks are put in place, again wipe the surface of the top laminate material as well as the entry material. Place the entry material on top of the stack and tape it in place. The entry material size should be such that it clears the pins and does not extend beyond the stack edges. Pinning the entry material to the stack is not recommended because it tends to constrict the movement of the material, causing separation between it and the stack and resulting in entry burrs and possible drill breakage.

24.5 HOLE QUALITY

24.5.1 Definition of Terms

The terms in Tables 24.1 and 24.2 are commonly used to describe drilled hole defects observed on copper and substrate surfaces. It is important to be able to identify these defects specifi-

TABLE 24.1 Copper Defects

Defect	Definition	Type
Burr	Ridge left on external surface	Mechanical
Debris	Drilling residues	Mechanical
Delamination	Separation of the copper from the substrate	Mechanical/heat-related
Nail-heading	Burr on internal copper layer	Mechanical/heat-related
Smearing	Thermomechanically bonded resin deposit	Heat-related

TABLE 24.2 Substrate Defects

Defect	Definition	Type
Debris pack	Drilling residues packed into voids	Mechanical
Delamination	Separation of the substrate layers	Mechanical/heat-related
Loose fibers	Unsupported fibers in the hole wall	Mechanical
Plowing	Furrows in the resin	Heat-related
Smear	Thermomechanically bonded resin deposit	Heat-related
Voids	Cavities due to torn-out supporting fibers	Mechanical

cally rather than in general terms. Using a general term such as *roughness* may imply voids or plowing. While voids are a mechanical type of defect, plowing is a heat-related type of defect. Therefore, excessive voids would lead one to examine the chip load (infeed rate) used; plowing would lead one to look at surface speed (spindle speed).

24.5.2 Examples of Drilled Hole Defects

Examples of typical drilled hole wall defects are shown in Figs. 24.9 and 24.10.

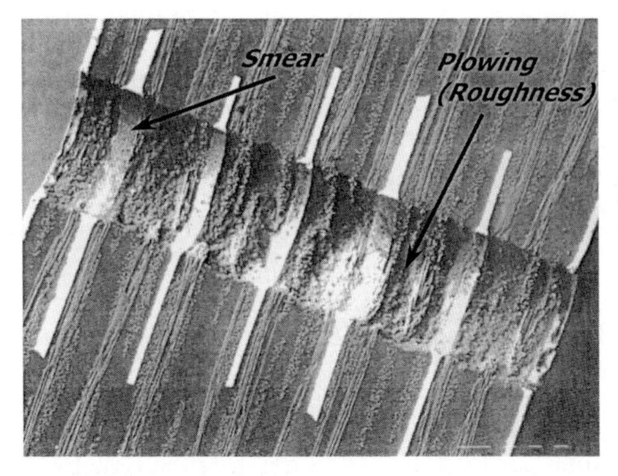

FIGURE 24.9 Cross section of drilled hole showing examples of smearing and plowing.

FIGURE 24.10 Cross section of drilled hole showing examples of nail-heading.

24.6 POSTDRILLING INSPECTION

A wealth of information is available by simply examining the materials from the drilled stack and the drill bits. For instance, inspecting the drill bits will allow one to determine if wear occurs at consistent rates (for drills of same diameter) or will reveal whether hit count maximums are excessive and the type of drilled hole wall defects to expect. Bonded debris and/or extensive wear to the drill corners imply high drilling temperatures (resulting in greater extents of defects such as smearing and plowing) or materials that are not fully cured and point to a problem with the materials (laminate, entry, or backup) or may suggest an excessive surface speed.

Extensive primary cutting edge wear indicates abrasive materials and may require lowering stack heights, reducing hit counts, or replacing entry or backup materials.

Burrs on the surfaces within the stack mean that there is a problem with the way panels are assembled and pinned or may be the result of warped panels. Entry or exit burrs on the outer laminates should cause one to question the entry and backup materials or the infeed rate.

The point of a postdrilling inspection process is that if one takes the time on a regular basis, to examine materials and drill bits after drilling, many drilling problems would be solved before they get out of hand.

24.7 DRILLING COST PER HOLE

Material and processing costs, as well as the resulting total drilling cost, may be determined by using an analysis matrix such as a cost model specifically designed for this purpose and generated with the use of a computer spreadsheet program. The advantage of using a spreadsheet is that it allows changes to be made in, for instance, specific material prices and processing times or parameters, and allows instantaneous viewing of the resulting effects on the total drilling costs, the cost per panel, and the average cost per hole. Knowing the cost per hole is important because it allows comparing different jobs or processing situations. Following is a step-by-step description of how to construct a drilling cost analysis matrix such as the one shown in Fig. 24.11.

24.7.1 Machine Time

Table A in the drilling cost analysis matrix is used to calculate the total time that is required to complete the job. First, the different drill sizes (a) and their respective total drilled holes per panel (b) as well as the total number of panels (c) to be drilled are determined and entered in the spreadsheet; this allows the spreadsheet to calculate the total number of holes for each size to complete the job (d).

Second, using the appropriate drilled stack height (e), the total number of drilled stacks (g) and the total number of drilled hits per drill size (f) can be calculated. The total number of drilled hits (drill strokes) is the total number of drilled holes per panel (b) divided by the number of panels per drilled stack (e).

Third, the number of total drilled stacks (g) is divided by the number of stations per machine (stacks per load [h]) to calculate the number of machine loads (i).

Fourth, the total drill time per load required per drill size (j) is entered to calculate the total machine time for each drill size (k).

Fifth, the total times of each of the drill sizes are simply added up to arrive at the total time required to finish the job.

An option is to enter the total drill time per load instead of entering the time for each of the drill sizes and multiplying the total drill time per load by the number of machine loads to determine total machine time.

TABLE A Machine Time

a	b	c	d	e	f	g	h	i	j	k
Drill bit size	Number of holes per panel	Number of drilled panels	Total number of drilled holes	Panels per drilled stack	Total number of drilled hits	Total number of drilled stacks	Stacks per machine load	Number of machine loads	Drill time per load (h)	Total machine time
0.0135	7,000	120	840,000	2	420,000	60	4	15.0	1.12	16.80
0.0160	5,000	120	600,000	2	300,000	60	4	15.0	0.80	12.00
0.0225	3,125	120	375,000	2	187,500	60	4	15.0	0.48	7.20
0.0350	1,500	120	180,000	2	90,000	60	4	15.0	0.19	2.85
0.0520	800	120	96,000	2	48,000	60	4	15.0	0.10	1.50
0.0700	250	120	30,000	2	15,000	60	4	15.0	0.04	0.60
	17,675		2,121,000						2.73	40.95

TABLE B Drill Bits

l	m	n	o	p	q	r	s	t	u	v
Drill bit size	Cost per new tool	Cost per repoint	Number of repoints per tool	Total cost per drill bit life	Number of uses per drill bit life	Average cost per drill bit use	Total number of drilled hits	Number of hits per drill bit use	Total number of drill bit uses	Total drill bit cost
0.0135	$1.25	$0.25	2	$1.75	3	$0.58	420,000	1,000	420.0	$245.00
0.0160	$1.20	$0.25	2	$1.70	3	$0.57	300,000	1,000	300.0	$170.00
0.0225	$1.15	$0.25	3	$1.90	4	$0.48	187,500	1,250	150.0	$71.25
0.0350	$1.15	$0.25	3	$1.90	4	$0.48	90,000	1,500	60.0	$28.50
0.0520	$1.15	$0.25	4	$2.15	5	$0.43	48,000	2,000	24.0	$10.32
0.0700	$1.30	$0.25	4	$2.30	5	$0.46	15,000	2,500	6.0	$2.76
										$527.83

TABLE C Entry and Backup Material

	w	x	y	z	aa
Material type	Cost per ft^2	Stack size ft^2	Cost per stack*	Total number of drilled stacks	Total material cost
Entry material	$0.53	3.00	$1.59	60	$95.40
Backup material	$0.65	3.00	$0.98	60	$58.50
					$153.90

* Backup material = cost/2 (uses per panel)

TABLE D Labor and Burden

	ab	ac
	Cost per hour	Cost total
Labor	$15.00	$614.25
Burden	$10.00	$409.50
		$1,023.75

TABLE E Total Costs

Cost variable	Cost per panel	Total cost	% of total
Drill bits	$4.40	$527.83	30.9%
Entry material	$0.80	$95.40	5.6%
Backup material	$0.49	$58.50	3.4%
Labor	$5.12	$614.25	36.0%
Burden	$3.41	$409.50	24.0%
Total drilling cost	$14.21	$1,705.48	100.0%

TABLE F Drilling Costs

Drilling cost per panel	$14.21
Average cost per 1000 holes	$0.804

FIGURE 24.11 Drilling cost analysis matrix. (*Courtesy of LCOA Technical Center.*)

24.7.2 Drill Bits

The cost of the drill bits needed to complete the job may be determined after the average cost per drill bit use has been calculated. To find the average cost per drill bit use, the typical number of repoints for the particular size (o) is multiplied by the cost of each repointing (n). The resulting cost, added to the new drill bit price (m), is the cost per drill bit life (p). By dividing the cost per life by the number of uses (q) per life (the number of times the bit is repointed + 1), you arrive at the average cost per drill bit use (r). Next, by dividing the total number of hits per drill size (s) by the number of maximum hits per drill bit use (t), the number of required drill bit uses (u) for each size may be determined. Then calculate the total cost per drill bit size (v) by multiplying the number of uses (u) by the average cost per use (r). The sum of the total costs of each of the drill bit sizes (v) brings you to the total cost of drill bits needed for the job. This cost, of course, is true only with the assumption of no drill bit breakage.

24.7.3 Entry and Backup Materials

Entry and backup material cost per stack (y) is determined by multiplying the stack size (square foot per panel [x]) by the cost per square foot. Remember to divide the backup cost by 2 since each backup panel may be used twice. Total material cost (aa) is calculated by multiplying cost per stack (y) by total number of drilled stacks (z).

24.7.4 Burden and Labor

Using typical burden and labor rates per hour (ab), these values are multiplied by the number of hours to complete the job (determined in Table A) in order to calculate the total burden and labor costs (ac).

24.7.5 Total Drilling Cost and Cost per Hole

After entering the required data in Tables A through D of Fig. 24.11, the total drilling cost and the cost distribution (see Table E) as well as the drilling cost per panel and the cost per hole (see Table F) can be calculated. Because the cost per hole typically ranges around $^1/_{10}$ of a cent, a more accurate and easier way to comprehend this value is by showing the average cost per 1000 holes, as is done in the cost model.

CHAPTER 25
PRECISION INTERCONNECT DRILLING

Terry Haney
Excellon Automation Co., Rancho Dominguez, California

25.1 INTRODUCTION

As circuit density continues to increase along with the demand for higher and higher accuracy in hole location, the drilling machines and the environment in which the machines operate must be tightly controlled to achieve success in the drilling operation.

High-density interconnect (HDI) has been defined as referring to holes with diameters of 0.006 in. or less. HDI holes were generally expected to be made by nonmechanical means, such as lasers, plasma, or photoimaging. However, most holes continue to be created by mechanical means, and as the hole size approaches or in some cases decreases to less than 0.004 in., special challenges face the mechanical drilling process. With the density increase, innerlayer registration is of utmost importance. Artwork generation and fabrication processes can introduce layer shift. This shift is particularly troublesome in high-layer-count printed circuit boards (PCBs). This chapter addresses these issues of small-hole drilling as well as layer-to-layer registration.

In this chapter, holes created by traditional mechanical drilling methods are referred to as "drilled," and the process is referred to as "drilling." Holes created by laser, plasma, or photoimaging are not really drilled, even though the term is often used to describe them. When this chapter references nonmechanical via-hole creation processes, these are referred to by their specific type.

25.2 FACTORS AFFECTING HIGH-DENSITY DRILLING

PCB technology requires holes as small as 0.002 in. (0.05 mm) to be drilled with extremely high accuracy, particularly when drilling dense hole patterns. The processes and machines used to drill these holes now constitute a highly developed science.

As the drilling process approaches these HDI dimensions, numerous factors become increasingly critical, such as:

- Hole location
- Predrilling process issues
- Drill room temperature/relative humidity

- Vacuum
- Drill bit condition
- Dynamic spindle run-out
- Backup and entry material (type and thickness)
- Maximum spindle speed
- Depth control

25.3 LASER VERSUS MECHANICAL

Although it is somewhat subjective, the decision to drill with mechanical or laser systems can be influenced by the following considerations:

- **Aspect ratio determination** If the via to be formed is more than 0.016 in. (6.3 mm) deep, it is better to use a mechanical drilling process.
- **Material type** Materials that tend to have excess smearing that clogs the flutes in a mechanical system are usually better on a hybrid laser system.
- **Size of holes** If the hole to be drilled is larger than 0.008 in., then use a mechanical system for speed and hole wall quality.

25.3.1 Lasers for PCB Processing

Several types of lasers can be used to fabricate PCBs, including:

- Infrared (IR) (CO_2) lasers (9.4 μm to 10.6 μm)
- Ultraviolet (UV) lasers for PCB
- YAG (yitrium-aluminum-garnet)
- YVO_4 (yitrium-lithium-floride)

See Fig. 25.1 for diagram of laser wavelengths by type.

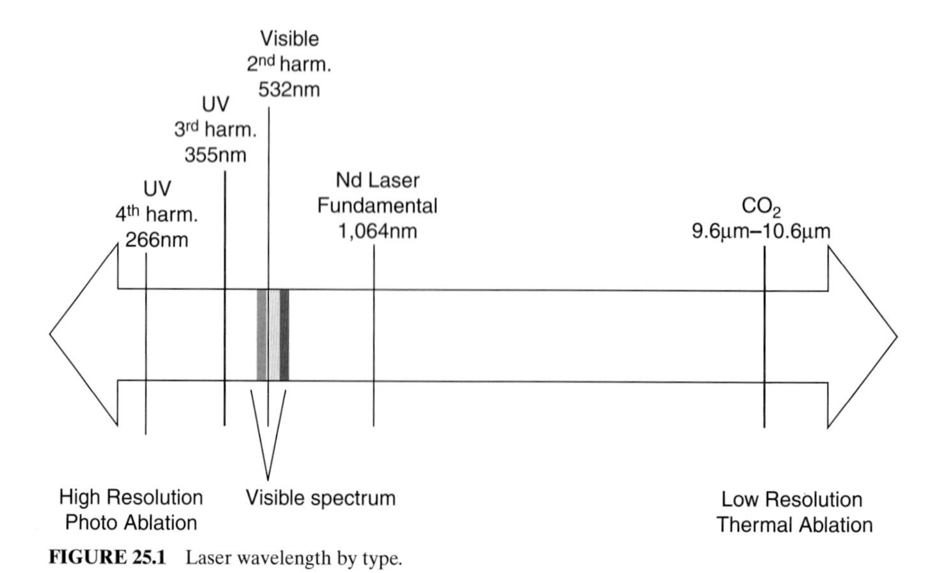

FIGURE 25.1 Laser wavelength by type.

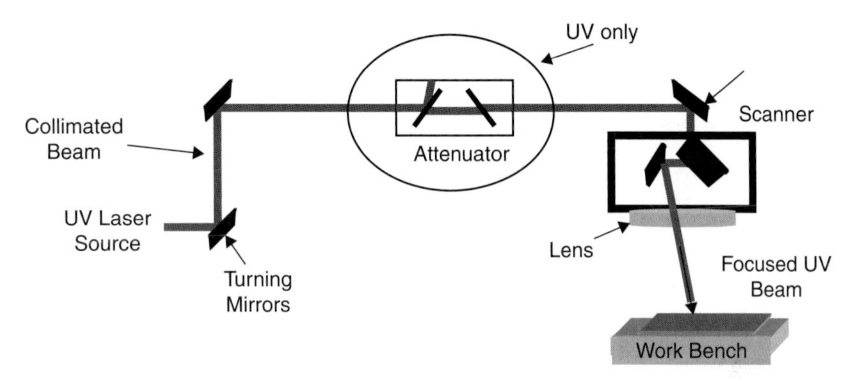

FIGURE 25.2 Laser beam delivery.

25.3.2 Beam Delivery

Laser source outputs a collimated beam. Turning mirrors (see Fig. 25.2) guide the beam to the scanners. The attenuator is used to control UV beam power. Scanners move the beam on the worktable. The lens focuses the beam to the worktable.

25.3.3 IR (CO_2) Drilling

IR drilling requires the following:

- **Thermal ablation** This ablation requires heating, melting, and vaporizing. CO_2 IR wavelengths are reflected by copper.
- **Large focus diameter** The focus should be ~200 μm.
- **Larger beam size** The beam is adjustable through defocusing or apertures.
- **Percussion drilling** Each pulse removes a portion of dielectric material.
- **High dielectric removal rate** Scanner speed limits throughput.
- **Postprocess clean** A desmear process is required.

See Fig. 25.3 for an overview of the CO_2 process.

25.3.4 UV Drilling

UV drilling requires the following:

- **Photo ablation** This process breaks molecular bonds. Minimal heat damage is inflicted on the surrounding area.
- **Precise cutting** The small focus diameter should be ~20 μm.
- **Percussion drilling** The hole diameter must be the same as the focus diameter.
- **Trepanned drilling** The hole diameter must be greater than the focus diameter.

UV light is absorbed well by most PCB materials, but each material absorbs at a different rate.

Conformal Mask Processing

A mask of holes made by drilling, etching, or laser-ablating the top layer of copper. CO_2 is used for dielectric removal and a high ablation rate.

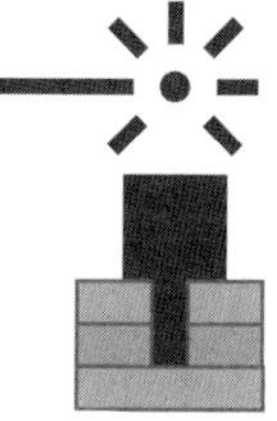

Direct Dielectric Drilling

No copper layer is required.
The mask in the laser beam defines the via diameter.

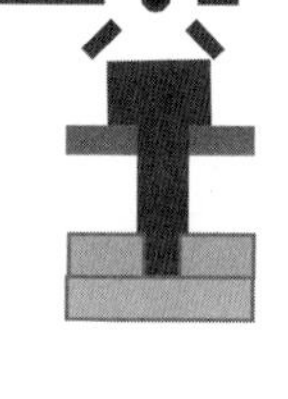

Via Formation with CO_2 Only

The copper layer is etched and oxidized.
CO_2 is used for copper and dielectric.

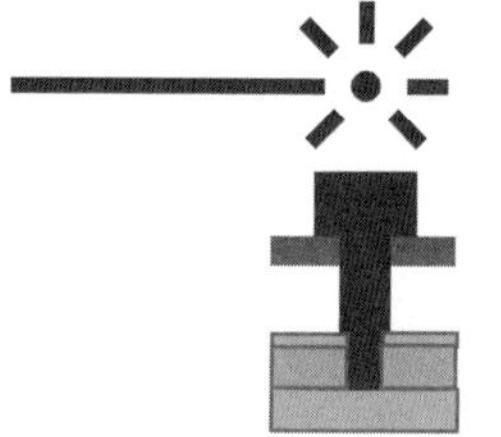

FIGURE 25.3 CO_2 process overview.

25.3.5 Creating Board Elements with Lasers

Vias are created by having the laser beam make concentric circles or spiraled cuts. Laser pulse trains can be used to create vias or lines. Hole desmear can also be done by laser.

25.3.6 UV Lasers

25.3.6.1 Yitrium-Aluminum-Garnet (YAG).

- The output pulse is ~120 nanoseconds.
- YAG was first used for laser PCB drilling.
- It provides high pulse energy.

25.3.6.2 Yitrium-Vanadate (YV04).

- YVO_4's output pulse is ~20 nanoseconds.
- Its repetition rate is ~100 KHz.
- Currently YVO_4 is most commonly used for PCB drilling.

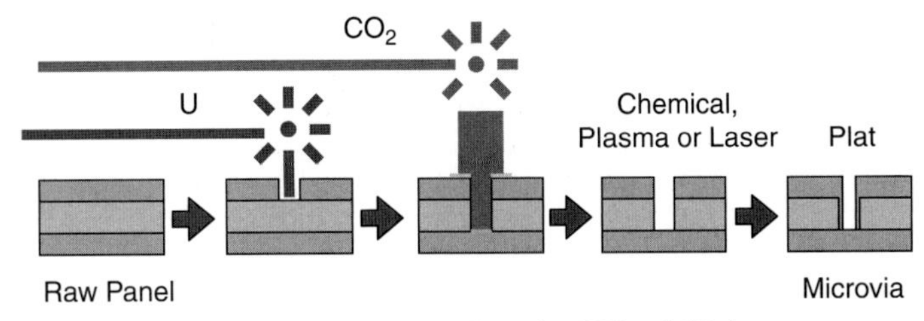

FIGURE 25.4 Hybrid laser via formation using UV and CO_2 lasers.

25.3.6.3 Yitrium-Lithium-Fluoride (YFL).

- The output pulse is ~50 nanoseconds.
- YFL is more efficient, providing five times the absorption coefficient of YAG at the pump wavelength.

25.3.7 Hybrid Lasers (UV and CO_2)

Figure 25.4 shows a flow of the use of hybrid lasers for via formation. This hybrid is the combined use of CO_2 and UV laser technology.

25.4 FACTORS AFFECTING HIGH-DENSITY DRILLING

The following are factors to be considered regarding mechanical drilling.

25.4.1 Positioning/Hole Location

The positioning system for conventional drilling machines consists of x-, y-, and z-axes. The x- and y-axes refer to the position of the circuit board under the drill spindle, and the z-axis to the plunging and retracting motions that allow the drill spindle to feed the drill bit into the substrate material. Machine accuracy is defined by the machine's ability to position the table under the spindles to the desired programmed x and y values. All process variables, such as tool design, stack height, feeds and speeds, and tooling methods, must be eliminated to evaluate machine accuracy.

The stability of the machine is critical when microdrilling dense drill patterns. When the drilling system is designed, stability is one of the most important factors taken into consideration. The foundation or base of the machine is usually constructed of granite or similar material. Granite is typically used because of its mass and low coefficient of thermal expansion (CTE) to temperature fluctuation. The upper structure of the machine is designed with a high natural frequency, which allows the structure to stabilize quickly, reducing the time the machine must pause before making a rapid movement to the next hole location.

Vibration caused by forces such as punch presses, drilling machines operating in close proximity, or any other type of equipment capable of causing severe vibration can have a detrimental effect on HDI drilling, such as drill bit breakage and hole mislocation. Each drilling machine should be isolated from these forces as thoroughly as possible. The ideal solution is to have each drill machine placed on its own isolated pad made of reinforced concrete. This setup serves to isolate the drill machine from a large percentage of the vibration transmitted through the shop floor.

25.4.2 Drill Room Temperature and Relative Humidity

All materials have coefficients that, because of the precision required, are critically important to the operation of drilling machines. Temperature has a profound impact on the accuracy of the drilling machines. With changes in ambient temperature, the materials used to build the machines may expand or contract, changing the position of the drilled hole in the printed circuit board.

Keeping a consistent room temperature is essential for maintaining accurate positioning of the machine. A stable temperature throughout the PCB process assists in decreasing the amount of potential growth and contraction of the substrate materials. The drill room should be kept at a constant temperature and humidity, typically 72 ± 2°F (22 ± 1.1°C) and 45 to 60 percent humidity.

25.4.3 Vacuum

Sufficient vacuum flow is imperative when drilling high-aspect-ratio holes in tightly drilled grids. The vacuum system serves two main purposes:

- To extract debris from the drilling surface
- To keep the tool cool by removing the debris, thus decreasing the amount of friction in the hole wall

Insufficient vacuum flow leads to hole wall quality problems such as gouging, nail heading, and excessive smearing. Table 25.1 provides a hole-quality troubleshooting matrix that addresses these problems. The minimum industry standard for vacuum strength is approximately 20 in. of water measured at the pressure foot.

There are numerous applications such as high-aspect-ratio drilling that require an increased amount of vacuum pressure to remove debris sufficiently, and the equipment manufacturer can provide specific guidelines for special situations. Increasing the z-stroke distance between the pressure foot insert and the top of the drilled stack also assists in cooling the drill bit.

25.4.4 Drill Bit

Drill bits (called *tools*) are critical in the drilling process. A wide variety of tools are available and your tool vendor can be very helpful with selecting feeds and speeds. Be sure to try several vendors to find the best tools for your particular process.

25.4.5 Drill Bit Condition

The condition of the used drill bit is a key indicator of the control and capability of the drilling process and provides an indication of the condition of the drilled hole wall. A wealth of process information can be extracted from the used drill bit. Examining drills before the

TABLE 25.1 Hole Quality Troubleshooting Matrix

Problem	Common causes	Parameter solutions	Process solutions	Possible equipment issues
Entry burrs	Excessive infeed rates	Lower infeed Decrease chip load Increase SFM	Change entry material type and/or thickness	Pressure foot lead set incorrectly; pressure foot insert damaged
Exit burrs	Warped panels Excessive infeed rate/SFM rate	Increase pressure foot, pressure value	Correct warpage issue in lamination; clean area between panel and backer	Verify pressure foot pressure is correct (45 psi)
Nailheading	Chipped/damaged tools Insufficient flute volume Poor drilling parameters	Modify feeds and speeds Decrease SFM	Remove defective drill bits Redefine minimum flute volume values	Machine not running at specified feeds and speeds
Rough holes	Chipped drill bits Poor drill concentricity Poorly cured laminate	Modify feeds and speeds	Inspect used tools for excessive wear patterns Replace worn drill bits Check laminate cure	Machine not running at specified feeds and speeds
Drill bit breakage	Poor drill concentricity Manufacturing defects Excessive drill depth Poor parameters	Check feeds and speeds Correct excessive runout Check z-axis offset values Clean and adjust collets	Check depth into backup Reset pecking depth values Replace drill bits Check program for double hits	Slop in lead screw Machine configured incorrectly Excessive spindle runout Poor machine maintenance
Excessive smear	Undercured laminate Excessive SFM Worn drill bits	Increase infeed rate Decrease SFM	Verify correct laminate cure	Machine not running at specified feeds and speeds

repointing process can be a warning indicator of drilling problems in a real-time mode soon after the process has been affected.

Drill machine operators and drill bit resharpeners should regularly evaluate the condition of used drill bits and report the findings to the process engineers. The areas of greatest concern when evaluating the condition of used drill bits are:

- Excessive margin wear
- Large amounts of burned substrate material attached to the relief areas
- Chips in the primary cutting edges of the tools

When tracked over time, this information can help alert the engineering team of any sudden changes in the drilling process. Although the used drill bit condition information is subjective, with properly trained personnel it can serve as a key indicator of the level of hole wall quality.

25.4.6 Dynamic Spindle Run-Out

The spindle houses the collet, which holds the drill bit that produces the hole in the printed circuit board. Modern spindles are capable of running at a wide range of rpm (15,000 to 200,000) with a power rating near one horsepower (1 hp). Due to the accuracy required when drilling with small drill bits, the drill bit must run true in the spindle. Excessive spindle run-out causes the drill bit to rotate eccentrically, which leads to inaccurate hole location, drill bit breakage, and hole wall quality problems.

The spindle run-out must not exceed 0.002 in. (0.005 mm) total indicated run-out (TIR) in a static situation. This trueness is achieved by using electrically powered spindles that rotate by the means of air bearings rather than mechanical ball bearings.

Other factors, such as poor collet condition and adjustment, can also lead to excessive run-out. Frequent collet cleaning and adjustment are necessary to ensure accurate mechanical hole location. Some modern drill machines are equipped with a tool metrology gauge (TMG). This feature measures the run-out for each drill bit during the tool change process, and if the run-out value measures over the predefined range, the tool is rejected. TMGs can help increase productivity and quality by identifying potential problems before they occur.

25.4.7 Spindle Speed

As via holes get smaller and smaller, maintaining an adequate cutting speed is essential, particularly with chip loads exceeding 0.001 in. per revolution. Increased spindle speed also allows for higher productivity because it corresponds to maintaining consistent chip load values while increasing z-axis feed rates. Spindle speed settings control the rotation of the drill bit during the drilling cycle. Measured in revolutions per minute, speed impacts hole wall quality as well as the condition of the drill bit's cutting edge. Excessive spindle speeds can result in premature wear on the cutting edges of the drill bit and may clog the margin relief area of the tool with burnt substrate (see Table 25.2). Spindle speeds are dependent on the process and the application. The maximum spindle speed capability is a factor when microdrilling, particularly when the desired surface foot per minute (sfm) values are above 300. Note the maximum achievable sfm for the microdrilling diameter range when utilizing the 180,000 rpm parameters for FR-4 material listed in Table 25.2. When the maximum allowable spindle speed is increased to 180,000 rpm, the 300 sfm value can be achieved when microdrilling, as shown in Table 25.2. As the table indicates , the upper spindle speed range accommodates microdiameter drill bits. Table 25.3 shaws sprindle parameters for FR-4 materials at various spindle speeds.

TABLE 25.2 Spindle Parameters for FR-4 Materials at 180,000 rpm

Spindle Speed 180,000 RPM

Desired Cutting Speed: 500 SFM

Tool size (inch)	Infeed rate (in/min)	Spindle speed (rpm)	Chip load	SFM
.0071	077	180,000	.00043	334
.0080	094	180,000	.00052	377
.0100	130	180,000	.00072	471
.0120	134	159,200	.00084	500
.0145	137	131,700	.00104	500

TABLE 25.3 Spindle Parameters for FR-4 Materials at Various Spindle Speeds

Spindle Speed 125,000 RPM
Desired Cutting Speed: 500 SFM

Tool size (inch)	Infeed rate (in/min)	Spindle speed (rpm)	Chip load	SFM
.0071	054	125,000.0	.00043	232
.0080	065	125,000.0	.00052	262
.0100	090	125,000.0	.00072	327
.0120	105	125,000.0	.00084	393
.0145	130	125,000.0	.00104	474

25.4.8 Chip Load

Chip load is calculated by dividing the feed rate by the spindle speed. The term identifies the amount of penetration the drill bit completes per revolution. As commonly used, the chip load accurately describes the proper cutting action needed to drill certain applications successfully.

25.4.9 Surface Speed

The surface speed in sfm is described as the distance that the drill bit's outermost cutting edge travels in 1 min. Cutting speed is frequently used in the metal-cutting industry, and the meaning is the same in the PCB drilling process. Particular substrate materials drill and fabricate favorably at certain cutting speeds. Cutting speed should remain consistent throughout the drill range (except when limited by spindle speed capability). For general guidelines, see Table 25.4.

TABLE 25.4 General Guidelines for Cutting Speed

Cutting speed (sfm)	Material type
200	Plastics, acrylics, soft materials
300	Teflon, Kapton
400	Polyimide, cyanate ester, BT epoxy, high-T_g laminates
500	FR-4, tetrafunctional, multifunctional multilayers
600	FR-4, difunctional rigid, double-sided

25.4.10 Retraction Rate

The z-axis return stroke can be programmed as well. This rate, known as *retraction*, should be set to the value that will minimize the time that the drill bit spends inside the drilled hole. The z-axis stability and drill bit diameter often influence the optimum retract values. An unstable or worn z-axis can lead to poor depth control, poor hole wall quality, and high levels of drill bit breakage. Another advantage of high retraction rates is the speed of the drill stroke.

Increasing production in this case through faster drill stroking is a high priority for the PCB manufacturer. Machine technology allows the user to set retraction rates upward to 1,400 in./min (35,560 mm/min).

25.5 DEPTH-CONTROLLED DRILLING METHODS

There are three common methods of depth control used in the drilling process:

- Manual through-hole drilling
- Machine depth-controlled drilling
- Controlled-penetration drilling

25.5.1 Manual Through-Hole Drilling

Manual through-hole drilling utilizes a down limit value to determine the bottom of the z-axis stroke. This method does not utilize the top of the stack or the top of the backing material as a reference point. Careful consideration should be given when a down limit is set. The operator needs to ensure that the drill bit is penetrating all the way through the stack of laminated panels. Shallow penetration or incomplete holes can result in drilling scrap. Excessively deep penetration leads to increased drill bit cutting edge wear, excess debris to be removed, higher drilling temperatures, and drill bit breakage. The down limit should be set at a value that enables the drill bit to clear the point and also avoids drilling too deeply into the backing material.

25.5.2 Machine Depth-Controlled Drilling

Depth-controlled drilling utilizes the top of the drilled stack as its reference point. Before this depth-control method can be utilized mechanically, the machine needs to know what the distance is from the pressure foot insert to the tip of the drill bit.

To determine this distance, the manufacturer performs a TMG check using standard measuring devices. Once this distance is established, any negative (–) values entered into the depth box will cause the z-axis to stop at the desired depth measured from the top of the stack.

25.5.3 Controlled-Penetration Drilling

Controlled-penetration drilling is utilized for drilling applications where the depth is referenced from the top of the backing material. This mode of depth control is not affected by stack height. The specified depth value or z-axis offset is treated as the desired amount of penetration relative to the top of the backing material. For the machine to be run in mode 3 depth control, a process called *mapping* must be performed prior to drilling any panels. During the mapping cycle, the machine touches down on the surface of the backing in 2 in. increments.

At each touchdown point, the thickness value of the backing is recorded. Once the mapping procedure is complete, the z-axis zero is established. Any negative values entered into the offset box represent the drill depth into the backing material. Controlling the penetration of the drill bit into the backing material assists in decreasing heat generation, improving hole wall quality by limiting the amount of debris, and decreasing drill bit breakage due to excessive depth.

25.6 HIGH-ASPECT-RATIO DRILLING

As circuits have gotten denser, not only have the traces gotten smaller, but the number of layers in a board has increased. The result is smaller holes penetrating thicker boards. To accommodate this increase in hole aspect ratio, two processes have been developed:

- Peck drilling
- Pulse drilling

25.6.1 Peck Drilling

Peck drilling is accomplished by dividing the total z-stroke into separate increments rather than completing the drill stroke in one action (see the example shown in Fig. 25.5).

25.6.1.1 Advantages. The advantages of peck drilling are as follows:

- Decreased drill bit breakage
- Lower aspect-ratio values
- Improved positional accuracy
- Decreased bottom-panel burring

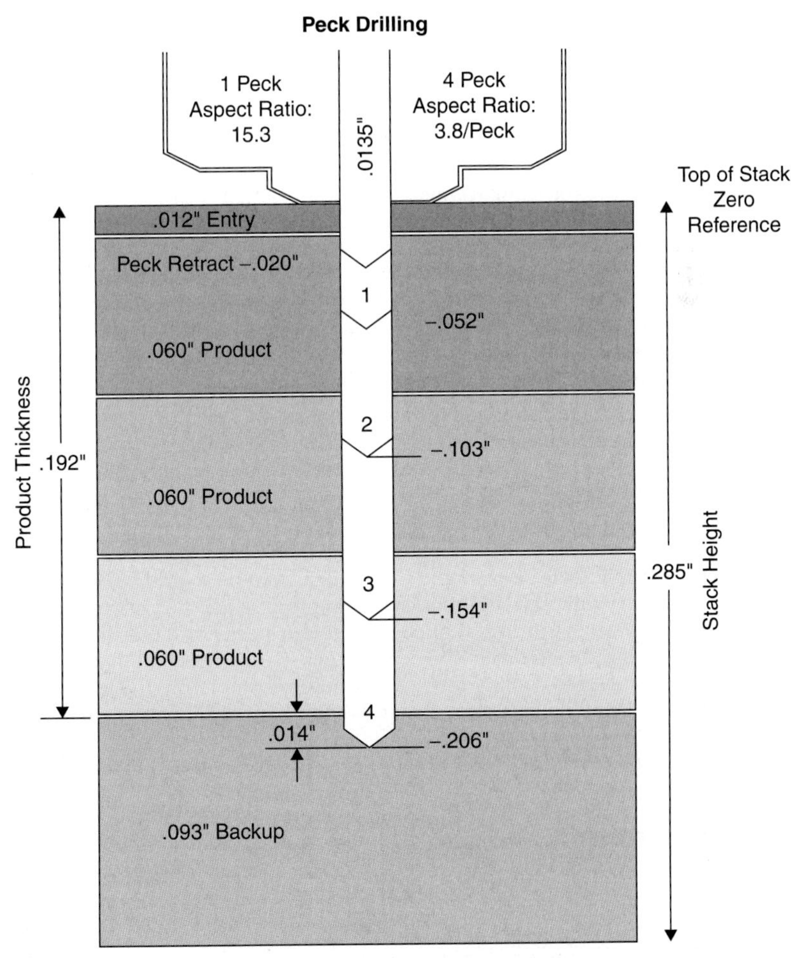

FIGURE 25.5 An example of through-hole peck drilling, showing a stack of three boards with entry and backup material creating a stack height of .285 in. and a total z-axis drill stroke of .206 in. The effective aspect ratio of 15.3 is reduced to 3.8 per peck when four pecks are made to complete the hole.

*25.6.1.2 **Effective Aspect-Ratio Reduction.*** The primary accomplishment of peck drilling is the decrease in effective aspect ratio value. Whereas aspect ratio is defined as the thickness of the board divided by the hole diameter, effective aspect ratio is defined as the total z-stroke divided by the smallest drill bit diameter. For example, if the total z-stroke equals 0.209 in. (the total thickness of the board) and the smallest diameter is 0.025 in., the aspect ratio and the effective aspect ratio would have the same value of 8.36. By taking a total z-stroke of 0.209 in. and dividing it into four pecks, the aspect ratio is decreased from 8.36 to 2.08. In general terms, an aspect ratio scale of 1 through 15 is commonly used, 1 being conservative and 15 being on the aggressive side of the scale.

*25.6.1.3 **Decision to Use Peck Drilling.*** The decision as to when to use peck drilling is very subjective.

Drill bit breakage and bottom-panel burring are ordinarily the main factors considered when electing to peck drill:

- **Decreased drill bit breakage** Decreased drill bit breakage is the primary benefit of peck drilling. If the z-stroke is reduced, the drill bit has less debris to extract per revolution, decreasing the chances of clogged flutes, which can cause the drill bit to seize in the hole and break. Setting pecking parameters is a very subjective science; at best, it is an approximation.
- **Hole wall quality** Analysis of the hole wall quality results should be completed before implementing any pecking procedures.
- **Hole location improvement** Another benefit of peck drilling is an improvement in hole location. Once a drill bit is deflected, it will continue at that deflection angle as it drills through the stack. Essentially, hole location is worse at the exit point than at the entrance point, particularly when drilling through thick panels with small-diameter tools (high-aspect-ratio drilling). Peck drilling can assist in achieving better hole location by lessening the effects of deflection by dividing up the total drill stroke, therefore decreasing the effective aspect ratio value.
- **Bottom-panel burring** Peck drilling can assist in decreasing the amount of bottom-panel burring by reducing the drilling temperatures resulting from drilling thick panels with small drill bits. Plowing, voids, and debris pack are reduced by peck drilling due to the decrease in the amount of debris the tool must extract per z-axis stroke.
- **Drill bit cooling** When peck drilling, each individual stroke increment returns to the upper limit value. For the example in Fig. 25.5, a value of 0.015 in. above the stack was used.

Figure 25.5 is an example of through-hole peck drilling, showing a stack of three boards with entry and exit material creating a total stack height (or z-axis stroke) of 0.285 in. The effective aspect ratio is reduced when four pecks of the drill are made to complete the hole. Increasing the distance above the stack (sometimes referred to as the second upper limit value) assists in cooling the drill bit before completing the next z-stroke. By cooling the drill bit, the manufacturer can substantially reduce nail-heading values and smearing.

*25.6.1.4 **Disadvantages.*** Peck drilling has some disadvantages. For example, it can cause the following:

- Increased nail heading
- Increased smearing
- Hole wall roughness
- Increased cycle times

25.6.1.4.1 Hole Wall Quality. Peck drilling increases the overall cycle time due to the greater number of z-strokes necessary to complete one drilled hole. Nail-heading values tend to be worse for peck drilling due to the certainty that some copper pads will make contact with the drill bit several times as opposed to making only one contact when using a single drill stroke.

Hole wall smearing can also be a problem when peck drilling. The increase in heat caused by drilling the same hole several times can cause the substrate material to become heated to its melting point, which can lead to deposits of resin along the hole wall. A similar effect on hole wall roughness can also be seen.

25.6.1.4.2 Reducing the Effect of Disadvantages. Methods for reducing the negative effects on the hole wall caused by peck drilling include the following:

- Increasing the feed rate decreases the amount of time the drill bit spends in the hole.
- Minimizing the number of pecks also decreases heat generation and reduces overall drilling cycle time.

The design of the tool has a major effect on hole wall quality when peck drilling. To decrease heat generation in the hole wall, the manufacturer must reduce the amount of friction caused by the drill bit making contact with the hole wall. To accomplish this reduced friction, the manufacturer can use an undercut tool design, sometimes referred to as a headed drill bit. The undercut tool design decreases the amount of contact by reducing the diameter of the tool, which creates a larger clearance area, consequently decreasing the amount of flute making contact with the hole wall. The undercut tool design reduces the amount of nail heading and smearing caused by peck drilling.

25.6.2 Pulse Drilling

Pulse drilling is designed primarily as a method to reduce problems related to "bird nesting," where excess debris builds up and remains attached to the flutes and shanks of the drill bit after it is retracted from the drilled hole. This condition also occurs in specialized drilling processes such as high-aspect-ratio drilling, drilling of test fixture materials, and drilling of abrasive materials used in high-temperature applications.

Pulse drilling acts as a chip breaker by driving the z-axis into the material using a short series of steps or pulses.

25.6.2.1 Advantages. Unlike peck drilling, where the tool is retracted completely out of the drilled hole, in pulse drilling the z-axis pauses for milliseconds between each drill stroke while remaining in the drilled hole. This action allows the drill bit to extract debris up the flute before continuing the next stroke. Due to the rapid rate at which debris accumulates, the duration of each pulse must be kept to a minimum, resulting in a decreased infeed rate. To reduce this effect, pulse drilling uses a depth sensor on the drilling machine to detect the top of the stack. Once the top of the stack is detected, the spindle begins the pulse-drilling process.

25.6.2.2 Disadvantages. Pulse drilling increases the amount of heat generated in the hole wall. It should be used only when hole wall quality is not the most significant aspect of the drilling operation.

25.7 INNERLAYER INSPECTION OF MULTILAYER BOARDS

Innerlayers of multilayer boards can shift during fabrication processes and create poor layer-to-layer alignment. During the artwork exposure or fabrication of the multilayer boards, the position of the copper shifts relative to the nominal stacked position. This can create breakout or missing of the copper pads when drilling the interconnection holes.

Two methods of inspecting the innerlayers of multilayer panels exist. One that has been used for several years is x-ray. This technology has progressed significantly and is widely used.

25.7.1 X-Ray

X-ray technology is used to inspect the innerlayers of multilayer PCBs and to verify that the drill pads are located properly. Innerlayer shift is identified easily in an x-ray image, and a simple technique can be used to measure the misalignment accurately. In developing the PCB artwork, a set of stacked pads are located in the coupon area. These are inspected and a "best-fit" centering is calculated.

X-ray inspection also is used to determine movement of each individual layer. In this case, a locating pad with a corresponding layer pad is on each individual layer. The system can measure the distance from the locating to layer pad and determine movement of each layer due to the location of the layer pads. PCBs can be inspected after tooling holes are drilled to verify that they are in the proper locations. In addition, x-ray inspection can be performed after final drilling to verify that the holes are centered within the pads.

25.7.2 Direct Vision

Using direct vision technology, the drilling machine opens a conical hole in the board's outer coupon area (or interior if space is available) and physically examines it with the help of an autofocus/autozoom camera at the copper traces exposed by the cut. These traces represent the various layers and can be measured to indicate their x and y variance from nominal. This is done for each internal layer, and algorithms are applied to generate a new drilling program that will provide for a best-fit drilling scenario.

25.7.2.1 Overview. The autofocus/autozoom camera looks at fiducial targets that are a cross pattern of traces spaced larger than the diameter of the tool used to drill the targets. The cross target should be repeated as a pattern shown in Fig. 25.6 to enable multiple depth cuts for thick panels.

25.7.2.2 Process. This pattern should be stacked on each internal layer and placed outside the circuitry. It is recommended that the patterns be as close to the circuitry as possible without interference to gather the best possible layer movement data.

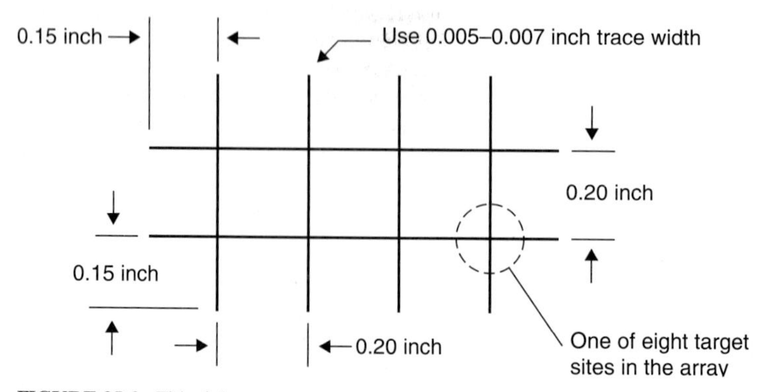

FIGURE 25.6 Fiducial target.

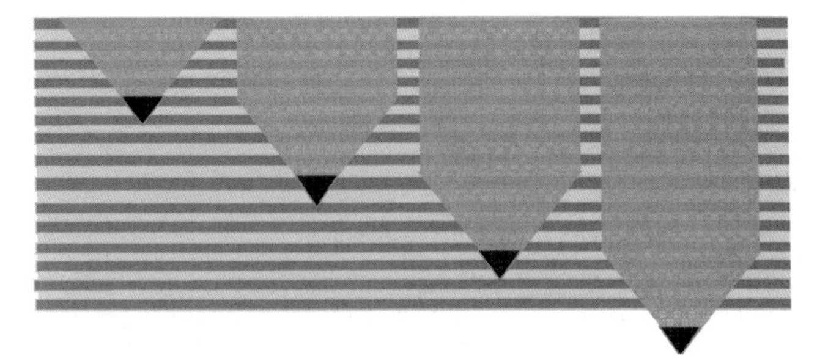

FIGURE 25.7 Multiple entries exposing all layers.

It is also recommended that this pattern be repeated around the circuitry in four places around the corners to enable the most flexible selection of analysis algorithms as described in the discussion of analysis later in this section.

With thick boards, all of the layers may not be visible with one countersink hole. The conical part of the drill bit is the only part where layers are exposed. The tip of the drill does not expose the layer adjacent due to alcohol residue during the cleaning process. Repeat entries at progressive depths expose all of the internal layers, as shown in Fig. 25.7.

After the conical holes are drilled, an automatic alcohol sprayer is activated to clean debris caused by the drilling process. The camera then activates its autofocus/autozoom mode to close in on the subject area.

Figure 25.8 shows the image created by the camera on the first conical hole. You can see the copper layers and some of the movement in the innerlayers.

FIGURE 25.8 Camera-generated conical hole image.

The operator sets the area in which the analysis software should attempt to locate the edges of the traces (see Fig. 25.9). These bounding boxes are mouse-controlled and should be wide enough to enclose all traces.

Gray circles represent the layer for the system to search. The area inside a boundary where the gray circle crosses a trace is searched for trace edges.

The analysis software then locates the trace edges and indicates them with hash marks (see Fig. 25.10). If any of these are incorrect, the operator can easily move the marks with the mouse.

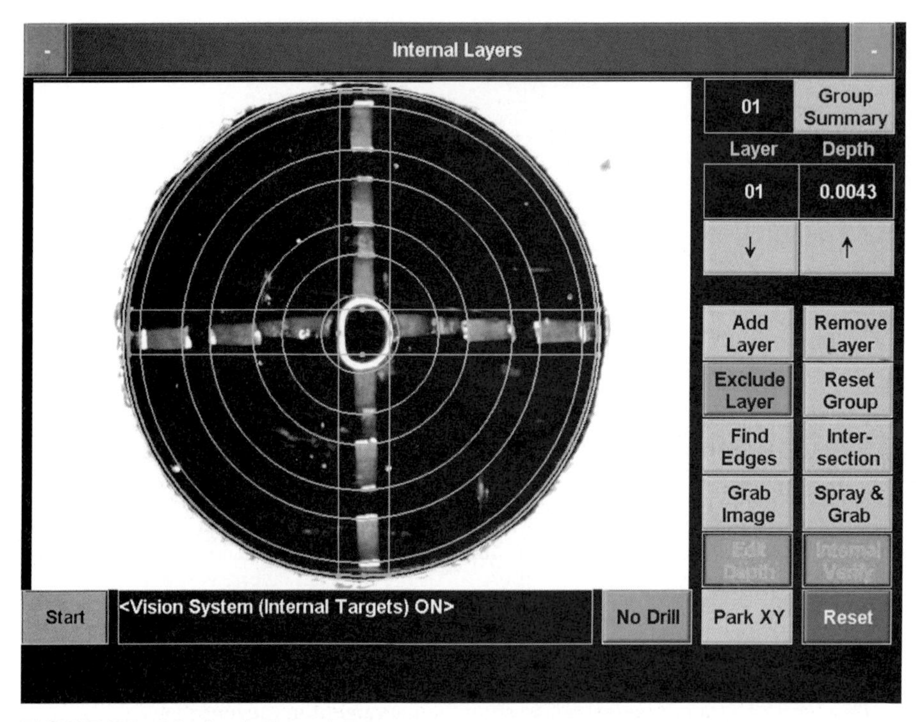

FIGURE 25.9 Selecting the inspection areas.

FIGURE 25.10 Trace edges located.

25.7.3 Correction for Registration Alignment (Direct or X-Ray)

After the data have been recorded by the x-ray or direct image system, the operator can select from several algorithms to perform several functions.

If there were only two corner fiducials inspected, the software can perform the following function:

- **Two-point offset and rotation** Two alignment fiducials are inspected (see Fig. 25.11) to calculate the offset and rotation values. The pattern is rotated to a point located about halfway between the two fiducials. The offset is calculated based on the average distance of the fiducials from their programmed location after the rotation is applied.

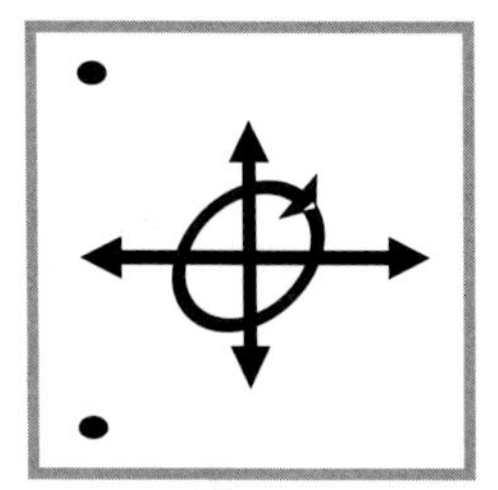

FIGURE 25.11 Two-point offset and rotation.

If three or four cornered fiducials are inspected, the analysis software can perform the following:

- **Three- or four-point offset, rotation, and scaling** Three or four alignment fiducials are inspected (see Fig. 25.12) to calculate the vision corrections. If only three alignment points are used, they may be located on any three corners. The offset is calculated based on the average distance of the fiducials from their programmed location after the rotation is applied.
- **Three- or four-point best-fit** This function is the same as the three- or four-point offset, rotation, and scaling algorithm, except scaling is not applied (see Fig. 25.13).
- **Three- or four-point alignment** The pattern is linearly adjusted in both axes to match the alignment fiducials. This generally transforms the pattern into a trapezoidal shape. Four-point alignment is used to adjust the pattern to match the visible landmarks precisely.

FIGURE 25.12 Three- or four-point offset rotation and scaling.

FIGURE 25.13 Four-point best-fit.

When the selected algorithm is applied, the operator has some choices. One is to elect to go ahead and drill the board on the same machine if it happens also to be an x-ray drilling machine or the Intelli-Drill, since such machines are also high-precision single spindle drillers with a 180,000 rpm spindle. This is the best choice because it does not require the operator to remove the panel from its tooling and all reference points are still aligned.

The operator can, however, elect to remove the panel and output the modified drill program for drilling on another machine, thereby freeing the Intelli-Drill or x-ray system to inspect another panel.

A third option is for the operator to inspect a number of panels and have the algorithms averaged to produce the best-fit program for the entire lot.

25.7.4 Saving Data

The data that are generated can also be stored in an optional database that allows searches to determine, by board type or part number, what is changing over time so that predrill processes can be modified to eliminate error as much as possible. Material types and vendors can also be tracked to verify performance. A utility for basic searches is included with the database option. For more detailed searches, any standard database acquisition program may be used.

25.7.5 Examples of Database Output

Examples of database output include the following:

- Artwork scaling data for each layer on each panel with an average and standard deviation for the order
- Shift data for each panel with an average and standard deviation for the order
- Front-to-back registration error for each layer with average and standard deviation for the order
- Layer-to-layer registration for each panel
- Optimized offset for each panel and average per order or a grouping of similar panels
- A part number history database for trend analysis
- Skew analysis by panel

Figures 25.14 and 25.15 show examples of analysis data provided by the utility.

25.7.6 Process Improvement for the Future

This method is a good tool to increase yields of complex multilayer panels as well as multilayer flex material. The actual layers are exposed and inspected to determine their position with

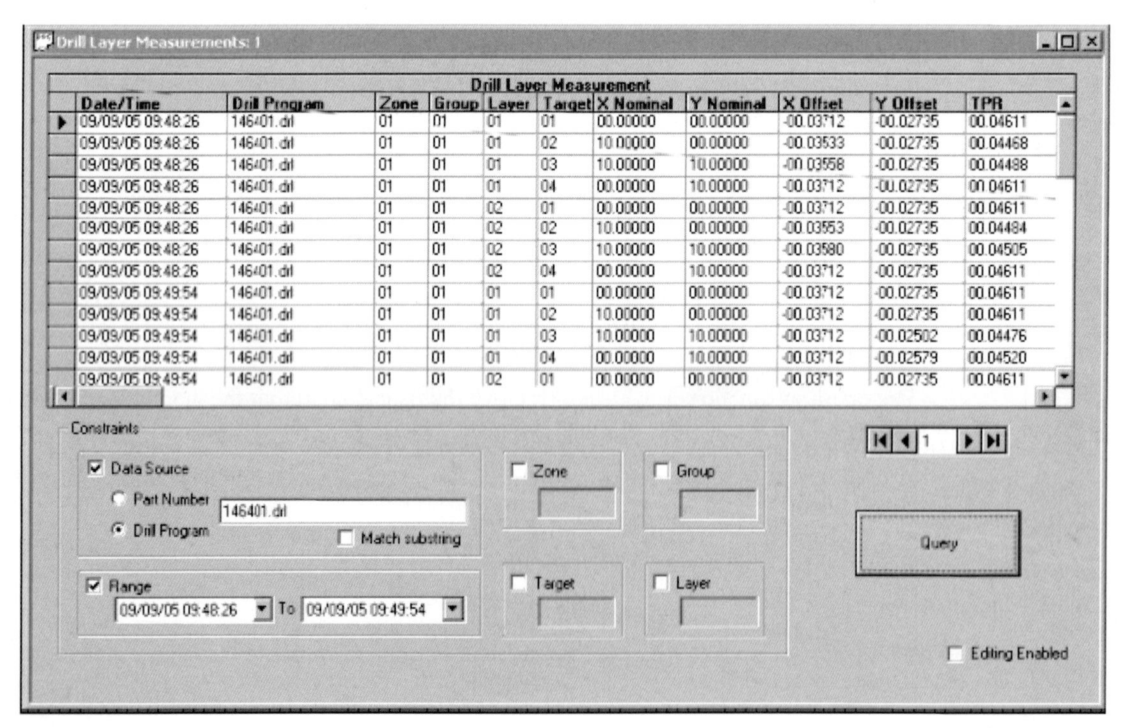

Drill Layer Measurement

Date/Time	Drill Program	Zone	Group	Layer	Target	X Nominal	Y Nominal	X Offset	Y Offset	TPR
09/09/05 09:48:26	146401.drl	01	01	01	01	00.00000	00.00000	-00.03712	-00.02735	00.04611
09/09/05 09:48:26	146401.drl	01	01	01	02	10.00000	00.00000	-00.03533	-00.02735	00.04468
09/09/05 09:48:26	146401.drl	01	01	01	03	10.00000	10.00000	-00.03558	-00.02735	00.04488
09/09/05 09:48:26	146401.drl	01	01	01	04	00.00000	10.00000	-00.03712	-00.02735	00.04611
09/09/05 09:48:26	146401.drl	01	01	02	01	00.00000	00.00000	-00.03712	-00.02735	00.04611
09/09/05 09:48:26	146401.drl	01	01	02	02	10.00000	00.00000	-00.03553	-00.02735	00.04484
09/09/05 09:48:26	146401.drl	01	01	02	03	10.00000	10.00000	-00.03580	-00.02735	00.04505
09/09/05 09:48:26	146401.drl	01	01	02	04	00.00000	10.00000	-00.03712	-00.02735	00.04611
09/09/05 09:49:54	146401.drl	01	01	01	01	00.00000	00.00000	-00.03712	-00.02735	00.04611
09/09/05 09:49:54	146401.drl	01	01	01	02	10.00000	00.00000	-00.03712	-00.02735	00.04611
09/09/05 09:49:54	146401.drl	01	01	01	03	10.00000	10.00000	-00.03712	-00.02502	00.04476
09/09/05 09:49:54	146401.drl	01	01	01	04	00.00000	10.00000	-00.03712	-00.02579	00.04520
09/09/05 09:49:54	146401.drl	01	01	02	01	00.00000	00.00000	-00.03712	-00.02735	00.04611

FIGURE 25.14 Table view.

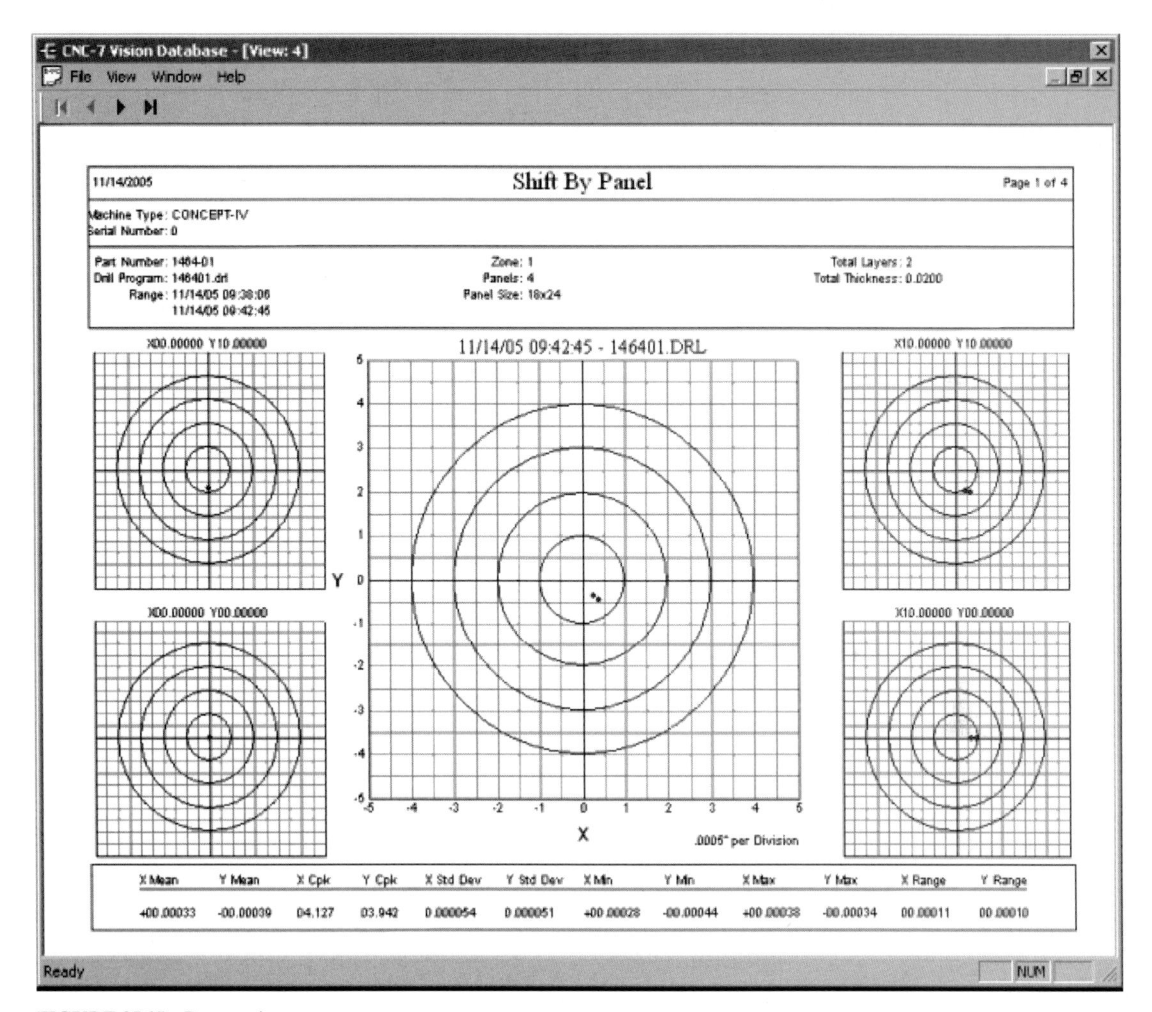

FIGURE 25.15 Report view.

respect to the programmed position. These data are then applied to various algorithms to produce new parts programs that locate the drilled holes to maximize the innerlayer connectivity. The data can be saved in an optional database and accessed later to provide correction data for the CAD, imaging, and fabrication processes.

CHAPTER 26
IMAGING

Brian F. Conaghan
Parelec Inc., Rocky Hill, New Jersey

26.1 INTRODUCTION

Imaging is the process that patterns the metal conductor to form the circuit. This process involves a multistep integration of imaging materials, imaging equipment, and processing conditions with the metallization process to reproduce the master pattern on a substrate. Large features (200 μm and greater) can be very economically formed by screen printing. Feature sizes smaller than 200 μm, however, are formed using a photolithographic process, which is the main discussion of this chapter. As circuit densities have increased over the years, the imaging process has continually evolved to enable commercial production of finer features. This need for high-density interconnect (HDI) is driving the industry to feature sizes of 10 μm or lower. Imaging equipment and materials have been developed to meet this challenge.

To achieve a reproducible high-yield/low-cost imaging process for a given feature size, various factors must be carefully balanced. This chapter outlines the chemistry and equipment options for the photolithographic imaging process, highlighting trade-offs that must be considered so that process engineers, designers, and procurers of printed wiring boards (PWBs) and HDI structures have an overview of the process considerations that can enable a manufacturable product.

Printing is an alternate process for imaging. Screen printing is used to print resist on foil and silver inks on polyester. Ink jet printing has been used to pattern resist, legend, and silver conductors. Recently an ink jet system has been introduced that prints a UV-curable catalyst layer to electroless plate copper at thicknesses of 0.05 to 5 microns.[1] Significant development work is underway on organic transistors patterned by ink jet and other printing methods. These techniques are beyond the scope of this chapter.

The photolithographic process sequence for imaging is given in Fig. 26.1. Basically the process

Clean Surface

⇓

Apply Photoresist

⇓

Expose Photoresist

⇓

Develop Photoresist Image

⇓

Pattern Transfer Image
(plating or etching)

⇓

Strip Photoresist

FIGURE 26.1 The photolithographic process sequence.

involves applying a light-sensitive polymeric material (photoresist) onto the substrate of interest, exposing this photoresist to light with the desired pattern, and developing the exposed pattern. This developed pattern is used for either subtractive (etching) or additive (plating) metal pattern transfer. Details of these metallization processes are given in subsequent chapters. After metallization, the photoresist is stripped from the surface and the panel is ready for further processing.

26.2 PHOTOSENSITIVE MATERIALS

The photosensitive polymeric systems used as photoresists in the interconnect industry are either liquids or dry films formed from a liquid solution. Dry-film photoresists are the industry standard, but liquid photoresists have become widely used, particularly for fine-feature circuits. Both types of photoresist can be used for a variety of processing requirements.

26.2.1 Positive- and Negative-Acting Systems

Photoresists may function in a photographic sense in either a positive or negative tone. The difference, illustrated in Fig. 26.2, is a result of the specific chemical reactions that occur on exposure to light. For the common negative-acting systems, exposure initiates cross-linking

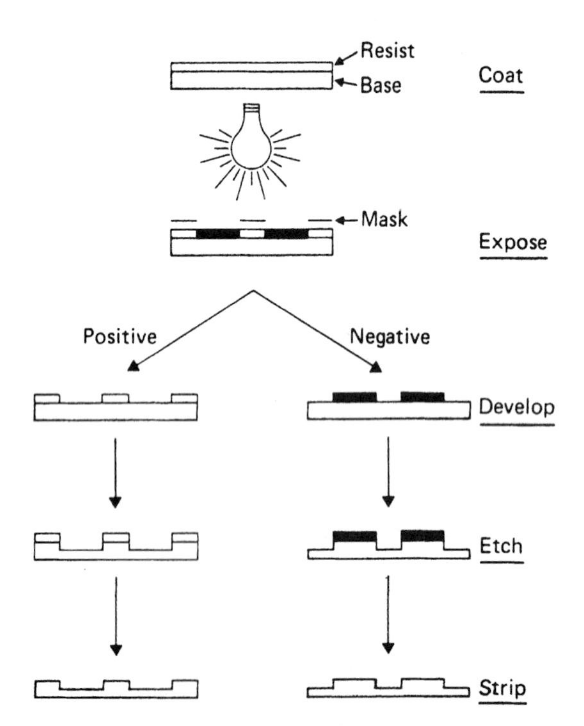

FIGURE 26.2 The photolithographic process sequence for positive and negative tone photoresists. *(Reprinted by permission of C.G. Willson,* Introduction to Microlithography, Theory, Materials and Processing, *ACS Symposium Series 219, Washington, D.C., 1983, p. 89.)*

between one or more of the components in the polymeric matrix, reducing the photoresist's solubility in the developing solution. The exposed regions remain after development of the image. For positive-acting photoresists, the chemistry is often the novalac resin-based materials utilized by the semiconductor industry. Upon exposure, an acid-catalyzed reaction occurs that increases the solubility of the photoresist in the developer solution. In this case, it is these exposed regions that are removed in the developer solution.

Photographic tone can have an effect on product yield caused by contamination between the phototool with the master image and the photoresist. For a negative-acting photoresist, contamination blocks the light and prevents cross-linking, leading to a mousebite (partial reduction in line width) or an open circuit in the conductor after etching. For a positive-acting photoresist, contamination blocks the light and prevents the acid-catalyzed solubilization, leaving excess metal and a possible short after etching. Since circuit patterns often have a greater area of spaces than lines, there is a lower probability of a contaminant causing an actual defect with the positive-acting photoresist. However, negative-acting photoresists are the most widely used since many other factors impact yield and contamination must be minimized in any photolithographic process..

26.2.2 Decision Factors

The large variety of commercially available photoresists makes it a challenge to determine which one is likely to perform best for a given application. Key technical and economic factors that must be considered are summarized in Fig. 26.3. The primary consideration is the end use of the pattern.

- The photoresist must be chemically compatible with the subsequent pattern transfer steps (i.e., can withstand the chemical environment such as the solution pH) for it to function as an accurate mask for pattern transfer.

Key Considerations in Photoresist Selection

Metallization Process
- Etching
  - Acid
  - Akaline
- Plating
  - Copper
  - Tin
  - Tin/Lead
  - Gold
  - Nickel
  - Full-Build Electroless Copper

Panel Design
- Substrate Topography
- Required Copper Thickness
- Number and Size of Holes
- Minimum Feature Sizes

Economic Factors
- Compatibility with Existing Application and Exposure Equipment
- New Capital Equipment Investment Required
- Productivity
- Process Latitude for High Yield
- Material Cost

FIGURE 26.3 Criteria for selection of photoresist materials.

- The nature of the substrate itself in terms of topography and surface features must also be considered. Is the photoresist required to cover or tent plated through holes or tooling features, and must it conform to existing circuitry? This consideration will dictate if dry-film or liquid photoresists are appropriate and, in some instances, the required tone.
- The required product feature dimensions are equally important. They are usually specified in terms of line width and spacing with a given allowable variation. Since these product specifications are for the final conductor features, the impact of photoresist processing on the critical dimensions must be established. Each photoresist has an inherent contrast (i.e., the rate of change of solubility on exposure to light). This contrast, combined with the resist thickness, phototool, light source used for exposure, and developing conditions, determines the finest feature that can be imaged. For photoresists used in the PWB industry, the minimum line width achievable in production is usually 10 to 25 μm greater than the photoresist thickness. Novel process sequences have demonstrated it is possible to overcome this traditional barrier.[2]
- The adhesion of the photoresist is another key attribute. The chosen photoresist must have good adhesion to the underlying substrate during pattern transfer but must also be able to release cleanly from the substrate during stripping.

26.3 DRY-FILM RESISTS

Dry-film photoresists are generally used for pattern formation prior to both plating and etching.

These films have a three-layer structure, with the photosensitive material sandwiched between a polyester coversheet and a polyethylene separator sheet (see Fig. 26.4). The photoresist thickness is typically 25 to 50 μm, but thinner and thicker films are used for special applications. The photoresist is applied to the printed wiring board (PWB) substrate using lamination: The polyethylene separator is removed, and the photoresist flows under the heat and pressure from the laminating rolls to adhere to the substrate. The polyester coversheet

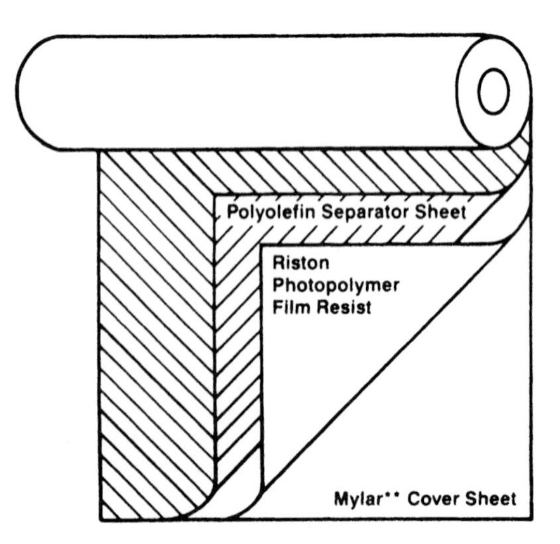

FIGURE 26.4 Dry-film photoresist components. *(Reprinted with permission of DuPont Electronics, DuPont Technical Literature.)*

performs several key functions. It protects the photoresist during handling, prevents sticking to the phototool during contact printing, and acts as an oxygen barrier during exposure. The optical properties of this polyester coversheet are important, especially for producing features less than 100 μm.

Dry-film manufacturers have increased the optical clarity and reduced the thickness of the polyester sheet to achieve the required image quality. Thin dry films of 25 μm thickness and less generally use this high-quality coversheet.

The chemical and mechanical properties of dry-film photoresists have been tailored to withstand various plating solutions and acid or ammoniacal-etching solutions. They can be grouped in terms of the processing chemistry used to develop the image:

- **Aqueous-developable** This group is the industry standard.
- **Semiaqueous-developable** These are used in special applications.
- **Solvent-developable** This is the original system (see Table 26.1). Since solvent development used chlorinated hydrocarbons, use of the solvent-developable photoresists has generally been abandoned.

TABLE 26.1 Summary for Dry-Film Photoresist Types and Chemistries

Photoresist type	Develop solution	Strip solution	Application
Aqueous-processable	Sodium/potassium carbonate	Sodium/potassium hydroxide	Acidic solution: etching, plating Limited alkaline solutions: alkaline etch
Semiaqueous-processable	Sodium tetraborate/ Butyl cellusolve	Ethanolamine or Butyl carbitol/Butyl cellusolve	Alkaline solutions: etching, precious metal plating
Solvent-processable	(Methyl chloroform) Alternate to CFCs	(Methylene chloride) Alternate to CFCs	Strong alkaline: full-build electroless Cu plating

26.3.1 Chemical Composition Overview

Dry-film photoresists vary in their exact chemical compositions, but all contain the following components:[3]

- **Polymer backbone**
- **Photoactive compound**
- **Monomer**
- **Dye**
- **Additives** The polymer backbone sets the fundamental solubility and chemical resistance of the system.

On exposure, the photoactive compound absorbs light of the appropriate wavelength and subsequently reacts with the monomer to alter the solubility of the film in exposed areas. The dye changes color on exposure to provide a visible latent image of the master pattern in the film, referred to as the printout image. Although the printout image decreases the efficiency of the desired photoreaction by consuming a portion of the exposed photoactive compound, it has proven to be a valuable manufacturing aid and is used in almost all dry-film photoresists.

Additives include adhesion promoters, flexibilizers, and other compounds that can improve a desirable property. All the ingredients function together to provide the process latitude needed for the photoresist's specific range of applications.

26.3.2 Aqueous-Processable Dry Films

The majority of materials used are based on an acrylic polymer with various forms of photoreaction initiation. The initiator's absorption is designed to coincide with the major emission wavelength of the mercury arc lamps in the ultraviolet region of the spectrum at 365 nm.

Many are based on the use of Michler's ketone and its subsequent sensitization of a triplet state promoted radical chain polymerization. This also explains the material's sensitivity to oxygen, an efficient triplet quencher. These materials are also very efficient at converting light energy into a chemical reaction due to the radical chain reactions; absorption of one photon of light results in many cross-links in the polymer matrix. Usual exposure doses measured as the irradiance from 330 to 405 nm to cross-link these materials functionally vary from 25 to 90 mJ/cm^2 depending on the exact chemistry used and the thickness of the materials, which compares favorably to positive-acting materials used for optical lithography in semiconductor integrated circuit (IC) production, where doses of 200 to 500 mJ/cm^2 are common.

The images in these materials are developed in alkaline solutions of 1 percent or less by weight of sodium or potassium carbonate. The complete removal of the photoresist after pattern transfer is accomplished in 1M or greater solutions of sodium or potassium hydroxide at elevated temperatures, often with an antitarnishing additive to limit oxidation of the copper.

Thus, in general, these materials exhibit excellent stability in acidic solution, but they do have varying stability to more strongly acid and alkaline solutions. In fact, there are three subclasses of materials: for acid etching of copper, for acid plating of copper, and for alkaline (ammoniacal usual) etching of copper. Increased acid stability is required for use in acid copper plating, in which the pHs of the solutions often are below 1. For ammoniacal copper etching at pH 8 to 9, increased alkaline stability is required. This latter capability illustrates the flexibility in tailoring these materials to their end use since these materials are imaged using similar alkaline chemicals: develop at pH 10.3 and strip at pH 13. Often these alkaline-stable materials are developed at high temperatures.

In the past, several photoresists were formulated for exposure at a visible wavelength, 450 nm, using specialized exposure equipment—either visible lasers or magnified projection printing. More recently, laser direct imaging (LDI) based on the Ar^+ laser output clustered around 360 nm has been commercialized. Conventional photoresists can be exposed on LDI equipment, but higher photospeed has been key to the productivity and economics of LDI.

Photoresists requiring only 10 mJ/cm^2 are now commercially available for the major applications. In general, these high-speed photoresist processes are similar to conventional ones, but improvements include photospeed, sensitivity to yellow safelights, and postlamination hold.[4]

Novel aqueous-developable photoresists have been formulated to process in dilute acidic media. A dry film based on electrophoretic depositable (ED) materials was shown to have excellent stability in strong alkaline solutions, potentially useful for etching polyimides or for full-build electroless copper plating.[5]

26.3.3 Semiaqueous- and Solvent-Developable Dry Films

Semiaqueous-developable photoresists are used when increased resistance to chemical attack is required, such as the highly alkaline solutions used in polyimide etching or gold and precious metal plating applications. Their composition is similar to the aqueous-developable photoresists, but the polymer backbone is less alkaline-soluble. This increased chemical stability provides accurate image transfer by maintaining the integrity of the surface and sidewalls and minimizing underplating. Semiaqueous-developer chemistry is slightly alkaline (sodium tetraborate) and contains an organic assist, typically butyl cellusolve. Process control for this mixture is more complicated than that used for aqueous development.

Solvent-developable dry-film photoresists were the first to be commercially produced. Since they are the only materials that can withstand the harsh conditions in some chemical processing sequences, they still have some interest. They are based on methyl methacrylate polymers and a Michler's ketone initiation system similar to that used for aqueous-developable photoresists. Although they can have very high contrast, they are prone to scumming, leaving residue in the spaces between exposed photoresist features after development. Slight etching of the underlying substrate is often required to have a clean image-transfer step, especially for pattern-plating applications. Traditionally these materials were developed in trichloroethane and stripped in methylene chloride. However, as the hazards associated with chlorinated hydrocarbons were recognized, use of these solvents has been practically eliminated. This has forced the use of alternative develop and strip chemistries for solvent development.[6]

26.4 LIQUID PHOTORESISTS

Liquid photoresists come in a variety of chemistries.[7] Many are simply coatable forms of the dry-film materials. They have been used primarily for patterning innerlayers and are applied to the substrate by roller, spray, curtain, or electrostatic coating techniques. These methods are discussed later in the chapter (with equipment). Applied thickness is typically 6 to 15 μm, giving superior resolution compared to dry-film resists that are generally thicker. Since resolution is approximated by the thickness, they have the ability to resolve features less than 25 μm. Conformation of liquid photoresist to the copper substrate is generally very good, and it is possible to achieve high yield at 50 μm features and less.[8] However, cleanliness of the substrate and the coating and drying operation is critical. Liquid photoresists are dried to allow for hard contact with the phototool during exposure, maximizing resolution. If soft or off contact is used, resolution and yield decrease, especially for features 100 μm or less. For the majority of liquid photoresists, the developing and stripping conditions are similar to those for aqueous-developable dry films: mild alkaline sodium carbonate and strong alkaline sodium hydroxide. Exceptions are positive-acting novolac-based photoresists, thermal photoresists,[9] and negative-acting acid-developable photoresists.

26.4.1 Negative-Acting Liquid Photoresists

The chemistry of the typical materials is similar to that of the aqueous-developable dry films. They depend on a radical chain reaction to promote cross-linking in the acrylic polymer. Their sensitivities differ depending on the initiator system chosen. Materials are available with minimal solvent content, and a few are diluted with water.[10,11] This reduces the emissions from coating and drying that must be contained and treated.

Acid-developable photoresists mentioned in the dry-film aqueous section can also be formulated as electrophoretic-depositable liquids. These materials have high contrast, good resolution, and chemical stability in strongly alkaline solutions since they are developed in aqueous acid solutions.

26.4.2 Positive-Acting Liquid Photoresists

The other alternative is a positive tone novolac resin-based system that finds its application most usually in the semiconductor industry for imaging integrated circuits. It has excellent resolution and chemical stability in acid and a large number of basic solutions (precious metal

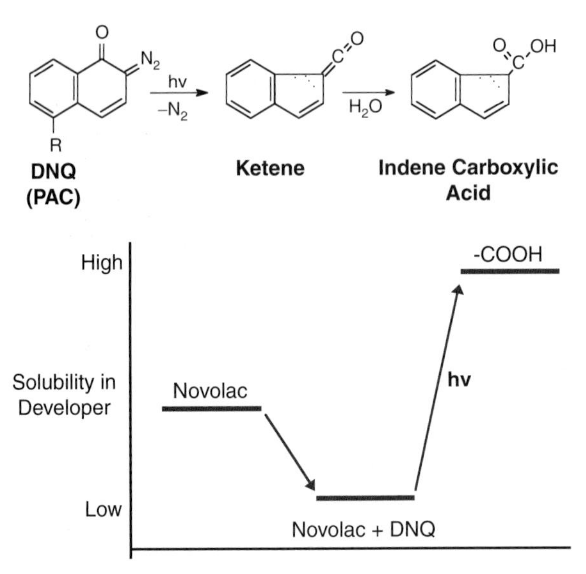

FIGURE 26.5 Chemistry mechanism of positive-acting liquid photoresists.

plating). The chemistry involves the conversion of a base-insoluble diazonaphthalquinone to the base-soluble indene acid. Thus, this photo-driven conversion alters the solubility of the base material in alkaline solutions, allowing for image formation (see Fig. 26.5).

26.5 ELECTROPHORETIC DEPOSITABLE PHOTORESISTS

Electrophoretic depositable (ED) photoresists[12,13] were introduced in the late 1980s with expectations of broad adoption by the interconnect industry. The early materials were modifications of those used for coating automobiles and appliances. These fast-coating processes provide a defect-free film of high durability on conductive substrates and are well suited to large-volume production. Anodically and cathodically deposited materials that are both positive and negative acting have been introduced but have achieved limited market penetration in the interconnect industry. Advances in dry-film photoresists and liquid photoresists reduced the interest in implementing a new photoresist process. (A detailed discussion of electrophoretic depositable photoresists is included in Sec. 17.1.6 of the fourth edition of this handbook.)

26.6 RESIST PROCESSING

Imaging is a sequential process, and the various steps are very interdependent. Equipment and processing conditions play a large role in the image quality, yield, and productivity that can be achieved. The choice among the numerous options for each step depends on the type of substrate to be imaged, the feature sizes to be produced, the existing equipment, the type of photoresist, and the productivity and economics of the production line. Alternatives for each process step are described and compared.

26.6.1 Cleanliness Considerations

Cleanliness is an issue throughout the entire photolithography process, including yellow room contamination level (cleanliness class), solution filtration, equipment cleanliness class, and product handling. Its importance increases as feature size decreases. The level of cleanliness needed to give acceptable yield for the finest features must be established and maintained.

Sticky rollers, either as free-standing equipment or as handheld items, are one type of equipment used to remove contamination from flat surfaces, either panels or equipment.

26.6.2 Surface Preparation

Good adhesion between the photoresist and the substrate is essential for the remainder of the process to be successful. All laminates arrive at the imaging area, dirty, and the exact process sequence for surface preparation is chosen based on the nature of the contamination. Epoxy dust generated during trimming and processing of the laminate is removed by mechanical cleaning. The copper surface is either foil or foil with electroless or electroplated copper. Foil typically has an antitarnishing agent of chromium and zinc that must be removed for reproducible imaging results. During cleaning, the texture or surface roughness of the copper is altered, promoting mechanical adhesion between the copper and the photoresist. The extent of this texturing is measured by the usual parameters for surface roughness, the height and spacing of the topography measured by profilometry (see Fig. 26.6).

R_a Arithmetic mean of all departures from the mean line

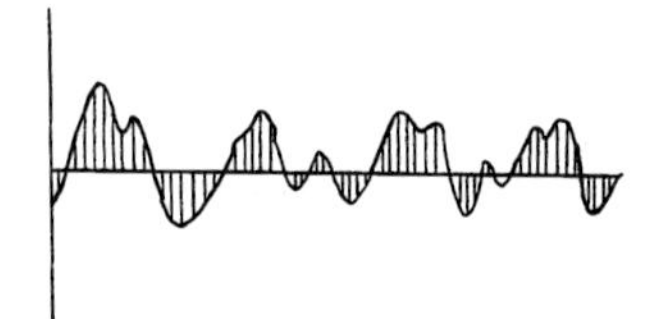

R_{max} Maximum peak to valley height within sampling area

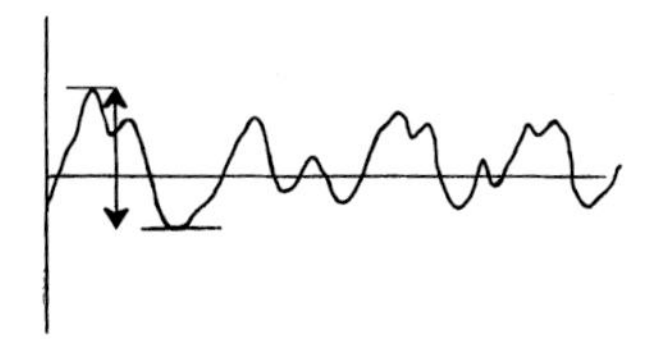

S_m Mean spacing between peak measured at the mean line

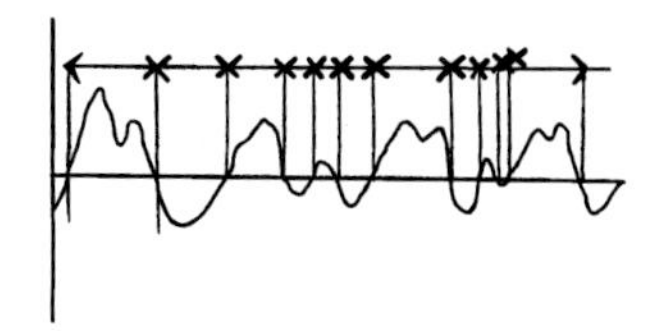

FIGURE 26.6 Surface roughness parameters.

TABLE 26.2 Copper Surface Chemical Composition as Measured by X-Ray Photoelectron Spectroscopy (XPS)

Sample	% Cu	% O	% C	% N	% Zn	% Cr
Initial		46	24		16	12
Preclean 1	12	19	68			
Preclean 2	6	10	70	14		

Chemical cleanliness can be determined in a variety of ways. Wetting is often tested using a *water break test* or more analytically in terms of contact angle, with a low value being desirable.

In addition, analytical techniques (Auger and x-ray photoelectron spectroscopies) can be used to evaluate the chemical composition of the surface (see Table 26.2).[14]

26.6.2.1 Mechanical Cleaning. Most mechanical cleaning equipment physically abrades the copper surface using a pumice or aluminum oxide slurry either directly (scrubbing or vapor blasting) or impregnated into brushes. Scrubbing provides an even final surface texture, whereas brushing creates grooves (see Fig. 26.7). The copper thickness must be sufficient to provide the desired final conductor height after cleaning. Mechanical cleaning is more difficult for very thin flexible panels.

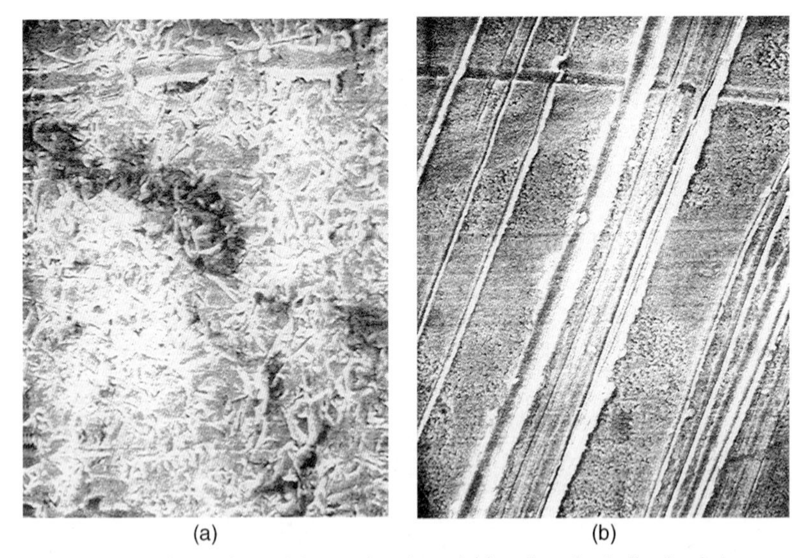

FIGURE 26.7 Comparison of the pumice cleaned (a) and mechanically abraded copper surfaces (b), 1200X. *(Reprinted with permission of D. P. Seraphim, R. C. Lasky, and C-Y Li,* Principles of Electronic Packaging, *McGraw-Hill, 1989, p. 383.)*

The texture of the mechanically cleaned surface affects the adhesion between the photoresist and the copper. The resist should conform to the topography, and thus, the extent and type of texture must be tailored to the type of photoresist. If the gouges are too deep, a dry film applied with traditional lamination will have difficulty conforming to the surface and create defects: near or full opens after etching, and underplating or shorts after pattern plating. Thus, the cleaning process, photoresist type, and application method must be matched to image at high yields.

Conveyorized equipment is used that passes the panels through the brushes or slurry and then to rinse chambers before exiting through a dryer section. If rinsing is insufficient, residual

TABLE 26.3 Cleaning Solutions Used to Remove Various Contaminants

Cleaning technique	Process chemistry	Contaminant removed
Abrasion	Pumice, Al oxide	All
Plasma	CF_4/O_2, O_2/H_2O	Organic
Chemical solutions (aqueous etchants)	H_2SO_4, HCl, etc.	Inorganic
Chemical solutions (nonaqueous)	Alcohols, etc.	Organic
Thermal	N_2	H_2O

Source: Adapted from *Introduction to Microlithography,* 1st ed., L. F. Thompson and M. J. Bowden, American Chemical Society Symposium Series 219, Washington, D.C., 1983, p. 184.

pumice can remain on the copper surface. Maintaining proper functioning of the mechanical parts that contact the product is essential since brushes and slurry deteriorate with use. Equipment designs that isolate the abrasive from the majority of the tool are more durable.

26.6.2.2 Chemical Cleaning. Chemical cleaning includes a variety of solutions since the solubility of contaminants can be quite different (see Table 26.3). Grease and fingerprints require a soap solution or solvent to dissolve them. Antitarnish treatment removal and copper roughening is done with mild etchants for the copper, such as ammonium persulfate, sodium persulfate, or peroxide/sulfuric acid. Depending on the oxidation level and the amount of antitarnish treatment, there is often an induction time for copper removal. Solutions used can be common chemicals or proprietary mixtures in which surfactants and other additives have been included. Alternatively, the oxide and antitarnish treatment can be removed initially, followed by a surface roughening. In this case, the oxide removal is accomplished with mild acidic cleaning solutions such as sulfuric acid.

The selection of the cleaning solution also depends on the copper thickness. Thin "seed" layers used for electrolytic pattern plating can tolerate very little etching, and extremely dilute solutions or dry methods are used. Thus, chemical adhesion can be more important than mechanical.[15] Selection of the proper process chemicals and sequence depends on the overall conductor formation process and the photoresist to be used.

Chemical cleaning sequences are often contained in conveyorized spray equipment with rinsing between each step. Batch processing in a tank system is also used, with a hoist for basket movement between solutions. Uniformity of the etching of panels within a basket must be measured, but is generally quite good and reproducible as long as the immersion time is short and the bath's chemical composition is consistent.

26.6.2.3 Electrolytic Cleaning. A unique inline conveyorized tool built by Atotech uses electrolytic cleaning.[16,17] The antitarnish is removed electrolytically along with oils and fingerprints in a uniform process with very little copper removal. The surface texture is then altered by microetching and passivated prior to dry-film photoresist lamination or other coatings.

26.6.3 Photoresist Application

The technique used to apply the photoresist to the substrate depends on the type of photoresist selected. The various techniques used with dry-film and liquid photoresists are discussed.

For any of these techniques, the cleanliness of the operation—both the equipment and the room—and the handling of the product are critically important, especially for high density. It is also advisable to minimize handling at this stage by using automated inline equipment for the preclean step and automatic loading and unloading.

26.6.3.1 Dry-Film Hot Roll Lamination. In this process, both temperature and pressure are used to laminate the dry-film photoresist to the panels. After stripping off the polyethylene separator sheet, the photoresist is contacted to the substrate in a nip between two heated rolls. Heat reduces the viscosity of the photoresist and pressure causes it to flow, conforming to the irregularities of the copper surface.

The weave of the laminate affects the topography of the substrate and how much flow is required to achieve good conformation, an essential condition for high yield. Dry-film photoresists are tailored to have rheology that maximizes flow during lamination while preventing flow in the roll prior to lamination, which could cause edge fusion. Panels are often heated with heated rolls and infrared heaters prior to lamination to improve photoresist conformation. The key process parameters affecting conformation are laminator speed, laminating roll temperature, and rigidity of the rolls with respect to the core material and rubber durometer. For aqueous-developable materials, wet lamination has proven to be effective in improving conformation and yield for innerlayers. By using a water-alcohol mixture, wet lamination can also be applied to tent and etch inner via holes to minimize the land diameter.[18]

Both automatic and manual tools are available for film lamination (see Fig. 26.8). With manual tools, the panel is placed in the nip and pulled through by the rotation of the rollers.

The continuous film of photoresist must be trimmed for each panel on at least two sides. This technique is labor-intensive and extremely dirty due to the chips of resist generated by trimming.

However, it is useful for very thin materials and nonstandard-size panels in small lot sizes. With automatic equipment, the basics are the same but the panel enters from an automatic loader or a conveyorized preclean tool directly into and through the dry-film laminator.

This is a relatively clean operation with little handling of the panels since the photoresist is automatically trimmed to the appropriate size and placed within the board edge.

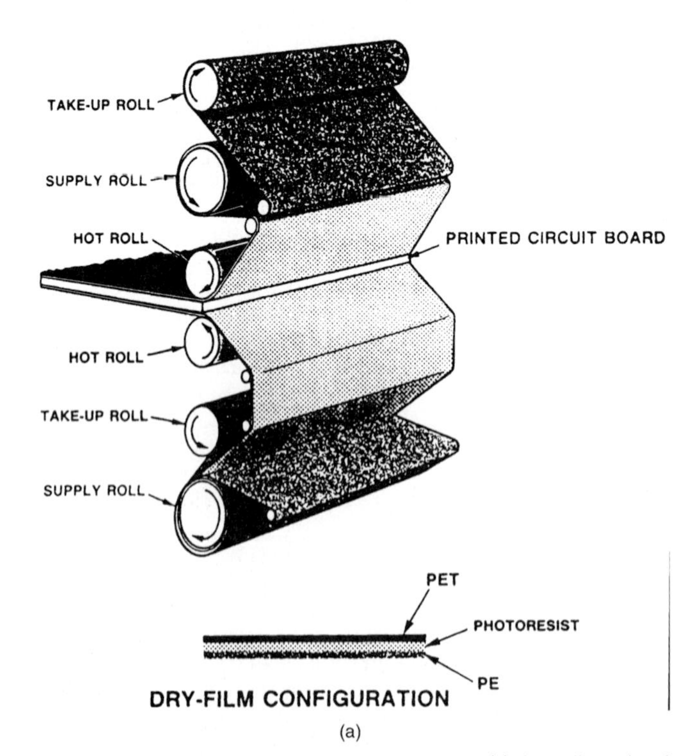

FIGURE 26.8 Configuration of a hot roll laminator: (a) three-dimensional schematic; (b) two-dimensional schematic. (*Reprinted with permission of E. S. W. Kong,* Polymers for High Technology, *ACS Symposium Series 346, Washington, D.C., 1987, p. 280.*)

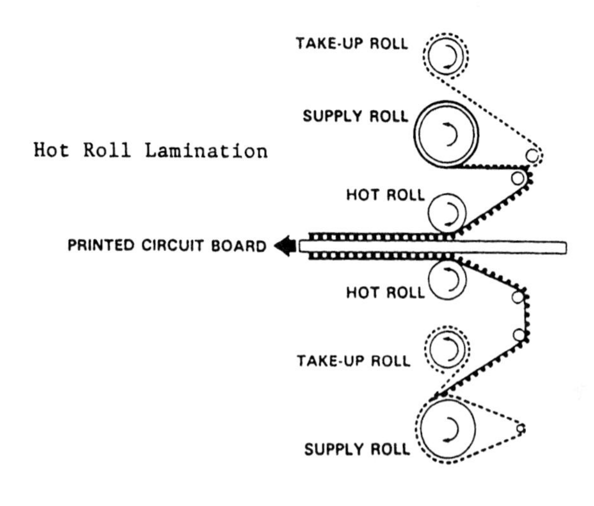

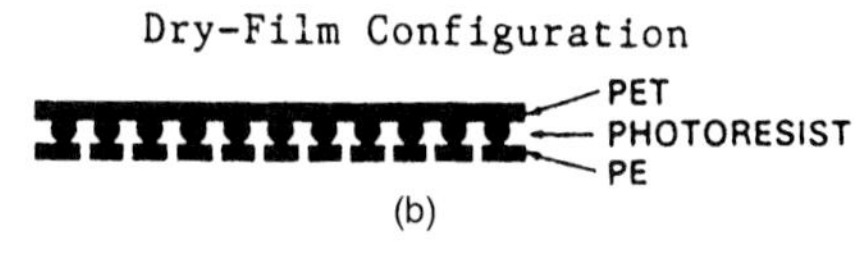

(b)

FIGURE 26.8 (*Continued*)

Equipment is available to convey even very thin panels. Panels are often heated, wetted (wet lamination), or cleaned before the actual lamination.

26.6.3.2 Dry-Film Vacuum Lamination. This method is similar to hot roll lamination but the temperature and pressure used to adhere the photoresist to the substrate is applied by a heated vacuum platen instead of laminating rolls. Vacuum lamination is useful for products with tall or closely spaced features that make it difficult for the photoresist to conform. The vacuum removes the air that would be trapped and pulls the photoresist into tight spaces, providing conformal coverage. This equipment is also used for material that will deform nonuniformly under the pressure of laminating rolls, such as thin unreinforced polyimides. With vacuum lamination, the pressure is evenly distributed over the panel, improving dimensional stability control.

26.6.3.3 Liquid Coating. A great variety of methods are used to apply liquid photoresist to panels. Roller, spray, electrostatic, and electrophoretic coating provide double-sided application and are well suited to high-volume manufacturing. Curtain coating is single-sided, requiring either two passes of the panel through the equipment with slight drying between coatings or an inline equipment layout with two coating and drying units. This causes a difference in the solvent content of the two sides. Screen-coating equipment is available for either single-side or double-side coating. In all instances, the cleanliness of the surroundings, equipment, and solutions is essential for high-yield processing. Most equipment is designed to minimize material consumption by collecting and filtering excess liquid so it can be recycled to the feed tank.

26.6.3.4 Roller Coating. This method places liquid photoresist on the board by transferring the liquid from one set of rollers to another (see Fig. 26.9). The exact number, physical configuration, and surface of the rollers vary, as well as how the material is metered. Often, gravure roller coating is used. Grooves or crosshatches in the rollers that contact the panel determine the amount of liquid deposited on the substrate and are set with a measured overlap or interference so that the board is squeezed between the rollers.[19] Photoresist viscosity also influences the coating thickness applied. Improvements in panel handling and in confining the solution to

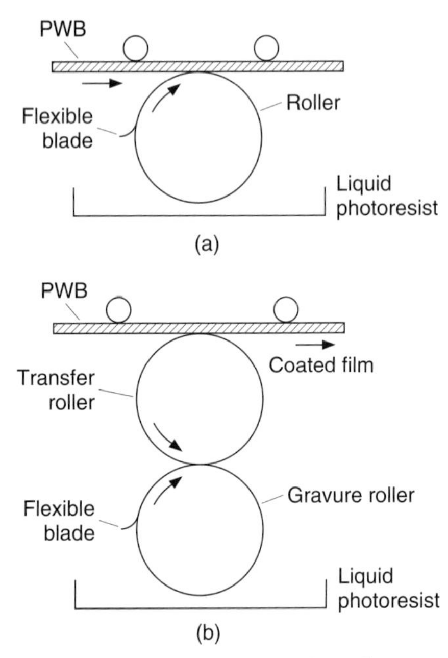

FIGURE 26.9 Configuration of a roll coater: (a) direct coating; (b) indirect coating.

the center of the panel allow for simultaneous two-sided "postage stamp" coverage similar to that obtained with dry-film photoresists, leaving the tooling and location holes clear of material. In commercial systems, the coater is conveyorized, with panels entering a clean oven immediately after coating. The throughput of these systems can be as high as 240 panels per hour.

26.6.3.5 Spray Coating. Although very thin and uniform coatings are possible with this method, its use in resist imaging has been limited due to concerns about waste caused by overspray.

The spray head traverses a larger area than the substrate to be coated, and the amount of liquid photoresist that becomes overspray can be significant. Exact characteristics of the coating in terms of thickness and topographical coverage depend on spray-head configuration, nozzle backpressure, droplet size, and conveyor speed.[20]

26.6.3.6 Electrostatic Coating. This method is similar to spray coating except that the rotation spray head is charged and the panel is grounded. Since the atomized photoresist is attracted to the substrate, there is less material waste than with spray coating.[21] Electrostatic coating is used for solder mask since it provides good coverage of conductor sidewalls. However, its use with photoresist is limited to innerlayers. It is unsuitable for outerlayers because it deposits a thicker coating on the rim of a plated through hole while leaving its interior uncoated.

26.6.3.7 Electrophoretic Coating. This method requires materials specially formulated for anodic or cathodic deposition. The equipment schematic in Fig. 26.10 illustrates the process sequence: preclean, coat, permeate rinse, rinse, and dry.[22] Panels are placed in the photoresist solution wet or are often sprayed with solution to ensure wetting, especially if the parts have blind or through vias. Voltage is applied, and within 20 sec. to 3 min. an insulating film forms.

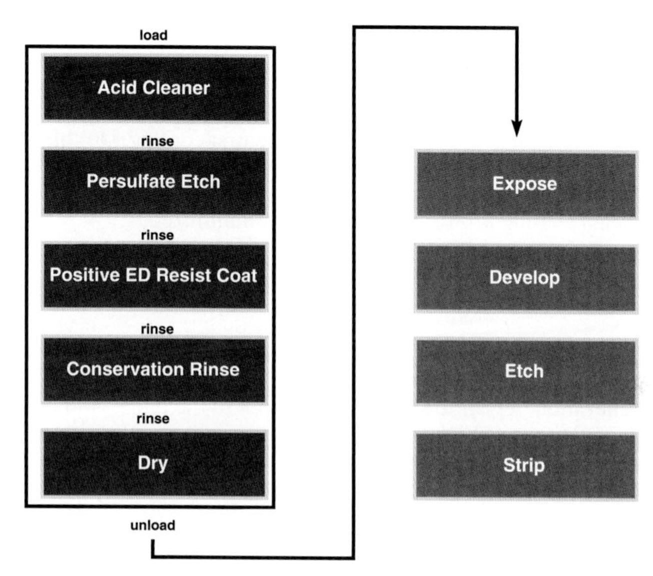

FIGURE 26.10 Configuration of an electrophoretic photoresist coater. *(Reprinted with permission of J. Dubrava et al., "Development of a Novel Positive-Working Electrodeposited Photo Resist Process for the Production of High Density PWB Outerlayers," presentation at the IPC Conference, San Diego, California, Spring 1995.)*

Rinsing removes loosely bound material that is returned to the plating cell after ultrafiltration to separate the photoresist and the counterion. This aids in maintaining the ionic balance.

After a final water rinse, the panel is dried to remove water and consolidate the film. The film is now tack-free and can be imaged.

The chemical composition of the photoresist determines whether the equipment components are made from polyvinyl chloride (PVC), polypropylene, or polyethylene. Electrodes are made of stainless steel and encased in an ion-selective membrane that is flushed to prevent the build-up of counterion migrating there during deposition. The permeate and ultrafiltration unit is important for long-term stability of the coating solution. The coating unit is often enclosed in a clean environment to reduce the amount of particles found in solution or resting on the panel as the process proceeds.

26.6.3.8 Curtain Coating. This method is commonly used to coat liquid solder mask materials. The panel is conveyed at high speed through a vertically falling curtain of the liquid photoresist. The coating width can be adjusted so that the edges are not covered. Temperature and viscosity are controlled for reproducible coatings, and coating thickness is determined by the conveyor speed and the solution flow rate.[23] Material in the curtain is returned to the sump and reused, giving very efficient material utilization. This method is one-sided, and the first-side photoresist coating must be partially dried before coating the reverse side. Due to the extra handling required to coat one side at a time and the inability to convey very thin panels through the curtain at high velocity, this method has seen only limited use with photoresist.

26.6.3.9 Screen Coating. Another common method for coating solder mask with either a pattern or flood coverage is screen coating. The screen mesh size and the liquid solution viscosity set the wet coating thickness. The screen is placed above the panel to be coated, and the material is forced through the openings in the screen onto the panel surface, where it forms a film.

Equipment is available for either single-side or double-side coating. With single-sided screen printers, the coating must be partially dried before the reverse side is coated. This causes a difference in the solvent content between the two sides that would result in nonuniformity for very precise lithography. In addition, the sequential nature of the coating allows debris and contaminants to be embedded into the first coating. Double-sided screen printers eliminate a drying step and minimize panel handling. The advantages of screen coating are that the equipment is relatively inexpensive to purchase and operate, "postage stamp" (small embedded areas) coating is possible, and waste is minimized.

26.6.4 Expose

Expose is the actual imaging step, reproducing the master pattern in the photoresist. The relief image is created after subsequent development. The expose process elements are phototool generation, registration of the phototool to the panel, and exposure through the phototool by the light source. Light-source alternatives include contact printing, either collimated or uncollimated; proximity printing; projection printing; and laser direct imaging. The noncontact methods separate the phototool from the substrate, reducing yield losses caused by contamination between the phototool and photoresist. With laser direct imaging, there is no phototool and the design data file of the master pattern drives the laser beam to expose the photoresist.

Phototools are generated on either film or glass substrates, depending on the feature sizes that are being patterned and the durability required. The panel is either mechanically aligned, with pins holding the phototool with respect to the product, or optically aligned, with alignment features dictating the movement of the phototool and product. The alignment requirements are interrelated to those achieved at composite lamination and drilling. The registration scheme is contained within the exposure equipment.

26.6.4.1 Conventional Imaging

26.6.4.1.1 Artwork Generation. Polyester and glass are the substrates used for phototools. They differ in optical properties, dimensional stability, and durability. Silver halide on polyester is used to create first-generation phototools on a laser plotter. This image is often contact printed onto diazo on polyester to make the copies used in production. Optical absorbance of the different phototool materials versus wavelength is plotted in Fig. 26.11.

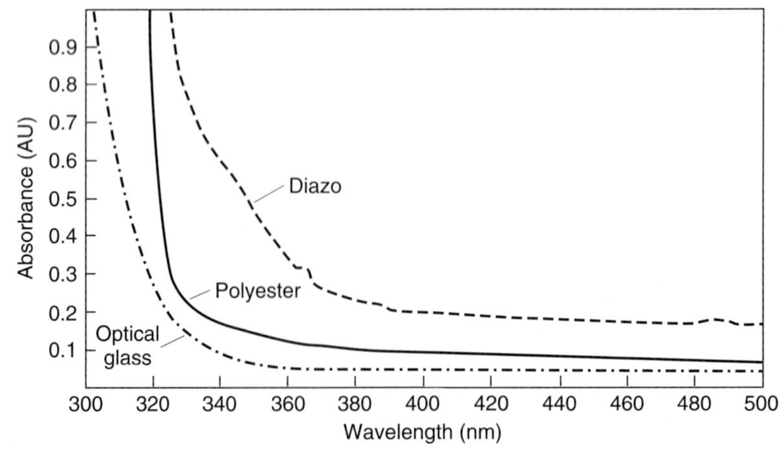

FIGURE 26.11 Optical absorbance of common phototool materials in the spectral region of PWB photoresist exposure.

Diazo phototools are the least expensive, and glass is the most expensive. Although the less-expensive materials have poorer transmission, this property is generally not limiting, except for applications requiring high intensity. Table 26.4 illustrates the impact on exposure time. Edge definition is a major difference among the options, as shown in Fig. 26.12.The photos illustrate that chromium on glass gives the sharpest edge definition. Edge definition is also affected by the pixel size of the phototool plotter and the spot's addressability (see the section on LDI). For applications that use a low-contrast photoresist and poor light collimation, these differences are not measurable on the printed photoresist image or in the patterned conductor. However, for high-density imaging, edge definition is important, as is the dimensional stability of the phototool.

TABLE 26.4 Comparison of the Optical Transmission of Various Phototool Materials

Property	Diazo	Polyester	Glass
Absorbance at 365 nm	0.271	0.107	0.047
Transmission at 365 nm	54%	54%	90%
% increase in exposure time vs. glass	40	13	0

Film dimensions change more rapidly with temperature than glass and are also affected by humidity. Careful control of both temperature and humidity is required throughout the life of a phototool (see Table 26.5).

TABLE 26.5 Dimensional Stability of Phototool Substrate Materials with Respect to Temperature and Relative Humidity

Substrate	Expansion coefficient, ppm/°C	Expansion coefficient, %RH/°C
Soda lime glass	9.2	0
Low-expansion glass	3.7	0
Pyrex	3.2	0
Quartz	0.5	0
Polyester	18	9

Source: Kodak Technical Literature (ACCUMAX 2000) for polyester film; *Tables of Physical and Chemical Constants,* Longman, London, 1973, p. 254; H. J. Fischbeck and K. H. Fischbeck, *Formulas, Facts and Constants,* Springer-Verlag, Berlin, 1987, for glass data.

The durability of the phototool is also important, especially for use in a manual contact printer. Damage depends on phototool handling, the exposure tool, the cleanliness level of the operation, and the surface of the product being contacted. Commonly, glass artwork is usable for 100 to 400 contacts with repair, while film artwork is usable for 20 to 50. For large numbers of the same part, number cost per contact may actually be lower with glass artwork, while for prototype parts, film artwork is ideal. However, the majority of production panels are exposed with film artwork. High-density applications such as chip scale packaging and multichip modules using laminate (MCM-L) may require glass artwork. The requirements of the product determine the type of phototool that should be used.

26.6.4.1.2 Registration. The image is placed on the panel with respect to a point of reference so that the image is well aligned to previous and future features. In multilayer construction, the innerlayer images are aligned front to back for successful lamination and drilling of the plated through holes (PTHs). The outerlayer image is then aligned to the drilled holes. The same considerations are relevant for single-layer boards. Thus, accurate alignment

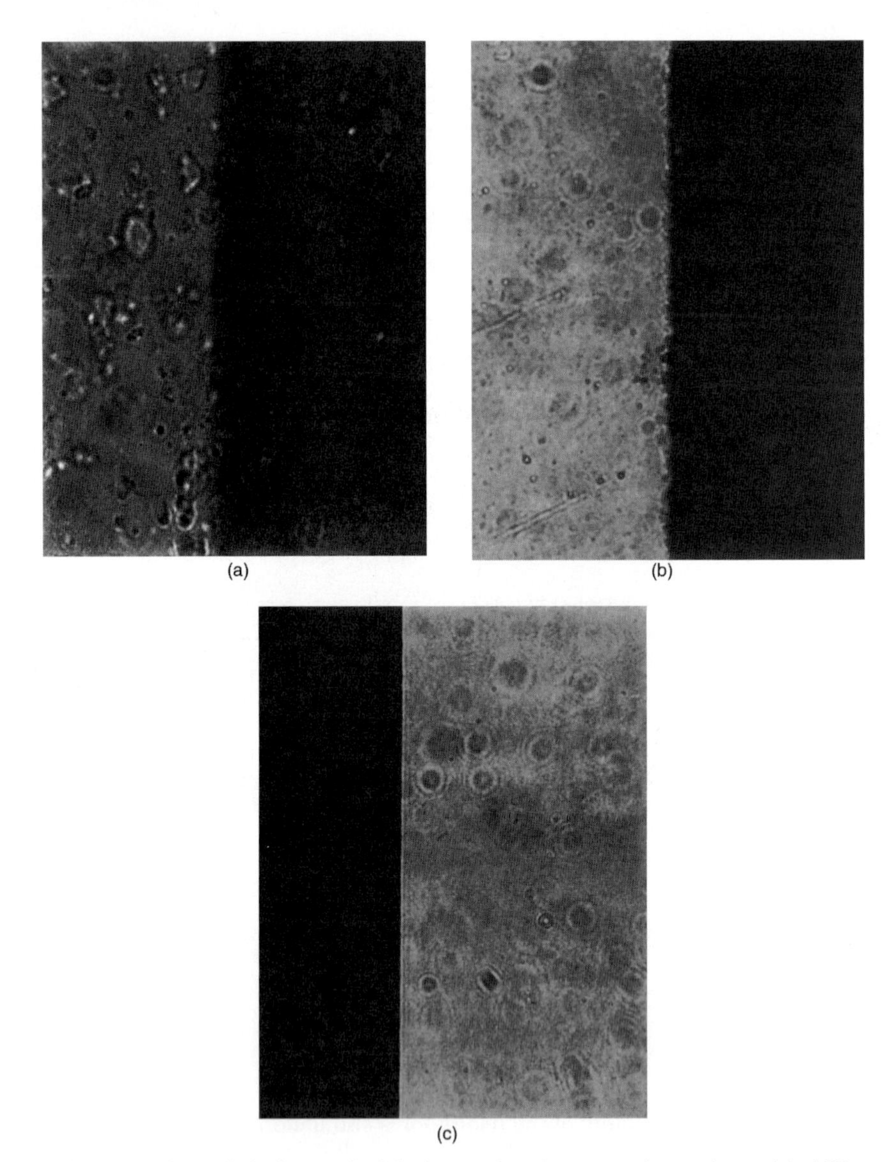

FIGURE 26.12 Optical micrograph of the image edge of common phototool materials, 1600×: (a) diazo on polyester; (b) silver on polyester; (c) Cr on glass.

is clearly part of the imaging process and also depends on knowledge and control of the dimensional stability of the product. The image must be scaled to match the dimension of the board at the exposure step; for multilayer structures with several photolithographic steps, a series of dimensional measurements are required prior to product manufacture.

Mechanical registration entails the use of fixed pins or other mechanical devices to hold the phototool and the panel in place during exposure. The actual configuration of the pins and their shape varies. Both two- and three-point systems are used. The three-point systems well define the center location since they are located at the extremes of the area to be patterned.

The shapes of the pins are either round or elongated with one flat side. In the latter case, the flat side defines the edge, while round pins center within the hole in the panel and phototool.

Film artwork alignment holes are punched with respect to the product pattern by plotting alignment targets as part of the product pattern and optically aligning and punching a slot or hole. The edges of the punched feature wear, and registration will deteriorate with use. For glass artwork, the glass is drilled at the alignment locations and a bushing is placed at the center of the hole. The pin is then inserted through the bushings in top and bottom artwork and through the panel. With use, the bushings do move and must be reset for maximum reproducibility.

Optical alignment systems operate either manually or automatically. In both instances, the phototool is plotted with alignment targets—usually an opaque dot that is smaller than a drilled or punched hole in the panel. With backlighting, the dot (phototool) is aligned to the center of the hole (product). Often, three locations at the edges of the panel are used. In manual systems, micrometers are used to move either the part or artwork. In automatic equipment, a vision system calculates the necessary movement, and motors then move either the phototool or the product into alignment. Since there is no wear on the phototool, when it is optically registered the accuracy is maintained with usage. The absolute accuracy achieved with optical methods is superior to mechanical registration, with the best accuracy and reproducibility obtained with automatic optical systems. The extra expense for automated equipment is warranted if it is needed to meet stringent product requirements.

26.6.4.1.3 Exposure Control and Measurement. The role of exposure is to change chemically the photoresist and its solubility in the developer solution. The appropriate energy dose is determined experimentally by measuring the combination of dose and development that is needed to produce features with straight sidewalls. The photoresist is coated on an optically clear substrate and exposed from the backside. Contrast curves, plots of the log of the exposure dose versus the thickness of the film remaining, are used to identify the functional cure point—for example, the dose that loses thickness less than 10 percent (see Fig. 26.13).

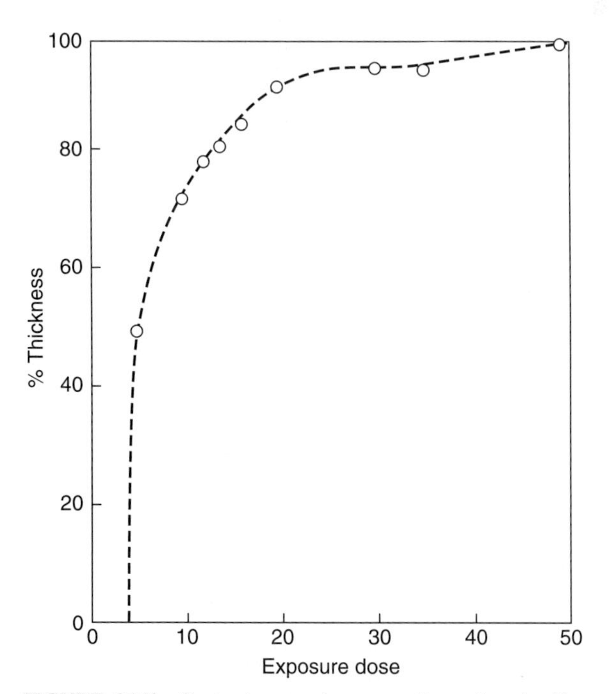

FIGURE 26.13 Contrast curve for a negative-acting dry-film photoresist, percent film thickness remaining versus exposure dose (mJ/cm^2).

Exposure in the region of stability ensures that the base of the material has reacted. Step wedges, film strips with a series of neutral density filters, are also used to determine appropriate exposure doses. The dose is varied to obtain photoresist residue on the area with the manufacturer's recommended step value. This technique depends on appropriate and consistent developer conditions.

Step wedges are often used to control the expose or expose-and-develop processes.

The energy incident on the photoresist is the product of the lamp intensity and the time of exposure:

$$\text{energy} = \text{intensity} \times \text{time} \tag{26.1}$$

Thus, the exposure dose can either be measured directly by using an integrating radiometer or indirectly by measuring exposure time and light intensity with a radiometer. The radiometer response must be matched to the spectral output of the light used so that representative measurements are made (see Fig. 26.14). Most exposure equipment provides the option of using either direct energy or time measurement. With a stable light source, both approaches are equally reproducible.

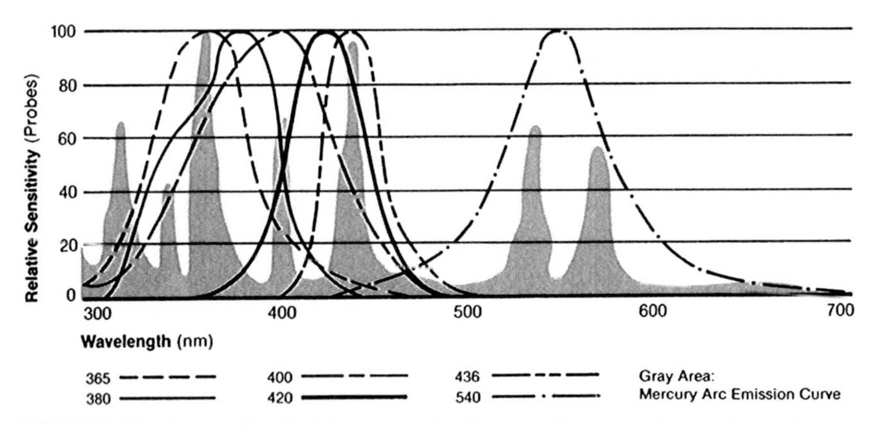

FIGURE 26.14 Output of Hg/Xe light source and radiometers. *(Reprinted with permission of Optical Associates, Inc., Milpitas, California, Technical Literature.)*

26.6.4.1.4 Contact Exposure Tools. This equipment places the phototool and the panel in direct contact and draws a vacuum between the pieces for hard contact. A range of light wavelengths is used, often from a mercury or mercury/xenon light source, which is stable and has a high intensity output at ultraviolet and visible (UV/Vs) wavelengths (see Fig. 26.14). The light is distributed over the area to be exposed by placing the board at a distance from the source, either noncollimated or collimated by the use of optical elements, as shown in Fig. 26.15. Collimation refers to the angle of incidence of the light onto the photoresist, and it is critically important for fine-line images. Defining fine spaces is the challenge for both additive and subtractive conductor formation, and exposure under the opaque areas cannot be tolerated. The lamp intensity for noncollimated sources is greater, and therefore less time is required to expose the panel. Process throughput is noticeably enhanced for photoresists requiring a large dose.

In addition to collimation, good contact between the phototool and the panel is the most important factor to control in fine-line formation. Any gap will allow exposure under the

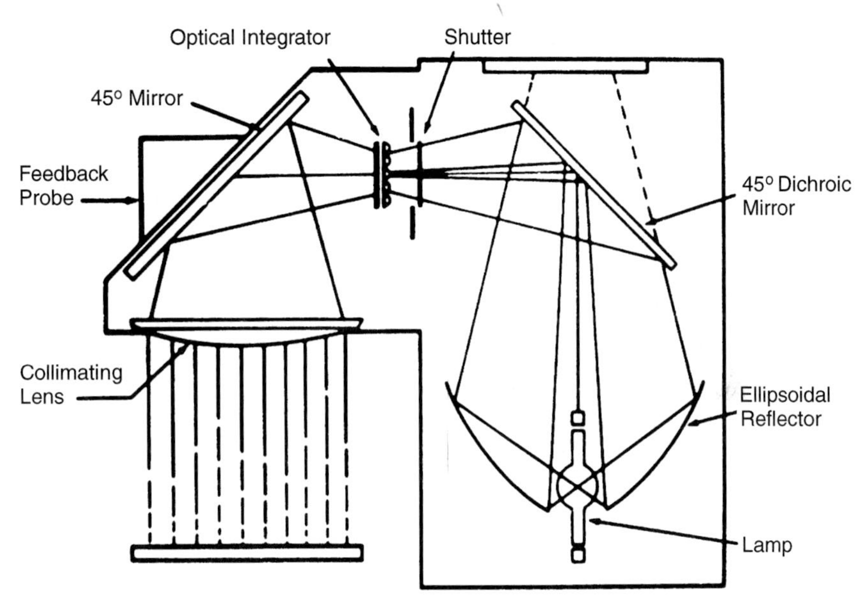

FIGURE 26.15 Configuration of lamp and optics for a collimated contact printer. A noncollimated source is identical except for the collimation lens in the lower left. (*Reprinted with permission of Optical Associates, Inc., Milpitas, California, Technical Literature.*)

opaque, resulting in poor linewidth control and reduced resolution. The sensitivity to off-contact exposure varies among photoresists, and newer resists are formulated with increased off-contact latitude.[24] Since the lamp's spectral output degrades over time, the lamp must be routinely replaced to maintain a stable photoprocess. The exact frequency for replacement depends on the line and space definition required, but the lamp life is typically 1,000 hours.

Failure to replace a lamp can result in it exploding and damaging the optical elements in the exposure unit, which is far more costly than routine maintenance.

26.6.4.1.5 Noncontact Exposure Equipment. Semiconductor imaging demonstrated the yield limitations in using contact-exposure equipment. Noncontact methods had to be developed to achieve acceptable conductor yield as critical dimensions decreased to 0.18 μm and below.

The PWB industry is following this evolution, driven by the increased demand for products with fine features of 25 μm and less. Semiconductor tool concepts cannot be used directly due to important differences between IC and PWB products, namely the physical size of the substrate and features to be imaged and the flatness of the substrate. The key to noncontact methods is the use of appropriate optics for clear image placement in the photoresist (Fig. 26.16), even if the image transfer function has degraded. For PWB applications, this includes the depth of focus to allow for panel warpage and the resolving power, given the contrast of the resist. The following approaches address these concerns and are viable alternatives to contact printing.

26.6.4.1.6 Proximity Printing. This is the oldest method for off-contact printing and requires no modification to the equipment optics. The phototool is held out of contact from 125 to 500 μm. The image in the photoresist suffers since the distance between the phototool and the part allows exposure under the opaque area and imaging of debris on the phototool. Nevertheless, resolution as fine as 75 μm has been reported for proximity printing with thin coatings of liquid resist.[25]

26.6.4.1.7 Projection Printing. Three approaches are possible for this technique: scanning, stitching, and magnified projection printing. Scanning and stitching evolved from semiconductor

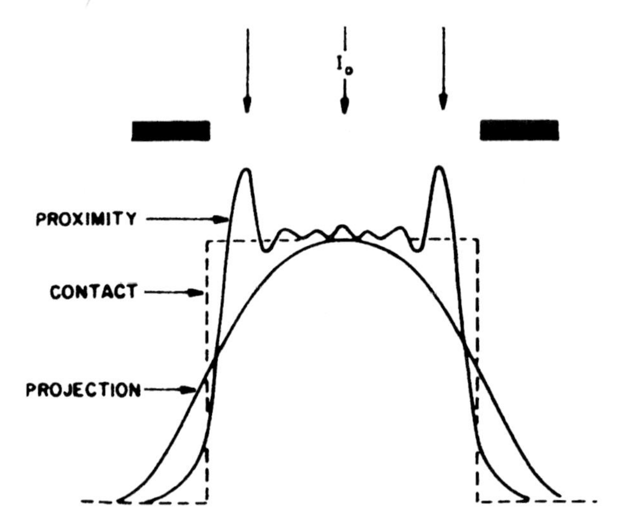

FIGURE 26.16 Image transfer for contact, proximity, and projection printing.

and MCM-D thin-film technology, where smaller, flatter substrates are used. They require movement of the phototool and/or the substrate during exposure. Magnified projection imaging is unique to the PWB application and has no moving optical components. A schematic of all three systems is given in Fig. 26.17.

- The scanning printer requires the synchronized movement of the part and the phototool during exposure. Precise alignment of the two pieces must be maintained during the motion. Periodic adjustment for flatness is usually required since this equipment has a small depth of focus. For typical PWB formats, a large phototool is needed, requiring expensive glass phototools for their flatness, rigidity, and sharp line definition. This exposure technique has demonstrated high resolution over a large, relatively flat area using a high-contrast resist.[26] Few PWB applications exist, but the technique is useful for flat panel display manufacture.
- Stitcher projection printers are step-and-repeat tools that use a repeating image or "stitch" together several images. The artwork generation and management in the equipment are complicated, which has hampered these printers widespread use for large panels. The usual active area size as defined by the optics is approximately 6 in. × 6 in. Thus, for a standard 18 in. × 24 in. panel, 12 artwork changes are required to expose one side. These systems have very high resolving capability and large depth of focus. Again, glass artwork is used, but these smaller plates are easier to generate and use than in a scanning projection printer.
- The magnified projection imaging concept was first demonstrated in the late 1980s with the SeriFLASH exposure tool, utilizing 436 nm illumination and a 5 in. × 5 in. liquid crystal display as the phototool. The image was magnified six times to expose an 18 in. × 24 in. panel, but demonstrated photoresist definition was limited to 125 μm since the liquid crystal features were one-sixth of the final conductor width. The equipment was subsequently modified for 365 nm exposure with a glass phototool to improve resolution. Fifty-μm line and 63 μm space resolution has been demonstrated.[27] The exposure time required is similar to that of contact exposure equipment. With optimized optics, this may become a high-resolution tool that utilizes conventional photoresist and phototools.

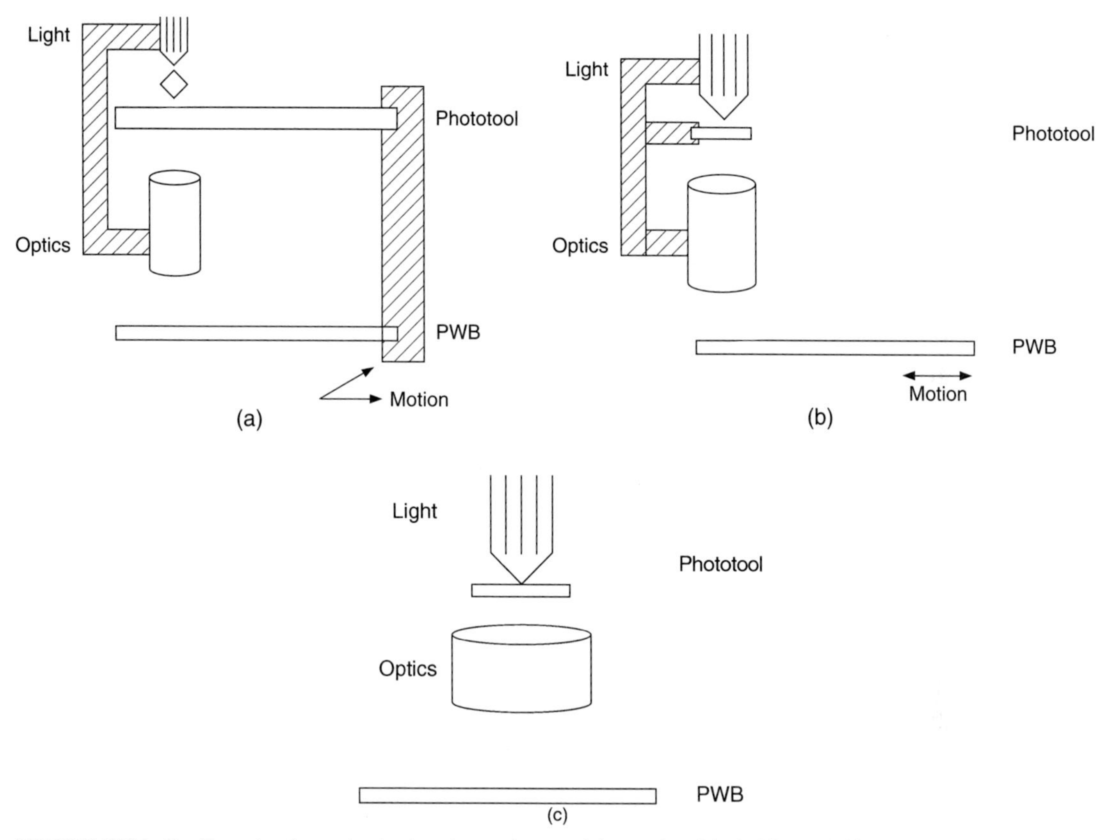

FIGURE 26.17 Configuration for projection imaging equipment: (a) scanning, (b) stitching, and (c) magnified.

26.6.4.2 Laser Direct Imaging. The LDI method of exposure provides the ultimate in flexibility for customized pattern compensation factors and for engineering changes to the product design since it does not require a phototool. Systems introduced in the 1980s used an Ar^+ laser with visible optics, requiring specialized photoresists that were sensitive to visible light. Improved Ar^+ systems relied on the UV wavelengths, making it compatible with conventional photoresists. However, high-speed photoresists were needed to achieve the productivity for high-volume production. Concerns about the lifetime (3,000 to 5,000 hours) and power consumption (60 to 80 kW) of gas lasers led to development of solid-state laser systems at various wavelengths.[28,29,30] The initial solid-state laser systems used 355 nm with a polygon mirror to scan the panel. Recently developed LDI systems have used digital mirror devices (DMD) instead of polygon mirrors and image at a wavelength of 405 nm. Filters are used to block 355 nm light, extending the lifetime of the DMD device. Due to the small size of the individual mirrors, about 1.5 microns, DMD technology provides better resolution than an 8,000 dpi polygon mirror system. However, the area coverage is smaller and several DMDs working in parallel are needed to expose a PWB panel. Systems have also been introduced that use no photoresist at all, laser ablating either tin for use as an etch resist or physical vapor deposition (PVD) copper on polyimide for use as a plating base.[31,32] In addition to varying in laser type and wavelength, specific equipment designs differ in optics, either single- or multiple-beam operation; platen design; single- or double-side exposure; pixel shape; and spot size and resolution.

Multiple-beam operation increases exposure productivity, a key requirement for economic feasibility in large-volume manufacturing. Target production rates for the various systems ranged from 60 to 180 panels per hour (18 in. × 24 in. with 50 μm features). However, installed LDI systems have been used primarily for prototype and small-lot, quick-turn production of high-density boards. For these applications, savings are achieved due to artwork elimination and reduced setup time. For volume production of medium- to high-technology boards, it is possible to achieve an economic return through yield savings and the ability to produce designs that had not been technically or economically possible with conventional imaging.[33] LDI technology makes it possible to scale the electronic data and and compensate for dimensional changes in sequential build-up (SBU) structures. Although the UV systems can expose conventional photoresist, use of high-speed photoresist is needed to maximize tool throughput. Most LDI systems use a flatbed platen design and can image both innerlayers and outerlayers.

Systems with an external drum architecture, similar in design to laser photo plotters, are limited to innerlayers because the panel must have sufficient flexibility to conform to the drum (Fig. 26.18).The pixel shape can be either square or gaussian. Square pixels are made up

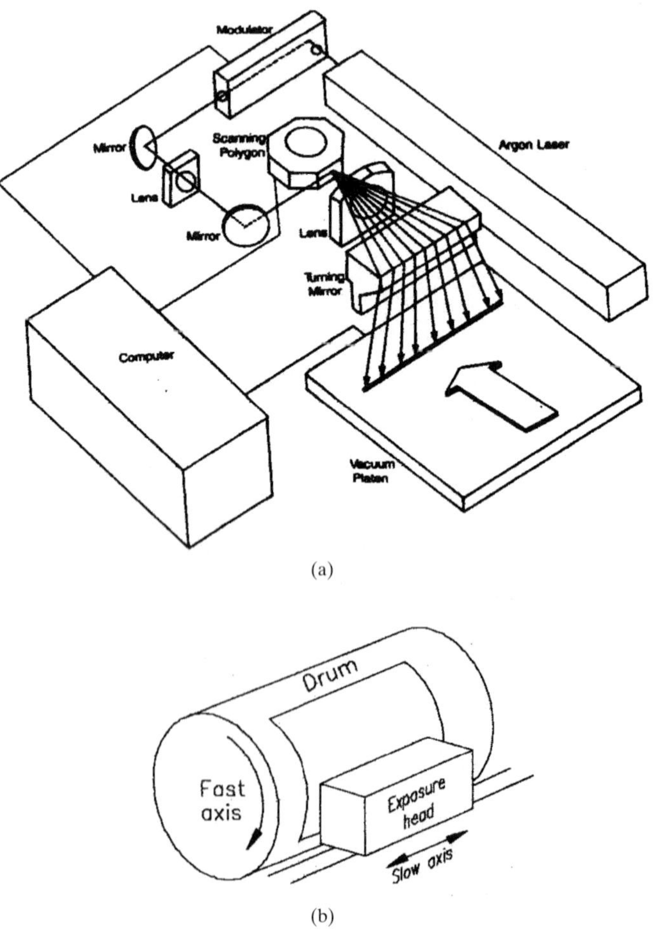

FIGURE 26.18 Schematics of laser direct imaging equipment: (a) flatbed platen; (b) external drum.

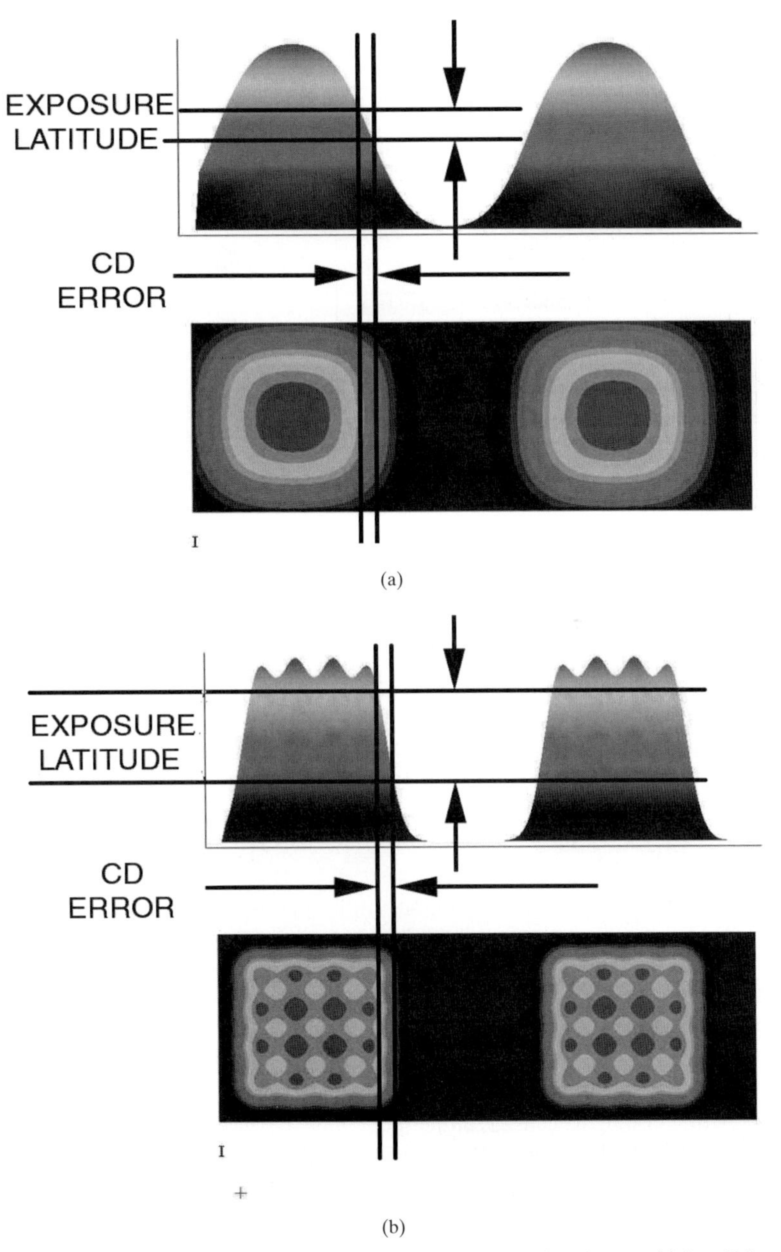

FIGURE 26.19 Computer simulation of aerial image for a 25 μm feature: (a) four 12.5 μm laser spots on 6.25 μm pixel spacing; (b) four 6.25 μm laser spots on 6.25 μm pixel spacing.

of many point sources of light and can theoretically achieve perfect stitching. However, in real systems there is intensity variation due to mechanical vibration and edge rounding. Gaussian pixels use a single-point source and may have less intensity variation. Due to the diffraction-limited nature of light, greater depth of focus can be achieved.[34] Figure 26.19 illustrates that

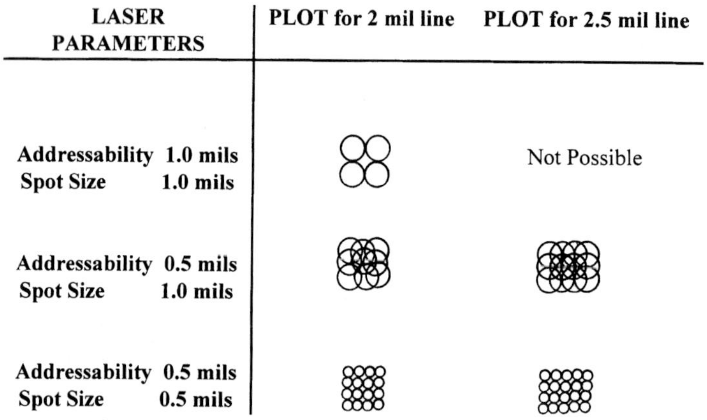

LASER PARAMETERS	PLOT for 2 mil line	PLOT for 2.5 mil line
Addressability 1.0 mils Spot Size 1.0 mils		Not Possible
Addressability 0.5 mils Spot Size 1.0 mils		
Addressability 0.5 mils Spot Size 0.5 mils		

FIGURE 26.20 Final image dimension as governed by spot size and addressability of a laser plotter or laser direct exposure tool.

imaging with a smaller laser spot size gives steeper aerial image sidewalls, increasing process latitude.[35] In addition to spot size, addressability of the beam is important in determining the system resolution, as shown in Fig. 26.20.

In selecting an LDI system, you must consider both the capital cost and operating cost. However, the critical factor is that the equipment combined with the photoresist material is capable of achieving both the resolution needed for the product mix and the productivity needed for the production volume. Photoresists for LDI are available that are compatible with the various metallization processes.[36]

26.6.5 Develop

In this process step, the solubility difference between the exposed and unexposed areas of the photoresist creates a relief image of the master pattern. The panel is immersed in an appropriate solvent, and the process conditions are adjusted to control the clearing time for dissolving the unexposed areas for negative photoresists or exposed areas for positive ones. Total dwell time is set to approximately double the time to clear, commonly called a 50 percent breakpoint. Solution concentration, temperature, and agitation are key variables. The resulting photoresist images should be distinct, with vertical sidewalls. Failure to achieve this indicates that one of the previous steps requires adjustment. For images larger than the phototool dimension, the cause is either incomplete development, overexposure, or poor contact during exposure. For images that are smaller than expected, either the exposure dose is too low or development is too aggressive. Distorted images can be caused by problems with preclean, application, or exposure. The common equipment for developing is spray conveyorized, either horizontally or vertically.

Additives are used in the developer solution to prevent foaming. The solution is filtered to remove resist particles and either replenished with fresh solution to maintain a consistent dissolved resist content and solution concentration or operated continuously for a certain amount of product and then replaced. Waste-developer solution is treated (aqueous and semiaqueous) or distilled and reused (solvent). Rinsing is important in stopping the dissolution and, for aqueous photoresists, water with a high-mineral content often improves the resist image and the conductor yield. Tank systems can also be used with photoresists that have a wide-process latitude. Ultrasonic agitation is often used to aid in the dissolution.

There are additional steps that improve the resist removal in the line channels and the conductor formation yield. Plasma treatment has been used effectively to improve product yield, especially with respect to shorts in a print-and-etch process. In addition, for some aqueous-developable dry films, a heat treatment after exposure has improved the space definition, and spaces equal to or smaller than the resist height have been resolved. Thus, these process steps ensure that tight resolution requirements can be met.

26.6.6 Strip

After the pattern transfer has been completed, the photoresist is removed from the substrate using equipment similar to that used for development. The stripping solution swells and dissolves the photoresist, stripping it either in sheets or as small particles. The equipment design must effectively remove and separate the skins. Often, brushes and ultrasonic agitation are added to aid in the resist removal. As with developing, filtration is important to keep the spray nozzles clean and keep fresh solution reaching the panel. For stripper chemistries that oxidize the copper, an antitarnishing agent is often added either to the stripping solution or as part of the rinsing.

26.7 DESIGN FOR MANUFACTURING

As design features are continually reduced to produce higher-density interconnects, tighter control of every step in the conductor formation process is required to achieve high yield. The maximum possible yield with an etching or plating process depends on the conductor dimensions such as the conductor pitch in terms of line and space, the conductor thickness, and the size and shape of the capture pads around plated through-holes (PTH) and vias.

26.7.1 Process Sequence: Etching versus Plating Considerations

For a specific circuit design, there is often a question of the appropriate process sequence. Although the capabilities of imaging and the pattern transfer step dictate the overall limitations, the decision depends on the unique capabilities of the manufacturing line to be used. Some general considerations can clarify the true issues for most production situations.

Photoresist patterning has different constraints for etching and plating. For etching, a thin photoresist is desirable to maximize the etchants' attack in the channel developed in the photoresist. Resolution does not limit the process, since very small spaces can be resolved in either liquid or thin dry-film photoresists. The etching process itself is key to conductor formation.

Etching the copper from the channel becomes increasingly difficult as feature size is reduced, particularly for thicker metal layers. Therefore, the key criteria defining the capabilities of the print-and-etch process is the minimum space cleared, constrained by the final conductor height and the etchant chemistry and equipment.

For pattern plating, the photoresist thickness must be greater than or equal to the final thickness of the conductor, and it must be possible to develop a photoresist channel equal to the final conductor width. As conductor thickness increases, it becomes more challenging to produce channels with a higher aspect ratio. Since photoresist resolution is on the order of its thickness, it would be difficult to resolve smaller than a 38 μm photoresist channel for 25 μm thick plating, irrespective of the final conductor spacing. Thus, the challenge for pattern plating is to resolve and develop fine channels in thicker photoresist materials and then ensure that the plating solution wets the bottom of the narrow photoresist channels.

Thus, in choosing the conductor formation process, the thickness of the conductor and the linewidth and spacing are the key parameters. A generalized relationship between them is

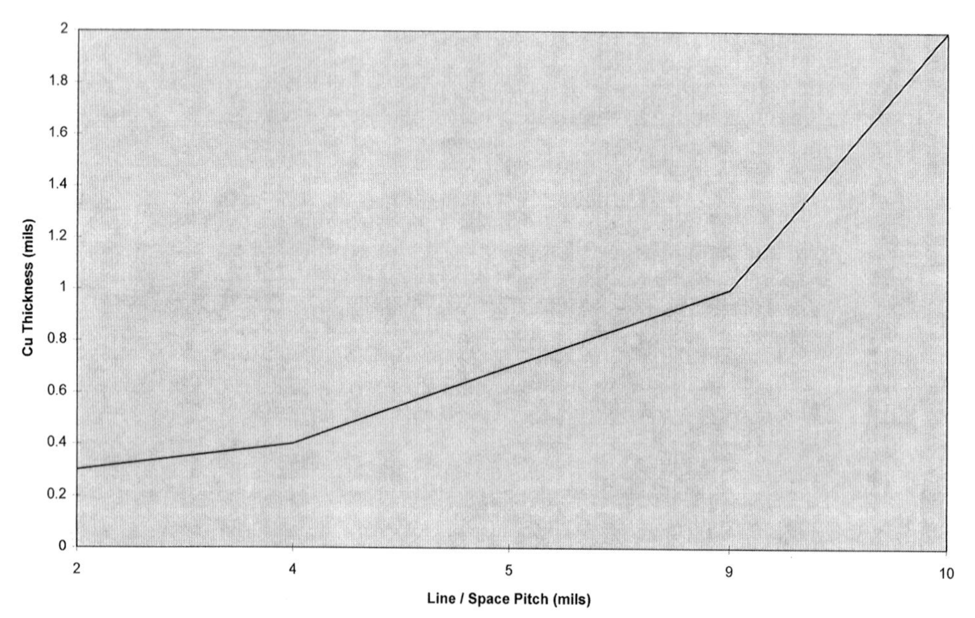

FIGURE 26.21 Sample plot of manufacturing line process capability to aid in processing sequence decisions.

found in Fig. 26.21. It is important that this type of plot is known for the production area before the process sequence is determined.

26.7.2 Line and Space Division for a Fixed Pitch

It is common for product designs to have a fixed pitch, whether it refers to the I/Os for direct chip or packaging attachment or to the spacing between PTHs. This space is often divided equally between conductors and spacing. For etching or pattern-plating processes, avoiding their respective resolution limitations can increase the pattern transfer yield.

In etching, the etchant undercuts the photoresist and an etch factor is used to obtain straighter sidewalls and improved linewidth control. If lines and spaces are equally allocated in the circuit design, then the photoresist must resolve a smaller space than a line. This is difficult for fine features, except for very thin photoresists. For fine pitch, having a larger space than line will give higher yield. For pattern plating, the spacing in the photoresist is the limiting factor. After pattern transfer, this will become the line. In this instance, equal line and space is more acceptable, but based on the photoresist concerns, a wider final line than space is preferred. At the same time, it is desirable to increase the spacing for reduced incidence of line shorting, which translates to a wider "line" in the photoresist. The former is usually more important. Therefore, for both processes there is preferably balance between linewidth and spacing.

26.7.3 PTH Capture Pad Size and Shape for Optimum Line Formation

Just as the conductor line and spacing can be optimized, the PTH capture pad shape and size can also be altered to increase the yield. The absolute dimension of the feature is dictated by the placement accuracy of the drilling process and the overall dimensional stability of the product. The feature is sized to ensure that the PTH and conductor are connected. There are varying specifications as to the extent of capture that is required.

Feature dimensions above the line would be processed with pattern plating, and those below the line with print and etch.

Depending on the direction and magnitude of the dimensional stability and the drill wander and accuracy, the shape required to capture the PTH can be changed. This would reduce the size of this feature in at least one direction. In consequence, the spacing between the line and pad increases. Since this location is a change to the nominal line-to-line spacing, the narrowing of the channel results in shorting between the features in both a print-and-etch process and additive processes. In the former case, it is more difficult to clear the space, and in the latter, the narrower resist width is often underplated. Thus, when possible, an elongated pad will benefit the final conductor yield.

REFERENCES

1. Woznicki, T., "Flex Circuit Manufacturing in a Box?" *Printed Circuit Design and Manufacture,* February 2006, p. 26.
2. Stoll, R., "High Definition Imaging," *PC Fab,* Vol. 17, No. 6, June 1994, p. 31.
3. Hayes, E., "An Overview of Dry Film Imaging Chemistry," *PC Fab,* May 1988, p. 74.
4. McKeever, M. R., "Laser Direct Imaging—Trends for the Next Generation of High Performance Resists for Electronic Packaging," *Proceedings, S12-1, 2000 IPC Expo and Technical Conference,* San Diego, California, 2000.
5. Choi, J. H., "Chemistry and Photoresist for Electroless Deposition," presentation at the IPC Spring 1992 Conference, Bal Harbor, Florida.
6. U.S. Patent 5,268,260, November 7, 1993.
7. Sutter, T.C., "Liquid Photoresist Systems—An Overview," *Board Authority* supplement to *CircuiTree,* Vol. 1, No. 3, October 1999, p. 22.
8. Gangei, J., "Pushing the Envelope in Innerlayer Primary Imaging," *Board Authority* supplement to *CircuiTree,* Vol. 1, No. 3, October 1999, p. 30.
9. Taff, I., and Benron, H., "Liquid Photoresist for Thermal Laser Direct Imaging," *Board Authority* supplement to *CircuiTree,* Vol. 1, No. 3, October 1999, p. 66.
10. Almond, R.J., Goewey, M. E., and Jobson, B. F., "Lowering Innerlayer Fabrication Cost with Liquid Resists," presentation at the IPC Spring 1995 Conference, San Diego, California.
11. Gurian, M., and Ivory, N., "Performance Requirements of Primary Liquid Resists," *Electronics Packaging and Production,* March 1995, p. 49.
12. Murray, J., "ED Processes Revisited," *PC Fab,* May 1992.
13. Nakahara, H., "Electrodeposition of Primary Photoresists," *Electronic Packaging and Production,* February 1992, p. 66.
14. Dietz, K.H., "Surface Preparation for Primary Imaging," presentation and paper at the IPC Spring 1992 Conference, Bal Harbor, Florida, p. TP-1025.
15. Moreau, W., *Semiconductor Lithography,* Plenum Press, New York, 1989, pp. 651–664.
16. Crum, S., "Surface Preparation Process Improvements," *Electronic Packaging and Production,* July 1993, p. 24.
17. Atotech technical information.
18. Yamada, H., "Wet Lamination Technology and New Application for Image Transfer," poster presentation at Electronic Circuits World Convention 8, September 1999, Tokyo, Japan.
19. Patel, R., and Benkreira, H., "Gravure Roll Coating of Newtonian Liquids," *Chemical Engineering Science,* Vol. 46, No. 3, 1991, p. 751.

20. Young, C., "The In Line Spray Conformal Coat Process," *Proceedings of the Technical Program, NEPCON West 89*, Vol. 2, Cahners, Des Plaines, IL, p. 1676.
21. Marks, D., and Lee, T-C., "Electrostatic Applications of Resists and Solder Masks," *PC Fab,* September 1991, p. 100.
22. Dubrava, J., Pai, D., Rychwalski, J., and Steper, J., "Development of Novel Positive-Working Electrodeposited Photoresist Process for the Production of High Density Outerlayers," presentation at the IPC Spring 1994 Conference, Boston, Massachusetts.
23. Nguyen, J., "Curtain Coating in Soldermask Application," *Proceedings of the Technical Program, NEPCON West '89,* Vol. 1, Cahners, Des Plaines, IL, p. 869.
24. Cox, S., "Wide Exposure Latitude Photoresists and Fine Line PWB Manufacturing," *Board Authority* supplement to *CircuiTree,* Vol. 1, No. 3, October 1999, p. 34.
25. W. R. Grace and Co., ACCUTRACE technical information.
26. Muller, H. G., Yuan, Y., and Sheets, R. E., "Large Area Fine Line Patterning by Scanning Projection Lithography," *IEEE Transactions on Components, Packaging and Manufacturing Technology Part B: Advanced Packaging,* Vol. 18, No. 1, 1995, p. 33.
27. Feilchenfeld, N. B., Baron, P. J., Kovacs, R. K., Au, D. T. W., and Rust, R. D., "Further Progress with Magnified Image Projection Printing for Fine Conductor Formation," presentation at the IPC Fall 1995 Conference, Providence, Rhode Island.
28. Dietz, K. H., "Alternatives to Contact Printing," *CircuiTree,* Vol. 12, No. 5, May 1999, p. 120.
29. Vaucher, C., "Solid or Gas?" *CircuiTree,* Vol. 14, No.1, January 2001, pp. 96–97.
30. Stone, D., "Use of a Solid State Laser to Expose High Speed LDI Resists in PCB Fabrication," presentation at EPC 2000 Conference, Maastricht, Belgium, October 4, 2000.
31. Siemens news release at IPC Expo 2000, San Diego, California.
32. Kickelhain, J., "New Excimer Laser Technology—Ultra Fine Lines (15 μm) without Etching," poster presentation at Electronic Circuits World Convention 8, Tokyo, Japan, September 1999.
33. Waxler, S., and Spinzi, S., "Direct Imaging Implementation: Real World Case Studies," presentation at IPC Printed Circuit Expo 2000, San Diego, California.
34. Kesler, M., "Direct Write for HDI Substrates," presentation at HDI Expo, Phoenix, Arizona, September 2000.
35. Tamkin, J. M., "The Impact of HDI on Fine-Line Lithography," *Board Authority* supplement to *CircuiTree,* Vol. 1, No. 3, October 1999, p. 59.
36. Liebsch, W., "New Dry Film Developments for Laser Direct Imaging," *CircuiTree*, Vol 19, No. 10, October, 2006, p. 32.

CHAPTER 27
MULTILAYER MATERIALS AND PROCESSING

C. D. (Don) Dupriest
Lockheed Martin Missiles and Fire Control, Dallas, Texas

Valerie A. St. Cyr
Teradyne, Inc., Semiconductor Test Division, North Reading, Massachusetts

27.1 INTRODUCTION

Printed circuit board (PCB) structures and processes continue to be driven by device miniaturization, functional densification, and speed. The choice of components, actives and passives, and connectors are primary determinants of layer count and layer options, such as copper foil weights. Reductions in chip voltages are matched by an increase in current demand; this in turn causes increases in the number and thickness of the plane layers. More device leads combined with decreasing pitch increases the number of vias and interconnecting layers, while miniaturization and weight reduction at the product level forces reduction in the physical package for the product.

New structures or advancements on existing structures are then needed to make these interconnections. These structures—such as blind and buried vias, multi-lamination (sub-laminations), and build-up technologies—directly impact multilayer processing. Deeper blind vias and buried vias have caused a major increase in the demand for reliable via filling materials and methods. Relatively new materials, equipment, and processes have and are being developed to address this need. A new section in this chapter, "Filled Via Processing and Sequential Lamination," addresses filled vias internal to the PCB.

Ever increasing frequencies continue to shift demand to materials with lower attenuation and propagation delay. To guarantee the electrical performance requirements of this product type and to benefit from the increased raw material and processing costs, the industry is specifying each element in the construction more so than in the past, and with less latitude. For instance, requirements for post-pressed thickness tolerance on a prepreg fill section, or surface roughness of the treatment side of the copper foil, or percent resin of the prepreg or core are much more likely to find their way onto fabrication prints.

Globalization of sourcing for PCBs continues to have a major impact on the physical design or the fabrication processes for the PCB construction. Mass lamination techniques have scaled to higher layer counts, and alternate tooling methods and panelization approaches have contributed to increases in material utilization and throughput. These technologies increase production volumes and reduce costs and are discussed in this chapter in Secs. 27.4.4 and 27.4.6.

The latest challenge to the physical PCB is coming from regulatory bodies and consumers who demand products that are less damaging to the environment. Materials developed to meet these demands are known colloquially as either lead-free or RoHS-compliant (that is, in accordance with the Restriction of the Use of Certain Hazardous Substances in Electrical and Electronic Equipment Directive). This chapter will not deal at length in the specific requirements of this directive, but will address the major impacts to the PCB materials (see Section 27.2.3) and subsequent processing (see Secs. 27.5.5 and 27.6.2).

Lead-free joints, which have no appreciable lead content (for what "appreciable" means in this context, refer to the appropriate directive; see also Chapter 10) reflow at temperatures 30 to 40°C higher than the melt point of eutectic solder. These temperature increases have necessitated the development of newer resins or blends, different curing agents, and fillers to provide a laminate system that can survive and remain reliable after these elevated heat excursions. The changes to the laminates themselves are covered elsewhere in more detail (see Chapters 6 and 7), but those properties directly applicable to the performance of the PCB are discussed in this chapter in Sec. 27.2.2.

In some market segments, particularly the consumer market, "green" technology is an important product differentiator or even a market requirement. This chapter also introduces halogen-free laminates in Sec. 27.2.

27.1.1 Relevant Specifications, Standards, and Test Methods

Industry standards groups such as the IPC have defined properties requirements and performance specifications to facilitate the choice of laminates. Design standards are available to inform the material selection process to achieve a desirable form, fit, or function. Standardized property tests have been developed to provide a common method to test for property values and to interpret results. Some of the most commonly used are:

- IPC-2221, "Generic Standard on Printed Board Design"
- IPC-4101, "Specification for Base Materials for Rigid and Multilayer Printed Boards"
- IPC-4104, "Specification for High Density Interconnect (HDI) and Microvia Materials"
- IPC-4652, "Metal Foil for Printed Wiring Applications"
- IPC-TM-650, "Test Methods Manual"

Table 27.1 lists some laminate tests covered in IPC-TM-650.

TABLE 27.1 IPC-TM-650 Laminate Tests

Test Method #	Title
2.4.24C	"Glass transition temperature and Z-axis thermal expansion by TMA"
2.4.24.1	"Time to delamination (TMA method)"
2.4.24.2	"Glass transition temperature of organic films—DMA method"
2.4.24.3	"Glass transition temperature of organic films—MA method"
2.4.24.4	"Glass Transition and Modulus of Materials Used in High Density interconnection (HDI) and microvias—DMA method"
2.4.24.5	"Glass transition temperature and thermal expansion of materials used in high density interconnection (HDI) and microvias—TMA method"
2.4.24.6	"Decomposition temperature (T_d) of laminate material using TGA"
2.4.25C	"Glass transition temperature and cure factor by DSC"
2.6.25	"Conductive anodic filament (CAF) resistance test: X-Y axis"

27.2 PRINTED WIRING BOARD MATERIALS

A basic understanding of material performance is necessary for both the designer and fabricator of the printed wiring board (PWB). Understanding the materials available is one of the primary tasks for designing for manufacturability and performance. It is necessary to match the product end-use performance requirement as well as the environmental exposure the board experiences in the fabrication and circuit card assembly (CCA) processes with the capabilities of the material. It is common for boards to experience as many as five thermal excursions in CCA.

With the introduction of the RoHS directive, the temperature at which the board is assembled is typically 30 to 40°C higher than usual eutectic reflow temperatures. Current lead-free pastes require peak temperatures from 230 to 250°C. The actual peak temperature for any CCA is a variable based on the specific alloy paste, the component lead metallurgy, and also the combined component and board mass. Many studies have reported data that document the plated through hole damage that accumulates from successive reflows as a function of the expansion of the laminate with increasing temperature. Other experiments have demonstrated that the peak temperature has a primary effect on the ability of the PCB to withstand assembly without delamination or ruptured vias. Materials have been developed to withstand these higher lead-free assembly (LFA) temperatures with residual life cycle reliability. However, quality and reliability of any unique part is codependent on processing and design variables in conjunction with the proper material selection.

27.2.1 IPC-4101 Specification

To facilitate the many choices of laminates and their associated properties, industry standards groups such as the IPC have defined minimum performance specifications and have issued several specifications to inform the selection process. Some of the most commonly used material specifications are those that deal with laminate, prepreg, and copper foil. IPC-4101, "Specification for Base Materials for Rigid and Multilayer Printed Boards," and IPC-4652, "Metal Foil for Printed Wiring Applications," are the primary specifications for clad laminates, prepregs, and foils. Another specification, IPC- 4104, "Specification for High Density Interconnect (HDI) and Microvia Materials," deals with many of the new materials for HDI, such as epoxy-coated microfoils, as discussed in this chapter.

The bare-board base materials are identified by using the classification system found in IPC-4101, which has specification sheets for various grades of material. The specification sheet is also known as a *slash sheet* since each material specification is preceded by a forward slash (/). Each sheet's heading typically identifies the resin system, the reinforcement, flame retardant, presence of fillers, as well as some performance parameters, such as glass transition temperature (T_g) or decomposition temperature (T_d). More detail is in Sec. 27.2.2.

27.2.1.1 Keywords. IPC-4101B includes "Keywords" to assist the user to select slash sheets that have applicability to a typical use. The specification cautions that Keywords are not specification requirements and "having a particular Keyword in the header section, does not guarantee the acceptability of this material for a specific application." Nor are Keywords meant to convey the "only" use for that material. The Keywords used in this chapter to point to particular materials by slash sheet are: "RoHS compliant," "Low Halogen Content," "CAF Resistance," and "High Decomposition Temperature."

27.2.1.2 Limitations of Specification. IPC-4101B also includes guidance on qualification assessment for base materials. The user of this specification should be aware, however, of the boundaries and caveats of this specification and the consequent responsibility on the stipulator to understand these limits. The tests proscribed in IPC-4101 are on either raw cores or prepregs laminated into simple cores. The manufacturer of base laminates must test its laminates for the inherent properties of the materials, before transformation by fabrication processes. This means that the data are relevant to compare materials to a baseline value and like materials to each other, but these are not tests for post-fabrication long-term reliability or time to failure in the end-use environment (with the exception of CAF resistance, which is an optional test), although they can be and are indicators of reliability for a particular use when properly understood.

27.2.1.2.1 Conductive Anodic Filament (CAF) Resistance. The Keyword "CAF Resistant" is more of a descriptive than quantitative term. CAF growth resistance is tested per test method 2.6.25, but the acceptability of the results is stated only as "Pass/Fail." Acceptability is as agreed upon between user and supplier (AABUS). Furthermore, the voltage applied during the test is a variable, although the specification recommends 100VDC for material evaluations. The "CAF Resistant" Keyword essentially means that the laminator has conducted some level of CAF testing on the material and/or has optimized some element of the system to reduce CAF propensity. When reviewing a laminator's CAF test results, the user should know what applied voltage was used and if that is sufficient for the user's product reliability requirements, the test population, and the distribution of the "failed" nets as a function of the interaction of distance and time.

27.2.1.2.2 RoHS-Compliant. The Keyword "RoHS Compliant" is also descriptive, intended to convey that the laminate is sufficiently thermally robust to withstand the rigors of lead-free reflow assembly temperatures. For a more quantitative approach, look for materials that comply with slash sheets that show results for time to delamination (T_d) by thermal mechanical analyzer (TMA). This test is optional and the results, given in minutes to delaminate at three benchmark levels of 260, 288, and 300°C, are minimums. The establishment of these levels and the minimum time to delamination puts a laminate in the class of the more thermally robust materials, but does not guarantee any minimum performance for a particular multilayer board (MLB) in its specific assembly temperature exposures.

Table 27.2 shows elements defined in the slash sheets in IPC-4104. Table 27.3 shows 20 slash sheets from the 55 in IPC-4101B, and some of the key elements from over a dozen listed properties.

27.2.2 Critical Properties

In addition to providing circuit interconnection, a multilayer printed wiring board (ML-PWB) provides the electrical and mechanical platform for the system. This means that the electrical and thermal properties of the ML-PWB material are very important for the proper functioning of the system. Among the properties of importance are dielectric constant, D_k (also known as *Er*); dielectric loss, D_f (or $\tan\delta$); glass transition temperature, T_g; time to delamination, T*xxx*; thermal decomposition temperature, T_d; coefficient of thermal expansion, CTE; and moisture absorption. The following sections discuss the importance of these properties to an ML-PWB substrate.

27.2.2.1 Electrical Properties. Electrical properties include the dielectric constant and dielectric loss.

27.2.2.1.1 Dielectric Constant (Dk). Impedance and transmission speed are both affected by D_k. Since impedance must be matched throughout a high-speed circuit, this means that the impedance of other circuit elements determines the required impedance of a circuit line. The designer can use standard equations to select values for line width, dielectric thickness, and dielectric constant that will achieve the desired impedance (47, 50, and 60 ohms are common requirements). However, factors such as interconnection density and ML-PWB imaging capability can constrain the line width, leaving the trade-off available to the designer between dielectric thickness versus dielectric constant. As the average number of layers in ML-PWBs increases, the

TABLE 27.2 Slash Sheet IPC-4104 Series

Slash sheet IPC-4104	T_g (°C)	Resin system	Reinforcement
4104/12	>140	Epoxy	N/A
4104/19	240	Polyphenylene ether	N/A
4104/20	180	Epoxy blend	N/A
4104/21	150	Epoxy blend	N/A
4104/22	120	Epoxy liquid, epoxy coated	N/A
4104/23	240	Epoxy	Non-woven aramid

TABLE 27.3 Header Information from Selected Specification Sheets in IPC-4101B

IPC-4101 slash sheet	Resin system	Reinforcement	Fillers	Flame retardant	T_g (°C) min.	T_d (°C) min.
/13	Polyester, vinylester	Woven E-glass	Inorganic fillers	Bromine	N/A	Not stated
/24	Multifunctional epoxy	Woven E-glass	No	Bromine	150	Not stated
/41	Multifunctional epoxy	Woven Aramid	No	Bromine	150–200	Not stated
/42	Polyimide	Woven E-glass	With or without	N/A	220	Not stated
/53	Polyimide	Non-woven Aramid paper	No	N/A	220	Not stated
/55	Multifunctional epoxy	Non-woven Aramid paper	No	Bromine	150–200	Not stated
/70	Cyanate Ester	Woven S2-glass	No	Bromine	230	Not stated
/71	Cyanate Ester	Woven E-glass	No	Bromine	230	Not stated
/92	Multifunctional epoxy	Woven E-glass	No	Phosphorus	110–150	Not stated
/93	Multifunctional epoxy	Woven E-glass	No	Aluminum hydroxide	110–150	Not stated
/94	Multifunctional epoxy	Woven E-glass	No	Phosphorus	150–200	Not stated
/95	Multifunctional epoxy	Woven E-glass	No	Aluminum hydroxide	150–200	Not stated
/96	Polyphenylene Ether	Woven E-glass	No	Phosphorus	175	Not stated
/97	Multifunctional epoxy	Woven E-glass	Inorganic fillers	Bromine	110	Not stated
/99	Multifunctional epoxy	Woven E-glass	Inorganic fillers	Bromine	150	325
/101	Di-functional epoxy	Woven E-glass	Inorganic fillers	Bromine	110	310
/121	Di-functional epoxy	Woven E-glass	No	Bromine	110	310
/124	Multifunctional epoxy	Woven E-glass	No	Bromine	150	325
/126	Multifunctional epoxy	Woven E-glass	Inorganic fillers	Bromine	170	340
/129	Multifunctional epoxy	Woven E-glass	No	Bromine	170	340

dielectrics between layers must decrease to keep the overall thickness constant. Thinner dielectrics require either lower D_k value material or thinner lines for constant impedance, sometimes both. Low D_k value materials are frequently the enabler to achieve higher layer count PWBs.

There are also applications that require higher D_k material. This allows finer lines to achieve a target impedance when dielectric thickness is already established, providing for greater densification.

Transmission speed is important because the signal transit time affects device timing and determines at what circuit length transmission line effects become important. The transmission speed of an electromagnetic wave in a dielectric medium is the speed of light divided by the square root of D_k. Air has a D_k of 1.0, and electromagnetic waves travel at the speed of light (12 in./ns). At 1 GHz, standard FR-4 ML-PWB materials have a D_k of 4.4 and the transmission speed is reduced to 5.7 in./ns. With a material having a D_k of 3.4, transmission speeds reach 6.5 in./ns, while even lower D_k materials can achieve speeds up to 8 in./ns. Although these are small improvements, the higher transmission speed provided by a low-D_k material may be important in some high-speed applications.

27.2.2.1.2 Dielectric Loss (D_f or Tan δ). The energy absorbed by the dielectric media is called dielectric loss. Attenuation is proportional to tan δ and signal frequency. For standard FR-4 ML-PWB materials, tan δ is 0.02, which translates into serious losses at frequencies above 1 GHz. For circuits operating above 1 GHz, a lower loss material is required. There are a number of material choices for lower dielectric attenuation, including blended epoxies, hydrocarbon ceramic, polytetrafluoroethylene (PTFE), and PTFE with ceramic. The tan δ values of these materials are approximately .01, .004, .002, and .001, respectively.

27.2.2.2 Thermal Properties. The important thermal properties of a laminate are the glass transition temperature, the coefficient of thermal expansion, the time to delamination, and the decomposition temperature.[1,2,3] These properties quantify the material's reactions to extreme temperatures and so are indicator's of the materials' suitability for a particular reflow profile and residual capacity for withstanding heat input (such as rework or hot use environments). T_g alone is insufficient to predict a materials response to LFA temperatures. In fact, since each test measures a different response to temperature, all the tests together provide a broad determination of suitability to a particular use.

27.2.2.2.1 Glass Transition Temperature (T_g). The T_g of a resin is the temperature at which the resin reversibly changes from a glassy state to a rubbery state. This loss of modulus creates an effective limit on the operating temperature of the system. T_g also affects the thermal fatigue life of the plated holes in the ML-PWB. Higher values of T_g translate into lower total z-axis expansion, which in turn means less stress on the plated through holes (PTHs), all other variables being held constant. There are three methods that are currently used to evaluate T_g: differential scanning calorimetry (DSC), thermomechanical analysis (TMA), and dynamic mechanical analysis (DMA). Results from the three methods vary since they measure different properties associated with the glass transition. Each has its rationale, strengths, and weaknesses. It is important to note not only the value of T_g given for a particular material, but the method by which the T_g was determined; only then can meaningful comparisons between materials be made.

T_g measurement by differential scanning calorimetry (DSC) defines the T_g as a change in the heat capacity of the material. The deflection in the heat rate absorption curve, W/g/°C, is used to identify the second-order thermodynamic change from a glassy solid to an amorphous solid that is the glass transition.

DSC is the method most commonly used by laminators for determining, and reporting, the T_g. The sample for DSC weighs between 15 to 25 mg and is tested with copper foil on both sides. This method is well suited to the testing of laminates and provides a precision measure for T_g on cores and cured prepreg (laminated to foils) for the laminator. These results are used for product acceptance and process control. DSC results are generally 5 to 10°C higher than when the test is conducted by TMA.

T_g measurement by thermomechanical analysis (TMA) determines the T_g by the change in thermal expansion as the polymer goes from glass state to rubbery state due to a change in free molecular volume. A change in the slope of the z-axis expansion rate indicates the glass transition. Typically, the data sheet will report the T_g and the CTE (discussed in more detail later in the chapter) in the z-axis above and below the T_g ("% CTE < T_g" and a "% CTE > T_g"), where the slope (viewed when graphed with temperature on the x-axis and percentage expansion on the y-axis) is much steeper above the glass transition. Alternately or additionally, the data sheet may report the percent thermal expansion (PTE), which is the total change in thickness over a temperature range of interest—for instance, from a temperature below T_g to the anticipated soldering temperature (e.g., 245 °C, 260 °C). IPC-4101B slash sheets 99, 101, 121, 124, 126, and 129 have a new requirement for reporting z-axis CTE in %TE from 50 to 260 °C.

Both the slope above T_g and the temperature at T_g are important to understanding the total expansion of the material. When the data sheet lists PTE, and the top temperature is sufficient to your application, then it is far easier to compare competing materials to each other and their expansion response to the assembly temperature.

For Test Method TM-2.4.24 (TMA), the sample is 0.25 × 0.25 in. × 0.20 in. minimum; for increased accuracy, the sample should be between 0.030 and 0.060 in. The sample has no internal or external copper, with only the resin, fillers (if applicable), and reinforcing system affecting the result. Test Method 2.4.24.5 (TMA) is used to determine the CTE ppm/°C values for the x- and y-axis, using Method B, a thin specimen (≤ 0.020 in.); it can also provide the T_g and PTE. TM 2.4.24.5,(TMA) Method A, will provide z-axis expansion (but not x- and y-axis expansion), T_g, and PTE for thicker specimens (≥ 0.020 in.).

Since TMA measures T_g as a function of the expansion of the material, and will also provide the z-axis CTE, it is the method fabricators use most commonly for T_g determination to infer a materials' thermal stability during assembly. For dielectric materials used for the formation of HDI layers (thinner layers, = 0.006 in., adjacent to or forming the outerlayer, and having microvias to provide interconnection with the balance of the construction), IPC-4104 states that the preferred method for determining T_g and CTE above and below T_g is Test Method 2.4.24.5.

T_g measurement by dynamic mechanical analysis (DMA) determines the elastic modulus (or storage modulus) as a function of temperature, thereby identifying the glass transition region in plastics. The T_g is accompanied by a rapid reduction of the flexural strength. DMA typically gives T_g values 10 to 15°C higher than DSC.

DMA has not traditionally been used by the board fabrication industry; however, some evidence suggests that this technique might be a better differentiator for and indicator of a materials' inherent thermal stability in the LFA reflow environment since it combines temperature with torsional oscillation. Additionally, this test has the advantage of being very sensitive and accurate with thin film materials.

27.2.2.2.2 Time to Delamination (Txxx). Time to delamination is a test to determine the elapsed time at an elevated temperature when a sudden and irreversible expansion, indicative of a delamination, occurs. This is at a temperature higher than the T_g of the material. Historically this test called for an isothermal temperature of 260°C (hence the abbreviated common test name, T260) until expansion due to delamination is detected by TMA. See Fig. 27.1 for an example of a TMA plot. For common materials, the T260 varies from a few minutes to hours. With the advent of LFA, higher temperatures 288°C and 300°C are now used to evaluate and differentiate materials. The test is the same as T260 with the exception of the isothermal temperature. IPC-4101 RoHS-compliant laminates pass T260 at 30 min. minimum, T288 at 15 min. minimum, and T300 at 2 min. minimum, when tested by TM 2.4.24.1.

27.2.2.2.3 Thermal Decomposition Temperature (T_d). The thermal decomposition of a laminate is a test to determine the temperature at which a set weight loss percentage occurs. This is indicative of an irreversible degradation of the chemical bonds within the system coincident with the alteration of the laminate's physical properties, or indicative of outgassing of volatile products. The mass of the sample is weighed; the sample is then placed in the thermogravimetric analyzer (TGA). The temperature at which there is a 2 percent weight decrease is

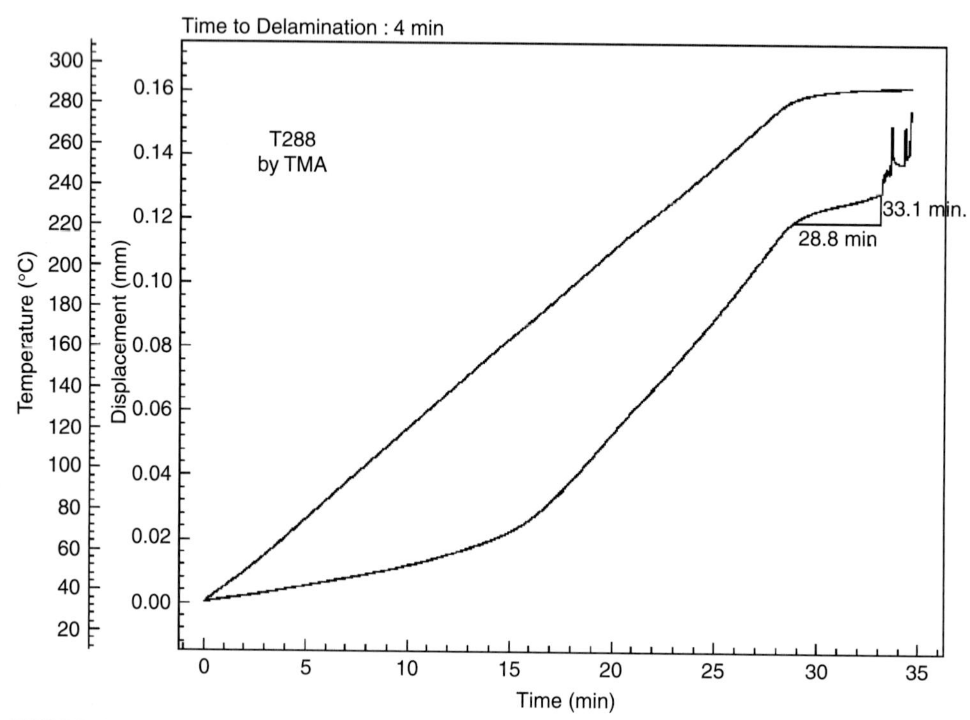

FIGURE 27.1 TMA plot: time to delamination at 288°C.

recorded (T_d 2 percent), and the temperature to a 5 percent weight decrease is subsequently recorded (T_d 5 percent). IPC-4101 RoHS-compliant laminates pass T_d 5 percent at 340°C minimum. Delta T_d (T_d 5 percent–T_d 2 percent) is considered by some as an indicator for the onset and rapidity of decomposition. For laminates with identical T_d 5 percent values but differing T_d 2 percent values, the lower the delta T_d the more thermally resistant the material. Note that T_g and T_d do not correlate. A low T_g material can have a high T_d and vice versa. Materials with identical T_g can have different T_d. See Fig. 27.2 for an example of a TGA plot.

27.2.2.2.4 Coefficient Of Thermal Expansion (CTE). The CTE of a laminate is comprised of values for x-y axis, or in-plane expansion, and z-axis, or vertical expansion (which is further comprised of the percent expansion below and above the glass transition temperature). The data are reported in ppm/°C; a typical plot is shown in Fig. 27.3.

The x-y CTE for standard FR-4 at 14 to 20 ppm/°C is higher than for ceramic or silicon. The resulting thermal expansion mismatch between an ML-PWB and the attached devices can lead to solder joint fatigue failures when the system undergoes multiple heat cycles during power-up and power-down. Packages with compliant leads can accommodate the CTE mismatch and so can be used with standard ML-PWB material systems.

The z-axis CTE for standard 130 to 135 °C T_g FR-4 is typically about 40 ppm/°C T_g and 250 ppm/°C > T_g. At LFA temperatures of 245 to 260°C, the material expansion is so dramatic that the traditional materials experience delamination and PTH ruptures. The new generation of Lead Free Assembly Capable(LFAC) FR-4 laminates generally utilizes one or both of the following approaches to minimize the CTE above T_g: a change to non-dicy curing agents, typically phenolic, and/or the dispersion in the resin of some amount of inorganic fillers to restrain the z-axis expansion.

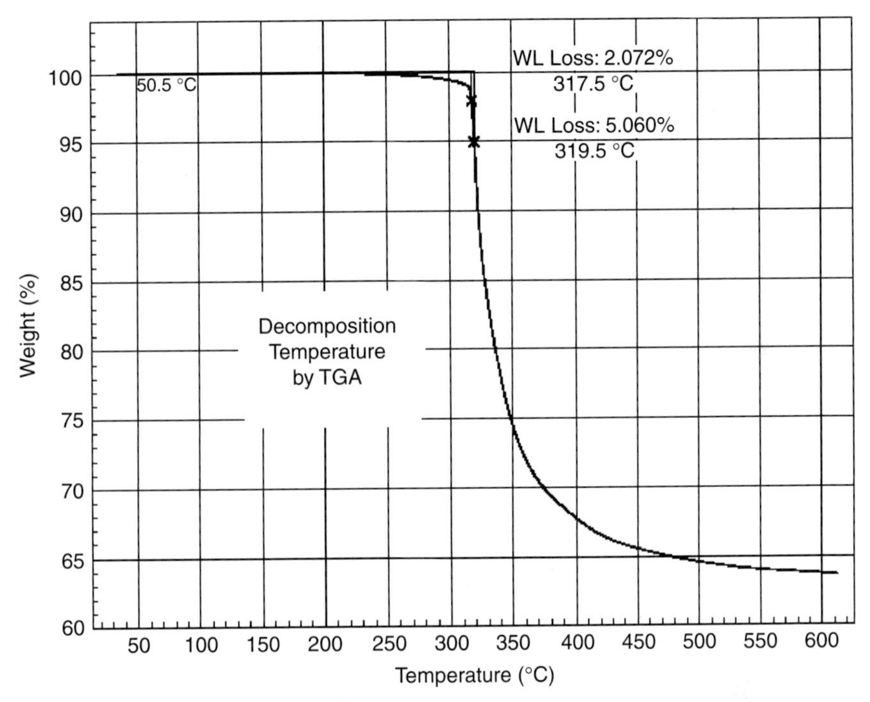

FIGURE 27.2 TGA plot: decomposition temperature.

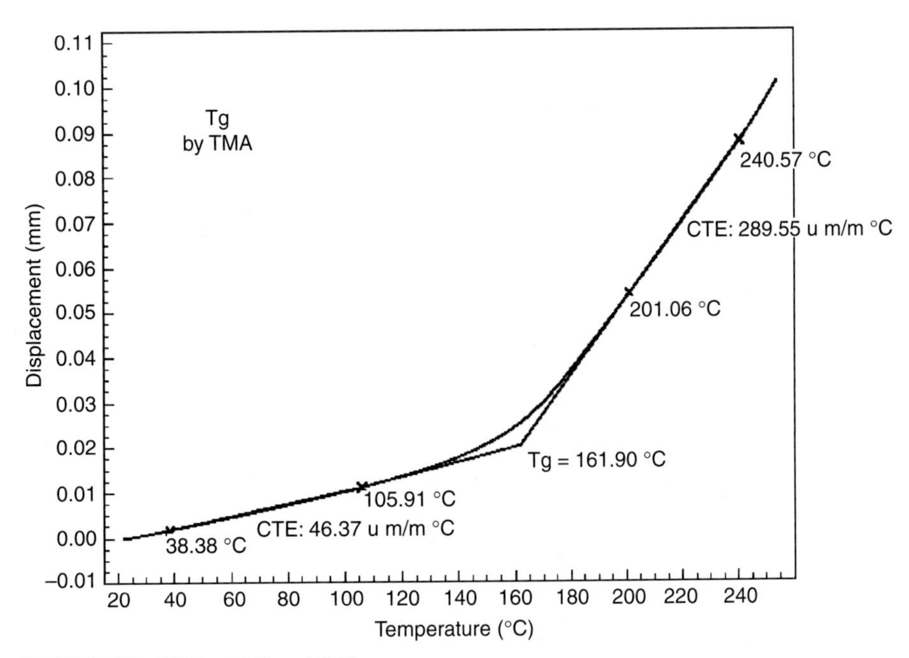

FIGURE 27.3 TMA plot: T_g and CTE.

27.2.2.2.5 Moisture Absorption. Moisture is the enemy of an ML-PWB. Absorbed water raises D_k somewhat and can raise tan δ significantly, hampering the functioning of the circuit at high frequencies. For applications in humid environments, a low moisture-absorbing material should be selected. Additionally, moisture absorption increases leakage current, which in turn degrades the ability of the material to resist CAF formation under bias.

Trapped moisture expands with temperature, so the board sees more thermal expansion (additive to the effects of the laminate CTE) at assembly, which causes thermal defects such as delamination, blistering, or barrel cracking. The severity of these problems depends on the storage environment prior to soldering and the peak temperature of the reflow. If storage times are short, humidity is low, and reflow temperature is moderate, moisture will not be a serious issue. However, if the ML-PWB material has high moisture absorption, or is subjected to high humidity for an extended period of time, or if the assembly temperature is especially high, special care must be taken. This can include storing with a desiccant, baking prior to assembly, or both.

27.2.3 Properties of the Resin Systems

Materials with highly cross-linked epoxy resin reinforced with woven fiber glass are the most common in use. Bromine is reacted with the epoxy matrix and is used to provide fire retardancy. Most epoxy-based materials satisfy the Underwriters Laboratories (UL) classification of V-0 for fire retardancy. The generic term for this family of epoxy resin materials is "FR-4," with "FR" standing for flame retardant and "4" an assigned number indicating "epoxy." Epoxy-fiberglass materials are sold by many suppliers and have become a commodity material.

27.2.3.1 Epoxies. A variety of epoxies are available for use, including halogen-free epoxies and materials with enhanced thermal, electrical, and mechanical properties.

27.2.3.1.1 Standard Epoxies. Two resin systems are used to make GF laminate: difunctional and tetrafunctional. These systems are distinguished by the nature of the epoxy cross-linking. In a difunctional system, the epoxide molecule has two cross-linking sites, and the cured epoxy contains long linear molecular chains. Pure difunctional laminates have excellent physical properties, and for many years were the mainstay of the industry. They have a T_g of 120°C, which is adequate for most use environments, but is low for some applications and too low for LFA.

The epoxide molecule of a tetrafunctional epoxy has more than two cross-linking sites. This allows a high cross-link density and a high T_g. A pure tetrafunctional system is expensive and difficult to work with. To meet the need for a T_g, above 120°C, laminators blend difunctional and tetrafunctional resins, producing a mixture referred to as multifunctional. Around 1985, some laminators began to sell a multifunctional epoxy blend with a T_g between 130 and 145°C. This blend was called tetrafunctional, even though it actually contained both difunctional and tetrafunctional epoxies. This blend is available at little or no price premium over a difunctional laminate. Laminates with a higher fraction of tetrafunctional resin are available. These systems, which are called multifunctional, have T_g values in excess of 170°C and are sold at a price premium of approximately 10 percent over difunctional systems. Multifunctional blends often have lower moisture absorption and higher thermal degradation temperature than difunctional systems. However, in some multifunctional systems, these properties are not improved. Care must be taken in selecting a multifunctional system to ensure that all properties of importance are enhanced.

Modern ML-PWBs often operate at elevated temperatures due to the heat output from semiconductor devices. As boards become thicker and holes smaller, these thermal cycles result in an increasing threat to the reliability of plated holes. For example, plated holes have been shown to fail when subjected to multiple thermal cycles up to temperatures near the T_g of the base material. These cycles can easily occur when a high-power device is turned on and off. One solution to this problem is to use materials with a higher T_g.

ML-PWBs often have surface-mount devices on both sides of the board, and they receive three or more solder operations during the assembly of connectors and devices. In addition, because of the value of a completed assembly, the board must be able to withstand additional soldering operations needed for occasional removal and replacement of defective devices. Boards made with difunctional GF epoxy can suffer from lifted lands, cracked PTH barrels, or substrate blisters during these multiple soldering operations. The solution is to use materials with low moisture absorption and high thermal degradation temperature.

27.2.3.1.2 RoHS-Compliant Epoxies (Non-Dicy Cure Agent). Conventional epoxy resin systems have commonly used dicyandiamide (dicy) as the curing agent. Experiments reported in the industry have concluded that dicy-cured FR-4 PWBs even with high T_g and optimized PWB fabrication do not withstand the thermal stress of lead-free assembly conditions. Phenolic-cured (novolac) epoxies had demonstrated better thermal stability than the dicy-cured systems. Many of the new RoHS-compliant LFAC laminates use a phenolic curing agent. Many of the new laminates, in addition to changing the curing agent, also incorporate some percentage of inorganic fillers to restrain the z-axis expansion of the system.[4]

These changes fundamentally impact the fabrication of the base materials into ML-PWBs especially in the lamination, drilling, and hole-preparation processes. The process impacts will be discussed in later sections.

27.2.3.2 Halogen-Free. Halogen-free laminates are a response to the European Union direction to require original equipment manufacturers (OEMs) to reclaim electronics to remove hazardous substances from the waste stream (leaving aside the discussion of which substances are hazardous and how hazardous those substances are considered to be). Waste management of PCBs containing halogens will be costly. Halogen-free (HF) technology is sometimes used to refer to laminates that have no tetrabromobisphenol A (TBBPA), which is the primary flame retardant used in FR-4 laminate. TBBPA is an organic molecule whose composition includes approximately 59 percent bromine. There is no legislation or regulation calling for the removal of TBBPA; however, the driver for halogen-free technology is the eco-friendly consumer market. For this reason, these laminates are also sometimes called "green" laminates. HF laminates can be identified by a non-bromine flame retardant in the composition.

The main chemical families used for nonhalogenated flame retardants are phosphorous compounds, nitrogen compounds, inorganic fillers, and compounds containing aluminum, magnesium, or red phosphorous. Curing agents are available that improve cross-linking with epoxy systems, completely curing the resin. These proprietary curing agents, when combined with fillers such as alumina trihydrate (ATH) up to 50 percent by weight, can achieve higher T_g values, comparable to multifunctional epoxies, and provide better flame retardancy. Various formulation recipes can lower CTE and decrease moisture absorption.

HF laminates, however, require different processing than brominated FR-4s, in lamination, wet processing, and drilling. Also, HF laminates differ from each other based on the flame-retardant system; they are not as similar to each other as one FR-4 is to another FR-4. The fabricator needs to work closely with the laminator to develop processes suited to the particular formulation. The engineer specifying the HF material for the product needs to understand both the properties of the specific brand of HF material, and the capabilities of the fabricator to produce quality MLBs given the design's properties and performance requirements.[5,6,7]

27.2.3.3 Materials with Enhanced Thermal Properties. There are three alternate resin systems commonly available for applications requiring thermal properties superior to those available from the enhanced multifunctional epoxies: polyimide (PI), cyanate ester blends (CE and BT), and polyphenylene oxide (PPO) blends.

27.2.3.3.1 Polyimide. Probably materials in the polyimide family provide the best thermal stability. This family of materials has a T_g in excess of 250°C, a high thermal degradation temperature, and a CTE less than that of epoxy. Polyimide resins can be coated on glass fabric to produce an ML-PWB substrate that processes like epoxy. With a high T_g and high thermal degradation temperature, polyimide ML-PWBs provide high-temperature reliability.

For systems that must operate at temperatures above 200°C, polyimide is a good choice. The combination of a high T_g and a relatively low CTE results in excellent fatigue life for plated holes. This makes polyimide a good candidate for very thick boards and for applications where the system must survive multiple thermal cycles over a wide temperature range.

Polyimides have several disadvantages. Early versions used a solvent called MDA, which is widely viewed as carcinogenic. Although there was no evidence of MDA release during ML-PWB fabrication or use, many fabricators refused to process MDA-containing polyimide. Fortunately material suppliers have been able to formulate MDA-free versions of polyimide, and most polyimide on the market today is rated as MDA-free. The second disadvantage of polyimide is fast moisture uptake. Generally, polyimide users combat moisture by including multiple bakes in the board fabrication and assembly processes. The third disadvantage of polyimide is cost. The cost of a polyimide laminate is up to two to three times that of an epoxy laminate. In addition, ML-PWB fabrication requires a high-temperature lamination.

Polyimide material has been used in low volume since the early 1980s. Broad acceptance was always limited by cost and fabrication issues. This material family should be specified only for systems that must operate with high reliability in an extreme environment, such as a high ambient temperature or extremes in thermal cycling. This includes some military applications and potentially a few consumer applications such as under-hood automotive electronics. For most commercial applications, lower-cost materials—such as conventional epoxy blended with CE or PPO resin—may be used.

27.2.3.3.2 Cyanate Ester Blends. The second most stable resin system is the triazine or cyanate ester family. In its pure form, the cyanate ester resin is brittle and is difficult to drill without cracking. Low peel strengths are typically associated with its use. In addition, cyanate ester is an expensive resin system. As a result, cyanate ester resin is often blended with epoxy and a small amount of polyimide. This blend, called BT (after its two ingredients, bismaleimide and triazine), can be coated on conventional glass cloth to produce a laminate.

BT laminates have a T_g of 180°C and a high degradation temperature. For most high-temperature applications, they are a direct substitute for polyimide. They have the added advantage that the moisture sensitivity and processibility of BT are much closer to those of conventional epoxy than those of polyimide. Other than a high-temperature post lamination bake required for a full cure, the BT process is the same as an epoxy process. The major drawback of the BT laminate is cost. Although BT is much less expensive than polyimide, it is still up to 1.5 times the cost of epoxies. The result is that BT is a popular replacement for polyimide, but its use is limited to specialty applications. The trend for most high-temperature commercial applications is either a multifunctional epoxy or an epoxy-PPO blend.

27.2.3.3.3 Polyphenylene Oxide (PPO)-Epoxy Blends. One of the cost-effective high-temperature laminates in use is a material based on a blend of PPO and epoxy resin. This material has a T_g of 180°C and a T260 of 60 minutes or more. When coated on glass, the PPO blend processes like epoxy. The only exception is the need for a high-temperature bake to achieve full cure. The major advantages of this laminate family are very low moisture absorption and a small cost premium. The disadvantage of PPO is that it has a broad T_g with softening beginning well below 150°C. This has an adverse effect on the ultimate fatigue life of high-aspect-ratio holes. Care should be taken in using this material in applications where the system must operate over a very large number of thermal cycles that include temperatures in excess of 130°C.

27.2.3.4 Materials with Enhanced Electrical Properties. The resin and the reinforcement determine the D_k and tan δ of a composite material. Standard FR-4, made up of epoxy resin and woven glass, has a D_k in the range of 4.0 to 4.4 and a tan δ of 0.02. This may be reduced somewhat by replacing some or all of the epoxy resin with PPO, cyanate ester, or PTFE. Further reductions require replacing the glass reinforcement, which contributes a D_k of 6.0 to the effective D_k of the system.

27.2.3.4.1 Cyanate Ester Blends. Both the tan δ and the D_k of the cyanate ester resin system are much lower than those of epoxy. The best results are achieved with pure cyanate ester. However, as mentioned earlier, this resin is brittle and difficult to drill. The BT blend provides an excellent compromise. BT has a D_k of 2.94 and a tan δ of 0.01. This improvement is useful in some applications; however, it is inadequate for many high-performance applications, and the approximately 100 percent cost penalty for BT has prevented its broad acceptance.

27.2.3.4.2 PPO Blends. As with cyanate ester, both the tan δ and D_k of the PPO resin are lower than those of epoxy. For processibility and cost reasons, a blend of approximately 50 percent PPO and 50 percent epoxy is generally used. This gives electrical performance similar to that of BT, but with a cost penalty of only 20 to 50 percent above standard epoxy. It is likely that PPO will find wide use in those applications such as high-performance workstations that need a small improvement in electrical performance. Unfortunately, the characteristics of a PPO-epoxy blend are often inadequate for very high-speed applications such as supercomputers and wireless (RF) applications.

27.2.3.4.3 PTFE-Based Laminates. This resin system is commonly known as Teflon. Of all resin systems in common use, the best electrical performance is provided by PTFE. The D_k of PTFE is close to 2.0 and the tan δ is less than 0.001. This is a significant improvement over all other materials. In addition to offering excellent electrical properties, PTFE has excellent thermal properties. It is a thermal plastic that can operate at temperatures above 300°C without softening, oxidation, or other forms of degradation. PTFE is naturally fire retardant, so it does not need bromine addition to achieve the UL rating of V-0. It has very low moisture absorption.

For ML-PWB applications, PTFE has three serious disadvantages:

- **Poor processibility** Because PTFE is a high-temperature thermoplastic, it is generally laminated in a high-temperature lamination cycle not achievable in conventional lamination presses. The fabricator can avoid this poor processibility by using a low-temperature adhesive, but this can lead to serious compromises with electrical and thermal properties of the laminate. The second major difficulty in processing PTFE is the hydrophobic nature of the material. This makes it difficult to clean and metallize the holes. Generally, plating must be immediately preceded by a special fluoride etch, TETRA-ETCH, that activates the hole wall and enhances wetting. As an alternative, some PTFE systems can be activated in plasma etch; however, this does not work with all PTFE materials.
- **High CTE** PTFE maintains full strength up to and above solder temperature. When this strength is combined with a high CTE, PTH reliability is a serious concern. This is a particular problem with glass-reinforced PTFE, where the x-y constraint of the glass magnifies the out-of-plane expansion. A second problem attributable to high CTE is excessive warpage whenever Teflon is combined with epoxy based laminate in an unbalanced hybrid structure.
- **Cost** PTFE-based laminates are often 100 times more expensive than epoxy-based laminates. This makes it difficult to justify the use of PTFE if any alternative exists.

At least three types of laminate systems are available with PTFE resin:

- **PTFE resin coated on a conventional glass mat** Although the construction of this laminate is most like standard FR-4, it is not recommended for ML-PWB applications. The large CTE of PTFE coupled with the in-plane strength of the glass reinforcement leads to a very high out-of-plane expansion that can easily fracture a plated hole during soldering. In addition, the glass cloth raises the D_k of the composite. This material is commonly used in applications such as antennas or RF microstrip cables, where the circuit is placed on one side of the substrate and a ground plane on the other. Plated holes are not needed, and a very low D_k is not critical.
- **Expanded PTFE film impregnated with cyanate ester or epoxy resin** This material has no glass cloth reinforcement. The resulting increase in in-plane CTE reduces the out-of-plane expansion, allowing the reliable use of plated holes. Although the non-PTFE resin

impregnating the PTFE film increases D_k and tan δ over those of a pure PTFE layer, this composite is at least as good as glass-reinforced PTFE. Because of the conventional resin surface of this material, it can be laminated using conventional methods and materials, allowing the fabrication of ML-PWBs containing a mixture of PTFE and epoxy layers. The resulting hybrid board can contain any symmetrical mix of epoxy and PTFE layers. The high in-plane CTE of the PTFE layer leads to severe warp in any nonsymmetrical stack-up. Hybrid boards are less expensive than pure PTFE boards because they are laminated in a conventional press and they minimize the use of costly PTFE layers. However, the high cost of PTFE and the need for a fluoride etch still makes these boards expensive. The major application for this material is in supercomputer designs that have high-speed circuits requiring enhanced electrical properties mixed with others that operate on FR-4. This approach does not work in applications where a mixture of RF signals and digital logic requires a nonsymmetrical design with low-loss material on one side and standard material on the other.

- **PTFE mixed with a low-D_k ceramic** A third approach, in which up to 60 percent ceramic (by weight) is combined with PTFE resin, results in a very interesting material. This material has a low CTE, low tan δ, and low D_k, and, by a proper choice of the ceramic, it is possible to have an in-plane CTE that matches that of epoxy. This allows the fabrication of an unbalanced hybrid structure without excessive warp. The high ceramic loading typical of this material minimizes PTFE use and reduces costs. A commercial version of this material is available with a cost in the range of four to five times that of standard FR-4, rather than the 100 times typical of other PTFE options. The only serious drawback of this material is the use of a special process such as plasma etch to ensure hole wall wetting. Success has been reported with using H_2 or He mixed with O_2 to activate the hole for plating.

27.2.3.5 Materials with Enhanced Mechanical or Conduction Properties. In cases where the device package (leadless area array with ball grids) is incompatible with the CTE of a standard material, a low-expansion substrate must be used. This can be achieved in several ways. In the past, leadless surface-mount technology (SMT) focused on replacing the woven glass in standard FR-4 with woven quartz or aramid fibers. Although this reduced the expansion of the substrate, both materials were expensive and difficult to process.

27.2.3.5.1 Aramid. A better solution for CTE match is to use a non-woven aramid mat material that is resin-impregnated; it is available with either modified epoxy or polyimide resins. Non-woven aramid has lower in-plane expansion under heat input, and is closer to the in-plane expansion of leadless ceramic chip carriers (LCCCs) or thin small outline packages (TSOPs). A low in-plane expansion reduces the strain on the solder joint, which in turn improves assembly yields and long-term field reliability.

As a random fiber, it also provides a more consistent environment for high-frequency signals since the material under the signal at any point along its length has the same effective dielectric constant. (Woven fiberglass-reinforced laminates impart discontinuous effective dielectric constant along the length of the conductor varying between the resin-rich and glass-dominant sections of the laminate.) In addition, the interface with the bottom of the conductor is smoother, reducing attenuation of the copper trace. The aramid layer is usually only the outermost layer where the chip is attached, whereas the balance of the layers consists of standard fiber-glass reinforced materials. Aramid-reinforced materials process much like traditional FR-4 and are compatible for mixed dielectric constructions. This material is also very compatible with laser drilling.

There are trades-offs to consider with aramid. The total thickness of the board is usually kept low, since the aramid has a high out-of-plane or z-axis expansion. Typically, a thickness of at least 0.015 in. is needed to offset the effective x-y CTE of the balance of the MLB, which in turn means that outerlayer line widths are wider for a 50-ohm impedance. Also to be considered is the lower copper peel strength and high moisture uptake.

27.2.3.5.2 Carbon Composite Laminates. Another past approach was to laminate a low-expansion metal such as Invar into the ML-PWB. The Invar plane could double as a heat sink or a power/ground plane. This approach, however, added much weight and thickness to the product and is in limited use.

New materials such as STABLCOR[8] are thinner and lighter. Epoxy glass systems have a thermal conductivity of approximately 0.3 W/m°K; the conductivity of copper is approximately 385 W/m°K; STABLCOR has products with a 325 and a 650 W/m°K conductivity capability. Carbon composite laminate has a very low negative CTE, approximately –1.10 ppm/°C, so it contracts very slightly as temperature rises. The user can tailor the surface CTE of a standard ML-PWB from 3 to 12 ppm/°C. With a very high tensile modulus of elasticity, this material adds considerable rigidity for applications needing high stiffness.

The material is used typically as a ground (GND) plane in the circuit. If 1 GND is used it must be in the center. If two are used then they do not need to be in the center, but they must be symmetrical within the stackup. Three can be used if one is in the center and the other two are symmetrical in relation to each other, and so forth, with even GNDs symmetrically placed and the last (odd) GND in the center. This does add height to the circuit. Also the D_k is higher at 13.36, but since the usage is as GND, that does not negatively affect the electrical performance of the circuit.

Holes not connected to the GND must be predrilled and back-filled to provide clearance from the plane (which is entirely conductive, as opposed to laminate cores, which are conductive only on the foil surfaces). The extra processing, extra layers, and cost of the material make this a specialty solution for specific applications needing high heat dissipation or very low CTE or very high stiffness.

27.2.4 Materials Summary

Material properties affect PWB operation in important ways. The preceding discussion focused on thermal and electrical properties. The use of small, high-aspect-ratio holes, thick boards, and high-power devices drives the need for materials with improved thermal properties. The transition to LFA requires materials with high thermal stability. Similarly, the use of high-speed devices, including the RF circuits in wireless applications, drives the need for improved electrical performance. Materials exist with improved properties, but the ideal material that combines low cost with improved thermal and electrical properties does not exist. Therefore, the designer must consider the needs of each design before selecting a cost-effective material.

The need for improved thermal properties includes a material with high T_g, low CTE, high thermal decomposition temperature, long time to delamination, and low moisture absorption. Polyimide meets the first four needs at the expense of increased moisture absorption and high cost. BT satisfies all these needs, but cost remains an issue. The most economical solutions are the multifunctional epoxies and the epoxy-PPO blends. Multifunctional epoxies are relatively inexpensive, provide a significant improvement in T_g, and, in some cases, provide small improvements in thermal degradation temperature and moisture absorption. The PPO blends provide significant improvement in all areas at a slightly higher cost. If moisture absorption is an issue, the PPO blend may be the best choice. For LFA the "RoHS-compliant" formulations may be necessary.

Both BT and PPO-epoxy boards provide small improvements in electrical properties. If the real need is to reduce D_k so that the board thickness can be reduced, these materials may be useful. However, if dielectric loss is a serious problem, the only real solution is to use PTFE. Unfortunately, PTFE is very expensive and difficult to process, so it should be specified only where absolutely necessary. For example, RF circuits will nearly always require PTFE.

The ceramic-PTFE blends used in a hybrid construction may be the most cost-effective solution for these applications.

27.3 MULTILAYER CONSTRUCTION TYPES

The construction of the rigid multilayer ML-PWB can take on many variations. To help categorize the various constructions, the IPC has developed industry PWB design specifications, defining them by class and type. Grouping the ML-PWB into categories facilitates the ability of designers and fabricators to communicate using a common format.

27.3.1 IPC Classifications

IPC classifications specify generic PBC design and rigid organic printed board structure.

27.3.1.1 IPC-2221: Generic Standard on Printed Board Design. There are many ML-PWB structures. This section discusses the methods and materials for the basic and several of the advanced structures. The IPC has two comprehensive standards for the design of rigid printed circuits, IPC-2222 and IPC-2226. The classification system within these standards for the ML-PWB by structure is shown. The distinction between the two is the focus on microvias in the later standard.

27.3.1.2 IPC-2222: Rigid Organic Printed Board Structure Design. This standard covers products with conventional feature sizes:

- **Type 3** This is a multilayer board without blind or buried vias (see Fig. 27.4)
- **Type 4** This is a multilayer board with blind and/or buried vias (see Fig. 27.5)

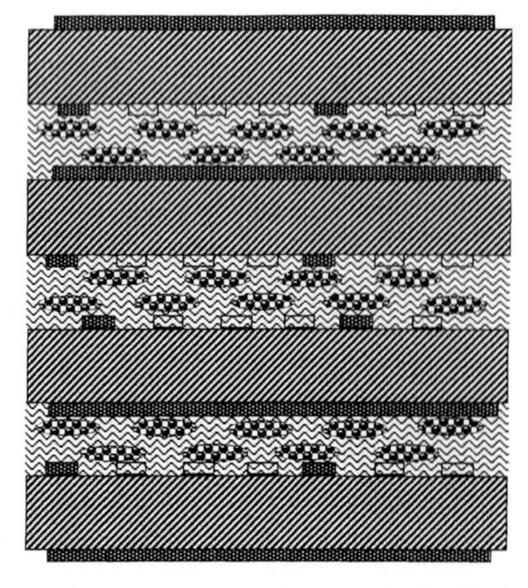

FIGURE 27.4 Type 3 (eight-layer ML-PWB) multilayer board without blind or buried vias.

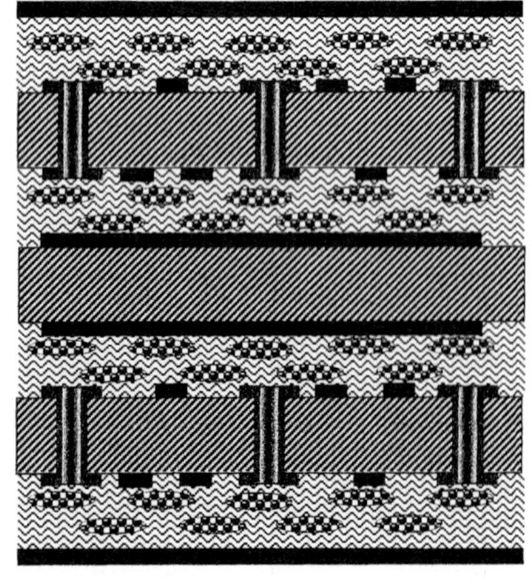

FIGURE 27.5 Type 4 (eight-layer-B/V MLPWB) multilayer board with buried vias.

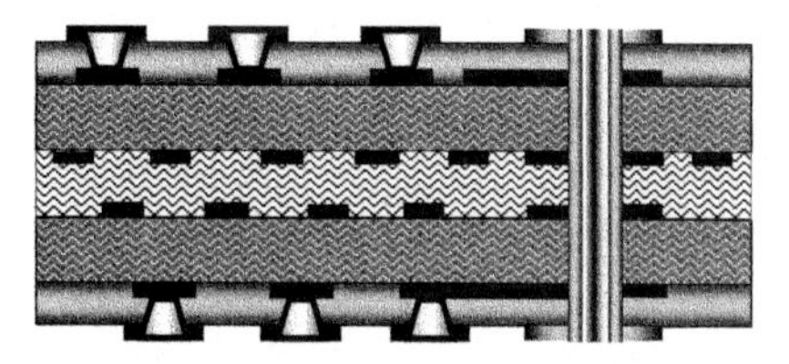

FIGURE 27.6 Type I (six-layer HDI MLPWB) high-density multilayer board with blind vias from top and bottom layers and through vias connecting the outerlayers.

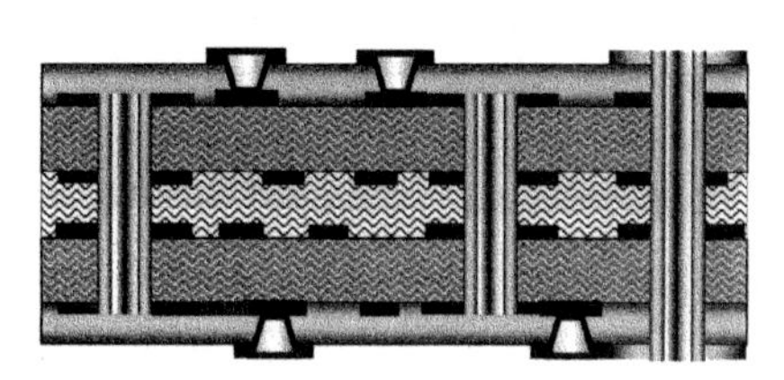

FIGURE 27.7 Type II (six-layer HDI MLPWB) high-density multilayer board with vias as well as buried vias in the core and through-vias connecting the outerlayers.

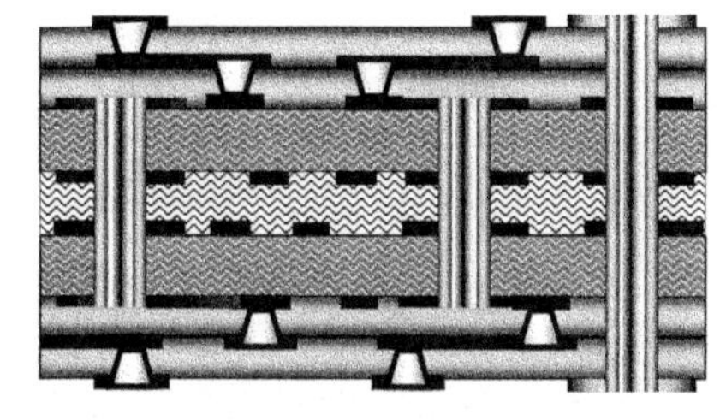

FIGURE 27.8 Type III (eight-layer HDI MLPWB) high-density multilayer board with blind vias as well as buried vias in the core and through vias connecting the outerlayers.

27.3.1.3 IPC-2226: Design Standard for High-Density Interconnect (HDI) Printed Boards. This covers products with high-density feature sizes:

- **Type I** 1[C]0 or 1[C]1 has through vias connecting the outerlayers (see Fig. 27.6).
- **Type II** 1[C]0 or 1[C]1 has buried vias in the core and may have through vias connecting the outerlayers (see Fig. 27.7).
- **Type III** >2[C]>0 may have buried vias in the core and may have through vias connecting the outerlayers (see Fig. 27.8).

27.3.2 Basic Type 3 ML-PWB Stack-Ups

The type 3 construction can be said to be the most basic of PWB multilayer technologies. An ML-PWB is fabricated by the bonding (laminating) of copper-clad details consisting of imaged and etched laminates (typically double-sided). The bonding medium, known as prepreg, is a B-staged (partially cured) reinforced resin. The imaged details consist of C-staged (fully cured) laminate. These material components are arranged by layering according to the design documentation. This layering method in fabrication, known as the stack-up, follows the layer numbering order of the design. The stack-up formation method is often loosely defined in the design documentation; therefore, a good understanding of the lamination options is necessary. The lamination options described in this discussion refer to methods used to form the outerlayers and to form layer pairs. The material resin system should be defined on the design documentation. Refer to IPC-2221/2222 for minimum suggested design documentation requirements.

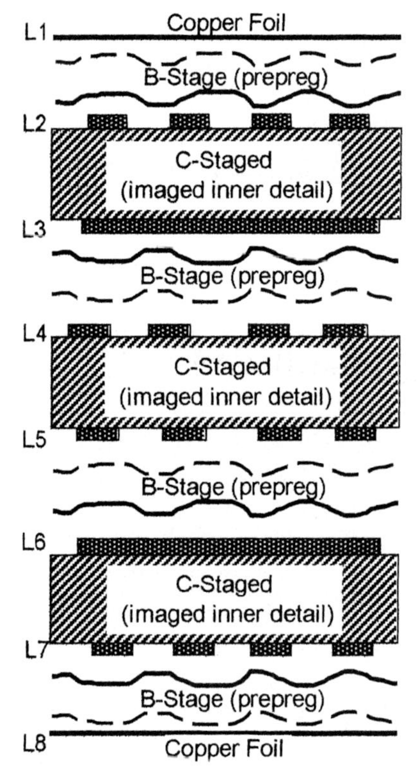

FIGURE 27.9 Foil outer stack-up (eight-layer ML-PWB), a typical stack-up for a foil board.

The basic ML-PWB stack-up can be constructed with two options to produce the outerlayers and to form layer pairs. A third option that employs single-sided "capped" clads is rarely used and is not discussed. Often, in the case of designs having an odd number of layers, a combination of these methods is employed. The following three-layer stack-up methods are discussed:

- **Foil outerlayer construction (foil outer)** In this stack-up, the outerlayers are formed by using a sheet of copper foil per side.
- **Clad outerlayer construction (clad outer)** A clad detail provides the copper for the outerlayer.
- **Odd-layer construction** This stack-up method balances a construction consisting of an odd number of layers.

27.3.2.1 Foil Outer Stack-Up. A board produced with copper foil outers is fabricated from one or more patterned innerlayer details and two copper sheets. The copper sheets form the outerlayers of the fabricated ML-PWB. This stack-up is the least expensive way to fabricate an ML-PWB and is by far the most popular design option. Figure 27.9 shows a typical stack-up for a foil outer board consisting of eight layers.

The stack-up shown contains three imaged clad details bonded with two sheets of prepreg at each opening and having two sheets of copper foil on the outside. When possible, the higher resin content prepreg ply should face the signal layer side. This is especially important when the signal is a heavyweight copper (2 oz. or more). The copper layers, numbered from L1 to L8, begin with the top foil layer (typically called the primary side). In the design pictured, the layers L2, L4, L5, and L7 represent signal layers. Layers L3 and L6 represent power/ground layers. When the final imaged patterned is produced, it can provide another signal layer pair or a set of ground layers, or a "pads-only" pattern to support vias and component holes, as well as the device footprints of the electrical components and their associated fan-out patterns. Some of the advantages of this bonding method are:

- **Lower raw material cost** Loose sheet copper foil and prepreg sheets are more economical than clad laminate.
- **Lower consumable material cost** The reduction in imaging resist and chemistry results in a cost savings.
- **Lower labor cost** The reductions in material handling and in the processes of imaging and pre-lamination result in less total labor.

27.3.2.2 Clad Outer Stack-Up. Figure 27.10 depicts the same eight-layer design as Fig. 27.9, except the outerlayers are formed with a clad laminate. This stack-up arrangement requires four clad details, as compared to only three for the foil-stacked board shown in Fig. 27.9. This makes the clad outer board more expensive than the more commonly used foil outer construction.

However, the clad outer design has one less B-stage opening and no copper foil. In addition, the clad outer boards are patterned on only one side prior to lamination. These factors partially offset the higher cost of this design. In Fig. 27.10, notice how the layer pairs have changed position placement. Layers L3 and L6 are now paired with a signal layer, which may be advantageous for controlling the dielectric thickness. See Sec. 27.3.3 for a discussion on controlled impedance. Some of the advantages of using the clad outer bonding method include:

- **Improved surface topography** When surface flatness is critical, the clad outer construction provides a smoother surface. When a signal layer of heavyweight copper is the first layer down, it is possible for weave texture imprinting to occur. This topography condition, sometimes referred to as "telegraphing," can sometimes imprint the circuit pattern when using a foil build-up.
- **Improved handling** Depending on the layer pairing, sometimes additional copper can be retained, as in a case where a signal-to-signal imaged detail can be avoided. This is mostly the case when the layer pair details are being produced on very thin core laminate.
- **Improved dielectric thickness control** Sometimes, it is necessary to control the thickness at a specific layer pair tightly. This could be due to high-voltage considerations or to aid in the precision spacing of a signal plane to a ground plane where controlled impedance is necessary. This is possible because the C-staged (fully cured) clad laminate typically has an improved thickness tolerance.

The chief disadvantage with a clad outer layer design is in the potential for lower yield. This is because the external copper foil must be protected by the image mask during all the internal layer image, etch, strip, and oxide processing steps; then the mask must be removed at a later point in the process so that the external details can be fabricated. The potential exists to scratch the photomask during all the handling steps, which will result in shorts or opens (depending at which point in the process the scratch occurs).

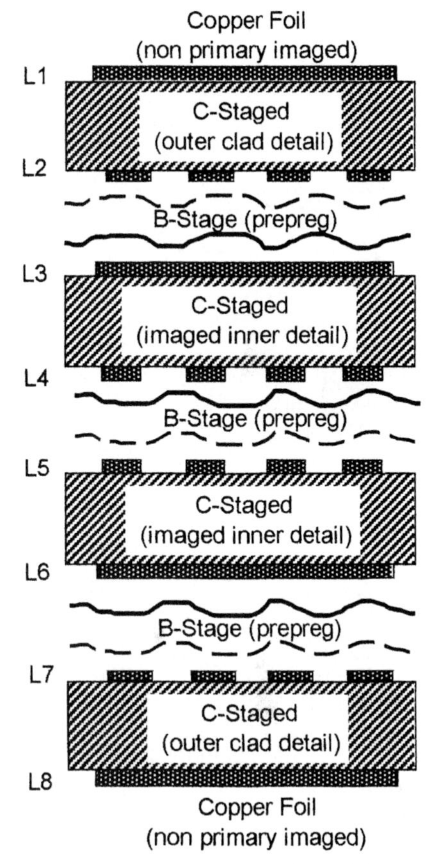

FIGURE 27.10 Clad outer stack-up (eight layer ML-PWB) multilayer board design with the outerlayers formed with a clad laminate.

27.3.2.3 Odd-Layer Stack-Up. Sometimes, due to signal routing or the need to have a greater dielectric space, an odd-layer circuit is formed. Similar in characteristics to the clad stack-up, the odd-layer construction technique employs a non-imaged single-sided clad laminate. Material can be procured with a single side, but a warning is associated with its use due to a lack of bondability. When this material is manufactured at the laminator supplier with only copper on one side, a release sheet is used against the non-copper side. The release sheet produces a slick, smooth surface. The non-copper surface thus requires an aggressive surface preparation to promote bondability. One alternative to gain the same benefit of the single-sided clad is to etch the copper off one side of a double-sided clad. The etched side becomes the bond surface, which has greater adhesion due to the copper tooth imprint left behind. See Fig. 27.11 for an example of a balanced odd-layer construction.

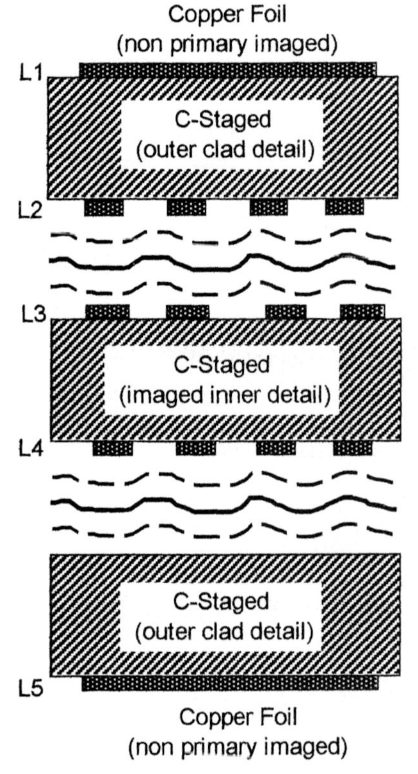

FIGURE 27.11 Odd-layer clad stack-up (five-layer ML-PWB), an example of a balanced odd-layer construction.

Some of the uses of the single-sided odd-layer stack-up are:

- **Odd-layer construction option** The single-sided construction technique allows an opportunity to supply a layer for circuit designs having an odd layer count.
- **Odd-layer core balance** One of the problems associated with a build-up of odd-layer circuits is the ability to maintain a balanced construction of the cores. Depending on the overall design thickness and layer count, this can become a major concern to reduce warping. Figure 27.11 depicts an example of an odd-layer stack-up with three equal core thicknesses.

27.3.3 Controlled-Impedance Stack-Up

The number of ML-PWBs containing controlled impedance signals is significant. The signal lines in an ML-PWB have a characteristic impedance value expressed in ohms (Ω). Active devices, connectors, and transmission lines also have characteristic impedance. In a high-speed circuit environment (typically starting at 2 to 3 MHz and above), if the signal energy is reflected rather than transmitted, the signal impedance mismatch is known to have discontinuity. This means that, to avoid signal losses due to reflections, all impedance characteristics must be matched. The physical relationship between a signal line and the surrounding power/ground layers determines the impedance of that signal line.

The controlled-impedance stack-up requires specific manufacturing discipline that must be adhered to in order to maintain the integrity of the designed circuit. An ML-PWB design requiring a controlled-impedance fabrication for a design for manufacturing (DFM) discussion between the designer and manufacturer in order to avoid producibility traps and induced error. Several resources are available to assist in developing the proper manufacturing parameters to ensure that the product meets the specified tolerance. IPC-2141, "Controlled Impedance Circuit Boards and High Speed Logic Design," and IPC-317, "Design Guidelines for Electronic Packaging Utilizing High-Speed Techniques," are two of the industry standard documents that contain formulas and examples of various impedance configurations.

Typical producible process tolerances for common impedance values are as follows:

- 50 Ω traces typically have a tolerance around 5 to 7 percent.
- 75 Ω traces typically have a tolerance around 12 to 15 percent.
- 100 Ω traces typically have a tolerance around 17 to 20 percent.

When tolerances are required to be tighter than this, it places more significance on the stack-up and etch of the circuit. The design requirement should be clearly specified in the ML-PWB documentation. The preferred way to specify an impedance requirement is in a drawing note rather than by applying dimensional boundaries. Firm dimensional boundaries do not allow the manufacturer freedom to meet the end requirement based on process knowledge.

It is recommended that during the design process software modeling be performed and then confirmed with the manufacturer before fabrication. Several excellent simulation software tools are available commercially. Some also exist as freeware and shareware. However, it is best to

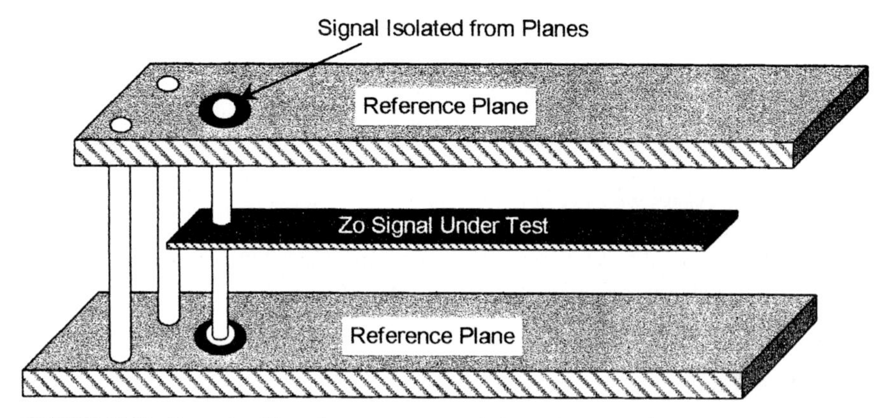

FIGURE 27.12 Example of impedance test coupon design and layout.

confirm the models' predicted values by actual process verification. The most versatile method of verification is with a test coupon. The test coupon should accurately represent the construction configuration of the stack-up, except it requires a few special prerequisites in order to yield reliable test data. Figure 27.12 shows a typical impedance test coupon stack-up. Note that the signal reference planes of the coupon are shorted together. Also, the simulated signal line is isolated and open-ended; the signal lines are usually at least 6 in. long. The footprint of the coupon via holes that feed through to the surface should be verified to match to the spacing of the time domain reflectometry (TDR) instrument probe. This will reduce the risk of rendering the specimen not testable or necessitating the use of expensive adjustable probes. The IPC test method IPC-TM-650 2.5.5.7 contains suggested spacing, but it is best to confirm this with the manufacturer's preference. Examples of some common impedance types are shown in Fig. 27.13.

The factors most influential on characteristic impedance are:

- **Dielectric separation (H)** The separation of the signal line between the reference planes has a significant influence on characteristic impedance. The variability of the dielectric layering must be reduced to minimize the effect on tolerance. This is where stack-up selection becomes critical when determining whether the signal-to-plane opening will be made with a clad laminate or within the bondable area of the B-staged (prepreg) resin. This is also where glass

Symmetrical Stripline

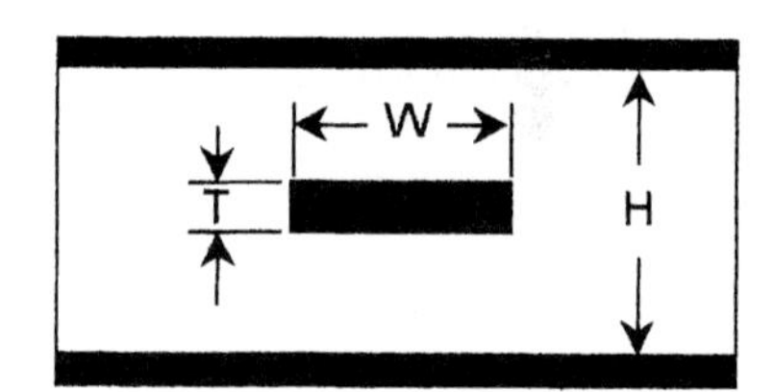

Offset Stripline

Edge-Coupled Symmetrical Stripline

FIGURE 27.13 Examples of common impedance stacks.

style selection becomes important. The glass style and subsequent resin content will have different effects on the nominal thickness obtained.

- **Conductor width (W)** The finished width of the conductor can produce variance, which is likely to occur from lot to lot. Therefore, process control measures are necessary when producing controlled impedance circuits. The density of neighboring circuits will have an effect on the final etch. Often it is advantageous to modify the artwork line width at phototool generation for predicted variance.
- **Dielectric constant (D_k)** Choosing a laminate resin system with a consistent dielectric constant has an influence on characteristic impedance over higher frequencies. The influence of dielectric constant becomes most critical when the ML-PWB design is a high-layer-count design. The lower the D_k value of the resin system, the thinner the overall board can be. Typically, the manufacturer has little opportunity to affect the dielectric constant, because the material type is specified by the design. Here, it is important that the designer/manufacturer know the D_k value range of the laminate supplier's resin system. Caution: Do not use the D_k value of the neat resin, but that of the composite laminate, which will vary somewhat with glass style. This is sometimes referred to as effective D_k (Eff D_k).
- **Conductor thickness (T)** The conductor thickness or copper weight can also affect the final impedance value. Here, as with conductor width, manufacturing process variations can have an inverse effect on the precision of the impedance value. Some modern software simulation tools, such as those from Polar Instruments, Inc., allow values for conductor profile (area) to be included in the simulation. This is more significant when lines are of heavy copper. It is best to avoid routing of impedance lines on plated subcores due to the added variability.

For external impedance tracks and internal plated-up tracks such as on subcores, autothieving patterns will minimize large variations in plating height between tracks in dense areas and those in uncongested areas. This will increase the fabricator's ability to produce impedance values consistently within a smaller range.

27.3.4 Sequential Laminations

When a design includes buried and/or blind vias, it typically requires a set of sequential lamination and plating cycles. These technologies are defined in the industry design standard IPC- 2221/2222 and are known as Type 4 ML-PWBs. These build types, when employing industry standard feature sizes, are mature in the industry. Complementing technologies, employing use of sequential processes for ML-PWB designs containing microvias less than 0.15 mm, are considered as build-up technologies. The terminology of build-up technologies encompasses many design stack-up variations; they can take on many forms and employ a multitude of methods. The build-up technologies defined in IPC-2315 and IPC-2221/2226 include categories named Type I, II, III, IV, V, and VI. The advanced build-up technologies use the materials found in IPC-4104. These include materials for layer forming, dielectric insulation, and interconnectivity. Included are photo-imageable and non-photo-imageable materials (liquid, paste, or dry-film nonreinforced dielectric); adhesive-coated dielectrics (reinforced and nonreinforced); and conductive foils and paste (coated and non-coated, photo-imageable). For a detailed discussion of these processes, see Chap. 22 and 23. This discussion will limit itself to addressing processes for standard technology Type 4 and advanced technology build-up Types I, II, and III, which can be manufactured with conventional processes.

27.3.4.1 The Buried Via Stack-Up. To avoid hopelessly complex routing, each signal net is generally routed using only one pair of layers with what is called *Manhattan geometry*. This means that diagonal routing is avoided and all signal lines run in a horizontal or a vertical direction. To avoid blockage and side-to-side cross-talk problems, horizontal lines are run on

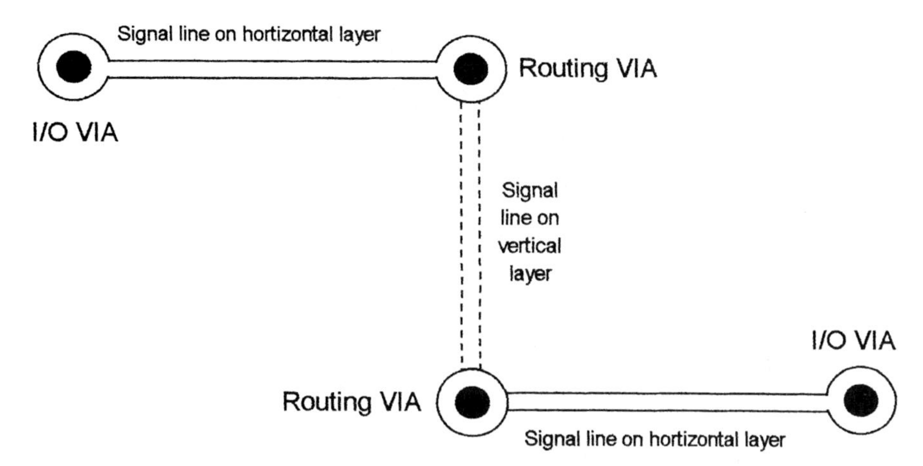

FIGURE 27.14 An example of a net routing using Manhattan geometry.

one layer and vertical lines on the other. This means that, in addition to a via at each end of the net (I/O via), most nets need one or two additional vias (routing vias) to change direction from horizontal to vertical. Figure 27.14 shows an example of Manhattan geometry.

The net shown in Fig. 27.14 has two I/O vias and two routing vias. The I/O vias connect the signal lines to the board surface, where the net connects to an input or output point of an active circuit. The routing vias are used to change directions from horizontal to vertical. With through-hole vias, the routing vias pass through all layers, consuming valuable routing space. A high-layer-count board with many signal layer pairs can run out of via sites. In this case, additional layers will not improve routing completion and the board is said to be via-starved. One solution to a via-starved board is to use buried vias. The buried via connects two adjacent signal layers and provides routing vias without affecting routing on other layers. Figure 27.15 shows an eight-layer board with two buried via layers.

Buried vias do not pass through the board, so they do not congest the routing on other layers. In addition, the same via site can be used simultaneously on different layer pairs. Since buried vias are drilled and plated in the thin laminate prior to lamination, they can be very small and positioned very accurately, saving additional routing space. In some applications, buried vias are placed where needed with no reference to a predetermined grid. This gridless routing approach gives very high computer-aided design (CAD) autoroute completion.

To benefit from buried vias, signal layer pairs must be routed on opposite sides of the same C-stage component. In Fig. 27.15, L4 and L5 are the only buried via candidates. In Fig. 27.10, no signal layers can benefit from buried vias. The design shown in Fig. 27.15 is able to use buried vias on two layer pairs because a power/ground layer pair separates them. In non-buried via designs, it is best to use one power/ground plane between each signal layer pair. This gives cross-talk isolation and impedance reference. A buried via design uses a second redundant layer to force the next signal layer pair onto the same imaged detail. In other words, for a high-layer-count design to use buried vias on all signal layer pairs, power/ground plane pairs must be inserted between each signal layer pair. This increases the number of layers in the board and increases cost. Another disadvantage of buried vias is the cost associated with the extra drilling and plating operations.

27.3.4.2 The Blind Via Stack-Up. Another via option is the blind via, which connects the surface layer to one or more internal layers but does not go through the board to the opposite outside layer. Figure 27.16 shows an example of an eight-layer ML-PWB with both blind and

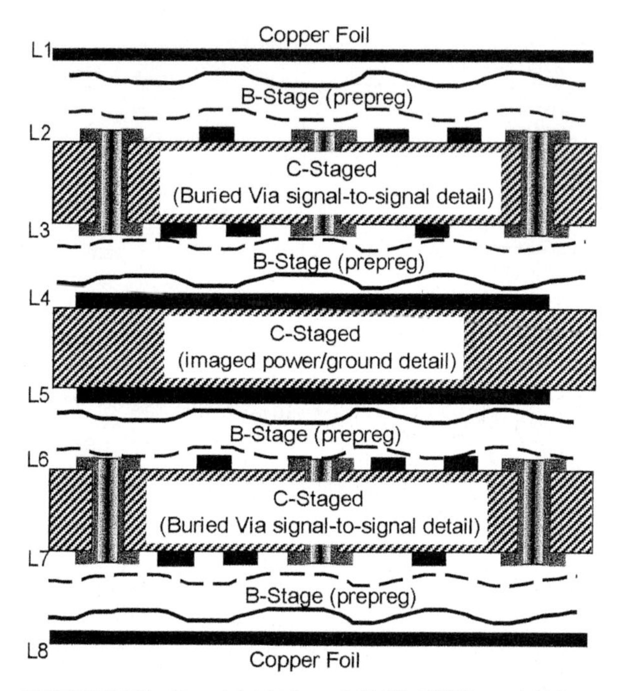

FIGURE 27.15 Type 4 (eight-layer B/V ML-PWB), an eight-layer board with two buried via layers.

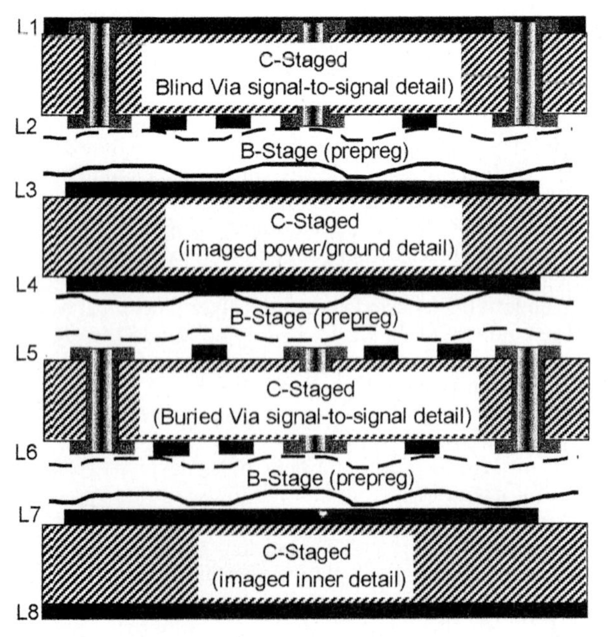

FIGURE 27.16 Type 4 (eight-layer blind and buried via ML-PWB). The design contains buried vias connecting L5 and L6. Blind vias connect layers L1 and L2.

buried vias. The design shown in Fig. 27.16 contains buried vias connecting L5 and L6. Blind vias connect layers L1 and L2. The buried vias can be fabricated in layer details as described earlier or can be fabricated by blind drilling. Blind vias may become very important in very dense double-sided surface-mount designs that have interference between I/O vias from opposite sides on the board. If this is particularly troublesome, a via-in-pad (VIP) approach or dog-bone escape pattern is used. The VIP approach places a blind hole directly in the device I/O pad. The dog-bone approach places a blind hole in an adjacent pad next to the solderable land.

A second application for buried vias is to ensure complete side-to-side electrical isolation. This is particularly important in wireless designs where the RF circuits must be shielded from other circuits. Through vias allow RF electric fields to escape from a shielded region. A blind via eliminates this problem and allows RF functions to be combined with logic and control functions on the same board.

The ultimate use of blind vias is effectively to convert a dense double-sided surface-mount design to a pair of less-dense single-sided surface-mount designs. To see how this is possible, visualize an ML-PWB with fine-pitch SMDs on both sides as two separate boards with some level of side-to-side interconnection. If the board is manufactured as two separate subassemblies, the I/O connections on one side will not interfere with the I/O connections on the opposite side. For example, consider a 16-layer board fabricated as two 8-layer subassemblies. The subassemblies are laminated, drilled, plated, and patterned like a standard ML-PWB, with the exception that the sides that will become the board exterior are blanket-metallized. After patterning, the two subassemblies are laminated together and then processed as a standard ML-PWB. This process is another form of sequential lamination. The only through holes required in this design are the relatively few that provide side-to-side interconnection. Since each via site is used twice, a 100 mil. grid may replace a 50 mil. grid, providing a significant increase in innerlayer routing resources. This type of ML-PWB structure is very costly and has mostly been replaced with the routing features of the high-density design.

27.3.4.3 The High-Density Stack-Up. The buried via stack-up design came about for a solution to gain signal routability when CAD routing solutions were pressed to the limit. Early on, Type 4 designs were considered specialty products and utilized standard-feature-size characteristics. With the advent of more sophisticated CAD routing tools, it became possible to auto-route with greater efficiency, sometimes avoiding the necessity and cost of buried vias. Soon high-I/O, full-matrix components created new interconnectivity demands. Today buried via sections defined as cores in IPC-2221/2226 are used for the central starting point of manufacture, and then build-up layers are applied to complete the design as an HDI structure. As noted earlier, the common discriminator that separates the high-density design from a conventional Type 4 is mainly the feature size. For example, the HDI Type II, as shown in Fig. 27.7, uses a conventionally processed ML-PWB Type 3 board as the core (L2 to L5) and the PTHs become buried after the high-density features of L1 and L6 are added. Microvias are used to form a connection to L2 and L5; then through holes are formed to tie L1 through L6. The microvias and holes are now metallized in one cycle. The build-up layers in the example use thin dielectrics to provide a close proximity to the adjacent layer. This is required to accommodate the microvia connection and maintain producibility for metallization.

Figures 27.17 to 27.19 show a comparison of the HDI build-ups for Types I, II, and III. In each case, complexity is added by building add-on layers sequentially. The build-up layers are typically for signal routing and contain low copper weight. Dielectric-coated (nonreinforced) microfoils (as thin as 9 to 12 μm) are employed to provide a low-profile thickness after copper plating. This is needed to facilitate the image of fine-line widths typically associated with HDI. These styles of circuits offer some economy through use of a variety of dielectrics. The use and style of construction type must be matched with the product's expected application environment and operating life. Often the need for CTE match of area array components compels the application of a nonwoven aramid layer at the surface. The resin system selection should be based on the expected CTE mismatch concerns, such as when a ceramic device is surface-mounted to the assembly. Thermal cycling of power on and off can cause earlier failure when a large mismatch is present.

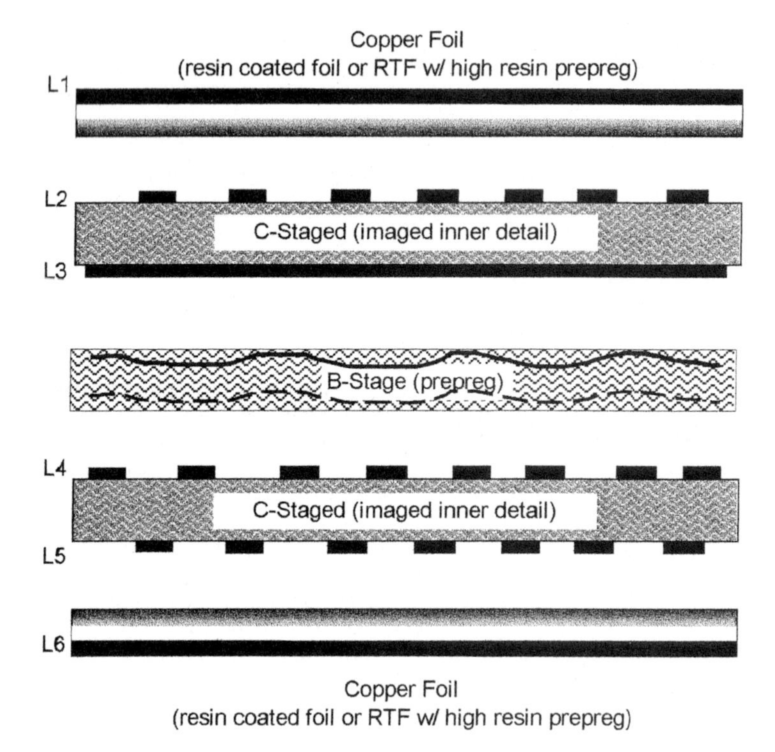

FIGURE 27.17 Type I (six-layer HDI stack-up) shows one example of how to stack up an ML-PWB with HDI features.

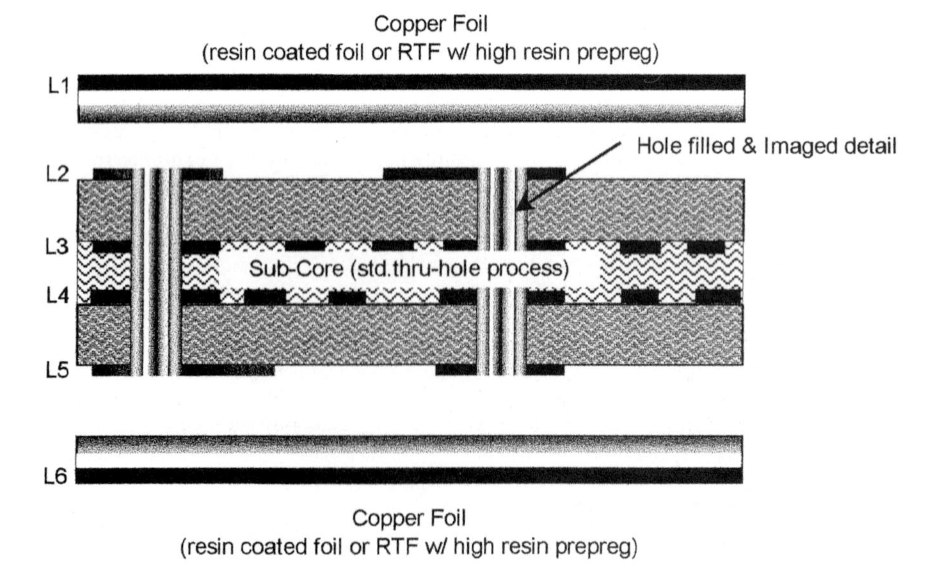

FIGURE 27.18 Type II six-layer stack-up having at its core a four-layer board with through holes, which later become buried.

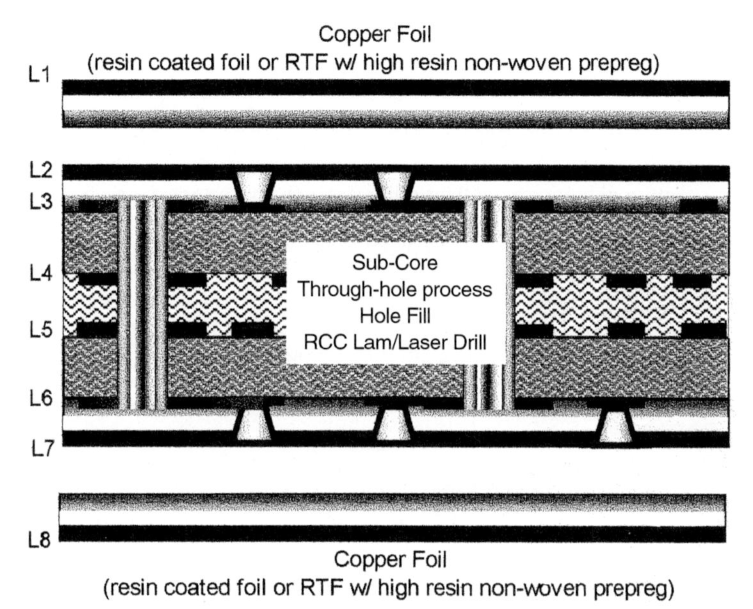

FIGURE 27.19 Diagram showing how an HDI Type III (eight-layer HDI stack-up) might be produced employing conventional lamination techniques.

Manufacturing capability becomes critical when targeting HDI features. The IPC has taken the responsibility of developing industry standards to assist in benchmarking and quantifying capability limits. Manufacturing features such as imaging, etching, hole formation, plating, and lamination registration are all strained and must be optimized. Other processes that are equally challenged are the end-product continuity testing to access the high-density features.

27.3.4.3.1 HDI Type I Stack-Up. Figure 27.17 shows one example of how to stack up an ML-PWB with HDI features. The economies of this design are obvious because it requires only one lamination cycle. The density push in this level of HDI focuses on imaging and microvia routing. By routing at a high circuit density, the surface area of the board is better utilized. The outerlayers are typically formed with a low height dielectric. When utilizing conventional lamination, this is usually a resin-coated foil or a microfoil bonded with a high-flow non-woven aramid prepreg. Special non-woven series of aramid fiber laminates have been developed to yield an equivalent dielectric thickness of 1.9 mil.

Special handling is required for the microfoils, which are usually 9 to 12 μm thick. To facilitate handling, a sacrificial carrier foil is sometimes cohered to the microfoil externally for added stiffness. After lamination, one option is to process the microvias through laser drilling and conventional through-hole drilling. Sometimes a clad outer construction may be best, depending on the laser technology chosen. For example, infrared (IR) (CO_2) laser ablation using an etched mask (small etched openings the size of the via) in the outer copper can be performed at primary print. This process helps reduce possible misregistration error. One metallization cycle is required to tie in all vias electrically. This reduces costs compared to sequential cycles.

27.3.4.3.2 HDI Type II Stack-Up. Density demands, when not met by the single lamination approach of Type I, must then address the use of sequential processing when employing standard ML-PWB techniques. The Type II stack-up shown in Fig. 27.18 has at its core a four-layer board with through holes, which later become buried.

27.3.4.3.3 HDI Type III Stack-Up. The use of a Type III structure may be employed when routing densities are greatly pressed. At this point of complexity and above, alternative

HDI approaches should be investigated because of the extra cost associated with performing repetitive cycles with conventional ML-PWB processes. Alternatives to lamination provide build-up of the dielectric and conductive layering through other means. These should be discussed with the manufacturer prior to selection. It is important to remember that the manufacturing method related to fabrication can have a great impact on the design rules chosen at CAD layout. Figure 27.19 represents how an HDI Type III might be produced employing conventional lamination techniques. Here, the core substrate is a four-layer ML-PWB detail similar to the Type II starting construction discussed earlier. Processing for the first pair of build-up layers follows the processing analogy of the Type II structure. Here, as in other foil laminated build-ups, the starting copper thickness should be kept to a minimum. Plating uniformity becomes critical when attempting to image/etch high density features. After L2 and L7 have been defined, the final copper layer pair (L1 and L8) is laminated. At this final build-up, since no through holes are present, the flow of the resin-coated foil or prepreg should be sufficient to fill the microvias and flush the circuits. Once the additional layer is laminated, another level of microvias may be produced.

27.3.5 Filled-Via Processing and Sequential Laminations

The requirement to fill vias is driven most often by routing density. When high-density-area array components are utilized, the quantity of vias per square inch greatly increases in the local area under the device. Buried vias or blind vias are frequently the solutions to through-via starvation.

Buried vias, unless otherwise prefilled, will fill with resin during final lamination. The volume of resin necessary to fill the buried vias is dependent on the diameter, length, and total quantity. Buried vias might straddle a single core, or might interconnect several cores and dielectric separations in a sub-lamination section. If there is insufficient resin in the prepreg to fill the buried vias, those vias can starve bonding resin from the local area where they are concentrated. To prevent the prepreg resin from entering the buried hole, the fabricator often is required to prefill the vias prior to a build-up lamination with a resin or paste formulation. Other design constructions require blind vias that are planar and within the land pattern at the surface mount attachment locations. These VIP structures free up real estate on the component attachment surfaces and provide enhanced signal integrity at high frequencies. Prefilling internal buried vias or blind VIPs with a hole-fill resin can provide a more robust interconnect structure, improved lamination integrity, and a planar surface in the case of the blind vias.

27.3.5.1 Fill Materials. Since the fill material is an additional fabrication material that becomes a part of the design construction, procurement documentation is required to specify a fill material type and thereby implement the via fill process. The selection and documentation of the fill material require the same consideration as the base laminate preference. This is especially critical when targeting a lead-free-compatible process. Currently, an industry-based material specification for via fill material does not exist. Therefore, specific fill-material brands may be named on the drawing, or some other form of user/supplier agreement must be established. The fabricator has preferences for the type of material used for via fill. Just as suppliers often have preferences for a specific solder mask brand, they also often prefer to use of a specific via fill material around which they have developed their principal processes. Supplier preferences can be driven by specific via fill material characteristics, such as accessibility, equipment compatibility, process supportability, plateability, and/or shelf/pot life. This may complicate source selection, or it might influence the use of a dedicated service center for the hole-fill process. The fabricator may not always know the reliability of its preferred material for a given via structure or end-use environment.

Determining the properties of the various fill materials may be difficult for the user. Obtaining properties from the material suppliers' data sheets may be possible for some properties, whereas others are more difficult to obtain; for example, many manufacturers' data

TABLE 27.4 Common Via Fill Manufacturers and Material Properties

Mgf. (Link)	Type	Material[3] (part number)	Pencil Hardness	Color & Plateability	Thermal conductivity (W/m/°K)	T_g °C	CTE under T_g (ppm)	CTE over T_g (ppm)
Peters	Non-conductive (Ceramic Filler)	PP 2795	9H	White & Light Gray[1]	N/A	140	40	150
Peters	Non-conductive (Ceramic Filler)	PP 2794[3]	N/A	White[1]	N/A	115	40	105
SAN-EI	Non-conductive	PHP-900 IR-10F[3]	N/A	White[1]	N/A	160	32	83
Taiyo	Non-conductive	TCHP-200 DB4 (2 part)	8H	Green[1]	N/A	130	24	78
DuPont	Conductive	CB-100 Copper /Silver	4H	Silver[2]	3.5	115	35	47
Tatsuta	Conductive	AE 3030 Copper/Silver[3]	N/A	Silver[2]	7.8	171	40	86
Taiyo	Conductive	SCHP-7901 Silver	4H	Silver[2]	6.7	110	45	120

Notes: Some via fill materials are not typically compatible with permanganate chemical processing. Confirm with the specific supplier before using. The denotation of N/A indicates that the data were not readily available.
[1]Compatible with Plasma desmear, requires electroless followed by electrolytic processing.
[2]Compatible with Plasma desmear, conventional electroless recommended but may be use direct electrolytic processing.
[3]Capability with a lead-free profile should be verified; typically the higher T_g materials with a lower ppm above T_g will be more compatible.

sheets omit the modulus. Table 27.4 provides a cross-reference for some of the more common material types with properties available. Additional properties and performance data may also be found at the specific manufacturer web page.

27.3.5.2 Concerns and Common Defects. Concerns have been associated with filled-via structures arising from a lack of process maturity and the potential impact on reliability. These concerns include plating adhesion, air entrapment (voids) created during the filling operation, CTE mismatch with the bulk resin, and the risk of overplanarization to remove excess resin. Entrapped voids are suspected to have an effect on hole wall integrity due to outgassing; CTE expansion mismatch could have an adverse interaction with the material and copper CTE properties. Overplanarization can grind down the plated hole wall knee, weakening the interconnect at the barrel plate. This is commonly called a reduction in wrap copper. A lack of wrap copper will leave a butt joint at the interconnect that is vulnerable to pad rotation (lift) knee crack. Limited reliability data are available within the industry for these structures, yet they are being widely deployed. The performance criteria for the via fill method cannot be properly quantified for a definitive requirement without tangible reliability data. Additionally, to apply the most cost-effective via fill method with confidence, the fabricator must know the method's compatibility with the assembly process and targeted mission criteria. For all these reasons, qualification testing to the end-use environment with a preproduction build is recommended.

Designers need to be aware of potential processing issues associated with specifying via fill. Although the risks are becoming better known, the manufacturing issues are often difficult to detect from standard coupon analysis. Construction integrity weaknesses often are not found until latent defects occur or become revealed by aggressive thermal stress screening. Figure 27.20 is an example of a design employing conductive via fill within two via structures (blind and through holes) sharing a common surface layer. This example is free of plating interconnect defects; the buried structures are fabricated without an internal cap plate; the voids in the conductive fill are typical for this type of material.

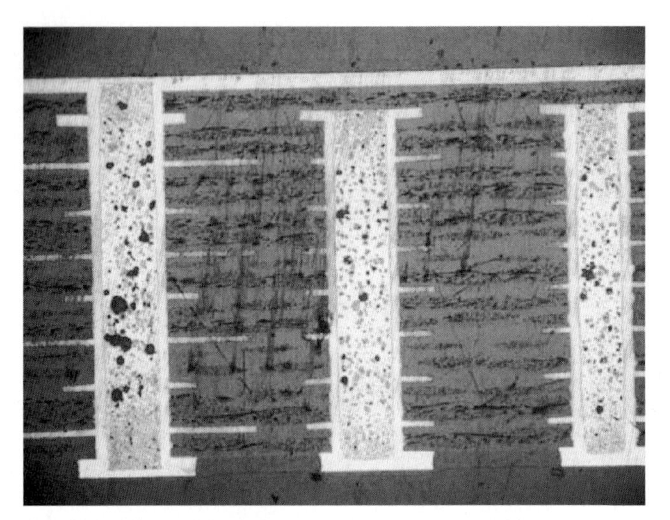

FIGURE 27.20 Micrograph of conductive via fill.

Figure 27.21 is a nonconductive fill with a close-up view of a micrograph illustrating the plating integrity and surface planarization that has maintained the wrap copper requirement. In this example, the full plating requirement was deposited without a separate button plate cycle. The wrap copper demarcation is noted and no separation has occurred at the cap plate.

Some common failure mechanisms, which illustrate the risk associated with fill technology, are shown in Fig. 27.22. Here, micrographs show a via fill with a lack of plating adhesion and wrap copper, allowing interconnect separation.

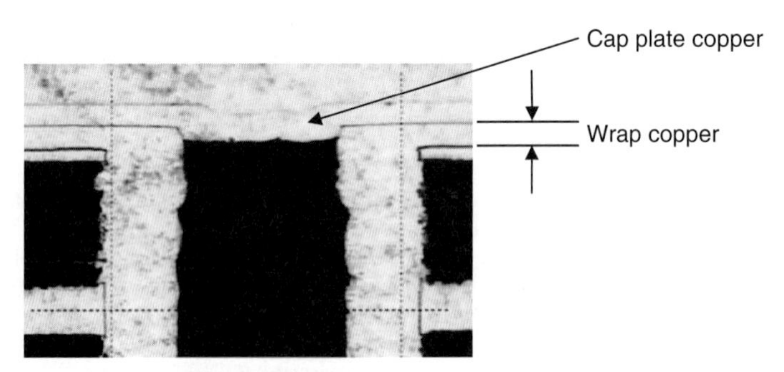

FIGURE 27.21 Micrograph of a nonconductive via fill with good plating features.

27.3.5.3 ***Via Fill Processing and Flow.*** A via fill process flowchart can be used to communicate the modified steps of a typical manufacturing flow. Two process maps are shown to illustrate two typical process flow variables of a plated-cap versus nonplated-cap via fill process. Reliability testing to segregate the performance of via fill materials with the cap-plate and noncap-plate features in the selected application should be performed prior to selecting this design feature.

When a plated-over via is required, plateability of the fill material should be a chief consideration. Fabricators should screen for this compatibility across the various via fill materials

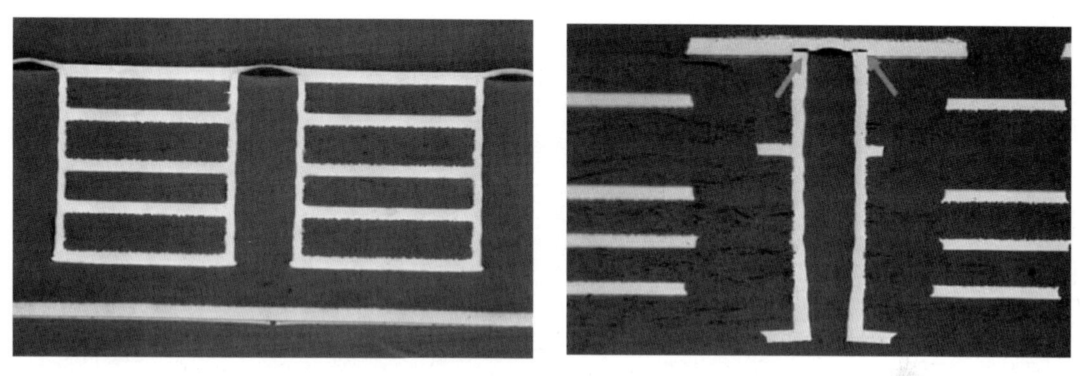

FIGURE 27.22 Micrograph of a nonconductive via fill with defective features.

within their plating process to determine the integrity of the metalization. The flowcharts illustrate the fabrication methods employing a panel plate scheme. The flow could also be modified for a pattern plate methodology. The two process flows are illustrated in Figs. 27.23 and 27.24. The key via fill process steps are highlighted for clarity.

27.3.5.3.1 Pre-Via-Fill Processing (Plating Variables). The plateability of via fill material is an important feature to understand due to industry trends that have shown that plated caps may be problematic in some design constructions. Outgassing of the fill material may separate the metallization during sequential lamination or thermal stress. Poor adhesion can also contribute to a separation but typically can be screened with standard tape testing. Figure 27.25 illustrates a cap-plated buried structure and a micrograph showing a cap-plated via with separation occurring.

When the main design attribute is to limit resin from flowing into to holes at sequential, lamination, the plating over the filled structure may not be required. Here, the filled panel after planarization can be staged for lamination. Figure 27.26 illustrates a noncap-plated buried structure, and the micrograph shows a noncap-plated via with lamination separation occurring.

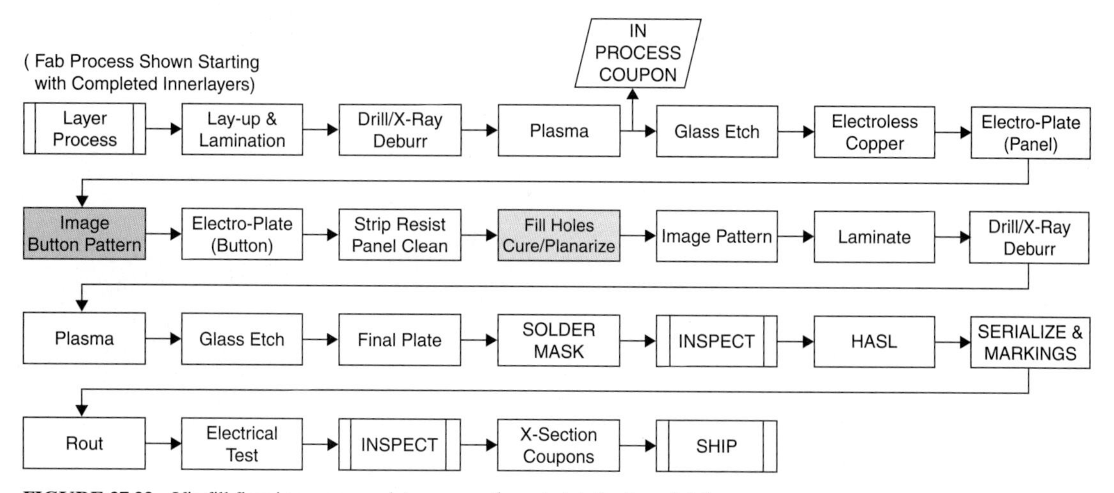

FIGURE 27.23 Via fill flow in a noncap plate process (panel plate/button plate).

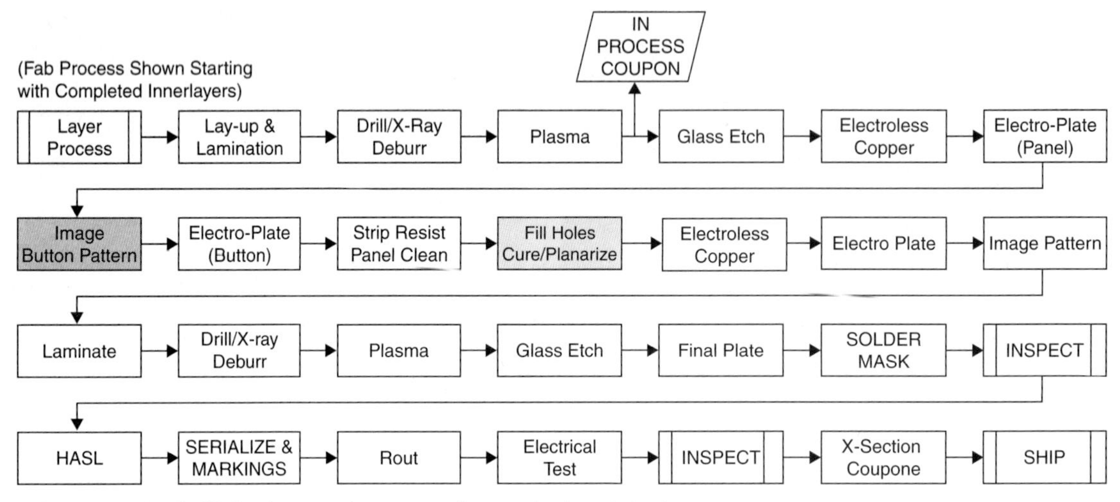

FIGURE 27.24 Via fill flow in a cap-plate process (button plate/panel plate).

FIGURE 27.25 Cap-plated filled structure.

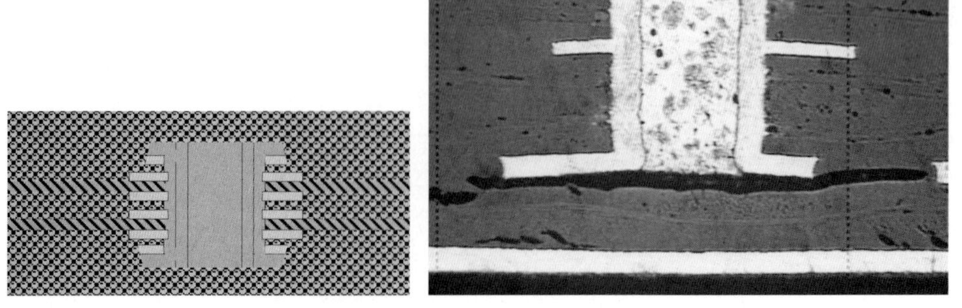

FIGURE 27.26 Noncap-plated filled structure.

Each of these techniques requires the fabricator to understand the fill material compatibility with their base laminate system and plating processes. Typically, process compatibility can be screened with a simple solder-float thermal stress analysis and cross section.

Prior to plating, the fabricator must understand the ability of the fill material to accept the electrolytic process. Some via fill materials have been reported to not be compatible with permanganate chemistry commonly found in electroless copper lines. Some cases of early failure of the via fill cap-plating integrity have been traced to this process. Elimination of the programming step to process the board through permanganate will avoid this incompatibility. Plasma desmear is the generally recommended preplating preparation step. Some resin chemistries that are loaded with ceramic are susceptible to excessive surface roughening if the plasma is too aggressive as is typical in a full etchback cycle. This becomes problematic when secondary nonfilled plated vias are required and are processed with a full etchback cycle. In this instance, it is best to choose a nonceramic filled material or modify the drill sequence so they are protected. Typically, the conductive fill materials may be electroplated directly, omitting the electroless process. If the final surface plate has secondary drilled holes that require metallization, all features may be exposed to the electroless process anyway. If directly plating the material, the fabricator should characterize the material to confirm peel strength because some cases of plating separation have been traced to poor adhesion in direct plate applications. Therefore, when choosing a via fill material, it is best to develop a process map of the required via structure processing and understand the process sequence of when drilling, filling, and plating occur.

Figure 27.27 illustrates a typical manufacturing process targeting plating allowances for the successive plating steps prior to via filling. The first plating step establishes the minimum wrap plating requirement, and deposit thickness should be adjusted for the procurement specification and the balance of final plating. Here the substrate has a panel plate followed by a button plate. One advantage of the button plate is that it serves as a visual guide for the planarization removal process. Acting as a gauge, the button, when completely sanded off, indicates any additional sanding can be assumed to reduce the wrap plating.

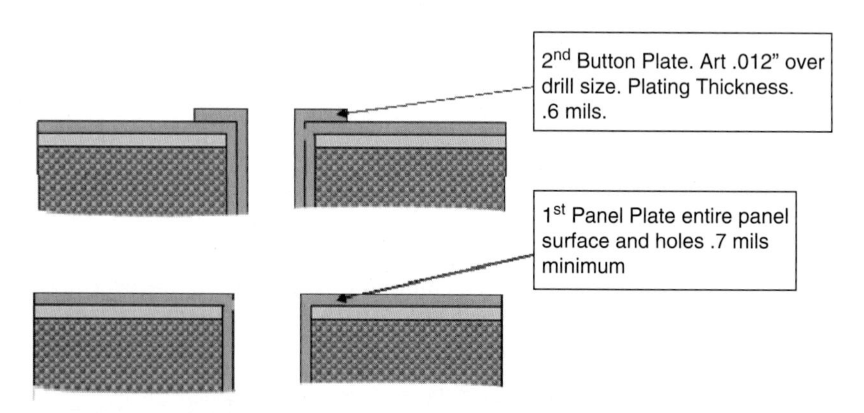

FIGURE 27.27 Pre-hole-fill plating.

27.3.5.3.2 Hole-Fill Process. Once the prefill-plating process is complete, the next major step is the actual hole-fill process. The fill materials can be divided into two major categories: conductive and nonconductive. A generic hole-fill process flow is shown in Fig. 27.28.

The specific equipment and technique can vary by the material type used. The method chosen to apply the material in the holes can be influenced by many variables such as material cost, pot life, viscosity, equipment resources, and the quantity of panels to be processed. Some hole-fill materials lend themselves to a variety of fill methods. Some of the common hole-fill

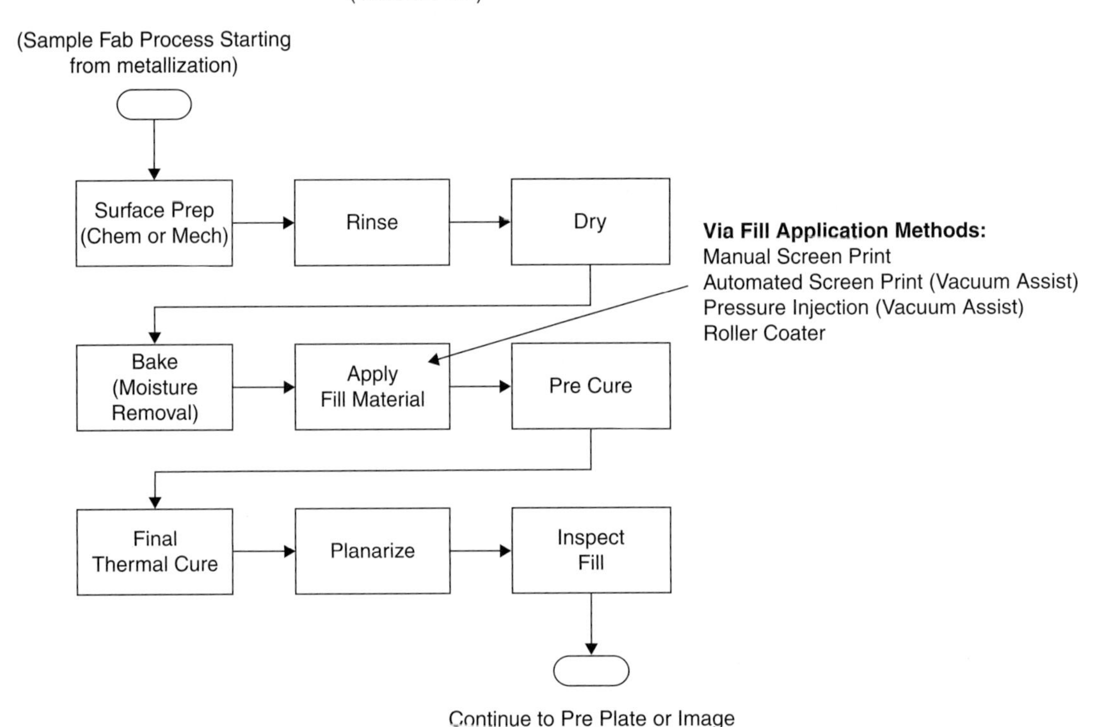

FIGURE 27.28 Generic via fill application flow.

application methods are noted in the flowchart in Fig. 27.28. Application guideline parameters should be followed as noted in the respective material supplier data sheets. Via-hole filling may be considered a finesse process due to its sensitivity to variables that occur from design to design and the availability of support equipment to perform the process. The various application methods each have advantages and disadvantages associated with the technique employed.

When a fabricator is screen printing, a mask sized just over the hole size keeps excess material from being applied to the surface. Screen printing with a dot mask using a back side receiving template with open holes sized just over the plated hole diameter yields a domed fill bump. These bumps can act as a visual indicator similar to a button plate pad to indicate when the planarization is flush to the surface. Other methods include roller coating modified to squeegee excess material off the surface of the panel. Commercially available equipment designed specifically to apply fill material is now available. One such machine uses a pressure-assisted injection method to force material into the holes under a vacuum. The principal advantage of this technique is the ability to reduce the opportunity for void entrapment.

Figure 27.29 illustrates an example of how the filled structure forms with successive steps for a noncap-plated fill process. Note in this example how the surface button plating is mostly removed at the planarization step. Process development should be performed to target the optimum plating build-up method in regard to the process control of available planarization equipment. This means that whatever plating method is employed, both the deposit thickness across the panel and the net removal of deposit thickness as a result of planarization must be

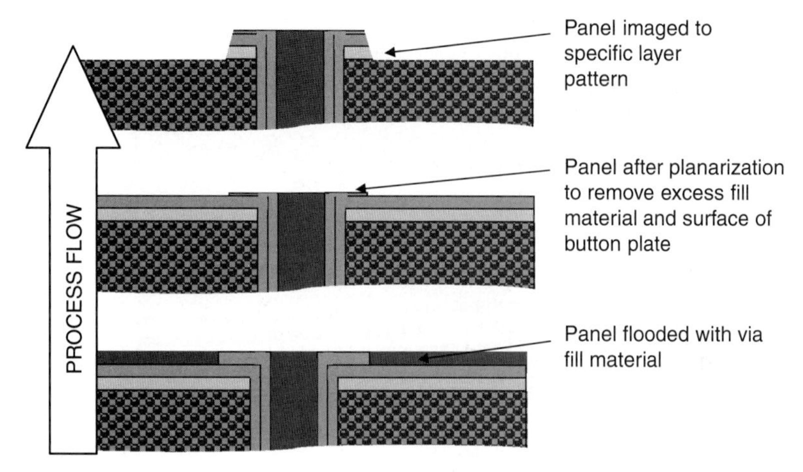

FIGURE 27.29 Noncap-plated fill and planarization.

understood. Simple eddy current measurements of multiple locations across the panel are recommended prior to and after planarization sanding passes.

Once the planarization is complete, the panels are ready for imaging and etching to allow typical process flow, depending on construction type. The plating method chosen should take into account the surface pattern and construction level of difficulty. Multiple plating cycles are often required if multiple filled structures occur on the same surface plane. Excessive copper plating complicates the imaging, so the starting foil thickness should be kept to a minimum. Figure 27.30 illustrates how the fill method forms from use of a method employing cap plating over the fill material. It should be noted how the copper thickness requiring etch can quickly build up, making pattern definition difficult.

27.3.5.4 Specifications. The industry has now released a design guideline document outlining via protection classifications, including the hole fill as described in this chapter. The IPC-4761, "Design Guide for Protection of Printed Board Via Structures," has established via-protection techniques, creating specific types identified as Types I through VII. IPC-4761 contains many combinations of techniques where an added material is utilized to plug, tent, cover, or fill a via structure. Many of the techniques established are used in Class 1 and 2 hardware or where circuit card assembly consideration is the primary focus. Many of the 4761 via-protection types are of lower cost and lesser complexity than are those used for hole fills. The designer is encouraged to consult the application guide included in Table 5-1 of IPC 4761 to narrow the selection prior to layout and specifying a method. The via fills described in this chapter fall under the Type V and VII classifications. These types have been the focus for developing performance-based requirements as reliability data have become available.

Amendment 1 of the rigid board performance specification IPC-6012B will introduce requirements for maintaining a minimum wrap copper. The specification limit is being established at 0.0005 in. minimum wrap as verified at cross section. This is considered a conservative value resulting from the nonuniformity of the process. Correlation data have revealed that coupons on the panel perimeter rarely match that of the actual product. Uniformity of the planarization across the entire panel is difficult, so a conservative value at cross section was established to ensure that the part maintains wrap copper. The reliability data gathered support this value due to the additional manufacturing risk that can occur from the added fabrication steps. However, some data collected in MRB action and assessment indicate that wrap copper as low as 0.0002 in. is acceptable for some end-use environments. It is likely that future revisions of

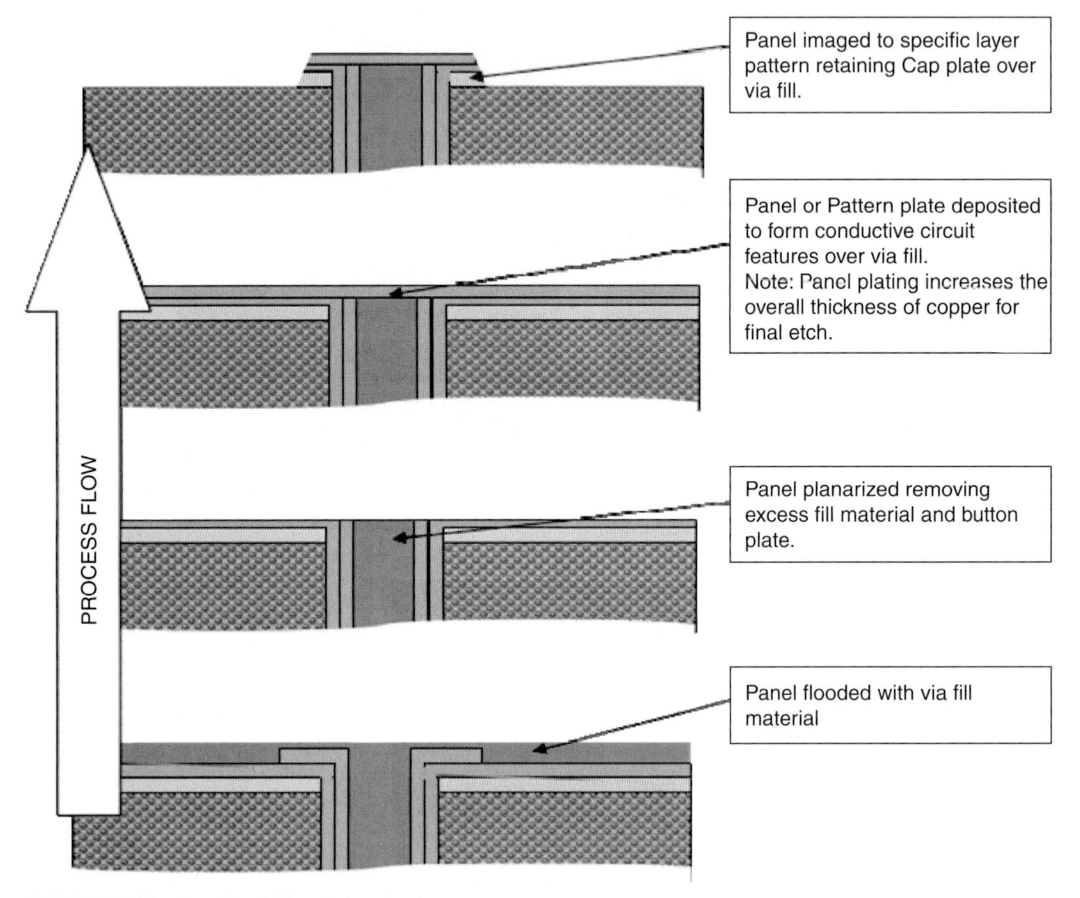

FIGURE 27.30 Cap-plated fill and planarization.

IPC 6012 will relax the 0.0005 in. wrap requirement as the industry matures in regard to controlling the planarization process and additional reliability data become available.

27.3.5.5 Via Fill Summary and Additional Considerations. The manufacturing of reliably filled vias should not be underestimated. Considerable process development is required to optimize the interaction of maintaining plating wrap to lower the risk associated with via fill planarization. The nature of reported failures due to compromised interconnect integrity have ranged from infant mortality at assembly to latent field failures. These failure mechanisms are very difficult to screen with standard performance verification methods. Even post-assembly environmental stress screening (ESS) may not exercise sufficient stress levels to screen out marginal product effectively. This has directed some users to specify thermal shock temperature cycling at the bare-board production lot level for acceptance testing. This level of lot acceptance becomes more critical for lead-free assembly thermal stress environments, requiring the user to be more cognizant of the fill material selection.

A number of commercially available material choices exist with a wide variety of physical properties. Matching a material to a design type is necessary prior to specifying it on a drawing. Industry dialogue indicates the desire eventually to establish a material performance specification for fill materials similar to the IPC SM 840 for solder mask. The absence of a material specification requires the user to specify the brand(s) of acceptable fill materials through a user/supplier agreement. The process of specifying a hole-fill material can add considerable cost and difficulty to the PWB manufacturing process. The process labor cost is relatively flat to apply a material, with the most significant cost variable being the material type. Some of the silver-filled conductive materials are roughly twice the cost of nonconductive materials. The added process time and material cost can be roughly estimated at $25 to $50 per panel not including setup fees or minimum lot charges.

The impact to reliability is judged to be weighted more toward the suppliers' ability to successfully process the panels without degrading plated interconnects rather than the fill material type. This is followed by the ability of the manufacturer to metallize a specific fill material. The greatest number of failures reported in industry have been related to the planarization process. To this end, commercially available planarization equipment has come to the marketplace. Due to the rapid introduction of hole fill and process immaturity, many suppliers were faced with establishing the sanding processes with modified deburr machines or fully manual methods. Even with automatic planarization equipment, careful consideration should be given when specifying via fill for designs with heavy-weight copper or rigid flex constructions. Any construction that lends itself to circuit pattern imprint at the surface will have difficulty in the planarization process. Panel thickness at via structures can often be several mil. thicker than open circuit areas. This change in height is not easily recognized and is problematic for producing a uniform sanding surface. Additionally, some softer materials or non-glass-reinforced materials may not be able to be planarized without distortion or gouging.

With more designs migrating to higher-density layout schemes, it is expected that the use of via fill will increase. The process of via fill is gaining maturity for Class 3 hardware but this has also been the market segment most affected by reliability issues. Many end users who learned difficult lessons quickly introduced their user-defined specifications for hole fill. User-defined specifications often do not lend themselves to producible limits, so consideration for via fill should be part of the DFM dialogue during the design layout cycle. The user and supplier must work closely together to understand the impact of specifying hole fill for each design construction method employed.

27.4 ML-PWB PROCESSING AND FLOWS

27.4.1 Process Flowcharts

Attempting to visualize the process flow of the manufacture of multilayer printed wiring products can be overwhelming. One way to help picture the multiple paths a board travels is through the use of flowcharts. The flowchart in Fig. 27.31a is a typical process flow for the beginning innerlayer process identified as 1 through 4. Figure 27.31b is a typical finish board flow process after lamination and drill. Three additional diagrams are provided in Fig. 27.32 that represent possible sequential flows for HDI products. They are identified as processes 5, 6, and 7, referring to how the different HDI types might flow. It should be noted that alternative flows and methods are possible in HDI. This discussion is limited to mostly conventional processing. Major aspects of these processes are discussed in detail in the following sections.

27.4.2 Innerlayer Materials

The multilayer process begins with the accumulation of the innerlayer clad dielectric laminate.

27.4.2.1 Documentation and Specifications. The ML-PWB design documentation should specify the specific material system to be used in the fabrication. Typically materials are identified

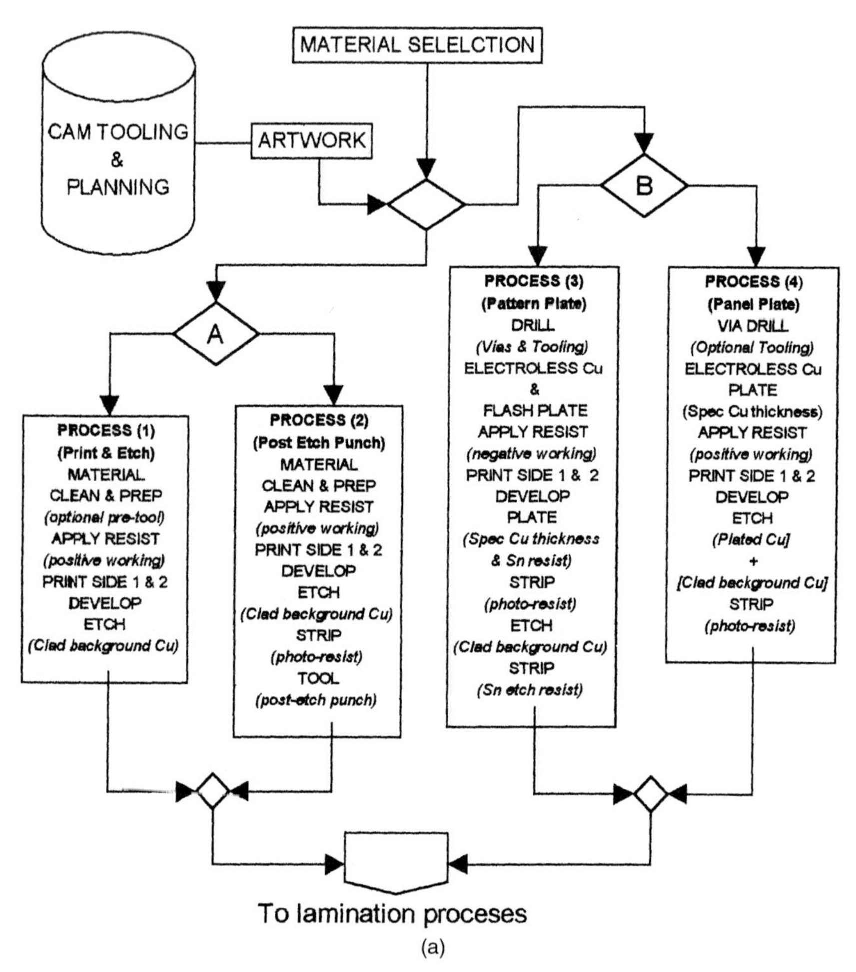

FIGURE 27.31 Process flowcharts: (a) A typical process flow for the beginning innerlayer processes identified as 1 through 4; (b) a typical finished board flow process after lamination and drill.

by IPC-4101 callout or slash sheet designation, for example, IPC-4101/24. Copper weights as well will be specified within the drawing documentation. The specification IPC-4562, "Metal Foil for Printed Wiring Applications," provides reference to the various weights of copper foils along with an applications use guide. The common terminology refers to foils less than oz./ft.^{2} as a metric thickness of micrometers. The most commonly used foil is STD-Type E electrodeposited (IPC-4562/1) with a weight of 1 oz./ft.^{2} and a nominal thickness of 1.35 mil. (35.5 μm).

27.4.2.2 Copper Foil. High-current applications use heavier (thicker) copper of 2 oz. or above. With copper weights above 3 oz., processing difficulty is increased. High-density circuit designs that have low voltage and that are mainly concerned with passing signals may use thinner copper weight foils, such as 18 μm or less. When sequentially processing clads as buried via pairs, the fabricator must use thin starting foil to promote line definition and image integrity and, where necessary, impedance-controlled tracks.

Copper foil is fabricated in an electroplating process on a rotating drum that produces a coarse-grained columnar copper. The economy of the foil process always yields copper at the minimum thickness tolerance. Due to the speed of the foil process, often a course grain structure can yield poor elongation. Elongation is a property of major importance in reducing trace fracturing. Standard copper foils typically fail at elongations around 3 percent. IPC-4562/3 high-temperature elongation (HTE) provides a slight, but important, improvement in elongation from 5 to 8 percent. High-ductility electrodeposited foil, called HD-Type E (IPC-4562/2), is specified to withstand an elongation of 10 percent minimum. Foil vendors also sell special fine-grain-structure, annealed, or wrought foils. The foils with higher elongation are advertised to have superior etch performance for fine-line densities.

Standard foils have a rough surface called *tooth* on one side (drum side out) and a smooth or shiny surface on the opposite side (drum side). The rough side, treated with an adhesion promoter, is laminated against the C-stage dielectric to ensure good adhesion. Since the shiny side of the copper has poor adhesion characteristics, the ML-PWB fabricator must include adhesion promotion steps prior to resist lamination and prior to final board lamination. Double-treated copper foil has an adhesion promoter applied to both sides and is attractive in high-volume applications. Although double-treated foil requires no further adhesion promoting treatments, it has several process disadvantages to overcome:

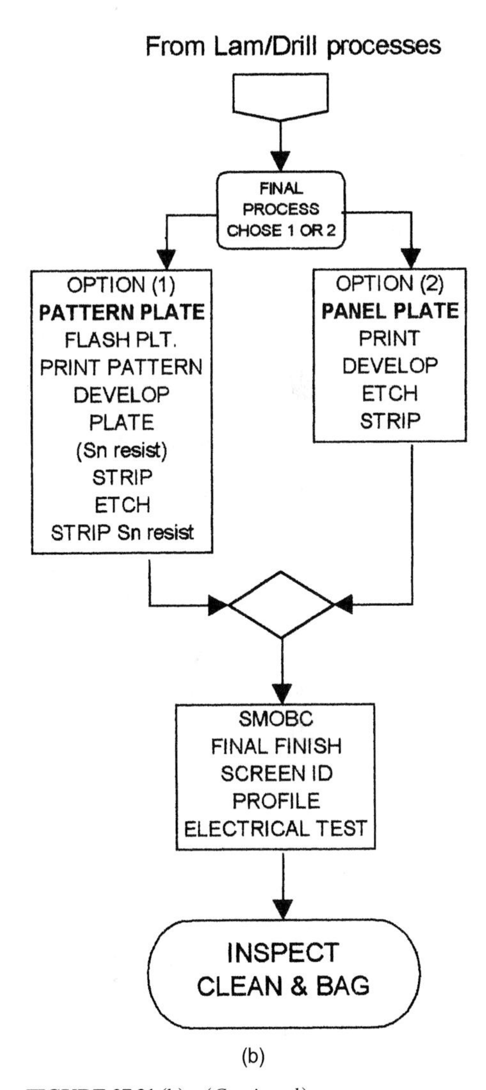

(b)

FIGURE 27.31 (b) (*Continued*)

- Cleanliness and material handling sensitivity is more critical.
- The cost often offsets the savings from the eliminated processes.
- It is somewhat fragile, so it is difficult to rework.
- Complete resist development is difficult, leading to a high incidence of shorts.
- It is not compatible with the plating processes used to make buried and blind vias.

Another foil type offered by laminate suppliers, reverse-treated foil (RTF), offers an advantage for producing fine lines. The RTF copper has adhesion promoter applied to both sides and is classified by 4562 as code R (reverse-treated bond enhancement [cathode side] stain-proofing on both sides). This approach provides advantages to imaging fine lines. When the copper tooth is reversed, the fabricator can improve line definition by allowing the etch chemistry process to stop at the surface of the laminate.

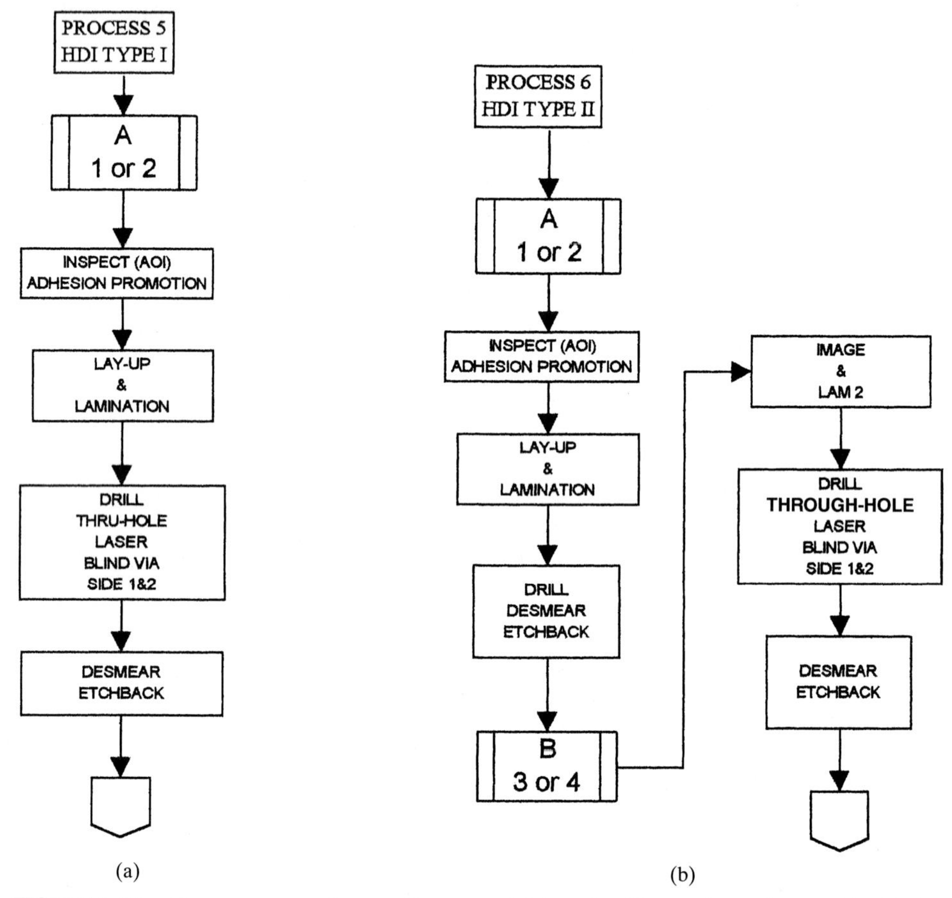

FIGURE 27.32 Diagrams representing possible sequential flows for HDI products. They are identified as processes 5, 6, and 7, referring to how the different HDI types mentioned in the text might flow.

27.4.3 Innerlayer Process

An innerlayer detail is essentially a thin double-sided printed circuit. The standard innerlayer process contains no plated holes because it is produced by using a print-and-etch process. Blind, buried via layers and laminated cores contain holes that must be plated in either a pattern-plate or a panel-plate process. Figure 27.31a shows a typical flowchart for four innerlayer process options. Processes 1 and 2 support standard innerlayers, and processes 3 and 4 can be used for buried via innerlayers or core subassemblies. All four processes start with a bare copper-clad laminate and end with a patterned double-sided circuit. The patterned circuit must be inspected and treated to enhance adhesion prior to further ML-PWB lamination. All four of these sequences work equally well with any of today's materials systems.

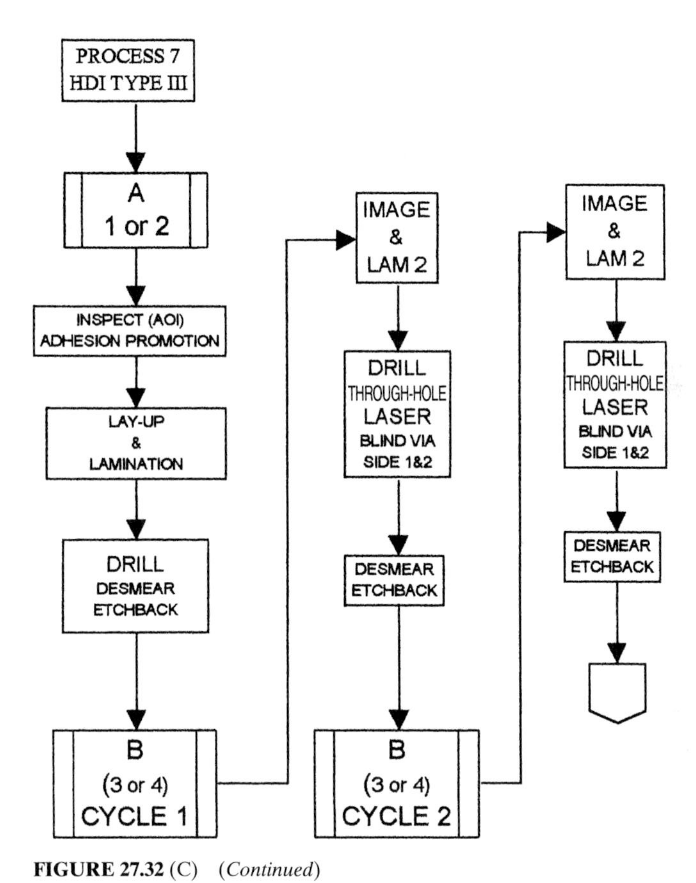

FIGURE 27.32 (C) (*Continued*)

27.4.3.1 ***Process Sequence Flows 1 and 2.*** Process sequence flows 1 and 2 are variations of how to produce a standard innerlayer with no plated holes by print-and-etch or post-etch-punch processes:

- **Process 1: print and etch** The laminate layers receive an application of photoresist. After photoresist application, the panel tooling holes are punched with the predefined pattern to match the lamination tooling. This is followed by the image printing (exposure) process, where the punched tooling holes align the artwork pattern. After print exposure, the inner panel is ready for development, which clears the non-image area for etching. The artwork is a negative image and produces a retained pattern of "what light sees stays." The resist that stays behind protects the copper, which becomes the circuit pattern. The etch process then removes the unprotected background copper. The stripping process then removes the resist material and the layer is ready for adhesion promotion.
- **Process 2: post-etch-punch** The difference in this sequence is the order in which the laminate material is punched (tooled). The innerlayer detail is resist-coated and placed in the exposure cabinet. The artwork layers have already been pre-aligned as the top and bottom

layer pattern pair within the exposure frame. This is defined as a pinless registration system. The process then is followed by etching and stripping. The final step is to punch the tooling holes, which are aligned optically to etched targets. The advantage of this process is improved innerlayer registration because it compensates for panel in-plane dimensional movement that occurs when stresses are relieved after etching.

27.4.3.2 Process Sequence Flows 3 and 4. Process sequence flows 3 and 4, as shown in Fig. 27.31a, are plating processes that may be used for blind and buried via layer pairs or laminated cores. The flows represent two distinctly different plating methods referred to as pattern plate and panel plate. Note that in these processes, the printed image ultimately is registered to tooling holes as well. Depending upon the tooling system, often the tooling holes are drilled at the same time as the via holes. This ensures optimum alignment between drilled holes and printed features due to dimensional movement of the materials.

- **Process 3: pattern plate** The distinguishing factor of this process is that final build-up plating is performed after imaging. Extra processing steps are required in this process to protect the primary metallization. Typically, a deposit of tin metal acts as etch resist, then is stripped after etching. Note that this process is suitable for defining small circuit patterns. The copper plating is selectively built up on the circuit pattern. The etch process need only remove the base-clad copper to define the circuit. Photoresist thickness must be matched with respect to the image resolution of the circuit.
- **Process 4: panel plate** The diagram of process 4 represents the flow for a panel plate substrate. Here the plating deposit covers the entire surface area of the substrate (panel). Photoresist is applied post-plating and provides the only protection to the defined circuit and the plated via holes. The via-hole protection is referred to as tenting. Photoresist materials chosen must have characteristics with tenting properties to avoid resist breakout and voiding of the plated hole. Note that this method places a burden on plating uniformity across the entire surface of the panel due to high current densities at the perimeter of the panel. Imaging of fine features is difficult because the base-clad copper and plated copper are etched together. Additionally, the panel plating of thin-core laminates can induce dimensional shrinkage of the substrate. Process parameters should be optimized through experimentation to applying a low-stress deposit.

27.4.4 ML-PWB Tooling

The tooling system employed for ML-PWB fabrication is one of the most critical aspects of the process and requires much forethought. The investment in tooling is significant and not easily changed. Planning is required to determine the degree of flexibility that is needed in a tooling system based on the types of ML-PWBs that will be produced in that facility. The ML-PWB tooling system can be broken down into four aspects: the panel size scheme, the front-end tooling, the method of generating tooling holes, and the hard tooling (plates).

27.4.4.1 Panel Size Scheme. The first decision is to choose from essentially three broad approaches to panel size: full sheet, or mass-lamination; standard panel sizes based on cuts from a full sheet; or customized (flexible) panel sizes based on cuts from a full sheet. Any particular facility may also employ more than one approach.

27.4.4.1.1 Mass-Lamination. Sheet sizes can be 24 × 48 in.; 36 × 48 in.; 48 × 52 in.; 1.5 meters × 2 meters; or other similar large formats. This approach is well suited to medium or mainstream technology requirements in lines and spaces, and hole-to-pad ratio. It works best with larger sized PWBs since too many small PWBs leave too much blank space between the circuits, wasting the principal advantage of full-sheet lamination. Typically, the copper foils are standard weights. The cores are .004 in. (1.0 mm) or thicker, although some mass laminators are capable at .002 in. (0.05 mm); and the lamination is a single-resin system, not a mixed dielectric build. The most common construction is from epoxy resin, since most of the enhanced electrical or thermomechanical resin systems are not available in large sheet sizes.

27.4.4.1.2 Standardized Size Cut Panels. This approach is dominant in high-end technology PCBs. Some very common panel sizes are 18 × 24 in. and 20 or 21 × 27 in. The facility strives to have the fewest number of standardized panels that will accommodate the majority of the MLB sizes typical of their customer base. Tooling is then either fully fixed or modular based on those panel sizes, but other new panel sizes are not possible without new hard tooling and die. Having four to eight panel sizes is common. These sizes are based on full yield cuts from the laminator's full sheets; for example, the laminator might offer a 36 × 48 or 42 × 54 in. sheet and the fabricator orders (or cuts from full sheets) 14 × 18, 12 × 24, 18 × 24, and 21 × 27 in. panels. This system standardizes tooling into the fewest number of different sizes of hard tooling (e.g., caul plates and separator plates; see section 27.5.1.1). This approach is also best suited to advanced feature sizes, where limiting the panel size to just over the MLB size has the advantage of reducing some degrees of difficulty such as the registration requirement over a larger area. Fixed panel sizes work well for mixed dielectric builds and sequential laminations. Very large cut panel sizes such as 24 × 36 in., approaching mass lamination size, are uncommon but available in very high-volume or backplane fabrication specialty facilities.

27.4.4.1.3 Custom-Cut Panels. Asian fabricators have pioneered flexible custom-cut panel sizes to maximize material utilization further and so reduce costs. The difference between standard and flexible cut panels is that standard-cut panels have a limited number of predetermined sizes and consequent hard-tooling configurations, while the flexible cut panels afford a continuous range of size because the tooling isn't fixed but highly configurable. Flexible cut panels are often paired with riveting instead of pinning to hold the details, prepreg, and foil together for lamination. Many Asian facilities will offer both approaches, selecting the best suited to the job at hand.

27.4.4.2 Front-End Tooling. The first aspect of the ML-PWB tooling is commonly called front-end tooling or simply computer-automated manufacturing (CAM) tooling. (See IPC-2514A, "GenCAM [BDFAB], Sectional Requirements for Implementation of Printed Board Fabrication Data Description." IPC-2514 is a standard developed by the Computerized Data Format Standardization Subcommittee [2-11] of the Data Generation and Transfer Committee [2-10] of the Institute for Interconnecting and Packaging Electronic Circuits. The GenCAM format is intended to provide CAD-to-CAM, or CAM-to-CAM, data transfer rules and parameters related to manufacturing printed boards and printed board assemblies. IPC-2514 is a section requirement within IPC-2511B, "Generic Requirements for Implementation of Product Manufacturing Description Data and Transfer XML Schema Methodology.") The front end is where manufacturing personnel using a CAM software package generate all phototools, the associated computer numerical control (CNC) electronic files (used by equipment such as drillers and routers), the construction stack-up, and the part manufacturing instructions, or "work order," also called the "traveler." The CAM station mirrors the tooling method chosen for the shop and the job's work order. The CAM software overlays the tooling pattern on each circuit layer to produce the master pattern alignment. The master artwork pattern is then photo-plotted to reproduce the artwork film tools for the imaging operations. Several important manufacturing functions fall within the bounds of the front-end tooling responsibility.

27.4.4.2.1 Design Rule Check (DRC). Design rule check (DRC) is where the electronic design data are analyzed against a specific set of fabrication rules, a virtual design for manufacturing (DFM). The capabilities and attributes of these rules are based on sound, industry-proven values that match the technology targeted for construction. The default values embedded into the CAM are input by the manufacturer as technology files. These files should reflect the capability of the fabricator's equipment and processes in light of how they are utilized against the ML-PWB type being produced.

27.4.4.2.2 Panelization. To gain economies in the manufacturing process, the individual board design is stepped and repeated within the artwork tooling. *Panelization* is the term used to describe this method. Here the elemental goal is to yield as many parts on one process panel as possible. The CAM software allows manipulation of the single design pattern to place or nest multiples within the confines of the chosen panel dimensions.

27.4.4.2.3 Artwork Generation (Photo-Plotting). Once the design has been formatted into the desired panelization, it is ready to be made into a working phototool. The CAM data are output to a server that converts the data into a language recognizable by the plotting equipment. The plotter exposes the circuit pattern on the film tool. Typically, the film tool is polyester-based with a silver emulsion and is exposed by means of a laser or fiber-optic light source. The resolution of the light source directly affects the ability to image the technology type. Note that when the technology type approaches the feature size of HDI, the use of an alternative approach to phototooling may be employed. This is known as laser direct imaging, and, as the name implies, the image is directly exposed onto the production innerlayer, thereby eliminating the use of a phototool.

27.4.4.2.4 CNC Drilling Routing Tools. The machine communication method commonly employed in the automation of the drilling and routing process is output from the CAM tooling as CNC programs (from the earliest computer numerical control machines, as differentiated from manual or template positioning systems). The front-end tooling software facilitates the organization and optimization of the CNC routine. The routine must match the flow of the technology process steps being produced. When sequential processing is required, the drilling routing data must be broken out into separate routines. When the technology type involves HDI features that require laser drilling, specific tooling accommodations have to be considered to match the discipline of the laser equipment—for example, the use of special imaged fiducial targets or a special dot mask aligned to the drill locations.

27.4.4.2.5 Automatic Optical Inspection (AOI). Another aspect of the tooling set is the data configuration to support the automated optical inspection (AOI) process. AOI is considered part of the detailed inspection process. When the AOI equipment supports what is known as CAD reference, the original design data (which were reproduced in the phototool) are now output as a configured file to compare the imaged circuit layer to the original designed circuit (see Chap. 53 for a complete discussion of inspection).

27.4.4.2.6 Electrical Testing. The final aspect of the tooling set is preparation of the electrical design data to support an electrical testing routine. The purpose of the electrical testing within the ML-PWB production process is to verify the integrity of circuit continuity. The electrical data required for the test must be extracted from the circuit pattern unless a separate net list is provided that contains the connectivity. The standard data format for electrical testing is IPC-356. (See also IPC-2515A, "Sectional Requirements for Implementation of Bare Board Product Electrical Testing Data Description [BDTST].")

Electrical testers can take on many forms, but are generally divided into two major categories, the bed-of-nails and the flying probe. Tooling for a bed-of-nails tester consists of outputting data to manufacture a fixture to accommodate the pins, which make contact with the circuit nets. When a flying probe tester is employed, the net data are fed into the machine, which uses its own software to configure the routine to probe continuity and perform isolation.

The chosen method of test tooling requires an examination of the economics based on the volume and technology of the ML-PWB. For small prototype lots required in short cycle times, flying probe may be more suitable than bed-of-nails testing. Although the throughput of a flying probe unit is less than that of a bed-of-nails tester, the flying probe does not require expensive, and time consuming to construct, fixtures. When the technology type involves HDI features, the only alternative is to use a flying probe due to the limitation of the grid spacing in a bed-of-nails tester. High-density test requirements continue to drive flying probe technology. (See Chaps. 54 and 55 for a complete discussion of electrical testing.)

27.4.5 Tooling Hole Formation

The shape, size, and location of tooling holes or slots are dependent on the investment tooling system determined by manufacturing engineering for a given facility, and there a variety of methods to form them. The tooling holes range in size and shape from 0.125 to 0.250 in. diameter to a slotted hole of 0.187 × 0.250 in. The common methods used to produce the mechanical alignment in innerlayers are punch and die or drilling, but the point in the process where they are formed can vary depending on the total tooling scheme.

27.4.5.1 ***Prepunch.*** A fixed punch and die set is used to generate tooling holes/slots prior to imaging/etching. To avoid having to cut away the photoresist material over the tooling hole, fabricators sometimes perform the punching after resist lamination. In this case, cleanliness has to be significantly refined. If a cut sheet resist laminator is used, it can be adjusted so the resist falls inside the boundary of the tooling locations. The punching occurs in a hydraulic punch press, where alignment is ensured by means of an edge stop on two axes of the panel to square the alignment. The die set penetrates the laminate in one stroke, punching all holes at one time while holding the material flat. Material is punched one piece at a time, yet high throughput can be achieved by this method. Punching of multiple laminates at once is not recommended because the quality and size of the tooling hole/slot are affected. Additionally, thicker laminates (greater than 0.032 in.) require the punch and die set to have a slightly larger tolerance to accommodate the spring-back of the hole within the composite material. Overall, the prepunch method is very accurate and repeatable. Materials with a high degree of movement (low-dimensional stability) are at a significant disadvantage in this method due to the drift of the tooling location after innerlayer processing.

27.4.5.2 ***Post-Etch Punch.*** An optically aligned punch is used to generate holes/slots after imaging/etching. The fully processed innerlayers are edge-aligned into the punch; a pair of video cameras locates on a diagonal pair of printed fiducials; a servo-system aligns the panel, minimizing alignment errors due to panel distortion; and tooling holes are punched. This process punches an optimized set of tooling holes/slots in the layer just prior to lamination. Post-etch punch can be automated so it has high productivity, but it is slower than a simple punch and die. However, the productivity gain achieved by printing without tooling reduces some of the productivity loss from using post-etch punch.

27.4.5.3 ***Drilled Tooling.*** When processing follows the process sequence 3 and 4 described earlier, tooling locations may be drilled. This allows the tooling locations to be formed at the same time as the via-drill cycle. To minimize distortion, which affects final registration accuracy, it is recommended that the drill program be scaled. Scaling factors must be characterized through experimentation. Scaling helps minimize the dimensional effect that processing has on the true positions of the holes. Alternatively, extra sets of tooling holes may be produced just for the imaging process, followed by a post-etch punch operation to optimize the lamination registration.

27.4.6 Tooling System

There are numerous rationales for innerlayer tooling schemes, and this has resulted in a wide assortment of tooling systems being used across fabricators. Over the years, many tooling systems have evolved, ranging from full-perimeter distribution of holes to as few as four holes located on the midaxis. One low-cost tooling system comprises use of rivets in a flexible layout scheme. The tooling system comprises a substantial investment and therefore merits considerable engineering conceptual judgment. Many shops become trapped with poorly conceived tooling schemes that are difficult to migrate from due to the cost of converting earlier designs.

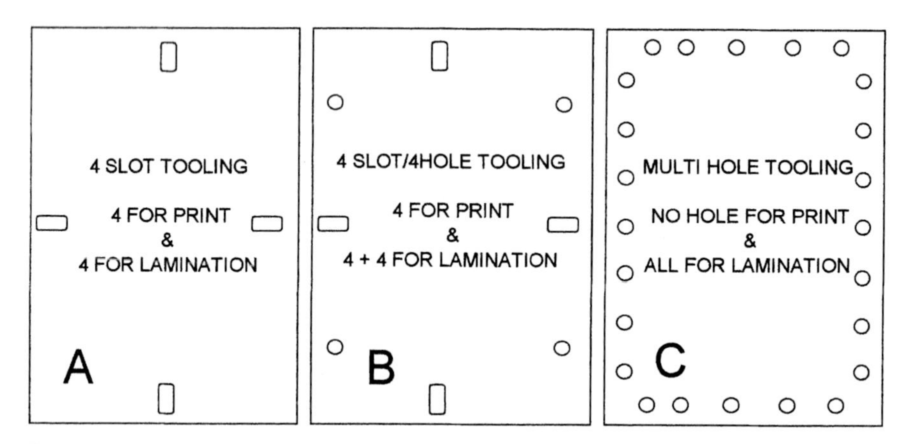

FIGURE 27.33 Three different tooling schemes: (a) four-slot tooling; (b) four-slot/four-hole; (c) a tooling system with a full perimeter of holes placed.

Therefore, careful analysis should be given to the product mix and technology types to be manufactured prior to choosing a scheme.

The primary purpose of the tooling system in ML-PWB manufacturing is to facilitate layer-to-layer alignment during the lamination cycle while maintaining a positional reference for subsequent processing. The tooling hole arrangement is mirrored within the front-end tooling routines mentioned earlier. The master alignment of each part keys off the tooling scheme. The primary tooling locations may be used at all process steps or sometimes used to introduce secondary reference holes.

Figure 27.33 shows three different tooling schemes. Figure 27.33a shows a system known as four-slot tooling. This is an excellent system, because the use of four slots allows for growth and shrinkage of the layers during processing. The four slots may be used for printing, inspecting, and lamination, or, if post-etch punch is used, they are used for inspection and lamination only. Figure 27.33b shows a system known as four-slot/four-hole. This is common in a prepunched system where the slots are used for image printing and the extra holes are engaged at lamination. Alternatively, punched slots can be used for innerlayer image and lamination, while post-lamination holes are drilled for plated holes' drilling and outerlayer imaging. In this method, targets imaged in the details under the outerlayers are exposed by controlled depth boring; a camera then locates the targets and a spindle drills the locating holes for the secondary operations. Figure 27.33c shows a tooling system with a full perimeter of holes placed. This style would be considered overdetermined. It has a disadvantage at lamination because imaged panels typically have to be stretched over the alignment pins. A fourth tooling system not shown uses low-profile rivets to align layers. The rivets can be arranged outside the pattern area and do not penetrate into the caul plate. Rivet height is chosen based on the appropriate panel thickness.

27.4.7 Imaging

27.4.7.1 Surface Preparation. The first step in the imaging operation is surface preparation to enhance photoresist adhesion. Double-treated foils (DTFs) and foils laminated with the tooth side up (RTFs) require minimum surface preparation. Generally, a process through a tacky roller machine to remove dust and foreign material is adequate. For cores with standard foil with the "shiny side up" (non-tooth or drum side), a more aggressive treatment is needed. Common options include a conveyorized chemical cleaning and/or an abrasive cleaning. Automated equipment is available to perform this function with pumice or aluminum oxide slurry. Scrubbing the layer by hand is acceptable but labor-intensive, and can provide inconsistent results.

However, acidic 4F grit pumice is required for single-step cleaning. Standard copper foils are coated with a chromate conversion to eliminate formation of heavy copper oxidation during storage. This coating, and any oxides, must be removed to provide a microroughening of the surface. When using any of the mechanical cleaning methods, care must be taken to avoid mechanical deformation (stretching) of the thin laminates. Stresses induced will be relieved after etching of the image, causing positional movement.

27.4.7.2 Photoresist. Photoresist is supplied in both dry-film and liquid forms. Dry films are a popular choice because of the simplicity of application. For buried via innerlayers, dry films are preferred because liquids are difficult to use with through holes. The major weakness of dry-film resists is sensitivity to surface flaws and a tendency to lift on poorly prepared surfaces.

27.4.7.3 Dry Film. Resist film is typically 1.5 to 2.0 mil. thick. Both thinner and thicker resist is available as a specialty product. Dry-film resist is applied with a hot roll laminator. To aid in production of fine lines and spaces, the use of a wet nip applicator aids in filling surface imperfections. Precision flatness and uniform temperature are equally important as well. Popular resists for high-density applications include films formulated with high-speed sensitivity for laser technology.

27.4.7.4 Liquid Photoresist. Liquid resists work well in a print-and-etch process. They have excellent adhesion and a tolerance for surface flaws and are relatively low in cost. The disadvantage of a liquid photoresist is the need to produce a perfect coating. Foreign material, skips, thin spots, and dewetting all cause serious image problems. The use of a wet coating can also cause problems due to resist contamination of the transport system.

27.4.7.4.1 Roller Coating. The least expensive and most popular way to coat liquid resists is with a pair of roller coaters consisting of pinch rollers in which one or both rollers is used for coating. The coating roller has a closely spaced, precisely cut spiral grove. Liquid is metered onto this roller and then transferred to the panel. A good roller coater produces an extremely uniform coating. By carefully selecting coater parameters, it is possible to achieve a coating thickness control of 0.1 mil. Some roller coaters use the top roller as a coating roller, allowing the panel to be transported on a clean conveyor system. In these systems, the resist can be dried or the panel can be printed wet. Other roller coaters use both the top and bottom rollers for coating. These roller coaters require a handling system that holds the panel by its edge until the resist is dry. The coating roller is the weak link in a roller coating. Alignment problems result in nonuniform coatings. Flaws produce repeating defects in the coating. Worn or improperly cut grooves result in a low-quality coating.

27.4.7.4.2 Curtain Coating. The curtain coater operates by pumping a waterfall of liquid through a narrow slot. By carefully controlling the slot width, pumping pressure, and viscosity, the fabricator can create a well-controlled curtain of liquid. When a panel moves through the curtain, a thin coating is applied to one side. Curtain coaters produce a good-quality coating, but not all photoresists have the proper viscosity for curtain coating.

27.4.7.4.3 Electrodeposition. With electrodeposition, a polymer is deposited on a biased metal surface submerged in a liquid medium by a process that is analogous to plating. It gives a well-controlled, high-quality coating. Both curtain coating and electrodeposition are expensive and are not widely used for innerlayer processing.

27.4.7.5 Photo Print Exposure. The third step in imaging is photo printing. The three types of photo-printing machines are flood, collimation, and direct image. Most flood printers consist of high-UV (5,000 W) light sources housed within large reflectors to distribute the light uniformly across the imaged surface. Hard contact by means of a vacuum is required between the film and the innerlayer. The silver emulsion side of the polyester film tool makes direct contact with the resist. This allows good image reproduction without collimated light. The drawbacks of contact printing are defects associated with poor contact, poor productivity due to the time required to establish hard contact, and the potential for artwork damage.

The alternative, collimation, an off-contact printing method, holds the film a small distance above the innerlayer surface and prints with collimated light. The collimated machine uses a high-UV lamp with mirrors to reflect intense light straight down on the innerlayer. This method permits a high degree of automation. Off-contact printing is absolutely necessary for printing wet layers. The disadvantage of off-contact printing is a loss of resolution due to incomplete collimation and a great sensitivity to dust and scratches on the artwork.

The other style of printing is referred to as direct imaging, and is the technology of choice where production of high-density line and space features is required. In this process, the electronic data are input into the direct image machine instead of being used to produce a phototool. The direct image, typically output through a laser, is used to expose the image by means of a series of on/off pulses manipulated within an x-y axis. The throughput of this process can be limited by the combination of resist speed (sensitivity) and mechanical positioning of the next panel.

27.4.8 Develop, Etch, and Strip

27.4.8.1 Developing. All modern innerlayer photoresist materials are aqueous-soluble. This means they develop in a mild caustic solution. Note that resists that use solvent developing are no longer in use due to significant health and environmental concerns. The developing process is easily automated by means of conveying the part through a sprayed caustic solution. The developing solution removes the non-imaged resist, which was not polymerized by the light source. Therefore, in effect, "what light sees stays." The quality of the developed image is often quantified by means of a density step chart that was previously exposed. By comparing the step held on both sides of the imaged panel, the fabricator can determine process control.

27.4.8.2 Etching. Etching chemistry is determined by the process flow sequence as outlined in Fig. 27.32. Etching can be classified as print and etch when using processes 1, 2, or 4 and using a cured photoresist as an etch resist. Etching is classified as print-plate-etch when using process 3 employing a sacrificial metal layer as an etch resist.

The preferred etch chemistry of the print and etch process is cupric chloride. Cupric chloride is easier to control than the ammoniacal etchant used with finished boards, and it can be easily regenerated. Etching occurs by spray erosion with the highly loaded (≤25 oz./gal.) copper-soluble solution. A high degree of automation is possible in the print-and-etch process. Since there is no plating step, it is possible to feed printed innerlayer panels directly into a conveyorized inline machine that develops the resist, etches the circuits, and strips the resists. These automated machine lines are called develop, etch, strip (DES) and improve productivity by reducing the defects caused by handling.

The print-plate-etch process follows the same general methods as used in the finished board steps (see Chap. 29). The use of a metal etch resist requires ammoniacal or acid etch. Resist developing is separated from etching by a plating step, so a stand-alone resist developer is required.

27.4.9 Inspection

Once innerlayer etching is complete, the innerlayer must be inspected. The innerlayer inspection criteria follow industry-adopted performance limits (IPC-600 and the IPC-6000 series). These limits must be confirmed prior to lamination into an expensive multilayer to prevent decreasing yields. Flaws are categorized to reflect processing defects related to laminate materials and circuit definition. Line width and spacing are of primary importance on dense signal layers. In the case of controlled-impedance layers, it may be necessary to gauge line widths. Power and ground layers are less likely to have defects and are sometimes excluded. Most modern circuitry precludes manual visual inspection due to the inability to accurately detect flaws and the associated intense labor. AOI is the industry standard for inspecting innerlayers.

AOI has the advantage of being able to detect circuit nicks and protrusions. False errors can be a problem and are time consuming to confirm. AOI technology using CAD reference (the comparison of the circuit image to the design data) has a lower escape rate. Cleanliness and copper oxidation are factors that can influence poor performance in most AOI technologies. Two common methods exist for imaging the circuit. In one, the innerlayer is scanned with a tiny laser spot that causes the substrate to fluoresce. The circuit is seen as a dark image against the light emitted from the substrate. The other method illuminates the circuit with a bright light so that its image appears bright against the dark background of the substrate. Both methods have strengths and weaknesses. The fluorescent method is blind to surface flaws, and substrate flaws can confuse it. The top-light method is oversensitive to the surface appearance of the copper. The AOI inspection of high-density features presents new challenges to the technology related to minimizing false errors. Modern machines employ a combination of image acquisition techniques while performing virtual reasoning against a background scan of design rules that are compared to the design image data.

27.4.10 Adhesion Promotion

Epoxy does not adhere well to untreated copper surfaces. This means that some type of treatment must be applied to the innerlayer before lamination. One option is to use double-treated copper, discussed earlier. Double-treated copper has a rough surface with a treatment supplied by the material vendor. Many ML-PWB fabricators report excellent results with double-treated copper. Others report problems with contamination and difficulty with rework. The alternative to using pretreated copper foil is to use a chemical treatment after etching.

27.4.10.1 Copper Oxide, Reduced Copper Oxide, and Alternative Oxides. There are many variations of the copper oxide treatment, but they all work by producing a rough surface topology that enhances adhesion. Depending on the exact nature of the treatment, the color varies from a light brown to a velvet-like black appearance. The differences among the treatments include the density of the oxide and the ratio of cupric oxide to cuprous oxide. To maximize bond strength, the fabricator should avoid oxide formulations that form tall vertical crystalline structures. Low-profile, self-limiting formulations generally provide the most consistent adhesion.

Most copper oxide treatments are applied by dipping the innerlayer in a hot (85 to 95°C) caustic bath for 1 min. or more. The oxide process generally uses a batch process that requires vertical racking of the layers. Rack design is critical to prevent problems with innerlayer damage and contamination. By minimizing the contact points, the fabricator can eliminate this concern. Racks require coating to insulate conductive contact to reduce electromotive effect in the oxide chemistry, which will polarize conductors and prevent oxide from being applied on random isolated traces. Since the oxide treatment is done after inspection and immediately prior to lamination, it can be the source of quality problems if improperly done.

The copper oxide surface treatment is very soluble in acids such as those found in the electroless copper line. If the epoxy-copper interface is fractured during drilling, the dark oxide will dissolve during the plating process, leaving a distinct pink ring around holes. This is not a serious reliability threat; it is generally considered a cosmetic problem but it often causes false concern. One way to minimize pink ring problems is to use an oxide reduction step after oxide formation. This step converts the copper-oxide crystals back to copper metal, preserving their topology. This reduced oxide surface has limited shelf life, so lamination should follow this process within 48 hours. Reduced peel strength should be expected as well.

Copper oxide treatments are not recommended for material systems that require processing temperatures significantly above 180°C. This includes Teflon and some forms of polyimide film materials. Oxide reduction begins to occur spontaneously above this temperature, reducing adhesion. The use of a thin brown oxide, or a reduced oxide, will reduce the risk of this failure mode. A second problem with polyimide film is that it is soluble in strong caustics, so if the treatment time in the oxidization process is excessive, significant substrate damage can occur. When processing circuits with microfoils having high-density features, the fabricator should reduce processing times because the oxide chemistries can reduce copper thickness.

The LFAC laminates are compatible with current brown oxides and reduced copper oxide treatments. The oxide treatment should be kept thin, not exceeding 0.4 mg/cm^2. Alternative oxide treatments, such as peroxysulfuric oxide, are strongly preferred for some HF laminates. The cores for ML-PWBs that will be subject to LFA should be post-oxide baked to remove moisture—generally 120°C for at least 30 minutes for signal layers and 60 minutes for power/ground layers.

27.4.10.2 Silane-Based Adhesion Promotion (Oxide Alternative Treatments). The silane-based process is an alternative to the copper oxide process. Silane can be used to bond epoxy to other materials. One end of the silane molecule bonds to the epoxy. If the other end of the molecule is modified to bond to the secondary material, the silane can serve as a bridge, greatly enhancing adhesion. It is commonly used in this mode to enhance the adhesion of epoxy to glass. It can be also used to enhance the adhesion of epoxy to copper.

This silane process attaches a thin silane layer to the copper surface. In lamination, the active epoxy molecules bond to the silane. This causes it to be a binding layer holding the epoxy to the copper. The objective is to achieve a stable coupling between the silane and the copper. This can be done by precoating the copper with a metal such as tin that reacts with the silane. When properly applied, the silane treatment is very stable and is resistant to chemical attack and delamination. The silane process has the advantage that it can be conveyorized in an inline process system. The major weakness of the process is that silane layers can absorb water and fail in some environments. The proper choice of silane chemistry minimizes the risk.

27.4.11 Drilling

The point at which the drilling process occurs is determined by the ML-PWB sequence chosen.

27.4.11.1 Standard Drilling of Innerlayers. Standard buried via innerlayers are drilled and plated prior to lamination using the same procedures as a finished ML-PWB. Buried via innerlayers are typically thin, and it is easy to drill very small holes. Whereas an 8 mil. hole is difficult to drill and plate in a 62 mil. thick ML-PWB, it is relatively easy in a 5 or 10 mil. thick innerlayer. Some manufacturers report success with holes as small as 4 mil. However, very small bits are expensive and easy to break. Special handling is required to load the bits and the drill machine must be vibration-free with very low run-out. For most ML-PWB shops, 8 mil. is a practical lower limit for mechanically drilled buried and blind vias. At 8 mil. it is generally possible to drill thin layers in stacks up to 100 mil. thick, greatly increasing drill productivity. However, a 10 mil. bit is less costly than an 8 mil. bit, so cost-sensitive products should try to design around a 10 mil. bit lower limit for buried vias.

LFAC laminates with fillers can be tougher and more brittle than regular FR-4s requiring different drilling parameters. As an example, the standard FR-4 might be drilled at up to 450 surface feet minute (SFM) with a chip load up to 2.0 mil/rev. and a maximum hit count of 1,500 holes; the LFAC epoxy-fiberglass-comparable laminate might be drilled at up to 350 SFM with a chip load up to 1.5 mil/rev. and a maximum hit count of 1,000 holes. Consult your laminate manufacturer's guidelines or applications engineer.

The major challenge in buried via processing is handling. Care must be taken in mechanical operations such as deburring to avoid mechanical damage or distortion. Often a frame is used to stiffen the layer during plating. Blind vias may be fabricated like buried vias and mechanically drilled in a top (or bottom) sub-lamination section prior to full board lamination, or controlled-depth drilling may be used after lamination.

Controlled-depth drilling has the advantage that standard innerlayer processing is used, including foil stacking. Controlled-depth drilling has several limitations:

- Blind holes cannot be stack-drilled, severely limiting drill productivity.
- It is difficult to plate a blind hole whose depth exceeds its diameter; this necessitates limiting the maximum hole depth or increasing the drill diameter such that a maximum aspect ratio of 1:1 is preserved.
- Drill depth tolerances make it easy to under- or overdrill a blind hole, producing a quality or reliability risk.

All of these limitations are avoided when the blind vias are drilled through a two-sided board used for the outerlayer and next innerlayer in a stack-up. These holes are drilled prior to lamination just as a thin two-sided plated through board. In this case, the process sequence for a blind via layer is identical to the sequence for a buried via layer. Such layers can be stack-drilled.

There is no aspect ratio limitation and the need for controlled-depth drilling is completely eliminated. To be able to drill blind vias prior to lamination, a clad outer stack-up is required. One side of the outer component layer is patterned prior to lamination, while the other is patterned after lamination. This means that when the blind via is metallized, the unpatterned side is blanket-metallized. This is true for either pattern plate or panel plate. This outside layer is metallized again when the holes in the finished ML-PWB are metallized. The result can be very thick plating on the exterior of a blind via board. To minimize this problem, the fabricator should plate the blind via innerlayer with the minimum possible current density on the blanket-metallized side.

27.4.11.2 Drilling Related to Laminated Subcores. The drilling process for laminated subcores usually follows the rules for finished board drilling and is design-dependent. When designs follow this methodology, they typically push the limits of plating aspect ratio. Density constraint typically drives this requirement due to routing availability. In this case, additional density can be gained through the use of laser blind vias.

27.4.11.3 Laser Drilling. Laser-drilled blind vias can be placed on either or both sides of the ML-PWB, and they have inherently accurate depth control. Productivity of the laser is very high. When employing laser blind drilling, it is necessary to use thin insulating dielectrics to produce a thin stack height to the sublayer. Depending on the type of laser employed, it is sometimes necessary to image fiducial targets the next layer down for alignment. This is the case when employing an UV or Nd:YAG laser, which can penetrate copper. Other laser-drilling alignment targets are required when employing a carbon dioxide (CO_2) laser, which will penetrate only the dielectric. Very high drill rates can be achieved on a scanning CO_2 laser. Here a mask pattern is etched in the copper to allow exposure of the laser beam to the dielectric. The laser removes the dielectric, stopping at the copper pad below. Multiple via stacks can be accomplished with this method by repeating the mask opening at different layer depths. Aspect ratio should always be kept in mind. Most success has been reported when the mask diameter is kept greater than 0.003 in. For best accuracy, a minimum of three etched targets is required and should be referenced to the CNC drill data.

27.5 LAMINATION PROCESS

The lamination process is an essential step in the fabrication of an ML-PWB. The process is also one of the longest cycle time operations. Therefore, when the process method calls for repeat cycles, or what is called sequential laminations, it can be a significant cost driver. (*Note:* Sequential lamination is often paired with distinct, separate drill and plate processes for the subcore that further ratchets up pricing, as they are also time-intensive operations.) The lamination process involves two distinct yet linked operations: lay-up and laminating.

27.5.1 Lay-Up and Materials

Lay-up occurs in a clean and controlled room environment. The level of environmental control depends on the circuit feature technology being fabricated. The lay-up process for standard multilayer processing is relatively forgiving. However, working with material that is hygroscopic will require additional measures. The innerlayer details require a bake cycle (typically 120°C for 1 hour minimum) to remove moisture. A reduced bake cycle duration is indicated when certain oxide reduction chemistries are employed. Once layers are readied for lay-up, they should be processed as soon as possible. Should extended hold times become necessary, storage in a nitrogen-purged dry box is recommended.

The lay-up operation is often referred to as "building up a book." The operation follows a guide, often referred to as a stack-up sheet. The guide sheet, which depicts the engineering design of the ML-PWB, is highly recommended to minimize error. The written and illustrated guide to the book build-up process is made a part of the job planning/traveler. Due to the complexity of some products, the operator follows this guide in a recipe-type manner to perform the systematic construction. The lay-up may consist of some four or more PWB material components or other subassemblies. The lay-up operation produces a large assembly when complete that consists of the tooling plates, consumable press materials, and the ML-PWB detail. Refer to Fig. 27.34 and 27.35 for illustrations of the following.

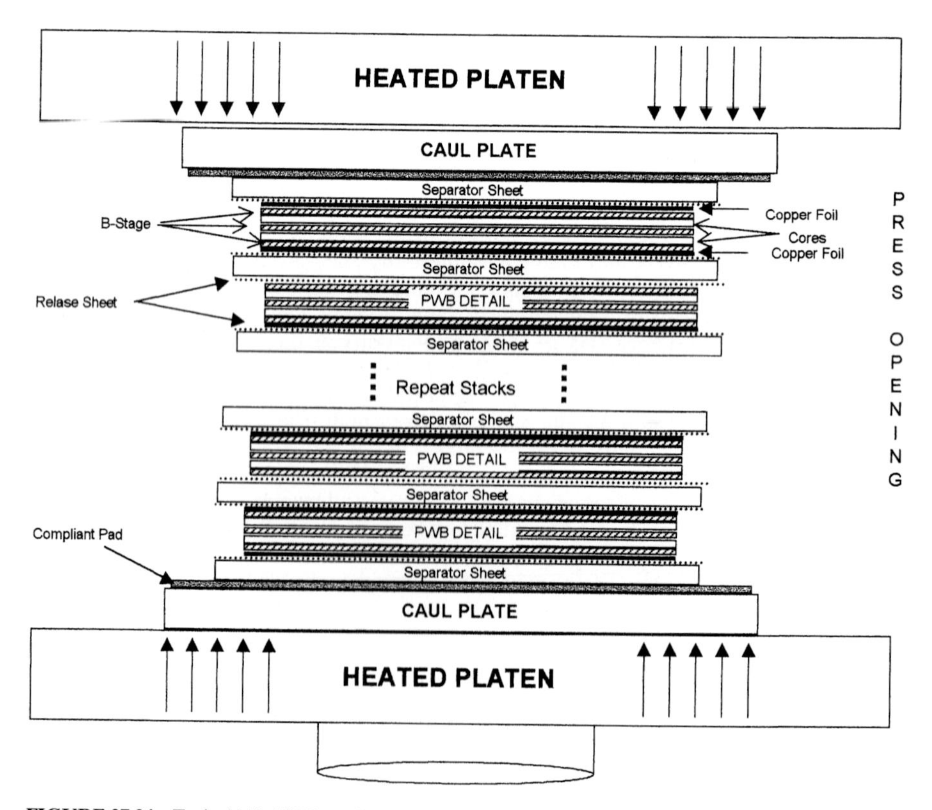

FIGURE 27.34 Typical ML-PWB stack-up for hydraulic lamination.

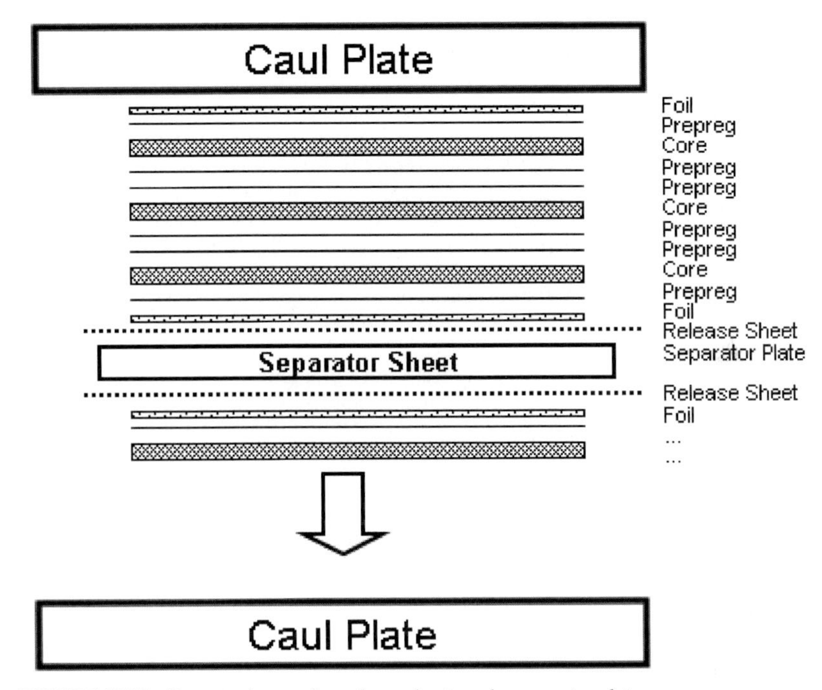

FIGURE 27.35 Press stack-up using release sheets and a separator plate.

27.5.1.1 Tooling Plates. The outermost item in the stack is termed a caul plate or carrier plate. These are thick, oversized metal plates, generally made of $^3/_8$ in. (9.525 mm/0.375 in.) thick steel. The 4130 alloy of steel is often chosen due to its precision machining capability for placement of the holing holes. Sometimes a hardened aluminum alloy is chosen, but is not highly recommended due to its high in-plane expansion. The purpose of the caul plate is to provide a stable base to transport the ML-PWB stack.

27.5.1.2 Separator Plates. The multilayer boards in the stack are isolated from each other by metal separator plates. The separator plates provide a mold surface for the laminated ML-PWB. It is extremely important that the separator plates be clean and free of debris. Both aluminum and steel separator plates are used. Separator plates should be cleaned regularly. The surface finish of the plates becomes very critical when laminating against microfoils that will be printed with fine lines. The most common type of steel is one of the 400 series stainless steels that have a very durable surface. Hardened 300 series steels are also occasionally used. Separator plate thicknesses range from as thin as 0.015 in. to as thick as 0.062 in. The thicker plates are more rigid and resist the tendency of internal layer features to print through from one ML-PWB to another. Thin aluminum plates have the advantage of being disposable, eliminating the need for cleaning.

27.5.1.3 Consumable Press Materials. The purposes of consumable press materials placed between the tooling plate and the top layer are to ensure that pressure is applied uniformly and to provide thermal lagging to correct heat rise. The purposes of consumable release sheets placed in contact with the foils are to protect the outerlayer foils, keeping them resin-free, and to allow easy separation of the foil from the separator sheet.

Among the materials used to provide for pressure and temperature uniformity are multiple sheets of Kraft paper, silicone rubber pads, expanded mat paper, and composite board. Kraft paper and composite board have the advantage of low cost, but produce an odor that some press operators find objectionable. Silicone rubber pads have the advantage of being reusable, but are limited by the number of heat cycles they can withstand. Silicone rubber pads, when close to the end of their life, go through reversion and leach out silicone oil, which can become a source of contamination. They also have limited success in controlling edge voids. Excellent results are reported with expanded paper mat products that are commercially available. Mat paper press pad materials come in different thicknesses. Some are even produced with a sandwich layer of release material cohered. Other products exist that flow and thus act as a stop for resin during lamination of subcores with predrilled blind vias.

Some form of release material is required against the outer foil surface of the ML-PWB. The release material acts as a nonstick slip sheet to keep the copper surface smooth, minimize the plate cleanup, and catch the resin run-out. Figure 27.35 shows the location of the release sheets. Alternatively, oversized foil can be used to catch the resin run-out. Or, a product called C-A-C (for copper-aluminum-copper) is used that consists of two sheets of foil laminated to either side of an aluminum carrier sheet. In this configuration, the one sheet of C-A-C is between two PCBs with the bottom copper foil over prepreg to form the top side of the bottom PCB, and the top copper foil under prepreg to form the bottom side of the top PCB. Figure 27.36 shows the location of the C-A-C sheet. The aluminum serves to replace the steel or aluminum separator plates and has the added benefits of reducing the stack height (which can sometimes translate to increased productivity by enabling another board to be made in the "book") and providing excellent protection to the surface of the outerlayer foils. Since they are not handled as foils but as a more rigid composite, there is no propensity to wrinkle, and since they are covered for the entire book-building operation, they are not subject to scratching or having errant debris introduction cause a pit during lamination. Finally, when the book is broken down, the aluminum between the ML-PWBs is easily separated from the foils, leaving the distinct ML-PWBs. The aluminum itself is sometimes sent to the drilling department to be used as an entry material, and then sent to reclamation.

27.5.2 Lamination Stack-Up

The lamination process is key to building a reliable ML-PWB. In the lamination process, the board is subjected to heat and pressure that melts the B stage (bonding sheets) and causes it to flow. This encapsulates the circuits and fills any buried vias. The B stage then cures, establishing a good mechanical bond to the inner detail layers. A variety of materials can be utilized in standard laminating press cycles. (Refer to the preceding section 27.3.5 for a detailed discussion of the materials and prefill process.) A typical ML-PWB stack-up is shown in Fig. 27.35. For productivity reasons, multiple stack heights are repeated within each opening of the press. The various components and their placement within the book are shown.

A standard lamination uses tooling pins that go from caul plate to caul plate, passing through each board and all of the separator sheets. Since the CTE of stainless steel roughly matches the in-plane CTE of a multilayer board, a tight fit to the pin is possible. Aluminum has a much higher CTE, so if aluminum is used, a loose fit to the tooling pin is required. Some fabricators use as few as four tooling pins. Others may use 20 or more. In addition to easy stacking, the four-slotted pin system minimizes problems arising from material growth and shrinkage in an overdetermined tooling system. On the other hand, users of systems with a large number of pins believe that firmly anchoring the ML-PWB to stainless steel obtains better dimensional stability.

27.5.3 Lamination Breakdown

Once the ML-PWB is fully cured and then cooled to room temperature per the manufacturer's guidelines, the book is taken apart; this process is called "breakdown." This is usually a

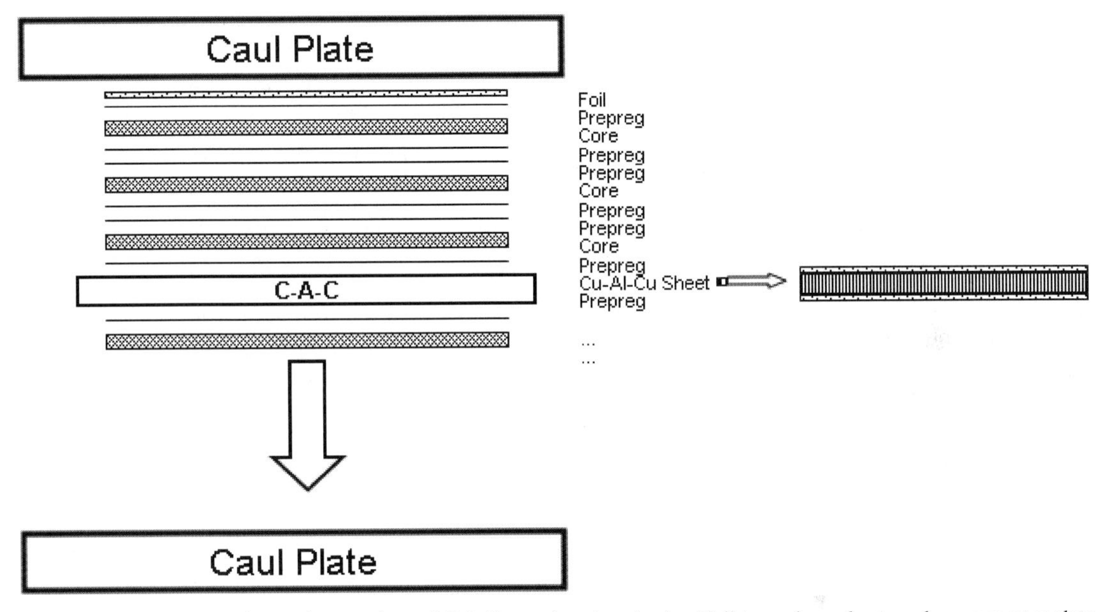

FIGURE 27.36 Press stack-up using one sheet of C-A-C to replace two sheets of foil, two release sheets, and one separator plate.

manual process which consists of de-pinning the book, then separating the individual ML-PWBs from each other at the separator sheets or C-A-C boundary as shown in Fig. 27.36. The panels are sent to an automatic thickness checker to ensure compliance with requirements, then on to an edge-finishing machine. The edge finisher removes the rough cured prepreg from the edges (called "ooze out") and any copper foil overhang by routing the four sides. The machine will then bevel and polish the edges so that they are not sharp and there are no protruding glass fibers or copper slivers. The finishing process might also include a bore to remove surface epoxy from around the tooling holes or slots.

27.5.4 Lamination Process Methods

*27.5.4.1 **Standard Hydraulic Lamination.*** A standard hydraulic press generally has a top or bottom ram and several floating platens that create multiple press openings. Typical presses have four to eight openings. The output of the press is based on the stack height in each opening. Low-layer-count designs can yield as many as 96 panels per press cycle based on a stack height of 12 by 8 openings. High-layer-count and thick laminated boards have the stack height reduced to ensure uniform heat rate from the outside to innermost stack. Steam, hot oil, or electrical resistance heaters may heat hydraulic-style presses. The steam and hot oil presses have the advantage of a fast heating rate, but the temperature of the heating fluid limits their maximum temperature. For steam, this is generally below the lamination temperature needed for high-temperature materials. If a steam press must be used, polyimide, PPO, and cyanate esters can be oven-baked to complete their cure. However, the thermal plastic adhesive layers used with Teflon are not compatible with a steam press.

*27.5.4.2 **Vacuum-Assisted Hydraulic Lamination.*** Many hydraulic presses use a vacuum to eliminate volatiles. Users of vacuum presses report a reduction of edge voiding and the ability to laminate at a lower pressure. Almost all hydraulic presses sold since 1990 are equipped with a vacuum chamber. In a typical process, the ML-PWB caul plate is loaded onto a carrier sheet, where a spring-loaded rail holds it off the press platen and limits the heat transfer to the ML-PWB stack. Next, the vacuum chamber is closed and a vacuum is drawn and held. A typical vacuum cycle may be from 15 to 60 min., depending on the nature of the materials being laminated. This gives the vacuum time to pull air, moisture, and other volatiles out of the ML-PWB stack. At the completion of this prevacuum process, the press is closed, compressing the spring-loaded rails and establishing good thermal contact with the press platens.

27.5.4.3 Autoclave. The autoclave is a popular tool in the composites industry and has found limited use by board fabricators. The autoclave is a sealed cylindrical chamber that subjects the ML-PWB stack to a high-pressure heated gas. The ML-PWB stack is sealed in a vacuum bag, and the hydrostatic pressure from the gas produces the force necessary for lamination. In principle, the autoclave is an excellent machine for producing a void-free lamination. Since the pressure is hydrostatic, it eliminates the problem of low pressure at the panel edge. This results in void-free panel borders and increases the usable area on a panel. In practice, an autoclave requires a long prevacuum cycle and has a slow heatup rate. It is also vulnerable to problems with vacuum bag failures. The result is that autoclaves have not found broad acceptance.

27.5.5 Critical Lamination Variables

Since hydraulic presses are by far the most common type of press, the following discussion focuses on that process. The quality of the ML-PWB lamination is affected by both the pressure ramp and the temperature ramp. There are almost as many unique cycles (also known as press recipes) as there are fabricators. However, in general the lamination cycle can be divided into four regions: B-stage melt; B-stage flow; B-stage cure; and cooldown. Figure 27.37 shows a typical lamination cycle with these critical variables identified.

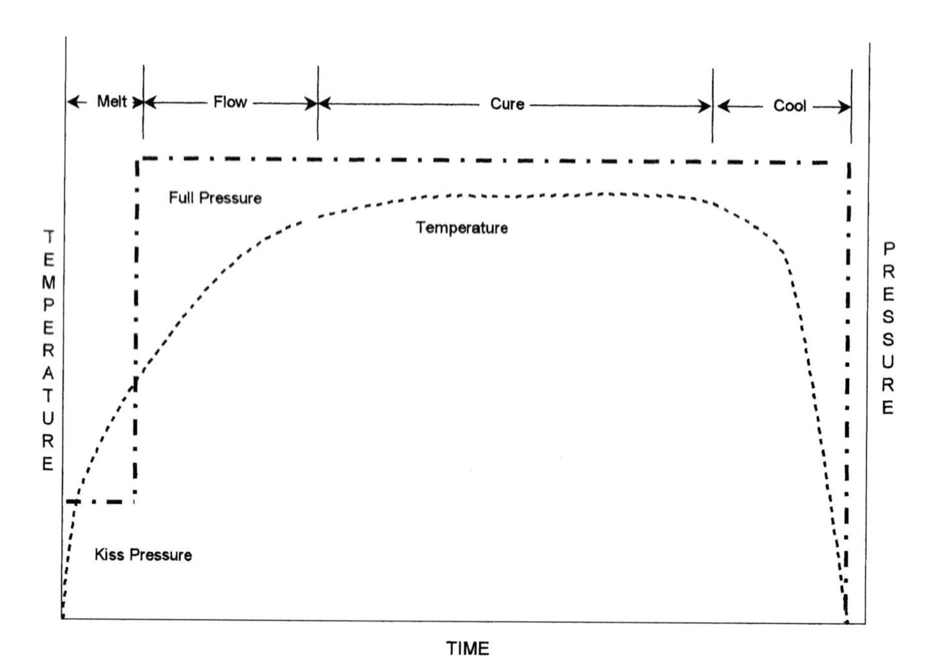

FIGURE 27.37 Typical temperature and pressure cycle for hydraulic lamination.

27.5.5.1 B-Stage Melt. During the melt cycle, the B stage is solid and the pressure should be low. Excessive pressure during this portion of the cycle will damage the glass cloth and exaggerate problems with image print-through, also known as "telegraphing." Most press cycles start with a low pressure known as a kiss pressure. The kiss pressure is high enough to ensure good thermal contact without damaging the ML-PWB. The length of the kiss cycle depends on the heating rates and the cure kinetics of the B stage. In a hot press cycle, the ML-PWB stack is loaded into a preheated press, and the heating rate may approach 20°C/min. In a hot press, the kiss cycle should be limited to a few minutes, or the B stage will begin to cure before flow is complete. At the opposite extreme, the press is loaded cold, and the heating rate is determined by the heating ramp of the press. For a slow heating rate of 5°C/min., a kiss cycle of 15 min. is appropriate.

Some LFAC laminates, however, have process guidelines where a best practice is to have a kiss cycle with 0 psi with a slow heat rise on the order of not more than 15°C/min. until approximately a 90°C internal temperature. This step is held for 30–40 min. to drive out moisture before pressure is applied and the flow stage initiates.

27.5.5.2 B-Stage Flow. The second portion of the cycle begins when the B stage liquefies, but before its viscosity begins to rise due to curing. In this region of the cycle, the liquid B-stage flows and encapsulates the circuitry. As long as the internal features are surrounded by liquid B stage, print-through is not an issue. The key to good results is to select a pressure that will allow reproducible resin flow but that does not squeeze out all available resin before the progressing cure stops the flow. Once again, the exact pressure depends on the temperature cycle, the B-stage viscosity characteristics, and the B-stage cure kinetics. For a fast-curing B stage and a fast temperature ramp, pressures as high as 600 psi may be needed to ensure complete circuit encapsulation. On the other hand, with a slow ramp and a long B-stage working time, high pressures give excessive flow and best results are obtained at 200 psi. Improper melt staging can result in "footballing," a term used to describe a panel whose center is significantly thicker than its outside. To determine proper melt, it is recommended to install a thermocouple in the center stack edge to chart the actual heat rise through the flow.

A typical dicy-cured FR-4 multifunctional epoxy has a flow range, called the critical range, at the temperatures between 70–130°C. HF and LFAC laminates have a critical range typically between 80–140°C. During the critical range, the heat rate input and pressure are critical variables. A typical dicy-cured FR-4 press recipe would be 4–8°C/min. rise at 200 to 300 psi pressure. HF and LFAC laminates' recipes are usually lower in the rate of rise and higher in pressure, such as 2–4°C /min. and 225–360 psi pressure.

Table 27.5 shows some press recipes from the processing guidelines posted on the laminators' web pages. This table is intended to illustrate that moving from standard FR-4 to other HF or LFAC laminates will require the development of new press recipes.

TABLE 27.5 Press Recipes for Selected Materials

	T_g °C (DSC)	Rise (°C/min.)	Critical range (°C)	Pressure (psi)	Cure time (min.)/Temp (°C)
Standard Dicy-Cured FR-4	175	4–7	70–130	200–300	60/182
Non-Dicy FR-4 Laminate	155	2–4	80–140	275–360	90/193
HF Laminate	160	2.3–5.6	>100	380	90/200
LFAC Laminate A	185	4.4–7	70–130	200–300	75/185
LFAC Laminate B	190	2.3–5.6	83–139	225–325	75/193
LFAC Laminate C	210	2–4	80–140	275–360	90/193
LFAC Laminate D	190	2–4.5	80–135	200–300	120/200
LFAC Laminate E	215	2	80–135	300	120/200
Polyimide	250	2–4	80–138	225–275	200/218

*27.5.5.3 **B-Stage Cure.*** In the third part of the cycle, flow has stopped and the resin cure is proceeding. The temperature is held at its maximum value to minimize the time to obtain full cure. For a typical epoxy system, this usually is approximately 180°C for 60 min. LFAC laminates, however, tend to require longer and hotter cure stages: up to 200°C internal temperature for 120 min., and longer for thicker boards. Some materials, such as polyimide, require a significantly higher cure temperature for even longer.

*27.5.5.4 **Cooldown.*** The last part of the cycle is the cooldown cycle. In Fig. 27.37, it is suggested that the pressure is released after some cooling has occurred, but before the stack reaches room temperature. In many modern systems, the ML-PWB stack is transferred hot to a low-pressure cooling press. It is important to control the cooling rate to minimize warpage. It is generally desirable to cool through T_g in a stress-free state without any significant thermal gradients present. A properly designed cooling press will meet these conditions.

27.5.6 Critical B-Stage Variables

During a typical lamination cycle, the B stage undergoes several significant changes. At the beginning of the cycle, the B stage is a solid with a low cross-link density and a melt temperature near 90°C. As the temperature rises, the B stage melts and becomes a high-viscosity liquid. As the press heats further, the viscosity of the liquid drops. When the B stage begins to cure, viscosity reaches a minimum and begins to rise. The region around the viscosity minimum is called the region of maximum flow. The wider this region and the lower the minimum viscosity, the more flow occurs. Figure 27.38 shows a schematic viscosity curve for a typical cure cycle. In a high-flow B stage, the initial cure level is low. This results in a longer time at temperature before the B-stage viscosity rises due to cure. This is often described as a long gel time. In terms of Fig. 27.38, a high-flow B stage has a low minimum viscosity and a wide region of maximum flow. A low-flow B stage has a higher degree of initial cure and may include flow restrictors to increase the minimum viscosity. High-flow B stages are useful

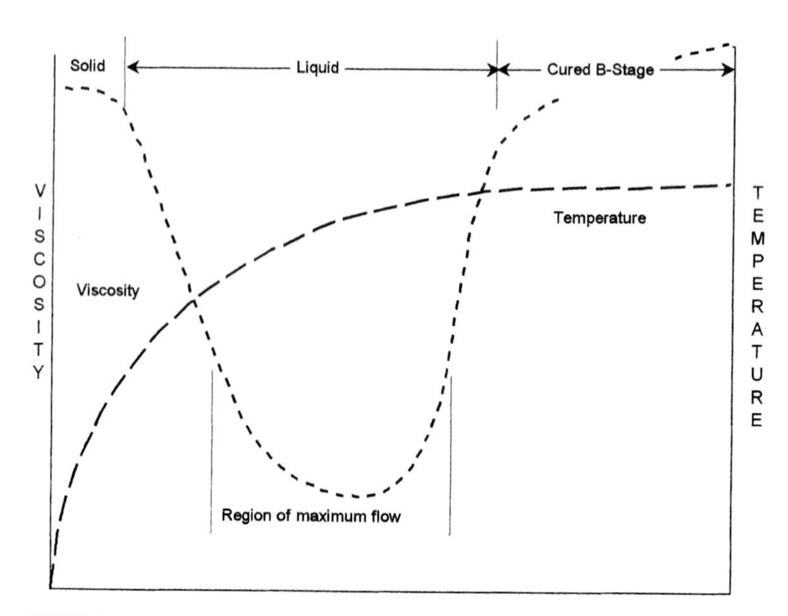

FIGURE 27.38 Typical viscosity curve for B stage during lamination.

in presses with a high-heating rate where the resin may begin to cure before flow is complete. They give excessive flow if used in a press cycle with a very slow heating rate.

27.5.7 Single-Ply versus Multiple-Ply B-Stage Fill Considerations

For dielectric separations that could be filled by either one "thick" ply of B stage—for instance, style 7628—or by two thinner plies of B stage—for instance, style 1080—the fabricator and engineer need to determine which approach is best on a product-by-product basis.

Building ML-PCBs with single-ply B-stage fill is common and has the following advantages:

- Costs are reduced significantly when sheets of prepreg are eliminated—the main advantage.
- Z-axis expansion is easier to control since there is usually less resin with a thicker style.
- Overall thickness has less variation since there are fewer elements in the construction.
- Single-ply fill allows for overall thickness reduction when two plies of any style are replaced by one ply of the same style or thinner.

Building with two plies is usually either preferred or required with the following conditions:

- The separation to be filled is between two planes generating a large voltage bias and additional dielectric separation is required to minimize the potential of dielectric withstanding breakdown.
- The separation to be filled is between two layers having thicker (70 micron) copper foil. In this case, there is frequently insufficient resin in a single ply to encapsulate the features and prevent air bubbles in etched-out regions.

27.6 *LAMINATION PROCESS CONTROL AND TROUBLESHOOTING*

A good lamination cycle produces a flat ML-PWB that is free of moisture or voids and has a fully cured substrate. All layers must be well registered. The ML-PWB must be free of warp and the thickness must be within the specification. Any controlled impedance layers must have the correct pressed dielectric separations above and below the signal layer. Each of these requirements puts special demands on the lamination process. To assist in setting control measures, the following process indicators (see Table 27.6) should be monitored with a Statistical Process Control (SPC) method.

TABLE 27.6 Process Variables and Limits

Process indicator	Specification limit	Gauge type or method
Thickness control	Per engineering drawing or internal spec requirement	Over-arm micrometer
Heat rise °/min. through melt	Per resin system requirement	Chart recorder
Cure	T_g of resin system	TMA
Post cure	Dwell time/temp per resin system requirement	Chart recorder
Registration	Per engineering drawing or internal spec requirement	Appropriate coupons or x-ray inspection equipment
Dielectric separation	Per engineering drawing	Cross section

27.6.1 Common Problems

27.6.1.1 Voids and Moisture. Substrate voids are a serious problem in many lamination processes. One source of these problems is moisture. B stage is very hygroscopic and must be stored in a low-humidity environment to preclude serious void problems. C-stage components also have a tendency to absorb water, and many fabricators use a bake to dry layers prior to lamination. However, for a fast innerlayer line with a good dryer in the oxide line, an innerlayer bake may not be necessary. Voids generally cluster in the low-pressure regions near the edge of the panel. This effect is minimized by the use of vacuum lamination. Increasing the lamination pressure can also reduce voids. However, the use of high pressures with high-flow materials can result in excessive flow, which leads to other substrate flaws such as resin starvation.

27.6.1.2 Blisters and Delamination. Blisters and delamination are also caused by trapped volatiles that collect in the low-pressure regions associated with print-through. If a board has wide copper borders on every layer, blisters are often found in the lower-pressure circuit areas adjacent to the borders. The best solution to this problem is to avoid areas of heavy copper adjacent to low-density circuit areas by replacing solid borders with a dot or stripe pattern. Also, the glass style and resin content of the prepreg should be matched to the weight of copper thickness adjacent to the prepreg.

27.6.1.3 Under Cure. The requirement of full cure is relatively easy to obtain if proper cure time and temperature are used. One measure of the cure level is T_g. Periodic measurements of T_g are an excellent check for material and process consistency. Another way to check cure is to make two consecutive measurements of T_g: If the epoxy is only partially cured, it will continue to cure during the first measurement and a higher T_g will be detected on the second measurement. A shift in T_g of more than 5°C is an indication of under cure. This measurement is typically performed on a TMA. Figure 27.39 shows an example of an epoxy TMA run; note that the second TMA run has a delta of 3°C.

27.5.1.4 Post-Lamination Bake. Many fabricators bake the ML-PWB at 150°C for up to four hours after lamination. One purpose of this baking is to ensure a complete cure.

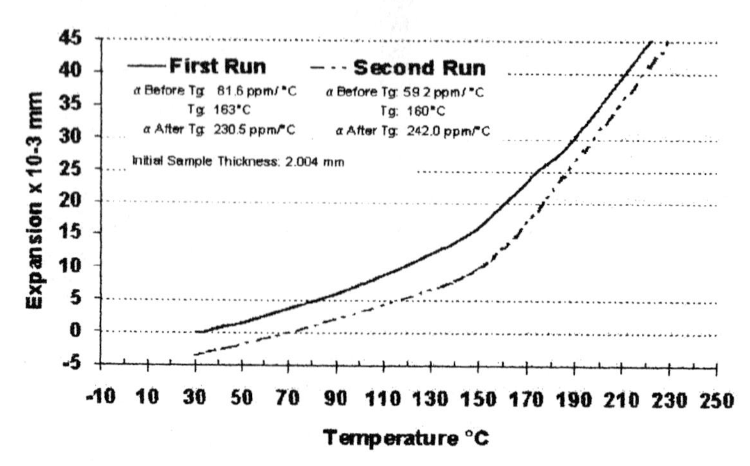

FIGURE 27.39 TMA analysis of epoxy with woven fiberglass. (*Courtesy of Microtek Labs, Anaheim, CA.*)

Although a bake will advance the cure, it is unnecessary if a proper lamination cycle is used (although polyimide is an exception; check the recommended processing guidelines). As discussed earlier, this can be verified by T_g measurements. However, additional baking beyond full cure degrades the material and reduces T_g.

A second purpose of the bake is to reduce the warpage often seen in the outer boards of a stack-up. Although a post-lamination bake will flatten the boards, it is more of a repair than a root solution. If the panels are cured in the press in a way that ensures they go through T_g in an isothermal stress-free state, warpage should not occur. The warpage seen in the outer boards of a stack is generally a symptom of nonuniform cooling.

The third purpose of a bake is to relax internal stress and improve registration. Internal stresses are a symptom of an overdetermined tooling system. If such a system is used, a bake may improve registration. If the more popular four-pin system is used, a bake is unnecessary. Always confirm the need of a post cure with the recommendations of the material supplier.

27.6.2 Special Consideration for Non-Dicy, Non-Bromine, and LFAC Laminates

For decades, the majority of PCBs have been built with epoxy resins, dicy-cured, and with bromine flame retardant. The industry has developed around this paradigm. Drills and drilling, chemistry, lamination, and circuit card assembly are all well understood with reference to standard FR-4. Although there are fabricators with processes developed to manufacture PCBs using laminates other than standard FR-4, this knowledgebase is not distributed industrywide. Especially problematic is the intersection of requirements for HF and/or LFAC with high thermal resistance and improved electrical characteristics, such as low attenuation in the material.

Many laminators recognized that the heart of their product line would need to be revamped from the entry-level cost-conscious material all the way through to the highest performance material. This was a large undertaking, requiring many different trials to find the right ingredients, in the right combinations, with the desired properties, especially the response to the new elevated reflow temperatures. As is frequently the case, there are trade-offs. For instance, reducing CTE by increasing filler is desirable to a point; the increase of filler increases the dielectric constant and loss tangent properties of the material, making it less suitable for high-speed digital products. The line-up from the bottom to the top of the performance ladder, with and without bromine, capable of withstanding high-temperature reflow, isn't complete in all cases.

These materials have not been in manufacturing and the field as long as the time-honed standard epoxy systems. Also, as they have been used, they have also been continuously refined—which has caused processes to be continuously redefined. The major processes that need to be developed and optimized specifically for nonstandard FR-4 laminates are innerlayer adhesion promotion, lamination, drilling, and hole wall preparation. The fabricator must work closely with its laminate and chemistry providers to develop the specific process parameters that will work with their lamination equipment and chemistries.

The major problems have been delamination (both adhesive and cohesive failures), voiding, and hole wall pull-away. Some of these quality problems can be discovered prior to shipment by cross-section analyses, but some do not manifest until after high-temperature reflow. Since laminators test laminates rather than finished PCBs, and fabrication and design features both effect the final product, a cautious fabricator will test completed PCB parts for quality and reliability, simulating the intended assembly soldering profile. Accordingly, submitting an actual final ML-PWB to multiple cycles through a production reflow oven with a lead-free soldering profile can be advantageous. After reflow exposure, the parts are visually inspected for blisters, then cross sections are taken at both high and low hole density areas to look for internal delamination or laminate cracks, voiding, and hole wall pull-away. Figures 27.40 through 27.43 show examples of these defects.

FIGURE 27.40 Blistering.

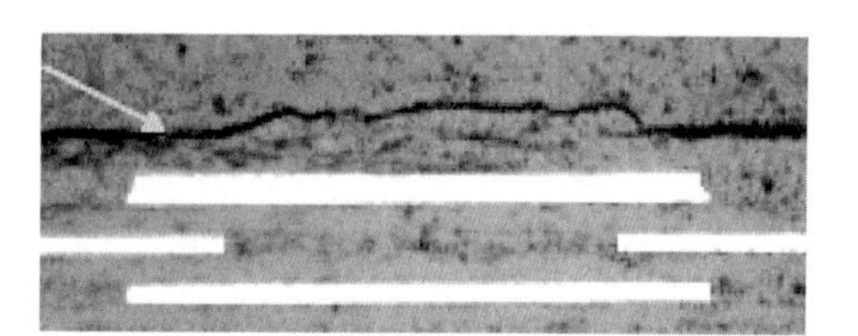

FIGURE 27.41 Laminate crack.

FIGURE 27.42 Void.

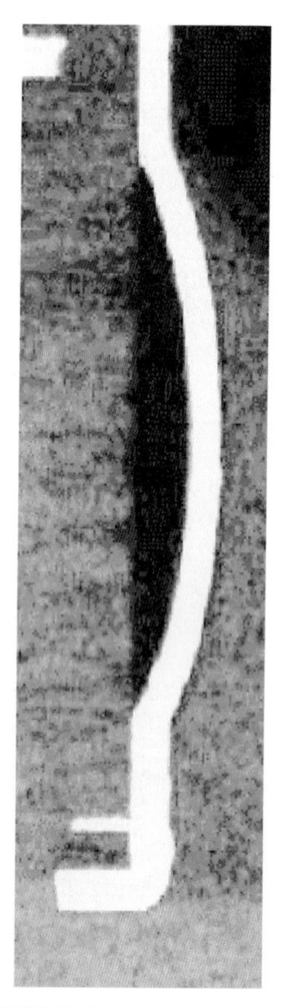

FIGURE 27.43 Hole wall pull-away.

27.7 LAMINATION OVERVIEW

The standard hydraulic vacuum press cycle is flexible and has very high output. Through the use of multiple openings and stacked lamination, high productivity is obtainable. Hydraulic lamination works effectively with all modern materials systems.

ML-PWB fabricators use many different laminate systems, press cycles, and B-stage formulations. The most significant differences among press cycles are in the heating rate, cure temperature, and cure time. At one extreme, the ML-PWBs are loaded into a cold press. This gives a very slow heating rate, and a B stage with a low flow is needed to avoid excessive flow. This material works well at low pressures, minimizing print-through and innerlayer distortion. A vacuum cycle is recommended to minimize voiding in low-pressure regions near the panel edge. The other extreme is a hot-loaded press with a very fast temperature rise. This cycle needs a high-flow material a high pressure to complete the resin flow cycle before the onset of cure. Although a vacuum is less important with this cycle, it will minimize edge voids. Some materials require a moisture drive-off prior to heat using a 0 psi low temperature hold.

Some fabricators use bakes both before and after lamination. The bake before lamination is designed to ensure moisture removal from cores after the application of the adhesion promotion chemistry; follow the recommendations of the laminate manufacturer carefully. The bake is also needed if innerlayers are stored at high humidity prior to lamination, or as a standard process for hygroscopic materials such as some of the LFAC laminates or polyimide. Other fabricators use a bake after lamination to complete the cure, reduce warp, and relieve stress. Although a post bake will achieve these goals, it is usually unnecessary in a controlled lamination process. In the case of high-temperature materials such as polyimide, cyanate ester, and PPO, a post bake is a useful way to achieve a full cure in a process where the maximum press temperature is limited.

27.8 ML-PWB SUMMARY

The printed wiring board manufacturing industry continues to be challenged with a high rate of change. Today's global market and added environmental considerations are driving investments and innovation in research and development of integrated components (e.g., laminates), subcomponents (e.g., resins and glass), and process consumables (e.g., oxide chemistry) for PCB fabrication. The manufacturing flow has only a backbone of standard methodologies, with many optional branches and staging through some process sequences multiple times (as in sequential lamination and HDI technologies). PCB fabrication has become much more complex, interactions between processes and components are compounded, and the user requirements are more formidable. A successful implementation of any of the high-density constructs or high-temperature-resistant materials should always be complemented with sound statistical process verification and testing appropriate to the end use.

REFERENCES

1. Bergum, Erik J., "Application of Thermal Analysis Techniques to Determine Performance Entitlement of Base Materials through Assembly," presented at IPC Printed Circuits Expo, 2003.
2. Plastics Technology Laboratories, Inc., "Dynamic Mechanical Analysis (DMA)," from http://www.ptli.com/.
3. Murray, Cameron, "Testing and Evaluation of HDIS Materials, Cameron Murray, presented at IPC Printed Circuits Expo, April 1998.
4. Ehrler, Sylvia, "Compatibility of Epoxy-based Printed Circuit Boards to Lead Free Assembly," presented at IPC Printed Circuits Expo, March 2003.

5. Luttrull, D., and Hickman, F., "New Halogen-Free PCB Materials for High-Speed Applications and Lead-Free Solder Processes," *Future Circuits,* March 2001.
6. Levchik, S., "New Phosphorous-Based Curing Agent for Copper Clad Laminates," IPC Printed Circuits Expo, February 2006.
7. Fisher, Jack, "The Impact of Non-Brominated Flame-Retardants on PWB Manufacturing," *IPC Review,* May 2000.
8. Burch, C., and Vasoya, K., "The Thermal and Thermo-mechanical Properties of Carbon Composite Laminate," presented at IPC Printed Circuits Expo, February 2006.

CHAPTER 28
PREPARING BOARDS FOR PLATING

Jim Watkowski
MacDermid Incorporated, Waterbury, Connecticut

28.1 INTRODUCTION

A major part of manufacturing printed circuit boards involves *wet process* chemistry. The plating aspects of wet chemistry include deposition of metals by electroless (metallization) and electrolytic (electroplating) processes. Topics to be described here are multilayer processing, electroless copper, direct metallization, electroplating of copper and resist metals, nickel and gold for edge connector (tips), tin-lead fusing, and alternative coatings. Specific operating conditions, process controls, and problems in each area will be reviewed in detail. The effects of plating on image transfer, strip, and etching are also described in this chapter. See the printed circuit plating flowchart in Fig. 28.1.

Two driving forces have had major influence on plating practices: the precise technical requirements of electronic products and the demands of environmental and safety compliance.

Recent technical achievements in plating are evident in the capability to produce complex, high-resolution multilayer boards. These boards show narrow lines (3 to 6 mil), small holes (12 mil), surface-mount density, and high reliability. In plating, such precision has been made possible by the use of improved automatic, computer-controlled plating machines, instrumental techniques for analysis of organic and metallic additives, and the availability of controllable chemical processes. Military-specification-quality boards are produced when the procedures given here are closely followed.

28.2 PROCESS DECISIONS

Process and equipment needs dictate the physical aspect of the facility and the character of the process, and vice versa. Some important items to consider are the following.

28.2.1 Facility Considerations

1. *Multilayer and two-sided product mix:* Lamination presses and innerlayer processes are required.
2. *Circuit complexity:* Dry film, photoimageable resist, and a clean room are needed.

3. *Level of reliability (application of product):* Extra controls and testing are required.
4. *Volume output:* Equipment sizing and building space are needed.
5. *Use of automatic versus manual line:* Productivity, consistency, and a workforce are required.
6. *Wastewater treatment system:* Water and process control capability must be available.[1]
7. *Environmental and personnel safety; compliance with laws.*
8. *Costs.*

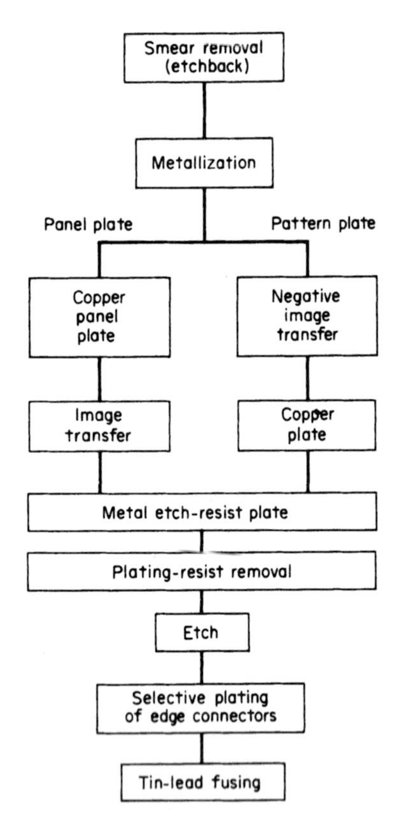

FIGURE 28.1 Printed wiring board plating flowchart.

28.2.2 Process Considerations

1. *Material:* The principal printed board material discussed will be NEMA grade FR-4 or G-10, that is, epoxyfiberglass clad with 1.2-, 1-, or 2-oz. copper. Other materials will be briefly mentioned because they can significantly alter plating and related processes.
2. *Standard:* Plated through hole (PTH) is the current standard of the industry. The following purposes, objectives, and requirements apply to both multilayer and two-sided boards:
   a. *Purposes:* Circuit density; double-sided circuitry
   b. *Objectives:* Side-to-side electrical connection; ease of component attachment; high reliability
   c. *Requirements:* Complete coverage; even thickness; a hole-to-surface ratio of 0.001 in. minimum; no cracks; no voids, nodules, or inclusions; no pull-away; no epoxy smear; minor resin recession; optimum metallurgical structure; M/L compatibility
3. *Image transfer:* Photoimageable, dry film, or screening of plating resists will depend on board complexity, volume, and labor skills.
4. *Electroless copper:* The type chosen will depend on the method of image transfer as well as on the need for panel plating. These processes are readily automated. The largest percentage (estimates are as high as 95 percent) of printed circuit board manufacturers worldwide rely on the electroless copper method for hole metallization.
5. *Direct metallization technology (DMT):* Some of the remaining printed circuit board manufacturers have eliminated the electroless copper step and converted to DMT. Developed in the 1980s, DMT methods produce a conductive surface on the nonconductive through-hole surfaces (see Chap. 30 for a detailed discussion). Electroless copper baths contain formaldehyde and chelators. In addition, the baths use large volumes of water, are difficult to control, and make it difficult to treat waste. DMT claims include increased productivity, ease of control, and lower hazardous material involvement. Because of these characteristics, DMT will probably gain significant acceptance as a primary hole metallization method.[2] Acceptance of the DMT process has been delayed by the high cost of conveyorized equipment and chemicals.

DMT primary technologies include:

a. Palladium
b. Carbon-graphite
c. Conductive polymer
d. Other methods (see Chap. 30 for further information)

6. *Electroplating processes.* Deposit requirements are as follows:

Electrical conductivity
Mechanical strength
Ductility and elongation
Solderability
Tarnish and corrosion resistance
Etchant resistance
Compliance with military specification (mil-spec)

Details emphasizing operation, control, and mil-spec plating practices are given elsewhere. Metal-plated structures of completed PC boards are as follows:

Copper/tin-lead alloy
Copper/tin (SMOBC)
Copper/tin-nickel (nickel)/tin-lead
Copper/nickel/tin-nickel/silver

7. *Strip, etching, and tin-lead fusing:* Methods required by these steps are determined by the preceding processes and by the need for automation.

28.3 PROCESS FEEDWATER

28.3.1 Water Supply

Printed circuit board fabrication and electronic processes require process feedwater with low levels of impurities. Large volumes of raw water must be readily available, either of suitable quality or treatable at reasonable cost. New facilities must consider water at an early stage of the site selection and planning process. Zero discharge, although a desirable goal, is very costly and difficult to achieve.

28.3.2 Water Quality

Highly variable mineral content causes board rejects and equipment downtime, as well as reduced bath life, burdened waste treatment, and difficult rinse water recovery. Many water supplies contain high levels of dissolved ionic minerals and possible colloidals that can cause rejects in board production. Some of these impurities are calcium, silica, magnesium, iron, and chloride. Typical problems caused by these impurities are copper oxidation, residues in the PTH, copper-copper peelers, staining, roughness, and ionic contamination. Problems in the equipment include water- and spray-line clogging, corrosion, and breakdown. The best plating practices suggest good-quality water for high yields. The need for water low in total dissolved solids (TDS), calcium hardness, and conductivity is well known. Good water eliminates the concern that the water supply may be responsible for rejects. Although water quality is not well defined for plating and PC board manufacturing, for general usage, some guidelines can be assigned as follows. Where high-purity water is required, see Sec. 28.3.3.

Typical quantities are:

TDS:	4 to 20 ppm
Conductivity:	8 to 30 μS/cm
Specific resistance:	0.12 to 0.03 M
Carbonate hardness ($CaCO_3$):	3 to 15 ppm

Somewhat higher values are acceptable for less-critical processes and rinses.

28.3.3 Water Purification

Two processes widely used for water purification are reverse osmosis and ion exchange. In the reverse osmosis technique, raw water under pressure (1.4 to 4.2 MPa or 200 to 600 lb./in.^{2}) is forced through a semipermeable membrane. The membrane has a controlled porosity that allows rejection of dissolved salts, organic matter, and particulate matter, while allowing the passage of water through the membrane. When pure water and a saline solution are on opposite sides of a semipermeable membrane, pure water diffuses through the membrane and dilutes the saline water on the other side (osmosis). The effective driving force of the saline solution is called its *osmotic pressure*. In contrast, if pressure is exerted on the saline solution, the osmosis process can be reversed. This is called the *reverse osmosis* (RO) process, and involves applying pressure to the saline solution in excess of its osmotic pressure. Fresh water permeates the membrane and collects on the opposite side, where it is drawn off as product.

Reverse osmosis removes 90 to 98 percent of dissolved minerals and 100 percent of organics with molecular weights over 200, as shown in Table 28.1.

A small quantity of dissolved substances also facilitates deionized (DI) water production, wastewater treatment, and process rinse water recovery, since it makes recycling less costly and more feasible. An RO system will result in lower costs for DI water preparation and for process water recycling. The setup for recycling requires additional equipment for polymer addition, filtration, and activated carbon treatment.

TABLE 28.1 Purified Water Supply Values
Typical in/out RO values

TDS, ppm	SiO_2, ppm	Conductivity, μS	Hardness-$CaCO_3$, ppm
170/4	30/1	130/8	24/1
240/7	45/2	200/14	35/2
300/10	60/2	250/20	45/3

DI water purification is used when high-purity water is required, for example, in bath makeups, rinses before plating steps, and final rinses necessary to maintain low ionic residues on board surfaces. Mil-spec PC boards must pass the MIL-P-28809 test for ionic cleanliness. This is done by final rinsing in DI water. Deionized water is made by the ion exchange technique. This involves passing water containing dissolved ionics through a bed of solid organic resins. These convert the ionic water contents to H^+ and OH^-. Deionized water systems are more practical when using feedwater low in ionic and organic content.

Other requirements are as follows:

pH: 6.5–8.0

Total organic carbon: 2.0 ppm

Turbidity: 1.0 NTU

Chloride: 2.0 ppm

28.4 MULTILAYER PTH PREPROCESSING

Two-sided printed circuit boards are usually processed by first drilling and deburring, then metallizing the through hole. Multilayers require treatment prior to any metallization process. Resin smear must be removed from the innerlayers, and texturing of the epoxy surface enhances hole wall adhesion. The preprocessing can be categorized as either smear removal or etchback.

Fresh solution replenishment is crucial for effectively attacking smear and achieving uniformity through the holes. Circulation pumps and 90°C work piece movement are required to prevent temperature stratification of the heated baths as well as assist in forcing fresh chemistry through the holes. Higher-aspect-ratio holes may require vibration to dislodge air bubbles from the holes and enhance uniformity of attack (see Chap. 29).

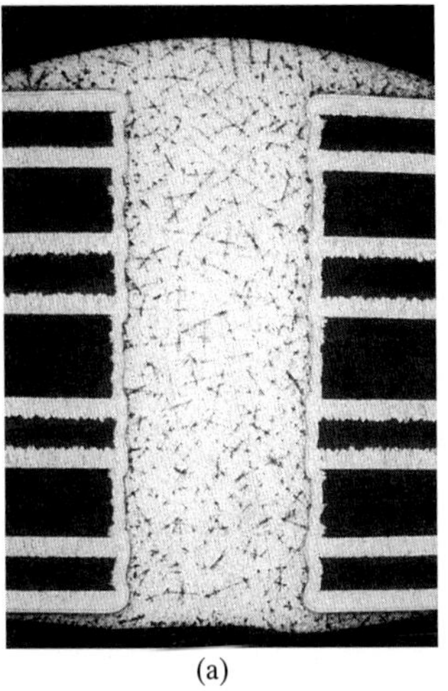

(a)

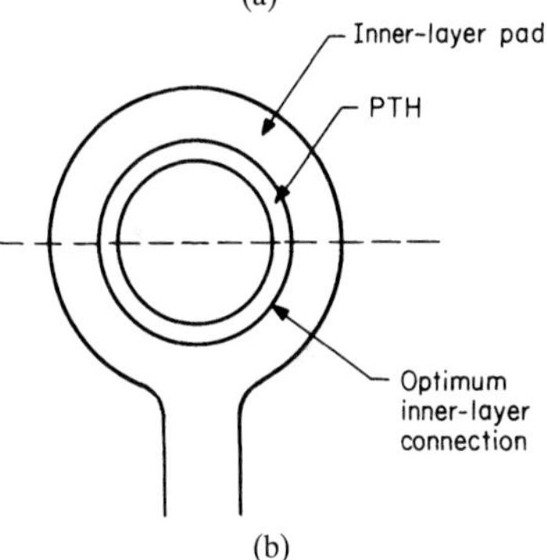

(b)

FIGURE 28.2 PTH vertical and horizontal cross sections illustrating optimum innerlayer connection and smear removal.

28.4.1 Smear Removal

High-speed drilling machines are used to drill holes into the PWB dielectric. During the drilling operation, sufficient heat is generated, causing the epoxy to melt. As the drill is withdrawn, the melted epoxy smears across and coats the innerlayer copper surfaces, causing a condition commonly known as *drill smear* (see Figs. 28.2 to 28.6).

28.4.2 Etchback

Etchback refers to the aggressive removal of the epoxy and glass that results in the protrusion of the innerlayer copper. When two surfaces of the innerlayer are exposed, it is referred to as two-point connection, and when three surfaces of the innerlayer are exposed, it is referred to as three-point connection. Physically, the innerlayer copper will protrude from the drilled-hole three-point connection for copper bonding, which is required on some mil-spec boards (see Fig. 28.7).

28.4.3 Smear Removal/Etchback Methods

The four methods commonly used utilize hole wall epoxy or dielectric oxidation, neutralization-reduction, and glass etching. Three of the four common methods employ technology that

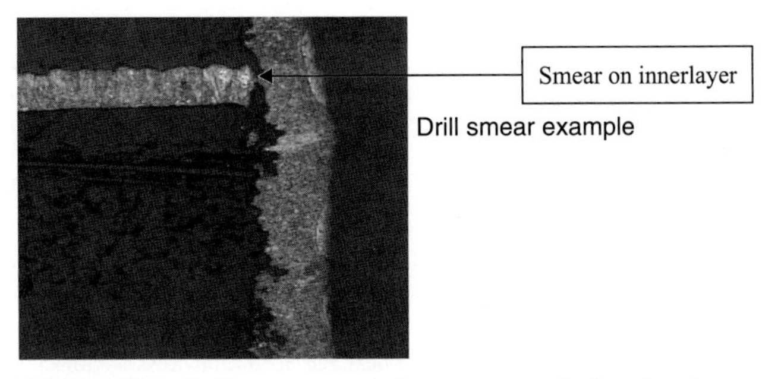

FIGURE 28.3 Drill smear example (note smear on the innerlayer).

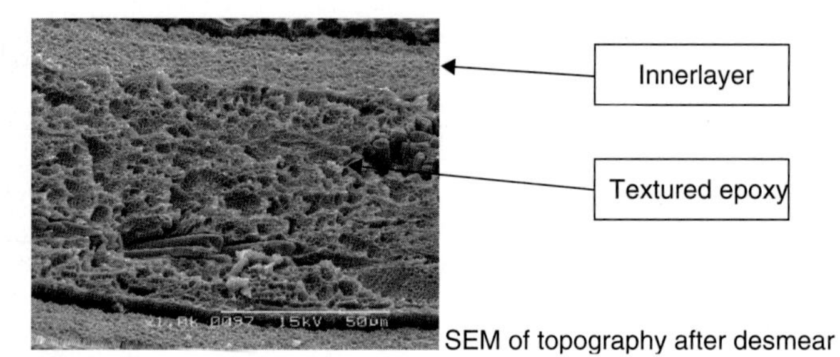

FIGURE 28.4 Scanning electron microscope (SEM) view of topography after desmear.

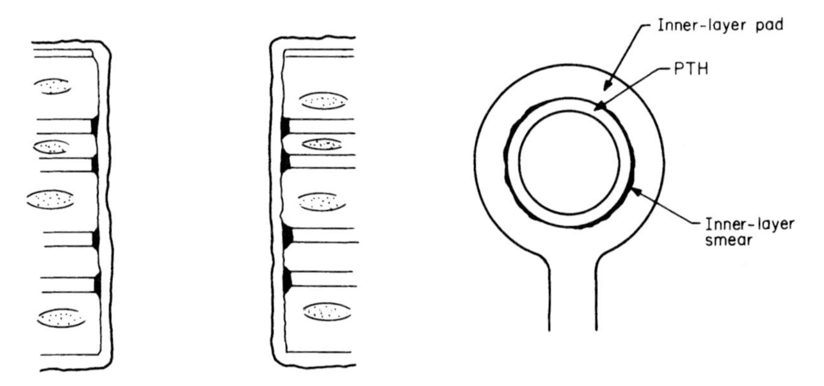

FIGURE 28.5 PTH vertical and horizontal cross sections illustrating innerlayer smear.

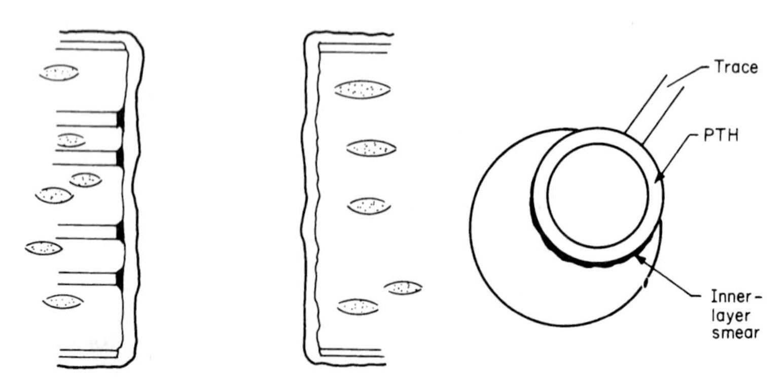

FIGURE 28.6 PTH vertical and horizontal cross sections illustrating innerlayer smear and misregistration.

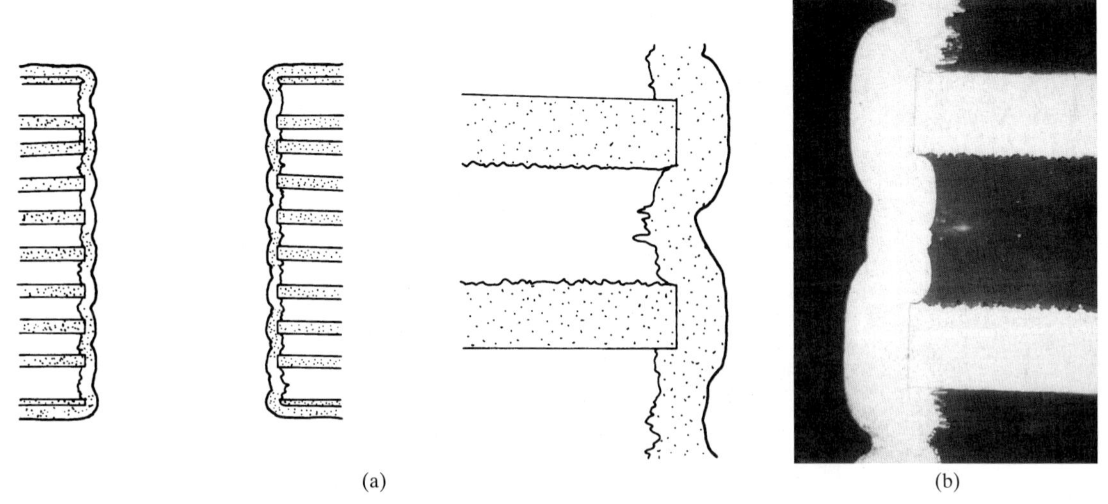

FIGURE 28.7 PTH vertical cross section illustrating optimum connection and etchback. (a) shows a drawing of expected cross section (b) shows an actual cross section with expected results.

oxidizes resin or dielectric, followed by a neutralization-reduction step. The fourth method—sulfuric acid—solvates or dissolves resin or dielectric. All methods remove smear, but the resultant surface of the epoxy affects hole wall pull-away and adhesion.

28.4.3.1 Sulfuric Acid. This process has been used extensively for many years because of its ease of operation and reliability of results. Disadvantages are lack of control, inability to attack certain resin systems, and the inability to create significant topography resulting in poor hole wall adhesion. Operator safety is crucial since concentrated sulfuric must be used.

28.4.3.2 Chromic Acid. Chromic acid is a more controllable method that imparts some topography for enhanced hole wall adhesion and has a longer bath life. However, insufficient neutralization of the Cr^{+6} leads to copper voids. The line cost of using chromic acid is appealing. However, the Restriction of Hazardous Substances (RoHS) and waste treatment regulations of chromic acid add substantial costs that make this method a nonviable option. Etchback is possible by double processing.

28.4.3.3 Permanganate. Permanganate is the most widely used and preferred method in the industry. This easy-to-control process incorporates electrolytic regeneration, which adds to consistency and increases bath lives. Permanganate effectively removes drill smear from most materials and imparts some texturing to the resin that increases hole wall adhesion.[3] (See Fig. 28.2) for an example of hole wall topography after etchback. Problems result from lack of chemical control or inconsistent cure of the multilayer package.

Permanganate is also used as a second step in conjunction with the other methods to enhance hole quality.[4]

28.4.3.4 Plasma. Plasma utilizes electrical current to generate a plasma from fluorocarbon gases. This plasma oxidizes the epoxy resin and removes drill smear. Longer cycles or double cycling will generate etchback. Plasma works on any material and is especially effective when processing polytetrafluoroethylene (PTFE)-based dielectrics. This process has few

TABLE 28.2 Smear Removal and Etchback Processes*

Sulfuric acid	Chromic acid
Rack panels	Rack panels
Sulfuric acid, 96%	Chromic acid 20-s dwell, 15-s drain, 3 min., 140°F room temperature
Neutralizer 8 min., 125°F	Reducer 3 min., room temperature
Ammonium bifluoride 3 min., room temperature	Ammonium bifluoride 3 min., room temperature
Unrack panels	Unrack panels
High-pressure hole cleaning	High-pressure hole cleaning
Release to electroless copper/DM	Release to electroless copper/DM
Permanganate	**Plasma Etch**
Solvent conditioner 90°F, 5 min.	Plasma Oxyget, CF_4, 30 min.
Alkaline permanganate 170°F, 10 min	Glass etch (optional)
Neutralizer-120°F, 5 min.	High-pressure hole cleaning
Glass etch, 4 min., room temperature	Release to electroless copper/DM
High-pressure hole cleaning (optional)	
Release to electroless copper/DM	

* Water rinses after each step are not shown.

steps and does not require the use of large quantities of chemicals. Disadvantages of plasma are nonuniformity of the treatment across the batch and across the panel. The texture generated on the epoxy is minimal, and an ash is commonly left behind that leads to poorer hole wall adhesion. A secondary cleanup, such as permanganate, is commonly used after plasma to remove the ash. The cost of the equipment and its maintenance is high. Controls must be provided to prevent air pollution.

28.4.4 Process Outline: Smear Removal and Etchback

The four common methods for smear removal and etchback are given in Table 28.2. Combinations of these methods are also in use because of added reliability they offer to both process and product.

Polyimide and polyimide-acrylic systems are processed in chromic acid or plasma. Teflon™, and R T Duroid™ materials are treated (before operations) in sodium-naphthalene mixtures to yield void-free, high-bond-strength copper in the PTH. The introduction of lead-free material and halogen-free material has challenged the desmear processes. The new materials are more chemically resistant, and thus more aggressive desmear cycles are required to remove any smear. These more chemically resistant resin systems are more difficult to texture. Alternative solvents, higher temperatures, and cycle optimization may be required to process these newer materials effectively.

28.5 ELECTROLESS COPPER [5,6,7,8,9,10]

28.5.1 Purpose

This second series of chemical steps (after smear removal) is used to make panel side-to-side and innerlayer connections by metallizing with copper. The process steps required include racking, cleaning, copper microetching, and hole and surface catalyzing with palladium and electroless copper. Typical steps are as follows:

1. *Cleaner-conditioner.* Alkaline cleaning is used to remove soils and condition holes.
2. *Microetch.* This slow acid etching is used for removal of copper surface pretreatments, oxidation, and presentation of uniformly active copper. Persulfates and sulfuric acid hydrogen peroxide solutions are commonly used.
3. *Sulfuric acid.* This acid removes persulfate residues.
4. *Predip.* This process maintains the balance of the next step.
5. *Catalyst (activator).* Neutral or acid solutions of palladium and tin are used to deposit a thin layer of the catalytic metal on the nonconductive surfaces. Some palladium is unintentionally absorbed onto the foil surface and innerlayers.
6. *Accelerator (postactivator).* This process removes colloidal tin on board surfaces and holes or change the charge of the palladium tin colloid.
7. *Electroless copper.* This alkaline-chelated copper-reducing solution deposits thin copper in the holes (20 to 100 μin) and surfaces.
8. *Antitarnish.* This neutral solution prevents oxidation of active copper surfaces by forming a copper conversion coating.

28.5.2 Mechanism

Equations 28.1 and 28.2 illustrate the process.

$$Pd^{+2} + Sn^{+2} \rightarrow Pd + Sn^{+4} \quad (28.1)$$

$$CuSO_4 + 2HCHO + 4NaOH\ Pd \overset{Pd}{\rightarrow} Cu + 2HCO_2Na + H_2 + 2\ H_2O + Na_2SO_4 \quad (28.2)$$

However, Eq. 28.2 is an old two-step catalyzation sequence that is no longer in use in the industry.

28.5.3 Electroless Copper Processes

New materials and increased reliability criteria have precipitated a newer generation of electroless copper baths. Lead-free assembly forces the PWB to undergo higher thermal excursions, thus increasing the stress on the plated through holes. Proprietary additives (which are not listed in Table 28.3) affect the integrity of the electroless copper-plated deposit. Selection from the several types available depends on the type of image transfer desired. Operation and control of three bath types and the function of constituents are given in Tables 28.3 and 28.4.

The problems encountered with this system are as follows:

1. *Uncoppered holes or voids.* Voided holes appear as black or dark color in the holes. Corrective actions : Ensure that the process is functioning correctly by verifying the times, temperatures and concentrations of all preceding process tanks, including the desmear process as specified. The electroless copper bath components should be verified as well as the dwell time. Low loading factor, low temperature, or high air agitation lead to low bath activity and poor plating characteristics. Voids may also be caused by an overaggressive electrolytic preclean line.
2. *Hole wall pull-away.* This is the separation of the copper pulling away from the dielectric. The defect is usually detected during the cross-sectioning inspection. If the hole wall pull-away is severe, large blisters can be seen in the hole with the naked eye. Hole wall

TABLE 28.3 Electroless Copper Processes
Operation and control

	Low deposition	Medium deposition	Heavy deposition
Copper	3 g/l	2.8 g/l	2.0 g/l
HCHO	6–9 g/l	3.5 g/l	3.3 g/l
NaOH	6–9 g/l	10–11 g/l	8 g/l
Temperature	65°–85°F	115° ± 5°F	115° ± 5°F
Air agitation	Mild	Mild/moderate	Moderate
Filtration	Periodic	Continuous	Continuous
Tank design	Static	Overflow, separate sump	Overflow, separate sump
Heater	Teflon	Teflon	Teflon
Panel loading	0.25–1.5 ft^2/gal	0.1–2.0 ft^2/gal	0.1–2.0 ft^2/gal
Replenish mode	Manual	Manual or continuous	Automatic
Idle time, control	70–85%	Turn off heat	Turn off heat
Deposition time	20 min	20 min	20–30 min
Thickness	20 μin	40–60 μin	60–100 μin

pull-away can be caused by insufficient texturing of the epoxy dielectric during the desmear process, overconditioning, overcatalyzation, or insufficient acceleration. The electroless copper bath may become overactive and plate a stressed deposit. Ensuring that the process is functioning as specified with regard to dwell times, chemical concentrations, and bath temperature are the corrective actions.

3. *Bath decomposition.* This is rapid plating out of the copper. Common causes are bath imbalance, overloading, overheating, lack of use, tank wall initiation, and contamination.
4. *Electroless copper–to copper-clad bond failure.* Copper-to-copper adhesion failures can occur on the surface or on the innerlayers. Surface copper adhesion failures arise from unclean copper. Dry-film and/or developing residues, overconditioning, insufficient rinsing, insufficient microetch, and overcatalyization are common causes. Innerlayer copper-to-copper adhesion failures, commonly referred as ICD, may be due to all the items mentioned. Further causes may be related to the electroless copper bath itself. A stressed deposit due to high bath activity or decomposition has poor integrity and may fail under reliability testing. Specific additives in supplier formulations may contribute to ICDs as well.
5. *Staining.* Copper oxidation is due to moisture or contamination on the copper surface. Corrective action involves dipping boards in antitarnish or hard rinsing in DI water.

TABLE 28.4 Electroless Copper
Function of constituents

	Constituent	Function
Copper salt	$CuSO_4 \cdot 5\,H_2O$	Supplies copper
Reducing agent	HCHO	$Cu^{+2} + 2e \rightarrow Cu^0$
Complexer	EDTA, tartrates, Rochelle salts	Holds Cu^{+2} in solution at high pH; controls rate
pH controller	NaOH	Controls pH (rate) 11.5–12.5 optimum for HCHO reduction
Additives	NaCN, metals, S, N, CN organics	Stabilize, brighten, speed rate, strengthen

28.5.4 Process Outline

This outline presents the typical steps in an electroless copper process:

1. Rack
2. Clean and condition
3. Water rinse
4. Surface copper etch (microtech)
5. Water rinse
6. Sulfuric acid (optional)
7. Water rinse
8. Preactivator
9. Activator (catalyst)
10. Water rinse
11. Postactivator (accelerator)
12. Water rinse
13. Electroless copper
14. Rinse
15. Sulfuric acid or antitarnish
16. Rinse
17. Scrub (optional)
18. Rinse
19. Copper flash plate (optional)
20. Dry
21. Release to image transfer

28.6 ACKNOWLEDGMENT

Significant portions of this chapter are drawn from Edward F. Duffek, Ph.D., "Preparing Boards for Plating," *Printed Circuits Handbook, 5th ed,* Clyde F. Coombs (ed.), 2001, Chap. 28.

REFERENCES

1. Carano, M., *Proceedings of 16th AESF/EPA Pollution Prevention & Control Conference,* 1995, p. 179.
2. Nargi-Toth, K., *Printed Circuit Fabrication,* Vol. 15, No. 34, September 1992.
3. Deckert, C. A., Couble, E. C., and Bonetti, W. F., "Improved Post-Desmear Process for Multilayer Boards," *IPC Technical Review,* January 1985, pp. 12–19.
4. Batchelder, G., Letize, R., and Durso, Frank, *Advances in Multilayer Hole Processing,* MacDermid Company.
5. Stone, F. E., "Electroless Plating—Fundamentals and Applications," G. Mallory and J.B. Hajdu (eds.), American Electroplaters and Surface Finishers Society, Inc., 1990, Chap. 13.
6. Deckert, C. A., "Electroless Copper Plating," *ASM Handbook,* Vol. 5, 1994, pp. 311–322.
7. Murray, J., "Plating, Part 1: Electroless Copper," *Circuits Manufacturing,* Vol. 25, No. 2, February 1985, pp. 116–124.

8. Polakovic, F., "Contaminants and Their Effect on the Electroless Copper Process," *IPC Technical Review,* October 1984, pp. 12–16.
9. Blurton, K. F., "High Quality Copper Deposited from Electroless Copper Baths," *Plating and Surface Finishing,* Vol. 73, No. 1, 1986, pp. 52–55.
10. Lea, C., "The Importance of High Quality Electroless Copper Deposition in the Production of Plated-Through Hole PCBs," *Circuit World,* Vol. 12, No. 2, 1986, pp. 16–21.

CHAPTER 29
ELECTROPLATING

Jim Watkowski
Mac Dermid Incorporated Waterbury, Connecticut

29.1 INTRODUCTION

The reality in printed circuit evolution is that the parts are continuing to increase in degree of difficulty and complexity. A new genre of high-density interconnect (HDI) boards is making the transition from leading edge to mainstream. These boards are characterized by a combination of a series of complex features that include buried and blind vias, high-aspect ratio plating, small-hole plating (as low as 6 mil diameter holes), and fine lines and spaces side by side with ground plane areas of different sizes.

The industry, however, still has a need to produce simpler products, such as single-sided and double-sided boards; these are still in demand. Multilayer boards with lower layer counts (four to eight) fall in this category also.

This chapter focuses on all aspects of electroplating. Emphasis will be on acid copper plating as the main process for providing interconnection. Tin, tin-lead, nickel, and gold electroplating will also be covered as important plated materials. In addition, the latest innovations in technology for meeting the challenges of product complexity, such as pulse plating and horizontal conveyorized plating, are discussed in detail.

29.2 ELECTROPLATING BASICS

Electroplating is the production of adherent deposition of conductive surfaces by the passage of electric current through a conductive metal-bearing solution. The rate of plating depends on current and time and is expressed by Faraday's law (Eq. 29.1):

$$W = (ItA)/(nF) \tag{29.1}$$

where W = metal, g
I = current, A
t = time, s
A = atomic weight of the metal
n = number of electrons involved in metal ion reduction
F = Faraday's constant (96,485°C/mol)

Plating occurs at the cathode (the negative electrode). Accordingly, deposition thickness is determined by time and by the current that is impressed on the surface being plated; for example, by using the preceding formula, one can readily calculate the weight of the deposited metal. The weight can then be converted to a specific thickness over a known area. The rate of deposition of the most common metals is shown in Table 29.1.

TABLE 29.1 Rate of Deposition of the Most Common Metals Grams

Metal	Grams deposited per amp.hr	Amp. hr. per sq.ft. to deposit 0.001 in	Amp. hr. per sq.dm. to deposit 25 μm
Copper	1.186	17.8	1.88
Tin	2.214	7.8	0.82
Lead	3.865	6.9	0.73
Nickel	1.095	19.0	2.00
Gold	7.348	6.2	0.65

To plate 1.0 mil of copper, one would need 17.8 amperes per square foot (ASF) or 1.88 amperes per decimeter (ASD) for 1 hour (60 min.). Properties of electrodeposits are shown in Table 29.2.

TABLE 29.2 Properties of Electrodeposits

Property	Cu	Ni	Au	SnPb	Sn
Melting point, °F	1980	2600	1945	361	450
Hardness, VHN	150	250	150	12	4
CTE X 10^{-6}/°F	9.4	8.0	8.2	12.2	12.8
Conductivity, %IACS	101	25	73	11.9	15.6
Electrical resistivity, $\mu\Omega$/cm	1.67	6.8	2.19	14.5	11.1
Thermal conductivity, CGS°C	0.97	0.25	0.71	0.12	0.15

29.3 HIGH-ASPECT RATIO HOLE AND MICROVIA PLATING

The combination of increasing layer counts (panel thickness) and decreasing hole size results in higher aspect ratios.

29.3.1 Aspect Ratios

Aspect ratios are calculated by dividing panel thickness by the hole size. (See Fig. 29.1 for example of a high-aspect hole with a 15:1 ratio.) Standard direct current (DC) plating bath formulations are not sufficient to plate these geometries. Modifying the concentrations of the electrolyte components is required. In general, lowering the copper concentration and increasing the acid concentration improves through-hole distribution, as indicated in Table 29.3. Proprietary additives play a role maximizing through-hole distribution as well.

Solution agitation is an integral part of successfully plating high-aspect ratio through holes. Ninety degree work rod agitation with a $^1/_2$ in. stroke is the most common method of replenishing

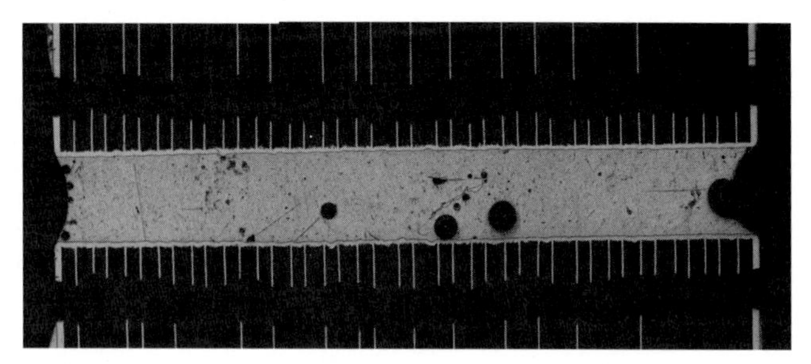

FIGURE 29.1 A 15:1 aspect ratio plated with MacDermid PPR 100.

solution in the holes. Other adjuncts such as high-flow circulation pumps or e-ductor systems aid in solution turnover and replenishment of solution inside the holes. A decrease in corresponding current density is required to achieve improved throwing power and to prevent copper burning.

For aspect ratios above 10:1, pulse plating yields the best microdistribution.

TABLE 29.3 Plating Solutions for Different Aspect Ratios

	High aspect ratio formulation	Medium aspect ratio formulation	Low aspect ratio formulation
Aspect ratio	>8:1	3:1 – 8:1	<3:1
Copper sulfate ($CuSO_4 \cdot 5H_2O$)	45 g/l	60 g/l	75 g/l
Free sulfuric acid 66°Be electronic grade	300 g/l	260 g/l	225 g/l
Chloride ion (Cl^-)	50 ppm	50 ppm	50 ppm

29.3.2 Blind Microvia Plating

Smaller holes and increasing circuit density are drivers that push technology for "via-in-pad" and "stacked microvia" designs that help minimize PWB real estate. It has been established that via-in-pad designs, where the via is not filled, can lead to voids in the BGA solder joint that can negatively affect reliability.

Stacked vias, or the sequential build-up process, is a technology where vias (either blind or through hole) are metallized, plated to required thickness, then circuitized. Another layer of dielectric is applied and the process repeated.

Electrodeposited copper (see Fig. 29.2) has become an effective method of filling microvias, replacing the alternative via filling techniques, especially as the conductor patterns become finer and the pitch density of the component packaging increases. Resin or conductive pastes have been used for via filling, but the reliability problems associated with these materials include subsequent polishing, entrapped air within the blind micro via, and void formation during the reflow process.

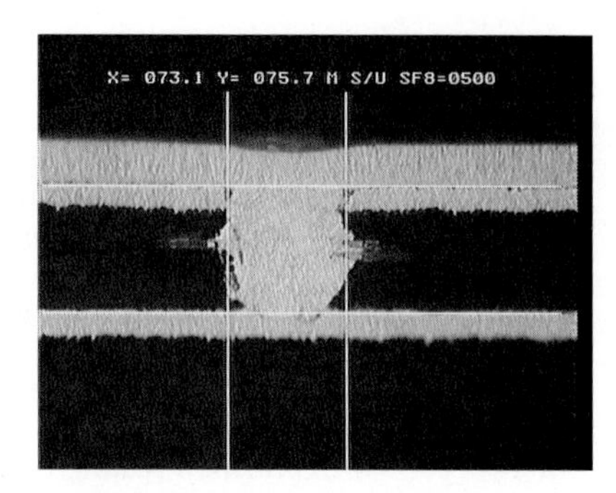

FIGURE 29.2 Example of a plated filled blind microvia.

All major suppliers to the PWB industry have developed processes for filling blind microvias.

The electrolyte compositions for via filling baths typically run with high concentrations of copper (up to 50 g/l as metal) and lower concentrations of acid (approximately 100 g/l). These concentrations are juxtaposed to high-aspect ratio plating. Thus, these process usually are utilized in a panel plating mode for filling vias only. This type of process usually requires some type of planarization step post-filling to decrease the amount of copper plated on the surface.

Proprietary additives are key for consistent filling and for filling vias of various diameters and depths. Typical systems will contain carriers, brighteners, and levelers.

Further developed processes are capable of plating through holes and filling vias simultaneously, although aspect ratio capability is limited and should be tested before committing to production.

29.4 HORIZONTAL ELECTROPLATING

As the degree of part complexity continues to increase, new innovative plating solutions are being put into place. Periodic pulse reverse (PPR) is an example of this, and another major development is the advent of horizontal acid copper plating.

Here, the panel is transported in a horizontal mode through the plating equipment with stationary anodes below and above it. The panel continues to plate as it travels, interfacing with one anode after the other. The panel comes off the end of the line plated.

The plating thickness is a function of time and current density. The dwell time is determined by the length of the plating module and the conveyor speed. Plating at 20 ASF for 60 min. in horizontal equipment running at 1 m/min. would require a plating module that is 60 m long. This length is not practical. Current density has to be increased to 40, 60, 80, and higher ASF. At 80 ASF, the equipment length needs to be only 15 m.

New chemical additives were formulated to accommodate high-current-density plating.

However, this current density offered new challenges: at high-current densities, most plating solutions lose throwing power. Another problem that was encountered early on was that the anode film particulates would land on the top side of the panel and eventually create nodulation. This did not occur on the bottom side.

Two major developments helped make horizontal processing popular:

- Pulse plating, coupled with solution impingement, resolved the throwing power problem of plating at high current densities.
- Nodulation was resolved by taking the copper dissolution out of the module and replacing the standard anode with an inert or insoluble anode. Copper dissolution or oxidation was achieved by different means, such as the use of rectification, the use of ozone, or the dissolution of copper oxide.

29.4.1 Advantages of Horizontal Processing

The most dramatic advantage of horizontal plating is the uniformity from panel to panel. All panels go through the equipment in an identical fashion. They all see the entire bank of anodes, and they are in a single plating solution. All panels coming off a horizontal line are alike. If for some reason one edge of the panel has higher plating, then all panels will have the same edge equally high. In vertical plating, different panels are plated in different parts of the cell, in different positions in the rack, and even in different tanks or baths. Although vertically plated panels may all be within specifications, there will always be inherent variability that the shop has to contend with in downstream processing beyond the plating line.

Horizontal plating machines can be easily automated. Loaders and unloaders are usually integrated in the system. Automation of chemical analysis and dosing is incorporated with minimum cost increase. The equipment is easily linked to pre- and post-processes. Horizontal processing lends itself well to continuous production flow and reduced cycle time.

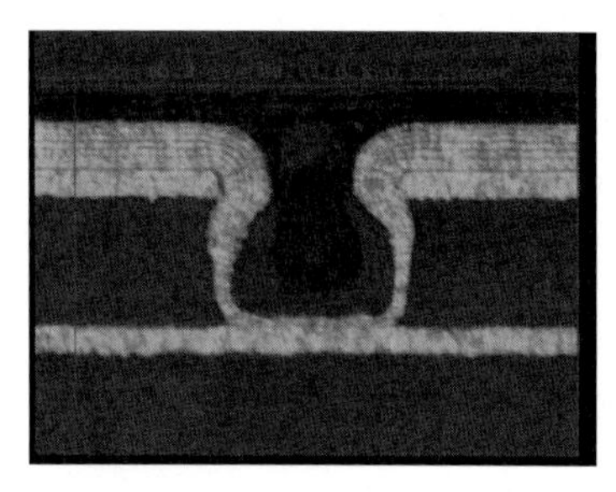

Horizontal DC	
Via Diameter	90 μm
Via Depth	72 μm
Surface Thickness	19 μm
Base Thickness	4 μm
Throwing Power	25%
Current Density	7 ASD

FIGURE 29.3 Horizontal plated through hole(PTH)/DC-plated resin-coated copper.

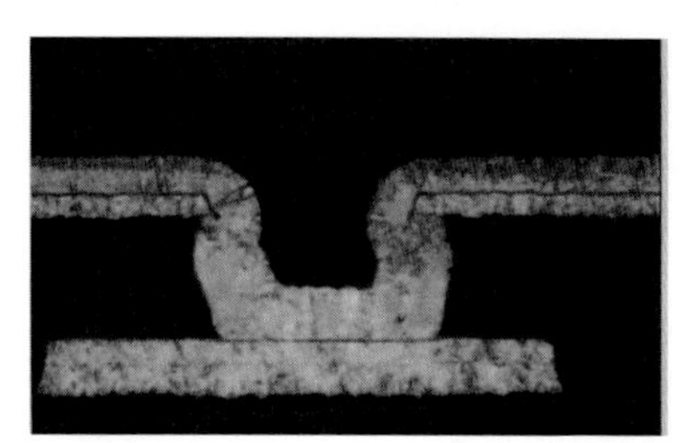

Horizontal Pulse	
Via Diameter	90 μm
Via Depth	72 μm
Surface Thickness	17 μm
Base Thickness	15 μm
Throwing Power	88%
Current Density	7 ASD

FIGURE 29.4 Horizontal PTH/pulse-plated resin-coated copper.

Anode uniformity is a major advantage in horizontal plating. The size and geometry of an insoluble inert anode do not change. Anode variability and maintenance are eliminated. The latter is a very labor-intensive and hazardous activity is associated with vertical plating.

Horizontal equipment is completely enclosed. This minimizes the operator and shop personnel's exposure to acid and chemical fumes and creates a better and safer working environment.

Horizontal plating gives superior surface thickness distribution. Variation of less than 10 percent across a panel-plated surface is normal. This is due to the anode's close proximity to the cathode; usually, the anode-to-cathode spacing is between 8 and 10 mm, or 0.3 and 0.4 in.

In addition, surface-to-hole thickness variation is minimized. Excellent throwing power (80 to 100 percent) is achievable in blind vias (aspect ratios greater than 1:1 [i.e., with depth exceeding hole diameter]) and in through holes (with aspect ratios of 10:1). This is the result of solution dynamics coupled with periodic pulse reverse rectification. This feature makes horizontal pulse plating an enabling technology for HDI applications. Figures 29.3 and 29.4 show the difference in throwing power of DC and PPR plating in a 1:1 via hole. Figure 29.5 shows an example of horizontal PPR plating of a via with an aspect ratio greater than 1, that is, the depth is greater than the diameter of the via.

Copper Gleam PPR-H	
Via Diameter	120 μm
Via Depth	160 μm
Surface Thickness	17 μm
Base Thickness	17 μm
Throwing Power	100%
Current Density	8 ASD

FIGURE 29.5 Horizontal PTH/pulse-plated microvia.

Horizontal equipment is capable of handling thin material (as thin as 10 mil). In the vertical mode, thin parts are a challenge to rack and to transport. Core plating requires the use of a metal frame to add rigidity to the part so that it can be handled and agitated in the bath during the plating cycle. This adds labor and creates handling defects.

29.4.2 Drawbacks of Horizontal Processing

To date, most horizontal equipment is designed around panel plating. Pattern plating has its advantages and is being worked on by a series of equipment suppliers and could be available in the near future.

The equipment requires a substantial investment and is a major deviation from standard vertical processing. There is both a learning curve and engineering know-how that will be needed to run this equipment smoothly. The equipment is integrated, and either it all works or it all doesn't. It requires a large parts inventory and trained equipment maintenance crew to ensure uninterrupted operation.

29.5 COPPER ELECTROPLATING GENERAL ISSUES

Copper is the most commonly used metal in the structure of a printed circuit board. Copper has high electrical conductivity, strength, and ductility, and low cost. It is readily plated from simple solutions, and it is readily etched. MIL-STD-275 states that electrodeposited copper shall be in accordance with MIL-C-104550, and shall have a minimum purity of 99.5 percent as determined by ASTM-E-53, and shall be 0.001 in. (1.0 mil). Requirements for good soldering also indicate the need for 1.0 mil of copper and smooth holes. Copper plating is generally regarded as the slow step in manufacturing PC boards.

Copper electrolytic plating of printed circuit boards is a key process step in circuit manufacturing. After all, it forms the traces that carry the signal. The physical properties of the deposit, specifically tensile strength and elongation, have to be controlled so the product can withstand, without cracking, the temperature excursions of assembly and of use. A successful copper-plating system must be able to provide a copper deposit with tensile strength between 40 and 50 kpsi and percent elongation between 10 and 25 percent. As demands for speed and impedance control continue to increase, the thickness uniformity of the plated copper becomes extremely critical.

In the late 1990s, significant developments occurred in HDI boards, usually for portable products. HDI has traditionally focused on the high-end telecommunications, imaging, and computing markets. Handheld electronics products present unique challenges in terms of size, weight, functionality, and cost. This poses a series of challenges to the board manufacturer in every processing area. Acid copper electroplating is quickly becoming the limiting factor in most board shops today. High-aspect ratio plating, 3 mil lines and spaces, laser-drilled microvias, as well as buried vias, are forcing the industry to look for innovative approaches to the traditional methods of plating.

29.5.1 Pattern Plating versus Panel Plating

Panel plating does not involve any patterning or imaging before plating and hence has a uniform surface geometry; with good plating practice, less than 10 percent thickness variation across the panel and from panel to panel is achievable. The challenge here is to get good throwing power so as not to overplate the surface to achieve 1.0 mil in the hole. An overplated surface limits the line width and spaces that can be circuitized by subsequent etching.

Pattern plating, as the name implies, occurs after pattern or image transfer and, accordingly, is the real challenge for the plating process. Here, the surface geometry is not uniform and may have extremes of isolated traces and pads, as well as ground plane areas. This gives rise to great disparity in primary current distribution, which can result in two to four times thicker copper in the isolated areas, compared with the ground plane areas.

29.5.2 Thickness Distribution

A formidable challenge facing acid copper plating of printed circuit boards is thickness distribution on the surface of the panel (whether pattern or panel plated) and thickness distribution in the barrel of the hole or the via, namely throwing power. If the throwing power is 100 percent, then plating 1 mil of copper in the hole would result in plating 1 mil on the surface.

At 50 percent throwing power, to achieve 1.0 mil in the hole would result in 2.0 mils on the surface, which would limit the etching capabilities in the panel plate and increase the thickness variation in pattern plate.

HDI compounds the plating challenges with microvias, small holes, and fine lines. Different approaches are being practiced to resolve the following four plating issues:

- Low-current-density plating
- Chemically mediated process
- Electrically mediated process
- Horizontal conveyorized systems

29.5.3 Additives in Acid Copper Plating

Additives used for bright acid copper plating are divided into three categories: (1) the carrier or suppressor, (2) the additive or brightener, and (3) a leveling component. Each component has a specific role in controlling the quality of the deposit. Polarization curves that show the ensuing current as the voltage is increased are used to study the effects of the various components on the plating system and how they affect the deposit.

Figure 29.6 shows a polarization curve where the current is plotted against the voltage. The data can be obtained from a potentiostat setup, shown in Fig. 29.7. Here, we have three electrodes in the plating solution. The first is a counter electrode and is usually made of copper or platinum; the second is the working electrode and is a rotating platinum disk; and the third is the reference electrode, like a double-junction calomel. Voltage is applied between the working and the reference electrodes, and the ensuing current is measured between the working and the counter electrodes. This is the principle upon which cyclic voltametric stripping (CVS)

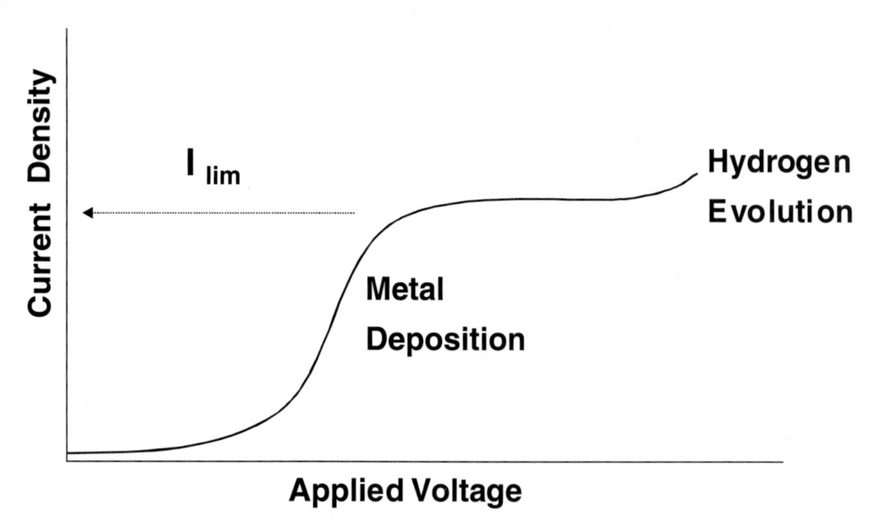

FIGURE 29.6 Acid copper: cathodic polarization curve.

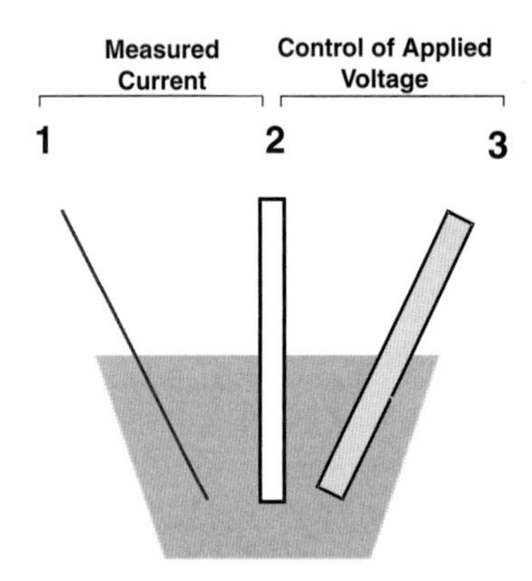

FIGURE 29.7 Acid copper: polarization study: three-electrode cell.

is based. CVS equipment is commonly used to assay the additives in the plating solution. A closer look at the polarization curve shows four distinct areas:

- **Low-current area** Here, there is a minimum increase in current as the voltage is increased.
- **Metal deposition area** This is an area where there is a significant increase of current with applied voltage.
- **Limiting-current-density area** This area is marked by a plateau, where there is no increase of current as the applied voltage continues to rise.
- **"Hydrogen evolution" part of the curve** This is the last area.

The area of most interest when plating, of course, is the area of metal deposition. The slope of this part of the polarization curve can be altered (suppressed or accelerated) by the addition of organic additives to the plating bath.

29.5.4 Carrier/Suppressor

Carriers are large-molecular-weight poly-oxyalkyl-type compounds. Figure 29.8 shows the effect of the carrier on the polarization curve. The addition of a carrier alone did not alter the

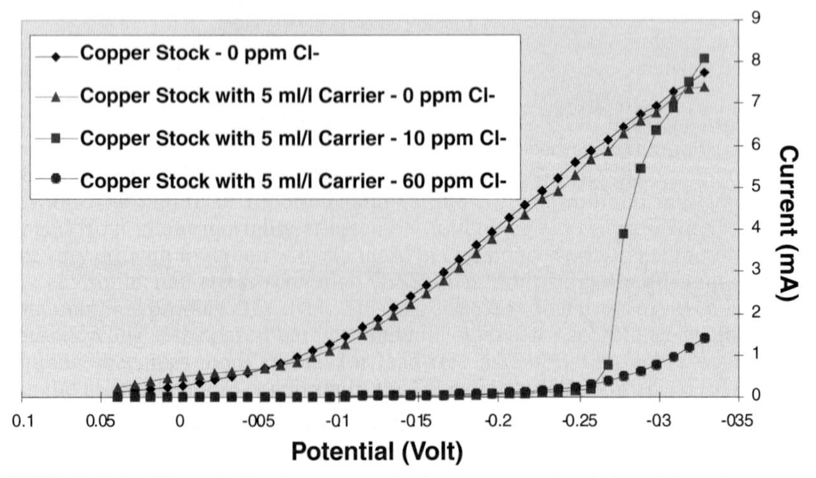

FIGURE 29.8 Effect of chloride concentration in a copper stock solution with carrier.

polarization curve; however, when this is done in the presence of 10 ppm of the chloride ion, a marked suppression is observed initially but is not sustained as voltage continues to increase.

At 60 ppm of chloride, the suppression is sustained; the result is that it now takes more voltage to produce the desired current.

The suppression is a result of the effect of the carrier on the diffusion layer (also referred to as the Helmholtz double layer). The carrier is adsorbed to the surface of the cathode; this results in increasing the thickness of the diffusion layer. The result is better organization.

This gives rise to a deposit with a tighter grain structure (see Fig. 29.9). The carrier-modified diffusion layer also improves plating distribution and throwing power, without burning the deposit.

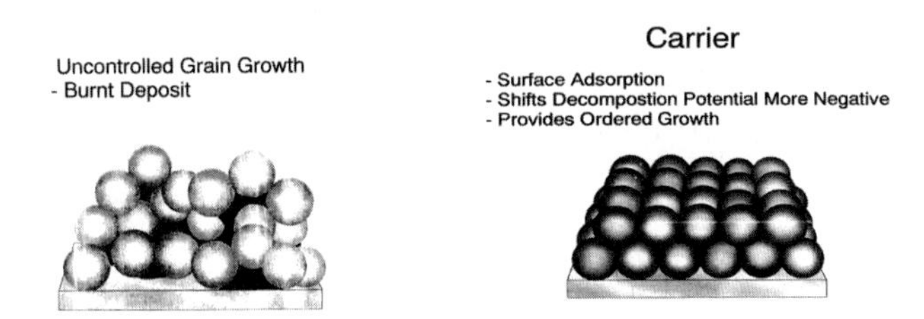

FIGURE 29.9 The effect of a carrier on the grain structure of the deposited copper.

29.5.5 Additive/Brightener

These are typically small-molecular-weight disulfide compounds, with the general structure R-S-S-R. The R groups are organic functional groups and vary from one brightener system to the other. Figure 29.10 shows the effect of a brightener on a suppressed polarization curve.

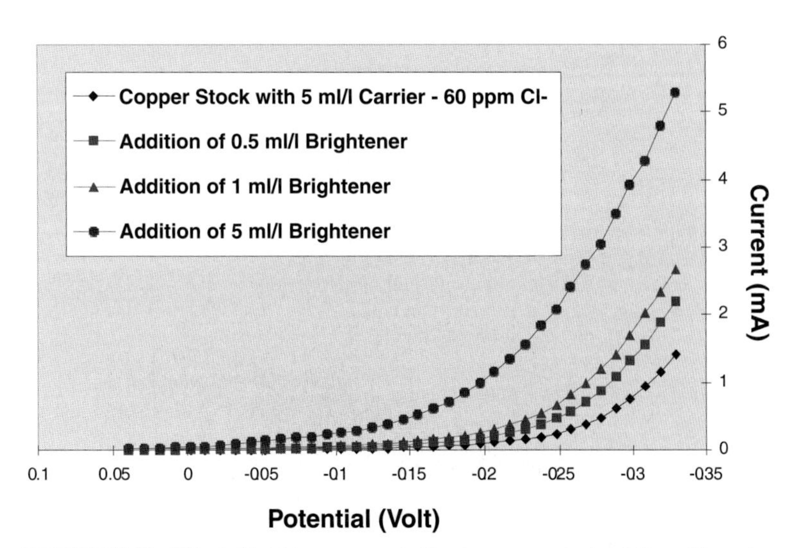

FIGURE 29.10 Effect of brightener concentration in a copper stock solution with carrier.

Basically, the brightener increases the current and reduces the suppression. The brightener plays a key role in determining the physical properties of the deposit. It is a grain refiner and, as such, it has a direct enhancing effect on the physical properties of the deposit, such as tensile strength and percent elongation.

29.5.6 Levelers

Levelers are used to offset localized high-current-density areas that result from high points or sharp edges in the plating surface. An example would be the dog-bone effect, occasionally seen at the knee of the hole.

Levelers are a class of compounds, typically aromatic, that adsorb preferentially at high points in the plating topography due to the increased mass transport possible at these sites.

The adsorption of the leveler at these sights creates localized suppression, allowing plating in the lower-current-density areas to catch up with high areas—hence, the leveling effect.

29.5.7 Low-Current-Density Plating

One way to achieve good distribution is the use of a low-current acid copper-plating system, which is designed specifically to produce a deposit with the required physical properties at reduced current density. Some shops are presently plating HDI boards at 6 to 8 ASF for a period of 4 to 6 hours. This will give excellent plating distribution, due to the increased suppression at these low plating voltages.

Low-current-density plating baths or electrolytes are characterized by lower copper concentration and increased sulfuric acid concentrations (refer to Table 29.4). The additive package of carrier and brightener is specifically formulated to operate at these low-current densities.

TABLE 29.4 Operation and Control

Operating variable plating	Low–current density plating	Conventional CD* plating	High-speed plating
Copper	1–2 oz./gal	2–3 oz./gal	7–8 oz./gal
Copper sulfate	4–8 oz./gal	8–12 oz./gal	28–32 oz./gal
Sulfuric acid	25–32 oz./gal	22–28 oz./gal	5–8 oz./gal
Chloride	40–80 ppm	40–80 ppm	40–80 ppm
Additives	As required	As required	As required
Temperature	75–85°F	70–85°F	75–100°F
Cathode CD	5–15 ASF	20–40 ASF	40–150 ASF
Anodes			
Type	Bars or baskets		
Composition	Phosphorized 0.04–0.06% P		
Bags	Closed-napped polypropylene		
Hooks	Titanium or Monel		
Length	2 in. shorter than rack		
Anode CD	25–50% of cathode CD		
Properties			
Composition	99.8% (99.5% min., ASTM-E-53)		
Elongation	10–25% (6% min., ASTM-E-8 or ASTM-E-345)		
Tensile strength	40–50 kpsi (36 kpsi min, ASTM-E-8 or ASTM-E-345)		

*CD, current density.

These types of products are in use throughout the industry and are tolerated as long as the numbers of HDI boards are a small percentage of the overall production. This practice, however, dramatically reduces the output of the plating line and increases the cost of the part. As the percentage of HDI boards increased, the need for a different answer became apparent.

29.5.8 Chemically Mediated Process

Here, the carrier is engineered for maximum chemical suppression, even at a high-current density.

This is demonstrated by the shift in the Tafel slope (see Fig. 29.11). These baths are capable of good throwing power and surface distribution while plating at relatively high-current densities. "Super-throw" baths are capable of 85 percent throw in a 5:1 aspect ratio board while plating at 30 and at 35 ASF. This kind of bath contrasts with the low-current bath and can improve productivity significantly, while still operating under conventional DC plating.

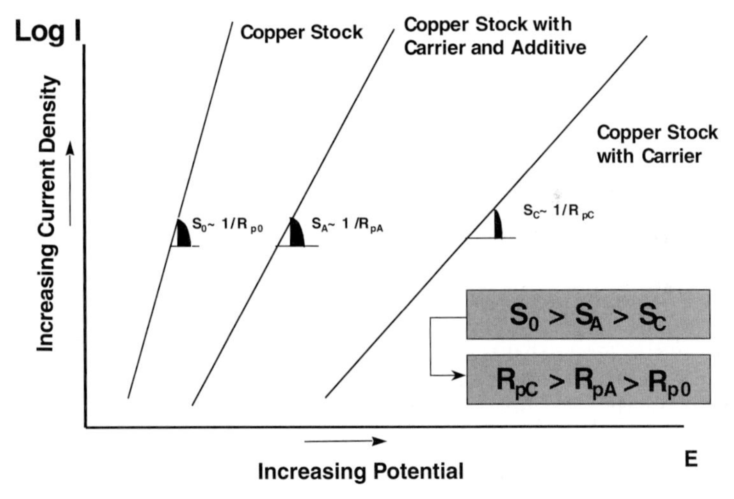

FIGURE 29.11 Tafel plots: direct current.

The system fits readily into existing in-house equipment and in-house know-how. Figure 29.12 shows an example of a blind via that is 5.0 mil in diameter and 3.0 mil deep, plated with this type of bath at 20 ASF for 90 min.

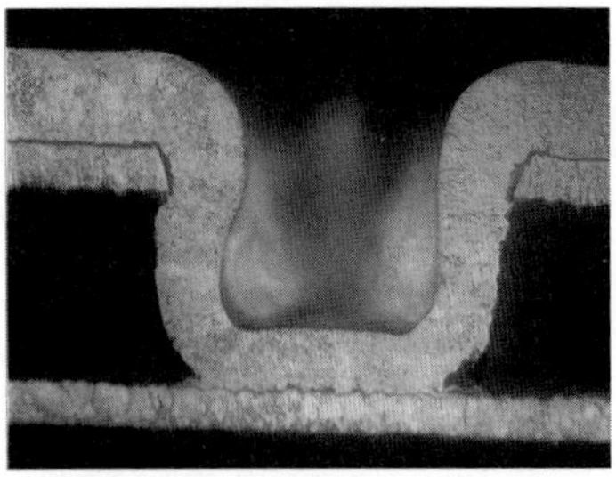

29.5.9 Pulse Plating (Electrically Mediated Process)

In this plating system, the bath is engineered to respond to a periodically pulsed reverse current. The required Tafel slope shift is accomplished through rectification (see Fig. 29.13). The rectifier produces a pulsed wave with a forward cathodic current that is perturbed by short anodic pulses. The forward current at 1X (e.g., 30 ASF) is maintained for 10 ms and the reversed at 3X (e.g., 90 ASF) for 0.5 ms, for example. The duty cycle may vary (e.g., 20 ms forward with 1.0 ms reverse).

FIGURE 29.12 Example of a blind via that is 5.0 mil in diameter and 3.0 mil deep, plated with chemically mediated bath.

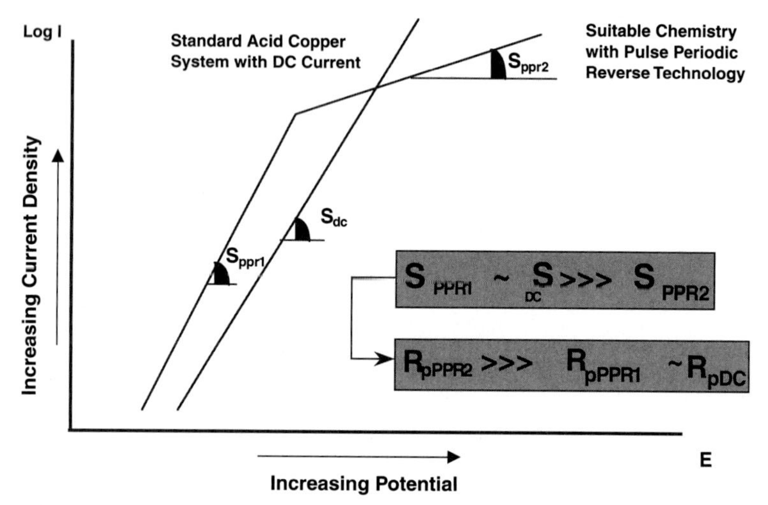

FIGURE 29.13 The pulse periodic reverse effect.

There is also room to optimize the forward-to-reverse current ratio; 1:3 is only one example. The shape of the wave is very important here, also. A square wave with minimum rise time gives the best results (see Fig. 29.14).

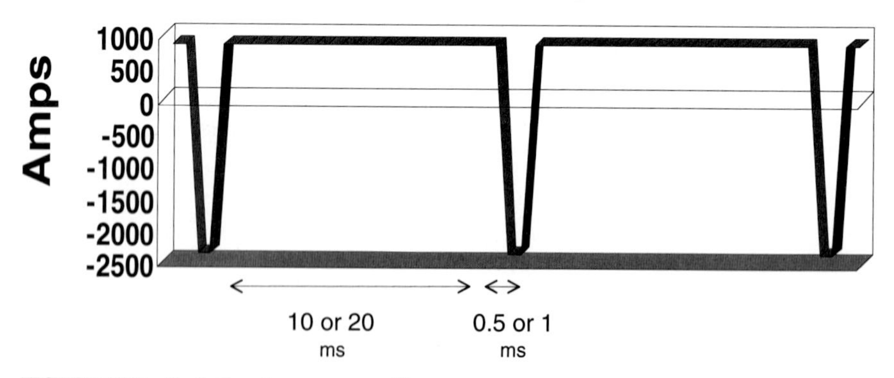

FIGURE 29.14 Periodic pulse reverse rectifier output wave.

Chemical suppliers have designed specific additive packages that ensure maximum response to the periodic pulse reverse wave. During the reverse cycle, the additive is preferentially desorbed from the high-current-density areas. This results in less plating acceleration due to the additive, and more suppression due to the carrier. Since the low-current-density areas of the panel will also receive a lower reverse pulse, the acceleration is reduced to a lesser degree than in the high-current-density areas. This leads to greatly improved plating thickness distribution.

Often, the difference in plating thickness between isolated surface features and ground planes will be no greater than 0.5 mil. This effect is so powerful that throwing power greater than 100 percent is commonplace.

Figure 29.15 shows the plating in a hole that is 0.013 in. in diameter in a 0.100 in. thick board (8:1 aspect ratio). The board was plated using Copper Gleam PPR (a proprietary additive) under the following conditions: forward current density, 30 ASF forward-to-reverse (F:R) CD ratio, 1:2.67 duty cycle, 20:1 ms plating time, and 60 min.

The results are an average thickness on the surface of 1.1 mil and an average thickness in the hole of 1.4 mil. Reverse pulse plating gives dramatic improvements in copper thickness distribution beyond the capabilities of the chemically mediated process. It is clearly the wave of the future as the HDI boards become more and more complex. However, it involves capital investment for the pulse rectifier, which may be five to six times the cost of an equivalent DC rectifier. It also involves a learning curve for matching the new parameters (F:R ratio, duty cycle, ASF, and the shape of the wave) to meet the specifications of the part number being plated.

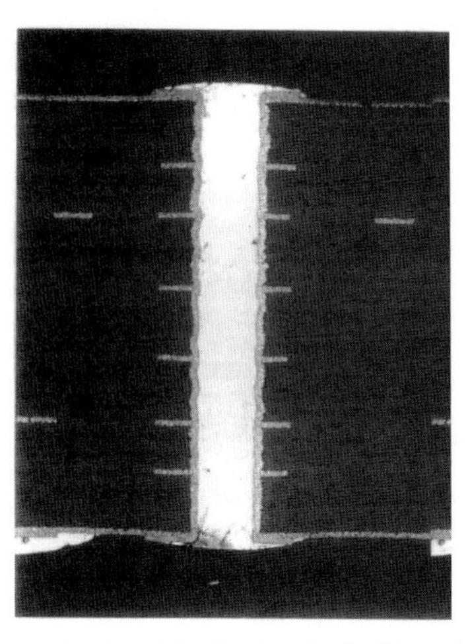

FIGURE 29.15 One hundred mil board with 13 mil diameter hole plated with Copper Gleam PPR for 60 min. Surface-to-hole ratio is 0.75.

29.5.10 Key Factors for Uniform Plating

To have day-to-day control and achieve ductile, strong deposits, and uniform copper thickness, you need the following controls:

1. Maintain equipment following best practices, such as uniform air agitation or e-ductors (also known as "ser-ducters") in the tank, equal anode/cathode distances, rectifier connection on both ends of tank, and low resistance between rack and cathode.
2. Maintain narrow-range control of all chemical constituents, including organic additives and contaminants.
3. Conduct batch carbon treatment as needed.
4. Control temperature at 70° to 85°F, or as specified by supplier.
5. Eliminate contaminants in the tank from preplate cleaners, microetchants, and impure chemicals.

The plating thickness distribution of your plating bath can be readily evaluated. Use noncopper-clad or bare epoxy laminate. First, plate 80 to 100 μin of electroless copper, then rack the panels and plate at the specified ASF to plate 80 to 100 μin of plated copper (usually 1/10 of the plating cycle time). Place the rack in a microetch solution until 60 to 80 percent of the copper is etched away. Remove and examine. The remaining copper is where the high-current density is. This is a useful tool in optimizing bath geometry, especially anode placement and panel placement on the rack. It is also useful in designing thieving and shielding for the best thickness distribution.

The plated-copper thickness distribution for a specific pattern can also be studied the same way. After plating the electroless copper on non-copper-clad bare laminate, image the pattern.

Plate for a limited time to get approximately 80 to 100 μin, or 0.2 to 0.3 μm. Strip the resist off the panel and place in microetch until 80 percent of all copper is removed. The remaining copper is where the high-current density is.

29.6 ACID COPPER SULFATE SOLUTIONS AND OPERATION

The preferred industrial process uses an acid copper sulfate solution containing copper sulfate, sulfuric acid, chloride ion, and organic additives.

29.6.1 Solution Makeup by Current Density

Using the proper additives, the resultant copper is fine-grained with tensile strengths of 50,000 lb/in.^{2} (345 MPa), a minimum of 10 percent elongation, and a 1.2 surface-to-hole thickness ratio (see Table 29.4).

29.6.2 Operation and Control

- **Agitation** Air (vigorous) is agitated from an oil-free source at 70° to 80°F. Or e-ductors (airless high-volume, low-pressure circulation).
- **Filtration** The operation is continuous through a 3 to 10 μm filter to control solution clarity and deposit smoothness.
- **Carbon treatment** New baths do not require activated carbon purification. Circulation through a carbon-packed filter tube is recommended to control organic contamination; for design and number, consult with supplier. The need for batch carbon treatment is indicated by corner cracking after reflow; dull, pink deposits; and haze, haloing, or comet trails around the PTH. Carbon-treat about every 1,500 (Ah) per gal. The following is the procedure for batch carbon treatment.

1. Pump to the storage tank.
2. Clean out the plating tank.
3. Rinse and clean the tank.
4. Leach with 10 percent H_2SO_4.
5. Adjust the agitators.
6. Clean the anodes.
7. Heat the solution to 120°F.
8. Add 1 or 2 qt of hydrogen peroxide (35 percent per 100 gal of solution). Dilute with 2 pt water, using low-stabilized peroxide.
9. Air-agitate or mix for 1 hour.
10. Maintain heat at 120° or 140°F.
11. Add 3 or 5 lb. powdered or granular carbon per 100 gal. of solution. For a specific carbon source, seek recommendations from your supplier. Mix for 1 to 2 hour.
12. Pump back to plating tank promptly or within 4 hours.
13. Analyze and adjust.
14. Dummy plate at 10 A/ft^2 for 6 hours. Panels should be matte and dull. Replenish with additive.
15. Follow supplier instructions for electrolyzing and start-up.

- **Contaminants** In general, acid copper tolerates both organic and metallic contaminants.

Organic residues may come from cleaners, resists, and certain sulfur compounds. Dye systems usually are more resistant than dye-free systems with respect to certain cleaner constituents.

Metals should be kept at these maximums: chromium, 25 ppm; tin, 300 ppm; antimony, 25 ppm. Nickel, lead, and arsenic may also cause roughness and other problems.

29.6.3 Process Controls

29.6.3.1 Bath Composition

- Copper sulfate is the source of metal. Low copper will cause deposit burning; high copper will cause roughness and decreased hole-to-surface thickness ratios or throwing power.
- Sulfuric acid increases the solution conductivity, allowing the use of high currents at low voltages.
- However, an excess of sulfuric acid lowers the plating rate, whereas low acid reduces hole-to-surface ratio (throwing power).
- Chloride ion (Cl^-) should be controlled at 60 to 80 ppm. Below 30 ppm, deposits will be dull, striated, coarse, and step-plated. Above 120 ppm, deposits will be coarse-grained and dull. The anodes will become polarized, causing plating to stop. Excess chloride is reduced by bath dilution or by dragout.
- Additive components analysis and control are critical for consistent product quality. The primary analytical tool is CVS, and the Hull cell continues to be a useful complementary tool. Excessive or insufficient additive will cause deposit burning and corner cracking. This condition can be judged by metallographic cross-sectioning and etching. Optimum-quality plated metal is fine-grained and equiaxed (nondirectional) and shows no laminations or columnar patterns.
- Concerning water quality, the use of DI water and contamination-free materials such as low chloride and iron will give added control and improved deposit quality.

29.6.3.2 Temperature. Optimum throwing power and surface-to-hole ratios are obtained by operating at room temperature (i.e., 70° to 80°F). Lower temperatures cause brittleness, burning, and thin plating. Higher temperatures cause haze in low-current-density areas and reduced throwing power. Cooling coils may be necessary during a hot summer or under heavy operation.

29.6.3.3 Deposition Rate. A thickness of 0.001 in. (1.0 mil) of copper deposits in 54 min. at 20 A/ft^2, in 21 min. at 50 A/ft^2.

29.6.3.4 Hull Cell. Operation at 2 A will show the presence of organic contamination, chloride concentration, and overall bath condition. However, an optimum Hull cell panel is only a small indication that the bath is in good operating condition, since test results are not always related to production problems. More reliable results are obtained by adjusting the bath before Hull cell testing. See Sec. 29.12.4 for procedures on Hull cell.

29.6.4 Cross-Sectioning Results—Troubleshooting

Sectioning with etching provides information on the plated copper that explains PTH quality in terms of processing factors. Besides showing the overall quality, cross-sectioning gives information on thickness and on possible problems, such as drilling, cracking, blowholes, and multilayer smear. Copper deposits with columnar or laminar patterns indicate inferior copper properties. Cross sections of optimum copper deposits are fine-grained and equiaxed (structureless) upon etching.

29.6.5 Inferior Copper Deposits

These may be caused by any of the following:

- Either low or excess additives
- Chloride out of range (i.e., too high or too low)
- Organic, metal, or sulfur (thiourea) contamination
- Excess DC rectifier ripple (greater than 10 percent)
- Low copper content with bath out of balance
- Roughness in drilling, voids in electroless steps, or other problems introduced in earlier processing

29.6.5.1 Cracking and Ductility. Resistance to cracking is tested by the following:

- **Solder reflow** Wave soldering and cross-sectioning are acceptable alternatives.
- **Elongation** Two mil copper foil should exceed 10 percent elongation. Acid copper elongation should range between 10 and 25 percent. Frequent testing gives more meaningful results.
- **Float solder test** This test includes prebaking and flux, using a 5 to 20 sec. float in a solder pot (60:40) at 550°F, followed by cross-sectioning for evaluation.
- **Copper foil bulge test** This test measures tensile strength by puncturing copper at high pressure.
- **DC ripple** High values of rectifier ripple (8 to 12 percent) may cause inferior copper deposits and poor distribution of thickness.

29.6.5.2 Visual Appearance. When plated, copper has a semibright appearance at all current densities. Unevenness, hazy or dull deposits, cracking, haloing around the PTH, and low-current-density areas indicate organic contamination. Carbon treatment is required if these conditions persist. Burned, dull deposits at high-current densities indicate low additive content, contamination, solution imbalance, or low bath temperatures. If dull, coarse deposits result at low-current densities, the chloride ion is not in balance.

When chloride is high or bath temperature too low, anodes may become heavily coated and polarized (the current drops). Decreased throwing power (surface-to-hole ratio), reduced bath conductivity, or poor-quality plating may also indicate contamination and are corrected as follows:

- Maintaining solution balance and chloride content at 60 to 80 ppm
- Circulating solution through filter continuously, passing through a carbon canister periodically, or by batch carbon treatment
- Analyzing organic additives by CVS or Hull cell
- Checking metal contaminations every three months
- Keeping the temperature between 70° and 85°F
- Checking anodes daily and replacing bags and filters (rinsed in hot water) every three to four weeks

29.6.5.3 Problems. Table 29.5 lists problems that appear after copper plating. Two groups are listed, with the first group readily correlated to the copper-plating process. Thin, rough copper plating in the PTH may also be exhibited by outgassing and blowholes during wave soldering. Figures 29.16 through 29.25 illustrate some of these effects.

TABLE 29.5 Printed Wiring Board Copper-Plating Defects

Defect in copper process	Cause
Corner cracking	Excess additive, organic contamination in solution
Nodules	Particulate matter in solution; also drilling, deburring residues
Thickness distribution	See Sec. 29.5.2
Dullness	Off-balance solution, organic contamination
Uneven thickness in PTH	Organic sulfur (thiourea) contamination
Pitting	Additive malfunction, defective electroless, or preplate preparation
Columnar deposits	Low additive, rectifier malfunction
Step plating, whiskers	Excess or defective additive
Defect, overall manufacturing process	**Cause**
Voids	Malfunction of electroless copper steps, also preplate cleaner
etching	
Inner–layer smear	Drilling or malfunction in smear removal
Roughness	Drilling or drilling residues
Hole-wall pullaway	Malfunction of smear removal or electroless copper steps
Copper-copper peeling	Surface residues from electroless and/or image transfer
Soldering blowholes	Drilling roughness, voids, and thin plating

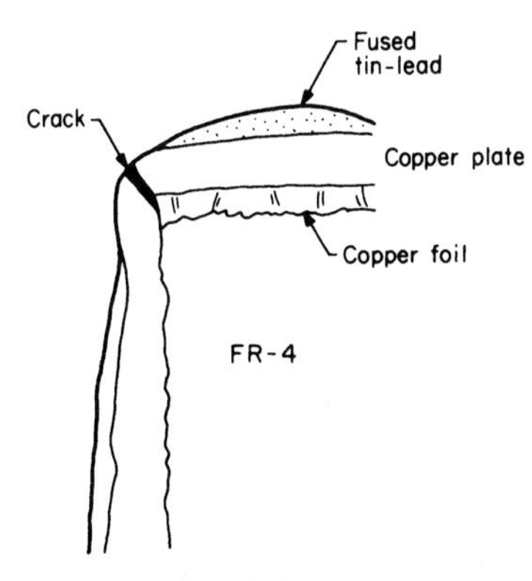

FIGURE 29.16 PTH gvertical cross section illustrating corner cracking.

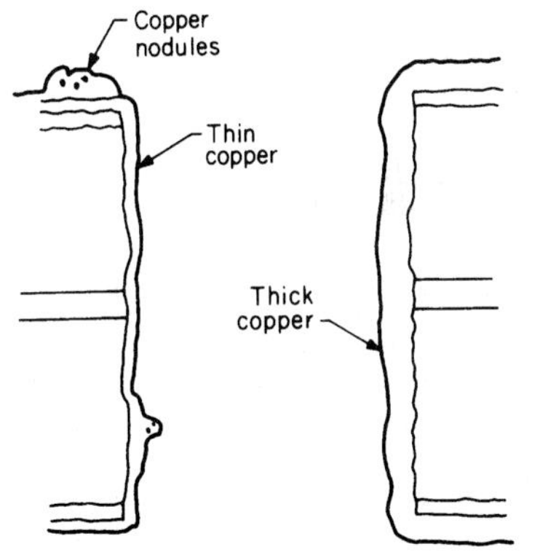

FIGURE 29.17 PTH vertical cross section illustrating uneven, thick/thin copper plating.

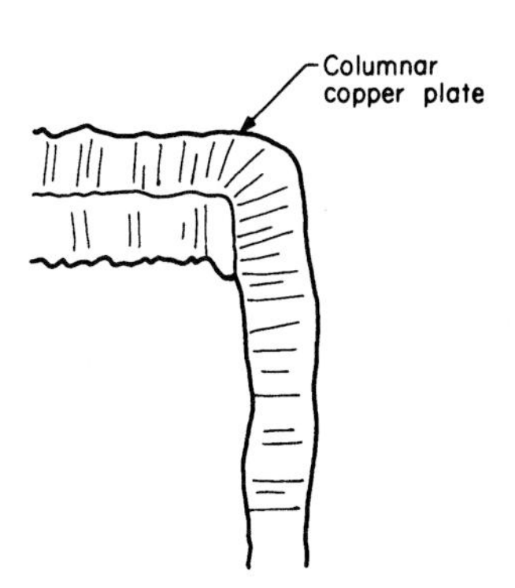

FIGURE 29.18 PTH vertical cross section illustrating columnar copper deposit structure.

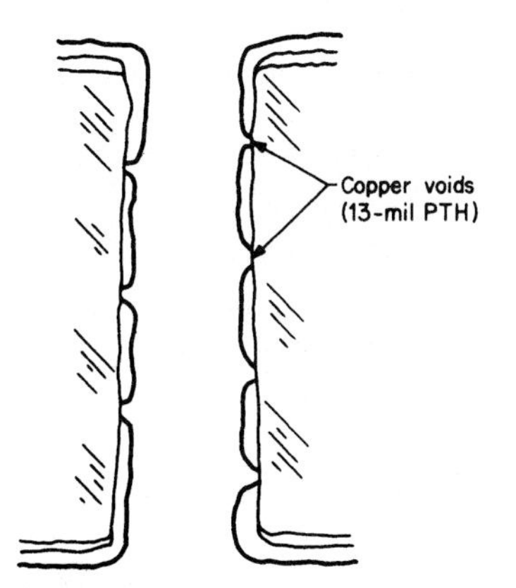

FIGURE 29.19 PTH vertical cross section illustrating copper voids.

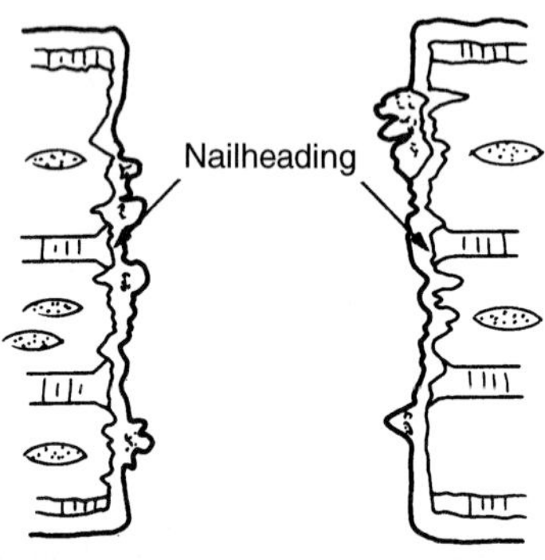

FIGURE 29.20 PTH vertical cross section illustrating rough, nodular copper plating due to drilling roughness. Nail-heading is also shown.

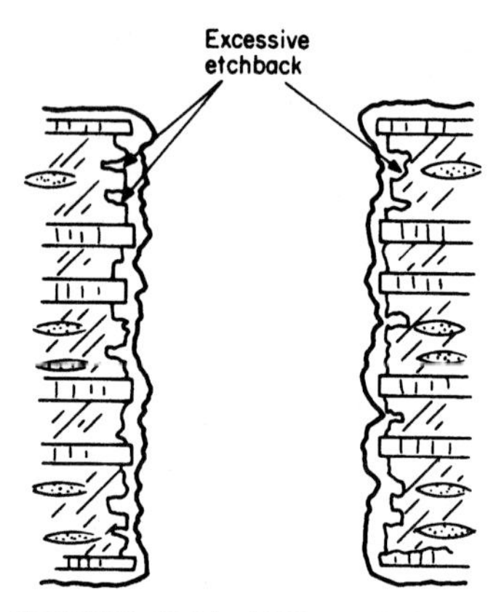

FIGURE 29.21 PTH vertical cross section illustrating hole roughness due to excessive etchback.

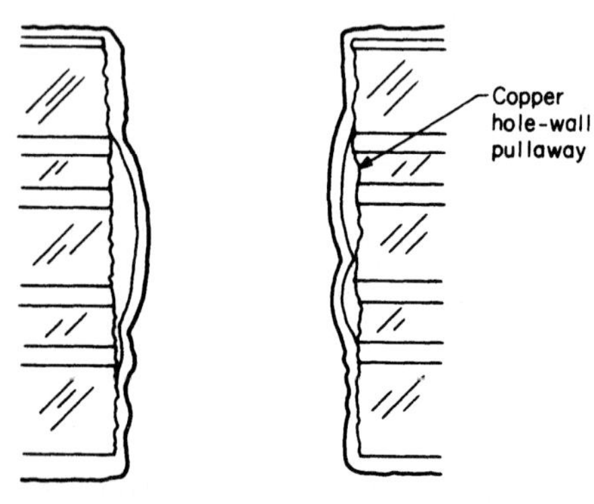

FIGURE 29.22 PTH vertical cross section illustrating copper hole wall pull-away.

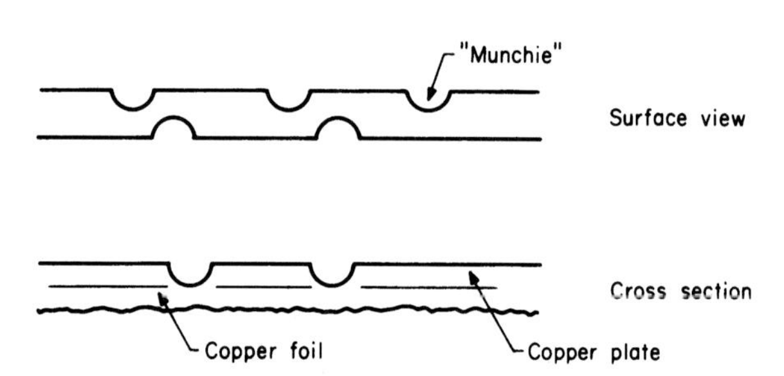

FIGURE 29.23 Trace surface view and vertical cross section illustrating "munchies" (or "mouse bites") and pitting.

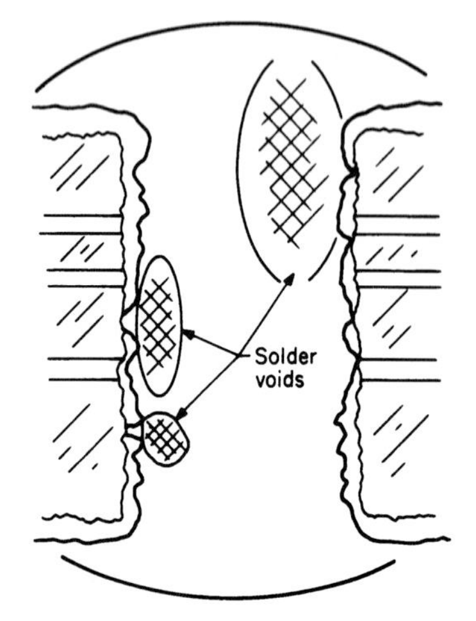

FIGURE 29.24 PTH vertical cross section illustrating wave-soldering blowholes and thin, rough copper plating.

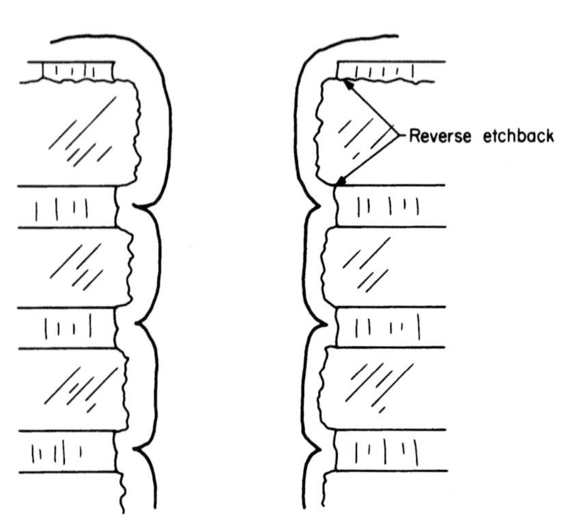

FIGURE 29.25 PTH vertical cross section illustrating reverse etchback.

29.7 SOLDER (TIN-LEAD) ELECTROPLATING

Solder plate (60 percent tin, 40 percent lead) is widely used as a finish plate on printed circuit boards. This process features excellent etch resistance to alkaline ammonia, good solderability after storage, and good corrosion resistance. Tin-lead plating is used for several types of boards, including tin-lead/copper, tin-lead/tin-nickel/copper, solder mask over bare copper (SMOBC), and surface-mount (SM). Fusing is required on all tin-lead-plated surfaces. Thickness minimums are not specified.

The preferred composition contains a minimum of 55 percent and a maximum of 70 percent tin. This alloy is near the tin-lead eutectic, which fuses at a temperature lower than the melting point of either tin or lead and, thus, makes it easy to reflow (fuse) and solder. (The composition of the eutectic is 60 percent tin, 37 percent lead, with a melting point of 361°F.)

Fusing processes include infrared (IR), hot oil, vapor phase, and hot-air leveling for SMOBC.

Plating solutions currently available include the high-concentration fluoboric acid–peptone system as well as low-fluoboric, nonpeptone, and a nonfluoboric organic aryl sulfonic acid process. These processes are formulated to have high throwing power and give uniform alloy composition.[1] The sulfonic acid process has the advantage of using ball-shaped tin-lead anodes. Table 29.6 gives details of operation and control of two high-throw tin-lead (solder) baths.

TABLE 29.6 Tin-Lead Fluoborate: Operation and Control High

	HBF_4/peptone	Low HBF_4/proprietary
Operating conditions		
Lead	1.07–1.88 oz./gal	1.4–3.0 oz./gal
Stannous tin (SN^{+2})	1.61–2.68 oz./gal	3.0–4.5 oz./gal
Free fluoboric acid	47–67 oz./gal	15–25 oz./gal
Boric acid	Hang bag in tank	Same
Additive	Use as needed by Hull cell and Ah usage.	Use as needed by Hull cell and Ah usage.
Temperature	60–80°F	70–85°F
Cathode current density	15–18 A/ft^2	10–30 A/ft^2
Agitation	Solution circulation	Mechanical and solution circulation
Anodes		
Type	Bar	
Composition	60% tin-40% lead	
Purity	Federal Specification QQ-S-571[30]	
Bags	Polypro	
Hooks	Monel	
Length	Rack length minus 2 in	
Current density	10–20 A/ft^2	

29.7.1 Agitation

The solution is circulated by a filter pump, without allowing air to be introduced.

29.7.2 Filtration

A 3 to 10 μm polypro filter is needed to control solution cloudiness and deposit roughness.

29.7.3 Carbon Treatment

The solution is batch-treated at room temperature every 4 to 12 months. If the solution is clear and colorless, additive is added based on Hull cell and analysis. Do not use Supercarb or hydrogen peroxide.

29.7.4 Contaminations

Several classifications of contaminations are possible:

- **Organic** Such contamination results from peptone or additive breakdown and plating resists. Periodic carbon treatment is needed.
- **Metallic** Copper is the most serious of these contaminations. It causes dark deposits at low-current densities (in the plated through hole) and may coat the anodes. The maximum levels of metallic contaminants allowable are copper, 15 ppm; iron, 400 ppm; and nickel, 100 ppm.
- **Nonmetallic** The maximum levels allowed are chloride, 2 ppm; and sulfate, 2 ppm.
- **Dummy** Copper is removed by dummy plating at 3 to 5 A/ft^2 several hours each week. (Dummy plating is plating at a low-current density, using corrugated metal sheets or scrap panels.) Other metals such as iron and nickel in high concentrations may contribute to dewetting and cannot be easily removed. Ensure that the copper is covered by a tin-lead coating before reducing the current density for dummying.

29.7.5 Solution Controls

Stannous and lead fluoborates are the source of metal. Their concentrations and ratio must be strictly maintained, as they will directly affect alloy composition. Fluoboric acid increases the conductivity and throwing power of the solutions. Boric acid prevents the formation of lead fluoride. Additives promote smooth, fine-grained, tree-free deposits. Excess peptone (three to four times too much) may cause pinholes (volcanoes) in deposit when reflowed. Testing by Hull cell and periodic carbon treatments is indicated. The peptone add rate is about 1 to 2 qt per week for a 400 gal tank. Only DI water and contamination-free chemicals should be used—for example, <10 ppm iron-free and <100 ppm sulfate-free fluoboric acid. A clear solution is maintained by constant filtration.

29.7.6 Deposition Rate

A layer of 0.5 mil tin-lead is deposited in 15 to 17 min. at 15 to 17 A/ft^2. The best practice is to plate at 10 to 25 A/ft^2 (2.3 copper current). Higher currents lead to coarse deposits and more tin in the alloy. Excessive current causes treeing and sludge formation.

29.7.7 Deposit Composition

The composition should be 60 percent tin to 40 percent lead. The composition should be confirmed with a deposit assay. Alloy composition is determined by the ratio of tin and lead in the solution; thus Sn (oz./gal) + [total Sn + Pb (oz./gal)] × 100 = %Sn (29.2).

29.7.8 Hull Cell

This test shows overall plating quality and the need for peptone, additives, or carbon treatment, as well as the presence of dissolved copper in the solution.

29.7.9 Visual Observation of Plating

When plated, solder has a uniform matte finish. The deposit should be smooth to the touch. A coarse, crystalline deposit usually indicates the need for additives or peptone or a current density that is too high. Peeling from copper is generally related to the procedure. The fluoboric acid predip should not be used as a holding tank (for more than 5 min.) prior to solder plating.

Acid strength and cleanliness are also important in this case. Load the tank with some residual current (5 to 10 percent). Rough, dark, thin, or smudged deposits may be due to organic contamination and may require carbon treatment. Dark deposits, especially in low-current-density areas and in the plated through hole, are due to copper contamination or to thin plating.

A Hull cell test will confirm deposit, and a dummy plate will remove copper.

29.7.10 Corrective Actions

- Keep the bath composition in balance.
- Use contaminant-free chemicals.
- Dummy-plate once a week at 3 to 5 A/ft^2.
- Circulate the solution through a filter continuously.
- Maintain additives by Hull cell and by analysis.
- Analyze for copper, iron, and nickel at least once a month.
- Carbon-treat on schedule.

29.8 TIN ELECTROPLATING

Tin is used extensively for plating electronic components and PC boards due to its solderability, corrosion resistance, and metal etch-resist properties. The current MIL-STD-275 does not include tin plating, although earlier versions stated a required thickness of 0.0003 in. Specifications covering tin plating are MIL-T-10727 and MIL-P-38510, which say that tin must be fused on component leads.

29.8.1 Acid Tin Sulfate

This is the most widely used system. Among the many processes available, some produce bright deposits for appearance and corrosion resistance; others give matte deposits, which can be fused as well as soldered after long-term heating. Tin sulfate baths are somewhat difficult to control, especially after prolonged use.[2,3] Table 29.7 describes their operation and control.

29.8.1.1 Process Controls

- **Agitation** A filter pump circulates the solution without allowing air to be introduced. Cathode rod agitation is also useful for a wider range of plating current densities.
- **Filtration** A 3 to 10 μm polypropylene filter is needed to control excess cloudiness and sludge formation.
- **Temperature** The preferred temperature for deposit luster and visual appearance is 60 to 65°F. Use cooling coils. Baths can operate up to 85°F, but may result in smoky, hazy deposits.
- **Carbon treatment** Light carbon filtration at room temperature removes organic contaminations. New baths are also made up if problems continue with deposit quality, solderability, thickness control, and cost-effectiveness.

TABLE 29.7 Acid Tin Sulfate: Operation and Control

Operating conditions:	
Tin	2 oz./gal
Sulfuric acid	10–12% by volume
Carrier, additives	Replenish by Ah usage and spectrophotometry
Temperature	60–65°F for bright, 65–85°F for matte
Cathode current density	10–30
Current efficiency	100%
Plating rates	0.3 mil @25 A/ft^2
Anodes:	
Type	Bars
Composition	Pure tin
Bags	Polypropylene
Hooks	Monel or titanium
Length	Rack length minus 2 in
Current density	5–20 A/ft^2

- **Contaminations** These include three groups of contaminants.
  - **Organics** These come from additives and breakdown of resist.
  - **Metallic** The effects of metallic contaminants on the plated deposits and the maximum levels allowable are as follows: copper darkness, 5 to 10 ppm; cadmium dullness, 50 ppm; zinc dullness, 50 ppm; nickel streaks, 50 ppm; iron dullness, 50 to 120 ppm; and chromium dullness, 5 ppm.
  - **Nonmetallic** The maximum level allowable is 75 ppm of chloride.
- **Anodes** To maintain tin content, remove anodes when the bath is idle.

29.8.1.2 Solution Controls

- **Bath constituents** Stannous sulfate and sulfuric acid are maintained by analysis, additives by spectrophotometry, Ah usage, Hull cell, and the percentage of sulfuric acid additions.
- **Electronic-grade chemicals** Such chemicals must be used to control metallic contaminations of cadmium, zinc, iron, and so on.
- **Control additive** Low levels of additives must be maintained.
- **Hull cell** This test is useful for control of additive levels and plating quality. See Sec. 29.12.4 for the procedure.
- **Rinsing after plating** Adequate rinsing after plating is important to control white or black spots on tin surfaces.
- **Visual observation** When plated, tin has a uniform lustrous finish. The deposit should be smooth to the touch.

29.8.2 Problems with Tin Electroplating

- **Dull deposits** These are due to an out-of-balance condition of the main solution constituents, that is, low acid (<10 percent), high tin (>3 oz./gal), improper additive levels, contamination by metals or chlorides, or high temperatures (>65°F).

- **Peeling tin** Tin comes off due to low acid (<10 percent) or organic contamination.
- **Slivers** These are caused by overetching. A review of etching practice is indicated, as well as the use of 1.2 oz. copper foil.
- **Pitting** If substrate is not the cause, check precleaning, solution balance and contaminants, current efficiency, and current densities. High-current densities may cause pitting.
- **Strip and etching residues** Tin is attacked by strong alkaline solutions. To control residues and spotting, use mild room temperature stripping and alkaline-ammonia etching.
- **Poor solderability** This may be caused by excess additives or contaminations in the bath, poor rinsing, bath age, or excessive thicknesses (>0.3 mil).

29.9 NICKEL ELECTROPLATING

Nickel plating is used as an undercoat for precious and nonprecious metals. For surfaces such as contacts or tips that normally receive heavy wear, the uses of nickel under a gold or rhodium plate will greatly increase wear resistance. When used as a barrier layer, nickel is effective in preventing diffusion between copper and other plated metals. Nickel-gold combinations are frequently used as metal etch resists. Nickel alone will function as an etch resist against the ammoniacal etchants.[4] MIL-STD-275 calls for a low-stress nickel with a minimum thickness of 0.0002 in. Low-stress nickel deposits are generally obtained using nickel sulfamate baths in conjunction with wetting (antipit) agents. Additives are also used to reduce stress and to improve surface appearance.

29.9.1 Nickel Sulfamate

Nickel sulfamate is commonly used both as an undercoat for through-hole plating and on tips. Conditions given in Table 29.8 are applicable for through-hole and full-board plating.

TABLE 29.8 Nickel Sulfamate: Operation and Control

Operating conditions:	
pH	3.5–4.5 (3.8)
Nickel	10–12 oz./gal
As nickel sulfamate	43 oz./gal
Nickel chloride	4 oz./gal
Boric acid	4–6 oz./gal
Additives	As required
Antipit	As required
Temperature	130 ± 5°F
Cathode current density	20–40 ASF
Anodes:	
Type	Bars or chunks
Composition	Nickel
Purity	Rolled depolarized, cast, or electrolytic; SD chips in titanium basket
Hooks, baskets	Titanium
Bags	Polypro, Dynel, or cotton
Length	Rack length minus 2 in.

29.9.1.1 Process Controls

- **pH** With normal operation, pH increases. To lower pH, use sulfamic acid (not sulfuric acid). A decrease in pH signals a problem; anodes should be checked.
- **Temperature** The preferred temperature is 125°F. Low temperature causes stress and high-current-density (CD) burning. Higher temperatures increase softness of nickel deposit.
- **Agitation** Solution circulation between panels is done by filter pump and/or cathode rod agitation.
- **Filtration** This is donc continuously through 5 to 10 μm polypro filter, which is changed weekly.
- **Deposition rate** For 0.5 mil, plate 25 to 30 min. at 25 A/ft^2. For 0.2 to 0.3 mil, plate 15 min. at 25 A/ft^2.
- **Contaminations** These include three groups of contaminants:
  - **Metals** The following are the maximum allowable metal content by atomic absorption: iron, 250 ppm; copper, 10 ppm; chromium, 20 ppm; aluminum, 60 ppm; lead, 3 ppm; zinc, 10 ppm; tin, 10 ppm; and calcium, 0 ppm. These metal contaminants lower the deposition rates and cause nonuniform plating. Cooper and lead cause dark, brittle deposits at low CD. Dummy-plate at 3 to 5 A/ft^2. Iron, tin, lead, and calcium cause deposit roughness and stress.
  - **Organics** These cause pitting, brittleness, step plating, and lesser ductility.
  - **Sulfates** These cause solution breakdown and should not be added.
- **Carbon treatment** Circulate through a carbon canister for 4 hours (more than one canister may be required) or batch-treat to remove organics. The basic six steps to batch-carbon-treat are as follows:

  1. Heat to 140°F.
  2. Transfer to the treatment tank. Do not adjust pH.
  3. Add 3 to 5 lb. carbon per 100 gal of solution. Premix outdoors.
  4. Stir 4 hours at temperature.
  5. Let the solution settle 1 to 2 hour.
  6. Filter the solution back to cleaned tank.

29.9.1.2 Common Problems. Pitting, stress (cracking), and burned deposits are common problems:

- **Pitting** This is caused by low antipit, poor agitation or circulation, boric acid imbalance, or the presence of organics. In testing for antipit, the solution should hold bubble for 5 sec. in a 3 in. wire ring and for less than 5 sec. in a 5 in. ring. In the case of severe pitting, cool to room temperature, add 1 pt/200 gal hydrogen peroxide (35 percent), bubble air, and reheat to 140°F. Carbon-treat as described in Sec. 29.9.1.1. Adjust pH before restarting.
- **Stress** This refers to the cause of deposit cracking. Although low nickel chloride causes poor anode corrosion (a rapid decrease in nickel content), high chloride causes excess stress. To prevent this, maintain pH, current density, and boric acid for control below 10 kpsi.
- **Burns** A low level of nickel sulfamate causes high-current-density "burns."

Common treatment includes the following:

- **Hull cell** Plate at 2 A for 10 min. with gentle agitation. This test is useful for bath condition and contaminants.

- **Visual observation** The plated metal has a matte, dull finish. Burned deposits are caused by low temperature, high-current density, bath imbalance, and poor agitation. Rough deposits are due to poor filtration, pH out of spec, contamination, or high-current density.

Pitting is due to low antipit, poor agitation, bath imbalance, or organic contamination. Low plating rates are due to low pH, low-current density, or impurities. Gassing at the cathode is a sign of low plating rates.

29.9.2 Nickel Sulfate

This is typically plated with an automatic edge connector (tip) plating machine. Table 29.9 gives the operating conditions that apply to these systems. For additional instruction, follow the details in Sec. 29.9.1.

TABLE 29.9 Nickel Sulfate: Operation and Control

Operating conditions:	
pH	1.5–4.5
Nickel	15–17 oz./gal
Nickel chloride	2–4 oz./gal (with soluble anodes)
Boric acid	5–7 oz./gal
Stress reducer	As required
Antipit	As required
Temperature	130 ± 5°F
Cathode current density	100–600 A/ft^2
Current efficiency	65%
Anodes:	
Composition	Nickel or platinized titanium

Nickel anodes are preferred in tip machines because the pH and the metal content remain stable. The pH will decrease rapidly when insoluble anodes are used. The pH should be maintained at 1.5 or higher with additions of nickel carbonate. Stress values are higher than in sulfamate baths, with values of about 20 kpsi.

29.10 GOLD ELECTROPLATING

Early printed circuit board technology used gold extensively. In addition to being an excellent resist for etching, gold has good electrical conductivity, tarnish resistance, and solderability after storage. Gold can produce contact surfaces with low electrical resistance. In spite of its continued advantages, the high cost of gold has restricted its major application to edge connectors (tips) and selected areas, with occasional plating on pads, holes, and traces (body gold). Both hard-alloy and soft, pure gold are currently used. Plating solutions are acid (pH 3.5 to 5.0) and neutral (pH 6 to 8.5). Both automatic plating machines for edge connectors and manual lines are in use.

29.10.1 Acid Hard Gold

To a large extent, acid golds are used for compliance to MIL-STD-275, which states that gold shall be in accordance with MIL-G-45204, Type II, Class 1. The minimum thickness shall be

0.000050 in (50 μin); the maximum shall be 0.000100 in. Nonmilitary applications require 25 to 50 μin. A low-stress nickel shall be used between gold overplating and copper. Type II hard gold is not suited for wire bonding. These systems use potassium gold cyanide in an organic acid electrolyte. Deposit hardness and wear resistance are made possible by adding complexes of cobalt, nickel, or iron to the bath makeup. Automatic plating machines are being used increasingly because of the enhanced thickness (distribution) control, efficient gold usage, productivity, and quality. A comparison of automatic versus manual plating methods for edge connectors is given in Table 29.10.

TABLE 29.10 Acid Gold-Cobalt Alloy: Operation and control

	Manual	Automatic
Gold content, troy oz./gal	0.9–1.1	1–3
pH	4.2–4.6	4.5–5.0
Cobalt content	800–1000 ppm	800–1200 ppm
Temperature	90–110°F	100–125°F
Solution density	8–15 Be	12–18 Be
Replenishment per troy oz. gold	8 Ah	6.5 Ah
Current efficiency	50%	60%
Agitation	5 gal/min.	50 gal/min.
Anode-to-cathode distance	2–3 in.	1.4 in.
Anodes, composition	Platinized titanium	Platinized titanium
Cathode current density	1–10 A/ft^2	50–100 A/ft^2
Thickness	40 ± 10 μin.	40 ± 2 μin.
Deposition rate for 40 μin.	3–6 min.	0.3–0.6 min.
Deposition composition	99.8% gold, 0.2% cobalt	99.8% gold, 0.2% cobalt
Hardness	150 Knoop	150 Knoop

Gold solution contaminants

Metal	Maximum ppm	Metal	Maximum ppm
Lead	10	Iron	100
Silver	5	Tin	300
Chromium	5	Nickel	300–3000
Copper	50		

Organics: Tape residues, mold growth, and cyanide breakdowns.

29.10.1.1 Process Controls

- **Analysis** Gold content, pH, and density should be maintained at optimum values. Operation at very low gold content causes early bath breakdowns with loss of properties, less current efficiency, and less cost savings. The pH is raised by using potassium hydroxide and is lowered with acid salts. Solution conductivity is controlled by density, which is adjusted with conductivity salts. Hull cell is not recommended for this purpose.
- **Anodes** Platinized titanium should be replaced when operating voltages are excessive or when thick coatings develop on the anode surfaces.
- **Recovery** Plating solutions should be replaced when contaminated, when plating rates decrease, or after about 10 total gold content turnovers.

29.10.1.2 Problems. Difficulties can be controlled by proper gold bath and equipment maintenance. Some typical situations follow:

- **Discolored deposit** This may be due to low brightener; to metal contaminants such as lead, low pH, low density, or organic contaminations; or to leaking from tape. Gold plate is stripped to evaluate nickel. For these problems, first try to lower the pH, raise the density, or increase the brightener before replacing the bath or using decreased current densities.
- **Gold peeling from nickel** This is generally due to inadequate solder stripping, leaky tape, or poor activation. Methods to increase nickel activation include increasing acid strength after nickel plating and using a fluoride activator, cathodic acid, or gold strike. A gold strike that is compatible with gold plate and has low metal content and low pH is available. Its main purpose is to maintain adhesion to the nickel substrate.
- **Wide thickness range** To narrow the thickness range, improve the solution movement between panels; clean, adjust, or replace anodes; and adjust or replace the solution.
- **Low deposition rate** This is characterized by excessive gassing and low efficiencies. To correct this condition, adjust solution parameters and check for contaminants such as chromium.
- **Pitting** Strip gold and evaluate nickel and copper by cross-sectioning.
- **Resist breakdown** Low current efficiencies cause this condition. Use solvent-soluble screened or dry film for best results.

29.10.2 Pure 24-Karat Gold

High-purity 99.99 percent gold processes are used for boards designed for semiconductor chip (die) attachment, wire bonding, and plating solder (leaded) glass devices, for their solderability and weldability. These qualities comply with Types I and III of MIL-G-45204. The processes are neutral (pH 6 to 8.5) or acid (pH 3 to 6). Pulse plating is frequently used. Table 29.11 gives typical conditions for a neutral bath.

TABLE 29.11 Neutral Pure Gold: Operation and Control

Gold content	0.9–1.5 troy oz./gal
pH	6.0–7.0
Temperature	150°F
Agitation	Vigorous
Solution density	12–15 Be
Replenishment	4 Ah/troy oz.
Current efficient	90–95%
Cathode current density	1–10 A/ft^2
Deposition rate for 100 μin., @5 A/ft^2	8 min.
Deposition composition	99.99% + gold
Hardness	60–90 Knoop

29.10.3 Alkaline, Noncyanide Gold

Various processes for alloy and pure gold deposits are available. Solutions are based on sulfite-gold complexes and arsenic additives and operate at a pH of 8.5 to 10.0. A decision to use this process is based primarily on the need for uniformity (leveling), hardness (180 Knoop), purity, reflectivity, and ductility. PC board use is limited to body plating, since wear characteristics of the sulfite-gold are not suitable for edge connector applications. The microelectronics industry uses these processes for reasons of safety and gold purity. Semiconductor chip attachment, wire bonding, and gold plating on semiconductors are possible applications and are enhanced by using pulse plating without metallic additives.

29.10.4 Gold Plate Tests

Several routine in-process and final quality control tests are performed on gold:

- **Thickness** Techniques are based on beta-ray backscattering and x-ray fluorescence. Thickness and area sensitivity are as low as 1 μin. with 5 mil pads.
- **Adhesion** Standard testing involves a tape pull test.
- **Porosity** Tests involve nitric acid vapor and electrographics.
- **Purity** Lead is a common impurity that must be controlled to <0.1 percent. Other tests included discoloration by heating, electrical contact, and wear resistance.

29.11 PLATINUM METALS

Interest in platinum systems usually soars in a climate of high gold prices, even though plating results are not always as dependable as those obtained when using gold.

29.11.1 Rhodium

Deposits from the sulfate or phosphate are hard (900 to 1000 Knoop), highly reflective, extremely corrosion-resistant, and highly conductive (resistivity is 4.51 μ/cm). Rhodium plate is used where a low-resistance, long-wear, oxide-free contact is required. In addition, rhodium as a deposit on nickel for edge connectors has been replaced by gold. This is due to difficulty in bath control problems with organic and metallic contaminants, as well as cost. Table 29.12 shows details of rhodium plating.

TABLE 29.12 Rhodium Sulfate: Operation and Control

Rhodium	4–10g/L
Sulfuric acid	25–35 mL/L
Temperature	110–130°F
Agitation	Cathode rod
Anodes	Platinized titanium
Anode-to-cathode ratio	2:1
Cathode current density	10–30 A/ft^2
Plating rate	At 20 A/ft^2, 10 μin. will deposit in 1.4 min., based on 70% cathode current efficiency

29.11.2 Palladium and Palladium-Nickel Alloys

Deposits of 100 percent palladium, 80 percent palladium with 20 percent nickel, and 50 percent palladium with 50 percent nickel find use as suitable deposits for edge connectors. Deposits are hard (200 to 300 Knoop), ductile, and corrosion-resistant. A palladium-nickel undercoat for gold shows good wear and electrical properties.

29.11.3 Ruthenium

Deposits of ruthenium are similar to rhodium but are plated with easier control, greater bath stability, and high-current efficiency at lower cost. Deposits are usually stressed.

29.12 SILVER ELECTROPLATING

Silver is not widely used in the printed circuit industry, although it finds applications in optical devices and switch contacts. Thicknesses of 0.0001 to 0.0002 in. (0.1 to 0.2 mil) with a thin overlay of precious metal are specified.

Silver plating should not be used when boards are to meet military specifications. The reason for this is that, under certain conditions or electrical potential and humidity, silver will migrate along the surface of the deposit and through the body of insulation to produce low-resistance leakage paths. Tarnishing of silver-gold in moist sulfide atmospheres also produces electrical problems on contact surfaces due to diffusion of the silver to the surface.

Another reason for the lack of acceptance of silver is that silver is plated from an alkaline cyanide bath, which is highly toxic. Bright plating solutions that produce deposits with improved tarnish are usually related to black anodes and are due chiefly to solution imbalance, impurities in anodes, or solution roughness and pitting. Most metals to be plated, particularly the less noble metals, require a silver strike prior to silver plating to ensure deposit adhesion.

29.13 LABORATORY PROCESS CONTROL

Laboratory processes available for electroplating control include conventional wet chemical analysis, advanced instrumental techniques, metallographic cross-sectioning, and Hull cell.

29.13.1 Conventional Wet Chemical Analysis

The traditional wet chemical methods for metals and nonmetal plating solution constituents are available from suppliers and in the literature.[5,6] These methods also make use of pH meters, ion electrodes, spectrophotometers, and atomic absorption. The composition of liquid concentrates for plating solutions is shown in Table 29.13.

TABLE 29.13 Composition of Liquid Concentrates

Chemical	Formula	Weight, lb./gal	Percent	Metal, oz./gal
Acids				
Sulfuric	H_2SO_4	15.0	96	—
Hydrochloric	HCl	9.8	36	—
Fluoboric	HBF_4	11.2	49	—
Alkaline				
Sodium hydroxide	$NaOH$	12.8	50	—
Ammonium hydroxide	NH_4OH	7.5	28	—
Metals				
Copper sulfate	$CuSO_4 \cdot 5H_2O$	9.7	27	9
Copper fluoborate	$Cu(BF_4)_2$	12.9	46	25.4
Stannous fluoborate	$Sn(BF_4)_2$	13.3	51	44.3
Lead fluoborate	$Pb(BF_4)_2$	14.4	51	65.0
Nickel sulfate	$NiSO_4 \cdot 6H_2O$	11.0	44	17.8
Nickel sulfamate	$Ni(NH_2SO_3)_2$	12.9	50	24
Nickel sulfamate	$Ni(NH_2SO_3)_2$	12.3	43	20
Nickel chloride	$NiCl_2 \cdot 5H_2O$	11.2	54	23.7

29.13.2 Advanced Instrumental Techniques

New techniques have been developed for the control of organic additives in copper plating. Continued development is in progress in the area of measurement of such additives in nickel, gold, and tin solutions. Methods used include liquid chromatography, ultraviolet/visible (UV/VIS) spectrophotometry, CVS, ion chromatography, UV persulfate oxidation, and polarography. These techniques can detect contaminations in various processes, and they show the need for an effectiveness of carbon treatment. Table 29.14 lists and references these techniques, which are having a major influence on plating process capabilities. Literature references list 136 entries on these techniques.[7]

TABLE 29.14 Advanced Instrumental Analysis Techniques

Technique	Constituent
Cyclic voltammetry stripping	Organics and inorganics
Liquid chromatography with UV/VIS	Organics and inorganics
Ion chromatography	Ionic species
Polarography	Organics and inorganics
Ion selective electrode	Ionic metals, nonmetals
Atomic absorption (AA)	Metals, nonmetals
UV oxidation	Total carbon

29.13.3 Metallographic Cross-Sectioning[8,9]

A method for cross-sectioning PC boards is as follows:

1. *Bulk cutting.* Remove a manageable-sized piece of board or assembly by shearing or abrasive cutting.
2. *Precision cutting.* Perform low-speed sawing with a diamond wafer blade to produce vertical sections about 1 × 1.2 in., and cut a horizontal section next to PTH pads.
3. *Mounting.* Encapsulate sections vertically and horizontally in epoxy resin.
4. *Fine grinding.* Hand-grind using 240-, 320-, 400-, and 600-grit silicon carbide papers.
5. *Rinsing.* Rinse the sample between grits.
6. *Polishing.* Diamond-polish (6 μm) on nylon cloth and alumina polishing (0.3 μm) on nap cloth using a rotating wheel. To polish the sample, place it on the rotating-wheel polisher and move it slowly in the opposite direction. Polish for 4 min. if using a 6 μm diamond on nylon, and 1 min. on 0.3 μm alumina on nap cloth. Clean and dry between polishing compounds.
7. *Etching.* Apply a cotton swab for 2 to 5 sec., soaked in a solution of equal parts of ammonium hydroxide and 3 percent hydrogen peroxide. Rinse in water and dry carefully.
8. *Documenting.* Observe and photograph the sample with a microscope at 30 to 1500 × magnification.

29.13.4 Hull Cell

Although the advanced techniques discussed previously provide precise control of plating solutions, Hull cell testing is still widely used in the industry. Its advantages are low cost, simplicity of operation, and its actual correlation with plating production. Its main disadvantage is that defects in copper plating frequently are not shown by this method. For example, Hull cell testing will not help in detecting dull plating, roughness, or pitting. The procedure starts with brass panel preparation, in the following order:

1. Remove the plastic film.
2. Treat with cathodic alkaline cleaner.
3. Soak in 10 percent sulfuric acid.
4. Rinse.

Repeat these steps until the panel is water-break free. Proceed with the Hull cell as follows:

1. Rinse with test solution.
2. Fill to mark.
3. Adjust temperature and agitation.
4. Attach panel to negative terminal.
5. Plate.

Agitation should be similar to tank operation—that is, vigorous air bubbling for copper, gentle stirring for tin and tin-lead, and none for tin-nickel. Plate copper and nickel at 2 A and other metals at 1 A. The effects of bath adjustment, carbon treatment, dummy plating, and so on, are readily translated from the Hull cell to actual tank operations. See previous sections on metal plating for Hull cell results, and consult the supplier for test equipment.

29.14 ACKNOWLEDGMENT

Significant portions of this chapter are drawn from George Milad, "Electroplating," *Printed Circuits Handbook, 5th ed,* Clyde Coombs (ed.), 2001, Chap. 29.

REFERENCES

1. Rothschild, B. F., "Solder Plating of Printed Wiring Systems," *Proceedings of the Printed Circuit Plating Symposium, California Circuits Association,* November 5–6, 1969, pp. 10–21, and November 12–13, 1968, pp. 61–65.
2. Jacky, G. F., "Soldering Experience with Electroplated Bright Acid Tin and Copper," *Circuit Technology Today,* California Circuits Association, October 1974, pp. 49–74.
3. Davis, P. E., and Duffek, E. F., "The Proper Use of Tin and Tin Alloys in Electronics," *Electronic Packaging and Production,* Vol. 15, No. 7, 1975.
4. Kilbury, R. G., "Producing Buried Via Multilayers: Two Approaches," *Circuits Manufacturing,* Vol. 25, No. 4, 1985, pp. 30–49.
5. Foulke, D. G., *Electroplaters' Process Control Handbook,* rev. ed., R. E. Krieger Publishing Company, 1975.
6. Langford, K. E., and Parker, J. E., *Analysis of Electroplating and Related Solutions, 4th ed.,* R. Draper Ltd., 1971.
7. See Ref. 64, Chap. 7, "Additives."
8. Nelson, J. A., "Basic Steps for Cross Sectioning," *Insulation/Circuits,* 1977.
9. Wellner, P., and Nelson, J., "High Volume Cross Sectional Evaluation of Printed Circuit Boards," IPCWC- 4B-1, 1981.

CHAPTER 30
DIRECT PLATING

Dr. Hayao Nakahara
N.T. Information Ltd., Huntington, New York

30.1 DIRECT METALLIZATION TECHNOLOGY*

For 40 years the PTH process of choice has been palladium followed by electroless copper, but there have been at least 12 DMT processes challenging that established process, with several hundred installations, a significant percentage of the total number of PCB shops. The basic idea for the palladium systems dates back to the Radovsky patent of 1963,[1] which claimed a method of using an electrically nonconductive film of palladium in a semicolloidal form to directly metallize through holes in printed circuit boards. Radovsky's invention was never commercialized. The basic idea for the carbon/graphite systems dates back to the very early days of eyelet boards when Photocircuits was experimenting with graphite, silver, and other media to turn their single-sided PCBs into reliable double-sided boards.

30.1.1 Direct Metallization Technologies Overview

From the many media and technical variations, some common elements have evolved.

30.1.1.1 Direct Metallization Technologies Common Elements. There are two elements common to all DMT:

1. Holes must be conditioned more specifically and thoroughly than for electroless copper.
2. Conductive media must be removed from the copper foil in a majority of the techniques (an exception is DMS-E).

It is understood that additional desmear steps are necessary or advisable when processing multilayer PCBs. Common elements to all horizontal conveyorized DMT systems are:

- Throughput is typically 6 to 15 min for a panel, with the next panel following 1 in behind
- Tremendous economies in rinse water use
- Lower consumption of chemicals

* All proprietary names are trademarked by their respective owners.

- Fewer steps than in its vertical mode
- Panel-to-panel uniformity

Many DMT processes work better than electroless copper on substrates with "difficult" resin systems such as PTFE, cyanate ester, or polyimide.

30.1.1.2 Direct Metallization Categories. DMT falls into four broad categories:

1. Palladium-based systems
2. Carbon or graphite systems
3. Conductive polymer systems
4. Other methods

30.1.2 Palladium-based Systems

30.1.2.1 Palladium/Tin Activator with Flash Electoplating. EE-1,[2] the first commercialized direct metallization technique was invented in 1982 at Photocircuits and PCK. It uses a palladium/tin activator followed by mandatory flash electroplating. The flash-plating bath contains a polyoxyethylene compound to inhibit the deposition of copper on the foil surface without inhibiting deposition on palladium sites on nonconductive surfaces (holes, edges, substrate). Deposition occurs by propagation from the copper foil and grows epitaxially along the activated surface of the hole. Coverage takes about 5 to 6 min. This flash is pattern- or panel-plated subsequently to full thickness in any electroplating bath. Microetching is incorporated in the accelerator to remove palladium sites and nail heads from innerlayers. A special cleaner/conditioner is used. This process has never been adapted to horizontal equipment. (See Fig. 30.1.)

30.1.2.2 Palladium/Tin Activator with Vanillin. DPS[3] was invented in Japan in the late '80s. This method uses a palladium/tin activator with vanillin, followed by pattern or panel electroplating. It employs a special cleaner/conditioner and a carbonate accelerator. The three key solutions—cleaner/conditioner, activator, and accelerator—all operate at elevated temperatures. DPS has recently been adapted to horizontal equipment, but works well in the vertical mode, both manual and automatic. Following the last step, the *Setter,* DPS yields a stable, grayish conductive palladium film in the holes. It is believed that the cleaner/conditioner slightly solubilizes the activator, attracting it to the nonconductive surface, and that the vanillin lines up the palladium molecules, and directs them towards the work, hence giving lower electrical resistance and better adhesion to nonconductive surfaces. It is also claimed that there is little palladium/tin left on the copper foil, so it is easily soft-etched away in the normal electroplating preplate cycle. (See Fig. 30.2.) DPS was the first DMT process to suggest an Ω meter as the standard quality assurance (QA) tool.

30.1.2.3 Converting Palladium to Palladium Sulfide. Crimson,[4] invented by Shipley, employs a conversion step after the activator where palladium is changed to palladium sulfide, which is claimed to be more conductive for subsequent electrolytic copper plating. The *enhancer* stabilizes the conductive film so that it is chemically resistant to imaging steps. The *stabilizer* neutralizes residues from the enhancer, thereby preventing contamination of subsequent steps. The *microetch* selectively removes activator from copper surfaces to achieve optimum copper-to-copper bond and reliable dry film adhesion. The process works best in conveyorized horizontal equipment and can be followed by pattern or panel electroplating. (See Fig. 30.3.)

30.1.2.4 Process Variations. ABC,[5] invented in Israel by Holtzman et al., is similar to EE-1. It has been adapted to conveyorized horizontal equipment, but must be followed by a flash electroplating in a proprietary bath.

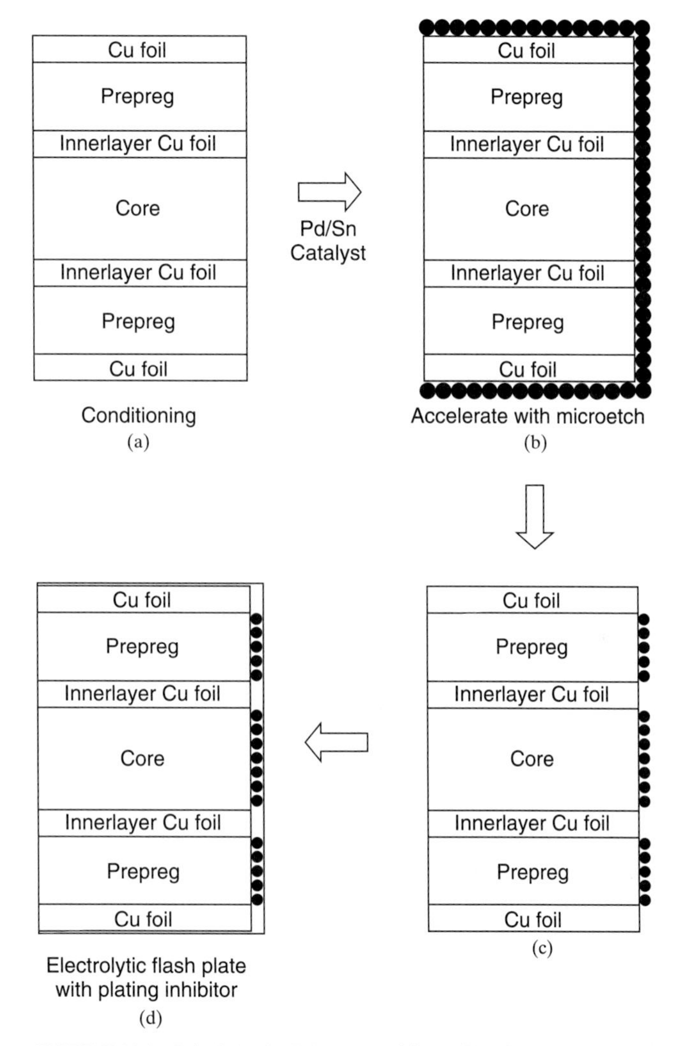

FIGURE 30.1 Principle of EE-1 process: (*a*) conditioning—making glass and epoxy surfaces receptive of Pd catalyst sites; (*b*) palladium catalyst sites; (*c*) acceleration removes excess tin and microetch removes catalysts from the surface of Cu foil; (*d*) flush Cu plate coppers enter surface in drilling hole walls.

Conductron, from LeaRonal, is similar to DPS with the addition of a special cleaner/conditioner and a glass-etch step. It has been adapted to conveyorized horizontal equipment and can be followed by pattern or panel electroplating.

Envision DPS, from Enthone-OMI, and Connect, from M & T (now Atotech), are fairly similar to each other, and to DPS, though each has a specific cleaner/conditioner and modified accelerator. No adaptation has been made to horizontal processing. Both processes can be followed by pattern or panel electroplating.

Neopact,[6] from Atotech, uses a tin-free palladium activator in colloidal form. The subsequent postdip removes the protective organic polymer from the palladium, leaving it exposed and with increased conductivity. It has been adapted to conveyorized horizontal equipment, works well in vertical, and can be followed by pattern or panel electroplating. (See Fig. 30.4.)

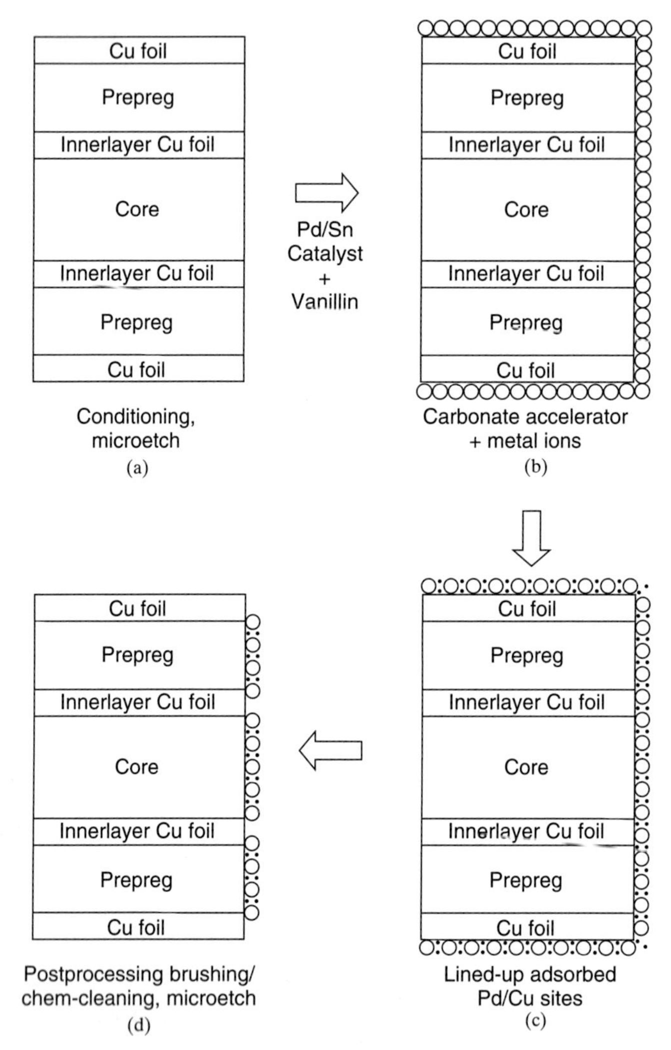

FIGURE 30.2 Principle of the DPS process: (*a*) conditioning for receptive surface; (*b*) catalysts adhering to the surface; (*c*) stronger adhesion; (*d*) ready to plate.

30.1.3 Carbon/Graphite Systems

30.1.3.1 Carbon Suspensions. Black Hole,[7] the second direct metallization technique, was patented by Dr. Carl Minten in 1988 and pioneered by Olin Hunt, who sold their technology to MacDermid in 1991. MacDermid improved the process considerably and called it Black Hole II. Instead of palladium activator, Black Hole II uses carbon suspensions as its conductive medium. Polyelectrolyte conditioned nonconductive surfaces absorb carbon sites, and they "line up" after heating. To ensure sufficient conductivity, the carbon treatment is performed twice. Residues of carbon sites must be removed from the copper foil surface by a

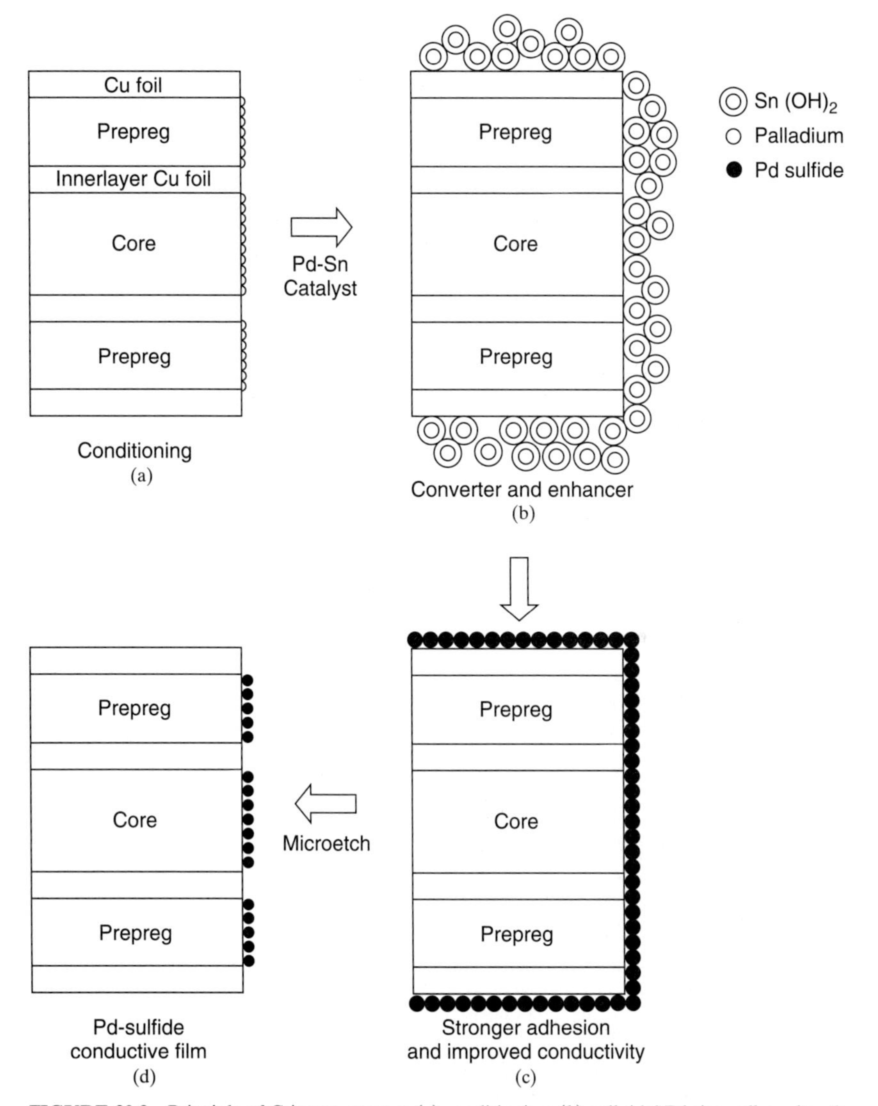

FIGURE 30.3 Principle of Crimson process: (*a*) conditioning; (*b*) colloidal Pd sites adhered to the surface; (*c*) Pd-sulfide (stronger adhesion of Pd catalysts); (*d*) microetch to remove Pd from Cu surface for better Cu-Cu adhesion.

microcleaning step. Black Hole II has been well adapted to conveyorized horizontal equipment and can be followed by pattern or panel electroplating. (See Fig. 30.5.)

30.1.3.2 Graphite. Shadow[8] is from Electrochemicals (division of LaPorte Industries, UK) and uses graphite as its conductive medium. The process sequence of Shadow is very simple and involves fewer steps than most DMTs. Electrochemicals and one of their fabricators, Eidschun Engineering, made the breakthrough in inexpensive, compact, conveyorized horizontal equipment, and the Shadow process is well adapted to this mode. It can be followed by pattern or panel electroplating.

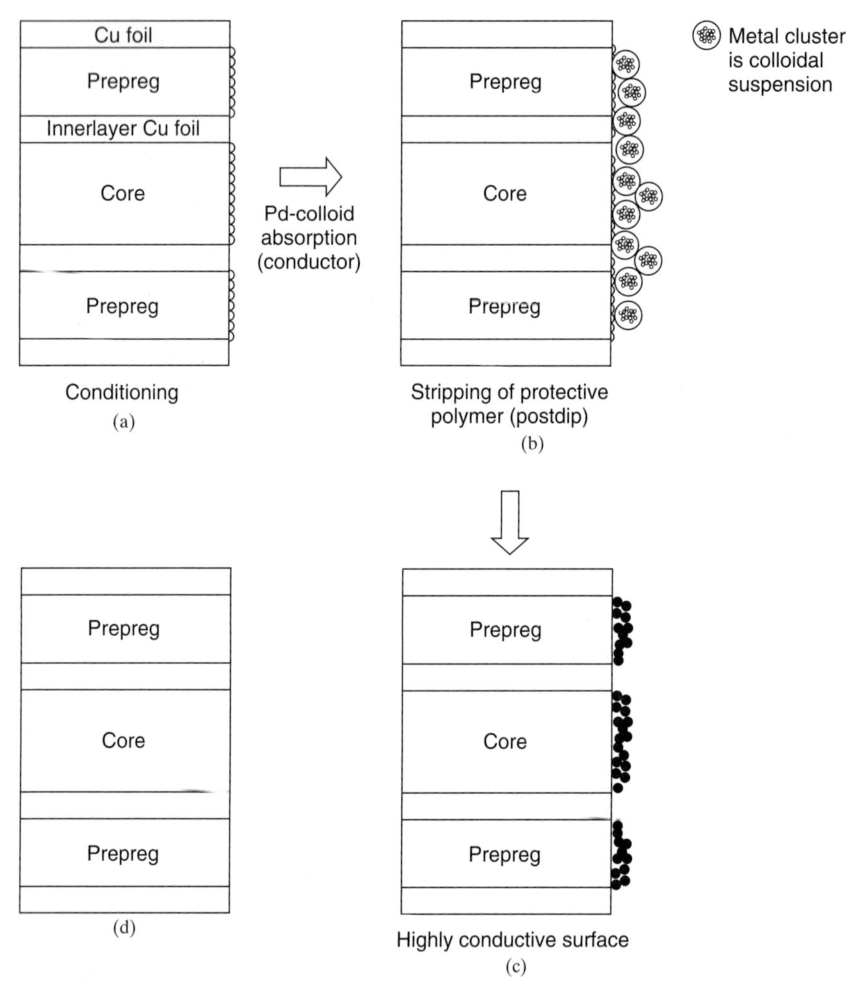

FIGURE 30.4 Principle of Neopact process: (*a*) conditioning surface; (*b*) metal cluster in colloidal suspension adhered to the hole wall; (*c*) stripped Pd adhering to the hole wall; (*d*) Cu reduced.

30.1.4 Conductive Polymer Systems

30.1.4.1 DMS-E. DMS-E[9] from Blasberg is a second-generation DMS-2 process with which they pioneered in this field. DMS-1 was similar to EE-1. After microetch and conditioning, a potassium permanganate solution forms a manganese dioxide coating in the holes which acts as an oxidizing agent during subsequent synthesis reaction. In the catalyzing step, an EDT* monomer bath wets the manganese dioxide surfaces especially well. During the sulfuric acid fixation step, a spontaneous oxidative polarization takes place, forming a black conductive poly-EDT film on the nonconductive areas of the PCB. This technique is very suitable for use in conveyorized, horizontal equipment, since the oxidative conditioning step is very hot (80 to 90°C), and there are solvents involved. It can be followed by pattern or panel electroplating. (See Fig. 30.6.)

* EDT = 3,4 Ethylendioxythiophene.

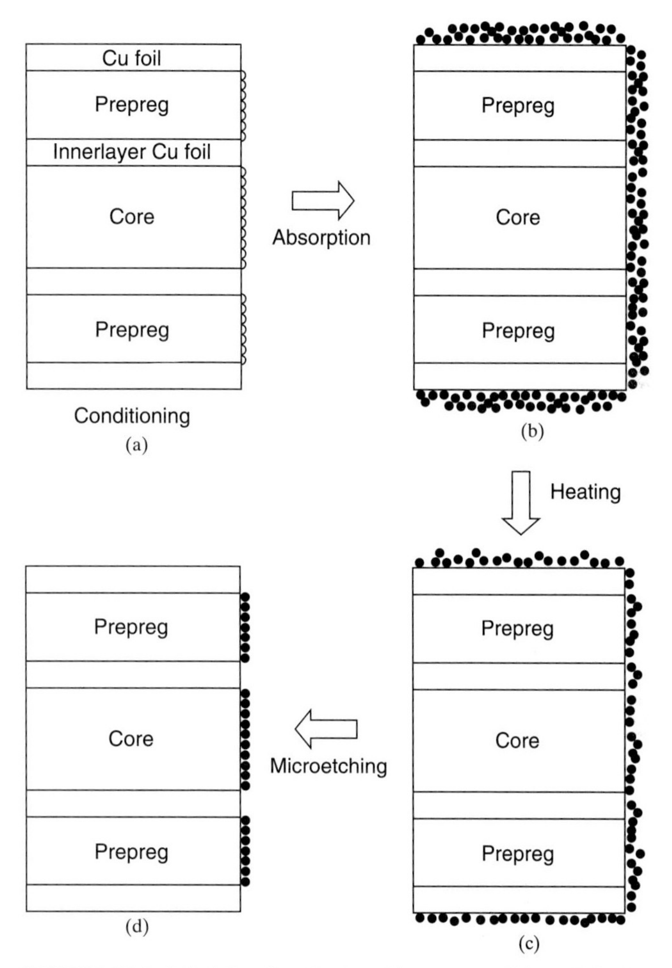

FIGURE 30.5 Principle of carbon/graphite process: (*a*) conditioning; (*b*) carbon particles adhered to the surface; (*c*) dense more conductive carbon conducting film; (*d*) removal of conduction film from Cu surface resulting in stronger Cu-Cu interface.

30.1.4.2 Compact CP. Compact CP[10] was developed by Atotech in 1987 and is essentially similar to DMS-E, except that it combines the catalyzing and fixation steps, it uses an acid permanganate, and the conductive film is a polypyrrole. The technique is very suitable for use in conveyorized, horizontal equipment. It can be followed by pattern or panel electroplating.

30.1.5 Other Methods

There are several other novel ways to metallize holes in PCBs, such as Phoenix and EBP from MacDermid, Schlötoposit from Schlötter. There are conductive ink and laser scribing/filling techniques; there are sputtering and sequential plating, to name a few, but they do not fall within the scope of this chapter.

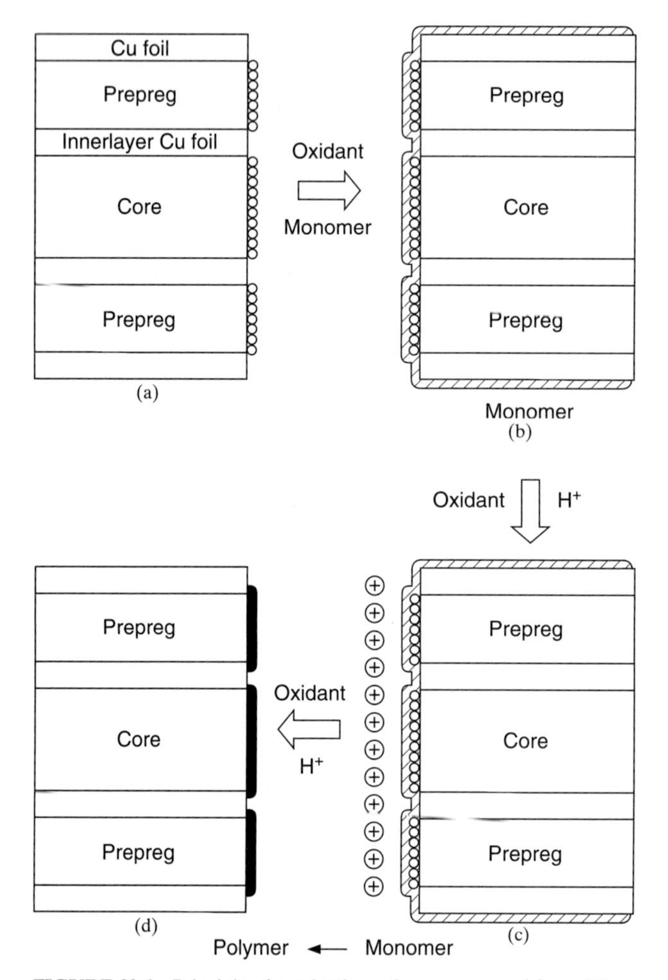

FIGURE 30.6 Principle of conductive polymer process: (*a*) conditioning; (*b*) monomer created on the surface; (*c*) oxidant reduces monomer to polymer yielding highly conductive surface; (*d*) further improved conductive surface.

30.1.6 Comparative Steps of DMT Process

30.1.6.1 DMT.1: Palladium-based Systems. All vertical modes only, for comparison purposes, are shown in Table 30.1.

30.1.6.2 DMT.2: Carbon/Graphite Systems—Conveyorized Horizontal. Conveyorized horizontal systems are shown in Table 30.2.

30.1.6.3 DMT.3: Conductive Polymer Systems. Conveyorized, horizontal systems with optional electrolytic flash are shown in Table 30.3.

TABLE 30.1 DMT.1
Palladium-based systems—all vertical mode only

EE-1	DPS	Crimson	ABC	Conductron	Envision DPS	Neopact
Cleaner conditioner	Cleaner conditioner	Cleaner conditioner	Cleaner conditioner etch	Glass conditioner	Conditioner	Etch cleaner
Rinse	Rinse	Rinse	Rinse	Rinse	Rinse	Rinse
Microetch	Microetch	Predip	Activator	Cleaner conditioner	Carrier	Conditioner
Rinse	Rinse	Activator	Rinse	Rinse	Activator	Rinse
Predip	Predip	Rinse	Salt remover	Microetch	Rinse	Predip
Activator	Activator	Accelerator	Rinse	Rinse	Generator	Conductor
Rinse	Rinse	Rinse	Dry	Predip	Rinse	Rinse
Accelerator	Accelerator	Enhancer	Microetch	Activator	Stabilizer	Postdip
Rinse	Rinse	Rinse	Rinse	Rinse	Rinse	Rinse
EE-1 electrolytic flash	Setter	Microetch	Reactivator	Accelerator	Microetch	Dry
Rinse	Rinse	Rinse	Rinse	Rinse	Rinse	
Dry	Dry	Dry	ABC electrolytic flash	Acid Dip	Dry	
			Rinse	Rinse		
			Dry	Dry		

TABLE 30.2 DMT.2
Carbon/graphite systems—conveyorized horizontal

Black Hole II	Shadow
Cleaner	Cleaner conditioner
Rinse	Rinse
Black Hole I	Shadow bath
Dry	Heated dry
Rinse	Inspection chamber
Condition	Microetch
Rinse	Rinse
Black Hole II	Antitarnish
Heated dry	Dry
Microclean	
Rinse	
Antitarnish	
Rinse	
Dry	

TABLE 30.3 DMT.3
Conductive polymer systems—conveyorized, horizontal, shown with optional electrolytic flash

DMS-E	Compact CP
Microetch	Microetch
Rinse	Rinse
Conditioner	Cleaner/conditioner
Rinse	Rinse
Oxidative conditioning	Permanganate
Rinse	Rinse
Catalyzing	Polyconductor
Fixation	Rinse
Rinse	Soft-etch
Acid dip	Rinse
Electrolytic copper	Acid dip
Rinse	Electrolytic copper
Dry	Rinse
	Dry

30.1.7 Horizontal Process Equipment for DMT

Although many DMTs fit easily into existing electroless copper lines, whether manual or automatic, and perform well in a vertical mode, DMT is becoming inextricably linked to horizontal processing in a conveyorized machine. Like so many inventions, necessity was its mother. Certain DMT processes are marginal in the vertical/basket mode, and handling each panel individually was the only solution. The advent of compact and inexpensive horizontal DMT equipment has been a catalyst. Although Atotech pioneered horizontal electroplating equipment with their Uniplate system, until the development of horizontal DMT machines it was not feasible to engineer a horizontal PTH machine. Now, however, for those who want only flash plating, or even full electroplating, the entire process—both PTH and galvanic—may be horizontal, conveyorized, and automated with the advantages of reduced chemical consumption; radically reduced use of rinse water; panel-to-panel uniformity, reliability, and quality; reduction in operating personnel; reduction of handling; a fully enclosed operating environment; and JIT delivery. Stackers/accumulators further minimize handling and optimize efficiency.

30.1.8 DMT Process Issues

Anything brand new experiences some "teething problems" at the beginning, and DMT is no different. But most of the processes referred to in this chapter are in at least their second generation and some are already in their third, so there are no serious contraindications to the use of DMT. However, there are a few caveats:

1. Esoteric base material does not run well with all DMT.
2. Not all DMT performs as well in the vertical mode as in the horizontal.
3. Certain DMT is better suited to the rigors of multilayer production.
4. Some perform better than others with very small holes.
5. Some processes are cleaner than others.
6. Certain DMT is more sensitive than electroless copper to organic contamination in electrolytic copper plating.

7. Rinsing is important, and some DMT needs special rinsing.

8. Rework is simple in most DMT, but it can be abused, i.e., it changes the cost structure.

9. Quality assurance tools for DMT (how do you know that holes will plate void-free?) are being developed, but at the moment, the only absolutely certain method is to flash electroplate, because, unlike electroless copper, there is not much to see in the hole after DMT.

10. Some DMT analyses and operating controls are rudimentary.

30.1.9 DMT Process Summary

Strong ecological and health reasons will drive the use of DMT for the manufacture of PTH boards. Since the cost of PTH metallization constitutes only 2 to 3 percent, at most, of the total process cost of a PCB, the direct saving resulting from DMT is very small. However, the fringe benefits of this technology are rather significant: water conservation, no noxious chemicals, minimal panel handling, reduced waste treatment, lower labor costs, fewer rejects, etc.

REFERENCES

1. Radovsky, U.S. Patent 3,099,608, 1963.
2. Morrissey, et al. (Amp/Akzo), U.S. Patent 4,683,036, July 1987.
3. Okabayashi (STS), U.S. Patent 4,933,010, June 1990.
4. Gulla, et al. (Shipley), U.S. Patent 4,810,333, March 1989.
5. Holtzman, et al. (APT), U.S. Patent 4,891,069, January 1990.
6. Stamp, et al. (Atotech), PCT WO 93/17153, September 1993.
7. Minten, et al. (MacDermid), U.S. Patent 4,724,005, February 1988.
8. Thorn, et al. (Electrochemicals), U.S. Patent 5,389,270, February 1995.
9. Blasberg, Europatent 0489759.
10. Bressel, et al., U.S. Patent 5,183,552, 1993.

CHAPTER 31
PWB MANUFACTURE USING FULLY ELECTROLESS COPPER

Dr. Hayao Nakahara
N.T. Information Ltd., Huntington, New York

31.1 FULLY ELECTROLESS PLATING

Fully electroless plating has been recognized as a viable technology for some time. It is especially useful for the formation of fineline conductors and considered excellent for plating small, high-aspect-ratio holes because of its high throwing power when compared with that of galvanic plating (see Fig. 31.1). However, its use was limited to the manufacture of double-sided and simple multilayer printed wiring boards (PWBs) for some time after the commercial introduction of the additive process called CC-4 began at Photocircuits Corporation in 1964. This was due to some relatively poor physical properties of electrolessly deposited copper, such as elongation of 2 to 4 percent, compared to 10 to 15 percent achieved by galvanically deposited copper.

The view on electroless plating technology began to change in the mid-1970s when IBM decided to utilize the technology for the fabrication of multilayer boards (MLBs) to package its then top-of-the-line mainframe computers.[1] IBM and other PWB makers using fully electroless copper plating have continuously improved the properties of electrolessly deposited copper since the early 1980s. IBM has continued to use the technology for the fabrication of more advanced MLBs for mainframe and supercomputers.[2,3] Stimulated by IBM's work, and because of technical necessity, NEC Corporation and Hitachi Ltd. of Japan also applied electroless plating technology for the fabrication of MLBs for their mainframe and supercomputers.[4,5]

Today, electrolessly deposited copper is considered as reliable as galvanically deposited copper, and fully electroless plating technology is finding its way to many applications.[6–20] In this chapter, we will discuss various methods of PWB fabrication by means of fully electroless copper-plating technology.

PWB fabrication technology using electroless plating is often referred to as *additive technology*. Therefore, throughout this chapter, the words *electroless* and *additive* will be used interchangeably. Electroless copper plating for through-hole metallization, which deposits a thin film of copper on the wall of plated-through-holes (PTHs), typically from 0.3 to 3 μm thick, is a technology different from the one under consideration and will not be discussed in this chapter (see Chap. 29, "Electroplating").

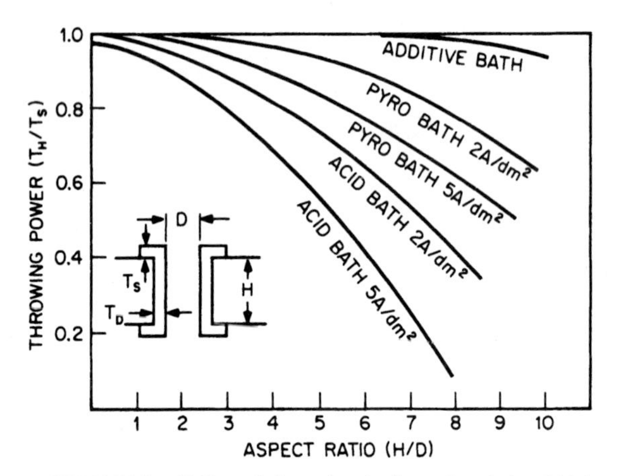

FIGURE 31.1 Ability of electroless bath to plate hole of high aspect ratio without reducing throwing power.

31.2 THE ADDITIVE PROCESS AND ITS VARIATIONS

There have been many variations to additive processes,[20,21] but there are three basic additive processes commercially practiced, as illustrated in Fig. 31.2:

1. Pattern-plating additive
2. Panel-plate additive
3. Partly additive methods

31.3 PATTERN-PLATING ADDITIVE

Pattern-plating additive methods can be classified further into three different approaches, as described in the following sections, depending on the base substrates used.

31.3.1 Catalytic Laminate with CC-4®*[22]

CC-4 stands for *copper complexer number 4,* EDTA, the fourth complexing agent successfully tried by Photocircuits in the early 1960s for full-build electroless copper-plating solution. Over time, the term CC-4 has been used frequently as an adjective, such as in "CC-4 process" or "CC-4 bath."

31.3.1.1 Process Steps. The CC-4 process starts with catalytic laminates coated with catalytic adhesive. The process sequence is as follows:

1. Catalytic base laminate coated with catalytic adhesive (both sides)
2. Hole formation

* CC-4 is a registered trademark of AMP-AKZO Corporation.

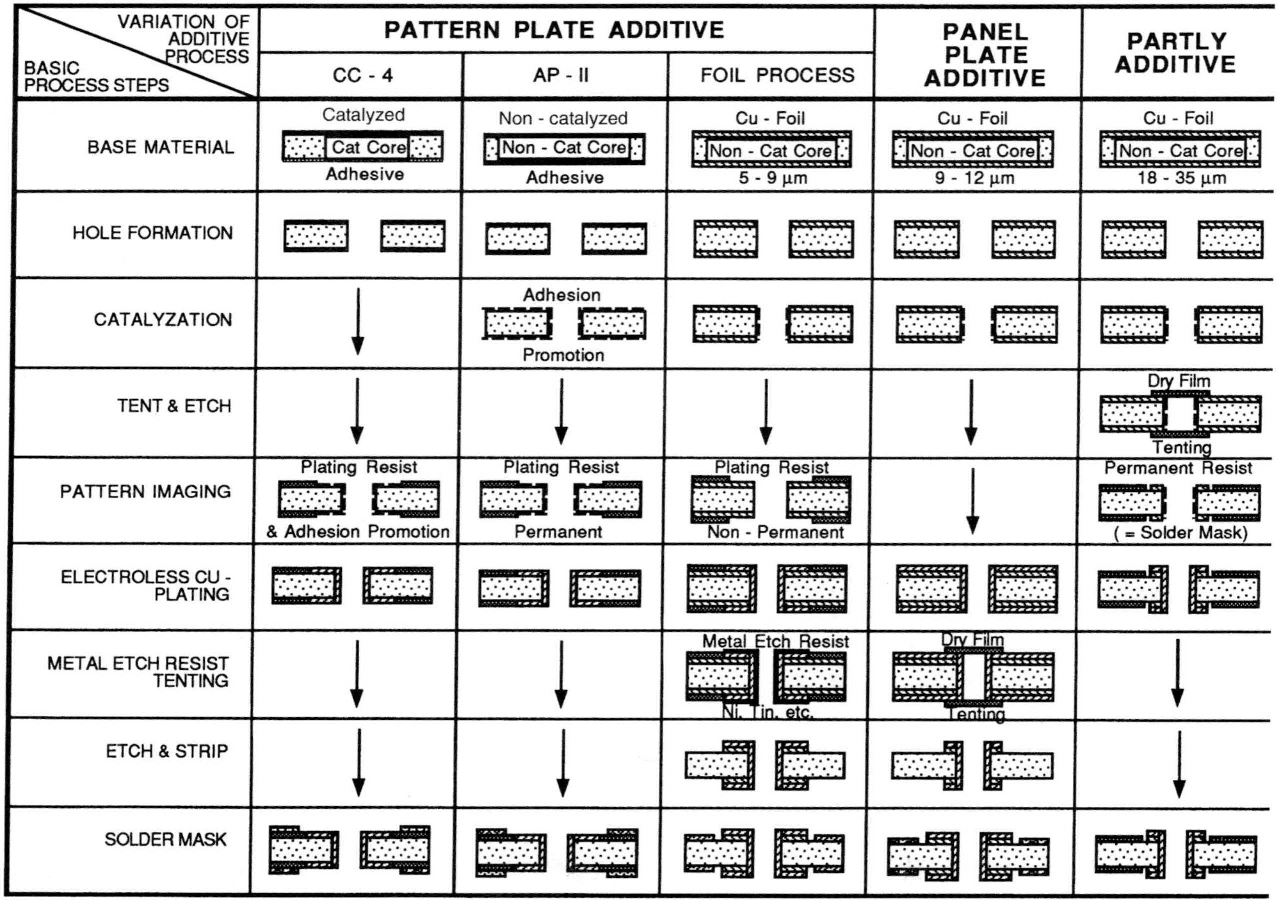

BASIC PROCESS STEPS / VARIATION OF ADDITIVE PROCESS	PATTERN PLATE ADDITIVE			PANEL PLATE ADDITIVE	PARTLY ADDITIVE
	CC - 4	AP - II	FOIL PROCESS		
BASE MATERIAL	Catalyzed Cat Core Adhesive	Non - catalyzed Non - Cat Core Adhesive	Cu - Foil Non - Cat Core 5 - 9 μm	Cu - Foil Non - Cat Core 9 - 12 μm	Cu - Foil Non - Cat Core 18 - 35 μm
HOLE FORMATION					
CATALYZATION	↓	Adhesion Promotion			
TENT & ETCH	↓	↓	↓	↓	Dry Film Tenting
PATTERN IMAGING	Plating Resist & Adhesion Promotion	Plating Resist Permanent	Plating Resist Non - Permanent	↓	Permanent Resist (= Solder Mask)
ELECTROLESS CU - PLATING					
METAL ETCH RESIST TENTING	↓	↓	Metal Etch Resist Ni, Tin, etc.	Dry Film Tenting	↓
ETCH & STRIP	↓	↓			↓
SOLDER MASK					

FIGURE 31.2 Variation of additive process.

3. Mechanical abrasion of the adhesive surfaces for better adhesion of plating resist
4. Application of plating resists (screening ink or dry film resist)
5. Formation of microporous structure of exposed adhesive surfaces by chemical etching in acid solution (CrO_3/H_2SO_4 or CrO_3/HBF)
6. Fully electroless copper deposition on the conductor tracks and hole walls
7. Panel baking
8. Application of solder mask and legends
9. Final fabrication and test

31.3.1.2 Resist Issues. When the conductor width is 8 mil (0.2 mm) or wider, image transfer (step 4) for a majority of CC-4 boards is still done by screen printing with thermally curable or UV-curable ink, since the cost of imaging by screening method is less than one-third that of dry film imaging. The plating resist is permanent; that is, it stays permanently as an integral part of the board. UV-curable ink is preferred because of its shorter curing time that tends to minimize lateral ink flow while being cured. Lateral ink flow narrows the conductor width since the conductor is formed in the trench between plating resists. See Fig. 31.3.

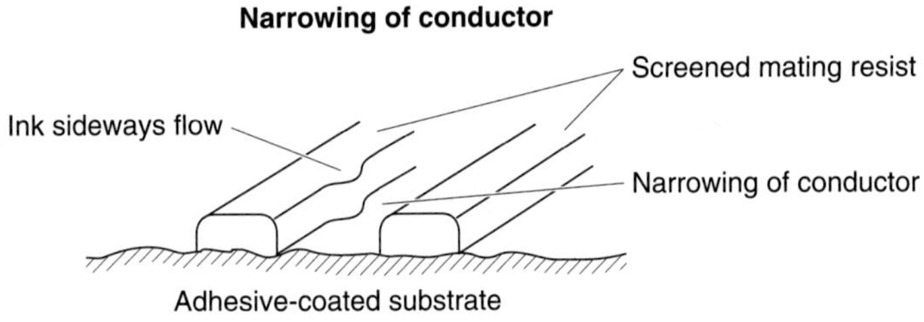

FIGURE 31.3 Narrowing of conductor due to lateral flow of plating resist after application.

31.3.1.3 Typical Plating Conditions. Although the practical limitation of screen printing conductor pattern is 8 mil (0.2 mm), in a mass production environment, 6-mil (0.15 mm) conductor patterns have been successfully screened, but such a pattern is confined to a limited area of the panel. Therefore, a conductor pattern which is less than 6 to 7 mil is normally formed by dry film resist. E.I. du Pont de Nemours & Co. and Hitachi Chemical Co., Ltd., are two major suppliers of permanent dry film plating resists which withstand the hostile plating bath environment. Typically, additive bath temperatures range from 68 to 80°C and pH from 11.8 to 12.3, and panels must stay in the bath for more than 10 h with plating speed of typically 2.0 to 2.5 μm/h.

Since the use of solvent such as 1-1 trichloroethane is being banned, semiaqueous-developable permanent dry film resist has been developed by Hitachi Chemical to withstand the plating condition as previously described and is used extensively in Japan.

Since paper-based laminates are still the dominant materials in Japan (although CEM-3 and FR-4 laminates have been used increasingly in recent years), some allowance for this material in the plating solution is made. Therefore, plating baths for paper-based laminates are usually maintained at the lower end of the operating temperature (68 to 72°C) in order not to delaminate them.

31.3.1.4 Permanent Plating Resist. The use of a permanent plating resist creates a flush surface, where the copper and resist are at the same height from the base laminate. This offers two distinct benefits. In the board fabrication process, after the conductors are formed, the

FIGURE 31.4 Flush surface resulting from additive plating with permanent plating resist. *(Photo courtesy of Ibiden Co. Ltd.)*

chance of their destruction due to scratch is minimized. (See Fig. 31.4.) In addition, the flush surface makes the application of solder mask easier and reduces the consumption of solder resist by as much as 30 percent when compared to that incurred in a conventional subtractive, etched foil process. In the board assembly process, a permanent plating resist acts as the solder resist and tends to minimize solder bridges during the soldering operation, particularly for fine-pitch devices.

31.3.2 Noncatalytic Laminate with AP-II[22]

AP-II is an abbreviation of Additive Process II developed by Hitachi Ltd. This process also makes use of permanent plating resist. However, unlike the CC-4 process, it starts with noncatalytic laminate coated with noncatalytic adhesive. Its process sequence is as follows.

31.3.2.1 Process Steps

1. Noncatalytic laminate coated with noncatalytic adhesive
2. Hole formation
3. Mechanical abrasion of adhesive surfaces followed by chemical roughening (adhesion promotion)
4. Catalyzation of the surfaces and hole walls
5. Application of plating resists

The rest of the process is essentially the same as the CC-4 process (steps 6 through 9).

31.3.2.2 Advantages/Disadvantages. This catalyzing version of the pattern-plating additive process has one advantage over the CC-4 process, that is, the use of noncatalytic base laminates which are cheaper than catalyzed laminates. On the other hand, this process has a disadvantage also in that pattern repair before plating is not possible because resist which protrudes sideways and makes the conductors narrower cannot be removed since the Pd catalyst underneath the protruding part of the resist tends to be removed as well if repair is attempted, resulting in plating voids. However, in a really fineline conductor case such as 4 mil (0.1 mm) or below, repair attempts will not be made; hence, this disadvantage does not usually impede this catalyzing version of the pattern-plating additive process.

31.3.2.3 AP-II Variation AAP/10. A variation of the AP-II process called the AAP/10 process has been developed and practiced by Ibiden Co., Ltd., of Japan to fabricate high-layer-count MLBs with fineline conductors down to 3 to 4 mil.[23,24] The AAP/10 process is essentially the same as AP-II except that the adhesive used for the AAP/10 process has better insulation characteristics, suitable for higher performance.

31.3.3 Foil Process

Although an adhesive-coated surface provides adequate insulation characteristics for most applications, some boards, such as ones used for mainframe computers, require higher insulation resistance than adhesives can render. When IBM decided to use fully electroless plating to form fineline conductors on innerlayers of high-aspect-ratio MLBs in the mid-1970s, it took an approach different from the adhesive system.[1]

31.3.3.1 Process Steps. The process developed at IBM is as follows:

1. Copper-clad laminate (5-μm copper foil).
2. Drill holes.
3. Treat copper surface with benzotriazol.
4. Catalyze the panel.
5. Dry film laminate for conductor pattern (reverse pattern).
6. Do fully electroless copper deposition on the conductor track.
7. Metal overplate as etch resist (tin-lead, tin, nickel, etc.).
8. Strip dry film and etch copper.
9. Strip metal etch resist.
10. Do oxide treatment for subsequent lamination.

31.3.3.2 Process Issues. This foil process requires strong adhesion of dry film plating resist on copper surface. Benzotriazol treatment of the copper surface improves adhesion of dry film. However, the most crucial step is cleaning the copper surface (virgin copper) to obtain an optimum pH for best reaction between copper and benzotriazol.

Hitachi Ltd. found that the use of compounds having amino and sulfide groups is effective in suppressing delamination between copper surface and plating resist.[21] It uses dry film resist which contains a trace of benzotriazol rather than treating the copper surface with the chemical. Baking the panel after exposure/development of dry film at about 140°C for 1 h is found to improve the adhesion between copper surface and dry film during plating operation and at the same time prevents this chemical from breaching out into the plating bath, which could poison the bath.

Table 31.1 shows representative high-tech MLBs fabricated by means of additive technology. Innerlayer patterns of all MLBs listed in the table are made by the foil process. The outerlayer patterns are formed by the panel-plate additive process, which is the subject to be discussed in the following section.

TABLE 31.1 Large-Scale MLBs

Computer	IBM ES-9000	NEC ACOS-3900	Hitachi M-880
Materials	Brominated epoxy glass	Polyimide glass	Maleimide styryl
Dielectric C	NA	4.6	3.6
Size, mm	600 × 700 × 7.4	446 × 477 × 8.0	534 × 730 × 7.1
No. layers	22	42	46
Cond. width, μm	81	70	70
Cond. height, μm	43	30	65
PTH dia., mm	0.46	0.6/0.3	0.56/0.3
Aspect ratio	16/1	27/1	23/1
No. of PTHs	42,000	17,000	100,000
No. of IVHs	8,000	11,000	NA
Innerlayer cond. formation	Additive pattern 5-μm copper	Additive panel 12-μm copper	Additive pattern 12-μm copper
Outerlayer formation	Additive panel	Additive panel	Additive panel
Impedance, Ω	80 + 10, −9	60 ± 6	NA

Source: N.T. Information Ltd.

31.4 PANEL-PLATE ADDITIVE

Maintaining uniform thickness on the panel surface in a galvanic plating operation is difficult. As the aspect ratio becomes higher, maintaining uniform thickness on the hole wall is also a problem.

Electroless plating offers two distinctively superior features in this respect:

1. Its excellent throwing power for high-aspect-ratio holes (Fig. 31.5)
2. Its ability to deposit copper film of even thickness over the entire circuit panel.

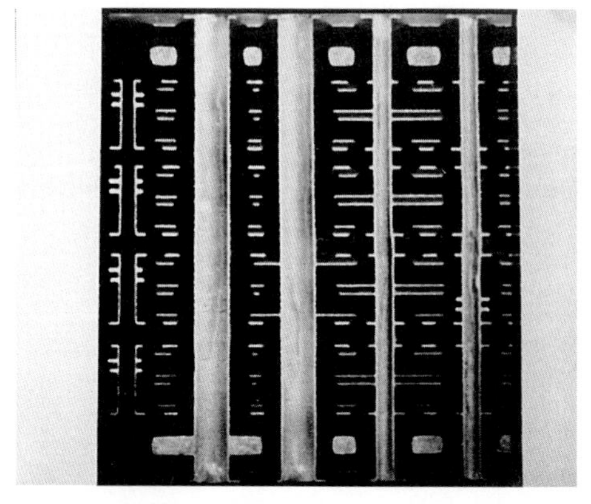

FIGURE 31.5 Excellent hole coverage is achieved by fully additive plating. This shows the cross section of a hole with a 27:1 aspect ratio. *(Photo courtesy of Hitachi Ltd.)*

These features of fully electroless plating are ideal for the fabrication of high-tech MLBs, as described in Table 31.1. All signal conductors of the MLBs are formed on innerlayers and the outerlayer surfaces provide pads only for through holes and some bonding pads for decoupling capacitors. The process steps of the panel-plate additive method are essentially the same as those of the galvanic panel-plate method:

31.4.1 Process Steps

1. Copper-clad laminate (thin copper foil is preferred).
2. Perform hole formation.
3. Catalyze holes.
4. Perform fully electroless copper deposition (panel-plate).
5. Perform dry film tent-and-etch.

31.4.2 Process Issues

Tenting with dry film becomes difficult when annular rings are very small or in some cases when there are no annular rings at all. In such cases, a positive electrodeposition system may be applied to overcome this difficulty.

It should be cautioned, however, that there is a limitation to this panel-plate additive approach, even though the thickness of deposited copper film is much more even across the panel. If very fine conductors are to be formed by this method, the conductor cross section becomes intolerably distorted from the ideal rectangular shape as the ratio of the width to the height of the conductor approaches unity, as shown in Fig. 31.6.

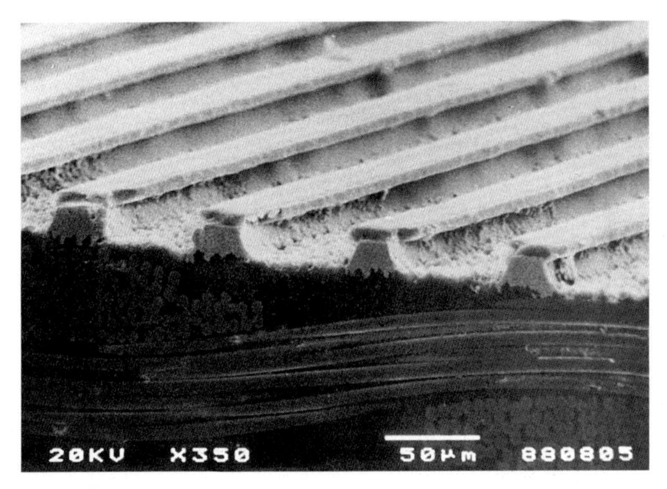

FIGURE 31.6 Formation of fineline conductors by etching process results in undesirable, nonrectangular conductor cross section. *(Photo courtesy of Ibiden Co., Ltd.)*

31.5 PARTLY ADDITIVE

As in the case of the panel-plate additive method, the partly additive process starts with copper-clad laminate.[11–14] The essence of this process is to minimize the problem encountered in etching fineline conductors through thick copper which results from through-hole plating (in the conventional etched foil process, through-hole is plated first), by forming conductors first, prior to through-hole plating.

31.5.1 Process Steps

1. Copper-clad laminate.
2. Perform hole formation.
3. Catalyze hole walls and remove catalyst sites from the surfaces by brushing.
4. Perform image transfer by screen print or dry film tenting.
5. Form conductor pattern by etching and stripping the resist.
6. Apply plating resist over the entire surface, but leaving pads and holes uncovered.
7. Perform electroless copper deposition onto the pads and hole walls.

31.5.2 Process Issues

By proper selection of catalysts and etching solutions, the catalyst on the hole walls survives during the etching operation, posing no problem in the subsequent plating step.

In step 6, plating resist can be applied by screen printing when the circuit density is not so high. When the density becomes higher, dry film or liquid photoimageable resist can be used. Liquid resist can be applied by open screen coating, roller coating, or curtain coating.

In some cases, step 6 can be bypassed when sideways line growth is not a problem (Fig. 31.3). In such a case, the plating speed is reduced by 20 to 30 percent, from the normal speed of 2.0 to 2.5 μm, in order to avoid copper deposition on unwanted areas (bare laminate areas). Conductors must be made narrower than the intended final width at the design stage to compensate for the side growth.

31.5.3 Fine-Pitch Components

As the lead pitch of multipin devices becomes narrower than 20 mil (0.5 mm), bridging between adjacent pads during plating can occur even when plating resist is placed between them. As a result, the practitioners of the partly additive process are abandoning this process and converting to the panel-plate additive method based on the experiences they gained from practicing partly additive technology.

31.6 CHEMISTRY OF ELECTROLESS PLATING

Chemical reactions which take place in electroless plating have been treated well in the literature.[4,21] However, they are discussed again here because some of the due comments are relevant to the chemistry of electroless plating.

31.6.1 Electroless Plating Solution Chemical Reactions

Main ingredients of most electroless plating bath formulations consist of

NaOH	pH adjustment
HCHO	Reducing agent (formaldehyde)
$CuSO_4$	Source of copper iron
EDTA	Cheleting agent

31.6.1.1 Copper Deposition. The chemical reaction of copper deposition may be represented by

$$\begin{gathered} Cu(EDTA)^{2-} + 2HCHO + 4OH^- \rightarrow \\ Cu^0 + H_2 + 2H_2O + 2CHOO^- + EDTA^{4-} \end{gathered} \tag{31.1}$$

Under the alkaline pH condition, the cupric ions would normally combine with OH^- to produce cupric hydroxide [$Cu(OH)_2$], a useless precipitate. When a cheleting agent such as EDTA is added, it prevents cupric hydroxide formation by maintaining the Cu^{2+} in solution. Once the deposition of metallic copper starts through catalytic sites, the reaction in Eq. (31.1) continues because of the autocratic nature of the reaction.

31.6.1.2 Cannizzaro Reaction and Formaldehyde Concentration. While this main reaction, represented by Eq. (31.1), takes place, other undesirable side reactions proceed, also in competition with the reaction. One of the major difficulties is the lowering of formaldehyde concentration in the solution due to its disproportionate consumption in alkaline solution. This is known as the *Cannizzaro reaction*, which may be characterized by

$$2HCHO + OH^- \rightarrow CH_3OH + HCOO^- \tag{31.2}$$

This reaction continues independently. Fortunately, methanol (CH_3OH) one of the by-products of the Cannizzaro reaction, tends to shift the equilibrium of the reaction in Eq. (31.2) to the left and thus prevents the decrease of formaldehyde concentration by an unproductive reaction. By controlling the plating conditions properly, the wasteful consumption of formaldehyde by the Cannizzaro reaction can be retained within 10 percent of the consumption in the reaction in Eq. (31.1).

31.6.1.3 Fehlings-Type Reactions. In addition to the Cannizzaro reaction, the following side reaction also competes for formaldehyde:

$$2Cu^{2+} + HCHO + 5OH^- \rightarrow Cu_2O + HCOO^- + 3H_2O \tag{31.3}$$

31.6.1.4 Other Formaldehyde-Related Reactions. Under conditions which would favor the Fehlings-type reaction [Eq. (31.3)], spontaneous decomposition of uncatalyzed plating solutions produces precipitation of finely divided copper with attended vigorous evolution of hydrogen gas. The finely divided copper produced is due to the disproportionate quantity of Cu_2O under alkaline conditions:

$$Cu_2O + H_2O \rightarrow Cu^0 + Cu^{2+} + 2OH^- \tag{31.4}$$

Furthermore, formaldehyde may act as a reducing agent for the cuprous oxide to produce metallic copper:

$$Cu_2O + 2HCHO + 2OH^- \rightarrow 2Cu^0 + H_2 + 2HCOO^- + H_2O \tag{31.5}$$

31.6.1.5 Filtering. The Cu^0 nuclei thus produced according to Eqs. (31.4) and (31.5) are not relegated to deposition on substrate but are produced randomly throughout the solution and become the catalytic sites for further undesirable copper deposition. Continuous filtration with a filter of pore size 20 μm or smaller can improve extraneous copper due to these catalytic sites.[25] The use of continuous filtering also improves the ductility of copper by eliminating codeposition of impurities. In most mass production facilities of PWBs by additive technology, two or three stages of filtering with filters of pore sizes down to 5 μm are common.

31.6.2 Use of Stabilizers

Since the reaction represented in Eq. (31.3) indicates the greatest degree of instability of the plating bath, various measures are taken to counteract this reaction. Alkaline cyanide has been a popular stabilizer used for the CC-4 bath, since cyanide forms strong complexes with Cu^+ ($Cu^+ + 2CN^- \rightarrow Cu(CN)^{2-}$), but relatively unstable complexes of Cu^{2+}. However, since cyanide also reacts with HCHO, it is difficult to control. 2,2′-dypiridil also chelates Cu^+ and does not react with HCHO. Hence, it is a more favored stabilizer, particularly for those baths used in Japan. On the other hand, a dypiridil bath deposit begins to have less ductility after several plating cycles and weekly bath makeup may be necessary.

Aeration of the plating bath is known to stabilize the solution, and vigorous aeration is commonly used to operate modern full-build electroless baths.

31.6.3 Surfactant

Reduction in surface tension assures that the plating solution is in intimate contact with the catalyst nuclei and permits thorough surface solution interaction. For this reason, a wetting agent or surfactant is included in bath formulations. Polyethelene oxide, polyethelene glycol (PEG), etc., are popular surfactants. Proper usage of surfactants also prevents the accumulation of gaseous material (air and hydrogen bubbles) on the surface of the board, particularly around the entrances of holes, which would also tend to lower the surface-solution interaction and thus seriously impair proper metal deposition.

31.6.4 Reliability of Deposited Copper and Inorganic Compounds

Generally, copper film with high elongation and reasonable tensile strength is preferred for through-hole reliability, although the relation between through-hole reliability and ductility and tensile strength is still not understood clearly.

The use of inorganic compounds containing vanadium or germanium as a part of chemical ingredients in plating baths enhances the quality of deposited copper.[13] Most full-build electroless baths used today contain either V_2O_5 or G_2O_5 to improve the physical properties of deposited copper.

31.6.4.1 Typical Copper Properties. Typical physical properties of reliable electrolessly plated copper are (as plated):

Elongation	8 to 10 percent
Tensile strength	35 to 45 kg/mm^2

After annealing the copper, elongation increases to 12 to 18 percent and tensile strength decreases slightly. These properties are equivalent to the ones resulting from acid copper. Since all panels are baked at about 140 to 160°C for 30 to 60 min during the course of the production process after plating, deposited copper gets annealed, thus assuring high through-hole reliability.

31.6.5 Removal of Impurities

Use of continuous filtering has been mentioned to remove some of the impurities. In addition there are other methods.

31.6.5.1 Electrodialysis. Electrodialysis[26,27] is effective to remove formate ions, sulfate ions, and carbonate ions which are by-products of reactions defined in Eqs. (31.1), (31.2), (31.3), and (31.5).

31.6.5.2 Overflow Method. Another very effective way to remove these unwanted by-products is the *overflow* method. Since essential components of the plating solution are consumed continuously, they must be replenished to maintain desirable plating conditions. In so doing, a considerable amount of the solution overflows because of the addition of water. The overflow also contains the reaction by-products as well as chelated copper ions.

After a certain amount of overflow solution is accumulated, sodium hydroxide and formaldehyde are added to the overflow solution and heated to promote spontaneous decomposition of the plating solution to precipitate copper metal. Addition of copper powder accelerates the precipitation. This so-called *chemical bomb-out* recovery method can be operated in bulk or continuously. The remaining solution can be treated with sulfuric acid to recover over 95 percent of the EDTA in the solution. Usually, the purity of EDTA thus recovered is higher than that of purchased EDTA.

Ninety-eight percent of the solution is recovered as permeate and is used in such processes as rinsing. The concentrate, the other two percent, is high in salt content (Na_2SO_4 and HCOONa) and may be evaporated to produce a small amount of solid waste or discharged into the ocean, if the plant is located near the ocean and locally approved.[6] This chemical bomb-out method can achieve two objectives at once: removal of unwanted impurities and recovery of chemicals. With this system, it is easy to maintain Baume at the desired 9°.

31.6.6 Environmental Issues of Formaldehyde

Formaldehyde is considered to be a carcinogen. If exposed to formaldehyde fumes of certain concentration for a long time, it can cause allergy conditions and perhaps cancer, although cancer due to long-time exposure to formaldehyde in the PWB industry has never been reported.

With a good exhaust system and scrubbing through water and peroxide solution, formaldehyde can be made quite harmless. It is a matter of economy. EDTA can be treated as previously mentioned.

Reducing agents alternative to formaldehyde such as glyoxylic acid[27] and hypophosphite (M_3PO_2)[28] have been suggested and tried, but they have not been able to replace formaldehyde as an effective reducing agent at this stage.

31.7 FULLY ELECTROLESS PLATING ISSUES

31.7.1 Efficiency of Electroless and Galvanic Plating

PWB manufacturers which are not familiar with electroless plating often have the misconception that electroless plating takes "too long." On the contrary, electroless plating is faster than galvanic plating, and its volumetric efficiency is twice as much as galvanic plating.

Electroless plating baths operate at the speed of 2.0 to 5.0 μm/h, 2.5 μm being the medium speed. Because of its high throwing power, a deposition thickness of 25 μm on the panel surface also assures 25 μm in the hole. That is, 10 h are needed to plate this thickness. The loading factor in electroless plating can be maintained at 4 dm^2/liter. A 2500-gal (10,000-l) tank can plate about 400 m^2 (4000 ft^2) of panels in one shot in 10 h. That is, 400 ft^2/h, or approximately 130 (18-in × 24-in) panels/h from a 10-m^3 tank. This is equivalent to twice the speed of galvanic plating from the same volume of plating solution. When small, high-aspect-ratio holes are involved, this figure becomes more favorable to electroless plating because the current density of galvanic plating must be reduced to 10 to 12 A/ft^2, which requires more than 2 h of plating for 25 μm in the hole.

Figure 31.7 shows the plating racks filled with panels waiting to be placed in electroless plating tanks. In large-volume operations, the panel-racking operation is fully automated.

31.7.2 Photochemical Imaging Systems

Various photochemical imaging systems have been developed and tried to be put into practice.[6] However, no photochemical imaging systems are used for real production today. They are all unstable and the lateral growth of conductors poses a problem for the fineline applications for which they are intended.

31.7.3 Molded Circuit Application

A handful of three-dimensional molded circuit manufacturers (or, the preferred identification, *molded interconnection device* or MID) have continued their quest for market with this

FIGURE 31.7 Plating racks filled with panels to be plated. *(Photo courtesy of Adiboard, Brazil.)*

product and have met with some success. (See Fig. 31.8.) There are several alternative methods to manufacture MIDs, of which the *two-shot* molding method can produce the most sophisticated three-dimensional interconnection devices.

A part which is to be metallized (circuitized) is molded first. Then, a portion of the first mold which is not desired for metallization is surrounded by the second molding, exposing

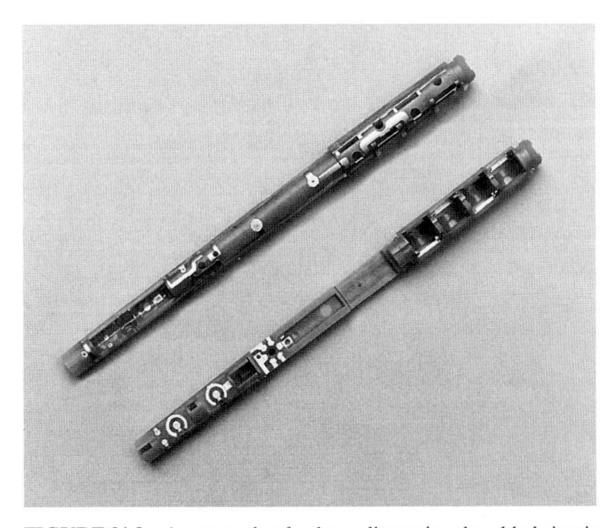

FIGURE 31.8 An example of a three-dimensional molded circuit application, in this case a light pen for computer input. *(Photo courtesy of Mitsui Pathtek.)*

only the portion desired to be metallized. This two-shot molding method can be classified further into two alternative methods, which are similar to CC-4 and AP-II.

One is to use catalyzed engineering plastic for the first molding and noncatalyzed plastic for the second molding. Adhesion promotion consists of a swell-and-etch step followed by chromic acid etch. The other alternative is to mold the first part with noncatalyzed plastic materials and expose the molded part to an adhesion promotion process and catalyze. The second part is then molded, as in the first case.

These parts are placed in an electroless plating bath for circuitization. A high rate of growth is expected for MIDs, with the automotive market being the most promising application area.

REFERENCES

1. W. A. Alpaugh and J. M. McCreary, IPC-TP-108, Fall Meeting, 1976.
2. R. R. Tummala and S. Ahmed, "Overview of Packaging for the IBM Enterprise System 9000 Based on the Glass-Ceramic Copper/Thin Film Thermal Conduction Module," *IEEE Trans. on Comp., Hybrids and Mfg. Tech.,* Vol. 14, No. 4, Dec. 1991, pp. 426–431.
3. L. E. Thomas, et al., "Second-Level Packaging for the High End ES/9000 Processors," *IBM J. of Research & Development.*
4. H. Murano and T. Watari, "Packaging Technology for the NEC SX-3 Supercomputers," *IEEE Trans. on Comp., Hybrids and Mfg. Tech.,* Vol. 15, No. 4, Aug. 1992, pp. 411–417.
5. A. Takahashi, et al., "High Density Multilayer Printed Circuit Board for HITAC M-880," *op. cit.,* pp. 418–425.
6. Clyde F. Coombs, Jr. (ed.), *Printed Circuits Handbook, 3d ed.,* McGraw-Hill, New York, 1986, Chap. 13.
7. U.S. Patent 2,938,805, 1960.
8. U.S. Patent 3,628,999, Dec. 1971.
9. U.S. Patent 3,799,802, Mar. 1974.
10. U.S. Patent 3,799,816, Mar. 1974.
11. U.S. Patent 3,600,330, Aug. 1971.
12. Japanese Patent 43-16929, July 1968.
13. Japanese Patent 58-6319, Feb. 1983.
14. H. Steffen, "Additive Process," Ruwel Werke GmbH, private communication.
15. K. Minten and J. Toth, "PWB Interconnect Strategy Using a Full Build Electroless Plating System: Part I," *PC World Conference V,* Paper No. b-52, Glasgow, Scotland, June 1990.
16. K. Minten, J. Seigo, and J. Cisson, "Part 2: Etch Characteristics," *Circuit World,* Vol. 18, No. 4, Aug 1992, pp. 5–12.
17. K. Minten and J. Cisson, "Part 3: Characterization of a Semi-Additive Naked Palladium Catalyst," *Circuit World,* Vol. 19, No. 2, Jan. 1993, pp. 4–13.
18. K. Minten, K. Kitchens, and J. Cisson, "Part 4: The Future of FBE Process," private communication.
19. N. Ohtake, *Electronics Technology* (Japanese), 27(7), June 1985, pp. 55–59.
20. N. Ohtake, "New Printed Wiring Boards by a Partly Additive Process," *PC World Conference IV,* Tokyo, Paper No. 44, June 1987.
21. S. Imabayashi, et al., "Partly-Additive Process for Manufacturing High-Density Printed Wiring Boards," *IEEE Trans.* 0569-5503/92/0000-1053, 1992.
22. H. Akaboshi, "A New Fully Additive Fabrication Process for Printed Wiring Boards," *IEEE Trans. on Comp., Hybrids and Mfg. Tech.,* Vol. CHMT-9, No. 2, June 1986.
23. R. Enomoto, et al., "Advanced Full-Additive Printed Wiring Boards Using Heat Resistant Adhesive," *PC World Conf V,* Paper No. B6-1, Glasgow, Scotland, June 1990.
24. K. Ikai, "Full Additive Boards," *Electronic Materials* (Japanese), Oct. 1993, pp. 58–63.

25. H. Honma, K. Hagihara, and K. Kobayashi, "Factors Affecting Properties of Electrolessly Deposited Copper," *Circuit Technology* (Japanese), Vol. 9, No. 1, 1994, pp. 31–36.
26. T. Murata, "Chemistry of Electroless Plating," *Circuit Technology* (Japanese), Vol. 8, No. 5, 1993, pp. 357–362.
27. J. Darken, "Electroless Copper—An Alternative to Formaldehyde," *PC World Conf V,* Glasgow, Scotland, Paper B/6/2, June 1990.
28. J. P. Pilato and Joe D'Ambrisi, "Process for Metallization by Copper Deposition Without the Use of Formaldehyde," MacDermid paper, private communication.
29. R. E. Horn, *Plating and Surface Finishing,* Oct. 1981, pp. 50–52.
30. A. Iwaki, et al., *Photographic Science & Engineering,* Vol. 20, No. 6, 1976, p. 246.

CHAPTER 32
PRINTED CIRCUIT BOARD SURFACE FINISHES

Don Cullen
MacDermid, Incorporated, Waterbury, Connecticut

32.1 INTRODUCTION

32.1.1 Purpose and Functions of Surface Finishes

The printed circuit board (PCB) conductor surface finish forms a critical interface between the component and the interconnection circuitry. As its most essential function, the final finish process is intended to provide exposed copper circuitry with a protective coating in order to preserve solderability. In a more expanded view, the final finish coating must meet dozens of functional criteria, including solderability, environmental, electrical, physical, and durability demands, as summarized in this chapter. Perhaps no other step in the PCB manufacturing process has undergone more change in the era of surface mount manufacturing than the final finish chemical process. A wide variety of functional, environmental, engineering, cost, productivity, and failure mode issues have contributed to this frequent change.

32.1.2 Impact of the Lead-Free Transition

The impact of lead-free manufacturing, as a result of the Waste Electrical and Electronic Equipment (WEEE) directive and Restriction on Hazardous Substances (RoHS), has changed surface finish specifications for PCBs and the soldering of those surfaces at assembly.

Lead-free assembly has changed just about every aspect of the surface finish technology. The elimination of lead (Pb) in most designs has rendered the standard surface finish hot air solder level (HASL) increasingly obsolete. At assembly, the use of new solder alloys affects the compatibility of the finishes with the metallurgy of the final solderjoint. Many of the assembly issues related to surface finishes, such as surface-mount technology (SMT) pad wettability, plated through hole (PTH) fill, and voiding are exacerbated by the Pb-free assembly process. Finally, the temperature of Pb-free soldering has a major impact on the shelf-life and solderability of the Pb-free finishes.

32.1.3 Technology Drivers

Each company in the manufacturing supply chain has needs specific to its manufacturing step. The methods used to achieve these goals have evolved with developments in PCB technology. Historically, technicians were able to solder components directly to copper using very active fluxes. When PCBs were first made in large production quantities, chemists devised ways to protect the copper using metal or organic coatings. Electroplated tin-lead became a popular coating as a metallic layer that could be subsequently reflowed. Design changes forced the era of solder mask over bare copper (SMOBC). With SMOBC, reflowed plated tin-lead was no longer possible because the tin-lead liquefies during the soldering process. HASL, the successor to plated tin-lead, became the surface finish of choice worldwide. Increasing PCB circuit density inspired the reemergence of organic coatings, such as new formulation organic solderability preservatives (OSPs) and metallic layers such as electroless nickel/immersion gold (ENIG.) The environmental initiative to eliminate ozone-depleting chemicals led to the next change in surface finish preference. Use of no-clean assembly operations prompted a decline in OSP and an increase in ENIG. Both ENIG and OSP continue to be widely used, but each has its own weakness, allowing for the proliferation of new low-cost, high-functional immersion metal finishes such as immersion silver and immersion tin.

32.1.3.1 Fabrication Requirements. The company producing the bare printed circuit board—that is, the PCB fabricator—has needs that are specific to that business, as follows:

- Low chemical and process cost
- Simple operation (low engineering demands)
- Reworkability
- Easily measured thickness
- Conveyorized equipment
- Electrically testable surface
- No sensitivity to poor handling
- Long shelf-life
- Inexpensive processing equipment and high throughput
- No attack of soldermask
- No toxic chemicals (waste should be easily treated)
- Specification by most original equipment manufacturer (OEM) customers
- Capability of being selectively coating with electroplated gold

32.1.3.2 Assembly Requirements. The assembly engineers, whether contract Electronics Manufacturing Services (EMS) provider, independent original design manufacturer (ODM), or captive OEM, have requirements that are very different from those of the PCB fabrication customer:

- Flat surface for stenciling paste
- Low contact resistance for in-circuit test (ICT)
- Tolerance for paste misprint cleaning
- Long incoming shelf-life (more than one year)
- Long midassembly shelf-life (more than two weeks)
- Compatibility with all pastes and fluxes
- Capability of bonding with Al and/or Au wire

- The ability to be visually inspected (prevents solder mask residue non-wetting)
- Ease of handling
- Excellent and predictable wettability
- Fiducial recognition
- No warp and twist of the PCB
- Good compliant pin and connector contacts
- Good Pb-free soldering with no need for nitrogen
- No solder bridging or plugged holes

32.1.3.3 OEM Requirements. The original equipment manufacturer has the following additional requirements:

- Best solderjoint reliability; strong intermetallic
- No impact on cost of bare board
- Prevention of defects (black-pad, voiding, whiskers)
- Corrosion resistance in uncontrolled field environments
- Usefulness as an electrical test point and EMI shield
- Functionality as a touchpad/keypad contact
- Good electrical performance of high-speed signals
- Adherence to industry-consensus specifications
- Complete documentation from the chemical supplier
- Established supply base of installed fabricators
- Audited fabricators

32.2 ALTERNATIVE FINISHES

Surface finishes available in the PCB industry are as follows:

- HASL
- Reflowed tin-lead
- ENIG
- OSP
- Antitarnish and preflux
- Immersion silver
- Immersion tin
- Electroless palladium (Pd)
- Electrolytic nickel/electrolytic gold
- Electroless gold
- Direct immersion gold
- Solid solder deposit

Table 32.1 shows these alternative surface finishes and their typical thicknesses.

TABLE 32.1 Major Surface Finish Options with Typical Thickness

PCB surface finish	Typical thickness	
Hot Air Solder Level		2–40 μm (80–1600 μin)
Organic Solderability Preservative		0.1–0.6 μm (4–24 μin)
Electroless Nickel/Immersion Gold	Nickel:	3–5 μm (120–200 μin)
	Gold:	0.05–0.15 μm (2–6 μin)
Immersion Silver		0.1–0.4 μm (4–16 μin)
Immersion Tin		0.6–1.2 μm (24–48 μin)

Figure 32.1 shows the estimated usage of these surface finishes, indicating that no single finish will satisfy all needs. These are discussed in the following sections.

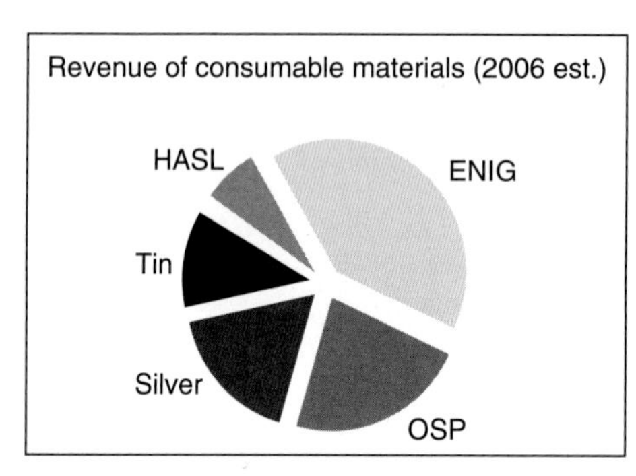

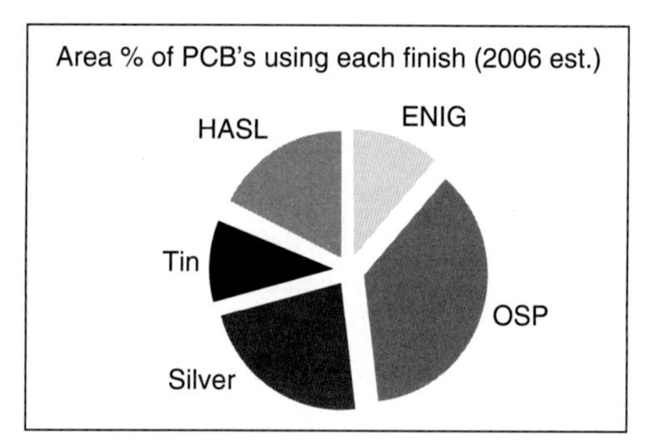

FIGURE 32.1 Worldwide surface finish usage estimates by (a) dollar amount and (b) area share of PCBs manufactured.

32.3 HOT AIR SOLDER LEVEL (HASL OR HAL)

HASL is the process of immersing a circuit board into molten solder. For the vast majority of HASL applications, the solder alloy is tin-lead. For Pb-free applications, the main alloys include tin-copper, tin-silver-copper, tin-copper-nickel, and tin-copper-nickel-germanium alloys. The thickness and texture of the HASL surface result from the surface tension of molten solder. As such, the thickness varies from 2 to 20 μm or more. HASL boards demonstrate excellent shelf-life, solderability protection, and resistance to poor handling, but they do tend to have higher levels of residual ions left over from fluxing. The HASL operation forms an intermetallic with copper circuitry at the PCB shop. This extra thermal step, with the associated loss in copper, is a concern for Pb-free assemblies. Expansion of the PCB substrate at high temperatures can warp and twist the board. The dimensional movement, along with uneven pad flatness and plugged vias, makes assembling dense circuitry (less than 0.5 mm pitch) difficult or impossible. The worldwide decline of HASL is partly due to assembly concerns and partly due to Pb restrictions.

32.3.1 Fabrication Process

HASL can be applied in vertical or horizontal equipment. As a pretreatment, the copper is cleaned, microetched, and fluxed. It is common for the HASL pretreatment steps to be carried out in conveyorized equipment, and then the panel to be transferred to a vertical machine that dips the individual panel into molten solder. As the panel is raised, air knives blow off excess solder. To handle the very high temperatures of the molten alloy, equipment is manufactured from stainless steel, titanium, and other alloys. Aggressive rinsing is needed to remove flux residues, as the HASL process is potentially much dirtier than other surface finish processes. Surface finishing engineers must be aware of safety concerns, such as exposure to Pb and the possibility of equipment fires.

32.3.2 Advantages and Disadvantages

Table 32.2 compares advantages and disadvantages of HASL.

TABLE 32.2 Advantages and Disadvantages of HASL

HASL Advantages	HASL disadvantages
Inexpensive relative to other finish options	Not flat
Cu/Sn solderjoint	Paste misprints
Widely available	Poor fine-pitch assembly
Reworkable	Thick intermetallics
	May contain Pb
	Thermal damage to board
	Solder bridging
	Plugged PTHs

32.3.3 Failure Modes

HASL involves a delicate balance between sufficient solder coverage and excess solder. Inadequate solder results in exposed copper intermetallics and reduced shelf-life. Excess solder (which is more common) forms bridging among fine-pitch features. If the solder is not removed sufficiently from through holes, pin components do not place at automated assembly. A chief failure mode is due to HASL's lack of flatness. At assembly, stencils do not gasket well to the board surface, so paste misprints and skips can be problematic. Assemblers find difficulty stenciling paste on HASL finished PCBs using technology below 0.5 mm pitch. Positioning a HASL board (see Fig. 32.2) in the fixture for stenciling can be challenging if the PCB dimensions changed during HASL thermal steps.

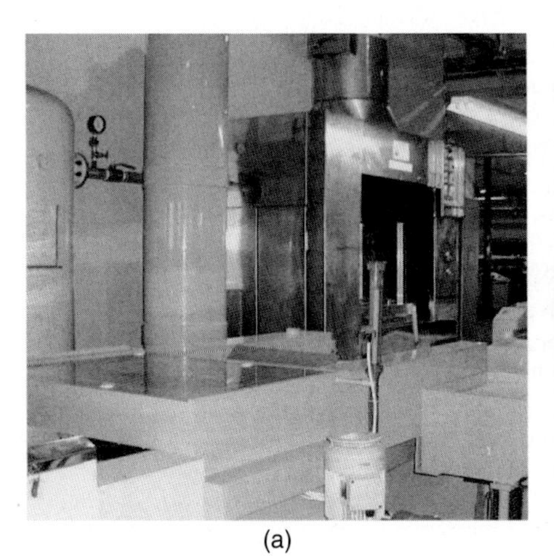

(a)

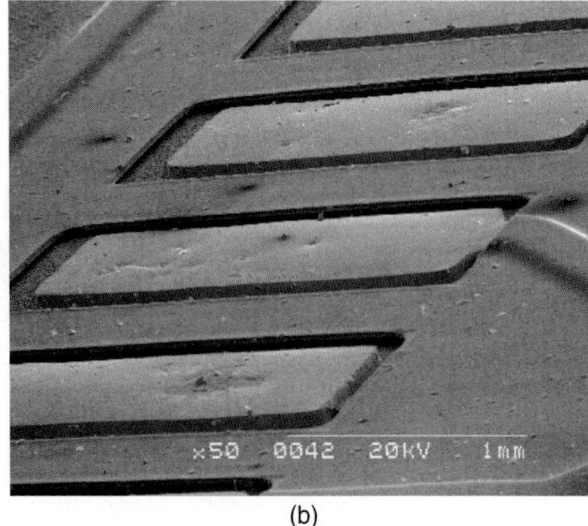

(b)

FIGURE 32.2 (a) HASL equipment (b) surface scanning electron micrograph (SEM).

32.4 ELECTROLESS NICKEL IMMERSION GOLD (ENIG)

Chemically, gold is the ideal element for the external coating of PCBs. Gold does not form an oxide, so it is virtually unaffected by temperature and storage conditions that might reduce the shelf-life of other finishes. In addition, gold dissolves nearly instantaneously into solder, promoting superior wettability.

Gold is limited, however, in other ways. Gold embrittles solderjoints when present in excess of about 3 percent by weight, so gold coatings of 0.3 μm maximum thicknesses should be used in soldering. Gold dissolves very quickly into copper as well. To prevent the mixing of gold with copper, and the eventual solderability problems caused by oxidized copper at the PCB surface, a layer of nickel is deposited to separate the metals. ENIG does not rely on electrolytic plating with outside power; in practice, nickel is deposited as a phosphorous-reduced "electroless" nickel.

32.4.1 Chemistry

ENIG deposition is dependent on the electromotive series of elements. Gold readily deposits directly on copper as an immersion bath, but as stated previously, a barrier of nickel is needed. In common practice, 3 to 5 μm of nickel is used in ENIG, with a thin gold coating of 0.05 to 0.15 μm (typically) to prevent nickel oxidation.

Nickel does not deposit directly on copper without prior catalysis. Palladium or ruthenium metal forms an immersion deposit on copper as the catalysis or activation step of ENIG. Catalysis baths employ the immersion (galvanic displacement) chemical mechanism.

As with all finishes, extra steps are required before the deposition of catalyst, Ni, and Au. All commercial ENIG processes rely on a cleaner and microetch prior to catalysis in order to remove trace residues from the solder mask and tin strip processes before final finish. Microetch is normally based on peroxide, persulfate, or monopersulfate systems.

32.4.2 Fabrication Process

Table 32.3 shows the process flow for ENIG surface finish.

TABLE 32.3 Process Flow for ENIG

Process	Chemicals	Time (min.)
Cleaner	Aqueous solvents, detergents, and emulsifiers. Acid can help undercut residues and remove thick copper oxides.	2–5
Microetch	In acidic environments, copper is oxidized by persulfate ($S_2O_8^{++}$), peroxide (H_2O_2), or proprietary monopersulfate acid mixes. Sulfuric acid is common to all types of microetch. 1–2 μm of Cu is oxidized to Cu^{++} and dissolved.	1
Catalyst	Immersion deposit (galvanic displacement) $Pd^{++} + Cu(0) \rightarrow Pd(0) + Cu^{++}$ (or) $Ru^{+++} + Cu(0) \rightarrow Ru(0) + Cu^{++}$	1–5
Electroless Nickel	Nickel is deposited through chemical (sodium hypophosphite) reduction. Rate control, by pH and temperature, determines phosphorous content, usually 8–11% by weight.	10–20
Normally, automated controllers are needed to maintain this dynamic bath. Nickel can be measured optically or with titration. Ammonia maintains pH.		
Immersion Gold	Immersion deposit (galvanic displacement) $Au^{+} + Ni(0) \rightarrow Au(0) + Ni^{++}$	8–15
Gold is added as $KAu(CN)_2$. Do not use gold salts containing Co or Ni.		

Due to the long dwell times, high operating temperatures, and complicated chemical reactions, the ENIG process is always conducted in vertical tanks. To achieve the throughput required in production, many ENIG systems make use of two or three electroless nickel (EN) tanks, which allows for heating new baths or nickel stripping. Using two active EN tanks greatly improves throughput, but the total process ordinarily exceeds 1 hour in cycle time.

In addition to the normal plastic materials used in tank construction, stainless steel can be used for EN tanks. This is possible only with the use of anodic passivation, which employs an electrical bias applied to the steel to prevent reduction of nickel ions on the steel. By using steel tanks, fabricators avoid the frequent stripping of plastic tanks required due to the slow plate-out of nickel on rough surfaces and localized hot spots.

32.4.3 Advantages and Disadvantages

Table 32.4 shows comparison of advantages and disadvantages of ENIG surface finishes.

TABLE 32.4 Advantages and Disadvantages of ENIG

ENIG advantages	ENIG disadvantages
Flat; fine-pitch assembly	Expensive
Surface contacts	Brittle Ni/Sn solderjoint
Widely available	Not reworkable
No copper dissolution	Solder mask attack
No Pb	Black-pad, black-line nickel
Strong PTH "rivet"	Signal loss at RF frequencies
Long shelf-life	Very complicated process

32.4.4 Failure Modes

ENIG failure modes at fabrication are mainly due to the activity of the chemical process steps. Underactive catalysis yields skip plating in the subsequent nickel bath. Nickel skips allow copper to migrate through gold and prevent solder wetting. Overactive catalysis can lead to extra metal between circuit features, known as extraneous plating. This can cause short circuits. Overactive nickel baths deposit low phosphorous content, leading to the interconnection failure mode known as black-pad (described in Sec. 12.2). Overactive gold chemistry attacks nickel along nodule boundaries and produces black-pad. Such gold conditions may be detected with frequent x-ray fluorescence (XRF) thickness measurement. Underactive nickel and gold baths pose skip-plate problems. Peeling of gold films can result from too much rinse time after nickel or from poor rinse quality. Tape testing of ENIG is mandatory. Very rough nickel deposits are possible when the electroless nickel bath is not closely maintained. Nickel baths can destabilize, plate onto the tank walls, and produce inferior coatings. Lastly, soldered ENIG surfaces result in a nickel-tin solderjoint. This structure has been demonstrated to be less tolerant of physical shock than the copper-tin solderjoint. The brittle intermetallic is unavoidable with ENIG, so mobile devices usually do not employ ENIG solderjoints.

32.5 ORGANIC SOLDERABILITY PRESERVATIVE (OSP)

As the name connotes, OSPs are very thin organic coatings used to preserve the solderability of PCB surface copper. Differing from alternative surface finishes that attempt to serve other PCB functions, OSPs are generally limited to solderability preservation only. The most widely used OSPs, benzimidazoles and phenylimidazoles, have evolved from earlier antitarnish coatings and can endure multiple soldering operations. OSP is the least expensive and simplest of all major surface finish options, so it enjoys wide use throughout the world. OSP use is limited, however, due to poor contact functionality for ICT and electrical test. Also, it suffers from degradation after multiple thermal operations, so solder wetting on Pb-free assemblies can pose problems.

32.5.1 Chemistry

OSP is the simplest chemical process of the surface finishes, but the chemical itself is actually quite complicated. In brief, a large organic molecule (see Fig. 32.3) is dissolved into a solution of water and organic acid. The circuit board is exposed to the solution, and the OSP molecule attaches to the PCB's exposed copper surfaces. A chemical bond is formed between copper and a nitrogen group in the OSP. Copper is present in the chemical bath, so multiple layers of an OSP-copper complex can deposit on the surface, reaching a thickness of 0.10 to 0.60 μm.

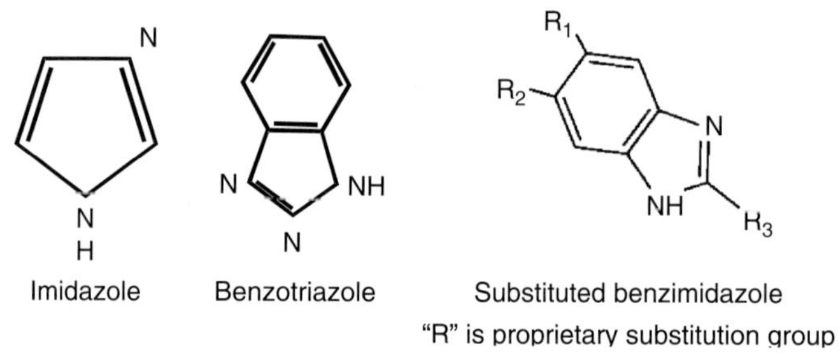

FIGURE 32.3 OSP molecule types.

Another generation of OSPs is designed to survive exposure to multiple high-temperature soldering operations. These chemicals are normally based on benzimidazole with attached functional groups to add molecular weight and thermal resistance. Substituted phenylimidazole and OSP systems with metal codeposits have been used as well. The intent of the new generation of OSP is to withstand Pb-free soldering temperatures. These "high-temperature" OSP coatings, made for use with Pb-free alloys and the associated fluxes, can be deposited at higher thicknesses and may be measured as high as 1.0 micron.

32.5.2 Fabrication Process

Table 32.5 show the fabrication process for OSP, including chemicals and dwell times.

Almost universally, OSP is applied in conveyorized equipment. The cleaner and microetch may use flood or spray application, but OSP is applied using flood. Because the OSP process is very mild to the PCB materials, rinsing is very important to ensure that contamination from all prior fabrication steps is removed. Often, OSP is applied when the PCB is in part form after the PCB has been electrical tested, routed, and depanelized. This is to preserve the

TABLE 32.5 Process Flow for Organic Solderability Preservative

Process	Chemicals	Time (min/)
Cleaner	Aqueous solvents, detergents, and emulsifiers. Acid can help undercut residues and remove thick copper oxides.	1–4
Microetch	1–2 μm of Cu is oxidized to Cu^{++} and dissolved.	1.0
Pre-Dip	Normally, the predip employs ingredients of the OSP bath without the OSP chemical itself.	0.5
OSP	OSP, acetic or formic acid, source of copper or other metals.	0.5–1.0

fragile OSP surface, which is sensitive to handling, contamination, and fingerprints. The OSP parts should stabilize at room temperature and controlled humidity before packaging, as any moisture in the package can turn the OSP surface black.

OSP thickness is normally determined by indirect sampling. No method exists for determining the OSP thickness during production on the actual PCB. Instead, a sample coupon is processed along with the production parts. The coupon is immersed in an acid/solvent solution that dissolves the coating. The solution is transferred to a UV-VIS cell, where the absorbance of light allows the estimation of the thickness of OSP on the coupon. That measurement is used as a quality control for the production boards.

The lack of a true thickness measurement control is a key issue to OEMs that require control metrics. If the OSP coating is too thin, the PCBs will have inadequate protection. If the coating is too thick, there is a possibility for poor solder wetting and poor PTH filling. The OSP coating is invisible to the eye, so there is no way to inspect for unprotected copper if the OSP coating was not applied correctly.

32.5.3 Advantages and Disadvantages

Table 32.6 compares the advantages and disadvantages of using OSP as a surface finish.

TABLE 32.6 Comparing the Advantages and Disadvantages of OSP

OSP Advantages	OSP disadvantages
Flat; fine-pitch assembly	No direct thickness measurement
Least expensive finish	High surface contact resistance
Simple process	Poor PTH soldering
Widely available	Shorter shelf-life
Cu/Sn solderjoint	Exposed copper on final assembly
No Pb	Not inspectable before assembly
Reworkable	Degrades with multiple reflow
High productivity at fabrication	Handling-sensitive
	Problems with paste misprint cleaning

32.5.4 Failure Modes

The OSP process is simple, so few failure modes exist at fabrication due to chemical stability. Aggressive microetch and poor rinsing can result in attack at the junction of solder mask and exposed PCB metal, especially if nickel-gold has been selectively deposited on edge fingers. Galvanic attack at this interface can cause electrical opens. PCBs using selective nickel-gold can also suffer from deposition of OSP on the gold surface. OSP on copper prevents good electrical test contact and in-circuit testing. OSP deposition on selective gold can prevent the intended contact functionality of the gold pad. Skip plating can be a concern. Very thin solder mask residues will prevent OSP coating. Because the OSP coating is invisible, it cannot be inspected, so skip plating will lead to non-wetting at assembly. OSP is the coating most sensitive to handling, storage, and shelf-life. Copper under OSP oxidizes with time and heat, so weak fluxes present soldering issues. The Pb-free temperature increase has narrowed the OSP process window to the point where PTH solderability may not meet IPC standards.[1–3]

32.6 IMMERSION SILVER

Silver was chosen as a PCB surface finish from the earliest days of PCB fabrication due to some inherent advantages. Silver is the best electrical conductor and exhibits the lowest contact resistance of any metal. As a precious metal, it is more resistant to oxidation than other more base metals. On early PCBs, electroplated silver was discovered to form metal dendrites between conductors. During the 1990s, new immersion silver was introduced as an HASL alternative, and users developed data showing that dendrite formation was not an issue. With the Pb-free transition, immersion silver gained widespread use. OEMs with specific needs specified silver to overcome issues with other finishes. Immersion silver is a galvanic displacement process with a typical thickness range of 0.1 to 0.4 μm. The fabrication process is easy, but the coating can suffer from tarnish when left exposed after assembly in field use. Heavy tarnish can be a predictor for corrosion and functional loss.

32.6.1 Chemical Coating

The PCB copper is prepared using separate cleaning and microetch steps. Over time, these steps can be taken for granted, but improper use can cause many defects. The predip bath removes excess acidity from the PCB and prevents chlorides from entering the silver bath, where silver chloride would quickly precipitate. The silver bath is quite simple chemically, relying on the galvanic displacement of less noble copper with two atoms of more noble silver. Proper rinsing and drying is essential.

32.6.2 Fabrication Process

Table 32.7 shows the fabrication process for immersion silver surface finish, including the chemicals used and dwell times.

TABLE 32.7 Process Flow for Immersion Silver

Process	Chemicals	Time (min)
Cleaner	Aqueous solvents, detergents, and emulsifiers. Acid can help undercut residues and remove thick copper oxides.	1–4
Microetch	1–2 μm of Cu is oxidized to Cu^{++} and dissolved.	0.5–1.0
Predip	Normally, the ingredients of the silver bath without the silver metal itself.	0.5
Silver	Silver, acid, grain refiners, chelation $Ag^{+} + Cu(0) \rightarrow Ag(0) + Cu^{++}$	0.5–3.0
Postdip (optional)	Can be an antitarnish or an extra cleaner for demanding ionic cleanliness applications.	

Most immersion silver in large production is applied in conveyorized equipment. The cleaner and microetch may use flood or spray application, but the silver solution is applied using flood. Because the silver process is very mild to the PCB materials, rinsing is very important to ensure that contamination from all prior fabrication steps is removed. Often, silver is applied in part form after the PCB has been routed, depanelized, and electrically tested. This is to preserve the silver surface, which can be sensitive to handling, contamination, and fingerprints. In smaller-scale production and prototype manufacturing, silver can be applied in vertical process tanks.

32.6.3 Advantages and Disadvantages

Table 32.8 compares the advantages and disadvantages of using immersion silver as a surface finish.

TABLE 32.8 Advantages and Disadvantages of Immersion Silver as a Surface Finish

Silver advantages	Silver disadvantages
Flat; fine-pitch assembly	Handling-sensitive
Inexpensive finish	Solderjoint microvoid history
Simple process	Exposed silver on final assembly can corrode
Widely available	Tarnish
Cu/Sn solderjoint	Silver had a history of dendrite formation
No Pb	Attack at the solder mask interface
Reworkable	
High productivity at fabrication	
Easy thickness measurement	
Excellent surface contact functionality	
No degradation with multiple reflow soldering	

32.6.4 Failure Modes

The silver surface readily shows tarnish from exposure to sulfur and chloride environments. Excessively tarnished parts have limited solderability. As a galvanic process, aggressive silver baths can attack copper circuitry. Attack can occur at the interface with solder mask or at the electroless copper layer on sidewalls. The attack is more severe at high thicknesses over 0.50 μm and if the PCB is reworked with multiple passes through immersion silver. Copper attack can result in electrical opens. Aggressive silver baths, when deposited on rough copper, can also attack on pad surfaces, causing "caves" in the copper. Studies show that caves can result in layers of very small voids at the copper-solder interface after soldering. These structures, termed *microvoids,* can dramatically reduce solderjoint strength and produce electrical opens. Each of the preceding failure modes is worsened when the silver bath plates nonuniformly. The chemicals need to be formulated and operated in such a way to provide uniform galvanic exchange metal deposition.

32.7 IMMERSION TIN

Immersion tin is a thin coating of pure tin, typically 0.6 to 1.2 μm thick, which protects the underlying copper from oxidation and provides a highly solderable surface. Tin is deposited using a galvanic displacement process and can be applied in vertical or conveyorized equipment. Tin boards are used primarily for solderability and have very good compliant pin connector functionality.

Tin readily forms an intermetallic with copper, so tin can pose challenges with shelf-life, multiple soldering operations, and contact functions. Tin is also restricted in some regions due to environmental concerns over the use of the bath ingredient thiourea.

32.7.1 Chemical Coating

As with all immersion metal systems, the copper needs to be perfectly clean and textured before the tin plating step. A predip is used to maintain proper chemical balance in the tin bath and prevent contamination. The tin deposition is not a direct galvanic displacement, since tin is more electronegative than copper. Thiourea is used in tin chemical formulation to create a copper-thiourea complex on the surface. This complex becomes more electronegative than tin, and then participates in the immersion reaction. Because of the need to drive the reaction, the tin bath contains relatively high concentrations of chemicals, which must be thoroughly rinsed. Waste treatment of the system is complicated due to high levels of thiourea.

32.7.2 Fabrication Process

Table 32.9 shows the fabrication process for the immersion tin surface process, including chemicals used and dwell times.

TABLE 32.9 Process Flow for Immersion Tin

Process	Chemicals	Time (min.)
Cleaner	Aqueous solvents, detergents, and emulsifiers. Acid can help undercut residues and remove thick copper oxides.	1–4
Microetch	1–2 μm of Cu is oxidized to Cu^{++} and dissolved.	0.5–1.0
Predip	Normally, the ingredients of the tin bath without the tin metal itself.	1.0
Tin	Tin, thiourea, acids, whisker inhibiting trace metals, chelation.	4.0–12.0
Postdip (optional)	Can be an extra cleaner for demanding ionic cleanliness applications.	

Few production installations use immersion tin in horizontal conveyorized applications due to the chemical process time and the higher temperatures used. Conveyorized lines are long and therefore expensive. The use of vertical tanks carrying baskets of panels for immersion tin plating makes better use of production floor space. Due to the high sulfur reactivity of the thiourea ingredient, tin lines should be dedicated to the tin process only and not mixed in with OSP, silver, or ENIG lines where sulfur can harm the surfaces.

32.7.3 Advantages and Disadvantages

Table 32.10 compares the advantages and disadvantages of immersion tin used as a surface finish.

TABLE 32.10 Advantages and Disadvantages of Immersion Tin Surface Finish

Tin Advantages	Tin Disadvantages
Flat; fine-pitch assembly	Handling-sensitive
Inexpensive finish	Use of thiourea (a carcinogen)
No Pb	Exposed tin on final assembly can corrode
Cu/Sn solderjoint	Whiskers
Excellent compliant pin functionality	Degradation with time/ heat and multiple reflow soldering
Less visible tarnish	Solder mask attack
Reworkable	Difficult thickness measurement

32.8 OTHER SURFACE FINISHES

Other surface finishes include reflowed tin-lead, electrolytic nickel/electrolytic gold, antitarnish and reflux, electroless palladium, electroless gold, direct immersion gold, and solid solder deposit.

32.8.1 Reflowed Tin-Lead

Before the widespread use of solder mask, a primary method for protecting copper circuitry was to electroplate a coating of tin or tin-lead on the copper. After component insertion, the tin plating could be liquefied with oven or vapor-phase reflow to form solderjoints. The use of solder mask over bare copper to take advantage of surface mount, wave soldering, and mixed assembly restricted the use of tin plate finishing. For simple technology product, tin or tin-lead plate and reflow remains a viable fabrication and assembly method.

32.8.2 Electrolytic Nickel/Electrolytic Gold

The electrolytic nickel-gold process employs galvanic electroplating. The circuit board is placed on a rack and electrically connected to a power supply (rectifier), When immersed in a solution of metal ions, the metals are reduced as a metal coating on the PCB by the supply of electrons. The metal thickness grows in a very predictable way depending on the time, current, area plated, and efficiency of the reaction. Relatively few PCBs are plated "full-build" with nickel-gold as the only finish. More frequently, the coating is applied selectively in combination with another finish such as OSP. For a large number of PCBs that employ edge tab connection, the PCB is electroplated in a special plating line where only the exposed edge features are immersed into the plating cell. Electroplated nickel-gold is expensive and difficult to use in high production, so it is commonly used only on features requiring high-force physical connections or gold wirebonding. Several micrometers of nickel are deposited, and 0.50 to 1.5 μm of gold protects the nickel from oxidation. Thick gold forms gold-containing phases in the solderjoint. The embrittlement of these phases has prevented the wide use of electrolytic nickel-gold as a solderable finish.

32.8.3 Antitarnish and Preflux

Antitarnish surface finishes are thin, organic coatings designed to prevent copper oxidation. Antitarnish chemicals include imidazole and benzotriazole. Preflux coatings are resins or rosins. These materials cannot endure long storage times and normally can be used only for single-sided soldering operations. Much of the use of simple organic coating was replaced by OSP as the cost of OSP declined, but antitarnish and preflux chemicals are still used in simple PCB fabrication.

32.8.4 Electroless Palladium

The original concept for electroless palladium was as an alternative to the high cost of gold. Palladium shows some of the oxidation resistance similar to precious metals, and can offer a good physically strong deposit for electrical contacts and wirebonding. However, palladium is a very difficult metal to deposit electrolessly, causing substantial engineering support and yield problems. The rise in the cost of palladium metal nearly removed it from any market share worldwide. Electroless palladium reemerged in OEM testing as a possible cure for ENIG black-pad, but it is not widely used. Palladium is sensitive to the environment, so a flash of immersion gold may be used over palladium. The coating is considered not reworkable. Palladium phases in the solderjoint pose an embrittlement hazard.

32.8.5 Electroless Gold

Gold can protect copper and resist oxidation like no other metal. But electrolytic nickel-gold and ENIG both suffer from certain limitations. Electroless nickel with a coverlayer of electroless gold is intended to overcome these limitations. As with ENIG, electroless nickel underlayer is needed as a barrier layer to prevent copper to gold interdiffusion. Electroless gold can achieve a sufficient thickness to allow gold wirebonding and some physical connection functions. It also deposits much more uniform thickness than electrolytic gold. Better thickness control allows somewhat thinner coatings, so gold embrittlement can be prevented. Cost and process control considerations prevent more widespread use. In addition, all the limitations of the use of electroless nickel still apply when considering this finish.

32.8.6 Direct Immersion Gold

Because electroless nickel is the cause for many of the disadvantages of ENIG, some designers have considered the use of immersion or electroless gold directly on top of underlying copper. This process works for boards that are assembled without a long shelf-life and do not require multiple thermal excursions. Without the barrier coating of nickel, however, the copper and gold soon dissolve into each other and form an unsolderable oxidized intermetallic. As direct gold costs more than immersion silver or tin coatings, its main advantage is its improved possibility for gold wirebonding.

32.8.7 Solid Solder Deposit

Technologies exist for the deposition of solder onto the bare PCB at the fabrication shop using non-HASL techniques. One method employs the deposition of tin-lead with a physical flattening of the solder meniscus. Other methods have been proposed as well. The use of these deposition methods has not penetrated widely into the production market.

32.9 ASSEMBLY COMPATIBILITY

Figure 32.4 shows a PCB as it progresses through the assembly and soldering process, showing the resulting intermetallics in cross section.

32.9.1 Solderability Test Methods

Refer to Chap. 42 for detailed information on solderability of PCB surface finishes. Testing methods are contained in the comprehensive joint industry specification ANSI/IPC J-STD 003. Surface finishes are commonly tested in production at the board fabricator using a solder float or solder dip method. When qualifying the finish, an OEM or assembler uses more equipment-specific tests such as wetting balance, rotary dip, or paste spread methods. Solderability should be consistent from day to day and lot to lot, with very little degradation after exposure to at least three assembly thermal exposures.

FIGURE 32.4 A surface finish progressing through assembly; the bare finish, with stenciled paste, as soldered, and in cross-sectional view of resulting intermetallics.

32.9.2 Misprints

For many years, a primary failure mode in PCB assembly was solderpaste misprints. Misprints occur when the solder stencil or screen does not form a close contact or good gasketing with the circuit features. If there is a gap between the stencil and the copper, or if the screen is slightly misregistered, solderpaste squeezes between the features, causing solder bridging and resulting in a lack of solder where the paste deposition was intended. The uneven surface of HASL is a chief cause of paste misprints. Flat surface finishes such as silver, tin, and OSP virtually eliminate solderpaste misprints. If misprints do occur for other reasons such as misalignment, the finish should be able to be cleaned, staged for reprinting, and soldered without any lack in wetting. OSP may be removed during misprint cleaning; the PCBs should be assembled immediately if this occurs.

32.9.3 Shelf-Life, Storage, and Handling

Each finish is vulnerable to solderability loss if the PCB is handled or stored in non-ideal environments. OSP, generally given a 6 to 12 month shelf-life, is degraded with time and temperature, as the organic and underlying copper can oxidize. Silver, with a shelf-life of 12 months, is affected by exposure to sulfur and chloride environments that can cause excessive tarnish. Tin, with a shelf-life of 6 to 12 months, forms intermetallics with copper over time and at elevated temperature. ENIG has a shelf-life exceeding 12 months and can withstand most environments, but is sensitive to steam aging and degrades more quickly than other finishes in mixed flowing gas testing. HASL is considered to have virtually unlimited shelf-life. The shelf-life of any PCB, with any surface finish, can quickly degrade if the parts are exposed to contaminating materials. Any contaminant film that prevents flux and solder from coming into intimate contact with the surface finish will result in non-wetting.

32.9.4 Solderjoint Metallurgy

The basic function of the circuit board is to act as a mechanical and electrical connection between components of an electrical device. Application of solder to a PCB surface results in the formation of an intermetallic compound. The metals in the solder and the metals on the PCB participate in the intermetallic formation. The intermetallic acts as the physical glue holding the solder to the circuitry. The type of intermetallic depends on the type of solder and the type of PCB finish.

The vast majority of solders used in electronics (SnPb, SnAgCu, SnCu, SnCuNi) is based on tin. Most of the time, the tin forms an intermetallic with copper (Cu_3Sn and Cu_6Sn_5). The primary exceptions are ENIG and electrolytic NiAu finishes. Soldering to nickel-based finishes results in Ni_3Sn_4 and $NiSn_3$ intermetallic compounds. The physical and electrical performance differences between copper and tin intermetallics do influence the design of electronic devices. In general, solderjoints formed using intermetallics of nickel are more brittle and lead to earlier failure in physical shock testing.

HASL begins to form copper-tin intermetallics when the HASL is deposited at the PCB shop. When HASL is soldered at the assembly house, the HASL coating liquefies and mixes with the molten solderpaste or wave solder. The intermetallic formed at the PCB interface deepens with each thermal excursion. HASL intermetallics can reach thicknesses of concern to reliability engineers. Intermetallic layers formed using high tin content solders (Pb-free) grow more quickly due to the increased availability of tin and the elevated temperatures. With the consumption of copper into intermetallics, there is a concern with the thinning of copper, especially at the knee of plated through holes.

The OSP coating is displaced by the flux during SMT and wave soldering. The exposed copper forms a copper-tin intermetallic with the solder. If the flux exposes more copper than the solder spreads, a ring of unprotected copper results. Some engineers consider this ring of exposed copper to pose a corrosion risk. For Pb-free soldering, best results are achieved by selecting a flux system that maintains activity throughout the entire soldering temperature range.

Immersion silver dissolves into the tin phase of molten solder. The silver does not melt, but rather forms a solid solution. The dissolution rate of silver is in the range of 0.5 to 1.5 μm per sec. at standard soldering temperatures. Once the silver is dissolved, the underlying copper forms intermetallics with tin as discussed previously. Flux has little effect on silver, but can help clean trace contaminates and reduce surface tension.

Immersion tin dissolves very quickly into SMT and wave solder. The tin may not melt (m.p. 232°C) but dissolves into the liquefied solder readily. Even before soldering, the immersion tin begins to form intermetallics with the copper beneath. In fact, with extended storage, the tin may be completely consumed by intermetallics before soldering. At soldering temperatures, the copper-tin intermetallics form at much higher rates. Areas of the PCB exposed to the assembly temperatures, but not soldered, will form thicker and thicker intermetallic layers until the pure tin is consumed.

ENIG is a special case. When exposed to solder, the top layer of gold very quickly dissolves into the tin phase at a rate of about 3 μm per sec. at standard assembly temperatures. Once the underlying nickel is exposed, it begins to form intermetallics with tin. Nickel dissolves very slowly in solder (<0.002 μm per sec.) so the nickel remains as a barrier between solder and copper. The intermetallics formed between nickel and tin are very thin and more physically brittle than the copper tin Intermetallic Compound (IMCs).

Electroless palladium can also act as a barrier layer if used in sufficient thickness. Automotive PCBs produced in the 1990s specified a maximum of 0.2 micron of Pd metal with the goal to incorporate the entire Pd layer into the solderjoint.

Figure 32.5 shows the interaction of four surface finishes with solder during the assembly process.

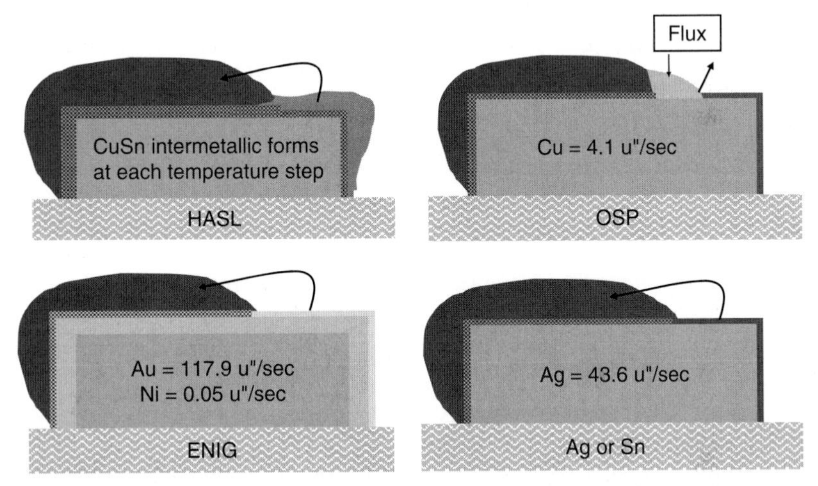

FIGURE 32.5 Displacement of PCB surface finishes during component assembly and soldering. (a) HASL: PCBs show some intermetallic formed during the HASL operation. Upon subsequent assembly, the HASL coating melts into the solderpaste or wave solder, and the intermetallic grows in thickness. (b) OSP: The OSP coating is removed by the flux during component assembly. The underlying copper forms the intermetallic with tin during soldering. (c) ENIG: Gold dissolves quickly into the tin phase of solderpaste or wave solder. The intermetallic is formed between tin and nickel because the rate of dissolution into nickel is slow. (d) Silver or Tin: Silver and tin coatings dissolve very quickly into the tin phase of solder. The intermetallic is formed with underlying copper.

32.10 RELIABILITY TEST METHODS

Reliability is determined using a variety of methods, usually used in combination. Often an unassembled PCB is exposed to a conditioning or aging environment. The conditioning may consist of time, temperature, humidity, and contamination metrics. The conditioning is normally at a higher level than the environment in which the device is intended to be used so that the test environment imparts an accelerated age condition.

Temperature exposure for PCBs is commonly conducted at 155°C for 4 or 8 hours to study shelf-life. To study the effect of soldering, simulation is conducted at the actual reflow temperature, without contact with flux or solder. Temperature cycling of assembled devices reflects the difficulty of the intended end-use environment. Although some cycling may occur at 0°C to 100°C, more severe cycling may require –55°C to 125°C. Assembled parts may also undergo physical testing conditions such as physical shock, vibration, bending, twisting, and impact.

Environmental exposure testing is common for PCB surface finish evaluation. Exposure to ambient air during reflow is common, as is long-term exposure to polluting environments such as mixed flowing gas. Physical contact with contaminating materials such as fluxes, assembly materials, fuels, and even diesel fuel and carbonated beverages has been documented.

Electrical testing is another class of reliability testing. The function of a solderjoint formed with the surface finish can be measured by burn-in testing, electrical functional tests, and/or high-current-density application for extended periods. The contact resistance of an unassembled surface finish is an important measurement. The PCB finish, when exposed to heat and humidity metrics, as well as environmental exposure, may need to pass a sensitive four-point probe contact resistance check.

32.11 SPECIAL TOPICS

32.11.1 High-Speed Signals

Electrons travel along the outer regions of a conductor. This phenomenon, known as the skin effect, is most pronounced at the high signal frequencies of Global Positioning Systems (GPS), aerospace, navigation, mobile telephony, computer servers, and crash avoidance devices. At high frequencies, the skin depth of electron travel is of a thickness similar to that of the PCB surface finish. For example, at 10 GHz, the skin depth is less than 1 micron. Therefore, the electrical conductivity of surface finish materials is important at those higher frequencies. Above 2 GHz, circuit designers find that the use of electroless nickel hinders signal integrity. Silver, the most conductive element, is used for devices operating at higher frequencies.

32.11.2 Inspectability

To ensure good soldering, it is necessary for the assembler to ensure that all copper is protected and no solder mask residue remains on solderable features. By inspecting the bare board, the engineer can reject parts that show bare copper due to skipped surface finishing or mask residue. A special case is OSP, where inspection of the protected copper is impossible because OSP-protected areas have the same visual appearance of uncoated copper with or without thin solder mask contamination.

32.11.3 Contact Functionality

Many circuit board assemblies rely on surface contact functionality to operate. Examples include touchpads, edge contact rails for grounding/shielding, test probe contact points, zero insertion force connectors, chassis and enclosure bolts, and surface contacts for mating with nonsoldered connectors/interposers. OSP does not meet the demands of surface contact, so selective plating or soldered contact designs must be used. Immersion tin may lose contact functionality as the intermetallic oxidizes. HASL will maintain contact ability, but suffers from material creep under load, which may allow loose connections after field use. Silver maintains contact functionality even when tarnished, but loses contact ability if heavily corroded by the environment. ENIG maintains contact functionality over long periods of time, but is also susceptible to excessive corrosion. Electroplated Ni/Au is the best performing finish for long-term contact functionality.

32.11.4 Wirebonding

Some device designs call for wirebonding directly to the PCB or laminate chip carrier. In general, most surfaces are wirebondable with aluminum wire. Al wirebonding may be impossible with OSP or HASL finishes. For gold wirebonding, only electrolytic Ni/Au is widely used. Good bond strengths have been achieved on ENIG, Ni/Pd/Au, electroless gold, direct gold, and silver, but most production uses "soft" electrolytic Ni/Au to meet gold wirebonding demands.

32.11.5 Compliant Pin, Press-Fit Connectors

Compliant pin, or press-fit, connectors rely on contact with the surface finish in plated through holes. The main concern is the dimension of the finished hole size. When using HASL, the PCB designer must allow for the extra thickness of solder that will reduce the diameter of the PTH. HASL alternative finishes are flat, so will have negligible effect on finished hole size, since the thickness of the finishes is less than the tolerance of the drilling and plating operations. The material properties of the finish have some affect on compliant pin capability. Soft materials such as HASL and immersion tin perform better by providing low insertion force. Hard materials, such as copper and nickel, yield higher insertion force. OSP and silver act like copper due to the thinness of the coating.

32.12 FAILURE MODES

See Table 32.11 for descriptions of failure modes of specific surface finishes

TABLE 32.11 Failure Modes Specific to PCB Surface Finishes

OSP	HASL	ENIG	Immersion Tin	Immersion Silver
PTH Hole Fill	Solder Bridging	Black-Pad, Black-Line Nickel	Tin Whiskers	Post-Assembly Corrosion of Exposed Surfaces
Exposed Copper after Soldering	Plugged Holes	Brittle Fracture	Ionic cleaning Failures	Tarnish
Post-Assembly Corrosion of Exposed Surfaces	Equipment Safety	Solder Mask Attack	Solder Mask Attack	Solderjoint Microvoids
Handling Defects	Paste misprints	RF Signal Loss	Post-Assembly Corrosion of Exposed Surfaces	Electrochemical Migration
		Post-Assembly Corrosion of Exposed Surfaces		

32.12.1 Whiskers

Whiskers form as a single crystal growth from the surfaces of mainly pure tin coatings (see Fig. 32.6). The whiskers are tiny, but can form lengths exceeding the spacing between electrical features on the circuit board or components. As pure tin is electrically conductive, reliability engineers fear that tin whiskers will form short circuits. In fact, there have been several high-profile failures attributed to tin whisker growth.

Whiskers form more quickly from pure deposits of tin over copper and brass. Although they have been reported in literature from other elements, the vast majority of whiskers in electronics have resulted from tin plating.

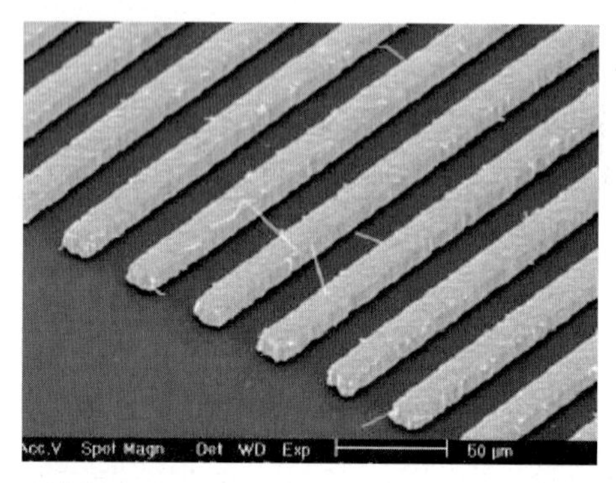

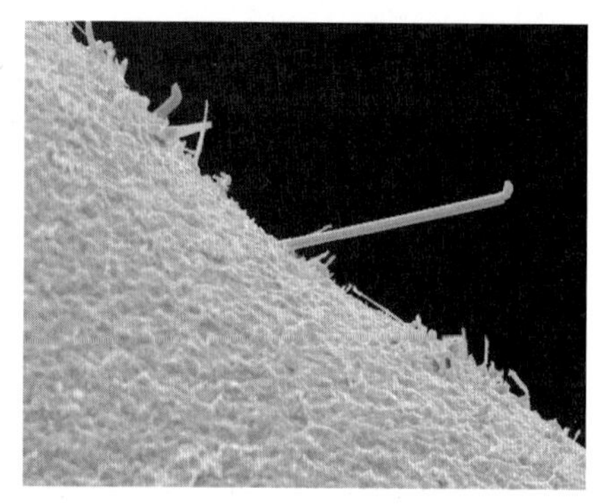

FIGURE 32.6 Tin whiskers on fine-pitch contacts.

The use of lead in tin plating and solder has been used to inhibit tin whisker growth. Rare reports of tin whiskers have been reported from tin-lead coatings. Whiskers evolve from tin surfaces as a stress-relief mechanism between tin crystals. Most stress in tin deposits arises from the growth of copper-tin intermetallics at the interface with underlying copper. With a much slower rate of nickel-tin intermetallic formation, tin whiskers form at a much lower rate than tin coatings on nickel underplate.

Several tin-whisker mitigation techniques have been proposed:

- Inclusion of lead, silver, and other metals in the tin deposit
- Use of a nickel tin coating as a barrier between copper and tin
- Annealing of tin as a stress relief
- Conformal coating as a barrier to whisker growth

In practice, the only effective whisker mitigation strategies involve nickel barriers and contamination of the tin deposit with a sufficient amount of metal, such as 2 percent Pb.

32.12.2 Interfacial Fracture

Interfacial fracture is a failure mode usually associated with ENIG, and is also known as black-pad, brittle nickel, and black-line nickel (See Fig. 32.7). Technically, interfacial fracture simply refers to the location of physical separation when a solderjoint is tested to failure. Black-pad (black-line nickel) is a phenomenon of ENIG that, when sufficiently severe, results in interfacial fracture. Brittle nickel refers to the inherent weakness of solderjoints formed from nickel-tin intermetallics, regardless of the mechanism of failure.

Black-pad is a name given to the surface of a circuitry feature after interfacial fracture. The separation occurs between the solder and an underlying surface of irregular, high-phosphorous, rough nickel. Aggressive gold displacement corrodes the rough nickel selectively along the nodule boundaries of the electroless nickel. The nickel is easily corroded due to the low phosphorous content of the as-deposited electroless nickel coating. Low phosphorous deposits (less than 9 percent) are far more easily corroded than higher phosphorous (10 to 12 percent by weight) EN coatings. The simple remedy in prevention of black-pad is the use of high phosphorous coatings, based on the operating parameters and chemical formulation of the nickel bath. However, any nickel bath can be destabilized with contamination such as divalent sulfur from undercured solder mask. Less stable EN baths plate at a higher rate and result in low phosphorus deposits.

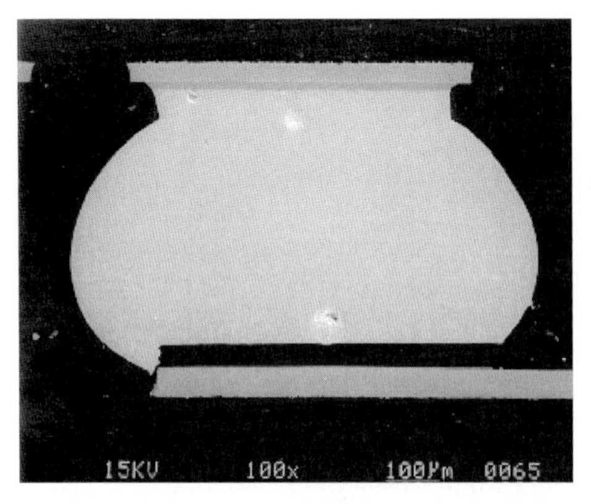

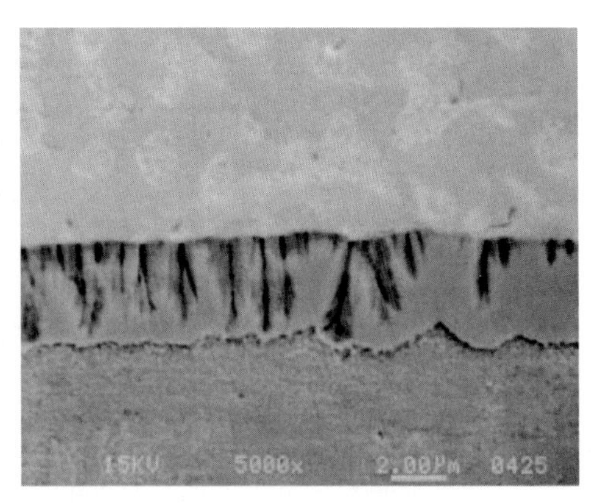

FIGURE 32.7 The result of black-pad and the microstructure of a black-pad nickel interface.

Even with good chemical formulation, correct operating conditions, and good inspection, black-pad still occurs in PCB assembly. Certain PCB designs are more vulnerable to black-pad due to galvanic effects set up in the plating bath in areas of extreme anode/cathode ratio. Users should be aware that there is no complete solution to black-pad, but adoption of the following measures is highly recommended:

- Cure the solder mask as much as practical to prevent photoinitiator contamination of the nickel bath.
- Use a cleaner before the nickel bath capable of removing material from the solder mask that may be leached out in the nickel bath.
- Operate the nickel bath to result in 9 percent phosphorous or higher.
- Test the corrosion resistance of the nickel deposit using nitric acid exposure.
- Use a mild gold bath that delivers a gold thickness target of 0.07 to 0.15 μm.
- Use function testing such as tape test and XRF to screen for black-pad.

32.12.3 Dendrites and Electrochemical Migration

Metals, when patterned as conductors and placed under electrical bias, can suffer from a vulnerability to electrochemical migration. When exposed to an electrolyte (such as dirty water), PCBs form an electrochemical cell. At the anode, metal is stripped of electrons (ionized) and can dissolve in the electrolyte. The corrosion, movement, and redeposition of metals is termed *electrochemical migration*, and can form interesting structures such as metal dendrites. All metals are susceptible to dendrite formation. Indeed, dendrites formed from PCB circuitry usually contain copper sourced from the patterned outerlayer copper. Groups such as IPC, Battelle, and Underwriters Laboratories have studied electrochemical migration. Documents issued by these groups show the causes of migration to include surface cleanliness, electrical bias, environmental humidity, and conductor spacing. The selection of surface finish material is not a leading factor in vulnerability to dendritic growth. However, a metal will corrode and migrate more quickly than others depending on the type of contaminating environment. In general, more noble metals such as gold resist corrosion better than copper. This is a main reason for the use of a protective surface finish. Metals such as silver and tin are attacked by sulfur, leading to corrosion. Electroplated silver, used to some extent in component terminations, does have a higher propensity for dendritic growth. Electroplated silver was abandoned as a PCB finish in the 1960s due to dendritic growth. Immersion silver does not behave in the same way as electrodeposited silver.

32.12.4 Solderjoint Voiding

Solderjoint voiding can result from various causes, including flux type, PTH blowholes, poor wettability, the Kirkeldall effect, and improper surface finish deposition. Surface finishes contaminated with organic materials have been shown to exhibit solderjoint voids after assembly. Silver has been connected with the void phenomenon termed *microvoids*, *champagne voids*, or *planar microvoids*. The mechanism for microvoiding has been determined to result from overaggressive silver bath deposition on improperly treated substrate copper. When the silver exchange reaction attacks the copper nonuniformly, a cave can be created in the copper from the localized galvanic attack. The cave entraps flux or moisture, resulting in a series of voids just above the intermetallic layer. If too many voids form at the pad interface, the reliability of solderjoint strength is compromised. Following the chemical suppliers' exact operating recommendation resolves this problem. Other types of voids, such as process voids, pinhole voids, and microvia voids, are not caused by the type of surface finish selected, but depend on the PCB design and assembly parameters.

32.12.5 Solder Mask Attack

The chemicals involved in depositing PCB surface finishes can attack the solder mask. Immersion tin directly attacks the mask by chemical dissolution. Thiourea in the tin bath acts as a solvent for the mask. Exposure time and temperature of contact with the bath should be carefully watched. The surface should be tape-tested for mask adhesion. ENIG processing creates an environment that can lift slivers of solder mask. The temperature of the nickel bath is not the culprit; rather, the generation of hydrogen bubbles as a by-product of nickel plating at the interface is alleged to cause lifting of the mask. Immersion metals such as gold, silver, and tin remove metal as part of the chemical reaction. If the immersion bath removes copper from underneath the edge of the solder mask, slivers of mask lose adhesion and show up in tape testing.

In all cases, the mask preparation steps are critical. If the copper is treated well to adhere to solder mask, the interaction from surface finish chemicals is greatly reduced. It is imperative that there is no wedge between the mask and the copper into which the surface finish chemicals can creep. In extreme cases of chemical entrapment, the chemical reactions are accelerated in the wedge, leading to nonuniform plating. Immersion silver, tin, and OSP can lead to excessive copper corrosion at the solder mask edge and result in electrical opens. This phenomenon is also known as *solder mask interface attack*.

32.13 COMPARING SURFACE FINISH PROPERTIES

Table 32.12 compares surface finishes by type and property and summarizes the material presented in this chapter.

TABLE 32.12 Comparison of Surface Finishes by Type and Property

	HASL	OSP	ENIG	Pd	Tin	Silver
Flat	No	✓	✓	✓	✓	✓
Solderjoint	Cu-Sn	Cu-Sn	Ni-Sn	Ni-Sn	Cu-Sn	Cu-Sn
Contact	E-test, ICT	No	E-test, ICT, keypad	E-test, ICT, keypad	E-test	E-test, ICT, keypad
Wirebond	No	No	Al	Au, Al	No	Au, Al
Cost	$	0.7 × $	3 × $	5 × $	0.8 × $	0.8 × $
OEMs	All	All	Most	Few	30	150
Fabs	Most	Most	200	Few	40	250
Reflows	6	2–4	6	6	2–3	6
Shelf-life (months)	18	6	24	24	6	12 mos.
Compliant Pin	+	−	−	−	++	+

REFERENCES

1. IPC-4552, "Specification for Electroless Nickel/Immersion Gold (ENIG) Plating for Printed Circuit Boards," October 2002.
2. IPC-4553, "Specification for Immersion Silver Plating for Printed Circuit Boards," June 2005.
3. IPC-4554, "Specification for Immersion Tin Plating for Printed Circuit Boards," January 2007.

CHAPTER 33
SOLDER MASK

David A. Vaughan
Taiyo America, Inc., Carson City, Nevada

33.1 INTRODUCTION

33.1.1 Definition and Terminology

The term *solder mask,* as used in this chapter, is the same as the terms *mask, soldermask,* or *solder resist,* which are frequently used interchangeably in the printed circuit board (PCB) fabrication industry. The Association of Connecting Industries (IPC) has standardized on the use of *solder mask,* since that is the term most widely used in the industry. The existing definition for "Solder Resist" from IPC-T-50 is, "A heat-resistant coating material applied to selected areas to prevent the deposition of solder upon those areas during subsequent soldering." With the evolution of PCB technology, solder mask coatings are now asked to provide much more functionality than simply a soldering aid.

33.1.2 Purpose

When solder masks were first used, their purpose was to protect the circuitry of the PCB from solder during the assembly operation. Eliminating short circuits was the main objective. As the use of solder mask became more prevalent and the material properties improved, solder mask was used for other functions, such as to provide environmental protection to the assembled board, and to serve as a plating resist for final finishes, dielectric protection, and other functions that take advantage of the properties of the cured mask.

33.1.3 History

Solder mask began with the use of two-part epoxy, thermally curable inks that were an offshoot of inks that were used in the printing industry. These inks were applied by screen printing with the use of a patterned screen, since these materials were not photoimageable. One-part inks that were ultraviolet (UV) curable were available soon after, but these also had to be applied by a patterned screen because they could not be imaged and developed.

Photoimageable solder masks were developed in the mid-1970s and were offshoots of photoresist materials used for patterning the circuitry. Photoimageable solder masks were developed in both liquid (LPI) and dry film (DFSM) and both were used widely. Today, more than 98 percent of the solder mask used is applied in liquid form, and most is photoimageable.

33.2 TRENDS AND CHALLENGES FOR SOLDER MASK

The use of solder mask continues to evolve with the technology. This section discusses some of the issues involved.

33.2.1 Circuit Density

Since the development of circuit boards, the dimensions of the holes and features have been decreasing. Smaller component attachment dimensions require corresponding decreases in solder mask features that challenge the solder mask materials and processes.

33.2.1.1 Feature Size. Smaller solder mask features have eliminated non-photoimageable solder masks from all but the lowest technology designs, and continue to require ongoing improvements in resolution and image reproduction. A very high percentage of solder mask being used is photoimageable.

33.2.1.2 Hole Size. Holes in boards that have been coated with LPI solder mask will have some solder mask material in the holes after application. Typically this material must be removed in the development process. There are very significant differences in how easily different solder mask materials develop from holes. There is also a difference in the amount of solder mask material that is in the holes between application methods, with spray and curtain coating leaving less material than screen printing.

33.2.2 Component Attachment in a Lead-Free Environment

As many PCB fabricators and assemblers transition to lead-free processing, they are faced with solder alloys that have an operating temperature up to 30°C (54°F) hotter than the traditional eutectic tin-lead solders. These increased soldering temperatures raise concerns about the ability of the solder mask (as well as all other materials) to resist embrittlement, discoloration, loss of adhesion, and cracking with repeated exposures to the higher temperatures. Some existing products may not be acceptable for lead-free processing.

33.2.3 High-Density Interconnection (HDI)

HDI technology creates small blind microvias that are plated to provide interconnects to buried layers of circuitry. These small holes can either be coated with solder mask or filled and plated over so that the real estate of the hole can be utilized for a circuit or a component mounting pad (via-in-pad).

If the hole is simply coated with solder mask and the solder mask is not developed from the hole, the air trapped in these small holes may expand and create a bubble or blister, or may erupt, exposing the copper in the hole to potentially corrosive chemistries. These holes should be developed cleanly so that they receive the final finish, should be protected before solder mask application with an inert final finish or be completely plugged.

Plating over microvias requires that they are totally filled, planarized, roughened, and catalyzed before the plating process. If a conductive plugging material is used, the catalysis step typically can be eliminated. Material selection for this application is particularly important because not many materials can withstand the roughening and plating and still maintain good adhesion of the copper to the plugging material after exposure to soldering temperatures. If the via is in a component mounting pad (via-in-pad), the plugging material must not shrink and cause a significant depression (dimple) in the pad. This is especially true for wire bonding applications that require a very flat planar surface.

33.2.4 Environmental

The Restriction of Hazardous Substances (RoHS) Directive 2002/95/EC places restrictions upon specific materials that may be included in PCB fabrication materials, including solder masks. These requirements may require some products to be reformulated. Some products and/or markets may require low-halogen materials to comply with company or industry specifications. Again, these requirements have required some solder masks to be reformulated or new products to be developed.

Many localities have strict air emissions regulations that limit the materials and the amount that may be exhausted to the atmosphere. Individual processes or pieces of equipment may require a permit to exhaust any volatile materials. Hazardous air pollutants (HAPs) and volatile organic components (VOCs) are of particular concern. Limits on these parameters can limit mask selection and application processes. For example, curtain coating and spray mask application are not viable in some areas due to the higher percentage of solvent needed for proper application. Screening utilizes solder mask at a much higher percent solids (less dilution), so air emissions may be only half that of other methods.

33.2.5 Technical Service and Troubleshooting

The solder mask process is quite complex. It relies upon all PCB fabrication processes that preceded it and plays a key role in all processes after it has been applied, through assembly and the life of the PCB. As a result, it is recommended that the specifying authority choose companies that can provide technical expertise and troubleshooting assistance. Companies that offer service and experienced engineering personnel in the PCB fabricators are becoming scarcer, making the supplier's ability to provide service a key issue in solder mask selection.

33.3 TYPES OF SOLDER MASK

33.3.1 Non-Imageable

Early solder masks could not be photoimaged—that is, placed in its exact location with a photographic process. They could not be hardened selectively in one area and then be removed from other areas. These masks had to be applied in the final pattern desired, typically through a patterned screen. The "silk screening" process (the screens are actually nylon) is limited in that it can neither resolve the fine features nor meet the registration requirements and tolerances needed for today's PCB designs. The use of these materials is limited to lower technology products.

33.3.2 Imageable

Higher-density circuitry, with its close tolerances and image reproduction requirements, has mandated the use of photoimageable solder mask. Assemblers need fine dams (or "webs") of

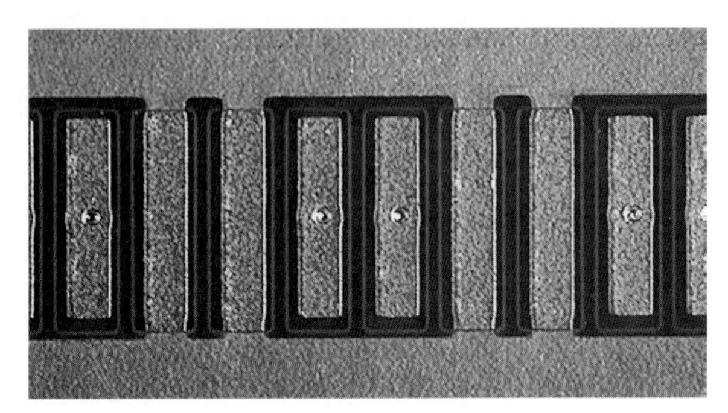

FIGURE 33.1 An example of fine solder dams, or "webs," imaged between fine pitch SMT pads.

solder mask between fine-pitch component leads (see Fig. 33.1) to enable assembly with low defect rates. These webs are frequently as narrow as 0.003 in. and can be as narrow as 0.001 in. A photoimageable mask is required to meet these needs.

Photoimageable masks are available in either liquid or dry-film forms. Liquids account for more than 98 percent of the products used, due to their lower cost, greater availability, ease of use, and acceptability in assembly processing. Dry-film masks require special application equipment, are expensive, are often not robust when used as a resist for final finishes such as ENIG or immersion tin, and are too thick for use in some assembly processes.

33.3.3 Temporary

A number of temporary masks are used in PCB fabrication and assembly processing. The properties of these masks vary considerably depending upon the purpose for which they have been used. The typical application for these masks is to protect circuitry from plating and/or soldering. Various products can be removed mechanically by peeling ("peelable mask") or by dissolving in cleaning chemistry. Since they do not become a permanent part of the finished assembly, there are no industry standards for these products. PCB fabricators and assemblers should be careful to verify the compatibility of any temporary mask with their solder mask(s) to avoid problems such as swelling, blistering, and so on.

33.4 SOLDER MASK SELECTION

33.4.1 Availability and Consistency

There are ongoing changes in the makeup of the supplier base of electronic materials, including solder mask. Many acquisitions have taken place and several products eliminated in the recent past, so users must have confidence that they will have a continued source of supply of solder mask that is the same lot to lot for the future. International Organization for Standardization (ISO) qualification for the manufacturing operation as well as for environmental considerations are good indicators of a supplier with a long-term view.

33.4.2 Performance Specifications

33.4.2.1 IPC-SM-840. This is the predominant solder mask material specification used globally. Most manufacturers test their products to verify that their products are capable of meeting the requirements of this document. The SM-840 includes physical, chemical, thermal, and electrical requirements for two classes of application.

33.4.2.2 Underwriters Laboratories UL94. This specification defines the degree of flammability of a material. Solder mask suppliers generally test their products for flammability according to UL94. The solder mask is applied to a specific type of base material or construction before burn testing. The main variables in this test are the specific base material and construction, the mask type, and the thickness of each. Since most solder mask has no flame retardant, thinner mask and thicker base material will give better results.

33.4.2.3 Telcordia GR-78-CORE. This specification, from Telcordia, is commonly referred to by its previous name, BellCore. With the introduction of the C Revision to IPC-SM-840, the Class T requirements were accepted by the BellCore as equivalent to meeting the GR-78-CORE specification. This specification is still available as part of a package of documents from Telcordia that is quite expensive. However, since Telcordia recognizes the IPC-SM-840 Class T specification for solder mask, it is typically not necessary to purchase a copy of the GR-78 specification.

33.4.2.4 Military Applications. The military specification that applies to solder mask is MIL-P-55110. Since the mid-1990s, the revisions of this document refer to the IPC-SM-840 specification for solder mask requirements.

33.4.2.5 NASA Outgassing. Materials that will be used in space applications are tested under extremely high vacuum conditions. To be acceptable, they must exhibit a maximum total mass loss (TML) of 1.0 percent and have a maximum collected volatile condensable material (CVCM) of 0.10 percent. A list of tested products is available from the NASA web site at http://outgassing.nasa.gov. This site features "Outgassing Data for Selecting Spacecraft Materials," which lists products that the Goddard Space Center has tested (and the tests' results). Not all materials tested have passed the TML and CVCM specifications. Independent test labs can also be used to test for this property.

33.4.2.6 Other Customer Industry Application Specifications (AABUS). Often PCB customers have special requirements for their solder mask. These can be a modification of the IPC criteria or some additional requirement not included in the SM-840 document. There is a term used to describe these requirements: as agreed between user and supplier (AABUS). The key word here is "agreed," since both parties must be completely aware of the requirement and what it takes to achieve the required result.

33.4.3 Environmental and Health Considerations

33.4.3.1 RoHS Compliance. Most PCB fabricators now insist that any solder mask products they use be compliant with Directive 2002/95/EC of the European Parliament and of the Council of 27 January 2003 on the restriction of the use of certain hazardous substances in electrical and electronic equipment, the RoHS directive, and Directive 2003/11/EC, the 24th amendment to Council Directive 76/769/EEC. This directive places requirements on specific materials and amounts of each that may be included in a formulation. Meeting this criterion is essential for most printed wiring board (PWB) fabricators.

33.4.3.2 Lead-Free. One of the materials eliminated with RoHS compliance is lead. The largest impact of eliminating lead is the higher soldering temperatures required for lead-free solders. Soldering temperatures for these alloys are typically 10 to 20°C hotter than for eutectic tin-lead solders. For example, one of the current alloys of choice is SAC-305 (tin-silver-copper solder) has a melting point 34°C higher than eutectic solder. Higher temperatures place additional stress upon the solder mask and could potentially lead to adhesion loss, color change, embrittlement, or cracking.

33.4.3.3 Low Halogen. The low-halogen requirement came about due to concerns about incineration of scrap PCBs generating dioxins under some conditions, and originally centered on some brominated flame retardants used in base laminates. It progressed from that into an additional concern over the chlorine content. This requirement is sometimes referred to as "halogen-free," but in reality, the true requirements place a limit on the amount of bromine, chlorine, and their sum in the content of the material. Current low-halogen requirements on the cured solder mask are as follows:

Organization	Spec-test method	Chlorine	Bromine	Total Halogen
JPCA	JPCA-ES-01-1999	<900 ppm	<900 ppm	
IEC	IEC 61249	<900 ppm	<900 ppm	<1,500 ppm

Although solder masks do not typically contain flame retardants, and hence no bromine, they often contain chlorine. The chlorine generally comes from pigments that give the mask its desired color, and also from residual catalyst from the resin (solder mask raw material) manufacturing process. A number of manufacturers offer low-halogen solder mask materials that meet the specified criteria.

33.4.4 Circuit Density Issues

As the pitch of circuitry and packages has decreased, it has also generated the demand for smaller holes and finer solder mask features.

33.4.4.1 Hole Size. As mentioned in Sec. 33.2.1.2, small hole sizes are a significant challenge for the solder mask process. All application processes introduce some solder mask into the holes, and this material must be removed (developed) from the holes before curing. There are two primary considerations in achieving finished PCBs with no mask in the holes. First, it is easier to achieve mask-free holes when less solder mask is introduced into the holes during the application process. In general, spray applications put the least mask into holes, and single-sided screening introduces the most. Curtain coating and simultaneous double-sided screen printing produce more moderate levels of solder mask in holes.

The ability to remove all solder mask from the holes is dependent upon the solder mask material itself, the tack drying, and the developing processes. Some solder masks inherently develop faster than others, so the choice of products has a significant impact on development. If the solder mask is subjected to an overly aggressive (too hot and/or too long) tack drying process, the mask may not develop at its optimum speed. The developing equipment and process should be set up according to the solder mask manufacturer's recommendations. This includes the chemistry, pressures, flow rates, nozzle type(s), process cycle times, and so on.

33.4.4.2 Solder Dams ("Webs"). To reduce soldering defects in assembly, designers have placed fine solder dams, or "webs," between attachment pads. The width of these dams is frequently in the 2.5 to 3.0 mil range, with some designs requiring dams as narrow as 1.0 to 1.5 mils wide. Producing features this fine is not possible with all solder mask materials, and the imaging process for very fine features will typically be modified from normal production with wider dams.

33.4.5 Assembly Considerations

33.4.5.1 Thickness. When PCBs had all through hole components, solder mask thickness was not of much concern. However, with the advent of surface-mount technology in the 1980s, excessive solder mask thickness caused assembly problems, including:

- Tombstoning of discrete devices in reflow soldering
- Excessive opens with discrete devices in wave soldering
- Solder paste printing problems (excessive paste, solder balls, paste on back of stencil, etc.)

Flip-chip assembly, with its finer pitch, has placed further restrictions on the maximum mask thickness. These thickness restrictions effectively eliminated dry film solder mask in most applications and relegated it to a specialty product for special applications.

33.4.5.2 Cleanliness. After PCBs have been assembled, they must meet specific ionic and visible cleanliness requirements. It is important that PCBs produced with the specific solder mask through the normal process be qualified through the assembly process to ascertain whether the production parts will meet expectations. Each solder mask and flux combination throughout the assembly process may give a different cleanliness result.

33.4.5.3 Solder Balls. Some assembly processes are more prone to generate solder balls than others. Prequalification of the solder mask with the assembly chemistries and processes is essential to predict production performance. If solder balls occur during assembly, in some cases they can be reduced by using a mask with a different surface finish. Satin or matte products sometimes have less of a tendency to have solder balls. Modifying the solder mask cure can have an effect on the level of solder balls also—consult with the solder mask supplier before deviating from its recommended cure process.

33.4.5.4 Assembly Material and Process Compatibility. When qualifying a new solder mask or new PCB supplier, it is critical that all existing materials and processes be evaluated for compatibility before production is scaled up. This should help avoid major issues where there is a large quantity of PCBs awaiting assembly that cannot be assembled with the existing assembly materials and processes. Care should be taken to include evaluation of:

- Soldering fluxes and pastes
- Cleaning agents
- Multiple heat exposures
- Adhesives
- Underfills
- Temporary solder masks
- Tapes
- Anything else in assembly that depends upon the surface of the solder mask

33.4.5.5 Conformal Coating Compatibility. Again, qualification of the solder mask, through all of the fabrication processes, is required to ensure that no problems with production boards arise. Since conformal coating is applied after assembly, a significant amount of processing and exposure to other chemistries will have happened since the solder mask has been applied. The conformal coating compatibility testing should be done on a board that has been through all of the assembly processing. The cleanliness of the PCB surface after the final assembly process has a great impact upon the adhesion of the conformal coating. One cannot expect conformal coating to adhere well to a dirty board. These coatings were not meant to be applied over a no-clean flux, but over a clean substrate.

33.4.6 Application Method for Installed Equipment

Choose solder mask that is compatible with the existing or proposed application method. Not all solder masks can be screen printed, sprayed, and curtain-coated. Consult with the solder mask supplier for the appropriate solder mask for your application method.

33.4.7 Gloss

Solder masks are available in different levels of gloss. Typical terminology for finishes includes gloss, semigloss, satin or semimatte, and matte. These finishes represent a certain level of reflectivity of light and microroughness. Gloss level is generally specified by end users for assembly or cosmetic reasons.

33.4.8 Color

Likewise, different levels of color may be available. Most solder mask is green, but many other colors, such as red, blue, black, yellow, and white, are often specified. Some solder masks are specified to be clear.

33.4.9 Packaging versus PCB

When solder mask is used on substrates that will become part of a semiconductor package. The JEDEC Solid State Technology Association (once known as the Joint Electron Device Engineering Council) specifications apply in addition to the IPC specifications. JEDEC specifications include additional, more aggressive testing that generally requires a more robust solder mask. Low moisture absorption is important for a mask used in packaging applications.

33.4.10 Surface Finish Compatibility

The choice of the final surface finish for the copper has an effect upon how the solder mask performs. And, with the elimination of lead in solder, the compatibility of new final finishes needs to be confirmed before their implementation. Most of these finishes are applied after the solder mask so the process for mask application can have a significant impact upon final quality of the PCB:

- **Hot air solder leveling (HASL)** This has been the most common final finish, utilizing 63/37 eutectic or 60/40 tin/lead (Sn/Pb) alloys. The PCB is coated with flux, immersed into molten solder, and withdrawn through air knives to remove excess solder. Lead-free solders are expected to require solder temperatures significantly hotter than what has been required for eutectic solder. The higher temperatures are expected to place higher stress on the solder mask and may also present more difficult post-HASL cleaning challenges.
- **Electroless nickel/immersion gold (ENIG)** ENIG is typically applied to the metal surfaces of the PCB after the solder mask process. Each plating chemistry supplier has its own ENIG bath, and each has a different level of aggressiveness to the solder mask. Consequently, the solder mask and ENIG chemistry must be qualified together to avoid breakdown of the solder mask over circuits or around openings on metal. A thicker coating of

solder mask may alleviate solder mask breakdown problems. Excessive microetching before plating can undermine the solder mask and induce tape test failure as the overhanging mask is broken off.

- **Immersion tin** The same comments from ENIG also apply to immersion tin.
- **Immersion silver** Immersion silver is relatively benign to the solder mask, and few, if any, issues are known as of the writing of this book.
- **Organic solderability preservative (OSP)** This is another process and chemistry that has little impact on the solder mask.

33.5 SOLDER MASK APPLICATION AND PROCESSING

33.5.1 Surface Preparation

Solder mask should be applied to a clean, dry surface. The cleaning step should remove all particulates, oils and greases, and unwanted oxides. In the case of bare copper, this step should provide some "tooth" to the copper to enhance the mechanical bond of the mask to the copper. The cleaning step is followed by a rinse and drying step that leaves the panels ready for coating with mask. The surface preparation technology chosen is dependent upon the metallization of the circuitry and, to a lesser extent, the base laminate material.

33.5.1.1 Copper Circuitry. A large majority of panels produced have solder mask coated over bare copper (SMOBC). The goals of the surface preparation process are to remove all unwanted oxides and contaminants from the copper and to impart a microroughness or "tooth" to the surface, and to leave a clean, dry surface for coating.

33.5.1.1.1 Incoming Copper. The quality of the surface of the copper is very important for achieving good solder mask adhesion. All traces of the etch resist metallization, typically tin or solder, must be removed prior to the solder mask surface preparation process. Solder mask does not adhere well to tin or solder residues or tin-copper intermetallic surfaces.

33.5.1.1.2 Preparation Methods. The most common surface preparation method is pumice scrubbing with either a silica or aluminum oxide media. Most equipment in use is brush pumice, but jet pumice may be used also. With this method, the main considerations are keeping the pumice concentration within recommended levels, periodic changing of the media, and good equipment maintenance. A prepared surface from a proper pumice cleaning will have a uniform rosy pink, matted surface that is free from stains. Considerations in pumice-based preparation techniques include the following:

- **Brush pumice versus jet pumice** Both techniques utilize a slurry of pumice in water. With brush pumice the slurry is pumped onto the panel surface under low pressure, and rotating brushes scrub the pumice against the panel surface to clean and roughen it. With jet pumice, the slurry is sprayed through nozzles at high pressure onto the panel surface; the impact of the particles against the panel surface provides the cleaning and roughening.
- **Silica versus aluminum oxide** The media used in pumice scrubbing degrades with usage. Silica pumice is softer than aluminum oxide and must be changed more frequently, but it is much less expensive; thus relative cost differences in operation may not be significant. Silica pumice particles degrade by fracturing into smaller pieces and not providing the needed impact to roughen the surface. Aluminum oxide particles become rounded with age and will "peen" the surface instead of roughening it. Both conditions lead to a less than optimum solder mask adhesion.

Chemical cleaning, usually in the form of a microetch, may be used to prepare the copper surface. Not all microetch chemistries will provide the proper tooth for best adhesion, so a product should be selected that provides some tooth (a matte surface is a good indicator) instead of a bright shiny surface. Good rinsing and drying, and possibly a passivation chemistry, are important to avoid rapid staining of the copper since it is very active after chemical cleaning.

Abrasive brush or compressed pad technology is generally acceptable for solder leveled panels, but if other final finishes such as ENIG or immersion tin are used, it may not provide sufficient adhesion. The abrasive in the brushes leaves directional scratches in the copper that can lead to the wicking of plating chemistries under the mask that will cause loss of adhesion around mask openings on copper. If this method is to be used, it should be followed by a chemical cleaning that will provide the microroughness needed for good adhesion.

A very good alternative to abrasive or chemical cleaning is to impart an oxide to the copper surface. If the copper has not been cleaned or if there are residual etch resist metals on the surface, the copper will not oxidize. The oxide provides a fine structure and microroughness for very good adhesion. When using this technique, the fabricator should use a "reduced oxide" so that cleaning chemistries in subsequent processes will not attack the oxide layer and cause adhesion loss at the mask-copper interface.

33.5.1.1.3 Rinsing and Drying. After the cleaning step, the panel surfaces must be thoroughly rinsed with good quality water. City or well water is often inadequate due to high levels of minerals, particulates, and other contaminants. Poor water quality can lead to blistering of the mask and/or discoloration of the copper under the solder mask. Deionized (DI) water is the best rinsing agent, but reverse osmosis (RO)-treated water is also used effectively.

After the panels have been rinsed, the drying step should remove the water from the panels—both from the surface and from the holes. Air knives should be incorporated to blow out the holes and sweep the water from the surface. It is undesirable to dry water on the boards since any droplets that are dried will leave behind any dissolved minerals and other contaminants that had been in the water.

33.5.1.2 Other Metals

33.5.1.2.1 Gold, Tin, or Solder. If the gold-plated surface is perfectly clean, no surface preparation is required. However, if there is any question about cleanliness, or if the panels have been exposed to any chemical vapors or particulates, or have been awaiting solder mask for any appreciable time, they should be cleaned. Depending upon the extent of surface contamination, the panels can be only rinsed and dried, or may be cleaned in a mild detergent solution before rinsing and drying. Abrasive cleaning is not recommended.

33.5.1.2.2 Mixed Metals. Two good examples of the use of mixed metals are for panels that have been selectively solder-stripped and for panels that have had fingers or contacts plated with gold before the solder mask process. The surface preparation process must be selected based upon the most delicate of the metals on the panel.

33.5.2 Solder Mask Application

33.5.2.1 Liquid Masks. Solder masks applied in liquid form can be either photoimageable or not. Most solder mask in use today is photoimageable, meaning that after application it will be patterned in a photolithographic process. A non-imageable product must be applied only to those areas to be coated, and areas where metal or base material must be exposed cannot be coated.

33.5.2.1.1 Application—Liquid Photoimageable (LPI) Solder Mask. LPI solder masks are applied typically by screen printing, spray, or curtain coating. Although each method has pros and cons, all are used successfully in the industry.

Screen printing is the most common technique of applying solder mask. This technique utilizes a non-imaged, tightly stretched screen of precise construction and a squeegee to apply the solder mask through the screen onto the panel. The screening technology includes a wide range of equipment and techniques, from single-sided hand screening to fully automated double-sided equipment:

- **Single-sided coating (hand or flat bed)** In this application, each side of the panel is fully coated (except for a border on each side) with solder mask, one side at a time. The panel may be tack dried between coating the two sides, or the second side may be coated immediately using a dimple plate to protect the coating on the first side. It is advantageous to coat both sides of the panel before tack drying so that both sides see the same tack-drying cycle. One concern about this technique is the amount of solder mask that is pushed into the holes. Since this is a single-sided process, there is nothing to restrain the amount of mask that gets pushed into the holes, so it may be more difficult to develop all of the mask out of the holes.
- **Double-Sided Coating** Vertical double-sided screen printing applies solder mask to both sides of the panel or panels simultaneously. The screening units used are available in various levels of automation, up to fully automated screening units that are integrated with tack drying ovens. Because they apply mask to both sides at the same time, the units put less mask into holes, making it easier to develop mask out of holes—a particularly important consideration when working with small holes that must be free of solder mask. Front-to-back squeegee alignment is critical for putting the least amount of solder mask into the holes.

A key variable in screen printing is the screen mesh. Most applications utilize a non-calendared, nylon mesh material. These products are designated with the number of threads per inch and the diameter of the threads in microns. Typical examples of commonly used mesh counts are 86-120, 83-100, 86-100, 92-100 or 110-80. Supplier literature will provide a theoretical ink volume for each type that can be used to determine the relative difference in coating weight to be applied to the panel.

Table 33.1 provides an overview of the screen process key variables and the effect of operating a process at the extreme of that variable.

Spray coating accounts for a limited percentage of the solder mask coating. There is renewed interest in spray coating since it can allow more latitude in removing solder mask from holes. As the holes become smaller, some solder masks and processes have difficulty achieving clean holes after development. At the same time, because the solder mask is sprayed at reduced viscosity, the amount of solvent (VOCs) that is exhausted to the environment is significantly higher than for screen printing, the use of spraying may be limited in some localities.

TABLE 33.1 Screening Process Variables and Effect Operating at the Extremes

Process variable	Effect of low	Effect of high
Mesh Tension	Mesh sticking to screen Poor quality printing Low coating thickness	Nonuniform coating across panel
Squeegee Pressure	Skips (areas with no solder mask)	Low coverage over circuits
Squeegee Speed	Lower circuit coverage	Higher coverage
Squeegee Shape	Very sharp edge, lower coverage	Slightly rounded edge, higher coverage
Squeegee Angle (from panel surface)	Lower coverage, Less ink in holes	Higher coverage, More ink in holes
Snap-off	Screen sticking to panels	Skips or difficult printing
Ink Viscosity	Lower coverage, More flow or sagging if too low	Higher coverage, Bubbles if too high

Two basic types of spray equipment have been used for spray coating PCBs with solder mask, HVLP (high volume, low pressure), and electrostatic. HVLP spray systems are most common. Single- and double-sided spray units are available.

The solder mask is sprayed at a reduced percent solids compared to screen printing, so circuit coverage and tack drying can be more of a problem, especially if the ink and/or spray guns are not heated. Heating of the ink allows more solvent to evaporate between the nozzle and the panel, resulting in a higher viscosity liquid on the panel. The higher viscosity theoretically allows greater circuit coverage with less ink.

The conventional HVLP spray process utilizes a low pressure stream of solder mask liquid that is atomized by a high volume stream of air. The atomized solder mask is directed to the PCB surface in a specific pattern so that as either the panel or the spray head(s) move, a uniform coating is applied to the panel. In some systems, the ink is heated just before spraying to lower its viscosity and aid in the evaporation of solvent between the spray nozzle and the panel. Typical process variables for HVLP spray include ink viscosity, ink (pot) pressure, atomization air pressure, conveyor speed, and ink and atomization air temperatures.

In a typical electrostatic spray application, the ink is atomized and given an electrostatic charge while the PCB panel is grounded to attract the ink particles. The intent is to limit the amount of overspray and provide a more even coating. In reality, with the infinitely variable PCB designs to be coated, there are often many ongoing adjustments required to minimize Faraday effects that cause uneven distribution of the spray. Often solder mask formulations need to be modified for effective use in electrostatic spray applications. The main process variables in electrostatic spraying are solder mask temperature and viscosity (dilution), bell speed and voltage, back plane voltage, and shaping air pressures. These variables are adjusted to provide a uniform coating of the desired thickness or weight.

Table 33.2 lists the process variables and the effect of operating at the extremes of that variable.

Curtain coating is an application method where the panel to be coated is rapidly passed through a continuous falling "curtain" of solder mask. The curtain is created by pumping the solder mask through a die with a slot of controlled opening. The curtain falls vertically through an opening between two conveyors that carry the panels horizontally through the curtain, thus coating the panel on its upper side. Solder mask that is not coated onto the circuit board panels falls into a trough between the conveyors and is recirculated. The pump and conveyor speeds and the gap in the die control the coating thickness. Solder mask used in curtain coating applications must be coated at reduced viscosity or percent solids (as in spray applications), so the increased VOCs emitted to the atmosphere may limit curtain coating's acceptability in some localities.

TABLE 33.2 Key Variables for Spray Coating and Effect of Low and High Levels for Each Variable

Process variable	Effect of low	Effect of high
Ink Viscosity	Low circuit coverage, Tack drying difficulties	Mottled or "orange peel" coating, excessive atomization, and overspray
Ink Temperature	Low circuit coverage	Possible clogging of lines or nozzle
Ink (Pot) Pressure	Thin coating, low coverage	Thick coating, drying problems
Atomization Air Pressure	Mottled surface	Excessive overspray
Wet Weight or Wet Film Thickness	Low circuit coverage	Thick coating, insufficient drying, and/or bubbling

TABLE 33.3 Key Variables for Curtain Coating Solder Mask Application and the Effect If the Process is Operated at the Extremes of That Variable

Process variable	Effect of low	Effect of high
Ink Viscosity	Low circuit coverage, dewetting, Streaks or breaks in the curtain ("blips"), Sagging of the coating	Unstable curtain, streaks, improper leveling, skips
Ink Temperature	Coat at incorrect percent solids, Sagging of the coating, Thin coverage	Unstable viscosity, high solvent evaporation, Inconsistent leveling
Pump Pressure	Thin coating, low coverage	Thick coating. drying problems
Conveyor Speed	Excessive coating	Insufficient coating, Ink in holes, Teardropping on bottom side
Wet weight or wet film thickness	Low circuit coverage	Thick coating, Insufficient drying and/or bubbling, Sagging of coating
Air flow around the curtain	None	Unstable curtain, "blips," Nonuniform coating

Curtain coating is a highly productive process that gives uniform coating across panels. It has good flexibility for coating panels of different sizes (above a minimum) and gives very high material utilization.

The quality of the coating from this process is highly dependent upon the ability to maintain a uniform, stable curtain of solder mask. The solder mask must be maintained at the specified temperature and viscosity to avoid coatings that are too thick or thin, streaks, skips, or breaks in the curtain that cause no coating in localized areas, commonly referred to as "blips." Curtain coating is sensitive to circuit height and typically requires process changes as height increases to avoid skips in the coating. Thin panels may become airborne at high conveyor speeds, and thick panels may slip on the conveyor and not accelerate to proper coating speed.

Typical process variables for curtain coating include solder mask viscosity and temperature, pump pressure, die (gap) opening, conveyor speed, and air flow. Table 33.3 lists the key variables for curtain coating and the effect of operating the process at the extremes of each variable.

33.5.2.1.2 Tack Drying—Photoimageable Solder Mask. LPI solder mask inks received from the supplier contain solvents to control the viscosity in the correct range for their intended application process. These solvents must be removed to harden the surface of the coating so that it may be handled and processed through the exposure step. Because the coated solder mask is wet and tacky, tack drying should closely follow the mask-coating process to reduce the possibility of handling damage and airborne debris becoming embedded in the wet ink. Care must be taken that panels do not touch each other in tack drying because they will mark each other and will not dry properly due to localized inadequate air flow across the panel surface. (Dry film, UV, and thermally cured solder masks do not require tack drying.)

Panels that have had insufficient tack drying will be sticky (phototools will stick to the solder mask); they may also exhibit "artwork marking," where the pressure from the exposure process causes the mask surface to be smoothed, changing the cosmetic appearance of the surface. If solder mask has excessive tack curing, it will become difficult or impossible to develop—a condition called "lock-in."

Oven types used for tack drying include batch ovens, tunnel ovens, and infrared (IR) ovens.

Batch ovens can be as small as a bench-top unit where panels may be inserted individually or as large as a walk-in oven that may be loaded with many panels in racks or carts. The wet panels are loaded through a door or doors that are gasketed or sealed to the operating area so that the vapors emitted are not exhausted into the work environment.

Total tack drying time in a batch oven is the sum of the ramp-up time plus the desired time at temperature. Ramp-up time of the panels is dependent upon the initial temperature of the oven, the heating capacity of the oven, the amount of makeup or purge air, and the loading (thermal mass) of the panels, racks, and carts placed in the oven at start-up.

Cure times are generally specified to be a specific time at temperature. Once the ramp-up time has been determined for the specific temperature and worst-case loading (generally a full oven load), the total time for the cure process is the ramp-up time plus the cure time. Oven doors should not be opened during a cure cycle because cure volatiles may escape into the environment and temperature will be lost and potentially cause incomplete curing.

Pros of using batch ovens include the ability to load large quantities of panels of any thickness or design at one time with minimal labor. Cons include having to wait for the oven to be empty before starting a new cycle. Opening of doors causes significant drop in temperature and may cause under curing if the time is not adjusted for the time below curing temperature.

In conveyorized *tunnel ovens,* panels are generally placed in slots in the conveyor or in racks that ride on the conveyor. The main controls are temperature and conveyor speed. Exhaust from the oven serves the same purpose as purge air and should be regulated per the oven manufacturer's specification. Curing is typically uniform and is independent of panel design and construction.

With IR ovens, panels typically lay flat and are processed one at a time, and cure times are very rapid. Multiple cure recipes are usually needed to compensate for differences in panel design, materials, and construction. Proper handling of the exhaust is important, as with all curing ovens.

Several issues must be considered in the tack drying process:

- **Time at temperature** Proper curing requires specified time at temperature. Ovens should be set up with a profiler to ensure that variables such as oven loading, heating capacity, ramp-up time, and so on, compensated for when establishing process conditions.
- **Air flow** Air flow in ovens must be parallel to the panel surface so that there is uniform temperature across the panels and the solvent is swept away from the coated surface.
- **Debubbling time** With some solder mask applications, it is important to have some time (usually 5–10 min. maximum) for the coating to debubble before the tack drying occurs. This allows time for the bubbles to break and for the coating to heal itself.
- **Purge air** All ovens need fresh air to be drawn into the oven to replace solvent-laden air that is exhausted. Without sufficient purge, air panels will not dry properly.
- **Condensate** Volatiles from the solder mask can condense on cool surfaces of the oven and ductwork. This material must be cleaned periodically so that it does not plug air passages. Proper safety equipment should be worn during cleaning.

See Table 33.4 for a list of oven types and their advantages and disadvantages.

33.5.2.1.3 Exposure. Exposing photoimageable solder mask requires equipment that provides the needed intensity of light of the proper wavelength to expose the solder mask to an extent that the exposed mask will withstand the developing process without washing away, lifting, or getting surface attack.

TABLE 33.4 Solder Mask Drying Ovens and Their Associated Advantages and Disadvantages

Oven type	Pros	Cons
Batch	Productivity—large quantities with minimal manpower Low investment Not dependent on panel design, materials, or construction Consistency	Potential for nonuniform temperature Opening oven doors causes temperature to drop Cure time may be oven-loading dependent
Tunnel	Consistency Uniformity Not dependent on panel design, materials, or construction	More handling Higher investment Thin panels may need a fixture to prevent them from bending
Infrared (IR)	Very fast cycle time Productivity	Multiple recipes usually required to compensate for different materials, designs, and constructions Cannot cure every PCB design/construction Requires automation or extra handling High investment

Exposure equipment must provide a means of holding a phototool in intimate contact with the surface of the panel being exposed and provide the required amount of light (millijoules/cm^2) of the needed wavelength distribution (which may depend on the solder mask) at the appropriate intensity (mW/cm^2). Unfortunately, many exposure devices that are used for exposing solder mask are substandard, including many units that were designed many years ago for exposing primary imaging photoresist. Modern units have high-intensity lamps (7 kW or higher) and sufficient lamp and exposure frame cooling to provide a high-quality image very quickly.

Lamp intensity and energy output should be monitored routinely with a radiometer sensitive to the appropriate wavelength to verify that the lamps are performing as expected. Intensity will decrease with lamp hours, making exposure times become longer. At the same time, the infrared output will increase, potentially causing frame cooling issues. Hot exposure frames cause artwork to stick to the mask, resulting in artwork marking or mask to transfer to the phototool.

Laser units are available for exposing photoresist that have been used for exposing solder mask. These solder masks may be specially formulated to be sensitive to the laser wavelength. Laser imaging eliminates solder mask registration problems, but is not as productive as conventional exposure. Equipment cost is very high compared to conventional lamp exposure equipment.

Underexposure will result is insufficient polymerization to withstand the development process properly without damage to the solder mask. Typical symptoms may include a dull or chalky surface, lifting of small features (e.g., solder dams), excessively undercut sidewalls, or lifted mask.

Overexposure causes image growth where the exposed feature is larger than on the artwork and openings in the solder mask will be smaller. If the growth is excessive, the mask may grow onto the pads.

33.5.2.1.4 Development. Development of the unexposed solder mask material must be sufficient to clean out completely the smallest holes that must be free of mask.

Photoimageable solder masks may be developed in either aqueous or solvent chemistries. The developer chemistry needed depends on the mask formulation. In the development process, solder mask that has not been exposed to UV light and polymerized is washed away, or developed, from the PCB panel to reveal the desired pattern of solder mask.

For aqueous development, the chemistry is a warm, mildly alkaline (pH 10.6–11.3) solution of sodium carbonate (Na_2CO_3) or potassium carbonate (K_2CO_3). Proprietary chemistries are also available. Antifoam is generally used to control foaming as the development solution loads with developed mask.

Solvent-developing chemistry typically uses gamma-butyrolactone (also known as "GBL" or butyrolactone) or butyl carbitol (2-(2-Butoxyethoxy)ethanol) solvent.

The development process must be set up to ensure that the smallest holes are free of undeveloped solder mask. This is very different from photoresist because the breakpoint will often be 10 to 15 percent of the development chamber instead of nearly 50 percent with photoresist.

Key process variables for the development process are the time in the developer and rinse sections; solution temperature, pressure, and nozzle type; flow rate; and configuration.

It is very important to have an inspection step immediately after development to check for registration, mask in holes, feature resolution, dam retention, lifting, and so on. Once the panels have been cured. it is significantly more difficult to strip and rework the solder mask, and the rework process is more likely to cause the panels to be scrapped.

33.5.2.2 Dry-Film Solder Mask (DFSM) Application. The process for DFSM differs from LPI solder mask only in the application step. Surface preparation, exposure development, and cure are the same as for LPI solder masks.

33.5.2.2.1 Application with Vacuum Lamination. DFSMs typically have no solvent in them to enable them to conform to circuits without entrapping significant amounts of air along or between circuit traces. To apply these films to circuit topography, a laminator must be able to laminate panels in a vacuum chamber. The predominant type of vacuum laminator has heated upper and lower platens. The panel or boards to be coated will have DFSM gently applied (without any pressure) to one or both sides and are placed into the vacuum chamber formed between the upper and lower platens. The platens are closed together and a seal is formed. Vacuum is drawn in the chamber. When a specified time has been reached (determined by the board type and thermal mass, DFSM type and thickness, and board/panel plated circuit height), the flexible diaphragm on the upper platen (which had been held tightly against the upper platen surface by vacuum) is released to slap down onto the panel surface. The heat will have made the DFSM plastic so that it will conform to the circuits, and the vacuum prevents poor lamination along circuits. The lamination is completed when the platens separate and atmospheric pressure forces the DFSM into all voids.

33.5.2.2.2 Roll Lamination. Lamination of DFSM with a standard hot roll laminator is unacceptable for all except the least demanding applications. It is impossible to laminate a flat film of solder mask to PCB topography without trapping air along circuits. There are vacuum hot roll laminators that will work satisfactorily, but these are not widely available.

33.5.2.3 Non-Imageable

33.5.2.3.1 UV and Thermally Curable Solder Masks. Because these products are not photoimageable, they must be applied with patterned screen printing in the exact pattern that is needed on the finished PCB. After their application, they are cured by the appropriate cure recipe as dictated by the supplier. There is no opportunity to develop off solder mask from areas where it is unwanted, so registration of the screen to the panel is very important.

33.5.2.3.2 Ink Jet Application. This technique utilizes digital data and ink jet technology to coat solder mask only in areas where it is wanted. This eliminates the photoimaging steps and should reduce or eliminate registration problems. As of the writing of this book, no such machines are being used in production for solder mask application.

33.5.3 Curing

33.5.3.1 Cure Purpose. All solder mask products require a cure step or process for the solder mask to achieve the desired material properties. Curing typically increases hardness and the thermal and chemical resistance. The desired curing process is determined by the specific solder mask used. Consult the technical data sheet for the product and follow the supplier's recommendations for the appropriate cure.

33.5.3.2 Types of Cure—UV, Thermal, IR.
33.5.3.2.1 Thermal Cure. Nearly all solder masks require a thermal cure. It is generally specified as a specific time at temperature. When calculating the correct time, the fabricator must add the temperature ramp-up time to the required time at temperature to arrive at the total time in the oven. For a conveyorized oven, a thermal profile will show the time above the minimum temperature.

If the thermal cure is to be accomplished with an infrared oven, the time will be much shorter but temperatures will be far greater than for a convection oven. Also, a number of recipes are usually needed to compensate for materials with different IR absorbance and panels with different thermal mass or thickness.

33.5.3.2.2 UV Cure. Some solder masks require high-intensity UV light to cure. This is typically applied in a conveyorized UV curing unit. Calibration of the curing unit is important to ensure the solder mask is cured properly.

33.5.3.3 Effects of Undercure and Overcure. Improperly cured solder mask can exhibit poor properties that will become apparent in subsequent processing. Care must be taken to ensure that the solder mask is cured according to the supplier's recommendations. The following are some of the potential problems caused by improperly cured solder mask.

Undercure problems include:

- Blistering or adhesion loss in final finish or assembly
- High ionic contamination
- Residues from assembly soldering
- Solder balling
- Handling damage and soft coating

Overcure problems include:

- Embrittlement and cracking
- Adhesion loss

33.5.4 Stripping Solder Mask

Stripping of solder mask is something to be avoided, but at times is required due to problems such as misregistration, incorrect or wrong sized artwork, handling damage, and so on. It is important to recognize problems as early in the process as possible since the stripping process must be more aggressive after each process step.

33.5.4.1 Stripping after Coating. At this stage, the solder mask may be removed easily. An LPI solder mask can simply be developed off. (*Note:* The mask should be tack dried first to avoid contaminating the developer with wet mask and/or solvent.). UV and thermal mask must be removed manually with an appropriate solvent.

33.5.4.2 Stripping after Photoimaging. The stripping task becomes more difficult. Aggressive chemistries, typically highly caustic and at high temperature, are required.

These chemistries are aggressive and the process, if not controlled, can damage the butter coat of the laminate and render the panel scrap. Small holes filled with solder mask may not be able to be cleaned out in the stripping operation without damaging the panel.

33.5.4.3 Stripping after Cure. Stripping solder masks is most difficult at this stage, and doing so in fact may be impossible for some solder masks. The time required to strip the solder mask may allow exposed portions of the laminate to become damaged and cause the panel to be scrapped. Every effort should be made to identify panels needing stripping before curing.

33.6 VIA PROTECTION

Via protection is an increasingly complex issue for PCB designers and fabricators. There are multiple options for protection—organic and conductive, platable or not, before or after final finish, and more. The materials and processes for each type are not the same, so there is no single answer, and few answers that are easily performed with high productivity. Inks for plugging come in a variety of processing types, including thermal cure, UV cure, photoimageable, conductive, platable, and so on. The panel design will narrow the choice for the specific application.

33.6.1 IPC-4761, "Design Guide for Protection of Printed Board Via Structures"

IPC-4761, published in July 2006, gives excellent guidance for how best to protect vias and what not to do. It does not give specifications for the plugging materials themselves, but a future document is expected to do so. This document calls out a number of classes of via protection that can be used by designers and fabricators.

One very important message from this document is that vias should be protected from both sides, not only one side. If single-sided protection is used before an "inert" final finish, a ring void of copper results at the barrel-plug interface that is subject to corrosion due to entrapped microetch, flux, or other corrosive chemistry. Also, some of these materials leach out in subsequent processes and can kill plating baths, cause high ionics, induce electrochemical migration, or present other long-term reliability concerns.

33.6.2 Material Specifications

Although there are no "official" specifications for the materials used for plugging, there are some specifications that apply to solder mask that should be considered when using a material other than solder mask to protect vias. UL94 flammability and IPC-SM-840 tests—including especially resistance to solvents and cleaning agents, as well as non-nutrient testing—may be necessary. Dependent upon the specific circuitry design, other specific tests from IPC-SM-840 or other specifications may be appropriate.

33.6.3 Material Selection Considerations—Solder Mask versus Specialty Inks

33.6.3.1 Shrinkage—Percent Solids. LPI solder masks have solvent in the formulation to control viscosity and tackiness. Drying the solvent causes the plug of mask in the via to shrink, typically causing a dimple and some thinning of the coverage on the knee of the hole. This may not be a problem if the plug is covered with another coating of solder mask or if the annular ring of the via does not pick up solder or other final finish. Specialty hole plugging inks are available that are very high in solids and shrink very little, as are 100 percent solid inks that eliminate this concern.

33.6.3.2 Platability—Cu Adhesion. More frequently, designers are specifying that the material used to plug the hole be platable so the real estate can be reclaimed as in via-in-pad designs or in subcomposite structures. Such applications require a material that can be roughened, catalyzed, and plated with good copper adhesion.

33.6.3.3 Image Definition. If the vias to be filled are in very close proximity to areas that must be solderable (e.g., surface-mount technology [SMT] pads) it may be desirable or, mandatory to utilize a plugging product that is photoimageable to ensure the ink does not encroach upon solderable surfaces.

33.6.3.4 Other Properties. One property important for vias needed to conduct thermal energy is thermal conductivity. Products used to fill such vias are generally highly loaded with a material with good conductivity such as silver or copper. Some designers believe that boards that must operate at high temperatures should have plugging materials with low coefficient of thermal expansion (CTE) and high glass transition temperature (T_g), but there is little data to substantiate benefits of such materials.

33.7 SOLDER MASK FINAL PROPERTIES

Final properties are often characterized into physical, electrical, mechanical, and chemical properties. Most people who specify solder mask utilize the criteria in the IPC-SM-840 document when specifying desired mask performance. There are other industry and regional specifications but the IPC specification is recognized worldwide.

33.8 LEGEND AND MARKING (NOMENCLATURE)

Many boards still require legend and marking to identify part numbers and component locations, for company logos, date codes, and a multitude of other reasons. These can be etched into the copper or, more typically, added in a separate process after solder mask. For good legibility, materials are of a color that contrasts with the solder mask color.

33.8.1 Types

33.8.1.1 UV. These materials are not photoimageable, but are applied in the specific desired pattern through a patterned screen and then are cured with high-intensity UV light.

33.8.1.2 Thermal. These materials are not photoimageable, but are applied in the specific pattern desired through a patterned screen and then are thermally cured, generally in a batch or conveyorized oven.

33.8.1.3 Photoimageable. These inks are applied to large areas of panels or over complete panels, as with an LPI solder mask. They are tack dried, imaged, developed, and cured, often at the same time as the solder mask is cured. The advantages of these products are that they eliminate expensive screen making and achieve excellent resolution, even over the topography of circuits.

Special inks have been formulated for LPI legend applications that have higher pigment loading. The advantage of these inks is that they can be coated much thinner while achieving good color contrast. This increases the productivity through the tack and developing processes and gives a more robust finished product since it is thinner and less likely to be chipped.

Standard solder masks of contrasting color may be used but they will have to be coated thicker to achieve similar contrast. The higher sidewall thickness makes the legend more susceptible to chipping.

33.8.1.4 Ink Jet. These are actually UV and/or thermal inks that are applied with an ink jet. A limited number of these machines are in production as of early 2007. Initial equipment cost is very high, as is the price of ink, and progress in productivity is expected in the future.

33.8.2 Legend Specifications

33.8.2.1 IPC. No approved IPC specification exists for legend inks. A committee is drafting a specification, IPC-4781, that should be issued in 2008.

33.8.2.2 Military. The military has an obsolete specification that was written years ago around a two-part epoxy ink that most MIL applications still specify. Many of the criteria (such as shelf life, pot life, composition, etc.) in this specification are irrelevant to the final legend performance, but they are still enforced as long as the specification is used. More recently the specification for legend ink called out for many military applications is CID A-A-56032D, entitled "Commercial Item Description—Ink, Marking, Epoxy Base." Although military specifications call out epoxy-based marking inks, a number of fabricators have had other inks approved on a waiver basis for military applications.

33.8.2.3 UL94. When areas of legend ink on the board are larger than a UL flame strip (1/2 × 5 in.), it is treated as if the legend is another coating that needs UL qualification. If all areas are smaller than the flame strip dimension, no separate UL qualification is needed.

33.8.3 Legend Performance

33.8.3.1 Legibility and Resolution. The legend must be readable to be useful. As circuitry becomes more dense, the feature sizes needed for legend become smaller also, causing processes to be reestablished to give the desired quality. Some fabricators have switched to a photoimageable (LPI) legend to give the desired resolution for the legend on high-density circuitry.

33.8.3.2 Adhesion. The legend must be applied to a clean, dry surface and be properly cured to achieve optimum adhesion. This can be a challenge when legend is applied after HASL or other final finish that could leave a residue (such as HASL flux) on the surface. Legend adhesion must be evaluated in combination with the solder mask and its processing. Different legend materials have different adhesion to different solder masks.

33.8.3.3 Final Finish Compatibility. Like solder mask, legend materials that are applied before the final finish must survive the final finish processes and chemistries. Not all legend materials will survive all finishes and these must be qualified together, the same as with solder mask.

CHAPTER 34
ETCHING PROCESS AND TECHNOLOGIES

Marshall I. Gurian, Ph.D.
Marshall Gurian Consulting, Tempe, Arizona

34.1 INTRODUCTION

One of the major steps in the chemical processing of subtractive printed boards is *etching,* or removal of copper, to achieve the desired circuit patterns. Etching is also used for surface preparation with minimal metal removal (microetching) during innerlayer oxide coating and electroless or electrolytic plating. Technical, economic, and environmental needs for practical process control have brought about major improvements in etching techniques. Batch-type operations, with their variable etching rates and long downtimes, have been replaced completely with continuous, constant-etch-rate processes. In addition, the need for continuous processing has led to extensive automation along with complete, integrated systems.

The most common etching systems are based on alkaline ammonia and cupric chloride. Other systems include peroxide–sulfuric acid, persulfates, and ferric chloride. Process steps include resist stripping, precleaning, etching, neutralizing, water rinsing, and drying. This chapter describes the technology for etching high-quality, fine-line (0.003 to 0.005 in) circuits in high volume at a practical cost, as well as continuous processing, constant-etch rates, and control at high dissolved copper capacities. Increasingly, the production of uniform and constant feature geometries calls for precision and statistically robust control of the circuitization processes and materials.

There remains the ever increasing need to balance the process selection according to the factors of cost, environmental and regulatory compliance, stable factory productivity, low worker intervention requirements, compatibility with board design, and construction innovations. It is significant that environmental concern has eliminated the use of chromic–sulfuric acid and ammonium persulfate etchants and chlorinated solvents. Limitations on chlorine gas and volatile organic emissions loom as a significant factor in determining future process choices, and pressure to eliminate brominated compounds and lead content may have consequences for processing mandated by material developments. More than ever, the simplification of process selection and vendor support appear to drive process selection in the industry.

Typical procedures are given for etching organic (i.e., dry film) and metal-resist boards, and for innerlayers. Strippers and procedures for resist removal are described based on resist selection, cost, and pollution problems. The properties of available etchants are also described in terms of resist compatibility, control methods, ease of control, and equipment maintenance. Other considerations include chemical and etchant effects on dielectric laminates, etching of thin-clad copper and semiadditive boards, solder mask on bare copper (SMOBC), equipment selection techniques, production capabilities, quality attained, and facilities.

34.2 GENERAL ETCHING CONSIDERATIONS AND PROCEDURES

Good etching results depend on proper image formation in both organic innerlayer print-and-etch and plated-metal etch resists. Etch personnel must be familiar with screened, photosensitive, and plated resists commonly used. The etching of printed boards must begin with suitable cleaning, inspection, and pre-etch steps to ensure acceptable products. Increasingly, metal foil or plating uniformity in structure thickness and freedom from coatings and defects must be maintained to achieve uniform small-feature definition. Plated boards also require careful and complete resist removal. The steps after etching are important because they remove surface contamination and yield sound surfaces. This discussion considers the various types of resists and outlines typical procedures used to etch printed boards using organic and plated resist patterns.

34.2.1 Screened Resists

Screen printing is a common method for producing standard copper-printed circuitry on metal-clad dielectric and other substrates. The etch-resist material is printed with a positive pattern (circuitry only) for copper etch-only boards or with a negative image (field only) when plated through-holes and metal resist are present.

The type of resist material used must meet the requirements for proper image transfer demanded by the printer and for stripping chemistry compatibility. From the metal etcher's point of view, the material needs to provide good adhesion and etch-solution resistance; be free of pinholes, oil, or resin bleed-out; and be readily removable without damage to substrate or circuitry.

Typical problems are excessive undercutting, slivers, unetched areas, and innerlayer shorts in multilayer boards. In addition, conductor line lifting may occur when the copper-to-laminate peel strength (or surface contamination) is below specification.

34.2.2 Hole Plugging

Plugged-hole, copper-only boards use alkaline-soluble screen resists in a unique manner. The technique, called *hole plugging,* makes the SMOBC board possible. This technique can be used with screened images and to augment *tent-and-etch* processing for small annular ring structures.

34.2.3 UV-Cured Screen Resists

Ultraviolet-cured solventless systems are available for print-and-etch and plating applications. These products are resistant to commonly used acidic plating and etching solutions. Stripping must be evaluated carefully.

34.2.4 Photoresists

Dry-film and liquid photoresist materials are capable of yielding the fine-line (0.003 to 0.005 in) circuits needed for production of surface-mount circuit boards. Like screened resists, photosensitive resists can be used to print either negative or positive patterns on the metal-clad laminate. Although dry-film and liquid materials differ in both physical and chemical properties, they are considered together for our purpose.

In general, both positive- and negative-acting resists offer better protection in acidic rather than alkaline solutions; however, negative-acting types are more tolerant of alkaline solutions. Negative resists, once exposed and developed, are no longer light sensitive and can be processed and stored in normal white light. The positive resists remain light sensitive even after developing and must therefore be protected from white light. Liquid photoresists, although less durable, are capable of finer line definition and resolution.

Positive-acting resists are subject to the same problems as are negative-acting resists, although they may be easier to remove cleanly, after exposure, from areas to be etched or plated. Problems related to plating photoresist-coated boards are reviewed in Chap. 29.

34.2.5 Plated Etch Resists

At present, the most extensive use of metal-plated resists is found in the production of double-sided and multilayer plated-through-hole circuit boards. The most commonly used resists are tin (matte and bright surface). Solder plate (60 percent Sn, 40 percent Pb) continues to be used on some reflow-treated solderable coating situations, but it is not favored for a sacrificial resist that is to be removed for SMOBC processing. Nickel, tin/nickel, and gold are occasionally used. Silver is used to some extent for light-emitting and liquid crystal applications. Details concerning the deposition of these metals are given in Chap. 29. The use of these resists is described in the following paragraphs.

34.2.5.1 Tin Plating. Thin tin deposits (0.0002 in) are used to make SMOBC boards where the tin etch resist is stripped after etching. Alkaline ammonia etchants are usually favored. Other etchants, such as sulfuric acid–hydrogen peroxide and ammonium persulfate–phosphoric acid, have been especially formulated for bright tin. Tin plating (directly over barrier layers of nickel or tin/nickel) has been used because of its optimum solderability. Cupric and ferric chloride etchants attack tin and are not used.

34.2.5.2 Solder Plate. Tin/lead solder (0.0003 to 0.001 in thick) is the plated etch resist for "fusing" into a solderable finish. The 60Sn/40Pb alloy offers good etchant resistance with few problems but must be treated with a brightening solution to retain solderability. Increased reliability may be achieved by the use of solder plate over tin/nickel.[1] Thin solder deposits (0.0002 in) can be used for the SMOBC process but have no benefit over tin plating for this purpose. The most suitable etchants are alkaline ammonia and sulfuric acid–hydrogen peroxide. Ferric and cupric chloride acid etchants cannot be used because of solder plate attack. Postetch neutralization rinses are needed, especially with alkaline systems, to rinse away etchant residues and to maintain optimum surface properties.

34.2.5.3 Tin/Nickel and Nickel. Tin/nickel alloy (65 percent Sn, 35 percent Ni) and nickel plate, used as is or overplated with gold, solder, or tin, are also capable metal resists for etching copper in alkaline ammonia, sulfuric acid–hydrogen peroxide, and persulfates.

34.2.5.4 Gold. With an underplate of nickel or tin/nickel, gold provides excellent resistance to all the common copper etchants. Some etchants have a slight dissolving effect on gold.

34.2.5.5 Precious Metals and Alloys. Rhodium has been described as being a suitable resist for edge connectors on boards; however, the plating process is difficult to control. When plated over nickel, rhodium tends to be thin and porous and to lift during etching. Because of varying surface properties, 18-karat alloy gold and nickel/palladium must be evaluated carefully when used as substitutes for pure gold systems.

34.2.5.6 Silver. Although silver is not used on most printed boards (MIL-STD-275 states that it shall not be used), it has found some applications for camera, light-emitting, and liquid crystal devices. Copper etching using silver as resist can be done with alkaline ammonia solutions. Silver loss is about 0.0001 in/mm.

34.2.5.7 Etching Procedures for Plated-Metal Resists. Etching of the metal-resist-plated boards begins with removal of the plating resist using commercial strippers. The stripping process must be sufficient to remove resist trapped under plating edges. Gold, solder, and tin resists must be handled carefully because they scratch very easily. Tin/nickel alloy and nickel plate, however, are very hard and resistant to abrasion.

The procedure after etching includes thorough water rinsing and acid neutralizing to ensure removal of etchant residue on the board surface and under the traces. Alkaline etchants are followed by treatment with proprietary ammonium chloride acidic solutions, ferric and cupric chloride with solutions of hydrochloric or oxalic acid, and ammonium persulfate with sulfuric acid. Alkaline cleaning is used for tin/lead boards etched in chromic–sulfuric acid. Etchant residue not removed before drying or reflow results in lowered electrical resistance of the dielectric substrate and in poor electrical contact and soldering on the conductive surfaces.[2] Care must be taken to minimize undercutting and overplating to eliminate slivers of plating breaking off and bridging circuitry.

A problem common to etching printed boards is not having the entire area etch clean at the same time. This occurs when etch action is more rapid at the edges of printed areas than in a broad expanse of copper. If very fine patterns and lines are required, the result can be loss of the pattern due to undercutting, especially when the board is left in the etcher until all the field copper has etched away. Modern etchers with reduced pooling of etchant have reduced this problem.

Fine-line boards in high-volume production may require special fine-line etchants, high-resolution photoresists, thin-clad laminates, controlled plating distribution, and thin base foil processing. Care must be taken to balance etching and thickness factors. In some cases, fine-line etchant additives have actually made cleanout more difficult in spaces of 0.003 in and narrower.

34.3 RESIST REMOVAL

The method used for resist stripping must be carefully evaluated when a resist is selected. The effect on board materials, cost and production requirements, and compliance with safety and pollution standards must be taken into account. Solvent-based stripping solutions are under significant environmental pressure. Chlorinated solvents and cyclic compounds (toluene, xylenes, etc.) have been banned, and many glycol ethers are restricted. In spray applications, care must be taken to capture or eliminate VOC emissions. For these reasons, aqueous or primarily aqueous stripping systems are necessarily widely used.

34.3.1 Screen Resist Removal

Alkali-soluble resist inks are generally preferred. Stripping in the case of thermal and UV-curable resists is accomplished in 2 percent sodium hydroxide or in proprietary solutions. The resist is loosened and rinsed off with a water spray. Adequate safety precautions must be taken, since caustics are harmful.

Conveyorized resist-stripping and etching machines use high-pressure pumping systems that spray hot alkaline solutions on both sides of the boards. Certain laminate materials such as the polyimides may be attached by alkaline strippers. Measling, staining, or other degradation is noted when strippers attack epoxy or other substrates. Control of concentration, temperature, and dwell time usually can prevent serious problems.

Screened vinyl-based resists are removed by a dissolving action in solutions of chlorinated, petroleum, or glycol ether solvents. Methylene chloride and toluene (both currently not usable) were used extensively in cold stripper formulations. The most common strippers are acidic formulations that contain copper brighteners and swelling, dissolution, and water-rinsing agents. The usual procedure for static tank stripping involves soaking the coated boards in at least two tanks of stripper. Excessive time in strippers is to be avoided because of attack on the butter (top epoxy) coat, especially on print-and-etch or single-sided boards. Water is a contaminant in most cold strippers.

34.3.2 Photoresist Removal

34.3.2.1 Dry Film. Dry-film resists have been formulated for ease of removal in aqueous-alkaline solutions. Strippers are available for both static tank and conveyorized systems, although conveyorized systems are preferred because the spray action aids in separating the resist from the copper surfaces and washing it off the panel. Prompt processing after etching avoids resist adhesion "lock-in," which can make removal difficult. Aqueous-alkaline stripping results in partially undissolved residues of softened resist films. These residues are captured in a filter system and disposed off in accordance with waste-disposal requirements. Design and availability of specific filters to match the character of the "skins" can be a significant factor in the selection of the proper resist. Environmental caution must be taken to properly evaluate and dispose of resist residues. Metal content (particularly lead) may lead to classification of these residues as toxic waste.

34.3.2.2 Negative-Acting Liquid Photoresist Removal. Negative-acting, liquid-applied photoresist can be readily removed from printed boards that have not been baked excessively. Baking is critical to removal because it relates to the degree of polymerization. Since overbaking is also damaging to the insulating substrates, processes should stress minimal baking—only enough to withstand the operations involved. The negative-acting resists are removed by using solvents and commercial strippers. In this case, the resist may not dissolve; instead, it softens and swells, breaking the adhesive bond to the substrate. Once that has taken place over the entire coated area, a water spray is used to flush away the film.

34.3.2.3 Positive-Acting Photoresist Removal. Positive-acting photoresists are removed in commercial organic and inorganic strippers if baking has not been excessive. Removal by exposure to UV light and subsequent dipping in sodium hydroxide, TSP, or other strong alkaline solutions is also effective. Overbaking makes removal difficult. Machine stripping is done in a solution of $0.5N$ sodium hydroxide, nonionic surfactants, and defoamers.

34.3.3 Tin and Tin/Lead Resist Removal

In SMOBC processing, the metal-plated resist is removed to present a flat, clean copper surface for solder mask definition. Tin/lead alloys can be stripped in oxidizing fluoride solutions such as fluoboric acid and hydrogen peroxide or ammonium bifluoride with hydrogen peroxide or nitric acid. (Caution: machine construction must be made compatible with fluorides by elimination of titanium and glass components.) Commercial formulations are available to be used inline after the etch machine rinses. Accumulations of spent solution or filtered lead-fluoride deposits must be treated as hazardous waste and have been accepted by solution vendors for treatment and disposal costs. Modern applications usually use lead-free tin plating resists, which can be fluoride containing as previously discussed, or compounds of ferric chloride followed by nitric acid. In either case, feed-and-bleed solution management with filtration and periodic chamber cleanout has been successful.

34.4 ETCHING SOLUTIONS

This section is a survey of the technology and chemistry of the copper etching systems in common use. Selection of practically available etchants has been limited by economic, operational, and (environmental) regulatory concerns. Fabricators have been forced into practical trade-off decisions to suit situations. Two etchants in particular, chromic–sulfuric acid and ammonium persulfate, are no longer practical considerations due to environmental pressures. Other formulations and choices have been modified to suit these pressures.

There are two basic etchant needs to be met. The first is traditional foil etching for print and etch, plate/tent and etch, and pattern plate and etch. Virtually all processes in the United States and Europe use constant-rate systems for alkaline ammonia or cupric chloride etchants for this purpose. The second need is developing technology for specific precision very-fine-line etching—including foil thinning and thin metallization clearout for HDI constructions and fine features. (See Sec. 34.7 for additional discussion and mention of additional chemistries that may be useful for these applications.)

Continuous constant-rate systems with process automation represent current practice in production etching. These systems feature feed and bleed of replenishment chemicals under control of process instrumentation that monitors and responds to real-time changes in properties of the working solution. The "bleed" stream of the working composition etchant is usually returned to suppliers for copper reclamation and chemical recycle. The resulting degree of constant etching rate allows for stable and repeatable performance to achieve practical manufacturing processes. However, as demands for higher precision increase, further sophistication in controls will be necessary.

Key factors for selecting an etching chemistry include:

- Board design requirements
- High yield
- Compatibility with resist
- Etch rate (speed)
- Equipment required for process control of etch rate, regeneration, and replenishment
- Ease of equipment maintenance
- By-products disposal and pollution control
- Operator and environmental protection

The preceding factors serve to evaluate copper etchants to be used. Introduction, chemistry, properties, and problems are given in this section, along with suggestions for selection and control.

34.4.1 Alkaline Ammonia

Alkaline etching with ammonium hydroxide complexing is increasingly used because of its continuous operation, compatibility with most metallic and organic resists, high capacity for dissolved copper, and fast etch rates. Continuous (open-loop) spray machine chemical control systems are universally used. This operation provides constant etch rates, high work output, ease of control and replenishment, and improved pollution control. However, rinsing after etching is critical, and the ammonium ion introduced to the rinses presents a waste-treatment problem. On-site closed-loop regeneration with chemical recycling is commercially available but not routinely practiced because of facility requirements, capital cost, fluctuating economics depending on copper commodity pricing, and worker requirements. The general economical and environmentally appropriate operating strategy is to recycle the by-product etchant products under contract to a supplier who reclaims or reconstitutes the copper contained and regenerates the ammoniacal constituents into a reformulated replenisher solution for return to fabricators.

34.4.1.1 Chemistry. The main chemical constituents function as follows:

1. Ammonium hydroxide (NH_4OH) acts as a complexing agent and holds copper in solution.
2. Ammonium chloride (NH_4Cl) increases etch rate, copper-holding capacity, and solution stability.
3. Copper ion (Cu^{2+}) is an oxidizing agent that reacts with and dissolves metallic copper.
4. Ammonium bicarbonate (NH_4HCO_3) is a buffer and as such retains clean solder holes and surface.
5. Ammonium phosphate [$(NH_4)_3PO_4$] retains clean solder and plated through-holes.
6. Ammonium nitrate (NH_4NO_3) increases etch rate and retains clean solder.
7. Additional additives are included in most formulations to enhance speed and/or sidewall protection. Thiourea or its derivatives are often used, although a newer thiourea-free formulation with improved undercut protection is available.
8. Continuous operations consist of single-solution makeup buffered to a pH of 7.5 to 9.5.

Alkaline etching solutions dissolve exposed field copper on printed boards by a chemical process of oxidation, solubilizing, and complexing. Ammonium hydroxide and ammonium salts combine with copper ions to form cupric ammonium complex ions [$Cu(NH_3)_4^{2+}$], which hold the etched and dissolved copper in solution at 18 to 30 oz/gal.

Typical oxidation reactions for closed-loop systems are shown by the reaction of cupric ion on copper, and air (O_2) oxidation of the cuprous complex ion:

$$Cu + Cu(NH_3)_4^{2+} \rightarrow 2Cu(NH_3)_2^{+} \tag{34.1}$$

$$2Cu(NH_3)_2^{+} + 2NH_4^{+} + 2NH_3 + \tfrac{1}{2}O_2 \rightarrow 2Cu(NH_3)_4^{2+} + H_2O \tag{34.2}$$

There is strong evidence that the etching rate is dependent on the diffusion of $Cu(NH_3)_2^{+}$ from the copper surface (Eq. 34.1) into the bulk of the active solution, where oxidation per Eq. 34.2 occurs.[3] Etching can be continued with the formation of $Cu(NH_3)_4^{2+}$ oxidizer from air during spray etching and as long as the copper-holding capacity supported by Cl^- ions is not exceeded.

34.4.1.2 Properties and Control. Early versions of alkaline etchants were batch operated. They had a low copper capacity, and the etch rates dropped off rapidly as copper content increased.[4,5,6] It was found to be necessary to add controlled amounts of dissolved oxidizing agents to speed up the rate and increase copper capacity at a constant temperature. Batch operation is no longer supported by commercial suppliers.

Etching solutions are operated at 120 to 130°F and are well suited to spray etching. Efficient exhaust systems are required because ammonia fumes are released during operation.[7] Etching machines must have a slight negative pressure and moderate exhausting to retain the ammonia necessary for holding dissolved copper in solution. Care must be taken that sufficient fresh air to supply needed O_2 is introduced to balance the extraction. Currently available solutions offer constant etching of 1 oz (35 μm) copper in 1 min or less, with a dissolved copper content of 18 to 24 oz/gal.

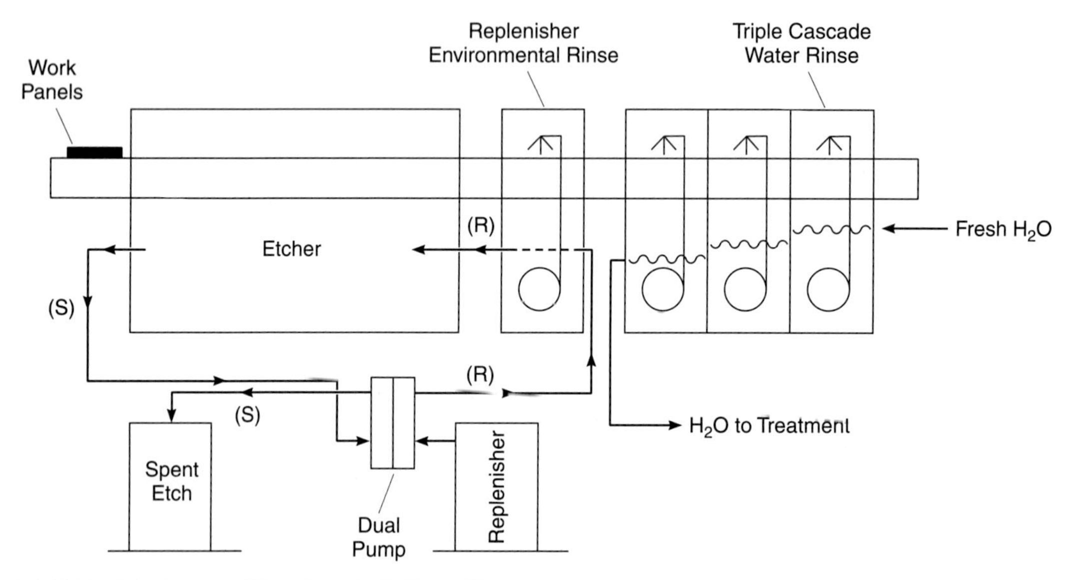

FIGURE 34.1 Automated flow-through alkaline etching system.

34.4.1.3 Continuous Systems. A practical method of maintaining a constant etch rate with minimal pollution uses automatic feeding controlled by specific gravity or density.[6] The process, which is generally referred to as *bleed and feed,* is illustrated in Fig. 34.1. As the printed boards are etched, copper is dissolved, and the density of the etching solution increases. The density of the etchant in the etcher sump is sensed to determine the amount of copper in solution. When the density sensor records an upper limit, a switch activates a pump that automatically feeds replenishing solution to the etcher and simultaneously removes etchant until a lower density is reached. An elegant by-product of this technique is the introduction of the replenisher into the first rinse stage following the etcher. The rinse captures the copper from the panel and reintroduces it to the etcher, where it is eventually eliminated to the by-product tank with the rest of the copper. When used with aqueous processing photoresists, it has been found that low pH (7.9 to 8.1) aids etch reliability. However, maintaining this pH by adjustment of exhaust extraction has been difficult. Measurement of pH and controlled addition of anhydrous ammonia into the etchant has improved consistency. It has been established that control of free ammonia as well as ammonium chloride and oxygen levels are needed to stabilize etching rates. Direct injection of oxygen has also been attempted to stabilize the etching rates between low and high copper etching demand process utilization periods.

Typical operating conditions are as follows:*

Temperature	120 to 130°F (49 to 50°C)
pH	8.0 to 8.8
Specific gravity at 120°F (49°C)	1.207 to 1.227
Baumé, Be°	25 to 27
Copper concentration, oz/gal	20 to 22
Etch rate, 0.001 in/mm	1.4 to 2.0
Chloride level	4.9 to 5.7 mol/L

* PA Hunt Chemical Corporation, West Paterson, NJ.

A study of the etching rate vs. dissolved copper content shows the following effects:

0 to 11 oz/gal	Long etching times
11 to 16 oz/gal	Lower etch times but solution control difficult
18 to 22 oz/gal	Etch rates high and solution stable
22 to 30 oz/gal	Solution unstable and tends toward sludging

All work must be thoroughly rinsed immediately upon leaving the etching chamber. Techniques such as replenisher rinsing and multiple cascade water rinsing assist in thorough rinsing with controlled effluent and limited water use.[8] Do not allow the boards to dry before rinsing. Etched circuitry with plated tin/lead solder resist also requires an acid application of solder brightener for reflow of the coating. Thiourea-containing brighteners have largely been eliminated due to use of sacrificial plating resist (SMOBC) and environmental pressures. Rinsing must be sufficient to remove etchant from under circuit edges and to completely clean the circuit surfaces and plated through-holes.[8] Multiple-stage cascading water rinsing and air knife drying result in clean, stain-free surfaces (see Fig. 34.1). Modern processes often follow the etching rinse with metal-resist stripping and rinsing chambers before exiting the machine conveyor.

34.4.1.4 Closed-Loop Regeneration. True regeneration requires the following:

1. Removal of portions of the spent etching solution from the etcher sump under controlled conditions, according to the amount of copper added to the etch bath.
2. Chemical restoration of spent etchant (i.e., removal of excess by-products and adjustment of solution parameters for reuse).
3. Replenishing of etchant in the etching machine, balancing the actual production use demands. Constant etching conditions are achieved when regeneration is continuous. Regeneration by these methods is expensive and is thus limited to large printed-circuit facilities. The principal methods of regeneration are crystallization, liquid–liquid extraction, and electrolytic recovery.

- Crystallization reduces the copper level in the etchant by chilling and filtering precipitated salts. This is followed by refortification and adjustment of operating conditions.
- Liquid–liquid extraction[9,10] is gaining acceptance because of its continuous and generally safe nature. The process involves mixing spent etchant with an organic solvent (i.e., hydroxy oximes) capable of extracting copper. The organic layer containing copper is subsequently mixed with aqueous sulfuric acid, which extracts copper to form copper sulfate. The copper-free etchant is restored, and the copper sulfate is available for acid copper electrowinning. Closed-loop regeneration systems reduce chemical costs, sewer contamination, and production downtime, but require floor space, resources, and technical attention. Economics are directly affected by copper market prices.
- Electrolytic recovery of copper directly and through membrane cells from an ammonia-complexed copper sulfate etchant yields possible benefits such as reduced waste shipments and cost savings.[11,12]
- These processes are usually employed by the large-scale vendor/recycling plants that offer a spent removal and replenisher replacement service as part of the product contract. This practice allows for economies of scale and environmental responsibility to be removed from the fabrication plant site.

34.4.1.5 Special Problems Encountered During Etching[13]

1. *Low etch rate with low pH, <8.0.* This is caused by excessive ventilation, heating, downtime, and spraying when the solution is hot, under conditions of adequate replenishment or low ammonia. The pH must be raised with anhydrous ammonia. Automatic replenishment equipment must be checked.

2. *Low etch rate with high pH, >8.8.* This is caused by high copper content, water in etchant, or underventilation.
3. *Low etch rate with optimum pH.* This is due to error of copper thickness, oxygen starvation in the etcher, or contamination of the etchant.
4. *Solder attack.* This is caused by excess chloride in the etchant, improper tin/lead deposits, or insufficient phosphate content.
5. *Residue on solder plate, holes, and traces.* These may be caused by etchant solution imbalance or by spent finisher.
6. *Under- or overetching.* This may be due to improper pH or equipment adjustments.
7. *Sludging of etch chamber with low pH, <8.0.* See special problem 1. Sludge will be gritty and dark blue. This may be corrected by the addition of anhydrous ammonia.
8. *Sludging of etch chamber with high pH, >8.8.* See special problem 2. Sludge will be light blue and fluffy. This may be due to the copper concentration exceeding the capacity of the chloride concentration. It may be corrected by adding ammonium chloride. It may also be due to water added to the etchant.
9. *Presence of ammonia fumes.* The cause of this is leaks in the etcher. Ventilate for operator safety.
10. *Pollution.* This results when water coming from the etchers contains dissolved copper. If so, it must be chemically treated and separated from ammonia-bearing rinses. Thin-clad copper laminates present a further problem because their faster movement through the etcher increases the transport of etchant into the rinses. Evaluate isolation roller placement and cleanliness, internal etcher overspray, and proper chemical and water rinsing.

34.4.2 Cupric Chloride

Cupric chloride is the second mainstream etchant of choice. (See Table 34.1 for characteristic cupric chloride solutions.) It is more compatible with photopolymer resists and is the etchant of choice for precision etching of small features. Because the cuprous ion buildup renders the etchant impractically slow, this etchant must always be used as a regenerative system with continuous oxidizer management and control. These systems have been the subject of many innovations to achieve continuous regeneration, lower costs, and a constant, predictable etch rate. Steady-state etching with acidic cupric chloride permits high throughput, material recovery, and reduced pollution. Regeneration in this case is somewhat complex but is readily maintained. Dissolved copper capacity is high compared with that of batch operation. Cupric chloride solutions are used mainly for fine-line multilayer inner details, print-and-etch boards, and panel-plate/tent-and-etch boards.[14] In addition to photoresists, screened inks, gold, and tin-nickel have been used. Solder and tin resists are not compatible with cupric chloride etchant.

TABLE 34.1 Characteristic Cupric Chloride Etching Solutions

Component	Solution 1[15]	Solution 2[16]	Solution 3[17]	Solution 4[18]
$CuCl_2 \cdot 2H_2O$	1.42 lb	2.2 *M*	2.2 *M*	0.5–2.5 *M*
HCl (20°Be)	0.6 gal	30 mL/gal	0.5 *N*	0.2–0.6 *M*
NaCl		4 *M*	3 *M*	
NH_4Cl				2.4–0.5 *M*
H_2O	*	*	*	*

* Add water to make up 1 gal.

34.4.2.1 Chemistry. The etching reaction is as follows:

$$Cu + CuCl_2 \rightarrow Cu_2Cl_2 \tag{34.3}$$

Chloride ions, when added in excess as hydrochloric acid, sodium chloride, or ammonium chloride, act to solubilize the relatively insoluble cuprous chloride and thereby maintain stable etch rates. The complex ion formation can be shown as

$$CuCl + 2Cl^- \rightarrow CuCl_3^- \tag{34.4}$$

The formation and diffusion of this chloro-complex form of the Cu(I) ion determine the etching rate for cupric chloride. Thus, the rate is affected by chloride ion concentration as well as the typical diffusion concentration and film thickness variables.

The etchant is regenerated by reoxidizing the cuprous chloride to cupric chloride by several chemical choices as illustrated by the following equations:

Air

$$2Cu_2Cl_2 + 4HCl + O_2 \rightarrow 4CuCl_2 + 2H_2O \tag{34.5}$$

This method of regeneration is not used because the oxygen reaction rate in acids is slow and the solubility of oxygen in hot solution is low (4 to 8 ppm). Spray etching induces air oxidation, but is not sufficiently reactive to support practical etching rates. An additional option of using ozone can accelerate the rate of this process, but ozone concentrations in an oxygen stream are difficult to maintain at above 3 percent, thus limiting utility.

Direct Chlorination

$$Cu_2Cl_2 + Cl_2 \rightarrow 2CuCl_2 \tag{34.6}$$

As the traditional regeneration choice, the chlorine reaction is rapid, positive, and easily controlled. Liquefied chlorine gas has been generally available, and standardized commercial addition systems have been available. Recently, pressure on chlorine distribution and use due to safety and environmental regulations have made availability and implementation more expensive and difficult. It is significant that the reaction simplicity indicates that all etched copper is converted directly to cupric chloride (combination of Eqs. 34.3 and 34.6). Thus, there is no effect on water or acid balances from the etching reaction itself.

Hydrogen Peroxide

$$Cu_2Cl_2 + 2H_2O_2 + 2HCl \rightarrow 2CuCl_2 + 2H_2O \tag{34.7}$$

This process is more complicated because both hydrogen peroxide and hydrochloric acid add water in the reaction process. Since hydrogen peroxide is unsafe to store and handle in concentrations above 35 percent, there is considerable water added with both the oxidizer and the hydrochloric acid. The result limits the copper-holding capacity of the etchant. The addition is reasonably and simply controlled as a direct ratio of HCl, and peroxide is added to the bath to maintain the copper oxidation. This formulation has been favored in Europe and to some extent in Asia.

Sodium Chlorate

$$3Cu_2Cl_2 + NaClO_3 + 6HCl \rightarrow 6CuCl_2 + NaCl + 3H_2O \tag{34.8}$$

Sodium chlorate has become one of the most widely used formulations. The oxidizer solutions can be mixed from powder and are also commercially supplied as 45 percent solutions. The water balance must also be carefully controlled with this oxidizing plan, and the amount of sodium chloride in the working bath can be boosted by addition with the oxidizer. This can be a rate enhancer by supplying chloride ions needed by Eq. (34.6), and since the chloride ion is separately added from the hydrochloric acid, the free acid can be maintained at very low levels (<0.1 normal).

Electrolytic

(Cathode) $$Cu^+ + e^- \rightarrow Cu \tag{34.9}$$

(Anode) $$Cu^+ \rightarrow Cu^{2+} + e^- \tag{34.10}$$

Electrolytic systems of direct and membrane cell design have been employed. However, this technology is not a major factor in current manufacturing practice.

34.4.2.2 Properties and Control. Early cupric chloride formations had slow etch rates and low copper capacity and were limited to batch operation.[15–21] Regenerated continuous operation using modified formulations has brought useful improvements. Etch rates as high as 55 s for 1 oz copper are obtained from cupric chloride–sodium chloride–HCl systems operated at 130°F with conventional spray-etching equipment. Copper capacities are maintained at 20 oz/gal and above. However, more recently, higher copper contents and adjusted chemistries at 125°F typically etch at 75 to 90 s for 1 oz copper.

34.4.2.3 Continuous Etching and Regeneration. Systems in use include chlorination, sodium chlorate, hydrogen peroxide, and electrolysis.

Chlorination. Direct chlorination has been the preferred technique for regeneration of cupric etchant because of its historically low cost, high rate, efficiency in recovery of copper, and pollution control. The cupric chloride–sodium chloride system (Table 34.1, no. 3) is suitable. Figure 34.2 shows a generalized process.[17] Chlorine, hydrochloric acid, and sodium chloride solutions are automatically fed into the system as required. Sensing devices include oxidation-reduction instruments (Cu oxidation state), density (Cu concentration), level sensors,

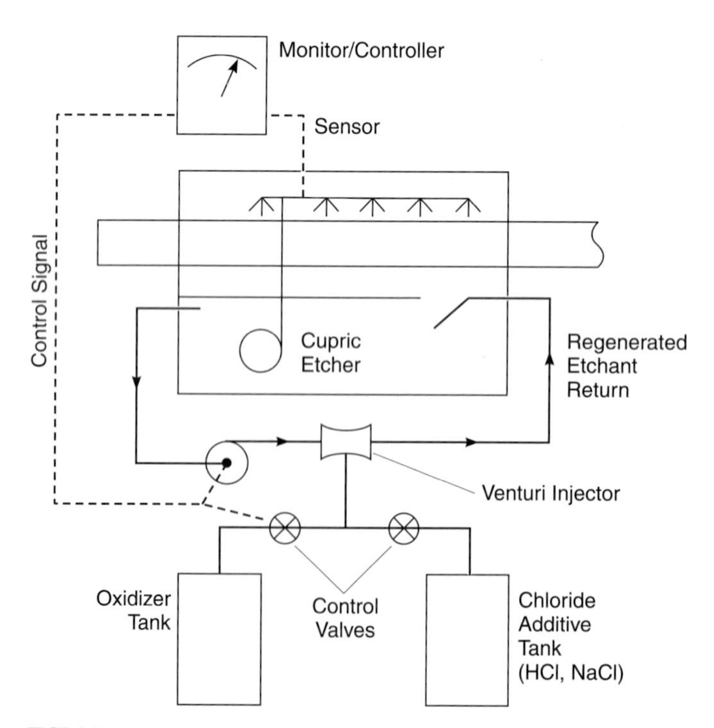

FIGURE 34.2 Cupric chloride chlorination regeneration system.

and thermostats. Chlorination is reliable and controllable. Other factors are safety and solution control. Optical colorimetric sensors can be used to detect Cu(I) levels. However, these cells are subject to fogging by organic contamination and crystalline buildup with the hazardous defect of continuing to add chemicals past the proper control points and liberating chlorine into the work area.

1. *Safety.* Use of chlorine gas requires adequate ventilation, tank and cylinder storage, leak-detection equipment, emergency protocols, personal protective equipment, operator training, and fire department approval and inspection of installations.

2. *Solution control.* An increase in pH will cause the copper colorimeter to give erroneous readings caused by turbidity in the solution. Organic deposits can also foul controls. Excess NaCl at 18 to 20 oz/gal copper causes coprecipitation of salts when the solution is cooled. Solution filtration and etch tank cleanliness must be maintained. Replacement of electrodes, instruments, and metering devices should be scheduled regularly.

Chlorate Regeneration. This method, currently widely supported, uses sodium chlorate, sodium chloride, and hydrochloric acid and is an alternate method similar to chlorination. A control sequence similar to the one shown in Fig. 34.2 may be used. Sodium chlorate is available in commercial solutions of several strengths with some degree of formulated agents (usually sodium chloride) added to the mixture. These solutions are added with hydrochloric acid either as a coordinated addition or in a special control logic sequence that allows very low free acid. This control system uses a multiple sensor color or turbidimetric optical system. The addition of one component (e.g., oxidizer) to a flowing sample cell occurs when the primary cell senses are triggered. A second cell allows this addition so long as there is a change from the first cell input. When there is no longer a change response and the first cell is still in "triggered" condition, the second component (e.g., HCl) is added. The logic continues to flip-flop until the first cell is satisfied. The etch sump conditions seem to have a partially delayed response to the chemical addition reaction such that oxidizing seems to occur over a longer period. Therefore, proper addition and mixing in the sump must be considered for stable control operation.

1. *Safety.* Sodium chlorate is a potent oxidizing agent that can support combustion. It is necessary to ensure that solution spills are completely cleaned up and that the rags or other materials are not allowed to dry out. Excess addition of reagents above amounts to support the actual copper etched may form chlorine gas that will be liberated into the etch machine and environment. Proper instrumentation and training for this eventuality must be implemented.

2. *Solution control.* If there is insufficient free acid content in the etch sump to react with the oxidizer and copper (Eq. 34.8), a dangerous imbalance can occur. First, the instruments and controls may not respond properly, and the etch reaction may slow down. Second, any subsequent addition of acid or mixing with acid-containing wastes can liberate chlorine gas in an uncontrolled and dangerous manner. Proper chemical balance is another issue. Water is added in all of the chemical streams and is produced in the chemical reaction. Therefore, any excess chemical addition can dilute the chemistry and change the copper-holding ability of the etchant.

Hydrogen Peroxide Regeneration. Hydrogen peroxide uses a two-part chemistry much like sodium chlorate. ORP instrumentation is identical in the control of adding proportional amounts of oxidizer and hydrochloric acid. Peroxide tends to decompose on sitting so that the spills are not as dangerous. Older application of colorimetric cells for the control of dosing has largely been eliminated.

1. *Safety.* Hydrogen peroxide solutions can become unstable and disassociate into oxygen and water with the liberation of heat. This can occur with explosive force in a confined vessel such as a drum stored in sunlight or heat. It can also occur in plumbing with closed valving.

The decomposition is catalyzed by many metals, including Cu, Ni, and Fe. If traces of metal or metal salts are allowed to contaminate piping runs, the piping may also rupture explosively. Facilities should be constructed with pressure relief capabilities.

Electrolytic Regeneration. The electrochemical reversal of the etching of copper shown in Eq. 34.9 and 34.10 is claimed to be effective and economical. Descriptions of this system are given in the literature.[15] On a large scale, electrolytic regeneration requires a high investment in equipment and materials, as well as high power consumption. The chemical formulation is relatively high in acid content and low in copper, a condition not optimized for etching effectiveness or plating efficiency.

The etchant is a solution of cupric chloride and hydrochloric acid (Table 34.1, no. 1). Etchant flows continuously between spray-etching machines and a plating tank. In the plating machine, two processes take place simultaneously: copper is plated at the cathode, and regeneration of the spent etchant occurs at the anodes. Copper recovery may not return copper value and may be expensive, inconvenient, and cause difficulty with recycling.

34.4.2.4 Problems with Cupric Chloride Systems

1. *Slow etch rate.* Depending on chemical formulation, cupric chloride is inherently slower than ammonia and must be properly evaluated before production expectations are established. Check to ensure no resist residues or chromate coatings remain on the panel surface before etching. A slower than usual rate is frequently due to low temperatures, insufficient sump mixing, or poor solution chemical control. If temperature and agitation are under control, slow etching with a dark green solution may result from a low cupric ion content, indicating insufficient oxidizer. Acid must also be added to clarify cloudy solutions. (Before acid addition, make sure excess oxidizer is not present.) In regenerative systems, the source of oxidation may be depleted. It is advisable to analyze specific gravity, free acid, and total chloride on a regular schedule and maintain control charts to ensure that bath controllers are consistent and operating properly.

2. *Sludging.* This occurs if acid is low or if water dilution occurs.

3. *Etcher goop.* This is a common floating accumulation of photoresist compounds leached into and precipitated from etch baths. Accumulations appear to increase with higher acid formulations. Formation can be limited by proper exposure of photoresists and continuous recirculation of the etch bath through carbon filtration. Periodic, complete drain-down of the etching machinery and thorough mechanical and chemical equipment cleaning with a sulfamic acid–type commercial cleaner can minimize problems.

4. *Yellow or white residues on copper surface.* Yellow residue is usually cuprous hydroxide. It is water insoluble and is left when boards are etched and alkali-cleaned. A white precipitate will probably be cuprous chloride, which can remain after etching in solutions that are low in chloride ion and acid. To eliminate both conditions, the solution in which the board is rinsed just before final water-spray rinsing should be 5 percent by volume hydrochloric acid.

5. *Waste disposal.* Spent or by-product etchant is usually sent off-site for copper reclamation. There is usually a fee for this service depending on copper content and distance to the reclamation facility. The solutions must be free of unreacted oxidizer (see previous chlorate system discussion). Etchant can contain traces of zinc, chromium, and arsenic from the foil treatments.

34.4.3 Sulfuric Acid–Hydrogen

Sulfuric-peroxide systems are used extensively for copper surface preparation by microetching the surface to provide texture and active surfaces. These formulations have been used in the past for primary circuitization etching but are no longer favored because of slow etch rate and needs for accompanying control of exotherm and copper crystallization. However, the increasing need for fine-line etching has renewed use possibilities for this and other etchants (see Sec. 34.7.4). In essence, these call for etching foils thinner than ½ oz and the control precision offered by slower etchants. They are compatible with metal resists and many organics.

34.4.3.1 Chemistry. Typical constituents of both immersion and spray etchants and their functions are as follows:

1. Hydrogen peroxide is an oxidizing agent that reacts with and dissolves metallic copper.
2. Sulfuric acid makes copper soluble and holds copper sulfate in solution.
3. Copper sulfate helps to stabilize etch and recovery rates.
4. Molybdenum ion is an oxidizing agent and rate exaltant.[22]
5. Aryl sulfonic acids are peroxide stabilizers.[23]
6. Thiosulfates are rate exaltants and chloride ion controllers that permit lower peroxide content.[24]
7. Phosphoric acid retains clean solder traces and plated through-holes.[25,26]

The etching reaction is as follows:

$$Cu + H_2O_2 + H_2SO_4 \rightarrow CuSO_4 + 2H_2O \tag{34.11}$$

34.4.3.2 Properties and Control. The earlier technical problems of slow etch rates, peroxide decomposition, and foaming in spray systems have been solved, but critical concerns still remain. Among these are process overheating, etchant composition balance with by-product recovery, etchant contamination, and the dangers of handling concentrated peroxide solutions. The specific problem of decomposition of peroxide during idle hours has caused disastrous equipment meltdowns, and thus requires thermal management while equipment is otherwise not in operation.

34.4.3.3 Closed-Loop Systems. Production facilities require continuous recirculation of the etchant through the etching tank or machine and the copper sulfate recovery operation. Etchant replenishment is controlled by chemical analysis and by additions of concentrates.

Copper sulfate recovery is based on lowering the solubility of $CuSO_4 \cdot 5H_2O$ by decreasing the etchant temperature to 50 to 70°F. Crystallizer equipment and conveyorized discharge has been available for this process.

34.4.3.4 Problems Encountered with Peroxide Systems

1. *Reduced etch rates.* This problem can be caused by operating conditions, solution imbalance, or chloride contamination.
2. *Under- and overetching.* A review of etching conditions, solution control, and the resist stripping process may show deviations from normal. In the case of immersion etching, the solution and panel agitation may need to be increased. When spray-etching, a check of nozzles and line clogging is indicated.
3. *Temperature changes.* Recirculating water rates and thermostats need to be examined regularly. Overheating may be due to high copper content, contamination, or rapid peroxide decomposition.
4. *Copper sulfate recovery stoppage.* Examine solution balance, heat exchanger, and other recovery equipment. Exit conveyor jamming difficulties from thermal excursions and varying crystal size and liquid content cause major shop floor problems.

34.4.4 Persulfates

Ammonium, sodium, and potassium persulfates modified by certain catalysts have been adopted for the etching of copper in PC manufacturing. Continuous regenerative systems and a batch system using ammonium persulfate are no longer used. Wide use is made of persulfates as a microetch for innerlayer oxide coating and copper electroless and plating processes. Persulfate solutions allow all common types of resists on boards including solder, tin, tin/nickel, screened inks, and photosensitive films. Formulations of ammonium persulfate catalyzed with mercuric chloride are no longer used, primarily because of environmental issues, costs, and improved alternatives. In general, persulfate etchants are unstable and will exhibit decomposition, lower etch rates vs. copper content, and lower useful copper capacity.

34.4.4.1 Chemistry. Ammonium, potassium, and persulfates are stable salts of persulfuric acid ($H_2S_2O_8$). When these salts are dissolved in water, the persulfate ion ($S_2O_8^{2-}$) is formed. It is the most powerful oxidant of the commonly used peroxy compounds. During copper etching, persulfate oxidizes metallic copper to cupric ion as shown:

$$Cu + Na_2S_2O_8 \rightarrow CuSO_4 + Na_2SO_4 \tag{34.12}$$

Persulfate solutions hydrolyze to form peroxy monosulfate ion (HSO_4^{1-}) and, subsequently, hydrogen peroxide and oxygen. This hydrolysis is acid catalyzed and accounts for the instability of acidic persulfate etching solutions.

Ammonium persulfate solution, normally made up at 20 percent, is acidic. Hydrolysis reactions and etchant use cause a reduction of the pH from 4 to 2. The persulfate concentration is lowered, and hydrated cupric ammonium sulfate [$CuSO_4 \cdot (NH_4)_2SO_4 \cdot 6H_2O$] is formed. This precipitate may interfere with etching.

Solid persulfate compounds are stable and do not deteriorate if stored dry in closed containers. Solution makeup composition includes various catalysts, including organic matter and transition metals (Fe, Cr, Cu, Pb, Ag, etc.). Materials for storage must be chosen carefully. Persulfates should not be mixed with reducing agents or oxidizable organics.

The useful capacity of the etchant is about 7 oz/gal copper at 100 to 130°F. Above 5 oz/gal of copper, it is necessary to keep the solutions at 130°F to prevent salt crystallization. The etch rate of a solution containing 7 oz/gal of dissolved copper is 0.00027 in/min at 118°F.

34.4.4.2 Batch Operation. Sodium persulfate is preferred because it has minimal disposal problems and somewhat higher copper capacity and etch rates. A composition of 3 lb/gal sodium persulfate with 15 ppm of $HgCl_2$, 1 gal of proprietary additive, and 57 ml/gal of H_3PO_4 has been successfully used for batch-type spray etching.[27] Etch rates vary throughout bath life and range from 0.0018 to 0.0006 in/min for copper content of 0 to 7 oz/gal. Prepared solutions must be aged for 16 to 72 h before etching when proprietary additives are used.

34.4.4.3 Problems with Persulfates

1. *Low etch rates.* Since the solution may decompose, it will be necessary to replace the bath. If solution is new, add more catalyst and check for iron contamination.
2. *Salt crystallization.* Salts crystallize on the board and cause streaks, damage the solder plate, and plug the spray nozzles or filters. When copper content is high, blue salts may precipitate.
3. *White films on solder surface.* This may occur normally or when the lead content in the solder plate is too high.
4. *Black film on solder.* This condition can result when the solder alloy is high in tin. If solder reflow or component soldering is to follow, activate by tin immersion or with solder brighteners. Adjust phosphoric acid content in etchant and solder-plating conditions.
5. *Spontaneous decomposition of etch solution.* This breakdown is due to contaminated, overheated, or idle solutions. Ammonium persulfate etchants are unstable, especially at higher temperatures. At about 150°F, the solution decomposes quickly. Use it soon after mixing.

6. *Disposal.* The exhausted etchant consists mainly of ammonium or sodium and copper sulfate with a pH of about 2. Two methods for disposal are:

- Electrolytic disposition of the copper on the surface of passivated 300 series stainless steel is one disposal method. The spent etchant is acidified with sulfuric acid prior to electrolysis. Once the copper has been removed, the remaining solution can be diluted, neutralized, checked, and discarded. The copper can be removed from the cathode. Spent sodium persulfate can be treated with caustic soda.
- Addition of aluminum or iron machine turnings to a slightly acidified solution is another practical but possibly more difficult means of removing the dissolved copper. The reaction, especially in the presence of chloride ions, will be violent, and considerable heat will be given off if the solutions are not diluted.

34.4.5 Ferric Chloride

Ferric chloride solutions are used as etchants for copper, copper alloys, Ni/Fe alloys, and steel in PC applications, electronics, photoengraving arts, and metal finishing. Current use of ferric chloride etchant in printed wiring fabrication is extremely limited in the United States because of costly disposal of the copper-containing etchant, and the much better commercial support for ammoniacal and cupric chloride etchants. There is still considerable use for alloy etching and photochemical machining applications.

Ferric chloride can be used with screen ink, photoresist, and gold patterns, but it cannot be used with tin or tin/lead resists. However, ferric chloride is an attractive spray etchant because of its ease of use, holding capacity for copper, and ability to be used on an infrequent batch application basis.

The composition of the etchant is mainly ferric chloride in water, with concentrations ranging from 28 to 42 percent by weight. Free acid is present because of the hydrolysis reaction and the need to maintain an acid environment. The natural acidity is usually supplemented by additional amounts of HCl (up to 5 percent) to prevent formation of insoluble precipitates of ferric hydroxide. Commercial formulations for copper alloy etching are usually 36 Be, or approximately 4.0 lb/gal $FeCl_3$, and may contain antifoam and wetting agents. Customary acid operating HCl content is 1.5 to 2.0 percent.

The effects of ferric chloride concentration, dissolved copper content, temperature, and agitation on the rate and quality of etching have been reported in the literature.[28,29]

34.4.6 Chromic-Sulfuric Acids

These etchants for solder- and tin-plated boards were preferred for many years. More recently, their use has been completely eliminated due to Cr(VI) listing as a critical environmental hazard. Other problems with chromic-sulfuric etchants are difficulty in regeneration, inconsistent etch rate, the low limit of dissolved copper (4 to 6 oz/gal), and dangerous degradation of PVC and polypropylene equipment. Chromic acid etchant is suitable for use with solder, tin/nickel, gold, screened vinyl lacquer, and dry or liquid film photoresists. Chromic-sulfuric mixtures etch copper slowly, and additives are needed to increase the etch rate. For example, sodium sulfate and iodine have been used for rate increase. Alkaline etchants have become so well controlled that there is no justification for the risks and costs of chromic acid formulations.

The following listing demonstrates some of the operational difficulties with chromic-sulfuric etchants.

1. *Solder attack.* The protective value of solder depends on the formation of insoluble compounds on the surface. Solder is attacked if the sulfate content of the bath becomes very low or contains chloride or nitrates. The solder plate composition can also cause etchant attack. When the lead content becomes low, the sulfate film protection is lowered, and protection is lost.

2. *Slow or no etching of copper.* This can be caused by low chromic content, low temperature, insufficient acid, or high copper content. The solution is maintained as close as possible to 30 Baumé (pH about 0.1, temperature 80 to 90°F) and should be discarded when copper metal content exceeds 5.5 oz/gal.

3. *Staining of board materials.* The surfaces of dielectric substrates such as paper-based phenolics are attacked by chromic acid etchants. Removal is difficult, and the boards are generally rejected.

4. *Disposal.* Spent chromic acid etchants present a serious disposal problem. Disposal must comply with pollution standards and approved practice.

5. *Safety hazards.* Chromic acid is an extremely strong oxidizing agent. It will attack clothing, rubber, plastics, and many metals. Safety measures require adequate ventilation to keep fumes out of room air, synthetic rubber gloves, aprons, face and eye shields, and storage away from combustible materials. Dermatitis and nasal membrane damage are possible dangers.[7]

34.4.7 Nitric Acid

Etchant systems based on nitric acid have not found extensive application in PC manufacture. Copper etching is very exothermic, which may lead to violent runaway reactions. Problems with this system include solution control, attack on resists and substrates, and toxic gas fuming. However, nitric acid has certain advantages. These include rapid etching, high dissolved copper capacity, high solubility of nonsludging products, availability, and low cost.

34.4.7.1 Chemistry. Reaction in strong acids is shown by the following equation:

$$3Cu + 2NO_3^- + 4H+ \rightarrow 3Cu^{2+} + 2NO_2 + 2H_2O \tag{34.13}$$

Recent work shows that process improvements are possible.[30,31] Controlled etching has been attained in solutions containing 30 percent copper nitrate, water-soluble polymers, and surfactants. Dry-film resists work well in this etchant. An important finding was that straight wall trace edges were achieved using nitric acid. This could result in higher yields and density of fine-line boards. The major difficulty in obtaining the sidewall results is the requirement for specific crystollagraphic grain structure to the copper foil that has been difficult to reproduce in production quantities.

34.5 OTHER MATERIALS FOR BOARD CONSTRUCTION

Printed board laminates are usually composed of copper foils bonded to organic dielectric materials, ceramics, or other metals.

1. *Organic dielectrics.* These are thermosetting or thermoplastic resins usually combined with a selected reinforcing filler. Thermoset-reinforced materials used for rigid and flexible boards provide overall stability, chemical resistance, and good dielectric properties. Thermoplastic materials are also used for flexible circuit applications. A factor in material selection is the effect of process solutions, etchants, and solvents on the material. In addition, the adhesives used in laminating metal to substrate can be softened, loosened, and attacked by some solutions. Flat-surface planarity is an aid to precision etching. This can be improved by selection of laminate with fine glass woven structures that do not express a waviness to the free copper foil surface.

2. *Thin-clad copper.* Etched printed boards with epoxy laminate of ¼ oz (9 μm) copper or less show minimal lateral etching, thus enabling higher fidelity to pattern-plated traces. The problem with very thin copper foils is pinhole density and fragility in the laminate

manufacturing process. One method of avoiding these problems is to start with a more conventional ½-oz (18-μm) foil and reduce its thickness by etching to a thinner state (3 to 9 μm).[32] This technique depends on a uniform starting foil and a very well-controlled etching process. Although the particulars of the published process were not disclosed, cupric chloride and sodium persulfate formulations have been used for this process.

3. *Reverse foil laminate.* This construction is receiving attention for improving precision and control of etched lines. In order to achieve this foil structure, the *tooth* or crystalline roughness must also be controlled in the foil. The tooth side is usually placed on the dielectric interface of the laminate to increase peel strength and present a flat surface for processing. However, one variation is to place the *drum* or flat side of the laminate toward the dielectric and leave the tooth side upward. This forms a flat interface so that the etching stops at this surface, instead of the conventional extra etching to remove the teeth that are buried in dielectric.

4. *Semiadditive copper.* A copper thickness of 0.000050 to 0.000200 in with subsequent copper and resist metal plating shows no overhang or sliver formation. These layers must be of sufficient thickness and uniformity to withstand the cleaning and preparation steps for plating the required conductors.

34.6 METALS OTHER THAN COPPER

34.6.1 Aluminum

Aluminum-clad flexible circuits find use in microwave stripline[33] and radiation-resistance applications. Aluminum and its alloys have good electrical conductivity, are lightweight, and can be plated, soldered, brazed, chem-milled, and anodized with good results. Laminate dielectrics include PPO,[33] polyimide,[34] epoxy-glass, and polyester.

Precleaning for resist application includes nonetch alkaline soak, water rinsing for 5 to 10 s in chromic-sulfuric acid, rinsing, and drying. Preferred etchants include ferric chloride (12 to 18° Baumé), sodium hydroxide (5 to 10 percent), inhibited hydrochloric acid, phosphoric acid mixtures, solutions of HCl and HF, and ferric chloride–hydrochloric acid mixture.

Screen-printed vinyl resists and dry-film photoresist are the most durable for deep etching or chem-milling. A dip in a 10 percent nitric or chromic acid solution will remove residues from the surface or edges of conductor lines that may be left on some alloys. Dilute chromic acid has also been used for this purpose. Spray rinse thoroughly with deionized water after etching.

34.6.2 Nickel and Nickel-Based Alloys

Nickel is increasingly used as a metal cladding, electroplated deposit, or electroformed structure for printed wiring because of its welding properties. Nichrome- and nickel-based magnetic alloys are other examples of materials requiring special etching techniques.

The methods previously described are adaptable to image transfer and etching of nickel-based materials. Etching uses ferric chloride (42° Baumé) at about 100°F. Other etchants include solutions made from one part nitric acid, one part hydrochloric acid, and three parts water, or one part nitric acid, four parts hydrochloric acid, and one part water.

34.6.3 Stainless Steel

Alloys of stainless steel are used for resistive elements or for materials with high tensile strength. Etching of the common 300 to 400 series can be done with the following solutions:

1. Ferric chloride (38 to 42° Baumé) with 3 percent HCl (optional).
2. One part HCl (37 percent) by volume, one part nitric acid (70 percent) by volume, one to three parts water by volume. Etch rate is about 0.003 in/min at 175°F, useful for high 300 to 400 series alloys.
3. Ferric chloride + nitric acid solutions.
4. One hundred parts HCl (37 percent) by weight, 6.5 parts nitric acid by weight, 100 parts water by weight.

34.6.4 Silver

Silver, the least expensive precious metal, has excellent properties, including superior electrical and thermal conductivity, ductility, visible-light reflectivity, high melting point, and adequate chemical resistance. As such, it is widely used throughout the electronics industry. Flexible circuit structures with silver are used in electronic cameras and LED products.

Standard image-transfer methods are suitable. Pre-etch cleaning should include a dip in dilute nitric acid. Mixtures of nitric and sulfuric acids are effective etchants. With silver on brass or copper substrates, a mixture of 1 part nitric acid (70 percent) and 19 parts sulfuric acid (96 percent) will dissolve the silver without attacking the substrate. The solutions should be changed frequently to prevent water absorption and the formation of immersion silver on the copper.

Etching can be done with a solution containing 40 g chromic acid, 20 mL sulfuric acid (96 percent), and 2000 mL water.[35] This is followed by a rinse in 25 percent ammonium hydroxide. Thin films of silver are etched in 55 percent (by weight) ferric nitrate in water or ethylene glycol. Solutions of alkaline cyanide and hydrogen peroxide will also dissolve silver. Use extreme caution. Electrolytic etching is also possible with 15 percent nitric acid at 2 V and stainless-steel cathode.

34.7 BASICS OF ETCHED LINE FORMATION

The basic process of chemical etching to form features has been studied and modeled extensively. However, the best scientific knowledge can only proceed to an understanding of factors influencing the process. The practical execution of the process to manufacture useful circuits contains elements of experience in the best practice of several processes and choices in the selection of materials and process variations to achieve technically sound, cost-effective, and manufacturable results. In order to make fine-line circuit products, it is necessary to understand the process fundamentals so that the limitations may be understood and then overcome. The intent of this section is to briefly review the underlying science and discuss factors and practices for precision etching.

34.7.1 The Image

34.7.1.1 Phototools. It seems obvious that artwork is the first defining step of the image. If the phototool lacks definition, integrity, or dimensional stability, there is no possibility of improving the image in subsequent processing. The edge definition and contrast are particularly important. Images of multiple pulses or *spots* must be overlapped so that there is minimum waviness of edges. As the feature to be resolved becomes smaller, the image must have better and cleaner integrity. It is desired that the artwork be at least an order of magnitude (factor of 10) better than the final tolerance required for the final product. Therefore, if the line tolerance is ±0.0001 in, then the artwork should be ±0.00001 in.

34.7.1.2 Image Integrity. The effectiveness of the image translated from the artwork/exposure process can be best achieved by optimization through all of the processing. Surface preparation, resist application, exposure, handling, development, and cleanliness must be maintained sufficiently to produce images with proper integrity. Image integrity may be gauged periodically with a resource such as the Conductor Analysis Technologies etch evaluation protocol. Defects such as shorts and opens have been determined to be related to resist integrity, and persistent linewidth repeating patterns may be image related. General concepts indicate that the thinner the resist, the shorter the light path; and the less optical interface layers between the light source and the copper surface, the better fidelity of the image to the phototool. However, constant improvement in film, exposure, and phototool products makes fine-image options dependent on proper optimization and constancy of technique and cleanliness as well as choice of technology.

34.7.2 Basics of Processing

34.7.2.1 Diffusion—The Controlling Mechanism. In Sec. 34.4, the chemical reactions of the etching process are discussed. In order to understand the formation of the actual shapes of the foil cross section, the influences of diffusion and fluid flow must be understood.[32] As a result of the etching reaction at the reaction point on the copper surface, a complex ion concentration is built up and active etching chemicals are depleted. In order to complete the reaction, this complex ion must move through a static *boundary layer* into a place where there is fresh etchant supplied to complete the reaction and carry away the etched product. (See Fig. 34.3.) This boundary-layer shape is dependent on the specific shape of the resist and etching wall, the fluid flow over the surface, and the critical fluid flow in the channels formed by the resist and the etched side wall.

34.7.2.2 Fluid Flow Contribution. Examination of Fig. 34.3 shows that the contour of the boundary layer and its thickness vary with the shape and narrowness of the channel. If the surface is flat with no resist image, the boundary layer is thin and only dependent on the fluid properties and velocity across the surface. On the other hand, with resist images forming channels with depth, and the etched copper forming further channels, it is easy to see how the flow of fluid in the channel can be much different than the surface flow. In order to make matters even more confusing, the circuit traces and resist images form actual channels in the surface.

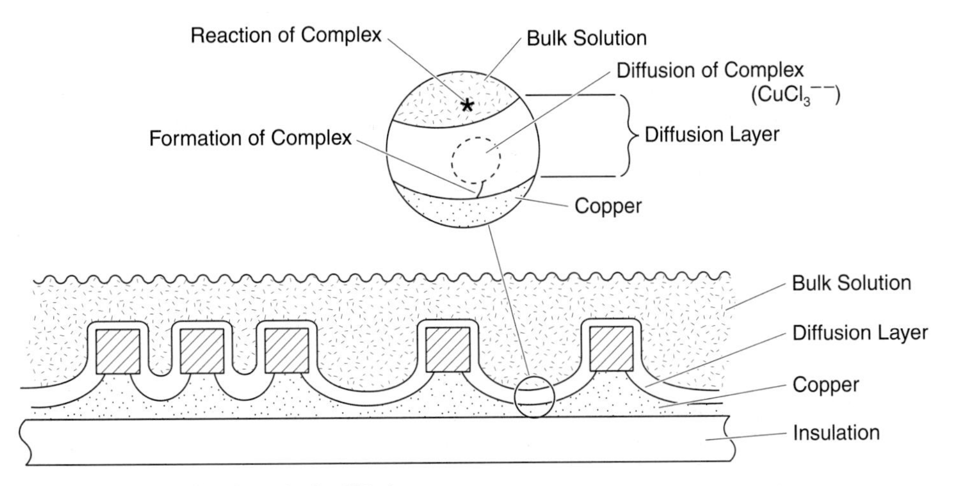

FIGURE 34.3 Etching dynamics for diffusion.

Much like a river bottom, the fluid flow in these individual channels is influenced by but different from the overall surface flow. In general, the narrower and deeper the channel, the slower the flow; and, critically, the diffusion boundary layer is more significant and etchant slows. In general terms, this is the reason that the middle section of a group of parallel, closely spaced lines etches slower than the outside lines. Similarly, this is the reason why a sharp 90° angle has a slower-etching section in the inside edge of the bend.

The concept of spray etching adds another dimension to the fluid dynamics. The surface of the panel is never really free of a liquid film. The top side of a panel has the etch liquid held up between delivery by the sprays and removal by flow off the edges. This flow is influenced by the number, flow, and pattern of the nozzles, the interaction of the flow between the sprays, the motion of the panels, and the effects of rollers and other contacting and shadowing objects. Even the bottom side retains fluid by surface tension and is influenced by the amount of liquid moving under the direction of the nozzles. The end result is complicated flow patterns that are the result of testing and optimization by the machine manufacturers.

34.7.3 Trace Shape Development

34.7.3.1 Etched Trace Shape Description. During etching, the gap between resist areas is removed evenly at first, and then progressively cup-shaped until the center area is broken through. After this, the area of exposed laminate opens progressively and the etching works on the side walls, etching in and under the resist as the side wall is relieved to expose more insulation and isolate the desired feature. (It has been noted that very little undercutting is apparent until the center is opened.) Figure 34.4 illustrates the attack with standard definitions for R (resist width), B (base of trace), T (top of trace), and t (foil thickness). There are two descriptive measures of the etching process—undercutting and etch factor—which are defined in Eqs. (34.14) and (34.15). These definitions are not universal, so it is necessary to clarify the definitions to correctly interpret discussions using these factors. Undercutting defines the average overhang of resist after top width reduction, and etch factor defines the average side wall inward taper of the etched feature per unit thickness. Occasionally, the etch factor is inverted or different measured values are used.

Undercut: $$U = (R - T)/2 \tag{34.14}$$

Etch factor: $$F = 2t/(B - T) \tag{34.15}$$

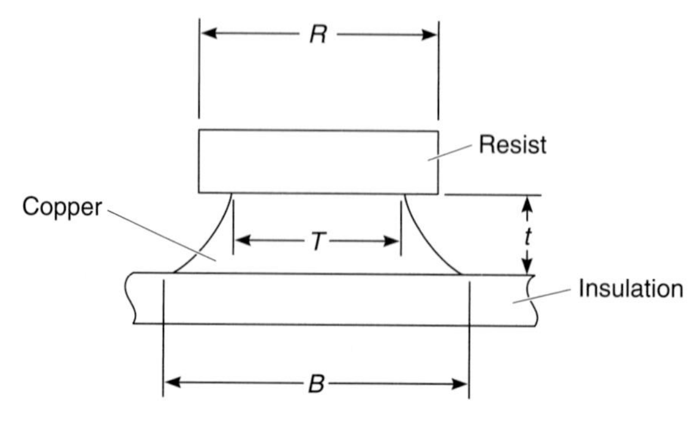

FIGURE 34.4 Trace cross-section measurement.

It is significant that these two units of measure are useful in describing the development of the shape of an etched part. It is desirable that U be minimized and F maximized for best feature definition.

34.7.3.2 Undercut and Etch Factor Development. The development of trace shapes has been studied by observing etched traces at progressive times.[36,37] This allows one more parameter to be developed, the extent of etch, which is the ratio of R/B. This factor equals 1.0 when the bottom of the trace equals the resist width, one measure of ideal etching. Note that if $R < B$, this implies that the trace is underetched, and $R > B$ indicates overetching. The traces were etched in cupric chloride using 3.0-mil line and space patterns of 1.0-mil resist on 1-oz (1.4-mil) copper foil. The progression of etch values is found in Table 34.2 and Fig. 34.5.

TABLE 34.2 Etch Progression

Etch time, S	Undercut, U mil	Etch factor, F	Extent of etch, R/B
90	0.05	0.90	0.5
110	0.30	1.75	0.75
125	0.45	2.33	1.1
140	0.525	2.67	1.0
165	0.75	3.11	1.25

There is an important conclusion to be drawn from these data. In order to compare or characterize an etch process (etcher to etcher, etchant to etchant, condition to condition, etc.), it is not sufficient to only use the undercut or etch factor alone. The data must be evaluated at a given R/B point for the same foil, resist, artwork, and panel size. If these factors are not held constant, the statement that "process A gives less undercut" has little meaning—additional etching gives more undercutting and straighter side walls.

34.7.3.3 Sensitivity Analysis. One of the extensions of the previous analysis is the ability to evaluate the sensitivity of the process. This is a study of the change in the result (R/B) due to the additional etching (relative time). Relative time is the ratio of the actual etching time to the nominal etch time taken to reach R/B of 1.0. In Table 34.2, it took 140 s to reach $R/B = 1$, but only 25 s (18 percent) more etching to achieve 25 percent overetching ($R/B = 1.25$). This sensitivity indicates that a small amount of overetch time could cause overetched 3-mil circuits. In this example, the etching process is slow enough to be controlled with the adjustment of the conveyor speed in the etcher. However, if the etchant were twice as fast (alkaline, for example) with the same sensitivity, it would be difficult to accurately "dial in" the proper result. Further evaluation of the cumulative variation from several variable properties on the overall product capability statistics can be found in the literature.[38]

34.7.3.4 Implication for Improved Processes. As a practical process application, it is often desirable to improve the etch factor to get a more square shape of the etched profile. One method for accomplishing this is to design the phototool for wider lines than desired, and then overetch to the desired endpoint. This practice increases F at the expense of also increasing U. However, as spaces get smaller, there is no space available for additional resist linewidth, because the fluid activity is restricted so that the etchant is stagnant between the traces. Therefore, resist width near the final line specification and precise control of the etch process are required to obtain desired shape profiles within the $R/B = 1$ degree of etching. As feature specifications tighten, the process to do this may need to be altered. One patented approach is to finish the etch with a specific slower-speed fine-tune etching chemistry to be able to stop the etching process at precisely the correct point.[39]

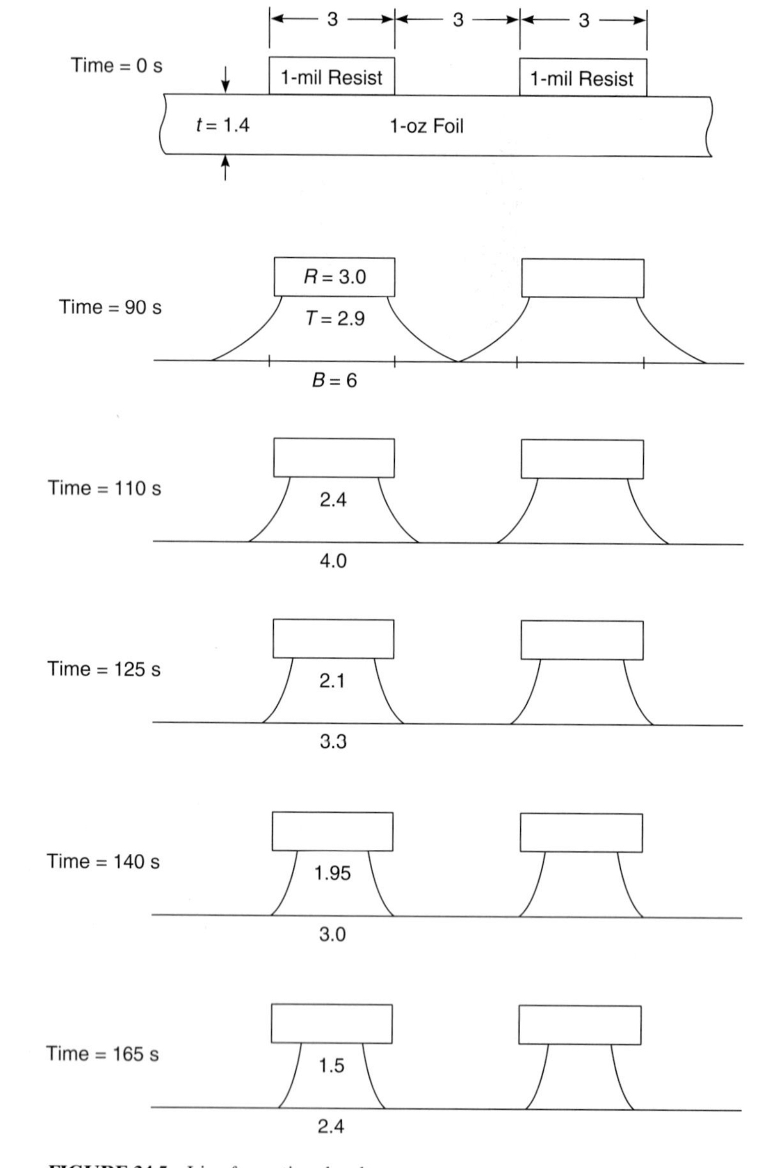

FIGURE 34.5 Line formation development.

34.7.4 Fine-Line Formation Etching Requirements

34.7.4.1 Fine-Line Definition. The term *fine line* is a relative one because the current state of processing by conventional available technology is constantly improving. Therefore, fine lines can be considered the next extension of capability that requires precision higher than that afforded by available technology. This is understood by means of statistics and standard gauging tools. The measure C_p (simple process capability) evaluates the variability of output results (6*s*, where *s* is the standard deviation of the linewidth distribution) to the specification limits of variation.[40] The data for the performance can be obtained by using a standardized test vehicle (IPC-9251) or a commercial testing program by CAT . In general, fine lines can then be defined as the linewidth point at which yield performance begins to drop below the run of production. Note that it is up to the customer and shop to determine what product line variation (usually a percent of line dimension) is allowed. It is apparent that this definition depends on the available technology and operation of the individual shop.

34.7.4.2 Limitations—Practical Rule of Thumb. It is useful to have a practical limit to express an understanding of where the technologies may limit performance. In the case of etched linewidth, it has been expressed that the total of the resist thickness plus the etched foil thickness would limit the gap between the etched features. Therefore, for 1.2-mil dry-film resist over 1-oz copper foil, the total thickness is 2.6 mil. This could be used as a practical limit for both the trace and gap. Further limits can be determined by the undercut and etch factor experienced for the same type of etched features. Therefore, using the $R/B = 1$ data from Table 34.2 ($U = 0.525$ mil), a 2.6-mil resist line would be 2.6 mil at the trace bottom and would have a 1.5-mil etched top reduction, leaving only a 1.1-mil top surface. It must then be determined if these dimensions (with allowance for variations) are sufficient for the design functionality.

34.7.4.3 Thinner Is Better. In general, it follows that thinner foil and thinner resist allow for smaller features. However, if thin foil (9 μm, ¼ oz) is used, the traces may be plated to achieve better current capability. Thin resist may be used, but the image may be susceptible to handling damage or damage in the process equipment. However, for coated resists of 0.4 mil and ¼-oz. foil (0.35 mil), the result could indicate a limit of 0.75 mil (19 μm). In fact, there has been a process using 3- and 5-μm foil to produce 30-μm line- and spacewidth patterns with 14 μm additional plated copper (total trace height of 17 and 19 μm). This method uses both panel and pattern plating.[41] The original foil was etched from thicker commercial foil laminate to the working thickness.

34.7.4.4 Changing the Microchannel Flow. As previously explained, the flow in the microchannels surrounding the features is critical in achieving uniform and accurate etching results. There is a unique process development using fibers to affect localized fluid mixing in these channels.[42] This approach requires specific patented equipment and processing methods.[43] Reported results down to 50-μm (2-mil) lines and spaces in 1-oz (34-μm) copper have been produced with 1.4-mil (37-μm)-thick resist. This is a very unique and important result because conventional foil and resist technology can be used to make features significantly smaller than conventional capability through improved fluid application mechanics.

34.7.4.5 Beyond Fine? HDI Impacts. High-density interconnect (HDI) technology involves several technology changes to form the circuitry. The highlight, as the name implies, is small via structures made in a layer-by-layer methodology with a built-up thin insulator and thin foil structure. To achieve the interconnection, thin and precisely controlled traces are required. The technologies, such as the two previous items, are becoming available to attack the needs of the products. As layer counts increase, the accumulating yield implications are significant, so that each process must produce minimum defects. For etching, one of the challenges is to make the resist conform to the surface—which forces the structural materials and processes to engage planarity of the end stage process as an issue. Etching capability requires improving the precision, stability, and capability of processing. The flowchart in Fig. 34.6 graphically illustrates the effects required for etching concerns and approaches to address these.[44] It is significant that there are no new concerns added to previous discussions.

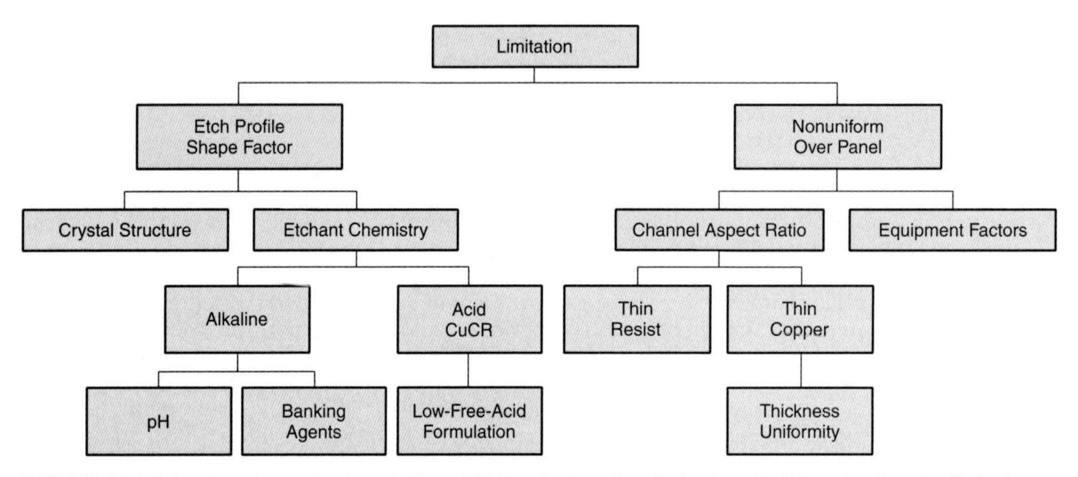

FIGURE 34.6 Flowchart for study of resolution of HDI etched product limitations. Profile and uniformity limitations may be studied for process improvement by progressing down the chart for contributing elements.

34.7.4.6 Stability and Control. It is important to measure and understand sources of variation in order to eliminate them. The gauging tools mentioned previously are very useful for collecting the data for analysis. One important feature is the positional and orientational variation in line geometry. Various conclusions can be taken from feature variation by location across the panel and orientation relative to the direction of motion. Repeating the gauging technique on a regular basis can allow observation of changing conditions within the process (such as plugged nozzles and other defects).

Another type of variation can be obtained by examining a series of gauge panels progressively processed with specific time-lag variations. If etching chemical controls are not adequate, the etch rate may vary in the sequence and therefore cause line variation with time. Usually, regeneration equipment can be adjusted or modified to minimize these variations, but the value of such instrument upgrades can be determined easily by data recorded about the type and extent of defects incurred.

34.8 EQUIPMENT AND TECHNIQUES

Etching equipment for circuit line formation has developed many variations on spray processing equipment. Often, as members of a coordinated automatic immersion plating process line, there are immersion etching tank varieties for microetching and surface preparation. However, for individual panels, conveyorized spray processing is clearly established because of efficient part handling and processing effectiveness.

34.8.1 Spray Equipment Basics

34.8.1.1 Construction. Etching equipment must be made from materials for long life, good dimensional stability, and resistance to chemicals and temperatures (130°F and higher). The basic construction includes rigid plastics (PVC, polypropylene, CPVC, and PVDF), metals (titanium, Hastalloy C), elastomers (vinyl, Viton A, Kel F, ethylene-propylene), specialized composites (epoxy-glass, graphite, and carbon-filled polymers, and other specialized formulations), and glass or clear and translucent polymers for visibility. The challenge is to select proper grades, compositions, and properties for mechanical strength, functionality, and durability. Equipment manufacturers perform these tasks in their design process.

34.8.1.2 Controls. Controls and electronics are significant because the increasing robustness and precision can have a large impact on cost. Therefore, the compromise on cost-effectiveness/technology trade-off often is made with a market-price primary goal. Therefore, cost-benefit analysis must be made by purchasers to evaluate available upgrades in instrumentation. The key controls are temperature, pressure, conveyor speed, and safety interlocks. Temperature is especially important because the etching reactions are usually exothermic and heat must be removed by cooling water coils to maintain stable rates. Conductivity, pH, density, ORP, and other process chemical control instruments may be added to monitor and control the processes. These instruments may be integrated into a PLC or computerized control system or may be individual control-panel stand-alone instrumentations. The wiring and the control cabinet itself must be suitably fume and hazard protected for water and chemical splashes as well as corrosive fumes. It is desired to have critical condition displays available to operators in the machine environment as well as some means of monitoring and recording process information.

34.8.1.3 Spray Strategies. Individual equipment designers have implemented very different types and mechanisms of spray attack on the panels. It is important to realize that the spray system design must be carefully integrated with the conveyor and mechanical design to minimize the effect of rods, wheels, panel control devices, and other mechanisms to achieve both uniformity and geometry of the features produced. It must be realized that the effect of the sprays is not merely the attack by droplets on the surface, but that the sprays are the prime drivers of the direction and effectiveness of the microchannel fluid flow on the panels. Therefore, separate controls and gauges must be provided for top and bottom spray arrays. For maximum precision, four quadrant controls are often beneficial. The mechanical spray arrays can be described as a variation of the following types.

Fixed Spray Array. This is a system of spray tubes containing multiple nozzles connected transversely across the conveyor to a header pipe on one or both sides. The spray nozzles are placed into the tubes in a pattern designed to optimize the effect. Care must be taken to install and orient the nozzles properly for designed effect. Mechanical intervention to selectively activate (or deactivate) portions of the design to effect the fluid dynamics of particular patterns should be included.

Oscillating Sprays. This system uses spray tubes that are connected generally to be aligned with the conveyor direction (often at a slight angle). The ends of the spray tubes are made with rotating seals so that the tubes can be moved back and forth by an oscillating crank or gearing mechanism. The nozzles move the liquid in a pattern generally across the conveyor travel. Individual spray tubes may be throttled and/or selectively activated.

Reciprocation. These nozzles are usually placed in an array similar to the fixed system, and the entire array is moved by a cranking mechanism in a direction across the conveyor. Maximum precision offerings include various patterns of nozzles that can be separately valved and selectively activated by computer controls.

Spray Nozzles. Spray nozzles are available in several types, sizes, and materials. For etching it has been established that PVDF nozzles are robust and consistent. The nozzle patterns are flat fan, full cone (round and squared), and open cone designs. Generally, flat fan nozzles have large openings and high impact for a given flow rate, while round cones have lower impact and orifice restrictions to ensure the pattern is filled. One of the principal failings of nozzles is simply clogging. The best array design can become dangerously nonuniform simply by plugging a small number of sprays. There is a commercially available nozzle that has a self-clearing fan design. Many are installed. For some applications, filtration between the pump and sprays has been installed to prevent plugging.

Selection. All of these combinations of nozzles, arrays, and mechanisms offer benefits and drawbacks that are often counterintuitive to the end user. It is necessary, therefore, to test and evaluate the performance of the spray systems for the work type and working environment. Testing of performance must be evaluated in a carefully controlled trial or by comparison with a standardized gauging tool.

34.8.2 Spray Equipment Options

34.8.2.1 Vertical Panel Orientation

Batch Operation. Some of the earliest equipment is still useful for prototype manufacture. This equipment holds a single panel on a workpiece that is generally rotated about an axis and sometimes oscillated as well. The sprays are addressed from one or both sides of the work and may be fixed or oscillating. This type of equipment usually is limited to small panels. The liquid all drains off the panel and falls into a collection sump. The panels must be removed after etching is complete and placed into a separate rinse vessel. (Some equipment has been designed to divert the drain and spray water onto the same work orientation.) This equipment is not capable of high capacity but has been able to etch high-quality work.

Conveyorized Operation. There are several versions of large machines that accept vertically oriented workpieces, convey them on a series of rollers from beneath the panel, and stabilize the work with rollers or wires. There are some machines that suspend the panels from the top surface. Vertically oriented nozzle arrays, either stationary or moving, spray the panel surfaces. These machines have been more successful in developing than etching because the depth of the image to be developed is formed by the exposure rather than completely depending on the etching fluid dynamics. All the liquid sprayed on the panels is removed by running downward off the bottom panel edge; therefore, the nozzles must be specifically designed to offset the effects of higher amounts and speeds of downflowing etchant progressively toward the panel's lower edge. It is obvious that very thin workpieces can have difficulty being conveyed through this design.

34.8.2.2 Horizontal Panel Orientation

General Operation. The most widely used configuration of etching equipment handles the panels horizontally on a roller table support. The panels placed on the rollers move constantly through arrays of sprays above and below the work. One of the principal design challenges is the fluid flow control of the liquid pool that builds up on (and runs off) the top surface of the panel. The various designs of the spray system must allow equally good and equally fast etching beneath the panel as on top. In a well-conceived design, the simple adjustment of spray pressures can achieve this balance.

Conveyance. The conveyor design is particularly critical. Rollers must turn at constant speed to ensure that the surfaces are not rubbed and damaged. Chains and gear designs have been successfully used. The potential problem areas are the entrance and exit squeegee rollers and the changes from one machine drive section to another. The roller wheels themselves need to be made of compliant materials that have flat area contact with the panels (rounded contact areas allow points that can cause resist damage, especially if the rollers harden with age). The wheels must be spaced far enough apart to allow the bottom spray to hit the panel with minimum shadowing, but close together enough to convey thin work without sagging or deforming.

Thin panels (0.0015- to 0.003-in insulation core) are difficult to convey because the leading edges must feed consistently from one roller to the next in spite of the downward force of the fluid. In addition, the deflection of the panels between rollers due to fluid loading can cause lack of uniformity in etching. Often, special mechanical guides such as clips or wires are used to overcome these difficulties. Problems occur when the guides stretch or are misplaced during production. If one panel jams, then the others behind it continue to jam and pile up, resulting in a major problem when discovered. The clearing of the jam often requires removal of rollers that are usually enmeshed in the mass of displaced guides and work. Not only are damaged panels costly, the downtime to renew the equipment and reestablish production causes significant production delay. One particular patented design features a combination of rollers and spray force that assures robust transport while simultaneously improving planarity for spray action.[45]

Truly flexible circuitry requires even more enhancements for reliable transport. Frames and screens have been used in the past, but these fixtures require extra handling and cause drag-over and cleaning problems. For sheet-cut panels, the best solution is a leading edge support attached to the panel by tape or mechanical clips. Usually these panels are attached and removed manually, but automatic systems could be provided. Flexible circuits can also be manufactured in continuous rolls that can be threaded through multiple processes. In addition to the previous difficulties with planarity and drag-over, the precise synchronization of roller speed throughout all the processes is mandatory. Therefore, machine and processing speeds must be carefully planned at original manufacture to allow proper dwell settings and latitudes. Rinsing and drag-over present problems.

34.8.3 Rinsing

Rinsing is a process technology that needs particular attention because of the multiple needs of uncontaminated panels, conservation of rinse waters, and environmental discharge limitations. It is no longer sufficient to flood the panel with large amounts of water that are then discharged to sewers. The capability of technical analysis of this process has been published.[46]

34.8.3.1 Cascade Rinsing. The principal tool to effect the required change is the concept of multiple-stage cascade rinsing. This process includes the concept that a robust flow recirculated on a panel can reduce the concentration to a level in equilibrium with a much lower circulated water stream. The lower stream passes through a series of chambers of increasing concentration, while the work passes through a series of chambers of increasing cleanliness (in reverse order from the water). The end result is a small use of water capturing a relatively high level of contamination. The contamination level is many times the discharge limit for direct environmental discharge. However, the fact that the stream is relatively small and highly concentrated allows for reasonable treatment and recovery of by-products by ion exchange, direct electrowinning, or membrane cell concentration. In certain circumstances, the waste stream may be comingled with a main process stream for recycling (see Fig. 34.1).

34.8.3.2 Drag-over Minimization. All of the contamination that eventually ends up in the water stream is that introduced by drag-over from the process itself (or from a recirculated environmental rinse). Therefore, the first step in successful rinsing is the reduction or prevention of materials from the panel. The conventional means of attempting this is by a set of pinch or squeegee rollers at the output of the chemical module. These rollers may be increased in effectiveness by softness or by pressure. Often, rollers made of PVA foam have been employed effectively. However, these rollers must be kept clean and moist to be effective. Other methods include improved baffling, ensuring capture of overspray, and low-velocity air blow-off. Proper management of drag-over on conveyorized equipment has reduced residues by a factor of 10 or better. Thin panels cause difficulties because the compromise made between conveying reliability devices and planarity and contact of squeegee rollers and surface tension reduces effectiveness of retention. Thick panels (over 0.125 in) also cause problems because squeegee rollers and baffles must lift to clear the panels, allowing gaps for solution escape. It is therefore a subject for engineering study and process management trade-off decisions that must be considered during the specification process.

REFERENCES

1. E. Armstrong and E. F. Duffek, *Electronic Packaging and Production,* vol. 14, no. 10, October 1974, pp. 125–130.
2. W. Chaikin, C. E. McClelland, J. Janney, and S. Landsman, *Ind. Eng. Chem.,* vol. 51, 1959, pp. 305–308.
3. Jieh-Hwa Shyu, "Electrochemical Studies of Etching Mechanisms in Ammoniacal Etchants," IPC TP-751, October 1988.
4. U.S. Patent 3,466,208, J. Slominski, 1969.
5. U.S. Patent 3,231,503, E. Laue, 1966.
6. U.S. Patent 3,705,061, E. King, 1972.
7. I. Sax, *Dangerous Properties of Industrial Materials,* rev. ed., Reinhold Publishing Corp., New York, 1957, p. 464.

8. Marshall I. Gurian, "Rinsing as a Process Technology," Paper T14, *Proceedings of Printed Circuit World Convention VI,* San Francisco, May 1993.
9. U.S. Patent 3,440,036, W. Spinney, 1966.
10. *Solvent Extraction Technology,* Center for Professional Advancement, Somerville, NJ, 1975.
11. Galvano Organo, *Printed Circuit Fabrication,* vol. 16, no. 1, January 1993, pp. 42–47.
12. Atotech USA, Inc., State College, PA.
13. K. Murski and P. M. Wible, "Problem-Solving Processes for Resist Developing, Stripping, and Etching," *Insulation/Circuits,* February 1981.
14. C. Swartzell, *Printed Circuit Fabrication,* vol. 5, no. 1, January 1982, pp. 42–47, 65.
15. G. Parikh, E. C. Gayer, and W. Willard, *Western Electric Engineer,* vol. XVI, no. 2, April 1972, pp. 2–8; *Metal Finishing,* March 1972, pp. 42, 43.
16. L. Missel and F. D. Murphy, *Metal Finishing,* December 1969, pp. 47–52, 58.
17. F. Gorman, "Regenerative Cupric Chloride Copper Etchant," *Proceedings of the California Circuits Association Meeting,* 1973; *Electronic Packaging and Production,* January 1974, pp. 43–46.
18. U.S. Patent 3,306,792, W. Thurmal, 1963.
19. L. H. Sharpe and P. D. Garn, *Ind. Eng. Chem.,* vol. 51, 1959, pp. 293–298.
20. J. O. E. Clark, *Marconi Rev.,* vol. 24, no. 142, 1961, pp. 134–152.
21. O. D. Black and L. H. Cutler, *Ind. Eng. Chem.,* vol. 50, 1958, pp. 1539–1540.
22. U.S. Patent 4,130,454, B. Dutkewych, C. Gaputis, and M. Gulla, 1978.
23. U.S. Patent 3,801,512, C. Solenberger, 1974.
24. U.S. Patent 4,130,455, L. Elias and M. F. Good, 1978.
25. U.S. Patent 3,476,624, J. Hogya and W. J. Tillis, 1969.
26. A. Luke, *Printed Circuit Fabrication,* vol. 8, no. 10, October 1985, pp. 63–76.
27. U.S. Patent 2,978,301, P. A. Margulies and J. E. Kressbach, 1961.
28. E. B. Saubestre, *Ind. Eng. Chem.,* vol. 51, 1959, pp. 288–290.
29. W. F. Nekervis, *The Use of Ferric Chloride in the Etching of Copper,* Dow Chemical Co., Midland, MI, 1962.
30. U.S. Patent 4,482,425, J. F. Battey, 1984.
31. U.S. Patent 4,497,687, N. J. Nelson, 1985.
32. Don Ball, "The Surface Mechanics of Fine-Line Etching," *Printed Circuit Fabrication,* vol. 21, no. 11, 1998.
33. F. T. Mansur and R. G. Autiello, *Insulation,* March 1968, pp. 58–61.
34. H. R. Johnson and J. W. Dini, *Insulation,* August 1975, p. 31.
35. P. F. Kury, *J. Electrochem. Soc.,* vol. 103, 1956, p. 257.
36. Marshall Gurian, "Process Effects Analysis for Fine Line Production," IPC Paper WCIV-28, *Printed Circuit World Convention IV,* Tokyo, 1987.
37. Marshall Gurian, "Reliable Fine Line Wet Processing," *Printed Circuit Fabrication,* vol. 10, no. 12, 1987.
38. Marshall Gurian, "Fine Line Processing: The '90's Are Here!" *Printed Circuit Fabrication,* vol. 13, no. 5, 1990.
39. U.S. Patent 5,904,863, Michael Hatfield and Marshall Gurian.
40. Michael Brassard and Diane Ritter, "The Memory Jogger II," Goal/QPC, Methuen, MA.
41. Yasuo Tanaka, Hireyuki Urabe, and Morio Gaku, "Three Micron Copper Foil Clad Laminate for 30/30 Micron Line/Space Circuit," *CircuiTree,* vol. 10, no. 11, 1997.
42. Igor Kadija and James Russel, "New Wet Processing for HDI's," *CircuiTree,* vol. 12, no. 5, 1999.
43. U.S. Patents 5,024,735, 5,114,558, and 5,167,747, Igor Kadija.
44. Karl Dietz, "Process and Material Adaptations for HDI Requirements," *CircuiTree,* vol. 13, no. 12, 2000.
45. U.S. Patent 4,607,590, Don Pender.
46. Marshall Gurian, "Rinsing as a Process Technology," Paper T14, *Proceedings of the Printed Circuit World Convention VI,* 1993; summarized in *Printed Circuit Fabrication,* vol. 20, no. 7, 1997.

CHAPTER 35
MACHINING AND ROUTING

Gary Roper
One Source Group, Dallas, Texas

35.1 INTRODUCTION

Laminate machining consists of the mechanical processes by which circuit boards are prepared for the vital chemical processes of image transfer, plating, and etching. Such processes as cutting to size, drilling holes, and shaping have major effects on the final quality of the printed board. This chapter will discuss the basic mechanical processes that are essential to producing the finished board.

35.2 PUNCHING HOLES (PIERCING)

35.2.1 Design of the Die

It is possible to pierce holes down to one-half the thickness of XXXPC and FR-2 laminates and one-third that of FR-3 (Fig. 35.1). Many die designers lose sight of the fact that the force required to withdraw piercing punches is of the same magnitude as that required to push the punches through the material. For that reason, the question of how much stripper-spring pressure to design into a die is answered by most toolmakers: "as much as possible." When space on the dies cannot accommodate enough mechanical springs to do the job, a hydraulic mechanism can be used. Springs should be so located that the part is stripped evenly. If the board is ejected from the die unevenly, cracks around holes are almost certain to occur. Best-quality holes are produced when the stripper compresses the board an instant before the perforators start to penetrate. If the stripper pressure can be made to approach the compressive strength of the material, less force will be required and the holes will be cleaner.

If excessive breakage of small punches occurs, determine whether the punch breaks on the perforating stroke or on withdrawal. If the retainer lock is breaking, the cause is almost certain to be withdrawal strain. The remedy is to grind a small taper on the punch, no more than 11.2 in. and to a distance no greater than the thickness of the material being punched. If the grinding is kept within those limits, it will have no measurable effect on hole quality or size. The other two causes of punch breakage are poor alignment, which is easily detected by close examination of the tool, and poor design, which usually means that the punch is too small to do the job required.

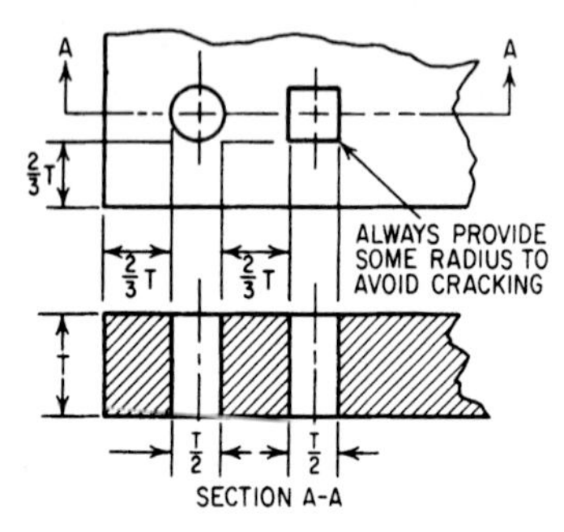

FIGURE 35.1 Illustration of the proper sizing and locating of pierced holes with respect to one another and to the edge of paper laminates. Minimum dimensions are given as multiples of laminate thickness *t*.

35.2.2 Shrinkage of Paper-Based Laminates

When paper-based laminates are to be punched, it must be remembered that the materials are resilient and that their tendency to spring back will result in a hole slightly smaller than the punch that produced the hole. The difference in size will depend on the thickness of the material. Table 35.1 shows the amount by which the punch should exceed the print size in order to make the holes within tolerance. The values listed should not be used for the design of tools for glass-epoxy laminates, the shrinkage of which is only about one-third that of paper-based materials.

TABLE 35.1 Shrinkage in Punched Hole Diameters, Paper-Base Laminates

Material at thickness	Material at room temp.	Material at 90°F or above
1.64	0.001	0.002
1.32	0.002	0.003
3.64	0.003	0.005
1.16	0.004	0.007
3.32	0.006	0.010
1.80	0.010	0.013

35.2.3 Tolerance of Punched Holes

If precise hole size tolerance is required, the clearance between punch and die should be very close; the die hole should be only 0.002 to 0.004 in. larger than the punch for paper-based materials (Fig. 35.2 and Table 35.2). Glass-based laminates generally require about one-half that tolerance. Dies have, however, been constructed with as much as 0.010 in. all-around clearance between punch and die. They are for use where inspection standards permit rough-quality holes.

A die with sloppy clearances is less expensive than one built for precision work, and wide clearance between punch and die causes correspondingly more break and less shear than a tight die will cause. The result is a hole with a slight funnel shape that makes insertion of components easier.

Always pierce with the copper side up. Do not use piercing on designs with circuitry on both sides of the board, because lifting of pads would probably occur.

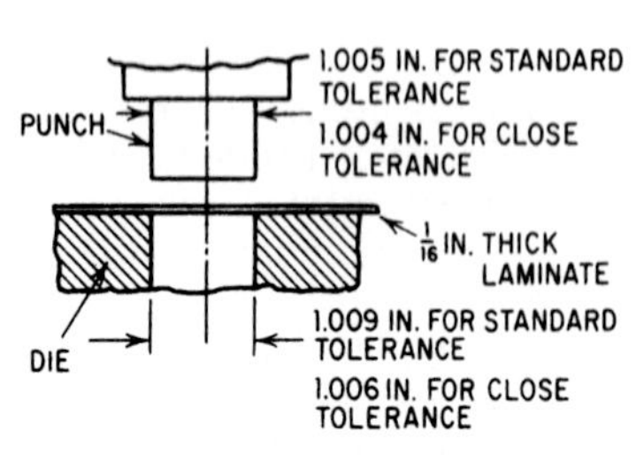

FIGURE 35.2 Example of proper tolerance of a punch and die.

35.2.4 Hole Location and Size

Designs having holes whose distance from the edge of the board or from other holes approaches the thickness of the material are apt to be troublesome. Such designs should be avoided; but when distances between holes must be small, build the best die possible. Use tight clearance between punch and die and punch and stripper, and have the stripper apply plenty of pressure to the work before the punch starts to enter. If the distance between

TABLE 35.2 Tolerances for Punching or Blanking Paper-Base Laminate

Material thickness	Base material	Tolerance on hole size, in.	Tolerances, in, on distance between holes and slots, 90°F				Tolerances for blanked parts, overall dimension, in.
			Up to 2 in.	2 to 3 in.	3 to 4 in.	4 to 5 in.	
To and including 1.16 in	Paper	0.0015	0.003	0.004	0.005	0.006	0.003
Over 1.16 in to and including 3.32 in.	Paper	0.003	0.005	0.006	0.007	0.008	0.005
Over 3.32 in to and including 1.8 in.	Paper	0.005	0.006	0.007	0.008	0.009	0.008

holes is too small, cracks between holes may result even with the best of tools. If cracks between holes prove troublesome, plan the process so that the piercing is done before any copper is etched away. The reinforcing effect of the copper foil will help eliminate cracks. Most glass-epoxy laminates may be pierced, but the finish on the inside of the holes is sometimes not suitable for through-hole plating.

35.2.5 Warming Paper-Based Material

The process of punching paper-based laminates will often be much more trouble-free if the parts are warmed to 90 or 100°F. That is true even of the so-called cold-punch or PC grades.

Do not overheat the material to the point at which it crumbles and the residue is not ejected as a discrete slug. Overheated material will often plug the holes in the die and cause rejects.

Opening the taper on the takeaway holes will reduce plugging, but the most direct approach is to pierce at a lower temperature. Glass-epoxy is never heated for piercing or blanking.

35.2.6 Press Size

The size of the press is determined by the amount of work the press must do on each stroke.

The supplier of copper-clad sheets can specify a value for the shear strength of the material being used. Typically, the value will be about 12,000 lb./in.^{2} for paper-based laminate and 20,000 lb./in.^{2} for glass-epoxy laminate. The total circumference of the parts being punched out multiplied by the thickness of the sheet gives the area being sheared by the die. If all dimensions are in inches, the value will be in square inches. For example, a die piercing 50 round holes, each 0.100 in. in diameter, in 0.062 in. thick laminate will be shearing, in square inches:

$$50 \times 0.100 \text{ in.} \times 3.1416 \times 0.062 \text{ in.} = 0.974 \text{ in.}^2$$

If the paper-based laminate has 12,000 lb./in.^{2} shear strength, 11,688 lb. of pressure, or about 6 tons, is required just to drive the punches through the laminate. Bear in mind that, if a spring loaded stripper is used, the press will also have to overcome the spring pressure, which ought to be at least as great as the shear strength. Therefore, a 12 ton press would be the minimum that could be considered. A 15 or 20 ton press would be considerably safer.

35.3 BLANKING, SHEARING, AND CUTTING OF COPPER-CLAD LAMINATES

35.3.1 Blanking Paper-Base Laminates

When parts are designed to have shapes other than rectangular and the volume is great enough to justify the expense of building a die, the parts are frequently punched from sheets by using a blanking die. A blanking operation is well adapted to paper-based materials and is sometimes used on glass-based ones.

In the design of a blanking die for paper-based laminates, the resilience, or yield, of the material previously discussed under Piercing applies. The blanked part will be slightly larger than the die that produced it, and dies are therefore made just a little under print size depending on the material thickness. Sometimes a combination pierce and blank die is used.

The die pierces holes and also blanks out the finished part.

When the configuration is very complex, the designer may recommend a multiple-stage die: The strip of material progresses from one stage to the next with each stroke of the die.

Usually in the first one or two stages, holes are pierced, and in the final stage, the completed part is blanked out.

The quality of a part produced from paper-based laminates by shearing, piercing, or blanking can be improved by performing the operation on material which has been warmed. Caution should be exercised in heating over 100°F because the coefficient of thermal expansion may be high enough to cause the part to shrink out of tolerance on cooling. Paper-base laminates are particularly anisotropic with respect to thermal expansion; that is, they expand differently in the x and y dimensions. The manufacturer's data on coefficient of expansion should be consulted before a die for close-tolerance parts is designed. Keep in mind that the precision of the manufacturer's data is probably no better than ±25 percent.

35.3.2 Blanking Glass-Based Laminates

Odd shapes that cannot be feasibly produced by shearing or sawing are either blanked or routed. Glass blanking is always done at room temperature. Assuming a close fit between punch and die, the part will be about 0.001 in. larger than the die which produced it. The tools are always so constructed that a part is removed from the die as it is made. It cannot be pushed out by a following part, as is often true when the material has a paper base. If material thicker than 0.062 in. is blanked, the parts may have a rough edge.

The life of a punch, pierce, or blank die should be evaluated with reference to the various copper-clad materials that may be used. One way to evaluate die wear caused by various materials is to weigh the perforators, or punches, very accurately, punch 5,000 pieces, and then reweigh the punches. Approximately 5,000 hits are necessary for evaluation, because the initial break-in period of the die will show a higher rate of wear. Also, of course, the quality of the holes at the beginning and end of each test must be evaluated. Greatly enlarged microphotos of the perforator can be used for visual evaluation of changes in the die.

35.3.3 Shearing

When copper-clad laminates are to be sheared, the shear should be set with only 0.001 to 0.002 in. clearance between the square-ground blades (Fig. 35.3). The thicker the material to

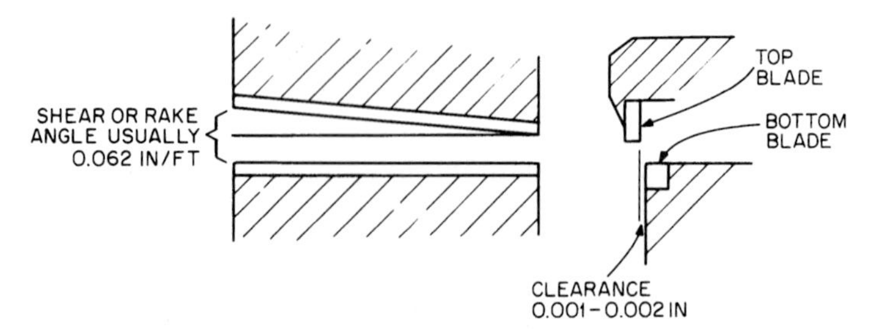

FIGURE 35.3 Typical adjustable shear blades for copper-clad laminates.

be cut, the greater the rake or scissor angle between the top and bottom shear blade. The converse also is true: The thinner the material, the smaller the rake angle and the closer the blades.

Hence, as in many metal shears, the rake angle and the blade gap are fixed; the cutoff piece can be twisted or curled. Paper-base material can also exhibit feathered cracks along the edge that are due to too wide a gap or too high a shear angle. That can be minimized by supporting both piece and cutoff piece during the shear operation and decreasing the rake angle. Epoxy-glass laminate, because of its flexural strength, does not usually crack, but the material can be deformed if the blade clearance is too great or the shear angle is too large. As in blanking, the quality of a part produced from paper-based laminates by shearing can be improved by warming the material before performing the operation.

35.3.4 Sawing Paper-Base Laminates

Paper-base laminates are much harder on sawing tools than are the hardest woods, and therefore a few special precautions are necessary for good saw life. Sawing paper-based laminates is best accomplished with a circular saw with 10 to 12 teeth per inch of diameter at 7500 or 10,000 ft/min. Hollow-ground saws give a smoother cut; and because of the abrasive nature of laminated materials, carbide teeth are an excellent investment. (See Fig. 35.4 for tooth shape.)

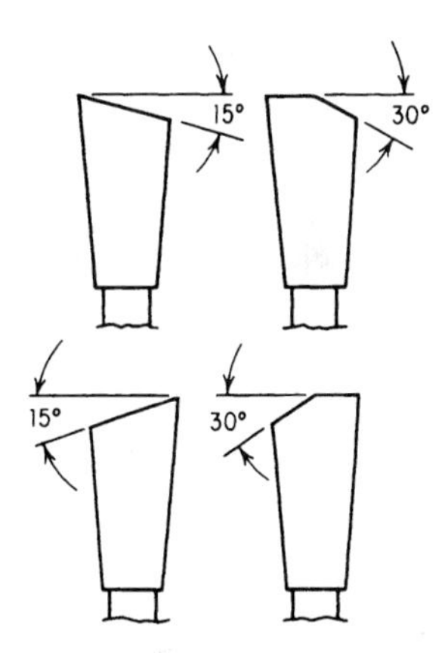

FIGURE 35.4 Commonly used saw tooth designs for paper and cloth laminates. At left, two successive teeth on a 15° alternate-bevel saw. At right, two successive teeth on a 30 ft alternate-corner-relieved (AC-30) saw.

When a saw does not last long enough between sharpenings, use the following checklist. (These steps could have a cumulative effect and change saw life by a factor of 4 to 5.)

1. Check the bearings for tightness. There should be no perceptible play in them.
2. Check the blade for run-out. As much as 0.005 in. can be significant.
3. When carbide teeth are used, inspect them with a magnifying glass to make sure a diamond tool no coarser than 180 grit was used in sharpening them.
4. If the saw has a thin blade, use a stiffening collar to reduce vibration.
5. Use heavy pulleys with more than one V belt. Rotating parts of the system should have sufficient momentum to carry the saw tooth through the work smoothly and without variation of speed.
6. Check the alignment of the arbor and the motor mounting.

All these steps are intended to reduce or eliminate vibration, which is the greatest enemy of the saw blade. If vibration is noticed, find the source and correct it.

35.3.5 Sawing Glass-Base Laminates

When glass-based laminates are to be sawn, carbide-tooth circular saws can be used; but unless the volume of work is quite low, the added investment required for diamond-steel-bonded saws will be paid for in future savings. The manufacturer's recommendation of saw speed should be followed; usually it will be for a speed in the neighborhood of 15,000 ft/min. at the periphery of the saw blade. When economics dictate the use of carbide-tooth circular saws for cutting glass, use the instruction previously given for paper-based laminates (see Fig. 35.4 for tooth shape) and remember that each caution regarding run-out, vibration, and alignment becomes more important when glass-reinforced laminates are sawn.

35.4 ROUTING

Modern circuit board fabricators rely principally on routing to perform profiling operations.

The high cost and extended lead times for blanking dies, combined with the problem of design inflexibility of hard tooling, limit the punching operation generally to very high volumes or designs specific to die applications. Shearing or sawing are limited to rectangular shapes and generally are not accurate enough for most board applications.

In the modern circuit board fabrication industry, rapid response to customer lead times and economies of universal process application are well served by routing, especially multiple-spindle computerized numerical controlled (CNC) routing. Routing consists of two similar, yet vastly different fabrication processes:

- CNC multiple-spindle routing
- Manual pin routing

The similarities consist of the use of high-speed spindles, utilizing carbide cutting tools, and generating high cutting rates.

35.4.1 Pin Routing

Pin routing is a manual routing process utilizing a template that has been machined of aluminum, FR-4 laminate, or a fiber-reinforced phenolic. The template is made to the finished board dimensions and has tooling pins installed to register to the board's tooling holes. The package (which can have up to four pieces in a stack) is routed by tracking the template against a pilot pin protruding from the router table. The pin height is less than the template thickness. Usually, the machine pilot pin is the same diameter as the router bit and can be offset adjusted to give the operator flexibility in optimizing dimensions. Work should be fed against the rotation of the cutter to prevent the cutter from grabbing.

Pin routing can be an economical process when a small generation of boards is profiled, or if the shapes required are relatively simple. For pin routing to be effective, generally a very skilled operator is required to fabricate the template and to route the boards. Outside machine shops can build aluminum route fixtures for each customer application; however, lead times and costs per order must be considered. Pin routing is usually used by small shops not able to invest in the CNC equipment and its associated support, or as a specialty process, off-line from CNC routing.

In the best pin-routing operations, the volumes produced cannot be compared with those of multiple-spindle routing.

35.4.2 CNC Mechanical Routing Applications

The applications of CNC routing extend well beyond merely cutting board profiles. The ability to produce boards in multiple-image modules reduces handling not only in the board shop, but in every subsequent operation, from packing, component assembly, wave solder, and test.

This is of special value when dealing with a postage-stamp-size part or wire-bondable gold surfaces. Where handling must be minimized, the module acts as a pallet throughout these operations. In addition, unusual or irregular shapes, small or large, can be palletized to simplify handling and conveyance. In Fig. 35.5, examples of tab-routed, or multiple-image, modules are shown.

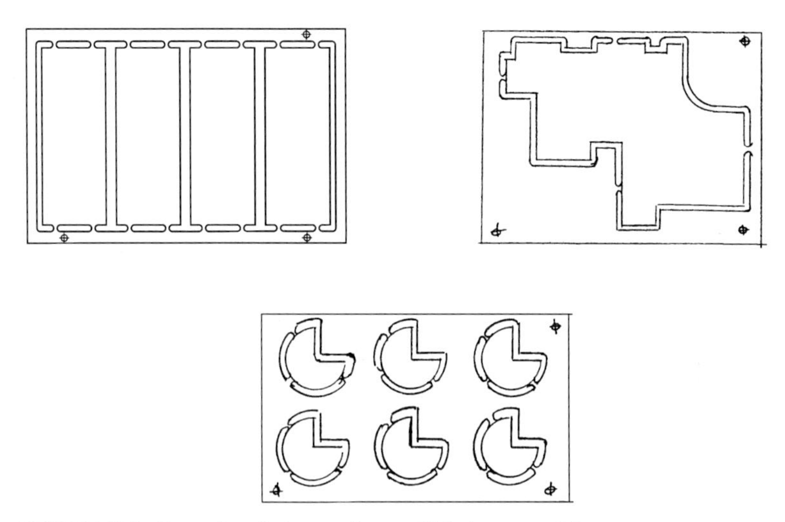

FIGURE 35.5 Examples of tab-routed, or multiple-image, modules.

In Fig. 35.6, each part is shown attached to the frame with breakable or removable tabs and can include features permitting tab removal below the board edge of the image periphery.

Note on Fig. 35.6 the use of a score line to ease tab removal. Scoring, alone or in concert with routing (as discussed in Sec. 35.5), can play a large role in board and pallet separation.

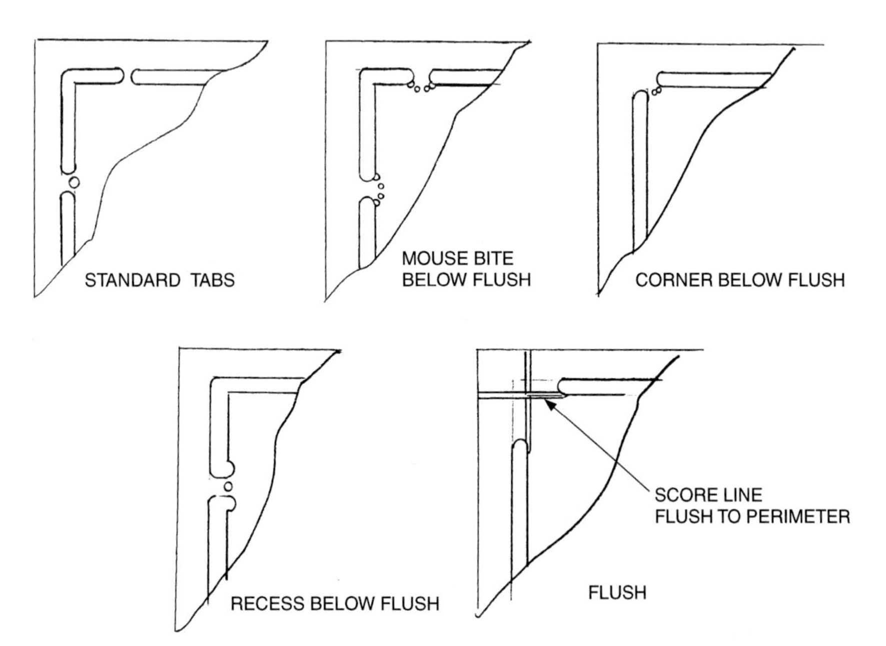

FIGURE 35.6 Each part is shown attached to the frame with breakable or removable tabs. A variety of breakable tabs is shown. Also shown is a score line for ease of tab removal.

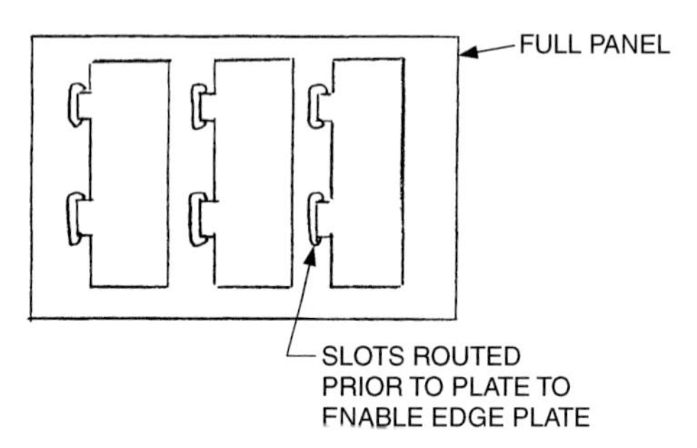

FIGURE 35.7 Multiple-image palletization.

Beyond multiple-image palletization, as shown in Fig. 35.7, CNC routing can provide a variety of board requirements including:

- Internal cutouts
- Slots
- Counter bores
- Board edge conditioning for plating

There are many benefits available through CNC routing beyond those of efficient part profiling. A little planning before the release to production can improve manufacturability as well as provide many no-cost benefits to assembly, soldering, and so on.

35.4.3 CNC Operation

CNC router equipment has the ability to process high volumes of circuit boards very accurately and economically, yet it is coupled with features to enable quick program and setup.

This coupling enables the same processing used for high volume to be utilized for prototypes and short-lead-time production. With circuit board data files so universally available, the part programming time has dropped to a few minutes, as opposed to the hours it once took, while setups remain at about 15 to 20 min., plus cutter labyrinth and first article routing.

Router operation consists of multiple spindles (two to five) capable of operating from 6,000 to 36,000 r/min. or more. The router path (x-y table movement and spindle plunge and retract) is determined by program. This permits any number of paths and any location.

The preferred method of registration of the panel to the machine table is to use the full panel and the tooling holes previously drilled. Tooling holes internal to the part provide for the highest accuracy, although manufacturing panel tooling may be used if considered earlier in the process.

35.4.4 Mechanical Routing with CNC Equipment

Mechanical routing is the most common technology used for rigid circuit board routing. The cuts are obtained with a metal cutter.

35.4.4.1 Cutter Offset. Since the cutter must follow a path described by its centerline, it must be offset from the desired board edge by an amount equal to its effective radius. This is the basic cutter radius, and it will vary with the cutter tooth form. Most newer-generation equipment will automatically adjust for the cutter radius. However, either manually or automatically, this is a basic element of routing planning. Since the cutters deflect during the

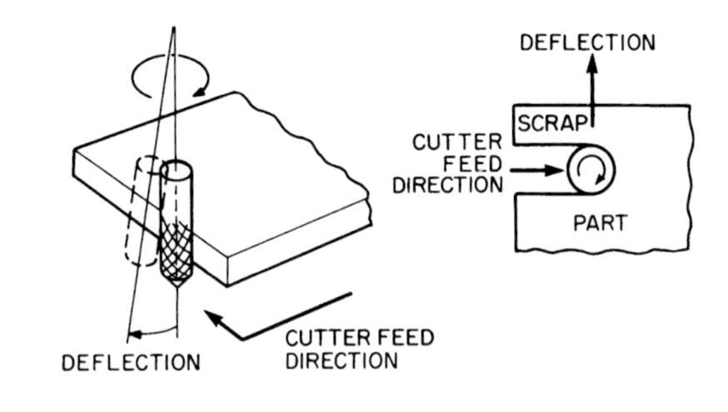

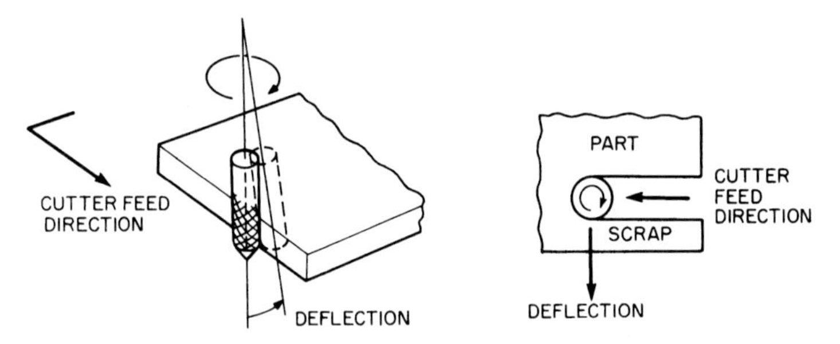

FIGURE 35.8 Effect of cutter deflection on part size and geometry: (a) Clockwise cutting (recommended for outside cuts) deflects cutter away from part. That leaves outside dimensions large on first pass unless compensated for in programming. (b) Counterclockwise cutting (recommended for inside cuts and pockets) deflects cutter into scrap. Therefore, inside dimensions of holes or cutouts will measure small unless compensated for in programming.

routing operation, it is necessary to determine the amount of deflection to be added to the basic cutter radius before expending large amounts on programming parts (Fig. 35.8). Cutter compensation values can be varied.

Variables that affect deflection are thickness, type of material, direction of cut, feed rate, and spindle speed. To reduce those variables, the manufacturer should:

- Standardize on cutter bit manufacturer, selected diameters, tooth form, and end cut.
- Fix spindle speed (24,000 r/min. is recommended for epoxy-glass laminates).
- Rout in clockwise direction on outside cuts, counterclockwise on inside cuts.
- Standardize on single or double pass.
- Fix feed rates for given materials. (Note that higher rates will increase part size, and slower feed rates will decrease part size.)
- Develop documented process controls after experimenting with varied parameters.

35.4.4.2 Direction of Cut. A counterclockwise direction of feed (climb-out) will leave outside corners with slight projections and inside corners with small radii. A clockwise direction of feed (rake cut) will give outside corners a slight radius, and perhaps give inside corners a slight indentation. You can minimize such irregularities by reducing the feed rate or cutting the part twice.

35.4.4.3 Cutter Speed and Feed Rate. The variables affecting cutter speed are usually limited to the type of laminate being cut and the linear feed rate of the cutter. A cutter rotation of 24,000 r/min. and feed rates up to 150 in./in. may be used effectively on most laminates, although cutter feed direction may require a lower feed rate. Teflon-glass and similar materials, the laminate binder of which flows at relatively low temperatures, require slower spindle speeds (12,000 r/in.) and high feed rates (200 in./in.) to minimize heat generation. The graph in Fig. 35.9 shows recommended feed rates and cutter offsets for most standard laminates at various stack heights. The cutter used is a standard 1.8 in. diameter burr type.

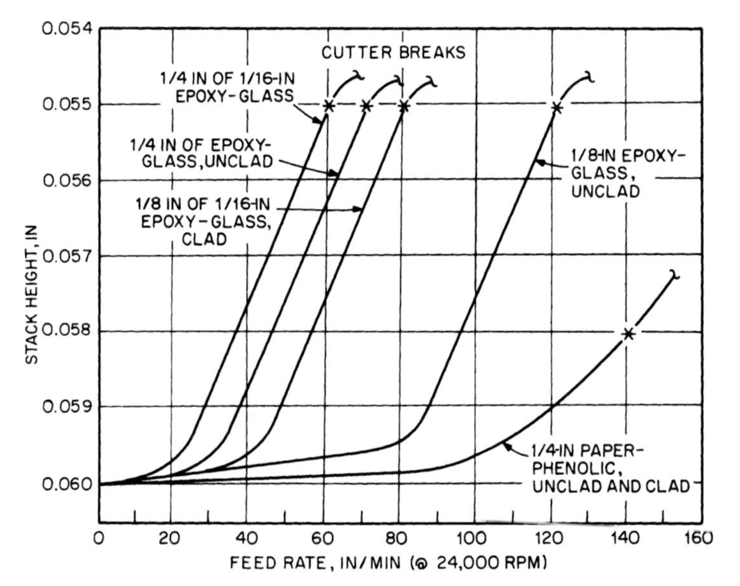

FIGURE 35.9 Recommended feed rate of 24,000 r/min. for varying stack heights of specific thickness of material using 0.124 in. diameter burr cutter.

35.4.4.4 Cutter Bits. Because of the precise control of table movement in CNC routing, cutter bits are not subjected to shock encountered in pin routing and stylus routing, and therefore small-diameter cutters (0.031 in. diameter) may be used successfully. However, the fabricator would do well to standardize on 0.124 in. diameter cutters because they are suitable for most production work and are readily available from a number of manufacturers in a variety of types. The resulting 0.062 in. radius on all inside corners is usually acceptable if the board designer is aware of it.

Cutter tooth form is more important in CNC than in other routing. Because of the faster feed rates possible, it is important that a cutter have an open tooth form that will release the chip easily and prevent packing. Many standard diamond burrs available on the market will load with chips and fail rapidly. The carbide cutting bit will normally cut in excess of 15,000 linear in. of epoxy-glass laminate before erosion of the teeth renders the cutter ineffective or too small.

If extremely smooth edges are required, a fluted cutter may be used. Single- or two-flute cutters with straight flutes should be used when cutting into the foil if minimum burring is desired. It should be noted that such cutters will be more fragile than a standard serrated cutter, and feed speeds should be adjusted accordingly. When a slightly larger burr can be tolerated, two- and three-flute left-hand spiral cutter bits should be used because of their greater strength. The left-hand spiral will force the work piece down rather than lift it, assuming a right-hand turning spindle.

35.4.5 Laser Routing with CNC Equipment

Laser routing can be an effective alternative to mechanical routing especially for flexible circuit board materials. Laser cutting is very accurate, allowing for the profiling of very small parts. There are several types of lasers in use, and some equipment uses more than one type in a single piece of equipment.

Some laser types include the following:

- Nd:YAG (neodymium-doped yttrium aluminium garnet; Nd:$Y_3Al_5O_{12}$), a crystal that is used as a lasing medium for solid-state lasers
- CO_2 cuts using heat only
- UV cuts using both thermal and chemical reaction

It is important to match your application and material with the laser type. Using the wrong application can result in burned material.

35.4.6 Tooling

To simplify tooling and expedite loading and unloading operations, effective hold-down and chop-removal systems should be provided as part of the machine design. Various methods may then be devised to mount the boards to the machine table while properly registering them to facilitate routing the outline. Some machine designs have shuttle tables available so that loading and unloading may be accomplished while the machine is cutting. Others will utilize quick-change secondary tooling pallets or subplots that allow rapid exchange of bench loaded pallets with only a few seconds between boards.

35.4.6.1 Tooling Plates. Tooling plates utilize bushings and a slot on the centerline of the active pattern under each spindle. They are doweled to the machine table (Fig. 35.10). The plates may be made by normal machine shop practice, or the router may be used to register and drill its own tooling plate. Mounting pins in the tooling plate should be light slip fit.

35.4.6.2 Subplates. Subplates should be made of Benelax, linen phenolic, or other similar material. The pattern to be routed should be cut into the subplates' surfaces. The patterns act

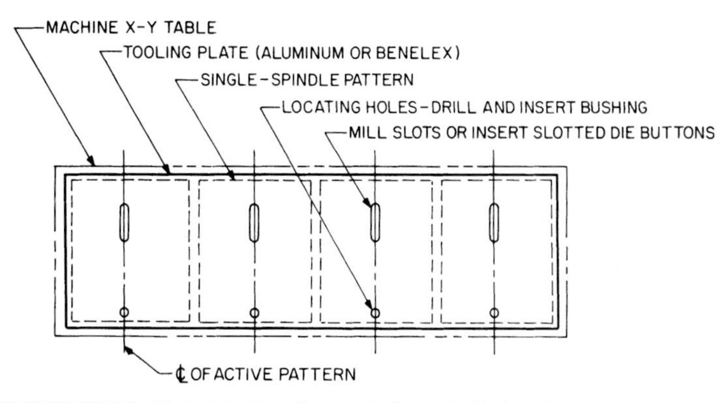

FIGURE 35.10 Typical tooling of numerically controlled routing.

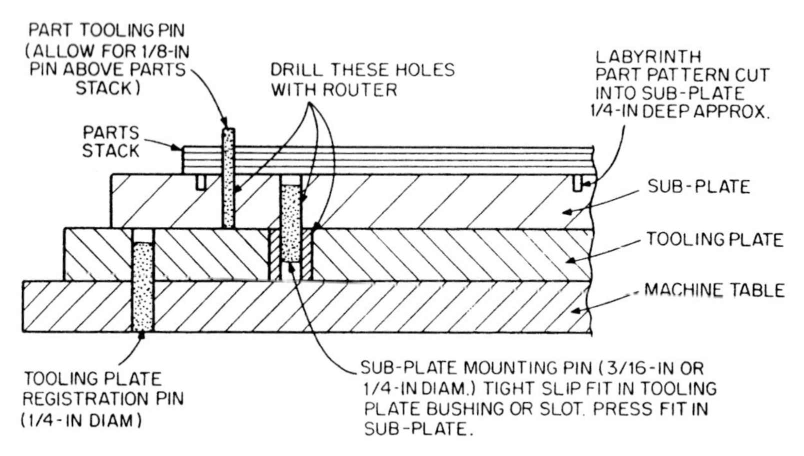

FIGURE 35.11 Tooling schematic for numerically controlled routing.

as vacuum paths and aid in chip removal. Part-holding pins should be an interface fit in subplates and snug to loose-fit in the part, depending on cutting technique used (Fig. 35.11).

It is recommended that the programmer generate the tooling and hold-down pinholes in addition to the routing program. That will provide absolute registration between the tooling holes and the routing program.

35.4.7 Cutting and Holding Techniques

Since the precision required for cutting board outlines, as well as the placement of tooling holes for registering boards, will vary, a number of different cutting and holding methods may be used. Three basic methods are illustrated here.

Experimentation will determine which method or combination of methods is most applicable to a particular job. With all methods, the minimum dimension for board separation with a 0.125 in. cutter is 0.150 in.

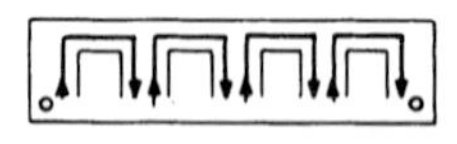

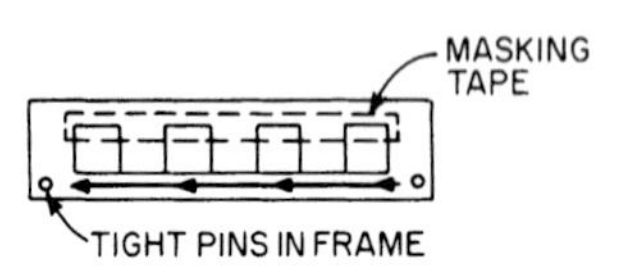

FIGURE 35.12 No-internal-pin method. Step 1: Cut three sides (a). Step 2: Apply masking tape (b). Step 3: Cut parts away.

- **No-internal-pin method** If no internal tooling pins are used, the procedure of Fig. 35.12 may be employed, but it is normally used only when no other method is possible. Characteristics of this method are as follows: accuracy, ±0.005 in.; speed, slow (best used with many small parts on a panel); load, one panel high for each station.
- **Single-pin method** The single-pin method is illustrated in Fig. 35.13. Characteristics of this method are as follows: accuracy, ±0.005 in.; speed, fast (quick load and unload); load, multiple stacks.
- **Double-pin method** In the two-pin method, there is a double pass of cutter offset; see Fig. 35.14. Make two complete passes around each board, the first pass at a recommended feed rate and the second at 200 in./min. Remove scrap after the first pass. Characteristics of this method are as follows: accuracy, ±0.002 in.; speed, fast (highest-accuracy system—loads and unloads slower than single-pin method due to tight pins); load, multiple stacks.

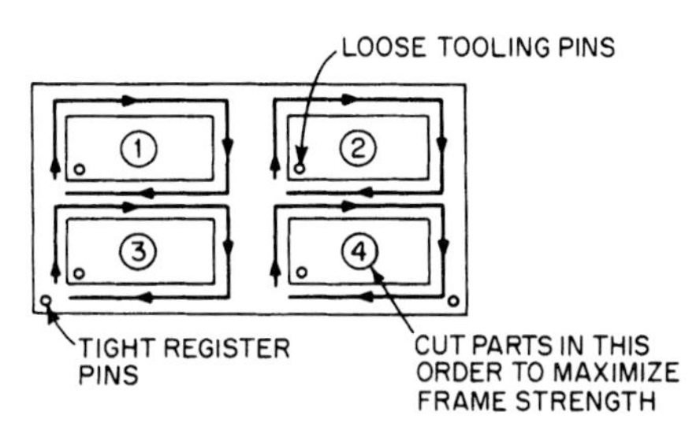

FIGURE 35.13 Single-pin method.

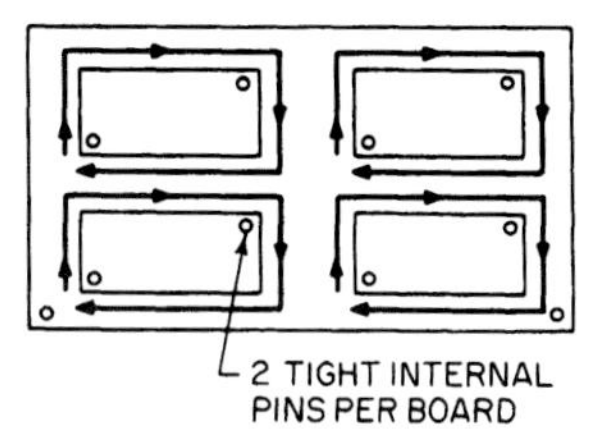

FIGURE 35.14 Double-pin method.

35.5 SCORING

Scoring is a circuit board fabrication method used to make long, straight cuts quickly, and therefore is often used to create rectangular profile board shapes. More commonly, however, it is used in concert with CNC routing for complex shapes, enabling each tool to be used to its unique advantage. When used with routing, scoring has a much wider application and can provide simple breakaway for complex profiles (Fig. 35.15).

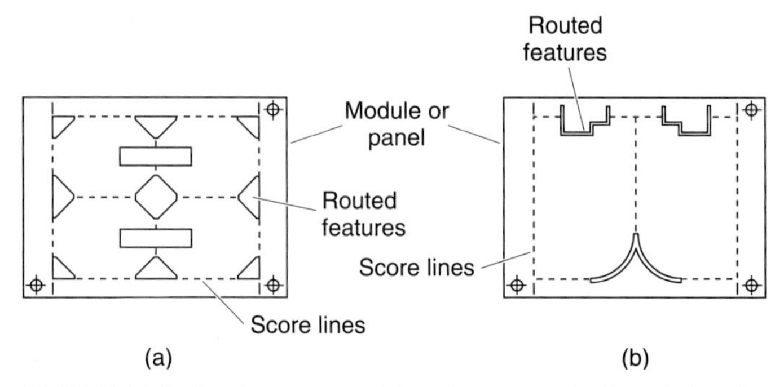

FIGURE 35.15 Scoring processes: (a) module or panel with typical scoring lines and routed corners; (b) scoring lines and routed complex features.

35.5.1 Scoring Application

Scoring is accomplished by machining a shallow, precise V-groove into the top and bottom surfaces of the laminate, generally with the use of CNC equipment. The two most significant elements of the score line are as follows:

- The positional accuracy from the reference feature (usually the registration hole)
- The depth of the score, which determines the web thickness

The final edges of a scored circuit board are yielded by breaking the panel, or border, at the score line (Fig. 35.16). The angle of the cutting tool is reflected in the V-groove geometry, and limiting this angle to 30° to 90° will minimize the score line intrusion into traces near the edges of the circuit board. The score line exposes the laminate glass fibers and resin. Measurements from these surfaces will vary greatly, even though the score line is precisely machined

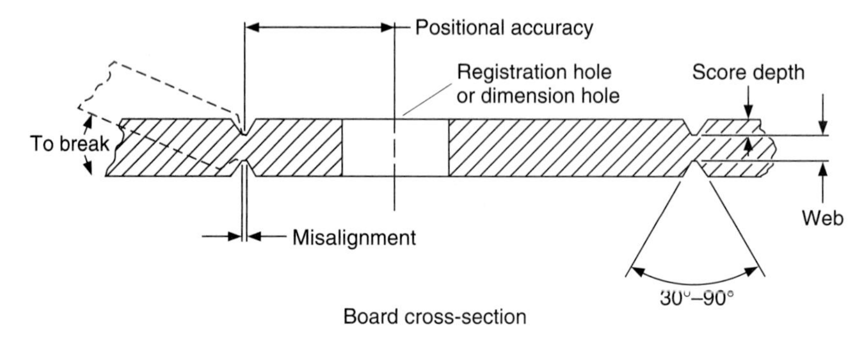

FIGURE 35.16 Cross section of board, showing finished V-groove and break.

(Fig. 35.13). These irregular surfaces will be noticed as dimensional growth and should be considered in design or planning when designated as a scored edge.

Positional accuracy Board cross-section Registration hole or dimension hole Misalignment Web Score depth To break 30°–90°

The dimensional accuracy of the final board is determined by the degree of precision with which the following are performed:

- Misalignment within ±0.003 in. of the score line from the desired location
- Web thickness within ±0.006 in. of the designated dimension

Typically, a nominal web thickness is 0.020 in. for 0.060 in. thick FR-4 boards and 0.014 in. for 0.030 in. thick FR-4 laminate. For CEM-1 or CEM-3 materials, 0.040 in. and 0.024 in. nominal web thickness values apply, respectively. These web thickness values enable sufficient module strength to avoid accidental or premature score separation while providing simple breaking efforts without excessive edge roughness or growth.

35.5.2 Operation

Two major types of panel scoring systems are available:

- Dedicated CNC scoring machines utilizing high-speed carbide or diamond-embedded cutter blades, operating as a pair, one on each side of the board. This generates the V-groove on each side simultaneously
- Drills or driller/routers equipped with scoring software and spade type carbide bits, generating score lines on one side of the panel at a time (Fig. 35.14)

35.5.2.1 Dedicated Scoring Equipment. The dedicated CNC scoring machines are high-production, precision computer-driven machines. With an exception or two, they utilize blade type cutting tools of all-carbide or with carbide inserts, as well as diamond-embedded varieties, and are designed to self-center the panel.

The panel feed rate is high due to the blade's ability to operate at high-surface-feet cutting rates. Scoring both sides simultaneously with one pass contributes greatly to elevated production processing. This equipment utilizes pin or edge registration, with positioning of score lines and steps by programmed instructions. The vertical adjustment of the cutter blades permits variation in V-groove depth, and on many models jump scoring is available. The ability to do jump scoring, or score/no-score segments along a simple line at desired points, is programmable.

35.5.2.2 Multiple-Role Machines. Scoring with CNC drillers or driller/router machines equipped with scoring software produces score lines only on one side of the panel per machine cycle, although each spindle can be used. The panel and program data must be flipped to score the second side. The panel registration method is similar to that of routing, using existing tooling holes to pin the panel to the machined tooling plates. The tooling plates must be machined flat to ensure uniformity of score depth. Brush-type spindle pressure foot inserts should be used to apply downward pressure during score line cutting. Spade-type carbide tools of various angles and configurations are used in the spindles. Typically, multiple passes (two or three) may be required to produce a clean, uniform score line.

35.6 ACKNOWLEDGMENT

Significant portions of this chapter are drawn from James Cadile, Leland E. Tull, and Charles G. Henningsen, "Maching and Routing," *Printed Circuits Handbook, 5th ed.,* Clyde Coombs (ed.), 2001, Chap. 35.

P · A · R · T · 7

BARE BOARD TEST

CHAPTER 36
BARE BOARD TEST OBJECTIVES AND DEFINITIONS

David J. Wilkie
Everett Charles Technologies, Pomona, California

36.1 INTRODUCTION

Advances in packaging technology resulting in finer board geometry, including the various forms of high-density interconnection (HDI), have combined with increasing data rates to put significant pressure on the electrical test area. Fixture construction is more expensive and requires improved process control. Advanced test methods, such as radio frequency (RF) impedance testing, are more often required. Global price competition demands reduced costs. Meanwhile, original equipment manufacturers (OEMs) demand that board manufacturers accept increased liability for defective product—thereby demanding improved fault coverage. This chapter is devoted to the why, what, where, when, and how of current electrical test methods useful in meeting these requirements.

36.2 THE IMPACT OF HDI

Test engineers are currently confronting significant changes in test requirements that derive from changes in the product itself. Notable among these changes are those driven by the growth of various HDI technologies. Examples of common HDI applications are direct chip attach (DCA), high-density ball grid array (BGA), and variants of these often referred to as chip scale packaging (CSP). In addition, higher-density input/output (I/O) connectors are commonly applied. In addition to the changes in physical geometry, HDI also implies increasingly high data rates and clock speeds. Advanced means are being applied that verify not only the ability of the product to interconnect electronic components, but to do so in a manner that guarantees signal fidelity.

An analogy is possible comparing the impact of HDI on board technology with the impact of the sound barrier on aerodynamic flight. This analogy is readily applied to final electrical testing. It was popularly imagined that the sound barrier was impassible, or could be passed only by Buck Rogers in his silver rocket—that beyond the speed of sound was no place for an airplane to go. As it turned out, passing the sound barrier *was* difficult, and required rethinking many aerodynamic principles. But the resulting vehicles are still recognizable as aircraft. They have wings, they burn fuel. They meet all of the objectives of earlier craft, and do so in a superior manner. The test processes that are resulting from the impacts of HDI upon final electrical testing are similarly recognizable evolutions of previous methods, employing familiar base technologies, but in new combinations and with added features that actually improve test coverage rather than diminishing it.

Proposals have circulated for radical new measurement approaches to electrical testing. These have included the use of electron beams, laser-stimulated photoelectric effects, and gas plasma techniques in configurations similar to existing flying probers. In each case, the board is scanned without use of a custom test fixture. To date, all of these methods involve compromise of fault coverage, or add little or no fine-pitch performance beyond conventional methods. Relaxation of test criteria is not a relief available to most users. HDI product types demand increased fine-pitch capability. These finer trace and via geometries increase the risk of latent defects and increase interest in measurement methods most sensitive to precursor symptoms of such defects. As a result, none of these approaches has seen widespread adoption.

Rather than witnessing radical breakthroughs, the market has seen a steady improvement in the availability of software and hardware tools that provide for the effective combination of the best features of conventional test systems and increased availability of test systems dedicated to specialty markets, in particular the small-format laminated chip carrier product type.

36.3 WHY TEST?

Why pay for testing? Why do we need to test? The answer has several components. The primary assumption is that not all boards produced are good. If we could look at the yields through the history of the industry in printed wiring boards using similar technology, we would see exponential reductions in the percentage of bad boards produced. Process improvements and a wide variety of quality improvement programs continue to make a difference. In spite of these improvements, the need for electrical testing remains.

36.3.1 The Rule of 10s

One reason to test is to block the addition of further value to defective product. Consider the printed wiring board as a component of a completed assembly. A commonly accepted relative measure of the cost of faults in a completed electronic assembly can be expressed by the rule of 10s (see Fig. 36.1). The idea is that the earlier a fault is caught, the less it will cost. An example might be an open in a bare board that is not found at bare board test. The faulty bare board is now loaded with the components, soldered, and tested. If the fault is found at the loaded board level, repair or scrapping of the assembled board is much more expensive than repair or scrapping at the bare board level. There are situations where the assembler of the board charges back to the manufacturer of the bare board some portion of the cost of the scrapped components, assembly labor, and/or production cost. This is increasingly true with the high-cost and high-capability integrated circuits (ICs) present in so many designs. A fault that passes a system-level test and makes it to the end user entails even higher cost. Some of these costs are very tangible in the form of field service labor, downtime, parts, and labor. Other costs are counted in lost business and reputation.

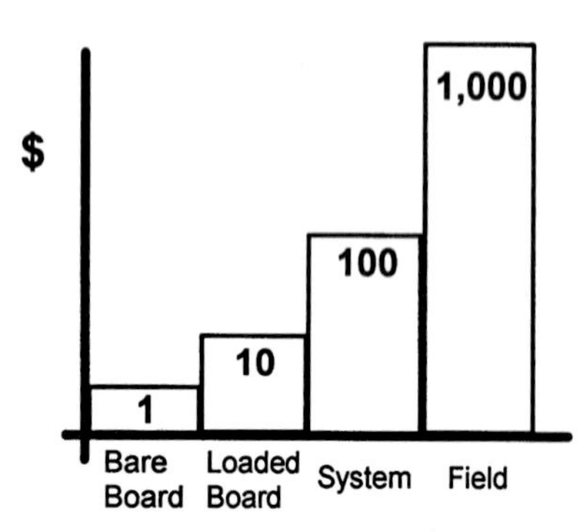

FIGURE 36.1 Rule of 10s: cost of a found fault.

36.3.2 Satisfying Customer Requirements

Where the rule of 10s is a logical big-picture approach to the necessity of testing, requirement from a customer is quite the opposite. The board user has already determined that electrical testing is appropriate and necessary and has made it a requirement as part of supplying the board. It is the obligation and responsibility of the bare board manufacturer to meet this requirement of purchase contracts. In practice, we find such customer requirements stated in various ways.

36.3.2.1 Generic or 100 Percent Board Test. Unfortunately, this requirement may be nothing more than the words "100 percent electrical test" appearing somewhere in a purchase order or a larger package of documentation. Not only is such a specification relatively meaningless, but the issuer of such a requirement is led to believe that the boards will be tested to the most stringent requirements—that it will not be possible for a single bad board to be delivered. Nothing could be further from the truth. Such a specification fails to define what a 100 percent test consists of. The purchaser of the bare board will certainly be dissatisfied when he or she receives a bad board. Although the specific fault on the bare board in question may have a perfectly viable explanation (might not be detected electrically, or may be undetectable by the specific methods employed), the 100 percent test specification implies that no explanation is acceptable. In many cases, application of this general sort of specification is combined with a customer-imposed limitation on test cost, which in turn requires cost-driven limits on the thoroughness of electrical testing performed by the board manufacturer.

36.3.2.2 Written Test Specification. It is the responsibility of both the bare board manufacturer and the bare board purchaser to agree upon a test specification that is specific in terms of test thresholds and method of testing while keeping in mind the appropriateness to the application for the bare board design. Some limits on test thresholds and test methodology may be due to equipment and capabilities of the manufacturer. It is the responsibility of the bare board manufacturer to educate the end user as to the implications of test specifications in thoroughness and cost of testing. A matrix of test options, fault detection capabilities, escape risks, and associated costs should be presented to the end user. It is the responsibility of the end user to work with the bare board manufacturer to develop specifications appropriate to the bare board being produced and to accept and plan for any testing compromises imposed by cost or specification limits. Areas that should be covered in the specification are:

- Data-driven testing, source of data, format, and integrity
- Test thresholds: maximum allowable for continuity verification
- Test thresholds: minimum allowable threshold and applied voltage for isolation verification
- Test fixture methods: single fixture with simultaneous access of all test points versus split-net fixtures versus flying probe
- Test point optimization
- Board marking indicating electrical test satisfaction
- Resolution of discrepancies in process (contacts and procedures)
- Requirements for special test methods (time domain reflectometry [TDR], embedded passives, etc.)
- Resolution of escapes

Such agreements make for good communication, clear understanding, happy customers, good reputation, and good business.

36.3.2.3 Using Available Standards. Industry standards or guideline documents such as IPC-9252, "Guidelines and Requirements for Electrical Testing of Unpopulated Printed Boards" (which replaces IPC-ET-652), are useful in developing agreed-upon written specifications for a particular product.

Standards documents must be applied in consideration of the target application. An excessively rigorous specification would result in unacceptable test cost with little quality improvement.

Arbitrarily high isolation voltages or continuity test currents can actually lead to product damage. Conversely, relaxing isolation resistance specifications on boards containing sensitive amplifiers would be inappropriate. These are but a few examples. In IPC-9252, the IPC has made efforts to distinguish between general classes of board applications, suggesting specific electrical test requirements that attempt to recognize different levels of criticality in testing.

Final determination of the suitability of a given set of requirements must be made in cooperation with technical representatives of the bare board purchaser.

36.3.3 Electrical Testing as a Process Monitor

A third benefit of testing is the improvement of processes and the resultant reduction in costs. To raise yields, reduce scrap, or generally improve quality, there must be a system of measurements able to express results of changes. One of the best areas to collect data is electrical testing. This usually requires an integration of testing and repair data. While the test system can identify a bad board and the location of a fault, a repair operator can further classify the fault to associate it with a specific process (imaging, plating, etching, solder masking, etc.).

These data can be quantified and analyzed in a variety of ways, and corrective action may then be directed to one or more of the following levels.

36.3.3.1 Faults Specific to the Operation. Fault types may be detected that are not typical of the industry or competition. It can be difficult to assess this type of situation accurately because data commonly available from the customer, alternative vendor, or competitor are likely to be narrow or biased. It may be better to analyze fault types specific to an operation in terms of cost or profitability. Are certain product types profitable to manufacture? What would be the cost to make them profitable?

36.3.3.2 Faults Specific to the Process. An analysis of fault data may lead to a specific process that regularly induces faults in a wide variety of part numbers, or in a variety of part numbers sharing key characteristics. This may point to requirements for new settings, procedures, materials, training, personnel, equipment, etc.

36.3.3.3 Faults Specific to a Part Number. Fault data on a specific part number are useful in correcting and optimizing processes for future runs of the same part number, and often for similar future part numbers. It may be possible to feed back electrical test data in real time using software systems, depending upon lot size, factory flow, and the type of test and repair analysis equipment available for use. In this situation, results from electrical testing may help drive some adjustments or quick changes in the process that will immediately improve the yield and reduce scrapping, repair, and retesting costs.

36.3.4 Quality System Improvement

Quality control analysis of fault data is only a piece of what needs to be in place already, that is, a quality control system or process. The value of electrical test (or other process) data for process monitoring is only as great as the quality control framework making use of it. It is possible for electrical test results to become a vehicle for improvements to a board manufacturer's quality system.

36.4 CIRCUIT BOARD FAULTS

For electrical test purposes, faults may be defined as test system measurement results other than those programmed to be representative of a good board. The faults detected may or may not impact the functionality of the circuit board, though in most cases they will. Some guidelines may prove useful when more specific guidance from the original specifier or designer is lacking. Obviously, gross shorts and opens in a circuit are likely to cause problems. Will an electrical test system find all the faults? No. The definition of "all the faults" is too subjective.

The electrical test system will not detect all faults related to aesthetics, annular rings, layer-to-layer registration, etc., unless they present an effect measurable by the test system. Further, the electrical measurement employed, type of test fixture, test program generation method, and end-use requirements vary too widely to state broadly that electrical testing will find all the faults.

For purposes of the following discussion, it is worthwhile to clarify the distinction between a defect and a fault. A fault is a test system designation for an item that does not meet the expected criteria. A defect refers specifically to the board and a defect in its design, fabrication, appearance, etc. Not all defects can be detected by the test system.

36.4.1 Fault Types

Table 36.1 presents common fault types. For those fault types that are commonly detected, it is valuable to distinguish types of tests from types of faults. It may be preferable to refer to isolation testing or continuity testing rather than to shorts or opens testing, as the latter meaning is often confused Shorts and opens are results, not test types. Test methods themselves are discussed in detail elsewhere in this chapter.

TABLE 36.1 Test and Fault Types

Test type	Fault type identified by test
Continuity	Open
Isolation	Short and/or leakage
TDR or network analysis	RF impedance fault
Hi-pot	Voltage breakdown

36.4.1.1 Shorts. Shorts, hard shorts, or short circuits are defined here as erroneous (undesired and unexpected) low-resistance connections between two or more networks or isolated points, typically exhibiting a fairly low electrical resistance value. Shorts are reported as failures of the isolation test of the product. Shorts are produced in a variety of ways, including exposure problems, underetching, contaminated phototools, poor alignment of layers, defective raw material, and improper solder leveling.

36.4.1.2 Opens. Opens represent an absence of expected circuit continuity, or in other words, a missing connection. This divides a circuit network such that the network is split or divided into two or more pieces. Opens are reported as failures of the continuity testing of the product. Opens are produced in a variety of ways, including overetching, underplating, contaminated phototools, contaminated raw material, layer registration errors, and mechanical damage. A common problem during electrical testing is "false open" errors, typically the result of localized contamination on the product or test probe that prevents proper connection to the test system. Of particular concern in testing substrates employing microvias are latent defects (those which may appear after test during subsequent substrate assembly processing). Examples are improperly formed conductors at stress points where cracks may form during thermal or mechanical stresses of assembly. Small HDI features are less tolerant of such defects. These may be completely undetectable at the time of test, or limited examples may be detected by especially sensitive ohmic measurements able to detect a limited conductor cross section where a future open circuit may occur. Frequent observation of such defects indicates the need for process changes if field failures are to be avoided.

36.4.1.3 Leakage. A leak or "leaking network" is essentially a type of short. Leaks are also referred to as high-resistance shorts, and differ from hard shorts in that they exhibit a higher resistance value. The precise division between the two types of error reports varies according to the type of equipment used, and some equipment does not attempt to distinguish them in fault reports. As in the case of hard shorts, leakage is a failure of the isolation testing of the product.

Common leakage causes are moisture, chemicals, or debris. Contamination can occur during innerlayer fabrication, lamination, plating, solder masking, or any stage due to handling. Chemical contaminants are often deposits of metal salts left as artifacts of the chemical processes used to manufacture the product. With sensitive test methods, even fingerprints can result in detectable leakage between networks. Such contaminants are often spread over an area of the board such that several networks become interconnected. Consideration of the potential for multiple network involvement is useful in selecting the particular isolation test algorithm or test method, as these methods differ in their sensitivity to this situation. Isolation test methods are discussed elsewhere in the chapter. Some circuits are not sensitive to the high-resistance loads immediately presented by such contaminants. But it is important to note that, in the presence of time, electric fields, and moisture, it is possible for the resistance of a high-impedance short to decrease greatly. Contamination sites may facilitate the growth of metallic crystals that reach out as thin metal threads between networks, forming hard shorts.

Thus an area of the product exhibiting unusual leakage may, at some future time, exhibit hard shorts between networks. This amplifies the need for effective high-impedance testing of high-reliability product types as a means of preventing latent field failures. Note that the nature of board materials is such that they can absorb moisture relatively easily, and they do so over time. Thus even a fairly "dry" contamination providing a very weak electrical path may eventually result in sufficient metal migration during product operation to cause a serious field failure.

36.4.1.4 RF Impedance Fault. Many circuits produced today are required to operate at very wide bandwidths. Examples include fast microprocessors, fast general digital circuits, RF amplifiers in wireless devices, etc. Just as we must use a proper type of cable to connect a television antenna to a TV receiver, it is important for specific RF characteristics to be maintained in interconnections between components of fast electronic circuits on printed wiring boards. One parameter commonly specified and measured is the RF transmission line impedance of the signal traces. This parameter is strongly affected by the materials used in fabricating the board, the trace thickness and width, and the spacing from ground planes and adjacent signals. A common method of measuring RF impedance is to employ TDR.

The TDR measurement provides a statement of RF impedance as a function of distance along the trace. (Distance and time are related here, as the electric signal flows through the board at velocities approaching the speed of light.) TDR testing is often performed on a test coupon attached to the product during manufacture and subsequently disconnected. TDR testing on actual product traces is also done on selected traces, but is complicated by the need for a trace length of several inches, uninterrupted by branches or other constructions. Common values for RF impedance on circuit boards range from the low tens of ohms to several hundred ohms.

RF impedance should not be confused with ordinary direct current (DC) resistance and cannot be measured with common ohmmeters, even though the same unit of measure—the ohm—is used.

RF parameters of interconnections can also be characterized in the frequency domain using instruments referred to as network analyzers, but this method is not common in bare board testing. Requirements for RF impedance testing are more commonly applied as signal frequencies exceed 100 MHz.

36.4.1.5 Hi-Pot Faults. Hi-pot or high–potential voltage breakdown testing is often confused with isolation testing. The tests are very similar, and, to the extent that high voltage is employed in an isolation test, they might achieve a similar result. But isolation testing of bare boards commonly occurs with 250 volts or less applied between networks, and hi-pot testing is often performed at values from 500 volts to several kilovolts. Hi-pot testing attempts to verify the strength of the insulating material between networks by subjecting it to so high a voltage that a catastrophic or avalanche-type voltage breakdown will occur if the insulator is subpar. In contrast, isolation testing on bare boards attempts to detect the small current flowing through contamination (or, for that matter, a hard short) before voltage breakdown occurs. Of course, if the insulator is very weak, then avalanche failures can occur at almost any voltage. Hi-pot testers tend to be benchtop devices with a pair of test leads and no switch matrix, and hi-pot test requirements usually specify that the voltage be applied for a sustained period of time.

Hi-pot testing is valuable for inspection of very thin insulating core material, before circuits are etched. This can serve to detect z-axis faults or contamination in the material before value is added in subsequent processes. It can be impractical to perform hi-pot tests between all conductors of finished fine-pitch boards, as the atmospheric environment (air) between conductors will break down before high voltage is reached. The slow speed at very-high-voltage hi-pot testing and the costs of suitable fixtures and electronics present problems. For final product inspection, it is arguable that a very-high-impedance isolation test provides the superior solution.

CHAPTER 37
BARE BOARD TEST METHODS

David J. Wilkie
Everett Charles Technologies, Pomona, California

37.1 INTRODUCTION

Although the main bare board testing technology is electrical, it is important to consider that nonelectrical methods are also important in the acceptance or rejection of bare printed wiring boards (PWBs). This chapter therefore includes detailed descriptions of both electrical and nonelectrical testing methods.

37.2 NONELECTRICAL TESTING METHODS

There are two nonelectrical acceptance/rejection methods, both based on inspection processes:

- Visual inspection
- Automatic optical inspection

37.2.1 Visual Inspection

Visual inspection is a very manual approach in that it makes use of people, good lighting, some type of training defining what is acceptable and what is not, and good operator judgment.

Usually a comparison to a known good product or the artwork is made. If the operator has seen the board often, he or she becomes more skilled at finding faults and looking for faults in likely locations. As product complexity has increased, we find that many modern products are not suited to this method. Many innerlayer defects are completely undetectable, and even the external layer complexity is visually overwhelming. Visual inspection often remains appropriate for detecting cosmetic defects, such as poor solder masking or physical damage. Such defects generally fall outside the realm of electrical testing as they are not detected by electrical means.

37.2.2 Automatic Optical Inspection

There are computer-based visual inspection methods, referred to as automatic optical inspection (AOI). AOI equipment compares the board or its innerlayers to expected data and/or design rules that have been programmed into the controlling computer. These can be generally accepted parameters or design-rule-based parameters, or windows of acceptable dimensions for each specific feature on the board. As with manual visual inspection, faults found with this method can imply that there may be an impact on the board's functionality, but the board's functionality and interconnect are not directly tested. Distinctions between the aesthetics of the board's features and its fitness for use are difficult to differentiate, and may result in false failures. Rather than being used in final testing, AOI is used for inspection of innerlayers prior to lamination, with the goal of increasing final yield by weeding out the majority of defective layers prior to the addition of further value. As such, AOI can achieve significant financial benefit. AOI can detect some defect types not readily detected by electrical means, particularly "mouse bites" (brief narrowing of the conductor cross section). AOI may also be used to inspect the outside layers after lamination, but is not generally accepted as a quality assurance substitute for final electrical testing.

37.3 BASIC ELECTRICAL TESTING METHODS

Electrical testing is the final test method frequently used to determine whether a board should be shipped. Electrical testing emulates the intended function of the board conductor and insulator patterns by passing currents through conductors and applying voltages across insulators. Such direct electrical measurement requires that the board come into physical contact with a measurement system. Two test types are almost universally performed: continuity and isolation testing. Some other tests may be applied selectively, depending upon the product and customer requirements.

Test order is usually such that the continuity test is performed first. This verifies that each network is intact within itself, and that contact is established between any test fixture and the product. The isolation test can then be performed using only a single test point per network. Some test methods attempt indirect inference of continuity and isolation without making direct current (DC) measurements. These methods are commonly employed in flying probe systems.

37.3.1 DC Continuity Test Method

Continuity testing checks for the expected continuous path within each electrical network. This is done in a series of point-to-point measurements within each network. The resistance found in each measurement is compared to the selected continuity resistance threshold. If the measured value is higher than this threshold, then a fault report is generated. For complex networks, multiple measurements are required in order to ensure that all extremities of the network are interconnected. For example, the network with test points labeled A through D, shown in Fig. 37.1, could be tested in the sequence illustrated in Fig. 37.2. For a network with four test points, the minimum number of tests would be three to determine whether all points are connected. If a board contains a total of N isolated networks, and contains a total of X test points, we may calculate the number of continuity measurements C as $C = X - N$. In assigning test points, the software system often is programmed to delete those that are unnecessary. This is referred to as test point optimization. In Fig. 37.1, a test point located along the path between D and the branch to C would not be useful. Various optimization rules may be applied, but should be applied with care to ensure that adequate test coverage remains in place.

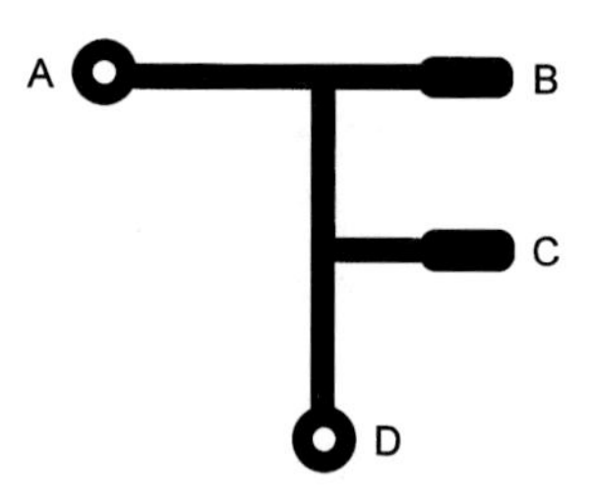

FIGURE 37.1 Sample network.

37.3.1.1 Continuity Test: Two-Wire versus Four-Wire Switching. Switch matrixes are constructed using either two-wire or four-wire circuits. A diagram comparing a single continuity measurement being performed on both two-wire and four-wire matrixes is shown in Fig. 37.3.

Grid test systems use solid-state switches to connect appropriate test points to an internal measurement system (which is effectively an ohmmeter). Current I is driven from the upper test point through the product network and returned to the measurement system through the lower test point. The resultant voltage V across the network is measured. The continuity resistance R is then determined using the relationship R = V/I.

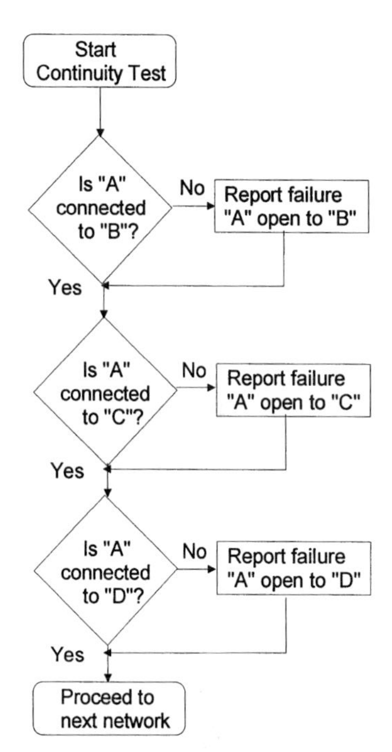

FIGURE 37.2 Continuity test algorithm.

37.3.1.2 Two-Wire Switch Matrix Construction. The simplest, most affordable, and most common construction of the switch matrix is to have a pair of switches for each test point that can connect the test point to either the high side or low side of the ohmmeter. This technique is illustrated on the right side of Fig. 37.3. However, as the solid-state switches exhibit a certain amount of on-resistance, they contribute error to the measurement. The resistance of the switches is added to the resistance we measure, increasing the likelihood of failing the test. These errors can be reduced in various ways:

- Use switches with low on-state resistance (and thus a small amount of variation in on-resistance). This means using physically larger transistor dies. Such devices are not expensive, but are harder to integrate into integrated circuits. Thus, most two-wire designs rely upon discrete output transistors as switch elements.
- Use software to subtract the estimated typical on-resistance. As switch resistance effects vary from device to device, and change with temperature, some error remains.
- Make multiple measurements for each continuity to be inspected, and mathematically subtract most of the switch resistance. However, this slows the measurement process somewhat by requiring extra measurements and requires high-quality fixture construction such that contact resistances do not vary during the required multimeasurement sequence.

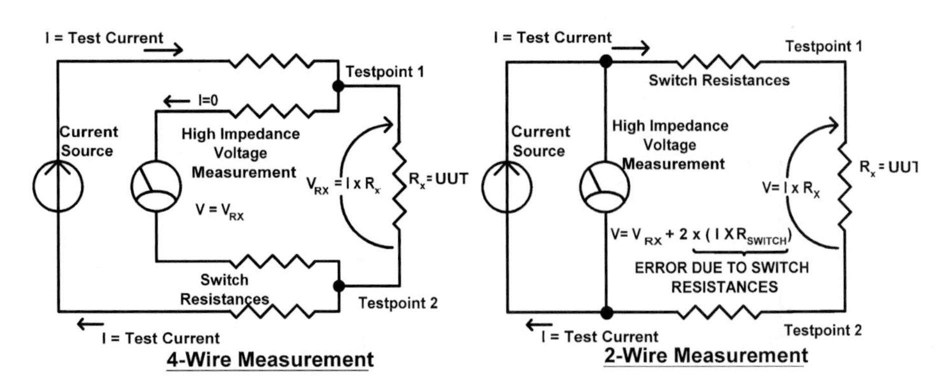

FIGURE 37.3 Continuity measurement circuits.

- Attempt to learn the resistance of each switch. Some systems require the operator to install a shorting plate over all test points, and the system then learns the sum of the switch resistances plus contact resistances to this plate. Note that this technique can actually add error if a high contact resistance value is learned for a particular test point, perhaps due to dirt on a probe. The excessive resistance value learned is subtracted from all subsequent measurements made with that probe, masking some high product resistances. The net result is a slightly increased possibility of passing a board that should be failed.

37.3.1.3 Four-Wire Switch Matrix Construction. The left side of Fig. 37.3 represents a Kelvin or four-wire switch matrix construction. Notice that each test point will require four switches, so that any test point can be connected to the high-side current drive, high-side voltage sensor, low-side voltage sensor, or low-side current return via separate paths. Because the voltage-sensing portion of the measurement system has an extremely high input resistance, almost zero current flows through the extra pair of switches that connect the test points to this circuit. The result is that there is zero voltage drop across these sense switches, and the measurement unit sees exactly the voltage across the unknown load. Because the measurement system knows exactly the current flowing through the load, and exactly the voltage across it, it can calculate the load resistance accurately. The dominant error term in this situation is generally the contact resistances in the fixture and fixture-product interface. Also, the smaller transistors leak less current in the off state, thereby permitting isolation testing at higher thresholds with smaller errors. The continuity accuracy benefit of this technique is realized only if high-quality fixtures with low contact resistance are employed.

37.3.1.4 Continuity Threshold. The continuity resistance threshold parameter is usually specified in the range from a few ohms to 1,000. Several standards useful in suggesting continuity thresholds are summarized in Table 37.1. IPC-9252 has actually weakened the recommendations of IPC-ET-652, which it replaces. As discussed earlier in the chapter, these must be applied with judgment. Generally, a lower continuity threshold provides a more stringent test of the board. Networks with resistances of 5, 10, or 25, although rare when using copper traces of moderate length, can significantly impact the functionality of precision measurement instruments or high-speed computer products. At the same time, it should be noted that there are practical and economic considerations in determining how low the continuity test threshold should be set. Part of the limitation comes from the test system's measurement and switch matrix capabilities and, to a greater part, from the type of test fixture used. At the time of this publication, a 10-continuity resistance test threshold is a common lower limit for production testing with good-quality systems and fixtures.

TABLE 37.1 Examples of Continuity Resistance Test Threshold Standards

IPC Board Class	IPC-9252	IPC-ET-652 (obsolete)	MIL-55110D (obsolete)
Class 1: General Electronic	< 50–100 Ohms	< 50 Ohms	< 10 Ohms
Class 2: Dedicated Service	< 50 Ohms	< 20 Ohms	< 10 Ohms
Class 3: High Reliability	< 50 Ohms	< 20 Ohms	< 10 Ohms

37.3.1.5 Continuity Test Current. Continuity test current is not addressed in IPC-9252 or in most other publications. Use of high current has been proposed as a means of burning out weak traces or mouse bites. But such currents may also damage good traces. If this occurs after the test system has already determined that there is a good connection, the result is a board that once tested as good but is now bad. It is preferred that the continuity test not be invasive or destructive. Typical test currents today are in the range of 5 to 50 mA.

37.3.1.6 Continuity Test False Opens. A common problem with continuity testing is a high incidence of false failures due to fixture and product contamination, poor product registration, or fixture damage. Dramatic improvement is often possible with the addition of product and fixture cleaning methods. Separation of the board-testing environment from such dust-producing processes as drilling/routing can be invaluable in increasing throughput. High-voltage pulses are sometimes used to overcome thin-film contaminants or oxides coating the surface of the board and preventing good contact with the test probe. As no current is initially flowing through the oxide, some test systems offer a feature delivering a high-voltage pulse of strictly limited current and duration, and therefore limited total energy. While brief, the energy level is higher than for a normal continuity test and a small risk of damage at defect sites remains. Adding test time for automatic or manual retesting of the product yields additional boards, but this common approach becomes expensive when large numbers of false opens occur. It may be advisable to correct the root cause.

37.3.2 DC Isolation Test Method

Isolation testing verifies the presence of adequate electrical isolation between networks that are not intended to be connected to one another. Typically a resistance measurement is made from a given network to another net (or group of nets). If the measured value exceeds the specified isolation resistance threshold while the specified voltage is applied, then the measurement is considered to have passed. Otherwise, a fault report is generated. So long as contact with the net has been ensured (during the continuity test), only a single test point per network is required to perform the isolation test. The actual number of isolation measurements required to test a given board can vary substantially with details of the algorithm employed, with subtle impacts upon fault coverage. These issues are discussed in a separate section.

37.3.2.1 Isolation Resistance and Voltage. As isolation testing is a means of evaluating the ability of product insulation to withstand voltage and prevent current flow, it is common for the isolation test specification to include not only a statement of minimum resistance, but also an applied voltage. This is the voltage that the insulator must withstand while exhibiting at least the minimum isolation resistance. Given the relationship $R = E/I$, increasing the applied voltage is also a means of increasing the measurement current level up out of the noise floor internal to the test system. Thus, high-voltage-capable bare board test systems usually are able to verify isolation at higher resistance levels and do so at reasonable speed. Typical values for isolation resistance can range widely, from as little as 1 k. to as much as 1000 M. Values of 2 to 10 M are common, but higher values are useful in detecting trace contamination. Excessive humidity in the test area may preclude the use of very high thresholds and may affect accuracy at lower thresholds. Values below 50 to 55 percent relative humidity are desirable. Common isolation test standards (see Table 37.2) have not always kept pace with the capabilities of new equipment to test at elevated and sensitive thresholds. In any event, assignment of specific test thresholds would ideally be based upon a competent analysis of the intended application of the specific circuits on the product to be tested.

There is somewhat less emphasis on test voltage than was true previously, probably as a result of increasingly fine geometry in the product. Older or simpler test systems may be limited to 10 to 40 volts, but most are able to apply 100 to 250 volts during the isolation test.

TABLE 37.2 Examples of Isolation Resistance Test Threshold Standards

IPC Board Class	IPC-9252	IPC-ET-652 (obsolete)	MIL-55110D (obsolete)
Class 1: General Electronic	> 500 K Ohms	> 500K Ohms	> 2 M Ohms
Class 2: Dedicated Service	> 2 M Ohms	> 2 M Ohms	> 2 M Ohms
Class 3: High Reliability	> 10 M Ohms	> 2 M Ohms	> 2 M Ohms

TABLE 37.3 Examples of Isolation Test Voltage Standards

	IPC-9252	IPC-ET-652 (obsolete)	MIL-55110D (obsolete)
Minimum isolation test voltage	High enough to provide sufficient current, but avoid arcover	High enough to provide sufficient current, but avoid arcover	All least 40 volts, or twice the rated voltage of the board, whichever is greater

A higher voltage increases the test current, providing a better signal-to-noise ratio and generally improving test speed at higher isolation thresholds. Excessive voltage is inappropriate for very fine-pitch substrates and may result in damage from arcing in normal environments (see Table 37.3). The insulating properties typical of modern board material suggest that it may be of little value in specifying elevated voltages while using low (relaxed) threshold resistances. Raise the resistance threshold first, then use enough voltage to get adequate speed and accuracy.

37.3.2.2 True Isolation Test Method. Several different algorithms have been developed for sequencing the switching state during the isolation test, and the choice may affect test coverage. In the most rigorous method, each network is individually tested to determine the total parallel leakage resistance to all other networks on the product. This requires one measurement per network, as illustrated in Fig. 37.4 for a board with three networks. Each network is, in turn, given a chance to charge to an elevated voltage. All other networks are connected together and to 0 volts at this same moment. Notice that only one test point is needed per network.

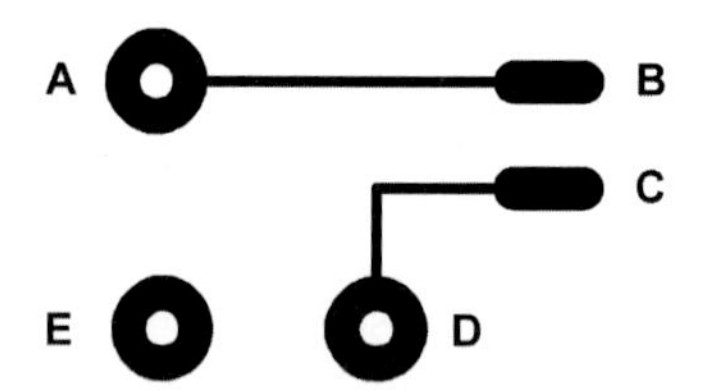

Test 1: Raise net AB positive while nets CD, E are Grounded.
Test 2: Raise net CD positive while nets AB, E are Grounded.
Test 3: Raise net E positive while nets CD, AB are Grounded.

FIGURE 37.4 True isolation test of three-network board.

If the network under test is shorted or leaking to any of the other networks, or to any combination of them, it will fail to charge adequately and the test will fail. The pattern continues until each network has been tested.

Figure 37.5 illustrates the status of the measurement system during the three tests. Notice that test points A and C are used to access their respective networks, and that test points B and D are not needed for the isolation test. These test point switches remain open through all measurements.

A key feature of this method is the ability to answer the rigorous question: "How well isolated is this network from the rest of the board?" Consider the example illustrated in Fig. 37.6.

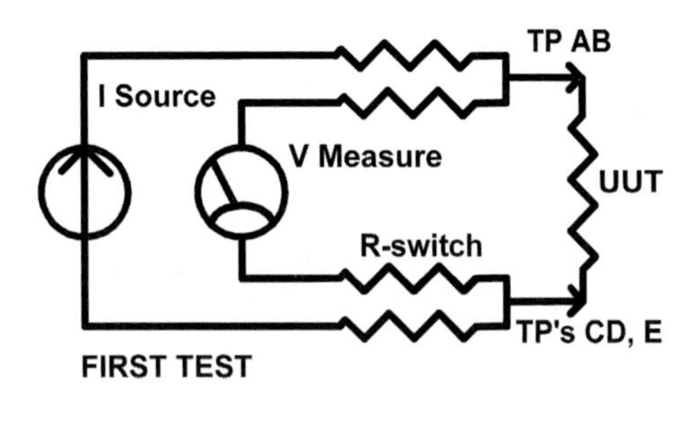

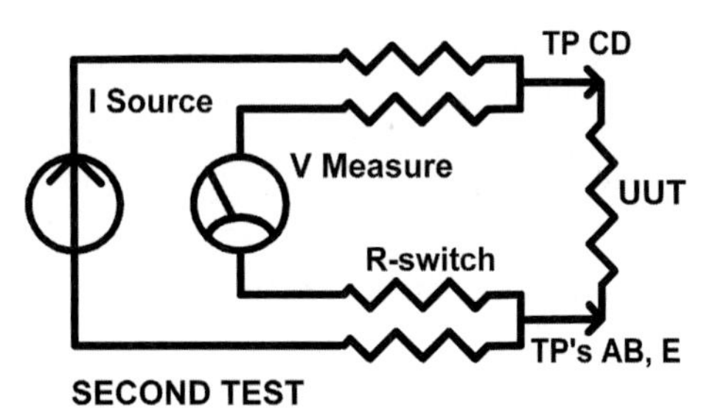

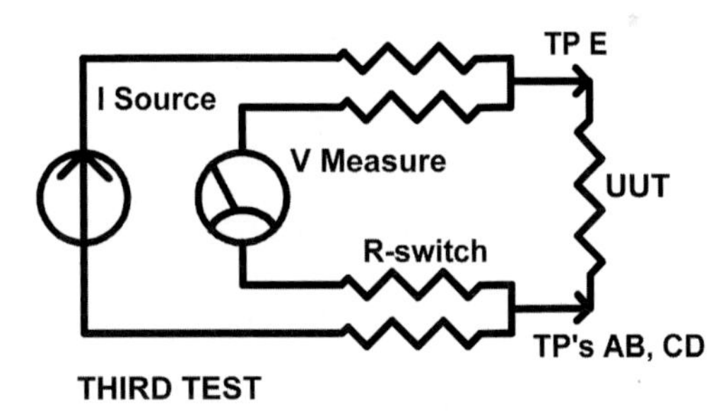

FIGURE 37.5 Measurement sequence for three isolations.

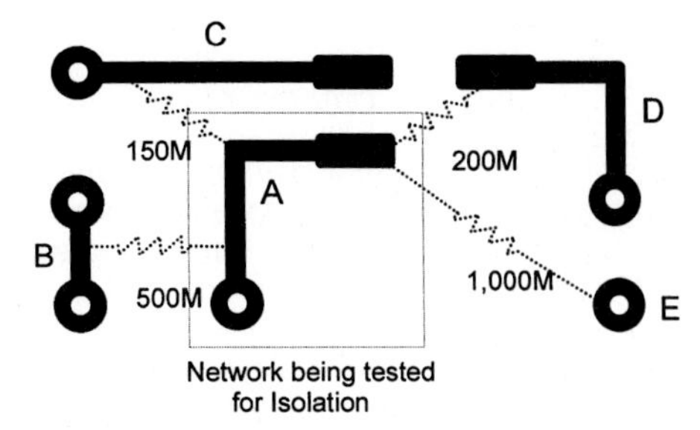

FIGURE 37.6 Parallel leakage detection.

Here the isolation threshold is 100 M. Four leakage paths exit network A, one each to networks B, C, D, and E. Each path, measured separately, is well above the 100-M. pass/fail threshold. But electrically we can only guarantee that network A is isolated by the parallel combination of these four resistances, given by:

$$RA = \frac{1}{(1/R_{AB}) + (1/R_{AC}) + (1/R_{AD}) + (1/R_{AE})} = 68 \text{ Mohms}$$

Thus, the true isolation test method will correctly fail this measurement. A not unlikely real-world situation would be a smear of contamination that touched A, B, and C.

37.3.2.3 Log of (N) Isolation Test Method. Table 37.4 illustrates the log of *N* isolation test method, and compares it to the true isolation method described earlier. The log method offers the powerful advantage that only a small number of measurements are required to test a complex board. That number is given by log2(*N*) rounded up to the next whole number, where *N* is the number of networks.

Thus, for a board with eight networks, we need only three measurements. The gain is even greater for large boards. For a realistic board with 4,000 networks, only 12 measurements would be needed.

This method involves a compromise of fault coverage for parallel leakage detection. Consider the example shown in Table 37.4. The top half of the table considers a very simple board with eight networks and illustrates the pattern of measurements required by the log method.

TABLE 37.4 Isolation Methods Compared

	Network name							
Measurement number	A	B	C	D	E	F	G	H
Log method								
1	+	+	+	+	–	–	–	–
2	+	+	–	–	+	+	–	–
3	+	–	+	–	+	–	+	–
True method								
1	+	–	–	–	–	–	–	–
2	–	+	–	–	–	–	–	–
3	–	–	+	–	–	–	–	–
4	–	–	–	+	–	–	–	–
5	–	–	–	–	+	–	–	–
6	–	–	–	–	–	+	–	–
7	–	–	–	–	–	–	+	–
8	–	–	–	–	–	–	–	+

Networks noted with a + in a given measurement are connected to the upper side of the measurement system. Networks with a – are connected to the lower side. Notice that in each measurement, about half the networks are positive and half are negative.

A hard short between any two networks in Table 37.4 will be detected by either method. In at least one measurement, a + will be at one end of the short, and a – at the other. Both methods work well for hard shorts, and parallel resistance effects play no significant role. But consider the effect of two 15–M. leaks, one from network A to network B and the other from network A to network C. If the threshold is 10 M., this fault will not be detected by the log method, but the true method will fail test measurement #1. The log method is ineffective because there is no one measurement that has the + end of the measurement system on network A while the—end is connected to networks B and C simultaneously. (Were the A–B resistor moved to A–D, then we would detect an error on measurement #2 of the log method.)

The log method may fail to detect certain combinations of high parallel leakage. For product types where a low isolation resistance threshold is acceptable, this may not be a significant issue. Note that this risk may be reduced by substantially increasing the test threshold, such that a larger proportion of individual leaks are likely to trigger a fault report.

37.3.2.4 Isolation Failures: Distinguishing Shorts and Leaks. Earlier discussion made a distinction between hard shorts and leakage failures, although both reports result from the isolation test. Consideration of either the true isolation test method or the log(N) method shows us that neither method immediately tells us the opposite end of the short or leak (refer again to Table 37.4). In the true isolation method, we know that the network under test is leaking or shorted to some other network, but we don't know which one. Initially the log(N) method tells us slightly less: We know only that the short is between that half of the networks switched positive and the half switched negative (or ground). Upon fault detection, either method must call a subroutine to search all network pairs systematically to localize the fault. Searching all networks in pairs for leakage can be slow as well as ambiguous. Therefore, some test systems allow the user to disable searching for leakage.

In the case of distributed leaks, a clear answer may be impractical because searching pairs of networks never connects all of the leakage paths in parallel. The effort to localize a high-resistance leakage may report the entire problem, part of the problem, or none of the problem. The exact result depends upon the actual leakage resistances present, the pattern of networks involved, and the isolation resistance threshold. Note that a flying probe system seeking to "verify" defects may be unable to detect some or all leakage for the same reason. It may be advisable to return such boards to a grid (or other fixtured tester) for final pass/fail.

37.4 SPECIALIZED ELECTRICAL TESTING METHODS

Certain special testing methods have evolved to provide for detection of defects more subtle than simple shorts, opens, or leaks. Additional methods have evolved as suitable to specific types of test equipment, notably the flying prober. These are discussed in the following text.

37.4.1 Hi-Pot Testing

Hi-pot or high-potential testing is very similar to isolation testing, but is commonly distinguished by the magnitude of the applied voltage and, to some extent, by the expected behavior of the detected failure. In the context of bare printed circuit boards, hi-pot testing usually refers to voltages over 250 volts, often 500 to 3,000 volts. The objective is to locate faults in the dielectric (insulating) layers of the board that may result in subsequent field failures at lower voltages.

When the dielectric is subjected to a voltage substantially in excess of the expected working voltage, certain types of material defects can result in an "avalanche" mode of failure of the insulation. Ionization of the intervening material or atmosphere occurs, resulting in a sudden increase in current flow to some relatively large value. Visible arcing and/or burning may result at the failure site. This contrasts somewhat with the isolation test, where a very subtle leakage current flow is usually detected using a somewhat lower voltage. Most often, the isolation test is less destructive as the total energy delivered is less. However, depending upon product conditions, arcing can occur at many voltages and in such cases an ordinary isolation test is essentially the same as a hi-pot test.

It is often impractical to perform hi-pot tests on finished products, except at limited numbers of test points. Equipment limitations are one factor, it being expensive to construct test equipment and fixtures capable of routing such high-voltage signals to a large number of test points. However, the product itself is often a poor target for such tests when evaluated in finished form. Modern products are often constructed with relatively fine spaces between conductors.

On the surface layers of the board, the exposed component connection sites are usually too close together to withstand very high voltage stress without surface arcing. Such arcing is destructive to perfectly good product. The voltage at which such arcing will occur is a function of product geometry and atmospheric conditions.

In consideration of this, the most common practical application of hi-pot testing is to inspect raw material for defects before etching. For example, a thin FR-4 core, clad with copper on both sides, might be evaluated by placing a high voltage across the opposing sides. Any crack or other defect in the insulating material may ionize, resulting in a large current flow. Thus the defective material is rejected before substantial value is added in subsequent processing.

37.4.2 Embedded Component Tests

Methods of embedding certain electronic components *within* the board have been developed. The most common example is the embedded or "buried" resistor. Such resistors are constructed by embedding a layer of partially conductive material within the board. By selectively removing (or adding) material, the resistance value is adjusted. Accuracy ranging from a few percent to many tens of percentage points is realized. Typical resistance values range from a few ohms to thousands of ohms. The most common use of this technology is within high-speed digital circuit designs to replace large numbers of termination resistors with a resistance of 200 ohms and lower. Measuring these values accurately is challenging, requiring good fixture construction and cleanliness. Another difficulty with testing of buried resistors is obtaining usable expected value data for the resistors. At the board shop, most resistors are specified as a shape and a type/thickness of resistance material rather than as a specific resistance value and tolerance.

It may be necessary to analyze the circuit pattern and compute the net resistance of series and parallel combinations, arriving at a measurable final resistance value for the tester.

Embedded capacitors may also be present. All boards exhibit capacitance between various traces, and especially between planes. The amount of capacitance is determined by the parallel surface area of the conductors, the thickness of the insulator between them, and the dielectric constant of the insulating material. Some product designs seek to maximize the desirable noise suppression benefits of capacitance between power and ground planes by making the intervening insulator thin. Some go further and use special insulating cores for these layers.

This is the common form of buried capacitance. Most bare board test systems are not well suited to measuring capacitance values, particularly small ones. In some cases, no measurement is specified and it is only important that the test system tolerate the presence of the capacitance. When measurements are specified, it is usually true that few test points are involved. It may be expedient to use benchtop equipment.

37.4.3 Time Domain Reflectometry

Time domain reflectometry (TDR) is a measurement method often used to verify the radio frequency (RF) impedance of a signal conductor on a circuit board. The RF impedance is important to the proper function of high-speed digital or RF applications. Examples include computer products, cellular telephones, radios, etc. The RF impedance of a signal path should not be confused with the DC resistance, as verified during the continuity test. It is common for a trace with seriously errant RF impedance to demonstrate a very solid DC connection (i.e., low DC resistance). The RF impedance of the trace is most strongly affected by the trace's width, thickness, z-axis spacing from the ground plane, the location of adjacent traces, and the relative dielectric constant of the type of insulating core used to build the board. These parameters are usually rather constant within a particular panel, justifying use of coupons as a means of monitoring the delivered product. Standard bare board testing systems (other than flying probes) do not incorporate TDR capability, as the signal paths through the fixture will not pass the fast-risetime TDR signal. TDR is commonly conducted on a test bench manually.

TDR test systems inject a very fast-risetime voltage step into one end of the conductor. Discontinuities in the RF impedance level along the conductor result in reflected voltage waves being returned to the driving point, where they are collected by the same probe that injected the signal. The result is usually presented as a graph of RF impedance versus distance from the point of injection. Because of reflections and disturbance occurring at the point of injection, a minimum trace length is needed to get a meaningful measurement. Extremely short networks are not ideal candidates for TDR measurement. A trace 2 in. or more in length is generally required to obtain a useful reading. Longer traces may provide improved accuracy of the result. A branch-off to a second signal path along the measured length will disturb measurement badly; thus an undisturbed signal path provides the best measurement target.

A typical TDR result graph is illustrated in Fig. 37.7. This illustrates a region of approximately 50 ohms RF impedance, rising toward infinity as the trace ends in an open circuit at the right, but falling between the upper and lower pass/fail bounds in the region of interest.

Sample testing using handheld probes is popular and practical, but the influence of hand and body position on the measurement can be significant, reducing the repeatability of the result. Several vendors offer flying probe solutions dedicated to TDR test. While requiring some additional setup they are faster at higher volumes and greatly improve repeatability.

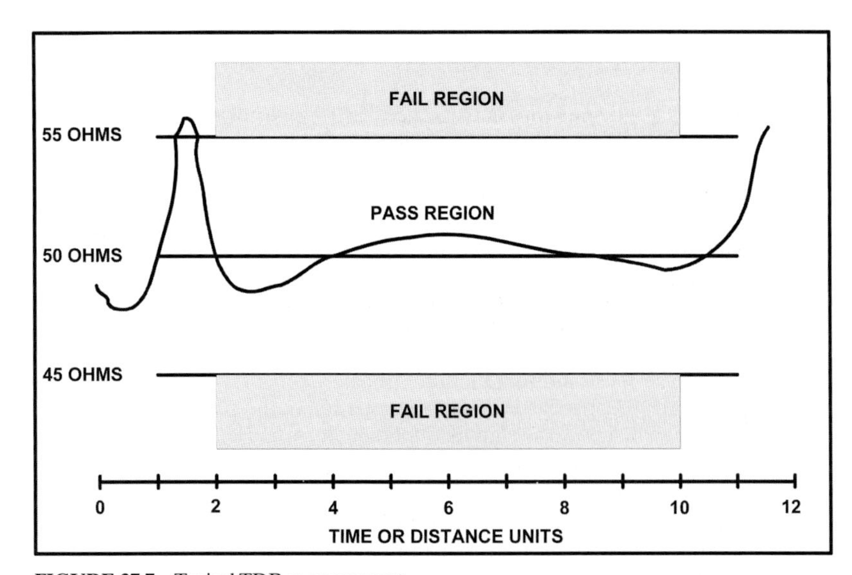

FIGURE 37.7 Typical TDR measurement.

37.4.4 Test Methods Unique to Flying Probe Systems

Because flying probe systems contact pairs (or other limited numbers) of points at any one time, they cannot directly perform precisely the same isolation measurements as can universal grid (and other fixtured) systems. DC isolation is performed between pairs of networks, with the number of measurements reduced by determining which networks are adjacent to the network being tested and therefore likely to cause isolation problems. Otherwise, flying probers are capable of ordinary DC continuity and isolation measurement, as discussed earlier. In addition, most flying probe vendors have developed alternative measurement methods that reduce the number of measurements and therefore reduce time lost to mechanical motion. These methods are discussed in the following text.

37.4.4.1 Indirect Measurement of Isolation or Continuity. Different vendors have developed various implementations of indirect measurement, but in general these methods share a general assumption that a given product network will display a certain amount of capacitive or other electromagnetic coupling to neighboring planes or traces. The amount of coupling is affected by the geometry of the traces involved. If a trace is broken, the remnant will display reduced coupling. Similarly, if a trace is shorted to another trace, the amount of coupling will increase substantially. If the test system measures the amount of coupling from each trace to a ground plane (or other electrical environment), a degree of confidence can be obtained that all traces are intact. No direct measurement of isolation or continuity is necessarily made.

Instead, the coupling signature is used to imply the correctness of the configuration. As no direct measurement of isolation or continuity is made, we refer to these methods as indirect measurements.

Typically these methods are highly reliable in detecting hard shorts and opens, but may be less effective in detecting distributed contamination or high-resistance connections of several meg ohms or more. Specific methods include constant-current capacitance measurement (charging or discharging variations), voltage-source measurement of Resistance-Capacitance (RC) time constant, alternating current (AC) capacitance measurement, and measurement of the electromagnetic coupling of an AC signal between adjacent networks.

The capacitive techniques using constant currents are generally based upon the approximation:

$$C = i\frac{dt}{dv} \text{ or, } C = i\frac{\Delta t}{\Delta v}$$

where C is the network capacitance, i is the charge or discharge current, and v is the change in voltage that occurs during the measurement time t. Related, though more complex, behaviors occur for RC time constant and AC coupling measurements.

Using the preceding example, note that if two networks are shorted, the resulting capacitance is the sum of their individual capacitances. The larger capacitance causes a slower charge rate than is expected for either net individually. The system may note that these two networks are adjacent and displaying suspiciously similar charge time behavior. The system tags the networks as suspicious, and either fails them or verifies the presence of a short with a DC measurement. Networks containing an open will have reduced capacitance and will charge or discharge too quickly. These conditions are illustrated in Fig. 37.8.

To use these indirect methods, most systems require testing of a first article board using standard direct measurement methods for both continuity and isolation. Once the board is known to be good, the signal-coupling behavior is learned for each network on this board. The learned values are saved and compared to measured results from subsequent boards as described previously. Tremendous speed increases are obtained by eliminating repetitive probing of each network at multiple locations, particularly in the case of isolation testing. Some users combine a traditional DC continuity measurement with the indirect method as a substitute for isolation.

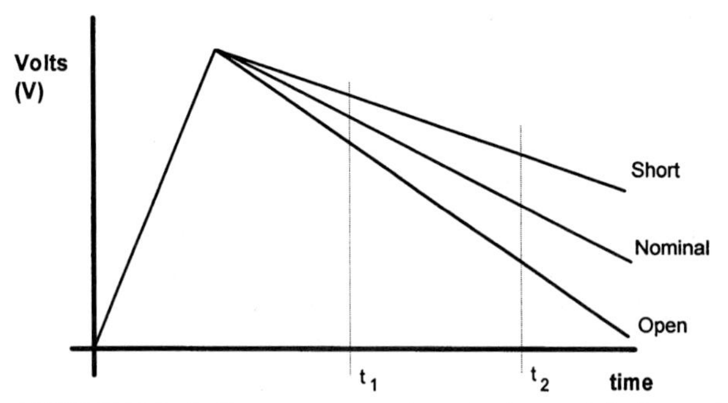

FIGURE 37.8 Indirect measurement: constant-current discharge.

37.4.4.2 Adjacency Analysis: Isolation Testing on Flying Probe Systems. Adjacency analysis simplifies isolation testing on flying probers by reducing the number of measurements, even for direct measurement methods. In adjacency analysis, a database is prepared listing each network and all traces found to be immediately adjacent to that network according to a set of geometric criteria. Such locations are assumed to represent the sole opportunities for shorts and leaks. Isolation testing is then performed only between adjacent pairs, this measurement being well suited to the nature of the flying prober. It is important that attention be paid to whether such analysis is three-dimensional. Networks may cross over a network horizontally, but on a different layer of the board. A hole or defect in the intervening insulating layer may allow a short or leakage to occur.

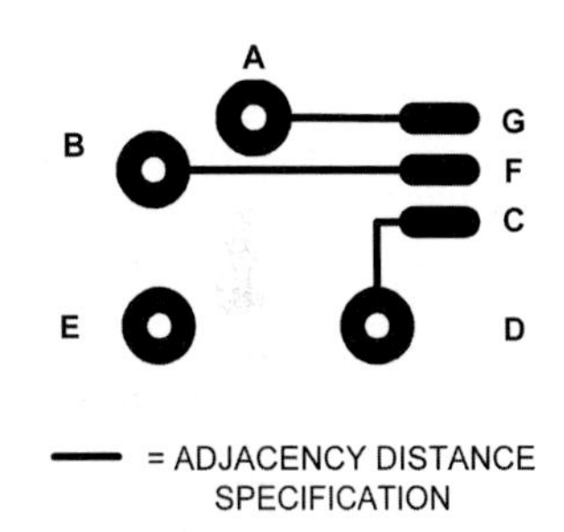

FIGURE 37.9 Network adjacency.

In the example illustrated in Fig. 37.9, network A would probably be judged adjacent to B, but not to C. B would be considered adjacent to A and D. Assuming that the qualifying dimension was a bit smaller than the distance from E to either B or D, E might not be judged adjacent to either.

37.5 DATA AND FIXTURE PREPARATION

An essential element in any test operation is the means of generating test programs, and, in the case of fixtured tests, the design of an appropriate test fixture. There are several methods for deriving this program and fixture information.

These methods vary from simple to complex and from escape-prone to very sound. The most common terms used in the language of bare board testing for the test program development are self-learning and netlist testing. *Netlist testing* is a misnomer that confuses computer-aided design (CAD) output with test inputs and implies that there is not much in between. In this discussion, we will use the term *data-driven test program* to indicate the program to be loaded into the test system is derived from the board's design data (which may be provided in any of several different forms). On the other hand, a self-learned program is derived by placing an assumed good board on the test system and causing the test system's internal computer to create a program automatically that can be used to compare other boards to the detected pattern. It is worth noting that the description "known good" is often used to describe this first board, when in fact our knowledge of its quality is often imperfect.

Flying probe test systems generally require no fixturing, except sometimes a frame to hold thin boards or to load multiple small boards simultaneously. Flying probers still require test program generation and optimization. Instead of a fixture, the product program contains probe position information for planned test. Data preparation processes are therefore similar to those of fixtured tests, but only test program data are output.

Note that the extractions of program and fixture information are quite naturally linked. In fact, the final test program cannot be created until the fixture is defined. The fixture determines how the individual test points of the test system are connected to specific product locations.

Knowledge of how the fixture has taken hold of the product is essential to determining which system test points to employ in performing specific measurements. Thus, the same software system generally outputs both the final test program and the fixture drilling/loading information.

37.5.1 Self-Learning

Self-learning has increasingly limited application and is subject to certain escape risks derived from assuming that the original circuit board is good. The self-learning method requires that a fixture be already available, as well as a known good board (preferably). A shorting plate is placed on the fixture in place of the product, and the test system uses this to identify which test points are employed in the test fixture. (All test points found shorted to the plate are considered active. This eliminates test points of no interest, shortening the test program and saving considerable execution time during both learning and testing operations. This set of active points is sometimes referred to as a mask.) The known good board is then substituted for the shorting plate, and the pattern of product interconnections is learned and saved as a test program.

A few boards are tested, and, if results appear reasonable, the program is considered valid. The key drawback of the method is that self-learned test programs determine that all the boards are the same, not that they are good. Moreover, for economic fabrication of the test fixture, it is necessary to process product data anyway; thus we may as well have output a data-driven test program. With virtually 100 percent of board designs being CAD-driven, there is little motivation to use self-learning today.

37.5.2 Data-Driven Testing

The preferred method of deriving program and/or fixture data is data-driven programming (DDP), sometimes referred to as netlist testing. The basic idea of data-driven test programming is to test the board using the same database used to specify its manufacture—in other words, the original design database. The fixture and DDP development process can be divided into two stages:

- Input/extraction, the preparation and processing of the various possible input sources into a format usable by the second stage
- Fixture data, test program, and repair file output, and subsequent application of the data.

Factors such as data quality and completeness, the format of data available, and fixture design can significantly reduce or extend software processing and engineering time. Board technology, size, and complexity tend to increase time requirements. While software tools are not always the most immediately visible tools in the testing department, the completeness and automation of software tools can have a large impact upon whether you are able to set up and test boards quickly and efficiently rather than struggling to build fixtures quickly—while testing very few boards.

Ideal fixture software readily accepts the product data, analyzes the data, recommends an optimal fixture design including all fabrication details, and outputs computerized numerical control (CNC) files for fixture fabrication, material requirements, special procedures required, and a test program in the proper format for the selected test system(s).

37.5.2.1 Selecting Fixture Software. A complication arises due to the mechanical incompatibility of many types of test systems. If the available equipment includes something of a grab bag of different equipment types, the test manager's task is made considerably more difficult in several ways. Aside from the obvious problems of manufacturing different fixture types, the manager is also faced with difficult load-leveling tasks. In such circumstances, the test manager may be forced to select general-purpose fixture software packages. By virtue of offering many configuration options, these require additional operator training and present increased opportunity for error. If the testing equipment uses a shared fixture type, it becomes possible to standardize the software process (and raw material). Optimized configurations may be locked into the software, minimizing error opportunities.

37.5.2.2 Test Data Extraction Methods for Fixtures and Programs. The initial stages of processing include the acceptance and inspection of the input data to ensure that they have been presented in a valid format. These data are then scanned to determine which sites on the board deserve test points, and what measurements are required among these test points in order to ensure proper continuity and isolation of the product. There are a variety of input sources, summarized in subsequent sections.

A usual goal of the extraction process is to eliminate unnecessary test points. According to predefined rules, the software identifies those end points of each network segment where test pins can and should be placed. For example, a T-shaped network probably requires three test points, located at the extreme extensions of each arm of the T. This provides sufficient connectivity to the product to permit a complete continuity test. One point per network is the minimum needed to attempt isolation testing, and two points per network make the isolation test more robust by permitting the tester to ensure that these points are in proper contact with the product.

The following is a generic description of the software process, where CAD or photoplot information is available.

- **Data orientation** The toughest side of the board in terms of test centers is usually oriented face down. It is often easier to manage a heavier, more complex fixture on the bottom side of the test system. This may require flipping or inverting the data.
- **Test site identification and optimization** Depending upon customer preference, it is often true that some test points can be deleted from the program. Figure 37.10 shows a network with a midpoint B that does not need to have a test point on it. The test points at A and C are all that are necessary to determine whether this network is isolated and contains no opens. Test point optimization can reduce the number of points in a fixture by 10 to 50 percent, thereby reducing the cost of the fixture and the time needed to test the board.

FIGURE 37.10 Optimizing a network.

Although optimization provides some economic and minor throughput benefits, this must be offset by any impact upon the repair process and by any fault coverage risk. Midpoints can provide failure data that help the repair station more easily pinpoint the actual fault site. Where an operator is required, or when flaws appear in the algorithm, the deletion of test points can add to the risk of test escape. Optimization is sometimes not performed, speeding data preparation and reducing escape risk but adding cost to the fixture.

- **Pin type assignment** Once the test sites have been identified, test pins can be assigned to the features to be tested. Different pin types may be required for different features. For example, a through hole might require a large head or larger-diameter pin than a surface-mount technology (SMT) pad on 0.025 in. centers. The software usually includes a table relating feature types to pin types.
- **Test system type identification** Mechanical details of the fixture must suit the target test system. The fixture in a universal grid system requires assignment to a grid of uniform centers. A more detailed description of fixture and system types is included elsewhere in this chapter, but obviously the fixture must be built in the correct size and grid density to match the specific test system to be employed. Figure 37.11 shows three of the most common grid configurations available. Single density (100 points per sq. in., or a 0.100 in. grid) is the most common, with double-density systems being added as product technology demands.

Of course, a few systems still use wired fixtures, and here the output will be a set of drill files, a wire list, and the test program. The balance of this discussion focuses on tilt pin translator–equipped grid systems:

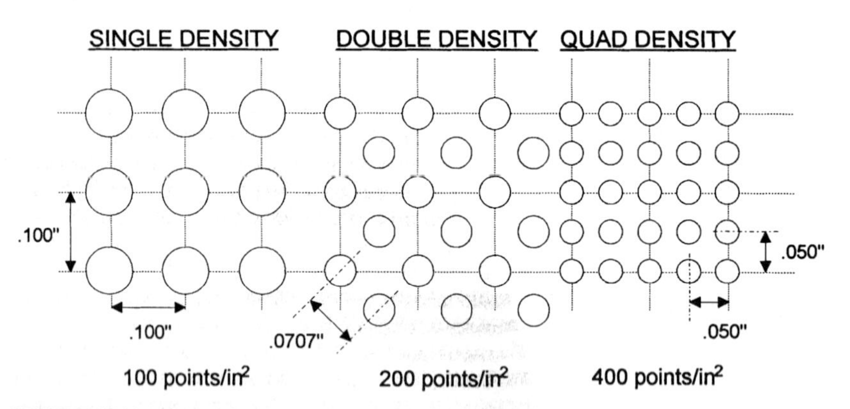

FIGURE 37.11 Single-, double-, and quad-density universal grid patterns.

- **Grid pin mapping** The fixture software determines which system grid location will be mapped to each of the previously determined on-product test targets. Several factors are considered:
  - No two targets can be assigned to the same grid location.
  - No assignments can allow the pins to touch each other as they tilt through the fixture (cross pins).
  - Drilled holes in fixture plates cannot break out into adjacent holes (adequate spacing on all plates).
  - No assignment can require a pin to tilt beyond the predetermined deflection limit (determined by the fixture and pin design).
  - Test sites that are very close together may require that on-product targets be staggered such that the test pins are more widely separated. (See Fig. 37.12.)

- Hole diameter and placement relative to the pin diameter and pin deflection must be optimized. A nearly vertical pin might be assigned a hole diameter just slightly greater than its shaft diameter. This clearance balances avoidance of friction against pointing accuracy during testing. Other pins of the same diameter that are highly tilted will require a slightly larger hole to avoid binding. The fixture adjusts drill sizes for each hole, depending on the amount the pin is deflected. This maintains pointing accuracy while eliminating pin binding. Hole locations may also be shifted slightly to compensate for the fact that the pin is pointed at an angle and that the point of contact will not be at the center of the hole. The result is a more reliable test fixture.

FIGURE 37.12 Test point staggering.

- **Test program output** With all pins fully mapped to grid locations, the test program file can now be constructed and saved.
- **Tooling** The last step in data preparation requires the marrying of the job being processed with the tooling. The tooling includes tooling to the test system and tooling for the specific fixture design utilized.
- **Fixture file outputs** The completed CNC files for fixture drilling and fabrication can now be created and the test fixture prepared.

37.5.3 Data Formats

Modern circuit boards are designed using CAD systems. These systems output data files used by the board shop to generate phototools, drill files, routing patterns, and other tools used to fabricate the boards. While several standards have been developed for data outputs from this process that can provide useful data input to the final test process, adoption has not been universal. As a result, test engineers and software providers have developed means of hijacking and converting the photoplot data stream, and can use this to drive the test process with certain compromises and a degree of inconvenience. Both approaches are discussed in the following text.

37.5.3.1 Gerber Format Data Extraction. The primary example of "hijacked" photoplot data arrives in Gerber photoplotter format. Even here a standard was eventually developed. RS-274X is a standard that affords a degree of consistency to the photoplotter files. This data stream is a series of drawing commands that direct the photoplotter in drawing the copper patterns for each layer of the board. The drawing is made with a selection of "apertures" that define different pen tip shapes and dimensions. Thus a fine line is drawn by choosing a small aperture, issuing a light-on command, and directing a move. These data are contained in a very large set of files, containing a picture of each layer of the board. Combined with a drill file and aperture definitions, the appropriate information for generation of fixture and test program files is present, but a considerable amount of processing is necessary to develop understanding of layer-to-layer connectivity and finally identify test targets.

Although the standardization of Gerber data has been improved with efforts such as RS-274X, it remains a data format originally optimized for the purpose of photoplotting and not for testing. CAD system developers and board designers have a great degree of freedom in how they design board features and in how these features are represented in Gerber. As a child will readily demonstrate, many different sequences can be used to draw even simple shapes. A rectangle might be created by "flashing" a square pattern, or might be created by an elaborate spiral using a much finer aperture, a zigzag pattern, or a sequence of overlapping flashes. This variety leads to occasional errors, because it is difficult for any Gerber extraction routine to be capable of handling all possible permutations. The software system must successfully interpret any of these examples (and many others) as defining the same intact square area of copper conductor pattern on the final product. Some methods result in huge file sizes.

Despite these problems, this method is far superior to self-learning, digitizing, or drill file extraction, and remains in common use in a variety of software packages. The data requirements to start from this format include:

- Photoplot files for each layer, solder mask layers, etc.
- Photoplot files for silk screen legend(s) (required for computer-aided repair)
- Aperture file
- Drill file
- Board outline

37.5.3.2 CAD/CAM Data Extraction Method. Recognizing the problems inherent in photoplot extraction, system vendors have agreed to add support for test-oriented output formats, in particular IPC-D-356 and 356A. These formats provide data in a much more readily digested format, and eliminate most ambiguities in processing the data. These formats include the following data:

- Signal ID, network name, and/or network number
- Reference designator or PIN number (e.g., U14, 12, R11) for the related component on board
- X-y coordinate of pad center (minimum data set requirement if grouped as connected)
- Pad dimensions relative to the center, and size of the hole (if any)
- Resistor or other component values (if appropriate and not usual)
- Board side (top or bottom)
- Mid-network flag suggesting that test point placement may not be required.

Often this data set is converted to a standard such as IPC-D-356 from the CAD/CAM system's internal data formats by means of an intermediate converter. These software converters, although simple, are usually customized for each individual CAD system. A converter must be updated or modified with any output changes made by a CAD vendor due to a CAD system software update or new product introduction. The number of converters required by an independent PWB manufacturer could be quite high, as data will likely be received from many different CAD system types. Fortunately, many of the CAD vendors now include direct output of IPC-D-356 data or readily provide converters. Acceptance of these standards is now such that in many cases even Gerber input data are first converted into IPC-D356 format prior to final processing.

37.5.4 Outputs from Data Extraction

Once all the preparation steps and processes have been performed, there are several outputs generated by the fixture software.

- **Test file** A data-driven test program is outputted in a format compatible with the test system type. This file informs the test system of the measurements expected and of pass/fail criteria. Some test system formats permit the data-driven test program to support a graphical representation of the fixture and/or board on the test system monitor, provided that sufficient data are included in the test program file. This graphical presentation is useful in fixture and/or program debugging.
- **Fixture fabrication files** Probably the most significant output—the drill files, one for each plate or pass required by the fixture design—is needed in drilling to start building the fixture. (In the case of wired fixtures, a wiring list is also required.)

- **Repair/verification files** Files can also be outputted that support the repair or verification function. These relate the graphical image of the board to the assigned test point locations in the program. Preparation often relies on inclusion of Gerber data in the input data stream to the software system. Extensions to IPC data formats are planned to better support the repair function without this recourse to photoplot data. Some test systems may carry trace image information in the test program file as well, for enhanced debugging support on the test system.

37.5.5 Setting Up a Fixture

With the fixture assembled and CAD data–derived program prepared, the next step is to set up the fixture on the test system and validate the fixture and program. Details vary with the fixture and system used, but in general the following steps are taken:

- Program data are loaded into the test machine from the disk or network.
- Test thresholds are set to the desired continuity and isolation values.
- Compression settings on the test system are adjusted to the proper values for the fixture type. In some cases, these values may already be included in the test program file.
- The new (or recalled to duty) fixture is compressed with a nonconductive material of similar thickness to the board being tested.
- An all-isolations version of the test program is executed to verify that the fixture does not contain any internal shorts.
- The fixture is then compressed again with a conductive plate in place of the product, intentionally shorting all fixture test points together. Often this plate is a simple piece of copper-clad G-10. Copper oxidation can prevent reliable contact, and users sometimes wrap the plate in fresh aluminum foil or use a flash-gold-covered plate. In this step, an all-shorts version of the test program verifies that all expected test points are continuous through the fixture, up to the shorting plate. The intent is to verify that the number of test points is correct and that, under multiple compressions or closures, all points are present and remain in contact with the shorting plate.
- Assuming proper alignment and cleanliness, the product can then be tested with the final test program. If certain errors immediately repeat on all boards, then it is reasonable to suspect a pin-loading error or other error in the fixture, and investigation is warranted.

The basic techniques just described are generally applicable to all fixture types, whether tilt pin, wired-dedicated, or other specialized types.

Do not underestimate the value of removing dust and debris from the product and fixture, both during setup and occasionally during operation. Target sizes on modern products are not tolerant of debris, and false open circuit reports are an immediate result. Tacky roller-type cleaning systems may be helpful in periodically removing debris from test fixtures, system grids, and the products themselves. Follow the test equipment manufacturer's recommendation regarding the use of electrostatic discharge (ESD)-safe cleaning materials near the test system.

Adjustment of fixture compression is also important, especially on press-type systems where the amount of compression stroke is controllable. (In vacuum fixtures, the dimensions of various fixture plates and components usually fix the amount of compression.) Undercompression usually leads to poor contact and false open results. Overcompression can cause excessive marking of the product by probe tips, probe damage, and fixture damage.

Overcompression is a very common problem, and, unfortunately, so is product damage in the form of excess probe marking. It seems intuitive to just press harder when you experience contact problems, and this is often the first step taken. But the actual change in force per spring probe is very small as it travels further, until the spring probe hits bottom—and at that point you begin damaging the product almost immediately. A typical spring probe in a grid system presents about 139 g of force when compressed 0.167 in. (two-thirds of travel). At zero travel (uncompressed), the spring inside is already preloaded to about 55 g. The force increases at a spring rate of 503 g/in. Just before bottoming, the spring force peaks at about 180 g. Thus, there is only modest force gain possible before serious danger of bottoming the probe occurs. Consult the spring probe manufacturer for details, but most probes function well at about two-thirds of full travel. If contact problems remain at this level of compression, leave further compression adjustment as a last resort and look elsewhere. Generally, you should never cause the probes to hit bottom. (Some older vacuum fixture probes are a notable exception.)

37.6 COMBINED TESTING METHODS

As product density and geometry become more challenging, the cost of fixture construction increases. In a large number of cases, the most challenging test sites occur at only a few areas of the product. It is possible to combine test techniques such that less expensive means are used to test the majority of product features, while more advanced and costly techniques are reserved for the most challenging test sites. Some examples follow.

37.6.1 Split Net Testing

This is perhaps the oldest and most primitive example of a sequential or combinational test. When the product complexity exceeds the capacity of a fixture design to solve the test requirement with a single fixture, it is possible to divide the test responsibility among multiple fixtures. The flip test method of dual-side access described earlier is one example of this. More typically, the test department has dual-side-access equipment, but the density of test sites exceeds the capability of the fixture and/or test system.

One way to split the responsibilities is to divide continuity and isolation between fixtures. For example, one fixture could be limited to two test points per network. These two would be checked for continuity with each other (to verify that there are no open pins in the fixture) and would then be used to perform a 100 percent isolation test of the product. A second fixture would complete the continuity test. In very difficult cases, multiple fixtures can be used.

Handling boards in this scenario is complicated. Boards that pass the first fixture must be stacked and prepared to run on the second fixture. But boards that fail the first fixture must be tagged with failures, and in many cases must be run through the repair process, retested, and combined with the boards being sent to fixture 2. Such failed boards must not be confused with any failed boards from fixture 2, which have already completed the first phase of testing.

Unless there is a large continuous flow of boards with two different test systems/fixtures set up, it is also likely that the two fixtures will have to be installed/removed from the test system a number of times.

37.6.2 Manual Combination of Methods

It is also possible to divide the test burden by testing the majority of ordinary test sites with a test fixture, followed by a flying probe system to test strictly limited portions of the product. Without special software, one has the same handling problems as in the split net case. But it is possible to automate the data handling between the two test systems using network resources, such that a single error tag results from both tests. All boards then flow through both test systems, and any rejects flow to repair. This flow is very similar to that of ordinary testing, resulting in less confusion. An interesting additional benefit of this combination is that the flying prober not only can test those sites that you wish to avoid fixturing, but can also retest any of the failures noted by the universal grid. This can eliminate the majority of false open reports caused by stuck pins and contamination, improving the total first-pass yield.

CHAPTER 38
BARE BOARD TEST EQUIPMENT

David J. Wilkie
Everett Charles Technologies, Pomona, California

38.1 INTRODUCTION

The largest volumes of bare boards are tested on fixtured systems. However, for smaller volumes or special purposes, "flying probe" systems may be preferable. This chapter discusses them both to enable the user to make the most effective decision based on testing objectives and volumes.

38.2 SYSTEM ALTERNATIVES

38.2.1 Fixtured Systems

Each dedicated, or hard-wired, test system presents some sort of standardized interface pattern of test points. This may consist of a universal grid's continuous array of points (bed of nails) or may be a connector pattern of some type. In the most primitive systems, a simple cable connector may be presented. Each of these types of equipment uses one or more types of customized test fixtures to connect this standard electrical interface pattern to the unique contact pattern of a particular product to be tested.

Each fixtured system includes a measurement unit that can be connected to any of some thousands of test points by a solid-state switching matrix. A central computer controls the measurement unit and switching matrix. This computer also controls the press (or vacuum mechanism), which can compress the product against spring-loaded test points presented to the product by a customized test fixture.

To imagine the system's operation, first consider the measurement unit to be an ohmmeter. The computer commands the switch matrix to connect the red lead of the meter to, say, test point 17. It then connects the black lead of the meter to test point 1027. The measurement system can then be commanded to measure the resistance of the product under test between test points 17 and 1027. The test program dictates a long series of such measurements organized to test the product completely. Because the switching and measuring occur at the high speeds possible with modern electronics, a complex board can be tested in seconds.

38.2.2 Dedicated (Hard-Wired) Fixture Systems

Originally the most popular system type, dedicated systems are generally being displaced by systems employing less expensive and more accurate fixturing methods. The word "dedicated" refers to the fact that it is generally impossible to salvage very much material from these fixtures for reuse in other fixtures. The cost of these fixtures derives from this fact, and from the high labor and material costs involved in the original construction.

38.2.2.1 Advantages of Dedicated Fixture Systems. A primary advantage of these systems is that, because each test point is routed within the fixture by a flexible wire, it is possible to position any test point at just about any location. If your product requires 8,000 active test points, then you need no more points than this in the test system. One hundred percent utilization of available test points is possible. This contrasts with the universal grid approach discussed later, wherein points can be shifted only a small distance from the original location on the grid. In that case, you must have enough electronic test points to cover the entire surface of the product at some constant density, and any given fixture will generally use only a portion of these points. Thus, the capital equipment cost of a dedicated system may be lower, but the ongoing fixture costs may be much higher. Dedicated systems are generally built with no more than about 10,000 test points, with smaller numbers common.

38.2.2.2 Construction of Dedicated Fixtures. The test fixture for such systems generally consists of a rigid box structure with one side facing the product to be tested and some other side or surface presented to the interface pattern of the test system. The side facing the product includes the probe plate. This plate of insulating material is drilled to mount an array of spring-loaded pins able to make electrical contact with the product. The reverse side of these pins is connected within the box by physical wires to the system interface connector on the opposite side of the fixture.

Fabrication of the fixture involves the usual data-processing steps, followed by spring probe type selection, drilling of the probe plate, installation of spring probes, drilling of the interface plate, and installation of system mating contacts or connectors. Many spring probe tip styles are available to suit different diameters of target pads and holes (see Fig. 38.1).

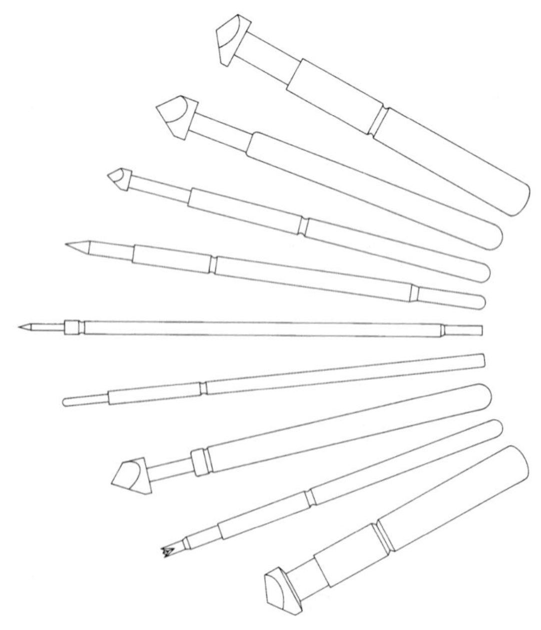

FIGURE 38.1 A variety of spring probes and receptacles. *(Courtesy of Everett Charles Technologies.)*

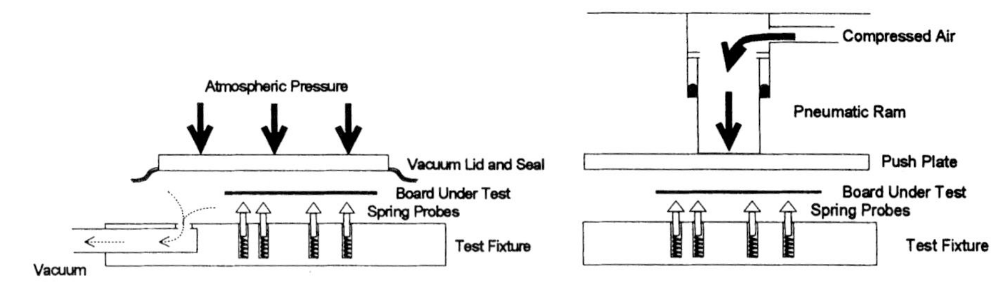

FIGURE 38.2 Vacuum versus pneumatic compression.

In most cases, the probe plate actually carries a socket or receptacle into which a replaceable spring probe is inserted, simplifying service. The underside of the receptacle protrudes into the fixture box and carries a wire-wrap tail. Each of these receptacle tails must be wired to an individual system interface contact on the bottom of the fixture.

Compression of the product onto the fixture spring probes may be accomplished with electric drives, pneumatic air cylinders, or vacuum (see Fig. 38.2). Vacuum compression can result in the lowest system cost, but adds cost to the fixture and limits the total number of probes that can be compressed.

Many forms of fixture interfaces have been provided, with many appearing as small universal grid patterns. Some very simple test systems have been constructed with a large number of ribbon (or other) cable connectors presented as the only connection to the fixture. The matching wired fixture may be troublesome to connect (because of the large number of cables), but is left connected for days or weeks at a time to a single fixture type due to the quantities of boards being tested.

38.2.3 Flying Probe–Type Test Systems

Smaller volumes and specialty parts are tested on flying probe–type test systems. These consist of a small number of robotic probes with independent measurement abilities. These are moved among the various product test target locations, and a sequence of measurements is made. Such systems offer the powerful advantage that no fixture preparation is required (although data must still be processed to prepare the test program). The best of these systems are also able to provide extremely high probe placement accuracy, generally exceeding that of fixtured systems. The primary drawback of such systems is low throughput, as a result of the need to reposition the probes mechanically between measurements.

38.3 UNIVERSAL GRID SYSTEMS

The most flexible and widely used electrical test solution is the universal grid test system (see Fig. 38.3). Most systems now include upper and lower grids, permitting simultaneous dual-side access for test of surface-mount technology (SMT)–type products. Grids offer very high test speeds, and moderate the cost of test fixtures by permitting the reuse of many fixture components. Thus, today, the majority of product is tested on universal grid systems and fixtures.

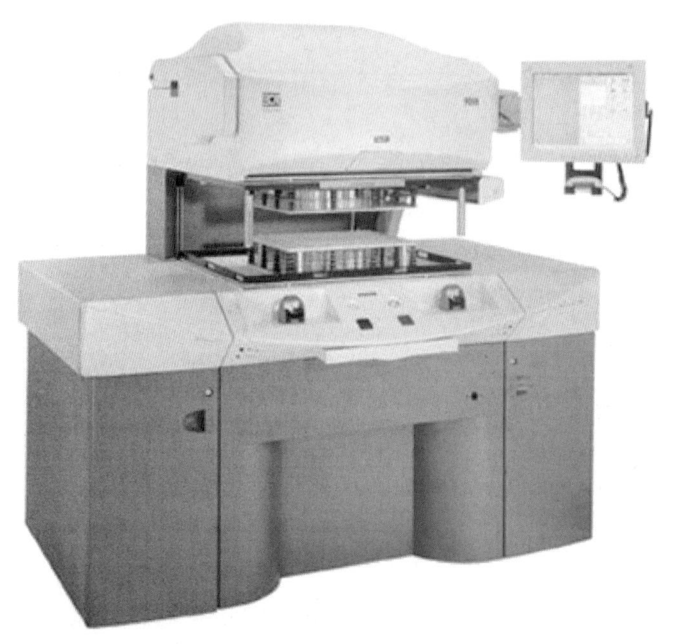

FIGURE 38.3 Universal grid test system. *(Courtesy of Everett Charles Technologies.)*

38.3.1 Universal Grid Test System Design

The universal grid test system presents a rectangular array of equally spaced test points. Generally this is chosen to be large enough to cover the test area of the largest product type to be tested (see Fig. 38.4). It is common to speak of the density of test points presented.

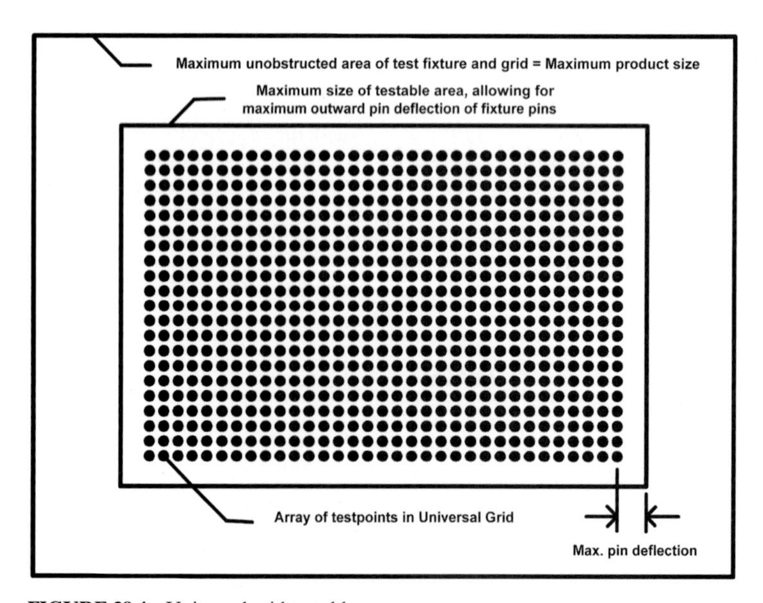

FIGURE 38.4 Universal grid testable area.

A single-density system presents points spaced at 0.100 in. (10 per in.). Thus, there are 100 points per sq. in. Similarly, a double-density system has test points spaced at 0.0707 in. and provides 200 points per sq. in., and quad-density spacing is 0.050 in. for a density of 400 points per sq. in. As grid cost is largely a function of the number of test points, larger sizes and/or higher-density configurations become expensive. With modern grid designs, the test system grid size can be upgraded in the field by addition of electronic modules. With older designs that use wire between the grid and the electronics, upgrades may be less practical.

38.3.2 Exclusion Mask Fixtures for Universal Grid Systems

Occasionally, some applications involve product whose test-point spacing exactly matches the grid pattern. This situation may permit use of a very simple fixture referred to as an *exclusion mask*. This is composed of a thin glass-epoxy sheet, perhaps 0.030 in. thick, drilled at those locations where test probes are desired to pass through from the grid to make contact with the product. The exclusion mask (see Fig. 38.5) prevents unused grid probes from contacting the product unnecessarily, preventing erratic contacts and marking of the product. To use such a mask, it is necessary that the grid be constructed with a pointed or chisel-tip probe. Unfortunately, few products exist that are so simple to test.

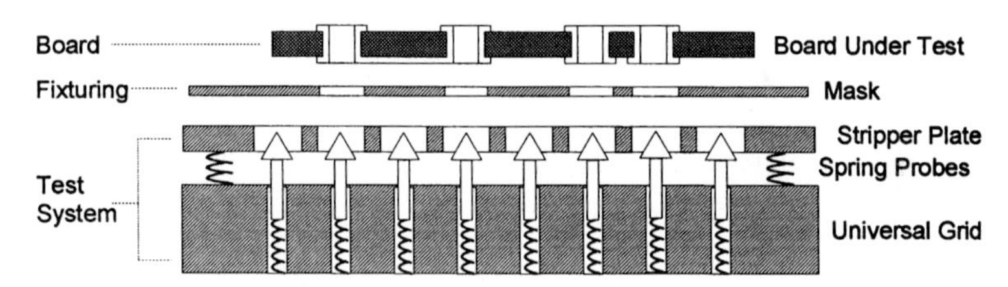

FIGURE 38.5 Exclusion mask fixture.

38.3.3 Pin Translator Fixtures for Universal Grid Systems

Most boards manufactured today are tested on universal grids through the use of pin translator fixtures. These fixtures are rapidly designed and produced using available software systems, and the most expensive material components (the contact pins and plate spacers) are generally reuseable. These fixtures are also referred to as *tilt-pin fixtures* or *grid fixtures*. Grid fixtures and related software remain the fastest-evolving area in electrical testing. Particular fabrication details will not remain current for long. Therefore, this discussion employs current practice only as means of illustrating key issues such as density and registration.

In a tilt-pin fixture, a rigid pin serves the dual roles of providing an electrical contact path from a grid test point to the product and translating the grid test-point location a small distance horizontally (in x- and y-coordinates) such that contact is made to the desired target on the product surface. Because the product targets are unlikely to be perfectly centered above grid test point locations, these pins are usually somewhat tilted from vertical. The amount of tilt or displacement varies according to the amount of x-y shift required. Several lightweight plates of plastic material support these pins. The individual hole locations in each plate are offset the amount required to achieve the desired pin tilt and spaced apart the required distances (see Fig. 38.6). These pins are variously referred to as *tilt pins, translator pins*, or *fixture pins*.

To enable grid test points to be translated to exactly the correct target location on the product, it is important that each translator pin fall into the correct sequence of holes while the fixture is being assembled. If sufficient intermediate guide plates are employed, then the geometry of drilled holes can be so arranged that a virtual tunnel is created for each

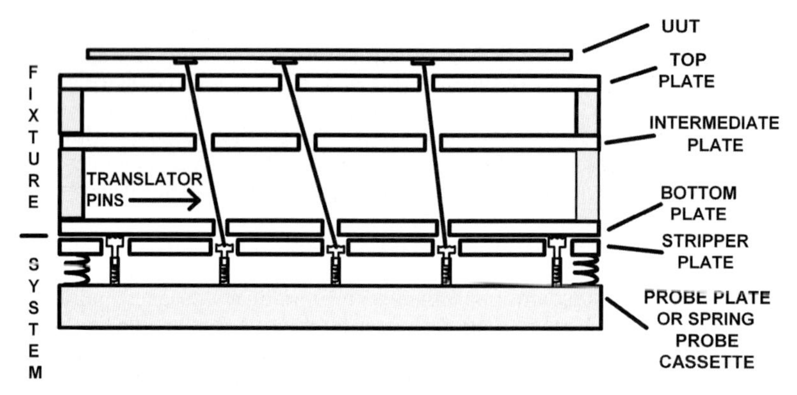

FIGURE 38.6 Single-side multiplate tilt-pin translator fixture.

translator pin. It becomes impossible for a pin to fall into an incorrect sequence of holes. More plates are required when the maximum degree of pin tilt is increased. Calculation of plate count, position, and drill hole locations is a key function of the data extraction and fixture software, whose process is described elsewhere in this chapter.

38.3.3.1 Test Pins for Pin Translator Fixtures. A wide variety of pins are available for use with different system types and for different applications (see Fig. 38.7). These vary in length, tip style, material, cost, and thickness. For a given angle of tilt, a longer pin is generally able to

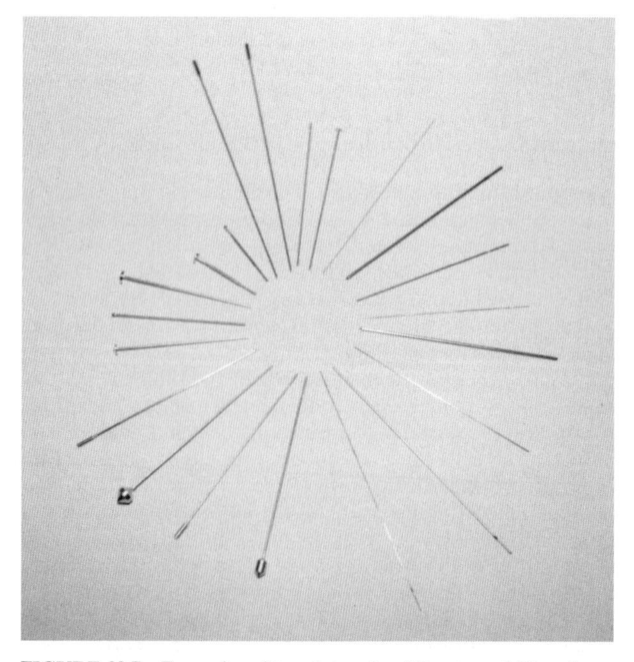

FIGURE 38.7 Examples of translator pins. *(Courtesy of Giese International.)*

translate the test point a larger horizontal distance (larger pin displacement). Pins with enlarged heads of various shapes are useful for probing large-diameter holes. Aggressive shapes such as chisels provide the pin with enhanced abilities for cutting through product contaminants, but may be unsuited to closely spaced test applications. At present, the vast majority of pins are headless music wire pins with lengths ranging from 2.5 to 3.75 in. When probing a large-diameter through hole, common practice now is to probe the annular ring around the hole rather than pay the penalty for an odd-headed pin type.

38.3.3.2 Pin Displacement in Grid Fixtures. Pin displacement capability is one of the more significant performance measurements of a pin-fixture design combination. A larger displacement ability provides greater freedom in allocating test points, resulting in improved ability to fixture complex products on a test system of given density. With an arbitrary fixed rule of thumb of 10° maximum tilt, we can see from Fig. 38.8 that the longer the pin, the further the pin can be displaced from its on-grid location to its on-product target location. The maximum pin displacement that can actually be obtained is a function not only of pin length, but also of fixture design and the resultant maximum tilt angle that can be tolerated in the design. The fixture software usually seeks to arrange pins such that the maximum tilt in the fixture is minimized. This decreases the number of guide plates required, minimizes the number of different hole sizes required in the plates, and tends to reduce friction and binding. Nonetheless, a design that permits a maximum tilt capability ultimately provides the greatest flexibility in assigning test points and generally permits the solving of denser products on grids of a given density.

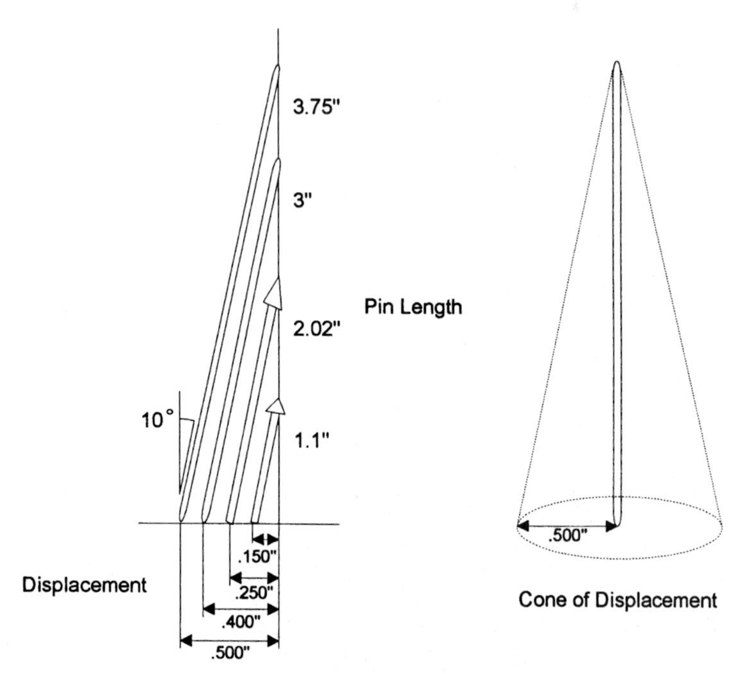

FIGURE 38.8 Approximate pin displacement as a function of pin length.

38.3.3.3 Pin Binding and Friction in Grid Fixtures. As noted, the grid fixture pins must lean some distance to connect the grid to the on-product target location. Thus, the relationship between the plate thickness, hole diameter, and angle of displacement is critical. If the plates

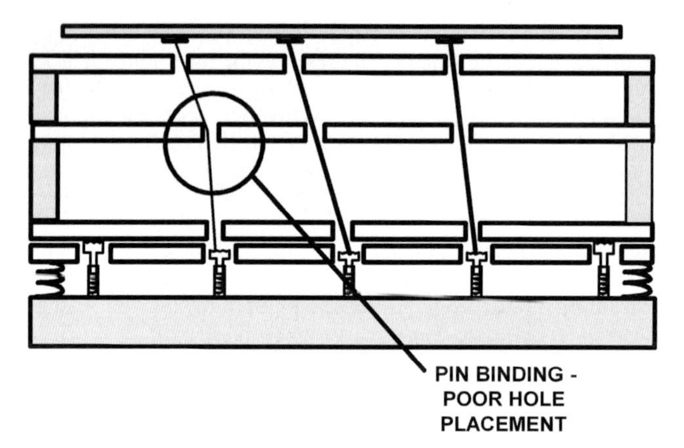

FIGURE 38.9 Pin binding in multiplate tilt-pin fixture.

capture the pin too tightly, binding and friction can result. A pin that is held down below other pins in the fixture by binding will fail to make contact with the product (see Fig. 38.9).

The spring force of the spring probe in the universal grid is intended to supply the pressure that presses the tilt pin against the product. Every bit of friction in the tilt-pin motion subtracts some of this force. If the friction is great, not only may the pin fail to contact, but if the pin is stuck high (toward the product surface), the product may press on it with undue force in shoving it back into the fixture, resulting in excess product marking. Binding can be caused within a single plate by hole diameters that are too small for the degree of pin tilt. Binding can result between plates if the plates are offset from one another, forcing the pin to bend as it snakes its way through the various plates. Binding can also result if the plate spacing is incorrect as compared to the design values.

38.3.3.4 Calculations for Tilt-Pin Fixtures. Modern fixture construction software should perform all of the necessary calculations automatically. Yet it is valuable to understand the basic decision processes.

38.3.3.4.1 Minimum Hole Diameter Calculation. The relationships between pin displacement, pin diameter, plate thickness, and drilled hole diameters are trigonometric. We calculate Eq. 38.1:

$$A = \sin^{-1}(PD/PL) \text{ and} \tag{38.1}$$

$$HS_{min} = [SD/\cos(A)] + [PT \times \tan(A)] + HT$$

where
HS_{min} = minimum hole size in a particular fixture plate
PT = fixture plate thickness
SD = pin shaft diameter as it passes through the plate
PL = overall pin length
PD = horizontal displacement distance of the pin tail relative to the pin head location
A = angle of pin tilt expressed as degrees from vertical. (i.e., a vertical pin has a tilt of 0°)
HT = minimum acceptable tolerance to ensure that the pin slides freely in the hole

The first term in the equation derives from the effective increase in shaft diameter in the horizontal plane as the pin tilts (see Fig. 38.10). The second term derives from the horizontal

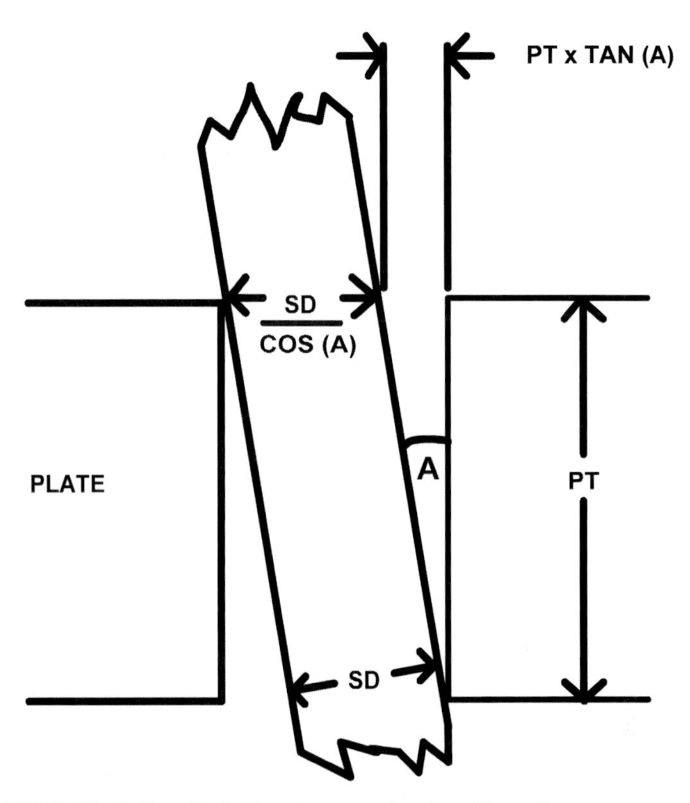

FIGURE 38.10 Illustration of hole diameter calculations for a tilt-pin fixture.

distance that the center line of the pin travels as it passes through the plate thickness. The third term, as mentioned earlier, is to provide a minimal clearance.

For angles of tilt below 15°, the value of the cosine is close to 1 (over 0.96), and for rough calculations the term may be omitted. (For a 20 mil diameter pin, the contribution of pin tilt to effective shaft diameter is less than 0.7 mil). Similarly, for the same small angles, the calculation of the tangent can be simplified to $\tan(A) \approx PD/PL$. The error incurred increases as the plate thickness increases. For a 0.1 in. thick plate, the error at 15° will again amount to something less than a 1 mil hole diameter.

With these simplifications, for low tilt angles and thin plates, the formula may be reduced to that shown in Eq. 38.2, with an extra mil or two added to the clearance.

$$HS_{\min} \approx SD + [PT \times (PD/PL)] + HT \tag{38.2}$$

For the highest-accuracy applications, use the exact calculation.

38.3.3.4.2 Drill Bit Size Selection. Note that the minimum hole size calculated previously is often not the ideal drill size. The most common material used for grid fixtures is polycarbonate.

Polycarbonate tends to have a finished hole size that is about 0.001 in. smaller than the drill size for plate thicknesses typically employed in fixtures. Therefore, 0.001 in. should be included in the tolerance figure HT to permit accurate drill bit selection.

38.3.3.4.3 Pin Lean and Pointing Accuracy. Pointing accuracy is defined as the pin's ability to hit an intended target, usually described as a radial dimension (the distance from the intended target to the actual contact location). Because the pin is typically not perfectly vertical, there are several possible sources of error, depending upon the type of pin employed.

With the older headed pin types, the head sat above the top plate of the fixture. With modern headless approaches, the pin tip is usually flush with the upper surface of the top plate while in contact with the product. As the pin tip projects less (nearly zero) above the top plate surface, its tip position in x-y coordinates is better controlled by the hole in the top plate. For this reason, most modern fixture designs operate with the probe tips arriving flush with the top plate surface while the product and fixture are compressed.

It is also preferable that the pins remain approximately flush when compression is released. If they project above the top plate surface, the tip moves out of position in x-y coordinates, as illustrated in Fig. 38.11. When a product is placed on such a pin, the tip initially contacts an incorrect target location. During compression, the pin must drag across the product surface as the pin retracts into the flush position. Thus some spring mechanism, which supports the fixture and pins above the array of grid spring probes, is commonly employed. The fixture may rest on a spring-loaded stripper plate over the grid of spring pins, or the fixture may be designed with spring feet on its underside. In some designs. the fixture kit itself is compressed.

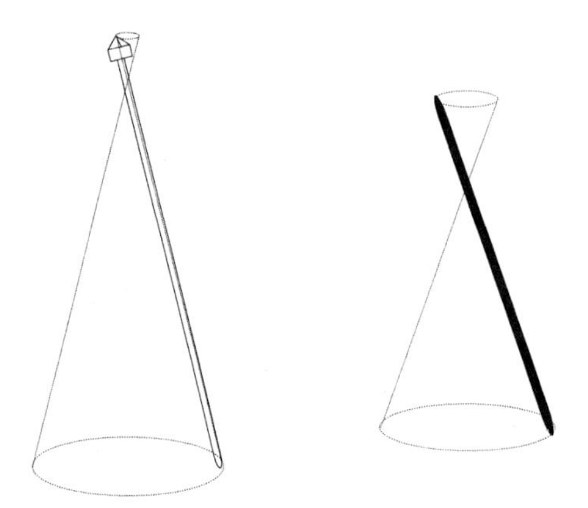

FIGURE 38.11 Pointing accuracy.

Some fixture designs have used bent or curved test pins. In such cases, the pin is bent by including additional fixture plates such that the tip of the pin as it comes through the top plate is perpendicular to the test pad. This allows the top plate holes to remain on the product and mirror the board under test and thus eliminates concerns about the effects of pin contact at an angle. Common examples of such fixtures were built as spring boxes, such that the overall thickness of the fixture decreases during compression. This eliminates the need for spring feet or a stripper plate, but complicates friction and binding issues as the entire array of test pins must bend and flex during every compression. Contact reliability, product marking, and ease of pin loading all suffer.

38.3.3.4.4 Offset Error Compensation. Figure 38.12 represents the problem associated with the deflection of a pin retained in the top plate and its ability to hit the center of the pad or target. As described earlier, the hole diameter must be sufficiently oversized to allow the pin to pass through the hole at its angle of deflection without binding. Vertical force on the pin from the product above tends to press the pin against one side of the hole, as illustrated. (This has the curious effect that tilted pins can display more consistent targeting performance than do perfectly vertical pins.) As the pin tilts, the center line of the pin becomes substantially shifted from the center line of the hole (note the offset distance *OS* in the illustration). Finally, the conical tip associated with most modern pins has a certain finite radius—the pin is not infinitely sharp. Thus, in the illustration, the product will not contact the pin on its center line. As illustrated, contact will come at the highest point of the tip radius, slightly to the right of the

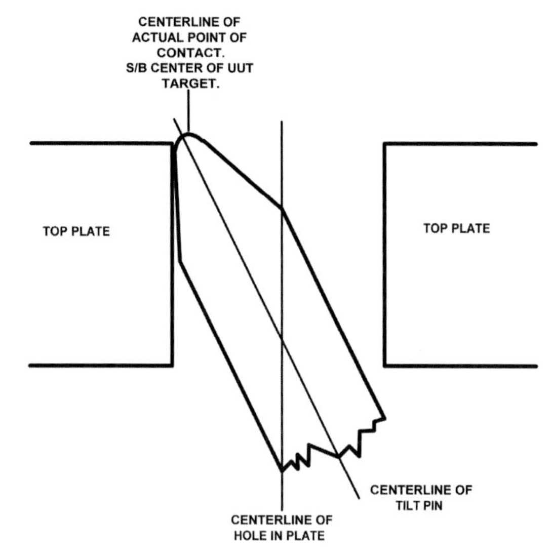

FIGURE 38.12 Pin tip offset error.

center line. Thus, the theoretical contact point will be at neither the center line of the pin nor the center line of the hole. Probing targets only a few mil wide with probes two or three times as wide requires high-quality probes displaying consistent tip geometry, and software able to calculate the actual point of contact carefully.

38.3.3.4.5 Countersink Drilling of Top Plate. The calculations and considerations just discussed illustrate that thicker plates and higher pin tilt present some interesting problems when combined. A highly tilted pin in a thick plate requires a substantially oversized hole if binding is to be prevented. Looking through the hole at a viewing angle equivalent to the tilt of the pin, the hole will appear to be oval, with the narrow axis considerably shorter than the drill bit diameter. Of course, the pin is thereby allowed somewhat more freedom of movement along the major axis of the oval. This bit of looseness can be a problem with very fine-pitch products.

A better solution is to use a very thin plate for the upper fixture plate, backed up by a thicker plate drilled oversize. Alternately, a single plate can be countersunk. The effect is the same, as illustrated in Fig. 38.13.

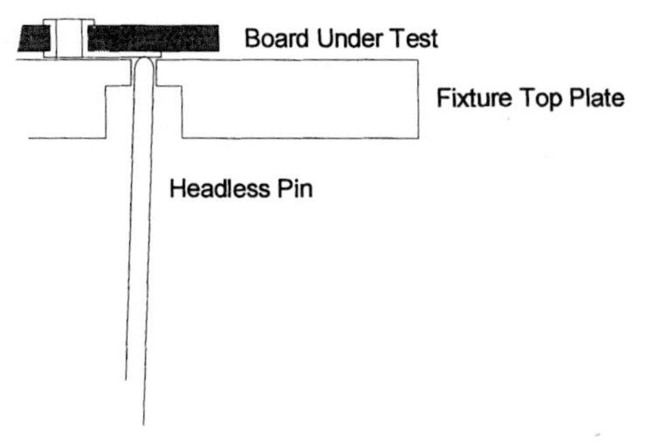

FIGURE 38.13 Countersink drilling of top plate to avoid binding or oversize holes.

38.3.3.5 ***Retaining Fixture Pins.*** It is necessary to retain the pins in a translator fixture such that they do not fall out of the fixture during ordinary use and handling. Older headed pin types were often used in single-sided fixtures. The oversized head protruded above the top fixture plate, and therefore the pin was unable to fall out the bottom of the fixture (although they fell out all too easily if one accidentally inverted the fixture). Early top-side fixtures attempted to duplicate this method for the upper fixture by employing a different pin in the upper fixture half, using an oversized "tail" on the fixture. However, the oversized geometry of the heads is inappropriate for fine-pitch testing, and therefore the use of such pins has fallen out of favor.

The desire to use thin, headless music wire pins necessitated the invention of new means of pin retention (see Fig. 38.14). These have included:

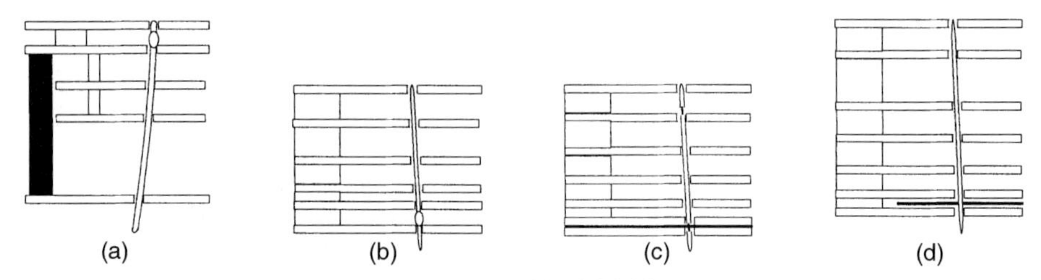

FIGURE 38.14 Alternative pin retention methods. (a) (b) (c) (d)

a. Crimping a wide flat into the pin, and trapping the oversize flat between two plates

b. Trapping the pin in a sheet of foam rubber such that friction retains the pins

c. Passing the pin through a sheet of spandex cloth that grips the pin

d. Passing the pin through a sheet of drilled latex rubber that grips the pin with flexible flaps

38.3.3.6 ***Examples of Real-World Fixture Designs.*** While many combinations of fixture techniques have been employed, it is worthwhile to compare several popular examples. No substantial discussion of older headed rigid pin types is included here, as their use is largely disappearing. Most fixtures today are built with headless pins that compress flush into the fixture without bottoming on the spring probe below, such that the contact force is entirely controlled by the spring (excluding friction or binding effects).

Figure 38.14a represents a common bent or curved pin design using 3.75 in. music wire pins. The upper and lower fixture plates are held apart by a spring mounting, and the entire fixture compresses during actuation. The pins hit the intended target perpendicularly, because they bend toward vertical as they reach the product area. But extra force is required, increasing pad marking. Hole diameter in the top plate can be tightened to improve pointing accuracy because all pins pass through perpendicularly, and a fairly thick top plate is possible. For a given pin length, the maximum possible horizontal displacement is more limited than in straight pin designs, because all displacement must occur in the lower portion of the pin length. The hanging middle plates do not contribute to torsional stability of the fixture design, increasing distortion under compression. This in turn can increase pin binding as the pins try to bend and slide through the compressing fixture envelope. The pins usually used have a flattened crimp trapped between the upper two plates and must be loaded or serviced with the upper plate removed. Pin loading is therefore difficult. The increased diameter in the crimp region limits close spacing of the probe. Finally, the crimp cannot be employed on very fine-diameter pins, as they would weaken excessively.

Figure 38.14b illustrates a significant evolution from Fig. 38.14a. Here the pins are not bent or curved. Shorter pins (3 in.) are sometimes used, providing about the same horizontal translation as in Fig. 38.14a. Deleting the bend significantly reduces pin binding. A retention crimp similar to the previous example is used, relocated toward the bottom of the fixture where the pins are more widely spaced. Pins are loaded from the bottom side, with the lower plate removed. This design still requires disassembly of plates for pin servicing, but the bottom plate is somewhat easier to reinstall due to the possibility of larger hole sizes. Fixtures are usually rigid as opposed to spring-loaded, reducing friction effects during compression.

Variations of the fixtures illustrated in Figs. 38.14a and 38.14b are produced using a foam-rubber pin retention method. In these versions, the crimp on the pin is deleted and a sheet of foam rubber material is sandwiched between the two lower plates. During loading, pins are forced through this foam, and friction holds them in place. This provides the important advantage that individual pins can be replaced without fixture disassembly. Unfortunately, high-density zones incorporating many pins stretch the foam rubber material. The stretched foam holds adjacent pins so tightly that a group of pins becomes reluctant to move except as an almost solid block. The resulting pin binding effect causes contact reliability and product marking problems. Also, the foam tends to deteriorate with time and/or heavy use. As densities have increased, this fixture has become difficult to use.

Figure 38.14c illustrates a retention technique employing a very thin sheet of drilled Mylar combined with a unique feature on the pin. The Mylar sheet is sandwiched between standard polycarbonate plates. The pins employed begin as straight-sided music wire pins 3 in. long. A grinding operation reduces the pin diameter in a short region (about 0.2 in.) of the pin length. The Mylar retaining sheet is drilled with a hole diameter that is slightly smaller than the original pin diameter. As the pin is inserted, it is forced through this undersized hole until the reduced-diameter zone reaches the Mylar sheet. At that point, the pin can slide freely over the 0.2 in. narrowed length. Only in this region are the pins able to move quite freely up and down. The height of the Mylar plate in the fixture is matched to the reduced-diameter portion of the pin, such that the pin is free to move a short distance up and down from its nominal position without falling out of the fixture. Pins are generally specified with the grinding operation performed at locations near both ends of the pin, such that the pin is symmetrical and can be loaded without regard to orientation. This pin demonstrates low friction and good contact, but a few problems remain. The Mylar sheet is thin and fairly brittle. It is not uncommon for a particular hole to fracture when extracting or reinserting a pin, such that the damaged hole can no longer retain a pin. Replacing the Mylar sheet requires unloading all pins and removing one or more plates. Pins for this fixture are expensive, as the grinding operation is difficult. The requirement to grind the pin undersize along a portion of its length limits the minimum pin diameter which may be used without causing the pin to become excessively weak at the narrowed section. As product pitch and density demands have increased further, these issues have become significant limiting factors.

Figure 38.14d represents techniques employed in more recently developed fixtures. These fixtures include nonoriented, nonfeatured pin designs, usually 3.75 in. long. A key feature is retention via drilled latex rubber or undrilled spandex cloth. The drilled latex is not sandwiched but rather floats in a gap between two plates, amounting to 0.1 in. or so. Drilling tends to cut or tear a slit in the rubber rather than to cut a round hole. The result upon pin insertion is that the pin is gripped by a pair of tiny rubber flaps, allowing short motions of the pin to occur without actually rubbing the pin against the rubber. (The flaps are able to move short z-axis distances with the pin.) This provides essentially frictionless pin motion over the short range of travel needed to conform to surface irregularities on the product. Properly sized to match pin diameter, the presence of the holes or slits tends to relieve any accumulation of stress in the rubber sheet in high-density areas, providing good contact reliability under such conditions. The lack of motion against the rubber sheet also promotes long life of the fixture. The presence of the hole or slit provides stress relief in the rubber in high-density situations. Molded one-piece plate spacers speed assembly and disassembly, lower cost, and provide accurate z-axis position control of the plates. Use of the longer 3.75 in. pin provides maximum pin displacement.

The featureless music wire pin is economical and widely available. A countersunk top plate hole or two-ply top plate is generally used for high-performance applications. Notice that no disassembly is required to service a pin.

A variation of this fixture employs a stretchable fabric web in place of the latex, generally referred to as *spandex*. Under close examination, this material appears as a very fine, stretchable net. The stretchable net requires no drilling operation—pins are just pressed through during the pin-loading operation. However, some stress accumulation is observed in very dense areas.

38.3.3.7 Intermediate Guide Plates in Fixtures. Early pin translator fixtures were constructed with only two plates, top and bottom. Modern fixtures add intermediate plates to speed fixture building, produce a more reliable fixture, and make the fixture determined. Determined fixtures ensure that a pin started into a fixture hole during pin loading is able to follow one and only one path through the fixture (without bending). This is required if data-driven test programs are to be used because the test program assumes a specific mapping of grid test points to product targets. The determined pin load must remain constant if the fixture is taken apart and later reloaded; otherwise, the test program will become invalid.

Intermediate plates simplify and shorten the pin-loading process because each pin must fall into the intended track. Assemblers do not have to pick between several possible holes or spend time checking for misloads and/or short circuits. Determined pin loads permit the fixture software to calculate the side loads imposed on fixture plates by the combination of tilted probes in each region of the fixture, adjusting the pin-loading plan until these side loads are balanced. This prevents the fixture from collapsing sideways under compression. The intermediate plates also support finer-diameter pins, ensuring that they do not buckle under load.

As density increases, holes in the various plates become more closely spaced. At some point, for a given plate stack-up, there is the risk that some pair of holes on the next plate will become close enough that a pin being loaded might fall into the wrong hole. At that density, the fixture software must add an additional intermediate plate to the design. Thus, denser fixtures generally require more plates. Fixture software should analyze the requirements of each fixture and delete unneeded plates.

38.3.3.8 Fixture Plate Spacing Techniques. Errors in the vertical (z-axis) position of fixture plates can result in pin binding. As a tilted pin travels through the plate stack-up, the hole position in each plate shifts slightly in the x and y directions, according to the amount and direction of tilt. If a plate were to be mounted either too high or too low in the plate stack-up, the drilled hole would no longer be at the correct x-y position, forcing the pin to bend in order to pass through the hole. Pins would become difficult or impossible to load; binding and friction would occur during operation. Sensitivity to this problem increases as the amount of tilt increases.

Early multiplate fixtures used a plate-spacing technique referred to as *post and doughnut,* which had mediocre control of z-axis plate position. A long screw passed through a stack consisting of the fixture plates and thick plastic washers. The washers were drilled in the center, and the plates were drilled with corresponding mounting holes. The screw passed through all of these in turn. The thickness of a given washer determined the space between two plates. The problem was that the errors in washer and plate thickness accumulated through the stack-up. At greater pin tilts, the result was binding. It turned out that the thickness of inexpensive fixture plate material was not terribly precise.

Alternative means of plate spacing have been devised in which a single molded (or machined) part is responsible for the entire fixture stack-up, and which eliminate intermediate plate thickness as an error contribution. One method is illustrated in Fig. 38.15.

FIGURE 38.15 Molded spacer prevents pin binding.

This molded spacer provides a series of flat surfaces upon which individual plates sit. In a top view, the shape of each surface is modified such that plates slip easily over the top of the spacer, but are locked into position when the spacer is rotated 90°. As long as the spacer is accurately molded, no cumulative plate spacing error results. For midfixture plate supports, a similar part of smaller diameter is used. Short screws at the top and bottom plate secure the support in place. Alternative methods use machined bars around the periphery of the fixture. Either method results in accurate fixtures, reduced parts count, and ease of assembly.

38.3.3.9 Headed Music Wire Pins. For very large through-holes, the preferred probing method is to probe the annular ring of the plated through-hole. If, however, this ring is insufficient, a headed pin may be required. Such pins are manufactured with a machined tip attached to a standard music wire pin. Various head shapes are available. These headed music wire pins work in the fixture design much the same as the headless pin, that is, flush in the top plate (see Fig. 38.16), with the exception that the headed pin must be loaded from the top individually.

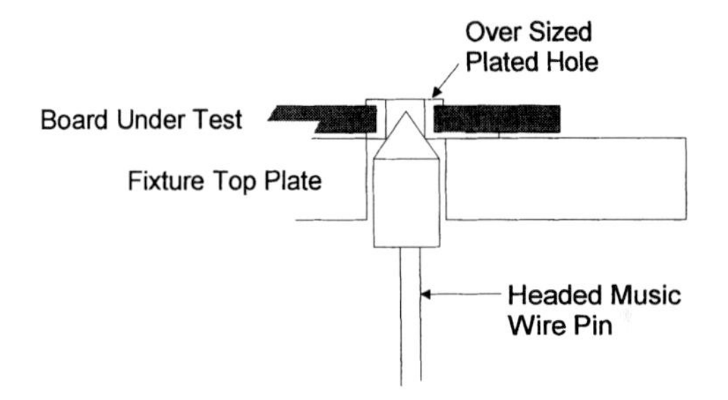

FIGURE 38.16 Headed pin in a tilt-pin fixture.

38.3.4 Dual-Side Testing Considerations

Modern SMT boards often have networks that terminate at component locations on both sides of the board. Continuity testing of such networks requires that the top and bottom terminations of such networks be probed simultaneously. There are three ways to test boards that require dual-side access:

- Flip test
- Clamshell upper fixture
- Dual-side universal grid fixture

These methods vary in terms of test coverage, fixture cost, labor cost, risk of test escape, and required infrastructure. Each method is discussed in the following subsections.

38.3.4.1 Flip Testing. Flip testing is a last-resort method employed where up-to-date dual-side equipment is not available and compromising in test coverage is acceptable. The flip test method requires building either two fixtures, one with the top-side image and one with the bottom-side image, or one fixture with both images on it. The board is then tested one side at a time. Although this does provide partial fault coverage, it does not test those plated holes that link the top-side half of a network to the half on the bottom of the board. This poses a significant risk that bad boards will escape the test process. The problem only becomes worse if these interconnecting vias are tented (that is, if they have solder mask over them), preventing probing. In this case, the traces that run from the tented via to the first pad on the network that can be probed are also untested at both sides of the board.

Flip testing is expensive. Not only are two fixtures built, but also many test points are duplicated between the fixtures. Much additional board handling and potential for confusion are involved. The cost and risk incurred would likely justify an equipment upgrade to dual-side test capability.

38.3.4.2 Clamshell Upper Fixture. Clamshell testing is another compromise method employed where dual-side equipment is not available. This method provides top-side access through the use of a wired (dedicated) top-side fixture. The method is superior to flip testing in that it can provide complete simultaneous access and full test coverage. However, fixture costs are quite high, especially for top-side test points. In this scenario, a more-or-less normal pin translator grid fixture is built for the bottom side of the board. The board should be oriented such that the busy side of the board faces down, as this minimizes the costly top-side fixture (see Fig. 38.17). Additional test points (called *transfer points*) are added outside the product area on both the top and bottom fixtures, conducting test points from the lower fixture and grid to the upper fixture assembly. Normal wired fixture construction methods are used within the upper fixture to relocate these points to desired top-side product target positions.

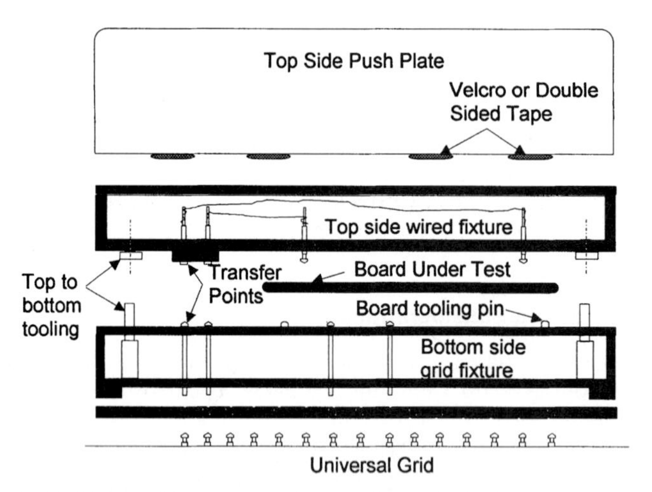

FIGURE 38.17 Transfer point clamshell fixture for dual-side access.

To use this method, it is necessary that the lower grid be larger than the product under test such that a number of transfer test points adequate to provide for the upper fixture can be placed outside the product area. Again, the cost of the extra contact points, spring probes, wiring, and other special material and labor processes associated with the topside fixture may well justify an equipment upgrade.

38.3.4.3 Dual-Side Access Universal Grid. The third and most widely accepted method of providing for volume testing of dual-side product is the double-sided universal grid. Such systems provide electronics and grid probe fields on both the top and bottom sides. Upper and lower translator fixtures are built for dual-side products using the standard techniques discussed earlier, providing economical fixtures, reusable pins and hardware, full fault coverage, excellent throughput, reliable contact, and accurate targeting.

38.3.5 Press Units

The major mechanical component of the grid is a press unit. The three major compression methods used to compress the product onto the fixture are hydraulic, pneumatic, and electric. The total force requirement can be quite large. The spring probes employed in grids usually require a force in the range of 4 to 10 oz. each. A large, complex board may have 20,000 active points per side. This amounts to 5,000 to 12,500 lb. of force, depending upon the probes used.

Hydraulic drive is probably the least used compression method. Although it provides tremendous power, it is difficult to maintain and fluid leakage is a problem. Pneumatic and electric compression are popular and successful. Electric drives are easy to control, but become expensive and/or slow when larger forces are required (as in the case of larger grids). Pneumatic systems are fast and powerful, but require a good supply of clean, dry, compressed air. With any method, accurate compression control is critical to reliable fixture performance.

As noted, universal grids present the test points as an array of spring-loaded contacts. A variety of contact tip shapes have been used, but most contacts used today offer waffle tips.

Waffle tips are nominally flat, but with an *H*-shaped pattern of grooves cut into the surface to provide a somewhat aggressive contact surface. Such a surface is better able to cut through thin contaminants on any surface it contacts.

38.4 FLYING-PROBE/MOVING-PROBE TEST SYSTEMS

Moving probe, flying probe, and x-y prober are all names for test systems that make use of two or more test points that can be accurately positioned anywhere on the board surface by means of a computer-controlled motion system (see Fig. 38.18). Probe tips can be retracted in a z-axis direction away from the board surface, then moved in the x and y directions to a new board

FIGURE 38.18 Flying-probe test system. *(Courtesy of atg Test Systems GmbH.)*

FIGURE 38.19 Two probe tips contacting board. *(Courtesy of Probe Inc.)*

location and extended again to make contact. Figure 38.19 shows two probe tips contacting board. Dual-side systems generally provide a minimum of two independently moving probes per side for a minimum total of four heads.

38.4.1 Advantages of Flying-Probe Systems

The major advantage of these systems is elimination test fixtures, making these systems ideal for small-to-moderate volume production. Advanced flying-probe systems provide highly accurate probe placement, and contact the board with minimal force, leaving no discernable witness marks on most surfaces. They are very well suited to testing the finest pad sizes. Although not subject to limitations due to test-point density, these systems do slow down as additional test points are added.

Direct current (DC) continuity test measurements are accomplished by placing one probe tip at each end of the continuity to be verified and performing a measurement. The probes then retract, move to the next measurement site of interest, make contact, and make the next measurement. Similarly, it is possible to perform DC isolation testing of a network by placing one probe (probe A) on the network, while another (probe B) checks all other networks in sequence for the presence of a short or leak. Probe A can then be stepped to the next network to be tested and the process repeated. This continues until each network has been checked against all others.

This requires a huge number of measurements. For a board with N networks, the total number of measurements M required would be:

$$M = (N2 - N)/2$$

For example, a fairly ordinary board with 1,000 networks would require 499,500 measurements. Fortunately, several means of reducing the measurement burden have been developed.

The program preparation software can analyze the conductor pattern on the board and, noting that short circuits should occur only between physically adjacent conductors, reduce the number of measurements required. This is termed *adjacency analysis*. To achieve further time savings, the testing department might use an indirect measurement method. Please refer to Sec. 38.4.4.

Some indirect methods require only a single probing action of each net, combining continuity and isolation testing with some extra probing to verify any suspect results. Theoretically, our example board with 1,000 networks might have isolation verified with only 1,000 probe placements.

38.4.2 Economics of Flying-Probe Systems

The primary limitation of flying-probe test systems is throughput. Although systems with up to 16 test heads are commonly available, these systems still lose a substantial fraction of the operating time due to mechanical positioning of the heads. Yet the cost savings derived from elimination of direct fixture costs, fixture support infrastructure, fixture debugging, and so on, are substantial.

Accurate comparison of the costs associated with grid and prober methods requires inclusion of all the costs associated with each approach. Common omissions include:

- Labor overhead
- Floor space allocated to fixture storage, materials, and assembly
- Costs outside the test department (such as drilling)
- Stranded investment in completed fixtures, pins, other fixture materials
- Stranded investment in drills and other equipment
- Lost throughput of test systems due to maintenance
- Lost throughput of fixtured systems during fixture setup, debugging, and maintenance
- Peak throughput demand

It is quite possible that, depending upon product mix, a testing department would be made more efficient with a larger investment in capital equipment (in the form of flying probe systems) and a smaller investment in fixtures. With smaller runs and/or smaller lots necessitating frequent fixture teardown/bring-up, this becomes more likely.

38.4.3 Flying-Probe Throughput Enhancement Systems

The tremendous advantages of flying probers for advanced substrates (such as the ability to hit the smallest targets even on large panels, the lack of fixture cost, witness mark reduction or elimination, and fractional ohm measurement) have traditionally been offset by the relatively slow speed of test compared to universal grid methods. As products have included increased amounts of high-density features, test managers find that flying probers are the solution that can reliably contact the smaller features without damaging them. This has provided tremendous motivation for flying-probe manufacturers to find ways to increase throughput.

Recognizing that many of the test targets are easily measured (only a fraction of the sites on most boards presents the majority of the challenge), one new approach uses an extra piece of equipment to prescreen the boards before they are placed on the flying prober, eliminating many of the measurements from the flying prober's task list. This approach is described in this section, with each of the two commercially available models summarized.

In the first case, a row of thousands of regularly spaced sliding contacts moves relative to the board surface, scanning the surface electrically as a row of scan lines. The contacts are designed to be individually flexible and to apply minimum force, eliminating witness mark concerns for most surfaces. Adjacent to the row of contacts is a planar section of a pliant conductive surface. This surface electrically grounds all exposed conductors on the board that lie beneath any part of the planar area. This conductive plane is fixed in position relative to the row of scanning contacts and extends to within a fraction of an in. of the row of scanning contacts. The relationship is illustrated in Fig. 38.20.

FIGURE 38.20 Flying-probe throughput enhancement system. *(Courtesy of Everett Charles Technologies.)*

As relative motion of the board to the scanning contacts/pliant conductive plane occurs, a high-speed measurement apparatus constantly makes measurements among the scanning contacts and between the scanning contacts and the pliant conductor. For example, if a scanning contact is at one moment making electrical contact at one point end of a network so that another end of the same network is being grounded by the compliant planar surface, then current injected into the scanning contact will fail to increase the voltage of (that is, fail to charge) the network. The pliant planar surface holds the network grounded. Similarly, if two adjacent networks are simultaneously contacted by different scanning contacts, and if at that moment no area of at least one of these networks is grounded by the compliant planar, then it will be observed that at least one of the two networks can be charged to a high voltage and that they must not be shorted together. Software in the prescreening system uses the real-time measurement results to "image" the board and constantly adapt to any shifts in geometry. Thus the system solution is useful for very-fine-pitch substrates appearing in large-panel form.

In a manner similar to the preceding examples, significant portions of most boards may be tested using this apparatus. Some portions of the board usually remain untested, and these measurements are reserved for the subsequent final test by the flying-probe test system. The flying prober also retests and verifies any failures reported by he prescreening system, effectively eliminating most false failures. The resultant speed increase of the combined systems is tremendous, so that true production volumes can be tested. It is important to note that no fixture is required and that both systems provide for true DC measurement of continuity and isolation. Moreover, because high-density interconnect (HDI) features that are too fine for the scanning contacts of the prescreener are simply passed on to the prober, this solution seems to be appropriate for very advanced substrates.

A second commercially available pre-screening system is very similar, but uses a capacitive measurement made at each time of the finely spaced row of scanning contacts. The overall function is very similar to that just described, except that the alternative of a true DC measurement of continuity and isolation is not provided. For those customers that do not perform such tests with their existing flying probers, this may not prove a significant limitation. For high reliability applications where DC continuity testing is required the utility of this method may be reduced.

38.5 VERIFICATION AND REPAIR

The test system identifies faults on the board. Assuming that repaired boards are acceptable to the end user and that it makes economic sense to repair boards, a repair process normally follows testing. Because fixture problems, contamination problems, product registration problems, and other issues often result in false error reports, a verification process usually is added between testing and repair. During verification, a technician reads the fault data and makes confirming measurements to determine whether the reported faults are genuine.

If verification of the fault finds that the fault report is valid, the board proceeds through the repair function. If the fault report is not valid, the board may be either retested or directly shipped, depending on customer requirements, in-house policy, etc.

When deciding on a ship-from-verification policy, be mindful of the possibility of multiple defects masking one another (an open hiding a short) or appearing only under the pressure of the fixture. If possible, it may be preferable to perform a complete retest.

In some cases, simple tabletop meter equipment is used for verification, and verification may actually be done at the repair station by the repair technician. More commonly, some form of computer-assisted search-and-display tool is employed. The computer has information concerning the product and test program available, and searches the artwork pattern for areas likely to contain the fault being considered. The resulting risk area is displayed to the operator or superimposed as a visual projection onto the suspect board, simplifying the task of placing meter leads.

Automatic verification may be accomplished by a flying-probe–type system. The flying probe performs a similar analysis, but does the retesting itself and displays a final result. Video camera systems may provide image capture of suspected fault sites, storing the image for subsequent recall at the repair station. Advanced systems may even suggest the repairability of particular fault sites according to user-defined rules.

Figure 38.21 shows an example of a fully automatic verification system and Computer Aided Repair (CAR) station. At the repair station, a display computer similar to that

FIGURE 38.21 Fully automatic verification system and CAR station. *(Courtesy of Everett Charles Technologies.)*

described for verification purposes is used to highlight the proposed repair site to the operator, who performs the appropriate cuts or welds.

38.6 TEST DEPARTMENT PLANNING AND MANAGEMENT

In most cases, a new manager inherits a selection of existing equipment, processes, and personnel. Nonetheless, over a period of time, the manager has the same opportunity to shape the test floor operation as does the unlikely individual starting from a clean sheet of paper. To the extent that the past suggests the future, the new manager may expect increased density, finer pitch, higher total test-point counts, decreased tolerance for test escapes, shorter product delivery times, shorter product lifetimes, and larger numbers of smaller batches of boards for just-in-time delivery.

38.6.1 Equipment Selection

Most equipment being purchased for general bare board testing consists of universal grids and flying probers. The increased capabilities of flying probers and increased demand for small lot and prototype testing have increased the number of probers selected. This trend seems likely to continue, but each shop must determine its appropriate solution individually.

A first consideration in selecting equipment lies in recognition of the type of business you anticipate. For very-high-volume production of a variety of part numbers, the universal grid is a likely participant. For medium volumes, considerable savings may result from the elimination of test fixturing costs and delays that can be achieved by combining flying-probe test systems with throughput enhancement systems, as described in Sec. 38.4.3. For prototype or small lot production, a quantity of flying-probe systems may be the best alternative due to the elimination of fixture expense. Dedicated (wired) fixture–type equipment is rarely demanded due to fixture cost, excepting very simple systems that remain devoted to an extremely high-volume run of a single product type for extended periods of time.

A further important concern is the level of technology that you anticipate. Verify that it does not outrun the capabilities of the fixture or test system types that you plan to use. Plan for change, as the equipment that you purchase must last for several years. Also consider the level of support available from the equipment vendor, including ordinary maintenance service, emergency service, employee training support, and applications support.

For high-volume operations, you can achieve substantial throughput and consistency of operation by adding automatic handling systems to a universal grid system configuration. For prototypes and moderate production volumes, the ability to place a stack of boards on a flying probe system and walk away for an hour or so, rather than wait for a board to come out every three minutes, can result in huge labor savings. Note that some very large, very thin, or very small product types may not be well suited to automatic handling without the use of special carriers or adapters. It may be best to size the automation equipment to handle the major portion of production and leave unusual product configurations on manually loaded equipment.

An often overlooked but practical consideration is maintaining some degree of commonality in equipment and fixtures. It is not uncommon for fixture problems to lead a test engineer to change equipment types several times, resulting in a motley assortment of incompatible (but expensive) test systems but failing to resolve the fundamental underlying issue. With data-driven test processes and automated fault data management in the mainstream, the ability of equipment to share data, fixtures, and operating procedures will decrease cost, increase throughput, simplify training, and enhance opportunities for combinational test solutions. This is harder to achieve with a random mix of equipment. If this seems to state the obvious, try this experiment: Visit some shops, stopping in drill rooms and in the electrical testing area. Do you see a difference in the assortment of equipment? Department planning should include not only new equipment, but also an exit strategy for old and inefficient equipment that you wish to retire.

38.6.2 Fixtures: Build or Buy, and What Type?

If you plan to use fixtures, the selection of fixtures is probably more important than the choice of equipment on which you choose to run the fixtures. Certainly you must have equipment at least adequate to the fixture, and with sufficient test-point density for the product. This is where a careful evaluation of product technology is crucial.

If the product includes even fairly simple SMT boards, then you should plan immediately for a dual-side capability. If the product includes fine-pitch SMT, dense BGA patterns, and such, then you must consider higher-density universal grids and modern tilt-pin fixtures consistent with the test-site density on these products.

If you plan to build fixtures in-house, recognize that you are creating a complete manufacturing operation. Measurement of complete material, labor, inventory, and overhead costs as well as operating efficiency is critical to effective management of this operation. Don't just measure how fast you build a fixture, but include measurement of the time it takes to get it running. Consider whether you would be better served by outsourcing fixture fabrication.

Avoid manufacturing multiple styles of fixtures. Ensure that you have an inventory management process that does not run out of pins at midnight, but does not have half a million dollars tied up in obsolete pin inventory. Consider outsourcing at least the more difficult fixtures to specialists. You may save time, plus gain a partner who provides exposure to alternative techniques and materials.

38.6.3 Selecting Fixture Software

Today's tilt-pin translator fixture is largely a software product. It should be possible for you to establish a flow of information from your computer-aided manufacturing (CAM) department to the front-end fixture software that automates the design of fixtures and the creation of test programs. (The same goal holds true for processing data to create flying-probe test programs.) A premium tool will pay for itself many times over in reduced fixture scrap, reduced late deliveries, and elimination of test escapes. If you can select a tool that is also used by several fixture outsourcing vendors, you will have a backup supply of fixtures in the event of equipment problems, personnel problems, or workload peaks. If you have special and specific demands for the fixture design, verify that the proposed software package readily supports your requirements. And, as in the case of the test system, evaluate the software vendor's ability to sustain and support your operations.

CHAPTER 39
HDI BARE BOARD SPECIAL TESTING METHODS

David J. Wilkie
Everett Charles Technologies, Pomona, California

39.1 INTRODUCTION

Printed circuit technology is now being applied to high-density interconnection (HDI) applications that require consideration of new testing techniques. Notable examples include many types of advanced integrated circuit (IC) packaging introduced in recent years, such as area-array devices, ball grid arrays (BGAs), laminate chip carriers (LCCs), and circuit boards employing direct chip attach (DCA) or chip-scale packages (CSPs). In some of these cases, semiconductor dies are being directly attached to a printed circuit surface, and in the case of CSP, the package size is approximately the same as a raw die. Flip-chip die (or CSPs) may be attached to a circuit board by an array of solder balls. Other devices may attach with a peripheral array of wirebond pad sites just outboard of the die area. For high-pin-count devices, this last example can produce the finest side-to-side test-point pitch as efforts are made to squeeze a large number of contacts onto the limited periphery of the die. The flip-chip or CSP bump array attachment parts generally produce the highest test-point density per unit area. The flip-chip or CSP approaches allow the designer to distribute the solder ball contacts over the full surface area of the die or package. For fixtured test systems, the ability of the tester to provide a sufficient density of test points and the ability of the fixture to achieve the needed pitch and accuracy must be considered.

The general goal of electrical testing for HDI substrates remains about the same—detection of opens, shorts, and leakage, although often with increased concern for latent defects, whose appearance can be made more likely by the fine geometry employed in traces and microvias. The measurement electronics are often the same, and are not further discussed here except for special cases. For some applications, time domain reflectometry (TDR) or other radio frequency (RF) tests are desired, but are complicated by the short signal runs on some of these substrates.

Sensitivity to pad marking is increased as there is not much pad area to mark. Marking that penetrates the metallization layers on the board can affect the chemistry of the solder joint, adversely impacting reliability. It is quite challenging to prepare test fixtures able to test such products on universal grids given the need to minimize witness marks, meet aggressive continuity test threshold expectations, and contact the finest geometries. When HDI parts are tested in panelized form, the very small target sizes combine with potential expansion/contraction issues in the panel construction to place a maximum demand upon test-point positional accuracy within the fixture. The ability of the latest camera-equipped flying-probe systems to locate these targets individually and probe them without marking combines with the fixture cost savings to make this method of test very attractive, especially as new techniques improve throughput (see Sec. 39.4.3). Because of the high costs associated with HDI fixtures, some fixtured solutions compromise test coverage such that the fixture can be simplified to a practical degree.

39.2 FINE-PITCH TILT-PIN FIXTURES

Many HDI structures can be tested with advanced examples of tilt-pin fixtures. Using such techniques as test pins that taper (see Fig. 39.1) to a very fine diameter at one end, fixtures have been built for devices with pitch down to about 0.010 in., with research and development (R&D) examples at about 0.008 in. These fixtures contain additional plates to support the very thin pins (so they do not buckle under pressure or short to adjacent pins). Those fixture plates very close to the product must be fairly thin, or the closely spaced holes will break out into one another. Such fixtures increase the demand for higher grid densities. These fixtures and pins are moderately more expensive to prepare than standard fixtures, but use similar technologies and processes. Excellent process control is important, as is excellent pin tip symmetry.

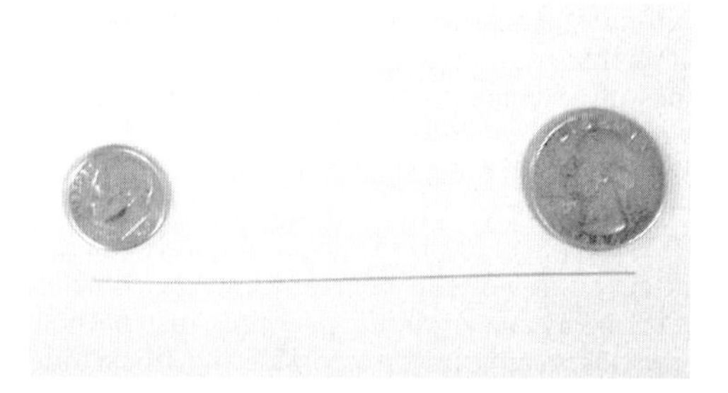

FIGURE 39.1 Tapered pin for a fine-geometry fixture.

Areas of the product requiring large numbers of probes are subjected to considerable total force. If no equal and opposing force is applied on the opposite side of the product, bowing or "potato chipping" of the product will occur. Ultimately this can collapse the opposing test fixture, but long before this, the probing accuracy of a fine-pitch fixture will be affected because the product no longer lies in a flat horizontal plane. To avoid this, the fixture software can add support within the opposing fixture, directly opposite the densely probed site. This additional support can be accomplished with a spacer affixed between the top and bottom fixture plates or by using blind pins. Blind pins are ordinary test pins located in the opposing fixture at sites where the top (product side) plate of the probing fixture is not quite drilled all the way through. If a quantity of blind pins equal to the quantity of dense product site probes is employed, then the opposing forces are perfectly balanced in this region.

When dealing with fine-pitch fixtures, the problem of product and fixture registration must be considered. Even if the fixture is manufactured perfectly, the product may not align. Because the tooling features of the product (edge or tooling hole) are added in a process separate from the artwork production, the artwork and tooling features are usually misaligned.

With HDI substrates, it doesn't take much for this error to move pads completely away from the assigned test pins. To overcome this, a variety of test systems are available with optical and/or electrical means of sensing misregistration. Movable plates in the fixture are controlled by servomotors, which reposition the product through motion of the tooling pins. As the product grows in size, or as the manufacturing process deteriorates, the registration errors may not be sufficiently constant over the product area to permit the use of motorized alignment. In this case, none of the probe-per-pad fixtured approaches may be effective.

39.3 BENDING BEAM FIXTURES

Bending beam fixtures are somewhat similar to tilt-pin fixtures, except that they use extremely thin test pins manufactured from a special alloy. As in the tapered pin fixture, a significant number of thin supports are required near the product, where the pin tends to be held fairly straight to avoid conflict with other nearby pins. These pins are not held rigid, but are expected to buckle under force. This buckling displaces the pin sideways. The buckled portion may be located at some distance from the product plane, where the probes can be more widely separated. Once the pin has buckled, the force applied to move through a substantial travel is constant. This is very different from the case for a spring probe, where the force increases at a constant rate determined by the spring constant. Bending beam fixtures provide good immunity to witness mark damage. The basic technique has been in use in specialty situations for many years.

At the interface side, these pins may mate to spring probes instead contacting a rigid contact surface. In some commercial examples, the pin continues away from the product as a piece of wire routed to widely spaced contacts on the interface portion of the fixture. These in turn interface with traditional spring probes. Within the fixture, the probe wire is bonded in place at some distance from the product, defining the separation between the probe and wiring portions of its overall length. Such fixtures are quite expensive, and replacing damaged probes can be challenging. Thus, this technique is usually restricted to small areas and limited test-point counts.

39.4 FLYING PROBE

The more precise examples of flying-probe systems are very well suited to HDI testing. With on-head optical pattern recognition guiding the probe tips, they can contact extremely small targets—even in the presence of significant product registration problems. They can probe small features without marking or disturbing the surface. These systems can be equipped with true Kelvin contact tips supporting effective single-measurement fractional ohm testing, well suited to the sort of fractional ohm continuity test that may be desirable in efforts to detect precursor phenomena to latent defects caused by cracks or other geometry problems. The principle drawback is, of course, test speed. Used alone, these systems are ideal for prototype or small volume production applications where fixture production would be quite expensive. Used with throughput enhancement systems as described in Sec. 39.4.3, these systems can accommodate moderate production volumes.

39.5 COUPLED PLATE

In a situation where most of the target product is relatively easy to fixture, but a DCA or CSP site is difficult, it may be possible to use coupled plate–type testing. A variety of proprietary products are available, but the basic technique is fairly similar. The primary assumption is that each signal network arriving at the DCA or CSP site is accessible by a traditional probe at some other site on the board. These probe sites are used to perform a normal isolation test of the product. Continuity testing of most of the board is performed in the usual manner also.

To verify the continuity of the signals to the DCA or CSP site, a small metal plate or antenna is suspended just over the site, perhaps insulated from the site only by a thin dielectric. One at a time, the standard probe sites are used to inject some sort of alternating current (AC) signal or pulse into each network in turn. If the network is properly continuous to the DCA or CSP site, then a signal of appropriate amplitude is detected at the antenna. Substantial deviation from the proper amplitude means that something is wrong. The different methods vary in terms of the applied signal, antenna characteristics, and so on.

This method eliminates the need for super-fine-pitch probing at the DCA or CSP site and also avoids marking of the product at this site. However, the test does not perform a true DC test of continuity and may miss high-resistance connections, which would be detected by a low-threshold DC continuity measurement. Signals that loop from one DCA or CSP pin to another and go nowhere else may be untestable as there is no outboard probe site at which to inject the test signal.

39.6 SHORTING PLATE

This method is employed in circumstances similar to the coupling plate method just described. In this case, however, the plate must be movable during the test. Generally, a pencil-sized pneumatic actuator is mounted within the test fixture to accomplish this motion. The plate employed here is a small flat metal plate that is the size of the DCA or CSP site and covered with an electrically conductive rubber (see Fig. 39.2). When pressed against the product, it shorts together all of the pads at the DCA or CSP site. Continuity testing is performed in this condition by using the outboard probe sites to confirm that all of these (otherwise isolated) networks are shorted together via the path to the DCA or CSP site. All other networks are tested normally for continuity. Then the shorting plate is removed and a normal isolation test is performed. Key advantages of this method are the accomplishment of true DC measurements for both isolation and continuity and the use of standard bare board test systems.

In some applications, there is concern about any trace chemicals that may be left behind by the conductive rubber, although outright marking is minimal. Cleanliness of the product is

FIGURE 39.2 Pneumatically actuated shorting plate.

critical, as the rubber or plate must be replaced when excessive dirt is embedded. As with the coupled plate, certain signal topologies are untestable or difficult to test. Signals that loop from one DCA or CSP pin to another and go nowhere else are untestable, as there is no external probe site with which to verify either continuity (to the DCA or CSP shorting pad) or isolation. Any signal that connects two DCA or CSP pads on the same package may be partially untestable, even if the signal continues to an external probe site. In this case, it is possible to verify that the external signal arrives at the device, and it is possible to verify isolation of this segment. But there is no simple way to discern whether the two device pads are joined to each other. (If they are already shorted together on the board and engaging, the conductive rubber has no effect.) This latter case may sometimes be mitigated by segmenting the shorting pad, such that each of the target device pads is in a separate segment. Then, if the connection on the board is good, the halves will be joined through the board.

39.7 CONDUCTIVE RUBBER FIXTURES

Several designs for fixture systems employing conductive rubber as the basic probe element have been offered commercially. In some cases, the rubber is a specialized material in sheet form and is conductive only in the z-axis (through the thin sheet vertically, but not sideways across its surface).The fixture is itself made from a circuit board, with slightly raised pads to compress the rubber tightly against desired product sites. The fixture board connects to the grid electronics at its reverse side. Other designs have included various types of locally deposited rubber dots, usually of conductive rubber that is not sensitive to orientation. Again, a rubber probe is formed. Problems with cost, complex manufacturing, complex repair, dirt sensitivity, and suitability to very small pad areas (which limit contact quality to the rubber) seem to have prevented widespread adoption.

39.8 OPTICAL INSPECTION

Optical inspection has been discussed elsewhere. It is generally applied early in the fabrication process as a yield improvement and data collection tool, not as a means of final product qualification. However, with improvements in resolution, the type of defect that may escape undetected becomes somewhat more limited, and optical inspection is argued as a possible means of final testing.

With complex multilayer product, optical inspection will not be able to identify assembly/contamination-induced defects internal to the board in any case, and may still be limited in cases of contaminants or very fine-geometry shorts and opens on external layers. The equipment is somewhat slower than universal grid test systems, particularly when run at very high resolution. For such reasons, it is still not common practice to employ optical inspection (by itself) as a substitute for electrical testing. This may be a method that develops further in the future or that finds acceptance in special circumstances.

39.9 NONCONTACT TEST METHODS

Quite possibly the greatest daily irritant (and cost) in the operation of a typical test area today revolves around the cost of building test fixtures. Customers still hate to pay for them, and the creation of fixtures burdens the board shop with an entire manufacturing process that seems to add no value and that distracts from the main productive purpose of the factory. The only fully effective commercially available systems that require no test fixtures are flying-probe test systems and the related throughput enhancement systems. In addition, extensive work has been done by several organizations to attempt to develop alternative noncontact means of fixtureless test. To date, no fully successful noncontact method has yet emerged with a capability for parametric measurement of the normal range of both continuity and isolation, yet some brief description of various techniques seems warranted.

39.9.1 Electron Beam Methods

When a test system contacts a product with an electromechanical contact, it uses this contact to inject or remove electrons. That is, current flows. The sort of electron beam common to the picture tube in a television might be used to do the same thing without contacting the product physically. Experimental systems have been built by several firms to date, but have shared some common problems. First and foremost, the amount of current delivered by the electron beam is so small that only a very poor continuity test is possible. Increased interest in fractional ohm and/or high-current testing to detect latent continuity defects runs counter to this limitation. Lab systems have been limited to continuity thresholds of 100,000 ohms or more, and this only very slowly. As most users wish to test at 10 to 100 ohms, this is a major compromise. (A smear of contaminant across an open circuit would appear to be a perfectly good conductor.)

Test speeds are affected by product capacitance, as a longer time is required for the limited current to achieve significant voltage effect in a highly capacitive environment. Also, such systems must operate in an extremely high vacuum of laboratory grade. Expensive pumping systems are required. Costly air-lock systems with multiple stages and robotic product handling are probably required to keep any reasonable flow of material through the system. (Staging product in and out through a series of chambers avoids the time delay of constantly pumping down the main chamber when loading new product.) At the present time, flying-probe systems seem to offer superior test and measurement coverage at more practical operating costs.

39.9.2 Photoelectric Methods

Subjecting a metal surface to an intense beam of light, such as that from a laser, can cause electrons to be ejected from the metal. As in the case of the electron beam, a very small current flow can be induced. Again a vacuum is required (so that the ejected electrons can be measured before they collide with air molecules), though the vacuum requirement is less stringent than that for the electron beam technique. The resistance at which continuity measurements can be made is quite limited, as in the case of the electron beam. (See discussion of electron beam method in Sec. 39.9.1.) Again, test speed suffers due to product capacitance. No practical system has resulted from investigative work.

39.9.3 Gas Plasma Methods

Fluorescent light bulbs emit light because a gas is subjected to an electric field, which adds energy to the electrons orbiting the gas molecules until some break free. As they attempt to reattach themselves to the gas molecules, they eject their excess energy as photons (light). The plasma consists of a mixture of gas molecules, ionized gas molecules (missing electrons), and free electrons. In this state, the gas can conduct an electric current. If a jet of plasma is directed from a small nozzle to the surface of a circuit board, the effect is that of building a "gas probe." In broad terms, the gas probe can be used just as any other probe or test is completely possible.

Generally a noble gas such as argon is used. The gas residue is nontoxic, only a tiny amount of gas is consumed, and substantial current can flow to the product. Several companies have developed experimental systems in the form of flying-probe systems.

Although gas probes offer the benefit of little or no product marking, it is difficult to achieve one that is as fine as the best mechanical probes. Adjacent jets tend to combine, producing a short circuit just as if two mechanical flying-probe tips shorted while testing. Thus, the commercial system offered to date includes completely traditional mechanical probes as well, using these to probe closely adjacent (i.e., HDI) sites. As a flying-probe mechanism is still used, there is little speed advantage to date. Eventually some benefit may be obtained due to the elimination of any wait time for z-axis travel of the prober head, although this is already the fastest motion axis of most flying-probe systems. The cost of this technique is modestly higher than that of an ordinary flying prober.

39.10 COMBINATIONAL TEST METHODS

One technique that offers immediate practicality in resolving difficult testing situations for high-density product—often using existing equipment—is generally described as combinational testing or sequential testing. As the name implies, this is testing in one or more stages, using a combination of test techniques. Combining techniques inevitably adds complication. The simplest example is the use of a universal grid to test the majority of a product, followed by a flying probe system to test HDI features and reverify the failures reported by the grid. Software tools have simplified the process of combining test methods.

P · A · R · T · 8

ASSEMBLY

CHAPTER 40
ASSEMBLY PROCESSES

Paul T. Vianco
Sandia National Laboratories, Albuquerque, New Mexico

40.1 INTRODUCTION

The electronics "revolution" has been sustained through the development of products having increased miniaturization, enhanced functionality, improved reliability, and reduced costs of manufacturing. In fact, innovations in electronics assembly methodologies have kept manufacturing costs low with each new generation of product.

40.1.1 Feature Density

Product designers have been able to explore new packages, materials, etc., so as to increase functionality further while, at the same time, reducing the size and weight of both consumer as well as high-reliability military and space electronics. The result has been greater challenges for process engineers. For example, there is a continued trend of decreasing component sizes. Leadless ceramic chip capacitors having sizes of 0804, 0603, and 0402 are commonplace, whereas smaller 0201 devices are being introduced into product lines, particularly handheld products. Manufacturing resources are now developing the means to tool up for 01005-size components. Fine-pitch and area-array packages—ball grid array (BGA), chip-scale packages (CSP), and flip-chip (FC) or direct-chip attach (DCA)—provide the means to increase device functionality significantly. Input/output (I/O) interconnections that are reaching several thousands on BGA packages require stringent solder paste printing, part placement, and reflow processes controls to minimize defects. At the same time, the need for more complex, multilayer substrates has also placed limitations on the process window so as to avoid damage to circuit board microvias and fine traces that are needed to support higher assembly densities. Lastly, these challenges are made more complicated by the use of Pb-free solders. Changing equipment parameters as well as alternative surface finishes impact solderability performance and thus process yields as well.

40.1.2 Printed Circuit Board Assembly Process

There are two basic steps in the assembly of a printed circuit board: (a) placement of the components (resistors, capacitors, etc.) on the substrate and (b) soldering those components into place. Although this is a fairly accurate description of a through-hole, hand-soldering operation, nearly all electronics assembly operations are, in fact, considerably more complex. Multiple-step assembly processes provide the versatility to incorporate different component package types and a wide variety of substrate configurations and materials as well as to accommodate frequently changing production volumes in order to meet prescribed defect levels and reliability requirements. A more accurate, albeit still relatively general, listing of assembly process steps consists of the following:

1. Preparation of the component and substrate surfaces to be soldered
2. Application of the flux and solder
3. Melting of the solder to complete the joint
4. Post-process cleaning of the soldered assembly
5. Inspection and testing

Some of these steps may be either combined together or eliminated, depending on the particular product line.

It is important that the manufacturing engineer and operator understand the critical steps in the printed circuit board assembly process to ensure the manufacture of a cost-competitive, reliable product. That understanding includes both the general function of the equipment as well as the activity taking place *inside* the machines. The following sections of this chapter describe in detail the printed circuit board assembly processes.

40.1.3 Assembly Process Categories

Assembly processes can be placed into the following three categories, which are described by the types of circuit board components:

- Through-hole technology
- Surface-mount technology
- Mixed technology, which is a combination of through-hole and surface-mount components on the same circuit board

Within each of these assembly technologies are different levels of automation that equipment resources offer. The degree of automation will be optimized, depending on the product design, bill of materials, capital equipment expenditures, and actual manufacturing costs.

It is important to remember that through-hole printed circuit boards and their assembly processes remain a critical technology in the electronics industry, albeit clearly not at the same production volumes as experienced before the advent of surface-mount technology (SMT). Through-hole technology may be used because it is the only format available for some components, particularly large devices such as transformers, filters, and high-power components, all of which require additional mechanical support that is offered by through-hole interconnections. A second reason for using through-hole technology is economics. It may simply be more cost-effective to use through-hole components, together with manual assembly (i.e., no automation) to produce an electronics assembly. Of course, through-hole technology is not limited to manual assembly. There are varying degrees of automation that can be used to assemble a through-hole circuit board.

40.1.4 Pb-Free Technology

The introduction of Pb-free technology has not changed, per se, available electronics assembly processes (reflow, wave, hand soldering, etc.). However, it has caused manufacturing engineers to reassess the parameters used in those processes because of two factors: (1) higher process temperatures required by Pb-free solders, and (2) the poorer solderability of these alloys (owing largely to the absence of Pb). With respect to the five general process steps noted previously, the use of Pb-free solders primarily affects steps 3, 4, and 5.

The need for higher processing temperatures limits the assembly "process window" of Pb-free solders. A higher *nominal* temperature is needed to accommodate the temperature variation at components across a circuit board to ensure melting of the solder and adequate wetting and spreading at each interconnection. On the other hand, the maximum temperature must be limited to prevent thermal damage to heat-sensitive devices and the circuit board.

The poorer solderability of the Pb-free solders presents several challenges as well. Although the longer time needed to heat the component lead or termination and circuit board is often cited as the underlying source of poor solderability for Pb-free solders, it is primarily the higher surface tension of the Sn-based alloys (in the absence of Pb) that restricts wetting and spreading behavior. The need for longer heatup times is a particular issue with "faster" assembly processes such as wave soldering and hand soldering. However, the intrinsically poorer solderability impacts all assembly processes because it can degrade the quality of hole fill and fillet development for both short and relatively long (e.g., reflow) assembly processes.

The intrinsic solderability performance of Pb-free solders is being improved by two means. First, new flux formulations are available that more effectively reduce the surface tension of the solder. Second, alternative surface finishes can be specified for the component I/Os and/or circuit boards that improve wetting and spreading activity exhibited by the Pb-free alloys.

Strictly from an assembly process point of view, the mixing of Pb-free and traditional Sn-Pb solder can be beneficial. The Sn-Pb solder can improve the wetting and spreading performance of the Pb-free solder by two phenomena. First, Pb contamination lowers the molten solder surface tension of the solder. Second, the Pb contamination reduces the melting temperature of the Pb-free alloy. However, concerns are raised by the mixing of Sn-Pb and Pb-free solders and its effect on the long-term reliability of interconnections under thermal-mechanical fatigue environments.

Lastly, the use of Pb-free solders impacts the post-assembly cleaning step (step 4) and the inspection step (step 5). The higher process temperatures can produce more tenacious flux residues that require more aggressive cleaning steps to ensure their removal. Also, the more tenacious residues affect the ability of test probes to contact test site pads on the circuit board. Poor contact can be responsible for detecting false opens on the assembly.

40.2 THROUGH-HOLE TECHNOLOGY

Through-hole technology refers to assemblies having components with leads that are inserted into holes in the circuit board and soldered into place. This technology has been used since the early days of electronics (the 1920s). The leap in automation came in the early 1960s with the advent of wave soldering. The particular drawback of through-hole printed circuit boards, which led to the introduction of surface-mount technology (see the next subsection), was low assembly densities. The relatively large devices, and the circuit board real estate required for holes, limited the density of components, which in turn restricted product functionality and further miniaturization. A photograph of a through-hole circuit board is shown in Fig. 40.1 along with an optical micrograph that shows the through-hole solder joint of the component lead extending through the circuit board.

An advantage of through-hole technology is reduced cost in some applications. That cost benefit may be realized by lower labor costs in some parts of the world that support hand assembly, which is relatively easy with the larger components and products, and lower board densities. Even when fully automated—either by wave soldering, selective soldering, or paste-in-hole/reflow—the capital equipment and manufacturing costs can still be lower than those required for surface-mount assembly.

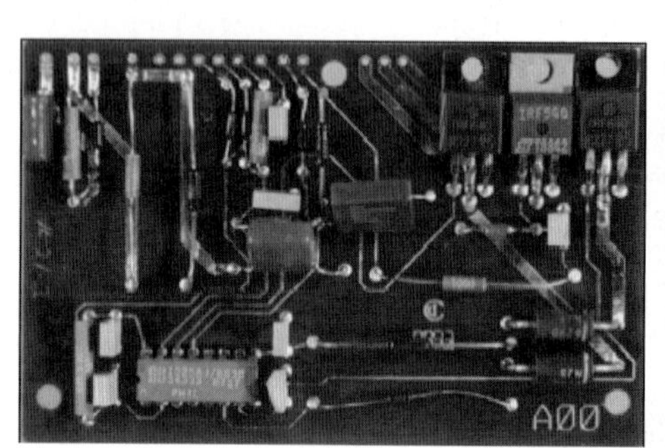

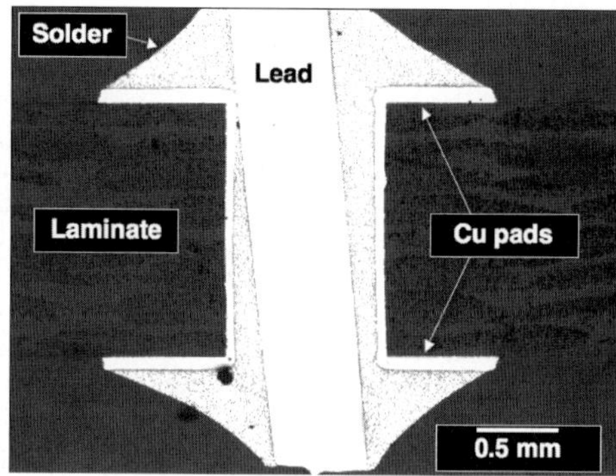

FIGURE 40.1 (a) Photograph of a through-hole printed circuit board. (b) Optical micrograph of the component lead as it is soldered into the circuit board hole. *(Courtesy of Sandia National Laboratories.)*

40.2.1 Impact of Pb-Free Soldering on Through-Hole Technology

The change to Pb-free solders affects through-hole assembly technology. First, there is equipment selection. Manual assembly operations may require purchasing higher-temperature soldering irons. At the other end of the cost spectrum, there are the expenses to replace several hundreds of pounds of Sn-Pb solder with a Pb-free alternative in the pot of a wave-soldering machine. Alternatively, it may be advantageous simply to purchase an entirely new machine to avoid cross-contamination with legacy Sn-Pb operations and/or to have the latest technology that mitigates the higher erosion activity of Pb-free solders on machine components.

The use of Pb-free solders also requires a review of the process parameters. The specific aspects are the higher processing temperature required to use Pb-free alloys and the reduced solderability of these alloys. It has been fortuitous that Pb-free solders, when used in manual (soldering iron) and wave processes, have not required a very significant increase in soldering iron temperature as was first anticipated in the early stages of Pb-free technology development. The added temperature margin was of least concern with manual soldering processes because the localized heating limits thermal degradation to either the component or substrate. Some concern arose in the wave-soldering process because the substrate is exposed to the molten solder bath. On the other hand, paste-in-hole technology that uses a Pb-free reflow process subjects through-hole components and substrates to temperatures higher than those to which they are normally accustomed, thereby possibly degrading them.

The soldering of a through-hole interconnection is also susceptible to the intrinsically poorer solderability of Pb-free alloys. The higher surface tension of these solders results in a slower flow of molten alloy between the pin and barrel, thereby potentially requiring a longer soldering time. Also, the poorer solderability can result in reduced wetting of the pad and limited fillet formation, particularly on the opposite side of the circuit board to which the joint is being soldered. Some of these solderability issues can be mitigated by changing flux and/or the use of alternative surface finishes (e.g., Au solderable finishes). Both options can reduce the solder surface tension as well as enhance the base material metallurgical reaction, thereby improving solderability.

40.2.2 Design Considerations

The design of a through-hole printed circuit board must necessarily consider currently available equipment and assembly practices as well as the potential for advances in future machines and processes. It is important that circuit board designs comply, as much as possible, with recommendations provided by industry standards, such as those of the IPC Association

Connecting Electronics Industries (IPC) and the Joint Electron Device Engineering Council (JEDEC). These recommendations include hole diameters, trace dimensions, and feature spacing. The overriding document is the specific product drawing. Any design deviations from industry standards must be fully considered with regard to how such changes will affect the assembly (process).

The following are important design considerations regarding the assembly process for a through-hole printed circuit board products:

- Tooling requirements (holes, edge clearances, etc.)
- Registration holes (manual alignments or vision systems)
- Component lead-hole sizes
- Circuit board dimensions (length, width, and thickness)
- Size and density of components
- Pb-free solders

Three of these factors are discussed below because of their general role on all assembly processes.

40.2.2.1 Component Lead and Circuit Board Hole Sizes. Designing the correct component lead hole diameters begins by referring to the appropriated industry standard(s) (e.g., IPC, Electronic Industry Association [EIA], etc.). Hole tolerances must take into account runout by the drill, etch-back, barrel-plating thicknesses, and the need for a nominal gap of 0.07 to 0.15 mm between the pin and hole to support the capillary flow of the molten solder. In addition, there are added tolerance considerations due to the variation in component lead diameters as well as the positioning accuracy of the equipment.

40.2.2.2 Circuit Board Dimensions (Thickness). The thickness of the circuit board is typically governed by the product design and the number of layers required for signal routing. Thickness has minimal direct impact on the ability of automated machines to accept a circuit board physically. (The same is true of hand soldering.) However, the board thickness affects the soldering process. As thickness increases, it becomes more difficult to supply a sufficient amount of heat to the joint area that allows the molten solder to fill the through-hole completely prior to solidification. Copper layers contained within the laminate, which are used for signal transmission, ground planes, radio frequency (RF) shielding, and power sources as well as layers for thermal management, act as additional heat sinks that can impede the flow of molten solder into the hole.

40.2.2.3 Pb-Free Solders. Design rules and industry standards that have been established are based primarily on experiences with eutectic Sn-Pb solder. Studies are investigating through-hole processes for Pb-free solders. Part insertion, equipment tooling considerations, even paste placement for paste-in-hole technology will be largely unaffected by the change of solder alloy. As noted previously, in spite of the higher liquidus temperature, hand-soldering iron tips and wave-soldering pots will use similar temperatures. The design engineer will have to address the generally poorer solderability of Pb-free alloys, which arises from the higher surface tension of the solder. Flux selection and alternative surface finishes can mitigate this discrepancy to some degree. Other measures may be necessary. For example, the designer may be required to increase the spacing between components in order to avoid solder bridges that cause short circuits during wave and selective soldering. The high Sn content of Pb-free solders results in a greater degree of erosion of the Cu features on the circuit board. This phenomenon is particularly severe at the knee of the through hole when the molten solder is highly agitated such as in a wave-soldering or selective soldering machine. Therefore, it may be necessary to reduce the time that the joint is exposed to the molten solder or to select an alternative Pb-free alloy (Sn-Ag-Ni or Sn-Ag-Ni-Ge) that is less prone to Cu erosion.

40.2.3 Assembly Process

The through-hole printed circuit board can be a cost-effective technology for many applications. One determining factor is the level of automation used to make the product, which can range from hand assembly to fully automated processes (inline or batch). The specific assembly steps include component insertion (also referred to as "board stuffing"), lead trimming, soldering, and post-assembly cleaning. Labor costs, capital expenditures, board design, and production volumes are contributing factors toward determining the details of these steps.

There are two general formats for assembly processes: the cell or batch process and the line process. Both methods are discussed in the following subsections.

40.2.3.1 Cell (Batch) Process. The cell process routes the circuit boards in batches between the different steps. The cells or workstations are not always in immediate proximity to each another and can be entirely manual, semiautomated, or fully automated in terms of the actual process step. For example, component placement may be fully automated, but require several machines for inserting the different component types. Circuit boards are typically loaded and unloaded by hand between machines. Table 40.1 lists the advantages and disadvantages of the cell process. The cell or batch process is best suited for a facility that assembles a high mix of low production volume products (e.g., prototype development or high-reliability circuit boards) where flexibility is necessary on the factory floor.

TABLE 40.1 Advantages and Disadvantages of the Cell Process

Advantages	Disadvantages
• Single machine outage does not immediately stop the entire process line.	• Higher product flow time due to the transfer of parts between cells is unattractive for high volume applications.
• Adds greater flexibility to the process by creating alternative workflow routes.	• Part transfer between cells increases the likelihood of handling damage
• Attractive for high-mix, low volume applications that require frequent equipment change-outs and re-tooling.	• Difficult to predict product line throughput in a multi-cell assembly process.

40.2.3.2 Line Process. The second approach is the line process, where the different insertion machines, as well as the soldering process in some cases, are linked together with automatic board-handling equipment. Table 40.2 lists the advantages and disadvantages of the line process line. The line process is best suited for a factory floor where high production volumes (e.g., consumer electronics) and a low mix of product types are typical of the manufacturing operations. Less versatility is required by assembly processes, which justifies the capital equipment expense.

TABLE 40.2 Advantages and Disadvantages of the Line Process Line

Advantages	Disadvantages
• Improved manageability of material inventory, product flow, and operator resources is attractive for high volume, low mix applications.	• Failure or maintenance outage of a singe machine failure can potentially halt an entire assembly line.
• Shorter process times by eliminating the transfer of parts between machines.	• Reduced equipment flexibility is unattractive for high mix assembly.
• Reduce the likelihood of handling damage to the assembled product.	• Capital equipment costs and fixed floor space requirements are critical considerations.

The configurations of most through-hole components fall into one of three geometries: axial leaded, radial leaded, and dual inline pin (DIP) packages. These traditional configurations are used for resistors, capacitors, transistors, crystals, and, in the case of active devices, the DIP package. There are also odd-form packages for devices such as transformers, switches, and relays. New package configurations such as the pin grid array (PGA) are being developed to accommodate the increased functionality and further miniaturization of active devices. Besides the actual component body size, shape, and lead configuration, the other factor that can impact the through-hole assembly process is the lead finish. First of all, the finish can potentially add significantly to the diameter of the lead, which must be taken into account when considering the tolerance budget for the hole in the board design. Secondly, for hot solder dipped leads, the potential accumulation of solder at the end of the lead can interfere with part insertion.

The Pb-free alloys can impact the hand-soldering assembly process. First, the higher melting temperature of these alloys requires a slightly longer soldering time. In the case of hand soldering, tip temperatures designated for Sn-Pb process can be used for the Pb-free soldering of traditional through-hole designs. However, hotter tips and/or irons with a higher power rating may be required for "borderline" designs such as those having bigger component leads or thicker circuit boards. Secondly, the Pb-free solders have a higher surface tension that tends to slow wetting and spreading on surfaces as well as capillary flow into holes. For example, the Pb-free solder may not fully coat pad surfaces on the opposite side of the circuit board. Third, the high Sn content of the Pb-free solder increases the rate of erosion of the soldering iron tip, wave-soldering machine parts, and the Cu features on the circuit board.

40.2.4 Hand-Soldering Process

The sequence of steps used to hand solder a through-hole circuit board varies somewhat between different applications. First, there is part insertion. If the insertion operation is fully manual, the parts will be "kitted" into specific groups: The first group is inserted and then soldered; the second group is inserted and soldered; and so on. The order of the groups is determined so as to maximize throughput as well as to take into account human factors in order to minimize part error, lead damage, and operator fatigue or inattention. In semiautomated processes, the operator may receive the circuit board for soldering that was partially or fully populated by machine.

Next, there is the soldering step. The location of the joint where soldering is performed depends on the architecture of the circuit board. For single-sided boards without a plated through-hole, soldering must be performed on the component side. On the other hand, in the case of double-sided and multilayer circuit boards with plated-through holes, soldering is typically performed on the bottom side in order to avoid potential heat damage to components by the soldering iron, particularly on densely populated circuit boards.

The hand-soldering process proceeds as follows:

1. The operator applies flux to the joint.
2. The soldering iron tip is contacted to one side of the component lead (see Fig. 40.2). The tip should not contact the circuit board pad, if possible. It may be necessary to contact the pad of thicker circuit boards.

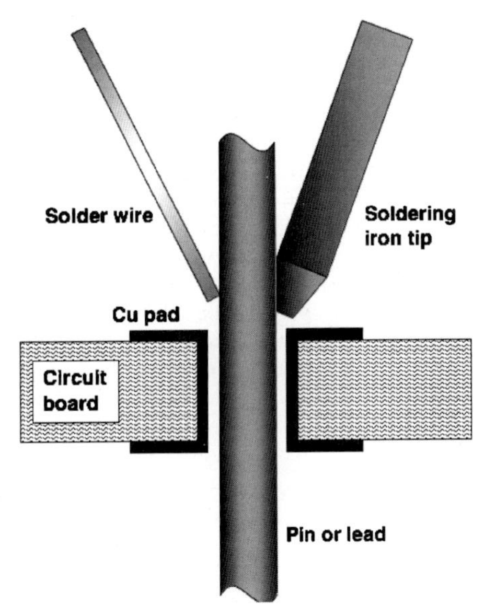

FIGURE 40.2 Diagram showing placement of the soldering iron tip for heating the lead and placement of the solder wire to complete the joint.

3. The solder wire is brought into contact at the side of the lead opposite to the soldering iron tip. The solder wire should not contact the tip in order to melt it. Once melting, the solder is allowed to wet and spread over the surfaces as well as to flow through the hole. In the case of flux-cored wire, the flux application step is omitted. A properly designed process—that is, one having a soldering iron with adequate power as well as tip temperature and geometry—should complete the joint in approximately 3 to 7 sec.
4. Once the soldering process is completed, the circuit board is cleaned of flux residues when required by the flux type or the product's long-term reliability requirements.

Hand soldering may be achieved in an assembly (or "slide") line process for larger volumes. In this case, each one of several operators solders only a few of the total components of the circuit board. In the case of single-person work cells, the operator may solder all of the components to the circuit board. Alternatively, the work cell may be used to add odd-form components as the final assembly step on a nearly complete circuit board.

40.2.5 Automated Soldering and Insertion Technology

In the case of automated assembly processes, the general sequence is part insertion followed by the soldering step. (Slightly different steps are used for the paste-in-hole process; that is, the paste may be applied before or after part insertion.)

40.2.5.1 Components. Part insertion is typically performed in the following order:

1. DIP package
2. Axial leaded components
3. Radial leaded components
4. Odd-form components

The following subsections provide a brief description of each of the component formats and the manner in which the insertion machine addresses those formats.

Choosing the appropriate insertion machine depends on production volume requirements and the mix of products (part density, board sizes, etc.) that must be assembled in the factory. These factors determine the machine's physical dimensions, placement speed, and part-handling (staging) features.

40.2.5.1.1 Dual Inline Pin (DIP) Package. The first auto-insertion process is placement of the DIP packages. The typical DIP package (see Fig. 40.3) is manufactured in one of two widths: 7.6 mm and 15.2 mm. Package lengths vary with the number of pins. For example, a 6-pin DIP package can have the same width and length of 7.6 mm. On the other hand, a 42-pin DIP package has a width of 15.2 mm and a length of 54.6 mm. The DIP packages are typically received in 61 cm length plastic tubes. The insertion machine carries these tubes vertically in magazines. The part shuttle selects the appropriate package and carries it to the center of the machine for insertion into the circuit board. Tooling on the shuttle head holds the package by the leads during the insertion process. The tooling is simply scaled for different DIP package dimensions.

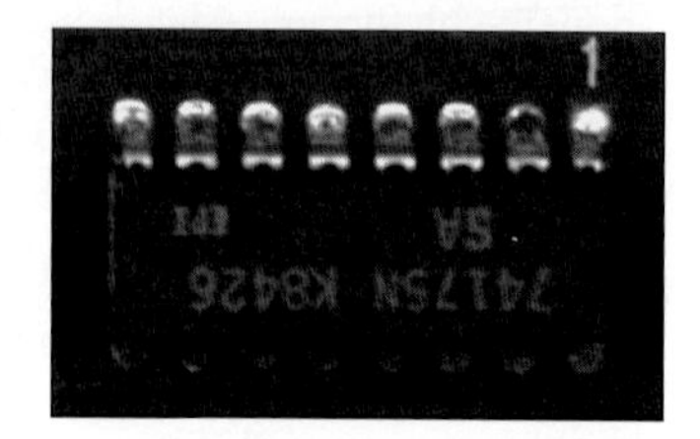

FIGURE 40.3 Dual inline package (DIP). *(Courtesy of Sandia National Laboratories.)*

Dual inline packages are either molded plastic (PDIP) or ceramic (CerDIP, or just CDIP). The surface finishes on PDIP packages are typically electroplated layers of Ni (solderable layer) and Sn-Pb (protective finish) over a Cu lead frame. The Sn-Pb finish is being replaced with 100 percent Sn layers to meet environmental requirements. The CDIP package has low-expansion Fe-based alloy leads that are either electroplated with Cu finish for solderability or the alternative lead finish of Ni (the solderable layer) followed by Au (the protective layer). Although the DIP package is rarely received from the manufacturer with a hot-solder-dipped coatings, these coatings may be applied, after-market, to prevent Sn whiskers with 100 percent Sn finishes (PDIP) or Au embrittlement with Au finishes (CDIP). Aside from potential hole interference with a hot-solder-dip coating, there is also the need to unpackage and repackage the components, which can potentially damage the leads as well as add to the overall cost of the assembly process.

40.2.5.1.2 Axial Leaded Components. Following the DIP packages is insertion of the axial leaded components, which are taped together at the leads by the manufacturer to prevent damage. The taping of the leads conforms to EIA specification number 296-E. Examples of axial leaded devices are shown in Fig. 40.4. During the insertion process, the machine cuts the axial components from their tape to ready them for sequencing into the shuttle head. The leads on axial devices are typically copper or a Cu-clad, Fe-based alloy material. Electroplated Sn-Pb or 100 percent Sn coatings are used to preserve lead solderability. Hot-solder-dipped coatings (Sn-Pb or Pb-free) may be increasingly encountered as a replacement for 100 percent Sn coatings, especially for high-reliability applications. The coating thicknesses must be considered when determining the tolerance stack-up between the hole and pin. Also, hot-solder-dipped coatings often lack uniformity, increasing in thickness near the end of the lead where molten solder accumulated prior to solidification.

Choosing the appropriate insertion machine depends on production volume requirements and the mix of products (part density, board sizes, etc.) that must be assembled in the factory. These factors determine the machine's physical dimensions, placement speed, and part-handling (staging) features.

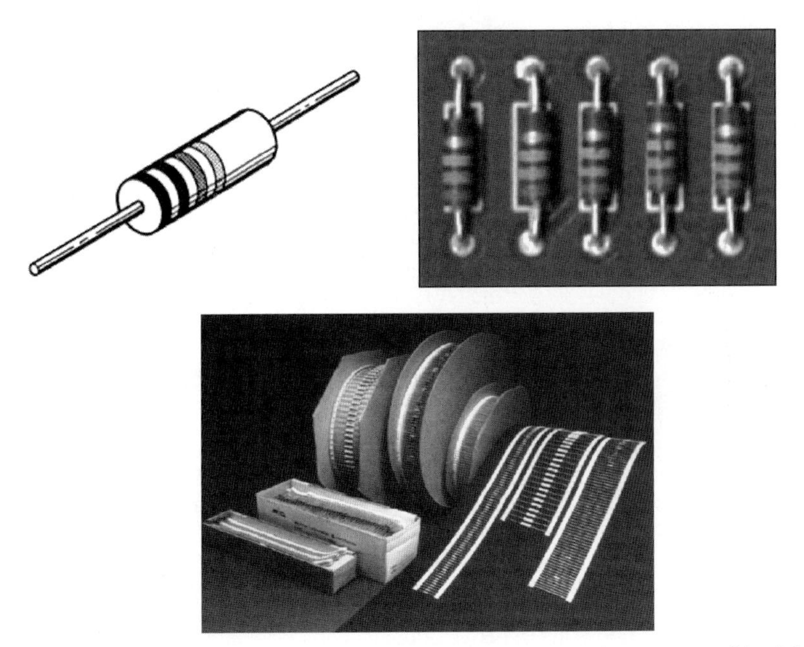

FIGURE 40.4 Examples of axial leaded devices: (a) axial leaded components; (b) axial leaded components in reels and boxes. *(Courtesy of Sandia National Laboratories.)*

Axial leaded components are inserted before radial leaded components because, typically, axial components are smaller overall than radial components. Because the associated tooling size is also smaller, inserting axial components prior to the radial components facilitates the stuffing of higher-density circuit boards.

40.2.5.1.3 Radial Leaded Components. After DIP and axial leaded components, the radial components are inserted into the circuit board. Radial components can vary greatly in size, shape, height, and weight (see Fig. 40.5). The lead finishes—types and thicknesses—are identical to those of axial leaded components. The possible exceptions are hermetically sealed components, such as active Si chip devices in TO-5 cans, optoelectronic packages, and relays. Such devices typically have low-expansion, Fe-based alloy leads that allow for them to be sealed into the header with a glass-to-metal joint. Although some device manufacturers use a Sn-Pb or 100 percent Sn finish over an electroplated Ni (solderable) layer, other suppliers prefer an Au (protective) finish over the Ni (solderable) layer to provide adequate solderability. In such cases, the Au layer is removed by hot solder dipping in order to prevent Au embrittlement of the circuit board interconnection.

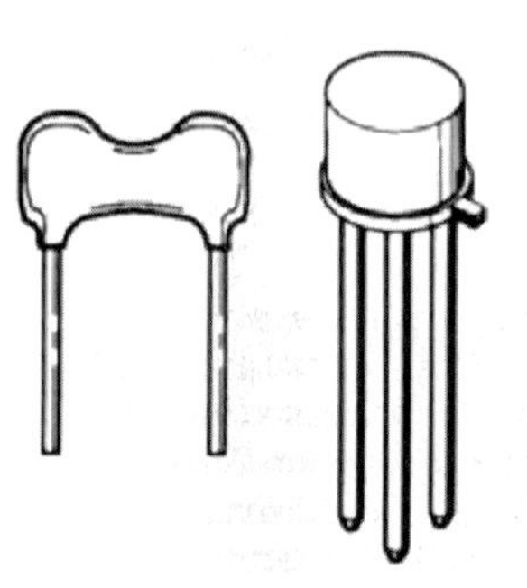

FIGURE 40.5 Radial leaded components: (a) individually; (b) in taped reels. *(Courtesy of Sandia National Laboratories and Uni ersal Instruments.)*

Radial leaded components are taped together at the leads. Radial insertion machines remove the components from the tape and insert them in sequence on to the printed circuit board using a shuttle system similar to that used for other components. The insertion machine tooling for radial components is designed to avoid previously inserted DIP and axial components.

40.2.5.1.4 Odd-Form Components. The odd-form components are the last devices to be assembled on the circuit board. Odd-form components, by definition, are those packages that are not readily addressed by automated assembly because (a) they are used at insufficient volumes on the product to justify machine "space" or (b) they are of a geometry (shape or size) that lacks the customer demand to justify the machine manufacturer to provide off-the-shelf tooling for them.

High-power applications often require a variety of odd-form components, including simply oversized DIP, axial leaded packages, or radial leaded packages. Also, odd-form components include transformers, switches, relays, and connectors. Several of these packages are shown in Fig. 40.6. Unusual package sizes and geometries are often accompanied by unusual lead configurations and materials. It is not uncommon to have Cu- and Ni-based alloy leads, or even leads made of refractory materials such as Mo and Ni. The leads may be round, square, or in the form of ribbons. Irrespective of the lead material or nontypical geometry (which must have the hole geometry addressed at the circuit board design step), it is *always* required that the leads have adequate solderability. Leads are typically plated with solderable layers (Cu or Ni) and

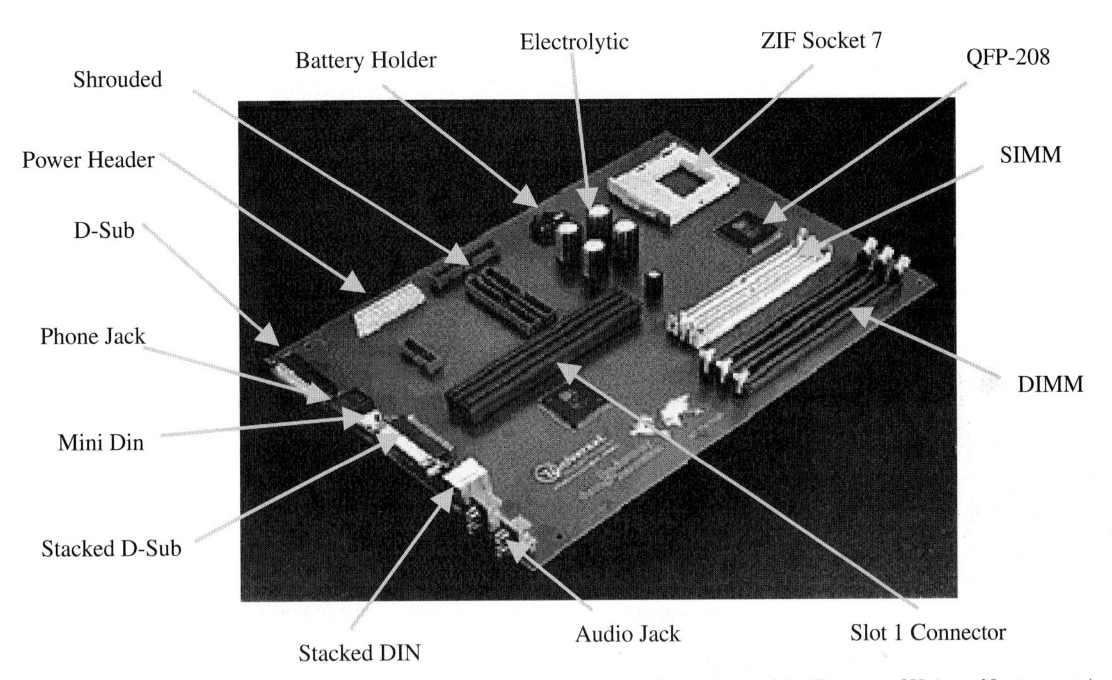

FIGURE 40.6 Examples of odd-form, through-hole components and a circuit board assembly. *(Courtesy of Uni ersal Instruments.)*

protective finishes (Au, Sn, Sn-Pb, etc.) very much like those used on traditional components to support the soldering process.

40.2.5.2 Insertion Equipment. Because the demand for through-hole circuit boards has remained strong, insertion equipment capabilities have steadily improved. Brushless servomotors and state-of-the-art motion controllers and sensors have replaced most pneumatic drive assemblies and bulky (and slow) mechanical switches and relays, respectively. Some equipment options include the capability to test parts electrically prior to insertion. At insertion rates of up to 40,000 components per hour (CPH), defect rates are a couple of hundred parts per million or lower for the common axial and radial leaded packages. Interchangeable tooling has allowed the automated insertion of many odd-form components, as well. The change to Pb-free solders does not have a direct impact on insertion equipment technology.

40.2.6 Automated Soldering and Wave Soldering

The most commonly used process for the soldering of through-hole and mixed (through-hole and surface-mount) circuit boards is *wave soldering.* The wave-soldering process is shown schematically in Fig. 40.7. The populated (or "stuffed") circuit board is secured to a conveyor belt. The conveyor belt carries the board through the fluxer, then the preheat stage, and lastly on to the molten solder wave.

Interestingly, one of the most critical steps in the wave-soldering process is the application of the flux. Manual fluxing has given way to more precise, automated equipment in order to improve yields. The advantage is better control of the *quantity* of flux on the circuit board as it enters the wave because the flux controls both the entry and exit geometries of the molten solder wave, which in turn are instrumental toward minimizing skips, bridges, or icicle defects. Flux can be applied by spraying techniques or by passing the circuit board through the foam or "suds" of the flux. The latter technique is provided by the "foam fluxer."

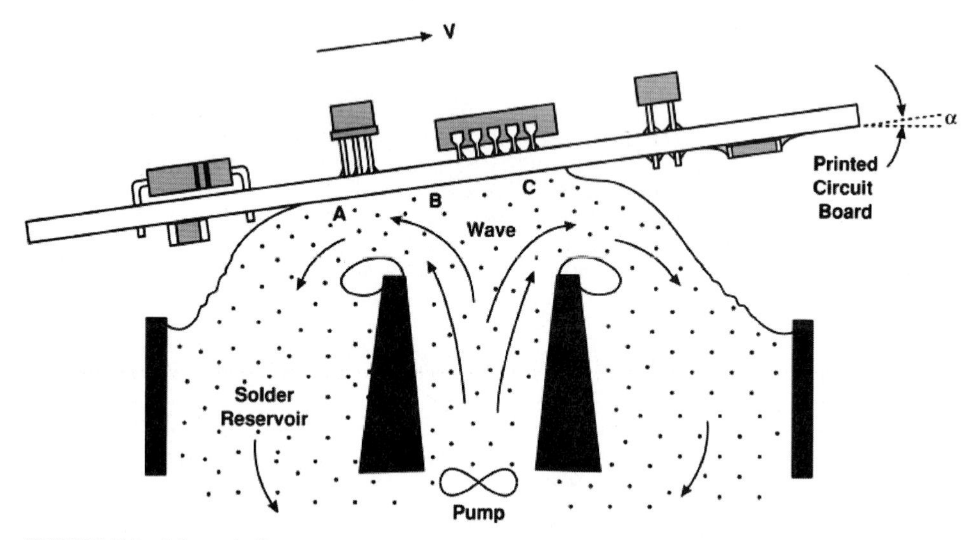

FIGURE 40.7 Schematic diagram showing the wave-soldering process.

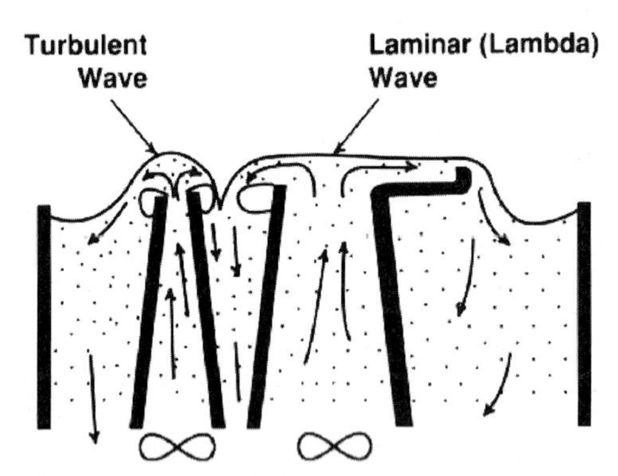

FIGURE 14.8 Schematic diagram of the dual solder wave configuration.

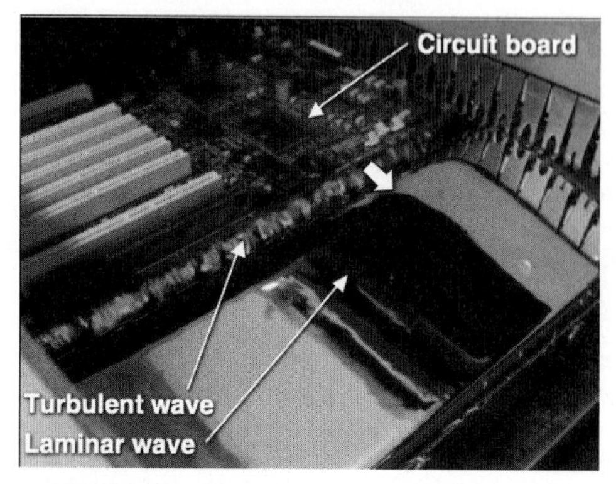

FIGURE 40.9 Photograph of the dual-wave configuration. *(Courtesy of Vitronics-Soltec.)*

After the fluxing step, the circuit board passes through the preheat stage, which is comprised of a set of radiant heaters. As the circuit board is heated, this step evaporates volatiles from the flux coating (which if not adequately removed can lead to voids and blow holes) and activates the chemical reactions between the flux and oxides on the component lead or termination and the circuit board conductor.

The solder wave is created by pumping molten alloy upward through a nozzle, where it exits to a certain height and then falls back into the bath. The solder bath temperature is 260°C for eutectic Sn-Pb solder. The printed circuit board is carried by the conveyor so that is passes over the surface of the wave. The bottom side of the circuit board contacts the wave, allowing the molten solder to wet the exposed pads as well as to flow upward through the holes by capillary action. The speed of the conveyor and the take-off angle—that is, the angle at which the circuit board approaches the wave—are critical parameters for minimizing wave-soldering defects.

Another critical parameter of wave soldering is the wave geometry. The schematic diagram in Fig. 40.8 shows the commonly used dual wave. The first wave is turbulent so as to counteract the surface tension of the molten solder, thereby literally forcing the molten solder into confined geometries in order to initiate the wetting process. The second wave is smooth or laminar. The laminar wave is located at the exit point because its geometry reduces the propensity for bridges and icicle defects to form as the molten solder retracts itself from the circuit board. The photograph in Fig. 40.9 shows a circuit board entering the turbulent wave of a dual-wave soldering system.

A variation of wave soldering is the so-called selective soldering process. Instead of using a long wave to accommodate the dimensions of an entire circuit board, a small fountain is created of molten solder. The reduced geometry allows for the soldering of individual components or several components on only selected areas of the circuit board. A manual operation is used to apply flux and preheat the circuit board prior to the actual soldering process.

The change to Pb-free solders has affected the wave-soldering process. Fortunately, the 260°C bath temperature used with eutectic Sn-Pb solder has proven to be adequate for most applications using the 99.3Sn-0.7Cu (wt. percent) and Sn-Ag-Cu alloys. Some users have preferred to raise the pot temperature to 260 to 270°C. The elements Ni and Ge have been added to the Pb-free solders in order to provide the shiny fillets that inspectors are accustomed to observing with Sn-Pb solders. The higher surface tension of the Sn-based, Pb-free solders increases the tendency for opens, bridges, and icicles to form as the circuit board exits the wave. A change of flux chemistry as well as adjustments to the take-off angle and conveyor speed have been used to minimize those defects. Lastly, a particular problem with Pb-free wave soldering is erosion of the machine parts by the high-Sn alloy compositions. Special coatings are put on the impeller, nozzles, baffles, and bath walls to mitigate the erosion process.

40.2.7 Paste-in-Hole (Pin-in-Paste) Soldering

Very rapid production volumes are realized with inline surface-mount assembly processes. In general, the surface-mount assembly process consists of three basic steps: (a) screen printing of solder paste, (b) pick and place of components, and (c) reflow of the solder paste. To obtain similar throughputs for through-hole circuit boards, paste-in-hole soldering is being investigated as an alternative to wave soldering. Solder is provided to the joint in the form of a paste, using screen or stencil printing, or dispensed from a needle. The component is then inserted into the hole. In some applications, the parts are inserted into the circuit board and subsequently the solder is dispensed at the hole using a needle and pump. Then the assembly is sent through the reflow furnace. The primary hurdle with paste-in-hole soldering is supplying a quantity of paste that is sufficient not only to develop a fillet on that side of the joint, but also to fill the hole and, optimistically, create a fillet on the bottom side of two-sided and multilayer circuit boards.

A second concern for paste-in-hole soldering is the temperature sensitivity of the through-hole components. Recall that in wave, selective, or even hand soldering, the temperature rise is confined largely to the leads and circuit board. At worst, the component package materials experience only the preheat temperature environment. However, in paste-in-hole soldering, components must withstand the soldering temperatures within the reflow furnace. Therefore, a conversion to this solder process requires an in-depth review of the allowable temperature range for all of the components.

40.2.8 Cleaning

The final step in the soldering process is the removal of flux residues from the completed circuit board. No-clean fluxes, as the name implies, produce residues that need not be removed from the assembly after soldering. The corrosive activators are locked up in a hard, polymerized residue. However, no-clean flux residues can potentially interfere with in-circuit testing (probes) and inspection, degrade cosmetic appearance, inhibit rework activities, and lessen the adhesion of conformal coatings. In the event that any one of these factors becomes critical, it is best to switch to a cleanable flux than to develop a cleaning process for a no-clean flux.

Low-solids fluxes (which are not to be confused with no-clean fluxes) are simply formulated with reduced solids content so that very little residue is left on the circuit board. Low-solids flux can be successfully used in wave-soldering and paste-in-hole processes, thereby eliminating the need for a cleaning step. However, they are somewhat more difficult to use in hand-soldering processes because the higher temperature of the soldering iron tip can cause premature loss of the activating agent, resulting in poor solderability.

The cleaning step can be performed in either a batch or inline machine. The optimum equipment is determined by production volumes, floor space, and capital expenditure costs. Smaller, batch cleaning (or "dishwasher") machines are very cost-effective for overall batch-type processes (e.g., hand soldering) as well as for circuit boards assembled in low production volumes. Inline cleaning equipment is placed at the end of the soldering assembly line to accommodate high production volumes. Whether batch or inline equipment is used, an equally important consideration is selecting the type of cleaning solution. Solvent-based, aqueous, and semi-aqueous cleaning materials are commercially available that meet environmental regulations and provide adequate efficacy for the removal of flux residues.

The impact of Pb-free soldering on the cleaning step depends on the particular assembly process. By virtue that the same pot temperatures are used in Pb-free wave soldering as are specified for Sn-Pb solder, there is no added thermal degradation to the flux residues that would cause them to be more difficult to remove from the circuit board surfaces. Such degradation may be observed, albeit to a limited extent, with hand assembly due to slightly longer soldering times. Thermal effects are potentially most significant with paste-in-hole soldering because of the increased temperature settings of the reflow furnace required to melt Pb-free alloys.

40.3 SURFACE-MOUNT TECHNOLOGY

Surface-mount technology refers to assemblies that have components soldered to pads on the surfaces of the circuit board. Components may populate only one side (single-sided) or both sides (double-sided) of the circuit board. Surface-mount technology dates back to the 1960s, when it was developed for hybrid microcircuit (HMC) assemblies for which it was difficult to put holes into the ceramic substrates. The advent of surface-mount technology for laminate substrates, though, is relatively recent (c. 1980). The advantages of surface-mount technology include smaller components and greater board densities. The large holes have been replaced by small vias for signal conduction between sides and internal layers. Finer traces and reduced component heights also contribute to increased circuit board miniaturization and functionality. Surface-mount circuit boards are shown in Fig. 40.10.

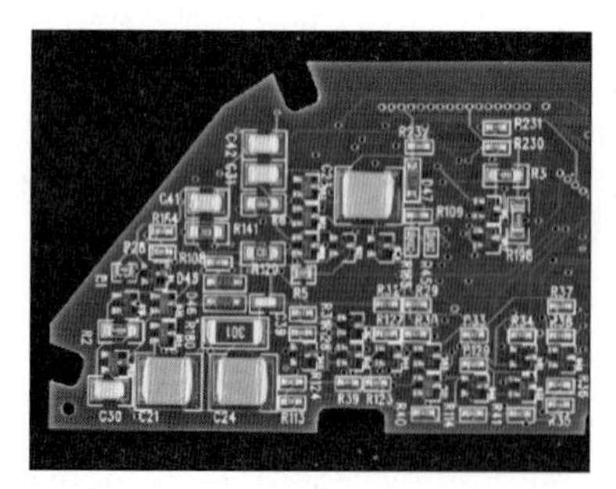

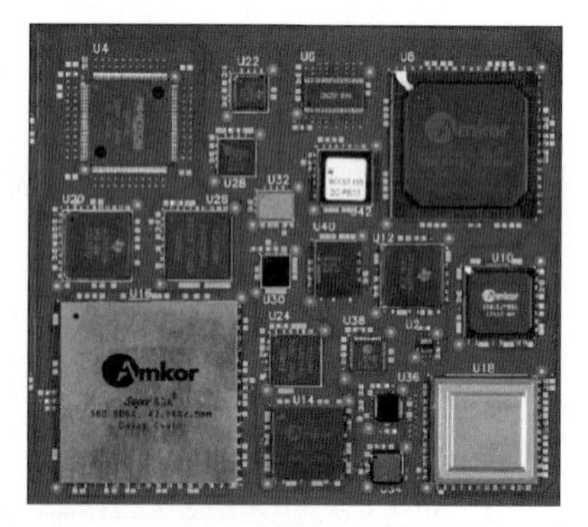

FIGURE 40.10 Surface-mount technology circuit boards showing the range of passive and active device sizes and geometries. *(Courtesy of Sandia National Laboratories and American Competitiveness Institute.)*

The trend in surface-mount technology is to use smaller passive devices such as capacitors, resistors, and inductors. Also, there is the use of embedded passive devices, that is, resistors and capacitors that are located within the circuit board laminate. Embedded passive devices free up additional surface area for larger, active components.

Active devices are experiencing two opposing trends. On the one hand, memory components (RAM, SDRAM, etc.) are becoming smaller as more transistors are being packed on to the silicon chip. On the other hand, microprocessors and application-specific integrated circuits (ASICs) are becoming larger because of increased functionality on larger chips. Both trends have seen a shift away from peripherally leaded packages to area-array packages. Area-array packages include BGAs and the smaller counterparts, those being CSPs and DCA/FC technology. Examples of peripherally leaded and area-array packages are shown in Fig. 40.11. The advantages of the area-array technology include a reduction of the component footprint by eliminating the leads that extend from the package. Also, fewer assembly defects results from damage to the fragile leads during packaging, transportation, and part placement on the circuit board.

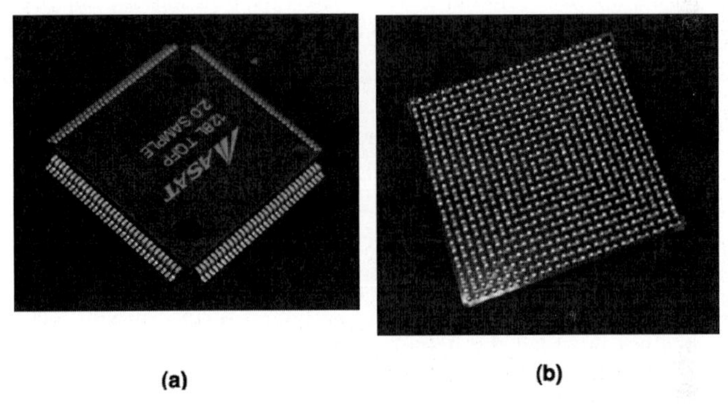

(a) **(b)**

FIGURE 40.11 Photographs of (a) peripherally leaded and (b) area-array surface-mount packages. *(Courtesy of Sandia National Laboratories.)*

At the inception of area-array technology, the size and pitch of the I/Os were initially higher when compared to what was then 0.4 mm and 0.5 mm fine-pitch, peripherally leaded packages. However, as I/O counts have increased with the functionality of area-array packages, solder ball size and pitch have decreased significantly, particularly when taking into account DCA technology.

Increased device functionality and further miniaturization, which have resulted in higher board densities, are placing stricter requirements on circuit board technology. Increased internal conductor layer counts result in the need for thicker laminates. Whereas once the 1.58 mm board thickness was the norm, now 2.29 mm is commonplace, with many products requiring thicknesses greater than 2.54 mm. The internal conductor layers are being designed to carry not only the electrical signal, but also contribute to thermal management by removing excess heat from active devices. Similarly, the vias within the laminate are used to transmit signals between layers and the surfaces as well as remove excess heat from large active components (e.g., microprocessors, ASICs, etc.). Vias are reaching aspect ratios (length/diameter) of 8:1 to 10:1, which is nearing the fabrication as well as reliability limits for these structures.

A particular advantage of surface-mount technology is the lower manufacturing cost resulting from automated assembly processes. Solder paste, which is the combination of solder metal powder, flux, and thixotropic agents, is applied in highly controlled amounts (thickness and area) using screen or stencil printing as well as dispensing techniques. Pick-and-place machines handle even the smallest components, locating them precisely on the solder paste deposits (or "bricks"). The tacky nature of the flux component in the paste keeps the components in place. The assembled (or "stuffed") printed circuit board then passes through the convection/radiation reflow furnace or vapor phase (or condensation) reflow oven to melt the solder. The machines responsible for the steps—paste printing, component placement, and reflow—are connected by conveyor belts to create the inline process. In fact, the final step—cleaning the circuit board—may also be a part of the assembly process sequence.

Of course, various levels of automation may be considered, depending on production volumes and capital expenditure costs. However, with continued miniaturization of surface-mount products as well as the strict requirements for the repeatable placement of specific solder paste volumes and accurate placement of components, it will be a foregone necessity to assemble surface-mount technology with a fully automated assembly process.

Mixed technology refers to the combination of surface-mount and through-hole components on the same printed circuit board. The lack of components in a surface-mount package is nearly always the reason for using a through-hole counterpart. In general, the surface-mount devices are soldered first to the top side of the circuit board using a convection/radiation reflow furnace or vapor-phase soldering machine process. (Surface-mount is performed first because the through-hole components would interfere with solder paste printing and pick-and-place steps.) The through-hole components are then soldered to the board. The actual soldering process is performed at the bottom side. For larger numbers of through-hole components, the board is wave-soldered. If there are surface-mount components located on the bottom side, they can likewise be wave-soldered; however, those components must first be secured in place by an adhesive. Hand soldering may be preferred if there are relatively few through-hole devices or if bottomside surface-mount devices cannot be exposed to the solder wave.

40.3.1 Pb-Free Soldering

Pb-free soldering technology has affected surface-mount technology. Fortunately, the fallout has not included the need for new automation equipment. Lead-free solder paste behaves essentially the same as its Sn-Pb counterpart in the printing process. Slight modifications have been suggested for the screen and stencil apertures to accommodate the reduce solderability of Pb-free alloys once the paste enters the reflow furnace. Pick-and-place equipment can remain unchanged by a Pb-free soldering process. Reflow furnaces must be able to accommodate the higher process temperatures required by Pb-free solders. Vapor phase reflow machines are available with fluids that provide the higher process temperatures needed for Pb-free assembly. (As noted earlier, significant equipment costs are associated with altering a wave-soldering machine for the Pb-free assembly, which would pertain to mixed-technology printed circuit boards. Those costs include the replacement of the solder as well as the replacement of some components that are susceptible to erosion by the high-Sn alloys.)

The most significant impact of Pb-free technology is in reflow process parameters, specifically the time-temperature profile. Longer times at higher reflow temperatures increase the likelihood of thermal damage to components and substrate materials. Also, in seeking to reach higher reflow temperatures, one must also consider the sensitivity of the solder paste to the temperature ramp-up rate. An incompatibility between the paste and time-temperature profile can lead to increased levels of defects such as voids, solder balls, short-circuits due to paste slump, etc. Similarly, excessively fast cooldown rates from the higher reflow temperatures can result in board warpage and cracked passive devices. The greater need to control the Pb-free time-temperature profile favors greater control capabilities for the reflow furnace.

The Pb-free solders have intrinsically poorer solderability performance than their Sn-Pb counterparts. The slower rate of wetting and spreading is not as critical of a factor in reflow surface-mount assembly as it is the faster wave- and hand-soldering processes. However, the higher surface tension of Pb-free alloys may limit the extent of solder spreading on a pad, particularly with bare Cu, leaving the corners uncoated by soldre. The choice of flux and use of alternative surface finishes can mitigate this artifact in some cases. A high surface tension also increases the likelihood of *tombstoning* by passive devices particularly for the smaller chip resistors and capacitors. The degree of tombstoning can vary from the device having one termination just lifting from the pad to the component standing up completely on one end.

40.3.2 Design Considerations

The design of a surface-mount circuit board requires the consideration of three factors pertaining to the assembly process. First, there is a wide variety of solder-joint configurations for the different package types. Second, higher board densities can lead to upward of several tens of thousands of solder joints being fabricated at one time. Third, component miniaturization, as well as the unexposed solder joints of area-array packages, further diminishes the viability of performing repair or rework activities to correct solder-joint defects. A number of design rules are printed in documents (e.g., IPC SM-782 and D-330) that govern the size and location of bond pads. Surface-mount circuit boards are at least double-sided and, in most cases, multilayer, so that board design also must consider the interior of the laminate where vias are placed to conduct signals between the layers and surfaces.

Assembly requirements also place constraints on the circuit board layout. A very high part density requires a large number of apertures in the solder paste stencil, which can cause the stencil to become locally too flimsy to control the solder paste deposit. A surface-mount circuit board with a very wide range of component sizes and package configurations may require multi-thickness stencils to properly control the paste deposit. Solder paste printing quality is a determining factor in solder-joint defects observed after the reflow process.

The layout design of the surface-mount circuit board is a factor in the optimizing of the time-temperature profile used in the reflow process. Two important factors are overall thermal mass of the circuit board, determined primarily by the board thickness and number of internal layers as well as the variety package sizes and materials that can lead to temperature gradients as large as 20°C between the smallest passive devices and larger ball grid array packages over the board surface. Also, double-sided circuit boards require two reflow processes. Therefore, larger components should be placed on the side of the circuit board that is soldered last. If placed on the first side to be soldered, large packages may fall off of the board when the circuit board is upside down for the second side's soldering step, unless those devices are secured in place with an adhesive or "staking" compound.

The design of a printed circuit board for Pb-free assembly must consider the same points as are considered for Sn-Pb technology, but with a greater emphasis placed on the effects of higher process temperatures – e.g., temperature gradients between components across the circuit board. Foremost, there are the enhanced temperature gradients that may be created across a large circuit board having a high mixture of different component sizes and shapes. In the case of a mixed-technology product, there may be limitations placed on the maximum substrate thickness due to the less-solderable Pb-free alloys. It may also be necessary to limit via aspect ratios (typically remaining less than 10:1) to prevent immediate damage (defects) by the reflow process or a loss of long-term reliability.

40.3.3 Assembly Processes

The basic assembly process sequence for surface-mount technology is (1) solder paste printing, (2) pick and place of components, and (3) reflow of the solder. The assembly process is complicated when through-hole components are combined with surface-mount devices on the same circuit board; this is the previously mentioned mixed technology. In general, hand assembly of surface-mount boards is not always practical given the small size of passive devices and hidden solder joints of area-array packages. The use of hand soldering for repair and rework activities is limited by these same factors.

This section lists the sequence of process steps for the different circuit board categories. The term *top side* has typically referred to the side with the greatest density of components and/or where the larger, active device packages are clustered together. As the functionality of printed circuit board products increases, these physical distinctions between the two sides have become less obvious. Therefore, the circuit board sides are simply given a designation: A or B.

1. Single-sided (top side), surface-mount only:
   a. Print paste
   b. Place components
   c. Reflow solder paste
2. Double-sided, surface-mount only:
   a. Bottom side, print paste
   b. Bottom side, dispense adhesive for larger components, if necessary
   c. Bottom side, place components
   d. Bottom side, reflow solder paste and cure adhesive

   Flip the board over to the top side:

   e. Top side, print paste
   f. Top side, place components
   g. Top side, reflow solder paste
3. Double-sided, mixed technology (with bottomside wave soldering):
   a. Bottom side, dispense adhesive for surface-mount components
   b. Bottom side, place surface-mount components
   c. Bottom side, cure adhesive

   Flip the board over to the top side:

   d. Top side, print paste
   f. Top side, place components
   g. Top side, reflow solder paste
   h. Top side, insert through-hole components

   Keep the top side facing up:

   i. Wave-solder the through-hole and the bottom-side surface-mount components

In case (3), for mixed-technology products, there is some flexibility regarding the sequence of staking (gluing) of bottomside components versus the topside soldering step, based largely upon the capabilities of the pick-and-place equipment.

In the case of area-array packaging technology having solder balls (BGA, CSP, and DCA), there is the option of *not* adding solder paste to the pads. Instead, only the flux is deposited on the pads, which serves to tack the package in place during transport of the board to the reflow oven. The solder balls provide the solder to complete the joint; however, the penalty is a reduced gap thickness that may affect self-alignment during soldering (as well as solder-joint reliability later on). Omitting the solder paste is most often considered for the DCA/FC assembly process because the solder balls, and the corresponding pads, are so small that it is not possible to print the acceptable quantity and dimensions of the paste consistently.

There is also the use of step-solder processes for double-sided, surface-mount-only circuit boards, that is, case (2). A high-temperature solder, often the 96.5 Sn-3.5 Ag (Sn-Ag) alloy ($T_{eutectic}$ = 221°C), is used to attach the bottomside components. These solder joints do not remelt when the topside joints are made using the Sn-Pb solder ($T_{eutectic}$ = 183°C). Somewhat tighter process control is required, since the 221°C Sn-Ag eutectic temperature is very near the optimum, peak process temperature for the Sn-Pb process (210–220°C). This approach provides the option of placing larger passive and active components on both sides of the circuit board without the need for an adhesive dispensing and curing step. Unfortunately, the use of the higher melting temperature, Pb-free Sn-Ag-Cu solders all but eliminates this step-soldering approach without identifying a higher melting temperature solder for the first step.

The following sections examine each of the steps in a surface-mount assembly process, beginning with an examination of component types. The different dispensing technologies are reviewed, including those for solder paste and adhesives. Component placement machines are examined, followed by a description of the use of the various heat sources such as convection/radiation reflow ovens, wave-soldering equipment, condensation reflow apparatus, hand soldering, and conduction soldering. Lastly, cleaning processes are described.

40.3.3.1 Components. A wide variety of components are available for surface-mount printed circuit boards. The form, size, and materials used for these components change constantly as new offerings are made by suppliers to meet miniaturization, functionality, and reliability requirements. The most common surface-mount components are the passive or "chip" devices—the resistors, capacitors, and inductors. Chip capacitors and resistors are often referred to by a four-digit number, such as 1825, 1210, or 0804. The first two digits refer to the component body length, which is the distance between the terminations, and are in hundredths-of-inches (0.*xx*). The second two digits refer to the body width in hundreds-of-inches (0.*xx*). (For passive devices, a similar numeric designation is given that is based upon the metric system [mm]. The actual values are very close to the English values and can be a source of confusion, especially when partnering with overseas companies.) Thus, an 1825 capacitor has a length of 0.18 in. and a width of 0.25 in. A stereo photograph appears in Fig. 40.12 that shows examples of passive components. Chip resistors tend to be very robust and thus relatively immune to damage during the assembly process. Multilayer chip capacitors are temperature-sensitive and therefore more prone to cracking during assembly, particularly under fast temperature ramp rates.

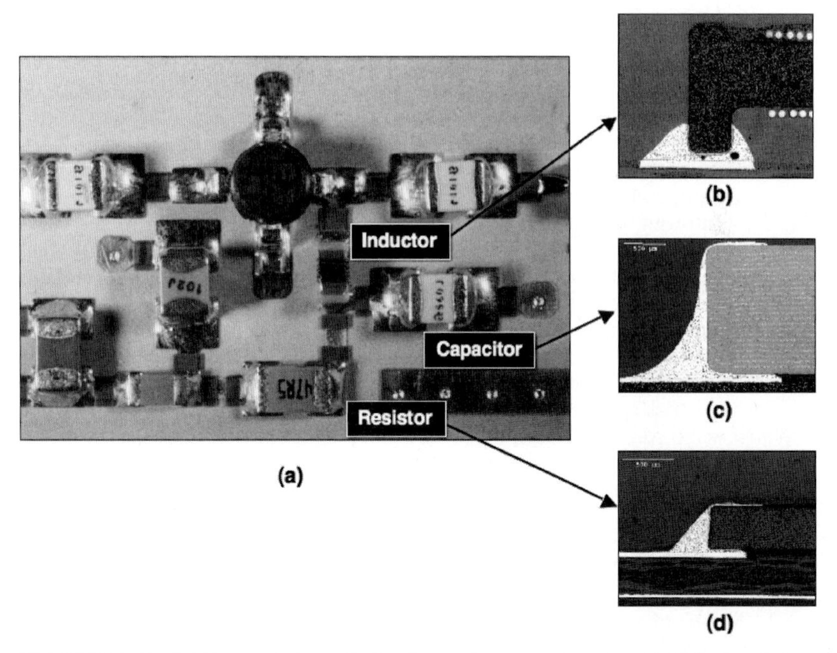

FIGURE 40.12 (a) Stereo photograph showing surface-mount passive (chip) devices: inductor, capacitor, and resistor; (b–d) cross sections of their respective solder joints. *(Courtesy of Sandia National Laboratories.)*

The chip resistor is constructed of a piece of alumina ceramic on top of which is deposited a thin film that serves as the resistive element. Attached to the resistive element at the top, on the ends, and partially at the bottom are the termination structures to which the solder joint is made. The termination structures are comprised of a fired, Ag-based thick film; a Ni or Cu barrier layer; and lastly an electroplated finish of Sn, Sn-Pb, or Au.

Chip capacitors are constructed from a special oxide-based ceramic that is built up of alternating layers of ceramics and thin film layers that provide the device capacitance value. This capacitor is the multilayer thin film (MLTF) type. The second capacitor type has electrodes on the top and bottom surfaces of a homogenous "block" of ceramic. The ceramics used to make chip capacitors tend to be more fragile than the alumina body of a chip resistor. The built-up layer structure of MLTF capacitors causes them to be particularly susceptible to both mechanical and thermal shock. Chip capacitors use similar metal layers to build up the terminations that are soldered to the circuit board pad as were described previously for resistors.

Chip inductors are of two types as well. Coil inductors are comprised of a fine Cu wire wound around the alumina body. The body size and number of windings determine the part inductance. The second type is the thin-film inductor. The winding is fabricated from a patterned thin film, which is deposited upon a piece of alumina. (Although the latter are considerably easier to fabricate, they have a limited range of inductance values.)

As much as 40 percent of passive components on a surface-mount product are chip capacitors. The miniaturization of these devices is critical for reducing the size and weight of electronic products. For handheld electronics (such as cell-phones, PDAs, and pagers), common capacitor sizes are from 0603 down to 0402 and 0201.

Some diodes and all active devices come in a wide variety of peripheral lead and area-array packages. The diodes and transistors are typically used in small-outline (SO) packages; they are the small-outline diode (SOD) and small-outline transistor (SOT) packages. The package body is a plastic overmolding compound. There are two leads for the SOD and three for the SOT. The leads are very robust, have a gull-wing shape, and are formed from Cu or an Fe-based alloy. Larger active devices require more I/Os. These devices use the small-outline integrated circuit (SOIC) package with gull-wing leads protruding from the two sides of the long dimension. The gull-wing leads are very robust with a pitch of 1.27 mm (50 mil) or 0.635 mm (20 mil). Pitch is the distance between the centerlines of two neighboring leads.

A further increase in the number of I/Os was realized by placing leads around all four sides of the package; the leads can have the gull-wing geometry or have the J-lead geometry. The J-configuration reduces the bond pad area by bending the lead inward, under the package. Like the gull-wing, the J-lead geometries are robust at 1.27 mm (50 mil) and 0.635 mm (20 mil) pitches.

Pitches below 0.635 mm, beginning with 0.5 mm and 0.4 mm pitches, are called fine-pitch (package) technology. The smaller leads are considerably more fragile, causing them to be more susceptible to damage during handling and pick-and-place activities. Also, the fine-pitch packages have a more stringent coplanarity requirement for the leads. Coplanarity specifies the degree to which the lead bottoms must be at the same level around the periphery of the package. A non-coplanar lead—one that is lifted high—is more likely to generate an open because of the reduced amount of solder used for these smaller I/Os. A lead that is too low will be damaged during package placement and also displace the solder paste deposit, causing a defective joint or short circuit to a neighboring lead after the assembly step.

A second type of peripheral I/O package is the leadless ceramic chip carrier (LCCC). This package is made of ceramic material; the I/Os are castellations on all four sides of the package. Solderability of the castellations is obtained by a Ni finish over which is deposited an Au layer. The Ni and Au layers extend down the castellation and under the frame to form a pad. This package can be used only on so-called matched or low-expansion substrates, that is, circuit board substrates with a thermal expansion coefficient that matches that of the ceramic package. Otherwise, the solder joints are quickly degraded by thermal mechanical fatigue (TMF) when exposed to even modest cyclic temperature environments.

Area-array packages include the BGA, CSP, the land grid array (LGA), DCA/FC, and the ceramic column grid array (CCGA). The common characteristic of these packages is that the solder joint is made to an array of solderable pads on the bottom side of the package rather than to peripheral leads or castellations. The difference between a BGA package and a CSP package is that the latter is stipulated to have molding compound dimensions that are less than 1.2 times the corresponding die dimension. There are no specified limits on a BGA package size.

The typical pitch size is 1.27 mm and 1.0 mm for BGA and CSP packages. Here, the pitch is the distance between the center points of any two balls or lands. Therefore, alignment requirements are not very stringent for these area-array packages. Also, there is a sufficient quantity of solder to allow self-alignment between the package and circuit board pads by the surface tension of the molten solder. However, as ball counts reach several thousands, reduced ball sizes and pitches are required, which in turn necessitate closer tolerances on part placement. This is also the case of DCA, where the solder ball size and pitches tend to be as small as 0.10 mm and 0.25 mm, respectively.

The CCGA is a variant of the BGA in which the solder balls have been replaced with solder columns. The columns allow the ceramic package to be assembled to an organic laminate circuit board of considerably larger thermal expansion coefficient by absorbing the higher strains created by the larger thermal expansion mismatch between the two materials. The columns are created from one of the high-melting-temperature, Pb-based alloys (e.g., 95Pb-5Sn or 90Pb-10Sn) that will not melt during a eutectic Sn-Pb reflow process. The columns may also have a Cu spiral wound around them to enhance durability as those columns are susceptible to damage during handling and part placement.

The accelerated development of surface mount components has generated packages and I/O configurations that have not yet been standardize, resulting in odd-form devices. Examples of odd-form components include surface mount switches and connectors as well as a variety of inductors, (Fig. 40.13) LEDs, and transformers. Typically, so-called surface-mount connectors may actually be mixed technology with through holes providing the mechanical strength needed for cable insertion and removal activities and surface-mount leads establishing the electrical connection. (The through-hole interconnections can be made by paste-in-hole techniques or be soldered manually.)

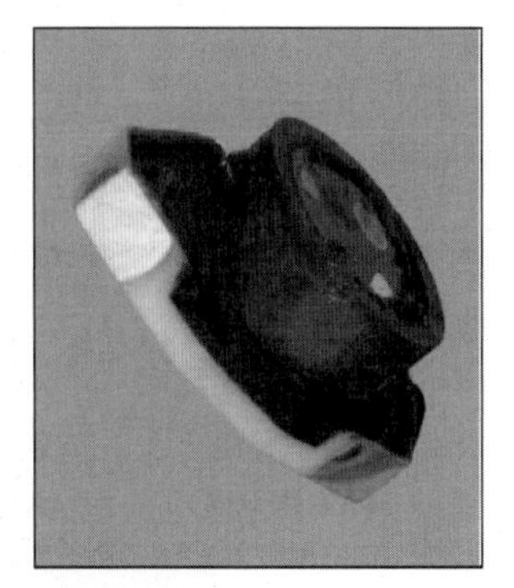

FIGURE 40.13 Odd-form components as exemplified by surface-mount inductors. *(Courtesy of Sandia National Laboratories.)*

A number of assembly-related issues must be addressed with odd-form components. First, it is necessary that correct pad dimensions be designed on the circuit board. Also, the stencil must have the correct aperture size to print an adequate quantity of solder paste. The pick-and-place machine may require custom tooling in order to handle these components. Lastly, odd-form parts are typically larger and heavier. Therefore, it is possible that they will not readily self-align while the solder is molten during the reflow process.

The conversion to Pb-free solders has significantly affected surface-mount components. In the case of leadless passive devices and peripherally leaded packages, the traditional electroplated Sn-Pb finish has been replaced with 100 percent Sn coatings. The Sn coatings have raised concerns regarding the development of Sn whiskers that can potentially short nearby conductors in service. The Sn-Pb solder balls of BGA, CSP, and DCA technologies, having a melting temperature of 183°C, are being replaced with Sn-Ag-Cu alloys with a melting temperature of 217°C. In the case of DCA/FC and CCGA applications, the high-Pb alloys used for the balls and columns, respectively, will not melt during a Sn-Ag-Cu solder process used to create the second-level interconnections.

40.3.3.2 Dispensing. In surface-mount technology, three classes of materials must be dispensed on to the circuit board: adhesives, fluxes, and solder pastes. Each of these three categories, which utilize very similar equipment options for the actual dispensing action, is discussed in the following subsections.

40.3.3.2.1 Adhesives. Adhesives are used to secure surface-mount devices to the circuit board. An adhesive may be required when, for example, exposing surface-mount components to a wave-soldering process used to assemble a mixed technology circuit board. Also, larger components on double-sided circuit boards may be adhesively bonded in place to prevent them from falling off when the board is turned upside down for the second reflow step. Under these circumstances, the package weight exceeds the surface tension force of the molten solder that keeps smaller components on the board. The adhesive must be able to withstand the temperature conditions of the wave or reflow soldering process as well as the chemical activity of the flux. Adhesives may also be required to anchor larger surface-mount devices to the circuit board. This additional strength is needed for service specifications that include mechanical shock and vibration environments.

(Adhesives are typically not used for through-hole components because clinching the leads provides a sufficient anchor to keep the component on to the board prior to and during the soldering process. After soldering, through-hole solder joints are sufficiently robust to withstand heavy shock and vibration environments. Nevertheless, in very severe environments, adhesives may be used to anchor through-hole components to the substrate.)

It is important to control the quantity of dispensed adhesive since there must be sufficient material to perform the attachment function. On the other hand, too much adhesive can cause run-out onto the solder pad or component I/O, resulting in poor solderability. Some adhesives are prone to run-out or "bleeding" caused by their separation into individual liquid components that together comprise the adhesive material. When bleeding by a component liquid occurs beyond of the adhesive deposit, that bleeding liquid can contaminate nearby solderable surfaces. In the case of very dense circuit boards, excessive migration of the whole adhesive migration, or the bleeding of one of its components, can contaminate the pads of other components, affecting their placement and solderability.

Although the function of the adhesive is to anchor components in place for the soldering process, the adhesive deposit remains as a part of the assembly after soldering, and as such must not interfere with next-assembly steps or negatively impact the long-term reliability of the circuit board. For example, some epoxies readily absorb moisture or other organic compounds. Those absorbed materials may outgas during subsequent temperature excursions that could contaminate critical surfaces (e.g., sensors) when the product is in service. Therefore, it is important to select only adhesive materials that are qualified for a particular application.

Adhesives materials that are used in electronics assembly processes are typically based upon epoxies or silicones. Adhesives can also be described by the following four functional/materials categories: thermosetting adhesives, thermoplastic adhesives, elastomer adhesives, and toughened alloy adhesives. Each group differs in its compositions, the type of curing cycle, and its pre- and post-cure material properties. The curing cycle, which typically requires an elevated temperature/time profile, must not degrade the components present on the board or the laminate itself. It is understood that the added step of a curing cycle slows the overall assembly process.

Thermosetting adhesives cure by heat or a catalytic reaction that cross-links the polymer chains. Once cured, these materials remain very strong and will not readily soften at elevated temperatures. Epoxies are one group of thermosetting adhesives, which are used widely in electronics assembly because they do not weaken at the high temperatures of wave and reflow soldering environments. Also, epoxies are resistant to attack by solvents and aqueous-based cleaning solutions. Epoxies can be one-part, in which the curing agent and resin are already mixed together, or two-part, in which the two chemistries are combined prior to application. Although the one-part epoxies are convenient from the assembly standpoint, their storage and handling must be rigidly controlled to prevent curing prior to use. Both epoxies are cured by exposure to elevated temperatures, which can range from less than 100°C to as high as 125 to 150°C, for time periods of 1 to 4 hours, depending on the specific product recommendations. There is very little outgassing associated with the curing of thermosetting adhesives. Higher residual stresses can be generated by these materials due to their rigidity at elevated temperatures when there is a significant thermal expansion mismatch among the package, the epoxy, and the circuit board substrate. The "permanency" of thermosetting adhesives complicates repair or rework functions. The removal of these adhesives usually requires mechanical scraping and abrasion that can damage the component and circuit board.

Thermoplastic adhesives soften when exposed to elevated temperatures. These adhesives are weaker than thermosetting epoxies. However, when tolerances are less restrictive, these materials may be preferred for assembly applications, particularly when excessive residual stresses are of concern during the soldering process temperature cycle. Thermoplastic materials are less resistant to solvents and aqueous-based materials. These adhesives tend to absorb these liquids more readily, resulting in dimensional changes (swelling) and more outgassing than is observed with thermosetting adhesives.

Thermoplastic materials have curing temperatures that are lower and time durations that are shorter than thermosetting materials. Some compositions will cure at room temperature, making them suitable for temperature-sensitive components or when thermal expansion mismatch, residual stresses are of concern. Another advantage of thermoplastic adhesives is that because they readily soften at elevated temperatures, they can be easily removed to allow for the rework of components.

Elastomeric adhesives are a subset of thermoplastic adhesives. These materials tend to be very tough, yet have a higher degree of elasticity. The silicone (rubber) adhesives are examples of this category. The lack of rigidity limits the application of these adhesives in the soldering assembly process. Curing temperatures are relatively low, as some compositions cure at room temperature. However, the curing cycle of some silicone adhesives include considerable outgassing and, moreover, the outgassing of vapors that can be corrosive to metal surfaces (e.g., acetic acid).

Toughened alloy adhesives are blends (or alloys) of elastomeric materials and epoxy resins that together form this special class of thermosetting-like adhesives. These adhesives are engineered to provide both high structural strength and sufficient toughness (ductility) to resist damage due to either thermal or mechanical shock. Examples of this type of material include epoxy-nylon adhesives.

All of these materials have been engineered to have properties that can accommodate one or more of the various dispensing techniques used in printed circuit board assembly (which are discussed later in this chapter). However, these properties do not remain optimal indefinitely. There are two stages of degradation. The first stage is the shelf-life of the material, which is the time frame during which the adhesive keeps its properties while the container remains unopened. Manufacturers date-code specify shelf-life based on changes to the mechanical properties (strength, ductility, etc.) and physical properties (glass transition, density, liquid viscosity, etc.) that occur to the adhesive. Density and viscosity directly impact dispensability.

The second stage of degradation occurs when the adhesive is removed from the container, mixed if required, and is loaded into the dispensing equipment. Exposure to air, even under room temperature conditions, causes the adhesive to begin curing on the assembly floor. The curing can alter density and viscosity and thus the dispensing properties of the adhesive. Indications of a significantly cured adhesive include a clogged dispensing machine, run-out or bleeding of the deposit, and "stringers" of material created as the dispensing tool moves from one location to another spot.

40.3.3.2.2 Fluxes. The dispensing of fluxes has more limited applications in surface-mount technology (aside from mixed technology that includes a wave-soldering step). Flux dispensing is performed when solder is not simultaneously delivered to the joint—for example, the wave soldering of a mixed technology circuit board or the attachment of an area-array package for which the solder balls provide a sufficient quantity of alloy for the joint. In fact, flux dispensing is most widely used for DCA/FC applications. The DCA joints often require only a very small quantity of solder that would be difficult to control or to place accurately with typical printing equipment and stencils. Thus, the quantity of solder present in the ball forms the joint, and only flux must be added to the assembly.

Soldering fluxes are generally available only as a liquid for electronics applications. The low viscosity prevents the flux from being precisely located on the circuit board using stencil printing. Therefore, flux is usually dispensed by spraying techniques. However, once flux is sprayed on the surface, the circuit board must be soldered very quickly as volatiles and other components begin to evaporate from the flux coating.

In the case of DCA applications, the flux is applied directly to the component. The process is shown schematically in Fig. 40.14. The die is placed into a bath containing a very thin layer of flux. The thickness of the layer controls the quantity of flux taken by the die when it is withdrawn from the bath. The flux must be sufficiently tacky to anchor the die to the substrate for transport to the reflow step. Because of the large exposed surface area of the flux bath, the flux volatiles readily evaporate so that the bath must be replenished on a regular basis.

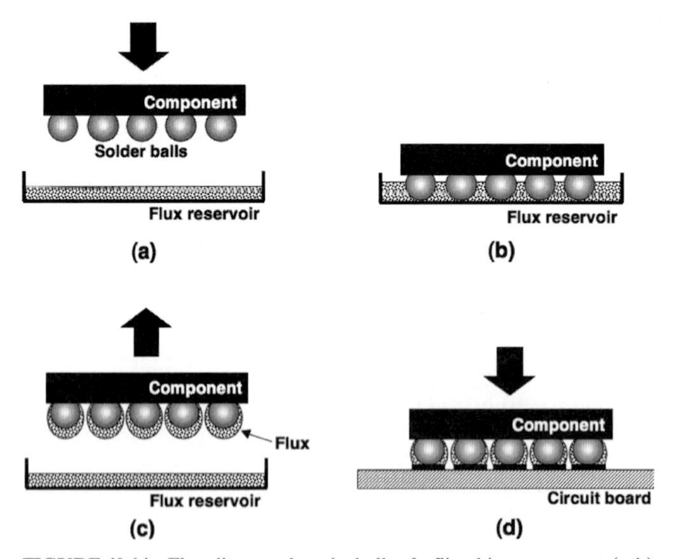

FIGURE 40.14 Flux dispensed on the balls of a flip-chip component: (a, b) The solder balls are immersed into the flux reservoir; (c) the component is removed from the reservoir with the solder balls coated with flux; (d) the component is placed on the circuit board.

40.3.3.2.3 Solder Paste. The dispensing of solder paste is the most widely used means of providing flux and solder metal to the joint for the reflow process. The primary components of the solder paste are the solder metal and the flux. The solder metal is typically 80–90 wt. percent of the paste. Aside from the In- and Zn-containing solders, there is very little sensitivity of the paste properties to the solder metal composition for Sn-based alloys. Important properties of the paste with respect to dispensing are the solder powder particle size and weight percent or "metal loading" of the paste.

The flux can be any one of the various compositions—rosin-based, no-clean, low solids, and water-soluble. The flux also provides the tack that keeps the components attached to the board prior to the reflow step. Other ingredients in the paste are the thixotropic agents. It is the thixotropic agents, together with the metal content and flux, that determine the viscosity of the paste.

Viscosity controls the dispensing properties of the paste, irrespective of the particular technique (needle, screen printing, etc.). The viscosity and hence the dispensability of the paste change with time, whether slowly in an unopened jar on the shelf or more quickly when the paste is exposed to air while awaiting the dispensing step. Care should be taken to monitor strictly the manufacturers' recommendations for shelf-life and lifetime on the assembly floor. Poor solder paste dispensing accounts for the majority of solder-joint defects observed in surface-mount technology.

40.3.3.2.4 Dispensing Methods. There are five primary methods for dispensing adhesives, fluxes, and solder pastes:

1. Pin transfer
2. Screen or stencil printing
3. Time-pressure pump dispensing
4. Archimedes screw pump dispensing
5. Positive displacement pump dispensing

The latter three methods generally produce a single deposit per dispensing step. The first two methods enable the application of the material at multiple sites in a single step. Of course, the five techniques do not perform equally well for the three materials. This section discusses the advantages and disadvantages of each technique.

Pin transfer is the simplest technique for dispensing adhesives and flux. Although a single pin may be used to deposit material one location at a time, a matrix of pins can do so at multiple sites, the pin transfer process is illustrated in Fig. 40.15. This technique is suitable for adhesives

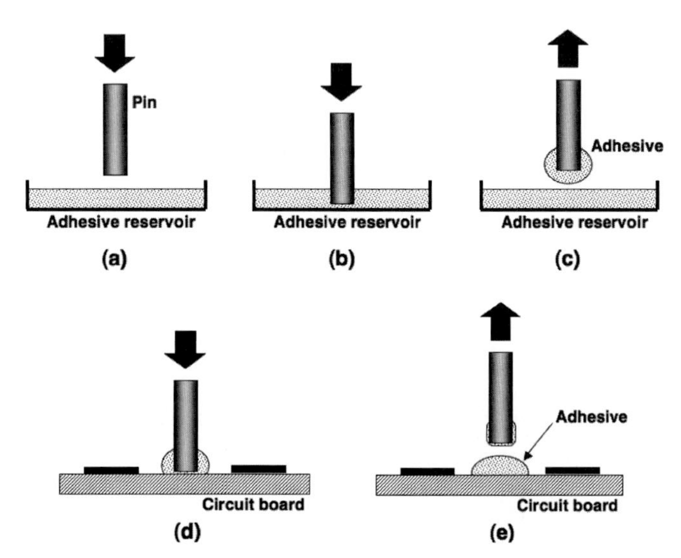

FIGURE 40.15 Pin transfer technique: (a) The pin is brought down to the adhesive reservoir; (b) the pin is immersed in the adhesive; (c) the pin is withdrawn from the reservoir with a repeatable quantity of adhesive; (d) the pin is positioned on the circuit board, allowing the adhesive to contact the surface; (e) the pin is withdrawn from the surface, leaving behind a well-controlled volume of adhesive.

and fluxes; it does not work as well with solder paste. A pin is dipped into a reservoir of the adhesive or flux. The length and diameter of the pin determine the quantity of material that is picked up upon its withdrawal from the reservoir. The pin is then lowered to a position just above the surface of the printed circuit board that allows the flux or adhesive to contact the circuit board. Surface tension causes a portion of the adhesive or flux to be deposited on the printed circuit board. It is very important that the pin does not touch the board because this will cause inconsistent dot sizes and shapes. This system requires the substrate to be relatively flat and free from distortion. The nature of the pin array can also allow for adhesive to be applied to the board even after through-hole parts have been put into place.

A similar principle to pin transfer is used for the application of flux on the solder ball of FC components for DCA (See Fig. 40.14). The die is immersed into a thin film bath of flux. The flux depth allows only the balls to be coated so that, in effect, the solder balls become the pin, taking up the flux. The flux on the solder balls is transferred with the die to the circuit board, where it provides the tack function as well as fluxing action for the solder balls during the reflow step.

An important consideration about the pin transfer technique is that it requires an open bath of the adhesive or flux. Adhesives readily absorb water from the air. Fluxes lose vehicle (water or alcohol) and possibly other constituents through evaporation. By either mechanism, the material properties change, which affects the quantity of fluid retained on the pin and deposited at the site (including the flip-chip process previously described). Adhesives must have sufficient "wet strength," and the fluxes must have enough tack to hold the component in place for the duration of component placement activity as well as subsequent handling of the circuit board on its way to the curing oven or reflow oven.

Screen or stencil printing can be used for adhesives as well as solder paste. The low viscosity of most flux solutions precludes their successful dispensing by this technique. The adhesive or solder paste is deposited through openings in the stencil or screen called *apertures*. The apertures are located over the locations on the circuit board where the adhesive or solder paste is required. The placement of material is performed by a squeegee pushing a quantity of adhesive or paste ahead of it, over the screen or stencil, as illustrated by Fig. 40.16a.

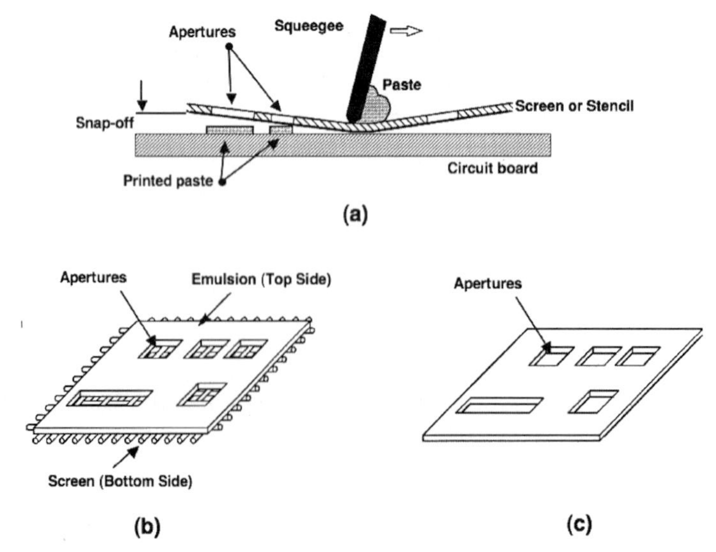

FIGURE 40.16 Screen and stencil printing: (a) the squeegee movement used to print adhesive or paste through a screen or stencil; the construction of a screen (b) and stencil (c) used for printing.

The difference between a screen and a stencil is in their respective structures, as shown in Fig. 40.16b, c. The screen is comprised of two layers: the emulsion layer and the actual screen that supports the emulsion layer. The apertures through which the adhesive or solder paste is deposited are created in the emulsion by photoimaging techniques. The adhesive or solder paste simply flows past the screen cross-hatched wires.

The stencil is simply a sheet of metal or alloy (commonly Mo, Ni, brass, or stainless steel) having the apertures formed in them. The apertures can be created by one or a combination of the following techniques:

- Photoimaging (photoresist definition) combined with etching by wet chemistry
- Laser cutting
- Build-up technology by electroplating processes

The choice of fabrication technique depends on the required sizes and densities of the apertures. The stencil has replaced the screen in most surface-mount printing applications, not only because of its simpler construction, but also because it can accommodate finer, denser circuit board features.

The thickness of the screen or stencil and the size of the individual aperture openings are the parameters that control the *quantity* of adhesive or solder paste deposited on the circuit board. Secondary factors are the aperture wall quality, the material viscosity, the hardness of the squeegee, and the speed of the squeegee. In the case of solder paste, the quantity that is actually deposited is usually *less than* the volume of the aperture, which is the product of the width, length, and stencil thickness. The degree of that discrepancy is called the *transfer factor* or *transfer coefficient*. Values can range from 60 percent for very small apertures to nearly 100 percent for larger aperture openings.

Screen or stencil printing is the most widely used means for depositing solder paste on a surface-mount circuit board. The preferred paste viscosity for screen printing is 250 to 550 kcps (kilo-centipoises) for an 80 mesh screen. In the case of stencil printing, the desired viscosity is 400–800 kcps. The ability to print consistently upward of tens of thousands of paste deposits per circuit board has been critical to the realization of high-volume electronics production. This process is being developed for through-hole circuit boards and is referred to as *paste-in-hole* or *pin-in-paste* technology.

Solder paste printing technology includes stepped stencils. Stepped stencils are made with two different thicknesses and are used when a circuit board has such a wide range of device pitches and joint configurations that a single stencil cannot yield the optimum paste deposits for all the components. The thinner sections are used for the very-fine-pitch packages, whereas the thicker sections deposit paste for the larger-pitch components. These stencils are more expensive to fabricate than the single-thickness products.

The printing of adhesive or solder paste with a screen or stencil has several limitations. It can only be performed in a single pass; if there is a fault during that pass, the board must be removed and cleaned prior to a second attempt. Second, the circuit board surface must be flat and have no obstructions that will interfere with the stencil or screen surface "sealing" against the board surface as the adhesive or paste is pushed into the apertures by the squeegee. As such, it is important that the screen or stencil be thoroughly cleaned of paste residues prior to use in order to minimize printing defects, which can subsequently become solder joint defects if not caught prior to reflow. Third, stencils and screens wear with time, resulting in an increased number of printing defects. The harder the metal or alloy, the longer is the lifetime. For example, brass stencils, which are relatively inexpensive, have a short service life. Stainless steel stencils have a longer lifetime, but are also considerably more expensive.

Referring to the solder paste printing process with a stencil, Pb-free solder pastes behave very similar to the Sn-Pb pastes for leaded and area-array pitches of greater than 0.5 mm. At the smaller pitches that are characterized by smaller apertures, it has been observed that the Pb-free solders have a slightly reduced transfer coefficient. The likely cause is the reduced density of the Pb-free solder particles, which becomes a significant factor when so few particles are passed through the aperture. Therefore, it may be necessary to slightly open up the apertures to ensure a sufficient quantity of Pb-free solder at the joint.

Time-pressure pump dispensing is a method to deposit adhesive or solder paste by applying a pressure pulse for a specific time duration on a reservoir of material (see Fig. 40.17). A precisely controlled quantity of adhesive or solder paste emerges from the selected orifice size, which is deposited on the circuit board. Often, the material comes prepackaged in a syringe that is inserted into the machine.

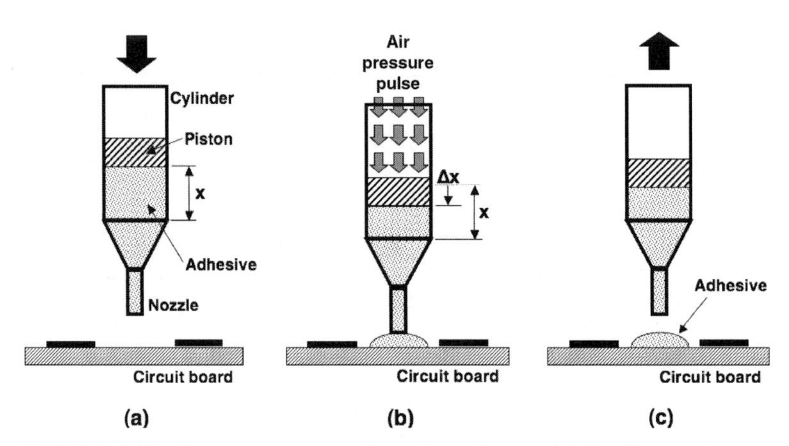

FIGURE 40.17 Time-pressure pump dispensing technique: (a) The dispenser is positioned at the circuit board site; (b) the dispenser is lowered to near the surface at which time, a pulse of air pushes the piston a distance Δx to dispense a set quantity of adhesive; (c) the dispenser is raised from the site, leaving the adhesive deposit.

As with all dispensing techniques, the flow properties of the adhesive or solder paste are important factors that determine the consistency of the deposit between the different sites. Shelf-life requirements should be strictly followed, particularly those pertaining to time allowed for the material to be present in the dispenser due to rapid degradation when open to the factory floor environment. The desired viscosity for pump dispensing through a nozzle is 100 to 400 kcps.

Machines utilizing the time-pressure dispensing technique can dispense different deposit sizes on a single board. One approach is to have multiple nozzles or syringes on a single head, using the same pulse profile. The second approach is to preprogram a different time-pressure pulse that alters the deposit quantity from a single orifice or syringe. This pump-dispensing technique (as well as those described in the following paragraphs) is slower than screen or stencil printing. However, it offers greater flexibility in terms of well-controlled deposit quantities and location.

Archimedes screw pump dispensing utilizes an Archimedes screw to push the adhesive or solder paste out of a nozzle (see Fig. 40.18.) The speed and duration of a turn, as well as the size of the opening, determine the quantity of adhesive or paste deposited on the site. As in the case of the time-pressure technique, different deposit quantities can be realized by multiple spindle (heads) or a computer program that alters the screw speed or turn duration to change the amount of material dispensed from the same orifice. All other considerations with respect to the roles of adhesive or solder paste viscosity and shelf-life also apply to this technique.

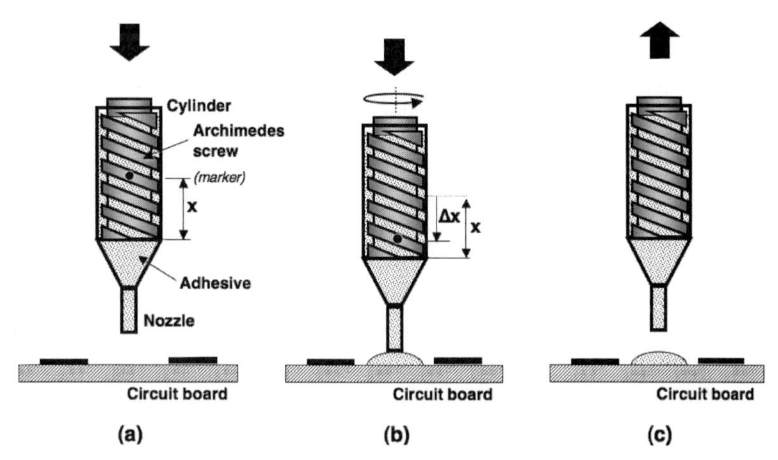

FIGURE 40.18 Archimedes screw pump dispensing technique: (a) The dispenser is lowered to above the circuit board surface; (b) the Archimedes screw is turned a set rotation (Δx), pushing adhesive out of the nozzle; (c) the dispenser is lifted from the circuit board.

Positive displacement pump dispensing uses a piston, rather than an air pressure pulse, to control the deposit quantity. This technique is used primarily to dispense adhesive (see Fig. 40.19). First, the nozzle is placed in a bath of adhesive, in which it retracts to draw a

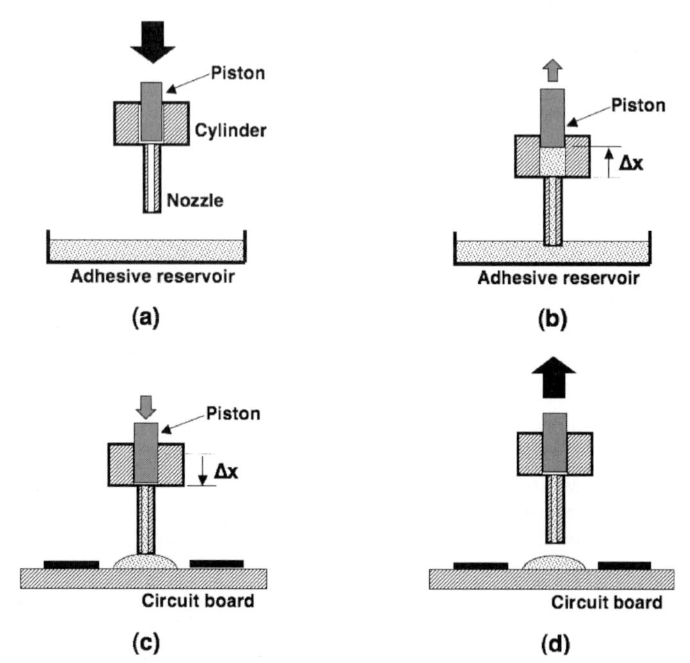

FIGURE 40.19 Positive displacement pump dispensing technique: (a) The dispenser is brought to the adhesive reservoir; (b) the nozzle is immersed into the adhesive and the piston withdrawn a distance Δx, resulting in a set quantity of material being pulled into the cylinder; (c) the dispenser is brought to just above the circuit board surface so that a reverse displacement of the piston pushes out a quantity of adhesive; (d) the dispenser is lifted from the circuit board.

quantity of adhesive that fills the orifice and small cylinder at the top of the orifice. This suction step works best with low-viscosity fluids. It is difficult to pull up relatively high-viscosity materials such as solder paste and some adhesives. Next, the piston moves downward into the cylinder, forcing an exact amount of adhesive out the nozzle and onto the printed circuit board. This technique consistently produces the same volume of material. Besides the dimensions of the cylinder as well as the speed that the piston displaces, the viscosity of the adhesive also affects the dispensed quantity. Shelf-life specifications should be strictly enforced.

The primary objective in each of the five dispensing techniques is to deposit *consistently* a *specific quantity* of adhesive or solder paste at each designated site. Too small of a quantity of adhesive, especially dot height, can fail to attach the part to the board. Too much adhesive causes it to run on to the solder pads, degrading solderability. In the case of solder paste, an insufficient quantity of paste will cause an incomplete joint or, in the worst case, an open circuit. An excess of solder paste results in a fillet that is difficult to inspect for solderability or risks formation of short circuits between neighboring interconnections.

40.3.3.3 Component Placement. The purpose of the component placement machine—also called the "pick-and-place" machine—is to select the proper component, orient it correctly, and then place it on the circuit board, all with degrees of accuracy and precision that minimize defects on the finished product. In addition, the component must be placed on the printed solder paste, the dispensed adhesive, or a combination of the two deposits with a controlled pressure or release distance that does not excessively spread out either material or damages the component package. Moreover, the placement machine must execute these tasks as quickly as possible in order to maximize the production volume. Lastly, the equipment must be sufficiently versatile to address continually changing electronic packages, specifically dimensions and I/O configurations.

The change to Pb-free solders does not have an explicit impact on component placement machine technology. Indirectly, however, the need for alternative surface finishes on both the components and circuit board fiducials, which have different reflectance characteristics, can affect the performance of the vision systems used to locate accurately both the circuit board and the tooling that delivers the component to the board.

Several machine types are available. The turret-style chip shooter and the gantry-style or flexible fine-pitch (FFP) machines have both been used extensively in the assembly of consumer electronics, telecommunications, mainframe and server computers, as well as for lower volume, high-reliability electronic products. However, the demands for even higher production volumes as well as the flexibility to change product lines rapidly have caused manufacturers to consider alternative machine architectures that include high-speed stepper motors and optical sensors as well as highly parallel methodologies that place multiple components at the same time.

40.3.3.3.1 Turret Systems (Passive Devices). The basic turret or "chip shooter" has been used to place passive components (i.e., capacitors, resistors, etc.) since the early development of surface-mount technology. A photograph of the turret head and schematic diagram of its operation are shown in Fig. 40.20. Multiple heads are positioned around a stationary, horizontally rotating turret. A moving feeder carriage positions tape feeders that deliver a component to each head. After the part is located in the head, the turret rotates it to a vision processing station where a charge coupled device (CCD) camera acquires an image of the part. This image is processed so that the part will be precisely located over the circuit board location. As the turret continues rotating, a moving table positions the printed circuit board so that the target location is in position under the turret head to receive the component. The part is lowered to the circuit board and released. The head rotates to acquire another component, and the cycle repeats itself.

Listed in Table 40.3 are general performance characteristics for turret chip placement technology. This technology is constantly addressing smaller passive devices (0101 and 01005) as

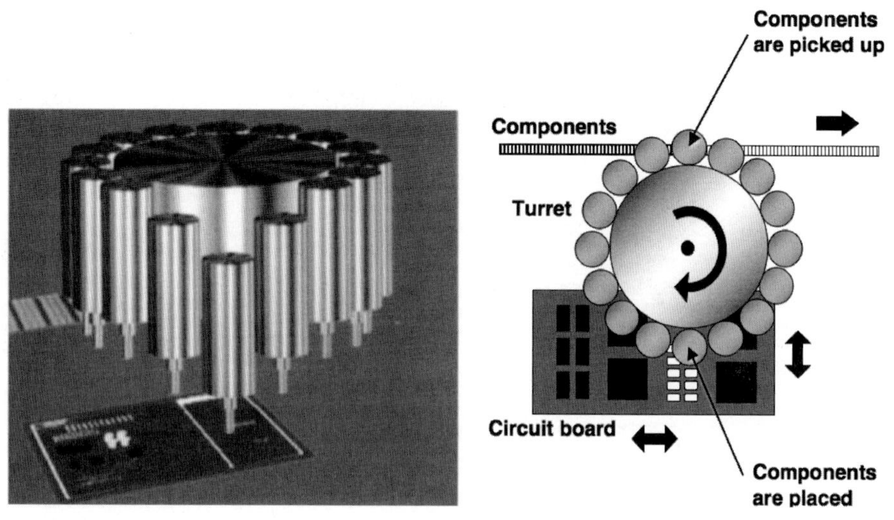

FIGURE 40.20 Turret head used to retrieve chip components from a tape and place them on the circuit board. *(Photo courtesy of Uni ersal Instruments.)*

well as an increased demand for placing bare die (flip-chip) components on more densely populated circuit boards. Equipment manufacturers and users must continually address new package configurations with different I/O geometries.

40.3.3.3.2 Gantry Systems (Active Devices). The gantry-style architecture differs from the turret because the printed circuit board is fixed in place and the moving gantry locates the component over the correct position (see Fig. 40.21). The part feeder is stationary. The gantry method is typically used to place larger components (e.g., SOICs, PLCCs, etc.). Several attributes of gantry technology are listed in Table 40.4.

TABLE 40.3 Capabilities of Turret Chip Placement Technology

Property	Range
Part range	0201 (English) passive chip devices to 10 mm area-array packages
Part capacity	Several hundreds
Placement speed	25,000–40,000 CPH*
Capabilities	1. Handle both chips and small area-array packages. 2. Moving turret and PCB to increase placement speed. 3. Flexibility of using either tape or bulk part supply.

*CPH = components per hour

Several equipment variations can be specified to increase the variety of parts placed on the circuit board. Each gantry is made to accommodate different parts when equipped with a multispindle placement head. The placement head is positioned at each feeder location to receive a part. Then, the head moves the component to the upward-looking vision station for inspection

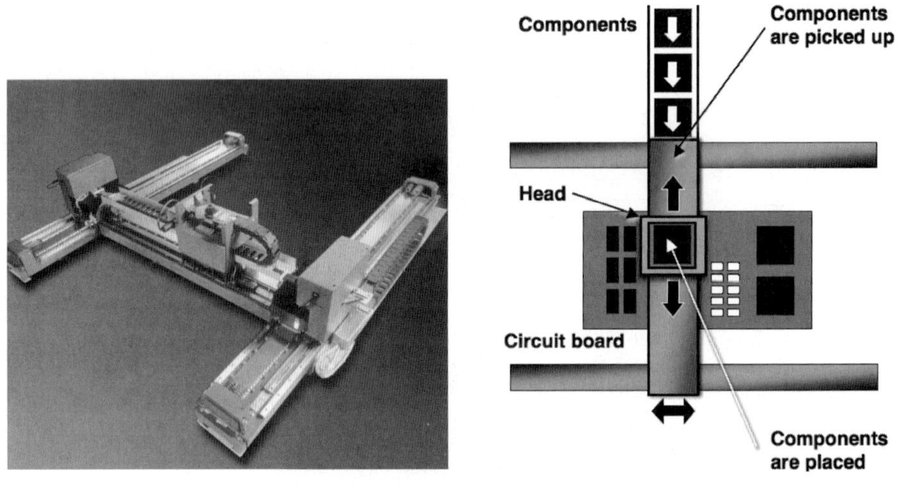

FIGURE 40.21 Gantry architecture used for placing larger components such as SOICs, PLCCs, and area-array packages. *(Photo courtesy of Uni ersal Instruments.)*

TABLE 40.4 Capabilities of Gantry Technology

Property	Range
Part range	Full SMT (SOD, SOT, SOIC, PLCC, CCGA, BGA), direct-chip attach (DCA), and odd-form components.
Part capacity	Several hundreds
Placement speed	5,000–15,000 CPH*
Capabilities	1. Oversized and odd-form parts placed with very high accuracy. 2. Moving gantry, fixed PCB and feeders. 3. Flexibility of using tape, tube, bulk, and tray part supplies.

*CPH = components per hour

followed by its placement on the printed circuit board. A second option is a placement machine made with dual gantries, each placing a single component type or multiple components with the spindle head option.

Several features in gantry-style equipment allow component placement speeds to approach those of turret systems in "high-speed" models. See Table 40.5 for the performance

TABLE 40.5 Capabilities of High-Speed Gantry Technology

Property	Range
Part range	Full SMT (Passives, SOD, SOT, SOIC, PLCC, CCGA, BGA), direct-chip attach (DCA), and odd-form components.
Part capacity	Several hundreds
Placement speed	15,000–21,000 CPH*
Capabilities	1. Oversized and odd-form parts placed with very high accuracy. 2. Oversized board capacity. 3. High volumes of tape, tube, bulk, and tray part supplies.

*CPH = components per hour

of some dual-beam (gantry) equipment. Machine options that enhance placement speeds include tape splicing for uninterrupted machine operation and simpler bank-changing capability to allow for the rapid changeover of components between different product lines. Also, part inspection can be made using a head-mounted camera while the gantry(ies) is (are) in motion, which eliminates the time interval required for part inspection with a stationary, upward-looking camera.

Further enhancements in capabilities of component placement equipment can be realized in *massively parallel* architectures. The attributes of this approach are listed in Table 40.6. Multiple placement modules are capable of picking up, inspecting, and placing components at the appropriate location on the circuit board. The circuit board is stepped by an indexing conveyor to locate the placement site precisely under the overhead component.

TABLE 40.6 Capabilities of Massively Parallel Gantry Technology

Property	Range
Part range	0201 (English) passive chip devices to 25 mm area-array packages
Part capacity	Several hundreds
Placement speed	60,000–100,000 CPH*
Capabilities	1. Chips and small area-array packages. 2. Large board size. 3. Tape part supplies in multiple bans.

*CPH = components per hour

40.3.3.3.3 Machine Vision Technology. A brief which is description is provided of one of the most critical advancements in component placement *machine vision technology*. The earliest pick-and-place function relied on mechanical stops (*detents*), switches and precision tooling to ensure that components were placed at the correct locations and with the proper alignment of I/Os to pads. As both board densities and component varieties increased, this technology was too slow to meet the requirements for higher production volumes and reduced placement defects. A critical incentive to move away from mechanical registration toward vision-based placement came from smaller I/Os on components. DCA/FC uses die bump pitches as small as 0.1 mm. Passive device sizes are commonly 0402 and 0201. Fine-pitch, quad flat pack (QFP) components have lead pitches as small as 0.3 mm. Each of these cases underscores the need for very high placement precision that can be realized only with machine vision technology.

In addition, there has been a steady increase of odd-shaped devices that include inductors as well as LEDs, surface-mount connectors, etc. The result has been circuit boards with a greater mix of package types and sizes. Consequently, it is considerably less expensive and time-consuming to reprogram a computer-based, machine vision system to recognize these components than it is to retool a machine based on mechanical relays, detents, and such for component placement.

Machine vision technology uses electronic cameras and optics together with specialized computer software to control the stepper motors responsible for positioning the component and the circuit board (site) relative to each other with the required accuracy and precision. To realize this objective, the placement machine must identify the component in the turret or gantry and establish the position of the turret or gantry. At the same time, the placement machine must know the position of the circuit board. The computer software ties these two requirements together by being programmed with the artwork (drawings) that identify the location of each component on the circuit board. Component recognition and circuit board recognition are discussed next, followed by comments regarding vision system limitations.

Component recognition is typically obtained from the configuration of the I/Os. The I/O configuration includes two attributes: the I/O's shape, whether it is a beam lead, gull-wing

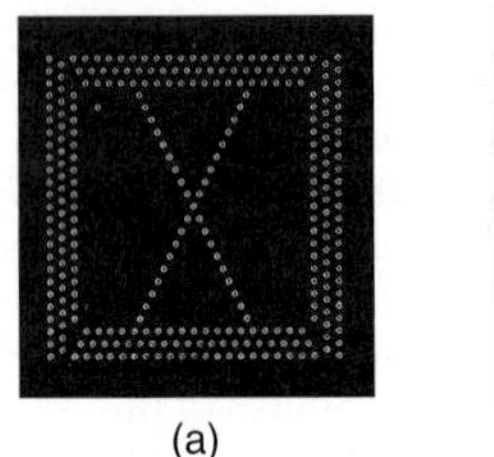
(a)

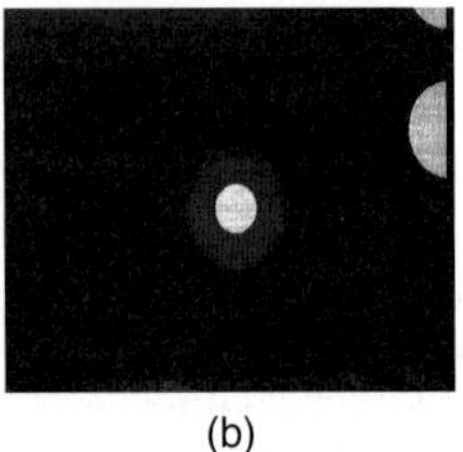
(b)

FIGURE 40.22 Vision system images: (a) a flip chip component; (b) a printed circuit board fiducial. *(Courtesy of Universal Instruments.)*

lead, or solder bump; and the I/O's layout, which may specify, for example, peripheral leads on two rather than four sides, or solder balls as full area-array packages rather than perimeter, area-array packages. Shown in Fig. 40.22a is the machine vision image of a flip-chip die with a perimeter array of solder bumps. The vision system determines the position of the package based upon the coordinates of the I/O locations, for example, two or four of the corner leads or solder bumps. Additional fiducials and/or nonsymmetries in the I/O layout (e.g., missing corner leads or bumps) are used to establish the rotational orientation of the component.

Besides determining component type and orientation, vision systems have also been programmed to recognize damage to components. For example, fine-pitch quad flat packages are prone to bending of the very small leads, particularly at package corners, due to worn die sets or improper handling. In the case of area-array packages (BGAs, CSPs, and DCA die), there can be missing balls or solder bumps. The damaged component is rejected into the "scrap bin" and a new unit retrieved for placement. It should be noted that component recognition vision systems are designed primarily for part placement. Although defect recognition can be programmed into the software, increasing the degree of inspection function slows the placement process significantly. Therefore, the optimum approach is to allow the vision system to identify only gross component defects, targeting those defects that would occur between incoming inspection and the component placement step. Otherwise, full, incoming component inspection should be performed *before* the parts are loaded into the placement machine.

Circuit board recognition requires that the machine (computer) be able to locate the circuit board site precisely for the component. First, the circuit board is secured on the conveyor by mechanical clamps, vacuum chuck, or other technique. The circuit board is then positioned under a camera that identifies the registration marks or fiducials on the surface (see Fig. 40.22b).

This process is repeated for two to three other fiducials on the circuit board. At this point, the machine "knows" the location and orientation of the circuit board and thus, through the artwork design stored in the software, correlates the fiducial locations with the location of each component site. Subsequently, the software matches the coordinates of the circuit board site to those of the component positioned in the turret or gantry. Then the software instructs the stepper motors to position the component over that site and lower it on to the circuit board.

Some applications, such as those using ceramic substrates that are prone to inconsistent degrees of shrinkage during fabrication (e.g., low-temperature, co-fired ceramic, or LTCC), may develop a discrepancy between the software design and the actual circuit board component positions (pads). In this case, the placement site of each component is determined directly using fiducials in close proximity to that component site. Although local fiducials may enhance component placement accuracy to some degree, especially when tolerance stack-up in the product causes it to deviate from the design files, this approach requires added processing time by the machine computer. The result is a decrease in placement speed that can develop into an appreciable process delay, particularly for large production volumes.

Vision system limitations are determined by the speed with which the computer can process information (e.g., circuit board coordinates, component geometries, defects). The more information to be processed, the slower is the component placement step. For products requiring the placement of thousands of parts per circuit board, even an additional few tenths of a second per component can add up to a significant loss of production throughput.

Similarly, there are operational limitations on vision systems themselves (cameras and optics). The trade-off is between resolution and the range of processable part sizes. The underlying premise is pixel count. The vision system requires a minimum number of pixels to recognize a feature (lead, solder bump, fiducial, etc.). A very small feature (e.g., a flip-chip solder bump) requires both high magnification and high resolution (i.e., pixels/length or pixels/area) in order for the system to recognize the feature. However, using the same system to recognize a large BGA package would require a magnification that may well be outside the capability of those same optics. Moreover, even if the package could be brought into the field of view, using the same high resolution, which is now not necessary, will overload the computer memory and bog down the software processing step, resulting in a slowing of the placement process. And conversely, the same camera that can efficiently process a large, 32 mm QFP (208 I/O) in a single image would generally not have sufficient resolution to process a flip-chip solder bump of 0.1 mm diameter. Therefore, the choice of optics is critical to maximize the efficiency of the pick-and-place function for a particular product. For optimum efficiency, it may be necessary to place the two components on separate machines. Alternatively, multiple cameras and optics may be used on a single piece of equipment. Cost becomes a critical factor.

Once the components have been placed, the circuit board is soldered using one of the techniques briefly described in the following sections. The component placement step is synergistic with the soldering step and, in particular, reflow soldering. That synergism arises from the fact that placement inaccuracies can be compensated by the self-alignment of packages arising from the surface tension (more accurately, the solder-flux interfacial tension) of the molten solder. This self-alignment phenomenon opens the placement window for passive devices larger than about 0603, smaller LCCCs, and 1.27 mm pitch area-array packages of up to several hundred balls. Unfortunately, for smaller passive devices and larger area-array packages, the self-alignment process becomes less capable of compensating for placement errors. In the case of smaller passive devices, there is the increased likelihood that an imbalance of surface tension forces can result in an asymmetric movement of the part, causing the tombstoning defect. In the case of larger area-array components, poor placement (which is becoming more critical with the finer pitches of smaller balls) cannot always be remedied by self-alignment simply because of the greater weight of the component.

The topic of soldering is described in greater detail in Chap. 42. Therefore, this chapter offers only a brief overview of each approach, with the discussion limited to its relevance to the overall assembly process.

40.3.3.4 Reflow Soldering. Reflow soldering is the technique in which a circuit board, which has the components placed on the solder deposit (either as paste or preforms) is passed through a furnace (oven) in order to melt the solder and form the joints. The furnace may be a batch type in which the circuit boards must be loaded and unloaded, one group at a time. The operator inputs the batch furnace time-temperature profile into the controller that alters the power to a set of heating coils as a function of time. The atmosphere can be very well controlled, including the use of vacuum. The batch furnace is advantageous for small production lots, including development work or when the time-temperature profile and environment must be carefully controlled for the application.

The second furnace has an inline configuration. The circuit boards continuously enter one end unsoldered and exit the other end soldered. Therefore, the inline furnace can be part of an overall assembly line, receiving stuffed circuit boards from the component placement machine via a conveyor without operator intervention. The temperature of the different zones along the length of the furnace and the conveyor speed determine the time-temperature profile. More zones (five to seven are typical) provide for better control of the soldering profile. Inert atmospheres, typically nitrogen (N_2) to minimize cost, can be maintained to better than 20 ppm oxygen (O_2). However, control of both the time-temperature profile as well as atmosphere cannot match that of a batch furnace. Also, vacuum conditions are not possible with most inline equipment. Nevertheless, inline furnaces are well suited for high production volumes and are the most widely used furnace type for electronics assembly.

Whether batch or inline, furnace selection is based not only on throughput rates, but is also determined by the type of product being assembled. A greater complexity to the circuit board requires more control of the time-temperature profile to ensure that all of the solder joints are completed at minimum defect levels. In some applications, the extra furnace zones are used to control the cooling rate of the soldered circuit board to prevent thermal shock damage to sensitive components or substrates.

The introduction of Pb-free solders has impacted reflow soldering, less so in terms of actual equipment temperature capabilities than in the development of a suitable time-temperature profile. The heat source technology (infrared, convection, or mixed) can provide the higher reflow temperatures of the Sn-Ag-XCu alloys (T_{melt} = 217°C versus 183°C for the traditional Sn-Pb solder). Of course, higher energy usage and maintenance costs are likely. Two generalized reflow profiles are used for Pb-free soldering. They are illustrated graphically in Fig. 40.23. The soak-reflow profile in Fig. 40.23a derives from the traditional Sn-Pb eutectic solder profile, but with the reflow "spike" increased for the higher melting Sn-Ag-Cu alloys. The soak step provides for activation of the flux as well as heatup of the circuit board and components. The continuous ramp or "hat" profile in Fig. 40.23b allows for a more rapid heatup rate, which reduces the time that heat-sensitive components and materials spend at elevated temperatures. On the other hand, the relatively faster heatup rate can increase the chances of thermal shock damage to some components—for example, larger plastic and ceramic devices—or to circuit board structures such as vias.

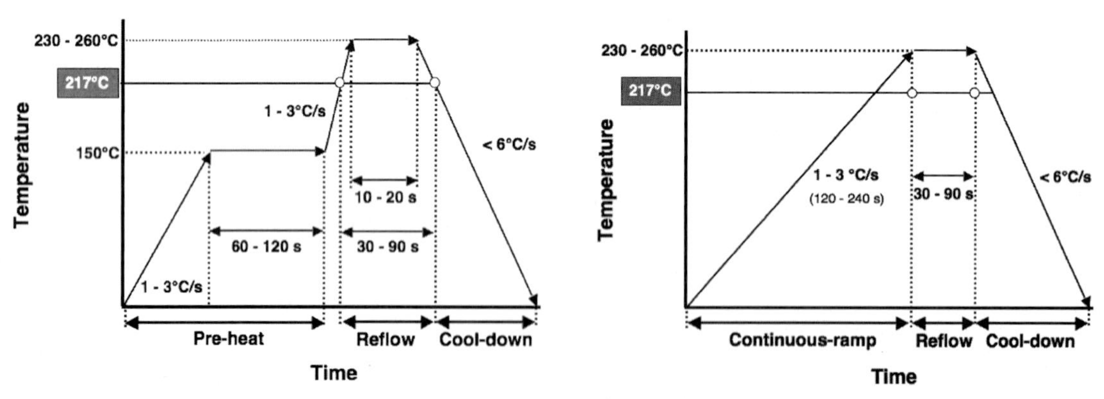

FIGURE 40.23 Generalized time-temperature profiles for Pb-free reflow soldering: (a) the soak-reflow profile; (b) the continuous ramp or "hat" profile. Peak temperatures will vary, depending on the particular circuit board product.

In terms of optimizing the time-temperature profile, a balance must be established between achieving sufficiently high temperatures that will reflow the solder of every component (size and shape) and preventing thermal damage to other components or the substrate. Therefore, the higher melting temperatures of Pb-free solders make it more challenging to develop a time-temperature profile that will successfully melt the solder paste of larger components without causing thermal damage to smaller devices or to the circuit board material.

Therefore, in the case of some very-high-mix products, it may be necessary to solder larger packages in a separate operation (e.g., hand soldering or selective soldering). Under this circumstance, the latter packages would be considered as odd-form components.

40.3.3.5 Wave Soldering. Wave soldering is used when surface-mount technology is mixed with through-hole components on the circuit board. (See Figs. 40.7–40.9 showing wave soldering as it is used for through-hole circuit boards.) Surface-mount devices that are present on the same side as the wave are soldered as well, being held in place by an adhesive. The surface-mount components must be resistant to thermal shock because of the temperature spike experienced upon entering *and* leaving the molten solder wave. The temperature on the topside surface of the circuit board typically remains well below the solder solidus temperature, thereby preventing the reflow of any solder joints present there.

Wave-soldering equipment can be used either in an inline or in a batch process since the equipment has generally the same construction. The inline approach is used to support high production volumes. Small systems that are used in a batch-like mode include those used for *selective soldering*. This equipment includes miniwaves or solder fountains and are used to solder only localized areas of the circuit board, such as when attaching through-hole connectors, transformers, or switches. The advantage of selective (wave) soldering is that the entire circuit board need not be exposed to an elevated temperature.

An important consideration in wave soldering is the supporting fixtures that hold the circuit board on of the conveyor. In reflow soldering, the conveyor can support the entire bottom side of the circuit board. However, in wave soldering, the conveyor does not offer such support, as it must allow the wave to contact the circuit board bottom side. Therefore, it may be necessary to provide an additional fixture to prevent warpage and sagging of larger circuit boards.

The impact of Pb-free technology on wave soldering has largely occurred in the equipment performance. It has been determined that the same solder bath temperatures that are used for Sn-Pb processes (250 to 270°C) are suitable for the Sn-Ag-XCu Pb-free alloys. Therefore, excessive dross formation and flux residue removal have not become a significant problem during equipment operation. The lack of shiny fillets with the Sn-Ag-XCu alloys has been addressed by modified alloys having Ni and Ge additions that alter the solidification process, which leads to shinier fillet surfaces.

The Pb-free alloy compositions are more prone to erode the wave machine's structures, such as the impeller, baffles, and pot walls. New wave-soldering machines address this problem through the use of alternative steel alloys and ceramic coatings.

40.3.3.6 Condensation (Vapor Phase) Soldering. Condensation soldering, also referred to a vapor phase soldering, uses a working fluid's heat of condensation to reflow the solder paste or solder preform to make the joints. Because the condensation reflow process had its origins in the early days of surface-mount technology when production runs were more limited in volume, these machines were primarily batch-type units. Subsequently, inline equipment was developed to "attach" the vapor phase reflow machine to the back end of the placement equipment for higher production volumes. Two particular attributes of condensation soldering are (a) the temperature of the product cannot exceed the vaporization (or condensation) temperature of the working fluid, thereby preventing overheating of temperature-sensitive materials; and (b) the temperature is very uniform over all of the components and the substrate, thereby minimizing temperature gradients that could warp or crack component or laminate materials.

This process fell out of favor in the early 1990s, for two reasons. First, the working fluid for Sn-Pb solder, Dupont's Freon TMF, was categorized as an ozone-depleting substance (ODS). Its use was initially restricted and then prohibited by the Montreal Protocols. Second, the increasingly more complex circuit boards that were being designed with surface-mount technology required more precise control of the time-temperature profile, which lead to the development of multizone reflow furnaces having convective and infrared heating capabilities.

There has been a resurgence in the use of condensation soldering. Alternative working fluids have been developed for both Sn-Pb and Pb-free processing that are compatible with environmental regulations. Preheaters are added to reduce the thermal shock when parts entered the working fluid, thereby providing a more controlled time-temperature profile. Lower capital cost and properties of the condensation heat source previously noted cause condensation or vapor phase reflow to be well suited for prototype development programs and small production volumes.

40.3.3.7 Hand Soldering. As noted earlier, in the highly automated surface-mount assembly line having automated paste printing and component placement capabilities, hand soldering would not be particularly advantageous in terms of maximizing production volumes. However, in some applications, the assembly process includes a hand-soldering operation. For example, often odd-form components cannot be incorporated in a pick-and-place machine, or temperature-sensitive devices cannot be exposed to the reflow furnace environment. Under these circumstances, the hand-soldering step is performed *after* mass soldering (reflow, wave, etc.). The implication is that hand soldering is now being performed on what is potentially a very-high-valued circuit board. Therefore, factors such as handling damage, electrostatic discharge (ESD) damage, and thermal damage to nearby components by the soldering iron tip, as well as flux residue contamination, must be thoroughly addressed at the development stage of the hand-soldering process.

The use of Pb-free solders has not impacted the hand-soldering process itself. There is a slight lengthening of the soldering time, typically from 3 to 4 sec. to 5 to 7 sec., due to the higher melting temperatures of the Pb-free solders. The same soldering iron equipment can provide the necessary tip temperatures. The high-Sn solder compositions, coupled with their higher melting temperatures, can more quickly degrade the tip. Lastly, the operator should not mistake a duller appearance to the fillet surface for a "cold" solder joint.

40.3.3.8 Conduction (Sikama) Soldering. Conduction soldering is the process in which heat arrives to the solder paste by means of conduction through the substrate. The process, which is illustrated in Fig. 40.24, is also termed Sikama soldering, being named after the primary

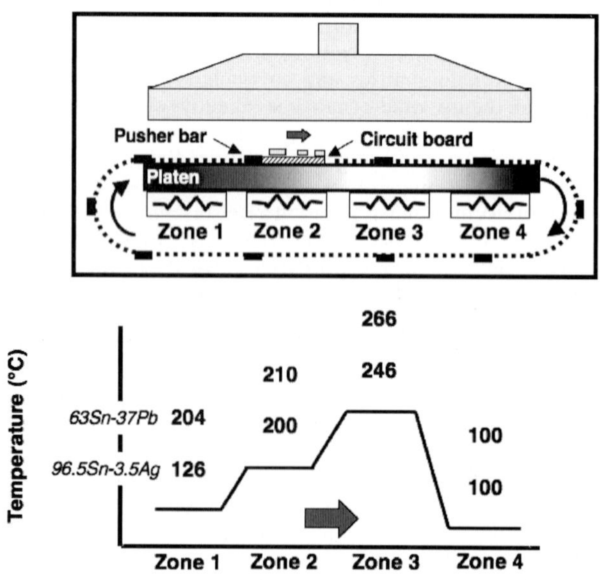

FIGURE 40.24 Schematic diagram depicting conduction soldering. The pusher bar presses the circuit board along the surface of platen. Heaters under the platen determine the time-temperature profile.

equipment manufacturer. The equipment is set up to accommodate an inline process, although it can be used in batch mode as well.

The circuit board is pushed along on top of the heated platen by rods attached to a conveyor. Different sets of heating elements under the platen heat the latter locally to different temperatures along its length, creating heat zones such as those in a reflow furnace. The soldering process can be performed in air or using an inert atmosphere. The fact that the substrate is fully supported along its entire footprint (which is needed to maximize heat input) precludes excessive warpage.

The conduction heating process has been used almost exclusively for ceramic substrate because these materials do not degrade when exposed to the high platen temperatures. Thermal degradation would be more likely for organic laminates. Best results are realized when this process is used with relatively thin substrates (<1.0 mm).

40.3.3.9 Cleaning. Cleaning is typically the final step in the circuit board assembly process. There is a variety of cleaning equipment, from hand-loaded, batch-type "dishwasher" machines for smaller production volumes, up to larger inline units that are integrated into the back end of the assembly line for large production volumes. In response to environmental regulations, cleaning solutions have moved away from the once popular organic solvents to aqueous and semi-aqueous cleaners based on water- and water-alcohol-based compositions. Similarly, the handling of the waste stream is also important. Closed-loop systems are preferable to once-through cleaning process because of regulations governing the disposal of the waste stream. In the case of assemblies requiring multiple soldering steps, the cleaning portion of the prior soldering step does not leave a residue on the surface that degrades the solderability required for the follow-on process.

The use of a Pb-free process does not necessarily have a significant impact on the cleanability of circuit boards. At one time, it was believed that flux residues from Pb-free processes would be more difficult to remove, due to the higher soldering temperatures. Subsequently, it has been determined that thermal degradation of those residues is not as severe, so current cleaning processes appear to perform adequately.

40.3.3.10 Pb-Free Solders. The equipment types described in the preceding sections are often combined to form an electronics assembly line. The term "line" implies that the equipment is physically located end to end, thereby allowing for the movement of circuit boards between the different functions or stations. Although this scenario is, in fact, typically the case, it need not always be so. Under some circumstances, it may be advantageous to locate some stations a distance away others on the shop floor, thus requiring operators to move product between the two sites on wheeled racks. This is usually the case with inspection and repair/rework functions as well as for some cleaning processes due to health or safety concerns. The setup and hence the flow of a product assembly process are referred to as the (assembly) line architecture.

Numerous variables are considered when developing any assembly line architecture. The underlying consideration is optimizing the *level of automation*. Some of these variables include the following:

- Available floor space and facilities (electricity, exhaust, computer networks, etc.)
- Equipment footprint and cost of ownership or lease
- Labor costs
- Production volumes and product changeover flexibility
- Circuit board technology (through-hole, surface-mount, or mixed)
- Soldering technology: Sn-Pb versus Pb-free
- Inspection and quality requirements

It is important for the manufacturing engineer to consider not only each of these points individually, but to also apply the proper weighting factors of their importance when integrating them together to develop a process for the product(s).

The introduction of Pb-free solder technology does not, in and of itself, alter the previously described principles with regard to assembly-line architectures. The equipment is very similar and the factors that determine an optimum line configuration remain unchanged. Instead, consideration must be given to the cost/benefit of having an *entirely separate* assembly line that is dedicated to Pb-free products. In the case of wave soldering, the need for a second line is intuitive; it is simply not practical to empty and refill the solder pot with different alloys from both a process time-management point of view as well as the metallurgical-based concern of mixing the two alloys. An interim approach is to exchange only solder pots in the machine, ensuring that all residues of the previous solder have been removed from all equipment surfaces.

However, even in the case of solder paste printing and reflow soldering, many manufacturers have determined that two separate assembly lines are required, even to the extent of physically locating them at separate locations in the plant. There are simply too many risks associated with potentially mixing the two assembly technologies, particularly in high-volume operations where a significant quantity of product may be built before an error is recognized. The potential mix-ups include:

- Use of the incorrect solder paste
- Placement of components with the incorrect surface finish—Pb-Sn (high-reliability electronics) versus 100 percent Sn (consumer electronics)
- Incompatible moisture sensitivity levels (MSLs) of plastic molded components
- Vision system settings for identifying fiducials of alternative surface finish
- Reflow furnace settings for the different time-temperature profiles

40.4 ODD-FORM COMPONENT ASSEMBLY

An accurate definition of an odd-form component is one that cannot be placed and/or soldered with equipment *available in the assembly line*. The likelihood of having to address an odd-form component is actually quite high. Because of the wide range of current component technologies as well as those expected in the future—whether through-hole or surface-mount—there is always the chance that a package cannot be placed or soldered in the assembly line. Certainly, if volumes warrant, new equipment can be purchased, or special tooling made, for adapting the placement and/or reflow equipment for odd-form components. However, there is a significant cost penalty to this approach, especially if it is performed for a variation of sizes and configurations. Each package variation would require separate tooling designs or equipment modifications. However, the advantages of automating odd-form assembly operations include a higher production volume as well as fewer defects resulting from the automated rather than manual operations. Therefore, it is not unusual for odd-form component assembly to be a mix of both manual practices as well as automated techniques.

Generally, the introduction of Pb-free soldering affects odd-form components in a manner not unlike that for other surface-mount and through-hole components. The two primary issues are I/O surface finishes and temperature sensitivity of each component's material set. High-reliability applications have to monitor components for 100 percent Sn finishes. Because odd-form components often fit this category due to a large size, hot Sn-Pb solder dipping is a more viable option to remove the 100 percent Sn finish. Otherwise, the tracking of incoming material and verification of the actual surface finish may be required.

Temperature sensitivity of devices will have to be documented in terms of surviving the Pb-free reflow profile. In fact, some devices that were previously processed on a Sn-Pb assembly line may have to be set aside as odd-form components because they cannot be exposed to the Pb-free process temperatures. Another risk of the higher-processing Pb-free reflow temperatures is the damage to moisture-sensitive components. The typical Pb-free time-temperature profile adds one or two MSLs to a component. Therefore, devices at MSL levels 5 or 6 per a Sn-Pb profile may have to be assembled to the circuit board in a separate operation to avoid damage, thus causing them to be odd-form components.

40.4.1 Components

Connectors comprise approximately 75 to 80 percent of odd-form components. See Fig. 40.6. For example, there are universal serial bus (USB) connectors, dual inline memory module (DIMM) connectors, phone jacks, and zero-insertion-force (ZIF) connectors, as well as a variety of surface-mount connectors. Other odd-form components include sockets, electrolytic capacitors, transformers or large inductors, LEDs, relays, switches, as well as speakers and vibrators for handheld communications devices. In surface-mount technology, odd-form components are most often through-hole packages, including both passive and active devices as well as connectors. When addressing such odd-form components, the first step is to check the availability of an equivalent surface-mount version since there has been a shift toward surface-mount versions of many of these devices. Of course, the materials used to construct the surface-mount replacement must survive the reflow furnace environment, and especially the Pb-free reflow temperatures, particularly connectors because of their complex internal construction. In some cases, it has been possible to replace the through-hole connector with one of a variety of solderless products.

Also, it is not necessarily size or weight that causes a component to be identified as "odd-form" per the particular assembly line. Temperature-sensitive and moisture-sensitive devices may also be excluded from a reflow step of the assembly process to prevent damage to them. The latter scenario is of particular concern when converting from a Sn-Pb to a Pb-free process.

DCA refers to the assembly of an *unpackaged* active semiconductor device directly onto the circuit board. The interconnections are made either by wirebonding or by solder bumps on the front side of the die. The latter application is referred to as FC attachment and can be implemented into a surface-mount assembly process as long as existing equipment can accommodate the placement accuracy required for the very small solder bumps and pads. Otherwise, a dedicated cell may be required.

The DCA process, which uses wirebond interconnections, is based on the following steps. First, an adhesive is dispensed on the board and the die placed on it. The adhesive, which is often thermally conductive for heat management, is cured. The I/O pads on the die are wirebonded to the appropriate pads on the circuit board. Then, the die and wire bonds are encapsulated in an epoxy to prevent damage to them. The encapsulant, or "glob top," is then cured. Critical to the glob top technology are the following factors: complete coverage of the die and wire bonds with minimal run-out beyond the circuit board pads; adequate adhesion between the die and circuit board; and good adhesion between the encapsulant and the circuit board beyond the perimeter of the die.

40.4.2 Manual Assembly

In the least complicated format, only an operator places the odd-form components on the circuit board. The soldering process is automated in that the operator picks components from bulk bins and repeatedly positions them onto the circuit board as it is carried along the conveyor line. For paste-in-hole assembly, the solder paste and other components (surface-mount or through-hole) may already be on the board, being held in place only by the tackiness of the paste. Therefore, the operator must not inadvertently displace other components off of their sites during handling.

Also, the operator must pay special attention to handling and ESD damage while placing components. Also, those components that incorporate snap-in features (e.g., connectors) can require that the operator exert relatively high forces on the circuit board. Thus, the operator must not damage either that component or other components and the circuit board. Often, special tooling is used to press snap-in components on the circuit board to prevent such damage.

A large number of applications require the operator to place and then manually solder the odd-form component into place, often after the other components have been soldered into place. The operator must be cognizant of handling and ESD damage as well as prevent thermal damage or the inadvertent reflow of solder joints for neighboring components. The impact of

Pb-free soldering would be most significant in this circumstance. It is necessary to verify that a longer hand-soldering time or a higher tip temperature does not cause thermal damage to the component in question or cause the reflow of nearby solder joints.

It should be noted that irrespective of whether manual assembly is performed on only stuffed boards that will be subsequently soldered or on circuit boards that have already been soldered, the units have a higher value to them because of this processing, which is performed beforehand. Therefore, *any* damage (handling, ESD, thermal, etc.) will result in a significantly higher yield loss to production volumes as well as higher costs associated with the repair or scrapping of damaged circuit boards. Therefore, manual assembly steps must be given due consideration during product design and process development activities.

40.4.3 Automated Assembly

The automated assembly of odd-form components uses the same placement equipment as is used for other components, but with special tooling or a dedicated machine (cell), which is already set up with the required tooling and component feeder systems, that can be introduced into the assembly process. The latter equipment typically places more than one type of odd-form component for increased versatility. For example, the close-up photograph in Fig. 40.25 shows the type of gripper tooling that can place different component packages (to within its design limits, of course). After accuracy and component-type requirements are met, speed is the critical performance metric for odd-form placement machines. Odd-form component placement is nearly always a bottleneck or slow point in the line; therefore, minimizing assembly time devoted to odd-form components will speed up the entire process flow.

FIGURE 40.25 Odd-form component placement machine having a flexible servo-driven gripper tooling that can place a wide range of different component configurations. *(Courtesy of Universal Instruments.)*

There are both advantages and disadvantages to the automated assembly of odd-form components. Advantages include the following:

- consistent as well as higher quality and faster throughput result from eliminating manual processes.
- Automated assembly saves floor space, as machines typically take up less area than operator work benches.
- Solder dispensing is integrated into the placement process flow, minimizing related defects.
- Automated assembly complements an overall, computer-integrated manufacturing infrastructure already in place with the other assembly steps that, in turn, facilitate process flow, quality control, and product changeover.
- Automation eliminates operator-related health and safety issues (fumes, repetitive tasks, etc).

The disadvantages of automation include the following:

- Capital equipment costs can be high for machines, dedicated tooling, and plant modifications.
- Prepackaging components are required for automated feeders.
- Equipment changeover for different product lines is somewhat slower due to the need for special component packaging and/or feeders.
- Higher production volumes can generate a greater quantity of defective product before a process problem is identified.

40.4.3.1 Component Packaging Formats. Odd-form components come in a variety of sizes, geometries, and I/O configurations that determine the ease with which the component can be "automated" into the assembly process line. Another important variable is the component placement machine. Typical component packaging formats for the placement machine include tape and reel (radial and axial leaded), extruded tube, tray, continuous strip, and bulk containers. Each of these formats has benefits and drawbacks, as noted in the following subsections. Industry specifications have standardized nearly all of these packaging schemes. For example, there is EIA-468-B for radial devices and EIA-296-E for axial components.

40.4.3.1.1 Tape and Reel. Tape and reel allows for the delivery of parts to the circuit board at very rapid rates. See Fig. 40.26. Tapes are available generally in three formats: flat-carrier tape, deep-pocket tape, and semi-deep-pocket tape. Flat-carrier tape is the traditional method of carrying both odd-form (as well as form-fit) radial and axial components. It is inexpensive and can provide a large quantity of parts for rapid placement rates. Deep-pocket and semi-deep-pocket tapes provide room for the larger component bodies. The packaging of multiple parts having a similar pitch is possible with semi-deep-pocket tape; the particular advantage is to conserve feeder space, particularly when only a few of each type of component are required on the circuit board.

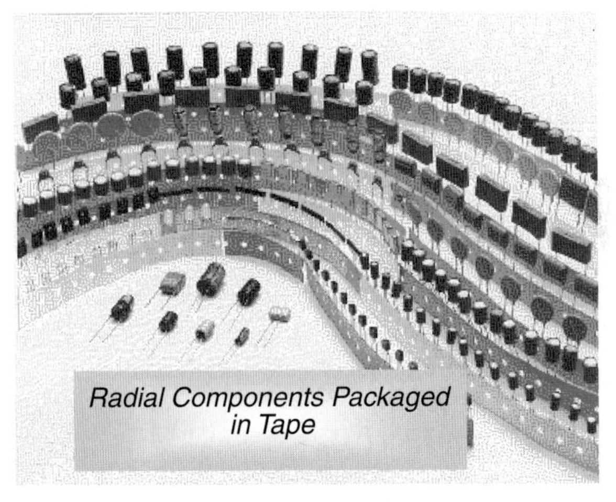

FIGURE 40.26 Radial components packaged in tapes that would come off of reels. *(Courtesy of Uni ersal Instruments.)*

Several drawbacks are associated with tape carriers. For example, tall or heavy parts may change position or break the cover of deep-pocket tape carriers. The potential consequence is that the tape and/or components will jam or damage the feeder mechanism of the placement machine, thereby interrupting the assembly process.

40.4.3.1.2 Extruded Tube. Extruded tube is a commonly used format for odd-form components. See Fig. 40.27. Components such as D-Sub connectors, phone jacks, transformers, relays, and shrouded headers are typically supplied in tube packaging. Recently, component manufacturers have begun to supply long connectors such as single inline memory modules (SIMMs), DIMMs, and other designs of long-aspect ratios in edge-stacked tubes.

Tube packaging is particularly beneficial for protecting the component at a reasonable cost. Also, tubes are less likely than tape to cause problems feeding components into the placement machine. Of course, it is necessary to verify that the tube geometry is compatible with the feeder mechanism on the placement machine. The tube material should be sufficiently thick so as not to bend readily. Bending of tubes is the primary cause for the jamming of devices inside

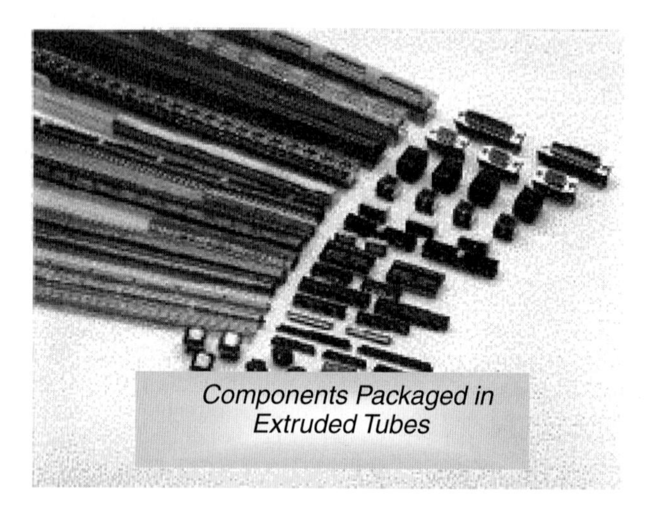

FIGURE 40.27 Components in tube packaging. *(Courtesy of Universal Instruments.)*

of them during the feeding process. Moreover, heavy components packaged in thin cross-sectional tubes may actually interfere with adjacent tubes when stacked in the automated feeder.

40.4.3.1.3 Matrix Trays. Matrix (or waffle) trays are an inexpensive means to supply components that are generally too large or heavy for tube or tape packaging and must be individually separated from one another. See Fig. 40.28. An important consideration is that the trays themselves may not be a suitable starting point for the automated placement processes especially for very small components. First of all, the components have too much room to move about in each bin to establish coordinates for initiating the pickup and insertion process. Also, component leads are not always oriented correctly in the insertion plane.

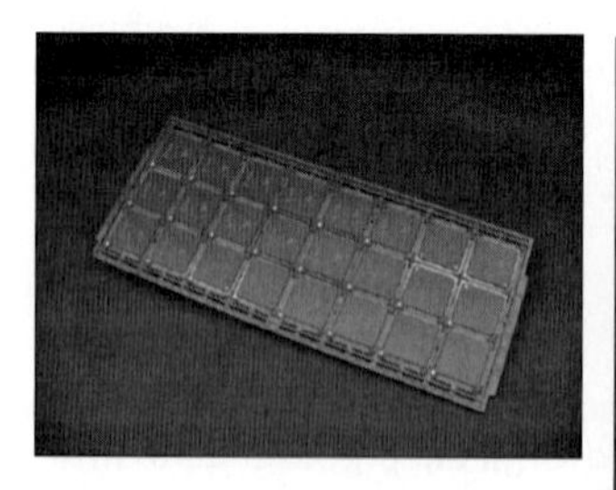

FIGURE 40.28 Components packaged in matrix trays. *(Courtesy of Sandia National Laboratories.)*

Several potential issues may need to be considered when choosing trays. Trays lack the rigidity required by many assembly machines for part pickup. Also, trays hold a more limited quantity of parts to the placement machine, and thus require more operator changeout during assembly. There are no standards for vacuum-formed trays that are used for odd-form components. Lastly, trays simply take up a greater amount of feeder space in a placement machine and on the floor.

40.4.3.1.4 Continuous Strip. Continuous strip packaging is designed for high-volume automation applications of odd-form components. This technique eliminates the need for separate component packaging formats because the individual parts are all attached in sequence to a single metal or plastic strip as part of the assembly process. The parts are located with a specified pitch that can be accommodated by the placement machine. Some components that are packaged in this manner include headers, battery clips, transformers, and motor brushes. Either the metallic part (e.g., brushes) or the leads that are part of the component can be stamped from a continuous strip of metal or thermosetting plastic. This type of packaging provides high levels of online parts inventory. However, such strips are application-specific and may carry a higher cost to automate due to the need for specially designed feed mechanisms.

40.4.3.1.5 Bulk. Bulk packaging eliminates the expense of component packaging altogether by simply placing the devices in a large container. A significant number of odd-form components are shipped in this format. Of course, the component manufacturer has likely established the cost/benefit of the fallout of parts due to handling and shipping damage versus the cost of added packaging by one of the previously mentioned formats. On the other hand, the degree of part damage may not be acceptable to the circuit board assembly house because of increased downtime due to machine jams and part replacement.

Bulk packaging is the most difficult packaging format to automate for odd-form components. Generally, bulk components require a specially designed bowl feeder to supply parts to the machine with some degree of consistent orientation, which increases cost and process implementation lead times. Also, the unusual configuration of bowl feeders may require a dedicated placement machine (cell) that may have a significant idle time. The feeders usually take up considerable space and are not easily retooled for product changeover. The placement machine must have then the versatility to orient each part consistently from a haphazard "pile" of devices so that the pickup device can obtain each unit and correctly place it on the circuit board. In spite of these numerous drawbacks, bulk packaging is a cost-effective approach for odd-form components in numerous product applications.

40.4.3.2 Equipment. Advances continue to be made in the capabilities of odd-form assembly equipment. Early vintage, single-task robotic systems are giving way to machines that perform all of the alignment and functions of modern placement equipment, and which can be integrated into existing assembly lines. New machines are designed for speed and the flexibility to handle different odd-form component geometries as well as a variety of circuit board configurations.

Typically, automated odd-form component placement machines use an overhead gantry-style positioning system and vision and optics systems underneath the head. The emphasis is on placement accuracy given the high component density of surface-mount boards and, in some applications, the use of solder paste-in-hole printing methods for a 100 percent reflow soldering process. The component must be placed accurately into the solder paste the first time so that the solder paste is not disturbed, which can result in potential defects. Interchangeable grippers and vacuum pickups as well as improved component feeding mechanisms provide greater flexibility for different product types, faster placement rates, and minimum equipment downtime.

40.4.4 Cleaning

The cleaning of assemblies having odd-form components, like the placement of those devices themselves, can occur at one of several points along the assembly process. If the components are placed prior to the principal soldering step, then they will be cleaned along with the rest of the circuit board. The higher temperatures of a Pb-free solder process can make this cleaning step somewhat more difficult, depending on the tenacity of the flux residues.

On the other hand, when the odd-form devices are soldered to the circuit board after the primary soldering process has taken place, the subsequent cleaning procedure must take into account the other components that are already on the circuit board, particularly potential handling damage. It may be preferred to use a no-clean or low-residue flux to minimize or eliminate the added cleaning step required for the later placement of odd-form components.

Because odd-form components are often larger and/or have more complex constructions (geometry and materials), there is an increased likelihood of flux residues becoming trapped in confined areas. Therefore, the cleaning process must be sufficiently thorough to remove residues from those locations on the component. Similarly, it is necessary to verify that the cleaning solution has likewise been removed from those locations as part of the process.

40.5 PROCESS CONTROL

Process control refers to the capability of repeatedly making product units that meet performance specifications (including long-term reliability) within an allowable frequency of repairable and nonrepairable defects. There are two underpinning premises of process control, which can be described specifically with regards to circuit board assembly: (1) The assembly process is nominally capable of making a circuit board that meets the performance specifications; (2) the equipment and materials set are capable of making those acceptable circuit boards repeatedly. The types of defects, and the frequency of their occurrence, are the metrics for establishing and monitoring the process using statistical process control (SPC). Defects can be detected by visual or machine inspection, or are identified by electrical performance of the assembly (referred to as in-circuit testing).

It is beyond the scope of this chapter to provide a detailed explanation of SPC. Rather, there will be a qualitative discussion of process control that addresses those factors affecting circuit board assembly, beginning with defect types in electronic solder joints. Then the discussion will turn to process control as it pertains to the basic assembly steps (e.g., dispensing, pick-and-place, etc.). The various material sets (e.g., circuit board, solder paste, flux, etc.) will be incorporated in the discussion. Also, the use of Pb-free soldering technology will be addressed where applicable.

The yield drops that result from an out-of-control process impose significant cost penalties. High production volumes can potentially generate large quantities of defective printed circuit boards before a defect trend is discovered. Therefore, it is critical, first of all, to establish process control *and* maintain it throughout production and, second, be able to identify product defects, determine the responsible process parameter, and quickly correct the faulty equipment, material, or operation.

40.5.1 Defects

Defects are defined as those artifacts that are outside the window of acceptable attributes for the finished circuit board assembly. Thus, defects are not limited to the solder joints, specifically, but can also include damage to the circuit board material as well as degradation to component structures (e.g., molding compound, leads or terminations, etc.). Defect types and their allowable frequencies (often expressed in parts-per-million solder joints or product units) vary with the different assembly processes and applications. Therefore, product drawings, in conjunction with industry standards (e.g., IPC-610), are used to establish accountable defect types for printed circuit boards.

Defects are most often detected by visual inspection or automated optical inspection (AOI). Other means of nondestructive evaluation (NDE) include electrical testing, x-ray inspection, and ultrasonic inspection. The preferred NDE inspection technique for BGA and CSP solder joints is x-ray inspection. Automated x-ray inspection equipment is often placed directly into the assembly process line for circuit board products having a large number of area-array components.

There are destructive evaluation techniques that can also be used to find defects. Metallographic cross sections provide a means to inspect the internal structures of the solder joints, components, and circuit boards, albeit in only a single plane at a time. Mechanical tests (die shear, pull tests, etc.) can also be used to identify defects. However, because these approaches are destructive, and because they require long lead times to obtain results, they are generally inapplicable for real-time process control. They are better suited for process development or failure analysis activities.

Defects serve two functions for the process engineer. First, defects provide the quantitative metric with which to ascertain process control, using SPC techniques. Second, the microstructural details of the defect(s) can be used in a failure analysis study to identify the root cause of an "out-of-control" process.

There is very little difference between the types of defects that occur during a Pb-free assembly process and those that have been documented for Sn-Pb assemblies. The two exceptions are grainy solder fillets and fillet lifting of Pb-free, through-hole interconnections. The grainy fillets resulting from the use of some Pb-free alloys are intrinsic properties of their solidification behavior. Because they do not necessarily represent a cold solder joint—that is, one that is only partially melted—the appropriate revisions have been made to industry specifications to account for this effect. The second phenomenon, fillet lifting, occurs on some Pb-free plated through holes. Although several studies have concluded that fillet lifting does not necessarily impact either the short-term performance or long-term reliability of these interconnections, they may still be categorized as defects, depending on the particular product requirements.

Finally, the frequencies of defects on a Pb-free product may differ from those for the same Sn-Pb assembly. Of course, proper process optimization will minimize the frequency of such defects. However, slightly different solder paste printing characteristics at finer pitches, as well as overall poorer wetting and spreading behavior of Pb-free solders, may result in an optimized process that simply cannot achieve the low defect rates of a Sn-Pb process under the capabilities of the equipment set. Additional mitigation steps that can be taken by the design and process engineers include altering the stencil design (e.g., slightly wider apertures); using alternative circuit board and component I/O surface finishes that enhance Pb-free solderability (e.g., one with a Au protective finish); changing the solder alloy composition (Bi-containing solder); replacing existing equipment to provide wider process window capabilities; or taking no action and simply accepting the reduced process yields, based on an accurate cost/benefit analysis.

40.5.2 Dispensing

Dispensing processes include those for an adhesive, solder paste, or both (e.g., surface-mount wave or selective soldering). Because the adhesive is dispensed prior to the solder paste, it is possible to inspect for defects—incorrect dot volume, run-out on nearby solder pads or lands, and stringers—prior to dispensing solder paste.

By and large, the greatest number of solder-joint defects can be traced to the solder-dispensing process and, specifically, the stencil or screen printing step. Therefore, process control is critical here. An absence of the control of any one of these attributes can significantly impact process yields. Important factors include the following:

- Have the appropriate stencil or screen design and materials (e.g., thickness, aperture geometry, wall finish, etc.) for the circuit board product. Design considerations must also address the need to widen fine-pitch aperture openings slightly for printing Pb-free solder paste.
- Properly design the circuit board to provide recognizable fiducials and an absence of burrs or other artifacts that can interfere with the dispensing machine and/or stencil performance.
- Ensure that the stencil is free of particles and other contamination.

- Establish the optimum dispensing equipment parameters and, in the case of solder paste printing, squeegee materials.
- Select the correct solder paste (powder size, metal loading, and flux type) for the dispensing process and adhere to shelf-life and shop-floor-life recommendations.
- Select and properly maintain dispensing equipment (e.g., squeegee replacement schedules, vision system calibration, cleaning, lubrication, etc.).

Dispensing defects can be detected by visual inspection or automated inspection techniques. Such automated techniques include those based on visible light images as well as laser profilometry that determines the actual volume of the adhesive or solder paste deposit. However, inspection slows the process assembly line. The more joints that are selected for inspection (that is, not all joints need to be inspected) and the greater the information detail required from of the inspection results (referring to the height profilometry data collection), the longer the delay in the process flow.

It may be preferred that, on the detection of a dispensing defect, the board is cleaned of all adhesive or solder paste and the dispensing process repeated again. However, in other cases, it may be possible to correct isolated dispensing defects, such as the manual assembly of odd-form components or when as many or more defects would be created by a redispensing step. The latter case is particularly pertinent for very dense circuit boards.

40.5.3 Component Placement

Successful control of the placement process, whether a highly automated chip "shooter" step or the manual placement of odd-form components, must meet the following objectives:

- Select the correct component for the designated circuit board location.
- Place the component at that location, in the proper orientation, and within the tolerances of the x, y, and z (height) coordinates as well as angular rotation specification.
- Minimize damage to the component being placed, the circuit board features, and nearby components.
- Minimize disturbing the dispensed adhesive or solder paste under the placed component prior to curing and reflow, respectively.

Important automated placement equipment factors are the stiffness of the support frame and positioning system, the geometry and size of the placement head (chip shooter or gantry architecture), as well as the capabilities of the camera and vision system. The machine frame must be sufficiently stiff to withstand the accelerations and decelerations of the placement heads, which are becoming heavier to accommodate more part types to reduce changeover intervals. The configuration of the placement head must not cause damage to neighboring components nor disturb already-dispensed adhesive or solder paste. Vision system capabilities —resolution and field of view—and associated software are a trade-off between information gathering and component placement speeds.

Linear motors can achieve placement accuracies of 5 to 10 μm. The required degree of accuracy is a function of the size of the I/O (either lead, ball, or termination geometry) and the size of the circuit board pads on which the part is to be placed. Although it is preferred that the entire termination be centered on the pad footprint, this is not a practical objective due to tolerance build-ups. Errors are designated by the individual x and y linear (offset) coordinates, together with the rotational offset. In general, industry standards have allowed I/O-to-pad offsets of between 50 percent and 25 percent depending on the class of electronics (i.e., consumer electronics to high-reliability military and space assemblies). Placement errors can also be expressed by a single parameter, the lead-to-pad (LTP) ratio, which is calculated by an equation based on the x, y, and rotational offsets. The LTP is typically computed for the one lead that is most impacted by the machine error, resulting in a single LPT value per component type. An industry standard is available (IPC-9850) that provides a methodology by which to compare the placement accuracies of different machines. Statistical process control can be used to establish the placement performance of a particular machine *and* the components for a specific circuit board product.

As noted previously, placement defects can be identified in the machine by automated inspection. Alternatively, such defects may be detected by a separate machine inspection station or through visual inspection. A trade-off must be made between the number of parts inspected and detail of that inspection on the one hand, and the assembly process flow on the other. The more information that must be processed by the inspection step, the slower the placement process. Finally, every attempt is made to correct placement defects prior to the soldering step.

40.5.4 Soldering

Control of a soldering process must address the two premises previously noted. First, the soldering process—be it reflow soldering, vapor phase soldering, or manual soldering—must nominally be capable of making every joint on the printed circuit board without damage to the components or the laminate materials. Second, the soldering step must perform the assembly process repeatedly to meet product quality (defect) specifications.

Although the first premise seems intuitive, it must be recognized that some circuit board products may simply be outside the process capabilities of some equipment or manual capabilities, no matter the degree to which the machine parameters or operator actions are optimized. For example, in the case of manual soldering, it becomes increasingly more difficult to make a satisfactory joint when a multilayer circuit board exceeds approximately 2.4 mm. The soldering iron cannot provide sufficient heat to the joint structure to support wetting and spreading by the molten solder without causing thermal damage to the component or the Cu pad. A similar circumstance can come to pass with automated soldering process equipment.

Generally, with the increased thickness and/or complexity of the circuit board as well as with the increased mix of components, the gap widens between the process requirements of the product and the process capabilities of the equipment or operator. There are two possible solutions to this discrepancy: Either a capital investment can be made to upgrade the equipment (if available) to assemble the product successfully, or the primary assembly process must be limited to those components having similar geometries and thermal characteristics. The remaining devices are then treated as odd-form components and are soldered to the circuit board using secondary processes (e.g., selective soldering or manual assembly).

There are ancillary measures that can be taken to enhance the capability of a soldering process. For example, in the case of manually soldering components to a thick circuit board, the latter may be preheated to augment the heat input of the soldering iron. Similar approaches have been used for selective soldering, particularly when both the component and the circuit board are thermally massive.

However, these ancillary steps must be thoroughly investigated prior to implementation. First of all, these added steps not only impact the production throughput, but they may also affect the defect rate of the next-assembly process. For example, excessive preheating of a circuit board can degrade the solderability of the other pad features or cause thermal damage to vias or the underlying laminate material. Second, to maintain minimum defect rates for the assembly, these ancillary process steps must be as tightly controlled as the primary soldering processes.

Lead-free assembly generates an increased concern for the nominal process to make all of the solder joints on a circuit board. The issue has less to do with the ability of the actual equipment and fluxes to accommodate the high process temperatures than it has to do with temperature gradients across the circuit board versus the temperature sensitivity of components.

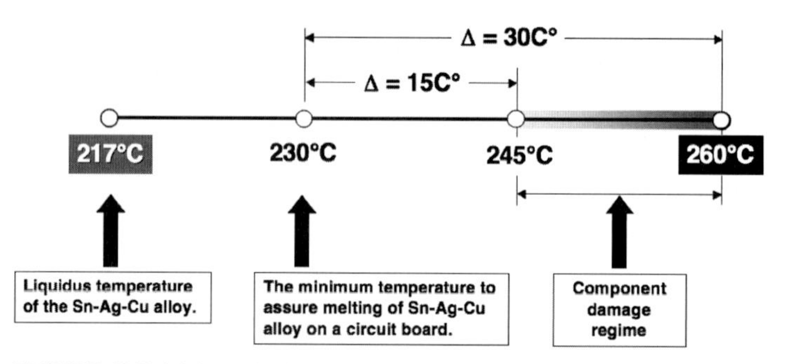

FIGURE 40.29 Schematic diagram illustrating the available process window for the Sn-Ag-Cu Pb-free solders (having melting temperatures of 217°C) referencing the typical reflow process time-temperature profile. The 230°C mark is the minimum process temperature to ensure adequate solderability. Temperatures exceeding 245°C increase the likelihood of thermal damage to larger plastic-molded area-array and flat-package devices. Temperatures above 260°C increase the chances of thermal damage to smaller passive devices (e.g., chip resistors, capacitors, inductors, and filters) and the circuit board structures (e.g., vias) and laminate material.

This point is illustrated by the schematic diagram in Fig. 40.29, which pertains primarily to the reflow process and typical time-temperature profiles.

A minimum temperature of 230°C is required to ensure adequate wetting and spreading by the Pb-free solder on circuit board pads as well as on component leads and terminations. Temperatures exceeding 245°C increase the likelihood of thermal damage to larger, plastic-molded packages (e.g., BGA and QFP devices). When temperatures exceed 260°C, there is the potential for thermal degradation of passive chip components (chip capacitors, inductors, or filters) as well as to circuit board structures (e.g., vias) and laminate materials. In the case of Sn-Pb eutectic solder that melts at 183°C, the minimum process temperature of 215°C ensures adequate solderability, which is 15°C lower than that of a nominal Pb-free process. The available Sn-Pb process window is 215°C to 260°C, or a ΔT equal to 45°C, rather than the ΔT of 30°C for the Sn-Ag-Cu Pb-free solders (T_{melt} = 217°C).

The important consequence of the smaller process window is that, as the mix of components increases on the circuit board, the likelihood increases that the solder joints on some of the larger components will not reach the minimum 230°C required for good solderability when the maximum circuit board temperature is required to be less than 260°C. Similar process window scenarios can be evaluated for wave soldering and manual processes. For example, in the latter case of manual soldering, the operator is unable to use the same soldering tip or iron to assemble all of the components. On the other hand, a soldering iron that is found to be effective for soldering large components may be too hot and/or too large for smaller devices, causing thermal and/or physical damage, respectively, to them.

The second premise of process control is the repeatability of making an acceptable solder joint on each assembled circuit board. With the exception of a major equipment failure that alters the time-temperature profile, the capability to solder the same joint repeatedly in an acceptable manner is more a function of the *consistent* solderability of the material set's properties. Those solderability properties include the efficacy of the flux, the properties of the solder metal (e.g., surface oxidation and particular to the case of solder paste, the solder powder particle size, and metal loading), and the solderability of the component I/O and associated circuit board feature. Moreover, the solderability of component I/Os and circuit board features is often identified as the primary factor in repeatedly making acceptable interconnections.

Unfortunately, because poor solderability is often the leading cause of poor reproducibility of acceptable solder joints, this does not bode well for Pb-free solders. Intrinsically, the Pb-free solders generally show poorer solderability than the eutectic Sn-Pb composition.

Therefore, to minimize assembly (solderability) defects, process engineers must be particularly vigilant to ensure that the surfaces of component I/Os and circuit board pads have optimum solderability.

40.5.5 Cleaning

The use of no-clean and low-solids fluxes has reduced the need to clean a greater number of circuit boards. Nevertheless, high-reliability electronics, as well as those used in harsh environments, may still require the removal of flux residues. The primary metric used to control the cleaning process is the quantity of residues that remain on the circuit board or components. Visual inspection (white light or ultraviolet light) and ionic testing are the two most commonly used methods to assess a cleaning step. In addition, in-circuit testing (ICT) provides a secondary means to indicate excessive residus presence. The residues prevent the electrical probe from reaching the conductor, which is registered as an "open" in the data collection routine. Unexpectedly large numbers of "open" defects may indicate excessive flux residues on the circuit boards as well as signify an actual electrical performance defect. Also, in the case of high-frequency circuit boards (radio frequency and microwave), ICT can identify flux residues as indicated by signal leakage and other parasitics when inspecting for electrical performance.

The cleaning step must be carefully considered for a Pb-free soldering process. The higher processing temperatures can increase the tenacity of residues, particularly those of fluxes designed for Sn-Pb processes. This point addresses fluxes used separately, as in manual or wave soldering, as well as the flux in solder pastes. The cleaning step may need to be enhanced to ensure that these residues are removed satisfactorily from the circuit board surface. On the other hand, flux formulations that have been developed specifically for the Pb-free soldering processes may create residues that are not compatible with cleaning processed developed specifically for Sn-Pb technology.

40.5.6 Network Communications

The generation, transmittal, and storage of process information is critical for real-time control of a circuit board assembly process, whether that process is for building consumer electronics in high volume or involves a low-volume assembly line for high-reliability military, space, or satellite electronics. The information that is generated by operators or automated machines on the shop floor typically includes commodity usage (i.e., circuit boards, components, etc.) as well as defects and machine malfunctions. Commodity usage information can then be made available to inventory control in the plant stockroom or even to the bill-of-material (BOM) suppliers across town and around the world, identifying the need for additional components, flux, and such at that particular site.

Process information is also accessed by the manufacturing engineer, who is monitoring the overall assembly flow, as well as by the process engineers or technicians who are directly responsible for equipment operations (e.g., furnace zone temperatures, flux levels, etc.). In addition, information may be shared between the different machines on an assembly line to prevent slowdowns in process flow due to equipment outage. Such slowdowns and stoppages can impact upstream steps. For example, there can be shelf-life issues with solder paste that is left out in a printer too long due to a stoppage created at a subsequent component placement machine or reflow oven.

It is often necessary to archive process-related information. Archived data are used to track defects as a means of process control. These data can assist in quickly determining the root cause of an out-of-control process. Process information can also be used to monitor long-term equipment performance, indicating the need for periodic preventive maintenance or machine replacement. Easy access to historical process information can also facilitate the reintroduction of a particular product line with a minimum investment in process development efforts.

Finally, archived process data are useful in an investigation of field failures. The traceability of the troubled product to the exact process parameters used on the assembly line at that time, material lots, operators, and so on can be accessed; these data are invaluable for the follow-up root-cause analysis, particularly in the case of high-valued, high-consequence electronics systems (e.g., military equipment, space and satellite hardware, as well as medical and automotive electronics). Understanding the circumstances under which the defective product was made can also provide insight into implications regarding long-term reliability and specifically warranty costs versus liability and the need to implement a product recall.

The need to access information quickly as part of assembly operations extends beyond process and inventory control. Peripheral information such as equipment software and operations manuals must also be quickly accessible. The ready availability of boot-up software and instructions minimizes the downtime required to reestablish a product line or effect repairs on a malfunctioning machine.

Once the need for a network has been determined, the locations of terminals, servers, and information storage equipment must be established. Local area networks (LANs) and the worldwide web (WWW, or Internet) can link access points located in adjacent rooms or halfway around the world. However, it is critical that the network uses consistent computer and data transmittal protocols in order to provide uninterrupted, information handling in an efficient manner. For example, serial communication via an RS-232 port is relatively common, but too slow for most machine applications. Ethernet communication via transmission control protocol/Internet protocol (TCP/IP) is a preferred method because most equipment suppliers offer this network option. The TCP/IP method allows computers to share resources, no matter whether the operating system is for an Apple®, a version of the Microsoft® operating system for PCs, or the UNIX operating system. Also, the Semiconductor Equipment Communication Standard (SECS)/Generic Equipment Model (GEM) is widely used as the interface between the host computer and assembly machines.

Lastly, the need for efficient network communications is particularly acute with the advent of Pb-free solders. In the case where OEMs as well as contract manufacturing service (CMS) companies are assembling both Sn-Pb and Pb-free technologies, it is essential to control all aspects of the assembly process with precision. In fact, process control extends beyond simply certifying that the reflow, wave, or soldering iron correctly reflect the higher soldering temperatures. It must also make certain that components and circuit boards have been delivered to the appropriate process line having the correct surface finishes and/or MSLs. Also, process control must establish that other factors, such as flux type and cleaning procedures that may be unique to the Pb-free process, are being used on the appropriate product line.

40.6 PROCESS EQUIPMENT SELECTION

The selection of assembly process equipment is an extensive undertaking for the manufacturing engineer. The three principle variables are equipment utilization, costs, and physical plant requirements. These three areas are listed in order of consideration, not in the order of importance. The analysis behind equipment selection can be very complex, given the wide variety of product designs, potential automation options, and access to worldwide labor markets. Unfortunately, the failure to consider thoroughly the details underlying any one of these three factors can result in a significant financial burden, lost time to market, and poor product quality/reliability for the OEM or CMS. The following sections provide relatively general discussions of these three areas. The reader is referred to many excellent texts on the topic of manufacturing engineering, as well as source books that specifically address assembly process equipment for electronics manufacturing.

40.6.1 Equipment Utilization

Equipment utilization refers to two aspects of the equipment. The first aspect is the operational time during which the machine is actually performing its intended function. The second aspect is production volume (expressed as parts or assemblies per unit time) provided by the machine while it is in operation. The consequences of operational time can be considered as follows: When the machine is performing its intended function, it is making product and as such is making a profit for the company. On the other hand, when a machine is idle, not only is it not making product (or profit), it may be losing money for the company. Those monetary losses result from wasted electricity and other utilities while the machine is on standby, labor costs for idle operating personnel, continued draw-down on the maintenance interval, and overall loss the equipment lifetime.

Next, equipment utilization is discussed in terms of throughput or product volume. It should be noted that the general approach is nearly the same for both Sn-Pb and Pb-free technologies. Only the specific cost factors may differ slightly between the two cases. A determination of equipment utilization requires that the appropriate metrics be defined that accurately reflect the machine activity. For example, in the case of component placement machines, the utilization metric may be components placed per unit time or it may be the number of stuffed circuit boards that are completed per unit time. Printed circuit boards or finished assemblies per unit time are the typical metrics for paste printing and reflow furnaces, respectively. For example, component placement rates are expressed as, components per hour (CPH). Some manufacturing engineers prefer to analyze utilization in terms of time per product unit, which is the inverse of production rate. Such an approach would be exemplified by measuring the time required to print the paste on, or to stuff, a single circuit board.

It is important to select a metric that is most appropriate for the machine and the products that are processed in it. Moreover, the metric should allow the process development engineer to monitor the performance of the machine process accurately; it can also be used to compare specifications between different equipment brands for a procurement activity.

Different methodologies can be used to exercise an equipment utilization analysis. The details of these techniques are many and beyond the scope of this text. Nevertheless, it is important to be familiar with several key attributes that are used to make utilization assessments. Descriptive definitions of several attributes are shown below:

- *Theoretical throughput* is the output of a machine based upon its underlying design.
- *Actual throughput* is the output of the machine for the particular product line or application. This value is always less than the theoretical value.
- *Utilization or utilization efficiency* is the ratio of actual throughput to theoretical throughput.
- *Operation time* is the total time during which the equipment may be required to operate. For example, it may be an eight-hour shift or it may be 24 hours if three shifts are running in the factory.
- *Equipment uptime* is the time that the machine is available to perform its task; it may not necessarily be performing the task. Therefore, this time period takes into account the various contributions to downtime, such as preventive maintenance, unexpected breakdowns, resupply of commodities, and so on.
- *Productive time* is the time during which the machine is actually performing its task.

These various attributes can be further broken down as the need arises; for example, several factors can account for loss of equipment uptime, including scheduled maintenance versus unexpected machine failures, commodity resupply, etc.

These attributes are reduced to mathematical variables that are used in equations to calculate the performance of a particular machine. However, the assembly process for electronic circuit boards is comprised of several machines or cells that are *in series*. Therefore, once the performance attributes are determined for a given machine, they must be integrated to predict the performance of the entire assembly line. For example, it is preferred that the throughputs of the individual machines be matched for the product application. A mismatch of machine capabilities can generate bottlenecks in the assembly process. The slowest machine or function then controls the production rate of the entire line. The obvious consequence of a bottleneck is an underutilization of the other equipment, which ultimately appears as an increase in manufacturing costs.

But there are also quality control reasons to smooth out the flow of bottlenecked assembly processes. For example, if the component placement machine has a throughput of 100 circuit boards per hour, but the reflow machine can only process 90 circuit boards per hour, then printed and stuffed boards will spend time in the open factory air awaiting entry into the reflow step. The consequence is a degradation of paste properties, resulting in the increased likelihood of solderability defects and a drop in product yield. Such technical ramifications must also be addressed when considering equipment utilization for an assembly process.

40.6.2 Costs

There are several cost factors to consider when establishing an assembly line: the initial capital investment for the machine(s); depreciation and lifetime (obsolescence) of that equipment; operational costs such as labor rates, electricity, water, or the use of inert atmospheres; and maintenance support costs. Many of these cost factors are determined on a "per product unit" basis so that an estimate of potential production volumes is often required for their calculations. Many OEMs and CMS providers have had to establish their own cost models to take into account labor rates, energy costs, shipping costs, and tax structures that are specific to the assembly plant location—on-shore as well as overseas. Because of the cost variables just mentioned, it is difficult to establish a universal model for these calculations.

When addressing the implementation of Pb-free soldering technology, one must consider two aspects regarding equipment costs. First are the direct costs of purchasing and installing equipment that can provide a Pb-free assembly capability. Second is the cost of simply developing a Pb-free assembly line in the factory.

Regarding the first point, equipment capability vis-à-vis Pb-free solder pertains primarily to the soldering process, including the reflow furnace, wave-soldering machine, soldering irons, etc. Printing and placement equipment is largely unaffected by the Pb-free soldering technology. Most reflow equipment used for Sn-Pb soldering is nominally capable of also supporting Pb-free soldering for moderately complex circuit boards. Changes are simply made to the time-temperature profile of a reflow furnace. On the other hand, for more highly complex products, a replacement reflow furnace may be required that has more than the typical five to seven zones. A larger number of zones allows more flexibility—and greater control—over the time-temperature profile. In the case of wave soldering, the exchange of Sn-Pb solder for a Pb-free alloy may also require the replacement of the pot, impeller, and baffles to prevent excess erosion by the higher Sn solders. Last, it is expected that erosion of soldering iron tips will be greater with the Pb-free solders. Tip technology has changed little from the Ni-coated or Fe-conted Cu core construction. Therefore, it may simply be a matter of stocking extra tips for those operators using Pb-free solders.

The second cost aspect addresses the manner in which a Pb-free capability is to be implemented on the factory floor. Specifically, the decision must be made whether to have separate Sn-Pb and Pb-free solder assembly lines or to mix the two technologies using the same assembly equipment. Critical factors include capital costs, factory floor space, infrastructure required to prevent inadvertent mixing of technologies, and projected product volumes. It is preferred to separate the two assembly capabilities physically. This separation is not limited to the actual machines, but also includes support functions such as solder material storage and component inventory control in order to prevent the use of the wrong solder or components and circuit boards with incorrect finishes. The extent of this separation of capabilities will determine the need for the capital investment in new machines as well as the cost of duplicating some support functions and personnel that are needed for adequate process control.

40.6.3 Physical Plant

The cost of having a soldering process line extends beyond the capital costs of the equipment or even the labor costs for operators and process engineers. There are also the costs of operating the machines themselves. The factory must have adequate floor space for the assembly line footprint, as well as the access required to move parts and finished assemblies to and from the line. There are also space requirements to perform maintenance functions safely and make repairs to equipment. Finally, there are assembly costs arising from utility usage: electricity, compressed air (pneumatic actuators), and water (cleaning).

If the reflow or wave-soldering machine is to have an inert atmosphere—typically nitrogen is cheaper than argon—then it becomes necessary to select between the options for the gas source. Those alternatives are compressed gas tanks, a liquid nitrogen tank, or an onsite nitrogen generation facility. The compressed gas tanks are usually used only for short production runs. For most moderately sized electronics assembly factories, the most cost-effective option is to use one or more liquid nitrogen tanks, off of which is bled the nitrogen gas. Only for very-large-scale manufacturing facilities is an onsite generation plant the best option. Ancillary costs include the installation of pipe lines that transport the gas to the specific machine sites on the factory floor as well as the provision for adequate ventilation of the used gas from the machine to the external environment.

The physical plant costs are generally similar for both a Pb-free assembly process and a Sn-Pb line. The two possible differences would be electricity usage, which may be greater with the Pb-free line due to the higher soldering temperatures, particularly with respect to a reflow process, and the need for an inert atmosphere capability for the soldering machine.

40.7 REPAIR AND REWORK

It is necessary to clearly define the terms *repair* and *rework*:

Repair, which is also referred to as "touch-up," describes the *single-step* procedure of applying flux, solder, and heat to a previously formed joint in order to improve it. Besides meeting acceptance criteria, improving the fillet geometry may also be required to bring the interconnection into a geometry "window" that is commensurate with long-term reliability predictions.

The benefits of repair operations have long been debated. On the one hand, optimizing the fillet geometry or hole fill will improve solder-joint consistency in order to accommodate long-term reliability prediction criteria. However, the opposing viewpoint is that the added reflow temperature "cycle" degrades the reliability of the other solder-joint structures, such as solder mask adhesion, the bond between Cu features and laminate, as well as damage to the laminate material itself. It is likely that any repair-induced degradation will increase with a lesser quality of the circuit board. Therefore, particular consideration should be given to the use of repair steps on consumer electronics circuit boards, which are typically made of lower quality materials. Fortunately, repair steps would be most likely used on high-reliability electronics (military, space, and satellite applications), which typically use higher-quality circuit board materials and construction.

A *rework* procedure is typically performed to replace a defective component on the circuit board. Therefore, the rework procedure has the following *multiple steps*:

1. Remove the previous solder joints in order to release the defective component from the circuit board.
2. Dress the circuit board solder pads by removing any excess solder while the latter is molten.

3. Clean the circuit board of solder metal (spatter or balls) and flux residue.
4. Place the new component on the circuit board site. Solder is supplied to the joint area by preplaced paste or preform as well as by solder wire (in a hand-soldering process). Flux must also be added to the joint area when using preform or when hand soldering with a wire, unless the latter is flux-cored.
5. Apply the heat source in order to re-create the joints.
6. Clean the circuit board of solder metal (spatter or balls) and flux residue.

The obvious distinction between the repair and rework processes is the number of heating cycles to which the printed circuit board is exposed in the latter case. There is one heating cycle in a repair operation that requires melting of the solder, whereas there is a minimum of three such cycles in a rework procedure. The circuit board laminate material, solder mask, and adhesion of the Cu conductor to the laminate must be sufficiently robust against degradation caused by these heating cycles. Poor thermal stability can result in manufacturing defects that reduce production yields, or aging defects that jeopardize the long-term reliability of the product once it reaches service. Often, preheating of the circuit board is used to reduce thermal shock and thermal gradients that may potentially damage the laminate material, vias, Cu pads, etc. Also, the replacement component must have adequate thermal stability to accommodate the typically faster heating and cooling rates of the more abbreviated rework time-temperature profile vis-à-vis the original assembly process.

A second distinction made between repair and rework procedures is the heating methodology. A repair operation addresses one solder joint at a time. Therefore, heating is localized to a relatively small area. On the other hand, in a rework procedure, the heat is supplied to the site in a more de-localized manner because of the need to de-solder multiple joints at one time. Similarly, in the re-soldering of a multiple I/O component, it is preferred to make all of the joints at one time in order to minimize the rework time. The heat source may be a hot-bar placed across a row of gull-wing leads or hot air that envelops the entire component site; for example, such a heat source might be used for the rework of a BGA package.

The de-localized heat flow raises concerns regarding thermal damage. For example, with respect to the component in question, there is potential thermal damage to the package material and internal components (e.g., the Si die) and interfaces. Also, there is possible heat damage to the Cu bond pads as well as hole and via structures in the underlying circuit board. Beyond the immediate component, there is the potential risk of remelting nearby solder joints or thermal damage to the neighboring component packages, particularly on very densely populated circuit boards. Thermal damage to nearby components can arise from any one or any combination of the three heat transfer modes: radiation (hot-bar or soldering iron), convection (hot air), or conduction (through holes or along other conductor features). During process development, strategically placed thermocouples should be used to determine the temperature rise of the reworkable component as well as those of neighboring structures prior to implementing a repair or rework procedure.

A third distinction between repair and rework is product time frame. Repair steps are typically performed on circuit boards that have been inspected immediately following the assembly process, prior to shipment to the customer. However, rework is not limited only to being performed immediately following a post-assembly inspection (visual or in-circuit testing). Rework activities can also be performed on electronics assemblies well after the products have entered service. The rework may be performed to replace a broken device or to upgrade the functionality of a legacy system. Although the nominal process steps are similar in the latter case, there are complicating factors that must be addressed in the process. The electronics must be taken apart to the level of the component or daughter-board in need of repair. The removal of potting material and conformal coatings must not damage the other components or their solder joints. Also, dirt, corrosion products, and other contaminants must be removed from the assembly prior to the first heating cycle. Following the rework procedure, it will be necessary then to reapply conformal coatings and encapsulants as well as to reassemble all hardware. Each of these steps must now be taken outside the scope of the original assembly process, yet achieve the same quality without ancillary damage to components, circuit boards, or other solder joints.

A further complication to the rework procedure performed on either post-assembled or field-return hardware is the use of adhesives or staking compounds to anchor the device to the circuit board. Because many adhesives used for this purpose are thermosetting, they do not readily soften with the application of heat. Care must be exercised when applying heat to soften adhesives so that the temperature rise does not damage the underlying laminate or neighboring components. Organic solvents such as those containing alcohols, acetone, methyl chloride, and so on (the use of some of which may be strictly controlled for environmental health and safety reasons) can successfully dissolve a number of adhesive compounds. However, it must be confirmed that these solvents do not attack the package materials of the components, the solder mask and laminate, or critical markings.

Lead-free soldering raises the following two concerns regarding repair and rework procedures: higher melting temperatures of the Pb-free solders, and the mixing of Sn-Pb and Pb-free solders. Clearly, the higher melting temperatures of the Pb-free solders increase the likelihood of thermal effects that include remelting of neighboring solder joints and thermal damage to component packages and the circuit board laminate. As noted previously, this concern can be readily addressed during process development efforts, through the use of thermocouples placed on the targeted component as well as on neighboring components, solder joints, and circuit board surfaces.

The second issue is that of mixing Pb-free and Sn-Pb solder. This situation would be less likely to occur with newly fabricated product, when repair and rework procedures are performed immediately after assembly inspection. The same solder would be used for both processes. Rather, the concern arises particularly with regard to the rework of legacy products that are returned from service. Older, field-return electronics were most likely assembled with a Sn-Pb solder. In those cases where Pb-free rework procedures are in full use, there is the likelihood of intermixing Sn-Pb and Pb-free solders in the interconnections. In terms of the rework process itself, the poorer solderability of Pb-free solders will be improved by Sn-Pb residues remaining on the circuit board pads. (The I/O finish of the replacement component will likely be Pb-free, unless the component was in storage as a life-of-program-buy [LoPB] for that system. Then, the Sn-Pb finish on the I/O will also improve Pb-free solderability.)

However, because there are concerns regarding the long-term reliability of mixed-solder interconnections, it may be necessary to remove more thoroughly any Sn-Pb solder residue from the circuit board pad or through hole. Care must be exercised to not overwork the Cu features, which may result in a loss of their adhesion to the laminate (e.g., pad lifting). Other degradation modes include excessive dissolution of the Cu features, particularly at the knee on the top and bottom of through-hole barrels as well as damage to the solder mask or underlying laminate material.

Repair and rework procedures differ between through-hole-mount and surface-mount interconnections. That difference is due to the size and geometries of the interconnections, the number of I/Os on the packages, the thermal sensitivity of the components, and the overall density of the printed wiring assemblies. The following subsections describe these factors for the two technologies.

40.7.1 Through-Hole Technology

By and large, the repair and/or rework of through-hole solder joints still uses hand-soldering procedures (see Fig. 40.30). In the case of a repair procedure, the first step is to add flux and solder to the joint. Then the soldering iron is contacted to the lead, thereby allowing the solder to reflow. The tip should be removed from the joint as soon as possible to prevent excessive heat conduction up the lead and into the component, where it may damage the active (Si chip) device(s), glass-to-metal seals, etc.

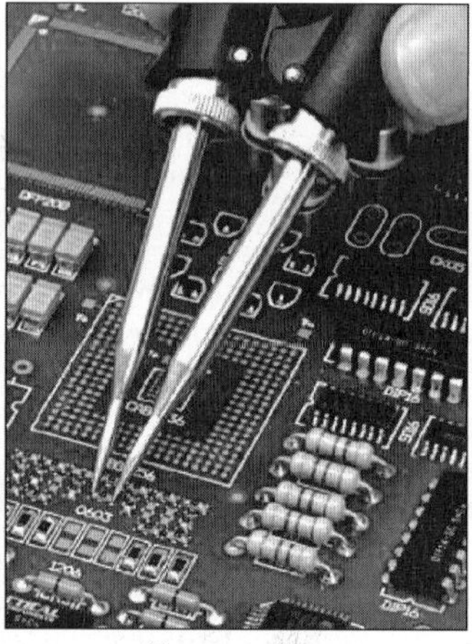

Point tip   Tweezer tips   Hot bar tip

FIGURE 40.30 Hand-soldering techniques using point, tweezers, and hot-bar tips to perform repair or rework procedures. *(Courtesy of OK International and Electronic Products and Technologies.)*

In the case of rework procedures, the most difficult step is often de-soldering the damaged component for the latter's removal from the circuit board. In the case of axial leaded devices, it may be possible to cut the leads of the damaged component. The component can then be removed from the circuit board, thereby increasing access to the joint area for the de-soldering operation. Lead cutting is more difficult for radial leaded and high-pin-count components such as DIPs, PGAs, and connectors due to their shorter lead lengths and limited clearances between the package and circuit board surface.

Typically, the operator must remove the solder from the joint, usually one joint at a time, to release the component. A light coating of flux is applied to the Cu braid; the braid is then placed between the fillet and the soldering iron, with preference given to resting the tip near the top of the fillet, against the lead. This technique limits possible damage to the Cu pad, solder mask, or laminate. Sufficient solder must be removed from each joint so that the component can be extracted with *minimum* force to prevent damage to the Cu barrel or the Cu pad.

The "dressing" of the through-hole and Cu pad is the next step. An excess of solder remaining in the hole will hinder lead insertion for the new component. It is difficult to extract physically all solder residue form inside a hole, even at modest aspect ratios (board thickness-to-hole diameter). This difficulty is further compounded with thicker, multilayer circuit boards because the soldering iron tip cannot compensate for the large thermal mass of the laminate or the heat-sink effects caused by the internal Cu layers. Finally, it is often necessary to contact the Cu pads of the circuit board briefly, with the braid between the pad and the tip, to remove any excessive solder residue.

Soldering of the replacement component on to the circuit board has fewer complications than does part removal. Using a hand-soldering process, after the device has been inserted into the circuit board, each joint is made one at a time. The operator provides real-time inspection of the joint for any defects. Finally, if necessary, flux residues can be removed from the circuit board using the appropriate cleaning procedures.

Thicker, multilayer boards are more difficult to solder due to their greater thermal mass and the heat-sink effects of the internal layers. Preheating of the circuit board can be used to augment the soldering iron heat source. Typically, the circuit board is heated to a temperature in the range of 100 to 125°C. However, because the preheating step is applied to a full-up assembly (minus the defective part), consideration must be given to the heat sensitivity of other components on the circuit board. Preheating with a hot plate may be acceptable for some applications. However, when temperature-sensitive components are nearby, other heating methods such as hot-air (gas) sources may be required to localize the temperature rise away from other components or structures.

Several techniques can be used to assist with the hand-soldering process (de-soldering or re-soldering), or even substitute for it altogether. For example, there is equipment that can heat multiple joints at a time. The so-called hot-bar technique is a soldering iron with a long bar for the tip. The bar simultaneously heats up a row of leads for the de-soldering or re-soldering step. Hot air or hot gas guns can provide more diffuse heat sources that can melt all of a component's solder joints at the same time, thereby facilitating the removal and replacement of larger components that have a greater number of I/Os.

Selective soldering, or solder "fountains" that perform like miniature wave-soldering machines, are an effective approach for the de-soldering and re-soldering of through-hole components. In fact, this technique is particularly well suited for connectors. The thermal energy provided by the molten solder fountain can overcome the heat-sinking effects of the connector structure and thick, multilayer circuit boards. However, it may be necessary to preheat particularly thick circuit boards (>3.0 mm) having multiple internal layers in order to melt the solder joints of a defective product successfully, as well as to achieve hole fill and adequate fillets on the replacement part. Again, care must be exercised to prevent thermal damage to neighboring components or the undesirable remelting of their solder joints.

It was noted previously that a Pb-free solder option raises two concerns: the higher soldering temperatures required by the new alloys, and the mixing of Sn-Pb and Pb-free solders, particularly when performing rework procedures on legacy hardware. Both concerns are generally of lesser magnitude for through-hole solder joints. The physical separation of axial or radial leads lessens the likelihood of thermal damage to the circuit board or replacement component. Potential processing and reliability effects resulting for the mixing of Sn-Pb residues with Pb-free solder are less significant due to the greater quantity of solder required in through-hole interconnections. In particular, the solder composition within the fillets, which provide most of the strength and service life of the joints, would be least impacted by Sn-Pb contamination.

40.7.2 Surface-Mount Technology

Solder-joint repair and the rework of surface-mount components pose a greater challenge due in large part to the reduced size of the devices and associated I/Os. (In fact, the difficulty of repairing or reworking surface-mount circuit boards was one of the leading drivers for the electronics industry, as a whole, to re-embrace more stringent process control techniques, including SPC, to minimize assembly defects.) Contributing hurdles included the higher board densities that accompany the small part sizes; an increased sensitivity of component materials to potential thermal damage including the Si die resulting from flip-chip attachment and thinner molded packages; as well as solder joints that are out of direct view from the operator, as in the case of area-array packages.

Another important consideration is the overall fragility of surface-mount Cu pads on the circuit board compared to through-hole structures. In the latter case, damage to the bond between the Cu pad and the laminate was less likely due to a greater robustness of the feature. When such damage in the form of lift-off occurred, it was of a lesser consequence because the hole barrel provided an additional mechanical attachment to the laminate. On the other hand, surface-mount pads are smaller in order to accommodate finer package I/O pitches and overall higher board densities, giving them less overall bond area with which to adhere to the board. Also, the pads are attached to the laminate without a hole or, at best, with only a small via, which also increases the sensitivity of Cu pads to thermal damage in the form of lift-off.

Repair procedures are typically performed on surface-mount interconnections using hand-soldering techniques. A stereomicroscope of 10 to 50 × magnification is essential at the workstation. The soldering iron tips are miniaturized to accommodate the smaller component sizes and higher board densities. In general, this technique does not necessarily reduce the time to repair the smaller solder joints, because the smaller tips also have a reduced thermal mass for supplying heat to the joint. As always, the soldering time should be minimized as much as possible. Also, in the case of ceramic chip resistors, inductors, and MLTF capacitors, direct contact between the tip and the component termination should be avoided to prevent de-lamination of the metallization. It is particularly important to avoid contacting multilayer chip capacitors with the soldering iron, as these materials are particularly sensitive to thermal shock, which can lead to cracking of the device body.

Rework procedures based on hand-soldering techniques are still used for surface-mount technology. Manual techniques include equipment innovations that can accommodate the smaller part sizes, multiple I/Os per package, and higher board densities. However, it is recognized that more precisely controlled procedures are required for reworking fine-pitch, peripherally leaded packages and area-array components (BGAs, CSPs, and flip-chip). Often, an operator simply does not have the physical ability to perform the necessary actions or to execute steps that must be done simultaneously. Therefore, rework "stations" are available that can perform some or all of the steps in a semiautomated or even fully automated fashion, depending on the equipment capabilities. Predetermined thermal profiles are used to remove a defective component and then to install the new part. Positioning of the replacement part is achieved manually or by stepper motors that can be executed manually or controlled by software commands.

The three steps in the rework process for surface-mount technology are as follows. First, there is removal of the defective part. The component removal step subjects the circuit board to the first of three solder reflow cycles. Different heat source methods are available to address the variety of package I/O configurations (e.g., leadless, gull-wing leads, etc.). The removal of the defective component, particularly high I/O fine-pitch leaded and area-array packages, is best done when all of the solder joints are simultaneously molten to prevent damage to the more fragile Cu pads. Soldering tweezers are used to melt both solder joints of chip resistors, capacitors, inductors, or diodes. The tweezers are then used to pull the component quickly from the site. The hot solder bar "tip" is used to melt, simultaneously, all of the solder joints in along a row as may be required for peripheral leaded packages, such as gull-wing leaded and J-leaded packages.

Hot air or hot gas (nitrogen) has been used to reflow, simultaneously, all solder joints of fine-pitch (high I/O count), peripheral leaded packages (e.g., QFPs) as well as those of area-array packages that include BGA components, CSPs, and FC components. An example of a hot-gas nozzle applied to a BGA package is shown in Fig. 40.31. Control of the hot-air flow maximizes heat delivery to the solder joints and minimizes the temperature rise in the component body. The shroud also limits the exposure of neighboring devices and their solder joints to elevated temperatures.

Unfortunately, unlike with one or two solder joints, it is not always possible for an operator to "sense" the presence of a few unmelted solder joints among the several hundreds or thousands of small interconnections on a fine-pitch peripherally leaded or BGA package. The consequence is damage to the Cu pad, which must then be repaired if possible. Under these circumstances, rework stations have become invaluable. Vacuum pickups or finger mechanisms attach to the component and then pull on in it with only enough force to release it from the circuit when all of the solder joints are molten, thereby preventing damage to the pads.

Once the defective component has been removed from the circuit board, the second step is to remove excess solder from the Cu pads. This step is also referred to as *dressing* the pads. Remaining solder bumps and spikes interfere with the placement of the new component. More importantly, excess residual solder, the quantity of which is not tightly controlled, causes variability to the replacement component's solder joints, which can lead to possible opens, shorts, or a loss of long-term reliability. Because the cleaning or dressing step requires a solder reflow heat cycle (the second of three), the time-temperature profile of that cycle must be

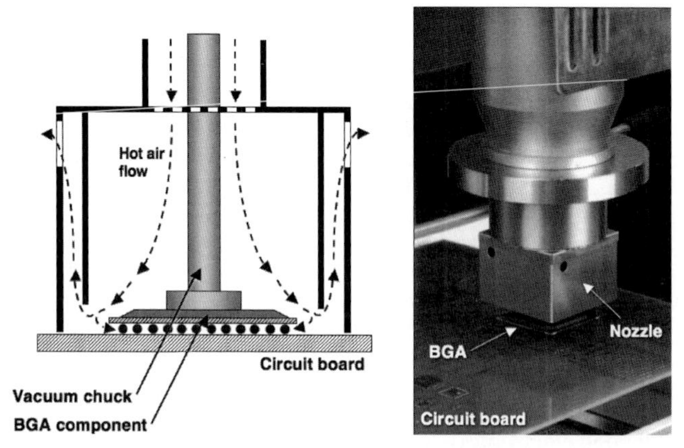

FIGURE 40.31 A hot-air shroud that is used to simultaneously reflow all of the solder joints on a BGA package. *(Courtesy of OK International.)*

minimized to prevent damage to the more fragile Cu bond pads and nearby traces as well as to the laminate itself. This procedure is typically performed manually.

The third step in the rework process is the assembly of the new component to the circuit board. Here, the circuit board site is exposed to the third of three solder reflow cycles. Also, it is at this point that semiautomated or automated rework capabilities can be particularly advantageous when compared to a manual operation. First, there is the step of providing solder to the joint pads. The presence of other components precludes the use of stencil or screen-printing dispensing techniques. Although solder preforms are a viable approach for components of larger I/Os, the most versatile and repeatable method for surface-mount components is dispensing paste through a syringe. This method may be performed manually when the number of I/O are relatively small. However, for particularly small passive components, fine-pitch peripheral leaded packages, and area-array devices, automated solder paste dispensing is the preferred technique to control both position and paste quantity.

In the case of some BGA and DCA/FC components, there may not be the need for additional solder. Only a flux is dispensed on the Cu pads prior to the component placement step. The tackiness of the flux anchors the part in place prior to reflow soldering.

Next, the new component is placed on the circuit board. The extent of rework placement automation depends on the mix of components. Specific variables include the package type (a passive chip device, a QFP with leads, an area-array package, etc.); the package size and its fragility, which determine handling requirements; and the number and pitch of the interconnections. At one extreme, manual assembly procedures are well suited for large leadless chip resistors and capacitors (generally sizes >1206 English) as well as LCCCs and peripherally leaded packages having pitches as small as 0.4 mm. Smaller components, particularly passive chip devices having sizes less than or equal to 0402 (English) are best placed by the mechanical handling and positioning capabilities of a rework station. At the other extreme are the 1.0 mm, high I/O count area-array packages. In these cases, the fine ball pitch, coupled with the inability to position the underside balls accurately over the circuit board pad, nearly always requires some level of automation to ensure the correct placement of the component.

After placement of the component, the soldering process is performed to make the joints. Irrespective of the exact method whether it is hot gas or a soldering iron, or the extent of automation the overall objective should be that the time-temperature profile resemble as closely as possible the process used to make the original solder joints. The solder paste is the same as that used in the initial assembly process. As a result, each step, from the preheat stage to melting of the solder and subsequent cooling sequence, must be controlled to minimize defects in the rework solder joints.

The cleaning step can be eliminated by the use of no-clean or low-solids fluxes in the rework procedure, whether the flux is used alone or is contained in the solder paste. However, if a flux was used that requires that the residues must be removed, some precautions must be followed. In the case of a circuit board product that was reworked immediately following the initial assembly process, one can use the same cleaning steps used in that original process again by simply passing the circuit board through the cleaning cycle a second time. On the other hand, in the case of hardware that is typically built-up further by next-assembly steps, additional materials compatibility concerns must be addressed with respect to the cleaning solutions. For example, it may be necessary to apply the cleaning materials only to the immediate location of the rework activity, avoiding otherwise sensitive materials elsewhere on the circuit board.

As noted previously, the two general concerns that must be addressed regarding Pb-free solder rework are higher soldering temperatures and the mixing of Sn-Pb and Pb-free solders for fielded electronics. By comparison to through-hole technology, both issues are amplified in surface-mount technology. Surface-mount components have greater temperature sensitivity due to the materials used in their construction as well as their smaller size. Multilayer chip capacitor ceramics are particularly prone to cracking under thermal shock conditions caused by too fast of a heating ramp or too fast of a cooling rate. In the case of area-array packages, excessively high temperatures cause the larger packages to warp or "potato chip." In the extreme cases, warpage can cause opens between the solder ball and pad due to an increased gap or result in short circuits caused by a shortened gap that compresses the solder balls, allowing neighbors to contact one another.

The second concern is that of mixing Pb-free solders with a Sn-Pb solder residue that may remain on the circuit board after removal of the defective component. Again, this issue pertains primarily to field-return hardware that had a legacy of Sn-Pb solder. The levels of Pb contamination are potentially higher in the smaller, surface-mount solder joints than they are in through-hole interconnections, so that the impact of Pb on melting point depression, solderability, and, more importantly, long-term reliability is similarly more significant. This concern is acute with Bi-bearing, Pb-free solders, especially those with greater than 5 wt. percent Bi, in which the formation of a low-melting-temperature (96°C) Sn-Pb-Bi ternary phase can potentially degrade long-term reliability. However, studies indicate that simply exercising due diligence with current pad-cleaning techniques will remove Sn-Pb residue to an extent that such effects are minimal. The consequence of unnecessarily subjecting the pad to additional heating cycles to remove all traces of Sn-Pb solder is damage to the Cu/laminate bond, which can result in a loss of electrical reliability.

When there is a likelihood that a significant amount of Pb contamination (generally >5 wt. percent Pb) will remain in the Pb-free interconnection, ensuring a minimal impact on long-term reliability requires that the Pb be completely intermixed throughout the interconnection. Under the circumstance when the Pb is not completely distributed in the joint, two distinct microstructures are created: one that is Pb-rich, and the other in which Pb is absent. This condition is of particular concern with BGA and CSP solder joints. The boundary between the two segregated microstructures is weakened with respect to thermal mechanical fatigue failure, thereby potentially reducing the reliability of the interconnection.

40.8 CONFORMAL COATING, ENCAPSULATION, AND UNDERFILL MATERIALS

When required, the final steps in the assembly of a printed circuit board are the applications of conformal coatings, encapsulants, and underfill materials. Conformal-coated circuit boards are shown on the bottom of the stereo photograph in Fig. 40.32a; the same circuit board,

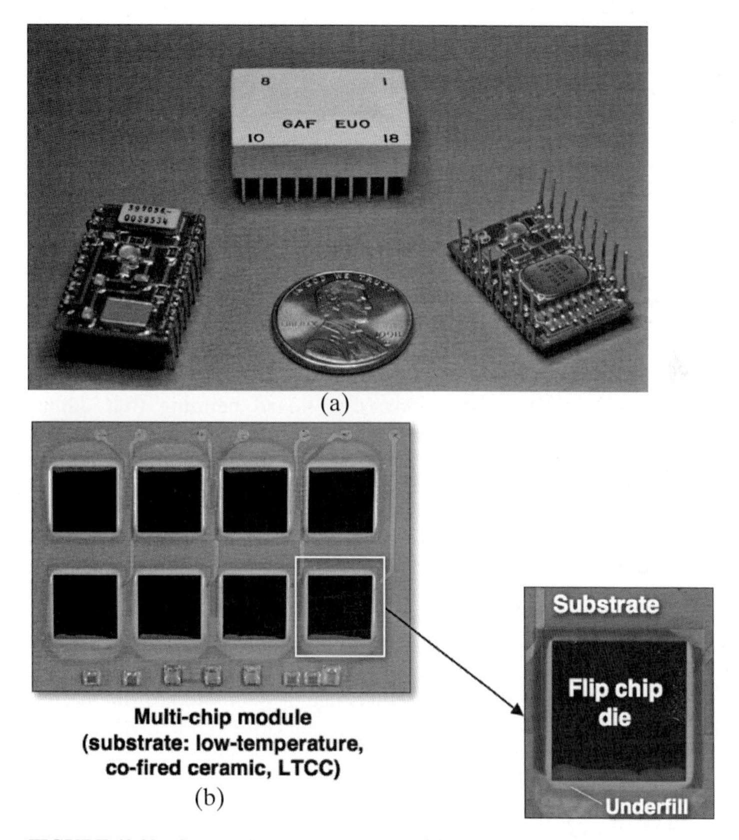

(a)

(b)

FIGURE 40.32 Stereo photographs showing (a) conformal-coated (bottom) and encapsulated (top) versions of a surface-mount circuit board and (b) underfilled flip-chip components. *(Courtesy of Sandia National Laboratories.)*

encapsulated, is shown in at the top of the photograph. An example is shown in Fig. 40.32b of underfilled flip-chip devices.

The role of the conformal coating is to protect the circuit board and components against large-scale contamination from dirt, water, liquid chemicals, corrosive gases in air pollution, mold, and fungus. Because the layers are very thin (<0.25–0.50 mm) and relatively soft, they do not offer significant protection against handling damage or mechanical loads in shock-and-vibration environments. Protection against the latter conditions is obtained from encapsulants, also called *potting* materials, and underfill materials. Underfill materials perform a function similar to adhesives by securing the component to the circuit board. The difference is that the underfill material is applied after soldering and must fill the entire gap between the package and the circuit board. In the case DCA/FC, the underfill also improves the thermal mechanical fatigue reliability of the solder joints.

The details of these materials are discussed in other chapters. However, it is important to understand their role in the overall circuit board assembly process. First of all, the functions of these three materials require that they have good adhesion to all surfaces. Therefore, the printed wiring assembly must be cleaned of any flux residues as well as those residues left behind from the cleaning procedure.

Second, the application of conformal, encapsulant, and underfill materials must be tightly controlled in order for them to function properly. The nominal thickness and its uniformity are critical for conformal coatings. A coating that is too thin is a less effective barrier against harsh environments. Thickness and uniform coverage have become even more critical as many designers are relying on conformal coatings to mitigate the consequences of Sn whisker growth on 100 percent Sn coatings. On the other hand, a conformal coating that is too thick can interfere with rework procedures as well as prevent fit-up of the circuit board into racks and backplane cabinets. In general, visual inspection is used as a nondestructive method to control the application processes.

In the case of encapsulant or potting materials, there must be complete filling of the mold, especially around the components, to prevent stress concentrations that can damage surface-mount solder joints. Potentially damaging stresses can also be generated when the density of the potting material varies excessively about the circuit board. The resulting variation of physical and mechanical properties of the foam encapsulant (thermal expansion coefficient, modulus, etc.) can lead to excessive board bending that damages solder interconnections, particularly those of large components (e.g., BGA devices). Therefore, it is necessary to control the encapsulant material itself as well as the "casting" process to prevent these defects.

Unfortunately, most nondestructive evaluation techniques are not effective at finding encapsulation defects. Voids and density variations are not readily detectable by visual inspection or by x-ray analysis. In the case of x-ray, the inherently low density of encapsulant foams limits the contrast between filled and unfilled (void) regions. Scanning acoustic microscopy (SAM) has had some limited success when the encapsulant layer is relatively thin, which is not often the case. Witness samples taken during the casting process are used to confirm foam density.

A number of studies have been performed that underscore the crucial requirement to control the underfill process used on flip-chip devices. Even small variations in underfill material properties (e.g., viscosity, percentage of filler, etc.) or the method for dispensing the underfill material between the die and substrate can significantly alter the long-term reliability of these interconnections.

The accepted method of nondestructive testing used to control the underfill process is SAM. The thin layer allows this technique to detect voids in the underfill material, which when located near the solder interconnections can be responsible for a significant loss of thermal mechanical fatigue reliability. X-ray techniques can be used to monitor the density of the underfill material, specifically, the distribution of filler material within the layer under the die. Density variations can indicate a larger distribution of underfill mechanical and physical properties, which may affect long-term reliability performance of the solder joints. Quantitative image analysis can be coupled into SAM and x-ray analysis data to provide valuable process control tools for the factory floor.

40.9 ACKNOWLEDGMENT

R. Boulanger is acknowledged for contributions to this chapter, which is based on "Assembly Processes," *Printed Circuits Handbook, 5th ed.,* Clyde E. Coombs (ed.), 2001, Chap. 40. The author also wishes to thank M. Dvorack of Sandia National Laboratories for his careful review of the chapter.

CHAPTER 41
CONFORMAL COATING

Jody Byram
Lockheed Martin Commercial Space Systems, Newtown, Pennsylvania

41.1 INTRODUCTION

Once a circuit card assembly (CCA) has been fully assembled and tested, it often needs to be protected from the environments in which it will ultimately be used. Moisture, salt, dirt, fungus, and a variety of other contaminants as well as mechanical shock and vibration will cause failure on an unprotected assembly. The CCA's last line of defense against the elements is a thin layer of coating that conforms to the shape of all the part leads, solder joints, and other complex features on the completed CCA. This conformal coating (or flow coating) is applied as liquid to the fully assembled CCA, and is cured in place to form a protective layer of insulation for the assembly.

This chapter presents:

- An overview of the five basic conformal coating types
- Preparatory steps necessary to ensure a successful coating process
- Various methods of applying conformal coating
- Repair and rework processes
- Design guidance on when and where coating is required, and which physical characteristics and properties are important to consider

41.1.1 Objectives and Applications for Conformal Coatings

Besides protecting the CCA from the elements, conformal coating also provides:

- Mechanical support for components during mild shock and vibration
- Electrical insulation for high-voltage components, especially at altitude
- Chemical resistance
- Prevention of short circuits due to tin whisker growth from pure tin-plated part leads and cases

Because of the extra cost involved with the cleaning, coating, curing, and touchup operations, conformal coating was usually reserved only for equipment that required high reliability, such as military and aerospace electronics, and equipment that needed to operate in severe environments. With the introduction of automated application robots and ultraviolet (UV) curable chemistries, conformal coating has become more common.

41.1.2 Overview of Types of Conformal Coatings

Conformal coating materials are available in five basic chemistries, each with multiple application methods and several drying/curing options. As such, no one material will fit every application, but the wide variety will allow the designer to select a material that meets most technical, budgetary, and manufacturing needs. The various coatings and their characteristics are summarized in Table 41.1.

The five basic chemistries are:

- **Urethane** This coating has good moisture and chemical resistance and good electrical insulation.
- **Silicone** Being soft and having good adhesion, silicone is useful over a wide temperature range.
- **Epoxy** This coating is tough, durable, and very chemically resistant.
- **Acrylic** Acrylic is easy to apply and remove.
- **Para-xylylene** This very thin, pinhole-free coating is vapor deposited instead of being applied as a liquid.

More information on the less common types of conformal coating may be found in IPC-HDBK-830.

41.1.3 Environmental Issues

The predominant environmental issue with coatings was the use of solvent thinners to aid the application process. Most coatings are available in low-solvent or zero-solvent formulations, so that the emission of solvents can be reduced or eliminated. The change to lead-free solders and plating has produced an increase in the use of conformal coatings, since conformal coating has been shown to prevent tin whiskers (see Chap. 29 on tin plating) from causing short circuits, and is used for that purpose in types of products that have not been coated before. The higher-temperature lead-free soldering processes are irrelevant to the conformal coating step, since the coating is applied after the soldering process.

41.2 TYPES OF CONFORMAL COATINGS

The most commonly used types of conformal coatings include:

- Urethane (Type UR)
- Silicone (Type SR)
- Epoxy (Type ER)
- Acrylics (Type AR)
- Para-xylylenes (Type XY)

41.2.1 Urethane (Type UR)

Urethanes are relatively easy to apply with dip, spray, or brush. Cure times (other than UV-curable) range from a few hours for two-part heat curable, to several days for a single-part room temperature cure. Typical application is .001 to .005 in. thick. They are often used for space applications due to their low outgassing characteristics.

TABLE 41.1 Conformal Coating Materials and Their Characteristics*

	Material type	Urethane			Silicone			Epoxy				Acrylic			XY
	Cure methods	Solvent evap.	Heat cure	UV Cure	RTV	UV Cure	Catalyzed	Solvent evap.	Heat cure	UV Cure	Catalyzed	Solvent evap.	Heat cure	UV Cure	Vapor deposited
Application method	hand paint	x	x	x	x	x	x	x	x	x	x	x	x	x	
	dip	x				x		x				x			
	hand spray	x	x	x	x	x	x	x	x	x	x	x	x	x	
	automatic spray	x	x	x	x	x	x	x	x	x	x	x	x	x	
	selective coating	x	x	x	x	x	x	x		x	x	x	x	x	
	vapor deposited														x
Removal	Solvent	x	x	x	x	x	x	x				x	x		
	Thermal	x	x	x	x	x	x	x	x	x	x	x	x	x	x
	Mechanical	x	x	x	x	x	x	x	x	x	x	x	x	x	x
	Plasma etch														x
Benefits	Easy to rework											x	x		
	moisture resistent	x	x	x	x	x	x					x	x	x	x
	simple cure											x	x	x	
	Biologically compatable														x
	High temperature use				x		x	x	x		x				
	Abrasion resistent	x	x	x				x	x	x	x				x
	Solvent Resistent		x	x					x	x	x				x
	Complete coverage														x
	Reversion resistent		x	x								x	x	x	x
	Cold temperature Use		x	x	x	x	x								x
Risks	High VOC	x	x					x				x			
	Requires strict viscosity control											x			
	flammable											x			
	aided by primer				x	x	x	x	x	x	x				x
	cure affected by thickness and mass		x						x				x		
	cure inhibition or reversion	x					x	x			x	x	x		
	High cure shrinkage								x				x		
	not for low temp	x						x	x				x		
	Incomplete cure in shadow			x		x				x				x	
	Has pungent odor			x		x				x	x			x	
	Brittle at high temp			x						x				x	
	rework difficult			x				x		x	x			x	x
	UV wavelength affects cure			x						x				x	
	potential for contamination							x							
	difficult process							x							x
	Poor edge cure								x						
	Short work life				x						x				
	Moisture affect cure	x													
	Long complete cure time	x													
	Health and safety concern	x													
	Violent reaction with water		x												
	Requires moisture for cure				x										
	Low abrasion resistance				x	x	x								
	High CTE				x										
	May contaminate other products				x	x	x								
	Intermittant solvent resistance				x	x	x								
	difficult adhesion						x								
	batch process														x
	Full masking required														x

*Disclaimer: Due to the widely varying nature of the different formulations even within a single column heading, the above information is general in nature and cannot reflect the nature of all available materials that fall under that heading.

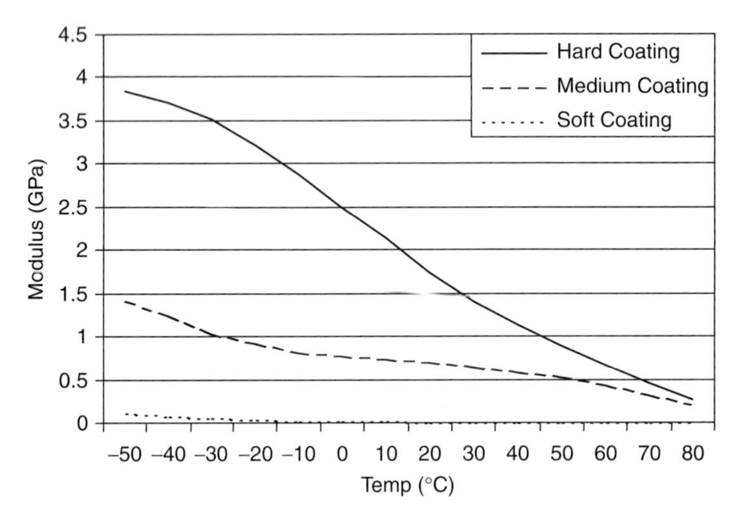

FIGURE 41.1 Elastic modulus versus temperature for three different type UR coatings.

41.2.1.1 Chemistry. Urethanes are based on a di-isocyanate and polyol backbone. They are available in solvent evaporative cure, heat cure, and UV cure formulations.

41.2.1.2 Properties. Since urethanes are polymerized and cross-linked in place, they have excellent resistance to chemical, moisture and solvents. They are available in hardnesses ranging from tough, abrasion-resistant varieties to low-modulus versions suitable for extreme temperature ranges (see Fig. 41.1).

41.2.1.3 Advantages and Disadvantages. Urethanes have good adhesion to most materials, including epoxy part bodies, metals, and ceramics. As such, the coating process is fairly robust. Since urethanes are chemically resistant, they are also difficult to remove except by thermal or mechanical means. Urethanes can be soldered through, although this often results in a brownish discoloration that must then be removed.

41.2.2 Silicone (Type SR)

Silicones are very flexible, and stay that way over a wide range of temperatures. They have good adhesion to a variety of surfaces, but contaminate the surface and, once applied, prevent other materials from adhering. Typical application is .002 to .008 in. thick. They are often used in automotive applications, where high temperature and moisture-resistance are needed.

41.2.2.1 Chemistry. The silicone polymer chain is based on an alternating silicon-oxygen backbone. SR coatings are available in three types, room temperature vulcanizing (RTV), UV cure, and catalyzing (addition) cure.

41.2.2.2 Properties. Silicones have relatively stable properties from –55°C to 200°C. Their coefficient of thermal expansion (CTE) is higher than that of urethanes, but this is mostly offset by their lower modulus, so that the level of stress exerted on parts is still relatively low. They have high resistance to moisture and humidity, as well as polar solvents.

41.2.2.3 Advantages and Disadvantages. Silicones are usable over a wide temperature range and are relatively easy to remove via mechanical or thermal means. If mishandled, silicones may contaminate the work area, causing adhesion problems on other CCAs. Careful process sequencing and process separation is necessary to prevent this.

41.2.3 Epoxy (Type ER)

Epoxies are chemically stable and very resistant to chemical attack. Typical application thickness is .001 to .005 in. Epoxies are useful in extreme environments where chemical vapors or high temperatures are present.

41.2.3.1 Chemistry. Epoxy coatings are based on epoxy resin systems and come in four types: solvent evaporation, heat curing, UV curing, and catalyzed.

41.2.3.2 Properties. Epoxies have a low CTE that matches well with printed wiring board (PWB) epoxy resin, since they share very similar chemistry. They have a higher T_g than most of the other coating materials. They are also very tough and abrasion resistant, so that rework is very difficult; epoxy coating can form the basis of an antitampering coating.

41.2.3.3 Advantages and Disadvantages. Epoxies are useful at moderately high temperatures, up to about 150°C. Because of their strength, they also provide mechanical support for components.

The disadvantages of epoxy coatings are their usually pungent odor and the possibility of skin irritation. They are difficult to rework. Some formulations of epoxies are chemically delicate and do not cure properly in the presence of inhibiting compounds. Cure shrinkage is also of concern for fragile components; a softer buffer coating should be applied locally before the epoxy, particularly for assemblies that need to withstand wide swings in temperature.

41.2.4 Acrylics (Type AR)

Acrylic coatings are applied .001 to .005 in. thick and are often used in military and consumer electronics.

41.2.4.1 Chemistry. Acrylic coatings are usually supplied as dissolved pre-polymerized acrylic chains. The acrylic chemistry does not cure by polymerization and cross-linking as the other coating materials do, but instead hardens gradually as the solvent evaporates. Acrylics are also available in heat-curable and UV-cured formulations.

41.2.4.2 Properties. Acrylics are easy to apply, and are the easiest coating to remove, since relatively mild solvents soften and dissolve the acrylic coating while leaving the epoxy encapsulated parts and PWB unharmed. They can be cured quickly.

41.2.4.3 Advantages and Disadvantages. The greatest strengths of acrylic coating are the ease of rework and the fast room temperature cure. Acrylics provide good moisture resistance and fluoresce easily, aiding inspecting under UV lamps.

Since they are so easy to rework, acrylics are also susceptible to inadvertent chemical attack from solvent splash during hand cleaning of solder joints elsewhere in the assembly. The high emission of solvent inherent in the acrylic process makes them less environmentally friendly than other materials. As the solvent evaporates, the coating shrinks and exerts stress on the components, so acrylics may not be suitable for all low-temperature applications.

41.2.5 Para-Xylylene (Type XY)

Para-xylylene coatings are unique, since they are applied by vapor deposition (see Sec. 41.4.6) rather than as a liquid. They are applied .0005 to .002 in. thick. Para-xylylene coatings are often used in biomedical devices due to the inert character of the coating.

41.2.5.1 Chemistry. XY coating is supplied as a dimer, a polymer chain only two units long. During the vacuum deposition process, the dimer powder is gradually vaporized and is recondensed onto the CCA, polymerizing with the material already deposited. Typical application is .0005 to .002 in. thick.

41.2.5.2 Properties. XY coating is chemically inert and moisture-resistant. Very thin, uniform layers are possible, with no pinholes or voids. XY coating also has very high dielectric strength. Due to the nature of the deposition process, no volatiles are generated.

41.2.5.3 Advantages and Disadvantages. XY coating is the highest performing coating in many regards. It has the highest dielectric strength, and is the lightest weight, lowest outgassing, mostly chemically inert, and most moisture-resistant. It provides the most uniform coating over, under, and inside parts where liquid coatings cannot be applied.

The coating process must be performed in batch mode, using specialized coating equipment. It may be impossible to mask adjustable components such as potentiometers adequately. Rework is difficult; since the thin coating cannot be peeled, a microabrasion process is usually required to remove the coating. Additional staking or adhesive bonding of parts is required, since the thin coating provides little mechanical support.

41.3 PRODUCT PREPARATION

Most coating problems can be traced to improper preparation. The four preparation steps are:

- Cleaning to remove flux, oils, and contaminants
- Masking to ensure coating is applied only where desired
- Preparing the surface or priming to promote adhesion of the coating
- Baking out to prevent moisture from being trapped under the coating

41.3.1 Cleaning

Thoroughly cleaning the CCA before conformal coating is the best way to ensure proper adhesion of the coating. The various cleaning chemistries and their effectiveness on different kinds of fluxes are treated in depth in the Chapter on, "Fluxes and Cleaning." Besides removing flux, the cleaning process removes any contamination such as mold release agents on plastic part bodies, fingerprints, etc.

No-clean fluxes reduce the need for waste water and waste solvent treatment. For boards that require conformal coating, the benefits of a no-clean system must be balanced with the risks of poor coating yields. A minor change in flux or coating chemistry could render the coating incompatible with the no-clean flux being used, causing delamination, cure inhibition, or other defects. In general, coating results are not as good with no-clean flux as with standard flux plus cleaning, yet in some cases may still be adequate to do the job.

41.3.2 Masking

To control the areas where coating is applied, some kind of masking is usually needed. Masking tapes or adhesive circles may be used to keep flat areas and holes free of coating. Since the tape needs to be removed completely after the coating operation, the tape adhesive must be capable of withstanding the cure temperature without peeling or becoming permanently bonded to the surface.

To mask irregular surfaces, latex masking material is used. This material is applied with automatic or manual dispensing equipment and is cured in an oven before the coating operation is performed. It then peels off easily after coating.

For high-quantity production, boots or covers can be molded to fit over the areas that do not require coating. These are reusable, easy to install, and produce well-defined repeatable results.

Rather than being masked, areas may be sealed with a permanent bead of a thixotropic (non-runny) adhesive. This eliminates the demasking step, and is useful for complicated geometries such as the back of connectors and the perimeter of ball grid arrays (BGAs).

41.3.3 Priming and Other Surface Treatments

The adhesion of some coatings—notably epoxies, silicones, and para-xylylenes—will be improved if the CCA surface is primed before coating. Primers are applied as a very thin coat using dip or spray (generally the thinner the better) and oven-dried to drive any remaining solvent from crevices, vias, and such. If the primer fails to wet all areas of the board, the coating will not wet either but will easily delaminate instead. Such surfaces are not compatible with the primer, and an adjustment in primer chemistry is required. For difficult surfaces, such as large fluoropolymer parts or substrates, a more aggressive measure such as plasma-etch or mechanical abrasion may be needed to promote adhesion.

Mechanical abrasion (grit blasting) may be used to roughen troublesome surfaces to promote coating adhesion. It may be easier to do this step as the parts are fabricated (in the case of mechanical parts) or as part of the make-ready process before assembly. Microabrasion can be used with proper care on the finished assembly as long as electrostatic discharge is controlled by the design and use of the abrasion equipment, and is monitored daily to ensure that it is working properly. Microabrasion is also a viable method for removing coating.

41.3.4 Bake-Out

Since the coating is intended to be waterproof, it is imperative that all moisture be baked out before the coating is applied, especially for hydroscopic substrates such as GI polyimide. If moisture is left in place, it can cause corrosion of traces and/or parts and will promote dendritic growth between conductors, as well as enabling the growth of conductive anodic filaments (CAF) along the glass fibers in the weave of the PWB. A bake of 93°C for 4 hours will be sufficient to drive out moisture from the CCA.

41.4 APPLICATION PROCESSES

The liquid coatings can be applied by a variety of methods, including:

- Hand painting the coating with a brush
- Dipping the board into the coating
- Hand spraying with an aerosol can or handheld spray gun
- Automatically spraying the board with coating as it passes by on a conveyor
- Selectively applying coating to certain areas of the board with a robotic nozzle
- Vapor depositing para-xylylene onto the board with specialized equipment

41.4.1 Hand Paint

Hand painting allows coating to be applied wherever desired with no setup time and low equipment cost. It is still the preferred method for touching up thin spots in a bulk applied coating and for coating replaced components. It can be used to reach areas where spray or automatic nozzles cannot reach, and works well for smaller, one-of-a-kind items. Hand painting is the most variable of all methods.

41.4.2 Dip

Dip coating works best for coatings that have a long pot life at room temperature, such as solvent evaporation cured coatings and one part materials. Since the unused coating remains in the dip tank, a long (or infinite) pot life keeps the waste down to a minimum.

The viscosity of the material in the bath should be checked daily, using a flow cup viscosimeter or other method, and solvent added to keep the viscosity of the material within the acceptable range. For a given dip speed, a lower viscosity will produce a thinner coating. The dip machine (see Fig. 41.2) will have bars or hangars on which to hang the board to be coated. Insertion speed should be slow enough to prevent the formation of bubbles as the coating flows around and under surface mounted parts, etc. Withdrawal speed should range between 1 to 6 in./min and will dictate the finished coating thickness; a slower withdrawal will produce a thinner coating.

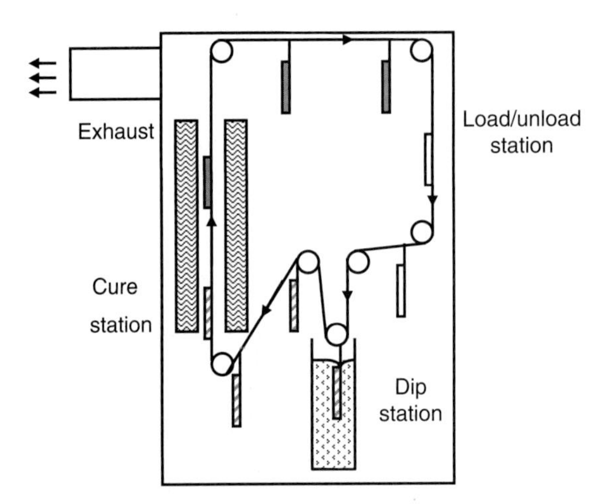

FIGURE 41.2 Dip coating machine.

Dip coating cannot be used multiple times to build up thicker layers of acrylic coating, since solvent in the bath will loosen or dissolve previously applied layers of acrylic.

Any areas that are not to be coated must be completely masked before dipping. Assemblies can be partially dipped to leave the top edge of the board uncoated if the board is designed so that all components that need coating are below the level of the bath.

41.4.3 Hand Spray

Hand spraying is used for all types of liquid coatings. The boards to be coated are sprayed several times; the angle is changed 90° each time so that complete coverage is achieved. Tall components may shadow or block the spray from lower profile components behind them, so care must be taken to ensure all areas of the board are fully coated. A witness strip is often sprayed at the same time for use in measuring the coating thickness.

As with the dip method, viscosity control is a key factor to reducing bubbles, cobwebs, and improper coating thickness.

Two part materials are mixed immediately prior to filling the spray gun container, or are mixed in line if a continuously fed mixing gun is used. Single-part materials (particularly acrylics) are also available in aerosol cans, eliminating mixing altogether.

41.4.4 Automatic Spray

For high-volume coating application, a semiautomatic spray machine is used to eliminate the dependency on the operator and allow in-line coating. The boards enter the machine on a paper-covered conveyer and are sprayed by a rotating or reciprocating spray head. The motion of the spray head(s) is designed to cover a given width of the belt completely from all angles. Shadowing is of greater concern with semiautomatic spray equipment, since special treatment cannot be given to any one area of the board.

Full masking is required to keep the coating out of areas that should not be coated.

Once the board is coated on one side, it is partially cured (or fully cured), turned over, and coated on the other side.

41.4.5 Selective Coating

The selective coating process utilizes a robot similar to the automated pick-and-place machine to dispense coating only where it is needed. Four or five axis machines can be used to coat the sides of tall parts and around corners.

The greatest benefit to selective coating is that it drastically reduces the amount of labor needed for masking and mask removal; areas that need to be free of coating simply have no coating applied there. Coating utilization is highest of all methods, since almost no coating is lost in continuous production.

Spray patterns may be either atomized, spraying droplets onto the board and providing a thinner coating (see Fig. 41.3), or continuous, applying a stream or sheet of liquid to the board

FIGURE 41.3 Selective coating using an atomizing nozzle. *(Courtesy of Asymtek.)*

FIGURE 41.4 Selective coating using a continuous flow nozzle. *(Courtesy of Chipco. Inc.)*

(Fig. 41.4), producing a well-defined boundary between the coated and uncoated areas and eliminating the need for masking.

In the same way as the semiautomatic sprayer, the board is tack-cured and then turned over and coated on the other side. An automated board flipper may be used to accomplish this to automating the entire coating process.

41.4.6 Vapor Deposition

Vapor deposition is only used with XY coatings, and requires specialized coating equipment. In the evaporator, a measured amount of dimer is placed in a crucible. The coating cycle is then started, and the chamber is pumped down to an almost complete vacuum (see Fig. 41.5).

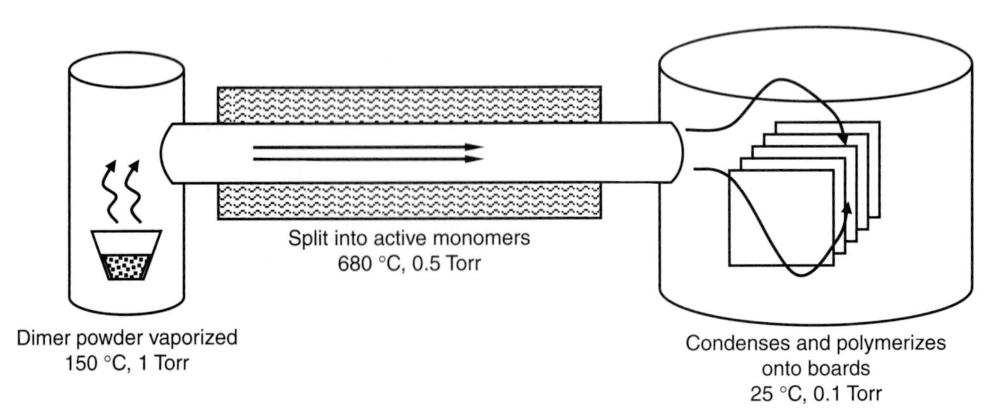

FIGURE 41.5 Para-xylylene vapor deposition process.

During the deposition process, the dimer powder is vaporized and recondenses onto the batch of CCAs, forming a uniform coating covering all corners and edges evenly.

The vapor deposition process drives the vapor into any crevices and openings. Vapor deposition has the most stringent masking requirements, since leaks in the maskant will still admit vapor that will contaminate the surface behind, and air pockets inside masking boots or tape may dislodge the boot in the vacuum and create leaks as the air escapes.

41.5 CURE, INSPECTION, AND FINISHING

The finishing steps require most of the labor involved in conformal coating process. Three finishing operations are:

- Curing the coating
- Inspecting under UV light to verify complete coverage and touch-up
- Removing mask materials

41.5.1 Cure

Each of the four liquid chemistries is available in a formulation offering different cure mechanisms, as shown in Table 41.1.

41.5.1.1 Solvent Evaporation. Systems that are cured by solvent evaporation can be air dried, or slightly heated to drive the solvent out faster. Care must be taken not to overheat the freshly coated assemblies; the coating will skin over and cause bubbles. Long-term exposure to the unevaporated solvents trapped in the coating may also be deleterious to the materials in the parts and PWB. Drying of solvent cure assemblies must be done under a ventilation hood or other properly ventilated area.

41.5.1.2 Room Temperature Vulcanization (RTV). Some silicone-based coatings use the room temperature vulcanization (RTV) cure process, which consumes moisture from the surrounding environment as part of the cure process. Some means of humidity control is required; if cure is attempted in a dry oven, the material will not cure properly, if at all.

41.5.1.3 Heat Cure. Oven-curable systems use an extended bake, ranging from 15 min. to several hours or longer, for the material to cure fully. Thermally activated systems require the increased temperature to start the curing process, whereas thermally accelerated systems will cure eventually at room temperature but are baked to speed up the process and improve material properties. Generally, the hotter the oven, the quicker the cure, and the harder the finished material.

41.5.1.4 UV Cure. For high-volume production, UV-curable materials are used. Immediately after the coating machine is a booth with high-intensity ultraviolet lamps to illuminate the assembly. After several seconds, the assembly has cured enough to be able to be handled at the next stage of assembly, such as coating the opposite side. UV-curable materials do not cure fully under components, where the coating is not illuminated. Therefore, most UV-curable materials also contain a secondary curing mechanism, such as heat cure, so that at the end of the line the CCA can be fully cured in a single baking step or naturally come to full cure in ambient conditions inside the finished product. Since different coatings are sensitive to different wavelengths of UV light, it is important to coordinate the purchase of the light source with the supplier of the coating.

41.5.1.5 Catalyzed. The catalyzed process involves mixing two parts together just before application, to initiate polymerization and cross-linking. This process occurs slowly at room temperature but may be accelerated by elevating the temperature in an oven. Catalyzed coatings may be affected by contaminants present in the PWB or parts being coated, resulting in uncured areas or peeling coatings. Since the pot life of catalyzed materials tends to be short, it is imperative that the mixed material be cleaned from sprayers, tubing, and so on, before it thickens and hardens.

41.5.2 UV Inspection

Coatings contain a fluorescent dye that will glow under UV light and show where the coating has (or has not) been applied. Areas without coating can be touched up at this time.

The areas most susceptible to insufficient coating are corners of parts, where the coating can flow away and leave little behind; the protruding ends of through-hole part leads; and shadowed areas under or behind larger components. If large flat areas are uncoated and show beads of coating, there is a wetting problem due to improper surface preparation or a material incompatibility.

The coating should also be inspected for improper thickness, bubbles, trapped material or debris, and incomplete curing (see Fig. 41.6).

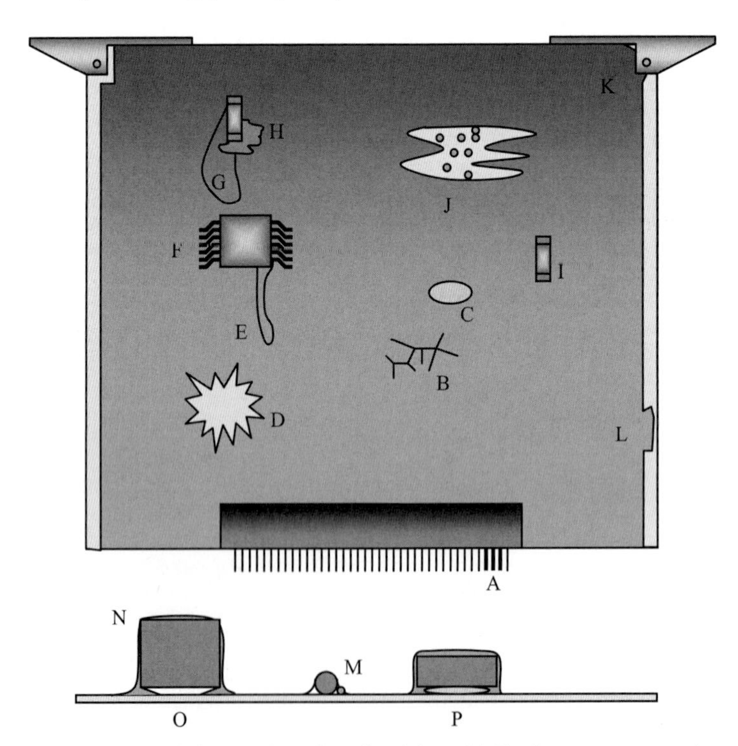

FIGURE 41.6 Various conformal coating defects. (a) Coating on connector pins (b) Cracked or crazed coating (c) Delaminated or blistered (d) Uncoated areas (e) Excess coating drips and runs (f) Excess coating bridging flat pack leads (g) Solvent runoff from rework affecting surrounding coating (h) Overheated, burned coating from rework (i) No buffer coating applied to Glass parts when used with hard coating at cold (j) Dewetting; coating beading up instead of wetting to board (k) Excess coating interfering with movable items (l) Excess coating in masked areas (m) Excess bubbles from trapped solvents (n) Insufficient coverage on corners and edges (o) No coating applied under parts (p) Uncured coating under parts.

41.5.3 Mask Removal

Mask materials should be removed carefully; to prevent peeling, it may be desirable to score or cut the coating at the edge of the mask so that the coating tears along the correct line. Ensure that the mask is removed entirely and that no residue has been left behind on the board. Latex mask that has been overheated will be sticky and difficult to remove completely, as will overheated masking tapes.

41.6 REPAIR METHODS

Since the different varieties of available coatings vary widely in their properties, removal methods must be evaluated and developed on practice boards first before being used on deliverable CCAs. The coating can then be removed as easily as possible, with the least risk to the CCA. Several methods to remove conformal coating are:

- Using a solvent to loosen or dissolve the coating
- Using a hot soldering iron or hot-air nozzle to solder through the coating while simultaneously removing the offending electrical part
- Mechanically cutting, abrading, or picking away the coating
- Etching the coating away using rarefied gas plasma (typically for XY coating only)

41.6.1 Solvent

The easiest coatings to remove are acrylics, since they do not react when they are applied. For local coating removal, the area to be repaired can be cotton swabbed several times with the proper solvent. Care must be taken to keep the solvent only in areas that are to have the coating removed. The entire board may be stripped of coating with a solvent soak. This is most useful on acrylic coatings, but is also possible for some urethanes and silicones, which swell and loosen when soaked. Note that this process may damage the components or solder joints on the board, which will be stressed as the expanding coating pushes up the components. The solvent may attack the components and PWB itself; epoxy coatings cannot be removed with solvent or chemical strippers, since many parts and the PWB are also made of epoxy.

41.6.2 Thermal

Thin coatings may be loosened from solder joints with a soldering iron or a hot-air reflow tool, while the faulty part is being desoldered. The area is then cleaned of charred and or damaged coating, and the repair completed. Although this method is simple, it usually produces toxic gases, and must be performed with adequate ventilation. The decomposed coating is also difficult to clean off of the board and soldering iron.

41.6.3 Mechanical

Mechanical means such as scraping and picking are necessary to remove all coatings, even if the bulk has been removed via one of the preceding methods. Softer coatings such as UR or SR can be scraped away with a wood implement, so as not to cause damage. A soft rotary tool may be used for larger areas. The board may be heated slightly to make the job easier. For harder coatings, a hot knife is used to cut through the material.

A microabrader can also be used to abrade the material away. The surrounding board areas must be protected from contamination by the abrasion media.

41.6.4 Plasma Etch

An extended plasma etch will remove XY coatings from large areas, since the plasma erodes the surface evenly. Since the plasma will chemically change the surface (leaving behind oxides, reduction products, and ash), compatibility of the plasma-stripped surface with the new coating must be verified.

41.7 DESIGN FOR CONFORMAL COATING

When designing a conformal coating system into the product, the designer needs to consider a number of questions:

- Is coating required for this product, or should it be omitted for cost savings or other reasons?
- What material properties are most important for this application, and which materials will provide the best balance?
- What areas should not be coated?

41.7.1 Is Conformal Coating Required?

One of the basic questions to consider is whether or not to coat the particular assembly. Generally this will be a trade-off between reliability and cost.

41.7.1.1 Why Use It? Customers may flow down requirements for conformal coating for industries where it is commonly used such as military or aerospace electronics. The following are some of the reasons to consider the use of conformal coating:

- The end products are used in humid or condensing atmospheres, such as electronics boxes for military and shipboard applications. It should be noted that although conformal coatings are moisture resistant, they are not waterproof, and should not be used for items that will see extended or continuously wet conditions. For those applications, use a watertight enclosure and conformal coat the boards inside. Conformal coating will protect against condensation, such as when the product is brought from a cold environment into a warm humid one.
- Clean conformal coating does not support mold and fungus growth.
- The end products can become contaminated by dirt and debris. Conformal coating is especially useful in space applications, since loose debris will float readily from place to place inside the electronics box.
- Conformal coating greatly improves the board's capability to withstand high voltage, especially at altitude or low pressure, since an electrical arc now has to travel through two layers of insulation rather than just jumping from one trace to its neighbor.
- Conformal coating protects against tin whiskers. Pure tin plating has the tendency to recrystallize and extrude long (>1 mm) whiskers of pure tin, which can short to neighboring part leads and cause short circuits. Conformal coating provides protection against short circuits caused by tin whiskers. Even if a tin whisker can force its way through a layer of coating on the way out, it cannot force its way back through the coating on the neighboring lead without bending and buckling.

- Conformal coating provides mechanical support for components during shock and vibration. Components weighing more that $^1/_4$ oz. per lead should be supported by other means as well, such as staking with a bead of adhesive along the side.
- Lastly, conformal coating may be made opaque, and thus provide a level of security. A thick opaque epoxy coating will prevent the casual observer from determining what is under the coating.

41.7.1.2 Why Not Use Coating? Most of the reasons for not coating on an assembly stem from the recurring cost of cleaning, masking, demasking, touch-up, etc. The target cost of the product may not support the extra cost incurred by the coating, and if reduced reliability of the uncoated assembly is still acceptable, the rational decision is not to coat the assembly.

Since the coating is a dielectric material, it changes the impedance of high-speed traces on the surface of the PWB. Boards with microstrip are usually not coated for this reason. Radio frequency (RF) PWBs for severe environments can still be coated if the traces are buried microstrip or stripline, which are unaffected by the additional coating on the board surface.

41.7.2 Desirable Material Properties

When selecting a coating, the entire suite of material properties must be considered in order to choose the best coating for the application. Unfortunately there is no "one size fits all" solution, but many different coatings, each suited to different applications, as shown in Table 41.1. The following are some of the properties to consider:

- **Ease of application and producability** The intended production environment may determine the cure process and application process that is best for the job, and will narrow the field of alternatives.
- **Compatibility** The coating must not be inhibited from curing by any material on the completed assembly. Often the best way to address this is to ask the manufacturer and try out the coating in question on a sample of the materials that you are going to use.
- **Moisture resistance**
- **Dielectric constant** For RF and high-speed circuits, a material with a low dielectric constant will change circuit performance less than a material with a higher one.
- **Dielectric strength** For high-voltage application, materials with a high dielectric strength should be used. For a valid comparison, tests must be performed on samples of equivalent thickness, since thin samples will break down at a higher volt/mil than thick samples of the same material.
- **Insulation resistance** Also known as *resistivity*, insulation resistance is of concern in high-voltage or high-impedance circuits, where the resistance in the circuit is very high (>10M ohm) and is significant compared to the resistance of the coating and PWB laminate. For these applications, choose a coating with high (>10^15 ohm-cm) resistivity and excellent moisture resistance.
- **Hardness** For products that need to withstand cold temperatures <–10°C, a soft coating is recommended so that at cold temperatures the coating will not exert excessive force on fragile components (see Fig. 41.7). For products that need good abrasion resistance and protection from external handling damage, a harder coating must be used.
- **Glass transition temperature (T_g)** For assemblies that need to withstand cold environments, select a coating that either has a T_g below the coldest temperature required or a T_g of 25°C and above. Coatings that have a low T_g in the range of operation will suddenly get hard at cold temperatures and impose great stress on the surrounding components.

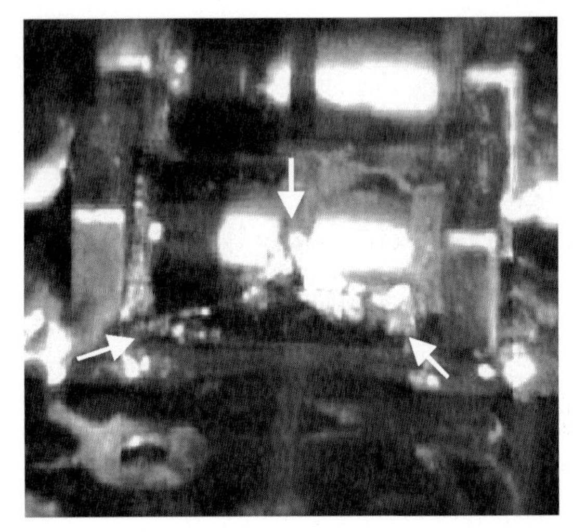

FIGURE 41.7 Glass diode cracked by high-modulus coating.

- **Elongation** Materials with high elongation are resistant to cracking and abrasion.
- **Reworkability** AR coatings are the easiest to rework, whereas XY and ER are the hardest.
- **Abrasion resistance** This is ability of the coating to withstand damage from handling and later assembly operations, as well as in the field.

41.7.3 Areas to Mask

During the design, the designer needs to consider all the areas and parts on the board that need to be uncoated and identify them on the assembly drawing, shop instructions, or other paperwork that the coating technician will use. Although it is obvious to the design engineer that an expensive infrared (IR) sensor element should not be coated, it may not be so obvious to the technician holding a drawing that says, "Coat entire board," and nothing else. Being specific on the drawing clarifies what should be coated and how.

Consider the following areas:

- Coating on connector pins can cause open circuits, and may interfere with the full mating of the connector. It is necessary to mask the back of the open back connectors as well as the front, to keep the coating from flowing along the back of the pin into the active area. Also, pin alignment depends on a certain amount of looseness on the pins, and filling the connector shell with coating will defeat this feature.
- Coating should be kept from the edges of plug-in CCAs where the card guides travel. The coating will interfere with proper operation of the card guide and may prevent proper grounding of the card.
- Mounting holes for screws should be masked to keep the coating from filling or reducing the size of the hole and to ensure that the PWB is properly grounded.
- Test points should be masked so that they can be reliably accessed during testing.
- The adjustment screws of adjustable components. If the adjustable components have openings in them, the openings should be masked also.

- If the product contains sensors for air temperature, pressure, humidity, etc., the coating should not block the air from entering the sensor.
- Optical devices, light-emitting diodes (LEDs), infrared sensors, and light sensors may have their operation affected by the coating. This is especially true for UV and IR devices since the coating may be opaque to UV and IR wavelengths of light. The coating may also darken, cloud, or turn brown as it ages, eventually affecting the operation of the circuit. Connectors for fiber-optic cable should also remain clear of coating for the same reason.
- Components mounted in sockets should have the sockets masked and the components inserted after coating.
- Areas under BGA packages should not be underfilled with conformal coating but with an underfill adhesive particularly suited for that purpose. If no underfill is used, the edges of the BGA should be masked, or dammed with a permanent adhesive barrier to prevent the coating from entering under the BGA. Due to the shrinkage of conformal coating, during thermal cycling the balls will be crushed and the reliability of the BGA will be adversely affected.

REFERENCES

1. Ritchie, B., Loctite Corporation, "Process Requirements for Solvent-Free UV Conformal Coatings," Adhesives in Electronics '96, Second International Conference on Adhesive Joining and Coating Technology in Electronics Manufacturing, Stockholm, Sweden, June 1996.
2. Dow Corning, "Conformal Coatings Tutorial." Dow corning "Conformal Coating Tutorial" is a web page: http://www.dowcorning.com/content/etronics/etronicscoat/etronics_cc_tutorial.asp.
3. Bennington, L., "Conformal Coating Overview," Loctite Corporation. 1995.
4. Wu, F., and Goudie, S., *IPC-HDBK-830*, IPC, 2002.
5. NASA-STD-8739.1, 1999.
6. MIL-I-46058C, 1972.
7. Woodrow, T., and Ledbury, E., "Evaluation of Conformal Coatings as a Tin Whisker Mitigation Strategy," IPC/JEDEC, 2005.

P · A · R · T · 9

SOLDERABILITY TECHNOLOGY

CHAPTER 42
SOLDERABILITY: INCOMING INSPECTION AND WETTING BALANCE TECHNIQUE

GERARD O'BRIEN
S.T. and S. Company Bayport, New York

42.1 INTRODUCTION

The power of solderability testing can be lost, even to people who routinely perform solderability tests, because of the function of the test method and a lack of understanding the process. Most solderability testing has been used as a means of testing incoming components or boards in the hope of catching defects and thus avoiding line-related issues. When a defect occurs on the line it is too late.

42.1.1 Dip and Look

Testing has most often used a protocol best described as "dip and look," in which a technician completely immerses the wettable area of a PCB or component into a solder bath for a prescribed period of time and then evaluates the sample visually once it has cooled down. Pictorial guidelines were used to help the technician to accept or reject the samples being tested. Some incoming inspection departments may have stressed the solderable deposit by exposing it to steam for eight hours and then performed the solderability test. Steam exposure for tin lead coatings is a well documented and applicable stressing technique, but the same isn't true for Pb-free surface finishes. In all cases, the result is basically subjective and the "data" produced from these tests provides little information either to the process engineer responsible for the assembly line or the plating line that is applying the solderable deposit. With the exception of the very extreme "it wet or it did not wet with solder" which is very rare, the remaining population of results achievable are lost in dip and look testing.

Plating, solderability, and soldering in the tin lead world has such a large process window that in the majority the dip and look testing suffices with defects normally in the parts per billion. The occasional disaster does occur but it is rare and attributable to an assignable cause. This process latitude and statistical process control has resulted in the demise of the incoming inspection departments at the CM or OEM. The technician who would have once performed the testing has been replaced by a clerk who only job is to reconcile the purchase order quantity with the received quantity.

42.1.2 Introduction of Lead-Free Assembly

The introduction of lead-free assembly as a direct consequence of the European Union as shown in "Reduction of Hazardous Substances (RoHS) and Waste Electrical and Electronic Equipment (WEEE) directives," changes everything. While lead-free assembly is nothing new, it has existed as a function of design requirements and has been a niche market in a lead-dominated one. Plating and assembly of these non lead-bearing platings and solders were learned and understood by the users in this unique market. However, with the RoHS and WEEE directives, this niche, highly-specialized market has become the norm overnight and the forgiving tin lead process is gone. Up and down the supply chain, it's a new game that the supplier base needs to learn and learn quickly if quality and costs are to be maintained. This "new game" can be learned quickly if real and useable data is available to the process engineers supporting the plating and assembly processes. This means the re-introduction of the "incoming inspection departments" once used for the testing of components and PCBs, but with more sophisticated tools than the tweezers and solder pots of old. Solderability testing, if performed as outlined and detailed in this chapter, will help with process development, aid with the choice of surface finish, distinguish marginal from excellent suppliers and if followed widely can short circuit the time to return to the process latitude that tin lead provided.

42.2 SOLDERABILITY

To understand the issues of solderability it is first important to understand what is meant by the term, including appropriate definitions, the physical properties of the materials involved, and the various standards developed to guide the user to ensure that the solder process will be successful.

42.2.1 Industry Standards

Two common solderability specifications for Components and Printed Circuit Boards are ANSI-JSTD 002 and ANSI-J-STD 003 respectively. In each specification, there are detailed test protocols as well as accept and reject criteria. While there are different specific test methods, the general test method can best be described as "dip and look" with the responsibility of accepting or rejecting the parts being tested being placed firmly into the hands of an inspector that evaluates by eye. With this test, the inspector uses the percentage of area that is wet with solder: with 95 percent as a pass and 94.999 percent as a failure. It should be stated that these "dip and look" tests do not usually represent what actually occurs during the assembly process. Both specifications have process simulation tests, but the action of immersing a component or PCB coupon into a static solder pot using tweezers remains the most common test method in use. Table 42.1 details the test protocols found in these J-STD documents.

TABLE 42.1 Solderability Test Methods with Accept/Reject Criteria

Test document	Test method	Comments
ANSI-JSTD-002	Test A, Test B, Test C, and Test S	All basically dip & look methods that have little to no possibility of meeting GR&R requirements
ANSI-JSTD-002	Test D	A vital test for Pb-free
ANSI-JSTD-003	Test A, Test B, Test C, Test D, and Test W	All basically dip and look methods that have little to no possibility of meeting GR&R requirements

In both these specifications are test methods that are *not* approved for accept/reject criteria, but that produce data that is useful for process development, shows what really happens during the assembly process, and if tested correctly meets the most stringent of evaluations to modern measurement analysis. These are tests E, F, and G for the J-STD-002 and test F for the J-STD-003. The test protocol detailed in this chapter is designed to change the test status of the wetting balance to one capable of producing accept/reject data, and more importantly, to provide accurate data for process improvements.

42.2.2 Definitions of Terms

For soldering to occur some obvious fundamentals must be met:

1. The part needs to be fluxed—see following definition.
2. The part must be solderable—see following definition.
3. There must be sufficient thermal energy to allow wetting to occur—typically 35° to 50°C above the Liquidus point of the alloy being used.

Soldering is defined as a metallurgical joining method using a filler metal (the solder) with a melting point below 842°F (450°C). Soldering relies on wetting for the bond formation, and is defined in the ANSI-JSTD-002 as the ability of a metal to be wetted by molten solder.

Wetting is defined as the formation of a relatively uniform, smooth, unbroken, and adherent film of solder to a basis metal.

42.2.3 The Choice and Impact of Fluxes for Solderability Testing

Flux is an integral part of soldering and solderability testing. A flux used for solderability testing should *never* produce false positives or negatives. It should demonstrate the true characteristic of the sample being evaluated. Some examples are whether it's old, composed of intermetallics, buried under a layer of oxides, contaminated with organics, hiding a poorly wettable basis metal, or showing a clean fresh readily solderable surface.

42.2.3.1 Flux Functions. A flux has three main functions:

1. To act as chemistry that reduces oxides on a wettable surface.
2. To provide the correct interfacial energy potential between the flux/solder interface and the solder substrate interface.
3. To prevent oxidation from forming during the soldering process and to a lesser extent in the case of liquid fluxes they may protect the surface on the PCB from the extreme temperatures. It may also act as a heat transfer medium.

Fluxes are necessary for soldering. While flux-free soldering is possible, in the presence of a reducing gas atmosphere, it is very uncommon. When attempting to check for solderability characteristics without the use of a flux, at best the solder will stick to the sample but will not wet the surface to form a true metallurgical bond. At the other extreme end of the scale are people using fluxes so strong that nothing will ever fail a solderability test. These fluxes do not represent anything used in the assembly industry. A great example of this is the use of the Hot Air Solder Level (HASL) flux for outgoing solderability testing. This type of flux typically uses 5 percent HCl or 5 percent HBr as the active ingredient which, when applied to a surface finish, will reduce any oxide layer present and reactivate any potentially passivated nickel layer and thus always produce a pass result. This is not what is supposed to be used and it does not provide any information pertaining to the actual solderability of the sample being tested because it passes all the time.

42.2.3.2 Fluxes and Evaluation of Solderability. There has been a fundamental change in how solderability is evaluated. The need to demonstrate a year of shelf life has been replaced with a test for how robust a deposit or plating is. The more robust the deposit, the greater the potential shelf life and the ability to survive the complicated higher temperature exposures found in Pb-free assembly. When making the choice in terms of solderability testing, it is important that the chosen flux is active enough to overcome finger oils and light surface oxidation and to promote wetting. At the same time, it should not be capable of reducing heavy oxides or wetting to exposed and oxidized intermetallic compounds. For many years the use of the R flux for solderability testing was the norm, and it was referenced in that venerable specification MIL STD 202, method 208. The weight consists of 25 percent colophony and 75 percent alcohol. The activation level is approximately 0.05 percent as measured by chloride content, since this is not very active and being a naturally occurring substance there is variation year to year and location to location.

The JSTD-002/003 committees working with NIST, NEMI, and Industry undertook a study to look into the use of this flux. Using an R type flux for solderability testing produced excessive noise in the data, which was generated from the flux alone. The inability of the flux to deal with simple soils, such as a finger print for example, could produce a false negative result. The outcome of the study was a recommendation to add a known quantity of an active ingredient, low enough to not produce false positives but strong enough to deal with normal sample contamination not related to the plating or surface preparation of the sample. Testing of OSP, ENIG, and Palladium surface finishes with an R flux inevitably produced failures that were never seen in production. These surface finishes, particularly ENIG and OSP, are in common use and a suitable flux needed to be created to test them. The added ingredient was diethylammonium hydrochloride and the flux specified for solderability testing became known as "a standard activated rosin flux (type ROL1 per J-STD-004)." This flux has been shown to be excellent in differentiating solderability issues from finger oils and sample issues. It is sold under many names. The list of current suppliers can be found in the appendix of both J-STD-002 and 003.

42.2.3.3 Lead-Free Solderability and Fluxes. For lead-free solderability testing, a further extensive study was undertaken by the J-STD-002/003 committees again. The outcome of the testing was to recommend and specify the use of a more active flux. The activation level has been increased to 0.5 percent as measured by chloride content. This flux and the increase in test temperature are specified in the J-STD-002C and 003B.

42.2.3.4 Wetting and Solderability. Soldering is governed by the laws of physics. In soldering, three phases are present:

1. The solid phase (parts to be soldered)
2. The liquid phase (molten alloy)
3. The vapor phase; flux evaporation (atmospheric air in the majority, could be nitrogen)

The molecular interactions of these three phases taken in pairs are surface tensions (see Figs. 42.1 and 42.2).

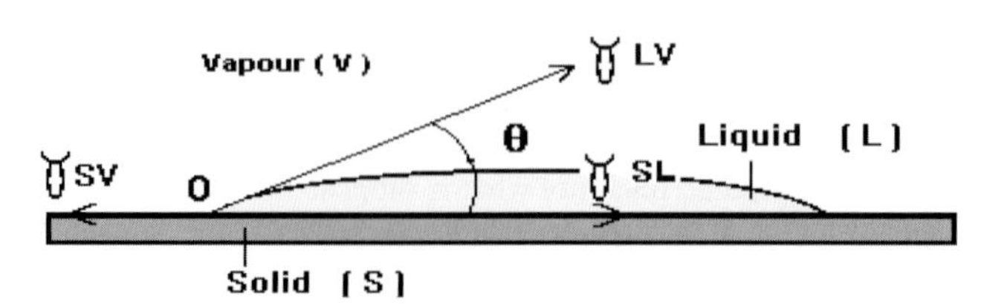

FIGURE 42.1 Theory of wetting.

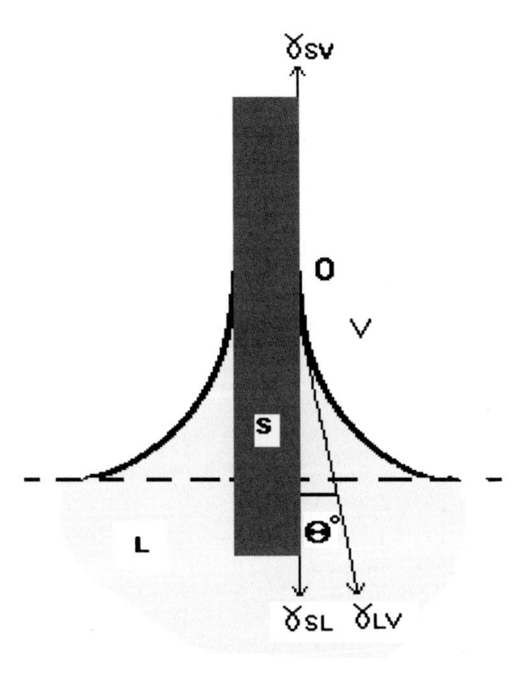

FIGURE 42.2 A graphical representation of a lead immersed in molten solder and wetting.

Figure 42.1 shows a soldering alloy pellet put onto the surface of a metal plate, which has been previously fluxed and heated to a temperature at least equal to the Liquidus point of the alloy, where :

1. γSL represents the solid—liquid phase
2. γSV represents the solid—vapor phase
3. γLV represents the liquid—vapor phase
4. Represents the junction between the solid, liquid (solder) and fluxed surface

When a lead, component, or PCB is immersed into solder during a solderability test, the following can be derived from the component solder interface Equation 42.1

$$\gamma SV + \gamma SL + \gamma LV = 0$$

$$\gamma SV = \gamma SL + \gamma LV.\cos\theta \text{ (YOUNG's relation)}$$

$$F = \gamma LV.\cos\theta.P - \rho \cdot v \text{ (LAPLACE law)}$$

where F = capillary forces
ρ = specific mass of Sn/Pb (molten alloy in the pot)
v = volume of the component part immersed in the molten alloy
$\rho \cdot v$ = Archimedean thrust or buoyancy generated by the component part immersed in the molten alloy.
γLV = alloy/flux surface tension
P = component wettable perimeter

Where θ is the angle formed by the surface of the solid and the liquid at their extreme point of contact—also known as the wetting angle.

It should be noted that

1. The relations are true only if the sample is immersed in a perpendicular direction at the surface of the alloy and the dimensions are constant.
2. The angle θ is directly linked to surface tensions and is thus representative of the wetting quality.
3. The smaller the wetting angle θ the better the solderability!

A rating of solderability, quality-based on contact with the wetting angle is listed in Table 42.2.

TABLE 42.2 A Relative Index of Solderability as a Function of Contact Wetting Angle

Angle value (°)	Quality rating
$\theta \leq 30$	Excellent
$\theta \leq 40$	Good
$\theta \leq 55$	Acceptable
$\theta > 55$	Rejectable

42.3 SOLDERABILITY TESTING—A SCIENTIFIC APPROACH

Solderability defects can result from poor line management, poor chemical management, incorrect deposit thickness (whether real or incorrectly measured), poor choice of chemical supplier, poor understanding of metallurgy and solderability, and/or a combination of all of these items. Being able to measure solderability correctly and accurately is necessary for fixing these problems and for producing acceptable parts for the end user. If the test always accepts the sample, or there is little to no discrimination between inputs to the line, the supplier will never develop a robust process and then will never supply parts with zero defects accordingly. However, if the solderability test is performed correctly, data can be generated to determine whether significant relationships exist between two or more variables that have a positive or negative impact on the final product performance. These are known as *analytic studies,* as defined by Deming, and are used to increase the knowledge about the system of causes that affect the process. These are considered among the most important uses of measurement data because they lead ultimately to better understanding of processes.

The benefit of using data-based procedures is largely determined by the quality of the measurement data used. If the data quality is low, the benefit of the procedure is likely to be low, such as from dip and look solderability testing. Similarly, if the quality of the data is high, the benefit is likely to be high also. The quality of measurement data is related to the statistical properties of multiple measurements obtained from a measurement system operating under stable conditions. For example, someone might be testing plated lead frames of varying deposit thicknesses and stressing the parts using steam exposure. If there is a reduction in solderability, as a function of thickness and time exposure, then the data produced would be considered of high quality. If there is no discrimination between any of the groups, and all pass or all fail, then the quality of the data would be considered low.

42.3.1 Commonly Used Statistical Properties

The statistical properties most commonly used to characterize the quality of the data are *bias* and *variance*. Bias refers to the location of the data relative to the true value and variance refers to the spread in the data. Variance is a major player in producing low-quality data.

Over the years, the science of evaluating pieces of test equipment and their suitability to accurately measure whatever they are being used for has evolved. By having a test that is scientifically and statistically believable, the results obtained can be trusted and thus correct judgments can be made on their results. Using a child's toy to measure the thickness of a lead frame is obviously silly, but attempting to fix a plating line problem for solderability using tweezers and a solder pot falls into the same category—just silly!

42.3.2 Gauge Repeatability and Reproducibility (GR&R) and Measurement System Analysis (MSA) Background

The first evolutionary step was the use of Gauge Repeatability and Reproducibility (GR&R) testing that took into account both the suitability of the test equipment as well as the operator/technician/engineer performing the test. Detailed worksheets were created by the likes of IBM and it became commonplace to speak of a gauge with a 10 percent GR&R rating being very acceptable for its task, with minimal negative influences imparted by the person conducting the test. A gauge with values between 10 and 30 percent was said to be acceptable, and anything above 30 percent required immediate attention to determine whether it was the piece of equipment, the operator, or the interaction between the two that was producing such unacceptable results.

Solderability testing by dip and look methods do *not* meet this minimum requirement.

The science of GR&R has since evolved into Measurement System Analysis (MSA), where GR&R is only one part of a system that evaluates the total measurement capability. MSA is an experimental and mathematical method of determining how much the variation within the measurement process contributes to overall process variability. It takes into account

- **Bias** The difference between the observed value and reference value.
- **Stability** The total variation in the measurements obtained with a measurement system, on the same master or parts, when measuring a single characteristic over an extended time period.
- **Repeatability** The variation in measurements obtained with one measurement instrument when used several times by an appraiser; all while measuring the identical characteristic on the same part.
- **Reproducibility** The variation in the average of the measurements made by different appraisers using the same measuring instrument when measuring the identical characteristic on the same part.
- **Linearity** The difference in the bias values through the expected operating range of the gauge.

All of these requirements can be attributed to solderability testing, even though at first glance it would appear to have nothing to do with it. Being able to test to a protocol that meets the previous requirements provides useable data for process development and improvement and hopefully will facilitate a much shorter development time than might otherwise be the case.

Classic solderability testing, using dip and look techniques, cannot meet any of the previous requirements.

42.3.3 The Wetting Balance and Measurement System Analysis (MSA)

For solderability testing there is only one method that has the potential to meet the requirements of MSA and that is testing by wetting balance. Not all wetting balances are, of course, created equal and the ease of use, or lack there of, can have a serious impact on the final MSA analysis and ultimate acceptance of the data. Part of the problem in previous evaluations of the wetting balance has been the samples used for testing the performance of the machine. Typically, with GR&R or MSA, samples are tested repeatedly by numerous people in a random order. When testing with a wetting balance, soldering a part is required. This means that it is a one use item that cannot be tested again. Testing components clearly is not an option, as the natural variation from a plating process (assuming it is one that is in control), may contribute enough variation to fail the evaluation. To get around this one-time test, the solution has been to place known weights that are calibrated and certified on to the wetting balance (which is measuring force as a function of displacement of a sensor) and then record the force measurement as a function of the weight applied. Calibrated weights in the range of 100 milligrams to 5 grams were used and this was thought to provide the necessary range and usability for running a GR&R test. The results obtained from numerous round-robin industry studies have been terrible, with no correlation test site to test site or machine to machine from the same manufacturer being obtainable let alone obtaining an acceptable MSA, until the methodology and equipment was understood.

A very detailed study was performed on the failed GR&R's of the past and an examination into the source of variation was undertaken. The results showed that slight misplacement of the test weight could have a huge impact on the result, especially in the lower weight ranges where small components are typically tested. Figure 42.3 shows normalized data from one round of testing with calibrated weights. The correct value should be 0.98 mN. It is clearly evident that the degree of spread and variation in this one-person test, which is typical for this type of test, would result in a rejection of the wetting balance as a scientifically acceptable measuring device for solderability force measurements.

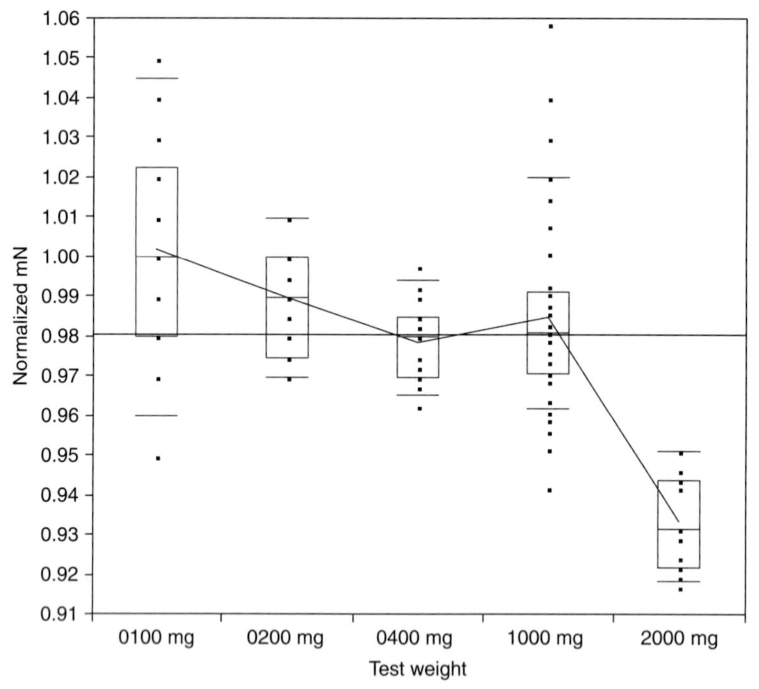

FIGURE 42.3 Normalized data for one individual from a "placing the test weights" wetting balance test—the correct value is 0.98 mN.

42.3.4 Wetting Balance Protocol to Meet the Requirements of an Acceptable GR&R

What was needed was a test protocol that allowed the wetting balance to be used for what it was designed for: measuring solderability. This required producing a test sample that is considered uniform and consistent, sample to sample. This "known good coupon" was created and subsequently a test protocol developed. This detailed test protocol maybe found in the Appendix sections of the ANSI-JSTD-002 and ANSI-JSTD-003 documents.

42.3.4.1 Wetting Balance Protocol Development. This protocol was run using three types of wetting balance. To demonstrate the power of both the protocol and the wetting balance, three individuals were requested to run the test at every location: an expert in wetting balance use, a technician who knew about the wetting balance, and someone who never even heard of solderability testing, let alone a wetting balance. Generally secretaries and customer sales people were used for this group. Figure 42.4 shows the results obtained for one wetting balance combining ten test per individual. A series of industry-wide, round-robin tests were conducted on three different manufacturers of wetting balance machines and among ten individuals. The results show that while there may be some performance differences from machine manufacturer to machine manufacturer, it is nothing that cannot be fixed now that a MSA-capable test protocol exists for the wetting balance. Based on the data produced, the test protocol was accepted by both the solderability specification committees and inserted into the next released revisions—ANSI JSTD-002C and ANSI-JSTD-003B.

Based on this work, the industry has a test protocol that stands up to any scrutiny pertaining to statistical GR&R and MSA evaluations or critiques.

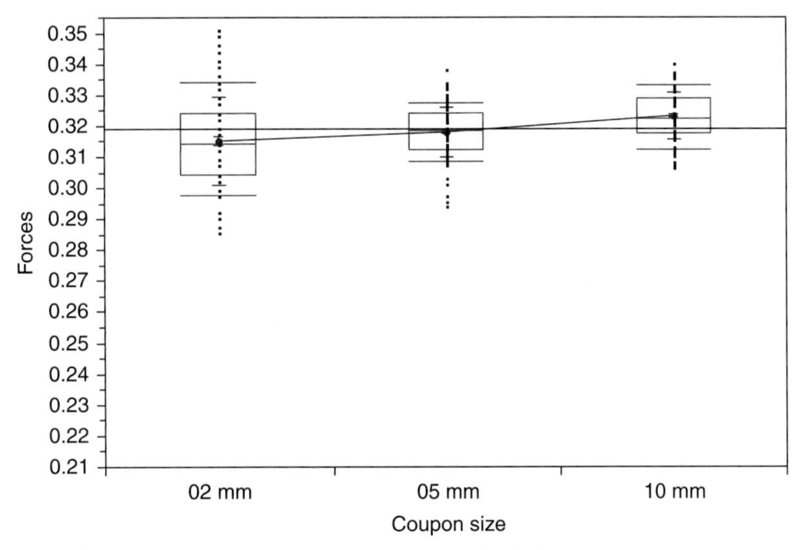

FIGURE 42.4 An example of the consistency of wetting balance using the test protocol—ten individuals participated in the generation of the previous data.

42.3.4.2 The Output from a Wetting Balance. The output from a wetting balance test provides detailed information in both graphical and numerical formats. Depending on the software, multiple tests may be displayed on one graph, showing consistency or lack there of in the performance of a test group. Average values may also be displayed; some softwares even displays the contact angle. It also clearly distinguishes between non-wetting (soldering never took place) and de-wetting (soldering initially occurred but due to basis metal issues the solder could *not* remain wetted to the substrate). Figure 42.5 depicts the action of soldering using a wetting balance, the same motion applies if using a globule or immersing at an angle. Figure 42.6 shows the typical outputs from a wetting balance.

a. Sample reaches the surface of the solder bath
b. Sample at end of immersion depth—(buoyancy)
c. Forces at equilibrium
d. Maximum wetting force
e. Sample lifted out of the solder bath
f. Sample is out of the solder bath

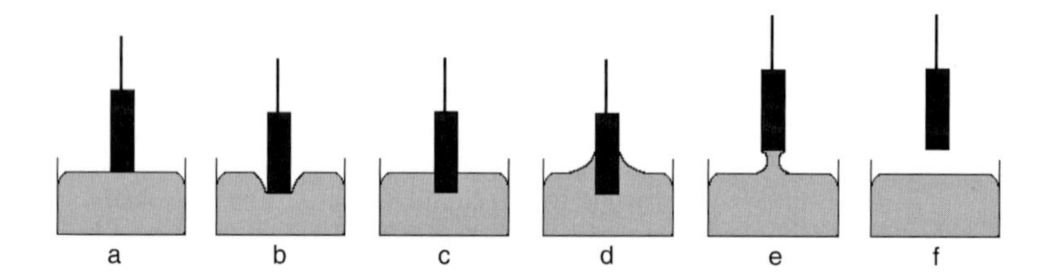

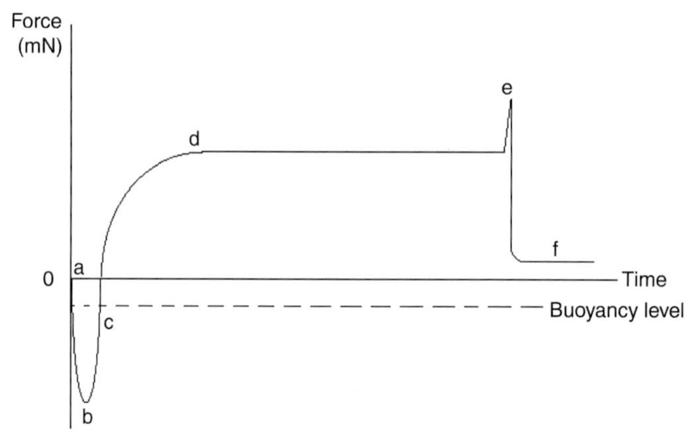

FIGURE 42.5 An example of the consistency of wetting balance using the test protocol—ten individuals participated in the generation of the data.

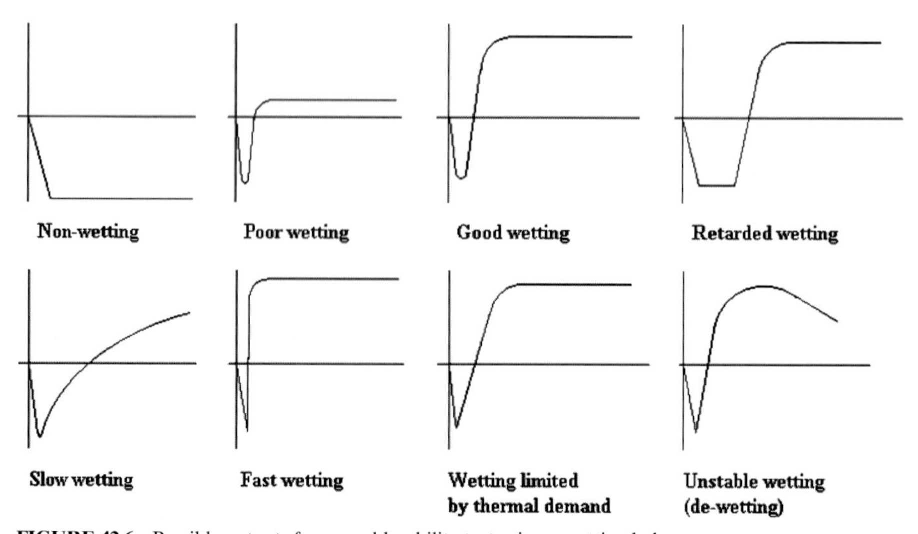

FIGURE 42.6 Possible outputs from a solderability test using a wetting balance.

42.3.4.3 Reporting the Results—mN or mN/mm. The output from a wetting balance is typically a force measurement. The force value in and of itself does not help interpret how good or bad the test specimen is, but obviously, the higher the force the better. The size of the test specimen and the wettable length of the specimen are key characteristics in how much force is produced, and the buoyancy of the device also is a key characteristic. To rationalize solderability performance over a myriad of components and PCB thicknesses, the use of normalized data is recommended. The wetting force is then reported as mN/mm and the ability to standardize on minimum acceptable values regardless of part size is now possible. Some wetting balance software does this normalization automatically, but other software that doesn't will need to be manipulated by exporting to Excel(tm) or similar programs to normalize the output data.

42.4 THE INFLUENCE OF TEMPERATURE ON TEST RESULTS

The test temperatures specified for solderability testing with SnPb is 235°C and 245°C for PCBs and components respectively. The test temperature for SAC305 testing should be 255° C for both the component and PCB solderability specifications. Testing at lower temperatures will produce slower wetting times that maybe reported as failures incorrectly. As the temperature increases, there is the possibility of the thermal degradation of the flux being used for the test, but unfortunately not everyone follows the recommendations of the solderability specifications. If this occurs, a false negative result may be obtained—check with the flux supplier before using temperatures above 260°C for testing regarding the stability of the flux.

Figure 42.7 is a comparison of average wetting times for ENIG, tested at 235°C as a function of solder alloy used. The SnPb alloy wet the coupon after 0.75 seconds, SAC305 took more than 4 seconds and SnCu did not wet during the 10 seconds of the test.

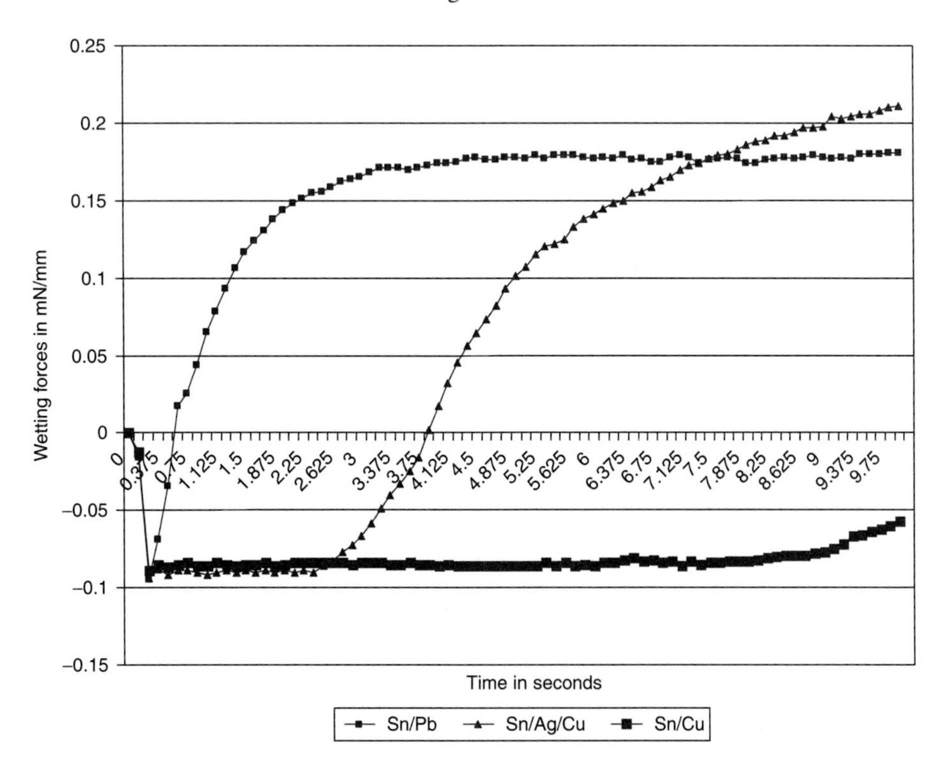

FIGURE 42.7 Wetting speeds for three common solder alloys at 235°C.

42.5 INTERPRETING THE RESULTS: WETTING BALANCE SOLDERABILITY TESTING

Solderability testing results should be useable in process development. The classic dip and look tests produce only go/no go data that has little value in troubleshooting or developing a plating line. It cannot differentiate if one pre-clean chemistry is better than another or the effect of the micro-etch rate on basis metal removal because it does not produce useable data. The part tested by dip and look is immersed into the solder bath and it either wets or it doesn't wet—with no shades of gray in between. Solderability testing with a wetting balance can detect subtle changes in many of the process steps that are part of the plating process of a PCB or a component.

The shape of the curve and the rate of rise give the informed user the vital information necessary to troubleshoot or refine their process. A slow rise of the wetting curve is indicative of oxides. That is, is the rinsing too aggressive, the pH too low, or the temperature too high? Initial good wetting, followed by low forces being produced, indicates poor surface preparation of the basis metal. Is the micro-etch able to attack the basis metal or is the preclean leaving a barrier film that prevents it from doing its work? Table 42.3 details some of the powers of a useable solderability test (as compared to a dip and look test) for the process of an immersion chemistry such as tin or silver. It is assumed that the process is in control and all process sequence steps are being followed.

TABLE 42.3 Comparison of the Power of a Wetting Balance Test to a Dip and Look Test for Process Development

I Sn or I Ag Process Flow and Solderability Testing		
Process step	Dip and look testing	Wetting balance
Acid Preclean	Changes in chemistry *not* detectable	Changes in chemistry detectable
Rinse		
Microetch	Changes in chemistry *not* detectable	Changes in chemistry detectable
Rinse		
Pre-dip	Changes in chemistry *not* detectable	Changes in chemistry detectable
I Sn or I Ag plating	Changes in chemistry *not* detectable	Changes in chemistry detectable
Rinsing (too short, too long, too much oxygen, incorrect pH)	Changes in chemistry *not* detectable	Changes in chemistry detectable
Drying (leaving parts wet or over-drying)	Not detectable	Detectable

Testing with a wetting balance will provide the following information:

1. The speed of wetting   This will show how solderable the part is. This will show if the part is old, how well or poorly the part has been stored, and/or how well the process control is for the plating process.
2. The rate of rise to maximum wetting force   This will show the level of oxides present on the surface in the case of a gold-plated part, and how thick the deposit is. This again will provide information on process control and/or age and storage of the parts being tested.
3. The maximum wetting force produced   The higher the force the better—provides information on the basis metal solderability, the surface preparation, or lack thereof. Low forces recorded on a test, even if the wetting time is fast, is of *major* concern.
4. Once the part wets and produces positive wetting forces, do they remain consistent (ideal), continue to increase (indicative of oxide reduction), or begin to de-wet (the worst

solderability defect)? It should be noted that some 100 percent tin surface finishes and other lead-free finishes have such a smooth surface that adhesion of sufficient flux for a five or ten second test is not possible and the de-wetting being recorded is a function of flux consumption rather than a fundamental solderability test.

5. The output of the wetting balance provides "hard data" which then gives the engineer the ability to perform a statistical comparison of both the population being tested and the other production batches from the same supplier or off the same plating line.
6. The impact of changing fluxes This is especially useful for lead-free assembly development, given the myriad of surface finishes the assembly engineer is now faced with.
7. The impact of stressing (steam exposure or dry bakes or temperature/humidity) on the robustness of the plating can easily be determined as a function of reduced wetting forces or increased wetting times.

The importance of the previous information cannot be over stated as the industry transitions to Pb-free assembly. A test that provides everything in the previous list as its output can improve a process. It can also be used to improve a supplier or disqualify a supplier, select a flux or a surface finish, and/or the best combination of both to minimize defects.

42.6 GLOBULE TESTING

All the previous information provided has been based on the use of a solder bath for testing when using a wetting balance. There is another test method possible for the wetting balance known as the *globule test method,* in which a pellet of solder of known weight, for example 25 mg, or for size, 4 mm is placed onto a solder iron tip of suitable size that is encapsulated in a non-wettable aluminum block. The solder pellet is fluxed, melts on the heated tip and then forms a globule into which the part to be tested is immersed. Historically, the globule was designed for testing wire or leads of a component that are circular in cross section. In that process, the sample to be tested was positioned horizontally to the globule and then, using a microscope and a stop watch, the time needed to reform the globule after wetting the wire was recorded. Similar to the mathematics explaining the physics of soldering in Fig. 42.1, using the globule for testing components of circular cross sectional areas can be logically explained.

The use of the globule test method for other components is very common. The output of the globule test is in mN and consistent placement of the sample, relative to its immersion point in the globule, is critical if variation due to the test is to be minimized. Three globule sizes are popular: 4mm, 2.5mm, and 1mm, and these are chosen based on the size of the component to be soldered. Placing a large chip endcap into the smallest globule will provide insufficient solder to totally wet the part causing artificially low values to be recorded. Placing the part offset to the globule will affect the true immersion depth that the machine has been preprogrammed to; again providing insufficient solder to wet the part resulting in low forces. Modern globule test machines use video and programmable XY stages to consistently position the sample in the same place relative to the solder globule. After each test, the globule needs to be replaced with a fresh pellet in order to maintain the same dimensions. This is necessary so that the immersion depth and solder available for wetting are consistent. Because of the need to prevent the globule from oxidizing so that it doesn't impact the solderability results, it must be fluxed with a rosin flux that does not evaporate. Testing with fluxes other than rosin-based results in a mix of flux types being used. If a VOC or VOC-free flux is used to prevent oxidation of the globule, it will evaporate rapidly and produce false readings. The ANSI-JSTD-002 has detailed component-orientation diagrams to help minimize variation due to placement error when using the globule test method.

42.7 PCB SURFACE FINISHES AND SOLDERABILITY TESTING

The PCB surface finishes all have their unique wetting curves and shapes. Understanding what a good curve should look like for a given surface finish is very important, especially when using the wetting balance to detect subtle changes that are process indicators of potential problems to come. A test that does not provide numerical data will never be able to achieve this.

It is very important that the solderability test, including any stressing that may be imparted on the deposit, be representative of the failure mode that the surface finish finally succumbs to naturally over time. The use of steam to stress a deposit other than a lead-bearing one is not to be used; the recommended stressing is eight hours of exposure to 72°C/85 percent R.H.

42.7.1 Typical Solderability Failure Modes By Surface Finishes

The typical failure modes for the most common PCB surface finishes are listed next. It should be noted that these wetting curves are for PWB's that were *not* packaged and purposefully exposed to the environment for a worst case scenario. PWB's correctly packaged and stored will last a lot longer in the majority:

- HASL (standard 63/37 Sn/Pb) This has been the standard by which all other surface finishes have been measured. It is said that, "nothing solders like solder." It is a very robust deposit if the solder is maintained for correct alloy ratio and the copper concentration is kept below 0.3 percent by weight. The thinner the deposit the shorter the mean time to failure, whereas the thicker the deposit the more problems with chip placement and paste transfer. A good solder level deposit for a 1206 chip pad should not have thickness readings below 50 microinches with an average value of 120 microinches being excellent! The ultimate failure of solder is that after many years in storage, the copper migrates through the deposit and ultimately becomes oxidized upon reaching the surface. This renders the deposit unsolderable. Before reaching the surface and oxidizing, the growing copper tin intermetallic is solderable. Surface oxides are typically not an issue for HASL. Wetting times are instantaneous for a new deposit, with rapid rise to the maximum wetting force –0.25 to 0.27 mN/mm is common when using a standard 0.2 percent activated test flux. As the part ages, there is a slight reduction in wetting forces and a slight increase in wetting times. (See Fig. 42.8)
- HASL (Pb-free) The majority of Pb-free HASL deposits use Cu. This is constantly added to as a function of the leveling process. In the tin-lead world, this copper is a source of contamination and needs to be controlled tightly. For Pb-free HASL it is a more important control point for the deposit. Additional copper in the bulk may mean a shorter shelf life as well, and may have an impact on the Liquidus of the surface finish. Be cautious of using this finish unless the supplier has excessive amounts of data and can show control of the alloy composition
- I Ag This finish is perhaps the most susceptible to correct storage and packaging. Packing materials that contain sulfur will destroy the solderability of the deposit in a very short period of time. The use of dessicants (in the majority containing sulfur, though sulfur-free is available), rubber bands, and slip sheet papers that are *not* pH neutral and sulfur-free should *never* be part of the packaging materials. Never take the PWB's out of their original packages unless to assemble the boards. If there are left over PWB's then place them inside a bag and fold it over to prevent air impingement. Exposure to the air will result in tarnishing of the surface (sulfate formation) that will destroy solderability. This can happen in as little as 28 days. (See Fig. 42.9) If stored correctly with no exposure to air, a gradual reduction in wetting force due to copper oxide growth will occur.

Figure 42.10 shows the natural reduction in wetting forces for the same I Ag deposit when stored in a bag. The deposit is porous by definition and the reduction in wetting forces is a

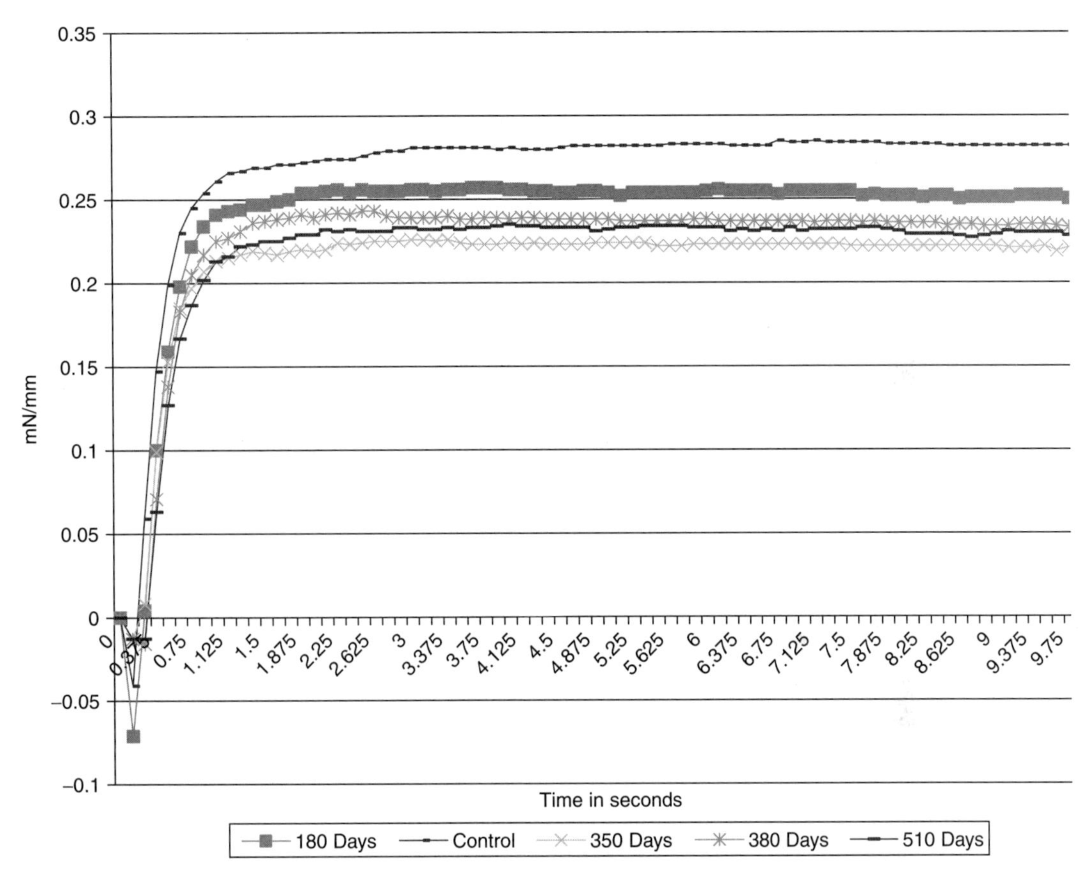

FIGURE 42.8 The normal deterioration of a 63/37 Sn/Pb HASL finish after 510 days in unprotected storage. Wetting times increase only slightly to 0.05 seconds.

function of copper oxide growth that is forming at the Ag/Cu interface. The speed of wetting is not impacted; even after 629 days of storage, Ag dissolves rapidly at the test temperature of 235°C (approx 50 microinches/second).

- ENIG the immersion gold deposit is porous by definition. It does offer very good protection to the underlying nickel, but over time the porosity of the deposit results in the passivation of the nickel surface and as the soldering is done to this layer and not to the gold, the wetting forces are reduced. (See Fig. 42.11) This process should take years to occur. Normal storage of the boards will not impact the initial wetting time, as the dissolution rate of the gold is not impacted—it dissolves at 153 microinches/second at 235°C. An example of the power of the wetting balance is the ability to detect excessive rinsing post nickel deposition: it will passivate the surface and produce slow wetting times similar to old parts. Depending on the flux activation levels used, the degree of nickel passivation can produce a total non-wet condition to an excellent wetting result. After 842 days of unprotected storage, the nickel deposit continues to wet under 0.5 seconds, but again the exterior metal—immersion gold—is porous by definition. So, while doing an excellent job at protecting the underlying basis nickel from passivation, time eventually takes its toll.

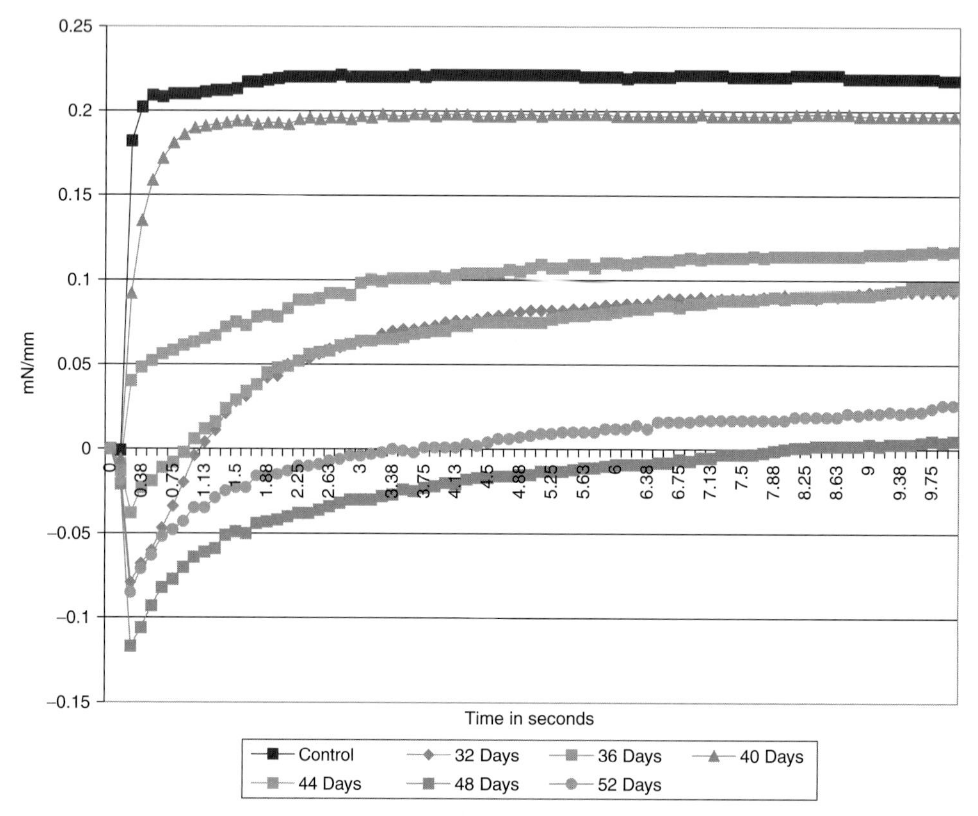

FIGURE 42.9 The rapid decrease in wetting forces for I Ag when stored incorrectly. While this may seem to be a problem, it is actually an advantage as I Ag is supposed to tarnish as a function of exposure to air. If it does not tarnish, there is too much "organic" contamination in the deposit which will produce its own series of problems.

- OSP With the exception of the first generation of organic coatings, which were mono layers of approx 40Å in thickness and were very sensitive to handling/storage, the current generation of OSP coatings are robust and have excellent shelf life. (See Fig. 42.12 Their ultimate failure is a gradual break down of the deposit and subsequent formation of copper oxide. Even as this occurs, the reduction in solderability performance is not as dramatic as with other surface finishes—naturally occurring copper oxide not being very tenacious and/or hard to reduce.
- I Sn This is one of the more difficult surface finishes to correctly apply and maintain from a chemical view point. As such, the possibilities for solderability defects are many. Both the basis metal (copper) and the I Sn deposit have a natural affinity for one another. The copper migrates through the tin as a function of time and temperature; the hotter the temperature the shorter the time to reach the surface. Once there, the intermetallic oxidizes and becomes non-solderable. As a function of plating, copper is displaced from the surface of the PWB and contaminates the plating bath; potentially reducing the mean time for copper to reach the surface of the part. The deposit thickness likewise plays a crucial role in determining shelf life—the thinner the deposit, the shorter the shelf life all things being equal. In addition to this, the catalyst used to make the plating chemistry work (Thiourea) can

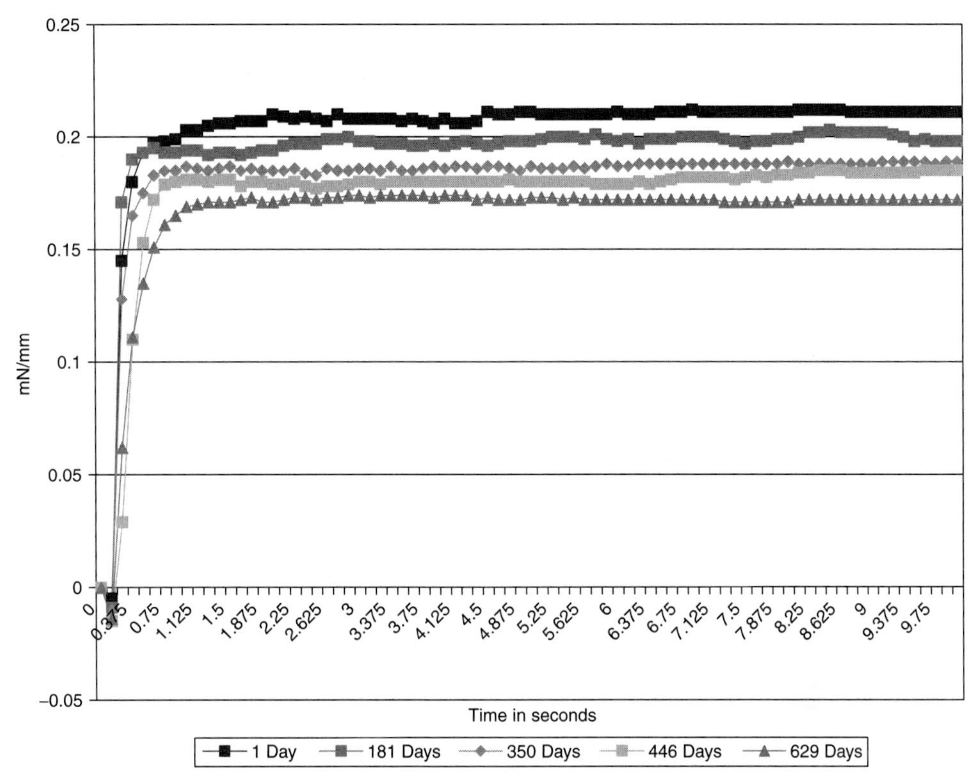

FIGURE 42.10 A gradual reduction in wetting forces over 629 days of "correct storage" for Immerssion Silver (I Ag) is shown.

break down and co-deposit sulfur into the tin, which will destroy solderability. This co-deposition may not be seen with a standard solderability test—even a wetting balance test—as it requires the heat from a first pass assembly process to move the sulfur to the surface, which then destroys the solderability of the deposit. It is recommended that thermal stressing of the deposit is used in conjunction with wetting balance testing to ensure that the deposit is robust and free of sulfur. Finally, I Sn has two oxides—SnO and SnO2—the former easily reducible and solderable, the latter *not*. SnO2 can occur due to poor bath maintenance or poor rinsing control. Again the use of stressing techniques, such as 8 hours of 72°C/85 percent R.H., is recommended to help detect the presence of such surface oxides. (See Fig. 42.13 for an example of I Sn gradual decline).

- Direct Immersion Gold (DIG) This is a relatively unknown deposit, but one with major potential. The application of a special immersion gold deposit does not corrode the underlying copper, and based on the density of the deposit it performs more like a 30 microinch deposit rather than the 3 microinch deposit it actually is. The underlying copper does migrate through the gold eventually, but it takes many years under normal storage conditions and the failure mode is a simple copper oxidation that is easier to reduce than a passivated nickel layer. (See Fig. 42.14).

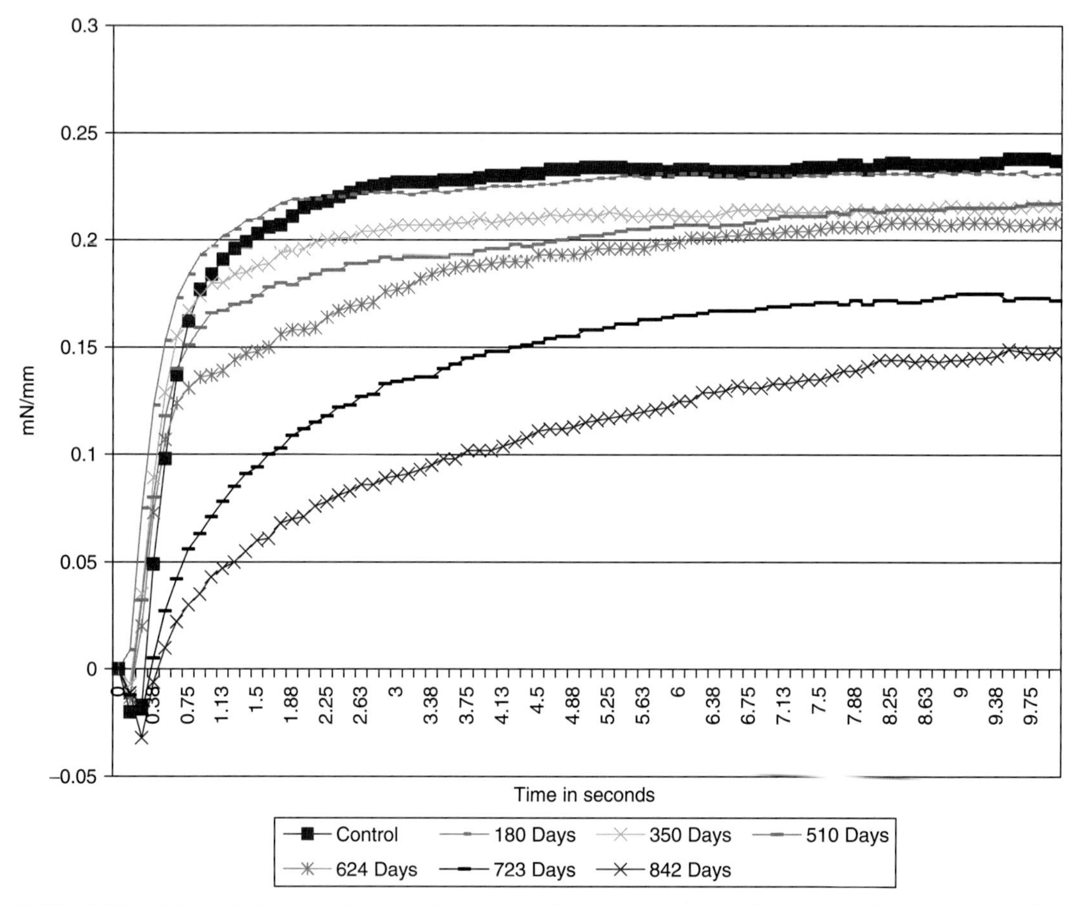

FIGURE 42.11 The gradual decrease in wetting forces for ENIG can be seen as a function of the passivation of the underlying nickel. The deposit, after 842 days of unprotected storage continues to wet under 0.5 seconds, but the exterior metal (immersion gold) is by definition porous and will eventually deteriorate.

42.7.2 PCB Surface Finishes and Solderability Testing for Plated Through Holes

The one area where dip and look type testing works better over a wetting balance is in the examination of solderability of plated through holes. While the majority of assembly is surface mount, there remain a large number of connectors that are through-holes as well as other devices—capacitors etc that are still assembled with wave soldering or intrusive paste reflow. Verification of through holes solderability is equally important if line defects are to be prevented. A solder float is still the preferred method as this forces the solder to rise against gravity to wet the plated through-hole. A number of points need to be made:

1. There are no requirements saying that vias need to be filled with solder—sometimes they do, sometimes they do not. The size of the via and of the solder mask clearance defining the

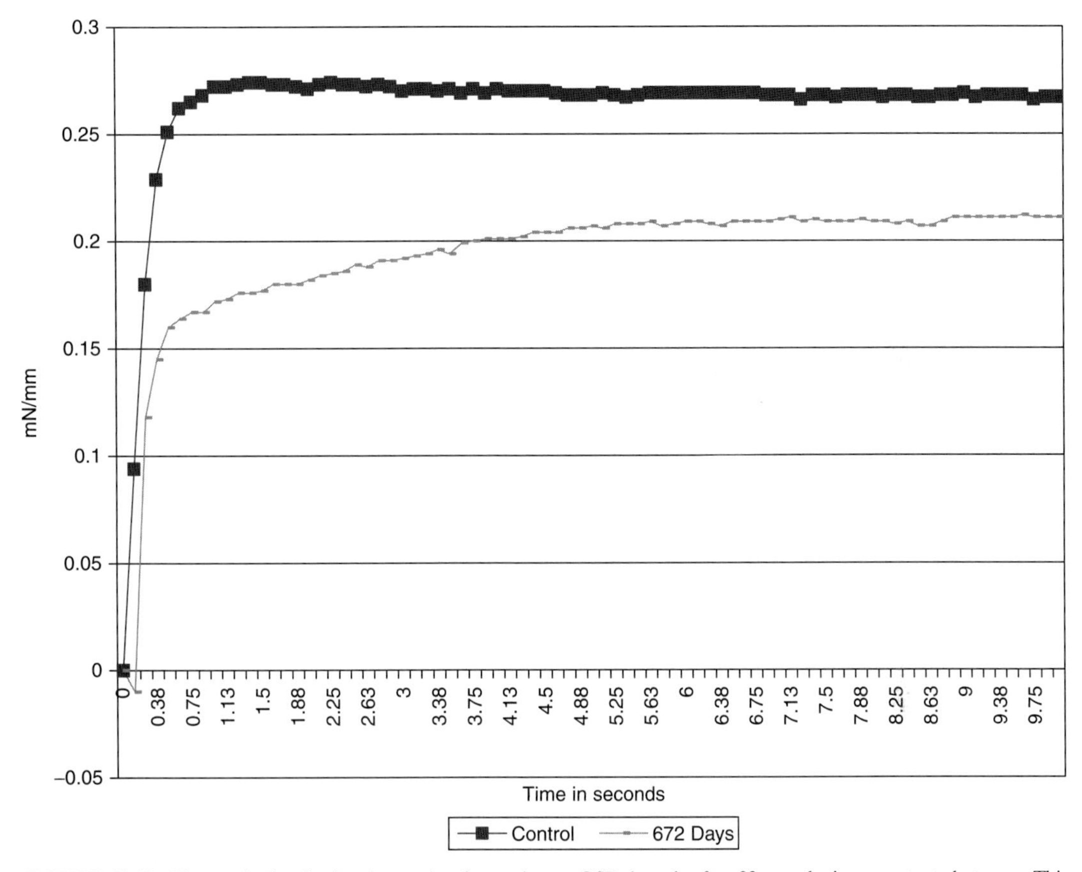

FIGURE 42.12 The gradual reduction in wetting forces for an OSP deposit after 22 months in unprotected storage. This shows the growth of the copper oxide layer which is obviously less solderable than an oxide-free copper surface.

via, are key factors that influence whether they wet or not. The surface finish, however; must be plated inside the vias if required by the specification or purchase order.

2. With the exception of tin-lead surface finish—HASL—there is *no* requirement to produce a positive meniscus for meeting class 3 acceptability. The solder should wet the hole completely, but it does not have to show positive meniscus.
3. The requirement to have flux wet the hole and drain, while may be seen as stating the obvious, cannot be overstated. The flux needs to be adjusted to the specified specific gravity to allow for wetting to occur. If the holes are small or the aspect ratio is high, additional agitation rather than a simple dunk will be required to ensure complete wetting of the hole wall with the flux. Additionally, the need to tap the sample to ensure complete drainage is also necessary. Failure to do both will result in either incomplete hole fill and/or non-wets: both of which are failures but a function of the test rather than the surface finish.
4. Adhere strictly to the 2×2 coupon-size specification and if the sample has many power or ground planes or a heavy overall copper-rich PCB, use *pre-heat*!
5. Hole fill with Pb-free alloys are typically more difficult and may require an increase in test time to allow for the final outcome of the test—hopefully complete hole fill. Experiment

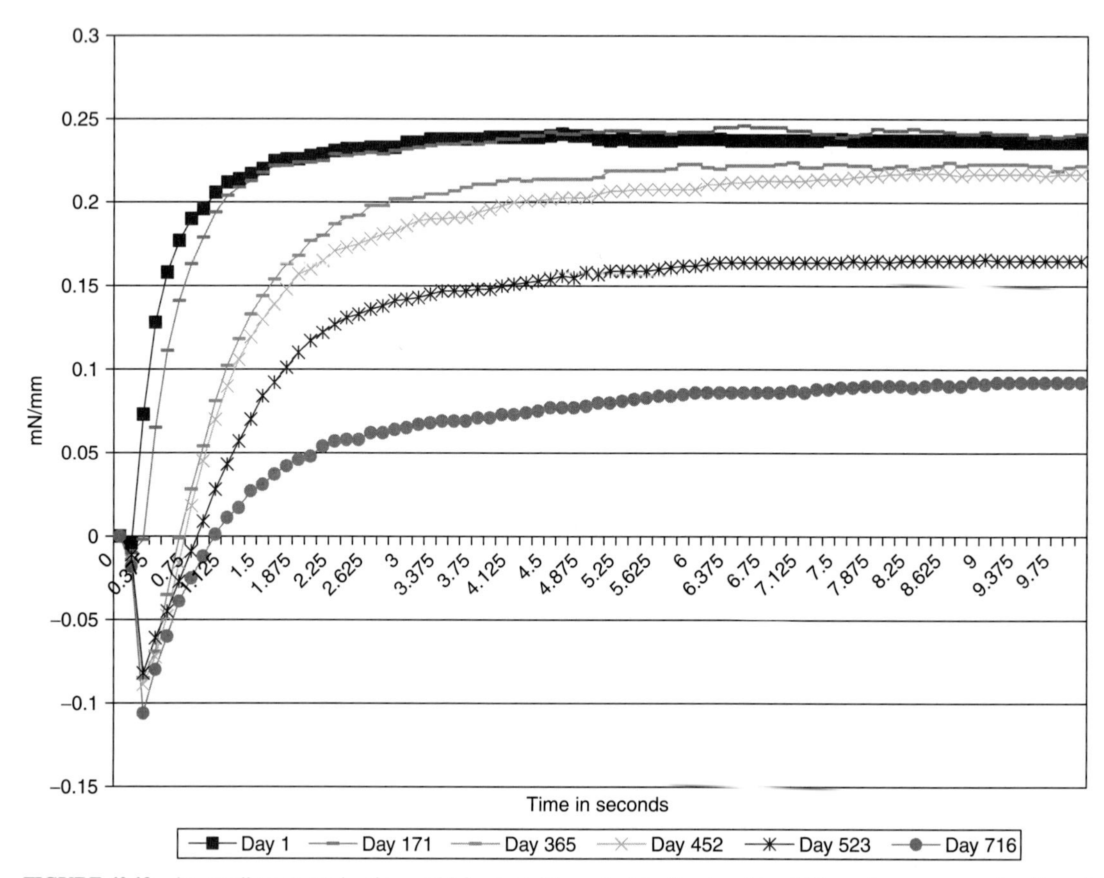

FIGURE 42.13 An excellent example of a good I Sn deposit's gradual decline in solderability performance as a function of intermetallic growth over 716 days. The wetting times increase slightly, which is also as a function of SnO growth on the surface of the deposit.

with test times as a function of board thickness and copper weight. Any data generated from such experimentation should be submitted to the ANSI-JSTD-003 committee.

42.8 COMPONENT SOLDERABILITY

The majority of components require the use of some form of plating to provide solderability and shelf life—very similar to what is needed for PCBs. There are myriad plating options, both from the metallurgy used to the application methods employed. Process control for component plating could be argued to be more critical than for PCBs due to the speed of plating and the number of components plated /hour as well as the various plating techniques and alloys. Prior to the RoHS directive, the component surface finish in the majority and of choice has been tin lead. Suitable for all plating applications from barrel to rack to strip, the plating of this surface finish was easily controlled; hence its solderability. The chemistry used in the majority was

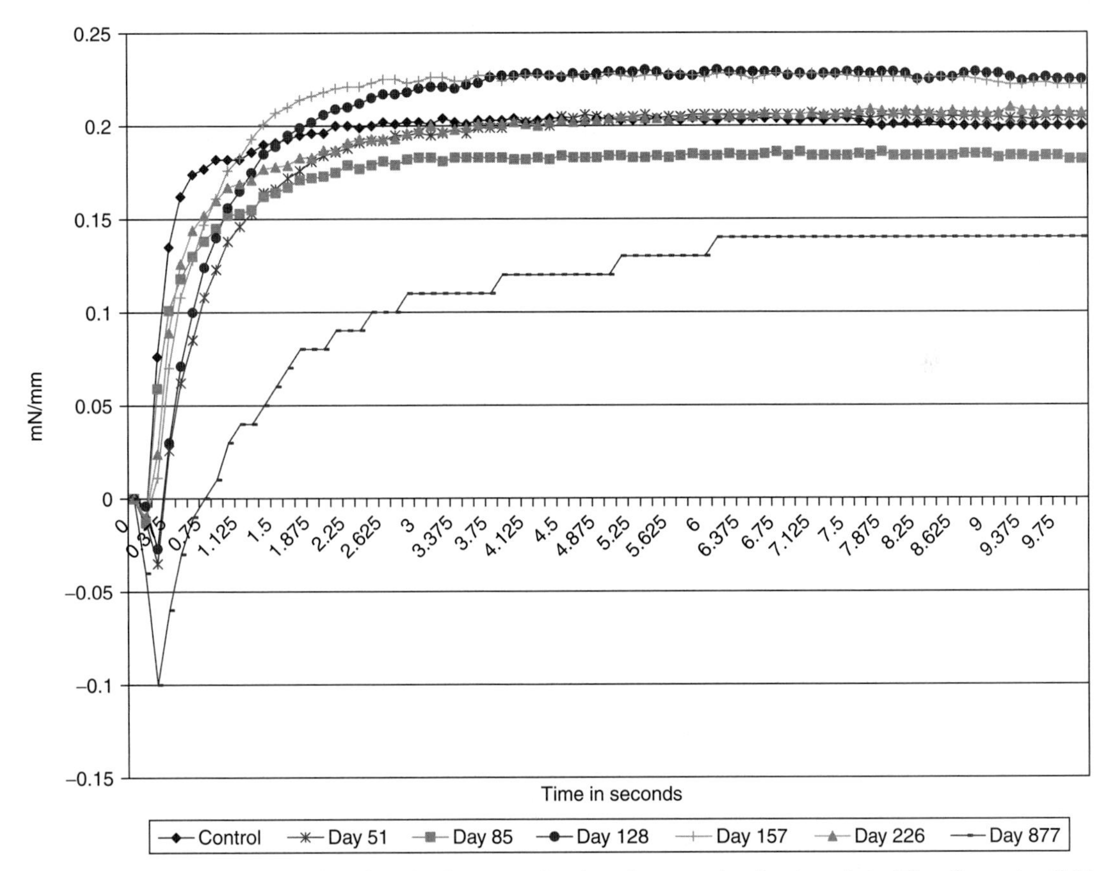

FIGURE 42.14 A gradual reduction of wetting forces as a function of copper migration through the Direct Immersion Gold (DIG) deposit after 877 days in storage.

initially formulated in the 1960s and may have seen some tweaking over the years for plating efficiency. In addition to the common place tin lead, exotic deposits such as nickel palladium, nickel gold, or palladium silver etc have also been used. While these deposits also offer extensive shelf life and excellent solderability performance, fear of gold embrittlement has resulted in the pre-tinning of gold plated devices prior to assembly to remove the gold from the component. The tinning process changes the solderability performance dramatically.

42.8.1 Solderability of Component Basis Metal

Unlike the PCB world where the basis metals used are inherently solderable, such as copper or nickel, the component world deals in basis metals that are non-solderable such as stainless steel or brass (solderable with fluxes not typically found in the electronics industry), marginally solderable (such as Alloy 42—a nickel iron alloy), to easily solderable (such as C194 copper alloy). This fact needs to be realized when designing with or choosing a metal or alloy for a given component and the simple question, "Will this part solder if the surface finish application or quality is marginal?" needs to be asked. If the answer is no, as when using stainless steel for example, then great care will need to be exercised in both the process control and metallurgy chosen for the part. If the answer is inherently yes then a greater process window should be expected. It should be noted that testing and soldering with Pb-free alloys only acts to reduce any process window.

While this is a chapter on solderability and not necessarily metallurgy, it is important to point out a few factors that may have been missed in designing for component solderability 101. If these fundamentals are followed then the results obtained from solderability testing of components may make more sense and help you deal with problems on the line. The following points should be followed to minimize solderability issues as a function of basis metal choice:

1. Is the basis metal solderable by itself when tested with the flux specified by the ANSI-JSTD-002? Brass, stainless steel, and beryllium copper—common basis metals for components—will not solder as tested to the JSTD-002. C194 or C197, phosphorous bronze, alloy 42, thick film Ag, or Pd Ag are solderable.
2. The ideal method to make a non-solderable basis metal solderable is to use either a barrier or intermediate metal that *is* solderable, such as nickel or copper. A less ideal method, and one commonly used, is to plate a heavy deposit of tin lead or 100 percent tin over it and hope that sufficient plating remains post soldering.
3. The use of thin deposits of a final finish, over a non-solderable basis metal, may result in initial wetting. This is followed by de-wetting when the surface finish is consumed by the tin in the solder.
4. If soldering to a surface finish, plated over a non-solderable basis metal, the reliability of the solder joint will be a function of the adhesion capability of the plating to the basis metal. Even if a good solder joint is formed between the assembly alloy and the component surface finish, the reliability will be poor if the plating peels from the surface. The reliability of the component will be placed firmly back into the plating process and the surface preparation by the plater. You should ask yourself if you trust your plater and the consistency of the process.
5. The preferred method for dealing with a non-solderable basis metal is to plate with a solderable metal. Adhesion of this plating to the basis metal is critical. Acid copper plating does not have great adhesion qualities to alloy 42 or alloy 52, so the use of a cyanide copper strike prior to acid copper plating is the correct protocol. By doing this, the adhesion is guaranteed and provides an excellent solderable surface beneath the surface finish.
6. The conversion to Pb-free has resulted in the dominance of 100 percent Sn as the solderable surface finish. Deposits as thin as 1.5 micron over non-solderable basis metals are being specified, possibly as a cost saving issue, but if the contact time with the solder is slightly longer or rework is needed, this very thin deposit is consumed and defects are inevitable.
7. One positive aspect of the RoHS directive is the use of defusion barrier metals. Nickel is the most common, but copper greater than 1.5 micron seems to work as well, acting as a tin whisker mitigation strategy. By requiring the defusion barrier, the possibility for solderability improvements exists. The quality of this diffusion barrier is likewise very important and its solderability should not be sacrificed for a whisker mitigation strategy.
8. Switching from a tin lead-plated component to a lead-free finish is something not to be taken lightly. Changing from tin lead to 100 percent tin may seem to be relatively easy and logical but a 100 percent tin deposit is more susceptible to line issues such as rinse pH control. Also, its impact on the surface finish is more than tin lead ever was. Similarly, changing from a base metal to an exotic such as nickel palladium gold will require the purchase of some expensive sophisticated measurement tools that were not necessary before. Plating a tri-metal deposit to an accurate thickness is of itself not easy and failure to do so has produced some interesting solderability challenges.

42.8.2 Solderability Testing and the JSTD-002

A very useful test is specified in the document called Resistance to Dissolution of Metallization. Parts are tested by immersing them in solder at a higher temperature (–260°C) for an extended period of time and once removed, examination of the deposit is undertaken. This test will certainly catch the thin deposits applied over non-solderable basis metals and will also catch poor surface preparation of solderable basis metals. This test should be used on *all* non-solderable basis metal components and for *all* thick film metallization parts and should be used before basic solderability testing is begun.

As per the JSTD, the test time for components is typically five seconds. The part is completely immersed into the solder bath and then removed, cleaned, and evaluated for dip and look testing. The part may have wet instantaneously or it may have wet at 4.999 seconds. This is something that can never be discovered via dip and look testing. This wetting time becomes critical if there are differential wetting times between endcap metallizations. When a chip with two endcap metallizations wets inconsistently endcap to endcap, the resultant forces of soldering will cause one end to rise up off the pad and the chip will exhibit the classic "tombstone" appearance, in which one end of the component becomes detached from the land area and the surface forces on the other end cause the component to stand on end, resembling a "tombstone" (See Fig. 42.15).

Testing of these parts with a wetting balance revealed the problem and, more importantly, gave information to the supplier about the potential root cause of the problem.

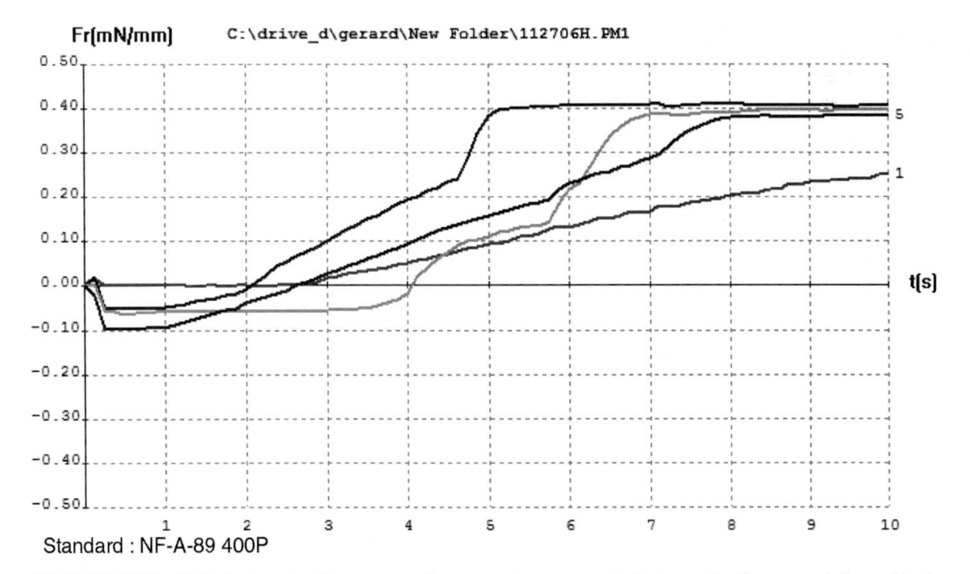

FIGURE 42.15 This shows a test for some surface mount components that pass the five second dip and look test, but clearly show an issue with surface oxides prior to reaching a very acceptable final wetting force. The supplier of these parts, using dip and look testing, certified them to meet solderability requirements but the end customer had line problems.

CHAPTER 43
FLUXES AND CLEANING

Gregory C. Munie
Kester Itasca, Illinois

Laura J. Turbini
University of Toronto, Ontario, Canada

43.1 INTRODUCTION

Cell-phones, iPODs, digitial cameras, laptops and desktop computers, DVDs, and indeed all electronic equipment in our world have one thing in common: They each contain an electronic assembly composed of integrated circuits and discrete components soldered to a printed wiring board (PWB) to create an electrically functional circuit.

In years gone by, printed circuit assemblies were made exclusively with through-hole components. Today's assemblies contain components that are both through-hole and surface-mount, and some contain unpackaged integrated circuits (ICs) attached directly to the PWB. Except in the case of direct chip attachment by wirebonding or conductive adhesive, the use of a soldering flux to ensure good joining is almost universally practiced. Thus, soldering flux choice has become an important consideration in the manufacturing of electronic assemblies. As electronic product designs have moved to more densely packed PWBs with fine lines and spacing, double-sided surface-mount technology, fine-pitch components, ball grid arrays (BGAs), chip-scale packages (CSPs), stacked packages, etc., the role of the soldering flux has become a critical link in the chain of events required for a successful manufacturing process.

In the past, the soldering process was usually followed by a cleaning step to remove flux residues and to ensure the reliability of the assembly. This practice was challenged in the early 1990s by the elimination of a major class of cleaning materials, which had been used for flux residue removal. The targeted cleaning agents include chlorofluorocarbons (CFCs) and methyl chloroform (1,1,1—trichloroethane) that have been shown to destroy stratospheric ozone.[1] These were replaced by aqueous detergent cleaners and saponifying agents, and in some cases cleaning was eliminated by the use of low solids or no-clean fluxes.

Up to the mid-1970s, the predominant flux chemistry was rosin-based. Today, although rosin fluxes are still in use, great advances have been made in flux chemistries and a large variety of formulations with very different chemistries are being used.

The implementation by the European Parliament of Directive 2002/95/EC on the restriction of the use of certain hazardous substances (Restriction of Hazardous Substances [RoHS]) in electrical and electronic equipment, including lead in solder, has had significant impact upon the manufacturing process. Traditional tin-lead solder melts at 183°C, whereas the most common lead-free alternatives have a much higher melting temperature, including tin-copper (227°C), tin-silver (221°C), and tin-silver-copper (217°C). This affects the substrate materials, the flux formulations, the cleaning materials and processes, among other things.

This chapter will review the assembly, soldering, and cleaning processes, briefly touch on choice of surface finishes for both printed boards and components, discuss the soldering flux in terms of its role in soldering, review the variety of flux chemistries available along with their advantages and disadvantages, and comment on the challenges for flux formulation and cleaning due to lead-free soldering temperatures.

43.2 ASSEMBLY PROCESS

The basic elements of an electronic assembly are the PWB, the components and the interconnecting metallization. The majority of PWBs are made of epoxy-glass substrate with copper or solder-coated copper metallization. In many cases, a solder mask covers most of the circuit traces, leaving exposed only those patterns where soldering will occur during assembly. Boards designed for through-hole components usually contain copper-plated through holes that may be solder-coated; those designed for surface-mount components will have metallized pads for interconnecting the surface-mount parts to form the assembly.

43.2.1 Assembly Process Issues

The assembly process for electronics includes wave soldering, reflow soldering, and vapor phase soldering as well as selective soldering. Components range from through-hole and surface-mount—BGA, μ-BGA, CSP, and stacked packages—to chip-on-board (COB). The high density and reduced size of today's electronics has led to system-in-a-package (SIP), where the integrated circuit (IC) package contains several active devices, and system-on-a-chip (SOC), where all the functions of an electronic system are integrated on a single chip. These package types increase the complexity of the electronic assembly process, requiring adhesive attachment for bottom-side discretes, underfill for CSP, wirebonding and encapsulation for COB, and solder paste, paste flux, and liquid flux for various applications.

When manufacturing complex printed wiring assemblies (PWAs), the thermal process chosen must take into account the thermal mass of the assembly, the component density, the solder flux and solder paste characteristics, and the maximum temperature limitation of the components. Most of the components assembled to the board in the soldering process have a maximum temperature of 240°C for traditional Sn/Pb soldering. Some electronic components, such as electrolytic capacitors and plastic-encapsulated components, are not rated to experience the high temperatures required to process with lead-free solders. The resulting heat-induced degradation can result in early field failures. Also, the higher temperatures required for lead-free solders are not compatible with many optoelectronic components. The increased heat can cause a variety of conditions with these components, among them electrical variances, changes in silver-epoxy die attach properties, delamination between plastic and lead-frame parts, deformation of plastic encapsulants and plastic lenses, damage to lens coatings, and changes in the light transmission properties.

43.2.2 Assembly Temperature Profiles

IPC/JEDEC[2] has developed a recommended reflow profile for nonhermetic packaged semiconductor components. These recommendations are based on the temperature taken at the top side of the packaged device, and also based on package volume, excluding external leads, or solder balls in the case of BGAs, and nonintegrated heat sinks. Table 43.1 lists the recommended range for Sn/Pb and Pb-free assemblies.

TABLE 43.1 Reflow Profiles Recommended for Sn/Pb and Pb-free Assemblies Based on Temperatures Taken on the Package Body

Profile feature	Sn/Pb assemblies	Pb-Free assemblies
Average ramp-up rate	3°C/sec. max.	3°C/sec. max.
Preheat		
Temperature min.	100°C	150°C
Temperature max.	150°C	200°C
Time	60–120 sec.	60–180 sec.
Time maintained above melting Temperature	60–150 sec.	60–150 sec.
Time within 5°C of peak temperature (T_p)	10–30 sec.	20–40 sec.
Ramp-down rate	6°C/sec. max.	6°C/sec. max.
Time 25°C to T_p	6 min. max.	8 min. max.

The maximum recommended temperature for the package depends on the package thickness and volume. Table 43.2 lists the recommended reflow temperatures for Sn/Pb processing, whereas Table 43.3 lists the recommendations for Pb-free processing.

TABLE 43.2 Recommended Maximum Package Reflow Temperatures for Sn/Pb Process

Package thickness	Volume <350 mm^3	Volume ≥350 mm^3
<2.5 mm	240 +0/–5°C	225 +0/–5°C
≥2.5 mm	225 +0/–5°C	225 +0/–5°C

TABLE 43.3 Recommended Maximum Package Reflow Temperatures for Pb-Free Process

Package thickness	Volume <350 mm^3	Volume 350–2,000 mm^3	Volume >2,000 mm^3
<1.6 mm	260 +0°C	260 +0°C	260 +0°C
1.6–2.5 mm	260 +0°C	250 +0°C	245 +0°C
≥2.5 mm	250 +0°C	245 +0°C	245 +0°C

43.3 SURFACE FINISHES

Surface finish is defined as the outerlayer metallization on a component or substrate being soldered. Both the type and quality of the surface finish have a major impact on surface solderability and on process soldering ability. Here, *solderability* is defined as the ability of the surface to be wetted by a solder, and the forming of an intermetallic layer between the solder and the base metal. *Soldering ability* is defined as control of the soldering process parameters, such as a heating profile that produces that wetting action. The preference is always to have a surface that exhibits good solderability—that is, ease of wetting and formation of high-quality solder interconnections—for all soldering processes and materials to which it might be exposed. Soldering ability is best described as the quality of a solder process such that the operating window is wide, that is, the process is not overly sensitive to minor shifts in process parameters such as time and temperature. Thus a process with good soldering ability would be easy to control and would produce high-quality interconnections given that the incoming materials have good solderability.

43.3.1 Substrate Finishes

The substrate surfaces of choice are usually one of the following: electroless nickel/immersion gold (ENIG), immersion silver (ImAg), and organic solderability preservative (OSP). Two other finishes, hot air solder leveling (HASL) and immersion tin, are also in use, but HASL has declined in volume and immersion tin has not reached significant volumes due to concerns over reliability.

- **ENIG** This finish is widely used because it provides a very flat surface for surface-mount component placement. It has good solderability and a long storage life but is somewhat expensive compared to other finishes. Careful control of the gold plating process during board fabrication is required to prevent a defect known as black-pad. Black-pad is caused by galvanic corrosion of the nickel layer, creating corrosion products that affect the ability of the surface to be wet with solder. If severe enough, black-pad will impact reliability of the final solder joint.
- **ImAg** Immersion silver provides a flat surface with good solderability and storage life. It is also less expensive than ENIG. However, the ImAg finish is prone to tarnish under certain atmospheric conditions; in the presence of sulfur, it forms sulfides that darken the surface and reduce solderability.
- **OSP** Organic solder preservatives are inexpensive and also provide a flat surface for surface-mount technology (SMT) placement. Storage life for OSP is less than for the metallized finishes, and multiple thermal cycles cause it to degrade.
- **HASL** One of the oldest surface finishes, HASL is based on exposing the PWB to molten solder and then removing excess solder using hot-air knives. In the lead-free world, the processing temperature is higher than for tin-lead. HASL is inexpensive, has good solderability if the solder is sufficiently thick, but has the disadvantage of not being a flat surface.
- **ImSn** This finish is similar to immersion silver in most characteristics. However, because tin tends to form metallic whiskers that can cause electrical shorts, it is not widely used.

43.3.2 Component Surface Finishes

Component surface finishes, like those of substrate finishes, are intended to provide good solderability. In the past, the majority of component leads had a tin-lead solder finish. This finish was either plated on the base metal of the component or was applied by dipping the leads in liquid metal. Given the push for lead-free assembly, most of these finishes containing lead are being replaced by alternatives. The main contenders in replacement of lead containing finishes are based on noble metals (e.g., palladium or gold) or some form of tin plating. Palladium finishes have been in use for many years and are a known commodity. However, as noted previously, the tendency of tin to form metallic whiskers as a result of stresses in the plated layer is a matter of serious concern for the electronics industry.

43.4 SOLDERING FLUX

Soldering[3] is defined as the process of joining metallic surfaces with solder without melting the basis metal. For this joining to take place, the metal surfaces must be clean of contamination and oxidation. This cleaning action is performed by the flux,[2] a chemically active compound that, when heated, removes minor surface oxidation, minimizes oxidation of the basis metal, and promotes the formation of an intermetallic layer between solder and basis metal. The soldering flux has several functions to perform. It must:

- React with or remove oxide and other contamination on the surface to be soldered
- Dissolve the metal salts formed during the reaction with the metal oxides
- Protect the surface from reoxidation before soldering occurs
- Provide a thermal blanket to spread the heat evenly during soldering
- Reduce the interfacial surface tension between the solder and the substrate in order to enhance wetting

To perform these functions, soldering flux formulations contain the following types of ingredients: vehicle, solvent, activators, and other additives.

43.4.1 Vehicle

The vehicle is a solid or nonvolatile liquid that coats the surface to be soldered, dissolves the metal salts formed in the reaction of the activators with the surface metal oxides, and, ideally, provides a heat transfer medium between the solder and the components or PWB substrate. Rosin, resins, glycols, polyglycols, polyglycol surfactants, polyethers, and glycerine are among the major chemicals used. Rosin or resins are selected when more benign chemicals are required since their residues are less apt to cause reliability failures. Glycols, polyglycols, polyglycol surfactants, polyethers, and glycerine are used in water-soluble flux formulations because they provide excellent wetting of the board surface and dissolve the more active materials used in these formulation.

43.4.2 Solvent

The solvent serves to dissolve the vehicle, activators, and other additives. It evaporates during the preheat and soldering process. The solvent chosen depends upon its ability to dissolve the flux constituents for a given formulation. Alcohols, glycols, glycol esters, glycol ethers, and water are common solvents used.

43.4.3 Activators

Activators are present in the flux formulation to enhance the removal of metal oxide from the surfaces to be soldered. They may be reactive at room temperature, but their activity is enhanced as the temperature is raised during the preheat step of the soldering process. Among the common activators found in flux formulations for traditional soldering fluxes are amine hydrochlorides; dicarboxylic acids, such as adipic or succinic; and organic acids such as citric, malic, or abietic. Higher molecular weight activators are needed for lead-free soldering. Activators containing halide and amine give excellent soldering yields but may cause reliability problems if not properly removed in a well-controlled cleaning step. The halides may be present in either ionic or non-ionic form depending on the level of activation desired. Ionic activators provide, in general, a higher level of activation for a given halide content.

43.4.4 Other Additives

Soldering fluxes often contain small amounts of other ingredients that serve a specialized function. For example, a surfactant may be added to enhance the wetting properties. This constituent can also assist in the foaming characteristics required for foam flux application. Other additives may be included to lower the interfacial surface tension between the molten solder and the PWB as it exits the solder wave, decreasing the chance for solder bridges to form. Solder paste formulations require the presence of additives to provide good viscosity or flow characteristics, low slump during the preheat step, and good tack characteristics for holding the component in place until reflow occurs. Finally, cored-wire flux used for hand-soldering contains a plasticizer to harden the flux ingredients that are present in the core of the wire.

43.4.5 Flux and Temperature Issues

The flux becomes active as it is heated. In traditional flux chemistry for Sn/Pb solder, the assembly is preheated to around 100 to 125°C to remove the solvent and begin to activate the chemicals used to remove the metal oxide. After this plateau, the temperature is increased above the melting point of the solder (183°C) to 240°C for sufficient time to reflow the solder paste, and then the assembly is cooled, solidifying the solder and creating a metallurgical bond between the board metallization and the components. For lead-free soldering, the preheat plateau temperature is higher—150 to 200°C—and the peak temperature is 245 to 260°C. This requires solvents that evaporate at a higher temperature and activators that become chemically active at a higher temperature. In addition, new activators are needed to address the new metallurgy on board surfaces and new lead-free solders containing Ag, Cu, and much higher levels of Sn. The decrease in pitch for area-array packages has resulted in a need for finer solder powders. This is driving a change in paste flux chemistry since the higher surface area of these fine powders results in increased oxide formation and decreased paste stability.

43.5 FLUX FORM VERSUS SOLDERING PROCESS

Although one usually thinks of soldering flux as a liquid, in fact the soldering flux can exist in several different forms. *Liquid flux* is commonly used for wave-soldering or hand-soldering applications. *Paste flux,* a thick, viscous flux, is used to hold components to the board prior to reflow of the already present solder. Direct attachment of a solder-bumped chip or BGA packages to solder pads on a PWB is an example of paste flux's usefulness. Also known as *tacky soldering fluxes,* these materials are useful in repair of area-array devices. Solder paste contains *solder paste flux,* which is the solder paste without the solder particles. Solder paste is used in surface-mount attachment of components such as chip carriers, quad flat packs (QFPs), BGAs, μ-BGAs, CSPs, leaded packages, and discrete resistors and capacitors. *Flux-cored solder wire* is used for hand-soldering, and *flux-coated or flux-cored solder preforms* are used as the solder/flux source in some applications such as backplane connector pin reflow soldering.

Fluxing underfills can also be considered as a new type of flux material. Many fine-pitch, low-profile array packages suffer from poor reliability due to their low standoff from the substrate, small area of solder interconnection and CTE mismatch with the substrate. Often, a reinforcing material is dispensed around the solder joints after assembly to improve attachment reliability. Fluxing underfills provide fluxing action for the solder process and cure into a hard supporting material during the soldering process, eliminating the need for a post-solder dispensing step.

43.6 ROSIN FLUX

Early flux formulations for electronics used rosin, which is a naturally occurring resin, obtained from the sap of a pine tree. Its exact composition varies depending on the part of the world from it originates as well as the time of the year. Rosin contains a mixture of resin acids, the two most common ones being abietic acid and pimeric acid (see Fig. 43.1). Rosin has been a favorite material for soldering because it liquefies during the soldering process, dissolves the metal salts, and then solidifies when cooled, entrapping and for the most part immobilizing the contaminants. In addition, rosin has some innate fluxing activity since its molecular structure contains a weak organic acid. Finally, rosin-based fluxes work well in hand-soldering and repair operations because rosin provides good heat transfer characteristics.

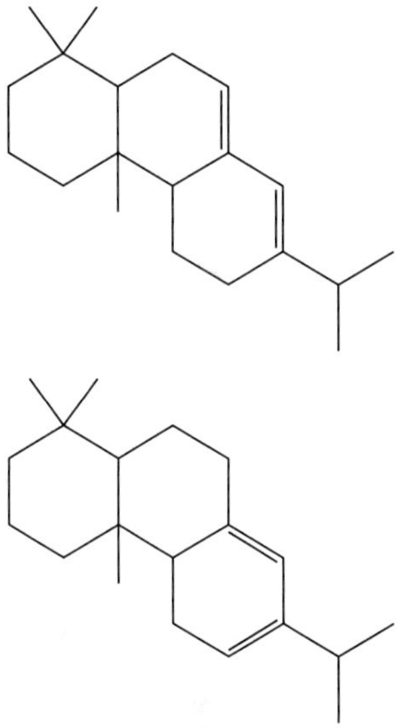

The activity of a rosin-based flux will be determined by the activators and surfactants, which are part of the formulation. Some activators help to remove metal oxide but also leave residues that are essentially noncorrosive. Halide-containing activators leave residues that could be corrosive if too much is left on the board.

Military specifications[4] in the past have required the use of rosin-based fluxes. These were defined as pure rosin (R), rosin mildly activated (RMA), rosin activated (RA), and rosin super-activated (RSA), based upon the level of halide activators they contained. Typically only R or RMA fluxes were approved for high-reliability military applications.

FIGURE 43.1 The two most common rosin isomers are abietic acid and pimaric acid.

In the 1970s, telecommunication companies in the United States and Europe predominantly used rosin fluxes for their wave-soldering requirements. These companies had their own internal set of test methods for selecting noncorrosive rosin flux formulations. Using their selection criteria, they believed the fluxes to be sufficiently safe. For their applications, they cleaned only the bottom side of the assembly, where rosin residues needed to be removed to ensure good electrical contact during bed-of-nails testing.

In the early 1980s, the IPC developed a flux characterization criteria based on the flux and flux residue activity. Thus, fluxes were classified as:

L = low or no flux/flux residue activity
M = moderate flux/flux residue activity
H = high flux/flux residue activity

These designators were determined by a series of tests, including the copper mirror test, a qualitative silver chromate paper test for chlorides and bromides, a qualitative spot test for fluorides, a quantitative test for halides (chloride, bromide, and fluoride), a corrosion test for flux residue activity, and a surface insulation resistance test at accelerated temperature and humidity conditions.

The industry standard "Requirements for Soldering Fluxes" (J-STD-004A) includes a test for electrochemical migration and updates the earlier IPC-SF-818 solder flux specification to include some international elements from the International Standards Organization (ISO-9454). In addition to defining the flux categories L, M, and H, it notes the absence or presence of halides by a 0 or 1 that is added to the descriptor. In parallel the international standards, industry standards list the basic chemical constituents as RO (rosin), RE (resin), OR (organic), or IN (inorganic) (see Table 43.4).[5] The test methods for this specification are contained in the Test Methods Document (IPC-TM-650).[6]

TABLE 43.4 Flux Constituents and Activity Levels

Flux materials of composition	Symbol	Flux activity level (% Halide)	Flux type
Rosin	RO	Low (0%)	L0
		Low (<0.5%)	L1
		Moderate (0%)	M0
		Moderate (0.5–2.0%)	M1
		High (0%)	H0
		High (>2.0%)	H1
Resin	RE	Low (0%)	L0
		Low (<0.5%)	L1
		Moderate (0%)	M0
		Moderate (0.5–2.0%)	M1
		High (0%)	H0
		High (>2.0%)	H1
Organic	OR	Low (0%)	L0
		Low (<0.5%)	L1
		Moderate (0%)	M0
		Moderate (0.5–2.0%)	M1
		High (0%)	H0
		High (>2.0%)	H1
Inorganic	IN	Low (0%)	L0
		Low (<0.5%)	L1
		Moderate (0%)	M0
		Moderate (0.5–2.0%)	M1
		High (0%)	H0
		High (>2.0%)	H1

43.7 WATER-SOLUBLE FLUX

Water-soluble fluxes have also been called "organic acid fluxes." This name is misleading since all fluxes used for electronic soldering contain organic ingredients and many contain organic acid activators. The term "organic acid flux" probably originated from the designation of water-soluble fluxes as "organic" and those activated with organic acid activators as "organic acid." Other activators for these fluxes include halide-containing salts and amines. Although the correct name for this category of fluxes is *water-soluble,* it should also be noted that the flux solvent is normally not water, but alcohols or glycols.

As the name implies, water-soluble fluxes are soluble in water and their soldering residues are also expected to be water-soluble. These fluxes are much more active than rosin fluxes, have a wider process window, and give a higher soldering yield with reduced defects. This means that the final assembly will require less touchup or repair. On the down side, water-soluble fluxes contain corrosive residues that, if not properly removed, will cause corrosion in the field and long-term reliability problems.

As indicated earlier, water-soluble fluxes usually contain glycols, polyglycols, polyglycol surfactants, polyethylene oxide, glycerine, or other water-soluble organic compounds as the primary vehicle. These provide good solubility for the activators, which are usually the more corrosive amines and halide activators. With the onset of highly efficient cleaning equipment, this type of flux became popular for computer and telecommunication applications.

F. M. Zado[7] raised concern in the late 1970s that water-soluble fluxes affect the electrical characteristics of the epoxy-glass laminate by reducing the insulation resistance. This reduction was due to the dissolution of the polyglycols of the flux formulation into the epoxy substrate during the soldering process. Later work by J. Brous[8] indicated that some polyglycols were much more deleterious than others. In general, polyglycol containing fluxes (and fusing fluids used by board manufacturers to fuse tin-lead plating) can cause an increase in moisture absorbance of the epoxy-glass substrate.

The user must consider several factors when determining whether to use water-soluble fluxes in a given application. One important factor is the operating environment. If the assembly will experience extremes of temperature while under power, it is possible that localized condensation will occur and dendrites will form, shorting out some of the circuit elements. Assemblies of this nature should be conformally coated. A second critical factor is the use of a board design and cleaning process, which ensures that corrosive residues are removed. A third consideration is the voltage gradient in the electrical design of the circuit. Within reduced lines and spacings, a failure mechanism called conductive anodic filament (CAF)[9] formation has been associated with high humidity and high voltage gradients. This will be described in detail in a later chapter.

43.8 LOW SOLIDS FLUX

Until the mid-1980s, liquid soldering fluxes were formulated in 25 to 35 percent (weight percent) solids or nonvolatile liquid. Then flux chemistries changed and new formulations that were lower in total solids content came on the scene. These fluxes are composed principally of weak organic acids, often with a small amount of resin or rosin. Early formulations had 5 to 8 percent solids, but today's low solids fluxes are 1 to 2 percent solids composition. For lead-free soldering, weak organic acids with higher molecular weight are being used.

Nomenclature of these fluxes moved from "low solids flux" to "low residue flux" to "no-clean flux." The thinking was that the amount of residue left by these fluxes after soldering was so minimal that it did not need to be removed. This is only true if the residues are noncorrosive.

One challenge of low solids fluxes is the processing window. Unlike water-soluble fluxes that give very low defect levels and have a wide processing window, the soldering process for low solids fluxes must be carefully designed. To begin with, the recommended preheat temperatures are different from those for rosin flux and the preferred solder wave temperature is lower than that for rosin. Additionally, the solderability of incoming boards and components must be ensured. Although water-soluble flux can cut through heavy metal oxide layers, the amount of fluxing ingredients in low solids fluxes is not sufficient to accomplish the task.

Lower cost manufacturing can be achieved with low residue fluxes if the cleaning step can be eliminated. This assumes that the incoming components and boards are clean, and that operators handling the boards are careful not to introduce contamination. This requires a flux with noncorrosive residues that will not impede or contaminate the electrical bed-of-nails test probes.

A new category of low solids fluxes was introduced in the early 1990s to meet the needs of localities where volatile organic compounds (VOCs) are regulated. These fluxes are marketed as VOC-free or low-VOC fluxes. The solvent in this case is 100 percent water or at least greater than 50 percent water. Use of these fluxes requires special care in the preheat step where the water (solvent) must evaporate before the assembly reaches the solder wave. Failure to do this will result in excessive solder ball formation.

Solder fluxes and pastes have gone through significant evolution since the early 1980s. As of 2006, in North America about 70 percent of the fluxes were not cleaned, 25 percent were water-soluble, and 5 percent were rosin-based for military applications.[10] However, in the lead-free soldering world, more cleaning is required.

43.9 CLEANING ISSUES

Flux removal is motivated by several factors, and the importance of these factors varies with the end use of the product. In general, the purpose of cleaning is (1) to remove corrosive residues from the soldering process or handling operations (some fluxes have high levels of halide activators that would cause corrosion were they not removed); (2) to remove rosin, resin, or other insulating residues that would interfere with bed-of-nails electrical testing; (3) to remove rosin, resin, or other residues that attract dust, dirt, and other airborne contaminants; (4) to ensure that the board is free of residues and contamination prior to conformal coating.

In the early days of soldering, cleaning was either not done, because the flux was mostly rosin and noncorrosive (e.g., the telecom fluxes), or cleaning of rosin flux residue was done with solvents. Typically the solvent was a halogenated material, as such solvents were considered relatively nontoxic and were nonflammable. However, in the late 1970s, the Environment Protection Agency (EPA) listed some of these solvents, such as perchloroethylene and trichloroethylene, as suspected carcinogens and lowered the allowed level of these in the operator-breathing zone. This led to the application of CFC azeotropes with methyl alcohol, ethyl alcohol, and methylene chloride as replacements.

However, the CFCs were not without their own environmental problems. In September 1987 at the United Nations Environment Program Conference in Montreal, 24 countries signed the Montreal Protocol on Substances that Deplete the Ozone Layer. The protocol was the beginning of an international agreement on the reduction of CFCs that were shown to have a deleterious effect on the stratospheric ozone layer that protects us from harmful ultraviolet radiation. The initial protocol targeted CFCs and halons (bromine-containing CFCs) and proposed a 50 percent reduction in production by 1998 based on 1986 levels. Revisions to the protocol led to a ban on the manufacture of CFCs and other ozone-depleting chemicals effective by the end of 1995.

43.9.1 Solvent Cleaning

With the general elimination of the halogenated solvents, manufacturers have changed their flux and cleaning choices. The new cleaning materials must be capable of dissolving the particular chemical residues from a given flux formulation. Several detailed cleaning references are available.[11,12,13,14,15,16]

Solvent cleaning options[17] have focused on semiaqueous cleaning, solvent defluxing using a blend of solvents, and vapor degreasing using a single or co-solvent mix. Semiaqueous cleaning usually involves a two-step process where the assembly is first cleaned in the organic solvent, then rinsed in a separate step. Terpenes have been widely used as the solvent with dibasic esters and non-ionic surfactants added to improve cleaning and rinsing. An alternative semiaqueous cleaning solvent is composed of high molecular alcohols. The solvent is usually sprayed under immersion onto the assemblies in its pure form. This step is followed by a rinse step that emulsifies the semiaqueous solvent and dissolves the ionic residues. Innovations[18] have shown that an aqueous emulsified solution of the solvent can also provide good cleaning.

Solvent cleaners may be inline or batch. Inline cleaning usually involves spray under immersion. The temperature of the solvent wash tank is set below its flash point, and several rinse tanks follow the wash tank to ensure good cleaning before the final drying step. Solvents such as glycol-ether blends are used for this process. Batch degreasers may use n-propyl bromide in specialized applications.

Traditional batch cleaning was performed in a vapor degreaser that has a wash sump of boiling nonflammable azeotropic solvent and a rinse sump. A vapor blanket forms above the sump and is contained using cooling coils at the mouth of the machine. The assembly is first placed into the wash sump, and then is transferred to a rinse sump that contains clean solvent from the condensation at the cooling coils. As the assembly is removed, it is held for a period in the vapor blanket for a final rinse and drying of the remaining solvent.

Co-solvent cleaning involves the use of a rinsing agent that is volatile and a solvating agent that is not volatile. The assembly is placed into the wash tank of boiling solvent where the solvating agent enhances the cleaning process. The vapor blanket above the wash sump contains only the rinsing agent. The assembly is moved from the wash sump into the rinse sump, which contains only the condensed rinsing agent. Finally, the assembly is removed from the rinse sump and held in the vapor blanket for a final rinse and drying step.

43.9.2 Aqueous Cleaning

Aqueous defluxing is performed using a high pH saponifying solution that is diluted to a 10 to 25 percent aqueous solution. The saponifier may be organic or inorganic, with additives such as solvents and surfactants to improve wetting and cleaning. The organic cleaning agent usually contains an amine, such as monoethanol amine, whereas the inorganic cleaners use buffered metal salts such as NaOH/sodium carbonate/sodium bicarbonate. The wash step is usually done in an inline spray machine, followed by a de-ionized (DI) water rinse with several spray sections to remove the more contaminated material, then by a final rinse and a final blow-dry step. The cleaning agents in the saponifier react with the rosin or resin from the flux to form a "soap" in the same manner that the detergent used to wash greasy dishes reacts with the grease and helps it dissolve, or at least be emulsified and removed in the rinse water. Once the rosin or resin is removed, the rinse water solubilizes the ionic residues that were on the surface of the board as well as those trapped within the solid rosin matrix.

Table 43.5 summarizes the possible cleaning agents for the various flux types.

TABLE 43.5 Soldering Flux/Paste Cleaning Options

Flux/paste type	Possible cleaning agents	Comments
1. Rosin/resin	a. Detergent cleaning	
	b. Semiaqueous cleaning	
	c. Solvent cleaning	
2. Low-residue flux	a. No-cleaning step	If flux is rated L0 or L1 in ANSI-J-Std-004
	b. Aqueous cleaning	If flux does not contain rosin or resin
	c. Detergent	If flux contains rosin or resin
	d. Solvent	
3. Water-soluble	a. Water	
	b. Detergent cleaning	

43.9.3 Flux and Cleaning Issues Related to Lead-Free Soldering

Solder flux residues are much more difficult to clean after lead-free soldering.[19,20,21] Fluxes formulated to remove oxidized metals (SnPb, Cu) for traditional tin-lead soldering must be modified to address the needs of different metallization for the solder, the board finish, and component terminations. Soldering flux formulations for lead-free soldering use higher molecular weight resins and activators for the more aggressive flux chemistry needed to achieve wetting. Also, the higher tin content of the alternative solders (SnAgCu, SnCu) requires more aggressive flux chemistries for good wetting.

The process window for lead-free soldering requires higher peak reflow temperatures and longer times above liquidus. Soldering in a nitrogen environment improves wetting and reduces oxidation of the flux residues and the solder. Due to the higher reflow temperature, greater polymerization of the flux constituents can occur—producing residues that are more difficult to remove.

The cleaning challenge represented by lead-free soldering must be addressed by modifying the cleaning process to increase the wash time, wash concentration, and wash temperature.

Cleaning for fine-pitch components soldered under lead-free conditions requires an increase in cleaning time, the use of increased solution agitation (e.g., ultrasonic), and the modification of spray nozzles and impingement angle. Flux residues are more difficult to remove when they set. Thus, it is important to minimize the time between reflow and cleaning. Aqueous cleaning chemistries should include wetting agents, solvating agents, and reactive agents. Water-soluble fluxes that can be cleaned with water after lead-based soldering conditions may require the addition of some cleaning agent to remove the higher molecular weight organic acids and the tin salts created when the flux decomposes during lead-free reflow.

43.10 SUMMARY

This chapter has reviewed the processing issues and materials used for the soldering and cleaning processes. In the lead-free era, thermal profiles and component maximum temperatures require different reflow profiles. The various types of soldering fluxes, their constituents, and their role in the soldering process have been discussed and the formulation changes required for the higher temperature soldering examined. Cleaning materials and processes have been identified and the challenges with higher temperature soldering discussed.

REFERENCES

1. Fisher, D. A., et al., "Relative Effects on Stratospheric Ozone of Halogenated Methanes and Ethanes of Social and Industrial Interest," *Scientific Assessment of Ozone: 1989*. Vol. II. World Meteorological Organization Global Ozone Research and Monitoring Project, Report No. 20, 1989, pp. 301–377.
2. IPC/JEDEC J-STD-020C, "Moisture/Reflow Sensitivity Classification for Nonhermetic Solid State Surface Mount Devices." July 2004.
3. ANSI/IPC-T-50G, "Terms and Definitions for Interconnecting and Packaging Electronic Circuits." December 2003.
4. MIL-F-14256, "Flux, Soldering, Liquid (Rosin Base)." June 15, 1995.
5. ANSI J-STD-004A, "Requirements for Soldering Fluxes," available from the IPC. January 2004.
6. IPC-TM-650, "IPC Test Methods Manual." January 2003.
7. Zado, F. M., "Effects of Non-Ionic Water Soluble Flux Residues," *The Engineer,* Vol. 27, No. 1, p. 40. (1983).
8. Brous, J., "Electrochemical Migration and Flux Residues Causes and Detection," *Proceedings of NEPCON West,* February 1992, pp. 386–393.
9. Mitchell, J. P., and. Welsher, T. L., "Conductive Anodic Filament Growth in Printed Circuit Materials," paper prepared for the Circuit World Convention II, Munich, published as IPC Technical Report WC-2A-5, June 1981.
10. Biocca, Peter, "Flux Chemistries and Thermal Profiling: Avoiding Soldering Defects in SMT Assembly," *Proceedings of SMTA International,* Chicago, September 30, 2001.
11. Cala, F.R., and Winston, A.E., *Handbook of Aqueous Cleaning Technology for Electronic Assemblies,* Electrochemical Publications, 1996.
12. "Post Solder Solvent Cleaning Guidelines," IPC-SC-60. August 1999.
13. "Post Solder Semiaqueous Cleaning Guidelines," IPC-SA-61A. June 2002.
14. "Post Solder Aqueous Cleaning Guidelines," IPC-AC-62A. January 1996.

15. "Guidelines for of Cleaning Printed Boards and Assemblies," IPC-CH-65A. September 1999.
16. Kanegsberg, B., and Kanegsberg, E., *Handbook for Critical Cleaning,* CRC Press, 2001.
17. Sanders, J. R., Chute, S., Soma, J., and Fouts, C., "A Comparison of Cleaning Technologies for New Lead-Free Solder Paste Formulatons," *Proceedings of SMTAI International,* 2005, pp. 871–875.
18. Breunsbach, R., "New Developments in Simplified, Low Cost, Semi-Aqueous Emulsion Cleaning Technology," *Proceedings of NEPCON West,* 1992, pp. 1217–1225.
19. M., Bixenman, Miller, E.,and Rued, F., "Lessons Learned and Best Practices Developed for Cleaning Pb-Free Flux Residues from Printed Circuit Assemblies and Advanced Packages," paper presented for the IPC/JEDEC International Conference on Lead-Free Electronic Components and Assembly, 2006, Singapore.
20. Tosun, Umut, Wack, Harald, Becht, Joachim, Schweigart, Helmut, Afshari, Sia, and Ellis, Drik, "Defluxing of Eutectic and Lead-Free Assemblies in a Single Cleaning Process," *Proceedings of SMTAI International,* Chicago, September 24–28, 2006, p. 160.
21. Davies, Matt, Chute, Susan, Sanders, John R., Soma, Jay, and Fouts, Christine, "A Comparison of Lead-Free Solder Assembly Defluxing Processes," *Proceedings of PC EXPO/APEX 2006,* Anaheim, February 5–10, 2006, S08-03-1.

P · A · R · T · 10

SOLDER MATERIALS AND PROCESSES

CHAPTER 44
SOLDERING FUNDAMENTALS

Gary M. Freedman
Hewlett-Packard Corporation, Business Critical Systems, Singapore

44.1 INTRODUCTION

Soldering is a technology that has been used extensively for utilitarian and decorative purposes for thousands of years. In the last 100 years, soldering has been raised from art to science for joining electrical assemblies. By the late twentieth century, the tin-lead solder joint and properties of its constituent materials had been studied well enough that it could be modeled for purposes of predicting reliability.

Circuit board soldering has traditionally been accomplished with eutectic or near eutectic tin-lead (Sn-Pb). The eutectic alloy is approximately 63 percent Sn and 37 percent Pb by weight and is the mainstay of printed circuit board assembly. A binary alloy (composed of two constituent metals), its melting and solidification are well understood and predictable. Being a eutectic alloy, it has a distinct melting point of 183°C. This temperature is low enough that it is compatible with a significant set of materials such as circuit board laminates, integrated circuit (IC) package encapsulants, ceramics used for packaging, silicon devices, molded connector bodies, etc. Materials for these components were in part codeveloped for use with Sn-Pb solder for joining. Now the electronics industry is faced with tremendous change spurred mainly by the European Union's (EU) environmental initiatives to limit electronics waste and restrict toxic materials used in electronic manufacturing. The main thrust of the legislation is to limit or ban a short list of materials that the EU deems harmful to the environment. Lead (Pb) is most prominent on the list of restricted materials, and limitations on its use have the biggest impact to electronics manufacturing. Now the industry is forced to adopt Pb-free solder alloys. Most viable Pb-free alloys have a higher melting temperature than Sn-Pb solder. Most Pb-free alloys are composed of three or more metals.

All of this comes at a time when circuit board component density, layer count, and performance are higher than ever before. The complexity of printed circuit assemblies continues to increase in every sector of the electronics industry. The number of solderable contacts of IC components is on the rise as circuit board components both miniaturize and increase in body size due to ever-increasing input/output (I/O) count. Through-hole technology, although on the wane, is still prevalent. Motherboards for personal computers are more powerful than computer room servers that are only a few years older. Consumer electronics are highly sophisticated and internally complex. Mobile telephones are multifunctional incorporating still- and video-cameras, personal digital assistants (PDAs), wireless Internet, and localized wireless connection to user devices such as hands-free devices, all in smaller and smaller packages. Portable music systems such as MP3 and MP4 players have more memory than many personal computers (PCs) in use today. They must be built small enough to fit in a pocket and be shock-resistant.

The ever-increasing demand for product performance drives increases in component density, the number of I/Os, increased board layers, and finer printed wiring board conductor trace spacing in order to make electrical routing possible. Higher density boards and higher layer counts result in more thermally massive assemblies, which are more difficult to solder during surface mounting and doubly difficult to assemble during wave soldering. When both surface mounting and wave soldering are used on the same assembly, it is referred to as *mixed-mount technology*. The proliferation of press-fit connections (force fitting compliant electrical contacts into a circuit board's plated through-holes) has alleviated some of the problems of mixed-mount assembly and has been driven largely by the difficulties faced in wave soldering large and thick circuit boards. See Chap. 49 in this book for an introduction to press-fit technology. Land grid arrays (LGAs) with separable pressure interconnect has become more prevalent especially for high-end motherboards for computing and telecommunications. LGA incorporation on a board has influence on layout and implications for soldering. This will be discussed in Chap. 50.

In the fifth edition of this handbook, there was a separate chapter on Pb-free soldering. Since Pb-free is now the primary focus of electronics manufacturing, its impact has been incorporated into the text of this chapter, which previously concentrated only on Sn-Pb soldering. The reader should understand that Sn-Pb soldering is still important, as many types of electronic assemblies will still be manufactured with lead-based solders until the year 2010 as per the new EU Restriction of Hazardous Substances (RoHS) legislation. Also, there may be problems associated with mixing Pb-free solders with leaded (Pb'd) packages and Pb-free packages with Pb'd solder. Since the latest EU legislative initiatives have profound effects on every aspect of electronics manufacturing, it is necessary to understand the impacts before delving into the technology of soldering.

44.2 ELEMENTS OF A SOLDER JOINT

There are three important constituents of a solder joint: the two materials or surfaces to be joined and the solder itself. Each has its own set of attributes and variables that contribute to the ease and quality of soldering. A delicate balance of material conditions and process parameters determine the solder joint appearance, resultant strength, and reliability of the soldered assembly. The composition of the solder, the surface finish on lead, pad or plated through-hole (PTH), environmental factors, chemical implications, and thermal conditions all impact the soldering process. Each will be discussed.

44.3 COMMON METAL-JOINING METHODS

Welding, brazing, and soldering are the three most widely used metal-joining techniques. Although this section does not provide a rigorous review of welding and brazing, a cursory comparison of these metal-joining processes is useful in setting the stage for an in-depth discussion of circuit board soldering.

Welding involves heating of two similar or dissimilar abutting metals to the point that they fuse into one another. Unlike brazing and soldering, there is considerable depth to the metallurgical changes at their interface that take place in the metals being joined during welding. Sometimes a filler metal is applied to lower the melting temperature of the weldment metals or for desired mechanical properties of the finished bond. As in brazing and soldering, sometimes chemical treatment is required to help dissolve surface oxides and also a cover gas may be applied to retard oxidation during the high temperatures of the bonding process.

There are three distinguishing characteristics that differentiate welding from brazing and soldering: the temperature of the bonding process, the depth of material dissolution from the bonding surfaces, and the reliance on surficial wetting and spreading of the filler material.

Brazing and soldering involve the use of a filler metal (brazing compound or solder), which has a significantly lower melting point than the basis metals associated with the two bonding surfaces. The filler material must be able to alloy with the surfaces to be joined. As the filler is heated to liquidus, there is near-surface metallurgical dissolution and mixing (alloy formation) with the contacted bonding surfaces. Continued heating above the liquidus temperature of the filler metal results in its further spreading and alloying via molten metal surface tension effects. As the bonding surfaces are cooled, the filler metal returns to solidus and a bonded assembly results.

There are only subtle differences between soldering and brazing, the primary distinction being the temperature required for joint formation. Brazing is generally accomplished at a relatively high temperature (>450°C) as compared to soldering. (There is no consensus in the literature, but values between 400°C and 500°C are commonly sited.) Brazing temperatures would be detrimental to most electronic components and PWBs. Since circuit board assembly is done almost exclusively by means of soldering at temperatures generally below 260°C, this chapter is devoted to that traditional joining method.

44.4 SOLDER OVERVIEW

Solder is the cement that joins lead to pad, imparts the mechanical robustness required for a reliable assembly, and also possesses the electrical conduction needed for circuitry. Generally composed of an alloy of metals, it is chosen to melt at a temperature compatible with other materials associated with the soldered assembly. Once molten, the solder must wet the component lead and bond pad. Upon solidification, the resultant solder joint must provide bond strength to survive differences in thermal expansion rates of the associated component assembly. There must be compatibility of the solder alloy and related materials to requirements for assembly and service at elevated operating temperatures as well as resistance to mechanical shock and vibration.

Since the start of electronics, Sn-Pb has been the dominant solder alloy choice. Its eutectic alloy possesses a relatively modest liquidus temperature (183°C) and a working range typically about 205 to 230°C. This overage in temperature by about 20 to 50°C is applied to ensure that all parts have achieved a temperature commensurate with good solder wetting. It also may be necessary to reach such temperatures on thermally massive boards where the lightest parts may heat quickly while the most thermally aggressive parts may just barely achieve a temperature adequate for good solder wetting. Everything in the electronics industry has been tailored to working in that temperature regime. ICs and passive components, laminates, and process equipment associated with soldering have been qualified for use at those temperatures.

Now, due to the EU's RoHS directive and similar legislation elsewhere in the world, Sn-Pb is being phased out from mainstream electronics manufacturing. Only a small fraction of printed circuit assemblies (PCAs) will remain as Pb'd solder assemblies as permitted by RoHS exclusions and exempted military applications. As time progresses, it will become increasingly difficult to source Pb-bearing parts for use with RoHS Pb-exempt circuitry.

There are no drop-in replacement solders for Sn-Pb. Although there are many candidates, the most widely considered alloys require higher process temperatures and necessitate changes to components, circuit board laminates, and even process equipment.

44.5 SOLDERING BASICS

Although these chapters will give great emphasis to Pb-free solders and soldering, it is best to demonstrate soldering process dynamics with Sn-Pb solder, for a couple of important reasons: first, it is our best understood metallurgical system, and second, it is a binary alloy with predictable bonding attributes. It will benefit the reader first to have a good understanding of this relatively simple alloy system. The case presented will be based upon eutectic Sn-Pb solder (63 wt percent Sn and 37 percent Pb) in contact with solderable and non-solderable coatings on a basis metal.

Primary soldering steps include:

- Intimate contact of the solder to materials being joined
- Slow application of heat to warm the parts to be soldered
- Oxide removal from the joining surfaces and solder metallurgies
- Application of heat sufficient to melt the solder
- Solder wetting to joining surfaces and intermetallic formation
- Quenching of the solder liquidus

There are a few other materials that need to be covered for this discussion. The first is solder paste. This is a mixture of minute solder beads, flux, and other materials to give it specific rheological characteristics for dispensing and chemical agents for metal surface preparation. For surface-mount applications, it is typically stenciled onto PWB bonding pads, and then the electronic component is placed upon the solder paste deposit. The paste holds the component in place during the reflow process. The second is flux, which, as mentioned previously, is a key component of solder paste. The flux is a heat-activated chemical agent used to clean solderable surfaces. Both paste and flux will be covered in subsequent sections of this chapter.

44.5.1 Intimate Contact of Solder to Surfaces to Be Joined

This is the most basic of requirements. The solder has to be in contact with the materials to be joined. The contact area of the solder is not wholly important as long as the solder is in point contact with the surface to be soldered when it reaches liquidus. Surface tension effects and metallurgical wetting will complete the spreading of solder contact.

44.5.2 Slow Heating of Boards and Parts to Be Soldered

This is important for three main reasons. First, overly rapid heating can cause certain parts to experience thermal shock and subsequent failure as in cracking or may cause degradation of device electrical characteristics. Second, solder paste may spatter if heated too rapidly. Last, appropriate heating rate is crucial to good surface preparation through fluxing. Balance between time and temperature has to be determined so that the flux has enough time to accomplish its cleaning step and does not dry out prematurely or spatter. Prolonged heating cycles can also cause re-oxidation of flux cleaned parts. The following subsection covers fluxing more thoroughly.

44.5.3 Oxide Removal from Bonding Surfaces and from Solder

Most materials, when in equilibrium with our oxygen-rich environment, develop an oxide coating. Upon heating, solder surfaces as well as the bonding surfaces will more thoroughly oxidize in a normal air environment. If a silver-bearing surface is exposed to a sulfur-containing ambient (sulfur-tainted air), sulfidation occurs and that tarnish also inhibits soldering. Generally, the higher the storage temperature, the more oxidation is present unless oxide growth is self-limiting, as it is in some materials. The same holds true for soldering process temperature or process time. The longer and hotter, the more oxidation or tarnish is a problem if there is no flux or if the process renders the flux ineffective such as by overheating.

Oxides and tarnishes act as a physical barrier preventing alloy formation between the solder and the metal to be soldered. In the case of gold (Au), known to remain largely oxide-free, there is insufficient oxidation to degrade soldering. But unless oxidation is removed from the solder itself, solder alloying with the Au may not be possible or will be incomplete. Note that in most soldering processes, the ambient can be altered to mitigate the detrimental effects of oxygen or other airborne contaminants. This will be discussed later.

The most generally applied remedy for the effects of oxidation and tarnish as relates to the soldering process is the application of a chemical agent, flux. It is formulated to react with specific metallurgies and removes tarnish and oxidation. It also acts as a barrier preventing fluxed metal surfaces from re-oxidizing prior to and during the joining process. The word *flux* comes from the Latin *fluxus,* which means flow or flowing. Flux ensures that the solder, once molten, will flow over the surfaces to be bonded, unconstrained by oxide skins on the solder or the metals to be joined. Certain materials form oxides very rapidly, and some oxides are rather chemical-resistant. Nickel is one such element. Copper, like nickel, can also form resistant oxides and may necessitate the use of strong fluxing agents to achieve a bondable surface. Tin and silver oxides are easily attacked by even weak organic acids. Gold, which is known to remain oxide-free, can be applied as a thin, nonporous barrier over less noble, oxide-stripped metals such as nickel. During reflow, the gold dissolves quickly into the solder and the solder bonds to the underlying oxide-free nickel. Figure 44.1 shows the difference between a solder wettable surface and a flux-resistant oxide-coated surface.

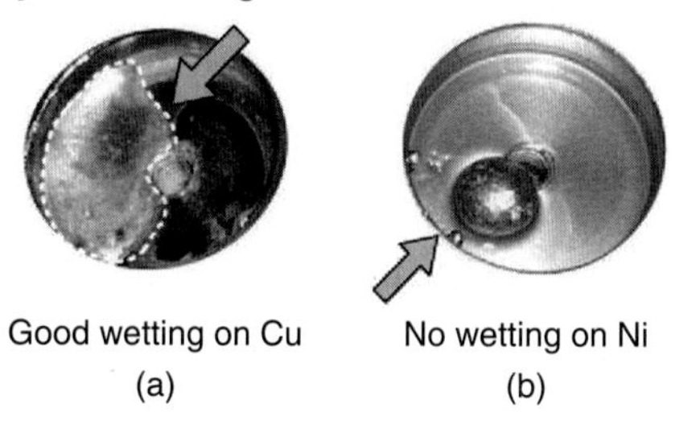

FIGURE 44.1 Comparison of soldering to copper (a) and nickel surface (b) with a weak organic solder flux. Note that in the case of the copper surface, the solder has wet and spread, characterized by a low wetting angle (the extent of wetting is indicated by the dashed outline). In the case of the nickel surface, the flux was ineffective in penetrating the oxide layer and solder was not able to wet to it. Instead, the solder beaded up on the surface of the nickel. *(Courtesy of Hewlett-Packard.)*

Several types of fluxes are available for soldering, but only two broad categories are in widespread use. The first is aqueous-clean, also known as water-clean flux. This is generally composed of relatively strong organic acids and may be fortified with halogens to increase its chemical activity. Aqueous-clean formulations are meant to be thoroughly washed from the surface printed circuit assembly after soldering. If left on the board, corrosion products will form and cause corrosion and electrical failure.

There are several electrical components that make aqueous-clean chemistry less attractive. These include dual inline package (DIP) switches, sealed switches (which are known to leak), high-density connectors, large area-array packages, micro-BGAs, and any other components with low headroom between the underside of the package and the surface of the PWB. It is for this reason, plus the opportunity to eliminate an expensive and vagarious process step (aqueous-cleaning) and associated equipment, that the vast majority of electronics manufacturers have widely embraced no-clean soldering fluxes.

The second is rosin-based, with additives to make it chemically active to varying degrees. Removal requires solvent, but it can be formulated so that the residue is inert and can remain on the board. This is termed no-clean flux.

44.5.4 Heating to Liquidus

Once heated to melting, solder begins to form a metallurgical bond with contacted metals if wettable by dissolution and alloy formation. After the process starts, it progresses slowly, causing the solder to spread as it dissolves surfaces and alloys with it. The molten solder is drawn by surface tension to fill fine capillaries, and surface tension causes the solder to flow across wettable surfaces to some extent, forming webs of solder known as *fillets*. The solder fillet acts as a mechanical gusset imparting strength to the resultant solder joint.

44.5.5 Solder Fillet Formation

The solder fillet is an overt manifestation of surface tension and wetting. Fillets are readily apparent in Fig. 44.2, which shows the fillet as a web of solder extending from the PWB bonding pad to the lead of the component.

Gull-wing leads rely on good heel fillets for solder joint strength. Toe fillets may be present or absent depending whether there is exposed copper or other oxidized or unsolderable exposed metal at the lead tip where the lead-frame was excised from its tooling strip. Toe fillets are of little importance, adding little strength to the soldered assembly. The bulk of the strength comes from the heel fillet and the solder wetted to the capillary between the bottom of the component lead and top of the bond pad. Similarly, side fillets may be present, but if the component lead width is on the order of the bond pad width, there may not be enough room to develop good side fillets. The wetting angle is also known as the *dihedral angle*. The lower the dihedral angle, the better the wetting.

For an example of a PTH solder fillet, see Fig. 44.2.

The fillet is a reasonable indicator of the degree of solder wetting and, therefore, of process goodness. It is generally believed that the higher solder fillet, the better the solder joint, but high fillets can be indicative of three problems: excessive overheating, which can result in brittle solder joints; excessive solder volume, and loss of component lead flexibility due to overly high fillet formation.

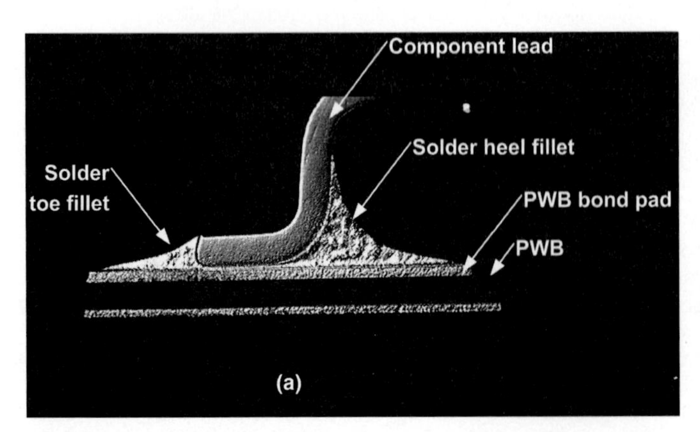

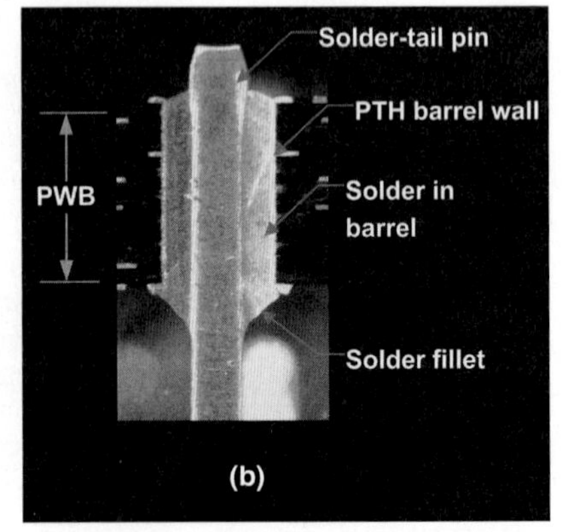

FIGURE 44.2 Solder fillet formation: (a) Cross-section of an SMT solder joint; (b) a PTH solder joint in cross-section. In the case of the SMT solder joint, solder has wet to the pad and lead. The low wetting angles and good fillet formations are indications of good solder wetting. The PTH solder joint shows good fillet formation on the secondary side of the PWB. At the top, the primary side, the fillets are not as well formed, probably due to lower temperature at the top side of the board. *(Courtesy of Hewlett-Packard.)*

If too much solder wicks up the lead, it may cause solder shorting, particularly near the package body, where the lead spacing may be finest. Therefore, fillet examination alone is not sufficient to accurately assess soldering process performance. But, for a fixed set of process parameters, fillet appearance may provide a good means of first-pass comparative assessment of solderability and soldering results.

44.5.6 Intermetallic Compound Formation

Intermetallic compound (IMC) formation is the key to soldering. IMC is the local alloying created at the boundary of the solder liquidus and the surfaces of the metal with which it is in contact. It is the essence of soldering. There is much misinformation about IMCs. One thing is certain: Without intermetallic formation, there is no solder joint. This is stressed because some erroneously claim lack of IMC in such processes as laser soldering. In the case of that technology, the IMC can be exceptionally thin immediately after soldering due to the short time of liquidus, but it is and must be present for soldering to occur. High temperature, time, and intimate contact of the liquid solder enhance the rate and volume of intermetallic compound formation.

The IMC is a crystalline intermediate alloy phase composed of some or all of the contacting metal constituents. This high-alloy concentration region generally has properties that are vastly different from those of either the solder or the contacted metals. It is a brittle material that is electrically conducting or even semiconducting. Solder is thought of as a soft, forgiving material of low melting-point and good electrical conductivity. It is ironic that the essence of soldering—the actual bond, or more correctly, the bond's anchor (IMC)—is in fact, a brittle, high melting-point composition of generally poor electrical conductivity.

44.5.7 Cooling and Phase Change to Solidus

As the solder approaches its solidus temperature, the rate of intermetallic compound formation decreases significantly and the solder joint is formed. It is critical that the solder freezes prior to circuit board handling for two important reasons. First, accidental movement of the PWB may cause the components to move in the molten solder, which could result in defects such as solder joint opens (disbonds) or solder shorts (unwanted bridging of solder from one bond pad to the next). Second, a solder joint disturbance at the critical transition from liquid to solid may result in a bond of degraded reliability. This type of bond has an inherently grainy look and can even have microfractures within caused by motion-induced interruption of orderly crystal growth during solidification.

The cooling rate at the tail end of the process is as important as at the onset of heating. Extreme cooling rates may result in component cracking or electrical performance degradation induced by differential contraction of materials.

CHAPTER 45
SOLDERING MATERIALS AND METALLURGY

Gary M. Freedman
Hewlett-Packard Corporation, Business Critical Systems, Singapore

45.1 INTRODUCTION

Numerous considerations are crucial for successful circuit board assembly and useful product life. Assembly processes and materials that result in solder-joint formation can also have a profound effect on long-term solder-joint reliability as well as a product's environmental exposure and handling during its lifetime. Design and assembly engineers need to take into consideration temperature extremes and thermal excursion ramps that the product may endure as well as anticipated mechanical shock and vibration, the number of power (on-off) cycles expected over its lifetime, exposure to airborne particulates, gaseous pollutants, humidity, and handling. Such exposures may impact the printed circuit assembly (PCA) during soldering, when being handled on the factory floor, or while in the hands of the product's end user. These are but a sampling of variables that influence product reliability and should be considered in defining design and assembly attributes. PCA designs are most often driven by electronic device availability, electrical performance requirements, affordability, and the product's end-use environment and operating conditions.

This chapter will strictly avoid offering up generalized rules for all soldering process applications. The materials and requisite solder-joint reliability properties for a desktop computer are markedly different from those for the ignition system printed circuitry associated with the engine of a motor vehicle. Similarly, mobile electronics (mobile phones, laptop computers, PDAs, etc.) have the added requirements of mechanical shock-resistance as well as high-volume manufacturing needs, including low cost, ease of assembly, and high yields. In some cases, the cost to repair a PCA or an entire product may exceed the value of the combined material and labor content of the first-pass assembled product. It is also well understood that the reliability of a repaired assembly is not as good as a properly soldered first-pass printed circuit assembly due to additional board handling and solder-joint degradation as a result of re-soldering. The solder flux chemistry and how it is applied and heated may also impact product reliability. These points reinforce the need to understand PCA processes thoroughly and to solder properly the first time. There are fundamental rules that will become evident in discussions of metallurgies, fluxes, and processes. If these rules are heeded, the results are optimal soldering yield and the best reliability. However, the only way to know whether an assembly will endure its intended use and environment is to develop appropriate accelerated testing methods that can accurately gauge product field reliability.

It is incumbent on the PCA process engineer to understand the materials properties and phenomena that influence the soldering process and the reliability of the finished assembly. A basic understanding of mechanics, thermodynamics, metallurgy, and chemistry and how they apply to a finished solder joint is required for good PCA process engineering. The process engineer should influence product design for best manufacturability and reliability and also assemble boards with the highest reliability possible given the board design, materials variables (surface finishes, solder, solder flux, etc.), and components that comprise the bill-of-materials. Additionally, there are certain critical rules one could follow for incoming printed circuit board cleanliness, solder purity, etc. These are best described in standards documents, such as those published by the IPC.

There are also abundant pitfalls in soldering. Too much gold in tin-lead solder will cause solder-joint embrittlement. Too high a process time or temperature can have the same effect on solder joints. Contamination of tin-bismuth solder joints with lead (Pb) can cause joint failure. The residues left by some fluxes can be corrosive. Electrochemical migration induced by flux corrosion products is a common mode of circuit failure that is difficult to pinpoint by electrical testing. Dendritic filaments can form between two board features of differing electrical potential. Because these microscopic dendrites have limited current-carrying capability, they may act as miniature fuse links that may break and re-fuse, making the root cause of electrical failure quite difficult to diagnose and locate. Thus, there are many variables associated with circuit board assembly, and the process engineer has to rely on science and engineering fundamentals in order to contribute to intelligent PCA design and defect-free assembly.

45.2 SOLDERS

The properties of an alloy are dependent on its constituent metals and the ratio of the number of atoms of one constituent to the atoms of the other. Variations in these atomic ratios result in alloys of vastly different properties that are exploited to suit a particular application. Melting points can be adjusted, as can hardness, tensile strength, shock resistance, etc. In trying to achieve a suitable melting point, there are always trade-offs, and other material properties may degrade to the point that the alloy may not be suitable for its end-use application. Such is the case in the search for lead-free alloy substitutes. To date, no direct replacements approximate the properties of the well-understood standard of eutectic Sn-Pb (Sn63) solder. In most cases, liquidus temperatures (the temperatures at which a material goes from solid to liquid) of the Pb-free solder alloys are higher than that of eutectic Sn-Pb. As mentioned previously, this imparts many changes to process, process equipment, printed wiring board (PWB) laminates, and electronic components.

Families of alloys that are created from two or more metals in different ratios are called *alloy systems*. Tin-lead (Sn-Pb) is one such system, available in any number of compositions (50Sn-50Pb, 60Sn-40Pb, 10Sn-90Pb, etc.). Some solder alloys are composed of two elements. These are binary alloys such as tin-lead Sn-Pb or tin-bismuth (Sn-Bi). Others are ternary systems such as tin-silver-copper (Sn-Ag-Cu), which is also commonly referred to as *SAC* (for the *S* in Sn [tin], the *A* from Ag [silver], and the *C* from Cu [copper]. SAC305 is an abbreviation for Sn-3.0 wt percent Ag-0.5 wt percent Cu. Similarly SAC3807 is an abbreviated form of Sn-3.8 wt percent Ag-0.07 wt percent Cu.

Quaternary alloys are composed of four elements (e.g., Sn-Ag-Bi-Cu) and pentanary alloys (e.g., Sn-Ag-Cu-In-Sb) are composed of five. The latter is least understood; binary systems are best understood and easiest to formulate. Some of the small differences in compositions (0.5 wt percent) as seen in ternary, quaternary, and pentanary systems are difficult for solder vendors to control accurately during formulation. Also, in wave soldering, where the solder pot contents are in constant contact with metals from component leads and circuit board surface finishes, materials will be leached into the molten charge in the solder pot and will effect solder composition over the long term. Minor constituents of a solder may be depleted more quickly during wave soldering.

45.2.1 Tin-Lead Solders

Since these alloys have been the staple of PWB assembly and since Restriction of Hazardous Substances (RoHS) legislation exempts the alloy for certain applications until 2010, it is useful to understand this simple alloy system. A sound understanding of Sn-Pb will help in the intelligent selection of lead-free (Pb-free) solders and the mechanics of manufacturing reproducible and reliable solder joints. Until now, Sn-Pb solder was the alloy family of choice for electronics manufacturing with a long history dating back to the origins of circuitry.

45.2.1.1 The Sn-Pb Eutectic Alloy. Sn and Pb can be formulated into many alloys but its eutectic alloy is the most familiar to the electronics assembler. It is a ductile metal exhibiting a melting point below 200°C, a temperature range that has proven benign for most of the materials associated with or attached to a printed circuit board. Its fatigue resistance is sufficient for most commercial applications. Sn-Pb is easily fluxed by very weak organic acids and its shelf-life in terms of solderability is long lasting and generally not an issue. There are many solder alloys to choose from, but few as compatible as eutectic tin-lead when it comes to production assembly soldering of PWBs. There are also several metallurgical properties that make this alloy attractive as a solder, which explains why it is so widely embraced as the paradigm of solders. That is not to say that other metal alloys cannot be substituted, but this composition has endured the test of time and many searches for alternatives.

A eutectic alloy has several useful and interesting characteristics. Like all alloys, it has a melting point below that of its constituent metals (see Table 45.1)

TABLE 45.1 Sn-Pb Constituent Melting Points Compared to Its Eutectic Melting Point

Material	Melting point (°C)
Sn	232
Pb	327.4
63wt% Sn:37wt% Pb*	183*

*Eutectic alloy composition.

The eutectic composition has the lowest melting point of any other members of its alloy family. It has a discrete melting point rather than a melting range and its constituents are combined in a single, specific alloy composition. At the eutectic temperature, the eutectic metals coexist in a liquefied alloyed state as opposed to a state where there are some solids of one metal and liquid of the other.

When non-eutectic Sn-Pb alloys melt, phases of Sn or Pb precipitate out of the liquid solution and give rise to a solid-liquid mixture with broad melting range. The further from the eutectic point, the greater the band of the melting range for the alloy. This region of broad melting temperature is known as the "pasty" or "plastic" range. Non-eutectic compositions do not solidify as quickly as eutectic alloys. Because eutectic solders nucleate and crystallize very rapidly at the solidus point, fine grain growth and high mechanical strength in the solidified solder joint result. Upon solidification, non-eutectic compositions result in internal stresses and coarse, dull solder joints with perturbed surfaces. These are all points in favor of using eutectic or near-eutectic solders for circuit board assembly.

45.2.1.2 High Pb-Content Solders. High Pb content Sn-Pb solders are known for producing compliant solder joints owing to elemental lead's high ductility. Due to high melting temperatures, these alloys are generally used to solder non-organic circuit board substrates such as ceramic. An organic laminate such as glass-epoxy is likely to incur degradation through a reflow cycle at such temperature extremes as required by alloys as 5Sn-95Pb (melting range = 301–314°C) and 10Sn-90Pb (melting range = 268–302°C). For these solders, the soldering process temperature is generally chosen to be about 20 to 40°C higher than the melting point or the upper bound of the melting range for reflow and even higher for wave soldering. High-temperature solders are difficult to work with and are relegated to niche applications such as ceramic hybrid circuits or as a contact material for ceramic ball grid array (CBGA) and ceramic contact grid array (CCGA) packages and hand-soldering operations. Even in these applications, the solder ball or column is often soldered to the package using a lower melting-point formulation.

45.2.1.3 Tin-Lead Additives

45.2.1.3.1 Antiscavengers. Occasionally a small fraction of silver (Ag) is added (up to 2 wt percent) to improve solder joint appearance and to retard Ag scavenging. Scavenging occurs when there is the propensity of one metal to dissolve rapidly and thoroughly into another upon reaching liquidus. A good example would be silver traces on a thin-film ceramic device and the use of Sn-Pb solder to bond to it. During soldering, the Ag would dissolve very rapidly into the Sn-Pb. If too much Ag is dissolved, then the bonding pad on the thin-film ceramic device may be rendered unsolderable due to the lack of a wettable solder surface. The addition of Ag to Sn-Pb solder would slow the dissolution rate of Ag from the thin-film ceramic bond pad and retard scavenging.

45.2.1.3.2 Brighteners. Sometimes Ag is added to solder to improve wetting, making the solder joint smooth and shiny. When not used as a scavenger, the addition of Ag should be avoided as it is not usually needed for most electronic joining applications with Sn-Pb.

45.2.1.4 Sn-Pb Intermetallic Compound. In addition to lowering the melting point of pure Sn (232°C), Pb retards Sn-Cu intermetallic formation by piling up at the intermetallic boundary, frustrating tin-copper intermixing. Sn-Cu intermetallic compound (IMC) is crucial for solder-joint formation, but if the intermetallic layer is too thick, the resultant solder joint will be brittle and subject to failure during thermal cycling or mechanical shock. Although Sn-Pb solder is known for its ability to wet well to a number of component lead platings and circuit board finishes, Pb actually inhibits wetting and keeps the solder localized to the targeted solder-joint area. Excessive solder spreading can be detrimental in three ways. First, if solder wicks away from the intended solder joint area, the resultant joint will be solder starved, and weaker than intended. Second, if the solder were too mobile, it could wick up connector leads and into the connector, decreasing inner contact flexibility, decreasing contact gap, and changing connector contact physics and resulting in a less reliable interconnect. Finally, if the solder wicks up too high on a gull-wing component lead, it inhibits the flexibility of the component lead and makes it more susceptible to mechanical failure.

45.3 SOLDER ALLOYS AND CORROSION

Just as some metals are more corrosion resistant than others, the same is true of solder alloys. When a solder corrodes on a circuit board, it is more than an aesthetic issue. Contaminants on the surface of the PWB between adjacent, oppositely charged conductors can result in corrosion dendrites. These are tiny conductive crystals—filament networks that may extend from one conductor to the other. They can have enough current capacity to cause electrical shorting or they may heat up to the point of fusion, melt, interrupt current flow, and return the assembly to normal operation. This can be a cyclical event with the dendrites growing, fusing, and regrowing, making diagnosis difficult. Finer-pitch surface-mount geometries are particularly susceptible to this phenomenon.

M. Abtew et al. published a chart of the electromagnetic force (EMF) of various metal couples present in some lead-free solder alloys.[1] As a rule, the lower the EMF, the more corrosion-resistant the alloy. Eutectic Sn-Pb, the basis of comparison, was by far the lowest value on the list at 0.010 volts. The next nearest, Sn-51In, was 0.201 volts or 20 times the voltage as that of Sn-Pb solder. Other values reported were Sn-57Bi at 0.323 volts, Sn-9Zn at 0.624 volts, Sn-3.5Ag at 0.937 volts, and Sn-80Au at 1.636 volts. So tin-lead solder is significantly more resistant to corrosion and dendritic growth than the other seven alloys studied. This will be an area to watch as the world moves forward with Pb-free soldering.

45.4 PB-FREE SOLDERS: SEARCH FOR ALTERNATIVES AND IMPLICATIONS

The European Union's RoHS legislation has spawned widespread changes in electronic assembly. The biggest among them is the required move to Pb-free solders for most electronic products. Although this topic is a hotly debated, the use of elemental lead (Pb) in the electronics industry and more specifically in solders is a small percentage of the total industrial use of Pb worldwide. Accurate and current numbers are difficult to come by, but estimates range from less than 1 percent to as much as 10 percent worldwide. There are some excellent reasons, though, to find suitable substitutes for Pb where possible. Pb is toxic and governments worldwide spend large sums on abatement programs, education, and health care assistance to those afflicted with toxic levels of lead, mostly from peeling Pb-based paint, which is already regulated. Pb salts are sweet to the taste but toxic. The Romans purportedly added Pb salts to sweeten their wine. Electronic assemblies are rarely linked to toxicity through oral ingestion, but may be the source of contamination through improper disposal of the assemblies, components, raw materials, or their by-products. Purity of the water table may be affected by inappropriate land-filling of large quantity of Pb-bearing circuit boards; solder dross; spent solder pot contents; trace concentrations released as part of aqueous cleaning processes; and Pb-bearing ion exchange filters from aqueous cleaning units, plating baths, etc.

Replacement solders, while perhaps nontoxic themselves, may require deleterious raw stock or result in noxious by-products from processing or from ore refinement. It is wise to proceed cautiously and look at the impact of Pb removal from the electronics industry from all perspectives. There are many nonPb bearing materials associated with printed circuit board assembly, each with a set of material components and process chemicals that may be problematic in a public health sense. Products, by-products, raw materials, and their derivatives must be handled and disposed of in a conscientious and environmentally considerate manner. Attention to better reclamation and disposal methods of industrial and household waste streams, as well as adequate education concerning the same, probably holds more significance to public health than defining a quick substitution for Sn-Pb solder. However, it is the requirement and, therefore, a knowledge of alternatives is mandatory for design and process engineers.

The database for the material properties of widely used Sn-Pb solders is still incomplete. Needless to say, for a new Pb-free solder system, it will be much less complete and take several years to characterize the qualities and performance of a new solder. Since so many new solders are being investigated, this will further slow down solder development and characterization since efforts will be spread out rather than concentrated on a common solder alloy. As of this writing, several corporations have had product recalls associated with the changeover to Pb-free solders where there were unknown interactions with soldering system materials or lower-than-expected field reliability.

45.5 PB-FREE ELEMENTAL ALLOY CANDIDATES

Of the 90 naturally occurring elements, only 13 can be practically combined with each other to form a practical solder (see Fig. 45.1).

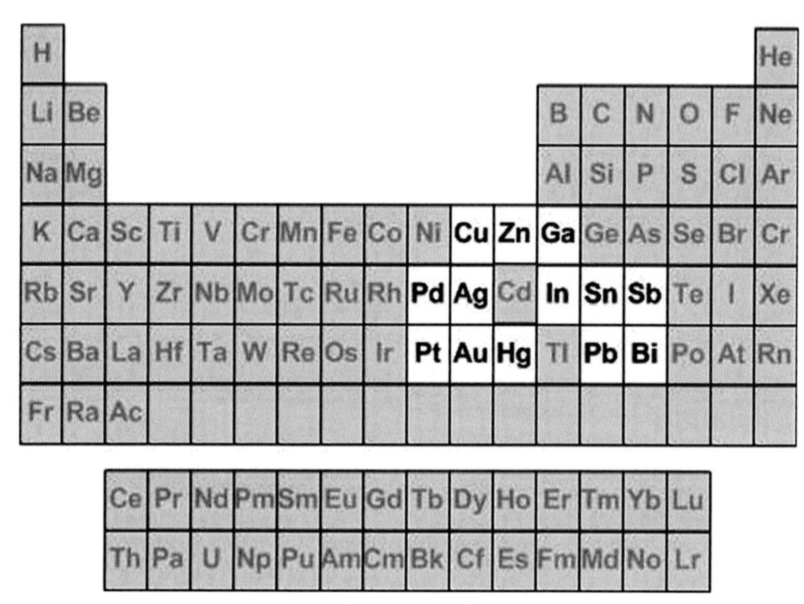

FIGURE 45.1 Highlighted elements can be combined to form solders useful for circuit board assembly.

Of those elements that can combine to form solders, several have limitations that restrict their practical implementation for solder use (see Table 45.2). Gallium, gold, indium, platinum, and palladium cannot be mined or refined in sufficient quantity to satisfy the needs of the electronics industry and are prohibitively expensive, blocking their consideration as a major constituent of a worldwide solder supply. The Bismuth (Bi) supply, a by-product of lead refining, would be marginally sufficient. Mercury and antimony are too toxic. Both mercury and Pb are already on the EU's RoHS list of restricted materials. Gallium and mercury possess too low a melting point to be used on their own. This leaves five metals for practical solder alloy consideration: bismuth (Bi), copper (Cu), silver (Ag), tin (Sn), and zinc (Zn). Due to

TABLE 45.2 Elemental Candidates for Solders

Element	Symbol	Melting point (°C)	Comments
Antimony	Sb	630.5	Toxic, used as a minor constituent of some solders
Bismuth	Bi	217.5	Already in use
Copper	Cu	1084.5	Already in use
Gallium	Ga	29.75	Too expensive
Gold	Au	1063	Too expensive
Indium	In	156.3	Too expensive
Lead	Pb	327.5	Banned for use in EU
Mercury	Hg	−38.83	Toxic, restricted for use in EU
Palladium	Pd	1550	Too expensive
Platinum	Pt	1768.3	Too expensive
Silver	Ag	960.15	Expensive but already used in solders in small quantities
Tin	Sn	231.89	Already in use
Zinc	Zn	419.6	Already in use

its value, silver, too, can be used only in small proportions for solder and is already in wide use that way today. Given the four remaining metals (Bi, Cu, Sn, and Zn), there are many possible metallurgical permutations available for solder alloying, but there are few that either possess low enough liquidus to be useful for circuit board assembly or have attractive physical properties as a solder. Although Pb-free solders have been in use for many years (jewelry making, plumbing, brazing, etc.), few have been studied sufficiently for electronics assembly, most have too high a melting point to be useful, and none come with the well-documented properties database as is the case for Sn-Pb solder.

45.5.1 Candidate Pb-Free Solders

Although many manufacturers completed the conversion to Pb-free soldering in time for the July 1, 2006, implementation of the RoHS directive in Europe, there was still no consensus across the industry for a single tin-lead solder replacement at the time. Pb-free alloy systems receiving the most attention include those listed in the Table 45.3.

TABLE 45.3 List of Lead-Free Solders and Their Temperature Ranges

Alloy families	Most common compositions[a] (wt %)	Solidus temperature[b] (°C)	Liquidus[b] temperature (°C)
Sn-Cu	Sn-0.07Cu	227[3]	227[c]
Sn-Ag	Sn-3.5Ag	221[3]	221[c]
	Sn-5.0Ag	221	240
Sn-Ag-Cu	Sn-3.0Ag-0.05Cu	217	221
	Sn-3.5Ag-0.9Cu	217[c]	217[c]
	Sn-3.5Ag-0.7Cu	217	220
	Sn-3.8Ag-0.7Cu	217	220
	Sn-3.9Ag-0.6Cu	217	223
	Sn-4.0Ag-0.5Cu	217	225
Sn-Ag-Bi	Sn-1.0Ag-57Bi	137	139
	Sn-3.4Ag-4.8Bi	200	216
	Sn-3.5Ag-5.0Bi	208	215
	Sn-3.5Ag-1.0Bi	219	220
	Sn-2.0Ag-7.5Bi	191	216
Sn-Ag-Bi-Cu	Sn-2.5Ag-1.0Bi-0.5Cu	214	221
	Sn-2.0Ag-3.0Bi-0.75Cu	207	218
Sn-Bi	Sn-58Bi	138[c]	138[c]
Sn-Zn-Bi	Sn-8.0Sn-3.0Bi	191	198
Sn-Zn	Sn-9.0Zn	199[c]	199[c]

Notes:
[a]There are too many non-eutectic formulations composed of the same elements in varying ratios to list comprehensively here. Each variation in composition will have its own melting range. The ones listed in this table are examples only to show approximate temperature range and composition.
[b]There are discrepancies in the literature as to the exact melting point or melting range of alloys listed, but the table will be useful for approximating phase-change characteristics of the alloys listed.
[c]Eutectic alloys have a distinct melting point; solidus and liquidus temperatures are the same. Non-eutectic alloy melting is characterized by a melting range where the alloy is present as both solid phase(s) and liquid phase simultaneously (pasty range). At the upper temperature of the melting range, only the liquid phases exist.

45.5.1.1 Characteristics of Pb-Free Solders. Pb-free solders generally exhibit poorer wetting and spreading characteristics than Sn-Pb alloys, but may offer an advantage in tensile strength and creep resistance. In terms of the soldering process, a whole new understanding of the various Pb-free alloys is required. Propensity for solder void formation, solder paste shelf-life, solder paste service-life (while on the stencil), printability and effects on stencil and squeegee life, fatigue characteristics, alloy interactions with board and component surface finishes, corrosion resistance, mechanical shock resistance, and numerous other properties will have to be understood, especially for the dense, high-end electronic assemblies that are expected to last many years in the field. It may be less an issue for consumer electronics that are considered obsolete within a couple of years. However, ruggedness, especially for portable consumer electronics, is a concern. Many of the lower melting point solders have precious elements, whereas some of the inexpensive alternatives wet poorly, are prone to oxidation during soldering, and tend to corrode. A few of the most favored alloy systems will be briefly discussed.

45.5.1.2 Bismuth Alloys. Bismuth (Bi), an elemental by-product of Pb smelting, is generally associated with lower melting point solder alloys. It is in short supply and Bi solders cost about twice that of Sn-Pb alloys. Bi is compliant like Pb and shares several of its properties, such as high specific gravity and ductility.[2,3,4] Pure bismuth melts at 271.3°C, about 50°C lower than that of Pb. Very easily oxidized, bismuth alloys are best soldered with the aid of highly activated fluxes or in a nitrogen environment.

Bi forms a binary eutectic with Sn in the proportions 58Bi:42Sn (m.p., 138°C). There are numerous bismuth alloys in use, many composed of two or more metals in addition to bismuth. The Sn-Bi eutectic, if contaminated with Pb, can be problematic as it is known to form a ternary alloy with a melting point of 96°C adversely affecting solder-joint fatigue characteristics. In some applications, solder joints will fall apart if the service temperature is high and the low melting point Sn-Bi-Pb alloy is formed. This becomes all the more critical with low-volume Sn-Bi solder joints. The Pb can come from solder predeposited on component leads, Sn-Pb hot-air solder leveled pads, or both.

Bismuth expands upon freezing, whereas tin contracts. A phenomenon called fillet lifting has been reported[5,6] (see Fig. 45.2). It is mostly associated with bismuth ternary alloys such as Sn-Cu-Bi and Sn-Ag-Bi used in plated through-hole (PTH) wave soldering, but has also been observed with Sn-Bi system.

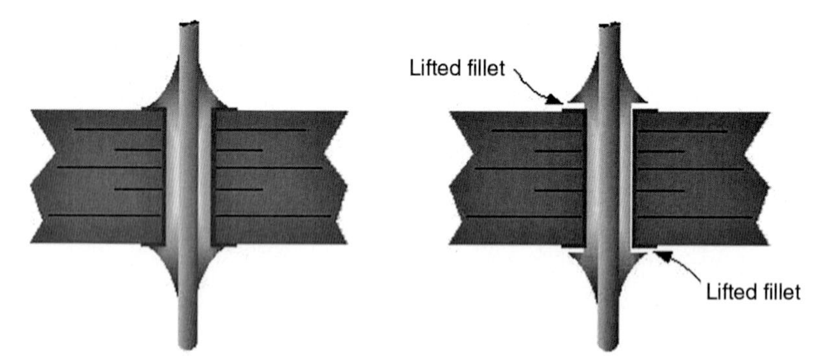

FIGURE 45.2 The expansion upon solidification in a bismuth-containing solder alloy can cause solder fillets to fracture and lift from the plated through-hole annular pads (right). Compare this to the schematic on the left, which is indicative of a conventional tin-lead solder joint.

As the board cools, it contracts at a different rate than the solder, and the fillets may lift, especially if cooled slowly or contaminated with Pb. Rapid cooling (~3°C/sec.) has been advised but may not completely eliminate this phenomenon. Care must be taken not to exceed the component manufacturer's specified heating and cooling ramp rate. Too high a cooling rate may result in component cracking or degraded component reliability. Fillet lifting may also result in lifted PTH annular ring (bonding pad) if solder-joint strength is high and bond strength of the ring to the board is low.

Some Bi alloys have too low a melting point for many applications. Despite its excellent tensile strength and thermal cycling endurance (better than Sn-Pb), Sn-Bi eutectic (m.p., 138°C) would not be practical for use in most automotive applications or even for high-end computer assembly, where the service temperature may approach the melting point of this alloy. Bi also oxidizes rapidly, making Bi alloys difficult to maintain during wave soldering due to copious oxide (dross) production.

Bismuth soldering alloys are noted for their brittleness[7,8]; nonetheless, several bismuth solders are used commercially, the most popular alloys containing Bi as a minor constituent (in the 2 to 14 percent range). Aside from bismuth alloys' inappropriateness for high-temperature applications and Pb contamination, safety is an issue. According to C. White and G. Evans,[9] safety is particularly a concern when using alloys with high concentrations of bismuth. Cadmium (Cd), a poisonous metal, is a commonly associated bismuth contaminant. Care is recommended in handling of this solder, and special venting requirements may also be mandated for its use. On the other hand, the RoHS legislation has reinforced prior EU restrictions on cadmium, limiting it as a constituent or contaminant to less than 100 ppm by weight.

Bismuth and indium alloys are useful for *step soldering,* which refers to the use of two different solder alloys with two different melting temperatures used on the same circuit board. The higher temperature alloy (e.g., Sn-3%Ag-0.5%Cu, melting onset, 217°C) is used to solder surface-mount components and an alloy of lower melting temperature (e.g., 58%Bi:42%Sn, m.p., 138°C) is used for wave-soldered components. The temperature experienced at wave solder will not cause the Sn-Ag-Cu to reflow, thus preserving the integrity of surface-mount technology (SMT) solder joints.

In the case of repair or replacement of components, the use of low melting temperature solder will have no adverse thermal effect on adjacent or reverse-side solder joints. In fact, use of low melting point solder reduces the possibility of pad delamination or through-hole barrel cracking; failure mechanisms associated with localized overheating, especially at repair.

45.5.1.3 Sn-Ag-Cu. As previously stated, the Sn-Ag-Cu (SAC) family is receiving the most attention worldwide. Its tensile strength is superior to that of Sn-Pb, but its shear strength is worse. It is easily fluxed and moderate in melting temperature roughly 35°C higher than eutectic Sn-Pb solder. The elevated temperature regime makes this alloy more difficult to flux since the flux tends to dry up and activate too early in the process. This problem is overcome with higher molecular weight resin fluxes and volatile organic compound (VOC)-free flux formulations since there are no alcohols or other low boiling organics present to aid in evaporation. The SAC family has good compatibility with most Pb-free board surface finishes and Pb-free component platings. Its process window is narrower than that of Sn-Pb. It has been demonstrated that SAC can be used with parts containing Pb, but the reliability of the solder joint is less than if soldering Pb'd to Sn-Pb or Pb-free to SAC. SAC solders sell at about two-and a-half to three times the price of Sn-Pb solder alloys. A comparison of Sn-Pb solder to SAC solder is offered in Table 45.4.

TABLE 45.4 Comparison of Sn-Pb Solder with Pb-Free SAC Solder

Solder alloy	Tin/Lead Sn63:Pb	Tin-Silver-Copper Sn-(3.0–4.0%)Ag-(0.5–0.7%)Cu
Melting temp.	183°C	~217°C → ~221°C
SMT peak temp.	~215±10°C	~240±10°C
Wave solder pot temp.	250°C → 260°C	260°C → 270°C (with higher preheats and longer soldering dwells)
Solder reliability	Adequate	Collecting data
Process compatibility	Adequate	Collecting data

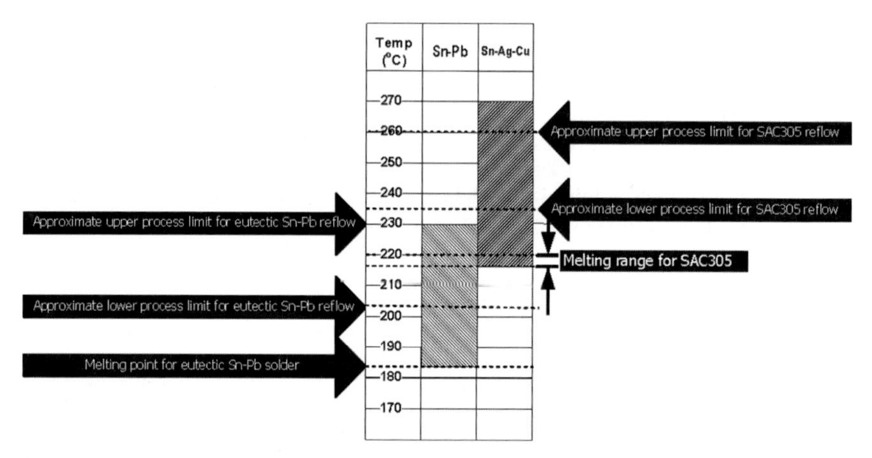

FIGURE 45.3 SAC solder requires higher process temperatures than Sn-Pb.

Figure 45.3 demonstrates the process temperature differences for Sn-Pb soldering compared to SAC.

A condition known as *sporadic brittle fracture* has been reported and may be related to the use of SAC alloy soldered to electroplate gold over electroplate nickel. It is most often associated with BGA solder ball-to-BGA package attach, but may also occur in board-level solder joints at the BGA ball-PWB interface. The root cause of failure is still unknown.

45.5.1.4 Sn-0.7Cu Alloy. Eutectic Sn-Cu (m.p., 227°C) is being used as a wave solder charge material. It is compatible with most board surface finishes and component lead-frame finishes. Its relatively low cost (about 50 percent more than Sn-Pb solder) makes it an attractive Pb-free alternative. It also has been demonstrated to be an effective alloy for Pb-free hot-air solder-leveled (HASL) surfaces on PWBs. One problem with Sn-Cu alloy, especially at wave soldering or fountain soldering, is that small amounts of dissolved copper will cause a dramatic rise in the melting point of the alloy. The high melting temperature of this alloy causes rapid material dissolution from PWB surfaces and component lead platings. Care must be taken to top off the wave solder pot with an appropriate amount of Sn or Sn-Cu to maintain proper Sn-Cu alloy proportions.

45.5.1.5 Indium Alloys. This metal and its effect on the solder joint is very much the same as for bismuth. Indium (In) is very soft and ductile in its elemental form and melts at about 157°C. In-based soldering alloys are much more expensive—about 25 times that of Sn-Pb—due to indium's scarcity, thus it is not suited as a dominant metal in solder. Price and supply make it unattractive even as a trace agent or a minor solder alloy constituent.[10] B. Allenby et al.[11] concluded that the world's indium supply would be depleted in about one year if indium were consumed at the same rate as Pb.

An In-Sn eutectic occurs at 58 wt percent In with a eutectic point temperature of 120°C. It forms intermetallic compounds with copper from the circuit board's bonding pad or copper-based component leads. Copper is highly soluble in indium.[12] Indium has been demonstrated to be of value in reducing the leaching of gold into solder. F. Yost[13] reported a lower Au dissolution rate for indium-based solders as compared to Sn solders. The use of indium allows soldering to gold films without thoroughly consuming the gold. The resulting solder joints are not brittle, as is the case of Sn-based solder joints exposed to too much gold.

Indium and some of its alloys can wet to glass, quartz, and other normally unsolderable inorganic substrates with the aid of an ultrasonic soldering tool. Ultrasonic soldering will be discussed in the chapter on soldering techniques.

45.5.1.6 Zinc Solders. Zinc (Zn) alloys, favored by some Japanese companies, oxidize rapidly. Solder paste shelf-life has been an issue even when these alloys are refrigerated at very low temperatures. Dross formation in wave soldering has also been problematic. Zn alloys are also known to have some corrosion problems.

45.6 BOARD SURFACE FINISHES

Most every circuit board consists of a stack-up of insulator material such as fiberglass and conductive layers composed of copper. Copper oxidizes quickly and its oxides are somewhat chemical-resistant, making it tough to solder without aggressive fluxing. Copper also corrodes in the long term from left-over process residues, fingerprints (body salts), or airborne contaminants. It is for these reasons that copper traces and solder lands on circuit boards are protected by solder mask, plating, or other surface treatments. The following section covers the most common board finishes. The process engineer needs to bear in mind that in addition to solderability, other surface finish–related issues may affect manufacturing or cost, including the following:

- **Probe-ability** This is the ease with which conventional in-circuit test probes are able to penetrate surface oxides or solder flux residue.
- **Tribology** The frictional properties of a material become important for press-fit connections.
- **Cost** Some surface finishes are significantly more expensive than others.
- **Shelf-life** This is the ability of a surface finish to resist oxidation or other surface conversion or corrosion and remain solderable.
- **Reliability** This refers to the resultant solder joint strength as it relates to long-term use and resistance to shock, vibration, and other environmental factors.
- **Corrosion resistance** Some surface finishes are more prone to corrosion than others. Immersion silver and copper are two such finishes.
- **Other surface finish–related defects** Immersion silver (Imm-Ag) sometimes exhibits linear arrays of microvoids at the intermetallic interface which detracts from solder joint strength. Electroless nickel/immersion gold (ENIG) sometimes results in brittle fracture if the plating chemistries are not maintained properly.

45.6.1 Hot Air Solder Leveled Surfaces

HASL coated boards have been the industry mainstay and the most common board surface finish worldwide. HASL is generally eutectic or near-eutectic Sn-Pb flowed onto exposed surface pads and traces on a finished PWB at the board shop prior to electronic assembly. It is inexpensive, eminently solderable, and easy to apply, and has excellent shelf-life/aging properties. There is the old adage in the industry that "nothing solders like solder," which has made HASL a safe and preferred finish for many years. Previously it was thought that the HASL process could not deliver flat enough pads for fine-pitch surface-mount. Now it is a well-proven surface finish for component lead pitches down to at least 0.5 mm. Reputable board shops no longer consider HASL difficult to apply.

RoHS for the most part marks the impending demise of the Sn-Pb HASL surface finish. There are efforts to find other metallurgies to extend the useful life of HASL equipment.

Sn-Cu has been demonstrated to be one such HASL coating material. Studies are underway to determine long-term reliability impact of this change in material on the board and resultant solder joints.

45.6.2 Organic Solder-Preservative-Copper

Known as OSP-Cu,[14,15] organic solder-preservative-copper is a treatment and organic coating applied to the copper lands on the PWB at the board shop. The lands are first given a microetch to remove oxides and to increase surface area for OSP coating adhesion. The protective coating is an organometallic compound such as benzotriazole or imidazole, that chemically bonds with the freshly etched copper bonding pads and PTHs of the PWB. The coating retards copper oxidation and is meant to retain copper solderability through multiple reflow cycles. OSP-Cu is widely used and best for "just-in-time" board assembly strategies since its useful life is time-, temperature- and humidity-dependent.

During the soldering process, the organic coating degrades with each reflow cycle, allowing increasing levels of copper oxidation. The chemical-resistant copper oxide inhibits solder wetting, requiring ever stronger solder fluxes. This is particularly the case with thick {2.36 mm (0.093 in.) or greater}, thermally aggressive boards with mixed-mount technology. Reflow results on the secondary side and primary side of such boards are generally satisfactory. But by the time such a board goes through wave soldering, the OSP coating has largely lost its effectiveness in preventing copper oxidation and incomplete barrel fill may result on PTH component soldering.

When solder paste printing goes awry and the PWB needs to be cleaned of the misprinted or smeared solder paste, most organic solvents used for removing the paste can remove, thin, or react with the OSP coating reducing its effectiveness in preventing copper oxidation. Care must be taken to ensure solvent compatibility with the OSP coating and its impact on the coating's ability to protect the copper after cleaning.

Another deficiency of OSP-Cu is found at in-circuit test (ICT). As the board is passed through top- and bottom-side reflow cycles and wave soldering, the unsoldered and unfluxed test points build up a layer of copper oxide. This oxide layer is difficult to pierce even with sharp pin- or blade-type ICT probes, resulting in open circuits and multiple fixture reseatings to attain probe penetration and electrical contact. Some have adopted coating the test points with solder; however, no-clean residues can also lead to ICT probe contact problems. The best strategy goes against modern PCA practices: reverting to the use of water-clean solder paste and wave solder flux and coating test points with solder paste. The post-soldering, aqueous cleaning cycle removes any flux residues, enhancing ICT probe contact on solder-coated test point on the finished PCA. Most large electronic manufacturers have worked diligently to remove aqueous cleaning from their factories—a move for improved process economics and significant environmental impact. Therefore, OSP-Cu is best used where ICT is not required, such as on low-end consumer products.

45.6.3 Electrolytic Nickel/Electrolytic Gold

This plating combination has been in use for many years and is considered a high-end surface finish but a highly reliable one in terms of solderability and product life. The amount of gold (Au) is small, but the plating process uses toxic and expensive materials that further increase the price of the plating. This plating system should not be confused with electroless nickel/immersion gold (ENIG). That surface finish will be discussed separately and does not have the reliable history of electrolytic nickel/electrolytic gold.

The electroplated nickel (Ni) serves as a barrier between the copper and the gold. Ni on its own is a poor candidate for a solder pad coating as it oxidizes rapidly and its native oxide is very stable—chemically tough to dissolve or react away. Unoxidized Ni can alloy with tin, which means that it can be soldered. Were the gold deposited over the copper directly, the gold layer would begin to interdiffuse with the copper even at room temperature forming an intermetallic compound. The reaction rate is temperature-dependent. This intermetallic, like most intermetallic compounds, is difficult to solder, requiring a higher process temperature, a more active fluxing agent, or both. The Au-Cu system has poor solderability and the brittleness of this intermetallic results in diminished solder-joint strength.

Ni-Au coatings are in widespread use worldwide. Ni coating thickness is generally on the order of $\geq 2.54\ \mu m$ ($>100\ \mu in$). Au is a more difficult metal to plate. It tends to coalesce into islets, resulting in a porous coating. Any pores in the gold overcoat allow oxidation of the underlying Ni, rendering it locally unsolderable. To preclude this, the gold is liberally plated with a targeted minimum thickness greater than about 0.127 μm (5 μin) to ensure complete Ni coverage.

There are problems with the Ni-Au metallurgical system. First and foremost, it is well known that gold can cause solder joint embrittlement. When gold is introduced to tin as in tin-lead solder, it forms alloys with both the Pb and Sn. The Au-Pb alloy has two predominant intermetallic compounds: Au_2Pb and $AuPb_2$. Of the two, the latter is of greatest concern. It begins to form at the Au-Pb eutectic (85 wt percent Pb, m.p., 215°C), a temperature that encroaches into the normal reflow solder processing regime. $AuPb_2$ is stable below 254°C and is characterized by brittle, plate-like structures. Regardless of the composition, size, and concentration of the intermetallic grains in the solder joint, they can either strengthen it (forming miniature reinforcing bars in a solder "cement") or they can embrittle it, detracting from the solder's inherent ductility.

Gold is non-oxidizing, like platinum and palladium, its noble metal neighbors in the periodic table. It readily alloys with most common solder alloy compositions, especially Sn-Pb and Sn-Ag-Cu. A. Korbelak and R. Duva[16] consider gold the most readily solderable material, but Sn-Au intermetallics are of even more serious concern than those of Au-Pb. The former is a characteristically brittle intermetallic that has been widely linked to solder-joint failure. Too much gold in a tin-lead solder imparts a dull, grainy look to the resultant joint—the only possible visual clue, although there are other conditions that can result in a similar appearance. There are marked differences in solder-joint integrity when it is loaded with Au. The embrittlement causes a reduction in the joint strength and fatigue life. Although there is much controversy over how much gold is too much, there is universal agreement that Au can be detrimental to the final assembly if its fractional weight-percent is not strictly controlled. The controversy is in part due to the type of product being soldered, its intended use, and its life cycle. If an electronic assembly is subjected routinely to high operating temperatures and/or frequent "on-off" power cycling, embrittled solder-joint fatigue problems will be evident. Although there is unanimous agreement that gold can be responsible for solder joint embrittlement and reduced solder-joint service lifetime, there is wide disagreement on the exact amount of gold permissible. Reported values generally fall between 2 wt percent and 10 wt percent, with most experts agreeing that less than 2 wt percent is a safe value.[17,18]

Ebneter[19] determined that Au as a thin, protective coating 0.762 μm to 5.84 μm (30 μin to 230 μin) does not pose an embrittlement problem. He also determined that the Au thickness required was a function of the plated grain size and porosity. Further, grain size is dependent on the type of plating bath (cyanide or acid) used. Hedrig[20] also concluded that >5 percent Au in Sn-40Pb solder was problematic. Foster[21] reported a small net increase in solder-joint strength with a 2.5 percent Au addition to the same alloy, but when the addition was as high as 10 percent, there was a marked degradation in solder-joint strength. C. J. Thwaites[22] reported that 4 percent gold in a Sn-Pb solder joint is the threshold value above which there is a marked embrittlement and reduction in fatigue resistance.

There are three important Au-Sn intermetallics: AuSn, $AuSn_2$, and $AuSn_4$. Eutectic points are at about 85 wt percent Sn (215°C) and 20 percent Sn (280°C). Gold can also form a ternary compound when joined with Sn-Pb solder, exhibiting a eutectic at about 175°C.[23] This can radically alter reflow characteristics and solder-joint performance. Au, like Pb, is poorly soluble in Sn, which means that it dissolves in hot Sn, Pb, or Sn-Pb solder and precipitates out of solution during the liquid-to-solid phase change of the solder. When it does, it forms brittle, plate-like crystals that appear needle-like in cross-section.[24]

When an intermetallic forms a distinct interfacial layer or otherwise high localized concentration, it is a rigid structural element, prone to cracking and brittle fracture failure on flexing. Therefore, even though the solder joint is composed of a ductile Sn-Pb composition, its overall strength and endurance is dominated by the brittle intermetallic composition and its thickness and structure within the solder joint.

As component lead pitch decreases, so does the amount of solder-per-joint contribution from the solder land metallurgy. Au concentration becomes increasingly difficult to control because its plated thickness cannot be easily regulated to accommodate the reduced surface area and lower solder volumes that accompany finer-pitch component requirements. After all, Au deposit thickness is dependent on its deposition grain size and plated porosity. The porosity must be overcome in order for the Au to serve its function in preventing oxidation of the underlying nickel on a Ni-plated substrate.

Another pitfall of Cu-Au or Cu-Au-Sn systems is that of Kirkendall void formation. Pores in the solder joint or at copper-gold interfaces occur through solid-state diffusion. Copper, which has a relatively high solubility and solid-state mobility in gold, diffuses into the gold. Transported by grain boundary diffusion at temperatures below 150°C and by bulk diffusion above that temperature, Au-Sn regions result along with atomic vacancies. When lattice vacancies are in profusion, voids can be observed, and this material depletion weakens the solder joint. Several references are available on this phenomenon.[25,26,27,28] Kirkendall voids should in no way be confused with voiding that results from entrapped gases, liquids, or residues from solder paste in the surface-mount process or from fluxes or from PWB laminate materials in wave soldering.

Although there is still much controversy on the subject of Au concentrations and effects on the solder joint, there is no doubt that this metal system is being widely exploited successfully in electronics manufacturing. It is, however, prudent to ensure that the Au levels are kept to a minimum via adequate vendor management and implementation of good process control methods for solder deposition. Note that because gold is a non-oxidizing material and therefore does not deplete the flux, more flux is available to work on the other components of the metallurgical system—namely, the underlying nickel (which must be kept from oxidizing), the solder, and the component lead.

The Ni-Sn intermetallic is more brittle than that of Cu-Sn, and this surface finish is generally not recommended for applications where the product may experience high mechanical stresses such as shock or vibration.

It has been stated that in the Ni/Au system, the soldering actually occurs on the Ni surface and the Au serves as a protective coating. Another intermetallic of this system, $NiSn_3$, has been cited as a problem in long-term solderability degradation.[29] This IMC rapidly develops a platelet structure at low temperatures, with the greatest growth rate in the vicinity of 100 to 140°C. When intermetallic platelets grow large enough, they may penetrate the surface of the overlying tin, oxidize, and become a material difficult to flux and solder. The presence of elemental Pb was determined to retard $NiSn_3$ IMC formation.

Ni/Au surface finish is in use with Pb-free soldering, but there is currently a watch in the industry as sporadic brittle fracture has been observed on occasion. The exact nature of this fracture mechanism is yet to be understood and it is not certain that Ni/Au surface finish is at fault or just coincidental with some failures.

45.6.4 Electroless Nickel/Immersion Gold (ENIG)

As the name implies, ENIG is not an electroplated surface finish. Still comprised of Ni with an overcoat of Au, ENIG is a lower-cost alternative than electrolytic Ni/Au. When the plating process is kept in control, it gives an excellent, solderable, and reliable surface finish. On the other hand, it is a finish that has caused many well-known reliability problems. Terms such as "black-pad," "black line nickel," and "mud cracks" describe some of the evidence of failed ENIG-related problems. Although these phenomena are most commonly observed on BGA packages plated with ENIG for package-to-ball attachment,[30] the failure has also been seen on solder joints on PWBs plated with an ENIG surface finish.

The main defect is brittle failure of the solder joint—a clean break beneath the intermetallic layer formed on the PWB bonding pad. Were the package completely pried loose from the board, a high surface area, dull finish on bonding pad would be evident—the well-known black-pad. Often the pad has an array of cracks visible on its surface reminiscent of mud cracks. Once the failed component is removed, the black-pad afflicted bonding pads is unsolderable or unreliable for further use and the board must be scrapped. Although the exact cause of ENIG failure is not well understood, it seems to be linked to low phosphorus concentration in the electroless nickel plating bath.

ENIG's appeal is its modest cost, flat topography consistent with all plated surface finishes, excellent solderability, excellent testability, and reasonable shelf-life if plated well. Unfortunately, the risk associated with this surface finish must be weighed alongside its benefits. In general, the assembler will find good results with ENIG. Unfortunately, there are no quick screening methods to weed out failure-prone ENIG plating lots.

45.6.5 Immersion Silver (Imm-Ag)

This electroless silver-plated surface finish is in common use for Pb'd assembly and is also compatible with the most common Pb-free solders. Although it is highly solderable and known to be a good material for probe contact during in-circuit test measurement, it does bring with it some negative attributes. It has reasonable storage shelf-life if stored properly (air-tight bags), but is known to combine with sulfur contaminants in the atmosphere and tarnish or sulfidize. The silver sulfide is somewhat resistant to weak organic fluxes as used for no-clean soldering.

A new failure mechanism associated with this surface finish has been reported.[31,32,33] Referred to as "planar microvoids" or "champagne voids," they are small voids in alignment atop the intermetallic layer of a solder-joint bond line (see Fig. 45.4).

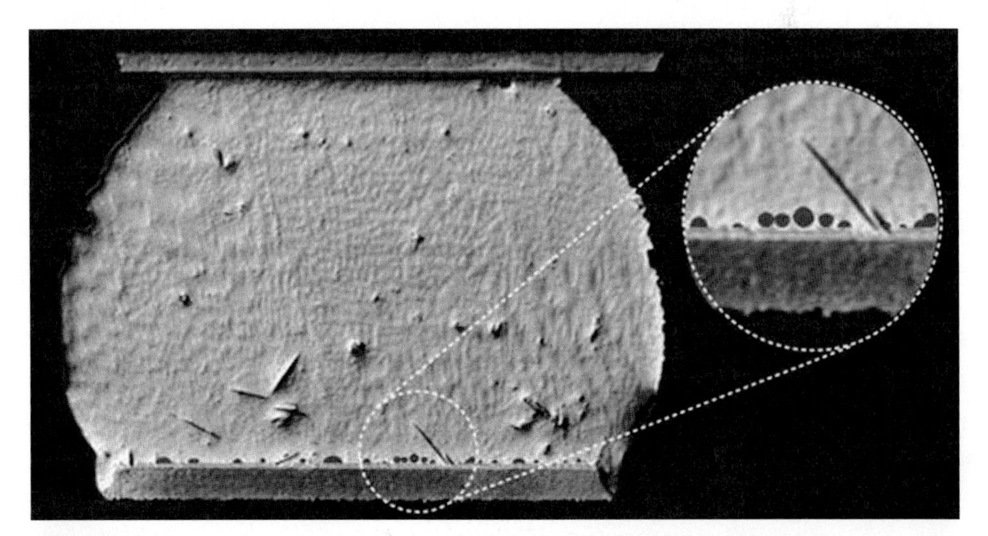

FIGURE 45.4 Cross-sectional micrograph of SAC BGA solder ball bonded to a silver surface finished PWB bond pad. Planar voids are clearly evident in the section and in the magnified inset. Voids in close adjacency above the intermetallic layer and the BGA ball can significantly detract from the solder-joint strength. *(Courtesy of Hewlett-Packard).*

Planar voids result in diminished bond strength and solder-joint fracture. This defect has been responsible for some product failures and recalls. The root cause of the microvoids is generally believed to be the result of volatile organics incorporated during Ag plating, but there also seems to be a correlation between Ag thickness and Cu topology on the bond pad.

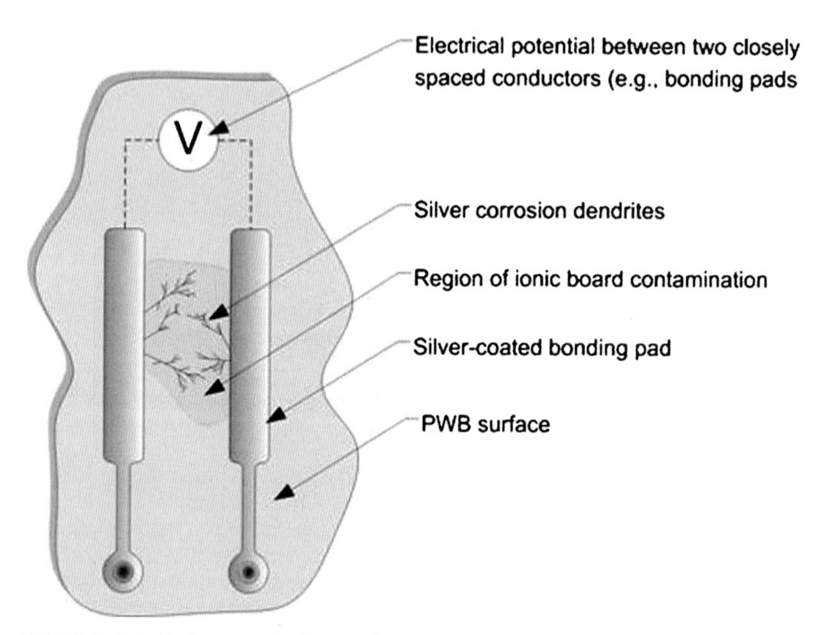

FIGURE 45.5 Ionic contamination on a board surface can adsorb moisture from the atmosphere. If an electrical charge is present, silver from bond pads may corrode and electrochemically migrate, resulting in a short circuit between bond pads.

Silver ions are known for their mobility in the presence of an electrical field (see Fig. 45.5). Ionic residues on a board with silver-coated surface finish together with adsorbed moisture from the atmosphere can result in electrochemical migration and corrosion dendrites originating from any exposed, unalloyed silver plating. These dendrites are most commonly seen on the board surface but can also occur under solder mask. If the dendrites are long enough, they may cause electrical shorting to one another or to adjacent traces. Since the corrosion dendrites are very thin, they are poor electrical conductors. If enough current is passed, the dendrites may fuse and the once-shorted circuit will become open again. If moisture and contaminant conditions are right, corrosion dendrites may form again and result in an electrical short. This cyclical growth-and-fuse phenomenon can complicate diagnosis of the source of the shorting.

Silver is an expensive metal. With its increased demand for solders (e.g., Sn-Ag-Cu), it is expected to rise in price, making it economically less attractive.

45.6.6 Tin (Sn)

Tin deposited directly over copper, whether by electrolytic or electroless methods, should be avoided except for low-end/low-cost, short-field-life products. Sn alloys with copper, even at room temperature, form a brittle intermetallic compound, albeit slowly. It has limited solderable shelf-life at room temperature and the Sn-Cu intermetallic can inhibit soldering and may result in incomplete solder wetting or solder joints of inferior strength. For tin to be used effectively, an appropriate non-porous barrier metal, such as Ni, should be deposited between the Cu and the Sn. Immersion Sn can be plated only as a very thin layer since its deposition is by a self-limiting replacement reaction. When there is no more exposed Cu on the surface of the PWB during the surface-plating operation, Sn deposition will cease. Therefore, it is impossible to plate enough Sn by this method to extend the solderable shelf-life or reduce the brittleness of the solder joint.

When fresh, tin has reasonable solderability, but also tends to oxidize quickly. There are two important phenomena that are associated with Sn platings: tin whiskers and tin pest.

45.6.6.1 Sn Whiskers. Not to be confused with corrosion dendrites, whiskers are crystallites (see Fig. 45.6) that grow from pure tin surfaces into free space whereas corrosion dendrites grow on ionic contaminated, hydrated surfaces between two charged conductors. Although whiskers have been studied since the 1940s, their root cause is yet to be fully understood.

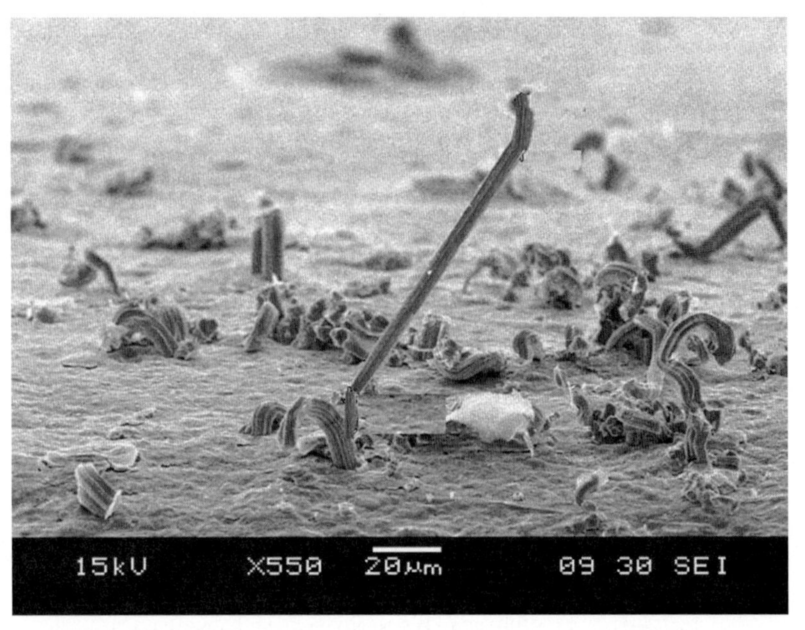

FIGURE 45.6 Scanning electron microscope (SEM) micrograph of Sn whiskers growing from a pure Sn surface. When whiskers break, they pose danger of shorting between fine-pitch conductors. Whiskers can also grow long enough to bridge to a nearby conductor. *(Courtesy of H. Hsu, Mitac Corporation, ShunDe, China).*

The concern over whisker growth is related to electrical short circuits. If a tin whisker is long enough, it may break due to mechanical shock or vibration. The broken whisker may bridge two closely spaced contacts or traces on a component or PWB. Whiskers need not break to cause a short circuit, as they can grow long enough to bridge two adjacent, closely spaced conductors. In 1998, the Galaxy IV communications satellite failure was traced back to tin whiskers. Spare boards for that satellite stored on Earth clearly showed whisker growth. It is ironic that the first whisker issue occurred in 1946 on Bell telephone equipment, causing another telecommunications disruption.

Theories abound with regard to tin whisker formation. There are lattice dislocation theories and grain boundary diffusion theories. But the most prevalent theory for whisker generation relates their origin to compressive stresses within the tin lattice and perhaps weak spots in the overlying natural oxide on the tin surface (see Fig. 45.7). When tin grains are compressed against each other, and particularly when there are large compressive forces against a small, near-surface grain, the small grain may start to recrystallize or extrude in the least stressful direction, outward through a weak spot in the native oxide layer. Growth will continue until the lattice energy reaches equilibrium. Tin whiskers can grow to a diameter of approximately 1 micrometer and their length can reach tens of micrometers.

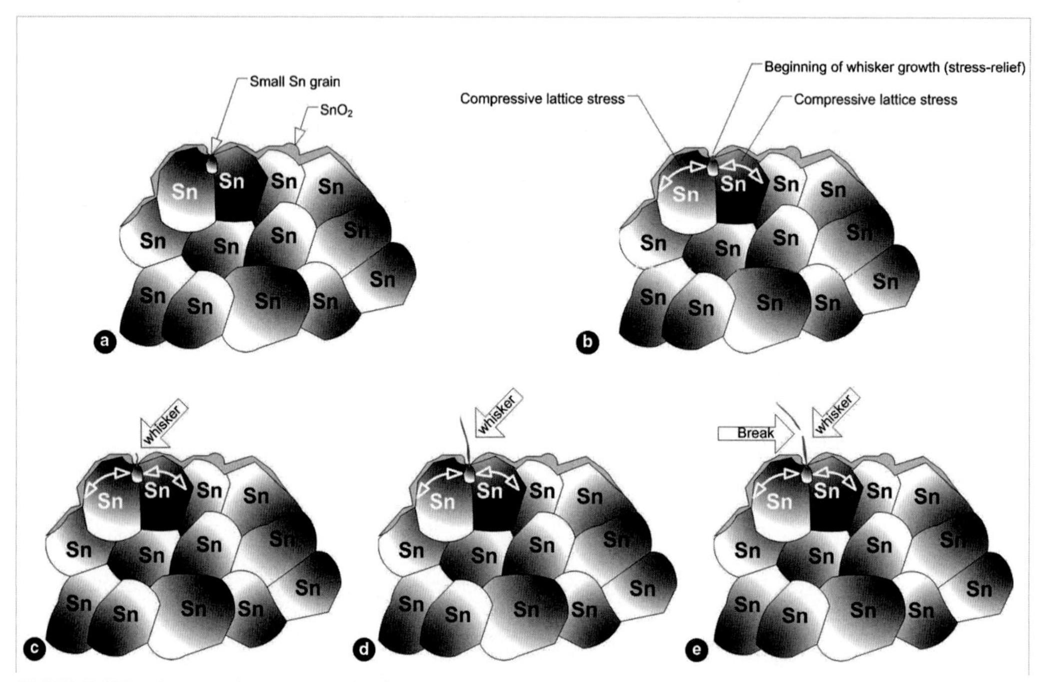

FIGURE 45.7 Compressive stress within the tin lattice may cause whisker formation. The picture sequence depicts a likely mechanism for tin whisker growth and risk to printed circuit assembly: (a) A tin lattice with small, near-surface grain between two large grains. (b) Compressive stresses cause recrystallization of grain. A thin spot in the oxide layer allows rudimentary whiskers to penetrate the surface. (c) The whisker continues to grow. (d) The whisker continues to grow. (e) The whisker breaks due to mechanical shock or vibration. Whisker fragment poses a shorting risk to the PCA.

Now that Europe's RoHS legislation restricts Pb to a level of less than or equal to 1,000 parts per million by weight in electrical and electronic assemblies, there is much more focus on tin whiskers. Traditionally, many electrical components had tin-lead surface finishes, as did printed wiring boards (HASL). In cases where pure tin finishes were used on components, small additions of lead were added. The resulting lead-doped tin lattice imparted some compliance to the structure and reduced the risk of tin whisker formation. For the vast majority of electrical and electronic products, that is no longer an option for many classes of components.

Where tin and copper are in contact with one another, copper is found to diffuse into tin faster than tin into copper. The disruption to the tin lattice results in compressive strains, which sets the stage for whisker initiation. The solution is the addition of a thick, non-porous nickel barrier layer over the copper. Tin is then plated over the nickel. The thick nickel deposit is generally under tension and the tin diffuses into the nickel faster than the nickel into the tin. Since the nickel exerts no compressive influence on the tin lattice, tin whiskers are less likely to form.

Note that an added RoHS exemption allows the addition of Pb to Sn plated component leads if the component lead pitch is <0.65 mm (<0.025 in.). Besides metallic dopants, there are other ways to mitigate the growth of tin whiskers. Brighteners in tin-plating baths should be avoided and matte tin is preferred to minimize whisker growth. Annealing tin coatings is known to reduce whisker growth as recrystallization minimizes lattice stress. Ensuring that the reflow temperature during PCA soldering is above the melting point of tin (232°C) will reduce whisker risk. Additionally, the use of a barrier metal between copper and tin is recommended whether for circuit board or component surface finish.

45.6.6.2 Tin Pest. In addition to risk from whiskers, tin also poses another problem in some applications. At about 13°C, tin undergoes a crystalline structure transformation. The tetragonal crystal of white (β) tin converts to the cubic allotrope, gray (α) tin. Since gray tin has less density than white tin, there is an expansion of the tin lattice upon transformation and the tin starts to fall apart, turning to a gray dust. So tin should be avoided as a surface finish for low-temperature duty. The transformation is most profound at –40°C but will occur, albeit more slowly, at temperatures below the allotropic transition threshold. As in the case of tin whiskers, tin pest has been combated with the addition of a dopant metal such as bismuth, antimony, or lead. Lead, however, is no longer an option due to the RoHS directive and similar mandates.

REFERENCES

1. Abtew, M., and Selvaduray, G., "Lead Free Solders for Surface Mount Applications," *Chip Scale Review,* September 1998, pp. 29–38.
2. Shuldt, G., and McKay, C., "Amalgams for Electronics Interconnect," *Proceedings of the Seventh Electronic Materials and Processing Congress,* 1992, pp. 141–145.
3. Strauss, R, and Smernos, S., "Low Temperature Soldering," *Circuit World,* Vol. 10, No. 3, 1984, pp. 23–28.
4. Marshall, J., Calderon, J., and Sees, J., "Microstructural and Mechanical Characterization of 43–43–14 Tin-Lead-Bismuth," *Soldering and Surface Mount Technology,* No. 9, October 1991, pp. 25–27.
5. NCMS Lead-Free Solder Project Final Report, August 1997.
6. Bath, J., Handwerker, C., and Bradley, E., *Circuits Assembly,* May 2000, p. 31.
7. Patanaik, S., and Raman, V., "Deformation and Fracture of Bismuth-Tin Eutectic Solder," *Proceedings of ASM International-Materials Development in Microelectronics Packaging Conference* August 1991, pp. 251–256.
8. Mei, Z., and Morris, J. W., "Characterization of Eutectic Sn-Bi Solder Joints," *Journal of Electronic Materials,* Vol 21, No. 6, 1992, pp. 599–607.
9. White, C., and Evans, G., "Choose the Right Alloy for Each Soldering Job," *Research & Development,* March 1986.
10. Allenby, B., et al., "An Assessment of the Use of Lead in Electronic Assembly," *Proceedings of Surface Mount International Conference,* 1992, pp. 1–28.
11. Freer, J., and Morris, J., "Microstructure and Creep of Eutectic Indium-Tin on Copper and Nickel Substrates," *Journal of Electronic Materials,* Vol. 21, No. 6, 1992, pp.647–652.
12. Yost, F., "Soldering to Gold Films," *Gold Bulletin,* No. 10, 1977, pp. 94–100.
13. Boggs, D., "Anti-Tarnish: One Alternative to HASL," *Electronic Packaging and Productions,* August 1993, pp. 31–35.
14. Murray, J., "Beyond Anti-Tarnish: An SMT Revolution," *Printed Circuit Fabrication,* February 1993, pp.32–34.
15. Korbelak, A., and Duva, R., *48th Annual Technical Proceedings of the American Electroplater's Society,* 1961, p. 142.
16. Foster, F. G., "Embrittlement of Solder by Gold from Plated Surfaces," Papers on Solders, American Society for Testing Materials, STP No. 319, 1962.
17. Foster, F. G., "Gold Plated Solder Joints," *Product Engineering,* August 19, 1963, pp. 50–61.
18. Ebneter, S. D., "The Effect of Gold Plating on Soldered Connections," NASA Technical Report, Accession Number: 65N36777; Document ID: 19650027176; Report Number: NASA-TM-X-53335, 1965.
19. Hedrig, G., "Soldering of Gold Thin Films," *Finomechanika* (Precision Mechanics), Vol. 9, No. 4, 1970, pp. 108–118.
20. Foster, F. G., "Embrittlement of Solder by Gold from Plated Surfaces," Papers on Solders, American Society for Testing Materials, STP No. 319, 1962.

21. Thwaites, C. J., "Soft Soldering," *Gold Plating Technology,* Electrochemical Publications, 1974, ch. 19, pp. 225–245.
22. Karnowsky, M., Rosenweig, A., *Trans. Met. Soc. Am. Inst. Min. Engrs,* 242, 2257 1958.
23. Fox, A., et al., "The Effect of Gold-Tin Intermetallic Compound on the Low Cycle Fatigue Behavior of Copper Alloy C72700 and C17200 Wires," *IEEE Transactions on Components, Hybrids, and Manufacturing Technology,* Vol. CHMT-9, No. 3, September 1986, pp. 272–278.
24. Prince, A., "The Au-Pb-Sn Ternary System," *Journal of Less-Common Metals,* 12, pp. 107–116, 1967.
25. Shewmon, P., *Diffusion in Solids,* McGraw-Hill, 1963, pp. 115–136.
26. Seitz, F., "On the Porosity Observed by the Kirkendall Effect," *Acta Metallurgica,* Vol. 1, 1953, p. 355.
27. Zakel, E., and Reichl, H., "Au-Sn Bonding Metallurgy TAB Contacts and Its Influence on the Kirkendall Effect in Ternary Cu-Au-Sn System," *Proceedings of the 42nd Electronic Components & Technology Conference,* May 1992, pp. 360–371.
28. Haimovich, J., "Intermetallic Compound Growth in Tin and Tin-Lead Platings," NWC TP 6896, EMPF TP 0003.
29. Moltz, E., "Use and Handling of Semiconductor Packages with ENIG Pad Finishes," Texas Instruments Application Report SPRAA55, August 2004.
30. Bryant, K., "Investigating Voids," *Circuits Assembly,* June 2004, pp. 18–20.
31. Yau, Y-H, et al., "The Properties of Immersion Silver Coating for Printed Wiring Boards," IPC/APEX, Anaheim, 2005.
32. Aspandiar, R., "Voids in Solder Joints," IPC/CCA PCB Assembly and Test Symposium, May, 2005.

CHAPTER 46
SOLDER FLUXES

Gary M. Freedman
Hewlett-Packard Corporation, Business Critical Systems, Singapore

46.1 INTRODUCTION TO FLUXES

In a simplistic sense, the soldering process model is generally treated as solid and liquid metals interfacing with a gas (the soldering ambient). However, realities dictate a modified view. Nearly all metals oxidize when exposed to air and these oxides, whether on the solder itself or on the metals to be joined, always impede soldering. To keep the solder and surfaces to be joined bondable, the oxide must be removed and bonding surfaces must be kept oxide-free until soldering is complete. A chemical fluxing agent is used to this end and is indispensable to the process of joining. Fluxes are tailored to both the metals to be joined and soldering process temperature. Some metal oxides such as aluminum and nickel have chemical-resistant oxides, which make soldering difficult. Even copper oxides can be tough to dissolve. Most fluxes used in printed circuit assembly are acid-based and are formulated to be strong or weak depending on the metal systems to be joined, post-soldering treatments, and reliability needs.

Fluxes are typically composed of acids and high molecular weight materials that are slow to dry, slow to evaporate, and slow to decompose. They are often so acidic that they can etch, dissolve, or disrupt oxides and tarnishes on solder and surfaces to be joined. Both the absence and presence of liquid flux have an effect on the extent of solder wetting. Some fluxes can transition from a runny liquid during preheat to a viscous liquid or gel and ultimately to a gummy solid as it dries out, polymerizes, and decomposes during the soldering cycle.

46.1.1 Flux Delivery Systems

Fluxes can be delivered in many different ways. For surface mounting, solder paste is used. The paste is a mixture of pure solder spheres, a fluxing agent, and other materials to adjust rheology for efficient stencil deposition of the solder paste on printed wiring board (PWB) solder lands. In wave soldering, a liquid flux is sprayed or otherwise applied to the secondary side of the board before the preheat and soldering steps. In hand-soldering, the flux can be an integral part of the solder wire, which has lumens within the wire filled with flux (see Fig. 46.1). As the solder melts, the flux is released to do its job.

It can also be applied as a flux paste with miniature artist's brush, with a pick, or as a liquid via fine-tipped flux pen. It has been shown that flux can also be delivered as a gas. This technology is rarely used, but will be discussed briefly.

46.1.2 Flux Functions

Selection of the flux type depends on the solder alloy, metal finishes on the board and components, condition of the surfaces to be joined, the type of soldering process selected, required solder-joint attributes, and the intended final use of the assembly. Some fluxes leave residues that are meant to remain on the board. Some fluxes are difficult to clean or may interfere with in-circuit test (ICT) probing or may hinder conformal coating or under-fill adhesion. Post-process flux deposits can cause corrosion, especially in the electric field between adjacent conductors on the surface of the printed circuit board (PCB). Fluxes may impart gas bubbles, or voids, which may be "frozen" into the solidified solder (see Fig. 46.2). The resultant voids may detract from the mechanical strength if they are large enough, are in abundance, intersect bonding surfaces, or are in close proximity to one another.

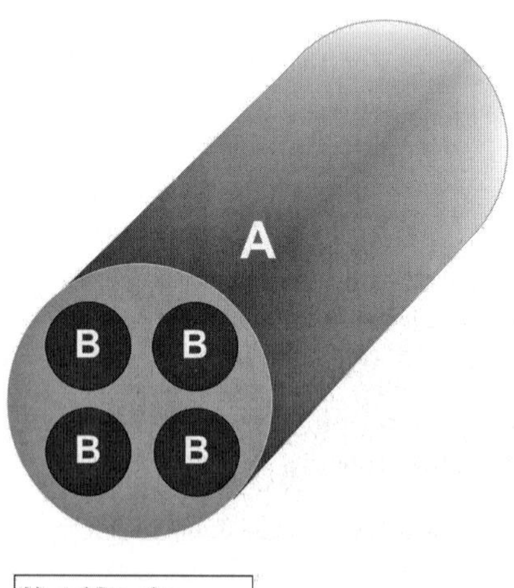

FIGURE 46.1 Solder wire for hand-soldering showing flux-filled lumens.

When soldering is conducted in an air environment, once the flux reacts with the oxide layer on the solder and the metals to be joined, the metal surfaces are rendered chemically unstable and are quite vulnerable to reoxidation as they once again strive for equilibrium within the oxygen-rich ambient atmosphere. The flux residue coats oxide denuded, solderable surfaces as well as the solder and retards oxidation. If the solder flux is exhausted or fully evaporated before or even after oxide stripping and before the onset of solder reflow, oxidation would recur and solder-joint formation would be impeded. Solder fluxes are generally formulated with high-boiling-point organic materials to slow evaporation, allowing overcoating of the cleaned metal surfaces. Further, if the flux residues were to become polymerized or charred prior to the solder melting temperature, this too could inhibit the flow of solder. It is therefore necessary to heed manufacturer solder paste or flux time and temperature guidelines from the flux or paste chemistry manufacturer, although some process optimization is recommended.

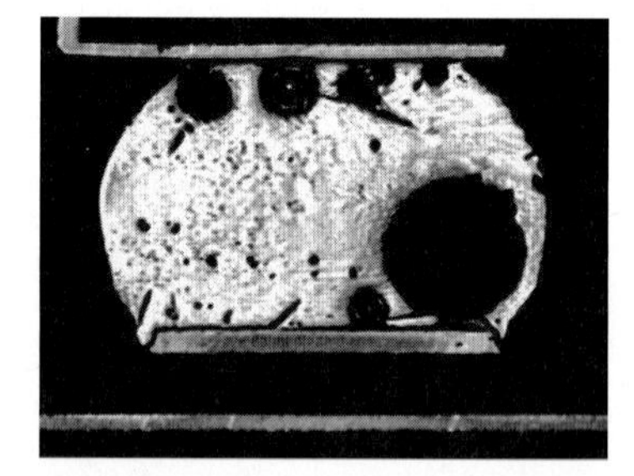

FIGURE 46.2 A BGA solder joint rife with voids from flux-derived gases and improper soldering profile. These voids can detract from the reliability of the solder joint. The IPC-A-610 workmanship standard for printed circuit assemblies (PCAs) should be consulted on void acceptability. *(Courtesy of Hewlett-Packard.)*

The presence of the hot liquid flux during soldering helps to promote thermal uniformity on the board. This is very important in mass reflow and may play an even more crucial role in some solder-in-place processes such as hot-bar laser. In some forms of laser soldering, the flow of hot flux may help to preheat leads and pads, readying them for solder joint formation. In the case of hot-bar soldering, a pool of liquid flux may help to even out thermal disparities over the length of the heated bar.

46.2 FLUX ACTIVITY AND ATTRIBUTES

A flux's ability to prepare a surface for soldering is dependent on its chemical strength or *activity*. Just like reagents on the chemistry lab bench, there are various concentrations available. A flux may be naturally strong or it may have additives to enhance its reactive power.

These additives are referred to as *activators*.[1] Activators accelerate the fluxing action either by direct chemical contribution or by catalysis. The activator can be an inorganic or organic material and combines with other ingredients in the flux, or decomposes with temperature to form an acid that is effective for oxide removal. Halogens, amino-halides, organic acids, and other materials are common activators. As the flux is heated, the activator begins its work. Flux heating for the soldering process has to be tempered for the particular type of flux formulation and solder system. Some activators may decompose if heated too quickly. Others may not be effective for a long enough time. Weaker fluxes (flux + activator) may require a longer preheat time and higher temperature in order to have sufficient chemical activity to remove oxidation and tarnish from parts to be joined as well as from the solder. Flux selection has to be appropriate for the solder alloy melting temperature as well as for the type of soldering process. The solder chemistry chosen will greatly influence the process time-temperature profile.

Like most chemical reagents, a flux's reactivity is in part a function of the exposure time and the temperature; many exhibiting Arrhenius-like behavior, where time-dependent chemical performance increases sharply with temperature. Some fluxes are weakly active at room temperature, but most require a significant thermal assist to enable them to be useful. Ultimately the only valid test of flux performance is whether the system works for the types and conditions of printed circuit boards and components encountered in the course of the required soldering process. A more active flux may be needed for low melting point solder, especially if manufacturing demands high throughput for this process step. In this case, the flux will not have the benefit of longer exposure times and higher process temperatures to assist in its job of oxidation removal. Higher-temperature alloys require fluxes that can withstand higher process temperature for reflow. It must not volatilize or decompose too rapidly nor polymerize or char at temperatures commensurate with soldering. Some metals are more prone to oxidation than others. Bismuth and zinc used in solder and copper on circuit boards are notable examples that may require a more active flux. A nitrogen-rich environment is helpful in augmenting flux performance. (See also Chap. 43, "Fluxes and Cleaning.")

46.3 FLUX: IDEAL VERSUS REALITY

The ideal flux would be one that is potent, requires only a small quantity, leaves little or no residue (and what residue does remain would be thin, transparent, nonreflective), easily penetrated by in-circuit test probes, electrically insulating, and chemically inert. If not inert, it should be easily and completely removable in an inexpensive solvent. However, this is far from actual materials and process reality.

Flux must be applied in sufficient quantity for thorough oxide removal and to avoid both metal reoxidation and flux exhaustion prior to solder wetting. Some post-soldering flux residues or form by-products may have deleterious effects on the circuit assembly since they may be chemically reactive and result in corrosion of exposed metals over the long term (see Fig. 46.3). The corrosive agents can be electrically conductive, resulting in electrical leakage (soft shorts) between closely spaced, oppositely charged conductors on the circuit board surface exacerbated by the residue's hygroscopic moisture uptake from the atmosphere. Corrosion may be augmented either by natural electromagnetic field (EMF) couples between two different metals or electrochemically driven by a charge between two adjacent conductors. The metallic ions, electrochemically driven, can plate out as metal dendrites on the surface of the board and produce intermittent or even persistent hard short circuiting (see Fig. 46.3).

Flux must be heated to a temperature to allow the best reactive conditions without drying out or denaturing. For the sake of economy and cleaning, a thin application is necessary and should be confined to the bonded site(s) of interest.

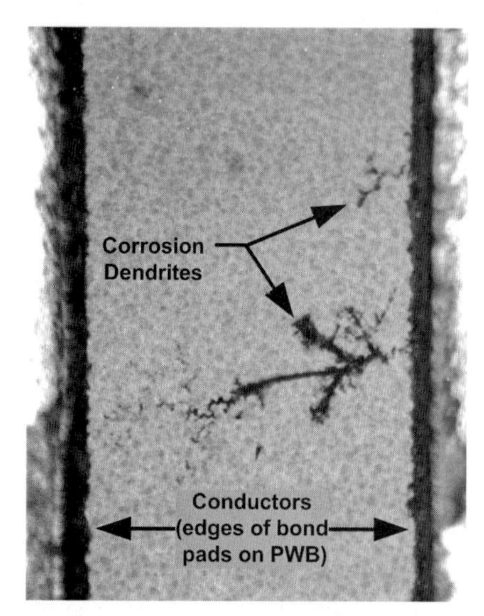

FIGURE 46.3 Corrosion dendrites caused by corrosive, hygroscopic flux residues. *(Courtesy of Hewlett-Packard.)*

46.4 FLUX TYPES

The 1989 adoption of the Montreal Protocol[2] imparted far-reaching changes to electronic manufacturing. In the circuit board assembly arena, it drove the elimination of chlorofluorocarbons and the reduction of volatile organic compounds for earth's ozone layer preservation. Many fluxes that used volatile organics such as alcohols were reformulated to use water as their solvent. Solvent-clean fluxes, which required non-aqueous cleaning chemicals (alcohols, chlorofluorocarbons [CFCs], hydochlorofluorocarbons [HCFCs], etc.), were practically eliminated from the marketplace.

Today only two flux categories are in widespread use for PCB assembly: water-clean and no-clean. The former leaves a residue that must be removed from the PCA; otherwise, corrosion will set in. The latter, as the name implies, leaves a benign residue that is meant to be left on the circuit board. In fact, some no-clean residues perform as a conformal coating shielding the board from moisture and atmospheric contaminants. The choice between water-clean and no-clean has a profound effect on the process and may also affect subsequent process steps or long-term reliability. Both can be effective oxide strippers for soldering, but there are several considerations for choosing and using the proper flux.

46.5 WATER-CLEAN (AQUEOUS) FLUXES

Also referred to as aqueous-clean fluxes or organic acid (OA) fluxes (a misleading term, because most fluxes are organic acid-based), this category contains the most aggressive fluxing agents used for large-scale commercial board assembly. It can consist of strong organic acids, inorganic acids, amines, amino salts, or often a weak organic acid fortified with reactive materials such as halogen salts (Br^-, F^-, I^-, Cl^-) that turn acidic during the flux decomposition process. For this reason, as the name implies, this type of flux requires an aqueous wash step to remove any reactive residue traces to avoid corrosion on the PCA.

Many of the water-soluble solder pastes are hygroscopic or leave hygroscopic, corrosive residues on the circuit board after soldering. This property can impact solder paste or flux shelf-life and the solder paste's physical attributes during printing and soldered assembly reliability. Most solder pastes are thixotropic; that is, they shear thin as they are being squeegeed onto the board, becoming thinner with use. Generally, pastes are formulated to limit the degree of thixotropy. In addition to thixotropic thinning, some solder pastes, when exposed to a humid atmosphere, pick up moisture, which further thins paste rheology. Paste stenciling becomes more difficult to control and tack, and slump characteristics can be adversely affected. Ironically, some of these same organic acid solder pastes are prone to dry out in low-humidity conditions. Check paste manufacturer recommendations for optimal temperature and humidity regimes for storage and use. Localized environmental control of air temperature and humidity within the printer enclosure or factory-floor air conditioning can help to preserve solder paste rheological characteristics.

46.5.1 Cleaning Process

The cleaning consists of more than a dip in a tank of water. Generally commercially available, conveyorized wash tunnels are used for effective removal of the reactive residue from the soldered PCA. Such equipment can be many meters long and may have numerous zones consisting of cold-water spray nozzles, hot-water spray nozzles, and hot-air drying. Some systems may incorporate sumps where the board is briefly conveyed underwater for a short time to enhance residue dissolution. High-purity (de-ionized) water is recommended to enhance washing to eliminate or minimize corrosion. Saponifiers are sometimes injected into the water stream to target organic residues and convert them to water-soluble entities. Following the saponification step, thorough rinsing with pure water is required to remove traces of the saponifying agent.

46.5.2 Component Compatibility

Not all components are compatible with the water-wash process. Circuit board mounted switches are notable in this regard. Even some switches sold as "environmentally sealed" or "water-wash–compatible" have been found to leak during the high-pressure wash spray or submersion steps of aqueous cleaning. This can result in failure of the switch to work properly (the water causes an electrical short within the switch) or, if there is enough ionic content in the wash water or in the switch itself, switch contact corrosion may occur in the long term.

Many circuit assemblies are very dense with parts. The higher the part density, the less effective aqueous cleaners are. Like stones in a stream, the water swirls around them and eddy currents of relatively stagnant water form behind the stones. Components on a circuit board act in the same way, as they block the free flow of water and impede proper rinsing even in the most turbulent water spray systems. To counteract this, aqueous wash machines have multiple spray nozzles directed at board surfaces in several directions and angles. Many components have low headroom spaces beneath them, inhibiting rinsing effectiveness. In the case of PGAs, ball grid assemblies (BGAs), ceramic column grid arrays (CCGAs), ceramic ball grid arrays (CBGAs), etc., the water's path is further frustrated by the presence of solder joints beneath the component. Increasing I/O count, increased package size, and decreasing space between the PWB and component underside make water washing all the more difficult. The same issues hold true for chip-scale packages (CSPs) where the headroom is little more than a capillary and good wash-water exchange is unlikely. High-density, multirow connectors are also difficult to rinse. All of the preceding are equally difficult to dry. Water beneath components or within components may interfere with in-circuit test or functional test results immediately following aqueous cleaning. Water from aqueous wash has been known to enter integrated circuit (IC) encapsulant vent holes and cause shorts internal to the IC (see Fig. 46.4).

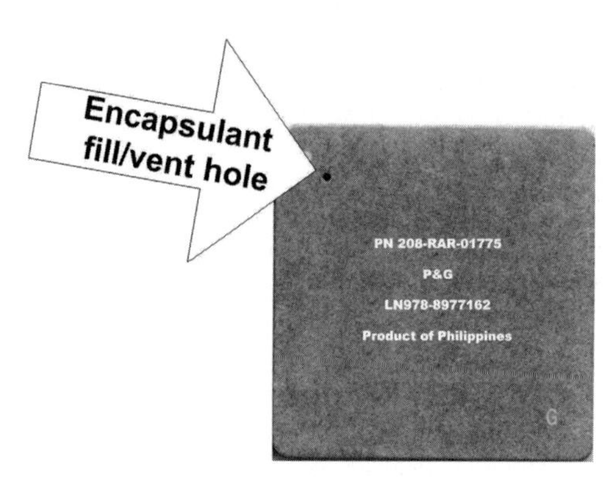

FIGURE 46.4 If water enters an improperly sealed encapsulant vent/fill hole on an IC package, shorting or corrosion can occur. *(Courtesy of Hewlett-Packard.)*

46.5.3 Manufacturing Impact of Water-Clean Process

There are significant issues to consider when using water-clean fluxes, and the decisions are dependent on the specific parts and final product:

- **Expense** Aqueous washers are expensive and require a large footprint on the manufacturing floor. Generally, high-purity, de-ionized water is used instead of tap water to enhance rinsing effectiveness and to minimize corrosion or shorts caused by tap water particulates or ionic species such as chlorine or fluorine as found in treated public water supplies.
- **Additional process steps** Washing represents several additional handling steps, and with each handling step comes risk to the board. Due to the force of the water jets in the cleaning machine, boards, especially smaller ones, are put in baskets or carriers to go through the water-wash process. This involves additional materials (baskets/carriers) and handling. There is ample opportunity to knock parts off the board during loading or unloading of these baskets. Occasionally tall parts (such as surface-mount capacitors) are blown off the board by the force of the water blast. Boards that somehow escape the water-wash process will develop corrosion in short order.
- **Additional labor** Besides the need for additional operators to load and unload the conveyorized wash machine, increased maintenance is required for the aqueous cleaning machine, de-ionized water system, and waste-water treatment.
- **Environmental impact** The de-ionized water system requires frequent monitoring and filter replacement. Solder fines washed off the PCA must be filtered from the waste-water stream, and dissolved metals and other species harmful to the environment must be monitored and removed if of a high enough concentration before releasing spent water to the drain. This process is not suited to geographies with water shortages, water restrictions, or poor-quality water, or where the cost of water is high.

46.5.4 Benefits of Water Washing

There are significant benefits to using water-clean fluxes, but the decision involves trade-offs and must be made based on the specific needs of the parts and the product:

- **Unimpeded inspection** PCAs that have gone through a proper aqueous cleaning process are easy to inspect because there is no flux residue on the surface to interfere with solder-joint inspection by eye or automated optical inspection devices.

- **Best adhesion for conformal coating** Some boards that are intended for harsh environments rely upon a conformal coating process to enhance their resistance to moisture and airborne contaminants. The coating is silicone, epoxy, or other material. For good adhesion of these coatings, aqueous cleaning is usually prescribed for the boards.
- **Recommended for use with under-fill compounds** For added solder-joint reliability, epoxies or other adhesives are sometimes dispensed under certain critical components to anchor them to the PWB surface. Most under-fill compounds require a clean substrate for best bonding, so the underside of the component and board to be bonded in this method should be free of flux residues.
- **Possible electrical implications for RF circuitry** Flux residues can interfere with some high-speed circuit conduction by changing top surface dielectric properties and increasing parasitic capacitance. It is for this reason that radio-frequency (RF) circuitry is often specified to be assembled using water-soluble fluxes.

46.6 NO-CLEAN FLUX

Boosted largely by cost, floor space, and environmental concerns, no-clean fluxes now dominate printed circuit assembly. They are generally composed of rosins (e.g., natural acidic plant residues such as pine pitch) and/or resins. Often, but not exclusively, the term *resin* is used to denote a manmade rosin analog.

Historically, pine sap—*colophony,* as it is referred to in older publications—is the most notable natural product used as a flux. Once extracted from the tree, the pine rosin is purified and neutralized to remove excess acidity imparted by the extraction process and dissolved in alcohol. Termed *water-white rosin,* the key fluxing agents are a mixture of abietic and pimaric acids. Water-white used to be the basis for many commercial flux offerings.

Rosin is a solid at room temperature and can be dissolved or partially dissolved in alcohols. During the soldering process, the alcohol or other volatiles in the flux begin to evaporate and the rosin-based solids precipitate. The solids soften at low temperatures (50 to 70°C) and begin to flow, covering the surfaces to be fluxed. With higher temperatures come higher reactivity and oxides and other tarnishes are stripped, readying surfaces for solder wetting. The most widely used fluxes today are resins but act in the same fashion. Generally weak organic acid activators such as succinic, malonic, and adipic acids are added to increase the reactivity of the flux. These, too, volatilize or decompose during the reflow cycle and only a clear polymerized, non-corrosive residue is left on the board. The need for cleaning this residue is absent, hence the term *no-clean*. The best additives are ones that will not evaporate or decompose too early in the heating cycle but will completely volatilize or decompose at or near the peak temperature of the soldering process.

Halogenated activators should be avoided with no-clean chemistries. Some high-temperature organic acids may not completely volatilize or decompose, so reliability testing of the flux residues should be conducted to ensure that corrosion will not impact product reliability. Besides adding activators, one can boost a flux's chemical reactivity by increasing the rosin or resin content, but the higher the rosin/resin content, the heavier the post-soldering flux residue. Pure rosinic residues, especially from high-rosin-content formulations, are benign, hydrophobic, and electrically insulating, but they tend to be sticky.

46.6.1 Benefits of No-Clean

No-clean benefits include lower manufacturing costs through fewer process steps, less board handling, less maintenance, and reduced floor space than is required for board assembly via water-soluble fluxing. There are also environmental benefits of no-clean, as there is almost no waste effluent other than from the solvent within the paste or flux that evaporates during the process and from solvents used to clean the stencil, squeegee blades, and any misprinted boards.

This represents an extreme reduction in waste stream to the environment as compared to the use of water-soluble paste and flux. There are volatile organic compound (VOC)–free no-clean formulations that make no-clean even more environmentally friendly and attractive. These are generally compounded with a water or water-soluble solvent instead of alcohol.

46.6.2 The Negatives of No-Clean

Before talking about the negatives of no-clean, it should be understood that the no-clean approach to circuit board assembly is widely embraced and is the mainstay of the industry. Any negatives of this technology are counterbalanced by major benefits. No-clean is impractical for only a small number of niche applications.

46.6.2.1 Difficulty of Inspection. No-clean residues, especially from surface-mount technology (SMT) soldering, are generally hard, transparent polymerized deposits that interfere to some degree with solder-joint inspection. The deposit can crack and sometimes operators will mistake the cracked flux residue deposit for solder-joint cracking. The residue is also reflective, which can interfere with visual inspection or automatic optical inspection (AOI) methods.

46.6.2.2 Decrease in Barrel Fill at Wave Soldering. Since no-clean fluxes are weaker fluxing agents than water-soluble formulations, wave soldering is a somewhat less robust process. Poorer barrel fill is generally found with no-clean fluxes, although in most cases subtle changes to the wave solder machine configuration and wave solder time-temperature profile are sufficient to gain excellent or acceptable plated through-hole (PTH) barrel fill.

46.6.2.3 ICT Probe Noncontact. Much of the no-clean flux residue associated with wave soldering of the secondary side of the board is washed off by the molten solder during the wave solder operation. Generally, ICT is accomplished by bed-of-nails testing where pointed spring probes contact dedicated test pads located on the secondary side of the board (see Fig. 46.5a). Although these test points are masked off during wave soldering using selective wave soldering pallets (shields), sometimes the liquid flux, when dispensed to the secondary side for the wave solder operation, finds its way between the pallet and the board (see Fig. 46.5b). Flux-covered test pads may be difficult or impossible to probe, depending on the thickness of the solder flux deposit and type of flux used. Without cleaning, these residues will persist. Adjustments to probe tip style and spring force may be necessary to overcome the noncontact issue.

There are different types of test probes and different spring forces available. Generally, single-point probes rather than crown probes will have the best chance to penetrate the hardened flux residue. With a single-point contact, point force is maximized rather than spread out over multiple points. Using probes with high spring rates may cause the board to flex and damage solder joints or pierce through pads.

46.6.2.4 Not Recommended for Conformal Coating or Under-Fills. No-clean residue may inhibit adhesion of under-fills or conformal coatings. In fact, the flux residue may wick under a component and completely block an under-fill from entering between the board and the IC package. There are some new developments in this area and there are under-fill/flux combinations that may be effective.

46.6.2.5 Corrosion from Repair. Very often, even water-clean boards are repaired with no-clean formulations to avoid another aqueous cleaning step. Sometimes after a board is totally assembled, components of the board may not be suited for exposure to water. Although no-clean residues after soldering are generally benign, care must be taken to minimize the flux dispensed during board repair. There is a chance that the pooled flux will not be

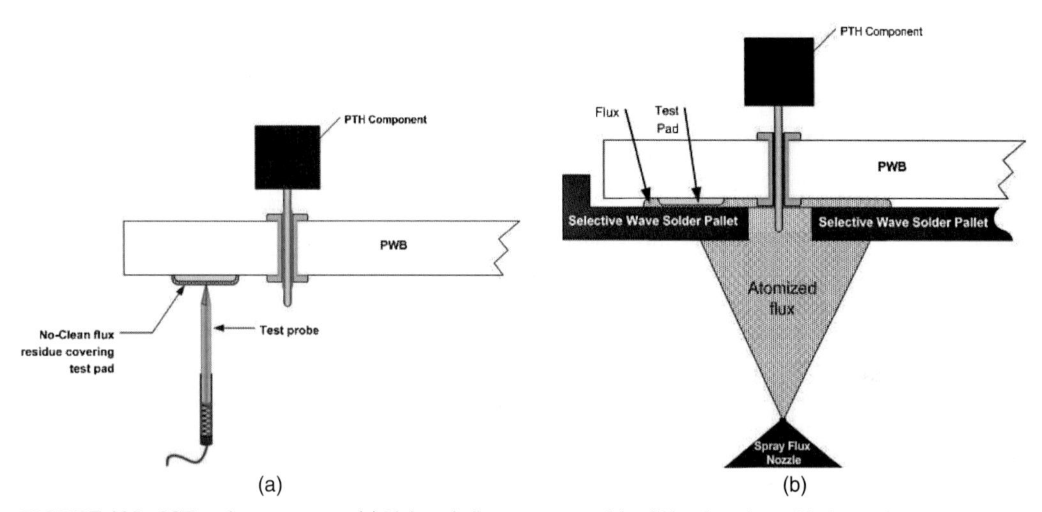

FIGURE 46.5 ICT probe noncontact: (a) If there is flux on a test pad, it will harden after soldering and may prevent the test probe from making electrical contact to the test pad. The test result may be interpreted as an "open" circuit. (b) As the board is fluxed in preparation for the wave solder process, the liquid flux may be drawn by capillary action between the wave solder pallet and the PWB. When this occurs, the in-circuit test pad (the target for bed-of-nails testing) may become fouled with solder flux. The flux residue inhibits probe contact.

fully heated and divested of its organic acids (activators). If this happens, the flux will remain acidic and promote corrosion. Therefore, it is important to ensure that:

- The smallest amount of flux is used for a repair
- Flux is confined to the area, that is, the lead(s) to be repaired
- The use of liquid fluxes is avoided, as they are prone to spreading
- The smallest diameter flux-core solder wire is used for repairs
- Ultra-fine, electrostatic discharge (ESD)–safe, felt-tip flux pens or paste flux is applied with an ultra-fine artist's brush, fine pick, or pin
- All the flux dispensed has reached a near soldering temperature

46.7 OTHER FLUXING CAVEATS

Several general issues regarding fluxes must be considered before selecting one for a process and product.

46.7.1 Solder Paste Misprint Cleaning

When solder paste is printed in a manufacturing setting, occasionally the paste deposit is misprinted or smeared, requiring that it be removed before reprinting. Although this is not truly a fluxing issue, it does have ramifications to the process and reliability and is an important flux chemistry–related topic. Here are some guidelines for accomplishing this.

46.7.2 Aqueous-Clean Solder Paste Misprint Cleaning

A standard aqueous board cleaner or an aqueous stencil washer can be used for solder paste removal. Here are some simple guidelines:

- Ensure that all parts on the board are compatible with aqueous cleaning. Some surface-mount switches may not be compatible and may need to be removed from the board or masked in some way to prevent water or flux residue leakage.
- If SMD placement has begun, remove any high-value components from the paste on the side that is misprinted or smudged.
- Do not wipe solder paste off the board, as this will force paste into vias, PTHs, and solder mask clearance areas. Once reflowed, any solder residues may interfere with PTH component insertion or press-fit operations.
- If tape is used to mask gold fingers or other features, remove it before the commencing misprint cleaning to avoid entrapping solder beads or paste residue beneath the tape.
- Allow the paste to wash off during the aqueous cleaning process using a standard aqueous cleaning machine or an aqueous-based stencil washing machine
- Direct immersion of the board should be avoided. Immersion may result in water and diluted flux residues leaking into already mounted switches—even into switches that are specified as hermetically sealed and water-wash–compatible.
- Do not use solvents other than water to clean the board.
- Do not use a saponifer or solvent other than water unless the flux chemistry has been tested and qualified for compatibility and unless saponification or other solvent is in use with standard post-solder board cleaning.
- Do not use ultrasonic cleaners. Even random-frequency/random-amplitude ultrasonic cleaners may impact certain devices that may be already mounted on the board (e.g., on the secondary side). Crystal oscillators are notoriously sensitive to ultrasonic cleaning. Other components have been known to fail as a result of ultrasonic agitation and the consequential damage to internal bonds or die attach.
- After cleaning, use a microscope with as high a magnification as is needed to see individual solder spheres from the paste. Ensure that solder spheres are not present, or, if they are, ensure compliance with the workmanship standard in use. Also ensure that paste beads are not agglomerated in plated through holes. Generally a $\geq 30 \times$ stereo zoom microscope with good lighting is recommended for this inspection.
- If excess solder beads are still apparent, the board should be recleaned. Try rotating the board 90° or 180° for the repeat cleaning cycle.

46.7.3 No-Clean Solder Paste Misprint Cleaning

Cleaning boards printed with no-clean paste requires a little more care, but many of the same rules apply as for cleaning a water-soluble printed board:

- If placement has begun, remove any high value components from the side that is misprinted or smudged.
- If tape is used to mask gold fingers or other features, remove it before commencing misprint cleaning to avoid entrapping solder beads or paste residue beneath the tape.
- Do not wipe solder paste off the board. Instead, allow the paste to melt off in a stream of solvent recommended for the solder paste in the stencil washer or board washer. The solder paste vendor can provide a list of recommended cleaning solutions.
- Rely on the solvent in stencil or board cleaner to wash off the printed solder paste instead of trying to wipe it off.

- Do not use water to clean a no-clean soldered board unless the solder paste manufacturer recommends it, as it may result in a sticky, white residue on top of flux residues. This may impact electrical surface insulation resistance and corrosion.
- Do not use saponified water with no-clean unless the process has been tested for materials compatibility and subsequent corrosion. Surface insulation resistance (SIR) testing and electrochemical migration (ECM) testing are in order.
- Direct immersion should be avoided. Immersion can cause cleaning solvents and diluted flux residues to enter switch and other components. This, in turn, can result in an electrical short internal to the switch (from cleaning solvent conductivity), an open (from insulating flux residue), or corrosion (also from flux residue).
- Do not use ultrasonic cleaning. Even random-frequency or random-amplitude ultrasonic cleaners may impact certain devices that may be already mounted on the board (e.g., on the secondary side). Crystal oscillators are notoriously sensitive to ultrasonic cleaning. Other components have been known to fail as a result of ultrasonic agitation, which can damage or weaken internal bonds or die attach.
- After cleaning, use a microscope with as high a magnification as is needed to inspect for individual solder spheres from the paste. Ensure that solder spheres are not present; if they are, they should be in compliance with the workmanship standard. Also ensure that paste beads are not agglomerated in plated through holes, as they may interfere with PTH component insertion or press-fit operation. Generally a ≥30 × stereo zoom microscope with good lighting is recommended for this inspection.
- If excess solder beads are still apparent, the board should be recleaned. Try rotating the board 90° or 180° for the repeat cleaning cycle.

46.7.4 Do Not Use Aqueous Cleaning on No-Clean Repaired Assemblies

Very often, completed boards that are prepared with a water-clean flux formulation are repaired using no-clean flux. This strategy precludes the need for another wash cycle. In another situation, there may be parts that are on the board or associated with an assembly that are not compatible with a wash cycle (electromagnetic interference [EMI] shields, enclosure pieces, etc.); in such cases, a no-clean repair is necessary. However, once a board is repaired with a no-clean flux, it is important not to subject the board to an aqueous cleaning process for the reasons previously described. Conversely, aqueous cleaning must be used if a water-wash flux chemistry is chosen for repair.

46.7.5 Test for Flux Compatibility

Certain no-clean fluxes are not compatible with each other. Their chemical interaction could result in corrosion on the PCA. To avoid corrosion issues, solder paste flux, wave solder flux, and repair flux should be tested separately and in combination for SIR and ECM with the board surface finish.

46.7.6 Ensuring Full Flux Activation

As mentioned previously, it is important to heat the flux sufficiently to activate it (make it chemically reactive) and to burn off the activators by the end of the soldering process. This is true even of no-clean solder chemistries. If an unactivated, insufficiently heated no-clean solder flux is left on a board, the weak organic acids in the flux residue could cause long-term corrosion. Usually temperatures between 80° and 150°C (depending on the flux composition) are sufficient to activate a flux fully. Avoid pooling flux. To minimize the possibility of corrosion, dispense liquid flux sparingly by using ultrafine artist paint brushes or fine-tipped, ESD-safe felt fluxing pens or a pick. Of course, the best practice will preclude the use of liquid fluxes. Instead, dispense paste flux with an ultrafine artist paint brush or pick, or use no-clean flux core solder wire to minimize flux deposition and for the best chance of reaching flux activation temperature.

46.7.7 Localized Solvent Cleaning

To remove excess flux on a board, cleaning solvents can be used locally. Sometimes these solvents are formulated specifically for flux cleaning and are sold as such. Sometimes pure chemical solvents, such as alcohols, are used. Very often solvents and flux cleaners leave gummy residues on the board. They can also dissolve the no-clean flux residue and redistribute it, leaving a messy-looking assembly. The dissolved residues can find their way into board mounted connectors via capillarity or direct dripping. This can result in electrically insulating flux residues on connector contacts or on board-edge gold fingers. Both can lead to interconnect electrical opens.

Methanol is a better cleaner than isopropyl alcohol (IPA) due to lower water content and it dries much quicker and more thoroughly than the latter. Care must be taken with methanol as it is flammable and its fumes are toxic. This is often true of other commercially available flux cleaners. Although sometimes flux cleaning is necessary or useful, it is best to avoid the need for cleaning by minimizing the flux dispensed and dispensing the flux only to the lead or leads to be repaired. Training, skill, patience, and practice are required for accurate flux dispensing and soldering.

46.8 SOLDERING ATMOSPHERES

The natural heat of the soldering process, combined with the oxygen in normal factory air, can inhibit soldering by creating oxides that compete with the action of fluxes. To avoid this problem, manufacturers can use inert or reactive gases for the local atmosphere at the solder site.

46.8.1 Nitrogen Soldering

Nitrogen (N_2) as a cover gas for soldering processes has been in use for many years and is considered commonplace in today's electronic assembly.[3,4,5,6,7,8,9,10,11,12,13,14,15] Its role in reducing oxidation during the joining process ensures best solder wetting. Soldering in a nitrogen atmosphere has become increasingly important in recent years, driven by the rise of no-clean fluxes and solder pastes and the upsurge in the use of area-array devices such as BGAs, CCGAs, CBGAs, CSPs in reflow, and PGAs, along with complex, finer-pitch connectors in wave soldering.

Area-array devices have one thing in common: their solder joints are hidden by the package body and it is impossible to repair an isolated open solder joint beneath the package. Nitrogen provides added insurance that best solder wetting will result if soldering conditions and materials are process-appropriate.

The most recent increase in N_2-assisted soldering comes with the onset of large-scale lead-free soldering implementation around the world. Higher process temperature translates to higher oxidation rates during soldering. Also, as previously discussed, certain solder alloys are more prone to oxidation than conventional Sn-Pb solders.

46.8.1.1 Use of N_2 for Increased Fluxing Effectiveness. The soldering environment is an important variable. Obviously the most common and least expensive soldering ambient is air, but many board assemblers introduce nitrogen to their reflow ovens, wave soldering machines, or repair equipment. N_2 is particularly effective when working with no-clean solder pastes or fluxes and it will gain increased importance with the conversion to Pb-free soldering. Most all of the Pb-free solders that are in widespread use for the electronics industry are slower to wet and require greater reflow temperatures than the old Sn/Pb standard. The introduction of nitrogen limits further oxidation of the metals to be joined, the solder, and even the flux. This in turn increases fluxing effectiveness and enhances solder wetting. Also, several of the Pb-free soldering alloys are prone to oxidation, such as solders containing Bi, Zn, or In. Reduced dross (oxides on top of liquid solder) production rates are generally seen when nitrogen is used to limit the oxygen content of the wave soldering atmosphere. Numerous citations have indicated that soldering can be effectively augmented by the use of an inert environment.

46.8.1.2 Use of N_2 for Lighter, More Transparent No-Clean Residue. The yellow or brown coloration of some no-clean residues is absent when boards are soldered in a nitrogen atmosphere. Clearer no-clean residues make for easier inspection of solder joints on finished PCAs. It should be noted that the use of nitrogen even keeps the PWB laminate coloration looking the same as before it went through the soldering process.

46.8.1.3 Nitrogen Negatives. There are some negative points that must be considered when working with N_2.

46.8.1.3.1 Cost. There are some negative points that must be considered when working with N_2. First and foremost, nitrogen is an added expense for manufacturing. For factory usage, it is impractical and dangerous to store large numbers of gaseous cylinders or liquid dewars onsite; for this reason, bulk storage is needed. This comes with facility capital costs including: cement pad for outside storage tank and vaporizers to convert stored liquid N_2 into usable room temperature gas. Pressure regulation is needed as well as piping and further pressure regulation for delivery of nitrogen to the soldering equipment. Then there is the incremental cost of nitrogen usage.

46.8.1.3.2 Tombstoning. It has been found that even modest flows of N_2 are beneficial. However, too much nitrogen during reflow leads to passive component tombstoning (see Fig. 46.6).

FIGURE 46.6 Passive component exhibiting tombstoning. Differential heating and surface tension effects have drawn the lightweight chip away from the surface of the PWB during reflow. On solidification, the component is frozen in the raised position and an open circuit results.

During reflow, solder may wet to one end of a chip before the other, usually due to one bonding pad heating before the other. As the solder continues to wet around the one end, the surface tension of the molten solder on that wetted end results in a lifting of the chip off the board before the solder paste goes molten and wets the opposite end. This results in an open solder joint. This defect, sometimes referred to as the "Manhattan Effect," "standing proud," or "drawbridging," is most often seen on the smallest passive devices. The defect is exacerbated by excessive oven atmosphere inerting. Oxygen monitors can be used to sense the amount of nitrogen in an oven. Maintaining a reflow oven oxygen concentration level of about 1,000 ppm to 1,500 ppm O_2 will benefit reflow and minimize tombstoning. Inerting to this level is also significantly more economical since nitrogen consumption is less than when running at lower O_2 concentrations.

46.8.2 Fluxless Soldering

There have been many discussions and investigations into fluxless soldering, but often this term is incorrectly applied. Disruption of the integrity of oxides on the parts to be soldered and on the solder itself is a requirement for solder wetting. If a thermal or thermomechanical method is solely employed (e.g., as by ultrasonics), then it is truly a fluxless method. If a chemical agent is applied for oxide removal, whether it is in solid, liquid, or gaseous phase, it is still a flux-based process. Using cover gases alone does nothing to clean oxidized leads, bonding pads, or solder, but will limit further oxidation.

46.8.3 Gaseous Fluxing

It has already been discussed that flux can exist as a liquid or solid, but it can also be dispensed as a gas. There have been several investigations into the realm of fluxing with various gaseous species, including hydrogen, carbon monoxide, carboxylic acids, and others. Some have been demonstrated to be effective fluxing agents, but either their efficacy or economic impact, or both, is not yet sufficient to displace current liquid flux methods. Investigations continue in this area.

46.8.4 Hydrogen

There has been talk in the industry of hydrogen as a fluxing agent, but at soldering temperatures, hydrogen is predominantly a diatomic, unreactive gas; in fact, even at 1,730°C, it is only 0.33 percent dissociated at atmospheric pressure.[16] Therefore, it is relatively stable chemically, serving ostensibly as an expensive cover gas—an inerting blanket rather than a reactive fluxing agent. H_2 explosively combines with oxygen, so it is typically mixed with nitrogen or another inert gas when used for processing. This mixture, called *forming gas,* is susceptible to separation. Since hydrogen is lighter than air, it rises and may pocket in the upper recesses of the soldering machine or in the room. Its lower explosive limit in air is only 4.65 percent[17] so precautions have to be taken either to keep below this value or to preclude exposure to an ignition source. Its upper explosive limit is 93.9 percent, above which hydrogen is once again considered safe, but this is impractical economically and a gas containment vessel would rule out a reasonable continuous process and affordable equipment.

46.8.5 Carbon Monoxide

Carbon monoxide is an interesting choice. Very dense, it readily displaces air from the soldering environment, but since it is a heavier-than-air toxin with considerable public notoriety, it is unlikely that it will ever be considered an attractive assistive agent for reflow. Also, carbon monoxide does not dissociate sufficiently well at soldering temperatures to be useful for effective fluxing, but does make an excellent, weakly reducing cover gas much as hydrogen does.

46.8.6 Methyl Bromide

This material has been shown to be an effective fluxing agent but is toxic. In fact, it is now a banned pesticide.

46.8.7 Carboxylic Acids

The idea of a gaseous fluxing agent has much merit. It could be inexpensive, eliminate post-solder cleaning, and leave no residue to impede inspection or test. Without residues, the prospect of corrosion would be minimized.

Carboxylic acids in gaseous phase have been demonstrated to be useful in this regard, but it has yet to undergo large-scale testing in a manufacturing environment. Of the carboxylic acids, formic and acetic have received some attention as fluxing agents. They are very simple materials, weak organic acids that decompose to water and carbon dioxide at temperatures above 160°C.[18] Acetic, in its dilute form, is vinegar. Both formic and acetic are widely used industrially and are inexpensive reagents. H. J. Hartmann[19,20] has shown that as formic decomposes, it is a good reducing agent for oxides of solder. The process produces meta-stable tin formate, which further decomposes, yielding elemental tin, water, and carbon dioxide. Acetic acid follows a similar reaction and decomposition cycle, forming meta-stable tin acetate and ultimately elemental tin. Especially attractive is the fact that there is no flux residue on the circuit board after the gaseous carboxylic acid reactive soldering process.

R. Iman et al.[21] have demonstrated formic and acetic to be especially effective for fluxing in the wave-soldering process in conjunction with a light application of adipic acid, a common food preservative and also a known fluxing agent. However, a water rinse was required to remove the adipic residue.

Work by G. Disbon and S. M. Bobbio[22] and K. Pickering, et al.[23] has demonstrated that a plasma can be used to strip oxides from circuit boards and components, and, if they are maintained in an inert environment, they can be soldered without the use of additional liquid fluxing agents. In these schemes, the plasma serves as the oxide-stripping flux.

REFERENCES

1. McKay, C. A., "The Role of Activators in Fluxes for Microelectronics Soldering," Microelectronics and Computer Technology Corporation Technical Report P-I I-405–91, 1991.
2. United Nations Environment Programme, "Montreal Protocol on Substances That Deplete the Ozone Layer," Final Act.
3. Hwang, J. S., "Controlled-Atmosphere Soldering: Principle and Practice," *Printed Circuit Assembly,* July, 1990, pp. 30–38.
4. Stratton, P. F., Chang, E., Takenaka, I., Onishi, H., Tsujimoto, Y., "The Effect of Adventitious Oxygen on Nitrogen Inerted IR Reflow Soldering with Low Residue Pastes," *Soldering and Surface Mount Technology,* No. 13, February 1993, pp. 12–15.
5. Aguayo, K., "Increasing Soldering Yields through the Use of a Nitrogen Atmosphere," *Journal of SMT,* November 1990, pp. 3–9.
6. Aguayo, K., and Boyer, K., "Case History-Utilization of Nitrogen in IR Reflow Soldering," paper presented at the Technical Proceedings of NEPCON West, Anaheim, 1990.
7. Mead, M. J., and Nowotarski, M., "The Effects of Nitrogen for IR Reflow Soldering," *Technical Proceedings of SMT IV-34,* 1998, pp. 34–1 to 34–20.

8. Fenner, M., "Solder Paste for No Clean Reflow," *Soldering and Surface Mount Technology,* No. 13, February 1990.
9. Bandyopadhyay, N., Marczi, and Adams, S., "Manufacturing Considerations for a No Clean No-Residue Soldering Process," *SMART VI,* Orlando, 1990, pp. 398–415.
10. Ivankovits, J. C., and Jacobs, S. W., "Atmosphere Effects on the Solder Reflow Process," *Proceedings of SMTCON,* Atlantic City, 1990.
11. Arslancan, A. N., "IR Solder Reflow in Controlled Atmosphere of Air and Nitrogen," *Proceedings of NEPCON West' 90,* Anaheim, 1990.
12. Keegan, J., Lowell, N. C., and Saxeena, N., "Solder Joint Defect Analysis for Inert Gas Wave Soldering," *Proceedings of NEPCON West,* 1992, pp. 672 690.
13. de Klein, F. J., "Open vs. Closed Reflow Soldering," *Circuits Assembly,* April 1993, pp. 54–57.
14. Morris, J. R., and Bandyopadhyay, N., "No-Clean Solder Paste Reflow Process," *Printed Circuits Assembly,* February 1990, pp. 26–31.
15. Lea, C., "Inert IR Reflow: The Significance of Oxygen Concentration of the Atmosphere," *Proceedings of Surface Mount International,* San Jose, August 1991, pp. 27–29.
16. *Van Nostrand's Scientific Encyclopedia, 4th ed.,* D. Van Nostrand Co., Inc., 1966, p. 870.
17. Weast, R. C., *Handbook of Chemistry and Physics, 49th ed.,* Chemical Rubber Co., p. D–58.
18. Arnow, L. E., and Reitz, H. C., *Introduction to Organic and Biological Chemistry,* C. V. Mosby Company, 1943, p. 182.
19. Hartmann, H. J., "Soft Soldering under Cover Gas: A Contribution to Environmental Protections," *Elecktr. Prod. und Prftechnik,* 1989, H. 4, s.37–39.
20. Idem, "Nitrogen Atmosphere Soldering," *Circuits Assembly,* January 1991.
21. Iman, R., et al., "Evaluation of a No-Clean Soldering Process Designed to Eliminate the Use of Ozone Depleting Chemicals," IWRP CRADA No. CR91–1026.
22. Disbon, G., and Bobbio, S. M., "Fluxless Soldering Process," U.S. Patent No. 4,921,157, May 1, 1990.
23. Pickering, K., Southworth, C., Wort, Parsons, A., and Pedder, D. J., "Hydrogen Plasmas for Flux Free Flip-Chip Solder Bonding," Journal of Vacuum Science and Technology, Vol. A, No. 8(3), May-June 1990, pp. 1503–1508.

CHAPTER 47
SOLDERING TECHNIQUES

Gary M. Freedman
Hewlett-Packard Corporation, Business Critical Systems, Singapore

47.1 INTRODUCTION

There have been many soldering methods devised over the millennia. The most ancient are the soldering torch and the soldering iron, but they are inadequate for assembling today's densely populated printed circuit boards. These are both directed-energy bonding methods, where the heating is localized to a specific small piece of board real estate to accomplish soldering of one or a localized set of joints. Oven reflow or wave soldering are examples of mass soldering techniques, where an entire circuit board is soldered simultaneously (or sequentially in the case of wave soldering). These techniques encompass a much larger area of heating than was possible by the iron. Thus, today's techniques can be divided into categories according to area of heating or number of joints formed simultaneously.

Soldering techniques must be judged by their applicability, cost, and end result. Mass soldering techniques are by far the most common for high-volume circuit board assembly. Some of the directed-energy soldering methods are now catching on in volume manufacturing.

47.2 MASS SOLDERING METHODS

The mass flow or reflow methods are suited for high-volume manufacturing. The entire board is heated and large numbers of components on the board are soldered simultaneously. The two most common of these methods are oven reflow soldering and wave soldering. A third technique, vapor phase reflow soldering, has dwindled in popularity due to environmental concerns regarding the use of the chlorofluorocarbon-based solvents that were key to this process. Now, however, perfluorocarbons are substituted and the technique is still in use.

The choice of soldering method is dictated by the types of components and boards being soldered, the required throughput rate, and requisite solder-joint properties. There are no clear-cut rules. Once the exclusive domain of wave soldering, some plated-through hole (PTH) components are being assembled along with surface-mount components in reflow ovens. Conversely, some surface-mount components are bonded by wave soldering.

47.3 OVEN REFLOW SOLDERING

Oven reflow is primarily used for surface-mount component soldering. To prepare a board for this process, solder paste is deposited and then forced through a metal stencil by either a metallic or polymeric squeegee. The paste contains both solder flux for preparing the metal surfaces for solder attachment and sufficient solder for joint formation. Components are placed in the solder paste on the board and the populated printed circuit board (PCB) is inserted into a conveyorized reflow oven. The oven is set to raise the circuit board and component temperatures gradually. Flux in the solder paste activates with the increase in temperature and strips metal surfaces on the components, board, and solder of oxides that inhibit solder joint formation. Finally, enough heat is imparted for the solder to flow (meaning to liquefy, also known as *reflow*). When built and implemented properly, the oven reflow process results in a controlled and predictable heating and cooling cycle and reproducible solder-joint formation.

Rapid heating of the paste is widely known to be a source of solder ball formation. Solder balls (isolated spheres of solder not necessarily connected to the solder mass of the joint) can be problematic for printed circuit board assemblies. They can induce electrical shorts, especially with finer-pitch components where solder ball diameters may be on the order of component lead or printed wiring board (PWB) pad spacing. Solder paste heating rates must be controlled to preclude solder ball generation. Similarly, the soldering time-temperature profile must be carefully adjusted to prevent excessively high temperatures that can cause flux charring or "caramelizing" and solder-joint degradation. The circuit board itself may also fall victim to the reflow process and deteriorate if temperatures are not maintained properly or evenly. It is for this reason that much equipment development has ensued.

47.3.1 Reflow Oven Subsystems

Even the simplest reflow ovens consist of several subsystems: insulated tunnel, board conveyor, heater assemblies, cooling, and venting (see Fig. 47.1).

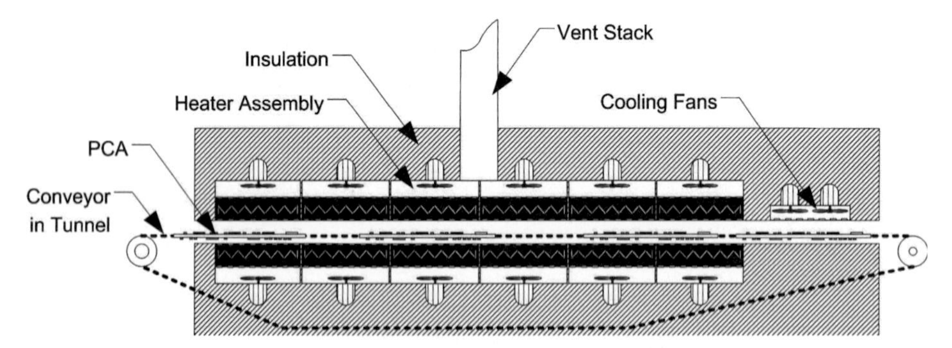

FIGURE 47.1 Cross-sectional view of reflow oven with top and bottom heater assemblies, cooling module, vent stack, insulated tunnel, and printed circuit assemblies (PCAs) atop motorized conveyor.

Reflow ovens have reached a high level of sophistication, and there are many other items that enhance oven suitability for the manufacturing floor. Those items, beyond the aforementioned reflow oven subsystems, are niceties, accessories, and gimmicks offered by oven manufacturers, but are not discussed in this section; however, those subsystems previously listed are reviewed to impart an understanding of the basics of oven construction and operation as well as the most advantageous configurations.

47.3.1.1 Tunnel. The tunnel is a thermally insulated passage through the length of the oven where the board is heated and cooled for a continuous reflow process. It serves to insulate the heaters and boards from the external (room) environment just as much as it is designed to preserve thermal conditions as prescribed by the process and demanded of the heaters. Boards are moved through the tunnel and past multiple heaters by an adjustable constant-speed conveyor permitting controlled and gradual preheating, reflow, and post-reflow cooling of the circuit board.

Consideration of tunnel dimensions is critical for the reflow application. Short tunnel ovens may not permit a profile adequate for attaining prescribed reflow temperature-versus-time profiles for larger, thermally massive PCB assemblies. Tunnel height dimensions must also be adequate to accommodate the tallest components or component heat sinks. Tunnel width will limit the size of the board that can be introduced to it.

47.3.1.2 Conveyors. There are two main conveyor systems used in reflow ovens: pin-chain and mesh belt. One is required and both are recommended for any reflow machine.

47.3.1.2.1 Pin-Chain Conveyor. The pin-chain conveyor, also known as an *edge-hold conveyor,* looks like a bicycle chain with a pin protruding inward from evenly spaced links, as shown in Fig. 47.2.

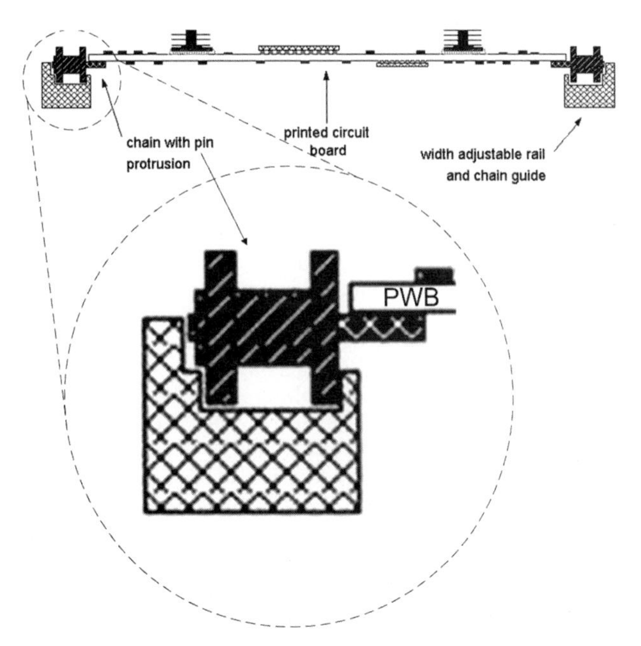

FIGURE 47.2 Cross-sectional view of circuit board carried on a pin-chain conveyor.

There is one such chain on each side of the reflow oven. The circuit board rests on these pins on the inner aspect of the chains and is transported through the oven during the soldering process. The two chains are driven from a common motor, appropriately geared to ensure that the circuit board is conveyed through the oven evenly and to preclude angling and jamming of the board in the oven. Conveyor speed is adjustable and is one of the major factors of reflow soldering in the thermal (time-versus-temperature) profile development.

Pin-chain conveyors are best for circuit boards with components on both sides since the circuit board is conveyed by its edges. This edge-hold method prevents misregistration of parts to be reflowed and also eliminates movement of or mechanical interference with previously reflowed components on the bottom side of the board during transport through the oven.

Of course, an edge keep-out area must be designed into the board such that there will be no pin-chain or component interference.

When reflowing thin, large circuit boards, they may sag when held by the edges on a pin-chain conveyor. The sagging is due to thermal excursion above the glass transition temperature (T_g) of the circuit board epoxy during the reflow process. T_g is the temperature at which a partially crystalline polymer will change from a hard structure to a rubbery, viscous state, as in the case of FR-4 and other such laminate materials. The T_g of most circuit board laminates is in the range of about 135°C to about 200°C, well below the peak reflow process temperatures of most solders. Sagging will be pronounced due to the fact that the board is unsupported at its middle as it surpassed its T_g during transport on the pin-chain conveyor. Copper ground and power planes within the board provide significant support, but may not impart enough stiffness to counter this problem. Sagging alone or together with rail-to-rail nonparallelisms at process temperature can even lead to boards falling off the conveyor. Most of the major vendors in the reflow oven market have conquered the rail twist problem, but hot testing an oven with precisely toleranced test vehicles or even aluminum plates should be performed prior to final acceptance of the equipment or reliance on it. Board sagging can be mitigated, to some extent, by means of mechanical stiffeners affixed permanently or temporarily to the circuit board or by reflowing the board on a pallet. It should be noted that stiffeners and pallets may affect the thermal mass of the board, making reflow more of a challenge. Care should be taken to ensure that the attachment of stiffeners does not interfere either mechanically or thermally with components that may be positioned close to the board edge. Design rules should preclude the use of reflow-challenging components or component fields too close to the circuit board edges. Additionally, some ovens are sold with an accessory steel cable strung the oven length that is meant to serve as an antisag support. An additional adjustable supporting chain is sometimes installed for the same reason and meant to be in contact with the bottom side of the board. The cable or support chain must not contact components on the bottom side of the board, as the chain could dislodge components; thus, clearance must be designed into the board or panel for the support cable or chain. The chain and its supporting rail can interfere with the reflow profile. The board will be cooler in the shadow of this additional rail as it blocks air flow from the bottom blowers.

In a properly designed oven, the pin-chain inflicts little thermal influence on the conveyor. However, some edge-hold systems have been known to impart a dramatic effect on the thermal transfer at board edges, resulting in overheating or heat sinking, depending on the oven's performance and rail position. Some vendors offer rail heaters to counteract this effect. It is best to avoid this type of system because it adds further complexity to the oven and makes the job of process control all the more difficult. This should not be a problem with the conventional pin-chain conveyor because in actuality the pin-chain is only in point-contact with the circuit board. Thermal transfer through these contact points is very poor. Further, most of the newest reflow oven manufacturers have replaced carbon steel chains with stainless steel chains. Stainless steel has better wear characteristics and is a poorer thermal conductor than other commonly available chain materials.

47.3.1.2.2 Mesh Belt Conveyor. Mesh belts used in reflow ovens are generally fabricated of stainless steel links. Some are wide open with large spaces, many centimeters from link to link. Others are more like chain mail armor (Fig. 47.3). The widely spaced links allow more air flow to the bottom side of the board.

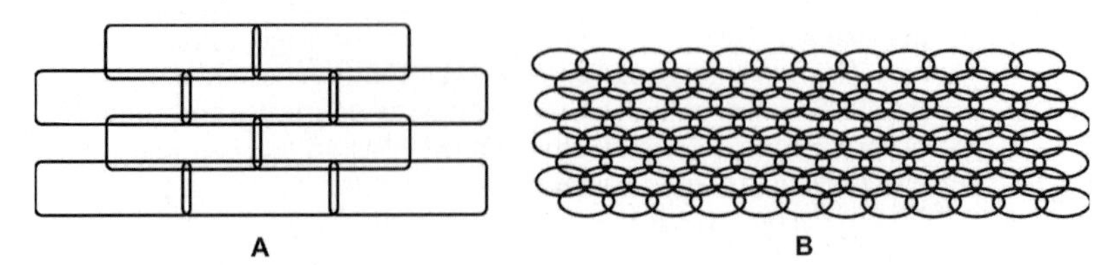

FIGURE 47.3 A large space mesh belt (a) allows more air flow than the "chain mail armor" mesh (b).

The mesh belt is close to the full width of the tunnel and traverses its entire length. Mesh belts offer exceptional versatility, as there is no need to adjust for various board widths and never a concern for board sagging or dropping.

The best situation, however, is a combination of conveyors with the pin-chain running above a mesh belt. In this case, redundancy can save boards, reduce maintenance time, and enhance personal safety. The mesh belt here serves as either an optional means of conveyance for boards populated on a single side or as a safety net to catch a board that may fall off the pin-chain conveyor. Dropped boards are a rare occurrence if the pin-chain conveyor is set properly and the equipment, process, and materials are properly specified and controlled. If the mesh belt were not used with the pin-chain, any dropped boards would "cook" on the heater assemblies, resulting in the release of noxious, irritating fumes. Additionally, the heater or heaters affected could become encrusted with the decomposed board laminate, which may impact their thermal performance. If a thermocouple, the sensing instrument for oven temperature control, is damaged or insulated by the laminate decomposition products, overheating of the zone could occur, causing oven damage and manufacturing downtime.

The biggest problem with mesh belt conveyors is that the bottom side of the board is in direct contact with the conveyor surface. Any components mounted on the bottom side of the board could be damaged or dislodged before or after reflow. While the board is at reflow temperatures, components could be forced into misalignment or nudged off bonding pads by contact with the conveyor mesh. Often, where there are components on both sides of the board, a carrier (pallet) is used to keep the board from contacting the mesh belt. The carrier adds thermal mass, so it is wise to clear unneeded materials or perforate it to allow good air flow during the soldering process. Time-temperature profile development should include such a carrier.

47.3.1.3 Oven Heaters. Each heater's thermal output is sensed via a thermocouple, which is used to close the loop to the heater controller. In larger, production-worthy systems, heaters are located both above and below the plane of the circuit board and are at least as wide as the conveyor. Thermal uniformity across a 60 cm tunnel width can be better than ±2°C on top-of-the-line ovens. This is a function of the tunnel insulation, heater performance, heater control, and convective mixing of the heated air or gas.

There are several heating schemes used in reflow ovens, the result of years of technological evolution. Focused infrared (IR) lamps have given way to secondary emission panel heaters and, finally, to the forced hot-air convective ovens that are the de facto industry standard today.

47.3.1.3.1 Infrared Heater Types. Alternative IR heater types are shown in Fig. 47.4.

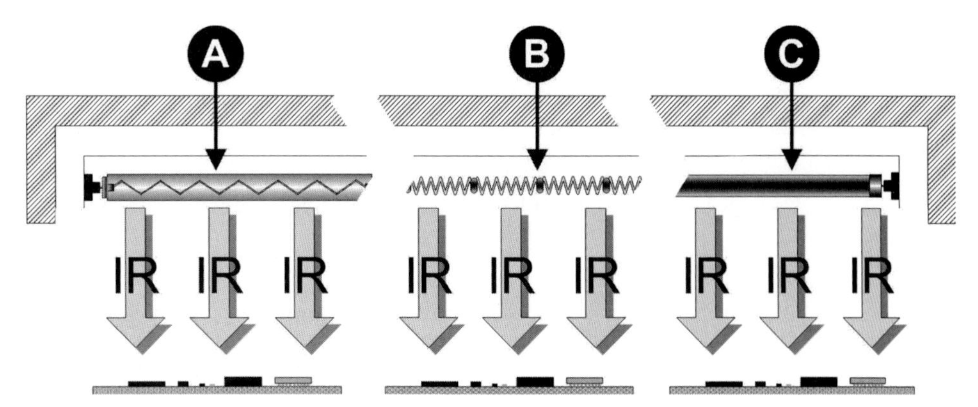

FIGURE 47.4 Three infrared heating schemes for reflow ovens: (a) IR lamp; (b) open resistance wire heater; and (c) resistance rod (calrod-type cartridge heater).

Early ovens utilized focused and unfocused IR lamps, mounted in the reflow oven tunnel. These bathe the solder paste-coated circuit boards and associated components with a broad spectrum of photonic energy heavily weighted to the IR end of the electromagnetic spectrum. The radiant energy absorbed by the printed wiring assembly (PWA) and related materials brings about the thermal increase needed to reflow solder. As the board travels beyond the last reflow heaters at the exit end of the oven, the absorbed heat is lost to the environment or the board is actively cooled by fans. The cooling results in resolidification of the molten solder and solder-joint formation.

Material such as plastic and ceramic component bodies, lead and pad metallurgies, solder, printed circuit board laminate, solder paste flux, and adhesives all absorb infrared radiation at different rates. Therefore, the direct radiative approach is known to cause overheating of some components and underheating of others. For this reason, direct IR heating is no longer the preferred method for reflow.

47.3.1.3.2 IR Panel Heaters. Better than direct IR heating is the secondary IR emitter panel (see Fig. 47.5). It is constructed out of metal or ceramic platens heated either conductively via attached resistance heaters or by direct IR irradiation on the back side of the panel.

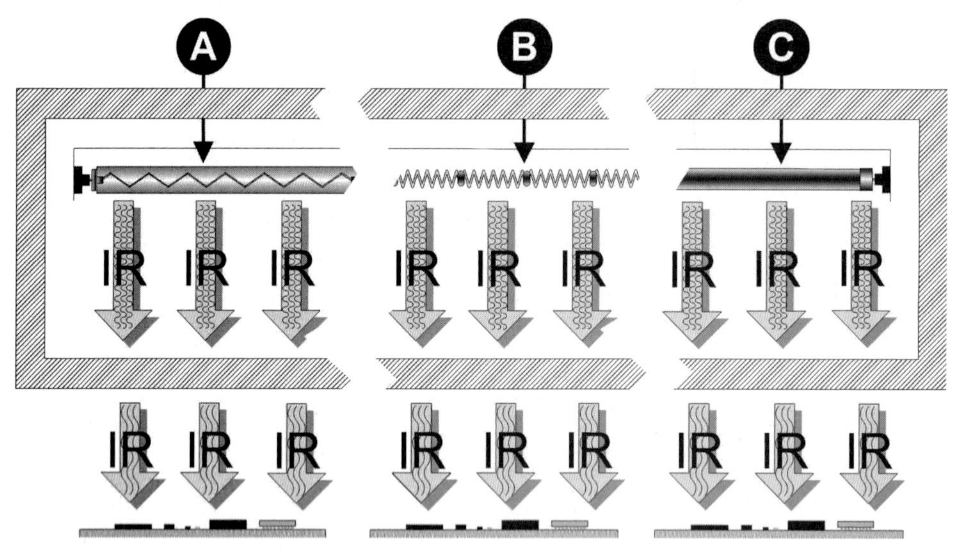

FIGURE 47.5 Secondary IR panels rely on radiant IR sources such as (a) lamps, (b) wire resistance heaters, or (c) heating rods/cartridge heaters. The primary IR is impinged on a baffle and the absorbed short-wave radiation is converted to longer wavelength radiation. This imparts more even heating of the PWB and components during soldering.

During reflow, the circuit board is shielded from direct short wavelength IR impingement from IR lamps or filament heaters. Instead, it is heated by blackbody emission of the heated platen. This results in much longer wavelength IR emission, slower heating rates, and more even heating—a significant improvement over the direct IR lamp oven reflow soldering method.

A variant of this method is the convective IR oven. This technique also relies on radiant or panel IR heating, but the air in the oven is stirred by fans to enhance uniformity of heating. However, techniques such as IR lamp, IR emitter, and combination convective units have been largely supplanted by a more favorable method: forced-air convection, where air is forced through the panel heaters at high velocity. This produces very even oven temperatures and rapid heating response and control. This method is the mainstay of oven reflow.

47.3.1.4 Venting. Oven venting is required to remove harmful fumes from the oven tunnel. If there is too little venting, fumes will not be removed effectively. Too much exhaust will pull in room air through the tunnel entrance, tunnel exit, or both, which can disturb thermal performance. If nitrogen is being introduced to lower oxygen levels, too strong an exhaust draw will negate the effect of the nitrogen or make it difficult to control the oxygen level within the oven. For a more detailed look at venting, see Sec. 47.3.2.

47.3.1.5 Cooling. Just as important as heating the solder, the solder must be brought back to solid state before the board is handled. Otherwise, components sitting in liquid solder could move. Also, as a material such as solder solidifies and crystallizes, mechanical motion can perturb the lattice structure. The resultant solder joints may be of poorer quality and lesser reliability. Some ovens rely upon passive cooling, whereas others have active cooling elements such as fans or even chillers (fans blowing over chilled radiators).

47.3.2 Forced-Convection Reflow Oven

This is the preferred tool for controlled and reliable reflow soldering dependent on heated air recirculated at high velocity in the oven tunnel. The only direct irradiation of the PCBs is by the longest IR wavelengths emitted by the hot surfaces within the tunnel.

47.3.2.1 Forced-Convection Reflow Module. Heating is accomplished primarily through heated air streams directed at the PWB's top and bottom surfaces. This method imparts greater thermal uniformity and controllability of the reflow process than any other reflow oven heating methods. It precludes overheating of the circuit board and/or components caused by preferential absorption common to direct IR heating methods. Although the combination IR-convective method has helped in this regard, it is no substitute for the hot-air forced-convection technology.

Top-of-the-line forced-air-convection ovens have heating modules mounted above and below the board conveyor. They accomplish heating by passing a high-velocity air stream drawn from the tunnel ambient over resistance heaters, through perforated platens. The heating elements are typically shielded from the board to mitigate the effects of any direct or indirect IR impingement on the circuit board. The high-velocity air flows are baffled and regulated to impart a uniform temperature across the tunnel width. Each module is quickly heated and highly controllable. Several heating modules are set side by side along the length of the oven. It is typical to duplicate the forced-convection module array on the bottom of the oven tunnel to impart controllability of the reflow time-temperature profile and to help in even heating of the board from both sides simultaneously. Fig. 47.6 depicts this arrangement.

47.3.2.2 Cooling. Some ovens rely upon passive cooling to lower the temperature of the PCB assembly below the solder liquidus point. In this case, the board traverses a region of the oven devoid of heaters cooling along the way. This is generally adequate for thin, low-thermal-mass boards. But the reality of multilayer, densely packed circuit boards necessitates some means of cooling to ensure that solder joints are solid prior to exiting the oven and before handling the board. To accomplish this, many ovens incorporate an area of active cooling after the last heating zone within the oven (see Fig. 47.7).

Active cooling comes in many forms. Forced-air cooling by fans is the most common. Fans can be deployed on top, on the bottom, or in combination for board cooling. Among the most exotic but the most efficient is the water-to-gas heat exchanger found in some ovens. This provides a stream of temperature-controlled cool air at the exit end of the oven for board cooldown and has distinct advantages in nitrogen atmosphere reflow systems. This method uses whatever process gas is in the oven, cooling it and redirecting it onto the circuit board. Thus, this method does not require additional nitrogen for cooling, does not draw any additional air into the oven that might degrade the nitrogen atmosphere, and precludes the need

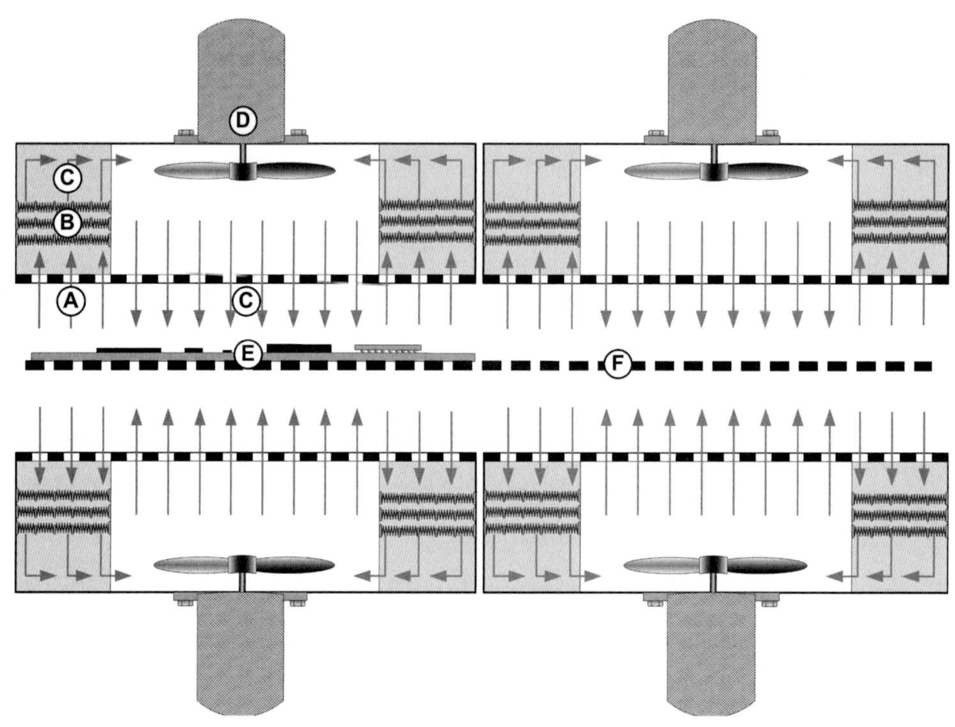

FIGURE 47.6 Forced-convection heating modules. Cooler air (a) is drawn across the resistance heater (b) by a fan (d). The heated air (c) is forced through a perforated plenum tuned for even heat distribution to the PWB (e), which is riding on the conveyor (f). A second set of forced-convection heating modules is directed at the underside of the PWB. Several such modules are arrayed the length of the oven for precise thermal profile control.

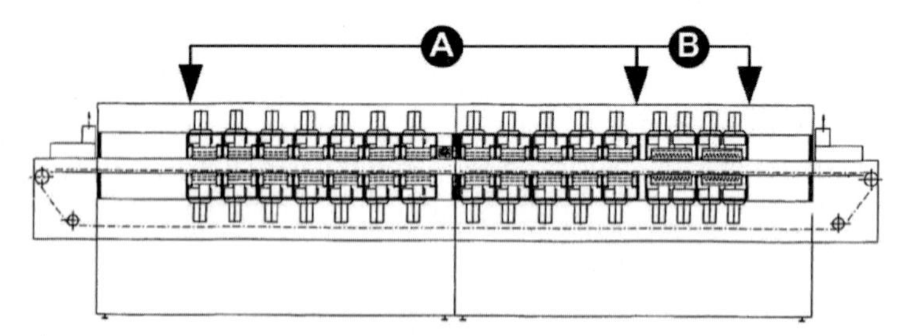

FIGURE 47.7 Forced-convection oven with 12 top and 12 bottom heating modules (a). Each module is independently heated and controlled. The last two top and bottom modules on the right (b) are for active cooling of the soldered circuit board to ensure that solder has reached solidus before exiting the oven. In this machine, cooling zones are aided by fans blowing over water-cooled radiators. *(Courtesy of Heller Industries).*

for cooling fans outside of the oven adjacent to the oven exit. This approach minimizes air turbulence at that tunnel exit and air entrainment into the oven's nitrogen environment. Water-to-gas cooling is efficient, but it must be included in the reflow profile to ensure that thermal ramp rate is correct. Some ovens use polymeric "muffin" fans. These must be kept on at all times to prevent overheating of the fan blades and the motors' plastic fan support structures. Other ovens use air rakes or even air amplifiers to direct a flow of cooling compressed air or nitrogen at the board.

In-oven cooling schemes often result in condensation of flux volatiles and flux decomposition products. As deposits build up on the coolest surfaces within the oven, they change the thermal transfer characteristics of the cooling system. This alters the thermal profile of the oven over a long period of time. Flux condensates may also drip onto circuit boards as they pass through the oven. Nitrogen atmosphere ovens running no-clean s face the biggest challenge in trying to control these condensates. No-clean pastes give off abundant volatiles during reflow. Ambient air entrainment is necessarily low for precise control of the nitrogen ambient within the oven, whereas nitrogen volume is minimized to keep board processing costs reasonable. The result is low gas exchange within the oven's tunnel and little dilution of condensable vapors. Condensates build up rapidly on the cooler surfaces and are baked on over time and difficult to remove. Oven manufacturers have devised many elaborate schemes for flux condensate management. Gas stream filtration in and around cooling zones, replaceable filters, cold traps, and cold fingers with self-cleaning burn-off cycles (kitchen oven–style) have been incorporated into some of the newest ovens. Several ovens have quick-change finned radiators that are swapped out, cleaned, and readied for the next change-out.

Many silicon integrated circuit (IC) packages or passive component devices have specifications for maximum thermal ramp rate. Exceeding this heating or cooling rate may damage die attach materials, crack packages, or otherwise alter electrical performance of the package. This is especially true of ceramic devices. Often the ramp is recommended in the range of 1–4°C per sec. Additionally, some solders like Sn-Ag-Cu (SAC) require cooling rates in this range to allow for quick solidification and to discourage solder alloy segregation. Cooling too slowly encourages redistribution of alloy constituents and silver platelets, or needles can form that may detract from solder-joint strength.

47.3.2.3 Venting. An often overlooked reflow oven subsystem is exhaust venting. This is most important from an industrial hygiene point of view, but it can also have a profound effect on the soldering process itself. During the soldering process, minute quantities of metal oxides accumulate in the oven atmosphere or coat oven tunnel surfaces. The dangers of prolonged exposure to microquantities of lead (Pb) or lead compounds are well documented. With the change to Pb-free materials, the risks are likely lower, but there should still be concern for inhalation of any metal particulates. As previously discussed, during reflow, solder paste volatilizes and releases paste reaction products. Since most solder pastes and fluxes are highly guarded proprietary formulations, the paste and flux vendors may neither accurately disclose the composition reagents nor report the decomposition products in the mandated Material Safety Data Sheet (MSDS). Additionally, if a PCB should fall off the pin-chain conveyor or a component should fall through the mesh belt, it may land on a high-temperature surface—either the heater itself ,or, in the case of forced-air convection reflow machines, the perforated baffles above the heater assembly. Either the board or a plastic component will overheat, decompose, and release unpleasant or even dangerous fumes. A properly designed and implemented exhaust system will mitigate these hazards.

A high-velocity exhaust may influence oven performance. Too much exhaust will cause a significant net flow of atmosphere through the oven. This can result in unwanted turbulence within the oven, upsetting intentional forced-convection patterns and making temperature regulation more difficult, reducing zone-to-zone influence (zone separation), decreasing ease of profile establishment, and resulting in process variability.

Some ovens are set with one central exhaust flue, whereas others may be designed to have ports at both the tunnel entrance and exit. A dedicated manometer or other vent sensor should be installed in each exhaust line to monitor setup and routine exhaust performance. Blast gates in the exhaust line permit flow adjustment. It is customary to mount vent sensors after the flow damping device. Follow oven manufacturer's guidelines for exhaust requirements. Since the exhaust gases are laden with volatile materials, mostly from the solder paste, active exhaust systems should be checked routinely for performance. Impeller blades can foul because of the condensates from these cooling volatiles. Exhaust stack monitors should help pinpoint such problems.

The smoke from a piece of smoldering cotton twine can be used to sense whether there is a net flow into or out of the oven. When venting is correct, there should be a lazy net flow into the oven at both the entrance and exit. Of course, the oven should be hot and fully operational for proper exhaust setup. Similarly, it would be instructive to perform this test using hottest and coolest reflow profiles to ensure that heater module settings do not affect the exhaust conditions for minimal net flow into the oven at entrance and exit. Once the exhaust properties are set, repeat the measurements with dummy boards in the oven and observe whether the properties' presence has an influence on net flow. Fine-tune the exhaust flows and repeat testing as needed. Proper exhaust setting will help achieve consistency in oven reflow soldering and protect plant personnel from exposure to noxious chemical or metal particles generated from the reflow soldering process.

47.3.3 Reflow Oven Requirements

47.3.3.1 Multiple Heating Zones. Reflow ovens are designed with multiple heating zones at the top and bottom along the tunnel length. Each heater or zone is independently settable to a different temperature and with sensor circuitry for stable temperature control. Note that a zone here is the top and opposing bottom heater pair, as shown in Fig. 47.6. Multiple heater zones allow the required adjustment and shaping of the time-temperature soldering profile necessary for successful heating, activating of flux, reflow of solder paste, and cooling of the soldered assembly. The number of heating zones required depends on the thermal mass of the product being assembled; the more heating zones, the more controllable the reflow process.

47.3.3.2 Heater Independence. Zone separation—the influence of one thermal zone upon another—is also essential for controllable and repeatable reflow soldering. One heating zone set to 250°C should have minimal affect on an adjacent zone set to 200°C. There are no set rules for zone separation requirements; however, the greater the zone separation, the more controllable the process. Testing this parameter prior to buying or deploying a reflow oven is necessary for proper process setup. To test, set one heater zone high and adjacent zones lower. Allow the oven to stabilize. Determine conditions that cause the higher temperature zone to influence the lower set-point zones. The lower set zones may rise in temperature above their set point; an indication that the hotter zone is causing uncontrollable temperature rise in adjacent zone(s). In setting the oven for circuit board assembly, avoid the large differential zone-to-zone thermal conditions and stay within the range of safe zone separation as previously characterized.

47.3.3.3 Temperature Distribution. Temperature distribution across the width of the tunnel determines the quality and uniformity of heating. Top-quality ovens have rail-to-rail thermal uniformity of about ±2°C, even for large board widths. Such tight thermal distributions are necessary for controllable and consistent soldering.

47.3.3.4 Heater Responsiveness. Another important characteristic of an oven is responsiveness of the oven's heaters and accompanying thermal controller circuitry. As a board or multiple boards pass through the oven, the oven's heat is absorbed by those cooler boards on their way to reflow. The oven has to respond to that net heat loss absorbed by the train of PWBs with their solder paste and components. The thermocouples built into the oven should sense the thermal impact on the tunnel's ambient and the net loss in temperature from the flow of product in the tunnel. It is important not to space the boards too closely together, as doing so will fully segregate the top from the bottom heaters and alter oven performance. The spacing between the boards in a forced-air convection oven must be sufficient to obviate this top-to-bottom divisional isolation of the heater zones. The spacing is dependent on line throughput requirements and thermal mass of the product design. Board separation may be as small as a few inches, but the effects of board loading and spacing need to be determined empirically and checked to detect any impact on the process time-temperature profile. Any assessment of new equipment ought to include load testing to ensure that the oven can handle anticipated process throughput and capacity. The oven's profile needs to be examined before, during, and after loading of the board train and compared to the performance with the same oven settings for a single board.

47.3.3.5 Ovens and RoHS. The onset of the Restriction of Hazardous Substances (RoHS) legislation and the move to Pb-free reflow has implications for reflow equipment. When using ovens set up for Sn-Pb soldering, there are five areas that the process engineer must pay attention to when making the move to Pb-free reflow.

47.3.3.5.1 Heater Robustness. Check with the reflow oven manufacturer to ensure that the heater and fan assemblies (heater modules) will be compatible with the rigors of the higher oven temperatures associated with Pb-free reflow. Although an oven may be able to reach the higher temperatures, the heater modules may degrade or fail as a result of running close to their maximum rated output.

47.3.3.5.2 Over-Temperature Sensors. Most good reflow ovens are equipped with bi-metallic switches or other over-temperature sensors to prevent catastrophic overheating of heater modules induced by a broken or faulty thermocouple or failing thermal control circuitry. When an oven is set up for Sn-Pb soldering, lower temperature bi-metallic switches are incorporated than when an oven is set up for Pb-free soldering. The lower temperature bi-metallic switches may trip the oven heater circuitry when used with the higher temperatures necessary for achieving Pb-free reflow. This will be most apparent in the highest temperature zone. Check with reflow equipment manufacturer that the bi-metallic switches in an older oven will be compatible with the higher-temperature regimes of Pb-free soldering.

47.3.3.5.3 Firmware. In some older reflow ovens, the computer control heater firmware does not allow the heater modules to be set high enough for Pb-free reflow. If the rest of the oven is compatible for Pb-free soldering, check that heater modules are settable for the higher peak temperatures required of Pb-free soldering. A firmware upgrade may be necessary.

47.3.3.5.4 Seals and Others. Check with the reflow oven manufacturer that soft seals and other oven components in older reflow equipment will be compatible with the higher temperatures of Pb-free soldering.

47.3.3.5.5 Vent Stack. Since vent gases are much hotter when running with most Pb-free solders (due to the higher melting point or melting range of these solders), ensure that all elements of the venting system are compatible with this increase in temperature. This is particularly necessary closest to the oven itself, where vent stack gases are hottest. Check flexible vent ducts, duct seals, flexible vent stack couplings, and stack monitors (manometers, velocimeters, etc.) to ensure that they are compatible or not overheating.

47.3.4 Elements of Reflow

The reflow time-temperature profile is of greatest importance in the oven soldering process. It is the relationship of temperature with respect to time required to bring a PCB assembly to solder liquidus and back to solidus before exiting the oven.

47.3.4.1 Profile Dependencies. The solder paste composition, circuit board lay-up (number of layers, board material), component types, component count, and the number of components per unit area on the board will dictate or influence reflow profile characteristics. Oven performance such as heating characteristics (number of heater zones and responsiveness of each module) will dominate the process and profile. Although the requirements of the in use determine each profile, how well the profile is maintained depends on the materials to be reflowed, the heating characteristics, repeatability health of process equipment, and the attention to detail by the process engineer.

47.3.4.2 Solder. The solder alloy chosen dictates the peak temperature and time required for reflow and wetting. It may also influence the post-reflow cooling rate chosen, as some solders, especially Pb-free compositions, have preferred cooling ramps.

47.3.4.3 Flux and Other Solder Paste Components. As discussed previously, solder paste is typically composed of three main ingredients: solder alloy, flux, and other materials added to help with rheological properties critical for solder paste printing on the PWB. As mentioned, the solder alloy dictates the peak temperature regime, whereas the flux and other constituents dictate the pre-reflow heating ramp rate. Heating too rapidly may prematurely dry out solder paste components. Overly rapid heating may cause the solder paste to spatter. Too little heat may not allow proper activation of the flux. There is no universal profile recipe for reflow for a given solder alloy. The solder paste manufacturer will provide guidance for the time-temperature profile for each solder paste formulation. The reflow characteristics should be consistent from batch to batch for a given paste formulation.

47.3.5 The Reflow Profile

There are four distinct steps in the reflow profile that must be accommodated by the reflow oven for successful soldering. Each step must be tailored for the solder paste composition and must be accomplished in proper, controllable, and repeatable reflow equipment. A misstep at any of the four stages of reflow can result in product loss. Refer to Fig. 47.8 for this discussion on thermal profile.

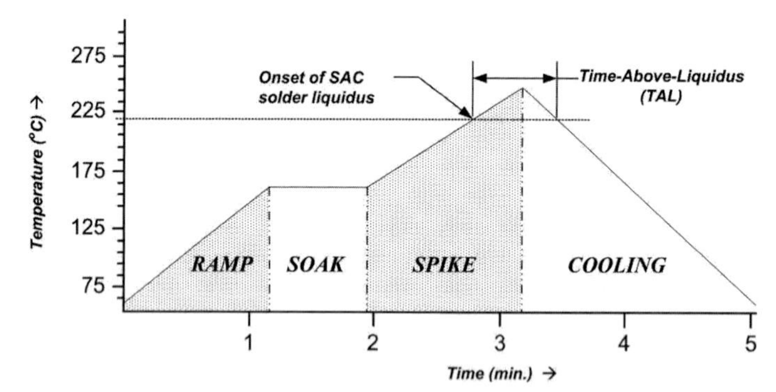

FIGURE 47.8 Schematic of a generic reflow profile for Sn-Ag-Cu solder paste. Time-temperature values are not precise and are shown for illustrative purposes for a discussion of reflow soldering.

47.3.5.1 Initial Ramp. In this step of the reflow time-temperature profile, preheating of the boards, components, and solder paste is initialized. The solder paste begins to lose some of its volatile components and the flux becomes chemically active (activates). If the ramp gradient is too steep, volatiles will be given up too rapidly, boiling will result, and the solder paste will spatter. This can cause explosive solder ball formation and decrease local solder volume, and thus impair bond reliability. Unattached solder balls may form solder bridges between two closely spaced conductors and produce an electrical short circuit.

Overly rapid heating is known to cause component cracking. This is particularly true of ceramic components. To be safe, restrict the ramp rate to between 2 to 4°C/sec., but investigate component manufacturers' specifications as well as the solder paste vendor's recommendation for the maximum allowable heating ramp.

47.3.5.2 Thermal Soak. During this step, the solder, board, and components are further heated. The flux from the solder paste flows onto all metal surfaces in which it is in contact, continues to react away surface oxides, and also acts as a barrier to prevent oxidation. The soak is designed to provide the requisite time and thermal energy for the flux's prolonged chemical reaction with oxides and tarnishes on metal surfaces and the solder. Were the flux to activate fully too early in the process, it could dry out or be spent too soon. If its potency is lost, fluxed metal surfaces will reoxidize in the critical moments before the onset of solder liquidus, and solder wetting (alloying) would be inhibited.

47.3.5.3 Spike

47.3.5.3.1 Onset of Solder Liquidus. In this step, the solder undergoes a phase change from solid to liquid. The liquid metal solder wets onto fluxed metallic surfaces and flows along the component lead-to-PWB pad interfaces drawn by surface tension, capillary forces, and convective flows within the liquid metal. This is the essence of soldering and effective solder-joint formation.

Characterized by a rapid thermal rise, the spike's peak temperature is chosen to be well above the solder's melting point to ensure that all parts of the board, all components, and all solder paste deposits have surpassed solder liquidus onset. As examples, for eutectic Sn-Pb solder, the peak temperature is generally 25 to 40°C above the melting point. For Sn-Ag-Cu solders, the peak is a more conservative 15 to 30°C above the onset of liquidus. The lower temperature overage for SAC solder as compared to Sn-Pb is due to the higher melting range of the SAC alloy family and the potential damage to components and the circuit board due to the higher-temperature regime needed for reflow. Also, since the temperature is higher than for Sn-Pb, oxidation of heated metal surfaces is more likely, as is solder paste decomposition or charring.

47.3.5.3.2 Surface Tension Effects. While the solder is molten, surface tension effects draw component leads into best registration with circuit board pads if the component is light enough to float upon the molten solder. Passive devices (resistors, capacitors, inductors, etc.), most components with formed leads, and plastic or laminate BGAs self-align during this step. Solder surface tension is typically not strong enough to effect self-alignment of heavy ceramic components such as ceramic ball grid arrays (CBGAs) and ceramic column grid array (CCGAs).

47.3.5.3.3 Intermetallic Compound Formation. As previously discussed, intermetallic compound (IMC) formation occurs when the molten solder wets to metal surfaces of component leads and PWB pads. IMC thickness is dependent on the temperature and time at or above liquidus. The longer the solder is heated, the thicker the intermetallic layer, and thick IMCs result in brittle bond lines and degraded solder-joint reliability. For this reason, overheating the circuit board during the spike should be avoided. Time-above-liquidus (TAL) should be in accordance with the solder-paste manufacturer's recommendation, which is typically 30 to 60 sec., although there are no hard and fast rules for TAL. A longer TAL for some components may be evidenced on large thermal mass boards that are slow to heat and heat unevenly. In this case, some devices may overheat while others barely reach soldering temperature. Similarly, some component or board metallurgies may be slow wetting and a longer TAL may be necessary.

Hot air solder leveling (HASL) or reflowed tin on copper bond pads possess a thin intermetallic layer between the applied surface finish and the pads. During surface-mount technology (SMT) reflow, the addition of molten solder on top of this preexisting intermetallic layer will further its growth. Therefore, the TAL should be limited to minimize IMC but sufficient for good solder wetting.

47.3.5.3.4 Materials Degradation. During spike, board laminates, component bodies, etc., can degrade, so trials should be performed to ensure that all materials are compatible with chosen reflow parameters. If connectors are being applied during SMT reflow, make sure that normal contact force has not been jeopardized by distortion or softening of the connector body during the spike. Inspection of connector contact gaps before and after reflow and comparison to the connector manufacturer's specifications will provide clues as to whether the connector is truly compatible with the reflow process. This is most important during the transition to Pb-free soldering due to the novelty of materials and questions of compatibility with higher reflow process temperatures.

47.3.6 Cooling

The fourth and final stage is cooling. It is here that the board's temperature is lowered beyond that of solder liquidus prior to exiting the oven. Once again, recommendations for device heating-and-cooling ramp rates should be heeded. Most electronic packaging and board materials are reluctant to shed heat quickly. If the board is thick, the laminate, a relatively poor thermal conductor, will remain hot. It must be cooled to below solder liquidus prior to exiting the oven to preclude dislodging of any soldered components or disturbing the solder-joint solidification process. On the other hand, with the trends toward thinner circuit boards, slimmer components, and hot, fast-soldering profiles (to keep up with high-volume product demand), it is wise to ensure that the negative thermal ramp requirements are considered for this step also.

As mentioned previously, the solder paste manufacturer will provide recommended profile guidelines, but those are only a starting point. The process engineer should optimize the reflow oven profile with product profile boards and analysis based on actual product runs. Board profiling will be covered in a subsequent section.

Older textbooks and industrial references cite completeness of fillet formation and solder reflectivity as typical hallmarks of a good reflow process, but these are subjective measures that may not be true indicators of solder-joint quality. As mentioned previously, overheated joints will be brittle although they may appear to be well wetted. SAC and other Pb-free solder alloys have a naturally grainy-looking solid structure as compared to that of Sn-Pb, so a dull solder joint is not necessarily indicative of a poor solder joint. Tests such as tensile pull-peel, shock and vibration, thermal cycling, x-ray for solder void content, and metallographic cross-sectioning are much better means of assessing solder-joint quality. There is yet another implication with the move to SAC solders. If not quenched quickly, this alloy is known to result in some constituent separation. Silver may precipitate out as platelets or needles. This can be minimized by using a fast-cooling ramp. Check with component manufacturer's specifications so as not to jeopardize component integrity.

47.3.7 Successful Oven Reflow

Successful soldering is dependent on several factors including suitable, well-maintained, and controlled reflow equipment; good-quality solderable parts; a thermally balanced board designed for the reflow process; a well-tested and reliable solder paste; a proven time-temperature profile; and good thermometry techniques.

47.3.7.1 Adequately Maintained and Controlled Reflow Equipment. The oven should have small thermal differentials across the width of the tunnel and should be capable of adequate heating and cooling ramps as well as adequate zone separation characteristics. It should be stable for a given zone setting, not varying by more than a couple of degrees centigrade. Similarly, the conveyor speed should be stable, not varying by more than 1 cm to 2 cm/min.

47.3.7.2 Good-Quality Solderable Parts. The best reflow equipment, profiles, and solder paste will not make up for inadequacies in part quality or restore solderability. Good solderable parts and boards are a must for successful reflow.

47.3.7.3 Coplanarity. Coplanarity of component leads, BGA balls, or CCGA columns is a must. All component leads should share a common seating plane within the specified tolerances associated with the component vendor's specification. The coplanarity requirement also depends on the dispensed solder paste height and paste deposit uniformity, as the solder paste accommodates some small differences in lead-to-lead coplanarity both as a paste and also as the solder assumes the characteristic dome shape upon melting.

47.3.7.4 Component Storage. Extended storage should be avoided and parts should be kept cool and dry prior to use. Sufficient quality control methods should be established to ensure that the solder and components are solderable. Plastic molded parts such as plastic quad flat packs (PQFPs) and plastic ball grid arrays (PBGAs) can absorb moisture from the atmosphere over time. When moisture-laden plastic packages are subjected to reflow soldering, the entrapped vapor may expand to the point where the package fractures, often damaging the silicon die and associated wires internal to the package. This phenomenon is commonly known as *popcorning* throughout the industry. Ensure that plastic packaged components are stored in unopened bags as supplied by the component vendor until time of use. Once the components are removed from the manufacturer's packaging, follow IPC/JEDEC guidelines[1] for useful shelf-life in the unbaked state and any bake-out requirements prior to soldering.

Parts must be suited for the temperature required by the reflow process. Lastly, ensure that component contact or surface finish is compatible with the solder being used, be it leaded or lead-free.

47.3.7.5 Board Design for Oven Soldering. Thermal balance of the board design is important for reliable and adequate reflow. Ensure that a design for manufacturing (DFM) review of the board has been done and that thermally massive components are not relegated to one portion of the board. Also, it is desirable to distribute smaller components rather than to create component fields to permit uniform thermal balance across the board. Components should be spaced adequately to prevent shadowing of smaller devices on the board by larger, neighboring components. This is particularly important in IR ovens without the forced-convection option. If the oven is characterized by poor uniformity across the tunnel width, the board should be oriented, if possible, to take advantage of the imbalance; i.e., the board edge with greatest thermal mass should be aligned to the oven edge with the highest recorded process temperatures. Board components should be designed far enough away from board edges to preclude any mechanical interference with the pin-chain conveyor. Since the oven rails can have an influence on the reflow or damage of bottom-side components, leave enough edge clearance on the board to accommodate the pin-chain pin length plus some additional margin in case the board is moved from one reflow oven to another.

47.3.7.6 Paste and Profile. The profile provided by the solder paste vendor is only a recommended starting point that needs to be optimized for the reflow conditions and process time requirements. Solder paste with finer solder particles are more troublesome to flux owing to their greater total surface area and corresponding greater surface oxide volume per unit volume of solder.

It is important to limit exposure to high temperatures and particularly time above liquidus, as it is in this regime that the intermetallic layer growth is most vigorous. The thicker the intermetallic layer, the more brittle and less reliable the solder joint will be.

47.3.8 Impact of Pb-Free on Reflow Soldering

Although the move to Pb-free soldering represents a major change for the industry, its impact on the reflow process and reflow equipment is minimal. In many cases, reflow ovens used for Pb-bearing solders can be used for Pb-free reflow. There has been concern that Pb-containing dust collected on the tunnel walls, fans, and such could contaminate Pb-free solder, but it is unlikely that the small number of Pb-bearing particles in a properly vented oven will provide enough contamination to exceed the RoHS limit for Pb of 0.1 percent by weight.

As mentioned previously, to permit Pb-free reflow, reflow equipment may require some oven modifications. There is little impact to the reflow process and most of the changes are material related. Because many of the Pb-free solders are slower to wet out on leads and pads, extended wetting times may be required. Many Pb-free solder pastes are more prone to solder void formation. Extended soak (pre-heat) times and slightly higher peak temperatures are known to help with void reduction.

47.3.9 Good Thermometry Techniques

Understanding the thermal impact of the oven on the board and the board on the oven is critical for a controlled and reproducible reflow process. Thermometry is the only practical method for validating these influences by means of oven profiling. Oven-and-board interactions are explored in this presoldering step and the oven is adjusted for proper time-temperature profile of the PWB with the parts to be soldered. Some of the most important process points determined through board and oven profiling are:

- Heating-cooling ramps and ensuring that they are within the recommended bounds prescribed in the surface-mount device (SMD) component and solder paste specifications
- Peak process temperature and duration
- Dependency on the heater types, quantities, cooling methods on board, and component materials
- Conveyor speed
- Heat distribution along the oven's width
- Heating characteristics of the PWB and the SMD components upon it
- Rate at which the oven reacts to the board's presence

47.3.10 Product Profile Board

All of the preceding and many more variables have dramatic impact on the soldering process. The only way to assess these is to gauge accurately the thermal energy distribution in the oven in the presence and absence of the board. Most process engineers trust their favorite profiling board to determine the health of their reflow oven or assess the performance of a new oven prior to purchase. The profiling board is simply a thermocouple-instrumented product board or test vehicle that is run through the oven for the purpose of determining the oven's effect on the board for given heater and conveyor speed set points. The thermocouples trace the temperature profile of the oven as seen by the component lead or package as the board is conveyed through the oven for reflow.

However, reliance on a product board to determine oven performance may mask some inherent problems with the reflow oven. If the board is not thermally balanced in terms of components or board innerlayers, it may not be able to detect a strong transverse (rail-to-rail) thermal differential. It is for this reason that a balanced oven diagnostic profiling device should be made and run to assess oven performance independent of a product board. That is not to say that the oven diagnostic board is a direct replacement for the product profiling board. The latter is to verify that the critical sectors of the board or, better yet, device leads are heating per set profile. Thermocouples for this board are deployed at the leading edge, trailing edge, center, and sides, with the thermocouple beads ideally embedded within the solder joints of disparate component types. Additional thermocouples can be placed on unusually massive components as well as on adjacent smaller components to determine whether a large component is thermally shadowing its diminutive neighbor. The number of thermocouples deployed depends on the package types, known oven characteristics, board thickness, board layout complexity, etc. The following sections describe good thermometry practices and the oven diagnostic board, as well as some of the thermocouple measuring innovations that are now a routine for large-scale mass reflow manufacturing operations.

47.3.10.1 Oven Diagnostic Board. The oven diagnostic board can be a very simple or a rather complex device. Its job is to probe the normal operating conditions of the oven and determine whether all the heaters and forced-convection fans are working properly. It is useful in assessing not only the oven's time-temperature profile parallel to the tunnel's long axis, but also perpendicular to it at the rail edges. The data recorded from this board are not necessarily be translatable to the establishment of a product time-temperature profile, but are more suited

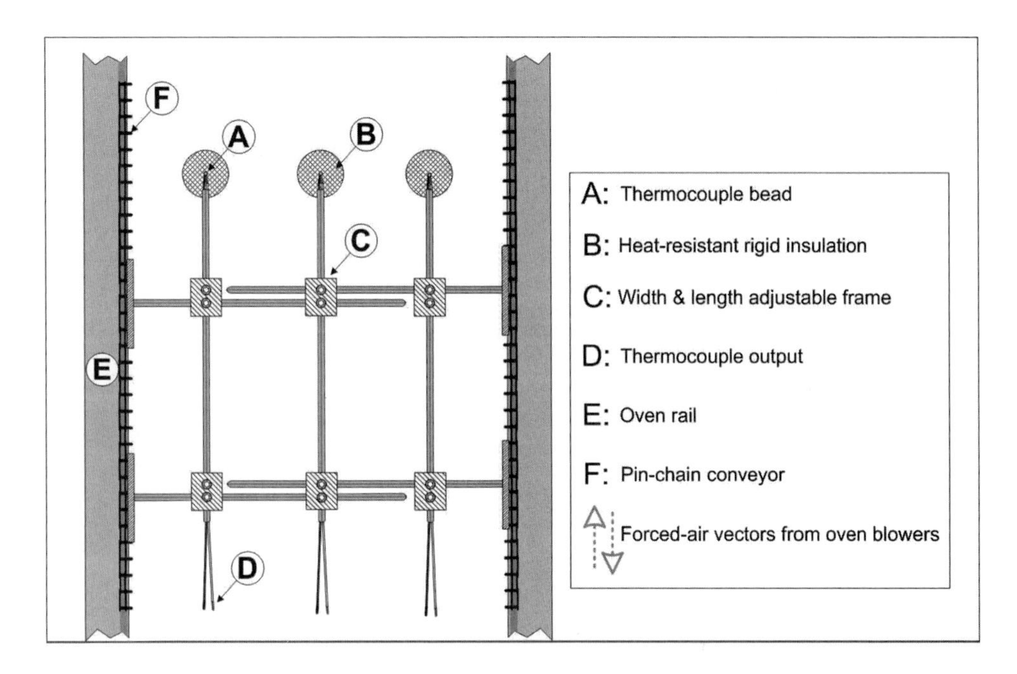

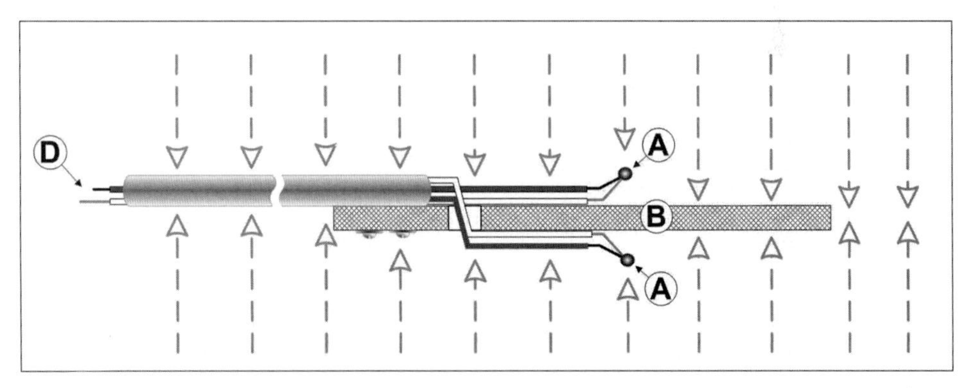

FIGURE 47.9 Schematic plan view and detailed side view of an adjustable oven characterization fixture (oven diagnostic tool) composed of pairs of thermocouples, shielded from one another by a rigid, heat-resistant insulator. The thermocouple positions are adjustable in width and length for investigating temperature distribution along the tunnel width.

to determining oven performance. The diagnostic board, whether a product board or a separate test board or other device, should be thermally symmetrical. It should be made from a low thermal conductivity material, or better yet, a thermal insulator. It should also be of low thermal mass. Attached to this are thermocouples mounted top-side and bottom in an evenly spaced array across the width of the material. They should be mounted close enough to the conveyor rails to gauge their influence on the thermal profile. Thermocouple beads should be identical in size and spacing from the insulating material. Furthermore, they should also be mounted slightly off the insulating material such that the heated oven environment can freely swirl around them. Measuring with such a device provides a true picture of the oven's thermal environment. Figure 47.9 shows a schematic of one such device developed for oven diagnostics.

47.3.10.2 Printed Circuit Board Thermometry. Knowledge of good thermometric practices and thermocouple use is critical to acquiring accurate and useful board time-temperature profiles and oven thermal performance information.

47.3.10.2.1 Good Thermocouple Practices. A thermocouple is composed of two atomically disparate metals or alloys in contact with one another. The wires are joined by twisting them tightly together or preferably by welding (Fig.47.10a,b). When both legs of the thermocouple are held at the same temperature and joined at a discrete point, an electronic difference between the two metals results in a measurable voltage. This electromagnetic field (EMF) is temperature-dependent and once calibrated can be used for accurate temperature measurement.

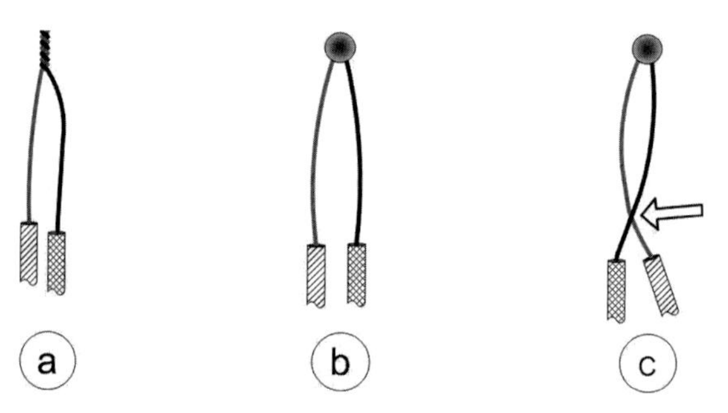

FIGURE 47.10 Thermocouples: (a) twisted couple; (b) welded couple; (c) welded couple with an inadvertent couple (see arrow).

Thermocouples are simple to make. The two different wires need only touch one another to form the couple. This simplicity in construction can also be problematic. If the two wires inadvertently contact (see Fig. 47.10c), thermocouple output will be related to the temperature at that accidental juncture.

In the case of a thermocouple sandwiched between a component lead and PWB pad (see Fig. 47.11) and an accidental junction formation nearby, the thermocouple will register the oven air temperature rather than the conducted temperature of the component lead, the solder, and the PWB. Since the oven air temperature is generally much hotter than the PWB traveling through it, the resultant soldering time-temperature profile (based on the false thermocouple reading) may be cooler than desired and cold (under-reflowed) solder joints may result.

47.3.10.2.2 Welded versus Twisted. A thermocouple wire pair should be welded to help preclude noise-induced measuring errors from intermittent wire pair contact. Small, inexpensive bench-top welders are available for this purpose. The two halves of the thermocouple pair should touch at only one point—that closest to the area desired for temperature measurement. The welded pair has a significant advantage in that the resulting weld bead can be made small and uniform in shape. As a rule of thumb, the weld bead diameter can be made as small as about 1.5 times the diameter of the individual conductors that compose the couple. The temperature will be sensed only at the weld bead. Regardless of whether the couple is made or bought, always check with a magnifier or microscope to ensure the integrity of the bead and that couple wires are only touching at the bead and not at any point below it. During manufacture of a thermocouple assembly, the usual practice is to twist the thermocouple wires together prior to welding. Upon welding, the twisted material will fuse and melt back. After welding, it may be necessary to untwist the pair just behind the bead. Of course, this is risky because many of the thermocouple materials are brittle before welding and are even more so afterward.

47.3.10.2.3 Most Common Couple. The most common thermocouple used for PWB soldering application is the "K" type, Ni-Cr/Ni-Al (chromel-alumel) couple, as it possesses a

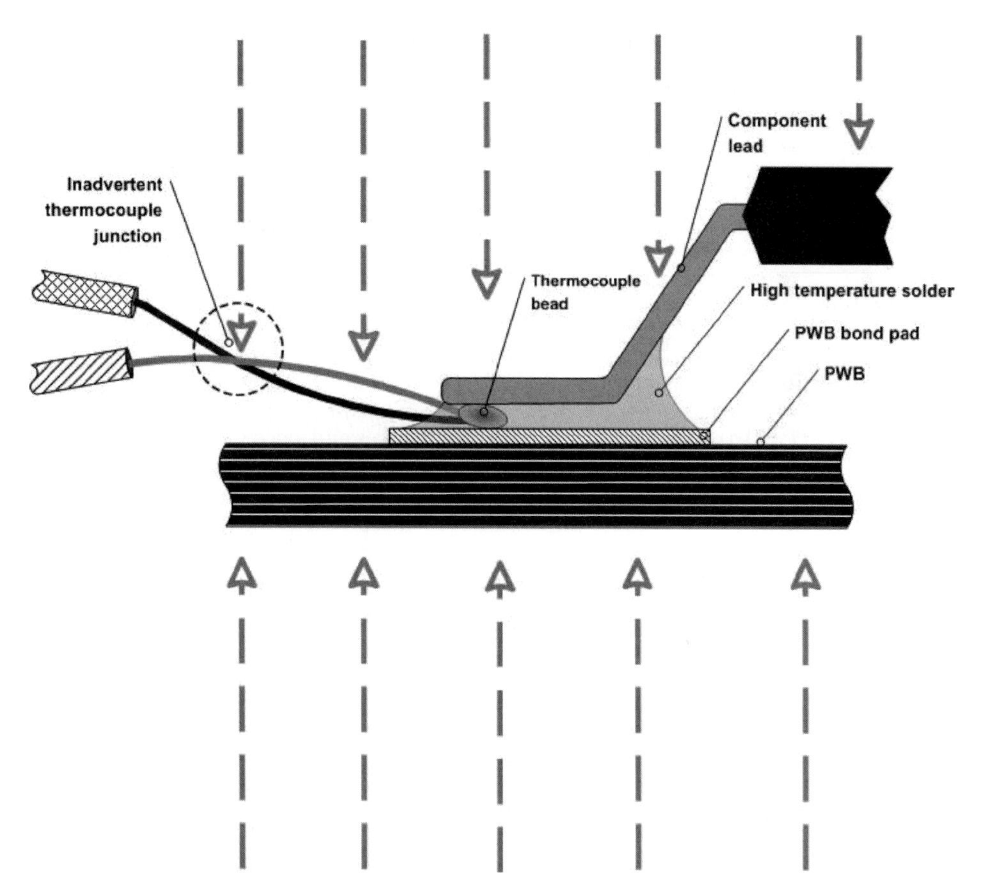

FIGURE 47.11 Arrows indicate impinging forced-convection flows from oven blowers. The inadvertent thermocouple junction (crossed wires) will record oven local air temperature rather than the intended measurement target—the solder-joint area.

thermal measurement range best suited for soldering of circuit boards regardless of alloy type. There are two main devices in use for measuring the output of the thermocouple: the reflow tracker and digital thermometer.

47.3.8.2.4 Measurement Instrumentation. One of the two most primitive measurement devices is the electronic (digital) thermometer with output hand-recorded or tied to a chart recorder. The other (though much more sophisticated) device is the reflow profiler or tracker, an electronic package that can not only output the time-temperature plot of multiple thermocouples but can also make certain assessments of the data, such as heating ramp rate, preheat time, time above liquidus, cooling ramp rate, and other attributes. Tracker systems are computer-based and battery-powered and can generate detailed reports and graphs of a reflow cycle. Most are sold with a thermal protective cover that enables them to travel through the oven, thus eliminating the resistive losses and thermal heat-sinking effects of excessively long thermocouple wires. Some even output temperature by wireless transmitter so that measurement data can be observed in real time.

Once a selected above-ambient threshold temperature has been reached, such as 30°C, the tracker begins recording the experienced thermal environment per unit time. It samples and records the circuit board's thermocouples at short, programmable time intervals (several times per second) for the entire trip through the oven's tunnel. When the tracker emerges from the exit end of the oven's tunnel, it is removed from its protective thermal barrier and docked to a computer. The data stored in the electronic tracker are uploaded to the computer, conditioned, displayed, reported, and printed for analysis. The user defines the reporting format, including the type of data sought, thermocouple plots, and other relevant, run-related data. Some systems are capable of making predictive corrections to the profile based on current measured conditions and recorded oven settings. This can further enhance the ease of adjusting a complex multizoned reflow oven to meet the requirements of a particular job. Whether a simple or complex model is selected, the reflow tracker is invaluable for setting up the oven and checking process repeatability.

When an electronic tracker is used, its presence in the oven may have an impact on the resulting profile data by affecting the local aerodynamics within the oven. The trackers also have thermal mass, but all trackers are sold with an insulating cover to protect their electronics. It is best to check the influence of the tracker's presence as it traverses through the oven tunnel. To do this, use long thermocouple wires and, once the oven has warmed to a steady-state operating condition, run the profile tracker with the tracker a fixed distance behind the profile board. Note that the tracker is usually placed to follow the board in reflow. Were the tracker to precede the thermocoupled profile board, it might cause the heaters to ramp up as the oven attempts to compensate for the thermal mass of the tracker or compensate from the disrupted or deflected air flow. Although that is what happens during a reflow cycle as a cooler board enters a heated zone of the oven, the oven would attempt to adjust for the board's thermal mass. The added thermal mass or disruption of the air flows from the profile tracker would not be a factor in the normal reflow process.

Rather than assessing the reflow tracker effect on the profile, if you maintain the tracker one board length away from the instrumented board, you can eliminate the likelihood of the tracker having any influence on the measured profile.

47.3.10.3 The Profile Board. The profile board is the keystone for any soldering process. It is a board loaded with its complement of appropriate components. Thermocouples are attached to critical locations to ensure that proper soldering temperatures are reached during the bonding cycle. Typically thermocouples are embedded under BGAs (innermost joint and outermost joint) and on fine-pitch and even some passive components. The idea is to ensure that the most difficult, thermally challenging parts are receiving enough heat and that the lightest components are not overheating at the same time. Sometimes it is impossible to achieve the perfect balance, and in such cases the thermal edge is given to the area array components to ensure good soldering beneath while slightly overheating the lighter components.

It is good practice to attach thermocouples to components on the leading edge of the board, the trailing edge of the board, the sides, and the middle. It may take more than one trip through the oven to collect data for all these locations. There is no substitute for a profile board. Some have tried weighing the board for an estimate of thermal mass. Generic profiles with overly hot peak temperatures have been applied and other supposed short-cuts have been attempted. None of these approaches can be 100 percent successful. There is no replacement for preparation and proper thermal assessment with a profile board.

Typically for oven reflow soldering, one profile board is prepared for the secondary-side reflow profile (in which only secondary-side components are mounted), then another for primary-side reflow profile (in which both secondary- and primary-side components are mounted). Another board is prepared for wave soldering (all SMDs loaded plus PTH parts). These boards also double as rework profile boards.

It is wise to attach thermocouples to power and ground pins if that information is readily available. These connection points are the most thermally challenging points for a particular device. This is especially true on thick motherboards where there may be multiply interconnected, thick ground, or power planes embedded within the board, which are slow to heat and slow to cool.

Always allow the profile board and the reflow profile tracking instrument to cool to room temperature before attempting another profile check. Cooling with a fan greatly decreases the wait time.

47.3.10.3.1 Thermocouple Deployment. The size and position of the thermocouple are critical for accurate and reliable thermometry. There are many methods for securing the thermocouple bead to the PWB and some simple rules for proper thermocouple use:

- The junction must be in contact with the object to be measured.
- Each of the thermocouple legs must be kept at the same temperature.
- There should be only one junction (the bead) between the two legs of the thermocouple.
- The measuring technique must not interfere with the outcome of the measurement.

47.3.10.3.2 Thermocouple Contact. The thermocouple bead must be in intimate contact with the lead and pad combination of the solder joint to be measured. Therefore, the thermocouple bead must be inserted between the device lead and the circuit board bonding pad. It is good practice to tailor the size of the thermocouple wire and, thus, the bead, for the scale of the solder joint to be measured. Flattening the thermocouple bead also helps in positioning it between lead and pad. Flattening can be accomplished with smooth-jaw pliers, a smooth-jaw vise, or a hammer. After flattening the bead, check to ensure that it is neither cracked nor broken (see Fig. 47.12).

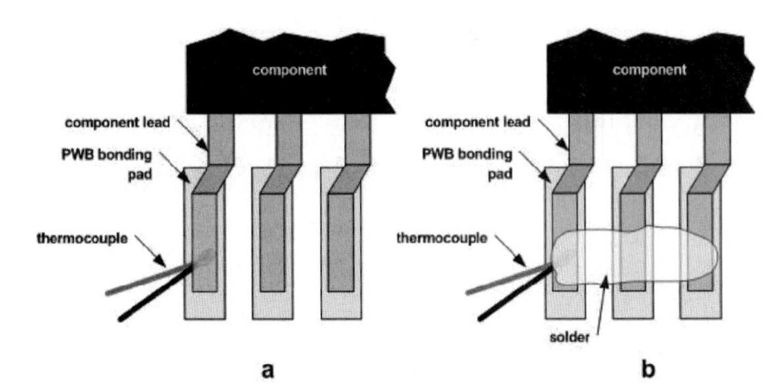

FIGURE 47.12 Checking the quality of the termocoupling: (a) A good thermocouple deployment. The bead is secured beneath a single lead and held in place with high-temperature solder. (b) The thermocouple is held in place by high-temperature solder but the solder has bridged to multiple leads. Temperature indicated will be that of the entire thermal mass of leads and solder rather than a single lead as in (a).

For BGAs and other area-array devices, it is necessary to drill a very small diameter hole through the board and into the BGA ball or column. This procedure is done with a very fine drill bit slightly larger than the BGA ball. The drill bit should be kept short (slightly longer than the board plus ball thickness) to avoid breaking the bit. It helps to drill the hole location before assembly and then redrill to remove solder for thermocouple insertion after reflow. An overly hot thermal profile can be used for assembly of the thermal profile board. Drilling into the ball at that time completes the path for thermocouple insertion. As long as the hole diameter is kept small, there is no need to add solder to retain the thermocouple within the BGA. If the hole is kept slightly larger than the thermocouple bead, the radiated air temperature within the hole and around the thermocouple bead should be the same as the ball temperature. A high-temperature adhesive or heat-resistant tape should be used to anchor thermocouple leads to

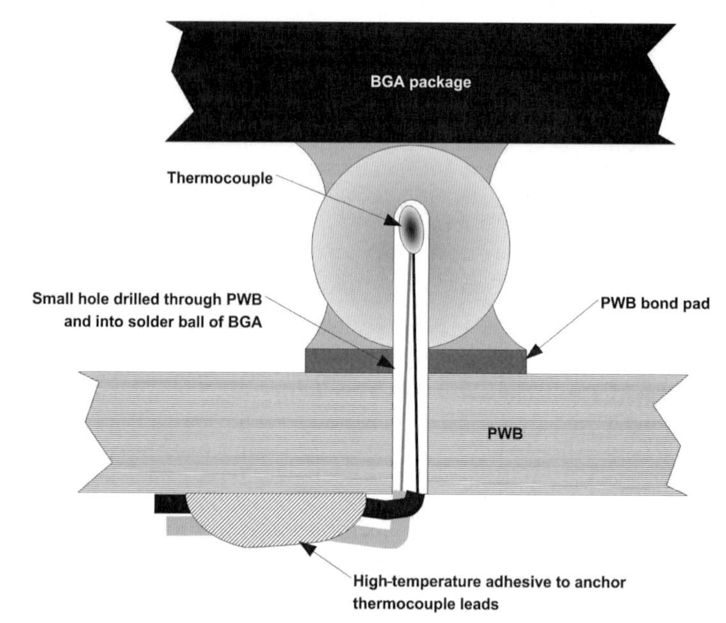

FIGURE 47.13 Thermocouple embedded into BGA ball. Leads are anchored to the board with a high-temperature adhesive or high-temperature tape.

the board to keep them in place and to prevent dislodging of the bead from its intended measurement target (see Fig. 47.13). Note that there is no need to seal the drilled hole into which thermocouple is inserted. It is a miniature blind hole largely filled with thermocouple wires. Air transfer within this hole will be negligible.

47.3.10.3.3 Thermocouple Size. Many factors can confound the solder-joint temperature measurement process, regardless of whether the thermocouple bead is properly deployed between lead and pad. Among these are the size of the bead and the thermal mass of the thermocouple assembly. In general, the use of finer-gauge thermocouples such as 30 to 36 American Wire Gauge (AWG) (see Table 47.1) is recommended. This will permit insertion between lead and pad even at very fine lead and pad pitches. Ideally the thermocouple bead will be completely within the solder joint to be measured.

TABLE 47.1 AWG versus Wire Diameter and Anticipated Thermocouple Bead Diameter

AWG	Wire diameter (in.)	Wire diameter (mm)	Bead diameter (in.)	Bead diameter (mm)
20	0.032	0.81	0.048	1.22
24	0.020	0.51	0.030	0.76
30	0.010	0.25	0.015	0.38
36	0.005	0.127	0.0075	0.19

Strive for the shortest possible length of thermocouple wire pair that is practical for the application. Long, large diameter wire may conduct heat away from the solder joint to be measured, altering its true heating nature.

47.3.10.3.4 Securing Thermocouples: Thermally Conductive Adhesives. If an adhesive is to be used to hold a thermocouple bead, its thermal conductivity must be taken into account. Most thermally conductive adhesives are not nearly as conductive as metal or solder. The best are filled with high thermal conductivity materials such as aluminum, boron nitride, or other thermally conductive ceramics. If the thermal conductivity is not high enough, a false thermocouple reading will result. Thermally conductive adhesives are meant to be applied thinly and conductivity depends on bond-line thickness. It is wise to baseline a conductive adhesive against a soldered thermocouple to ensure that it is not insulating and affecting measurement results. Also ensure that it survives multiple reflow cycles and still maintains it thermal conductivity. Were the adhesive to release the thermocouple bead during the process, then measurement data would be useless. Many adhesives will not tolerate the high-temperature environment of soldering or the continuous thermal cycling encountered in trying to adjust a soldering profile. This is especially true with the onset of Pb-free soldering and the higher soldering temperatures associated with the move away from lead-bearing alloys. High-conductivity adhesives are also expensive.

47.3.10.3.5 Securing Thermocouples: Adhesive Tapes. Tape should be avoided for holding the thermocouple bead to the board as it is known to be an unreliable technique. Problems arise as the tape heats up and stretches, expands, or even releases from the board, allowing the thermocouple bead to float off its measurement target. The tape also shields the thermocouple and target solder joint from the oven ambient, which is contrary to the task at hand—measuring the effect of the oven environment on the intended solder-joint location.

47.3.8.3.6 Securing Thermocouples: High-Temperature Solder. Best practice is to use a high-temperature solder to secure the thermocouple bead to its measurement site. The bead is not really soldered, as the K-type thermocouple wire is not easily wettable. Instead, the solder bead is positioned under a component lead, and the lead is soldered with the high-temperature alloy, merely encasing the bead in the solder.

The chosen high-temperature solder should have melting range or melting point well above the intended peak reflow temperature. An alloy such as 10Sn/Pb (melting range near 300°C) is appropriate as SAC solder melts in the vicinity of 220°C and the peak soldering temperature is generally less than 255°C. Before applying the high-temperature solder, it is advisable to use a fluxed copper wicking braid to remove any solder associated with normal manufacturing assembly of the board, such as SAC or Sn/Pb, as these may depress the melting range of the high-temperature solder.

47.3.10.3.7 Profile Board Components. The profile board can be populated with dummy or electrically failed or reject components. The board can be used over and over and should be saved. Once the oven profile has been established and data recorded, a quick check of the oven with the profile board before a soldering run is useful to ensure that the oven is working properly and reproducibly.

47.3.10.4 Profile Board Spacing. In ovens with pin-chain conveyors, it is common practice to place the profile logging device (reflow tracker) on a dummy board that is trailed behind the profile board. The distance between the profile board and the dummy board or data tracker carrier should be at least one board length. Thermocouple wire length beyond the thermal profile board should slightly longer than one board length (see Fig. 47.14).

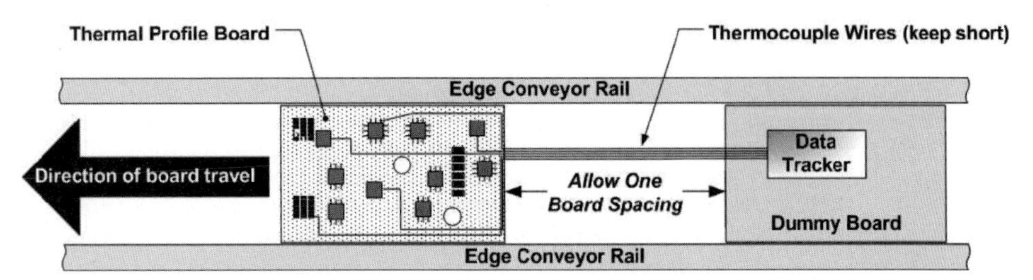

FIGURE 47.14 The thermal profile board should be kept a distance of about one board length from any dummy board or carrier for the data tracker. This enables the heaters to function properly for a board of a given thermal mass.

Were the two boards abutting, the heaters would try to compensate for a much larger thermal mass and overheat. Heater performance would be much different than during steady-state manufacturing and not give a good picture of heater performance during thermal profile. This rule also applies to thermal profiling at wave soldering.

47.3.11 Pin-in-Paste Soldering (Intrusive Reflow Process)

An adjunct to the surface-mount process, this method, sometimes referred to as *intrusive reflow,* allows the soldering of some through-hole (solder-tail) parts into plated-through holes on the circuit board during SMT oven reflow. This process can eliminate or reduce the need for wave soldering—a step prone to defects.

The through-hole components (axially leaded parts, pin-grid arrays, solder-tail connectors, etc.) are inserted into their respective PTHs before or after surface-mount component placement. Once the surface-mount components and the solder-tail parts are placed, the paste-bearing board is then passed through the SMT reflow oven. During the reflow process, the molten solder coalesces around the through-hole pins wetting between the pin and barrel. Surface tension and capillary action draws the solder down the barrel to complete the solder joint.

47.3.11.1 Solder Paste Deposition for Pin-in-Paste Soldering. To prepare a board for pin-in-paste soldering, you deposit generous amounts of solder paste over or adjacent to the circuit board's targeted PTH sites. This is done during solder-paste stenciling in preparation for SMD component placement and reflow (see Fig. 47.15).

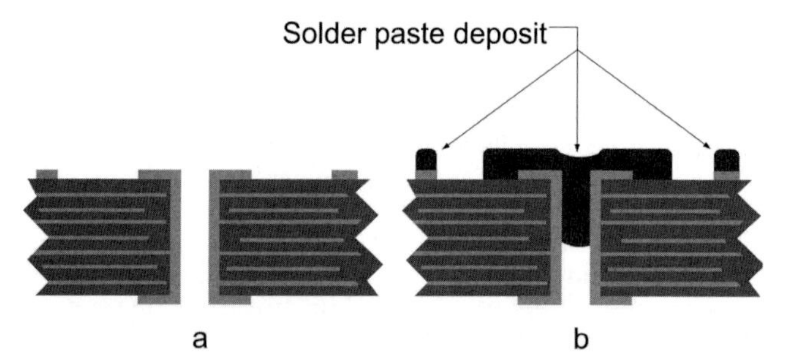

FIGURE 47.15 Solder paste stenciling in preparation for SMD component placement: (a) a PWB cross-section prior to solder paste deposition; (b) the same PWB after solder paste deposition. Note that SMD pads and PTHs have received solder paste deposits. In the case of (b), paste has been forced into the PTH by the squeegee during solder paste printing.

Of course, stencil apertures have to be created in the surface-mount stencil to accommodate paste deposition on PTH sites for pin-in-paste soldering. If solder volume is not a concern, it is desirable not to dispense paste into the PTH. This is discussed later in this chapter.

47.3.11.2 Solder Volume. The inability to apply enough solder paste to meet standard through-hole solder-joint acceptability criteria is one of the major shortcomings of pin-in-paste reflow. That is why this technique is usually relegated to boards ≤1.6 mm (0.063 in.) thick. Requisite solder volume is dictated by component pin pitch and the available printing space between component leads; stencil thickness, which is generally limited by the smallest or finest pitch components on the PWB; and PTH volume and associated component lead displacement volume.

Since requisite solder volume for pin-in-paste reflow is a function of the ratio of PTH volume to component lead displacement, it follows that reducing PTH barrel size is advantageous, especially in the case of thick PWBs. But as the volumetric annulus between lead and barrel is reduced, there is a tendency for excessive void formation. Solder pastes formulated for surface-mount applications consist of solder spheres in a creamy matrix of organic chemicals composed of soldering flux and other materials to help with printing and component placement as required for SMT processing. Voids in the solder are caused by vaporization of solder paste organic components during the reflow process. In an unlikely twist, enlarging a plated-through hole in a thicker board may also benefit the pin-in-paste technique. The paste printing process drives solder paste into the PTH, effectively increasing the solder volume available for joint formation. A double-print cycle forces more paste into through holes. If solder paste is in the PTH, some quantity of it will be displaced as the component's lead is inserted (see Fig. 48.16a).

Upon reflow, the solder generally wets back up the pin (see Fig. 47.16b). Sometimes, though, the solder will form a bead on the pin tip, as seen in Fig. 47.16c). Although electrically it is of no

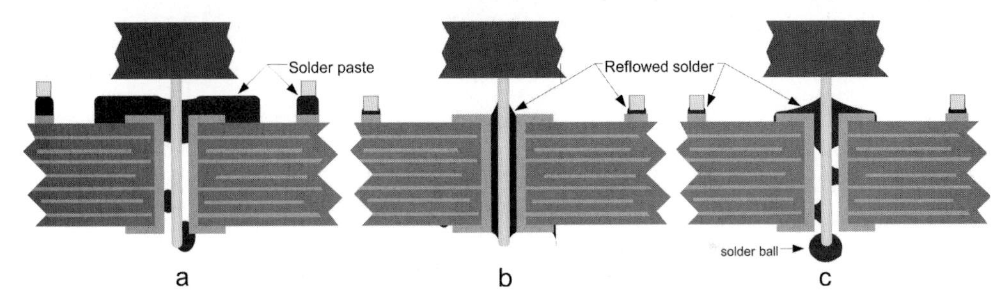

FIGURE 47.16 Board sections illustrating pin-in-paste process: (a) A PTH component pin inserted through solder paste squeegeed in and around PTH. (b) At reflow, the solder has melted and wet to pin and component leads. If solder volume is correct, surface tension effects will draw the solder between pin and barrel and a good solder joint will form. (c) The solder paste displaced by the pin insertion has melted into a ball around the pin tip. So much solder was displaced that voids are apparent between the pin and the PTH barrel.

consequence and well adhered since it is soldered, it can interfere with critical clearances (such as aboard to chassis) or impede proper board seating at in-circuit test or in other fixtures.

A solder paste stencil can be made to occlude solder deposition on top of or in the board's PTH, as can be seen in Fig. 47.17.

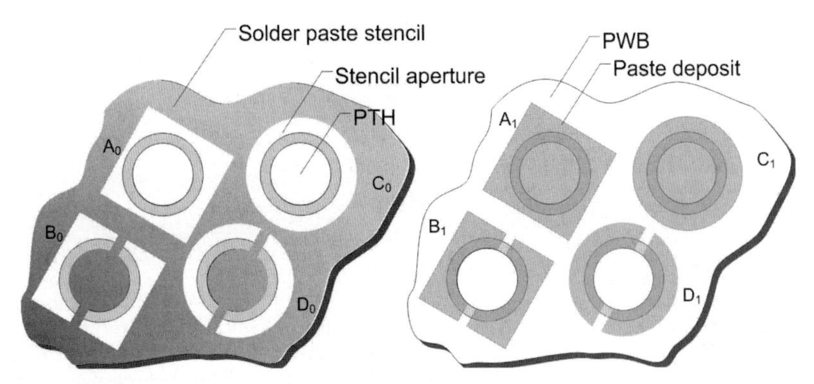

FIGURE 47.17 A_0, B_0, C_0, and D_0 are solder paste apertures cut into a conventional stainless steel stencil. Apertures B_0 and D_0 are designed to prevent solder paste from being forced into through holes. Corresponding paste deposits are indicated as A_1, B_1, C_1, and D_1. C_0/C_1 and A_0/A_1 are usual round and square apertures and deposits, and solder paste is allowed to enter corresponding PTHs.

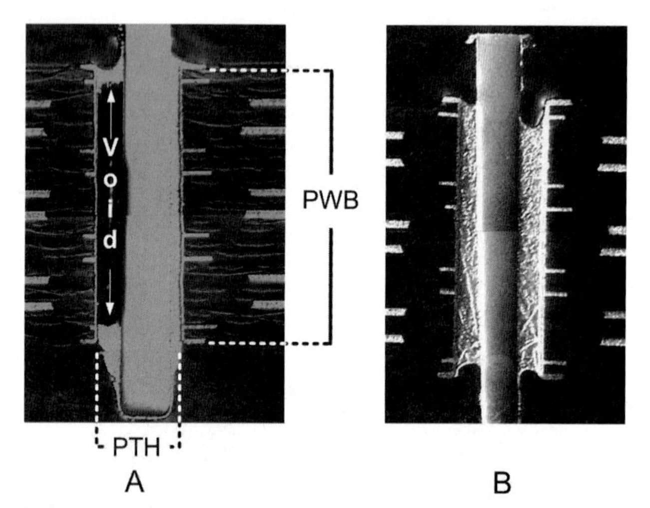

FIGURE 47.18 Solder-joint defects can result from uneven solder distribution: (a) A solder-starved pin-in-paste solder-joint showing large void; (b) a good pin-in-paste solder joint showing nearly 100 percent barrel fill and no voids. *(Courtesy of Hewlett-Packard).*

Stencil designs that preclude solder paste from entering the PTH also result in a significant reduction of solder paste on the top surface and in the PTH. These reductions may make the pin-in-paste technique impractical for some boards. If the solder volume is too low, the available solder distributes unevenly around the PTH pin and the PTH barrel, resulting in voids and inferior solder joints (see Fig. 47.18a, b).

A normal incident x-ray will highlight circumferential voids in PTHs, as evidenced in Fig. 47.19.

47.3.11.3 Pin-in-Paste on Thick Boards. Since the solder volume increases proportionately with increasing board thickness, pin-in-paste soldering is best suited for thin PWBs (≤1.6 mm). The technique is limited by available solder and required solder volume for proper barrel fill. There are two ways to deal with the added volume.

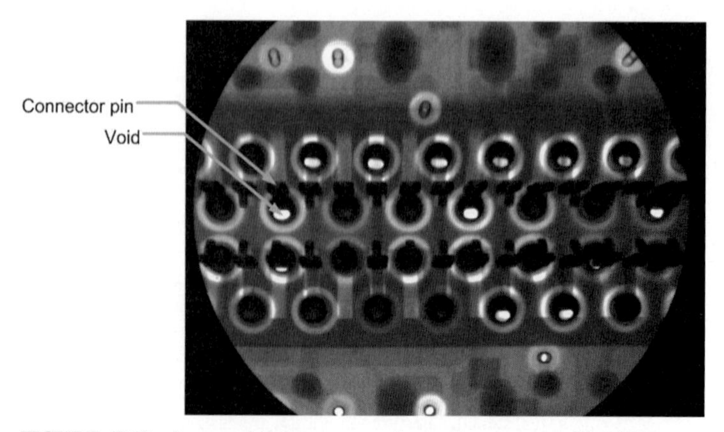

FIGURE 47.19 A normal incident x-ray photo of a connector. The light spots around the pins are areas of no solder or voids (reduced solder thickness). *(Courtesy of Hewlett-Packard).*

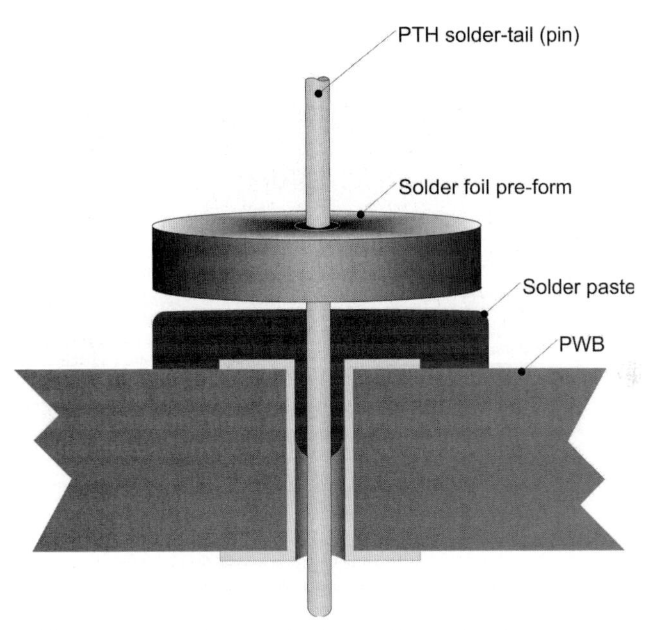

FIGURE 47.20 Application of solder pre-form foil washer to augment pin-in-paste solder volume. Note that the flux from the solder paste is sufficient for pre-form soldering.

47.3.11.4 Addition of Solder through Pre-Forms. Solder foil pre-forms (stamped solder foil) can be added to through-hole parts to augment solder volume (see Fig. 47.20). This approach is expensive and application of the pre-forms is tedious and time-consuming.

47.3.11.5 Buried Intrusive Method. There is a new method for formation of a somewhat nonstandard solder joint in thick PWBs.[2] A foreshortened pin, on the order of that used for a board 1.6 mm in thickness, is used in the thick board. So, too, solder volume is treated as if for a board 1.6 mm thick. When the solder melts, it coalesces and wets to the pin and barrel as in any through-hole soldering process. Conventional thinking would have the molten solder running through the barrel and out the bottom side of the board. Instead, nearly all the solder is retained by surface tension forces around pin and barrel. If the solder volume is calculated properly and aptly delivered, there is 100 percent pin wetting both longitudinally and circumferentially (see Fig. 47.21). Accelerated thermal cycling followed by tensile testing has shown that resultant solder-joint reliability is sufficient and equivalent to conventionally formed pin-in-paste or wave-soldered joints of twice the pin-wetted length.

47.3.11.6 Temperature Compatibility. If pin-in-paste soldering is to be used, check that components are temperature-compatible with the oven reflow process. The high temperatures and long exposures associated with oven reflow soldering may cause unsuited molded component bodies to melt or warp. Connector contact normal force may be affected if the molded connector body softens or distorts. Solder joints or wire bonds internal to some devices may become disbanded, and some, such as electrolytic capacitors, may leak or even explode as a result of an oven reflow cycle. Check the component manufacturer's specification for thermal limits and compatibility with oven reflow soldering.

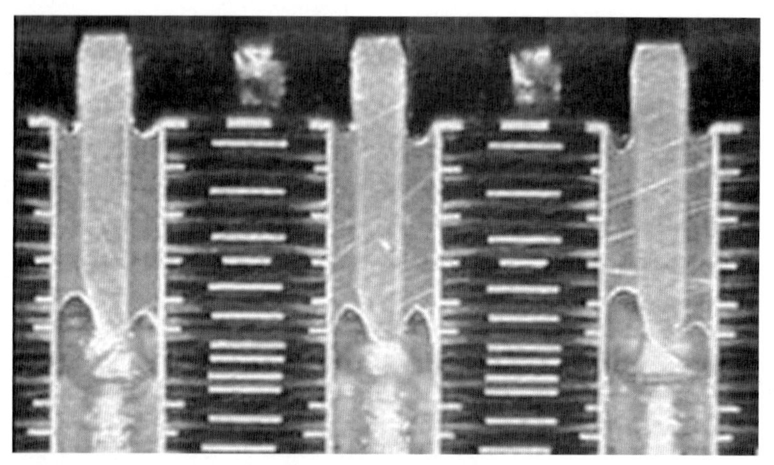

FIGURE 47.21 Cross-sectional micrograph showing pin-in-paste buried intrusive soldering. Note that very little solder has drained.

47.4 WAVE SOLDERING

Once the predominant method for mass assembly of circuit boards, wave soldering has taken a back seat to oven-based reflow soldering. The ease of surface mounting led to a rapid rise in popularity of oven-based technology, and proliferation of surface-mountable packages has led to a decreased reliance on wave soldering. Nonetheless, through-hole componentry persists and mixed-mount (surface-mount plus through-hole) technology still may be the only alternative for some assemblies, and it is unlikely that wave soldering will disappear from PWB manufacturing in the short term. Since wave soldering, compared to reflow soldering, is fraught with defects, it behooves the designer and assembler to minimize the number of PTH components on a board.

47.4.1 Wave-Solder Process Basics

Wave soldering utilizes a reservoir of molten solder pumped and circulated to form a standing wave. The circuit board is prepared with devices for wave soldering in one of three ways:

- Coarse-pitch surface-mount components, especially passive devices, are affixed to the bottom side of the PWB using surface-mount adhesive. The adhesive is cured prior to wave soldering. The adhesive is in contact with the body of the component, which is aligned with its respective pads on the circuit board in preparation for wave soldering.
- Solder-tailed components—such as connectors, PGAs, or other through-hole devices—are inserted into PTHs from the top side of the board.
- Solder-tailed components such as axially leaded devices are inserted from the top side of the circuit board, whereas the leads are clinched on the bottom side of the board.

The circuit board is placed on a motorized, edge-hold conveyor where it is fluxed, then preheated both to activate the flux and give the PWB a thermal boost. The board is next skimmed over the crest of the molten solder wave. Only the bottom of the circuit board is exposed to the molten solder (see Fig. 47.22).

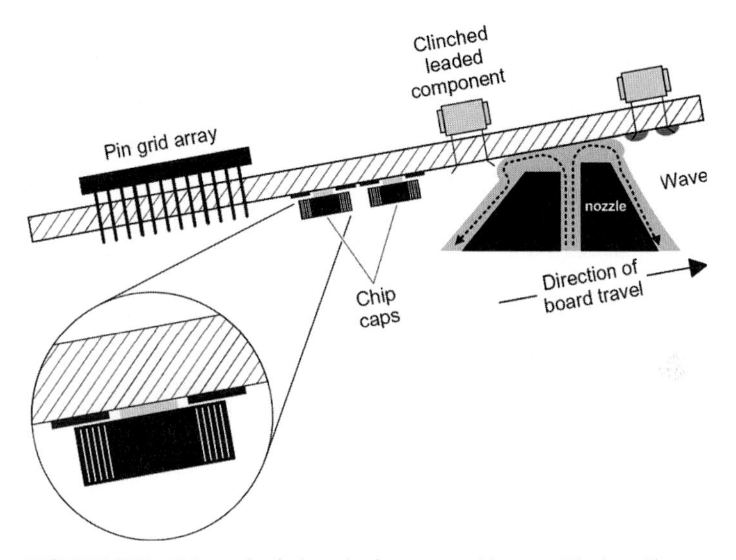

FIGURE 47.22 Schematic of a board going over a solder wave. The board is conveyed over the pumped standing wave of molten solder. Only the secondary side of the board is exposed to the wave. The wave height is set to contact the bottom of the board very slightly.

When exposed to the wave, the adhesive-attached surface-mount devices pick up solder on their metal contacts and the solder bridges from the contact to its corresponding bonding pad on the bottom side of the PWB. In the case of solder-tail components, the molten solder is drawn by capillary action between the lead and the PTH barrel. If the barrel and lead are hot enough and well fluxed, the solder fills up the barrel and wick to form fillets from pin to barrel. As the board continues past the wave, it cools, solder solidifies, and joint formation is complete.

One attribute of this process is the speed at which solder joints are formed. Much quicker than oven reflow soldering, the wave-solder process allows little time for preheating, fluxing, and solder-joint formation, which also explains the variability of this process.

47.4.2 Wave-Solder Machine Subsystems

There are five basic subsystems of a wave soldering machine, including conveyors, fluxers, preheaters, solder pots with pumps and heaters, and ventilation. If alcohol-based fluxes are being used, it is wise to install a fire suppression system also.

47.4.2.1 Conveyor. As for the reflow oven, the wave solder machine is generally equipped with a mechanized, speed-controlled conveyor to transport the board through the soldering process. In the case of wave soldering, the board is always held by a width adjustable edge-finger conveyor, which grabs the board by its edges and neither occludes solder exposure nor interferes with wave dynamics. The stainless-steel edge fingers need to be cleaned routinely to avoid flux build-up.

47.4.2.2 Fluxer. The fluxer is responsible for delivering a uniform application of soldering flux in a sufficient quantity for solder-joint formation. The short preheat and liquidus times characteristic of the wave-soldering process dictate the use of slightly stronger fluxing agents than are used in reflow soldering. Although this is the case, the vast majority of wave solder applications have migrated to the use of no-clean fluxes. For the flux to do its job, it must be heated to an ample temperature to permit best reactive conditions without drying or denaturing it. For the sake of economy, the thinnest flux application is deposited. The flux quantity has to be sufficient to remove oxides from the PTH barrels, solder, and component leads. On the other hand, since post-wave solder no-clean flux residues can confound in-circuit test probe contact, the thinnest application that still permits adequate fluxing is recommended.

Flux quantity may also be of concern in the wave-soldering process for yet another reason: fire hazard. Flux-laden boards are preheated going into the wave. If the flux application is too heavy, the flux may drip onto preheater elements. This may cause the flux to volatilize rapidly, combine with oxygen in the atmosphere, and provide the right conditions for flame initiation. Even if there is not direct exposure of the liquid flux to preheaters, if the quantity of volatile, flammable components is high enough to be an ignition source, then an explosive condition may develop. With the advent of more eco-friendly, water-based fluxes, fire hazard is less of a concern.

Foam and wave fluxing have given way to spray fluxing as the predominant flux application method. All three techniques are discussed here.

47.4.2.2.1 Foam Fluxer. Foam fluxing is accomplished by pumping and aerating a stream of liquid flux through a porous metal nozzle, a fritted glass, or porous stone. The nozzle, also called a *chimney,* shapes the flow of the aerated flux. The board to be soldered is run over the foaming flux and then heated to activate the flux before reaching the solder wave. As it moves past the solder wave, the solder wets to solderable metals and solidifies to complete the soldering process. Foam fluxing is particularly effective for soldering of PTH assemblies. Surface tension draws the foam into the PTH barrels, just as is the case with solder. The resulting flux deposit on the pin and barrel is thin and uniform.

47.4.2.2.2 Wave Fluxer. The wave fluxer works much as the solder wave itself. The printed circuit board is moved over a standing wave of solder flux. The height of the wave and depth of the board penetration into the wave are adjusted to allow for proper flux application thickness. Capillary action, as in all the fluxing methods for through-hole components, draws the flux into interstice between component lead and the barrel. As with foam-fluxing, it is difficult to control the amount of flux delivered. For the most part, flux applied for wave soldering is largely removed by the turbulence of the solder wave; however, it is possible to bake on the flux if the preheater temperatures are set to too high. This may impede fluxing, and the residues can even preclude otherwise solderable contacts from the solder during initial contact with the wave.

47.4.2.2.3 Spray Fluxing. Because of its economy and accuracy, spray fluxing has grown to be the predominant technology. Precise amounts of a low-solids flux can be applied generally to a board or can be selectively delivered to small areas of the board to be soldered. There are two principal methods of spray fluxing: air-spray and ultrasonic. Both are effective and well tested. Ultrasonic methods minimize flux volumes consumed in manufacturing.

If not set properly, spray fluxing may result in a flux-laden airborne mist that deposit wherever the air or process gases carry it. Flux contamination of edge-card gold fingers and connectors is most notable. This can disrupt proper electrical contact when a daughter-card is inserted into a connector. It is wise to cover any exposed gold fingers with an acrylic adhesive polyimide tape[3] or other shielding to prevent contact fouling.

47.4.2.2.4 Fluxer Maintenance. As fluxes are exposed to the atmosphere, they are vulnerable to evaporation of the volatile constituents. This is the case even with water-based fluxes. Evolution of volatiles is significantly enhanced in systems such as wave-fluxer or foam-fluxer, where the flux is open to the atmosphere or processing environment, has significant exposed flux surface area, and is constantly being recirculated. Therefore, the flux needs to be monitored and maintained. Although some automatic systems are now available, most require routine measurement and adjustment of the flux's specific gravity with a hydrometer. Flux thinner must be added to restore the specific gravity to compensate for evaporative losses. In addition, the volume of the flux in the system must be adjusted to the proper level. The fluxer must be maintained to prevent impact on process yield. The flux manufacturer can provide information as to the specific gravity-per-flux target and also recommend an appropriate thinning formulation.

The flux reservoir should be emptied, cleaned thoroughly, and refilled with a fresh charge of flux periodically. Sometimes the flux develops a residue that changes the flux's surface tension characteristics, clogs nozzles, or dispenses pores in foaming systems. In addition, flux may become contaminated with debris conveyed by the PCB to be soldered, which may impact assembly quality. Since the PTH relies on a minute capillary to be filled first by flux and ultimately by solder from the solder-wave process, any small particulates entrained in the flux or the solder may impede PTH barrel fill. Also, while the flux reservoir is empty, it is a good idea to inspect it to ensure that the materials of construction are holding up to the rigors of system operation and prolonged contact with fluxing agents that may prove corrosive in the long term. An inspection of the materials of construction for the entire system should be made prior to committing to the purchase of a wave-soldering machine. Unfamiliar or untested materials should be avoided unless there is sufficient literature, test results, or customer experience to ensure compatibility with fluxes that will be used in that system.

47.4.3 Preheating

Preheating the PWB and components serves three purposes in the wave-soldering process:

- Helps the board and components reach a temperature sufficient for the flux to activate and clean surfaces to be joined in preparation for wave soldering
- Lessens thermal differentials and decreases the likelihood of components cracking due to thermal shock when impacted by the intense heat of the molten solder wave
- Allows the board to ramp quickly to solder-melting and -wetting temperatures when it comes in contact with the solder wave

A component's resident time in the solder wave is 10 to 30 times shorter than for a comparable solder joint formed during reflow oven soldering, so everything in wave soldering has to happen very rapidly. On the other hand, heating time-temperature ramps must be observed in accordance with the component manufacturer's specifications. Components and PWB bonding pads must be fluxed sufficiently to allow adequate solder-joint formation, yet there must be enough flux left on the circuit board after preheating to protect the newly fluxed surfaces until the assembly reaches the solder wave.

Just as in the reflow soldering oven, there are multiple solder-wave preheater styles, but only two are in prevalent use: radiant preheaters (direct and indirect IR) and forced-air convective preheaters. Both are effective and both have their advantages. In fact, the best configuration is a combination of the two. Some wave-soldering machines can be equipped with both top and bottom preheaters. This can be advantageous for thermally massive boards.

47.4.4 The Wave

Numerous solder-wave configurations are available. Rather than provide a detailed discussion of each, this section offers a basic overview of the elements of the process in relation to the wave. As previously mentioned, the molten solder is pumped to form a standing wave. This is accomplished by a spinning impeller on the bottom or side of the solder reservoir. Once the solder is molten, the impeller motor is activated and the solder wells up between baffles and nozzle that reside within the solder reservoir. A combination of impeller speed and baffles-and-nozzle configuration dictates overall standing-wave characteristics. The nozzles and baffles are generally adjustable, as are impeller speed, molten solder temperature, board introduction angle, and board conveyor velocity. These, along with preheater settings, define the profile parameters or process variables that must be tamed to accomplish high-yield wave soldering.

Since the solder is a molten, turbulent liquid, theoretically it can make intimate contact with the underside of the circuit board. Therefore, thermal uniformity is generally not difficult to control as long as the wave contact area along the width of the PWB's underside is uniform. Since the solder is in contact with the underside of the PWB and since the underside of the PWB is coated with flux, the molten solder has the opportunity to wet to PWB lands and component leads. As the process progresses, the solder wicks between the pin and barrel. After leaving the wave, the board cools rapidly, liquid solder solidifies, and solder joints result.

47.4.5 Wave-Soldering Accessories

Solder-wave manufacturers have devised a number of options that may be helpful to the process. Some of these have extended wave-soldering capability to allow finer-pitch soldering or have permitted wave soldering of thicker PWBs.

47.4.5.1 Gas Knife. The gas knife—more commonly called the air knife—directs a high-velocity stream of heated air or nitrogen at a glancing angle to the bottom side of the board directly after the board emerges from the wave. It can be effective in relieving solder bridges from tight interstitial pin fields or closely spaced passive device fields. When the gas knife is used with air instead of nitrogen, increased rates of dross formation may result. Also, if the angle is wrong and the velocity is too high, the exit gases could disrupt the solder wave, resulting in solder opens or shorts. If it is set too cool, the knife will have little effect or may exacerbate solder bridging. Since the advent of the air knife, there has been less dependence on the teardrop-shaped trailing pads used to minimize trailing edge solder bridges.

47.4.5.2 Sonic Assist. Although not in common use, another accessory imparts high-amplitude sonic pulses into the solder wave. This can help drive solder into PTHs and increase wetting, especially on thick printed wiring boards. Care must be taken during setup, as very high amplitudes may pump so much solder up the barrel that it may result in topside solder bridging or solder splashing. This feature can help with barrel fill on thick PWBs. Its value is limited, since it does not affect board heating, which is the primary factor in barrel fill.

47.4.6 Wave-Solder Diagnostics

As in all cases of mass soldering, the use of a product-specific, fully populated time-temperature profile board is necessary to ensure that critical areas of the board to be soldered are maintained at proper temperatures for each stage of the process. Topside SMT component solder joints must be kept well below the melting point of the solder used to join them to the PWB. At the same time, PTH barrels and PTH components must surpass liquidus temperature to guarantee sufficient wetting and capillary rise of the molten solder. With the complexity of today's mixed-mount boards, this is more of a challenge than ever. A profile board should be constructed and instrumented just as for oven reflow soldering, but thermocouples should be attached to both sides of the board and monitored simultaneously during the wave-solder profile run. Topside SMT solder joints should be monitored to ensure that they stay beneath the reflow temperature of the solder alloy being used. Connector bodies should be monitored to ensure that they are not overheated. Generally, for 63Sn/Pb solder, a temperature of approximately 120°C is chosen for maximum preheat to preclude reaching reflow temperature (183°C) on the top side of the board when the board touches the molten wave. Of course, this thermal limit varies greatly with board size, thickness, component type, number of board layers, number of connections to inner planes, thickness of inner planes, component layout density, etc. In the case of Pb-free solder, the preheat temperature should be about 50 to 70°C less than the liquidus temperature of the soldering alloy used for SMT. The use of wave pallets (shields) in the context of selective wave soldering will be discussed later.

In addition to the profile board, there are numerous useful diagnostic tools commercially available for dynamically measuring the wave height and wave contact area and assessing resident time in the wave for a given conveyor speed.

47.4.7 Dross

Hot solder is prone to rapid oxidation at the air-liquid solder interface. Although the wave is in constant motion, the solder is actually flowing beneath a thin, stationary, plastic film of solder oxides. In Sn/Pb, Sn-Cu, or Sn-Ag-Cu systems, the skin is composed chiefly of tin oxide but also contains oxidized traces of other alloy constituents or contaminants. The oxide skin on the surface of the wave and atop the solder in the reservoir is called *dross*. The dross has a beneficial aspect in that it helps to limit oxidation of the recirculating solder in the solder pot, but it also interferes with solder-wave dynamics.

When adjusted properly, the board meets the crest of the wave and disrupts the oxide skin. In doing so, the fluxed components and board are immersed in the flowing, oxide-free molten solder. If all steps are carried out properly, the solder alloys with the fluxed, oxide-free component leads, component pads, and PTH barrels. Upon exiting the wave solder machine, the assembly cools below the solder liquidus temperature and solder joints are formed. The more the system is used to solder boards, the faster the contamination and dross build-up.

47.4.7.1 Impact of Dross. Dross is of concern from three points of view:

- **Dross's impact on the process** Dross is solder lost from the manufacturing process.
- **Economics of dross** In high-volume manufacturing, the amount of solder lost to dross can mean hundreds of dollars of lost solder per week per machine. Of course, the actual cost depends on machine use, solder alloy, and quantity of boards soldered. However, the dross can be returned to some solder founders for recycling.
- **Sn drift** Excessive dross on the surface of the solder can disrupt normal wave dynamics. In the case of Sn/Pb solder, Sn oxidizes more easily than Pb. The solder can become Sn depleted over the long term. This is known as Sn drift. The same is true of Sn-based Pb-free solder alloys, but since some alloys are nearly all Sn (SAC305 with 96.5 percent Sn by weight), this effect will be less dramatic. However, other Pb-free alloy perturbations such as dissolution of Cu become much more important. For example, in the case of the eutectic Sn-Cu alloy, the copper content is only 0.7 wt percent. Small changes in the copper content by means of copper dissolution of copper pads on the board can have a dramatic effect on the melting temperature of the solder.

47.4.7.1.1 Dross-Related Process Defects. Dross can generally change the dynamics at the wave-land-lead interface, discouraging adequate pull-back of the solder and encouraging defects. Molten solder droplets can become entrapped in the dross as a result of wave turbulence and mixing of the oxides with the molten solder. Once oxidized, they are unable to rejoin solder in the reservoir. As the dross passes through a PTH pin field intended for soldering, it is raked by the pins; the pins acting as a strainer. The dross containing liquid solder can hang up on the pins and cause shorting (an effect known as *drossy shorts*). This dross-thickened layer can also block the solder wave from effectively contacting PWB land and component lead, and opens (skips) result.

47.4.7.1.2 Taming Dross. Various schemes have been devised to control dross, such as co-mixing the pumped solder with a mineral oil, which floats to the surface, blanketing the solder from the atmosphere. Liquid reducing agents can be added to the solder as well as fluxes. In the long term, none of these has proven popular. The most popular method of dealing with dross is manually scooping dross from the surface of the solder reservoir with a ladle. Machines that vacuum dross are also available.

Inerting is effective in reducing dross formation and ensuring best fluxing from no-clean or other weakly activated fluxes. Nitrogen as a cover gas has become the most effective method of dross management. R. Iman et al.[4] have explored the use of formic acid in gaseous form as an additive to the nitrogen atmosphere for the soldering process. The fundamentals of this process were demonstrated by H. H. Hartmann,[5,6] who had shown formic acid to be an effective gaseous fluxing agent for soldering. Dross forms when the solder is exposed to the atmosphere, so limiting atmospheric contact limits dross production. As mentioned previously, nitrogen blanketing of the solder pot and the wave will help in this regard. Another simple but effective method is to turn off the wave pump(s) when the solder wave is not in use.

47.4.8 Metal Contaminants

Metal contaminants can also affect wave soldering. These stem from two sources: contaminants from the bar solder used to charge the solder pot, and, to a greater degree, from the materials joined during the soldering process. As the hot, circulating wave washes over component leads and PCB through-hole lands and pads, materials are dissolved into the solder. Even if leads and pads are solder-plated or solder-dipped, there is the opportunity for adulteration of the solder in the reservoir, dependent on coating composition, thickness, and underlying basis metals and impurities within surface finishes as well as resident time in the wave and wave temperature. Copper, gold, silver, additional tin or lead, and intermetallic compound precipitates are all common contaminants derived from the slight dissolution of component lead and pad or coatings thereon during soldering. See Chap. 46 for discussion of these metals in the solder process.

Solder reservoir contamination can have significant process impact, eventually altering melting range of a solder, injecting particulates into solder joints, and altering wetting characteristics. This may lead to an increase in shorts or opens or even result in solder-joint embrittlement. The composition of the wave reservoir can be assayed by a testing service to determine its impurity content by scooping a small amount of solder and checking its solidus and liquidus points (melting range) against a pure standard. This method is less accurate than chemical analysis for contamination assessment and will not be adequate for detecting the presence of some impurities such as intermetallic precipitates that may not affect the solder alloy composition.

Excessively high temperatures in the solder reservoir and low speeds through the wave should be avoided to limit dissolution of component leads, PWB pads, and PTH barrel materials.

Note that the solder wave is maintained at fairly high temperatures, generally 245 to 265°C, depending on the solder alloy, but board exposure is short (2 to 10 sec.). Compare this to oven reflow, where time above liquidus is typically 30 to 90 sec. As in oven reflow, thermal shock in wave soldering can lead to component cracking or degradation problems, so the preheating rate should be tempered to match component vendor recommendations—sometimes to as little as 2°C per sec.

Bar solder impurities such as aluminum, gold, cadmium, copper, and zinc can increase surface tension of the solder and make the process more prone to bridging. D. Bernier[7] reviews some of these contaminant effects and describes empirically derived impurity limits.

47.4.9 Hygiene

Airborne tin, lead oxide fumes, and other metal particulates are unhealthy to breathe. Risks associated with lead oxide intake are well documented, but inhalation of any particulate should be considered potentially hazardous. The health risks associated with working with Pb-free alloys are not as well known as those of the Sn-Pb system. With any alloy, precautions should be taken, especially during wave-solder system maintenance procedures, to preclude particulate inhalation. Donning a personal particulate mask and washable or disposable outer garments (contamination suit) is recommended. Proper hygienic venting of the work area, not only during maintenance procedures but for normal soldering operation, is necessary.

47.4.10 Design for Wave Soldering

In most factories, defect levels at the wave step are higher than those for the oven mass reflow process. The defects are related to poor process setup, poor process control, inadequate PWB design, or any combination of the three. Although wave soldering has been around for a long time, it is still not very well understood, due mainly to varying machine configurations and number of process variables.

Many wave designs are available from the various wave-solder machine manufacturers. There are wave machines that provide multiple smaller, turbulent wave(s), which are best for leadless components such as surface-mount passive devices (resistors, capacitors, etc.). Smoother-flowing waves are recommended for components with leads and through holes as well as coarse-pitch surface-mount devices. Wave dynamics are dictated by process values as well as the materials in contact with the wave. As the solder wets to the circuit board materials, solder wetting contact angle and solder viscosity imposes wave peel-off characteristics (see Fig. 47.23).

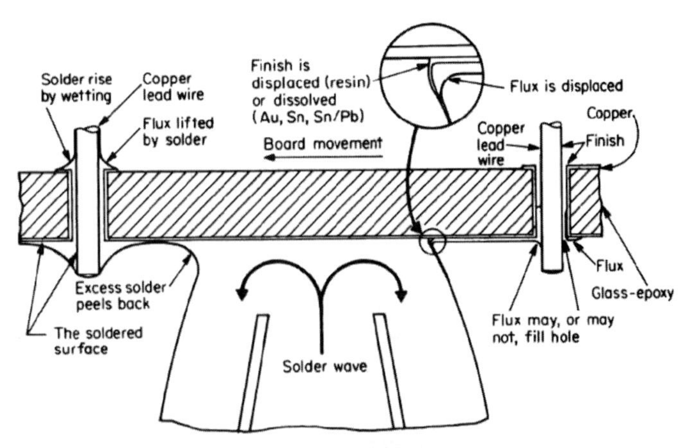

FIGURE 47.23 PWB passing over a solder wave. Note the tendency to pull a web of solder from the wave as the board passes over. The solder-wettable lead-pad combinations become one with the wave as the board passes through it. Good soldering relies on flux-aided wetting and the ability of the solder web to break cleanly from the wave when surface tension forces are exceeded as the board moves away from the wave. *(Courtesy of Alpha Metals, Inc).*

The hot-air knife, as previously described, is sometimes useful in removing unwanted solder bridges from between component leads while the solder is still molten. This is especially useful in connectors and PGAs with interstitially arrayed pin fields (see Fig. 47.24).

As the wave moves through this frustrated interstitial pin field, areas in which fluid flow stagnates may occur and opens or insufficiently soldered pins may result. In addition, the interstitial pin field may not drain solder well enough in other areas, causing solder bridge formation. Numerous wave designs are aimed at improving the soldering yields of these large packages, but there is no clear choice in wave style for all applications.

In designing circuit boards for the wave-soldering process, it is important to anticipate the flow of the wave over the bottom surface of the board. Placing tall components in front of short component fields should be avoided and component spacing should be maximized; otherwise, solder from the wave may be blocked from washing over the targeted joining area. The taller component may shadow the flow of the wave, causing areas of flow eddying or stagnation in the vicinity of shorter components behind it. Also, via placement should be as far away from the leads as is practical to avoid the aforementioned problems of re-reflow.[8]

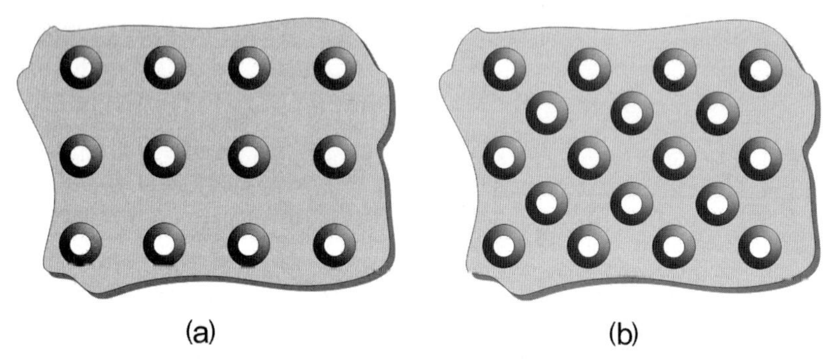

FIGURE 47.24 Circuit board segments with (a) an orthogonally arrayed connector PTH pattern and (b) an interstitially arrayed connector PTH pattern.

Component spacing from the edge of the board is dictated by wave flows and the surrounding structure. F. W. Kear's review[9] of the thermal aspects of through-hole solder-joint formation offers a glimpse of the physical phenomenon of the wave-soldering process.

47.4.11 Wave-Solder Pallets

The pallet, selective soldering pallet, or shield is a masking device meant to shield certain secondary (bottomside) components or features of the PWB during the wave soldering process. On its topside are a recessed nest for the PWB and retention features to hold the PWB within the nest. There are cutouts that expose those areas to be soldered to the flux and the standing solder wave (see Figs. 47.25 and 47.26).

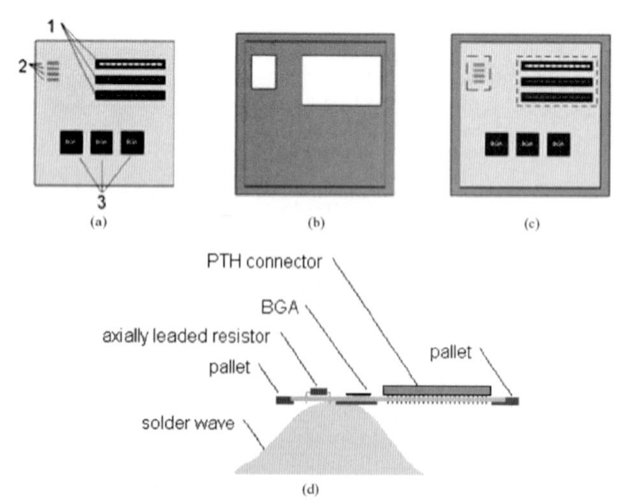

FIGURE 47.25 (a) A wave-solder pallet: (1) a PWB with PTH connectors; (2) PTH axially leaded resistors; (3) BGAs. (b) A pallet with cutouts for PTH components. It also shields BGA via pattern and other components and features on the bottom side of the PWB. (c) A PWB nested in a pallet. Dashed lines indicate areas in the pallet that were cut out to expose the bottom side of the board to the wave. (d) A masked board passing over the solder wave, showing open and shielded areas.

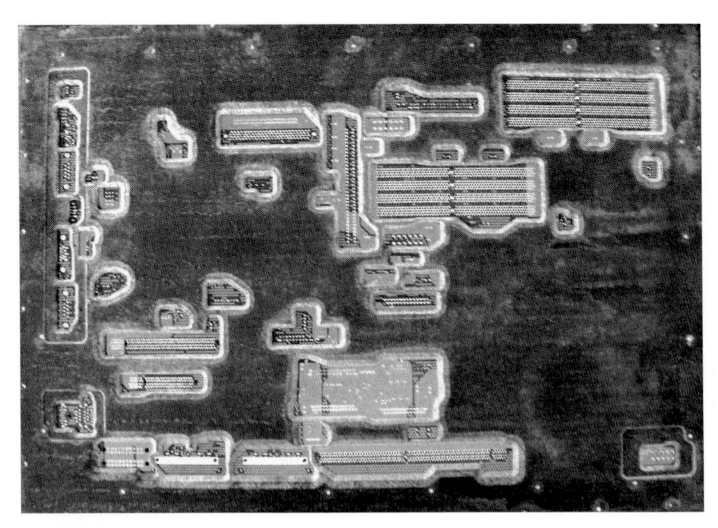

FIGURE 47.26 A wave-solder pallet for a PWB. The light areas are cutouts that allow the component solder-tails to come into contact with the solder wave. The dark areas shield the board, component vias, and critical components from the wave. *(Courtesy of Hewlett-Packard).*

The typical wave solder pallet is made of a nonwetting, electrostatic discharge (ESD)–safe, thermally insulating material such as Delmat or another high-temperature epoxy-glass composite.

47.4.11.1 Pallet Uses. A pallet may serve several purposes, including the following:

- Bottomside SMD solder-joint protection
- Via and via field shielding from solder[10]
- Edge card finger shielding from solder
- Press-fit connector site shielding
- PWB distortion prevention
- Test-point protection
- Plated chassis hole protection

47.4.11.2 Wave-Solder Pallet Design. Although the wave solder pallet can prevent damage to a board, if not designed properly it can also interfere with soldering. There are only a few rules governing wave pallet design:

- **Limited pallet thickness** The pallet must not be too thick. The design must not hold the board so high as to preclude the molten solder wave from contacting areas to be soldered.
- **Largest possible pallet openings** Pallet openings should be as large as possible for good PCA preheating and to permit free access of the molten solder to the targeted areas of solder-joint formation.
- **Clearance between cutout walls and component pins** Clearance must be provided between the edge of the pallet cutout and the component pins to be soldered. If the pins are too close to the pallet opening wall, the wave may not contact the pins. Further, the pallet is somewhat of a thermal sink and inhibits the thermal rise necessary for soldering.
- **Board locks and overclamps** Board locks to clamp the board into the pallet are useful. If none are included, the PWB will float up when contacted by the wave. Overclamps may be

needed to keep the board tight against the pallet during the wave-solder process. Overclamps are also useful to hold connectors and other nonclinched parts from buoying up when contacted by the wave. Care must be taken that the overclamps and board locks are not so massive or otherwise interfere with preheating of the board.

- **Pallet support** The pallet design should include enough areas of support under the board so that a thin, large PWB will not sag as it traverses the wave.
- **Bevel cutouts** Cutout openings should be beveled to aid in the hydrodynamics of the wave, ensuring that the wave smoothly enters and exits each cutout.

47.4.12 No-Clean Flux Residue and In-Circuit Test

During wave soldering, most of the flux from that process is consumed or washed off the bottom side of the board when in contact with the molten solder wave. However, if no-clean flux residue seeps between the pallet and the board, it will not be removed by the wave process. Any remaining residue, if covering in-circuit test pads, may be an impediment to proper in-circuit test probe contact. Care should be taken to minimize flux deposition for wave soldering. In terms of board design, test pads should be moved as far from intended wave-solder pallet openings as possible.

47.4.13 Designing PWBs for Wave Soldering

Complex PCAs with BGAs and fine-pitch SMT components on boards thicker than about 2.36 mm (0.093 in.) may be difficult to wave-solder if care in layout is not observed.

- **Use appropriate heat relief** It is good design practice to use appropriate heat reliefs (e.g., spoked pads) to provide electrical conduction but limit thermal conduction from the PTHs to ground or power planes internal to the board (see Fig. 47.27). This will enhance heatup of the PTH and limit thermal sinking to the ground or power planes and thus also enhance PTH barrel fill if process parameters are correct.

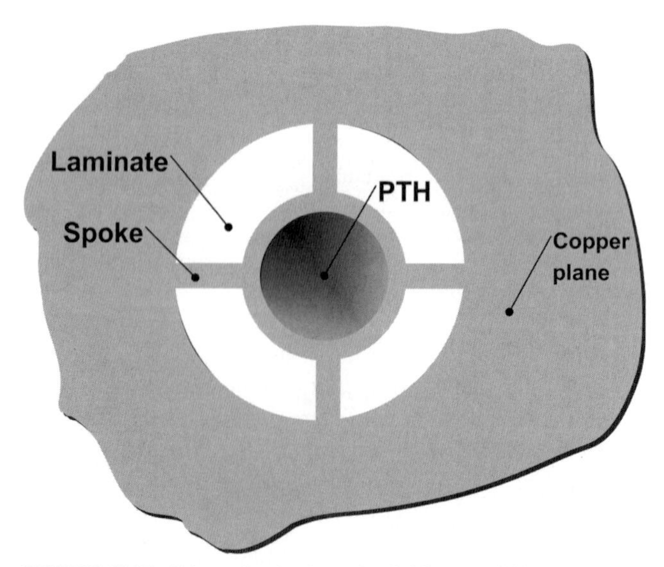

FIGURE 47.27 Schematic of a thermal relief feature within a PWB. The spokes connect to the PTH to the conduction plane and limit thermal transfer. The limited thermal path enhances heating of the PTH for soldering.

- **Limit the number or plane attachments** Where possible, limit the number of ground or power plane attachments to a PTH barrel. Each attachment point will result in thermal dissipation into the copper plane. This lost heat will detract from the soldering process making the job of pre-heating and solder wetting all the more difficult.
- **Cluster components and provide clearance** Group wave-soldered PTH components together. Avoid interspersing nonwave soldered SMD components within PTH fields. Conversely, avoid designs that place an isolated PTH component amid a field of SMD components. Such a placement requires a small wave pallet opening that inhibits solder-wave fluid flow, pre-heating, and wave contact.
- **Place the components properly in relation to edges and use selvage edges** When possible, avoid placing PTH components directly against the edge of the board. There must be clearance from the board edge to the pallet to allow the board to nest within the pallet. A selvage edge (a removable and disposable board edge extension that is part of the board itself) can be designed into the PWB (see Fig. 47.28). After wave soldering, the selvage edge can be removed by rotary cutter or routing. Note that post-solder routing is not preferred, especially for no-clean soldered assemblies. The abrasive router dust may remain within connectors and scratch connector contact surfaces or mating connector contacts. The vibration from routing may also affect the reliability of certain components.

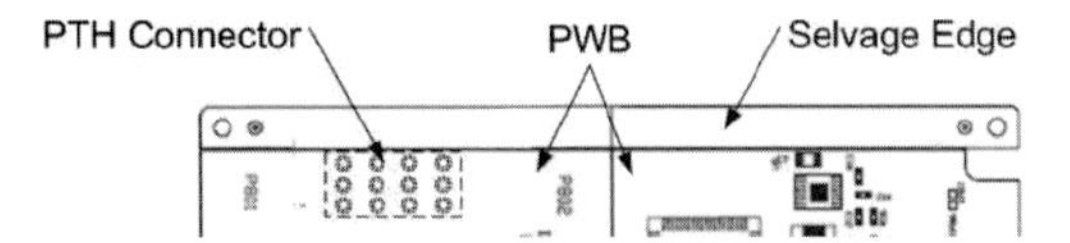

FIGURE 47.28 Board schematic showing a removable selvage edge, a disposible edge to enable wave soldering of the PTH connector near the true edge of the board.

- **Avoid double-sided wave-soldered PTH components** Double-sided wave soldering is generally fraught with defects. Determine whether SMT or press-fit technology can be used to replace a second wave-soldering operation. The industry-standard preference is to restrict all wave-solderable components to one side (components inserted through the primary side) of the PWB.
- **Do not place components near gold fingers** Clearance from gold fingers is required so as not to contact the gold fingers with solder. The gold finger should be away from the pallet nest edges. Covering the gold fingers with an acrylic-based high-temperature adhesive tape will help to keep solder and flux from the gold fingers. The tape is not an effective mask if exposed directly to the solder wave.

47.5 WAVE SOLDER DEFECTS

There are many types of wave-soldering systems, each with its unique advantage as claimed by the manufacturer. The soldering engineer has to assess these improvements as they relate to the type of assembly being soldered. The technology has matured significantly, but the degree of equipment complexity attests to the complexity of the process. There are many process variables associated with this operation. If these are not understood or properly controlled, wave-soldering defects such as skip soldering (electrical opens) and bridges (electrical shorts) will occur as in any other soldering operation. Another important defect is the influence of secondary-side wave soldering on primary-side surface-mounted components. If parameters are not controlled properly, it is possible to cause primary-side SMDs to reach reflow temperatures. This can induce solder opens, shorts, or solder-starved joints. This phenomenon is most commonly associated with thin, densely populated, double-sided PWBs with fine-pitch surface-mount devices such as quad flat packs and BGAs soldered to the primary side of the board. Thermal conduction from the wave through the board itself, through electrical vias and along electrical traces in and on the board, may provide enough heating to cause previously soldered parts to reflow again. Solder can be drained away through vias or wicked-up surface-mount component leads, resulting in opens or weakened soldered joints with an extreme hourglass-shaped attachment of solder from the bottom of the lead to the bonding pad.

A common defect associated with no-clean wave soldering is the seepage of flux between the pallet and the board. The resulting flux residue may inhibit in-circuit test probe contact. Care should be taken to ensure that the board fits well into the pallet nest, is rigidly indexed within the nest of the pallet, and is adequately retained against the pallet surface. As a pallet ages, it may shrink, take on twist, or bow and even delaminate. All of these may interfere with proper board seating.

If a component is seated tight against the PWB, it may block the flow of solder into the PTH. A gas pocket may form at the primary side of the hole. Gas pressure may force liquid solder out of the hole, preventing adequate PTH hole fill. This can be remedied by applying a small spacer under the component as shown in Fig. 47.29.

The spacer must be made of a heat-resistant material such as silicone.

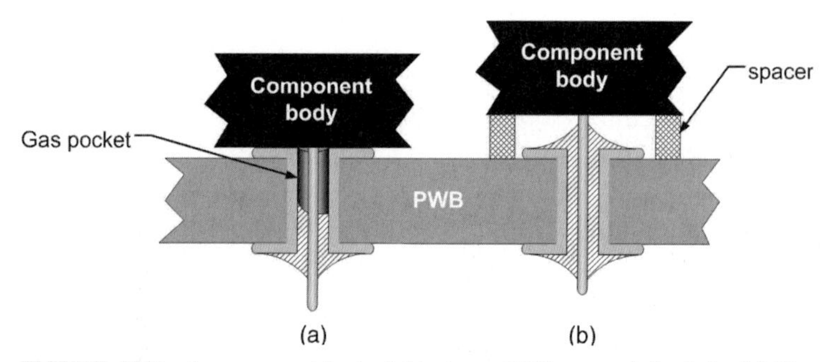

FIGURE 47.29 A component body tight atop a PTH can seal the hole: (a) The resulting gas pocket in the PTH can exclude liquid solder; (b) this situation can be remedied with silicone or other heat-resistant spacers applied beneath the component body.

47.5.1 Wave Soldering in the Pb-Free Context

The advent of lead-free solders has brought with it several issues:

- **Solder pot temperature** In wave soldering with eutectic Sn-Pb, the solder pot temperature is typically in the range of 240 to 260°C. Surprisingly, the move to Pb-free soldering has not necessitated a big change in solder pot temperature. For SAC305 (Sn-3 percent Ag-0.5 percent Cu), solder pot temperature is typically maintained between 255 and 265°C. It is wise to use the lowest practical solder pot temperature to reduce solder oxidation (dross formation) and to limit dissolution of PWB or component materials. In many cases, major assembly houses are using the same solder pot temperature for Pb-free solder as was used for Pb-containing solder, and conveyor speed is slightly reduced for better preheating and to compensate for the lower wetting rate characteristic of most Pb-free solders.
- **Oxidation** Some of the Pb-free solder constituents—such as copper, zinc, and bismuth—are notorious for oxidizing. When using alloys in which such constituents are major components, nitrogen inerting is mandatory to ensure good soldering and to conserve solder composition by retarding oxidation and excessive dross formation.

- **Flux** Both alcohol-based and water-based (volatile organic compound [VOC]-free) fluxes are compatible with most Pb-free alloys. The longer preheat times and slightly higher wave temperature are more of a challenge for alcohol-based fluxes, as dry-out is a concern. Higher flux solids content will help in this regard. Water-based fluxes are the direction in which the industry is moving, and a higher boiling point is widely used for Pb-free soldering.
- **Pb-free wave-solder joint appearance such as duller solder joints** Just as in reflow soldering, wave-soldered joints tend to be duller than those obtained from eutectic Sn-Pb. This is not a defect but a reality.
- **Fillet lifting** This has been previously covered, but it is worth noting that the effect is more prevalent in Pb-free alloys than in Sn-Pb alloys. Fillet lifting occurs when the solder cools at a different rate than the board. The Sn-Bi system is notorious for resulting in lifted leads. Bi expands upon freezing and can rip itself off the bonding pad or rip the bonding pad from the laminate. Fillet lifting is generally controllable by optimization of the cooling ramp rate after soldering.

When associated with the primary side of the board, fillet lifting is not considered a defect. If found on the secondary side (bottom) of the PWB, it is considered a workmanship defect with reliability consequences.

- **Hot tears and shrink holes** Most solders shrink upon solidification, and this is true of SAC solder. This alloy is known to form microcracks (see Fig. 47.30) on its surface during the phase change from liquid to solid. These cracks are called *hot tears* or *shrink holes*. They are generally benign as long as the crack is shallow (that is, the bottom of crack can be seen) and the crack neither extends to contact component lead nor PTH features (annular pad or barrel).

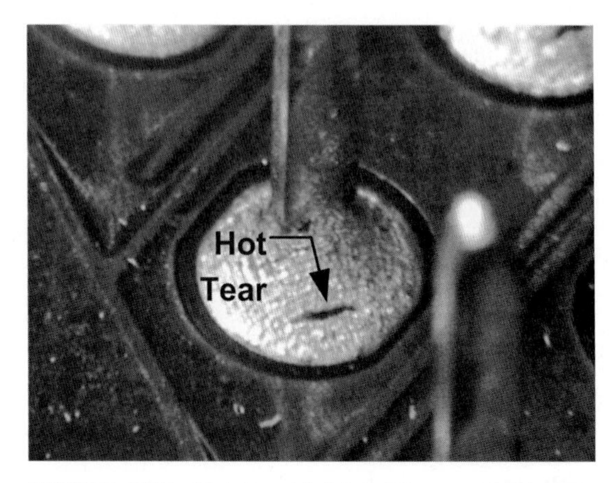

FIGURE 47.30 Hot tear (shrink hole) on an SAC wave-soldered PTH joint. *(Courtesy of Hewlett-Packard).*

47.5.2 Equipment

There is little difference between wave-solder equipment set up for Sn-Pb and that configured for Pb-free soldering. There are two points to consider before converting equipment. First, the level of Pb contamination will be high in the Sn-Pb equipment. This can be remedied by draining the system of Sn-Pb solder and purging the system with a charge of pure tin. The tin scavenges any Pb from the solder pot, delivery lines, and nozzles. Good cleaning of the wave machine will take care of any trace Pb contaminants, but proper protective gear (respirator and clothing) should be donned for this cleanup.

Sn is known to dissolve many metals. With the high Sn content of the SAC solders (SAC305 is 96.5 wt percent Sn), check to ensure the machine is compatible with the higher tin content. The solder pot, impeller, delivery lines, and nozzles are all subject to iron leaching. This can interfere with the process by changing the composition of the solder. It can also be dangerous if a leak develops in the system allowing molten solder to leak from the solder pot. Some coated steels (such as ceramic and nitride) and cast iron are generally compatible. Changing over older systems to Pb-free compatibility can be an expensive proposition.

47.6 VAPOR-PHASE REFLOW SOLDERING

Because of safety and environmental concerns and compliance with the Montreal Protocol for the reduction of ozone-depleting chemicals, this soldering technique had fallen out of favor. With the advent of Pb-free soldering, there is increased interest in vapor-phase soldering for SMT applications, but it is likely that it will remain a niche application. Due to its continued diminished status, vapor-phase reflow is covered only briefly.

47.6.1 Basic Process

As in any other reflow technique, the board must be supplied with sufficient solder for joint formation. Most commonly, solder paste is stenciled onto circuit board pads. Components are placed onto the solder paste in preparation for reflow. The board is conveyed into the reflow chamber, where it is exposed to the vapor-phase of a boiling liquid. This liquid is inert with respect to the solder and board. It is a dense synthetic of high boiling point (slightly higher than that of solder liquidus but not high enough to damage the circuit board or components). Historically, chlorofluorocarbons were used, but now only hydrofluorocarbons (HFCs) are acceptable due to environmental concerns. Material costs are high, at several hundred dollars per pound.

When conditions are optimized, the hot vapors begin to condense on the cooler PWB, heating it. As the process progresses, sufficient energy for sustained solder reflow results. Solder wets to lead and pad and, as the board is removed from the hot vapor, the molten solder solidifies, bonding component leads to circuit board pads.

47.6.2 Machine Subsystems

The vapor reflow soldering machine is composed of three main subsystems: conveyor, reservoir vessel, and heaters. The vessel has cooling coils surrounding it, placed well above the level of the liquid in the reservoir. These condense the vapors, returning the majority to the reservoir, as shown in Fig. 47.31.

47.6.3 Advantages and Disadvantages of Vapor-Phase Reflow Soldering

The process is exceptionally uniform in heating and is also very precise in temperature. This method of reflow negates problems associated with varying topologies and areas of high thermal mass that confound other soldering processes. Since the soldering occurs in an inert atmosphere, fluxes of lower activation level can be used. In the absence of air, solder-joint quality is generally excellent.

Although vapor phase reflow has the appearance of being a fast process, solder paste reflow recommendations must still be followed. If the paste is heated too rapidly, paste volatiles may boil, resulting in explosive solder ball formation. As in other soldering methods,

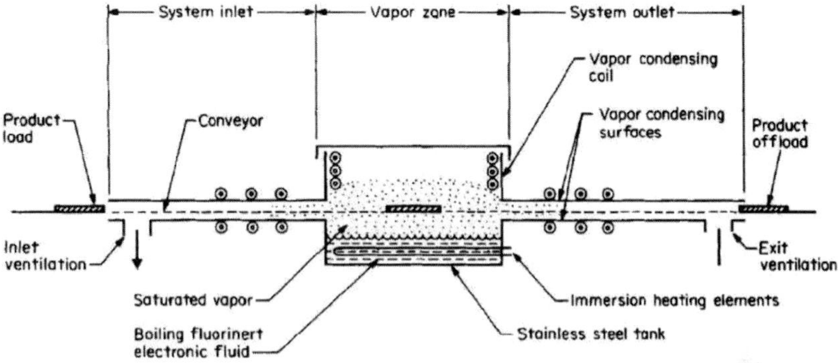

FIGURE 47.31 In-line single-vapor heating system schematic. *(Reprinted with permission from* Electronic Packaging and Production, *November 1982, p. 63, Fig. 1).*

the maximum heating ramp rate must be commensurate with component manufacturer recommendations. A slow preheat is required to keep components from cracking. Plastic package popcorning is pronounced with this method of reflow.

Besides legislated environmental issues, there are health risks associated with vapor-phase reflow soldering. With the continued recycled use of HFCs, toxic materials such as hydrofluoric acid and perfluorisobutylene are formed. These must be neutralized and their by-products must be disposed of properly.

When massive, densely populated boards are introduced to the vapor reflow oven, vapor collapse can occur—a condensation rate that outpaces that of vaporization. The result is that the internal atmosphere of the oven thins dramatically to the point that it cannot sustain adequate reflow. Vapor-phase machines that rely on immersion heaters are prone to this phenomenon. More recent machines include massive heating element housings that provide sufficient thermal inertia to preclude this problem.

Increased incidence of tombstoning, solder ball formation, and component displacement has been noted in this method of reflow. The condensing vapor transfers heat directly to the best thermal conductors, component leads, and pads. Excessive solder wicking can occur, transporting solder from between the lead and pad up the lead, where it is not needed. This can cause solder bridges proximal to the component body. Excess solder on the component lead decreases the lead's flexural compliance, making it less reliable. Furthermore, if the solder is wicked up the lead, the lead-pad interface will be solder–starved, resulting in an inferior solder joint. Additional information can be gleaned from older reference books and other authoritative publications, including previous editions of this book.[11,12]

47.7 LASER REFLOW SOLDERING

The laser has proven to be a versatile tool in many industries. For circuit board assembly, it can be used for marking, single-point soldering, or rework. Lasers can accommodate bonding of the finest or coarsest peripherally leaded surface-mount packages, and its performance and applicability are independent of device population density on the board, the thickness of the printed circuit board, or the presence or absence of package heat sinks. It can be used with most any component lead or PWB pad surface finish and any solder, including Pb-free alloys.

When laser soldering is implemented properly, high soldering yields and exceptional bond characteristics are possible[13]. Its use is usually relegated to low-volume soldering applications.

For successful laser soldering, it is necessary to use tooling to hold the component—or, more specifically, the component leads—down to circuit board pads.

Laser-based reflow is used as a rework tool, and laser systems are packaged expressly for this purpose and sold commercially. Boards are preheated to about 100°C within the system by means of a bottomside resistance heater. A diffuse laser beam is then scanned around the package by a rapidly moving galvanometer. The impinging energy heats surface-mount leads or an entire BGA to the point of solder reflow for package removal. This system has the advantage of such localized heating that only the package targeted for rework reaches reflow temperature. Everything around it stays close to the system preheat temperature with almost no thermal spill-over from the laser. This is a distinct advantage over the hot-air repair method, where the hot gas stream used for package heating and reflow can inadvertently melt solder joints on adjacent components or components on the opposite side of the board.

Laser soldering requires no hot zones to profile or maintain, does not induce substrate warpage, and requires no heated bonding head that can degrade or vary with usage. Components can be spaced exceptionally close to one another—closer than currently permitted for other techniques, especially when one considers rework.

47.7.1 Lasers

The word *laser* derives from the acronym for *light amplification via stimulated emission of radiation.*

47.7.1.1 Laser Subsystems. All lasers are comprised of three main subassemblies: power supply, cavity/oscillator, and beam delivery optics (see Fig. 47.32 and Table 47.2). Although there are many configurations and additional accessories that may enhance a laser's output, only the basics are covered here.

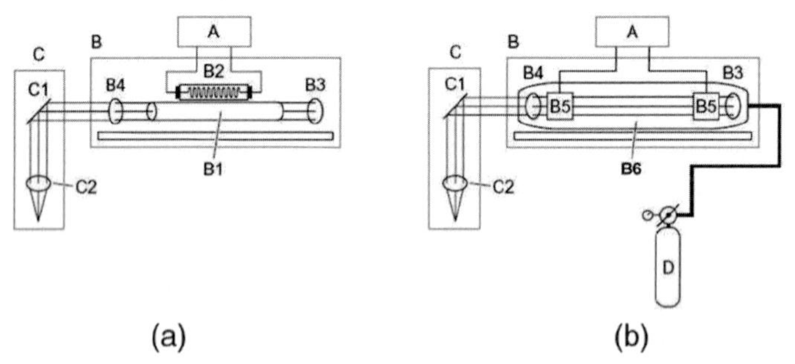

FIGURE 47.32 Comparison of the two most popular lasers used for soldering: (a) Nd:YAG; (b) CO_2.

47.7.1.1.1 Cavity/Oscillator. *Lasing,* which is the laser photon generation and amplification process, occurs in the cavity, also known as the oscillator. The cavity contains the lasing medium, a material that is both the photon originator and amplifier of the laser beam. Photons, electrons, or other high-energy sources are used to raise certain atoms or ions in the lasing medium to a temporarily stable electronic transition, a meta-stable state. When the medium is returned to its stable, ground state, energy is released in the form of heat and fluoresced photons. These photons are, for the most part, reflected at the ends of the cavity and focused back through the lasing medium by the cavity optics, generating more photons with each additional pass.

TABLE 47.2 Laser Systems and Subsystems

System	Subsystems	Name
A		Excitation power supply
B		Cavity
	B1	Nd:YAG crystal (lasing medium)
	B2	Flash lamp (excitation source for Nd:YAG)
	B3	Retroreflector
	B4	Partial reflector (laser beam exit optic)
	B5	Electrodes (excitation source for CO_2 laser)
	B6	Quartz tube
C		Delivery optics
	C1	Mirror
	C2	Final objective
D		Lasing gas CO_2, N_2, He

There are always two types of reflectors that bound a laser's cavity. The rear or retroreflector focuses and returns nearly 100 percent of the emission wavelength photons back through the cavity, further stimulating the lasing medium to emit photons. The front reflector is a leaky mirror lens, one that is reflective but somewhat transmissive also. It reflects most of the emission beam back into the oscillator for further stimulation of the lasing medium while allowing a small fraction of the laser beam to pass through it. The component of the beam that is transmitted by the front partial reflector is the working laser beam. It is monochromatic by virtue of the lasing medium, which limits photon emission to a set of discrete wavelengths, or laser lines, which are characteristic of the medium material, and by virtue of the optical coatings on the front reflector, which allow only one laser line through it.

49.7.1.1.2 Delivery Optics. Delivery optics, placed in the path of the laser beam, direct it to the work piece. In the case of laser soldering, the work piece is the circuit board or, more specifically, the leads, pads, or lead-pad combinations. The beam can be easily steered by means of wavelength-appropriate mirrors and can be focused to the required spot size via a final objective lens to accomplish soldering. The delivery optics can be held fixed and the circuit board moved beneath or, conversely, the optics can be moved to direct the beam as needed. Either can be done with great precision sufficient for any PCB soldering task. Moving the beam and moving the board both have their own advantages. Moving the optics in any direction is very simple and will not interfere with any PCB or surface-mount device fixturing requirements. Also, moving the circuit board may cause misalignment of component leads and circuit board pads. Fixed optics are more stable and require fewer adjustments, although, if engineered properly, moving optic beam delivery systems can be exceptionally stable and accurate.

Fiber-optics are in vogue for laser beam delivery, but it should be noted that, when dealing with high-energy-density laser beams as required for soldering, fibers are subject to damage if not precisely maintained. Fiber delivery has the advantage of allowing for easier steering of a beam in a complex machine that may preclude the use of orthogonal mirror and lens beam steering.

47.7.1.2 Beam Characteristics. There are many key differences between a laser beam and other light sources. (The term *light* is generally reserved for those wavelengths that are part of the visible spectrum and detectable by the human eye, but for the purposes of this discussion, *light* and *photonic emission* will be used interchangeably.) Laser beams have several distinctive characteristics. They are generally set to emit monochromatic radiation, which is also coherent radiation. Coherence here relates to the synchronized propagation of photons; that is, all waves of the emission radiation are in phase with one another.

47.7.1.3 Laser Criteria for Soldering. The choice of a laser for soldering is predicated on emission wavelength, required power, required beam diameter, machine reliability, and price. The optical properties of the target material are of primary importance in the laser selection process. In the case of surface-mount soldering, the absorption, reflection, and transmission characteristics of the PWB laminate as well as the reflectivity and absorptivity of the metallics (leads, pads, and solder) involved in the soldering process must be considered. The majority of circuit boards produced are composed of fiber-reinforced epoxy resin, although there are others. Each laminate has its own unique optical properties and characteristic laser damage threshold (LDT). LDT can be defined as the energy required to alter or damage the target material. In the case of an organic-based laminate PWB, it would be the laser energy necessary to char the board or cause the bonding pads to lift from the surface. Most any lead-pad surface finish on most any circuit board material can be soldered. For the purpose of this section's discussion, the focus will be on industry-standard glass-epoxy laminate and most any solder, be it lead-bearing or lead-free.

47.7.1.4 Laser Choices. There are few practical choices for laser selection, since only a few possess the characteristic energy, production-tested reliability, and economics necessary for PWB assembly. The most common are the neodymium drifted yttrium aluminum garnet (Nd:YAG) laser, an example of a solid-state laser, and the carbon dioxide (CO_2) laser, which has a gaseous lasing medium. These two types of laser are among the most common of the industrial lasers. Each has been on manufacturing floors in various industries since the early 1980s

They are versatile in terms of applications and capable of the output required to weld, braze, solder, cut, and mark. Although other lasers could be used for soldering, the Nd:YAG and CO_2 types are the most commonly available. They have proven track records and are also the most commonly reported in terms of application to soldering.

47.7.1.5 Carbon Dioxide (CO_2) Lasers. In the case of the CO_2 laser, there are mirrors at both ends of the tube that contains the lasing gas—a mixture of carbon dioxide, nitrogen (N_2), and helium (He). Each of the components of the mixture helps in the lasing process. The CO_2 is the lasing medium. Electricity is discharged directly into the CO_2, resulting in its ionization and establishing a meta-stable state that decays, resulting in fluorescence (photon generation) and heat. These photons are reflected back through the laser cavity to stimulate the production of additional photons further. As this process continues, the beam intensifies and emissions from the cavity compose the working laser beam. The introduction of N_2 helps in the transfer of energy to excite the CO_2 molecules into their meta-stable state. The helium, with its high thermal conductivity, helps to transfer some of this thermal energy to the laser's cooled cavity walls. Were the cavity gases to become too hot, the CO_2 would dissociate and would be less effective as a lasing medium.

CO_2 lasers are known for their reliability and stability in manufacturing. Their emission is well into the IR spectrum at 10.6 μm (10,600 nm). It has limited use in PWB soldering since that wavelength is well absorbed by most organic materials (such as epoxy laminates) and is well reflected by most metallurgies. This is disadvantageous, since to get sufficient energy into a lead-pad combination, the CO_2 beam has to be kept to a small size to prevent it from spilling onto the PWB laminate. If the PWB were irradiated either directly or by errant reflection off a specular surface, it would char and the resultant carbon-rich residues might be electrically conductive enough to cause an electrical short circuit. There are three other points that detract from the use of the CO_2 laser as the preferred tool for circuit board soldering, each of which is discussed in the following subsections. Despite these drawbacks, the use of CO_2 lasers for soldering has been widely reported in the literature.

47.7.1.5.1 Materials Compatibility. The long wavelength of the CO_2 laser is readily absorbed by organic materials (e.g., glass-epoxy laminate, polyimide, and molded encapsulants).

47.7.1.5.2 Fiber-Optic Delivery. The CO_2 laser's output is not compatible with fiber-optic delivery. The wavelength is well absorbed by the most common fiber-optic material, fused silica.

47.7.1.6 Nd:YAG Lasers. The neodymium drifted yttrium-aluminum-garnet (Nd:YAG) laser, commonly referred to as the YAG laser, has a solid-state medium. It relies on lamp excitation of a cut and polished yttrium-aluminum-garnet crystal that has been doped with neodymium to stimulate this solid-state lasing medium. The photons released by the solid-state lasing medium are used to stimulate more lasing medium-derived photons. Otherwise, the same lasing principles apply to the operation of that laser as to the gaseous CO_2 laser. Nonlinear optical materials can be used with the YAG laser to double the output frequency, halving the operating wavelength. This can be advantageous when working with highly reflective materials such as gold and copper, which are more optically absorptive at lower wavelengths. This, however, detracts from laser performance by decreasing available operating power, adding complexity to the system as well as increasing maintenance requirements.

47.7.1.7 Spot Size. First, the minimum practical spot size is relatively large. The theoretical diffraction-limited spot size of a focused beam is directly proportional to the beam wavelength and lens diameter, as described in Eq. 47.1.

$$S = f\lambda/D \tag{47.1}$$

where S = diffraction limited spot size
f = focal length of the lens
λ = wavelength of the laser
D = lens diameter

Thus, for a lens of 25 mm diameter and 100 mm focal length, in conjunction with CO_2 or YAG lasers, the spot sizes shown in Table 47.3 are theoretically possible.

TABLE 47.3 Theoretical Spot Size versus Laser Type

Laser type	Emission frequency (μm)	Theoretical spot size (μm)
Co_2	10.6	42
Nd:YAG	1.06	4.2
Nd:YAG (Frequency doubled)	0.532	2.1

Imperfections in the lens or beam shape and other factors prevent practical achievement of these minimal spot sizes. Generally, the attainable focused beam diameter on the factory floor is about two to three times larger than the ideally calculated spot size. Note that for an Nd:YAG laser, the spot size is at least 10 times smaller than that of the CO_2 laser, providing a fine, high-energy-density spot for soldering.

47.7.1.8 Lasing Operating Modes. Both YAG and CO_2 lasers can be operated in a variety of modes, each of which can be used to advantage in soldering:

- **Continuous wave (CW) operation** CW is a constant emission analogous to the continuous output of a light bulb that is powered by a direct current source.
- **Pulsed laser output** Furthering the light bulb analogy, this mode is akin to a bulb operated by an alternating current or pulsating power supply, although more like a strobe with intense bursts. The pulsing can be attained through a variety of methods, including switched power supply, capacitive discharge, or mechanical shutter, or by means of an optically manipulated shutter, as by acoustooptic, electrooptic, or magnetooptic methods.

47.7.2 Laser-Soldering Fundamentals

There are relatively few variables associated with the laser in its application to soldering. This is one of the advantages for laser processing. Beam wavelength, irradiation time, and beam power are important to the process, as are the properties of the materials being joined. Reflectivity, thermal conductivity, and laser damage threshold must be understood before soldering is attempted. The wavelength will be fixed by the laser of choice, and the shorter the wavelength, the smaller the theoretical spot size, as discussed. Generally, the reflectivity of a metal is lower at shorter wavelengths, thus a metal is more easily heated by an Nd:YAG laser than with a CO_2 laser. The converse is true for many polymeric materials, which are more absorptive at longer wavelengths and therefore more prone to damage with increasing wavelength. Many polymeric materials are also absorptive at the UV end of the spectrum. A carbon dioxide laser beam is more likely to impart damage to a circuit board than the beam of an Nd:YAG laser, but either laser can inflict damage if the energy density is high enough. In fact, either laser can easily cut through a component lead or drill through a circuit board if parameters are not adjusted properly.

The reflectivity of metals varies widely with composition and surface condition. Every metal can be heated with a laser as long as the energy density of the beam and dwell time are sufficient for heating. This is also the case with laser-irradiated component leads, board pads, and solder during the laser bonding process. Measurements of the reflectivity of Sn-Pb solder show that a eutectic alloy can be as high as 74 percent at 10.6 (m or as low as 21 percent at 1.06 μm. Therefore, in the case of the CO_2 laser at 10.6μm on a Sn or Sn-Pb-plated lead and solder-coated pad, much energy will have to be directed at the metals to start the absorption process, since so little laser energy is absorbed by the solder and converted to heat; the rest of the energy is lost to reflection. The reflected or multiply reflected beam may impinge on adjacent components and damage package bodies or even the circuit board itself, causing charring. Therefore, Nd:YAG is preferable for circuit board soldering.

47.7.3 Through-Lead and Through-Pad Bonding

Generally, the laser beam is directed at the component lead to accomplish soldering, but when the lead material is highly reflective, as would be the case with a gold-plated finish, heating is slow and irradiation times impractically long. An alternative has been demonstrated whereby the beam is directed at solder-coated circuit board bonding pad. If there is leeway in design, the bonding pad can be extended to aid in this method, dubbed *through-pad bonding* (see Fig. 47.33b). In contrast to the usual beam impingement on the component lead, or *through-lead bonding* (see Fig. 47.33a), the beam is directed at the more absorptive solder or on the circuit board land, increasing process efficiency and locally melting the solder in the vicinity of beam impingement. Since the molten solder is in intimate contact with the lead and pad, the heat from the bonding process is efficiently transferred. The process can result in rapid soldering along the entire length of the component lead and circuit board pad with rich solder fillets evident if the process is conducted properly.

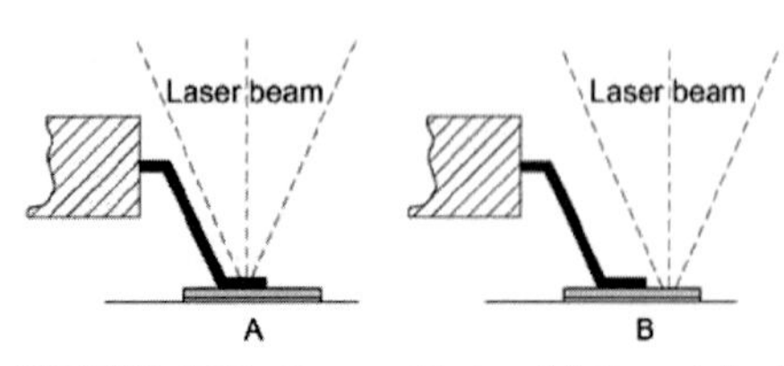

FIGURE 47.33 Laser soldering: (a) through-lead laser soldering; (b) through-pad laser soldering (B).

47.7.4 Single-Point Laser Soldering

This method requires a laser beam that is smaller in diameter than the length or width of the component lead or bonding pad. The beam is stepped to each lead-pad combination and delivered with sufficient energy to result in solder reflow. The beam can be CW, pulsed, or multiply pulsed to accomplish soldering as long as there is enough radiation delivered to cause the solder to undergo the phase change required for soldering and wetting. Due to the small beam diameter, as required by this technique for fine-pitch components, the energy density can be exceptionally high. In fact, it can be so high that if the lead is not in good contact with the bonding pad, then the laser's beam can damage the lead, perhaps cutting through it rather than soldering it. Overly intense irradiation can also cause a bonding pad to delaminate from the PWB.

In any laser soldering technique where very small spot size is required, the beam diameter must be precisely controlled. A small variation in beam diameter, either through change in laser parameters or in working distance (focus), can dramatically affect the delivered energy density. In the cases of a 10 W Nd:YAG laser beam focused to a spot of 0.1 mm (0.004 in.) and a spot of 0.2 mm (0.008 in.), this seemingly small change in spot size will result in a large change in power density (see Eq. 47.2).

$$\mathrm{P} = p/\mathrm{d} \tag{47.2}$$

where P = power density
p = average power
d = focused beam diameter.

For this example, the power density would vary from 1,273 W/mm^2 for the 0.1 mm diameter beam to 318 W/mm^2 for the 0.2 mm beam diameter, a fourfold reduction emphasizing the criticality of strict maintenance and process control when working with a small spot size.

In most cases, a Gaussian distribution of the beam's energy distribution across the focused diameter is assumed. It is customary to measure the beam at the 1/e^2 point (13.5 percent of the peak height), so, in fact, the beam is impinging on a slightly larger area. However, the most intense portion of the laser's beam is confined to the region bounded by the 1/e^2 points in two dimensions, as shown in Fig. 47.34.

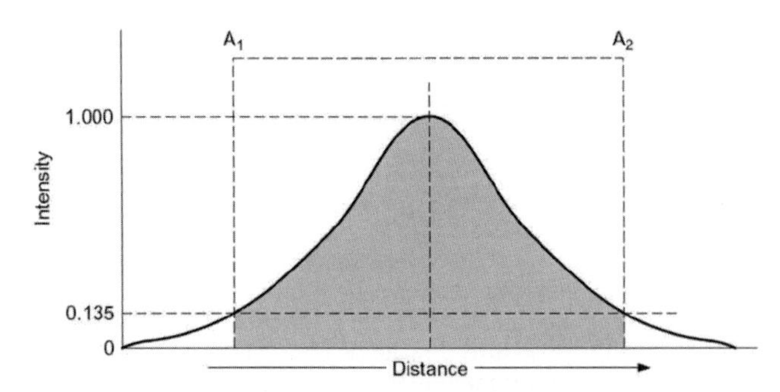

FIGURE 47.34 Laser energy density is measured from points on the distribution curve where the beam's intensity is 1/e^2 (13.5 percent) of peak intensity or between points A1 and A2 in this plot.

Single-point laser bonding has been applied to soldering tape automated bonding (TAB) lead frames to silicon die, TAB leads to PWB, or multichip modules as well as for soldering gull-wing leaded components to circuit boards. This technique is not suited for bonding area-array devices such as BGAs or CBGAs to PWBs.[14]

47.7.5 CW Scanning

In sharp contrast to single-point laser soldering, this technique relies on a larger, defocused moving beam to accomplish the heating. The beam may be many times the size of the narrowest lead or pad dimensions. In this technique, the laser beam generally spills over onto the substrate material, irradiating the lead, pad, or lead-pad combinations as well as interpad board surfaces. Because a larger beam diameter is used, a long focal-length lens and generous delivered laser spot size can be employed. This results in a wide process window in terms of final beam diameter and working distance of lens to the circuit board surface. The laser spot is moved at a rate such that the beam exposure to the board is below the laser damage threshold limit. It heats the lead-pad combination, the solder flux, and the board sufficiently to cause the solder to change state.

A variation of this technique utilizes orthogonally mounted galvanometers to move an Nd:YAG laser beam around the periphery of a surface-mount device sitting atop solder lands on a circuit board. The beam is driven at high velocity repetitively around the surface-mount device until all leads are heated to the point of solder liquidus. As the beam is turned off, the solder solidifies and solder joints result. This same galvanometer-driven technique has gained commercial acceptance as a BGA rework tool. A slightly defocused laser beam is aimed at the body of the BGA. The beam is rapidly moved around the package body at a rate below the laser damage threshold of the packaging material. With each pass, the component heats up a little more, eventually melting the BGA solder joints. The package is removed with an automated vacuum pickup tool while the solder is still molten. A new BGA can be attached using this same technique. This method can be applied to any type of rework, although component leads and pads would be targeted rather than the body, as was the case for the BGA.

47.7.6 Multiple-Beam Laser Soldering

One of the attractions of working with a laser is the fact that its beam can be split for multiple use within one station or even shared between two or more stations. The split can be accomplished by the use of bifurcated fibers or beam splitting mirrors. The beam of a single laser cavity can be duplexed to solder two sides of a surface-mount component simultaneously. It is entirely possible to share a common beam between two or more laser soldering stations either simultaneously or in a time-shared manner.

47.8 TOOLING AND THE NEED FOR COPLANARITY AND INTIMATE CONTACT

In all modes of laser bonding, the component leads must be in contact with solder lands on the PCB to accomplish joint formation. It is therefore necessary, in most cases, to use a specialized hold-down tool to ensure that leads contact pads. This detracts from the ideal of noncontact laser bonding. Several techniques have been developed and reported. One such method uses a transparent hold-down medium such as glass, quartz, or transparent high-temperature plastics. There are several significant problems associated with this approach, which can result in an inconsistent soldering process. First, these materials are rigid, making it impossible to hold a lead down to a low-lying pad abutted by normal height pads or, conversely, a high pad abutting lower pads (see Fig. 47.35).

During the soldering cycle, the glass can accumulate spattered flux and flux by-products that may change the delivered laser beam intensity, adding variability to the process.

Comb or pin arrays have been applied that match the lead configuration of a package. This type of fixture is expensive and easily damaged, especially if the soldering flux glues the comb teeth to the bonded component lead. Probably the most common method for rigid-leaded

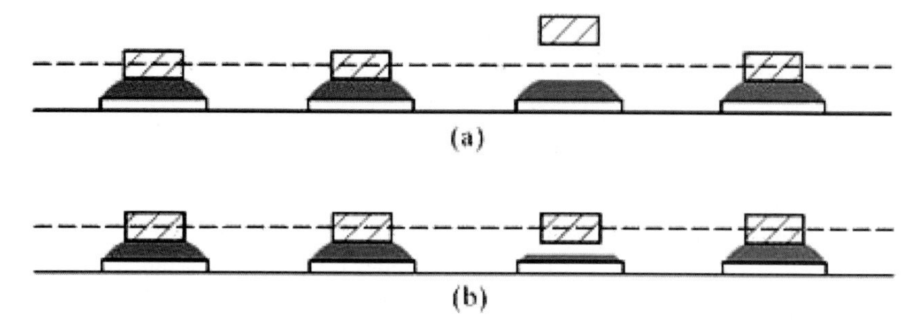

FIGURE 47.35 For laser soldering to be effective, it is necessary to have lead-to-pad, pad-to-pad, and lead-to-pad coplanarity: (a) noncoplanar component leads on constant-volume (coplanar) solder pads; (b)a coplanar lead array on noncoplanar PWB solder lands.

surface-mount packages is the body-push method (see Fig. 47.36), in which a force is applied to the component body, springing the leads just slightly. As the solder melts below the lead, the lead drops down to board level, where it is frozen in the cooling solder. It is necessary to control the amount of push such that the solder joint is not stressed by a lead frozen in compression. This is particularly the case for coarse-pitched component leads, where much force may be required to overcome a coplanarity problem. When board pads and component leads are reasonably coplanar, then this method is adequate for high-yield, high-reliability soldering as long as the noncoplanarity is in a negative direction, as indicated in Fig. 47.36a.

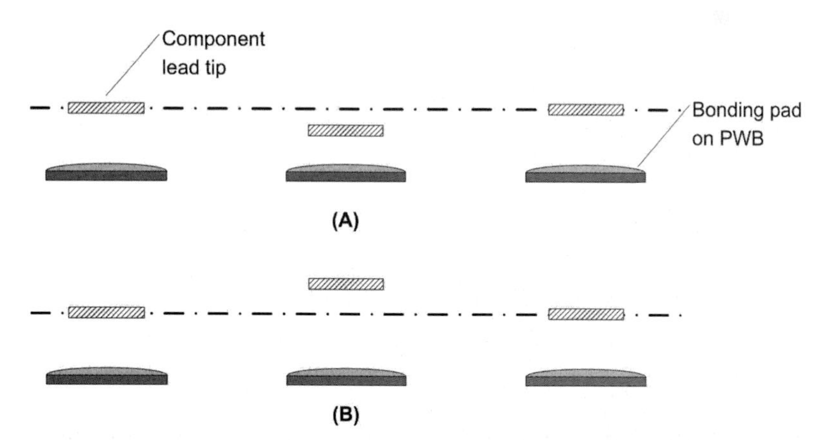

FIGURE 47.36 The body-push method would be effective in (a), where a lead is below the level of adjoining leads. In the case of (b), too much force would be needed to push the component body down to make the high component lead coplanar with adjacent leads. This could result in too much stress in the solidified solder joints.

A compliant hold-down method has been reported that is inexpensive to fabricate and easy to implement. This consists of a silicone rubber foot and an aperture that allows the beam unobstructed access to the lead-pad combination. This method has been proven effective for TAB as well as for rigid leaded surface-mount device bonding. It obviates the problems associated with transparent hold-down methods in terms of flux build-up. The compliant foot holds leads to pads and can accommodate lead-to-pad noncoplanarity differences.

47.8.1 Flux for Laser Bonding

Requirements are much the same as for other reflow methods. Flux must be active enough to remove oxides from the component lead, bonding pad, and the solder. Its optical characteristics in terms of reflection, absorption, and transmission must not interfere with the solder process. If too absorptive, the flux could char, becoming more absorptive, and then overheat and wreak damage to the underlying circuit board. Also, if denatured, it will lose its chemical effectiveness as a fluxing agent. In laser soldering, particularly in the scanned methods, the flux adds a path of heat transfer to the board and preheats the next joint to be soldered. In the absence of a liquid flux, the PWB is more likely to char if the laser impingement is not stringently controlled. During soldering, the laser beam's output is quick and intense. The high energy of the process results in exceptionally high processing temperatures for very short durations. The high processing temperature aids in flux activation, making even the mildest of no-clean fluxes an effective soldering agent.

47.8.2 Fluxless Laser Soldering

There have been many reports of fluxless laser soldering. As used for inner-lead bonding to silicon die, these utilize a high-peak-energy, short-duration pulse. Under these conditions, when the mass of the lead and the bonding pad is small enough, the process is more akin to welding with the component lead melting and alloying with an underlying metal such as tin. To accomplish conventional reflow soldering by means of a laser, or any other technique, it is necessary to employ a liquid solder flux or gaseous analogue pretreatment to bond effectively. Plasma cleaning and inert storage may be one such method. Ultrasonic assisted laser soldering can be effective, but is slow, requiring precise placement of an ultrasonically agitated head and running the laser beam either through the ultrasonic tool or precisely adjacent to it. The ultrasonic agitation breaks up the oxides surrounding the molten solder and the component lead, allowing joining of the liquid solder to the component lead metallurgy. The use of gaseous phase carboxylic acid fluxing with laser bonding has been demonstrated.[15] This precludes the need for conventional liquid fluxes and leaves no visible residues on the circuit board. No cleaning is required and the assembly reliability is not degraded from the use of a gaseous fluxing approach.

47.8.3 Laser Solder-Joint Characteristics

There are very few differences between solder joints prepared by laser as compared to those formed by other methods. Laser soldering results in intermetallic formation, but the layer is extremely thin if the laser soldering cycle heating is kept short—much thinner than that found in solder joints manufactured by more conventional methods. Since the heating is highly localized, cooling is rapid resulting in exceptionally fine solder grain growth. The fine grain growth leads to greater joint strength initially. The strength advantage, although significant in magnitude, tapers off as the solder joint ages and metal grain coarsening occurs. Coarsening rate depends on ambient temperature and time.

47.8.4 Solder Sources and Defects Associated with Laser Reflow

Solder requirements are the same as for any other process. There are no alloy composition requirements specific to laser soldering, as this soldering method is compatible with leaded or lead-free solder alloys. Even high-temperature alloys can be soldered by this technique. When single-point laser reflow is applied, the board quality and integrity are not compromised if parameters are chosen and adequately controlled.

Laser soldering is not prone to solder bridges and if component leads are held down to solder-coated PWB lands, then soldering defects are rare. Perhaps the most common characteristic defect unique to laser soldering is charring or burning of the circuit board, which occurs if too high an energy density is used and the laser damage threshold of the PWB is exceeded. Charring or burning of the PWB can also occur if the circuit board is grossly contaminated with grease or other organic contaminants. This is not to say that boards for laser soldering have extraordinary requirements for cleanliness. Requirements should be considered the same for this technique as for any other reflow process.

47.8.5 Laser Safety Issues

Lasers are categorized by their safety hazard potential. A full review of these will not be provided, but suffice it to say that Class 1 is an intrinsically safe laser, posing no intra-ocular danger, while Class 4 lasers pose the greatest hazard. All lasers considered for soldering use are of the Class 4 variety. Because of this, they are generally embedded in appropriate interlocked cabinets with laser-safe viewing ports or a video camera incorporated to preclude direct ocular exposure to the laser's intense beam. When the Class 4 system is embedded in an interlocked cabinet, it is considered a Class 1 system and poses no hazard to the immediate area; laser-safe eyewear need not be worn except when interlocks are overridden for system maintenance. As mentioned previously, lasers, especially Nd:YAG, are known for their high uptime and lack of required expendables.

47.9 ADDITIONAL INFORMATION SOURCES

There are numerous excellent supplemental texts available for additional detail and instruction in laser technology. Those by S. Charschan,[16] Hecht,[17] and Ready[18] are particularly useful. Additional information on CO_2 lasers in the realm of soldering can be found in references.[19,20,21,22,23,24]

47.10 HOT-BAR SOLDERING

Specifically suited to surface-mount assembly of packages with lead frames, hot-bar soldering has been in use for several years. The technique relies on a resistance-heated element to push component leads into contact with solder and bonding pads, simultaneously reflowing the solder. Compression of the leads onto the circuit board lands is continued as the heat is ramped down. Upon cooling, the solder solidifies and the heating element is withdrawn from the newly formed solder joints. The heated element is commonly referred to as the *hot-bar,* although the term *thermode* is also in widespread use.

Hot-bar soldering is best suited for low-thermal-mass, single-sided surface-mount assemblies. Each component type requires its own hot-bar bonding head assembly. These can be expensive, with the price dependent on the complexity, materials of construction, precision required, and overall size of the bonding head.

47.10.1 Solder Application

The use of solder paste is discouraged because blade heatup is fast and may result in explosive solder ball formation from rapidly volatilized paste constituents. Also, the paste is likely to squeeze out from between the component lead and circuit board pad, which can cause the solder to bridge to adjacent conductors. In fact, even with solid solder coatings on the circuit board, bridging can be problematic in hot-bar bonding. This is usually a function of the volume of solder on the pad, the quantity of flux, its degree of activity, and lead-pad pitch. As the solder is melted and displaced by the hot-bar, it bulges laterally to the point that the solder masses of two or more adjacent pads may touch one another, forming a solder bridge. Once the bridge has formed, the forces associated with lead-pad wetting and capillarity may not be strong enough to overcome the surface tension conditions established during bridge formation. If that is the case, the bridging defect(s) will persist (Fig. 47.37).

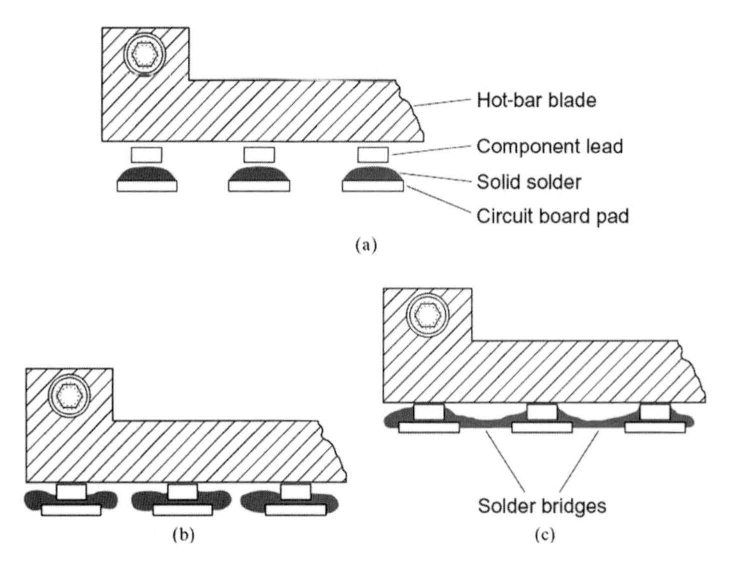

FIGURE 47.37 As the hot-bar is forced into contact with component leads and circuit board pads (a), the solder may bulge laterally as shown in (b), resulting in solder bridges (c). Careful control of solder volume, applied pressure, and temperature profile during hot-bar soldering is required to preclude this defect.

Solid solder coatings such as hot air-leveled pads, solder-plated boards, or stenciled solder paste reflowed prior to hot-bar bonding are preferred for this bonding method.

47.10.2 Fluxes and Fluxing for Hot-Bar Soldering

Soldering flux is applied prior to component placement and soldering. Flux choice should be tailored to the hot-bar process. The flux can be either liquid or paste. The flux chosen should be tested to resist charring and the development of polymerized decomposition products, or lacquers, which adhere to both circuit board and hot-bar. Residue build-up on the bar can adversely inhibit hot-bar performance by diminishing thermal transfer. The residue can also become thick and uneven enough to prevent the hot-bar from squarely contacting component leads. Baked-on aqueous-clean flux residues can make flux cleaning difficult. With no-clean flux, overheated residues may detract from the visual appearance of the printed circuit assembly.

47.10.3 The Hot-Bar Soldering Operation

Generally, the board is preheated on a hot stage beneath the hot-bar, off to the side or even in a box oven prior to soldering. Sometimes hot air is used for preheating. This step is generally done after flux application, so the thermal profile has to be adjusted to prevent flux dry-out. Once component leads are aligned with solder-coated PWB pads, the hot-bar is pressed against the lead-pad combinations and heated to reflow. Precise control of the heating process is required to prevent flux spattering or solder bridging. A mechanical stop is employed to keep the bonding head from exerting too much force on the lead and pad combination. Excessive force can cause the component leads to slide off pre-reflowed (domed) solder-coated pads. It can also enhance solder bridge formation and solder-starved joints by physically displacing solder from between leads and pads.

One approach to preventing bridged or solder-starved joints is to retract the hot-bar by several hundredths of a millimeter once the solder starts to reflow. This results in a surface tension coupling of solder between lead and land, reduced propensity for solder bridge formation, and robust joint formation upon cooling. As in any other soldering process, it is important to maintain proper reflow time-temperature characteristics. In hot-bar soldering, it is of paramount importance that the blade seating plane is maintained normal to the surface place of the PWB. On the other hand, it should be noted that the coplanarity of solder on the PWB pads does not have to be exact. The hot-bar will reflow this solder and it will self-planarize, as does any liquid, and remain planarized into the solid phase.

47.10.4 Hot-Bar Construction

The hot-bar bonding head can be composed of one or several blades. They are designed to solder one side, two sides, or all four sides of a gull-wing leaded component simultaneously. Generally the hot-bar assembly is configured to accommodate the maximum span of a lead set, permitting simultaneous bonding of all leads on a package side. In the case of some very long connectors or other large packages, the thermal uniformity of a single blade may not be adequate for the process. This is sometimes remedied by using a smaller hot-bar and stepping it along the length of the lead set until all leads are bonded. Another approach uses multiple hot-bars side by side, each individually controlled, to achieve the span and required thermal uniformity. Bar configuration is designed for the lead form to be bonded. It should sit flatly on the foot of the lead, neither contacting the radiused area of the lead form nor significantly overhanging the lead toe, as shown in Fig. 47.38.

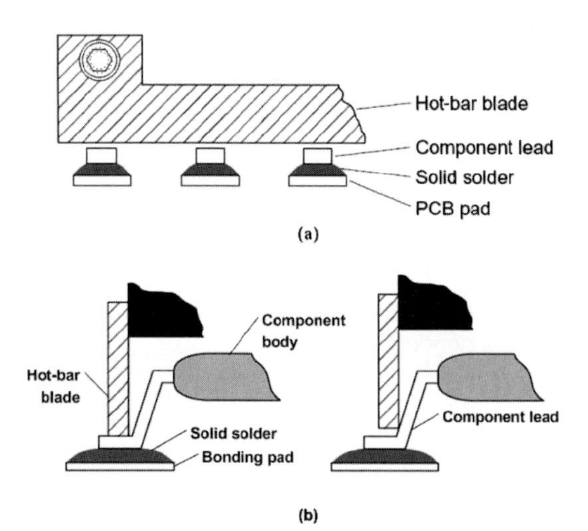

FIGURE 47.38 A hot-bar in contact with leads: (a) transverse view; (b) lateral view. The hot-bar must be positioned so as not to interfere with the leg of the formed component lead. It should sit squarely on the foot of the lead.

The bar itself can be manufactured to nearly any dimension, but there are limitations on its size; the longer the blade, the worse its longitudinal thermal uniformity. Variations in blade temperature may cause it to distort due to differential thermal expansion and contraction. Thermal uniformity is critical for consistent lead-to-lead soldering and joint quality per component side. The allowable variation in bar temperature is limited by the component specification, the damage threshold of the board to be soldered, the solder, and product reliability requirements. As in any other soldering process, the variation in process temperature must be understood and controlled to ensure highest board assembly yields and best-quality solder joints. Most hot-bar systems are configured with a fine-gauge thermocouples welded to the bar, an integral part of a closed-loop, temperature-controlled hot-bar heater system. There may be more than one thermocouple per bar.

Numerous bar designs and materials of construction are available; tungsten, titanium, and molybdenum are commonly chosen for their electrical resistance, durability, thermal conductivity, immunity to flux damage, and solder nonwetting characteristics. Some ceramics are also used for blade construction. The bar must be designed for uniform heating across its length and it must be able to shed heat rapidly to allow the solder to resolidify within a reasonable time. The blade has to expand and contract uniformly throughout the soldering cycle. Some blades have been seen to develop a "frowning" or "smiling" profile in their z-axis during heating due to differential thermal expansion or built-in stress in the metal, as shown in Fig. 47.39. For the same reasons, transverse warping of the blade is also commonplace.

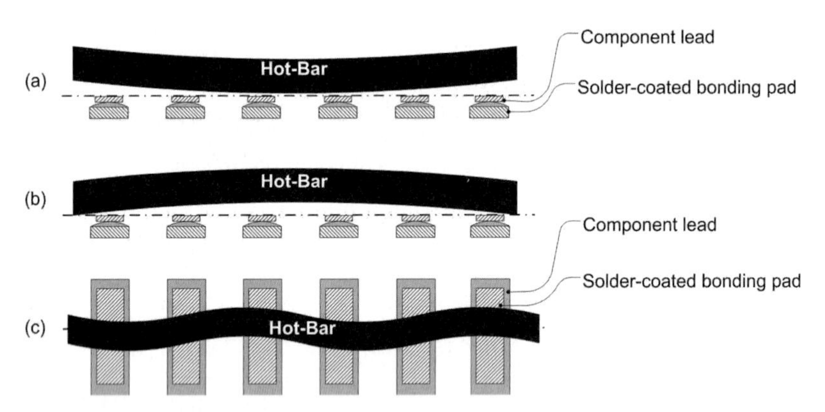

FIGURE 47.39 If not constructed properly, the hot-bar can develop a "smiling" profile (a), a "frowning" profile (b), or may warp along its length (c). (a) and (b) are front views whereas (c) is a top view. The dashed line in (a) and (b) shows the desired seating plane.

These conditions detract from proper blade contact with lead and pad combinations and are indicative of poor hot-bar design or poor choice of materials. Blade warpage, if the curvature is sufficiently large, may result in solder opens. There have been many solutions proposed in the realm of structure, materials of construction, and electrical input to permit highest uniformity of blade heating and contact to the bonding pads. D. Waller et al.[25] have patented a rigid molybdenum truss blade that has been found to be stable dimensionally and thermally uniform over long spans (in excess of 76 mm).

Usually the hot-bar blade is attached to a self-leveling, spring-loaded bonding head that allows the blade to self-planarize in relation to the underlying component leads and circuit board surface. Nonetheless, variations in board contour, such as a localized high or low spot, can render the concept of self-planarization useless and solder opens can result. Rigid-member hot-bars are excellent for accommodating differences in lead coplanarity because the blade forces lead to pad. Driving a heavily sprung lead onto its corresponding bonding pad can, however, lock stress into the resultant solder joint because the solder is working to resist the spring force of the lead. This can detract from the joint reliability and cause premature joint failure.

47.10.5 Maintenance and Diagnostic Methods

As with any process equipment, it is important to maintain the hot-bar properly. Check it frequently for distortion. Ensure that it remains flat and perpendicular to the circuit board to be bonded. Scrub the hot-bar on a ceramic flat to remove any baked-on flux residue. This may need to be done every few bonding cycles, depending on the flux and the criticality of the assembly.

Several diagnostic tools can be applied to assess hot-bar conditions. The two most important characteristics that must be understood and monitored are thermal performance and blade planarity.

47.10.6 Thermal Monitoring

One of the most common methods of assessing hot-bar performance is to use a thermocouple-instrumented board, as is the case with any other soldering method. Preparing such a board for thermode use has its own requirements. Fine-gauge thermocouples are attached to leads of the component or the bonding pad of the circuit board to be soldered. Positioning thermocouples at both ends of the lead set and also near its center helps quantify the longitudinal thermal uniformity of hot-bar performance during soldering. It is also useful to deploy thermocouples on surrounding components to ensure that the hot-bar soldering process does not jeopardize the integrity of adjacent, previously soldered joints.

When preparing the thermal profile board for hot-bar soldering, avoid placing the thermocouple bead(s) between component lead and bonding pad. The added height of the bead would prevent the bar from contacting adjacent component leads and results in point-contact heating that is not indicative of the bar's normal operation. Instead, place the thermocouple bead at the pad extension area in front of the lead tip or at the lead heel as indicated in Fig. 47.40.

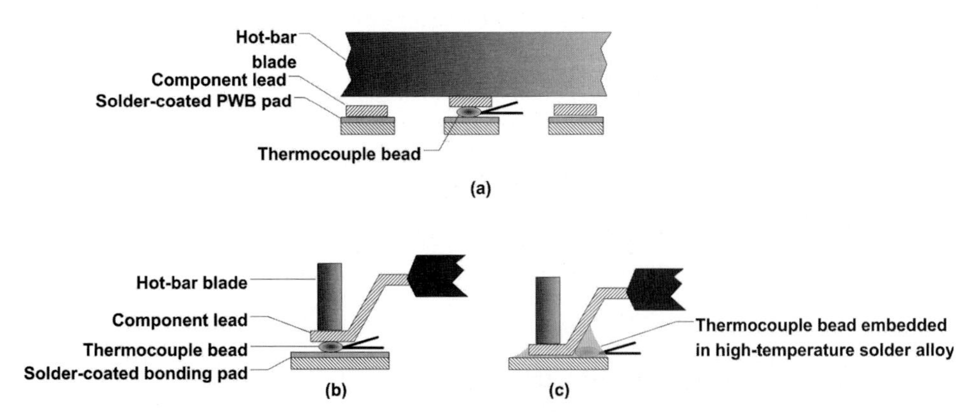

FIGURE 47.40 Just as in any other soldering method, a thermal profile board is needed for hot-bar soldering. Care must be taken that the thermocouple bead does not interfere with seating of the lead to pad or with the hot-bar blade to the lead and pad combination (a, b). Instead, the thermocouple can be attached with a high-temperature solder alloy in front of or behind the component lead (c) so that a common lead, pad, and bar seating plane is maintained.

47.10.7 Blade Planarity

As stressed throughout this section, the coplanarity of the hot-bar blade or multiblade assembly with respect to the bonding surface is of extreme importance to preclude solder opens during hot-bar soldering. There are numerous measurement methods that can help diagnose this; unfortunately, though, they are meant for evaluation of the blade(s) at room temperature, not at bonding temperature. As previously mentioned, a blade may distort temporarily or permanently during heating, but planarity measurements on hot blades are impractical. Therefore, the majority of techniques encountered are performed at room temperature.

J. A. Wilkins[26] suggests the use of a colorant, such as from a marking pen, applied to the cold bonding surface of a freshly cleaned thermode. Once the ink is dry, the blades are then scrubbed over a clean, flat ceramic plate. Low spots on the bar are indicated by the presence of colorant remaining on the bar after several circular swipes on the ceramic flat. Single-bar and two- or four-sided blade assembly planarity can also be evaluated using an array of ground, leveled, rigidly mounted pressure transducers. Bar pressure differential, an indicator of blade planarity, can be adjusted so as to be uniform from end to end and from blade to blade on the two-up or four-up hot-bar assembly. Of course, some hot-bar assemblies are self-leveling, but even these should be checked for planarity and force per blade to ensure best uniformity during soldering. The use of pressure transducers has poor spatial resolution.

47.10.8 Process-Induced Defects

Solder bridging is the most prevalent problem associated with this method of soldering. Recall that the solder can be squeezed out of the solder joint and that conditions may favor solder bridge formation. Solder opens result from lack of coplanarity between the hot-bar and the plane of the circuit board surface. Lead misalignment during the bonding cycle is another defect that detracts from this method. As pressure is applied to component leads by the hot-bar prior to the onset of solder liquidus, leads are sometimes forced to slide down any domed pre-reflow the solder deposits. This displacement causes misregistration of component leads to bonding pads, and may build stress into soldered joints. If the forces are great enough, it may also cause the whole package to move and misalign the entire lead set.

Because the heating is rapid, the thermode temperature is necessarily well above the solder liquidus temperature. If the time-temperature cycle is not carefully controlled, overly thick intermetallic compound formation can be a problem. This is especially true in this process, where the solder may be largely displaced from between lead and pad, and bond lines are excessively thin. Within the joint, the volume of intermetallic compound (hard and brittle) may be large compared to the remaining solder (soft and compliant). If this is the case, solder joints will be less reliable and more susceptible to brittle fracture. All of these obstacles have prevented the widespread acceptance of hot-bar bonding in manufacturing except in some niche applications where soldering by other methods may be difficult. Hot-bar bonding is most useful for low-volume, fine-pitch surface-mount soldering and rework.

47.11 HOT-GAS SOLDERING

Hot-gas soldering relies upon a heated stream gas heated to effect reflow. This noncontact, directed-energy method is most suited to bonding surface-mount components. Although hot-gas soldering has been around for years and after numerous machine offerings, it is not a popular method for soldering despite its evolutionary improvements. Instead, it has made its mark in the area of component rework—removal of previously soldered devices from a circuit board and replacement of same.

One disadvantage of this soldering method is that the thermal energy is not well localized. Most machines typically emit a hot-gas jet too large to be isolated to reflow only the device of interest. The gas jet, once impinged upon the board and component leads, is deflected and its backwash can be problematic. It may cause unwanted reflow of previously formed joints, especially on closely spaced adjacent components. This problem is typically overcome by the use of baffles that are either applied to adjacent components or by a singular baffle that confines the gas jet to the component to be soldered.

Hot-gas soldering nozzles are available in a variety of forms. The simplest of nozzles is the single orifice, which can be translated around the entire periphery of a component (see Fig. 47.41a). Some machines offer a double translatable nozzle assembly that can solder two opposing sides of a component simultaneously (see Fig. 47.41b).

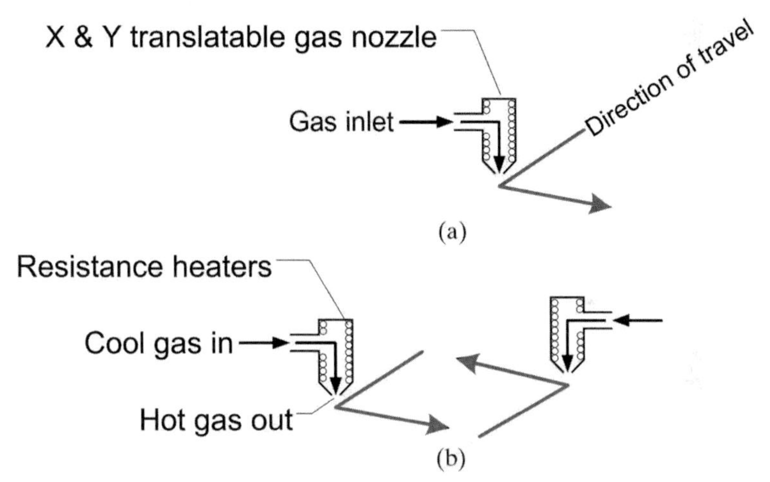

FIGURE 47.41 Hot-gas soldering nozzles: (a) A programmable moving nozzle translates hot gas in x-y axes; (b) double nozzles, each programmable, allow simultaneous soldering of all four sides of a component.

Solder can be applied to the board as a paste, solid pre-form, or solder-coated pad. In all cases, the component must be held down during the soldering process to ensure contact of component leads with bonding pads. Proper gas pressure, temperature, nozzle translation speed, and flux are required to effect joint formation. Otherwise, the same reflow considerations are required for this technique as for any other. Heating ramp rate, solder paste preheating, peak temperature, liquidus duration, etc., must be observed for successful joint formation and joint reliability considerations.

A thermocouple-instrumented profile board populated with components is needed to determine the proper soldering profile. Adding thermocouples to adjacent solder joints is recommended to ensure that the soldering profile will not inadvertently re-reflow neighboring components.

47.12 ULTRASONIC SOLDERING

This method relies on a heated, ultrasonically vibrated soldering tip that simultaneously melts and agitates the solder (Fig. 47.42). The ultrasonic energy is transferred from the soldering tip through the molten solder droplet beneath it and ultimately to the component lead and circuit

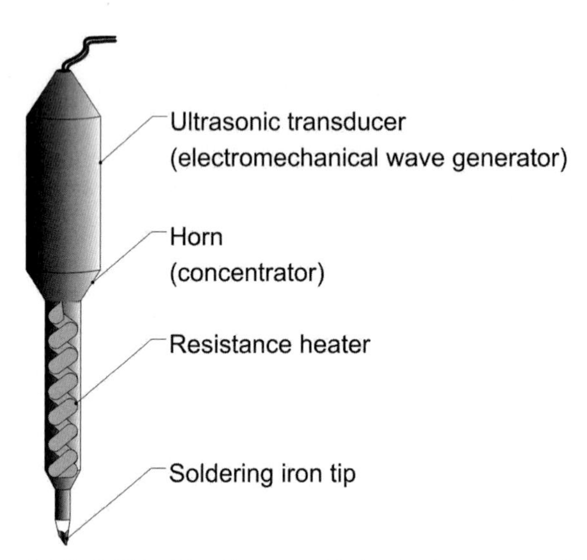

FIGURE 47.42 The ultrasonic soldering iron is composed of four main components: an ultrasonic transducer; a horn for concentrating and directing ultrasonic energy; a resistance heater; and the soldering tip, which emits both thermal and ultrasonic energy.

board pad. The high-energy agitation of the solder droplet helps to cleanse bond-inhibiting materials from the solder and solder-metal interfaces (component lead and bonding pad). If soldering temperature is appropriate, any exposed solderable metal surfaces are rendered wettable by this technique. Ultrasonic soldering precludes the need for chemical fluxing agents. Aluminum and other difficult-to-join metals can be soldered by this method. The viability of this technique has been well proven on a commercial scale in the manufacture of air-conditioner heat exchangers.[27,28]

Ultrasonic soldering has also been applied as a batch or continuous mass reflow process. In these instances, the molten solder is ultrasonically agitated while the assembly to be soldered is immersed in it. Similar arrangements have been made for ultrasonically vibrating the part while dipping it in a molten bath or wave of solder. These mass processes are more common for non-electronic assembly.

Care must be taken in ultrasonic soldering to tune the tip amplitude and/or frequency to the mass of the system being soldered. Overagitation results in excessive cavitation and splashing of the liquid solder. This generates solder balls, which could short finer-spaced leads or pads. Additionally, ultrasonic agitation increases the dissolution rate of any soluble metals into the solder at any given temperature, which may degrade solder-joint strength.

This technique can be useful for the repair of opens or the installation of new or change-order components onto completed circuit assemblies. Because no flux is required, a previously cleaned board will stay clean through repair or upgrading operations. This technique is applicable to all peripherally leaded surface-mount components that are not susceptible to ultrasonic energy damage. It can also accommodate through-hole device soldering. Equipment availability is limited, with only a few manufacturers worldwide. Several past and recent publications provide a comprehensive review of this technology's applicability and attributes.[29,30,31,32]

REFERENCES

1. IPC/JEDEC J-STD-033, "Standard for Handling, Packing, Shipping and Use of Moisture/Reflow Sensitive Surface Mount Devices."
2. Freedman, G. M., Patel, K. B., Batchelder, R. G., "Method and Process for Soldering Reflow-Compatible Plated Through-Hole Components into Printed Wiring Board in Order to Circumvent the Wave Soldering Process (Buried Intrusive Reflow Soldering and Rework)," *Research Disclosure*, May 2006, pp. 543–545.
3. Freedman, G. M., Baldwin, E. A., "Method and Material for Maintaining Cleanliness of High Density Circuits during Assembly," *Proceedings of the SMTA Pan-Pacific Conference*, February 2003.
4. Iman, R., et al., "Evaluation of a No-Clean Soldering Process Designed to Eliminate the Use of Ozone Depleting Chemicals," IWRP CRADA No. CR91-1026, Sandia National Laboratories, 1991.
5. Hartman, H. H., "Soft Soldering under Cover Gas: A Contribution to Environmental Protection," *Elektr. Prod. Und Prftechnik*, H.4, 1989, pp. 37–39.
6. Idem, "Nitrogen Atmosphere Soldering," *Circuits Assembly*, January 1991.
7. Bernier, D., "The Effects of Metallic Impurities on the Wetting Properties of Solder," *Proceedings of the First Printed Circuit World Convention*, Uxbridge, UK, Vol. 2, 1978, 2.5–1.5.
8. Hallmark, C., Langston, K., and Tulkoff, C., "Double Reflow: Degrading Fine Pitch Joints in the Soldering Process," *Technical Proceedings of NEPCON West*, 1994, pp. 695–705.
9. Kear, F. W., "The Dynamics of Joint Formation," *Circuits Assembly*, October 1992, pp. 38–41.
10. Manko, H., *Solders and Soldering, 2nd ed.*, McGraw-Hill, 1979.
11. Wassink, R. J. K., *Soldering in Electronics*, Electrochemical Publications, Ltd., 1984.
12. Hutchins, C. L., "Soldering Surface Mount Assemblies," *Electronic Packaging and Production*, Supplement, August 1992, pp. 47–53.
13. Hartmann, M., et al., "Experimental Investigations of Laser Microsoldering," *SPIE*, Vol. 1598, Lasers in Microelectronics Manufacturing, 1991, pp. 175–185.
14. Spletter, P. J., and Goruganthu, R. R., "Bonding Metal Electrical Members with a Frequency Doubled Pulsed Laser Beam," U.S. Patent No. 5,083,007, January 1992.
15. Freedman, G., "Atmospheric Pressure Gaseous Flux-Assisted Laser Reflow Soldering," U.S. Patent No. 5,227,604, July 1993.
16. Charschan, S., *Lasers in Industry*, Laser Institute of America, 1972, p. 116.
17. Hecht, J., *Understanding LASERs,* Howard W. Sams, 1988.
18. Ready, J. F., *Industrial Applications of LASERs,* Academic Press, 1978.
19. Burns, F., and Zyetz, C., "Laser Microsoldering," *Electronic Packaging and Production*, May 1981, pp. 109–120.
20. Chang, D. U., "Analytical Investigation of Thick Film Ignition Module by Laser Soldering," *Proceedings of ICALEO '85*, 1985, pp. 27–38.
21. Hartmann, M., et al., "Experimental Investigations of Laser Microsoldering," *SPIE*, Vol. 1598, Lasers in Microelectronics Manufacturing, 1991, pp. 175–185.
22. Hernandez, S., "Wire Bonding with CO_2 Lasers," *Surface Mount Technology*, March 1990, pp. 23–26.
23. Wright, E., "Laser vs. Vapor Phase Soldering," *Proceedings of the 30th Annual SAMPE Symposium*, March 1985, pp. 194–201.
24. Lish, E. F., "Laser Attachment of Surface Mounted Components to Printed Wiring Boards," paper presented at the Sixth Annual Soldering Technology Seminar, Naval Weapons Center, China Lakes, CA, February 1982.
25. Waller, D., Colella, L., and Pacheco, R., et al., "Thermode Structure Having and Elongated, Thermally Stable Blade," U.S. Patent No. 5,229,575, July 20, 1993.
26. Wilkins, J. A., "Heat Transfer Control for Hot-Bar Soldering," *Proceedings of Surface Mount International*, 1993, pp. 186–192.

27. Gunkel, R., "Solder Aluminum Joints Ultrasonically," *Welding Design and Fabrication*, Vol. 52, No. 9, September 1979, pp. 90–92.
28. Shuster, J. L., and Chilko, R. J., "Ultrasonic Soldering of Aluminum Heat Exchangers (Air Conditioning Coils)," *Welding Journal* (U.S.), Vol. 54, No. 10, October 1975, pp. 711–717.
29. Hosking, F. M., Frear, D. R., Vianco, P. T., and Keicher, D. M., "Sandia National Labs Initiatives in Electronic Fluxless Soldering," *Proceedings of the 1st International Congress on Environmentally Conscious Manufacturing*, Santa Fe, September 18, 1991.
30. Vianco, P. T., Rejent, J. A., and Hosking, F. M., "Applications-Oriented Studies in Ultrasonic Soldering," *Proceedings of the American Welding Society Convention and Annual Meeting*, Cleveland, 1993.
31. Antoncvich, J. N., "Fundamentals of Ultrasonic Soldering," *American Welding Society, 4th International Soldering Conference, Welding Journal Research Supplement*, Vol. 55, July 1976, pp. 200-s to 207-s.
32. Shoh, A., "Industrial Applications of Ultrasound," *IEEE Transactions on Sonics and Ultrasonics*, Vol. SU-22, March 1975, pp. 60–71.

CHAPTER 48
SOLDERING REPAIR AND REWORK

Gary M. Freedman
Hewlett-Packard Corporation, Business Critical Systems, Singapore

48.1 INTRODUCTION

As printed circuit assemblies (PCAs) become more densely populated, the probability of rework increases. So does the complexity of rework steps. Rework deserves consideration both in terms of board design as well as in conducting the repair. Deep preheating cycles, careful thermal profiling, and thermal masking of adjacent components become critical. PCAs are often mix-mount (surface-mount technology [SMT] and plated through-hole [PTH]). Additionally, they may contain press-fit components or other subassemblies that may be sensitive to overheating. Aside from hand repair with a soldering iron, there are two techniques that are the mainstays of repair: solder fountain and hot-gas. The former is a variant of wave soldering and is tailored to removal or post-soldering addition of PTH (solder-tail) components. Hot-gas (also known as hot-air) soldering is used primarily for removal and replacement of SMT components.

Because repair of dense assemblies is difficult, this chapter covers some additional descriptions and tips for repair. The onset of Pb-free soldering has increased the difficulty of repair processes due to generally higher melting temperatures and slower wetting times that are characteristic of the most prevalent Pb-free solders. Otherwise, repair involving Pb-containing solders and Pb-free solders is the same.

48.2 HOT-GAS REPAIR

This technique consists of directing a heated stream of air or other process gas such as nitrogen at the component to be removed. Modern hot-gas repair stations are sophisticated, with computer control, predictive reflow profiling, vision systems to assist in component alignment to bonding pads on the printed wiring board (PWB), preheating stations, and accommodations for component removal. Hot-gas repair is most often used for SMT components but can be adapted for PTH component removal.

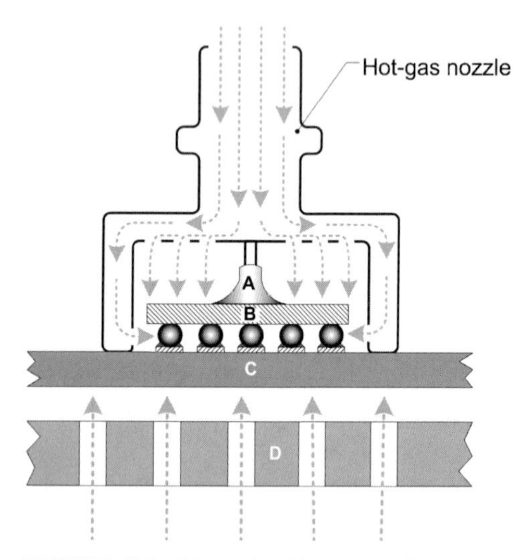

FIGURE 48.1 Schematic of hot-gas repair of a ball grid array (BGA) (B) soldered to a circuit board (C). Hot-gas (air or nitrogen) is injected through a baffled nozzle. In this diagram, the heated gas is directed under and on top of the BGA. A heat-resistant suction cup (A) is used to remove the BGA once all joints have reached reflow temperature. Gas injected through a heated platen (D) is used to preheat the circuit board, lessening the thermal load and shortening the rework cycle.

48.2.1 Hot-Gas Repair Equipment

48.2.1.1 Hot-Gas Nozzle. The heated gas is generally delivered through a component-specific nozzle. The gas stream is directed at the top of the component, under the component, or a combination of the two, depending on the nozzle design (see Fig. 48.1).

48.2.1.2 Preheater. Preheaters are used to help overcome the thermal inertia of the board. Since the nozzle exposes only a small amount of the board to a high-temperature gas stream, heating through this mechanism is limited. The preheater uses heated gas or infrared heating but heats a much larger area of the board. The preheating is set to attain a temperature much lower than the reflow temperature of the solder, but high enough that when nozzle heating is applied, reflow can take place in a reasonable period of time without overheating the component to be replaced or other components adjacent to the repair site.

Most hot-gas repair machines have a built-in bottom-side preheater. An alternative to preheating would be to bake the board in a box oven. The problem with this approach is rapid cooling of the board from the time of removal from the oven and during setup of the board on the hot-gas repair station.

48.2.1.3 Vacuum Pickup Device. This is generally a silicone rubber suction cup or a solid metal tube nested within the nozzle and connected to a vacuum pump. Silicone is chosen for its heat resistance and compliance. The vacuum can be toggled on and off to pick up and release a component. The vacuum pickup device serves two purposes. First, it is used to remove a component from the circuit board once the component has been heated by the hot-gas to the point of reflow. Following dressing of the repair site, the suction cup is used to hold the replacement component rigidly during lead-to-pad alignment and placement steps.

48.2.1.4 Adjustable Stage. This permits alignment of the nozzle to the PWB in the x-, y-. and θ-axes. It is important for component removal but even more critical for component replacement and alignment of component leads to PWB bonding pads.

48.2.1.5 Vision System. This is used to align component leads to PWB bonding pads manually. Usually a split-image video system is used to view both component leads and board bonding pads simultaneously. The images are superimposed and the operator manually adjusts them for the best lead-to-pad alignment.

48.2.1.6 Computer Control. This is the heart of most hot-gas machines. It monitors and controls the various operations of the system, such as preheating and thermal profile. The rework engineer programs the profile manually.

48.2.2 Hot-Gas Repair Process

48.2.2.1 Thermal Profiling. As with any reflow process, whether for first-pass build or for repair, it is necessary to construct a thermal profile board as previously described to assess the impact of heating and to ensure that the heating is adequate, in accordance with the thermal limits described by the solder paste manufacturer, the component specifications for thermal exposure, best solder wetting, and minimized intermetallic formation. In the case of repair, a dummy board is used with thermocouples attached to the component to be removed and critical adjacent components to ensure that reflow conditions are correct.

48.2.2.2 Component Removal Preheat. As discussed previously, preheating may be necessary to overcome the thermal mass of the assembly and to ready the board for the reflow step. Preheat temperature is set well below the reflow temperature of the solder alloy in use to confine reflow to the device being removed or replaced and to discourage widespread reflow in the vicinity of the removal or repair.

48.2.2.3 Reflow and Removal. The same considerations apply for this as for any reflow step. The ramp, soak, peak temperature and time above liquidus are dictated by the solder alloy used and should be roughly the same profile parameters that were used for the initial assembly of the board. Once the component has reached complete reflow, the vacuum pickup tool is engaged and the component is lifted vertically. Care should be taken to minimize the time above liquidus to minimize intermetallic compound thickness.

48.2.2.4 Site Dressing. Following component removal and while the board is still hot, the solder lands on the PWB can be prepared for component replacement by means of a hand soldering iron and a braided copper solder wick or a solder vacuum machine that simultaneously heats the pad by hot gas and vacuums up any excess solder. During the component removal step, solder surface tension capriciously allows some of the solder from the joint to be removed with the package leads and some to be left on the bonding pads. In the case of area-array packages (BGA, ceramic ball grid array [CBGA], etc.), the ball may remain with the package or with the land. This can make solder land dressing a labor-intensive operation. Care at this step is critical for success in resoldering a fresh component and for reliability. The dressed bonding pads should be smooth so as not to interfere with fresh component placement.

48.2.2.5 Part Bake-Out. As with first pass soldering, it is important to observe rules concerning component bake-out to remove any absorbed moisture and preclude part cracking during the reflow heating cycle. Guidance for part baking can be found in IPC/JEDEC J-STD-033.[1]

48.2.2.6 Solder Paste or Flux Deposition. Although this is not part of the hot-gas station, it is part of the hot-gas work cell and needs to be discussed. For most SMT components, solder paste is added to either the board or the component to effect a reliable solder joint for component replacement. This is generally done by ministencil similar to what is used for SMT solder paste deposition, but on a smaller scale; the stencil is only slightly larger than the component footprint. The paste is pushed through the stencil apertures with a razor blade or a miniature stainless-steel squeegee. Programmable syringe-dispenser machines are sometimes used, but solder paste is known to separate when under pressure, i.e., solder alloy separates from the solder flux. Choosing the largest needle bore, slowest dispense rates, and lowest pressures will help to mitigate paste separation. Also, use of direct drive auger pump rather than an air-pressure-driven dispenser will help to reduce paste separation.

Sometimes in BGA repair, solder paste is not used. Instead, solder from the BGA ball is used to make the solder joint, and solder flux must be added for this bonding step. The flaw in this methodology is that lowered solder volume results as compared to the original solder joint and the BGA reliability may not be as robust as when first soldered. In fact, any repaired joint is less reliable than a first-pass solder joint due to thickening intermetallic layer formation that comes with each reflow cycle. Addition of solder paste results in a better process because the paste compensates for small differences in BGA ball sizes. It domes upon reflow, adding some compliance to make up for non-coplanarity solder balls. Opens are less likely to occur, although more process precision is required to dispense the paste and place the BGA without causing solder bridges. Eliminating solder paste is not an option for area-array devices with noncollapsible interconnects, such as ceramic column grid arrays (CCGAs) or CBGAs. Solder has to be added for rebonding of this type of component since the interconnects (balls or columns) are of a high-temperature solder alloy and will not reflow. Most commonly, the necessary solder is added by the aforementioned ministencil method.

48.2.2.7 Soldering Preheat. After flux deposition, the board is preheated to a temperature about 50 to 70°C below solder liquidus. Generally, the hot-gas rework station has its own preheater and component and board preheating is part of the reflow profile.

48.2.2.8 Component Pickup. As the board is preheating, the component is picked up and readied for placement. After the component is placed on a pickup station, the nozzle with vacuum pickup is placed and lowered over the component and the vacuum is activated. The component is raised up and readied for the alignment step.

48.2.2.9 Component-to-PWB Alignment, Placement, and Reflow. Generally, a split vision optics system is used and the semitransparent image of the component to be placed is superimposed over the video image of the PWB bonding pads for that component. When the two are manually aligned, the component is lowered on to the solder paste- or flux-coated bonding pads and the thermal cycle begins.

Once reflowed and the solder reaches solidus, the rework process is complete. The board is then ready for inspection.

48.2.2.10 Inspection. Visual inspection should be conducted as practical to ensure integrity of any visible solder joints and also the area surrounding the rework. Ensure that solder mask is not charred or delaminated. Also check whether bonding pads have delaminated (not possible on inner rows of area-array packages).

For BGAs and other area-array devices, following a visual check, use transmission x-ray to inspect for solder bridges. Sometimes this method also reveals lack of solder. Lastly, if in-circuit testing or functional testing is done on the first-build product, then it should be applied to the reworked product also.

48.2.3 Hot-Gas Rework Guidelines

Generally the number of SMT solder rework cycles per component per board is kept to two or three. Beyond that, the board should be scrapped to avoid reliability problems. The key in all reflow steps is to limit the amount of time at or near solder liquidus in order to minimize intermetallic compound (IMC) formation. IMC thickness increases with each reflow cycle and is temperature-dependent also. IMC is brittle—the weakest link in terms of solder joint structure and resultant joint strength. During component rework and replacement, minimizing thermal excursion is all important. Too often, overly high temperatures are used during rework and dwell at liquidus are prolonged as compared to a normal mass reflow soldering cycle. When a component is removed and replaced, IMC from previous soldering operations or other thermal excursions remain on the pad since the IMC becomes an integral part of the pad.

For illustrative purposes, consider the case of a surface-mount component with a copper lead-frame and copper bonding pad and either a leaded (Pb'd) or lead-free (Pb-free) solder alloy on the bottom side of the PWB. During the first-pass assembly, the bottom side of the board is built first. During that soldering process, an intermetallic layer forms at the copper lead-to-solder interface and also at the solder-to-bond pad interface. During topside soldering, the components on the bottom side of the board undergo a second reflow. Assume that a part soldered during the bottomside assembly fails and requires removal. The removal and replacement processes necessitate additional reflow cycles to the affected component site, including component removal; site repair, and component replacement. With each soldering cycle, the intermetallic layer builds and reliability generally degrades with increasing

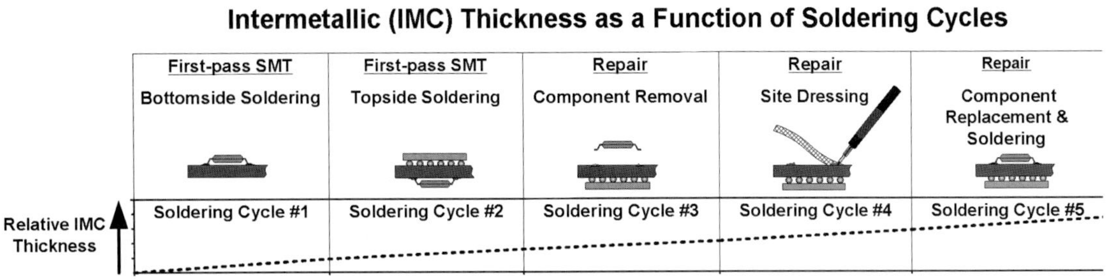

FIGURE 48.2 Soldering steps required for first-pass soldering of a bottomside component on a PWB and success reflow steps encountered during first-pass soldering and repair. Note that the intermetallic compound formation thickness is not truly linear with each step. Thickness depends on materials of the soldering system, time above solder alloy liquidus, peak temperature, etc. The illustration is meant to show that with each reflow cycle, there is increased intermetallic layer thickness.

intermetallic layer thickness, Therefore, careful process control is needed to minimize time above solder liquidus. Figure 48.2 depicts soldering and repair steps and a hypothetical impact to intermetallic layer thickness common to any soldering and repair cycles.

The thermal profile used for component replacement should be predicated on the four things: (1) the solder paste manufacturer's recommendations for reflow, (2) component manufacturer's specification for peak temperature and heating ramp rate, (3) optimization of the reflow profile, and (4) results of thermal profile established from trials with the instrumented product profile board.

For no-clean boards, the same solder paste used for the initial assembly should be used for the repair. For aqueous-cleaned boards, often a no-clean solder chemistry is used to preclude the need for another wash cycle. If solder paste is not added, then a paste flux (identical to that comprising the solder paste) should be used. This will obviate solder chemistry interactions and reduce likelihood of long-term corrosion.

48.3 MANUAL SOLDER FOUNTAIN

PTH components are generally repaired by solder fountain, which is akin to a wave-soldering system with a vertically pumped jet of solder. It is significantly smaller than a wave-soldering machine and generally lacks a conveyor, a fluxer, and preheaters. Since the equipment is much smaller, often benchtop-mounted, it permits an operator to manually remove parts and replace them. The classic solder fountain is a manually controlled system. Nozzles are chosen to shape the molten solder jet to match closely the size of the area to be soldered. Nonetheless, a shield or pallet is often used to localize the soldering and to prevent contact of the molten solder with previously soldered components on the board.

The board requiring repair is shielded, except in the area of component replacement, and positioned over the solder fountain. The shaped solder fountain is directed at the repair site. Once the component solder joints go molten, the component can be pulled out with gloved hand or pliers, some flux can be added, and a new component can be inserted to complete the rework within one thermal excursion. Care must be taken to ensure that all solder joints have gone molten on the component being reworked. If not, PTH barrels can be ripped out of the board or barrel annuli can be damaged.

To aid the solder fountain process, dense assemblies can be batch-heated in a box oven to make the repair process more efficient and limit exposure of the board to the turbulence of the molten solder wave. Limiting the soldered assembly's time near or above solder liquidus is key for solder joint and product reliability. As mentioned previously, time above solder liquidus along with peak temperature will define the intermetallic thickness for a given metals system, and intermetallic thickness will in part define the solder joint reliability. The turbulence of the solder fountain is known to enhance dissolution of certain metals into the molten solder of the fountain. Copper PTH barrels can be thinned and PTH annuli (bonding pad atop the PTH) can be thinned or completely dissolved, impairing the degrading solder joint either to make a connection or to ensure long-term reliability. The high turbulence, high temperature, and long dwell times all hasten dissolution in this fashion. For this reason, precise process control, including strict exposure time limits and minimum solder pot temperature, must be imposed for this largely manual operation. This process is best suited for repair and should be avoided, if at all possible, for first-pass soldering of printed circuit assemblies.

48.4 AUTOMATED SOLDER FOUNTAIN

This version of the solder fountain incorporates a programmable head and is used mostly for first-pass soldering, but can be an effective soldering repair tool once the PTH component is removed, the site is prepped, and a new component is placed. A programmable nozzle emitting a precise jet of molten solder is directed at the leads of the component to be soldered. The nozzles allow point soldering of very fine PTH pitches.

48.5 LASER

Laser is a recent innovation in rework and repair. The fundamentals of laser for initial soldering or for repair have already been discussed in Chap. 47. In the commercial incarnation of this technique, a laser beam is quickly scanned around component leads or, package surface, in the case of area-array devices such as BGAs and chip-scale packages (CSPs). It is most effective for plastic packaged components rather than the thermally massive CBGAs, CCGAs, and so on. In this technique, the component body is heated. Otherwise, there is little difference between this and alternative rework or repair techniques.

Since the energy of the laser beam is more tightly confined to the area it irradiates, the heating is more localized. This can be advantageous when trying to remove or replace a component on a densely populated, double-sided PCA. Otherwise, the same caveats must be kept in mind regarding IMC formation and pulling components before they are fully reflowed. Care must be taken to assess carefully the assembly for laser damage threshold so as not to char components or PCAs during either removal or component resoldering.

48.6 CONSIDERATIONS FOR REPAIR

48.6.1 Mixing Aqueous With No-Clean Chemistries

Care must be taken when using fluxes for the rework and repair processes. If an aqueous-clean solder chemistry is used for building the PCA, then either no-clean or aqueous flux formulations can be used. If an aqueous chemistry is used for the repair, then the board must be subjected to another aqueous cleaning cycle. Be sure to check that all components at that stage are compatible with the aqueous cleaning process.

No-clean solder flux formulations can be used for rework and repair even if the PCA was manufactured with an aqueous clean chemistry, although the reverse is not recommended. Sometimes when a no-clean board is subjected to an aqueous cleaning, the no-clean flux residue takes on a white, gummy characteristic that is conducive to dendritic corrosion, which can result in soft or hard electrical shorting. Saponified aqueous cleaning can be used, but must be tested for effectiveness in removing the polymerized flux residue from the no-clean process and for compatibility with the selected no-clean flux to avoid generation of corrosive byproducts. This is particularly important under connectors, area-array devices, and other components with low headroom between the underside of the body and the PWB surface.

48.6.2 Pooled Flux

Another consideration with no-clean solder chemistries is that of pooled, unactivated flux residue. No-clean solder flux is generally harmless after it has gone through the thermal cycle associated with solder reflow. It is activated and denatured by the heat. If the flux is applied generously and the flux pool does not see the high temperatures associated with solder joint formation, acids remaining in the unheated flux residue can cause reliability problems. These acids are normally denatured, evaporated, or sublimed through the high-temperature excursions of the soldering process. If the flux acids remain, corrosion can occur in the long term. There are commercial flux solvents that can be used to dilute and wash away excess flux; otherwise, the assembly can be baked briefly to about 120°C to denature the flux. Check with the flux manufacturer for specific recommendations regarding flux residue management. The problem can be avoided by eliminating the use of liquid flux and using small amounts of paste flux instead. In the case of core wire, use the smallest diameter practical to limit flux residue.

48.6.3 Component Bake-Out

Components that have been sitting out may be subject to water absorption. When components are heated during rework, where the heating cycle is more localized and perhaps more extreme than in the slow heating experienced during mass oven reflow soldering, moisture-induced popcorning may be a more serious problem. Follow IPC or component manufacturer recommendations for presoldering moisture bake-out. Refer to Sec. 50.2.2.5 for further discussion on this topic.

48.6.4 Gold Finger Protection

There is ample opportunity to ruin gold edge connector fingers on a circuit board during the repair operation. Care should be taken to keep both solder and flux away from gold fingers during repair. Mechanical shields or acrylic-adhesive polyimide tape can be used to protect fingers from smeared solder paste, loose solder debris, or flux. If solder comes into contact with gold, the gold will rapidly dissolve into the solder mass and the solder will spread on the finger. The solder cannot be removed from the gold finger without scraping and replating, a costly process. Solder is not a reliable connector contact material, especially with today's high-density, low normal-force electrical connectors.

Keep solder away from gold fingers. Keep gold finger temperature as low as possible. Ensure that gold fingers are never handled, even with gloved hands. Avoid lint and fibers. After rework, clean gold fingers with a suitable solvent. Use a soft brush, lint-free cloth, or other nonabrasive material for this cleaning. Clean gold fingers on each side of the board separately.

Avoid folding the lint-free cloth over both edges and wiping repeatedly; the cloth will wear against the glass-epoxy expunge and shred, leaving lint on the board. A better alternative is to use a soft, ESD-safe brush with solvent to clean the gold fingers and follow with a filtered-air blow-off to dry. Avoid any materials that will scratch the gold surface finish on the fingers. The underlying nickel, if exposed to the atmosphere will oxidize and, is known to be problematic for separable contact reliability.

REFERENCE

1. IPC/JEDEC J-STD-033, "Standard for Handling, Packing, Shipping and Use of Moisture/Reflow Sensitive Surface Mount Devices."

P · A · R · T · 11

NONSOLDER INTERCONNECTION

CHAPTER 49
PRESS-FIT INTERCONNECTION

Gary M. Freedman
Hewlett-Packard Corporation, Business Critical Systems, Singapore

49.1 INTRODUCTION

Electronic connectors are attached to printed circuit boards in four ways (see Fig. 49.1):

- Oven reflow surface-mount technology (SMT) soldering
- Wave soldering (for solder-tail components)
- Pressure interconnect, a solderless method relying on mechanical forces to hold interconnect elements together to make contact
- Press-fit, another mechanical, solderless method (which is the subject of this chapter)

Press-fit, also referred to as *press-pin, compliant-pin,* and a number of other trade names, has been in use for many years and is a proven and reliable interconnect method. Once used exclusively for passive backplanes, more recently press-fit connectors have gained in popularity and are commonly incorporated on complex motherboards and daughter-cards. Just as circuit boards have become more complex (with thicker, higher layer counts, densely routed circuit traces, and high component counts), so too have press-fit connectors. They are available with high pin density (pin pitch down to 1 mm) and pin counts in the thousands per connector and even strip-line shielding for enhanced high-speed signal performance. Press-fit connectors can be applied to either or both sides of a PCA and are repairable, although first-pass assembly yields are always near 100 percent. Press-fit is easier and more reliable than soldering and does not subject the printed circuit assembly (PCA) to additional thermal or chemical processes—an advantage for reliability, and especially advantageous for today's dense circuitry. Since it requires very little energy (no soldering), minimizes materials (no solder), and calls for no chemicals for its application, it is environmentally beneficial. The press-fit process is simple in theory and generally so in practice, relying on oversized connector leads forced into in plated through-holes in the PWB. The plated through-holes in the PWB are sized specifically for the press-fit connector to be applied. As the connector pin is forced into the board, there is slight deformation to the press-fit pin and to the PTH barrel. The result is a mechanically stable, electrical contact.

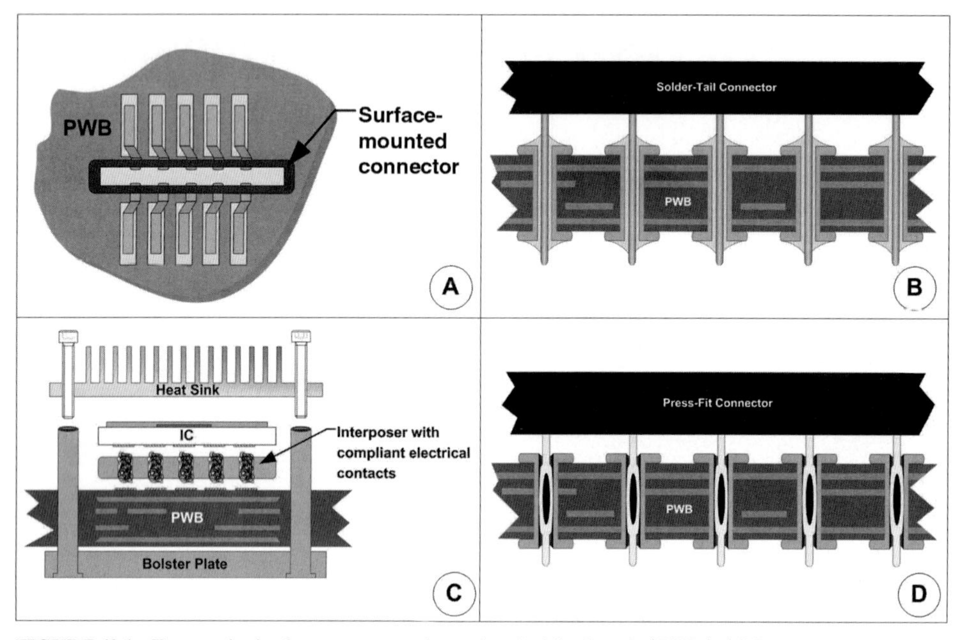

FIGURE 49.1 Four methods of connector attach to printed wiring boards (PWBs): (a) SMT applied by reflow soldering; (b) plated through-hole (PTH) solder-tails that are wave-soldered; (c) pressure-interconnect, which relies on mechanical forces for electrical contact; and (d) press-fit, which relies on mechanical deformation of the connector lead and PTH barrel to make intimate electrical contact.

49.2 THE RISE OF PRESS-FIT TECHNOLOGY

There is a resurgence in press-fit connector popularity driven by increasing board complexity, the component density of today's PCAs, and the appearance of backplanes populated with active components. Press-fit connectors are typically used on very thick boards that would be difficult, or impossible, to wave-solder. Elimination of wave soldering increases manufacturing yield and allows a denser component layout on the bottom side of the board, contrary to what is dictated by wave soldering. Another advantage of using press-fit connectors is that it is a means of lead (Pb) abatement since no solder is needed for press-fit installation. Given the climate of increasing importance of environmental responsibility and the potential for lead abatement legislation passing in several countries, use of press-fit components will likely increase. Since Restriction of Hazardous Substances (RoHS) legislation allows Pb for certain products such as two-processor servers and telecommunications equipment until 2010, there is no need to solder the same connectors that are press-fitted and, thus, press-fitting significantly limits the need for Pb.

Connector bodies may or may not be compatible with the thermal rigors of reflow or wave, so press-fit connectors are typically applied after all mass soldering of the PCA is completed. Further, press-pin connectors are not generally soldered into place since rework would be rendered impractical. Rework of press-fit connectors will be discussed later in this chapter.

49.3 COMPLIANT PIN CONFIGURATIONS

There are many different press-fit pin configurations—far too many for this chapter to review each one. Instead, three commonly used styles will be described as an introduction to press-fit technology. The first has a lead of square or rectangular cross-section that is sized to deform

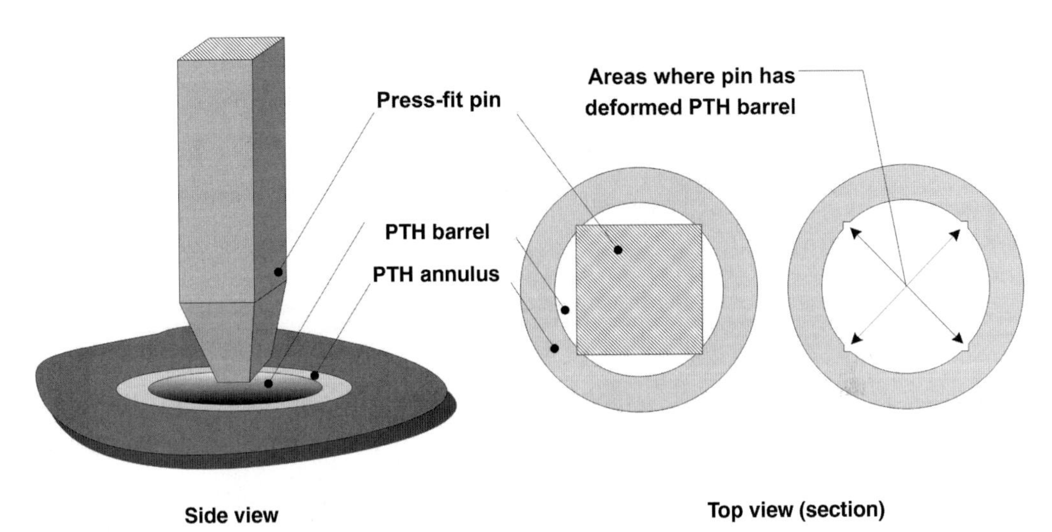

FIGURE 49.2 A square or rectangular pin forced into round hole. This is the most primitive of press-fit techniques.

the board's plated through-hole slightly upon insertion. This is the proverbial "square pin in a round hole." The lead is tapered at its end to facilitate initial PTH insertion (see Fig. 49.2).

Second is the square-pin variant sometimes referred to as the "H" section, which consists of a sculpted rectangle, an "H" in cross-section, whose corners are meant to deform (see Fig. 49.3).

The third and most prevalent configuration is the collapsible pin known as "eye-of-the-needle" (EON) (see Fig. 49.4).

Upon insertion and pressing into the PTH, the EON pin folds upon itself axially in a controlled manner, as depicted in Fig. 49.5.

The plated through-hole barrel is slightly deformed from the mechanical interference of the press-fit pin against it. The resultant forces from the collapsed press-fit lead against the PTH barrel wall result in stable, long-term mechanical and electrical contact.

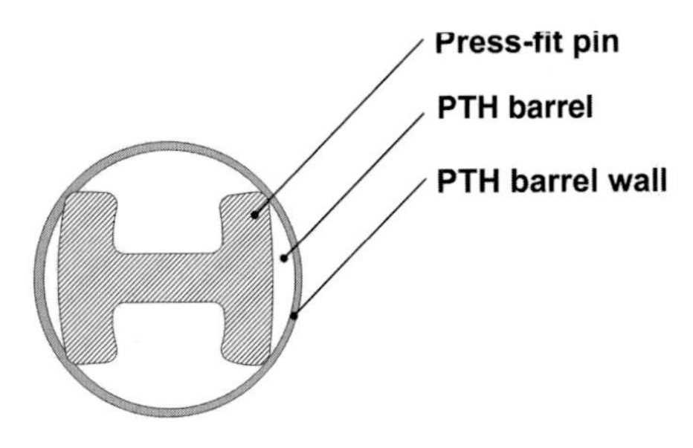

FIGURE 49.3 "H"-section press-fit pin. The sculpted cross-section of the pin allows the pin to deform slightly. This deformation along with some deformation of the PTH barrel upon insertion allows for reliable mechanical and electrical contact.

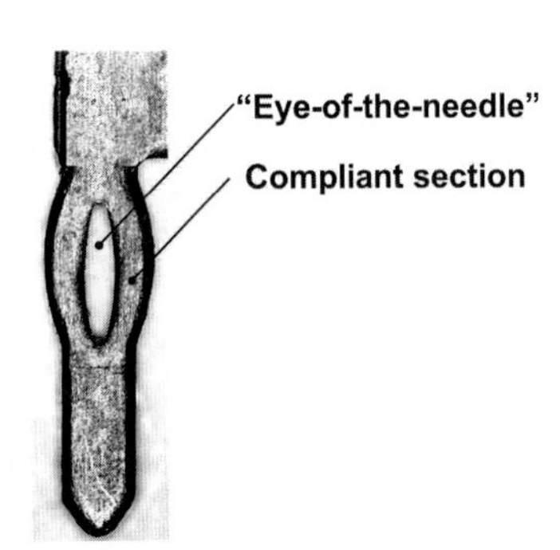

FIGURE 49.4 "Eye-of-the-needle" configuration press-fit pin, shown before compression.

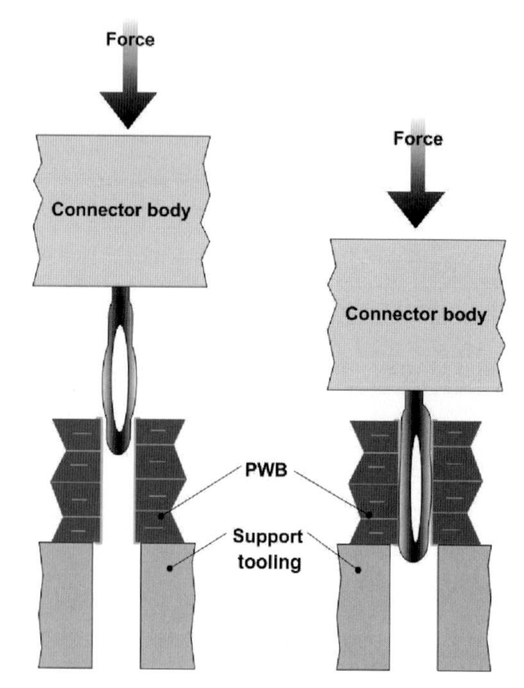

FIGURE 49.5 As the EON pin is driven into the PTH barrel, its compliant section collapses inward upon itself. The resultant forces between the pin and the PTH barrel wall result in a stable, intimate mechanical and electrical contact.

49.4 PRESS-FIT CONSIDERATIONS

49.4.1 Proper Loading

The load necessary to apply a press-fit connector depends on a variety of factors, including pin design, pin material, pin size, pin surface finish (plating), PTH surface finish, plating thickness, pin and hole size, and the number of pins per connector. The connector manufacturer should provide the required loading per pin for each connector for proper insertion and connector seating.

49.4.2 Correct Design and Size

Press-fit connector leads are designed to work within a tight range of PTH barrel diameters. The compliant pin normal force exerted against the PWB plated through-hole barrel wall must be maximized to anchor the connector into the PWB reliably, both for mechanical stability and electrical contact. The connector manufacturers will recommend proper PTH diameter and tolerances for their products. The resultant press-fit coupling is analogous to the solder joint. If the pin and barrel are sized properly and materials are appropriately finished, the anticipated joint life will be better than that of soldered assemblies. Also if the pin and barrel are sized correctly, the pin-to-barrel normal force will be high enough to form a gas-tight seal for long-term electrical and environmental contact reliability. The pin must be strong enough longitudinally to prevent buckling during the pressing operation. Pin normal forces must be optimized so as not to damage the PTH barrel—high enough for good contact, but light enough that the pin insertion will not damage the barrel or the PWB.

49.4.3 Reproducible Connector Press

Since pressing force generally ranges from some 10s of grams to kilograms per lead, the mechanical advantage of a reproducible press machine is required for press-fit connector application.

49.5 PRESS-FIT PIN MATERIALS

49.5.1 Base Material

For purposes of affordability, mechanical integrity, electrical properties, and materials conservation, all electrical contacts, regardless of assembly methodology, are made from inexpensive base metals and plated with a minimum thickness of a more precious or more practical material. Many press-fit connectors have been made of beryllium-copper alloy, but due to beryllium dust toxicity, a factor in press-fit lead preparation, the switch to safer phosphor-bronze or other copper alloys has occurred throughout the industry.

49.5.2 Surface Finishes: Pin and Barrel

Since most materials oxidize readily upon exposure to the atmosphere, gold or other noble metals are often used to prevent or retard oxidation of SMT or PTH component leads or PWB solder lands and plated though-hole barrels. However, most circuit boards and component leads are covered with less noble materials such as tin (Sn) or tin-lead (Sn-Pb) and instead rely upon a fluxing agent to remove naturally occurring oxides prior to and during the soldering process. In a solder joint, the component lead and pad interface materials are wetted by and sealed in solder. The solder gives the interconnection mechanical rigidity and seals the contact surfaces from the environment. This is not the case with press-fit technology.

The press-fit process is compatible with all surface finishes on the pin and the PTH barrel. The frictional properties of some materials are more favorable to the press-fit process than others. Sn-Pb is known to have favorable properties for the press-fit process. Pb is known for good tribologic properties and acts as a lubricant for the press-fit process. Copper is less favorable but is not a problem if the PWB hole is sized properly for the press-fit pin.

49.5.3 Materials in Light of RoHS

With the onset of the European Union's (EU) RoHS legislation, there is additional consideration for the press-fit process. The original directive banned Pb for most PWB materials and assembly. This resulted in two problems. First, press-fit connector contacts converted to Sn instead of Sn-Pb would risk Sn-whisker formation. Unlike Sn-coated SMT component leads, there is no reflow associated with press-fit technology. Without the reflow step, Sn-whisker formation is much more likely to occur. Second, increased frictional forces from Pb-removal on currently designed products made pressing difficult and in some cases resulted in damage to the PWB. EU's RoHS technical advisory committee was petitioned and an exemption to the RoHS legislation was enacted permitting the use of Pb in surface finishes for "compliant pin connector systems."[1]

49.5.4 The Gas-Tight Seal

In the case of the press-fit connector pin, there is neither a physical wetting of material nor metallic encapsulation to protect the interconnection. Additionally, no chemical fluxing agents are used.

Instead, the press-fit pin is anchored solely by the mechanical interference between connector lead and the PTH barrel of the PWB. Oxides on the surface of the press-fit pin and mating PTH barrel are broken loose by the high frictional forces associated with the press-fit assembly operation, so nobility of materials, oxide build-ups, and material shelf-life are less critical considerations than in soldering. If conditions are proper in terms of sizing and materials, the press-fit process results in a gas-tight contact between the connector pin and the PTH barrel wall.

The notion of a gas-tight seal as the product of a proper press-fit is crucial to connection reliability. If an electrical contact is to be reliable, the interface between the two mating surfaces must remain chemically and mechanically stable. The gas-tight, smeared, metal-to-metal contact attained during the press-fit operation mitigates oxidation of either contact (PTH barrel or press-fit pin) and prevents fretting corrosion, which is a common failure mechanism for mechanically mated contacts, especially those subjected to vibration. Press-fit interconnection is considered at least as reliable as a soldered connection and there is no solder joint to degrade or crack with time.

49.6 SURFACE FINISHES AND EFFECTS

Surface finish of the PWB's PTH has a significant effect on the force required to seat a press-fit connector. Changes in surface finish and PTH finish quality can have a profound effect on pressing force and reliability.

49.6.1 Tribology

Every material has frictional (tribologic) properties. Some are naturally lubricating, such as Sn-Pb alloys, and others are notably nonslippery, such as tin or copper. Copper is one of the least slippery of the surface finishes and typically requires more pressing force than Sn-Pb or gold finishes for a comparably sized hole. Tin's crystalline structure is also rather resistant to sliding. In some cases, extra attention to PTH finished hole size (FHS) is required. The PWB vendor should be alerted to the fact that holes are being placed in the board for press-fit, so FHS and surface finish conditions are appropriate for the press-fit component. In some cases, for some materials such as copper, the holes could be specified for the upper end of its FHS as expressed by the connector vendor to aid in insertion. FHS should be just that: the hole size after surface finish is applied.

If tribologic forces are too great, then the PTH barrel can be pushed out of the board or otherwise damaged. Trial boards and pressing parameters should always be assessed before full manufacturing commences. Only very small changes or imperfections can interfere with the typically high yields of press-fit connections.

As previously mentioned, recently the European Union exempted press-fit connectors from its RoHS legislation, allowing the use of Sn-Pb coatings on compliant tails. The Pb adds a small amount of lubricity, allowing for easier pressing into notably difficult surface finishes such as organic solderability preservative (OSP)–copper.

49.6.2 PTH Barrel Finish

Barrel finish can have a profound impact on the press-fit operation. If a PTH barrel is too thin or poorly sized for the compliant pin, the pin may pierce the barrel wall or break internal connections to it. OSP-copper finish—since it is a watery, thin topical coating—adds no appreciable thickness to the PTH lands or barrel, so its control in high-aspect-ratio holes is not an issue in terms of press-fitting. Electroless/immersion surface finishes are generally thin and self-limiting in thickness, and as with OSP-copper, there no appreciable change in PTH hole diameter adding little impact to the press-fit process.

At the other end of the spectrum are hot air solder leveling (HASL) and electroplated finishes, which may add significant material to PTH barrel walls and annular rings. Too thick a HASL coating can lead to PTH barrel diameter reduction. During pressing, the press-fit pins may be overcompressed, causing them to yield and collapse beneath the connector rather than go into their respective through-holes. Sometimes solder from the HASL surface finish can redistribute during SMT reflow cycles, causing a bulging meniscus within the PTH. This condition is depicted in Fig. 49.6a. Compare this illustration to that in Fig. 49.6b, which shows a normally finished PTH with even wall coating, a relatively uniform barrel diameter, and an evenly coated primary-side annular ring. With the advent of RoHS, HASL finishes are less common, but press-fit is commonly used in high-end servers and telecommunications equipment where the use of Sn-Pb solder is currently exempt from RoHS Pb-free requirements.

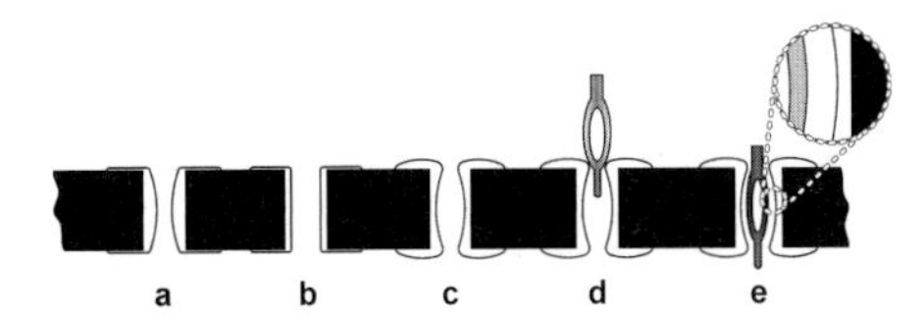

FIGURE 49.6 Barrel profile as a result of surface finish coating: (a) A PTH barrel with excess surface finish from HASL, which has formed a meniscus upon solder reflow. Other plating defects can result in such a meniscus, although the problem is uncommon. (b) A normal PTH barrel with evenly coated walls and annular rings. (c) Excessive plating on PTH barrel walls and annular pads. (d) Excess plating on annular ring causing pin interference and premature EON pin collapse. (e) An overly compressed pin resulting from excess plating on PTH annular ring. This condition poses a reliability risk.

Excessive plating on PTH barrel walls and annular pads (see Fig. 49.6c, d) can cause premature collapse of the compliant section of the press-fit pin. This condition may result in excessive pressing force to get the pin into the hole as well as insufficient normal contact force from the collapsed press-fit pin section to the barrel once fully seated in the hole. This condition poses an electrical contact reliability risk, as shown in Fig. 49.6e, where the pin is not actually in contact with the PTH barrel walls.

Figure 49.7a shows a precollapsed press-fit pin that is making only intermittent contact due to the PWB condition as depicted in Fig. 49.7b.

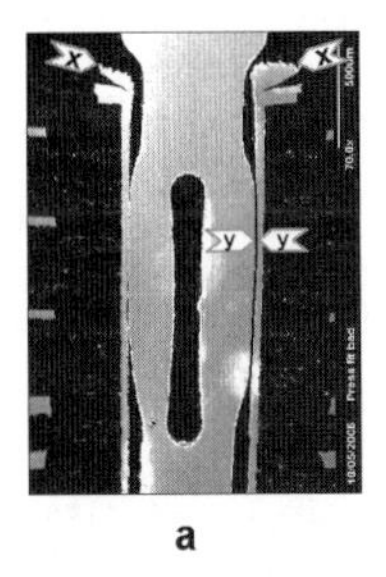

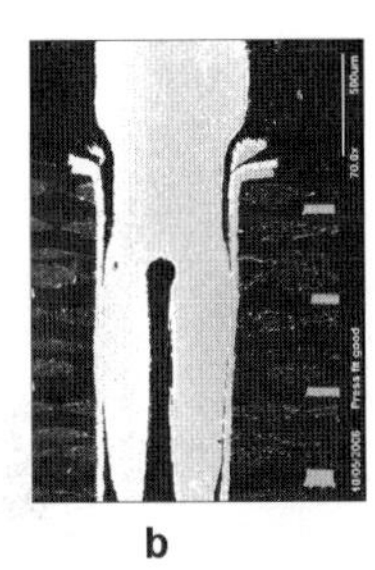

FIGURE 49.7 (a) Excess nickel plate on a PTH annular ring has caused two problems. First, it has shaved material from the press-fit pin (arrows labeled "x"). Additionally, it caused compression of the compliant section of the pin. When fully seated in the PTH, the pin was too loose for reliable contact. The "y" labels demark a gap between the press-fit pin and the PTH wall. Compare this to (b), where pin shaving is evident but the pin is in intimate contact with the PTH barrel. *(Scanning electron microscope (SEM) micrographs courtesy of A. Alexander, Jabil Circuits, Penang, Malaysia.)*

High-aspect-ratio holes (PWB thickness: PTH diameter) may have irregular plating where the thickness closest to the board's outer layers is within acceptable thickness limits, whereas plating thickness in the middle of the PTH may have unusually thin deposits. These thin deposits are caused by plating bath stagnation and exhaustion—a zone of poor electrolyte circulation within the high-aspect-ratio barrel. This condition may also lead to poor contact of the press-fit pin against the PTH barrel wall and contact reliability issues. The opposite can be detrimental also, where the PWB vendor strives to attain the proper plating thickness within the PTH barrel, but plates too much material at the barrel openings and on annular rings.

49.6.3 Slivers

The Pb of Sn-Pb solder-coated barrels is known to reduce the force required for press-fit connector insertion and seating. Although Sn-Pb coating is one of the best surface finishes for pressing, if it is thick, then the press-fit connector pin may skive a thin sliver of solder and push it out the secondary side of the PWB (see Fig. 49.8). Very often these slivers

FIGURE 49.8 A solder sliver (dashed outline) pushed out of a HASL-coated PTH.

remain attached to the pin or the PWB, or they become jammed between the pin and barrel. The sliver may break free of its source and cause an electrical short locally or elsewhere in the system.

49.7 EQUIPMENT

Several types of mechanical presses are used for applying press-fit connectors to PWBs. The lever-activated arbor press is the most primitive but suffices for simple, low-lead-count, coarse-pitch connectors. One step up is the pneumatically driven press. Although it is easier to operate than the lever-activated arbor press, it does little for process control or assembly repeatability. Higher on the evolutionary scale is the pneumatic/hydraulic combination (the so-called "air-over-oil" press), which offers a somewhat more controllable process. High-precision pressing cycles are mandated by the latest press-pin connectors with their delicate, molded body features, fragile electrical shielding, and fine pin pitch, so it is advisable to use a machine that is computer-controlled. Aside from force and speed precision for process reproducibility, press cycle data can be stored along with process recipes for ram force, speed of compression, and component location if equipped with programmable axes. Data logging is critical for statistical process control and is useful for press cycle root-cause defect analysis or machine troubleshooting.

The state-of-the-art connector press is electromechanically actuated, relying on a motor-driven coarse-pitch lead screw with tachometer speed control, z-axis positional encoders, and load cell feedback for an accurate pressing cycle (see Fig. 49.9).

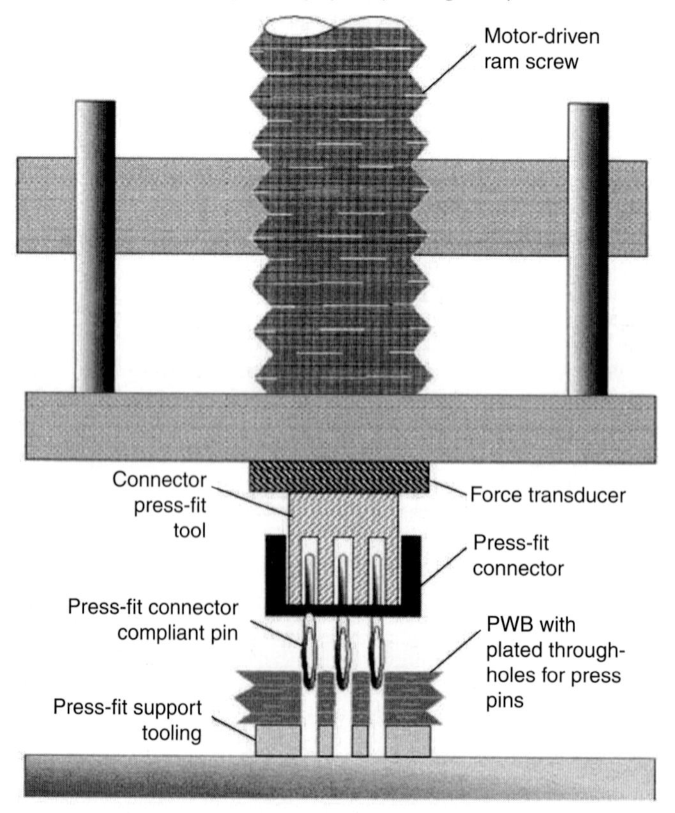

FIGURE 49.9 An electromechanical connector press.

Some commercially available machines are automated even to the point that they are able to shuttle boards into the press, pick and place connectors, select a connector-appropriate pressing tool and supporting tool anvil, rotate the tools to the proper orientation, and press multiple connectors sequentially on a single board or multiple nested boards. There are several ways to control the press-fit process. Methods and merits of each are discussed in the remainder of this chapter, but first a more detailed review of the mechanics of the press-fit process is in order.

49.8 ASSEMBLY PROCESS

Press tool and support plate are fabricated to match connector requirements and PCA layout. The PCA is placed on the press-fit tooling fixture support plate. The operator manually places the connector into the appropriate plated through-holes in the PWB. The ram and tool are aligned to the connector to be pressed. When press-fit connector leads are forced into PWB plated through-holes during a pressing cycle, there is a characteristic fingerprint of force versus vertical distance traveled for each connector and PWB type. The various events and slopes that comprise the fingerprint are shown in Fig. 49.10.

As pressing progresses, connector leads are forced into the plated through-holes and the compliant section of its pins deform elastically, then plastically. Continued resistance is encountered due to the friction of the collapsed pins sliding along the PTH barrel walls. Another inflection in the plot occurs where the connector fully seats against the PWB. Further force affects little or no change in distance and the press cycle ceases. Were the cycle to continue, damage to the PWB would result.

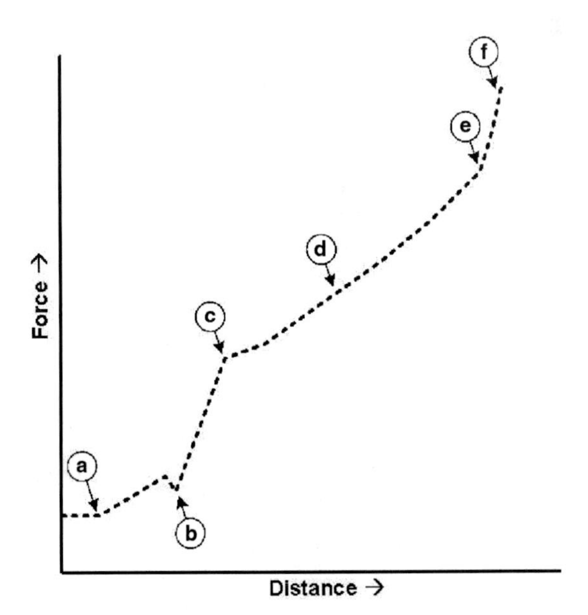

FIGURE 49.10 Force versus distance plot of a press-fit cycle. Inflection points and regions are as follows: (a) initial force is applied to the press-fit connector; (b) the compliant section of the pin collapses; (c) the pin collapse is complete; (d) the pin slides along PTH barrel wall, creating friction; (e) the connector is fully seated; (f) continued pressing makes little change in press distance. The pressing cycle is complete.

49.9 PRESS ROUTINES

For best results, a method that allows real-time measurement and control of the pressing process is preferred and is available on today's commercial presses. There are four commonly used pressing routines, but only three take advantage of force-sensing, distance-sensing, or real-time feedback control of the connector press. Connector complexity and equipment availability dictate press-fit assembly methodology.

49.9.1 Uncontrolled Pressing

This was the most prevalent technique for a long time, but it rapidly lost ground to more sophisticated pressing methods, as described later in this section. It commonly relies on an arbor press and human muscle to force the press-fit connector into the PWB. There is neither force sensing nor speed control. It is the least reliable way to press connectors into PWBs and is not recommended for complex connectors or fragile PWBs. If too high a ram speed is used on a delicate connector, its pins may buckle. This method may not provide enough force for some high-lead count connectors.

Although this type of pressing is discouraged, it may be adequate for low lead-count, coarse-pitch connectors in noncritical assemblies. As with any press routine, a suitable support fixture must be used beneath the board to keep the assembly from breaking or flexing. It should incorporate support to counter the forces of the ram, and relieved areas to accept protrusion of press-pins below the bottom side of the board and clearances for bottomside components as shown in Fig. 49.11.

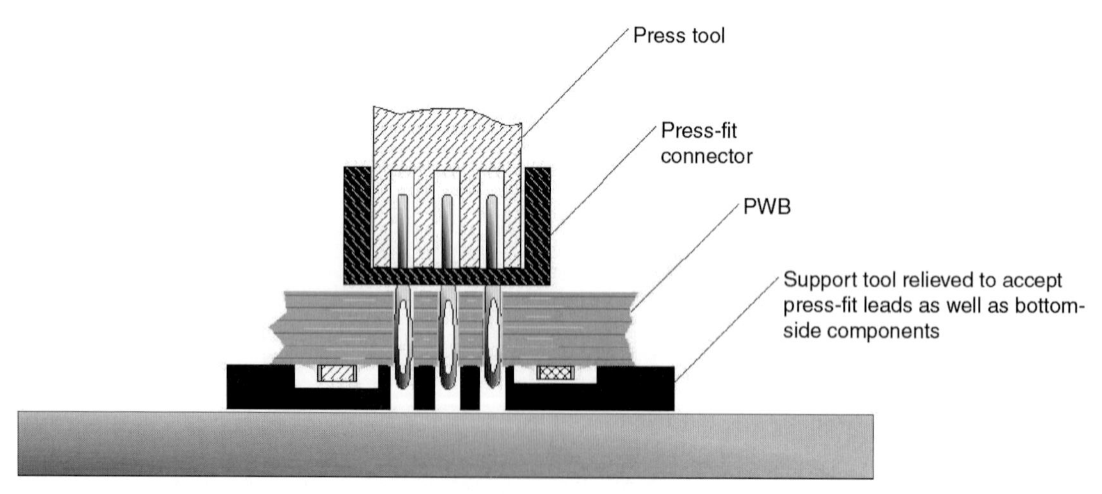

FIGURE 49.11 The supporting fixture is critical for good press-fit application. It must have clearances for press-pin protrusion and any bottom-side components.

49.9.2 Height-Limited Pressing

As previously described, connector presses have matured into sophisticated machinery; improved greatly over the arbor press. Some are capable of precisely determining ram height above a datum such as the top surface of the PWB. If board-to-board thickness is consistent, then the connector can be forced into the board to a predetermined standoff height as required by the connector specification. Just as connector pin dimensions vary from connector to connector, so do plated through-holes in the PWB. Pressing to a predetermined height (that is, *press to height*) ensures that the connector will be seated the same way each time, regardless of the pin or hole condition and the resultant force required to seat the connector (see Fig. 49.12). Variations in PWB thickness, variability in the molded cross-section of the connector body, or molding flash can result in imprecision and negate some of the benefits of this approach.

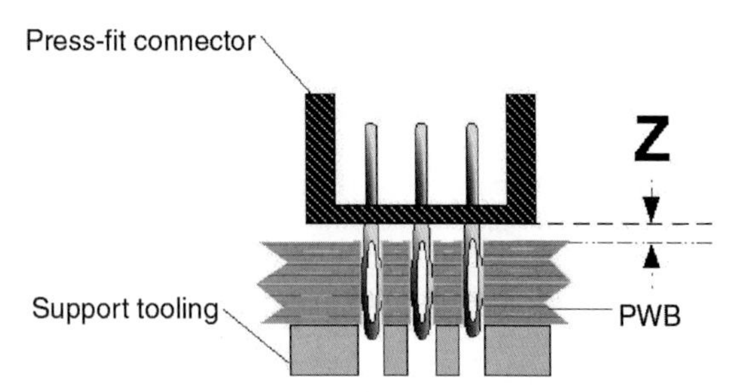

FIGURE 49.12 Height-limited pressing forces the connector to a predetermined standoff distance, "Z," from the surface of the PWB.

49.9.3 Force-Limited Pressing

Force-limited pressing (or *press to force*) relies on the intelligence of an instrumented connector press to sense that a predetermined force limit has been reached, at which time the pressing cycle is terminated. The force limit can be roughly set to coincide with the average force required per compliant pin (available from the press-fit connector manufacturer's specification) multiplied by the number of pins per connector. Application of upper and lower tolerances of specified force per pin coupled with knowledge from empirical trials will help refine the acceptable press force envelope for a given set of materials. This technique is highly sensitive to the material conditions of the compliant pins and the plated through-hole diameters.

49.9.4 Gradient-Limited Pressing

The slope or gradient of the final portion of the force-versus-distance plot (refer to Fig. 49.10, and the inflection labeled "e") can be used to trigger the end of the press cycle and retract the ram. This routine ensures that regardless of pin size, hole size, board or connector variations, the connector will not be overpressed. This is most reliable when a very steep slope is chosen and refined empirically. Gradients in the range of ≥75 percent are generally favored. This routine is also known as *press to gradient*.

49.9.5 Rework for Press-Fit Connectors

The press-fit operation is inherently high yielding, but problems with the process or a bad press-fit connector may necessitate the removal and replacement of a connector. As a result of the press cycle, the compliant pins of a press-fit connector are plastically deformed. Therefore, once used, the same connector cannot be removed and reinserted in a circuit board. However, most press-fit connectors are designed to be reworked—that is, repaired or replaced. In some cases, individual leads or contacts can be replaced. In other cases, banks of leads or contacts can be substituted. Some press-fit connectors require complete removal of the damaged connector for replacement with a new one. Many different press-fit connectors are on the market, and each has its own manufacturer-recommended repair strategy.

Press-fit assemblies are generally designed to accommodate at least two connector replacement cycles. The two cycles are predicated on (1) the normal force exerted by the press-pin's compliant section on the PWB's plated through-hole barrel wall, and (2) the quality and condition of the PWB's plated through-hole barrel surface after subsequent press-fit connector removals and replacements.

Were the normal force of the compliant section on the PTH barrel of a press-pin excessive, it might cut through the barrel and even damage surrounding innerlayer interconnects or traces. Instead, the normal force is optimized to permit gas-tight seal without excessive deformation or deep skiving of the plated through-hole barrel. Once a connector or connector pin is replaced, the new connector/pin must result in a gas-tight pin-to-barrel seal for reliable contact. Each time that a new press-pin is inserted into a plated through-hole, there is opportunity for the barrel to deform; the more insertions into the barrel, the deeper the deformation. It is for that reason that the industry standard design goal for compliant pin connectors is generally targeted at only two replacement cycles. Beyond two replacements, there is the possibility that the barrel will be thinned and damaged, impairing press-pin contact reliability or compromising the trace-to-barrel interconnection.

Press-pin rework tools vary greatly. In some cases, they are designed to remove and replace a single pin. In other cases, the tool may be designed to help dislodge an entire connector from a circuit board. Some tools are designed to remove the molded connector body only. Then individual pins are pulled by pliers one by one in preparation for replacement. Sometimes multiple press-pin ground or power leads are stamped into discrete wafers, and then the wafers are inserted into the plastic molded connector body to form a single press-fit connector. In this case, individual wafers can sometimes be removed and replaced rather than having to replace single leads or the entire connector. Repair tools and repair strategies are generally available from the connector manufacturer.

To avoid damage to the board, components, or solder joints, the PWB must be supported so that it is not flexed during press-fit component removal or replacement. It is a good practice to verify circuit integrity by electrical test before and after repair to ensure that it is not degraded as a consequence of the repair cycle. If a connector is replaced because of pin stubbing on the surface of the circuit board upon the original installation, ensure that the surface and underlying traces of the circuit board have not been damaged by the errant lead (see Fig. 49.13).

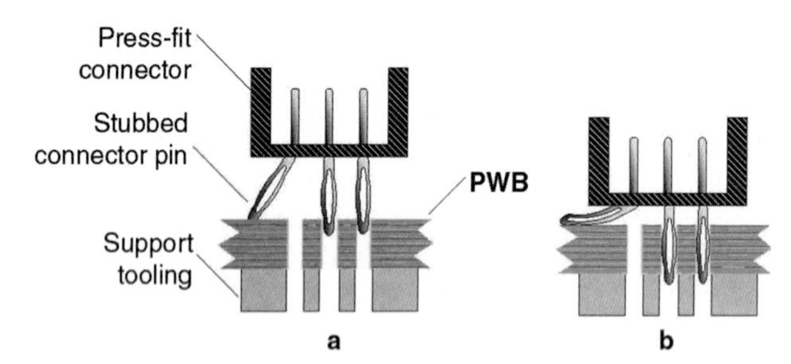

FIGURE 49.13 (a) Misaligned pin stubbing on the surface of the PWB. (b) The stubbed pin has folded under the connector. Stubbing may damage the PWB surface or subsurface traces.

49.10 PWB DESIGN AND BOARD PROCUREMENT TIPS

49.10.1 Clearance for Press-Fit Tools for Assembly and Rework

To avoid process problems and damage to adjacent components, there should be sufficient clearance in the PCA design to permit room for the pressing tool, board supports, and use of any associated rework tools.

49.10.2 Long Pins Preferred

When possible, use press-pin contacts slightly longer than the board thickness so that connector pins will protrude on the bottomside of the board after the pressing cycle. Pin protrusion permits easy inspection of the finished assembly. A missing pin is evidence of pin stubbing or buckling.

49.10.3 PTH Size and Finish

Follow the press-fit connector manufacturer's recommendation for PTH finished hole size, tolerances, and acceptable or tested surface finishes. Use cross-section sample boards with high-aspect-ratio (depth:diameter) holes to qualify the PWB vendor and determine typical PTH barrel profile. Ensure that wall thickness of plated through-holes are uniform throughout their depth and meet connector specifications. Make certain that the surface finish on the PTH annular ring does not interfere with insertion of the compliant press-fit pin.

49.11 PRESS-FIT PROCESS TIPS

Several things can be done to ensure adequacy of the press-fit operation in manufacturing.

49.11.1 Use a Proper Press

Use of an electromechanical press is recommended for best precision and process feedback. Its precision and feedback mechanisms will minimize the impact on the PWB, or connector.

49.11.2 Check Pins for Straightness

Prior to insertion into the board, check press-fit connector pins for straightness and integrity. An inexpensive go/no-go gauge (see Fig. 49.14) can be machined to check that connector pins are not bent. It should be constructed so that it neither compresses the compliant section of the press-fit pin nor damages its surface coating. This is especially helpful for fine-pitch press-fit connectors (1 mm centers) and is better than relying on visual inspection for pin field integrity.

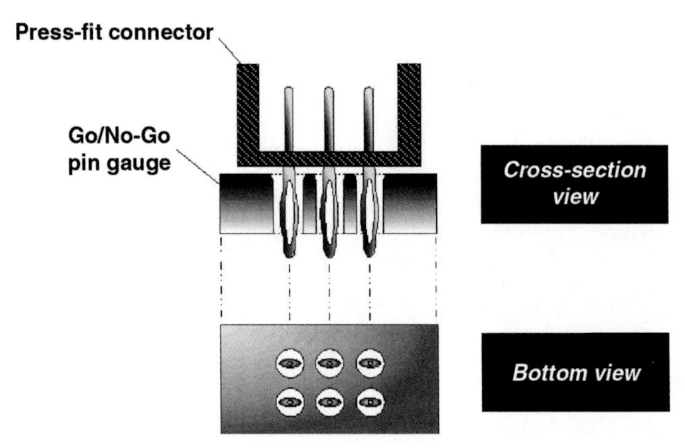

FIGURE 49.14 Gauge for checking connector pin position prior to pressing. This is best used for high lead count, fine-pitch pin press-fit connectors.

49.11.3 Keep the Press-Fit PTH Clear of Foreign Material

Protect press-fit PTH holes from extraneous solder, flux, surface-mount adhesive, spot mask, or other contaminants from surface-mount, wave-solder, or hand-solder processes. Use of a wave pallet or high-temperature masking tape (e.g., polyimide tape) to shield press-fit locations during other processing is advisable. If tape is used, inspect first articles to ensure that high-temperature masking tape residue does not foul press-fit holes. Polyimide tape with acrylic-based adhesive is best for this application, as residues will be absent or minimal.

49.11.4 Use the Proper Press Tooling

Use hardened-steel pressing tools as sold or recommended by connector manufacturcr to avoid damage to the connector or the PCA.

49.11.5 Provide Proper Board Support

Adequately support the PCA during the pressing operation so that connectors can be fully homed and that the board is not overflexed. Excessive board flexure may cause solder-joint cracking, broken PWB circuit traces, and board delamination. This is especially important for the characteristically inflexible ball-or column-based solder joints of area-array devices such as BGAs and CCGAs.

The best support will be localized directly under the connector to be pressed. Ensure that the support includes clearance for backside pin protrusion; otherwise, pins will stub and bend on the supporting anvil. Support tooling design should carry appropriate relief features for bottomside components.

49.12 INSPECTION AND TESTING

49.12.1 Connector Inspection

Check the connector after pressing for even bottomside pin protrusion, topside connector body seating height, damage to the connector body, inaccurate mating pin positions (within the connector as opposed to within the board), damage, or movement of electrical shields if the connector is so equipped.

49.12.2 PCA Inspection

Check the PCA for surface damage in the area of the pressing operation. A mispositioned or damaged press tool or board support can damage components or overly compress the laminate causing outer- or innerlayer trace damage. Look for errant tooling marks on the PWB surface after the press-fitting operation. Overcompression of the board can lead to compressed or severed traces within the board. A compressed trace may impact current capacity and result in localized circuit overheating or otherwise impact signal integrity.

49.12.3 Solder Sliver Inspection

Check the bottomside of the board for solder sliver formation. This is especially relevant in the case of HASL boards. Brush these out if required and inspect. Loose slivers may cause electrical shorting on the board or other boards within the system enclosure. Work with the

PWB vendor to ensure that drilled and plated holes are sized appropriately for the press-fit process and that excess solder from the HASL process does not adversely alter the finished hole size or shape.

49.12.4 Electrical Testing

In the case of high-friction surface finishes, such as OSP-copper, forces applied during the press-fit process may be so great as to cause damage to the PTH barrel or adjacent inner- or outerlayer circuit traces. Check the board visually and by electrical test as part of a first article inspection.

49.12.5 Check Tooling Frequently

Check for excessive tool wear or damage. A damaged tool can ruin a connector or PWB and replicate the damage over many boards.

49.13 SOLDERING AND PRESS-FIT PINS

Apply press-fit connectors after all SMT and wave-soldering operations. Do not try to solder press-pin connectors in place. Press-fit connector bodies may not be reflow-compatible. Soldering a press-fit connector will also make rework difficult or impossible.

REFERENCES

1. EU RoHS Commission Decision of 21 Oct 2005, Annex to Directive 2002/95/EC of the European Parliament and of the Council on the restriction of the use of certain hazardous substances in electrical and electronic equipment.
2. Parenti, D., and Mitchell, J., "Validating Press-Fit Connector Installation," *Circuits Assembly*, April 2003, pp. 26–29.
3. Ocket, T., and Verhelst, E., "Lead-Free Manufacturing: Effects on Press-Fit Connections," paper presented to the International Center for Electronic Commerce (ICEC) Proceedings, Zurich, September 2002.
4. Hilty, R. D., "Effect of Lead Elimination on Press-Fit Interconnects,"paper presented to the Institute of Electrical and Electronic Engineers (IEEE) HOLM Conference Lead Free Workshop, September 9, 2003.

CHAPTER 50
LAND GRID ARRAY INTERCONNECT

Gary M. Freedman
Hewlett-Packard Corporation, Business Critical Systems, Singapore

50.1 INTRODUCTION

The land grid array (LGA) device is a solder-less package that relies upon pressure contact to interconnect it to the board mechanically and electrically. Interconnection is accomplished through a socket or interposer that is sandwiched between the LGA integrated circuit (IC) package and the printed wiring board (PWB). The socket has compliant, conductive contacts that match the LGA and PWB contact land patterns on each. A system of insulators, stiffeners, and fasteners keeps the LGA in intimate mechanical and electrical contact with the PWB. Care in assembly and cleanliness is of utmost importance for successful pad-to-land contact and subsequent assembly reliability. The reader should be aware that there are leadless soldered packages also called land grid arrays, but this chapter will deal with the pressure-interconnect variety of leadless package.

Most often, an LGA-type interconnect is used for mounting a central processor unit (CPU) chip or other active device to a PWB. It is also used for some connectors. An advantage of this type of interconnect is that it makes it possible to exchange a chip without the need to unsolder and re-solder, facilitating rapid rework, engineering changes or field upgrades without soldering and without degradation in reliability. The LGA concept has been applied to connectors also. Although the technique has been in use for several years, it is still a niche technology but will likely increase in importance as input/output (I/O) counts continue to increase and package size limitations for soldering are met.

50.2 LGA AND THE ENVIRONMENT

Since the LGA connection to the PWB is solderless, it is environmentally friendly, requiring no heating for solder, soldering, or rework. It is easy to move into compliance with the European Union's (EU) Restriction of Hazardous Substances (RoHS) requirements since solder considerations (at least external to the package) are absent.

50.3 ELEMENTS OF THE LGA SYSTEM

There are five key elements and several subelements that make the LGA system practical and reliable. These are pictured in Fig. 50.1, but be aware that there are many variations. Key elements will be described in a generic sense in this chapter.

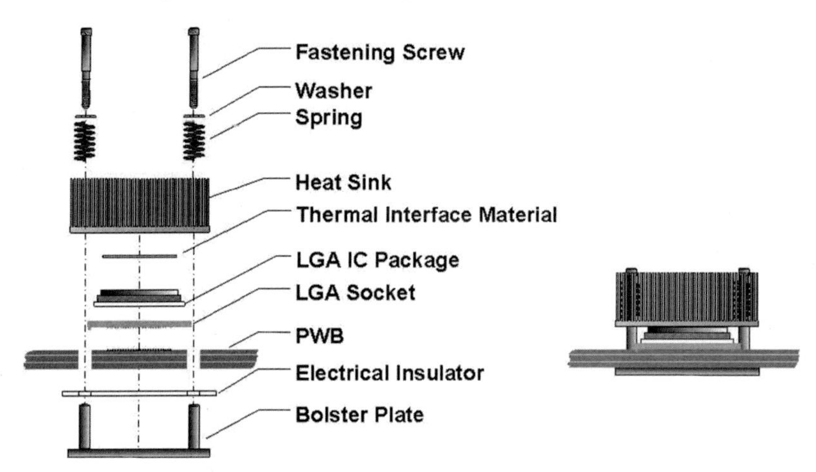

FIGURE 50.1 Elements of an LGA stack.

50.3.1 PWB and LGA Issues

50.3.1.1 Contact Lands. The PWB is much like any other board. It has lands to accommodate electrical interconnect of the LGA socket to the board. The lands correspond in size and shape to socket requirements. Most often, the LGA lands on the PWB are electrolytic nickel covered by electrolytic gold. Hardened gold is often used for added contact durability and reliability. Other surface finishes can also be used, but should be tested for reliable and long-lasting contact.

50.3.1.2 Drilled Holes. LGA sockets have the requirement for two sets of drilled holes in the PWB. One set is used to accommodate fastening of the bolster plate through the PWB. The second set is used to orient and retain the socket on the PWB. The closely toleranced orientation holes correspond to molded locating pins on the LGA socket and they are meant to locate the socket contacts properly on the LGA lands of the PWB. The socket manufacturer will provide guidelines and drill pattern recommendations. It is good practice to make the locating pins different sizes and to drill the PWB accordingly to polarize the socket so it cannot be misoriented on the board.

50.3.1.3 PWB Layout Considerations. There are some layout considerations such as avoidance of components in the vicinity of the LGA socket, bolster plate, and pressure plate. PWB flatness should be considered so that best contact can be made to the board once the LGA interconnect system is fully assembled.

50.3.2 Bolster Plate

LGA technology relies on a pressure interconnect system, and the bolster plate is the rigid backside element that serves to anchor the LGA system to the PWB. It is inserted through predrilled holes in the circuit board and provides the means for fastening the clamping mechanism

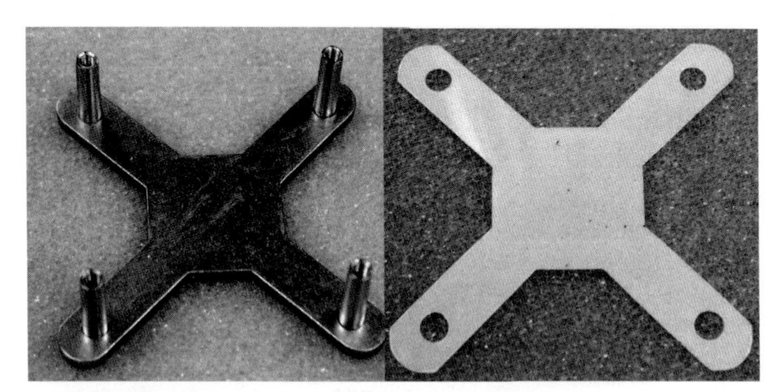

FIGURE 50.2 Stainless steel bolster plate (l) and insulator (r). *(Courtesy of Hewlett-Packard.)*

that holds the LGA stack together. It is typically fashioned from a stiff metal such as stainless steel or thick aluminum (see Fig. 50.2).

The bolster plate may be orthogonal or sculpted to accommodate components in the vicinity of the bolster plate or to permit access to test points for in-circuit testing (ICT) or diagnostics. The design of the bolster plate must be such that it is rigid upon fastening and must flatten out any local bow or warpage in the vicinity of the LGA contact field on the PWB to ensure that all electrical contacts are made and mechanical stability is imparted to the LGA stack.

Usually an insulator is applied over the surface of the bolster plate and positioned against the underside of the PWB to prevent electrical shorting of the metal bolster to any electrical features on the secondary side of the PWB (through-hole vias, test pads, and so on). It is best not to include any components on the secondary side of the board beneath the bolster plate. If components must be included, the bolster plate will have to be relieved for clearance so that these components will not be crushed during LGA assembly.

50.3.3 Active Component (LGA IC)

The LGA is an IC package usually made from a rigid substrate such as ceramic. It looks like a ceramic ball grid array (CBGA) without solder interconnects (balls) on its underside (see Fig. 50.3). In all other respects, it is like any other ceramic package. It typically has gold pads on its underside that serve as the contact points for the LGA socket. The contacts are usually electrolytic nickel overplated with electrolytic gold. Sometimes a hardening agent such as cobalt is co-plated with the gold to enhance contact durability for improved long-term reliability.

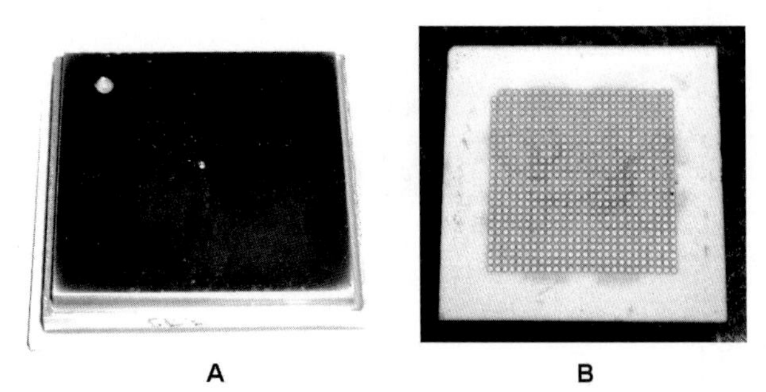

FIGURE 50.3 LGA package topside (a) and underside (b). The package is similar to a CBGA without any solder interconnects. *(Courtesy of Hewlett-Packard.)*

50.3.4 Socket

Many types of LGA sockets (also referred to as *interposers*) are available. There are interposers with traditional dimpled leaf-spring contacts, Fuzz Button contacts, and even metal-filled polymer contacts. The contacts are held on a molded insulating substrate. The contacts and socket frame are customized to fit the LGA package and corresponding PWB footprint (see Fig. 50.4).

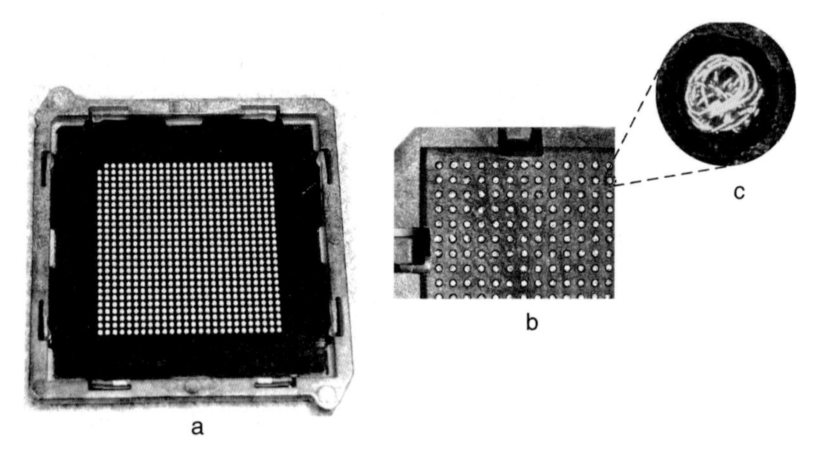

FIGURE 50.4 LGA sockets: (a) a socket with metal-filled polymer contacts; (b) a socket with Fuzz Button contacts; (c) closeup of a single wire-wound Fuzz Button.

Each contact is interconnected through the molded substrate in some fashion, and the socket can be considered double-sided for interconnecting both the LGA IC and the PWB (see Fig. 50.5).

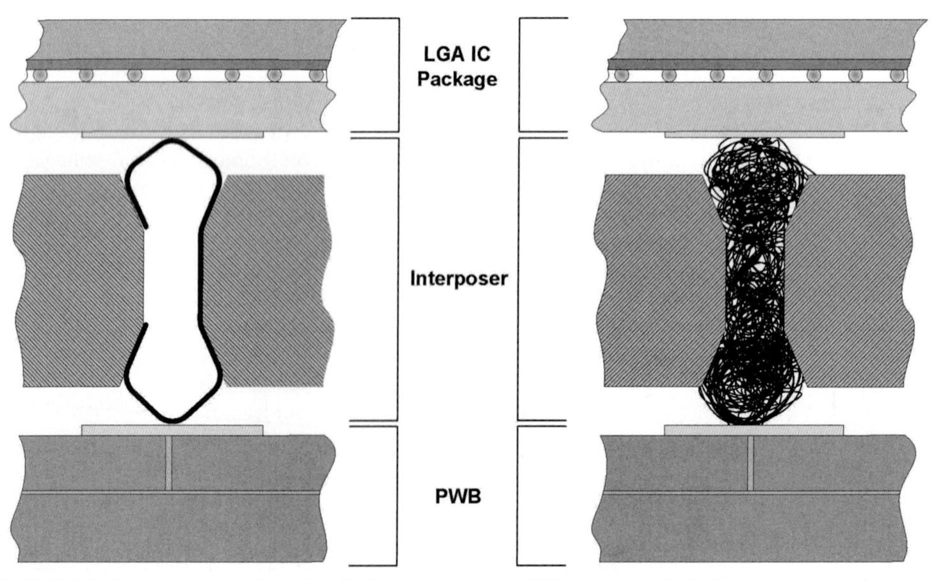

FIGURE 50.5 Interposer sockets electrically connecting the PWB with an LGA IC: (l) spring contact; (r) Fuzz Button contact.

Additional subelements such as fasteners and springs may be required to enable or enhance fastening, shock and vibration resistance, and thermal management.

Although Fuzz Button technology is among the most reliable of LGA interconnect schemes, some defects are common to this socket style. Pulled wires (see Fig. 50.6) as well as leaning and missing contacts have been seen, although the interconnect reliability is quite high if the socket is in good condition prior to assembly and if the LGA stack is assembled correctly.

If too much pressure is applied to metal-filled polymer contacts, they can deform and flow over time, lessening contact pressure, and result in open or intermittent contact failures.

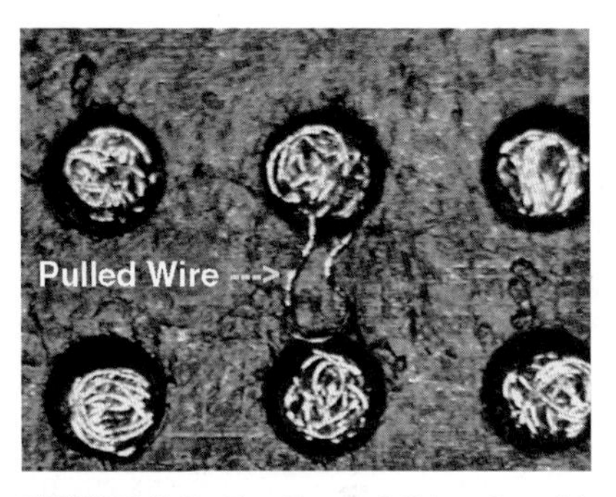

FIGURE 50.6 Fuzz Button LGA socket with a stray strand of wire, which can cause electrical shorting.

50.3.5 Pressure Plate

The pressure plate is fastened atop the LGA package and couples to the bolster plate. It forces the LGA package in contact with the socket and the socket to the lands on the PWB for completion of the electrical circuit. Besides being responsible for interconnecting the socket to package and board, coupled with the bolster plate, it imparts the necessary mechanical rigidity to the LGA stack for shock and vibration resistance and reliable operation. Of course, shock and vibration resistance will be dependent on bolster plate and pressure plate designs, including any springs, fasteners, and such. Force on the stack has to be such that it damages neither the LGA IC nor the PWB. Guidance is needed from the LGA package manufacturer and from the LGA socket manufacturer as to how much force can be applied to the stack without damaging the board, socket, or IC package.

50.3.6 Thermal Management

Frequently, thermal management solutions are built into the LGA stack. The heatsink, if constructed properly, can be used as the pressure plate. A thermal interface material such as graphite sheet or other thin thermal transfer material can be used between the heatsink and the IC package to enhance heat dissipation.

50.4 ASSEMBLY

In addition to good design, proper assembly strategy and precise assembly are crucial to LGA interconnect reliability. The following steps should help the engineer to achieve good results, but as stated previously, if the design is flawed, interconnect reliability will be as well.

50.4.1 Order of Assembly

LGA-mounted component assembly should be done after all other soldering, press-fit, and mechanical assembly steps are completed. Most LGA sockets are not rated for high-temperature excursions. The stress of other assembly may be enough to interfere with good LGA interconnection.

Gold pads on the PWB must remain as clean as possible—free of solder, flux deposits, chemicals, and contaminations. Once assembled, all chemicals must be kept away from the LGA socket. The heat of rework can also cause interposer damage or LGA device failure, so rework is best completed, where feasible, before LGA assembly.

50.4.2 Prior to PWB Processing

A high-temperature masking tape can be used to cover LGA lands on the PWB. This will protect it from flux residue, spattered solder, and so on. The tape should be capable of sustaining multiple reflow cycles. A polyimide tape with acrylic adhesive is recommended to minimize tape adhesive residue. Silicone-based adhesives have been found to leave residues that can interfere with electrical interconnect. Since silicone is largely insoluble in most common solvents, attempts to clean any residue will be futile.

50.4.3 PWB and LGA Land Cleaning

It is good practice to clean LGA pads on the IC and the PWB prior to application of the socket. Even if a masking tape is used, there is the opportunity for materials to condense under the tape. A wipe with isopropyl alcohol, methanol, or Vertrel-based cleaners is useful for normal flux condensates and other process residues. Do not flood the area with solvent. Instead, use a solvent-saturated lint-free wipe or sponge. Be aware that even lint-free materials can snag on PWB pads, edges, and drilled holes and may leave fibers or debris. Tests should be made to find the most appropriate and abrasion-resistant cleaning material. Ensure that the cleaned lands are perfectly dry before assembly. A Vertrel-based cleaners is ideal for this due to its low, sub–room-temperature boiling point. It can cause some local moisture condensation during the boiling process, which may leave a slight, localized, innocuous discoloration on land surfaces.

50.4.4 LGA Socket Cleaning

Although it is important for the socket to be clean, never wipe it to clean as it may damage the fragile contacts. Even the use of compressed air to remove dust may dislodge or damage a contact. It is best to work with the socket supplier to ensure that interposers arrive clean and ready for use.

50.4.5 Inspection

Inspect LGA pads on the PWB and on the LGA prior to assembly. Inspect both sides of the interposer to ensure that contacts are present and in good condition.

50.4.6 Socket Application

It is important to seat the socket onto the PWB such that it is maintained parallel to the PWB. Applying it at an angle can cause damage to the delicate contacts of the socket and compromise contact reliability. The same can be said about sliding the socket on the board; it is likely to result in socket contact damage.

The socket must be oriented properly on the board so that the contact pattern of the socket matches the land pattern on the PWB. There are molded guide pins on the socket that help with this, with one peg larger than the other to match a similar drill pattern in the PWB. The socket usually has a missing corner contact to match the LGA IC package and corresponding LGA land pattern on the PCA. It is important not to drop the LGA IC package on the socket, as doing so could damage delicate socket contacts.

50.4.7 Fastening

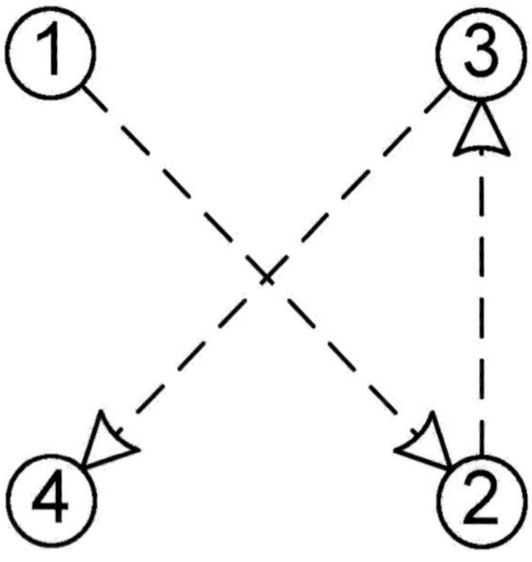

So many mechanical designs are available that listing any particular method for LGA assembly is impractical. Some rely on spring clips whereas others use traditional screw fasteners, as shown previously. If held by four screws, as often is the case, then a tire lug torque pattern should be employed to ensure even tightening (see Fig. 50.7).

Slow and even tightening is important to avoid damage to either the LGA IC device or the LGA socket and to ensure proper compression for reliable electrical contact.

The number of revolutions per fastener depends on the pitch of the screw and socket requirements. Ideally, a four-headed torque driver assembly should be used with each head rotating at the same speed and a calibrated slip-clutch on each driver to prevent overtightening of the stack.

A tooling plate with strategically located board supports should be created to hold the PCA during LGA stack assembly in order to minimize board flexure and to prevent damage to soldered parts on the PCA.

FIGURE 50.7 Manual torque pattern for even fastening of the LGA. Each fastener should receive the same number of turns or partial turns to ensure even seating of the LGA stack.

50.4.8 LGA Socket Rework and Reuse

Ease of IC replacement for upgrade or repair is one of the most important benefits of using pressure interconnect technology. Most LGA sockets are designed for multiple reuses. Check with socket manufacturer for the number of reuse cycles; some are specified for as many as 20. Reliability and reuse depend on a number of factors, including mechanical robustness of the socket, thermal exposure, proper assembly, and contamination from process chemicals such as flux, flux cleaners, alcohol, and so on. Note that there are usually no reuse limits on other parts of the system (LGA IC, printed circuit assembly, bolster plate, heatsink, or pressure plate), but these should be inspected to ensure that contact platings are in good condition and that mechanical fasteners are not showing signs of mechanical wear out.

When removing the LGA IC package component or the socket itself, ensure that the socket contacts have not stuck to either the IC or to the PWB lands. If there is contact sticking, the contacts can be bent, broken, or pulled out of the socket. Any of these will compromise interconnect integrity. Sticking is most prevalent with filled-polymer or Fuzz Button–style contacts, although there can be enough adhesion of plated spring contacts to LGA IC or PWB-plated lands to result in damage to the fragile socket contacts. It is also important to remove the LGA IC package straight out of the socket to avoid damage to the LGA socket if it is to be reused.

50.5 PRINTED CIRCUIT ASSEMBLY (PCA) REWORK

If soldered joints on the PCA need to be reworked in the vicinity of the LGA stack, use a thermal profile board, perform a mock repair, and measure the thermal effect on the socket. Ensure that the LGA IC and socket temperatures are maintained within their respective manufacturer's thermal guidelines. Also make certain that process chemicals, flux, cleaners, and such do not run under the LGA socket.

50.6 DESIGN GUIDELINES

50.6.1 Traces Beneath LGA Elements

It is advisable to keep active traces off the surface of the PWB where they could get crushed by the bolster plate or the socket. Since the bolster plate is flat and large, the forces are spread out. The molded socket frame exerts a more concentrated force on the PWB surface due to its narrow outline and could compress any surface traces running beneath it.

50.6.2 Clearance for Tall Components

The pressure plate or heatsink usually overhangs the LGA IC package. For this reason, tall components should be kept outside of the pressure plate/heatsink footprint to prevent mechanical interference with the LGA stack.

50.6.3 Components Under Bolster

As previously mentioned, it is unwise to put components under the bolster plate. If components are required in this area, then the bolster plate should be relieved to allow clearance of these components. This may require the use of an extra-thick bolster plate to preserve its stiffness after relief machining.

REFERENCE

1. Freedman, G. M., and Baldwin, E. A., "Method and Material for Maintaining Cleanliness of High Density Circuits during Assembly," *Proceedings of the SMTA Pan-Pacific Conference,* February 2003.

P · A · R · T · 12

QUALITY

CHAPTER 51
ACCEPTABILITY AND QUALITY OF FABRICATED BOARDS

Robert (Bob) Neves
Chair/Chief Technical Officer Microtek Laboratories, Anaheim, California, and Changzhou, China

51.1 INTRODUCTION

51.1.1 Reasons for Appropriate Acceptance Criteria

Other than the financial tie between the user and supplier, nothing affects the relationship between these two more than product acceptability and quality. These issues define the relationship. As the printed circuit board (PCB) is the foundation for almost all electronic products, the user cannot tolerate defective boards. At the same time, cost is always a critical issue and imposing expensive and unnecessary acceptance criteria can be a major disadvantage. Failure to provide acceptable and quality product is the largest factor that affects the deterioration of the relationship between user and supplier. All of the components in an electronic product rely on the PCB to get information and power from place to place. An interruption of this process because of a defective PCB results in a failed product. In other words, a poor-quality product at a low price is worthless.

51.1.1.1 Customer Satisfaction. Brand name is one of the most valuable things that the original equipment manufacturer (OEM) possesses. A product's success or failure is quite dependent on the perceived quality that the brand name suggests. OEMs rely on getting products quickly from the design programs to market, and a slowdown due to quality issues can significantly impact a product's introduction and thereby affect their brand. The negative publicity from field failures can also have a significant affect on a brand. Agreed upon acceptance and quality criteria reduce the likelihood of nonconforming PCBs.

51.1.1.2 Reduction of Cost. If the PCB buyer clearly states the required acceptability and quality criteria to the manufacturer, the manufacturer can more clearly understand the buyer's costs. Cost reductions are a permanent fixture in the manufacturing environment. For companies to survive in this competitive market, they must address costs. Unacceptable, low-quality product that cannot be sold is a direct and significant cost.

51.1.1.3 Liability and Litigation. Showing that the PCB was purchased to mutually agreed upon "standards" goes a long way to minimizing litigation and financial liability.

51.1.2 Printed Circuit Board Functions

A PCB is required to do three things for the buyer when it is received:

- The PCB physical form should match the intended design. The dimensions and placement position of the interconnection points along with the quality of the coating on these interconnection points must be acceptable and of high quality to facilitate the attachment of components.
- The PCB should provide proper interconnection between components. This allows the assembly to function properly as intended.
- The PCB should provide insulation between interconnection points that should not be connected.

These three items should be "acceptable and of high quality" at the time of purchase and continue to remain "acceptable and of high quality" throughout the expected life of the product. There are a lot of details between the few preceding lines defining those three things. The rest of this chapter will define the details of "acceptability" and "quality" for the PCB in order to fill in the spaces created by the three previously listed requirements.

51.1.3 Industry versus Customer Standards

Properly implemented acceptability and quality criteria give both user and supplier a clear picture of what is expected. Without clear instructions on how the acceptability and quality of the PCB should be implemented at the supplier, the user is unlikely to get a product that meets his or her needs. A clear definition of what the user wants can help make the relationship between the two much more successful, especially as complexity of the PCB makes acceptability and quality criteria more critical and challenging. In general, PCB manufacturers have gotten larger and produce many more PCBs. This makes the user's individual PCBs a much smaller part of the big picture and their uniqueness may be lost. Therefore, it is essential to define what is required for acceptability and quality clearly and early in the relationship between the user and supplier.

51.1.4 Objectives of Acceptability and Quality Criteria

The PCB needs to be supplied at a quality level that is capable of withstanding the assembly process, as it will undergo several operations to mount and rework components and connectors. This assembly and rework process has gotten more complicated and damaging to the PCB, especially with the introduction of lead-free products. The acceptability criteria should attempt to anticipate the conditions of the soldering process to ensure the PCB will survive the assembly and rework process and still function as intended.

51.2 SPECIFIC QUALITY AND ACCEPTABILITY CRITERIA BY PCB TYPE

Many unique inspection properties for PCBs are dependent on their type of manufacture. There are two major classifications of printed circuits: rigid PCBs and flexible PCBs.

51.2.1 Rigid PCBs

Rigid PCBs make up the bulk of what is built around the world. Rigid PCBs are used in a variety of applications and products. The materials and testing procedures for each type of PCB can be dramatically different. These manufacturing details are thoroughly covered elsewhere in this book and are only touched on here. The processing for multilayer PCBs is a multiple of that required for single- and double-sided PCBs, plus surface preparation, lay up and lamination processes are needed. The additional processes required for each additional layer introduce opportunity for anomalies to creep into the PCB.

51.2.1.1 Single-Sided PCBs. These contain no plated-through holes (PTH), eliminating the need to inspect the attributes associated with PTHs. These PCBs are much simpler to manufacture and PCB shops specializing in these types of PCBs have fewer processes to manage. Many single-sided PCBs are manufactured using a punching process to make part mounting holes and create the outline of the PCB. Single-sided PCBs are typically the least expensive type of PCBs and are usually used in simple and inexpensive applications.

51.2.1.2 Double-Sided PCBs. These differ from single-sided PCBs in that the circuitry is on both sides of the PCB and there are PTHs connecting the top to the bottom. Moving from single-sided to double-sided PCBs requires the manufacturer to keep consistency of the image from top to bottom, and also requires drilling and through-hole-plating processes. These additional processes also require increased inspection, as the attributes they represent can significantly affect acceptability and quality of the PCB.

51.2.1.3 Multilayer PCBs. These are much more complex than double-sided PCBs as they require that circuitry be embedded within the PCB. The PTHs must also make connection to internal layers as well as top and bottom layers. Multilayer PCBs also introduce the lamination process, which is where the manufacturer actually creates the composite PCB from layers of circuitry in a lamination press under high pressure and heat.

51.2.1.4 Backplane PCBs. These multilayer PCBs have connector arrays installed that hold other PCBs. These arrays must be rigid enough to ensure that insertion of PCBs does not affect their performance and reliability. They usually have many layers and tend to be large and thick. Here, again, the manufacturing processes differ from those of traditional multilayer PCBs, and additional requirements for the connector area and PTH areas are necessary.

51.2.1.5 High-Frequency PCBs. These can be any of the preceding types of PCBs and have unique properties that allow high-speed electrical signals to travel within or on the PCB without significant degradation. The materials used and geometry of the circuit are critical to produce the correct impedance for these signals to maintain integrity. Additional testing is typically performed to verify the more precise nature of the requirements for these types of PCBs, such as controlled impedance.

51.2.2 Flexible PCBs

Flexible PCBs must be capable of bending and flexing, either for installation or continuously throughout the product's life, and come in single-sided, double-sided, multilayer, and high-speed versions just like the rigid PCBs previously described.

51.2.3 Rigid/Flexible PCBs

Rigid/flexible combination PCBs combine the benefits of the rigid PCB with integrated flexible parts. This type of PCB requires evaluation of both the rigid and flexible sections for acceptability.

51.3 METHODS FOR VERIFICATION OF ACCEPTABILITY

51.3.1 Production PCB

Most acceptability criteria are evaluated on the production PCB by visual or dimensional means. Visual inspection is typically done under low magnification by human operators. Although the human eye is very capable of catching inconsistencies in the circuit patterns and surface materials, it is not always consistent and can allow nonconforming attributes to slip through. Because of this, many companies have turned to using automated optical inspection (AOI) machines to check pattern integrity, consistency, and dimensions. AOI provides a much more consistent evaluation of circuit geometry than the human eye, but is a slower and more costly option. Many visual inspection attributes have dimensional attributes attached to them. It is important to note that magnification for inspection of dimensional attributes should be sufficient to make an accurate measurement to determine acceptability.

51.3.2 Test Coupons and Patterns

For testing purposes, there are several reasons to utilize test coupons and patterns that are representative of the PCBs rather than the PCBs themselves. The downside of using coupons is that they take up space on the manufacturing panel and therefore increase the overall cost of the PCBs. Also, the test coupons are not actually part of the PCB and are typically placed at the edges of the production panel, where production attributes may be different from the production panel. This can cause the coupons and patterns to reflect different attributes than the PCBs they are attempting to represent. It is very important to take steps to ensure that the test coupons and patterns are actually manufactured in a manner that is representative of all the attributes of the PCBs associated with them. It is important to ensure that any holes in the coupon are drilled with the same tools and parameters as those on the corresponding PCB. Coupons and patterns are typically used in situations where using a PCB for testing would either not provide the information required or the PCB would have to be destroyed in order to get the information. Coupons and patterns can be used for microsectional analysis, electrical measurements, environmental simulation, and reliability evaluation. See Fig. 51.1[1] for an example of test coupons taken from IPC-2221. (Unless noted otherwise, figures in this chapter are courtesy of the Association of Connecting Electronics Industries [IPC], and are taken from the publication IPC-600, "Acceptability of Printed Circuits." These figures can be found in that document under the headings noted in this chapter for specific topics. Figures credited to the author of this chapter are taken from presentations made by him. Other figures are from the sources noted in the captions. For further details on design and placement of coupons, see IPC-2221.)

51.3.3 Microsection

Microsectioning the holes from a PCB provides a cross-sectional view of the construction of the PCB. Microsectioning a PCB with a design that removes nonfunctional pads makes it virtually impossible to inspect all the interconnecting layers in the hole, so microsections are typically done on associated and optimized coupons containing PTHs that contain pads on all the layers. Microsectioning PCBs requires great skill, as the PCB contains both hard and soft materials and the area of interest is very small. Improper microsection technique can either create or hide anomalies. For example, interconnection separation can be created or hidden by improper techniques.

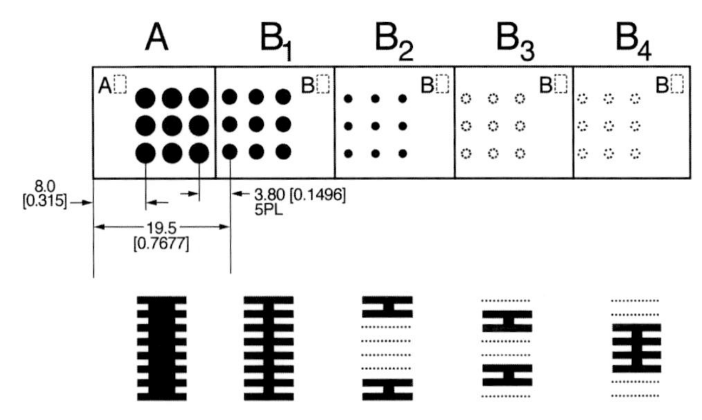

A = Component holes; solderability & rework. (as required)

B_1 = Through vias thermal stress. (most complex through-hole)

B_2 = Blind vias thermal stress; separate sequential plating cycle.

B_3 = Buried vias thermal stress; separate sequential plating cycle.

B_4 = Buried vias thermal stress; separate sequential plating cycle.

FIGURE 51.1 Quailty conformance test coupons taken from IPC-2221. *(Courtesy of IPC.)*

51.4 INSPECTION LOT FORMATION

An inspection lot consists of all the PCBs fabricated from the same materials, using the same processing procedures and constructions and produced under the same conditions. Differing PCBs can be combined to form an inspection lot, but then they must all be submitted together for inspection. It is the PCB manufacturer's responsibility to maintain traceability for all the PCBs in an inspection lot along with any corresponding test coupons that were separated from the production panels. The success of inspections greatly depends on how the inspection lot is formed. The goal of forming the inspection lot for a particular attribute is to allow, if applicable, a smaller sampling of the inspection lot to be representative of the entire lot. In forming inspection lots and sampling groups, the methodology for producing the attribute and the consistency of that methodology or process must be taken into consideration. Some PCB processes are done in batch form and others are done on a panel-by-panel basis. Understanding process techniques is critical in forming inspection lots and sampling groups that are representative of the product.

51.4.1 Sampling Inspection

For some attributes, 100 percent inspection is not necessary or prudent. In these cases, a sampling of the PCBs from the inspection lot is allowed to represent the entire inspection lot. Most performance documents specify which attributes can use sampling inspection and to what extent sampling may be performed. There are many sampling plans available; the most widely known, and historically used, is MIL-STD-105, which describes an Acceptable Quality Level (AQL) for samples of an inspection lot. However, it allows a small amount of defective material to be considered "acceptable" and has typically been replaced by sample plans where defective parts that are found during inspection cause the lot to be termed "nonconforming." This style of plan is defined as Defective Parts (C) Equals Zero (C = 0). If the inspection lot is considered "nonconforming" during a sampling procedure, the supplier may elect to screen 100 percent for the nonconformance, removing only the nonconforming PCBs from the inspection lot.

51.4.2 Inspection Tool Calibration

It is important that the all inspections be carried out with equipment that is calibrated and traceable to internationally recognized standards. Two of the documents that govern inspection equipment and facilities are ISO-17025 and IPC-QL-651.

51.5 INSPECTIONS CATEGORIES

51.5.1 Materials Inspection

This inspection usually consists of certifications from the material supplier, supported by data from the supplier's own inspections. The materials used must be inspected by the PCB manufacturer to ensure compliance with customer requirements.

51.5.2 Qualification Inspection

The PCB supplier must provide proof to the user of its qualifications to build the product specified. The user must then review this proof in order to minimize the risk associated with buying from the PCB manufacturer. This process usually includes the PCB manufacturer supplying a self-declaration of capabilities such as those found in the IPC-1710, and then supplying some kind of physical sample or testing data to support the assertions made in the self-declaration. It is up to the PCB user to determine the level and detail of qualification required.

51.5.3 Quality Conformance Inspection

These product verification inspections use specified acceptability criteria and are performed prior to shipment of the PCBs to the customers. These inspections can be defined either by industry or customer standards, or a combination of both.

51.5.4 Reliability Testing

This testing consists of looking for imperfections in the PCB that may show up some time during the product's field life. They include accelerated life testing for both the interconnection and insulation attributes of the PCB. Please see Chaps. 57, 58, and 59 for a detailed discussion of reliability testing.

51.6 ACCEPTABILITY AND QUALITY AFTER SIMULATED SOLDER CYCLE(S)

The user expects the PCB to survive the assembly and rework process. Temperature excursions that approach or exceed the base material's glass transition temperature (T_g) can significantly affect some attributes of the PCB. For these attributes, it is important that the PCB be examined after solder cycling that simulates the type and quantity of soldering operations (rework included) that the PCB is expected to undergo during the assembly process. It is

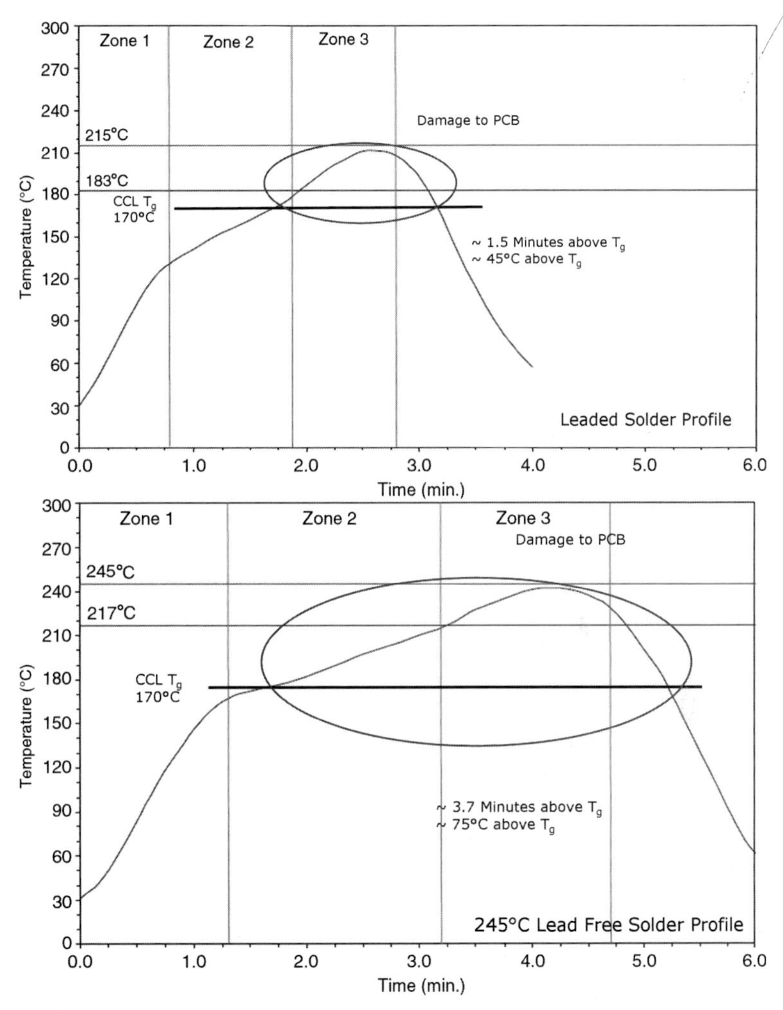

FIGURE 51.2 Damage done to PCBs during typical leaded and lead-free solder cycles. *(Courtesy of Microtek Laboratories.)*

important to note that each rework process usually includes two soldering cycles: one to desolder the affected part, and one to reattach a new part.

51.6.1 The Effects of Lead-Free Assembly on PCB Acceptability and Quality

The soldering process has a significant and detrimental effect on the PCB's reliability and quality. The lead-free assembly process increases both the soldering temperature and the time the PCB is exposed to that temperature. This will accelerate degradation of the base material and can have significant impact on the PCB. Figure 51.2 shows damage related to time and temperature above T_g. It is essential to establish that the PCB was capable of being assembled and was acceptable when it was delivered to the assembly organization, as poor assembly techniques can severely compromise the integrity of the PCB. It is almost impossible to look at a PCB after assembly and definitively state that the fault lies entirely with the PCB manufacturer or with the assembly organization.

51.6.2 PCB Issues Created by Lead-Free Assembly

- **Copper reduction** Lead-free solder, when molten, rapidly dissolves copper into it. One must be very careful when using lead-free solder on copper surfaces without a protective barrier of some type (e.g., nickel); otherwise, the copper conductors may be reduced to a point where they are not adequate for the design.
- **Base material degradation** Base materials begin to degrade and decompose during soldering and can reduce the insulating properties of the resin system in the base material.
- **Conductive anodic filament (CAF)** As the base material system degrades, the resistance to CAF formation decreases.
- **Delamination** In extreme cases, the resin system can lose cohesive or adhesive properties, causing catastrophic failure.
- **Increased expansion causing premature PTH failures** Base materials expand when exposed to soldering temperature. This expansion becomes pronounced above the T_g of the resin system in the base material. The increased temperature of lead-free soldering can cause premature failure of the PTHs due to this increased expansion. See Fig. 51.3 for PTH expansion characteristics with temperature.
- **Solderability** Lead-free solder does not wet as well to solderable surfaces as leaded solder does, so pad size, solder quantity, and flux type and quantity must be taken into account.

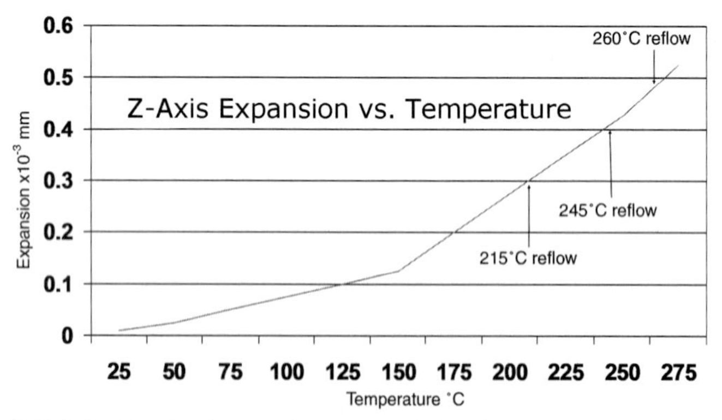

FIGURE 51.3 Z-axis expansion characteristics of 150°C T_g PCBs versus temperature. *(Courtesy of Microtek Laboratories.)*

51.7 NONCONFORMING PCBS AND MATERIAL REVIEW BOARD (MRB) FUNCTION

Standards and customer documentation establish minimum quality and acceptability requirements for the inspections performed. If the PCB does not meet these requirements, it becomes nonconforming. This does not necessarily mean the PCB is scrapped, although this is often what happens to nonconforming PCBs. What it does mean is that with a generic standard, samples that do not meet the stated requirements must be "dispositioned" outside the standard. To determine whether a nonconforming PCB is or is not "scrap," a material review board (MRB) must be established to evaluate the nonconformance(s). The MRB is usually made up of one or more representatives from the departments of quality, production, and design. The purpose of the materials review board is to effect positive corrective action within a short time to eliminate the cause of recurring discrepancies and prevent recurrence of similar discrepancies. The MRB's responsibilities include:

- Reviewing questionable PCBs or materials to determine conformance with acceptability, quality, and design requirements
- Reviewing nonconforming PCBs for effects on design functionality
- Authorizing repair or rework of nonconforming PCBs when appropriate
- Establishing responsibility and/or identifying causes for nonconformance
- Authorizing the scrap of excessive quantities of materials

When all else fails and the PCBs do not meet the intended end product use, they need to be scrapped.

51.8 THE COST OF THE ASSEMBLED PCB

The cost of the assembled PCB continues to increase because of density, complexity, and rising material and component costs. The value added to the PCB can be 10 to 20 times the cost of PCB. If it is found that a nonconforming PCB is the cause of failure, the total cost of the assembly process is likely to be passed on to the PCB manufacturer.

51.9 HOW TO DEVELOP ACCEPTABILITY AND QUALITY CRITERIA

51.9.1 Using Industry Standards

There are both advantages and disadvantages to using industry standards for acceptability and quality requirements.

51.9.1.1 Advantages

- International standards represent a consensus of what is possible and common for acceptance.
- They increase the likelihood of PCB users getting what they expected.
- The PCB manufacturer is unlikely to become as expert in complying with a "customer" specification as it would an industry standard.
- The more experience the quality personnel at the PCB manufacturer obtain in acceptability criteria, the better they will be at spotting nonconformance.
- Industry standards are a great place to start developing company-specific requirements for acceptability and quality.

51.9.1.2 Disadvantages

- International standards are developed by a cross section of the industry. The resultant standards often represent the "least common denominator" requirements in order to satisfy the broad range of participation.
- Although these standards may meet most of a PCB buyer's needs, they often leave deficiencies that necessitate additional requirements being imposed upon the PCB supplier.

- By nature, international standardization efforts always lag the state-of-the-art in the industry.
- Standards can over- or underspecify requirements for an industry segment's needs.

51.9.2 Industry Specifications

The following are lists of the most commonly used specifications for PCBs. IPC standards are the most widely used in the industry and therefore are referenced in this chapter as examples.

51.9.2.1 Defence Supply Center Columbus (DSCC)

- MIL-PRF-55110, "Printed Wiring Board, Rigid, General Specification for"
- MIL-PRF-31032, "Printed Circuit Board/Printed Wiring Board, General Specification for"
- MIL-P-50884, "Printed Wiring Board, Flexible or Rigid-Flex, General Specification for"

51.9.2.2 International Electrotechnical Commission (IEC)

- IEC-61188, "Design of Printed Boards and Printed Board Assemblies"
- IEC-62326, "Printed Board Performance for Capability Approvals"
- IEC-61189, "Test Methods for Printed Boards and Printed Board Materials and Printed Board Assemblies"

51.9.2.3 Institute for Interconnecting and Packaging Electronic Circuits (IPC)

- ANSI/J-STD-003, "Solderability Tests for Printed Boards"
- IPC-1710, "OEM Standard for Printed Board Manufacturer's Qualification Profile (MQP)"
- IPC-4552, "Specification for Electroless Nickel/Immersion Gold (ENIG) Plating for Printed Circuit Boards"
- IPC-4553, "Specification for Immersion Silver Plating for Printed Circuit Boards"
- IPC-A-600, "Acceptability of Printed Boards"
- IPC-6012, "Qualification and Performance Specification for Rigid Printed Boards"
- IPC-6013, "Qualification and Performance Specification for Flexible Printed Boards"
- IPC-6016, "Qualification and Performance Specification for High Density Interconnect (HDI) Layers or Boards"
- IPC-6202, "IPC/JPCA Performance Guide Manual for Single- and Double-Sided Flexible Printed Wiring Boards"
- IPC-TM-650, "Test Methods Manual"

51.9.2.4 Japan Printed Circuit Association (JPCA)

- JPCA-PB01 "Printed Wiring Boards"
- JPCA-HD01 "HDI Printed Wiring Boards"
- JPCA-DG01 "Design Guide for Multilayer Printed Wiring Boards"
- JPCA-DG02 "Performance Guide for Single- and Double-Sided Flexible Printed Wiring Boards"

51.9.2.5 Underwriters Laboratories (UL)

- UL-94, "Flammability Testing"
- UL-796, "Standard for Safety for Printed Wiring Boards"
- UL-796F, "Flexible Materials Interconnect Constructions"

51.9.3 Using Customer Standards

Customer standards should reflect the unique nature and needs of the company's products. Unfortunately, company history often can affect these criteria and overemphasize criteria for failures that were previously experienced. It is recommended that customer standards be used to augment, not replace, industry standards. Consider the following when creating a customer standard:

- Understanding where and how the product will be used can influence the use of extended or reduced criteria in specific requirement areas.
- Understanding the stresses to which the assembly process exposes the PCB is fundamental in defining the acceptability criteria. Process elements such as leaded or lead-free, reflow or wave solder, double-sided reflow, hand-soldering, and rework all significantly impact the requirements of the PCB.

51.10 CLASS OF SERVICE

Functionality through operational life should be the ultimate criteria for acceptance. To account for the fact that a PCB will see different end-use environments and conditions, acceptance criteria are typically broken into three classes of service. Each company must establish the level of acceptance as it is dependent on the functional criteria to which the PCB will be subjected. The IPC has established recommended guidelines for acceptance, based on end use, categorized as Classes 1, 2, and 3. The following subsections describe the classes.

51.10.1 Class 1: General Electronic Products

Class 1 includes consumer products and some computer and computer peripherals suitable for applications where cosmetic imperfections are not important and the major requirement is the functioning of the completed PCB.

51.10.2 Class 2: Dedicated Service Electronic Products

This class consists of communication equipment, sophisticated business machines, and instruments for which high performance and extended life are required, and for which uninterrupted service is desired but is not critical. Certain cosmetic imperfections are allowed.

51.10.3 Class 3: High-Reliability Electronic Products

Equipment and products for which continued performance or performance on demand is critical belong in Class 3. Equipment downtime cannot be tolerated, and the equipment must function when required (e.g., life support systems, flight control systems). PCBs in this class are suitable for applications in which high levels of assistance are required and service is essential.

51.11 INSPECTION CRITERIA

When inspecting PCBs, the requirements can be separated into externally observable attributes (those where acceptability can be determined from the outside of the PCB) and internally observable attributes (those that require microsectioning to determine acceptability). The actual requirements vary by specification, but the requirements stated in this chapter generally follow IPC document guidelines.

51.11.1 Externally Observable Attributes

Visual examination for applicable attributes is typically conducted at 3 diopters (~1.75 × magnification). If it is not possible to determine whether the condition is acceptable at this magnification, progressively higher magnifications should be used to confirm acceptability. Dimensional requirements such as conductor spacing or width may require different magnifications and devices with measurement scales or reticules as part of the inspection tool to allow accurate measurement(s) of the specified dimensions. The PCB users may also specify other magnifications and inspection tools. Note that the use of higher magnifications for visual-type attributes can producc false results due to illumination and attribute contour effects. Visual inspection criteria are difficult to define grammatically. An effective way to define visual inspection criteria is to use line illustrations and/or photographs. The IPC utilized this method in the publication IPC-A-600, "Acceptability of Printed Boards," to "visually standardize the many individual interpretations to specifications on printed boards." Of the different types of inspections specified, surface visual inspection is the least expensive. Visual inspection is performed on 100 percent of the PCBs or on a sampling based on an established sampling plan. Inspection for visual imperfections, following a sampling plan, is done on the premise that the PCBs were 100 percent visually screened during the fabrication process. Externally observable imperfections can be divided into 10 groups:

- Surface imperfections in the insulating material such as burrs, nicks, scratches, gouges, cut fibers, weave exposure, and voids
- Subsurface imperfections such as foreign inclusions, measling/crazing, delamination, pink ring, and base material voids
- Conductive pattern imperfections such as loss of adhesion, reduction of conductor width or thickness due to nicks, pinholes, scratches, surface plating, or coating imperfections
- Hole characteristics such as diameter, misregistration, foreign material inclusions, and plating or coating imperfections
- Marking imperfections affecting such characteristics as location, size, readability, and accuracy
- Solder resist surface coating, coverlayer, and coverfilm imperfections such as misregistration, blisters, bubbles, separations, adhesive squeeze out, delamination, adhesion, physical damage, and thickness
- Flawed dimensional characteristics affecting such characteristics as PCB size and thickness, hole size, pattern accuracy and registration, conductor width and spacing, and annular ring (dimensional inspection is usually performed using a sampling plan)
- An unclean PCB surface due to such materials as ionic and organic contaminates
- Poor Solderability of the PCB including wetting, de-wetting, and non-wetting
- Electrical discontinuity, circuit shorts, and if required, impedance

51.11.1.1 Surface Imperfections in the Insulating Material. Surface imperfections include burrs, nicks, scratches, gouges, cut fibers, weave exposure, and voids. The imperfections, when minor, are normally considered to be cosmetic issues, and usually have little or no effect on functionality. However, they can be detrimental to function in the edge board contact area and if they reduce the insulating properties of the material below acceptability limits.

51.11.1.2 Subsurface Imperfections. Subsurface imperfections are those that are visible from the surface of the PCB but are below the surface of the PCB. (see Fig. 51.4)

51.11.1.2.1 Inclusion. An *inclusion* is a foreign particle, metallic or nonmetallic, in a conductive layer, plating, or base material. Inclusions in the conductive pattern, depending on degree and material, can affect plating adhesion or circuit resistance. Inclusions in the base material are put into the categories of translucent/nonmetallic and opaque/metallic. Translucent particles are acceptable. Opaque particles are acceptable as long as they do not reduce the minimum conductor spacing below the minimum requirement.

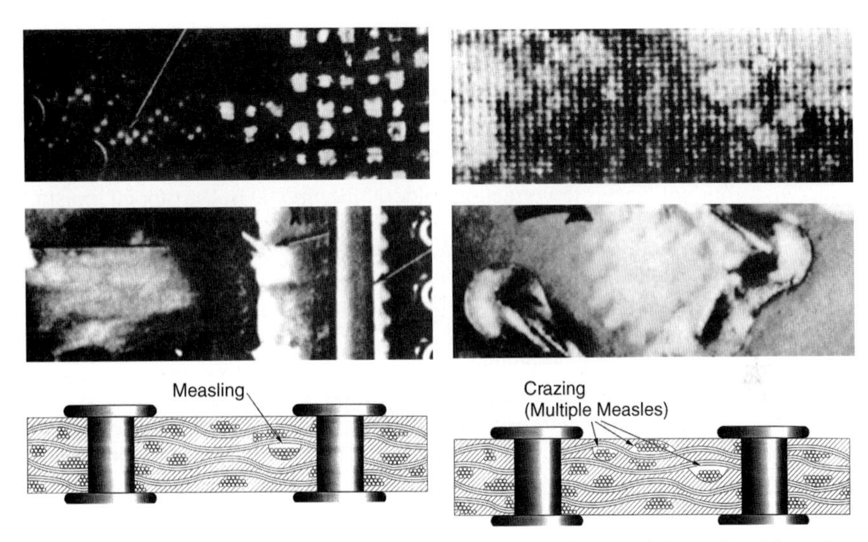

FIGURE 51.4 Base material defects: (a) blistering, (b) fiber exposure, (c) measling, (d) crazing, (e) crazing, (f) measling. *(Courtesy of Sandia Laboratories and IPC.)*

51.11.1.2.2 Measling/Crazing, Delamination, Blistering, and Weave Texture/Exposure. The IPC formed a special committee in 1971 to consider base material imperfections and to define them better with illustrations and photographs. The conditions are defined and discussed here:

- **Measling** This internal condition occurs in the woven-fiber-reinforced laminated base material in which the bundles are separated at the weave intersection. Measling manifests itself in the form of discrete white spots or "crosses" below the surface of the base material. A report compiled by the IPC, "Measles in Printed Wiring Boards," was released in November 1973. The report stated that "measles may be objectionable cosmetically, but their effect on functional characteristics of finished products are [sic], at worst, minimal and in most cases insignificant." The IPC Acceptability Subcommittee readdressed the subjects of measles and crazing in 1994 and verified the 1973 findings. IPC standards allow measles in PCBs with the exception of "high-voltage" applications. As hole spacing has continued to decrease, new attention has focused on measles. There is a concern that measling could contribute to the formation of CAF on designs with close hole spacing, and the requirements for measles will likely adjust to accommodate these designs.
- **Crazing** An internal condition occurring in the woven fiber-based base material in which the glass fibers within the yarn are separated, crazing can occur at the weave intersections or along the length of the yarn. This condition manifests itself in the form of connected white spots or crosses below the surface of the base material, and is usually related to mechanical and/or thermally induced stress. Limited crazing is an acceptable condition as long as it does not reduce the space between conductive patterns below the minimum conductor spacing, the distance of crazing does not span more than 50 percent of the distance between adjacent conductive patterns, and no propagation results from thermal testing that replicates the manufacturing process. If crazing is present at the edge of the PCB, it should not reduce the minimum distance between PCB edge and conductive pattern.
- **Delamination** This is a separation between plies within the base material, between a material and the conductive foil, or between any other planar separations within a PCB.

Blistering and delamination are considered to be major imperfections. Whenever a separation of any part of the PCB occurs, a reduction in insulation properties and adhesion occurs. The separation area could house entrapped moisture, processing solutions, or contaminants, and could contribute to electrochemical migration or produce other detrimental effects in certain environments. The delamination or blister area may also propagate to the point of complete PCB separation, normally manifesting during assembly operations. Last but not least is the effect on solderability in PTHs. Entrapped moisture, when subjected to soldering temperatures, has been known to create steam that blows holes through the plated side walls, exposing the resin and glass of the PTHs and creating large voids in the solder fillet.

- **Blistering** A localized swelling and separation between any of the layers of a laminated base material or between a base material and a conductive foil or protective coating, blistering is a form of delamination.
- **Weave texture** In this surface condition of base material, a weave pattern of glass cloth is apparent although the unbroken fibers of the woven cloth are completely covered with resin. Weave texture is an acceptable condition and should not be confused with weave exposure.
- **Weave exposure** In this surface condition of base material, the unbroken fibers of woven glass cloth are not completely covered by resin. Weave exposure is considered a major imperfection. The exposed glass fiber bundles allow wicking of moisture and entrapment of processing chemical residues into the base material.

51.11.1.2.3 Haloing. A mechanically induced fracturing or delamination on or below the surface of the base material, haloing is usually exhibited by a light area around holes, other machined areas, or both. Haloing should not penetrate more than 50 percent of the distance to the nearest conductor and should not reduce electrical spacing below the given requirement.

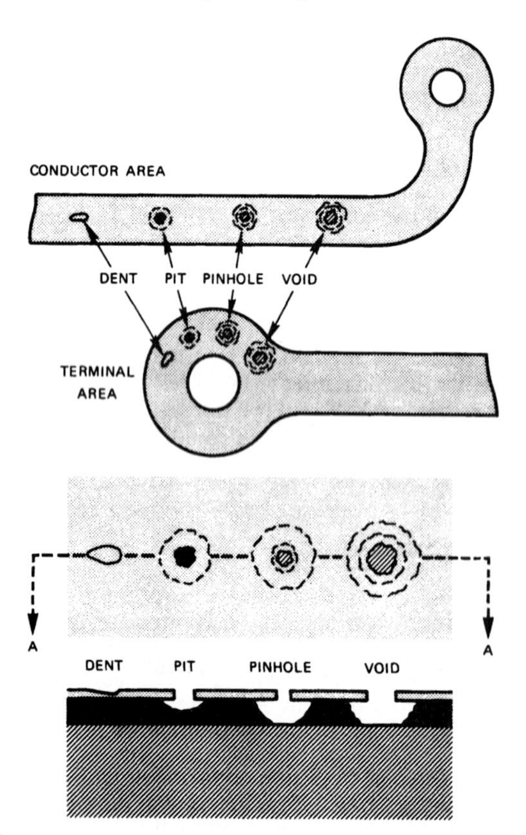

FIGURE 51.5 Dents, pits, pinholes and voids. *(Courtesy of Sandia Laboratories.)*

51.11.1.3 Conductor Pattern Integrity. Several methods can be used to determine conductor pattern integrity. Comparison equipment, overlays, and AOI machines are routinely used to compare the pattern to the digital image of the pattern. The most common conductor pattern inspection tool is the aided human eye using low (<3 ×) magnification. The loss of adhesion and imperfections that reduce the conductor volume or spacing make up this category of imperfections. Voids in conductors and lands can be detrimental to function, depending on the degree of imperfection. Voids or pinholes reduce the effective conductor width, reduce current-carrying capacity, and can affect other design electrical characteristics, such as inductance and impedance. Voids in lands are also detrimental to solderability. Pinholes or voids in the conductor or land can undermine the finish coating. The presence of conductive elements in areas in which they were not intended can reduce spacing to a point where the insulation between conductors can be jeopardized. The degree that these imperfections affect the PCB depends on the intended design of the circuit. The imperfections in this group are defined as follows. (see Fig. 51.5).

- **Dent** This is a smooth depression in the conductive foil that does not significantly decrease foil thickness.

- **Pit or dish-down** This is a depression in the conductive layer that does not penetrate entirely to the base material but does reduce the cross-sectional area of the conductor. These imperfections are almost impossible to detect as they rarely give visual cues to the inspector.
- **Scratch** This is a slight surface mark or cut that may or may not go all the way through to the base material.
- **Void or Pinhole** This is the absence of conductive material in a localized area going all the way to the base material.

51.11.1.3.1 Conductor Width. The conductor width is the observable width of a conductor at any point chosen at random on the PCB, normally viewed from vertically above unless otherwise specified. The conductor width affects the electrical characteristics of the conductor. A decrease in conductor width decreases current-carrying capacity and increases electrical resistance. Although the conductor width definition is very basic, there are two different interpretations as to where on the conductor the measurement is performed. The minimum conductor width (MCW) is measured at the minimum observed width of the conductor, and the overall conductor width (OCW) is measured at the observable width. Plating outgrowth of the top of the conductor during pattern plating can prevent the minimum width from being seen unless cross-sectioning is performed. The OCW is nondestructive and easily measured. The difference between the MCW and the OCW (see Fig. 51.6), which is measured from vertically above, can have an effect on the current-carrying capacity, inductance, and impedance of the circuit. IPC-6012, "Qualification and Performance Specification for Rigid Printed Boards," addresses conductor width requirements and allows added width reductions for edge roughness, nicks, pinholes, and scratches of 30 percent (Class 1) and 20 percent (Classes 2 and 3) from the minimum specified on the procurement documentation. The document states that this reduction also applies to the MCW reflected on the procurement documentation, and this allowance should be carefully considered against design needs.

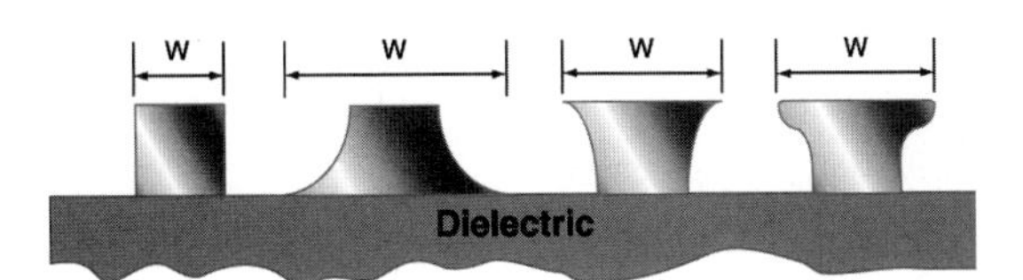

FIGURE 51.6 Overall conductor width (OCW) measurement "W." *(Courtesy of IPC.)*

51.11.1.3.2 Conductor Spacing. Conductor spacing is the distance between adjacent edges (not centerlines) of isolated conductive patterns or features. The spacing between conductors and/or lands is designed to allow sufficient insulation between circuits. A reduction in the spacing can cause electrical leakage or affect the capacitance of the circuit. The cross-sectional width of conductors is usually nonuniform. Therefore, the spacing measurement is taken at the closest point between the isolated conductors and/or external *annular ring* (the portion of conductive material completely surrounding a hole).

51.11.1.3.3 External Annular Ring. The portion of conductive material completely surrounding a hole is called the annular ring (see Fig. 51.7). The primary purpose of the annular ring is that of a flange surrounding the hole; it provides an area for the attachment of electronic component leads or wires and acts as an anchor for the circuit. The minimum annular ring dimension needs to be defined by the PCB user and will differ based upon the function of the

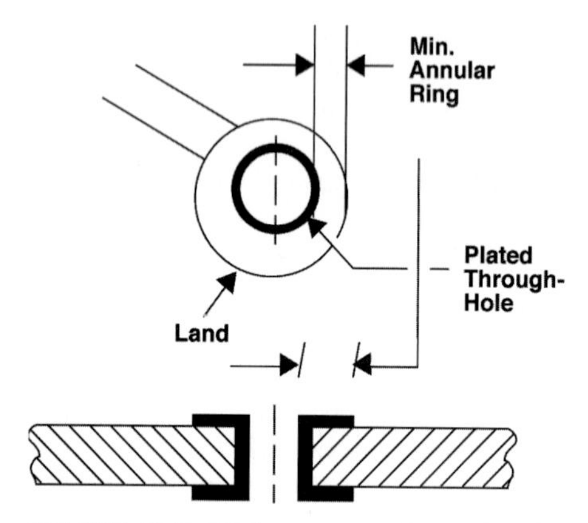

FIGURE 51.7 Minimum external annular ring (includes plating). *(Courtesy of IPC.)*

holes and whether they are plated or not. Annular ring dimensions that are less than those specified can interfere with attachment of the component, and a pad breakout (zero annular ring) condition associated with the area where the circuit enters the pad can reduce the current-carrying capacity of the circuit. After verifying the dimensions of the PCB and then verifying that the minimum widths of the annular rings on all other lands are within the procurement documentation requirements, the pattern is found to be in registration with the drilled hole location. Front-to-back pattern registration can also be inspected in this manner.

51.11.1.3.4 Plating Adhesion. A common method of inspecting for plating adhesion is the tape adhesion test, which is described in detail in the IPC test method IPC-TM-650, 2.4.1. The basic test method is performed by adhering tape of known properties onto the plating, rapidly removing the tape, then inspecting the tape for removed plating on the tape adhesive.

51.11.1.3.5 Peel Strength. Peel strength is the force per unit width required to peel the conductor or foil from the base material. It is usually associated with the acceptance testing of copper-clad base material upon receipt, but it is sometimes used to test the adhesion of the conductors to the finished PCB. Peel strength tests of conductors are usually performed after the specimens have been dip or reflow soldered. Peel strength tests of conductors are a good method of ensuring that the conductor-to-base material adhesion is sufficient to withstand assembly and rework operations.

51.11.1.4 Hole Characteristics. PTHs provide circuit interconnection from the surface of the PCB to internal layers and to circuitry on the opposite side. For this reason, much attention is correctly focused on the acceptability and quality of these holes. Nonconformance in the PTHs is a serious concern.

51.11.1.4.1 Roughness and Nodulation. *Roughness* is an irregularity in the side wall of a hole; *nodulation* is a small knot or an irregular lump found in or on the plating. Although roughness and nodulation are not desirable, they are allowable in small amounts. Roughness and nodulation create one or more of the following nonconforming conditions:

- Reduced hole diameter
- Impaired lead insertion
- Impaired solder flow through the hole
- Voids in solder fillet

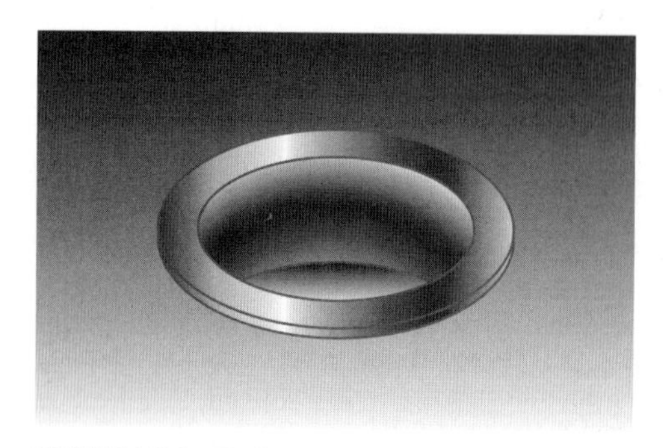
FIGURE 51.8 Plating void. *(Courtesy of IPC.)*

51.11.1.4.2 Voids in the Plating and Finished Coating. A void is the absence of material where it is intended to be. Copper-plating voids are evaluated again during microsectional analysis, but this sampling is very small and this attribute should be thoroughly evaluated during visual inspection of the PCB. When visually evaluating voids in the PTH copper, no more than 5 percent of the holes on the PCB should exhibit evidence of voids. In holes that have voids (see Fig. 51.8), there should be no more than one void that is less than 5 percent of the hole length. Voids in the finish coating of the PCB (see Fig. 51.9) that do not extend to the copper plating are allowed in larger quantities than voids in the copper plating. There should again be surface finish voids in no more than 5 percent of the PCB holes, and these holes should have no more than three voids that are each less than 5 percent of the hole length.

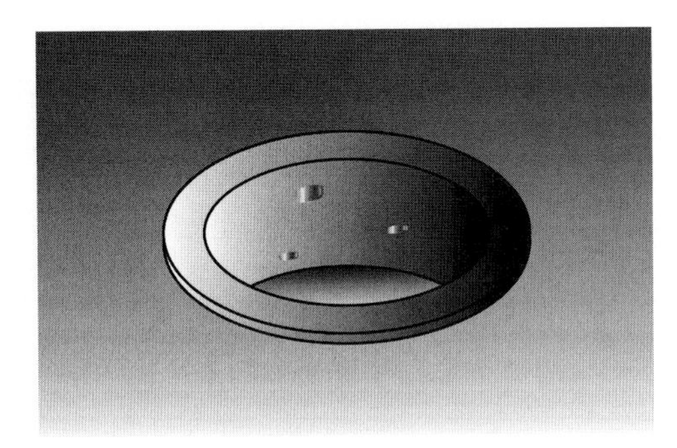

FIGURE 51.9 Voids in finished coating. *(Courtesy of IPC.)*

51.11.1.4.3 Hole Diameter. Hole size is the inner clearance diameter of the finished plated-through or unplated hole. A PTH is a hole in which electrical connection is made between internal or external conductive patterns, or both, by the deposition of metal on the wall of the hole. An unplated hole, termed an *unsupported hole*, is a hole containing no conductive material. Hole size measurement is performed to verify that the hole meets the intended design requirements. This size requirement is usually associated with the fit of component leads, mounting hardware, etc., plus adequate clearance for attachment metal. PTHs providing layer-to-layer interconnections, where no components are soldered into the hole, are called "via" holes. Via holes do not have a fit requirement, so only the plating integrity is critical. Drill blank plugs and optical magnifiers are used to verify hole diameter. The latter method is utilized when soft coatings over the copper are used. The optical method prevents deformation or removal of the soft coatings within the hole. When drill blank plug gauges are used, the inspector should use a soft touch to prevent damage to the hole. The drill blank gauges should be cleaned prior to use to prevent solderability degradation. Plating nodules are sometimes present in the hole and restrict the penetration of the gauges. Forcing the gauges into the holes causes the nodules to be dislodged, resulting in voids in the PTH side wall.

51.11.1.4.4 Lifted Lands. During visual examination there should be no evidence of the lifting of lands from the base material surface. Lifted lands can trap contaminates during the assembly process and are considered undesirable for assembly.

51.11.1.5 Marking (Legend/Silkscreen) Imperfections. The marking of part number, revision letter, part mounting information and orientation, etc., on a PCB is evaluated for location, size, readability, and accuracy. Marking imperfections are usually considered minor, but markings that are missing, incorrect, or partially obscured could have an effect on functionality. Marking that encroaches into the soldering area can also be a significant issue and should not reduce the solderable areas below what is stated in the procurement documentation.

51.11.1.6 Solder Resist Coating Imperfections. A solder resist is a coating material used to mask off or to protect selected areas of a pattern from the attachment of solder. A solder resist can also be used as electrical insulation between isolated conductors or to define the soldering area for ball grid array (BGA) patterns. Attributes inspected are registration, blisters, bubbles, adhesion, wrinkles, tenting, soda strawing, and delamination. Misregistration at land areas reduces or prevents adequate solder fillet formation or can expose adjacent conductors. The minimum land area to obtain adequate solder filleting should be reflected by the minimum acceptance criteria. Wrinkles, flaking, peeling, and delamination provide sites for moisture absorption and contaminants to settle on or into the PCB and can reduce the insulation between adjacent conductors, increasing the potential for electrochemical migration (ECM). When solder resist is used to define BGA patterns, the shape and position on or around the land is important to proper soldering of the BGA (see Fig. 51.10)

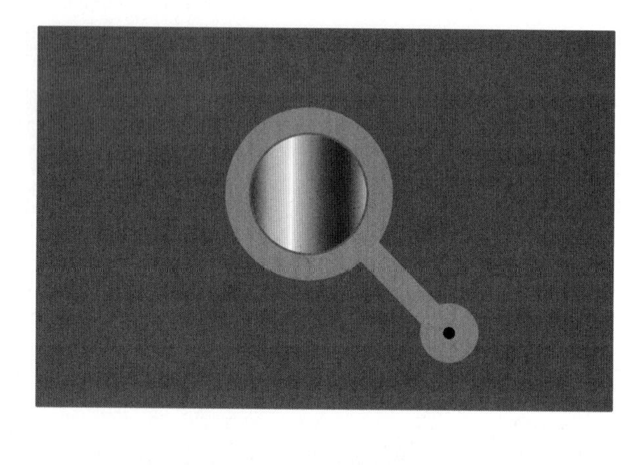

FIGURE 51.10 Soldermask defined BGA land. *(Courtesy of IPC.)*

51.11.1.6.1 Tenting of Vias. Tenting of via holes is the process of covering holes with solder resist where soldering is not required or desired. This is done to prevent the potential for shorting during the assembly process and to allow for vacuum-based test fixtures to draw the assembly down onto test pins for functional test. The requirement here is that all vias that are required to be tented are covered by resist.

51.11.1.6.2 Soda Strawing of Solder Resist. A soda straw is a long tubular void along the edges of conductive patterns where the solder resist is not bonded to the base material surface or edge of the conductor (see Fig. 51.11). If this void is allowed to reach the point where the solder resist meets the outside environment, fluxes, fluids, and other contaminates may end up trapped in the soda straw, leaving the potential for future ECM or corrosion.

FIGURE 51.11 Soda strawing between copper trace and soldermask. *(courtesy of IPC.)*

51.11.1.7 Coverlayer Characteristics. Flexible and rigid/flexible PCBs are covered with either a coating similar to solder resist or a coverfilm adhered to by acrylic adhesive. Covercoat acceptability requirements are the same as those for solder resist.

51.11.1.7.1 Coverfilm Separations. When a film is adhered to the flexible PCB surface, separations between the film and PCB surface can occur. There should be no more than three small separations that do not reduce the spacing between conductors by 25 percent or below the minimum stated in the procurement documentation. Separations along the edge of the PCB are not allowed, as they can trap contaminates.

51.11.1.7.2 Adhesive Presence on Attachment Surfaces. The presence of adhesive from the coverfilm should not reduce the solderable surfaces below the minimum reflected on the procurement documentation.

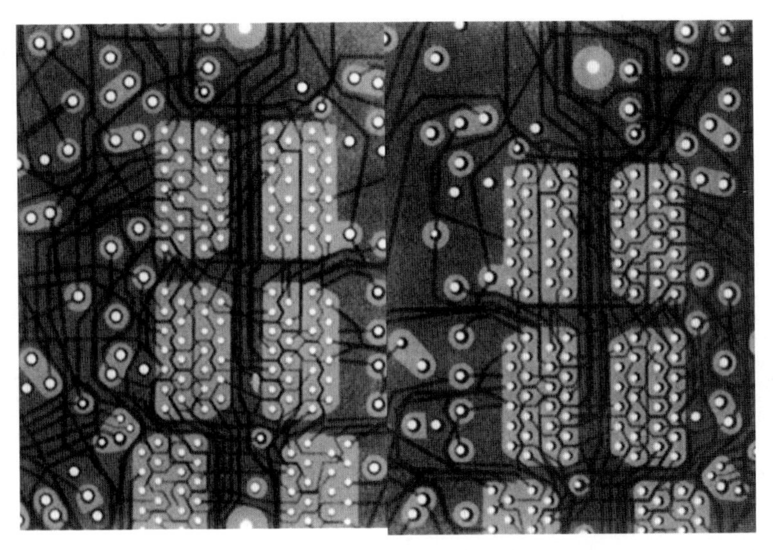

FIGURE 51.12 Layer-to-layer registration, x-ray method. *(Courtesy Sandia Laboratories.)*

51.11.1.8 Dimensional Characteristics. Contour dimensional inspection verifies that the outside border dimensions, cutouts, and key slots meet the procurement documentation requirements. Contour dimension requirements can be considered fit requirements. Both undersized and oversized PCBs can affect functionality, depending on the degree of requirement violation. Incorrect placement or size of key slots prevents the proper mounting of the PCB. Measuring methods vary from the use of a ruler or calipers to sophisticated coordinate measurement machines (CMMs). The sophistication of the method is naturally dependent on the required dimensions and tolerances.

51.11.1.8.1 External Dimensions. The external dimensions of a PCB are usually measured by calipers and micrometers. They can also be measured with a CMM or comparison gauges and templates. Tolerances for these dimensions should be reflected on the PCB procurement documentation. It is important to ensure that the measurement tools being used have sufficient accuracy to perform the intended measurements.

51.11.1.8.2 Registration, Layer to Layer, X-Ray Method. The x-ray method provides a nondestructive way to inspect layer-to-layer registration of internal layers of multilayer PCBs. It utilizes an x-ray machine and either Polaroid film or camera sensors to image the internal layers. The multilayer PCB is x-rayed in a horizontal position. The x-ray images are then examined for evidence of hole breakout of the internal lands. The lack of an internal annular ring denotes severe misregistration of layers (see Fig. 51.12).

51.11.1.8.3 Flatness (Bow and Twist). *Bow* is the deviation from flatness of a PCB characterized by a roughly cylindrical or spherical curvature such that, if the PCB is rectangular, its four corners are in the same plane (see Fig. 51.13). *Twist* is the deformation parallel to a diagonal of a rectangular sheet such that one of the corners is not in the plane containing the other three corners (see Fig. 51.14). Excessive bow and/or twist on a PCB are detrimental to the assembly process when the PCB must fit into card guides or packaging configurations in which flatness is critical. It can also interfere with the proper mounting of the assembly into the intended case. The method for bow and twist measurement can be found in IPC-TM-650 method 2.4.22. The maximum

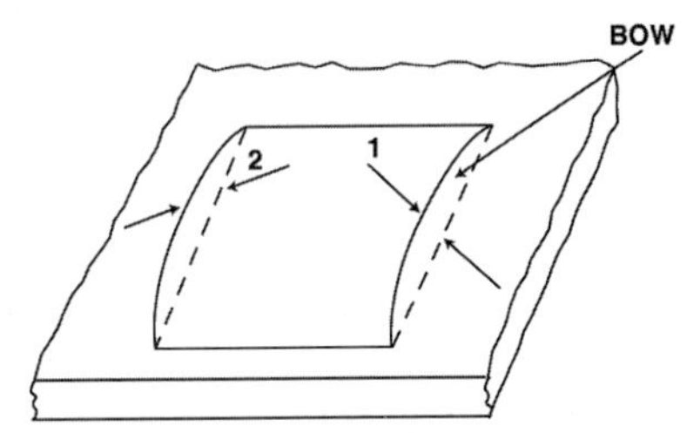

FIGURE 51.13 Bow measurement of PCB. *(Courtesy of IPC.)*

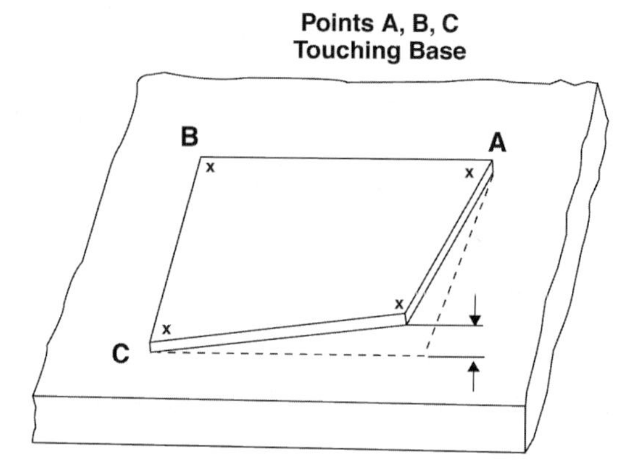

FIGURE 51.14 Twist measurement of PCB. *(Courtesy of IPC.)*

bow or twist allowed is 1.5 percent unless the PCB has surface-mounted devices (SMDs) attached where the requirement is reduced to a maximum allowable of 0.75 percent.

51.11.1.8.4 Cleanliness (Solvent Extract Methods). PCBs are fabricated using wet chemistry and mechanical techniques. Some of the chemical baths contain metallic salts and organic materials. These chemicals, if left on the surface, could affect the functionality of the PCBs by reducing electrical insulation resistance and promote corrosion or ECM on the conductor pattern. This condition is usually associated with low operating voltage (10 volts or less) and requires three components: moisture, metallic contaminants, and voltage. The ionic cleanliness test measures the conductivity of a de-ionized water and alcohol solution used to wash the PCB. Commercial equipment is available to perform ionic cleanliness testing. The organic cleanliness test uses an organic solvent wash (e.g., acetonitrile) to remove "loose" organics from the surface of the PCB. This wash solution is collected on a glass plate and the solvent is allowed to evaporate. Any organic contamination in the wash solution leaves a cloudy residue on the glass that can be subsequently analyzed for composition. These two tests are mutually exclusive in that the organic test cannot detect ionic contamination and vice-versa. Note that when a solder mask is required on PCBs, an additional cleanliness test is performed on the uncoated PCBs before coating.

51.11.1.9 Solderability Inspection. Solderability inspection measures the ability of the attachment points to be wetted by solder in the areas for the joining of components to the PCB. Three terms are used to describe acceptability of solderability:

- **Wetting** The formation of a relatively uniform, smooth, unbroken, and adherent film of solder to a basis material, wetting is the preferred and acceptable condition.
- **De-wetting** This condition results when molten solder has coated the attachment surface and recessed, leaving irregularly shaped mounds of solder separated by areas covered with a thin film of solder that fully covers the basis metal. De-wetting is not acceptable for attachment points but is allowed on ground or voltage planes for Class 1 and 2 products.
- **Non-wetting** This nonconforming condition is the partial adherence of molten solder to a surface that it has contacted, leaving some of the basis metal exposed.

51.11.1.9.1 Solderability Testing and Coating Durability. Many methods for inspecting PCBs for solderability both quantitatively and qualitatively have been established. However, the most meaningful method is that used in the actual assembly soldering operation (hand, wave, drag, reflow soldering, etc). ANSI/J-STD-003, "Solderability Tests for Printed Boards," addresses the recommended test methods and subsequent storage conditions that have no adverse effect

FIGURE 51.15 "Wetted" solder joint. *(Courtesy of IPC.)*

on the solderability of those portions of the PCB intended to be soldered. Test coupons or PCBs are subjected to accelerated aging for a length of time defined by a coating durability requirement reflected on the procurement documentation, followed by solderability testing. Solderability is acceptable when the surfaces to be soldered have fully wetted. See Fig. 51.15 for acceptable criteria for PTHs, and Fig. 51.16 for examples of non-wetting.

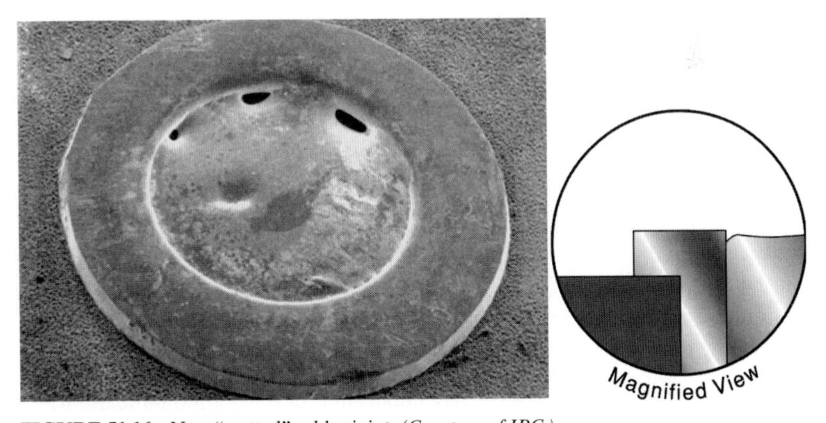

FIGURE 51.16 Non-"wetted" solder joint. *(Courtesy of IPC.)*

51.11.1.9.2 Coating Durability Requirements. The coating durability requirements are classified into three categories:

- **Category 1** Minimum coating durability is required. The PCB is intended to be soldered within 30 days from time of manufacture and likely to experience minimum thermal exposure.
- **Category 2** Average coating durability is required. This classification is intended for PCBs likely to experience storage of up to 6 months from the time of manufacture and moderate thermal or solder exposure.
- **Category 3** Maximum coating durability is required. This classification is intended for PCBs that are likely to experience long storage (over 6 months from time of manufacture) and that may experience severe thermal or solder processing steps, etc. It should be recognized that there may be significant difficulty involved, a cost premium, or delivery delay associated with PCBs specified to meet this durability level.

51.11.1.9.3 Surface Finishes. Tin-lead solder alloys (e.g., 63/37) are the most popular alloy used for surface finishes on PCBs. Other surface finishes are rapidly finding their way onto the PCBs. IPC-6012 lists more than 20 different surface finishes that are now in use for PCBs. Lead-free alloys are also appearing on PCBs. It is important to understand that the composition and type of the surface finish influence solderability. The procurement documentation must state specifications for surface finish. Methods available for analyzing the alloy composition on the plated PCB include wet analysis, atomic absorption, and x-ray fluorescence (XRF). XRF is popular because of the ease of obtaining the alloy composition and thickness nondestructively.

51.11.1.10 Electrical Inspection. These inspections are performed to verify circuit integrity after processing and substantiate that the electrical continuity, signal integrity, and isolation characteristics of the PCB meet the design specifications. Electrical inspection methods are typically performed on the production PCBs, but evaluation of signal integrity is routinely performed on specially designed test coupons. Two popular nondestructive electrical tests are insulation resistance and circuit continuity tests. These are usually performed on 100 percent of complex PCBs, especially multilayer ones. Automated electrical test equipment with either stationary or moving test probes allows the probes to make contact with the all parts of the circuit pattern in order to verify the continuity and isolation characteristics of the PCB.

51.11.1.10.1 Circuit Continuity. Continuity tests are performed on PCBs to verify that the printed circuit pattern is interconnected as designed. Testing can be performed with an inexpensive multimeter for simple PCBs or with more elaborate automated equipment such as a computer-enhanced bed-of-nails or flying probe tester. These types of automated testers can either be preprogrammed with the circuit interconnection information derived from the computer design files or can use a known good "golden" PCB as a test template to learn the interconnection information. The preferred method is to perform the continuity test on all PCBs submitted for acceptance. This method is especially recommended for multilayer circuitry, where the internal patterns and interconnections cannot be inspected visually after fabrication.

51.11.1.10.2 Insulation Resistance (Circuit Shorts). The purpose of this test is to measure the resistance offered by the insulation system of a PCB to an applied voltage. Low insulation resistances can prevent proper operation of circuits by permitting the flow of leakage current. This test can also reveal the presence of contaminants from processing residues. The test is performed by applying voltage to isolated nets and measuring the current flow between these nets. Test voltages of 40 to 500 VDC and minimum insulation resistances of 100 to 500 meg-ohm are popular. As with continuity testing, insulation resistance testing typically employs automated test equipment using stationary or moving probes.

51.11.1.10.3 Impedance. Impedance (see Fig. 51.17) is the amount of resistance a circuit offers alternating current (AC) signals. High frequency AC circuits require that the impedance of PCB circuits be matched to that of the components in order to transfer the most signal between the components on the assembly. Proper impedance matching also reduces radio wave emissions (EMF) from the assembly. Impedance is measured using Time Domain Reflectometry (TDR) and is typically measured using a specially designed test circuit on a coupon. Requirements for impedance as well as test circuit design must be supplied to the PCB manufacturer. Variations in impedance of up to 10 percent around the nominal value are expected and are not unusual do to standard process variations in circuit geometry and insulation make up. Tighter control of impedance is possible, but special attention must be made to processes and materials to achieve reductions in variation.

51.11.2 Internally Observable Imperfections

These are imperfections where evaluation from the outside of the PCB is not possible. They require the preparation of a microsection in order to view and evaluate conformance. Preparing a microsection of the PCB requires a great deal of skill. The method of sample removal, the microsection materials used, and the grinding/polishing techniques all significantly affect

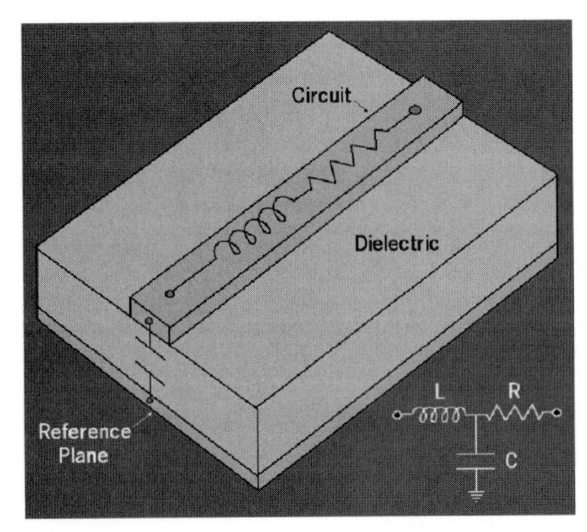

FIGURE 51.17 Impedance diagram of a PCB. *(Courtesy of Microtek Laboratories.)*

the ability to produce a microsection that will reveal and not hide imperfections. The basic methodology for the preparation of microsections can be found in IPC Test Method 2.1.1, but numerous papers and articles have been written on the art of the microsectional preparation process. The use of test coupons to perform microsections is optimal due to the fact that internal layer connections do not always appear for every layer in holes on the PCB, whereas they always appear in a properly designed microsection test coupon. See Fig. 51.18 for a microsection of a multilayer PTH.

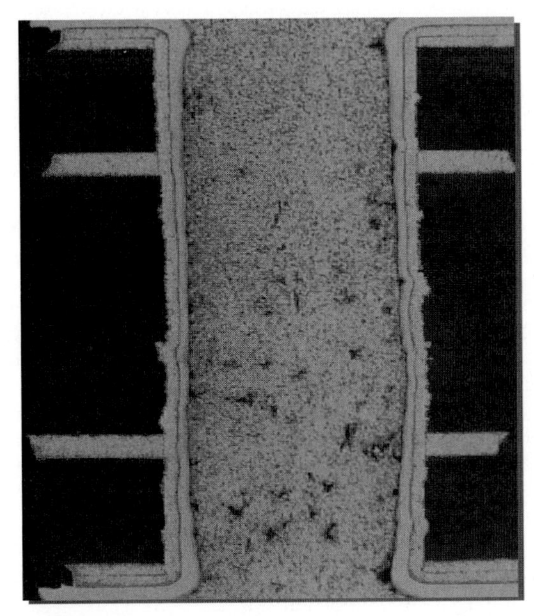

FIGURE 51.18 Vertical PTH microsection of a multilayer PCB. *(Courtesy of Microtek Laboratories.)*

51.11.2.1 Microsectioning. Samples for microsectioning are removed from the PCB or test coupon. The removed specimen should contain at least three of the smallest holes in a line that will allow microsectioning. Because shearing of microsection samples from a PCB or coupon can result in creating failure mechanisms that are not truly present in the production PCB, the practice is not recommended. Routing or sawing microsection samples is the preferred method for sample removal. The samples should be mounted in a resin-based medium that does not produce too much heat during curing and does not shrink away from the surface of the holes being evaluated. Holes should be fully filled, as partially or nonfilled holes result in poor mount conditions, difficult sanding/polishing processes, and possibly erroneous thickness results. Examination under a microscope for internally observable attributes should be performed at 100× magnification. If it is not possible to determine whether the condition is acceptable, progressively higher magnifications should be used until accurate determination of acceptance is determined. Polishing vertical PTH cross sections to the center of the hole +/– 10 percent is critical. If the hole is polished less than or more than the center of the hole, the measured plating thickness will be artificially high. Horizontal plane, PTH cross sections are recommended as references for internal-layer interconnection quality and are prepared by grinding from the top down through the hole to the inspection area. See Fig. 51.19 for an example of a horizontal microsection.

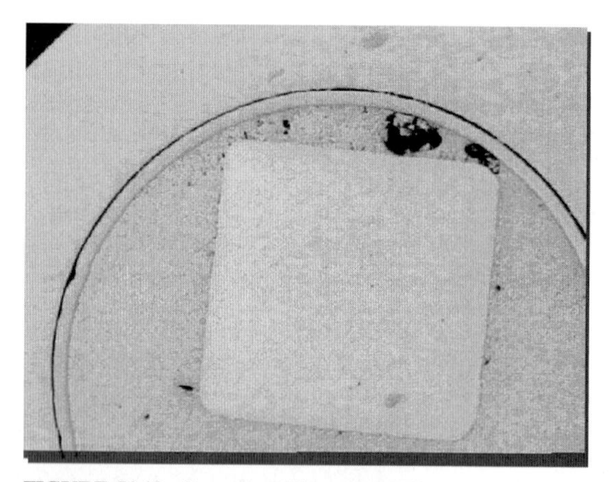

FIGURE 51.19 Example of interconnection separation in a horizontal microsection. *(Courtesy of Microtek Laboratories.)*

Although horizontal PTH cross sections are usually more accurate for plating thickness, interconnection separation, and annular ring measurements, they do not allow adequate inspection of other attributes such as voids, plating uniformity, adhesion, etchback, and nodules. Surface plating thickness measurements are taken from vertical cross sections of conductor areas. Microsection mounts are usually etched with an appropriate etchant to show grain boundaries between the copper-clad foil and copper plating and to remove copper smearing introduced during the polishing process.

51.11.2.1.1 Thermal Stress Solder Float Test. Soldering temperatures expose PCBs to thermal and mechanical stress and may deform the PCB shape and cause premature degradation of the base material. Thermal stress inspection is performed to predict the acceptability of the PCBs after the assembly and rework process. PTH degradation, separation of platings or conductors, and base material delamination are induced or amplified by the thermal stress test. The PCB specimen is oven-conditioned to reduce moisture, placed in a dessicator on a ceramic plate to cool, fluxed, floated in a solder bath, and placed on an insulator to cool. Visual inspection for imperfections is followed by microsectioning of the PTHs and inspection under a microscope for integrity. Test coupons are typically used for thermal stress inspection.

51.11.2.2 Microsection Evaluation. The evaluation of internally observable attributes in the microsection consists of the following set of inspections:

- Subsurface attributes in the PCB material such as delamination, blistering, cracks, ground plane clearance, and layer-to-layer spacing
- Internal conductor attributes including over- and under-etch, conductor cracks and voids, and foil thickness
- PTH attributes including size, annular ring, plating thickness, plating voids and nodules, plating cracks, resin smear, etchback, wicking, innerlayer (post) separation, and solder thickness

51.11.2.3 Subsurface Imperfections in the PCB Material

51.11.2.3.1 Base Material Voids after Thermal Stress. Heat from soldering operations will transfer quickly through the copper holes and pads in a PCB to the base material in close proximity to the copper. This rapid heat movement can easily cause imperfections to occur in the base material and this area near the holes is exempted from evaluation for base material imperfections. This area is defined as the "thermal zone" and described in detail by IPC-A-600. Outside this thermal zone, base material voids should not exceed 0.08 mm for rigid PCBs and 0.5 mm for flexible PCBs or flexible portions of rigid-flexible PCBs. Base material voids found in the same plane of flexible or rigid-flexible PCBs should not have a combined dimension greater than 0.5 mm (see Fig. 51.20).

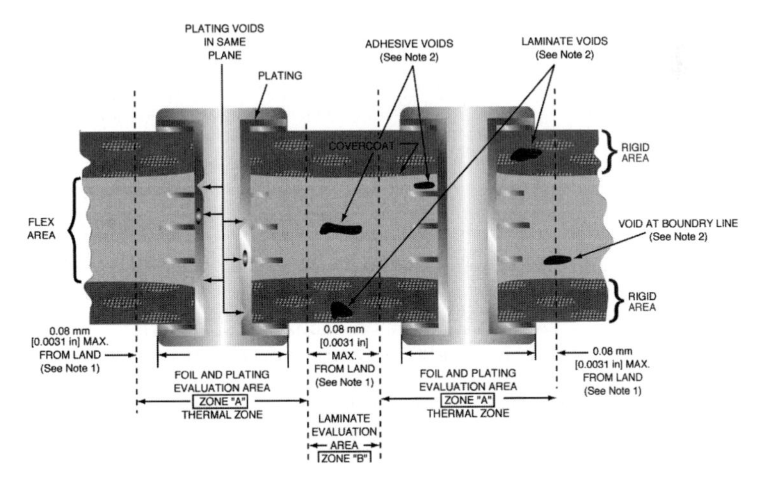

FIGURE 51.20 The thermal "A" zone. (*Courtesy of IPC.*)

51.11.2.3.2 Clearance of Holes to Power/Ground Planes. Electrical isolation between power/ground planes and nonconnected holes is necessary to ensure that shorting does not occur. This is especially true with holes that are nonplated and used for mounting of the PCB to a frame, as without proper clearance, the mounting hardware can cause shorting to the frame or between power and ground. The minimum distance of this clearance should reflect the minimum electrical spacing requirement on the procurement documentation (see Fig. 51.21).

51.11.2.3.3 Delamination and Blistering. Delamination or blistering observed in the microsection is a nonconforming condition and the inspection lot should be rescreened in accordance with the externally observable visual requirements for delamination and blistering set forth earlier in this chapter.

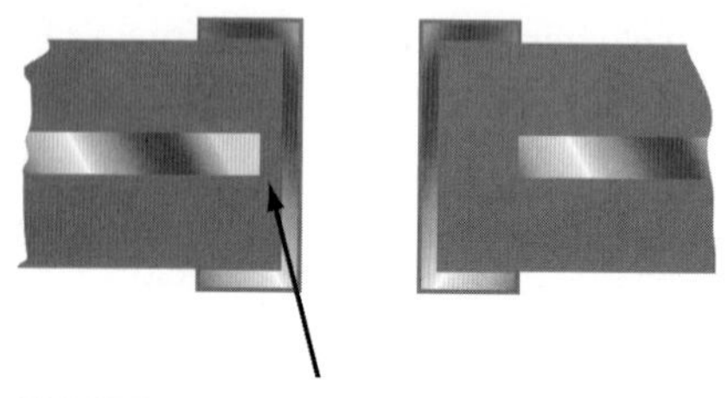

FIGURE 51.21 Clearance between PTH and power/ground plane. *(Courtesy of IPC.)*

51.11.2.3.4 Resin Smear. Resin smear is transferred from the base material onto the surface or edge of the internal conductive pattern during the drilling process. Excessive heat generated during drilling softens the resin in holes and smears it over the exposed internal copper areas. This condition creates a resin insulator between the internal land and subsequent plating in the holes, and the result is open or poorly connected circuits. The imperfection is removed from the holes by chemical or plasma cleaning prior to metallization. Inspection for resin smear is performed by viewing vertical and horizontal microsections of PTHs and any evidence of it is nonconforming. (See Fig. 51.22 for an example of resin smear prior to etchback.)

FIGURE 51.22 Resin Smear in PTH, vertical microsection. *(Courtesy of IPC.)*

51.11.2.3.5 Etchback. Etchback is a process for the controlled, positive removal of nonmetallic materials from side walls of holes to a specified depth. It is used to remove resin smear and then to remove the reinforcement material laterally to expose additional internal conductor surfaces. The degree of etchback is critical to function. Too much etchback creates excessively rough hole side walls and causes weak PTH structures. Etchback requirements range between 0.005–0.08 mm. The degree of etchback is measured by evaluating vertical PTH cross sections of multilayer PCBs and is determined "effective" if at least one of the top or bottom surfaces of the internal conductor is attached to the plated copper on both visible sides of every conductor layer. (See Fig. 51.23 for an example of acceptable etchback for all classes.)

51.11.2.3.6 Negative Etchback. This is a condition where the internal copper conductors rather than the insulating material in the hole have been laterally removed. This condition is not allowed when positive etchback is required. In all other cases, negative etchback should be minimal or nonexistent.

51.11.2.3.7 Layer-to-Layer Spacing. The minimum spacing between two consecutive layers in the PCB is determined by drawing an imaginary plane between the two most eccentric points of each of the two layers in the field of view and measuring the distance between these two planes. (See Fig. 51.24 for an example of minimum dielectric thickness between layers)

51.11.2.4 Internal Conductor Imperfections in the Microsection. In theory, circuits should be rectangular. Circuits are usually designed with this rectangular shape in mind. In practice, circuits are trapezoidal. Depending on how the circuit is formed, this shape could be close to

FIGURE 51.23 Effective etchback in PTH, vertical microsection. *(Courtesy of IPC.)*

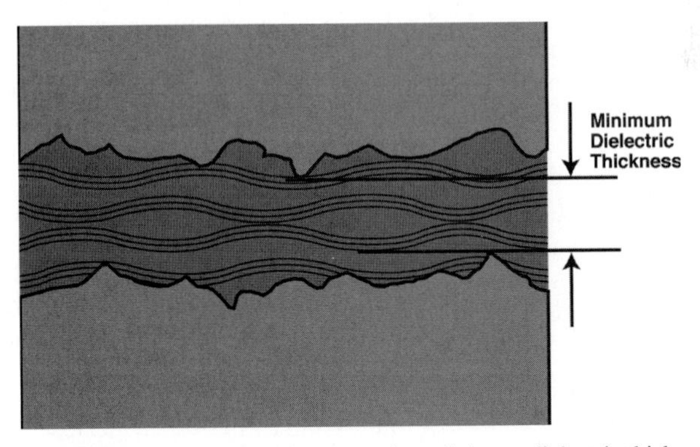

FIGURE 51.24 Layer-to-layer spacing, minimum dielectric thickness, vertical microsection. *(Courtesy of IPC.)*

a rectangle or differ significantly from it. The further from a rectangle that the shape is, the less volume the circuit has to conduct current compared to a corresponding rectangular circuit. Terms like *outgrowth, undercut, overhang,* and *etch factor* are used to define the attributes of these shapes (see Fig. 51.25 and 51.6).

51.11.2.4.1 Conductor Width. This attribute is measured at the widest point of the circuit in the microsection. The procurement documentation should state the minimum allowable width. (See Fig. 51.6)

51.11.2.4.2 Conductor Thickness. Conductor thickness depends on the thickness of the base foil used and the additional plating added to the surface during any plating processes. Allowances for expected reduction of both foil and plating thicknesses due to manufacturing processes must be considered. A detailed table of these minimum acceptable conductor thicknesses with allowances for manufacturing processing can be found in IPC-A-600.

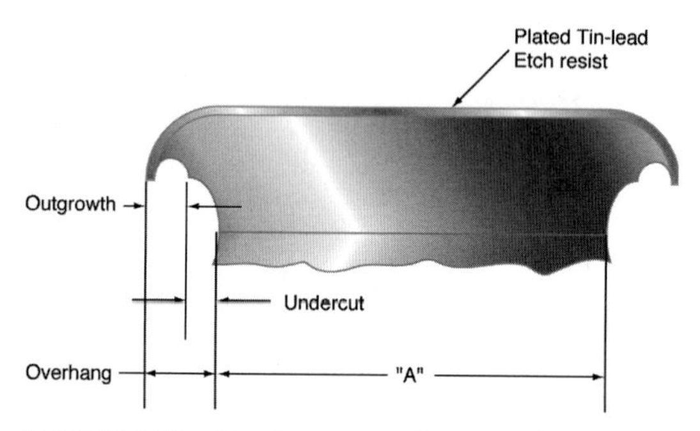

FIGURE 51.25 Circuit geometry. *(Courtesy of IPC.)*

51.11.2.5 PTH Imperfections. Nothing affects the function of the PCB more than imperfections in the PTH. These imperfections are usually not observable from the surface, and microsectional sampling of holes is limited to almost statistical insignificance. Despite this, evaluations of the microsection represent one of the most important evaluations that can be done on a PCB.

51.11.2.5.1 Annular Ring. Minimum internal annular ring is evaluated on the internal copper land that exhibits the shortest distance between its outer edge and where it ends in or at the plating in the hole. It should be noted that a microsection identifies layer shift in only one direction and that microsections in both x and y directions should be evaluated to determine accurately the actual shift that may be present. Horizontal sections from the top of the PCB down to the layer in question give a 360° view of the pad and can be used to exactly identify minimum annular ring and layer shift. The procurement documentation should specify the requirement for minimum annular ring (see Fig. 51.26), and in cases where the allowable annular ring is zero or "breakout," the breakout cannot be allowed in the area where the circuit enters the pad, as this will reduce the circuit's current-carrying capacity.

51.11.2.5.2 Lifted Lands. Lifted lands on the PCB surface prior to thermal stress are not allowed and if found, the entire inspection lot should be reevaluated to the externally observable visual requirements for lifted lands described earlier in this chapter. Lifted lands after thermal stress are allowed.

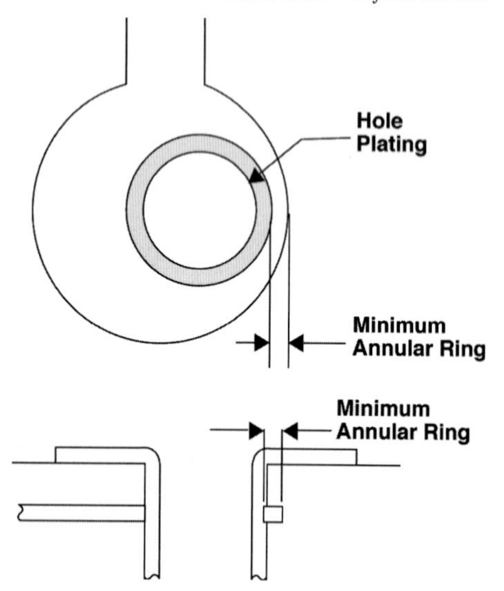

FIGURE 51.26 Internal annular ring measurement, horizontal and vertical microsection (does not include plating). *(Courtesy of IPC.)*

51.11.2.5.3 Cracks. Cracks in the foils and platings in the microsection are broken into the following six categories (see Fig. 51.27):

- "A" type cracks occur in the external foil. They do not extend into the plating and are acceptable for all classes of PCBs.
- "B" type cracks occur in the external foil. They go completely through the foil and extend into, but not completely through, the plating and are acceptable only for Class 1 PCBs.
- "C" type cracks occur in the internal foil and are nonconforming for all classes of PCBs.
- "D" type cracks completely penetrate the external foil and plating and are nonconforming for all classes of PCBs.
- "E" type cracks occur in the barrel plating of the hole. They are non-conforming for all classes of PCBs.
- "F" type cracks occur in the corner of the plating of the hole and are nonconforming for all classes of PCBs.

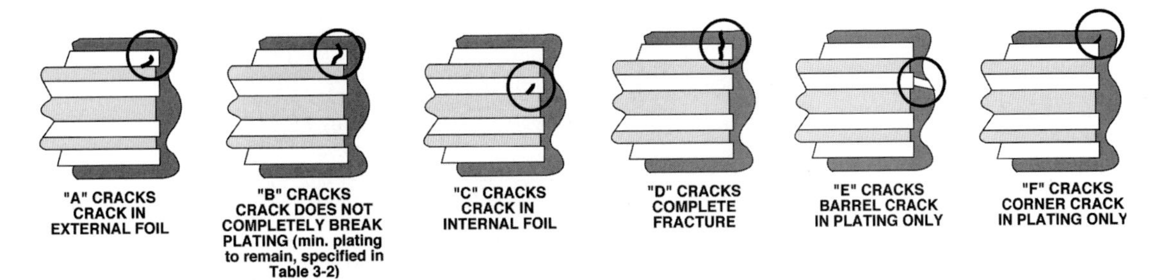

FIGURE 51.27 Types of cracks in the PTH, vertical microsection. *(Courtesy of IPC.)*

51.11.2.5.4 Plating Nodules. Plating nodules should not reduce the PTH diameter below the minimum requirement specified in the procurement documentation.

51.11.2.5.5 Plating Thickness. Plating thickness should be measured at three places in the plated hole and averaged. The absolute minimum plating thickness observed should also be recorded. The minimum plating thickness requirements for the three-area average along with the absolute minimum thickness observed are defined in IPC-6012 (see Fig. 51.28).

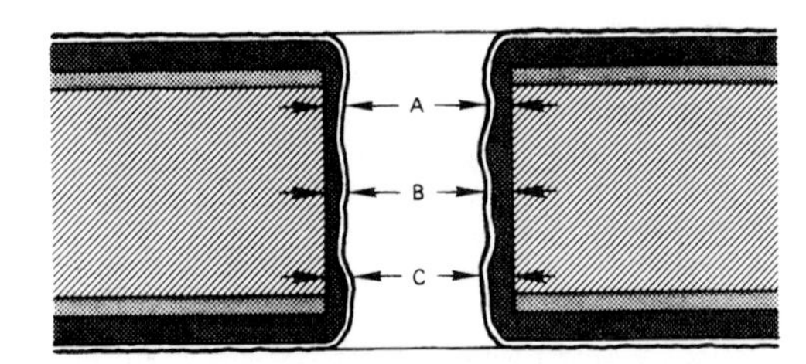

FIGURE 51.28 Measurement of plating thickness in PTH, vertical microsection. *(Courtesy of Sandia Laboratories.)*

51.11.2.5.6 Plating Voids. When found in the microsection, plating voids should be limited to one per microsection. No void should span more than 5 percent of the PCB thickness or be present at the interface of internal copper foil to the plating.

51.11.2.5.7 Solder Resist Thickness. When the procurement documentation specifies this thickness, it should be measured at the thinnest area observed between the copper and the surface of the solder resist. This is usually at the corner of the copper trace.

51.11.2.5.8 Wicking. During the drilling and hole cleaning process, glass fibers from the base material can be removed inward from the hole surface. During the copper-plating process, the copper will follow this glass removal topography, creating pockets of plating that extend laterally from the hole into the base material. This wicking can significantly reduce the electrical conductor spacing between holes or between holes and internal circuits. The maximum allowances for wicking can be found in IPC-A-600. (see Fig. 51.29).

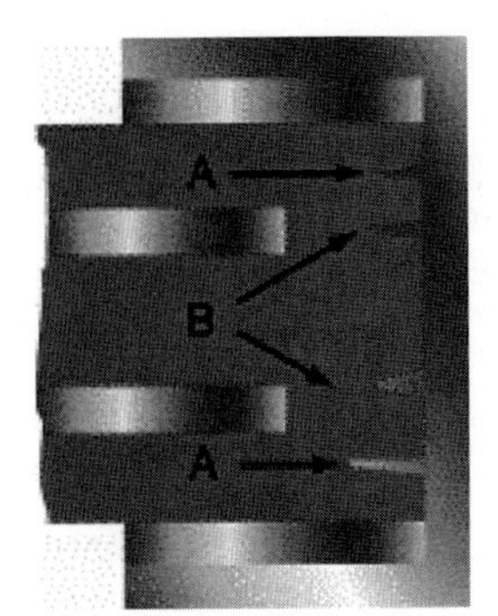

FIGURE 51.29 Wicking, vertical microsection (*Courtesy of IPC.*)

51.11.2.5.9 Interconnection Separation. This is defined as the separation between the internal copper foil and the copper plating in the hole. It is also sometimes referred to as *post separation*. Any evidence of this imperfection in the microsection makes the associated PCBs nonconforming. This imperfection is a severe issue that will likely reduce the service life of the PCB (see Fig. 51.30).

FIGURE 51.30 Interconnection separation, vertical microsection. *(Courtesy of Microtek Laboratories.)*

51.11.2.5.10 Material Fill of Blind and Buried Vias. When blind or buried vias are present in a PCB, it is important that they be microsectioned and evaluated in addition to any through holes evaluated. Buried vias are typically filled with resin, and this resin should fill a minimum of 60 percent of each buried via.

51.12 RELIABILITY INSPECTION USING ACCELERATED ENVIRONMENTAL EXPOSURE

Reliability inspection consists of performing specific tests to ensure that the PCB will function under the influence of the climatic and/or mechanical forces to which it will be subjected during use. Environmental tests are performed on preproduction PCBs or specifically designed test coupons to verify design adequacy and manufacturing process control. Specific tests are sometimes specified as part of the PCB acceptance procedure to expose a prospective failure situation. Specific environmental testing, as part of the acceptance procedure, is prevalent in high-reliability programs. Note that it is recommended that a simulation of the cycles required for component assembly and rework be performed as a preconditioning prior to environmental exposure. Test results then relate to the actual condition of the PCB after assembly. Specific test methods for performing these tests can be found in IPC-TM-650.

51.12.1 Thermal Shock

The thermal shock test is particularly efficient in identifying (1) PCB designs with areas of high mechanical stress and (2) the resistance of the PCB to exposure of high- and low-temperature extremes. This test is periodically required in IPC-6012 for product acceptance and is also used for PCB supplier qualification inspection. A thermal shock is induced on a PCB by exposure to severe and rapid differences in temperature. The test is usually performed by transferring the PCB from one temperature extreme (e.g., 125°C) to the other (e.g., –55°C) rapidly, usually in less than 2 min. A resistance difference in excess of 10 percent between the 1st and 100th cycles is considered nonconforming, but requirements in the thousands of cycles are sometimes specified for high-reliability PCBs. The thermal shock test can produce cracking of plating in the holes and base material de-lamination (refer to IPC-TM-650, method 2.6.7.2, for the procedure). Note that continuous electrical monitoring during thermal shock cycling will detect intermittent electrical circuits that may not be detected with periodic measurement techniques. Intermittent circuits usually occur at the high-temperature extreme.

FIGURE 51.31 Basic SIR test pattern. *(Courtesy of Microtek Laboratories.)*

51.12.2 Surface Insulation Resistance (SIR)

SIR testing evaluates the electrical resistance between two surface electrical conductors separated by dielectric material. This test detects the bulk conductivity and electrical leakage through surface electrolytic contaminants when the samples are exposed to a high-humidity environment. The test conditions are a relative humidity of 85 to 92 percent at a temperature range of 35 to 45°C without the use of an electrical forcing potential. (See Fig. 51.31 for example of a SIR test pattern.)

51.12.3 Moisture and Insulation Resistance (MIR)

This is an accelerated method of testing the PCB for the deteriorative effects of condensing high humidity and heat conditions typical of tropical environments. The test conditions are a relative humidity of 90 to 98 percent at a cycling temperature range of 25 to 65°C with a forcing voltage potential of 10 to 100 VDC applied to the test circuit(s). After the required test cycles are completed, the PCB is subjected to insulation resistance testing. Test specimens should exhibit no blistering, measling, warp, or delamination after the moisture resistance test.

51.12.4 Dielectric Withstanding Voltage

This test is used to verify that the component part can operate safely at its rated voltage and withstand momentary overpotentials due to switching, surges, and similar phenomena. It also serves to determine whether insulating materials and spacings on the PCB are adequate. The test is performed on either an actual PCB or a test coupon. Voltage is applied between mutually insulated conductors on the specimen or between insulated conductors and power/ground planes. The voltage is increased at a uniform rate until the specified value is reached. The voltage is held for 30 sec. at the specified value and then reduced at a uniform rate. Visual examination of the specimen is performed during the test for evidence of flashover or breakdown between contacts. The test can be either destructive or nondestructive, depending on the degree of overpotential used.

51.12.5 ECM and CAF Testing

ECM and CAF tests are intended to detect the growth of conductive metal filaments on a PCB under the influence of a direct current (DC) voltage bias. This filament growth may occur at either an external or internal interface. Growth of the metal filaments is by electrolytic deposition from a solution containing metal ions that are dissolved from the anode in the test circuit and then transported by the electric field toward the cathode. ECM comes in two forms. First are surface dendrites, which form on the internal or external surfaces between the two test electrodes. The second form of ECM is CAF, in which filaments form along the surface of the glass-fiber reinforcement between a hole and another hole or conductor along the surface of the fiber. ECM and CAF testing is done at a relative humidity of 85 to 92 percent at a steady-state temperature range between 35 to 85°C using forcing potentials ranging from 10 to 100 volts DC during the test.

CHAPTER 52
ACCEPTABILITY OF PRINTED CIRCUIT BOARD ASSEMBLIES

Mel Parrish
STI Electronics, Inc. Madison, Alabama

52.1 UNDERSTANDING CUSTOMER REQUIREMENTS

When determining the acceptance criteria to use in building or manufacturing a printed circuit board assembly (PCBA), one first and foremost must look at the contract between the supplier and the purchaser of the product. Always determine the needs and/or wants of the customer through contractual communications or, less formally, by talking to the customer if a contract is not detailed enough or does not even exist.

In some situations, contracts are not the norm, such as for consumer electronics built for retail sale or captive production efforts where the customer and producer are one and the same. In these cases, the company must determine the level of quality desired based on company culture and goodwill, reputation for quality products, and the desired life cycle of the product. Consequences for an unintended or unexpected failure may be a significant driver in this consideration. An electronic medical device used for life support has much greater consequence attached to continued successful operation than a disposable electronic device such as a toy that is intended to be discarded after a short term of use. Additionally, the intended operational environment may have an even greater impact on what should be expected for the acceptability of production. Consider that a product used for military operations such as space travel, arctic conditions, or tropical environments is subjected to much harsher conditions than an office product that sits on a desk in a temperature-controlled environment where the temperature will not vary more than 10°F during its expected life cycle. The range of these expectations is significant and is a paramount consideration for what can be defined as acceptable or nonconforming.

Most standards that define acceptability have roots traceable to the various Department of Defense (DoD) military standards established in the 1960s and 1970s. The advent of complex electronics systems required a standardized approach to acceptability based on lessons learned through testing and product failure analysis that was accomplished by laboratory or research activities. The resulting definitions for acceptance became very restrictive, and although the requirements were initially intended only for harsh environment exposure or products that have a significant consequence for failure, they were used successfully to increase performance of all soldered electronic assemblies. Most commercial production did not share the need for critical continued performance and the accompanying significant impact on production cost. Despite an effort to approach acceptance from a cost-prevention or value-added viewpoint, the military standards are no longer maintained and the DoD research sources for maintaining technology currency were eliminated. The demise of military standards allowed the industry to develop consensus standards for soldering processes and also acceptability. Undoubtedly the most significant of these is IPC-A-610, which creates varying levels of acceptability for broad application ranging from medical and military products to disposable electronics. Not all products need to survive after an explosion or operate for 30 years without interruption.

Each company should decide whether to build its product to an industry standard such as IPC-A-610, "Acceptability of Electronic Assemblies," or develop its own workmanship manual that would closely match IPC-A-610 levels for quality. The advantage of adopting a common standard is the universal understanding of expectations it supports between the customer and producer and thus circumvents the need for lengthy negotiations over individual interests. Where individual custom standards are used, the production processes must be adjusted for each customer or product as it is addressed for production. On the other hand, common standards can be transitioned from one project to the next without significant change. The use of a common standard can take into account the perspective of collective experiences of the consensus participants who are involved in the development. Creating an individual standard can also be a rather daunting and expensive task that can provide limited return without the broad application to bear the cost. In any case, the workmanship manual decision must support the company's objectives to achieve customer expectations for quality with regard to its product.

Detailed specifications may be called out in a drawing or contract, or there may not be any documentation that identifies requirements. In the latter case, the recommendation is to revert to accepted industry standards or an internal quality requirement such as a workmanship manual.

52.1.1 Lead-Free Soldering

Any acceptance criteria decision should define lead-free materials and processes acceptability. Until the advent of the European Union's adoption of the Directive on Restriction of Hazardous Substances (RoHS), which went into effect in July 2006, all electronic design and also production control was based on the precedent that soldering processes would be accomplished with tin-lead (SnPb) soldering materials. Tin-lead solder has specific properties that were instrumental in the multitude of component attachment methods that evolved for production of electronics. Had the solder of choice over the years been other than tin-lead, many of these common configurations might have evolved differently. The current reality is that component designs intended for tin-lead attachments are being soldered with various lead-free alloys, driven by the requirement to remove lead from electronic products. Acceptability for these designs has not changed that much, as reflected in current IPC acceptability standards. The attachments still need to have solder that flows over the connection interface surfaces, forms a fillet, and wets to the constituents that form the attachment. What has changed significantly is the surface appearance of the resulting solder connection. Most lead-free solders exhibit surface roughness that would cause concern for those accustomed to the normal appearance of tin-lead. In fact, if solder finishes created by tin-lead had the same surface appearance, they would have rightly been considered contaminated or overheated. The current IPC-A-610 standard provides a side-by-side contrast of similar solder connections accomplished with both tin-lead and lead-free solder alloys for appearance perspectives.

Outside of the considerations of acceptability appearance, there are additional concerns for materials compatibility issues and adverse properties such as hot tear, tin whisker, tin pest, accelerated and increased process temperature exposure, and others. These considerations should be developed elsewhere in this reference.

This section presents common possibilities for acceptance criteria.

52.1.2 Military, Space, and Commercial Specifications

Reference to these types of specifications is common in the applicable markets. No great amount of detail will be discussed; however, they are all worthy of some discussion since they can provide a great deal of insight into the rationale of "product acceptability" and its origin.

52.1.2.1 Military Specifications. Military specifications have been widely accepted for use in many markets outside of their intended military products. The one with the most relevance to workmanship or acceptability criteria at the assembly level was MIL-STD-2000, a unified standard created from several individual services and production standards. It was created to combine the various requirements into a single standard to make it easier for military contractors to implement a single performance requirement. Many lower-level supporting specifications for components, designs, materials, and printed circuit boards (PCBs) were part of the family of documents to support performance and acceptance requirements as a package. Each specification was designed to be compatible with the eventual success of the end product. In an effort to increase understanding by all of the parties involved, and to ease compliance issues, the services provided training and certification at various locations throughout the country, and even at some international locations. The typical training audience often included customer representatives as well as producers. The title of MIL-STD-2000 was "Standard Requirements for Soldered Electrical and Electronic Assemblies." However, all of the military soldering requirements standards, including MIL-STD-2000, have been rescinded for future production through DoD acquisition reform initiatives. Nevertheless, some products still use the requirements because they have been adopted internationally and because they provide relative benefit for the product's specific industry segment.

MIL-STD-2000 was an extensive document that defined the nonconformance conditions that required disposition as rework, repair, or use as is. This of course required some action by the manufacturer to evaluate a given defect anomaly and make a decision regarding its impact on the form, fit, or function of the intended use. It also allowed some level of customer involvement concerning these decisions if the contract prescribed it. MIL-STD-2000 also limited the assembly materials for a manufacturer in terms of the solder alloys, flux activity levels, cleanliness, conformal coating, solder mask, and their acceptance conditions, either through a flow down to a supporting specification or internal requirements. Any deviation from standardized process and materials as specified in these documents required qualification and/or customer consent before use. However, the document allowed flexible contractual application dependent upon the product demands for performance. "General Requirements" for standard production were much less stringent than "Detail Requirements," which were used for demanding performance or consequence products such as those intended for missile or space applications. Contracts were required to specify "Detail Requirements" before they were imposed; otherwise, only the "General Requirements" were required as a default. The intent of flexible requirements was to allow the use of a single standard for everything from complex weapon systems to office equipment. Unfortunately, they were not commonly applied that way, and the overly stringent requirements originally intended for critical systems were attached to applications such as office or communication products that did not benefit from the more stringent and costly production effort.

Solder connection characteristics for both plated through-hole and surface-mount devices were defined by solder finish, fractures, voids, solder coverage and quantity, and wetting and filleting. The PCB requirements after assembly were detailed in terms of conductor finish and condition, conductor separation from the board, cleanliness of the assembly, weave exposure, delamination, measles, haloing, and bow and twist. In addition, part markings were expected to remain legible even after assembly processing.

If specifically imposed by individual contract, the services provided, through established training resources, traceable training and certification of the people making decisions. This training provided a cadre of qualified instructors who in turn could train people for their company or organization. Its interesting that prior to elimination of these schools, the majority of students attending the certification-training program were from commercial production companies with no contractual performance for DoD. In addition to establishing product acceptance criteria, the specification also detailed methods for process control and defect reduction to be used in assembly. One hundred percent inspection was required unless specific conditions were met, so that sample-based inspection might be used with confidence.

As a result of elimination of the various military standards, most DoD and termed high-reliability production is defined through contractual implementation of the commercial IPC equivalent standards, such as J-STD-001, and IPC-A-610.

52.1.2.2 Space Production Requirements. Standards associated with space applications have significantly different purposes than process control documents commonly used for large-volume production electronics. For the purpose of space applications, there is seldom a large enough production volume to benefit from process development, and each individual product must be acceptable based on the environments experienced with space applications. These could be quite similar to the initial developments for DoD acceptance discussed earlier since they also involved space considerations. Current space application standards have evolved from NHB 5300 through to NASA-STD-8739.3, "Soldered Electrical Connections." Additionally, the space user community has created a space addendum to the commercial standard J-STD-001 for the opportunity to employ consensus standards used by many of their contractors for various other programs. The addendum modifies requirements to a degree consistent with space interests and in effect makes the performance and acceptance requirements similar to that of the NASA-generated standard.

52.1.2.3 J-STD-001 and IPC-A-610 Industry Standards. These consensus industry association standards are widely recognized as the most common standards for requirements and workmanship for PCB assemblies, both nationally and internationally. They allow flexible definition of requirements based on three distinct classes of product, and the acceptability and performance requirements change based upon the contract provisions for the individual product as these classes are applied.

52.1.2.3.1 J-STD-001, "Requirements for Soldered Electrical and Electronic Assemblies." Since "acquisition reform" and elimination of the very commonly adopted military standards, IPC standards have filled the resulting void and have also brought with them additional valuable practices and perspectives from commercial producers. J-STD-001, "Requirements for Soldered Electrical and Electronic Assemblies," was first released in April 1992 as a modification of the previous IPC-S-815, which had limited application, likely due to the competition from contractually imposed military and other industry-unique standards that were available to the users and managers at the time. IPC-S-815 had very similar criteria to that of the military baseline or "General Requirements" discussed earlier.

The resulting J-STD-001 defines three levels of electronic assemblies based on the intended end-item use. The three classifications were established to reflect differences in consequences of failure, harshness of the operating environment, and expected term of operation. The three classes are summarized as follows:

- **CLASS 1: General Electronic Products** This class includes products suitable for applications where the major requirement is function of the completed assembly.
- **CLASS 2: Dedicated Service Electronic Products** Class 2 includes products where continued performance and extended life is required, and for which uninterrupted service is desired but not critical. Typically the end-use environment would not cause failures.
- **CLASS 3: High-Performance Electronic Products** These include products where continued high performance or performance-on-demand is critical, equipment downtime cannot be tolerated, end-use environment may be uncommonly harsh, and the equipment must function when required. Examples include life support or other critical systems.

J-STD-001 addresses many of the same subjects as MIL-STD-2000, and in most cases for Class 1 and 2 production, the requirements are again similar to the "General Requirements" of the previous standards. The Class 3 requirements for the majority of the subject areas are similar to MIL-STD-2000, "Detail Requirements." Since the deactivation of MIL-STD-2000 and the associated revisions, many contract performance requirements have been transitioned to this document and family of IPC standards. As with the predecessor documents, it should be recognized that a successful soldering process requires consistent materials and the ability to select the materials for a given process that will perform well in combination with compatibility for the application. As with acceptability, users may have the flexibility to establish their own requirements for these materials, but it would be difficult to find suppliers willing to create materials qualifications for anything other than a standardized definition. The following discussion provides a brief overview of the "family of standardized criteria" for materials associated with acceptable production standards.

52.1.2.3.2 J-STD-002, "Solderability Tests for Component Leads, Terminations, Lugs, Terminals, and Wires." This standard prescribes the recommended test methods, defect definitions, acceptance criteria, and illustrations for assessing the solderability of electronic component leads, terminations, solid wire, stranded wire, lugs, and tabs.

Solderability evaluations are made to verify that the solderability of component leads and terminations meets the requirements established in J-STD-002 and that subsequent storage has had limited adverse effect on the ability to solder components to an interconnect. Determination of solderability can be made at the time of manufacture, at receipt of the components by the user, or just prior to assembly and soldering.

Similar to J-STD-001, J-STD-002 has its origin with military standards, which in this case was MIL-STD-202, Method 208. It defines standardized methods of evaluating the ability of a surface to accept solder and in addition defines the durability of the solderable surface with three categories of product and related conditioning to determine the product performance before soldering. Category 1 is the least demanding performance, where the product would be used soon after testing, and Category 3 is applied where the product may be stored for extended periods of time before soldering. The intent is to identify solderability problems before the expense of production is added or the production schedule is delayed while waiting for solderable replacement parts.

For a more detailed discussion of solderability and the related issues, refer to Sec. 52.5.

52.1.2.3.3 J-STD-003, "Solderability Tests for Printed Boards." J-STD- 003, "Solderability Tests for Printed Boards," was released in April 1992 to complement the requirements of J-STD-001. This standard prescribes the recommended test methods, defect definitions, and illustrations for assessing the solderability of printed board surface conductors, attachment lands, and plated-through holes.

The solderability determination is made to verify that the PCB fabrication processes and subsequent storage have had no adverse effect on the solderability of those portions of the PCB intended to be soldered. This is determined by evaluation of the solderability specimen portion of a board, or representative coupon, that has been processed as part of the panel of boards and subsequently removed for testing per the method selected.

The objective of the solderability test methods described in J-STD-003 is to determine the ability of printed board surface conductors, attachment lands, and plated-through holes to wet easily with solder and to withstand the rigors of the PCB assembly processes.

For a more detailed discussion of solderability issues, refer to Sec. 52.5.

52.1.2.3.4 J-STD-004, "Requirements for Soldering Fluxes." J-STD-004 prescribes general requirements for the classification and testing of soldering fluxes for high-quality interconnections. This standard is a flux characterization, quality control, and procurement document for solder flux and flux-containing material.

This standard defines the classification of soldering materials through specifications of test methods and inspection criteria. These materials include liquid flux, paste flux, solderpaste flux, solder-preform flux, and flux-cored solder. It is not the intent of this standard to exclude any acceptable flux or soldering aid material; however, these materials must produce the desired electrical and metallurgical interconnection.

52.1.2.3.5 J-STD-005, "Requirements for Soldering Paste." This standard prescribes general requirements for the characterization and testing of the solderpastes used to make high-quality electronic interconnections.

52.1.2.3.6 J-STD-006, "Requirements for Electronic Grade Solder Alloys and Fluxed and Non-Fluxed Solid Solders for Electronic Soldering Applications." This standard prescribes the nomenclature, requirements, and test methods for electronic grade solder alloys; for fluxed and non-fluxed bar, ribbon, and powder solders, for electronic soldering applications; and for "special" electronic grade solders.

52.1.2.3.7 Other Standards. There are many standards that comprise the necessary content for successful production practices that lead to acceptability. Each represents an element of the foundation that supports a complete and successful production effort. One example is moisture susceptibility and the procedures to ensure that the components are not damaged by popcorning during reflow. J-STD-020 and J-STD-033 provide classification and control for moisture-sensitive components. Others, such as IPC-CC-830, "Qualification and Performance of Electrical Insulating Compound for Printed Wiring Assemblies," deal with material acceptability of conformal coating. Each of the elements is important and should be considered as a package for overall acceptability.

Many of the material specifications reference test methods. Rather than include the actual test method with the publication of the specification, the IPC posts them on its web site, where they can be easily accessed and kept current without extensive revision of the various reference standards. This prevents possible conflict with outdated procedures. Criteria for the acceptance levels, however, are provided by reference standards such as J-STD-001, J-STD-004, and so on.

52.1.2.4 IPC-A-610, "Acceptability of Electronic Assemblies". Many companies use IPC-A-610, "Acceptability of Electronic Assemblies," as the standalone workmanship standard for their products. J-STD-001 establishes the requirements for soldering processes, whereas IPC-A-610 depicts the pictorial acceptability criteria for production, including soldering as identified in the J-STD-001 document. IPC-A-610 also addresses additional, broad criteria to define handling and mechanical workmanship requirements, among others. A large percentage of the acceptability criteria defined in this chapter is that which is illustrated in the IPC-A-610 standard.

The popularity of IPC-A-610 is apparent considering that it is available in many different languages and used throughout the world as the acceptance reference of choice for electronics.

The IPC-A-610 standard describes the acceptability criteria for quality electronic assemblies. It does not define process requirements, although the methods used must produce a completed solder joint conforming to the acceptability requirements as defined. Consistent with other IPC standards, the IPC-A-610 document details the acceptance criteria for each of three classes.

The three classes of product allow flexible application of the conditions for which the standard provides acceptance criteria. The customer should provide the inspector with the intended class for acceptability before the product is considered for acceptance. The three classes are 1, 2, and 3 and are the equivalents of the definitions provided by J-STD-001.

In addition, this standard discusses and visually illustrates degrees of compliance, such as "Target," "Acceptable," "Defect," and "Process Indicator," for each condition relative to the three classes of product. The importance of the condition varies with the expected performance of the class based on the preceding class criteria.

Acceptance definitions are summarized as:

- **Target** This condition is close to perfect or preferred; however, this desirable condition is not always achievable and may not be necessary to ensure reliability of the assembly in its service environment. Although this condition is not a requirement for performance based on the standard, it is provided to illustrate to users the desired condition that the creators of the criteria thought they should strive to achieve.

- **Acceptable** This condition, although not necessarily perfect, will maintain the integrity and reliability of the assembly in its service environment.
- **Defect** This condition may be insufficient to ensure the form, fit, or function of the assembly in its end-use environment. Defect conditions need to be dispositioned by the manufacturer based on design, service, and customer requirements. Disposition may be to rework, repair, scrap, or use as is. Repairing or using the assembly as is may require customer concurrence. A defect for Class 1 automatically implies a defect for Class 2 and 3. A defect for Class 2 implies a defect for Class 3.
- **Process indicator** A process indicator is a condition (not a defect) that identifies a characteristic that does not affect the form, fit, or function of a product. Such condition is a result of material, design, and/or operator- or machine-related causes that create a condition that neither fully meets the acceptance criteria nor is a defect.

As established by the predecessor military standards, process indicators are not intended to be reworked or even dispositioned, and they can be included in product delivery without correction of the product. They are, however, not desirable and indicate the relative success of the process and materials; if the process or materials could lead to defect conditions, the process may need to be adjusted or corrected for future production. Process indicators should be monitored as part of the process control system. Should the number of process indicators indicate abnormal variation in the process or identify an undesirable trend, the process should be analyzed. This analysis may result in action to reduce the variation and improve yields before there is a defect trend that may require rework or scrap or, worst of all, result in customer dissatisfaction.

52.1.3 Workmanship Manuals

Many companies use a workmanship document of some sort to describe what the operations should achieve during production. IPC-A-610 is used frequently by manufacturing and also quality personnel to determine acceptable quality levels of their product. Some companies have also developed very good workmanship manuals for their own use, especially to cover unique or uncommon design requirements. If a need exists to develop a unique workmanship manual, then this manual should reflect the intended contract requirements. It is certainly desirable to use a manual already in existence since it can become extremely costly to develop your own, and because customer review is unnecessary since the customer should already be aware of the common requirements definition. If minor changes are necessary to standard performance due to process, product or material limitations, they can often be addressed as a contractual waiver, addition, or exception to the standard definition.

52.2 HANDLING TO PROTECT THE PCBA

The handling of PCBs both before and after the soldering operation can be very important in terms of inducing possible damage to the board or contaminating the board such that subsequent operations are affected. There are three subjects that should be well thought out and controlled in an assembly operation for PCBAs: electrostatic discharge (ESD) protection, contamination prevention, and physical damage prevention.

52.2.1 ESD Protection

ESD is the rapid discharge of a voltage potential into an electronic assembly from a static energy source. The presence of a static discharge in the area of a sensitive device can damage the device even without actual contact. ESD-sensitive (ESDS) components found on the assembly and the amount of current generated by the discharge will determine whether immediate failure takes place. All types of ESD damage can be difficult to troubleshoot, but the most costly scenario is one in which the component is assembled and passes functional test only to fail later due to latent conditions resulting from ESD. This could cause the loss of a customer and even in some cases recall of the product.

Some electronic devices are more sensitive to electrostatic overstress (EOS) damage than others. The degree of such sensitivity within a particular device is related directly to the manufacturing technology employed. The trend for electronic components is smaller size, wider band pass, and lower voltage. These are all contributors to the ESD protection problem and increased sensitivity of the component.

Where sensitive components are handled, protective measures must be taken to prevent component damage. Improper and careless handling accounts for a significant portion of ESD damage to components and assemblies. Before handling ESD-sensitive components, equipment should be carefully tested to ensure that it does not generate damage-causing spikes. The preferred workstation to be utilized in electronic assembly is shown in Fig. 52.1.

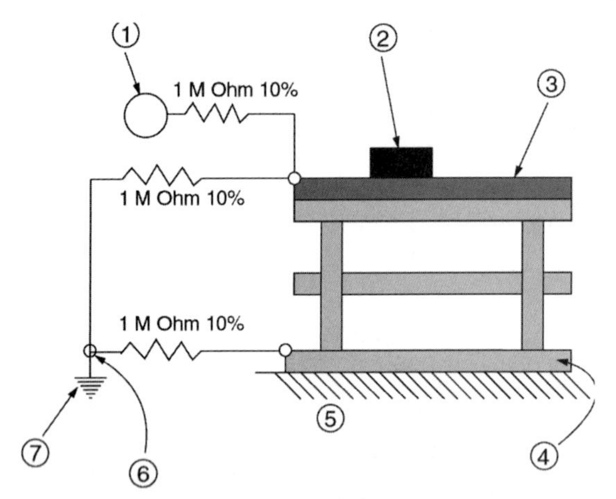

FIGURE 52.1 Target condition for EOS/ESD workstation (1) Personal wrist strap (2) EOS protective trays, shunts, etc. (3) EOS protective table top (4) EOS protective floor mat (5) Building floor (6) Common ground point (7) Ground. *(IPC)*

Static charges are created when nonconductive materials are separated. Destructive static charges are often induced on nearby conductors, such as human skin, and discharged as sparks passing between conductors. This can happen when a person having a static charge potential touches a PCBA. The electronic assembly can be damaged as the discharge passes through the conductive pattern to a static-sensitive component. Static discharges may be too low to be felt by humans (less than 3,500 volts) and still damage ESD-sensitive components.

Sensitive components and assemblies must be enclosed in conductive, static-shielding bags, boxes, or wraps when not being worked on, unless otherwise protected. ESD-sensitive items must be removed from protective enclosures only at static dissipative or antistatic workstations.

For ESD safety, a path to ground must be provided for static charges that would otherwise discharge on a device or board assembly. Provisions are made for grounding the worker, preferably via a wrist strap or a heel strap, provided conductive flooring is used.

52.2.2 Contamination Prevention

The key to any contamination problem is to prevent it from happening. The cost of subsequent operations required to clean a product or rework a product is orders of magnitude more than any cost associated with preventing the contamination from occurring. These contaminants can cause soldering and solder mask or conformal coating problems. There are any number of possibilities for inducing contamination in an assembly environment, such as dirt, dust, machine oils, and process residues; however, in many cases, contamination is caused by the human body, specifically the salts and oils on the skin.

In a high percentage of assembly areas, good housekeeping will take care of contamination from the environment. It should be a common event on each shift to clean workstations, sweep floors, dust fixtures, empty trash, and so on. This not only prevents contamination, but it also helps to keep personnel morale at a higher level. After all, we all desire to work in a nice, clean environment. It will also serve as a positive point if visitors or possible customers visit the assembly area.

To prevent contamination from the human body, every individual must be aware of the possibility of contaminating the PCBA. Printed circuits should be touched only on the edges away from any edge connectors prior to the soldering operation (see Fig. 52.2). Where mechanical assembly procedures require a firm grip on the board, gloves or finger cots should be worn. If the PCBA is to be conformal coated after the soldering operation, the handling of the assembly should be prevented. In some cases, using finger cots, gloves, or fixtures can reduce the contamination. Many producers utilize cleanliness testing to ensure that contamination levels remain within an acceptable range before the coating process, and to ensure adhesion and eliminate localized contaminants that could impact circuit operation and long-term performance. J-STD-001 provides levels for cleanliness acceptability.

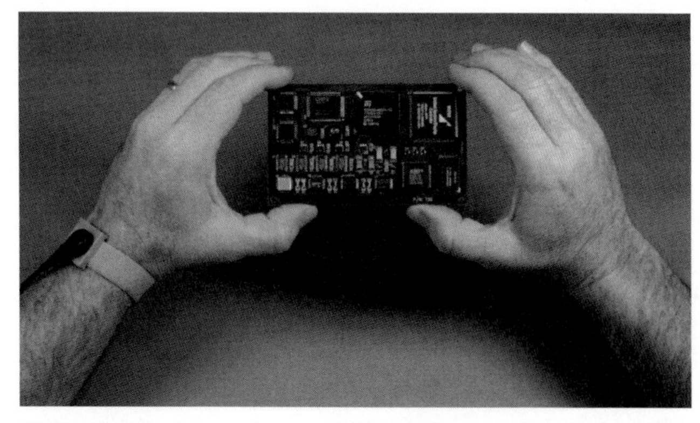

FIGURE 52.2 Appropriate method of handling board to avoid contamination and electrostatic discharger damage. Note position of fingers and the use of a grounding wrist strap.

52.2.3 Physical Damage Prevention

Improper handling can damage components and assemblies. Typical defects associated with handling are cracked, chipped, or broken components; bent or broken terminals; scratched board surfaces; damaged traces or lands; fractured solder joints; and missing surface-mount technology (SMT) components.

Physical damage caused by handling can ruin assemblies and cause a high scrap rate of components or assemblies. Scrap is costly and must be avoided to achieve an efficient and high-quality operation.

Well-maintained handling equipment is also very important in preventing physical damage. One good example is conveyor systems. PCBAs can be caught in conveyors and damaged beyond rework or repair capability and, unless the area of operation is staffed, the conveyor system can damage many assemblies in a very short period of time.

Care must be taken during assembly and acceptability processes to ensure product integrity at all times. The following guidelines provide general guidance:

- Keep workstations clean and neat. There must not be any eating, drinking, or use of tobacco products in the work area.
- Minimize the handling of electronic assemblies and components to prevent damage.
- When gloves are used, they need to be changed as frequently as necessary to prevent contamination from dirty gloves.
- Solderable surfaces are not to be handled with bare hands or fingers. Body oils and salts reduce solderability and promote corrosion and dendritic growth. They can also cause poor adhesion of subsequent coatings or encapsulant.
- Do not use hand creams or lotions containing silicone, because they can cause solderability and conformal coating adhesion problems.
- Never stack electronic assemblies, because physical damage may occur. Special racks need to be provided in assembly areas for temporary storage.
- Always assume the items are ESDS even if they are not marked.
- Personnel must be trained and follow appropriate ESD practices and procedures.
- Never transport ESDS devices unless proper packaging is applied.

52.3 PCBA HARDWARE ACCEPTABILITY CONSIDERATIONS

Most electronic assembly designs include a small percentage of mechanical assembly that requires hardware of various types. Some of the more common component types and the acceptability criteria associated with each are discussed in the following sections.

52.3.1 Component Types

52.3.1.1 Threaded Fasteners. Hardware stack-up for all threaded fasteners must be identified on engineering documentation. The stack-up can be critical, depending on the types of material used for both the hardware and the PCB. Any missing hardware must be found or replaced. Any damage to hardware that prevents it from accomplishing what it was designed to do is unacceptable. A good example of this is any screw or nut that has been stripped, cross-threaded, or damaged to the point that a screw or nut driver is no longer able to tighten or loosen the part (see Fig. 52.3).

A minimum of $1^1/_2$ threads should extend beyond the threaded hardware unless the hardware could interfere with other components. The maximum extension of threaded hardware is 3 mm plus $1^1/_2$ threads for hardware up to 25 mm long and 6.3 mm plus $1^1/_2$ threads for hardware greater than 25 mm long.

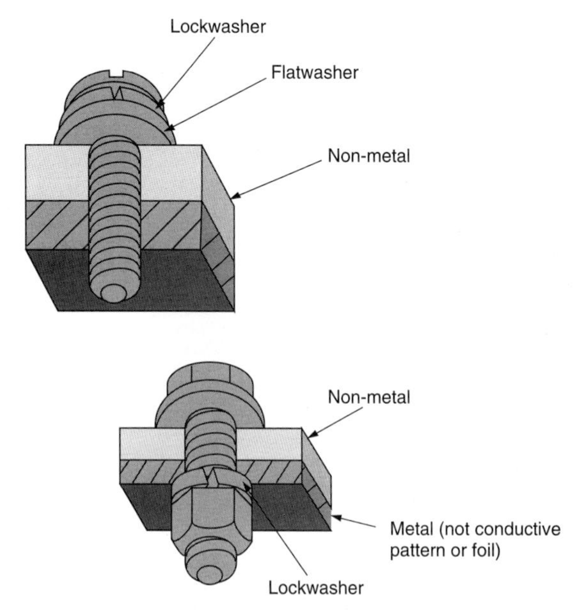

FIGURE 52.3 Hardware mounting for threaded fasteners (1) Lock washer (2) Flat washer (3) Nonconductive material (base laminate) (4) Metal (not conductive pattern or foil). *(IPC)*

Threaded fasteners should be tight to the specified torque on engineering documentation. If torque is not specified on engineering documentation, a generic torque table should be available for use in the assembly environment. Such a table is sometimes included in a workmanship manual.

52.3.1.2 Mounting Clips. Uninsulated metallic components, clips, or holding devices must be insulated from underlying circuitry. Minimum electrical spacing between land and uninsulated component body must not be violated (see Fig. 52.4).

The clip or holding device must make contact with the component sides on both ends of the component. The component must be mounted with its center of gravity within the confines of the clip or holding device. The end of the component may be flush or extend beyond the end of the clip or holding device if the center of gravity is within the confines of the clip or holding device (see Fig. 52.5).

52.3.1.3 Terminals. Terminals that are to be soldered to a land may be mounted such that they can be turned by hand, but should be stable in the z-axis. Terminals may be bent if the top edge does not extend beyond the base and no other mechanical damage such as fractures or breakage to the terminal or the solder joint have occurred (see Fig. 52.6). Common terminals utilized are turret, bifurcated, hooked, and pierced or perforated terminals. See Fig. 52.7 for an example of a damaged turret terminal and a cross section of an assembled board.

52.3.1.4 Swaged Hardware, Flared Flange. The shank extending beyond the land surface is swaged to create an inverted cone, uniform in spread and concentric to the hole. The flange should not be split, cracked, or otherwise damaged to the extent that mechanical strength is compromised or allows contaminating materials to be entrapped in the rivet or funnel (see Fig. 52.8).

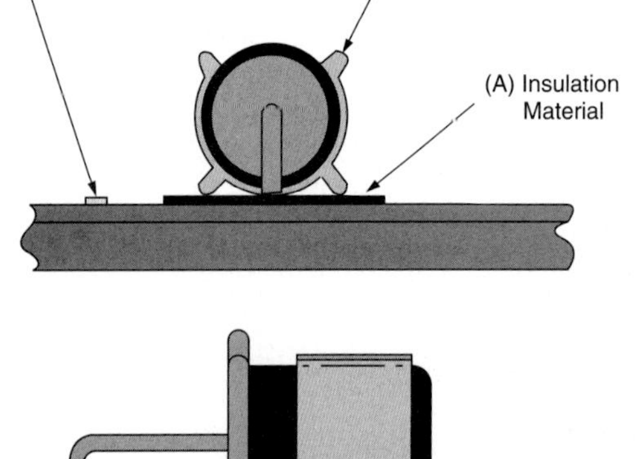

FIGURE 52.4 Component mounting clip insulaton requirements (1) Conductive patterns (2) Metallic mounting clip (3) Insulation material (4) Clearance. *(IPC)*

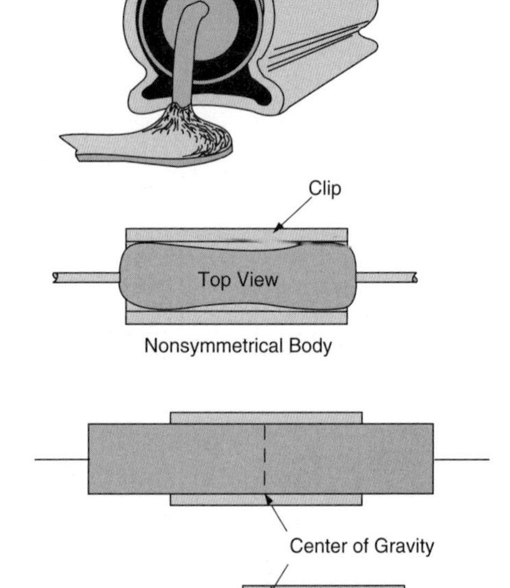

FIGURE 52.5 Component mounting requirements (1) Clip (2) Nonsymmetrical body (3) Top view (4) Center of gravity. *(IPC)*

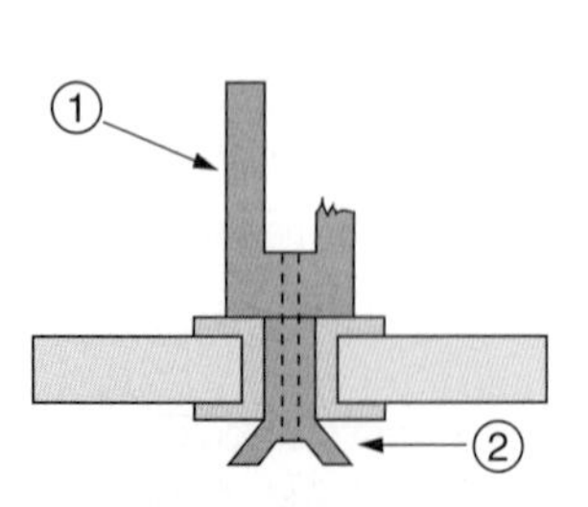

FIGURE 52.6 Bifurcated terminal (1) top edge and (2) base, swaged. Post is broken, but sufficient material is left to attach the specified wires/leads for class I. If both posts are broken, it is unacceptable. *(IPC)*

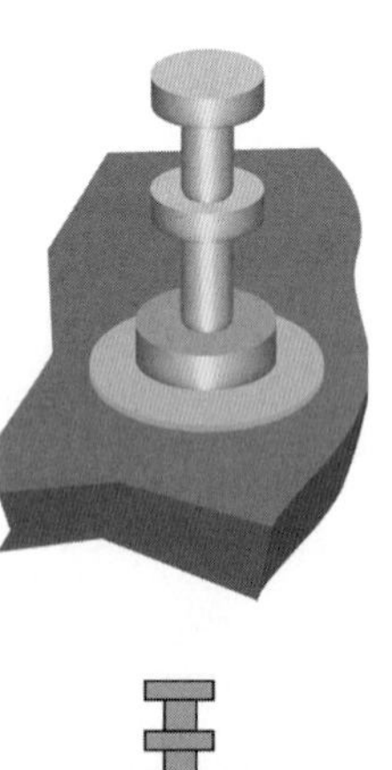

FIGURE 52.7 Acceptable installed turret terminal and cross section of assembled board. Terminal is intact and straight. *(IPC)*

FIGURE 52.8 Flared flange. Acceptable for all Classes.

After swaging, the area should be free of circumferential splits or cracks. It may have a maximum of three radial splits or cracks, provided that the splits or cracks are separated by at least 90° and do not extend into the barrel of the rivet or funnel (see Fig. 52.9).

52.3.1.5 Connectors, Handles, Extractors, and Latches. Ejectors, handles, and connectors should not exhibit any cracks in the component material emanating from the roll pin or rivet that mounts the component to the PCBA.

Roll pins should not protrude more than 0.015 in. from the surface of the ejector, handle, or connector. The major consideration for roll pins should be to ensure that any protrusion does not mechanically interfere with any other assembly. Damage to the part, PCB, or securing hardware is unacceptable.

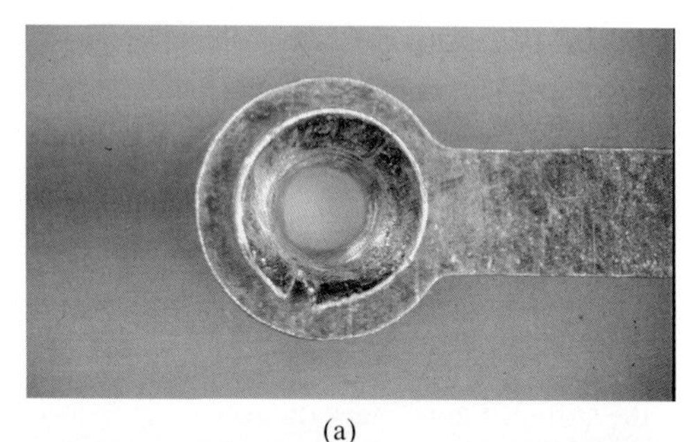

(a)

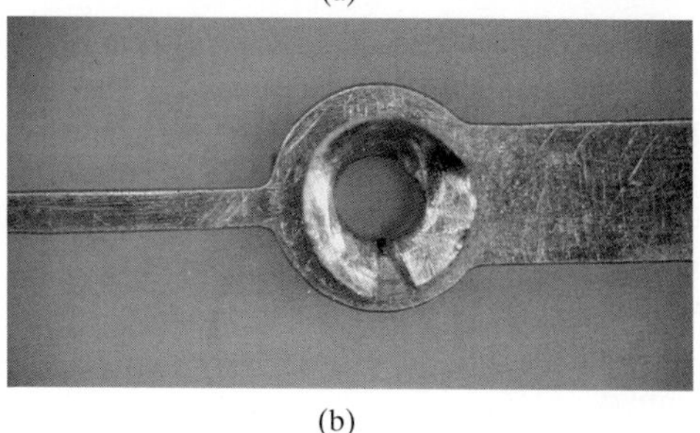

(b)

FIGURE 52.9 (a) Shows an example of acceptable flared swage-split while (b) shows an unacceptable split into barrel.

Connector damage can also affect the pins of the connector being pushed or bent out of specification. When connectors are mated, there is the chance that the female portion of the connector pin will be pushed backward and bent. This prevents adequate contact area between the male and female pins of the connector, if they will mate at all. This circumstance is an unacceptable condition.

The male connector pins might also be bent. When the male pin is bent significantly, the mating of the female and male connectors can cause mechanical damage to the connector casing due to the male pin being forced into the wrong female connector slot. In most cases, the damage is deformation, cracking, or breaking of the connector casing material, all of which are unacceptable.

When connector pins are installed (such as compliant pin, press fit), the pins must be straight within 50 percent of the pin thickness from the perpendicular. For Class 1 and 2 equipment, the PCB land may be lifted less than or equal to 75 percent of the annular ring width. Any land that is lifted across more than 75 percent of the annular ring or has been fractured is unacceptable. For Class 3 equipment, no lifted or fractured lands are acceptable. For all classes, visibly twisted pins, damaged pins, or pins inserted such that there is a nonstandard pin height beyond a specified engineering tolerance is unacceptable.

52.3.1.6 Faceplates. Faceplates must be clean and free of scratches or damage on the front surface. This is important since this is the surface that is commonly viewed by the customer and presents the cosmetic appeal to the product being purchased. An exaggerated example of this is the purchase of an automobile. Few if any consumers would be willing to accept a new automobile with noticeable scratches in the painted surface.

A good rule of thumb for scratch criteria that some companies have adopted as a workmanship standard includes the following:

1. A scratch must be visible when viewed under the following conditions to be considered a defect:
   - **a.** From a distance of 18 in.
   - **b.** With no magnification utilized
   - **c.** With normal room lighting used in the assembly area
2. A scratch must be visible from more than one angle to be considered a defect.
3. Metal or plastic faceplate scratches that exceed 0.125 in. in length are considered defects.

Surfaces exposed to frequent viewing should be free of blisters, runs, nicks, gouges, blemishes, or other abrasions that would detract from the general appearance of the finish. Faceplates must also be securely fastened to the PCB, that is, tightly enough to prevent physical movement.

52.3.1.7 Stiffeners. Stiffeners are commonly used on PCBA designs that are large in dimension to prevent warpage of the PCBA before, during, or after assembly operations. If a stiffener has kept a PCBA from warping outside acceptability criteria, it has done its job; however, the stiffener must meet the following criteria to be acceptable:

- Any marking or color coatings must be permanent. Loss of marking such that it is not legible, or fading or loss of color outside the color standard being used, is unacceptable.
- The stiffener must be properly seated and mechanically fastened. If the soldering operation is used to mechanically fasten the stiffener to the PCBA, then good wetting to the stiffener such that the stiffener is mechanically sound is required.
- It is unacceptable to have a loose stiffener on a PCBA. The stiffener must be securely fastened to the PCBA.

52.3.2 Electrical Clearance

Electrical clearance of hardware to components or electrical traces must be controlled by the design based on the expected voltage and environment's exposure. As a default, design standards such as IPC-2221 create a spacing requirement unless it is specifically defined for the particular PCBA (see Fig. 52.10 for examples). Minimum electrical clearance can be referenced throughout the acceptability requirements such as with contaminants or board laminate defects related to the conductors. Any violation of minimum electrical clearance is cause for a defect disposition consideration for any class, since it will directly impact the electrical performance of the product. Any condition that compromises the clearance should be verified by assemblers or inspectors to ensure that a shorting condition does not occur.

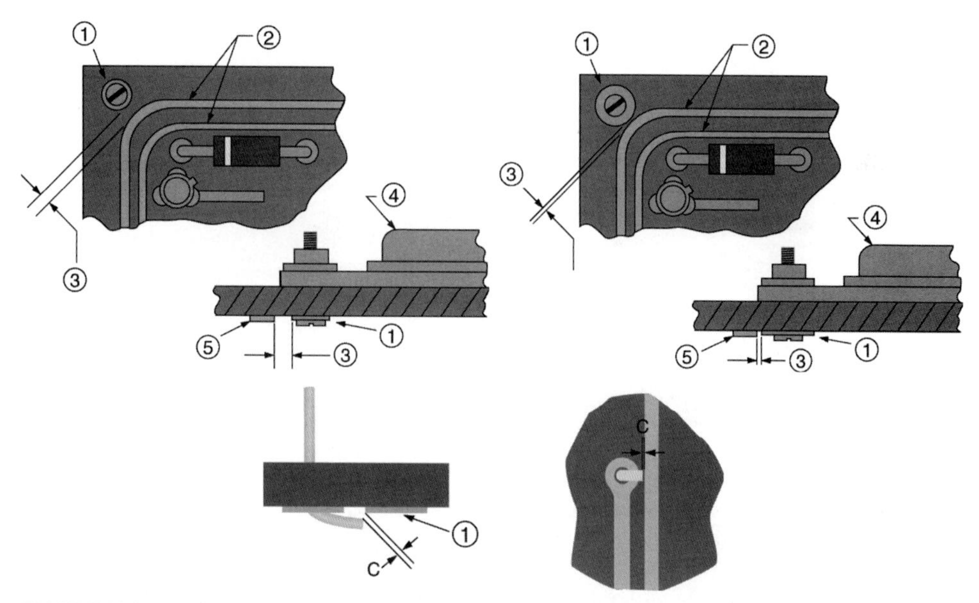

FIGURE 52.10 Examples of acceptable and unacceptable electrical clearance-hardware. (a) Acceptable where: (1) Metallic hardware; (2) Conductor pattern; (3) Specified minimum electrical clearance; (4) Mounted component; (5) Conductor; (b) Unacceptable where hardware reduces clearances to less than specified electrical clearance; (1) Metallic hardware; (2) Conductor pattern; (3) Spacing less than minimum electrical requirements; (4) Mounted component; (5) Conductor; (c) Unacceptable as lead is clinched toward a noncommon electrical conductor (1) and violates clearance; (C) requirement.

52.3.3 Physical Damage

Physical damage to hardware components in most cases refers to enough damage to a component to render it unusable in its application—for example, threads that are damaged to the point that mechanical fastening cannot be performed or a component that is fractured or broken such that it cannot perform the function for which it was intended.

Subjective judgments are common in this area since it is very much dependent on the application of the hardware in the given design, the environment in which the final product will be used, and life cycle expectations of the product.

52.4 COMPONENT INSTALLATION OR PLACEMENT REQUIREMENTS

Component installation or placement is the first step in the assembly of the PCBA. It may begin by preparation of the leads of a given package type in order to provide the correct lead protrusion, form the leads to fit the PCB plated-through holes or pads, or put a bend in the leads that will serve as a standoff from the component to the PCB.

52.4.1 Plated-Through Hole (PTH) Lead Installation

Several requirements apply to all PTH components. When using a polarized component, it must be oriented correctly. In all cases, if the orientation is incorrect, the board is unacceptable (see Fig. 52.11). When lead forming is required, the lead form must provide clearance and

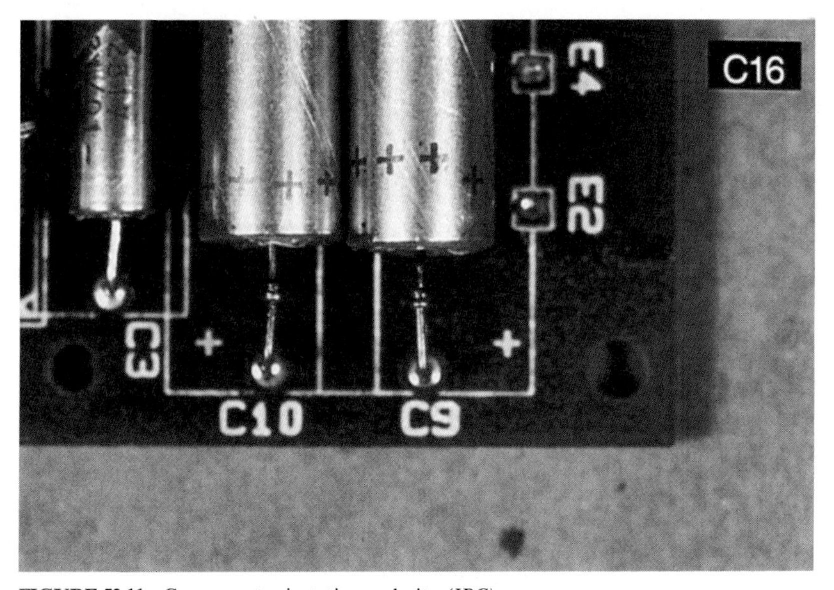

FIGURE 52.11 Component orientation–polarity. *(IPC)*

stress relief (see Fig. 52.12). Physical damage to the lead itself cannot exceed 10 percent of the diameter of the lead. Exposed basis metal as a result of lead deformation is a process indicator but does not render the board unacceptable.

52.4.1.1 Axial Leaded Components. The target condition for axial components is that the entire body length of the component should be parallel and in contact with the board surface (see Fig 52.13 for tolerances). If the component is required to be spaced above the board to avoid overheating the laminate, it must be mounted a minimum of 1.5 mm above the PCB surface. The maximum space between the component and the PCB surface should not violate the requirements for lead protrusion and cannot be greater than 3.0 mm for Class 1 and 2 and 0.7 mm for Class 3 equipment (see Fig. 52.13).

Leads must extend from the component body at least one lead diameter (L) or thickness, but not less than 0.8 mm from the body or weld, before the start of the lead bend radius (see Fig. 52.14).

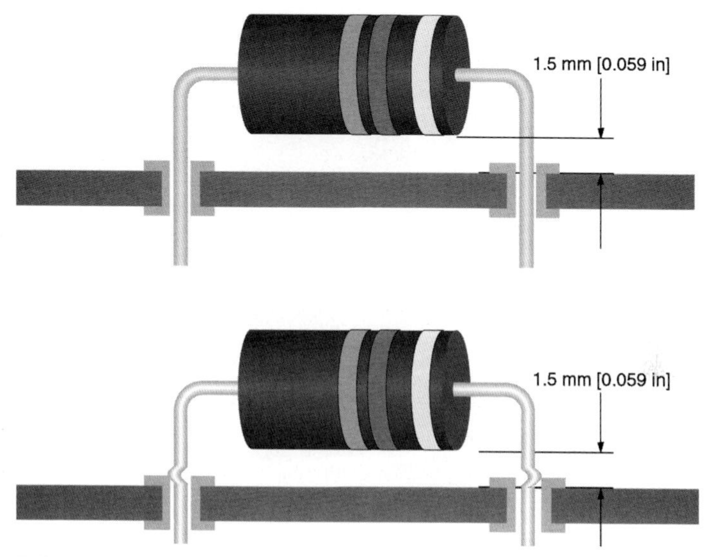

FIGURE 52.12 Examples of acceptable clearance and stress relief for PTH components. *(IPC)*

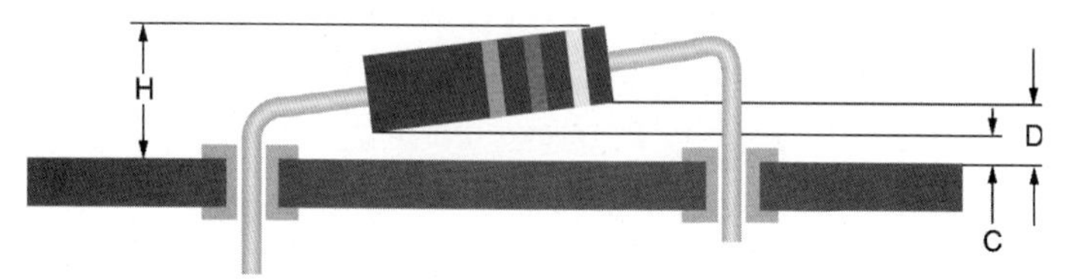

FIGURE 52.13 Amount of nonparallel allowed varies by class but is measured: (H) is component height, (C) is clearance between component body and board, and (D) is the farthest distance between the component body and the board.

No physical damage, such as chips or cracks to axial components, should be allowed; however, it is acceptable to allow minor damage. It becomes unacceptable if the insulating cover is damaged to the extent that the metallic element is exposed or the component shape is deformed (see Fig. 52.15).

52.4.1.2 Radial Leaded Components. The target condition for vertical radial components is that the body is perpendicular to the PCB and the component base is parallel to the PCB. It is an acceptable condition to have the component tilt from the perpendicular as long as it does not violate electrical clearance. The space between the component base and PCB should be between 0.3 and 2.0 mm (see Fig. 52.16) but it is considered a process indicator only if it exceeds that dimension. Some components, such as light-emitting diodes (LEDs) and switches, may need to be perpendicular due to mating requirements.

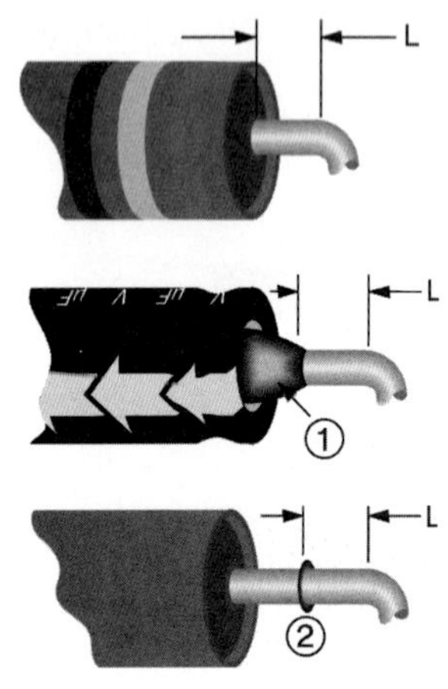

FIGURE 52.14 Lead extension from component body, where (1) is solder bead, (2) is weld. L is the distance from the effective component body to the bend. *(IPC)*

FIGURE 52.15 Example of physical damage to component after insertion. This is a "defect" condition and cause for rejection.(a) Axial lead component, (b) Chip component. *(IPC)*

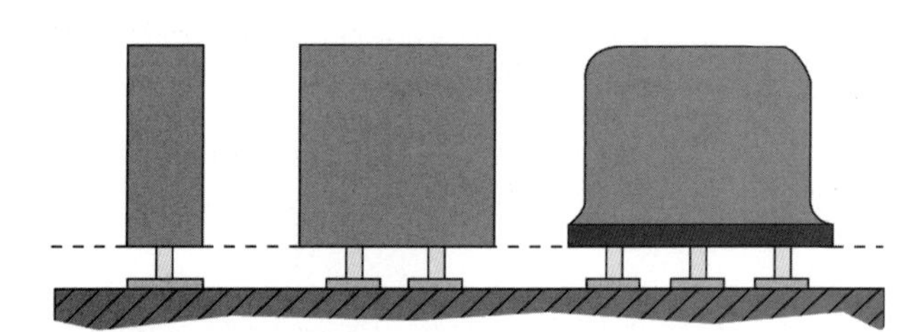

FIGURE 52.16 Radial leaded components spacing. Component is perpendicular and base is parallel to board. Acceptable clearance varies with Class. *(IPC)*

52.4.2 SMT Placement

Coplanarity is very important in the placement of SMT components with leads. Occasionally, lead preparation to form the leads of SMT components is necessary to achieve the proper coplanarity for manual or automatic placement of components on the PCB. Most times, however, the components are purchased packaged with leads prepared and ready to be automatically placed by pick-and-place equipment. Another critical parameter for all surface-mounted components is the accuracy of placement of the components onto the pads of the PCB.

52.4.2.1 Chip Components. Side overhang is acceptable up to one-half (one-quarter for Class 3) the width of the component end cap or PCB pad. End-cap overhang is not acceptable for all classes. The end-cap solder joint width is acceptable with a minimum solder joint length of one-half (three-quarters for Class 3) the component end cap or PCB land, whichever is less.

A side solder joint length is not required; however, a properly wetted fillet must be evident.

The maximum solder fillet height may overhang the land or extend onto the top of the end-cap metallization; however, the solder should not extend farther on the component body.

The minimum solder fillet height must cover one-quarter the thickness, or height, of the component end cap for Class 3 but is required to be evident only for Classes 1 and 2. The component end cap must have overlap contact with the PCB pad. There is no minimum contact length specified see (see Fig. 52.17 and Table 52.1 for an example and list of specifications).

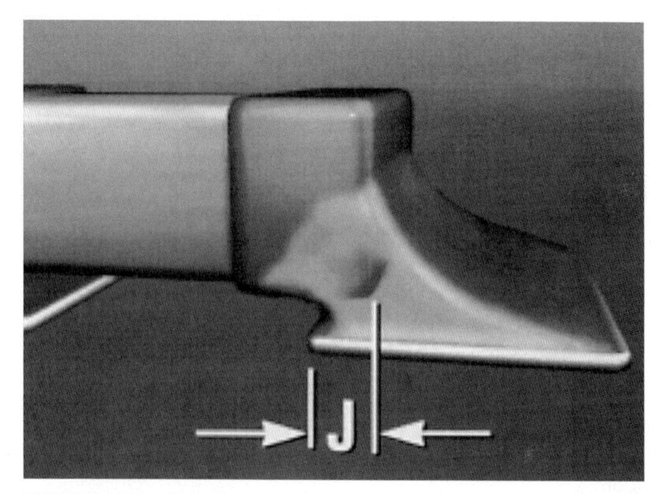

FIGURE 52.17 Example placement and solder fillet for surface mount chip component. Evidence of overlap of contact (J) between the component termination and the land is required. *(IPC)*

52.4.2.2 Metal Electrode Leadless Face (MELF) or Cylindrical Termination. Side overhang is acceptable up to one-quarter the diameter of the end cap. This dimension is different from the other SMT components due to the cylindrical feature of the component. End-cap overhang is not acceptable for all classes. The end-cap solder joint width is acceptable with a minimum solder joint length of one-half the diameter of the component end cap for Classes 2 and 3 but is required to be discernable only for Class 1. A side solder joint length must be a minimum of one-half (three-quarters for Class 3), but again is required to discernable only for Class 1. The maximum solder fillet height may overhang the land or extend onto the top of the end-cap metallization; however, the solder should not extend farther on the component body. The minimum solder fillet height must cover one-quarter the thickness, or height, of the component end cap for Class 3, and for Classes 1 and 2 the minimum solder fillet must exhibit proper wetting. The component end cap must have 75 percent of the cap dimension overlapping the land for Class 3 and 50 percent for Class 2. There is no minimum contact length specified for Class 1 (see Fig. 52.18 and Table 52.2).

52.4.2.3 Castellated Terminations. Side overhang is acceptable up to one-half (one-quarter for Class 3) of the castellation width. End overhang of the castellation is not acceptable for all classes. The castellation end solder joint width is acceptable with a minimum solder joint length of one-half (three-quarters for Class 3) of the castellation width (see Table 52.3). A minimum side solder joint length is required to extend beyond the edge of the component for all classes. The maximum solder fillet height is not a specified parameter for this type. The minimum solder fillet height must be 50 percent of the castellation height for Class 3 and 25 percent for Class 2. The only height requirement for Class 1 is that at least one fillet is visible.

TABLE 52.1 Dimensional Criteria for Chip Components, Rectangular or Square End Components—One-, Three-, or Five-Side Terminations

Feature	Dim.	Class 1	Class 2	Class 3
Maximum Side Overhang	A	50% (W) or 50% (P) whichever is less[a]		25% (W) or 25% (P), whichever is less[a]
End Overhang	B	Not permitted		
Minimum End Joint Width	C	50% (W) or 50% (P), whichever is less[b]		75% (W) or 75% (P), whichever is less[b]
Minimum Side Joint Length	D	Wetting is evident		
Maximum Fillet Height	E	The maximum fillet may overhang the land and/or extend onto the top of the end-cap metallization; however, the solder does not extend further onto the top of the component body		
Minimum Fillet Height	F	Wetting is evident on the vertical surface(s) of the component termination[c]		(G) + 25% (H) or (G) + 0.5 mm (0.02 in.), whichever is less[c]
Solder Thickness	G	Wetting is evident		
Termination Height	H	Unspecified dimension, determined by design		
Minimum End Overlap	J	Required		
Width of Land	P	Unspecified dimension, determined by design		
Termination Width	W	Unspecified dimension, determined by design		
Side mounting/billboarding[d,e]				
Width to Height Ratio		Does not exceed 2:1		
End Cap and Land Wetting		100% wetting land to end metallization contact areas		
Minimum End Overlap	J	100%		
Maximum Side Overhang	A	Not permitted		
End Overhang	B	Not permitted		
Maximum Component Size		No limits		1206
Termination Faces		Three or more faces		

Notes:
[a] Does not violate minimum electrical clearance.
[b] C is measured from the narrowest side of the solder fillet.
[c] Designs with via in pad may preclude meeting these criteria. Solder acceptance criteria should be defined by the user and the manufacturer.
[d] These criteria are for chip components that may flip (rotate) onto the narrow edge during assembly.
[e] These criteria may not be acceptable for certain high-frequency or high-vibration applications.

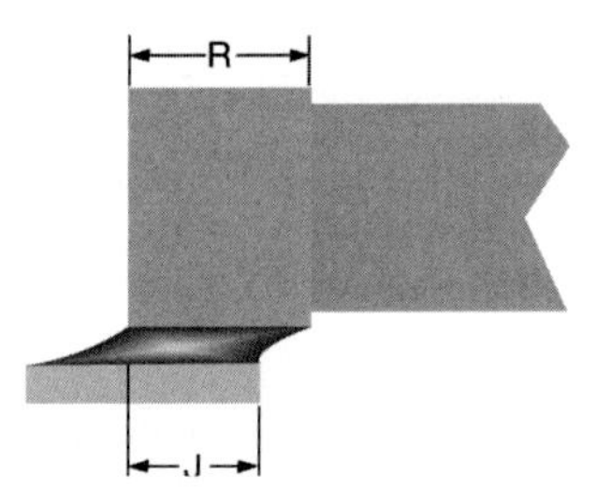

FIGURE 52.18 Cylindrical end cap (MELF) termination. "J" represents the overlap between component termination and land, "R" is length of component termination. *(IPC)*

52.4.2.4 Flat Ribbon, L, and Gull-Wing Leads. Side overhang is acceptable up to one-half (one quarter for Class 3) of the width of the component lead. Lead toe overhang is acceptable for all classes provided it does not violate minimum electrical clearance. The end-joint solder width is acceptable for Class 2 with a minimum solder-joint length of one-half (three-quarters for Class 3) of the component lead width. The lead-side solder-joint minimum length is one-time the lead width for Class 1. It must be three times the width when the lead length is greater than three times its width. If the lead length is less than three times its width, the side joint width must be equal to the length. For Classes 2 and 3, the maximum solder fillet height on components such as small-outline integrated circuits (SOICs) or small-outline transistors (SOTs) may touch the package body or end seal, but not that of ceramic or metallic components. The minimum solder heel fillet height must cover one-half of the lead thickness for Class 2 and must be equal to the lead thickness for Class 3 (see Table 52.4).

TABLE 52.2 Dimensional Criteria: Cylindrical End-Cap MELF Termination

Feature	Dim.	Class 1	Class 2	Class 3
Maximum Side Overhang	A	25% (W) or 25% (P), whichever is less[a]		
End Overhang	B	Not permitted		
Minimum End Joint Width[b]	C	Wetting is evident	50% (W) or 50% (P), whichever is less	
Minimum Side Joint Length	D	Wetting is evident[c]	50% (R) or 50% (S), whichever is less[c]	75% (R) or 75% (S), whichever is less[c]
Maximum Fillet Height	E	The maximum fillet may overhang the land or extend onto the top of the component termination; however, the solder does not extend further onto the component body		
Minimum Fillet Height (End and Side)	F	Wetting is evident on the vertical surface(s) of the component termination[d]		(G) + 25% (W) or (G) + 1.0 mm (0.0394 in.), whichever is less[d]
Solder Thickness	G	Wetting is evident		
Minimum End Overlap	J	Wetting is evident[c]	50% (R)[c]	75% (R)[c]
Land Width	P	Unspecified dimension, determined by design		
Termination/Plating Length	R	Unspecified dimension, determined by design		
Land Length	S	Unspecified dimension, determined by design		
Termination Diameter	W	Unspecified dimension, determined by design		

Notes:
[a] Does not violate minimum electrical clearance.
[b] C is measured from the narrowest side of the solder fillet.
[c] Does not apply to components with end-only terminations.
[d] Designs with via in pad may preclude meeting those criteria. Solder acceptance criteria should be defined by the user and the manufacturer.

TABLE 52.3 Dimensional Criteria: Castellated Terminations

Feature	Dim.	Class 1	Class 2	Class 3
Maximum Side Overhang	A	50% (W)[a]		25% (W)[a]
End Overhang	B	Not permitted		
Minimum End Joint Width	C	50% (W)		75% (W)
Minimum Side Joint Length	D	wetting is evident	Depth of castellation	Depth of castellation
Maximum Fillet Height	E	G + H		
Minimum Fillet Height	F	wetting is evident	(G) + 25% (H)	(G)+50% (H)
Solder Thickness	G	Wetting is evident.		
Castellation Height	H	Unspecified dimension, determined by design		
Land Length	S	Unspecified dimension, determined by design		
Castellation Width	W	Unspecified dimension, determined by design		

Note:
[a] Does not violate minimum electrical clearance.

52.4.2.5 J-Leaded Components. Side overhang is acceptable up to one-half (one-quarter for Class 3) of the width of the component lead. Lead toe overhang is not specified for any class.

The end lead solder-joint width is acceptable with a minimum solder-joint length of one-half (three-quarters for Class 3) of the component lead width. A side solder-joint length must be a minimum of $^{11}/_{2}$ times the lead width for both Class 2 and 3, but Class 1 requires only a visible fillet. The maximum solder fillet height is not specified; however, the solder fillet may not touch the component package body. The minimum heel solder fillet height must cover one-half of the thickness of the component lead or the lead thickness for Class 3 (see Table 52.5).

TABLE 52.4 Dimensional Criteria: Flat Ribbon, L, and Gull-Wing Leads

Feature		Dim.	Class 1	Class 2	Class 3
Maximum Side Overhang		A	50% (W) or 0.5 mm (0.02 in.), whichever is less[a]		25% (W) or 0.5 mm (0.02 in.), whichever is less[a]
Maximum Toe Overhang		B	Does not violate minimum electrical clearance		
Minimum End Joint Width		C	50% (W)		75% (W)
Minimum Side Joint Length[b]	When (L) Is ≥3 W	D	(1W) or 0.5 mm [0.02 in.], whichever is less	3 (W) or 75% L, whichever is longer	
	When (L) Is <3 W			100% (L)	
Maximum Heel Fillet Height		E	See Section 52.4.2.1		
Minimum Heel Fillet Height		F	Wetting is evident	(G) + 50% (T)[c]	(G) + (T)[c]
Solder Thickness		G	Wetting is evident		
Formed Foot Length		L	Unspecified dimension, determined by design		
Lead Thickness		T	Unspecified dimension, determined by design		
Lead Width		W	Unspecified dimension, determined by design		

Notes
[a] Does not violate minimum electrical clearance.
[b] Fine-pitch leads require a minimum side fillet length of 0.5 mm (0.02 in.).
[c] In the case of a toe-down lead configuration, the minimum heel fillet height (F) extends at least to the midpoint of the outside lead bend.

TABLE 52.5 Dimensional Criteria: J–Leads

Feature	Dim.	Class I	Class 2	Class 3
Maximum Side Overhang	A	50% (W)[a]		25% (W)[a]
Maximum Toe Overhang	B	Unspecified dimension, determined by design[a]		
Minimum End Joint Width	C	50% (W)		75% (W)
Minimum Side Joint Length	D	Wetting is evident	150% (W)	
Maximum Fillet Height	E	Solder does not touch package body		
Minimum Heel Fillet Height	F	(G) + 50% (T)		(G) + (T)
Solder Thickness	G	Wetting is evident		
Lead Thickness	T	Unspecified dimension, determined by design		
Lead Width	W	Unspecified dimension, determined by design		

Note:
[a] Does not violate minimum electrical clearance.

52.4.2.6 *Ball Grid Array (BGA) Components.* BGA package components are a very popular design due to significant packaging density. This allows designers to get more function in a smaller space. The component package is virtually impossible to inspect visually as traditionally allowed with other components since the solder spheres are located beneath the component. If cost justification exists, either due to volume or reliability provisions such as could be the case on Class 3 equipment, x-ray equipment can be used to verify the position quality and integrity of the solder joints. This could be costly in terms of the upfront capital investment of the x-ray equipment, but may pay for itself many times over in customer satisfaction or the prevention of defective product. Other options may use the x-ray as a process development tool where once the process is stabilized, the inspection is reduced to sampling or eliminated completely.

X-ray inspection can provide a visual depiction of the solder connection that must not contain solder bridges, incomplete reflow, attachment fractures, misalignment that violates minimum electrical clearance, or missing or damaged spheres. It's also considered defective for solder voids to displace more than 25 percent of a sphere (see Fig. 52.19).

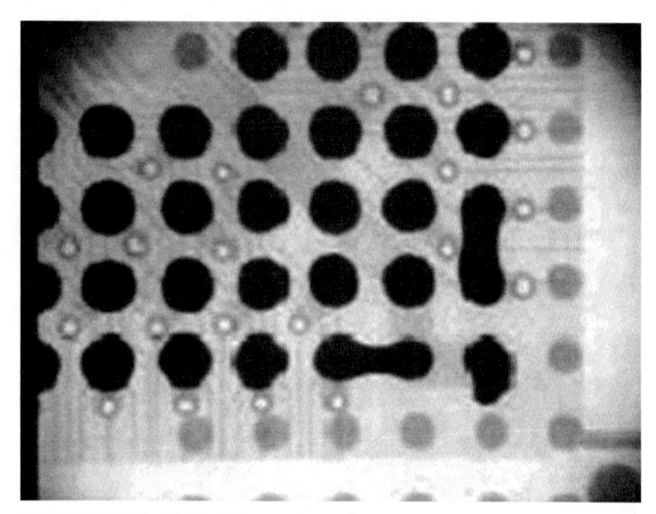

FIGURE 52.19 Ball grid array (BGA), bottom view showing solder ball connection. *(IPC)*

52.4.3 Use of Adhesives

Adhesives can often be used in SMT and PTH applications.

For SMT components, adhesive may be used to attach components to the bottom side of the PCB for mixed technologies (those containing both through-hole and SMT components). In some process flows, the SMT components are placed with adhesive between the solder pads; the PCB is processed through an adhesive curing cycle; then the components are wave soldered or reflowed onto the PCB along with the PTH components. In some processes, the adhesive is a temporary (soluble) attachment to stabilize the components during soldering and is to be removed after the solder is attached. In some extreme cases, the adhesive is used for both through-hole and SMT components to support the physical attachment of the components for severe operational conditions such as physical shock or vibration. Where intended to support the component physically, the attachment is specifically defined by drawing or individual requirements. The SMT components are acceptable as long as the adhesive has not contaminated the solder joint. If adhesive has contaminated the component solderable surface, lead, end cap, or PCB pad such that an acceptable solder joint is not achieved, the PCBA is not acceptable (see Fig. 52.20).

Normally in PTH component applications, adhesive is used to give large-profile and/or heavy components more mechanical stability. When adhesives are used in this manner, the following adhesive acceptance criteria apply:

- The adhesive must adhere to a flush-mounted axial component for 50 percent of the component length and 25 percent of its diameter on one side. The build-up of adhesive should

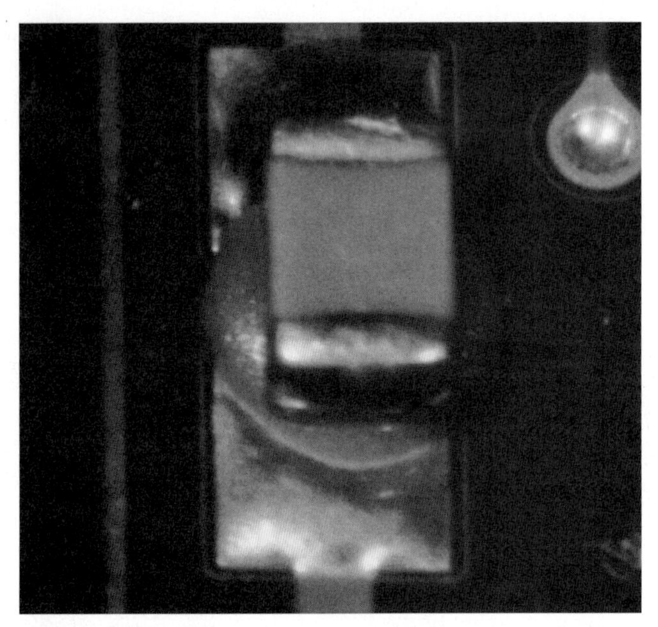

FIGURE 52.20 Adhesive visibly extended from under the component and visible in the terminal area. *(IPC)*

not exceed 50 percent of the component diameter, and adhesion to the mounting surface must be evident (see Fig. 52.21).

- The adhesive must adhere to a vertically mounted axial component for 50 percent of the component length and 25 percent of its circumference, and adhesion to the mounting surface must be evident (see Fig. 52.22).

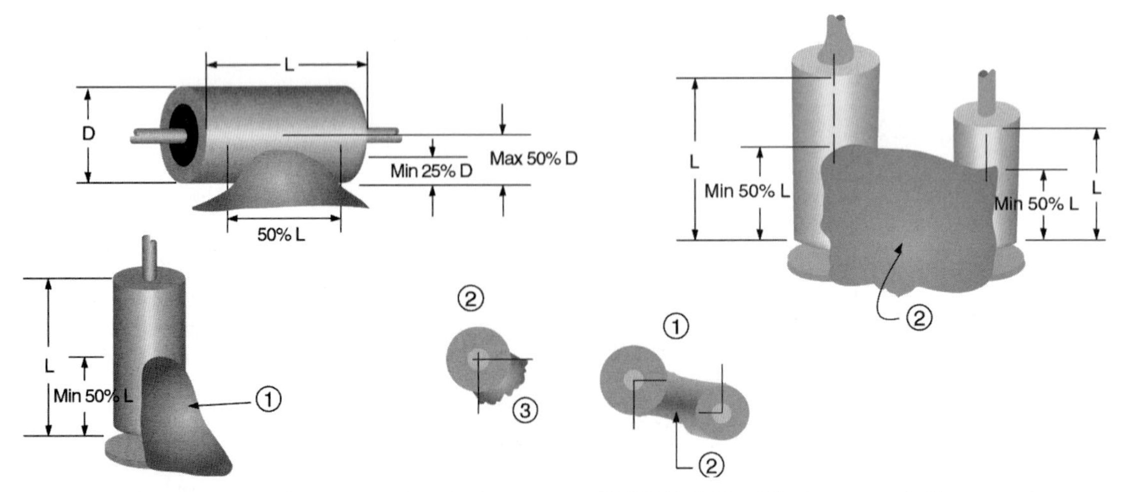

FIGURE 52.21 Adhesive on horizontal and vertically mounted components. (1) Adhesive, (2) Top view, (3) 25 percent circumference.

FIGURE 52.22 Adhesive on vertically mounted components. (1) Top view, (2) Adhesive. *(IPC)*

52.5 COMPONENT AND PCB SOLDERABILITY REQUIREMENTS

Component solderability, which includes the solderability of the PCB, is probably the most important single characteristic to consider when building PCBAs. To be successful in soldering PCBAs, good wetting must be achieved. It is an ongoing effort to ensure that all boards and electrical components purchased and received from suppliers maintain good solderability.

The key to success in this all-important task is to have an excellent working relationship with the suppliers and manufacturers of the components to ensure good, solderable components.

Until partnership relationships are achieved, there should be a sample of each lot or batch of components received that are subjected to solderability testing. This testing should be conducted in accordance with J-STD-002 and J-STD-003. These documents were written specifically to address solderability requirements for electronic assembly components and PCBs. The intent of the solderability testing is to qualify the surface to accept solder with less-than-optimum conditions. Additionally, artificial conditioning (aging) is used to simulate a normal aging period that the components might experience before use. Identification of weaker components can reduce the possibility of a solderability problem or production issues later on in the production cycle. Based on this lack of lead solderability, the components should be rejected and returned to the supplier, before they can be placed in the assembly process.

Packaging and handling of components are also important to maintain good solderability since many contaminants that could affect solderability can be transferred to solderable surfaces as a result of handling by equipment or personnel or the storage environment. Storage contamination may result from airborne contaminants, moisture, temperature, and also time. As tin-lead or lead-free component lead finishes age, they can oxidize, and this oxidation will affect solderability of the component. A rule of thumb used by many companies is to suspect the solderability of components two or more years old. In some cases, this age can be tracked easily for components marked with a date code. Other components too small to mark or that for some other reason are not marked with a date code cannot be traced as easily. Often companies keep records on the date of receipt and base the age of the component on this date. Although these records may not be completely accurate, they still serve the purpose. If components are known to be over two years old, a new solderability sample may be pulled to determine whether the components are still usable before production. Often they are not usable due to oxidation, intermetalic growth, or some other contaminant that the component has been exposed to over time.

52.6 SOLDER-RELATED DEFECTS

All solder joints should exhibit wetting as shown by a concave meniscus between the component and PCB being soldered. The outline of the component being soldered should be easily determined. The most common solder defects produced by the soldering process are discussed in the following sections.

52.6.1 PTH Solder-Joint Minimum Acceptable Conditions

Table 52.6 shows the minimum acceptable PTH solder-joint criteria. Fig. 52.23 shows the basic elements of solder fill for all classes.

TABLE 52.6 PTHs with Component Leads: Minimum Acceptable Solder Conditions[a]

Criteria	Class 1	Class 2	Class 3
A. Vertical fill of solder[b,c] Note: Less than 100 percent older fill may not be acceptable in some applications: e.g., thermal shock. The user is responsible for identifying these situations to the manufacturer	Not Specified	75%	75%
B. Wetting on primary side (solder destination side) of lead and barrel Target for Classes 1,2, and 3 is 360° wetting present on land and barrel	Not Specified	180°	270°
C. Percentage of land area covered with wetted solder on primary side (solder destination side) The land area does not need to be wetted with solder on the primary side	0	0	0
D. Fillet and wetting on secondary side (solder source side) of lead and barrel	270°	270°	330°
E. Percentage of land area covered with wetted solder on secondary side (solder source side)	75%	75%	75%

Notes:
[a] Wetted solder refers to solder applied by the solder process.
[b] The 25 percent unfilled height includes both source and destination side depressions.
[c] Class 2 may have less than 75 percent vertical hole fill.

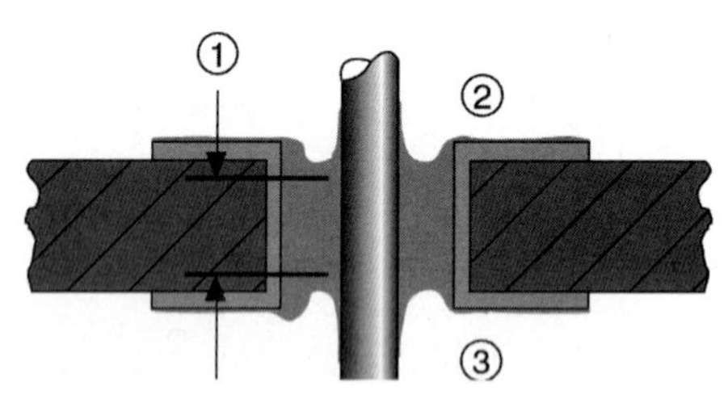

FIGURE 52.23 (1)–Vertical fill meets requirements of Table 52.6 (2)–Solder distination side (3)–Solder source side. *(IPC)*

52.6.2 Solder Balls or Solder Splash

Solder balls and splashes that violate minimum electrical design clearance, are not encapsulated in a permanent coating, or are not attached (soldered) to a metal surface are unacceptable (see Fig. 52.24).

52.6.3 Dewetting and Nonwetting

Dewetting is a condition that results when molten solder coats a surface and then recedes to leave irregularly shaped mounds of solder that are separated by areas that are covered with a

FIGURE 52.24 Solder ball causing a short circuit between pins in SOIC.

thin film of solder and with the basis metal not exposed (see Fig. 52.25). Nonwetting is the partial adherence of molten solder to a surface that it has contacted, while the basis metal remains exposed (see Fig. 52.26).

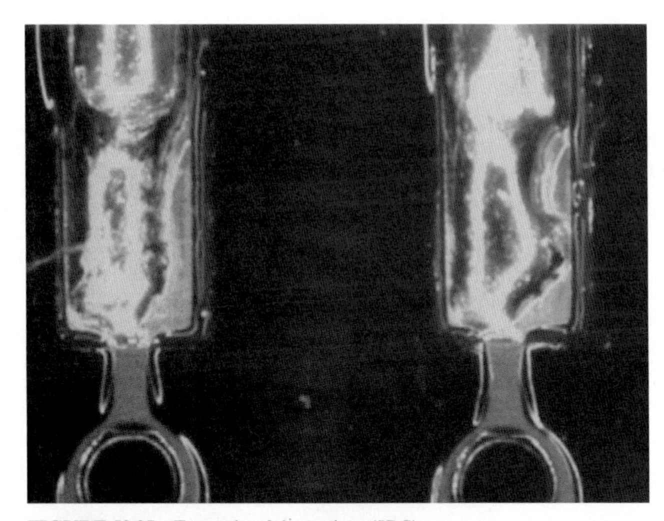

FIGURE 52.25 Example of dewetting. *(IPC)*

FIGURE 52.26 Example of non wetting. *(IPC)*

Dewetting and nonwetting of solder joints is generally caused by contaminants on either the component leads or on the PCB PTH or pads. A minimum of dewetting is allowed on solder joints assuming that the solder joint meets minimum requirements as defined in Table 52.6 and Sect. 52.4.2 by component package type, and good wetting is evident on the portion of the solder joint that does not display dewetting. Nonwetting is unacceptable since adequate wetting is not achieved and indicates a serious solderability problem on the component or the PCB.

52.6.4 Missing and Insufficient Solder

Missing solder is an obvious unacceptable condition since solder provides electrical continuity and some measure of mechanical attachment of the component to the PCB.

Insufficient solder becomes an unacceptable condition for SMT components when the minimum solder fillet requirements defined in Sec. 52.4.2 are not achieved. Insufficient solder becomes unacceptable for PTH components when the minimum requirements of section 52.6.1 and Table 52.6 are not achieved.

52.6.5 Solder Webbing and Bridging

Solder bridged between electrically noncommon conductors creates a shorting condition and is an unacceptable condition (see Fig. 52.27). Solder webbing is a continuous film of solder that is parallel to, but not necessarily adhering to, a surface that should be free of solder. Solder webbing is also an unacceptable condition (see Fig. 52.28).

FIGURE 52.27 Solder connection across conductors that should not be joined. *(IPC)*

For components with leads, with the exception of criteria stated in sec. 52.4.2.4, solder cannot come into contact with the component body or end seal. Solder contacting the component body or end seal is generally unacceptable.

52.6.6 Lead Protrusion Problems

Measurement of lead protrusion is defined as the distance from the top of the PCB land to the outermost part of the component lead, which can include any solder projection from the lead.

FIGURE 52.28 Examples of solder splashes and webbing. *(IPC)*

Solder projections (icicles) are unacceptable if they violate lead protrusion maximum requirements or electrical clearance, or if they pose a safety hazard; otherwise, solder projections are an acceptable condition on both SMT and PTH components (see Fig. 52.29).

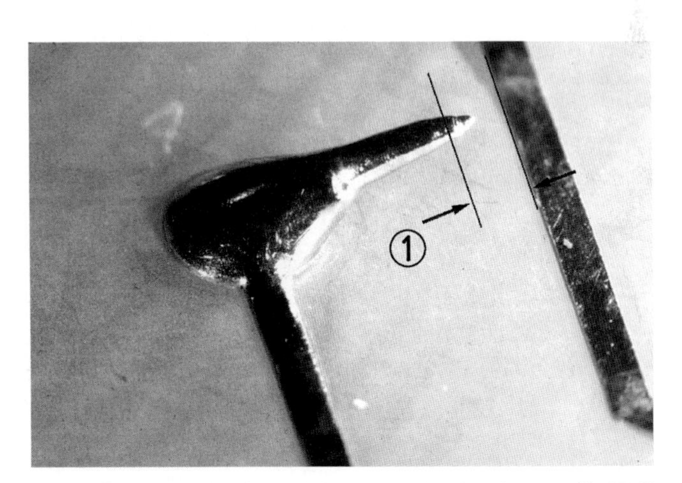

FIGURE 52.29 Projection violates minimum electrical clearance (1). *(IPC)*

For single-sided PCBAs, lead or wire protrusion must be visible in the solder for Classes 1 and 2, and must allow a clinch for Class 3. For double-sided and multilayer PCBAs in all classes, the minimum lead protrusion is for the lead end to be visible in the solder. The maximum lead protrusion for Class 1 is that there be no danger of shorts when the PCBA is used in its assembly application. For Class 2, the maximum lead protrusion is 2.5 mm, and for Class 3 the maximum lead protrusion is 1.5 mm (see Fig. 52.30).

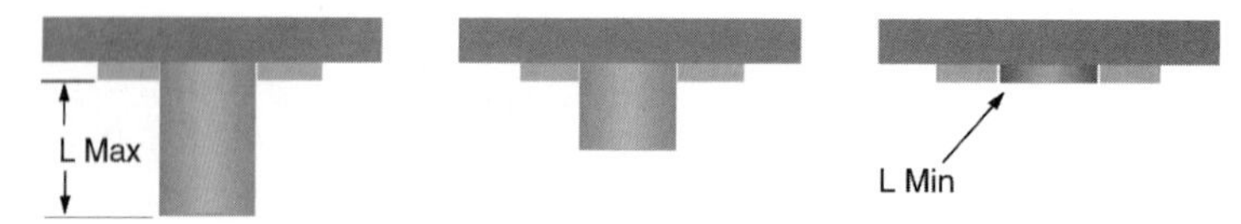

FIGURE 52.30 Lead protrusion violates minimum clearance requirements. Lead protusion exceeds maximum design height requirements. *(IPC)*

52.6.7 Voids, Pits, Blowholes, and Pinholes

Solder cavities (voids, pits, blowholes, pinholes) are acceptable for Class 1 and process indicators for Classes 2 and 3, provided the lead and land/pad are wetted and solder fillets meet requirements specified in Table 52.6.

52.6.8 Disturbed or Fractured Solder Joints

Solder joints may not be disturbed due to movement of the solder connection during solidification. However, a rough, granular, or uneven appearance, provided the wetting coverage criterion in Table 52.1 is met, is acceptable. Solder connections accomplished using many of the lead-free solder alloys and surface preparations will normally exhibit the rough or granular appearance more than eutectic tin-lead. Solder joints that are fractured or cracked are unacceptable for all classes of equipment.

Cutters cannot damage solder joints in which leads have been trimmed after soldering, due to physical shock transmitted through the lead. If lead cutting is required after soldering, the solder joint must be reflowed or visually inspected at 10 × magnification to ensure that the cutting operation did not damage the solder connection. No fractures or cracks are allowed between the lead and solder (see Fig. 52.31).

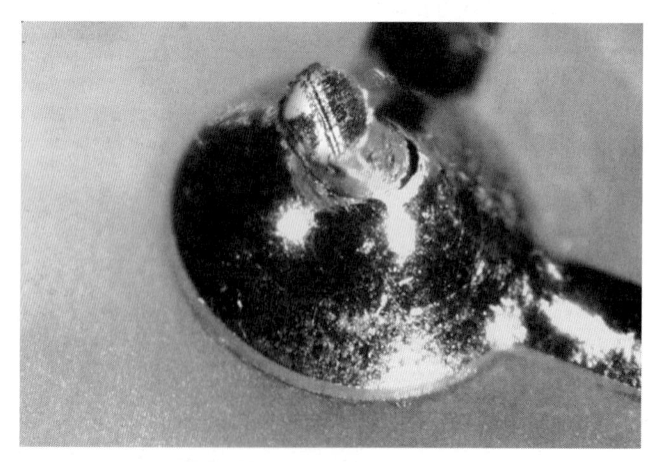

FIGURE 52.31 Example of disturbed solder joint on through-hole component joint. Note evidence of fracture between lead and solder fillet. *(IPC)*

52.6.9 Excess Solder

Excess solder conditions that produce a solder fillet that is slightly convex or bulbous and in which the lead is no longer visible are considered defects for unsupported hole lead termination.

The solder should not flow onto the body of SMT components or touch the component body for through-hole components.

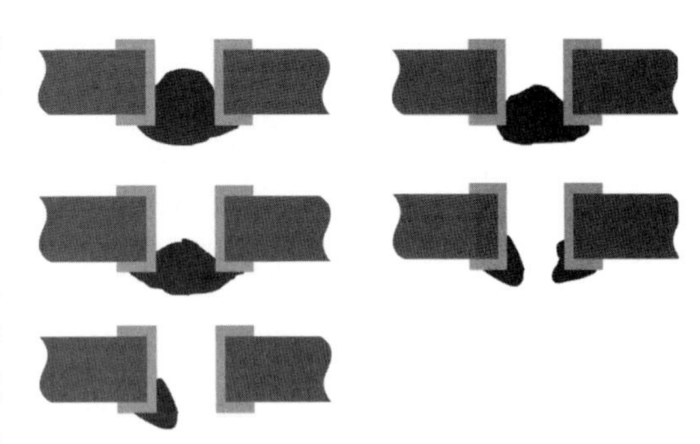

FIGURE 52.32 Via hole solder fill requirements. *(IPC)*

52.6.10 Solder Requirements for Vias

Plated-through via holes used only for interfacial connection do not need to be filled with solder. To prevent solder fill, manufacturers usually place temporary or permanent masks over the vias during the soldering process. PTHs or vias without leads when exposed to solder should meet the following acceptability requirements:

- The target condition is to have the holes completely filled with solder and the top of the lands showing good wetting.
- The minimum acceptable condition is to have the sides of the PTH wetted with solder.
- Solder that has not wet the sides of the PTH is considered a process indicator and product is not rejected (see Fig. 52.32).

52.6.11 Soldering to Terminals

When soldering wires to terminals, the lead outline must be visible and good wetting between the wire and terminal must be evident to be acceptable. The wire insulation gap can be near zero if the insulation has not melted into the solder joint. Slight melting of the insulation is acceptable. If the insulation gap is too large and allows potential shorting of wire to an electrically noncommon conductor, the joint is unacceptable. If the wire insulation is severely burned and the melt by-product intrudes into the solder joint, the joint is unacceptable

For round post terminals, the contact of wrap must be 180° for Class 3 and 90° for Classes 1 and 2. Most of the other terminal types, such as pierced terminals, require 90° of wire wrap (see Fig. 52.33).

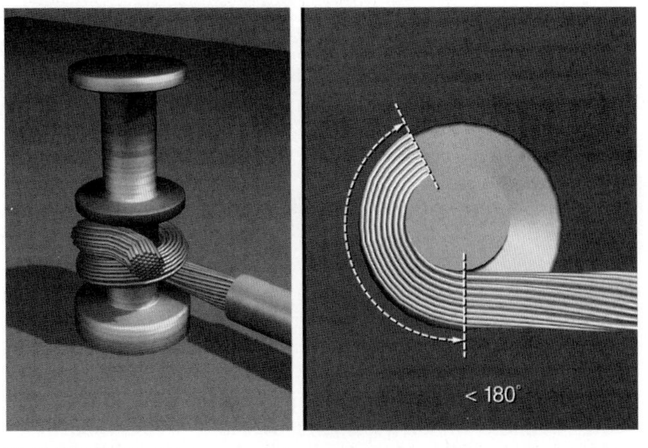

FIGURE 52.33 Minimum requirements for wrap of exposed wire on terminal for acceptable solder joint by class.

For solder cup terminations, the wire should be fully inserted to the bottom of the cup and solder fill must be greater than 75 percent of the milled edge.

52.7 PCBA LAMINATE CONDITION, CLEANLINESS, AND MARKING REQUIREMENTS

52.7.1 Laminate Conditions

Laminate defect conditions may be caused by the laminator, PCB fabricator, or assembly of the PCB. The major laminate conditions seen are measling, crazing, blistering, delamination, weave exposure, and haloing.

52.7.1.1 Measling and Crazing. Measling is an internal condition occurring in laminated base material in which the glass fibers are separated from the resin at the weave intersections. This condition manifests itself in the form of white spots or crosses below the surface of the base material and is usually related to thermally induced stress.

Crazing is an internal condition occurring in laminated base material in which the glass fibers are separated from the resin at the weave intersections. This condition manifests itself in the form of connected white spots or crosses below the surface of the base material and is usually related to mechanically induced stress (see Fig. 52.34).

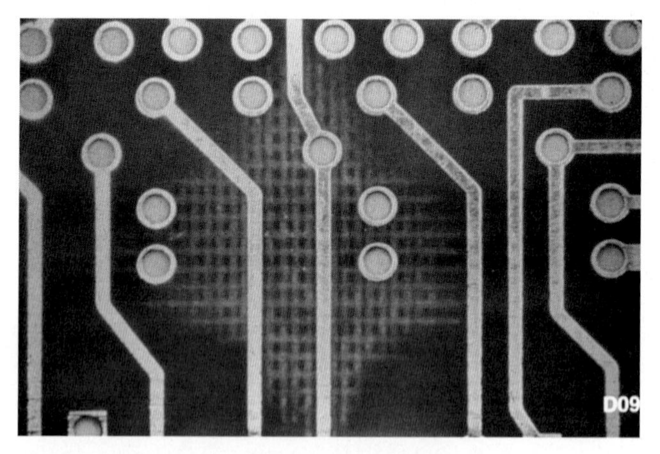

FIGURE 52.34 Crazing in base laminate. Crazing exceeds 50 percent of the physical space between internal conductors. *(IPC)*

Measling or crazing that occurs as a result of an inherent weakness in the laminate is a warning of a potentially more serious problem. If measling or crazing occurs in the assembly process, it usually will not propagate further. Measling is considered a defect for Classes 2 and 3 if the measle condition extends more than 50 percent of the span between internal conductors. The concern is in regard to the ability of the laminate to maintain dielectric integrity after exposure to high humidity and active chemistries and environments.

By the time that a PCB enters an assembly process, the operator who observes measling or crazing cannot determine the source of the problem. This fact places even more importance on getting high-quality PCBs from suppliers. This can be done through increased receiving inspection, source inspections, or a comprehensive qualification program and partnership with suppliers. In such a program, process control is exercised in the PCB fabrication, and parametric data from the supplier are sent to the buyer to review for compliance with requirements. This kind of information can also be used to justify a dock-to-stock program if desired.

52.7.1.2 Blistering and Delamination. Blistering is a localized swelling and separation between any of the layers of the base material or between the material and the metal cladding. Delamination is a separation between any of the layers of the base material or between the base material and the metal cladding. In contrast to measles that do not propagate, blisters and delamination conditions can increase in size with use.

Blistering and delamination cannot exceed 25 percent of the distance between PTHs or subsurface conductors (see Fig. 52.35).

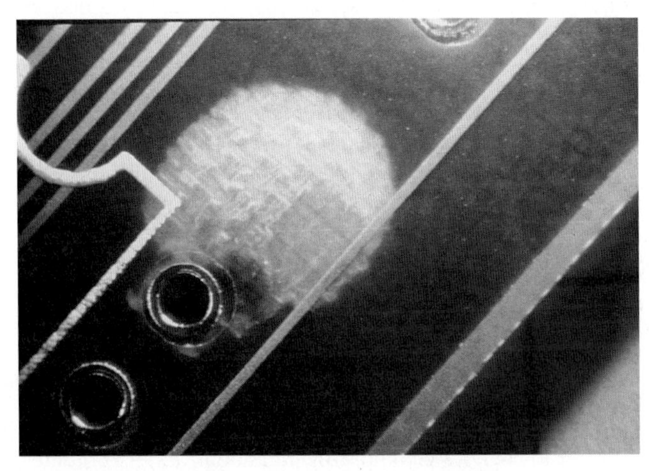

FIGURE 52.35 Example of blister in base material. *(IPC)*

52.7.1.3 Weave Exposure. Weave exposure is a surface condition of base material in which the unbroken fibers of woven glass cloth are not completely covered by resin. Weave exposure is acceptable if it does not reduce dielectric spacing below the minimum electrical clearance.

52.7.1.4 Haloing and Edge Delamination. Haloing occurs on or below the surface of the base material, appearing as a light area around holes or other machined areas. This condition is acceptable if the penetration of the haloing or edge delamination does not reduce edge spacing more than 50 percent of that specified in the drawing or other documentation or 2.5 mm maximum.

52.7.2 PCBA Cleanliness

Board cleanliness is needed to ensure sufficient removal of contaminants that could affect functionality at present or in the future. Some contaminants can actually promote growth of undesirable substances on the PCBA that can cause shorting or corrosion that would affect the PCBA's electrical functional integrity and dielectric properties over a period of time. These contaminants can be in the form of surface dendrites, internal conductive anodic filaments, and so on.

No visible residue from cleanable or any activated fluxes is allowed. Production operations may not be required to remove cleanable residues if qualification testing is performed that demonstrates no need for cleaning the assembly. No-clean or low-residue flux chemistries may be allowed during the soldering process. However, it is critical that the cleanliness of the elements such as the bare boards and components be specified, controlled, and closely monitored; otherwise, the contaminant build-up can far exceed allowable limits for the end item's functional performance.

For processes utilizing potentially corrosive rosin based fluxes, simple test options to determine cleanliness such as resistance of solvent extract (ROSE) tests can be used. Other more significant tests may be required to determine the residue properties of organic residue fluxes, such as those commonly used during hot air solder leveling (HASL) of boards. The most common of these are ion chromatography tests that can characterize the residuals and their potential hazards to the product.

In the event of test failure, immediate process corrections for the cleaning process or elimination of the source of residues must be accomplished before any additional product is assembled.

Particulate matter such as dirt, lint, dross, lead clippings, and such is not acceptable on PCBAs. Metallic areas or hardware on the PCBAs may not exhibit any crystalline white deposits, colored residues, or rusty appearance.

52.7.3 PCBA Marking Acceptability

Marking provides both product identification and traceability. It aids in assembly, in-process control, and field repairs. The methods and materials used in marking must serve the intended purposes and must be readable, durable, and compatible with the manufacturing processes as well as the end use of the product.

Fabrication and assembly engineering drawings should be the controlling documents for the locations and types of markings on PCBAs. Marking on components and fabricated parts should withstand all tests, cleaning, and assembly processes to which the item is subjected and must remain legible. Acceptability of marking is based on whether it is legible. If a marking is legible and cannot be confused with another letter or number, it is acceptable. Components and fabricated parts do not have to be installed so that reference designators are visible after installation. Missing, incomplete, or illegible characters in markings are unacceptable.

52.8 PCBA COATINGS

It should be noted that not all PCBA designs use conformal coating; however, when used, conformal coating must meet the following acceptability criteria. All PCBAs that utilize top- or bottom-layer etch (trace) runs have a solder mask over them if a solder wave or static solder bath process is used to solder components onto the PCB. Without the solder mask, the solder would bridge between many solder wettable surfaces, causing short circuits.

52.8.1 Conformal Coating

Conformal coating is an insulating protective covering that conforms to the configuration of the objects coated when it is applied to a completed PCBA. The purpose of conformal coating is to provide a temporary barrier to environmental exposure of chemicals, humidity, and condensation. It is also a barrier from surface contact of contaminants caused by handling, and can provide physical support for components from shock and vibration. It is not, however, a hermetic seal and is eventually permeable. To be most effective, conformal coatings should be homogeneous and transparent. Most coatings will contain an ultraviolet (UV) tracer to make inspection easier through the use of a black light. The conformal coating should be properly cured and not exhibit tackiness. Defects associated with conformal coating are as follows:

- Applied where not required
- Not applied where required
- Bridging of conductors due to:
  - Mealing (adhesion)
  - Voids
  - Dewetting
  - Cracks
  - Ripples
  - Orange peel/fisheye
- Inclusions
- Discoloration
- Incompletely cured
- Wicking into connectors or mating surfaces

Conformal coating thickness requirements for five types of coatings are listed in Table 52.7. The thickness is commonly measured on a coupon that has been processed with the assembly.

TABLE 52.7 Coating Thickness

Type	Material	Thickness
AR	Acrylic Resin	0.03–0.13 mm (0.00118-0.00512 in.)
ER	Epoxy Resin	0.03–0.13 mm (0.00118-0.00512 in.)
UR	Urethane Resin	0.03–0.13 mm (0.00118-0.00512 in.)
SR	Silicone Resin	0.05–0.21 mm (0.00197-0.00827 in.)
XY	Paraxylyene Resin	0.01–0.05 mm (0.00039-0.00197 in.)

52.8.2 Solder Resist

Solder resist or solder mask is a heat-resistant film coating used to provide mechanical shielding during soldering operations. Solder resist material may be applied as a liquid or a dry film.

Solder resist would be ineffective and considered a defect when there are blisters, scratches, or voids that bridge noncommon circuits, or if the exposure has the potential to allow solder bridges. It could be considered a process indication if blisters or flaking expose bare copper.

52.9 SOLDERLESS WRAPPING OF WIRE TO POSTS (WIRE WRAP)

Many applications of wire wrap are still utilized in equipment design, and standards of acceptance should be consistent. Standards of acceptance in this area are shown in IPC-A-610 and in other standards.

52.9.1 Wrap Post

The wrap post cannot be bent or twisted before or after the wire is wrapped to the post. The wrap straightness may not exceed approximately one post diameter or thickness from its perpendicular position.

52.9.2 Wire Wrap Connection

Table 52.8 shows the number of wire wrap turns of insulated and uninsulated wire used on the wrapped connection. The connection is made using automatic or semiautomatic wire wrapping devices. For this requirement, countable turns are those turns of bare wire or insulated wire in intimate contact with the corners of the terminals starting at the first contact with a terminal corner and ending at the last contact with a terminal corner (see Fig. 52.36).

TABLE 52.8 Minimum Turns of Bare Wire

Wire Gauge	Turns
30	7
28	7
26	6
24	5
22	5
20	4
18	4

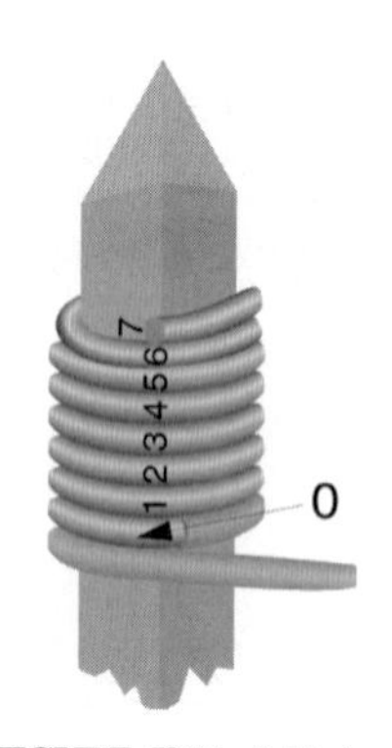

FIGURE 52.36 Solderless wrap acceptability. *(IPC)*

Maximum turns of bare and insulated wire are governed by the tooling configuration and space available on the terminal. The minimum countable turns must be as stated in Table 52.8. Electrical clearance between terminals as specified by engineering documentation must be maintained. Wire ends should in no case project to the extent that the required electrical clearance is compromised. The wire end should never project more than 0.12 in. (one wire diameter for Class 3) away from the terminal.

The wrapped conductors should be free of gaps (that is, each wrap should be in contact with the previous wrap), with turns not overlapping. The first and last one-half turns may have a space between turns, provided that the space does not exceed one diameter of the conductor wire. Excluding the first and last one-half-turns, the wrapped conductors may have a single space between them, if the opening does not exceed one-half the nominal diameter of uninsulated wire (see Fig. 52.37).

52.9.3 Multiple Wire Wrap Spacing

Typically no more than three wires are wrapped to a single post. When more than one wrap is used on a single post, there should be visible clearance between consecutively wrapped wires. The final wire wrap turn on a post must not extend beyond the working length of the post (see Fig. 52.38). The first insulated turn of a higher-level wire wrap may overlap the last turn of uninsulated wire on a lower-level wrap by a maximum of 1 turn for Class 3.

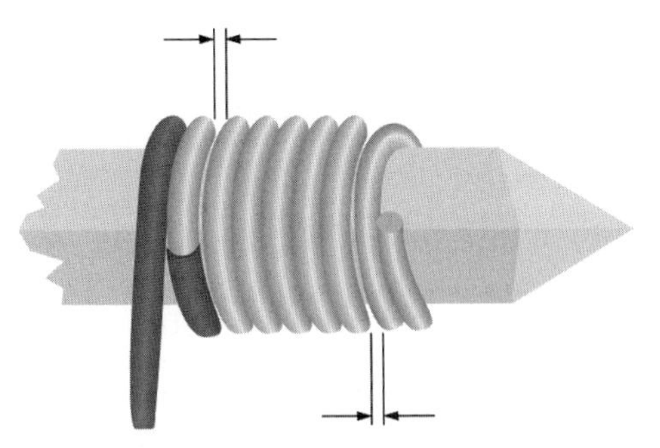

FIGURE 52.37 Solderless wrap with more than one space. Acceptability depends upon Class of product. *(IPC)*

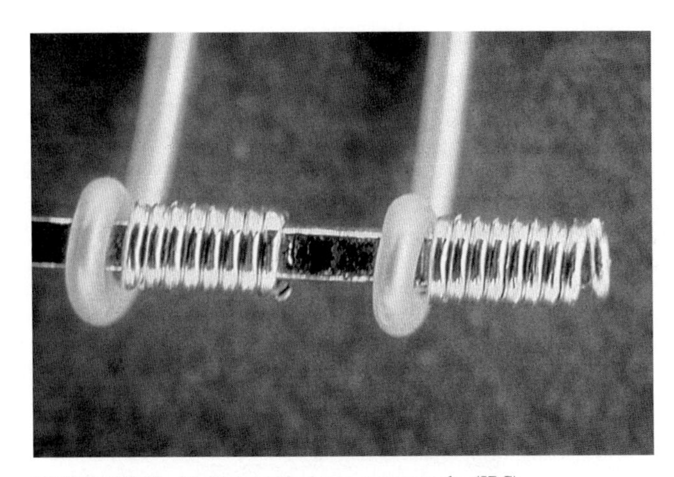

FIGURE 52.38 Multiple solderless wrap example. *(IPC)*

52.10 PCBA MODIFICATIONS

All modifications to PCBAs should be defined and detailed in approved engineering and/or methods documents. Jumper wires are considered components and are defined by engineering documentation for routing, termination, staking, and wire type.

52.10.1 Cut Traces

Trace cuts should be at least 0.030 in. wide, with all loose material removed. They should also be sealed with an approved sealant to prevent absorption of moisture.

Care should be taken when removing etch from PCBs to prevent damaging the laminate material.

52.10.2 Lifted Pins

Lifted pins should be cut short enough to prevent the possibility of being shorted to the pad from which they were lifted should they be pushed back down. If the component hole from which the pad was lifted does not contain a jumper wire, it should be filled with solder.

52.10.3 Jumper Wires

Jumper wires may be used on all classes of equipment electronic assemblies and in thick-film hybrid technology. The wire may be terminated in PTHs, on terminal standoffs, on circuit lands, or on component leads. It should be noted for Class 3 equipment that a wire cannot be placed into the same PTH with a component lead.

Recommended jumper wire is solid, insulated, plated copper wire. It should be the smallest wire diameter that can reliably carry the required current, and the insulation should be capable of withstanding soldering temperatures.

Jumper wires should be routed in the shortest x-y route possible (see Fig. 52.39). Wire routing must be documented for each part number. Assemblies having the same part number must be routed in the same pattern.

FIGURE 52.39 Example of X-Y routing of jumper wires. *(IPC)*

When jumper wires are used on the primary side (component side) of the PCBA, no wire is to pass over or under any components for Class 2 or 3. The wire may pass over lands, provided the wire can be moved away from the land for component replacement. To prevent damage to the wire from excessive heat, care must be taken to avoid running wires near heatsinks or components that generate high temperatures.

When jumper wires are used on the secondary side of the PCBA, the jumper wire should not pass through component footprints unless the layout of the assembly prohibits routing in other areas. If this condition occurs, it should be considered a process indicator. There is an exception to this is rule for edge connectors on the PCBA. Jumper wires should not pass over test points or vias used as a test point.

Jumper wires should be staked to the base material with an approved adhesive or tape dots or strips intended for this purpose. If adhesives are used, they must be fully cured as part of the completed PCBA. The wire should be spot-bonded along its route and must not be applied to lands, pads, or components. The interval of staking must be defined in the applicable engineering documentation, but the wire must be staked at all changes in direction of the wire. No jumper wires may be staked to components that are socketed or any moving part. Jumper wires must be staked or taut enough to prevent the wire from rising above the height of adjacent components. No more than two jumper wires may be stacked on a given route.

When a jumper wire is attached to leads on the secondary side of the PCBA or to axial components on the primary side of the PCBA, it must form a full 180 to 360° loop around the component lead. When a jumper wire is soldered to other component package styles, the wire should be lap-soldered to the component lead.

Jumper wires may be installed into a PTH with another component lead for Class 1 and 2 equipment; however, this is unacceptable for Class 3 equipment. Jumper wires may also be installed into via holes (see Fig. 52.40).

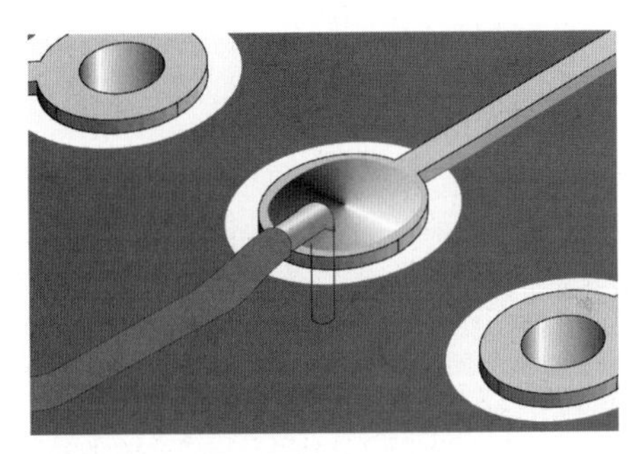

FIGURE 52.40a Example of jumper wire connected to plated through-holes (IPC).

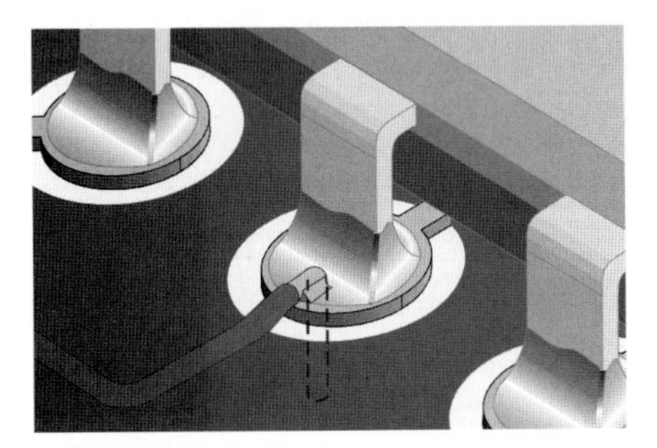

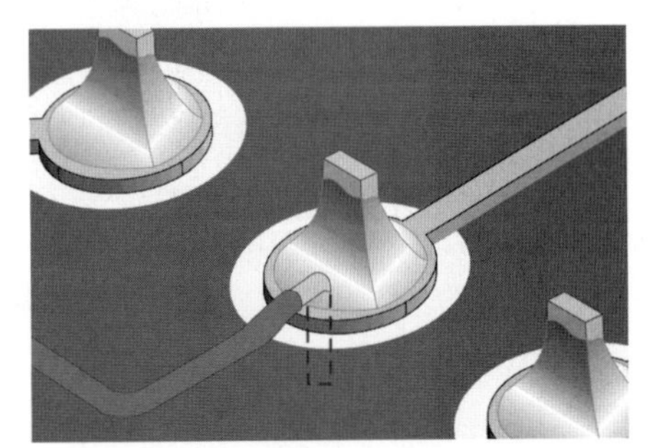

FIGURE 52.40b Example of wire connected to a via hole with a lead installed.

REFERENCES

1. ANSI/J-STD-001, "Requirements for Soldered Electrical and Electronic Assemblies."
2. MIL-STD-2000, "Standard Requirements for Soldered Electrical and Electronic Assemblies."
3. ANSI/J-STD-002, "Solderability Tests for Component Leads, Terminations, Lugs, Terminals and Wires."
4. ANSI/J-STD-003, "Solderability Tests for Printed Boards."
5. IPC-A-610, "Acceptability of Electronic Assemblies."

CHAPTER 53
ASSSEMBLY INSPECTION

Stacy Kalisz Johnson
Agilent Technologies Gilbert, Arizona

Stig Oresjo
Agilent Technologies Loveland, Colorado

53.1 INTRODUCTION

This chapter covers why manufacturers inspect printed circuit assemblies, how they have implemented and enhanced visual inspection, what automated inspection systems they are using, and how they have implemented these automated systems. The scope of this chapter includes only inspection of printed circuit assemblies during the assembly process, as typically shown in Fig. 53.1. Thus, it includes inspection of solder paste after the paste printing process step, components after the component placement process step, and solder joints after the solder reflow process step. Not included, however, is incoming inspection of components and the bare printed circuit board (PCB). The focus of this chapter is on production use of inspection, not the collection of measurements during process development in a research and development (R&D) environment.

53.1.1 Visual Inspection

Manufacturers of printed circuit assemblies have always visually inspected their boards at various points in the assembly process. There are a variety of reasons for visual inspection of assemblies, including eliminating obvious process defects quickly and meeting military specifications. With the advent and growth of surface-mount technology (SMT), visual inspection of printed circuit assemblies has also grown in importance and prevalence. SMT solder joints must carry a much bigger burden for mechanical or structural reliability than do plated-through-hole (PTH) solder joints. The pin-in-through-hole solder joint carries much of the mechanical burden, helping to keep the component attached to the printed circuit board. However, with SMT solder joints, it is often only the solder that keeps the component attached to the board. In some cases, only visual inspection could judge the mechanical reliability of an SMT solder joint.

As SMT geometries have continued to shrink and solder joints have become denser on printed circuit boards, visual inspection has become more difficult and thus its results have become less consistent and reliable. In addition, components such as pin grid arrays (PGAs)

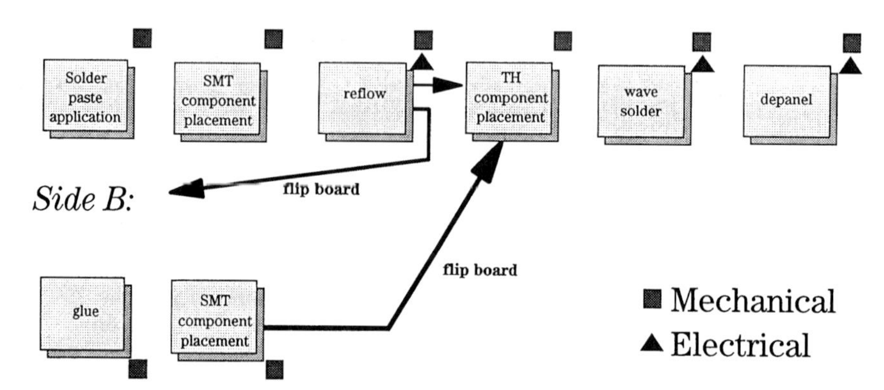

FIGURE 53.1 Generic manufacturing process for surface-mount technology printed circuit assemblies, including the possible locations within the process for inspection or testing of mechanical or structural attributes and electrical characteristics of an assembly.

or ball grid arrays (BGAs) completely hide their solder joints from view, which makes visual inspection not possible at all for those joints. As Fig. 53.2 indicates, achieving a high assembly process yield is more important as the number of solder joints per assembly increases.

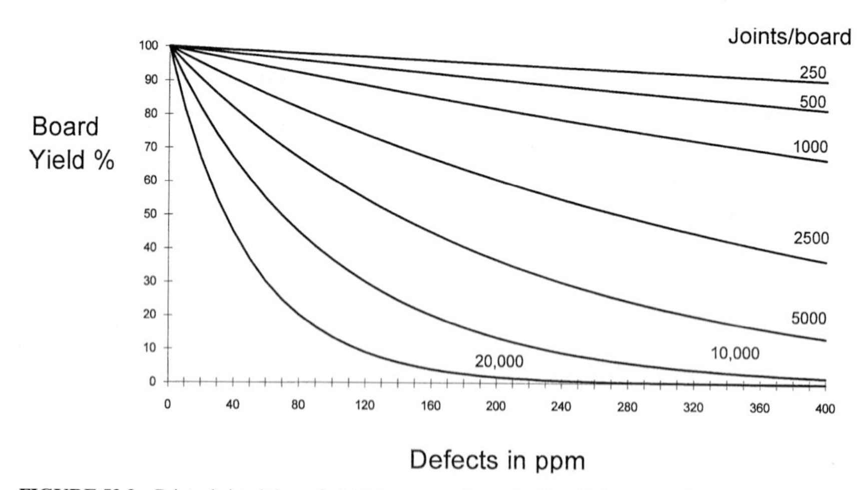

FIGURE 53.2 Printed circuit board yield decreases dramatically with increases in the number of solder joints per assembly if the defect rate per solder joint remains constant. For instance, at a 40 defects per million opportunities (DPMO) defect rate, the yield drops from 96 percent at 1,000 joints per assembly to 45 percent at 20,000 joints.

53.1.2 Automated Inspection

Inspection is an important source of process information without which high yields are difficult to obtain. Consequently, manufacturers have employed the following range of techniques either simply to enhance visual inspection or automate inspection fully:

- Microscopes, mono- and stereoscopic, with magnification from 4× to 10×
- Real-time video images created using simple light, x-rays, thermal imaging, or acoustics
- Fully automated inspection systems using light, laser, or x-ray imaging

The automation of inspection has evolved into systems that resemble the automated test equipment used to make electrical measurements and find electrical defects (faults). These automated inspection systems acquire real-time images, process the images to find and measure features within the image, and make accept/reject decisions based on this image processing. Additionally, they provide data that can be used for long-term statistical process analysis. Thus these automated systems remove the human—and human judgment—from the inspection process altogether.

53.2 DEFINITION OF DEFECTS, FAULTS, PROCESS INDICATORS, AND POTENTIAL DEFECTS

Faults, *defects*, *process indicators*, and *potential defects* are terms that are important when discussing test and inspection, so this section will define and explain them:

- **A *fault* is a manifestation of a defect** An example is a digital device output pin that does not toggle correctly. For simplicity, think about a two-input OR-gate whose output is stuck high. This is a fault and is a manifestation of a defect. The causing defect can be any one of several, including a defective component, an incorrectly placed component, an open input pin, or an open output pin. The fault class is a subset of the defect class. Electrical tests such as in-circuit test (ICT), boundary-scan, built-in-self-test (BIST), functional test (FT), and system test mainly detect faults.
- **A *defect* is, at the end of the manufacturing process, an unacceptable deviation from a norm** There may be defects that do not show up as faults. Examples are insufficient solder, a misaligned component, a missing bypass capacitor, and an open power pin. Inspection systems such as automated optical inspection (AOI) and automated x-ray inspection (AXI) detect many of the defects and also some of the same faults as electrical test. The definition includes the phrase "at the end of the manufacturing process," which is important when compared with "potential defect."

All defects, which also include the faults, need to be corrected before the product is shipped, if high quality standards are to be followed.

- **A *process indicator* is, at the end of the manufacturing process, an acceptable deviation from a norm** Good examples are insufficient solder or misaligned components. The insufficient solder may not be so lacking that it requires a repair action. However, if many of these conditions exist, a process improvement action may be required.
- **A *potential defect* is, in the manufacturing process, a deviation from a norm, that may or may not be a defect at the end of the manufacturing process** This category needs to be understood. An example is a pre-reflow misaligned chip component. This component may or may not self-align in the reflow oven. Another example is an insufficient paste volume that may not end up as a defective solder joint at the end of the manufacturing process.

Test and inspection engineers at the end of the line are mainly interested in finding faults and defects. Process engineers responsible for improving the manufacturing process are mainly interested in potential defects, process indicators, and systematic defects. Systematic defects are typically defects that occur on a larger number of boards and are due to some systematic problem, such as a bent nozzle on a pick-and-place machine, soldering issues with one type of component, or a design for manufacturing (DFM) issue.

It is important to realize that electrical test at the end of the manufacturing line only finds faults. It is therefore important to complement the test strategy with steps that find all defects. Finding all defects is critical if shipping high-quality, reliable products is a priority. Identifying potential defects and process indicators can also help in adjusting the manufacturing process to result in lower defect levels, higher yields, and lower cost. Inspection is therefore a very important complement to electrical test.

53.3 REASONS FOR INSPECTION

Manufacturers inspect printed circuit assemblies during production for a variety of reasons. Most of these reasons fall into the following categories:

- Improvement of process defect coverage
- Improvement in ability to meet customer specifications
- Detection of process defects as quickly as possible after they occur
- Ability to decrease process defect rates through statistical process control (SPC)

53.3.1 Process Defect Coverage

The goal of higher defect coverage is to prevent any defective printed circuit assemblies from reaching assembly of the board and making it into the final product, whether final assembly is at the same site or at a separate customer site. Figure 53.3 shows a typical spectrum of SMT process defects. Some of these defects, such as misaligned solder joints and solder balls, can be discovered only by inspection, not by electrical, functional, in-circuit, or boundary-scan test. For instance, marginal solder joints are those that pass electrical tests just after assembly completion, but eventually fail because of inferior mechanical strength. Insufficient solder and components partially misaligned with printed circuit pads are the two most common causes of low mechanical strength.

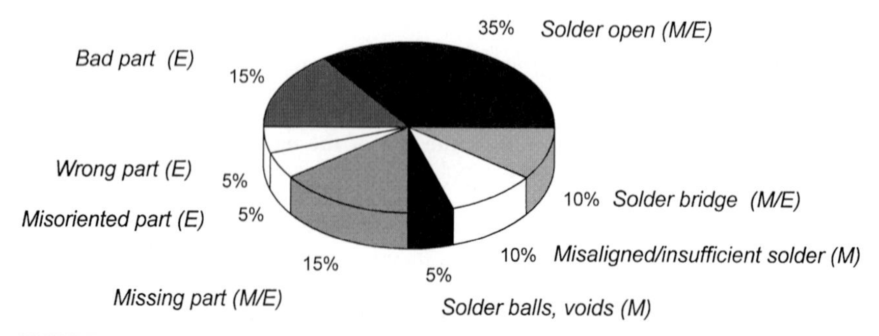

FIGURE 53.3 Typical process defect spectrum for surface-mount assemblies. Process defects marked "*E*" can be detected by electrical tests, those marked "*M*" can be detected by mechanical or structural inspection or testing, and those marked by "*E*" and "*M*" can be detected by either method.

Voids in solder joints, solder balls away from solder joints, and "cold" or dull solder are other examples of defects that can be found only by some kind of inspection.

53.3.2 Customer Specifications

Many printed circuit assembly workmanship specifications—including MIL-STD-2000A, IPC/EIA J-STD-001C, and, more recently, IPC-A-610D—require inspection of solder joints to ensure conformance to these specifications. For instance, MIL-STD-2000A prescribes, for all defense and aerospace government contractors, the attributes of solder joints to be inspected and describes the conditions that result in nonconformance. Most commercial customers of contract manufacturers also have very specific solder joint specifications that require some type of inspection of printed circuit assemblies. Although process control is theoretically preferred over 100 percent inspection, practically speaking, a majority of printed circuit production lines include some kind of inspection to ensure conformance to specifications. As the inspection speeds of the automated inspection systems have improved in recent years, this has become even more prevalent. It is likely that this trend will continue in light of the continued usage of finer pitch devices and the dawn of material challenges, such as lead-free implementation.

53.3.3 Quick Defect Detection and Correction

Quick detection of process shifts or defects can lower rework costs.

53.3.3.1 Preventing Defects. If a particular process step has drifted out of its control limits or specification limits, discovering this fact as soon as possible will prevent more defects from occurring. If the problem is found quickly enough and corrected immediately, perhaps no actual defects will occur.

53.3.3.2 Lowering Rework Costs. Defects found earlier in the process are often easier to repair. For instance, if a defect in solder paste deposition is found before a component has been placed in the paste, it is fairly inexpensive to wipe the solder paste off and start over. If this same defect is found after the solder is reflowed, the solder joint itself will have to be touched up, a more difficult and expensive rework step. The same is true for repairing component placement defects before the solder is reflowed, particularly for missing or misaligned components. Additionally, when organic solderability preservative (OSP) boards are being utilized for lead-free soldering, for example, they cannot be reworked as easily. Using process control and defect prevention can prevent a large number of non-reworkable boards in this case.

53.3.3.3 Making Defect Diagnosis Easier. Defects found earlier in the assembly process are often more easily diagnosed, shortening the overall repair time. An example is the inspection of solder joints. Inspection of solder joints detects defects specific to the solder joints and quickly determines each defect's exact location and characteristic. Waiting until a later electrical test stage could make diagnosis more difficult because at that point there could be several causes of the defect, such as a defective component and defects in other connections, in addition to the actual defective solder joint.

53.3.4 Statistical Process Control

SPC requires reliable data that can be analyzed either in real time or historically. Visual inspection collects defect data, such as the number of solder joint defects per assembly right after the solder reflow process (either reflow or wave soldering). Some manual and automated inspection techniques also take quantitative measurements of key assembly parameters, such as solder paste volume or solder joint fillet height. To the extent that these data are repeatable, manufacturers use defect data or measurements to characterize the amount of process variation from assembly to assembly or from solder joint to solder joint. When the amount of variation starts to drift outside its normal range or outside its control limits, manufacturers can assess the assembly process and monitor or choose to take action until the process is adjusted to eliminate this drift. Historical analysis of the defect or measurement

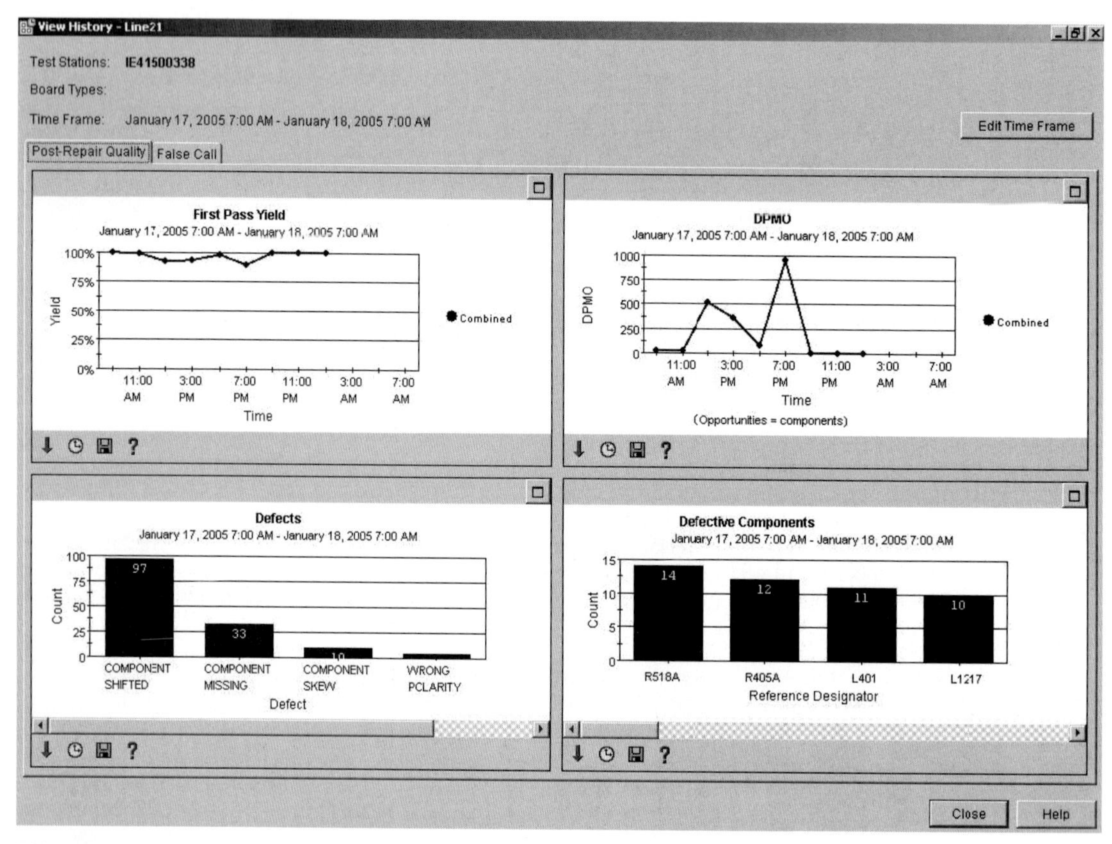

FIGURE 53.4 Sample charts to display data from automated inspection.

data also helps discover the cause of process variation. Eliminating the sources of process variation reduces the process defect rate, thus creating savings in rework costs and increasing product reliability. As shown in Figure 53.4, various charts to display available data from the automated inspection can be made available for corrective action and process control.

53.4 LEAD-FREE IMPACT ON INSPECTION

Other chapters of this book have covered the lead-free process, but this section covers the lead-free process from an inspection point of view. In general, the lead-free process is a less forgiving one because the wetting forces for lead-free alloys are not as strong as for tin-lead, which means that defect levels are likely to increase and the need for inspection for process control and defect containment will increase. The figures illustrate wetting issues of the lead-free process. Figures 53.5 and 53.6 show x-ray close-ups of two gull-wing joints on the same type of quad flat pack (QFP) device.

A key issue for manufacturers is the lead-free alloy melting point of 217°C, higher than the 183°C for lead alloys. The higher melting point will have significant impact on manufacturing processes and potentially on component reliability.

During the transition to a lead-free process, it is important to understand all of the materials in use, as mixing lead-free and traditional tin-lead components will create many issues during the transition to a lead-free process. Ideally, all lead-free components and materials will be utilized, but this is not always possible.

Figure 53.5 is an x-ray image of a tin-lead example. A good heel formation and toe formation can be seen. The side fillets are approximately the same, indicating that the component pins are centered on the pad. Note also that very little solder appears on the pad below the toe in the photo; all solder is around the component pins.

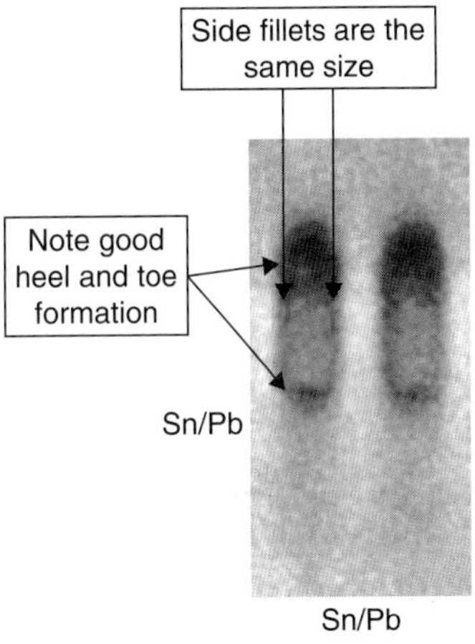

Side fillets are the same size

Note good heel and toe formation

Sn/Pb

Sn/Pb

FIGURE 53.5 An x-ray image of a tin-lead example.

Figure 53.6 is an x-ray image of a lead-free example and reveals a couple of wetting issues. Solder is over the full pad. The pad sizes on both boards (Figs. 53.5 and 53.6) are approximately the same. Also, note that the side fillet to the right is stronger than to the left, indicating that the component is misaligned on the pad. The toe fillet is insignificant. The heel fillet has formed correctly; a void can be seen in one of them. A void is also visible on the same pin further down on the pad.

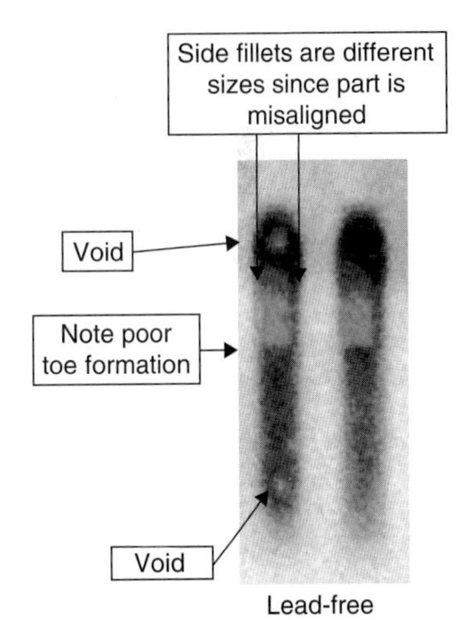

FIGURE 53.6 An x-ray image of a lead-free example.

During the transition to a lead-free process, expect to see many of these types of defects. Defect levels are expected to increase. The following types of defects are typically noticeable during lead-free transitions: opens, bridges, misalignment, tombstones, and voids. Because of the higher reflow temperatures, component defects are also expected to increase. Large variations in defect levels should be expected. For some board types, small increases or no increase will be seen, whereas for others a significant (more than 10 times) increase in DPMO values will be seen. These defect levels will vary from one manufacturing site to another.

Inspection and especially automated inspection will be a significant help in addressing the tighter process window acceptable for a lead-free process. For process control, pre-reflow inspection will provide the biggest benefit. The obvious benefit is that solder paste inspection (SPI) and pre-reflow automated optical inspection (AOI) systems provide early defect and potential defect detection and prevention specifically by identifying the defects that are expected to increase with the use of lead-free solder (opens, bridges, and so on, as previously mentioned in this chapter). An additional benefit—and perhaps the most valuable one—is the insight that the process engineer can gain by evaluating the results from SPI and AOI systems as lead-free materials are adopted. By integrating SPI and AOI systems in the manufacturing process as they move to lead-free materials, and then using the data for process optimization, the manufacturer can eliminate most of the process difficulties that a lead-free process creates. This implementation will in turn allow users to bring lead-free lines up to speed more quickly and efficiently, saving on labor, materials waste, and so on.

After reflow, the main objective is to identify defects so that repair action can take place. For this manual visual inspection, AOI post-reflow, or automated x-ray inspection, can be used. Even with good process control, defect levels are likely to increase, especially for higher-complexity boards. If defect levels are higher, then a good inspection strategy becomes more valuable.

53.5 MINIATURIZATION AND HIGHER COMPLEXITY

Printed circuit board real estate continues to be coveted. Boards are getting smaller and smaller as more and more capabilities are built on them. To keep up with this trend, complex packages and devices and miniature devices are cropping up. Many would argue that a 0402 chip component is small, but "small" becomes subjective when you consider 0201s and 01005s. The 01005 component's sizes typically range from 0.10 mm (0.004 in.) × 0.304 mm (0.012 in.) to 0.20 mm (0.008 in.) × 0.40 mm (0.016 in.), depending on the supplier. Note that the 0201 typical is 0.60 mm (0.0236 in.) × 0.30 mm (0.0118 in.), thus it's quite a jump. With devices on this size order, the design considerations are large and the propensity for defects is higher. Inspection must catch the defects, but at these sizes, approaching the eye of a needle, it is not practical to conduct inspections visually, and thus such automated test and inspection become paramount. Automated test and inspection continues to spur the development of camera and algorithm technology to keep up with the throughput, coverage, and size requirements demanded by the industry. Miniaturization in itself is quite a driver for these technologies. Add this to the material changes that are going on with materials changes such as lead-free materials, and the need for automated test and inspection only increases. AOI systems including three-dimensional (3-D) and two-dimensional (2-D) solder paste and pre- and post-reflow component inspection are typically lead-free-ready, as are x-ray test systems, with a few minor programming updates.

The advancement of the miniaturization movement also drives technologies such as multichip modules (MCM), system-in-a-package (SIP), and so on. These devices or boards, depending on how one refers to them, have many interconnects among themselves in addition to the requirements that they pose when the entire MCM, SIP, etc., is adhered to a printed circuit board. The number of hidden joints increases, the types of hidden joints increase, and the complexity of the attachment to the SMT board in general increases. All of these are contributors to the ongoing need for increase defect containment and defect coverage that only automated inspection and testing bring.

53.6 VISUAL INSPECTION

Visual inspection is the visual comparison by a human of some attribute of the printed circuit assembly with specified standards that describe the acceptable range for that attribute. The inspector normally picks up the printed circuit assembly or places it under a microscope and carefully observes particular attributes, such as the condition or bend radius of component leads or the wetting of solder joints to a lead. Visual inspection always involves human judgment in comparing the attribute to its conditions of conformance to standards.

53.6.1 General Inspection Issues

As Fig. 53.1 indicates, visual inspection can occur after each of the several printed circuit assembly process steps. Visual inspection often has different purposes, depending on where in the assembly process it occurs. These purposes fall into the following major categories:

- Quick detection of an assembly process step that is not operating within its normal range
- Detection of process defects as specified by the customer or industry or internal standards

53.6.2 Solder Joint Inspection Issues

Visual inspection after the reflow and wave-soldering process steps can also be just a quick scan for obvious defects to detect a process condition outside of control limits. In this case, the operator visually inspects for solder bridges, large solder balls or solder splashes, lifted leads, and a number of other improper conditions. Often, however, visual inspection after the soldering process steps is aimed at finding solder joints that do not meet specifications.

Visual inspection of solder joints against specification can cover 100 percent of the visual solder joints on an assembly. Inspecting samples of solder joints in conjunction with a documented process control system is also done. Inspecting for this purpose can be a lengthy process, taking as long as a half-hour for assemblies with 4,000 solder joints being measured against military specifications. As a rule of thumb, this kind of visual inspection has a throughput of about 5 joints per sec. compared to automated inspection techniques, which have capabilities of over 100 joints per sec. Thus, inspection of solder joints against specifications normally requires visual inspectors dedicated to this function with no other responsibilities. Visual inspection may still provide a low-cost alternative to automated inspection in some low-cost regions, which is one major factor contributing to its continued usage.

The visual inspector must be very familiar with the specifications of the attributes for each solder joint type. Each solder joint can have as many as eight different criteria for defects such as insufficient solder, misalignment, etc. Each assembly typically has more than six different solder joint types such as BGAs and QFPs, corresponding to different component packages with individual requirements for each. As an example, Fig. 53.7 shows the specification attributes for one component type (rectangular passive chips). Table 53.1 gives the corresponding conformance criteria for each attribute for this component type. More importantly, the visual inspector must be highly trained to make an accurate judgment of conditions on the borderline between good and bad. For instance, accurately determining whether a fillet height is one-quarter of the way up a component side that is only 0.05 mm high to begin with takes a lot of practice. Visual inspectors typically do not use any tools to help make these judgments. Rulers or calipers are very difficult or impossible to use to measure solder joint dimensions or thicknesses. Using reticules in microscopes in conjunction with coordinate-measuring machines is possible, but is usually much too time consuming to be done on a regular basis.

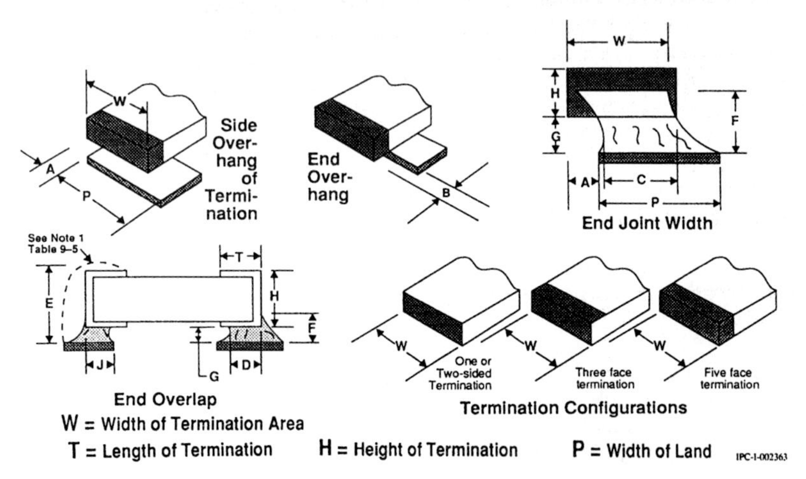

FIGURE 53.7 Inspection attributes for solder joints of surface-mounted rectangular passive chips.

Visual inspection often occurs at the end of the assembly process or even after all of the electrical tests on the assembly have been completed. It is not possible, however, for invisible or hidden joints. The visual inspector looks for scratches, partial delaminations, solder splashes or solder balls away from solder joints, and any other condition that does not affect assembly performance but does make the assembly look as though it might not be a high-quality product.

TABLE 53.1 Dimensional Criteria: Rectangular or Square End Components (Dimensions in Millimeters)

Feature	Dim.	Class 1	Class 2	Class 3
Maximum side overhang	A	50% (W) or 50% (P), whichever is less; Note 1	50% (W) or 50% (P), whichever is less; Note 1	25% (W) or 25% (P), whichever is less; Note 1
Maximum end overhang	B	Not permitted	Not permitted	Not permitted
Minimum end joint width	C	50% (W) or 50% (P), whichever is less	50% (W) or 50% (P), whichever is less	75% (W) or 75% (P), whichever is less
Minimum side joint length	D	Note 3	Note 3	Note 3
Maximum fillet height	E	Note 4	Note 4	Note 4
Minimum fillet height	F	Note 3	Note 3	(G) + 25% (H) or (G) + 0.5 mm (0.02 in), whichever is less
Solder fillet thickness	G	Note 3	Note 3	Note 3
Height of termination	H	Note 2	Note 2	Note 2
Minimum end overlap	J	Required	Required	Required
Width of land	P	Note 2	Note 2	Note 2
Width of termination	W	Note 2	Note 2	Note 2

Note 1. Shall not violate minimum electrical clearance.
Note 2. Unspecified parameter or variable in size as determined by design.
Note 3. Properly wetted fillet shall be evident.
Note 4. The maximum fillet may overhang the land or extend onto the top of the end cap metallization; however, the solder shall not extend further onto the component body.

53.6.3 Standards for Visual Inspection

Many standards cover printed circuit board assemblies. Most major electronics manufacturers have their own internally developed workmanship standards. Several industry and military standards also exist. However, the joint industry standard IPC-A-610D, "Acceptability of Electronic Assemblies," is the standard most often referenced for criteria defining reliable solder connections. This standard was developed by the Institute for Interconnecting and Packaging Electronic Circuits (IPC) (www.ipc.org).

The standard is a collection of visual quality acceptability requirements for electronic assemblies. It covers criteria for both pin through-hole and surface-mount technology solder connections. It also reflects requirements for three different classes of end products:

- **Class 1, general electronic products** These include consumer products and some computers and computer peripherals.
- **Class 2, dedicated service electronic products** This class includes communications equipment, critical business machines, and instruments where high performance is required and for which uninterrupted service is desirable.
- **Class 3, high-performance electronic products** These products include commercial and military equipment for which continued performance or performance on demand is imperative.

53.6.4 Capabilities of Visual Inspection

Visual inspection serves a number of important purposes well, but it also has several important limitations.

53.6.4.1 Advantages. Visual inspection can be an accurate and cost-effective method of defect detection for nonsubjective defects, such as larger missing components, larger misoriented or reversed components, or solder bridges. When low volume or lack of technical resources prevents the use of automation or more sophisticated tools, visual inspection is one alternative that finds many of the more subjective defects, such as solder joints with insufficient solder, lifted leads, or poor wetting. Finally, visual inspection can quickly detect when a process step has drifted significantly out of its control limits.

53.6.4.2 Disadvantages. Visual inspection also has many limitations, including the following:

- Low coverage of solder joint defect detection repeatability, particularly for fine-pitch parts, which results in high false accepts or defect escapes and high false reject rates
- Low coverage of component defect detection repeatability for smaller components such as 0402, 0201, or 01005 passive components
- Inability to see hidden solder joints in component types such as some connectors, PGAs, BGAs, chip-scale packages (CSPs), SIPs, or multichip modules (MCM)
- Inability to collect quantitative measurements in addition to defect data

53.6.4.3 Repeatability Limitations. Several studies have documented the low repeatability rate of visual inspection of solder joints. One such study was conducted by AT&T at its Federal Systems Division.[1] This study showed that even the same inspector inspecting the same assembly twice had a defect call repeatability rate of only about 50 percent. Two different inspectors inspecting the same assembly had a defect call repeatability rate of only about 28 percent. This study did not include any very-fine-pitch SMT solder joints or 0603 passive components, which are more difficult to inspect visually.

To alleviate this severe limitation somewhat, manufacturers have implemented the use of microscopes with a magnification level of 10×. Often, stereoscopic microscopes are used to provide visual inspectors a better three-dimensional view. Not as frequently, manufacturers have implemented light sources and cameras to capture real-time magnified video images of the assembly being inspected. The lighting and high-resolution video images can make it easier for visual inspectors to see the acceptance criteria for which they are searching. These enhancements do improve the repeatability rate of visual inspection. But the requirement for subjective human calls and the tedium of carefully inspecting thousands of connections per hour still results in repeatability rates much lower than desired. Low repeatability means many missed defects that escape to final assembly or to customers, impacting reliability and wasting and possibly damaging rework of good connections.

53.6.4.4 Hidden Solder Joints. Several component types used in printed circuit assemblies do not provide visual access to solder joints. These components include packages with bump arrays where the solder joints are distributed in a matrix under the entire component body, such as ball grid arrays and chip-scale packages. For these components, all of the joints except those on the edges of the array are completely hidden from view. Other packages, such as J-leaded components, where the connections are under the component body at the component edges, and 0.5 mm-pitch gull-wing components, where the solder joint heel is only 0.08 mm high and behind the component leads, also make visual inspection more difficult. Some manufacturers have attempted to solve this problem by using a penetrating imaging technique, such as x-ray, acoustics, or thermography, to acquire a live, magnified video image that the visual inspector can look at to make a defect call. However, the images resulting from these techniques are either less consistent or less well defined than a visual image, still requiring the inspector to make a difficult judgment call. Therefore, a low repeatability rate persists.

53.6.4.5 High-Complexity Boards. Many printed circuit boards today are entirely too complex to allow accurate visual inspection. Components as small as 0402, 0201, or even 01005 passive components are simply too small for accurate visual detection of missing parts and particularly misalignment defects. Add to this the fact that many printed circuit boards have hundreds of these component types per board, and the opportunity for the visual inspector to make a mistake becomes enormous. Further obstacles to accurate inspection are "families" of boards, where the basic layout of each board type is the same, but specific board types vary by purposely not mounting specific components. In these cases, a missing component can be completely acceptable even in places where the board layout may provide for a component to be mounted.

53.6.4.6 No Quantitative Measurements. Accurate quantitative measurements of dimensions ranging from 0.05 to 0.5 mm are not possible with visual inspection. Quantitative measurements provide much more information about a process step, allowing much tighter process control and providing insight into the causes of process variation, without which process improvement is a hit-or-miss proposition. Some manufacturers have implemented semiautomatic measurement tools to allow operators to take quantitative measurements. Examples include optical-focusing microscopes and semiautomatic laser triangulation equipment. These tools do collect useful quantitative measurements, but are typically limited to sampling. Sampling is required because the tools are very slow. Measurement of solder paste height, volume, and registration with pads is possible because of easy visual access to solder paste depositions as well as their simple rectangular shape.

Automated inspection techniques using automated inspection equipment significantly overcome the repeatability and measurement limitations of visual inspection. It is for this reason that many manufacturers are implementing automated inspection.

53.7 AUTOMATED INSPECTION

The automation of inspection has evolved into systems that resemble the automated test equipment used to make electrical measurements and find electrical defects (faults).

Automated inspection systems generate images of the item to be inspected (normally solder paste, components, or solder joints), digitally analyze the image to locate and measure key features, and, based on these measurements, automatically decide whether a defect exists or not.

Just like visual inspection, automated inspection systems do not require physical contact with the printed circuit assembly to generate the desired images. Unlike visual inspection, however, automated inspection removes human subjectivity from defect detection, thereby increasing repeatability rates typically by an order of magnitude. Many of the automated inspection systems also provide accurate, repeatable, quantitative measurements that directly correspond to process parameters, thus providing the means for process control and improvement.[2]

53.7.1 Measurements by Automated Inspection Systems

Figure 53.8 shows examples of measurements made by automated inspection systems.

53.7.1.1 Solder Paste Measurements. Typical solder paste measurements, shown in Fig. 53.8(a), are volume, area of pad covered, height, and misalignment with the pad. These quantitative measurements provide information about the paste viscosity, stencil registration, cleanliness, snap-off, and squeegee speed and pressure that can lead to improvement in the paste printing process.

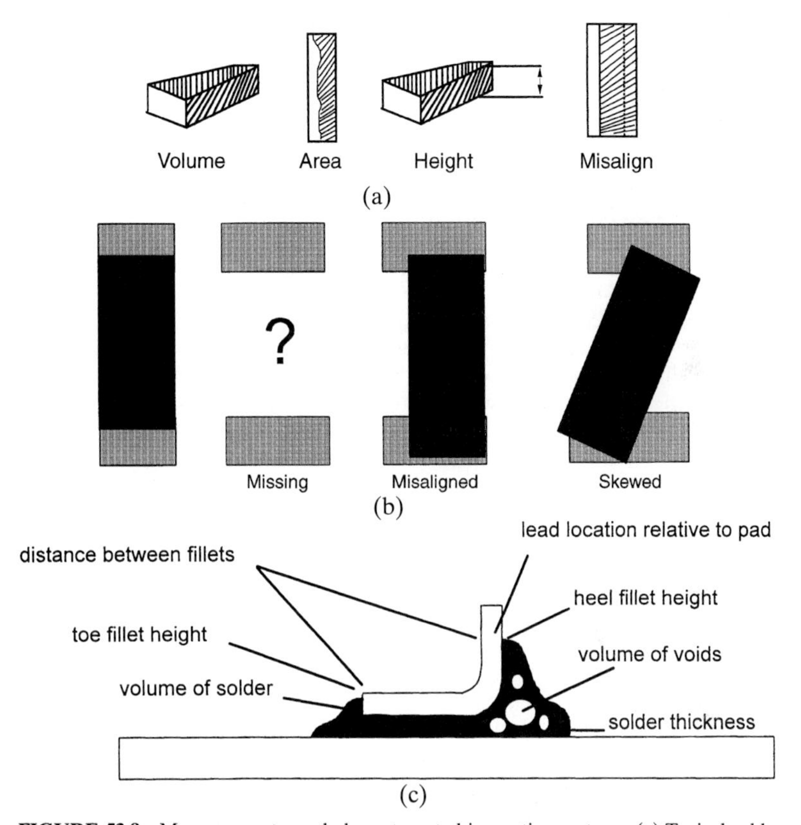

FIGURE 53.8 Measurements made by automated inspection systems: (a) Typical solder paste measurements made by automated inspection systems, including the volume of paste deposition, area of the pad covered by paste, height of the paste deposition, and offset of the paste from the pad; (b) typical component placement attributes inspected by automated inspection systems, including missing component, components misaligned along or across the pads, and skewed components; (c) typical solder joint measurements made by automated inspection systems, including volume of solder, toe and heel fillet heights, distance between fillets, void volume, average solder thickness across the solder joint, and offset between the solder joint and pad.

53.7.1.2 Pre-Reflow Measurements. Typical pre-reflow component placement measurements, depicted in Fig. 53.8(b), include whether or not the component is missing, misaligned, or skewed. These measurements are usually attribute measurements, just separating good from bad. But quantitative dimensional measurement of the amount of misalignment, just after the placement process step, is also possible. Component placement measurements provide information about the rate of placement accuracy. Optical character recognition (OCR) and optical character verification (OCV) techniques allow for additional levels of verification, including correct part.

53.7.1.3 Post-Reflow Solder Joint Measurements. Solder joint measurements, such as fillet heights, average solder thickness across the pad, void volume, and pin-to-pad offsets, as shown in Fig. 53.8(c), provide information about the paste printing process, the component placement process, and the solder reflow process steps. Attribute measurements, such as solder bridges, opens, or insufficient solder, are most common. Quantitative measurements of solder joints are also possible and provide greater insight into process parameters. For instance, the variation of heel fillet heights and void volume across an assembly provides insight into the oven temperature profile and spatial distribution. The variation of average solder thickness provides insight into squeegee speed and pressure and stencil cleanliness. And the pin-to-pad offsets measure the component placement accuracy.

53.7.2 Types of Inspection Systems

Inspection systems normally are dedicated to one type of measurement capability: solder paste, pre-reflow, or post-reflow inspection. For example, systems for solder paste measurements do not normally also make component placement measurements. The cost of combining different measurement capabilities into one system would typically make that system prohibitively expensive. More importantly, to reduce manufacturing costs, manufacturers want to implement linear, sequential production lines where an assembly always flows in one direction and goes through each machine only once per assembly side. So automated inspection systems fall into three major categories:

- Those that make three-dimensional solder paste measurements
- Those that make quantitative pre-reflow two-dimensional solder paste and/or component placement measurements
- Those that make post-reflow solder joint measurements and attribute component measurements

For all three types, the automated inspection system compares the measurements taken against a specified conformance range to accept or reject automatically a solder paste brick, component placement, or solder joint as being within specification.

53.8 THREE-DIMENSIONAL AUTOMATED SOLDER PASTE INSPECTION

53.8.1 Operating Principles

Automated inspection systems designed for 3-D solder paste measurements use structured light sources and cameras or digital light-emitting diode (LED) sensors to generate "three-dimensional images" of solder paste depositions. The structured light is normally some variation of a sheet of light or a laser line or spot. As Fig. 53.9 shows, the structured light sweeps across a solder paste brick, creating discontinuities and other features in the image that are calibrated to actual physical dimensions, such as height and area. In essence, the proportional nature of the deflection and height is turned into usable measurement results.

53.8.1.1 View Magnification. To achieve adequate accuracy, only a small portion of the assembly, called a *view*, is imaged at a single time. Magnification increases and digital quantization error decreases as the size of the view decreases. Views normally vary in size from 10 to 25 mm in diameter. So the automated inspection system may use more than 100 views to cover a typical assembly. The system moves the assembly or the image sensor on a positioning stage from view to view. The sum of move time plus image acquisition time normally makes up most of the total inspection time for a particular assembly.

53.8.1.2 System Throughput. The test speed for solder paste inspection systems varies anywhere from 2 to 22 cm^2/sec. The high end of this speed range can be fast enough to keep

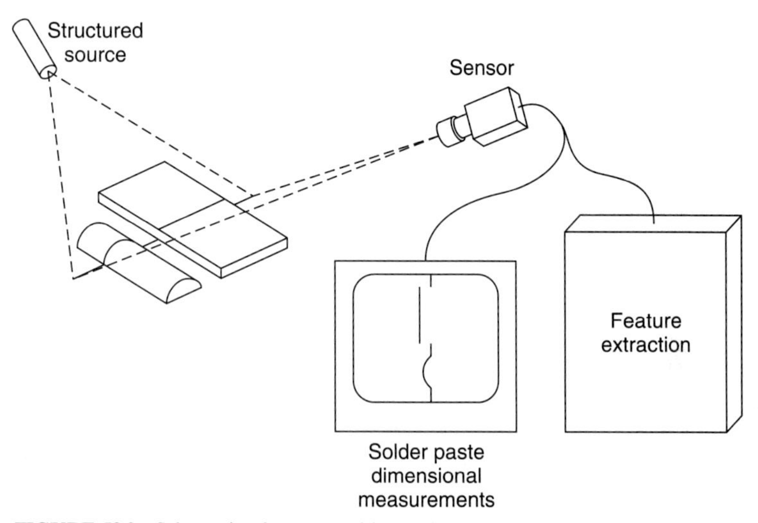

FIGURE 53.9 Schematic of automated inspection system for solder paste measurement. The camera or LED sensor obtains images with discontinuities in the laser line scan; the image-processing software finds these discontinuities, measures them, and calibrates them to real physical dimensions.

up with many automated printed circuit production lines. But for the slower systems, or even for the fastest systems on very-high-volume production lines, manufacturers inspect a specific subset of paste depositions on every assembly coming down the production line.

53.8.2 Applications

Automated solder paste inspection systems typically reside in the fastest speed ranges on the system throughput scale and therefore are utilized as process control tools in addition to defect detectors. Manufacturers typically use these systems to monitor the process by tracking the key quantitative solder paste measurements, normally height and volume, against control limits. The systems then alarm any condition where control limits are exceeded or definite drifts are detected, allowing the manufacturer to assess the production line and monitor or choose to take action until appropriate adjustments have been made. (The alarm generated by the system is normally an obvious flashing message on a computer monitor.) Because the paste printing process step causes a large percentage of all production defects, tight production control of this process step using quantitative measurements can significantly lower the process defect rate. In fact, automated solder paste inspection systems have been linked through software to paste printing systems, allowing not only process control, but also closed-loop feedback for semiautomatic adjustment of paste printing parameters based on the paste measurements.

53.8.3 Advantages and Disadvantages

Automated 3-D inspection of solder paste depositions has the following major advantages:

- Real-time process control of paste printing to lower defect rates and rework costs
- Quantitative measurements, including volume, to help permanently eliminate the causes of paste defects

- Defect detection where rework is easiest, before component placement and solder reflow
- Process characterization during the lead-free conversion with minimal program tuning

Automated inspection of solder paste depositions has the following limitations:

- It is sometimes too slow to cover all paste depositions on an assembly.
- It does not measure or detect defects in component placement or solder reflow.

53.9 PRE-REFLOW AOI

53.9.1 Operating Principles

Automated inspection systems designed for 2-D component placement and solder paste quantitative measurements typically use optical techniques consisting of multi-angle light sources and charged-couple device (CCD) cameras to generate images. As Fig. 53.10 indicates, these systems extract specific features, such as edges of the components or solder paste deposits, in the images and then use these features to determine quantitative measurements of component and deposit misalignment or deposit area.

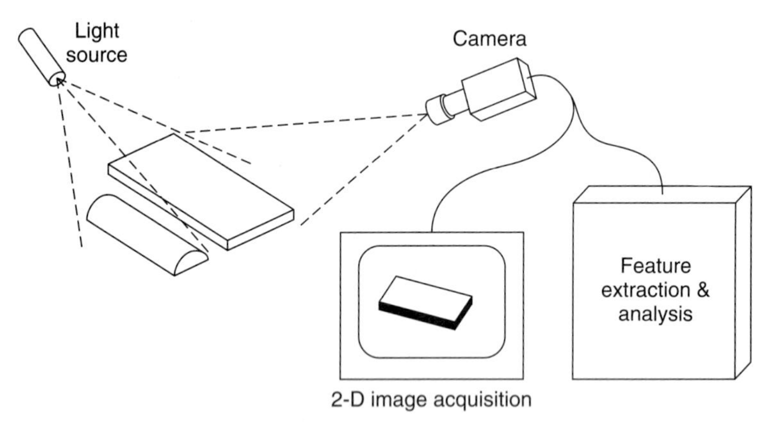

FIGURE 53.10 Schematic of automated inspection systems for component placement defect detection. The camera sensor obtains images of component positions relative to the printed circuit board. The image processing software extracts features from the image and compares them to present position limits to flag a placement as defective.

Automated optical inspection systems also image only a small portion, or *view*, of the assembly at a time. These systems normally can use somewhat bigger views than the 3-D solder paste systems because the features being extracted often do not require as much magnification as do measurements of solder paste depositions. However, inspection of components, such as 0402, 0201, and 01005 passive components, or very-fine-pitch deposits, can require the same level of magnification and therefore views as small as the 3-D solder paste inspection systems.

In general, pre-reflow AOI systems are typically two to three times faster than 3-D systems and range in inspection speed between 10 and 40cm^2/sec. The prices of 2-D component and solder paste placement automated inspection systems are typically somewhat less than those of the 3-D solder paste systems with the fastest inspection speed capability.

53.9.2 Applications

The majority of pre-reflow AOI systems are placed directly following the pick-and-place process step in the production line. These systems therefore only measure the amount of misalignment of components and are generally fast enough to cover all components on a printed circuit board and keep up with the production line's cycle time. So manufacturers often use these systems to detect all misalignment defects as well as missing components. But the most important purpose of these systems is process control, just as it is for the 3-D solder paste inspection systems. The intent is to discover process drift early enough to prevent defects by tracking the key quantitative misalignment measurements against control limits. The systems then alarm any condition where control limits are exceeded or process drifts are detected, allowing the manufacturer to assess the production line and monitor or choose to take action until appropriate adjustments have been made. Some system suppliers in conjunction with pick-and-place equipment suppliers can develop closed-loop feedback software links allowing semiautomatic adjustment of the pick-and-place equipment.

Pre-reflow AOI systems also have the ability to inspect 2-D solder paste; however, this ability is utilized only to inspect a small percentage of the solder paste deposits combined with the component misalignment measurements. Component misalignment measurements cover the passive components, whereas the solder paste measurements cover deposits for BGA, CSP, or fine-pitch QFP devices. Therefore, these systems are placed within production lines after the pick-and-place systems for passive devices but before the pick-and-place systems for the larger area-array and leaded devices. These systems serve the same purpose as those only meant for component placement measurement, both detecting defects and monitoring measurements within control limits to discover process drift as early as possible.

53.9.3 Advantages and Disadvantages

Pre-reflow AOI has the following major advantages:

- Real-time detection of systematic process defects where equipment is out of adjustment
- Detection of defects such as misalignment where rework is easier, before solder reflow
- Low-cost option for reasonable test coverage
- Process characterization during the lead-free conversion with minimal program tuning

Automated inspection of component placement has the following limitations:

- It does not measure or detect defects in solder reflow, and normally does not detect defects in solder printing.
- It does not measure 3-D for solder paste.

53.10 POST-REFLOW AUTOMATED INSPECTION

Solder joints have much more complex shapes than do solder paste depositions and components, so taking measurements of solder joints normally requires more complex imaging techniques than does measuring solder paste and components. Automated inspection systems for solder joints have used a variety of imaging technologies, including optical and x-ray imaging, thermography, cooling profiles of laser-heated solder joints, and ultrasonic imaging. But three technologies have dominated in these systems:

- Optical imaging using multiple light sources and cameras
- Transmission x-ray imaging
- Cross-sectional x-ray imaging

53.10.1 Post-Reflow AOI Systems

53.10.1.1 Operating Principles. Post-reflow AOI systems for solder joints are similar to, but more complex than, those for component placement. (Note that these systems also detect component defects such as missing, misaligned, or misoriented components, but their complex design is required for solder joint inspection.) Multicolor, multi-angled light sources are normally required to provide enough information in the images to allow sophisticated image-processing algorithms to detect the required features in the images accurately. If the light source is not multicolored or multi-angled, then many different cameras arc usually used instead, each camera mounted at a different angle relative to the board being inspected.

The multiple light sources and cameras create and detect shadows from various angles to detect features of solder joints of all types oriented in different directions on the printed circuit assembly. These optical systems often use higher magnification, particularly for the smallest passive components and fine-pitch components, and therefore capture a smaller portion of the assembly at any one time. The smaller views allow more image pixels per solder joint feature, allowing more accurate image processing and corresponding defect calls. The use of smaller views renders the throughput of these systems slower than that of component placement inspection systems. However, in general these systems are faster than 3-D solder paste measurement systems. Their throughput typically falls in the range of 10 to 40 cm^2/sec.

The systems' image-processing software uses sophisticated algorithms to extract specific features of solder joints, such as edges and areas of the solder joint in a specific range of angles relative to the board. Analysis of these extracted features then determines whether a defect exists or not. Defects detected include absence of solder, bridges, and grossly insufficient or excessive solder. These systems typically do not make quantitative measurements.

53.10.1.2 Application. These post-reflow AOI systems generate only attribute data. For instance, these systems detect the existence of a solder bridge between two joints or the absence or presence of a toe fillet on solder joints. But they do not measure the height of the heel fillet or the amount of solder in the solder joint. These systems typically do not measure how far a component is misaligned from its proper placement, but instead simply determine whether or not the component is misaligned more than a predetermined amount. These attribute data are not as useful for process control, and thus manufacturers use these systems strictly to detect defects. Normally these systems do, however, warn of a condition where the same defect has occurred several times consecutively or within a specific number of assemblies, indicating that some part of the process needs adjustment.

Post-reflow AOI systems cannot inspect hidden solder joints, such as those for ball grid arrays, pin grid arrays, and in some cases J-leaded devices, and can have high false accept or reject rates for some component types, such as fine-pitch components at or below 0.5 mm pitch or small outline transistor (SOT) devices. If tall components are placed on the board very close to smaller components, post-reflow AOI systems can also have difficulty inspecting these smaller components.

53.10.1.3 Advantages and Disadvantages. Post-reflow AOI systems of solder joints have the following advantages:

- They eliminate visual inspection by automating solder joint defect detection, thereby also reducing unnecessary rework due to false reject calls.
- They reduce rework analysis time by pinpointing defects to the exact solder joint.
- They afford real-time process control of all three process steps—paste printing, component placement, and solder reflow—to lower defect rates and rework costs.
- They can be used during the lead-free conversion with minimal program tuning.
- They are a low-cost option for reasonable test coverage.

Post-reflow AOI systems have the following limitations:

- Test throughput is not always fast enough to allow inspection of all solder joints within the manufacturing cycle time for the printed circuit assembly.
- A significant learning curve is required to become expert at developing solder joint tests with both low false accept and false reject rates.
- The ability to inspect hidden (or invisible) joints is not possible with post-reflow AOI.

53.10.2 Transmission X-Ray Systems

53.10.2.1 Operating Principles. Transmission x-ray systems radiate x-rays from a point source perpendicularly through the printed circuit assembly being inspected, as depicted in Fig. 53.11. An x-ray detector picks up a varying amount of x-rays depending on the thickness of metals that the x-rays are penetrating and converts the x-rays to light photons for a camera to create a grayscale image. The x-ray source is filtered so that metals of only a certain density range—lead, tin, gold, and silver—will absorb the x-rays. The copper leads and frames of components sitting on top of solder joints do not absorb the x-rays and are therefore practically invisible to the x-ray detector. Thus, x-ray systems can easily see the entire solder joint, no matter what component material may be on top of the joint blocking its optical or visual access. In other words, x-ray is the only automated inspection method equipped to inspect hidden joints such as those found in a BGA or CSP. The resulting x-ray image will be darker wherever the lead or tin solder or lead-free solder is thicker in the solder joint. The image processing capability of the system then searches for features, such as the heel and toe fillets, the sides of the solder joint, and even voids internal to the joint based on grayscale readings of the solder joint x-ray image. The system then uses predetermined decision rules to compare

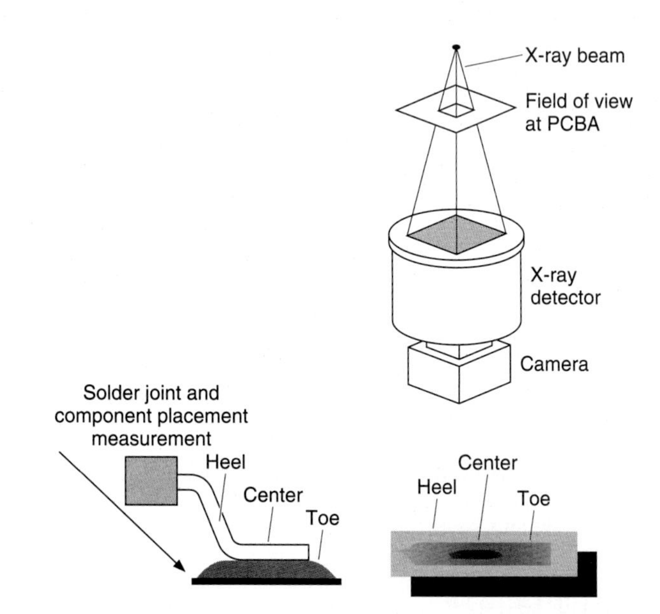

FIGURE 53.11 Schematic of transmission x-ray automated inspection system for solder joint defect detection. The x-ray detector converts a varying amount of x-rays to light, based on how much various parts of the solder joint absorb. The camera converts the light photons to an image, which is then processed to find solder joint features and flag defects accordingly.

the grayscale readings to acceptance criteria to accept or reject a solder joint automatically. For example, the system might compare the relative grayscale reading for the heel fillet region, the center of the solder joint, and the toe fillet region. The acceptance criteria might state that the heel fillet reading should be twice that of the center and that the toe fillet reading should be 50 percent higher than that of the center. If the actual readings do not meet these criteria, then the solder joint is reported as being defective.

The bottom of Fig. 53.11 shows an x-ray image of a gull-wing solder joint that shows the center of the joint as much darker than the heel fillet region. This solder joint is clearly defective, as the heel fillet region should always be darker and have a higher grayscale reading than the center of the joint, where the solder is thinnest for mechanically good solder joints. (The system's image processing capability is able to detect much more subtle changes in grayscales than can the human eye, allowing very accurate relative readings from one solder joint to the next.)

53.10.2.2 Application. Transmission x-ray technology works well for single-sided surface-mount assemblies. These automated inspection systems accurately detect solder joint defects such as opens, insufficient solder, excess solder, bridges, misalignment between pin and pad, and voids for most surface-mount solder joint types, including J-leads, gull-wings, passive chips, and small-outline transistors in hidden and nonhidden joints alike. These systems also detect missing components and reversed tantalum capacitors. Based on trends in grayscale reading, these systems also can accurately detect process drifts through real-time process control charting.

For double-sided assemblies, however, the transmission x-ray images of solder joints on the top side will overlap with the images of solder joints on the bottom side. The x-rays are absorbed by any solder in their path through the printed circuit assembly from the source to the detector. These overlapping images make accurate solder joint measurement impossible.

Transmission x-ray imaging also cannot easily distinguish between the top, bottom, and barrel of PTH solder joints or the bottom of BGA solder joints. So transmission x-ray systems cannot be used for accurate measurement of and defect detection in solder joints on double-sided assemblies or for PTH and BGA solder joints.

53.10.3 Cross-Sectional X-Ray Systems

53.10.3.1 Operating Principles. Cross-sectional x-ray systems radiate x-rays at an acute angle from vertical through the printed circuit assembly being inspected. As Fig. 53.12 indicates, images from all around the particular view being inspected are added together or integrated essentially to create an x-ray focal plane in space. This focal plane creates a cross-sectional image, approximately 0.2 to 0.4 mm thick, right at the focal plane, by blurring everything above and below the focal plane into the background, or *noise*, of the image. By moving the top side of an assembly into the focal plane, cross-sectional images of only the solder joints on the top side are created. By moving the bottom side of an assembly into the focal plane, cross-sectional images of only the solder joints on the bottom side are created. Separate images of top and bottom sides are always created, preventing any image overlap from the two sides.

53.10.3.2 Application. Cross-sectional x-ray automated inspection systems work well for all types of printed circuit assemblies, including single-sided and double-sided, surface-mount, through-hole, and mixed-technology assemblies. These systems accurately detect the same solder joint and component defects as do transmission x-ray systems, but, in addition, the cross-sectional x-ray systems accurately detect insufficient solder conditions for BGA and pin-through-hole solder joints.

Some cross-sectional x-ray automated inspection systems go beyond just grayscale readings of specific solder joint features. By carefully calibrating grayscale readings to actual solder thickness, it is possible to generate real-world measurements, in physical units rather than grayscale numbers, of fillet heights, solder and void volume, and average solder thickness for

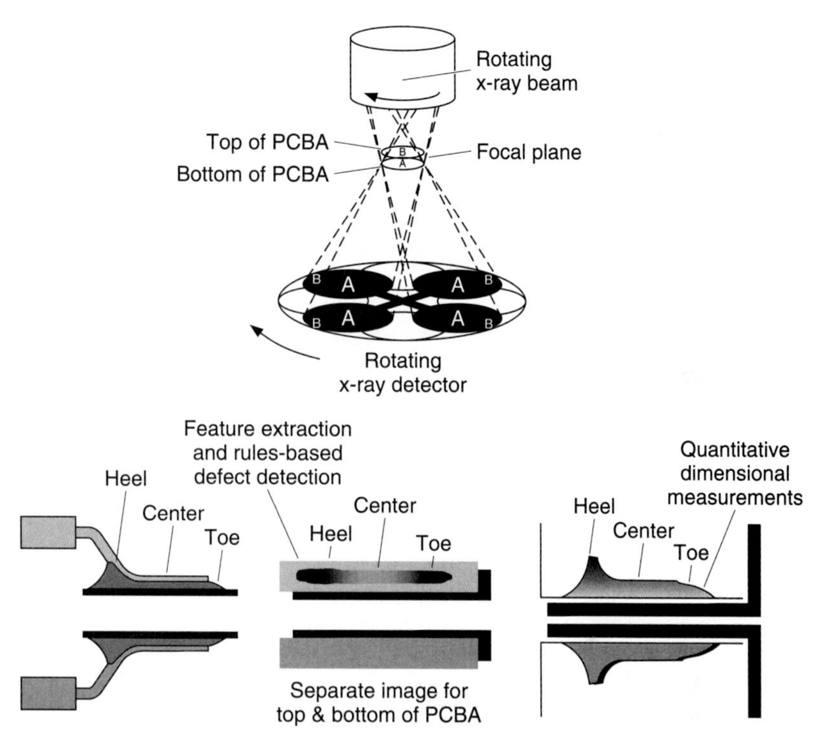

FIGURE 53.12 Schematic of cross-sectional x-ray automated inspection system for solder joint measurement. Adding images around a circle from a rotating x-ray beam and detector creates a focal plane that captures just the solder joints of interest and nothing else below or above.

the entire joint. Figure 53.13 shows an example of these calibrated measurements. This figure includes the actual cross-sectional x-ray image of tape automated bonded (TAB) solder joints.

The profile shown at the top of the x-ray image is generated by the system in physical dimensional units by interpreting and calibrating the grayscale readings of pin 193 in the x-ray image.

The table below the x-ray image includes example measurements for both pin 193 and pin 194.

Analysis of these physical thickness measurements of solder joints provides the information required for process characterization and improvement. For instance, variations in average solder thickness or volume for the solder joints across a single assembly or from assembly to assembly provide insight into the quality level of the paste printing process as well as sources of defects.

The image-processing software then finds and measures solder joint features and flags defects accordingly.

53.10.4 Advantages and Disadvantages of X-Ray Inspection

X-ray solder joint inspection systems can reach average inspection speeds of around 50 to 150 joints per second. X-ray solder joint inspection systems also have higher prices, typically about 50 to 100 percent more than the price of the optical solder joint systems with the fastest inspection speed capability.

Tape-automated bond (TAB)

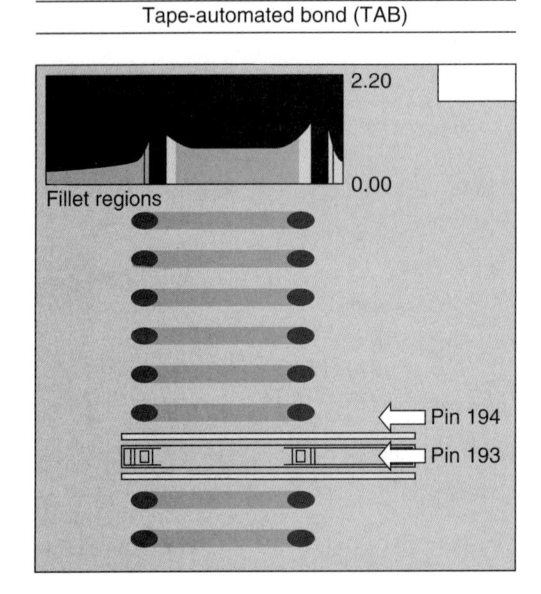

Good board (pin 6)

Reference designator	Inspection point	Thickness (in 0.001")
U1 pin 193	Pad	0.59
	Heel	1.18
	Center	0.69
	Toe	1.34
U1 pin 194	Pad	0.58
	Heel	1.20
	Center	0.68
	Toe	1.30

FIGURE 53.13 Cross-sectional x-ray image of TAB solder joints. Image-processing software converts the grayscale readings of the image for pin 193 into the side profile of solder thickness shown above the image. The actual calibrated measurements of average solder thickness across the pad, heel fillet height, center thickness, and toe fillet height processed from the images of pins 193 and 194 are shown in the table below the x-ray image. These measurements indicate that both of these solder joints are good.

Automated x-ray inspection of solder joints has the following major advantages:

- Its defect detection capability is extremely high.
- It eliminates visual inspection by automating solder joint defect detection, thereby also reducing unnecessary rework due to false reject calls.
- It reduces rework analysis time by pinpointing defects to the exact solder joint.

- It affords real-time process control of all three process steps—paste printing, component placement, and solder reflow—to lower defect rates and rework costs.
- It provides quantitative measurements to help permanently eliminate the causes of defects from all three process steps.
- It reduces failures at final assembly and in the field due to defective hidden solder joints and marginal solder joints due to insufficient solder or misalignment or excessive voids.
- During lead-free conversion, it can be used with minimal program tuning.

Automated inspection of solder joints has the following limitations:

- Test throughput is not always fast enough to inspect all solder joints within the manufacturing cycle time for the printed circuit assembly.
- A significant learning curve is required to become expert at developing solder joint tests with both low false accept and false reject rates.

53.11 IMPLEMENTATION OF INSPECTION SYSTEMS

Successful implementation of automated inspection systems into printed circuit assembly production lines requires a significant investment in training, process analysis, and system integration.

The implementation can be a lengthy process that requires concerted effort by engineers or skilled technicians. Listed here are highlights of what several manufacturers have learned are key aspects of successfully implementing automated inspection systems:

- **Assess requirements carefully** Start by carefully assessing the requirements for automated inspection in the particular production environment into which the system will be integrated. Determine exactly what kinds of defects are most important for the inspection system to detect, which measurements will most help with process improvement, and what benefits will generate the quickest financial return on investment.[3] This assessment must consider the testing and measurement capability that already has been implemented as well as new requirements arising from future printed circuit assembly designs.
- **Evaluate a select set of systems thoroughly** Select a small number of automated inspection systems to evaluate thoroughly and compare them against the system requirements. The evaluation should include a benchmark using printed circuit assemblies from production to determine the system's capabilities to detect accurately the important defect types within the required false reject rate, repeatedly make the required measurements, and not exceed the required test time. Elements of cost of ownership should be well understood, including test development time, maintenance skills and cost, expected system downtime, supplier support infrastructure, and supplier maintenance services and prices.
- **Plan for factory system interfaces** Consider and plan carefully for interfaces to other factory systems. These systems include board-handling equipment, bar code–reading systems, computer-aided design (CAD) systems for automatic download of board layout and component package information, and quality data management systems for SPC and historical quality tracking.
- **Focus on SPC measurements** Start with a focus on SPC measurements instead of defect detection. Until the process variation is reduced, most manufacturers will encounter either a false reject rate or a false accept rate that is higher than desired. Allowing one rate or the other to be too high while focusing on reducing the process variation first avoids time-consuming, unproductive tweaking of acceptance thresholds. Reducing process variation requires correlating measurements to the process parameters causing the variation and defects, and then properly adjusting these process parameters.[4]

- **Define defects carefully** With an understanding of the selected system's capability, carefully define the defects that must be detected for product quality and reliability. Many of the visual inspection criteria used in the past are not appropriate for automated inspection systems because the system takes objective and different measurements.
- **Invest enough resources** Do not underestimate the initial resource investment required to obtain optimum benefit from an automated inspection system. The implementation plan should include dedicated technical support for the first six months of operation and test development. Developing a thorough understanding of the measurement results and correlating the data to process parameters are key to successful use of the system. Implementation should also address the fact that production personnel will have to be convinced of the accuracy of the system's test results before full benefit can be obtained from the system.

53.12 DESIGN IMPLICATIONS OF INSPECTION SYSTEMS

Automated inspection systems in general do not require many changes or limitations in the design of printed circuit assemblies. Because these systems use noncontact measurement techniques, fixturing requirements present very few design limitations, for instance. However, the following requirements will facilitate automated inspection if they are considered during printed circuit design.

53.12.1 Automated Board Handling Requirements

- Parallel edges of the assembly or panel must have adequate clearance (typically at least 3 mm), to allow board-handling clamps or belts to grab the assembly.
- Three corners of the assembly or panel must include alignment fiducials (or solder joints in post-reflow configurations).
- Bar code identification of assembly number and serial number must be present at a predefined location on each printed circuit assembly.
- The board must be adequately rigid without a fixture to prevent excessive board vibration during movement.

Panels with prerouted breakaway boards or bare boards less than 30 mil thick present the biggest challenge: Component, heat sink, or daughter-board height above or below the bare board must not exceed the height clearance of the targeted automated inspection system.

53.12.2 Test Development Ease of Use Requirements

- There should be as few suppliers as possible (ideally one) for each component type. Variation in lead and component package dimensions from supplier to supplier for the same component forces longer and more difficult development of inspection programs for each printed circuit assembly type.
- Pad shapes and sizes must be uniform, particularly for each component package type. Variation in pad size and shape within a component package type forces longer and more difficult development of inspection programs for each printed circuit assembly type.
- Solder joints for optical and structured light automated inspection systems must be clearly visible.

- For transmission x-ray automated inspection systems, no components should be opposite or under dense structures, such as transformers, large capacitors, or thick steel heat sinks.
- There should be no silk-screened outlines around components. Although such outlines may be useful for visual inspection, they just confuse automated optical inspection systems.

REFERENCES

1. Donnel, A. J., et al., "Visual Soldering Inspection Inconsistencies—Interpretation of MIL-SPEC Visual Acceptance Criteria," AT&T Bell Laboratories, 1988.
2. Lancaster, Michael, "Six Sigma in Contract Manufacturing," *Proceedings of Surface Mount International Conference,* San Jose, CA, 1991.
3. Baird, Dennis L., "Using 3D X-Ray Inspection for Process Improvements," *Proceedings of Nepcon West Conference,* Hughes Aircraft Company, Cahners Publishing, 1993.
4. Sack, Thilo, "Implementation Strategy for an Automated X-Ray Inspection Machine," *Proceedings of Nepcon West Conference,* IBM Corporation, Cahners, 1991.

CHAPTER 54
DESIGN FOR TESTING

Kenneth P. Parker
Agilent Technologies, Loveland, Colorado

54.1 INTRODUCTION

In the latter part of the 1970s, it became clear that forces of technology were causing an evolution in board complexity that was quickly outstripping testing technology. It is quite possible to design boards that essentially cannot be tested in an economical sense. This will doom projects and products to failure.

The author has personal experience here, having consulted in the mid-1970s with a Silicon Valley start-up company that had created one of the first dedicated word processors. This product was based on a customized processor that had been designed and sent into production with absolutely no thought about how it was to be produced in volume. There was no test strategy. When it was determined that volume production was impossible, the design team was brought onto the production floor in a last-ditch effort to produce shippable units. High-priced design engineers could only produce one or two units a day, and they were no longer available for developing the next product. The company soon failed, a casualty of untestable design. The marketplace never got a chance to determine whether their product idea was a winner, because their design process was a loser!

In the 1970s and into the 1980s, it was very common for a design department to be physically and organizationally removed from the test department. The nature of product life cycles dictates that by the time a test department starts ramping up test program development for a board, the design team is off on the next project and finds it distracting to go back and help the test engineers with testing problems. Thus, in those years, the designers were unknowing contributors to difficult test problems. But what is different today, with the emphasis on outsourcing and contract manufacturing? Now, a design may be outsourced to one contractor, board layout to a second, test development to a third, and actual production to a fourth, all scattered about the globe.

Even if boards can be tested, the question bears asking, "Is there something a designer can do that will make testing easier, cheaper, more thorough, etc.?" The answer is that there is a great deal a designer can do (or fail to do) that will affect the testability of a board. The technology that addresses this problem is called design for testability (DFT).

By the middle of the 1980s, testing became a bottleneck in product development. The situation ultimately became severe enough that attention was paid to the effects that design has on testing. A landmark survey on DFT technology by Williams and Parker[1] brought DFT out of common lore and into the design lexicon of the electronics industry. That paper is still remarkably current* nearly 20 years later. In it were coined the terms *ad hoc testability* and

* This survey does not cover the topic of standards-based testing (Sec. 56.5), because the testability standardization work promulgated by the IEEE began in the later 1980s.

structured testability. Before jumping into the DFT discussion, however, some definitions are crucial.

54.2 DEFINITIONS

These definitions are fully discussed in Chap. 55 and are repeated here in an abbreviated form for reference.

1. A *defect* is an unacceptable deviation from a norm.
2. A *fault* is a physical manifestation of a defect.
3. A *fault syndrome* is a collection of measured deviations from an expected good outcome.
4. A fault is *detected* when an operation with an expected outcome is conducted and this outcome is not observed.
5. A fault is *isolated* when an operation with an expected good outcome and one or more failing syndromes is conducted and the outcome matches a member of the set of failing syndromes.
6. A *test* is one or more experiments that are specifically constructed to detect (and possibly isolate) failures.
   - A *detection test* has an expected good outcome.
   - An *isolation test* has an enumeration of possible fault syndromes indexed to specific failures.

The technology of testing, as covered in Chap. 55, is highly influenced either positively or negatively by the design of boards being tested. If the preceding definitions are not clear, then the discussion of loaded board testing should be digested first.

54.3 AD HOC DESIGN FOR TESTABILITY

Ad hoc design for testability consists of a set of simple design rules of the form "Do this, don't do that," where "this" and "that" are often not motivated with reasons. For example, when designing a board with ICs that have preset or clear pins, a rule might read as follows:

> "Tie unused preset and clear pins off through a 100-Ω resistor to a power rail; do not tie them directly."

The first-level reason for this is that a test engineer might want to access the preset/clear functions during testing, even though the designer did not use these functions. If these pins are tied through a resistor, a test engineer may still be able to manipulate them by applying a tester resource to them that can drive a signal in spite of the resistor. If these pins are tied directly to a power rail, the test engineer will never have that option. What might the difference be? Well, in a deeply sequential circuit, controlling the preset and clear functions might make the difference between a test that runs in milliseconds versus hours. Clearly, hours of testing (per board) are impractical, so the bottom line may be the difference between a thorough test and one that lacks significant fault coverage, affecting quality.

The real reason for the various ad hoc DFT rules is that, to effectively and economically test a circuit, one must be able to control and observe the circuit's behavior. Most rules are related to controllability and/or observability of the circuit. The rule just cited is a controllability rule. Observability rules typically suggest ways one might be able to monitor signals that are deeply embedded in combinatorial circuitry, or that are activated only rarely by complex sequential events.

Ad hoc DFT is essentially the only way many products can have improved testability when those products are constructed with off-the-shelf merchant parts. Large, vertically integrated companies have the advantage of being able to customize testability into the heart of a design, including the very ICs themselves. Application-specific ICs (ASICs) allow more of this as well.

54.3.1 Physical Access

Testing is performed once the board to be tested is connected by some adaptor mechanism to the test system. This may be accomplished via the edge connector(s) of the board, where the tester is given the same access that the board gets in its end application. But far more common is "bed-of-nails" in-circuit test access, where the board to be tested is physically mounted on a platen and depressed into a field of precisely positioned spring-loaded probes ("nails") that contact hundreds or perhaps thousands of internal board nodes. (See Sec. 55.4.2 in Chap. 55.) This can be a challenging mechanical proposition to implement, particularly when high-volume, reliable manufacturing is the goal.

Board designers must consider physical attributes of their boards early in the design process. They have a size target and then often find that there are density issues that may require fine-line geometries and two-sided component mounting to solve. The fact that in-circuit bed-of-nails access may be needed for testing should also be considered very early. See Ref. 2 for an excellent discussion of how test target pads need to be provided, and how artifacts of the board layout (particularly vias) may be used to satisfy some of these needs. However, due to the density revolution, full nodal access, which has been the holy grail of in-circuit testing, is often impractical. This leads to the question of what access is most important when full access is impossible. The answer to this question comes from the domain of circuit design.

54.3.2 Logical Access

Sometimes it is impractical to gain physical access to all nodes of a circuit. For example, Fig. 54.1(*a*) shows an IC containing a large amount of complex logic, much of it deeply buried and effectively inaccessible from the I/O pins. Figure 54.1(*b*) shows the same IC with two additional gates added. The first is an exclusive OR gate that collects three buried signals,

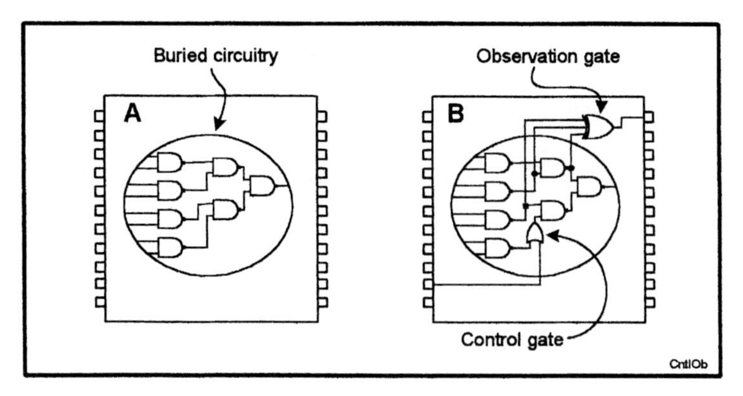

FIGURE 54.1 (*a*) An integrated circuit with deeply buried logic. (*b*) Same IC with controllability and observability logic added.

adds them together,* and brings the result to a spare I/O pin. This makes the states of those three signals much more observable. The second gate is an OR gate that allows us to insert a 1 from a spare input pin† into a deeply buried portion of the circuit. The assumption here is that the modified signal rarely reaches the 1 state in normal operation, so this extra controllability enhances our ability to manipulate the circuit for test purposes. Adding these extra gates can make it much easier to control and observe the performance of the IC. However, the downside is that we consumed a small amount of silicon area for the two gates, and, perhaps more importantly, we utilized two additional I/O pins. For many ICs, the extra cost of the pins could be a negative factor.

The choice of the optimal places to insert controllability and observability is not always clear, but rather is a trade-off of extra circuitry and pins vs. the difficulties that may be experienced in creating tests. There may be additional concerns as well with respect to circuit performance. For example, adding the controlling OR gate in Fig. 54.1 could add some signal propagation delay into the affected pathway, with detrimental results on system performance. When ad hoc circuit modifications are made for the sake of testability, it is likely that "seat-of-the-pants" decisions will be made rather than critical analysis. One person's decisions might be remarkably different than another's, with resulting great variability in testability improvements.

54.4 STRUCTURED DESIGN FOR TESTABILITY

Structured DFT was born inside companies that had vertical control over their designs, from custom ICs through systems. They also were well aware of their testing costs and realized that initial design decisions had a large impact on these downstream costs.

These companies studied the controllability and observability problems and instituted design rules into their design processes that, when followed, would guarantee that a circuit was testable. In the test department, where they also had complete control, they could utilize these added features with customized test development processes, gaining greatly enhanced levels of automation.

One of the earliest and most prominent structured DFT schemes was IBM's Level Sensitive Scan Design (LSSD), which was developed in the 1970s.[1] It is the precursor to what is called full internal scan technology now. In (greatly simplified) summary, LSSD design discipline requires every memory element (flip-flop or latch) to be constructed such that it obeys a testability protocol. This protocol allows two modes of operation: first, the normal operation of being a memory element in a design; and second, an operation for testing purposes, in which all memory elements can be connected into a serial shift register that can be loaded and unloaded by serial shifting. This makes every memory element a control point and an observation point within a circuit. No other memory elements are allowed in the design, e.g., no asynchronous feedback is allowed. This guarantees that circuitry between any control/observation point is combinatorial, not sequential (Fig. 54.2).

As might be imagined, these rules were looked upon by designers as restrictions of their creativity. Ed Eichelberger of IBM, a major proponent of LSSD, happens to stand about 5.5 feet tall. When asked in 1977 how designers received the LSSD rules, he quipped that, at the start, he was over 6 feet tall. Structured testability is not easy to implement. It requires commitment from the whole organization, starting with management.

The next piece of the puzzle was IBM's test generation software, which was able to automatically construct complete tests for combinatorial circuits (known as the D-algorithm and

* The addition is modulo-2 yielding a single bit. Any single-bit error delivered to the three inputs of the exclusive OR will cause the output to change from its expected state.

† This spare input pin should be held to a 0 value when the IC is performing its normal function. A pull-down resistor to ground could assure this, whereas during test, a tester signal could assert a 1 value when needed for testing.

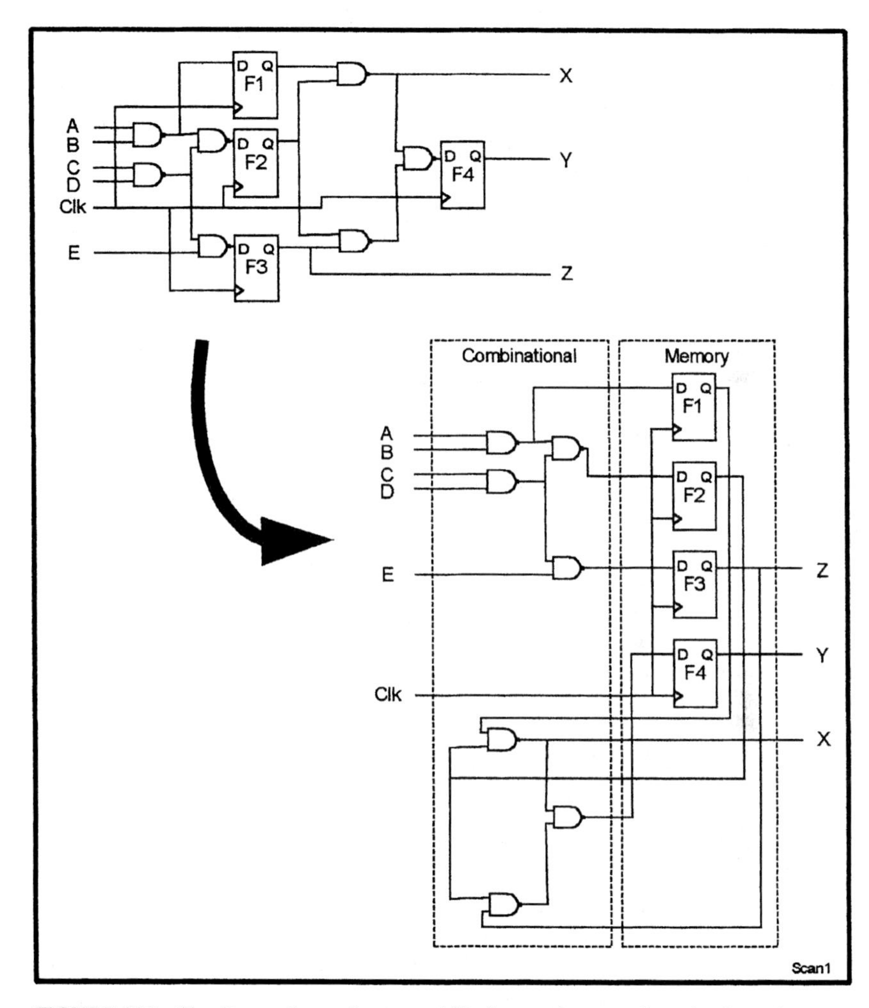

FIGURE 54.2 Circuitry made up of gates and flip-flops, redrawn to show that it can be represented as a bank of memory elements and a combinatorial circuit.

its derivatives). These tests could be shifted into the circuit and applied, and the results shifted out. This takes a lot of bookkeeping, but computers are good at that.

By using the LSSD discipline, IBM could verify that its designs would be completely testable, and those tests could be created by a computer program. Other companies such as Sperry Univac, Amdahl, Hitachi, etc., had similar proprietary structured approaches. Most smaller, nonintegrated companies were not able to participate in structured DFT—that is, until nonproprietary industry standards came into play.

54.5 STANDARDS-BASED TESTING

In the closing years of the 1980s, it became apparent that some sort of structured testability technology had to become accessible to the electronics industry at large. A small group of

European companies led by Philips formed the Joint European Test Action Group (JETAG) and began work on a testability standard. The effort quickly attracted the notice of North American companies, giving rise to the JTAG standard. As the proposal took shape, it was turned over to the IEEE, which ultimately produced IEEE Standard 1149.1-1990, Standard Test Access Port and Boundary-Scan Architecture,[3] in 1990.* Soon after 1149.1 came into being, a companion effort created IEEE Std 1149.4-1999, Standard for a Mixed-Signal Test Bus.[4] A complete coverage of these standards is beyond the scope of this chapter (see Ref. 5), so a brief summary is given here for an overview.

54.5.1 IEEE 1149.1, Boundary-Scan for Digital Circuits

The IEEE 1149.1 standard is a design discipline for digital ICs. It is a set of rules impressed primarily on the I/O structures of a device that allow two modes of operation, normal mode and test mode. In normal mode, the device performs its intended function. In test mode, the device obeys a protocol that has mandatory, optional, and customizable elements. The mandatory elements must exist, with the others being left as design options. The principle mandatory element of interest is a test mode dedicated to external test or EXTEST. When an 1149.1-compliant device is in EXTEST mode, its I/O pins are divorced from their normal operation and all internal functions of the device. Instead, the inputs become observation resources and the outputs become control resources† for test purposes. These resources are under control of the 1149.1 serial scan protocol. One can think of the I/O pins of the device being connected to shift register cells; states can be shifted in that will finally appear on all output pins (control) and the states of all input pins can be captured and shifted out (observe). This gives 1149.1-cognizant software a powerful tool for controlling and/or observing board-level node states. Figure 54.3 shows a simplified overview of the architecture.

Boundary register cells interposed between the IC pins and the internal logic surround the normal content of the IC called the mission logic. A small state machine called the test access port (TAP) is used to control the test functions. Four mandatory test pins (test clock [TCK], test mode select [TMS], test data in [TDI], and test data out [TDO])‡ give standardized access to the test functions. All 1149.1 devices have a 1-bit BYPASS register used to bypass the (much longer) boundary register if it is not needed in a given testing activity. Figure 54.3 also shows an optional IDCODE register that can be shifted out to uniquely identify the IC, its manufacturer, and its revision.

It is intended that collections of ICs (called chains, as shown in Fig. 54.4) with 1149.1 be connected TDO-to-TDI so that they may form a long, shiftable register structure. The primary use for the 1149.1 EXTEST capability is to conduct board-level tests for shorts and opens. This is an example of how resources included in an IC design may be used to help with the testing problem at other levels in the manufacturing process.

Briefly, interconnections between ICs are tested as follows. Consider the circuitry in Fig. 54.5. Some circuit nodes (also known as *nets* or *traces*) are accessible with a bed of nails and some are not. (To avoid clutter, the TCK and TMS signals are not shown.) Boundary scan can be used to test all the nodes; those with nails can be tested by coordinating nails with boundary scan resources and those without nails are tested solely with boundary scan.

---

* An IEEE standard has as a suffix the year of its creation or last update. A standard must be updated and/or reaffirmed every 5 years. Up to two supplements to a standard may be issued within the 5-year cycle. Users of a standard should keep up to date with it.

† Bidirectional signal pins can both observe and control the nodes to which they are connected.

‡ An optional fifth pin called Test Reset (TRST*) is an asynchronous active low reset for the 1149.1 circuitry. Because any TAP can be reset by five clock pulses to TCK while TMS is held high, TRST* is not actually needed for resetting an 1149.1 device. It is often included as a fail-safe measure with a board-level pull-down resistor providing a constant reset to the TAP. Many 1149.1-compliant ICs do not include the TRST* pin because the extra pin required may be too costly.

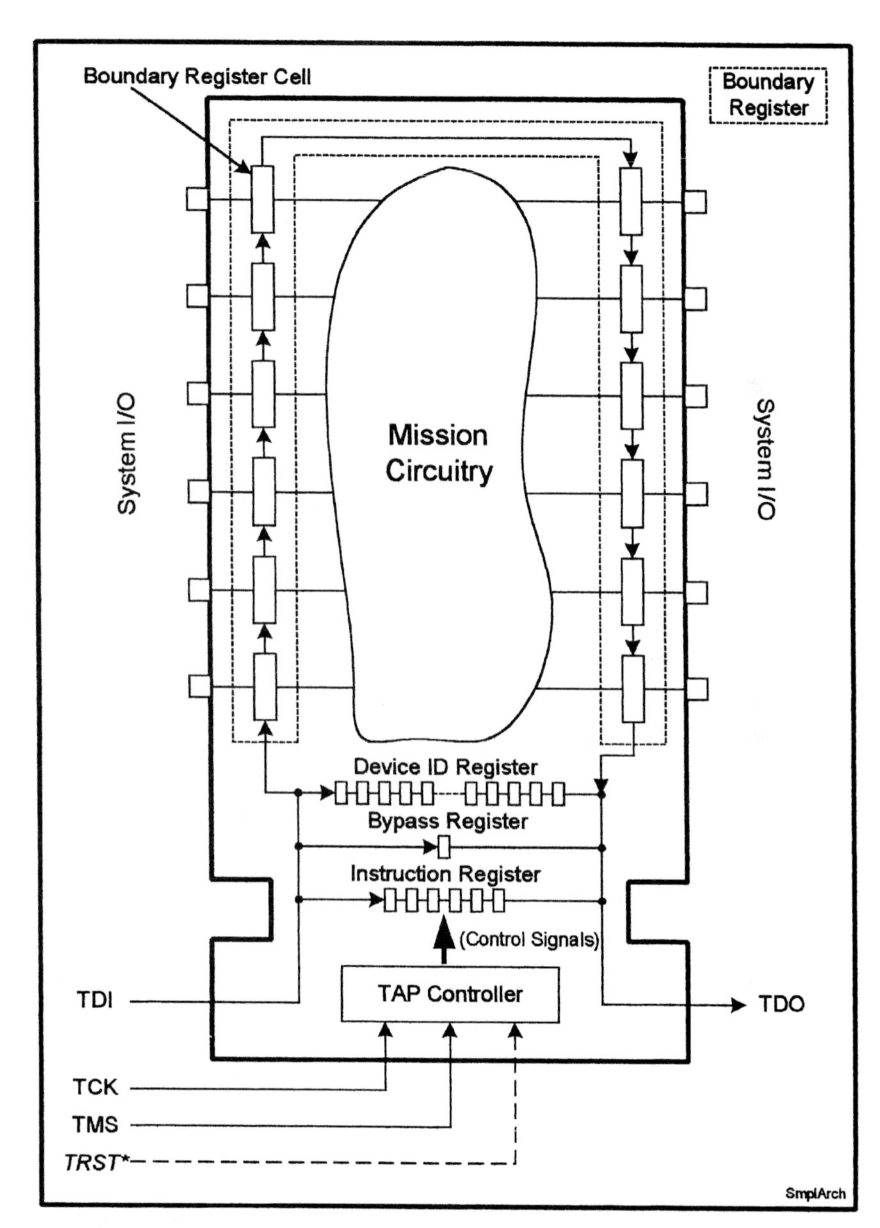

FIGURE 54.3 General architecture of an 1149.1-compliant integrated circuit.

As an example of this, boundary scan can be used to supply test patterns to nodes through a serialization process that a human would find laborious, but that is simple for a computer. When a set of patterns has been delivered (to control) and monitored (to observe) by the appropriate boundary register cells, we have uniquely identified each node with a "signature." A defect such as a short will cause two nodes to have deviant signatures, as shown in Fig. 54.6. Software can correlate observed deviations with the known boundary scan structure of each IC and the board netlist to yield a diagnostic message.

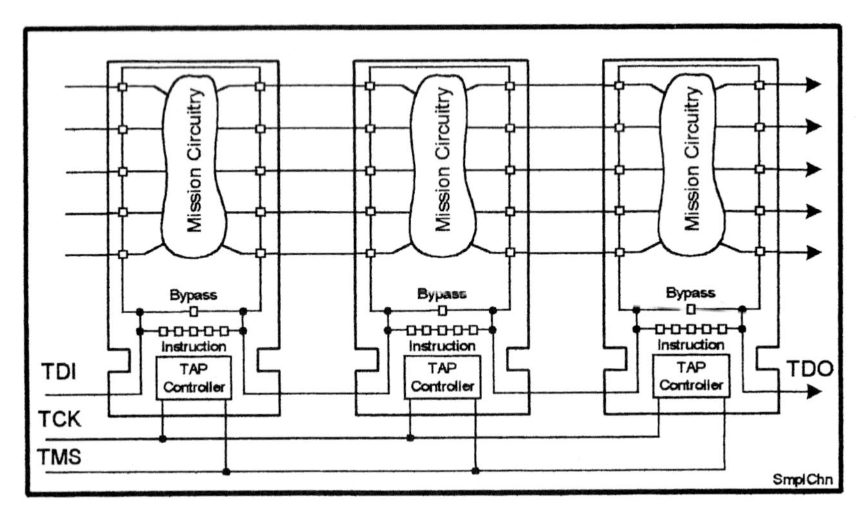

FIGURE 54.4 A collection (chain) of boundary scan devices can be used to test interconnections between ICs.

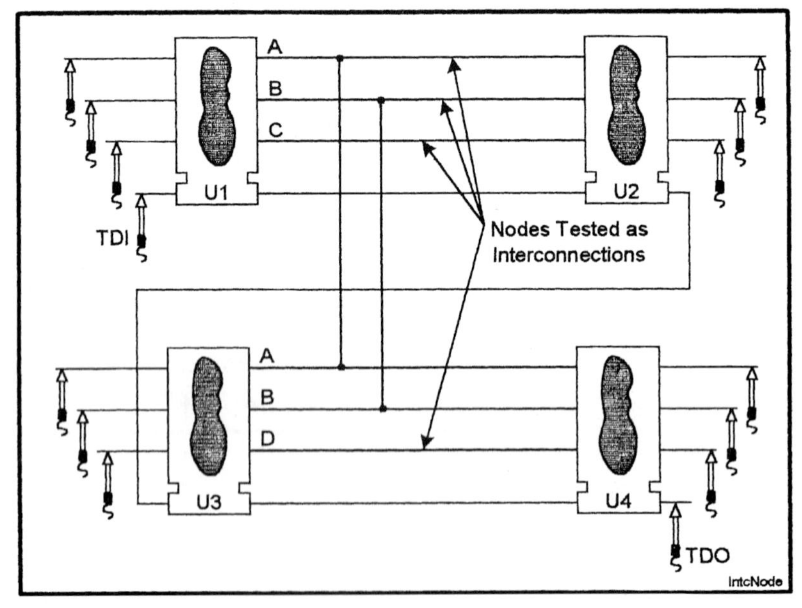

FIGURE 54.5 A set of boundary scan ICs with interconnections. Note that four nodes do not have bed-of-nail access.

The EXTEST capability can also be used during system testing to see if there are any system integration problems such as bad connections in backplanes and cabling. An IC designer may not see much attraction in EXTEST, but the 1149.1 standard offers other test modes that will allow a designer to access internal scan paths or built-in self-test functions. The 1149.1 standard's name has two parts, *Standard Test Access Port* being the first and crucially

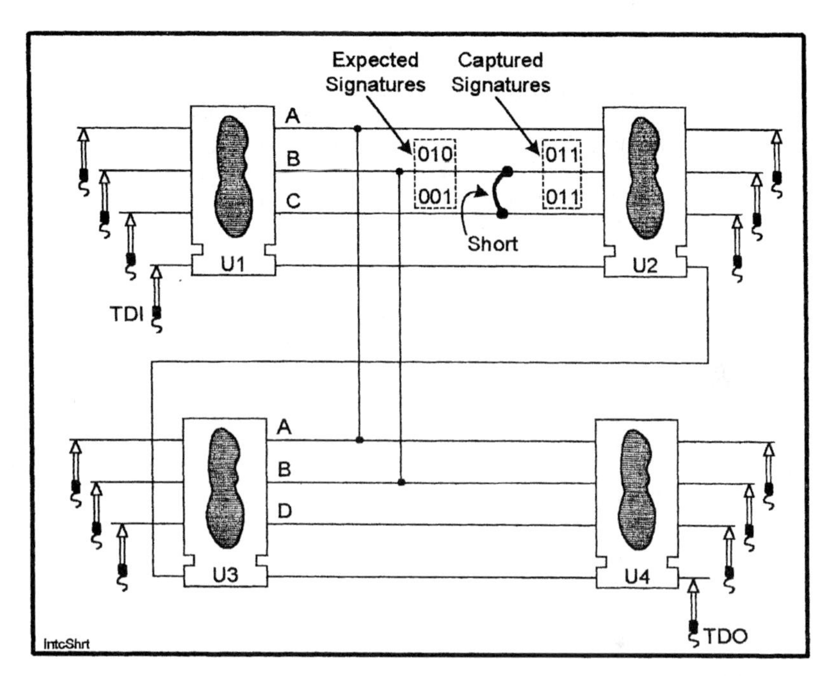

FIGURE 54.6 Interconnect testing drives unique patterns assigned to each node from drivers to receivers. In this case, a short is shown that creates a wire OR result.

important one. It signifies that the standard anticipates being used as a standardized protocol for accessing any on-chip, board-level, or system-level testability scheme. In support of this, the standard is deliberately extensible, allowing clever designers to implement additional operational modes that can be used to solve unique testing problems.

The 1149.1 standard has proven itself to be quite useful and it has several contributions. First, it allows the creation of software that can automatically write tests for boards where in the past the same level of test effectiveness was nearly impossible to achieve, and then only with weeks or even months of skilled labor. It is not uncommon to see a boundary scan board test prepared in a single day that otherwise might have taken weeks. Second, 1149.1 ICs can "read" their input pins and scan out the result. This allows diagnostic software to pinpoint the location of open solder problems where in the past an IC might have been falsely indicted as faulty. Third, it allows tests to be performed on digital circuits without 100 percent accessibility to board nodes. With the trend toward miniaturization of components making it difficult to provide full nodal access, boundary scan is allowing the elimination of many access points. Of course, not all points may be eliminated, so one must understand which are still necessary. Finally, because many industry segments are affected by the test problem, a standard offers a way for everyone to benefit. One can find a large number of applications and tools available to solve testing problems that would not have been possible without a standard.

54.5.2 IEEE 1149.4, Boundary-Scan for Mixed-Signal Circuits

Boundary scan (1149.1) is a digital testability standard. However, there is also a trend in the superintegration of circuitry toward higher mixed-signal digital-analog content in our designs. The IEEE has developed a mixed-signal testability bus with a new standard 1149.4.[4–6]

This standard is constructed as a superset of 1149.1 boundary scan, adding two additional analog test pins to the definition. The goal of the standard is to support opens and shorts testing of mixed-signal boards and to provide the capability of making analog value measurements of discrete analog components such as resistors, inductors, and capacitors without direct nodal accessibility, that is, a full bed of nails. (See Fig. 54.7.) It has been likened to in-circuit testing without a bed of nails, which is again not without caveats. (The elimination of test access points will still have to be done with thoughtful deliberation.)

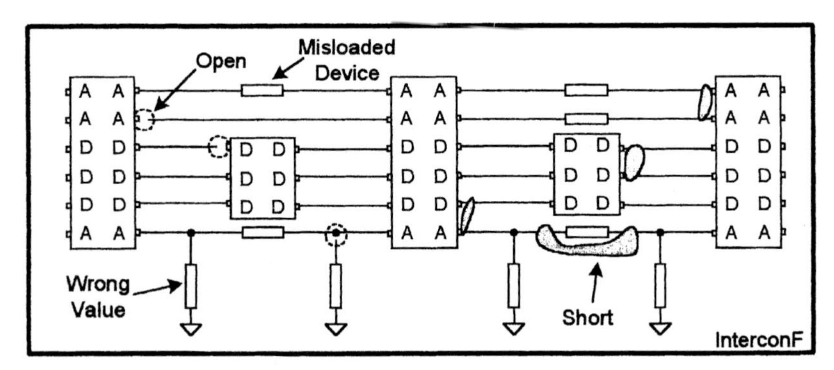

FIGURE 54.7 A mixed-signal circuit with some possible defects. IC pins marked A and D are analog and digital, respectively.

A mixed-signal device constructed with 1149.4 has the general architecture shown in Fig. 54.8. It is in many ways identical to an 1149.1 IC, but it has an additional analog test access port (ATAP) that is used to facilitate the control and observability of analog signals at device pins. The ATAP brings into the IC two additional analog signals that are used during testing.

With 1149.4, digital device pins are treated exactly as done with an 1149.1 boundary register. Analog pins have an augmented structure called an analog boundary module (ABM) as part of the boundary register. The ABM allows an analog pin to be tested for simple shorts and opens (this is called 1149.1 interconnect test emulation) as well as allowing the injection and/or observation of analog signals via the ATAP.

An ATE system can utilize the test resources in an 1149.4 IC as shown in Fig. 54.9. This requires the coordination of a digital test sequencer with analog test resources—in this case, a current source and voltmeter. Pathways from these resources and through the 1149.4 IC can be used to make measurements on discrete analog components on a board, even with no bed-of-nails access to the components.

The pathways needed for a measurement are provided by silicon switches in the ABM. Because switches implemented this way have significant nonlinearity and nonnegligible impedances, these must be accounted for in the measurement processes. Figure 54.10 shows how an analog device can be tested with two measurements. First, the tester's current source is connected such that current can flow along AT1 into the IC. There it flows on the AB1 bus inside to the ABM connected to pin 1 of the IC. The current flows through Z and then into pin 2 of the IC. Pin 2's ABM directs the current to ground, completing the current path. In Fig. 54.10(*a*), the ATE system's voltmeter is connected via the path AT2-AB2 to the ABM on pin 1, where it can observe the voltage at the top of Z. In Fig. 54.10(*b*), the voltmeter is switched to measure the voltage at pin 2, the bottom of Z. Subtracting the two voltage measurements gives the voltage drop across Z for a known current. Ohm's law gives us the value of Z, which we can check for the right value. This process works even with suboptimal silicon switches

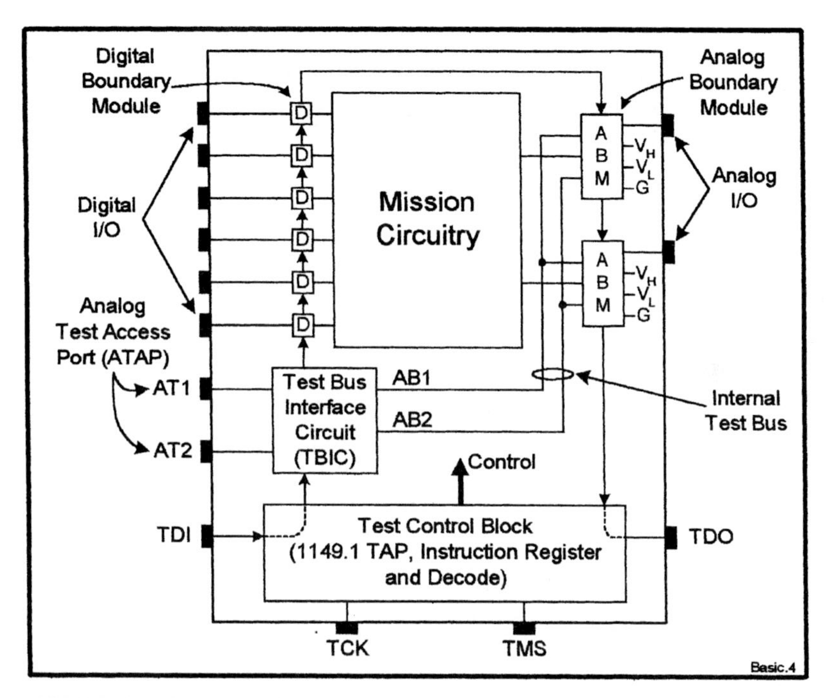

FIGURE 54.8 General architecture of an 1149.4-compliant IC.

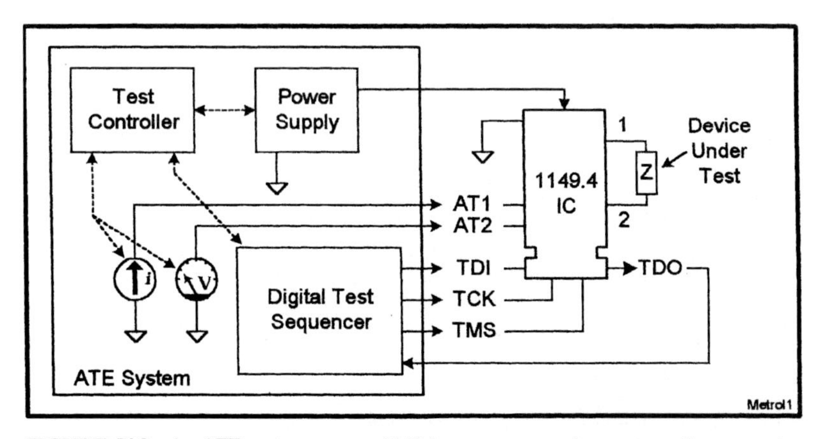

FIGURE 54.9 An ATE system can use 1149.4 resources to gain access to discrete analog components.

because we are using a current source to provide the known current and because the voltmeter pathway consumes a vanishingly small current and thus does not affect the accuracy of the measurement.

In the future, with these and other standards, test engineers may be able to do complex tests on superdense circuitry with far less nodal access than they were afforded in the past. This will be an enabling technology because without it the electronics industry may find it uneconomical to produce superdense designs except in very high-end applications.

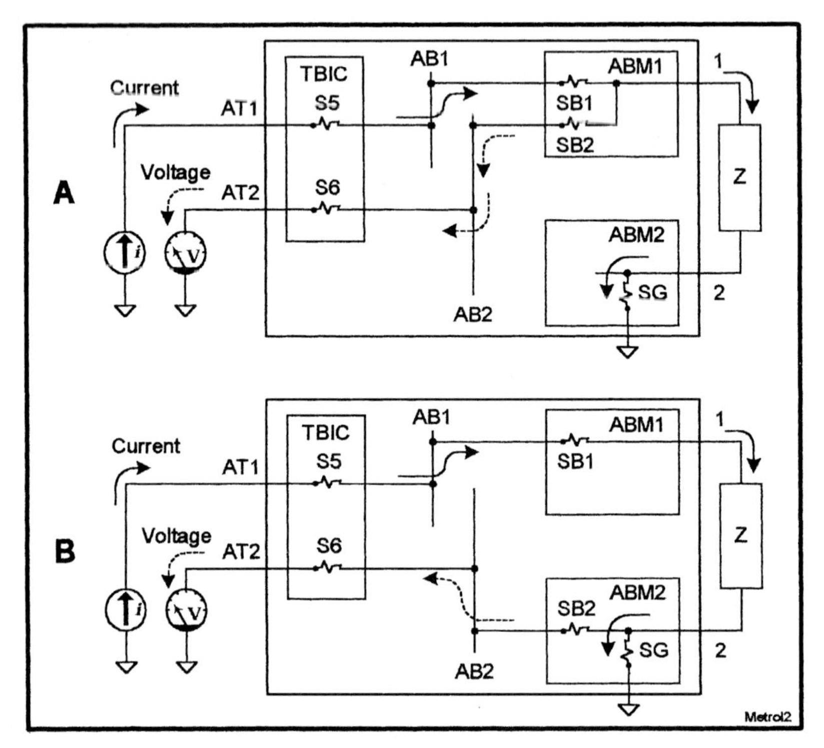

FIGURE 54.10 Two measurements (*a*) and (*b*) can be used to measure the voltage across Z for a known current.

REFERENCES

1. T. W. Williams and K. P. Parker, "Design for Testability—A Survey," *Proceedings of the IEEE,* vol. 71, no. 1, January 1983, pp. 98–112.
2. M. Bullock, "Designing SMT Boards for In-Circuit Testability," *Proceedings of the International Test Conference,* Washington DC, September 1987, pp. 606–613.
3. IEEE Standard 1149.1-1990, Standard Test Access Port and Boundary-Scan Architecture (includes IEEE Standard 1149.1a-1993), IEEE Inc., 345 E. 47th St., New York, NY 10017, USA.
4. IEEE Standard 1149.4-1999, Standard for a Mixed-Signal Test Bus, IEEE Inc., 345 E. 47th St., New York, NY 10017, USA.
5. K. P. Parker, *The Boundary-Scan Handbook: Analog and Digital,* 2d ed., Kluwer Academic Publishers, Norwell, MA, 1998.
6. K. P. Parker, J. E. McDermid, and S. Oresjo, "Structure and Metrology for an Analog Testability Bus," *Proceedings of the International Test Conference,* Baltimore, MD, October 1993, pp. 309–322.
7. R. W. Allen Jr. et al., "Ensuring Structural Testability of High-Density SMT Circuit Packs," *AT&T Technical Journal,* March/April 1994, pp. 56–65.
8. K. P. Parker, *Integrating Design and Test: Using CAE Tools for ATE Programming,* Computer Society Press of the IEEE, Washington, DC, 1987.
9. K. P. Parker, "The Impact of Boundary-Scan on Board Test," *IEEE Design and Test of Computers,* August 1989, pp. 18–30.

CHAPTER 55
LOADED BOARD TESTING

Kenneth P. Parker
Agilent Technologies, Loveland, Colorado

55.1 INTRODUCTION

Printed circuit boards, like everything else in the electronics industry, have been undergoing rapid technological evolution. This is only natural because everything, from the boards themselves, to the CAD systems that create them, to the components that populate them, to the assembly methodologies used to fabricate them, has been undergoing similar changes. These changes have common themes: greater functional density, better performance, improved reliability, and lower cost.

In the early 1990s, the move toward surface-mount technology (SMT) accelerated to the point that SMT designs became the rule. SMT supplanted, in large part, the familiar 100 mil centered through-hole package technology. The change came slowly, embraced by leading-edge applications that needed the density improvements that came with SMT. Many held back because they did not have a need for higher densities and could not justify the risk of putting new processes in place to manufacture with SMT. The process of perfecting SMT brought to light a surprising fact—it was more efficient once the necessary automation was put in place and perfected. There are now applications that use SMT not because of the density improvements it affords, but rather because of the efficiency it provides. This is confirmed by the fact that many new devices can no longer be obtained in the old-style through-hole packages.

Along with SMT came increases in lead pitch density. First there was 50-mil pitch; soon came 25-mil pitch, then came 15-mil pitch, and so on. Other technologies such as tape automated bonding (TAB), chip-on-board (COB), multichip modules (MCM), and ball grid arrays (BGAs) are gaining acceptance. The industry has for some time been in the midst of a packaging revolution. This revolution has applied to boards as well. The average board, by virtue of high-density interconnect (HDI), now has more layers, finer lines and spaces, buried vias, devices mounted on both sides, and so forth. The net result is that loaded printed circuit boards are becoming incredibly densely packed with highly sophisticated components. We are in fact experiencing a density revolution. Figure 55.1 shows a range of common electronic devices superimposed on a penny to show scale.

Testing has been impacted by all these changes. If perfect components were fed through perfect processes utilizing perfect machines run by perfect employees—testing would not be needed! Unfortunately, nearly perfect components are fed through processes that are subject to drift in many of the hundreds of variables that govern them, using machines that require careful calibration and preventive maintenance, operated by people who tire and err. For these reasons, testing is still an important part of loaded board manufacturing. However, as a result of the miniaturization brought on by the density revolution, our ability to gain the physical and

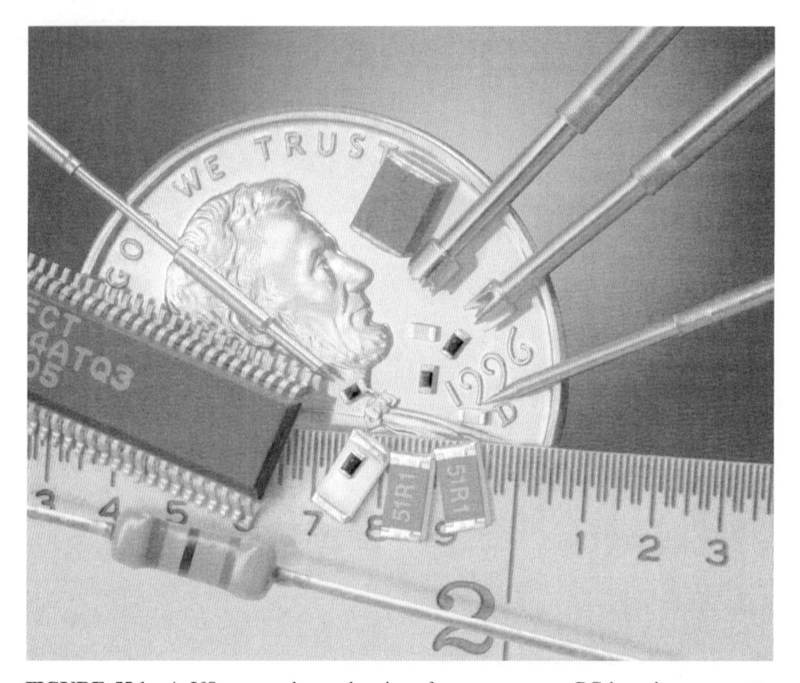

FIGURE 55.1 A US penny shows the size of some common PC board components, some older, some recent. The small black devices near Lincoln's bow tie are 04-02 (40 × 20 mils) surface-mount resistors. (Now 02-01 [20 × 10 mils] devices are being used.) Some 100-, 75-, and 50-mil test probes are shown for comparison. Note the lettering on the penny is done with 10-mil lines and the width of the circular part of the 9 is 35 mils, a common probe target size. *(Photograph courtesy of Agilent Technologies.)*

electrical access needed for testing purposes is increasingly hampered. Because access to a circuit is crucial for testing, accessibility difficulties make testing steadily more difficult to accomplish. On top of this, the electronics industry is expected to provide continued improvements in cost, reliability, and quality. Testing plays a crucial role in these improvements, as will be seen.

55.2 THE PROCESS OF TEST

In the earliest years of the electronics industry—indeed, before anyone may have called it an industry—there was no concept of test. A product was put together and shipped, because inherent in the putting together was a basic appreciation for how the product was supposed to look and behave. This crafting quickly gave way to mass production* of somewhat more complicated items by workers who were not themselves experts in the craft. Of course, today, workers with little or no fathoming of what it is they are producing turn out items of unbelievable complexity. If they had to know everything about the process, it would undoubtedly be too expensive to make the product.

* Perhaps the reader remembers the TV ads for Zenith televisions from about 25 years ago. They showed individual axial lead components being hand-wired onto lug strips and then soldered. The voice-over mentioned "*hand-crafted*" quality as if to imply that automated manufacturing (being used by Japanese manufacturers like Sony) was suspect.

It is also important to appreciate that boards are often themselves components of systems. Boards might have a respectable yield from a board test process of, say, 97 percent, which is to say, 3 of 100 still contain defects.* However, if 20 such boards are used in a system, the probability that the system will turn on is only 54 percent.† Because debugging a system is usually far costlier and more difficult than testing a board, there is much interest in higher post-test yields. This is why yields are being pushed into the 99 percent and higher levels.‡

Test has gone through three stages of evolution during this time. It has been used as a sorting process, then as a repair driver, and finally as a process monitor.

55.2.1 Test as a Sorting Process

Test can be used to sort boards into two piles, good and bad. Essentially, such testing provides only one bit of information about a board. This one bit of information provides little clue as to what the problem is or how to repair it. If all we intend to do with a bad board is discard it, then one bit of information may be enough. In some applications, for example making $2 digital watches by the million, it may be economically impossible to attempt to repair bad units. However, as will be seen soon, discarding bad boards may also lead to discarding of valuable information about the health of our manufacturing process.

55.2.2 Test as a Repair Driver

It is very often economically justifiable to repair a bad board, if it can be done quickly and with little skilled analysis needed. This is where modern test systems begin to come into their own. More bits of information are needed per bad board; a diagnostic test is needed that accurately resolves failures into defect reports that can be acted upon to effect repairs. Because a freshly repaired board will need retesting to ensure it is now defect free, one also would like to find as many defects as feasible in each pass across a tester to avoid a glut of work in progress.

55.2.3 Test as a Process Monitor

When repairing boards, valuable information about the health of our board manufacturing process is at our fingertips. Indeed, the repair process may have the ultimate defect resolution because faults detected by a tester may still not be perfectly correlated with defects (see Sec. 55.3). If one views test as a process monitor, one can gain valuable insight into what is happening upstream in many of the subprocesses that come together to produce boards. For example, one may see solder opens as a chronic problem. Further examination may show that they are happening in one area of a board more than in any other. That may lead to the examination of the solder application process, which shows that a solder screen squeegee has an uneven amount of solder paste on it. The root cause of this uneven distribution of paste can then be sought and corrected.

---

* It is important to differentiate test yield from defect coverage of a test. The yield of the manufacturing process before test might be (say) 50 percent, and the yield after test may be 97 percent. This may be accomplished with a test that detects (covers) 90 percent of the important defects.

† From basic probability theory, $(0.97)^{20} = 0.54$ is the chance that 20 independently manufactured boards with a probability of 0.97 of being fault free will all be fault free.

‡ Measuring yields in percentages is telling. In the IC industry, yields are measured in parts per million. One ppm corresponds to 99.9999 percent, and 250 ppm, a good IC yield, equals 99.975 percent yield. Will the board industry ever get into these ranges? Will it make sense to try?

It is quite important for the tester to do a good job of resolving faults into defects. Using an example of solder opens on digital devices, digital in-circuit test (detailed in Sec. 55.5.2) alone may indict devices as failing when it is really more of an open solder problem. The repair technician may notice this. However, it may well be overlooked because replacing the device "fixes" the problem. This pollutes the information about our process, causing one to call the IC vendor with complaints when one should be examining the solder process. If tests are focused more closely on defects than on faults, then one may be able to better trust tester-derived process information. This requires test engineers to be students of their manufacturing process and available test technologies so that test techniques are kept in balance with the defects that are most important.

55.3 DEFINITIONS

Testing is a word often used in a vacuum. What are tests actually looking for? The answer to this question has a huge impact on how to go about testing effectively. It makes little sense to test meticulously for problems with low likelihood of occurring, just as it makes little sense to inadequately test for likely problems. However, many of the terms used in the testing industry are overloaded such that surprisingly little agreement on their definitions may be found. Worse, this confusion is often not realized, which complicates communications between process people, test people, and equipment vendors.

55.3.1 Defects, Faults, and Tests

55.3.1.1 Defects. A *defect* is an unacceptable deviation from a norm. As an example, a bond wire could be missing in an IC. This defect may in turn be due to a problem (the "root cause") in a wire-bonding machine, such as it misfeeding wire. Other deviations may not be considered defects, but rather acceptable variations. For example, the plastic used to encapsulate a component may have significant variations in color that are still considered acceptable.

Defects require remedial action of some sort. Most often, defects lead to a repair operation that removes the defect and restores the product to acceptable status. In some cases, a defect may not be economically repairable and thus causes scrap. It is also possible that a defect will be overlooked and thus be delivered within a product to an end customer. In some cases, such as the manufacture of $2 wristwatches, some amount of defective product may be deemed acceptable. However, for many products, defects perceived by end customers can have grave economic consequences and are strenuously avoided.

55.3.1.2 Faults. A *fault* is a physical manifestation of a defect. (The word *fault* is often used synonymously with *failure.*) Thus a fault reveals the presence of a defect. A single defect may cause several faults (i.e., have several different manifestations), and a single fault could be the manifestation of several different defects.

A defect shows up as a fault. For example, the aforementioned missing bond wire on an input to a logic gate may cause it to see a permanent logic 1 rather than a time-varying signal. Several other defects can produce this same fault behavior. For example, missing solder between the input pin and the board, an electrostatically damaged input buffer within the IC, or a broken printed circuit trace between the upstream driver and the input will all exhibit the same faulty behavior. Similarly, one defect may cause several fault manifestations. For example, an open solder joint on an input pin (particularly a reset input) may cause an IC to show incorrect results, and these incorrect results vary in time.

An observed fault is not always a reliable pointer to a defect. For example, if an IC loaded onto a board has defective solder on one input pin causing an open circuit, this may appear to the IC to be a permanent, stuck-at-1 fault on that input. This faulty behavior may not be readily apparent because the effect of the erroneous logic 1 must propagate through the internal workings of the IC before its effects (improper output behavior) are seen.* When this faulty output behavior is finally observed, it can be a highly challenging task to relate this observed behavior to the input stuck-at fault caused by the defective solder. This is illustrated in Fig. 55.2.

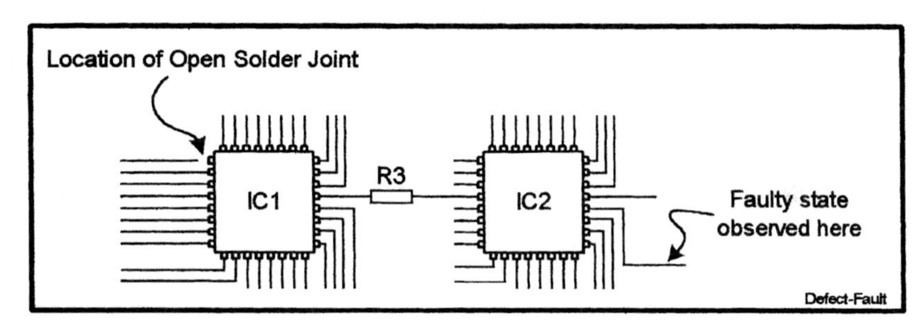

FIGURE 55.2 A defective solder joint on an input pin to IC1 causes a fault on an output of IC2. Note that IC1, R3, and IC2 are suspect, as well as six solder joints, two IC input buffers, two IC output buffers, and four IC bond wires.

55.3.1.3 Fault Syndromes. A *fault syndrome* is a collection of measured deviations from an expected good outcome. For example, a digital device may produce an incorrect response on several of its outputs when its inputs are stimulated in a certain sequence. The incorrect pins, their states at the point where they were incorrect, and the point in time where each failure was noted form a fault syndrome.

55.3.1.4 Fault Detection. A fault is *detected* when an operation with an expected outcome is conducted and this outcome is not observed. For example, when you turn on the power to a personal computer, you expect to see it boot up. If it doesn't, then you have detected a failure of some kind. There could be a plethora of reasons for this result, but you have little information to act upon. In other words, the underlying defect may not be easily resolvable.

55.3.1.5 Fault Isolation. A fault is *isolated* when an operation with an expected good outcome and one or more failing syndromes is conducted and the outcome belongs to the set of failing syndromes. For example, as your personal computer boots up, it may do a series of self-test operations. If any of these fail, a diagnostic message may appear on the screen that specifically identifies a failure. Again, the underlying defect may not be easily resolvable, but there is more information to act upon than you had when the fault was simply detected.

55.3.1.6 Tests. A *test* is one or more experiments that are specifically constructed to detect (and possibly isolate) failures. A *detection test* has an expected good outcome. An *isolation test* has an enumeration of possible fault syndromes indexed to specific failures. Thus, when a test fails, it provides us with information: a single bit of detection information, or a fault syndrome that may match an enumerated outcome, pointing to a fault.

A well-constructed detection test will detect a large number of potential failures. A well-constructed isolation test will accurately resolve a failure from a large list of potential failures. Note that an isolation test may encounter a failure that it was not constructed to isolate. If it fails (that is, if it detects the fault), it may not produce a syndrome that matches one from its

* See Sec. 55.6 for a discussion of how a new testing methodology may pinpoint a defect that conventional methodologies have difficulty isolating.

list, or it may produce a match that is erroneous. In either case, it has been reduced to a detection test.

Remember that an isolation test may point to a fault, but the fault may not be a good indicator of the actual defect that is present. For this reason, it is important to construct tests that target expected faults and that accurately resolve underlying defects. When defects are resolved correctly, it is then much easier to find and correct the causal problems.

There are, in general, three categories of faults being tested: performance faults, manufacturing defects, and specification failures.

55.3.2 Performance Faults

A *performance fault* is a fault in the performance of a system that occurs due to a mismatch of important parameters among the system's components. This mismatch is the defect. As a common example, the path delay seen by a digital signal as it passes through several components may exceed the intended design value, causing a malfunction. No single component in the path is defective, but the cumulative contributions of several components cause the performance fault. The repair for this defect is to replace one or more components in the path with new components specifically selected to give the proper delay.

There are several problems with testing for performance faults. First, the test developer must know about the circuit design in great detail. Second, it is difficult to set up a test that can resolve faults into specific defects (for example, the mismatch of parameters in several components). Third, it is difficult to avoid being confused by unanticipated defects that produce behaviors similar to the defects of interest.

Solving these problems implies great knowledge and understanding of a board design. Indeed, in some instances in a carefully designed board, the designer may have specific knowledge of some critical parameter that has to be precisely managed and can alert those responsible for test. However, much of the testing for performance faults carried out in the past was not done with this knowledge, but rather because of a lack of this knowledge. In the past, tools that could help control the key parameters of a design were unavailable, or designers were using components that were incompletely specified, or perhaps they were too trusting of their "seat-of-the-pants" instincts. It was expected that performance testing would give adequate coverage of any problems that might occur. In effect, performance testing was used to validate a design after the fact.

The expectation that performance testing will somehow protect products from the effects of poor design is now obsolete. It amounts to wildly shooting in the dark against a well-hidden, stealthy enemy force of unknown size and distribution. Given the ever increasing complexities of boards, it is simply not reasonable to expect that test engineers could stumble at random across effective tests for all possible design problems, and certainly not within the lifetime of the design. With the increase in effectiveness of design tools, designers should no longer be relying on test for design validation. Testing for performance faults will still be important, but it must be used in its proper role—to verify that critical parameters identified in the design process are properly controlled.

55.3.3 Manufacturing Defects

A *manufacturing defect* is a defect resulting from a problem in the manufacturing process. Manufacturing defects tend to be fairly gross in nature. Table 55.1 gives a list of potential manufacturing defects.

Manufacturing defects are the result of the havoc inherent in manufacturing processes. These defects result in faults that may be very easy to detect and to correlate with their root cause(s), but this is a function of the test approach. Some manufacturing defects can still be difficult to detect and resolve, as the example given of a solder open on a device input pin shows (Fig. 55.2).

TABLE 55.1 Examples of Manufacturing Defects Seen at the Board Level, Sources of the Problems and Causes

Defect	Source(s)	Cause(s)
Shorts between solder pins	Wave/reflow soldering	Too much solder, solder screen defect, pin misregistration, bent pins
Solder open	Solder application, wave/reflow soldering	Too little solder, solder screen defects, tombstoning, bent pins
Missing component	Placement, soldering	Shock, abrasion, too little glue
Wrong component	Placement setup, inventory, handling	Handling error, mismarked packages, operator error, wrong specifications
Misoriented component	Placement setup	Handling error, operator error
Dead component	Placement, soldering	Dead on arrival, handling damage, electrostatic damage

55.3.4 Specification Failures

Specification failures are similar to performance faults. Performance specifications are checked against requirements for the full range of operating conditions expected, such as temperature, humidity, vibration, electronic noise, etc. Specification test is often a regulatory or contractual requirement. One may argue that these specifications are largely unnecessary because, if a circuit design is robust, makes use quality components, is accurately assembled, and is thoroughly tested for manufacturing defects, one does not have to make it perform all of its functions to know that it works. This is the "proof by construction" argument, used by successful manufacturers of kitchen matches and bomb detonators. Full specification test may also be impossible in practical terms because there may be too many combinations of circuit functions versus operating ranges. If only a subset of combinations is to be checked, this begs the question of which subset.

If specification test is required nonetheless, such testing is usually carried out with custom-tailored test equipment that can simulate the range of operational environments of interest. Such testing may be quite time consuming, and the cost of the supporting test equipment may range from trivial to hyperexpensive. At one extreme, a manufacturer of I/O cards for personal computers may simply plug each one into a computer and see if it performs a simple loopback test. At the other extreme, a manufacturer of guidance computers for missiles may require a missile shot on a test range, full telemetry, and support from the Air Force and Navy. The bottom line is that you should seriously question the motivations for and expectations of testing for specification faults.

55.4 TESTING APPROACHES

We have seen a spectrum of faults, performance faults, manufacturing defects, and specification faults. Each has evolved a test technology.

55.4.1 Testing Boards for Performance Faults

Performance testers are often known as edge connector functional testers. They connect to the edge connectors of the board under test with mating connectors that are then adapted to the tester resources by a customized fixture called an adapter. In most cases, no other connections are made by the test fixturing to any internal nodes of the circuit. In essence, the board is tested in an environment that resembles its application environment to some degree.

Some performance testers have guided probe capability.[1,2] A guided probe is a manually positioned test probe with a measurement (and sometimes stimulus) capability and supporting software. It is used to temporarily gain access to internal nodes of a circuit, one at a time, where observations of internal circuit behavior during a test can be made and processed by software to provide enhanced fault resolution.

How does this software know what the nodes of a circuit are supposed to be doing during a test? One could enter all these data by hand if one knew the design of a circuit well, but this is impossible for any circuit that consists of more than a handful of gates. The most popular way to get these data is by logic simulation of the circuit.*[2] Good circuit simulation is often used by designers to prove the circuit behaves the way they expect it to under various input conditions (input vectors) they are interested in. Fault simulation is used to study how a circuit behaves when that circuit is perturbed by a modeled fault. An example of a modeled fault is a stuck-at-0 on a gate output, meaning no matter what the gate's inputs are, the output is 0. The single stuck-at model is the most prevalent fault model in existence. It assumes that only one stuck-at (either 0 or 1) may exist in a circuit at any one time. A fault simulator can be used to predict, for a given modeled fault, how the circuit will behave when stimulated with input (test) vectors. If the modeled fault causes one or more observation points (like circuit output pins) to deviate from normal behavior, we can declare the fault detected by that vector. In the common case of sequential circuits, it is more accurate to say all vectors detected the fault up to and including the one that caused the observable deviation.

Another technology used by some performance testers is the fault dictionary.[2] Fault dictionaries are prepared with fault simulators. A fault dictionary is a three-dimensional data structure of Boolean true/false bits (see Fig. 55.3). The first dimension is an enumeration of all test vectors. The second dimension is a list of modeled faults that are to be simulated. The third dimension is an enumeration of the circuit's output pins. A given bit in the fault dictionary is true if the corresponding output pin fails on the corresponding vector for a given fault.

* Logic simulation implies the treatment of the digital nature of the circuit, typically omitting any interaction with any analog portions of the circuit. Analog simulators do exist, but are implemented in a very different technology from digital simulators. Mixed-signal simulation is not anywhere near as viable as digital simulation, which presents a problem for test engineers.

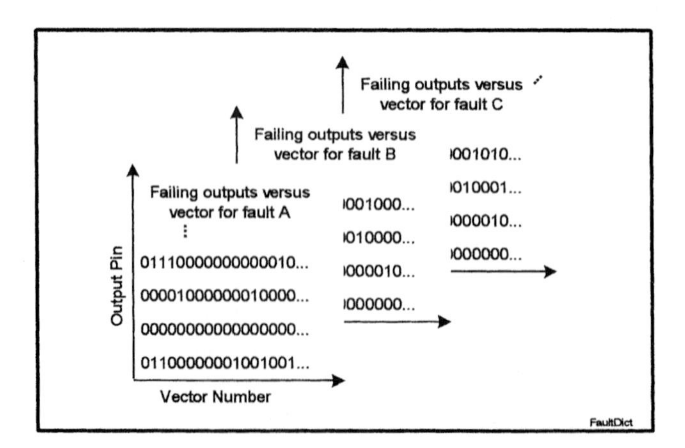

FIGURE 55.3 A fault dictionary. For each "page" corresponding to a modeled fault, bits set to 1 indicate that a given output pin fails for a given vector number. When testing is performed, the failing output(s) and failing vector number are noted and then each page is searched at those coordinates to look up a matching fault. (Note that more than one fault may produce a match.)

For a given vector and fault, the output pins that are expected to fail can be looked up in this dictionary. In a testing scenario, we take a failing test vector and list of failed output pins, and attempt to look up a fault (or faults) that match the failed pin behavior for that vector. Fault dictionaries work well when actual defects that cause a circuit to fail closely match the fault model. When defects occur that vary from the modeled faults, then a dictionary lookup may come up empty (no matching fault), or matches may be found but point to the wrong faults, or so many matches may be found that it is impractical to examine them all to see which (if any) are the actual problem. Dictionary generation is computationally intensive, dictionaries may consume huge amounts of storage space,* and dictionary lookups may not work. This technology was invented along with small-scale integrated bipolar technologies such as early TTL or ECL. The advent of large-scale integration and CMOS technology† has made fault dictionary technology nearly useless in board testing practice.

Performance testers are intended to emulate the environment the board would encounter in its native application. This means the tester itself needs to be carefully customized to supply this environment, or it must be a general-purpose tester with a great degree of flexibility. Flexibility brings cost, so commercial functional testers tend to be among the most expensive. Programming such machines is a complex task, due both to the flexibility of the tester and to the nature of the test requirements. It is difficult at best to get automated support for functional test programming. Typically, such programming takes extreme patience and a high level of skill—again, two more costs. To compound this, any last-minute design changes to the board may cause expensive and time-consuming test modifications or invalidate the tests altogether.

55.4.2 Testing Boards for Manufacturing Defects

Manufacturing defects can be detected by a functional test, but because functional testers are expensive and difficult to program, other techniques have arisen to test for them. These testers have one or more of the following advantages:

1. They are much easier to program, often requiring less than 10 percent of the time of functional test development.
2. Automatic test program generators that analyze the circuit design database do most of the work and relieve the skill required of the programmer.
3. Their programs are much less sensitive to design changes because they take a divide-and-conquer approach; the effects of design changes are localized, requiring that only a portion of the test be reprogrammed.
4. They offer much better defect resolution for many defects because localized portions of the circuit are being analyzed, reducing diagnostic complexity.

The principal technology is the in-circuit tester. This tester utilizes a high degree of nodal access—that is, connection to printed circuit traces—to perform its work. Nodal access is provided by a unique fixture called the bed of nails. (See Fig. 55.4.) This fixture is made up of a platen that supports the board under test. The platen is drilled with holes below each target point for nodal access. In the holes are spring-loaded "nails" that contact the target points on

* For example, say a circuit has 100,000 gates. The number of stuck-at faults to be simulated may easily be 200,000. A set of test vectors for these may easily exceed 500,000. If the circuit has a fairly normal complement of 250 outputs, then a fault dictionary for this example would consume the product of faults times vectors times outputs (2.5×10^{13} bits, or about 3.125 terabytes, give or take a gigabyte). Fault simulation times lasting months, even on very fast computers, have been reported. Of course, larger circuits will need more!

† CMOS VLSI circuits have several defect modes that are inconvenient to model in popular fault simulators, and thus fault models are more likely not to match defects. Two of the offending defect modes are bridging faults (intermetal shorts) and metallization opens that may actually introduce capacitive memory into a circuit. These are difficult to represent with traditional fault-modeling techniques.

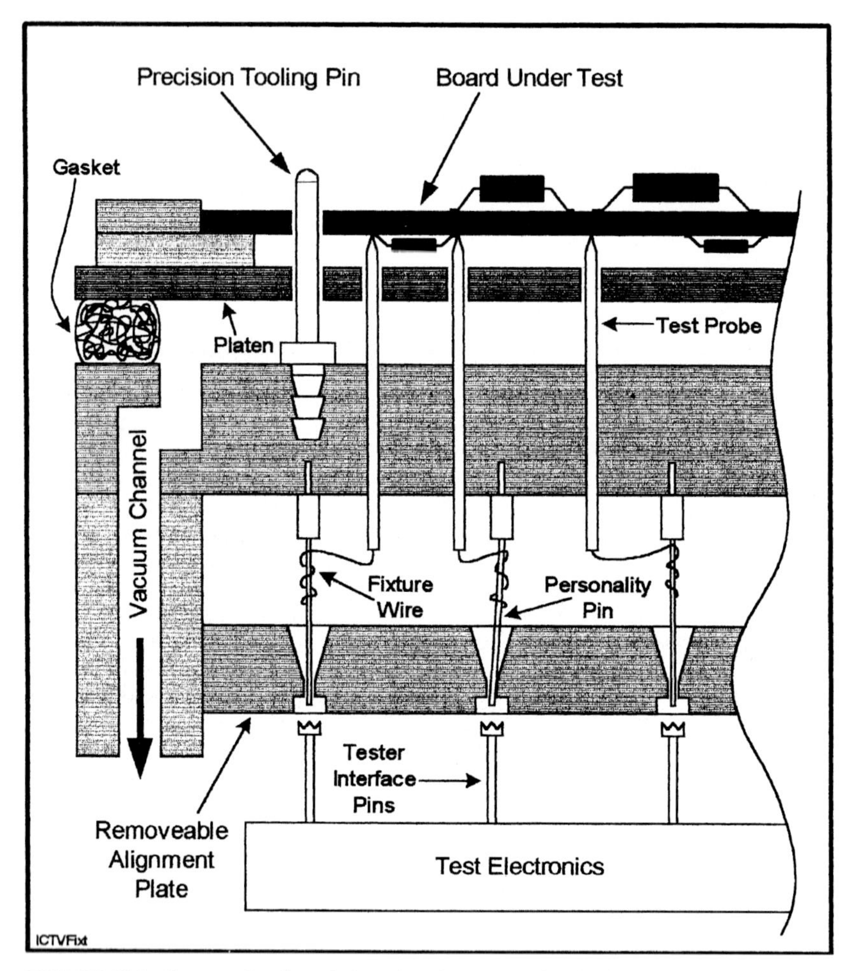

FIGURE 55.4 Cutaway drawing of a board resting on top of an in-circuit vacuum-actuated fixture, the bed of nails. The interface pins (the mechanical interface to the ATE pin electronics) are placed very close to reduce path lengths.

the board. (Special clamshell fixtures allow nail access to both sides of a board.) The platen with board forms the movable top layer of a vacuum chamber. When vacuum is actuated, the platen with board moves down, depressing the nail springs and causing nail contact with board nodes. (Sometimes mechanical actuation is used in place of or to assist vacuum.) The nails are wired (typically with wire-wrap technology) to the stimulus/measurement resources of the tester. These resources contain mechanical relays that allow connection of various tester functions to a given nail.

Once a tester has access to the nodes of a board, it can perform in-circuit (also called in situ) tests. The idea is to test components as if they are standing alone, while they in fact are part of a board circuit. The actual electrical processes used in in-circuit testing are explained in Sec. 55.5.

55.4.3 Testing Boards for Specification Faults

There is not much to add here because often, testing for specification defects is very similar to testing for performance faults. However, if for (say) contractual reasons, you are required to show your boards working as specified, you may be required to construct "live" situations where they are performing in a realistic system setting. Again, the extreme case of a missile shot on a test range could be involved. The question must be, is there an easier way?

55.5 IN-CIRCUIT TEST TECHNIQUES

The problem of in-circuit testing can be subdivided into two main categories, analog in-circuit testing and digital in-circuit testing.

55.5.1 Analog In-Circuit Test

Analog in-circuit test addresses testing for shorts in the printed wire circuitry; analog components, passive devices such as resistors, inductors, and capacitors; and simple semiconductor components such as diodes and transistors.* Analog in-circuit testing is conducted without applying power to a board; that is, it is an unpowered test methodology.

Shorts (unwanted connections between nodes) are tested for first because subsequent testing will profit from the assumption that shorts are not a factor, and subsequent testing may need to apply power to the board. Unpowered shorts testing may be accomplished by applying a small DC voltage† to a node while all others are connected to ground. If current flow is observed that is below a computed threshold, then the node cannot be shorted to any of the grounded nodes. If the current flow is above the programmed threshold, there may be an unexpected path to at least one of the grounded nodes. The destination node(s) can be determined by linearly searching the grounded nodes for the current flow (this is slow) or by using half-splitting techniques (which are fast‡) to determine the other node(s) the current is flowing to. When the algorithm has finished stimulating all nodes sequentially, it can declare which nodes are shorted and use x-y position data to show where to look for the problems. Typically, shorts are repaired before continuing with other tests because they may confuse the resolution of defects and may cause physical damage when the power is later turned on.

Unpowered tests on analog components such as resistors, capacitors, inductors, etc., are performed next. Again, low stimulus voltages keep semiconductor junctions turned off. AC voltages are applied and phase-shifted currents are measured in order to deduce the values of reactive components (capacitors, inductors). For a simple measurement of a lone§ impedance R, apply a voltage to one terminal of the component (through the bed of nails) and connect a

---

* Complex analog devices such as analog or mixed-signal ICs are not very amenable to analog in-circuit testing because they require power to be applied to the board. Simple diodes and single transistors can be tested by in-circuit stimulus that essentially examines the characteristics of their semiconductor junctions.

† The reason for using a small stimulus voltage (typically less than 0.2 V) is to prevent current flow through semiconductor junctions that may exist between nodes. This voltage will not turn on a junction. These junctions may exist in parasitic form within ICs and are not always documented.

‡ A half-splitting technique (also called a binary chop) is a fundamental algorithm of computer science. It works by successively considering half of a set of items (grounded nodes that are receiving current in this case) while removing the other half from consideration. It recursively divides sets repeatedly until only one item is left to consider. It has a complexity related to the logarithm (base 2) of the size of the original set of items. By comparison, a linear process has a complexity linearly related to the size of the original set.

§ This component may not be alone in a physical sense, but because low stimulus voltages are used, other connected devices such as ICs may be electrically quiescent so that the component is electrically alone.

current-measuring device (conceptually, an ammeter with zero impedance to ground) to the other component terminal. The current flow observed due to the known stimulus voltage is related to *R* by Ohm's law ($R = V * I$). See Fig. 55.5(*b*).

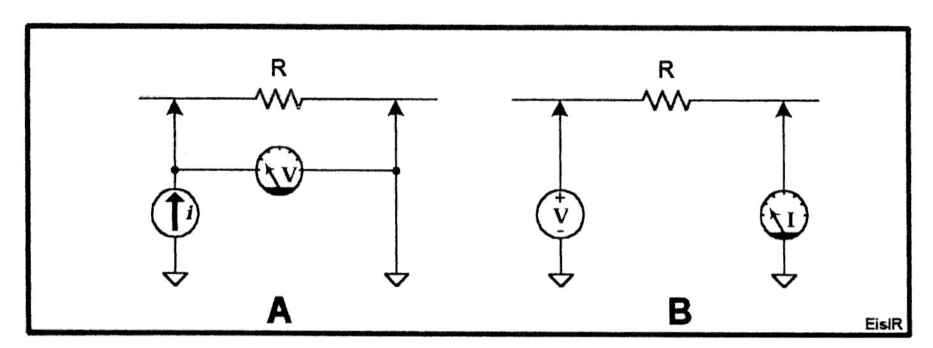

FIGURE 55.5 An analog component on a loaded board can be accessed by two in-circuit probes that connect its terminals with tester resources. There are two approaches: force a known current and measure a voltage (*a*) and force a known voltage and measure a current (*b*).

However, some discrete analog components may be connected to one another in ways that prevent simple measurement of the current flowing through the component due to parallel pathways that sidetrack some of the current. In Fig. 55.6, when applying a voltage to one terminal of *R*, a current also flows in a parallel path through R_a and R_b. The ammeter does not measure the true current through *R* because current from the parallel path also gets to the ammeter.

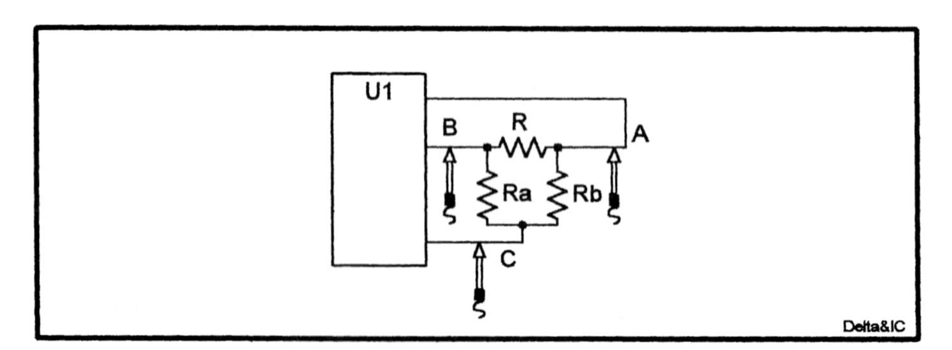

FIGURE 55.6 Three analog devices interconnected with an integrated circuit (U1).

The parallel path problem can be solved with a process called guarding. Guarding is accomplished by using a third nail to connect ground to the node marked *C* as shown in Fig. 55.7. When *C* is grounded, all the sidetracked current goes to ground because there is no voltage drop across R_b to attract current to or from the ammeter. All the current seen by the ammeter comes through *R*, so again Ohm's law gives the value of *R*. This is a classic three-wire measurement. It turns out that, for general component topologies, multiple guards (ground connections) may be required, but these are still considered three-wire measurements. In some cases where enhanced accuracies are needed or where there are extreme ratios of component

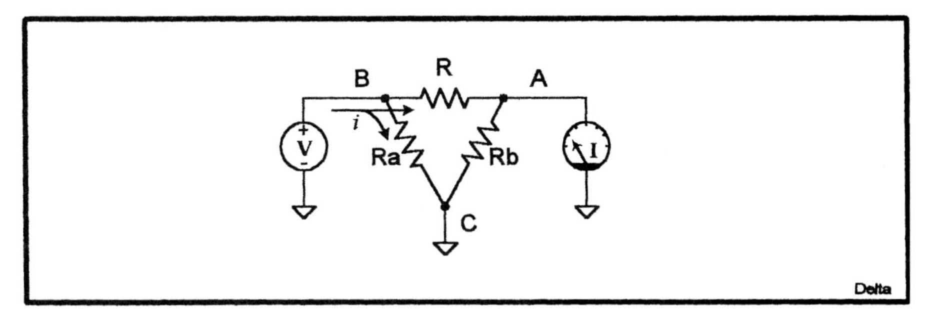

FIGURE 55.7 Equivalent circuit for Fig. 55.6 when making a guarded measurement.

values in a network, additional sense wires are used to eliminate errors that are due to small voltage drops in fixture nail, wire, relay, and trace contributions. (See Fig. 55.8.) See Ref. 3 for a discussion of enhanced measurement accuracies.

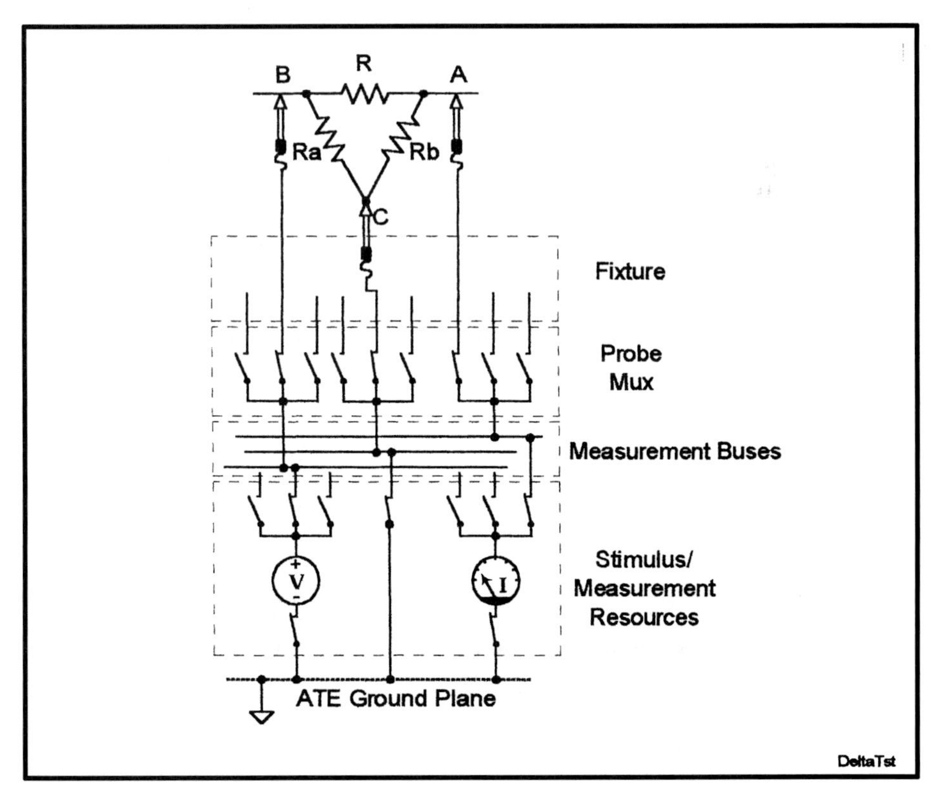

FIGURE 55.8 Routing of tester resources to the circuit to be tested.

55.5.2 Digital In-Circuit Test

Digital in-circuit test focuses on the digital components residing on a board, and requires that power be applied to activate the digital logic contained within the ICs. Just as analog components can be tested without removing them from a board, digital components can be tested the same way. The key technology is backdriving. When applying digital stimuli to a digital device's inputs (via the bed of nails), a tester's driver must overcome the voltage levels that connected upstream devices are producing. This is done by equipping the digital in-circuit tester with powerful, low-impedance drivers that can backdrive upstream drivers with enough current to create the desired signal voltage in spite of their state. With tester receivers connected by nails to the device's outputs, the tester can monitor these outputs for expected responses to the stimuli. (See Fig. 55.9.)

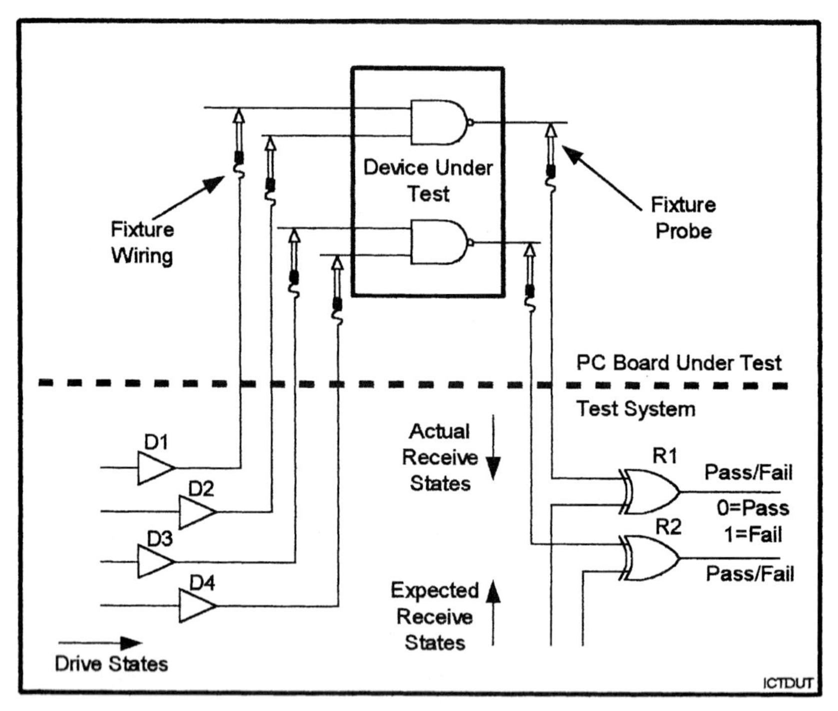

FIGURE 55.9 In-circuit digital test setup with full nodal access. The device under test is most often connected to other devices.

A problem with backdriving is that the tester drivers are abusing the upstream drivers in other devices* while testing the device of interest. Studies[4] have shown that this is a legitimate concern and that damage to upstream devices may occur. For example, overdriven silicon junctions or device bond wires may heat up enough to be damaged. Sometimes this damage may occur surprisingly fast (within milliseconds). This problem can be solved by careful application of tests with an eye toward their duration. If backdrive testing can be done quickly, and/or with appropriate cooling intervals, then damage can be successfully avoided.[5] This has allowed digital in-circuit testing to become the dominant testing technology.

A great advantage of digital in-circuit testing is that it is performed directly on the inputs and outputs of a targeted device. If the device should fail testing, this is seen directly, rather

* This abuse can be mitigated by "conditioning" upstream drivers. One method is to disable the upstream drivers by devoting additional tester drivers to driving states on disabling inputs (when they exist) such as output enable pins. Of course, the problem may simply be moved upstream, because the disabling inputs may also be connected to upstream drivers that are now being overdriven in place of the original drivers. This conditioning process may be carried out to several levels, if desired.

than having its identity masked by interactions with other devices. This is a major differentiation from functional testing. Faults typically can be resolved to two categories of defects: failed ICs or solder opens on I/O/power pins.

Another major differentiator is the ease—indeed, automation—of test programming that is possible with digital in-circuit testing. Tests for ICs can be prepared as if the ICs were standing alone,* stored in a library, and recalled from the library when needed. Modern digital in-circuit testers may have library tests for tens of thousands of devices. For custom, one-of-a-kind ICs for which a library test may not exist (e.g., ASICs), it is still substantially easier to create a test for just the one device than it is for a collection of ICs.

55.5.3 Manufacturing Defect Analyzer (MDA)

A manufacturing defect analyzer (MDA) is essentially a very low-end analog in-circuit tester. One way it maintains low equipment cost is by not having power supplies to power up a board. Another cost savings comes by having only rudimentary programming and operating software. Some amount of test accuracy and yield must be traded for these savings.

55.5.4 General-Purpose In-Circuit Tester

The workhorse of the electronics industry is the general-purpose in-circuit tester that merges support for analog and digital in-circuit tests. An example of a widely used system is shown in Fig. 55.10. It contains power supplies for powering boards and often contains sophisticated

* This assumes that the IC does not have any topological constraints on its I/O pins, such as having an input pin connected directly to ground or an output pin fed back to an input pin. In such cases, the prepared test may be incompatible with these constraints and may only be used after incompatible segments of the test are deleted, reducing fault coverage.

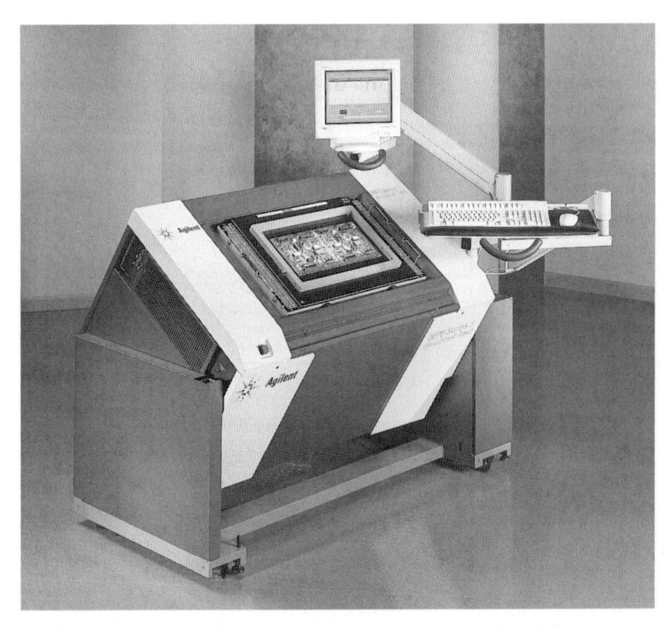

FIGURE 55.10 Example of a commercial in-circuit test head and operator terminal, with a printed circuit board mounted in the testing position on top of the bed-of-nails fixture.

analog in-circuit programming tools and extensive libraries of digital tests. The typical test and repair flow for this tester is shown in Fig. 55.11. Note the early exit to repair for boards that fail shorts testing. This avoids applying power to boards that contain shorts because these may present hazards to the board and human operator, plus they also confuse the diagnosis of faults later in testing.

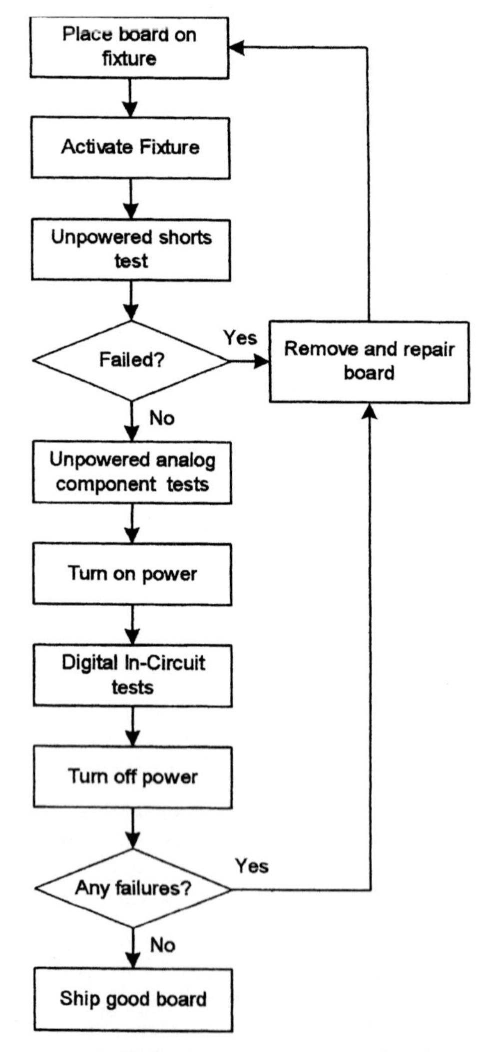

FIGURE 55.11 Typical test and repair flow for an in-circuit tester.

55.5.5 Combinational Tester

In situations where a manufacturing line has a variety of technologies in production, a need may exist for both functional and in-circuit testing. Thus, hybrid functional/in-circuit testers (commonly called combinational testers) exist. These machines give test engineers a full complement of tools to address testing problems. They also allow "one-stop" testing, where manufacturing defects and functional performance faults can be detected at one site in the production flow.

A combinational tester utilizes a bed of nails to perform in-circuit analog and digital tests. It may also utilize edge connector access to perform functional tests. In some cases these approaches are hybridized by constructing a two-stage bed-of-nails fixture that has two lengths of nails used in the bed, and a platen that has two stations of depression during operation. The first station is full depression, which brings all nails into contact with the circuit board for standard in-circuit access. The second station is partial depression, where only the longer nails still contact the board, perhaps at the board edge, for functional testing. The removal of the shorter nails removes the electrical loading they present to internal circuit nodes and allows the board to operate in a more natural environment.

55.6 ALTERNATIVES TO CONVENTIONAL ELECTRICAL TESTS

Alternative tests use radically different approaches to specifically address the resolution of defects. These may be needed to address blind spots in traditional electrical test methodologies. Two specific examples of defects that are difficult to resolve by conventional tests are given here.

First, consider a board with a large number of bypass capacitors. All of these capacitors are connected between power and ground so their parallel capacitances are summed. Using an analog in-circuit test, it is possible to test for this summed capacitance. However, if one capacitor is missing or tombstoned, the tester most likely will not notice because the resulting decrement in capacitance will likely fall well within the summed tolerances of the summed capacitances. The performance of the board at higher operating frequencies may be adversely affected, however, due to a loss of noise immunity. This could be detected (possibly) by a performance test. However, if this performance test failed, it could be very difficult to resolve which capacitor was missing. One non-traditional way to test for this problem is to use human visual inspection to look for missing capacitors. This may not find solder opens, however. Another approach would be to use an automated x-ray laminographic tester to check for the existence and solder integrity of each capacitor. Even the orientation of polarized capacitors can be verified. (See Chap. 53 for a discussion of x-ray inspection.)

Second, consider again the open solder problem on a digital device input. This defect may manifest itself as improper device behavior, but it would be wrong to replace the device (which would also fix the solder open) because the device is not at fault. Digital in-circuit testing has trouble resolving solder opens on input pins from bad devices.* Final resolution can be obtained either by visual inspection or by using a handheld probe to see if board signals reach IC legs, but this is becoming increasingly difficult as packaging dimensions continue to shrink (e.g., TAB) and newer attachment technologies (e.g., ball grid arrays) are used that prohibit visual inspection or probing altogether.

An alternative approach to the open solder problem uses a capacitive coupling technique to look for opens. The technique exploits the fact that many ICs have a leadframe that forms the conductive path from the legs of the device to the die bond wire pads. Using the bed of nails, all but one node attached to the IC can be grounded (this is an unpowered technique) and a small AC signal can be applied to the node that remains. An insulated metal plate pressed against the top of the IC forms the top plate of a capacitor and the stimulated IC leg and lead frame conductor form the bottom plate.

* Ignore for the moment the possibility of using boundary scan to solve this problem, which is discussed in Chap. 54.

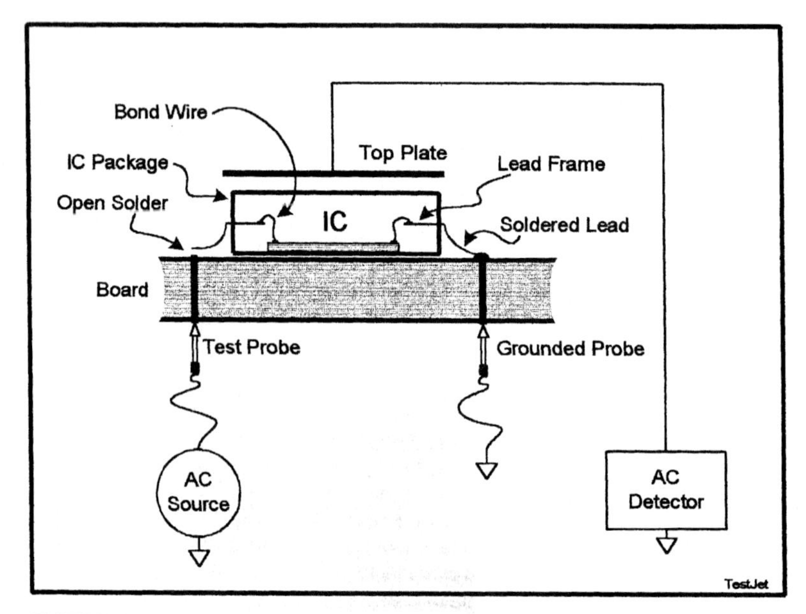

FIGURE 55.12 Capacitive leadframe testing used to find open solder joints.

Figure 55.13 shows an equivalent circuit for capacitive opens test for a properly soldered IC lead and for an open solder condition. The capacitance C1 may be on the order of 100 fF (100×10^{-15} F), which is small enough to require sophisticated detection electronics to measure in the face of environmental noise. Now, if the IC leg is soldered to the stimulated board node, the correct capacitance will be measured. If the solder joint is open, then a second small capacitance C2 now exists in series with the first. This reduces the measured capacitance by a factor of 2 to 10. Capacitive leadframe test allows testing of complex ICs for solder opens without knowing what the ICs actually do and without applying power. The technique requires no complicated programming, and gives accurate resolution of solder defects. The technique has been extended to allow testing of solder integrity for connectors and switches. A fixture for capacitive leadframe testing is shown Fig. 55.14.

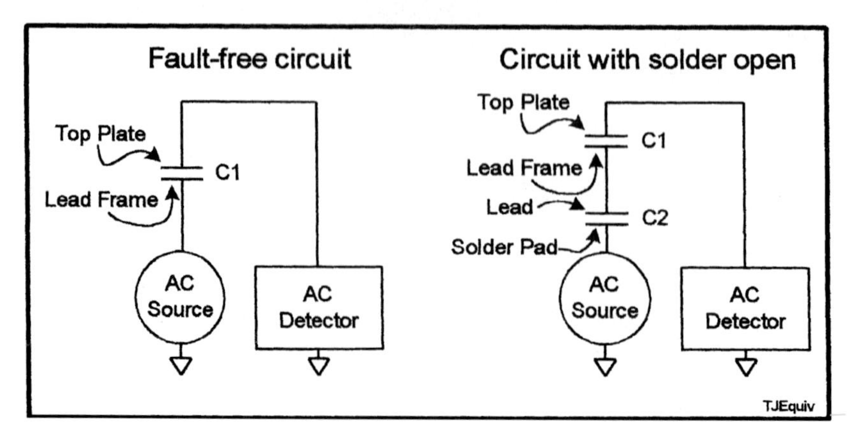

FIGURE 55.13 Equivalent circuits for fault-free and open solder conditions.

FIGURE 55.14 A test fixture with board on a bed of nails, with clamshell top fixture containing capacitive sensor plates. Notice eight closely spaced sensor plates (right) that test for opens in eight (white) connectors on the board.

55.7 TESTER COMPARISON

Table 55.2 summarizes the types of tester for comparative costs and capabilities. A majority of manufacturers value good diagnostic resolution and fault coverage, so it is not surprising that in-circuit and combinational testers are in widespread use. Next in prevalence are the MDA testers, which are typically used where coverage and diagnostic resolution may be sacrificed

TABLE 55.2 Costs and Capabilities of Various Testers

Tester type	Typical cost ($)	Programming time	Diagnostic resolution	Fault cover	Comments
MDA	10^4–10^5	1–2 days	Fair	Poor	No digital coverage; requires known good board for programming.
In-Circuit	10^5–10^6	5–10 days	Best	Good	Fixturing is a major portion of preparation time.
Combo	10^5–10^6	10–30 days	Best	Best	Functional test programming is a major portion of preparation time.
Functional	10^5–10^6	1–4 months	Fair	Fair	Very high skill required for test preparation and interpreting results.
Specification	10^3–10^8	Weeks to years	Poor	?	Very high skill required for test preparation and interpreting results.

for ease and speed of programming, usually in low-cost, high-volume products. Functional and specification testers are becoming rare, and are often only justified by the existence of contractual or regulatory requirements.

REFERENCES

1. W. A. Groves, "Rapid Digital Fault Isolation with FASTRACE," *Hewlett Packard Journal,* vol. 30, no. 3, March 1979.
2. K. P. Parker, *Integrating Design and Test: Using CAE Tools for ATE Programming,* Computer Society Press of the IEEE, Washington, DC, 1987.
3. D. T. Crook, "Analog In-Circuit Component Measurements: Problems and Solutions," *Hewlett Packard Journal,* vol. 30, no. 3, March 1979.
4. G. S. Bushanam et al., "Measuring Thermal Rises Due to Digital Device Overdriving," *Proceedings of the International Test Conference,* Philadelphia, PA, October 1984, pp. 400–407.
5. V. R. Harwood, "Safeguarding Devices Against Stress Caused by In-Circuit Testing," *Hewlett Packard Journal,* vol. 35, no. 10, October 1984.

P · A · R · T · 13

RELIABILITY

CHAPTER 56
CONDUCTIVE ANODIC FILAMENT FORMATION

Dr. Laura J. Turbini
University of Toronto, Ontario, Canada

56.1 INTRODUCTION

The failure mode, conductive anodic filament (CAF) formation was first observed in the mid-1970s by two different research groups, Bell Laboratories and Raytheon. Since then, further studies have led to an understanding of the factors that affect CAF formation. These include substrate choice, conductor configuration, voltage and spacing, processing, humidity and the storage, and use environment. CAF has been identified as a copper hydroxy chloride salt that causes catastrophic failure when it bridges because it has semiconductor properties. A CAF test method for multilayer boards has been developed by Sun Microsystems, and then introduced as an IPC test method. This chapter will explore these topics in more detail. It then identifies the printed wiring board (PWB) manufacturing tolerance limitations and makes recommendations for the design of CAF test coupons.

56.2 UNDERSTANDING CAF FORMATION

In the mid-1970s, Bell Labs' researchers were concerned about potential failures of printed wiring boards intended for high-voltage switching applications. They reported[1,2] on accelerated life testing of flexible PWBs coated with ultraviolet (UV)-cured resin, and identified two new failure modes: conductive bridges between conductors on the surface, and conductive shorts through the substrate.

AT&T Bell Labs' test vehicle (see Fig. 56.1) was a flexible epoxy-glass PWB, 0.005 to 0.007 in. thick with comb patterns of 0.008 in. lines and 0.009 in. spaces. Some combs were biased on the surface and some were biased through the substrate. Processed boards, coated with conformal coating, were tested from 35 to 95°C, 25 to 95 percent relative humidity (RH), and direct current (DC) voltages up to 400 V. For accelerated testing at 85°C, 80 percent RH, and 78 V bias, failures occurred within two to five days.

Bell Labs' technicians identified two major failure modes that they described as causing "catastrophic loss of insulation resistance due to the formation of conductive bridges between conductors." The first failure mode—*through-substrate shorts*—only occurred above 75°C and 85 percent RH and thus was not considered to be a problem at use conditions (see Fig. 56.2).

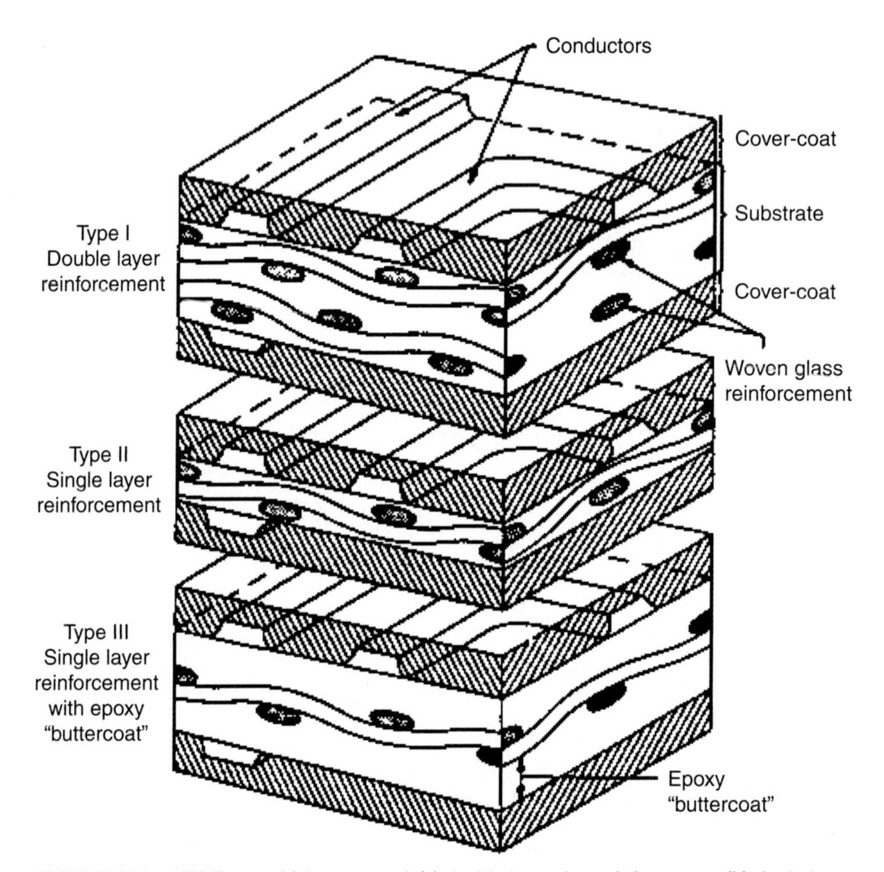

FIGURE 56.1 AT&T test vehicle compared: (a) double-layer glass reinforcement; (b) single-layer glass reinforcement; (c) single-layer glass reinforcement with extra "buttercoat."[2] ©1976 IEEE

The second mode involved shorts between conductors on the same side of the board in which conductive material accumulates between the glass bundles and the epoxy (see Fig. 56.3). R. H. Delaney and J. N. Lahti[2] noted that the thicker the buttercoat, the less this failure was observed. They also detected a failure, which they termed *anodic eruption failure mode,* in which corrosion

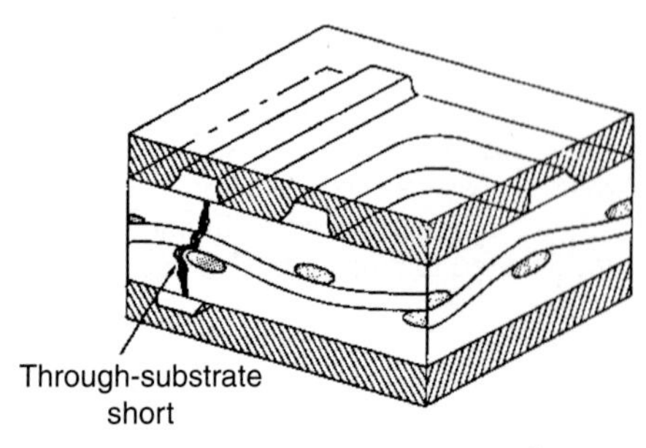

FIGURE 56.2 Through-substrate short.[2] © 1976 IEEE

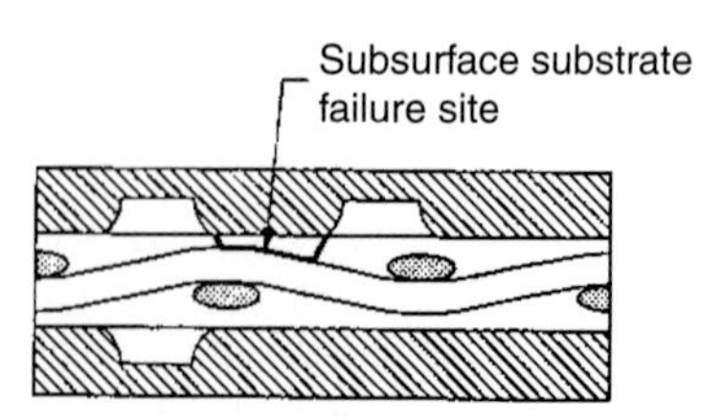

FIGURE 56.3 Subsurface substrate failure.[2] © 1976 IEEE

products emerged from the anode to the covercoat surface, charring the surface, and then growing back through the covercoat to the cathode, where it shorted (see Fig. 56.4).

In 1976, Aaron Der Marderosian at Raytheon, while studying measling, crazing, and delamination, examined the reliability of multilayer PWBs by applying a DC voltage between ground planes and conductor traces (see Fig. 56.5).[3] Test coupons were biased at 100 V and aged at 65°C and 95 percent RH for 10 days. Based on the results, Der Marderosian reported a failure that he termed the "punch thru" phenomenon. This is similar to the "thru substrate shorts" reported by Delaney.[2] To study this failure further, Der Marderosian obtained test coupons from three different vendors biasing some at 100 V DC, others at 100 V AC, and a third group was unbiased. Since "punch thru" was observed only when DC bias was applied, Der Marderosian concluded that this failure was due to electrochemically initiated metal migration. He reported that the number of incidents of "punch thru" decreased as aging voltage was decreased. The addition of a urethane conformal coating appeared to accelerate, not suppress, the problem.

"Punch-thru" is an electrical failure that eventually manifests itself in a rupture of the insulation between two layers of copper metallization. In the early stages, Der Marderosian observed conductive copper-containing deposits that he assumed were CuO along the glass fibers. Eventually these deposits shorted to the cathode, carbonizing the epoxy, which caused

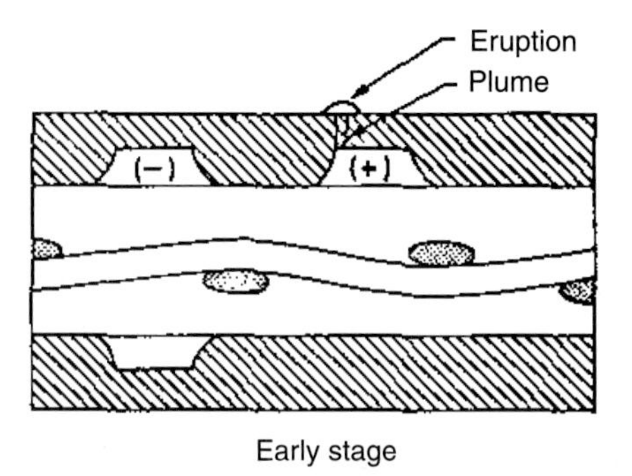

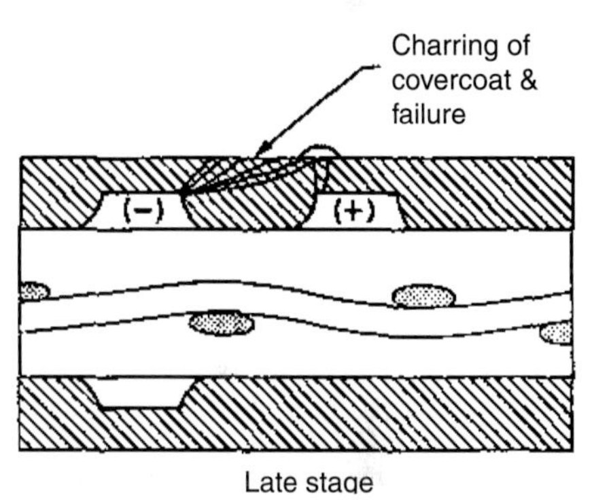

FIGURE 56.4 Anodic eruption failure mode.[2] © 1976 IEEE

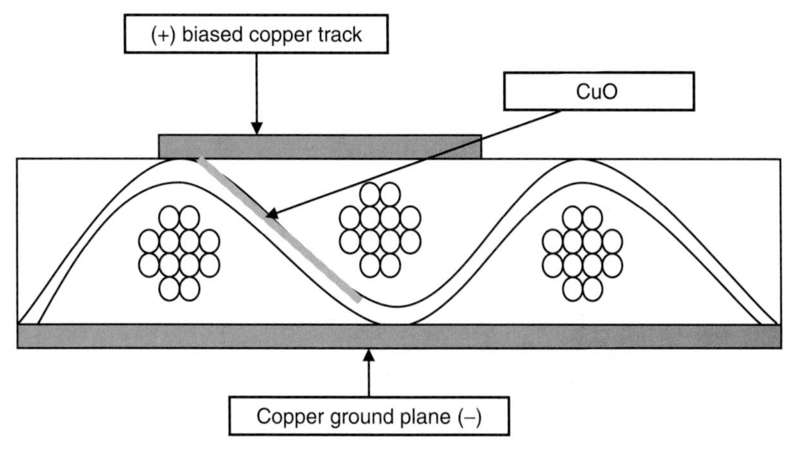

FIGURE 56.5 Raytheon test coupon with "punch thru" from anode trace to copper ground plane.[3]

the epoxy to be more conductive. Epoxy blowout then ruptured the glass fibers. In the later stages, Der Marderosian observed melting of the metal traces.

Equation 56.1 represents Der Marderosian's concept of the production of CuO and $Cu(OH)_2$ at the anode. At the cathode, reduction takes place yielding copper and hydrogen gas. He noted that copper hydroxide decomposes to copper oxide above 60°C.

$$\underset{\text{anode}}{3Cu + 3H_2O} = \underset{\text{cathode}}{CuO + Cu(OH)_2 + 2H_2 + Cu} \tag{56.1}$$

D. J. Lando et al.[4] first used the term conductive anodic filament (CAF) to describe this failure in 1979. They defined the mechanism of CAF formation as a two-step process: degradation of the epoxy/glass interface, followed by the electrochemical reaction (see Eq. 56.2).

Anode:

$$Cu = Cu^{n+} + ne^-$$
$$H_2O = {}^1\!/_2\,O_2 + 2H^+ + 2e^-$$

Cathode:

$$2H_2O + 2e^- = H_2 + 2OH^- \tag{56.2}$$
$$H_2O + {}^1\!/_2\,O_2 + 2e^- = 2OH^-$$
$$Cu^{n+} + ne^- = Cu$$

In 1980, T. L. Welsher et al.[5] reported that CAF was potentially a serious reliability problem for closely spaced conductors and proposed a mean time to failure (MTTF) based on Eq. 56.3:

$$\text{MTTF} = a(H)^b \exp\left(\frac{E_a}{RT}\right) + d\frac{L^2}{V} \tag{56.3}$$

where
H = humidity
E_a = activation energy
R = gas constant
T = temperature in Kelvin
L = conductor spacing
V = voltage
a, b, E_a, d = material-dependent

Welsher et al. noted that additional work was needed to determine the exact dependence of CAF on conductor spacing and humidity. They also reported that glass-reinforced triazine is CAF-resistant.

In 1981, they expanded their view of the MTTF and reported it as shown in Eq. 56.4:

$$\alpha\left(1 + \beta\frac{L^n}{V}\right) \cdot H^\gamma \exp\frac{E_a}{kT} \tag{56.4}$$

where
α, β = material-dependent constants
γ = humidity-dependent constant
n = related to the orientation of the conductors
L = spacing
V = voltage
H = humidity
E_a = activation energy
k = Boltzman's constant
T = temperature in Kelvin

W. J. Ready and L. J. Turbini[6] studied the effect of voltage and spacing on CAF failures for a hole-to-hole test pattern (see Fig. 56.6). Using two different spacings (0.50 mm and 0.75 mm)

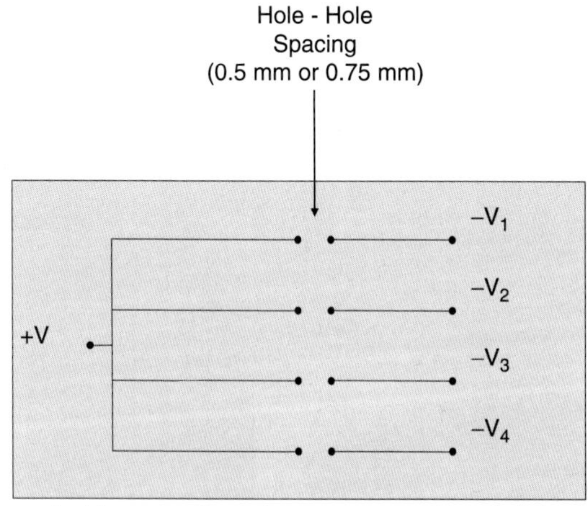

FIGURE 56.6 Hole-to-hole test vehicle used by Ready and Turbini.[6]

and two different bias voltages (150 V and 200 V), they were able to show the relation is L^4/V^2 (see Eq. 56.5):

$$\text{MTTF} = c \cdot \exp\left(\frac{E_a}{kT}\right) + d\left(\frac{L^4}{V^2}\right) \tag{56.5}$$

In addition, they provided evidence for different CAF morphologies when different flux constituents were used. Coupons that were aged without flux exhibited only a copper salt at the fiber and epoxy interface, whereas those processed with water-soluble flux showed the presence of a copper-containing material within the polymer matrix. Ready and Turbini also identified the chemical nature of CAF as atacamite: $Cu_2Cl(OH)_3$. An examination of the Pourbaix diagram (see Fig. 56.7) for the copper-chlorine-water system reveals that copper hydroxy chloride is insoluble below pH4 and thus this salt will grow from the anode, which is acidic. They also explained that the high conductivity of this salt, which causes catastrophic failure when it bridges to the cathode, suggests that it conducts through electrons and holes, that is, it has semiconductor properties.

CAF filaments are typically no larger than 50 μm in diameter and 0.2 mm in length.

56.3 ELECTROCHEMICAL MIGRATION AND FORMATION OF CAF

CAF is a conductive copper-containing salt created by electrochemical migration. Thus, it is important to review the electrochemical migration process in more detail. It begins with a substrate material, which serves as an insulator, preventing current flow between adjacent traces. This resistance to current flow is termed *insulation resistance*.

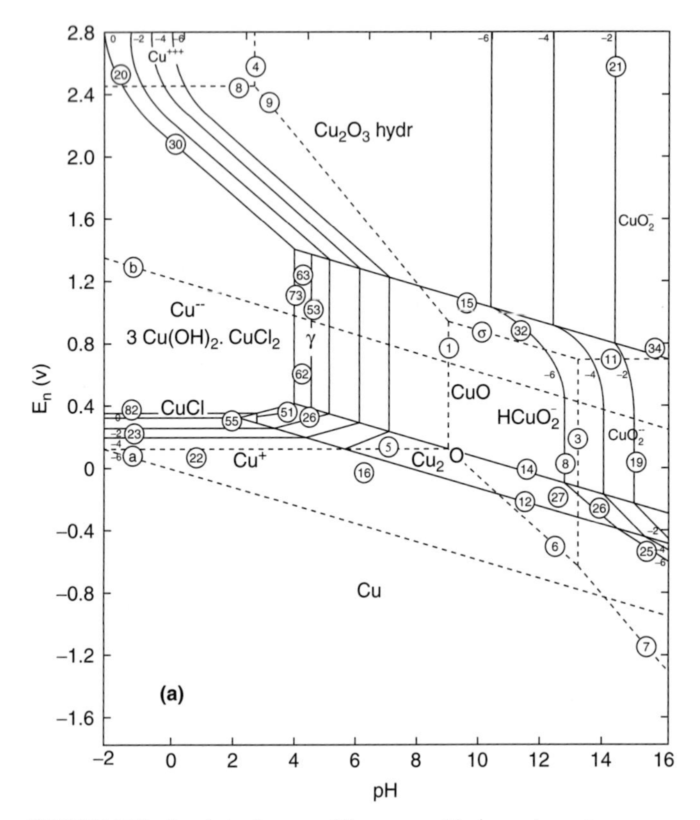

FIGURE 56.7 Pourbaix diagram of the copper-chlorine-water system.

Ohm's law (Eq. 56.6) describes the linear relationship between voltage, current, and resistance:

$$V = IR \tag{56.6}$$

where V = voltage drop
I = current in amps
R = resistance in ohms

Resistance is not an intrinsic property of a material. Rather, it depends on some geometric factors as well as the intrinsic resistivity of the substrate. For surface resistance, use Eq. 56.7:

$$R = \rho\left(\frac{d}{A}\right) \tag{56.7}$$

where ρ = resistivity of the material
d = the separation between the conductors traces
A = the cross-sectional area

Conductivity (σ) is the inverse of resistivity:

$$\sigma = \frac{1}{\rho}$$

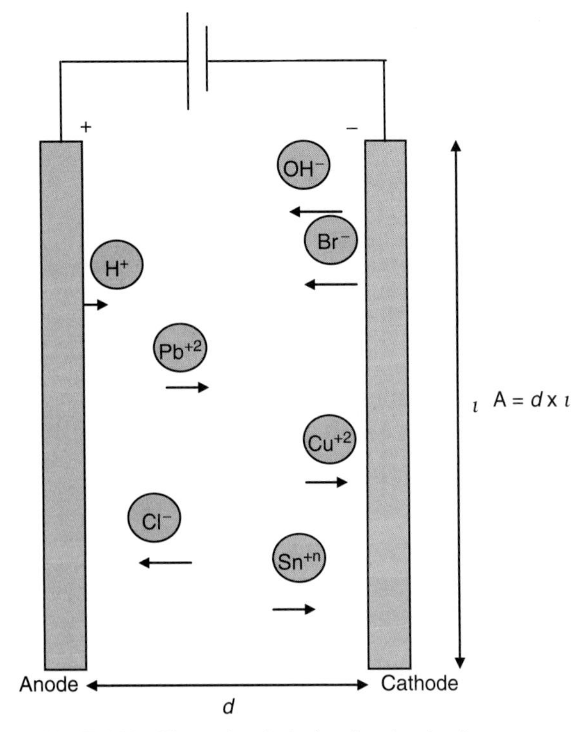

FIGURE 56.8 Electrochemical migration showing the movement of positive ions toward the cathode and negative ions toward the anode.

Both resistivity and conductivity can be affected by processing chemicals.

Electrochemical migration (see Fig. 56.8) is the movement of an ionic species under the influence of a DC voltage. Electrochemical corrosion increases as circuit spacing decreases. This is due to the increase in the electric field, which is inversely proportional to the spacing between the conductors (see Eq. 56.8).

$$E = \frac{V}{d} \tag{56.8}$$

where E = the electric field
V = the voltage
d = the spacing between the conductors

Moisture is essential to electrochemical migration. In the presence of moisture, metal ions form at the anode and migrate toward the cathode, where they plate out to form dendrites. When a dendrite bridges the gap, a short occurs that may burn out quickly due to the high current within the fragile dendrite.

In the case of CAF formation, copper ions are created electrochemically at the anode. Residual chloride ions from the reaction of epichlorhydrin with Bisphenol A (forming a precursor to the epoxy) exist at low levels in the FR-4 substrate. These react with the copper ions and water to produce the CAF—$Cu_2Cl(OH)_3$—which grows from the anode since it is insoluble in acid. The favored path for this is at the epoxy-glass interface.

Ionic contaminants, which dissolve in the moisture on the surface, increase the conductance (lower the resistance) of the insulating layer between the anode and cathode. Their presence enhances electrochemical migration. How much the migration is enhanced depends on several factors: the solubility of the ion, the mobility of the ion, the effect of pH on solubility, the reactivity of the ion, temperature, and relative humidity.

The solubility of the ion and its mobility in water are important because conductivity depends on both of these factors (see Eq. 56.9).

$$\sigma = \Sigma N_i q_i \mu_i \tag{56.9}$$

where N_i = the number of charge carriers of a particular type (e.g., Na^+ cations, or Cl^- anions)
q_i = the charge of the carrier
μ_i = the mobility of the charge carrier

The mobility of a given ion is dependent on temperature and its diffusion coefficient per Eq. 56.10:

$$\mu_{\text{ion}} = \left(\frac{q}{kT}\right) D_{\text{ion}} \tag{56.10}$$

where k = Boltzmann's constant
T = degrees Kelvin
D_{ion} = Diffusion constant for a particular ion.

Tables of ionic conductance in aqueous solution as well as diffusion constants can be found in the *Handbook of Chemistry and Physics*.[7]

Most contaminants from processing do not exist in isolation from one another. Rather, there may be several ionic species present, some of which will interact to create new ionic species.

Solubility of the contaminant salts also depends on the pH. Water ionizes under a bias voltage, creating an acidic medium at the anode and a basic medium at the cathode, as shown in the following reactions:

Anode: $H_2O = {}^1/_2O_2 + 2H^+ + 2e^-$

Cathode: $2H_2O + 2e^- = 2OH^- + H_2$ or
$O_2 + H_2O + 4e^- = 4(OH)^-$

Temperature is another important factor to consider. An increase in temperature increases ion solubility, mobility, and reactivity.

In summary, several factors affect electrochemical migration. These include the nature of substrate and metallization, the presence of contaminants, the voltage gradient, and the presence of sufficient moisture. This last element is critical, for without sufficient monolayers of water on the surface, ion mobility is impossible.

56.4 FACTORS THAT AFFECT CAF FORMATION

A number of factors affect CAF formation. These include substrate material, conductor configuration, processing, voltage and spacing, soldering flux, and humidity in the storage and use environment.

56.4.1 Substrate Material

Lando et al.[4] compared FR-4 with several substrates: G-10 (a non-fire-retardant epoxy/woven glass material), polyimide/woven glass (PI), triazine/woven glass, epoxy/woven kevlar, and finally polyester/woven and chopped glass. Similarly, B. Rudra et al.[8] performed an extensive experimental comparison among the substrates: bismaleimide triazine (BT), cyanate esters (CE), and FR-4. In addition, Ready[9] compared the CAF susceptibilities of FR-4 with CEM-3 (a substrate similar to G-10 except with chopped glass) and MC-2 (a blended polyester and epoxy matrix with woven glass face sheets and a chopped glass core). Of all materials tested by these investigators, the BT material proved to be most resistant to CAF formation (due to its low moisture absorption characteristics). Conversely, the MC-2 substrate proved to have the least resistance to CAF formation. The susceptibility of the materials follows the trend below:

MC-2 > Epoxy/Kevlar > FR-4 ~ PI > G-10 > CEM-3 > CE > BT

To ensure immunity to CAF, the laminate of preference is BT. In recent years, laminate suppliers have developed other CAF-resistant substrate materials. However, there is a cost penalty to consider.

56.4.1.1 Conductor Configuration. Lando et al.[4] evaluated several different conductor configurations:

- Line-to-line (L-L)
- Hole-to-line (H-L)
- Hole-to-hole (H-H)

This evaluation showed that susceptibility to CAF is H-H > H-L > L-L. Lahti et al.[10] demonstrated that the smaller the spacing between conductors and the greater the proximity of the glass fibers to the copper conductors, the faster the growth of CAF. They noted that for a multilayer board, failure initiated in the most deeply buried layers.

56.4.2 Processing

The epoxy-glass bond can be weakened by the drilling or depaneling processes in PWB manufacture. It will also be weakened during the soldering process due to the difference in coefficient of thermal expansion (CTE) of the epoxy and glass. Data from Turbini et al.[11] demonstrated that the increased soldering temperature associated with lead-free alloys significantly increases the incidence of CAF.

Another processing factor is the diffusion of polyglycols from water-soluble flux into the PWB substrate that occurs during soldering. Since the diffusion rate is temperature-dependent, the length of time that the board is above the glass transition temperature will affect the amount of polyglycol absorbed into the epoxy, and that will in turn affect its electrical properties. J. A. Jachim et al.[15] reported on water-soluble flux-treated test coupons that were prepared using two different thermal profiles. Those which experienced the higher thermal profile exhibited a Surface Insulation Resistance (SIR) level that was an order of magnitude lower than those processed under less aggressive thermal conditions.

56.4.3 Voltage and Spacing

Two other critical factors used in determining CAF susceptibility are voltage and spacing. The CAF growth is fueled by the electric field (V/d), and failure occurs more quickly with smaller spacing between conductors.

56.4.4 Effect of Soldering Flux

It has been shown that polyglycols[12] (which are found in water-soluble fluxes) diffuse into the epoxy during soldering.[13] This absorption has been shown to reduce electrical performance by increasing moisture uptake by the substrate.[14] Jachim et al.[15] were the first to link the use of polyglycols in soldering fluxes and fusing fluids to increased susceptibility to CAF formation.

Ready et al.[16] showed that the use of certain water-soluble fluxes or fusing (hot air solder level [HASL]) fluids could increase CAF formation. In examining a catastrophic field failure (see Fig. 56.9), Ready et al.[16] demonstrated the presence of a copper-bromide containing salt between an innerlayer power plane and ground pin, separated by 0.005 in. Using a test coupon manufactured in the same lot as the failed product, they extracted the flux residues from an innerlayer of the multilayer board and used ion chromatography to match the residues from the board's innerlayer with the constituents of the flux.

FIGURE 56.9 Catastrophic field failure of military hardware. The conductive filament grew from the +20 V ground plane to the –20 V ground pin. Flux residues enhanced this failure rate.[16]

The morphology of CAF is different when certain fluxes are used.[6] In Fig. 56.10(a), when no flux is used, CAF forms only as a crystalline filament at the epoxy-glass interface; in Fig. 56.10(b), the flux contained polyethylene propylene glycol (1800) and there appears to be copper-containing compounds in stratified layers within the matrix in addition to the filament at the epoxy-glass interface; and in Fig. 56.10(c), the flux contained a linear aliphatic polyether and one sees a copper-containing compound appearing in a striated morphology as well as in the filament at the interface.

56.4.5 Humidity in the PWB Storage and Use Environment

J. A. Augis et al.[17] determined that there is a humidity threshold below which CAF formation will not occur. They found that this relative humidity threshold depends on operating voltage and temperature. It is important to remember that this relative humidity need not be present in the operating environment. Moisture absorption can occur during any part of the assembly's lifetime.

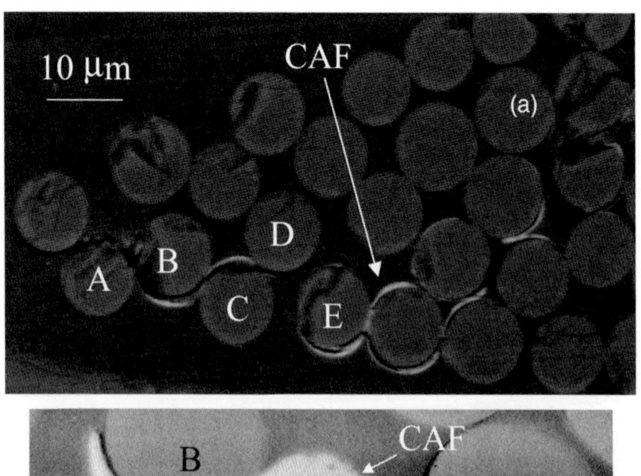

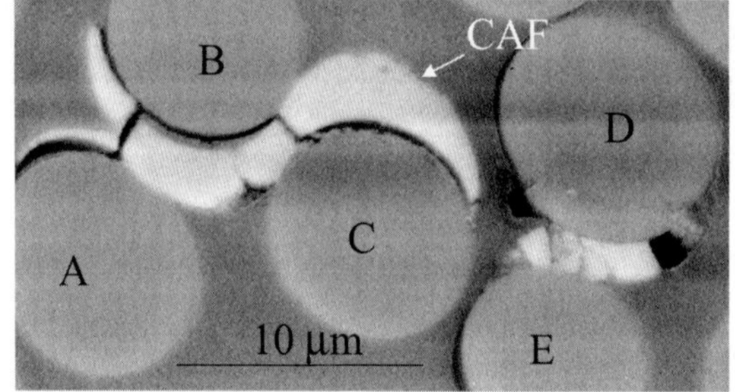

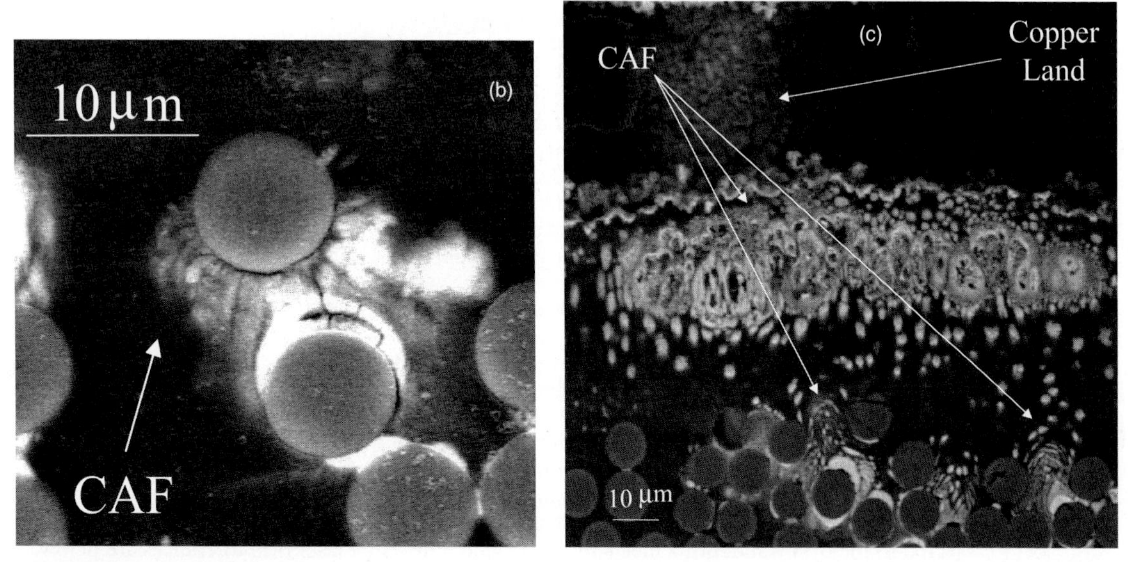

FIGURE 56.10 Morphology of CAF differs when different flux constituents are used: (a) CAF only at the glass-epoxy interface; (b), (c) an additional copper-containing compound within the polymer matrix.

This is particularly critical during transportation or storage, when the assembly may experience harsh environmental conditions.

56.5 TEST METHOD FOR CAF-RESISTANT MATERIALS

In the late 1990s, laminate suppliers began to develop new materials that they marketed as CAF-resistant. To evaluate these materials, K. Sauter[18] developed a CAF test vehicle, which consisted of a multilayer board with daisy-chained hole-to-hole spacing of 0.25 mm, 0.375 mm, 0.50 mm, and 0.625 mm. The holes were either inline with the glass fiber direction or staggered. Accelerated aging was done at 65 °C and 85 percent RH for 500 hours. Sauter's results show variations based on laminate material, manufacturer, and diagonal versus inline holes, with the former being more CAF-resistant. He also defined a "readily conductive region" around the plated-through holes, which must be considered in establishing design rules. This test vehicle and procedure has been developed as an IPC Test Method (IPC-TM-2.6.25), "Conductive Anodic Filament (CAF) Resistance."[19]

56.6 MANUFACTURING TOLERANCE CONSIDERATIONS

Test board designs for performing statistical studies of CAF-resistant materials must take into account the limitations of the PWB manufacturing process.[20] As hole-to-hole spacing becomes closer, the limits of the drilling process must be considered. Drill wander can be caused by splay when the drill enters the board at an angle. These two factors are exacerbated with smaller drill size. Typical manufacturing tolerances for hole-drilling equipment are specified at ±0.075 mm. Based on this, test board design for statistical CAF studies should have the minimum edge-to-edge spacing of 0.375 mm, since the manufacturing tolerance should be no more than 20 percent of the hole-to-hole distance.

CAF-resistant substrates are frequently more expensive than traditional FR-4, and therefore they are reserved for products that require high reliability, optimizing both cost and performance. Designs are multilayer rather than double-sided rigid boards. An important difference between a multilayer board and a double-sided rigid board is the glass-to-resin ratio. In drilling the holes, more damage will occur at the epoxy-glass interface for boards with higher glass content.

Typical double-sided PWBs use eight sheets of 7628-glass cloth with a thickness of 0.175 mm (0.007 in.) to create a 1.5 mm (0.062 in.) thick board. The portion of resin in this board would be 30 to 40 percent. For a multilayer board, the glass cloth thickness may range from 0.035 mm to 0.10 mm (0.001 in. to 0.004 in.) per layer. In this case, the amount of resin ranges from 45 to 50 percent in the case of the 0.10 mm glass cloth thickness, it ranges from 55 to 65 percent in the case of the 0.035 mm thickness. In the drilling process, the board with a higher resin content creates less hole damage than one with a lower resin content.

Microvias that are created by laser drilling can be located at a reduced pitch for CAF studies because one does not expect drill damage from a laser-ablated hole and it will not be subjected to the drill wander of a mechanically drilled hole. Thus, laser drill tolerances are below 0.025 mm (0.001 in.).

The catastrophic failure presented earlier in this chap. (see Fig. 56.9) was caused by a power plane (anode), which created a CAF that shorted to the hole (cathode). In designing a test board, the power plane must always be the anode in order to reduce the current density for corrosion. A large cathode in contact with a small anode creates a high-current density at the anode and increases the corrosion rate. Present BGA board designs have microvias of 1 mm (0.040 in.) or even 0.8 mm (0.032 in.). These become problematic when the plane-to-hole separation is below 0.2 mm (0.008 in.).

REFERENCES

1. Boddy, P. J., et al., "Accelerated Life Testing of Flexible Printed Circuits: Part I: Test Program and Typical Results," *IEEE Reliability Physics Symposium Proceedings*, Vol. 14, 1976, pp. 108–113.
2. Delaney, R. H., and Lahti, J. N., "Accelerated Life Testing of Flexible Printed Circuits: Part II Failure Modes in Flexible Printed Circuits Coated with UV-Cured Resins," *IEEE Reliability Physics Symposium Proceedings*, Vol. 14, 1976, pp. 114–117. © 1976 IEEE. Reprinted with permission.
3. Der Marderosian, A., "Raw Material Evaluation through Moisture Resistance Testing," paper presented at in IPC 1976 Fall Meeting, San Francisco, IPC-TP-125.
4. Lando, D. J., Mitchell, J. P., and Welsher, T. L., "Conductive Anodic Filaments in Reinforced Polymeric Dielectrics: Formation and Prevention," *IEEE Reliability Physics Symposium Proceedings*, Vol. 17, 1979, p. 51–63.
5. Welsher, T. L., Mitchell, J. P., and Lando, D. J., "CAF in Composite Printed Circuit Substrates: Characterization, Modeling and a Resistant Material," *IEEE Reliability Physics Symposium Proceedings*, Vol. 18, 1980, pp. 235–237.
6. Ready, W. J., and Turbini, L. J., "The Effect of Flux Chemistry, Applied Voltage, Conductor Spacing, and Temperature on Conductive Anodic Filament Formation," *Journal of Electronic Materials*, Vol. 31, No. 11, 2002, pp. 1208–1224.
7. Linde, D. R. (ed.), *Handbook of Chemistry and Physics, 80th ed.*, CRC Press, 1999.
8. Rudra, B., Pecht, M., and Jennings, D., "Assessing Time-to-Failure Due to Conductive Filament Formation in Multi-Layer Organic Laminates," *IEEE Transactions on Components, Packaging and Manufacturing Techniques—Part B*, Vol. 17, No. 3, 1994, pp. 269–276.
9. Ready, W. J., *Factors Which Enhance Conductive Anodic Filament (CAF) Formation*, Master Thesis in Materials Science and Engineering, Georgia Institute of Technology, 1997.
10. Lahti, J. N., Delaney, R. N., and Hines, J. N., "The Characteristic Wearout Process in Epoxy-Glass Printed Circuits for High Density Electronic Packaging," *IEEE Reliability Physics Symposium, Proceedings*, Vol. 17, 1979, p. 39.
11. Turbini, L. J., Bent, W. R., and Ready, W. J., "Impact of Higher Melting Lead-Free Solders on Reliability of Printed Wiring Assemblies," *Journal of Surface Mount Technology*, Vol. 13, No. 4, 2000, pp. 10–14.
12. Zado, F. M., "Effects of Non-Ionic Water Soluble Flux Residues," *Western Electric Engineer*, No. 1, 1983, pp. 41–48.
13. Brous, J., "Electrochemical Migration and Flux Residues: Causes and Detection," *Proceedings of NEPCON West 1992*, pp. 386–393.
14. Brous, J., "Water Soluble Flux and Its Effect on PC Board Insulation Resistance," *Electronic Packaging and Production*, Vol. 21, No. 7, 1981, p. 80.
15. Jachim, J. A., Freeman, G. B., Turbini, L. J., "Use of Surface Insulation Resistance and Contact Angle Measurements to Characterize the Interactions of Three Water Soluble Fluxes with FR-4 Substrates," *IEEE Transactions on Components, Packaging, and Manufacturing Technology, Part B*, Vol. 20, No. 4, 1997, pp. 443–451.
16. Ready, W. J., Turbini, L. J., Stock, S. R., and Smith, B. A., "Conductive Anodic Filament Enhancement in the Presence of a Polyglycol-Containing Flux," *IEEE International Reliability Physics Symposium Proceedings*, Dallas, 1996, pp. 267–272.
17. Augis, J. A., DeNure, D. G., LuValle, M. J., Mitchell, J. P., Pinnel, M. R., and Welsher, T. L., "A Humidity Threshold for Conductive Anodic Filaments in Epoxy Glass Printed Wiring Board," *Proceedings of 3rd International SAMPE Electronics Conference*, 1989, pp. 1023–1030.
18. Sauter, K., "Electrochemical Migration Testing Results: Evaluating PWB Design, Manufacturing Process, and Laminate Material Impacts on CAF Resistance," *Proceedings IPC Printed Circuits Expo 2002*, Long Beach, CA, 2002, EX02-S08-4.
19. IPC-TM-2.6.25 (IPC Test Method), "Conductive Anodic Filament (CAF) Resistance."
20. Parry, G., Cooke, P., Caputo, A., and Turbini, L. J. "The Effect of Manufacturing Parameters on Board Design on CAF Evaluation," *Proceedings of the International Conference on Lead-free Soldering*, Toronto, 2005.

CHAPTER 57
RELIABILITY OF PRINTED CIRCUIT ASSEMBLIES

Judith Glazer*
Hewlett-Packard Company, Electronic Assembly Development Center, Palo Alto, California

This chapter describes the response of functional printed circuit board assemblies (PCAs) to environmental stresses—that is, their reliability in service—and the influence of design, materials, and manufacturing decisions on this behavior. A variety of stresses may be present in the service environment of the assembly. *Thermal* stresses come from fluctuations in the ambient temperature in the service environment of the assembly or from power dissipation of high-power devices mounted on the printed circuit board (PCB). There are also thermal stresses associated with assembly and rework. *Mechanical* stresses may be due to bending and flexing of the assembly during later assembly steps or in service, mechanical shock during transportation or use, or mechanical vibration, for example, from cooling fans. *Chemical* sources of environmental stresses include atmospheric moisture, corrosive gases (for example, smog or industrial process gases), and residual chemically active contaminants from the assembly processes (for example, from flux). These environmental stresses may act singly or in concert with one another and the electrical potential differences that exist when the assembly is functioning to cause electrical failures in the printed circuit assembly (PCA). This chapter will focus on the reliability of the PCB and the interconnects to it. Reliability of the electrical components themselves is beyond the scope of this chapter (see Fig. 57.1).

By defining reliability as the response of functioning assemblies to environmental stresses, we have excluded the large class of production defects that are detected in the testing processes immediately after manufacturing or that will cause the assembly to be nonfunctioning from the outset. This chapter will focus on the delayed effects of manufacturing defects and the wear-out mechanisms of properly manufactured product.

The remainder of this chapter is organized into six major sections:

57.1 Fundamentals of Reliability

57.2 Failure Mechanisms of PCBs and Their Interconnects

57.3 Influence of Design on Reliability

57.4 Impact of PCB Fabrication and Assembly on Reliability

* Significant portions of this chapter are drawn from T. A. Yager, "Reliability," chap. 30, *Printed Circuits Handbook,* 3d ed. (Coombs, ed.), 1988.

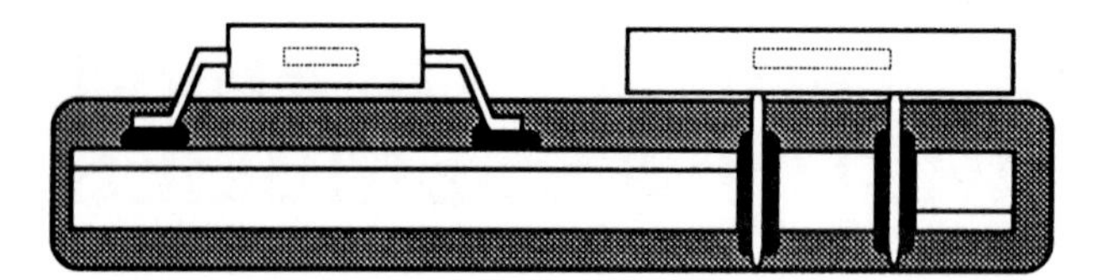

FIGURE 57.1 Schematic drawing of a printed circuit board assembly. This chapter focuses on the reliability of the printed circuit board and the interconnect between the printed circuit board and the components (shaded area in the drawing).

57.5 Influence of Materials Selection on Reliability
57.6 Burn-in, Acceptance Testing, and Accelerated Reliability Testing

Where applicable, each section covers printed circuit boards, printed circuit board assemblies, and components and their packages, in turn. Section 57.2 is the core of the chapter; it covers the fundamental of the failure mechanisms and is the assumed underlying basis of the subsequent sections. The breadth of this chapter, the complexity of the failure mechanisms involved, and the rapid evolution of the field mean that this chapter can provide only a brief overview of important topics in PCB and PCA reliability, many of which are the subject of books in their own right. The reader is encouraged to refer to the references and suggestions for further reading given at the end of the chapter before attempting quantitative reliability predictions.

57.1 FUNDAMENTALS OF RELIABILITY

57.1.1 Definitions

The reliability of a component or system can be defined as the *probability* that a functioning product at time zero will function *in the desired service environment* for a *specified amount of time.* Without these three parameters, the question "Is x reliable?" cannot be answered yes or no. Since reliability describes the probability that the product is still functioning, it is related to the cumulative number of failures. Mathematically, the reliability of an object at time t can be stated as

$$R(t) = 1 - F(t)$$

where $R(t)$ is the reliability at time t (i.e., the proportion of parts still functioning) and $F(t)$ is the fraction of the parts or systems that have failed at time t. Time may be measured in calendar units or some other measure of service time such as on/off cycles or thermal or mechanical vibration cycles. The unit of time that makes sense depends on the failure mechanism. When several failure modes are present, it is often helpful to think in terms of several time scales.

A plot of the failure rate of a product as a function of time typically takes the shape of a "bathtub" curve (see Fig. 57.2). This curve illustrates the three phases that occur during the lifespan of a product from a reliability perspective. In the first, infant mortality phase, there is an initially high but rapidly declining failure rate caused by infant mortality. Infant mortality is typically caused by manufacturing defects that went undetected during inspection and testing and lead to rapid failure in service. Burn-in can be used to remove these units before shipment. The second phase, the normal operating life of the product, is characterized by a period of stable, relatively low failure rates.

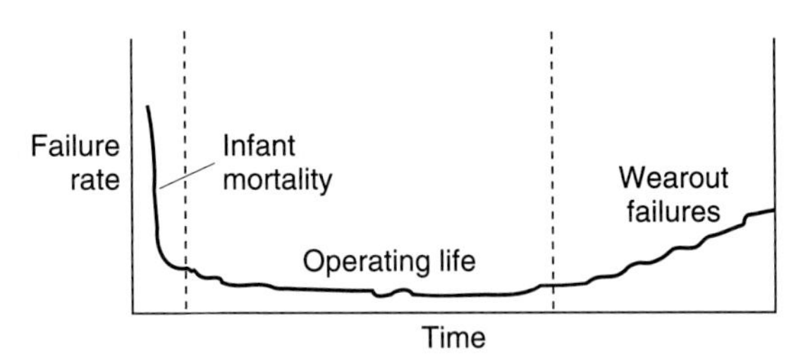

FIGURE 57.2 Classic bathtub reliability curve showing the three stages during the life of a product from a reliability perspective: infant mortality, steady-state, and wear-out.

During the operating life, failures occur apparently randomly and the failure rate r is roughly constant with time. An exponential life distribution is often assumed to describe the behavior in this region. In that case,

$$r = \left(\frac{N_t}{N_o}\right)\left(\frac{1}{\Delta t}\right)$$

and

$$R(t) = e^{-rt} = e^{\frac{-t}{\text{MTBF}}}$$

where N_t = number of failures in time interval Δt
N_o = number of samples at the beginning of the interval
MTBF = mean time between failures

During the third phase, the wear-out period, the failure rate increases gradually due to wear-out phenomena until 100 percent of the units have failed. For some systems, the second steady-state region may not exist; for solder joints, the wear-out region may extend over most of the life of the assembly. Understanding wear-out phenomena, which manifest themselves in properly manufactured parts after a period of service, and predicting when they will significantly affect the failure rate, are the primary focuses of this chapter.

Most wear-out phenomena can be characterized by cumulative failure distributions governed by either the Weibull or the log-normal distribution. Weibull distributions have been successfully used to describe solder joint and plated-through-hole fatigue distributions, while log-normal distributions are generally associated with electrochemical failure mechanisms. While these distributions may be quite narrow in some cases, their use should serve as a reminder that even with nominally identical samples, failures will be statistically distributed over time.

A practical use of fitting a distribution to reliability data is to extrapolate to smaller failure rates or other environmental conditions. To simplify the equations, the expressions in the text refer to the mean life of the relevant portion of the assembly. If the constants that define the failure distribution are known, the time to reach a smaller proportion of failures may be readily calculated. For example, for failure modes that are described by a Weibull distribution, the time t to reach $x\%$ failures is given by:

$$t(x\%) = t(50\%)\left[\frac{\ln(1 - 0.01x)}{\ln(0.5)}\right]^{1/\beta}$$

where β is the Weibull shape parameter, usually between 2 and 4 for solder joint failures.

57.1.2 Reliability Testing

Almost every reliability test program must solve the problem of determining whether an object is reliable in a calendar time period that is much shorter than the expected use period. Obviously, one cannot spend 3 to 5 years testing a personal computer that will be marketed for an even shorter time span or 20 years testing a military system. Depending on the failure mechanism, there are two approaches, which may be combined: (1) accelerate the frequency of the occurrence that causes failure and test the ability of the object to survive the expected number of events, or (2) increase the severity so that fewer occurrences are needed. Drop tests that simulate shock during transportation are an example of the first approach. Since the time between drops does not affect the amount of damage caused, a lifetime of drops can be conducted in rapid succession. However, the effect of temperature and humidity on corrosion over the lifetime of the product can be tested only by increasing the temperature, the humidity, or the concentration of contaminants, or some combination of these. The difficulty is ensuring that the test reproduces and/or correlates to the failure mechanism in service.

To use this data for making true reliability predictions—that is, *the probability of failures at a given time under given conditions*—testing must be continued until enough parts fail that a life distribution can be estimated. Unfortunately, this process can be time consuming and qualification tests are often substituted. Qualification test protocols specify a *maximum number of failures that may be observed in a specified period* in a sample of specified size. If few or no failures occur, a qualification test provides almost no information about the failure distribution; for example, the probability of failure during the next time interval is unknown. This limitation of qualification testing is minimized when the life distribution for properly manufactured samples is already known or can be estimated based on experience with similar designs. Many of the "reliability" tests described in Sec. 57.6 are actually qualification tests.

Many reliability or qualification testing schedules follow neither of these schemes. Instead, they test the ability of the product to survive a sequence of tests under extremely severe conditions for a short time or small number of exposures. Again, this type of testing may be adequate when it is supported by long experience with both the product type and its use environment; however, it is risky because it is not based on ensuring that probable failure modes will not occur in the life of the product. When new technologies or geometries are introduced, the old tests may not always be conservative. By the same token, irrelevant failure modes that would not occur in service may be introduced by the harsh test conditions.

57.2 FAILURE MECHANISMS OF PCBs AND THEIR INTERCONNECTS

This section will discuss the most important failure mechanisms of PCBs and the interconnects between PCBs and the components mounted on them. The discussion of PCB failure mechanisms will be more detailed since interconnect failures have been described far more extensively elsewhere. Whatever the environmental stress or the material response, these failures ultimately manifest themselves in terms of the functionality of the assembly, first as a change in electrical resistance between two points and then as electrical shorts or opens.

57.2.1 PCB Failure Mechanisms

PCB failure mechanisms fall into three groups: thermally induced failures, of which plated-through-holes are the most important example; mechanical failures; and chemical failure mechanisms, of which dendritic growth is the most important example.

57.2.1.1 Thermally Driven Failure Mechanisms. PCBs are exposed to thermal stresses in a variety of situations. These may be either prolonged exposure to an elevated temperature or isolated or repeated temperature cycles. These temperature cycles can cause various PCB failures. The most important sources of thermal stress are:

- *Thermal shocks and thermal cycles during PCB manufacturing.* Thermal shocks are usually defined as temperature ramps faster than ~30°C/s, but include any ramp fast enough that temperature differentials play an important role. Examples include solder mask cure and hot-air solder leveling.
- *Thermal shocks and cycles during printed circuit assembly.* Examples are glue cure, solder reflow, wave soldering, and rework using a soldering iron, hot air, or molten solder pot.
- *Ambient thermal cycles in service.* Examples are going from inside to outside temperatures or ground to upper atmospheric temperatures, and elevation in box temperature due to heat dissipation from functioning electronic components.

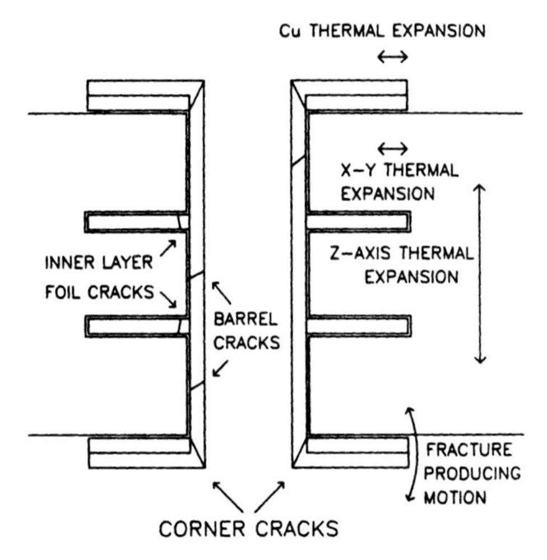

FIGURE 57.3 Schematic diagram of a plated-through-hole in a cross section of a four-layer printed circuit board showing common failure locations under thermal stress.

The primary PCB failure mechanisms accelerated by these thermal stresses are plated-through-hole cracking and delamination of the laminate.

Plated-Through-Hole Failures Due to Thermal Shocks or Cycling. Plated-through-holes (PTHs) are the most vulnerable features on PCBs to damage from thermal cycling and the most frequent cause of printed circuit board failures in service. PTHs include holes for through-hole (TH) components and vias that make electrical connections between layers. Figure 57.3 shows the common failure locations. Most organic resin-matrix substrate materials are highly anisotropic, with a much higher CTE above the glass transition temperature T_g in the through-thickness (z) direction than in the plane of the woven matrix cloth (the x-y plane of the board). Since above T_g the CTE climbs sharply, aggressive thermal cycles can result in large strains in the z direction and, consequently, on the PTHs (see Fig. 57.4[1]). The PTH acts like a rivet, which resists this expansion, but the Cu barrel is stressed and may crack, causing electrical failure. Figure 57.4 also illustrates the increasing strain on the barrel associated with a high temperature excursion. Failure may occur in a single cycle or may take place by initiation and growth of a fatigue crack over the course of a number of cycles. For high-aspect-ratio through-holes subject to repeated thermal shocks from room temperature to solder reflow temperatures (220 to 250°C) during board fabrication (e.g., hot-air solder leveling) and assembly (reflow, wave soldering, rework), it is not unheard of to encounter failures after 10 or fewer of these thermal cycles.

On a physical level, the number of thermal cycles to failure is affected by the strain imposed on the Cu in each cycle and the fatigue resistance of the copper. These factors are in turn controlled by a number of environmental, material, and manufacturing parameters. Low-cycle metal fatigue, in which most of the strain is plastic strain, can be treated approximately with the Coffin-Manson relation:

$$N_f \propto \frac{1}{2}\left(\frac{\varepsilon_f}{\Delta\varepsilon}\right)^m$$

where N_f = number of cycles to failure
$\Delta\varepsilon$ = strain
ε_f = strain ductility factor, which correlates closely with tensile ductility
m = constant near 2.

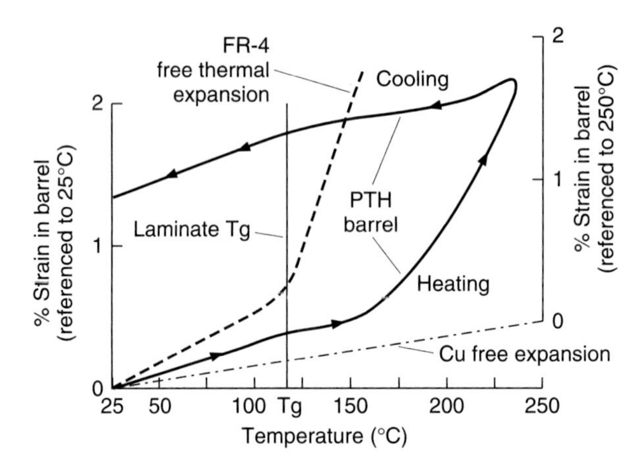

FIGURE 57.4 Strain vs. temperature for FR-4 (epoxy-glass), copper, and a PTH barrel in an FR-4 board during a single thermal cycle from 25 to 250 to 25°C. While the thermal expansion of the individual materials is fully reversible, much of the strain in the Cu PTH barrel is plastic, so most of the strain is not reversed during cooling. Note that the rate of thermal expansion of the FR-4 increases sharply at T_g. Results from Ref. 1.

This relation will significantly underestimate life for high cycle fatigue which can occur after repeated thermal cycling in service. The strain $\Delta\varepsilon$ can be estimated by finite element modeling or analytically. If no other data are available, ε_f for electroplated Cu can be approximated as 0.3.

The number of cycles to failure can be increased by increasing $\varepsilon_f/\Delta\varepsilon$, primarily by decreasing $\Delta\varepsilon$ by:

- Decreasing or eliminating thermal shocks by preheating the board before hot-air leveling, wave soldering, rework with a solder pot, etc.
- Decreasing the size of the thermal cycle (see Fig. 57.5[2]). Decreasing the size of the thermal cycle is the single most effective measure for increasing the life of the PTH, especially if the thermal cycle exceeds T_g.
- Decreasing the free thermal expansion of the laminate over the thermal cycle. The free thermal expansion can be reduced primarily by choosing a laminate material with a higher T_g, but also by choosing a laminate material (e.g., with Aramid fibers) with a low CTE below T_g (see Fig. 57.6).
- Decreasing the PTH aspect ratio (usually quoted as board thickness divided by finished hole size) by decreasing the board thickness or increasing the hole diameter (see Fig. 57.7). Aspect ratios tend to be higher in boards with eight or more layers because of their thickness and via density; aspect ratios greater than 3:1 require good-quality plating and aspect ratios higher than 5:1 are not recommended, in part because of the difficulty of achieving adequate plating thickness in the center of the barrel.
- Increasing the Cu plating thickness (see Fig. 57.8[2]). Increasing the plating thickness also increases the distance a fatigue crack must propagate to cause an electrical failure.
- Using Ni plating over the Cu (see Sec. 57.5.1.3 for more discussion).

The ratio $\varepsilon_f/\Delta\varepsilon$ can be increased by:

- Increasing the Cu ductility (increases ε_f) and yield strength (decreases ε). Cu strength and ductility are often inversely related, so these two factors must be balanced against one

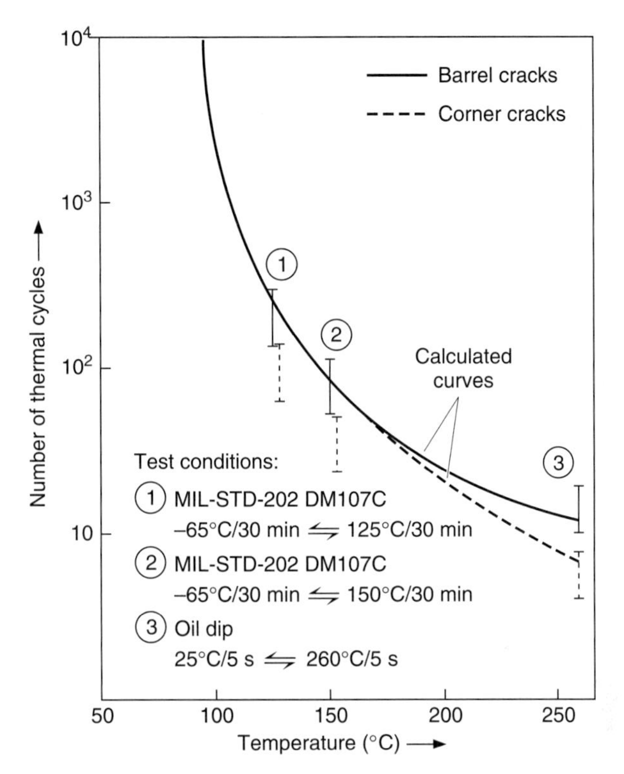

FIGURE 57.5 Peak temperature vs. number of cycles to failure by PTH barrel (solid bars) or corner (dashed bars) cracking in three different tests. Calculated lines are shown for comparison. Results are for pyrophosphate copper and FR-4. Other parameters are total strain energy required to cause fracture, 50 J/cm^3; hole radius, 0.45 mm; distance from hole center to free end, 0.8 mm; plating thickness, 0.02 mm; distance from hole center to pad edge, 0.8 mm; board thickness, 2 mm. Results from Ref. 2.

another. However, the strength-ductility relationship can be altered by the choice of plating bath and plating conditions.

The number of cycles to failure can be dramatically decreased by defects in the hole wall or Cu plating in the hole or PTH knee that act as stress concentrations (increasing the local stresses and strains) and/or facilitate crack initiation. Because of the importance of this failure mode, it has been extensively studied experimentally and with analytical modeling techniques and more quantitative models are available.[3,4,5]

Laminate and Cu/Laminate Adhesion Degradation. When a PCB is exposed to elevated temperatures for long periods of time, the adhesion between the Cu and the laminate and the flexural strength of the laminate itself will gradually degrade. Discoloration is usually an early symptom.

Several standards tests are used to compare the thermal resistance of different laminate materials. Cu adhesion is measured using a peel test.[6] Adhesion at elevated temperatures or after elevated temperature exposure gives some insight into the ability of the material to withstand rework and other high-temperature processes. Flexural strength stability is compared by measuring the times at 200°C before the flexural strength decreases to 50 percent of

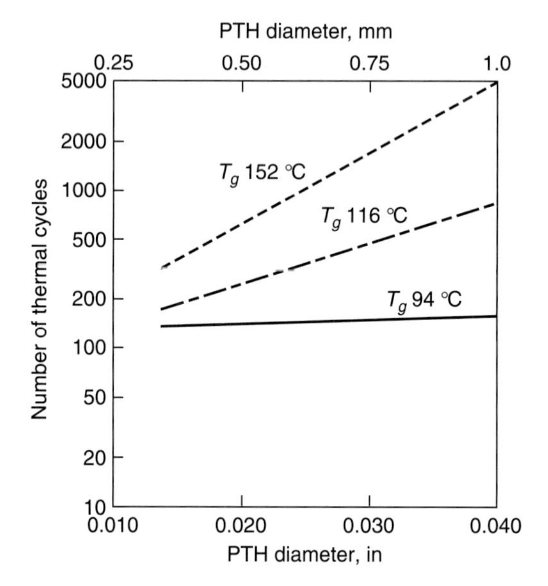

FIGURE 57.6 Effects of substrate T_g and PTH diameter on mean number of cycles to failure. The thermal cycle was 2-h cycle with extremes at –62 and +125°C. Multilayer printed circuit board thickness 0.10 in (2.5 mm); Cu in unfilled PTH is 1.2 mil (30 μm) thick. Results from Ref. 33.

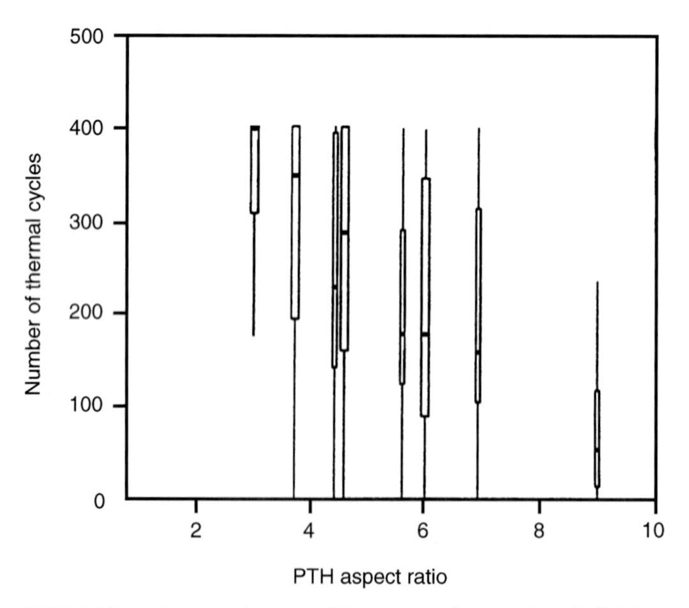

FIGURE 57.7 Cycles to failure vs. PTH aspect ratio for –65 to +125°C thermal shock cycles. Various hole diameters, board thicknesses, and board constructions. *(After Ref. 3.)*

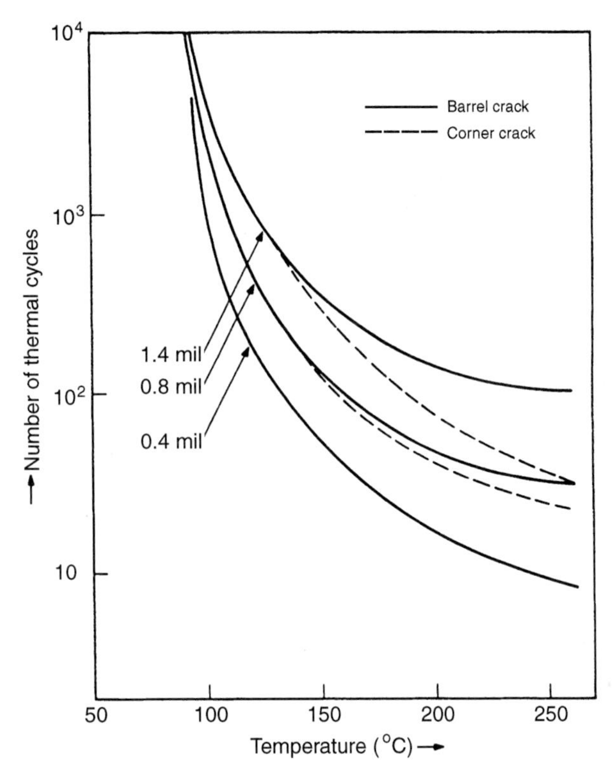

FIGURE 57.8 Effect of PTH plating thickness on number of thermal cycles to failure for thermal cycles with the indicated peak temperatures. For acid sulfate copper and FR-4 boards. Other hole parameters are the same as Fig. 57.5. *(After Ref. 2.)*

its original value. The quality of the bond between the resin and reinforcement is compared by measuring the time it takes for copper-clad laminate to blister during a solder float test at 290°C.[7]

57.2.1.2 Mechanically Induced Failures. PCBs may be mechanically loaded by test fixtures or processing equipment, when PCAs are loaded into card cages or fixtured into place with brackets, or when the assembly experiences mechanical shock or vibration in use. In general, once the PCA has been assembled, the interconnects to the components are the weak links in mechanical loading situations, not the boards themselves.

57.2.1.3 Electrochemical Failure Modes. The primary function of a printed circuit board is to provide electrical connections with the desired, stable, low-impedance and high-impedance insulation between them. A high surface insulation resistance (SIR) value is usually assumed by the circuit designer. Exposure to humidity, especially when ionic contaminants are present, is a common cause of insulation resistance failures that is accelerated by elevated temperatures and electrical bias. The impedance often decreases slowly over a long period of time. If the SIR value falls below the designed level, there will be cross talk between circuit elements that should be isolated, and the circuit may not function properly. Insulation resistance deterioration is particularly harmful for analog measurement circuits. If these circuits are used to measure low-voltage, high-impedance sources, changes in circuit impedance can result in the deterioration of instrument performance. Medical products that use sensors attached to a patient also pose special concerns, because deterioration of insulation resistance has the potential to cause electrical shock. For general applications, the surface resistivity is usually specified to exceed 10^8 Ω/□, but for these specialized applications higher values may be required. Electrochemical failures are usually accelerated by temperature, humidity, and applied bias.

High humidity is a significant cause of reliability problems because many corrosion mechanisms require water to operate. A humid environment is an excellent source of water, even when it is not condensing. Polymers commonly used in PCBs are hygroscopic; that is, they absorb moisture readily from the environment. This phenomenon is reversible; the moisture can be driven out of the PCB by baking it. The amount of moisture absorbed and the time to reach equilibrium with a humid environment depend on the laminate material, its thickness, the type of solder mask or other surface coating, and the conductor pattern.

The moisture absorbed by the PCB and ionic contaminants on or in the PCB play a role in a number of failure modes. Because the permittivity of water is much higher than that of most laminate materials, the increased water content can significantly affect the dielectric constant of the laminate, and thus may affect the electrical functionality of the board by increasing the capacitive coupling between traces. Absorbed and adsorbed water can lower SIR values, especially in the presence of ionizable contaminants (often from flux residues) and DC bias. The introduction of no-clean assembly processes has significantly increased the importance of measuring SIR values because contaminants left on the PCB remain there after assembly. Industrial pollutants can also be a source of ions that accelerate corrosion. In addition, typical industrial pollutants such as NO_2 and SO_2 can damage many materials used in PCAs, particularly elastomers and polymers. Some important mechanisms that cause failures due to low insulation resistance include dendritic growth and metal migration, galvanic corrosion, and conductive anodic filament growth. Whiskering can also cause electrical shorts, but neither electrical bias nor moisture is required.

Conductive Contaminant Bridging. Bridging of circuits by conductive salts may occur if plating, etching, or flux residues remain on the board. These ionic residues are good conductors of electricity in a moist environment. They tend to migrate across both metallic and insulating surfaces to form shorts. Corrosive byproducts, such as chlorides and sulfides formed in industrial environments, are chemically similar and can also cause shorting. An example of this type of failure is shown in Fig. 57.9.[8]

Dendritic Growth. Dendritic growth occurs by the electrolytic transfer of metal from one conductor to another; consequently, it is also termed *electrolytic metal migration.* It is also referred to as *electromigration,* although it should not be confused with the process that occurs in aluminum conductors in integrated circuits, which has a different mechanism. An example of a dendritic growth failure is shown in Fig. 57.10. Dendrites form on surfaces (including the interior surfaces of cavities) when the following conditions are met:

- Continuous liquid water film, a few molecules or more in thickness
- Exposed metals, especially Sn, Pb, Ag or Cu, that can be oxidized at the anode
- Low-current dc electrical bias

It is significantly accelerated by the presence of hydrolyzable ionic contaminants (for example, halides and acids from flux residues or extracted from polymers). Delaminations or voids that promote the accumulation of moisture or contaminants can promote dendritic growth. Conductive anodic filament growth (discussed later) is a special case of dendritic growth. Time to failure is inversely proportional to spacing squared and voltage. The failure mechanisms in accelerated tests have been reviewed.[9]

Dendritic growths usually form from cathode to anode. Metallic ions formed by dissolution at the anode are transported along a conductive path and reduced and deposited at the

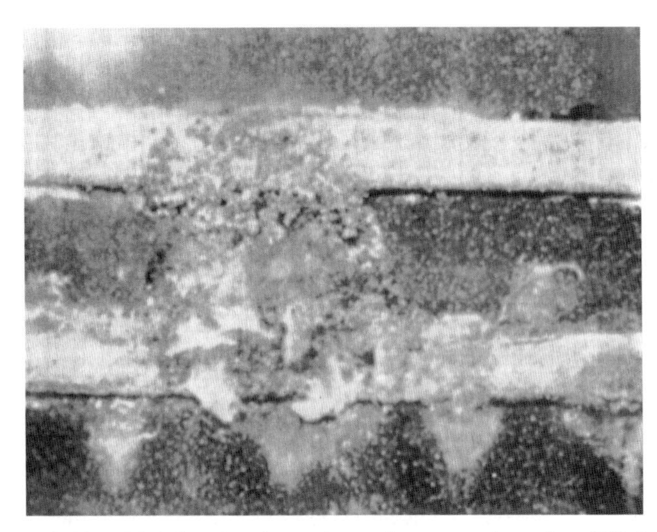

FIGURE 57.9 Migration of corrosion products across the surface of FR-4 bridging two conductors. From Ref. 11. *(From IPC-TR-476. Courtesy of Department of Defense.)*

FIGURE 57.10 Transmitted light micrograph through a PCB that failed in the field. Dendritic growth has formed at the interface of a UV-cured screened solder mask and the FR-4 surface.

cathode. The dendrite resembles a tree, since it consists of a stalk with branches. When the growth touches the other conductor there is an abrupt rise in current, which sometimes destroys the dendrite but may also cause an electrical circuit to temporarily malfunction or damage a device.

It has been proposed that the absorption of moisture produces an electrochemical cell. The following electrode reactions for Cu are an example:

At the anode: $Cu \rightarrow Cu^{n+} + ne^-$

$H_2O \rightarrow \frac{1}{2}O_2 + 2H^+ + 2e^-$

At the cathode: $H_2O + e^- \rightarrow \frac{1}{2}H_2 + OH^-$

where the majority of the leakage is due to the electrolysis of water. Copper metal is dissolved at the anode and migrates to the cathode, where it is no longer soluble. The dendrite that forms follows the resulting pH gradient.[10] The voltage difference between the cathode and anode also affects the rate of dendrite growth. When the cathode and anode are the same metal (e.g., Cu), the voltage difference is determined primarily by the applied bias, although the access of moisture and air also has an effect. Corrosion can be accelerated in a crevice because an oxygen concentration differential between the anode and the cathode develops. When the metals are dissimilar, galvanic corrosion may occur without a bias voltage.

If there is an applied bias, dendritic growth will occur almost instantly if the cathode and anode are under water. A simple laboratory experiment can prove the point. A 6-V bias across two conductors is sufficient to induce rapid growth (readily observable with a low-power microscope) even when distilled or deionized water bridges the conductors, although growth will occur faster with tap water.[11]

Galvanic Corrosion. Galvanic corrosion occurs between dissimilar metals because they have differing affinities for electrons (i.e., they are more or less electronegative). Galvanic series have been compiled for many common metals and alloys (see Table 57.1). Metals near the top of the series (noble metals) do not corrode; those near the bottom corrode easily. When these metals are near each other, the more noble metal becomes the cathode, the less noble the anode. Moisture is required to couple the two metals electrically. Applied bias is generally not required, but may accelerate the reaction if the polarity is correct. When the anode is very small compared to the cathode, its corrosion can be very rapid. Conversely, if the anode is much larger than the cathode, corrosion is unlikely to be serious, particularly if the difference in electronegativity is small.

Conductive Anodic Filament Growth. Conductive anodic filament growth (CAF) causes electrical shorts when a metal that dissolves anodically is redeposited at the interface between the glass (or other) fibers and the resin matrix of a printed circuit board. Conductive anodic filament growth is promoted by delamination at the glass-polymer interface, which may in turn be promoted by various environmental stresses including high temperatures (greater than about 260°C for FR-4) and thermal cycling. Shorts seem to occur most rapidly when a single fiber bundle connects two pads. Once delamination has occurred, the metal migration that causes shorts to occur is promoted by increasing temperature, relative humidity, and applied voltage. Small conductor spacings also significantly decrease times to failure.[12] In multilayer boards, failures occur faster on outerlayers than innerlayers because the surface layer absorbs moisture more readily. By the same reasoning, solder mask and conformal coating both increase the time-to-failure because they slow the absorption of moisture from the atmosphere into the board.

Whiskers. Whiskers are faceted filament-like structures that grow spontaneously on the surface of a plated metal and can cause shorts between closely spaced conductors (see Fig. 57.11). Whiskering can be differentiated from other causes of shorts such as dendritic growth, because neither an electrical field nor moisture is required for whiskers to form. Whiskering is a particular problem with pure tin. The whiskers grow in response to internal stresses in the plating or external loads. Sn whiskers are commonly 50 μm long and 1 to 2 μm in diameter.

TABLE 57.1 Standard Electromotive Force Potential (Reductions Potentials) for Elements Commonly Found in Electronic Assemblies

Reaction		Standard potential (Volts vs. standard hydrogen electrode)
Noble	$Au^{3+} + 3e^- = Au$	+1.498
	$Cl_2 + 2e^- = 2Cl^-$	+1.358
	$O_2 + 4H^+ + 4e^- = 2H_2O$ (pH 0)	+1.229
	$Pt^{3+} + 3e^- = Pt$	+1.2
	$Ag^+ + e^- = Ag$	+0.799
	$Fe^{3+} + e^- = Fe^{2+}$	+0.771
	$O_2 + 2H_2O + 4e^- = 4OH^-$ (pH 14)	+0.401
	$Cu^{2+} + 2e^- = Cu$	+0.337
	$Sn^{4+} + 2e^- = Sn^{2+}$	+0.15
	$2H^+ + 2e^- = H_2$	0.000
	$Pb^{2+} + 2e^- = Pb$	−0.126
	$Sn^{2+} + 2e^- = Sn$	−0.136
	$Ni^{2+} + 2e^- = Ni$	−0.250
	$Fe^{2+} + 2e^- = Fe$	−0.440
	$Cr^{3+} + 3e^- = Cr$	−0.744
	$2H_2O + 2e^- = H_2 + 2OH^-$	−0.828
	$Na^+ + e^- = Na$	−2.714
Active	$K^+ + e^- = K$	−2.925

Source: A. J. deBethune and N. S. Loud, *Standard Electrode Potentials and Temperature Coefficients at 25C,* Clifford A. Hampel, Skokie, Ill., 1964.

Once started, they may grow as fast as 1 mm per month. The tendency toward whisker growth is influenced by a variety of factors including plating conditions and the characteristics of the substrate. Growth may be inhibited by a Cu or Ni barrier layer beneath the tin coating. Pb seems to suppress whisker growth; eutectic Sn-Pb solder is considered almost immune. Whiskers do not cause the corrosion resistance or solderability of the tin coating to deteriorate, so tin may be used as a temporary finish. To avoid whiskering, plated pure Sn should not be used on closely spaced conductors that could short during service, such as connector terminations or component leads.[13,14]

57.2.2 Interconnect Failure Mechanisms

57.2.2.1 Thermally Driven Failure Mechanisms.

Thermal Fatigue of Solder Joints. Thermal fatigue of solder joints has been extensively researched in the last decade. The mechanism of fatigue, accelerated testing methods, and methods for predicting life have all been described at length, although there is still much controversy about many of the details.[15–17] These references also illustrate how modern finite element methods can be used to model the strains in the solder under both operating and accelerated test conditions. This section briefly reviews some of the important principles underlying solder joint thermal fatigue.

The focus of the discussion is on surface-mount solder joints, which have been extensively researched; however, many of the same principles also apply to through-hole solder joints. Through-hole joints are generally less prone to solder joint fatigue failures, so long as the through-hole barrel is full of solder. Ideally, complete fillets should be observed on both surfaces of the board. Reviews of this topic can be found in Refs. 18 and 19.

Thermal fatigue in solder joints occurs because of the thermal expansion (CTE) mismatch between the PCB and the component interconnected by the solder joint (see Fig. 57.12).

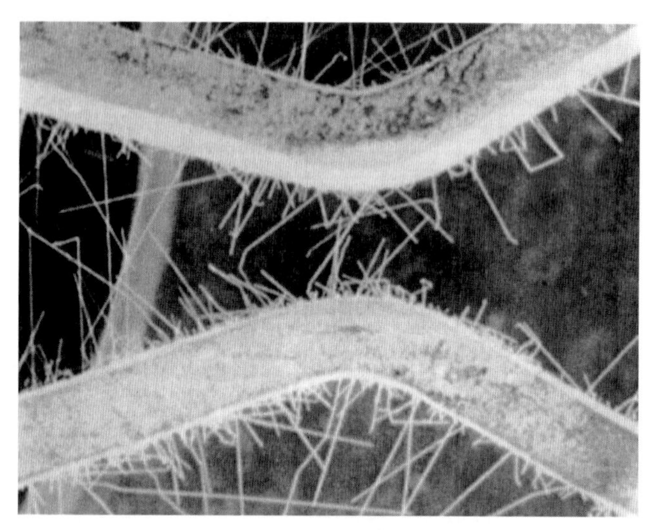

FIGURE 57.11 Tin whisker growth on a tin-plated surface. From Ref. 11. *(From IPC-TR-476. Courtesy of Burndy Corporation.)*

The imposed thermal cycle ΔT results in an imposed cyclic strain $\Delta\varepsilon$ of the solder joint, which is generally the weakest part of the system. The relationship is simple under the assumptions that the part and substrate are rigid, the solder joints are relatively small, and that homogeneous shear deformation caused by the global CTE mismatch predominates:

$$\Delta\varepsilon = \frac{(\Delta T)(\Delta\alpha)l}{h} \tag{57.1}$$

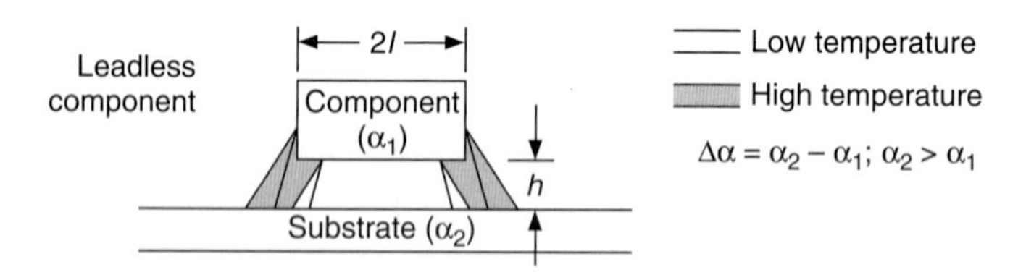

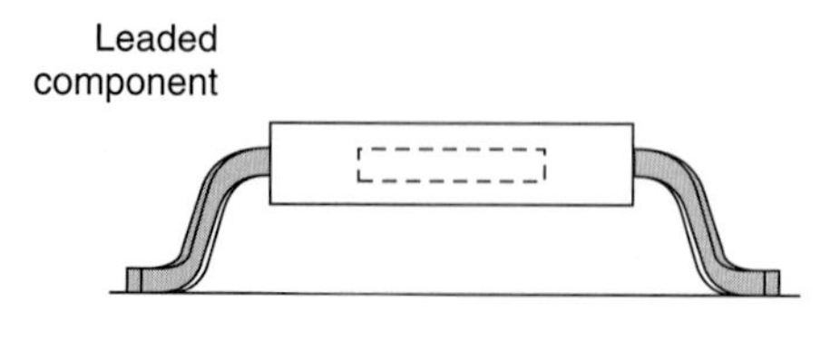

FIGURE 57.12 Schematic illustration of strains imposed on solder joints to leadless and leaded surface-mount components during a thermal cycle. Although the relative displacement of the substrate and component body is the same, the strain on the joint is reduced in the leaded case by the deflection of the lead.

where $\Delta\alpha$ = difference in thermal expansion coefficients of component and substrate
l = distance between center of component and joint
h = height of solder joint

If the component has leads or if the substrate is flexible, there will be some compliance in the system that will reduce the strain imposed on the solder joints. The local mismatch between the solder and the component lead or the pad or via metallization on the substrate can also contribute to the strains imposed on the solder.

Like plated-through-holes (PTHs), solder joints fail by a low-cycle-fatigue mechanism which can be crudely approximated by the Coffin-Manson relation

$$N_f = \frac{1}{2}\left(\frac{\varepsilon_f}{\Delta\varepsilon}\right)^m \tag{57.2}$$

where, again, N_f is the number of cycles to failure, ε_f is the fatigue ductility, and m is an empirical constant near 2. However, unlike the PTH case, the number of cycles to failure also depends on the frequency at which the cycles are imposed and the hold time at each temperature extreme. The reason for this dependence is that, for solders, the primary deformation mechanism causing thermal fatigue failures is creep.

The phenomenon of creep and its connection to fatigue are fundamental to understanding thermal fatigue of solder. Creep is time-dependent deformation that occurs gradually in response to a fixed imposed stress or displacement (see Fig. 57.13). Creep occurs by a variety of thermally activated processes. These processes play an important role only when the temperature exceeds half the melting temperature (in degrees Kelvin) of the material and, even then, the rate of deformation increases strongly with increasing temperature. For electronic solders, even room temperature is well above half the melting temperature; consequently, creep is the most important deformation mechanism of solder. When a displacement is first imposed, the strain is a combination of elastic and plastic strain. The elastic deformation is reversible and damages the microstructure relatively little, while the plastic deformation is permanent and contributes more significantly to the initiation and propagation of fatigue cracks in the solder (see Fig. 57.14). Given time, the creep process relieves some or all of the elastic stress through further permanent deformation. This additional deformation does further microstructural damage and increases the amount of plastic strain imposed when the thermal cycle is reversed. Because there is less time for creep to occur, rapid thermal cycles are less damaging than slow cycles or cycles with long hold times at the temperature extremes, a fact which is important in designing accelerated reliability tests as well as in service. The importance of creep makes the fatigue behavior of solder different from structural metals, such as copper, aluminum, or steel.

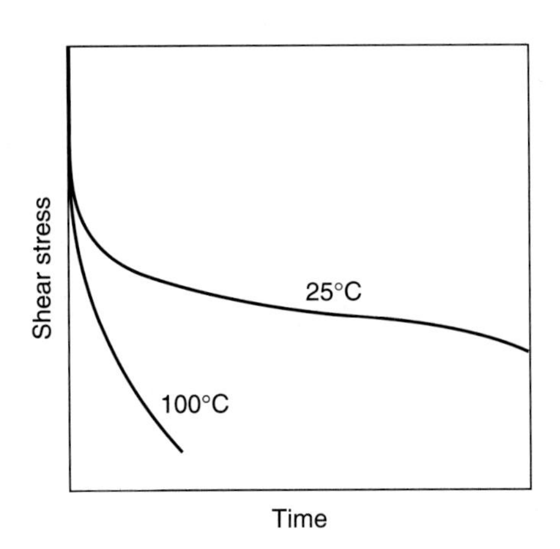

FIGURE 57.13 Typical behavior of solder in response to a constant applied displacement (for example, due to thermal expansion). The initial stress is relaxed over time as the solder elongates.

In summary, the effects of the thermal cycling profile on solder joint thermal fatigue life are as follows:

- *Temperature extremes:* Decreasing the size of the thermal excursion is the single most effective way to increase the life of the solder joints. Since creep occurs more rapidly at higher temperatures, decreasing the peak temperature of the thermal cycle further decreases the amount of creep deformation that occurs during the hold at high temperature.
- *Frequency:* The thermal fatigue damage per cycle is greater at lower cycling frequencies because there is more time for creep to occur, increasing the amount of permanent deformation. (Recall that most of the damage is

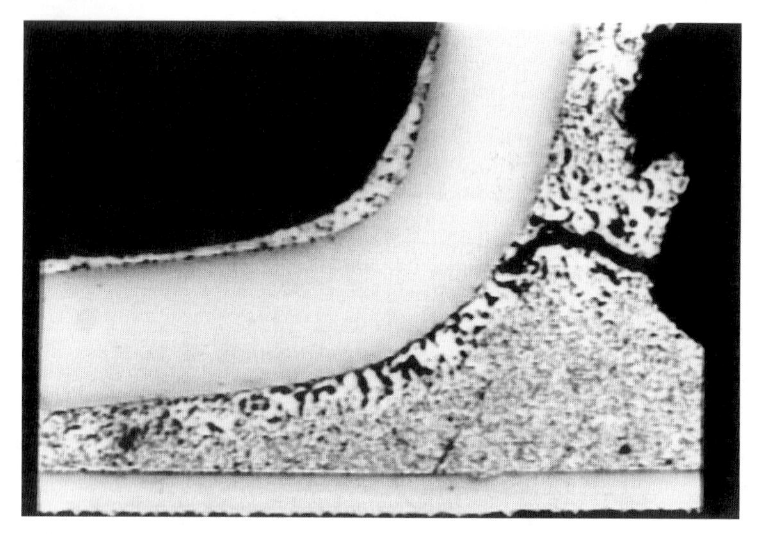

FIGURE 57.14 Thermal fatigue failure in eutectic Sn-Pb solder for a TSOP component. *(Photo courtesy K. Gratalo.)*

caused by the plastic deformation that occurs during each cycle, not by the cyclic stressing of the joint.)

- *Hold time:* As long as the stress on the solder joint remains nonzero, the thermal fatigue damage per cycle is increased if the hold time is lengthened, again because there is more time for creep to occur. Once the stress relaxation process has gone to completion, no further damage occurs and increasing the hold time further has no effect.
- *Thermal shock:* If the thermal cycle is extremely rapid, the components of the PCA may not be at the same temperature; consequently, the imposed strains may be larger or smaller than at slower rates.

Although the designer may be able to influence the peak temperature through cooling schemes, the thermal cycling profile and the frequency of thermal cycles in service are largely fixed by the application.

Solder joint fatigue life may be increased by decreasing the strain $\Delta\varepsilon$ imposed on the solder joint by:

- Choosing a package with a compliant attachment scheme. In this case, part of the strain is taken up by deflection of the lead, reducing the amount of strain in the solder. For these packages, the joint life can be further extended by decreasing the lead stiffness and increasing the joint area.
- Decreasing $\Delta\alpha$, the difference in the package and substrate thermal expansion coefficients, by carefully selecting the CTE of the package and substrate (see Secs. 57.5.3.1 and 57.5.1.1, respectively).
- Decreasing the size of the package, decreasing Δl.
- Increasing the height h of the solder joint.

Solder joint fatigue life may also be increased by:

- Decreasing the local CTE mismatch that occurs at the interfaces between the solder and the component lead and substrate metallization. While the substrate metallization is usually

Cu, which is moderately well matched to solder (17 vs. 25 ppm/°C), the leads may be made of a low-expansion metal such as Alloy 42 (~5 ppm/°C) or Kovar, as well as of Cu.

- Decreasing the mean stress imposed on the solder joint (for example, due to residual stress after assembly).
- Increasing ε_f or decreasing the creep rate of the solder by controlling the solder joint microstructure or by selecting an alternative solder. (Finer microstructures, which may be achieved by faster reflow cooling rates, have significantly greater fatigue lives because they are resistant to fatigue crack initiation and propagation. Unfortunately, solder microstructures coarsen over time, even at room temperature. Some solders, such as Sn-4Ag and 50In-50Pb, seem to have significantly improved fatigue lives over eutectic Sn-Pb; however, their higher reflow temperatures are not necessarily compatible with FR-4. See Sec. 57.5.2.2.)

Thermal Shock. Thermal shocks (>30°C/s) cause failures because differential heating or cooling rates introduce large additional stresses into the assembly versus thermal cycling. Under thermal cycling conditions, it is a good assumption that all components of the assembly are at approximately the same temperature. (High-power components may be an exception.) Under thermal shock conditions, different portions of the assembly are temporarily at different temperatures because their heating or cooling rates are not the same. These temperature transients are caused by differences in thermal mass and thermal conductivity across the assembly; they are caused by component selection and placement decisions and by differences in the physical properties of the materials used in the assembly. The temperature differences across the assembly and any resulting warpage can enhance the stresses normally imposed during temperature changes due to differences in CTE. Thermal shocks can cause reliability problems, such as solder joint failures in overload and crazing in conformal coatings leading to corrosion failures, as well as a range of component failures. Because of the differential thermal stresses that may be induced, thermal shock can cause failures that do not occur during slower thermal cycling between the same temperature extremes. On the other hand, rapid thermal shock cycles actually cause less solder joint fatigue than slower thermal cycles; because little creep occurs, more cycles are needed to cause solder fatigue failure.

57.2.2.2 Mechanically Induced Failures. PCAs can also fail in response to externally imposed mechanical stresses, for example, due to mechanical shock or vibration in shipping or in use. These failures can be divided into two categories: overload failures and mechanical fatigue failures, which are caused by mechanical shocks and vibration, respectively. Susceptibility to mechanical failures is closely linked to design of the PCA and the housing in which it is installed. The design determines the resonant frequency of the board, which in turn determines its response to external mechanical stresses. Cantilevers with low natural frequencies, such as an edge-mounted PCB with an unanchored large mass in the center, are particularly prone to failure. Depending on connector design and mounting scheme, solder joints to surface-mount connectors can also be vulnerable, particularly if there are many connector insertion cycles. See Ref. 20 for a more detailed discussion of design methodologies for mechanical durability.

Overload and Shock Failures of Solder Joints. When a printed circuit assembly is flexed, jolted, or otherwise stressed, solder joint failures may occur. In general, solder is the weakest material in the assembly; however, when it connects a compliant structure such as a leaded component to the board, the lead flexes and the solder joint is not placed under much stress. Solder joints to leadless components will see large stresses since the board can bend and the components themselves are usually rigid. These stresses can occur if the assembly is mechanically shocked, for example, if the unit is dropped, or during further assembly processes if the PCB is bent through a significant radius. The primary method of eliminating this failure mode is through package selection; however, other factors also play a role, including PCB design, process control during printed circuit board fabrication and assembly, and the shear strength, tensile strength, and ductility of the solder. Solder joints are particularly prone to failure in tension because the brittle intermetallics at the interface between the solder and the substrate are stressed. Joints with thick intermetallic layers are more susceptible.

Mechanical (Vibration) Fatigue. Vibration (a common source is an improperly mounted fan) can cause solder joint fatigue by repeatedly stressing the interconnects. Metal fatigue can occur even when this stress is well below the level that causes permanent deformation (that is, the yield stress). As for thermal fatigue of solder joints, the number of cycles to failure in mechanical vibration fatigue can be described by the Coffin-Manson relation. However, in contrast to thermal fatigue, failures usually occur after very large numbers of small, high-frequency cycles in which most of the strain in the solder is elastic ($\varepsilon = \sigma/E$, where σ is the stress and E is the elastic modulus of the solder). Consequently, creep does not play an important role in vibration fatigue. Although the damage per cycle is minimal, the number of cycles can be extremely high; they are often imposed at 50 or 60 Hz. Over time, a crack can nucleate, and subsequent cycles serve to propagate this crack. Again, the risk is much higher for joints to large leadless parts, since there is no compliant structure to take up part of the stress. The amount of damage to the solder joint depends on the imposed strain in each cycle, which depends largely on whether the excitation frequency is close to the natural frequency of the board. The mass of the component (including any heat sink) also plays an important role.

57.2.2.3 Electrochemically Induced Failures. The electrochemical failure mechanisms accelerated by temperature, humidity, and electrical bias that were described in Sec. 57.2.1.3 for printed circuit boards also apply to the remainder of the PCA. The solder used for the interconnects and the metal component terminations and lead frame finishes can also be involved in the reactions. The large number of dissimilar metals increases the complexity of the situation and the possibility of galvanic corrosion in a humid environment. In addition, contaminants introduced during printed circuit assembly such as flux residues can contribute to the failures.

57.2.3 Components

Although the failure mechanisms of electronic components in general have been described in detail elsewhere[21–24] and are beyond the scope of this chapter, there are a few failure mechanisms that are specifically associated with electronic assembly. In addition, component derating for high-temperature service should be evaluated if the unit will be exposed to a severe service environment. Component failures due to thermal shock, exceeding the maximum allowable component temperatures, and plastic package cracking can occur during reflow or wave soldering. These assembly-related failure mechanisms are described briefly as follows.

57.2.3.1 Thermal Shock. Multilayer ceramic capacitors may crack if exposed to thermal transients exceeding 4°C/s. These cracks are usually invisible, but may be the sites for dendritic growth in service, when the assembly is exposed to moisture under applied bias. Capacitors with high values and larger thicknesses are most susceptible. These failures can be avoided by following the manufacturer's requirements for maximum temperature excursion and rate of temperature change.

57.2.3.2 Overtemperatures. Many components, including connectors, inductors, electrolytic capacitors, and crystals, cannot survive the SMT reflow process, although most will survive wave soldering. Problems can include melting of internal soldered connections, melting or softening of polymeric capacitor dielectrics, and expansion of elastomeric materials. These failures can be prevented by carefully following the manufacturer's recommendations for maximum processing temperature.[25]

57.2.3.3 Molding Compound Delamination in Plastic SMT Packages. Plastic-packaged integrated circuits are generally transfer molded using a filled epoxy-based compound.

The plastic can absorb moisture, which tends to accumulate at interfaces within the package, such as the die attach paddle. Subsequent heating can cause the moisture to vaporize, causing delamination at the interface, eventually leading to package failure. This delamination phenomenon is also termed package cracking or "popcorning." The newer thin SMT components (e.g., TSOPs and TQFPs) are more susceptible because the distance the moisture must diffuse through the plastic to reach internal interfaces is shorter. Delamination when these components are exposed to high temperature may be prevented by ensuring that the components are dry through dry storage and/or baking.[23]

57.3 INFLUENCE OF DESIGN ON RELIABILITY

Design has a major influence on the reliability of any product. The implications of the demands of the application and expected service environment on product design should be considered as early as possible, since they can influence a wide range of decisions, including integrated circuit partitioning, package and substrate selection (which will impose specific design rules and electrical performance characteristics), component layout and box design, and heat sinking and cooling. IPC Standard D-279, Design Guidelines for Reliable Surface Mount Technology Printed Board Guidelines, is a good place to start in considering these issues. Section 57.2 has already described how design can promote or hinder certain failure mechanisms. Section 57.4 discusses the influence of materials, which are selected during the design process, on PCB and interconnect failures. This section highlights the importance of good thermal and mechanical design.

The size of the thermal cycles imposed on the PCA during power on/off cycles can have an overwhelming influence on the reliability of integrated circuits, solder joints, and plated-though-holes, especially if the external service environment is not particularly severe; consequently, good thermal design is critical to reliability. The thermal cycles imposed on the assembly can come from joule heating from high-power components and from ambient heating. Integrated circuit reliability depends on maintaining low enough junction temperatures, usually below 85 to 110°C, depending on the IC technology. During continuous operation, solder joint temperatures should be kept below about 90°C to avoid the extensive intermetallic growth and grain coarsening that occur during long-term exposure to higher temperatures. As described under Sec. 57.2.1.1 and 57.2.2.1, the size and number of thermal excursions directly affect the fatigue life of both solder joints and plated-though-holes. Component spacings, orientations, air velocities, and enhancements such as thermally enhanced packages, heat sinks, and fans can all have major effects on the thermal cycle experienced by the assembly. PCBs can also be enhanced with metal cores to improve heat dissipation.

As mentioned in the earlier descriptions of specific failure modes, package selection and via and PTH specification can have a major effect on reliability. Although small holes can be desirable from a design density perspective, use of the smallest holes (aspect ratios of 5:1 or greater) should be minimized to minimize the risk of PTH failures. This is especially true if the design includes large through-hole parts which are likely to be reworked frequently (for example, due to test escapes). Similarly, some package styles are more susceptible to solder joint fatigue than others. The effect of integration on reliability can depend on the difference in package styles for the options under consideration. Integration can have a positive effect by reducing the total number of connections that could fail. On the other hand, if it requires a large package with a large CTE mismatch to the substrate, integration may reduce the reliability of the assembly.

The effect of externally imposed mechanical shock and vibration on PCA reliability is largely determined by design factors, although substrate and package selection also play a role. Component placement and PCA mounting in the box determine the natural frequency of the board and, consequently, the extent to which the board deflects. High-mass packages, often due to large heat sinks, are particularly susceptible, especially if there is a large lever arm.

57.4 IMPACT OF PCB FABRICATION AND ASSEMBLY ON RELIABILITY

57.4.1 Effect of PCB Fabrication Processes

57.4.1.1 Laminate and Lamination. Delamination in PCBs may occur between the laminate materials or between the laminate material and the Cu foil. One cause of delamination is defective laminate material. Defects such as incomplete bonding at the resin/fiber interface can result in delamination due to formation of voids at these interfaces. Other common causes of delamination are excessive lamination pressure and/or temperature, contamination at interfaces, heavily oxidized copper foil surfaces, and lack of oxide treatment to enhance adhesion between copper innerlayers and prepreg. Debonding increases the risk of conductive anodic filament growth because it provides a place for moisture to accumulate. It can also result in increased stresses on the plated-through-holes (PTHs) during thermal cycling.

Laminate voids and resin recessions are separations of the laminate material from the copper conductor that may occur during multilayer PCB lamination. Most acceptability specifications prohibit voids larger than 0.076 mm (0.003 in); however, smaller voids are not generally considered to be detrimental to reliability. Some of the causes of laminate voids are entrapped air during lamination, improper flow of resin, and improper epoxy cure, perhaps due to improper lamination pressure and/or temperature, inappropriate heating rate, or too little prepreg.

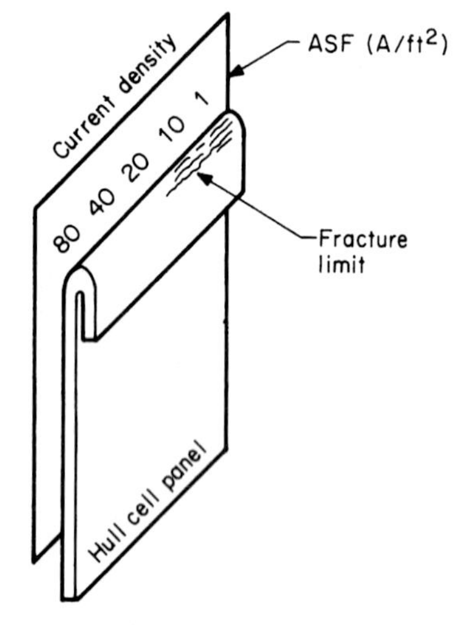

FIGURE 57.15 Ladwig panel ductility test. *(After Ref. 26.)*

57.4.1.2 Cu Foil. The major cause of innerlayer foil cracks seems to be poor ductility of the Cu. Poor foil ductility can have a more significant effect on PTH reliability than such well-known culprits as insufficient plating thickness and excessive etchback. A minimum of 8 percent elongation is required for 1-oz foil to eliminate this problem. Foils plated in a Hull cell can be easily evaluated for room-temperature ductility using a 180° bend test. This technique is illustrated in Fig. 57.15, in which the sample panel is bent flat parallel to the axis along which current is varied.[26] Fractures occur at current densities producing low-ductility copper. This test can also be used to evaluate the influence of bath chemistry on ductility or as a bath monitor. Poor copper ductility can be correlated to the microstructure observed in metallographic cross sections.[27]

57.4.1.3 Drilling and Desmear. Poor drilling and desmear (etchback) can cause PTH failures by providing stress concentrations that cause fatigue cracks to initiate. They can also cause voids and cracks at the interface to the copper plating, which can trap chemicals during plating and then contribute to conductive anodic filament growth. The following paragraphs describe the effects of poor desmear and some drilling defects which can cause poor plating, such as resin smear, rough walls, loose fibers, and burrs.

Resin smear can cause weak connections between plated-through-holes and innerlayer copper that fail under environmental stress. There is always some resin smear, which is removed by the desmear (etchback) process. If the desmear process is not effective, or if the resin smear is excessive, poor interconnection to the innerlayers can result. Possible causes of excessive smear are a dull drill or the wrong feed rate or drill speed, all of which can cause increased drill heating, resulting in more smear.

Similar errors in drilling setup can cause rough hole walls, loose fibers, or burrs. These defects are not serious in and of themselves, but can lead to rough plating or copper nodules, which introduce stress concentrations. Rough walls are typically associated with an incorrect feed rate or drill speed, or insufficiently cured material. Loose fibers may be caused by incorrect drilling parameters or improper cleaning. Burrs are usually associated with too fast a drill feed or a dull drill.

Poor drill registration can also decrease reliability of innerlayer via connections or the soldered connection to through-hole components. Poor registration can cause breakout on innerlayers; i.e., the drill hole may fall outside the pad on the innerlayer it is intended to connect to. Breakout increases the probability of PTH barrel failures. Breakout on outerlayers means that the solder fillet for a through-hole component will be partially missing, resulting in decreased reliability for some critical components.

Whether caused by excessive resin smear or not, poor etchback can result in a weak connection between the plating in the hole and the innerlayer copper. Etchback (see Fig. 57.16) removes laminate resin and woven glass in the hole so that the internal copper projects slightly into the hole, permitting the plating to make contact with the innerlayer foil on three sides. This strength is important to prevent cracking at the interface under thermal shock conditions. A review of innerlayer cracking in or around the electroless copper at the junction between the innerlayer foil and the electroplated Cu in the hole suggests that *negative* etchback, in which the electroplated Cu projects into the laminate may also give good results. Zero etchback, when innerlayer foil is flush with the hole wall, is the most dangerous case because the bond line between the foil and the plated copper is located at the point of maximum stress.[28] Causes of insufficient etchback include improper lamination and curing, hardened epoxy smear, a depleted smear removal bath, or a host of process control issues, including improper bath temperature, agitation, or time exposure.

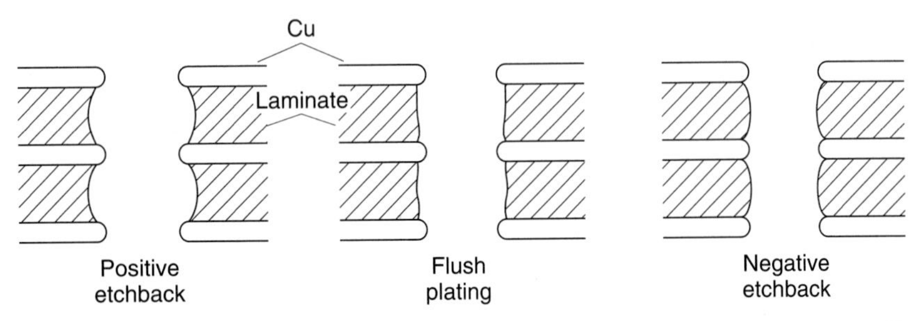

FIGURE 57.16 Schematic illustrations of positive, flush, and negative etchback. *(After M. W. Gray.[28])*

57.4.1.4 Plating. Defects from the plating process can be responsible for a variety of PTH reliability problems; in addition, as previously described, problems in earlier process steps such as drilling and desmear often show up as plating defects.

Uniform coverage of the hole with electroless Cu is critical to the strength of the through-hole and the adhesion of the metallization to the laminate. Oxidation of the innerlayer copper prior to electroless Cu plating is one source of poor plating adhesion. Poor control of bath composition can have the same effect.

The adhesion of the electroplated Cu to the electroless Cu and the ductility of the electroplated Cu also strongly affect PTH reliability. If the adhesion between the layers is poor, this interface may be the weak point where failure initiates when the PTH is subjected to thermal stresses. Causes can include tarnished electroless copper that is insufficiently microetched, burning the electroless copper with too much current in electrolytic plating, and film contamination in the electrolytic copper.[28] Susceptibility to innerlayer cracks may be identified by looking for cracks in microsections after a solder float test. The fatigue life of the copper is directly related to its ductility. Plating process parameters and plating additives can strongly affect the plating ductility. For example, Mayer and Barbieri[29] found that good thermal shock resistance of electrodeposits of acid copper depended on proper concentrations of three additives:

1. A leveling agent, to smooth over surface imperfections. (Without the leveling agent imperfections are reproduced in the deposit.)
2. A ductility-promoting agent, which functions to produce the equiaxed grain structure.
3. A carrying agent, to guide the other two components to create the equiaxed structure. (Striations occurred with insufficient carrying agent.)

Additive levels below a certain threshold make the bath more susceptible to impurity effects. For example, iron contamination of 100 mg/L without the recommended concentration of ductility-promoting agents was found to produce a columnar grain structure at the hole corners. Similarly, organic contaminants such as photoresist can produce laminar deposits.

Insufficient plating thickness in the barrel also directly reduces PTH reliability because the stress and, consequently, the strain in the Cu are increased. Overall insufficient plating thickness can be caused by a depleted bath or insufficient plating time, among other things. Insufficient plating thickness in individual holes can also occur as a result of nonuniformities in plating current caused by nonuniform copper feature density. It is particularly difficult to obtain adequate plating thickness in the center of high-aspect-ratio PTHs; good process control is important for aspect ratios greater than 3:1. Good coverage is difficult to obtain for aspect ratios greater than 5:1 by electroplating.

What constitutes "sufficient" plating thickness in PTHs is a subject of some controversy. Specifications range from 0.5 to 1 mil (12 to 25 μm) Cu thickness in the barrel. There are at least two reasons why there is no one right specification. First, different applications provide different levels of thermal stress and demand different levels of reliability. Second, design factors such as the aspect ratio of the plated holes determine the susceptibility of the PTHs to thermal fatigue. The IPC recommends an average minimum copper-plating thickness of 0.5 mil for consumer products (Class 1) and 1.0 mil for general industrial and high-reliability applications (Classes 2 and 3).

Poor coverage at the PTH knee can significantly accelerate PTH failures because it means the plating is thin at a point of high stress. It can be caused by excessive concentration of the organic leveling agents added to an electroplating bath.

57.4.1.5 Solder Mask Application. If it is properly applied, solder mask plays an important role in reducing the possibility of insulation resistance failures on PCBs. The solder mask protects the substrate from moisture and contaminants, which would otherwise promote shorting under electrical bias. The ability of the solder mask to perform this function depends on good conformity and adhesion of the solder mask to a clean, dry substrate. If solder mask conformity or adhesion is poor, moisture and other contaminants may accumulate at crevices or delaminations between the solder mask and the substrate. Substrate cleanliness is particularly critical, because in addition to causing poor solder mask adhesion, it can also provide the ionic species needed for rapid electromigration. When the laminate material absorbs moisture readily (e.g., polyimide, aramids), baking before reflow may be required to prevent solder mask delamination (as well as reinforcement/resin delamination). Other causes of poor adhesion or conformity include moisture on the board when the solder mask is applied, improper solder mask lamination or coating parameters, and improper solder mask cure parameters. Incomplete solder mask cure can create local soft pockets, which are common sites for delamination or contaminant entrapment. Solder mask over solder should also be avoided because delamination caused by reflow of the solder may allow contaminants to be entrapped.

57.4.2 Effects of Printed Circuit Assembly Processes

57.4.2.1 Stencil Printing and Component Placement. Stencil printing and component placement generally do not cause reliability problems; however, poor stencil design or stencil printing or placement process control can create issues with solder volume and component cracking. Very low solder volume can result in weak solder joints that fail rapidly in thermal fatigue or by overload. In some cases, excessive solder volume can also accelerate solder joint fatigue failures because the compliance of the lead is reduced. Assuming that the stencil was designed and manufactured correctly, low solder volumes are usually due to small or missing paste bricks caused by a clogged stencil aperture, a stencil that needs cleaning, or improper stencil-printing parameters. Paste bridging can cause low solder volumes on some joints and high volumes on others because one joint may rob solder from another. Paste bridging may be caused by improper stencil design or stencil-printing parameters or by excessive force when placing chip carriers (e.g., PLCCs) and quad flat packs (e.g., PQFPs). Excessive placement force can also cause component cracking, particularly for small leadless ceramic components.

57.4.2.2 Reflow. The reflow process attaches SMT and some TH components to the PCB by melting solder paste to form solder joints using an oven with a controlled thermal profile (see Fig. 57.17) and, in some cases, a controlled atmosphere (often N_2). Reliability problems that can arise due to improper reflow parameters can be grouped into three categories: damaged components, poor solder joints, and, for no-clean assemblies, cleanliness issues.

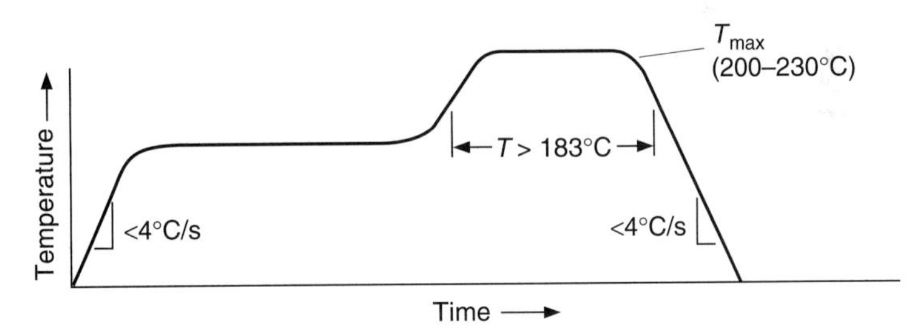

FIGURE 57.17 Schematic reflow profile for eutectic Sn-Pb solder paste, FR-4 substrate and typical surface-mount components illustrating key features from a reliability perspective.

Component Damage. The reflow process is responsible for most of the assembly-process-related component failures described in Sec. 57.2.3. These failures include molding compound delamination in plastic packages that have absorbed moisture (popcorning) and component failures due to overheating or thermal shock caused by excessive heating or cooling rates. All of these problems are preventable with good procedures and process controls.

Package cracking can be prevented by storing components in the unopened dry bags in which they are shipped and by baking the moisture out of components that have been exposed to ambient conditions for too long. The manufacturer's recommendations regarding bake-out conditions and maximum exposure times before reflow should be followed, but a good general guideline is that packages that have been exposed to the atmosphere for more than 8 h should be baked to a moisture content below 0.1 percent by weight immediately prior to use; a bake of 125°C for 24 h is usually safe, although shorter times may be acceptable. Note that the same concerns apply for rework and second-side reflow of double-sided boards; for example, if the boards are stored for several days between reflow steps, bake-out before the second reflow step may be required.

Component failures due to overheating or thermal shock can be prevented by monitoring the reflow temperature profile in several locations on the PCB to ensure that it meets the manufacturer's specifications for temperature-sensitive components. Measuring the board profile is important because temperatures on the board can differ significantly from oven panel temperatures and from the ambient temperature in each oven zone. Temperatures may also vary significantly across the board if there are large differences in the thermal mass of the components or in component density. Areas of the assembly that are devoid of components are particularly sensitive to overheating, which can damage the laminate as well as any small components in the area. Temperature variations across the assembly tend to be much smaller for ovens with predominantly convection heating than for ovens with predominantly infrared heating. A poor reflow profile can also cause a variety of other problems; some of the others that also affect reliability will be mentioned here.

Poor Solder Joints. A sound solder joint wets both the component termination and the substrate well, does not contain large or numerous voids, and does not have excessively thick intermetallic layers at the interfaces. When using solder paste, the reflow profile is the dominant factor in achieving these goals. Good wetting requires solderable incoming materials, but it also requires a reflow profile that gives the flux sufficient time to act in the right temperature range. In addition, the profile should ensure that all parts of the board are at least 15°C over the melting temperature of the solder for at least several seconds. "Cold," improperly formed solder joints can occur if the solder does not fully melt or if oxidation prevents the solder balls in the paste from melting together. The latter problem can be caused by an improper reflow profile or the wrong reflow atmosphere. Voiding is generally caused by a reflow profile that does not permit enough time for the solvents in the paste to boil off before the solder melts. All of these problems can be avoided by ensuring that the reflow profile of the board and reflow atmosphere (e.g., O_2 level) correspond to the manufacturer's recommendations for the solder paste.

Excessively long reflow times (time above the solder liquidus) can cause thick intermetallic layers to form at the interface between the solder and the component termination or substrate. Formation of an intermetallic layer at the solder interface indicates good metallurgical bonding, but thick intermetallic layers are undesirable because intermetallics are brittle and prone to fracture, especially if the joint is stressed in tension rather than shear. Because solder joint fatigue takes place in the solder rather than in the intermetallics or at the solder/intermetallic interface, the basic mechanism is unaffected. Nonetheless long reflow times and the accompanying thick intermetallic layers should be avoided. Cross-sectioning can be used to judge the extent of intermetallic growth; as long as the intermetallic layer thickness is relatively small compared to the joint thickness, reliability should not be adversely affected.[30] (Note however that minimizing reflow time is still a good thing; the reliability of all the components on the board is adversely affected by time at elevated temperature, both during processing and in service. Unfortunately, in developing a reflow profile, there is often a tradeoff between reflow time and peak temperature.)

Cleanliness Issues. An improper reflow profile can also cause solder balling and increase the amount of flux residue remaining on the board after reflow. The reliability concerns associated with these process issues are discussed in Sec. 57.4.2.4. Solder balls can be caused by a combination of improper paste storage or handling, incompatibility between the flux and reflow atmosphere, and a reflow profile that does not conform to the manufacturer's specification.

57.4.2.3 Wave-Soldering Process. Improper wave-soldering practice can cause reliability problems. The root cause is generally thermal shock, overheating of the top side of the board, or contamination of the solder bath.

Component Cracking. Ceramic components such as resistors and capacitors will crack under thermal shock conditions. When they are located on the bottom of the board they can be heated rapidly by the solder wave. Prevention is relatively simple; the assembly must be preheated before it hits the solder wave. A temperature difference between the component and the solder wave of less than 100°C is recommended; a typical preheat temperature is 150°C.

Hot Cracking. Hot cracking, also known as partial melting, can cause previously sound solder joints to fail during the wave-soldering process. A typical mixed TH/SMT assembly is manufactured by assembling the surface-mount components to the top side, inserting the through-hole components, and wave soldering these components to the board from the bottom side. The first step of the wave-soldering process usually involves preheating the entire board. During the wave-soldering process, the SMT joints on the top side will be further heated due to conduction of heat through the board, particularly if there are many vias. If these solder joints reach the melting temperature of the solder (usually 183°C), the joints will begin to melt. If the joints melt completely, the assembly may be intact after reflow; however, if they only begin to melt, the surface tension of the solder is insufficient to prevent cracks from forming between the portions that are still solid. This type of failure is often detected as an intermittent in the field, since in-circuit test fixtures may bring the two halves of the joint into mechanical contact, causing the joint to appear electrically good.

Solder Bath Contamination. Solder bath contaminant levels should be regularly monitored and limited to levels found in IPC-S-815. Many metals found on component terminations will dissolve into molten eutectic Sn-Pb solder. High Cu concentration is a relatively common occurrence that is associated with a rough solder surface and causes poor solderability. High Au concentrations can embrittle solder joints (see Sec. 57.5.1.3 for a discussion of this phenomenon).

57.4.2.4 Cleaning and Cleanliness. Improper handling procedures and improper selection and application of solder paste and wave-solder fluxes and their associated cleaning processes can cause ionic residues to be left on the board that result in low surface insulation resistance. Low SIR values can cause failures in and of themselves for some sensitive circuits and in other cases set up the conditions for further corrosion that eventually result in short circuits. Sodium and potassium ions and halide ions are the most commonly quoted culprits for these failures. The major source of sodium and potassium ions is handling, i.e., fingerprints. The primary sources of halide ions are soldering fluxes.

The elimination of chlorofluorocarbons (CFCs) mandated by the Montreal Protocol has caused most SMT manufacturers to switch to water cleaning or a no-clean process. Water cleaning has been used by most printed circuit board manufacturers for some time, but outgoing cleanliness was not carefully monitored since the boards were cleaned again after assembly. Both the no-clean and water-clean assembly approaches must meet certain criteria to provide reliable assemblies.

In a no-clean assembly process, there is no cleaning step after SMT or TH assembly. The finished assembly has whatever contaminants were present on the incoming board and components, plus any additional contaminants added during the assembly process. These contaminants are generally flux residues, both from the solder paste and the flux applied for wave soldering, although adhesives and fingerprints are other potential sources. A no-clean flux should have a low solids content so that it leaves little residue and be free of ionic contaminants such as halides that promote corrosion. Use of a flux that contains halides will result in low SIR readings and may result in shorting due to corrosion, particularly if the assembly is exposed to a humid environment. However, the incoming components and boards are cleaned, it is important that they are also free of halides when they arrive for assembly. Although SIR testing provides the best correlation with reliability, an ionic contamination test may be used for statistical process control. The measurement method may be found in MIL-P-28809.

Solder balls may also be a problem on no-clean assemblies. Solder balls are formed during reflow of solder paste when some solder is left behind when the solder melts and beads up and by spattering during wave soldering. These solder balls are usually washed off by solvent or water cleaning; however, in a no-clean process they remain on the board. Solder balls can cause shorts by bridging the pads of small capacitors or resistors or the leads of fine-pitch quad flat packs.

In a water-clean assembly process, the assemblies are cleaned with jets of deionized or saponified water after SMT and TH assembly. This process will work only if the flux residues and other contaminants are sufficiently soluble in either water or saponified water. It also depends on good access to the residues; consequently, a minimum component standoff that permits cleaning is required if it is possible for flux to get underneath the component body during assembly. It is almost as important that the board be thoroughly dried because water is an excellent medium for galvanic corrosion. Proper drying can be quite difficult even with substantial air flow since water has a much lower vapor pressure than CFCs do. If the component standoff is low, capillary action holds water in the small gap. If water cleaning is done in midprocess (e.g., before a reflow or wave-soldering step), plastic components may absorb moisture; in this case, the board must be baked out to prevent package cracking in subsequent high-temperature processes (see Secs. 57.2.3 and 57.4.2.2).

The rework process should not be overlooked in planning a flux and cleaning strategy. Compared to the automated processes that proceed it, it is typical to use a more aggressive flux and more of it to do rework. Use of a halide-free flux or proper cleaning after rework is essential to prevent cleanliness-related reliability problems.

Finally, the cleaning process itself can damage the PCA. Ultrasonic cleaning can damage components with internal wire bonds or die attach. It has also been observed to cause fatigue cracking of solder joints to LEDs and SOT-23s when the energy density was high because these components have terminations that are mechanically resonant near the generator frequencies. Solvent cleaning can attack the polymers used in solder masks, PCBs, conformal coatings, and components. D-limonene (terpene)-based solvents should be tested carefully for compatibility with exposed plastics and metals.

57.4.2.5 Electrical Test and Depanel. The electrical testing and depanel processes can impose large mechanical stresses on the PCB and its components. In-circuit electrical test utilizes a bed of nails or two beds in a clamshell arrangement to contact each electrical node on the board. The probes must contact the board with sufficient force to make good electrical contact. If the board is not properly fixtured or if the loading in a clamshell fixture is unbalanced, the resulting deflections can cause solder joint or component cracking. These cracks may cause electrical failures immediately or after some period of service. Depaneling, the process of separating individual images from a larger panel, is done by a variety of methods. The associated mechanical deflections or vibration can cause component cracking or solder joint fatigue.

57.4.2.6 Rework. Rework, whether repair of open or shorted solder joints or replacement of defective components, has a significant negative effect on component reliability. If there were not enough other incentives for low process defect rates, the effect on product reliability would be enough. Reworking the quality in does not bring the board to the quality level that would have been reached if the boards were built right the first time. Some of the ways rework processes can adversely affect reliability are described here.

Thermal Shock During Rework. Thermal shock to components is a concern during rework as it is during reflow. The maximum heating or cooling rate is driven by the requirements for ceramic capacitors and should not exceed 4°C/s.[31]

Rework of large through-hole components, such as pin grid arrays (PGAs) and large connectors, poses special problems. If it is improperly done it can result in PTH failures. Because the damage during these large thermal cycles is cumulative, the number of rework operations at a given site should be monitored and limited to a safe number. The number of cycles that will cause a fatigue crack to initiate in the copper in the barrel and propagate to failure depends on the aspect ratio of the PTH, the type and thickness of plating in the hole, the substrate material, etc.

Due to the large number of joints that must be melted at once and the large thermal mass of the components, rework of large TH components is often done with a solder pot. The thermal shock caused when the molten solder hits the board can cause PTH cracking due to z-axis expansion. A preheat step (to about 100°C for FR-4) helps to reduce the damage. The time the board is in contact with the solder fountain should also be minimized since dissolution of the copper plating inside the PTH occurs during this time. Thinning the plating in the PTH tends to increase the strain in the Cu during thermal cycling, further accelerating failure. If the total time for part removal and replacement is kept under 25 s, little dissolution is measured.[32] Weakening of the PTH by copper dissolution during PGA rework can be essentially eliminated by using NiAu plating. Although the thin Au coating that protects the Ni dissolves almost instantly during soldering, Ni dissolves quite slowly and effectively prevents thinning of the PTH metallization.

Damage to Adjacent Components. Rework can also damage components adjacent to the one being repaired or replaced. The hot cracking phenomenon during wave soldering can also occur in solder joints near the rework site if they reach the melting temperature of the solder. At slightly lower temperatures, rapid intermetallic growth can occur. Temperature-sensitive components can also be damaged. To prevent these problems, localized heating and shielding should be used and the temperature of adjacent components monitored. The generally recommended maximum temperature is 150°C. There are wide variations in the amount of heating of adjacent components among different types of rework equipment and between process protocols.[33]

Other Rework Concerns. Rework can cause a host of moisture-related problems including measling and package cracking. Both of these problems can be prevented by baking the PCA beforehand to drive out moisture and by minimizing the peak temperature and time at elevated temperature during rework. Rework temperatures also weaken the adhesive bond between the copper conductors on the PCB and the laminate material; use of force to remove components when the solder is not completely molten can cause the pad to lift off the board. The latter can be a particular problem when using a soldering iron.[21,31]

57.5 INFLUENCE OF MATERIALS SELECTION ON RELIABILITY

57.5.1 PCB

57.5.1.1 Substrate. Difunctional FR-4 is the workhorse material for high-reliability PCBs because its moderate z-axis expansion and moisture uptake characteristics are available at relatively low cost. Alternative substrate materials (see Table 57.2) are generally selected for more favorable properties in one or more of the following three areas: thermal performance, including maximum operating temperature and glass transition temperature; thermal expansion coefficient; and electrical properties, such as dielectric constant. Thermal performance characteristics and thermal expansion coefficient can have a significant effect on PCB and solder joint reliability. Other characteristics of these materials, such as moisture absorption, can also affect reliability.

PTH reliability can be improved by selecting a laminate with a lower z-axis CTE or a higher T_g. The damage caused to the PTH during a thermal cycle depends on the total z-axis expansion during the temperature change. Since the CTE is much lower below T_g than above,

TABLE 57.2 Physical Properties of Some Printed Circuit Board Laminate Materials

Material	CTE, x,y ppm/°C	CTE, z ppm/°C	T_g, °C
Epoxy glass (FR-4, G-10)	14–18	180	125–135
Modified epoxy glass (polyfunctional FR-4)	14–16	170	140–150
Epoxy aramid	6–8	66	125
Polyimide quartz	6–12	35	188–250

Source: After IPC-D-279.

the PTH strain can be reduced by increasing the T_g so that more or all of the cycle is below T_g (see Fig. 57.18*a*). Figure 57.18*b* shows that the increase in life can be quite significant. The strain imposed on the PTH can also be reduced by decreasing the CTE at temperatures below T_g, but the effect on total *z*-axis expansion is much smaller.

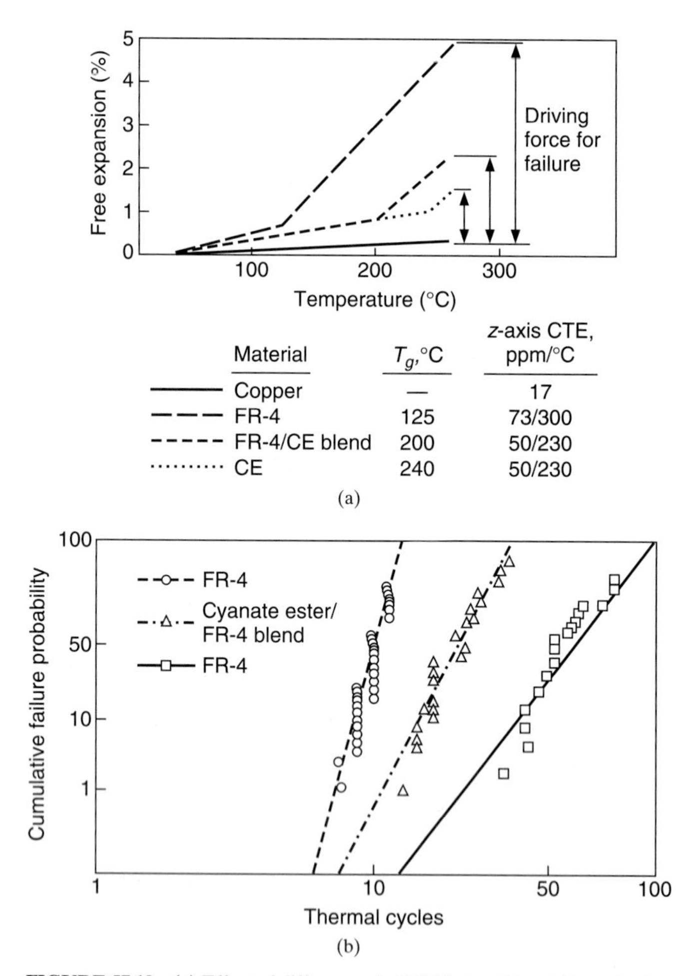

FIGURE 57.18 (*a*) Effect of differences in CTE below T_g and T_g on the free *z*-axis expansion of FR-4 (epoxy-glass), cyanate ester (cyanate ester-glass) and a cyanate ester/epoxy blend. T_g and CTE for each material below/above T_g are indicated. Cyanate ester is abbreviated CE in the figure. Cu is shown for comparison. (*b*) Weibull plot of PTH failures for these substrates during thermal shock cycling between 25 and 260°C. PTHs are 0.029-in diameter on 0.100-in grid on 0.125-in-thick laminate. *(After Fehrer and Haddick.[4])*

A variety of specialty resins with increased T_g is available, albeit at higher prices. Modified FR-4 materials with higher functionality offer the best combination of improved T_g at a reasonable price. Further improvements in T_g and other characteristics can be obtained with bismaleimide triazine (BT), GETEK, cyanate ester, and polyimide, but at greater price penalties.

Interconnect failures due to thermal fatigue of solder joints can be reduced by closely matching the *x-y* plane thermal expansion properties of the substrate to at-risk components. Large leadless ceramic components that are used because of their hermeticity pose a particular risk. Possible approaches include altering the laminate reinforcement material, adding constraining metal cores or planes, and switching to a ceramic substrate. The first two approaches are discussed here. A more extensive discussion of these options can be found in Ref. 33.

A lower *x-y* plane thermal expansion coefficient laminate can be obtained by replacing the continuous-filament E-glass used in most FR-4 PCBs with an alternative material. The CTE decreases as the fraction of silica dioxide (SiO_2) decreases and the level of quartz (as well as the cost) increases in the progression E-glass, S-glass, D-glass, and finally quartz, which has a CTE about one-tenth of E-glass. Aramid (Kevlar) actually has a negative CTE, but it is available in only a few glass styles. Some of the disadvantages of aramid fibers are higher *z*-axis expansion and higher moisture absorption relative to glass fibers that can result in decreased susceptibility to PTH failures and corrosion-related insulation resistance failures, respectively. Aramid fiber is also used to make nonwoven paper fabric which has a lower modulus, but also a much smoother surface because there is no weave pattern. This form has better dimensional stability and reduced microcracking during thermal cycling.

Low-thermal-expansion metal cores or planes can also lower the overall substrate CTE because they constrain the expansion of the polymer material they are laminated to (see Fig. 57.19). Copper-Invar-copper (CIC) is the most widely used material for constraining metal cores (also termed polymer-on-metal or POM construction), followed by copper-molybdenum-copper (CMC). The PCB and core are bonded with a rigid adhesive, usually in a balanced construction to minimize warping. Other special processing is also required. The CTE of the assembly can be estimated using a simple model for composite structures most often written as

$$\text{CTE(overall)} = \frac{\Sigma E\alpha t}{\Sigma Et}$$

where E, α, and t are the elastic modulus, CTE, and thickness, respectively, of the various layers. A more sophisticated model can be found in Ref. 34.

An example of the low overall CTE that can be obtained using a CIC core is shown in Fig. 57.20. Unfortunately, the constrained *x-y* axis expansion results in increased *z*-axis expansion that can reduce PTH reliability to dangerously low levels, especially in an environment in which the full mil-spec thermal cycle of –55 to +125°C is imposed. Consequently, use of polyimide is recommended with CIC cores. Because of its high T_g and low CTE below T_g, polyimide imposes much lower strains on the PTH for a given thermal cycle than other dielectrics.

Constructions utilizing constraining low-CTE metal planes usually use CIC layers in place of ground and power planes in a standard multilayer board. The same PTH reliability concerns that hold for CIC core boards apply to these PCBs as well. PTH reliability can be improved by using polyimide resin and by using CuNiAu or CuNiSn metallization in the PTHs. These substrates are easier to manufacture than metal core boards because standard PCB fabrication techniques can be used for the most part.

Resin material can affect fiber/resin delamination, one of the prerequisites for conductive anodic filament growth. Measling occurs at about 260°C for FR-4, but may occur at lower temperatures for boards with more hygroscopic resins.

57.5.1.2 Solder Mask. The three major types of solder mask—liquid screen-printed, dry film, and liquid photoimageable (LPI)—come with different benefits and concerns from a reliability perspective. The solder mask material should be selected for its compatibility with the heat and solvent characteristics of the assembly process, its capability to provide good conformity over surface features on the PCB, and its ability to tent vias if required. Since many of these characteristics are product-specific, only a few general guidelines can be provided here. Where tenting of vias is required to keep solder, moisture, or flux from wicking up

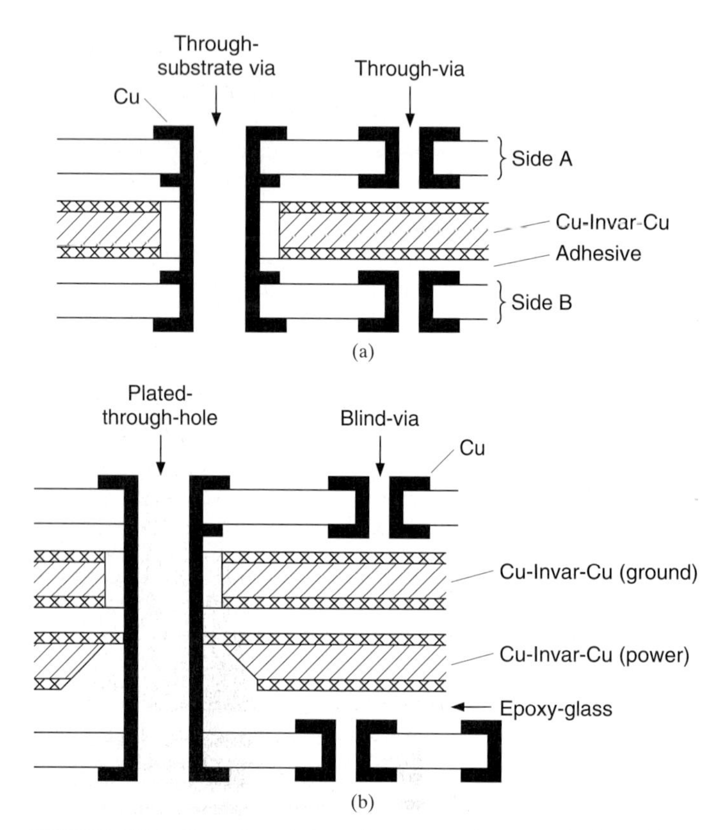

FIGURE 57.19 Examples of alternative construction using Cu-Invar-Cu to achieve low-*x-y* CTE: (*a*) metal core (sides A and B may be multilayer boards) and (*b*) metal plane constructions. *(After F. Gray.[34])*

under components, dry film solder mask should be used. However, excessively thick solder mask, particularly dry film over closely spaced traces, can result in crevices. If the solder mask cannot flow enough to adhere closely to the board, the resulting crevices can entrap contaminants such as flux that can accelerate corrosion later. LPI solder mask provides excellent coverage, resolution, and alignment to other features, but it generally cannot be used to tent vias. IPC-SM-840 defines the performance and qualification requirements for solder mask.

57.5.1.3 Metal Finish. The metal finish on the SMT and TH pads can have an impact on PTH reliability and on the reliability of the solder joints made to these pads. Common metal finishes for solder-mask-over-bare-copper (SMOBC) boards include hot-air solder leveling (HASL or HAL), organic-coated copper (OCC), and electroless NiAu. Galvanically plated CuNiAu and CuNiSn made by another processing route are also available. These finishes provide a solderable finish for later printed circuit assembly. The pros and cons of the various finishes are discussed in turn.

Of the common metal finishes, HASL is the only one which can directly reduce reliability of the board. In a typical HASL process, the board receives a severe thermal shock when it is dunked into a bath of molten eutectic Sn-Pb solder. The PTHs can survive only a certain number of solder shocks without failure; this process uses up one of these thermal cycles before the board leaves the fabricator.

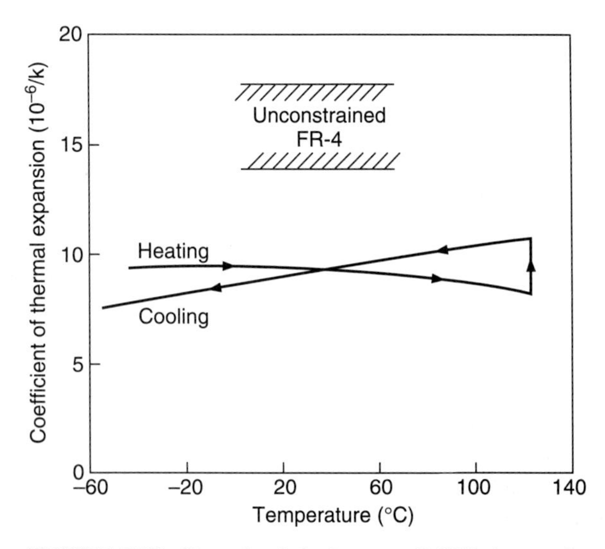

FIGURE 57.20 Example of the low overall CTE that can be achieved for a metal core construction similar to Fig. 57.19*a*. Two 0.055-in (1.4-mm)-thick multilayer FR-4 boards were bonded to a 0.085-in (2.2-mm)-thick copper-Invar-copper core. Data shown are for the third thermal cycle. *(After F. Gray.)*

Organic-coated copper provides a consistent, flat, solderable metal finish. Exposed copper after printed circuit assembly has been a persistent reliability concern, because it is generally not permitted on HASL boards. While exposed Cu on HASL boards is associated with poor solderability, which may be due to contaminants that were not removed before the HASL process, there is little evidence that exposed Cu on a properly processed OCC board causes reliability problems. Surface insulation resistance (SIR) testing shows that OCC boards have comparable or better performance than HASL boards in high-temperature, high-humidity storage tests.

CuNiAu boards fabricated either with the NiAu as the Cu etch resist or by the SMOBC process followed by electrolessly plating Ni and Au can confer improved PTH reliability. There are two mechanisms for the observed improvement: the enhanced rivet effect provided by the Ni and the elimination of Cu dissolution during solder shocks such as wave soldering or PGA rework. For high-aspect-ratio holes, electroless Ni confers an additional benefit because the plating thickness in the barrel is more consistent than for conventional electroplating.

In the simple picture of PTH failure shown in Fig. 57.3, the comparatively low CTE metal-plated PTH acts as a rivet that resists the *z*-axis expansion of the PCB. Because Ni has a higher elastic modulus than Cu, it strains less under the stress imposed by the expanding PCB. Consequently, adding Ni plating lowers the strain imposed on the Cu and lessens the amount of fatigue damage. In this model, the Ni protects the Cu, increasing PTH life.

The ability of the Cu to withstand the forces imposed on it by thermal expansion of the PCB is also dependent on the thickness of the Cu in the PTH. Unfortunately, in the SMOBC process all subsequent steps after pattern or panel plating reduce the Cu thickness from the plated amount. Nickel plating is resistant to the etches and developers used in later processing steps, so it protects the underlying copper from thinning due to dissolution. The HASL process and rework of large through-hole connectors or pin grid arrays (PGAs) can have particularly negative effects. Cu dissolves rapidly into molten eutectic Sn-Pb solder. During the HASL process or component removal and replacement with a solder fountain, large amounts of Cu can dissolve from the knee of the PTH. Nickel barrier plating minimizes this effect because Ni dissolves far more slowly in eutectic Sn-Pb solder than Cu does.

Use of Au plating can embrittle the eutectic Sn-Pb solder most commonly used in electronic assembly. Gold plating of various thicknesses is used for a variety of reasons, including as a solderability preservative over nickel plating, for connector contacts, and to provide wire-bondable pads. Reliability problems can arise because Au has a high solubility in eutectic Sn-Pb solders at reflow temperatures and dissolves extremely rapidly. In most cases, the Au finish on a PCB or component termination will be completely dissolved into the solder. In a wave-soldering process, the Au will be washed into the bath, requiring monitoring and bath changes to maintain the Au concentration at a low level that does not affect the process. However, in a reflow process, this Au remains in the finished solder joint. To avoid embrittlement of the solder by the $AuSn_4$ and $AuSn_2$ intermetallics that can form, the Au concentration should be kept below a critical level that most authors set at 3 to 5 percent by weight.[35,36]

For most components used today, a nominal 5-µin (0.1-µm) thickness Au flash to preserve solderability is harmless. However, if thicker Au plating (e.g., for connector contacts or wire bonding) is used, if the component lead pitch is less than 0.5 mm, or if the component lead terminations are also Au-plated, care should be taken to ensure that the Au concentration remains below the 3 to 5 percent by weight limit. For some applications in which use of thick Au is unavoidable, 50In-50Pb solder, in which Au dissolves very slowly, has been used to get around this problem.[37] Selective thick Au plating is another option.

The Au concentration in the finished reflowed solder joint can be estimated using the following equation:

$$\text{Wt. \% Au} = \frac{(\text{Au volume})}{[(\text{Au volume}) + (\text{solder volume})(\rho_{\text{solder/Au}})]}$$

where $\rho_{\text{solder/Au}}$ is the ratio of the density of the solder to the density of Au (0.4552 for 63Sn-37Pb solder). If the Au is less than 1 µm thick, it is usually valid to assume that all of the Au on the joined surfaces has dissolved into the joint. The solder volume should include any solder plated on either the component or board termination as well as that applied by stencil printing. It is common to specify only a minimum Au plating thickness; it is important to use a representative value to calculate the expected Au content in the solder joint.

57.5.2 Interconnect Material

57.5.2.1 Eutectic Sn-Pb Solder. Eutectic Sn-Pb solder, 63Sn-37Pb, and near-eutectic Sn-Pb solders, including 60Sn-40Pb and 62Sn-36Pb-2Ag, are used in the overwhelming majority of soldered electronic assemblies. From a reliability perspective, the most important characteristics of these solders are their susceptibility to creep and fatigue because ambient temperatures are so close to the metal temperature of the solder, their ability to dissolve common termination metals rapidly and in large amounts, and their tendency to form thick intermetallic layers with termination metals.

Although solder joint thermal fatigue is a major source of PCA field failures, the industry has used the same solder alloy for several decades. At present, there is no generally agreed-on alternative to eutectic Sn-Pb solder that has improved fatigue resistance as well as the favorable processing characteristics of eutectic Sn-Pb solder. However, there has been a tremendous surge in research into alternative solders, especially Pb-free solders, in the last decade and one can expect improved alloys in the future. There is some evidence that solders containing 2 percent Ag have improved properties in thermal cycling to high temperatures.

Many common termination metals dissolve rapidly in eutectic Sn-Pb solder, including Ag, Au, and Cu.[36] The dissolved metals can alter the properties of the solder. Reliability can also be impacted if the termination metal is dissolved entirely, the most notable case being Ag (or AgPd with less than 33 percent Pd) terminations on ceramic resistors and capacitors. If the entire termination thickness is dissolved, the solder will dewet from the ceramic part, leaving an open joint if the entire termination is dissolved, or a substantially weakened one if it is dissolved only in some areas. Use of 63Sn-36Pb-2Ag substantially reduces this problem, since the presence of Ag in the solder reduces the dissolution rate of the Ag forming the termination.

Finally, eutectic and near-eutectic Sn-Pb solders form intermetallics with the termination metals that influence the properties of the finished joints. On the most common termination metals, Cu and Ni, continuous intermetallic layers form: Cu_3Sn and Cu_6Sn_5 on copper, and Ni_3Sn_4, Ni_3Sn_2, and Ni_3Sn on nickel. These intermetallic layers are composed of compounds that are hard and brittle in comparison to both the solder and the termination metals. Although there have been few systematic comparisons made, it is general wisdom that when the intermetallic layers are very thick, solder joint reliability is reduced. While solder joint thermal fatigue mostly involves cracking in the solder, the intermetallics can affect their ability to withstand mechanical stresses, particularly in tension. The Ni-Sn intermetallics are particularly brittle. In every case, an effort should be made to minimize the total time the solder is molten during processing and to minimize the time above about 150°C once the solder joint is formed.

57.5.2.2 Other Solders. A number of other solders are used in specialized applications, including 80Sn-20Pb for lead finishes, 50In-50Pb for solder on thick Au, high-Pb solders such as 95Pb-5Sn and 97Pb-3Sn for flip-chip assemblies (usually on ceramic substrates), and low-temperature solders such as 58Bi-42Sn and 52In-48In, where hierarchical soldering is desirable. Further information can be found in Refs. 23, 24, and 38.

57.5.2.3 Conductive Adhesives. Electrically conductive adhesives are used today for specialized applications such as connections to LCD displays and attachment of small resistors and capacitors. These materials consist of conductive particles, usually silver flakes or carbon, suspended in a polymer matrix, most commonly epoxy. The electrical resistance of the contact to the PCB tends to be unstable over time, so these materials are not suitable for applications requiring a constant, low-resistance contact. The primary failure mechanism is moisture migration through the epoxy to the interface, resulting in oxidation of the contact metal. Adhesion strength is also a reliability concern. New materials suitable for a broader range of applications are under development. Further information can be found in Ref. 39.

57.5.3 Components

Components and their packages influence many of the field failures of electronic assemblies. Packages are primarily designed and selected for their ability to protect the electronic components inside; for example, ceramic packages may be selected over plastic ones for their greater hermeticity. This section will discuss the ways in which package selection can influence solder joint and cleanliness-related failures.

57.5.3.1 Package Selection to Minimize Solder Joint Thermal and Mechanical Failures. Minimizing solder joint thermal and mechanical fatigue failures means minimizing the global and local mismatches in the system and introducing compliance that minimizes the stresses and strains transmitted to the solder joints. Figure 57.21 illustrates the important features of these systems. The following describes how component parameters can influence the incidence of solder joint failures.

Surface-Mount Technology vs. Through-Hole Components. Although there is little compliance in the system, the reliability of through-hole solder joints generally exceeds that of surface-mount joints in thermal fatigue (assuming good solder fillets are present in both cases), because the loading geometry makes it difficult for a crack to propagate far enough to cause an electrical failure. However, the PTHs themselves may be susceptible to failure if they are exposed to even a few thermal cycles well above T_g.

Plastic vs. Ceramic Package. The global mismatch between the component body and substrate is minimized for most printed circuit boards if the package is plastic rather than ceramic. Most electronic ceramics have CTEs in the neighborhood of 4 to 10 ppm. Since the printed circuit board CTE is 14 to 18 ppm in-plane below T_g, the match to plastic packages which usually have average CTEs of 20 to 25 is better. The overall CTE of a plastic package can be significantly below that of the plastic if the die is large compared to the total package body. For example, TSOP components can have overall CTEs as low as 5.5 ppm. It is also worth recalling that component-level reliability must be considered; plastic packages suffer from other disadvantages versus ceramic packages, such as moisture absorption.

Leaded vs. Leadless Surface-Mount Components. Leadless surface-mount components with peripheral solder joints (e.g., leadless ceramic chip carriers, LCCCs) are more susceptible to solder joint failures due to thermal and mechanical stresses than leaded components because there is no compliance in the system (see Fig. 57.21). A compliant lead can take up relative displacement between the component body and the substrate during mechanical or thermal stressing. In doing so, it minimizes the stress and strain imposed on the solder joint, thus reducing the likelihood of failures. Large leadless components should be avoided whenever possible. If they must be used, the substrate must have as close a CTE mismatch as possible and be protected from mechanical stresses. A conformal coating should be considered.

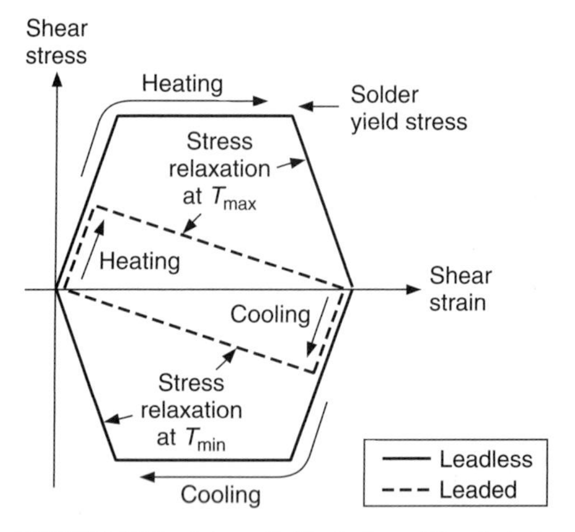

FIGURE 57.21 Schematic illustration of the strain in the solder during a thermal cycle for leaded and leadless surface-mount components. *(After W. Engelmaier)* (IPC-TP-797, Surface Mount Solder Joint Long-term Reliability: Design, Testing, Prediction.)

Ball grid arrays (BGAs) are a new style of leadless SMT component with an areal array of solder joints. The reliability of these components has come under intensive study. Plastic BGAs are less susceptible to solder joint fatigue failures than ceramic BGAs because the laminate and plastic body match the CTE of the PCB much better. At this time, it seems that there will be size and power limitations to ensure solder joint reliability.

Lead Compliance. As described, leadless components cause more reliability problems than leaded components because the entire displacement is imposed on the solder joint; however, there are also large differences in compliancy among leaded surface-mount components (see Fig. 57.22). Body height plays an important role because it determines the length of the

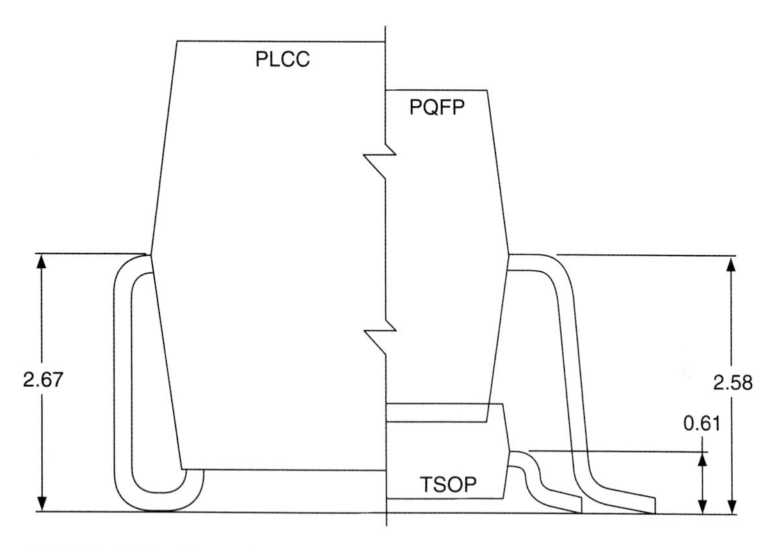

FIGURE 57.22 Schematic illustration of different surface-mount component lead types with widely differing compliance: (*left*) J-lead, (*right*) gullwing leads.

compliant beam. Other important lead characteristics that affect compliance are lead shape (e.g., J-lead vs. gullwing) and lead thickness (stiffness is proportional to thickness cubed).

Lead frame material also plays a role in determining solder joint life, although it is not as important as lead frame geometry. Common lead frame materials are Cu and Alloy 42 (Fe-42Ni); Alloy 42 has a better CTE match to silicon (and a greater mismatch to the solder), but it is much stiffer than Cu (see Table 57.3)

Thin small-outline packages (TSOPs), which are becoming increasingly common memory packages, pose the greatest solder joint reliability risks of any packages commonly used today. These packages generally have Alloy 42 lead frames and very low standoff from the PCB, resulting in a very stiff lead that transfers most of the relative displacement between the component and the substrate to the solder joint. The situation is exaggerated for this component because the overall CTE of the package is quite low. While adequate solder joint reliability can be achieved with TSOPs in many situations, some vendors have opted to encapsulate the solder joints with a filled epoxy to better distribute the stresses.[40]

57.5.3.2 Component Selection for Cleanliness. If fluid cleaning is used to remove flux residues after assembly, then a minimum component standoff (distance between the component body and the seating plane of the leads) is critical to ensuring proper cleaning and drying, and, therefore, resistance to corrosion and moisture-related failures. Industry standards for low standoff components permit the components to have 0–0.25 mm standoff from the board. These low standoffs permit corrosive residues and the cleaning fluid to be trapped

TABLE 57.3 CTE and Elastic Modulus at Room Temperature of Some Important Packaging Materials

	Cu	Alloy 42 (Fe-42Ni)	63Sn-37Pb Solder	Si
CTE, ppm/°C	17	5	25	3.5
E, GPa	130	145	~35	113

under the component. Poor fluid access can mean that flux residues are not removed. Drying is almost as important; it was not a major issue with highly volatile CFCs, but when the cleaning fluid is water, leaving it behind can promote moisture-driven failure mechanisms. Components which meet the standards for high standoff have a minimum standoff of 0.20 or 0.25 mm, which should be sufficient for cleaning and drying with water and most other cleaning fluids in use today. Using a no-clean process essentially eliminates these concerns about component standoff; however, an added concern is that any contamination of the external surface will remain. SIR testing should be conducted to ensure that the flux residues and other contaminants left on the board are not harmful.

57.5.3.3 Component Termination Selection for Joint Integrity. Surface-mount component termination finishes are generally tin-lead or tin, although other finishes are occasionally used. Good solderability is the foundation needed for forming a strong solder joint. Incoming cleanliness is another requirement that should go without saying, but has taken on increased importance with the advent of no-clean processes. With copper leads, formation of excessive quantities of Cu_3Sn and Cu_6Sn_5 intermetallics can occur if reflow times and temperatures are excessive. With pure tin, whisker growth can be a concern. When Au is used as a metal finish, the Au can be expected to dissolve rapidly into the solder joint. To avoid embrittling the solder, the finished joint should contain less than 3 to 5 weight percent Au (see Sec. 57.5.1.3).

Ceramic and ferrite components such as multilayer ceramic capacitors, chip resistors, and chip inductors are generally terminated with a fired-on silver or silver palladium paste. Because silver dissolves easily into molten Sn-Pb solder, a Ni/Sn or Ni/Au overplate is recommended.

57.5.4 Conformal Coatings

Conformal coatings are used for PCAs that require exceptional resistance to moisture, solvents, or abrasion. A wide variety of polymers are used to perform this function, including phenolic, silicone and urethane lacquers and silicone rubber, polystyrene, epoxy, and paraxylylene coatings. Epoxy and polyurethane-based coatings are the most commonly used. If conformal coatings are used in conjunction with a solder mask, they must be chemically compatible.[21,23]

Conformal coatings work by keeping contaminants away from the circuitry and preventing moisture from accumulating on the surface of the assembly. Since all conformal coatings are permeable to moisture, interfacial adhesion is essential to their function. Contaminants on the board that reduce the adhesion of the coating or trap moisture can cause the coating to fail, as can thermal stresses. When contaminants trap moisture, the coating will bubble up (vessicate), providing gaps where corrosion can occur. An ionograph is not always the right tool to detect the harmful level of ionic contamination; cleaning with both polar and nonpolar solvents before application is recommended.

A conformal coating that is not well matched to the use environment can actually promote new failure mechanisms that would not be found in the uncoated assembly. If it fills the gap beneath the component, the coating may place additional stresses on solder joints in thermal cycling by reducing or eliminating the compliance in the leads. It may also generate excessive stresses on the components if the service temperature drops below the T_g of the coating. Some coatings are not stable in hot, humid conditions.

57.6 BURN-IN, ACCEPTANCE TESTING, AND ACCELERATED RELIABILITY TESTING

This section reviews environmental stress testing procedures used to identify unacceptable parts or to estimate part life. These tests can be categorized by their intended goals: 100 percent screening to eliminate early failures (burn-in), acceptance testing on a sampling basis, and life distribution estimation (accelerated reliability estimation). Burn-in, more generally known as environmental stress screening (ESS), is an important subject in its own right that has been extensively written about; only its goals are described here. Acceptance or qualification testing and accelerated reliability testing are discussed together in sections covering testing of PCBs and PCAs under various environmental stresses.

Burn-in, also known as *environmental stress screening* (ESS), is used to eliminate early failures when infant mortality due to latent defects is a problem by exposing all parts to worst-case, but realistic, conditions. The conditions should not be too severe because burn-in decreases the useful life of the parts. On the other hand, in addition to increasing the reliability of the parts that are shipped, burn-in provides rapid feedback on process defects that can cause field failures. The most common use of burn-in in electronics is for integrated circuits, particularly memory chips, designed into leading-edge integrated circuit manufacturing processes.

Accelerated reliability tests are designed to cause failures that would occur by wear-out sometime during the service life of the part and provide the data for estimating the *life distribution* of the part. Estimating the life distribution requires that the test be continued until a large percentage of the parts fail. *Qualification tests* may be conducted under similar or even more severe conditions, but they are essentially pass/fail tests that are terminated after a specified period of time. Because there are few if any failures in a successful qualification, little new reliability information can be gained from this type of test. These tests should not be used routinely on all parts because they will greatly shorten the life of the parts.

Unfortunately, there is no standard suite of reliability tests in the industry, nor is there likely to be in the near future. There are several reasons for this. First, there are numerous service environments. The IPC has identified seven major application segments for electronic assemblies. Second, within these categories, the environment experienced by the assembly may differ from the external service environment, depending on product-specific design parameters, such as power dissipation and cooling efficiency, which influence the temperature and humidity in the vicinity of the assembly. Third, the desired product life and acceptable failure rates vary widely among applications and manufacturers. Finally, but of equal importance, as technology evolves, the tests must evolve, too. Use of reliability tests that once made sense but now are either unduly conservative or do not bring out potential failure modes of new designs has a high price in overdesigned assemblies that are larger or more costly than necessary and field failures that could have been predicted and prevented. This section describes some commonly used tests and methodologies for designing tests for new technologies or new applications.

As stated at the outset, reliability can be defined only after the service environment has been identified, and the acceptable failure rate over a specified service life is specified. If this environment proves unacceptable, the design can be modified by improving cooling, package hermeticity, cleaning, etc. If it is unclear whether the PCA will achieve the designed reliability goals, accelerated reliability tests should be designed to estimate the life distribution.

57.6.1 Design of Accelerated Reliability Tests

There are seven steps in accelerated reliability test design.

1. *Identify the service environment and the acceptable failure rate over a specified service life.*
2. *Identify actual environment of the PCA (modified service environment).* The service environment should be translated into the ambient environment actually experienced by the PCA. For example, the temperature experienced by the PCA is influenced by both power dissipation and cooling. The mechanical environment is influenced by shock-absorbing material, resonances, and so on.
3. *Identify probable failure modes* (e.g., solder joint fatigue, conductive anodic filament growth). Accelerated reliability tests are based on the premise that the frequency and/or severity of the environmental exposure can be increased to accelerate the incidence of the failure that

occurs in service in a known way, i.e., that the data can be used to predict the life distribution in the in-service PCA environment. This assumption makes sense only if the same failure modes occur in the test as in real life. It cannot be overemphasized that the accelerated tests must be designed around the real failure modes. Probable failure modes may be identified from past service experience, the literature, or preliminary testing or analysis.

4. *For each failure mode, construct an acceleration model.* An acceleration model that allows test data to be interpreted in terms of the expected service environment is crucial to life distribution estimation. It is also extremely helpful in designing good tests, so ideally the acceleration model should be developed before the accelerated reliability tests are carried out. Equation 57.2 for solder joint reliability plus Eq. 57.1 for strain in solder joints to rigid components is an example of an acceleration model. It predicts that increasing strain will decrease the number of cycles to failure in a specific way. Within a certain temperature range, increasing the temperature cycling range is a way of increasing the strain.

   In general, the acceleration model should be based on the rate-controlling step in the failure process. In some cases, the rate will be determined by an Arrhenius type equation; for example, if diffusion is the rate-controlling process:

$$D = D_o \exp\left(\frac{-E_a}{kT}\right) \quad \text{and} \quad x \propto \sqrt{Dt} \tag{57.3}$$

$$t_2 = \left(\frac{D_1}{D_2}\right) t_1 = t_1 \exp\left(\frac{-E_a}{k}\left[\frac{1}{T_1} - \frac{1}{T_2}\right]\right)$$

where D = diffusion rate
D_o = diffusion constant
E_a = activation energy for the process
k = Boltzmann constant

and T_1 and T_2 and t_1 and t_2 are two temperatures and corresponding equivalent diffusion times

   Note that even when temperature is an important factor, *an Arrhenius relationship may not exist;* in the preceding thermal cycling example, the failure rate is roughly proportional to $(\Delta T)^2$. Some acceleration models will be explored in the following sections.

   The limits of applicability of an acceleration model are as important as the model itself. Increasing or decreasing the temperature too much may promote new failure modes that would not occur in service or invalidate the quantitative acceleration relationship. For example, if the temperature is elevated above the T_g of the board, the z-axis CTE increases sharply and the modulus decreases, which may actually lessen the strains imposed on solder joints, but may also promote PTH failures.

   Finite element modeling (FEM) can be invaluable in developing and/or applying acceleration models for thermal and mechanical tests. Two-dimensional nonlinear modeling capability will usually be required in order to get meaningful results. Models can be constructed to estimate the stresses and strains in the material (e.g., the Cu in a PTH barrel or the solder in a surface-mount or through-hole joint) under operating conditions as well as under test conditions. These estimates will be far more accurate than the simple models provided in this overview because they can account for the interactions between materials in a complex structure and both elastic and plastic deformation.

5. *Design tests based on the acceleration models and accepted sampling procedures.* Using the acceleration model and the service environment and life, select test conditions and test times that simulate the life of the product in a much shorter period of time. The sample size must be large enough that it is possible to determine whether the reliability goal (acceptable number of failures over the service life) has been met.[41] Ideally, the life distribution in the accelerated test should be determined, even when the test period must be extended to do so.

6. *Analyze failures to confirm failure mode predictions.* Since an accelerated test is based on the assumption that a particular failure mode in the accelerated test is the same one that occurs in service, it is important to confirm by failure analysis that this assumption is valid. If the failure mode in the accelerated test is different from the one expected, several possibilities should be considered. (1) The accelerated test is introducing a new failure mode different from the one that will occur in service. Usually this means that the acceleration of one parameter (e.g., frequency, temperature, humidity) was too severe. (2) The initial determination of the dominant failure mode was incorrect. In this case, to understand the significance of the test results, a new acceleration model must be developed for this failure mode. The new failure mode may be promoted more or less effectively by the test conditions than the mode originally assumed. (3) There may be several failure modes. In this case, the two failure distributions should be considered separately, so that life predictions will be meaningful. The difficulty in determining which of the above scenarios holds is that for genuinely new technologies or service environments, the failure mode in service may not be known. In these situations, it is desirable to conduct a parallel test with less aggressive acceleration for comparison.
7. *Determine life distribution from accelerated life distribution.* The accelerated life distribution should be determined by fitting the data with the appropriate statistical distribution, such as the Weibull or log-normal distribution. The life distribution in service can be determined by transforming the time axis of the life distribution using the acceleration model. This predicted life distribution in service can then be used to estimate the number of failures in the specified service life.

The following discussion of testing for some specific failures will provide examples of this methodology.

57.6.2 Printed Circuit Board Reliability Tests

57.6.2.1 Thermal. PTH failures are the predominant source of PCB failures in service and predicting them is the primary goal of PCB testing at elevated temperatures. PTH reliability testing should simulate the thermal excursions of a PTH throughout its life. Generally, the most severe thermal cycles are experienced during assembly and rework.

Two basic types of tests are conducted: thermal stress or solder float tests, and thermal cycling tests. Both of these tests are intended to be accelerated tests for the PTH, not for the laminate; the thermal stress test, in particular, is expected to severely degrade the laminate. The delamination test is similar to a solder float test, but is conducted at a lower temperature specified by the laminate manufacturer; typically, a different fluid is required.

The most commonly accepted thermal stress test is MIL-P-55110 (also found in IPC-TM-650). Following baking at 120 to 150°C (250 to 300°F), the specimens are immersed in an RMA flux and floated in a eutectic (or near-eutectic) Sn-Pb solder bath at 288°C (550°F) for 10 s. Other investigators use a bath at 260°C. Following the test, the samples are cross-sectioned and the PTHs are examined for cracks. This is a severe test that ensures that the sample will survive a single wave-soldering or solder pot rework cycle.

Most thermal cycling tests for PCBs cycle the PCB repeatedly over a wide temperature range; many are actually thermal shock tests using liquid-liquid cycling. The results of five accelerated tests with different temperature extremes, ramp rates, and dwell times have been compared by the IPC, which also provides a simplified analytical model to estimate PTH life.[3] The results of all tests suggest the same approaches for maximizing PTH reliability, but they do not all correlate well quantitatively. Two of the most common tests are (1) oven cycling from –65 to +125°C, and (2) thermal shock cycling between oil or fluidized sand baths at +25 to 260°C. Figure 57.23 shows a suitable test coupon that contains 3000 PTHs interconnected in series, several PTH sizes, and varying annular ring sizes. The PTHs can be monitored during the testing. Figure 57.18*b* shows the type of data that can be collected in this type of test.

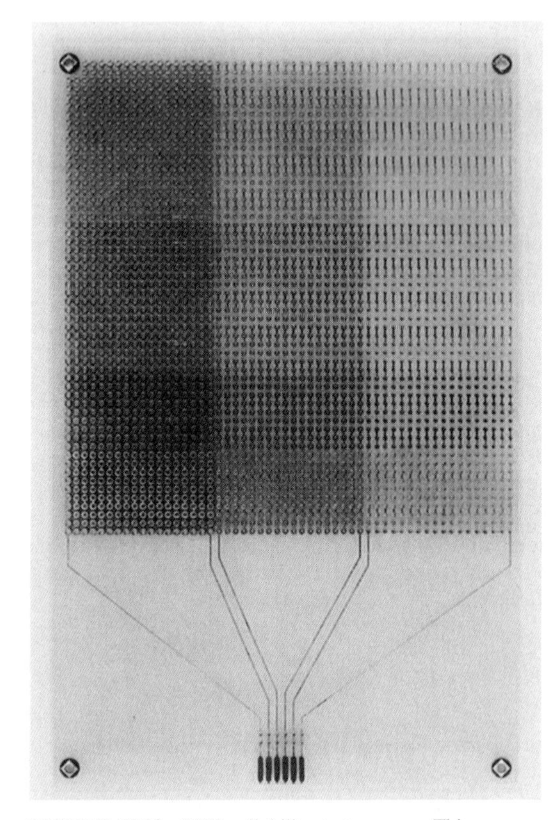

FIGURE 57.23 PTH reliability test coupon. This coupon contains three sets of 1000 PTHs interconnected in series on four layers. Each set is a different hole size. The pad size is also varied. Similar designs are available from the IPC.

57.6.2.2 Mechanical. Printed circuit boards are rarely subjected to mechanical tests that could cause electrical failures; however, adhesion of both Cu and solder mask to the laminate is critical and is often tested. Loss of solder mask adhesion can provide a place for corrodants and moisture to accumulate, which can be the cause of electrical failures when the board is exposed to temperature and humidity.

Adhesion is commonly tested using the peel test described in IPC-TM-650, Method 2.4.28. The simplest version of this test is conducted by scribing the adherent and dividing it into small squares. If the Cu or solder mask pulls off with a piece of tape with strong adhesive, the adhesion is inadequate. More quantitative tests that measure the actual peel strength are performed primarily by laminate and solder mask suppliers.

57.6.2.3 Temperature, Humidity, Bias. These tests are designed to promote corrosion on the PCB surface and conductive anodic filament growth, either of which can cause insulation resistance failures.

Surface insulation tests utilize two interleaved Cu combs with an imposed dc bias across the combs. These combs may be designed into existing boards or a coupon such as the IPC-B-25 test board shown in Fig. 57.24 may be used. The measured resistance (ohms) from the comb pattern can be converted to surface resistivity (ohms per square) by multiplying the measured

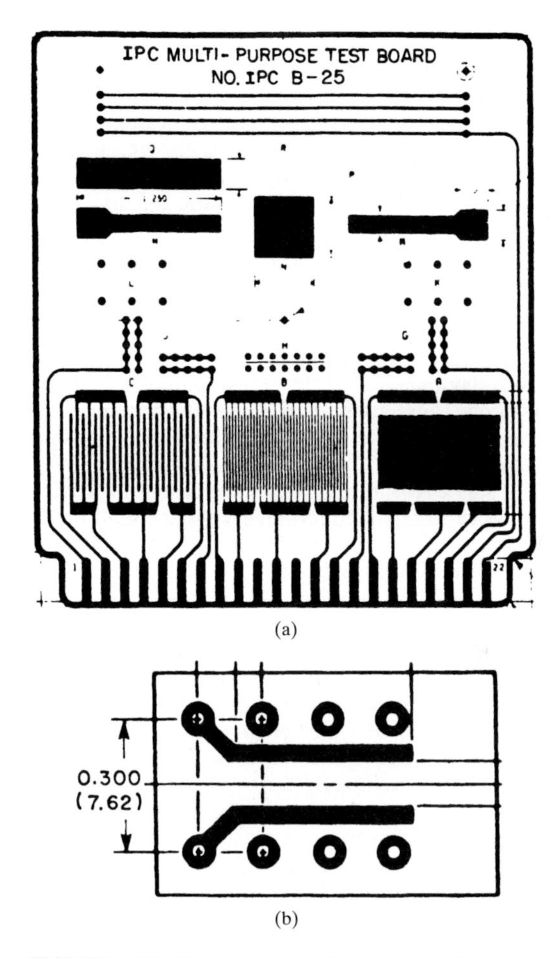

(a)

(b)

FIGURE 57.24 Test coupons used to check moisture, insulation, and metal migration resistance: (*a*) the IPC-B-25 test board, used to qualify the process; (*b*) The Y coupon, designed to be incorporated into production boards for statistical process control. From IPC-SM-840.

resistance by the square count of the pattern. The square count is determined geometrically by measuring the total length of the parallel traces between the anode and cathode and dividing by the separation distance. Special precautions are needed to make accurate measurements of insulation resistance.[42] Measurements of resistance above 10^{12} are very difficult and require careful shielding. Measurements of resistance below 10^{12} can be conducted in most laboratory environments if certain precautions are taken.

The actual tests are usually conducted at elevated temperature and humidity with an applied dc bias. A test for moisture and insulation resistance of bare printed circuit boards is included in IPC-SM-840A. The severity of the test depends on the intended use environment; for typical commercial products (Class 2), the test is conducted at 50°C, 90% RH, and 100 V_{dc} bias for 7 days. The minimum insulation resistance requirement is 10^8 Ω. The military test procedure for moisture and insulation resistance is specified in MIL-P-55110.[43] The moisture resistance test should be conducted in accordance with MIL-STD-202, Method 106, with applied polarization voltage (100 V_{dc}) and Method 402, Test condition A.[44] IPC-SM-840A also includes a test for electromigration resistance. The test is conducted at 85°C/90% RH at a 10 V_{dc} bias with a limiting current of 1 mA for 7 days. A significant change in current constitutes a failure. The samples are also microscopically inspected for evidence of electrolytic metal migration. A common test for dendritic growth due to flux residues is 85/85/1000 h at –20 V_{dc} bias. These tests are empirically based; however, several investigators have attempted to develop acceleration factors for these and similar tests.[45,46]

57.6.3 Printed Circuit Assembly Reliability Tests

57.6.3.1 Thermal. Most thermal cycling of PCAs is intended to accelerate solder joint thermal fatigue failures. In spite of the existence of an IPC standard, there is no standard accelerated test today that is suitable for all component and substrate combinations and all service environments. There are several acceleration models in the literature, each of which seems to fit the data well in at least some situations. All are based on a combination of empirical observations and fundamental arguments under simplifying assumptions. This topic remains a subject of active research since in some cases the predictions are significantly different. There is also a move to replace thermal cycling tests with mechanical cycling tests, which could be conducted in a shorter period of time; however, these tests are even further from standardization. Finally, for some components which dissipate a significant amount of power (usually 1 W or more), cycling the ambient temperature (which heats from the outside) may give quite different results from power cycling (in which heating occurs from the inside); for example, the failure location may shift from corner joints (which see the largest displacements) toward solder joints located near the chip (because they are hotter). Consequently, while thermal cycling will be adequate for most ASICs, memory chips, etc., power cycling should be considered for microprocessors, particularly those that dissipate more than a few watts.

Thermal shock testing is commonly used to test components, but it is not necessarily a substitute for thermal cycling. Because the temperature ramp is extremely rapid and the dwell at the extremes is generally short, there is little time for creep; consequently, the number of cycles to failure is increased. Furthermore, the rapid temperature change can induce differential thermal stresses that may be larger than those experienced during thermal cycling. These stresses can induce early failures, particularly if the failure is not in the solder.

There are some principles for designing thermal cycling tests to accelerate solder fatigue that seem to be generally agreed on. The following guidelines apply to gradual temperature cycling due to ambient heating inside a unit (e.g., due to power dissipation). If the unit will be subjected to extreme temperatures or thermal shock in service, these generalizations may not apply. A sample cycling protocol is shown in Fig. 57.25.

- The maximum test temperature should be below the T_g of the printed circuit board, for FR-4 below about 110°C. At T_g, the CTE of the board increases rapidly, but many other properties also change; for example, the elastic modulus of the board decreases. To avoid approaching the melting temperature of the solder and changing the mechanism of solder creep, the maximum temperature should also be kept below about $0.9T_m$, where T_m is the melting temperature of the solder in Kelvin. For eutectic Sn-Pb solder, T_m is 137°C, well above T_g. But for printed circuit board materials with high T_g values or low-melting-temperature solders, this restriction may take precedence. Using a peak temperature above these limits results in unpredictable acceleration.
- The minimum temperature should be high enough that creep is still the primary deformation mechanism of the solder, that is, at least 0.5 T_m, or –45°C for eutectic Sn-Pb solder. Many investigators prefer a higher minimum temperature (–20 or 0°C) to ensure that creep occurs fast enough to relieve the imposed shear stress during the allowed dwell time. Using too low a minimum temperature may seem to increase the acceleration factor (increased ΔT) while actually decreasing it (decreased $\Delta\varepsilon$), resulting in an overly optimistic life prediction.

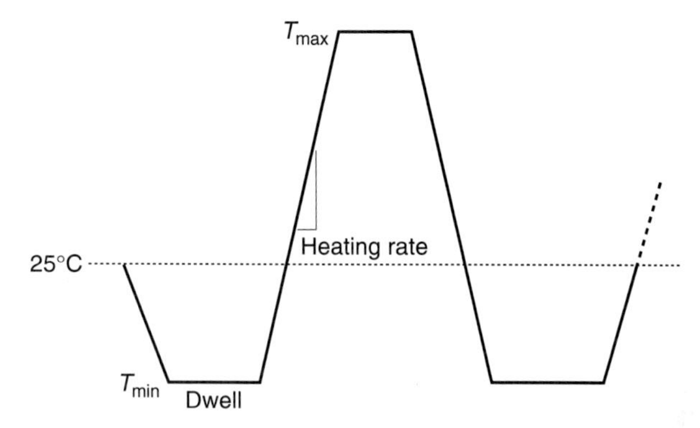

FIGURE 57.25 Schematic thermal cycling profile for testing solder joint thermal fatigue.

- The rate of temperature cycling should not exceed 20°C/min and the dwell time at the temperature extremes should be at least 5 min. The purpose of controlling the cycling speed is to minimize thermal shock and the stresses associated with differential heating or cooling. The dwell time at the temperature extremes is an absolute minimum needed to permit creep to occur. A longer dwell time is recommended, particularly at the minimum temperature.

As described here, there are several acceleration factors in the literature that may be applied to make life estimations from the accelerated test data that may be obtained using a thermal profile that fits the preceding criteria. One of the simplest expressions is due to Norris and Landzberg:

$$\frac{N_{op}}{N_{test}} = \left(\frac{\nu_{op}}{\nu_{test}}\right)^{1/3} \left(\frac{\Delta T_{test}}{\Delta T_{op}}\right)^2 \left(\frac{\phi_{test}}{\phi_{op}}\right)$$

where N_{op} and N_{test} = life under operating and accelerated test conditions, respectively
ν = cycling frequency
ϕ = mean temperature[47]

Another widely used expression may be found in IPC-SM-785. This expression tries to account for the effect of power cycling and the dwell time in the thermal profile; it also makes it possible to make predictions for one component or solder joint geometry based on data for another similar one. In its most simplified form, the acceleration factor for tests meeting the preceding criteria for FR-4 and eutectic Sn-Pb solder joints may be *approximated* as

$$\frac{N_{op}}{N_{test}} = \frac{\Delta T_{op}^{\ 2.4}}{\Delta T_{test}} \quad \text{for leadless surface-mount attachments}$$

$$\frac{N_{op}}{N_{test}} = \frac{\Delta T_{op}^{\ 4}}{\Delta T_{test}} \quad \text{for compliant-leaded surface-mount attachments}$$

assuming the test assemblies are nearly identical to the assemblies that will be put into service.

Mechanical fatigue cycling is increasingly being used as a quick way to induce solder joint failures. The goal is to *simulate* the thermal fatigue failure process in a much shorter test. The validity of this approach is still being investigated; although the imposed strain in each cycle is intended to be the same, the mechanical test eliminates thermomechanical effects (including creep) because the cycles are about two orders of magnitude faster. Nonetheless, mechanical cycling can certainly provide useful comparisons between different designs or package styles. The tests are usually conducted at a constant temperature with fixturing that puts the solder joints in shear when bending or tensile displacements are applied.

57.6.3.2 Mechanical. Mechanical vibration and shock can cause solder joint failures, particularly for large, rigid components or components with large, heavy heat sinks. Mechanical shock tests are usually modeled after drops that may occur during transportation or use. The test drops are generally quite severe, but few in number, since the system is not expected to be subjected to repeated drops in service. One common test uses a maximum acceleration of about 600 g, a maximum velocity of about 300 in/s, and a shock pulse of about 2.5 ms duration. The test setup is shown schematically in Fig. 57.26.

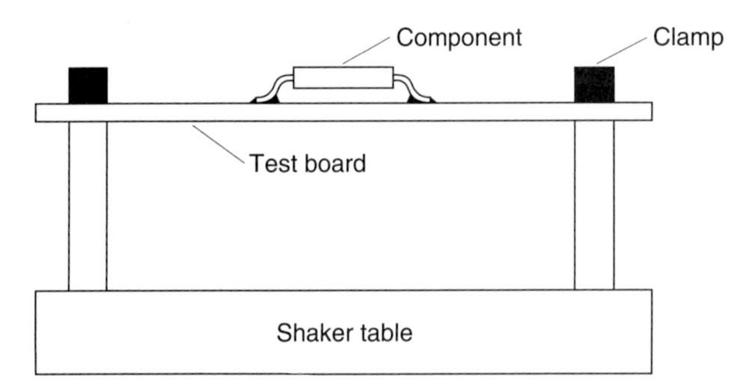

FIGURE 57.26 Schematic of setup used to test the resistance of printed circuit assemblies to out-of-plane mechanical vibration and shock.

On the other hand, the PCA may be exposed to millions of mechanical vibration cycles during its lifetime.[48] Depending on the application, both in-plane and out-of-plane vibration may play an important role. The damage done by these cycles depends primarily on whether the cycling frequency is near a natural frequency of the board, where large deflections can take place. For in-plane vibration, random vibration over a wide excitation frequency range is performed at a constant power spectral density. Most surface-mount components have high natural frequencies, and solder joint failures are rarely observed. For out-of-plane vibration, the following procedure is recommended:

- Design and fixture the test board so that each specimen is clamped at both ends and contains a single component at the center (see Fig. 57.26).
- Impose the vibration using a shaker table with sinusoidal excitation.
- Find the natural frequency of the sample by sweeping the frequency at a low amplitude (to prevent inadvertent damage before the start of the test). This natural frequency is near, but not the same as, the natural frequency at the larger amplitude that will be used for testing.
- Conduct the test by sweeping the frequency in a narrow range around the natural frequency previously determined. The amplitude can be set to correspond to a certain power spectral density or to achieve a desired deflection of the PCB.

57.6.3.3 Temperature, Humidity, Bias. The primary purpose of these tests is to identify surface insulation resistance (SIR) degradation due to corrosive materials left on the board from the assembly process or due to galvanic couples set up in the assembly process. The usual test procedure is to use SIR comb patterns on the PCB, and to subject the assembly to 85°C/85 percent relative humidity/–20V_{dc} for 1000 h. The bias voltage is dependent on the test device or test vehicle chosen.

One acceleration model that is applied is the modified Eyring model, which was developed for moisture-induced corrosion in plastic packages:

$$t_{50\%} = \left[A \exp\left(\frac{E_a}{kT}\right)\right]\left[\exp\left(\frac{C}{H_r}\right)\right]\left[D\exp\left(\frac{-V}{B}\right)\right] \tag{57.4}$$

where
- $t_{50\%}$ = time at which 50 percent of parts have failed
- A, B, C, and D = empirical constants
- E_a = thermal activation energy
- k = Boltzmann constant
- T = temperature in degrees Kelvin
- H_r = relative humidity
- V = reverse-biasing voltage[41,46]

The time to failure is also dependent on the concentration of ionic contaminants. The default industrial ionic contamination limit comes from MIL-STD-28809A; it is the equivalent of 3.1 $\mu g/cm^2$ of NaCl. For small plastic packages, enough data have been collected to show that an empirical acceleration factor for temperature and relative humidity $AF_{T,Hr}$ applies:

$$AF_{T,H_r} = 2^{(T+Hr)\text{test} - (T+Hr)\text{service}} \tag{57.5}$$

where
- AF = acceleration factor
- T = temperature in °C
- H_r = relative humidity in percent[46,49]

The rule of thumb for the effect of reverse bias is very device-specific; the following relationship was established for 20-V Schottky diodes in SOT-23 packages:[46]

$$AF_V = 7700 \exp\left(\frac{-V}{12.32}\right)$$

When both acceleration factors apply, the total acceleration factor is:

$$AF_{\text{total}} = (AF_{T,H_r})(AF_V)$$

57.7 SUMMARY

Reliability of electronic assemblies is a complex subject. This chapter has touched on only one aspect of the problem: understanding the primary failure mechanisms of printed circuit boards and the interconnects between these boards and the electronic components mounted on them. This approach provides the basis for analyzing the impact of design and materials choices and manufacturing processes on printed circuit assembly reliability. It also provides the foundation for developing accelerated testing schemes to determine reliability. It is hoped that the fundamental approach will enable the reader to apply this methodology to new problems not yet addressed in mainstream literature.

REFERENCES

1. M. A. Oien, "Methods for Evaluating Plated-Through-Hole Reliability," *14th Annual Proceedings of IEEE Reliability Physics,* Las Vegas, Nev., April 20–22, 1976.
2. K. Kurosawa, Y. Takeda, K. Takagi, and H. Kawamata, "Investigation of Reliability Behavior of Plated-Through-Hole Multilayer Printed Wiring Boards," IPC-TP-385, IPC, Evanston, Ill., 1981.

3. IPC-TR-579, "Round Robin Reliability Evaluation of Small Diameter Plated Through Holes in Printed Wiring Boards," Institute of Interconnecting and Packaging Electronic Circuits, Lincolnwood, Ill. The IPC recently initiated a second round robin study, the results of which should be available in about 1996.
4. F. Fehrer and G. Haddick, "Thermo-mechanical Processing and Repairability Observations for FR-4, Cyanate Ester and Cyanate Ester/Epoxy Blend PCB Substrates," *Circuit World,* vol. 19, no. 2, 1993, pp. 39–44.
5. D. B. Barker and A. Dasgupta, "Thermal Stress Issues in Plated-Through-Hole Reliability," in *Thermal Stress and Strain in Microelectronics Packaging,* J. H. Lau (ed.), Van Nostrand Reinhold, 1993, pp. 648–683.
6. IPC-TM-650, Method 2.4.8.
7. L. D. Olson, "Resins and Reinforcements," in *ASM Electronic Materials Handbook, Vol. 1: Packaging,* ASM International, Materials Park, Ohio, 1989, pp. 534–537.
8. D. W. Rice, "Corrosion in the Electronics Industry," *Corrosion/85,* paper no. 323, National Association of Corrosion Engineers, Houston, 1985.
9. J. J. Steppan, J. A. Roth, L. C. Hall, D. A. Jeannotte, and S. P. Carbone, "A Review of Corrosion Failure Mechanisms during Accelerated Tests: Electrolytic Metal Migration," *J. Electrochemical Soc.,* vol. 134, 1987, pp. 175–190.
10. D. J. Lando, J. P. Mitchell, and T. L. Welsher, "Conductive Anodic Filaments in Reinforced Polymeric Dieletrics: Formation and Prevention," *17th Annual Proceedings of IEEE Reliability Physics Symposium,* San Francisco, April 24–26, 1979, pp. 51–63.
11. "How to Avoid Metallic Growth Problems on Electronic Hardware," IPC-TR-476, Sept. 1977.
12. B. Rudra, M. Pecht, and D. Jennings, "Assessing Time-of-Failure Due to Conductive Filament Formation in Multi-Layer Organic Laminates," *IEEE Trans. CPMT-Part B.,* vol. 17, August 1994, pp. 269–276.
13. J. W. Price, *Tin and Tin Alloy Plating,* Electrochemical Publications, Ayr, Scotland, 1983.
14. D. R. Gabe, "Whisker Growth on Tin Electrodeposits," *Trans. Institute of Metal Finishing,* vol. 65, 1987, p. 115.
15. C. Lea, *A Scientific Guide to Surface Mount Technology,* Electrochemical Publications, Ayr, Scotland, 1988.
16. J. H. Lau (ed), *Solder Joint Reliability: Theory and Applications,* Van Nostrand Reinhold, New York, 1991.
17. D. R. Frear, S. N. Burchett, H. S. Morgan and J. H. Lau, eds., *Mechanics of Solder Alloy Interconnects,* Van Nostrand Reinhold, New York, 1994.
18. J. H. Lau (ed.), *Thermal Stress and Strain in Microelectronics,* Van Nostrand Reinhold, New York, 1993.
19. S. Burchett, "Applications—Through-Hole," in *The Mechanics of Solder Alloy Interconnects, op. cit.,* pp. 336–360.
20. E. Suhir and Y.-C. Lee, "Thermal, Mechanical, and Environmental Durability Design Methodologies," in *ASM Electronic Materials Handbook, Vol. 1: Packaging, op cit.*
21. IPC-D-279, "Design Guidelines for Reliable Surface Mount Technology Printed Board Assemblies," to be published.
22. L. T. Manzione, *Plastic Packaging of Microelectronic Devices,* Van Nostrand Reinhold, New York, 1990.
23. *ASM Electronic Materials Handbook, Vol. 1: Packaging, op. cit.*
24. R. R. Tummala and E. J. Rymaszewski (eds.), *Microelectronic Packaging Handbook,* Van Nostrand Reinhold, New York, 1989.
25. For a more comprehensive list of components that may be at risk, see IPC-D-279, "Design Guidelines for Reliable Surface Mount Technology Printed Board Assemblies," App. C, to be issued.
26. L. Zakraysek, R. Clark, and H. Ladwig, "Microcracking in Electrolytic Copper," *Proceedings of Printed Circuit World Convention III,* Washington, D.C., May 22–25, 1984.

27. G. T. Paul, "Cracked Innerlayer Foil in High Density Multilayer Printed Wiring Boards," *Proceedings of Printed Circuit Fabrication West Coast Technical Seminar,* San Jose, Aug. 29–31, 1983.
28. M. W. Gray, "Inner Layer or Post Cracking on Multilayer Printed Circuit Boards," *Circuit World,* vol. 15, no. 2, pp. 22–29, 1989.
29. L. Mayer and S. Barbieri, "Characteristics of Acid Copper Sulfate Deposits for Printed Wiring Board Applications," *Plating and Surface Finishing,* March 1981, pp. 46–49.
30. J. L. Marshall, L. A. Foster, and J. A. Sees, "Interfaces and Intermetallics," in *The Mechanics of Solder Alloy Interconnects,* D. R. Frear, H. Morgan, S. Burchett, and J. Lau (eds.), *op. cit.*, pp. 42–86.
31. M. Economou, "Rework System Selection," *SMT,* February 1994, pp. 60–66.
32. J. Lau, S. Leung, R. Subrahmanyan, D. Rice, S. Erasmus, and C. Y. Li, "Effects of Rework on the Reliability of Pin Grid Array Interconnects," *Circuit World,* vol. 17, no. 4, pp. 5–10, 1991.
33. F. L. Gray, "Thermal Expansion Properties," in *ASM Electronic Materials Handbook, Vol. 1: Packaging, op. cit.*
34. P. M. Hall, "Thermal Expansivity and Thermal Stress in Multilayered Structures," in *Thermal Stress and Strain in Microelectronics Packaging,* J. H. Lau (ed), *op. cit.,* pp. 78–94.
35. J. Glazer, P. A. Kramer, and J. W. Morris, Jr., "The Effect of Gold on the Reliability of Fine Pitch Surface Mount Solder Joints," *Circuit World,* vol. 18, 1992, pp. 41–46.
36. G. Humpston and D. M. Jacobson, *Principles of Soldering and Brazing,* ASM International, Materials Park, Ohio, 1993, Chap. 3.
37. F. G. Yost, "Soldering to Gold Films," *Gold Bulletin,* vol. 10, 1977, pp. 94–100.
38. J. Glazer, "Metallurgy of Low Temperature Lead-free Solders: A Literature Review," *International Materials Review,* vol. 40, 1995, pp. 65–93.
39. H. L. Hvims, *Adhesives as Solder Replacement for SMT,* vols. I and II, Danish Electronics, Light and Acoutics, Hoersholm, Denmark, January 1994.
40. A. Emerick, J. Ellerson, J. McCreary, R. Noreika, C. Woychik, and P. Viswanadham, "Enhancement of TSOP Solder Joint Reliability Using Encapsulation," *Proceedings of the 43d Electronic Components and Technology Conference,* IEEE-CHMT, 1993, pp. 187–192.
41. P. Tobias and D. Trindade, *Applied Reliability,* Van Nostrand Reinhold, New York, 1986.
42. ASTM D 257-78, Standard Test Methods for DC Resistance or Conductance of Insulating Materials.
43. Military specification for Printed Wiring Boards.
44. Military standard test method for Electronic and Electrical Component Parts.
45. P. J. Boddy, R. H. Delaney, J. N. Lahti, E. F. Landry, and R. C. Restrick, "Accelerated Life Testing of Flexible Printed Circuits: Part I and II," *14th Annual Proceedings of the IEEE Reliability Physics Symposium,* Las Vegas, Nev., April 20–22, 1976.
46. K.-L. B. Wun, M. Ostrander, and J. Baker, "How Clean is Clean in PCA," *Proceedings of Surface Mount International,* San Jose, Calif., 1991, pp. 408–418.
47. K. C. Norris and A. H. Landzberg, *IBM Journal of Research and Development,* vol. 13, 1969, p. 266.
48. For example, see J. H. Lau, "Surface Mount Solder Joints Under Thermal, Mechanical, and Vibration Conditions," in *The Mechanics of Solder Alloy Interconnects,* D. R. Frear, H. Morgan, S. Burchett, and J. Lau (eds.), *op. cit.,* pp. 361–415.
49. E. B. Hakim, "Acceleration Factors for Plastic Encapsulated Semiconductors," *Solid State Technology,* Dec. 1991, pp. 108–109.

FURTHER READING

ASM Electronic Materials Handbook, Vol. 1: "Packaging," ASM International, Materials Park, Ohio, 1989.

Frear, D. R., H. Morgan, S. Burchett, and J. Lau, *The Mechanics of Solder Alloy Interconnects,* Van Nostrand Reinhold, New York, 1994.

IPC-A-600D, Acceptability of Printed Boards, August 1989.

IPC-A-610A, Acceptability of Electronic Assemblies, Feb. 1990.

IPC-D-279, Design Guidelines for Reliable Surface Mount Technology Printed Board Assemblies, to be issued.

IPC-SM-785, Guidelines for Accelerated Reliability Testing of Surface Mount Solder Attachments, Nov. 1992.

de Kluizenaar, E. E., "Reliability of Soldered Joints: A Description of the State of the Art," *Soldering and Surface Mount Technology,* Part I: no. 4, pp. 27–38; Part II: no. 5, pp. 56–66; Part III: no. 6, pp. 18–27, 1990.

Lau, J. H. (ed.), *Solder Joint Reliability,* Van Nostrand Reinhold, New York, 1991.

Lau, J. H. (ed.), *Thermal Stress and Strain in Microelectronics Packaging,* Van Nostrand Reinhold, New York, 1993.

Lea, C., *A Scientific Guide to Surface Mount Technology,* Electrochemical Publications, Ayr, Scotland, 1988.

Tummala, R. and E. J. Rymaszewski (ed.), *Microelectronic Packaging Handbook,* Van Nostrand Reinhold, New York, 1989.

CHAPTER 58
COMPONENT-TO-PWB RELIABILITY—THE IMPACT OF DESIGN VARIABLES AND LEAD FREE

Mudasir Ahmad
Cisco Systems, San Jose, California

Mark Brillhart
Palo Alto, California

58.1 INTRODUCTION

The mandated conversion to lead free assemblies has led to the convergence of several critical issues in the microelectronics industry. The conversion has further compounded several other factors that have been pushing the envelope in Component to Printed Wiring Board (PWB) reliability, including

- Radical increases in package-to-board input/output (I/O)
- Rapid reduction in interconnect pitch
- Overall rise in surface-mount component density
- Higher speeds and correspondingly higher heat dissipation

All of these factors pose tremendous challenges with respect to reliability, and they cause complex design tradeoffs that must be considered while maintaining product reliability at required levels. Time to market and aggressive product life cycle pressures also require PWB designers to consider the impact that alternate designs and material sets may have on reliability, in a rapid fashion.

The PWB designer is faced with myriad packaged device types, field use environments, connectors, PWB materials, cost considerations, and real estate restrictions, which all must be optimized without degrading system reliability. Designers need tools to assess experimental reliability data, transform laboratory results into actual field loading conditions, and rapidly assess the reliability of different PWB layouts, material sets, and package types. Engineers must also have the capability for addressing complex loading conditions such as mini- and power-cycles and the knowledge of how they impact interconnect performance.

This chapter focuses on the design variables that impact the reliability of PWB-to-package interconnect (second- and third-level interconnect) of grid array devices and the impact of lead-free conversion on these design variables.

Some of the critical variables discussed in depth include:

- Double-sided construction
- PWB stiffness effect
- Multiple packages in a small area
- Package and PWB warpage; assembly and reliability challenges
- Solder joint geometry; assembly and reliability challenges
- Package and PWB pad size
- Heatsink design and attachment
- Package and PWB surface finish; Black Pad, Brittle Failure, and Kirkendall voiding
- PWB electrical test architecture
- Carrier chassis design
- Package design parameters such as overall size, silicon technology, ball array, ball size/pitch, heat spreader/stiffener, material selection, die/package aspect ratio, substrate material, low-*k* dielectrics and so on

The chapter is broken up into the following sections.

Section 58.2, "Packaging Challenges," outlines several of the performance challenges faced by next generation packages, and the key technical drivers behind these challenges.

Section 58.3, "Variables That Impact Reliability," contains a discussion of the common design variables that could impact the reliability of a product; ranging from PWB design rules to package design parameters. The failure modes relating to the choice of different package design parameters are outlined, along with recommendations on ways to mitigate different manufacturing and test-related failure mechanisms.

58.2 PACKAGING CHALLENGES

The reliability of many types of leadframe components (see Figs. 58.1 and 58.2) has been assessed in great detail.[1] Figure 58.3 shows the typical definitions of interconnect levels in assemblies. While leadframe devices are by no means infinitely reliable at the first and second level interconnects, they have historically posed a lesser risk than grid-array packaged components, such as ceramic ball grid arrays (CBGAs), plastic ball grid arrays (PBGAs), ceramic solder column carriers

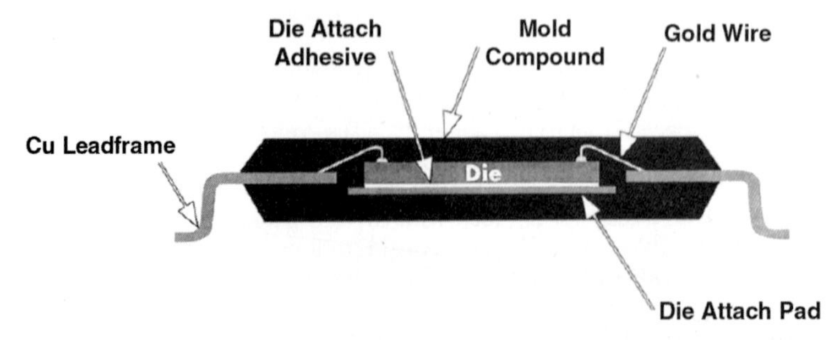

FIGURE 58.1 Diagram of a quad flat pack. Note the copper (Cu) leads that are attached by gold wire bonds to the die. This is the primary (or first-level interconnect) between the die and the package. The second-level interconnect will be created when the Cu leadframe is soldered to a motherboard via some type of surface-mount attach process. *(Courtesy of Amkor Technology, Inc.)*

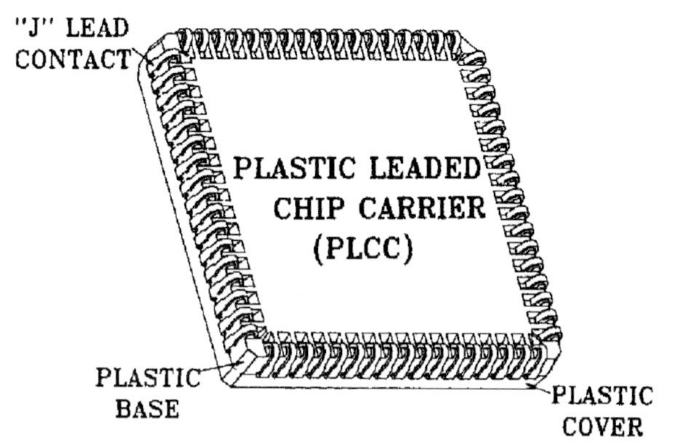

FIGURE 58.2 J-leaded device oriented bottom side up. *(Reprinted with permission from J. Lau, Ball Grid Array Technology, McGraw-Hill, New York, 1995, p. 20.)*

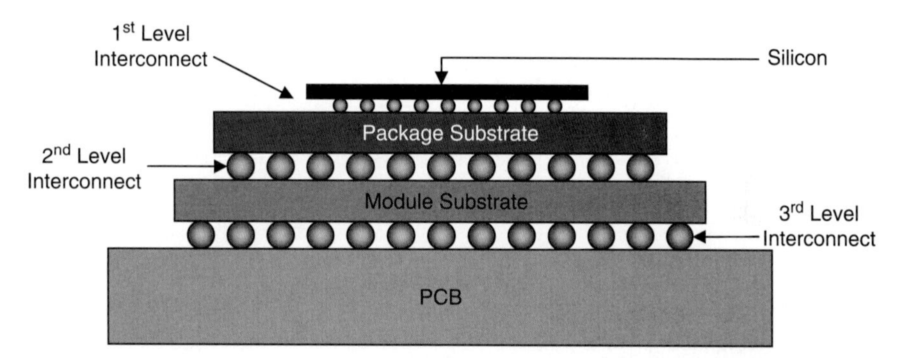

FIGURE 58.3 First-, second- and third-level interconnects defined, employing a flip-chip assembly as an example. The first-level interconnect is the primary connection between the silicon die and the package substrate. In this example it is created by the solder bumps between the die and the package. The second-level interconnect in this example is the "next" level of connection between the package substrate and the module substrate. The third-level interconnect in this example is created when the solder balls on the bottom side of the module substrate are attached via SMT to the PWB.

(CSCCs), flip-chip ball grid arrays (FCBGAs), and chip scale packages (CSPs). This is because the compliant leadframe in leadframe packages can absorb the differential strain induced in the solder joints, due to the mismatch in thermal expansivity between the silicon chip and the PWB. Ball grid array packages, on the other hand, rely entirely on the discrete solder balls absorbing the differential strain. Depending on the package construction, the solder joints at critical high strain locations could fail early, resulting in lower long-term reliability of the assembly. Figure 58.4 shows various grid array packaged devices.

To meet increasing I/O density requirements, many of these types of packages are either increasing in body size, decreasing in pitch, or both, to a point where the standard reliability envelope is being pushed to extreme levels. Additionally, most grid array packaged devices are high in cost (in many instances the grid array packaged devices are the most costly component on a printed circuit assembly) and thus cannot be deployed in a redundant design. Lack of redundancy creates critical path single points of failure (SPOFs). Failure of one of these devices can lead to catastrophic system-level failures. The proliferation of high-I/O, small-pitch grid array packaged devices throughout the electronics industry impacts almost all PWB engineers because any design has a high probability of containing one or more of these types of components.

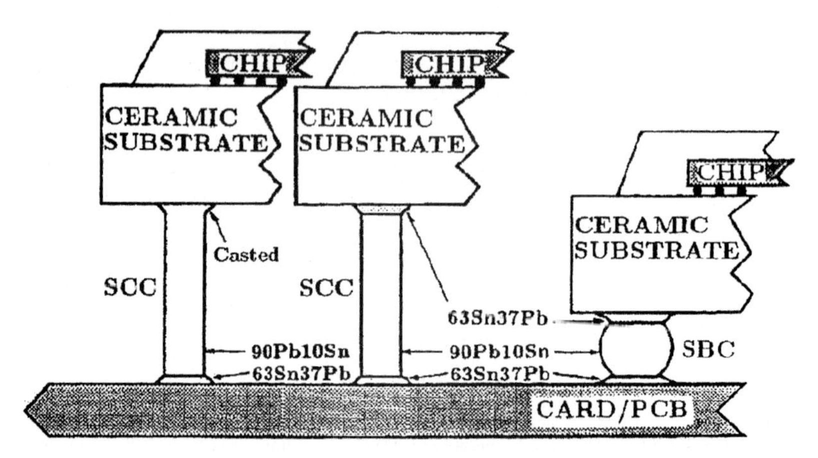

FIGURE 58.4 Cross-section of a ceramic solder column carrier and ceramic ball grid array (CBGA). Solder column carriers (SCC) and a solder ball carrier (SBC) are shown. The SBC is often referred to as a CBGA package due to the fact that it is a ceramic package substrate that employs solder balls to create a second-level interconnect between the package substrate and the PWB. The solder column carrier employs a column of high-lead solder (which is either casted onto the ceramic package substrate or attached by a layer of eutectic 63Sn/37Pb solder paste) to create a second-level interconnect between the ceramic package substrate and the PWB. A lead-free version of the SCC, made of copper columns is also being developed.[2] Typical ceramic ball grid array packages cannot achieve required reliability when body sizes exceed 32 mm per side. Ceramic packages having larger body sizes either employ SCC-type technology or connectors to create a second-level interconnect. *(Reprinted with permission from J. Lau, Ball Grid Array Technology, McGraw-Hill, New York, 1995, p. 27.)*

Moreover, at higher speeds, there is an ever increasing need for reduction in signal delay between processors and the memories they communicate with. These processors could be microprocessors, network processors, programmable logic devices (PLDs) and application-specific integrated circuits (ASICs). This has given rise to the prominence of System-in-Package (SiP) modules which significantly reduce the delay between on-PWB memory and ASICs (Fig. 58.5).

FIGURE 58.5 System-in-Package (SiP) Module. SiP modules help reduce signal delay between a processor and the memory devices it communicates with. This SiP has one processor in the middle, surrounded by four discrete memory packages, all on one module. The module is soldered to the PWB via the solder balls on the back side of the module substrate.

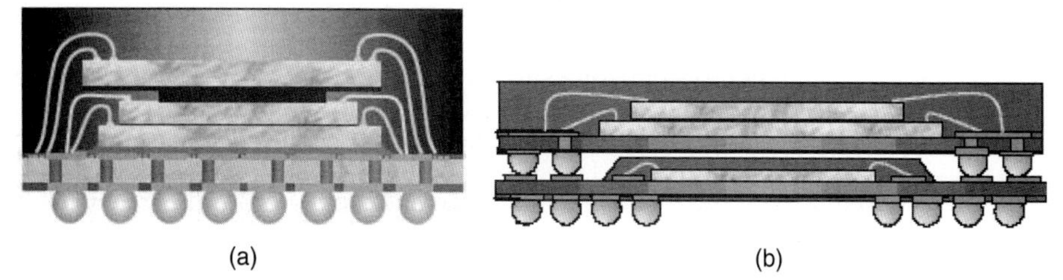

(a) (b)

FIGURE 58.6 (a) Stacked-die package. Several dies are stacked within a single package and wirebonded to a single substrate. (b) Stacked package. Several wirebonded packages are stacked on top of each other with solder balls. *(Courtesy of Amkor Technology, Inc.)*

SiP modules can have more than eight memories communicating with a single ASIC. The ASIC package is typically a flip chip, whereas the memories could be CSPs, stacked packages or PBGAs.

Stacked die and stacked package memories have also become more prevalent in applications where space is a constraint (Fig. 58.6). Stacking helps compact more memory within the same foot print and in applications where height is not a critical constraint, they help conserve valuable PWB real estate while at the same time providing greater integration and performance. But stacking technology raises its own unique set of reliability challenges. These challenges include heat dissipation from the multiple dies, reworkability and the reliability of the interconnects between the individual stacked dies or stacked packages.

58.3 VARIABLES THAT IMPACT RELIABILITY

Electronic packages are composed of complex materials that are assembled via intricate processes and then subjected to a wide range of service environments. It is crucial to note that both the die-packaging processes and surface-mount technologies can impart stresses and create residual stresses in the final packaged assembly that impacts its interconnect reliability. Assessing the reliability of package devices requires that all variables that can hinder or enhance reliability be considered.

58.3.1 Actual Product Environment

The actual product environment is what drives the reliability of the assembly. Printed circuit assemblies are deployed in a wide range of environments. Packages can be subjected to extreme temperatures and humidity. Automotive applications, for example, can see temperature ranges from –55°C to well over 95°C[3] along with up to 100 percent humidity. The rate at which temperature and humidity changes occur can be quite severe under the hood, causing components to transition from extreme cold to heat in a matter of minutes. This creates a very difficult environment to model, as the rate of occurrence of thermal and humidity cycles must be thoroughly understood. Severe shock and vibration can also be imparted to assemblies in this type of service environment, which adds to the complexity of reliability assessment. Great care must be taken when assessing these types of aggressive environments and their impact on packaging reliability. A thorough experimental analysis that employs thermocouples (to assess temperature conditions) and accelerometers (to assess shock and vibration loading conditions) should be employed to determine the actual end-use environmental conditions. These stressful environments are the primary driver for automotive applications employing extreme protection measures such as full encapsulation/potting for electronic devices placed under the hood.

Computer room environments represent the other extreme, with temperatures and humidity controlled to very tight tolerances. While servers, switches, hubs, routers, and other devices are housed in a controlled environment, their extremely large packages, coupled with power and mini cycles (discussed in the next chapter), create a set of loading conditions that are different from those found in automotive applications, but that are equally challenging. Worst-case loading conditions for a variety of end-use environments can be found in Table 58.1. Note the broad variation in temperature extremes, frequency of thermal cycles, and expected service life.

TABLE 58.1 Examples of Worst-Case Environments for Different Use Categories[3]

Use category	T_{min} (°C)	T_{max} (°C)	t_D (hrs)	Cycles per year	Typical years of service	Approximated acceptable failure risk (percent)
Consumer	0	+60	35	12	1–3	1
Computers	+15	+60	2	1460	5	0.1
Telecom	–40	+85	12	365	7–20	0.01
Commercial aircraft	–55	+95	12	365	20	0.001
Industrial and automotive passenger compartment	–55	+95	12	20–185	10	0.1
Military ground and ship	–55	+95	12	100–265	10	0.1
Space	–55	+95	1–12	365–8760	5–30	0.001
Military avionics	–55	+95	1–2	365	10	0.01
Automotive under hood	–55	+125	1–2	40–1000	5	0.1

T_{min}, minimum temperature; T_{max}, maximum temperature; t_D, dwell time at the operating temperature.

Prior to initiating any reliability assessment, it is crucial to understand the thermal, humidity, and mechanical loading conditions that a packaged device will be exposed to throughout its expected life time. Temperature maximums and minimums as well as the frequencies and rates at which the component cycles between extremes must be thoroughly understood. Complex thermal cycles (such as an outdoor product experiencing daily temperature changes coupled with a more local effect such as under-the-hood heating and cooling due to engine usage) must also be considered. Humidity and vibration loading conditions must also be evaluated as part of an overall package qualification process.

End-use conditions can be estimated by attaching thermocouples, humidity, and shock/vibration sensors in the final product because it is used in field operating conditions. This data needs to be analyzed over time and different expected field conditions to get useful estimates of the end-use environment as illustrated in Table 58.1.

Once this data is generated, it can be used in typical acceleration factor models relating accelerated stress tests to field data. For example, if it is determined that it takes x hours or cycles for an assembled package to fail in a laboratory test, the acceleration factor model can be used to extrapolate expected life of the device in typical field operating conditions. Details on acceleration factor models and sample calculations are provided in the next chapter.

Although environmental impact has a large effect on interconnect reliability, there are some basic design variables that can have a positive or negative impact on solder joint reliability. These critical design variables and their impact on reliability are discussed in the following sections.

58.3.2 Double-Sided (Mirrored BGA)

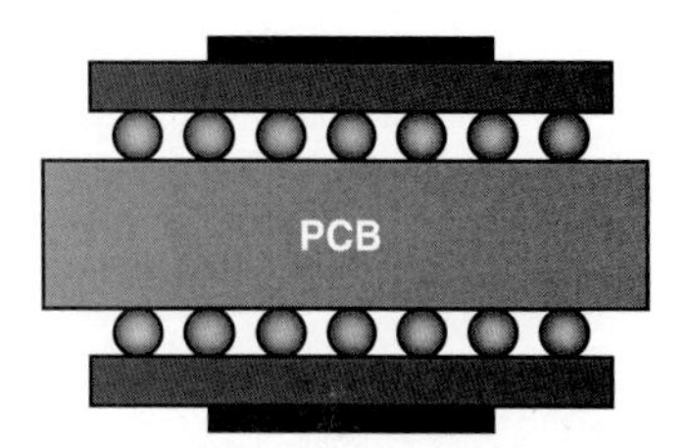

FIGURE 58.7 Double-sided BGA in mirror configuration. Note that the PCA is an exact mirror about the center line of the PWB. In some instances the top- and bottom-side packages share common vias.

The high-density revolution has forced PWB designers to place increasingly complex surface-mount devices in smaller and smaller spaces. This PWB density increase (where density refers to the number of components per unit area of PWB) may force designers to consider mirror BGA placement (see Figs. 58.7 and 58.8) or to place multiple devices on one side of a PWB.

The beginning of the density revolution occurred when designers began implementing double-sided technologies to allow for the placement of surface-mount components on both sides of a PWB. The continued push toward device density has resulted in placement of BGA components in mirror and quasi-mirror configurations, shown in Figs. 58.7 and 58.8.

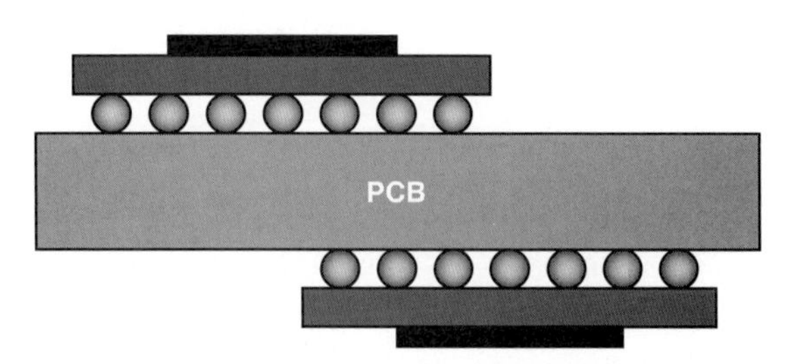

FIGURE 58.8 Double-sided BGA in quasi-mirror configuration. Note that although the packages are not right over each other, there is some overlap when looking from the top down or from the bottom up.

While the double-sided configuration certainly reduces signal delay, the primary disadvantage is a significant reduction in the thermomechanical reliability of the solder joint interconnects between the packages and PWB. Placing BGA packages in mirror or quasi-mirror configurations tends to rigidize the PWB assembly such that it does not bend as much during temperature cycling. The increased stiffness of the PWB due to double-sided mounting means the solder joints have to absorb more differential strain between the packages and PWB, resulting in as much as a 50 percent reduction in solder joint fatigue life (Figs. 58.9 and 58.10).[4] In terms of solder joint strain, the effect is the same as having a thicker or stiffer PWB. The primary failure mode is cracking in the solder joints near the package/joint interface.

Due to the dramatic decrease in assembly reliability when attempting mirrored configurations, these types of configurations should be avoided whenever possible. When it is impossible to avoid a mirror configuration, a quasi-mirror configuration without common through-hole vias is recommended (see Fig. 58.8). Common through-hole vias in this configuration tend to rigidize the assembly and contribute to further reduction in solder fatigue life. A standard rule of 50 percent reduction in fatigue life when employing a mirror configuration may be taken as

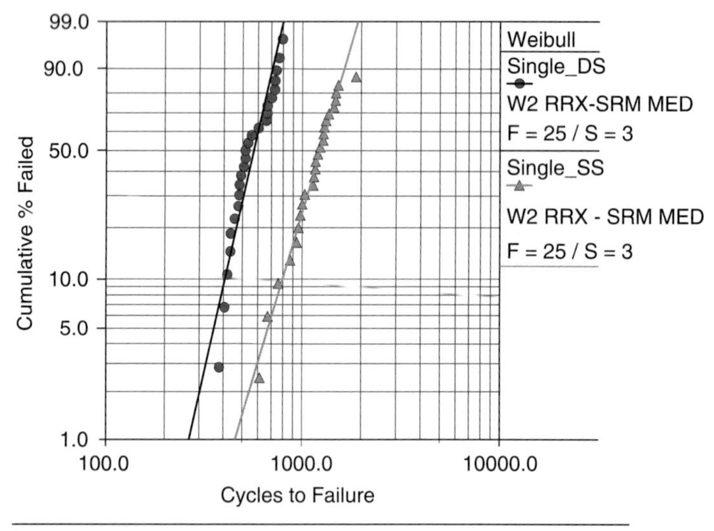

$\beta1 = 5.5657, \eta1 = 610.0411, \rho = 0.9484$

$\beta2 = 4.2934, \eta2 = 1347.6534, \rho = 0.9858$

FIGURE 58.9 Weibull plot of single and double-sided mounting of single lead-free extremely thin chip scale package (PS-etCSP®). *(Reprinted with permission from Amkor Technology, Inc./© [2004] IEEE ECTC[4])*

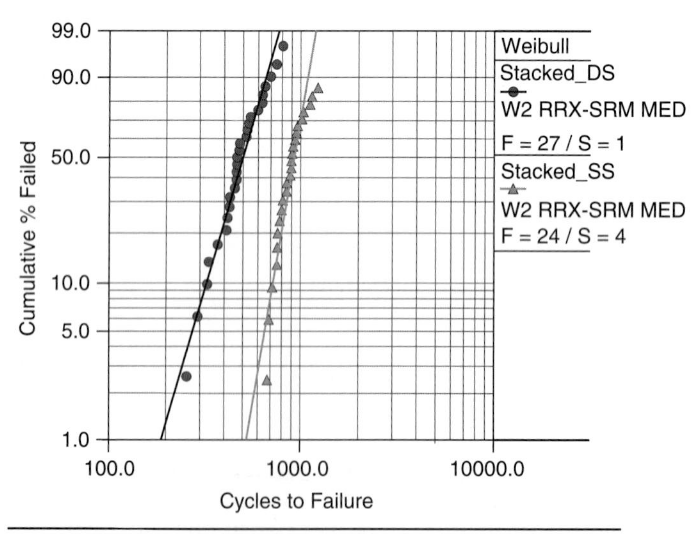

$\beta1 = 4.3555, \eta1 = 544.1006, \rho = 0.9837$

$\beta2 = 7.4308, \eta2 = 976.0282, \rho = 0.9395$

FIGURE 58.10 Weibull plot of single and double-sided mounting of stacked lead-free extremely thin chip scale package (PS-etCSP®). *(Reprinted with permission from Amkor Technology, Inc./© [2004] IEEE ECT[4])*

a first-pass approximation. Thermal cycle testing should be performed to ensure that the reliability degradation will still keep service life at an acceptable level when either a mirror or quasi-mirror configuration is considered.

58.3.2.1 Effect of Lead-Free Conversion. The reduction in lead-free package reliability when employing a mirrored double-sided configuration has been experimentally assessed.[4] Figure 58.9 and 58.10 show failure rates for single and stacked packages in single-sided and double-sided mirror positions.[4] Note the almost 50 percent reduction in fatigue performance when a single device is placed in a mirror configuration. The reduction is slightly less for the stacked packages, but the baseline reliability of the stacked package is 10 percent lower than that of the single package. The same trends have been shown with tin-lead eutectic assemblies, so it appears that the conversion to lead free would not mitigate this reduction in solder joint fatigue life due to double-sided mounting.

For details on Weibull distributions and failure life analysis, please refer to Sec. 59.2.2 in Chap. 59.

58.3.3 Board Stiffness Effect

As with the double-sided case described earlier, solder joint reliability generally decreases with PWB thickness. The thicker the PWB, the higher its stiffness, and the higher the differential strain that the solder joints have to absorb. However, the percent reduction in solder joint fatigue life due to board thickness varies for different package constructions (for example, CCGA, CSP, FCBGA, and PBGA).[5] The reduction is more pronounced in ceramic packages as compared to organic packages because of the higher mismatch in the coefficient of thermal expansion (CTE) between the package and PWB. Both thickness and number of layers of copper contribute to the stiffness of a PWB. As a result, the PWB thickness and layer count used to perform qualification testing should be representative of the end-use conditions in which the package would be used.

58.3.3.1 Effect of Lead-Free Conversion. The negative effect of PWB stiffness on solder joint reliability also applies to lead-free assemblies, as evidenced in Fig. 58.11. However, the amount of degradation could be different depending on the package type. Best practice recommendation is to use two different board thicknesses to perform temperature cycling qualification. This helps generate data for a specific package type over a range of board thicknesses. This recommendation is in line with the guidelines in the industry standard IPC 9701.[6]

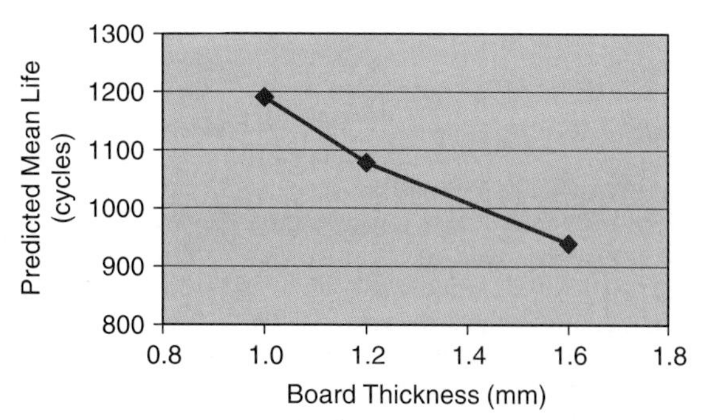

FIGURE 58.11 Effect of board thickness on solder joint reliability of stacked lead-free extremely thin CSP (PS-etCSP). *(Reprinted with permission from Amkor Technology, Inc./© [2004] IEEE ECT[4])*

58.3.4 High Density Due to Multiple Packages in a Small Area

A second source of reliability degradation due to density increase can be observed when a designer attempts to place many BGA packages in a small area on a thin PWB (see Fig. 58.12). The effective CTE of the packaged die or module tends to be much lower than that of the actual PWB. During reflow, the PWB and packages are in a stress-free state. Upon cooling, however, the packaged die contracts at a reduced rate (due to the lower effective CTE) than the underside of the PWB, causing a convex-type curvature to be imparted to the PWB. Moreover, if the thermal mass of the packages on the top side is higher than that on the bottom side, this can lead to a skew in the temperature distribution across the assembly, which in turn exacerbates the warpage of the assembly. Thus, care should be taken to distribute the packages as uniformly as possible across the PWB on both sides, and also to balance the copper in the PWB itself.

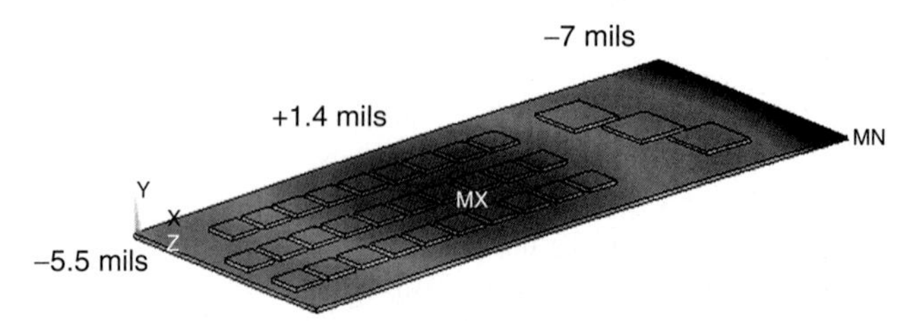

FIGURE 58.12 High-density placement of BGAs on top side of a PWB. This diagram illustrates the convex-type curvature of a PWB assembly where multiple BGA devices are all placed on one side of a board.

58.3.4.1 Effect of Lead-Free Conversion. Typical lead-free reflow temperatures (235 to 260°C) are higher than tin-lead reflow temperatures (220°C). As a result, the residual stresses built into the assembly after reflow could be higher in lead-free assemblies. These higher residual stresses are also a function of the cooling rate. Thus, it is important to not only control the heating side of the reflow profile, but the cooling side as well. The use of multiple zones for cooling is preferable, as it ensures better control. Moreover, forced convection ovens would be preferable over IR ovens for better control. Due to significant variations in the thermal mass of typical boards, it is difficult to provide universally applicable recommendations on reflow profiles.

58.3.5 Warpage Issues

The impact of warpage on a PWB assembly is multifaceted. First, warpage must typically be overcome when a PWB is placed in a card cage. Card cage placement of warped boards results in a rapid change in the loading of the solder joints that may lead to micro-crack formation or, in severe instances, solder joint fracture. Severe PWB warpage may create a card that physically cannot be placed in a card cage, resulting in a non-manufacturable product. Warped boards should be clamped in a controlled fashion to reduce the stress imparted to solder joints upon installation in a card cage.

58.3.5.1 Effect of Lead-Free Conversion. With the conversion to lead-free, it is imperative to ensure that the base material used in the PWB is compatible with the higher reflow temperatures. While more industry wide research is needed, the use of materials such as Dicyandiamide (DICY) cured FR-4 appears to be limited to applications with less than 240°C peak reflow temperature, whereas Phenolic cured PWBs are preferred for applications higher than 240°C. This is because of the better resistance to delamination observed for the phenolics after multiple reflows.[7,8] The thermal mass of the components and the amount of copper in large panel boards could make it difficult to maintain temperatures below 240°C. Detailed temperature measurements and reflow profiling should be performed on large boards to ensure that the temperature rise across the PWB is compatible with the PWB material.

58.3.6 Die and Package Stress Issues

A second issue associated with board warpage is die and package stress. The package experiences an amount of stress as it constrains the PWB from contracting during cooling after reflow. If this stress is severe enough, the silicon may crack. Also, severe stress may create multiple types of package delamination and/or cracking. The extreme danger here is that a device could pass full first-level qualification (first-level qualification typically consists of thermal cycling, humidity and temperature exposure, extreme-temperature storage, and shock and vibration experiments performed on packaged devices not attached to a PWB) yet suffer from delamination/die cracking due to stresses imparted by a warped PWB. This is one of the primary reasons that many packaged IC suppliers and consumers require that second-level qualification test vehicles include daisy chains at the first-level interconnect. Inclusion of first-level joints or wire bonds allows for the impact of second-level attach on first-level reliability to be assessed simultaneously with second-level solder joint reliability during a second-level qualification.[9]

58.3.6.1 Effect of Lead-Free Conversion. Lead-free conversion exacerbates these potential silicon-level failures. Since the assembly is subjected to a higher temperature range than with tin-lead assemblies, there is a higher propensity for early failures due to residual stresses during cool down. In addition, there are potential concerns over the effect on low dielectric constant (k) materials in the silicon during second level temperature cycling. This is discussed in more detail in Sec. 58.3.13.9. The effect of mold compound and underfill material selection on first-level interconnects is also discussed in more detail in Secs. 58.3.13.5 and 58.3.13.6 respectively.

58.3.7 Back-Side Component Issues

A third challenge related to warped PWBs concerns the back-side components in double-sided designs. During cool down after reflow, the top of the PWB is restrained from contraction due to the smaller effective CTE of the packaged device. This results in the bottom side of the PWB over contracting during cool down. The state of stress on the bottom solder joints due to warpage may be severe enough to fail solder joints or crack the die in bottom-side components. Additionally, the back-side components may be subject to residual stresses imparted after the second reflow (when the top-side components are attached via SMT) that may impact the long-term reliability of the devices on both sides and their second-level interconnects. As a result, it is important to characterize the thermomechanical reliability of the components that are mounted on the back-side by performing temperature cycling qualification.

58.3.8 Solder Joint Geometry Issues

Another consequence of severe warpage is the shape of the post-assembly solder joints. Joints can become excessively deformed, such that the cross section is too small to withstand fatigue. Joints can also become compressed or elongated, causing shorts or opens in some cases.

Moreover, with significant increases in body sizes of packages, there is a higher propensity for opens or bridges during reflow. For example, a 20 mm body-size package (1mm pitch)

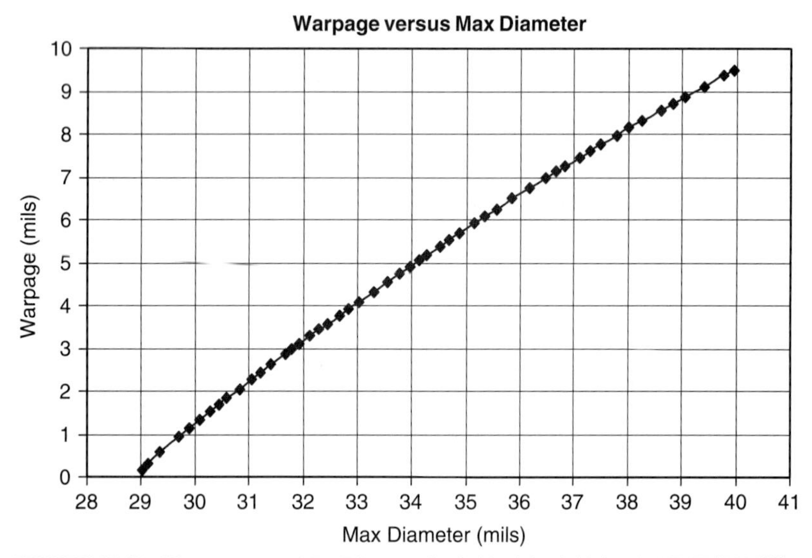

FIGURE 58.13 Warpage versus Max Diameter for Solder joint bridging © ASME [2005].[11]

would be easier to assemble on a relatively thin PWB, but with a 50 mm organic package, warpage is much more difficult to control, and it is a challenge to assemble it on a thick (≥125 mil, 20 + layer) PWB.

Tolerances are also difficult to control on large-panel, thick PWBs. Typical tolerances on high end PWBs are ±4 mils over the package site. Ensuring tolerances less than this is cost prohibitive. Current JEDEC guidelines[10] require ±8 mil coplanarity across the package. However, there are no specifications on the maximum acceptable warpage of the package at or above the reflow temperature. The typical relationship between maximum diameter and warpage is shown in Fig. 58.13 and 58.14.

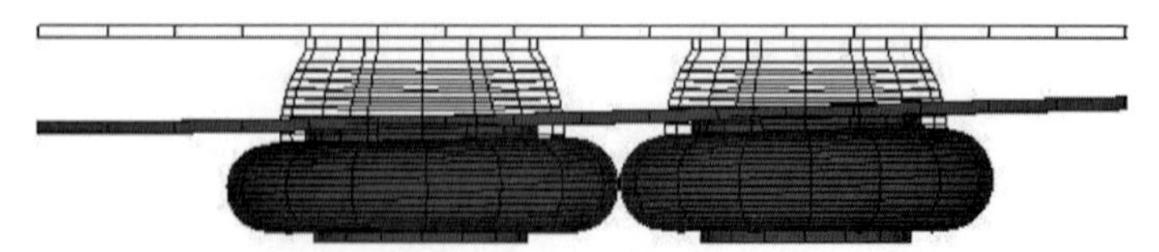

FIGURE 58.14 Bridging of solder joints–wireframe (unwarped), contour (warped) © ASME [2005].

The joint parameters used in the analysis are:

Paste volume (m^3)	4.909576e-011
Solder ball volume (m^3)	1.130972e-010
Upper pad diameter (mm)	0.60
Lower pad diameter (mm)	0.55
Viscosity of solder (Pa-sec)	2e-3
Surface tension (Nm^{-1}) (Tin-lead)	0.386

For a 1 mm pitch joint array, if the maximum diameter exceeds 40 mils (1.016 mm), bridging will occur. Based on the analysis results, if the effective warpage exceeds ~9 mils (0.23 mm) over a 7 mm distance along the package, the joints will bridge (Figs 58.13 and 58.14). Due to misregistration, bridging could occur at lower warpage levels. But, for the sake of clarity, the base of the solder joints has been assumed to have no warpage. Thus the predicted value (9 mils) is the *effective* warpage of the assembly. In practice, if the board has a concave warpage of 5 mils and the package has a convex warpage of 4 mils over 7 mm (above solder melting temperature), it could result in bridging.[11]

Given these values, there are a few critical points to note:

a. Simply maintaining the JEDEC recommended 8 mils coplanarity could be insufficient in some applications (for example, large body size packages on thick PWBs).

b. It is important to characterize the *effective* warpage between the package and PWB in addition to the warpage of the free standing package alone.

c. This warpage characterization should be performed over the entire reflow temperature range, as opposed to at room temperature. Of critical value is the warpage at or above the melting temperature. The data should be recorded in accordance with the JEDEC high temperature package warpage measurement specification.[12]

d. Where possible, the characterization should be performed on PWBs that are representative of actual PWBs used in volume production.

One useful way to mitigate solder bridging is to depopulate the corner-most balls of the package. In large body-size organic packages, a significant amount of warpage occurs at package corners, so depopulating the 6 corner-most balls at each corner could help absorb about 2 mils of effective warpage (Fig. 58.15). This helps minimize the likelihood of solder bridging/opens. Many package suppliers already depopulate the four corner balls to improve long term reliability of the solder joints.

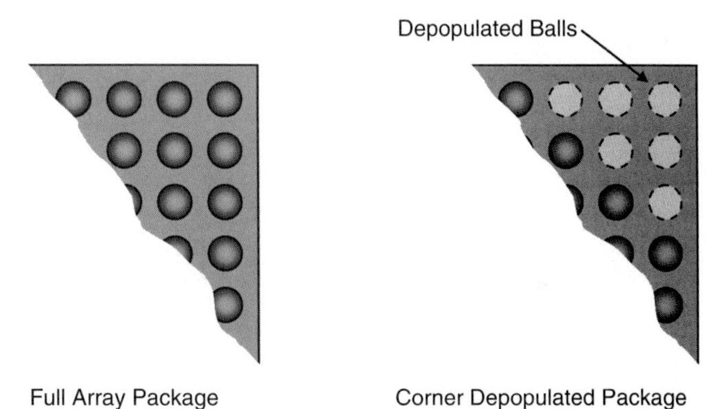

FIGURE 58.15 Depopulated corners to improve warpage-related assembly yield.

58.3.8.1 Effect of Lead-Free Conversion. It is important to note that the data presented in Figs. 58.13 and 58.14 is based on the surface tension of tin-lead solder. The surface tension[13] and contact angle of lead-free solders is higher than for tin-lead solders, but analytically, for the solder joint parameters analyzed, the warpage versus ball diameter curve is not significantly different from the data presented in Figs. 58.13 and 58.14. This is because the wetting area is presumed to be fixed in the analysis. For other solder geometries (gullwing, J-lead, QFN, and so on), the behavior of lead-free solders can be different.

58.3.9 PWB Pad Design

The size and shape of the pad used on the PWB has a significant impact on the reliability of BGA solder joints. In general, the reliability is optimal when the pad size/opening on the package side is the same size as that on the PWB side. Deviation from the same pad size could result in a reduction of up to 25 percent in solder joint fatigue life.[14] The side with the smaller pad size tends to fail earlier because the stress concentration is higher near the smaller pad. Generally, non-solder mask defined (NSMD) pads are used on the PWB for packages with solder pitches in the range 1.27mm to 0.8mm. As opposed to solder mask defined (SMD) pads, NSMD pads do not have a localized stress concentration around the solder joint, which tends to degrade the thermomechanical fatigue life of solder joints. (Fig. 58.16)

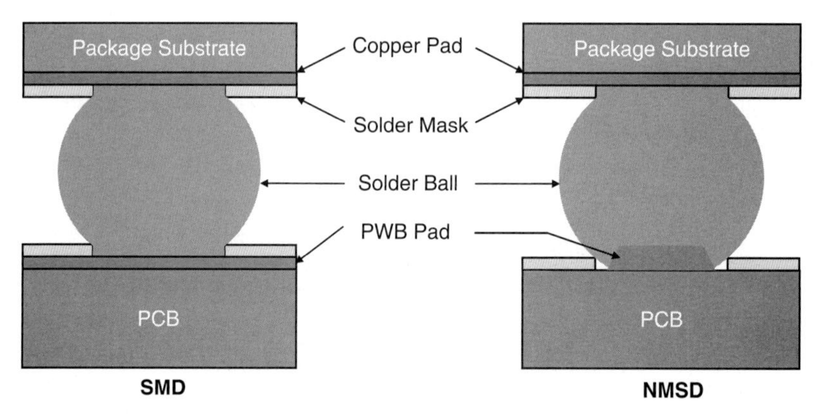

FIGURE 58.16 Solder-mask defined (SMD) and non-solder mask defined (NSMD) pad construction. For optimum reliability, the pad size on the PWB should be kept within 80 to 100 percent of the wetted pad area on the package side. This allows the stresses to be vectored equally between both ends of the solder joint, and if the PWB pad is slightly smaller, it allows a slightly higher stand off height, which also tends to enhance solder joint reliability.

For packages with solder joint pitches less than 0.8mm, a mix of NSMD and SMD pads is sometimes inevitable. At small pitches, it is difficult to route the traces from the pads in the typical "dog bone" fashion. Thus, to allow routability, the common power and ground pads are "bussed" together to form one large pad. The connected pads are SMD, whereas the individual signal pads are kept NSMD.

In general, it appears that while SMD pads tend to improve the resistance of solder joints to mechanical shock loading conditions, they also tend to degrade the thermomechanical fatigue life of the joints. Therefore, when the choice of pad construction is made, the impact on both thermomechanical fatigue and shock conditions should be characterized. (Details of typical shock testing conditions are outlined in Chap. 59).

It is important to note that the pad size specified in the fabrication drawings is not always the pad size that actually comes on all the fabricated PWBs. Actual measurements of pads specified at 12 mils NSMD could range anywhere from 9 to 15 mils. This is particularly critical for small pitch packages, because the solder volume is smaller and such variations in pad sizes could significantly alter the shape and reliability of the solder joints. A tolerance analysis such as worst case (WC) or root sum squares (RSS) should be conducted and a pad size specified such that it would not significantly exceed the pad size specified on the package side. Refer to the IPC-7351 specification for details of such tolerance analysis.[15]

Finally, it is clear that one common theme in the foregoing recommendations is that the PWB pad size should be pegged to the package side solder wetted area. Across the industry, the challenge with this recommendation is that the package pad size/opening is not always provided by package suppliers in relevant datasheets and mechanical drawings. To address this issue, JEDEC revised the JEDEC Publication 95[10] in 2004 to include nomenclature showing the pad size and construction on the package mechanical drawings. Adoption of this JEDEC nomenclature by package suppliers would go a long way in ensuring that the PWB design used by end users does not deleteriously impact the reliability of the package.

58.3.9.1 Effect of Lead-free Conversion. The impact of un-optimized pad size selection is detrimental to both tin-lead and lead-free assemblies. Fine pitch components (<0.8 mm pitch), regardless of solder metallurgy are much more prone to pad size-related failures.

58.3.10 Heatsink Design Issues

As PWB designs get more complex and more densely populated, the power dissipation from single cards can be on the order of 1000 W. A lot of heat needs to be transferred from individual cards, and from the chassis as a whole. Consequently, heatsinks are becoming larger and heavier. Heatsinks as large as 10 sq. in., weighing more than 1 lb are not unheard of. If not designed correctly, large heatsinks could transfer a significant amount of compressive load to the packages they are meant to cool. This compressive load could result in cracks in the package/silicon, or cause the solder joints to deform and bridge.[16] On the other hand, smaller heatsinks that are anchored to the package substrate could impart additional bending moments on the substrate that could negatively impact the reliability of the package.[17]

For bolt-down heatsinks, a significant amount of the compressive load comes not from the heatsink weight but from the bolt down mechanism itself. Tolerance variations in package height, warpage, heatsink stand offs, and so on only make the applied load worse. The attachment mechanism should be such that it can accommodate these variations in tolerance without transferring excessive load to the package.

Some of the attachment methods that could help accommodate tolerance variations and minimize the amount of loading imparted on to the package are shown in Fig. 58.17. The trade offs of using each method are summarized in Table 58.2.

Pressure-indicating film can be used to verify the amount of load applied on the package. The film contains tiny microcapsules of colored dye. These microcapsules rupture depending on the amount of pressure applied on the film, producing a pressure footprint. The results are not extremely precise, but can be used to estimate the amount of load applied, as a first pass approximation.

To verify the impact of heatsink loading on package solder joint reliability, temperature cycling can be performed on packages with heatsinks and compared with control samples. However, there are two caveats:

1. The heatsinks tend to cool the packages during temperature cycling, thus producing improved reliability data. The improvement may simply be because the solder joints are cycled over a lower temperature range. Thus it is important to either thermally shield the heatsink fins or calibrate the thermal cycling chamber such that the solder joints run at the same temperature range as control samples; in spite of the heatsinks.
2. Packages thermal cycled under compressive load could undergo two failure mechanisms: cyclic loading (fatigue induced opens) and static compressive loading (bridges due to solder collapse). Temperature cycling can accelerate the cyclic loading portion, but not the static loading portion. Thus, simply performing temperature cycling and comparing with control samples may not suffice. The static-loading component should also be accelerated, for instance, by dwelling at a high static temperature.

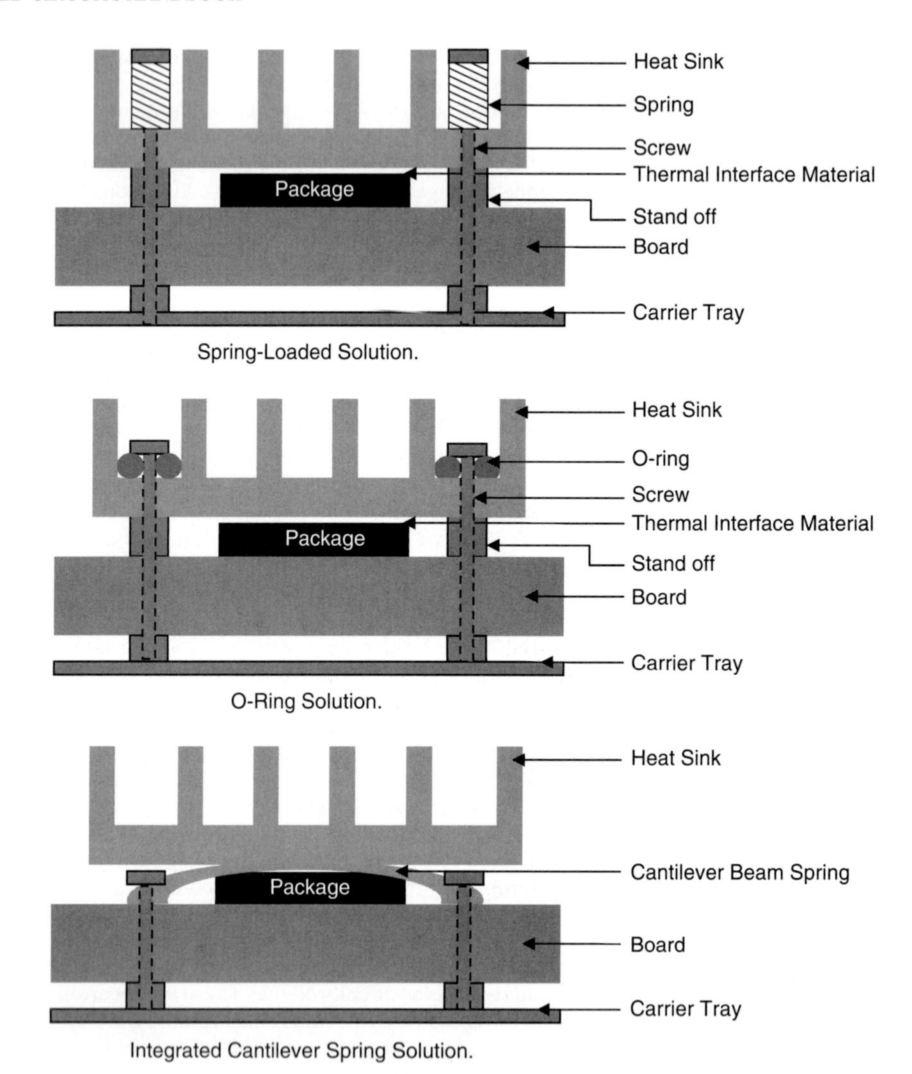

FIGURE 58.17 Heatsink attachment methods.

TABLE 58.2 Comparison of Different Heatsink Attachment Methods

Attachment method	Advantage	Disadvantage
Spring-loaded solution	Useful for absorbing significant tolerance variations by varying spring constants.	Relatively expensive. Bulky
O-Ring solution	Less expensive. Less bulky.	Marginal control in tolerance stack-up induced pressure.
Integrated cantilever spring solution	More versatile. Could be designed to absorb a wide range of expected loads.	Requires custom heatsink design. Could be more cumbersome to install.

58.3.11 Surface Finish Issues

In recent years, the role of package surface finish in the robustness of BGA packages has gained more prominence. Several different surface finishes are used across the industry, and there are trade offs with the use of each one. The surface finish used could impact intermetallics formed in the solder joints, which in turn can impact the thermomechanical and mechanical reliability of the package.

In recent years, a number of surface finishes have become more prevalent. The surface finishes with which the effect on package reliability is relatively well known, are outlined in this section:

58.3.11.1 Electroless Nickel Immersion Gold (ENIG). This has been the surface finish of choice for most high-end, flip-chip BGA packages. It enables the routability of fine pitch traces in high pin count packages. Historically, ENiG produces excellent board level solder joint reliability.

However, there are two primary concerns with ENiG:

1. *Brittle Fracture* This failure mode is seen in high strain/strain rate conditions, such as shipping, testing, and handling. A typical signature of this failure mode is a clean separation in the intermetallic between the pad and the solder joint. This failure typically has a low defect rate in production, but the ensuing line stops, customer returns, and root-cause analyses are quite expensive. Moreover, the unpredictability of the failure and the susceptibility to mechanical handling/testing/shipping conditions makes it quite undesirable.

   There are a few failure mechanism theories proposed to explain brittle fracture,[18] but no conclusive failure mechanism has been agreed upon in the industry. The predominant theory is based on the formation of Kirkendall voids during reflow, in the phosphorus-rich Ni layer and Sn-Ni intermetallic interface. Kirkendall voids are voids that form at the interface of two dissimilar materials. They form because the two materials diffuse into each other at different rates. Other factors such as the concentration of P in the Ni-P+ layer and the density of mudflat cracks in the Ni-P+ layer have also been shown to have an influence on the propensity of brittle fracture. The intermetallics formed in the joint during the assembly process are shown in Fig. 58.18.

The presence of pervasive Kirkendall voids could reduce the strength of the solder joint. A mechanically applied strain could result in the sort of brittle solder joint failure shown in Figs. 58.19 and 58.20.

- *Ni-Cu-Sn Intermetallic* Another phenomenon reported to cause a similar failure is based on the formation of a Ni-Cu-Sn ternary intermetallic in the solder joint.[20] The Ni-Sn intermetallic forms on the component side during ball attach. When the part is mounted on the board, copper migrates from the board pad side, across the BGA ball, to form a Ni-Cu-Sn intermetallic on the component side. The thickness of this ternary intermetallic, typically 3 to 5 μm thick, could grow with additional reflows.[20] Clearly, an important factor contributing to this phenomenon is the surface finish on the PWB side. If the surface finish has a barrier like Ni to prevent copper from migrating across the BGA ball, this ternary intermetallic is less likely to form on the component side.
- *Strain/Strain Rate* It has also been observed that the strain to failure of solder joints is a strong function of the strain rate. The higher the strain rate, the lower the strain-to-failure. One theory to explain this strain rate dependency is that the stiffness of bulk solder itself is strain rate dependent: at low strain rates, solder tends to deform and absorb some of the applied strain. Thus, less strain is actually transferred to the Kirkendall void-rich region. At high strain rate levels (typically above 5000 uε/sec.), bulk solder behaves more like a linear elastic material and transfers more of the applied strain to the voids, causing brittle solder joint fracture. A graph illustrating the relationship between strain and strain rate for solder joint failure is given in Chap. 59.

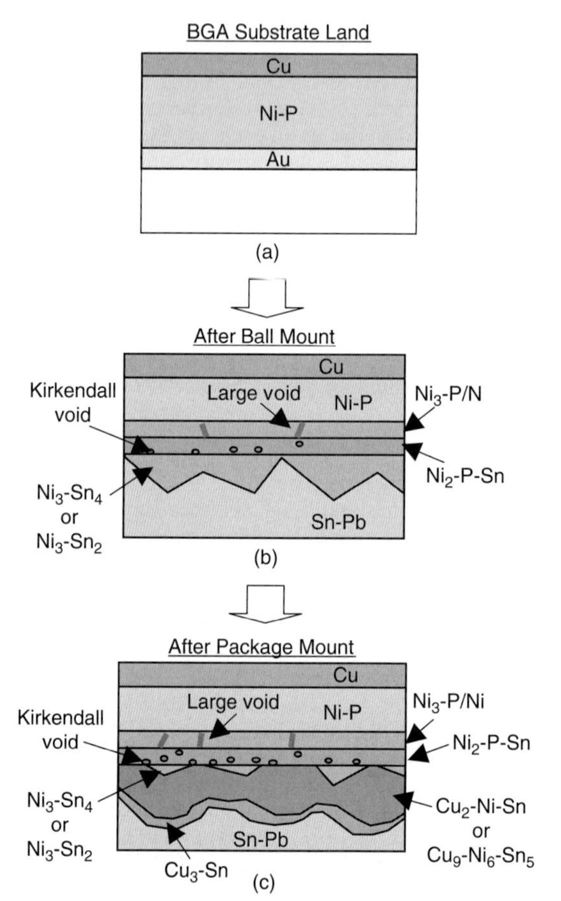

FIGURE 58.18 Solder joint intermetallic formation. (a) The constituent material stack up in a BGA substrate land pad. (b) After the ball is attached, a number of intermetallics are formed between the copper pad and the SnPb solder joint. Kirkendall voids are known to form at the phosphorus-rich Ni layer and Sn-Ni intermetallic interface. (c) After board level reflow, the interface region thickens and several more intermetallics are formed.[19] *(Reprinted with permission from Renesas Technology Corp./© [2003] IEEE ECTC.)*

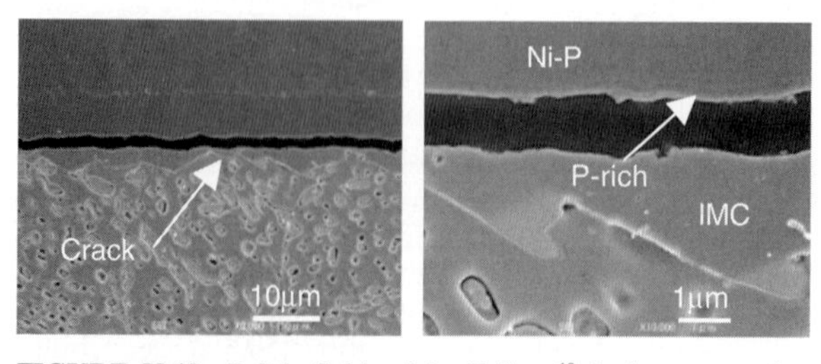

FIGURE 58.19 Brittle Solder Joint Failure.[19] A clear separation between the P-rich phase and IMC on the package side is shown. *(Reprinted with permission from Renesas Technology Corp./© [2003] IEEE ECTC.)*

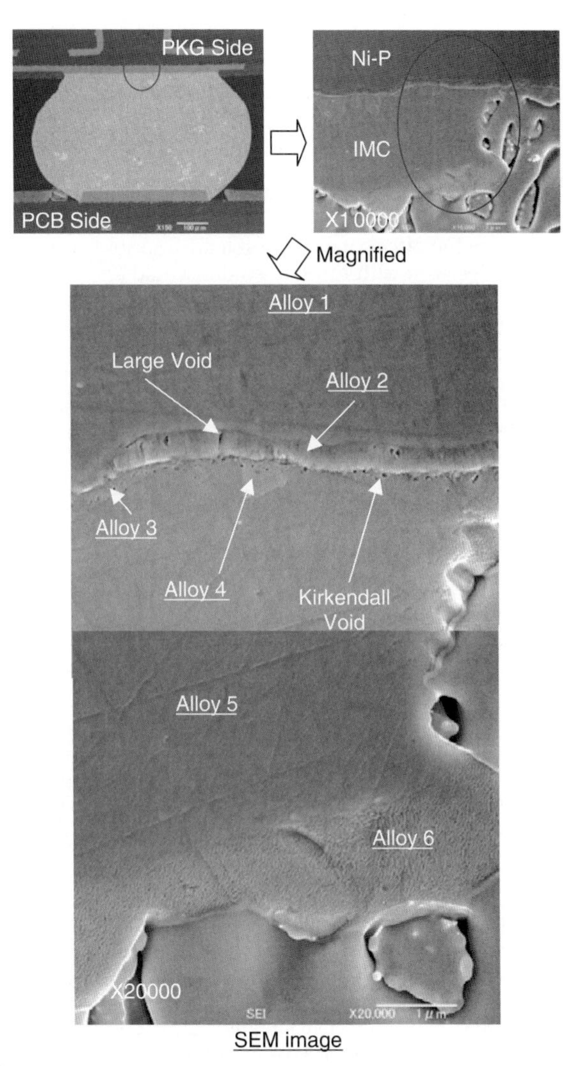

FIGURE 58.20 Intermetallics Formed In Solder Joint with Brittle Fracture.[19] *(Reprinted with permission from Renesas Technology Corp./© [2003] IEEE ECTC.)*

It is important to note that brittle solder fracture does not always occur exclusively. Frequently, it can be accompanied with pad lift, substrate cracking, or other intermetallic-related failure.

2. *"Black Pad" Corrosion Phenomenon*[21] This is a discoloration phenomenon caused by excessive oxidation of the Ni. This leads to increased contact resistance, which causes intermittent opens during electrical testing. It is believed that the "black pad" phenomenon is caused by impurities in the plating bath during the ENIG plating process. The impurities create pores in the gold plating, thus exposing the Ni to moisture and other contaminants. Consequently, the Ni oxidizes and manifests in varying extremes as a discoloration in the pad. Tighter process controls during the plating process and regular

purging of the plating bath can significantly mitigate this corrosion phenomenon. Failures stemming from black pad can be distinguished from brittle fracture by the following features: visible discoloration of the pads and a lot-to-lot dependency in failures due to impure plating baths.

58.3.11.1.1 Effect of Lead-Free Conversion. Given these known issues with ENIG, many package suppliers have moved away from it for lead-free applications. However, several package suppliers still use ENIG with standard tin-lead and lead-free solder. There isn't enough data to show the effect of ENIG on lead-free solder, but preliminary results indicate that the failure modes seen with tin-lead solder would also be seen with lead-free solder. In general, since lead-free solders are stiffer than tin-lead solders, they have higher strain/strain rate strength than tin-lead solders. More detail on tests to evaluate failure strength and failure modes is provided in the next chapter.

58.3.11.2 Solder-On-Pad (SOP). This is also known as "pre-applied solder" because it involves pre-applying a small amount of solder on the package substrate pad before the ball-attach process. Care must be taken to ensure that the height of solder applied on the pad does not exceed the thickness of the solder mask (Fig. 58.21).

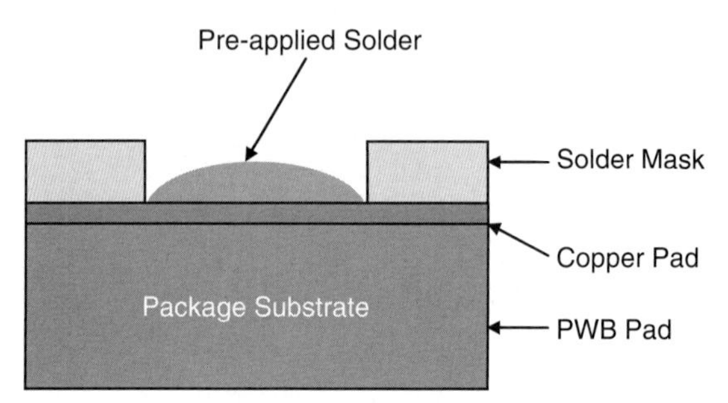

FIGURE 58.21 Solder-on-pad schematic.

Exceeding the height of the solder mask would result in alignment/registration issues during the assembly process. As with ENiG, there are concerns with the mechanical robustness of SOP solder joints. The Ni in ENIG serves as a diffusion barrier for the copper metallization. In the absence of Ni, copper diffusion occurs much faster and two intermetallics are known to form in SOP solder joints: Cu_6Sn_5 and Cu_3Sn. Under high temperature aging conditions, these intermetallics are known to grow in thickness. Thus, if tin-lead or lead-free parts with SOP are aged at high temperature (either in lab testing or in field conditions), they could have a higher propensity for brittle fracture.[22,23] By extension, one concern is that if solder joints can become brittle when aged over time, what would happen if the product has been operating in the field at high nominal operating temperatures, and it is subjected to shock conditions? Would it fail early? For instance, what if the product is a cell phone with an SOP based BGA used in it? If the user drops the cell phone after sustained usage, is there a propensity for brittle fracture? According to some publications on shock testing of lead-free assemblies (Cheng et al.),[22] aged SOP joints subjected to shock conditions tend to fail relatively early. However, this has not been universally demonstrated. Moreover, the shock conditions to which hand held products are tested (~1500 G max), are much more aggressive than typical shock conditions for server environments (~500 G max). In tests performed on aged tin lead SOP samples subjected to lower shock conditions, the parts were not shown to fail due to brittle fracture, even though the presence of some Kirkendall voiding was observed (Mei et al.).[24] However, the amount of Kirkendall voiding observed was not as pervasive as that reported by Cheng et al.[22] (Fig. 58.22)

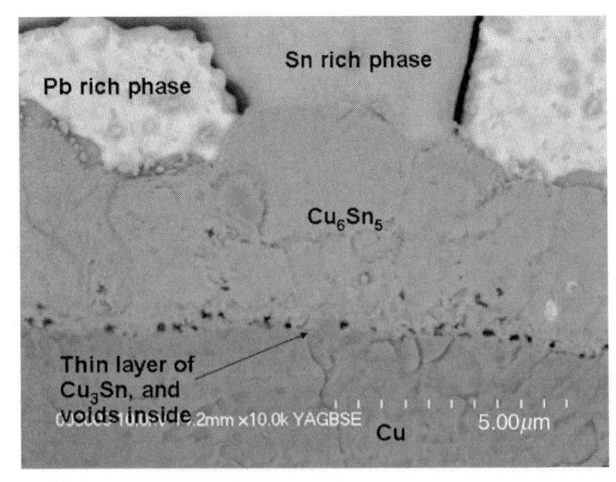

FIGURE 58.22 SEM Backscattering Electron Image of Tested Sample with Solder-On-Pad Surface finish – Package/Ball Interface. The image shows the intermetallics formed at the joint/package interface. These intermetallics include a thick layer of Cu_6Sn_5 and a thinner layer of Cu_3Sn. Small voids are seen at the Cu_6Sn_5 and Cu_3Sn interface.[24] *(Reprinted with premission from IEEE © [2005].)*

A curve fit of the data by Cheng[22]et al. reported by Mei et al.[24] as a function of void area and aging time for different temperatures is shown in Fig. 58.23.

The fit equation from the Cheng et al.[22] data is given as[24]:

$$A = Ct^{0.5} \exp\left(\frac{-Q}{RT}\right) \tag{58.1}$$

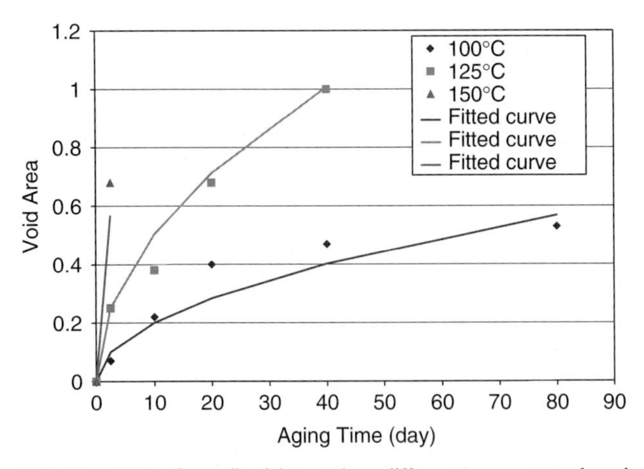

FIGURE 58.23 Curve fit of data at three different temperatures based on the data generated by Cheng et al.[24] The values are based on experimental measurements of void area versus time for different temperatures. *(Reprinted with premission from IEEE © [2005].)*

Where A = ratio of voided area to the joint interface area
t = time in days
Q = activation energy
T = absolute temperature in K
R = gas constant.
C = 145590 ($day^{-0.5}$)
Q = 45413 J/mol (0.47eV)

With this approximate curve fit, extrapolating to end-use conditions can give an estimate of the time it would take for the voiding to reach 50 percent of the interface area. For instance, at 50°C nominal operating temperature, it would take 6000 days (16.5 years) to reach 50 percent of the interface area. It is important to note that the mechanisms that control the propensity and pervasiveness of Kirkendall voiding are not clearly known at this time. There isn't enough experimental data in the industry at this time to conclusively validate the aforementioned relations.

The occurrence of Kirkendall voids is not limited to the package/solder joint interface. If OSP is used as the PWB surface finish, the same phenomenon is observed to occur on the PWB side.[24]

In terms of strain/strain rate strength, un-aged, SOP-based joints have been shown to have as much as twice the strength of ENiG joints. More details on comparison tests are outlined in the next chapter.

There are a number of other surface finishes used in the industry, such as Electroless Nickel Electroless Palladium Immersion Gold (NiPdAu), Immersion Silver, Immersion Gold, Immersion Tin, OSP and Electrolytic Nickel Gold. There are reliability and process trade offs with each surface finish. That is why it is recommended that strain/strain rate characterization and thermal cycling be performed for each set of surface finish before it is selected for the specific end-use conditions in which it will be used. The industry test methods used to evaluate different surface finishes are outlined in detail in the next chapter.

58.3.12 Final Electrical Test Architecture

During in-circuit testing (ICT), assemblies are subjected to high strains due to the number and distribution of test points applied underneath BGAs. Assemblies are also subjected to high strain rates in ICT because the speed of engagement of test points helps reduce the electrical contact resistance between the test probes and probe pads. Thus, from an electrical test perspective, high strain rates are desirable during ICT. However, there is a strain/strain rate range beyond which the likelihood of solder joint failure increases.

The actual strains induced in a printed circuit board assembly (PCBA) is a function not only of the number and layout of test points, but also the fixture design, package and PWB warpage, the surface finish used on the package and PWB, and the fixture actuation rate. During final test, it is very difficult, if not impossible to reduce the number of test points. Thus, if they are correctly designed-in early in the design process, it would minimize the likelihood of strain-induced failures in final testing.

A schematic showing the forces applied in the immediate vicinity of a BGA package in an ICT fixture is shown in Fig. 58.24. The distribution of test points underneath a BGA should be

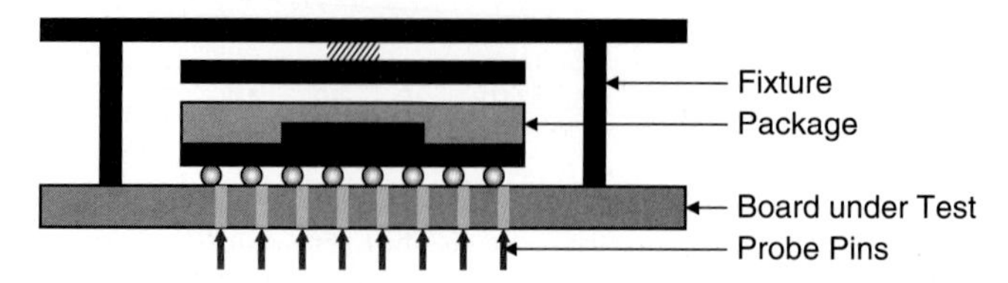

FIGURE 58.24 Schematic of ICT loading conditions.

made as uniform as possible. Several ICT fixtures have push fingers on the top side, which push down on the board, countering the force of the test probes. The use of "push-pins" or "push-fingers" on the top side of the board is not recommended. Push fingers induce localized stress concentrations, resulting in high strain in their immediate vicinity. If used, push fingers should not be placed close to package corners. For some generic recommendations on ICT test point layout, refer to IPC7351.[15] An alternative to push fingers is the use of a Milled City or zero-flex top plate on the top side.

This is basically a rigid metal plate which has a negative image of the entire board milled out. Thus, as opposed to having several push fingers, the entire board is pressed against the Milled City plate. This significantly reduces local stress concentration, and could potentially be used to enable PCBAs to withstand more test points. Milled City plates are expensive, but considering the cost of line stops and customer-related issues due to strain-induced failures in ICT, the non-recurring cost of the Milled City is a worthwhile investment.

58.3.12.1 Effect of Lead-Free Conversion. The conversion to lead-free solder has given even more prominence to this issue. With lead-free solders, a higher temperature organic solderability preservative (OSP) could be used. As a result, it could take more force to establish contact with the test pads. This translates to higher force applied to the PCBA. Moreover, the effect of mechanical bending on lead-free solders is yet to be fully characterized. Thus, there is a greater need to characterize the mechanical force applied on lead-free assemblies.

For the same applied force, the strain on thinner PWBs is higher than that on thick PWBs. Thus, this is more critical on assemblies with relatively thin boards (<93 mils thick). However, depending on the package, warpage, surface finish, and number of test points, it could also be seen on thick boards.

Strain gage characterization should be performed on fixtures with actual boards, in order to verify that excessive strains are not applied on critical components. In an effort to standardize strain characterization, the JEDEC/IPC 9704[25] test method was published in June 2005. This test method outlines the procedures and best practices relating to strain measurement in a manufacturing environment.

58.3.13 Package Design Parameters

A key factor that impacts second-level interconnect reliability is the package itself. Although the primary function of the package is to pass signals, power, and ground from the silicon device to the PWB, the package also provides protection for the silicon and enables a more robust interconnect than a direct chip attach. The critical package design parameters which can have a significant impact on the reliability of the solder joints are discussed in this section.

58.3.13.1 Overall Dimensions and Distance from Neutral Point (DNP). Package size continues to grow as the number of I/Os increase to meet the demands of ever increasing IC functionality. As IC complexity increases, memory is embedded in devices, while as silicon feature sizes decrease, I/O counts continue to be driven upward. Package dimensions have surpassed 50 mm. Package layer counts for both ceramic and plastic material sets continue to increase rapidly.

In addition to the solder joint geometry issues outlined in Sec. 58.3.8, increases in overall size tend to coincide with increases in the distance from the center of the package to the outer ball (see Fig. 58.25). This dimension is referred to as the Distance from the Neutral Point (DNP). As DNP increases, the amount of stress and subsequent damage inflicted on solder joints increases, especially during thermal cycles encountered in field environments. Thus packages with larger DNPs (and with all other material and design parameters equivalent) will have lower interconnect reliability. This is crucial when assessing second-level reliability data.

Most packaged IC suppliers qualify their products by thermal cycle experiments. The most common board level temperature cycling methodology used across the industry is outlined in IPC 9701.[6] The resistance in each solder joint is typically monitored in real time by incorporating a daisy-chain package and some type of data acquisition system. The end results of

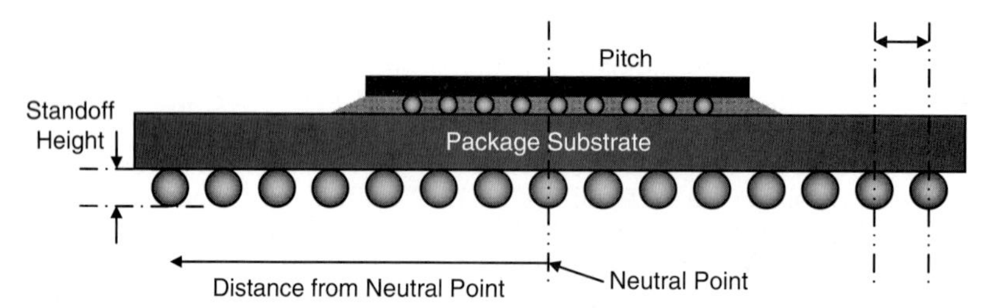

FIGURE 58.25 Distance from Neutral Point (DNP) for a BGA-type package. The DNP increases when moving away from the center (center line) of the package. Pitch is also indicated in this diagram. Pitch is defined as the center-to-center distance between two adjacent interconnects (in this example, between two adjacent solder balls).

these experiments are failure distributions typically presented as Weibull plots (for details on Weibull distributions and failure life analysis, refer to Sec. 61.2.2 in the next chapter) that show failure percentage as a function of time.[26] Note that if a package with a larger DNP than was qualified is considered, then there could be severe degradation in reliability and a full assessment should be conducted. However, if a package with a smaller DNP (10 to 20 percent smaller) than the package qualified (with all other material and design parameters being equal) is considered, then the proposed package should be at least as reliable as the qualified package.

58.3.13.2 Ball Pitch and Ball Size. Ball pitch is the center-to-center distance between two adjacent solder balls (see Fig. 58.25). In theory one could change the ball pitch and keep the same solder joint size. In practical applications though, decreases in ball pitch also require a reduction in solder ball diameter and pad size for physically accommodating the decrease in distance between adjacent balls. This decrease in ball size, with decreasing pitch, results in the formation of smaller balls with smaller stand-off between the package and the PWB. Smaller solder balls and decreased stand-off have been shown to decrease solder joint reliability (see Fig. 58.26).

The dashed line in Fig. 58.26 represents the finite element analysis results for a 0.889mm (35 mil) ball. The solid line represents finite element analysis results for a 0.762 mm (30 mil) ball. Note that the 0.889 mm ball is 1.3 times more reliable than the 0.762 mm ball. Thus reduction in pitch can result in decreases in reliability due to the subsequent decreases in solder ball size necessary to accommodate adjacent balls and the reduced stand-off heights due to smaller solder balls.

The PWB designer is therefore faced with a dilemma when considering a reduction in ball pitch. Not only is there a risk of reduced assembly reliability, but there is also a design challenge. A direct consequence of decreased pitch is added complexity in the ability to route all traces out from the package to the PWB. Designers are also sometimes forced to add additional layers to PWBs in order to accommodate reduced-pitch parts. This layer increase results in increased price and in some instances, drastic reductions in capacity due to the increased complexity associated with fabricating high-layer-count PWBs.

58.3.13.3 Ball Array (Full or Depopulated). Ball array is another variable that has an indirect impact on solder joint reliability. There is a coupled effect between the location of the edge of the packaged die, the solder ball array, and the state of stress in the solder joints. The silicon has a much lower CTE than plastic substrates. Thus the die can create a state of overconstraint, producing excessive stress in the solder joints under and near its edge. This overconstraint tends to be worst in joints near the edge of the die.

Figure 58.27 contains schematics of one configuration where a full array is employed and the edge of the die falls over the solder joints. Fig. 58.28 illustrates an alternative design where a region near the perimeter of the die is depopulated of all solder balls (the depopulated configuration).

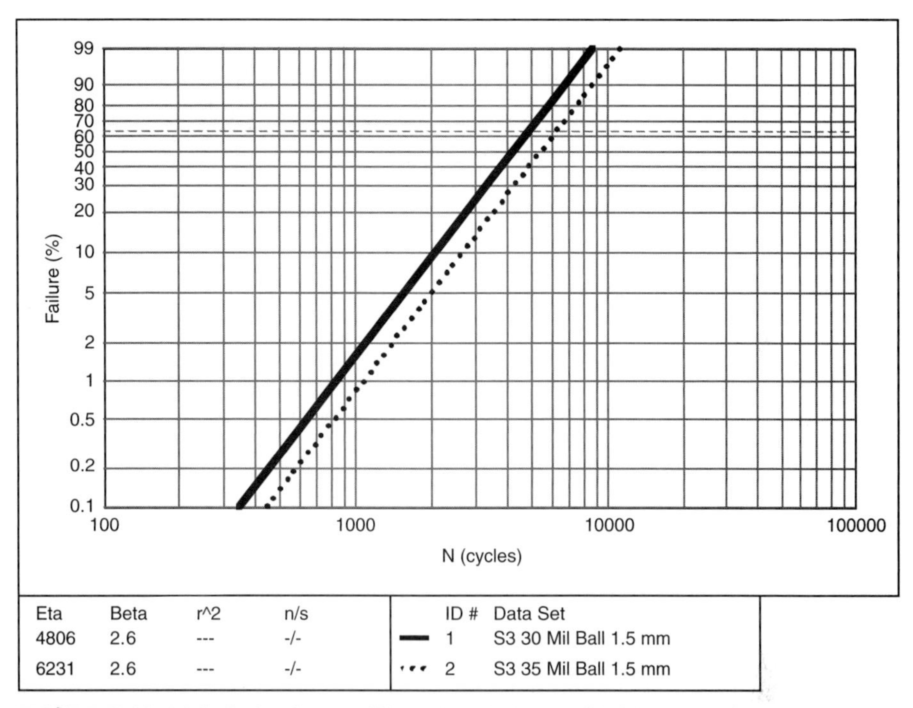

Eta	Beta	r^2	n/s		ID #	Data Set
4806	2.6	---	-/-	▬	1	S3 30 Mil Ball 1.5 mm
6231	2.6	---	-/-	•••	2	S3 35 Mil Ball 1.5 mm

FIGURE 58.26 Weibull plots for two different ball size/stand-off heights. The solid line indicates the finite element analysis (FEA) predicted failure distribution for an interconnect based on a 30-mil stand-off height. The dashed line indicates the FEA predicted failure distribution for an interconnect based on a 35-mil stand-off height. Note that a reduction in stand-off height correlates with a decrease in fatigue life.

Many plastic (laminate) package suppliers offer a depopulated option to enhance reliability by not placing solder joints in the worst-case regions under the perimeter of the die. Additionally, many package designers further reduce risk of CTE-induced joint failure by only placing redundant power and ground or nonfunctional thermal balls under the core. If redundant balls or thermal

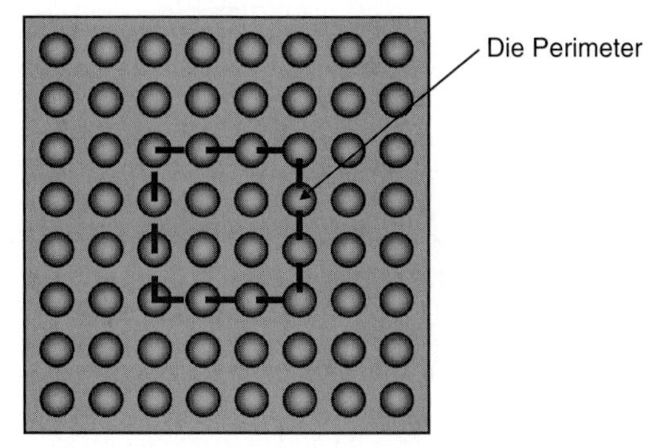

FIGURE 58.27 Full array with solder joints under the die perimeter. Solder ball locations are indicated by the array of circles. The die perimeter is indicated by the black dashed rectangular outline. Note that the die edge falls directly over solder joints.

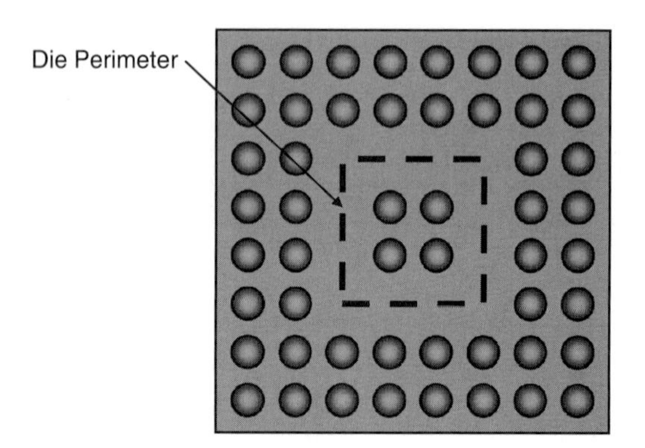

FIGURE 58.28 Depopulated array with no joints under the perimeter of the die. Solder ball locations are indicated by the array of circles. The die perimeter is indicated by the dashed rectangular outline. Note that the die edge falls in an area of depopulated solder balls.

balls do crack, the impact on device performance is negligible. Finally, in large laminate packages, the corner-most balls could also fail early. As a result, many package suppliers depopulate the corner balls of the package to enhance solder fatigue life.

58.3.13.4 Heat Spreader/Stiffener. A cross-sectional schematic of a flip-chip assembly, with a heat slug and stiffener, is shown in Fig. 58.29. The primary reason for employing a heat spreader lid is for thermal management. A secondary benefit is that the stiffener and heat slug reduce the potential for out-of-plane warpage occurring during reflow and thermal cycling. An additional benefit of incorporating a heat slug and stiffener into a laminate package design is improved ability to handle the package due to increased robustness of the packaged die.

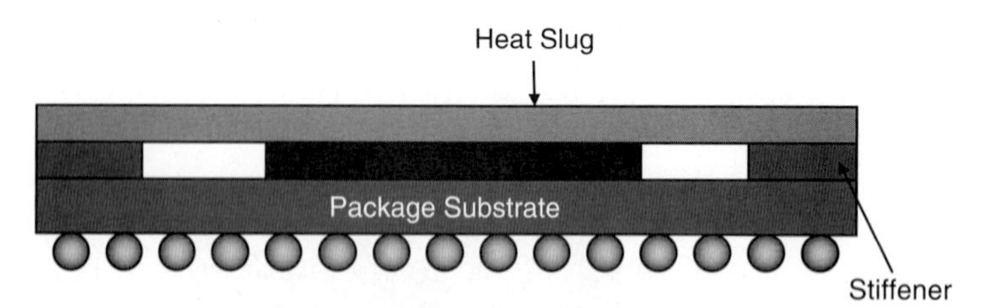

FIGURE 58.29 Flip-chip BGA with stiffener and heat slug. The stiffener provides added rigidity to the package and also provides mechanical support for the heat sink.

A variety of heat spreader materials have been used: copper, aluminum, steel, and aluminum silicon carbide (AlSiC). Copper provides the best thermal conductivity, but it has a high CTE. AlSiC has a slightly higher modulus than copper, but it has a lower CTE. As a result, in several applications, AlSiC is gaining more prevalence. Its thermal conductivity is lower than copper and comparable to aluminum (180 to 200 $Wm^{-1}K$).

For cost reasons, several package suppliers also use an alternative to the heat spreader/stiffener design, namely a single piece lid design to act as a rigid support and provide heat transfer at the same time. This design is easier to manufacture and also reduces assembly steps (See Fig. 58.30 for more details).

The thickness of a single piece lid is typically less than that of the heat spreader/stiffener combination. Changing the stiffness of the package could alter its out-of-plane warpage, which

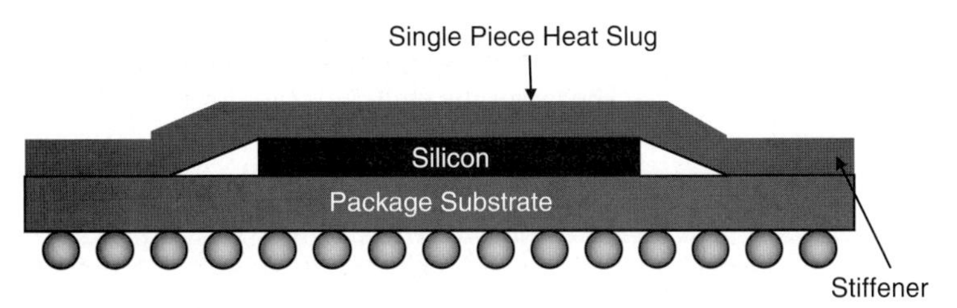

FIGURE 58.30 Single piece lid design.

in turn could impact the reliability of the solder joints. Therefore the effect on the overall stiffness should be characterized when using a single piece lid. One characterization method is to perform Moiré measurements on the free standing package, as a function of temperature.

Another variation of the standard heat spreader/stiffener design is the use of just a stiffener alone. This is usually for bare die applications where the wattage of the chip is so high that the thermal resistance between the chip and the heat spreader is unacceptable. In such cases, the stiffener is retained for stiffness and robustness, but the heat spreader is removed.

58.3.13.5 Mold Compound Selection. The overmold compound used in plastic ball grid array (PBGA) packages could significantly impact the assembly yield and reliability of solder joints. If the glass transition temperature (T_g) of the mold compound is significantly below the peak reflow temperature and if its modulus and CTE drastically change above its T_g, the package could warp significantly as it is heated. As a result, when the package reaches liquidus, the effective warpage between the package and PWB could be such that solder joint opens/bridges could occur. In large PBGA packages (>40 mm), it would be more effective to use an integrated heat spreader in the molded package to rigidize it and minimize excessive warpage. Integrated heat spreaders are primarily used to improve thermal performance, but they also help rigidize the package for improved assembly yield.

With the conversion to lead-free materials, the mold compound selected needs to be such that it can withstand the higher reflow temperatures of lead-free assembly. If the mechanical material properties of the mold compound (elastic modulus, CTE, and T_g) are not appropriately selected, the package could warp significantly at reflow temperatures and result in solder joint opens.

58.3.13.6 Underfill Material Selection. The underfill material selected in flip chip BGA (FCBGA) packages could significantly impact the reliability of the board mounted package. If the underfill material is too stiff, it would rigidize the assembly and thus transfer more strain to the silicon die. This could result in delamination at any of the critical interfaces: underfill/passivation, passivation/die, underfill/substrate, or within low-*k* layers.

On the other hand, if the underfill material is too soft, it could result in first level interconnect failure. A less stiff underfill could also help reduce board level assembly warpage, which in turn helps the second-level interconnects. Thus, the stiffness of the underfill selected should be optimized such that it does not lead to delamination or first-level interconnect failures.[9] In addition to stiffness, CTE, and T_g, there are several other factors that dictate the underfill material selected, such as thixotropy, viscosity, shrinkage, curing temperature, and curing time.

The choice of underfill material is also influenced by the conversion to lead-free. For lead-free packages, even though there is a European exemption on the first level interconnects,[27] many package suppliers are looking at using lead-free first level interconnects. As a result, the underfill material selected has to be such that it can withstand the higher reflow temperatures, and has low shrinkage for the increased temperature range.

58.3.13.7 Die/Package Aspect Ratio. The size of the silicon die relative to the package body size could significantly impact the reliability of the package. This is especially applicable to PBGA packages. The larger the die relative to the package, the higher the CTE mismatch-driven strains imparted to the second level solder joints.[28] The degree of impact varies with package body size and absolute die size.

58.3.13.8 Package Substrate Material. The choice of package substrate material can have a large impact on solder joint reliability. Some substrates require employing solder ball alternatives (such as the ceramic solder column carriers shown in Fig. 58.4) or even the elimination of a soldered solution due to interconnect reliability considerations. A wealth of information regarding ceramic and plastic substrates can be found in the literature.[26,29,30] This section focuses on the impact of those materials on solder joint reliability.

The two key material properties of a substrate that impact interconnect reliability are modulus of elasticity (E) and coefficient of thermal expansion (CTE). Modulus of elasticity is defined as the ratio of stress to strain and can be used to describe a material's ability to resist deformation when subjected to an applied load. CTE relates the change in length of material when heated or cooled and can be used to describe a material's deformation when subjected to temperature changes. Reference 31 contains detailed discussions of modulus of elasticity and coefficient of thermal expansion. A tremendous amount of material property data can be found in the literature.[26,29,30] Fig. 58.31 contains a sketch of a simplified solder joint interconnect. Note that the driving force for solder joint fatigue during field use is the coefficient of thermal expansion mismatch between the assembled package and the PWB. Additionally, the modulus of elasticity of the substrate can impact the fatigue life of solder joints by increasing (or decreasing) the driving force that is imparted to the solder interconnect.

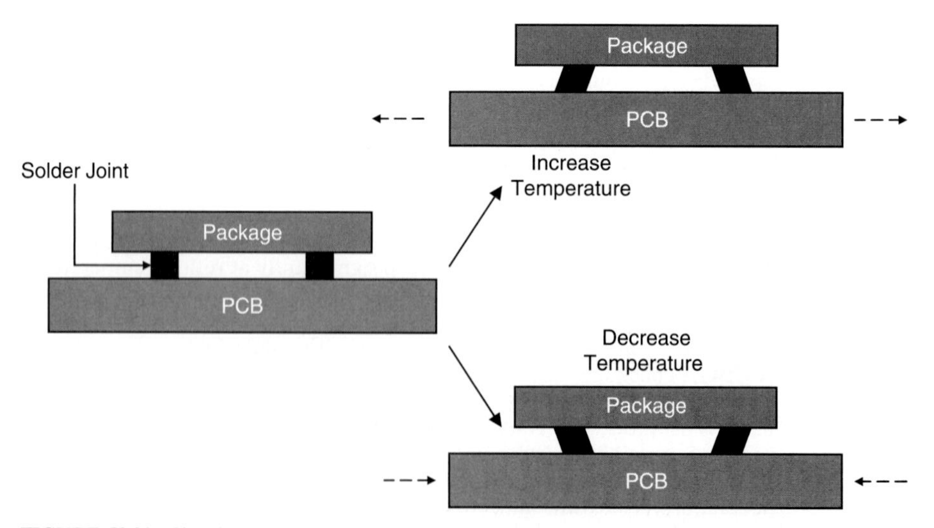

FIGURE 58.31 Simplified CTE mismatch effects. As the temperature increases, the PWB expands at a higher rate than the package due to the package having a lower effective CTE. As the temperature decreases, the opposite effect is observed.

Ceramic materials provide excellent CTE match between the die and package, which then reduces the risk of first-level solder bump failure in flip-chip applications. Unfortunately, this superb CTE match between the ceramic and the silicon die creates a large CTE mismatch between the ceramic package and the PWB. This CTE mismatch creates severe stress on the second level interconnects (see Fig. 58.31). This CTE mismatch, combined with DNP issues (previously discussed), limits ceramic BGA body sizes to approximately 32 × 32 mm. Ceramics with larger body sizes typically employ solder columns (Fig. 58.4) or compression-type sockets for second-level interconnect.

The CTEs of plastic packages tend to match the CTEs of PWBs fairly well. This CTE match allows for larger package body sizes to be employed when using a laminate-based BGA than with a ceramic package. It is crucial to note that many other variables independent of second-level reliability must be taken into consideration when selecting a package. Power requirements, signal integrity, first-level reliability (interconnect between the die and package), and cost must all be balanced when selecting the correct package for a specific application.

58.3.13.9 Impact of Low-k Dielectrics. Continued miniaturization and integration has led to the need for a multilevel interconnect layout to minimize the time delay due to parasitic resistance (R) and capacitance (C). The improvement in gate level device speed is negated by the propagation delay at the metal interconnects because of the increased RC time constant.[32] For instance, the RC delay of long interconnections contributes up to 50 percent of the time delay in devices with a gate length of 250 nm or less.[33] A reduction in the RC time constant requires the use of low resistivity interconnect materials and low capacitance interlayer films. The capacitance is a function of the material dielectric permittivity (ε_i), area (A), and thickness (d). ($C = \varepsilon_i\ A/d$). Increasing the area would require increased interconnect lengths, which in turn would result in high resistance. The thickness can also not be increased because it also increases resistance and poses manufacturing challenges in consistently filling the interlayer layers. Consequently, the dielectric constant (k) needs to be reduced. Hence the need for low-k materials. In addition to having a low dielectric constant, ideal low-k materials need to have low residual stress, high planarization capability, high gap-filling capability, low deposition temperature, and ease of process integration.[32] A list of possible candidates for low-k materials is outlined in Table 58.3.[33]

TABLE 58.3 Typical Low-k Materials[33]

Determinant	Materials	Dielectric constant
Vapor-phase deposition polymers	Fluorosilicate glass (FSG)	3.5–4.0
	Parylene N	2.6
	Parylene F	2.4–2.5
	Black diamond (C-doped oxide)	2.7–3.0
	Fluorinated hydrocarbon	2.0–2.4
	Teflon-AF	1.93
Spin-on polymers	HSQ/MSQ	2.8–3.0
	Polyimide	2.7–2.9
	SiLK (aromatic hydrocarbon polymer)	2.7
	PAE [poly(arylene ethers)]	2.6
	Fluorinated amorphous carbon	2.1
	Xerogels (porous silica)	1.1–2.0

However, several of the new low-k materials are porous and structurally weak. This poses several challenges in the back-end assembly process: poor adhesion, residual stresses in the inner layer dielectric (ILD) films, and so on. In addition to optimizing and better monitoring the wafer-sawing process, the effect of chip-to-package interactions also needs to be characterized. The choice of material set, underfill, substrate, lid material, and so on could impact the reliability of the low-k dielectric.

In flip-chip BGA (FCBGA) packages, the choice of underfill material in particular has been the subject of considerable interest.[34–36] In addition to a new material set and design rules, several packaging assembly manufacturers have evaluated the use of low glass transition temperature (T_g) underfills to reduce the stresses imparted to the silicon. However, while using a low T_g underfill may minimize the stresses on the low-*k*, it could in turn transfer the stresses elsewhere, such as in the first-level interconnect.[37] Moreover, the effect of the material set on ILD delamination and solder joint reliability may be different when the free standing package is stressed, as opposed to when it is mounted on a printed circuit board and then stressed.

In wirebond BGA packages, the effect of the wirebonding process on the low-*k* dielectric needs to be characterized.[38] In addition, the choice of mold compound material (modulus and CTE) could also impact the long term reliability of the low-*k* dielectric in the fully assembled package. The package with the selected material set should be subjected to component level and second level temperature cycling, followed by scanning acoustic microscopy (SAM) to ensure that there are no delamination-induced failures in the low-*k* dielectric or other critical interfaces.

REFERENCES

1. J. H. Lau (ed.), *Solder Joint Reliability: Theory and Applications*, Van Nostrand Reinhold, New York, 1991.
2. S. B. Park, Rahul Joshi, and Lewis Goldmann, "Reliability of Lead-Free Copper Columns in Comparison with Tin-Lead Solder Column Interconnects," *Electronic Components and Technology Conference*, 2004, pp. 82–89.
3. IPC-SM-785, Guidelines for Accelerated Reliability Testing of Surface Mount Solder Attachments, 1992. www.ipc.org.
4. Jin-Young Kim, Won-Joon Kang, Yoon-Hyun Ka, Yong-Joon Kim, Eun-Sook Sohn, Sung-Su Park, Jae-Dong Kim, Choon-Heung Lee, Akito Yoshida, and Ahmer Syed, "Board Level Reliability Study on Three-Dimensional Thin Stacked Package," *Electronic Components and Technology Conference*, 2004, pp. 624–629.
5. Rocky Shih, Sam Dai, Francois Billaut, Sue Teng, Mason Hu, Ken Hubbard, "The Effect of Printed Circuit Board Thickness and Dual-sided Configuration on the Solder Joint Reliability of Area Array Packages," *International Microelectronics and Packaging Society (IMAPS)*, 2004.
6. IPC 9701, "Performance Test Methods and Qualification Requirements for Surface Mount Solder Attachments," *IPC*, January 2002. www.ipc.org.
7. S. Ehrler, "Compatibility of Epoxy-based PCBs to Lead-Free Assembly," Circuitree, 1st June 2005.
8. S. Ehrler, "Comparison of High-Tg-FR-4 Base Materials," *IPC-Expo & Conference*, Long Beach, CA, 24–26th March 2003.
9. Mudasir Ahmad, Sue Teng, Jie Xue, "Effect of Underfill Material Properties on Low- *k* Dielectric, First and Second Level Interconnect Reliability," *IMAPS Conference on Device Packaging*, Scottsdale, AZ, 2006.
10. JEDEC Design Standard, "Design Requirements For Outlines Of Solid State And Related Products," *JEDEC Publication 95*, Design Guide 4.14, Ball Grid Array Package (BGA), Issue D, December 2002. www.jedec.org.
11. Mudasir Ahmad, Ken Hubbard, Mason Hu, "Solder Joint Shape Prediction Using a Modified Perzyna Viscoplastic Model," *ASME Journal of Electronic Packaging,* 2005, 127(3), pp. 290–298.
12. JEDEC Standard, JESD22b112, High Temperature Package Warpage Measurement Methodology, May 2005. www.jedec.org.
13. Benlih Huang, Ning-Cheng Lee, "Conquer Tombstoning in Lead-Free Soldering," *IPC Printed Circuits Expo*, SMEMA Council, APEX Designers Summit 2004, pp. S27-1–S27-7.
14. Lei L. Mercado, Vijay Sarihan, Yifan Guo, and Andrew Mawer, "Impact of Solder Pad Size on Solder Joint Reliability in Flip Chip PBGA Packages," *IEEE Transactions On Advanced Packaging*, Vol. 23, No. 3, August 2000, pp. 415–420.
15. IPC-7351, "Generic Requirements for Surface Mount Design and Land Pattern Standard," *IPC*, February 2005. www.ipc.org.
16. Marie S. Cole, J. Jozwiak, E. Kastberg, G. Martin, "Compressive Load Effects On CCGA Reliability," *SMTA International*, September 2002.

17. Michael L. Eyman, Gary B. Kromann, "Investigation of Heat Sink Attach Methodologies and the Effects on Package Structural Integrity and Interconnect Reliability of the 119-Lead Plastic Ball Grid Array," *Electronic Components and Technology Conference*, 1997, pp. 1068–1075.
18. Deepak Goyal, Tim Lane, Patrick Kinzie, Chris Panichas, Kam Meng Chong, Oscar Villalobos, "Failure Mechanisms of Brittle Solder Joint Fracture in the Presence of Electroless Nickel Immersion Gold (ENIG) Interface," *Electronic Components and Technology Conference*, 2004, pp. 732–739.
19. Kozo Harada, Shinji Baba, Qiang Wu, Hironori Matsushima, Toshihiro Matsunaga, Yasumi Uegai, Michitaka Kimura, "Analysis of Solder Joint Fracture under Mechanical Bending Test," *Electronic Components and Technology Conference*, 2003 pp. 1731–1737.
20. Shelgon Yee, Lodgers Chen, Justin Zeng, Roger Jay, "Ternary Intermetallic Compound - A Real Threat To BGA Solder Joint Reliability," *Journal of SMT*, 2004, Vol. 17, Issue 2, pp. 29–36.
21. Chong Kam Meng, Tamil Selvy Selvamuniandy, and Charan Gurumurthy, "Discoloration related failure mechanism and its root cause in Electroless Nickel Immersion Gold (ENIG) Pad metallurgical surface finish," *Proceedings of 11th IPFA,* Taiwan, 2004, pp. 229–233.
22. Tz-Cheng Chiu, Kejun Zeng, Roger Stierman, Darvin Edwards, Kazuaki Ano, "Effect of Thermal Aging on Board Level Drop Reliability for Pb-free BGA Packages", *Electronic Components and Technology Conference*, 2004 pp. 1256–1262.
23. Kejun Zeng, Roger Stierman, Tz-Cheng Chiu, Darvin Edwards, "Kirkendall void formation in eutectic SnPb solder joints on bare Cu and its effect on joint reliability," *Journal of Applied Physics*, 2005, 97, 024508.
24. Zequn Mei, Mudasir Ahmad, Mason Hu, Gnyaneshwar Ramakrishna, "Kirkendall Voids at Cu/Solder Interface and Their Effects on Solder Joint Reliability," *Electronic Components and Technology Conference*, 2005, pp. 415–420.
25. IPC/JEDEC-9704, "Printed Wiring Board Strain Gage Test Guideline," June 2005. www.jedec.org.
26. J. H. Lau (ed.), *Ball Grid Array Technology*, McGraw-Hill, New York, 1995.
27. Commission Decision of 21 October 2005, "Amending for the Purposes of Adapting to Technical Progress the Annex to Directive 2002/95/EC of the European Parliament and of the Council on the Restriction of the Use of Certain Hazardous Substances in Electrical and Electronic Equipment *(Notified Under Document Number C(2005) 4054)* 2005/747/EC, Official Journal of the European Union, EN, 25.10.2005, L280/18–19. http://eur-lex.europa.eu/LexUriServ/site/en/oj/2005/l_280/l_28020051025en00180019.pdf.
28. Yuan Li, Anil Pannikkat, Larry Anderson, Tarun Verma, Bruce Euzent, "Building Reliability into Full-Array BGAs," *26th IEMT Symposium PackCon*, 2000, pp. A-1-A-7.
29. R. R. Tummala, E. J. Rymaszewski, and A. G. Klopfenstein (eds.), *Microelectronics Packaging Handbook*, Chapman & Hall, New York, 1997.
30. J. H. Lau (ed.), *Thermal Stress and Strain in Microelectronics Packaging*, Van Nostrand Reinhold, New York, 1993.
31. J. M. Gere and S. P. Timoshenko, *Mechanics of Materials*, 2d ed., PWS Publishers, California, 1984.
32. Gary S. May, Simon M. Sze, *Fundamentals of Semiconductor Fabrication*, John Wiley & Sons (Asia) Pte Ltd., 2004, pp. 162–163.
33. T. Homma, "Low Dielectric Constant Materials and Methods for Interlayer Dielectric Films in Ultralarge-Scale Integrated Circuit Multilevel Interconnects" *Mater. Sci. Eng.*, 1998, 23, 243.
34. M. Jimarez, *et al*, "Technical Evaluation of a Near Chip Scale Size Flip Chip/Plastic Ball Grid Array Package," *Electronic Components and Technology Conference*, 1998, pp. 219–225.
35. K. Chen, *et al*, "Effects of underfill materials on the reliability of low-*k* flip-chip packaging," *Microelectronics Reliability*, 2006, Vol. 46, No. 1, pp. 155–163.
36. Pei-Haw Tsao, *et al*, "Underfill Characterization for low-k Dielectric/Cu Interconnect IC," *Electronic Components and Technology Conf*, 2004, pp. 767–769.
37. E. Hayashi, S. Baba, S. Idaka, A. Maeda, M. Satoh, M. Kimura, "Realization of Pb-free FC-BGA Technology on Low-k Device," *Electronic Components and Technology Conference,* 2005, pp. 9–13.
38. F. Keller, J. W. Brunner, T. Pan, "Optimization of Wire Bonding over Cu-Low K Pad Stack," *36th International Symposium on Microelectronics*, IMAPS, 2003.

CHAPTER 59

COMPONENT-TO-PWB RELIABILITY: ESTIMATING SOLDER JOINT RELIABILITY AND THE IMPACT OF LEAD-FREE SOLDERS

Mudasir Ahmad
Cisco Systems, San Jose, California

Mark Brillhart
Palo Alto, California

59.1 INTRODUCTION

In Chap. 58, the critical design variables that could impact the reliability and manufacturability of printed circuit assemblies (PCAs) have been outlined. The use of various assessment tools to quantify the expected life of PCAs is an important aspect of design for reliability. Data collected from standard tests can be used with acceleration transforms and finite element analysis to assess the expected service life of any interconnect. Experimental data play a crucial role in reliability assessments. Poor data or data obtained under incorrect experimental conditions can lead to grossly inaccurate reliability assessments. "Garbage in equals garbage out" is nowhere more true than in interconnect reliability assessments. The history and experience gained over years of use of tin-lead solders are no longer applicable to lead-free assemblies because lead-free solders behave differently under the same end-use conditions and relatively little data are available on the characteristics of lead-free solders. The formulas for predicting the field life of solder joint interconnects would have to be revisited and validated with extensive experimental data. Moreover, the introduction of lead-free materials is accompanied by new failure mechanisms that are yet to be fully understood and mapped into field-life prediction models.

59.1.1 Experimental Tests

The different levels of solder joint interconnects in a typical PCA are illustrated in Fig. 59.1.

A myriad of experimental tools have been developed to assess solder joint reliability rapidly. These include:

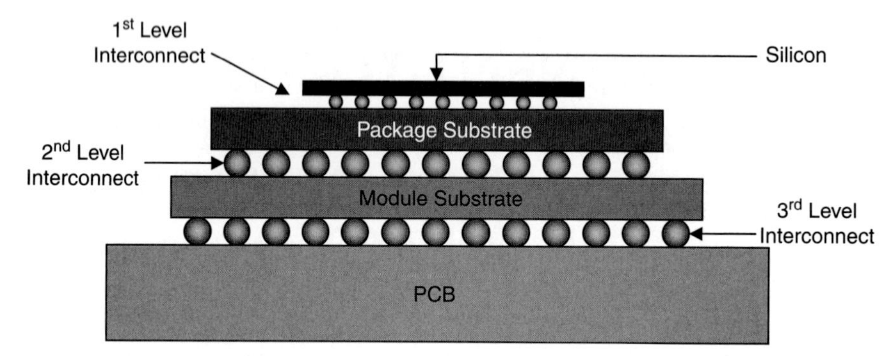

FIGURE 59.1 First-, second-, and third-level interconnects defined, employing a flip-chip assembly as an example. The first-level interconnect is the primary connection between the silicon die and the package substrate. In this example, the connection is created by the solder bumps between the die and the package. The second-level interconnect in this example is the next level of connection between the package substrate and the module substrate. The third-level interconnect in this example is created when the solder balls on the bottom side of the module substrate are attached via surface-mount technology (SMT) to the printed wiring board (PWB).

- Thermal cycling
- Thermal shock
- Air-to-air cycling
- Liquid-to-liquid cycling
- Mechanical bending
- Mechanical deflection
- Hyper-Peltier cooled thermal cycling

It is crucial for the PWB designer to understand the techniques employed, the applicability of the results, and the quality of the data when assessing the reliability of any package that will be placed on his or her PWB. Although rapid assessments are desirable, their benefit diminishes if the failure modes are fictitious and not indicative of expected field failures.

59.1.2 Impact of Lead-Free Materials on Reliability

The conversion to lead-free materials requires a reassessment of these experimental tools and field-life assessment techniques. New materials and failure modes need to be adequately captured in numerical and experimental test methods to model correctly the sort of thermal and mechanical conditions that an interconnect would be exposed to in field operating conditions.

The following key aspects of quantitative reliability assessment are discussed at length in this chapter:

- Thermomechanical reliability:
  - Temperature cycling tests used to determine the thermomechanical reliability of tin-lead and lead-free assemblies
  - The role of Weibull plots in estimating the failure distribution and characteristic life of solder joints from temperature cycling data
  - The acceleration transforms that can be used to estimate the field life of tin-lead and lead-free solder joints based on the results of temperature cycling tests
  - The role of power and minicycles in predicting the field life of solder joints

- Examples illustrating the use of acceleration transforms to compare the fatigue life of packages tested in different thermal cycling and end-use conditions
- Mechanical reliability:
  - Bend test methodologies and test methods that could be used to perform a relative comparison between different package types and surface finishes
  - Sample experimental data showing the comparative difference between different surface finishes and solder metallurgies
  - Shock test methodologies that could be used to perform a relative comparison between different packages and end-use conditions
  - Ball adhesion test methodologies: high-speed shear, pull testing, and impact testing

In addition, this chapter outlines the role of numerical analysis techniques in analyzing and improving the reliability of solder joint interconnects. Detailed procedures illustrating the use of finite element analysis (FEA)[1] in estimating the thermomechanical fatigue life of solder joint interconnects and monotonic bend testing are also discussed.

59.2 THERMOMECHANICAL RELIABILITY

In field operating conditions, ball grid array (BGA) solder joints are continuously subjected to strains due to the coefficient of thermal expansion (CTE) mismatch between the package and the PWB. These strains, which are caused by small but frequent fluctuations in temperature, accumulate as the device is powered up or down. As a result, cracks initiate in the solder joint interconnects and can eventually propagate, leading to solder joint opens. The ability of the solder joint interconnects to withstand these differential strains during field operating conditions for a desired lifetime is generally referred to as *thermomechanical reliability*. The most common method for accelerating this failure in the lab is by thermally cycling the assembled device through extreme temperatures while recording any electrical opens that may occur during the test.

59.2.1 Temperature Cycling

Thermal cycle experiments have been the current industry standard for assessing second-level interconnect reliability. These types of tests tend to produce solder joint and solder joint-pad interfacial fractures that are typically seen in field failures. The fact that thermal cycle tests produce the same physical failures allows for detailed acceleration transformations and finite element-based life assessments to be made employing thermal cycle data as input.

The industry test method that is widely used for this testing is IPC-9701.[2] IPC-9701 provides detailed guidelines for the recommended temperature cycling test methods for evaluating the reliability of surface-mount solder joints. In addition, IPC-9701 also provides guidelines for estimating the performance of solder joints in different field-use conditions from the recommended accelerated thermal cycling tests.

An important benefit of performing industry standardized testing as recommended by IPC-9701 is that device suppliers can perform this test routinely for their devices, irrespective of where the device would be used. The results of the tests can then be used to determine whether the device will survive in a variety of end-use conditions. This saves significant testing cost and design time while providing a baseline for comparison of devices from different suppliers.

Where there are minor differences between the conditions in which the device is to be used, finite element analysis can be performed to bridge the differences between the tested

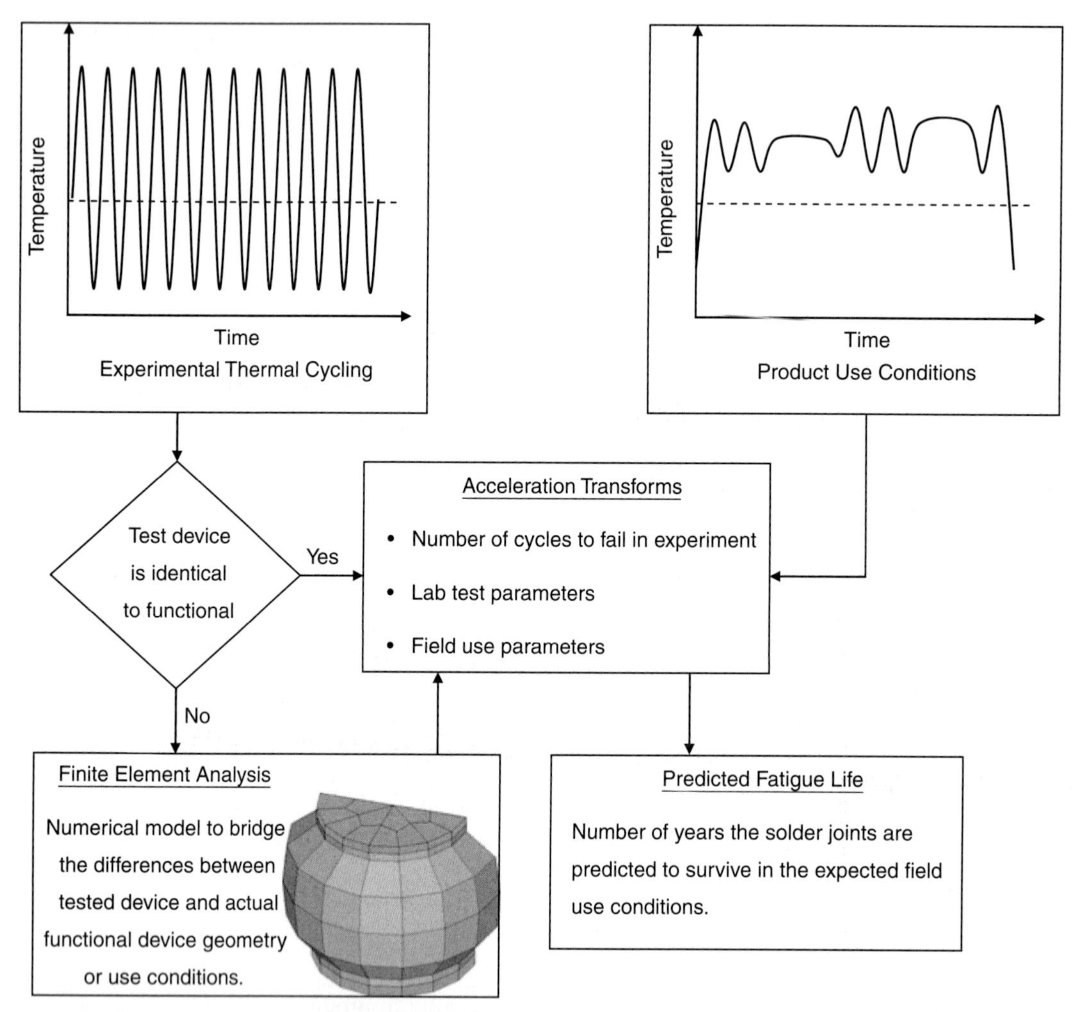

FIGURE 59.2 Flowchart of the solder joint thermomechanical reliability qualification process. Experimental data generated in the lab environment is compared with the expected product end-use conditions. If the package tested is identical to that used in the product, the acceleration transforms are used to map the fatigue life in the lab to the fatigue life in the product end-use conditions. If the package tested is similar but not identical to that used in the product, FEA is performed to account numerically for the differences between the tested and actual package before using the acceleration transforms.

device and the actual environment in which a very similar device would be used. The entire process by which the reliability of a device can be assessed is illustrated in Fig. 59.2.

Thermal cycle experiments are typically performed in single-chamber ovens. Dual-chamber systems are sometimes employed, but single-chamber equipment tends to dominate the test space. The main drawback of a dual-chamber oven is the lack of control over the thermal loading profile when transferring from one chamber (or thermal zone) to the other. Regardless of whether a single or dual chamber is used, temperature is cycled from a maximum (T_{max}) to a minimum (T_{min}) value. Temperature is ramped from one temperature extreme to another at a controlled rate, referred to as the *ramp rate*. Temperature is also held at the maximum and minimum values for a predetermined time known as the *dwell time*. Table 59.1 lists typical

TABLE 59.1 Typical Thermal Cycle Test Parameters for Second-Level Reliability Qualification

Parameter	Value
Maximum temperature (T_{max})	100°C
Minimum temperature (T_{min})	0°C
Ramp rates from T_{min} to T_{max} and T_{max} to T_{min}	10°C/min.
Dwell time at T_{max} and T_{min}	10 min.

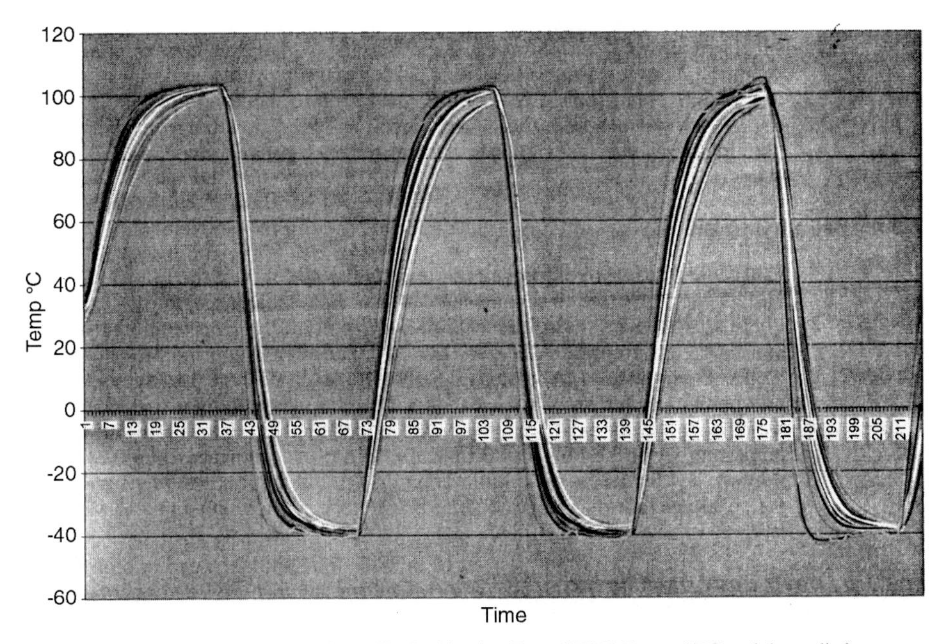

FIGURE 59.3 Typical thermal cycle profile. In this plot, $T_{max} = 100°C$, $T_{min} = -40°C$, and the cyclic frequency is 60 min.

experimental parameters for second-level reliability qualification. Figure 59.3 contains a plot of temperature versus time along with explanations of the key experimental parameters.

Data acquisition systems and/or event detectors (commonly referred to as *glitch detectors*) are employed to monitor resistance and/or scan for intermittent opens in real time. One of the early techniques for assessing solder joint integrity during thermal cycling was to remove parts periodically from the test chamber and measure resistance manually at room temperature.

The major risk associated with this technique is that opens are observed only at temperature extremes when the deformations due to CTE mismatch are largest. As the package cools to room temperature, most of the CTE-based deformations that pull cracked surfaces of solder joints are gone and there is a potential for cracked surfaces to be in enough physical contact to provide acceptable continuity. In some cases, the cycle number at which opens are observed at temperature extremes is much shorter than the cycle at which the same opens are observable at room temperature.

A schematic of a typical thermal cycling chamber, with the boards installed and connected to a data acquisition system, is shown in Fig. 59.4.

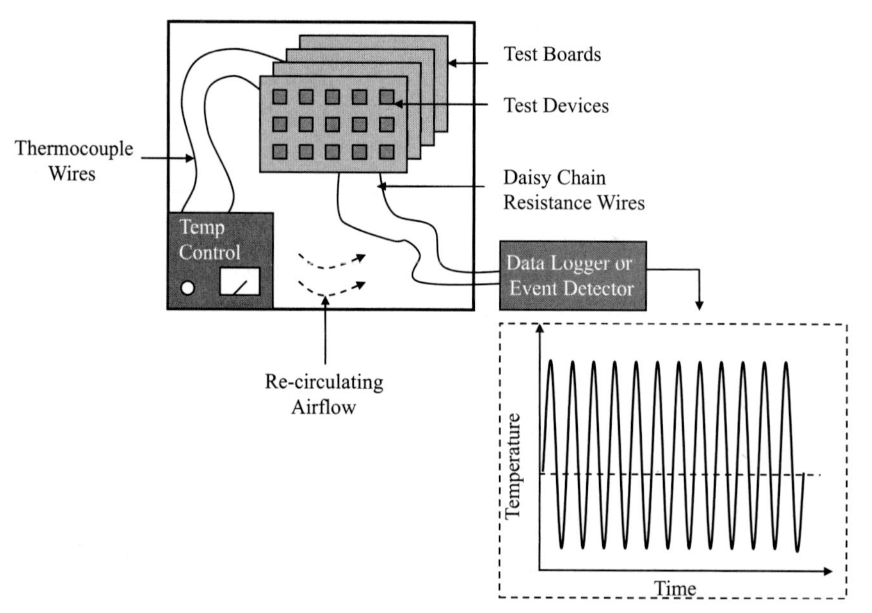

FIGURE 59.4 Schematic of a typical thermal cycling chamber and data acquisition system.

59.2.1.1 Effect of Lead-Free Conversion. The thermal cycling profile listed in Table 59.1 is also applicable to lead-free solders. Most of the damage accumulation occurs during the high-temperature dwell phase. If the high-temperature dwell cycle is shortened, the damage accumulated per cycle would be reduced. It is desirable to maintain a high-temperature dwell long enough to allow creep saturation. However, practical limitations prevent this; as a result, the recommendations in IPC-9701-A[3] (Appendix B) are that a 10 min. dwell time be used, and where possible a 30 min. (or higher) dwell time could be used. Finite element analysis could be used to calibrate the experimental data and determine the effect of dwell time on long-term reliability.

59.2.1.2 Thermal Shock. Thermal shock experiments continuously cycle parts from a high to a low temperature very rapidly. These types of experiments are typically used as screening tools or to run parallel experiments on a known good part and a proposed new component to assess the relative performance of each part. Thermal shock experiments are quite rapid, with thermal shock test durations on the order of days rather than the months associated with thermal cycle test times.

The main drawback of thermal shock experiments lies in the fact that these types of tests tend to induce failure modes not observed in the field. Additionally, even if solder joint fatigue can be induced, it is very difficult to map the high ramp rates associated with thermal shock loading conditions to the more gradual thermal loading conditions that occur in the field. Consequently, this test technique is limited to relative comparisons between packages in which the failure modes have been shown to be comparable under the same thermal shock test conditions.

59.2.2 Weibull Distributions

The results of the temperature cycling tests described in Sec. 59.2.1 can be used to estimate the field operating life of the solder joint interconnects by using the acceleration transforms outlined in detail in Sec. 59.2.3.

However, to determine the field operating life, one needs to know the life of the parts in lab testing. The raw data obtained from the lab test could be fitted with a failure distribution to determine the mean life of the parts. Typical failure distributions include Weibull, Normal, Lognormal, and Exponential. For wear-out type of failures, the Weibull and Lognormal distributions are usually used, with Weibull being the most common. Weibull distributions are lowest value distributions derived from the weakest-link theory.[4,5] Solder joint interconnects can be considered as connected in series. Usually, the failure of one joint at a critical location could cause the entire device to fail. The joints that fail early are usually located at the highest stress locations in the package. Devices with more resilient joints would not fail early. A Weibull distribution captures the minimum solder joint life, and the shape parameter captures the quality of the joints as a function of their construction and the applied stress.[4] There are different types of Weibull distributions: one-parameter, two-parameter, and three-parameter. The three-parameter Weibull Probability Distribution Function (PDF)[6] is as shown in Eq. 59.1.

$$f(t) = \frac{\beta}{\eta}\left(\frac{t-\gamma}{\eta}\right)^{\beta-1} e^{-\left(\frac{t-\gamma}{\eta}\right)^{\beta}} \tag{59.1}$$

where $f(t)$ = Probability Distribution Function
β = shape parameter (which defines the shape of the distribution)
η = scale parameter (which defines the characteristic life of the distribution; this is the time in which 63.2 percent of the tested samples would fail)
γ = location parameter (which defines the location of the distribution in time)
t = time

Two-parameter and three-parameter distributions are commonly used to fit the temperature cycling data. The two-parameter distribution fits a straight line though the data, whereas the three-parameter distribution fits a non-linear curve through the data. In general, the two-parameter distribution is more conservative than the three-parameter distribution. Best practice is to perform regression analysis to determine which distribution best fits the data.

Another important factor in the statistical assessment of test data is the *confidence interval.* Confidence intervals (or *confidence bounds*) take into account the quality of the data based on the number of samples tested.[7] When two separate experimental data sets are being compared, double-sided confidence bounds are useful in normalizing the data based on the number of samples tested in each data set. An example of upper and lower confidence bounds set at 90 percent is illustrated in Fig. 59.5.

59.2.3 Material Properties

A tremendous number of researchers have explored the reliability of Pb/Sn solder alloys.[8] A wealth of data, material properties, and proposed constitutive relationships have been captured in the literature.[8,9,10] However, there is relatively little information in the literature on the material properties of lead-free solders. Moreover, the material property information available is on slightly different compositions of lead-free solders, with tests performed in different ways.

One of the challenges with both Pb/Sn and lead-free solders is that they undergo viscoplastic deformation (creep) as a function of time, temperature, strain rate, and applied stress. A variety of creep deformation models have been used to model the viscoplastic behavior of lead-free solders. The Anand model has been successfully used to model the viscoplastic behavior of Pb/Sn solders. The model allows for the simultaneous incorporation of time-independent plastic deformation as well as time-dependent creep deformation.

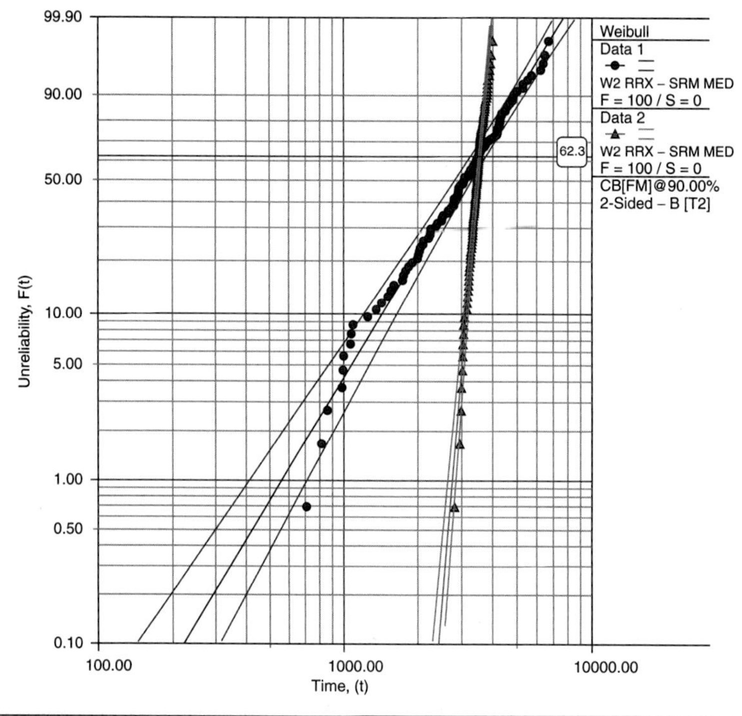

FIGURE 59.5 A typical two-parameter Weibull distribution. The plot shows two data sets with different shape parameter (β) values but comparable characteristic life (η). Extrapolating to 1 percent failure free life shows that the 90 percent lower-bound confidential interval for data set 1 is lower than that for data set 2, even though the characteristic lives are comparable.

The functional form of the Anand model is given as shown in Eqs. 59.2 through 59.5.[11,12]

$$\dot{\varepsilon}_p = A\left[\sin\text{h}\left(\xi\frac{\sigma}{s}\right)\right]^{\frac{1}{m}}\exp\left(\frac{-Q}{kT}\right) \tag{59.2}$$

where $\dot{\varepsilon}_p$ = inelastic strain rate
A = experimentally determined empirical constant
Q = activation energy
R = gas constant
ξ = stress multiplier
σ = equivalent stress
s = internal state variable
m = strain rate sensitivity parameter

$$\dot{s} = \left\{h_o\left(|B|\right)^a\frac{B}{|B|}\right\}\frac{d\varepsilon_p}{dt} \tag{59.3}$$

$$B = 1 - \frac{s}{s^*} \tag{59.4}$$

$$s^{*} = \hat{s}\left\{\frac{\frac{d\varepsilon_p}{dt}}{A}\exp\left(\frac{Q}{kT}\right)\right\}^{n} \tag{59.5}$$

where h_o = hardening/softening constant
a = hardening/softening strain rate sensitivity coefficient
s^{*} = saturation value of s for a given set of temperature and strain rate data
$\hat{s}$ = saturation coefficient
n = strain rate sensitivity coefficient for the saturation value of deformation resistance

The relevant material property information on the solder composition close to that recommended by iNEMI[13] (Sn3.9Ag0.6Cu) is listed in Table 59.2.

TABLE 59.2 Material Properties: Anand's Constants for Lead-Free Solder

Reference	Pei & Qu[14]	Rodgers[15]	Reinikainen[16]
Solder material	95.5Sn3.8Ag0.7Cu	95.5Sn3.8Ag0.7Cu	95.5Sn3.8Ag0.7Cu
Strain rate range	25–180°C	20–125°C	—
Specimens used	1 mm dog bone	4 mm dia. dog bone	—
S_0 (MPa)	21.57	24.04	1.3
Q/R (K)	10041	11049	9000
A (sec-1)	9450.6	8.75e+06	500
ξ	1.1452	4.12	7.1
m	0.1158	0.23	0.3
h_o	133.8025	9537	5900
$\check{S}$	13.3372	90.37	39.4
n	0.0402	2.26e–10	0.03
a	0.1082	1.2965	1.4

One disadvantage of the Anand model is that it mainly captures steady-state creep of solder, but it does not capture primary creep (see Fig. 59.6). Since a significant portion of solder joints' field life is spent in the primary creep phase, it is difficult to predict fatigue life driven by end-use conditions. In other words, although the Anand model can be successfully used in predicting the fatigue life of solder joints tested in laboratory conditions, it may not be very accurate in predicting the fatigue life of solder joints in field use conditions. What may be needed is a time-hardening creep model that can better capture primary, secondary (steady-state), and tertiary creep.

Another model that proposes the incorporation of primary and tertiary creep strain rates is the A-Ω model.[17] This model proposes that the primary creep strain rate be represented as an exponential decay function of increasing creep strain, and the tertiary creep strain rate be represented as an exponential growth function of increasing creep strain. However, this model is relatively new and has not been validated with independent experimental reliability data across the industry.

Other models available in the literature for different lead-free solder metallurgies are listed in Table 59.3.

A more comprehensive list of mechanical material properties of lead-free solders and other materials used in electronic packages can be found in Reference 22.

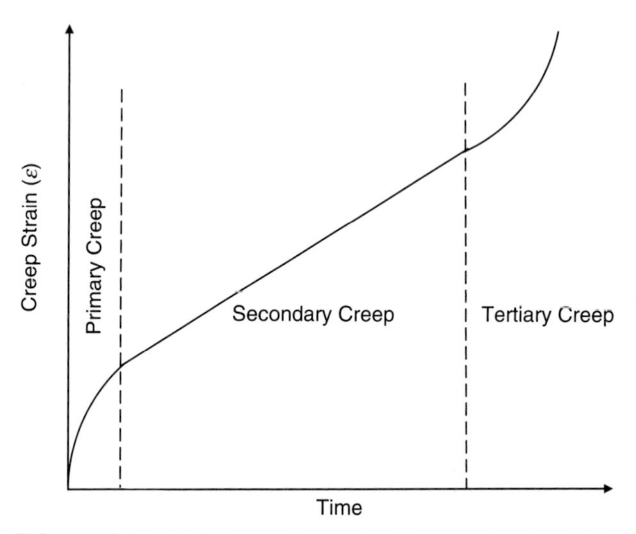

FIGURE 59.6 A typical creep strain versus time curve for viscoplastic materials. The curve shows that creep strain starts out as a non-linear curve, becomes steady for a significant period, and then increases significantly into tertiary creep. Most of the life of a solder joint is spent in the primary and secondary creep strain range of the curve.

TABLE 59.3 Material Properties for Lead-Free Solder[18]

Creep model	Metallurgy	Constant	Value	Ref.
$\dot{\varepsilon} = C_1[(C_3\sigma)]^{-C_2}\exp\left(\frac{-C_4}{T}\right)$	Sn4.0Ag0.5Cu	C_1 (s^{-1})	8×10^{-11}	Weise[19]
		C_2	12	
(Norton creep model)		C_3 (MPa^{-1})	1	
		C_4 (K)	8,996.57	
$\dot{\varepsilon} = A_1\left(\frac{\sigma}{\sigma_n}\right)^{n_1}\exp\left(-\frac{Q_1}{RT}\right)$	Sn4.0Ag0.5Cu	A_1 (s^{-1})	10^{-6}	Weise[20]
		n_1	3	
$+A_2\left(\frac{\sigma}{\sigma_n}\right)^{n_2}\exp\left(-\frac{Q_2}{RT}\right)$		Q_1/R (K)	34.6	
		A_2 (s^{-1})	10^{-12}	
		n_2	12	
(Double power law)		Q_2/R (K)	59.1	
$\frac{d\gamma}{dt} = C_1[\sin h(C_2\sigma)]^{C_3}\exp\left(-\frac{C_4}{T}\right)$	Sn3.8Ag0.7Cu	C_1 (s^{-1})	32000	Pang[21]
		C_2	37×10^{-9}	
		C_3 (MPa^{-1})	5.1	
(Garofalo creep)		C_4 (K)	6524.7	

59.2.4 Norris-Landzberg Acceleration Transformation

PWB designers typically receive a failure distribution plot (see Fig. 59.5) as part of a second-level package qualification report. This failure distribution plot is typically obtained in a thermal cycle oven under controlled temperature profiles that include the maximum and minimum temperatures (T_{max} and T_{min}, respectively) of the thermal cycle along with the frequency (f) of the experimental thermal cycle. The designer must then relate these experimental data obtained under ideal conditions to a different thermal cycle condition (or conditions) encountered in the product end-use environment. Although relative comparisons of two packages thermally cycled under identical conditions can be made in a ranking sense, this type of analysis yields no information regarding the actual lifetime of the solder joints under field loading conditions. This section focuses on the mapping of a laboratory thermal cycle to different end-use thermal profiles using strain as a damage metric.

59.2.4.1 Fatigue Behavior. The fatigue behavior of many metals (including Pb/Sn and lead-free solder alloys) can be described by the Coffin-Manson relationship (Eq. 59.6),[9] which relates the fatigue life of a metal to the plastic strain the material experiences during loading. Elastic strain is the reversible or recoverable deformation that occurs during a loading cycle. Plastic strain is the amount of irreversible (or permanent) deformation that occurs during a loading cycle.

$$\varepsilon = \varepsilon_e + \varepsilon_p = AN_f^{-c} + BN_f^{-m} \tag{59.6}$$

where
ε = total strain per cycle
ε_p = plastic portion of the cyclic strain
ε_e = elastic portion of the cyclic strain
N_f = number of cycles to failure
A,B,c,m = experimentally determined empirical constants

Most solder joints fail in the low cycle fatigue region (less than 10,000 cycles to failure),[9] where strain is primarily plastic. Thus, neglecting the elastic component of strain, Eq. 59.6 can be simplified and rearranged to yield a relationship between the number of cycles to failure and the plastic strain (see Eq. 59.7).[9]

$$N_f = (A/\varepsilon_p)^{1/m} \tag{59.7}$$

where
ε_p = plastic portion of the cyclic strain
N_f = number of cycles to failure
A,m = experimentally determined constants

This can be further simplified to the most common form of the Coffin-Manson relationship shown in Eq. 59.8.[9]

$$N_f \propto \varepsilon_p^c \tag{59.8}$$

where
ε_p = plastic portion of the cyclic strain
N_f = number of cycles to failure
c = an experimentally determined empirical constant

When any metal is cycled at a temperature above 50 percent of its melting temperature (homologous temperature), the strain rates, dwell times, and maximum temperature influence its fatigue life.[9] This is due to the occurrence of creep, along with plastic and elastic deformation, within these temperature regimes.[9] Norris and Landzberg proposed a modified version of the Coffin-Manson relationship that considers the effects of frequency (f) and maximum temperature (T_{max}) on solder joint fatigue life (Eq. 59.9).[9,10] This relationship is based on detailed physical and empirical analysis of the relationships between temperature and frequency on the thermal fatigue behavior of Pb/Sn solder alloys employed in interconnects.[10]

$$N_f = (A/\varepsilon_p)^{1.9} f^{1/3} \exp\left(\frac{0.123}{kT_{max}}\right) \tag{59.9}$$

where ε_p = plastic portion of the cyclic strain
N_f = number of cycles to failure
k = Boltzmann's constant (eV)
f = cyclic frequency
T_{max} = maximum temperature of the thermal cycle (K)
A = an experimentally determined empirical constant

59.2.4.2 Acceleration Factors. One of the challenges that must be addressed when attempting to employ Eq. 59.9 directly to assess fatigue life is the determination of the plastic strain and the value of the constant A. It is crucial to remember that the goal of this analysis is to develop an acceleration transform (or acceleration factor) that can be employed to use known fatigue-life data obtained under controlled laboratory conditions to estimate the number of cycles to failure under field conditions. The acceleration factor AF is defined as the ratio of the number of cycles to fail in the field N_{fF} to the number of cycles to fail in the lab N_{fL} (see Eq. 59.10).

$$AF = \frac{N_{fF}}{N_{fL}} \tag{59.10}$$

where AF = acceleration factor
N_{fF} = number of cycles to failure in the field
N_{fL} = number of cycles to failure in the lab

Substituting Eq. 59.9 for lab and field conditions into Eq. 59.10, one obtains 59.11.

$$AF = \frac{N_{fF}}{N_{fL}} = \frac{(A/\varepsilon_{pF})^{1.9} f_F^{1/3} \exp\left(\dfrac{0.123}{kT_{maxF}}\right)}{(A/\varepsilon_{pL})^{1.9} f_L^{1/3} \exp\left(\dfrac{0.123}{kT_{maxL}}\right)} \tag{59.11}$$

where AF = acceleration factor
N_{fF} = number of cycles to failure in the field
N_{fL} = number of cycles to failure in the lab
ε_{pF} = plastic portion of the cyclic strain in the field
ε_{pL} = plastic portion of the cyclic strain in the lab
f_F = cyclic frequency
f_L = cyclic frequency
T_{maxF} = maximum temperature of the thermal cycle in the field (K)
T_{maxL} = maximum temperature of the thermal cycle in the lab (K)
k = Boltzmann's constant (eV)
$_L$ = lab conditions
$_F$ = field conditions

Equation 59.11 can be algebraically rearranged to yield Eq. 59.12. Note that the "max" subscripts on the maximum temperatures have been dropped to simplify the notation.

$$AF = \frac{N_{fF}}{N_{fL}} = \left[\frac{\varepsilon_{pL}}{\varepsilon_{pF}}\right]^{1.9} \left[\frac{f_F}{f_L}\right]^{1/3} \exp\left[1414\left(\frac{1}{T_F} - \frac{1}{T_L}\right)\right] \tag{59.12}$$

where AF = acceleration factor
ε_{pF} = plastic portion of the cyclic strain in the field
ε_{pL} = plastic portion of the cyclic strain in the lab
f_F = cyclic frequency in the field
f_L = cyclic frequency in the lab
T_F = maximum temperature of the thermal cycle in the field (K)
T_L = maximum temperature of the thermal cycle in the lab (K)

When considering package interconnects of the same geometry and material set, one can reduce the strain term ε_p in Eq. 59.9 to the change in temperature.[10] This is a key feature that enables the development of an acceleration transform. The goal of this analysis is to develop a transform that allows one to map laboratory thermal cycle loading conditions to field loading conditions for the same package. Thus it is valid to assume that the strain term can be reduced to the change in temperature, as shown in Eq. 59.13.

$$\varepsilon_p \propto \Delta T \tag{59.13}$$

where ε_p = plastic portion of the cyclic strain
ΔT = difference between maximum and minimum temperature of the thermal cycle

Substitution of Eq. 59.12 (including $_L$ and $_F$ subscripts for lab and field, respectively) yields the final form of the acceleration factor relationship, which can be seen in Eq. 59.14.[10]

$$AF = \frac{N_{fF}}{N_{fL}} = \left[\frac{\Delta T_L}{\Delta T_F}\right]^{1.9} \left[\frac{f_F}{f_L}\right]^{1/3} \exp\left[1414\left(\frac{1}{T_F} - \frac{1}{T_L}\right)\right] \tag{59.14}$$

where AF = acceleration factor
f_F= cyclic frequency in the field
f_L= cyclic frequency in the lab
T_F = maximum temperature of the thermal cycle in the field (K)
T_L = maximum temperature of the thermal cycle in the lab (K)
ΔT_F= difference between maximum and minimum temperature of the thermal cycle in the field
ΔT_L = difference between maximum and minimum temperature of the thermal cycle in the lab

Figure 59.7 contains experimental failure data for a ceramic ball grid array (CBGA) cycled under the same two conditions. The predicted acceleration factor was calculated to be 3.2 employing Eq. 59.14. This predicted value correlates quite well with the CBGA data (acceleration factor ranges from 2.9 to 3.6). Note that this approach tends to be overly conservative when mapping large lab ΔT into small field ΔT (in the case of minicycles, to be discussed in a subsequent section).

59.2.4.3 Effect of Lead-Free Conversion. The acceleration factor equation (Eq. 59.14) is specifically derived from experimental data for Pb/Sn solder joints. What remains to be conclusively demonstrated is an equivalent acceleration transform for lead-free solder joints. Several studies have been performed, but there are no acceleration transforms accepted industrywide for lead-free solders. One of the acceleration transforms proposed is discussed in this section. A more comprehensive list of other field life prediction studies can be found in the appendix of IPC-9701-A.[3]

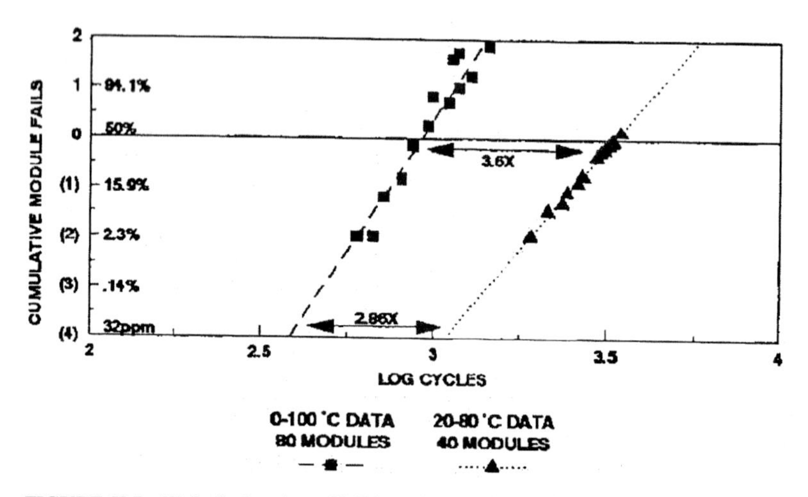

FIGURE 59.7 Weibull plots for a CBGA package cycled under two different loading conditions. Solid triangles and the dotted line are employed to indicate failures and associated failure distribution of a CBGA package cycled from 20 to 80°C (the least severe case presented in this figure). Solid squares and the dashed line indicate the failures and associated distribution for the identical package cycled from 0 to 100°C. Notice that an increase in the severity of the thermal cycle loading conditions (from 20 to 80°C to 0 to 100°C) creates an associated decrease in fatigue life. (*Reprinted with permission from J. Lau,* Ball Grid Array Technology, *McGraw-Hill, 1995, p. 162.*)

59.2.4.4 Modified Norris-Landzberg Acceleration Transform for Lead-Free Solder. Based on experimental data generated on a select variety of packages, Pan et al. proposed an updated form of the Norris-Landzberg acceleration transform (see Eq. 59.15).[23]

$$AF = \frac{N_{fF}}{N_{fL}} = \left[\frac{\Delta T_L}{\Delta T_F}\right]^{2.65} \left[\frac{t_F}{t_L}\right]^{0.136} \exp\left[2185\left(\frac{1}{T_F} - \frac{1}{T_L}\right)\right] \tag{59.15}$$

where AF = acceleration factor
t_F = dwell time at high temperature in the field
t_L = dwell time at high temperature in the lab
T_F = maximum temperature of the thermal cycle in the field (K)
T_L = maximum temperature of the thermal cycle in the lab (K)
ΔT_F = difference between maximum and minimum temperature of the thermal cycle in the field
ΔT_L = difference between maximum and minimum temperature of the thermal cycle in the lab

The primary differences between Eq. 59.14 and Eq. 59.15 are that the frequency term is replaced by the dwell time at high temperature, and the constants have been revised. This equation has been shown by Pan et al. to predict well within the temperature range of 0 to 100°C. However, predictions are less accurate for temperatures outside of this range.[23]

Although the aforementioned proposed acceleration transform is a first-order estimate, much more data would be needed to validate or improve this estimate. Several additional variables must be evaluated to study their impact on the acceleration transform. Some of the variables that impact the acceleration transform include the following:

- **Effect of different package types** The differential strain imparted on solder joints in a ceramic flip-chip ball grid array (FCBGA) is quite different from that imparted on a thin small outline package (TSOP) package, for example. As a result, the acceleration factors from one may not necessarily apply to the other. The construction and effective CTE mismatch of the packages to be qualified must be as close as possible to the packages tested.[23]
- **PWB thickness effect** A thick board is likely to constrain a package more rigidly than a thin board. Similarly, the number of layers in the board could significantly alter its overall stiffness. Tests are frequently done on thinner boards because they are less costly and easier to fabricate. However, if the end-use condition entails the use of a thicker board, IPC-9701 recommends performing the test on at least two different board thicknesses so that the data can be extrapolated to a range of board thicknesses. The effect of board stiffness on lead-free packages appears to be similar to that on Pb/Sn packages, as shown in Sec. 58.3.3 of Chap. 58.
- **Cycling frequency** As evidenced in the study performed by Pan et al. [23] and by Clech,[24] the temperature cycling frequency could have a significant effect on the reliability of the packages being tested. Both the ramp rate and the dwell time affect the results. Most of the damage accumulation during temperature cycling occurs in the hot dwell part of the cycle. A reduction in dwell time reduces the amount of damage accumulated per cycle, which in turn means longer solder fatigue life. A long hot dwell would result in almost complete creep stress relaxation in the solder joints. Consequently, some studies have indicated that a longer dwell time could result in eventual creep saturation.[25,26,27] Increasing the dwell time has shown some reduction in fatigue life in some studies, but the reduction was not very significant in some of those.[28] In addition to hot dwell time, ramp rate of heating and cooling also impacts the effective fatigue life of lead-free solder joints. The amount of reduction in lead-free solder fatigue life due to ramp rate is a function of the package type. Packages with higher effective CTE mismatched substrates (e.g., CBGAs) are more sensitive to the ramp rate than packages with a less differential strain (e.g., plastic ball grid arrays [PBGAs]).
- **Temp cycling range** Clearly, the higher the temperature cycling range, the lower the fatigue life of the solder joints. Some studies have also indicated that an increased average temperature could also result in a reduction in the fatigue life of lead-free solder joints. In other words, both factors—high maximum temperature and high average temperature—could contribute to a reduction in solder fatigue life.[24]
- **The Distance from neutral point (DNP) effect** This is the effective distance along the diagonal of the package, from the center to its outermost corner. Some studies have indicated that the DNP plays a significant role in altering the acceleration transforms of a lead-free package.[24]
- **Size and Orientation of tin (Sn) grains**[29] Depending on the cooling rate, Pb free solders containing a high percentage of Sn tend to have very large Sn grains with varied orientation. Sn has very anisotropic material properties, and consequently, the bulk properties of the Pb free solder joints could change significantly depending on the size and orientation of the Sn grains in any given solder joint. Consequently, the strength of the solder joints within a single BGA could vary significantly. In other words, the joints at the high stress regions may not be the first to fail. This random orientation of Sn grains within a solder joint poses a significant challenge in determining the fatigue life and acceleration factors of Pb free solder joints.[29]

It has also been observed that while the ductility of Pb/Sn increases with aging, the ductility of lead-free solders tends to decrease with aging.[30] In other words, the strain/strain rate characteristics of lead-free solders could change with aging. Further studies are also needed to quantify fully the aging effect and incorporate it into long-term field life prediction models.

59.2.4.5 Backward Compatibility. On small PWBs with relatively few components, it is possible to convert all the components to lead-free assemblies at once. However, in the transition to lead-free assemblies, there will be instances where not all the components on a complex

PWB can be converted to lead-free solder. This would inevitably create situations where both tin-lead and lead-free parts are attached on the same board.

Consequently, the use of Pb/Sn paste to assemble lead-free components (backward compatibility) would have to be studied. Component suppliers do not recommend the use of mixed solder metallurgies for a variety of reasons:

- The peak reflow temperature of lead-free solders is higher than that of tin-lead solders. Assembling lead-free components with Pb/Sn paste would involve limiting the peak reflow temperature to that of Pb/Sn components (~220°C). Consequently, the lead-free solders may not fully melt during reflow, resulting in unpredictable long-term solder joint reliability.
- The thermal gradients across components and the PWB could vary significantly, and as a result, it is difficult to ensure that the microstructure formed across the solder joints is consistent.
- There are no known and validated acceleration transforms to predict the field reliability of lead-free components assembled with Pb/Sn paste. Thus, even if the assembled parts pass temperature cycling relative to their Pb/Sn and lead-free counterparts, it is difficult to extrapolate to field conditions and predict how they will perform when subjected to actual end-use conditions.
- There is very little data on the bend/shock resistance of mixed-metallurgy solder joints. The intermetallics formed at the package-joint and PWB-joint interfaces can be quite complex and different. The resistance of these intermetallics to mechanical loading has not been characterized.

There is very little information in the literature on the process consistency of backward-compatible assemblies and their consequent long-term reliability. Some information can be found in References 28, 31, 32 and 34. Some reliability data have been shown to indicate that the reliability of backward-compatible assemblies can be significantly improved if the SnAgCu ball melts fully and good dissolution with the Pb/Sn paste is achieved.[31] However, the factors that influence the degree of dissolution, such as the peak temperature and time above liquidus (TAL), are generally difficult to control consistently. Moreover, the amount of dissolution obtained is a function of the solder ball volume relative to the volume of paste used. Consequently, packages with smaller solder ball pitch could also be more sensitive to assembly process variations in backward-compatible assemblies.

In view of all the unknowns and process dependencies outlined, the assembly of lead-free components with Pb/Sn paste is generally not recommended. If it has to be performed, detailed studies should be conducted to benchmark the consistency of the assembly process as well as the consequent effect on the long-term reliability (both thermomechanical and mechanical) of the assemblies.

59.2.5 Power and Minicycles

Consider a processor in a server. The server may only be powered down and back up one time per month (a conservative estimate). Power-downs and -ups are referred to as *power cycles*. A reliability estimate could be made assuming that one thermal cycle occurred per month from the operating temperature within the region of the package to ambient. This would yield a tremendously high lifetime estimate that would unfortunately be misleading. It is important to consider what happens to the packaged device during typical usage of the server. The device temperature will probably increase during usage and decrease when not in use. These smaller temperature rises can occur multiple times throughout a day. These types of thermal conditions are referred to as *minicycles* and can have a dramatic impact on the long-term reliability of an interconnect. Section 59.2.6.3 presents a technique for incorporating minicycles and power cycles in a lifetime assessment.

59.2.5.1 Miner's Rule. Miner's rule[33] is employed when attempting to assess fatigue life under multiple fatigue loading conditions. It is commonly used when addressing the combined power and minicycle thermal loading conditions described in the previous section. Miner's rule is based on the premise that fatigue damage accumulates over each cycle. When a critical quantity of damage has been imparted on a system, the system will fail.

Miner's rule states that the sum of the ratios of the actual number of cycles to fail of a given fatigue loading condition to the number of cycles to fail if only that fatigue loading condition occurs is equal to 1. Equation 59.16 contains a general expression of Miner's rule.[9]

$$\Sigma \frac{N_{ai}}{N_{oi}} = 1.0 \tag{59.16}$$

where N_{ai} = actual number of fatigue cycles that occur
N_{oi} = number of cycles to fail if only that specific fatigue cycle occurred

Miner's rule can be expanded and rearranged for power cycle and minicycle considerations. Equation 59.17 contains that expression.

$$N_{poweractual}/N_{poweronly} + N_{miniactual}/N_{minionly} = 1.0 \tag{59.17}$$

where $N_{poweractual}$ = actual number of power cycles that occur before failure
$N_{poweronly}$ = number of power cycles to fail if only power cycles occur
$N_{miniactual}$ = actual number of power cycles that occur before failure
$N_{minionly}$ = number of minicycles to fail if only minicycles occur

The Norris-Landzberg acceleration transformation can be combined with Miner's rule to estimate the fatigue life of a package. Note that the Norris-Landzberg method tends to be conservative when assessing minicycles whose changes in temperature are much smaller than the experimental temperature change applied when qualifying the part.

59.2.6 Practical Examples

59.2.6.1 Comparing the Reliability of Two Packages Qualified Under Different Thermal Loading Conditions

Given: Thermal cycle qualification data and test conditions for packages A and B (Table 59.4)

TABLE 59.4 Package A and Package B Qualification Data

	Package A	Package B
T_{max}	110°C	100°C
T_{min}	–10°C	0°C
Frequency	1 cycle/hr	1 cycle/hr
N_{50} (number of cycles to 50% failure)	2,291	3,575

Solution: It is necessary to decide which thermal profile will be kept as a constant and which one will be mapped. This problem can be solved both ways—each temperature cycling condition could be mapped into the other. Either method requires utilizing Eq. 59.10 and 59.14 to transform the number of cycles to failure. For clarity, let condition A refer to the thermal conditions that package A was qualified under, and let condition B refer to the qualification test conditions to which package B was exposed.

59.2.6.2 Map Package A Lifetime into Package B Thermal Loading Conditions. Using Eq. 59.10, one would derive Eq. 59.18.

$$AF = \frac{N_{fAA}}{N_{fAB}} \tag{59.18}$$

where AF = acceleration factor
N_{fAA} = number of cycles to failure in package A cycled from –10 to 125°C (condition A)
N_{fAB} = number of cycles to failure in package A cycled from 0 to 100°C (condition B)

Equation 59.14 can be rearranged to yield Eq. 59.19.

$$AF = \left[\frac{\Delta T_B}{\Delta T_A}\right]^{1.9}\left[\frac{f_A}{f_B}\right]^{1/3} \exp\left[1414\left(\frac{1}{T_A} - \frac{1}{T_B}\right)\right] \tag{59.19}$$

where AF = acceleration factor
f_A = cyclic frequency for condition A (1 cycle/hr)
f_B = cyclic frequency for condition B (1 cycle/hr)
T_A = maximum temperature for condition A (110°C, 383 K)
T_B = maximum temperature for condition B (100°C, 373 K)
ΔT_A = difference between maximum and minimum temperature of condition A (120)
ΔT_B = difference between maximum and minimum temperature of condition B (100)

Substitution of the values in Table 59.4 into Eq. 59.19 yields an acceleration factor of 0.641. Package A would be expected to survive 3,574 cycles ($N_{fAB} = N_{fAA}/AF$ = 2,291/0.641) if subjected to condition B thermal loading conditions. The packages appear to be comparable.

59.2.6.3 Assessing Expected Field Life of a New Package Based on Thermal Cycle Data

Given: Thermal cycle qualification data for a new ball grid array package (use package B in Table 59.4) and estimates of field loading conditions (see Table 59.5)

TABLE 59.5 Field Conditions for Example

	Power cycle	Minicycle
T_{max} (°C)	65	80
T_{min} (°C)	25	65
ΔT (°C)	40	15
Frequency	1 cycle per 30 days, 30 min. duration	4 cycles per day, 60 min. duration

Solution: The process for solving this problem is as follows. The experimentally obtained qualification data (presented in Table 59.4) must be mapped into number of power cycles to failure and number of minicycles to failure. These estimates assume that only power cycles occur and that only minicycles occur (that is the underlying assumption when mapping the experimental data to the end-use power cycle and minicycle conditions). After mapping the experimental data into field conditions, Miner's rule is employed to assess the number of cycles to fail based on the simultaneous occurrence of both power cycles and minicycles throughout the package lifetime.

When employing Eq. 59.14 to map power cycles and minicycles, it is important to consider the implications of how the *frequency* term is interpreted. The Norris-Landzberg approach assumes that the frequency corresponds to the duration of the cycle, not how often that cycle occurs.

Equation 59.14 can be employed to determine the acceleration factors. Equations 59.20 and Eq. 59.21 contain the expressions for power cycles and minicycles, respectively.

$$AF = \left[\frac{\Delta T_L}{\Delta T_P}\right]^{1.9} \left[\frac{f_P}{f_f}\right]^{1/3} \exp\left[1414\left(\frac{1}{T_P} - \frac{1}{T_L}\right)\right] \tag{59.20}$$

where AF = acceleration factor
f_P = cyclic frequency for a power cycle (2 cycles/hr)
f_L = cyclic frequency in the lab (1 cycle/hr)
T_P = maximum temperature of a power cycle (65°C, 338 K)
T_L = maximum temperature of the thermal cycle in the lab (100°C, 373 K)
ΔT_P = difference between maximum and minimum temperature of a power cycle (40)
ΔT_L = difference between maximum and minimum temperature of the thermal cycle in the lab (100)

$$AF = \left[\frac{\Delta T_L}{\Delta T_M}\right]^{1.9} \left[\frac{f_M}{f_L}\right]^{1/3} \exp\left[1414\left(\frac{1}{T_M} - \frac{1}{T_L}\right)\right] \tag{59.21}$$

where AF = acceleration factor
f_M = cyclic frequency for a minicycle (1 cycle/hr)
f_L = cyclic frequency in the lab (1 cycle/hr)
T_M = maximum temperature of a minicycle (80°C, 353 K)
T_L = maximum temperature of the thermal cycle in the lab (100°C, 373 K)
ΔT_M = difference between maximum and minimum temperature of a minicycle (15)
ΔT_L = difference between maximum and minimum temperature of the thermal cycle in the lab (100)

Substitution of the data in Tables 59.4 and 59.5 yields a power cycle acceleration factor of 10.6 and a minicycle acceleration factor of 45.6. Equation 59.10 can be rearranged to yield Eq. 59.22.

$$N_{fi} = AF_i \times N_{fL} \tag{59.22}$$

where N_{fi} = number of cycles to fail under minicycle (i) or power cycle (i)
N_{fL} = number of cycles to fail in the lab
AF_i = acceleration factor for minicycle (i) or power cycle (i)

Substitution of the lab number of cycles to fail from Table 59.4 into Eq. 59.22 for minicycle and power cycles yields 37,895 power cycles to fail (assuming only power cycles occur) and 163,020 minicycles to fail (assuming only minicycles occur).

Miner's rule (Eq. 59.17) can now be applied to determine the actual number of cycles to fail. It is important to note that at this stage the number of minicycles to fail if only minicycles occur and the actual number of power cycles to fail if only power cycles occur is known. The actual number of minicycles and power cycles to fail remains unknown; thus Eq. 59.17 is only one equation, whereas two unknowns still exist: $N_{poweractual}$ and $N_{miniactual}$.

Therefore, it is important to consider the number of minicycles that occur per power cycle. In this case, there is one power cycle per 30 days and 4 minicycles per day. This equates to 120 minicycles per power cycle. Equation 59.17 can now be written as shown in Eq. 59.23.

$$N_{poweractual}/N_{poweronly} + K^* \, N_{poweractual}/N_{minionly} = 1.0 \tag{59.23}$$

where $N_{poweractual}$ = actual number of power cycles that occur before failure
$N_{poweronly}$ = number of power cycles to fail if only power cycles occur
$N_{miniactual}$ = actual number of minicycles that occur before failure
$N_{minionly}$ = number of minicycles to fail if only minicycles occur
K = number of minicycles that occur per power cycle

Equation 59.23 can be solved for the number of power cycles to fail, which yields a value of 1,311 power cycles or 107.8 years. It is important to employ a confidence factor (safety factor). Taking 2 as a standard safety factor, the final fatigue life is estimated at 53.9 years.

59.3 MECHANICAL RELIABILITY

As discussed in Sec. 58.3.11 of Chap. 58 the surface finishes used on packages have been shown to have a significant impact on the mechanical robustness of BGA packages. With increased I/O density, finer pitch, and new substrate technologies (e.g., flex substrates), BGA packages tend to be more fragile and susceptible to mechanically induced failures. These failures are mostly catastrophic in nature, but the loading conditions could vary from bending during production assembly to shock due to shipping or accidental drops. Both monotonic and cyclic conditions could result in catastrophic failure either in manufacturing or field operating conditions. This section focuses on the test methodologies used in quantifying and benchmarking different mechanically induced loading conditions on printed circuit assemblies (PCAs). In general, the mechanical loading conditions that an assembly undergoes could be classified by the strain and strain rate level, as illustrated in Fig. 59.8. As Fig. 59.8 shows, bend and shock tests each cater to different ranges of the strain/strain rate spectrum.

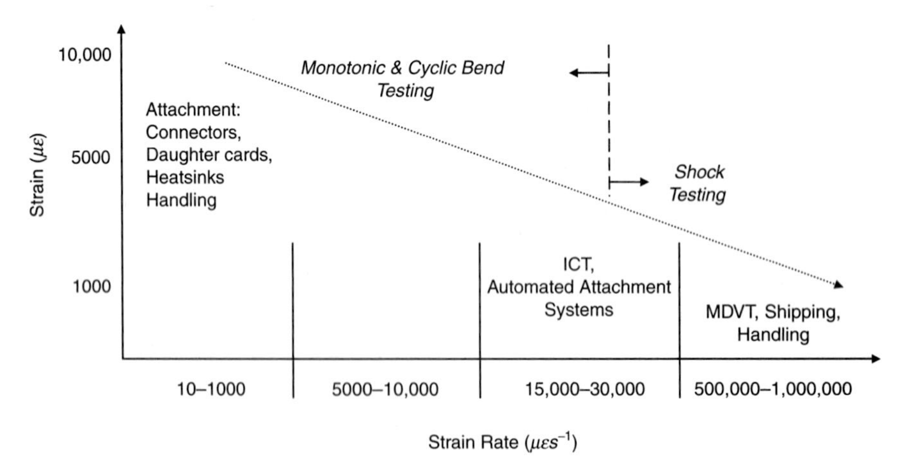

FIGURE 59.8 Spectrum of strain/strain rate in which mechanical testing fits with the assembly process. The monotonic and cyclic bend testing focuses on high-strain, low-strain rates, whereas the shock testing focuses on low-strain, high-strain rates.

59.3.1 Monotonic Bend Testing

Monotonic (bend-to-failure) testing is designed to simulate the assembly conditions to which a PCA is subjected in manufacturing. A variety of assembly steps impart loading conditions that are similar to a monotonic bend test, such as heatsink attachment, connector attachment, in-circuit test (ICT), handling, and card-cage installation. These assembly steps could mechanically induce several different failure modes. Monotonic bend testing in a controlled condition serves two critical purposes:

- A comparative assessment can be performed between different package types and material sets.
- The failure modes induced at different strain levels can be catalogued. Production operations can then be characterized with strain gauges and compared with the monotonic bend test data to determine whether excessive strain was applied on the PCA during any given assembly step.

To ensure consistency in results and to allow for ease of comparison, an industry standard test method has been developed and published jointly by IPC and JEDEC (IPC/JEDEC-9702).[35] This test method outlines in detail the requirements of the monotonic bend test method. The test setup is illustrated in Fig. 59.9. The typical failure modes observed at the solder joint interconnects are shown in Fig. 59.10.

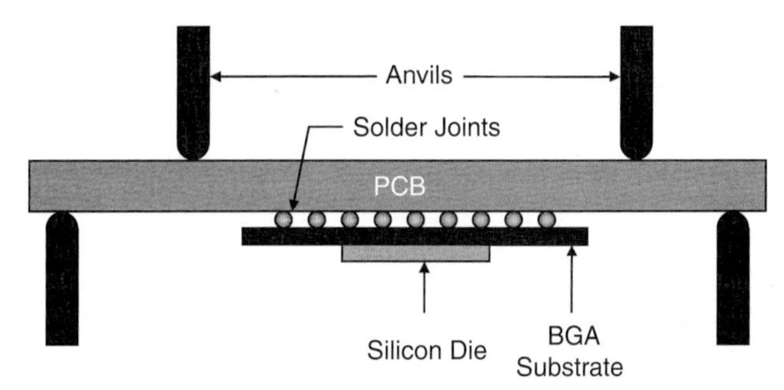

FIGURE 59.9 Schematic of four-point bend test setup. A four-point bend test setup is used to determine the strains-to-failure rate of the BGA solder joints.

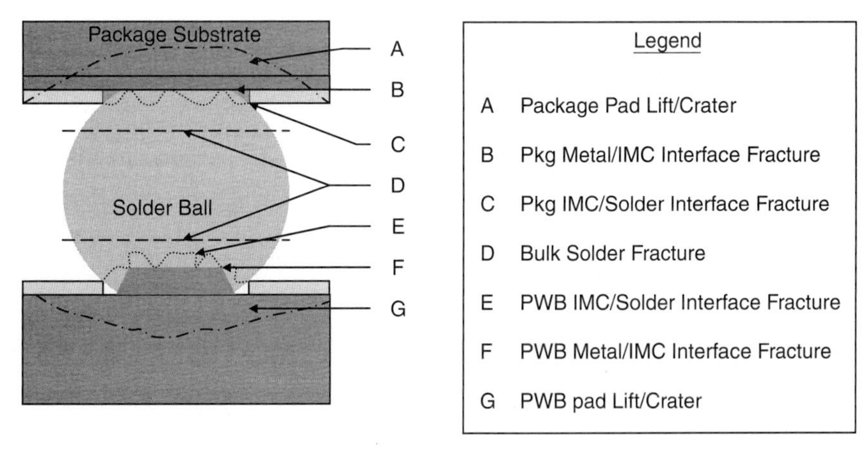

FIGURE 59.10 Different interconnect failure modes.

59.3.1.1 Strain Measurement in Production. A common procedure for performing strain measurements on live PCAs during production is outlined in IPC/JEDEC 9704.[36] This provides a comprehensive guideline on ways to perform strain measurements and reporting the recorded data. It also includes sample guidelines on the amount of strain/strain rate considered acceptable as a function of board thickness. One assembly process that is particularly susceptible to

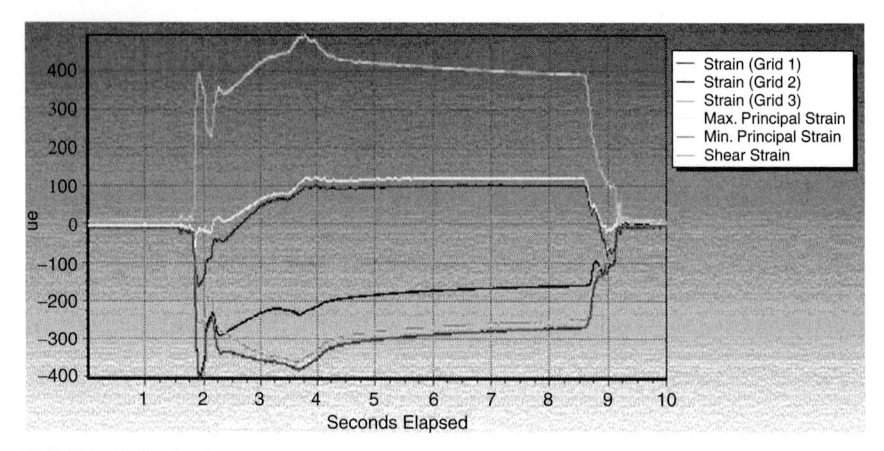

FIGURE 59.11 Strain versus time plot of an actual ICT fixture as it is engaged, the test is performed, and the fixture is subsequently disengaged. The sharp increase in strain during engagement and disengagement is the source of the high strain rate. The fixture is engaged at high speed to improve electrical contact between the probe pins and the PWB test pads.

high strain/strain rates is ICT. Once the ICT fixture is engaged, there is a sudden spike in strain (high strain rate), which is again repeated when the fixture is disengaged, as shown in Fig. 59.11.

Based on experimentally generated data on the production floor and in controlled four-point bend tests described earlier, guidelines can be developed on the maximum range of strain for a given strain rate that would be acceptable for a package in different production operations. A typical example of such a guideline is given in Fig. 59.12. It is important to note that this is just a sample guideline.

59.3.1.2 Effect of Lead-Free Conversion. Because of the conversion to lead-free solder, the intermetallics formed at the package-joint and PWB-joint interface would be different from those in Pb/Sn assemblies. Consequently, the response to bending of a lead-free assembly would be significantly different from that of a Pb/Sn assembly. There are relatively little experimental data in the industry comparing the mechanical reliability of lead-free soldered assemblies with Pb/Sn assemblies. One data set showing a relative comparison between Pb/Sn and SnAgCu for two different surface finishes is shown in Fig. 59.13.[37]

The results in Fig. 59.13 show the following:

- For ENIG surface finish packages, the average force-to-failure value for the SnAgCu assemblies is almost three times that of the Pb/Sn assemblies. This is because the SnAgCu joints are stiffer than the Pb/Sn joints.
- Failure mode A is significantly more pervasive in the Pb/Sn assemblies but not as much in the SnAgCu assemblies. Thus, the propensity for brittle solder joint fracture is higher in the Pb/Sn assemblies with ENIG.
- The introduction of a small amount of copper in the Pb/Sn assemblies tends to increase the bending strength of the Pb/Sn assemblies by more than two times.
- The average force to failure with a solder-on-copper (SOC or SOP) surface finish is higher than that with ENIG, especially for the Pb/Sn solder joints. For the SnAgCu joints, there is a marginal increase in relative strength.

Thus, mechanical bend strength is a function of the solder metallurgy and the surface finish used. These results indicate that SnAgCu solder joints are stiffer and have a higher average

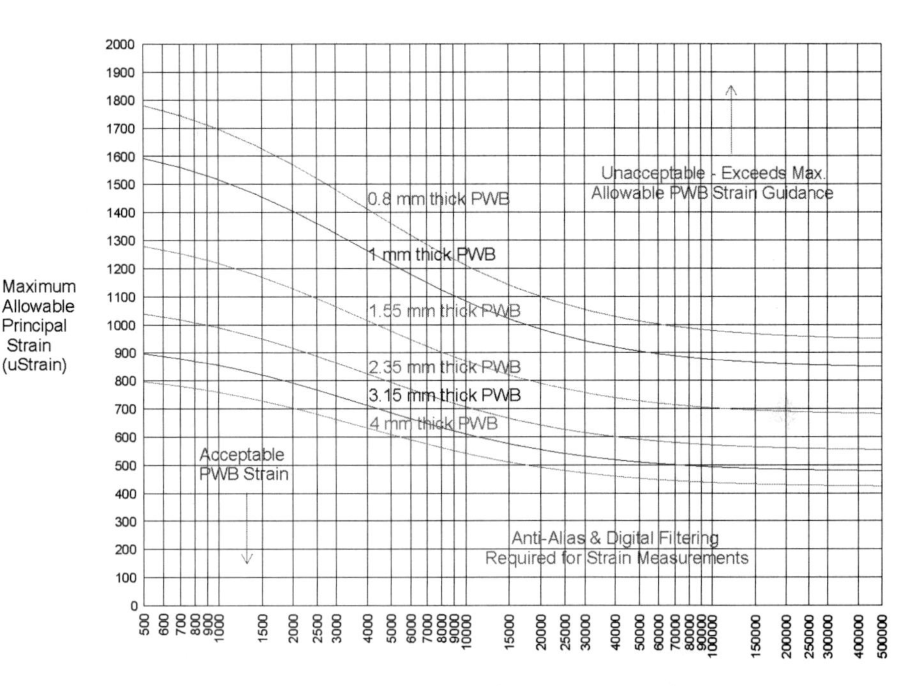

PWB strain = max. principal strain (absolute value) measured immediately adjacent to cornermost solder joint
Strain rate = change in strain (absolute value) between consecutive readings

Max. allowable strain = sqrt[2.35/(PWB thickness)]*[550 + 2.2e6/(strain rate + 4000)]

Units:
Strain [uStrain]
Strain rate [uStrain/sec]
PWB thickness [mm]

FIGURE 59.12 Typical example of strain/strain rate limits as a function of PWB thickness for BGA solder joint interconnect failure. Note that the acceptable strain limit decreases with increasing strain rate. Note also that for a given strain rate, the maximum acceptable strain limit decreases as the PWB thickness decreases. *(Reprinted with permission from Sun Microsystems, Inc.)*

force-to-failure value compared to Pb/Sn joints with the same surface finish. Another important factor in the bend strength of solder joints is the time after reflow, as shown in Fig. 59.14[37].

The results in Fig. 59.14 show the following:

- For ENIG surface finish packages, Pb/Sn joints are more sensitive to time after reflow than SnAgCu joints. The average force-to-failure value of Pb/Sn joints with ENIG increased by more than 50 percent, whereas the average force-to-failure value of SnAgCu with ENIG did not change noticeably over time.
- For solder pre-coat (SOP) packages, the strength of Pb/Sn joints increased by slightly more than 10 percent over three weeks, whereas the strength of SnAgCu joints with SOP does not change noticeably over time.

Thus, mechanical bend strength is a function of the time after reflow along with solder metallurgy and surface finish. The underlying failure mechanisms have not been understood clearly enough to explain the dependence on time after reflow.

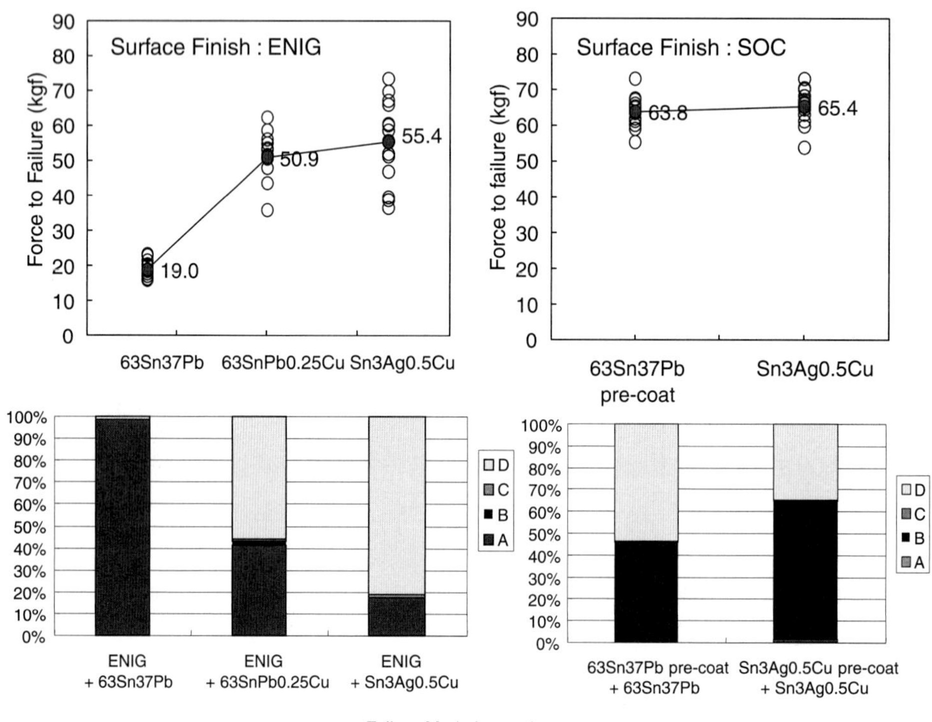

Failure Mode Legend:

Mode A: within the solder joint on the substrate side

Mode B: pad lift on the substrate side

Mode C: within the solder joint on the board side

Mode D: pad lift on the board side

FIGURE 59.13 Data set comparing the force-to-failure rate of Pb/Sn, Pb/Sn/Cu and Sn/Ag/Cu assemblies for two different surface finishes (ENIG and SOC). The histograms show a comparison of the different failure modes obtained in the bend tests.[37] All samples were tested within one day of board assembly. *(Reprinted with permission from Kyocera SLC Technologies. /© [2005] IEEE.)*

59.3.2 Cyclic Bend Testing

In addition to monotonic loading conditions, certain repetitive operations could also lead to mechanical failures. A typical example is a PWB used in cell-phone applications. The PWB could be subjected to multiple small deflections each time the cell-phone keypad is pressed. This can cause an incremental deterioration in the mechanical response of the PWB and lead to solder joint interconnect failures in BGAs mounted on the PWB. This sort of failure mechanism is more prevalent in relatively thin PWBs used in cell-phone applications. However, similar failures could also occur if, for example, multiple runs of ICT are performed on PCAs in high-end server/network applications. An industry standard test method has been developed by JEDEC to characterize the response of a PCA to cyclic loading conditions (JESD22-B113).[38] The layout of the test board as specified in the test method is similar to that for the monotonic four-point bend test. Some data on the cyclic bend response of SnAgCu solder joints can be found in Reference 39.

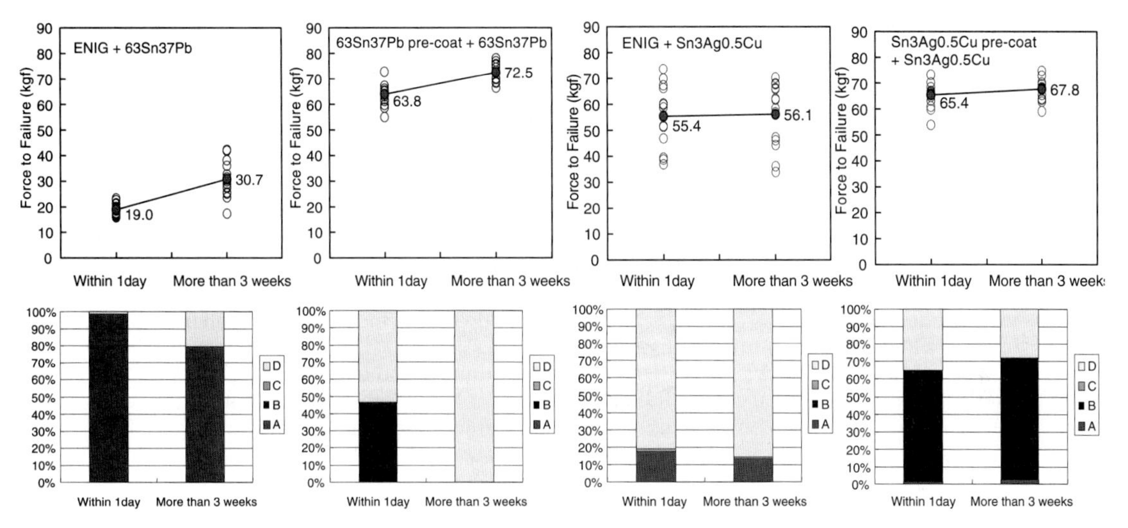

Failure Mode Legend:
Mode A: within the solder joint on the substrate side
Mode B: pad lift on the substrate side
Mode C: within the solder joint on the board side
Mode D: pad lift on the board side

FIGURE 59.14 Data set comparing the force-to-failure value of Pb/Sn and Sn/Ag/Cu assemblies for two different surface finishes (ENIG and SOP) as a function of time after reflow. The histograms show a comparison of the different failure modes obtained in the bend tests. *(Reprinted with permission from Kyocera SLC Technologies. /© [2005] IEEE.)*

59.3.3 Shock Testing

As with bend testing, a shock test methodology is needed to simulate the shock conditions that a PCA may undergo during assembly, shipping, and handling conditions. Shock test conditions usually induce strain rates on the order of 100,000 $\mu\varepsilon$/sec. (microstrain per second), whereas typical monotonic bend test strains are on the order of 10,000 $\mu\varepsilon$/sec. Thus, the strains to failure under typical shock test conditions are lower than the strains to failure under typical bend test conditions. However, the failure modes could be different. Hence the need to characterize both bend and shock conditions. There are different classifications of typical shock conditions. For example, those experienced by handheld products (cell-phones, personal digital assistants [PDAs], etc.) are different from those experienced by typical network/server products. The former requires resistance to more stringent shock levels (~1500 G),[40] whereas the latter requires resistance to shock levels on the order of ~500 G. An example of the typical shock test conditions applicable to these product categories is summarized in Table 59.6.

TABLE 59.6 Examples of Typical Shock Conditions Used for Different End-Use Environments (the Specific Conditions Tested Depend on the End Users)

	Typical field conditions	
Test parameters	Handheld products	Network/server products
Acceleration peak (G)	1,500–2,900	100–400
Pulse duration (ms)	0.3–0.5	1.2–2

In addition to shock testing, several component suppliers also perform drop testing. For handheld products in particular, some suppliers prefer to perform board level drop testing instead. Details of the drop test method are outlined in JESD22-B111.[41] Several component suppliers perform vibration testing in addition to shock testing to simulate shipping conditions.

The design of the carrier tray or chassis in which the PWB is installed could significantly impact the reliability of printed circuit board assemblies (PCBAs). If the resonant frequency of the PCBA matches that of the chassis, and the mode shapes corresponding to the resonant frequency are such that they induce a strain above the critical solder joint failure strain, brittle fracture can occur. Accelerometers and strain gages should be used to characterize both mode shapes and strains in the PWB assemblies mounted on carrier trays or in chassis. The chassis design should be optimized to ensure that the strains that might impact the PCBAs are minimized, especially at the critical locations. Ways to do this include adding stiffeners, frames, and struts in the chassis at critical, high-strain locations.

Finally, an additional way to mitigate shipping failures is to use shock indicators in the shipped product. A typical shock indicator changes color if the applied G values exceed a preset threshold. The use of a shock indicator in products can help in root cause analysis of failed parts.

59.3.4 Ball Shear/Pull Testing

Ball shear testers are employed to quantify the strength of the interface between a solder ball and its package prior to second-level attach. Ball shear experiments can also be employed to assess the solder ball-package pad interface qualitatively. This is accomplished by assessing the failure mode observed during a ball shear experiment.

However, certain failure modes are a function of the speed of the test. The failure modes observed at relatively low speeds (0.0001 to 0.0006 m/sec.) are different from those observed at high speeds (0.01 to 1.0 m/sec.). In general, solder shear strength tends to increase with shear speed. It is therefore important that failure mode characterization as a function of shear speed be performed before this technique can be used effectively as an inline monitoring tool to control the quality of substrate lots. To address this, JEDEC has published a revised version of the standard ball shear test to include high speed shear testing procedures (JESD22-B117A[42]).

In addition to speed, a number of other factors could impact the fracture strength of solder joints, including surface finish, pad size, ball diameter, ball metallurgy, and pad construction. All these need to be taken into account when comparing ball shear data across different packages. Another important factor that could impact the force-to-failure and the failure modes induced is the time after reflow. It is preferable to perform the test as early as possible after reflow.[43]

In addition to the high-speed shear/pull test, other methodologies are also being explored to characterize the interface strength of solder joints. One is a Charpy impact type of test.[44] This involves impacting the ball with a pendulum. The amount of energy imparted into the ball is estimated and the failure modes as a function of impact toughness are catalogued. This test is a sort of mini-shock test designed to produce an effect comparable to the high-speed shear test. Another methodology is the use of a hot pin pull test in which a heated pin is inserted into the solder ball, cooled, and then pulled at high speed. Efforts are being made to measure the force to cause failure, so that pull/shear force or impact energy could be used as a metric to assess the strength or weakness of a joint.

59.3.4.1 Effect of Lead-Free Conversion. Depending on the surface finish, the intermetallics formed in SnAgCu solders are different from those formed in Pb/Sn joints. Refer to Sec. 60.3.11 in Chap. 60 for more details. Preliminary results on high-speed pull/shear testing of SnAgCu joints indicate a higher interfacial fracture rate than Pb/Sn joint. However, the SnAgCu joints exhibited moderately higher fracture strength.[43]

59.4 FINITE ELEMENT ANALYSIS (FEA)

FEA is a numerical technique used to solve complex boundary value problems. It involves discretizing a complex geometry into smaller simpler geometries that can be represented by a set of differential equations. These equations can then be solved numerically to determine the deformation of the overall structure subjected to structural, thermal, or other loading conditions. FEA is a powerful numerical tool that has been used to assess the reliability of solder joints. FEA can be employed to:

- Estimate the fatigue life of a new package
- Compare reliabilities of different packaging options
- Assess impact on reliability due to material and design changes to the package
- Assess impact on reliability due to PWB material and design changes
- Determine changes in warpage due to package density and location

FEA has been used effectively in simulating thermomechanical, bending, and shock conditions. This section focuses on the use of FEA to determine the lifetime of packages subjected to thermomechanical strains, and the use of bend test simulations to determine the factors that influence the bend strength of a package.

59.4.1 FEA for Thermomechanical Lifetime Assessment

FEA lifetime assessments are typically broken down into four phases:

- Phase 1: global model
- Phase 2: local model
- Phase 3: single-joint lifetime assessment
- Phase 4: statistics-based package reliability estimate

Figure 59.15 shows the interrelationships between these four phases.

59.4.1.1 Phase 1: Global Model. Phase 1 involves building a global geometric model of the package. Models in FEA can be two-dimensional, 2.5-dimensional (strip), or three-dimensional. Two-dimensional models (plane strain or plane stress) are computationally very efficient as only a two-dimensional cross section of the assembled package is modeled. However, they are based on the assumption that the out-of-plane deformation and stresses in the assembly are negligible, which is not the case with discrete solder joints. Thus, although they are useful for providing qualitative deformation and stress comparisons, they are not useful for quantitative fatigue-life estimates. A 2.5-dimensional or strip model has a thickness, but it is very small—just enough to capture the round shape of the solder joints. While it is an improvement over a two-dimensional model, it does not capture the increased stress concentration at package corners. A three-dimensional model is a much closer representation of an assembled package. Although it is computationally challenging, it is the best way to estimate the deformation and stresses induced in an assembled package. A global model of one-eighth of a BGA package can be found in Figs. 59.16.

Note that only one-eighth of the package is modeled (Fig. 59.16). This is due to the fact that the package is octant symmetric, and thus symmetry boundary conditions can be employed to reduce the model size, which in turn increases the speed at which a solution can be obtained.[46] Wherever possible, the deformation results of the global FEA model should be verified with experimental warpage measurement (using tools like Shadow Moiré[45]).

A unit temperature rise is applied to the global model, the worst-case joint is determined, and the displacements of the worst-case joints are calculated. Worst-case joint determination

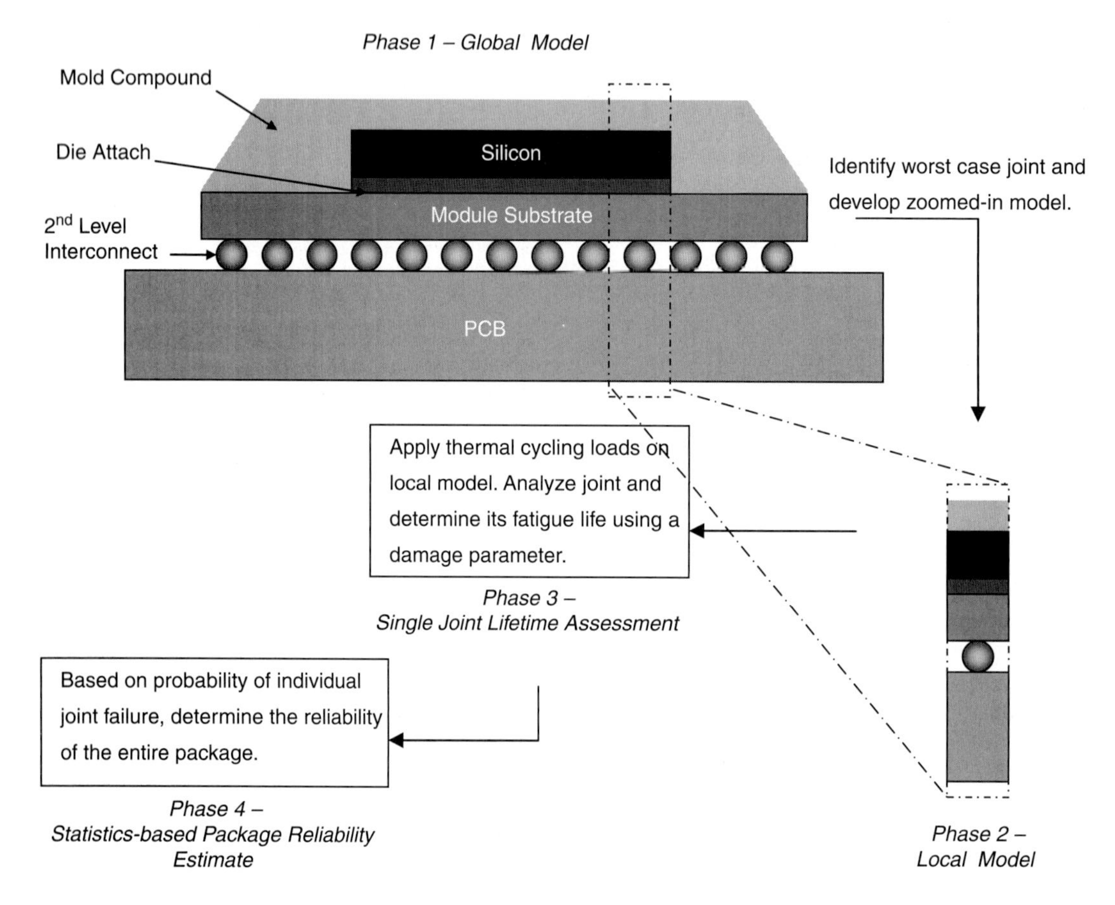

FIGURE 59.15 FEA flowchart. Phase 1 involves creating the global model to determine global strains and displacements across the entire package. The global model also helps identify the locations of maximum stress. Phase 2 entails zooming into the location with the maximum strain. The boundary conditions applied around the local model are extracted from the global model. Phase 3 involves applying thermal cycling loads on the local model and determining the fatigue life of the worst-case solder joint in the local model. Phase 4 entails the use of statistical assessment tools to determine the reliability of the other solder joints relative to the worst case and thus estimating the solder joint reliability of the entire package.

is typically based on a maximum stress or strain in the solder joints. It is essential to obtain the displacements of the worst-case joint because those displacements can now be incorporated in a local model that will simulate multiple thermal cycles on a worst-case joint.

Only a unit temperature rise is applied to the global model to reduce the computational run time. An additional assumption employed in global models to reduce run time is that all materials are linear-elastic and temperature-independent. Because the goal of the global model is to determine the worst-case joint and the corresponding displacement, this approach is typically valid. Reference 46 contains a series of experimental cases and corresponding FEA predictions that validate the global/local approach and the linear elastic, temperature-independent unit temperature rise technique for the packages studied.

59.4.1.2 Phase 2: Local Model. A local model of the worst-case joint is created and subjected to simultaneous temperature cycle and temperature-scaled displacement-loading conditions obtained from the global model. The displacement field of the worst-case joint obtained

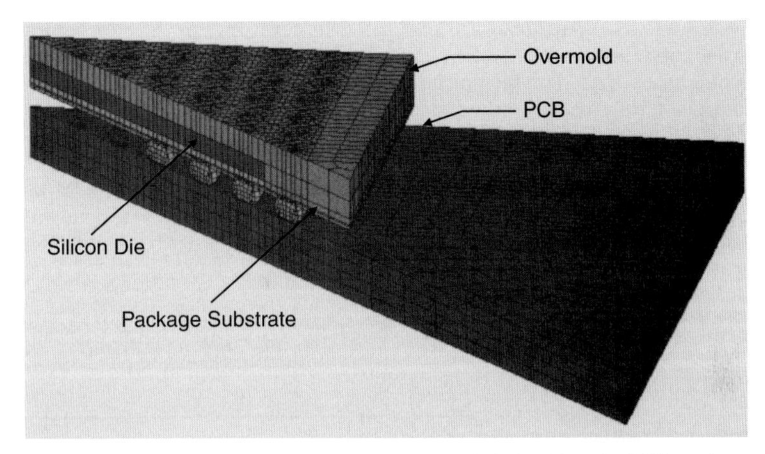

FIGURE 59.16 Global FEA model of one-eighth of a BGA. Note the PWB, package substrate, die, and overmold are all created by a series of elements.

in the global model when subjected to a unit temperature rise is scaled to match the expected displacement when the local model is subjected to actual thermal cycles. The advantage of this approach is that the local model can incorporate time and temperature material dependence because a reduced number of elements are employed when compared to the global model.

A local model of a solder joint can be found in Fig. 59.17. The local model is subjected to multiple thermal cycles and a final damage parameter is extracted. A typical thermal cycling loading profile to which the model is subjected is similar to that shown in Fig. 59.3. About two to four cycles are run to achieve stability of the hysteresis loop.

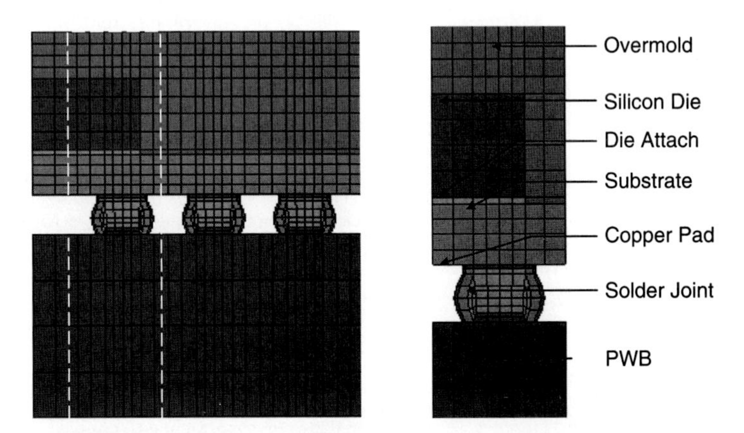

FIGURE 59.17 Close-up of three solder joints. The local model is created from one of the solder joints and its associated PWB, package substrate, die, and overmold. In this figure, the leftmost ball and associated material for the local model are indicated by the area inside the dashed rectangle.

Plastic work and plastic strain are two of the most common local damage parameters employed in solder joint lifetime estimations. The type of damage parameter selected is usually dictated by the experimental data available for a specific joint type (BGA, CCGA) that correlate damage with fatigue life. A methodology that employs plastic work and a correlation between plastic work and crack initiation and propagation can be found in References 10 and 46.

59.4.1.3 Phase 3: Solder Joint Fatigue-Life Assessment. The local model damage parameter is then employed to determine the fatigue life of a given joint. Techniques for assessing lifetime based on a damage parameter range from employing plastic strain (Coffin-Manson relationship, Eq. 59.8) to complex considerations of crack initiation and crack growth rates correlated with plastic work.[10, 46] Regardless of the technique, the output of this phase is an estimate of the reliability of one joint: the worst-case solder joint.

Some of the fatigue-life prediction models that have been employed for lead-free solder joints are outlined in Table 59.7. In addition, other models for predicting fatigue life from the finite element models and performing relative comparative analysis can be found in References 16 and 47.

TABLE 59.7 Different Fatigue-Life Models for Predicting Fatigue Life from the Finite Element Model Results[18]

Fatigue-life model	Constants	Test conditions	Reference
$N_f = C_1(\varepsilon_{cr})^{C_2}$ N_f = Number of cycles to fail ε_{cr} = Average equivalent creep strain	$C_1 = 4.5$ $C_2 = -1.295$	Flip-chip bumps with and without underfill. Wirebonded PBGA packages mounted on PWB	Schubert[48]
$N_f = C_1(\Delta W_{cr})^{C_2}$ W_{cr} = Average viscoplastic strain energy density/cycle	$C_1 = 345$ $C_2 = -1.02$	Flip-chip bumps with and without underfill. Wirebonded PBGA packages mounted on PWB.	Schubert[48]
$N_f = C_1(\varepsilon_{cr})^{C_2}$	$C_1 = 0.0468$ $C_2 = -1$	Several different BGA packages: CABGA, CBGA, FlexBGA, PBGA, TSCSP, etc. Double power law creep constitutive relations.[20]	Syed[49]
$N_f = C_1(\Delta W_{cr})^{C_2}$	$C_1 = 0.0015$ $C_2 = -1$	Several different BGA packages: CABGA, CBGA, FlexBGA, PBGA, TSCSP, etc. Double power law creep constitutive relations.[20]	Syed[49]
$1/N_f = 1/N_{fp} + 1/N_{fc}$ $N_{fp} = (W_p/W_{p0})^{1/c}$ $N_{fc} = (W_c/W_{c0})^{1/d}$ N_{fp} = Number of cycles to failure due to plastic damage N_{fc} = Number of cycles to failure due to creep damage W_p = Plastic work density calculated per load cycle W_c = Creep work density calculated per load cycle	$W_{p0} = 106.45$ $W_{c0} = 30.025$ $c = -0.51$ $d = -0.44$	Solder joint samples attached to copper coupons, subjected to cyclic loading. One-dimensional analytical model to derive constants.	Zhang[50]

A comparison of the results obtained by Schubert's[48] model with those obtained by Syed's[49] model can be found in Reference 51. It is important to note that these models are a function of the way that the material properties of the solder joints are modeled. In other words, the solder material properties used to derive these models play a significant role in the accuracy of the models. If a particular fatigue-life model is selected, the corresponding solder constitutive model used to derive the model constants should be used in the analysis.

In addition, several other factors could significantly alter the accuracy of a model, such as time step size, mesh density, material properties, reference temperature, element size, element type, and even software. All these need to be taken into account when using a model to predict fatigue life. Best practice recommendation is to run not just one, but a set of fatigue-life models and calibrate with existing experimental data before using the models to make fatigue-life predictions on packages on which no experimental data are available. Finally, it is important to note that these models only provide guidance on trends and relative comparisons. They should not be used as a substitute for experimental thermal cycle data for a given package under consideration.

59.4.1.4 Phase 4: Statistics-Based Package Reliability Assessment. This final phase considers the reliability of all of the solder joints when determining the reliability of the package. Techniques employed here range from simply assuming that all joints are of equal reliability to performing a series reliability calculation to solve simultaneously an assumed distribution of failures (such as a Weibull distribution; see Sec. 59.2.2) considering each joint. It is also common to weigh the reliability of joints other than the worst-case joint by considering the displacement of all joints with respect to the displacement of the worst-case joint.[46] This allows for the impact of DNP to be incorporated in the potential distribution of failures among joints at different locations.

59.4.1.5 Temperature-Dependent Material Properties. As mentioned earlier, the aforementioned procedure is based on the assumption that the mechanical properties of the package material set do not change significantly with temperature. There are certain packages, however, for which this technique is not applicable. A typical example is an FCBGA package in which the underfill material has a glass transition temperature (T_g) around 70 to 100°C and undergoes several orders of magnitude change in modulus above the T_g. Clearly, the properties of such a material are strongly dependent on temperature. In such a case, the following procedure can be followed:

1. The global model is cooled from its stress-free temperature (which is usually the adhesive cure temperature) to the lowest temperature to which the package has been thermal cycled (0°C if a 0 to 100°C thermal profile is used).
2. Cut boundary interpolation is performed at small temperature steps (such as 5 or 10°C), recording the deformation of the nodes at each temperature step. This information is stored in a set of data files.
3. A submodel is created of the worst-case solder joint (the one that shows the most stress or deformation in the global model). The files stored in step 2 are applied as load boundary conditions on the nodes forming the submodel boundary at each temperature step. Temperature-dependent material properties should be used in both models.
4. On completing the analysis, one can use the plastic work or strain accumulated to determine the expected fatigue life of the package based on the worst-case solder joint, as outlined in Phase 3 of the previous section.

If computer resources are available, a complete global model could be used to determine the plastic work or strain per solder joint. However, this could be very time-consuming depending on the complexity of the package under consideration.

59.4.1.6 Impact on First-Level Interconnect. In FCBGA-type packages using temperature-dependent material properties, failure resulting from solder joint fatigue may occur not

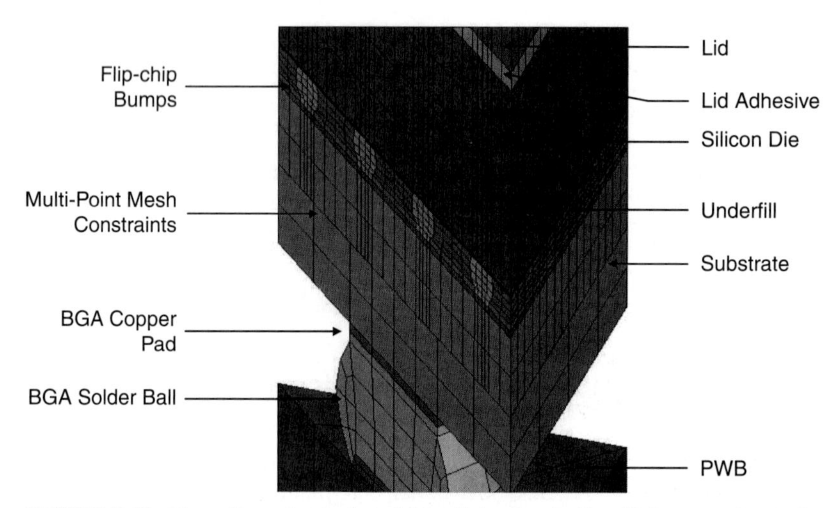

FIGURE 59.18 Three-dimensional submodel detail showing the flip-chip bumps and underfill included in the model. The submodel can be used to determine the fatigue life of the flip-chip bumps and the BGA solder joint underneath at the same time.

only at the board-level interconnect but also at the first-level interconnect (between the silicon die and substrate). The flip-chip bumps are encapsulated by the underfill to improve their thermomechanical reliability by evenly distributing the CTE-mismatch-induced strains between the die and substrate. If the underfill material properties are such that the underfill softens drastically above its T_g, then there is a possibility that the flip-chip bumps may fail early during thermal cycling.[52] To simulate this, the global-submodel approach outlined earlier can be used. In this case, the flip-chip bumps would also be incorporated in the submodel as shown in Fig 59.18.

59.4.2 FEA for Monotonic Bend Testing

FEA simulations of monotonic bend testing involve a quarter-symmetry model of the assembly placed between anvils and deformed while monitoring the strains induced in the package, solder joint interconnects, and the PWB. The details of the quarter-symmetry model are shown in Fig. 59.19.

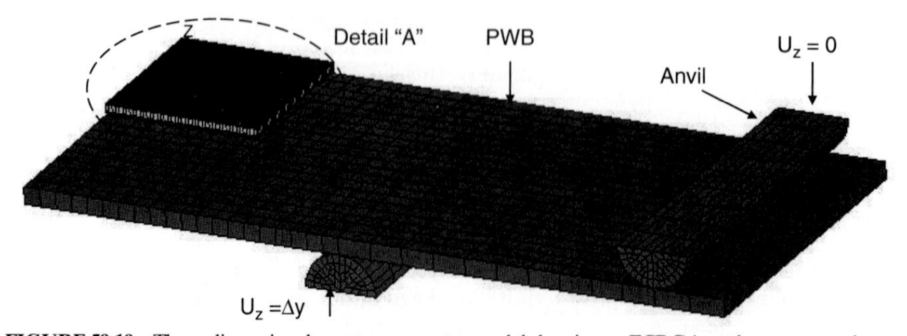

FIGURE 59.19 Three-dimensional quarter-symmetry model showing an FCBGA package mounted on a PWB between support anvils. Contact elements between the anvils and the PWB simulate the flexure of the PWB as one anvil set is moved vertically relative to the other. Detail A of the model is shown in Fig. 59.20.

Bend test FEA simulations achieve two primary objectives:

- To determine the relationship between PWB strain and critical joint strain. This in turn helps bridge differences between the controlled conditions in which the bend testing is performed and the conditions in which bending occurs during assembly.
- To perform a comparative assessment of different packages. FEA can be used to derive solder joint strains from experimental PWB strain data for different test conditions.

The FEA model is shown in Figs. 59.19 and 59.20. Solid structural elements are used to model the PWB, solder joints, and package details. Based on the premise that at high strain rates, the solder material becomes brittle and behaves almost like a linear elastic material, the solder joint elements are modeled as linear elastic. Contact elements are used at the interface between the PWB and anvils to simulate contact between the PWB and the anvils as the assembly is deformed. The solder joint geometry is modeled as closely as possible to the solder joints' actual shape after reflow.

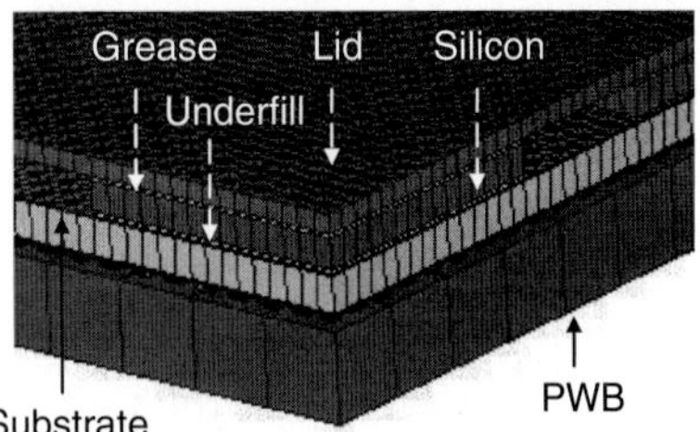

FIGURE 59.20 Detail A of the image model shown in Fig. 59.19. The image on the left shows the package (quarter-symmetry). The zoomed-in image on the right shows all the details of the FCBGA package construction.

A variety of different parametric analyses can be performed to determine the effect of different geometric variables on the critical joint strain. A few examples of such analyses are described in this section.

59.4.2.1 Strain Distribution Around the Package. Per IPC/JEDEC-9702, it is recommended that three strain gauges be placed around a package during four-point bend testing (see Fig. 59.21). These gauges are to be used to characterize the strain response of the given package

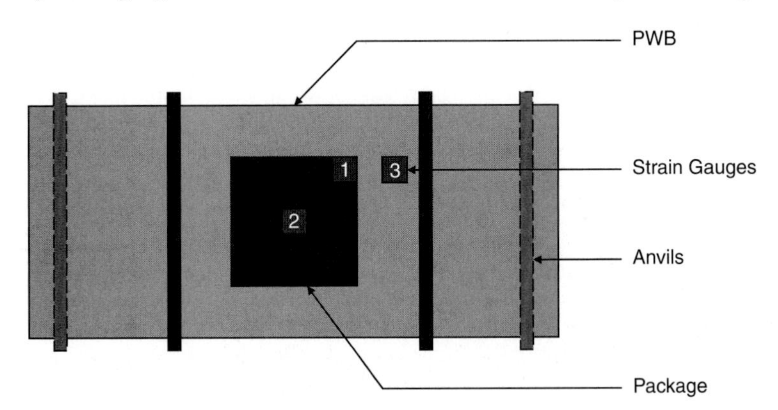

FIGURE 59.21 Nomenclature for the four-point bend test set up as outlined in IPC/JEDEC-9702. Three strain gauges are placed on the board to measure the strain distribution around the package. Gauge 1 is placed under the package, directly below the corner most solder joint. Gauge 2 sits directly below the package center. Gauge 3 is placed next to the package on the PCB topside, midway between the package edge and the anvil.

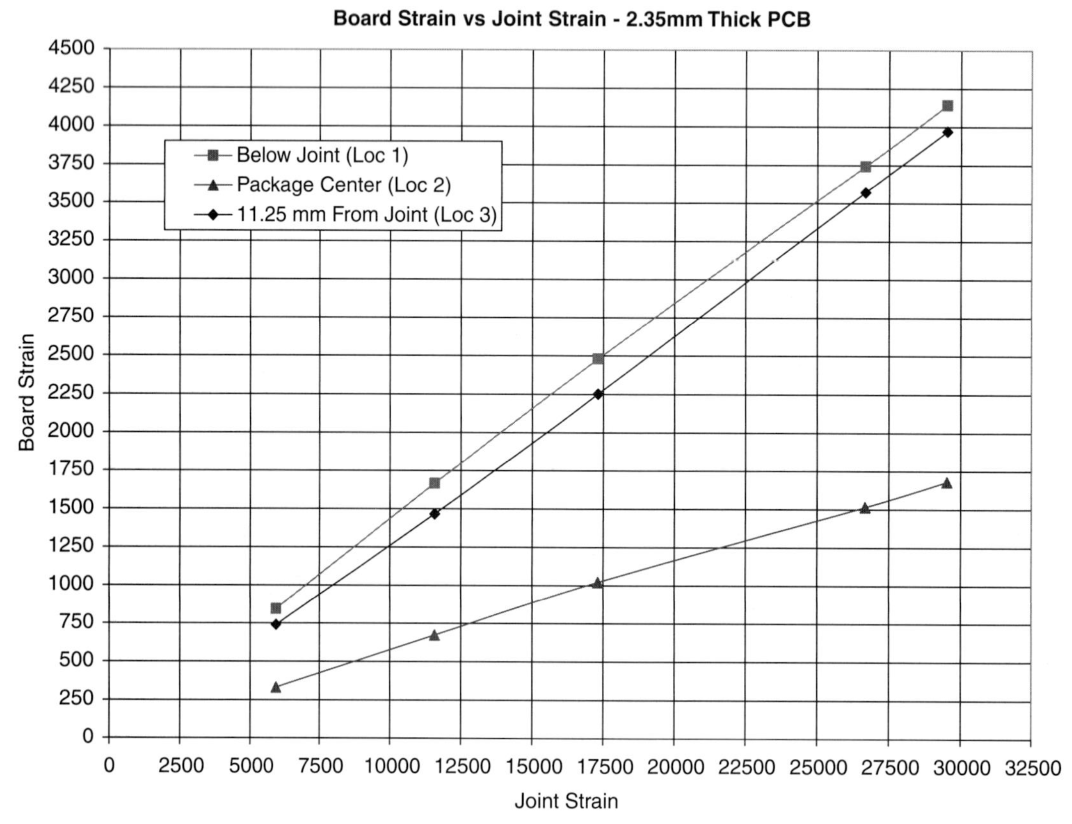

FIGURE 59.22 Solder joint-to-PWB strain relationship for three gauge locations. The results indicate that the gradient at locations 1 and 3 are relatively close. The gradient at location 2 is quite low. This indicates that the PWB strain at location 2 is least sensitive to variations in critical solder joint strain. The nomenclature for the strain gauge locations is given in Fig. 59.21.

mounted on a board of a certain thickness. FEA was used to determine the joint-to-board strain ratio for each case, which in turn can be used to determine which strain gauge location is most or least sensitive to board flexure. The results of the simulation are shown in Fig. 59.22.

The results indicate that the highest sensitivity to solder joint strain is obtained at locations 1 and 3 (see Fig. 59.21). Location 2 reflects the global bending of the entire assembly. These results are in line with measured experimental data. The absolute strain value and the change in strain at location 2 is least sensitive to the critical solder joint failure strain.

59.4.2.2 Effect of Package Pad Size on Joint-to-Board Strain. FEA can also been used to estimate the effect of changes in solder joint pad size on the joint-to-board strain ratio. Two different package pad sizes were used: 0.53 mm, solder mask defined, (pad design 1) and 0.4 mm, solder mask defined, (pad design 2) while the PCB pad design was kept constant (0.5 mm, non-solder mask defined) The results in terms of the joint-to-board strain at location 3 are shown in Fig. 59.23.

The results in Fig. 59.23 indicate that pad size change could have a significant impact on the strain in critical solder joints. As a result, care should be taken in using data generated with one pad size to qualify or accept the bending strength with another pad size, unless the absolute strain-to-failure value is much higher than the required strength.

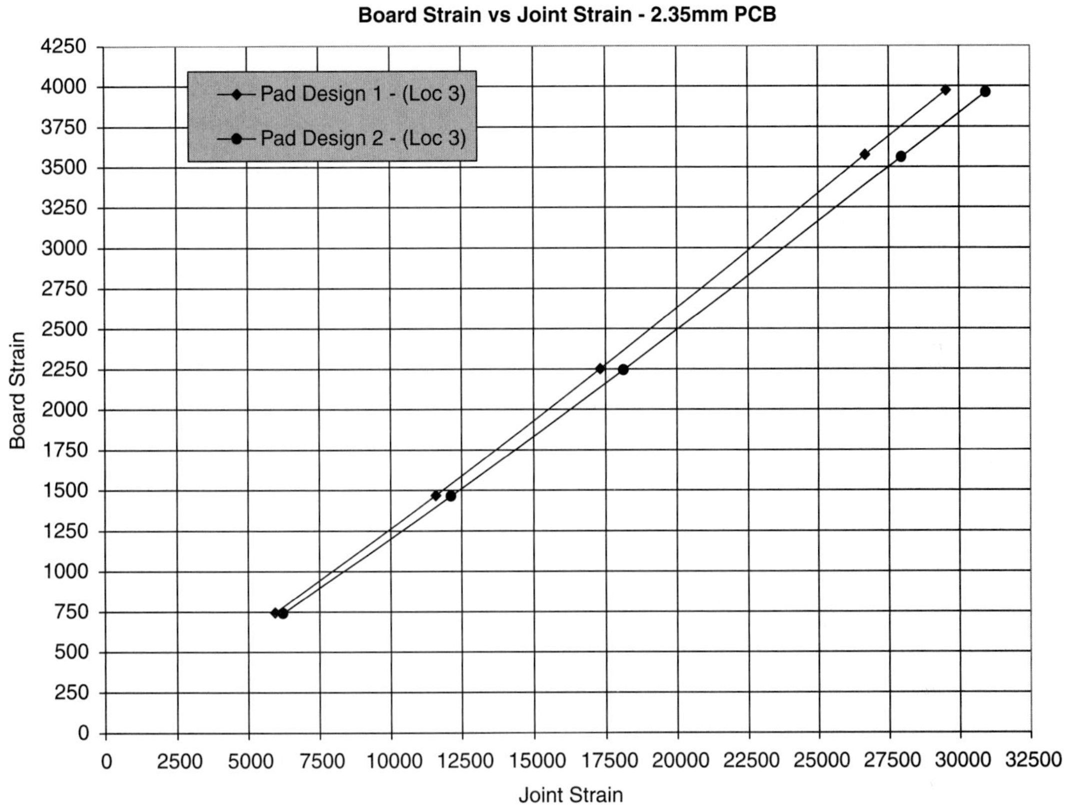

Board Strain vs Joint Strain - 2.35mm PCB

FIGURE 59.23 Two different solder joint pad sizes analyzed. The results indicate that for the same PWB strain, the solder joint strain is ~18 percent higher for pad design 2 as compared to pad design 1.

REFERENCES

1. Zienkiewicz, O. C., Taylor, R. L., and Zhu, J. Z., *The Finite Element Method: Its Basis and Fundamentals, 6th ed.*, Butterworth-Heinemann, 2005.
2. PC-9701, "Performance Test Methods and Qualification Requirements for Surface Mount Solder Attachments," IPC, January 2002.
3. IPC-9701A, "Performance Test Methods and Qualification Requirements for Surface Mount Solder Attachments," IPC, February 2006.
4. Clech, J-P., Noctor, D. M., Manock, J. C., Lynott, G. W., and Bader, F. E., "Surface Mount Assembly Failure Statistics and Failure Free Time," *Electronic Components and Technology Conference*, 1994, pp. 487–497.
5. Weibull, W., "A Statistical Distribution Function of Wide Applicability," *Journal of Applied Mechanics*, September 1951, pp. 293–297.
6. Reliasoft Corporation, "Weibull Probability Density Function," 1996–2006, available online at http://www.weibull.com/LifeDataWeb/weibull_probability_density_function. htm.

7. Idem, "What Are Confidence Bounds?" 1996–2006, available online at http://www. weibull.com/LifeDataWeb/what_are_confidence_intervals_or_bounds_.htm.

8. Lau, J. H. (ed.), *Thermal Stress and Strain in Microelectronics Packaging*, Van Nostrand Reinhold, 1993.

9. Tummala, R. R., Rymaszewski, E. J.,and Klopfenstein, A. G. (eds.), *Microelectronics Packaging Handbook*, Chapman and Hall, 1997.

10. Lau, J. H. (ed.), *Ball Grid Array Technology*, McGraw-Hill, 1995.

11. Darveaux, R., "Effect of Simulation Methodology on Solder Joint Crack Growth Correlation," *Electronic Components and Technology Conference*, 2000, pp. 1048–1058.

12. Anand, L., "Constitutive Equations for Hot-Working of Metals," *International Journal of Plasticity*, Vol. 1, 1985, pp. 213–231.

13. Bradley, Edwin (Motorola), Handwerker, Carol (NIST), and Sohn, John E., "NEMI Report: A Single Lead-Free Alloy is Recommended," *SMT*, January 2003, cover story.

14. Pei, M., and Qu, J., "Constitutive Modeling of Lead-Free Solders," *Proceedings of IPACK 2005*, 73411, 2005.

15. Rodgers, Bryan, Flood, Ben, Punch, Jeff, and Waldron, Finbarr, "Determination of the Anand Viscoplasticity Model Constants for SnAgCu," *Proceedings of IPACK 2005*, 73352, 2005.

16. Ng, Hun Shen, Tee, Tong Yan, Goh, Kim Yong, Luan, Jing-en, Reinikainen, Tommi, Hussa, Esa, and Kujala, Arni, "Absolute and Relative Fatigue Life Prediction Methodology for Virtual Qualification and Design Enhancement of Lead-Free BGA," *Electronic Components and Technology Conference*, 2005, pp. 1282–1291.

17. Clech, Jean-Paul, "An Extension of the Omega Method to Primary and Tertiary Creep of Lead-Free Solders," *Electronic Components and Technology Conference*, 2005, pp. 1261–1271.

18. Shangguan, Dongkai, *Lead-Free Solder Interconnect Reliability*, ASM International, July 2006, pp. 185–198.

19. Weise, S., and Muesel, E. "Characterization of Lead-Free Solders in Flip Chip Joints," *Journal of Electronic Packaging*, Vol. 125, December 2003, pp. 531–538.

20. Weise, S., Muesel, E., and Wolter, K. J., "Microstructural Dependence of Constitutive Properties of SnAg and SnAgCu Solders," *Electronic Components and Technology Conference*, 2003, pp. 197–206.

21. Pang, J. H. L., Xiong, B. S., and Low, T. H., "Creep and Fatigue Characterization of Lead-Free 95.5Sn-3.8Ag-0.7Cu," *Electronic Components and Technology Conference*, 2004, pp. 1333–1337.

22. NIST, "Materials for Microelectronics," , March 2004, available online at http://www. metallurgy.nist.gov/solder/.

23. Pan, N., Henshall, G. A., Billaut, F., Dai, S., Strum, M. J., Lewis, R., Benedetto, E., and Rayner, J., "An Acceleration Model for Sn-Ag-Cu Solder Joint Reliability under Various Thermal Cycle Conditions," *SMTA International*, 2005, pp. 876–883.

24. Clech, Jean-Paul, "Acceleration Factors and Thermal Cycling Test Efficiency for Lead-Free Sn-Ag-Cu Assemblies," *SMTA International*, 2005.

25. Bartelo, J., "Thermomechanical Fatigue Behavior of Selected Lead-Free Solders," *IPC SMEMA Council, APEX*, San Diego, January 14–18, 2001, LF2-2.

26. Sahasrabudhe, S., Monroe, E., Tandon, S., and Patel, M., "Understanding the Effect of Dwell Time on Fatigue Life of Packages Using Thermal Shock and Intrinsic Material Behavior," *Electronic Components and Technology Conference*, 2003, pp. 898–904.

27. Setty, Kaushik, Subbarayan, Ganesh, and Nguyen, Luu, "Powercycling Reliability, Failure Analysis and Acceleration Factors of Pb-Free Solder Joints," *Electronic Components and Technology Conference*, 2005, pp. 907–915.

28. Bath, Jasbir, Sethuraman, Sundar, Zhou, Xiang, Willie, Dennis, Hyland, Kim, Newman, Keith, Hu, Livia, Love, Dave, Reynolds, Heidi, Kochi, Ken, Chiang, Diana, Chin, Vicki, Teng, Sue, Ahmed, Mudasir, Henshall, Greg, Schroeder, Valeska, Nguyen, Quang, Maheswari, Abhay, Lee, M. J., Clech, Jean-Paul, Cannis, Jeff, Lau, John, and Gibson, Chris, "Reliability Evaluation of Lead-Free SnAgCu PBGA676 Components Using Tin-Lead and Lead-Free SnAgCu Solder Paste," *SMTA International*, 2005, pp. 891–901.

29. Bieler, T. R., Jiang, H., Lehman, L. P., Kirkpatrick, T., Cotts, E. J., "Influence of Sn Grain Size and Orientation on the Thermomechanical Response and Reliability of Pb-free Solder Joints," *Electronic Components and Technology Conference*, 2006, pp. 1462–1467.
30. Darveaux, Robert, "Shear Deformation of Lead Free Solder Joints," *Electronic Components and Technology Conference*, 2005, pp. 882–893.
31. Hua, Fay, Aspandiar, Raiyo, Rothman, Tim, Anderson, Cameron, Clemons, Greg, and Klier, Mimi, "Solder Joint Reliability of Sn-Ag-Cu Bga Components Attached with Eutectic Pb-Sn Solder Paste," *Journal of SMT*, Vol. 16, No. 1, 2003, pp. 34–42.
32. Nelson, Dave, Pallavicini, Hector, Zhang, Qian, Friesen, Paul, and Dasgupta, Abhijit, "Manufacturing and Reliability of Pb-Free and Mixed System Assemblies (SnPb/Pb-Free) in Avionics Environments," *Journal of SMT*, Vol. 17, No. 1, 2004, pp. 17–24.
33. Miner, M. A., "Cumulative Damage in Fatigue," *Journal of Applied Mechanics*, Vol. 12, *Transactions of the ASME*, Vol. 67, 1945, pp. A159–A164.
34. Hoffmeyer, M., Farooq, M., "Reliability and Microstructural Assessment of Hybrid CBGA Assemblies", paper presented at the 39th International Symposium on Microelectronics, IMAPS 2006, San Diego, October 8–12, 2006.
35. IPC/JEDEC-9702, "IPC/JEDEC Monotonic Bend Characterization of Board-Level Interconnects," June 2004.
36. IPC/JEDEC-9704, "Printed Wiring Board Strain Gage Test Guideline," June 2005.
37. Nakamura, Tomoko, Miyamoto, Yoshimasa, Hosoi, Yoshihiro, and Newman, Keith, "Solder Joint Integrity of Various Surface Finished Build-Up Flip Chip Packages by 4-Point Monotonic Bending Test," *Electronics Packaging Technology Conference*, 2005, pp. 465–470.
38. JEDEC Standard, JESD22-B113, "Board Level Cyclic Bend Test Method for Interconnect Reliability Characterization of Components for Handheld Electronic Products," March 2006.
39. Kim, I., and Lee, S-B, "Reliability Assessment of BGA Solder Joints under Cyclic Bending Loads," *International Symposium on Electronics Materials and Packaging, EMAP2005*, December 11–14, 2005, Tokyo Institute of Technology, Tokyo, Japan, pp. 27–32.
40. JEDEC Standard, JESD22-B110A, "Subassembly Mechanical Shock" November 2004.
41. JEDEC Standard, JESD22-B111, "Board Level Drop Test Method of Components for Handheld Electronic Products," July 2003.
42. JEDEC Standard, JESD22-B117A, "Solder Ball Shear," October 2006.
43. Newman, Keith, "BGA Brittle Fracture: Alternative Solder Joint Integrity Test Methods," *Electronic Components and Technology Conference*, 2005, pp. 1194–1201.
44. Ou, Shengquan, Xu, Yuhuan, Tu, K. N., Alam, M. O., and Chan, Y. C., "Micro-Impact Test on Lead-Free BGA Balls on Au/Electrolytic Ni/Cu Bond Pad," *Electronic Components and Technology Conference*, 2005, pp. 467–471.
45. Hassell, Patrick B., "Advanced Warpage Characterization: *Location* and *Type* of Displacement Can Be Equally as Important as Magnitude," *The Proceedings of Pan Pacific Microelectronics Symposium Conference*, February 2001.
46. Riebling, J. C., and Brillhart, M. V., "FEA Reliability Assessment Methodology Investigation to Improve Prediction Accuracy," *SMTA International*, 2000.
47. Lau, John H., Shangguan, Dongkai, Lau, Dennis C. Y., Kung, Terry T. W., and Lee, S. W. Ricky, "Thermal-Fatigue Life Prediction Equation for Wafer-Level Chip Scale Package (WLCSP) Lead-Free Solder Joints on Lead-Free Printed Circuit Board (PCB)," *Electronic Components and Technology Conference*, 2004, pp. 1563–1569.
48. Schubert, A., Dudek, R., Auerswald, E., Gollhardt, A., Michel, B., and Reichl, H., "Fatigue Life Models for SnAgCu and SnPb Solder Joints Evaluated by Experiments and Simulation," *Electronic Components and Technology Conference*, 2003, pp. 603–610.
49. Syed, Ahmer, "Accumulated Creep Strain and Energy Density Based Thermal Fatigue Life Prediction Models for SnAgCu Solder Joints," *Electronic Components and Technology Conference*, 2004, pp. 737–746. Corrected version available online at http://www.amkor. com/products/notes_papers/asyed_ectc2004_corrected.pdf.

50. Zhang, Qian, Dasgupta, Abhijit, and Haswell, Peter, "Viscoplastic Constitutive Properties and Energy-Partitioning Model of Lead-Free Sn3.9Ag0.6Cu Solder Alloy," *Electronic Components and Technology Conference*, 2003, pp. 1862–1868.

51. Stoeckl, Stephan, Yeo, Alfred, Lee, Charles, and Pape, Heinz, "Impact of Fatigue Modeling on 2d Level Joint Reliability of BGA Packages with SnAgCu Solder Bails," *Electronics Packaging Technology Conference*, 2005, pp. 857–862.

52. Ahmad, Mudasir, Teng, Sue, and Xue, Jie, "Effect of Underfill Material Properties on Low k Dielectric, First and Second Level Interconnect Reliability," paper presented at the IMAPS Conference on Device Packaging, Scottsdale, AZ, 2006.

P · A · R · T · 14

ENVIRONMENTAL ISSUES

CHAPTER 60
PROCESS WASTE MINIMIZATION AND TREATMENT*

Joyce M. Avery
Avery Environmental Services, Saratoga, California

Peter G. Moleux, P.E
Peter Moleux and Associates, Newton Centre, Massachusetts

60.1 INTRODUCTION

In the past, manufacturers of printed circuit boards have relied on end-of-pipe treatment and disposal for hazardous wastes generated in the fabrication process. These technologies are no longer optimal strategies for managing waste for two reasons. First, the potential liabilities involved with the handling and disposal of waste have increased and will continue to increase, and second, waste disposal costs have gone up significantly due to restrictions placed on land disposal. As a result, the industry is faced with the challenge of finding alternative methods for managing hazardous waste. This chapter presents a brief overview of some of the alternatives available to address this challenge, as well as a summary of some of the issues involved in implementation.

60.2 REGULATORY COMPLIANCE

Fabricators of printed circuit boards today are faced with a complex set of environmental requirements. In the United States there are three basic environmental statutes impacting the fabrication and assembly of printed circuit boards.

- Clean Water Act
- Clean Air Act
- Resource Conservation and Recovery Act (RCRA)

* This chapter is reprinted from the 4th edition. The basic issues, regulations, and processes are considered accurately stated and relevant to the 6th edition. For specific actions, however, it is recommended that waste treatment engineering and legal professionals be consulted to ensure that latest government expectations are understood at all levels of jurisdiction and that appropriate technology is applied to the resolution of specific issues.

60.2.1 Clean Water Act

The goals of the Clean Water Act are to "restore and maintain the chemical, physical, and biological integrity of the nation's waters." To accomplish these goals, discharges of industrial wastewater are subject to pretreatment requirements of federal, state, or municipal regulations. Industrial waste discharges are typically directed to a sewage treatment plant. Most sewage treatment plants use bacteria to biodegrade the organic matter present in the waste stream. Toxic materials such as copper, nickel, and lead from industrial discharges can pose a problem in two ways. These materials end up in the sludge from the sewage treatment process and can lead to disposal problems. Secondly, in high concentrations they can kill the bacteria in the treatment process, resulting in significant pollution of the receiving water. As a result, fabricators of printed circuit boards are required to pretreat their wastewater to specified levels prior to discharge to the sanitary sewer. The stringency of the requirements is ultimately determined by the use of the receiving water, as even minute amounts of toxics have been shown to have a negative impact on the aquatic environment. While the federal Clean Water Act specifies minimum pretreatment standards for fabricators of printed circuit boards, in most cases, state and local requirements may be more stringent. See Table 60.1 for an example of pretreatment requirements.

TABLE 60.1 Typical Pretreatment Requirements

Parameter	Limit, mg/L
pH	6.5–9.0
Copper	1.0
Nickel	0.5
Chromium	1.0
Silver	0.05
Cadmium	0.07
Zinc	0.5
Lead	0.2
Mercury	0.05
Aluminum	1.0
Selenium	0.2
Iron	2.0
Manganese	2.0
Tin	5.0
Cyanide	0.01
Phenol	0.05

60.2.2 Clean Air Act

The Clean Air Act established National Ambient Air Quality Standards (NAAQS) to achieve two goals:

1. Improve air quality in areas which fail to meet the standards.
2. Prevent significant deterioration of the air quality in clean air areas.

The states are responsible for achieving these standards by setting emission limitations and establishing timetables for compliance by sources. Printed circuit fabrication and assembly involves several processes that have an impact on air quality. Drilling, routing, sawing, and sanding create dust or airborne particulates. The plating process creates acid fumes and the etching process can generate ammonia if an ammoniacal etchant is used. Volatile organic compounds (VOCs) and lead particulates from the assembly process can pose a potential air pollution problem as well.

The technologies available for control of air pollutants include the following:

1. Electrostatic precipitators, baghouses (a type of dust collector with fabric bags mounted on frames), and cyclone separators are available to control airborne particulates.
2. Wet scrubbers containing a packed bed to provide surface area and water sprays are utilized for removing acid fumes. Addition of an acidic feed to the scrubbing liquid allows for the removal of ammonia. Addition of a caustic feed improves the scrubbing efficiency for other materials. Wet scrubbers are often used to prevent entrance of fumes back into the building from fresh air intakes.
3. Activated carbon filtration systems are utilized for removing chlorinated solvents and volatile organic compounds (VOCs). These can be regenerated on- or off-site.

60.2.3 Resource Conservation and Recovery Act

The Resource Conservation and Recovery Act (RCRA) goals are to protect human health and the environment by reducing or eliminating the generation of hazardous waste. To achieve this, the regulation mandates a system for managing hazardous wastes from cradle to grave. Anyone who generates, stores, treats, transports, or disposes of hazardous waste is subject to this regulation. The specific definitions of hazardous waste are spelled out in the federal statutes (40CFR Parts 260-280). Typically, however, the states have more stringent definitions. The states are responsible for the implementation and enforcement of RCRA. Wastes captured under this regulation in a printed circuit board fabrication facility include aqueous solutions with metals, acid or alkaline solutions with metals, sludges containing metals, etc. As a result, fabricators must comply with the following requirements.

1. Obtain an EPA identification number and apply for permits to generate, treat, store, or dispose of hazardous waste as appropriate.
2. Use appropriate containers for storage and disposal and approved manifests and labels for shipping.
3. Comply with technical requirements for on-site treatment of hazardous waste, including tank integrity standards, labeling, and secondary containment.
4. Keep records as appropriate for reporting to regulatory agencies.

A critical part of the regulation is to reduce and, where possible, eliminate the generation of hazardous waste. Waste minimization was specifically mandated in the 1984 Hazardous and Solid Wastes Amendments to the Resource Conservation and Recovery Act. This has had an enormous impact on the way waste is handled by printed circuit board facilities. Prevention of pollution has become the overriding goal in design with recycle and reuse technologies implemented only where pollution prevention is not feasible for technical and/or economic reasons. Chemical treatment of wastes should be utilized only where no other options exist.

60.3 MAJOR SOURCES AND AMOUNTS OF WASTEWATER IN A PRINTED CIRCUIT BOARD FABRICATION FACILITY

60.3.1 Major Sources of Waste

The following table indicates the major operations in the production of printed circuit boards with the waste streams they generate.

Source	Waste stream	Composition
1. Cleaning and surface preparation	Spent acid/alkaline baths Waste rinse waters	Metals, acids, alkalis
2. Electroless plating and deposition	Spent electroless copper bath Waste rinse water	Acids, palladium, complexed metals, chelating agents, formaldehyde
3. Pattern printing and masking	Spent stripper and developer solutions Waste rinse waters	Vinyl polymers, chlorinated solvents, organic solvents, alkalis
4. Electroplating	Spent plating bath Waste rinse waters	Metals, cyanide, sulfate
5. Etching	Spent etchant Waste rinse water	Ammonia, chromium, copper, iron, acids
6. Assembly	Aqueous and semiaqueous wastewaters	Lead, organics

60.3.2 Typical Amounts of Waste Materials

It is important to evaluate the waste generated in printed circuit board manufacturing in terms of the relative amounts of copper wasted from different processes in order to prioritize efforts at waste minimization. A waste audit should be performed to determine the types and amounts of waste generated. Figures 60.1 and 60.2 present two comparisons of copper-bearing wastes. The data presented are from two different printed circuit board fabrication facilities.

In Fig. 60.1, approximately 93 percent of the total amount of copper discharged was from the innerlayer and outerlayer etching process. The amount of copper discharged from the microetch baths (using sodium persulfate) was approximately the same as the copper contained in the rinses following acid copper and the microetch baths. That exact relationship is not always valid for every printed circuit board factory.

Figure 60.2 presents the results of an environmental audit at another printed circuit board facility in grams per hour of waste copper, excluding the large amounts of waste from I/L and O/L etching. The results presented in Fig. 60.2 closely resemble a "common" PCB factory.

The facility used to prepare Fig. 60.2 produced up to 1500 completed multilayer panels (18 in × 24 in) per 20-h day of production.

60.4 WASTE MINIMIZATION

60.4.1 Definitions

Waste minimization includes the procedures, operations, and equipment required to minimize the amount of waste produced. It includes anything that reduces the load on hazardous waste treatment or disposal facilities by reducing the quantity or toxicity of hazardous waste. Waste minimization should be approached as a hierarchy of options. The first step is *pollution prevention* or source reduction.

60.4.1.1 Pollution Prevention. Pollution prevention is the use of materials, processes, or practices that reduce or eliminate the creation of pollutants or wastes. Pollution prevention is often cost-effective because it may reduce raw material losses; reduce reliance on expensive end-of-pipe treatment technologies and disposal practices; conserve energy, water, chemicals, and other inputs; and reduce the potential liability associated with waste generation. Pollution prevention opportunities are intended to

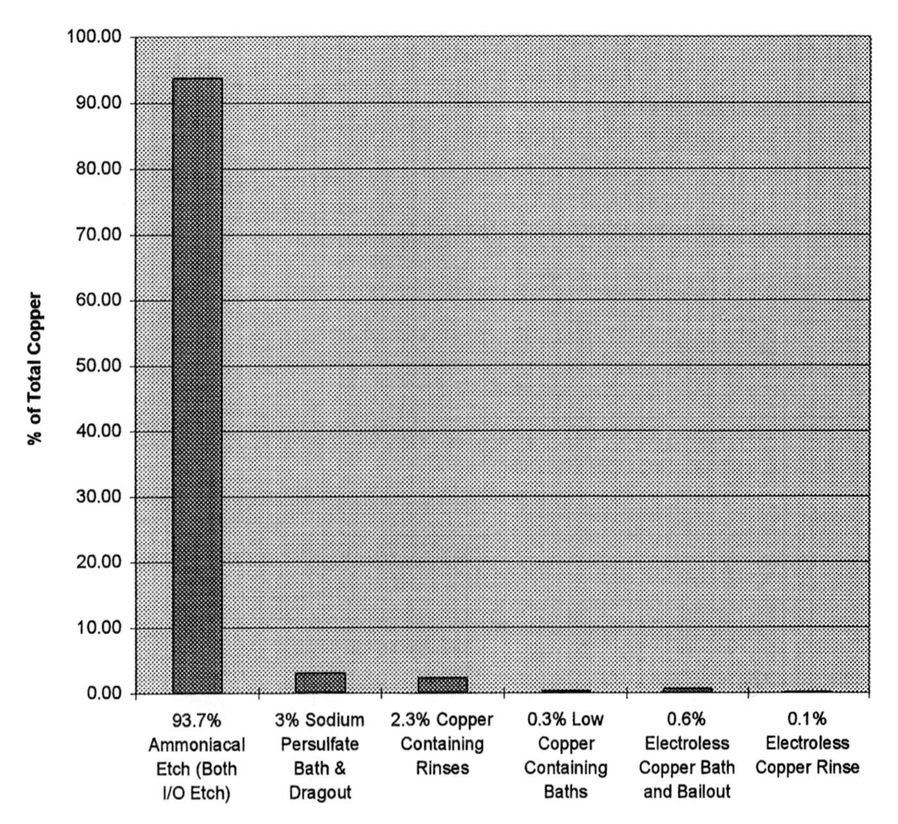

FIGURE 60.1 Amount of copper wasted for various waste streams expressed as a percent of total copper discharge.

1. Reduce the rinse water flow rate and volume.
2. Extend process bath life and reduce the dragout of a concentrated liquid (containing a contaminant such as copper).
3. Substitute materials where possible to eliminate or reduce the hazard of a particular waste.

60.4.1.2 Recycling and Recovery. The next step in the hierarchy of waste minimization options is *recycling* or *recovery* of any waste that cannot be reduced or eliminated at the source. It includes reuse of waste in the process or recovery of metals from a waste before disposal. Recycling and recovery can occur both on- or off-site.

60.4.1.3 Alternative Treatment. The last step in the waste minimization hierarchy is *alternative treatment.* Alternative treatments are selected to minimize the volume or hazard associated with a particular waste.

The benefits for implementing a waste minimization program can include savings in equipment and operating cost, recovery of natural resources, and significant reduction in risk of the liabilities associated with the disposal of hazardous waste. One measure of the effectiveness of a waste minimization project is the project's effect on the organization's cash flow. These projects should pay for themselves through reduced waste management and raw material costs.

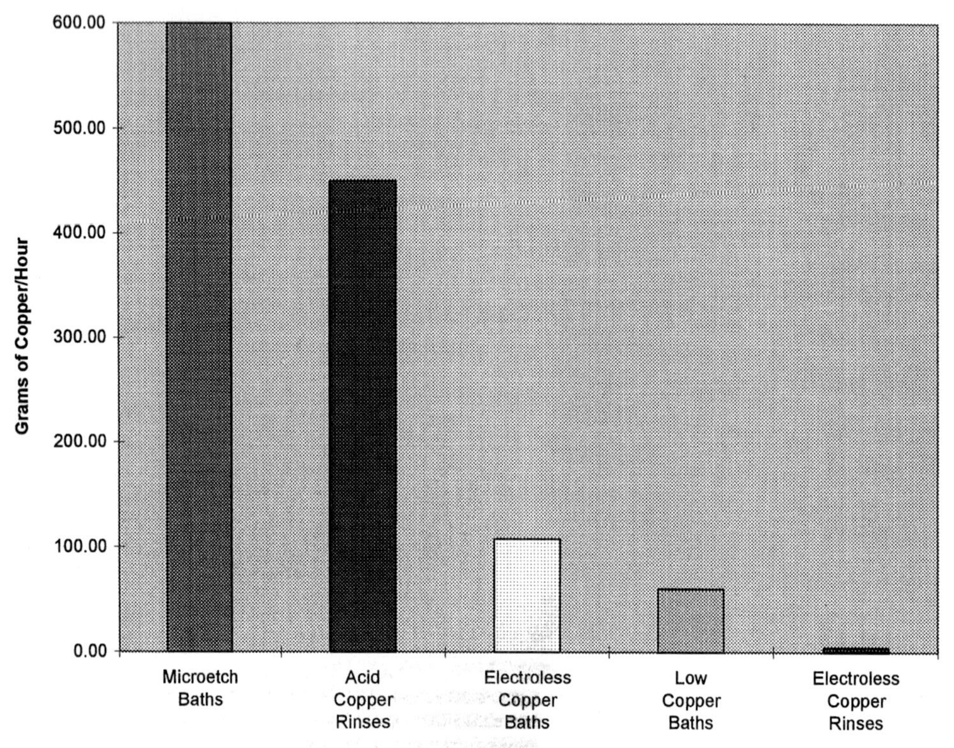

FIGURE 60.2 Major sources of copper expressed in grams per hour discharged. Inner- and outerlayer etching is excluded from this figure since it is so large in relation to the other contributors.

60.5 POLLUTION PREVENTION TECHNIQUES

Some common techniques available for pollution prevention in a printed circuit board fabrication facility are listed as follows. While this list is not all-inclusive, it provides an overview of the types of things that are important to consider. A brief description of each option is presented in the next section.

Rinse Water Reduction

1. Particulate filtration on deburr and panel-scrubbing operations with total or partial rinse water recycling
2. Etcher and conveyorized equipment design modifications
3. Immersion-type counterflow rinses
4. Alternating side spray rinses
5. Flexible orifice–type flow restrictors
6. DI and soft water for rinsing

Extend Bath Life

1. Filtration
2. Acid purification systems

3. In-tank electrolytic permanganate regeneration
4. Proper rack design and maintenance
5. Dragout recovery tanks
6. Monitor solution activity

Dragout Reduction
1. Etcher design
2. Automation
3. Rack design
4. DI and soft water for rinsing

Material Substitution
1. Acid tin as an etch resist
2. Eliminate thiourea from tin/lead stripping baths
3. Permanganate desmear
4. Direct metallization to eliminate electroless deposition and the use of formaldehyde
5. Nonaqueous waste resists
6. Use of nonproprietary chemistry to avoid chelating agents

60.5.1 Rinse Water Reduction

Most of the waste generated in the manufacture of printed circuit boards is from cleaning, plating, stripping, and etching. This section describes some of the techniques available for reducing the volume of rinse water used.

60.5.1.1 Particulate Filtration on Deburr and Panel-Scrubbing Operations. Deburrers are used to remove stubs of copper formed after the drilling of holes in double-sided and multilayer panels before they enter the copper deposition process where copper is deposited within the holes recently drilled in the panel. Scrubbers are used to remove oxides from printed circuit laminates, clean the surface prior to a surface coating to provide better adhesion, and remove residuals after etching or stripping. In deburring and board scrubbing, particulate materials are added to the water and are removed by various methods based on size and the weight of the copper particle such that the wash water becomes suitable for up to 100 percent recycle. The types of filtration available for this operation are cloth, sand, centrifugation, and gravity settling with filtration.

60.5.1.2 Etcher and Conveyorized Equipment Design Modifications. Use of recirculating rinse modules in the etcher and other conveyorized equipment will decrease the required flow rate of rinse water for that process step by about 50 percent without requiring significantly more floor space, compared to single-station spray rinse chambers (without recirculating rinses). In this application, fresh water is used for the final top and bottom nozzles in a rinse module. This water is collected in a sump located below the rinsing compartment. A pump recirculates this water through the first set of top and bottom nozzles (instead of using fresh water). As more fresh water enters the sump, the excess water overflows through a pipe fitting to drain. See Fig. 60.3.

Conveyorized equipment can also be used for, at least, innerlayer and outerlayer photoresist stripping and innerlayer and outerlayer photoresist developing, deburring, and panel scrubbing. Similar techniques to reduce water flow can be applied to these operations.

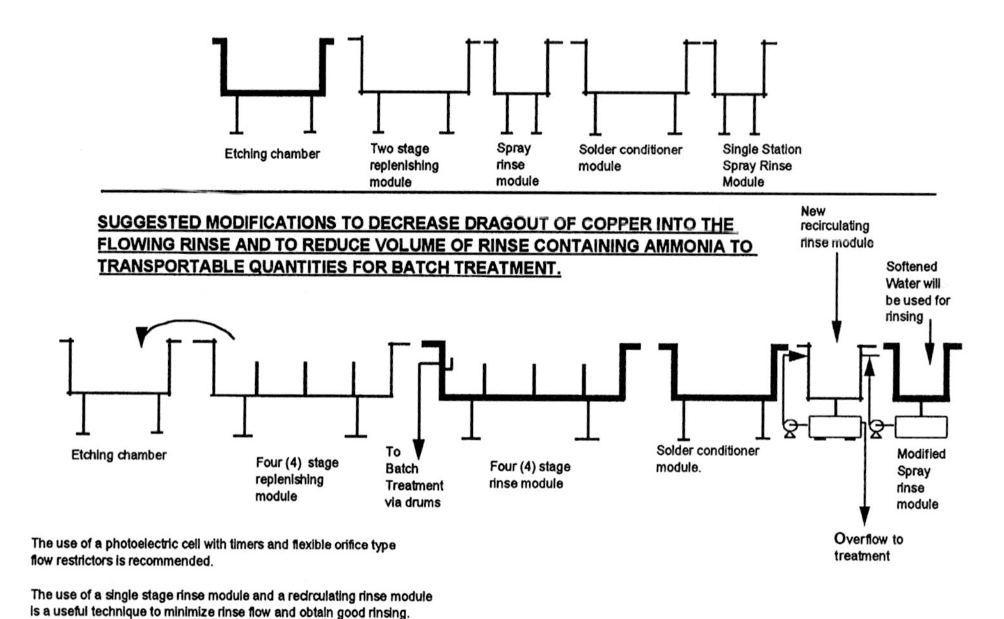

FIGURE 60.3 Suggested modifications to inner/outerlayer etching equipment.

60.5.1.3 Immersion-Type Counterflow Rinses. Counterflow rinsing, employing several single-stage rinse tanks in series, is one of the most powerful waste reduction and water management techniques for innerlayer processing and in the electroless copper process. In counterflow rinsing, the plated part, after exiting the process bath, moves through several rinse contact stages, while water flows from stage to stage in the opposite direction. Over time, the first rinse reaches a steady-state concentration of process dragout contaminants that is lower than the process solution. The second rinse (away from the process bath) stabilizes at even a lower concentration. This enables less water to be used to produce the same cleanliness compared to a single-station rinse tank. The higher the number of rinse stations connected in series, the lower the rinse rate needed for adequate removal of the process solution from the panel. A reduction in rinse water use for a given process step of up to 90 percent over what would be used without counterflow rinsing can be achieved with this technique.

A multistage counterflow rinse system allows greater contact time between the panels and the rinse water, greater diffusion of the process chemicals into the rinse water, and more rinse water to come into contact with each panel. The disadvantage of multistage rinsing is that more steps are required and additional equipment and work space is required.

60.5.1.4 Alternating Side Spray Rinses. Alternating side spray rinses can be used in place of or with immersion rinses in acid copper or solder plating lines. With this technique, rinse water pulsates on each side of a panel at different times so that penetration through holes can be achieved. The reduction in water consumption using an alternating side spray rinse for any given process step is about 85 to 90 percent of that consumed using single-stage immersion-type rinse. Another advantage of spray rinsing is that it will eliminate the need to dump the entire contents of an immersion-type rinse tank, for cleaning purposes, on a periodic basis.

60.5.1.5 Flexible Orifice–Type Flow Restrictors. Flow restrictors are available with flexible diaphragms and are rated for a specific flow rate in a specific pipe size. Typically a 2- to 3-gpm flow restrictor is recommended for a two-stage counterflow rinse tank. As long as a minimum water pressure exists in the water inlet line (about 20 lb/in^2 is considered minimum), the water flow into a rinse tank should not vary, although the water pressure will most likely vary. Flow restrictors are not adjustable after they are installed. By providing pipe union fittings both up and downstream from the flow restrictor fitting, the flow rate can be changed if necessary.

60.5.2 Extended Bath Life

Although process bath solutions are discarded infrequently, when spent or contaminated, they must be replaced. The resulting waste contains high concentrations of hazardous constituents. Bath life extension is thus an important means of minimizing waste in a printed circuit board manufacturing facility.

60.5.2.1 Filtration. Baths should be continuously filtered to remove impurities. Most manufacturers utilize spiral-wound cylindrical polypropylene-type filters with nominal pore-size ratings. Some manufacturers have decreased or actually eliminated the need to dump selected process baths, such as the brown oxide bath, by using a more expensive absolute-type mechanical filter.

The filter, in the brown oxide case, removes black copper metallic particles and other solids. Proper filtration will eliminate the cloudiness that may presently build up in the bath. A side benefit for filtering the brown oxide bath is the increased mechanical agitation in the bath that may improve the coating uniformity.

The filtration requirements for every bath should be reviewed with an objective to clean the bath and to decrease or eliminate its dumping frequency.

60.5.2.2 Acid Purification Systems. In the circuit board industry, mineral and semiaqueous acid solutions are used for activating metal and plastic surfaces and for the etching, cleaning, and stripping of racks and defective parts. These solutions become contaminated and lose strength (lowering the acid concentration) as a function of usage due to the increasing concentration of contaminants such as metals and the continuous dilution of the acid concentration. These less active acidic solutions are customarily disposed of by periodically dumping the bath at once, or slowly withdrawing a portion of, the bath.

From a process point of view, if the concentration of contaminants can be maintained low and the acid concentration maintained at the desired level, an improvement in performance of the acid solution will result (in other words, better quality control is provided using continuous bath purification).

Finding an economic method to continuously reuse 90 to 95 percent of an acid will (1) reduce incoming chemical consumption (for the factory), (2) reduce the cost to process parts (fewer rejects), (3) provide better quality control, and (4) reduce the chemicals (for example, caustic soda) required to neutralize the spent acid before discharge. Eliminating the need to handle spent acids will decrease the exposure of hazardous materials to employees and will reduce the wastewater treatment burden. This type of bath purification system may eliminate the need to dump selected concentrated wastes.

One technology for acid purification uses diffusion dialysis. Diffusion dialysis units utilize a membrane which allows anions to pass (i.e., Cl^- in the case of hydrochloric acid) while metal cations remain behind. This is the dialysis part of the process. At the same time, water molecules diffuse through the membrane in the other direction.

Each recovery unit produces two streams: a reclaimed acid stream, which is high in acid content and low in metal content, and a waste stream which contains a high concentration of metals and a low concentration of acid. The only required utilities are soft water and single-phase power.

The following commonly used acids found in the printed circuit industry can be purified using this technology:

Hydrochloric	Fluoboric
Hydrofluoric	Methane sulfonic
Nitric	Ferric chloride-hydrochloric acid
Nitric-hydrofluoric	Sulfuric acid

Methane sulfonic acid (MSA) is a relatively expensive acid used to strip (remove) solder (tin/lead) and copper metals. This acid is also used to condition metal surfaces prior to fluoborate free-solder electroplating when applying a final etch resist of solder onto multilayer printed circuit boards. By continuously recirculating the acid bath through a diffusion dialysis unit, a 95 percent reduction of chemical purchases (required by the process steps indicated in the first paragraph of this section) may result, producing less than a six-month return on the initial capital investment (ROI). Typically, 80 to 90 percent of the acid is recovered with 70 to 90 percent of the metals removed in an equal volume of dilute acid.

Another system uses only water in its process. The heart of this product is a shallow bed of special adsorption resin. This resin has the unique ability to adsorb strong mineral acids yet exclude metal salts. This is not an ion exchange process since only water is used, in a countercurrent fashion, to desorb the acid from the resin.

60.5.2.3 In-tank Electrolytic Permanganate Regeneration. Because permanganate is a powerful oxidizing agent only in basic conditions, desmear solutions usually contain 5 to 10 percent caustic soda. However, the alkalinity at high temperatures causes the permanganate (MnO_4^{-1}) ion to break down to manganate. The manganate reacts with water to form manganese dioxide (MnO_2), eventually forming a black sludge of MnO_2 which settles to the bottom of the desmear tank. This requires that the bath be dumped.

Oxidizing electrolytic regeneration systems are available to inhibit this reaction. A typical electrolytic regeneration system employs an in-tank probe composed of a copper rod as the cathode which is connected to a power supply. When current is applied, oxygen is generated in the permanganate solution which functions as the regenerate for the manganate. The efficiency of the system depends on the rate of manganate buildup, the amount of amperage generated, and the power supply to the in-tank regenerator. These probes attach to the top rim of the desmear tank and are always operating (as the permanganate ions will break down even when the bath is not active). In most cases, the use of this equipment has eliminated the need to dump this bath.

60.5.2.4 Proper Rack Design and Maintenance. By using plastic-coated racks, only the tips and contactors (which electrically and physically connect and support the panel to the carrying rack), have exposed conductive metal and must be stripped. If one of these is removable, then the amount of metal to be chemically or electronically removed from the rack is also reduced. This should lengthen the useful life of the stripping bath and further reduce the cost for hauling away or on-site treatment (if nitric acid is used) of the spent rack stripper. Use of racks with disposable contactors eliminates the need to strip the racks and could eliminate the use of nitric acid.

60.5.2.5 Dragout Recovery Tanks. The dragout tank's primary purpose is to collect most of the solution dragout before the panel enters the rinse tank. The dragout (recovery) tank is a rinse tank that operates without a continuous flow of feed water. It is positioned between a process bath and its rinse tank. The dragout tank can be either an immersion application (for the electroless and innerlayer operations) or a spray type application (on conveyorized equipment or for pattern- or panel-plating operations). After the panels have been removed from the process bath, they are immersed or sprayed in the dragout tank before they are rinsed in another tank. The concentrated rinse water can be collected separately from its following rinse.

Eventually, the concentration of the tank solution will increase to a point such that it provides an opportunity to

1. Recover metals.
2. Recycle a process bath since it is the same chemistry.

The disadvantage is that this opportunity may require additional production floor space.

A drip pan (also called a drain board) is one of the simplest methods for dragout recovery. The drip pan is located either under the rack while in transit or between tanks, and will capture drips of process solution from racks and panels as these are transferred between tanks. Drip pans not only save chemicals and reduce rinse water requirements, they also improve housekeeping by keeping the floor dry.

60.5.2.6 Monitor Solution Activity. Bath life can be extended by monitoring the bath activity and replenishing stabilizers and reagents. This will reduce dump frequencies and result in a reduction in waste.

60.5.3 Dragout Reduction

Dragout reduction is desirable as a waste minimization step because it results in the saving of process chemicals as well as a reduction in sludge generation.

60.5.3.1 Etcher Design. Etching machines can be the single largest source of copper waste in the discharge from a PCB facility. The amount of copper discharged and the rinse flow rate from that machine are a function of the machine design. An older etching machine often contains a single-stage etchant replenishing module positioned between the etching chamber and its continuously flowing single or multiple-stage water rinse chamber. Fresh etchant is fed to the replenishing module to wash the panels and that etchant (now containing copper washed off of the freshly etched panels), then flows (in a direction opposite to the direction of the panel movement) into the etching chamber. The continuous water rinse can contain from 100 to 500 mg/l or ppm of copper. Companies that use this type of equipment, with single-station rinse module, normally have floor space restrictions in their production area. See Fig. 60.3.

By way of comparison, adding a second stage replenisher station will reduce the range of copper concentration in the following rinse from 50 to 300 mg/l. Using a four-stage replenisher module and combination recirculating and then a single-stage rinse modules, will produce a rinse containing from less than 1.0 up to 2.0 mg/l of copper.

60.5.3.2 Automation. Computerized process control systems can be used for panel handling and process bath monitoring to prevent unexpected decomposition of a process bath, controlled rinse flow, and uniform panel withdrawal from each process bath. Since these systems require a significant capital expense for initial installation, typically only large printed circuit board companies will find this to be a cost-effective alternative.

60.5.3.3 Rack Design. Some plating rack designs allow the panels to be tilted to one side and the panel mounted at an angle (relative to the horizontal) to allow better solution drainage off of the panel. This will decrease the dragout.

60.5.3.4 DI and Soft Water for Rinsing. Natural contaminants found in water used for production purposes can contribute to the volume of sludge produced in waste treatment because they precipitate as carbonates and phosphates. Another advantage is that pretreated rinse water is cleaner; therefore, the normal diffusion of contaminants from the panel into the rinse water occurs more quickly. As a result, water requirements for each rinse may be reduced since rinse efficiency is improved.

60.5.4 Material Substitution

Material substitution is a waste minimization technique that should result in a reduced hazard associated with a particular waste or a simplied means of waste treatment. It is important that all process impacts are thoroughly understood prior to instituting any material changes.

60.5.4.1 Acid Tin as Etch Resist. Acid solder baths are often used as an etch resist before final etching of an assembled multilayer printed circuit board. The lead and fluoride present in this bath must then be removed in the waste treatment process. Acid tin can be used as the etch resist in place of solder under certain circumstances. An advantage of this bath is that it does not contain lead ions or fluorides. Use of this bath will eliminate these materials from the discharge.

60.5.4.2 Eliminate Thiourea. Thiourea is a known carcinogen which has been used as a chelating agent in Sn/Pb stripping baths. A peroxide-based bath can be substituted when the etch resist is acid tin, thus eliminating this concern.

60.5.4.3 Permanganate Desmear. To reduce the amount and toxicity of waste, most printed circuit board factories in the United States use the alkaline permanganate process instead of chromic acid for desmear. This technique has gained favor due to the wastewater treatment equipment requirements to reduce the chromic acid and the control problems (especially in humid environments) of 96 percent sulfuric acid. The plasma desmear process produces the least amount of hazardous waste, but that technique is used primarily for small batch operations and not high-volume continuous operations.

60.5.4.4 Direct Metallization. Alternatives to the electroless copper process have emerged to avoid the use of formaldehyde, a toxic and carcinogenic material. The use of formaldehyde has recently (as of June 1993) required extensive in-plant air monitoring and will require employers to have their affected employees periodically checked by a medical doctor. This is an additional unwanted expense for each PCB factory owner.

One alternative to the electroless process utilizes conductive polymers. Polymer thick film (PTF) technology is a method of screening polymer conductors, resistors, dielectrics, and protective coatings on a substrate or printed wire board to create basic circuitry or interconnections. The use of screened-on conductive films in place of the formaldehyde-containing electroless copper process is gaining acceptance in Asia (more so than in the United States at this time).

Other technologies are being applied commercially which use palladium or graphite to provide conductive pathways in the holes of multilayer printed circuit boards.

60.5.4.5 Nonaqueous Waste Resists. One of the most exciting waste minimization techniques currently being tested is the elimination of the need for rinsing from both photoresist developing and stripping operations. This new process may eliminate spent developer wastes and will eliminate the spent resist stripper bath dump.

The photoresist, being demonstrated at Circuit Center (Dayton, Ohio) and at E. G. & G. Mound, is applied in a production process as a thin film from a liquid solution onto a circuit board and air dried. When exposed to a high-intensity ultraviolet light source, these photoresists will decompose and vaporize as inert gas. There appears to be no rinsing required for these operations. In fact, these operations, as we currently know them, will be eliminated, thus reducing the cost of waste treatment. The existing 30-month demonstration project is a cooperative effort between the University of Dayton Research Institute, the Air Force, the U.S. Environmental Protection Agency, and the Edison Materials Technology Center (EMTEC).

Depending on the actual chemistry used, the use of this new process may reduce the hazardous air emissions from these processes commonly produced with conventional semiaqueous and aqueous chemistries containing butyl cellosolve or butyl carbitol.

60.5.4.6 Use of Nonproprietary Chemistry to Avoid Chelating Agents. Chelating agents are molecules that form a charged complex with a metal ion such as copper. They are used to enhance solubility and keep the metal ions in solution. Typical chelating agents used in printed circuit board fabrication include ferrocyanide, ethylenediaminetetraacetic acid (EDTA), phosphates, and ammonia. Chelated baths are intended to enhance etching, cleaning, and electroless plating but make waste treatment much more difficult because the metals are tightly bound in the complex, which inhibits precipitation. Often ferrous sulfate must be added to wastewaters to break the chelators prior to precipitation of metal hydroxides. The iron is precipitated as well as other metals, thus increasing the amount of sludge produced. Use of nonchelated chemistry where possible will eliminate this problem. If some chelating chemistry is required in the process, proper waste segregation should be implemented to minimize the volume of any chelated waste streams.

60.6 RECYCLING AND RECOVERY TECHNIQUES

60.6.1 Copper Sulfate Crystallization

60.6.1.1 Theory of Operation. The sulfuric acid/hydrogen peroxide etching reaction is:

$$Cu + H_2O_2 + H_2SO_4 \twoheadrightarrow CuSO_4 + 2H_2O \qquad (60.1)$$

Systems employing crystallizers are designed to

1. Lower the solution temperature to lower the solubility of copper.
2. Then remove copper in the form of copper sulfate pentahydrate crystals ($CuSO_4 - 5H_2O$).
3. Then reheat the bath (to prevent copper from clogging the return pipe line if the bath temperature is lowered further) before transferring the bath back to its working tank.

An advantage, from a waste treatment point of view, is that the bath is recycled and not routinely dumped. Other microetchants must be dumped when the copper concentration increases. The most significant advantage, from a process point of view, is that a low copper concentration can be maintained and the printed circuit board factory will minimize its purchase of microetchant chemicals.

A disadvantage to this process is that, at the time of this writing, there are no outlets available to receive and recycle the crystals, so the only alternative for manufacturers using crystallizers is to dissolve the copper sulfate crystals in spent dilute (10 percent) sulfuric acid. After the copper has dissolved, this material can be pumped to another holding tank for recirculation through an electroplating cell for reclaim of the copper.

60.6.2 Rinse Water Recycling

Rinse waters can be treated and reused in the process. This can be accomplished in two ways. Rinse water recycling can be accomplished with a point-of-source system. In this system, the flow from selected rinses is recirculated through cation and anion exchange columns and then returned to the point of origin. It may be necessary to add a carbon filtration step for rinses containing organics.

The second method would involve a central system. Selected rinses could be run through an ion exchange system for copper removal. Following this step, these rinses would pass through a general resin for cation removal, a general resin for anion removal, systems for organic removal such as activated carbon, ozonation or UV peroxide, and final filtration. In some cases, reverse osmosis could be substituted for the general ion exchange steps. Given the complexity of this treatment process, it would be cost-effective only for large printed circuit board facilities with high water use rates.

60.6.3 Copper Recovery via Electrowinning

60.6.3.1 Theory of Operation. Electrowinning is the reduction of copper ions to solid metallic copper at a cathode. The following are the chemical reactions governing copper recovery in an acidic solution. The reduction reactions that occur at the *cathode* are:

$$Cu^{+2} + 2e^{-} \longrightarrow\!\!> Cu\ (metal) \quad (60.2)$$

$$2H^{+} + 2e^{-} \longrightarrow\!\!> H_2\ (gas) \quad (60.3)$$

The reaction that occurs at the *anode* is:

$$H_2O \longrightarrow\!\!> \tfrac{1}{2}(O_2) + 2H^{+} + 2e^{-} \quad (60.4)$$

The first reaction describes the principal objective of electroplating (the reduction of copper to its solid form on the cathode). As electroplating proceeds, the waste becomes more acidic, as noted in Eq. 60.4.

The speed of these reactions will be inhibited if oxidizing chemicals are present. Typical oxidizing agents include peroxides and persulfates, which are normal components of microetching baths. If these oxidizers are not chemically reduced in the solution entering an electroplating module, additional reaction time must be anticipated to reduce the oxidizers electrically. Following that, the reduction of copper will occur. For this reason, a chemical reduction step (addition of sodium bisulfite ($NaHSO_3$) or sodium metabisulfite ($Na_2S_2O_5$) is recommended before the waste is recirculated through an electroplating module.

60.6.3.2 Central Systems. Copper can be recovered in a central system from the following sources:

1. Sulfuric acid regeneration of selective cation columns
2. Spent microetch baths
3. Dragout tanks
4. Copper sulfate electroplating bath bailout
5. Dissolved copper sulfate crystals from a crystallizer

The typical concentration of the preceding mixture (or any component of it) may range from 5 to 30 g of copper per liter. A central electrowinning system is shown in Fig. 60.4.

60.6.3.3 Parallel Plate Electrowinning Systems. The purpose of parallel plating, in a central system, is to recover copper as a 99.9 percent pure metal sheet (for resale) and to reduce the copper concentration in the liquid being recirculated to 1.0 g/l, or less. The efficiency of the reaction drops significantly below 1.0 g of copper per liter, as observed by the generation of heat in the bath by the electroplating process.

Many factors affect the ability of an electrolytic cell to recover copper. Important design parameters center around improving the mass transport. These include solution agitation and cathode agitation. Air agitation of the waste within the electroplating cell is used to increase the efficiency of the cell. Control of the air inlet rate, bubble size, current density and distribution are all critical to maintaining high efficiency in each parallel plate cell. Particles (so-called *dendrites*) or fines of metal can be dislodged and accumulate in dead spots with the plating cell tank. Electrical shorting out of the anode to cathode can occur when particles accumulate, creating the possibility of isolated locations where burning of the cell may occur.

60.6.3.4 High-Surface-Area Electrowinning Systems (HSA). HSA systems can be used to recover copper from a variety of solutions, including electroless copper concentrates and rinses. The use of high-surface-area cathodes improves the mass transport characteristics over flat plate cathodes. The HSA cathodes reduce electrode polarization potential and improve

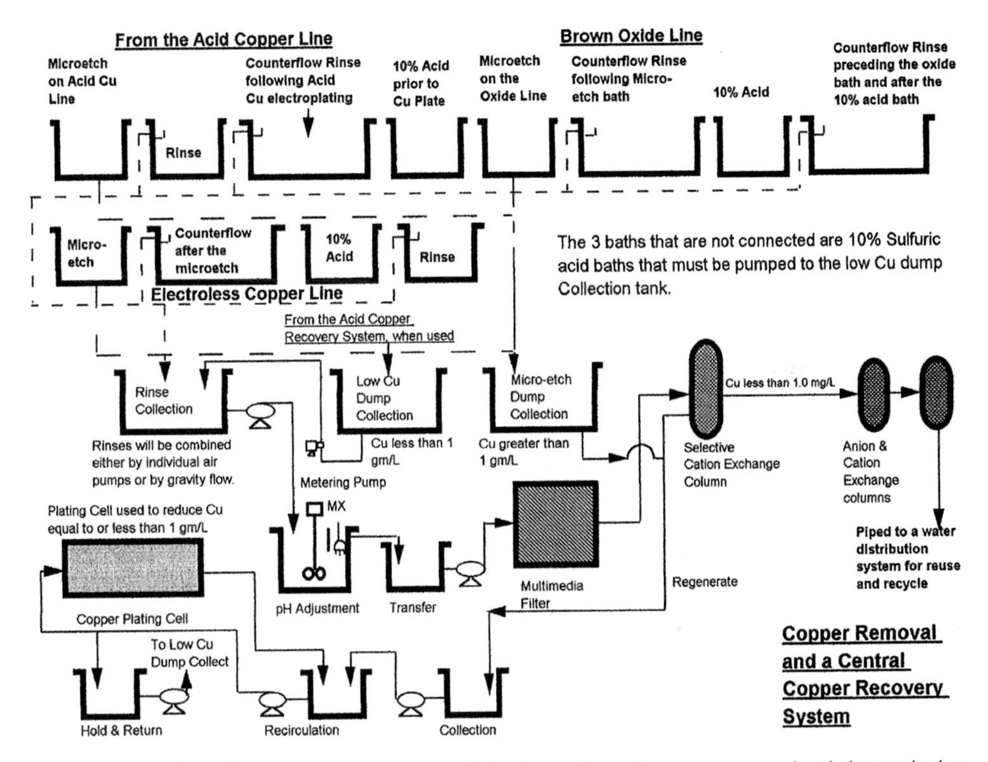

FIGURE 60.4 Central copper recovery system utilizing ion exchange for copper removal and electrowinning for copper recovery.

ion diffusion potential, allowing the copper to deposit rapidly on an HSA cathode from wastes with both high and low copper concentrations. HSA cathodes promote solution agitation by creating numerous bonding sites for copper. Expanded mesh carbon fibers or catalyzed foam are typical materials of construction in HSA cathodes. Some HSA systems require the purchase of new cathodes, while the cathodes in some systems can be regenerated for a period of time. To regenerate these cathodes they must be stripped of copper metal.

Stripping the cathode takes place in a second plating tank. The cathodes from the first tank must be manually removed and placed into the anodic position within the stripping tank. The copper is then recovered as a metal by electroplating the copper onto either a stainless steel or a copper laminate in a copper pyrophosphate electrolyte. The *reusable cathodes* are then available to be repositioned back to the first cell.

HSA systems can reduce the copper levels to low levels in an incoming waste stream, but it is important to note that copper is not recovered unless the stripping step takes place in the second plating tank.

60.6.3.5 Point-of-Source Systems. Dragout tanks can be located between a process bath (for example, a copper-plating bath in the pattern-plating line) and its following rinse tank. Typically, dragout tanks are positioned *only* adjacent to the process tanks containing the most difficult wastes. These include the etching baths (if not conveyorized), the microetch baths (which produce the most copper), and the copper electroplating (usually sulfate-based) baths.

The contents of each dragout bath can be recirculated through its own plating cell for copper recovery, then back to the dragout tank. In this way, the concentration of copper is maintained at a low level. The plating cell, for this application, could be either a parallel plate or an HSA system.

If dedicated electroplating cells are not used to recover copper directly from a dragout bath, the dragout bath can periodically be discharged to a dilute (or concentrated) copper collection tank for recovery through an ion exchange system (or a central electroplating system, respectively).

60.7 ALTERNATIVE TREATMENTS

60.7.1 Selective Ion Exchange

Ion exchange is a process in which ions, which are held by electrostatic forces to charged functional groups on the surface of the ion exchange resin, are exchanged for ions of similar charge in a solution in which a resin is immersed. Ion exchange is classified as an adsorption process because the exchange occurs on the surface of a solid (a resin bead), and the exchanging ion must undergo a phase transfer from the solution phase to the surface of the solid. A single-pass selective cation exchange column can reduce the copper concentration in the influent stream to less than 1.0 mg/l. If two cation exchange columns are in series, as indicated in Fig. 60.4, the expected copper effluent concentration should be less than 0.5 mg/l. Organics present in the incoming waste can foul the resin and render it unusable; therefore, a carbon filtration step occurs ahead of the ion exchange column. Wastes with significant levels of organic material are not treated with this technology.

60.7.1.1 Theory of Operation. Copper ions are normally present in their divalent (+2) state. The removal mechanism of divalent copper from an aqueous waste by ion exchange resins is:

$$Cu^{+2} + 2R\text{-}H \longrightarrow\!\!> 2H^{+} + R_2\text{-}Cu \tag{60.5}$$

where R = a cation exchange resin. Here the copper is "attached" to the resin beads.

Ion exchange equipment suppliers will usually furnish a water meter that totalizes the quantity of waste processed through the column. After the system has been calibrated by the equipment supplier, the water meter will activate an alarm after a specified volume of waste has been processed. An alarm will sound if the system requires manual regeneration. If the system is supplied with automatic regeneration, the water meter will deactivate a column while regeneration takes place. Where multiple columns are included within a system, valves must be adjusted to accommodate regeneration. This will be accomplished by manual or automated (pneumatic-, electric-, or hydraulic-actuated) techniques.

60.7.1.2 Regeneration. To recover the copper, the cation resins are first rinsed with water and then washed with 5 to 15 percent by volume sulfuric acid. This proceeds by the following reaction:

$$R_2\text{-}Cu + 2H^{+1} \longrightarrow\!\!> Cu^{+2} + 2R\text{-}H \tag{60.6}$$

where R = a cation exchange resin.

Copper is displaced from the resin by a proton (or hydrogen atom). The resulting sulfuric acid solution containing copper can be directed to an electroplating system to recover copper from the acidic solution. The remaining resin is washed with water to remove the residual sulfuric acid.

60.7.1.3 Spent Baths. Certain spent baths can be bled into the ion exchange system. These typically include the copper sulfate electroplating dragout, acid cleaners, predips, microetch and rinses, rinses following cupric chloride and ammoniacal etchants, and copper waste from electrowinning after reduction to 1.0 ppm or less.

60.7.2 Removal of Copper from the Electroless Copper Bath

60.7.2.1 Theory of Operation. It is well known that copper metal can be deposited onto metal (usually copper or palladium) within a PCB hole. Dissolved copper is automatically reduced (using formaldehyde as its reducing agent) to a solid metal and deposits itself onto the palladium inside of the hole while the panel is immersed in the electroless copper bath. The chemical reaction of copper deposition may be represented by

$$Cu\,(EDTA)^{-2} + 2HCHO + 4OH^{-1} \longrightarrow\!\!> $$
$$Cu^0 + H_2 + 2H_2O + 2CHOO^{-1} + EDTA^{-4} \tag{60.7}$$

This process can be utilized to remove copper from the electroless copper baths and bailout. Products on the market include canisters or modules which contain a proprietary spongelike material deposited with palladium and copper. The electroless copper solution passes through the canister where the copper is autocatalytically reduced, trapped by the spongelike media and filtered out of the waste stream. See Fig. 60.5

Typically, a minimum of two canisters are connected in series. As the copper concentration in the first canister approaches 1.0 mg/L, the sponge material is replaced by the material in the second canister. New sponge material is added to the second canister.

60.7.3 Sodium Borohydride Reduction

Although not as common or as inexpensive as the autocatalytic method, one of the simplest methods for the treatment of electroless copper is by the use of another strong reducing

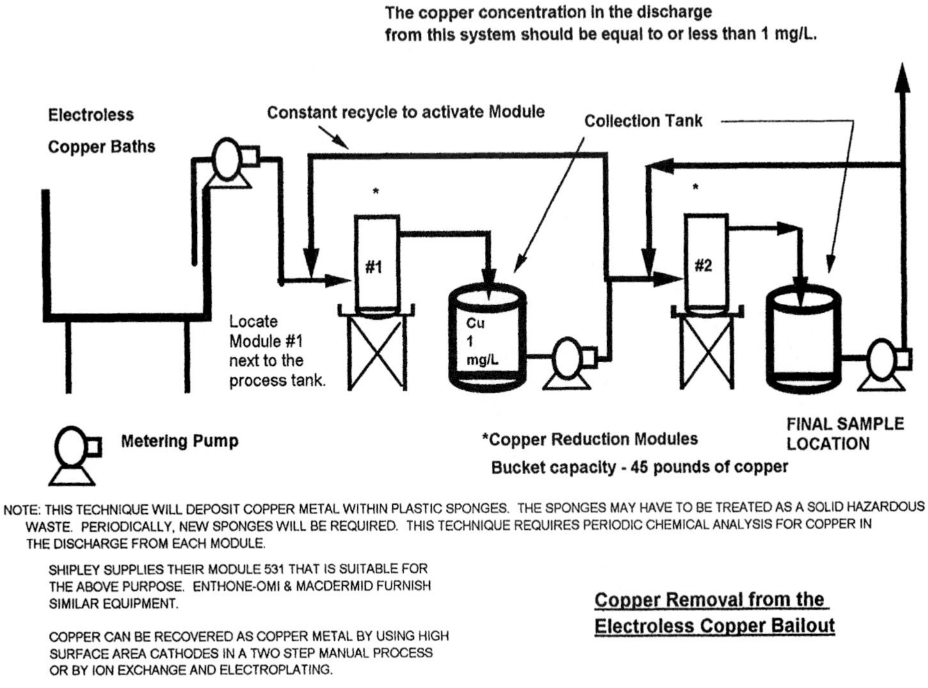

FIGURE 60.5 Copper removal from electroless copper bailout using copper reduction.

agent, sodium borohydride ($NaBH_4$). This chemical is commercially available as a solution of 40 percent caustic soda containing 12 percent by weight of sodium borohydride. The chemical reaction can be represented by

$$8MX + NaBH_4 + 2H_2O \Rightarrow 8M + NaBO_2 + 8HX \tag{60.8}$$

where M is the cation with +1 valence and X is the anion.

The spent electroless bath has to be in an active state for this method to work efficiently. Adding sodium hydroxide (NaOH) and/or formaldehyde (HCHO) to the bath will shift the equilibrium in the bath to this reactive stage. When this stage is reached, a small addition of sodium borohydride will catalyze the reaction, and reduced copper will precipitate out as metallic fines.

A typical process for using sodium borohydride requires the use of an open-top batch tank that is located in a well-ventilated area. Since $NaBH_4$ can react to release hydrogen as a side reaction, any source of flames or sparks must be eliminated. A 0.5 to 1 percent, by volume, $NaBH_4$ solution is added slowly to the batch reaction tank that contains the spent electroless copper solution. After a few minutes, the solution will effervesce vigorously and fine, pink copper particles will begin to form and become visible to the naked eye. Small bubbles of flammable hydrogen gas will also evolve from the solution, and these must be rapidly removed by appropriate exhausts. Usually within about one hour, the solution will turn clear and copper will precipitate from the solution. These fines will eventually settle to the bottom of the batch tank and the copper fines should be removed as soon as the reaction has been completed. Otherwise the copper fines may reoxidize and dissolve back into solution. More $NaBH_4$ will then be necessary to remove the copper. After the copper fines have been removed, hydrogen peroxide (H_2O_2) can be added to the clear solution to oxidize any excess formaldehyde to the less toxic formic acid.

This process can be used for other metal-bearing waste streams besides electroless copper. However, it is typically not a cost-effective means of treating nonchelated waste.

60.7.4 Aqueous and Semiaqueous Photoresist Stripping Bath Treatment

A significant volume of waste produced from PCB factory is from the stripping of resists. Resist stripping machines can utilize on-line or off-line equipment to remove semisolid photoresists on a continuous basis from the operating bath. Removing the suspended solids within the solution provides a longer life and more efficient usage.

In any case, there will be times when the bath becomes spent and must be dumped. Waste from this process should be segregated for treatment so that the sludge produced can be classified as nonhazardous. In addition, good engineering practice dictates that spent photoresist stripping baths not enter a continuous precipitation system since organic photopolymers may not always settle in a clarifier. Under some conditions (at certain temperatures or with certain organics), stratification of the organic material in the clarifier can occur. As a result, the organic solids can float and any metal hydroxide particles nearby may become attached to the floating material.

The treatment procedure for this waste uses chemical precipitation (caused by lowering the pH) and filtration (to remove the photoresist and residual metals). The treated wastes can be slowly discharged into the existing final pH adjustment tank. The specific treatment procedure follows.

The untreated pH of this waste is approximately 12.0 pH units. Sulfuric acid is slowly added to lower the pH to about 9.0. After a stable pH is achieved, the pH will be lowered to about 6.0 by the addition of acidic proprietary chemicals available at the time of this writing. At this low pH, the dissolved resist will precipitate out of solution and the solids formed can be removed by pumping the waste through a filter press.

60.8 CHEMICAL TREATMENT SYSTEMS

60.8.1 Definition

Chemical treatment of wastes involves the addition of chemicals to precipitate metals out of solution. This results in a sludge which must be disposed of. The sludge is then typically sent to a reclaimer for recovery of the metals or licensed disposal site. The latter option is becoming increasingly difficult and expensive. It should be avoided due to the unknown risks associated with the landfill of hazardous wastes. Chemical treatment of wastes should be utilized only when pollution prevention or recycling is not technically or economically feasible.

60.8.2 Treatment Process

Nonchelated metal rinses and dumps are typically treated in a two-stage process. In the first stage, rough pH control is accomplished with the addition of caustic soda and sulfuric acid. In the second stage, the pH is elevated with caustic to precipitate the metals in the hydoxide form. Lime has been used in this step in the past, but its use has been discouraged since it increases the sludge volume significantly.

Alternatively, the metals can be precipitated as in the sulfide form with the addition of FeS at pH 9. This process is more difficult to control and presents a potential hazard due to the possible evolution of H_2S. However, the lower solubility of the metal sulfides provides an advantage where discharge limits are more stringent.

Chelated metal rinses and dumps such as those from the electroless copper process provide more challenge in chemical treatment since the chelating agents bind the metals and keep them in solution. This problem is often addressed by the addition of ferrous sulfate to break down the chelated metal complexes, allowing the metal to precipitate as the hydroxide. It is often necessary to add large quantities of this reagent such that the ratio of iron to copper is 8:1. The disadvantage of this method is that iron will precipitate out as well, thus significantly increasing the volume of sludge produced.

Regardless of the specific treatment chemistry used, there are common components to all chemical treatment systems.

60.8.3 Collection System

The collection system is the tank or group of tanks where wastes are collected prior to treatment. The collection tanks should be sized to provide a minimum residence time of 20 min. Segregation of wastes can improve the efficiency of the waste treatment system. For example, process bath dumps can be collected in separate tanks and metered into the rinse waste stream at a low flow rate. Also, chelated wastes can be collected and treated separately from nonchelated streams, thus minimizing the use of ferrous sulfate.

60.8.4 pH Adjust

This is typically a dual-tank system where the pH is elevated and metals are precipitated as either the hydroxide or the sulfide, as previously described. Each stage must be adequately sized to provide adequate retention time (usually a minimum of 30 min) to ensure the completion of the reaction. Each tank is equipped with a mixer, pH probe, and controller. In addition, pH recorders should be provided to effectively monitor system performance.

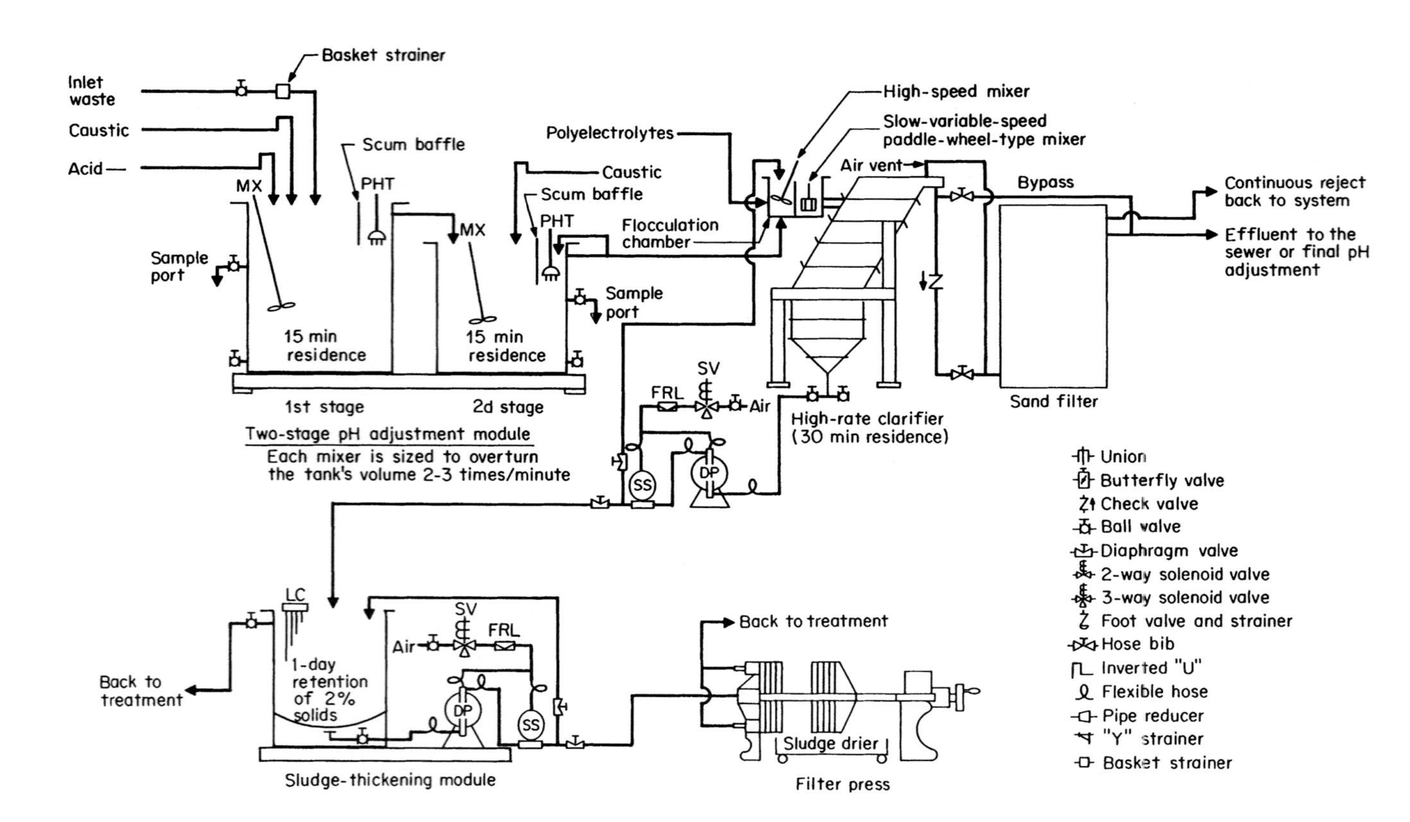

FIGURE 60.6 Treatment of metal-bearing wastes. MX = mixer; PHT = pH transmitter; FRL = filter, regulator, lubricator; SV = solenoid valve; SS = surge suppressor; DP = diaphragm pump; LC = level control. *(Courtesy of Baker Brothers/Systems.)*

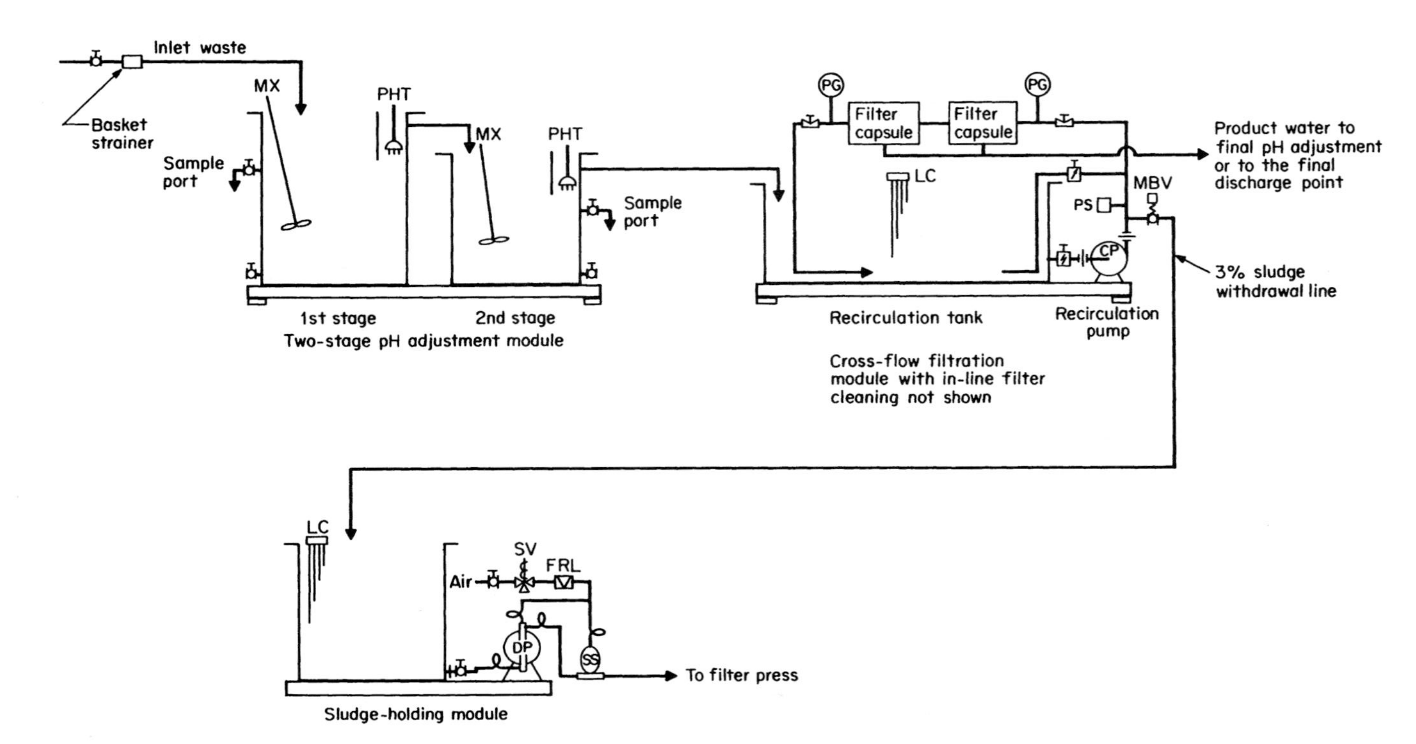

FIGURE 60.7 Cross-flow filtration. MX = mixer; PHT = pH transmitter; PG = pressure gauge; LC = level control; PS = pressure switch; MBV = motorized ball valve; CP = centrifugal pump; SV = solenoid valve; FRL = filter, regulator, lubricator; DP = diaphragm pump; SS = surge suppressor. *(Courtesy of Baker Brothers/Systems.)*

TABLE 60.2 Advantages and Disadvantages of Various PCB Wastewater Treatment Alternatives

Categories of major wastes	Category of waste minimization or treatment	Advantages	Disadvantages
1. Copper-containing acidic rinses.	Selective rinse water recycling with copper recovery.	Recovered copper metal can be sold. Recycling will reduce the quantity of waste being discharged. Reduced sewer use charges.	Most expensive. Requires flow control. Requires thorough and periodic bath analysis. Must address spent process baths.
	Copper removal from preselected rinses using selective ion exchange with copper recovery by electroplating.	Recovered copper metal can be sold. Minimizes the quantity of copper discharged.	More expensive equipment than used for chemical treatment. Requires expensive selective cation resins. Must address spent baths.
	Chemical treatment to precipitate copper.	Least capital cost. Requires clarification with filtration and proper pretreatment chemistry.	Produces sludge. Increasing liability. Requires intensive effort for labor. Must address spent process baths.
2. Copper-containing acidic spent process baths.	Bath purification with partial or total reuse using crystallization, ion exchange, filtration,and/or diffusion dialysis.	Reduce or eliminate the need to dump a bath. Recovered copper can be sold.	Requires through and periodic bath analysis. Optimum application is copper removal from rinse waters.
	Bleed spent bath into rinses for copper removal with copper recovery.	Recovered copper can be sold. Minimizes quantity of copper discharged.	More expensive equipment than used for chemical treatment.
	Allow a higher contaminant concentration before dumping.	Less frequent need to dump a process bath.	Requires through and periodic bath analysis.
	Copper recovery from microetchants.	Most copper recovery using least floor space and cost.	Requires that these wastes be collected and processed separately.
	Chemical treatment to precipitate copper.	Used to concentrate copper into sludge.	Produces sludge. Increasing liability. Requires intensive effort for labor.
3. Ammonium chloride etchants and rinses.	Recycle 95 percent etchant and rinse water.	Reduces fresh chemical and hauling purchases. Recovered copper can be sold. Reduces waste quantity.	May be justified only when both inner and outer layers are etched with this chemistry. Ammonium sulfate etchant requires a different method for recycling.
	Pipe rinse into copper removal with copper recovery system.	Recovered copper can be sold.	More expensive equipment than used for chemical treatment. Must continue to haul spent etchant.

4. Electroless copper bath growth.	Process chemistry substitution using palladium-based and/or graphite-based alternatives.	Eliminates formaldehyde.	May produce excess organic wastes. May not be suitable for all applications.
	Copper recovery.	Reduces copper discharge.	Expensive compared to other methods.
	Autocatalytic copper removal onto a sponge.	Effective removal of copper metal complexed with EDTA. Possible copper recovery.	Produces a waste to be transported. Continuous activation, heat, and monitoring are required.
	Chemical treatment.	Least capital cost.	Produces sludge. Chelating agents are still present in the treated waste.
5. Develop/Resist Strip.	Use resists which do not emit wastes.	Eliminates hauling solids.	May not be universally applicable.
	Chemical treatment to reduce volume compared to hauling spent process baths off-site for disposal or dumping waste to the sewer.	Reduces quantity of this material to be hauled for off-site disposal.	Solids handling could be a problem. Expensive proprietary chemicals. Minimal organic load reduction in treated waste. Requires segregation.
6. Other wastes including tin, tin/lead, nickel, and gold.	Recover metal and chemistry from dragout tank.	Bath and metal recovery and recycling are possible.	Requires space to install dragout tanks. Requires more waste segregation and can be expensive as more recycling and recovery is required.
	Chemical treatment and membrane filtration.	Least capital cost.	Produces sludge. Increasing liability. Requires intensive effort for labor.
7. Printed circuit assembly cleaning wastes.	Use of water-soluble solder mask and flux.	Eliminates toxic chemicals. Allows rinse recycling.	Not universally applicable.
	Use of semiaqueous or saponified cleaners for RMA fluxes.	Eliminates use of CFCs.	Possible chemical treatment of bath VOC emission from semiaqueous baths.

60.8.5 Settling Process

Following the second-stage pH adjust, waste flows to the flocculation chamber in a clarifier where polyelectrolytes and recycled sludge are added and rapidly mixed. The waste then enters a second stage where a slow-speed mixer is used to enlarge the hydroxides to make settling more effective. Finally, the flocculated waste will flow down the clarifier. The precipitated solids settle to the bottom of the clarifier and the clarified liquid is discharged to a final pH adjust system or in some cases to a sand filter prior to final pH adjust if required to meet discharge standards. (See Fig. 60.6.)

60.8.6 Cross-Flow Microfiltration

Cross-flow microfiltration systems are used in place of a clarifier and sand filter where the maximum discharge limit for copper is below 1.0 mg/l. Treated wastes are collected in a recirculation tank and pumped at turbulent flow and 10 to 35 psig through a series of capsules containing tubular filters. The majority of the recirculated wastewater plus all the suspended solids returns to the recirculation tank. The suspended solids free water will pass through the side walls of the tubular filter. Sludge is withdrawn near the discharge of the recirculation pump. The 3 percent (total suspended solids) sludge is pumped to a sludge holding tank and then pumped to a filter press. (See Fig. 60.7.)

60.8.7 Sludge Thickening and Dewatering

The purpose of the sludge thickening tank is to increase the concentration of the 1 to 2 percent sludge from the clarifier to approximately 3 percent. This is accomplished by continuously decanting the water from the sludge thickening tank and directing it back to the rinse collection tank. The solids from the sludge thickening or holding tanks are then pumped to a filter press for dewatering to reduce sludge volume.

The total suspended solids in the sludge can be increased to about 35 percent with a filter press. In addition, sludge dryers are available which increase the solids concentration to about 70 percent by adding heat to the sludge. This reduces the volume of sludge to be hauled away.

60.9 ADVANTAGES AND DISADVANTAGES OF VARIOUS TREATMENT ALTERNATIVES

All of the techniques described in this chapter have advantages and disadvantages. None of them are appropriate in all situations. Each potential application requires specific analysis and a thorough understanding of the technical and economic issues involved prior to implementation. Table 60.2 presents a summary of the advantages and disadvantages of some of the technologies discussed in this chapter.

P · A · R · T · 15

FLEXIBLE CIRCUITS

CHAPTER 61
FLEXIBLE CIRCUIT APPLICATIONS AND MATERIALS

Dominique K. Numakura
DKN Research, Haverhill, Massachusetts

61.1 INTRODUCTION TO FLEXIBLE CIRCUITS

Flexible circuits are a form of printed wiring interconnect structure built on thin, flexible substrates. They are also bendable to complete three-dimensional (3-D) wiring that rigid circuit boards cannot achieve. Because of this flexibility, flexible circuits have many advantages over other wiring methods and have many applications in electronic equipment that requires high-density wiring in a small space. As the samples in Fig. 61.1 show, most flexible circuits have cabling functions rather than the mounting functions of electronic components.

Tape-automated bonding (TAB) has not been considered as a type of flexible circuit because of the slightly different manufacturing processes. However, the basic construction and materials of the final products are the same; therefore, TAB is categorized as a type of flexible circuit in this book (see Fig. 61.2).

61.1.1 Advantages and Disadvantages of Flexible Circuits

The basic advantages of the flexible circuits are its thinness and ability to bend. However, a thin flexible circuit generates many supplemental advantages not available with other wiring methods (see Table 61.1). A 3-D wiring with component assembling in a small space and a long-term dynamic flexing with small radius are typical examples of what other wiring materials cannot replace. On the other hand, flexible circuits also have many disadvantages due to unstable thin constructions. It is necessary to consider how to avoid the disadvantages when using a flexible circuit in a packaging system. Otherwise, overall process yield and productivity will be low and the final cost will be relatively very high.

61.1.2 Economics of Flexible Circuits

The largest disadvantage of flexible circuits is that they cost more than rigid circuit boards or flat cables of the same size. Usually, flexible circuits are larger than rigid circuit boards to allow 3-D wiring capability; however, although the total circuit cost increases, using flexible circuits is still often cheaper than using multiple rigid boards and associated connectors.

FIGURE 61.1 Various examples of flexible circuits.

FIGURE 61.2 Example of fine TAB circuits.

That is, the cost of using flexible circuits must be compared to the total cost of rigid circuit boards in terms of the following:

- Connectors
- Wires
- Additional assembly costs

TABLE 61.1 Advantages and Disadvantages of Flexible Circuits

Advantages	Disadvantages
Thin	Fragile
Light	Unstable dimensionally
Bendable	Low reliability—fatigue
Flexible	Difficult to design
Long flexing endurance	Complicated manufacturing process
3-D wiring	Difficult to handle
High-density wiring in a small space	Need special tools for assembling
Avoid connectors and soldering	Low yield in assembling
Avoid missed wiring	Difficult to rework
Low total cost, including all elements	High cost of circuit fabrication

Because the total cost of a flexible circuit is greater than that of an equivalent rigid circuit board alone, designers should not choose flexible circuits over other wiring methods unless three-dimensional advantages are needed, especially for consumer applications that require low-cost wiring. A flexible circuit should be used only when it provides the solution for the application's wiring requirements (for example, when the application requires bendable wiring in a small space, lightweight, long-flex endurance, and so on).The critical question is whether the value of a flexible circuit's performance is greater than its cost.

Miniaturization of electronic products requires extremely lightweight wiring in a small space, and therefore there have been many opportunities for flexible circuits. Figure 61.3

FIGURE 61.3 Flexible circuits applied for digital camera. *(Courtesy of Canon).*

shows examples of flexible circuits applied to a digital camera in which their advantages make them the best technical solution, and also the least expensive. In this example, the entire circuit was divided into one multilayer rigid board and more than 10 flexible circuits to optimize the advantages and minimize the total wiring costs.

61.2 APPLICATIONS OF FLEXIBLE CIRCUITS

As described, a flexible circuit should be adopted only when other wiring or packaging methods cannot satisfy the application's requirements. Table 61.2 shows how features of flexible circuits relate to appropriate applications. The table shows typical examples of flexible circuit applications in electronic products. Portable electronic products such as cellular phones, digital cameras, automobile instrument panels, and handheld calculators require reliable 3-D wiring in a small space. In some cameras, more than 10 flexible circuits are squeezed into a small space. Flexible circuits are used for the hinge wiring in many electronic products that have mechanical moving parts, such as notebook computers, palmtop computers, and personal digital assistants (PDAs). Many new electronic products, such as hard disk drives, small printers, and clamshell type cellular phones, require a long-life dynamic flexible endurance. Multilayer rigid/flexible circuits have been effective in reducing wiring space and increasing wiring reliability in aerospace electronic equipment and small portable electronic products such as cellular phones. Only flexible circuits could satisfy all these wiring requirements. Efforts to minimize the cost of flexible circuit wiring for consumer applications are ongoing. Often, the miniaturization of a consumer electronic product assumes a wiring system with multiple flexible circuits, especially if the equipment has mechanical moving portions.

TABLE 61.2 Features and Applications of Flexible Circuits

Features	Applications
High-density wiring in a small space	Digital camera, cellular phone, flat panel display, video camera, facsimile, ultrasound probe
Lightweight	Calculator
3-D wiring	Automobile instrument panel, cellular phone, camera, notebook computer
Reliable wiring	Aerospace applications, industrial applications
Long flexing endurance	Printer, hard disk drive, floppy disk drive, CD drive, VCR

61.3 HIGH-DENSITY FLEXIBLE CIRCUITS

Serious progress in the density of flexible circuits was made during the second half of the 1990s, owing to the extreme miniaturization of electronic products. These products incorporate a new technical concept in the design of the circuits. They have a much higher density than traditional flexible circuits, with dimensions for traces, spaces, and via holes equivalent to those of rigid high-density interconnect (HDI) boards. (Details are described in the Chapter on Special Construction of Flexible Circuits.) They also have supplemental constructions on the circuit to allow higher termination capabilities. This type of construction is called an *HDI flexible circuit* (that is, a flexible circuit for high-density interconnects).

61.3.1 Specifications

Table 61.3 shows typical requirements for interconnection in the end product, and the preferred specification of circuit parameters for the use of high-density flexible circuits.

TABLE 61.3 Common New Requirements for High-Density Flexible Circuits

Requirements for end use	Preferred specification for flexible circuits
High-density wiring	Pitches smaller than 150 μm, via holes smaller than 150 μm in diameter
Thin and light	Base 25 μm or thinner, conductor thinner than 12 μm
High-density SMT	Small opening on coverlay, tin or lead-free plating
High-density connection	High dimensional accuracy ($<$0.3 mm diameter)
Direct bonding	High-density flying lead, pitch $<$100 μm
Chip-on-flex, flip-chip	High-density microbump array, pitch $<$100 μm
High reliability for vibration	High bond strength of conductors
For mechanical shock	High bond strength between layers
For wide-range heat cycle	High heat resistance
For salty moistures	High insulation resistance

Portable electronic products are exposed to more severe conditions than ever before. As a result, they need higher reliability for the circuits as listed in Table 61.3. In the case of portable telephones, they are often carried in the owner's pocket, where they may be exposed to a variety of conditions, including continuous vibration, mechanical shock, temperature, humidity, pocket lint, and even perspiration. Other portable electronic products, such as digital cameras, PDAs, and notebook computers, could be exposed to similar conditions.

61.3.2 Applications

Due to usage and assembly technologies, many applications require increased performance for high-density flexible circuits. The manufacturing processes for these circuits must adapt to new requirements imposed by additional process conditions. Table 61.4 lists major applications and the associated advantages of using high-density flexible circuits for a given purpose.

TABLE 61.4 Major Applications of High-Density Flexible Circuits

Applications	Reasons to use high-density flexible circuits
Wireless suspension of disk drives	High-density wiring in a small space Flexible, thin, light, 3-D wiring Direct connection to magnetic heads
Interconnections of disk drive actuators	High-density wiring in a small space Flexible, thin, light, 3-D wiring Repeated reparability
Interposer of CSP	High-density wiring in a small space Thin, light, 3-D wiring
Flat-panel displays	High-density wiring in a small space Flexible, thin, light, 3-D wiring Chip-on-flex capability High-density connection
Ultrasound probes	High-density wiring in a small space Flexible, thin, light, 3-D wiring Direct connection to ultrasound device
Cellular phones	High-density wiring in a small space Flexible, thin, light, 3-D wiring surface-mount technology (SMT) and COF capability
Inkjet printers	High-density wiring in a small space Flexible, thin, 3-D wiring Direct connection to printer head device

Without high-density flexible circuits, these applications could not achieve miniaturization or support valuable new functions. For example, specially developed high-density flexible circuits can provide highly reliable connections at a low cost for new tiny magnetic heads of hard disk drives with a vertical recording system. Chip-scale package (CSP) technologies can be developed with high-density flexible circuits. Driver integrated circuits (ICs) for flat panel displays in personal computers and large flat panel TVs are assembled with a set of chip-on-flex (COF) substrates. Due to the limited space within the camera bodies, small liquid crystal display (LCD) monitors for video cameras and digital cameras are assembled using long-tailed high-density flexible circuits. The wiring of the "next generation" of portable electronics cannot be completed without high-density flexible circuits.

61.4 MATERIALS FOR FLEXIBLE CIRCUITS

The basic difference between flexible circuits and rigid circuit boards is the thin and flexible materials used in the substrates of flexible circuits. Furthermore, because the flexible circuits have complicated constructions, supplemental materials other than copper-clad materials have been required to build a whole circuit.

61.4.1 Traditional Flexible Circuit Materials

Table 61.5 shows traditional major materials and typical examples of flexible circuits. A broad variety of the materials are employed to build traditional flexible circuits. Film materials, such as polyimide (PI) films and polyester (PET) films are the specially adapted for use in flexible circuits. For greater flexibility, rolled annealed (RA) copper foil is used as the major conductor material.

Coverlays and stiffeners are typical examples of the special components needed to build flexible circuits. Many types of materials have been introduced to accommodate the designs of the end applications.

TABLE 61.5 Traditional Major Materials Used to Build Flexible Circuits

Application	Typical materials
Base substrates	Polyimide film (PI), polyester film (PET), polyester telephthalate Thin glass-epoxy, fluorinated ethylene propylene (FEP), resin-coated paper
Conductors	Electrodeposited (ED) copper foil, RA copper foil, stainless steel foil, aluminum foil, etc.
Copper-clad laminates	Epoxy-based, acrylic-based, phenol-based (adhesive-based)
Coverlay	PI film, PET film, flexible solder mask
Adhesive layer	Acrylic resin, epoxy resin, phenol resin, pressure-sensitive adhesives (PSAs)
Stiffener	PI film, PET film, glass-epoxy, metal boards, etc.

61.4.2 HDI-Oriented Flexible Circuit Materials

Since 1990, many new materials for flexible circuits have been developed in conjunction with design and manufacturing processes to satisfy new requirements in HDI applications. These include:

- High-performance films
- Adhesiveless laminates
- Direct cast processes of liquid polyimide resins
- Photoimageable coverlay

TABLE 61.6 Materials for High-Density Flexible Circuits

Application areas	New materials for high-density flexible circuits
Base dielectrics	Polyimide films (Kapton K, E, EN, KJ; Apical NP, FP; Upilex S) Liquid polyimide resin, PEN film, LCP films
Conductor materials	Ultra-thin copper foils, sputtered copper, copper alloys, stainless steel foil
Copper-clad laminates	Adhesiveless laminates (cast type, sputtered/plated type, laminated type)
Coverlay	Photoimageable coverlay (PIC) (dry film type, liquid ink type)
Adhesive sheets	Hot-melt polyimide film

See Table 61.6 for a list of new materials for HDI flexible circuits. There is a significant difference between traditional flexible circuits and HDI flexible circuits. For example, more than 80 percent of traditional flexible circuits were covered by adhesive-based copper-clad laminates that use the traditional polyimide films, Kapton H™ or Apical AV™. Conversely, the majority of high-density flexible circuits used in large-volume applications utilize all new materials.

HDI applications require finer traces and microvia holes in severe manufacturing and application conditions. For manufacturing convenience, flexible circuit producers prefer thinner conductors and substrates. However, physical performances of thinner materials may not be optimal. Thinner materials impact both the performance and manufacturing yield of the final flexible circuit.

Dimensional stability of the materials in high-density flexible circuits during manufacturing is a key to good process yields. To produce reliable high-density flexible circuits, conflicts between materials, constructions, and manufacturing processes must be resolved. Measures to accomplish this objective have been undertaken. For example, new construction designs and manufacturing technologies now require use of the materials shown in Table 61.6.

61.5 SUBSTRATE MATERIAL PROPERTIES

Many kinds of materials have been developed as the major dielectric layers of flexible circuits; however, polyimide films and polyester films have been the major core material for both substrate films and coverlay films with adhesive for traditional flexible circuits. The polyimide films have been the major substrate material for the soldering process and other high-temperature terminations processes, such as wirebonding and flip-chip, because of their high-temperature resistance and good balance of physical properties. Polyester films are not available for standard soldering, but they are a low-cost solution for large circuits such as those used in automobile instrument panels and long cable circuits of printers.

61.5.1 Comparison of Substrate Materials

There have been examples of other materials, such as thin glass-epoxy, that have not been widely used. To satisfy the new concept of HDI flexible circuits, innovative uses of raw materials such as polyethylene naphthalate (PEN) film, liquid crystal polymer (LCP) film, and poly ether ether ketone (PEEK) film have been introduced. Comparisons are shown in Table 61.7.

61.5.2 Polyimide Film

Polyimide film continues to be the major substrate material for flexible circuits because only it can support high-temperature processes such as soldering and wirebonding. Both Kapton H film by Du Pont and Apical AV by Kaneka have been used in the flexible circuit industry for

TABLE 61.7 Comparison of Substrate Materials

	Polyimide	Polyester	Thin glass-epoxy	LCP	PEEK
Maximum operating temperature	>200°C	<70°C	~105°C	~90°C	>200°C
Standard thickness	12.5, 25, 50, 75, 125 μm	25, 50, 75, 100, 125, 188 μm	100, 150, 200 μm	(50 μm)	(50 μm)
Soldering	Applicable	Difficult	Applicable	Possible	Possible
Wire bonding	Possible	No	Difficult	Difficult	Possible
Color	Brown	Transparent	Transparent	Transparent	Milky
Moisture absorption	High	Low	Low	Low	Low
Dimensional stability	Acceptable	High	High	Good	High
Flexibility	High	High	Low	High	High
Cost	High	Low	Medium	Low	High

a long time. Both of these materials have a good property balance, and they are still the major substrates in traditional flexible circuits.

61.5.2.1 Properties of Polyimide Films. Table 61.8 shows major properties of typical polyimide films, Kapton H and Apical AV. These have good mechanical properties and provide excellent performance as base substrate films and coverlay films in dynamic flexing applications. They have flame-retardant characteristics, and it is not difficult to achieve Underwriters Laboratories (UL) flame class 94-V-0 or 94-VTM-0. The largest disadvantage of polyimide film is its higher cost compared to other common plastic films such as PET.

There are several barriers to using traditional polyimide films as the major substrate materials in HDI flexible circuits. Dimensional stability is the largest issue. Both Kapton H film and Apical AV by Kaneka have coefficients of thermal efficiency (CTEs) higher than 30 ppm, which is not acceptable for the large-volume production processing of HDI flexible circuits. Relatively high moisture absorption is another major barrier to using these films as the base

TABLE 61.8 Properties of Traditional Polyimide Films and New Polyimide Films

		Kapton H™	Apical AV™	Kapton E™	Apical NP™	Apical HP™	Upilex S™
Manufacturer		DuPont	Kaneka	DuPont	Kaneka	Kaneka	Ube
Thickness (μm)		12.5, 25, 50, 75, 125	12.5, 25, 50, 75, 125	12.5, 25, 50, 75	12.5, 25, 50, 75	12.5, 25, 50, 75	12.5, 25, 50, 75, 125
Tensile strength (kg/mm²)	MD	25.2	25.0	28.3	30.0	28.6	39.4
	TD	22.3	27.0	25.4	32.0	29.0	40.2
Elongation (%)	MD	85	119	16	82	40	22
	TD	83	114	32	73	38	21
Tensile modulus (kg/mm²)	MD	336.1	305.9	785.5	407.9	654.0	897.3
	TD	321.4	312.2	622.4	428.3	661.0	912.1
CTE (ppm) (100–200°C)	MD	27	35	3	17	12	14
	TD	31	31	12	13	11	15
CHE (ppm) (50°C, 35–75% RH)	MD	15	16	5	13	8	9
	TD	16	15	8	12	6	10
Heat (%) shrinkage (200°C, 2 h)	MD	0.18	0.08	0.03	0.06	0.06	0.07
	TD	0.20	0.03	0.02	0.02	0.02	0.10
Water absorption (23°C, 24 h)		2.8	2.9	2.2	2.5	1.1	1.9
Alkaline resistance		17.4	25.9	5.5	22.9	4.2	~0
Chemical etching		Possible	Possible	Possible	Possible	Possible	Difficult

material of the flexible circuits. High moisture levels in the films cause a lot of supplemental problems, mostly in increased temperature processes. Higher-performance polyimide films are required as the substrate material of HDI flexible circuits.

The primary materials used in industry to optimize cost and performance have been 50 micron thick polyimide film, used for industrial and avionics applications, and 25 micron thick films, used for consumer applications. However, since late 1990s, films as thin as 12.5 micron have been specified to reduce thickness and increase flexibility.

61.5.2.2 High-Performance Polyimide Film. Various manufacturers have made several aggressive measures to commercialize high-performance polyimide films. Upilex S by Ube Industry has excellent chemical stability, and it is a barrier for chemical processes such as adhesion with the other materials or chemical etching. Strong chemicals, such as hidrasine, are required to do the chemical etching. Upilex-S could be the major material as the substrate of TAB, which requires very high dimensional stability at high heat resistance. Newly developed polyimide films such as Kapton E, EN, and Kaneka's Apical NP and FP have higher dimensional stability and lower moisture absorption rates. It is possible to etch these films using mild alkaline chemistry. Basic properties of the materials are summarized in Table 61.8.

Hot-melt-type (thermoplastic) polyimide films have been developed as high-heat-resistant adhesive materials. Mitsui Chemical's TPI, DuPont's Kapton KJ, Ube's Upicel, and Kaneka's Pixeo are typical examples. These are coated on dimensionally stable polyimide films to ensure good physical performance.

61.5.2.3 Liquid Polyimide Resin (Photoimageable). Many new HDI applications require special capabilities and properties on the substrates that are not provided by commercial polyimide films. In these cases, circuit manufacturers initiate production by manufacturing polyimide substrates. Sometimes manufacturers have to start from the synthesis of polyimide resins.

Several liquid polyimide resins have been developed as the base materials of high-density flexible circuits. Some of these liquid polyimide resins can be photoimaged, and have been used in high volume as the dielectric layer and coverlay in the wireless suspension of hard disk drives.

These liquid polyimide resins could be the major dielectric materials of special high-density flexible circuits that demand extremely high density, down to 5 micron pitches with 10 micron via holes. The cost of these materials is higher than that of polyimide film. However, they have broader capability for meeting nonstandard requirements such as ultrathin substrates with microvia holes. The properties are very dependent on the manufacturer.

61.5.3 Polyester Film

PET film (polyester film or polyethylene telephthalate) has been produced as a common film for general use, for the following reasons:

- It can provide a low-cost solution.
- It has excellent mechanical performance at room temperature.
- It has low moisture absorption and its dimensional stability is excellent.

Unfortunately, PET film cannot keep its high performance at higher temperatures than 70°C, and it is available for only nonsoldering applications. (Several soldering technologies have been developed for PET-based flexible circuits, but they are very specialized.) Also, PET films are flammable and it is difficult to achieve UL flame class certification. Not many available PET films can satisfy the flammability requirements of UL-94. Typical properties are shown in Table 61.9.

TABLE 61.9 Basic Properties of PET Films (Nonflammable Grade)

Test Items	Properties	
	MD	TD
Tensile strength (kgf/mm²)	16	16
Elongation	110%	100%
Edge tear resistance (kgf/mm)	51	58
Folding endurance	37,000 times	39,000 times
Heat shrinkage	1.4%	–0.2%
CTE	30 ppm	
CHE	10 ppm	
Water absorption	0.7%	
Volume resistivity	4 × exp 15 Ω*cm	
Dielectric constant	3.0 for 1 MHz	
Dielectric loss	0.03 for 1 MHz	
Breakdown voltage	13.6 kV	
Light transmission	46%	
Chemical resistance	Stable in weak acid and alkaline solution Stable in organic solvent	
Flammability	UL-94-VTM-0	

Data courtesy of Mitsubishi Plastic.

61.5.4 Other Materials Used for Flexible Circuits

Thin glass-epoxy sheets less than 200 μm thick can be flexible and can be used as a base substrate of low-cost flexible circuits. Their physical properties are basically the same as those of rigid circuit boards, and standard manufacturing processes and assembling processes for rigid circuit boards are available for these materials. Standard soldering processes are applicable. But glass-epoxy sheets do not have high flexibility, and they are not available for repeated flexing use. The basic properties of these materials are listed in Table 61.10.

TABLE 61.10 Properties of Thin Glass/Epoxy Substrate

Items	Risho industrial		Matsushita electric
	MD	TD	(R-5766 series)
Tensile strength (kg/mm²)	24.5	17.0	—
Elongation (%)	3.2	1.6	—
Thickness (μm)	120		50, 70, 100, 130, 150, 180, 200
Dimension change by heat (%)	–0.05		—
Dimension change by humidity (%)	0.03		—
Surface resistance (Ω)	1 × exp 12–13		5.2 × exp 14
Volume resistivity (Ω*cm)	1 × exp 12–13		
Dielectric constant	—		4.7
Dissipation factor (at 1 MHz)	—		0.015
Water absorption (%)	1.0		0.18
Breakdown voltage (kV/0.1 mm)	3.6		—
Bond strength (kg/cm)	—		1.39
Soldering resistance	—		260°C × 120 s
Flammability	—		UL-94-V-0

These materials are supplied as copper-clad laminates.

Fluorized carbon polymer films have been applied as a low-loss substrate material for flexible circuits. However, they do not have good dimensional stability or adhesion characteristics. High cost of the materials is the major reason that they cannot be standard in flexible circuits. In the last 20 years, several heat-resistant films, such as polypalabalic acid film and polysulfon film, have been developed as alternative substrate materials in flexible circuits instead of polyimide films. Unfortunately, there was no successful material from a business standpoint. LCP films and PEEK have been considered as the new materials of high-speed flexible circuits based on the low dielectric constant and loss tangents.

61.6 CONDUCTOR MATERIALS

As flexible circuits have various mechanical stresses, their conductors also have requirements for more flexibility and toughness than traditional ED copper foils of standard rigid circuit boards. An RA copper foil is the solution for high flexing endurance. It is necessary for dynamic flexing use. RA copper foils are more expensive than ED copper, especially at thicknesses of less than 18 micron. Therefore, high-ductility electrodeposited (HD-ED) copper foils have been developed that have greater flexibility than ED copper foils and a lower cost than RA copper foils. The differences between the copper foils are shown in Table 61.11.

TABLE 61.11 Basic Properties of Copper Conductors

	ED foil	HD-ED foil	RA foil
Manufacturer	Furukawa Electric	Furukawa Electric	Japan Energy
Grade	STD foil	HD foil	RA
Thickness (μm)	9, 12, 18, 35, 70	18, 35	12, 18, 35, 70
Tensile strength (kg/mm^2)	TD 34	TD 32	MD 21.5 TD 18.9
Elongation (%)	TD 9	TD 23	MD 12.8 TD 9.5
Surface roughness (μm)	8	10	<3.5
Cost	Low	Middle	High

(Data measured for 35 μm Thick Foils).

Due to the stringent requirements of fine-line etching, thinner copper foils have been developed with small surface profiles. RA copper foil 12 μm thick is a standard product. Low-profile 12 μm thick ED copper foils have also been commercialized for fine-line etching. A special treatment has been developed to ensure reliable bond strength with each adhesive material. In the next five years, these materials will be standard and utilized in large volume. RA copper foils thinner than 10 μm are available; however, processing and high material costs need to be addressed before large-volume production can commence.

Copper foils thinner than 5 micron will be demanded for the semi-additive process to generate 10 micron wide traces or finer since the beginning of 2000s. Two types of technologies have been developed to have ultrathin copper foils. The first one is the "etch-down" process used to reduce the copper thickness from standard thick RA copper foils. Through this process, foils 3–5 micron thick can be obtained from 12 micron RA copper foils. The second is the use of ultrathin foils. ED copper foils 1–5 micron thick have been commercialized with the carrier copper foils. Ultrathin copper layers have been built on the smooth surface of other copper foils with standard thickness. The carrier copper foils are removed mechanically after the lamination with the base layers. Ultra thin copper foils have been applied to adhesiveless laminates through casting or lamination processes.

In addition, sputtering and plating technologies have achieved conductors thinner than 5 μm with semi-additive processes. Among others, these processes can generate copper, nickel, and gold conductors. Also, they can produce fine traces at less than 10 μm pitch on flexible substrates. Plating baths and chemicals have been developed for these technologies.

Special metal and alloy foils other than copper ones have been developed for specific applications.

Aluminum conductors have been developed as the low-cost solution. They have been applied for volume production of the keyboards of calculators, antennas of wireless devices, and so on. But they could not be universal conductor materials of standard flexible circuits because of difficulty in soldering and the special chemistries needed for etching. The wireless suspension of disk drives has consumed a large volume of special stainless steel and copper alloy foils because of the special mechanical performance of the circuits. A high-resolution printer head has also utilized thin tungsten foils as the conductor of thermal printer head circuits. Nickel-chromium alloy foils have been developed as the conductor materials of flexible heater circuits.

61.7 COPPER-CLAD LAMINATES

The majority of flexible circuit manufacturers start the process with copper-clad materials. The properties of these materials depend on the capabilities of laminate manufacturers, even though the same base films and copper foils are used. To choose the right materials, manufacturers must carefully consider the basic properties of each laminate material.

61.7.1 Adhesive-Based Laminates

Historically, copper-clad laminates with acrylic or epoxy adhesives have been the major materials for flexible circuits. Each manufacturer has developed a special resin grade or special additives to ensure reliable flexibility and bond strength. Other adhesive materials such as phenol resin or silicon resin have been developed; however, they have not become standard adhesive materials in flexible circuits.

The adhesive-based copper-clad laminates still represent more than 50 percent of the traditional flexible circuit market. The major properties of the materials are shown in Table 61.12.

The manufacturing process is illustrated in Fig. 61.4. Mostly, these films are processed and supplied in roll form. The surfaces of polyester films and polyimide films undergo special processes such as sandblasting and plasma treatment to achieve a reliable bond strength. A specially blended adhesive resin is coated on the film and dried. Then a copper foil is laminated continuously under appropriate temperature and pressure. The surfaces of the copper foils receive a specific treatment according to the requirements of each laminate's manufacturer.

TABLE 61.12 Basic Properties of Adhesive-Based Copper-Clad Laminates (Polyimide-Based)

Items		Properties	
Manufacturer	DuPont	Nikkan	Shin-Etsu
Grade	Pyralux LF	Nikaflex	RAR Series
Adhesive layer	Acrylic	Epoxy	Epoxy
Peel strength (kgf/ccm)	1.4	1.3	1.8
Dimensional stability MD	–0.08	–0.09	–0.09
(%) TD	–0.07	+0.03	–0.04
Flexing endurance MD	N/A	3200	2650
(MIT, 2.0R) TD		2950	2850
Insulation resistance (Ω)	1.0 × exp 11	2.5 × exp 13	1.0 × exp 13
Surface resistivity (Ω)	1.0 × exp 13	2.7 × exp 14	1.0 × exp 14
Volume resistivity (cm)	1.0 × exp 14	2.0 × exp 16	1.0 × exp 16
Soldering resistance	288°C × 5 min	280°C × 10 sec.	280°C × 10 sec.
Flammability	No	UL-94-VTM-0	UL-94-VTM-0

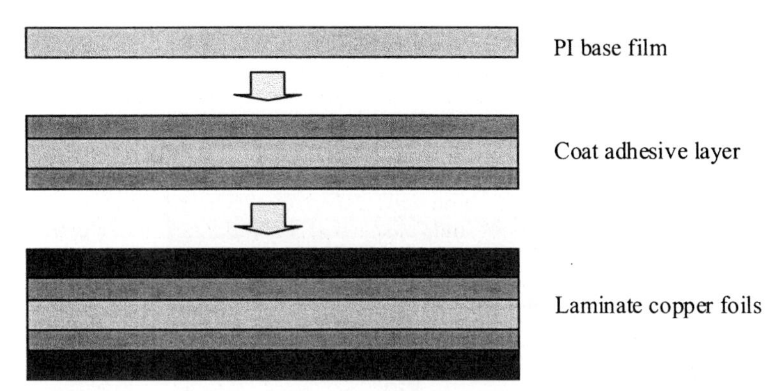

FIGURE 61.4 Manufacturing process for adhesive type laminate, showing a polyimide-based film combined with an adhesive layer coat and laminated to copper foils.

The same process is repeated for double-sided copper-clad laminates. (Some laminate manufacturers, however, have developed simplified manufacturing processes that can make the double-sided laminate in one single process to reduce manufacturing cost.) A well-conditioned aging process is important to achieving reliable bond strength and flexible characteristics of the laminates as the raw material of flexible circuits.

Most of the adhesive resins have lower heat resistances than polyimide films, and are the bottleneck for high-temperature processing of flexible circuits (e.g., lead-free soldering and wirebonding).

The flame-retardant properties of the laminates depend on the composition of the adhesive materials used by each manufacturer. Usually, a normal flame-retardant component in an adhesive resin has a negative effect on bonding. Several adhesive resins contain organic bromine molecules as the flame-retardant components, and will be eliminated for the ecological concerns. (For a discussion of this issue, see Chap. 6.) As soldering temperatures increase, heat-resistant adhesives or adhesiveless laminates systems are required.

61.7.2 Adhesiveless-Based Laminates

Several laminates without adhesive layers have been developed as the advanced materials for the next generation of flexible circuits. Lamination technology using epoxy resin or acrylic resin has been almost eliminated from HDI flexible circuits even though it uses new high-performance polyimide films as the substrates. Three types of adhesiveless copper-clad laminates have been developed (see Fig. 61.5):

- Cast
- Sputtering/plating
- Lamination

Each has different manufacturing processes and advantages as the major materials of HDI flexible circuits. Table 61.13 shows basic performances of typical adhesiveless laminate materials.

Currently, there is no perfect solution that satisfies all requirements including cost. Suitable adhesiveless laminate materials should be chosen according to the specifications of the final flexible circuits and the convenience of manufacturing processes.

61.7.2.1 Cast-Type Adhesiveless Laminates. The cast-type adhesiveless laminates have a good cost/performance balance. They have a high bond strength between substrate layers and conductor foils. They can also use special conductors such as copper alloy, nickel, stainless

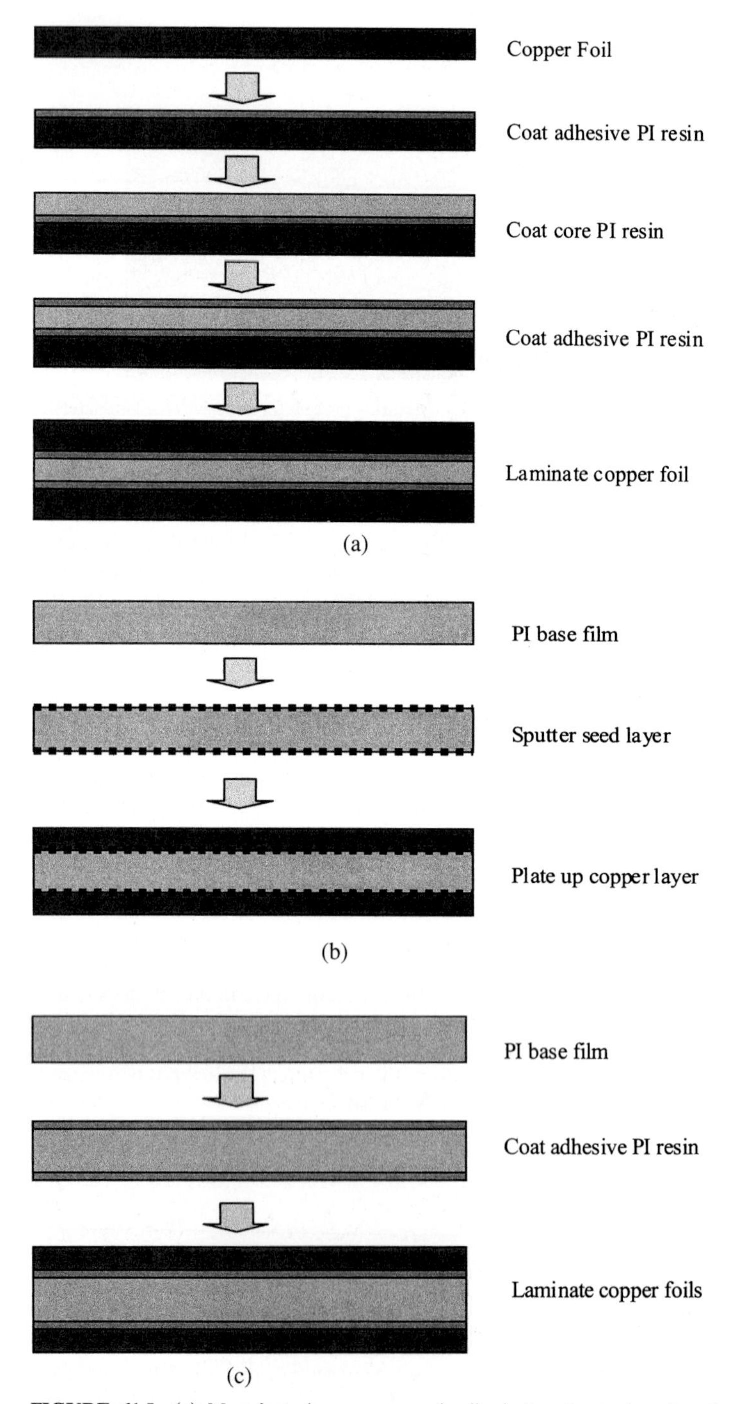

FIGURE 61.5 (a) Manufacturing process of adhesiveless laminate using the cast process, (b) Manufacturing process of adhesiveless laminate using plating process, (c) Manufacturing process of adhesiveless laminate using lamination process

TABLE 61.13 Comparison of Adhesiveless Copper-Clad Laminates

Items	Sputtering/plating process	Cast process	Lamination process
Manufacturer	Gould Electronics	Nippon Steel Chemical	DuPont
Grade	GouldFlex	Espanex	Pyralux AP
Choice of dielectric	Wide	Small	Small
Base etchability	Fair	Difficult	Difficult
Dielectric thickness (μm)	12.5–125	12.5–50	12.5–150
Choice of conductors	Small	Wide	Wide
Conductor thickness (μm)	~35	12–70	12–70
Bond strength	1.2	1.5	1.4
Double-sided	Available	Available	Available
Flexing endurance	N/A	180 (MIT 0.8R)	N/A
Dimensional change after etching	MD: –0.05%	MD: –0.02%	MD: –0.05%
	TD: 0.03%	TD: –0.02%	TD: –0.05%
Insulation resistance (Ω)	>1 × exp 10	>1 × exp 13	>1 × exp 14
Volume resistivity (Ω cm)	>1 × exp 14	>1 × exp 15	>1 × exp 17
Flammability	UL-94-VTM-0	UL-94-VTM-0	UL-94-VTM-0
Roll clad	Standard	Standard	Not available

steel, etc. In addition, they have good flexibility to generate very thin substrate layers down to 12 micron thick. They can have a wide range of copper thicknesses because of the process. There is no difficulty in making 70 or 105 μm copper laminates. Because it has been around longer, the cast-type laminate has a relatively broader availability than other adhesiveless laminates. The process can produce both single-sided and double-sided laminates.

The liquid polyimide resins are coated on a metal foil, directly controlling mechanical stress on the materials. Air-floating conveyors have been employed to reduce the stress. A thermal treatment process is the key to ensuring good physical performance of the laminates. Generally, two or more polyimide layers are coated on the metal foil to ensure both good dimensional stability and high bond strength. Usually, it is difficult to apply the chemical etching process to adhesive polyimide resins using standard alkaline solutions.

Producing double-side laminates through casting requires a special process because the chemical imidation reaction generates a lot of moisture and results in delamination during the lamination process. One special process is to coat hot-melt-type polyimide on the top of the substrate layer, and another is to laminate another copper foil at high temperature and pressure. Physical performance varies depending on the combination of the polyimide resins and lamination conditions. Usually, higher temperatures than 300°C are required.

61.7.2.2 Sputtering/Plating-Type Adhesiveless Laminates. A two-step process is required to produce sputtering/plating-type adhesiveless laminates. The first step is a kind of sputtering process to produce seed layers on substrate films. The seed layers have two functions: to generate conductive layers on the nonconductive plastic films, and to increase the bonding strength between base films and conductor layers. Generally, the sputtered layers are thinner than 100 nm. Therefore, the second step builds the thick conductor layers by electrical plating. To produce double-sided laminates, the process is usually repeated twice.

Sputtering/plating-type adhesiveless laminates can generate the thin conductors required for high-density circuits. Theoretically, these laminates provide stable conductor layers thinner than 10 μm at low cost. Technically, it is very possible to have 0.1–1.0 μm thick conductors. Although plating-type laminates provide more substrate material choices, their basic physical properties depend heavily on substrates. The majority of manufacturers introduced the sputtering process to generate a seed layer for good bond strength between the substrate and the conductor. Two or three steps of sputtering processes with nickel, chromium, and their alloys are employed to strike a good balance of chemical, physical, and electrical performance. Sometimes seed layers cannot be etched.

Few choices are available for conductor materials. Thick conductors are not optimal because they requires a longer plating process. Micron-size pinholes in the conductor layer are the biggest issue with this material. Several manufacturers have introduced additional chemical processes to eliminate these pinholes. Generally, the bond strength of the laminate is lower than that of other types.

61.7.2.3 Lamination-Type Adhesiveless Laminates. Thermoplastic-type polyimide resins with high-temperature melting points have been developed to serve as the hot-melt adhesives between copper foils and dimensionally stable polyimide films. The adhesiveless flexible laminates are produced by a heat lamination of the copper foils and base polyimide films that have thin hot-melt polyimide layers coated on the surface of the core polyimide films. The polyimide films, with hot-melt polyimide layers, are supplied by the major polyimide film suppliers. Temperatures higher than 330°C are required to produce reliable laminates. Materials processed at temperatures lower than 200°C are available for several hot-melt polyimide resins; however, their physical properties are inferior to those of standard laminates.

Producing double-sided laminates is not tremendously difficult. A conventional vacuum heat press with a higher processing temperature than 350°C is available to produce the laminate for small-volume production. A high-temperature roll laminator or belt laminator is used for high volume production.

Lamination-type adhesiveless laminates have advantages in small-volume production. Special polyimide films that have hot-melt polyimide resin on the surface of dimensionally stable polyimide film are available commercially. If a manufacturer has high-temperature heat press equipment, it is not difficult to produce the laminates. A compact roll-type heat laminator was developed to produce the rolled laminates.

Many choices of conductor materials are available for lamination-type adhesiveless laminates. Due to the special combination of conductor materials with stainless steel and copper alloy foils, this material has a variety of uses in multilayer rigid/flexible aerospace applications and in wireless suspension of disk drives. Generally, the adhesive polyimide is difficult to etch using standard alkaline etching solutions, so strong chemistry or a physical process such as plasma etching is required for the process.

Figure 61.6 shows the thickness ranges available for the substrates and conductors for each type material. All of the laminates have expanded their capabilities, and several thicknesses are overlapped. Therefore, circuit manufacturers have more choices for specific thicknesses and can consider other performance variables, including cost.

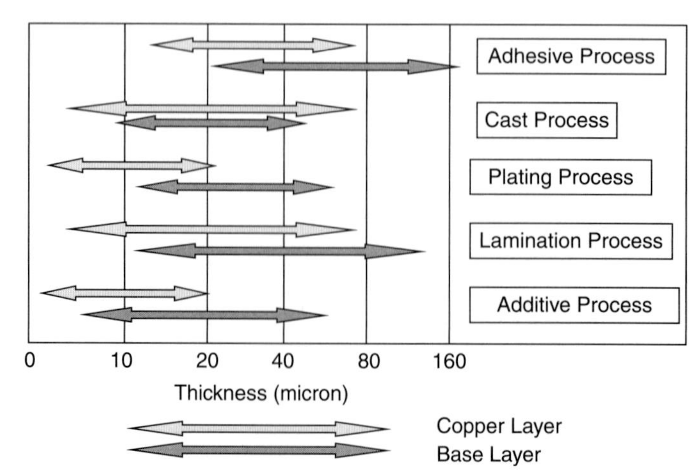

FIGURE 61.6 Thickness range of the copper clad laminates.

61.8 COVERLAY MATERIALS

One of the major differences between flexible circuits and rigid circuit boards is coverlay. In addition to the solder mask used for assembly, coverlay is the mechanical protector for the fragile conductors on flexible circuits. Film-based coverlay and flexible solder mask have been standard materials for traditional flexible circuits. Several types of photoimageable coverlay materials have been developed to satisfy the fine-resolution requirements of HDI flexible circuits:

- Film coverlay
- Screen-printable coverlay (flexible soldermask)
- Photoimageable coverlay (dry-film type and liquid type)

Several types of the materials are available for each coverlay type.

61.8.1 Film Coverlay

Usually the same films are chosen for the coverlay films as are used for the substrate layers. The films are coated with semicured acrylic or epoxy resin, which is used as an adhesive layer. The surfaces of the adhesive layers are covered with release films when the materials are delivered to circuit manufacturers. Semicured adhesive resins are unstable at room temperature, so film coverlay must be kept in refrigerators. The adhesive resins of the coverlay usually have only weeks of pot life even when they are kept in low-temperature environments, and exact conditioning is required for processing.

The process of manufacturing with film coverlay is very complicated, which makes process automation difficult and increases the cost. The manual registration process and the use of unstable film material are major issues when considering film coverlay for HDI flexible circuits due to small hole capability and lower dimensional accuracy. Basic properties of the typical materials are shown in Table 61.14.

TABLE 61.14 Basic Properties of Film Coverlay (Polyimide Film Base)

Manufacturer	DuPont	Nikkan Industry
Grade	Pyralux LF	CISV
Adhesive	Acrylic	Epoxy
Lamination temperature	180°C	160°C
Storage condition	Room temperature	Refrigerator
Bond strength (kg/cm)	1.3	1.0
Surface resistivity (Ω)	1.0 × exp 13	3.0 × exp 14
Volume resistivity (Ω*cm)	1.0 × exp 14	1.2 × exp 16
Soldering resistance	288°C × 5 min	280°C × 10 s
Flammability	No	UL-94-V-0

61.8.2 Screen-Printable Coverlay Ink (Flexible Solder Mask)

A screen-printing process similar to that used for rigid circuit boards is available for flexible circuits as the low-cost coverlay process. However, specially conditioned liquid solder mask materials are required to produce appropriate flexibility for the circuits. Standard solder mask materials designed for the rigid boards have small cracks and de-laminations when bending.

TABLE 61.15 Basic Properties of Screen-Printable Coverlay

Manufacturer	Taiyo Ink	Nippon Polytech
Grade	S-222	NPR-5
Base resin	Epoxy	Epoxy
Color	Green	Green
Hardness	5H	H
Peel strength	100/100 (crosshatch)	100/100 (crosshatch)
Flexibility	N/A	0.5 mm diameter
Soldering resistance	260°C × 20 sec.	260°C × 20 sec.
Insulation resistance (Ω)	>1 × exp 13	>1 × exp 12
Flammability	UL-94-V-0	UL-94-V-0

Generally, because of poor resolution, a screen-printing process does not satisfy the requirements of high-density SMT assembling on flexible circuits. Also, such as process does not provide good mechanical performance for dynamic flexing. Typical property examples are shown in Table 61.15.

61.8.3 Dry-Film–Type Photoimageable Coverlay

The basic idea of the photoimageable coverlay is to satisfy both of fine resolutions of the opening and physical flexibility by one single material. Also, its simple manufacturing process should reduce costs.

From a processing standpoint, there are two types of PIC materials: dry film and liquid. The dry-film type is easily processed when manufactured with vacuum laminators, which provide good encapsulation with the fine conductors. A vacuum press is also available for a small-volume production. However, the cost per unit area is higher for the dry-film type than for liquid materials.

Dry-film–type and liquid ink–type photoimageable coverlay materials have been developed with different resin matrixes to satisfy different technical requirements of HDI flexible circuits. A comparison of the materials is given in Table 61.16. Properties of the materials are shown in Tables 61.17 and 61.18.

TABLE 61.16 Comparison of Photoimageable Coverlay Materials

	Dry film type		Liquid ink type	
	Acrylic-base/epoxy-base	Polyimide-base	Epoxy-base	Polyimide-base
Applicator	Vacuum laminator	Roll laminator	Screen print Spray coat Curtain coat	Screen print Spray coat Roll coat
Thickness range (μm)	25–50	25–50	10–25	10–20
Minimum opening (μm)	70	70	70	50
Flexibility	*	Good	*	Good
Heat resistance	Acceptable	Excellent	Acceptable	Good
Electrical properties	*	Good	*	*
Chemical properties	Acceptable	Good	Good	Excellent
Handling	Easy	Difficult	Fair	Difficult
Technical hardle	Low	High	High	Very high
Storage condition	Refrigerator	Room temperature	Refrigerator	Freezer
Quick turn	Easy	Easy	Slow	Slow
Material cost	Medium	High	Low	High
Process cost	Medium	High	Low	Low

*Depends on manufacturer.

TABLE 61.17 Properties of Liquid/Epoxy-Type Photoimageable Coverlay

Manufacturer	PolyTech	Coates	Asahi	Peters
Grade	NPR-80	Image Flex	Focus Coat	Elpemer
Pencil hardness	5H	H, HB	4H	6H
Bond strength on copper	100/100	100/100	100/100	100/100
Flexibility	MIT test 0.5 mm R >500 times	1-mm mandrel: pass	180° bend:pass	3-mm mandrel: pass
Insulation resistance (Ω)	2 × exp 12 (IPC)	5 × exp 11 (IPC)	6 × exp 13 (JIS)	2 × exp 14 (VDE)
Volume resistance (Ωcm)	—	—	5 × exp 14	1 × exp 15
Dielectric strength	28 kV/mm	91 kV/mm	(2.1 kV/0.16 mm)	88 kV/mm
Soldering resistance	260°C × 10 s	260°C × 10 s	260°C × 1 min	288°C × 20 s
Flammability	UL-94-V-0	UL-94-VTM-0	N/A	UL-94-V-0
Resistance in Ni/Au plating	Pass	Pass	Pass	Pass
Exposure intensity (mJ/cm²)	150	400	300	150
Developer	Sodium carbonate	Sodium carbonate	Sodium carbonate	Sodium carbonate

61.8.3.1 Acrylic or Epoxy/Dry-Film Type. Acrylic or epoxy/dry-film types were the first materials developed to act as the flexible photoimageable coverlay for HDI flexible circuits. They have the same product concept as the photoimageable solder mask of rigid circuit boards. The same vacuum laminators and imaging equipment are available for these materials.

Managing the conditioning of the process is not difficult. Acrylic or epoxy/dry-film types are good for small-volume production because of the flexible manufacturing process. They may need some more improvements in electrical performance and chemical resistance if they are to be used in the general application of high-density flexible circuits. The present version

TABLE 61.18 Properties of Different Photoimageable Coverlay Materials

	DuPont	NSCC	Nitto Denko	Toray
	Puralux PC	Espanex SFP	JR-3000	Photoneece
Material	Acrylic/dry film	Polyimide/dry film	Polyimide/liquid ink	Polyimide/liquid ink
Pencil hardness	3H	—	—	—
Tensile strength (MPa)	N/A	226	120	148
Elongation	>55%	—	11%	36%
CTE (ppm/K)	130	23	35	16.1
Bond strength on copper	100/100	0.8 kg/cm	100/100	100/100
Flexibility	MIT 0.38R 100 times	MIT 0.38R 1200 times	N/A	N/A
Surface resistance (Ω)	>1.0 × exp 12	>1.0 × exp 13	—	>1.0 × exp 16
Volume resistance (Ω*cm)	3.4 × exp 16	1.0 × exp 16	5 × exp 15	>1.0 × exp 16
Dielectric strength (kV/mm)	>80	—	240	>300
Dielectric constant	3.5–3.6	3.5	3.3	3.2
Loss tangent	0.03	0.007	0.6	0.002
Soldering resistance	260°C × 10 s	350°C (JIS)	>300°C	>300°C
Flammability	UL-94	UL-94-VTM-0	Self-distinguishable	Self-distinguishable
Resistance in Ni/Au plating	Pass	Pass	Pass	Pass
Exposure intensity (mJ/cm²)	200	400	500	150
Developer	Sodium carbonate	Lactic acid solution	Alkaline solution	Special chemical

The materials were developed assuming different applications; therefore measured data are not exactly equivalent.

of the materials, however, yields a good balance of performances for both high-density soldering and flexing endurance.

61.8.3.2 Polyimide/Dry-Film Type. The physical performance of the polyimide/dry-film type is excellent. When applied to polyimide-based adhesiveless laminates, it provides very high heat resistance and good dimensional stability. Also, it has high electrical properties. It has been applied in high-density wiring of interconnection in the head suspension of disk drives. The major issues with this material are the complicated pattern-generation process and high material cost. It requires multiple chemical processes using special chemical solutions. A high-temperature baking of over 250°C is another difficulty. This is required to complete the standard manufacturing process of flexible circuits.

61.8.4 Liquid-Based Photoimageable Coverlay

Eventually, photoimageable coverlay could be the practical solution for HDI flexible circuits in large-volume production. It is similar to the solder mask of rigid circuit boards; however, the materials for flexible circuits must have high flexibility. None of the materials developed as solder masks for rigid boards are available for flexible circuits, even though dynamic flexing is not required. The liquid PIC requires special coating equipment such as a screen printer or sprayer. It can, however, provide a low-cost coverlay for large-volume production.

61.8.4.1 Epoxy/Liquid Ink. This material combination has been providing the best cost/performance combination for large-volume applications. Many manufacturers have tried to develop a good photoimageable coverlay with this combination, and Table 61.17 shows several successful examples. They provide good performance balances in mechanical, electrical, and chemical properties. A specialty is that these materials yield higher flexibility, without serious curling during the thermal processing, than other materials offer.

These materials are available by screen printing, spray coating, and other coating methods. UL-certified grades are available for commercial use. The flexibility of the materials allows several cycles of bending; however, the materials are not available for the long-term flexing that applications such as hard disk drives and printers require.

61.8.4.2 Polyimide/Liquid Ink. This material concept was generated as the reliable thin dielectric layer of flexible circuits for both substrates and coverlay. Polyimide/liquid ink is highly reliable for a wide range of thicknesses. A 10 μm thick layer could have sufficient performance to be a coverlay of a high-density flexible circuit. An appropriate combination with an adhesiveless copper laminate allows long-term dynamic flexing. The major issues are short pot life and high cost. Some of the ink materials must be stored at a temperature lower than 0°C. The materials require severe process conditions, especially for the final curing, where they require a temperature higher than 300°C. The cost of polyimide-based photoimageable materials is 5 to 10 times higher than for epoxy-based materials.

61.9 STIFFENER MATERIALS

All kinds of sheet or board material could be used as stiffener materials for the flexible circuits; however, several materials are commonly used. Typical stiffener materials for the traditional flexible circuits are listed in Table 61.19. Paper phenol boards and glass-epoxy boards are employed for relatively thick requirements. Polyimide films and polyester films are employed for relatively thin requirements. Aluminum plates and stainless steel plates are commonly used as the stiffener materials of flexible circuits. A specialty of the metal stiffeners is their forming capabilities after they are bonded on flexible circuits. Paper phenol and polyester are not available for thermo setting adhesives because of the low heat resistance.

TABLE 61.19 Comparison of Stiffener Materials

	Paper/phenol	Glass-epoxy	Polyimide	Polyester	Metal
Thickness range (μm)	500–2,500	100–2,000	25–125	25–250	Very wide
Soldering	Possible	Possible	Possible	No	Possible
Adhesion by thermoset resin	No	Possible	Possible	No	Possible
Cost	Low	High	Medium	Low	Medium

61.10 ADHESIVE MATERIALS

Several different types of adhesive materials can be used in flexible circuits depending on the application, except for the adhesive layer of copper laminates and film coverlays. The major applications are the bonding of stiffeners and multilayer constructions.

Table 61.20 compares the materials. Pressure-sensitive adhesives (PSAs) provide a low-cost solution due to their simple application process. A problem of PSA is creep characteristics under mechanical stress, as PSA cannot maintain the exact positions of the stiffener boards under the stress. Thermosetting adhesives provide a reliable bonding between the layers, but they require a long processing time in a heat press with suitable press pad materials and impose a higher processing cost.

Usually, the adhesive materials are supplied on or between separation sheets by rolled form. They should be kept in the refrigerators. The mechanical pretreatments are conducted with the separation sheets.

TABLE 61.20 Comparison of Adhesive Materials for Flexible Circuits

	Thermosetting epoxy and acrylic resin	Hot-melt-type polyimide resin	Pressure-sensitive adhesives (PSAs)
Reliability	High	High	Acceptable
Bond strength	High	High	Acceptable
Creep	Small	Small	Large
Soldering	OK	OK	OK
Application process	Complicated	Complicated	Simple
Process temperature	160–180°C	>330°C	Room temperature
Material cost	Low	High	Low
Total cost	High	Very high	Low

61.11 RESTRICTION OF HAZARDOUS SUBSTANCES (ROHS) ISSUES

Two issues have arisen regarding RoHS requirements for the flexible circuit materials: flame-retardant molecules with bromine in adhesive resins, and heat resistance for high-temperature processing with lead-free soldering. Although the issue of bromine is not actually a part of the RoHS requirement, it has been linked to the general environmental issues of printed circuit materials and processes.

Adhesiveless copper laminates can be the solution for the both of these problems because they do not use adhesive layers and have much higher heat resistance than adhesive-based laminates. However, because some of the applications need adhesive-based copper laminates, manufacturers of flexible laminates have been developing alternative materials without bromine-based flame-retardant reagents.

There have been technical barriers to developing new materials with a good balance of performances for coverlay eliminating bromine compounds. Film coverlay have to have an adhesive layer with some flame-retardant reagent. Few new epoxy resins have good balance for all of the physical performances without bromine compounds.

New materials of photoimageable coverlay have been developed without bromine molecules. Not many choices offer a balanced performance, especially high flexibilities for dynamic flexing. Polyimide-based photoimageable coverlays are the solution to satisfy the all of the physical requirements, but these polyimide materials are remarkably more expensive.

CHAPTER 62
DESIGN OF FLEXIBLE CIRCUITS

Dominique K. Numakura
DKN Research, Haverhill, Massachusetts

62.1 INTRODUCTION

Because of the broad wiring capabilities now available, it is possible to cover all 3-D wiring of an electronic product by one flex circuit, or by a set of multilayer rigid/flex circuits, without the need for connectors, extra cables, and soldering. However, to achieve this, the shape of the flex circuit may be very complicated and the size may be large. As a result, the final cost could be extremely high. If so, the use of flexible circuits is not an acceptable solution, especially for large-volume consumer applications. It may be better to divide the wiring into two or more parts to make each circuit simpler. A combination of simple circuits can be less expensive than one flex circuit, even though it mandates several connections between different circuits. A circuit designer should also consider the manufacturing convenience of flex circuits. There are serious differences among rigid circuit boards. A small design modification can sometimes reduce the total cost significantly.

62.2 DESIGN PROCEDURE

To achieve the best cost-performance combination for a wiring design with flex circuits, an appropriate design procedure is recommended, as illustrated in Fig. 62.1. The steps in the process are as follows:

1. A design concept for an electronic product is determined according to the basic specifications for the final product.
2. A logical circuit design and electronic circuit design are formulated.
3. A design for the equipment housing is made according to the final product concept. (This should sometimes be done independently in the case of consumer applications, especially for portable electronic products.)
4. The circuit design is formulated. Appropriate circuit selection is conducted depending on total conditions such as allowed space, sizes of components, operating temperature, possible assembling process, required reliability, and so on.

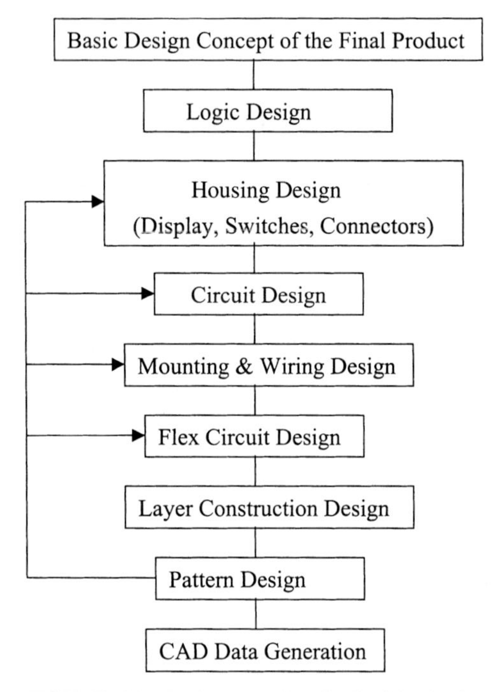

FIGURE 62.1 Design procedure for flexible circuits.

5. A flex circuit or a combination with a flex circuit is considered to be the solution for a small wiring space.
6. If the first circuit design does not satisfy the original specifications, the previous design steps are repeated to solve the conflicts.
7. A suitable value analysis is done to minimize the total wiring cost.
8. Conductor trace patterns and coverlay opening patterns are designed to minimize the mechanical stress.
9. Numerical data for photo plotter, NC drilling, punching dies, and other tooling is generated.

62.3 TYPES OF FLEXIBLE CIRCUITS

Flexible circuits allow for more variety in their basic structures and supplemental structures than do rigid circuit boards, as shown in Table 62.1. Flexible circuits have single-side or double-side constructions with through holes similar to rigid boards. However, multi-layer construction has significant differences from rigid boards; this is because the flexibility of the composite board will be very low when the circuit has more than two conductor layers. Therefore, the flexing parts of the circuit have to have special layer constructions to keep flexibility. The construction is called rigid/flex when the rigid part includes rigid board materials such as glass/epoxy. A simple single-side or double-side flexible circuit is mechanically fragile, so coverlay and stiffener boards are generally added as the supplemental structures of the flexible circuits. Further special constructions, such as flying leads and dimples, are used for specific applications.

TABLE 62.1 Construction Types of Flex Circuits

Classification	Structure types
Basic structure includes	• Single-sided flexible circuits • Double-sided flexible circuits with or without through-holes • Multilayer flexible circuits with through-holes • Multilayer rigid/flexible circuits with through-holes • Blind via holes, inner via holes • Flying-lead structure
Supplemental structure includes	• Coverlay structure • Stiffener structure • Dimple structure • Micro bump structure

An appropriate combination of a basic structure and supplemental structures should be chosen according to the requirements of the application. Also, appropriate termination technologies and suitable circuit constructions must be considered together.

62.3.1 HDI (High-Density Interconnect) Flexible Circuits

There are significant differences in design, materials, and manufacturing technologies between traditional flexible circuits and new high-density flexible circuits developed for high-density interconnect (HDI). Figure 62.2 shows one way of defining the areas of high-density flexible circuits by using the technical hurdles of manufacturing for both trace densities and via hole sizes. Advanced etching technology is needed to give the high-density flexible circuits fine traces. Also needed are new micro via hole generation technologies other than traditional mechanical drilling, such as laser, plasma, and chemical etching.

Ultra-high-density flexible circuits have been developed for special applications. They have extremely fine circuits that are smaller than 20 μm in pitch, as well as micro via hole connections

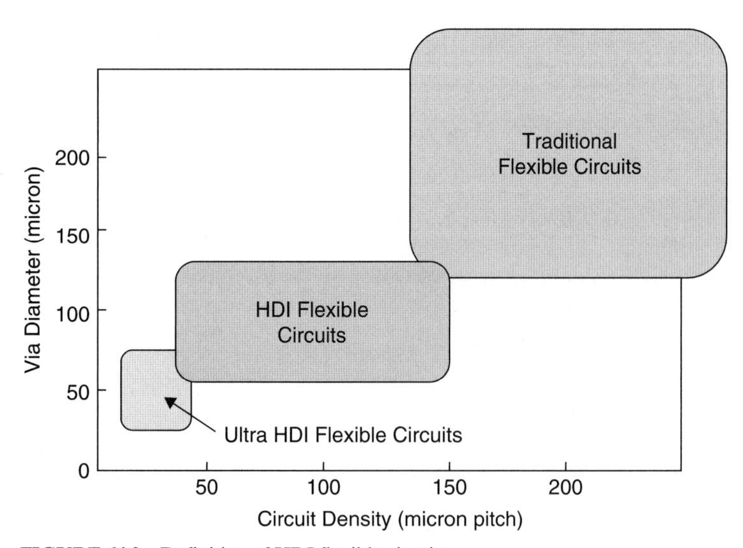

FIGURE 64.2 Definition of HDI flexible circuits.

smaller than 25 μm in diameter. Product concepts for ultra-high-density flexible circuits are very different from those for traditional flexible circuits in construction and manufacturing technologies. They need completely new concepts for the manufacturing processes, such as an additive process with new materials and new equipment. (See Chaps. 22 and 23 for detailed discussions of HDI design and fabrication.)

New high-density flexible circuits have supplemental structures such as flying leads and micro bump arrays to complete high-density terminations. As volume production is limited, the designers should consider the manufacturers' process capabilities seriously.

62.3.2 Single-Sided Circuits

A bare single-sided flex circuit has a construction very similar to those of the single-sided rigid circuit boards, except for the thickness of the substrate layers. On the other hand, the protection of the conductors of the flexible circuits is very different from the rigid boards. A film coverlay system is introduced for the flexible circuits instead of solder masks (see Fig. 62.3). As introduced in Chap. 61, there are several choices of substrate (see Table 62.2). A 25-μm-thick polyimide film substrate is the most common material for consumer applications, with large volume due to its low cost. For higher reliability thicker film should be chosen, while for lower reliability, a thinner film can be used.

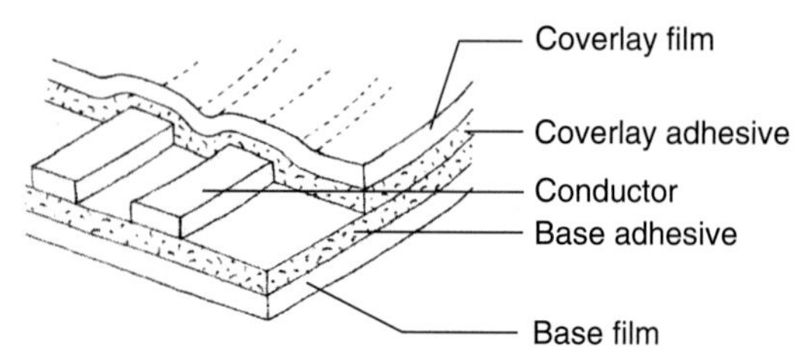

FIGURE 62.3 Basic structure of single-sided flexible circuit.

The thickness of the adhesive layers must be counted when total thickness is considered.

TABLE 62.2 Thickness of Base Substrates

	Base film (μm)	Adhesive (μm)
Polyester base/adhesive type	25, 38, 50, 75, 100, 125, 188	25–50
Polyimide base/adhesive type	12.5, 25, 38, 50, 75, 125	25–50
Polyimide base/adhesiveless type	12.5, 25, 30, 35, 40, 50	No
Glass-epoxy	100, 200	~0

Adhesive layer thickness is doubled for double-sided laminates.

Possible fine-line capability is illustrated for each copper thickness in. Fig. 62.4 Currently 18- and 35-μm copper foils are standard. Copper foils 12 μm and 9 μm thick are becoming the new standard to enable finer pattern etching. Thinner copper foils are available in sputtered/plated adhesiveless laminate materials. The fine-trace etching capability depends very much on the manufacturer, especially the exposure process and etching process.

62.3.3 Coverlay System of Flexible Circuits

The coverlay system of flexible circuits is one of the big differences from rigid printed circuit boards, as shown in Fig. 62.3. A coverlay has to have more functions than the solder mask of rigid boards. It should not only have solder dam capability, but also mechanical protection capability for the fragile conductors of flexible circuits. Also it is required to have high toughness to survive long flex endurance and to optimize the major advantage of flexible circuits.

Appropriate coverlay material and thickness including adhesive layers have to be chosen to have the best balance between the layers. Generally, the same film material laminated with appropriate adhesive as base substrate is selected as the coverlay materials. Polyimide film coverlay should be selected for the high temperature assembling processes such as soldering.

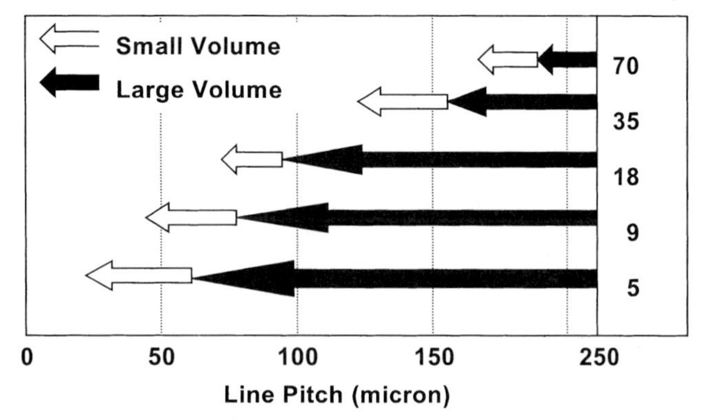

FIGURE 62.4 Fine line capabilities. Potential line pitch is noted for copper thickness alternatives along with whether a given thickness leads itself to high-volume production or requires too a high level of technical interaction.

The manufacturing process for the film coverlay is very complicated compared to the solder masks of the rigid boards, so the final cost of the circuits cannot be low. Flexible, screen-printable inks are a low-cost solution; however, most of the ink materials made of epoxy resins do not have good mechanical properties and cannot be applied for dynamic flexing usage. A combination of polyimide-based screen-printable coverlay on adhesiveless copper-clad laminate provides the smallest total thickness with very long flexing endurance life. A comparison of cross sections is shown in Fig. 62.5.

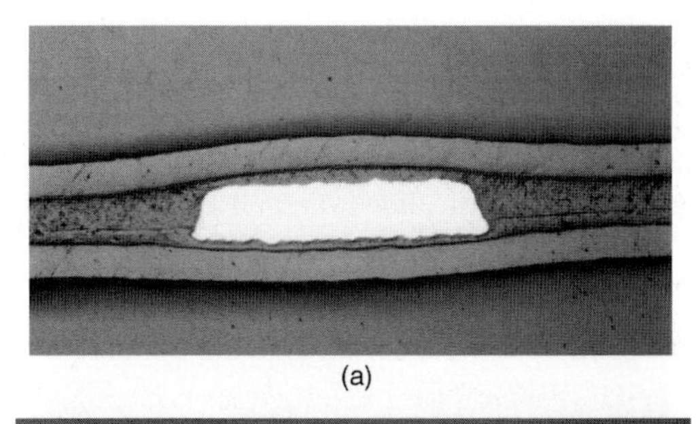

(a)

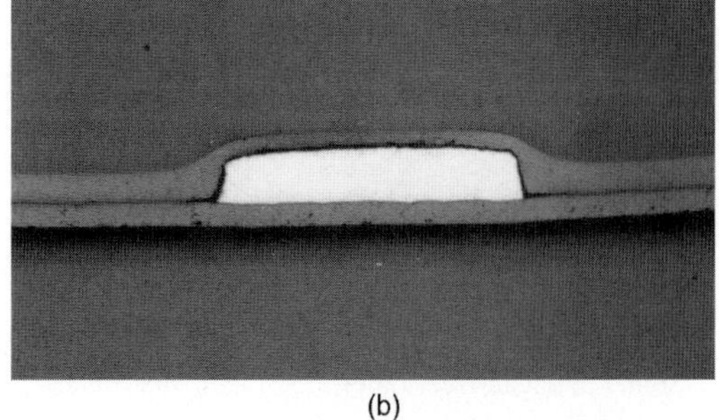

(b)

FIGURE 62.5 Cross section of a single-sided flexible circuit showing (a) adhesives-based laminate and film overlay; (b) adhesiveless laminate with liquid coverlay.

TABLE 62.3 Varieties of Coverlay

	Material selection	Thickness (μm)	Dimensional accuracy (minimum opening)	Reliability (flexing endurance)
Traditional film coverlay	PI, PET	30–100	±0.3 mm (800 μm)	High (long)
Film coverlay + laser drilling	PI, PET	30–100	±50 μm (50 μm)	High (long)
Screen print liquid ink	Epoxy, PI	10–20	±0.3 mm (600 μm)	Acceptable (short)
Photoimageable (dry film type)	Epoxy, PI, acrylic	25–50	±50 μm (80 μm)	Acceptable (short)
Photoimageable (liquid ink type)	Epoxy, PI	10–20	±50 μm (80 μm)	Acceptable (short)

High-density flexible circuits that have been developed since 1990 need a small pitch opening with high accuracy on the coverlay. Because traditional coverlay technologies have no capabilities to meet these requirements, new technologies were required. Many new technologies and new materials have been developed to satisfy the new requirements of high-density flexible circuits. The capabilities of the major technologies are shown in Table 62.3.

For example, laser drilling is capable of generating openings smaller than 100 microns on the traditional film coverlay; therefore it satisfies the requirements of physical performance and high density SMT assembling. However, the processing costs of the laser abrasion are much higher than other coverlay processes, and it is difficult to apply for the high volume production of consumer applications. A screen-printing process provides a slightly higher resolution than traditional pre-punching process of the film coverlay, but the process can not satisfy the recent requirement for high density SMT assembly. A practical solution is the combination of film coverlay for the flexing areas and solder mask for the SMT areas. It needs supplemental processing to complete the construction, accordingly the total process yield will be lower and the final cost of the circuits could be higher.

The idea of the photo-imageable coverlay could be the final solution to solve the all requirements, including the processing cost. The photo-imageable solder mask materials developed for rigid boards are too brittle, however, and a lot of cracks can be created easily by small radius bending—requiring flexible materials. Several types of flexible photo-imageable coverlay materials have been developed with different base resin systems. Most of the photo-imageable coverlay materials provide finer resolutions than 200 micron pitches, but the physical performances very depend on the base resin systems. Suitable material should be chosen for the applications.

Appropriate surface treatments for the bare conductors after the coverlay construction should be chosen according to termination technologies. Several examples are listed in Table 62.4. Details are explained in Chap. 64.

TABLE 62.4 Surface Treatments and Termination Technologies

Surface treatment	Termination technologies
Solder and lead-free solder	Lead-free soldering, FFC connector
Lead-free Hot-air leveling	Lead-free soldering
OSP (pre-flux)	Lead-free soldering
Hard Ni/Au plating	Connector, pressure contact
Soft Ni/Au plating	Wire bonding, direct bonding
Tin	Lead-free soldering

62.3.4 Surface Treatment Alternatives Stiffener boards

Thinness and bendability are the major advantages of flexible circuits. However, flexible circuits can lose mechanical strength when it is necessary as part of a composite assembly. The addition of stiffeners on flexible circuits is a practical solution to solve the conflicts, but appropriate stiffener board materials and adhesive materials have to be chosen based on the applications purposes.

Figure 62.6 illustrates typical examples of the stiffener systems. The first one shows a stiffener construction for an insertion part into a connector. A stiffener adjusts the total thickness of the insertion part. Heat resistant materials, such as polyimide films, are not required for the construction. PET films with PSA (pressure-sensitive adhesive) could provide the low cost combination. A second example shows a case of the assembling of the traditional lead components. A relatively thick stiffener board, such as glass/epoxy, is bonded on the component side to support heavy parts. The soldering for the leads is conducted at the other side. Heat resistant materials including adhesives have to be chosen to survive the high temperature soldering process. Thermo-setting type adhesives should be chosen instead of PSA for the high reliability requirements. The third example indicates the case of SMT assembling. A thin heat resistant stiffener material such as polyimide film is bonded on the other side of SMT assembling.

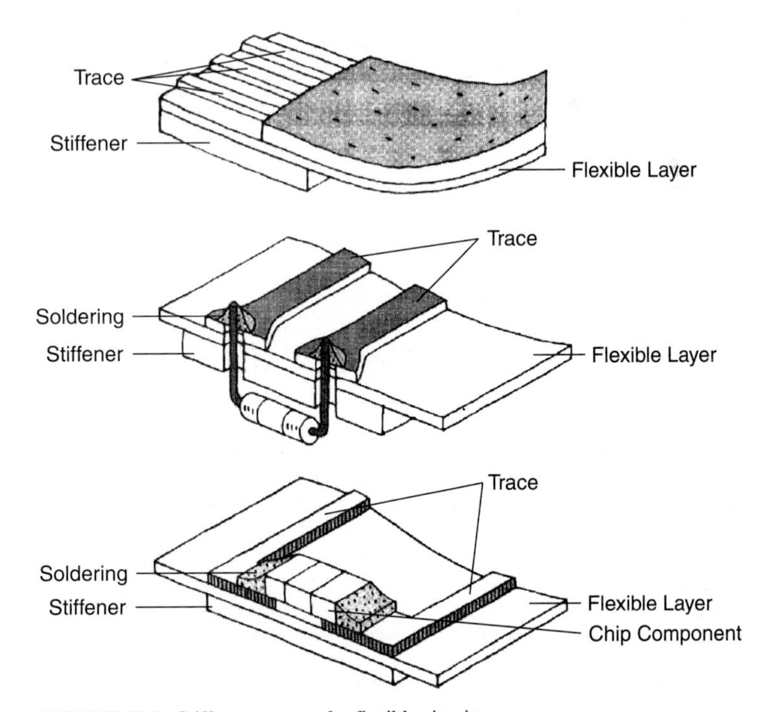

FIGURE 62.6 Stiffener systems for flexible circuits.

62.3.5 Double-Sided Circuits with Via-Holes

A double-sided flexible circuit can have the same through-hole structures for the traditional designs as rigid circuit boards. However, there are several critical differences in micro via holes on flexible circuits. Most of the differences come from a thin-film base substrate such as

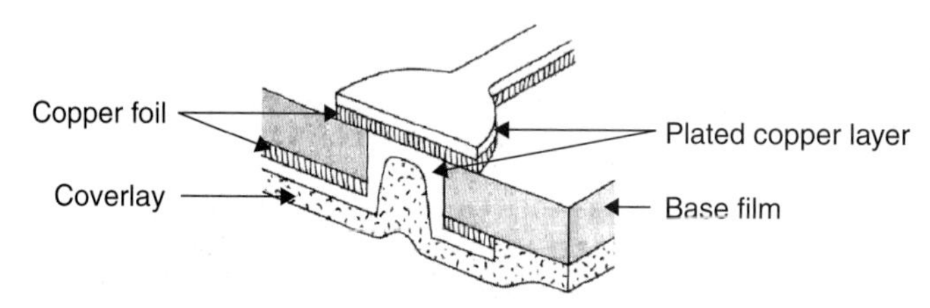

FIGURE 62.7 Blind via structure of double sided flexible circuits.

polyimide film. Electrically plated copper conductors are much more brittle than rolled-annealed copper foils; therefore, thin copper plating is preferred to maintain the flexibility of the circuits. Generally, sufficient through-hole reliability is established with 15-μm-thick copper plating. Copper plating should be eliminated from the dynamic flexing area by using a suitable masking method.

Adhesiveless, copper-clad laminates that have simple and uniform substrate layers and smooth copper surfaces have become popular in high-density flex circuits. Because of thin film materials, there are several choices in small hole generation and copper plating processes. Flex circuits can contain both through-holes and blind via holes for double-sided circuits as shown in Fig. 62.7. Micro-size blind via holes on double-sided flex circuits confer the ability to make SMT assembly density much higher.

The micro via hole capability depends on the process used by the manufacturer; however, general ideas are introduced in Chap. 61. Certainly the smaller via holes are available for the thinner substrates. Using the latest technology, such as excimer laser drilling, micro via holes smaller than 40 μm are available for 25-μm-thick adhesiveless polyimide base substrates.

Microvia hole capabilities for polyimide substrates are shown in Fig. 62.8.

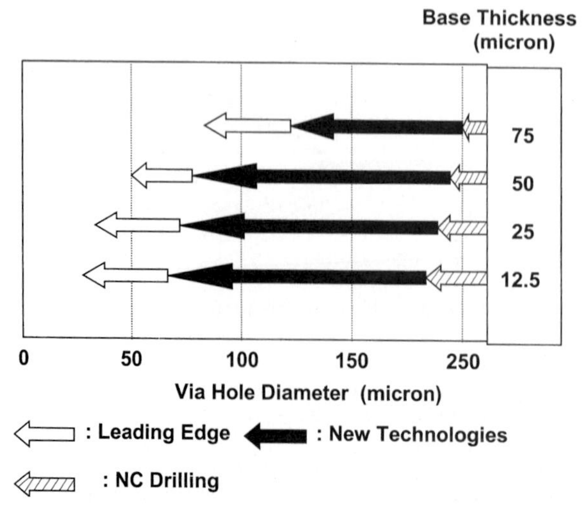

FIGURE 62.8 Microvia capabilities shown by base thickness and hole diameter. Also shown is the range of technologies from HDI to mechanical drilling required to create the vias.

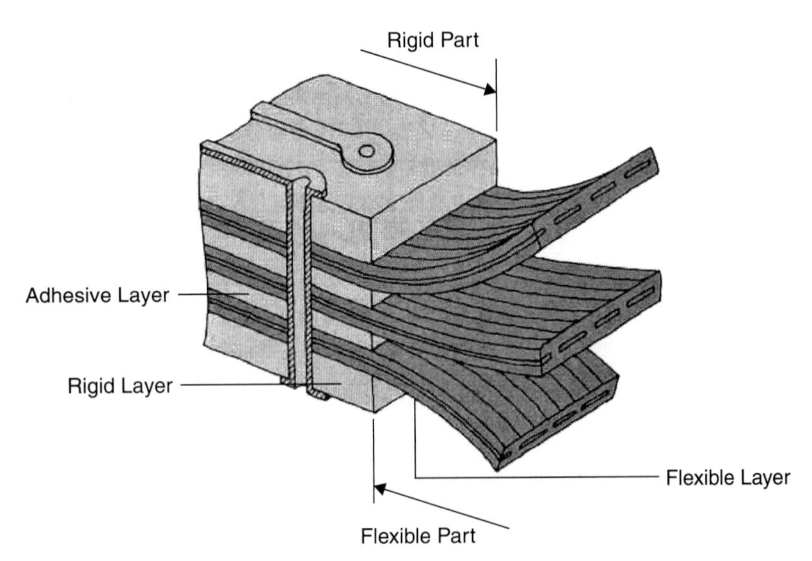

FIGURE 62.9 Multilayer rigid flex construction.

Basically, the same construction systems of coverlay, stiffeners, and surface treatments as single-side flexible circuits could be available for double-side flexible circuits.

62.3.6 Multilayer Rigid/Flex (Multilayer Flex)

A simple, multilayer, flexible circuit does not make sense other than for its small thickness, because these circuits will not have any more flexibility. Therefore, rigid/flexible constructions that contain both rigid parts and flexible parts have been developed. A typical layer construction is illustrated in Fig. 62.9. It is not a simple combination of flexible circuits and rigid circuits. Generally, there are common flexible layers that are separated in the flexible parts. A flexible layer of the flexible parts can have only one or two conductor layers to keep high flexibility. At the rigid parts, the flexible inner layers are sandwiched by cap layers that consist of rigid circuit materials such as glass epoxy or glass polyimide. The constructions of rigid parts are very similar to those of rigid multilayer circuits except for the thin polyimide dielectric layer inside. Thin polyimide films were used as the cap layer at the rigid part to reduce the total thickness required in portable electronics such as cellular phones. The multilayer construction in these cases, the circuits are called multilayer flex.

Through-hole and blind via hole constructions similar to those of multilayer rigid circuit boards are available using coated copper foil or photo-via processes for multilayer flexible and rigid/flexible circuits. However, there are more varieties from which to choose in both materials and manufacturing processes. Basically, the same assembly technologies are available on the rigid parts. The latest portable electronic equipment, such as palmtop computers, requires high-density SMT assembling on the rigid parts.

There are no limits to the number of flex layers. More than 30 layers are built into one rigid/flex circuit for aerospace applications. However, the circuits require very complicated manufacturing processes—more so than the standard rigid multilayer circuits—and therefore their costs are much higher than costs for rigid boards of the same size. Design should be considered seriously prior to manufacturing to reduce the total cost based on the capability of the manufacturer. A higher circuit density can reduce the layer counts and the total manufacturing cost.

62.3.7 Flying-Lead Construction (Double access construction)

A specialty construction for flexible circuits is the double-electrical access capability on even single conductor layer circuits, achieved by removing the base insulation layer under the conductors as shown in Fig. 62.10. It can be an alternative construction instead of double-side circuits with through holes for the low cost 3-D wiring. When the insulation layers are removed from both sides of the conductors, they can then be partially free from the organic materials. Unsupported conductors, also known as flying leads or flying fingers, are not a new idea in the construction of flexible printed circuitry. (Fig. 62.10) The biggest advantage of flying-lead technology is its extremely high heat resistance and thermal conductivity when compared to standard conductors insulated by dielectric substrate materials. Based on the lack of insulation, flying leads can be treated as a bare conductor material similar to bare wire. There are many high-density flexible circuit applications that require flying-lead terminations. Details are discussed in Chap. 64.

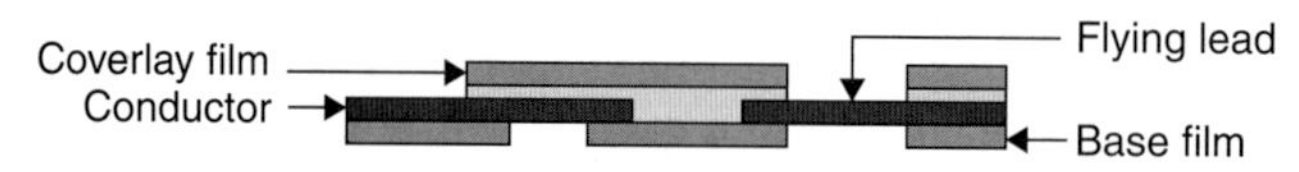

FIGURE 62.10 Flying lead construction (Double sided construction).

Consideration must be given to specific surface treatments required on the flying leads to make each of the various application terminations. Previously, tin/lead plating was the most common treatment for various soldering processes; however, it is expected to be replaced by lead-free technologies, which need higher temperatures for the soldering. Soft gold plating with a nickel under-plating is the standard surface treatment for the wire bonding and direct bonding to the other devices. The purity and the thickness of the plated gold, along with the bond strength of the plated layer, are important factors in bond performance.

Copper thickness is a key consideration when developing a new circuit design involving flying leads. Although it is easier to manufacture high-density circuits utilizing thinner copper, the resultant flying lead may be extremely fragile and will not have the mechanical strength of the thicker copper circuits (see Fig. 62.11). Manufacturing yield plays an important role in the overall circuit costs, and thinner copper foils typically result in more damage during processing. It is important to understand this correlation as poor circuit design can affect manufacturing yields.

Because the flying leads are basically bare conductors, suitable surface treatments should be conducted according to termination technologies.

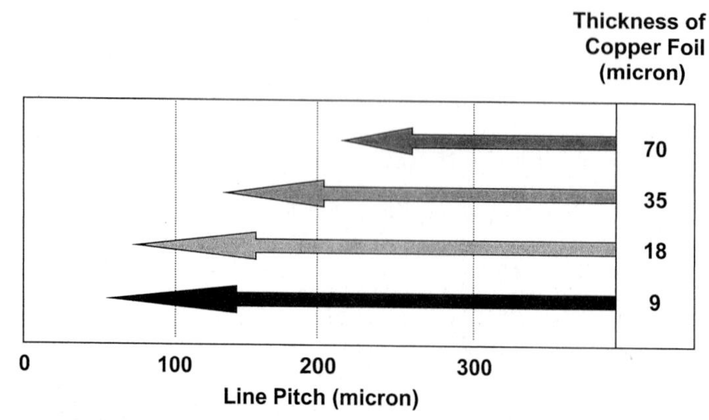

FIGURE 62.11 "Flying lead capability of copper foil, showing potential line width versus thickness of the copper foil."

62.3.8 Micro Bumps and Dimples

Many kinds of micro bump arrays and dimple arrays have been developed for high-density terminations of flexible circuits other than solder ball grid arrays. Recent technologies allow for a wide variety of shapes, materials, sizes, and pitches; down to 50 microns. The solder ball arrays are built by similar processes of the rigid BGA substrates. The combinations of the electrical and electroless-plating process provide a lot of choice for the shapes and sizes of the micro bump arrays on flexible circuits. The double-access capabilities of flexible circuits provide more choices for the locations of the micro bumps, as shown in Fig. 62.12, other than through coverlay. The micro bumps can be built on the bottom side of the conductors through base layers. They can be also built on both sides of the flying leads. An appropriate design should be chosen according to the requirements of the terminations. Whether it's a permanent connection or non-permanent connection is one of the key points needed for designing the basic constructions of the micro bumps. Bump density and reliability are other factors needed to design the micro bump structures. Details of the manufacturing processes are described in Chaps. 64 and 67.

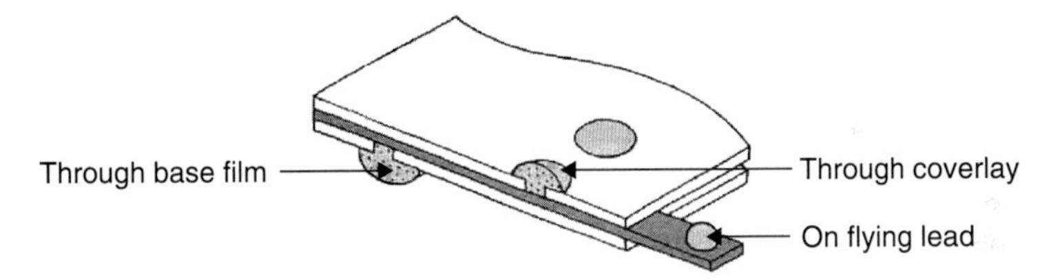

FIGURE 62.12 Locations of microbumps.

Dimple structure of flexible circuits is a low cost solution for the non-permanent termination. Multiple dimple arrays can be formed on the polyimide base or PET base flexible circuits by simple pressing processes. (Details are shown in Fig. 62. 13) A typical example of the dimple termination is the connection of the disposable cartridges of ink jet printers. Over 60 dimples with hard gold plating are formed at the end of flexible circuit for the connection with the ink cartridge. The dimple arrays can make reliable connections over one thousand times.

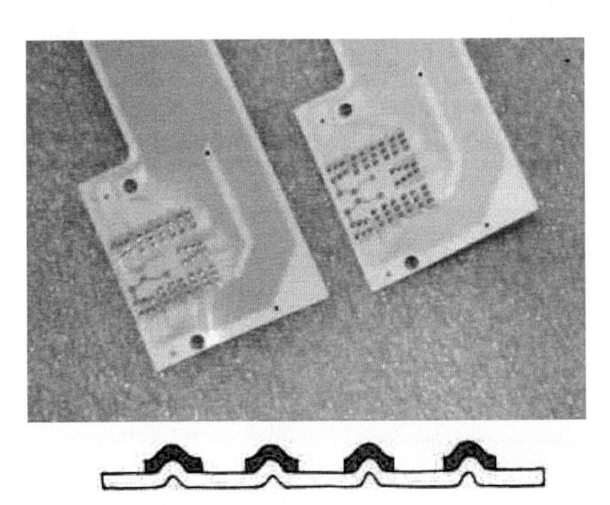

FIGURE 62.13 Dimple structures (Ink jet printer).

62.4 CIRCUIT DESIGNS FOR FLEXIBILITY

One of the major features of flexible circuits is their long flexing endurance. A suitable layer construction and materials survives over billion times of flexing with a small radius. A basic idea of the mechanical stress during the bending is illustrated in Fig. 62.14(a). To have the longest flexing life, the conductor layer should be the center of the layer construction. It should have symmetrical layer constructions for both sides of the conductor layer. The flexing life will be remarkably shorter when the conductor layer is out side of the bending center as shown in Fig. 62.14(b). A rolled annealed copper foil should be used instead of ED copper foil for a longer flexing life. But there is no exact relation between roll direction and flexing endurance; the exact performance should be evaluated for each material. Generally, 35 microns thick RA copper is more expensive than the same thick ED copper foil. The cost difference between RA copper foil and ED copper foil will be larger at thinner thickness. A high ductility ED copper foil could be alternative choice for middle class dynamic flexing. Its flexing life is shorter than RA copper foil, but it is less expensive and it can survive one million flexing instances with an appropriate layer construction and flexing radius.

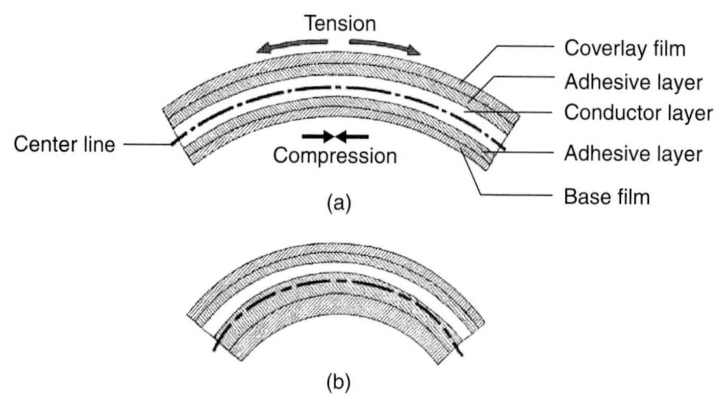

FIGURE 62.14 (a) and (b) Mechanical stress at bending, (a) Symetrical, (b) Nonsymetrical.

Another principle is that for dynamic flexing, the thinner the layer, the better. An 18-μm-thick copper foil has a longer flexing life than a 35-μm-thick copper foil. A 12.5-μm-thick polyimide film provides a longer flexing life than a 25-μm-thick polyimide film (Fig. 62.15). Adhesive layers should be as thin as possible for dynamic flexing. But thin adhesive layers do not have good bond strength and good encapsulation for traces, and therefore a suitable thickness should be determined based on experimental data. Generally, polyimide films have a longer life than polyester films. The combination of a thin, adhesiveless, polyimide base laminate and a thin coverlay of liquid polyimide resin could be the best solution to have the longest flexing life.

Many kinds of bending modes on flexible circuits could be practical, but in IPC-TM-650 it is recommended to evaluate the flexing life using the movement shown in Fig. 62.16.

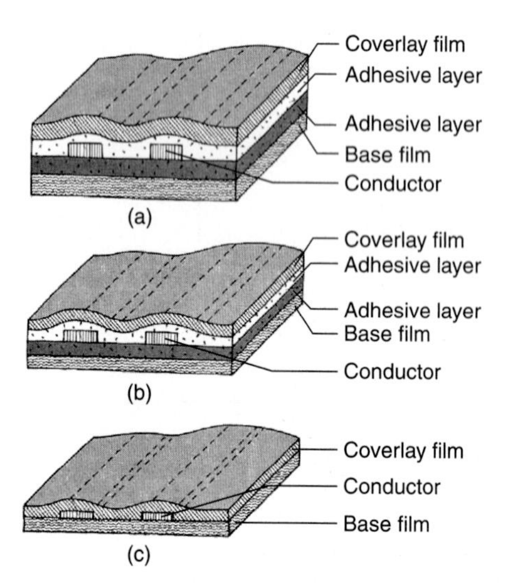

FIGURE 62.15 "Layer construction for dynamic flexing, (a) Standard layer construction, (b) Dynamic flexing construction with thin adhesive layers, (c) Dynamic flexing construction without adhesive layer."

The larger flexing radius provides a longer flexing life, as shown in Fig. 62.17.

There are several ideas for reducing the risks of flexing failures in actual cases for electronic products; Fig. 62.18 shows some examples.

Basically, double-sided flexible circuits with through holes are not available for dynamic flexing. The flexibility depends on the configuration of conductors in both sides of the base substrate, as shown in Fig. 62.19 shows examples. Bending the circuit with thin copper circuits on the outside of the flexible circuits may cause them to break, because the mechanical stresses are concentrated on the small traces.

A dynamic flexing area of a double-sided flexible circuit should have only one conductor layer. Coverlay on the other side of conductor should also be eliminated to enable symmetrical layer constructions. Copper plating for the through-holes of double-sided flexible circuits should be eliminated from dynamic flexing areas using a plating mask.

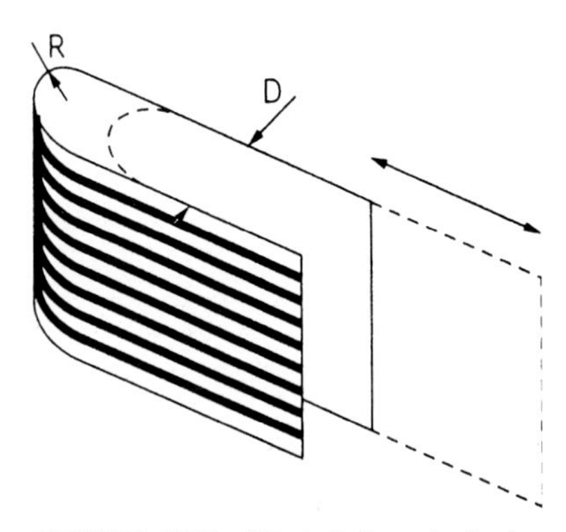

FIGURE 62.16 "Typical dynamic flexing mode of flexible circuit specified by IPC-TM-650."

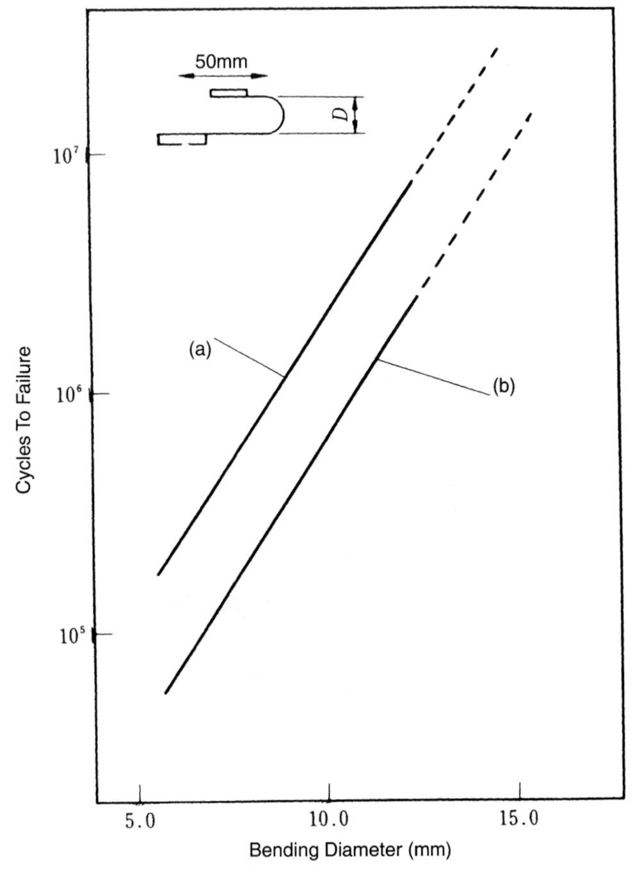

FIGURE 62.17 Flexing radius and life as specified by IPC-TM-650 test method. Bending Diameter (mm) a) RA copper foil, b) ED copper foil.

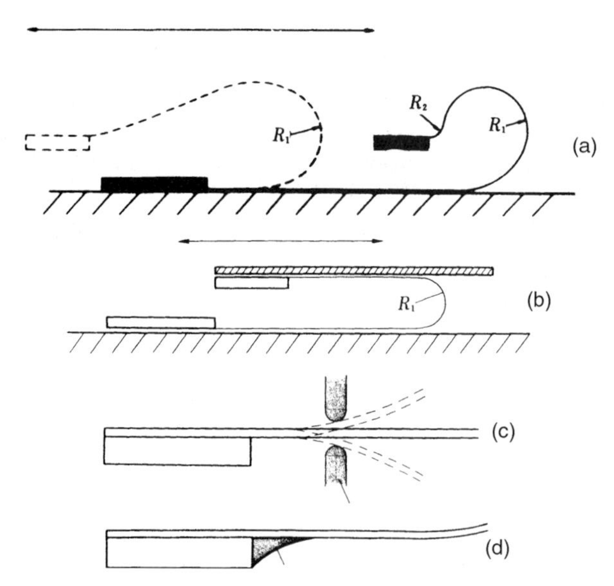

FIGURE 62.18 Design of dynamic flexing part of flexible circuits (a) Unacceptable technique, (b) guide board, (c) loose guide, (d) silicone rubber.

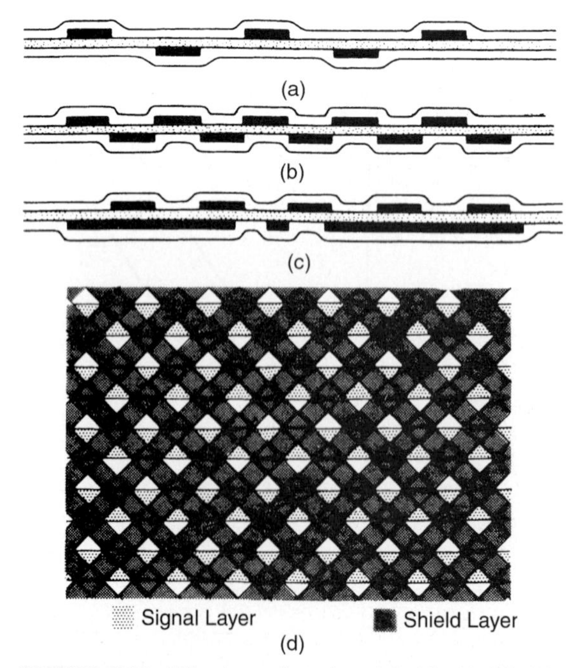

FIGURE 62.19 "Circuits configuration of double-sided flexible circhits at flexing parts, (a) Preferred, (b) Acceptable, (c) Unacceptable, (d) Meshed shield layer, (e) Crossing traces, (f) Separated shield layer for dynamic flexing, (g) Fan folded. (h) Coiled."

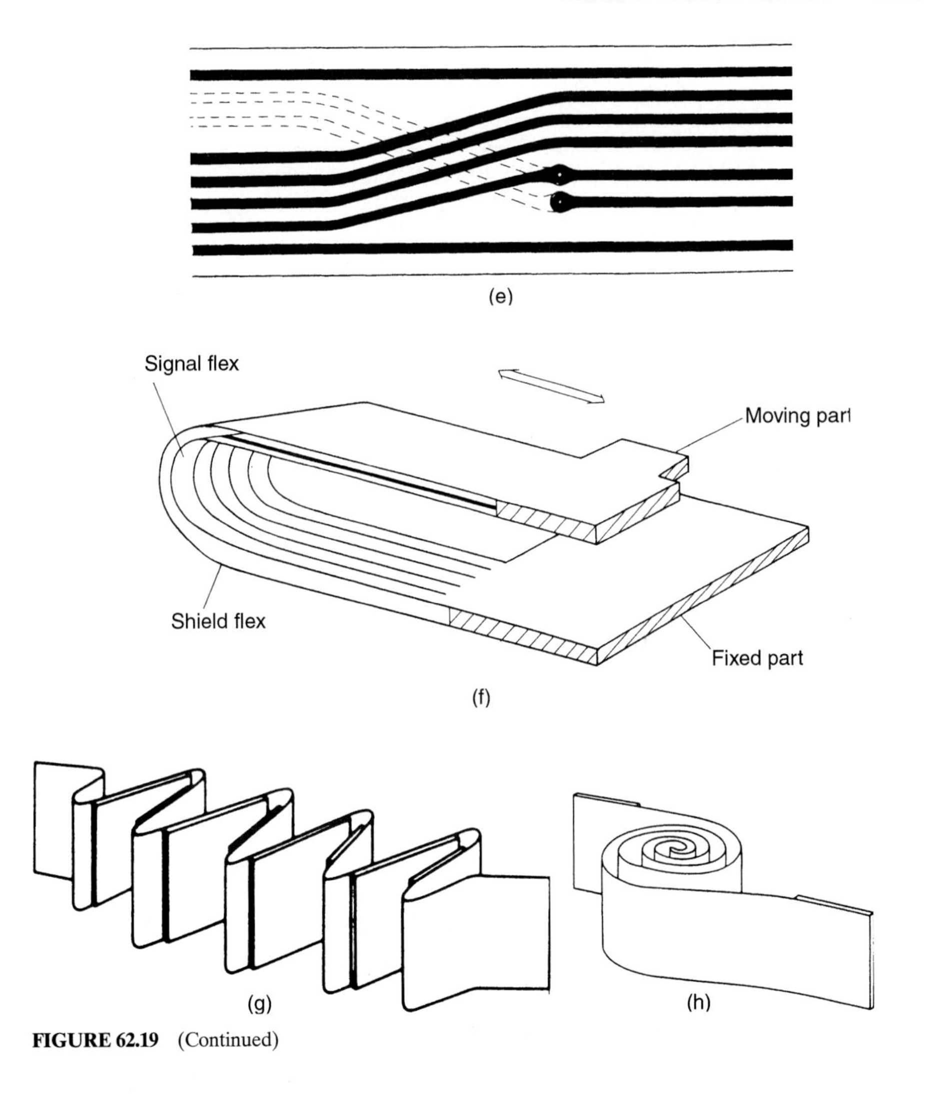

FIGURE 62.19 (Continued)

62.5 ELECTRICAL DESIGN OF THE CIRCUITS

There is no significant difference between the electrical performance of flexible circuits and that of rigid circuit boards. But there are small differences in conductivity and insulation resistance depending on the materials. Also, there are small differences in dielectric constant and tangent delta. Therefore, similar calculations can be applied to determine electrical performances for a flexible circuit. However, flexible circuits have a relatively long parallel circuit for cabling capability, and impedance should be managed carefully, especially for high-speed circuits with high frequencies.

62.6 CIRCUIT DESIGNS FOR HIGHER RELIABILITY

Because of the thin and fragile materials used, flexible circuits have lower mechanical reliability than rigid circuit boards. They have a low conductor bond strength and low base substrate tear strength. Nevertheless, most of the flexible circuits are subjected to more mechanical stresses due to movement. This means that special care is required in the circuit design to gain higher circuit reliability.

There are several ways to make the reliability higher. Figures 62.20 and 62.21 show several common ideas for increasing the reliability of flexible circuits by modifying conductor patterns.

The trace patterns should be smooth slopes between different widths. The pad size should be as large as possible. Figure 62.22 shows an example applied for automobile instrument panels. The conductors are designed to be as wide as possible, and the spaces between conductors are

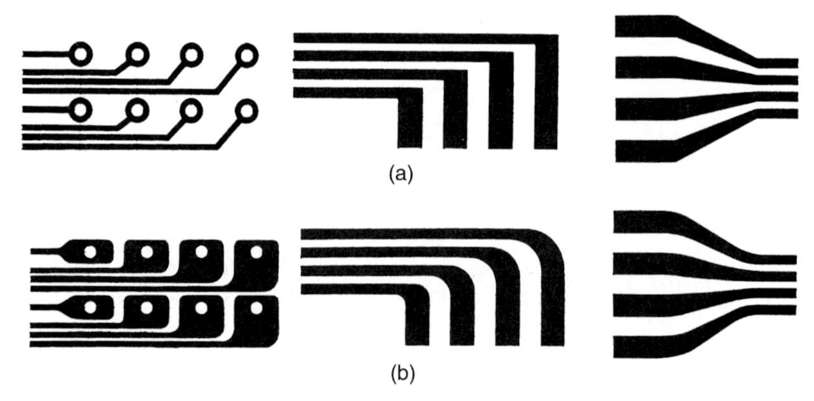

FIGURE 62.20 Reliability design for flexible circuits (a) Unacceptable, (b) Preferred.

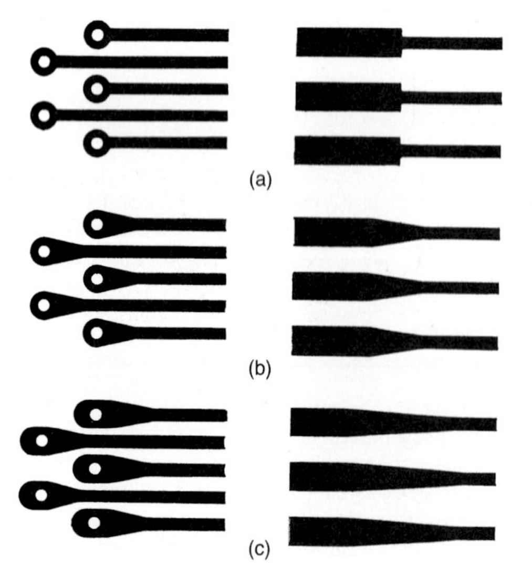

FIGURE 62.21 Reliable pattern design for flexible circuits, (a) Unacceptable, (b) Acceptable, (c) Preferred.

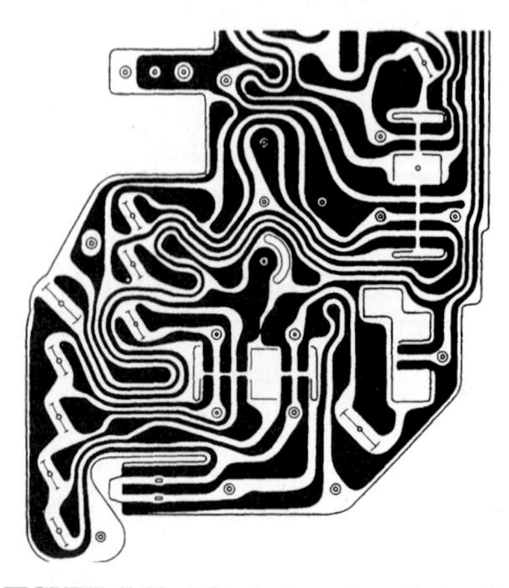

FIGURE 62.22 "Conductor pattern designed for the instrument panel of an automobile."

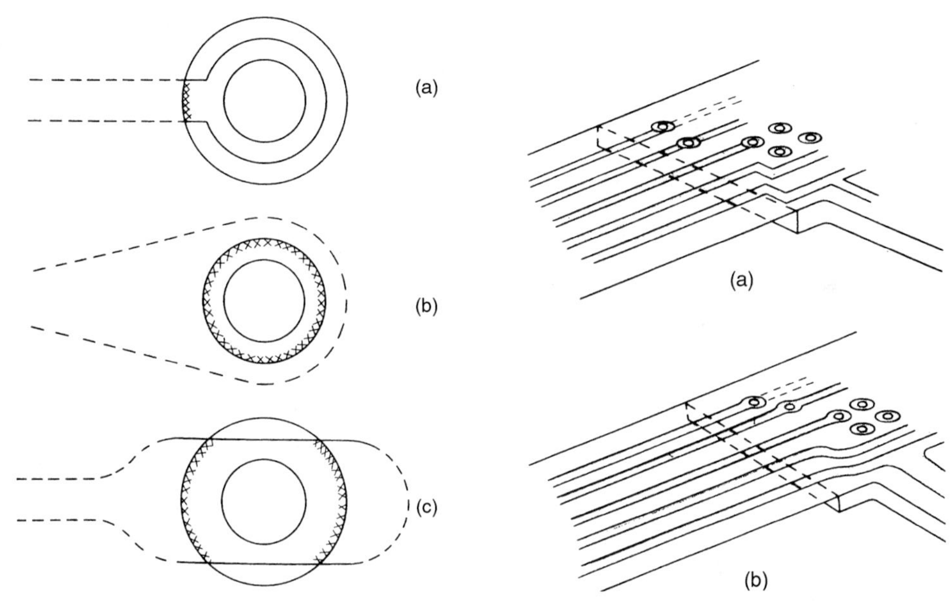

FIGURE 62.23 Coverlay opening for reliable solder pads. (a) Unacceptable, (b) Acceptable, (c) Preferred.

FIGURE 62.24 Circuit design at stiffener edge, (a) Unacceptable, (b) Preferred.

kept in the safety range. A suitable coverlay opening could increase the reliability, as shown in Fig. 62.23. The amount of squeeze-out of coverlay adhesives should be minimized, however. The edge areas of stiffener boards are dangerous places in which mechanical stresses are concentrated. Figure 62.24 shows ways of reducing these risks.

62.7 CIRCUIT DESIGNS FOR ROHS COMPLIANCE

As it was described in Chap. 61.11, there are two issues involved in satisfying RoHS compliance and Halogen-free requirements. They are the eliminations of bromine molecules and lead components. A suitable material combination should be selected to survive during the high temperature process of lead-free soldering. The adhesiveless copper laminates could be the suitable materials for this purpose. OSP (Organic Surface Protection) treatment or Tin-Copper plating could be the alternative solution for the surface finishing of the bare conductors instead of the eutectic solder and hot air leveling.

CHAPTER 63
MANUFACTURING OF FLEXIBLE CIRCUITS

Dominique K. Numakura
DKN Research, Haverhill, Massachusetts

63.1 INTRODUCTION

Basically, it is possible to produce the standard constructions of single- and double-sided flexible circuits and multilayer rigid/flex for small volume—if there is a set of manufacturing facilities for multilayer rigid circuit boards. However, it is not easy to achieve high productivity, with high process yield, for volume production of flexible circuits using these facilities, as they are not designed to deal with the unique issues of flexible circuits, such as their thin, flexible material. Therefore, manufacturing facilities for high volume flexible circuits must be designed to take advantage of these differences, including specific equipment and process conditions, as discussed in this chapter.

Table 63.1 shows the specialties for processing the flexible circuit materials. Most of the issues are caused by their complicated structures and fragile materials. The coverlay structures and the stiffener constructions require supplemental manufacturing processes, making process yields lower. The varieties of outer shapes make material yields and productivity lower. Thin, fragile materials can easily incur serious mechanical damage due to rough handling. Thin materials can also undergo significant dimensional changes in the manufacturing processes. They cause a pattern shift between the processes and make the process yields lower. High-level automation has basic barriers, discussed later, when applied to the manufacturing process for flexible circuits, because of its unstable materials. Many processes must be conducted by manual methods, which are, of course, labor intensive.

Process elements particular to flexible circuits are required to achieve a high manufacturing yield. Also, an appropriate total process design is required to achieve high productivity with high yield.

A roll-to-roll (RTR) manufacturing system is a high-productivity solution for volume production of flexible circuits. However, the RTR lines are not flexible for a non-standard construction, and they are available for only the early steps of the long manufacturing processes. Also, RTR processes have many limitations to apply.

63.2 SPECIAL ISSUES WITH HDI FLEXIBLE CIRCUITS

High-density flexible circuits, which have been developed since the mid-1990s, have special constructions and therefore require supplemental processes. In addition, high-grade manufacturing technologies are required to generate finer circuits. Sometimes, they are called HDI

TABLE 63.1 List of Specialties for Processing Flexible Circuits

Areas	Issues for processing flexible materials
Special constructions	Coverlay, stiffeners, and so on Complicated shapes
Materials	Thin, low stiffness Low dimensional stability Fragile for mechanical forces
Automation	Difficult for automation RTR available only for beginning

TABLE 63.2 New Manufacturing Technologies for High-Density Flexible Circuits

Processes	New technologies
Raw materials	Stable polyimide film (~20 ppm CTE) Thin copper foil (9, 12 μm) Adhesiveless copper-clad laminates
Microvia hole drilling	Laser (excimer, UV:YAG, carbon dioxide) Microhole punching High-accuracy *x-y* table Plasma etching, chemical etching
Copper plating	Direct plating Seed layer sputtering
Surface cleaning	New cleaning reagent
Etching resist	Thin dry film (~15 μm) Liquid resist
Exposure	Collimated light source, auto aligner Laser direct imaging
Etching	High-resolution etchant
Coverlay	Laser drilling Photoimageable coverlay (dry film type and liquid type)
Surface finishing	Electroless plating for Ni, Au, solder High-speed plating
Flying leads	Laser/chemical etching Chemical etching
Guide holes	Automatic high-accuracy punching machine
Inspection	Universal AOI Noncontact electrical tests
Roll-to-roll system	Low-tension EPC system Automatic aligner with CCD camera

flexible circuits because of new termination technologies. Major items are listed in Table 63.2. Details of the processes are described in each section.

Furthermore, ultra high-density flexible circuits have been considered recently as the necessary substrate materials for the next generation electronics packaging. Their circuit densities will be under 20 micron pitches with 20 micron diameter via holes. They need completely new manufacturing technologies, such as an additive process instead of traditional subtractive process. Ultra high-density flexible circuits will be considered separately.

63.3 BASIC PROCESS ELEMENTS

Figure 63.1 shows the basic idea of the standard manufacturing process flow for double-sided flexible circuits, with both through-holes and stiffener boards. Basically, most of the manufacturing processes for single-sided flexible circuits are included in these processes. The first half of the manufacturing process (through pattern etching) has few choices, and the automation of the processes are relatively easy. However, the second half has many options, depending on the design of the construction. The second half consumes more labors because of the difficulties of the automation, and its manufacturing cost per unit area is much higher than that of the first half. An appropriate consideration of process design should be conducted to increase the productivity and reduce the total cost. Good process design can reduce the total cost significantly, especially for complicated constructions. Typical ideas are introduced in the following sections.

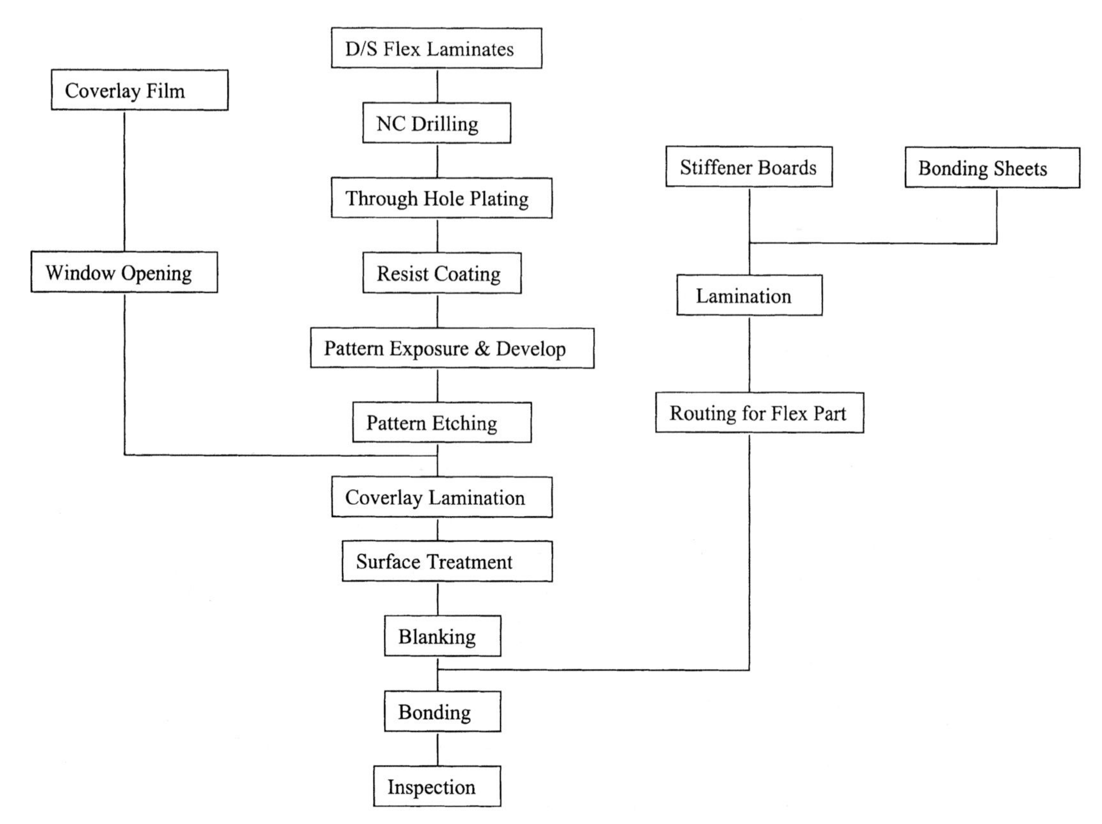

FIGURE 63.1 Standard manufacturing flow for double-sided flexible circuits.

63.3.1 Preparation of Materials

Most of the major materials for flexible circuits, such as copper laminates and coverlay films, are supplied in rolled form. The first process of manufacturing is to slit and cut the rolled materials into workable sizes. It is possible to cut with a manual sharing tool; however, an automatic machine is recommended to get an exact size without handling damage.

Cleaning processes for the copper foil surfaces of single-sided circuits should be conducted prior to cutting if roll-to-roll processing is available. A chemical etching system and soft brushing systems are recommended to use to reduce the scratches on the copper foils and mechanical stress on the thin substrate films. This could reduce handling damage and increase the manufacturing yield significantly.

63.3.2 Direct Cast Processes

For special substrate constructions such as ultra thin substrates or ultra-high-density flexible circuits, a direct casting process of polyimide resin has been developed instead of polyimide films. A photosensitive polyimide varnish is sometimes coated on a carrier stainless foil. Then an electroless and an electro copper layer are plated as the thin conductor layer. The casting process with photo-sensitive polyimide varnish is also available as a reliable coverlay of the HDI flexible circuits. The technology of the direct casting is valuable to build multi-layer constructions by semi-additive process. (See Sec. 63.4.2.) A special surface treatment on the polyimide is required for good bond strength. The thin copper layer is then etched to generate fine traces. Finally, the carrier tape is removed for flexibility. Sometimes the carrier tape is partially etched and kept as a stiffener.

63.3.3 Mechanical Generation of Through-Holes

There may be several different processes for each via hole construction in the case of flexible circuits. Suitable processes, including material combination, should be designed according to final requirements (for example, construction, circuit density, reliability, manufacturing volume, available equipment, urgency, and so on)

Figure 63.2 shows standard manufacturing processes of the through holes for the double-sided flexible circuits. A mechanical numerically controlled (NC) drilling process on double-sided flexible circuits can be done similarly to that used for rigid double-sided printed circuit boards. More laminates can be stacked up in the setup because of the thin materials.

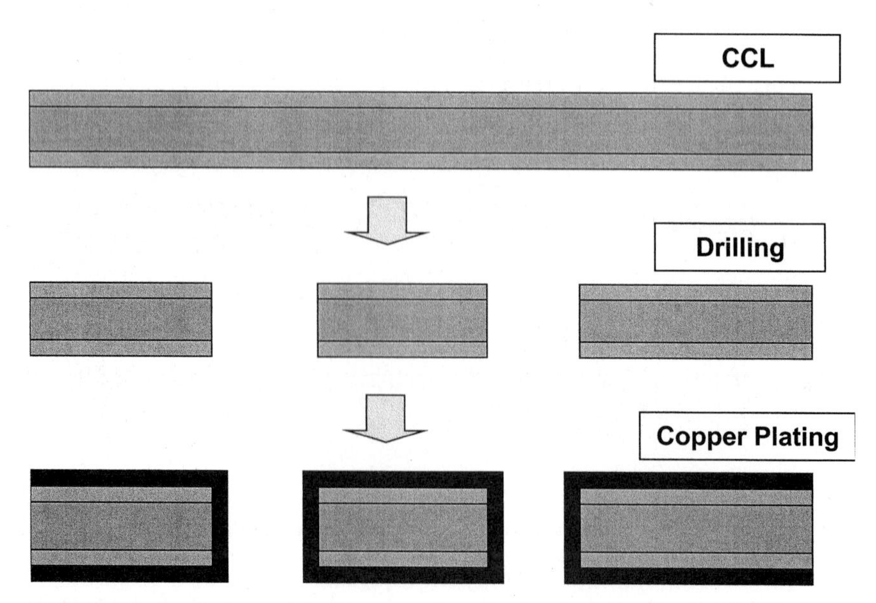

FIGURE 63.2 Standard manufacturing process for microvia holes in double-sided flexible circuits starting with copper clad laminate (CCL). The hole is created by drilling and then copper plated.

However, appropriate backing boards and drilling conditions should be determined to achieve good hole quality. A suitable combination of the NC drilling machine and drilling bit can generate smaller holes than 100 microns diameter.

A punching process with a suitable die set is available for the thin, flexible copper laminate. It can provide a low-cost solution, but it needs a large-volume production to compensate for the high-cost die. A small-diameter punching process controlled by an NC system is available for thin, flexible copper-clad laminates instead of using the NC drilling process. A sharp punching die set can generate smaller holes than 100 microns on thin flexible copper laminates with high though-put. Another advantage of the punching process is that it has a relatively easy applicability to RTR process. These mechanical hole-generation processes cannot make blind via holes on double-sided flex circuits.

63.3.4 Through-Hole Plating

Through-hole plating processes similar to those used in rigid board fabrication are available for thin, flexible circuits. Usually, a mild cleaning process is applied prior to the chemical plating process because of thin base layers and relatively small smear amounts. Both standard electroless (chemical), copper-plating processes, and direct copper-plating processes with conductive carbon paste have been established for the seed layer in the holes. Then, suitable thickness copper is plated electrically for whole surface of the panels. Generally, a thinner plated copper conductor than a typical 20 micron version is recommended to maintain the flexibility of the circuits. A thicker copper plating than 25 micron is recommended to have good reliability. Selective copper plating with suitable masking is required for dynamic flexing circuits. Precise thickness control of the through-hole copper is required, because the electrically plated copper does not deposit uniformly on the surface of the original copper foil if the raw material sheet is not fixed exactly on a fixture. Good design of fixtures and electrodes in the plating process is critical for achieving uniform copper thickness, because a thin, flexible sheet flops by agitation in the plating bath. As a result, it is not easy to achieve very uniform copper thickness for large panels. Non-uniform plated copper will cause a lower etching yield and reduce through-hole reliability, especially for small pitch traces.

As an adhesiveless copper-clad laminate does not have low-heat-resistant layers such as epoxy or acrylic resin, it provides excellent through-shapes without de-smearing processes.

Figure 63.3 shows a comparison of a hole plated with an adhesive-based laminate and a hole plated with adhesiveless laminate. The adhesiveless laminate provides a sharp through hole shape compared to the traditional laminate with adhesives.

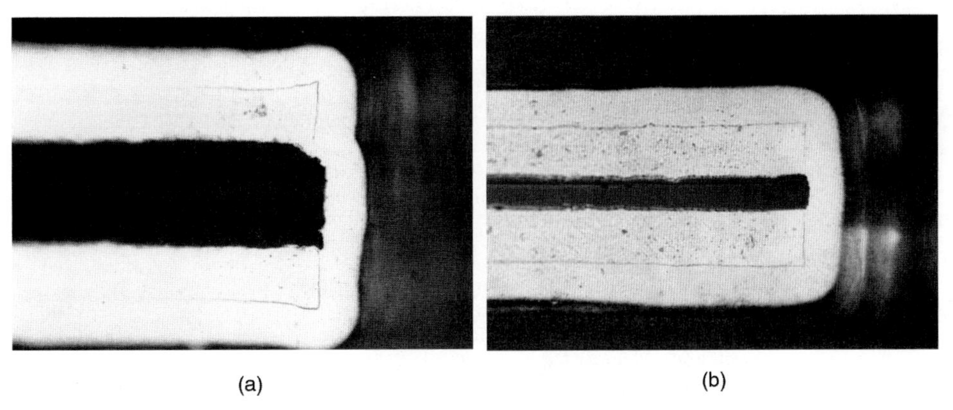

FIGURE 63.3 (a) Cross section of through-hole of a double-sided circuit with adhesive-based laminate. (b) Cross section of through-hole of a double-sided circuit with adhesiveless base laminate.

63.3.5 Microvia Hole Processes

Micro via hole generation can be one of the critical items in the building of high-density flexible circuits. Figure 63.4 shows standard manufacturing processes for micro via holes, including blind via holes for double-sided flexible circuits. Several new technologies have been developed for micro hole generation on thin polyimide dielectric layers, as shown in Table 63.3.

63.3.5.1 Mechanical Microvia Hole Creation. The newly developed NC drilling machine can generate micro via holes smaller than 100 μm in diameter using special drill bits. Technically, the newest NC drilling system can generate 50-μm-diameter holes on 50-μm-thick flexible copper laminates. However, it can drill fewer sheets in a stack than larger holes with slower speeds, and its productivity becomes smaller with smaller hole sizes. An NC micro hole-punching system can also generate micro holes smaller than 70 μm in diameter on 25-μm-thick polyimide base flexible copper laminates. It has lower productivity issues to the NC drilling machines; however, multiple punch die systems can increase productivity significantly. A roll-to-roll (RTR) capability is another benefit of the NC micro hole punching.

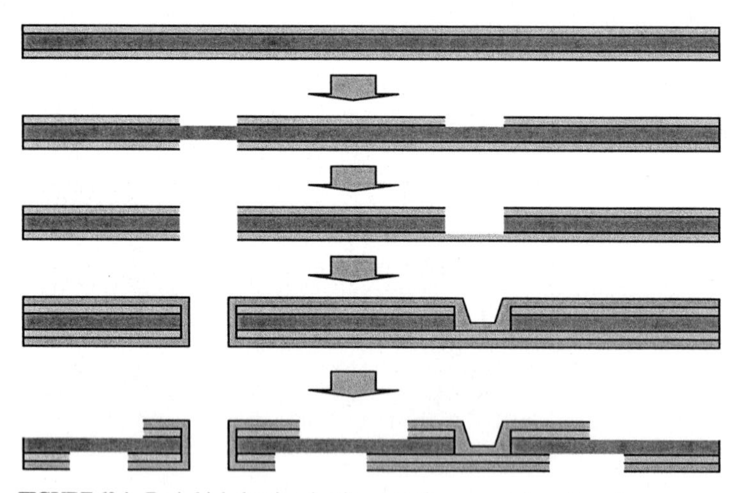

FIGURE 63.4 Basic high density circuit generation process by subtractive process.

TABLE 63.3 Comparison of Technologies for Hole Generation on Dielectric Layers

	Drilling	Punching	Laser drilling	Plasma etching	Chemical etching	Photo polymer process
Hole diameter (μm)	~50	~70	~10	~70	~70	~70
TH for D/S	Yes	Yes	Yes	Yes	Yes	Yes
TH for multilayer	Yes	No	Possible	No	No	No
Blind via hole	No	No	Yes	Yes	Yes	Yes
Material choice	Wide	Wide	Wide	Fair	Small	Small
Technical flexibility	Small	Small	High	Fair	Small	Small
Chemical waste	No	No	Small	Small	Serious	Fair
RTR availability	No	Possible	Possible	Difficult	Possible	Possible

63.3.5.2 Laser Microvia Hole Creation. Several laser systems have been developed to generate microvia holes for flexible circuits. The excimer laser has a wide capability to generate small holes on most organic substrates. It can generate holes 10 μm in diameter on 25-μm thick polyimide film, as shown in Fig. 63.5. The largest issue with the excimer laser is its slow speed. Appropriate process design is necessary to achieve a good productivity.

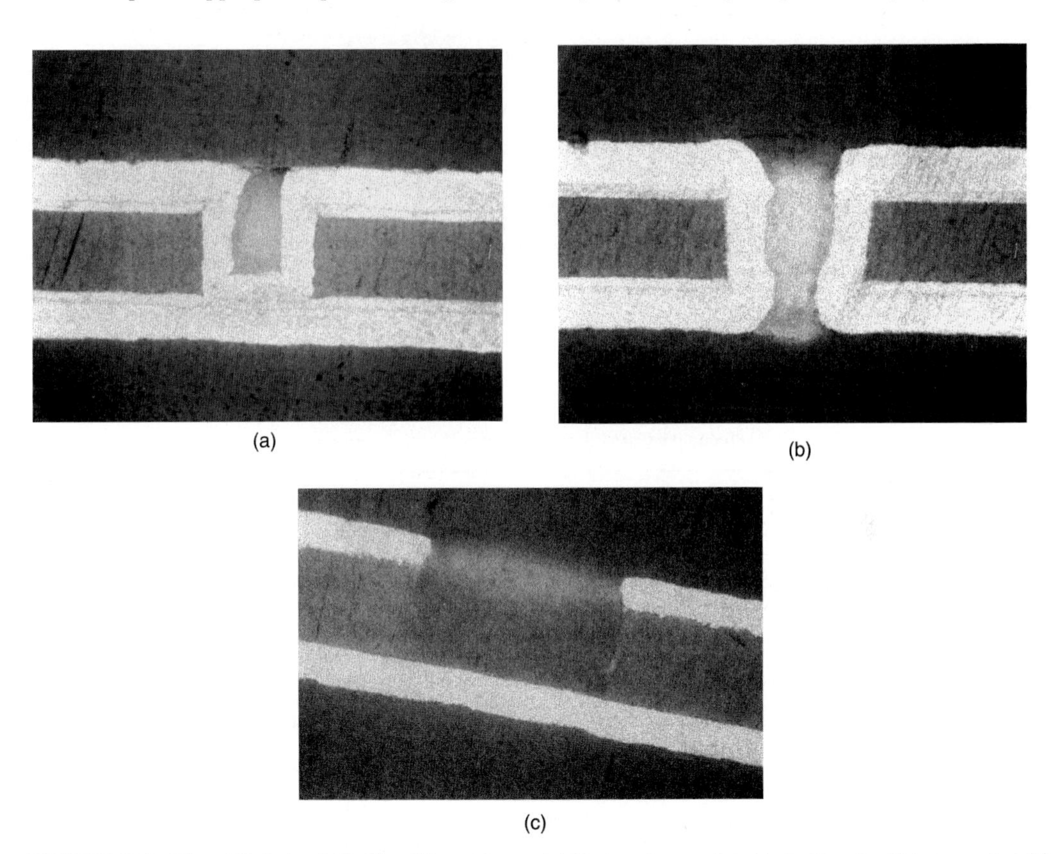

FIGURE 63.5 Micro-blind vias drilled by different lasers. (a) Excimer laser, plated hole diameter 25 'micrometer' (b) YAG laser, plated hole diameter 20 micrometer, (c) Carbon dioxide laser, plated hole diameter 100 micrometer. *(Source: Photo Machining.)*

The UV:YAG laser is another choice for generating microvia holes smaller than 50 μm in diameter on flexible materials. It has a higher productivity rate than the excimer laser. It can drill through both copper foil and flexible substrates. An issue with the UV:YAG laser is that it takes a long time to generate large holes.

The carbon dioxide laser has greater productivity than the excimer laser or the UV:YAG laser when used to generate via holes larger than 60 μm in diameter; however, it cannot drill through copper foil directly. As a result, a black surface treatment is required on the thin copper foils before the laser operation. (See Fig. 63.6) Comparisons of the technical capabilities of these micro via processes of the laser systems are shown in Table 63.4.

Another major advantage of the laser via generation systems is RTR capability compared to the traditional mechanical NC drilling systems for the volume production of the double-side flexible circuits. It needs some modifications and loader and un-loader for the RTR systems for the standard laser equipment. Several laser systems are available for RTR process.

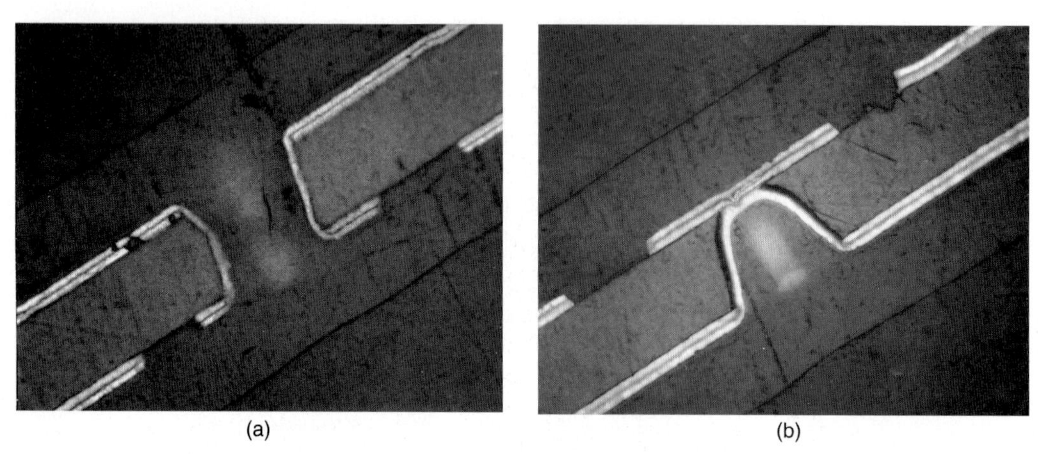

FIGURE 63.6 Microvia holes 50 micrometer in diameter created by carbon dioxide laser. (a) through hole, (b) blind via. *(Source: Photo Machining.)*

TABLE 63.4 Technical Capabilities of Microvia Hole Technologies for Double-Sided Flex Circuits

	Excimer	UV:YAG	Carbon dioxide	Plasma etch	Chemical etch
Hole size (μm)					
For small volume	~10	~15	~50	~70	~50
For large volume	25–150	25–75	75–250	~100	~75
Hole quality	Excellent	Good	Fair	Fair	Fair
Additional cleaning	Necessary	Necessary	Necessary	No	No
Technical hurdles	Low	Low	Fair	High	High
Cost	High	High	Fair	Low	Low

For only specific polyimide resins.

63.3.5.3 Conformal masks for Microvia Generation. Mostly, the laser systems are capable for the drilling for both of base films and copper foils of the flexible laminates. It is possible to make microvia holes through copper foils by the same machine in one process. However, the drilling speeds on the copper foils by lasers are remarkably lower compared to organic base films. Therefore, the conformal mask systems with copper foils are recommended to use for the volume productions with a large numbers of the holes in a unit area. The basic process is illustrated in Fig. 63.7. Firstly, small holes are generated on the copper foils by standard photolithography and etching process. The holes generated on the copper foils work as the masks for the laser drilling process. A slightly larger laser beam is irradiated on each opening of the copper foil, and the organic layers under the opening are engraved effectively.

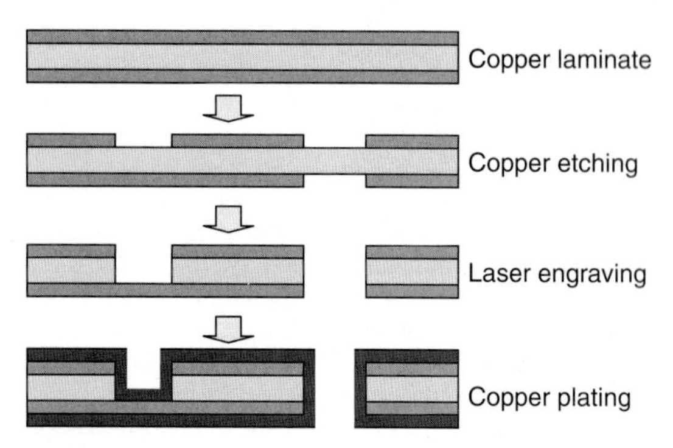

FIGURE 63.7 Conformal mask system for laser drilling.

To optimize the productivity of the laser drilling process, blind via hole structures are recommended for the high density flexible circuits.

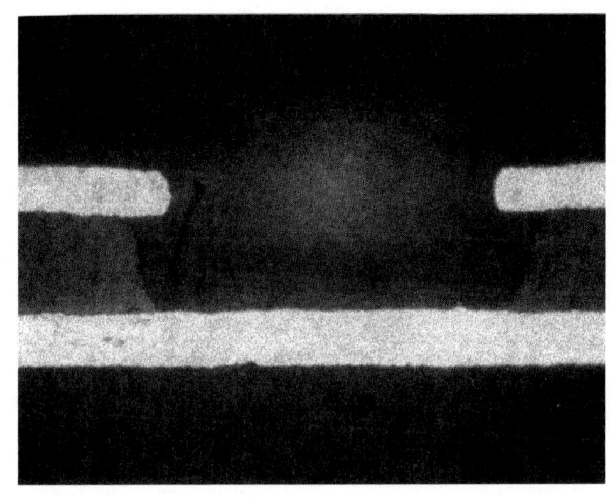

FIGURE 63.8 Microvia genterated by plasma etching. The hole diameter is 200 micrometers.

The basic idea of the conformal mask is applicable for the plasma and chemical etching introduced in the following section.

63.3.5.4 Plasma and Chemical Microvia Hole Creation. Plasma etching and chemical etching are the unique processes used to generate micro via holes on flexible circuit materials (see Figs. 63.8 and 63.9). Both can generate micro via holes smaller than 100 μm in diameter on 50-μm-thick polyimide substrates. Plasma processes can etch all kinds of organic materials, but standard alkaline etching can etch only specific polyimide materials such as Kapton™ or Apical™. Special strong chemicals are needed to etch Upilex™-type chemically stable polyimide substrates, which have high dimensional stability. Processing costs of plasma processes and chemical processes do not depend on the number of holes or sizes in a unit area.

FIGURE 63.9 Microvia holes made by chemical etching. The hole diameter is 40 micrometers. *(Source: Asahi Fine Technology.)*

They have a remarkably lower cost than laser processes, if there are many holes in the same area. On the other hand, they have several disadvantages. Generally, there are many parameters that affect process capabilities. An exact process conditioning is required for each construction of the flexible circuit.

Chemical etching is a convenient technology to apply to a reel-to-reel manufacturing system. It takes a large investment to introduce a reel-to-reel system for plasma etching system.

Waste treatment is another issue, especially in the case of wet chemical etching. The used etching solution for this process should be treated separately from the other chemical waste used for the wet process in the manufacturing lines.

Capabilities to process materials of each technology are shown in Fig. 63.10 and Table 63.5.

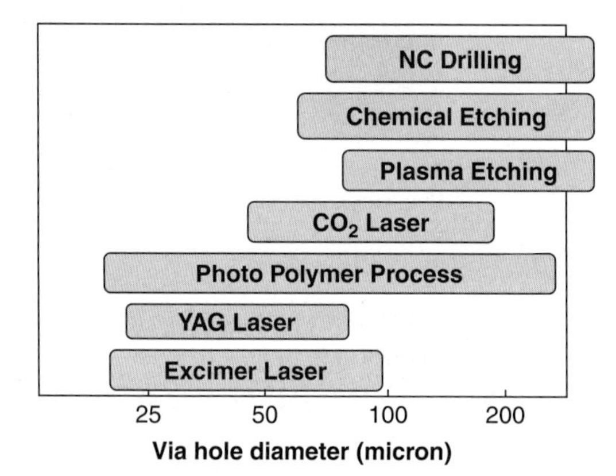

FIGURE 63.10 Microvia hole capabilities of the technologies.

TABLE 63.5 Material Availability

	Punching and drilling	Excimer laser	UV:YAG laser	Carbon dioxide laser	Plasma etching	Chemical etching
Copper	Yes	Slow	Slow	No	No	Yes
Epoxy	Yes	Yes	Yes	Yes	Yes	Difficult
FR-4	Yes	Slow	Slow	Yes	No	No
Kapton/Apical	Yes	Yes	Yes	Yes	Yes	Yes
Upilex	Yes	Yes	Yes	Yes	Yes	Difficult
Espanex	Yes	Yes	Yes	Yes	Yes	Difficult
NeoFlex	Yes	Yes	Yes	Yes	Yes	Difficult

Figure 63.11 shows examples of the cost comparison of the technologies for micro-through holes on 50-μm polyimide substrates. Figure 63.11(a) shows cost comparison of micro-size NC drilling versus the excimer laser and UV:YAG laser. NC drilling cost is inversely proportional to hole size. A conventional NC drilling process has small advantages over laser processes for holes smaller than 200 μm in diameter. Figure 63.11(b) shows a cost comparison of different laser systems. The UV:YAG laser is cost-advantageous in the 25- to 75-μm hole diameter range. However, its cost increases exponentially with hole size, especially for holes larger than 75-μm in diameter. The cost of the excimer laser process is more stable. It is proportional to

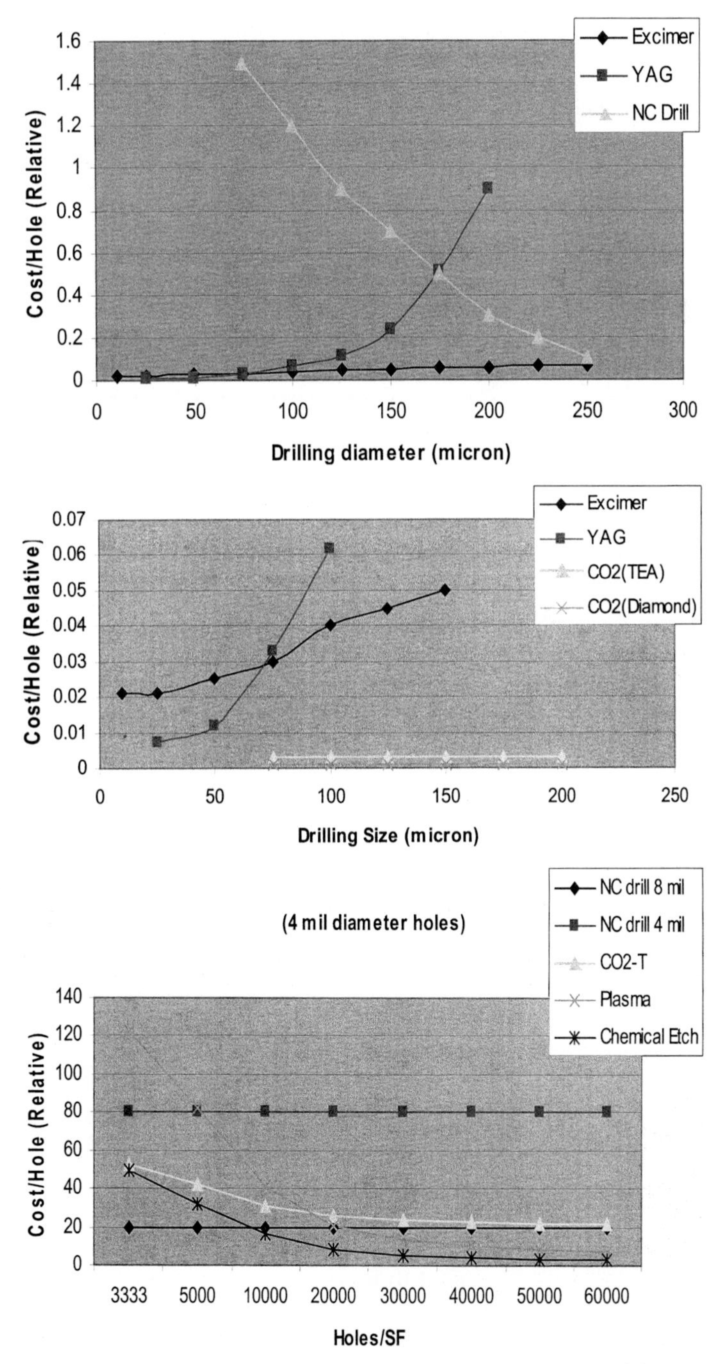

FIGURE 63.11 Relative cost comparison by hole creation technology, (a): Comparison of NC drilling cost in Excimer and YAG lasers, (b): Comparison of alternative laser processes (c): Comparison of NC laser, plasma, and chemical etch.

hole sizes in the 20- to 200-μm diameter range. Carbon dioxide lasers show extremely lower drilling cost than the excimer laser and UV:YAG laser for the holes larger than 75 μm in diameter. It is very possible to generate more than 10,000 holes, 100 μm in diameter, in 1 min with the Galvano mirror. This is a Diamond Carbon Dioxide Laser System™ that has higher productivity than a traditional carbon dioxide TEA laser system.

Figure 63.11(c) shows a drilling cost analysis of a plasma etching system and a chemical etching system as compared to a laser system or an NC drilling system. A unique cost feature of the etching procedure is that it does not depend on the number of holes or shapes, but on area sizes. Therefore, there are big cost advantages for circuits that have a large number of holes in a small area.

63.3.6 Wet Processes for the Flexible Materials

The conveyer systems of the wet processes designed for rigid circuit boards do not work properly for thin flexible materials. They can't make a high process yield generating a lot of wrinkles or scratches, even though the flexible sheet is fixed on a leader board or carrier frame. Appropriated conveyer system should be introduced for the volume production lines of the thin flexible circuits.

Several new ideas have been introduced to convey the thin flexible materials in the wet manufacturing processes. Large diameter conveyer rings with smooth surface have been employed to reduce the scratches on the soft thin copper foil surface. Smaller ring pitches, for both line direction and transverse direction, results in an overlapped conveyer system, which avoids the drop down between the rings and winding of the thin materials on conveyer rolls. Strings and thin bars between the conveyer rings are not recommended to use. These devices make small scratches on the soft thin copper surface, as shown in Fig. 63.12.

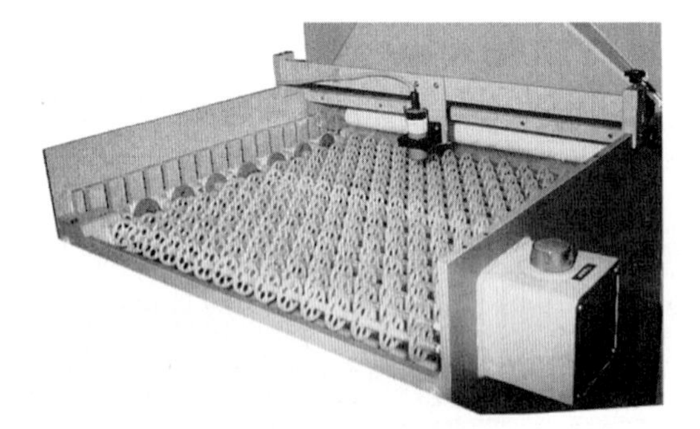

FIGURE 63.12 Special wheel conveyor system for thin, flexible materials showing high density and smooth wheels. *(Source: Camellia.)*

It looks like a conflict to have a stable conveyor and an effective spray of the liquids, therefore synchronized spray array with mild pressure are required to optimize the condition. A high spray pressure makes fluttering and partial distortion for the thin materials especially after etching.

Standard wet processing machines are designed to process both single side and double side circuits. Well-controlled soft rubber rolls are recommended for the conveyers of the RTR wet manufacturing process to avoid damage to the circuit surface.

63.3.7 Surface Cleaning

A pre-cleaning process is required to remove the stains and oxidations on the surface of the copper foils prior to the coating of photo resist materials. Several adhesiveless laminates have extremely shiny copper surface, therefore the chemical cleaning has an important function in making a better affinity with the etching resist, especially for the fine trace etching.

A mild chemical cleaning process, such as organic acid solution, should be applied for thin flexible materials for eliminating serious mechanical damages. Soft scrubbing or soft brushing must be used when a mechanical process is necessary to remove tough stains or oxidation on the copper surface. A severe mechanical brushing will produce mechanical stress on the thin materials and will cause the loss of dimensional uniformity. A soft brushing process on a belt conveyor has been developed to minimize mechanical damage to flexible materials, This process works only for one side of the laminates, therefore a two-step process is required for the double-side laminates.

63.3.8 Resist Coating

The same etching resist inks used for rigid circuit boards and screen printers are available for flexible circuits. These inks allow for relatively rough pattern etching with large volume. Also, similar dry films and laminators, designed for rigid boards, could be used for flexible circuits. RTR is can be available adding rolled loader and unloader. However, flexible dry films with a high etching yield should be selected. A tenting method with relatively thick dry films is used for the double-side circuits with standard size through holes.

A thinner dry film is recommended for the single-side circuits, especially for fine patterns. Dry films 15 or 20 μm thick can generate 30- to 40-μm lines/spaces on 12-μm-thick copper foils with a high yield. A liquid photo resist should be used for very-high-density traces with pitches smaller than 40 μm.

63.3.9 Pattern Generation

Similar exposure equipment is available for flexible circuits. A collimated light source machine is recommended for high-density traces with pitches smaller than 50 μm to achieve a high yield.

An appropriate matching tool for flexible materials should be developed in order to achieve good pattern alignment with high productivity. Suitable auto alignment systems with CCD cameras have been introduced to minimize the pattern shift from the via holes generated previously. Because of the dimensional changes in the previous mechanical and wet processes, appropriate dimensional corrections are required for the photo masks. A glass mask is required for tracers finer than 50 microns pitches instead of polyester base film masks. A set of reliable mask holder is required to keep the heavy glass masks in the machines. An exact alignment between top mask and bottom mask is required for the high density double-side flexible circuits.

Sometimes, partial dimensional corrections are required to adjust the distortions of the thin flexible materials. A higher alignment accuracy than +/–10 microns is required for the finer pitched than 50 microns. The easiest way for the dimensional correction is to reduce the panel sizes. But it reduces the process productivity. An alternative method is a step-and-repeat imaging with a small exposure area on a large panel.

It is possible to make a corrected photo mask for the imaging process in order to minimize the misalignment. Unfortunately, the distortion of the flexible laminate is unstable, sometimes, it needs to generate the corrected masks for each job lot. It asks the material to stay beside the exposure machine until the right mask is ready.

Laser direct imaging equipment (LDI) is capable of making the dimensional corrections for the serious distortions of the unstable flexible materials promptly. It provides a high productivity for the prototype or middle volume productions. On the other hand, it is not very capable for traces finer than 25 microns lines and spaces. Lower throughput compared to traditional exposure machines, especially for double-side imaging, is the barrier for the volume productions.

63.3.10 Etching

Ferric chloride or cupric chloride solution is recommended for use as the etching chemicals for flexible circuits, especially for high-density circuits. An alkaline solution has a higher process speed, but is not good on fine circuits because of unstable etching rates. As an alkaline solution makes some chemical attacks on polyimide surfaces, it should not be used for fine line requirements.

There are appropriate correction factors for the etching rates of each combination. Between conductors, patterns, thickness, and etching, solutions should be introduced for the photo masks, based on the actual trials, especially for the multiple parallel fine lines very common to high density flexible circuits. Usually, there are remarkable differences in etching factors between middle conductors and edge conductors, as shown in Fig. 63.13.

A set of dummy conductor patterns helps to minimize the difference of the etching factors.

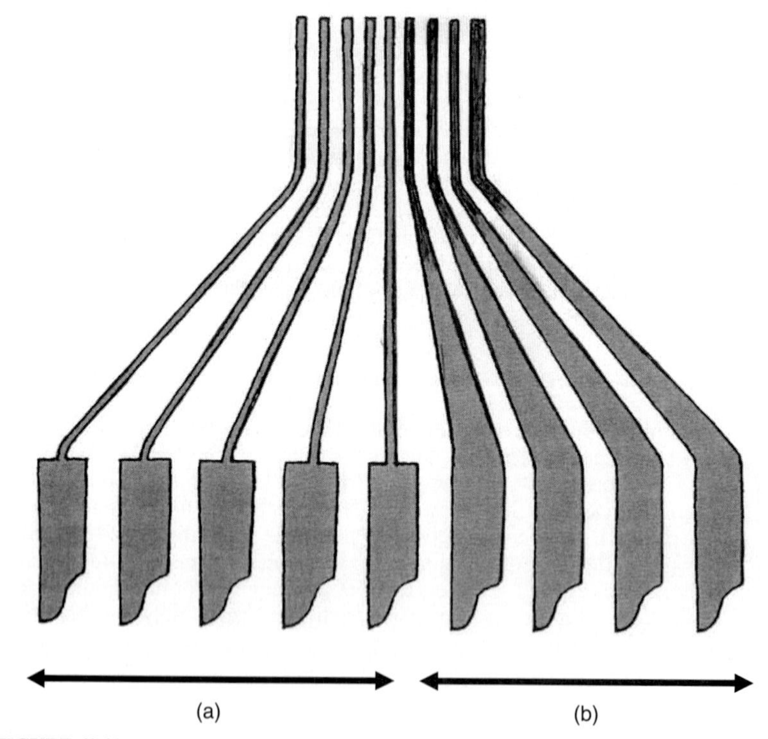

FIGURE 63.13 Pattern correction for the etching factors (a) Unacceptable (b) Preferred.

63.4 NEW PROCESSES FOR FINE TRACES

The generation of fine lines is the most basic process in building high-density flexible circuits. Three basic manufacturing concepts have been developed:

1. Advanced subtractive
2. Semi-additive
3. Full-additive

These methods have many variations to cover different construction concepts and specifications.

Innovative technologies have been developed for each unit process, such as the collimated light source exposure machine and the laser direct imaging system.

63.4.1 Advanced Subtractive Process

The subtractive (etching) process is the most common technology for generating copper-based circuitry on flexible film substrates. It is basically the same method used for rigid boards, but there are several differences in the process. An example is that it may be automated for use with a roll-to-roll process.

Subtractive technology with traditional equipment and materials cannot produce high-density traces with pitches smaller than 150 μm. However, many new technologies have been introduced for each key item, as shown in Table 63.6. The most advanced subtractive process with adhesiveless laminates can generate 30-μm-pitch traces on 9-μm copper foil for the single-side circuits using roll-to-roll manufacturing systems. A 50 micron pitch trace circuit is shown in Fig. 63.14 .

TABLE 63.6 New Technologies Needed to Generate Fine-Line Traces on Flexible Substrates

Processes	New technologies
Surface treatment	New soft etching chemicals (organic) Sputtering of seed layer
Resist coating	High-resolution liquid etching resist and plating resist (spray coater, dipping coater, etc.) Thin dry film (~15 μm) High-sensitivity dry film with high resolution (10 mJ/cm^2) Wet laminator
Exposure	Collimated light source with glass mask (~15-μm line/space) Autoalignment system for double-sided circuits (with accurate dimension compensation system, ~5 μm) Roll-to-roll conveyor system with low tension Laser direct imaging system (30–40 μm line/space)
Etching	High-resolution etchant (cupric chloride, ferric chloride)

It is possible to produce 20-μm-pitch traces by subtractive process (as shown in Fig. 63.15), but the process will have relatively low yields.. The advanced subtractive process can produce 60-μm-pitch traces on double-sided flexible circuits with thin copper foil plated to 20-μm-thick.

Processes New technologies Surface treatment New soft etching chemicals (organic) Sputtering of seed layer Resist coating High-resolution liquid etching resist and plating resist (spray coater, dipping coater, etc.)

Thin dry film (~15 μm) High-sensitivity dry film with high resolution (10 mJ/cm^2) Wet laminator Exposure Collimated light source with glass mask (~15-μm line/space) Autoalignment system for double-sided circuits (with accurate dimension compensation system, ~5 μm) Roll-to-roll conveyor system with low tension Laser direct imaging system (30–40 μm line/space) Etching High-resolution etchant (cupric chloride, ferric chloride)

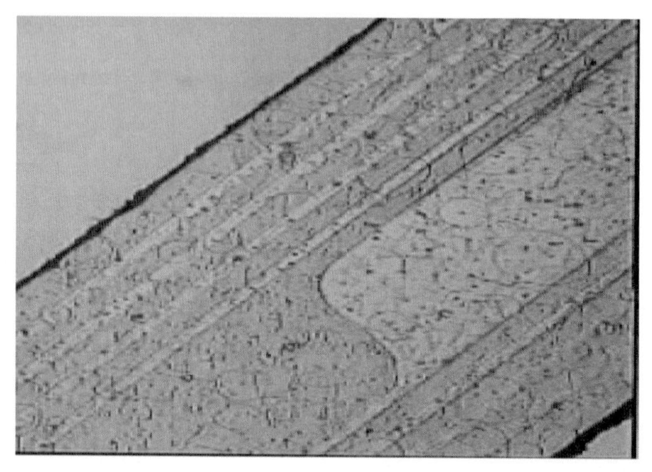

FIGURE 63.14 Fifty micron pitch trace produced by advanced subtractive process. *(DKN Research.)*

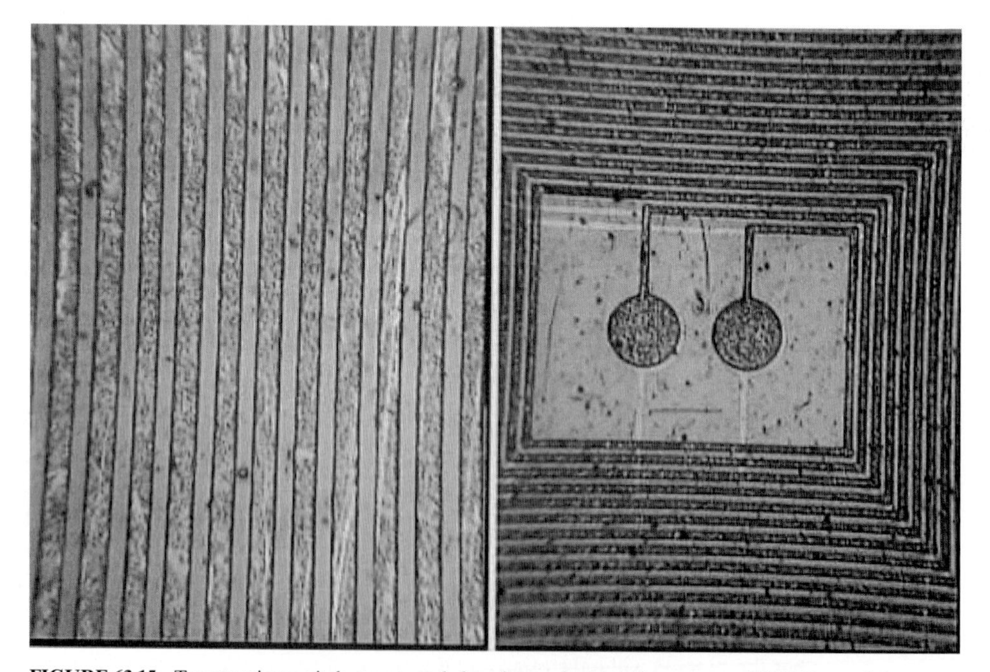

FIGURE 63.15 Twenty micron pitch traces made by advanced subtractive process. *(DKN Research.)*

63.4.2 Semi-Additive Processes

The first candidate to produce fine traces instead of subtractive process could be the semi-additive process. Several kinds of semi-additive processes—including thin copper laminates and sputtering—were developed to generate fine-line traces on specific polyimide substrate materials. An advantage of these processes is that the conductor layers can have varying thicknesses. Both processes create suitable electroless plating processes to achieve reliable bond strength on the substrates.

The first idea of the semi-additive process is that the chemical and electrical deposition makes the copper layers after the drilling on the plain base film followed by traditional imaging and etching process as illustrated in Fig. 63.16.

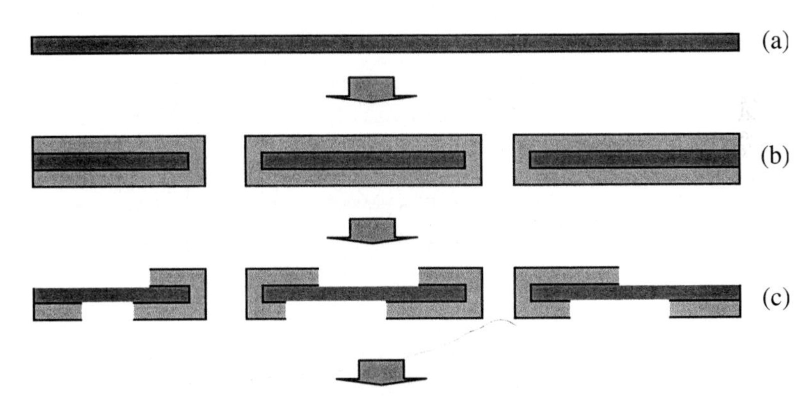

FIGURE 63.16 Basic high density circuit generation process flow for semi-additive method, (a) Plain film substrate, (b) Drilling via holes and plating of copper layers, (c) Etching of copper layer.

The thin copper laminate process does not need special technology such as sputtering, therefore it is relatively easy to design the manufacturing process. Figure 63.17 illustrates a typical manufacturing process for the double-side circuits.

1. It starts from double-side laminates with very thin copper layers.
2. The first step is drilling to form the via holes. Most of the drilling technologies are available to generate the small holes. Mechanical process including micro punching works for larger holes than 60 microns diameters. Chemical etching and plasma etching could be the low cost solutions for the high hole counts designs. But they need supplemental photolithography and copper etching process. Excimer laser and UV YAG laser are capable to generate smaller holes than 50 micron diameters for both through holes and blind via holes.
3. Copper layers are plated on the whole surface to build the conductive layer in the via holes, by both electroless and electrolytic processes, using, standard copper plating process. It is important that the processes are controlled well enough to make the thin copper layers with uniform thickness.
4. A negative pattern of conductor traces for the plating resist is formed on the copper surface by a high-resolution photolithography process.
5. Further electro plating is conducted to build the required copper thickness. The plating conditions should be well controlled to have a uniform thickness for whole circuits. A smaller panel size is recommended for an easy control of the plating conditions.
6. The plating resist is removed by standard stripping process.
7. As the final process, the seed layers between the conductor traces are removed by appropriate etching process. A supplemental metal plating is conducted as the protection layer for the process, prior to the stripping process of the plating resist.

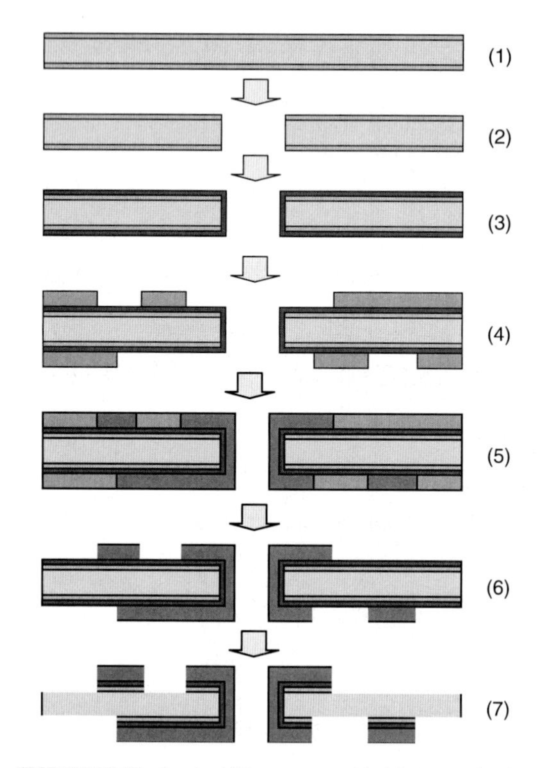

FIGURE 63.17 Semi-additive process with thin copper laminates.

FIGURE 63.18 Twenty micron traces on double sided flexible circuits made by semi-additive process.

Figure 63.18 and Fig. 63.19 show examples made by the semi-additive process.

The sputtering process has been developed to have thin seed layer for the semi-additive process with finer trace densities as shown in Fig. 63.20.

FIGURE 63.19 Fifteen micron traces with 20 micron holes for 10 layer circuits. *(Source: MicroComnex.)*

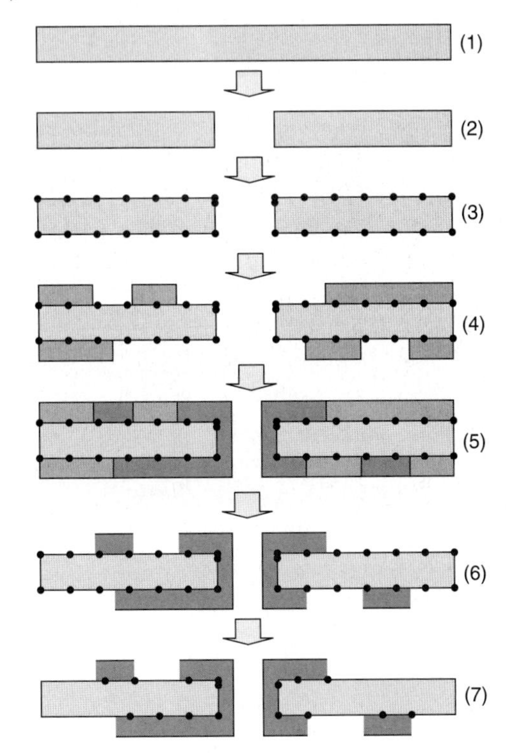

FIGURE 63.20 Semi-additive process with sputtering.

1. Because it's a plain plastic film, polyimide film is typically the starting material for the manufacturing process.
2. The small size holes are generated by mechanical processes or laser process. It needs supplemental photolithography process to use chemical etching and plasma etching for this process.
3. A set of sputtering processes is employed to form the electrically conductive thin seed layer, typically thinner than one micron. To have a reliable bond strength between organic layers and conductors, appropriate conditioning including surface treatment should be conducted.

The following processes are used for plating resist formation:

- Electro-forming of the copper conductors
- Stripping of the plating resist
- Etching of the seed layers
- A process that is basically the same as the previous semi-additive process, but is able to provide finer pitches with higher aspect ratios because of extremely thin seed layers. An optimized condition of the process can produce smaller than 5 micron pitches and larger aspect ratios than 1:5.

Figure 63.21 shows two examples of fine trace produced by sputtering/electro-forming process.

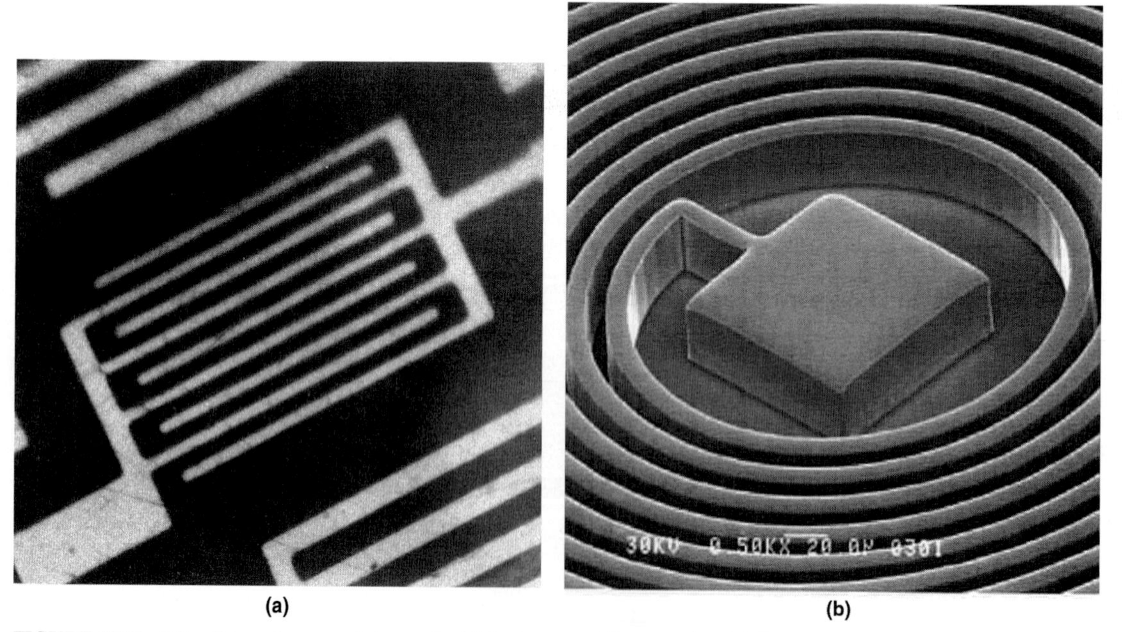

FIGURE 63.21 Ultra thin patterns made by semi-additive process with sputtering, (a) Ten micron pitch traces with one micron thick trace, (b) Ten micron wide traces with 25 micron thick traces. *(Dynamic Research.)*

A combination of the semi-additive process with a casting technology of the polyimide resin provides a capability to build multi-layer constructions with ultra fine traces and micro via holes. But the layer constructions and the manufacturing process could be very different from the traditional flex circuits. The manufacturing process starts from the preparation of the base layer. A commercialized polyimide film is available as the starting material for the process.

A casting of the liquid polyimide on a suitable carrier plate is another choice to have a thinner base layer than 12 microns. A glass or a stainless steel plate is recommended to use to manage the dimensions stable.

1. A seed layer is formed on the base layer by a series of sputtering processes.
2. A negative pattern of conductor traces is formed with plating resist by a high resolution photolithography process. The selection of the photo resist could be the key of the technology. A spin coating of the liquid photo resist is recommended for the high resolution.
3. An electrical forming of copper or other suitable metals is conducted to generate the fine conductor traces with necessary aspect ratios.
4. A set of process for stripping and etching of the seed layer between conductor traces is conducted to have isolated circuits. A gold plating on the conductors prior to the stripping process is recommended for the chemical protection of the traces.
5. A polyimide resin is coated on the circuits as the insulation layer for the next conductor layer. A photo-sensitive polyimide resin is recommended to coat to generate via holes by photolithography process. A combination of non-photo sensitive polyimide and Excimer laser is another choice. It has a flexible capability to form reliable insulation layers. But its productivity is low and processing cost per unit area is relatively high.
6. The second seed layer is formed by the same sputtering process as shown previously in process 2. A difference is that the thin seed layer have to cover the different surfaces in the via holes.
7. The same processes of photolithography of plating resist, electro forming of the conductors, stripping of the resist and etching of seed layer are conducted to build the second conductor layer.

An example of the processes is illustrated in Fig. 63.22.

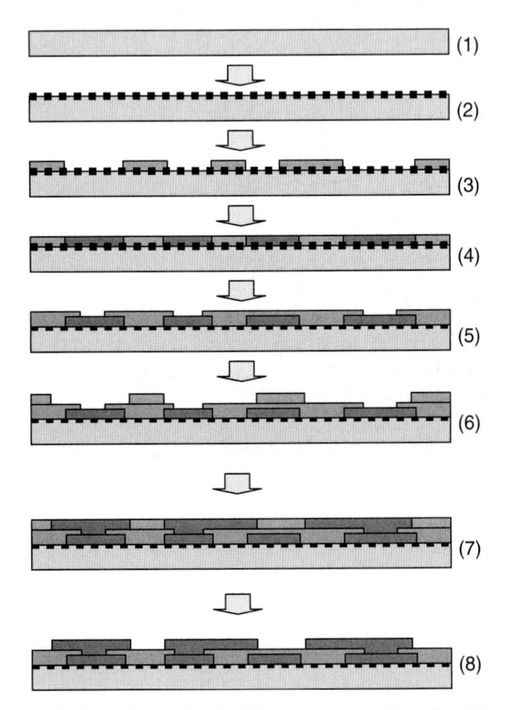

FIGURE 63.22 Semi-additive process with polyimide (PI) casting.

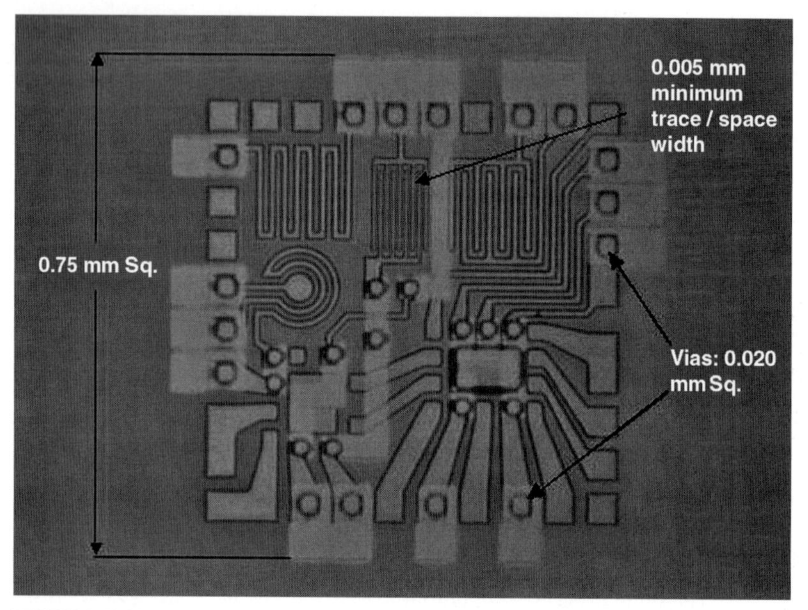

FIGURE 63.23 Ten micron pitch with 20 micron diameter via holes on the three layer flexible circuits made by semi-additive process with casting. *(Source: Dynamic Reaserch.)*

The same processes can be repeated to form more conductor layers. Figure 63.23 shows an example of ultra high density multi-layer flexible circuits with 10 micron pitch and 20 micron diameter via holes.

63.4.3 Fully Additive Processes

Several full-additive processes that can produce extremely fine traces on flexible substrate materials have been developed.

Several combinations of chemical deposition process of the conductor materials and photo-sensitive resists have been tried. Unfortunately, there is no good combination that satisfies the needs for both high-resolutions and physical performances. Further improvement is required to have practical high density flexible circuits.

Screen-printing has had a remarkable progress to generate fine resolutions. An appropriate combination of the screen printer and screen mask can provide smaller pitch traces than 20 microns as shown in Fig. 63.24. It could be the low cost solution for the volume production for the high density flexible circuits. The process is capable for both conductor materials and insulation materials. A screen printable conductive material could be the key of the process.

63.4.4 Process Comparison

A comparison of the fine line generation technologies is shown in Fig. 63.25. The advanced subtractive process can provide the best cost-performance solution when building a high-density circuit at pitches larger than 30 μm for single-side circuits and 50 μm for double-side circuits. The process does not need very different technologies from the traditional etching process, therefore the technical hurdles are relatively low and it minimizes the investment for the manufacturing equipment.

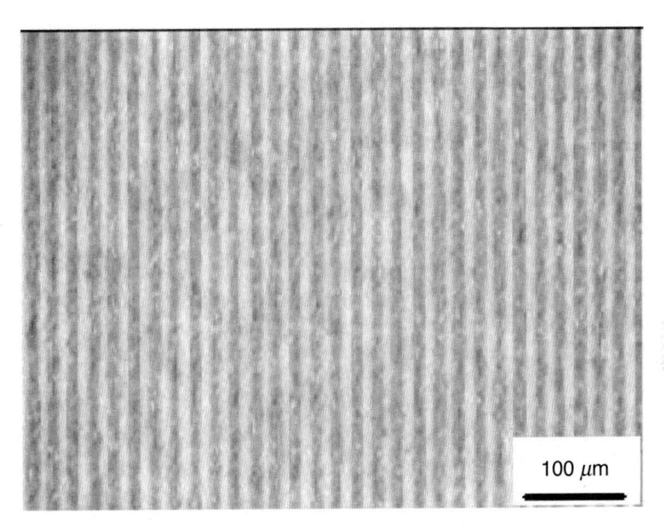

FIGURE 63.24 Twenty micron pitch traces made by advanced screen-printing process. *(Source: Micro Tech.)*

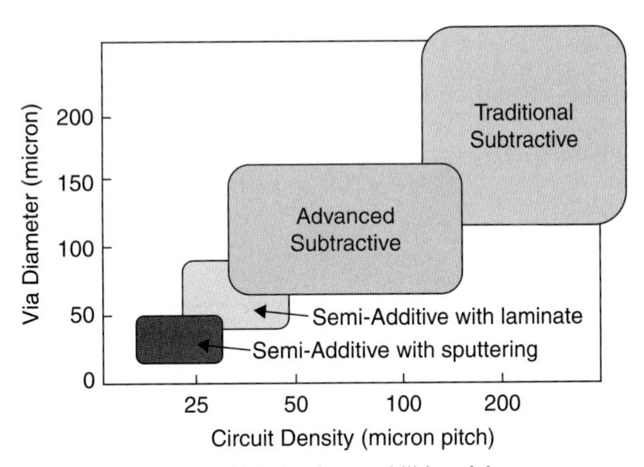

FIGURE 63.25 Total high density capabilities of the process.

Semi-additive process with thin copper laminates will provide reliable solutions to produce the fine line generation down to 20 micron pitches with 40 microns via holes for both single and double-side circuits. Technical hurdles are relatively low other than the control of the copper plating. The copper thickness of the original laminate is the key to producing finer traces.

Semi-additive process with sputtered seed layers is able to generate the finest traces, all the way down to 5 micron pitches with 10 micron via holes, on a thinner polyimide layers than 10 microns. A well conditioned sputtering is the key to have reliable bond strength for the fine traces. Additionally, a photolithography process with photo-sensitive polyimide resin provides the multi-layer capability with micro-size via holes for the ultra-fine flexible circuits.

It needs some more technical improvement to make the full-additive process with chemical deposition practical. Another full-additive process with high resolution screen printing will be a low cost solution for the 20 micron pitch single-side circuits when appropriate conductor materials are ready to apply. Combinations with the other technologies will create double-side and multi-layer capabilities.

63.5 COVERLAY PROCESSES

The coverlay construction of flexible circuits demands another special process. It is very different from the solder mask used for rigid circuit boards. Furthermore, recent high-density flexible circuits need new coverlay technologies. Table 63.7 summarizes the comparison of traditional coverlay technologies and new technologies. Traditional film coverlay has an excellent balance of physical properties and can provide high reliability, especially for long dynamic flexible endurance.

TABLE 63.7 Coverlay Process Options

	Accuracy (minimum opening)	Reliability (flexing endurance)	Material selection	Equipment/ tooling	Technical hurdles	Cost
Traditional film coverlay	Low (800 μm)	High (long)	PI, PET	NC drill, heat press	Need long experience	High
Film coverlay + laser drilling	High (50 μm)	High (long)	PI, PET	Heat press, laser	Low	High
Screen print liquid ink	Low (600 μm)	Acceptable (short)	Epoxy, PI	Screen printer	Medium	Low
Photoimageable, dry film type	High (80 μm)	Acceptable (short)	Epoxy, PI, acrylic	Laminator, exposure, developer	Medium	Medium
Photoimageable, liquid type	High (80 μm)	Acceptable (short)	Epoxy, PI	Coater, exposure, developer	Relatively high	Low

Unfortunately, it requires very complicated, labor-intensive processes, and it is difficult to introduce automatic manufacturing systems. Another big issue is that the standard film coverlay process cannot make small openings with high dimensional accuracy because of the manual handling of thin film materials.

Fine openings smaller than 200 μm in diameter with high dimensional accuracy and better than 100 μm positioning have been required for flexible circuits that need high-density interconnect (HDI). Examples include SMT of CSP or chip type passive components. Screen printing of flexible ink can provide a low-cost solution for volume production. However, it cannot provide a better solution for small openings with high dimensional accuracy. The technology of photo-imageable coverlay could be the right solution to the problem of having fine openings with high dimensional accuracy along with keeping the cost down by automation. A suitable photo-sensitive material has been in demand to satisfy all of the requirements.

63.5.1 Film Coverlay

A typical coverlay manufacturing process is shown in Fig. 63.26. Generally the same film materials are chosen for the base film and the coverlay film. A B-stage adhesive such as epoxy

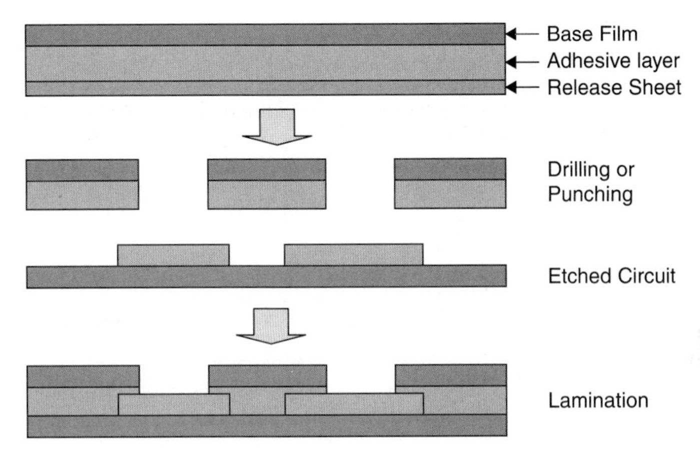

FIGURE 63.26 Lamination process of film coverlay.

or acrylic resin is coated on one side of the film covered with a release sheet. The epoxy-based coverlay films have a short shelf life in the room temperature and must be kept in a refrigerator. Openings for access windows are made by NC drilling or punching with a release sheet. The surface of traces must be cleaned before coverlay lamination takes place. It is difficult to introduce an automation system for the lamination process because the materials are very unstable after window openings are made. Manual alignment is the solution for this process unless the circuit needs an accurate registration higher than +/–0.5 mm. After the coverlay film is temporarily fixed on the circuit, the adhesive layer is cured in a heat press or autoclave similar as multi-layer rigid boards. A suitable curing condition should be chosen carefully based on the circuit design and material conditions. Epoxy resins change flow properties significantly by thermal aging, even though they are kept at a low temperature; therefore, a detailed conditioning should be performed based on the flow properties of the adhesives. A high flow of the adhesive resin makes good encapsulation of the conductors, but it also causes a squeeze out of the resin in the openings and makes serious stains on the copper traces. On the other hand, a low resin flow causes a poor encapsulation and results low bond strength and air voids beside the traces. The selection of the press pad materials and separation films is another key factor to have good encapsulation and minimum squeeze out of the glue from the edges. A vacuum press or a vacuum autoclave helps to make good encapsulation of the conductors, eliminating voids or air traps beside the traces, especially for a thick copper foil.

It takes more than one and a half hours for the whole heat press cycle to complete the chemical curing reaction of the adhesive resins, therefore the batch size and the lamination equipment have been becoming larger to make the productivity higher. But it makes the process flexibility lower. A quick press system has been developed to have a better flexibility. One set of coverlay films and an etched flexible circuit panel is hold in a small vacuum press for less than two minutes as the preliminary lamination. After that, several numbers of panels are baked in a thermal oven to complete the chemical curing of the adhesive resin. This process needs a special epoxy resin for the adhesives, but it provides a high flexibility for the small volume and prototype productions.

63.5.2 Screen-Printing Coverlay

Basically, the same process and the same screen printer are available for applying liquid coverlay materials as a solder mask on rigid circuit boards. The solder mask materials have to have appropriate flexibility to avoid the cracks during the bending. The only difference in the processes is the handling of the unstable flexible materials. Any vacuum used to fix the material on the printing table should be set to a low value to make the circuit flat without causing damage. A strong vacuum makes many small spots on the coverlay. The post-baking should be conducted with a suitable carrier that can keep the material flat to avoid supplemental damages.

An appropriate ink material and a suitable baking condition must be chosen to minimize the shrinkage of the material and have a uniform flatness (See Fig. 63.27).

63.5.3 Photoimageable Coverlay

As photoimageable solder mask systems for rigid circuit boards were developed in the 1980s, similar ideas were tried for flexible circuits. Ideally, fine openings and high dimensional accuracy were expected. Unfortunately, the materials for rigid circuit boards were too brittle for flexible circuits, and therefore new materials were required. Du Pont developed a dry film–type photoimageable coverlay (Pyralux PC™) and Nippon PolyTech developed a liquid ink–type photoimageable coverlay (NPR-80™) in the early 1990s. These were practically the first materials available for flexible circuits. They had volume applications in automobiles, disk drives, cameras, and so on. There were several problems with early versions, but there have been remarkable improvements, and application areas have expanded. More material manufacturers have been commercializing different materials, and therefore there are more choices nowadays. Reliability performances depend very much on materials and manufacturers. Material selection should be conducted carefully.

There are significant differences in the manufacturing processes between the dry film type and the liquid ink type, as shown in Fig. 63.28. The dry film-type materials are laminated by a vacuum laminator to ensure good encapsulation of conductors. The same laminators designed for dry film-type solder mask are available. Roll-to-roll vacuum laminators have been developed, but have not gained wide acceptance. Some materials are applicable by a simple heat roll laminator, but supplemental processes are required for good encapsulation. These machines are capable to manage both single sided and double-side in one processing. A range of 15 to 30 minutes holding time are recommended to stabilize the photo-imageable resins.

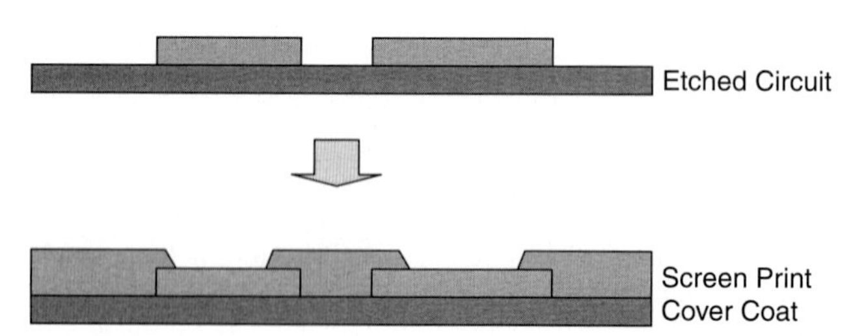

FIGURE 63.27 Screen print process of liquid ink coverlay.

There are several choices for application processes of liquid ink photoimageable coverlay.

Screen printing could be an easy method for most circuit manufacturers because they have been using the processes and machines for other purposes for a long time. Newly-introduced spray process or curtain coating could be a low-cost solution for volume production. These technologies are relatively easy to apply for RTR process. For pattern imaging, drying is necessary. Therefore, it needs to conduct the process two times for double-side circuits. Several automatic coating machines are designed for the double-side circuits, but they have become very big and have lost the flexibility for the small volume productions.

UV exposure with traditional equipment can be conducted to generate the pattern of window openings. The same chemicals, generally one percent sodium carbonate aqueous

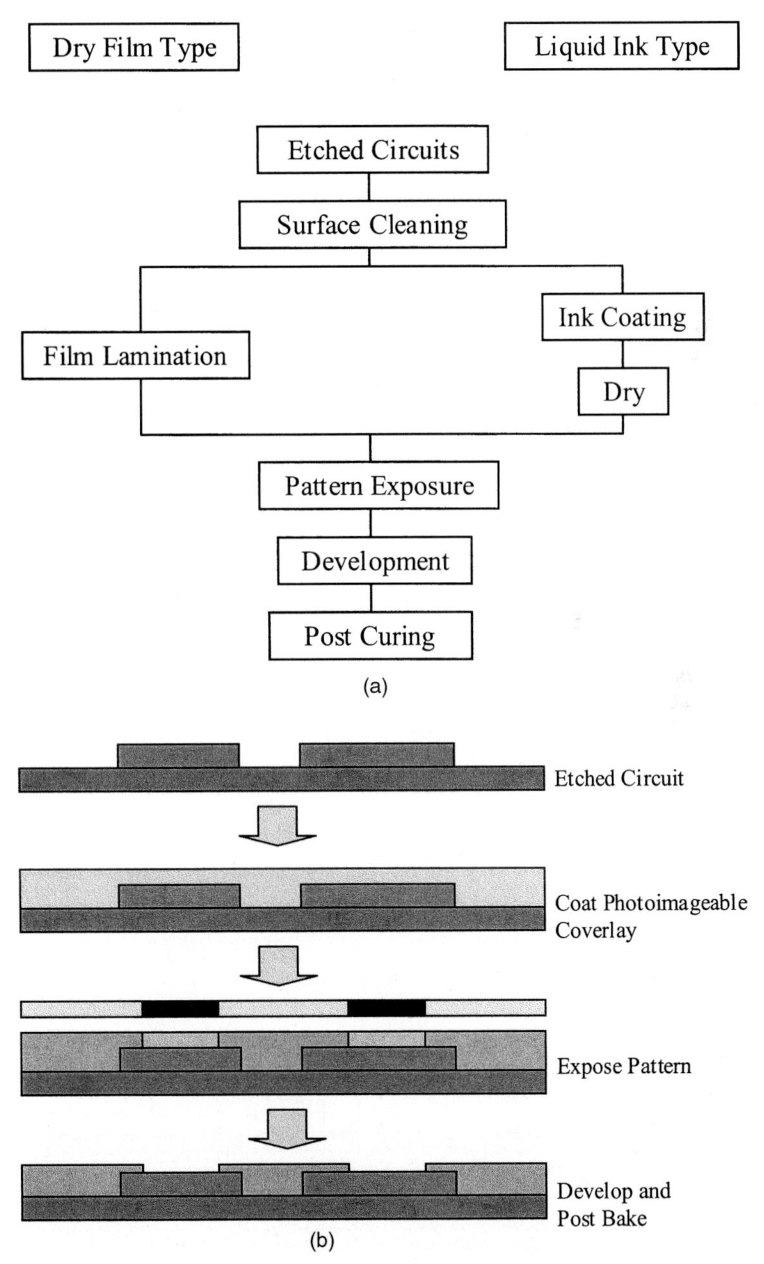

FIGURE 63.28 An example of a manufacturing process for photoimageable coverlay (a) A schematic flowchart of a generic process using either dry film or liquid resist, (b) A graphic flowchart starting with the etched circuit, showing the manufacturing process for a photoimageable coverlay.

solutions and the same wet lines as etching resist, are available to develop the photo-imageable coverlay.

A half hour post-baking processes for curing is often used for important for good physical performances. Usually, a temperature higher than 150 degrees C is required to achieve 100% curing.

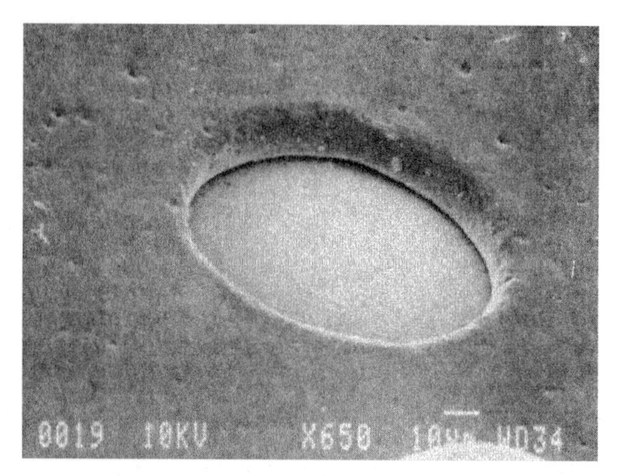

FIGURE 63.29 Small opening with photoimageable coverlay. Hole diameter is 100 micromieter.

Conventional air ovens or infrared furnaces with appropriate conveyers can be used for the post baking. Figure 63.29 shows a 100 micrometer hole made by photoimageable process.

63.5.4 Laser Drilling on Film Coverlay

Excimer laser drilling of traditional film coverlay can provide a perfect solution for both physical performances and fine openings with high dimensional accuracy. A traditional film coverlay is laminated on the etched traces without pre-punching or drilling. (See Fig. 63.30 for a manufacturing flowchart for laser drilling on film coverlay.) The latest excimer laser system can make sharp 50-μm openings on film coverlay without serious changes of physical

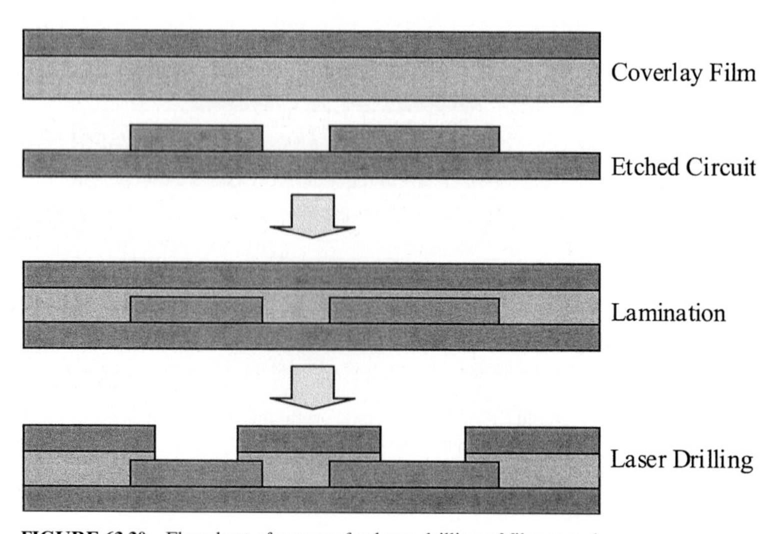

FIGURE 63.30 Flowchart of process for laser drilling of film coverlay.

FIGURE 63.31 Small coverlay opening drilled by excimer laser. Hole diameter is 100 micrometer. *(Source: Shinozaki.)*

properties (Fig. 63.31). It keeps the good mechanical, electrical, and chemical properties of the film coverlay, and the circuits are available for dynamic flexing. The largest issue with this technology is the low productivity and high processing cost. The processing time and cost are proportional to the total size of the opening areas; therefore it could be expensive when there are high-density SMTs. Several new laser systems such as the YAG laser and carbon dioxide laser have been developed to achieve a higher processing speed and higher productivity. A carbon dioxide laser with a Galvano mirror has a speed one order higher than that of an excimer laser, even though engraving quality is worse than that of the excimer laser (Fig. 63.32). It cannot remove organic residue perfectly from copper surfaces, and it requires a supplemental cleaning process such as plasma etching or chemical etching.

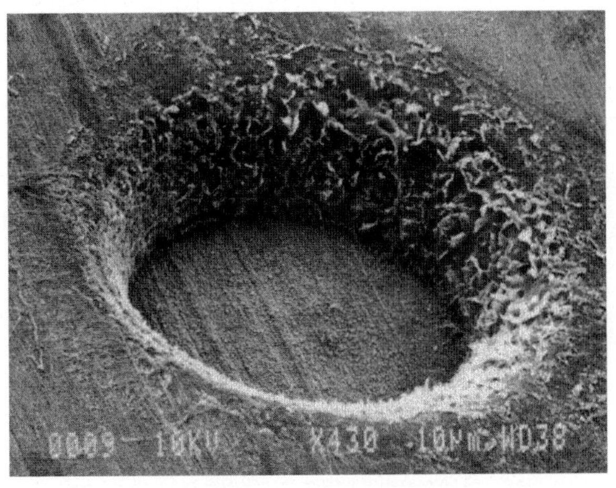

FIGURE 63.32 Small coverlay opening drilled by carbon dioxide laser. Hole diameter is 200 micrometer. *(Source: Shinozaki.)*

A laser process does not require supplemental tooling such as punching die or photo mask.

Also, its initial conditioning is significantly simpler than that for other technologies. Therefore, a laser process could be very valuable for prototype jobs and small-volume quick turnaround jobs.

63.6 SURFACE TREATMENT

Electro- and electroless plating of tin, nickel, gold, and solder are common surface treatments for flexible circuits. Solder plating by a hot-air leveler could be a low-cost solution. However, it entails severe high-temperature process conditions, and the poor uniformity of the solder is not acceptable for several applications.

Several new surface finishing technologies have been developed to satisfy the requirements of new termination technologies. A soft nickel/gold plating is the major process for the wire-bonding and flip chip bonding processes. A tin plating is also available for flip chip bonding. A hard nickel/gold plating is applied for the fine pitch pads of ACF termination (Anisotropic Conductive Film) and insertion parts of the FFC connectors. Tin- and lead-free solder plating will be an alternative treatment to traditional solder plating. OSP (Organic Surface Protection) will be a low-cost solution for next-generation soldering eliminating the lead components.

Prior to the surface treatment for the opened conductors, a suitable cleaning process is required to have good plating performances. Sometimes, it needs a strong chemical with mechanical brushing to remove the tough residues and oxidation on the copper surface made by previous processes. Strong alkaline solutions should be eliminated as the cleaning reagents. Several adhesive resins and photo-imageable coverlays have been attacked seriously and they cause de-lamination of the coverlay.

Electro plating is capable to provide a reliable metallic surface treatment controlling the thickness and the surface conditions. It is not difficult to manage the plated metal surface shiny or non-shiny. An issue is the all of the metal pad must be connected to the electrodes. It is not a big issue for the simple cable type flexible circuits, which have mostly parallel lines. But it is an serious issue SMT type circuits what have electrically isolated termination pads. They need supplemental lines for only electro plating, and they should be cut after the process.

Electroless plating has less limitation in its application. The plated metals are deposited uniformly for all bare terminals even though they are electrically isolated. On the other hand, electroless plating process has several limitations to control the thickness and the quality of the plated metals. A long processing time and high cost solutions cause the thick plating expensive. Basically, electroless plating is a kind of chemical reaction in an aqueous solution. Some of the plating solutions are high pH alkaline and they make serious damages for the adhesive layers and coverlay materials.

Micro bump arrays will be constructed by advanced processes. Optimized combinations of plating processes can build many kinds of bump shapes at a reasonable cost (see Table 63.8).

TABLE 63.8 Surface Treatments of Flexible Circuits

	Materials	Process conditions
Electroplating	Ni, Au, Sn, solder, etc.	Wet process
Electroless plating	Ni, Au, Sn, solder, etc.	Wet process
Hot-air leveling	Solder	>260°C
Solder roll coating	Solder	>260°C
OSP	Organic molecules	Wet process

63.7 BLANKING

A punching press similar to those used in the processing of rigid boards could be used for the blanking of flex circuits. Smaller tons are required to cut thin materials. Steel rule dies are the specialty for flex circuits (see Fig. 63.33). They can be prepared in a few days, and their costs are much smaller than those of hard tools. Their weight is also much smaller than that of hard tools and operators can manage them by hand. A typical cross section is shown in Fig. 63.33(*a*), a finished steel rule die is shown in Fig. 63.33(*b*), and some limitations on the configurations that can be used are illustrated in Fig. 63.33(*c*). Steel rule dies have less dimensional accuracy, lower productivity, and a shorter life than hard tools, and they need maintenance more frequently.

They are good at quick-turnaround jobs and small- to medium-volume productions. Table 63.9 shows a comparison.

A hard punching die is recommended for use for high dimensional requirements and large volume production. Zero clearance between the male and female dies is required for thin materials.

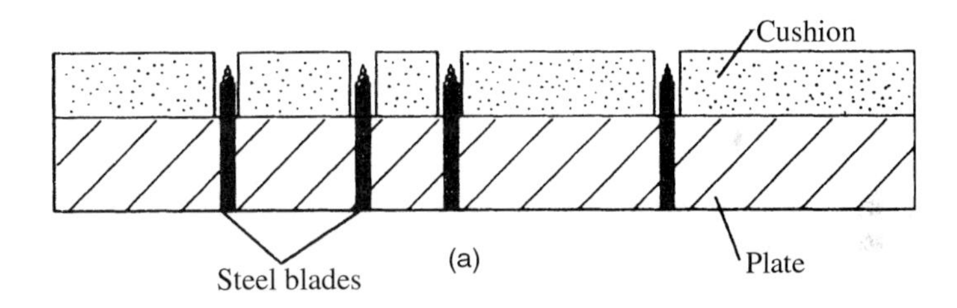

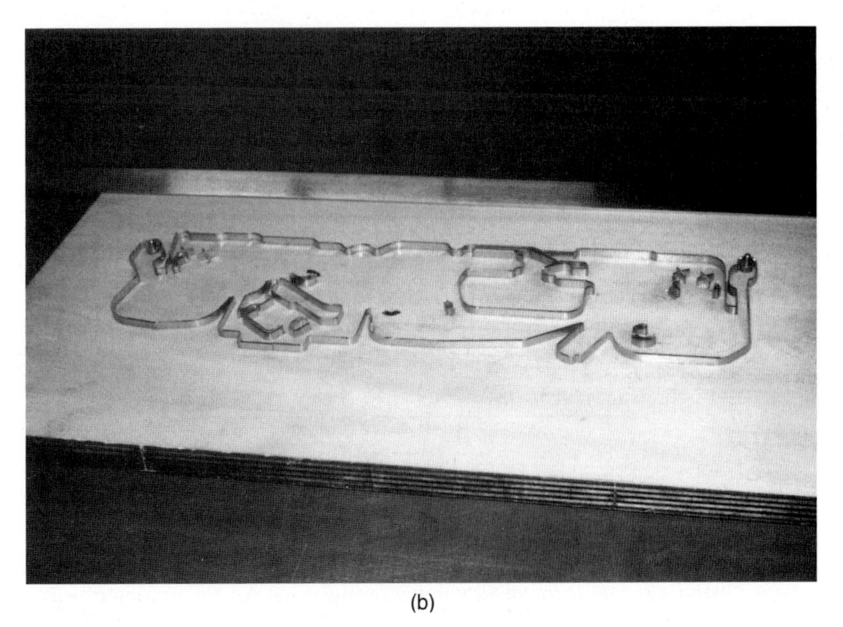

(b)

FIGURE 63.33 Steel rule die for blanking flexible circuits (a): A cross section of the configured die, (b): A sample finished die (c): Limitations on shapes that can be designed into a steel rule die.

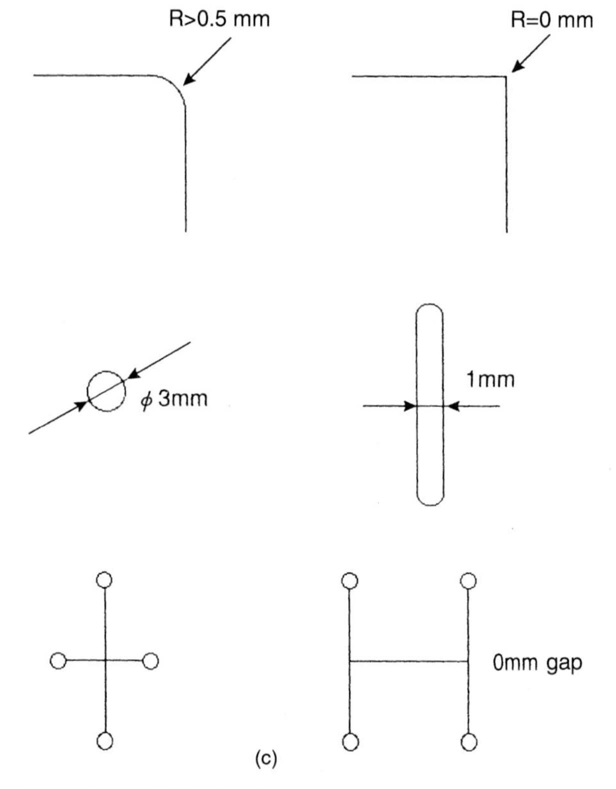

FIGURE 63.33 Continued.

TABLE 63.9 Comparison of Punching Dies

Items	Steel rule dies	Hard tool dies
Accuracy	~0.2 mm for whole	~0.01 mm for whole
Minimum hole size	3 mm diameter	0.4 mm diameter
Minimum slit width	Zero	0.5 mm
Minimum corner radius	0.5 mm	0.1 mm
Delivery time	2–4 days	More than 3 weeks
Weight	Light, manage by hand	Heavy, need a lift
Tooling cost	Low	High
Processing cost	High	Low

Special technology and periodical maintenance are needed when zero-clearance punching die sets are used.

Combinations of steel rule dies and hard tool dies are the trend for achieving high productivity and high dimensional accuracy at a low cost. Quick-change die systems are the new trend for high productivity for small-volume productions. An NC routing process is also available for the blanking of flex circuits. However, special backing boards and conditions are required to process unstable flexible materials.

It is very important to have accurate punching guide holes to create good perimeters for flex circuits. Unfortunately, a standard NC drilling machine does not provide accurate positions for each hole in a work sheet because of distortions during the previous manufacturing processes. An appropriate compensation for dimensional changes is necessary for each guide hole. Several punching and drilling machines that have self-alignment systems that use CCD cameras are available to ensure accurate positioning with guide holes. The latest machine has a level of accuracy better than +/–50 μm and high productivity based on automatic x-y table. Recently, higher perimeter accuracies have been required for specific areas, such as insertion parts for the small pitch FFC connectors. The auto alignment punching machines designed for the guide holes are modified for the trim punching.

Very different ideas are required to achieve a severe perimeter that allows tight spaces from conductors (smaller than 50 μm). A laser router that has self-alignment capability could be a possible solution. However, the processing speed is slow and expensive. A chemical etching process for the base could be a low-cost solution for volume production; however, the manufacturing process is very different from a standard process and a special process design is required. Its applicable construction and materials are limited, and suitable design changes are recommended.

63.8 STIFFENER PROCESSES

Flexible circuits have many supplemental constructions other than electrical traces to optimize their capabilities. The stiffener boards are the major one of them. An issue is that it is difficult to apply full automation to the process of making stiffener boards because of the many varieties. Most of the processes are conducted manually and are labor intensive. Sometimes stiffener boards account for a big percentage of the total manufacturing cost.

An adhesive film is laminated on a rigid board with a release sheet first. The boards are then routed by punching or by NC router. It is a very simple process if the adhesive material is pressure-sensitive adhesives (PSAs). Each piece of stiffener is placed on the flex circuits with appropriate pressure, mostly by hand. The process is more complicated when a thermo-set type adhesive material is required. A temperature higher than 160°C with a pressure higher than 20 kg/cm^2 is required for more than 30 min. A similar heat press used for multilayer circuit boards or film coverlay is necessary. Dummy boards should be prepared to make the pressure uniform.

A heat press with a vacuum chamber or autoclave could provide uniform adhesion for irregularly shaped stiffener boards.

63.9 PACKAGING

Many ideas have been introduced for packaging fragile HDI flex circuits. The packaging should provide easy handling capability, including the processing of the assembling side. A reeled tape modified from TAB technology is a valuable solution for volume production. A sheet carrier is another possible solution for medium- and large-scale production. Reusable cartridge or container designed for each circuit is recommended for volume production.

An appropriate packaging design results a lot of benefit for both flex circuit manufacturers and customers.

It needs a total design of the circuits including their manufacturing and assembling to optimize the productivities.

63.10 ROLL-TO-ROLL MANUFACTURING

Roll-to-roll (RTR) manufacturing is a specialty developed for rolled materials of flex circuits.

It can reduce manual handling and defects caused by human errors; therefore, when running effectively, it will increase productivity significantly and reduce manufacturing cost. On the other hand, RTR manufacturing has several serious disadvantages and a lot of applied limitations. It requires a large investment and advanced engineering capabilities. It is not appropriate for small-volume production, and can create a lot of scrap when not set up correctly.

Table 63.10 shows advantages and disadvantages of RTR manufacturing systems for flex circuits. Table 63.11 shows available manufacturing processes and typical examples. Most RTR systems are developed for single-sided flex circuits.

TABLE 63.10 Advantages and Disadvantages of RTR Systems

Advantages	Disadvantages
• High productivity	• Small flexibility for small volume
• High process yields	• Much scrap caused by small mistakes
• Low manufacturing cost	• Large investment cost
	• Limited for quick turnaround
	• Low flexibility for design changes

TABLE 63.11 Availability of RTR Processes

Process	Technical difficulty
Surface cleaning	Standard machines available
Via hole generation	Possible, but difficult
Via hole plating	Big investment
Resist coating	Standard machines available
Pattern generation	Standard machines available
Developing	Standard machines available
Etching/strip resist	Standard machines available
Coverlay	Possible, but difficult
Surface treatments	Standard machines available
Blanking	Possible, but difficult
Inspection	Possible, but difficult
Stiffener boards	Difficult

63.10.1 RTR Drilling

There have been significant technical barriers to applying the RTR system for double-sided flex circuit with through-holes. The NC drilling process is the bottleneck in introduction of the RTR system. A punching process, which is basically a continuous-flow process element, is an alternative technology to NC drilling of the RTR system to be used for double-sided flex circuits. However, it requires an expensive punching die and is not flexible for a design change.

An RTR base NC drilling machine has been developed, as shown Fig. 63.34. The machine is capable to generate smaller holes than 100 micron diameters processing multiple rolls in a same time.

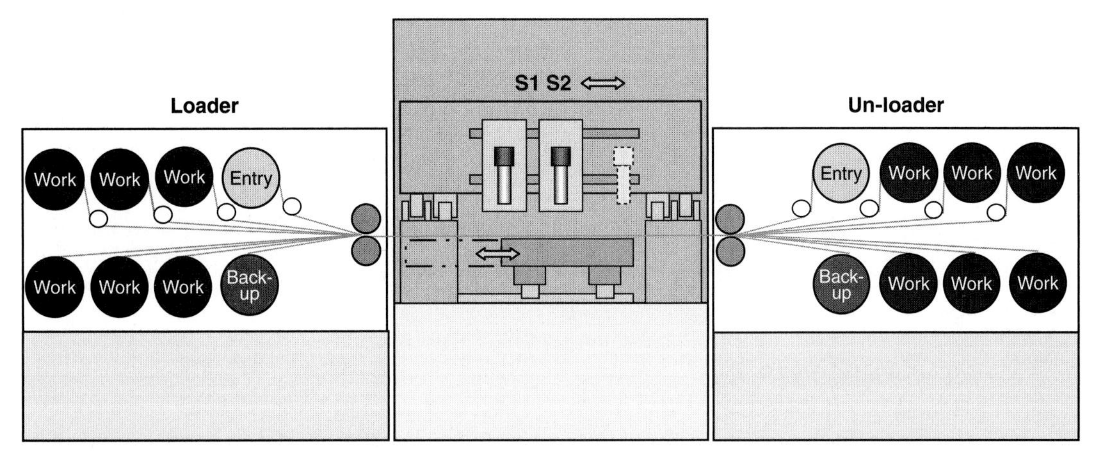

FIGURE 63.34 Reel-to-reel base numerical control drilling machine. *(Source: Hitachi Via Mechanics.)*

The new via hole technologies such as laser drilling and chemical etching are an innovative solution for generating small holes in rolled materials. A blind-via hole design is preferred for these RTR manufacturing processes.

63.10.2 RTR Copper Plating

Technically, RTR copper plating is possible. But the line will be very long, and it is not easy to have uniform quality for the wide rolls. There is a difficult choice which is better vertical system or horizontal process.

63.10.3 Pattern Generation

The pattern generation processes of resist coating, photo-print, development, etching and resist stripping are well established for single-side circuits. There are several key points such as straightness of the whole line, EPC (Edge Position Control), etc. Stability of loader/un-loader could be an example. Appropriate stabilities are required for the one-side holding systems of the rolled materials, especially for the wide materials.

There are not many technical difficulties with double-side circuits for we processes such as development or etching. On the other hand, special alignment systems are required between top/bottom and holes/traces.

63.10.4 RTR Coverlay Process

After etching process, technical hurdles to apply RTR become extremely higher compared to previous processes because of the difficult alignments between the patterns.

The coverlay process could be the typical example. RTR lamination of film coverlay is almost impossible. The screen-printing for RTR processing has developed for relatively rough circuit patterns. The photo-imageable coverlay system could be the right solution for the RTR process of the coverlay of the flexible circuits. However, there have been several limitations to apply.

63.10.5 Other Process

The other processes such as blanking and stiffeners are very difficult except special designs with large volume production.

63.10.6 Tape Processing

Manufacturing process of TAB (Tape Automated Bonding) has been developed by reel to reel from the beginning for specific applications. The details of the technology will be considered in Chap. 66.

63.11 DIMENSION CONTROL

Due to increasing density requirements and inherently unstable film substrate characteristics, dimensional control has become a key issue for both circuit fabricators and their end customers.

In order to achieve such densities while maintaining high production yields, careful dimensional consideration in both the design and manufacturing processes is necessary. An exact concept for dimension control of the whole manufacturing processes is required for high-density flex circuits. There are many factors that impact the dimensional stability of high-density flex circuits. These factors can be categorized as follows:

- Materials
- Circuit design
- Manufacturing processes

63.11.1 Materials

Substrate material selection may be the most important item to consider for dimensional control of high-density flex circuits. (For a more detailed discussion of flexible circuit materials, see Chap. 61.)

63.11.1.1 Polyimide. Kapton H™ and Apical AV™ are standard substrate materials for traditional flexible circuits. Unfortunately, these films have high moisture absorption rates, which create excessive dimensional instability, making them unsuitable for high-density flexible circuits.

Several newer polyimide films such as Upilex S™, Kapton E™, and Apical NP™ and FP™ have been commercialized to satisfy the requirements for increased dimensional stability.

A comparison of these base materials is shown in Table 63.8. Although these new polyimide materials have lower moisture absorption and CTEs than traditional polyimide films, their dimensional change is still not zero. In addition, process variations of 0.02 to 0.05 percent are common from batch to batch during film production.

Generally, material manufacturers fabricate polyimide film with a wide web. Special conditioning is necessary to ensure uniform properties for the whole web. Unfortunately, there are usually deviations from one side to another. Because the master roll is slit in halves or thirds according to the customer's convenience, each roll has different mechanical properties depending on its location in the master roll. If the master roll is slit into three rolls, a roll in the middle may have more uniform properties than the others. If the master is slit into two rolls, relatively wide edges will be wasted.

63.11.1.2 Copper-Clad Laminate. Most manufacturers of flexible circuits initiate their processes with copper-clad laminates. During production of these laminates, mechanical stresses are generated (by heat and pressure) that can become locked into the material. These stresses are partially released during the circuit manufacturer's etching operation and a certain amount of dimensional reversion takes place. In actuality, these stresses (or dimensional stability characteristics) differ in the machine direction (MD) vs. the transverse direction (TD). It is therefore important to review all the mechanical stresses in a laminated material and to consider some form of stress relieving prior to circuit fabrication.

IPC-TM-650 recommends a method of testing dimensional stability in a 9 × 10-in rectangle for both MD and TD. One standard-size test sheet is too small to represent the whole web.

Several samplings are required to evaluate the whole web. Yaw direction changes and local changes of dimensions that cannot be measured by the IPC test method are sometimes critical for high-density flexible circuits. Detailed investigation should be conducted.

63.11.1.3 Coverlayer Selection. Coverlayer selection should be considered the second important material factor for controlling dimensional stability. Generally this layer is adhered to the circuitry using heat and pressure, which can induce additional dimensional distortion.

Careful consideration should be taken if effective dimensional stability is to be achieved.

63.11.2 Circuit Design

As discussed earlier, most laminate materials undergo mechanical stresses, which generate non-uniform dimensional changes during the etching process. After etching, an etched area with substrate only has different physical properties from a non-etched area with substrate and copper foil. During manufacturing, the laminate materials generate different dimensional changes. A uniform circuit pattern design is necessary to control dimensional stability. A dummy copper pattern in an empty area can help to maintain optimal dimensional stability.

63.11.3 Critical Processes for Dimension Control

The critical manufacturing processes for dimensional control of flexible circuits are shown in Tables 63.12 and 63.13. In general, a high temperature or a wet chemical process can alter the dimensions of flexible circuits. Increased mechanical tensions can also impact dimensional control.

The major manufacturing steps for standard double-sided flexible circuits with through-holes are listed here. Each step should be considered carefully to determine its contribution to the overall distortion:

1. *NC drilling* Although NC drilling does not significantly impact the final dimension, certain scale factors based on assumed movements during the following operations may be factored into the drill pattern.
2. *Through-hole plating* Because the electric field in a through-hole plating bath is generally non-uniform, the copper thickness becomes uneven. An extremely thickly plated copper conductor will create mechanical stresses. During the etching process these stresses are released, generating some dimensional changes. In an extreme case, a copper-plated sheet cannot lie flat. An appropriate set of work fixtures and plating electrodes should be designed to create a uniform plating thickness.
3. *Dry film lamination and pattern printing* Due to the relatively low temperature and lack of wet processing (moisture), dry film lamination and pattern printing have little effect on dimensional control. Pattern imaging is, however, an opportunity to apply appropriate dimensional corrections for the following processes.
4. *Etching and stripping of resists* The most critical wet process in flexible circuit manufacturing is etching. Dimensional changes can occur when the by-products of the etching process are combined with the various materials. It is important to identify how to correct

TABLE 63.12 Key Items for Dimension Control in Flex Circuit Manufacturing

Categories	Key items for dimension control
Materials	Base film, copper, coverlay, bonding material
Design	Layer constructions, circuit density, circuit balance
Manufacturing process	Wet processes, heat processes, mechanically stressed processes

TABLE 63.13 Critical Processes for Dimensional Control

Process	Category	Importance
NC drilling	Mechanical	Medium
TH plating	Wet and heat	High
Cleaning	Wet and heat	Medium
DF lamination and print	Mechanical	High
Development	Wet and heat	High
Etching and stripping of resist	Wet	High
Coverlay, film lamination	Heat and pressure	High
Coverlay, screen printing	Heat	High
Coverlay, photoimageable	Heat	High
Lamination for multilayer	Heat and pressure	High
Plasma etching	Heat	Medium
HASL	High heat	High
Electro- and electroless plating	Wet and heat	Medium
Baking	Heat	High
RTR process	Mechanical	Medium

Heat means temperatures higher than 80°C.

the various dimensional alterations, particularly as additional factors are incorporated into the process.

5. *Coverlay processes* There are several coverlay processes that have been developed with varying process conditions for different concept materials. Film coverlay undergoes a manufacturing process with a temperature greater than 160°C and a pressure over 20 kg/cm^2.

   Cover coat ink withstands a screen-printing process with a drying temperature higher than 130°C. A newly developed photoimageable coverlay must be baked at over 150°C. The majority of the heat and pressure processes impact dimensional stability significantly. Additional factors, such as press pad heating, make dimensional changes even more complicated. Conditioning should be reviewed in detail for each construction.

6. *Lamination pressing for multilayers* Like the coverlay process, lamination pressing involves high temperature and pressure. Due to the number of layers, however, dimensional changes are more complicated because each level needs to be considered.

CHAPTER 64
TERMINATION OF FLEXIBLE CIRCUITS

Dominique K. Numakura
DKN Research, Haverhill, Massachusetts

64.1 INTRODUCTION

Major termination technologies for flexible circuits are listed in Table 64.1. Most termination technologies developed for rigid printed circuit boards are available for flex circuits with small modifications. Surface-mount technology (SMT) components can also be assembled through the same processes used for rigid circuit boards, such as standard pick and place and reflow soldering without significant conditioning. The card-edge connectors designed for rigid circuit boards are also available for flex circuits with the addition of suitable stiffener boards. In addition, because of the unique wiring capabilities and special construction of flexible circuits, many specialized termination technologies have been developed for them. These include solder fusing, anisotropic conductive film (ACF) connections, flying-lead direct bonding, dimple contacts, and flat flexible cable (FFC) connectors.

To accommodate the miniaturization of electronic products, many termination technologies have also been developed for high-density interconnect (HDI) flexible circuits. These technologies are key points in the assembly of various electronic components in small spaces, such as cellular phones, miniature liquid crystal displays (LCDs), keypad switches, antennas, connectors, microphones, speakers, battery cases, etc. It is common for several different termination processes to be used in each electronic product. Examples of applications include wireless suspension of disk drives and chip-scale packaging (CSP). These technologies use high-density flexible circuits as the interposers. Multiple special termination capabilities are required for these interposer flexible circuits.

All of the termination technologies of flexible circuits have both advantages and disadvantages.

No technology can satisfy all of the requirements at the same time. In general, flexible circuits have numerous wiring options, and therefore a thorough review of the technologies and the wiring requirements is required to design the appropriate termination.

64.2 SELECTION OF TERMINATION TECHNOLOGIES

To determine the optimal cost-performance combination, one first must establish the requirements from the assembly side and the capability of the termination technologies. Table 64.2 lists the critical items that are required to design a suitable termination.

TABLE 64.1 Termination Technologies for High-Density Flexible Circuits

Rigid PWB-based	Flexible circuit-based
• SMT • COB(chip-on-flex [COF]) • Wirebonding • Conductive adhesive • Card-edge connectors • Pair connectors • Flip-chip bonding	• Solder fusing • anisotropic conductive material (ACM) (ACF) • Direct bonding of flying leads • Dimple contact • Microbump contacts • FFC connectors • Nonconductive paste (NCP) bonding • Pressure contact

TABLE 64.2 Critical Items in the Design of High-Density Interconnects

Critical Item	Description
Objectives	What kinds of devices?
Termination density	Points per cm or cm^2 Number of connections in the allowed space
Permanent or nonpermanent	Repeated connections or not?
Circumstances used	Temperature, humidity, etc.
Condition processed	Temperature of the application
Reliability	Total performance
Cost	Material and handling

64.2.1 Termination Objectives

Several types of electronic components can be attached to flexible circuits. Table 64.3 categorizes these from the termination standpoint, assuming the use of high-density flexible circuits.

TABLE 64.3 Categories of Electronic Components

Category	Components
SMT devices	Discrete chip components, (BGA), CSP, multichip module (MCM), bare IC chips
Imaging devices	Large flat-panel display (FPD), small FPD, charge coupled device (CCD), linear sensors, 2-D sensors, printers
Wiring devices	Rigid PWB, ceramic-based circuits, flex circuits, cable, wires, connectors
Specialty devices	Keyboards, switches, printer heads, antennas, actuators, sensors, microphones, speakers, chimes

Because of the nature of their use, there are a limited number of locations where imaging devices such as LCDs and CCDs can be placed in electronic products. As a result, there are fewer choices for wiring materials other than flexible circuits. Newer imaging devices have numerous pixels in a limited area, which requires more connections with smaller pitches. In addition, it is often important to place the driver integrated circuit (IC) close to the device. Therefore, a flexible circuit is required for both wiring and chip-on-flex (COF). A large LCD, typically used in a personal computer and flat-panel TV, often needs several tape-automated bonding (TAB) circuits to create the wiring between a rigid board and the LCD device. A flat-panel display can be wired by a single high-density flexible circuit, which reduces the cost of the entire package.

It is common for a flexible circuit to be connected to other circuit boards. However, more traces and a higher density of the connections are then required. The basic design of the end products automatically determines the placement of specialty components such as keyboards, switches, antennas, actuators, printer heads, sensors, microphones, speakers, and buzzers in small electronic applications. Although these components do not require numerous wires, they do need high-density connections due to the extreme miniaturization of portable electronic products.

64.2.2 Nonpermanent or Permanent Termination

The permanence of a termination is determined by connection repeatability. For example, if connections are changed regularly, a nonpermanent termination should be utilized. Due to the number of reconnections, inkjet printers and electrical test probes should have nonpermanent terminations. Conversely, if a connection is to remain relatively constant, a permanent termination method should be employed. Termination method capabilities are profiled in Table 64.4.

TABLE 64.4 Capabilities of Termination

Methods	Permanent or nonpermanent	Termination density
SMT soldering	Permanent	350 μm pitch (2-D)
SMT wire bonding	Permanent	150 μm pitch (4 directions)
Solder fusing	Permanent	150 μm pitch
ACM connection	Permanent	30 μm pitch (100 contacts in 1 mm^2)
Direct bonding	Permanent	50 μm pitch (1- or 2-D)
Pressure contacts	Non-permanent	20 contacts in 1 cm^2 (1- or 2-D)
Dimple contacts	Non-permanent	20 contacts in 1 cm^2 (1- or 2-D)
FFC connectors	Non-permanent	300 μm pitch (1-D, 1 line)
Pair connectors	Non-permanent	500 μm pitch (1-D, 2 line)
Film connectors	Non-permanent	100 μm pitch (2-D, 100 contacts in 1)

64.2.3 Termination Density

To design a suitable termination solution with flexible substrates, one must consider the density and the number of terminations. An overview of each method's termination density is profiled in Table 64.4. Due to higher connection reliability, a contact array can produce more connections than a contact line in the same space with a larger pad pitch. Numerous wires must fit together in a limited space.

64.2.4 Wiring of Imaging Devices

A high-density flex circuit could be the best solution for the wiring of a small imaging device module designed for consumer electronics. In general, these devices have hundreds of pads that need to be connected with COF for the driver IC close to the device. The flex circuit usually requires a long tail to be connected to the other device. Only a high-density flex circuit can satisfy all the requirements as a single circuit.There are several choices in the design of the constructions of the COF area and the bonding area. An adhesiveless clad laminate having a stable substrate with polyimide resin is required to create a high connection yield for a high-resolution device. An anisotropic conductive material (ACM) connection is a low-risk solution because of the low processing temperature. A direct bonding process with fine flying leads can provide reliable connections for COF and wiring with the imaging device Due to increased temperature, the design requires a higher dimension control.

64.3 PERMANENT CONNECTIONS

If there is no functional reason to remove the connection during normal operation, a permanent connection is recommended because of the small mounting space and lower costs. Examples of this type of connection include:

- General soldering
- Through-hole leaded components
- General SMT components
- Solder fusing
- Wirebonding
- Direct bonding
- ACM connection
- Flip-chip connection
- NCP connection

64.3.1 Soldering

Soldering is still the most popular and common termination technology for flexible circuits. All kinds of solder and termination materials and equipment are available with suitable tooling guides. Usually, the soldering parts of the flexible circuits have stiffeners on the bottom side of the flexible substrates to make them tough enough to handle high-temperature soldering. However, the other parts of the circuits are still fragile, and appropriate frames or carrier boards have to be prepared to make the flexible circuits stable during the assembling processes, including screen printing, SMT mounting, solder reflow, and cleaning.

If adhesiveless laminates are used as the major materials, high temperatures can be used with lead-free soldering. In the case of SMT assembling, the flexible circuit should be fixed on a carrier frame. This allows it to be processed by the same mounting and soldering equipment as used for rigid boards.

64.3.2 Through-Hole Leaded Components

Often, through-hole leaded electronic components are necessary for specific applications.

Processes similar to those used for single-sided rigid boards are used to attach the components to the flexible material. However, because through-hole leaded components require

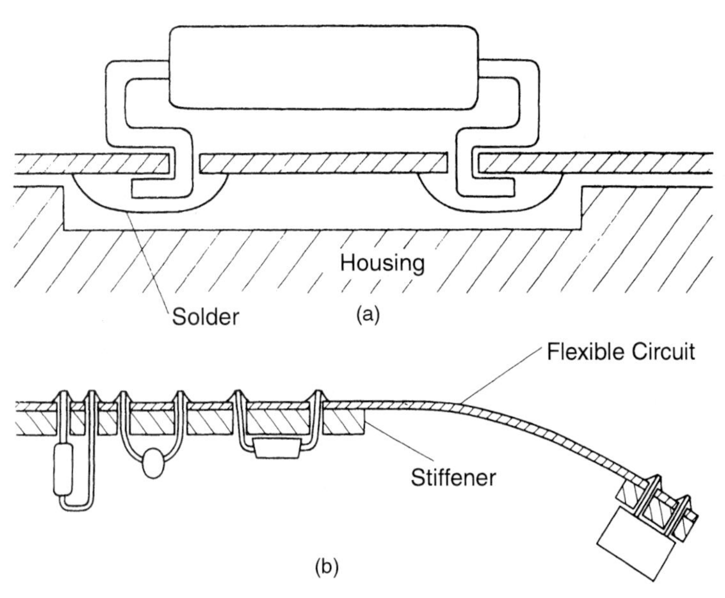

FIGURE 64.1 Examples of direct soldering of flexible circuits (a) Without stiffener, (b) With stiffener.

clinching of the leads and are heavier than SMT components, special fixtures and control are required when placing them on a fragile flexible circuit. This is to minimize mechanical stresses, especially when components are connected by soldering, as shown in Fig. 64.1 (a) and (b). A complicated stiffener system is required for the double-sided assembly of the leaded components on the flexible circuits.

64.3.3 SMT General Issues

Most SMT technologies can be used for flexible circuits without significant alteration, as shown in Fig. 64.2. In general, the same circuit design rules can be applied to SMT for a high-density

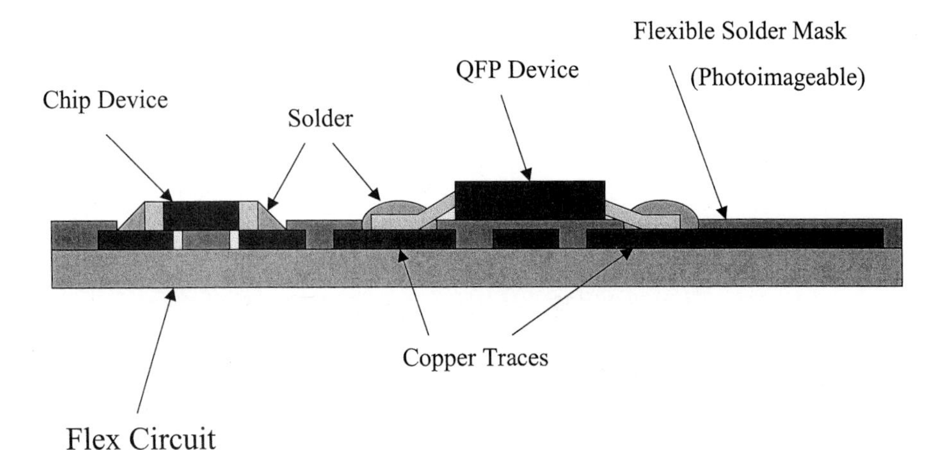

FIGURE 64.2 SMT soldering of high density flexible circuits.

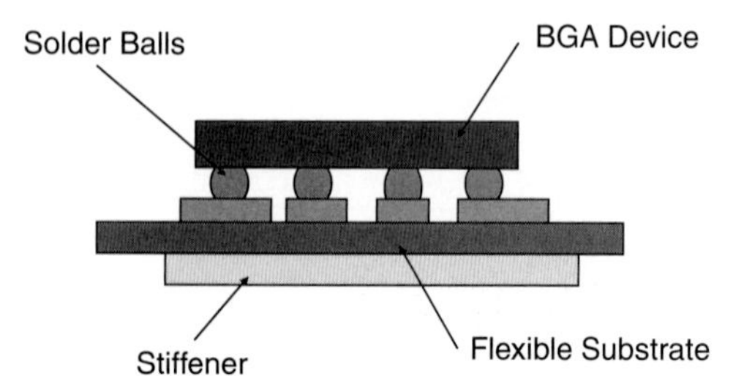

FIGURE 64.3 Mounting of BGA device on flexible circuit.

flexible circuit. A suitable thin stiffener sheet such as 125 μm thick polyimide film should be bonded to support SMT components. Figure 64.3 shows the case of the ball grid array (BGA) component with a thin polyimide stiffener. Adhesiveless copper-clad laminates need to be used as the raw material to withstand the high temperatures of typical lead-free soldering processes. A flexible photoimageable solder mask needs to be applied to cover the fine traces. A special dynamic flexible-grade photoimageable coverlay should be used if the flexible circuit has a portion of repeated flexing. A stiffener of suitable thickness needs to be bonded on the back side of an SMT area of a thin, single-sided flexible circuit. A thicker substrate layer should be used for double-sided assembly.

An appropriately designed stiffener board can increase assembly productivity without a big cost increase. A larger stiffener board and adhesive sheets are punched beforehand for the preparation of the assembling processes. The stiffener board has extra frame space that contains appropriate guide holes for the convenience of part mounting and soldering. The flexible circuits fixed on the stiffener boards can be managed the same as rigid printed circuit boards in the assembling processes. A standard mounting machine places small SMT components without additional tooling. In addition, the same high-temperature soldering equipment used for reflow soldering in an SMT line can be used here. In-circuit tests are conducted with an aluminum board. As the final process of the assembly, unnecessary parts of the stiffener boards are cut off and the whole circuit is folded to make a part of the module casing. An appropriate seal is made with the other casing parts to protect the module from the outside stress.

64.3.4 Solder Fusing

Although solder fusing (Figs. 64.4 and 64.5) is not a new technology for making connections for flex circuits, it is still a low-cost and reliable solution for multi trace connections. There has also been remarkable progress in fine-pitch capability. The newest applicator, coupled with well-controlled solder plating on fine traces of flex circuits, can provide a 150 μm pitch connection

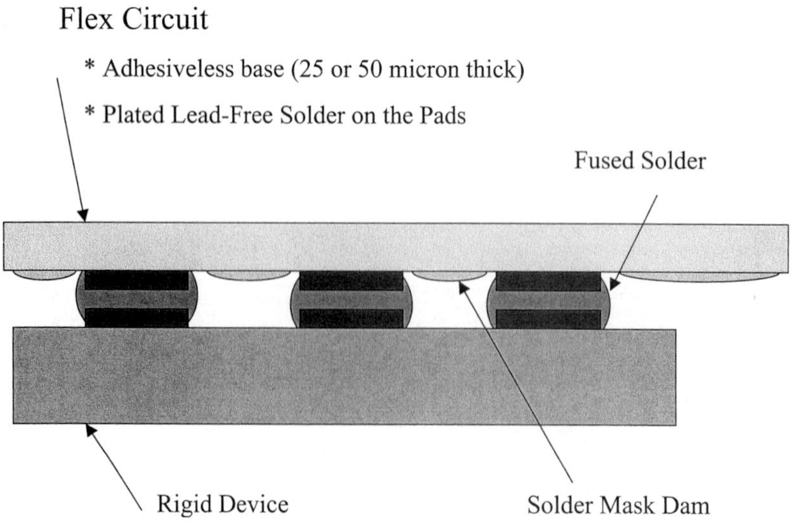

Flex Circuit
* Adhesiveless base (25 or 50 micron thick)
* Plated Lead-Free Solder on the Pads
Fused Solder
Rigid Device
Solder Mask Dam

FIGURE 64.4 Flexible termination by solder fusing.

FIGURE 64.5 Flexible circuits connected to a rigid board by solder fusing.

with the other circuit pads. A big advantage of the connection is a very low height even though the part is covered with an appropriate protection film. It is also possible to produce 30 connections in a 5 mm square, but high fusing temperatures with lead-free soldering are required. Because traditional adhesive materials such as epoxy or acrylic resin cannot survive the process, adhesiveless laminates need to be used as the major materials.

64.3.5 Wirebonding

A wirebonding process similar to COB has been applied for high-density flexible circuits without serious changes. Due to the high heat resistance of new adhesiveless laminates, standard bonding materials and equipment for flexible circuits can be used (see Fig. 64.6).

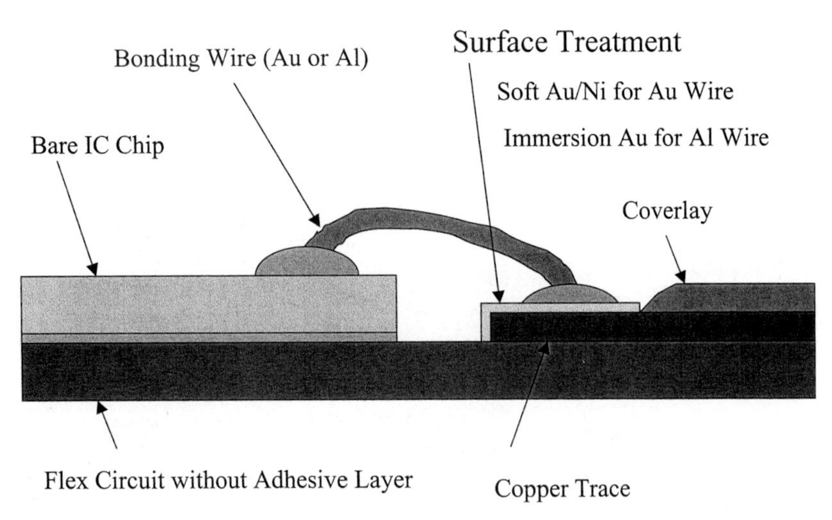

FIGURE 64.6 Wire bonding on and HDI board.

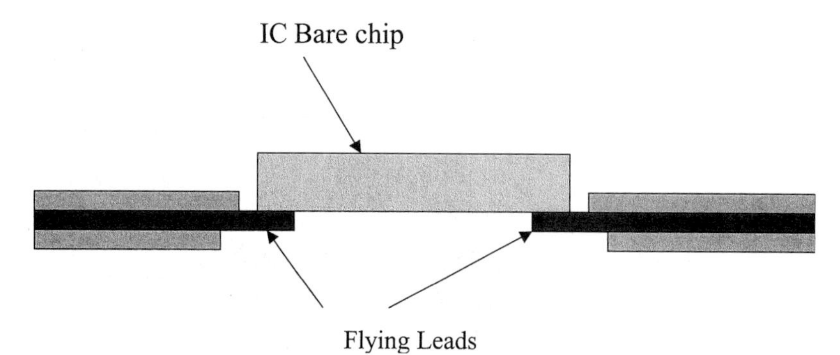

FIGURE 64.7 Direct bonding of flying leads.

Relatively thick soft gold plating with nickel base is required on the copper pads of the flexible circuits. The best combination of the bonding machine and bonding wire with a suitable process condition can make a 40 micron pitch connection between the high-density flexible circuits and the chip device. A relatively thick polyimide film is recommended to attach on the bottom side of the wirebonding area to make it stiff. The wirebonding process could be one of the standard assembly technologies for the COF system. Numerous driver ICs for LCD devices have been assembled by these methods.

64.3.6 Direct Bonding

Although flying-lead construction is not a new technology, fine-line traces created direct bonding. When a flying lead has a fine line smaller than 100 μm wide with nickel-gold plating on it, it can be processed using processes similar to wirebonding of the semiconductors. This can produce high-density, reliable connections between a flexible circuit and another circuit or device. (See Fig. 64.7). The basic idea was developed with TAB, but flexible circuits expand the capabilities and applicable areas. The small pitch direct bonding is used to assemble specific devices such as the magnetic heads of the hard disk drives and the semiconductor chip of inkjet printers (see Fig. 64.8).

The thickness of the soft gold plating on the nickel base is specified carefully based on the trials.

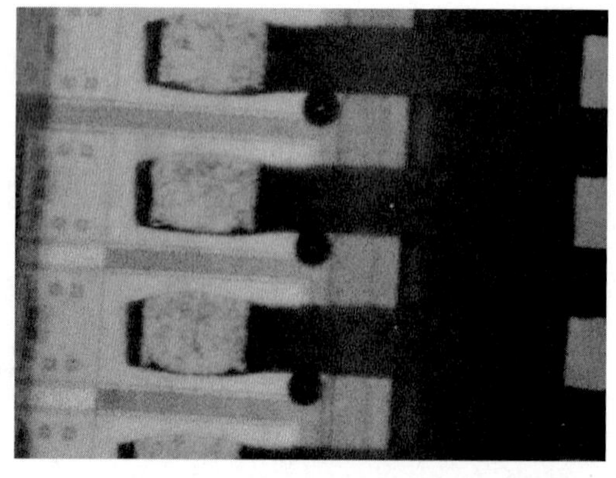

FIGURE 64.8 Direct bonding of fine flying leads. *(Ink cartridge of H.P.)*

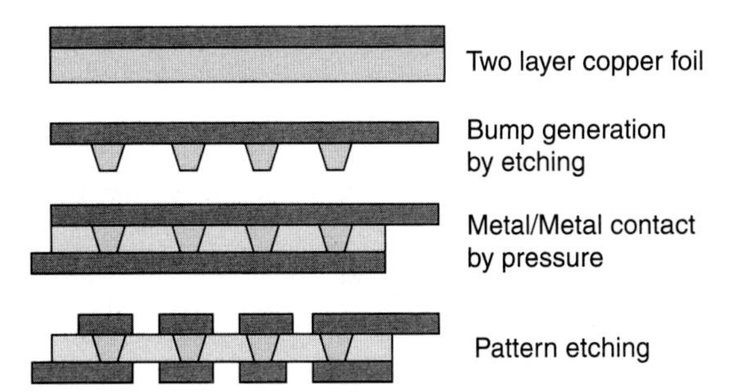

FIGURE 64.9 Flex interconnection through bump array.

64.3.7 Bump Connections

The microbump constructions are valuable for making reliable connections of the high-density flexible circuits with high-pin counts. Several ideas have been created to make practical termination technologies. Figure 64.9 shows an idea modified from the Bit^2 process developed by Toshiba. The copper bump arrays are formed by an etching process of the two-layer copper laminates on the heat-resistant polyimide film, and they have suitable plating such as nickel and gold. The microbump array is placed on the other circuit or devices through an adhesive layer of the insulation resin. A suitable combination of temperature and pressure makes a reliable metal-to-metal connection between the microbumps and the metal pad of the other printed circuits.

Supplemental etching processes after the heat press provide a special practice for building a low-cost rigid/flex construction.

64.3.8 Flip-Chip Bonding

The flip-chip connection is one of the termination methods with microbump arrays. The technology consumes a very small space for the package because all of the connections are conducted under the IC chips, as shown in Fig. 64.10. Arrayed gold bumps make reliable metal-to-metal connections with the tin- or gold-plated pad on the other devices through the combined process of the pressure and heat with ultrasound vibration. The process can provide smaller-pitch connections than 200 micron for large pin counts. Flip-chip mounting of bare IC chips is becoming more common in the memory devices of the portable electronics. The same technologies as used in rigid circuit boards are available for flexible circuits. Adhesiveless copper-clad laminate should be used when high-temperature processes are applied. Cast-type or lamination-type adhesiveless laminates could provide high-reliability because of high bond strength.

Applying underfill materials between the bare chip and the flexible substrate is recommended for higher reliability.

64.3.9 Anisotropic Conductive Material (ACM)

The connection of flexible circuits with anisotropic conductive materials is not new idea. An organic adhesive material that has conductive particles becomes conductive only in the z direction when sandwiched between a flexible circuit and the other device under a pressure

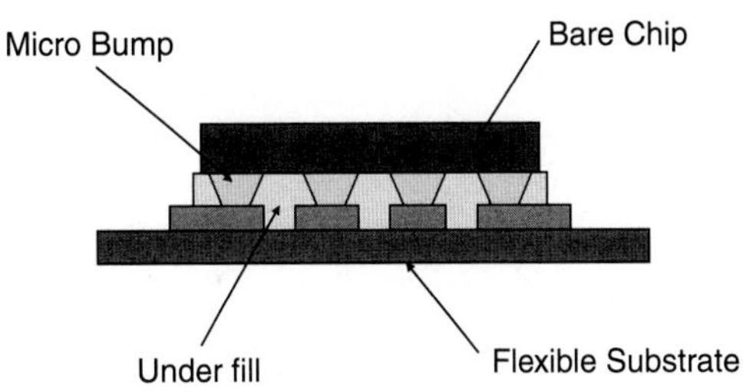

FIGURE 64.10 Flip chip mounting of bare chip.

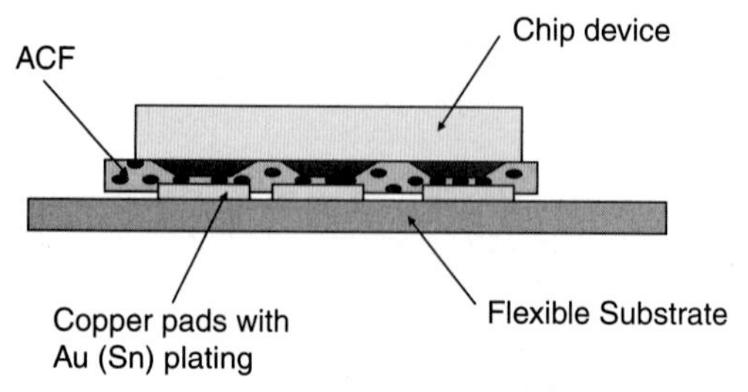

FIGURE 64.11 ACF (M) connection on HDI flex 30 micron pitch.

(see Fig. 64.11). The connection technology was widely applied for the simple displays of low-cost products such as calculators and clocks. Usually the conductors of the flexible circuits are built by screen printing of the carbon paste or silver paste, so the connection can be applied only for pitches larger than 1 mm with small currents.

Newly developed anisotropic conductive resins have generated several new termination technologies for high-density flexible circuits. Film-type and paste-type anisotropic materials are developed with suitable applicators. The film-type material, called anisotropic conductive film (ACF), has been widely used for the mounting of driver IC chips on the flexible substrate and connections between the flexible substrates and the glass substrates of the LCD devices. The latest material has 30 μm pitch connections in one direction for the IC chips on the gold-plated pads of the flexible circuits. Examples of the basic properties are provided in Table 64.5

The connections have relatively high resistance with a high noise level. The low-heat resistance resulting from the organic matrix of the anisotropic conductive materials is another barrier to apply as a reliable termination under sever operating temperatures.

TABLE 64.5 Basic Properties of the ACF Connection *(Source: Sony Chemical)*

Grade	CP9631SB	CP9221FS	CP7652K
Applications	FOG, TCP	FOG, COF	FOB, TCP
Min. pitch	70 microns	50 microns	200 microns
Thickness	25 microns	20 microns	45 micron
Matrix	Epoxy	Epoxy	Epoxy
Particle size	5 micron	5 micron	6 micron
Surface treatment	Gold plating	Gold plating	Gold plating
Process temp.	170 degree C	190 degree C	170 degree C
Process time	15 sec.	10 sec.	20 sec.
Repair	Possible	Good	Possible

FOG: Film on glass, FOB: Film on Board, TCP: Tape Career Package
COF: Chip on Film

64.3.10 Nonconductive Paste (NCP) Connection

An NCP connection looks similar to the flip-chip connection described in Section 64.3.7. The NCP connections do not need such severe conditions for both temperature and pressure. The nonconductive paste filled between the bare chip and flexible substrate works as an adhesive resin (see Fig. 64.12). It has good bond strength for both the bare chip and the substrate. The adhesive resin shrinks under a thermal treatment and generates negative pressure between the chip and flexible substrate. The negative pressure is remarkably strong at the contact points of the microbumps, so the metal-to-metal bonding is made between the microbumps and the pads of the flexible substrates (see Fig. 64.12.). A smaller contact of the microbump is recommended to generate a higher pressure at the bonding points. The gold bumps or soft gold-plated copper bumps and tin-plated copper pads usually are

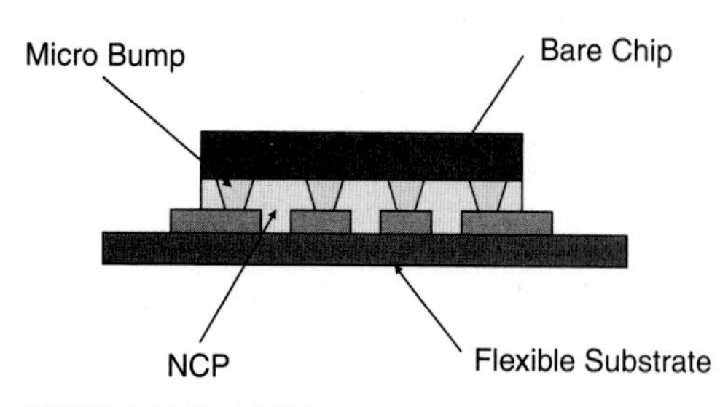

FIGURE 64.12 NCP bare chip packaging.

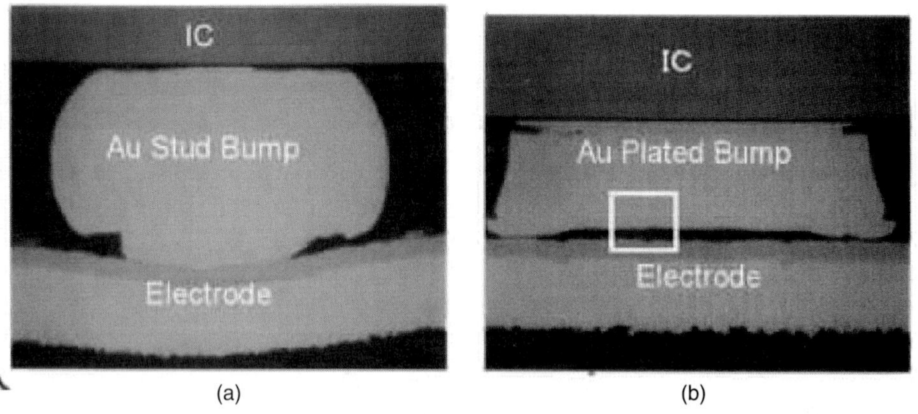

FIGURE 64.13 Cross section of NCP packaging, (a) Stud bump made by wire bonding machine. (b) Flat bump made by platting process.

used for this bonding. Stud bumps made by the similar machines as gold wirebonding provide needle shapes at the top of the bumps to make good bonding (see Fig. 66.13).

NCP connection provides a higher reliability than ACM connections with a pitch smaller than 100 micron. The density of an NCP connection looks lower than that of an ACM connection, but it covers more pin counts in a small space, utilizing the area connection capability.

64.3.11 Automatic Equipment for the Permanent Connections

Many kinds of equipment have been developed for the high-density permanent terminations of flexible circuits. However, the basic process and equipment look similar. A mounting machine places the devices on the reeled flexible circuit with accurate alignment by CCD cameras. The bonding head adds pressure and heat from the top of the electronic devices. An ultrasonic vibration of the bonding head could help to increase the reliability of the connections.

64.4 SEMIPERMANENT CONNECTIONS

Semipermanent connections have been required for rework, replacement of components and modules, and repeated connections. Realistic assembly processes for most electronic products are not perfect, so semipermanent terminations are required for rework and replacement of electronic components or modules in standard assembly processes to increase final manufacturing yields. Also, temporary connections are required for testing and adjusting electronic products and modules. These often need 5 to 10 reconnections, with as many as 100 reconnections required in extreme cases. Basic data on typical semipermanent termination technologies are summarized in Tables 64.6 through 64.8.

TABLE 64.6 Semipermanent Interconnects with Flexible Circuits

	Objectives	Application examples
Pressure contacts	Flex/rigid boards	Facsimile
Anisotropic conductive rubber	Flex/flex or rigid boards	Flat panel displays
Dimple contacts	Flex/rigid parts (flex-flex)	Inkjet printers
Microbump array	Flex/rigid boards and parts	Board testers

TABLE 64.7 Performance of Semipermanent Technologies

	Connection density	Repeatability	Trace density of flex circuits
Pressure contacts	150–μm pitch	~20	150–μm pitch
Anisotropic conductive rubber	100–μm pitch, 200 contacts per cm^2	~50	100–μm pitch

TABLE 64.8 Items to Be Considered in Semipermanent Termination Technologies

	Supplemental parts	Applicators	Remarks
Pressure contacts	Rubber strip and fixture	Simple tooling	Good dimensional control required
Anisotropic conductive rubber	Fixture and ACR strip	Simple tooling	Not available for high currents High contact resistance High noise level

64.4.1 Pressure Contact Termination

The basic idea of pressure contact termination technology is very simple and is not new. Pad patterns of the same pitch plated with gold are generated on a flexible circuit and opposite circuits, and they are attached to each other with uniform pressure using a suitable rubber strip (see Fig. 64.14). There is no high-temperature process to complete the connection; polyester base materials are available. The maximum connection number and density depend on the capability of dimension control and uniformity of the circuits. It is possible to make 200 connections with 150 μm pitch because of the processes all take place at room temperature. Neither special application equipment nor special constructions on flexible circuits are

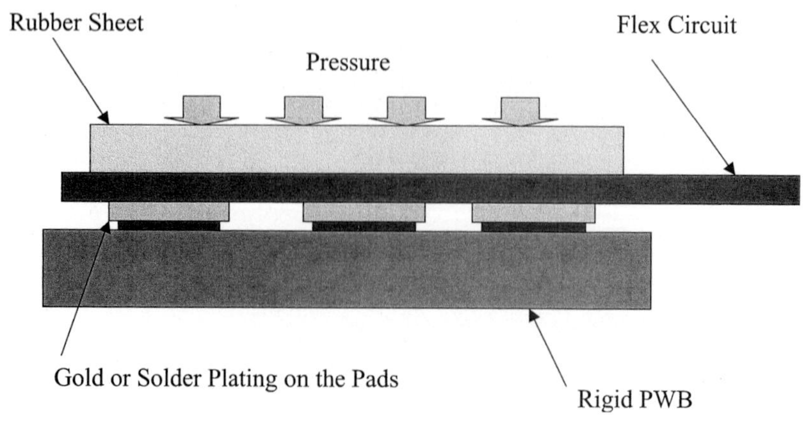

FIGURE 64.14 Pressure contact of flexible circuits.

required except simple tooling, so pressure contact termination can be a low-cost solution. Unfortunately, connection and disconnection take time, which makes this technology unsuitable for multiple repeated connections.

64.4.2 Anisotropic Conductive Elastic Materials

Newly developed anisotropic conductive rubber has generated several new termination technologies for high-density flexible circuits (see Fig. 64.15). The latest material has 200 μm pitch resolution in one direction. Another type of sheet rubber can have more than 200 contacts in 1 cm^2. These types of rubber can make more varieties for traditional pressure contacts. A disadvantage of the technology is high contact resistance, which makes it unavailable for large-current circuits. Also, it generally has a relatively high noise level. Table 64.8 gives the basic properties.

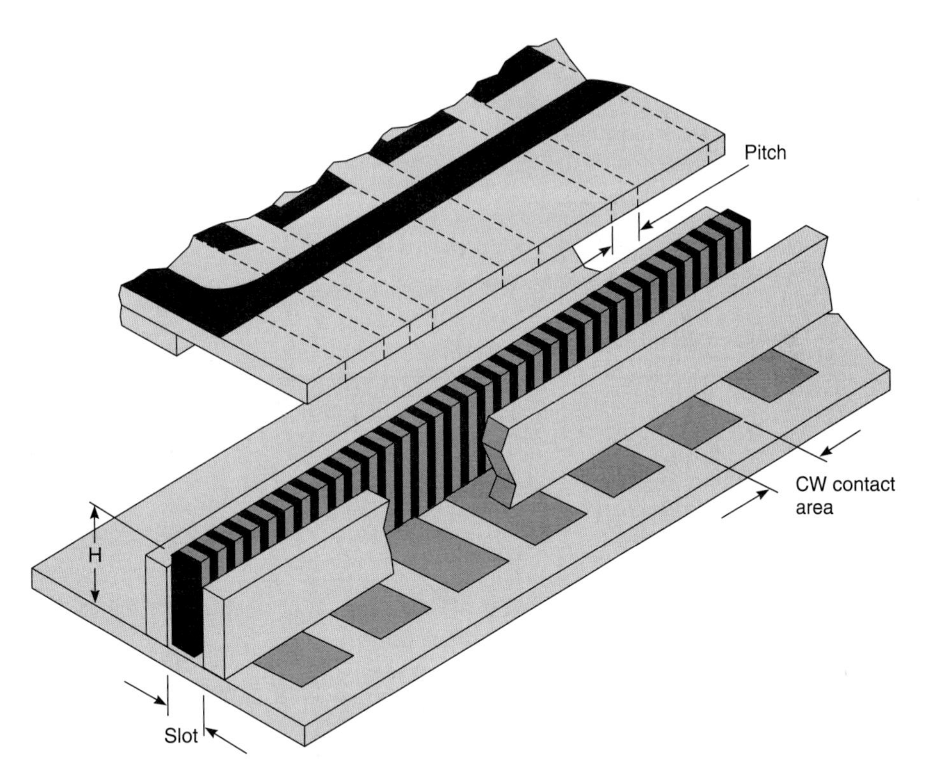

FIGURE 64.15 Pressure contact with anisotropic conductive rubber.

64.5 NONPERMANENT CONNECTIONS

Many repeated connections are required for replacement of disposable components such as inkjet printer cartridges and biological sensor devices. These require more than 1,000 repeated connections for a flexible circuit cable. More than 1 million repeated reliable contacts on flexible circuits are required for switching circuits and test probe circuits. Flat-panel switches and electrical test probe fixtures for microcircuits are typical examples. As flex circuits have high-density cabling capabilities other than the mounting capabilities of electronic components, there have been serious requirements for high-density nonpermanent termination.

There have been two kinds of approaches to realizing high-density nonpermanent flex circuits. The first is improvement in connection densities of traditional termination technologies. Crimp tabs, pressure contacts, key contacts, and FFC connectors could be categorized in this group.

Another approach is a new concept. Anisotropic conductive rubber, dimple contacts, and bump contacts are typical examples. The technologies are compared in Tables 64.9 through 64.11.

TABLE 64.9 Nonpermanent Interconnects with Flexible Circuits

	Objectives	Application examples
Crimp tabs	Flex-flex	Tach panel switches
Key pad contacts	Flex internal	Keyboards
FFC connectors	Flex-flex or rigid boards	Computer peripherals
Area array connectors	Flex-rigid boards	Industrial equipment
Dimple contacts	Flex-rigid parts (flex-flex)	Inkjet printers
Microbump arrays	Flex-rigid boards and parts	Board testers

TABLE 64.10 Performance of Nonpermanent Technologies

	Connection density	Repeatability	Trace density of flex circuits
Crimp tabs	1.27 mm pitch	~100	1.27 mm pitch
Keypad contacts	10 in 1 cm^2	10 million	200 μm pitch
FFC connectors	0.3 mm pitch	~100	150 μm pitch
Area array connectors	100 contacts per cm^2	~500	100 μm pitch
Dimple contacts	20 contacts per cm^2	~1000	0.5 mm pitch (150 μm pitch)
Microbump arrays	400 contacts per cm^2	~1 million	60 μm pitch

TABLE 64.11 Items to Be Considered for Nonpermanent Termination Technologies

	Supplemental parts	Applicators	Remarks
Crimp tabs	Tab and housing	Applicator	Low connection density.
Keypad contacts	Conductive keypads or a sheet	No	Density depends on mechanical key sizes.
FFC connectors	ZIF connectors	SMT soldering	Soldering on the other circuit.
Area array connectors	Male and female connectors	Special soldering	High-density traces are required on flex circuits.
Dimple contacts	(Housing)	Simple tooling	Suitable forming process shall be developed.
Microbump arrays	(Housing)	Special fixtures	Suitable processes to build microbumps must be developed.

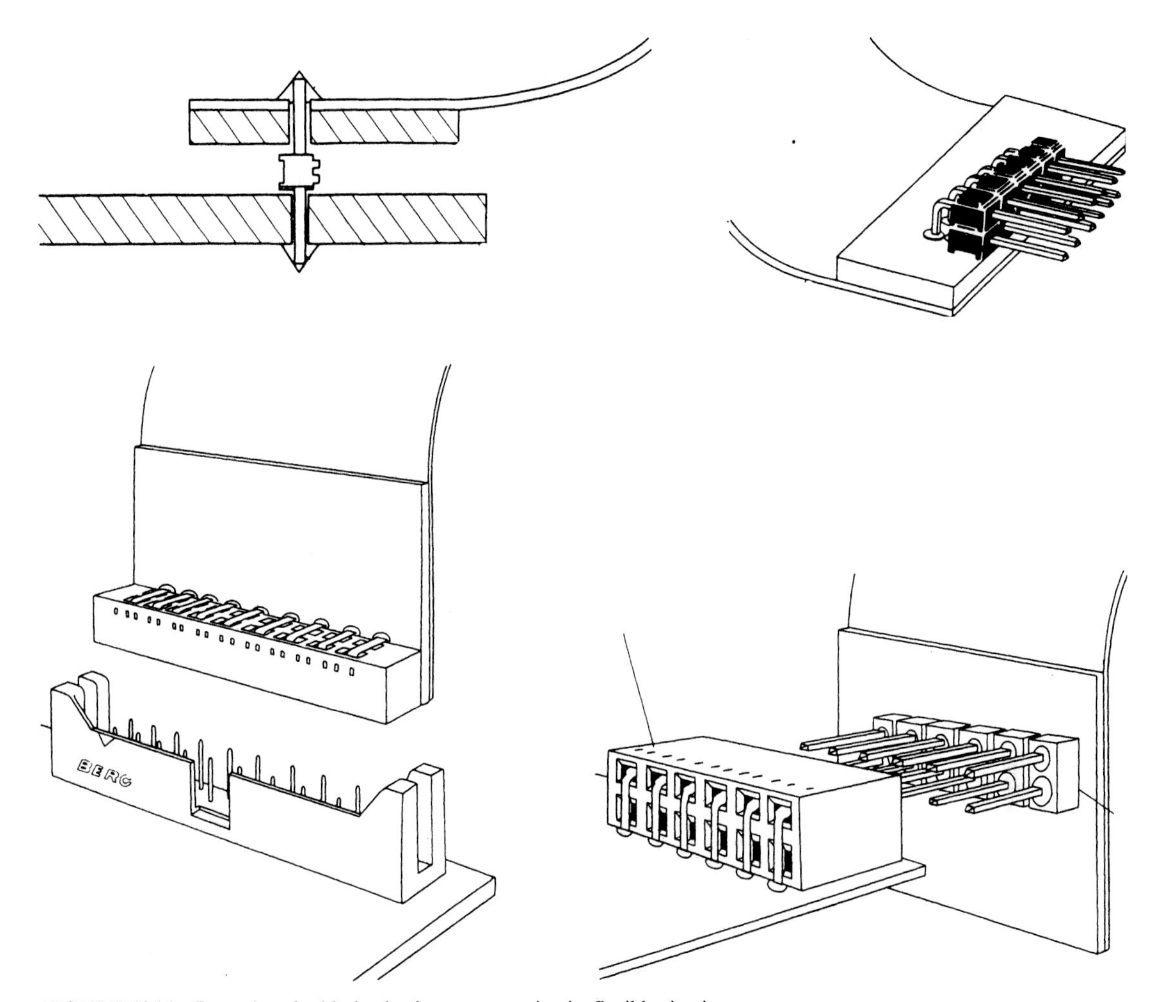

FIGURE 64.16 Examples of soldering lead-type connection in flexible circuits.

64.5.1 General Connectors

Most of the general connectors designed for other wiring technologies are applicable without serious changes. (Figure 64.16 shows a set of alternative designs.)

64.5.1.1 Card-Edge Connectors. All kinds of connectors designed for rigid circuit boards are available for flexible circuits. They can be mounted and soldered on flexible circuits supported with stiffener boards of suitable thickness. Card-edge connectors designed for rigid circuit boards are also available. A flexible circuit with a suitable stiffener board on the bottom can be inserted into the connector. However, the newest FFC connectors designed specifically for flexible circuits have a much higher density in a smaller housing and should be replaced by FFC connectors for large volume.

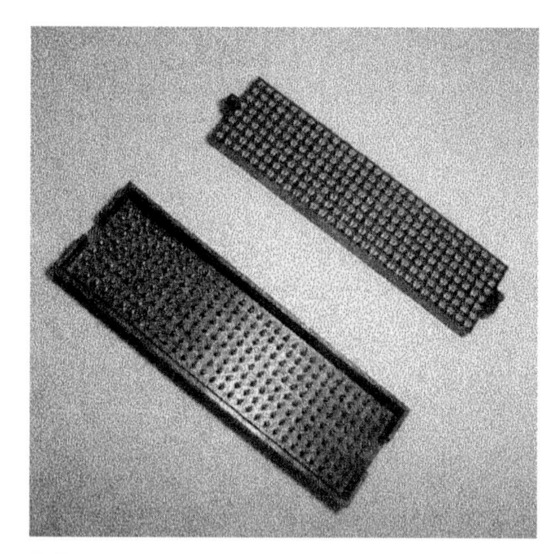

FIGURE 64.17 High-pin-count area array connectors available for high density flexible circuits.

64.5.1.2 Area Array Connectors. Density higher than 150 μm pitch on double-sided flexible circuits with through holes is required for high-pin-count connections. All kinds of connectors designed for rigid printed circuit boards are available for flexible circuits with some modification on flexible construction. A newly designed array connector for high-density flexible circuits can make 300 connections in a 10 × 15 mm area (see Fig. 64.17). Special constructions are not necessary, other than 0.5 mm pitch array soldering. However, traces with pitches smaller than 150 μm are required for high pin counts. They can survive 1,000 repeated contacts.

64.5.1.3 Crimp Tabs. Crimp tabs were originally developed for connections of flat conductor cables; however, they are also applied for flexible circuits because of the similar construction.

An applicator firmly staples female tabs mechanically through dielectric layers (see Fig. 64.18). Male tabs are stapled on the other cable. Both are covered with plastic housings. Because there is no soldering process, polyester-based materials are available. For a long

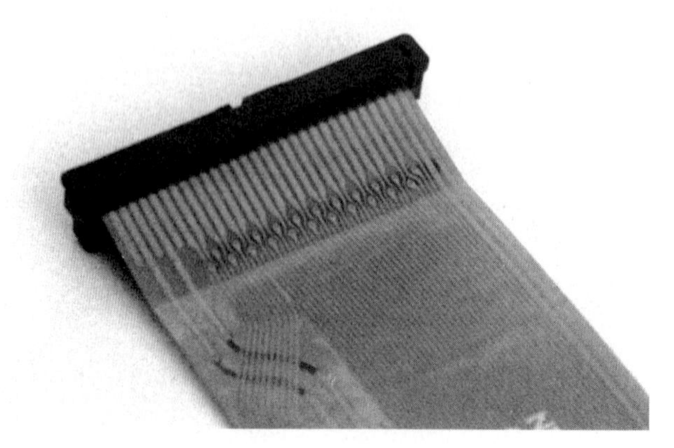

FIGURE 64.18 Clamp type connector for flexible circuits.

time, the standard pitch was 2.54 mm. A 1.25 mm system was introduced recently, doubling the connection density. The connectors can make reliable repeated connections more than 100 times. Total cost per connection is relatively high.

64.5.2 FFC Connectors for Flexible Circuits

Originally, FFC connectors were developed for laminated cable with flat conductors to connect to another circuit board. A similar construction of flexible circuits was developed to attach the connector. At this time, the standard pitch of the connectors was 2.54 mm. Due to increased trace densities of flexible circuits, small-pitch FFC connectors are now required. The newest connector can make 0.3 mm pitch connections for 30 traces. Zero in force (ZIF) mechanisms were introduced for easier handling. Due to limited height requirements for thin board assembly, a 1.0 mm high connector from the board surface is now available.

64.5.3 Pair Connectors for Flexible Circuits

Originally, the board-to-board connectors were designed for the connection between rigid circuit boards, and they have been also used for the connection of flexible circuits to rigid circuit boards. However, portable electronic products demand thin connectors with small pitches for the flexible circuit connections, so connector manufacturers have developed special connectors called *pair connectors* specifically for connecting high-density flexible circuits. Basically, they are a pair of male and female connectors separately mounted on the flexible circuit and rigid board by SMT soldering (see Fig. 64.19).

The smallest pitch of the pair connector is 0.5 mm. It is larger than the smallest pitches of FFC connectors; however, because the pair connectors have dual line contacts and soldering leads, they can manage more pin numbers in a limited space. A higher reliability for the repeated connections is another advantage of the pair connectors. Accordingly, they have been used as the standard methods for connecting high-density flexible circuits in portable electronics such as cellular phones and digital cameras (see Fig. 64.19).

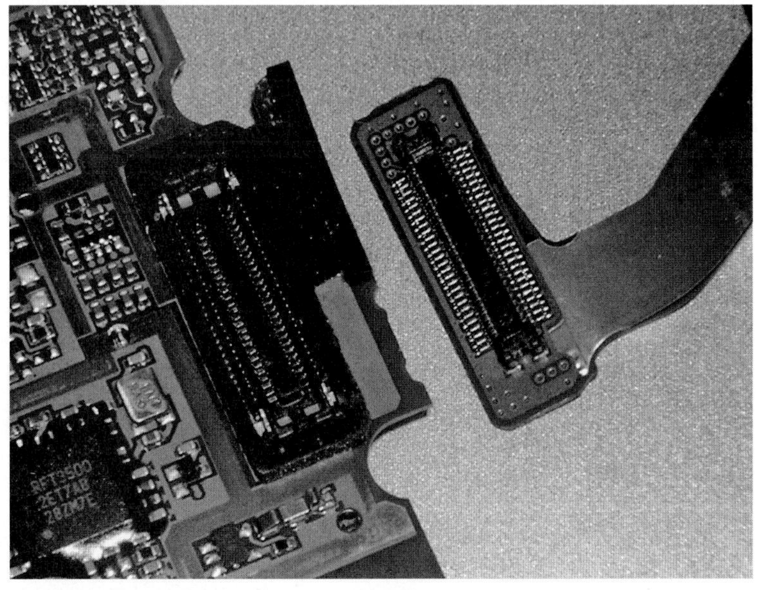

FIGURE 64.19 Pair connectors for flexible circuits. Example shown is for connection in cellular phone.

64.5.4 Bump and Dimple Array Contacts

Small bump arrays are valuable for the nonpermanent connections of high-density flexible circuits. Various bump constructions on flexible circuits have been developed through different manufacturing processes that are driven by the needs of specific applications.

64.5.4.1 Dimple Contacts. Dimpled pads of flexible circuits can be a low-cost solution for high-density connection of flexible circuits. One hundred contacts in 12.5 mm square between flexible circuits and the other circuit boards are possible. Appropriately shaped dimples are formed by pressure and temperature for a gold-plated pad. Dimpled pads with hard gold plating have enough life for several thousand connections and disconnections (see Fig. 64.20). The dimples are supported with bumped rubber to create long and reliable contacts. The pitches of the dimples cannot be smaller than 1.5 mm, but high-density termination can be realized with the area dimple array.

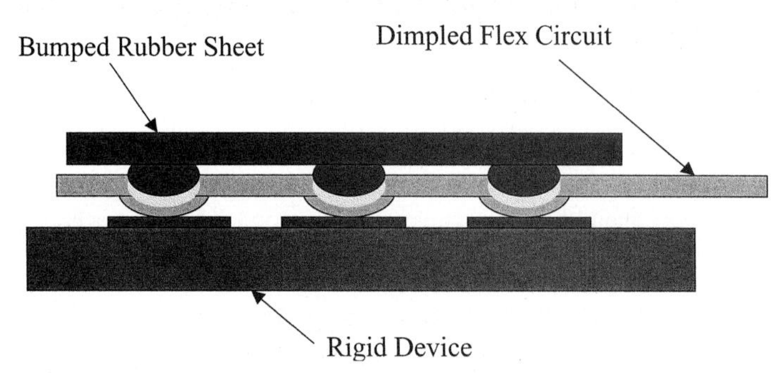

FIGURE 64.20 Dimple contacts for flexible circuits.

The dimpled connection was adopted for the disposable ink cartridges of the inkjet printers, and huge numbers of flexible circuits are consumed as the partner of the dimpled flexible circuits (see Fig. 67.21).

FIGURE 64.21 Dimple connections for ink jet printer. *(Source: LexMark.)*

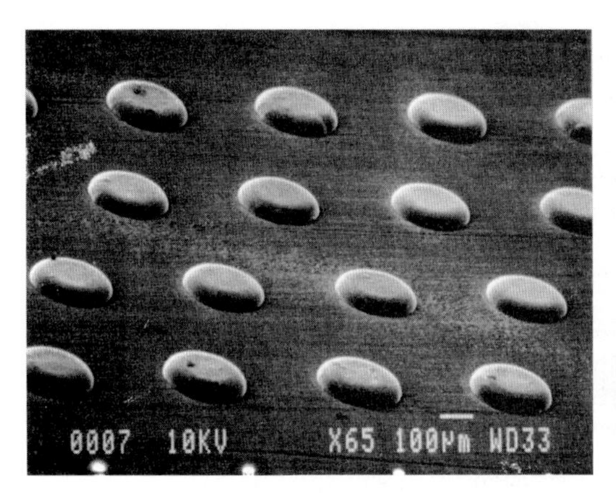

FIGURE 64.22 Flat disk-shaped microbump array on flexible circuit fir area contacts. Contact metal is copper-nickle-gold at 0.5 mm pitch. *(Source: Asahi Fine Technology.)*

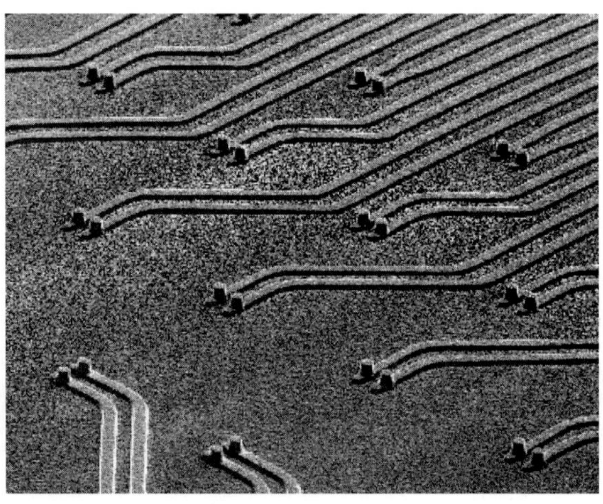

FIGURE 64.23 Column shape Cu/Ni/Au bump array on a flexible circuit substrate.

64.5.4.2 Microbump Array. Because of thin substrates, various constructions of microbump arrays have been developed on flexible substrates. A plating combination of nickel and hard gold on a copper bump provides reliable contact performances. A microbump array built on flexible circuits by a plating process enables high-density contacts with the other circuit boards. The process has a range of wide flexibility to generate various shapes of microbumps. It is possible to build more than 1,000 microbumps in 10 mm^2 if 30 μm line/space circuits are available. A mushroom-head copper bump plated with hard nickel and hard gold has enough life for millions of contacts (see Fig. 64.22).

64.5.4.3 Film Connectors. High height bumps built on the flexible substrates as shown in Fig. 64.23 enable various repeatable interconnections between flexible circuits and the other devices.. A flexible circuit with a microbump array makes made a connection thinner than 0.3 mm from the surface of the other circuits. Over 100 connections could be made in 5 mm square by a microbump array. The column-shaped microbumps work as the male parts of the connection. An appropriated design of the female parts of the flexible circuits enables over 1,000 repeated connections for the high-pin-count circuits.

Figure 64.24 shows the concept of a film connector made with bumped flexible circuits. The column-shaped microbump array is built on both sides of a thin, flexible substrate, and it works as the film connector between two flexible circuits. The film connector could generally be used with standard designs because it manages high-pin-count connections of flexible circuits without special constructions. Many varieties for the constructions could be possible depending on the combinations of flexible circuits and other devices.

64.5.5 Keypad Contacts

There are several different constructions for keypads on flexible circuits. Thick-film circuits on polyester film called *membrane switch* could be a low-cost solution with many millions of reliable contacts. The basic mechanism of the switching is illustrated in Fig. 64.25. Generally, the membrane switches are connected through FFC connectors to the main circuit board. Polyimide-based circuits are necessary with soldering of components.

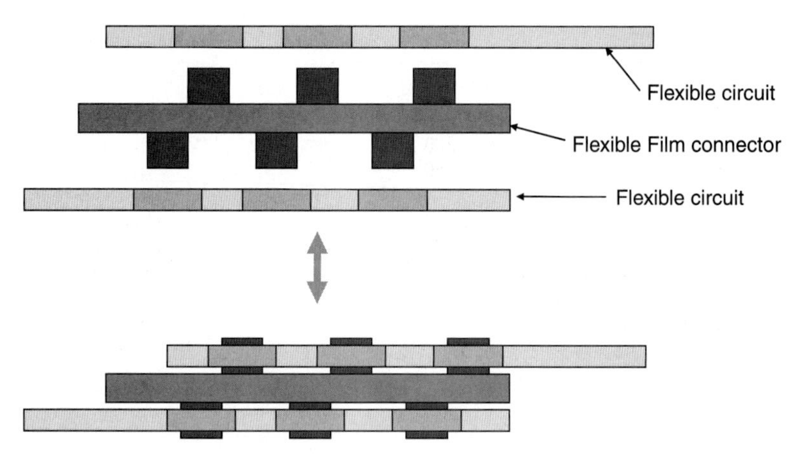

FIGURE 64.24 Thin connection by film connector.

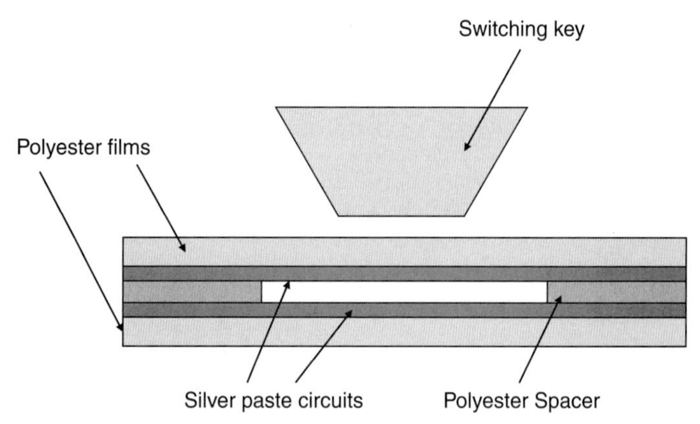

FIGURE 64.25 Membrane switch mechanisms.

The membrane switches built on polyester film are widely used as the standard keyboard circuits of personal computers. They are also used as the switching components of touch panel boards for home appliances such as microwave ovens and office machines such as copy machines. They are also used with the touch screens of display devices.

64.6 HIGH-DENSITY FLEXIBLE CIRCUIT TERMINATION

No perfect termination technology can satisfy all requirements, especially for high-density connections with high pin counts. On the other hand, higher performances are required for the new portable electronics to complete reliable terminations in limited spaces. Accordingly, many new termination technologies with flexible circuits have been developed to satisfy many requirements. Nevertheless, it is still difficult to use one termination technology to satisfy

TABLE 64.12 New Termination Technologies for High-Density Flexible Circuits

Termination technology	Termination density (μm pitch)	Repeatability	Requirement for flex circuits
Lead-free reflow	300	~2	Higher heat resistance (260°C for 30 sec.)
Lead-free hot bar reflow	150	~2	High heat resistance (400°C for 10 sec.) High dimensional accuracy (~0.02%)
Wire bonding	150	1	High heat resistance (450°C for 3 sec.)
Direct bonding	50	1	Flying-lead construction (50 μm pitch)
High-density BGA for flip-chip	200	1	Microbump array (200 μm pitch)
Anisotropic conductive materials	60	~5	High dimensional accuracy (~0.01%)
Small-pitch ZIF connectors	300	100	High dimensional accuracy (~0.05%) uniform tin plating
Contact probe array	200	~1000	Microbump array hard gold plating on nickel

all requirements. Therefore, suitable combinations are recommended to complete all terminations in an electronic product or module. Table 64.12 shows several examples of the major termination technologies used for high-density flexible circuits.

CHAPTER 65
MULTILAYER FLEX AND RIGID/FLEX

Dominique K. Numakura
DKN Research, Haverhill, Massachusetts

65.1 INTRODUCTION

Originally, the basic design concepts and manufacturing processes for the multilayer rigid/flex had been developed for aerospace equipment because of the need to manage reliable wiring in limited space. More than 30 conductor layers were built on a rigid/flex for complicated applications,. On the other hand, consumer portable products such as cellular phones and digital cameras have long been demanding high-density, low-cost wiring technologies and new design concepts and new manufacturing processes have been developed for this task.

A combination of rigid and flexible circuits can be called *rigid/flex, flex/rigid, or rigid/flexible.* However, they are called *multilayer flex* when flexible materials are used instead of glass/epoxy. The term *rigid/flex* is used throughout this chapter.

65.2 MULTILAYER RIGID/FLEX

Multilayer rigid/flex is basically a combination of rigid multilayer and flexible circuit processes.

However, the successful accomplishment of this combination requires a high level of skill in both areas, and it is important to have a clear understanding of the abilities and limitations of the chosen fabricator before designing this type of circuit for a product.

65.2.1 Basic Constructions

There are many different concepts in the construction design of multilayer rigid/flexible circuits.

Figure 65.1 shows the basic construction of a multilayer rigid/flex, both from a plane view and in cross section. Figure 65.2 shows an example of this construction with air gaps between the flexible layers, in order to have high flexibility. More than 30 layers are built in extreme cases, such as for aerospace applications. Because of the requirement for high-reliability, fine pattern designs with micro via holes cannot be used. In addition, leaded components are

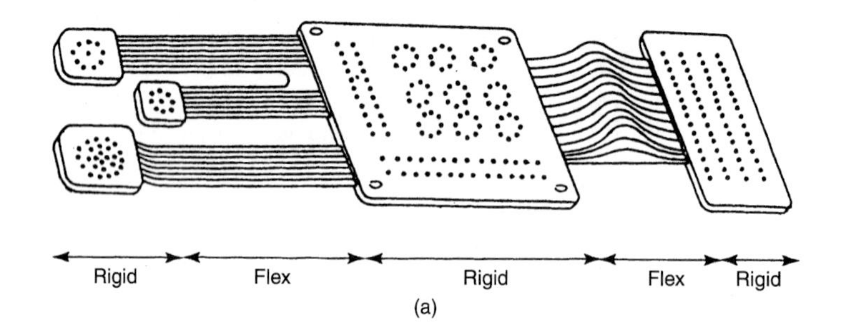

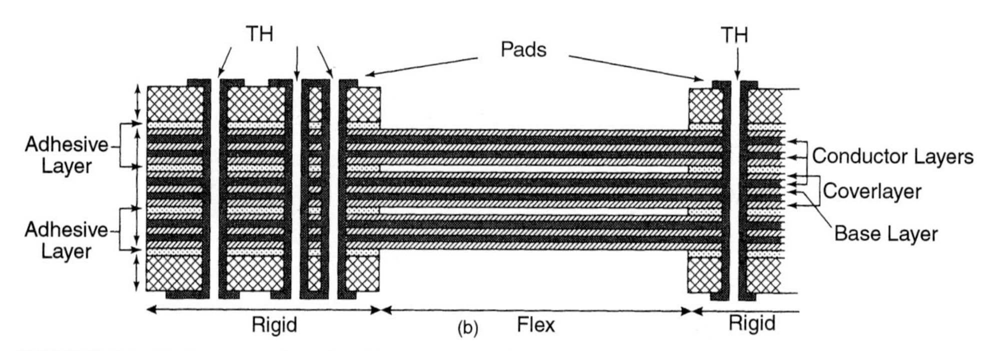

FIGURE 65.1 Basic construction of multilayer circuits (a) Plane view, (b) Cross section.

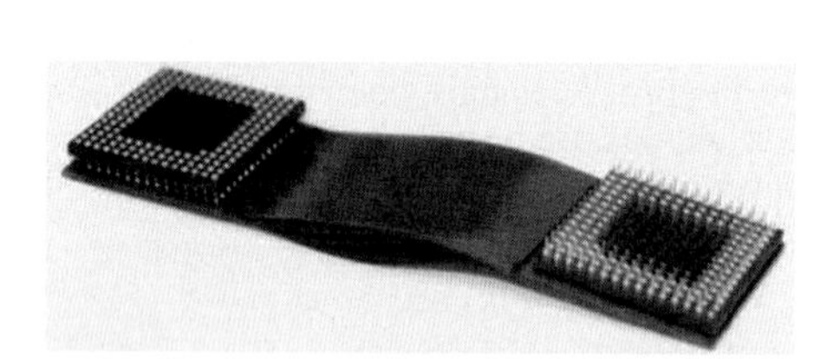

FIGURE 65.2 Air gap structure of multilayer rigid-flex board.

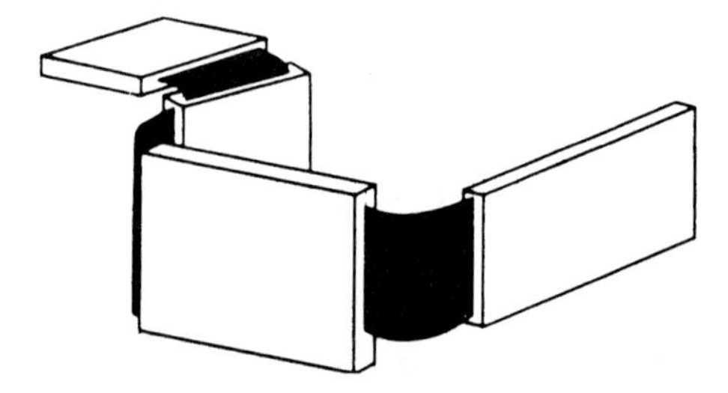

FIGURE 65.3 Folding type rigid/flexible circuit.

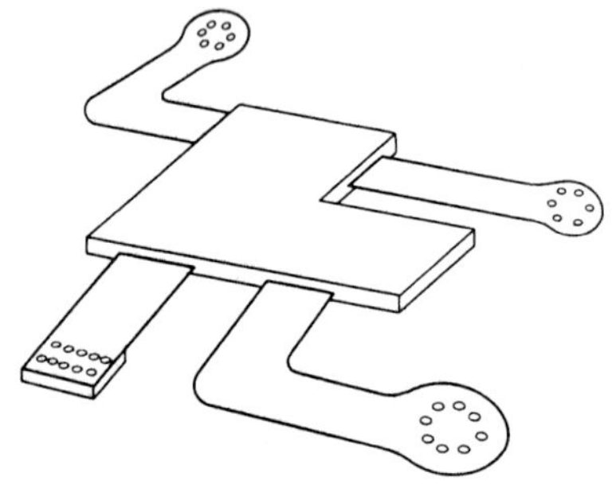

FIGURE 65.4 Flying tail type rigid/flexible circuit.

specified instead of SMT components. The design usually calls for relatively large lines/spaces and large-diameter through-holes with thick copper plating. There are several types, designated the folding type (Fig. 65.3), the flying tail type (Fig. 65.4), the bookbinder type, and so on, according to their shapes.

65.2.2 Materials

Several supplemental materials are necessary for building a multilayer rigid/flex, as shown in Table 65.1. It should be noted that there has been significant technical progress in the development of these high-performance materials.

TABLE 65.1 Materials for Multilayer Rigid/Flexible Circuits

Materials required	Traditional materials	High-performance materials
Flexible substrates	Traditional polyimide films (Kapton H™, Apical AV™)	New polyimide films (Kapton E™, Apical HP™, Upilex S™)
Copper-clad laminates (double-sided)	Polyimide film substrate Acrylic adhesive (epoxy adhesive)	Polyimide film substrate Adhesiveless laminate (cast materials or laminated materials)
Coverlay	Traditional polyimide films coated with acrylic or epoxy adhesives	New polyimide films coated with hot-melt polyimide adhesives
Bonding sheets	Acrylic resin films Epoxy resin films Polyimide films coated with acrylic adhesives on both sides	New polyimide films coated with hot-melt poyimide resin on both sides
Rigid substrates	Glass/epoxy boards	Glass–BT resin boards Glass-polyimide boards

The materials must have high heat resistance and high dimensional stability to survive during several high-temperature processes. Thicker polyimide films (50 μm thick or thicker) are recommended because the basic dielectric materials are required to have good stability in the manufacturing processes. An acrylic adhesive system in copper-clad laminates, as well as coverlay film and bonding sheets that have higher heat resistance than epoxy systems, are also recommended for use in manufacturing. Adhesiveless, copper-clad laminates, made by casting or lamination processes, usually perform better in high-temperature processes. They are also valuable because they reduce the total thickness of the finished board. Hot, melt-type, polyimide adhesive systems, including coverlay films and bonding sheets, have been developed to have very high reliability and smaller thicknesses. They can reduce smear level in drilled holes significantly. But they must be processed at temperatures higher than 300°C, and therefore require special facilities and conditioning.

65.2.3 Manufacturing Process Flow

There are many varieties of manufacturing processes for multilayer rigid/flexible circuits because of their complicated structures. Figures 65.5 shows a typical layer construction using

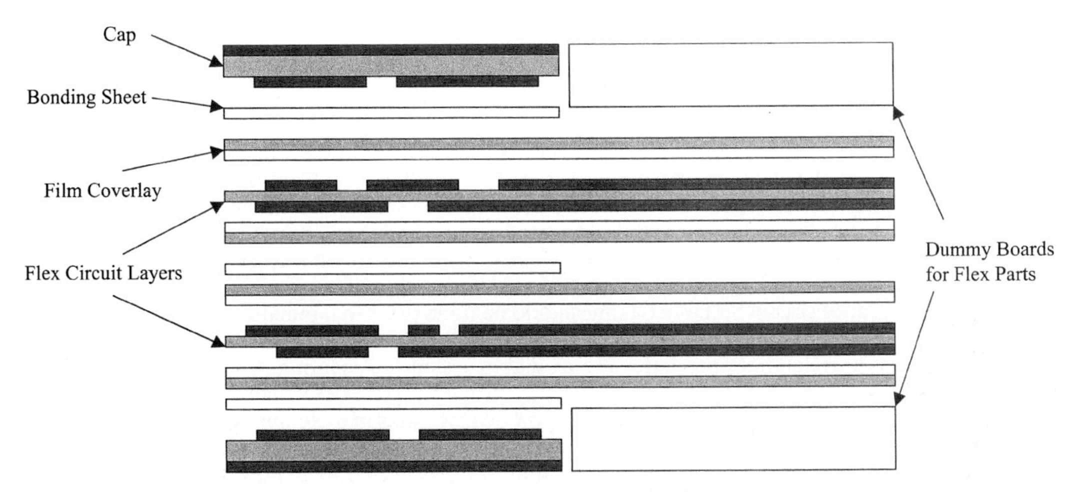

FIGURE 65.5 Standard manufacturing flow for multilayer rigid/flexible circuits.

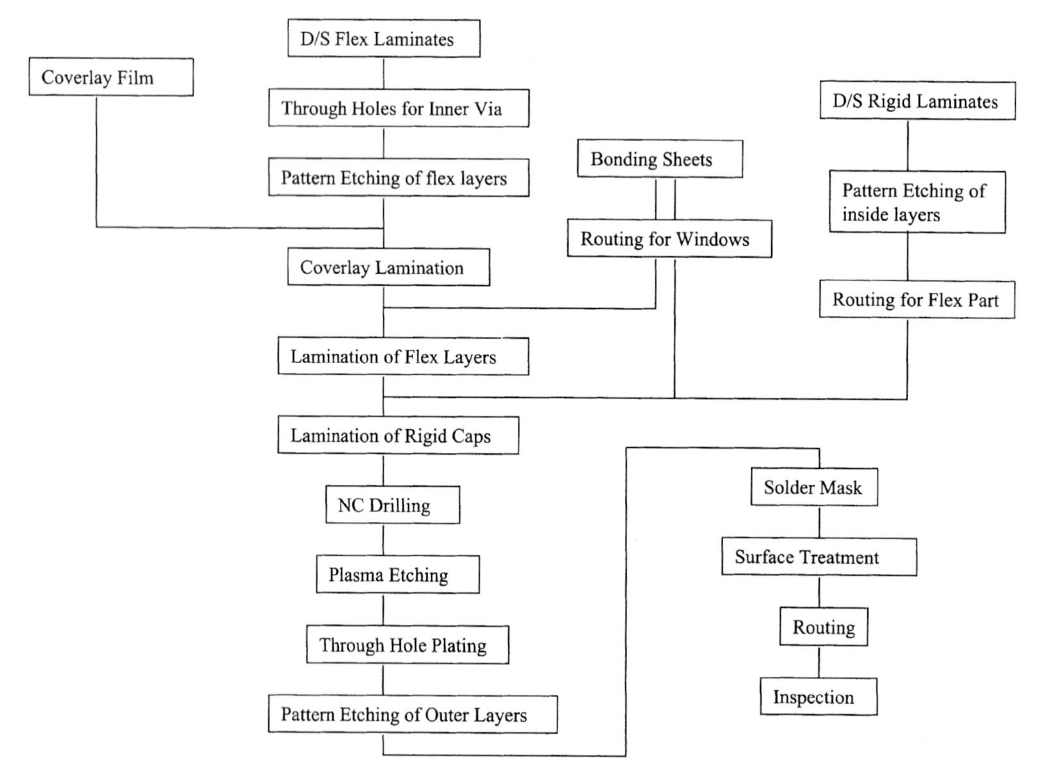

FIGURE 65.6 Material constructin for multilayer rigid/flexible circuits.

the standard manufacturing process for multilayer rigid/flexible circuits shown in Fig. 65.6. As seen in Fig. 65.6, the process starts with flexible, double-sided copper-clad laminates.

The conductive patterns are generated by the traditional etching process without through holes. All traces are covered with plain coverlay films, which have no openings. The multiple covered flexible circuits are bonded with bonding sheets, which are already routed to make openings for nonbonding flexible areas.

Caps (the outside layers of rigid parts) are made with rigid double-sided copper-clad laminates.

Traces are printed and etched only on the inside surfaces of the laminates as the first process. Flexible parts of the boards are removed through NC routing or punching. Both flexible layers and caps are bonded with bonding sheets that have been routed for flexible parts.

Appropriate dummy boards are prepared for the flexible part during the bonding process. A vacuum press is preferred for good bonding quality. An autoclave system is recommended for complicated circuit constructions, because the autoclave can create uniform pressure on the entire circuit. Suitable baking should be done prior to lamination or bonding processes.

As the laminated board can be handled in the same way as multilayer rigid boards, the same through-hole processes are available, except desmearing, which is dependent on the combination of materials used (see later). The drilling process is conducted after enough baking in basically the same way as for rigid boards. A plasma etching process should be applied to remove smears of acrylic resins in the holes. The etchback conditions should be determined carefully. The same through-hole plating process is available; however, detailed conditions must be determined based on reliability test data.

FIGURE 65.7 Example of typical multilayer rigid/flexible circuit with via holes. *(Source: Toshiba.)*

The rest of the processes are very similar to those used for multilayer rigid boards. Outer layer etching, coverlay (solder mask), and surface treatment are conducted in the same way.

When the dummy boards are removed after routing, the circuit becomes a rigid/flexible circuit.

65.2.4 Through-Hole Process

When inner via holes are required in flexible layers, the same through-hole process should be conducted prior to etching of inner-layers as for rigid board inner-layers. Figure 65.7 shows an example of inner via holes in flexible parts of a multilayer rigid/flex, and Fig. 65.8 shows a cross section of a multilayer rigid/flex plated through-hole. The through hole drilling can be accomplished by the same machine as used for rigid multilayer boards. However, the drilling conditions should be determined carefully for each material's construction.

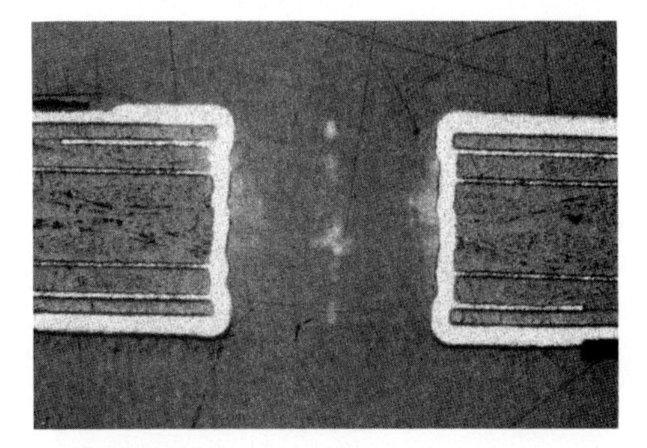

FIGURE 65.8 Cross section of a through-hole in a multilayer rigid/flexible circuit.

Baking conditions prior to drilling could be critical in creating reliable through-holes.

Desmearing of the rigid/flex is different from that of standard rigid multilayer boards. The acrylic adhesive materials will undergo serious swelling during standard permanganate desmearing processes, lowering through-hole reliability. A plasma etching process is recommended to remove the smears of acrylic adhesive materials in the drilled holes. Use of a polyimide adhesive system can reduce the smear level significantly, but this also requires a high-temperature process in special press equipment.

65.2.5 Reliable Through-Holes

Aerospace and industrial equipment require a higher reliability for the through holes than consumer electronic products. Copper plating thicker than 25 microns is recommended.

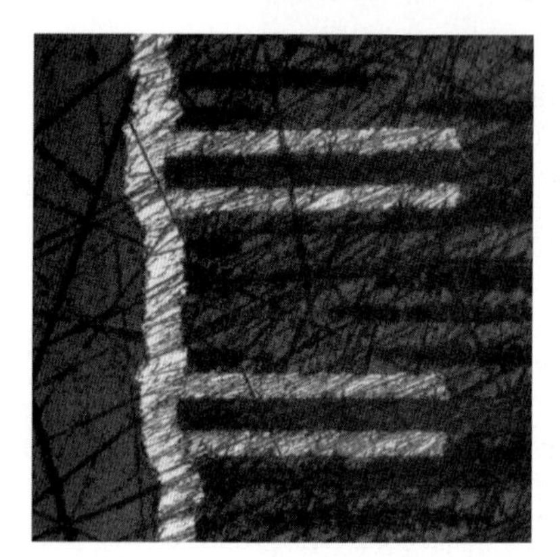

FIGURE 65.9 Etchback of through-hole in multilayer flexible circuit.

Several adhesive resins, especially acrylic-based adhesives developed for flexible circuits are not stable with the permanganate solution that is common for rigid multilayer boards, which means that a very narrow window is allowed for the desmearing of the rigid/flex process. If the materials are dipped too long in the solution, the adhesive layers swell and the reliability of the through holed will be damaged.

A plasma etching process has been introduced to remove the smear inside of the drilled holes as a substitute for using potassium permanganate solution for the polyimide base multilayer rigid/flex. The plasma gas removes the smear in the holes effectively without chemical damages on the adhesive resins.

The use of supplemental plasma etching, called the Etch Back Process, is recommended to achieve high reliability of through-hole multilayer rigid/flexible circuits. The plasma gas provides further etching on both of polyimide films and adhesive layer in the drilled holes. The plasma gas does not etch metal materials and it clears the surface of the copper foils for reliable copper plating as shown in Fig. 65.9.

65.2.6 Bookbinder Construction

A multilayer rigid/flex that has multiple conductor layers in its flexible (parts fixed by rigid caps at both sides) will not bend effectively if all the layers are of the same length. Special designs are not required while the flex areas have long distance and they use thin films and copper foils. They are sometimes called Air Gap Construction.

The Bookbinder Construction is recommended for the reliable interconnects of the multiple flexible layers in a rigid/flex with a short distance. A longer length of the flexible layer should be provided at the outer side of the flexible area. When they are bent, the flexible layers have uniform gaps between the flexible layers as shown in Fig. 65.10.

Special tooling and equipment are required to build the bookbinder construction, and the productivity is comparatively extremely low. As a result, this process is not used for consumer or other applications that require low cost.

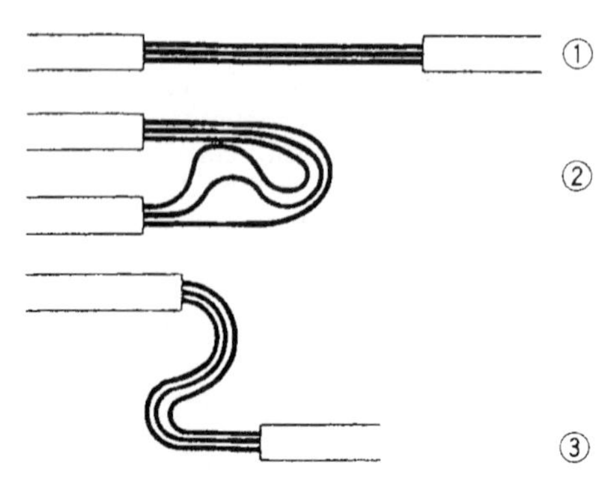

FIGURE 65.10 Design of bookbinder construction of multilayer rigid/flex circuit.

65.2.7 High Density Multi-layer Rigid/Flex for Industrial Use

High-density designs with fine traces and micro via holes for rigid/flex are not recommended for use in aerospace applications because they are not reliable enough under extreme conditions. Only traditional soldering processes are used for the terminations of the rigid/flex.

On the other hand, recent designs for industrial and medical applications have been using both high density and high reliabilities, therefore, new design concepts and termination technologies have been designed to meet this need.

FIGURE 65.11 Micro blind via hole on a multilayer rigid/flexible circuit made by a build-up process using resin-coated foil. The hole diameter is 150 micrometer. *(Source: Photo Machining.)*

Smaller trace pitches than 100 microns and smaller via-hole diameters than 100 microns have been introduced for the high density multilayer rigid/flex. To achieve the new design concepts, 25 micron or thinner polyimide films were introduced without adhesive layers for the thin copper laminates, mostly 18 microns or less. Several laser processes are used to generate small via holes, especially with blind via holes of the build-up process. (Fig. 65.11)

Several high density, but reliable termination technologies such as Ball Grid Array (BGA) and Flip Chip Bonding have been introduced for the high density rigid/flex.

65.2.8 Build-Up Process and Blind Via Holes

To build the high-density rigid/flex, the build-up process with blind via holes has been introduced as the new standard manufacturing process. Similar build-up processes are used with multilayer rigid and rigid/flex circuits; however, instead of the traditional manufacturing process used for multilayer rigid/flex circuits, more layers are built on the rigid part with resin-coated foils. Figure 65.12 shows a cross section of a blind micro via hole in a multilayer rigid/flexible circuit made by a buildup process with adhesiveless copper laminate and prepreg sheet. Holes smaller than 100 micron diameters were drilled by carbon dioxide laser and a thin copper plating was plated in the via hole conductors. Figure 65.13 also shows a

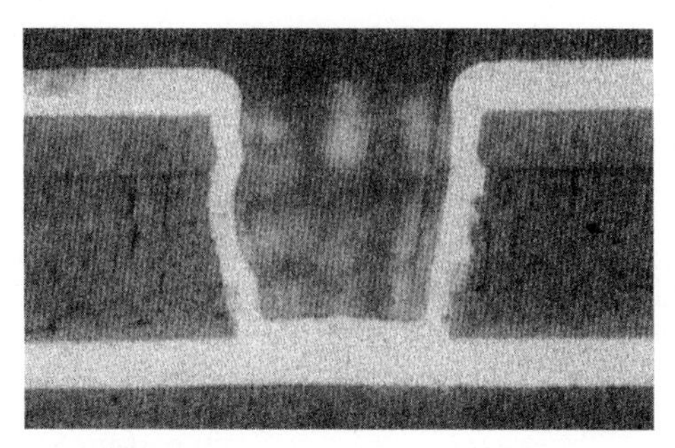

FIGURE 65.12 Micro-blind via hole on a multilayer rigid/flexible circuit made by a build-up process using adhesiveless laminate prepreg. *(Source: Photo Machining.)*

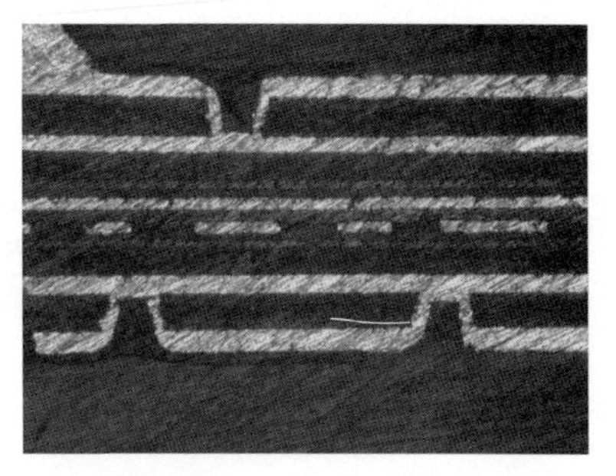

FIGURE 65.13 Micro blind vias on two sides of a multilayer rigid/ flexible cirucuit.

micro-blind via hole in a multilayer rigid/flexible circuit, made by a build-up process using an adhesiveless laminate, but with prepreg sheet and with hole diameter of 150 μm

65.2.9 High-Density Multilayer Rigid/Flex for Consumer Use

Consumer applications, such as digital cameras and cellular phones, often require high-density rigid/flex to complete complicated wiring and packaging in small spaces. They often use traces finer than 75 μm and blind via holes smaller than 150 μm in diameter with high-density SMT. Many small chip components are mounted on the rigid part of the circuits and the traces are connected to other circuits by soldering to have relatively large current capacities (See Fig. 65.14 for example of multilayer flexible circuit used in cellular phones.

Figure 65.15 shows another examples of rigid/flex applied for an LCD display module of a cellular phone. High density SMT including flip chip mounting of controller IC and small CMOS camera were conducted on a small rigid part of the rigid/flex. Onc of the flex parts was connected to the glass substrate of the LCD by ACF (Anisotropic Conductive Film) connection with 150 microns pitch. Another side of the flexible parts was to be inserted to the FFC connectors with 0.5 mm pitch.

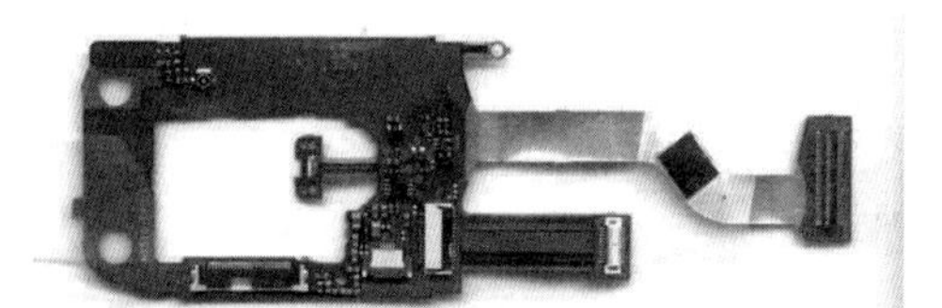

FIGURE 65.14 Multilayer flexible board for cellular phone application.

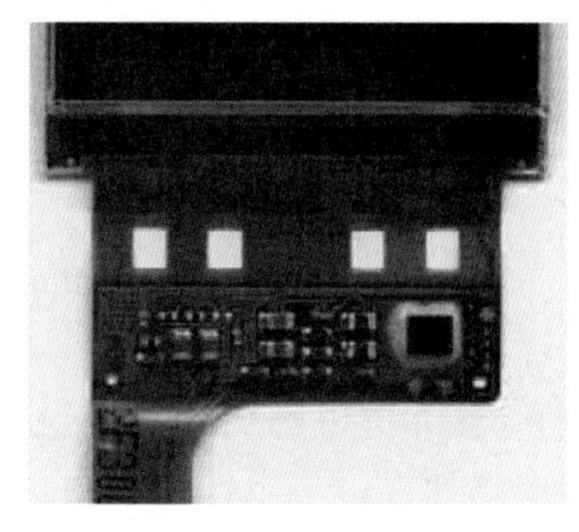

FIGURE 65.15 Rigid/flexible circuit for LCD driver.

Many technologies have been developed to satisfy these requirements. The use of a build-up process technology could be one of the solutions. The specialty of the consumer application is the large volume with extremely low cost, therefore a new design concept and manufacturing processes have been required.

65.2.10 Design Concept of Rigid/Flex for Consumer Use

To eliminate the conflicts of the requirements between high density and low cost, new design concepts for the consumer rigid/flex have been developed. The basic design concepts are listed in Table 65.2. Generally, thinner materials have been selected compared to those for aerospace and industrial use in order to reduce the total thickness significantly. Copper plating for the via hole connections should be less than 15 microns, but the high reliabilities should be kept.

Polyimide films or adhesiveless copper laminates can be used as the cap layer instead of thick glass epoxy materials for reducing the thickness of the rigid/area that is sometimes called multi-layer flex. The total thickness of the 6 layer rigid part for SMT can be less than 400 microns.

The numbers of the conductor layers should be 10 or fewer microns to keep total thickness small and reduce the manufacturing cost. Finer traces than 100 microns pitches should be

TABLE 65.2 Design Concepts of Rigid/Flex for Consumer Use

Base film	Polyimide, 25 micron thick or thinner
Laminate	Adhesiveless type by cast or lamination process
Conductors	Copper foil, 18 micron thick or thinner less than 10 layers
Type of via hole	Through hole, inner via hole, blind via hole
Min. via hole size	Down to 50 micron diameter
Trace density	Inner layer: down to 25 microns line/space Outer layer: down to 50 microns line/space
Copper plating	15 micron or thinner
Cover film	Polyimide, 25 micron or 12.5 micron
Bonding material	Epoxy or acrylic resin, 50 micron or thinner
Cap materials	Glass Epoxy, thinner than 250 micron Polyimide Film, 50 micron thick or thinner
Solder mask	Photo-imageable
Surface treatment	Ni/Au, lead-free solder or OSP
Shape	As smaller as possible

used instead of additional conductor layers. Figure 65.14 shows an example of multilayer flex designed for the display control module with hinge cables connected to key pad module in a cellular phone. It does not use glass epoxy for the cap layers and could minimize the thickness of SMT assembling ar.

The shape of the rigid/flex is a key to minimizing the cost. The whole rigid/flex should be in a small square or rectangular toz maximize the material efficiencies and processing efficiencies. The size of the rigid/flex should be minimized as much as possible to minimize the manufacturing costs.

65.2.11 Manufacturing of High Density Multi-Layer Rigid/Flex

To satisfy the requirements for high density with low cost for the consumer rigid/flex, several new manufacturing technologies have been introduced.

The buried, or inner, via hole system has been introduced to optimize the space factors of the inner conductor layers as shown in Fig. 65.16. The inner layer circuits with via holes can be produced as the same process of the double-sided flexible circuits. Adhesiveless copper laminates made by the cast process or lamination process are recommended for use in minimizing the smear during the drilling and optimizing the heat resistance during the multiple high temperature processes.

The build-up system with blind via holes has been introduced to generate fine outer layers and utilize the spaces for the via holes as shown in Fig. 65.16. The adhesiveless copper laminates with heat resistant epoxy bonding sheets are also recommended to use for the build-up layers.

FIGURE 65.16 Micro inner vias in multilayer rigid/flexible circuit construction.

Carbon dioxide lasers are used to generate micro blind via holes for the build-up layers. The conformal mask system is effective for optimizing the drillings speed and efficiencies of the laser system. The optimized condition with carbon dioxide laser can generate more than ten-thousand blind via holes per minute. Appropriate cleaning process such as plasma etching should be introduced before the copper plating to ensure high via hole reliabilitiy.

The roll-to-roll process could be available for high productivities in the inner layers if there are no inner via holes. However, it needs sever dimension control to use RTR process for inner layers with via holes.

The similar semi-automatic machines, such as guide hole punchers and punching presses used single- and double-sided, should be used because they have high productivities for the finishing processes. Steel rule dies are effective, and have higher productivity compared to NC routing machines, for the partial and whole blanking, especially for the half cutting of flexible parts of the thin multilayer flex. But the shape of the steel rule dies should be well designed to avoid the cracks or nicks at the boarders of the flexible parts.

Dimensional control is one of the major keys to have a high process yields. Even though dimensional stable laminates are selected, the thin flexible materials will have remarkable shrinkages during the multi-layer rigid/flex's multiple thermal processes of the. The correction factors for the each thermal or chemical process should be measured carefully. Appropriate work sizes should be chosen carefully to optimize the process yields.

CHAPTER 66
SPECIAL CONSTRUCTIONS OF FLEXIBLE CIRCUITS

Dominique K. Numakura
DKN Research, Haverhill, Massachusetts

66.1 INTRODUCTION

Because of thin base layers, the flexible circuits can have many special constructions that are impossible to build by traditional manufacturing processes of the rigid circuit boards. These special constructions generate further functional values for the flexible circuits. This chapter introduces the major constructions in use, along with their manufacturing processes.

66.2 FLYING-LEAD CONSTRUCTION

Flying-lead construction is a special case of the double access flexible circuit. The base layer under the traces are removed and the conductors are completely free from the organic materials. New HDI flexible circuit applications have arisen involving flying-lead constructions. Applications such as wireless suspension of hard disk drives, interposers for chip-scale packaging, and ultrasound probes depend on the reliability of this high-density interconnect. In competitive markets such as these, it is imperative that flying-lead applications offer low cost and high volume. The TAB (Tape Automated Bonding) is one of the major application of the flying lead costructions.

Traditional manufacturing processes used to create flying leads, such as pre-punching or pre-drilling, are not conducive to the quality levels and manufacturing yields required to achieve low cost, and alternative processes are needed, as described later.

66.2.1 Basic Design

A flying-lead structure is composed of a single layer of high-density-pitch copper conductors accessed from both sides of the substrate. Generally, the substrate side of a conductor on a circuit board is not electrically accessible because the conductive foil is typically laminated to the substrate board in sheet form. In the case of a flex circuit, however, the substrate can be

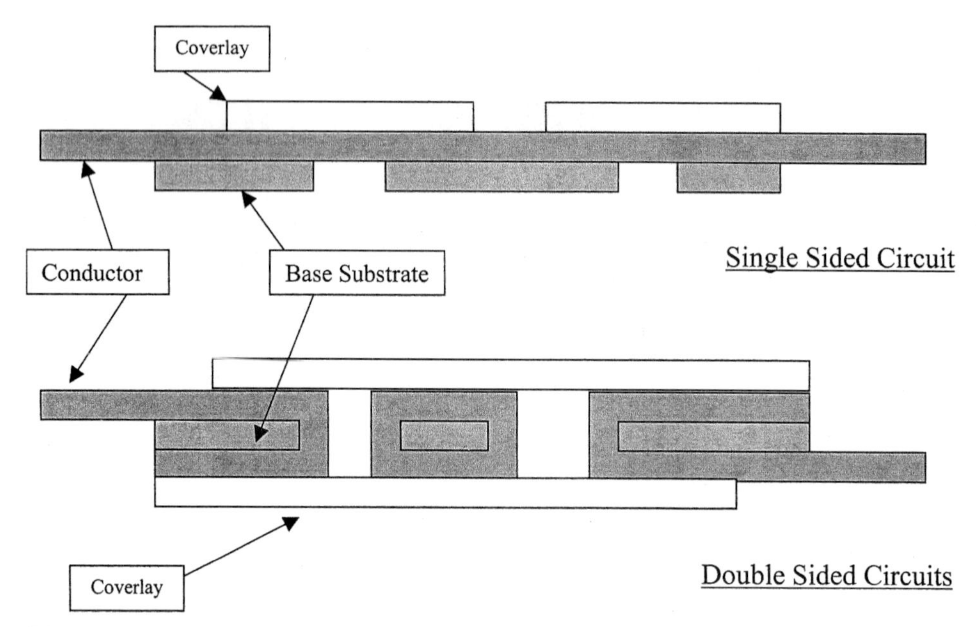

FIGURE 66.1 Basic constructions for Flying-leads for flexible circuits.

removed by various methods., Dual access is created by removing the coverlayer in the same area on both sides.

Figure 66.1 indicates two basic structures of dual-access conductors, in both the single sided and double-sided types of flexible circuits. Figure 66.2 shows actual examples of the fine flying leads built on the single side circuits and double side circuits. Basically, a coverlayer is applied to the side of the conductive layer opposite the substrate.The coverlayer can be processed easily to create the top-side exposure. The flying lead is created when the base substrate is also removed in the same area.

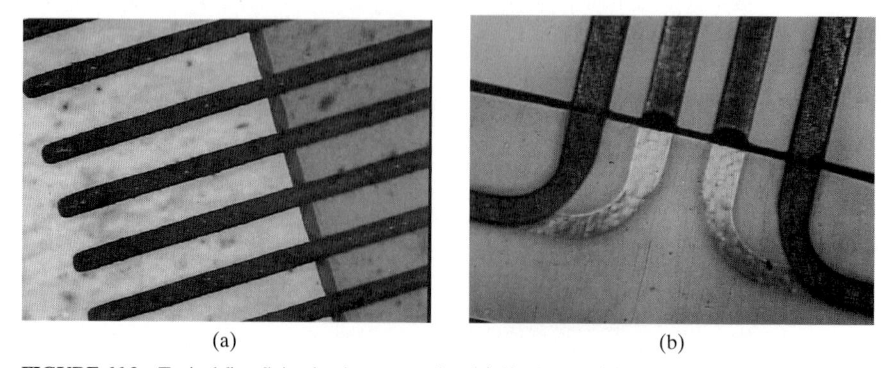

FIGURE 66.2 Typical fine flying leads construction (a) Single side, (b) Double side.

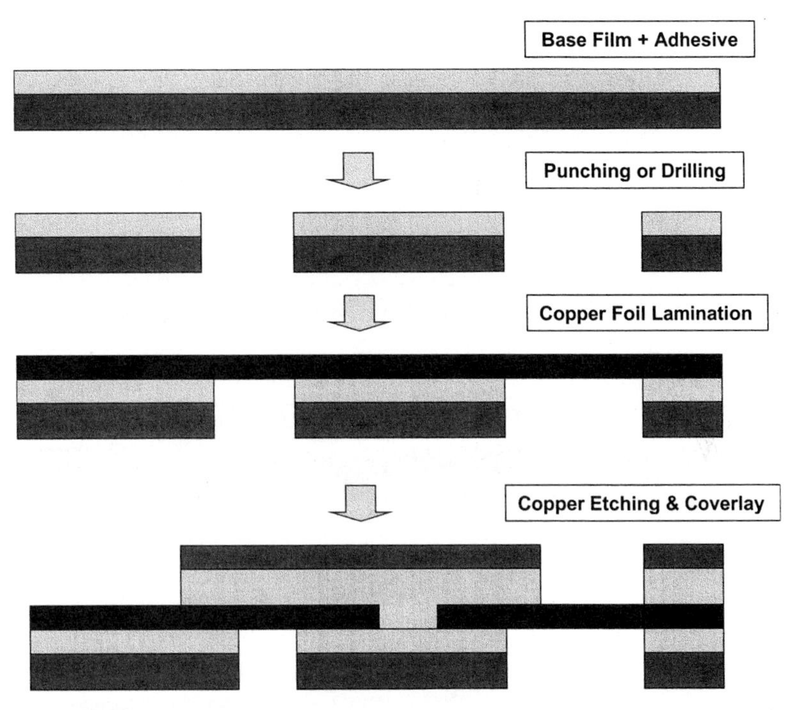

FIGURE 66.3 "Manufactuirng process for flying-lead flexiblse circuit construction using prepunching method."

66.2.2 Manufacturing Processes for Flying Leads

The fragility of flying leads, caused by the lack of dielectric support, is probably the largest disadvantage in the manufacturing process. Proper circuit design and layout is the key to the manufacturing process yield and subsequent cost and physical performance of the circuit. The most conventional manufacturing process for generating a flying-lead construction involves a pre-punching process similar to the device hole process of tape automated bonding (TAB). Figure 66.3 shows this process schematically. There is no proprietary technology or special tooling required for manufacturing the pre-punched window in the base substrate. This process, however, offers the manufacturer very little flexibility and is fairly labor intensive, which makes it difficult to maintain high productivity rates.

The handling of unstable punched films, along with thin, fragile copper foils, makes it difficult to achieve the high manufacturing yields required to support the cost model associated with high-density flexible circuits. These unstable film substrates also make it difficult to maintain dimensional accuracy between the fingers and the window, a key element for HDI flexible circuits. Adhesive flow or squeeze-out from the edge of the punched film along the conductor can also be an issue when dealing with small window exposures. Openings of less than 1.0 mm are difficult to create with high dimensional accuracy and yield.

Alternative technologies have been developed to generate high-density flying-lead designs with higher accuracy. Laser ablation, plasma etching, and chemical etching are just a few of the possibilities and are illustrated schematically in Fig. 66.4. A comparison of small opening capabilities of each of the processes is illustrated in Fig. 66.5, while a comparison of the technical capabilities of the processes is shown in Table 66.1. With these manufacturing technologies, many kinds of flying-lead constructions became possible.

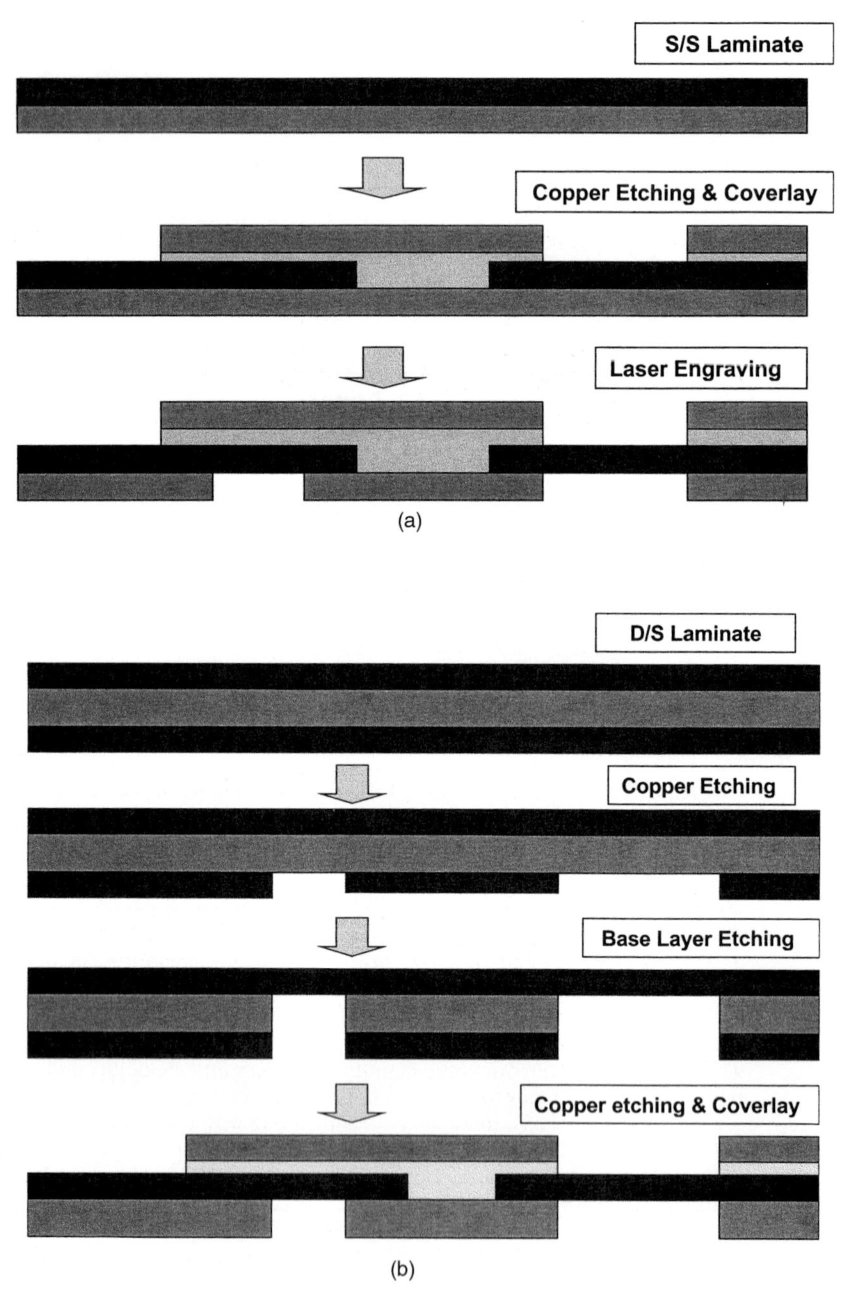

FIGURE 66.4 Manufacturing process for flying lead flexible circuit construction using (a) laser ablartion and (b) plasma etching or chemical etching.

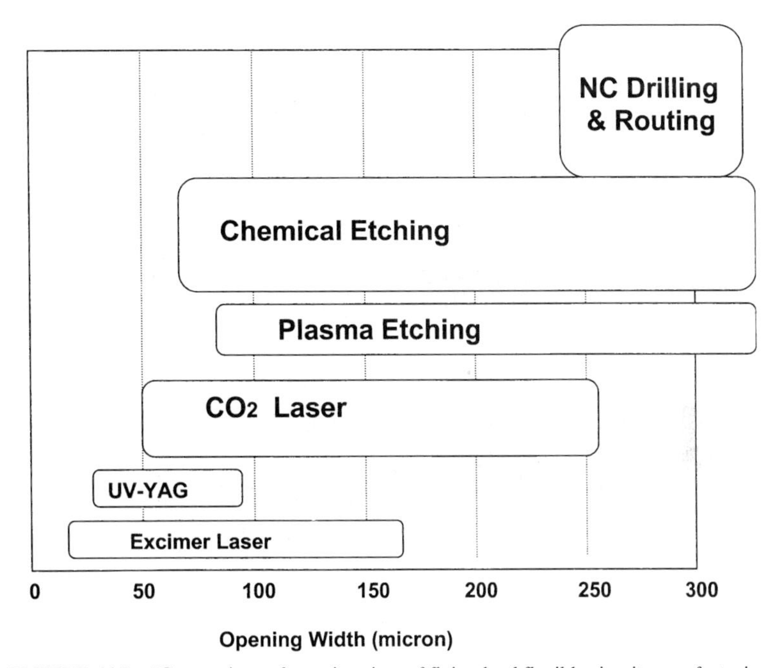

FIGURE 66.5 "Comparison of opening sizes of flying-lead flexible circuit manufacturing processes."

TABLE 66.1 Comparison of Flying-Lead Manufacturing Technologies

	Prepunching (NC routing)	Laser abrasion	Plasma etching	Chemical etching
Minimum slit width (μm)	800	50	100	70
Material availability	Fair	Wide	Wide	Limited
Design flexibility	Limited	Wide	Limited	Limited
Capability for high-density flying leads	Limited	Wide	Fair	Fair
Technical hurdles	Low	Fair	High	High
Registration	Difficult	Fair	Fair	Fair
Damage on flying leads	Serious	Small	Fair	Fair
Process cost	High	High	Low	Low
Investment for large volume	Small	High	High	Medium

66.2.3 Laser Abrasion

The excimer laser process has the capability to generate small openings with clearly defined edges on high-density flexible circuits. The narrow slit openings can be made smaller than 50 μm wide. The excimer process is available for all types of flexible circuit materials with high accuracy and fine openings. A carbon residue, typically removed by a suitable wet cleaning process, is found around the edges of the openings. One of the biggest disadvantages of excimer laser system is its slow speed, which accounts for the high processing cost when dealing with larger openings.

The UV:YAG laser process offers a higher productivity rate than the excimer laser for very small openings. It has the ability to cut both copper foil and organic substrate materials with good quality, which makes it suitable for cutting small openings on a wide range of high-density flexible circuits. Unfortunately, the productivity rate becomes extremely slow when cutting larger openings. Actually, it becomes impractical from a cost standpoint to utilize a UV:YAG laser system for openings larger than 10,000 μm^2.

Carbon dioxide TEA lasers, as well as diamond carbon dioxide lasers, have a much higher productivity rate when producing openings wider than 70 μm. The definition of the opening and quality of the edges, however, is somewhat less desirable than for other laser systems. In order to facilitate surface plating, a suitable cleaning process such as plasma is required to remove the thin residue remaining on the copper surface. A copper mask process helps to improve both quality and speed of the carbon dioxide laser process.

The excimer laser process, like the UV:YAG laser system, minimizes the potential for thermal damage on very thin flying leads. The carbon dioxide laser generates a remarkable amount of heat. The beam power density of the laser must be controlled carefully in order to process 18-μm copper with high-density flying leads narrower than 100 μm wide.

66.2.4 Plasma Etching and Chemical Etching

Plasma and chemical etching processes are other suitable choices for generating high-density flying-lead structures. Although their capabilities are somewhat less than those of laser processing, these processes can provide a lower-cost model of high-volume processing. A copper foil mask is required in order to define the exposure area, and specific process conditions are required to remove organic substrates effectively. These processes are quite effective for single-sided substrates. Double-sided substrates can be processed in this manner given very specific materials and construction.

A big advantage of these processes is the unlimited number of holes or openings that can be processed simultaneously. Performance value is high, while manufacturing costs remain fixed when larger sizes or higher numbers of holes are processed. The plasma process is capable of etching all kinds of organic materials; however, it must be performed in a vacuum chamber.

Capital investment is relatively large for high-volume production. Another issue involves uniform quality over a large working area/panel size in a vacuum chamber of limited size.

Depending on the process conditions, plasma etching creates a wall slope of 30 to 60° at the edge of the openings, and it is difficult to make holes smaller than 100 μm in diameter on a 50-μm-thick polyimide layer.

Chemical etching also offers a low-cost process for creating access openings in high volume on polyimide-based high-density flexible circuits. This process offers the capability of generating openings smaller than 100 μm in diameter or slit width on 25-μm substrates, while providing a very small slope on the wall definition of the opening. The issue in this process is finding the appropriate chemicals and creating the suitable conditions to etch the various flexible substrates.

Sodium hydroxide and potassium hydroxide are popular chemicals for etching Kapton® type polyimide materials. Dangerous chemistries and tightly controlled process conditions are required to etch dimensionally stable polyimide substrates such as Upilex™. Serious consideration must be given to choosing the appropriate circuit design, materials, and manufacturing processes. On the other hand, chemical processing such as this requires no specialized capital equipment, and it is extremely suitable for high-volume roll-to-roll manufacturing systems.

Typical high-density flying leads produced by new processes are shown in Fig. 66.6

66.2.5 Economical Comparison

Each technology has its own cost structure, which is dependent on the conditions. Figure 66.7 shows a comparison from an economic study of the various technologies. This study was conducted on 25-μm-thick polyimide substrate with a 300-mm^2 working size. The opening size is assumed to be 0.1×3 mm, with flying leads at a 150-μm pitch processed on 18-μm-thick copper.

FIGURE 66.6 "An example of a tab-shaped flying llead flexible circuit construction." *(Source: Asahi Fine Technology).*

Plasma and chemical etching processes have little dependency on the size or numbers of openings because they are based on batch processing of the materials in sheets. In contrast, the relative costs of the excimer and carbon dioxide laser process are proportional to the size of the area being processed as well as the opening sizes. This is primarily based on the small size of the laser beam as compared to the size of the opening, and on the need to overlap with each pass of the laser in order to remove the substrate in larger areas. The carbon dioxide

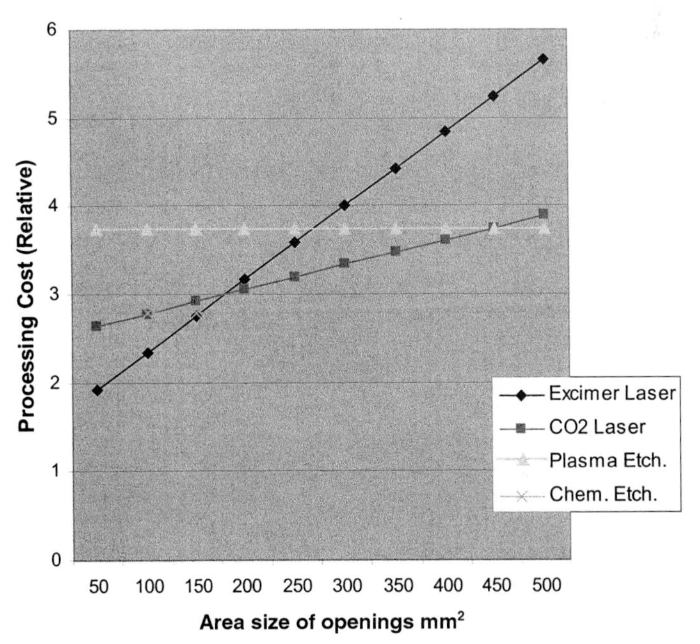

FIGURE 66.7 Cost comparison of flying-lead flexible circuit manufacturing processes.

laser, however, has higher manufacturing costs than the excimer laser process because of the additional chemical processes required to remove the organic residue left on the copper surface, but it has a lower associated manufacturing cost for larger opening sizes due to faster abrasion speeds.

66.3 TAPE AUTOMATED BONDING

TAB has been developed as the special wiring material for IC chips and liquid crystal displays (LCDs). Although it was developed separately from standard flexible circuits, they have similar constructions and they use similar materials, and can be categorized as one kind of flexible circuits.

66.3.1 Basic Concepts

Basic construction of TAB is the same as that of the flying-lead structure of flex circuits except for the web size used in manufacturing and the sprocket holes made for reel-to-reel conveyor systems which give TAB an appearance similar to that of cinema film. A 35-mm-wide web is the most common standard width for TAB. Webs 70 and 155 mm wide are available for larger circuits, nowadays. Because of the small working areas, TAB is good at producing fine traces with flying leads. The reeled circuits are convenient for automation of circuit manufacturing and termination. A short strip circuit made by standard manufacturing processes for flex circuits is also called a TAB circuit because it has the same functions. (Examples of TAB are shown in Fig. 66.8.)

FIGURE 66.8 "Examples of TAB high-density flexible circuits."

66.3.2 Manufacturing Processes

The basic manufacturing flow of single-sided TAB is the same as that of the traditional prepunch manufacturing process of flying-lead construction. Basically, all manufacturing processes of TAB are designed for automatic reel-to-reel equipment. Actual processing steps are slightly

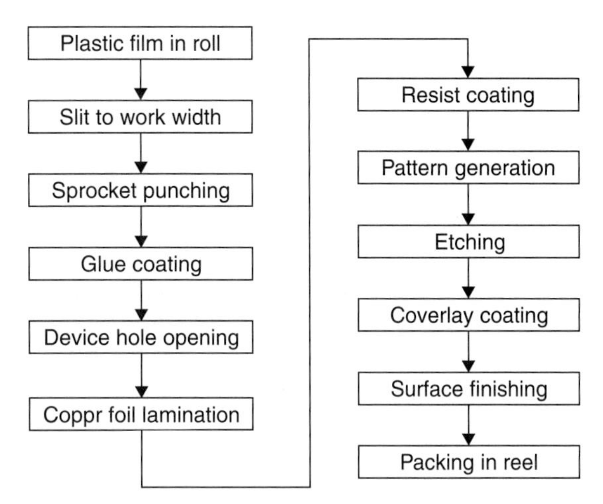

FIGURE 66.9 Manufacturing process flow of TAB.

different as shown in Fig. 66.9. The process starts from a plain film, mostly polyimide film. The first parts of the manufacturing are glue coating and punching of the hole generation. Secondary, a copper foil is laminated and the fine traces are generated by photolithography and chemical etching. Thirdly, final finishing such as cover coating and surface finishing is conducted. All of the processing equipment has to have loader and un-loader by reels,. The machines should be free from manual handling to eliminate the mechanical damages on the fragile fine flying leads. The whole process has to be kept in clean rooms to avoid the contamination for the fine lithography processes.

Because of the narrower material width, TAB constructions have high capability for fine-line generation with good dimensional accuracy. The chemical etching process is also available to generate device holes. The technology can generate 40-μm-pitch traces as the flying leads.

The manufacturing process for the double side TAB with fine flying leads and micro via holes is much more complicated. Very few manufacturers are capable to produce the products with high yield.

66.3.3 Technology movement of TAB

The TAB manufacturing technology has been experiencing a great progress since 1980s. However, it has been facing a turning point to change the basic process drastically for the finer circuitries.

TAB's manufacturing capabilities are very different from traditional flexible circuits and they cover different application area as shown in Fig. 66.10. TAB has been focusing single side fine traces with flying lead constructions. It has advantages for the volume productions of the small size circuitries because of special reel-to-reel manufacturing.

In early usage, only 35 mm wide web was available to have high dimensional stabilities. But the latest RTR manufacturing equipment can manage 155 mm wide materials to generate 40 microns pitch flying leads. Several new processes such as laser drilling and through hole plating have been introduced to produce double side TAB with fine flying leads.

Forty microns pitch of traces is almost the technical limit to manage the fragile fine flying leads for both of manufacturing and termination. Therefore, the technology is moving to COF (Chip on Film), that does not have device holes and flying leads, to have finer traces than 40 microns pitches. The manufacturing process and materials do not have significant differences from the

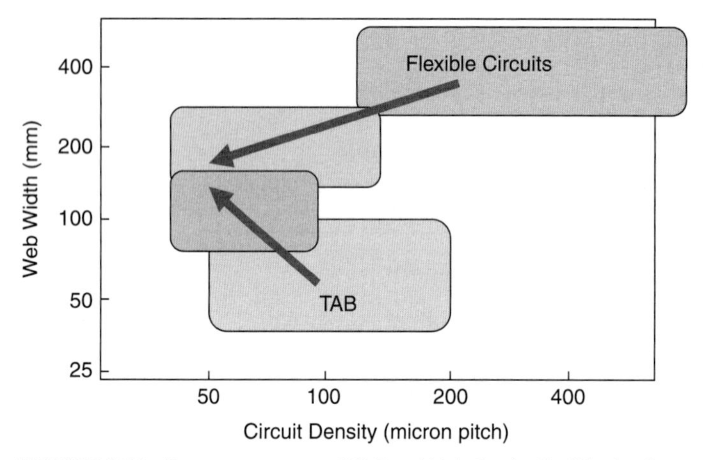

FIGURE 66.10 Recent movement of TAB and high density flexible circuits.

roll-to-roll manufacturing process of the fine flexible circuits. On the other hand, the fine line capabilities of the flexible circuits have been getting better reducing their RTR web size.

TAB and flexible circuits had different capabilities in the past. But they have been moving to the same fine circuit area, and there have not bee remarkable differences in their manufacturing processes as shown in Fig. 66.10.

66.4 MICROBUMP ARRAYS

A variety of shapes and materials for micro bump arrays on flexible circuits have been developed to accommodate specific applications.

66.4.1 Varieties of the Bump Shapes and Applications

There are several choices of places to build micro bumps on flexible circuits, as described in Chap. 64 (see Sec. 64.3.8). The same processes used to build ball grid arrays on rigid circuit boards are appropriate for flexible circuits with flexible photo-imageable solder mask. It is possible to build micro bumps on bare conductors using plating masks. Many micro bumps are built on fine flying leads without solder masks. A valuable construction of a flexible circuit is micro bumps built on the other side of the circuit through the dielectric layer. This can provide a reliable micro bump array because of the stable construction.

Many different shapes have been developed for micro bump arrays. The fusing solder process can make a spherical bump shape easily. Electrical plating of copper, nickel, gold, and combinations of these elements can provide many varieties of bump shapes. Changing the combination ratios and other factors can generate flat disks, flat domes, straight columns, etc. (Fig. 66.11). The best combination can provide 100- to 150-μm-pitch micro bump arrays on 50-μm-thick dielectric layers.

Most of the micro bump array constructions of flexible circuits have been developed for high-density termination (e.g., flip-chip, CSP, test probes, etc.). A suitable design shape and materials should be chosen according to the requirements of the termination. Several examples are shown in Fig. 66.12

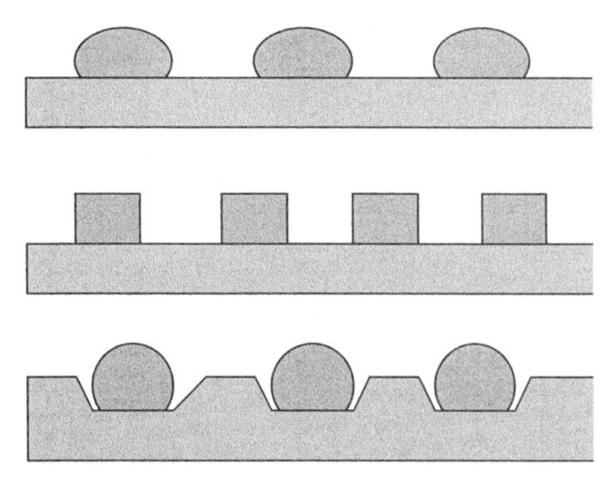

FIGURE 66.11 Microbump arrays on flexible circuits: examples of bump shapes.

(a) (b) (c) (d)

FIGURE 66.12 Various micro bump arrays on flexible circuits (a): Column shape copper bump array on the bare traces (b): Column shape Ni/Au bump array built through covelay (c): Solder ball array built though base substrate(d): Combination of flying leads and solder ball array.

66.4.2 Manufacturing

There are two kinds of manufacturing processes for the micro bump array on flex circuits:

1. The same technology as for ball grid arrays may be used. Solder paste is dispensed on the pads of the circuits with a liquid dispenser or screen printer. The solder paste will form spherical shapes for a refusing process. The process is available for relatively large solder balls.
2. Electroplating processes are used for small solder balls and non-spherical micro bumps.

Suitable combinations of different metal plating can provide wide varieties of micro bumps. There are several factors that determine the shape and size of the bumps. The thickness of the dielectric layer and the size of the opening are the major factors in determining the bump shapes. The bump shapes will be more spherical when the aspect ratio of the opening is large.

The bump shapes will be like flat discs when the aspect ratio is small. An appropriate plating mask will help to build straight, high bumps.

When the micro bumps need to be built through the insulation layers, it requires a suitable hole generation process. Several laser drilling processes are recommended for small sizes but exact dimensions. Plasma etching and chemical etching are recommended for low cost processing with large numbers of the holes.

66.5 THICK-FILM CONDUCTOR FLEX CIRCUITS

A thick-film circuit is a type of additive manufacturing process. If an organic resin matrix is used, the conductor can be flexible, and this process has been applied in flexible circuits.

66.5.1 Materials

A silver paste conductor with an acrylic resin matrix is the most popular conductor material as the major thick-film conductor material for flexible circuits. It can provide very flexible conductor layers via a simple screen-printing process. Copper-based and carbon-based paste materials have been developed as the low-cost materials, but their conductivity is very low and unstable, and therefore their applicable areas are limited.

The screen printing process with conductive paste provide low cost solutions for the large size flexible circuit employing economical plastic films such as PET films.

66.5.2 Applications

A silver paste thick-film flexible circuit built on polyester film can provide the lowest-cost solution for large circuits. Unfortunately, the conductivity of the traces is much lower than for copper foil conductors, and this material is not available for power circuits or signal layers of high-speed circuits. Standard soldering is not available because of the acrylic-based matrix.

The largest application of thick-film-based flexible circuits is in membrane switches, which need large circuit spaces but do not carry large currents. The majority of membrane switches are covered by silver paste thick-film flexible circuits. Major switching circuits of keyboard switches are also covered by the same constructions. Most of the key board circuits are covered by the silver paste base membrane switches. Most of the touch panel switches designed for micro waves, office appliances and medical equipment also use PET based membrane switches. Several examples are shown in Fig. 66.13.

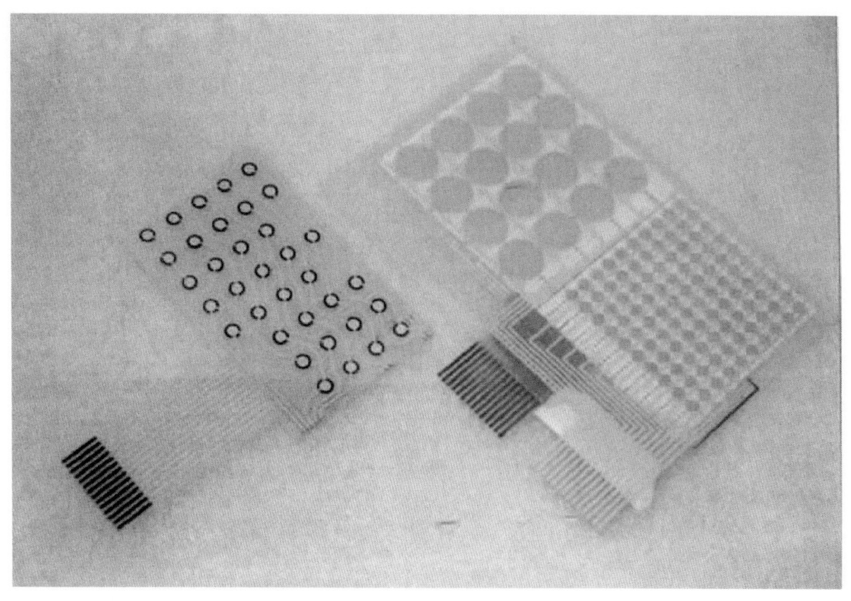

FIGURE 66.13 Membrane switch made by thick film process.

PET based thick film flexible circuits have been consumed for the touch panel switches of the electronic terminals such as PDA because of their transparency.

66.5.3 Manufacturing Process

It is a very simple process to build a thick-film flexible circuit. There are two steps:

1. Screen printing on a flexible film followed by baking.
2. Blanking.

This still requires that the process conditions be controlled carefully. The basic properties of these circuits change significantly depending on these conditions. A pre-baking step is recommended for polyester film substrate to eliminate thermal shrinkage during baking for printed conductor paste. It is not difficult to apply the roll-to-roll manufacturing system for large volume because of the simplicity of the processes.

The latest technologies of the printers and screens are able to produce 10 to 15 micron line space with low cost. The electrical properties and trace densities are very depend on the conductor paste. A combination with screen printable nano conductive paste will provide fine conductive traces by a simple manufacturing process.

66.6 SHIELDING OF THE FLEXIBLE CABLES

When the flexible circuits are used as the cables for the high-speed telecommunications, appropriate shielding systems are required. Generally, the cable will be bent frequently, therefore the shielding layers should also have high flexibilities.

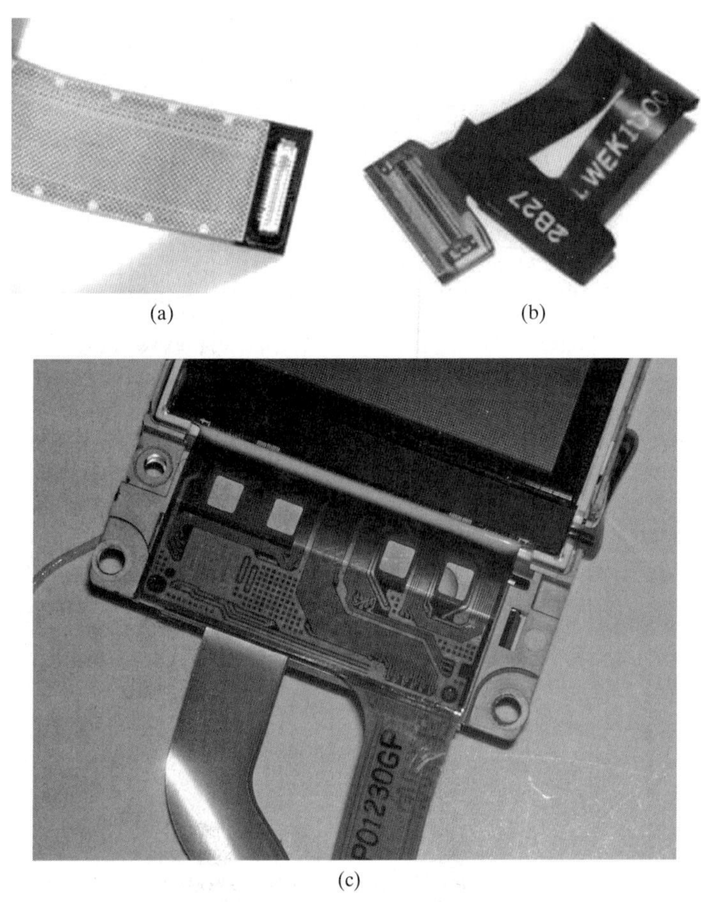

FIGURE 66.14 Shielding of the flexible cables (a): Screen-printing of silver paste (b): Screen-printing of carbon paste (c): Lamination of thin aluminum foil.

Several materials and constructions are available for the shielding of the flexible cables as shown in Fig. 66.14. Screen-printing of silver paste provides both high shielding effect and high flexibility. Screen-printing of carbon paste provides low cost solution. But its shielding effects are limited. Thermal lamination of a thin aluminum foil with thin flexible adhesive provides a reliable solution for both shielding efficiencies and flexibilities.

66.7 FUNCTIONAL FLEXIBLE CIRCUITS

Various new processes have been developed to build functional elements and devices on the flexible substrates. They have been generating more functions than wiring or assembling board of the discrete components. They are the similar ideas as embedded passive technologies of multilayer boards, but they generate more values with flexible substrates. Table 66.2 shows several material examples and applications.

TABLE 66.2 Functional materials for flexible circuits

Materials	Functions
NiCr, W	Embedded resistors
	Heater
	Mechanical sensors
BaTiO	Embedded capacitance
Tin oxide	Transparent electrode
Poly silicone	Embedded active devices
	Transistors and diodes
	Sensor devices
Organic EL	Flexible displays

Figure 66.15 shows examples of sputtered nickel/chromium alloy. The first one is embedded resistors and the second one is a small strain gage built on the flexible substates.

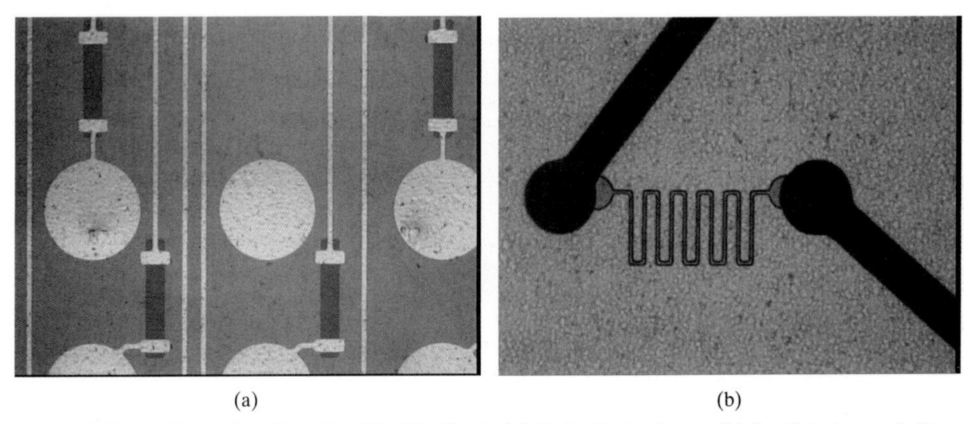

(a) (b)

FIGURE 66.15 "Examples of functional flexible circuits (a): Embedded resistance (b): Small strain gage built on flexible substrate." *(Source: MicroConnex).*

CHAPTER 67
QUALITY ASSURANCE OF FLEXIBLE CIRCUITS

Dominique K. Numakura
DKN Research, Haverhill, Massachusetts

67.1 INTRODUCTION

Because of the different design concepts and materials, there are several significant differences in quality assurance for flex circuits vs. that for typical rigid circuit boards. In addition, special test methods are required to guarantee the finished quality of high-density flex circuits, and new inspection technologies have been developed for this purpose.

67.2 BASIC CONCEPTS IN FLEXIBLE CIRCUIT QUALITY ASSURANCE

The basic concept of quality assurance for flexible circuits is the same as that for rigid boards, in the sense that any critical defect must be eliminated before shipping. This includes failures due to opens and shorts; dimensional degradation of pad arrays, serious defects on conductors, substrates, coverlay, and so on. These defects need to be inspected for each circuit, whether it is flexible or rigid. Technically, however, there are several differences in quality assurance for flexible circuits because of their additional functions, such as dynamic flexing capability.

A common issue is damage incurred during the manufacturing process because of the fragile materials. Even though the fine traces are inspected by a high-resolution automatic optical inspection (AOI) system during processing, it is possible that subsequent processes may cause new damage to the inspected circuits. This is a special issue for flying leads, which have no mechanical support under the traces. Therefore, a final inspection is necessary at the end of the manufacturing process.

The basic quality assurance concept for high-density flexible circuits is the same as for other circuit types. However, a difference for high-density flexible circuits is the acceptable defect size, which is one order smaller than that for traditional flexible circuits, necessitating higher inspection capabilities. Furthermore, new high-density flexible circuits contain additional structures such as flying leads and microbump arrays that require additional inspection capabilities for reliable termination. Exact 3-D accuracy and uniform surface conditions are required. Dimensional allowances are smaller than 0.3 percent, and sometimes 2-μm accuracy is required.

This is true for trace pitches, but it is also true for linewidths and spaces. Three-dimensional measurements are required to guarantee the quality of microbump arrays. Usually, it takes time to measure exact dimensions for each circuit, and this is one of the major costs of the circuits when the specification is very tight compared to the process capability.

There have been several basic technical issues regarding quality assurance for high-density flex circuits, as listed in Table 67.1. Because of the fine circuits and the severe acceptable level of the defects, low-magnification scopes and low-resolution AOI do not work effectively. Magnification of 20 × to 50 × is required for visual inspections of fine circuits, and higher resolutions are required for AOI systems. Low contrast between copper and polyimide base creates another difficulty for traditional AOI equipment. It is not easy for common AOI systems to distinguish the copper traces from the circuits on the other side of a double-sided circuit built on adhesiveless substrate, because the base layer is very transparent and the sensor detects both sides of the circuits as one.

TABLE 67.1 Key Inspection Issues for High-Density Flex Circuit

Item	Issues
Finer traces	Resolution of AOI
Micro–blind via holes	Low contrast for AOI
	Resolution of AOI
Unstable materials	Dimensional accuracy
Photoimageable coverlay	Low contrast for AOI
Fragile traces	Damage during processes
	Damage from test probe pins
Fragile flying leads	Damage after AOI
Microbump arrays	3-D constructions
Cosmetic defects	Serious failures caused by small defects

67.3 AUTOMATIC OPTICAL INSPECTION SYSTEMS

AOI is the most valuable inspection equipment for detecting dimensional defects in the traces and spaces of the printed circuits, especially for high-density traces. It's also very useful for the detection of cosmetic defects on the surfaces. Further sensitivities are required for AOI systems to detect small defects on coverlay, flying leads, microbumps, narrow space perimeters, and so on. These small defects have low contrast, different kinds of brightness, and different defect shapes, among other things. Traditional AOI systems are usually not appropriate for inspection in these circumstances.

Traditional AOI systems were designed for rigid circuit boards, and they were originally only good for the detection of copper trace qualities. They could work for 50-μm-line/space fine traces, but their inspection speed could be very slow. Therefore, remarkable progress has been made in the resolution and productivity of the newer AOI systems; and as a result, systems can now detect 1.5-μm defects on 15-μm traces. Also, the systems have other capabilities for detecting various defects on polyimide films and coverlay films. A careful review of AOI capabilities, compared to the specified design requirements, should be done prior to committing to a particular equipment.

67.4 DIMENSIONAL MEASUREMENTS

It seems to be an impossibility to require accurate dimensions on an unstable, thin, flexible material. However, high-density interconnects in particular demand very high accuracy in the termination area of the circuits, even though the dimensional allowance is smaller than the dimensional stability of flexible substrate materials. Therefore, high-accuracy 2-D or 3-D dimension measurement technologies are required.

There has been continuous progress in dimension-measuring equipment. The newest 3-D measurement equipment has a resolution higher than 0.2-μm and accuracy higher than 1 μm, with statistical data in a 200-mm square 100-mm high. It has automatic measuring capabilities to detect each measuring point, producing statistical data. It can provide standard deviation data for each point with *Cpk* value.

Usually, there are several critical points that require exact dimensional accuracy in a flexible circuit. They are guide marks and termination areas, and customers ask exact tolerance for each dimension. The stable dimensions, such as punched perimeters processed by blanking dies, can be represented with the first article. However, unstable dimensions such as line width and pad pitch should be checked for each circuit. A statistical method is available. The measurement can be conducted by sampling when *Cpk* is larger than 1.3, but 100 percent measurement should be done when *Cpk* is smaller than 1.0. See MIL-STD-414 for help in reducing the number of samples.

67.5 ELECTRICAL TESTS

Electrical tests of high-density flex circuits could be one of the largest issues. Electrical open/short performance must be guaranteed as the minimum function of flex circuits. A traditional contact electrical testing method presents two basic barriers in inspection of high-density flex circuits. The first is a geometric limit of probe pins. The costs of test fixtures are in inverse proportion to pitches of pin arrays, and 150-μm-pitch arrays could be the physical limit of pin probes. The second problem is physical damage caused by probe pins on fine traces, especially on fragile flying leads. A bus bar trace for a plating process is a problem in detecting shorts in circuits with the traditional pin-contact electrical-test method. A new additional electrical test method is required.

A noncontact electrical test system can detect all open/short failures for 50-μm-pitch fine traces without pin contacts. It does not make any dents or cause damage on the fine traces. It works for fragile, fine flying leads because it is a non-contact inspection system. The basic principle of the method is illustrated in Fig. 67.1.

An electric field sensor or a magnetic field sensor prepared for each circuit will detect all opens and shorts in branched circuits. Another big advantage of the non-contact test system is that it works for general circuits that have a common bus bar for electrical plating. The circuits must be isolated to conduct the electrical short test when traditional contact test methods are being used.

A non-contact electrical test system can inspect more than 5000 traces in a few seconds, eliminating physical damage to the circuits. It is also possible to use it in reel-to-reel processes. On the other hand, it still requires electrical contact on the other side of the fine traces, and therefore appropriate extra leads are required for pin or conductive rubber contacts.

67.6 INSPECTION SEQUENCE

Table 67.2 shows major quality assurance items for the final high-density flexible circuit products.

It is difficult to check all items after each manufacturing process. Therefore, an appropriate inspection sequence in the total manufacturing flow should be designed.

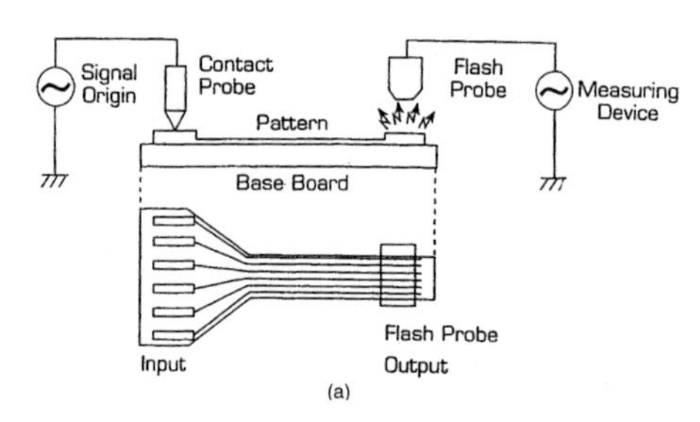

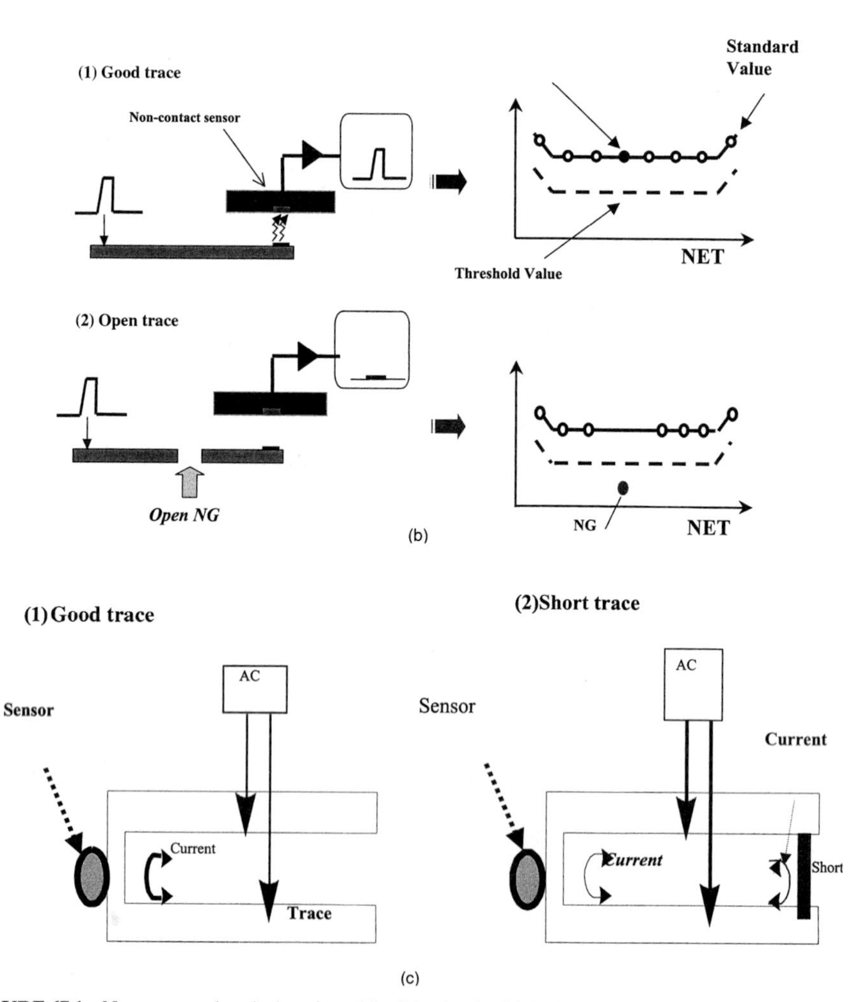

FIGURE 67.1 Noncontact electrical testing of flexible circuits (a) General principle of noncontact electrical testing showing contact only at the input. (b) Principle of noncontact testing for open circuits where NG means "no go" (rejected). (c) Principle of noncontact testing for short-circuits, using a bus bar circuit. *(Source:OHT.)*

TABLE 67.2 Quality Assurance Items for High-Density Flex Circuits

Item	Requirements
Physical properties	Sampling tests of raw materials
	Sampling tests of coupons
Opens/shorts	100% electrical inspection
Reliability of micro via holes	Cross sections obtained by sampling
	Daisy-chain test by coupon
Dimensional accuracy	Sampling by MIL-414 or 100%
Trace quality	100% inspection
Coverlay opening	Visual inspection, 100% or by sampling
Flying leads	Sampling test for dimensions
	100% for straightness
Cosmetic defects	100% visual inspection, audit by MIL-105

On the other hand, it is a basic concept that failures should be eliminated as early as possible to ensure a high process yield, from a quality control standpoint, especially for long manufacturing processes. Figure 67.2 shows the general manufacturing flow of high-density flexible circuits, and a recommended inspection procedure.

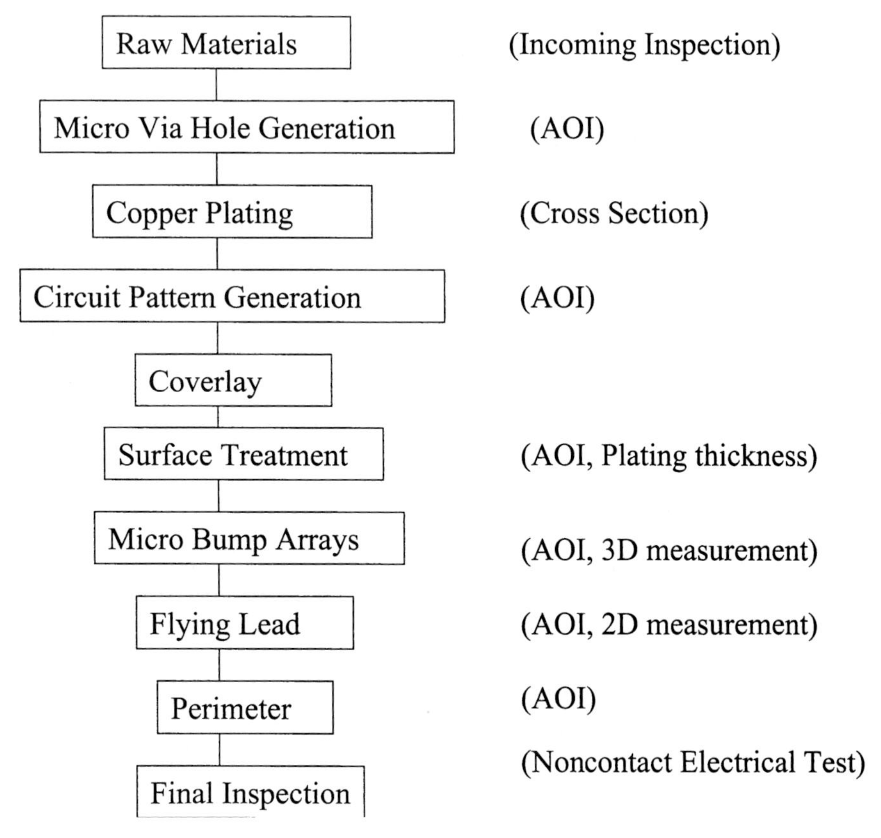

FIGURE 67.2 Manufacturing process for HDI flexible circuits and key inspection steps.

67.7 RAW MATERIALS

The basic properties of the raw materials should be guaranteed by material vendors; however, it is still necessary to conduct incoming inspections by circuit manufacturers. There is less experience with the new materials developed for high-density flex circuits than with traditional materials. Physical properties and cosmetic defects should be inspected by sampling.

67.8 FLEXIBLE CIRCUIT FEATURE INSPECTION

Flexible circuit features that require special consideration include:

- Micro-size holes
- Trace quality
- Coverlay and surface treatments

67.8.1 Micro-Size Hole Inspection

Several new technologies have been developed to generate blind holes smaller than 100 μm.

These include lasers, plasma etching, and chemical etching. Unfortunately, the reliability of micro-hole generation processes has not been established exactly, and a suitable inspection system should be used. Traditional hole checkers do not work on blind holes. An AOI system is sensitive to the low contrast of copper foil surfaces.

There is no perfect, nondestructive test method for evaluating the reliability of micro via holes. It is not easy to create suitable conditions for uniform plating of copper in very small holes. As a result, cross-sectional inspections should be done periodically. Also, a daisy-chain coupon arranged beside the circuits in the work should be tested periodically.

67.8.2 Trace Quality

As there is no perfect circuit generation process for fine traces, a high-resolution AOI system is required for high-density flex circuits. All serious defects larger than one-fourth of conductor widths or spaces must be detected by AOI. This means the AOI system must have a resolution higher than 5 μm for 25-μm traces. An appropriate sensitivity should be set in the AOI system to distinguish the copper traces from the polyimide base film and the traces on the other side of the film. It makes sense to apply AOI before etching to avoid a material loss. An algorithm of PC-Micro II can reduce the inspection time by applying high-resolution capabilities focusing on specific areas.

67.8.3 Coverlay and Surface Treatments

Coverlays and surface treatments are very difficult for current AOI systems to inspect because of the small optical contrast between coverlay materials and conductors or substrate materials. It may be better to apply AOI after a surface treatment, which will increase the contrast.

Gold plating and tin plating create a clear contrast, and a small defect can be detected by AOI. The thickness and quality of surface treatments should be inspected immediately after the processes.

67.8.4 Microbump Arrays

Both cosmetic inspections and dimensional measurements are required for microbump arrays.

Also, new ideas for inspecting 3-D constructions are needed. The uniformity of the bumps and cosmetic defects could be inspected by a suitable AOI system. The 3-D size of the bump and array pitches should be measured by a high-accuracy 3-D measuring system.

67.8.5 Flying Leads

The appropriate places for inspection of flying leads depend on the manufacturing process of the circuits. Flying leads should be inspected just after they are generated by copper etching or by engraving of the substrate. Both AOI inspection and dimensional measurement are required.

Usually, flying leads have a severe dimensional allowance for both line-width and pitches for reliable termination.

67.8.6 Electrical Testing (Opens/Shorts)

The electrical function is the basic and most important performance of the circuit. Unfortunately, the reliability of manufacturing processes for high-density flex circuits has not been established perfectly. There are many possibilities to cause further damage to the traces after the etching process and the AOI process, and therefore a 100 percent electrical test is necessary at the end of manufacturing to guarantee the circuit's performance. This is similar in objective to rigid, bare-board testing, and Chaps. 36 to 38 provide a detailed discussion of these issues.

Certainly, final electrical tests must not cause any additional damage to the fragile, fine traces.

The best solution is a non-contact electrical test. An appropriate design of non-contact sensors combining flash probe, flash shock, common flash, and surface probe of OHT can reduce the total inspection cost. A suitable circuit pattern should be arranged at the beginning. The inspection should be conducted before blanking for handling convenience when final circuits are very tiny.

67.8.7 Physical Properties (Final Properties)

The properties of raw materials do not represent final properties of the circuits. They undergo change during the manufacturing process, and therefore final properties should be measured at the end of the manufacturing process. Generally, most of electrical, mechanical, and chemical properties are different from those of raw materials. Critical properties such as flexing endurance, insulation resistance, and heat resistance should be measured carefully.

IPC-TM-650 indicates a flexing endurance test method based on a flexing mode, as shown in Fig. 67.3. Unfortunately, the test method takes a very long time (sometimes more than a month) to give a final result by conductor failure. JIS-C5016 indicates an alternative test method called the MIT test, as shown in Fig. 67.4. This method does not require more than 1 hour for one test. It is not specified by a public industrial standard, however, a new flexing endurance test method for the spiral wiring illustrated in Fig. 67.5 has been required by the mobile equipment manufacturers such as notebook computers or clamshell type cellular phones.

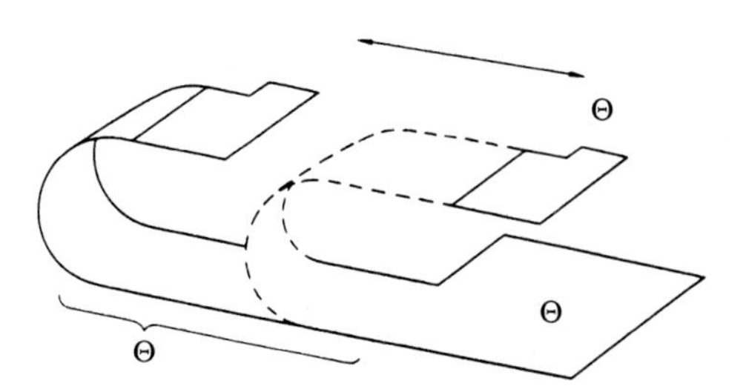

FIGURE 67.3 IPC test method for flexing endurance.

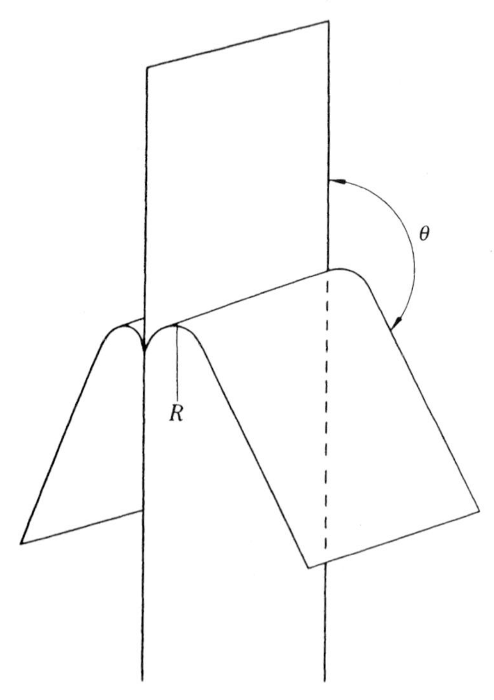

FIGURE 67.4 JIS-C5016 MIT test method for flexing endurance. R is the radius of the bend during the test symbol for "theta" is the angle from vertical.

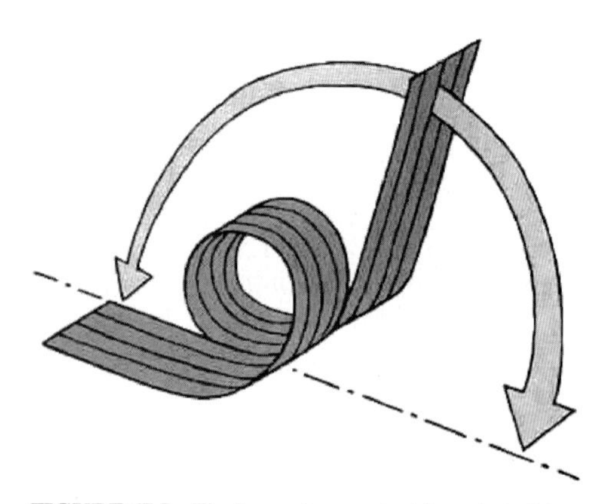

FIGURE 67.5 Flexing endurance test for spiral wiring.

67.8.8 Cosmetic Quality

There is no perfect automatic inspection system that can detect all cosmetic defects. Visual inspection at high magnification is the most reliable inspection method of eliminating all kinds of defects in high-density flex circuits. A stereomicroscope with high magnification helps the inspection procedure, but inspectors need training with exact criteria sample books. A final audit to confirm the level of the inspections is also recommended. MIL-STD-105 could help to reduce number of samples for the audit.

67.9 STANDARDS AND SPECIFICATIONS FOR FLEXIBLE CIRCUITS

Most of the major standards associations have generated new standards or upgraded existing ones for flexible circuits. Major flexible circuit standards and relating standards are listed in the following sections.

67.9.1 IEC

- IEC 249-2-15: Flexible copper-clad polyimide film of defined flammability grade
- IEC 326-7:Specification for single- and double-sided flexible printed boards without through-hole connections

- IEC 326-8: Specification for single- and double-sided flexible printed boards with through hole connections
- IEC 326-9: Specification for multilayer flexible printed boards with through-hole connections
- IEC 326-10: Specification for flex/rigid double-sided printed boards with through-hole connections
- IEC 326-11: Specification for flex/rigid multilayer printed boards with through-hole connections

67.9.2 IPC

- IPC-FC-231C: Flexible bare dielectrics for use in flexible printed wiring
- IPC-FC-232C: Specification for adhesive coated dielectric films for use as cover sheets for flexible printed wiring
- IPC-FC-234: Pressure sensitive adhesives assembly guidelines for single-sided and double-sided printed circuits
- IPC-FC-241C: Flexible metal-clad dielectrics for use in fabrication of flexible printed wiring
- IPC-RF-245: Performance specification for rigid-flex printed boards
- IPC-D-249: Design standard for flexible single and double-sided printed boards
- IPC-FC-250A: Specification for single and double-sided flexible printed wiring
- IPC-FA-251: Guidelines for assembly of single and double-sided flex circuits
- IPC-6013A: Qualification and performance specification for flexible printed boards

67.9.3 JIS

- JIS-C 5016: Test methods for flexible printed wiring boards
- JIS-C 5017: Flexible printed wiring boards, single-sided and double-sided
- JIS-C 6471: Test methods for copper-clad laminates of flexible printed wiring boards
- JIS-C 6472: Copper-clad laminates for flexible printed wiring boards (polyester film, polyimide film)

67.9.4 JPCA

- JPCA-BM 01: Copper-clad laminates of flexible printed circuits, polyester films, and polyimide films
- JPCA-FC 01: Single-sided flexible printed circuits
- JPCA-FC 02: Double-sided flexible printed circuits
- JPCA-FC 03: Visual defect criteria for flexible printed circuits

67.9.5 MIL

- MIL-STD-2118: Flexible and rigid-flex printed-wiring for electronic equipment
- MIL-C-28809: Circuit card assemblies, rigid, flexible and rigid-flex
- MIL-P-50884C: Printed wiring, flexible and rigid-flex

67.9.6 UL

- UL796F: Flexible materials interconnect constructions

APPENDIX
SUMMARY OF KEY COMPONENT, MATERIAL, PROCESS, AND DESIGN STANDARDS

David W. Bergman
(IPC)-Association Connecting Electronics Industries, Bannockburn, Illinois

The following documents center on surface-mount technology. These documents have been developed by standards organizations in the United States and internationally. The letters in each document designation indicate the organization that has responsibility for the document:

ECA: Documents prepared by the Electronic Components, Assemblies, Equipment and Supplies Association (ECA) of the Electronic Industries Alliance (EIA).

DoD: Documents prepared by the Department of Defense.

IEC: Documents prepared by the IEC — International Electrotechnical Commission.

IECQ: Documents prepared for Quality Assessment System for Electronic Components supported through the IEC.

IPC: Documents prepared by the IPC—Association Connecting Electronics Industries.

JEDEC: Documents of the Solid State Technology Association of the Electronic Industries Alliance (EIA).

JESD: Standards prepared by JEDEC.

J-STD: Joint industry standards (standards followed by more than one association).

MIL: Documents prepared by the military.

SMC: Documents published by the Surface Mount Council.

SURFACE MOUNT COUNCIL PUBLICATIONS

SMC-TR-001: An Introduction to Tape Automated Bonding Fine Pitch Technology

SMC-WP-001: Soldering Capability

SMC-WP-002: An Assessment of the Use of Lead in Electronic Assemblies

SMC-WP-003: Chip Mounting Technology

SMC-WP-004: Design for Excellence

SMC-WP-005: PWB Surface Finishes

COMPONENTS, GENERAL

EIA-186-E: Standard Test Methods for Passive Electronic Component Parts—General Instructions and Index

EIA-481-2-A: Embossed Carrier Taping of Surface Mount Components for Automatic Handling (16, 24, 32, 44, and 56 mm)

EIA-944: Surface Mount Ferrite Chip Bead Qualification Specification

EIA-945: Surface Mount Inductor Qualification Specification

EIA/IS-47: Contact Termination Finish Standard for Surface Mount Devices

EIA/IS-760: Surface Mount Wirewound Inductor Qualification Specification

EIA-PDP-100: Registered and Standard Mechanical Outlines for Electronic Parts

EIA-JEP-95: JEDEC Registered and Standard Outlines for Solid State and Related Products

EIA-JESD30B: Descriptive Designation System for Semiconductor Device Packages

JESD-625A: Requirements for Handling Electrostatic-Discharge-Sensitive (ESDS) Devices

PASSIVE COMPONENTS

Capacitors

ECA SP 4984: Surface Mount Aluminum Electrolytic Chip Capacitor with Polymer Cathode

EIA-198-E: Ceramic Dielectric Capacitors Classes I, II, III, and IV

EIA-469-C: Standard Test Method for Destructive Physical Analysis of High Reliability Ceramic Monolithic Capacitors

EIA-479A: Film-Paper, Film Dielectric Capacitors for 50/60 Hz Voltage Doubler Power Supplies EIA-535: Series of Detail Specifications on Fixed Tantalum Capacitors

EIA-595: Visual and Mechanical Inspection Multilayer Ceramic Chip Capacitors

EIA/IS-35: Two-Pin Dual In-Line Capacitors

EIA/IS-36: Chip Capacitors, Multi-Layer (Ceramic Dielectric)

EIA/IS-37: Multiple Layer High Voltage Capacitors (Radial Lead Chip Capacitors)

EIA/IS-717: Surface Mount Tantalum Capacitor Qualification Specification

IEC-384-3: Sectional Specification, Tantalum Chip Capacitors

IEC-384-10: Sectional Specification, Fixed Multilayer Ceramic Chip Capacitors

IECQ Draft: Blank Detail Specification, Fixed Multilayer Ceramic Chip Capacitors

IECQ-PQC-31: Sectional Specification, Fixed Tantalum Chip Capacitors with Solid Electrolyte

IECQ-PQC-32: Blank Detail Specification, Fixed Tantalum Chip Capacitor

Resistors

EIA/ECA-575 Resistors, Rectangular, Surface Mount General Purpose

EIA/ECA-576 Resistors, Rectangular, Surface Mount Precision

EIA/IS-30 Resistors, Rectangular Surface Mount Thick Film

EIA/IS-34: Leaded Surface Mount Resistor Networks, Fixed Film

ECA SP 5052 Resistors, Rectangular, Surface Mount, General Purpose

ECA SP 5053 Resistors, Rectangular, Surface Mount, Precision

ACTIVE COMPONENTS

EIA-JEP-95: JEDEC Registered and Standard Outlines for Solid State and Related Products

EIA-JESD11: Chip Carrier Pinouts Standardized for CMOS 4000, HC, and HCT Series of Logic Circuits

EIA-JESD21-C: Configurations for Solid State Memories

EIA-JESD22-B: Test Methods and Procedures for Solid State Devices Used in Transportation/ Automotive Applications (series format—consists of over 16 different test procedure documents)

EIA-JESD-26A: General Specification for Plastic Encapsulated Microcircuits for Use in Rugged Applications

EIA-JESD-30: Descriptive Designation System for Semiconductor Device Packages

EIA SP 4251 Detail Specification for Production Ball Grid Array (BGA), High Pin Count (1089 Pins and Greater) Socket for Use in Electronic Equipment

ELECTROMECHANICAL COMPONENTS

Connectors

EIA-364-C: Electrical Connector/Socket Test Procedures Including Environmental Classifications

EIA/IS-47: Contact Termination Finish Standard for Surface Mount Devices

IPC-CI-408: Design and Application Guidelines for the Use of Solderless Surface Mount Connectors

Sockets

EIA-5400000: Generic Specification for Sockets for Use in Electronic Equipment

EIA-540H000: Sectional Specification for Burn-In Sockets Used with Ball Grid Array Devices for Use in Electronic Equipment

Switches

EIA-448-23: Surface Mountable Switches, Qualification Test

EIA-520aaad: Detail Specification for Single Pole, Single Throw, Surface Mount Pushbutton Switches

EIA-520daad: Detail Specification for Single Pole, Double Throw, Surface Mount Subminiature Size Toggle Switches

EIA-5200000-C: Generic Specification for Special-Use Electromechanical Switches of Certified Quality

EIA-SP-2186: Detail Specification for Single Pole, Double Throw, Surface Mount Subminiature Size Toggle Switches

EIA-SP-3162: Detail Specification for Single Pole, Single Throw, Surface Mount Pushbutton Switches

IECQ-PQC-41: Detail Specification, Dual-in-Line Switch, Surface Mountable, Slide Actuated US0003

Printed Boards

IPC-6011: General Performance Requirements for Printed Boards

IPC-6012: Performance Specification for Rigid Printed Boards

IPC-6013: Performance Specification for Flexible Printed Boards

IPC-6015: Performance Specification for Organic Multichip Module Structures (MCM-L)

IPC-6016: Qualification and Performance Specification for High Density Interconnect (HDI) Layers or Boards

IPC-6018: Microwave End Product Board Inspection and Test

IPC/JPCA-6202: Performance Guide Manual for Single- and Double-Sided Flexible Printed Wiring Boards

MIL-PRF-31032: Printed Circuit Board/Printed Wiring Board Manufacturing, General Specification

MIL-P-50884: Military Specification for Printed Wiring, Flexible, and Rigid Flex

MIL-PRF-55110: Performance Specification, Printed Wiring Boards, Rigid General Specification

Materials

IPC-3406: Guidelines for Electrically Conductive Surface Mount Adhesive

IPC-3407: General Requirements for Isotropically Conductive Adhesives

IPC-3408: General Requirements for Anisotropically Conductive Adhesive Films

IPC-4101: Specification for Base Materials for Rigid and Multilayer Boards

IPC-4103: Specification for Plastic Substrates, Clad or Unclad, for High Speed/High Frequency Interconnection

IPC-4110: Specification and Characterization Methods for Nonwoven Cellulose Based Paper for Printed Boards

IPC-4121: Guidelines for Selecting Core Constructions for Multilayer Printed Wiring Board Applications

IPC-4130: Specification and Characterization Methods for Nonwoven "E" Glass Mat

IPC-4202: Flexible Base Dielectrics for Use in Flexible Printed Wiring

IPC-4203: Adhesive Coated Dielectric Films for Use as Cover Sheets

IPC-4204: Flexible Metal-Clad Dielectrics for Use in Fabrication of Flexible Printed Circuitry

IPC-4411: Specification and Characterization Methods for Nonwoven Para-Aramid Reinforcement

IPC-4412: Specification for Finished Fabric Woven form "E" Glass for Printed Boards

IPC-4552: Specification for Electroless Nickel/Immersion Gold (ENIG) Plating for Printed Circuit Boards

IPC-4553: Specification for Immersion Silver Plating for Printed Circuit Boards

IPC-4562: Metal Foil for Printed Wiring Applications

IPC-CC-830A: Qualification and Performance of Electrical Insulation Compounds for Printed Board Assemblies

IPC-CF-148: Resin Coated Metal for Multilayer Printed Boards

IPC-CF-152A: Metallic Foil Specification for Copper/Invar/Copper (CIC) for Printed Wiring and Other Related Applications

IPC/JPCA-4104: Specification for High Density Interconnect (HDI) and Microvia Materials

IPC-SM-840C: Qualification and Performance of Permanent Solder Mask for Printed Boards

J-STD-004: Requirements for Soldering Fluxes

J-STD-005: General Requirements and Test Methods for Electronic Grade Solder Paste

J-STD-006: General Requirements and Test Methods for Soft Solder Alloys and Fluxed and Non Fluxed Solid Solders for Electronic Soldering Applications

DESIGN ACTIVITIES

IPC-C-406: Design and Application Guidelines for Surface Mount Connectors

IPC-D-279: Reliability Design Guidelines for Surface Mount Technology Printed Board Assemblies

IPC-1902: Grid Systems for Printed Circuits (equivalent to IEC 60097)

IPC-2141: Design Guide for High-Speed Controlled Impedance Circuit Boards

IPC-2221: Generic Standard on Printed Board Design

IPC-2222: Sectional Design Standard for Rigid Organic Printed Boards

IPC-2223: Sectional Design Standard for Flexible Printed Boards

IPC-2224: Sectional Standard for Design of PWBs for PC Cards

IPC-2225: Sectional Design Standard for Organic Multichip Modules (MCM-L) and MCM-L Assemblies

IPC-2226: Design Standard for High-Density Array or Peripheral Leaded Component Mounting Structures

IPC-2251: Design Guidelines for Electronic Packaging Utilizing High Speed Techniques

IPC-2252: Design and Manufacturing Guide for RF/Microwave Circuit Boards

IPC-2615: Printed Board Dimensions and Tolerances

IPC-7351: Generic Requirements for Surface Mount Design and Land Pattern Standard

IPC-7525: Guidelines for Stencil Design

IPC-7530: Guidelines for Temperature Profiling for Mass Soldering (Wave and Reflow) Processes

IPC/JPCA-2315: Design Guide for High Density Interconnects (HDI) and Microvia

J-STD-026: Semiconductor Design Standard for Flip Chip Applications

J-STD-027: Mechanical Outline Standard for Flip Chip or Chip Scale Configurations

MIL-STD-2118: Design Standard for Flexible Printed Wiring

COMPONENT MOUNTING

EIA-CB-11: Guidelines for the Surface Mounting of Multilayer Ceramic Chip Capacitors

IPC-CM-770D: Guidelines for Printed Board Component Mounting

IPC-SM-780: Electronic Component Packaging and Interconnection with Emphasis on Surface Mounting

IPC-SM-784: Guidelines for Direct Chip Attachment

IPC-MC-790: Guidelines for Multichip Module Technology Utilization

IPC-4104: Specification for High Density Interconnect (HDI) and Microvia Materials

IPC-6016: Qualification and Performance Specification for High Density Interconnect (HDI) Layers or Boards

IPC-7095: Design and Assembly Process Implementation for BGAs

IPC/JPCA-2315: Design Guide for High Density Interconnects (HDI) and Microvias

J-STD-012: Implementation of Flip Chip and Chip Scale Technology

J-STD-013: Implementation of Ball Grid Array and Other High Density Technology

J-STD-032: Performance Standard for Ball Grid Array Bumps and Columns

SOLDERING AND SOLDERABILITY

EIA-448-19: Method 19: Test Standard for Electromechanical Components, Environmental Effects of Machine Soldering Using a Vapor Phase System

EIA-534: Applications Guide, Soldering and Solderability Maintenance of Leaded Electronic Components

EIA-638: Surface Mount Solderability Test

EIA/IS-46: Test Procedure for Resistance to Soldering (Vapor Phase Technique) for Surface Mount Devices

EIA/IS-86: Surface Mount Solderability Test

EIA SP 3408: Surface Mount Solderability Test

IPC-AJ-820: Assembly and Joining Handbook

IPC-HDBK-001: Handbook and Guide to the Requirements for Soldered Electrical and Electronic Assemblies

IPC-S-816: Troubleshooting for Surface Mount Soldering

IPC-TR-460A: Troubleshooting Checklist for Wave Soldering Printed Wiring Boards

IPC-TR-462: Solderability Evaluation of Printed Boards with Protective Coatings over Long-Term Storage

IPC-TR-464: Accelerated Aging for Solderability Evaluations

J-STD-001: Requirements for Soldered Electrical and Electronic Assemblies

J-STD-002: Solderability Tests for Component Leads, Terminations, Lugs, Terminals and Wires

J-STD-003: Solderability Tests of Printed Boards

JESD22: B102C Solderability Test Methods

QUALITY ASSESSMENT

IPC-A-20/21: Stencil Pattern for Solder Paste Slump Test

IPC-A-24: Flux/Board Interaction Board

IPC-A-36: CFC Cleaning Alternatives Artwork

IPC-A-38: Fine Line Round Robin Test Pattern

IPC-A-48: Surface Mount Airforce Mantech Artwork

IPC-A-600: Acceptability of Printed Boards

IPC-A-610: Acceptability of Printed Board Assemblies

IPC-7912: Calculation of DPMO and Manufacturing Indices for Printed Wiring Assemblies

IPC-9191: General Guideline for Implementation of Statistical Process Control (SPC)

IPC-9194: Implementation of Statistical Process Control (SPC) Applied to Printed Board Assembly Manufacture Guideline

IPC-9199: SPC Quality Rating

IPC-9201: Surface Insulation Resistance Handbook

IPC-9251: Test Vehicles for Evaluating Fine Line Capability

IPC-9252: Guidelines and Requirements for Electrical Testing of Unpopulated Printed Boards

IPC-9261: In-Process DPMO and Estimated Yield for PWAs

IPC/JEDEC J-STD-020: Moisture/Reflow Sensitivity Classification for Plastic Integrated Circuits Surface Mount Devices

IPC/JEDEC-033: Standard for Handling, Packing, Shipping, and Use of Moisture Reflow Sensitive Surface Mount Devices

IPC/JEDEC J-STD-035: Acoustic Microscopy for Non-Hermetic Encapsulated Electronic Components

JEDEC JEP-150: Stress-Test-Driven Qualification of and Failure Mechanisms Associated with Assembled Solid State Surface-Mount Components

JEDEC JEP-179: Stress-Test-Driven Qualification of and Failure Mechanisms Associated with Assembled Solid State Surface-Mount Components

JEDEC JESD-51-3: Low Effective Thermal Conductivity Test Board for Leaded Surface Mount Packages

JEDEC JESD-51-5: Extension of Thermal Test Board Standards for Packages with Direct Thermal Attachment Mechanisms

JEDEC JESD-51-7: High Effective Thermal Conductivity Test Board for Leaded Surface Mount Packages

JEDEC JESD-51-9: Test Boards for Area Array Surface Mount Package Thermal Measurements

JEDEC JESD-51-10: Test Boards for Through-Hole Perimeter Leaded Package Thermal Measurements

JEDEC JESD 51-11 Test Boards for Through-Hole Area Array Leaded Package Thermal Measurements

J-STD-028: Performance Standard for Flip Chip/Chip Scale Bumps

MIL-STD-883: Methods and Procedures for Microelectronics

SURFACE-MOUNT PROCESS

IPC-SC-60: Post Solder Solvent Cleaning Handbook

IPC-SA-61: Post Solder Semi Aqueous Cleaning Handbook

IPC-AC-62: Post Solder Aqueous Cleaning Handbook

IPC-CH-65: Guidelines for Cleaning of Printed Boards and Assemblies

IPC-5701: Users Guide for Cleanliness of Unpopulated Printed Boards

RELIABILITY

IPC-D-279: Design Guidelines for Reliable Surface Mount Technology Printed Board Assemblies

IPC-SM-785: Guidelines for Accelerated Surface Mount Attachment Reliability Testing

IPC-9151: Printed Board Capability, Quality and Relative Reliability ($PCQR^2$) Benchmark Test Standard and Database

JEDEC JESD 22-A113: Preconditioning Of Nonhermetic Surface Mount Devices Prior to Reliability Testing

JEDEC JESD 22-B113: Board Level Cyclic Bend Test Method for Interconnect Reliability Characterization of Components for Handheld Electronic Products

NUMERICAL CONTROL STANDARDS

IPC-2511: Generic Requirements for Implementation of Product Manufacturing Description Data and Transfer Methodology

IPC-2512: Sectional Requirements for Implementation of Administrative Methods for Manufacturing Data Description

IPC-2513: Sectional Requirements for Implementation of Drawing Methods for Manufacturing Data Description

IPC-2514: Sectional Requirements for Implementation of Printed Board Manufacturing Data Description

IPC-2515: Sectional Requirements for Implementation of Bare Board Product Electrical Testing Data Description

IPC-2516: Sectional Requirements for Implementation of Assembled Board Product Manufacturing Data Description

IPC-2517: Sectional Requirements for Implementation of Assembly Circuit Testing Data Description

IPC-2518: Sectional Requirements for Implementation of Bill of Material Product Data Description

EIA-267-C: Axis and Motion Nomenclature for Numerically Controlled Machines

EIA-274-D: Interchangeable Variable Block Data Contouring, Format for Positioning and Contouring/Positioning Numerically Controlled Machines

EIA-358-C: ANSI X3.4 American National Standard Code for Information Interchange for Numerical Machine Control Perforated Tape

EIA-408: Interface between Numerical Control Equipment on Data Terminal Equipment Employing Parallel Binary Data Interchange

EIA-431: Electrical Interface between Numerical Control and Machine Tools

EIA-441: Operator Interface Function of Numerical Controls

EIA-484-A: Electrical and Mechanical Interface Characteristics and Line Control Protocol Using Communication Control Characters for Serial Data Link between a Direct Numerical Control System and Numerical Control Equipment Employing Asynchronous Full Duplex Transmission

EIA-494-B: 32 Bit Binary CL (BCL) and 7 Bit ASCII CL (ACL) Exchange Input Format for Numerically Controlled Machines

TEST METHODS

EIA-469-C: Standard Test Method for Destructive Physical Analysis of High Reliability Ceramic Monolithic Capacitors

EIA-638: Surface Mount Solderability Test

EIA IS 46: Test Procedure for Resistance to Soldering (Vapor Phase Technique) for Surface Mount Devices

IPC-9501: PWB Assembly Process Simulation for Evaluation of Electronic Components

IPC-9502: PWB Assembly Soldering Process Guidelines for Non-IC Electronic Components

IPC-9503: Moisture Sensitivity Classification for Non-IC Components

IPC-9504: Assembly Process Simulation for Evaluation of Non-IC Components

IPC-9701: Qualification and Performance Test Methods for Surface Mount Solder Attachments

IPC-9850: Surface Mount Equipment Performance Characterization

IPC/JEDEC-9702: Monotonic Bend Characterization of Board-Level Interconnects

IPC/JPCA-6801: Terms & Definitions, Test Methods, and Design Examples for Build-Up/High Density Interconnection

IPC-TM-650: Test Methods Manual

JEDEC Standard 22 Series: Test Methods

JEDEC JESD 22-A111: Evaluation Procedure for Determining Capability to Bottom Side Board Attach By Full Body Solder Immersion of Small Surface Mount Solid State Devices

JEDEC JESD 22-A112: Test Method A112, Moisture-Induced Stress Sensitivity for Plastic Surface Mount Devices

JEDEC JESD 22-B105-B: Test Method Lead Integrity

JEDEC JESD 22-B106: Test Method B106c, Resistance to Soldering Temperature for Through-Hole Mounted Devices

JEDEC JESD 22-B108: Coplanarity Test for Surface-Mount Semiconductor Devices

JEDEC JESD 22-B117: BGA Ball Shear

JEDEC JESD 51-8: Integrated Circuit Thermal Test Method Environmental Conditions Junction-to-Board

JESD 22-B108: Coplanarity Test for Surface Mount Semiconductor Devices

REPAIR

IPC-7711A/7721A: Rework, Repair and Modification of Electronic Assemblies

TERMS AND DEFINITIONS

IPC-T-50F: Terms and Definitions for Interconnecting and Packaging Electronic Circuits

JEDEC JESD 12-1: Terms and Definitions for Gate Arrays and Cell-Based Integrated Circuits

JEDEC JESD 77: Terms, Definitions and Letter Symbols for Discrete Semiconductor and Optoelectronic Devices

JEDEC JESD 99: Terms, Definitions, and Letter Symbols for Microelectronic Devices

JEDEC JESD 100: Terms, Definitions, and Letter Symbols for Microcomputers, Microprocessors, and Memory Integrated Circuits

JEDEC JEP 99: Glossary of Microelectronic Terms, Definitions, Symbology

HOW TO OBTAIN THESE DOCUMENTS

Following are the addresses of the IPC and EIA, as well as other sources for documents shown in this SMT listing:

IPC—Association Connecting Electronics Industries
3000 Lakeside Drive #309S
Bannockburn, IL 60015-1249
Electronic Industries Alliance (EIA)
2500 Wilson Boulevard
Arlington, VA 22201-3834
Phone: (703) 907-7500

Global Engineering Documents (Part of IHS)
15 Inverness Way
East Englewood, CO 80112
Phone: (800) 854-7179

Military documents are available from the following address:

Standardization Documents Order Desk
Building 4D, 700 Robbins Avenue
Philadelphia, PA 19111-5094

Central office address of the IEC is as follows:

International Electrotechnical Commission (IEC)
3 Rue de Varembe
1211 Geneva 20, Switzerland

IEC documents are available from the following address:

American National Standards Institute (ANSI)
11 West 42nd Street
New York, NY 10036

GLOSSARY*

ACCELERATOR: A chemical that is used to speed up a reaction or cure, as cobalt naphthenate is used to accelerate the reaction of certain polyester resins. It is often used along with a catalyst, hardener, or curing agent. The term "accelerator" is often used interchangeably with the term " promoter."

ACCURACY: The ability to place the hole at the targeted location.

ADDITIVE PROCESS: A process for obtaining conductive patterns by the selective deposition of conductive material on an unclad base material.

ADHESIVE: Broadly, any substance used in promoting and maintaining a bond between two materials.

AGING: The change in properties of a material with time under specific conditions.

ANNULAR RING: The circular strip of conductive material that completely surrounds a hole.

ARC RESISTANCE: The time required for an arc to establish a conductive path in a material.

ARTWORK MASTER: An accurately scaled configuration used to produce the production master.

BACKUP MATERIAL: A material placed on the bottom of a laminate stack in which the drill terminates its drilling stroke.

BASE MATERIAL: The insulating material upon which the printed wiring pattern may be formed.

BASE MATERIAL THICKNESS: The thickness of the base material excluding metal foil cladding or material deposited on the surface.

BLIND VIA: Conductive surface hole that connects an outerlayer with an innerlayer of a multilayer PWB without penetrating the entire board.

BLISTERING: Localized swelling and separation between any of the layers of the base laminate or between the laminate and the metal cladding.

BONDING LAYER: An adhesive layer used in bonding other discrete layers during lamination.

BOND STRENGTH: The force per unit area required to separate two adjacent layers by a force perpendicular to the board surface; usually refers to the interface between copper and base material.

BOW: A laminate defect in which deviation from planarity results in a smooth arc.

B-STAGE: An intermediate stage in the curing of a thermosetting resin. In it a resin can be heated and caused to flow, thereby allowing final curing in the desired shape.

B-STAGE LOT: The product from a single mix of B-stage ingredients.

B-STAGE RESIN: A resin in an intermediate stage of a thermosetting reaction. The material softens when heated and swells when in contact with certain liquids, but it may not entirely fuse or dissolve.

BURIED VIA: Conductive surface hole that connects one innerlayer to another innerlayer of a multilayer PWB without having a direct connection to either the top or bottom surface layer.

BURR: A ridge left on the outside copper surfaces after drilling.

CAPACITANCE: The property of a system of conductors and dielectrics which permits the storage of electricity when potential difference exists between the conductors.

---

*Some terms may not be included in the glossary as they are treated in detail in the text. Please also see the index and subject chapters.

CAPACITIVE COUPLING: The electrical interaction between two conductors caused by the capacitance between the conductors.

CARBIDE: Tungsten carbide, formula WC. The hard, refractory material forming the drill bits used in PWB drillings.

CATALYST: A chemical that causes or speeds up the cure of a resin but does not become a chemical part of the final product.

CERAMIC LEADED CHIP CARRIER (CLCC): A chip carrier made from ceramic (usually a 90–96% alumina or beryllia base) and with compliant leads for terminations.

CHIP CARRIER (CC): An integrated circuit package, usually square, with a chip cavity in the center; its connections are usually on all four sides. (See *leaded chip carrier* and *leadless chip carrier.*)

CHIP LOAD (CL): The movement of the drill downward per revolution; usually given in mils (thousandths of an inch) per revolution.

CHLORINATED HYDROCARBON: An organic compound having chlorine atoms in its chemical structure. Trichloroethylene, methyl chloroform, and methylene chloride are chlorinated hydrocarbons.

CIRCUIT: The interconnection of a number of electrical devices in one or more closed paths to perform a desired electrical or electronic function.

CLAD: A condition of the base material, to which a relatively thin layer or sheet of metal foil (cladding) has been bonded on one or both of its sides. The result is called a metal-clad base material.

CNC: Computer numerically controlled. Refers to a machine with a computer which stores the numerical information about location, drill size, and machine parameters, regulating the machine to carry out that information.

COAT: To cover with a finishing, protecting, or enclosing layer of any compound.

COLD FLOW: The continuing dimensional change that follows initial instantaneous deformation in a nonrigid material under static load. Also called creep.

COLLIMATION: The degree of parallelism of light rays from a given source. A light source with good collimation produces parallel light rays, whereas a poor light source produces divergent, nonparallel light rays.

COMPONENT HOLE: A hole used for the attachment and electrical connection of a component termination, including pin or wire, to the printed board.

COMPONENT SIDE: The side of the printed board on which most of the components will be mounted.

COMPOUND: A combination of elements in a stable molecular arrangement.

CONDUCTIVE FOIL: The conductive material that covers one side or both sides of the base material and is intended for forming the conductive pattern.

CONDUCTIVE PATTERN: The configuration or design of the electrically conductive material on the base material.

CONDUCTOR LAYER 1: The first layer having a conductive pattern, of a multilayer board, on or adjacent to the component side of the board.

CONDUCTOR SPACING: The distance between adjacent edges (not centerline to centerline) of conductors on a single layer of a printed board.

CONDUCTOR THICKNESS: The thickness of the copper conductor exclusive of coatings or other metals.

CONDUCTOR WIDTH: The width of the conductor viewed from vertically above, that is, perpendicularly to the printed board.

CONFORMAL COATING: An insulating protective coating which conforms to the configuration of the object coated and is applied on the completed printed board assembly.

CONNECTOR AREA: The portion of the printed board that is used for providing external (input–output) electrical connections.

CONTACT BONDING ADHESIVE: An adhesive (particularly of the nonvulcanizing natural rubber type) that bonds to itself on contact, although solvent evaporation has left it dry to the touch.

CONTROLLED IMPEDANCE: The matching of substrate material properties with trace dimensions and locations to create a specific electric impedance as seen by a signal on the trace.

COPOLYMER: See *polymer.*
CORE MATERIAL: The fully cured inner-layer segments, with circuiting on one or both sides, that form the multilayer circuit.
CORNER MARKS: The marks at the corners of printed board artwork, the inside edges of which usually locate the borders and establish the contour of the board.
COUPON: One of the patterns of the quality conformance test circuitry area. (See *test coupon.*)
CRAZING: A base material condition in which connected white spots or crosses appear on or below the surface of the base material. They are due to the separation of fibers in the glass cloth and connecting weave intersections.
CROSS-LINKING: The forming of chemical links between reactive atoms in the molecular chain of a plastic. It is cross-linking in the thermosetting resins that makes the resins infusible.
CROSS TALK: Undesirable electrical interference caused by the coupling of energy between signal paths.
CRYSTALLINE MELTING POINT: The temperature at which the crystalline structure in a material is broken down.
CTE: Coefficient of thermal expansion. The measure of the amount a material changes in any axis per degree of temperature change.
CURE: To change the physical properties of a material (usually from a liquid to a solid) by chemical reaction or by the action of heat and catalysts, alone or in combination, with or without pressure.
CURING AGENT: See *hardener.*
CURING TEMPERATURE: The temperature at which a material is subjected to curing.
CURING TIME: In the molding of thermosetting plastics, the time in which the material is properly cured.
CURRENT-CARRYING CAPACITY: Maximum current which can be carried continuously without causing objectionable degradation of electrical or mechanical properties of the printed board.
DATUM REFERENCE: A defined point, line, or plane used to locate the pattern or layer of a printed board for manufacturing and/or inspection purposes.
DEBRIS: A mechanically bonded deposit of copper to substrate hole surfaces.
DEBRIS PACK: Debris deposited in cavities or voids in the resin.
DEFINITION: The fidelity of reproduction of the printed board conductive pattern relative to the production master.
DELAMINATION: A separation between any of the layers of the base laminate or between the laminate and the metal cladding originating from or extending to the edges of a hole or edge of the board.
DIELECTRIC CONSTANT: The property of a dielectric which determines the electrostatic energy stored per unit volume for a unit potential gradient.
DIELECTRIC LOSS: Electric energy transformed into heat in a dielectric subjected to a changing electric field.
DIELECTRIC LOSS ANGLE: The difference between 90° and the dielectric phase angle. Also called the dielectric phase difference.
DIELECTRIC LOSS FACTOR: The product of dielectric constant and the tangent of dielectric loss angle for a material.
DIELECTRIC PHASE ANGLE: The angular difference in phase between the sinusoidal alternating potential difference applied to a dielectric and the component of the resulting alternating current having the same period as the potential difference.
DIELECTRIC POWER FACTOR: The cosine of the dielectric phase angle (or sine of the dielectric loss angle).
DIELECTRIC STRENGTH: The voltage that an insulating material can withstand before breakdown occurs, usually expressed as a voltage gradient (such as volts per mil).
DIMENSIONAL STABILITY: Freedom from distortion by such factors as temperature changes, humidity changes, age, handling, and stress.
DIRECT IMAGING: The exposure of photo resist material with a laser without the use of positive or negative photo tool.

DISSIPATION FACTOR: The tangent of the loss angle of the insulating material. Also called loss tangent or approximate power factor.

DRILL FACET: The surface formed by the primary and secondary relief angles of a drill tip.

DRILL WANDER: The sum of accuracy and precision deviations from the targeted location of the hole.

DUMMY: A cathode with a large area used in a low-current-density pulsating operation for the removal of metallic impurities from solution. The process is called "dummying."

DWELL POINT: The bottom of the drilling stroke before the drill bit ascends.

EDGE-BOARD CONTACTS: A series of contacts printed on or near an edge of a printed board and intended for mating with a one-part edge connector.

EDX: Energy dispersive x-ray fluorescent spectrometer

ELASTOMER: A material which at room temperature stretches under low stress to at least twice its length but snaps back to its original length upon release of the stress. Rubber is a natural elastomer.

ELECTRIC STRENGTH: The maximum potential gradient that a material can withstand without rupture. It is a function of the thickness of the material and the method and conditions of test. Also called dielectric strength or disruptive gradient.

ELECTROLESS PLATING: The controlled autocatalytic reduction of a metal ion on certain catalytic surfaces.

EMULSION SIDE: The side of the film or glass on which the photographic image is present.

ENTRY MATERIAL: A material placed on top of a laminate stack.

EPOXY SMEAR: Epoxy resin which has been deposited on edges of copper in holes during drilling either as a uniform coating or as scattered patches. It is undesirable because it can electrically isolate the conductive layers from the plated-through-hole interconnections.

ETCHBACK: The controlled removal of all the components of the base material by a chemical process acting on the sidewalls of plated-through holes to expose additional internal conductor areas.

ETCH FACTOR: The ratio of the depth of etch to lateral etch.

EXOTHERM: A characteristic curve which shows heat of reaction of a resin during cure (temperature) versus time. The peak exotherm is the maximum temperature on the curve.

EXOTHERMIC REACTION: A chemical reaction in which heat is given off.

FIBER EXPOSURE: A condition in which glass cloth fibers are exposed on machined or abraded areas.

FILLER: A material, usually inert, added to a plastic to reduce cost or modify physical properties.

FILM ADHESIVE: A thin layer of dried adhesive. Also, a class of adhesives provided in dry-film form with or without reinforcing fabric and cured by heat and pressure.

FLEXURAL MODULUS: The ratio, within the elastic limit, of stress to corresponding strain. It is calculated by drawing a tangent to the steepest initial straight-line portion of the load deformation curve and using the equation $EB = L3m/4bd3$, where EB is the modulus, L is the span (in inches), m is the slope of the tangent, b is the width of beam tested, and d is the depth of the beam.

FLEXURAL STRENGTH: The strength of a material subjected to bending. It is expressed as the tensile stress of the outermost fibers of a bent test sample at the instant of failure.

FLUOROCARBON: An organic compound having fluorine atoms in its chemical structure, an inclusion that usually lends stability to plastics. Teflon* is a fluorocarbon.

GEL: The soft, rubbery mass that is formed as a thermosetting resin goes from a fluid to an infusible solid. It is an intermediate state in a curing reaction, and a stage in which the resin is mechanically very weak.

GEL POINT: The point at which gelation begins.

GLASS TRANSITION POINT: The temperature at which a material loses properties and becomes a semiliquid.

GLASS TRANSITION TEMPERATURE: The temperature at which epoxy, for example, softens and begins to expand independently of the glass fabric expansion rate, usually symbolized as T_g.

---

*Trademark of E. I. du Pont de Nemours & Company.

GLUE-LINE THICKNESS: Thickness of the fully dried adhesive layer.

GRID: An orthogonal network of two sets of parallel lines for positioning features on a printed board.

GROUND PLANE: A conducting surface used as a common reference point for circuit returns, shielding, or heat sinking.

GULL WING LEAD: A surface mounted device lead which flares outward from the device body.

HALOING: A light area around holes or other machined areas on or below the surface of the base laminate.

HARDENER: A chemical added to a thermosetting resin for the purpose of causing curing or hardening. A hardener, such as an amine or acid anhydride for an epoxy resin, is a part of the chemical reaction and a part of the chemical composition of the cured resin. The terms "hardener" and "curing agent" are used interchangeably.

HEAT-DISTORTION POINT: The temperature at which a standard test bar (ASTM D 648) deflects 0.010 in under a stated load of either 66 or 264 psi.

HIGH-DENSITY INTERCONNECT (HDI): Ultra fine-geometry multilayer PWB constructed with conductive surface microvia connections between layers. (Microvia is usually defined as a hole with a diameter less than 0.006 in.) These boards also usually include buried and/or blind vias and are made by sequential build-up lamination.

HOLE PULL STRENGTH: The force, in pounds, necessary to rupture a plated-through hole or its surface terminal pads when loaded or pulled in the direction of the axis of the hole. The pull is usually applied to a wire soldered in the hole, and the rate of pull is given in inches per minute.

HOOK: A geometric drill bit defect of the cutting edges.

HOT-MELT ADHESIVE: A thermoplastic adhesive compound, usually solid at room temperature, which is heated to fluid state for application.

HYDROCARBON: An organic compound containing only carbon and hydrogen atoms in its chemical structure.

HYDROLYSIS: The chemical decomposition of a substance involving the addition of water.

HYGROSCOPIC: Tending to absorb moisture.

I-LEAD: A surface mounted device lead which is formed such that the end of the lead contacts the board land pattern at a 90° angle. Also called a butt joint.

IMPREGNATE: To force resin into every interstice of a part, as of a cloth for laminating.

INHIBITOR: A chemical that is added to a resin to slow down the curing reaction and is normally added to prolong the storage life of a thermosetting resin.

INORGANIC CHEMICALS: Chemicals whose molecular structures are based on other than carbon atoms.

INSULATION RESISTANCE: The electrical resistance of the insulating material between any pair of contacts, conductors, or grounding devices in various combinations.

INTERNAL LAYER: A conductive pattern contained entirely within a multilayer board. IPC: Institute for Interconnecting and Packaging Electronic Circuits. A leading printed wiring industry association that develops and distributes standards as well as other information of value to printed wiring designers, users, suppliers, and fabricators. IR: Infrared heating for solder-reflow operation.

J-LEAD: A surface mounted device lead which is formed into a "J" pattern folding under the device body.

JUMPER: An electrical connection between two points on a printed board added after the printed wiring is fabricated.

KIRKENDALL VOIDS: In Cu-Au or Cu-Au-Sn systems, pores in the older joint, or at copper interfaces occur through solid-state diffusion. Copper which has a relatively high solubility and solid-state mobility in gold diffuses into the gold. Transported by grain boundry diffusion at temperatures below 150°C, and by bulk diffusion above that temperature. Au-Sn regions result along with atomic vacancies. Then lattice vacancies are in profusion . vpods can be observed and this material depletion weakens the solder joint.

LAMINATE: The plastic material, usually reinforced by glass or paper, that supports the copper cladding from which circuit traces are created.

LAMINATE VOID: Absence of epoxy resin in any cross-sectional area which should normally contain epoxy resin.

LAND: See *terminal area*.

LANDLESS HOLE: A plated-through hole without a terminal area.

LASER PHOTOPLOTTER (LASER PHOTO GENERATOR, OR LPG): A device that exposes photosensitive material, usually a silver halide or diazo material, subsequently used as the master for creating the circuit image in production.

LAYBACK: A geometric drill bit defect of the cutting edges.

LAYER-TO-LAYER SPACING: The thickness of dielectric material between adjacent layers of conductive circuitry.

LAY-UP: The process of registering and stacking layers of a multilayer board in preparation for the laminating cycle.

LCCC: Leadless ceramic chip carrier.

LEAD-FREE: Referring to solder alloys made without lead, to conform with the requirements of the European Union directive on the Restriction of Hazard Substances (RoHS) the most important of which, to the printed circuit industry, is lead. Often used to refer to any process that is designed to be compatible with "lead-free" alloys.

LEADED CHIP CARRIER: A chip carrier (either plastic or ceramic) with compliant leads from terminations.

LEADLESS CHIP CARRIER: A chip carrier (usually ceramic) with integral metallized terminations and no compliant external leads.

LEGEND: A format of lettering or symbols on the printed board, for example, part number, component locations, or patterns.

LOOSE FIBERS: Supporting fibers in the substrate of the laminate which are not held in place by surrounding resin.

MAJOR WEAVE DIRECTION: The continuous-length direction of a roll of woven glass fabric.

MARGIN RELIEF: The area of a drill bit next to the cutting edge is removed so that it does not rub against the hole as the drill revolves.

MASTER DRAWING: A document that shows the dimensional limits or grid locations applicable to any or all parts of a printed wiring or printed circuit base. It includes the arrangement of conductive or nonconductive patterns or elements; size, type, and location of holes; and any other information necessary to characterize the complete fabricated product.

MEASLING: Discrete white spots or crosses below the surface of the base laminate that reflect a separation of fibers in the glass cloth at the weave intersection.

MICROSTRIP: A type of transmission line configuration which consists of a conductor over a parallel ground plane separated by a dielectric.

MICROVIA: Usually defined as a conductive hole with a diameter of 0.006 in or less that connects layers of a multilayer PWB. Often used to refer to any small-geometry connecting hole the creation of which is beyond the practical capabilities of traditional mechanical drilling processes.

MINOR WEAVE DIRECTION: The width direction of a roll of woven glass fabric.

MIXED ASSEMBLY: A printed wiring assembly that combines through-hole components and surface mounted components on the same board.

MODULUS OF ELASTICITY: The ratio of stress to strain in a material that is elastically deformed.

MOISTURE RESISTANCE: The ability of a material not to absorb moisture either from air or when immersed in water.

MOUNTING HOLE: A hole used for the mechanical mounting of a printed board or for the mechanical attachment of components to a printed board.

MULTILAYER BOARD: A product consisting of layers of electrical conductors separated from each other by insulating supports and fabricated into a solid mass. Interlayer connections are used to establish continuity between various conductor patterns.

MULTIPLE-IMAGE PRODUCTION MASTER: A production master used to produce two or more products simultaneously.

NAILHEADING: A flared condition of internal conductors.

NC: Numerically controlled. Usually refers to a machine tool, in this case a drilling machine. The most basic type is one in which a mechanical guide locates the positions of the holes. NC machines are usually controlled by punched tape.

NEMA STANDARDS: Property values adopted as standard by the National Electrical Manufacturers Association.

NOBLE ELEMENTS: Elements that either do not oxidize or oxidize with difficulty; examples are gold and platinum.

OILCANNING: The movement of entry material in the *z* direction during drilling in concert with the movement of the pressure foot.

ORGANIC: Composed of matter originating in plant or animal life or composed of chemicals of hydrocarbon origin, either natural or synthetic.

PAD: See *terminal area.*

PADS ONLY: A multilayer construction with all circuit traces on inner layers and the component terminal area only on the surface of the board. This construction adds two layers but may avoid the need for a subsequent solder resist, and since inner layers usually are easier to form, this construction may lead to higher overall yields. PH: A measure of the acid or alkaline condition of a solution. A pH of 7 is neutral (distilled water); pH values below 7 represent increasing acidity as they go toward 0; and pH values above 7 represent increasing alkalinity as they go toward the maximum value of 14.

PHOTOMASTER: An accurately scaled copy of the artwork master used in the photo fabrication cycle to facilitate photo processing steps.

PHOTOPOLYMER: A polymer that changes characteristics when exposed to light of a given frequency.

PINHOLES: Small imperfections which penetrate entirely through the conductor.

PINK RING: The appearance of a halo of copper around the hole of a multilayer.

PWB PITS: Small imperfections which do not penetrate entirely through the printed circuit.

PLASTICIZER: Material added to resins to make them softer and more flexible when cured.

PLASTIC LEADED CHIP CARRIER (PLCC): A chip carrier packaged in plastic, usually terminating in compliant leads (originally "J" style) on all four sides.

PLATED-THROUGH HOLE: A hole in which electrical connection is made between printed wiring board layers with conductive patterns by the deposition of metal on the wall of the hole.(See *PTH.*)

PLATING VOID: The area of absence of a specific metal from a specific cross-sectional area: (1) When the plated-through hole is viewed as cross-sectioned through the vertical plane, it is a product of the average thickness of the plated metal times the thickness of the board itself as measured from the outermost surfaces of the base copper on external layers. (2) When the plated-through hole is viewed as cross-sectioned through the horizontal plane (annular method), it is the difference between the area of the hole and the area of the outside diameter of the through-hole plating.

PLOWING: Furrows in the hole wall due to drilling.

POLYMER: A high-molecular-weight compound made up of repeated small chemical units. For practical purposes, a polymer is a plastic. The small chemical unit is called a mer, and when the polymer or mer is cross-linked between different chemical units (e.g., styrene-polyester), the polymer is called a copolymer. A monomer is any single chemical from which the mer or polymer or copolymer is formed.

POLYMERIZE: To unite chemically two or more monomers or polymers of the same kind to form a molecule with higher molecular weight.

POTLIFE: The time during which a liquid resin remains workable as a liquid after catalysts, curing agents, promoter, etc., are, added. It is roughly equivalent to gel time.

POWER FACTOR: The cosine of the angle between the applied voltage and the resulting current.

PRECISION: The ability to repeatedly place the hole at any location.

PREPRODUCTION TEST BOARD: A test board (as detailed in IPC-ML-950) the purpose of which is to determine whether, prior to the production of finished boards, the contractor has the capability of producing a multilayer board satisfactorily.

PRESS PLATEN: The flat heated surface of the lamination press used to transmit heat and pressure to lamination fixtures and into the lay-up.

PRESSURE FOOT: The tube like device on the drilling machine that descends to the top surface of the stack, holding it firmly down, before the drill descends through the center of the pressure foot. The vacuum system of the drilling machine separates through the pressure foot to remove chips and dust formed in drilling.

PRINTED WIRING ASSEMBLY DRAWING: A document that shows the printed wiring base, the separately manufactured components which are to be added to the base, and any other information necessary to describe the joining of the parts to perform a specific function.

PRINTED WIRING LAYOUT: A sketch that depicts the printed wiring substrate, the physical size and location of electronic and mechanical components, and the routing of conductors that interconnect the electronic parts in sufficient detail to allow for the preparation of documentation and artwork.

PRODUCTION MASTER: A 1:1 scale pattern used to produce one or more printed wiring or printed circuit products within the accuracy specified on the master drawing.

PROMOTER: A chemical, itself a feeble catalyst, that greatly increases the activity of a given catalyst.

PTH: Plated-through holes. Also refers to the technology that uses the plated-through hole as its foundation.

QUADPACK: Generic term for surface mount technology packages with leads on all four sides. Commonly used to describe chip carrier-like devices with gull wing leads.

QUALITY CONFORMANCE CIRCUITRY AREA: A test board made as an integral part of the multilayer printed board panel on which electrical and environmental tests may be made for evaluation without destroying the basic board.

RAW MATERIAL PANEL SIZE: A standard panel size related to machine capacities, raw material sheet sizes, final product size, and other factors.

REGISTER MARK: A mark used to establish the relative position of one or more printed wiring patterns, or portions thereof, with respect to desired locations on the opposite side of the board.

REGISTRATION: The relative position of one or more printed wiring patterns, or portions thereof, with respect to desired locations on a printed wiring base or to another pattern on the opposite side of the base.

REPAIR: The correction of a printed wiring defect after the completion of board fabrication to render the board as functionally good as a perfect board.

RESIN: High-molecular-weight organic material with no sharp melting point. For current purposes, the term "resin," "polymer," and "plastic" can be used interchangeably.

RESIST: A protective coating (ink, paint, metallic plating, etc.) used to shield desired portions of the printed conductive pattern from the action of etchant, solder, or plating.

RESISTIVITY: The ability of a material to resist passage of electric current through its bulk or on a surface.

RIFLING: Spiral groove or ridge in the substrate due to drilling.

RIGID/FLEX: A PWB construction combining flexible circuits and rigid multilayer PWBs, usually either to provide a built-in connection or to make a three-dimensional form that includes components.

ROCKWELL HARDNESS NUMBER: A number derived from the net increase in depth of an impression as the load on a perpetrator is increased from a fixed minimum load to as higher load and then returned to minimum load.

RoHS: Acronym for "Restriction of Hazardous Substances," The name given to a Directive of the European Union meant to reduce certain materials considered detrimental to the Environment.

ROUGHNESS: Irregular, coarse, uneven hole wall on copper or substrate due to drilling.

SCHEMATIC DIAGRAM: A drawing which shows, by means of graphic symbols, the electrical interconnections and functions of a specific circuit arrangement.

SEM: Scanning electron microscope.

SEQUENTIAL BUILD-UP: A process for making multilayer PWBs in which already finished multilayers are laminated together to form a higher-layer-count final board, or in which additional outerlayers are added to finished multilayer PWBs.

SHADOWING: Etchback to maximum limit without removal of dielectric material from conductors.

SINGLE-IN-LINE PACKAGE (SIP): Component package system with one line of connectors, usually spaced 0.100 in apart.

SMC: Surface mounted component. Component with terminations designed for mounting flush to printed wiring board.

SMD: Surface mounted device. Any component or hardware element designed to be mounted to a printed wiring board without penetrating the board.

SMEAR: Fused deposit left on copper or substrate from excessive drilling heat.

SMOBC: Solder mask over bare copper. A method of fabricating a printed wiring board which results in the final metallization being copper with no other protective metal; but the non-soldered areas are coated by a solder resist, exposing only the component terminal areas. This eliminates tin-lead under the solder mask.

SMT: Surface mount technology. Defines the entire body of processes and components which create printed wiring assemblies without components with leads that pierce the board. Sn-Au-Cu: An alloy used as a replacement for eutectic tin-lead (Sn-Pb) in "lead-free" solders (often referred to, and pronounced as "snack")

SOI: SOIC package with J-leads rather than gull wing leads.

SOIC: Small-outline integrated circuit. A plastic package resembling a small dual-in-line package (DIP) with gull wing leads on two sides for surface mounting.

SOT: Small outline transistor. A package for surface-mounting transistors.

SPINDLE RUNOUT: The measure of the wobble present as the drilling machine spindle rotates 360°.

STORAGE LIFE: The period of time during which a liquid resin or adhesive can be stored and remain suitable for use. Also called shelf life.

STRAIN: The deformation resulting from a stress. It is measured by the ratio of the change to the total value of the dimension in which the change occurred.

STRESS: The force producing or tending to produce deformation in a body. It is measured by the force applied per unit area.

SUBSTRATE: A material on whose surface an adhesive substance is spread for bonding or coating. Also, any material which provides a supporting surface for other materials used to support printed wiring patterns.

SURFACE RESISTIVITY: The resistance of a material between two opposite sides of a unit square of its surface. It may vary widely with the conditions of measurement.

SURFACE SPEED: The linear velocity of a point on the circumference of a drill. Given in units of surface feet per minute—sfm.

TERMINAL AREA: A portion of a conductive pattern usually, but not exclusively, used for the connection and/or attachment of components.

TEST COUPON: A sample or test pattern usually made as an integral part of the printed board, on which electrical, environmental, and micro sectioning tests may be made to evaluate board design or process control without destroying the basic board.

TETRA-ETCH*: A nonpyrophoric (will not ignite when exposed to moisture) proprietary etchant.

TETROFUNCTIONAL: Describes an epoxy system for laminates that has four cross-linked bonds rather than two and results in a higher glass transition temperature, or T_g.

T_g: Glass transition temperature. The temperature at which laminate mechanical properties change significantly.

THERMAL CONDUCTIVITY: The ability of a material to conduct heat; the physical constant for the quantity of heat that passes through a unit cube of a material in a unit of time when the difference in temperatures of two faces is 1°C.

THERMOPLASTIC: A classification of resin that can be readily softened and resoftened by repeated heating.

---

*Trademark of W. L. Gore and Associates, Inc.

THERMOSETTING: A classification of resin which cures by chemical reaction when tested and, when cured, cannot be resoftened by heating.

THIEF: An auxiliary cathode so placed as to divert to itself some current from portions of the work which would otherwise receive too high a current density.

THIXOTROPIC: Said of materials that are gel-like at rest but fluid when agitated.

THROUGH-HOLE TECHNOLOGY: Traditional printed wiring fabrication where components are mounted in holes that pierce the board.

THROWING POWER: A measure of the degree of uniformity with which metal is deposited on an irregularly shaped cathode. Often refers to the ratio of amount of plated metal on the surface of a copper clad board to the amount plated on a side of a hole through the same board.

TWIST: A laminate defect in which deviation from planarity results in a twisted arc.

UNDERCUT: The reduction of the cross section of a metal foil conductor caused by the etchant removing metal from under the edge of the resist.

VAPOR PHASE: The solder-reflow process that uses a vaporized solvent as the source for heating the solder beyond its melting point, creating the component-to-board solder joint.

VIA: A metallized connecting holc that provides a conductive path from one layer in a printed wiring board to another. (1) *Buried via*—connects one inner layer to another inner layer without penetrating the surface. (2) *Blind via*—connects the surface layer of a printed wiring board to an internal layer without going all the way through the other surface layer.

VOID: A cavity left in the substrate.

VOLUME RESISTIVITY: The electrical resistance between opposite faces of a 1-cm cube of insulating materials, commonly expressed in ohm-centimeters. The recommended test is ASTMD 256 51T. Also called the specific insulation.

VULCANIZATION: A chemical reaction in which the physical properties of an elastomer are changed by causing the elastomer to react with sulfur or some other cross-linking agent.

WATER ABSORPTION: The ratio of the weight of water absorbed by a material to the weight of the dry material.

WEAVE EXPOSURE: A condition in which the unbroken woven glass cloth is not uniformly covered by resin.

WEAVE TEXTURE: A surface condition in which the unbroken fibers are completely covered with resin but exhibit the definite weave pattern of the glass cloth.

WETTING: Ability to adhere to a surface immediately upon contact.

WICKING: Migration of copper salts into the glass fibers of the insulating material.

WORKING LIFE: The period of time during which a liquid resin or adhesive, after mixing with catalyst, solvent, or other compounding ingredients, remains usable. (See *potlife.*)

INDEX